Animal Physiology

Animal Physiology

Richard W. Hill *Michigan State University*

Gordon A. Wyse *University of Massachusetts, Amherst*

Margaret Anderson *Smith College*

Sinauer Associates, Inc. *Publishers*
Sunderland, Massachusetts U.S.A

About the cover:

Snow geese above Basque del Apache National Wildlife Refuge, New Mexico.
Cover photo credit: Erwin Nielsen/Painet, Inc.

Animal Physiology

Address editorial correspondence and orders to:
Sinauer Associates, 23 Plumtree Road, Sunderland, MA 01375 U.S.A.
FAX: 413-549-1118
Email: publish@sinauer.com
Internet: www.sinauer.com

Library of Congress Cataloging-in-Publication Data

Hill, Richard W.
Animal physiology / Richard W. Hill, Gordon A. Wyse, Margaret Anderson.-- [1st ed.].
p. cm.
ISBN 0-87893-315-8 (harcover)
1. Physiology, Comparative. I. Wyse, Gordon A. II. Anderson, Margaret, 1941- III. Title.
QP33.H54 2004
571.1--dc22
2004005345

To Sue, from RWH

To Mary, from GAW

To Anita and Andy, from MA

Preface

For all of us, a commitment to teaching has been a top professional priority throughout our careers. We wrote this book out of that commitment.

Thomas Kuhn wrote that a textbook is more than anything a means of communicating to students the paradigms of their time. We have had three principal goals in preparing the content of this book. One, in accord with Kuhn's dictum, has been to articulate the central paradigms of contemporary animal physiology. A second goal for us has been to provide our readers with a resource where they will find accurate information about physiological systems and lucid explanations of physiological concepts. Our third goal has been to draw attention to the cutting edges, the places where the onward progress of research is challenging old paradigms and potentially creating footholds for new ones.

We have also had goals for presentation. We have wanted this book to be attractive and intriguing. To this end, each chapter starts with a vivid example of the application of the chapter to the lives of animals in their natural habitats. Moreover, five entire chapters, which we term the "At Work" chapters, break out of the usual textbook mold to discuss exciting topics such as diving and navigation with emphasis on experiments, theory maturation, integration, and prospects for future research. A third major element in our effort to make the book attractive and intriguing is extensive use of animal images. As students read this book, they will often be reminded by photographs and drawings of the animals we discuss.

In these pages we richly recognize the modern growth of animal physiology as a discipline integrated with other disciplines such as molecular biology, evolutionary biology, and ecology. As animal physiology has moved on from its old cataloging days to become more cosmopolitan and interdisciplinary, its scope has expanded seemingly exponentially. We have resisted the temptation to let pages expand in proportion. Instead, we have balanced competing demands for space so as to write a complete physiology at a length that can be read in a semester. Our intended audience is sophomores through beginning graduate students. To make the book accessible to as wide an audience as possible, we have included an extensive glossary and appendices on basic mathematical, chemical, and physical concepts. Our approach to the writing has been to seek advice and review from dozens of colleagues but opt for the pedagogical consistency of a three-author work. Margaret Anderson was principal author for Chapters 14, 15, 17, and 19; Gordon Wyse, for Chapters 10–13, 16, and 18; and Richard Hill, for the others.

We have tried to keep animals front and center. At the end of our production, as the orchestra goes silent and the klieg lights dim, we hope that animals leading their lives in their natural habitats will be the enduring image and memory left by this work—animals now better understood, but still with much to attract the curiosity of upcoming generations of biologists.

Hats off to Andy Sinauer, who took this project on, personally helped us to think big, and provided the resources to realize the ambitious goals that we set after having been so encouraged. We have all worked with many editors and publishers in our careers, and Andy is tops: an entrepreneur dedicated to putting the life of ideas on the printed page. Hats off too to Chris Small, Chelsea Holabird, Jefferson Johnson, David McIntyre, and the others at Sinauer Associates whose talents and dedication were indispensable to the production of this book. We feel privileged to have had Elizabeth Morales execute the art, which makes such a contribution to our pages.

Thanks to our students, who have challenged us, encouraged us, taught us, and—if nothing else—listened to us over our many years of classroom teaching. Our classes with our students have been our proving ground for teaching physiology and our source of reinforcement to take on ultimately a project of this magnitude. Thanks also to Michigan State University, the University of Massachusetts, and Smith College for their professional encouragement.

Of the many colleagues who have made contributions, Richard Hill would like in particular to thank Kjell Johansen, one of the greats, who way back at the beginning said without a moment's hesitation, "This is good." Energy still gets pulled from those words a quarter century later. Others who have encouraged us, taught us, reviewed chapters, provided photos or data, or prodded us by their criticism include Kellar Autumn, Lise Bankir, Bob Barlow, Al Bennett, Rudolf Billeter-Clark, Eric Bittman, Jeff Blaustein, Batrice Boily, Beth Brainerd, Eldon Braun, Klaus Bron, Bruce Byers, Don Christian, Barbara Christie-Pope, Corey Cleland, Randal Cohen, Dan Costa, Joseph Crivello, John Dacey, Bill Dawson, Hans-Rainer Duncker, Linda Farmer, Jane Feng, Milton Fingerman, Stephen Gehnrich, Fritz Geiser, Christopher Gillen, Judy Goodenough, Greg Goss, Edward Griff, John Harley, Bernd Heinrich, Gerhard Heldmaier, Ian Henderson, Raymond Henry, Dave Hinds, Richard Hoffman, Kay Holekamp, Charles Holliday, Henry John-Alder, Kelly Johnson, M. A. Q. Khan, Peter King, Leonard Kirschner, Rosemary Knapp, Rick Lee, John Lepri, Daniel Luchtel, William Lutterschmidt, Don Maynard, Kip McGilliard, Stephen McMann, Tim Moerland, David O'Drobinak,

Sanford Ostroy, Christine Oswald, Dan Otte, R. J. Paul, Linda Peck, Steve Perry, Chuck Peterson, Sandra Petersen, Richard Petriello, Nathan Pfost, Robert Rawding, Heinrich Reichert, Larry Renfro, David Richard, Bob Robbins, R.M. Robertson, Robert Roer, Stylianos Scordilis, William Seddon, Malcolm Shick, Laura Smale, George Somero, Tony Stea, Philip Stephens, Heather Thompson, Terry Trier, Walter Tyler, Kay Ueno, Joshua Urio, Mark Wales, Ewald Weibel, Susan Whittemore, Steve Wickler, Eric Widmaier, Robert Winn, Judith Wopereis, and Tom Zoeller. While acknowledging the many ways others have helped, we of course accept full responsibility for the product and invite readers' opinions on how we could do better.

Finally, we each have particular thanks to offer to the people in our personal lives whose support and patience have been indispensable. Richard Hill thanks Sue, Dave, and Chrissie, who have always been there even though the hours of writing have often meant long waits between sightings of their husband and father. Sue in particular has been a major contributor to this work by reading almost everything and offering the benefits of her knowledge and judgment as a biologist. Gordon Wyse thanks Mary for her editorial talents, support, and willingness to keep planning around this long project, and Jeff, Karen, and Nancy for inspiration. Likewise, Margaret Anderson expresses gratitude to her family and friends and to her students whose boundless enthusiasm and idealism provide great inspiration.

RICHARD W. HILL
East Lansing, Michigan

GORDON A. WYSE
Amherst, Massachusetts

MARGARET ANDERSON
Northampton, Massachusetts

To Our Readers

If you've ever been to a show and one of the producers stepped out on stage before the curtain went up to offer remarks about the upcoming event, you will understand the nature of these two pages. We, your authors, want to say a few words about the way we approached writing this book.

One of our primary goals has been to create a book in which you will find the fascination of physiology as well as its content. Thus we have started each chapter with an intriguing example that illustrates the application of the chapter to understanding the lives of animals. Flipping through the photographs that open the 28 chapters of the book, you will gain a quick sense of the many ways in which the study of physiology relates to biology at large.

Besides our desire to emphasize the fascination of physiology, we have also wanted to take time to emphasize the importance of integrating knowledge across physiological disciplines – and the importance of integrating physiology with ecology, behavior, molecular biology, and other fields. Moreover, we have wanted to discuss how concepts are tested and revised during research in physiology and to focus on the cutting edges in physiological research today. To help meet these goals, we have included five "At Work" chapters, which appear at the ends of five of the book's six parts and demonstrate how material learned in each of the parts can be applied to a topic of particular interest. You will find that the At Work chapters are written in a bit different style than the other chapters because they give extra emphasis to the *process* of discovery. We believe the topics we have chosen for these At Work chapters are especially interesting and important: They include diving by seals and whales, animal navigation, the function of human muscle in health and disease, mammals in the Arctic, and desert mammals. Each At Work chapter emphasizes the integration of scientific disciplines, the ways in which knowledge advances, and the current cutting edges. We hope you will find these chapters to be particularly enjoyable to read as well as informative.

One of the thrills of science today, besides the extraordinary pace at which new knowledge is being generated worldwide, is the revolution in how readily each of us can track down information. The first step in learning more about a field is to gather references. Even as recently as 15 years ago, the reference-gathering stage could easily require days or weeks. Today, however, the search engines available to find references in the scientific literature permit rapid assembly of information sources. Specialized search engines such as the Web of Science ®—which are similar to internet search engines but far more effective for exploring the scientific literature—will enable you to glean references rapidly from the thousands of scientific journals in which research is reported. Then they will enable you to read the abstracts of dozens of papers in a few hours of time, so you can identify the papers you want to read in full. Today *is* the information era. And indeed, knowledge *is* power. We encourage you to place a priority on mastering the tools available for information-gathering from the scientific literature.

You might wonder, if information is so easy to find, why should I take the course in which I am enrolled and why should I read this book? The answer in a few words is that extraordinary quantities of information create extraordinary challenges for synthesis. The more information each of us can locate, the more vulnerable we are to being befogged by it and the more we need frameworks into which we can assimilate new knowledge to organize it in a useful way. Sometimes science is presented as being the objective observation of facts. All commentators on the actual practice of science emphasize, however, that the mere accumulation of facts leads quite literally nowhere. Successful scientists require concepts that organize the facts, and in truth the progress of science is in its most essential form the testing of such concepts. For an analogy that may help clarify the usefulness of raw information, think of the last time you were at a well-attended, energy-filled party. In the midst of all the many animated conversations, if you had simply tried to assimilate the totality of the raw information your eyes and ears provided, you would likely have assimilated nothing. Instead, what we do on such occasions is organize the raw information, focusing on parts of it while ignoring other parts. Scientists create concepts that organize raw information. Then, in science, it is these concepts that we test for their accuracy and utility.

A good course taught with a good textbook provides a conceptual framework into which raw information can be fitted so that it becomes part of the life of ideas and concepts. We hope we have provided you with a conceptual framework that is "good for the future." By this we mean we have not tried merely to organize the knowledge already available. We have tried in equal measure to articulate a conceptual framework that is poised to grow and mature as new knowledge becomes available.

Just briefly we need to comment on units of measure. For 20 years there has been a revolution underway focused on bringing all human endeavor into line with a single system of units called the Système International (SI). Different countries have responded differently, as have different fields of activity. Thus, if you purchase a box of cereal in much of the world, the energy value of the contents will be quoted on the box in kilojoules, but elsewhere it will be reported in kilocalories. If you go to a physician in the United States and have your blood pressure measured, you will have it reported in millimeters of mercury, but if you read a recent scientific paper on blood pressures measured in a research laboratory in the United States, the pressures will probably be in kilopascals. The current state of transition in units of measure presents challenges for book writers just as it does for you. We have tried in each chapter to familiarize you with all the units of measure you might encounter in your reading or research on the subject of the chapter. Moreover, you will find in Appendix B a more-extensive discussion of the Système International, its relations to other systems of units, and the alternative ways presently used to express units.

All three of us who wrote this book have been dedicated teachers throughout our careers, and we have been fortunate to develop professional relationships and friendships with many dedicated learners. This book is a product of that two-way interaction. In the big universities today, there are many forces at work that encourage passivity and anonymity rather than two-way interactions. Our personal experience leads us to urge the opposite. We encourage you to talk science as much as possible with each other and with your instructors, whether in classroom discussions, study groups, office hours, or other contexts. This sort of active learning will contribute in a unique way to your enjoyment and mastery of the subjects you study. We have tried to write a book that will give you a lot to talk about.

RICHARD W. HILL GORDON A. WYSE MARGARET ANDERSON

Brief Contents

Contents

PART I Fundamentals of Physiology 1

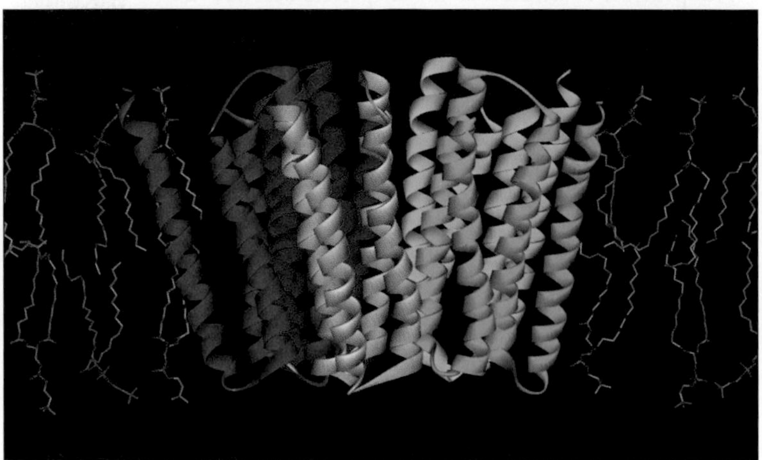

PART II Food, Energy, and Temperature 91

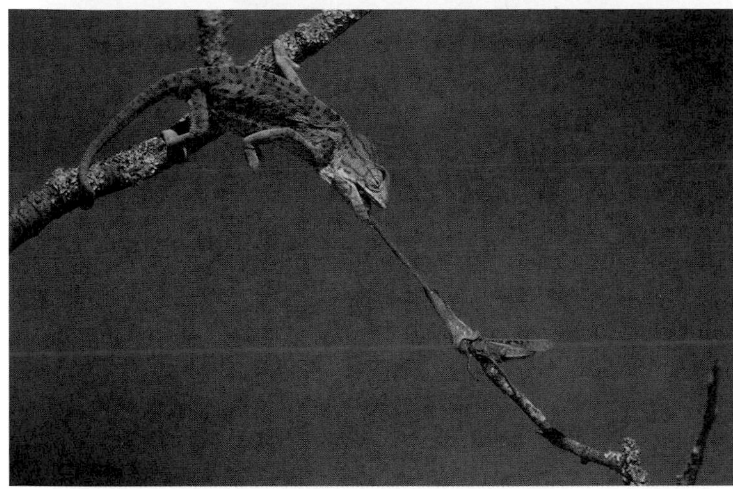

PART III Integrating Systems 257

PART V Oxygen, Carbon Dioxide, and Internal Transport 525

PART VI Water, Salts, and Excretion 663

25 Water and Salt Physiology: Introduction and Mechanisms 665

26 Water and Salt Physiology of Animals in Their Environments 685

PART I

Fundamentals of Physiology

Part I Fundamentals of Physiology

Previous page: The species of animals on any landscape each have a story to tell regarding how they function and how they relate to the environment they occupy. Each species has unique features. Related species, nonetheless, exhibit patterns because of their shared evolutionary heritage, and each fits in its own way within the patterns of its group.

1 Animals and Environments: Function on the Ecological Stage

Animal physiology is the study of animal function—the study of "how animals work." Physiology brings a special perspective to scenes like the salmon leaping a waterfall. We all know of the basic marvels of Pacific salmon—that they migrate from rivers to the open ocean and, years later, return to the very rivers of their conception to procreate the next generation. Physiologists seek to learn about the functional challenges the salmon face and determine how those challenges are met. When a salmon is hundreds of miles from land in the ocean and its time to breed approaches, what sensory and guidance mechanisms does it use to swim at high speed in a straight path to the correct coastline, as is often the case? And when it arrives at the coast, what sensory and integrative processes does it use to select the river where its life began years ago? These are the sorts of questions that physiologists ask.

Before the returning salmon enters freshwater, its blood is far more dilute than the seawater in which it swims. Freshwater, however, is itself far more dilute than the salmon's blood. What devices does the fish use to maintain its blood more dilute than seawater and more concentrated than freshwater, and how do its nervous and endocrine systems orchestrate the dramatic switch from one state to the other? Once in its natal river, a salmon no longer eats. Yet it may swim for many weeks and travel against the river current for hundreds of miles before reaching its spawning grounds. During its trip, the fish starves, gradually breaking down the substance of its body to supply its energy needs. How does the salmon control which body parts are broken down first, and which later? How does it generate swimming power and guide its course up the river so that its energy reserves last until it reaches its destination? When a waterfall is encountered, which muscles are used—and how do they generate enough power—to leap to the top? These are just a few of the questions that come to mind when physiologists ask how a salmon works.

As you start your study of physiology, we—your authors—believe you are at the beginning of a great adventure. We ourselves feel privileged to have spent our professional lives learning about how animals work, and we are eager to be your guides. If we could hop with you into a fantastic machine that would enable us to tour Earth in the realms we are about to travel, we would point out sperm whales diving an hour or more to depths of almost a mile, electric fish using modified muscles to generate lethal voltages, newborn reindeer calves standing wet with amniotic fluid in the frigid Arctic wind, reef corals

Migrating Pacific salmon Having spent several years feeding and growing in the Pacific Ocean, these fish have once again found the river in which they were conceived. Now they must power their way back to their birthplace to spawn, even though they ate their last meal at sea and will starve throughout their upriver journey.

growing prolifically because algae within their tissues permit internal photosynthesis, and moths flying through the night with bodies as warm as mammals' bodies. Each of these scenes draws the interest of physiologists and continues to spark new physiological research.

The Importance of Physiology

Why is the study of animal physiology important to you and to people in general? Not the least of the reasons is the one we have already emphasized—namely, that a full understanding and appreciation of all the marvels and other phenomena of the animal world depends on an analysis of how animals work. Although salmon, for instance, are interesting merely because of their life histories, they become even more interesting when studied physiologically. A physiological approach reveals, for example, that despite outward appearances, half or more of the body substance of a salmon is replaced with water as the fish migrates upriver, destroying its tissues to obtain energy for swimming, and the fish must manage this changeover carefully to have energy reserves remaining to breed at journey's end. The study of physiology leads one beyond surface impressions into the inner workings of animals, and nearly always this venture is not only a voyage of discovery, but one of revelation.

The study of physiology also has enormous practical applications because physiology is a principal discipline in the understanding of human health and disease. The analysis of many human diseases—ranging from aching joints to heart failure—depends on understanding how the "human machine" works. A physician who studies heart disease, for instance, needs to know how blood is made to flow into the heart chambers between one heartbeat and the next, how pressures are developed to eject blood into the arteries, how the cells of the heart muscle coordinate their contractions, and how the nutrient and oxygen needs of all parts of the heart muscle are met. We discuss these and other aspects of mammalian physiology extensively in this book. Even when we turn our attention to other types of animals, our study will often have application to human questions. One important reason is that nonhuman animals are often used as "models" for research that advances understanding of human physiology. Research on squids, for instance, has been indispensable for advancing knowledge of human neurophysiology because squids

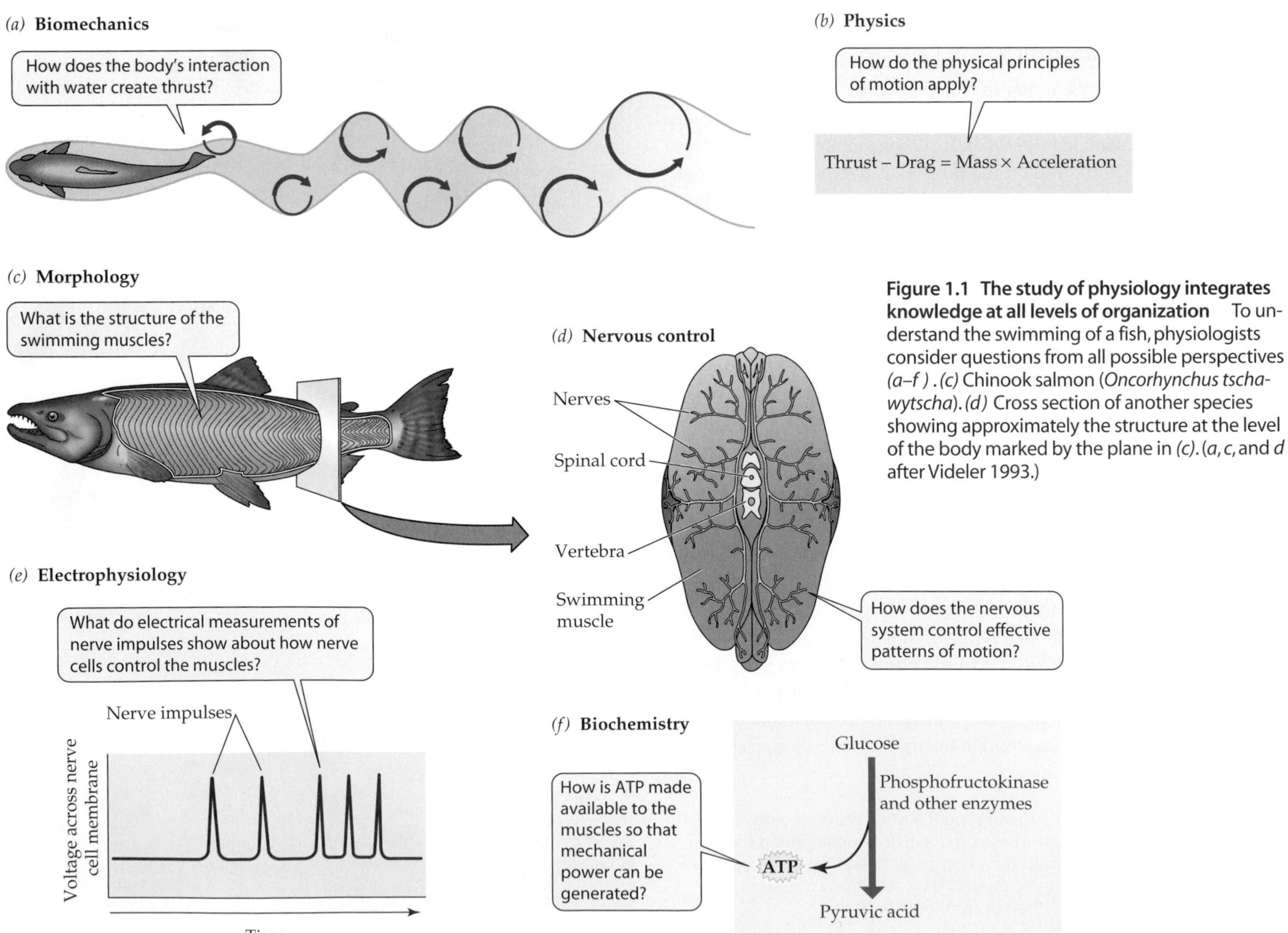

Figure 1.1 The study of physiology integrates knowledge at all levels of organization To understand the swimming of a fish, physiologists consider questions from all possible perspectives *(a–f)*. *(c)* Chinook salmon (*Oncorhynchus tschawytscha*). *(d)* Cross section of another species showing approximately the structure at the level of the body marked by the plane in *(c)*. (*a*, *c*, and *d* after Videler 1993.)

have nerve cells with cell processes so large that they can be studied in ways that would be impossible on small, mammalian nerve cells.

Physiology is as important for understanding the health and disease of nonhuman animals as of people. You might have wondered where funds are obtained to study salmon. The Canadian, American, and other governments have invested in salmon research in good part so that threats to the health of these important animals can be better understood. For instance, although individual dams might be designed to let the salmon pass, a series of dams on a river might so increase the overall cost of migration that the fish—living just on their stores of energy—would run out of energy before reaching spawning grounds. With studies of the energetics of swimming and leaping, managers can make rational predictions of the cumulative effects of dams, rather than simply altering rivers and waiting to see what happens. The effects of water pollutants, such as heavy metals and pesticides, are other important topics in salmon physiology.

In brief, physiology is one of the key disciplines for understanding

- The fundamental biology of all animals
- Human health and disease
- The health and disease of pets, farm animals, harvested animals, and other animals in human affairs

Physiology is also important because it is one of biology's most integrative disciplines, as Figure 1.1 illustrates. Physiologists study all the levels of organization of the animal body. In this respect, they are much like detectives who follow leads wherever the leads take them. To understand how an organ works, for instance, might require information about how certain enzymes work, and the study of the enzymes might lead to questions about the endocrine-mediated activation of genes that code for enzyme synthesis. Physiology pursues all these levels of biological organization. The study of physiology also employs the principles of physics and chemistry. Animals must obey the laws of physics and chemistry, and in fact animals often *exploit* those laws. Thus a physiological study often builds on a foundation of physical and chemical principles. Another significant point in discussing physiology's integrative nature is that the study of function typically goes hand and hand with the study of form; a review of anatomy is frequently needed to set the stage for the study of physiology, and as function is clarified, it typically helps account for anatomy. Also, physiologists must draw on knowledge of ecology and evolution because often the only way to comprehend the physiological differences between species—such as the differences between salmon and goldfish, or camels and polar bears—is to recognize that they evolved in different environments. Students often especially enjoy their study of physiology because the discipline is so integrative, bringing together and synthesizing many concepts that otherwise might seem independent.

Mechanism and Origin: Physiology's Two Central Questions

Physiology seeks to answer two central questions about how animals work: (1) What is the mechanism by which a function is accomplished, and (2) how did that mechanism come to be? To understand why there are two questions, consider the analogous problem of how a car works. In particular, how is an engine-driven wheel made to turn? To understand this function, you could disassemble a car and experiment on its parts. You could study how the pistons inside the cylinders of the engine are made to oscillate by forces released from exploding gasoline, how the pistons and connecting rods turn the drive shaft, and so forth. From studies like these, you would learn how the car works.

At the conclusion of such studies, however, you would have only half the answer to the question of how the car works. Presuming that you investigated a routine design of modern car, your experiments would tell you how a routine internal combustion engine turns a wheel by way of a routine transmission. Let your mind run free, and you may quickly realize that there are alternative designs for a car. The engine could have been a steam engine or a fuel cell based engine, for example. So when you ponder how a wheel turns, you see that you really face two questions: the *immediate* question of how a particular design of car makes a wheel turn, and the *ultimate* question of how that particular design came into being. Physiologists also face these two questions of *mechanism* and *origin*.

The study of mechanism: How do modern-day animals carry out their functions?

If you examine a particular car and its interacting parts to understand how it works, you are learning about the *mechanisms* of function of the car. Likewise, if you study the interacting parts of a particular animal—from organs to enzymes—to learn how it works, you are studying the animal's *mechanisms*. In physiology, **mechanism** refers to the components of actual, living animals and the interactions among those components that enable the animals to perform as they do.

Curiosity about mechanism is what inspires most physiologists to study animals, and studies of mechanism dominate physiological research. Physiology, in fact, is most clearly distinguished from other biological disciplines with which it is related, such as morphology or ecology, by its central focus on the study of mechanism. A physiologist typically begins an investigation by observing a particular capability that excites curiosity or needs to be understood for practical purposes. The capability of the human visual system to distinguish red and blue is an example. Another is the ability of certain types of nerve cells to conduct nerve impulses at speeds of over 100 m per second. Whatever the capability of interest, the typical goal of physiological research is to discover its mechanistic basis. What cells, enzymes, and other parts of the body are employed, and how are they employed, to enable the animal to perform as it does?

For a detailed example of a mechanism, consider how the light organ of a firefly produces light. Possibly our knowledge of this question will have advanced by the time you read this page because the subject is under active investigation. The biochemistry is clear (Figure 1.2*a*). A chemical compound named *firefly luciferin* first reacts with adenosine triphosphate (ATP) to form luciferyl-AMP. Then, if molecular oxygen (O_2) can reach the luciferyl-AMP, the two react to form a chemical product in which electrons are boosted to an excited state, and when this electron-excited product returns to its ground state, it emits photons. This sequence of reactions requires a protein catalyst, an enzyme called *firefly luciferase*. When a firefly is

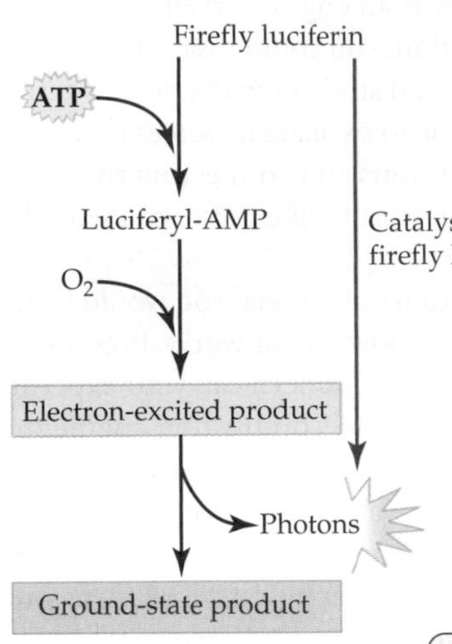

Figure 1.2 The mechanism of light production by fireflies (*Photinus*) *(a)* The chemistry of light production. *(b,c)* In the light cells, the luciferin reactions are spatially separated from mitochondria. When a light cell is not flashing *(b)*, the mitochondria intercept O_2. When a cell is flashing *(c)*, however, O_2 gets through to the luciferin reactions. Understanding of firefly flashing is a work in progress; thus, parts of this scenario are hypothetical. AMP = adenosine monophosphate; ATP = adenosine triphosphate.

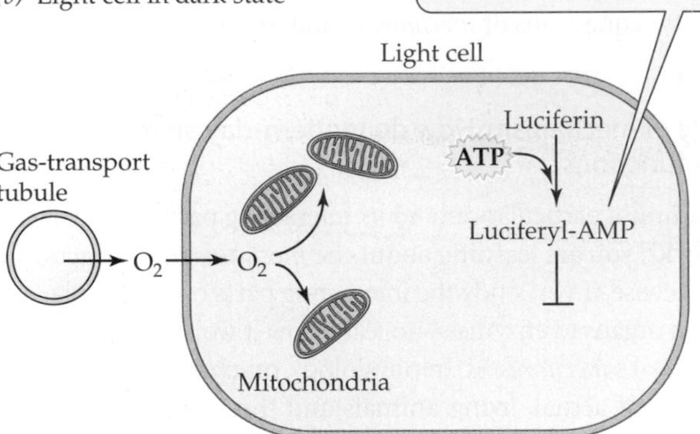

(c) Light cell in flashing state

In the flashing state, nitric oxide is produced under nervous control and bathes the mitochondria, preventing them from intercepting O_2.

Luciferin

ATP

Luciferyl-AMP

O_2 O_2 O_2 O_2

Nitric oxide

Pulses of O_2 reach the luciferin reactions, resulting in pulses of light.

not producing light (Figure 1.2*b*), any O_2 that reaches the insect's light cells via its gas-transport tubules is intercepted (and thereby prevented from reacting with luciferyl-AMP) by mitochondria that are positioned between the gas-transport tubules and the sites of the luciferin reactions. The light cells produce light (Figure 1.2*c*) when, because of stimulation by the nervous system, the mitochondria become bathed with nitric oxide (NO). The NO is hypothesized to block mitochondrial use of O_2, allowing O_2 through to react with luciferyl-AMP. Facts like these form a description of the *mechanism* by which fireflies produce light.

The study of a mechanism may become so intricate that decades or centuries are required for the mechanism to be fully understood. By definition, however, the complete mechanism of any given function is present for study in the here and now. A scientist can in principle fully describe the mechanism of a process merely by studying existing animals in ever-finer detail.

The study of origin: Why do modern-day animals possess the mechanisms they do?

Suppose a youngster observes a firefly produce a flash of light and asks you to explain what he has seen. One way you could interpret the request is as a question about mechanism. Thus you could answer that the brain of the insect sends nerve impulses that cause the light cells to become bathed with nitric oxide, resulting in the production of excited electrons through the reaction of O_2 with luciferyl-AMP. However, the youngster who asks you to explain the flashing of a firefly is probably interested in something else. The *reason* the firefly makes light is probably what is on your young friend's mind, rather than the mechanism. That is, the youngster is probably wondering *why* the firefly possesses a mechanism to make light.

For biologists, the answer lies in *evolutionary origins*. The mechanisms of modern-day animals are products of evolution, and thus the reasons for the existence of mechanisms lie in evolutionary processes. The study of evolutionary origins is a central aim of modern physiology because it promises to reveal the *significance* of mechanisms. If we can learn why evolution produced a mechanism, we will better understand what (if anything) animals gain by having the mechanism.

Because modern-day mechanisms evolved in the past, the question of origins is fundamentally historical. The origins of a mechanism, unlike the mechanism itself, cannot usually be observed directly in the here and now. Instead, origins must usually be studied indirectly, through inferences about the past from observations we can make in the present. The reliance on indirect reasoning means that evolutionary origins are rarely understood with the same certainty as mechanisms.

Natural selection is a key process of evolutionary origin

Natural selection is just one of several processes by which animals acquire traits during evolution, as we discuss later in this chapter. Natural selection, however, holds a place of special importance for biologists because, of all the modes of evolutionary change, natural selection is believed to be the principal process by which animals become fit to live in their environments.

Natural selection is the increase in frequency of genes that produce phenotypes that raise the likelihood that animals will survive and reproduce. During evolution by natural selection, such genes increase in frequency because animals with the genes are differentially successful relative to other members of their species. If we find that a physiological mechanism originated by natural selection within the prevailing environment, we can conclude that the mechanism is an asset; that is, it improves an animal's chances of survival and reproduction within the environment the animal occupies.

Adaptation is an important sister concept to natural selection. Because we discuss adaptation at length later, here we simply state that an adaptation is a physiological mechanism or other trait that is a product of evolution by natural selection. Adaptations are assets; because of the way they originated, they aid the survival and reproduction of animals living in the environment where they evolved. When we speak of the **adaptive significance** of a trait evolved by natural selection, we refer to the reason *why* the trait is an asset; that is, the reason *why* natural selection favored the evolution of the trait.

The light flashes of fireflies usually function to attract mates. The males of each species of firefly emit light flashes in a distinctive, species-specific pattern as they fly, thereby signaling their species identity to females (Figure 1.3). Using various sorts of evidence, students of fireflies infer that the firefly light-producing mechanism evolved by natural selection because light flashes can be used to bring the sexes together. Thus the mechanism of light production is an adaptation, and its adaptive significance is mate attraction.

Mechanism and adaptive significance are distinct concepts that do not imply each other

Why have we stressed that physiology faces *two* central questions? We have emphasized both that physiology studies mechanism and that it studies evolutionary origins to understand adaptive significance. Why *both*? Physiologists must seek answers to both questions because *mechanism and adaptive significance do not imply each other*. If you know the mechanism of a process, you do not necessarily know anything about its adaptive significance. If you know the adaptive significance, you do not necessarily know anything about the mechanism. Thus, to understand both mechanism and adaptive significance, you must study both.

As an example, consider light production by fireflies once again. Physiologists know of several mechanisms by which organisms can produce light. Thus even if fireflies were *required* to attract their mates with light, their mechanism of making light would not be limited theoretically to just the mechanism they use. The mechanism of light production by fireflies cannot be deduced from simple knowledge of the purpose for which the mechanism is used. Conversely, light flashes could be used for purposes other than mate attraction, such as luring prey, distracting predators, or synchronizing biorhythms. The significance of light production cannot be deduced from the simple fact that light is made or from knowledge of the mechanism by which it is made.

Figure 1.3 Male fireflies employ their mechanism of light production for an adaptive function, mate attraction Representative flashing patterns and flight paths of males of nine different species of fireflies of the genus *Photinus* from the eastern and central United States. Each line of flight represents a different species. For instance, the uppermost line represents *Photinus consimilis*, a species that flies well above the ground. To understand the format, imagine that you are watching the uppermost firefly as it leaves the tree and follows the numbered sequence. The differences in flashing and flight patterns among species allow males to signal their species to females. (From a drawing by Dan Otte in Lloyd 1966.)

François Jacob (b. 1920), a Nobel laureate, asked in a famous article whether evolution by natural selection more closely resembles engineering or tinkering. An engineer who is designing a machine can start from scratch. That is, an engineer can start by thinking about the very best design and then build that design from raw materials. A tinkerer who is building a new machine starts with parts of preexisting machines.

Evolution is like tinkering, Jacob argued: A population of animals that is evolving a new organ or process rarely starts from scratch; instead, it starts with elements that it already has on hand for other reasons. Think, for example, of the vertebrate lung, evolved as an outpocketing of the digestive tract. Would animals have made lungs from guts if they had been capable of starting from scratch? Throughout evolution, animals in a sense have had to remain capable of living in their old ways even as they have developed new ways. Thus, design from first principles—the engineering approach—has not been possible.

The tinkering aspect of evolution is a key reason why mechanism and adaptive significance do not imply each other. The mechanism employed to perform a particular process is not an abstraction but rests on what has come before in any particular evolutionary line. Consider, for example, the eyes of two groups of active aquatic animals: the cephalopod molluscs (squids and octopuses) and the fish. Both groups have evolved sophisticated eyes that permit lifestyles based on excellent vision. However, these eyes are built on very different retinal designs (Figure 1.4). The photoreceptors in the retinas of fish point away from the light; those of squids and octopuses point toward it. Moreover, whereas visual signals from the fish photoreceptors are extensively processed by networks of nerve cells with-

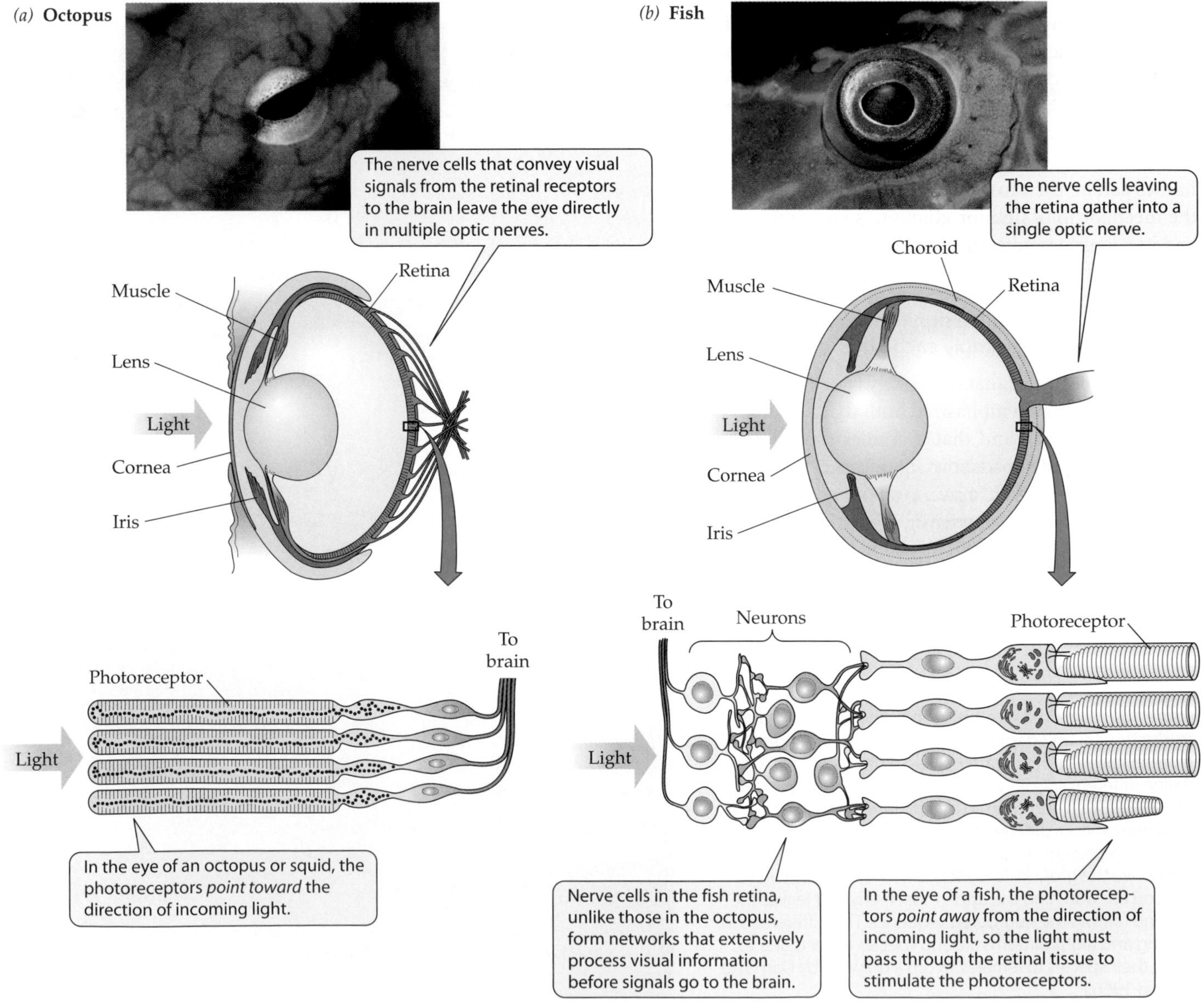

Figure 1.4 Structures that are remarkably similar in functional performance and adaptive significance can differ dramatically in details of how they are assembled and work Both cephalopod molluscs and fish have evolved excellent vision, but they see using different detailed mechanisms. (*a* after Wells 1966 and Young 1971; *b* after Walls 1942.)

in the retina before visual information is sent to the brain in a single optic nerve, in squids and octopuses the photoreceptors send their visual signals more directly to the brain in multiple optic nerves. The adaptive significance of excellent vision in the two groups of animals is similar, but the detailed mechanisms are not. A major reason for the differences in mechanisms seems likely to be that the two groups built their eyes from different preexisting structures.

This Book's Approach to Physiology

Mechanistic physiology, which emphasizes the study of mechanism, and **evolutionary physiology**, which emphasizes the study of evolutionary origins, have become recognized as major approaches to the study of animal physiology in recent years. The two approaches share the same overall subject matter: They both address the understanding of animal function. They differ, however, in the particular aspects of physiology they emphasize. The viewpoint of this book, as stressed already, is that both approaches are essential for physiology to be fully understood.

Comparative physiology and **environmental physiology** are additional approaches to the study of animal physiology. These approaches overlap mechanistic and evolutionary physiology, and they overlap each other. Comparative physiology is the synthetic study of the function of all animals. It contrasts, for example, with *human physiology* or *avian physiology*, each of which addresses only a limited set of animals. Comparative physiology is termed *comparative* because one of its major goals is to compare systematically the ways that various sorts of animals carry out similar functions, such as vision, breathing, or circulation. Environmental physiology (also called *physiological ecology*) is the study of how animals respond physiologically to environmental conditions and challenges, or—more briefly—"ecologically relevant physiology."

Our viewpoint in this book is mechanistic, evolutionary, comparative, and environmental. In other words, we stress

- The mechanisms by which animals perform their life-sustaining functions
- The evolution and adaptive significance of physiological traits
- The ways in which diverse phylogenetic groups of animals resemble each other and differ
- The ways in which physiology and ecology interact, in the present and during evolutionary time

Overlapping with the classifications already discussed, physiology is divided also into various branches or disciplines based on the *types of functions* that are performed by animals. The organization of this book into parts and chapters is based on the types of function. As Table 1.1 shows, the book consists of six major subdivisions, Parts I through VI, each of which focuses on a particular set of functions. The chapters within each part discuss particular aspects of the part topic. The last chapter in each part (except Part I) is an "at work" chapter that takes a synthetic approach to a prominent, curiosity-provoking topic in the part. *The principal goal of the "at work" chapters is to show how the material in each of the parts can be used in an integrated way to understand animal function.*

Now, as they say in theater, "Let the play begin." As we consider the principal subject of this chapter—function on the ecological stage—the three major players are animals, environments, and evolutionary processes. We address each in the remaining parts of this chapter.

Animals

Because animals are the focus of this entire book, the animal properties that deserve emphasis in this initial discussion are those that are of overriding importance for understanding animal physiology. One

TABLE 1.1 The organization of this book The "at work" chapters, printed in italics, exemplify how the material studied in a part can be used synthetically to understand a problem in animal physiology.

Parts	Topics within parts
Part I: Fundamentals of physiology	• Animals and environments (Chapter 1) • Molecules and cells (Chapter 2) • Transport of solutes and water (Chapter 3)
Part II: Food, energy, and temperature	• Nutrition, feeding, and digestion (Chapter 4) • Energy metabolism (Chapter 5) • Aerobic and anaerobic metabolism (Chapter 6) • Exercise (Chapter 7) • Thermal relations (Chapter 8) • *Mammals in frigid places (Chapter 9)*
Part III: Integrating systems	• Nervous systems and biological clocks (Chapter 10) • Neurons (Chapter 11) • Synapses (Chapter 12) • Sensory processes (Chapter 13) • Hormones and neuroendocrinology (Chapter 14) • Reproduction (Chapter 15) • *Animal navigation (Chapter 16)*
Part IV: Muscle and movement	• Muscle (Chapter 17) • Control of movement (Chapter 18) • *Muscle in human health and disease (Chapter 19)*
Part V: Oxygen, carbon dioxide, and internal transport	• Basics of oxygen and carbon dioxide physiology (Chapter 20) • Breathing (Chapter 21) • Gas transport in blood (Chapter 22) • Circulation (Chapter 23) • *Diving by seals and whales (Chapter 24)*
Part VI: Water, salts, and excretion	• Mechanisms of water and salt physiology (Chapter 25) • Water and salt physiology of animals in their environments (Chapter 26) • Kidneys and excretion (Chapter 27) • *Mammals in deserts and dry savannas (Chapter 28)*

of these properties is that animals are *structurally dynamic*. Another is that animals are *organized* systems and *require energy* to maintain their organization. Finally, both *time* and *size* are of fundamental significance in the lives of all animals.

One of the most profoundly important properties of animals is that the atoms of their bodies—their material building blocks—are in dynamic exchange with the atoms in their environments throughout life. This structural dynamism is a crucial way in which animals differ from inanimate objects like telephones. After a telephone is manufactured, the particular carbon and iron atoms that are built into its substance remain as long as the telephone exists. One might think by casual observation that the composition of a person, lion, or crab is similarly static. This illusion was abruptly dispelled, however, when scientists began using chemical isotopes as research tools.

Isotopes proved to be revealing because they permit atoms to be labeled and therefore tracked. Consider iron as an example. Because most iron atoms in the natural world are of atomic weight 56 (^{56}Fe), an investigator can distinctively label a particular set of iron atoms by substituting the unusual (but stable) alternative isotope of iron having an atomic weight of 58 (^{58}Fe). Suppose that we make a telephone in which all the iron atoms are of the unusual ^{58}Fe isotope, so that we can distinguish those iron atoms from the ones generally available. Years later, all the iron atoms in the telephone will still be of the unusual ^{58}Fe type. Suppose, however, that over the course of a year, we feed a person the unusual ^{58}Fe isotope, so that isotopically distinctive iron atoms are built into hemoglobin molecules and other iron-containing molecules throughout the person's body. If we then stop providing the unusual iron isotope in the person's diet, thereafter—as time goes by—the isotopically distinctive ^{58}Fe atoms in the body will leave and will be replaced with atoms of the ordinary isotope, ^{56}Fe, from the environment. Years later, all the unusual iron atoms will be gone. Thus although the person may outwardly appear to be structurally constant like a telephone, the iron atoms in the substance of the person's body at one time differ from those at another time.

The mechanistic reason for the turnover of iron atoms in an animal is that the molecular constituents of an individual's body break down and are rebuilt. A human red blood cell, for example, typically lives for only 4 months. When a red blood cell is discarded and replaced, some of the iron atoms from the hemoglobin molecules of the old cell are excreted into the environment, and some of the iron atoms built into the new cell are acquired from food. In this way, even though the number of red cells remains relatively constant, the iron atoms of the red cells are in dynamic exchange with iron atoms in the environment.

Essentially all the atoms in the substance of an animal's body undergo similar dynamic exchanges. Calcium atoms enter an animal's skeleton and later are withdrawn; some of the withdrawn atoms are replaced with newly ingested calcium atoms from the environment. Proteins and fats throughout an animal's body are continually broken down at substantial rates, and their resynthesis is carried out in part with molecules newly acquired from the environment, such as amino acids and fatty acids from foods. Adult people typically resynthesize 2% to 3% of their body protein *each day*, and about 10% of the amino acids used to build the new protein molecules are acquired from food.

Have you ever wondered why you need to worry *every week* about whether you are eating enough calcium, iron, magnesium, and protein? The principles we are discussing provide the explanation. If you were an inanimate object, enough of each necessary element or compound could be built into your body at the start, and you would then have enough forever. Instead, because you are alive rather than inanimate, you lose elements and compounds every day and must replace them.

As this discussion has illustrated, the *material boundaries* between an animal and its environment are blurred and dynamic, not crisp and static. Atoms cross the boundaries throughout life, so an atom that is part of an animal's tissues on one day may be lying on the forest floor or drifting in the atmosphere the next day. Possibly the most profound implication of these facts is that *an animal is not a discrete material object.*

The structural property of an animal that persists through time is its organization

If the atomic building blocks of an animal are transient, by what structural property is an animal *defined*? The answer comes from imagining that we can see the individual molecules in an adult animal's body. If we could, we would observe that the molecular structures and the spatial relations of molecules in tissues are relatively constant over time, even though the particular atoms constructing the molecules change from time to time. A rough analogy would be a brick wall that retains a given size and shape but in which the bricks are constantly being replaced, such that the particular bricks present during one month are different from those present a month earlier.

The structural property of an animal that persists through time is the *organization* of its atomic building blocks, not the building blocks themselves. Thus an animal is defined by its organization. This characteristic of animals provides the most fundamental reason why they require inputs of energy throughout life. As we discuss in detail in Chapter 5, the second law of thermodynamics says that for organization to be maintained in a dynamic system, use of energy is essential.

Most cells of an animal are exposed to the internal environment, not the external environment

Shifting our focus now to the cells of an animal's body, it is important first to stress that the conditions experienced by most of an animal's cells are the conditions inside the body, not those outside. Most cells are bathed by the animal's tissue fluids or blood. Thus the *environment* of most cells consists of the set of conditions prevailing in the tissue fluids or blood. Claude Bernard (1813–1878), a Frenchman who was one of the most influential physiologists of the nineteenth century, was the first to codify this concept. He coined the term **internal environment** (*milieu intérieur*) to refer to the set of conditions—temperature, pH, sodium (Na^+) concentration, and so forth—experienced by cells within an animal's body. The conditions outside the body represent the **external environment**.

The internal environment may be permitted to vary when the external environment changes, or it may be kept constant

Animals have evolved various types of relations between their internal environment and the external environment. If we think of the organization of the body as being hierarchically arranged, the relations between the internal and external environment represent a second potential level at which animals may exhibit organization. At a primary level, all animals consistently exhibit structural organization of their atoms and molecules. At this second level, animals sometimes—but only sometimes—exhibit additional organization by keeping their internal environment distinct from their external environment.

Animals display two principal types of relations between their internal and external environments. On the one hand, when the conditions outside an animal's body change, the animal may permit its internal environment to match the external conditions and thus change along with the outside changes. On the other hand, the animal may maintain constancy in its internal environment. These alternatives can be illustrated with temperature (Figure 1.5). If the temperature of an animal's external environment changes, one option is for the animal to let its internal temperature change to match the external temperature (Figure 1.5*a*). Another option is for the animal to maintain a constant internal temperature (Figure 1.5*b*). If an animal permits internal and external conditions to be equal, it is said to show **conformity.** If the animal maintains internal constancy in the face of external variability, it shows **regulation**. Conformity and regulation are extremes; intermediate responses to changes in the external environment are common.

Animals frequently show conformity with respect to some characteristics of their internal environment while showing regulation with respect to others. Consider a salmon, for example (Figure 1.6). Like most fish, salmon are temperature conformers; they let their internal temperature match the surrounding water temperature. Simultaneously, salmon are excellent chloride regulators; they maintain a nearly constant concentration of Cl^- ions in their blood, regardless of how high or low the outside Cl^- concentration is.

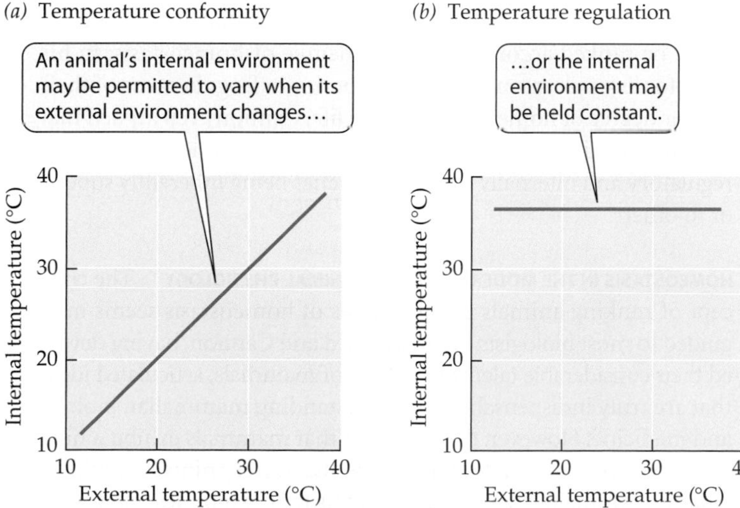

Figure 1.5 Conformity and regulation These examples from the study of temperature illustrate the general principles of conformity *(a)* and regulation *(b)*.

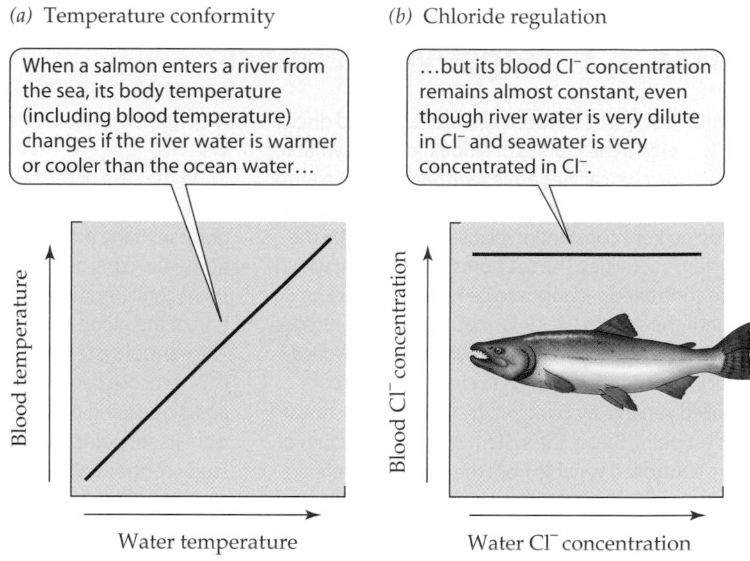

Figure 1.6 Mixed conformity and regulation in a single species Salmon are temperature conformers but chloride regulators, as shown diagrammatically here.

Regulation places greater demands on an animal's energy resources than conformity because regulation represents a form of organization; during regulation, organization is represented both by the maintenance of *constancy* inside the body and by the maintenance of a *distinction* between inside and outside conditions. A familiar analogy for the energy costs of regulation in animals is provided by home heating. A great deal of energy is required to keep the inside of a house at 22°C (72°F) during the cold of winter. This energy cost is entirely avoided if the inside temperature is simply allowed to match the outside temperature.

Homeostasis in the lives of animals: Internal constancy is often critical for proper function

Homeostasis is an important concept regarding the *nature* and *significance* of internal constancy. Soon we will define homeostasis using the words of Walter Cannon (1871–1945), who coined the term. To fully appreciate the concept, however, we must first recognize its historical roots in medicine. The two men who contributed the most toward developing the concept of homeostasis, Claude Bernard and Walter Cannon, were physicians and medical researchers, concerned primarily with human physiology. Healthy humans maintain remarkable constancy of conditions in their blood and tissue fluids. The notion of homeostasis was thus conceived during studies of a species that exhibits exceptional internal constancy, and later the notion was extrapolated to other animals.

BOX 1.1 Negative Feedback

The type of control that Claude Bernard discovered in his studies of blood glucose is what today we call **negative feedback.** In any control system, the **controlled variable** is the property that is being kept constant or relatively constant by the system's activities. The **set point** is the level at which the controlled variable is to be kept. **Feedback** occurs if the system uses information on the controlled variable itself to govern its actions. In *negative feedback*, the system responds to changes in the controlled variable by bringing the variable back toward its set point; that is, the system *opposes deviations* of the controlled variable from the set point. There are many detailed mechanisms by which negative feedback can be brought about in physiological systems. Negative feedback, however, is virtually synonymous with homeostasis and occurs in all homeostatic systems.

In the case of the blood glucose level that so intrigued Claude Bernard, the control system adds glucose to the blood if the blood glucose concentration—the controlled variable—falls below its set-point concentration, thereby opposing the deviation of the blood concentration from the set point. The control system removes glucose from the blood if the glucose concentration rises too high, thereby again opposing the deviation of the concentration from its set point. Biologists and engineers who study control systems have established that no control system can maintain perfect constancy in a controlled variable; putting the case roughly, a controlled variable must be a moving target for a control system to act on it. Thus the blood glucose concentration is not kept perfectly constant by the glucose control system, but during normal health it is kept from varying outside a narrow range. Box 8.1 will provide greater detail about control systems based on negative feedback.

In ***positive feedback***, a control system *reinforces deviations* of a controlled variable from its set point. Positive feedback is much less common in physiological systems than negative feedback. It occurs, however, and is not always pathological; for example, a functionally advantageous sort of positive feedback occurs when action potentials (nerve impulses) develop in nerve cells (see Chapter 11).

Claude Bernard was the first to recognize the impressive stability of conditions that humans maintain in their blood and tissue fluids. One of Bernard's principal areas of study was blood glucose in mammals. He observed that the liver takes up and releases glucose as necessary to maintain a relatively constant glucose concentration in the blood. If blood glucose rises, the liver removes glucose from the blood. If blood glucose falls, the liver releases glucose into the blood. Bernard stressed that as a consequence, most cells in the body of a mammal experience a relatively constant environment with respect to glucose concentration (Box 1.1). Bernard's research and that of later investigators also revealed that most cells in a mammal's body experience relative constancy of temperature, O_2 level, osmotic pressure, pH, salt concentrations, and so on because various organs and tissues regulate these properties at consistent levels in the body fluids bathing the cells.

Claude Bernard devoted much thought to the *significance* of internal constancy in humans and other mammals. He was greatly impressed with how freely mammals are able to conduct their lives regardless of outside conditions. Mammals, for example, can wander about outdoors in the dead of winter, seeking food and mates; fish or insects, on the other hand, are often driven into a sort of paralysis by winter's cold. Bernard reasoned that mammals are able to function in a consistent way regardless of variation in outside conditions because the cells inside their bodies enjoy constant conditions. He thus stated a hypothesis that remains probably the most famous in the history of animal physiology: *"Constancy of the internal environment is the condition for free life."* A modern translation might go like this: Animals are able to lead lives of greater freedom and independence to the extent that they maintain a stable internal environment, sheltering their *cells* from the variability of the outside world.

Walter Cannon, an American winner of the Nobel Prize who was born in the same decade that Claude Bernard died, introduced the concept of homeostasis to refer to internal constancy in animals. In certain ways, Bernard's and Cannon's views were so similar that Bernard might have invented the homeostasis concept, but the implications of internal constancy were clearer by Cannon's time. Because animals dynamically interact with their environments, the temperature, pH, ion concentrations, and other properties of their bodies are incessantly being drawn away from stability. Cannon emphasized that for an animal to be internally stable, vigilant physiological mechanisms must be present to correct deviations from stability. Thus when Cannon introduced the term *homeostasis*, he intended it to mean not just internal constancy, but also the existence of regulatory systems that automatically make adjustments to maintain internal constancy. In his own words, Cannon at one point described homeostasis as "the coordinated physiological processes which maintain most of the [constant] states in the organism."

An essential aspect of Cannon's perspective was his conviction that homeostasis is good. Cannon argued in fact that homeostasis is a signature of highly evolved life. He believed that animal species could be ranked according to their degree of homeostasis; in his view, for example, mammals were superior to frogs because of their greater degree of homeostasis. Late in life, Cannon even argued that human societies could be ranked on a scale of homeostasis—self-regulatory and internally constant societies being inherently superior to others.

HOMEOSTASIS IN THE MODERN STUDY OF ANIMAL PHYSIOLOGY The concept of ranking animals using degrees of homeostasis seems misguided to most biologists today. Bernard and Cannon, having devoted their considerable talents to studies of mammals, articulated ideas that are truly indispensable for understanding mammalian biology and medicine. However, the mere fact that mammals exhibit a high degree of homeostasis does not mean that other animals should be held to mammalian standards. Animals that exhibit less-complete homeostasis than mammals coexist in the biosphere with mammals. Indeed, the vast majority of animals thriving today do not achieve "mammalian standards" of homeostasis. Thus most biologists today would argue that a high degree of homeostasis is merely one of several ways to achieve evolutionary and ecological success. In this view, Bernard and Cannon did not articulate universal requirements for success, but instead they clarified the properties and significance of *one particular road* to success.

Recent research has clarified, in fact, that organisms sometimes achieve success in the biosphere by letting their internal environment *vary* with the external environment: the antithesis of homeostasis. Consider, for example, insects that overwinter within plant stems in Alaska. They survive by ceasing to be active, allowing their internal temperatures to fall to below –40°C, and tolerating such low tissue temperatures. Any attempt by such small animals to maintain an internally constant temperature from summer to winter would be so energetically costly that it would surely end in death; thus the tolerance of the insects to the *change* of their internal temperature in winter is a key to their survival. Even some mammals—the hibernators—survive winter by *abandoning constancy* of internal temperature; hibernating mammals allow their body temperatures to decline and sometimes match air temperature.

Both constancy and inconstancy of the internal environment—regulation and conformity—have disadvantages and advantages:

- *Regulation.* The chief disadvantage of regulation is that it costs energy. The great legacy of Bernard and Cannon is that they clarified the advantage that animals reap by paying the cost: Regulation permits cells to function independently of outside conditions.
- *Conformity.* The principal disadvantage of conformity is that cells within the body are subject to changes in their conditions when outside conditions change. The chief advantage of conformity is that it avoids the energy costs of maintaining organization of the internal environment relative to the external environment. Thus conformity is energetically cheap.

Neither regulation nor conformity is categorically a defect or an asset. One cannot understand mammals or medical physiology without understanding homeostasis, but one cannot understand the full sweep of animal life without recognizing that physiological flexibility is sometimes advantageous.

Time in the lives of animals: Physiology changes in five time frames

Time is a critical dimension for understanding the physiology of all animals because the physiology of animals invariably changes from time to time. Even animals that exhibit homeostasis undergo change. Details of their internal environment may change. Moreover, the regulatory processes that *maintain* homeostasis must change so that homeostasis can prevail, much as day-to-day adjustments in the fuel consumption of a home furnace are required to maintain a constant air temperature inside the home during winter.

As an example of change in an organism, consider yourself. If you read parts of this chapter in a warm room and parts in a cold room, your metabolic rate will likely be higher in the colder room. If you go for a run on a hot summer day, your rate of sweat production as you run may be high or low, depending on whether your run is your very first in the heat of summer or you have been running a lot under hot conditions in recent weeks. The hemoglobin in your red blood cells today is quite different in its O_2-transport properties from the hemoglobin you possessed as a newborn.

An important organizing principle for understanding the role of time in the lives of animals is to recognize *five major time frames* within which the physiology of an animal can change. The time frames can usefully be placed in two categories: (1) responses of physiology to changes in the external environment and (2) internally programmed changes of physiology. Table 1.2 lists the five time frames classified in this way. We will recognize these five time frames throughout this book as we discuss various physiological systems. If you keep the time frames in mind, you will find that they provide a useful tool for organizing your growing knowledge of time-dependent processes, so that you can recall the information you have learned and relate similar information studied on different occasions.

The concept of the five time frames overlies other ways of organizing knowledge about animal function. For example, the concept of time frames overlies the concepts of regulation, conformity, and homeostasis we have just discussed. When we speak of regulation, conformity, and homeostasis, we refer to *types* of responses that animals show in relation to variations in their external environments.

TABLE 1.2 The five time frames in which physiology changes

Type of change	Description
Changes in physiology that are responses to changes in the external environment	
1. Acute changes	Short-term changes in the physiology of individual animals: changes that individuals exhibit right after their environments have changed; acute changes are reversible
2. Chronic changes (acclimation and acclimatization)	Long-term changes in the physiology of individual animals: changes that individuals display after they have been in new environments for days, weeks, or months; chronic changes are reversible
3. Evolutionary changes	Changes that occur by alteration of gene frequencies over the course of many generations in populations exposed to new environments
Changes in physiology that are internally programmed to occur whether or not the external environment changes	
4. Developmental changes	Changes in the physiology of individual animals that occur in a programmed way as the animals mature from conception to adulthood and then to senescence
5. Changes controlled by periodic biological clocks	Changes in the physiology of individual animals that occur in repeating patterns (e.g., each day) under control of the animals' internal biological clocks

When we speak of the time frames, we address *when* those responses occur.

PHYSIOLOGY RESPONDS TO CHANGES IN THE EXTERNAL ENVIRONMENT IN THREE TIME FRAMES Individual animals subjected to a change of their external environment exhibit acute and chronic responses to the environmental change. **Acute responses** are exhibited during the first minutes or hours after an environmental change. **Chronic responses** are expressed following prolonged exposure to new environmental conditions. You might wonder why an individual's immediate responses to an environmental change differ from its long-term responses. The answer is that the passage of time permits biochemical or anatomical restructuring of an animal's body. When an animal suddenly experiences a change in its environment, its immediate responses must be based on the "old," preexisting properties of its body because the animal has no time to restructure. A morphological example is provided by a person who suddenly is required to lift weights after months of totally sedentary existence. The sedentary person is likely to have small arm muscles, and his immediate, *acute* response to his new weight-lifting environment will likely be that he can lift only light weights. However, if the person lifts weights repeatedly, as time goes by restructuring will occur; his muscles will increase in size. Thus his *chronic* response to the weight-lifting environment will likely be that he can lift heavy weights as well as light ones.

A familiar physiological example of acute and chronic responses is provided by human reactions to work in hot weather. We all know that when we are first exposed to hot weather after a period of living in cool conditions, we often feel quickly exhausted; we say the heat is "draining." We also know that this is not a permanent state: If we experience heat day after day, we feel more and more able to work in the heat.

Figure 1.7 shows that these impressions are not merely subjective illusions. Twenty-four physically fit young men who lacked recent experience with hot weather were asked to walk at a fixed pace in hot, relatively dry air. Their endurance was measured as a way of quantifying their physiological ability to sustain moderate exercise under the hot conditions. None of the men had sufficient endurance to walk for 100 minutes on the first day. However, as the days passed and the men had more and more experience with hot conditions, their endurance increased, as indicated by a steady increase in the number of men who could keep walking for 100 minutes.

From research on the physiology of human work under hot conditions, physiologists know that endurance under hot conditions changes because as people have increased experience with heat, their rates of sweat secretion increase, their sweat glands are able to maintain high rates of sweat secretion for dramatically lengthened periods of time, their sweat becomes more dilute (so they lose less salt), the blood flow to their skin becomes more vigorous (permitting improved delivery of internal heat to the body surface), and their heart rates during exercise in the heat become lower. Thus human physiology is restructured in many ways by repeated exposure to heat. For a person who has been living in cool conditions, the *acute* physiological responses to heat exposure are low exercise endurance, a low rate of sweat production, and so forth. Heat training poises a person to express *chronic* physiological responses to heat, such as high exercise endurance and a high capacity to sweat.

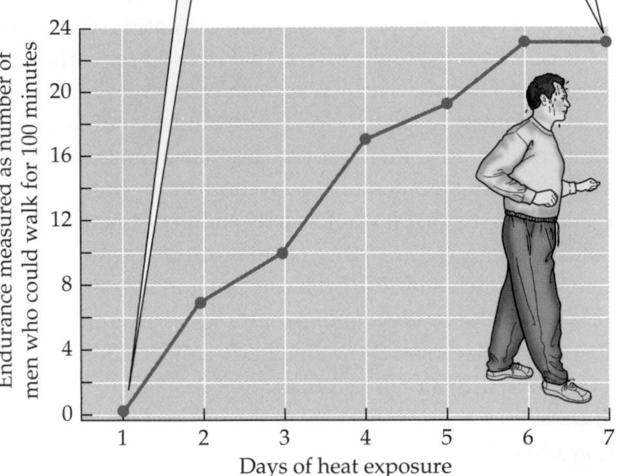

Figure 1.7 Heat acclimation in humans as measured by exercise endurance Twenty-four fit young men without recent heat experience were asked to walk at 3.5 miles per hour in hot, dry air (49°C, 20% relative humidity). Their endurance was used as a measure of their physiological capability to engage in moderate work under hot conditions. The acclimation illustrated by the chronic response is reversible; if heat-acclimated men return to a life of no heat exposure, they gradually revert to the endurance evident on day 1. (After Pandolf and Young 1992.)

The acute and chronic responses are, by definition, phenotypic responses of *individual* animals to environmental change. *Populations* may exhibit a third category of response to environmental change: **evolutionary responses** involving changes of genotypes. Collectively, therefore, animals display responses to environmental change in three time frames:

1. Individuals exhibit immediate, *acute* responses.
2. Individuals exhibit long-term, *chronic* responses. The length of time that an individual must be exposed to a new environment for chronic responses to be fully expressed is usually a few days to a few weeks.
3. Populations exhibit *evolutionary* responses.

Chronic responses by individual animals to environmental change are so common, diverse, and important that their study forms a special discipline with its own terminology. For many physiologists, the concepts of **acclimation** and **acclimatization** provide an important way to classify the chronic responses of individuals to environmental change. A chronic response to a changed environment is called *acclimation* if the new environment differs from the preceding environment in just a few highly defined ways.[1] Acclimation is thus a laboratory phenomenon. *Acclimatization* is a chronic response of individuals to a changed environment when the new and

[1]Some authors restrict use of the word *acclimation* to cases in which just one property differs between environments.

BOX 1.2 The Evolution of Phenotypic Plasticity

When animals express different genetically controlled phenotypes in different environments—when they acclimate and acclimatize—they require controls that determine which particular phenotypes are expressed in which particular environments. As an illustration, suppose that an animal has four possible phenotypes, *P1* through *P4*, and there are four environments, *E1* through *E4*. One option is that the individual could express phenotype *P1* in environment *E3*, *P2* in *E4*, *P3* in *E1*, and *P4* in *E2*. This set of *correspondences* between phenotypes and environments constitutes the individual's **norm of reaction**; that is, if we think of the phenotypes as one list and the environments as a second list in a matching game, the *norm of reaction* is like the set of lines that we would draw between items on the two lists to indicate which item on one matches which on the other.

Because it is genetically determined, *the norm of reaction itself can evolve and is subject to natural selection*. To see this, suppose that an individual other than the one just discussed expresses phenotype *P1* in environment *E1*, *P2* in *E2*, *P3* in *E3*, and *P4* in *E4*. In this case, the two individuals would differ in their norms of reaction. Suppose, now, that there is a population living in a variable environment, composed half of individuals with the first reaction norm and half of individuals with the second. If individuals of the first sort were to survive and reproduce more successfully as the environment varied, natural selection for the first reaction norm would occur. In this way the reaction norm itself would evolve in ways that would better adapt the animals to the variable environment in which they live.

A simple example is provided by tanning in people with light complexions. Suppose that there are two possible integumentary phenotypes: *high melanin* and *low melanin*. Suppose also that there are two environments: *high sun* and *low sun*. One possible norm of reaction would be to express high melanin in low sun and low melanin in high sun. Another norm of reaction would be to express high melanin in high sun and low melanin in low sun. If both of these reaction norms once existed, it is easy to understand why individuals with the second reaction norm would have left more progeny than those with the first, leading to the sort of reaction norm we see today among people with light complexions. Thus, phenotypic plasticity *itself* can evolve, and norms of reactions can themselves be adaptations.

old environments are different *natural* environments that can differ in numerous ways, such as winter and summer, or low and high altitudes. Thus animals are said to *acclimatize* to winter, but they *acclimate* to different defined temperatures in a laboratory experiment.

Acclimation and acclimatization are types of **phenotypic plasticity**, the ability of an individual animal to express two or more genetically controlled phenotypes. Phenotypic plasticity is possible because an individual invariably possesses the genetic code to adopt multiple phenotypes (Box 1.2). Increasing size of the biceps during weight training provides a simple example of a change in phenotype under control of genetically coded mechanisms, but there are countless other, less obvious examples. For instance, an adult individual's suite of active enzymes may change from one time to another because the genes for one suite of enzymes are expressed under certain environmental conditions, whereas the genes for another suite are expressed under different conditions.

PHYSIOLOGY UNDERGOES INTERNALLY PROGRAMMED CHANGES IN TWO TIME FRAMES The physiological properties of individuals sometimes change even if their external environment stays constant. As mentioned earlier, for instance, the hemoglobin in your blood today is different from the hemoglobin you produced as a newborn. This change in hemoglobin is internally programmed, occurring even if your external environment stays constant. Sometimes internally programmed changes interact with environmental changes. For instance, an internally programmed change might occur sooner, or to a greater amplitude, in one environment than in another. However, the internally programmed changes do not require any sort of environmental activation. There are two principal types of internally programmed change: developmental changes and changes controlled by periodic biological clocks.

Development is the progression of life stages from conception to senescence in an individual. Different genes are internally programmed to be expressed at different stages of development, giving rise to **developmental changes of an animal's phenotype.** Metamorphosing species provide the most dramatic examples of internally programmed developmental change, as when frogs lose their gills and develop lungs. Although developmental change is often more subtle in other sorts of animals, it is universal. Humans, for instance, not only change their type of hemoglobin but also alter their sets of hormones and digestive enzymes as they mature from infancy to childhood. Puberty is a particularly dramatic example of internally programmed developmental change in humans. The environment may change the timing of puberty—as when menarche is delayed by malnutrition—but puberty always occurs, no matter what the environment, illustrating that internally programmed changes do not require environmental activation.

Biological clocks are mechanisms that give organisms an internal capability to keep track of the passage of time (see Chapter 10). Most biological clocks resemble wristwatches in being periodic; that is, after they complete one timing cycle, they start another, just as a wristwatch starts to time a new day after it has completed timing of the previous day. These sorts of biological clocks issue signals that cause cells and organs to undergo internally programmed, repeating cycles in their physiological states, thereby giving rise to **periodic, clock-controlled changes in an animal's phenotype.** An enzyme under control of a biological clock, for instance, might increase in concentration each morning and decrease each evening, not because the animal is responding to changes in its outside environment, but because the internal timekeeping mechanisms program a daily oscillation in the enzyme concentration. The changes in enzyme concentration might mean that the animal is inherently better able to digest a certain type of food at one time of day than another, or is better able to destroy a certain type of toxin in the morning than in the evening. Biological clocks typically synchronize themselves with the external environment, but they go through their timing cycles inherently, and they can time physiological changes for days on end without environmental input.

Size in the lives of animals: Body size is one of an animal's most important traits

How big is it? is one of the most consequential questions you can ask about any animal. This is true because within sets of related species, many traits vary in regular ways with their body sizes. The length of gestation, for example, is a regular function of body size in mammals (Figure 1.8). Brain size, heart rate, the rate of energy use, the age of sexual maturity, and hundreds of other physiological and morphological traits are also known to vary in systematic ways with body size in mammals and other phylogenetically related sets of animal species. The study of these relations is known as the study of **scaling** because related species of large and small size can be viewed as *scaled up* and *scaled down* versions of their type.

Knowledge of the statistical relationship between a trait and body size is essential for identifying specializations and adaptations of particular species. To illustrate, let's ask if two particular African antelopes, the bushbuck and mountain reedbuck, have specialized or ordinary lengths of gestation. Answering this question is complicated precisely because there is no single norm of mammalian gestation length to use to decide; instead, because the length of gestation is a regular function of body size, a biologist needs to consider the sizes of the species to know what is average or ordinary. The line in Figure 1.8 is the statistically best-fitted line through all the data points relating gestation length to body size (see Appendix E: Fitting Lines to Data). Thus the line shows the length of gestation *expected* of an ordinary species at each body size. Notice that the length of gestation in the bushbuck is very close to what the line predicts for an animal of its size; the bushbuck therefore has an ordinary gestation length when its size is taken into account. The mountain reedbuck, however, is far off the line. According to the line, as shown in Table 1.3, an animal of the reedbuck's size is expected to have a gestation lasting 26.5 weeks, but actually the reedbuck's gestation lasts 32 weeks. Thus the reedbuck has evolved a specialized, exceptionally long gestation. Similarly, the gray duiker has evolved an exceptionally short length of gestation for *its* size (see Figure 1.8).

Body-size relations are important for analyzing almost all sorts of questions in the study of physiology, ecology, and evolutionary biology. If all one knows about an animal species is its body size, one can usually make useful predictions about many of the species' physiological and morphological traits by consulting known statistical relationships between the traits and size. Conversely, there is always the chance that a species is specialized in certain ways, and as soon as one has actual data on the species, one can identify potential specializations by the type of scaling analysis we have discussed.

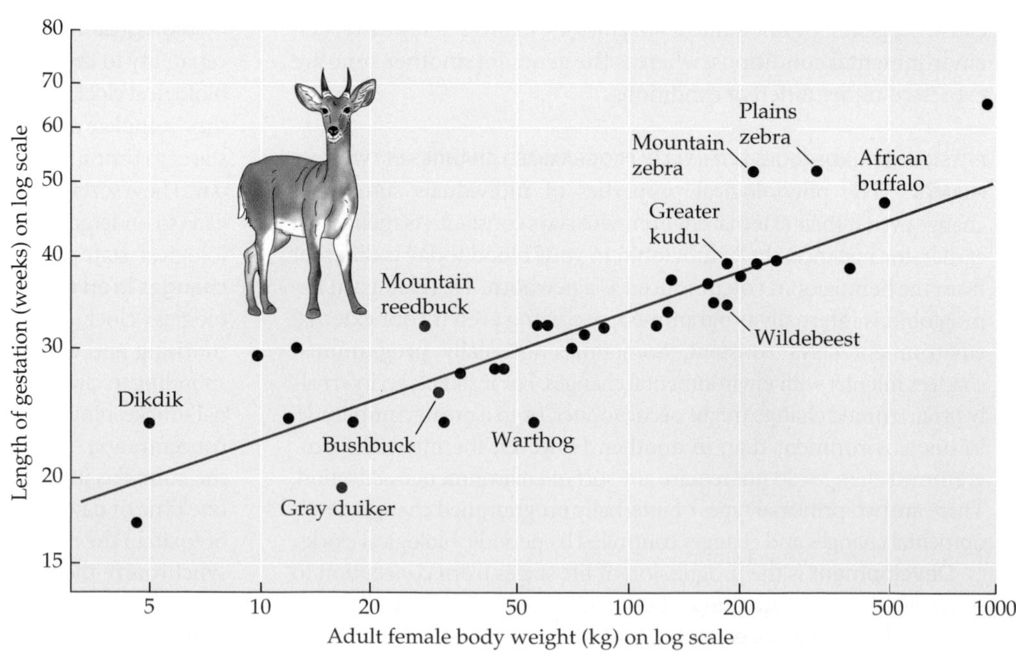

Figure 1.8 Length of gestation scales as a regular function of body size in mammals The data points—each representing a different species—are for African herbivorous mammals weighing 5 to 1000 kg as adults. The line is the best description of the overall trend based on statistical procedures. Both axes use logarithmic scales, which is why the numbers are not evenly spaced. (After Owen-Smith 1988.)

Environments

What is an environment? An important starting point in answering this question is to recognize that an *animal* and its *environment* are interrelated, not independent, entities. They are in fact defined in terms of each other, as noted over 100 years ago by none other than Claude Bernard. The environment in any particular case cannot be specified until the animal is specified. A dog, for instance, is an animal from our usual perspective, but if the animal of interest is a tapeworm in the dog's gut, then the dog is the environment. All animals, in fact, are parts of the environments of other animals. The birds in the trees around your home are part of your environment, and you are part of theirs. The interdependence of animal and environment is reflected in standard dictionary definitions. A dictionary defines an animal to be a living organism. An **environment** is defined to be all the chemical, physical, and biotic components of *an organism's* surroundings.

Earth's major physical and chemical environments

The physical and chemical environments on our planet are remarkably diverse in their features, providing life with countless challenges and opportunities for environmental specialization. Temperature, oxygen, and water are the "big three" in the set of physical and chemical conditions that set the stage for life. In this part we discuss their ranges of variation across the face of the globe and some highlights of how animals relate to these features.

TEMPERATURE The **temperature** of the air, water, or any other material is a measure of the intensity of the random motions that the atoms and molecules in the material undergo. All atoms and mole-

TABLE 1.3 Predicted and actual gestation lengths for two African antelopes of about the same body size

Species	Predicted gestation length (weeks)[a]	Actual gestation length (weeks)
Bushbuck (*Tragelaphus scriptus*)	27	26
Mountain reedbuck (*Redunca fulvorufula*)	26.5	32

[a] Predicted lengths are from the statistically fitted line shown in Figure 1.8.

cules ceaselessly move at random on an atomic-molecular scale. A high temperature signifies that the intensity of this atomic-molecular agitation is high; a low temperature signifies that the intensity of the random motions is low. Although some animals are temperature regulators and (at a cost of energy) keep their tissue temperatures relatively constant whether they experience high or low environmental temperatures, most animals are temperature conformers. As we discuss temperature here, the conformers are our principal interest because the level of atomic-molecular agitation in their tissues matches the level in the environments where they live.

The lowest temperature inhabited by active communities of macroscopic temperature-conforming animals is –1.86°C, in the polar seas. The very lowest temperature at which *any* active communities of temperature-conforming animals live occurs *within* the sea ice near the poles; minute nematodes and crustaceans, as well as algae, live and reproduce within the sea ice at temperatures that are not well defined but are a few degrees colder than in the surrounding water. The macroscopic fish and other animals living in the open waters of the unfrozen polar oceans (Figure 1.9) are far better known than the sea-ice animals, however. The open waters of the polar oceans remain perpetually within a few hundredths of a degree centigrade of –1.86°C, the lowest temperature at which seawater is liquid. Thus the fish, the shrimplike krill, the sea stars, and the other invertebrates of these oceans have tissue temperatures near –1.9°C from the moment they are conceived until they die. They do not freeze, either because their normal freezing points are similar to the freezing point of seawater,[2] or because they have special protections against freezing. Because they are very cold, one might imagine they live in some sort of suspended animation. Quite the contrary: The communities of temperature-conforming animals in the polar seas are active and thriving. In the ocean around Antarctica, for example, a sure sign of the vigor of the populations of fish, krill, and other invertebrates is that they reproduce and grow prolifically enough to meet the food needs of the famous Antarctic abundance of whales, seals, and penguins.

Figure 1.9 Fish in the sea around Antarctica spend their entire lives at body temperatures near –1.9°C These rock cod (*Pagothenia borchgrevinki*) lay their eggs in holes or depressions in ice sheets. They hatch, grow, feed, and mate at body temperatures near –1.9°C. Metabolically synthesized antifreeze compounds keep them from freezing.

Are the low tissue temperatures of polar fish and invertebrates really a challenge to them, or do those temperatures merely *seem* challenging to us because they make us cringe? One way to obtain an answer is to compare polar species with related nonpolar species. Tropical species of fish clearly find low temperatures to be challenging. Many, in fact, die if cooled to +6°C, even if they are cooled very gradually. Observations like this emphasize that the polar species have had to evolve special adaptations to thrive with their tissues perpetually at –1.86°C. The polar species themselves often die if they are warmed to +6°C, indicating that the tropical species also have adaptations to live at tropical temperatures. The evolutionary physiological divergence of these fish is dramatized by the fact that a single temperature can be lethally *cold* for the tropical species and lethally *warm* for the polar species.

Far greater extremes of cold are found on land than in aquatic environments. In Antarctica, the air temperature can drop to –90°C (–130°F); in the Arctic, it can descend to –70°C (–90°F). The extremes of animal adaptation to low tissue temperature are represented by certain extraordinary species of Arctic insects that spend the winter inside exposed plant stems or on the surface of pack ice. The insects are quiescent, not active, at these times. Nonetheless, it is impressive that some endure tissue temperatures of –60°C to –70°C, either in a frozen state (which they have adaptations to tolerate) or in an unfrozen supercooled state. The diversity of terrestrial animal life declines as latitude increases from the temperate zone toward the poles, indicating that the very cold terrestrial environments are demanding places for animals to occupy, despite evolutionary adaptability (Figure 1.10).

The temperature of the air or water on Earth usually does not go higher than +50°C (+120°F). Animals on land may experience even higher heat loads, however, by being exposed simultaneously to hot air and the sun's radiation. Some temperature-conforming animals from hot environments—such as certain desert insects and lizards—can function at *tissue* temperatures of 45°C to 52°C (Figure 1.11).[3] These are the highest tissue temperatures known for

[2] Dissolved salts and other dissolved compounds lower the freezing points of solutions. Most marine invertebrates have total concentrations of dissolved matter in their blood similar to the concentration in seawater. Consequently, their blood freezing points are about the same as the freezing point of seawater, and they do not freeze, provided that the seawater remains unfrozen.

[3] Normal human body temperature is 37°C.

Figure 1.10 Butterfly biogeography The diagram shows the number of species of swallowtail butterflies (family Papilionidae) at various latitudes. The reason there are relatively few species of animals at high latitudes may not be simply the low temperatures there, but may in part be a relay effect from the effects of cold on plants. Plants decrease in diversity and annual productivity toward the poles, affecting the food supplies of animals. (After Scriber 1973.)

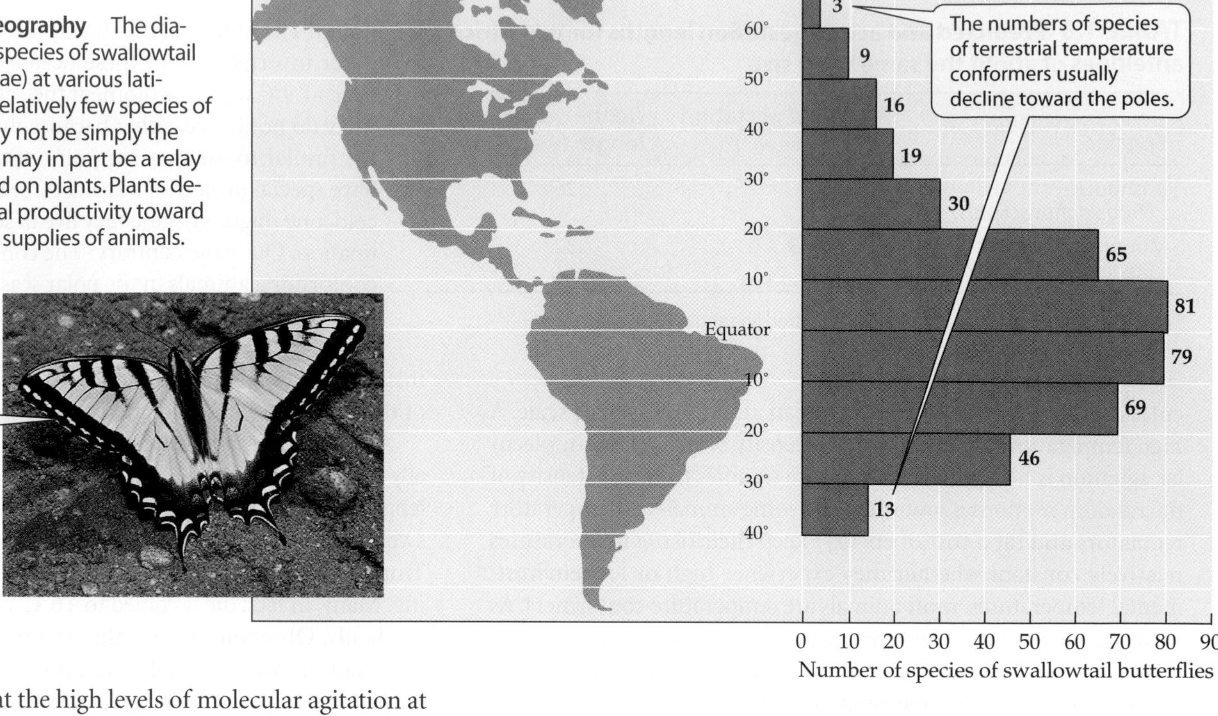

animal life, suggesting that the high levels of molecular agitation at such temperatures pose the greatest challenge that can be met by evolutionary adaptation in animal systems.

The hottest places in the biosphere are the waters of geothermally heated hot springs and underwater hot vents. These waters are often far above the boiling point when they exit Earth's crust. Although aquatic animals typically stay where the waters have cooled to 35°C to 45°C or lower, many prokaryotic microbes—bacteria and archaea—thrive at much higher temperatures than animals can. Some prokaryotes even reproduce at temperatures above 100°C.

Figure 1.11 A thermophilic ("heat-loving") lizard common in North American deserts The desert iguana (*Dipsosaurus dorsalis*) can often be seen abroad as the sun beats down on hot days. Although it does not usually expose itself to body temperatures higher than 42°C, it can survive 48.5°C, one of the highest body temperatures tolerated by any animal.

OXYGEN The need of most animals for oxygen (O_2) is a consequence of their need for metabolic energy. The chemical reactions that animals use to release energy from organic compounds remove some of the hydrogen atoms from the compounds. Each adult person, for example, liberates about half a pound of hydrogen every day in the process of breaking down food molecules to obtain energy. Hydrogen liberated in this way cannot be allowed to accumulate in any animal's cells. Thus an animal must possess biochemical mechanisms for combining the hydrogen with something, and O_2 is the usual recipient. O_2 obtained from the environment is delivered to each cell, where it reacts with the free hydrogen produced in the cell, yielding water.

The suitability of an environment for animals often depends on the availability of O_2. In terrestrial environments at low and moderate altitudes, the open air is a rich source of O_2. Air consists of 21% O_2, and at such altitudes it is relatively dense because it is at relatively high pressure. Thus, animals living in the open air typically have a plentiful O_2 resource. Even within burrows or other secluded places on land, O_2 is often relatively freely available because—as counterintuitive as it may sound—O_2 diffuses fairly readily from the open atmosphere through soil to reach burrow cavities, provided the soil structure includes gas-filled spaces surrounding the soil particles.

High altitudes, like the polar land masses, are among Earth's most challenging places, where the numbers of animal species are sharply reduced. Although high altitudes present animals with several stresses, their single greatest challenge is the decline in O_2 concentration with increasing elevation. Air at the top of Mount Everest—8850 m above sea level—is 21% O_2, like that at sea level; but the total air

Figure 1.12 Performance in an oxygen-poor environment Because of the difficulty of acquiring O_2 from rarefied air, the rate at which energy can be released from food molecules for use in work by humans is reduced at high altitudes, and the simple act of walking uphill becomes extremely arduous. Well-conditioned mountaineers are slowed to a walking rate of 100 to 200 m per hour near the tops of the world's highest mountains if they are breathing from the air rather than from oxygen tanks.

pressure is much lower, and gas molecules within the air are so widely spaced that each liter of air contains only about a third as much O_2 as at sea level.

The maximum rate at which animals can acquire O_2 at high altitudes is often much lower than the rate possible at sea level, and functions are consequently limited. At elevations above 6500 m, for example, humans breathing from the atmosphere are challenged just to walk uphill because of the limitations on their level of exertion imposed by the low availability of O_2 (Figure 1.12). Some animal species have evolved adaptations to succeed in the dilute O_2 of rarefied air in ways that humans cannot. One of the most remarkable species is the bar-headed goose (*Anser indicus*), which—in ways that physiologists still do not fully comprehend—is able to fly (without an oxygen mask!) over the crests of the Himalayas at 9000 m.

Water-breathing animals typically face a substantially greater challenge to obtain O_2 than air-breathing animals do because the supply of O_2 for water breathers is the O_2 dissolved in water, and the solubility of O_2 in water is not high. Because of the low solubility of O_2, water contains much less O_2 per liter than air does even when the water is fully aerated. For example, aerated stream or river water at sea level contains only 3% to 5% as much O_2 per liter as air at sea level does.

A common problem for animals living in slow-moving bodies of water such as lakes, ponds, or marshes is that the O_2 concentration may be even lower than in aerated water because dissolved O_2 may become locally depleted by the metabolic activities of animals or microbes. Density layering of water—which prevents the water from circulating freely—is a frequent contributing factor to O_2 depletion in the deep waters of lakes and ponds. Density layering occurs when low-density water floats on top of high-density water, causing distinct water layers to form. When this happens, there is often almost no mixing of oxygenated water from the low-density surface layer (where photosynthesis and aeration occur) into the high-density bottom layer. Thus, O_2 in the bottom layer is not readily replaced when it is used, and as microbes and animals in the bottom layer consume O_2, the O_2 concentration in the bottom waters may fall very low.

Differences in either temperature or salinity (or both) can cause density layering of water masses. In lakes during summer, density layering occurs because of temperature effects; sun-heated warm water tends to float on top of colder and more dense bottom water. The lake studied by a group of university students in Figure 1.13 provides an example of this sort of density layering. The bottom waters of this lake contained essentially *no* dissolved O_2 on the July day when the data were collected. Deep-water O_2 depletion has become more common in recent decades in lakes, ponds, and estuaries as human populations have enriched waters with organic matter. The organic matter supports the growth of microbes that deplete dissolved O_2. For deep-water animals to survive, either they must be able to tolerate low O_2 levels, or they must be capable of living elsewhere until O_2 is resupplied to the deep waters.

Animals have faced the challenge of low O_2 concentrations in certain sorts of water bodies from time immemorial. Unlike animals confronted with new, human-induced O_2 depletion, the animals living in primordially O_2-poor waters have been able to undergo long-term evolutionary adaptation to low-O_2 conditions. Examples of such bodies of water are many tropical rivers that are naturally very rich in organic matter, as in the Amazon basin. The warmth of these rivers lowers O_2 solubility and promotes rapid multiplication of microbes that use O_2. In addition, thick forest canopies may create deep shade over the rivers, impeding algal photosynthesis that otherwise could replenish O_2. Tropical rivers of this sort and their associated wetlands have experienced low levels of dissolved O_2 over eons of geological time.

Among the animals living in such waters, the evolution of air breathing is one of the most remarkable features. Hundreds of species of fish in these environments are air breathers, for example. Some take up inhaled O_2 across well-vascularized mouth linings or lunglike structures. Others swallow air and absorb O_2 in their stomachs or intestines. In one of nature's twists of fate, physiologists have discovered that it can be important for these fish to restrain gas exchange across their gills because the O_2 they get from breathing air can escape across their gills into the O_2-depleted waters they inhabit.

For animals confronted with challenges of short-term or long-term O_2 deficiency, whether in O_2-depleted freshwater environments or elsewhere, a potential solution over evolutionary time is to adopt a biochemistry that can attach hydrogen to molecules other than O_2. Many species—both air breathers and water breathers—have temporary options of this sort. Certain tissues in our own bodies, for example, can live without O_2 for 10 minutes at a time, and in fact most animals have certain tissues that can survive O_2 deprivation for minutes or tens of minutes. Suppose, however, that an animal's entire body must live without O_2 for many hours, days, weeks, or months. Doing so is possible, but as the period without O_2 lengthens, ever-fewer species have evolved biochemical specializations that enable

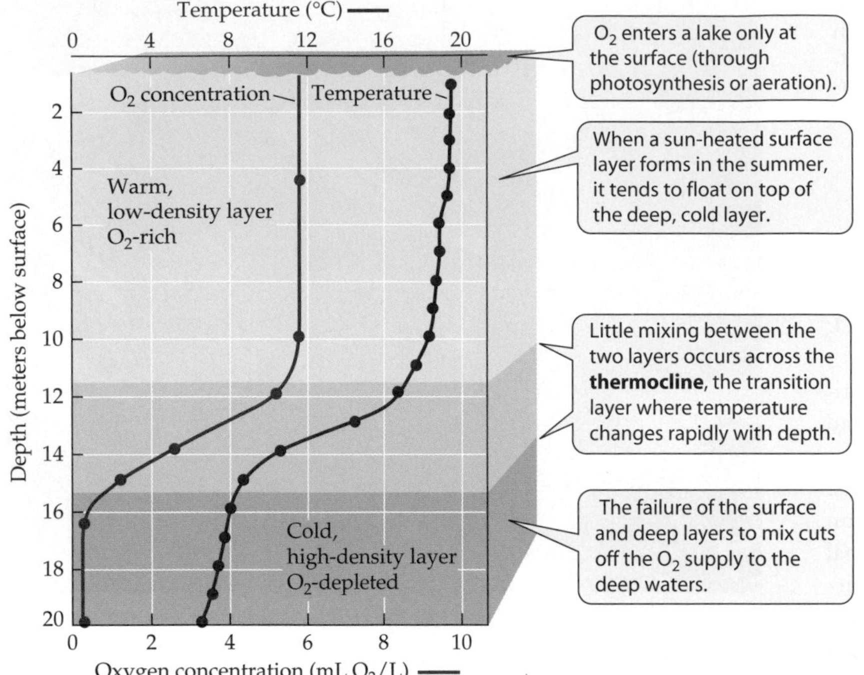

Figure 1.13 Density layering can cut off the oxygen supply to the deep waters of a lake Different densities of water do not mix readily. The O_2 concentration in the deep waters may fall to near zero because the animals and microbes living there consume O_2 that is not replaced. (From data gathered by a group of animal physiology students on a lake in northern Michigan in July.)

them to survive. Some exceptional animals are able to meet the ultimate challenge of living indefinitely in O_2-free environments. Intestinal parasites, such as nematodes and tapeworms, that live their entire adult lives in the O_2-free environment of the vertebrate gut cavity are the best-known examples.

WATER Water is the universal solvent in biological systems. Thus, it is required for the proper composition of the blood and all other body fluids. It is also important for animals because H_2O bound to proteins and other macromolecules as *water of hydration* is often required for the macromolecules to maintain their proper chemical and functional properties.

Animal life began in Earth's greatest watery environment, the oceans. Most invertebrates that live in the open oceans today—sea stars, corals, clams, lobsters, and so forth—are thought to trace a continuously marine ancestry. That is, their ancestors never left the seas from the time that animal life began, and thus the salinity of the oceans has been a perennial feature of their environment. The blood of each of these invertebrates, although differing a bit from seawater in composition, is similar to seawater in its total salt concentration (Figure 1.14). These animals therefore do not tend much to gain H_2O from their environment by osmosis, nor do they tend to lose H_2O from their blood to the seawater. Because this situation is almost universal among these animals, we believe it is the primordial condition of animal life. Thus we believe that for much of its evolutionary history, animal life lived in a setting where H_2O was abundant in the environment and little danger existed for an animal to be either dehydrated or overhydrated.

This benign situation was left behind by the animals that migrated from the oceans into rivers during their evolution. Freshwater has a very low salinity compared to seawater. When animals from the oceans, with their salty blood, started to colonize freshwater, they experienced a severe challenge: H_2O tended to seep osmotically into their bodies and flood their tissues because osmosis carries H_2O relentlessly from a dilute solution into a more-concentrated one. Today there are freshwater fish, clams, crayfish, sponges, hydras, and so forth, all descended from ocean ancestors. The process of invading freshwater from the sea is probably still continuing, in fact; some freshwater crabs, for example, are thought to be geologically recent immigrants to freshwater because they must return to the sea to breed. Over evolutionary time, freshwater animals have reduced their tendency to gain H_2O osmotically from their environment, but they have not eliminated it. A 100-g goldfish, for example, osmotically gains enough water to equal 30% or more of its body weight every day.

Vertebrates and several groups of invertebrates invaded the land from freshwater, and in so doing they confronted the most severe of all the water challenges on Earth. On land, evaporation of water into the atmosphere tends to dehydrate animals rapidly, and some terrestrial habitats are so dry that replacing lost H_2O borders on impossible.

When animals first invaded terrestrial habitats, they probably possessed integuments (body coverings), inherited from aquatic ancestors, that provided little or no barrier to H_2O loss. Water often evapo-

Figure 1.14 Marine invertebrates such as sea stars and corals have body fluids similar to seawater in their total concentration of salts, although marine bony fish are more-dilute than seawater Most scientists believe that the reason for the difference between the invertebrates and fish is evolutionary history. The invertebrates' ancestors always lived in the sea, but the fish's did not.

rates across these sorts of integuments as fast as it evaporates from an open water surface! Thus animals with such integuments can dehydrate to death in just a few hours if they are exposed to dry, warm air. Leopard frogs, wood lice, and earthworms are examples of modern-day animals that have integuments permitting free evaporation of water. Although living on land, they cannot possibly live steadily in the open air. Instead, they must stay in protected places where the humidity of the air is high, or if they venture into the open air, they must return often to places where they can rehydrate. Thus the danger of dehydration severely constrains their freedom of action.

For a terrestrial animal to lead a fully exposed existence in the open air, it must possess highly effective barriers to H_2O loss across its integument—barriers of a sort that never existed when animal life was entirely aquatic. Only a few major groups of animals have evolved such novel water barriers: reptiles (see Figure 1.11), mammals, birds, insects, and spiders. In each of these groups, excessive evaporation is prevented by integumentary lipids: thin layers of greasy or waxy hydrocarbons in the integument. These hydrocarbons liberate the animals to move freely about in the open.

Some terrestrial animals have adapted to land in part by evolving exceptional tolerance of dehydration. Although most terrestrial animals die if they lose half or more of their body H_2O without replacing it, the exceptional types can dehydrate more. The most extreme cases are certain invertebrates that can lose essentially all their body H_2O and survive in a dormant, crystalline state until H_2O returns. Certain tardigrades, for example, dry completely when deprived of water and then can blow about like dust, all the time ready to spring back to active life if water becomes available.

Contrary to what intuition might suggest, even some aquatic animals are threatened with dehydration. The bony fish of the oceans, such as the reef fish in Figure 1.14, are the most important example. These fish have blood that is only about half as salty as seawater, probably because they are descended from freshwater ancestors rather than from ancestors that lived always in the sea. The ocean is a desiccating environment for animals with dilute blood because H_2O is carried by osmosis from the blood to the more-concentrated seawater. These desiccating fish have an advantage over terrestrial animals desiccating in a desert—namely, that H_2O to replace their losses is abundant in their watery environment. To incorporate H_2O from seawater into their dilute bodies, however, they must in essence possess mechanisms to "distill" the seawater; that is, they must be able to separate H_2O from the salty seawater solution.

The environment an animal occupies is often a microenvironment or microclimate

In a forest, lake, or any other large system, small places inevitably exist where physical or chemical conditions are significantly different from the average in the system. For instance, when the average temperature of the open air in a woodland is 30°C, the temperature under a pile of brush on the forest floor might be 24°C. Although large-bodied animals are often, by necessity, exposed to the statistically average conditions where they live, small-bodied animals may enter the small places—the nooks and crannies—where they may find conditions that are far from average. Places within an environment that potentially differ from the environment at large in their physical or chemical conditions are called **microenvironments.** A related concept is that of **microclimates.** A microclimate is the set of climatic conditions (temperature, humidity, wind speed, and so forth) prevailing in a subpart of a system.

Because we humans are large organisms, our perception of the prevailing conditions in a place may bear little relation to the microclimates that smaller creatures find by entering distinctive subparts of the place. George Bartholomew (b. 1919), one of the founders of environmental physiology, expressed this important point well:

> Most vertebrates are much less than a hundredth of the size of man . . . , and the universe of these small creatures is one of cracks and crevices, holes in logs, dense underbrush, tunnels and nests—a world where distances are measured in yards rather than miles and where the difference between sunshine and shadow may be the difference between life and death. Climate in the usual sense of the word is, therefore, little more than a crude index to the physical conditions in which most terrestrial animals live.[4]

Desert environments nicely illustrate the point that Bartholomew makes (Figure 1.15). At head level (about 2 m aboveground), a human or a horse standing in the Arizona desert may experience daytime air temperatures that reach almost 50°C during the summer, combined with intense solar radiation. Humans and horses typically have no choice but to cope physiologically with these conditions be-

[4] From Bartholomew, G. A. 1964. The roles of physiology and behavior in the maintenance of homeostasis in the desert environment. *Symp. Soc. Exp. Biol.*, no. 18: 7–29.

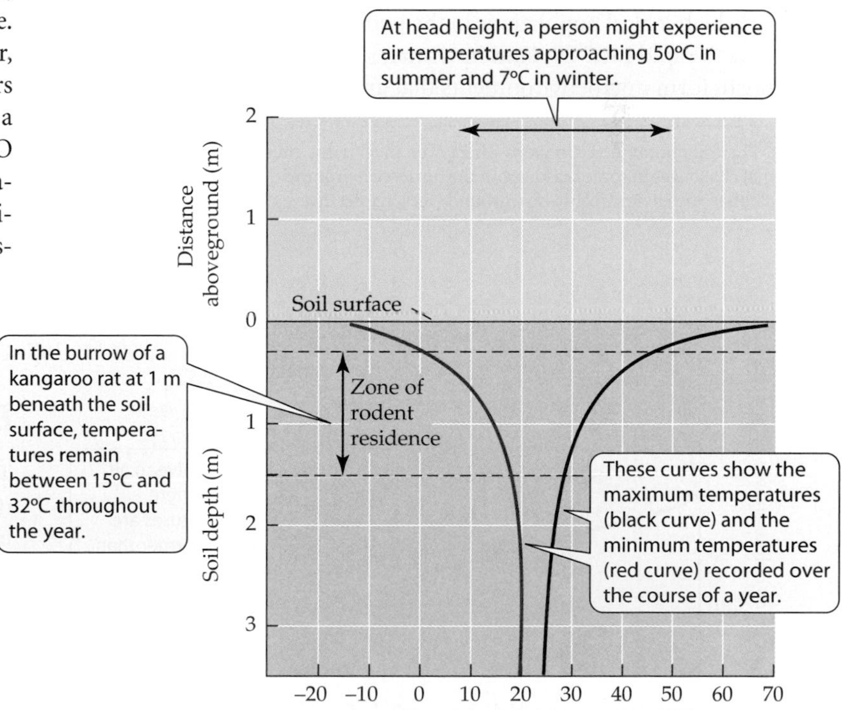

Figure 1.15 Microenvironments in a desert Annual range of temperatures in the soil and air and at the soil surface at a spot in the Arizona desert near Tucson. (After Misonne 1959.)

cause they are too large to escape by going underground or squeezing into patches of shade cast by cacti or desert bushes. Small desert rodents such as kangaroo rats and pocket mice are in a very different situation, however, because they can burrow deep into the soil, where thermal conditions are far different from those that humans associate with deserts. On the surface of the desert soil, the annual temperature range is actually greater than in the air above (see Figure 1.15); the soil surface becomes hotter than the air during the day as it absorbs solar radiation, and it becomes cooler than the air at night because it radiates infrared energy to the cold nighttime sky (see page 197). Beneath the soil surface, however, the annual range of temperature decreases dramatically with depth. At a depth of 1 m, the temperature remains well below the maximum aboveground air temperature during summer and well above the minimum air temperature during winter. In fact, in certain desert regions, such as that shown in Figure 1.15, the rodents never face significant heat or cold stress throughout the year when they are in their burrows![5]

Microenvironments must be considered in the study of virtually all the physical and chemical features of the places where animals live. In tall grass, for example, the wind speed is likely to be lower than in adjacent open areas, and because the weak wind permits moisture evaporating from the soil and grass to accumulate in the air rather than being blown away, the humidity also tends to be higher than in adjacent open air. Animals that enter tall grass are thus likely to find a less-desiccating microenvironment than in open fields nearby. Along a river, any side pool cut off from free mixing with the flowing water is likely to form a distinctive microenvironment; for instance, the concentration of dissolved O_2 in such a pool might be lower than that in the river at large if there is a lot of microbial decay occurring in the pool, but on the other hand, the O_2 concentration might be particularly high if algal populations in the pool photosynthetically produce O_2 at a rapid rate. On land in winter, spaces under snow in the far north form microenvironments that are not only windless but may also be 20°C (or more) warmer than the open air above (Figure 1.16). Thus a lemming burrowing under the snow experiences a very different environment from a reindeer standing above.

[5] There are hotter desert regions where even the burrow environment presents thermal challenges in some seasons, but the burrow environment is still far more moderate than the environment aboveground (see Chapter 26).

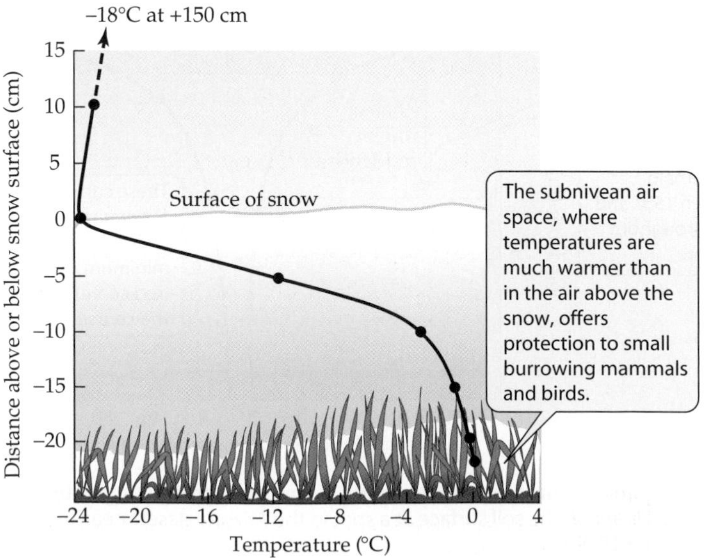

Figure 1.16 Winter microenvironments in the far north When lemmings, ptarmigans, or other small mammals or birds burrow under deep snow into the air space trapped beneath the snow (the subnivean air space), they enter a windless environment where the temperature may be 20°C warmer than in the air above the snow, and where they are protected by the overlying snow from losing heat radiatively to the frigid nighttime sky. The temperatures shown were measured on a March night in Sweden. (After Coulianos and Johnels 1963.)

Animals often modify their own environments

An important way in which *animal* and *environment* are interdependent is that animals modify their own environments. In the simplest case, animals behaviorally select the actual environments that they experience from the many that *could* be experienced. A frog stressed by desiccation on open terrain, for example, can raise the humidity and lower the wind speed of its environment by hopping into tall grass. On a larger scale, animal migrations typically provide new environments, as when birds from the Arctic enjoy balmy temperatures in winter by migrating to tropical latitudes. The environment of an animal is the animal's surroundings, and the surroundings depend on where the animal places itself.

A more subtle but equally important point is that the very presence of an animal in a place often alters the physical and chemical features of the place. The environmental alteration caused by the animal may then change that selfsame animal's behavior or physiological processes. Consider, for instance, a squirrel in a small cavity in a tree. In winter, the squirrel will warm the cavity to a higher temperature than would prevail in the squirrel's absence. The squirrel will then respond physiologically to the elevated temperature in the cavity, by cutting back its rate of metabolic heat production. In a like manner, a school of fish can deplete water of dissolved O_2 and then must cope with low "environmental" O_2 levels.

Because of phenomena like these, the analysis of an animal–environment interaction often requires dynamic calculations that take into account that the interaction is of a two-way, back-and-forth sort. After an animal has initially altered an environment, the animal may function differently because it is in a changed environment, and thus the animal's future effect on the environment may be different from its original effect.

Global warming represents a planet-sized example of this phenomenon. Most scientists believe that the use of fossil fuels by the global human population is changing Earth's atmosphere toward a composition that increases planetary heat retention, thereby raising the temperatures in which human societies will exist in the future.

Evolutionary Processes

The evolutionary origins of physiological traits form the subject matter of evolutionary physiology, one of the two most important branches of the modern study of physiology, as stressed earlier. Physiologists have long recognized that the traits of species are often well matched to the environments they occupy. For example, polar bears are well suited to deal with cold, and dromedary camels, with heat. Evolution by natural selection is believed by modern biologists to be the primary process that has produced this match between species and the environments they inhabit.

Carefully defined, **evolution** is *a change of gene frequencies over time* in a population of organisms. Suppose a population of animals contains a gene that codes for the oxygen affinity of hemoglobin (the ease with which hemoglobin combines with O_2). The gene has two alleles (alternative forms), one coding for high oxygen affinity and one for low oxygen affinity. At one time in the history of the population, 30% of all copies of this gene were of the high-affinity allele and 70% were of the low-affinity allele. After 1000 generations have passed, however, 60% of all copies of the gene are of the high-affinity allele and 40% are of the low-affinity allele. In this case, gene frequencies have changed. Therefore, evolution has occurred.

A more-complex question by far is whether *adaptation* has occurred. There are several known processes by which gene frequencies can change. Only one, natural selection, leads to adaptation.

Some processes of evolution are adaptive, others are not

Returning to the hypothetical example just discussed, suppose that in a population of animals occupying a particular environment, individuals with hemoglobin of high oxygen affinity are more likely to survive and reproduce than those with hemoglobin of low oxygen affinity. By this very fact, an allele that codes for high affinity will tend to increase in frequency in the population from one generation to the next (and an allele that codes for low affinity will tend to decrease). After many generations, the allele for high oxygen affinity might become so common that essentially all individuals born into the population have it. You will recognize this as the process of *natural selection*. Natural selection creates a better match between animals and their environments.

The concept of **adaptation**, which is intimately related to that of natural selection, has a specific meaning in the study of evolutionary biology. By definition, a trait is an adaptation if it has come to be present at high frequency in a population because it confers a greater probability of survival and successful reproduction in the prevailing environment than available alternative traits. Thus, adaptations are products of the process of natural selection. An adaptation is not necessarily an *optimum* or *ideal* state, because constraints on the freedom of natural selection may have precluded the optimum state from being an option (the optimum state, for instance, might never have arisen through mutation). As the definition stresses, an adaptation is the trait favored by natural selection from among the *available* alternative traits.

Now let us repeat our thought exercise but substitute different assumptions. Consider a population, in the same environment as analyzed before, in which the alleles for hemoglobin of high and low oxygen affinity are both at high frequencies. Suppose that the population experiences a drop in size, so that it contains fewer than 100 individuals. During this low point, a catastrophe strikes, killing individuals *at random*, regardless of whether they possess the high- or low-affinity allele. In a small population of 100 or fewer animals, random deaths could *by sheer chance* eliminate all individuals possessing one of the alleles. If all copies of the allele for high oxygen affinity are eliminated, then when the population later grows in size, it will have only the low-affinity allele, the *less-adaptive* allele. In this case the process of gene frequency change is a process of **nonadaptive evolution**. Because of chance, an allele that provides a lower probability of survival and reproduction than an available alternative comes to be the predominant allele in the population.

Processes in which chance assumes a preeminent role in altering gene frequencies are termed **genetic drift**. We have just described one scenario for genetic drift, that gene frequencies may shift in chance directions because of deaths or other blocks to individual reproduction in populations transiently reduced to small size. Another scenario is that when a species enters a new area and founds a new population there, the new population may exhibit changed gene frequencies from the preexisting population simply because the founding individuals are by chance genetically nonrepresentative of the population from which they came (a so-called *founder effect*).

Students of allele frequencies in natural populations believe that they observe many examples of genetic drift. For example, two populations of mice living 5 miles apart in seemingly identical woodlots usually exhibit many differences in allele frequencies thought to be produced by drift of one sort or another. Often the genes affected by drift seem to be ones that have little or no fitness effect; drift, in other words, seems to have its greatest influence on genes not subject to strong natural selection. That is not always the case, however.

Additional processes are known by which evolution may lead to nonadaptive outcomes:

- A trait may be common in a population simply because it is closely correlated with *another* trait that is favored by natural selection. For instance, a trait that itself diminishes the fitness of animals (their ability to survive and reproduce) may occur because it is coded by a gene that is subject to positive selection because of other, fitness-enhancing effects. The control by an allele of a single gene of two or more distinct and seemingly unrelated traits is called **pleiotropy**. An example is provided by a recently discovered allele of an enzyme-coding gene that simultaneously increases the resistance of mosquitoes (*Culex pipiens*) to organophosphate insecticides while decreasing the physiological tolerance of the mosquitoes to the cold of winter. When a population of mosquitoes is sprayed with insecticides, the population *may evolve toward a diminished physiological ability to survive winter* because of pleiotropy and selection for alleles that confer insecticide resistance.[6]
- A trait may have evolved as an adaptation to an ancient environment, yet persist even though the environment has changed. In the new environment, the trait may no longer be beneficial and thus may not be an adaptation. Traits of this sort are thought to be quite common by some evolutionary biologists because animals often colonize new places, and even if they stay in one place, climates often change radically over relatively short periods of ge-

[6] Besides pleiotropy, traits may evolve in tandem because of *linkage disequilibrium*, in which alleles of two or more genes on the same chromosome tend to be inherited together to a nonrandomly great extent.

ological time.[7] The need of many desert amphibians for pools of water to breed, and the possession of eyes by numerous species of insects and other arthropods that live obligatorily in caves, are two examples of traits that seem to exist today not because they are adaptations to present environments, but because they persist from the past. Similarly, the dilute blood of the bony fish of the oceans is probably a legacy of life in a different environment, not an adaptation to life in seawater.

A trait is not an adaptation merely because it exists

In the decades before 1979, many physiologists referred to *all* traits of organisms as adaptations. Traits were called *adaptations* merely because they existed, and stories (now sometimes recalled as "just-so" stories) were concocted to explain how the traits were beneficial. This habit ignored the possibility of genetic drift and other forms of nonadaptive evolution. The habit, in fact, reduced adaptation to a nonscientific concept because no empirical evidence was required for a trait to be termed an adaptation.

A major shift in the use of the concept of adaptation occurred in 1979, with the publication of a critique by Stephen J. Gould and Richard C. Lewontin. They stressed that natural selection in the present environment is just one of several processes by which a species may come to exhibit a trait. Thus a trait is not an adaptation merely because it exists, and when physiologists call a trait an *adaptation*, they are really making a *hypothesis* that natural selection has occurred. Moreover, just as is true with any other hypothesis, data must be gathered to assess whether the hypothesis is true or false. Because scientists usually cannot actually witness the evolution of a trait, the data that can be gathered are often indirect. Nonetheless, scientists must attempt to gather the soundest evidence possible. With this objective in mind, the study of adaptation is rapidly becoming an empirical (i.e., data-based) science.

Adaptation is studied as an empirical science

Biologists today are giving a great deal of attention to the question of how to obtain *data* that will guide a decision on whether or not a trait is an adaptation. Sometimes the biosphere presents a "natural experiment" that permits scientists to *observe* evolution taking place over multiple generations in a natural setting. Scientists cannot depend entirely on such natural experiments to study adaptation, because the natural experiments are uncommon and may not speak to questions of greatest interest. Nonetheless, a natural experiment may provide particularly useful insights into adaptation because it may allow the adaptiveness of a trait to be judged from all the angles that matter.

Industrial melanism is a famous phenomenon, with which you are likely familiar from your study of general biology, that exemplifies a natural experiment for adaptation. *Melanism* refers to a genetically coded dark body coloration. *Industrial melanism* is an evolutionary increase in the frequency of melanism in a population of animals living in an environment modified by human industries. A species of moth in the industrial regions of England has two genetically determined color states: light and dark. The moths were predominantly light-colored prior to the industrial era, when light-colored lichens covered the tree trunks on which they rested during the day. With increasing industrialization, the lichens on the trees were killed by pollutants, and soot from factories darkened the tree trunks. Within 50 years the moth populations in the industrial areas had become predominantly dark-colored because, from generation to generation, genes for dark coloration increased in frequency. Studies demonstrated that on dark tree trunks, the dark-colored moths were less likely than light-colored ones to be seen by avian predators.

From the *direct observation* of this natural experiment, we can say the following: In an industrialized environment, dark coloration became common in the moth populations by way of natural selection because it increased an individual's likelihood of survival in comparison with the available alternative coloration. Dark coloration thus met all the standards of our formal definition of adaptation and was indisputably an adaptation to the sooty environment.

Usually biologists are not able to observe evolution in action in this way. Thus, to study adaptation empirically, they must adopt other approaches. Several techniques have been developed—or are being developed—to study the question of adaptation when nature fails to provide an ideal natural experiment:

- *The comparative method.* The **comparative method** seeks to identify adaptive traits by comparing how a particular function is carried out by related and unrelated species in similar and dissimilar environments. *This method is based on the premise that although we cannot see evolution occurring in the past, the many kinds of animals alive today provide us with many examples of outcomes of evolution, and patterns we identify in these outcomes may provide insights into processes that occurred long ago.* Figure 1.17 illustrates a simple example of the use of the comparative method. Terrestrial vertebrates have lungs for breathing. If we were to look *only* at terrestrial vertebrates, we would have just that single isolated bit of knowledge about breathing mechanisms. However, if we also examine other unrelated terrestrial organisms, we discover a pattern: In insects, in land snails, *and* in terrestrial vertebrates, the breathing surfaces are parts of *invaginated* structures that hold the air rather than projecting into the air. This pattern is striking because *evaginated* breathing surfaces, which project into the water, are nearly universal among aquatic animals (note the gills of fish or crayfish). The occurrence of invaginated structures in multiple independent lines of modern terrestrial animals suggests that if we could see into the distant evolutionary past, we would witness individuals with invaginated breathing organs outcompete ones with alternative breathing structures on land. The pattern suggests that natural selection was at work, and that the invaginated breathing organs are adaptations to life on land.
- *Studies of laboratory populations over many generations.* Changes in gene frequencies can be observed over multiple generations in laboratory populations of fast-breeding animals such as fruit flies. By exposing such populations to specific, controlled condi-

[7] Just 18,000 years ago, the arid, warm deserts of Arizona and New Mexico were far more moist than today, and they were on average about 6°C colder, because of the last ice age.

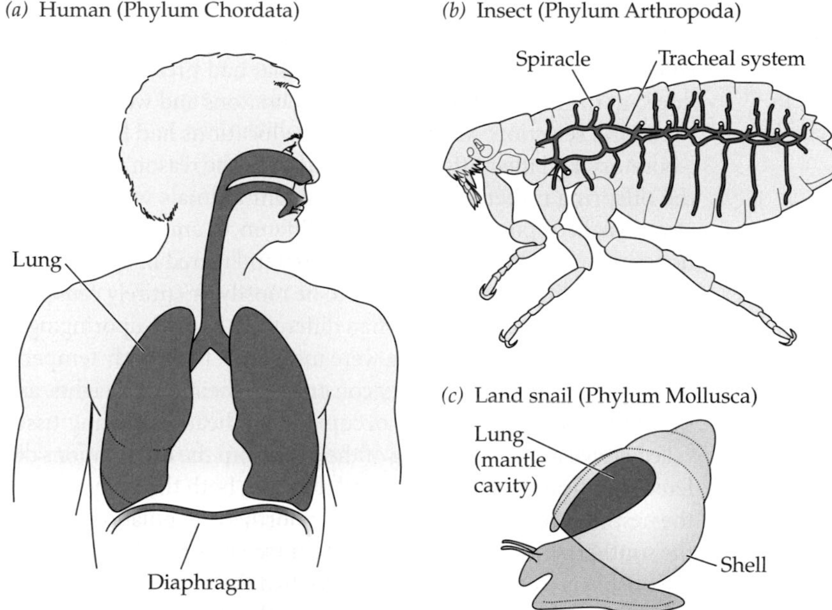

Figure 1.17 The comparative method Terrestrial vertebrates *(a)*, insects *(b)*, and land snails *(c)*—representing three phyla that separately colonized the land—have independently evolved breathing organs that are invaginated into the body. This convergence in the type of breathing organ suggests that invaginated breathing organs are adaptive for living on land.

tions (e.g., high or low temperatures), physiologists may observe which alleles are favored by natural selection when a particular condition prevails.

- *Single-generation studies of individual variation.* Individuals in a natural population of a single species typically vary in their physiological properties. For example, various individuals of a species of mouse vary in the maximal rate at which they can take in and use O_2; this rate is a key determinant of how long and hard the various individuals can sustain metabolic effort (Figure 1.18). Such natural variation among individuals of a species can be exploited to carry out single-generation experiments to determine which traits are most advantageous. To illustrate, suppose we trap several hundred mice in a wild population and measure the maximal rate of O_2 consumption of each, and then we release all the mice back into their natural population, where we monitor them until they die. If we find that individuals with particular O_2 consumption capabilities produce more young before dying than individuals with other O_2 consumption capabilities do, we will have insight into which capabilities are adaptive.
- *Creation of variation for study.* Biologists may be able to create variation in a trait that shows little or no natural variation among individuals of a species. Then competitive outcomes in natural or laboratory settings may be observed. Years ago, the principal application of this approach was morphological; for instance, the size of the ear cavities of desert rats was morphologically altered to assess which ear-cavity dimensions allowed the surest detection of predators. A newer approach is to use genetic mutants. Suppose that the vast majority of individuals of a species have a certain allele for a digestive enzyme but that an unusual mutant allele is found that produces a different molecular form of the enzyme. By controlled breeding, one can create a population rich in both alleles and then observe the relative advantages of the two enzyme forms under various conditions. A permutation of the genetic approach is to employ genetic engineering to create **knockout animals**, individuals that lack functional copies of a gene of interest. Such individuals are unable to synthesize the protein coded by the affected gene and thus can be used to assess the functional significance of the protein.
- *Studies of the genetic structure of natural populations.* Natural populations may be genetically structured in revealing ways. Genetic **clines** provide excellent examples. A genetic cline is a progressive change in allele frequencies or gene-controlled phenotype frequencies along an environmental gradient. Investigators find, for instance, that in a certain species of fish of the East Coast of the United States, alleles that are common in Georgia become progressively less common toward the north and are almost absent in New England (see Figure 2.21). Genetic patterns like this often are clues to how natural selection differs in its effects from place to place.
- *Phylogenetic reconstruction.* The goal of phylogenetic reconstruction is to determine the structure of the family tree (the ancestry) of groups of related species, often using molecular genetic data. The family tree is useful in two major ways. First, a family tree often facilitates the estimation of exactly *when* in evolutionary history each trait evolved; thus, for example, we might learn from a family tree whether the evolution of one trait preceded or followed that of another—knowledge that helps us understand the context of the

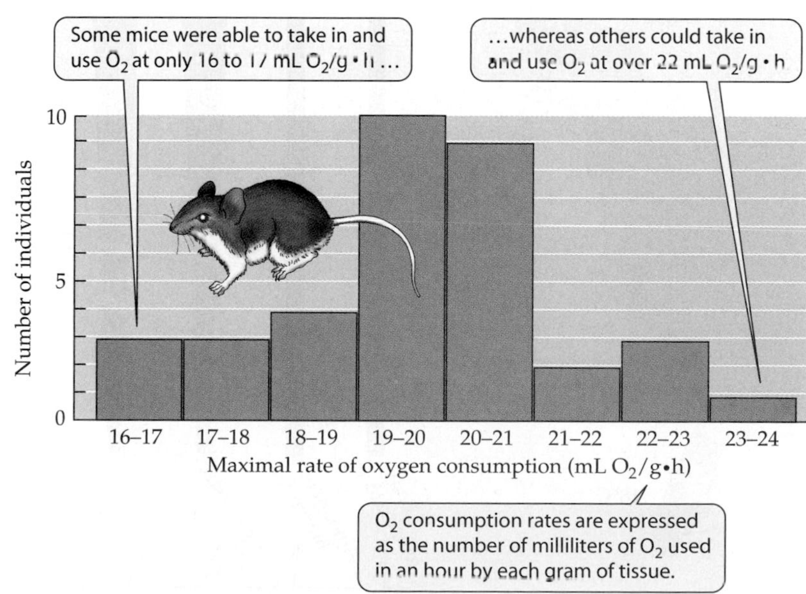

Figure 1.18 Physiological variation among individuals of a species This histogram summarizes the measured maximal rates of oxygen consumption of 35 deer mice (*Peromyscus maniculatus*) from a single natural population.

evolution of each trait. Second, a family tree clarifies whether a trait evolved independently more than once and helps us estimate the number of times a trait evolved; several independent origins in one environment suggest that a trait is adaptive to the environment. In this book we discuss several analyses of adaptation based on family trees, such as Figure 2.20 in the next chapter.

Evolutionary potential can be high or low, depending on available genetic variation

A key determinant of the course of evolution of a trait in an animal population is the amount of genetic diversity for the trait in the population. If there is no genetic diversity for a trait—that is, if all the individuals in a population are homozygous for a single allele—then evolutionary change in the trait is impossible (at least for the present). An example is a population of mammals in which all individuals are homozygous for an eye-color allele that codes for brown eyes. In this population as a whole, there is no genetic diversity for eye color, and thus natural selection cannot possibly alter eye color. By contrast, if the individuals in a population collectively have several different alleles of the gene for eye color—some alleles coding for brown, others for blue or olive—then the frequencies of the various alleles could be modified by natural selection, and eye color could evolve.

Physiologists are just beginning to take full account of the importance of genetic diversity in understanding evolutionary potential. Figure 1.19 provides a model example of the sorts of insight obtained through consideration of genetic diversity in studies of evolution. House mice were collected from five locations in eastern North America, chosen to represent a progression in winter severity, from mild winters in the south to severe winters in the north. The mice collected were from wild populations that had presumably reproduced at the five locations for many generations and were subjected to natural selection. Mice from the five locations had litters in the laboratory, and their offspring were studied. The reason for studying the offspring rather than the wild-caught animals was to gain as clear insight as possible into *genetic* differences among the populations; because all the offspring were born and reared in one environment, their differences were likely to be mostly or entirely caused by differences in genetics, rather than differences in early upbringing.

Four traits of the offspring were measured: their body temperatures, the sizes of the nests they constructed, their body weights, and the fractions of their bodies occupied by a heat-producing tissue called *brown fat.* The offspring of the mice from the five locations differed significantly in body size and nest size; both the body size and the nest size were higher in the colder, northern populations than in the southern ones, as one might expect (see Figure 1.19). However, the offspring from all five sets of mice had the same body temperatures and same amounts of brown fat. In the abstract, one might expect animals in a cold climate to evolve a lower body temperature and a larger quantity of heat-producing tissue than ones in a warm climate, but neither of these expectations is fulfilled in reality. Why has adaptation occurred in only two of the four traits studied?

Genetic diversity provides an important part of the answer. From prior studies, biologists know that house mouse populations exhibit relatively high genetic diversity in the genes that control body size and nest size; these two traits have responded to natural selection. However, house mouse populations exhibit little diversity in the genes that control body temperature and the amount of brown-fat they have; these two traits have failed to respond to natural selection in the very same mice. We do not know why genetic diversity is high for some traits and low for others. Examples like this show, however, that evolution by natural selection depends on the underlying genetic structure of populations and can be only as effective as genetic diversity permits it to be.

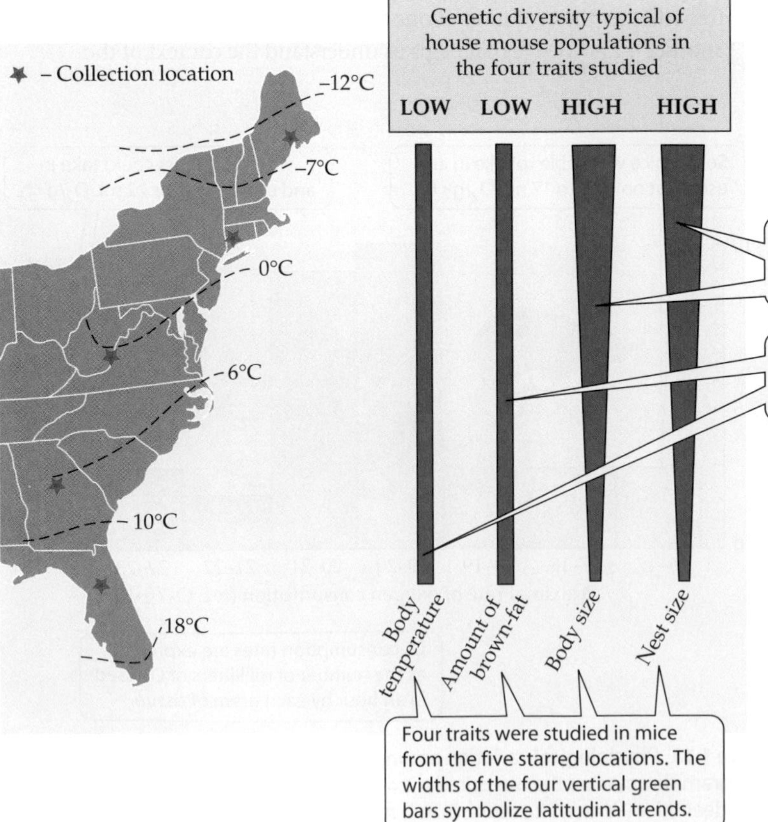

Figure 1.19 The effects of genetic diversity on evolutionary potential Tested in a constant laboratory setting, mice (*Mus domesticus*) from the more-northern locations attain larger body sizes and build bigger nests than mice from the more-southern populations. In contrast, despite having evolved in different climates, mice from the five locations show no differences in body temperature or in the amount of brown fat they have relative to body size. *Independent* studies reveal that in house mouse populations, there is high diversity in the genes that control the body sizes of the mice and the sizes of the nests they build, whereas the diversity of genes controlling body temperature and the amount of brown fat is low. The results support the hypothesis that the effectiveness of natural selection depends on how much genetic diversity exists in populations. The dashed lines on the map show average winter temperatures. (After Lynch 1992.)

Study Questions

1. There is a chance that a calcium atom or carbon atom that was once part of Caesar's or Cleopatra's body is now part of your body. Part of the reason is that most calcium and carbon atoms that were parts of these rulers' bodies did not go to their graves with them. Explain both statements.

2. Animals do not keep all their detoxification enzymes in a constant state of readiness. Thus they depend on phenotypic plasticity to adapt to changing hazards. An example is provided by an enzyme called *alcohol dehydrogenase*, which breaks down ethyl alcohol. People who do not drink alcoholic beverages have little alcohol dehydrogenase. Alcohol dehydrogenase increases when people drink alcohol, but it requires many days to increase to its maximum, leaving people undefended against alcohol's effects when they first start drinking after a period of not drinking. Propose reasons why animals do not maintain all their detoxification systems in a maximum state of readiness at all times.

3. Whereas the larvae of a particular species of marine crab are bright orange, the adults of the species are white. An expert on the crabs was asked, "Why are the two different life stages different in color?" She replied, "The larvae accumulate orange-colored carotenoid pigments, but the adults do not." Did she recognize all the significant meanings in the question asked? Explain.

4. Referring to Figure 1.8, do zebras, warthogs, and greater kudus have normal or exceptional gestation lengths? Justify your position in each case.

5. At least three hemoglobin alleles in human populations alter hemoglobin structure in such a way as to impair the transport of O_2 by the blood but enhance resistance of red blood cells to parasitization by malaria parasites. Explain how such alleles exemplify pleiotropy, and discuss whether such alleles could lead to nonadaptive evolution of blood O_2 transport in certain situations.

6. What are some of the microclimates that a mouse might find in your professor's home?

7. Figure 1.13 seems at first to be simply a description of the physical and chemical properties of a lake. Outline how living organisms participate in determining the physical and chemical (i.e., temperature and O_2) patterns. Consider organisms living both in the lake and on the land surrounding the lake. Consider also a recent research report showing that dense populations of algae sometimes change the temperature structure of lakes by raising the thermocline and thus increasing the thickness of the deep, cold layer; how could algal populations do this, and what could be the consequences for deep-water animals?

8. Do you agree with François Jacob that evolution is more like tinkering than engineering? Explain.

9. Explain how the comparative method, knockout animals, and geographical patterns of gene frequencies might be used to assess whether a trait is adaptive. As much as possible, mention pros and cons of each approach.

10. Certain species of animals tolerate body temperatures of 50°C, but the vast majority do not. Some species can go through their life cycles at very high altitudes, but most cannot. What are the potential reasons why certain exceptional species have evolved to live in environments that are so physically or chemically extreme as to be lethal for most animals? How could you test some of the ideas you propose?

Key References

Bennett, A. F. 1997. Adaptation and the evolution of physiological characters. In W. H. Dantzler (ed.), *Comparative Physiology*, vol. 1 (Handbook of Physiology [Bethesda, MD], section 13), pp. 3–16. Oxford University Press, New York. A thorough, compact review of adaptation and modes of evolution as seen through the eyes of a physiologist.

Feder, M. E., A. F. Bennett, and R. B. Huey. 2000. Evolutionary physiology. *Annu. Rev. Ecol. Syst.* 31: 315–341. A review of the evolutionary approach to physiology, emphasizing empirical methods to assess adaptation.

Gould, S. J., and R. C. Lewontin. 1979. The spandrels of San Marco and the Panglossian paradigm: A critique of the adaptationist programme. *Proc. Roy. Soc. London [Biol.]* 205: 581–598. The article that launched the modern reconsideration of adaptationist thinking. Deliberately provocative, it excites critical thought about key biological concepts that are often treated as truisms.

Lynch, C. B. 1992. Clinal variation in cold adaptation in *Mus domesticus*: Verification of predictions from laboratory populations. *Amer. Nat.* 139: 1219–1236. A fine example of the modern application of quantitative genetics to the study of physiological and behavioral evolution.

Mangum, C. P., and P. W. Hochachka. 1998. New directions in comparative physiology and biochemistry: Mechanisms, adaptations, and evolution. *Physiol. Zool.* 71: 471–484. An effort by two leaders to project the future of animal physiology, emphasizing the integration of mechanistic and evolutionary approaches.

Somero, G. N. 2000. Unity in diversity: A perspective on the methods, contributions, and future of comparative physiology. *Annu. Rev. Physiol.* 62: 927–937. A brief, modern, and stimulating view of why comparative physiology has been important in the past and will continue to be so in the future, written by a prominent comparative biochemist.

Selected References

Alexander, R. M. 1985. The ideal and the feasible: Physical constraints on evolution. *Biol. J. Linnean Soc.* 26: 345–358.

Bartholomew, G. A. 1986. The role of natural history in contemporary biology. *BioScience* 36: 324–329.

Brett, J. R. 1983. Life energetics of sockeye salmon, *Oncorhynchus nerka*. In W. P. Aspey and S. I. Lustick (eds.), *Behavioral Energetics: The Cost of Survival in Vertebrates*, pp. 29–63. Ohio State University Press, Columbus.

Brown, J. H., and M. V. Lomolino. 1998. *Biogeography*, 2nd ed. Sinauer, Sunderland, MA.

Calder, W. A., III. 1984. *Size, Function, and Life History.* Harvard University Press, Cambridge, MA.

Chevillon, C., D. Bourguet, F. Rousset, N. Pasteur, and M. Raymond. 1997. Pleiotropy of adaptive changes in populations: Comparisons among insecticide resistance genes in *Culex pipiens*. *Genet. Res.* 70: 195–204.

Crespi, B. J. 2000. The evolution of maladaptation. *Heredity* 84: 623–629.

Endler, J. A. 1986. *Natural Selection in the Wild.* Princeton University Press, Princeton, NJ.

Feder, M. E., A. F. Bennett, W. W. Burggren, and R. B. Huey (eds.). 1987. *New Directions in Ecological Physiology*. Cambridge University Press, New York.

Futuyma, D. J. 1998. *Evolutionary Biology*, 3rd ed. Sinauer, Sunderland, MA.

Garland, T., Jr., and S. C. Adolph. 1991. Physiological differentiation of vertebrate populations. *Annu. Rev. Ecol. Syst.* 22: 193–228.

Garland, T., Jr., and P. A. Carter. 1994. Evolutionary physiology. *Annu. Rev. Physiol.* 56: 579–621.

Gibbs, A. G. 1999. Laboratory selection for the comparative physiologist. *J. Exp. Biol.* 202: 2709–2718.

Gotthard, K., and S. Nylin. 1995. Adaptive plasticity and plasticity as an adaptation: A selective review of plasticity in animal morphology and life history. *Oikos* 74: 3–17.

Gould, S. J., and E. S. Vrba. 1982. Exaptation: A missing term in the science of form. *Paleobiology* 8: 4–15.

Hastings, J. W. 1996. Chemistries and colors of bioluminescent reactions: A review. *Gene* 173: 5–11.

Hoffmann, A. A., and P. A. Parsons. 1997. *Extreme Environmental Change and Evolution.* Cambridge University Press, Cambridge.

Jacob, F. 1977. Evolution and tinkering. *Science* 196: 1161–1166.

Jorgensen, C. B. 1983. Ecological physiology: Background and perspectives. *Comp. Biochem. Physiol. A* 75: 5–7.

Kingsolver, J. G., and R. B. Huey. 1998. Evolutionary analyses of morphological and physiological plasticity in thermally variable environments. *Amer. Zool.* 38: 545–560.

Levins, R., and R. Lewontin. 1985. The organism as the subject and object of evolution. In R. Levins and R. Lewontin, *The Dialectical Biologist*, pp. 85–106. Harvard University Press, Cambridge, MA.

Lewontin, R. C. 1978. Adaptation. *Sci. Amer.* 239(3): 212–230.

Liao, J. L., D. N. Beal, G. V. Lauder, and M. S. Triantafyllou. 2003. Fish exploiting vortices decrease muscle activity. *Science* 302: 1566–1569.

Madigan, M. T., and B. L. Marrs. 1997. Extremophiles. *Sci. Amer.* 276(4): 82–87.

Mangum, C., and D. Towle. 1977. Physiological adaptation to unstable environments. *Amer. Sci.* 65: 67–75.

Mayr, E. 1983. How to carry out the adaptationist program? *Amer. Nat.* 121: 324–334.

Mazumder, A., W. D. Taylor, D. J. McQueen, and D. R. S. Lean. 1990. Effects of fish and plankton on lake temperature and mixing depth. *Science* 247: 312–315.

Parsons, P. A. 1992. Evolutionary adaptation and stress: The fitness gradient. In M. K. Hecht, B. Wallace, and R. J. MacIntyre (eds.), *Evolutionary Biology*, vol. 26, pp. 191–223. Plenum, New York.

Parsons, P. A. 1996. Stress, resources, energy balances, and evolutionary change. In M. K. Hecht, R. J. MacIntyre, and M. T. Clegg (eds.), *Evolutionary Biology*, vol. 29, pp. 39–72. Plenum, New York.

Peterson, C. C. 1996. Anhomeostasis: Seasonal water and solute relations in two populations of the desert tortoise (*Gopherus agassizii*) during chronic drought. *Physiol. Zool.* 69: 1324–1358.

Prosser, C. L. 1986. *Adaptational Biology: Molecules to Organisms.* Wiley, New York.

Rose, M. R., and G. V. Lauder (eds.). 1996. *Adaptation.* Academic Press, New York.

Scharloo, W. 1989. Developmental and physiological aspects of reaction norms. *BioScience* 39: 465–471.

Schoenheimer, R. 1942. *The Dynamic State of Body Constituents.* Harvard University Press, Cambridge, MA. [The seminal book that introduced the concept of structural dynamism in animals.]

Spicer, J. I., and K. J. Gaston (eds.). 1999. *Physiological Diversity and Its Ecological Implications.* Blackwell, Malden, MA.

Stanley, S. M. 1989. *Earth and Life through Time*, 2nd ed. Freeman, New York.

Tracy, C. R., and J. S. Turner (eds.). 1982. What is physiological ecology? *Bull. Ecol. Soc. Amer.* 63: 340–347.

Travis, J. 1994. Evaluating the adaptive role of morphological plasticity. In P. C. Wainwright and S. M. Reilly (eds.), *Ecological Morphology*, pp. 99–122. University of Chicago Press, Chicago.

Trimmer, B. A., J. R. Aprille, D. M. Dudzinski, C. J. Lagace, S. M. Lewis, T. Michel, S. Qazi, and R. M. Zayas. 2001. Nitric oxide and the control of firefly flashing. *Science* 292: 2486–2488.

Waterlow, J. C., P. J. Garlick, and D. J. Millward. 1978. *Protein Turnover in Mammalian Tissues and in the Whole Body.* North-Holland, New York.

Waterman, T. H. 1999. The evolutionary challenges of extreme environments (part 1). *J. Exp. Zool.* 285: 326–359.

Waterman, T. H. 2001. Evolutionary challenges of extreme environments (part 2). *J. Exp. Zool.* 291: 130–168.

Wiebel, E. R., C. R. Taylor, and H. Hoppeler. 1991. The concept of symmorphosis: A testable hypothesis of structure-function relationship. *Proc. Natl. Acad. Sci. U.S.A.* 88: 10357–10361.

Williams, G. C. 1966. *Adaptation and Natural Selection.* Princeton University Press, Princeton, NJ.

2 Molecules and Cells in Animal Physiology

The two slow-moving animals pictured here are able to consume fast-moving prey because they have evolved ways to defeat the function of essential molecules and cellular structures in their prey. The first of these, the puff adder, is one of the slowest-moving snakes. It feeds on fast-moving rats, however, because it needs only a moment's contact with its prey to set in motion processes that will destroy key molecular-cellular properties on which a rat depends for proper function. Like rattlesnakes and other adders, the puff adder sits and waits for an unsuspecting animal to come close enough for a strike. It then lunges at its victim, sinks its fangs in, and in less than 1 to 2 seconds, injects a complex mix of compounds that attack critical molecules and cells. The snake then immediately releases the rat and tracks the victim as the rat's molecular-cellular mechanisms fall apart. Some of the injected compounds, for instance, strip the outer membranes from the rat's muscle cells, while other compounds make tiny holes in the rat's blood capillaries, permitting widespread internal hemorrhaging. When finally the molecular-cellular damage is so great that the victim can no longer function, the slow-moving adder moves in to eat.

The second example of a slow-moving animal that consumes fast-moving prey—the cone snail—feeds on fish. The snail lures fish to its vicinity by waving a wormlike body part that deceives the fish into coming close to eat. The snail then harpoons the fish with a hollow barbed tooth. The fish could easily tear itself loose if permitted just a moment's time to do so. The snail preempts such escape by injecting the fish through the tooth with compounds that rapidly disrupt the function of proteins that are essential for the function of the fish's nerve and muscle cells. In

Two slow-moving predators that use molecular weapons to capture fast-moving prey The puff adder is a notoriously sluggish, but deadly, African snake that resembles rattlesnakes in its strategy of hunting small mammals. The cone snail sits virtually motionless in the coral reef ecosystems it occupies yet feeds routinely on fish. Both predators produce venoms that disrupt vital molecular structures or mechanisms.

this way the fish's most promising defense, its ability to swim rapidly away, is immediately defeated. With the cells in its nervous system in disarray and its muscles paralyzed, the fish is ingested by the sedentary snail.

The actions of venoms and poisons remind us that all the higher functions of animals depend on molecules and on the *organization* of molecules into cellular structures and cells. An animal as spectacular as a racehorse or a mind as great as that of Socrates can be brought down in a moment if the function of key molecules is blocked or the normal organization of cells is disrupted.

A case can be made that the study of molecules and the cellular organization of molecules is the most fundamental study of biology, because molecules and cells are the building blocks of tissues and other higher levels of organization. Some scientists believe that all the properties of tissues, organs, and whole animals will eventually be fully predictable from knowledge of molecules and cells alone. Other scientists, however, believe that animals have **emergent properties**: properties of tissues, organs, or whole animals that will never in principle be predictable from mere knowledge of molecules and cells because the properties *emerge* only when cells are assembled into interactively functioning sets. Regardless of the resolution of this important debate, molecules and cells are critically important. Thus, here in our second chapter we turn our attention to the essential attributes of molecules and cells that bear on the study of animal physiology.

The cell membrane and the intracellular membranes of a cell form an appropriate starting point. These membranes, although exceedingly thin (6–8 nm), physically compartmentalize systems and thereby play crucial roles in organizing cellular and subcellular function; the cell membrane, for instance, separates the inside of a cell from the cell's surroundings, permitting the inside to have different properties from the outside. In addition, far from being inert barriers, the membranes are dynamic systems that *participate* in cellular and subcellular function. For example, cell membranes control the transport of materials between the insides and outsides of cells, and they mediate the responses of cells to changes in cell surroundings.

Cell Membranes and Intracellular Membranes

Each animal cell is enclosed in a **cell membrane** (*plasma membrane*) that is ordinarily composed primarily of a bilayer (double layer) of phospholipid molecules in which protein molecules are embedded (Figure 2.1). Each cell also includes many sorts of **intracellular membranes** (*subcellular membranes*), such as the endoplasmic reticulum, the inner and outer membranes of each mitochondrion, and the two closely associated membranes that form the nuclear envelope. The fundamental structure of all these intracellular membranes is also a bilayer of phospholipid molecules with protein molecules embedded in it. Given the ubiquity and importance of phospholipids in cell and intracellular membranes, it is not surprising that they are targets of venoms. A principal weapon in the complex venom of a puff

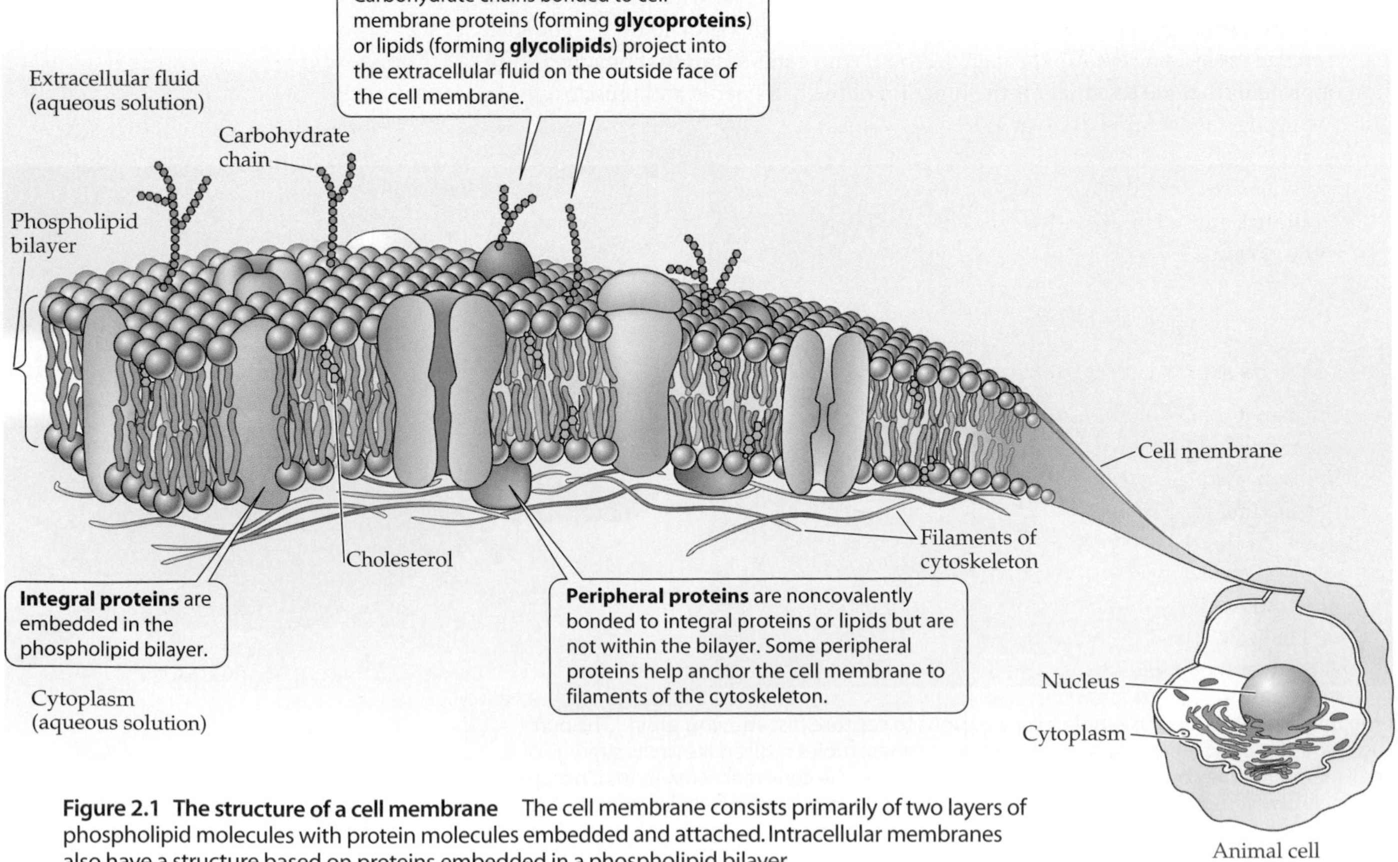

Figure 2.1 The structure of a cell membrane The cell membrane consists primarily of two layers of phospholipid molecules with protein molecules embedded and attached. Intracellular membranes also have a structure based on proteins embedded in a phospholipid bilayer.

adder or a rattlesnake is a set of enzymes known as *phospholipases*, which break up phospholipids. Among other effects, these enzymes destroy the phospholipid matrix of the cell membranes of a victim's skeletal muscle cells, thereby exposing the insides of the cells, setting membrane proteins adrift, and wreaking other havoc.

To understand the molecular logic of the structure of cell membranes and intracellular membranes, it is necessary to consider the *polarity* of molecules and the attendant attributes of **hydrophilic** and **hydrophobic** interactions. Consider vinegar-and-oil salad dressing as an everyday example of the effects of molecular polarity. Vinegar consists of acetic acid and water. Thus the dressing has three principal components: oil, acetic acid, and water. If the dressing sits still for a while, the acetic acid remains in solution in the water, but the oil forms a separate layer. This outcome occurs because the acetic acid is *hydrophilic* ("water-loving"), whereas the oil is *hydrophobic* ("water-hating"). Why do the two substances behave in these different ways? A principal reason is the polarity of the molecules. Acetic acid is polar and because of its polar nature is attracted to water; oil is nonpolar and therefore repelled from water.

The distribution of electrons in a molecule is what determines whether the molecule is polar or nonpolar. Within a **polar molecule**, electrons are unevenly distributed; thus some regions of a polar molecule are relatively negative, while others are relatively positive. Water is a polar molecule. Other polar molecules, such as acetic acid—and ions—intermingle freely with polar water molecules by charge interaction, forming solutions. Within a **nonpolar molecule**, electrons are evenly distributed and there are no charge imbalances between different molecular regions. Nonpolar molecules, such as the oil in salad dressing, do not freely intermingle with polar water molecules. Instead, water molecules tend to assemble *with each other* into arrays that surround nonpolar molecules. Because these arrays are thermodynamically less stable than a complete separation of the water and the nonpolar molecules, the nonpolar molecules separate into a distinct layer when not forcibly mixed with water by physical agitation. These principles help explain the structure of the phospholipid bilayer in cell membranes and intracellular membranes, and they also help explain the positioning of other chemical constituents within the bilayer.

The lipids of membranes are structured, diverse, and fluid

Phospholipids, which are the principal constituents of the matrix in which proteins are embedded in cell membranes and intracellular membranes, are lipids that contain phosphate groups (Figure 2.2*a*). They are **amphipathic**, meaning that each molecule consists of a polar part (within which there are regional differences of charge) and a nonpolar part (which lacks regional differences of charge). A membrane phospholipid consists of a *polar head* and two *nonpolar tails* (Figure 2.2*b*). The polar head is composed of the phosphate group, which forms a region of negative charge, bonded to another group that forms a region of positive charge, such as choline (see Figure 2.2*a*). Each nonpolar tail consists of a long-chain hydrocarbon derived from a fatty acid.

Whereas the polar part of a phospholipid molecule or any other amphipathic molecule is hydrophilic, the nonpolar part is hydrophobic. When phospholipid molecules are placed in a system of oil layered on water, they collect at the interface of the oil and water in a predictable way, with their polar, hydrophilic heads in the water layer and their nonpolar, hydrophobic tails in the oil. Of greater importance for understanding living cells is the fact that when phospholipid molecules are placed simply in an aqueous solution, they spontaneously assemble into bilayers, adopting the same bilayer conformation they take in cell membranes and intracellular membranes (Figure 2.2*c*). This bilayer conformation forms because it is thermodynamically stable. All the hydrophobic regions (the hydrocarbon tails) get together in the interior of the bilayer (away from the water), whereas the hydrophilic heads associate with the water on either side of the membrane. The energy barrier to mixing polar and nonpolar regions in the membrane is so great that in a cell membrane, it is nearly impossible for a phospholipid molecule to "flip" its polar head through the nonpolar interior and move from one side of the bilayer to the other (unless specifically catalyzed to do so).

A striking attribute of membrane phospholipids is their chemical diversity. Many different types of phospholipid molecules are possible because the two tails and part of the head, as shown in Figure 2.2*a*, can differ widely in their specific chemical composition. The cell membranes of human red blood cells contain over 150 different chemical forms of phospholipids, and similar diversity is seen in other cell membranes. The two layers of phospholipid molecules in any particular membrane, known as the two **leaflets** of the membrane, typically are composed of different mixes of phospholipid molecules.

The phospholipids in a cell membrane or intracellular membrane are **fluid**. Individual phospholipid molecules are not covalently bound to one another, and they move relative to each other. They are able to move about rather freely by diffusion *within each membrane leaflet*. The rate of this diffusion is great enough that a particular phospholipid molecule can travel around the entire circumference of a cell in a matter of minutes. The ease of motion of the phospholipid molecules in a membrane—their **fluidity**—depends in part on the degree of chemical saturation of the hydrocarbons that make up their tails.

The chemical saturation that helps determine membrane fluidity is defined by the number of double bonds in the hydrocarbons. A hydrocarbon is **saturated** if it contains no double bonds; it is **unsaturated** if it includes one or more double bonds. As shown in Figure 2.2*a*, a double bond may impart a bend to the part of the tail where it occurs. Bent tails of membrane phospholipids prevent tight, crystal-like packing of the tails in the hydrophobic interior of the membrane. This disruption of tight packing helps keep the phospholipid molecules free to move. Thus a greater proportion of unsaturated fatty acids in the tails of phospholipids results in a membrane with more fluidity.

In addition to chemical composition, temperature affects the fluidity of membranes; just as butter and other household lipids stiffen when they are chilled, the phospholipids in cell membranes tend to become stiffer at lower temperatures. During evolution, one important way in which cells have become adapted to different temperatures is alteration of the numbers of double bonds (the degree of unsaturation) in their membrane phospholipids. This is evident in fish of polar seas, for instance. The fish experience tissue temperatures so low that their cell membranes could be overly stiff. This problem is avoided, however, because they have cell membranes constructed of

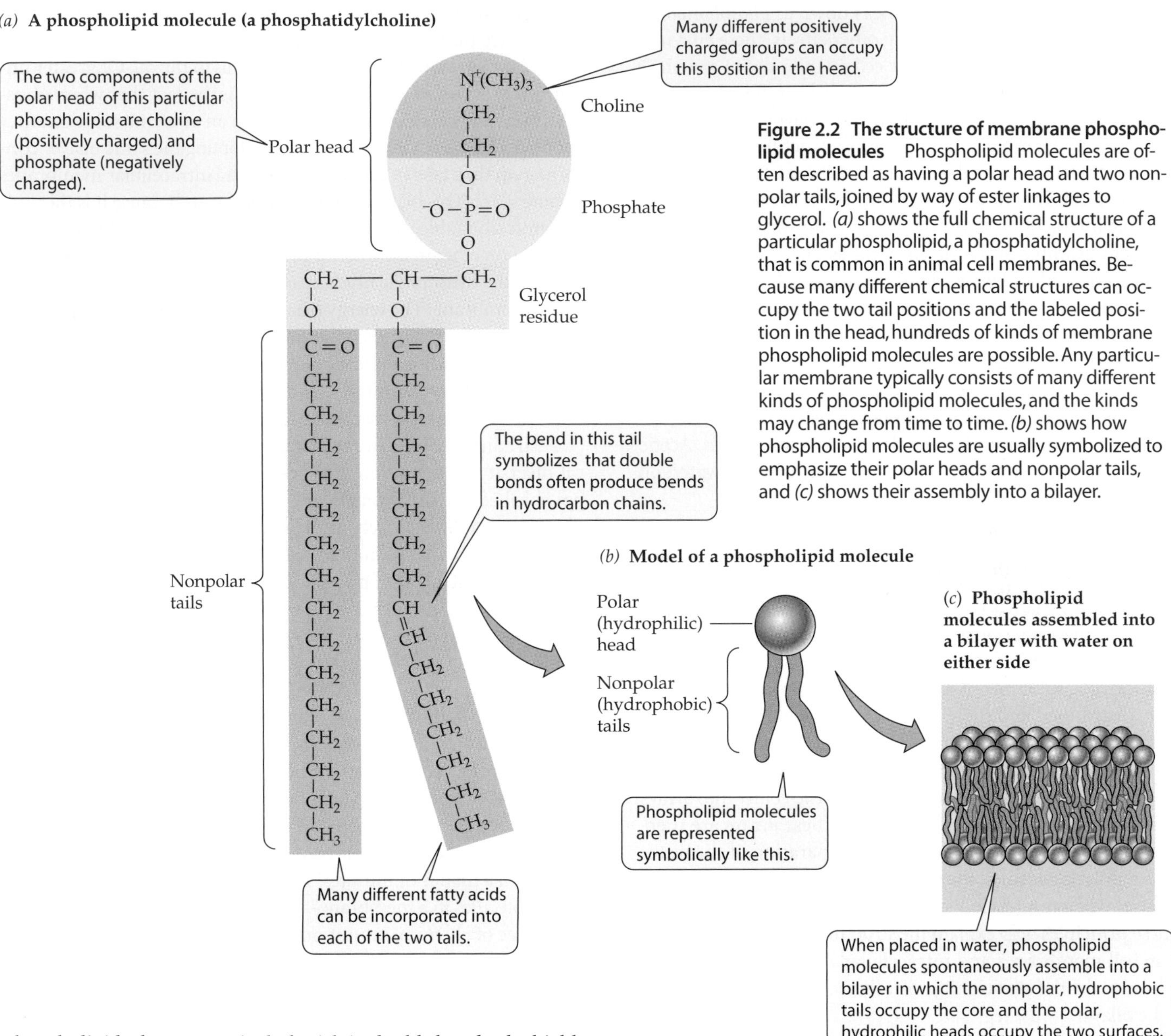

Figure 2.2 The structure of membrane phospholipid molecules Phospholipid molecules are often described as having a polar head and two nonpolar tails, joined by way of ester linkages to glycerol. *(a)* shows the full chemical structure of a particular phospholipid, a phosphatidylcholine, that is common in animal cell membranes. Because many different chemical structures can occupy the two tail positions and the labeled position in the head, hundreds of kinds of membrane phospholipid molecules are possible. Any particular membrane typically consists of many different kinds of phospholipid molecules, and the kinds may change from time to time. *(b)* shows how phospholipid molecules are usually symbolized to emphasize their polar heads and nonpolar tails, and *(c)* shows their assembly into a bilayer.

phospholipids that are particularly rich in double bonds; the highly unsaturated phospholipids are inherently quite fluid and thus less likely than other phospholipids to become detrimentally stiff at low temperatures. Recent research on the cell membranes of brain cells in fish demonstrates that the degree of phospholipid unsaturation depends in a regular way on the environmental temperatures to which various species are adapted (Figure 2.3). Tropical species of fish, which face little risk of having their membranes rendered too stiff by low temperatures, have evolved relatively saturated phospholipids, but as the temperature of the habitat falls, the degree of unsaturation of the phospholipids increases.

Besides phospholipids, cell membranes and intracellular membranes contain other classes of lipids, one of which is **sterols**. The principal membrane sterols are *cholesterol* and *cholesterol esters*. In cell membranes, which are typically much richer in sterols than intracellular membranes, sterols collectively occur in ratios of 1 molecule per 10 phospholipid molecules, up to 8 per 10. Cholesterol is mildly amphipathic and positioned within the phospholipid leaflets (see Figure 2.1), where it exerts complex effects on membrane fluidity.

Proteins endow membranes with numerous functional capacities

Proteins are the second major constituents of cell membranes and intracellular membranes. According to the **fluid mosaic model** of membranes, a membrane consists of a mosaic of protein and lipid molecules, all of which move about in directions parallel to the membrane faces because of the fluid state of the lipid matrix.

Membrane proteins are structurally of two principal kinds: integral and peripheral. **Integral membrane proteins** are parts of the membrane and cannot be removed without extraction procedures that take the membrane apart. Most integral proteins span the mem-

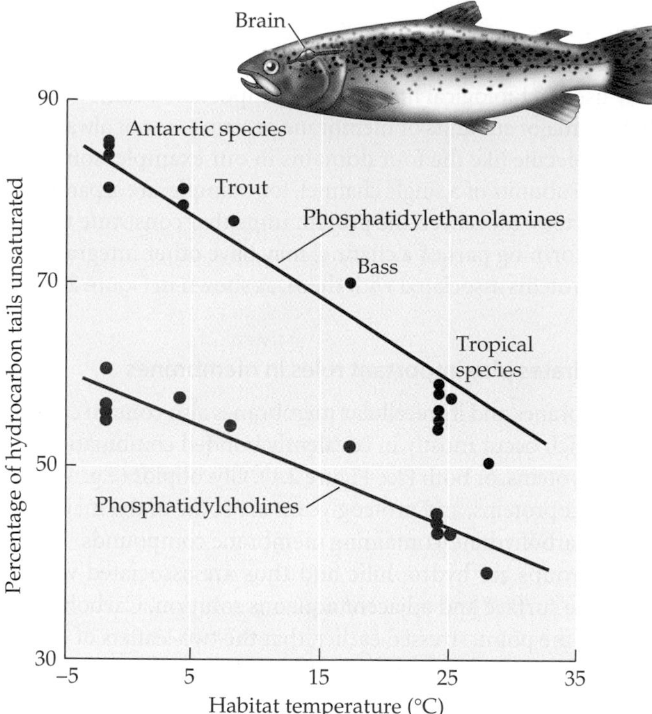

Figure 2.3 The degree of chemical unsaturation of the hydrocarbon tails of brain phospholipids in fish varies with habitat temperature Brain synaptic membranes of 17 species of teleost (bony) fish were studied. Measurements were made of the composition of the hydrocarbon tails of two categories of cell-membrane phospholipids, the phosphatidylcholines and the phosphatidylethanolamines, which differ in whether the group at the top of the head in Figure 2.2*a* is choline [—CH_2—CH_2—$N(CH_3)_3$] or ethanolamine (—CH_2—CH_2—NH_2), respectively. Each plotted symbol corresponds to the average value for one species. (After Logue et al. 2000.)

brane and thus are called *transmembrane* proteins (see Figure 2.1). These molecules have both hydrophobic and hydrophilic regions. As we will see in detail shortly, each hydrophobic region typically has an amino acid composition and a molecular geometry that allow it to associate with the hydrophobic hydrocarbon tails of the membrane interior. The hydrophilic regions of transmembrane proteins, on the other hand, typically protrude into the aqueous solutions bathing the two sides of the membrane.

Peripheral membrane proteins are associated with the membrane but can be removed without destroying the membrane. They are bonded noncovalently to membrane components (e.g., integral proteins) and are positioned on one side of the membrane or the other (see Figure 2.1). Their positioning means that the two leaflets of a membrane differ in protein composition, as well as phospholipid composition.

The proteins of cell membranes and intracellular membranes endow the membranes with capabilities to *do* many things. Five *functional* types of membrane proteins are recognized: **channels, transporters** (**carriers**), **enzymes**, **receptors**, and **structural proteins.** Because these types are classified by function, the *actions* listed in Table 2.1 *define* the five types. The categories are not exclusive: A membrane protein can be both a receptor and an enzyme, or a receptor and a channel, or a transporter and an enzyme, for example.

The molecular structures of membrane proteins are complex and are diagrammed in a number of ways, depending on the degree of chemical detail to be shown. To illustrate, let's focus on a channel, which is a type of membrane-spanning integral protein. Channels provide paths for ions or other materials in aqueous solution to pass through membranes. In our example the channel is formed by a single protein molecule, the secondary structure of which is shown in Figure 2.4*a*. (The structural hierarchy of proteins is discussed in Box 2.1.)

Each cylinder in Figure 2.4*a* represents a sequence of amino acids that forms a helix-shaped subunit, called an α-*helix*, within the protein structure. This channel protein exemplifies a common property of membrane proteins in that the subparts of the molecule exhibit repeating structural patterns known as **domains.** To identify the domains, review the molecule from left to right. You will note five α-helices linked closely together, then a sixth helix separated from the others by a longer string of amino acids; then you will note that this pattern of five closely spaced helices followed by a sixth more-separated helix is repeated three more times. On the basis of this repeating pattern, this molecule is said to show four domains, numbered I to IV, as illustrated in Figure 2.4*a*. The α-helices of the molecule in Figure 2.4*a* are predominantly hydrophobic and span the membrane

TABLE 2.1 The five functional types of membrane proteins and the functions they perform

Functional type	Function performed (defining property)
Channel	Permits simple or quasi-simple diffusion of solutes in aqueous solution (page 70)—or osmosis of water (page 87)—through a membrane; a simplified view of a channel is that it creates a direct water path from one side to the other of a membrane (i.e., an aqueous pore) through which solutes in aqueous solution may diffuse or water may undergo osmosis
Transporter (carrier)	Binds noncovalently and reversibly with specific molecules or ions to move them intact across a membrane; the transport through the membrane is *active transport* (page 74) if it employs metabolic energy; it is *facilitated diffusion* (page 74) if metabolic energy is not employed
Enzyme	Catalyzes a chemical reaction in which covalent bonds are made or broken (page 41)
Receptor	Binds noncovalently with specific molecules and as a consequence of this binding, initiates a change in membrane permeability or cell metabolism; receptor proteins mediate the responses of a cell to chemical messages (signals) arriving at the outside face of the cell membrane (page 56)
Structural protein	Attaches to other molecules (e.g., other proteins) to anchor intracellular elements (e.g., cytoskeleton filaments) to the cell membrane, create junctions between adjacent cells (Figure 2.7), or establish other structural relations

by associating with the hydrophobic interior of the phospholipid bilayer. The strings of amino acids that connect successive helices are hydrophilic and protrude from the membrane into the aqueous solutions on either side. In its natural state in a membrane, this protein is believed to be shaped into a closed ring in which the four domains form cylinder-like structures surrounding a central pore, as diagrammed in Figure 2.4*b*.

The three remaining representations of the membrane protein that are shown in Figure 2.4 are progressively simpler. The sort of representation in Figure 2.4*c*, which still shows that there are four domains or subunits, is a simplified way to represent the chemical structure of the molecule. The diagrammatic, semirealistic representation in Figure 2.4*d*, which leaves one guessing about the number of domains, is more simplified yet, and in Figure 2.4*e*, the channel is represented schematically (without any intention of resembling the actual molecule).

The interrelations of the presentations in Figure 2.4*a–e* are important to note because all of these sorts of presentations are commonly used in biological literature. An important additional detail is that the major subunits of membrane proteins are not always parts of one molecule like the four domains in our example; sometimes the major subunits of a single channel, for example, are separate protein molecules. Moreover, the protein units that constitute the central pore-forming part of a channel may have other integral or peripheral proteins associated with them, as shown in Figure 2.4*f*.

Carbohydrates play important roles in membranes

Cell membranes and intracellular membranes also contain carbohydrates, which occur mostly in covalently bonded combination with lipids or proteins, or both (see Figure 2.1). Glycolipids (e.g., gangliosides), glycoproteins, and proteoglycans are some of the major categories of carbohydrate-containing membrane compounds.[1] Carbohydrate groups are hydrophilic and thus are associated with the membrane surface and adjacent aqueous solution. Carbohydrates reinforce the point, stressed earlier, that the two leaflets of a membrane are typically different. In cell membranes, for example, the carbohydrate groups always project from the outer, extracellular face, not the inner, cytoplasmic face (see Figure 2.1). These carbohydrate groups are thought to serve as attachment sites for extracellular proteins and as cell recognition sites.

[1] The word fragment *glyco* refers to carbohydrates (after the Greek *glykeros*, "sweet").

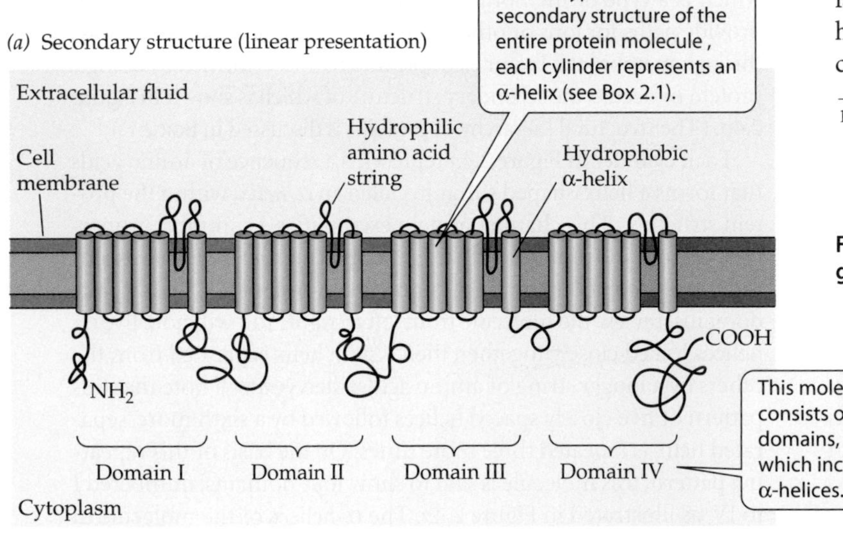

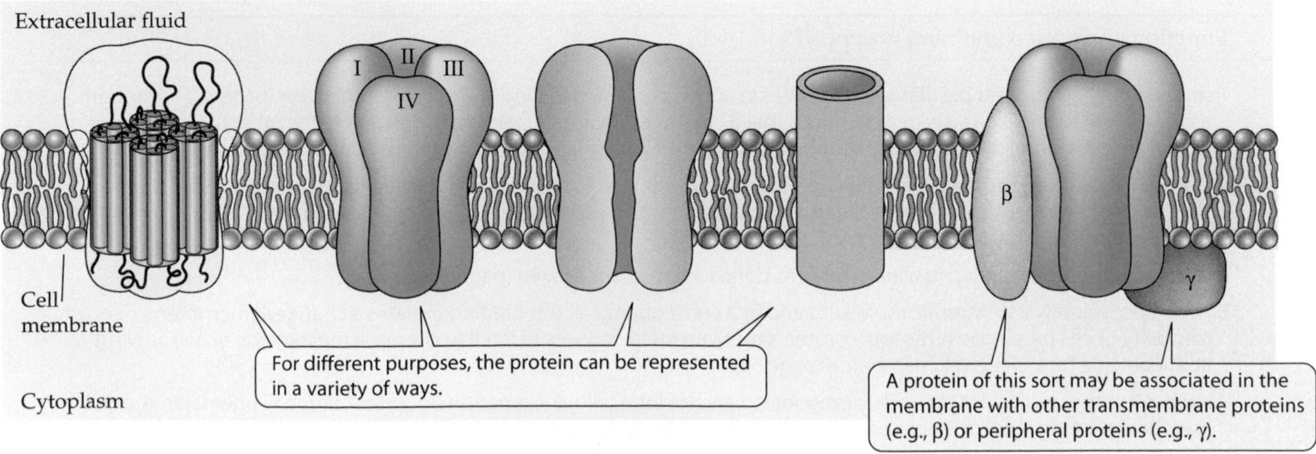

Figure 2.4 The structure of a transmembrane protein—a voltage-gated Na^+ channel—illustrating several modes of presentation

BOX 2.1 Protein Structure and the Bonds That Maintain It

Any protein molecule consists of a linear series of amino acids, coded by a linear series of bases in the genetic material (DNA). The series of amino acids is termed the **primary structure** of the protein (Figure A1). The primary structure is important but is inadequate to account for the functional properties of proteins. Instead, each protein's functional properties depend on the intricate *three-dimensional shape,* or **conformation,** of the protein molecule.

In the conformation of protein molecules, two increasingly complex levels of three-dimensional organization are recognized. The **secondary structure** of a protein molecule refers to subregions within the molecule in which the amino acids arrange themselves in highly regular geometric shapes; the two most common types of such highly ordered arrays are the α-helix and the β-sheet (pleated sheet) (Figure A2). The **tertiary structure** refers to the natural arrangement of an entire protein molecule in three dimensions, including its secondary structure and the other patterns of folding that give the molecule its particular conformation. Figure A3 shows two examples of tertiary structure. On the left is the three-dimensional structure of a subunit of the enzyme pyruvate kinase; it consists of several α-helices and β-sheets connected by strings of amino acids (the arrowheads on the sheets are merely artificial aids to help you trace the molecule from end to end). On the right is the three-dimensional structure of an O_2-binding protein (myoglobin) that includes several α-helices in its structure; this protein is drawn as a tube that follows the natural contours of the molecule. Multiple formats are used to diagram tertiary structure (Figure B).

The primary structure of a protein is maintained by **covalent bonds** between successive amino acids in the amino acid chain. Covalent bonds depend on sharing of electrons between atoms. A relatively great amount of energy is required to break a covalent bond; thus covalent bonds are **strong bonds**.

The secondary and tertiary structures of a protein are maintained for the most part by **noncovalent bonds** that do not entail sharing of electrons. These sorts of bonds are **weak bonds** because they can be broken with only relatively small amounts of energy. At least four mechanisms of noncovalent, weak bonding occur (Figure C). One is **ionic bonding,** which is electrostatic attraction between oppositely charged ionic regions of molecules. A second is **hydrogen bonding,** which occurs when a hydrogen atom covalently bonded to one atom (e.g., oxygen) is partially shared with another atom (e.g., another oxygen); for the most part, oxygen and nitrogen are the only sorts of atoms that associate by hydrogen bonding in biological systems. A third type of noncovalent bonding is **van der Waals interaction,** a type of attraction between nonpolar molecular regions that are close enough to each other to induce mutually attractive electric dipoles in each other's electron fields or otherwise attract in a similar way. The fourth type of noncovalent bonding is **hydrophobic bonding,** the tendency of nonpolar regions to associate with each other within an aqueous solution because their close association is more thermodynamically stable than an alternative configuration that would permit water molecules in between. The secondary structure of a protein (e.g., a β-sheet) is maintained mainly by hydrogen bonds. The tertiary structure is maintained by all four sorts of weak bonds.

One of the most important properties of the tertiary structure of proteins is that it is *flexible* because of the weak bonding that maintains it; small

Figure A The structural hierarchy of proteins

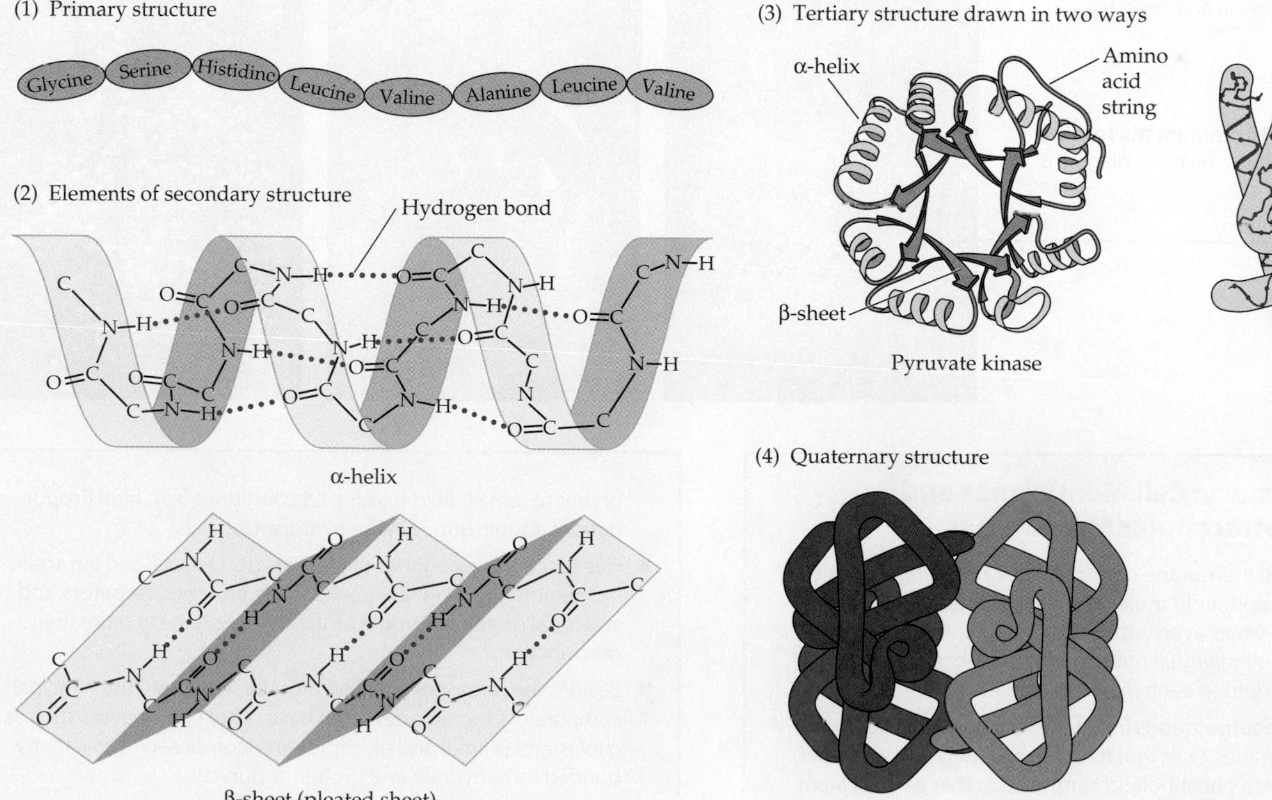

BOX 2.1 *(Continued)*

changes in three-dimensional shape are often essential for proteins to function properly. The conformation of a protein cannot, however, deviate too far from its native state. We are all aware of the radical change that occurs in the albumin protein of a hen's egg when we boil it; the protein changes from a transparent fluid to a white gel. As surprising as it may seem, the primary structure is not altered in this case. Instead, boiling corrupts the normal tertiary structure—an example of **denaturation**. Denaturation is a change in the three-dimensional structure of an intact protein that renders the protein nonfunctional. Cells have molecules called **molecular chaperones** that help ensure that a normal, functional three-dimensional conformation is assumed by proteins during their initial synthesis, and that sometimes are able to guide reversibly denatured proteins back to their functional conformations, thereby preventing them from becoming permanently denatured.

Some proteins have **quaternary structure** (see Figure A4). In such proteins, each finished molecule consists of two or more individual protein molecules bonded together into a multisubunit complex by weak (flexible) bonds. The quaternary structure refers to the three-dimensional arrangement of the subunit protein molecules in such an assembly. The example shown in Figure A4, which resembles mammalian hemoglobin, consists of four nearly identical molecules bound together in one assembly.

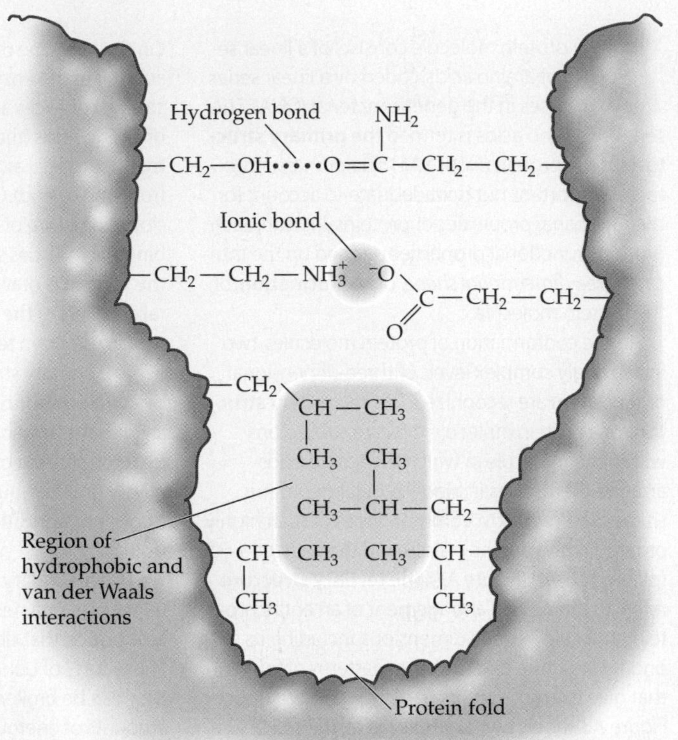

Figure C Types of weak, noncovalent bonds that are important in protein structure The bonds are illustrated where they stabilize a hairpin fold in a protein molecule.

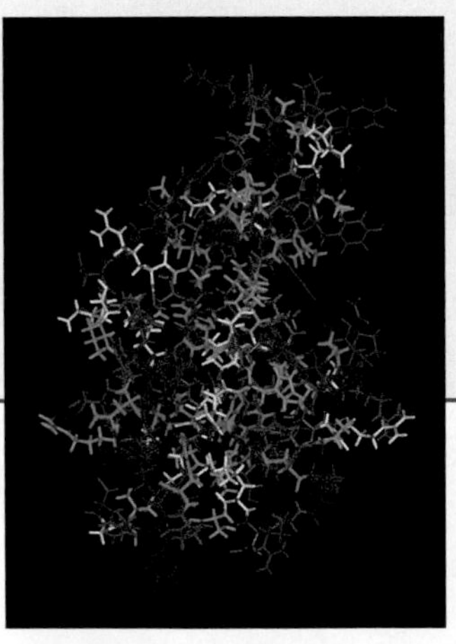

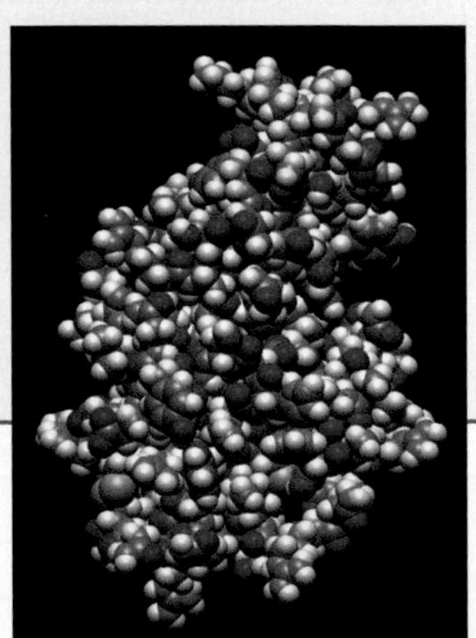

Figure B Three ways to diagram the tertiary structure of one protein All three diagrams represent lysozyme.

Summary: Cell Membranes and Intracellular Membranes

- The matrix of a cell membrane or intracellular membrane consists of a bilayer of phospholipid molecules. The phospholipids are chemically very diverse, even within a single membrane. They are fluid, meaning that individual molecules move about relatively freely by diffusion within each membrane leaflet.
- Animals exhibit adaptive trends in the phospholipid compositions of their cell membranes. Cells that function routinely at low temperatures tend to have a phospholipid composition that permits membranes to remain fluid under cold conditions (e.g., high proportions of double bonds in the hydrocarbon tails).
- Five functional categories of proteins occur in cell and intracellular membranes: channels, transporters, enzymes, receptors, and structural proteins. A single protein may engage in more than one function.
- Besides the phospholipids and proteins, which are the principal components, membranes often have other components such as cholesterol (a lipid) and glycoproteins (composed of covalently bonded carbohydrate and protein subunits).

Epithelia

When we consider the organization of tissues and organs in an animal's body, one of the first attributes worthy of note is that parts of the body are often bounded or separated by specialized epithelial tissue. An **epithelium** (plural *epithelia*) is a sheet of *cells* that covers a body surface or organ, or lines a cavity, thereby forming a boundary between the animal and the external environment or between functionally different regions of the body. Like the cell membranes of cells, epithelia are not simply physical boundaries. They also have numerous functional capacities and play major functional roles in animal physiology.

A **simple epithelium** consists of a single layer of cells (Figure 2.5*a*). Simple epithelia are exceedingly common; in the human body, for instance, the intestines, kidney tubules, blood vessels, and sweat glands are just some of the structures lined with simple epithelium. Each cell in a simple epithelium has an **apical surface** (*mucosal surface*) facing into a cavity or open space, and a **basal surface** (*serosal surface*) facing toward the underlying tissue to which the epithelium is attached. An epithelium typically rests on a thin, permeable, noncellular, and nonliving sheet of matrix material called the **basement membrane** (*basal lamina*), composed of glycoproteins and particular types of collagen. The basement membrane is secreted mostly by the epithelial cells, but the underlying cells also contribute. Simple epithelia are classified as *squamous*, *cuboidal*, or *columnar*, depending on how tall the cells are. The cells in a squamous epithelium are low and flat, whereas those in a columnar epithelium are high relative to their basal dimensions; the epithelium in Figure 2.5*a* is classified as cuboidal because the cells are about as tall as they are wide. Blood vessels usually do not enter epithelia. Instead, epithelial cells exchange O_2, CO_2, and other materials through the underlying basement membrane with blood capillaries located on the opposite side of the basement membrane.

The epithelium that lines the small intestines (midgut) of a mammal (Figure 2.5*b*) is an example of a simple epithelium that will be featured prominently in this book (e.g., in Chapters 3 and 4) and that introduces additional aspects of epithelial morphology and function. The apical surfaces of the cells in this columnar epithelium face into the lumen (open central cavity) of the intestine. Absorbed food molecules pass through the epithelium and basement membrane to reach blood vessels and lymph passages that transport them to the rest of the body. The intestinal epithelium—which includes absorptive-digestive cells and endocrine cells—illustrates that a simple epithelium can consist of two or more cell types. The intestinal epithelium also illustrates **microvilli** (singular *microvillus*), which are a common (although not universal) feature of epithelial cells.

Microvilli are exceedingly fine, fingerlike projections of the apical cell membrane. In the intestinal epithelium, the microvilli greatly increase the area of contact between the epithelial cells and the

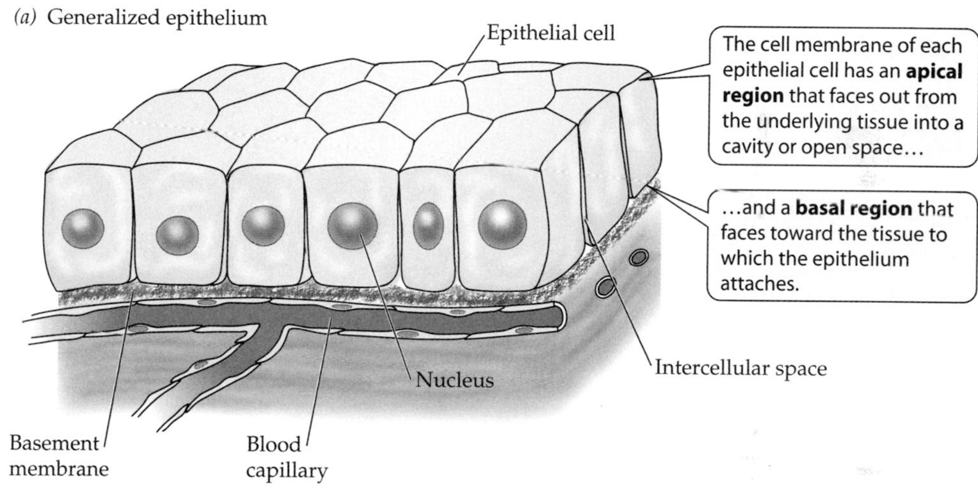

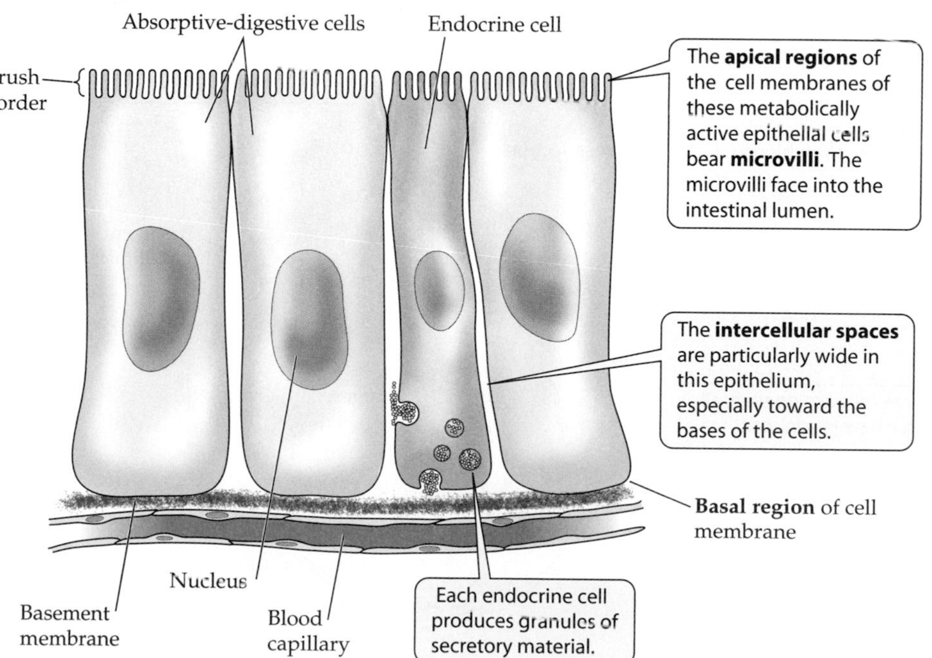

Figure 2.5 Simple epithelia *(a)* A generalized simple epithelium covering a free surface of a tissue. *(b)* The specific simple epithelium lining the mammalian small intestine (midgut). This epithelium consists of several cell types. Most cells are the absorptive-digestive cells emphasized here. Scattered among these cells are mucin-secreting cells (not shown) and at least ten types of endocrine or endocrine-like cells. Each endocrine cell produces granules of secretory material; the granules move to juxtapose themselves with the basal or near-basal regions of the cell membrane and disgorge into the spaces outside the cell, after which the secretions enter the blood for transport elsewhere. Endocrine-like cells termed *paracrine cells* also produce secretions, but their secretions affect nearby cells rather than acting on distant cells by way of the circulation.

(a) Epithelial cells can form tubules and follicles

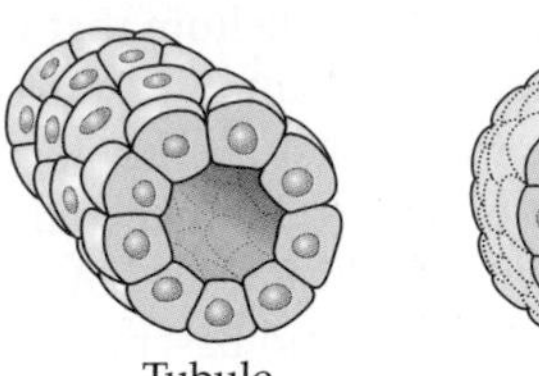

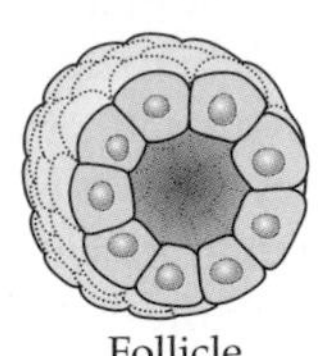

(b) Proximal tubule of mammalian kidney in cross section

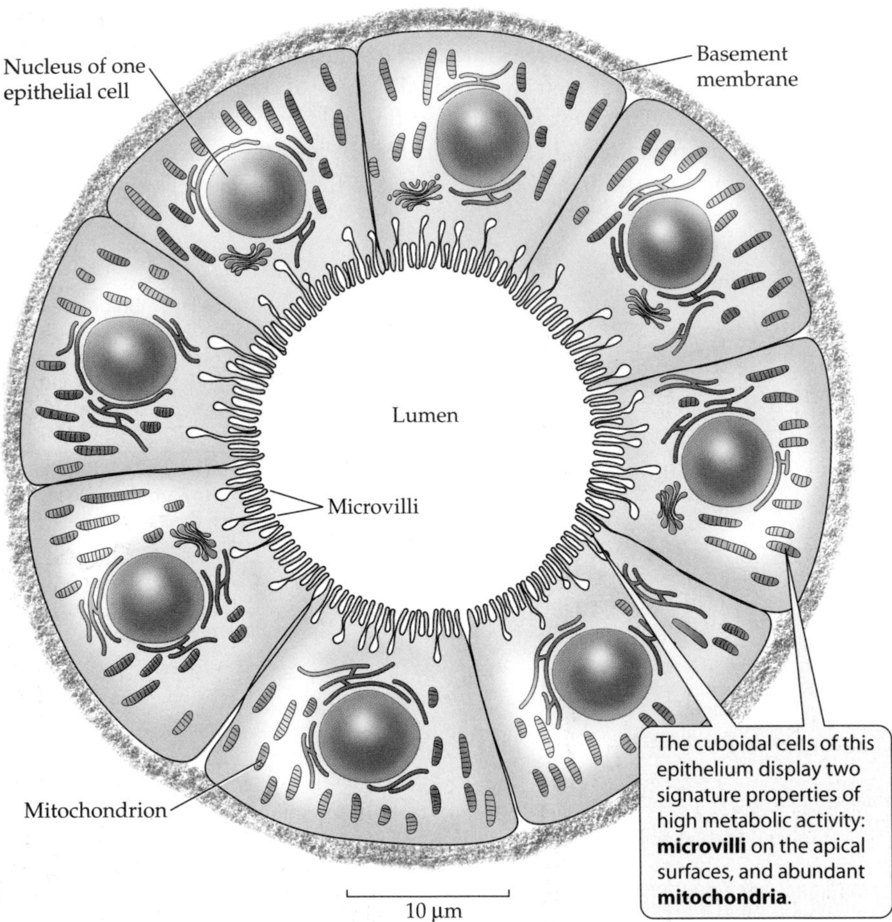

(c) Mammalian blood capillary in cross section

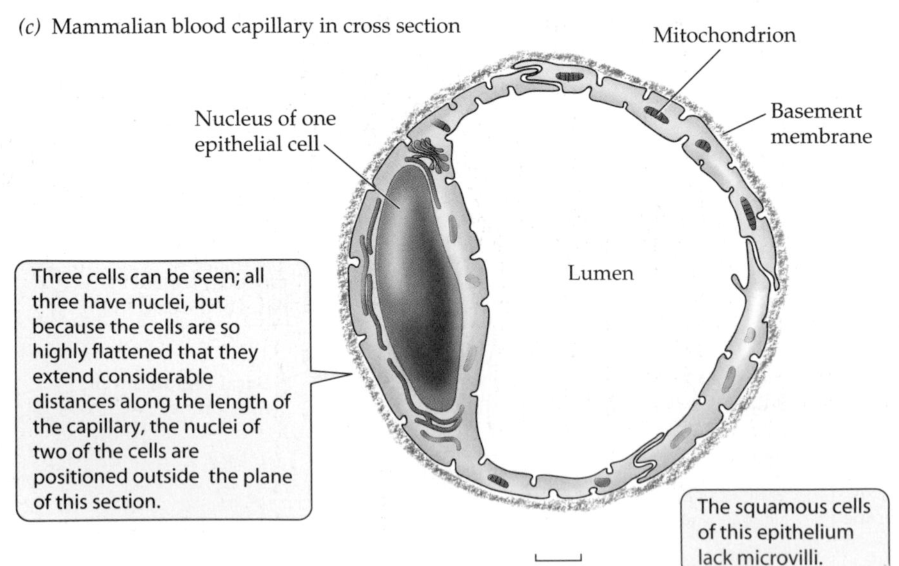

Figure 2.6 Tubules and follicles formed by simple epithelia *(a)* Both tubules and follicles are formed by simple epithelia wrapping into closed curves. Cross sections of two important tubular structures are shown in *(b)* and *(c)*; in each case the basal cell surfaces and basement membrane of the epithelium are on the outside. For historical reasons, the cells of blood capillaries are usually called *endothelial* cells, but they are a form of epithelium.

contents of the gut. Microvilli are most often found in epithelia that are very active in secreting or absorbing materials, such as the epithelia of certain kidney tubules and the pancreatic ducts, as well as the gut epithelium. Microvilli are often described as a **brush border** because they look like bristles on a brush when viewed microscopically.

Another significant aspect of diversity in simple epithelia is that tubules or follicles (hollow globes) are often formed by the wrapping of a simple epithelium into a closed curve (Figure 2.6*a*) supported by the basement membrane on the outside. The proximal parts of the kidney tubules, for example, consist of a circular layer of cuboidal epithelium bearing microvilli (Figure 2.6*b*). The blood capillaries are an especially important example. Each blood capillary consists of a single layer of highly flattened epithelial cells (lacking microvilli) supported by the epithelial basement membrane (Figure 2.6*c*). The basement membranes of capillaries are one of the important biochemical targets of the venoms of puff adders and rattlesnakes. The venoms cause internal hemorrhaging because they contain enzymes (metalloproteases) that break down the basement membranes, destroying the integrity of blood capillaries.

Adjacent cells in an epithelium are physically joined by cell-membrane junctions of several sorts; the four most important of these are tight junctions, septate junctions, desmosomes, and gap junctions (Figure 2.7). In the paragraphs that follow, we will look at each of these types of junction in turn.

A **tight junction** is a place were the cell membranes of adjacent cells are tightly joined so that there is no extracellular space between the cells; adjacent cells are perhaps 10 to 20 nm apart for the most part, but at tight junctions the cell membranes meet or fuse. Tight junctions typically occur between the sides of adjacent cells just a short distance away from their apical surfaces (Figure 2.8). Any given epithelial cell has tight junctions with adjacent epithelial cells in a continuous ring around its entire perimeter. *The ring of tight junctions demarcates the apical surface of the cell from its lateral and basal surfaces, giving rise to one of the most important distinctions in the physiological study of epithelia, the distinction between the* apical *and* basolateral *regions of cell membranes.* Many invertebrate groups have **septate junctions** instead of tight junctions. Septate junctions differ from tight junctions in their fine structure (see Figure 2.7), but they resemble tight junctions in their position and in the fact that they fully encircle each cell. Tight and septate junctions are sometimes aptly called **occluding junctions** because they block or occlude the spaces between

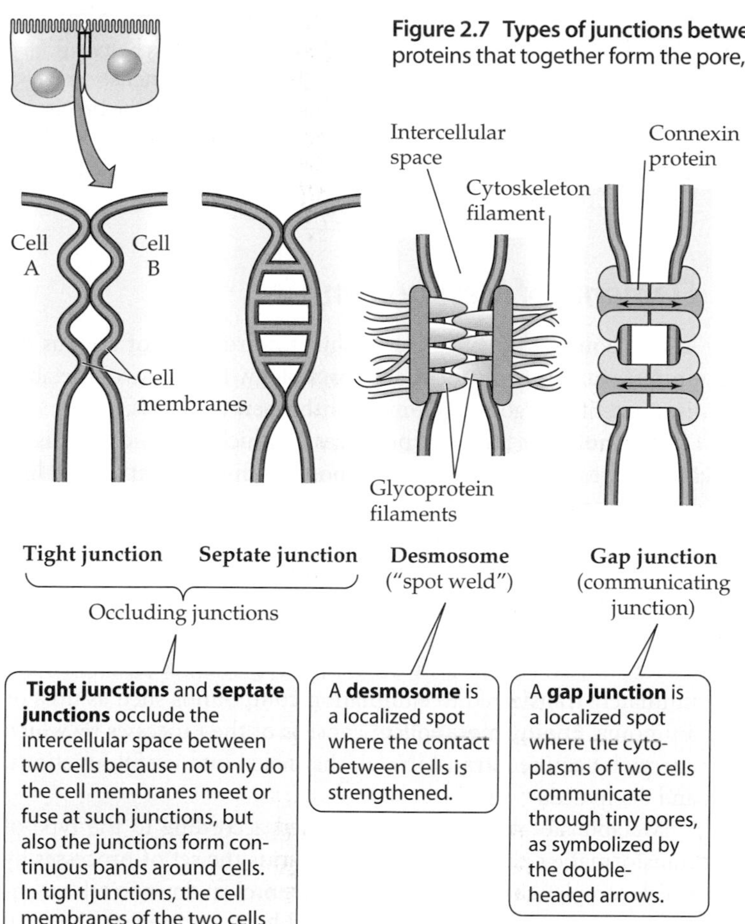

Figure 2.7 Types of junctions between cells At a pore in a gap junction, each cell has a ring of six connexin proteins that together form the pore, and the rings of the two cells line up to create continuity between cells.

adjacent epithelial cells, preventing open passage between the fluids on either side of an epithelium.

A **desmosome** (see Figure 2.7) is a junction at which mutually adhering glycoprotein filaments from two adjacent cells intermingle across the space between the cells. Desmosomes are often likened to rivets or spot welds because they occur as tiny isolated spots, not continuous bands, and their principal function is believed to be to strengthen and stabilize contacts between adjacent cells.

Gap junctions are like desmosomes in that they occur at discrete spots, but otherwise they are very different from all the other junctions we have discussed because within a gap junction are places where two adjacent cells lack cell-membrane boundaries, meaning that there is continuity between the cytoplasms of the cells. Molecules (such as simple sugars) and ions smaller than 1000 to 1500 daltons in molecular mass are able to pass between cells at gap junctions, although large solutes such as proteins cannot. Gap junctions are important in electrophysiological interactions between cells and are discussed in detail in Chapter 12.

A central feature of epithelia is that each epithelial cell is *functionally asymmetric*. As the membrane proteins of a cell diffuse about within the fluid phospholipid matrix of the cell membrane, they are unable (for reasons only poorly known) to diffuse through tight junctions. That is, the ring of tight junctions around each epithelial cell acts as a fence that keeps proteins from crossing between the apical and basolateral regions of the cell membrane. Thus the two regions of the cell membrane have different sets of channels, transporters, membrane enzymes, and other classes of membrane proteins, and they are functionally different in many ways. Differences also exist between the apical and basolateral regions in the membrane phospholipids composing the outer (but not inner) leaflet of the cell membrane.

One of the important functions of epithelia is to control and mediate the transport of substances between their two sides and, thus, between different body regions. Substances such as ions, nutrient molecules, or water pass through a simple epithelium by two types of paths (Figure 2.9). They may pass *through* cells by **transcellular** paths. Alternatively, they may pass *between* cells, in

Figure 2.8 The organization of an epithelial cell into apical and basolateral regions *(a)* The cell membrane of an epithelial cell is divided into apical and basolateral regions by a band of tight junctions. The apical and basolateral regions typically differ greatly in their protein components and functional properties. In this book, we will often use the schematic format shown in *(b)* to represent an epithelium. Microvilli do not always occur, but when they do, they are on the apical side only.

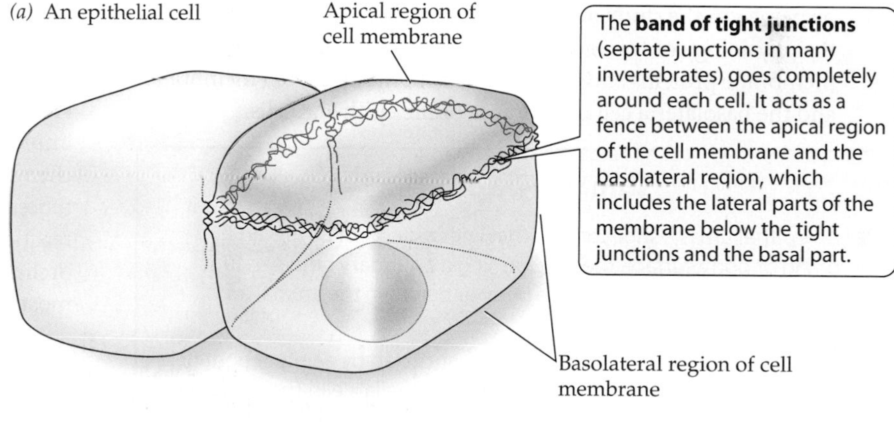

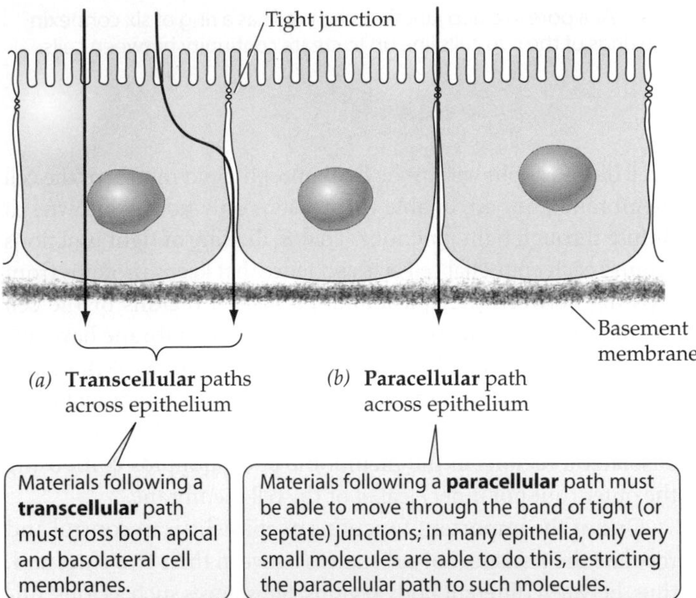

Figure 2.9 Transcellular and paracellular paths across an epithelium

paracellular paths. Tight junctions interfere with or block the paracellular movement of substances across an epithelium. In some epithelia the tight junctions prevent almost all paracellular movement. In others, however, the tight junctions permit extensive paracellular movement of certain sorts of molecules or ions, and the epithelia are described as **leaky**.

A substance that crosses an epithelium by a transcellular path must pass through two *cell* membranes. One of the most important principles in the study of epithelia is that for scientists to understand the physiology of transcellular transport, they must understand the membrane proteins and functions of both the apical cell membranes and the basolateral cell membranes of the epithelial cells.

Summary: Epithelia

- An epithelium is a sheet of cells that lines a cavity or covers an organ or body surface, thereby forming a boundary between functionally different regions of the body or between the animal and the external environment.
- In a simple epithelium, each cell is fully encircled by a ring of tight or septate junctions formed with adjacent epithelial cells. These occluding-type junctions seal the spaces between adjacent cells. Moreover, the ring of junctions around each cell divides the cell membrane into chemically and functionally distinct apical and basolateral regions.
- An epithelium rests on a nonliving, permeable basement membrane secreted by the epithelial cells and underlying tissue. The apical membranes of metabolically active epithelial cells often bear a brush border of microvilli, greatly enhancing their surface area. Besides the occluding junctions, adjacent epithelial cells are joined by structurally reinforcing "spot welds" called desmosomes and sometimes by gap junctions at which continuity is established between the cytoplasms of the cells.
- Materials pass through epithelia by paracellular paths between adjacent cells and by transcellular paths through cells. Materials traveling through a cell, must pass through both the apical and the basolateral cell membranes of the cell.

Elements of Metabolism

At this point in the chapter, we shift toward more of a focus on processes and less of a focus on morphology. The basics of metabolism constitute a good beginning for this new emphasis.

Metabolism is the set of processes by which cells and organisms acquire, rearrange, and void commodities in ways that sustain life. Metabolism involves myriad chemical and physical processes. To give order to their research, animal physiologists subdivide the study of metabolism. One way of doing this is according to specific commodities. For example, **nitrogen metabolism** is the set of processes by which nitrogen is acquired, employed in synthetic reactions to create proteins and other functional nitrogenous compounds, and ultimately transferred to elimination compounds such as urea or ammonia. **Energy metabolism** consists of the processes by which energy is acquired, transformed, channeled into useful functions, and dissipated.

Metabolism also may be subdivided according to the type of transformation that occurs. **Catabolism** is the set of processes by which complex chemical compounds are broken down to release energy, create smaller chemical building blocks, or prepare chemical constituents for elimination. **Anabolism**, on the other hand, consists of the processes that synthesize larger or more complex chemical compounds from smaller chemical building blocks, using energy. Catabolism is destructive; anabolism is constructive.

Metabolism depends on *sets* of biochemical reactions, such as the 30 or so reactions that cells employ to oxidize glucose into CO_2 and H_2O. The prominence of biochemistry in metabolism can give the impression that cells are just like test tubes: merely places where chemicals react. There is a massive distinction between cells and test tubes, however. Whereas test tubes are simply places where chemical reactions occur, cells orchestrate their own chemistry. The cellular orchestration of metabolism is directed by genes and mediated mostly by enzymes.

Enzyme Fundamentals

In his story "The Celebrated Jumping Frog of Calaveras County," Mark Twain appealed to the imagination of his readers by extolling the awesome jumping abilities of a frog, probably a common leopard frog (*Rana pipiens*) (**Figure 2.10*a***), named Dan'l Webster. Anyone who has ever tried to catch leopard frogs knows that when first disturbed, they can hop away at lightning speed. Thus it's hard not to smile in knowing admiration as Twain describes Dan'l Webster's celebrated jumping feats. Muscles can work only as fast as they are supplied with adenosine triphosphate (ATP). Amphibians, however, have only modest abilities to make ATP using oxygen (O_2) because (*a*) A fast-jumping amphibian: the leopard frog (*Rana pipiens*)

(*b*) A slow-jumping amphibian: the western toad (*Bufo boreas*)

Figure 2.10 Two amphibians with different jumping capabilities based in part on different levels of a key enzyme, lactate dehydrogenase

they have relatively simple lungs and can supply their cells with O_2 only relatively slowly. For leopard frogs to hop along as fast as they do when fleeing danger, they need to make ATP faster than the O_2 supply to their muscles permits. That is, they must make ATP by *anaerobic* mechanisms not requiring O_2. A crucial reason they can do this is that their leg muscles are well endowed with an enzyme called *lactate dehydrogenase.*

Compared to leopard frogs, toads such as the common western toad of North America (*Bufo boreas*) (Figure 2.10*b*) are not nearly as well endowed with lactate dehydrogenase. Thus they cannot make ATP to a great extent without O_2, and the slow rate of O_2 delivery to their muscle cells means a slow rate of ATP production, explaining why they cannot hop along as fast as frogs. Mark Twain could not have known it because the study of enzymes was just beginning during his life, but when he searched his mind for an amphibian that could inspire his readers as a "celebrated" jumper, he chose a frog rather than a toad in major part because frogs have more of the enzyme lactate dehydrogenase.

Enzymes are protein catalysts that *speed*, and often *regulate*, reactions. To appreciate the roles of enzymes, it is important to recognize that the vast majority of the biochemical reactions that occur in animals *do not take place spontaneously* at significant rates under physiological conditions. Cells are biochemically complex enough that, in principle, tens of thousands of reactions might occur in them. However, because reactions in general require catalysis to occur at significant rates, the particular reactions that do take place in a cell—out of all those that *could* take place—depend on the cell's own biosynthesis of enzyme proteins. *Enzymes represent one of the foremost means by which cells take charge of their own biochemistry.*

When we say that an enzyme is a **catalyst**, we mean that it is a molecule that accelerates a reaction without, in the end, being altered itself. The reaction catalyzed by lactate dehydrogenase (LDH) that is important for escape by frogs is the reduction of pyruvic acid to form lactic acid, a reaction in which each pyruvic acid molecule is combined with two hydrogen atoms (Figure 2.11*a*). Although the presence of LDH speeds this reaction, LDH is not itself altered by the reaction. Thus a molecule of LDH persists as it catalyzes the reduction of numerous molecules of pyruvic acid.

Enzymes are described as having *substrates* and *products*, and often there are two or more of each. To be exact about the substrates and products of LDH, a chemically complete presentation of the LDH-catalyzed reaction is needed (Figure 2.11*b*). The hydrogen atoms that reduce pyruvic acid are taken from a molecule we symbolize as $NADH_2$. NAD is an enzyme cofactor (nicotinamide adenine dinucleotide) found in all animal cells; and $NADH_2$ symbolizes the reduced form of this cofactor, the form that is combined with

(*a*) Simplified reaction

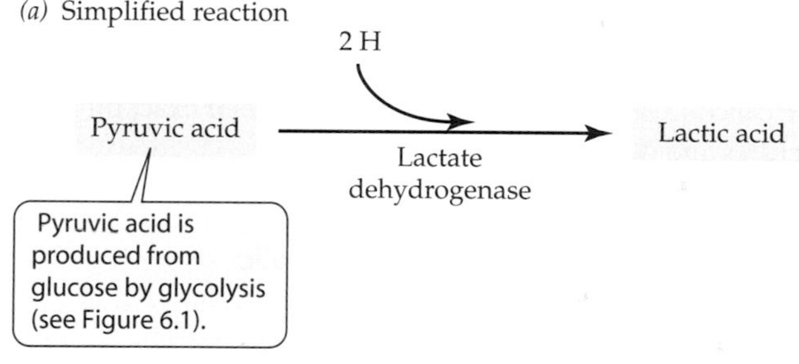

(*b*) Full reaction

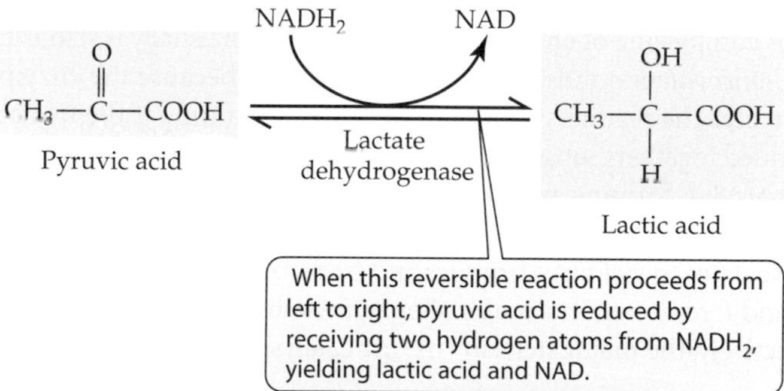

Figure 2.11 The reaction catalyzed by lactate dehydrogenase The enzyme cofactor nicotinamide adenine dinucleotide (NAD) acts as an electron (or hydrogen) shuttle by undergoing reversible reduction (forming $NADH_2$) and oxidation (forming NAD). As *(b)* shows, when the reaction catalyzed by LDH proceeds from left to right, $NADH_2$ produced elsewhere is converted to NAD, renewing the supply of NAD. The reaction catalyzed by LDH is reversible, but the NAD reaction involved in the reverse direction is not shown. Chapter 6 discusses the important role of the LDH-catalyzed reaction in ATP production.

hydrogen. The **substrates** of an enzyme are the initial reactants of the reaction that the enzyme catalyzes; the **products** of the enzyme are the compounds produced by the reaction. Thus, in the reaction we are discussing—the left-to-right reaction in Figure 2.11*b*—the substrates of LDH are pyruvic acid and $NADH_2$, and the products are lactic acid and NAD. Chapter 6 discusses how this reaction aids rapid jumping by frogs, as well as other forms of sudden, intense vertebrate exercise, such as sprinting by people. Put simply, the way the reaction helps is precisely that it produces NAD, an essential compound for ATP synthesis by glycolysis.

There are many kinds of enzymes; mammalian cells, for instance, typically synthesize several thousand. Because enzymes are so numerous, their nomenclature presents challenges. Each described enzyme is given a unique number and formal name by an international body called the Enzyme Commission (EC). Scientists usually do not use these numbers and formal names in everyday speech, however. Instead, informal "common" names are used. *Lactate dehydrogenase*, for example, is a common name and refers to an enzyme more exactly known as EC 1.1.1.27. The names of enzymes usually end in *-ase*. Thus, when you see a biochemical term that ends in *-ase*, it usually refers to an enzyme. Later we will see that a single enzyme may exist in multiple molecular forms in different tissues or different animal species, creating additional nomenclatural challenges. When we speak as we have in this paragraph about a "kind" of enzyme, we are referring to the specific reaction catalyzed by the protein. Lactate dehydrogenase—EC 1.1.1.27—is defined to be the particular kind of enzyme that catalyzes the reaction in Figure 2.11*b*. There is not, however, just one single molecule that can do this. Many related molecules catalyze the reaction, and therefore all of them are called LDHs, even though they vary in their exact molecular structures and detailed functional properties.

Enzyme-catalyzed reactions exhibit hyperbolic or sigmoid kinetics

For an enzyme molecule to catalyze a reaction, it must first combine with a molecule of substrate to form an **enzyme–substrate complex**. (Here, for simplicity, we assume there is only one substrate.) This complexing of enzyme and substrate, which usually is stabilized by *noncovalent* bonds, is *essential* for catalysis because the enzyme can alter the readiness of the substrate to react only if the two are bonded together. Substrate is converted to product *while united with the enzyme*, forming an **enzyme–product complex**, also usually held together by *noncovalent* bonds. The enzyme–product complex then dissociates to yield free product and free enzyme. Symbolically, if *E*, *S*, and *P* represent molecules of enzyme, substrate, and product, respectively, the major steps in enzyme catalysis are

$$\mathrm{E + S \rightleftharpoons E\text{–}S\ complex \rightleftharpoons E\text{–}P\ complex \rightleftharpoons E + P} \quad (2.1)$$

Note that, as stressed earlier, the enzyme emerges unaltered.

An enzyme-catalyzed reaction occurs at a rate that is affected by the relationship between the available number of enzyme molecules and the concentration of substrate. The **reaction velocity** (*reaction rate*) is the amount of substrate converted to product per unit time. At relatively low substrate concentrations, the reaction velocity increases as the substrate concentration increases (Figure 2.12). However, an increase in the substrate concentration does not increase the reaction velocity beyond a certain point. The reason for this overall behavior is precisely that substrate must combine with enzyme molecules to form product. As shown in Figure 2.12*a*, at low substrate concentrations (❶), the amount of substrate available is the limiting factor in determining the reaction velocity. Because all of the available enzyme molecules are not occupied by substrate when the substrate concentration is low, raising the substrate concentration (as from ❶ to ❷) increases the reaction velocity by utilizing more of the available enzyme molecules. At high substrate concentrations (❸) however, the amount of enzyme is the limiting

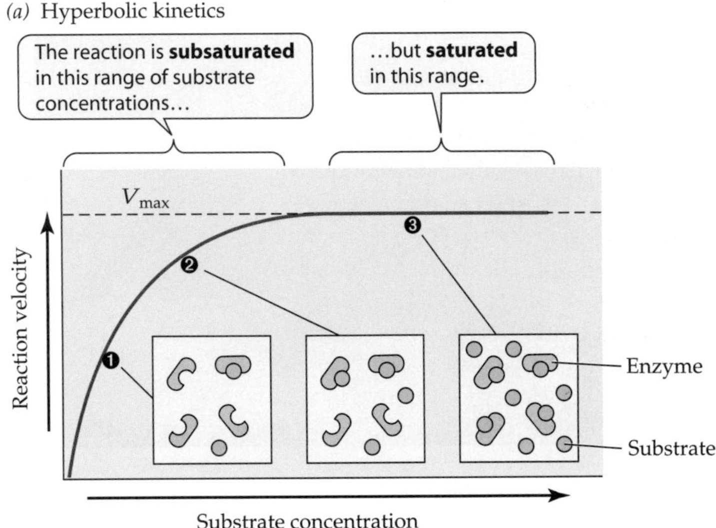

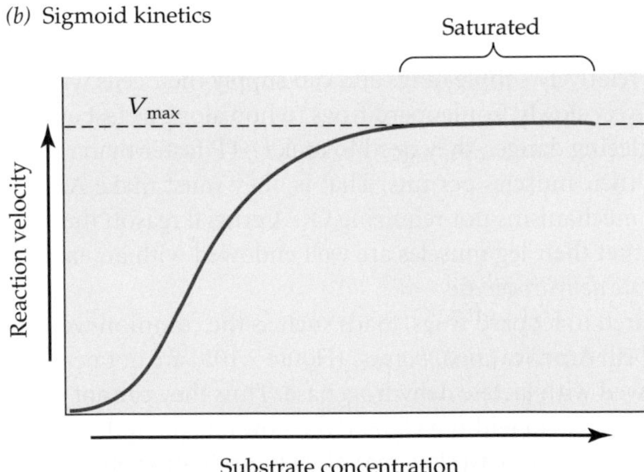

Figure 2.12 Reaction velocity as a function of substrate concentration *(a)* Some enzymes exhibit hyperbolic kinetics, in which the reaction velocity increases as shown, asymptotically approaching a maximum velocity, called V_{max}. The reaction velocity increases from ❶ to ❷ because the increase in availability of substrate allows a greater fraction of enzyme molecules to be engaged in catalysis at any given time. At ❸, however, the reaction velocity cannot increase further, because substrate is so abundant that all the enzyme molecules are engaged to the fullest extent possible. *(b)* Some enzymes exhibit sigmoid kinetics, in which the approach to V_{max} follows an S-shaped (sigmoid) trajectory; the reaction velocity cannot exceed V_{max} for the same reason as in hyperbolic kinetics.

factor in determining the reaction velocity. When the substrate concentration is high, the population of available enzyme molecules becomes **saturated**, meaning that each enzyme molecule is occupied by a substrate molecule nearly all of the time. Increasing the substrate concentration therefore cannot increase the reaction velocity further.

Enzyme-catalyzed reactions are one of the types of reactions that exhibit **saturation kinetics. Kinetics** refers to the velocity properties of reactions. A reaction exhibits *saturation* kinetics if it is limited to a maximum velocity because there is just a limited supply of a molecule (the enzyme in the case of enzyme-catalyzed reactions) with which other molecules must reversibly combine for the reaction to take place.

Two types of saturation kinetics are exhibited by various enzyme-catalyzed reactions: **hyperbolic kinetics** (*Michaelis–Menten kinetics*) (Figure 12.12*a*) and **sigmoid kinetics** (Figure 2.12*b*). Whether the kinetics are hyperbolic or sigmoid depends in major part on the chemical properties of the enzyme involved. Hyperbolic kinetics occur when each enzyme molecule has just one substrate-binding site for the particular substrate of interest or when there are multiple sites but the sites behave independently. Sigmoid kinetics occur when each enzyme molecule has multiple substrate-binding sites and the multiple sites influence each other by way of ripple effects within the enzyme molecule (discussed later) so that catalytic activity at any one site depends on whether or not binding has occurred at other sites.

A mathematical description of hyperbolic kinetics was first provided by Leonor Michaelis and Maude Menten in 1913. Their equation, after being revised by other chemists about a decade later, is called the **Michaelis–Menten equation:**

$$V = \frac{V_{max}[S]}{[S] + K_m} \tag{2.2}$$

where V is the reaction velocity at any given substrate concentration $[S]$, V_{max} is the maximum reaction velocity (assuming a certain fixed amount of enzyme to be present), and K_m is a constant that is usually termed the **Michaelis constant.**[2] This equation describes the curve plotted in Figure 2.12*a*.

Maximum reaction velocity is determined by the amount and catalytic effectiveness of an enzyme

Two properties determine the **maximum velocity** (**V_{max}**) at which a saturated enzyme-catalyzed reaction converts substrate to product (see Figure 2.12). One is the number of active enzyme molecules present. The second is the catalytic effectiveness of each enzyme molecule.

The catalytic effectiveness of an enzyme is expressed as its **turnover number** (**k_{cat}**), the number of substrate molecules converted to product per second by each enzyme molecule when saturated. Different enzymes vary enormously in turnover number. Indeed, even the molecular variants of a single enzyme can vary substantially in this crucial property. Some enzymes are so catalytically effective that when they are saturated, each enzyme molecule converts 10,000 substrate molecules to product each second, whereas others convert only one substrate molecule to product per enzyme molecule per second.

The catalytic effectiveness of an enzyme depends in part on the **activation energy** of the enzyme-catalyzed reaction, but it does not depend only on this property. To understand the implications of activation energy, it is necessary to recognize that a substrate molecule must pass through an intermediate chemical state termed a **transition state** to form a product molecule. Thus, one can think of any reaction, whether or not it is enzyme catalyzed, as involving first the conversion of the substrate to a transition state, and second the conversion of the transition state to the product. For a substrate molecule to enter the transition state, its content of energy must increase. The amount by which it must increase is the activation energy of the reaction. Molecules gain the energy they need as a result of random collisions with other molecules. Any particular substrate molecule has a continuously fluctuating energy content as it gains and loses energy through intermolecular collisions; as its energy content rises and falls, it undergoes reaction when its energy content is boosted by an amount at least equal to the activation energy. An enzyme accelerates a reaction by lowering the activation energy (Figure 2.13). The *extent* to which it lowers the activation energy is one factor that determines the enzyme's catalytic effectiveness.

According to modern theories of how enzymes work, catalytic effectiveness also depends critically on the rates at which enzyme molecules can go through molecular conformational changes required for catalysis. As discussed later, enzyme molecules change shape when they bind with substrate and again when they release product. There is reason to believe that different enzymes vary in the rates at which they can go through these necessary conformational changes, and differences in these rates may be as important as differences in activation energy in determining the relative turnover numbers of enzymes.

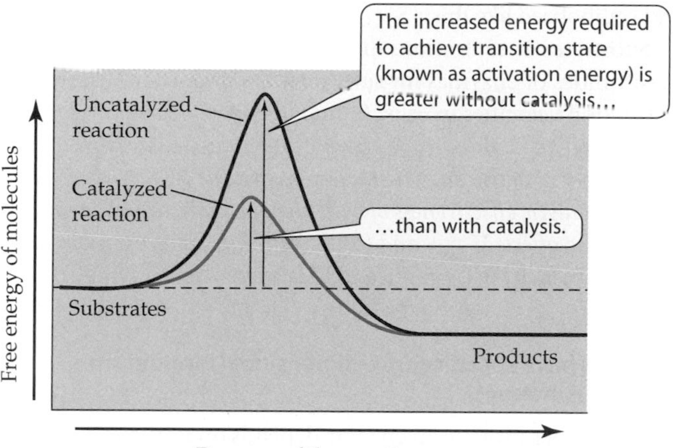

Figure 2.13 Enzymes speed reactions by lowering the needed activation energy Starting at their average energy level, substrate molecules must gain sufficient energy to enter a transition state before they can react to form product. The amount of energy required, the activation energy, is lowered by enzyme catalysts. Catalysts do not, however, alter the average free energy of either substrates or products; nor do they affect the relative concentrations of substrates and products at equilibrium.

[2] Square brackets signify concentration. Thus, $[S]$ is the concentration of compound S.

Enzyme–substrate affinity affects reaction velocity at the substrate concentrations that are usual in cells

In a cell, a collision between an enzyme molecule and substrate molecule does not necessarily result in the formation of an enzyme–substrate complex. The two molecules may instead collide and separate. The outcome of a collision depends on a property of the enzyme called **enzyme–substrate affinity**, which refers to the proclivity of the enzyme to form a complex with the substrate when the enzyme and substrate meet. An enzyme that is highly likely to form complexes with substrate molecules it contacts has a high enzyme–substrate affinity. Conversely, an enzyme that is unlikely to form complexes has a low enzyme–substrate affinity.

The affinity of an enzyme for its substrate affects the shape of the velocity–concentration relation at *subsaturating* concentrations of substrate, as illustrated in Figure 2.14*a* by enzymes with hyperbolic kinetics. Curve *x* in the figure represents an enzyme having a relatively high affinity for its substrate; curve *z* represents one having a relatively low affinity. All three enzymes represented in the figure have the same maximum velocity. The key difference among them is that at subsaturating substrate concentrations, the reaction velocity more closely approaches V_{max} if the enzyme has high substrate affinity (*x*) than low (*z*).

A convenient numerical expression of enzyme–substrate affinity for reactions showing hyperbolic kinetics is the **apparent Michaelis constant** or **half-saturation constant, K_m**, defined to be the substrate concentration required to attain one-half of the maximal reaction velocity. Figure 2.14*b* shows how K_m is determined for both the high-affinity enzyme *x* and the low-affinity enzyme *z* from Figure 2.14*a*. Note that the low-affinity enzyme has the greater K_m value. Thus, *K_m and enzyme–substrate affinity are related inversely*: A high K_m means low affinity, and a low K_m means high affinity. K_m is one of the parameters in the Michaelis–Menten equation (Equation 2.2). For enzyme-catalyzed reactions that follow sigmoid kinetics, the measure of affinity is once again the substrate concentration required to half-saturate the enzyme, but it is calculated in technically different ways and symbolized like this: $(S_{0.5})_{substrate}$.

Substrate concentrations in cells are usually subsaturating. Thus the affinities of enzymes for substrates are important determinants of reaction velocities in cells. *In sum, therefore, reaction velocities in cells depend on all three of the enzyme properties we have discussed: (1) the number of active enzyme molecules present (which affects V_{max}), (2) the catalytic effectiveness of each enzyme molecule when saturated (which also affects V_{max}), and (3) the affinity of enzyme molecules for substrate (which affects how close the velocity is to V_{max}).*

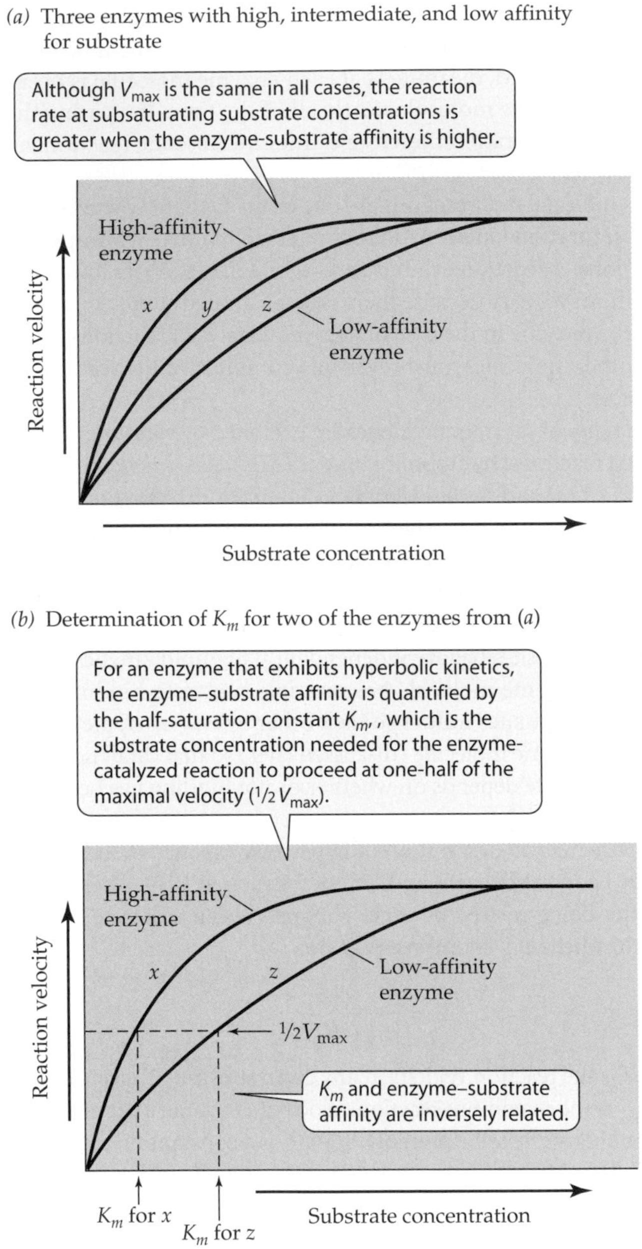

(*a*) Three enzymes with high, intermediate, and low affinity for substrate

(*b*) Determination of K_m for two of the enzymes from (*a*)

Figure 2.14 The approach to saturation depends on enzyme–substrate affinity

Enzymes have specific three-dimensional binding sites that often interact

Like any other protein, an enzyme depends on its three-dimensional molecular shape—its conformation—for its functional properties. One of the single most important attributes of enzymes and other proteins is that *their three-dimensional structure is stabilized mostly by weak, noncovalent bonds* (see Box 2.1). Weak bonds create flexible links between molecular regions that allow an enzyme's three-dimensional structure to change its detailed shape (while retaining its overall organization). Such shape changes, as already suggested, are crucial for proper enzyme function—for example, the binding of an enzyme to its substrate (Figure 2.15).

A substrate molecule binds with an enzyme molecule at a particular molecular region, at or near the surface of the enzyme, called the **active site** or **substrate-binding site**. The three-dimensional shape of the active site and the peculiarities of its chemical constituents complement a particular three-dimensional part of the substrate molecule such that the substrate molecule and the enzyme molecule fit together. The binding of the substrate and enzyme molecules is typically stabilized entirely by weak bonds, not covalent bonds. If an enzyme requires two or more substrates, the enzyme molecule has an active site specific for each. The active site(s) are said

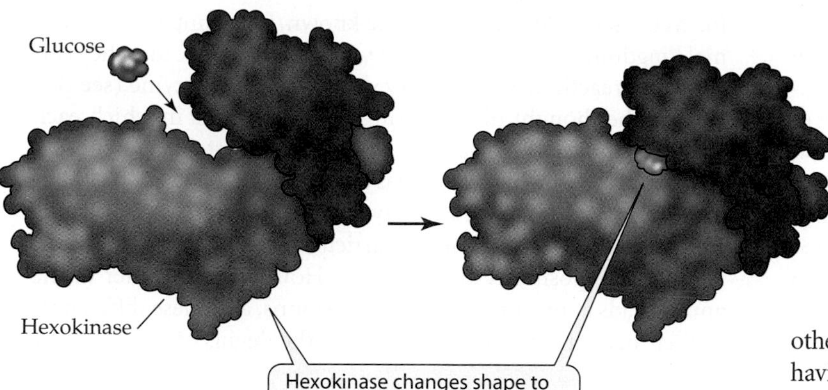

Figure 2.15 Molecular flexibility is important for enzyme function Hexokinase is the enzyme that ushers glucose into glycolysis by catalyzing glucose phosphorylation using a phosphate group from ATP (see Figure 2.18). The enzyme molecule is a single protein; the different shading of the two parts of the molecule is for visual clarity only.

to exist within a **catalytic vacuole**, a part of the enzyme molecule that provides a suitable setting for catalysis to occur.

Enzyme–substrate binding is sometimes said to resemble a lock and key fitting together, but this analogy is flawed in two important respects. First, the binding between the substrate and the corresponding active site on an enzyme is chemical and electrochemical in nature, not purely mechanical. The geometric array of specific chemical groups at the active site matches a complementary array of chemical groups on the substrate, facilitating the formation of weak chemical bonds that stabilize the enzyme–substrate complex. Second, the lock-and-key analogy erroneously suggests mechanical rigidity. In fact, the active site and other regions of an enzyme molecule are flexible and, as already stressed, change conformation when enzyme–substrate binding occurs (see Figure 2.15) and when product is released.

Many enzyme molecules consist of two or more noncovalently bonded proteins, and these often interact in important ways to determine enzyme properties. Enzyme molecules composed of two, three, or four protein subunits are called *dimeric*, *trimeric*, and *tetrameric*, respectively. All the subunits in a multisubunit enzyme may be chemically identical, or they may consist of two or more types. Multisubunit enzymes typically have multiple binding sites. The simplest version of this property is that a multisubunit enzyme may have an active site on each subunit. However, multisubunit enzymes often also have specific binding sites for molecules other than the substrate. These nonsubstrate-binding sites have important similarities to the active (substrate-binding) sites: They are at or near the surface of the enzyme molecule; they bind *noncovalently* and *reversibly* with specific molecules; and their specificity arises because they are complementary in three-dimensional shape and chemistry to parts of the molecules they bind. The substrates of enzymes and the molecules that bind to specific nonsubstrate-binding sites are collectively known as enzyme **ligands**. A ligand is any molecule that selectively binds by noncovalent bonds to a structurally and chemically complementary site on a specific protein; not just enzymes but also certain other sorts of proteins (e.g., transporters and receptors) are said to bind or combine with *ligands*, as we will see.

When an enzyme molecule has multiple binding sites, the binding of any one site to its ligand may facilitate or inhibit the binding of other sites to their ligands. Such interactions between the binding behaviors of different sites are termed **cooperativity**, whether they are facilitating or inhibiting. In **positive cooperativity**, ligand binding at one site facilitates binding of other sites on the same molecule to their ligands; in **negative cooperativity**, binding at one site inhibits binding at other sites on the same molecule. Besides being classified as positive or negative, cooperativity is also categorized as homotropic or heterotropic. In **homotropic cooperativity** the binding of a particular type of ligand facilitates or inhibits the binding of other molecules of the *same* ligand to the same enzyme molecule; homotropic cooperativity occurs, for example, when the binding of a substrate molecule to one of the active sites on a multisubunit enzyme molecule facilitates or inhibits the binding of other substrate molecules to other active sites (causing the kinetics to be sigmoid). In **heterotropic coooperativity** the binding of one type of ligand to an enzyme molecule influences the binding of other types of ligands.

A point deserving emphasis is that when cooperativity occurs, the interactions between binding sites on a molecule are interactions *at a distance*. The various binding sites on a multisubunit enzyme—whether they are sites for substrates or nonsubstrates—are usually not immediately next to each other. Instead, they are found at separate locations in the multisubunit molecular structure. Cooperativity occurs because the binding of a ligand to its particular binding site causes the conformation of the enzyme molecule to change in a way that ripples through the whole molecule, affecting the shapes and binding characteristics of all its other binding sites. The binding of a substrate molecule to an active site, for example, can cause shape changes that increase the affinity of all other active sites for the substrate.

A type of cooperativity that has great importance in the *control* of multisubunit enzymes is **allosteric modulation** (*allosteric modification*). Allosteric modulation means the modulation of the *catalytic* properties of an enzyme by the bonding of *nonsubstrate ligands* to specific nonsubstrate-binding sites, which are called **regulatory sites** or **allosteric sites**. The nonsubstrate ligands that participate in this sort of modulation are called **allosteric modulators**. In **allosteric activation** the binding of an allosteric modulator to its binding site on an enzyme molecule increases the affinity of the molecule's active sites for the substrate or otherwise increases the *catalytic* activity of the enzyme. In **allosteric inhibition** the binding of an allosteric modulator impairs the catalytic activity of an enzyme, as by decreasing its affinity for substrate. Allosteric modulation, as we will discuss, opens up vast regulatory possibilities.[3]

[3]Although the term *allosteric* was originally used only in the context of allosteric modulation, its meaning has evolved. Today *allosteric* is often used to refer to *any* form of enzyme conformation change that results from the noncovalent bonding of ligands to ligand-specific sites, not just allosteric modulation.

Enzymes catalyze reversible reactions in both directions

Like all other catalysts, enzymes accelerate reversible reactions in both directions. LDH, for example, can accelerate either the reduction of pyruvic acid (in Figure 2.11*b*, the reaction going from left to right) or the oxidation of lactic acid (in Figure 2.11*b*, the reaction going from right to left). Although all the reactions that take place within animals are reversible in principle, only some are reversible in practice. This is true because some reactions—for reasons unrelated to the enzymes that catalyze them—always proceed significantly in just one direction under the conditions that prevail in the body.

The direction of a reversible enzyme-catalyzed reaction is determined by the principles of **mass action**. Consider the following reversible reaction:

$$A + B \rightleftharpoons C + D \tag{2.3}$$

If the four compounds A, B, C, and D are mixed and then left alone, they will react until they reach equilibrium. The reaction equilibrium is characterized by a particular *ratio of concentrations* of the four compounds. At equilibrium, this ratio—$[C][D]/[A][B]$—is always the same. The principles of mass action state that if compounds are out of equilibrium, the reaction will proceed in the direction of equilibrium as dictated by the ratios of concentrations. Thus, for example, if the reactants on the left, A and B, are collectively too concentrated, relative to C and D, for the equilibrium state to exist, the reaction will proceed to the right, thereby lowering the concentrations of A and B and raising those of C and D.

An enzyme does not alter the principles of mass action. The catalytic effect of an enzyme on a reversible reaction is to increase the rate of approach to equilibrium from either direction. Note, nonetheless, that the substrate or substrates are different from the two directions. For instance, when LDH catalyzes the reaction in Figure 2.11 going from left to right, its substrates are pyruvic acid and $NADH_2$; whereas when it catalyzes the reaction going from right to left, its substrates are lactic acid and NAD. The enzyme–substrate affinity of an enzyme and its other kinetic properties are typically different for the substrates of the reaction going in the left-to-right direction than for those of the reaction in the right-to-left direction. Thus although an enzyme always catalyzes a reversible reaction in both directions, its catalytic behavior in the two directions may be very different.

Reversible reactions in cells are typically directional at any given time because they operate dynamically in a state that remains far from equilibrium. In a test tube, if A and B in Equation 2.3 were initially concentrated enough for the reaction to proceed to the right, the reaction itself would draw down the concentrations of A and B and create an equilibrium state. In a cell, however, the substrates of any one reaction are typically the products of other reactions; thus, A and B are likely to be replaced as they are converted to C and D, maintaining a state of disequilibrium in Equation 2.3. In this way, the enzyme-catalyzed reaction in a cell can display directionality even though the enzyme itself catalyzes both directions of reaction.

Multiple molecular forms of enzymes occur at all levels of animal organization

A single enzyme often exists in multiple molecular forms, which catalyze the same reaction, as stressed earlier. Dozens of described forms of lactate dehydrogenase are known, for example, in the animal kingdom. All are called *lactate dehydrogenase* because they catalyze one reaction. The primary structure of an enzyme (see Box 2.1) can be thought of as a string of amino acids in which each amino acid occupies a specific position in the string; an enzyme composed of 300 amino acids has 300 positions, for example. Multiple molecular forms of an enzyme typically have similar string lengths and are identical in the particular amino acids that occupy many of the positions on the string. However, they differ in the amino acids at one or more of the positions, and these differences in their primary structures often alter the details of their tertiary structures.

You might guess from what we have said about the existence of multiple molecular forms of enzymes that they are often related by evolutionary descent—that is, that certain forms evolved from others by mutations causing changes in the amino acid sequence. As we will see, biochemists in fact know enough about the exact structures of many different LDH molecules to be almost certain that the various forms of LDH are related by evolutionary descent, and probably it is a general rule that the multiple molecular forms of enzymes are families of evolutionarily related molecules.

LDH provides a good example for understanding multiple molecular forms of enzymes in greater detail. Individual vertebrates have two or three different gene loci that code for LDH proteins. Thus, two or three different forms of LDH protein—called LDH-A, LDH-B, and LDH-C—are synthesized in any one individual. The various gene loci are not, however, expressed equally in all tissues of an individual. An additional complexity is that each "finished" LDH molecule is a tetramer, consisting of four LDH protein molecules that are independently synthesized but linked together as subunits of the mature enzyme. The A and B forms are produced in all or nearly all vertebrates. Usually, skeletal muscle cells express the A genetic locus strongly and the B locus weakly. Thus, although some of the LDH tetramers produced in skeletal muscle consist of mixed A and B subunits, the principal type of LDH tetramer in the skeletal muscles is composed of all A subunits and is symbolized LDH-A_4. In contrast, the cells of heart muscle express the B genetic locus strongly, and their principal type of LDH tetramer is composed entirely of B subunits: LDH-B_4. In mammals, LDH-C is expressed in just a single organ, the mature testis; mammalian sperm LDH is mostly LDH-C_4.

Both finer-scale and larger-scale variation occur in the forms of LDH. At a finer scale than we have already described, two or more alleles may exist at each genetic locus within a species. Thus, for instance, a species might have two alleles for the A locus, meaning that two different types of the A protein can be synthesized; the skeletal muscles of the species would then exhibit multiple molecular forms of the finished LDH-A_4 enzyme (i.e., LDH-A_4 in which all four A subunits are of the sort coded by one allele, LDH-A_4 in which all four subunits are coded by the other allele, and LDH-A_4 in which some of the A subunits are coded by one allele and some by the other allele). At a larger scale, it is quite common for differences to exist between different species in the A, B, and C proteins. For example, although the A proteins synthesized by laboratory rats and by people are similar, they are not identical, so rat LDH-A_4 differs from human LDH-A_4.

Researchers have developed a complex terminology to describe all the possibilities. For the purposes of an introduction to animal physiology, a simple dichotomy between **isozymes** (*isoenzymes*) and **interspecific enzyme homologs** is sufficient. Isozymes are the different molecular forms of an enzyme produced by a single species; thus the LDH-A_4, LDH-B_4, and LDH-C_4 produced in people constitute three isozymes of LDH. Interspecific enzyme homologs are the different molecular forms of an enzyme coded by homologous gene loci in different species; human LDH-A_4 and rat LDH-A_4, for instance, are interspecific enzyme homologs (Figure 2.16).

Isozymes and interspecific enzyme homologs often differ not only in their *catalytic* properties but also in their *regulatory* properties. An important insight emphasized by recent research is that for differences in the amino acid structure to alter the catalytic or regulatory properties of an enzyme, the amino acid differences do *not* need to be located within the active site or within a nonsubstrate-binding site. In fact, amino acid substitutions seem typically to be outside the ligand-binding sites. Substitutions outside the ligand-binding sites can alter the catalytic or regulatory properties of an enzyme molecule because they can change the flexibility of the molecule (which is of fundamental importance), and they can cause shifts in the three-dimensional conformation of the whole molecule that indirectly affect the conformation of the binding sites.

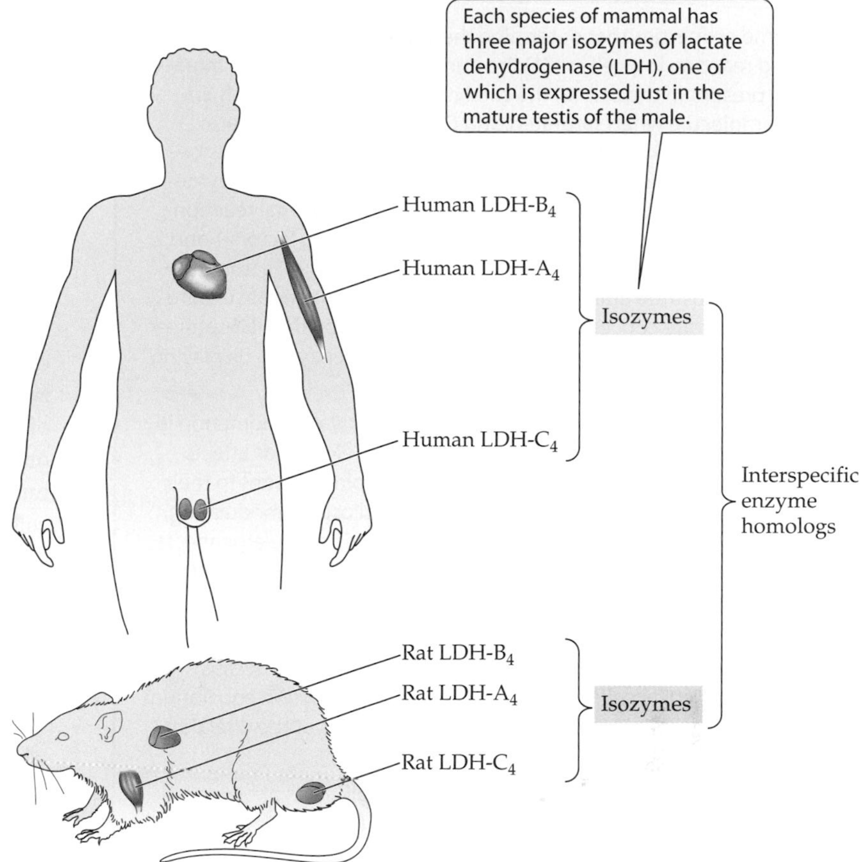

Figure 2.16 Isozymes and interspecific enzyme homologs The forms of LDH in rats are interspecific enzyme homologs of the forms in people.

When functional differences exist between isozymes or interspecific enzyme homologs, they often seem to be adaptive differences; that is, they often seem to *assist* the proper functioning of the animal. For an example that pertains to isozymes, consider LDH-A_4 and LDH-B_4, the two isozymes of LDH usually found in the skeletal muscles and heart muscle of a vertebrate. Of these two isozymes, LDH-A_4 is much more effective in using pyruvic acid as a substrate. As noted earlier, the reduction of pyruvic acid to lactic acid is an essential part of the mechanism by which skeletal muscles can at times make more ATP than their O_2 supply permits. Whenever people sprint for sport, cheetahs sprint for food, or frogs sprint for prizes in the Calaveras County races, the ability of their skeletal muscles to make ATP without O_2 allows the muscles to work exceptionally vigorously. Because this ATP production depends on the reduction of pyruvic acid, the particular isozyme of LDH found in the skeletal muscles—the isozyme that is superior in reducing pyruvic acid—plays a key role in the performance of intense exercise. The heart muscle, in contrast, rarely makes ATP without O_2, and its isozyme, LDH-B_4, is more suited to other functions than to the rapid reduction of pyruvic acid.

Interspecific enzyme homologs are often instrumental in the adaptation of species to different habitats. For example, two closely related species of West Coast thornyhead rockfish (*Sebastolobus*) possess different genetically coded homologs of LDH-A_4 in their skeletal muscles. One of the species lives at depths in the ocean above 500 m; the other lives 500 to 1500 m down. A problem with the forms of LDH-A_4 found in shallow-water species of marine fish, including the shallow-water rockfish, is that their affinity for substrate is highly sensitive to water pressure; these enzyme forms lose their affinity for substrate as pressure increases—so much so that at the high pressures in deep oceans, the forms become ineffective as enzymes. The homolog of LDH-A_4 synthesized by the deep-water species of rockfish is distinctive. It is relatively insensitive to pressure and retains a suitably high affinity for substrate even when the water pressure is high, helping to account for the adaptation of the species to live in the deep ocean.

Summary: Enzyme Fundamentals

- Enzymes are protein catalysts that accelerate reactions by lowering the activation energy required for reactants to reach transition state. To occur in cells, most reactions must be catalyzed by enzymes. Thus a cell controls which reactions occur within it by the enzymes it synthesizes.
- An enzyme must bind with its substrate to catalyze the reaction of substrate to form product. This binding, which is usually stabilized entirely by noncovalent bonds, occurs at a specific active site on the enzyme molecule, which is complementary in its three-dimensional chemical and electrochemical configuration to a portion of the substrate molecule. Enzyme molecules change shape when they bind to substrate or release product. These changes are permitted because the tertiary structure of a protein is stabilized by weak bonds.

- Enzyme properties that determine the velocity of an enzyme-catalyzed reaction in a cell are (1) the number of active enzyme molecules present in the cell, (2) the catalytic effectiveness of each enzyme molecule when saturated, and (3) the enzyme–substrate affinity. Enzyme-catalyzed reactions exhibit saturation kinetics because the reaction velocity is limited by the availability of enzyme molecules at high substrate concentrations. The maximal reaction velocity (V_{max}) that prevails at saturation depends on factors 1 and 2: the amount and catalytic effectiveness of the enzyme. Factor 3, the enzyme–substrate affinity, affects reaction velocity only at subsaturating substrate concentrations and is measured by the half-saturation constant (i.e., the Michaelis constant, K_m, for enzymes displaying hyperbolic kinetics).
- Multisubunit enzymes often exhibit cooperativity, a phenomenon in which the bonding of certain binding sites to their ligands affects (positively or negatively) the bonding of other binding sites to their ligands. An important type of cooperativity is allosteric modulation, in which a nonsubstrate ligand called an allosteric modulator affects the catalytic activity of an enzyme by binding noncovalently with a specific regulatory (allosteric) binding site. Both allosteric activation and allosteric inhibition are possible.
- Enzymes catalyze reversible reactions in both directions because their action is to accelerate the approach toward reaction equilibrium determined by principles of mass action, regardless of the direction of approach.
- Multiple molecular forms of enzymes occur at all levels of biological organization. Isozymes are multiple molecular forms of an enzyme within a single species; different isozymes may be coded by different alleles of one gene, or they may be coded by different genetic loci. Interspecific enzyme homologs are homologous forms of an enzyme in different species. Isozymes and interspecific enzyme homologs are often functionally different in adaptive ways.

Regulation of Cell Function by Enzymes

The catalytic nature of enzymes often receives such exclusive attention that enzymes are viewed merely as molecules that speed things up. At least as important, however, is the role that cellular enzymes play as agents of *regulation* of cell function. The biochemical tasks in a cell are typically accomplished by sequences of enzyme-catalyzed reactions called **metabolic pathways**. Enzymes participate in the regulation of cell function in two principal ways. First, the types and amounts of enzymes synthesized by a cell determine the metabolic pathways that are functional in the cell; any particular pathway is functional only if the cell synthesizes (through gene expression) all the enzymes the pathway requires. Second, the catalytic activities of the enzyme molecules that actually exist in a cell at any given time can be modulated as a way of controlling the rates at which the functional metabolic pathways operate.

The types and amounts of enzymes present depend on gene expression and enzyme degradation

Essentially all cells in an animal's body have the same genome, and the genome includes the genetic code for all enzymes that the animal can produce. Cells of different tissues differ, however, in their suites of enzymes. Moreover, any one cell typically differs from time to time in the types and amounts of enzymes it contains. A gene that codes for an enzyme is said to be **expressed** in a cell if the cell actually synthesizes the enzyme. Cells of various tissues differ in their enzymes—and one cell can differ from time to time—because only some genes are expressed in each cell at any given time. Gene expression is not all-or-none. Thus, for enzymes that are being synthesized by a cell, the rate of synthesis can be varied by modulation of the degree of gene expression.

The amount of a particular enzyme in a cell depends on the rate of degradation of the enzyme, as well as the rate of synthesis. All enzymes are broken down in specific and regulated ways by the action of proteolytic ("protein-lysing") enzymes. The rates at which various enzymes are degraded vary greatly, but all are degraded. Thus, unless an enzyme is synthesized in an ongoing manner, the enzyme will disappear from a cell. The amount of an enzyme present in a cell can be increased if synthesis is accelerated or degradation is slowed; conversely, the amount can be decreased if synthesis is slowed or degradation is accelerated. Animal cells are commonly able to increase or decrease the amounts of some of their key enzymes within hours or days. *Acclimation* and *acclimatization*—the long-term adjustments of individual animals to changes in their environments (see page 14)—are often dependent at the cellular level on changes in the amounts of key enzymes.

Variation in the rate of enzyme synthesis is the best-understood way that animal cells modify the amounts of their enzymes. The synthesis of an enzyme molecule in a cell requires several steps. First the stretch of DNA that represents the genetic code for the enzyme protein is transcribed to form pre-messenger RNA (pre-mRNA). Then the pre-mRNA is edited by posttranscriptional processing to form mature mRNA, and the latter moves from the nucleus to ribosomes. Next the mature mRNA on ribosomes is translated into the amino acid sequence of the protein. Finally, translation is followed by posttranslational processing that transforms the immature polypeptide chain into a mature protein.

Each of the steps in the synthesis of an enzyme, which begins with DNA and ends in a finished protein, is potentially modulated by a cell to control the rate at which the enzyme is synthesized. The first step, the transcription of DNA, for instance, is typically modulated by two types of specific regulatory regions of the DNA molecule that control whether, and how fast, transcription occurs in the relevant coding region of DNA. One type of regulatory region is the **promoter**, a DNA sequence located just upstream (toward the 5′ end) from the site where transcription starts. The second type of regulatory region consists of one or more **enhancers**, DNA sequences that may occur at various locations, even thousands of nucleotide bases away from the promoter. Proteins called **transcription factors** bind with the promoter and enhancer regions of DNA by way of DNA-matching subparts of their molecular strucures to control the extent to which RNA polymerase attaches to and transcribes the DNA-coding region responsible for a given enzyme. Transcription factors are highly specific and often work in sets, permitting different genes to be independently and finely controlled.

The processes that control the rates of synthesis of enzymes act on a variety of timescales to determine which metabolic pathways are functional in a cell. The **inducible enzymes** exemplify the modulation of enzyme synthesis on relatively short timescales. Such enzymes are present at low levels (or not at all) in an individual animal unless their synthesis is activated by specific inducing agents.

The cytochrome P450 enzymes found in the liver, kidneys, and gastrointestinal tract of vertebrates (and also found in most or all invertebrates) provide excellent and important examples of inducible enzymes. The function of the P450 enzymes is to help detoxify foreign compounds by oxidizing them. The foreign compounds themselves often serve as inducing agents for these enzymes. An individual animal that has not been exposed in the immediately preceding weeks or months to suitable inducing compounds has low levels of P450 enzymes because little or no enzyme synthesis occurs in such an individual and any preexisting molecules of the enzymes are removed by being degraded. The metabolic pathways that require P450 enzymes are thereby inactive. However, even a single exposure to an inducing compound will strongly induce increased synthesis of certain P450 enzymes in the individual, thereby activating some of these metabolic pathways. The P450 enzymes are a complex family of enzymes; over 30 kinds occur just in humans, for example. Many specialists hypothesize that the P450 enzymes evolved principally to help animals break down toxic compounds in their natural foods. Today, P450 enzymes are induced by certain human-made pollutants, such as halogenated aromatic hydrocarbons (HAHs), and the P450 enzymes help detoxify the pollutants. When an animal is exposed to HAHs, the HAHs enter cells and activate an intracellular receptor, which then acts as a transcription factor, causing expression of the P450-coding genes. Levels of liver P450 enzymes in fish are often employed today as indices of water pollution.

On an entirely different scale of time, the differentiation of tissues in an animal's body during development also depends on the regulation of the expression of genes that code for enzymes. Tissues become different in their sets of functional metabolic pathways, and they remain different throughout life, because of long-term controls on gene expression established during development. For example, bone marrow cells and skin cells differ in whether they synthesize hemoglobin because the two sorts of cells express different genes on a long-term basis; only the marrow cells have a functional metabolic pathway for hemoglobin synthesis.

Recently a major new technique for the study of gene expression, a technique based on DNA microarrays, was invented (see Box 2.2). This technique in principle permits the simultaneous measurement of the expression of hundreds or thousands of genes. Thus, molecular biologists expect that they will soon have means of understanding the roles of gene expression on all timescales in far greater detail than ever before.

Modulation of existing enzyme molecules permits fast regulation of cell function

Cells require speedier mechanisms of regulating their functions than are provided by even the fastest inducible enzyme systems. They achieve speedier regulation by modulating the catalytic activity of their *existing* enzyme molecules. A control mechanism that depends on changing the amounts of enzymes in cells usually requires many hours to be even minimally effective because increasing the synthesis or the degradation of enzymes cannot alter enzyme amounts more rapidly; if such mechanisms were the only controls, a cell would be like a car that could be accelerated or decelerated only once every several hours. Changes in the catalytic properties of already existing enzyme molecules, however, can often be brought about almost instantly, providing for moment-by-moment acceleration and deceleration of cell functions.

Some enzymes are better positioned in metabolic pathways than others to serve the requirements of rapid metabolic regulation. One category of well-positioned enzymes consists of those that catalyze **rate-limiting reactions.** In a linear metabolic pathway, it is possible in principle for the rate of one of the reactions to set the rate of the whole pathway. Suppose, for example, that of all the reactions in the pathway in Figure 2.17*a*, the conversion from B to C is inherently the slowest. Because the rate of the *whole pathway* would then be limited by the rate of that reaction, the conversion of B to C would be the

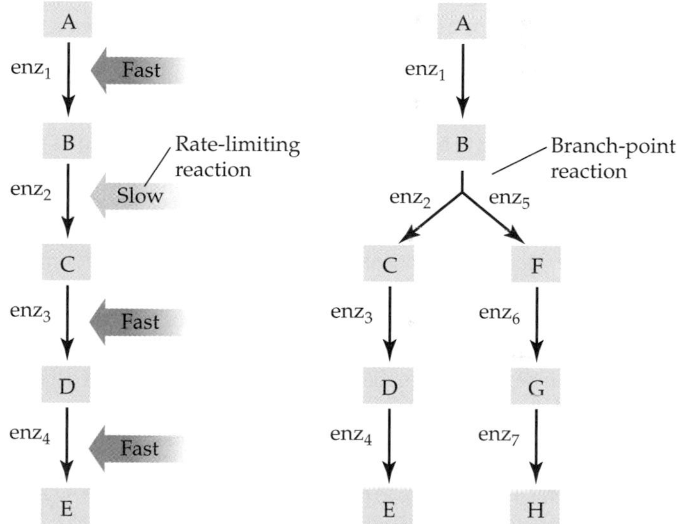

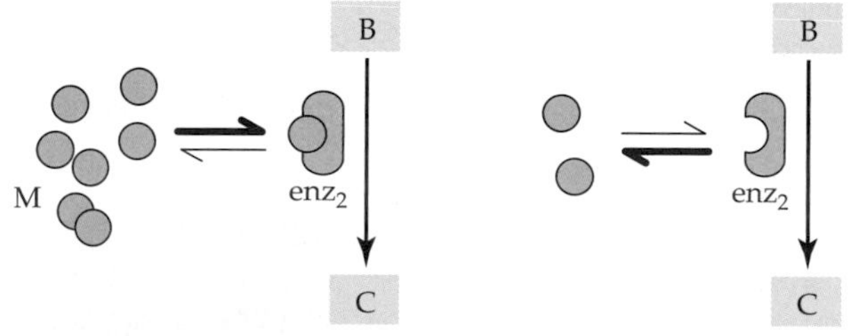

Figure 2.17 Enzymes that catalyze rate-limiting reactions and branch-point reactions are well positioned to exert control over metabolism The two reaction sequences in *(a)* and *(b)* are independent. A–H are reacting compounds; enz_1–enz_7 are enzymes. For example, B is the substrate of enz_2, and C is its product. *(c)* shows how an allosteric modulator (M) could interact with enz_2.

BOX 2.2 Gene-Expression Studies Using DNA Microarrays

A new and promising method for the *comparative study of gene expression* is the use of **DNA microarrays** (*DNA microchips, gene chips*). In its simplest form, a study using this technology compares the mRNAs (messenger RNAs) produced by a tissue under two different conditions, or possibly the mRNAs produced by two different tissues. Because expressed genes are transcribed to produce mRNAs, the mRNAs in a tissue provide information about which genes are being expressed by the tissue.

A DNA microarray consists of a set of spots of nuclear DNA. Each spot ideally contains DNA from just a single gene or presumptive gene in the species of interest. Enough DNA is placed in each spot that two mRNA samples could potentially hybridize to the spot without interfering with each other. The spots are laid out in a checkerboard pattern on a solid substrate, such as a glass plate, and are fused to the substrate. Because the spots consist of just short lengths of DNA, each is very tiny, and a grid of 2500 spots—representing 2500 different genes—might fit in a square 2 cm on a side, or even smaller. If the genome of the species of interest has been fully sequenced, defining the various genes to be included in spots is straightforward. The technique can be used with species that have not been fully sequenced, but defining the genes in the array is then less certain.

A typical experiment entails obtaining tissues from two sets of individuals that differ in a manner of interest. For instance, in a study of the role of gene expression in mammalian hibernation, tissue from individuals that are hibernating might be compared to that from individuals that are not. In a study of the responses of animals to a stressful environment (e.g., a stressfully high temperature), tissue from individuals exposed to stress might be compared to tissue from control individuals living under normal conditions. For purposes of illustration, let's assume that we are carrying out a study of the latter sort. Messenger RNA is extracted from the tissue of control animals and the tissue of animals exposed to the stressful environment (❶ in the figure). Here we will describe the subsequent procedure as if the mRNA itself were used, although in reality, some technical steps must be taken to stabilize the mRNA. Each of the two mRNA preparations to be compared is labeled with a fluor, a compound that has the potential to emit light by fluorescence (❷). Two different fluors are used to label the two mRNA preparations. A common approach is to label one with a fluor that will emit green light after it is fully processed, and the other with a fluor that will emit red light.

Next the two fluor-labeled mRNA preparations are permitted to hybridize with the microarray, the grid of DNA spots representing genes of the species being studied. Each mRNA hybridizes with its specific gene (❸). Thus, at this step each gene in the microarray potentially becomes labeled. A gene is labeled with mRNA from both preparations if it was being expressed in both tissues. A gene is labeled with mRNA from only one preparation if it was expressed in only one of the two tissues. A gene that was not expressed in either tissue does not become labeled at all. After hybridization, excess mRNA is washed away, leaving just the DNA array and the molecules of mRNA that have hybridized to it.

Finally, the fluors bonded to the microarray are visualized (❹). Typically visualization involves one or two additional technical steps. However, the essence of the procedure for visualizing the fluors is that the microarray is scanned with lasers that excite the fluors. The two distinctive fluors used to label the two tissues then emit their distinctive fluorescent wavelengths: green or red. Because green and red together produce yellow, each spot on the microarray emits green, red, yellow, or nothing. If the gene at a spot was not expressed in either tissue, it emits nothing. If a gene was expressed approximately equally in both tissues, it emits yellow. The most interesting genes are those that emit green or red because those are the genes that were expressed under only one of the two conditions studied.

Returning to our example, suppose we are comparing tissues taken from control animals and tissues from animals exposed to a stressful environment. On the basis of the microarray study, we would know which genes become expressed in response to the stressful environment, and which have their expression turned off in the stressful environment. Analysis of such gene-expression patterns promises to be an extraordinary tool for learning how animals modify their metabolic pathways under various conditions.

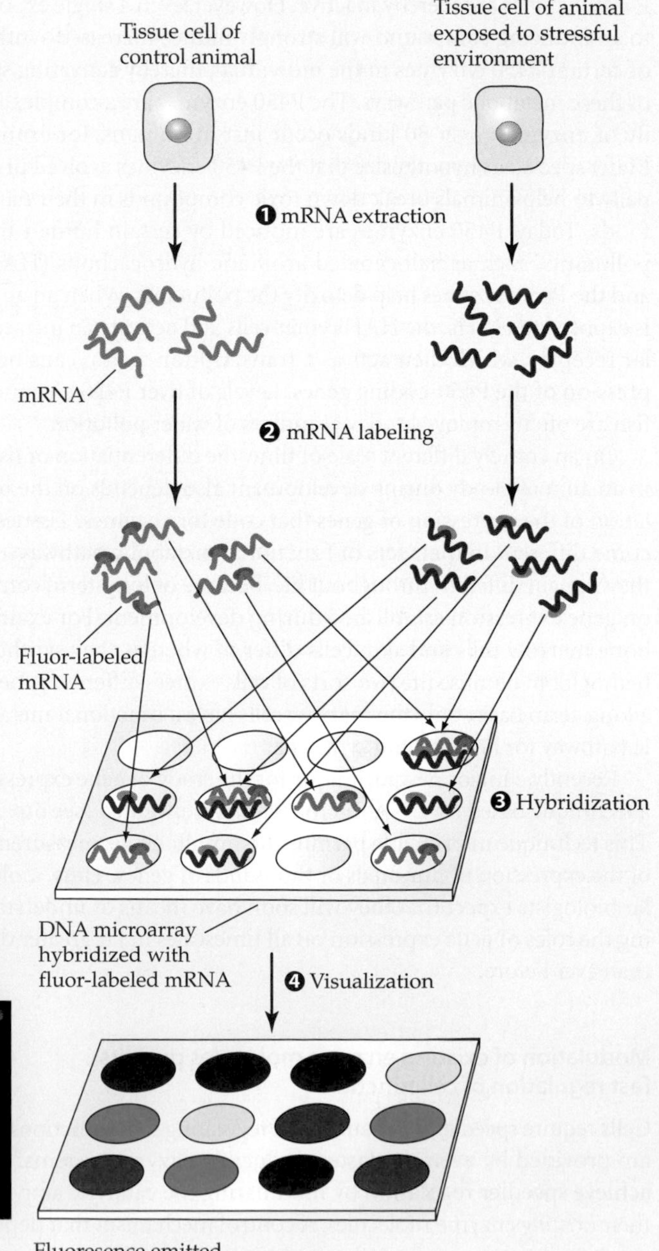

The steps in a comparative DNA microarray study of gene expression in two tissues The tissues in this study were taken from individual animals exposed to two different environmental conditions: a control environment and a stressful environment.

An actual microarray

rate-limiting reaction of the pathway. The catalytic effectiveness of enzyme enz_2 would then be crucial.

Enzymes that catalyze **branch-point reactions** in metabolic pathways are another category of enzymes that are well positioned to effect rapid metabolic regulation. A branching metabolic pathway permits two or more final products to be made from a single initial reactant. For instance, in Figure 2.17*b* either E or H could be made from A. The relative activities of the enzymes at the branch point—enz_2 and enz_5—determine which product is favored.

ALLOSTERIC MODULATION OF EXISTING ENZYMES Although binding sites for allosteric modulators do not occur in all enzymes, they are a common feature of enzymes that play regulatory roles. Allosteric modulation, in fact, is a principal mechanism by which cell function is regulated. Recall that allosteric modulators are nonsubstrate molecules that bind noncovalently with specific sites, termed *allosteric sites* or *regulatory sites*, on enzyme molecules and that thereby affect the catalytic activities of the enzymes. The binding of an allosteric modulator with a regulatory site is *reversible* and follows the principles of *mass action*. To illustrate, suppose that enz_2, the rate-limiting enzyme in Figure 2.17*a*, is allosterically modulated by a compound M. The reaction between M and the regulatory site on enz_2 would then be

$$\text{M} + \text{enz}_2 \rightleftharpoons \text{M-enz}_2 \text{ complex} \qquad (2.4)$$

Increasing the concentration of M shifts the reaction to the right by mass action, causing more enz_2 molecules to form M-enz_2 complexes (Figure 2.17*c*). Decreasing the concentration of M shifts the reaction to the left, causing fewer enz_2 molecules to be in complexes with the allosteric modulator. These adjustments, being driven by mass action, occur almost instantly, and they almost instantly affect the catalytic activity of the enzyme. Thus, allosteric modulation can occur *very rapidly*.

As stressed already, when an allosteric modulator binds with (or dissociates from) an enzyme, it alters the enzyme's ability to catalyze the conversion of *substrate* to *product*. This outcome occurs because the binding of the modulator to its *regulatory site* induces changes in the conformation of the enzyme molecule that ripple through the enzyme's molecular structure, affecting the *catalytically important properties* of the molecule, such as molecular flexibility or the conformation of the active site. An enzyme that has its catalytic activity increased by a modulator is said to be **upregulated**; conversely, one that has its catalytic activity decreased is said to be **downregulated**.[4] A single enzyme molecule may have two or more regulatory sites, each specific for a different allosteric modulator. In this case the individual modulators can exert reinforcing or canceling effects on the catalytic activity of the enzyme, offering elaborate regulatory possibilities.

If an allosterically modulated enzyme is the rate-limiting enzyme in a metabolic pathway, then the *entire pathway* may be upregulated or downregulated by allosteric modulation The downregulation of a whole pathway occurs, for example, during **feedback inhibition** (**end-product inhibition**), a common process in which a product of a metabolic pathway decreases the catalytic activity of a rate-limiting enzyme earlier in the pathway. Feedback inhibition would occur in the pathway in Figure 2.17*a*, for example, if enz_2, the rate-limiting enzyme, is downregulated by allosteric combination with compound E, the final product of the pathway. In this case an abundance of E in the cell would diminish the further formation of E by slowing the whole pathway. Conversely, if E were scarce, the rapid dissociation of E–enz_2 complexes by mass action would accelerate the reaction sequence that produces E. A metabolic pathway of this sort would act to stabilize levels of E in the cell by negative feedback (see Box 1.1).

The potential complexity of allosteric modulation is illustrated in the first three reactions of glycolysis, the metabolic pathway that converts glucose into pyruvic acid (Figure 2.18). The third reaction in glycolysis is catalyzed by *phosphofructokinase*, an enzyme of pivotal regulatory significance. The form of phosphofructokinase that occurs in most mammalian tissues (PFK-2) is allosterically modulated by over a half-dozen different substances, of which citrate and adenosine monophosphate (AMP) are particularly influential. Bonding of citrate to a citrate-specific regulatory site on the phosphofructokinase molecule inhibits catalysis. This modulation by citrate is essentially a case of feedback inhibition because in a cell with plenty of O_2, the pyruvic acid produced by glycolysis forms citric acid in the tricar-

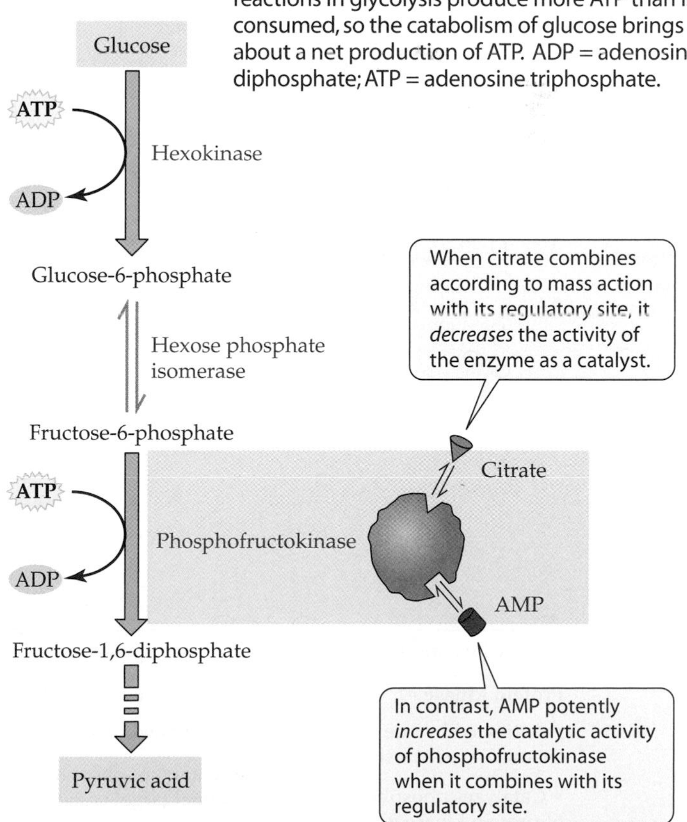

Figure 2.18 Phosphofructokinase, an allosterically modulated enzyme, is a key regulatory enzyme for the glycolytic pathway The first three reactions of glycolysis are shown. Citrate and adenosine monophosphate (AMP) are allosteric modulators of the mammalian phosphofructokinase known as PFK-2. Although the three reactions shown consume ATP, subsequent reactions in glycolysis produce more ATP than is consumed, so the catabolism of glucose brings about a net production of ATP. ADP = adenosine diphosphate; ATP = adenosine triphosphate.

[4] The concepts of *upregulation* and *downregulation* are used in additional contexts as well.

boxylic acid cycle; if the citrate concentration in a cell is high, allosteric downregulation of phosphofructokinase tends to restrain further entry of glucose into the glycolytic pathway that would produce more citrate. AMP very potently upregulates phosphofructokinase. A high concentration of AMP in a cell indicates that the cell has depleted its ATP (because AMP arises from the use of ATP). Under such circumstances, the allosteric modulation of phosphofructokinase by AMP can increase the catalytic activity of the enzyme 100-fold, accelerating the use of glucose to make more ATP.

COVALENT MODULATION OF EXISTING ENZYMES Besides allosteric modulation, **covalent modulation** (*covalent modification*) is the second major way that the function of cells is regulated by changes in the catalytic activity of existing enzymes. Covalent modulation occurs by way of chemical reactions that make or break *covalent* bonds (strong bonds) between modulators and enzymes. Although allosteric modulators are chemically very diverse, just a few principal chemical entities are employed in covalent modulation, of which the most common is phosphate.

Phosphorylation and **dephosphorylation**—the covalent attachment and removal of orthophosphate groups (HPO_4^{2-}), respectively—constitute the most important processes of covalent modulation. Phosphate groups (often symbolized P_i or PO_4^{2-}) are added to and removed from specific parts of modulated enzyme molecules, usually bonding with units of serine, threonine, or tyrosine in the enzyme structure. When a phosphate group bonds with an enzyme that is covalently modulated, the enzyme's activity is modulated because the shape of the protein changes, leading to changes in the catalytically important properties of the molecule. Often phosphorylation and dephosphorylation act as a very rapid type of *on–off* switch. An enzyme molecule may be completely inactive ("turned off") when it lacks a phosphate group, for example, and become activated ("turned on") when it bonds with a phosphate group. The transition between the downregulated "off" form and the upregulated "on" form can occur almost instantaneously because the enzyme protein is always present in the cell in this sort of regulation, rather than needing to be synthesized or degraded.

A crucial property of covalent modulation is that, unlike allosteric modulation, it *requires the action of enzymes* to catalyze the

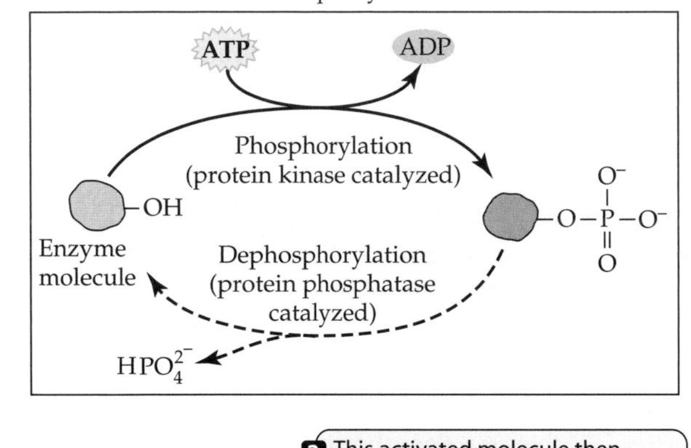

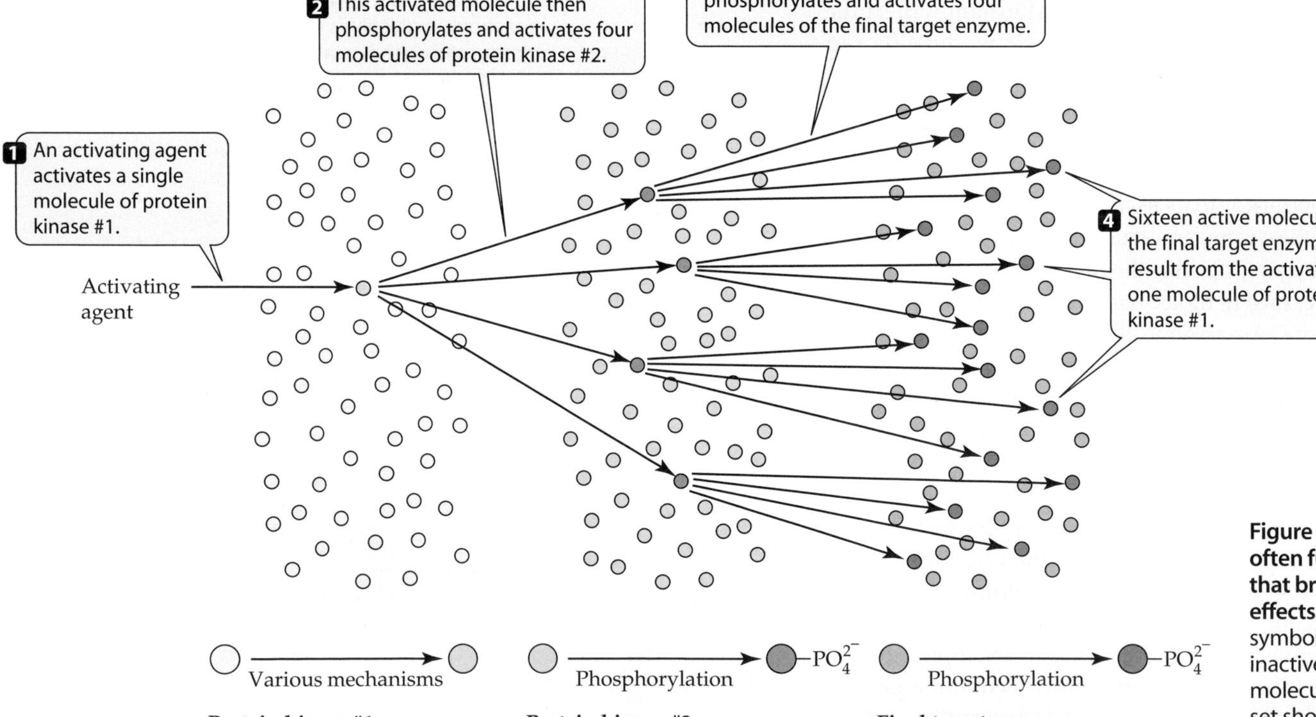

Figure 2.19 Protein kinases often function in sequences that bring about amplifying effects The light and dark symbols of each color denote inactive and active enzyme molecules, respectively. The inset shows greater detail of the phosphorylation of an enzyme catalyzed by a protein kinase (and later dephosphorylation catalyzed by a protein phosphatase).

making and breaking of covalent bonds. The enzymes that catalyze phosphorylation belong to a large class called **protein kinases**, which are enzymes that covalently bond phosphate to proteins using ATP as the phosphate donor (see the inset of Figure 2.19). The enzymes that catalyze dephosphorylation are *protein phosphatases*, which break covalent bonds between proteins and phosphate, liberating phosphate in the simple form of inorganic phosphate ions. Here we emphasize the protein kinases because they typically play a far larger role than the phosphatases in *controlling* the phosphorylation status of covalently modulated enzymes.

A significant question with regard to covalent modulation that you may have already wondered about is this: If phosphate is nearly always the modulator in covalent modulation, how does a cell prevent the simultaneous modulation of all of its covalently modulated enzymes? A key to the answer is that the protein kinases required for phosphorylation are *specific to the enzymes being modulated*. Hundreds of major types of protein kinases are known. Two different enzymes that are modulated by phosphorylation require the action of two different protein kinases to bind with phosphate, meaning that each can be controlled independently of the other. Some protein kinases phosphorylate proteins other than enzymes.

In carrying out their control function, protein kinases often act in multiple-enzyme sequences. That is, one protein kinase often activates another protein kinase, which then may activate a different sort of enzyme or possibly even a third protein kinase. The principal advantage of having multiple enzymes work in sequence in this way is believed to be **amplification** of the final effect. Amplification occurs because each molecule of an activated protein kinase may itself activate many molecules of the enzyme following it.

Figure 2.19 presents a simple example of an amplifying sequence of enzymes, consisting of two protein kinases and a final target enzyme that controls a critical metabolic process. The sequence is set in motion by an activating agent that activates a single molecule of the first protein kinase. This molecule then catalyzes the phosphorylation—and thus activation—of four molecules of the second protein kinase. Each activated molecule of the second protein kinase then catalyzes the phosphorylation of four molecules of the final target enzyme, so in total, 16 target-enzyme molecules are activated. Without the multiple-enzyme sequence, the initial activating agent itself would need to activate all 16 of the target-enzyme molecules. Initial activating agents are often signaling compounds that arrive at cells in amounts so minute that they could not by themselves exert large effects. The sequence in Figure 2.19 allows tiny quantities of a signaling compound to have 16-fold greater effects than otherwise would occur.

Summary: Regulation of Cell Function by Enzymes

- The metabolic pathways active in a cell depend on the enzymes present in the cell, as determined by the processes of enzyme synthesis (dependent on gene expression) and enzyme degradation. The presence or absence of enzymes in a cell is regulated on long and short timescales. During development (an example of a long timescale), tissues adopt tissue-specific patterns of gene expression that establish tissue-specific suites of enzymes and metabolic pathways. Inducible enzymes, such as the cytochrome P450 enzymes, exemplify shorter-term regulation of the presence or absence of enzymes and metabolic pathways.
- Very fast regulation of enzyme-catalyzed metabolic pathways is achieved by the modulation (upregulation or downregulation) of the catalytic activity of enzyme molecules already existing in a cell. Enzymes that catalyze rate-limiting or branch-point reactions are well positioned to mediate the rapid regulation of whole metabolic pathways in this way.
- Allosteric modulation and covalent modulation are the two principal types of modulation of existing enzyme molecules. Allosteric modulation occurs by way of the noncovalent binding of allosteric modulators to regulatory sites, governed by the principles of mass action. Covalent modulation requires the enzyme-catalyzed making and breaking of covalent bonds—most commonly with phosphate. Phosphorylation is catalyzed by enzyme-specific protein kinases, which usually are the principal controlling agents in covalent modulation.

Evolution of Enzymes

One of the great achievements of modern molecular biology is that the evolution of proteins can now be studied at the biochemical level. Two principal evolutionary scales are of interest, and they are studied with different scientific objectives and techniques. We discuss these here with an emphasis on enzymes.

One scale of interest in the study of enzyme evolution is the *evolutionary relationships of the enzyme forms found in large sets of related species*. The goal of research on these relationships is to reconstruct the family tree of the enzyme forms, so as to clarify enzyme evolution over geological scales of time, including the origins of enzymes and their later diversification. In research of this sort, scientists extract enzymes from each species of interest and determine the sequence of amino acids in each enzyme. After the amino acid sequences are known, the evolutionary relationships among the enzymes are estimated on the basis of similarities and differences in the sequences, as illustrated in Figure 2.20*a* by a set of five simplified enzymes. The results of such an analysis of 21 vertebrate lactate dehydrogenases (LDHs) are shown in Figure 2.20*b*.

The researchers who carried out the analysis in Figure 2.20*b*—and others who study LDH proteins—have concluded that there was just a single LDH gene when vertebrates first arose. All the modern LDH proteins thus belong to a single family tree. According to the amino acid sequence data, all the A forms of LDH in modern vertebrates are relatively closely related to each other (and thus all fall on one major branch—the upper branch—of the family tree in Figure 2.20*b*). Moreover, all the B forms are also relatively closely related to each other (and fall on a second major branch of the tree). The pattern indicates that the single original LDH gene duplicated (i.e., gave rise to two genes) early in vertebrate evolution at the point marked *. After the duplication, each individual vertebrate animal had two LDH genes. The two diverged evolutionarily to give rise to two lineages (two evolutionary families) of genes: one coding for the A forms of LDH, and the other for the B forms. The analysis also indicates that the A gene itself duplicated prior to the appearance of

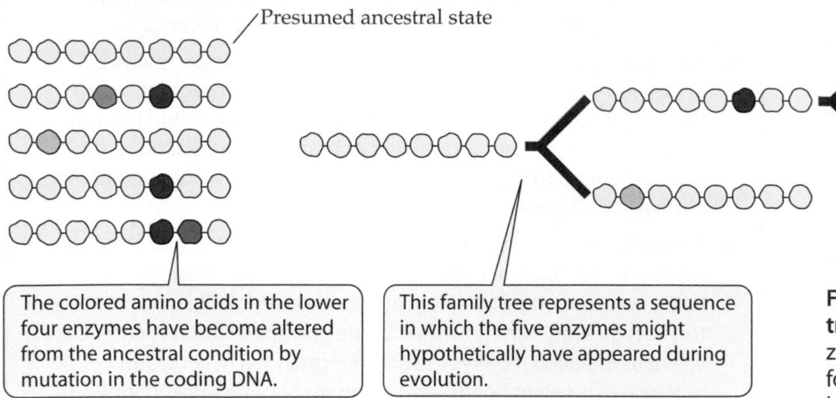

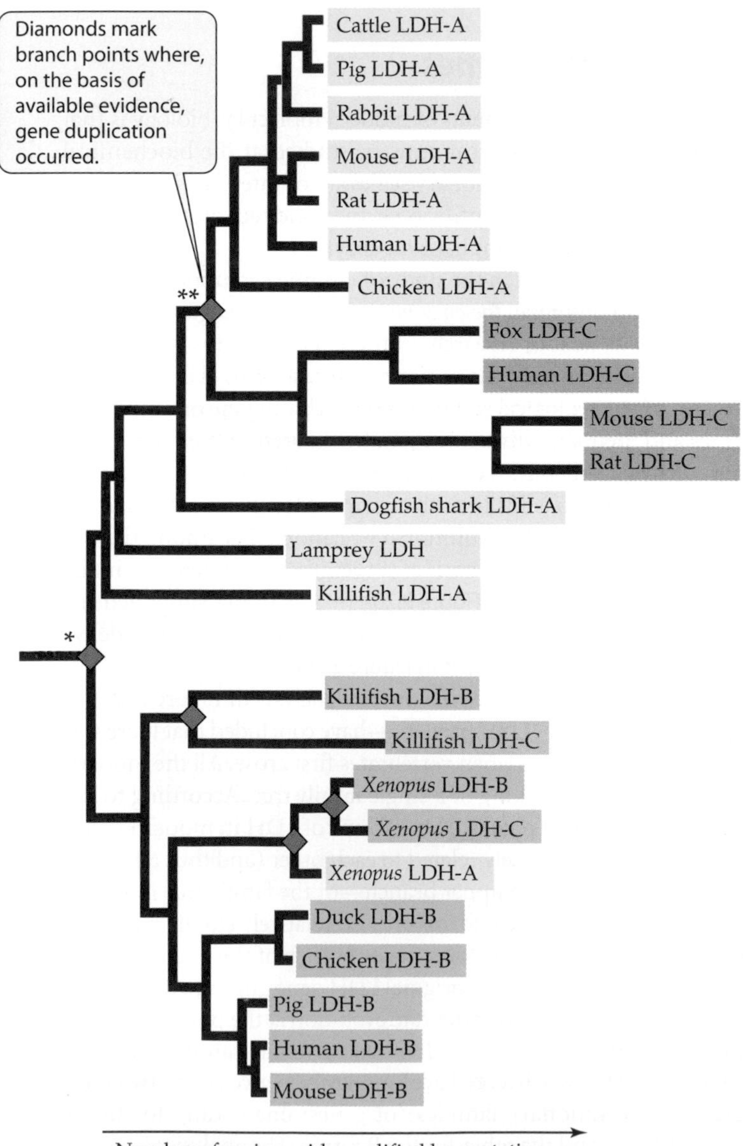

Figure 2.20 Enzymes and other proteins can be arranged into family trees based on their amino acid sequences *(a)* Five simplified enzymes, each consisting of a string of eight amino acids, and a proposal for how they could logically be arranged into a family tree. *(b)* The most logically cogent arrangement of 24 vertebrate LDH proteins into a family tree. The tree was determined by a computer program that took account of the primary chemical structures of the 24 proteins (300+ amino acids per protein) but that was provided no information on the presumed relationships of the animals sampled. Each horizontal line segment is proportional in length to the number of amino acids altered in that segment of the tree. Asterisks are referred to in the text. At the time this analysis was carried out, 24 LDH proteins had been sequenced, and all were included. (*b* after Stock et al. 1997.)

mammals at the point marked **. Following this duplication, one copy of the gene continued to code for the A form of LDH, and the other diverged to produce the C form. Thus, all three gene loci for LDH enzymes in modern mammals are descended from a single primordial LDH gene. According to the amino acid sequence data, the C forms of LDH in fish and amphibians are only distantly related to the C forms in mammals (these were all named "C" forms long ago, before their relationships were known). In sum, the study of the 24 LDHs illustrates how important features of the family tree of enzymes can be elucidated by study of the amino acid sequences of the enzyme forms in modern animals.

The second scale of interest in the study of enzyme evolution concerns the *evolution of alleles within single species*. An important goal of this microevolutionary scale of research is to study evolution in action. A case can often be made that changes in allele frequencies from place to place, or time to time, within the populations of a living species are sufficiently dynamic that they would not occur without the present-day action of natural selection or other evolutionary mechanisms. Thus scientists hope to obtain windows on the *present-day action* of evolutionary mechanisms by studying spatial and temporal patterns of allele frequencies.

A famous example of research focusing on the evolutionary mechanisms that determine allele frequencies is provided by studies of the killifish *Fundulus heteroclitus*, a small fish (5–10 cm long) found commonly in estuaries along the Atlantic seaboard. The waters along the coast from Georgia to Maine represent one of the sharpest marine temperature gradients in the world; in fact, the killifish living in Georgia experience body temperatures that, averaged over the year, are about 15°C higher than those of their relatives in Maine. In killifish, there are two major alleles of the B form of LDH (the form that occurs in the heart, red blood cells, liver, and red swimming muscles of fish). Killifish in the coastal waters of Georgia have mostly the *a* allele (symbolized B^a), whereas those in Maine have only the *b* allele (B^b). Moreover, the *a* allele becomes *progressively* less frequent from Georgia to Maine (**Figure 2.21**).

Several sorts of studies indicate that *modern-day natural selection maintains this geographical gradient of allele frequencies.* For example, individual killifish have been found to travel substantial distances; for this reason it is clear that interbreeding would rapidly even out the frequencies of the *a* and *b* alleles along the whole Atlantic seaboard if simply left to its own devices. The fact that different allele frequencies persist from place to place indicates that fish with different alleles undergo differential survival and reproduction; those with the *a* allele survive and reproduce better than those with the *b* allele in Georgia, and conversely, those with the *b* allele do better in Maine. Evidently we are witnessing natural selection right before our eyes in these killifish.

Trying to gain a better understanding of the natural selection that is involved in the distribution of allele frequencies in killifish, physiologists have explored how the two isozymes of the B protein, coded by the *a* and *b* alleles, differ in their functional properties as enzymes. They have found that the isozyme coded by the *b* allele has several functional advantages at low temperatures, and that the one coded by the *a* allele has advantages at high temperatures. Thus, mutation has given rise to two forms of this one enzyme protein, and both forms are retained because each is superior to the other in certain environments where the fish live.

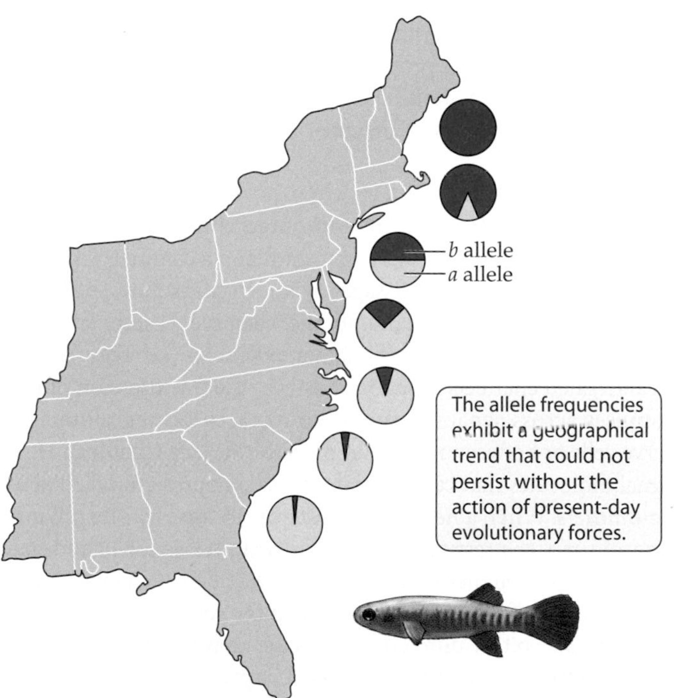

Figure 2.21 Contemporary evolution of an enzyme in the killifish *Fundulus heteroclitus* The pie diagrams show the frequencies of the *a* and *b* alleles of the gene for the B form of lactate dehydrogenase in fish of this one species collected at seven latitudes from Georgia to Maine. The green region in each diagram is the frequency of the *b* allele; the yellow region, the frequency of the *a* allele. (After Powers et al. 1993.)

Enzymes Are Instruments of Change in All Time Frames

Enzymes are primary instruments of physiological change in all five of the time frames that we first studied in Chapter 1 (see Table 1.2). Three of the time frames refer to changes in the physiology of animals that are *responses to changes in the environment.* The first of these three time frames is *acute physiological responses by individuals,* the responses that occur rapidly after the environment changes and are mediated by mechanisms already present in an animal's cells and tissues. Allosteric and covalent modulation of existing enzymes are major mechanisms of acute animal responses. For instance, if an animal runs rapidly away because it is frightened, the allosteric upregulation of phosphofructokinase by accumulation of adenosine monophosphate (AMP) in its muscle cells will immediately increase the rate that glucose is processed to manufacture more ATP to sustain muscular work (see Figure 2.18).

The second major time frame of response to the environment, the *chronic (long-term) physiological responses of individuals,* depends on reconstructions of physiological systems requiring hours, days, or longer periods to complete. Changes in the expression of enzyme-coding genes constitute a major mechanism of chronic responses. An example is provided by the cytochrome P450 enzymes. A fish or other animal acclimated to a toxin-free environment has low quantities of P450 enzymes in its cells. If the animal encounters toxins, it cannot increase its P450 enzymes immediately, because the enzymes must be synthesized, a process requiring many hours or days. In the long term, however, the animal acquires a greatly increased ability to detoxify harmful chemicals in its environment because of the induction of its P450 enzymes.

The third time frame of response to the environment, the *evolutionary change of physiological properties,* depends on changes of gene frequencies in whole populations over multiple generations. Genes that code for enzymes are frequently known to evolve by mutation, natural selection, and other mechanisms on both long and short scales of evolutionary time (see Figures 2.20*b* and 2.21). In this way, populations of animals take on new catalytic and regulatory phenotypes by comparison to their ancestors.

Besides the responses of animals to their environments, there are two time frames in which the physiology of individual animals is internally programmed to change independently of environmental change, and enzymes are primary participants in these time frames as well. One time frame of internally programmed change consists of the *developmental (ontogenetic) changes in an animal's physiology,* the changes that occur in a programmed way as an animal matures from conception to adulthood. Enzymes play essential roles in mediating changes of this sort. In young rats, for example, developmental programming of gene expression leads various enzymes to start to be synthesized at distinctive times during maturation (**Figure 2.22**). As new enzymes appear in cells (and old ones disappear), their collective catalytic and regulatory properties alter cell metabolism. For instance, although early fetal rats cannot synthesize liver glycogen because they lack glycogen synthetase, newborn rats are able to synthesize glycogen because the enzyme is suddenly expressed during the 5 days before birth.

Individual animals also undergo *periodic (e.g., day–night) physiological changes under control of internal biological clocks.* Enzymes

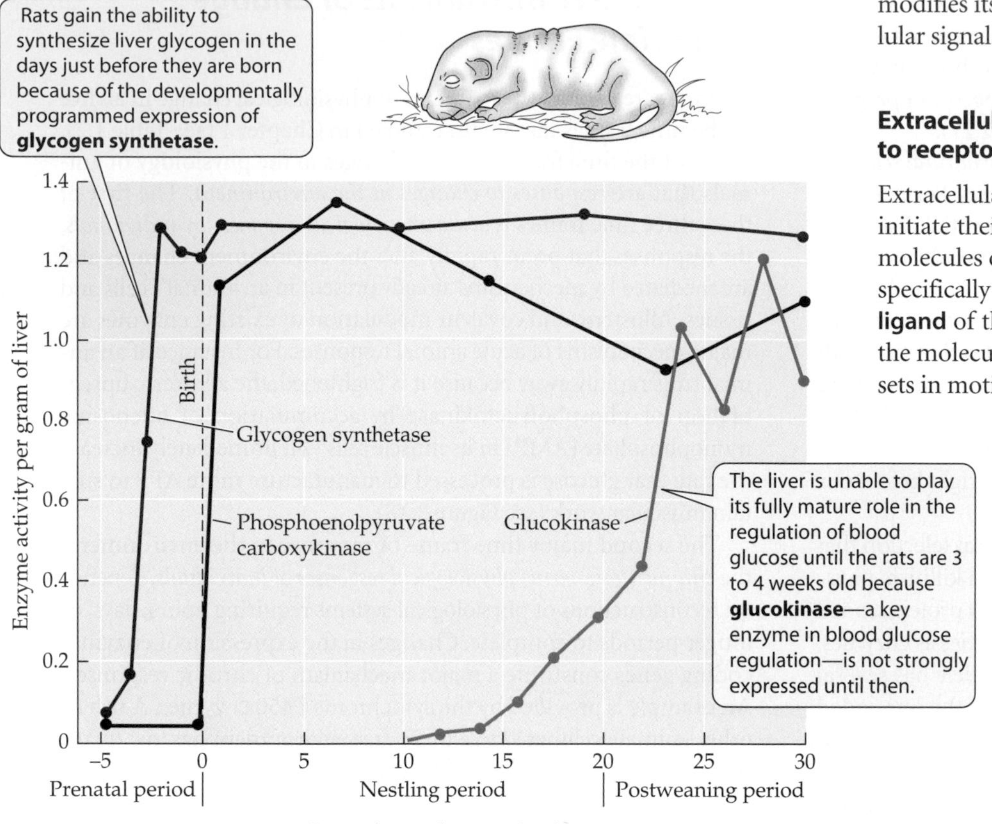

Figure 2.22 Developmentally programmed changes in the activities of three enzymes of carbohydrate metabolism in young rats The data shown are for liver enzymes in laboratory rats. Day 0 on the *x* axis is the day of birth. The time marked –5 is 5 days before birth. Weaning begins about 20 days after birth. Many enzymes resemble these in showing relatively sudden transitions from low to high activity at programmed times in development. (After Walker 1983.)

often mediate these changes, as shown by the fact that the catalytic activities of many enzymes rise and fall in rhythms that parallel the daily day–night cycle even when animals have no external information on the prevailing time. Some of these enzymes affect the abilities of animals to metabolize particular foodstuffs or detoxify particular foreign chemicals. Thus, food metabolism and responses to toxins vary between day and night because of internally programmed enzyme changes.

Cell Signaling: Signal Reception and Cell Signal Transduction

Cells send signals to each other that serve to coordinate cell functions throughout the body. Nerve cells, for example, signal muscle cells to contract, and certain endocrine cells signal liver cells to release glucose into the blood. When a signal arrives at a target cell, the cell must have mechanisms of **signal reception** to detect the signal. It must also have mechanisms of **signal transduction** by which it modifies its intracellular activities in response to the extracellular signal.

Extracellular signals initiate their effects by binding to receptor proteins

Extracellular signals such as neurotransmitters or hormones initiate their actions on a cell by binding with certain protein molecules of the cell, called **receptors**. A molecule that binds specifically and noncovalently to a receptor protein is called a **ligand** of the receptor. Ligand binding results in a change in the molecular conformation of the receptor protein, and this sets in motion a further response by the cell.

Receptors may be categorized into four functional classes: ligand-gated channels, G protein–coupled receptors, enzyme/enzyme-linked receptors, and intracellular receptors (**Figure 2.23**). Receptors in the first three categories reside in the cell membrane. This prevalence of receptors at the cell surface reflects the fact that *most signaling molecules cannot enter cells*. For the most part, signaling molecules are proteins or other, *hydrophilic* molecules that are unable to pass through the hydrophobic interior of the lipid bilayer of the cell membrane. Instead of entering cells, these signaling molecules bind to membrane surface receptors, which initiate their intracellular effects. Only hydrophobic or very small signaling molecules can enter a cell at meaningful rates through the cell membrane; once inside, such molecules bind to intracellular receptors.

Now let's discuss the properties of the four principal classes of receptors that we have identified: ligand-gated channels, G protein–-coupled receptors, enzyme/enzyme-linked receptors, and intracellular receptors.

LIGAND-GATED CHANNELS A **ligand-gated channel** is a cell-membrane protein that acts as both a receptor and a channel. This sort of channel opens to create a passageway for specific solutes, notably inorganic ions, through a cell membrane when the receptor site or sites on the protein bind to specific signal molecules (see Figure 2.23*a*). Ligand-gated channels function mostly in the transmission of nerve impulses across synapses, the narrow spaces between interacting nerve cells or between nerve and muscle cells (see Chapter 12). The signal molecules that carry signals across synapses are called *neurotransmitters*. When a neurotransmitter is released by one cell into a synaptic gap, the response of the receiving cell is for ligand-gated channels in its cell membrane to open. The opened channels permit increased flux of inorganic ions through the cell membrane, thereby changing the voltage difference across the membrane.

An example of synaptic transmission is provided when a nerve cell stimulates a skeletal muscle cell to contract. In this case, the specific neurotransmitter *acetylcholine* is released by the nerve cell and binds noncovalently to the receptor sites of acetylcholine receptors—which are ligand-gated channels—on the surface of the muscle cell. The channels then open and allow sodium (Na^+) and potassium (K^+) ions to flow through the cell membrane of the muscle cell, initiating

(a) Ligand-gated channel

Extracellular fluid
Ligand
Na^+
Cell membrane
Cytoplasm
K^+
In their typical mode of functioning, ligand-gated channels open to permit ions to pass through, thereby altering membrane electrical charge, when they bind to their ligands.

(b) G protein–coupled receptor and associated G protein system

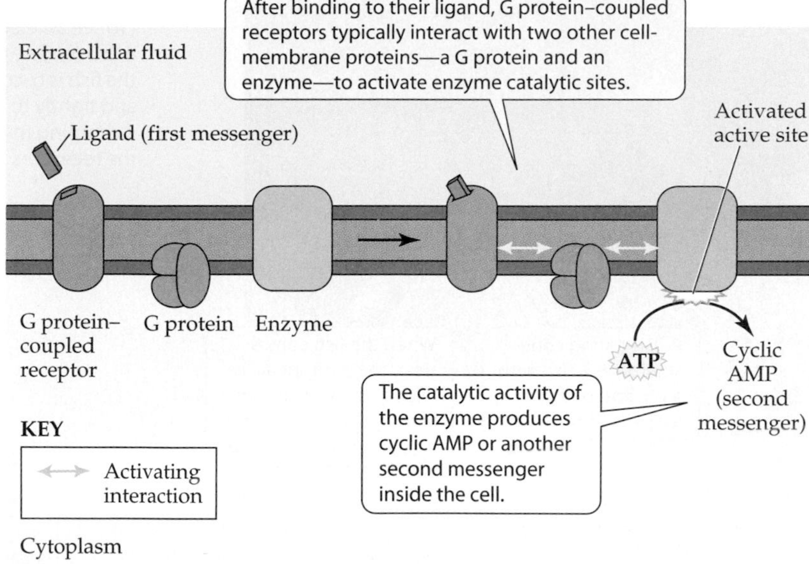

(c) Enzyme/enzyme-linked receptor

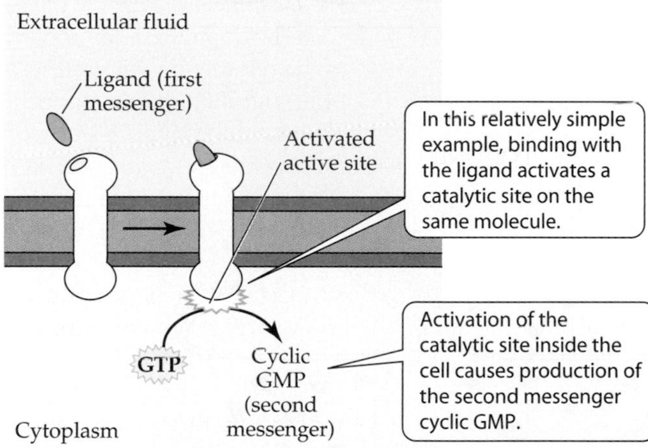

(d) Intracellular receptor

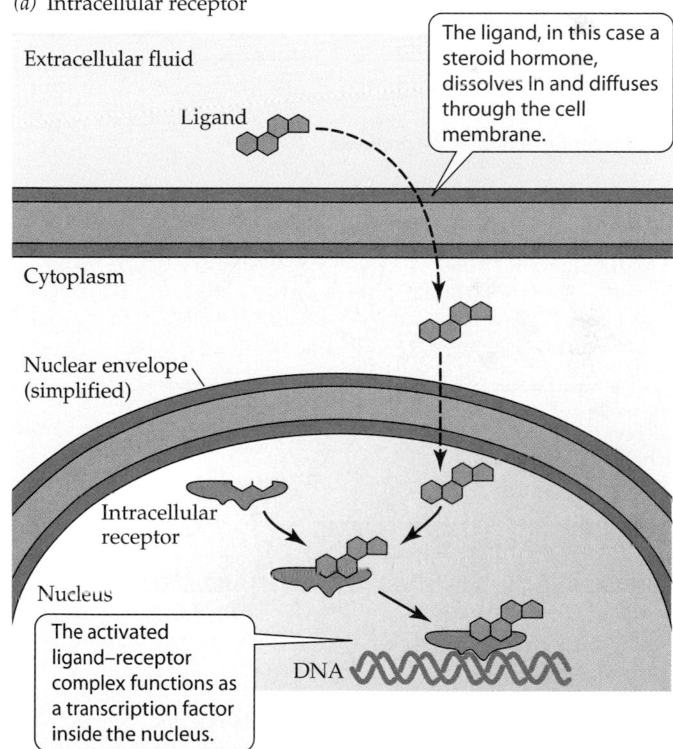

Figure 2.23 The four types of receptor proteins involved in cell signaling *(a)* A ligand-gated channel. The particular example shown, a muscle cell acetylcholine receptor, must bind a ligand molecule at two sites for the channel to open. *(b)* A G protein–coupled receptor. Details of the molecular interactions symbolized by double-headed arrows are discussed later in this chapter. *(c)* Enzyme/enzyme-linked receptors are themselves enzymes or, when activated, interact directly with other membrane proteins that are enzymes. One way or the other, binding with the ligand activates an enzyme catalytic site inside the cell. The example shown is the atrial natriuretic peptide receptor, which is particularly simple because it consists of just a single protein with both a ligand-binding site and a catalytic site. *(d)* Intracellular receptors are effective only for ligands that can dissolve in and diffuse through the lipid bilayer of the cell membrane. After a ligand enters the cell, it forms a complex with the receptor to initiate cellular responses. The example shown symbolically is a steroid hormone receptor, a type of receptor protein that is composed of a hormone-binding region and a region capable of binding with DNA. Binding with the hormone activates the receptor, and the activated hormone–receptor complex functions as a transcription factor. ATP = adenosine triphosphate; cyclic AMP = cyclic adenosine monophosphate; cyclic GMP = cyclic guanosine monophosphate; GTP = guanosine triphosphate.

a change in voltage across the membrane and a series of subsequent changes culminating in muscle contraction. Fish-eating cone snails incapacitate their prey in part using toxins that block these ligand-gated channels. One of the most potent of a cone snail's *conotoxins* is α-conotoxin, which specifically binds to the receptor sites on muscle-cell acetylcholine receptors, preventing the receptors from binding with or responding to acetylcholine (Figure 2.24). Because α conotoxin binds to the receptors very rapidly and tightly, the swimming

(a) A fish-hunting cone snail in action

(b) An example of an α-conotoxin

Different species of snails synthesize different forms of α-conotoxin, but all are very small molecules. This one, for example, consists of only 13 amino acids.

Disulfide bond

Glutamic acid – Cysteine – Cysteine – Asparagine – Proline – Alanine – Cysteine – Glycine – Arginine – Histidine – Tyrosine – Serine – Cysteine

Disulfide bond

(c) Block of receptor action by α-conotoxin

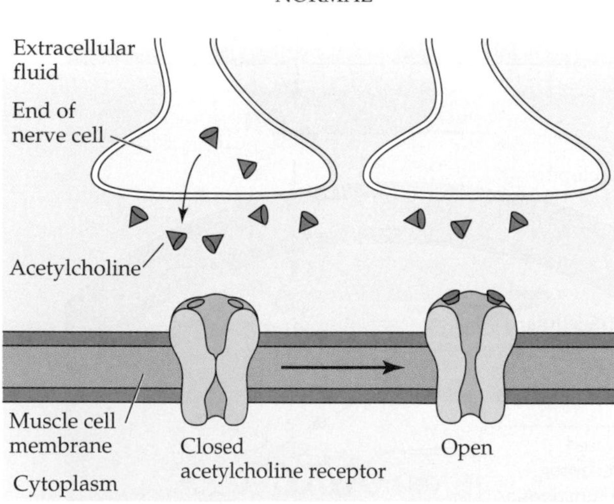

In a normal fish, when a nerve cell releases acetylcholine, the muscle cell receptors bind the acetylcholine, causing the receptors, which are ligand-gated channels, to open, thereby stimulating the muscle cell to contract.

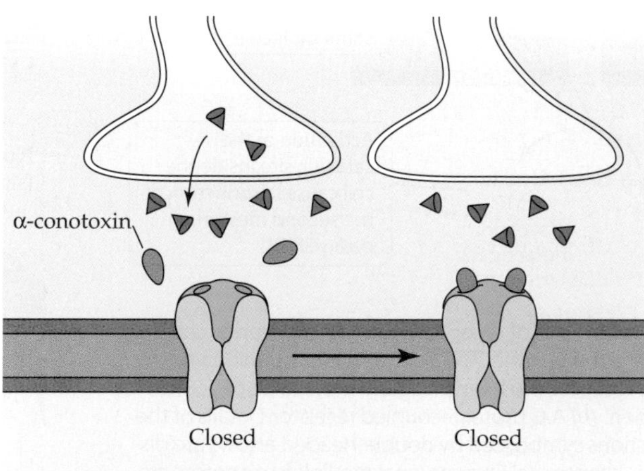

In a poisoned fish, the muscle cell receptors are unable to bind the acetylcholine because the receptor sites are blocked by the α-conotoxin. The receptors thus fail to open in the normal way, so the muscle cell is not stimulated and does not contract, meaning that the fish is paralyzed.

Figure 2.24 The defeat of a vital molecule by a venom *(a)* Speed of action in this cone snail's capture of a fish is imperative because the slow-moving snail could not pursue a fish that had even seconds to swim away. (To see an electron micrograph of the harpoon the snail uses, see Figure 4.8*d*.) *(b)* One of the most important toxins for the quick immobilization of the fish is α-conotoxin, a small polypeptide. *(c)* α-conotoxin binds quickly and tightly to the receptor sites on the acetylcholine receptors of the fish's swimming muscles. Thus, as shown in the "poisoned" part of the diagram, the receptors become incapable of binding acetylcholine.

muscles of a fish attacked by a cone snail are promptly blocked from responding to nervous stimulation, and the fish becomes paralyzed. Thus the whole fish is defeated by the incapacitation of a crucial ligand-gated channel protein in its body. Certain Asian krait snakes and the poison-dart frogs of Latin America have independently evolved toxins that also block this channel in their prey.

G PROTEIN–COUPLED RECEPTORS **G protein–coupled receptors** in cell membranes mediate cellular responses to many hormones and neurotransmitters, as well as many responses of sensory neurons. When a G protein–coupled receptor in the cell membrane of a cell is activated by binding its ligand, it activates a separate cell-membrane protein termed a **G protein**. The activated G protein may then directly exert an intracellular effect, or more commonly, it interacts with still another cell-membrane protein, usually an enzyme, and activates it so that a distinctive intracellular signaling compound is synthesized in the cytoplasm of the cell by the catalytic activity of the enzyme (see Figure 2.23*b*).

A major difference between reception based on G protein–coupled receptors and reception based on ligand-gated channels is that in general, *no sort of chemical passes through the cell membrane in the case of G protein–mediated reception*. In the most common type of such reception, one type of chemical brings the cell-signaling message to the extracellular side of the cell membrane; and a second, different chemical is produced on the intracellular side to carry the signal onward to the interior of the cell. The molecules, such as hormones or neurotransmitters, that bring signals to the cell membrane from the outside are called **first messengers**; the *intracellular* signaling molecules that carry the signals to the interior of the cell are called **second messengers**. The action of the G protein–mediated apparatus in the cell membrane is analogous to a relay race in which the first messenger brings the message to a certain point but then can go no farther and must activate a second messenger for the message to go on. An example is provided by the action of epinephrine (adrenaline) on a liver cell. Epinephrine, the first messenger, binds to a G protein–coupled receptor in the cell membrane, which initiates steps resulting in intracellular synthesis of the second messenger **3′-5′-cyclic adenosine monophosphate** (**cyclic AMP** or **cAMP**). Cyclic AMP then activates the intracellular responses to the epinephrine signal. Shortly we will return to this and other second-messenger systems in greater detail.

ENZYME/ENZYME-LINKED RECEPTORS **Enzyme/enzyme-linked receptors** are cell-membrane proteins that either are enzymes themselves or interact directly with enzyme proteins when activated. They are a more structurally and functionally diverse class of receptors than the two types of cell-membrane receptors we have already discussed. As in the case of G protein–mediated reception, molecules or ions do not pass through the cell membrane in this sort of reception, and enzyme/enzyme-linked receptors often activate the formation of second messengers. The simplest sort of enzyme/enzyme-linked receptor is a protein that is itself an enzyme; such a protein is composed of an extracellular receptor region, a membrane-spanning region, and an intracellular catalytic region (see Figure 2.23*c*). Binding of the extracellular signaling molecule to the receptor site activates the catalytic site at the other end of the molecule. The hormone atrial natriuretic peptide (ANP) acts on target cells in the kidney of a person to increase Na^+ excretion by way of this sort of receptor. When ANP binds to the receptor region on the outside of a cell, the receptor molecule catalyzes the formation of a second messenger, **3′-5′-cyclic guanosine monophosphate** (**cyclic GMP** or **cGMP**), inside the cell.

INTRACELLULAR RECEPTORS **Intracellular receptors** are the only class of receptors not localized at the cell surface. As noted earlier, most signaling molecules do not enter cells. Those that do are typically relatively small, hydrophobic molecules that can dissolve in and diffuse through the core of the lipid bilayer of the cell membrane. These signaling molecules include steroid hormones, thyroid hormones, retinoic acid, vitamin D, and nitric oxide. The receptors for these substances are located intracellularly, in the cytoplasm or nucleus. The usual pattern for intracellular receptors is that, after they are activated by binding with their ligands, they interact with DNA (see Figure 2.23*d*) to activate specific primary-response genes, the products of which may secondarily activate other genes.

When the steroid hormone estrogen arrives at a cell, for example, it passes through the cell membrane and binds to an estrogen-specific intracellular receptor, forming a hormone–receptor complex. The complex is itself a transcription factor that activates specific promoter and enhancer regions of the nuclear DNA, causing the expression of specific genes. The resulting effects can alter much of a target cell's metabolism, promoting female cellular phenotypes.

RECEPTORS ARE RELATED BY EVOLOUTIONARY DESCENT The receptors of any particular chemical type in modern-day animals are commonly related by evolutionary descent. For example, most of the ligand-gated channels in modern-day animals have very similar chemical structures and are coded by a single lineage of genes that has given rise to various specific channel proteins by mutation during the evolution of animal life. Similarly, all the G protein–coupled receptors belong to a single family tree, as do the intracellular steroid receptors.

Cell signal transduction often entails sequences of amplifying effects

When signaling molecules bind to cell-membrane receptors, *sequences of amplifying effects*—analogous to a chain reaction—are often involved between the moment that signal reception occurs and the moment that the final intracellular response occurs. The process by which epinephrine leads to the activation of glycogen breakdown to produce glucose in vertebrate liver cells provides a classic example of this widespread pattern (Figure 2.25).

When a person or other vertebrate experiences stress, such as the stress that occurs in anticipation of physical conflict, the adrenal glands secrete epinephrine into the blood. The circulation carries the epinephrine to the liver, where the hormone bathes liver cells, which contain abundant supplies of glycogen, a glucose-storage compound. The epinephrine itself cannot cross the cell membranes of the liver cells. "News" of its arrival in fact reaches the inside of each cell by way of a G protein–coupled receptor system.

The receptor system itself has important amplifying properties. To set the stage for discussing these, we need to note some details of G-protein function. Recall that a G protein–coupled receptor activates a cell-membrane G protein. G proteins get their name from the fact that they are modulated by binding with guanine nucleotides. A G protein bonded with **guanosine diphosphate** (**GDP**) is inactive. A G protein is activated when it is induced to exchange GDP for **guanosine triphosphate** (**GTP**). However, G proteins exhibit intrinsic GTP-destructive activity: When bonded with GTP, they tend to break down the GTP to GDP by hydrolysis. In this way, a G protein that has been activated by binding with GTP *tends to inactivate itself* by reverting to the inactive GDP-bonded form. The membrane G proteins, which are our focus here, are trimers in their inactive state. They dissociate into two parts when activated by GTP binding.

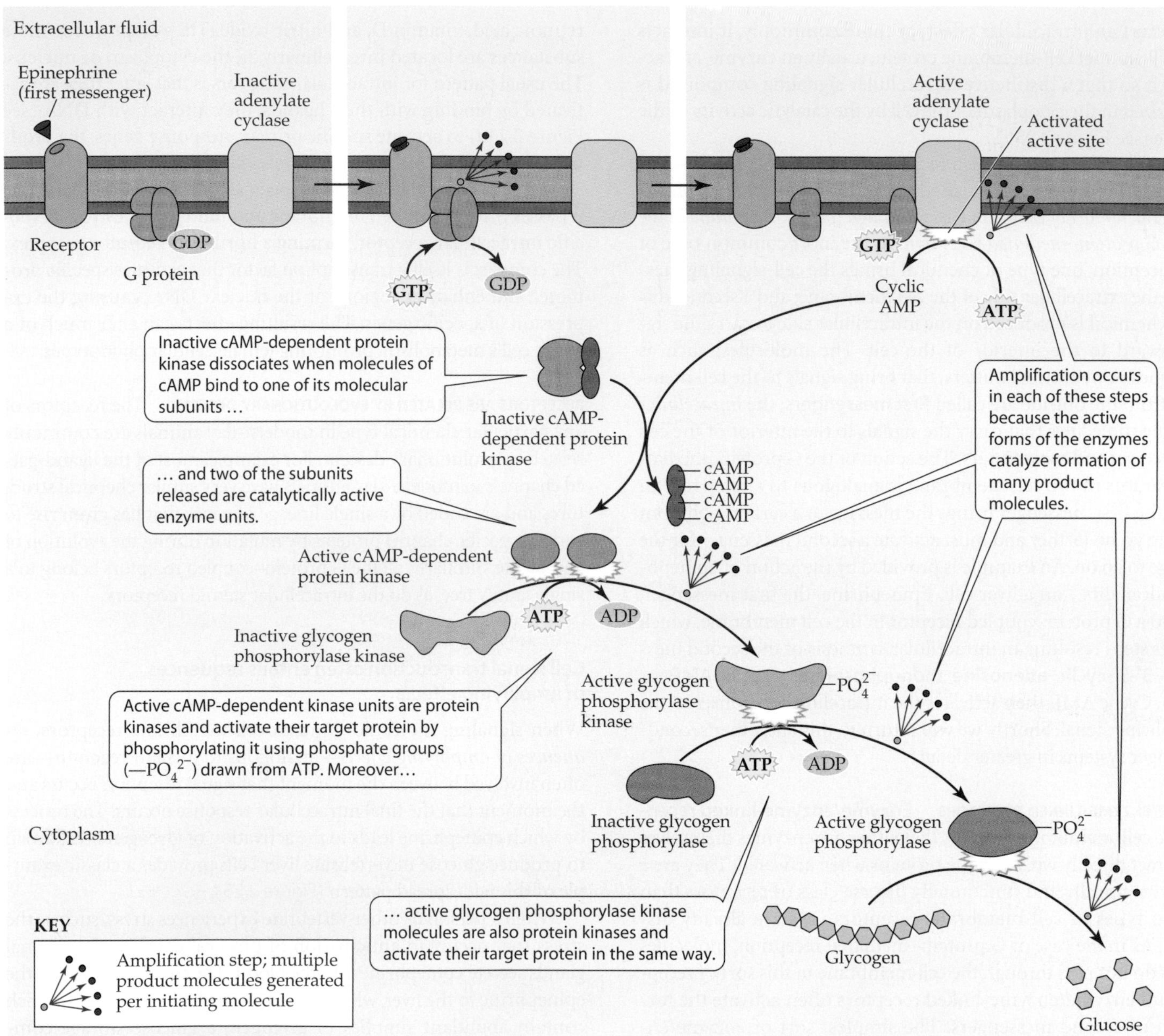

Figure 2.25 Amplifying signal transduction of an extracellular epinephrine signal results in enzymatic release of glucose inside a liver cell Because five steps in the epinephrine signal transduction pathway are amplifying, a very low epinephrine concentration can trigger a very large increase in glucose concentration. The cyclic AMP signal is ultimately terminated by the action of a cytoplasmic enzyme, phosphodiesterase. ATP = adenosine triphosphate; cyclic AMP = cAMP = cyclic adenosine monophosphate; GDP = guanosine diphosphate; GTP = guanosine triphosphate.

When epinephrine binds to its specific G protein–coupled receptor in the cell membrane of a liver cell, what first occurs is a series of amplifying reactions *within the cell membrane* (see Figure 2.25). The activated receptor interacts with molecules of G protein in the membrane to activate them by promoting loss of GDP in exchange for GTP. The G protein–coupled receptor and the G protein are separate, and both are free to diffuse in the fluid mosaic of the cell membrane. This means that an activated receptor must bump into a G-protein molecule to activate it, a situation that sounds inefficient until one realizes that it makes amplification possible. An activated receptor can bump into and activate many (perhaps 100) G-protein molecules. Each G-protein molecule remains active for a period of time that depends on how long it takes to inactivate itself (tens of seconds to several minutes), and while it is active, it can activate a cell-membrane enzyme, **adenylate cyclase**, which it bumps into by diffusion in the membrane; probably each activated G-protein molecule activates just one adenylate cyclase molecule because the activation requires steady linkage of the two proteins. Adenylate cyclase has an active site on the cytoplasmic side of the cell membrane, and when it is activated, it catalyzes the formation of the second messenger cyclic AMP (cAMP) from ATP inside the cell. Further amplification occurs at this step because a single activated molecule of adenylate cyclase can catalyze the formation of multiple molecules of cAMP during its active life.

The cAMP signal inside a liver cell triggers the activation of a series of intracellular enzymes (see Figure 2.25). Multiple amplifying steps occur in this series, which starts with two protein kinases and is a classic example of the type of amplification described in Figure 2.19. The cAMP second messenger activates a protein kinase named *cAMP-dependent protein kinase (cAPK)* by causing it to dissociate, forming two active enzyme units. The activated cAPK units phosphorylate, and thereby activate, a second protein kinase called *glycogen phosphorylase kinase (GPK)*. Finally, the active GPK molecules phosphorylate and activate the ultimate target enzyme, *glycogen phosphorylase (GP)*. The action of GP is to cleave glucose units from the glycogen polymers stored in the cell, and the glucose is then released into the blood for distribution throughout the body.

A small quantity of epinephrine can cause a flood of blood glucose because of the cumulative effect of the amplifications that occur in this cell signal transduction pathway. A cumulative amplification of about 10,000 can occur between the binding of an epinephrine molecule to a G protein–coupled receptor molecule and the formation of cAMP. Thus a blood epinephrine concentration of 10^{-10} *M* can result in an intracellular concentration of cAMP of 10^{-6} *M*. The protein-kinase cascade within the cell can then result in a further amplification of perhaps 1000, so that the concentration of activat-

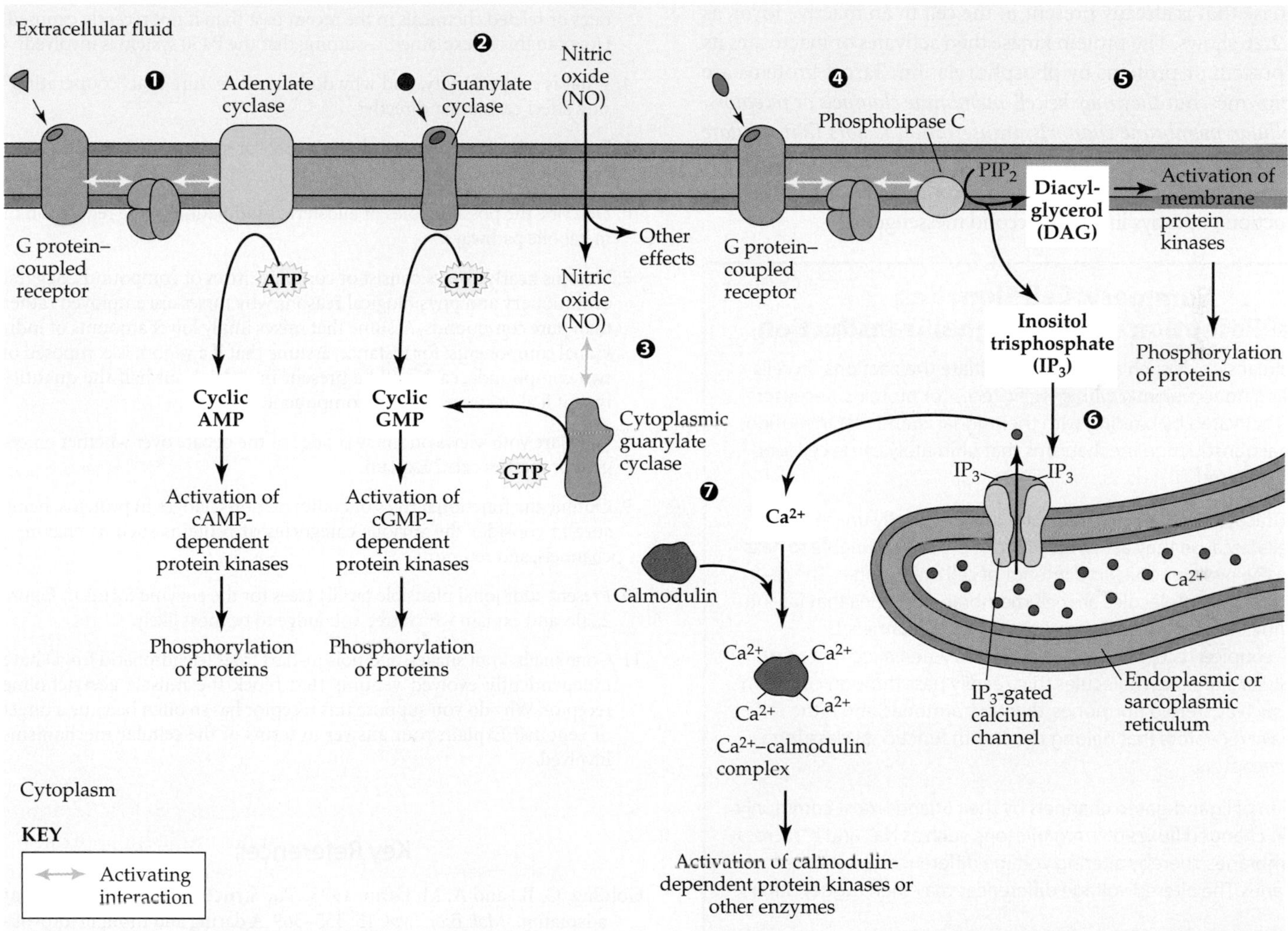

Figure 2.26 Second messengers in overview The production and the actions of five important second messengers are shown: cyclic AMP, cyclic GMP, diacylglycerol, inositol trisphosphate, and calcium ion. This diagram includes only some of the major cell signal transduction pathways that employ second messengers. (❶) Some G protein–coupled receptor systems employ cyclic AMP as a second messenger, as seen already in Figure 2.25. (❷) Receptor systems based on guanylate cyclase enzymes employ cyclic GMP as a second messenger. When activated, a guanylate cyclase produces cyclic GMP from guanosine triphosphate (GTP). In some cases, as in the atrial natriuretic peptide receptor system, the guanylate cyclase is a cell-membrane enzyme. (❸) Some guanylate cyclase enzymes are cytoplasmic. Many of the actions of nitric oxide (NO) are mediated by cyclic GMP produced by the activation of a cytoplasmic guanylate cyclase. (❹) Some G protein–coupled receptor systems employ diacylglycerol and inositol trisphosphate as second messengers. When such receptor systems are activated, the two second messengers are synthesized simultaneously from a cell-membrane phospholipid, PIP_2 (phosphatidylinositol 4,5-bisphosphate), by the catalytic action of activated phospholipase C, a membrane-associated enzyme. (❺) Diacylglycerol stays in the cell membrane because it is hydrophobic. (❻) Inositol trisphosphate, which is hydrophilic, enters the cytoplasm, where its major action is to open ligand-gated channels that release Ca^{2+} from intracellular stores such as the endoplasmic reticulum. (❼) Ca^{2+} ions sometimes act as second messengers, as when Ca^{2+} released by action of inositol trisphosphate activates the cytoplasmic protein calmodulin, which then can activate protein kinases or other enzymes.

ed GP is 10 million times the blood concentration of epinephrine that initiated the cell signal transduction process.

Several second-messenger systems participate in cell signal transduction

Several compounds—not just cyclic AMP—act as second messengers. The common second messengers, besides cyclic AMP, are cyclic GMP (cGMP); inositol 1,4,5-trisphosphate (IP_3); 1,2-diacylglycerol (DAG); and Ca^{2+} ions. Figure 2.26 provides an overview of some prominent second-messenger systems in which these compounds participate. For the most part, all the second messengers share with cyclic AMP the property that their immediate intracellular effect is to activate a protein kinase that is already present in the cell in an inactive form, as Figure 2.26 shows. The protein kinase then activates or inactivates its target protein or proteins by phosphorylation. Target proteins are often enzymes, but *they may be cell-membrane channels or receptors, intracellular-membrane channels, transcription factors that regulate gene expression, or virtually any other sort of protein.* Sequences of multiple signal-amplifying reactions are a common feature of the signal transduction pathways involving second messengers.

Summary: Cell Signaling
Signal Reception and Cell Signal Transduction

- Extracellular signals such as hormones initiate their actions on cells by binding noncovalently with specific receptor proteins. Receptor proteins activated by binding with their signal ligands set in motion cell signal transduction mechanisms that ultimately cause cell function to be altered.
- Most extracellular signaling molecules are chemically unable to enter cells because they are hydrophilic or otherwise unable to pass through the hydrophobic, lipid interior of cell membranes. The receptors for these molecules are cell-membrane proteins that fall into three principal functional classes: ligand-gated channels, G protein–coupled receptors, and enzyme/enzyme-linked receptors. Extracellular signaling molecules that readily pass through cell membranes, such as steroid hormones, thyroid hormones, and nitric oxide, have receptors that belong to a fourth functional class: intracellular receptors.
- Activation of ligand-gated channels by their ligands most commonly results in changed fluxes of inorganic ions, such as Na^+ and K^+, across cell membranes, thereby altering voltage differences across the membranes. The altered voltage differences may then trigger other effects.
- Activation of G protein–coupled receptors and enzyme/enzyme-linked receptors by their extracellular signaling ligands typically initiates the formation of second messengers, such as cyclic AMP or cyclic GMP, on the inside of the cell membrane. The second messengers, in turn, often trigger sequences of additional intracellular effects in which preexisting enzymes, most notably protein kinases, are modulated. A function of these sequences is amplification of the ultimate effect.
- Intracellular receptors, when activated by their ligands, usually bind with nuclear DNA and directly activate specific primary-response genes.

Study Questions

1. It is becoming possible for molecular biologists to synthesize almost any protein desired. Suppose you use a phylogenetic tree of modern-day enzymes (e.g., Figure 2.20b) to predict the amino acid sequence of a now nonexistent ancestral enzyme form. What insights might you obtain by synthesizing the ancestral enzyme protein?
2. Using lactate dehydrogenase as an example, explain why it is true to say that "multiple molecular forms of enzymes occur at all levels of animal organization."
3. One of the earliest pieces of evidence for the existence of the P450 enzymes was the recognition by physicians and veterinarians in the 1950s that barbiturate anesthetics interfere with their own effects when they are employed repeatedly. A person or other mammal requires a higher dose to be anesthetized if the individual has been exposed to barbiturates or related chemicals in the recent past than if not recently exposed. How can this be explained, assuming that the P450 system is involved?
4. What is cooperativity, and why does it not require that "cooperating" sites affect each other directly?
5. Explain why G protein–mediated receptor systems depend on membrane fluidity.
6. Describe the possible roles of allosteric modulation in the regulation of metabolic pathways.
7. Venoms nearly always consist of complex mixes of compounds. Suggest evolutionary and physiological reasons why mixes are employed rather than pure compounds. Assume that mixes imply lower amounts of individual components; for instance, assume that if a venom is composed of two compounds, each will be present in only about half the quantity than if it alone were the only component.
8. What are your views on the two sides of the debate over whether emergent properties exist? Explain.
9. Outline the functional roles of conformation changes in proteins, being sure to consider the various categories of proteins such as enzymes, channels, and receptors.
10. Present additional plausible family trees for the enzyme forms in Figure 2.20a, and explain which tree you judge to be most likely.
11. Cone snails, krait snakes, and poison-dart frogs (dendrobatid frogs) have independently evolved venoms that block the muscle acetylcholine receptor. Why do you suppose this receptor has so often become a target of venoms? Explain your answer in terms of the cellular mechanisms involved.

Key References

Golding, G. B., and A. M. Dean. 1998. The structural basis of molecular adaptation. *Mol. Biol. Evol.* 15: 355–369. A daring and thought-provoking introduction to the new field of paleomolecular biochemistry.

Hardison, R. 1999. The evolution of hemoglobin. *Amer. Sci.* 87: 126–137. An accessible treatment of research on the evolution of a protein, including studies of the evolution of the relevant promoter and enhancer regions of DNA, as well as coding regions.

Hochachka, P. W., and G. N. Somero. 2002. *Biochemical Adaptation: Mechanism and Process in Physiological Evolution.* Oxford University Press, New York. A peerless review of the modern study of biochemical adaptation written by two of the foremost scientists in the field. Pedagogically exceptional.

Powers, D. A., and P. M. Schulte. 1998. Evolutionary adaptations of gene structure and expression in natural populations in relation to a changing environment: A multidisciplinary approach to address the million-year saga of a small fish. *J. Exp. Zool.* 282: 71–94. This challenging paper records one of the most successful efforts to understand the molecular physiology and evolutionary biology of the ecological relationships of a species. There is probably no better exemplar of the future of environmental physiology.

Selected References

Alberts, B., A. Johnson, J. Lewis, M. Raff, K. Roberts, and P. Walter. 2002. *Molecular Biology of the Cell*, 4th ed. Garland, New York.

Bennett, A. F. 1974. Enzymatic correlates of activity metabolism in anuran amphibians. *Amer. J. Physiol.* 226: 1149–1151.

Eisenmesser, E. Z., D. A. Bosco, M. Akke, and D. Kern. 2002. Enzyme dynamics during catalysis. *Science* 295: 1520–1523.

Fields, P. A., and G. N. Somero. 1998. Hot spots in cold adaptation: Localized increases in conformational flexibility in lactate dehydrogenase A_4 orthologs of Antarctic notothenioid fishes. *Proc. Natl. Acad. Sci. U.S.A.* 95: 11476–11481.

Gibbs, A. G. 1998. The role of lipid physical properties in lipid barriers. *Amer. Zool.* 38: 268–279.

Goodsell, D. S. 1992. A look inside the living cell. *Amer. Sci.* 80: 457–465.

Hartl, F. U., and M. Hayer-Hartl. 2002. Molecular chaperones in the cytosol: From nascent chain to folded protein. *Science* 295: 1852–1858.

Hawkins, J. D. 1996. *Gene Structure and Expression*, 3rd ed. Cambridge University Press, New York.

Jones, C. T. (ed.). 1982. *The Biochemical Development of the Fetus and Neonate.* Elsevier Biomedical Press, New York.

King, J., C. Haase-Pettingell, and D. Gossard. 2002. Protein folding and misfolding. *Amer. Sci.* 90: 445–453.

Livingstone, D. R., and J. J. Stegeman (eds.). 1998. Forms and functions of cytochrome P450. *Comp. Biochem. Physiol. C* 121: 1–412. [Numerous individual articles.]

Lodish, H., A. Berk, S. L. Zipursky, P. Matsudaira, D. Baltimore, and J. Darnell. 2000. *Molecular Cell Biology*, 4th ed. Freeman, San Francisco.

Logue, J. A., A. L. DeVries, E. Fodor, and A. R. Cossins. 2000. Lipid compositional correlates of temperature-adaptive interspecific differences in membrane physical structure. *J. Exp. Biol.* 203: 2105–2115.

McIntosh, J. M., A. D. Santos, and B. M. Olivera. 1999. *Conus* peptides targeted to specific nicotinic acetylcholine receptor subtypes. *Annu. Rev. Biochem.* 68: 59–88.

Michal, G. (ed.). 1999. *Biochemical Pathways: An Atlas of Biochemistry and Molecular Biology.* Wiley, New York.

Mitic, L. L., and J. M. Anderson. 1998. Molecular architecture of tight junctions. *Annu. Rev. Physiol.* 60: 121–142.

Olivera, B. M. 1997. E. E. Just Lecture, 1996. Conus venom peptides, receptor and ion channel targets, and drug design: 50 million years of neuropharmacology. *Mol. Biol. Cell* 8: 2101–2109.

Palmer, T. 1995. *Understanding Enzymes*, 4th ed. Prentice Hall/Ellis Horwood, London.

Pierce, V. A., and D. L. Crawford. 1997. Phylogenetic analysis of glycolytic enzyme expression. *Science* 276: 256–259.

Price, N. C., and L. Stevens. 1989. *Fundamentals of Enzymology*, 2nd ed. Oxford University Press, New York.

Rasmussen, H. 1989. The cycling of calcium as an intracellular messenger. *Sci. Amer.* 261(4): 66–73.

Sotaniemi, E. A., and R. O. Pelkonen (eds.). 1987. *Enzyme Induction in Man.* Taylor & Francis, New York.

Spiegel, A. M., T. L. Z. Jones, W. F. Simonds, and L. S. Weinstein. 1994. *G Proteins.* Landes, Austin, TX.

Stock, D. W., J. M. Quattro, G. S. Whitt, and D. A. Powers. 1997. Lactate dehydrogenase (LDH) gene duplication during chordate evolution: The cDNA sequence of the LDH of the tunicate *Styela licata*. *Mol. Biol. Evol.* 14: 1273–1284.

Urich, K. 1994. *Comparative Animal Biochemistry*. Springer, New York.

Walker, R. 1983. *The Molecular Biology of Enzyme Synthesis: Regulatory Mechanisms of Enzyme Adaptation.* Wiley, New York.

Weiss, T. F. 1996. *Cellular Biophysics.* MIT Press, Cambridge, MA.

Williams, E. E., B. S. Stewart, C. A. Beuchat, G. N. Somero, and J. R. Hazel. 2001. Hydrostatic-pressure and temperature effects on the molecular order of erythrocyte membranes from deep-, shallow-, and non-diving mammals. *Can. J. Zool.* 79: 888–894.

3 *Transport of Solutes and Water*

Hummingbirds are dietary specialists that feed almost exclusively on flower nectars, which are principally solutions of sucrose (table sugar), plus other sugars, notably glucose and fructose. When sucrose itself is digested, it yields glucose and fructose. Thus a hummingbird has an abundance of glucose and fructose in its intestines after digesting a meal, and these simple sugars represent its immediate reward for feeding. The sugars, however, are in solution in the lumen—the hollow central core—of the intestines, and for them to do the bird any good, they must move into the bloodstream to be carried throughout the body.

You might at first imagine that glucose and fructose in the lumen of a hummingbird's intestines could simply diffuse across the intestinal epithelium into blood vessels. However, the hydrophilic or hydrophobic nature of the molecules needs consideration (see Chapter 2). Glucose and fructose are hydrophilic, and because hydrophilic molecules tend not to mingle with hydrophobic ones, the sugars can be expected to have difficulty dissolving in the highly hydrophobic interior of the apical and basolateral cell membranes of the epithelial cells lining the intestine. In fact, the glucose and fructose from a meal cannot diffuse into a hummingbird's bloodstream at significant rates. How then do they move into the blood? Questions of this sort are among the most common and important faced in the study of animal physiology.

In the life of any animal, many materials must move across cell membranes and epithelia in substantial quantities to resupply cells or tissues with needed raw materials, void wastes, maintain the proper composition of body fluids, and otherwise maintain the animal's integrity. The **solutes**—that is, the dissolved materials—that move across cell membranes and epithelia are chemically very diverse, and the mechanisms by which solutes and water cross are numerous. Thus a question that arises at the very start of our study of this complex topic is this: What term should be used to describe the movements globally? In this book we will use the term **transport** to refer in an entirely general way to any and all movements of solutes or water across cell membranes or epithelia, regardless of the mechanisms of movement.

Three examples in which transport is important to animals will help define the subject of this chapter and provide a focus for our study of transport. Here we describe the examples. Later, as the chapter unfolds, we will return to all three to discuss the specific transport processes involved.

1. One of our focal examples is *a single cell inside an animal's body*. A typical animal cell is bathed by extracellular fluid. Thus we can think of the cell membrane as separating an *intracellular fluid* on the inside from an *extracellular fluid* on the

Mechanisms must exist to transport sugars from a hummingbird's intestines into its blood following a meal Nectars are unusually rich in sugars. We may well wonder if sugar-transport mechanisms are exceptionally developed in hummingbirds because of the birds' nectar diets.

outside (Figure 3.1*a*). The solutions on the inside and outside of a cell are normally similar in their *sum-total* concentrations of dissolved matter; more exactly, they have similar osmotic pressures.[1] Consequently, the solutions inside and outside a typical cell do not tend to exchange water by osmosis to a great degree. However, there is a dramatic difference in the *solute compositions* of the two solutions. The concentration of potassium ions (K^+) is much higher inside an animal cell than outside, for instance, whereas sodium (Na^+) and chloride (Cl^-) ions are typically less concentrated inside than outside. These differences in solute composition suggest that the intracellular fluid of a cell is not at equilibrium with the extracellular fluid bathing the cell. What are the transport mechanisms by which ions like Na^+ and K^+ move toward equilibrium across a cell membrane, and which transport processes keep the intracellular fluid different in composition from the extracellular fluid?

2. Our second focal example is *the outer gill membrane of a freshwater fish living in a pond or stream*. The fluid portion of the blood of a freshwater fish—known as the fish's *blood plasma*—is much more concentrated in Na^+, Cl^-, and other inorganic ions than the surrounding pond or stream water (Figure 3.1*b*). In the fish's gills, the concentrated blood plasma is separated from the dilute pond water by just a thin epithelium. Because the gill epithelium is a permeable rather than impenetrable barrier, the blood plasma tends to lose ions such as Na^+ and Cl^- to the pond water across the epithelium, and H_2O tends to enter the plasma from the pond water across the epithelium. What are the transport mechanisms by which inorganic ions and water cross the gill epithelium? Are there transport mechanisms in the epithelium that help to maintain the difference in composition between the fish's blood plasma and the environmental water?
3. Our third focal example, continuing the illustration with which we started this chapter, is *the intestinal epithelium of the small intestine of a bird or mammal*. As we saw in Chapter 2 (Figure 2.5*b*), this epithelium consists of a single layer of cells bearing microvilli on their apical cell membranes (Figure 3.1*c*). Not only dissolved sugar molecules (such as glucose and fructose), but also dissolved amino acids and other compounds, must cross this epithelium from the intestinal lumen to the blood after a meal if they are to be of use to an animal. What are the transport mechanisms that move each of these materials across the epithelium?

[1] Osmotic pressures and osmosis are discussed later in this chapter.

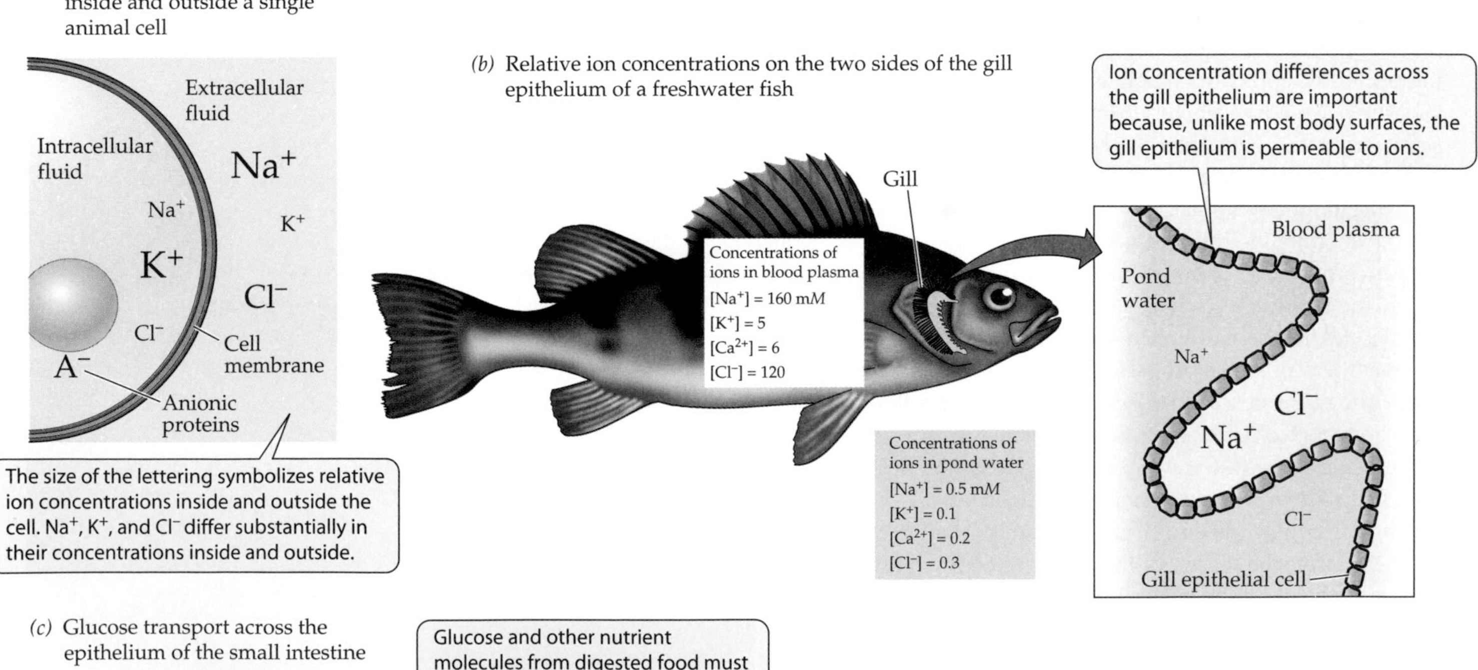

Figure 3.1 Three focal examples in which transport occurs Study of the three examples shown in this figure will bring to light most of the basic principles of solute transport. *(a)* A typical cell within an animal's body, surrounded by extracellular fluid. A^- represents anionic (negatively charged) proteins and other solutes that cannot cross the cell membrane and thus are trapped inside the cell. *(b)* A freshwater fish, showing representative ion concentrations in the blood plasma and surrounding pond water. A thin epithelium is all that separates the blood plasma and water in the gills of the fish. *(c)* The small intestine of a bird or mammal, emphasizing the intestinal epithelium lining the lumen.

As we start our study of transport mechanisms, an important organizing principle is the distinction between *passive* transport mechanisms and *active* ones. To define *passive* and *active* in this context, we first need to define **equilibrium**. Be cautious of the term, because it is commonly used in nonscientific ways, such as when people say that an angry child has "recovered his equilibrium" or that a city has reached "equilibrium size." These sorts of uses tend to confuse rather than clarify the scientific meaning of the term.

An accurate definition of *equilibrium* is based on the second law of thermodynamics, which describes the behavior of systems, termed **isolated systems**, which have no inputs or outputs of energy or matter.[2] The second law of thermodynamics states that when an isolated system undergoes change, it is not able to change in all conceivable ways. Instead, an isolated system can change in only certain, limited ways. Equilibrium is, by definition, the state toward which an isolated system changes; that is, it is the state toward which a system moves—internally—when it has no inputs or outputs of energy or matter. A system is *at equilibrium* when internal changes have brought it to an internally stable state from which further *net* change is impossible without system inputs or outputs. *The state of equilibrium is a state of minimal capacity to do work under locally prevailing conditions.* Thus a change toward equilibrium is always in the direction of decreasing work potential.[3]

Passive-transport mechanisms are capable of carrying material only in the direction of equilibrium. **Active-transport mechanisms**, on the other hand, *can* carry material in the direction opposing equilibrium. The word *can* is emphasized for an important reason: An active-transport mechanism does not *necessarily* carry materials in the direction opposing equilibrium, but it is *capable* of doing so.

Passive Solute Transport by Simple Diffusion

Simple diffusion is the most straightforward form of passive solute transport. To understand simple diffusion, consider a beaker divided by a nonliving membrane that is penetrated by many microscopic pores through which the sugar glucose, but nothing else, can pass. A solution of glucose in water is placed on each side of the membrane, with that on the left having a higher glucose concentration than that on the right. Viewing this beaker *macroscopically* (Figure 3.2*a*), we know from everyday experience what will happen: Glucose will move in net fashion from left to right across the membrane until ultimately the concentration of glucose is equal on both sides. However, what is the *mechanism of motion* that carries glucose from left to right? In addition, why do the concentrations of glucose on the two sides *stay equal once they have become equal*?

Answering these questions will reveal the nature of simple diffusion. To answer them, we need to look at the beaker *microscopically*. At temperatures above absolute zero, all atoms and molecules undergo ceaseless random motions. Imagine that you're taking a microscopic look at the membrane and the solutions just on either side, and you fix your eye on a particular glucose molecule to the left of the membrane (Figure 3.2*b*). This molecule, like all the others, will be moving constantly, sometimes colliding with and bouncing off other molecules in random directions. There is a chance that one of these random motions will carry the molecule through a pore in the membrane to the right-hand side. Over a period of time, more glucose molecules will move in this way from left to right than from right to left, merely because there are more glucose molecules per unit volume on the left. Macroscopically, these individual molecular events will result in net glucose transport from left to right. Once the concentrations on the two sides have become equal, glucose molecules will continue to move at random from left to right and from right to left. The numbers of molecules going in the two directions will then be equal, however, because of the equal numbers of glucose molecules per unit volume on the two sides. Thus, at a macroscopic level the concentrations on the two sides will remain equal once they have become equal.

Figure 3.2 Simple diffusion viewed macroscopically and microscopically *(a)* A beaker is divided in half by a porous artificial membrane that permits passage of glucose molecules, but nothing else. The solutions on either side of the membrane differ in glucose concentration (symbolized by different dot densities), and glucose diffuses through the membrane. *(b)* Looking at the same system microscopically, random motions are more likely to carry glucose molecules from left to right than from right to left. (Glucose pores of the sort in this example occur only in artificial membranes, not living biological membranes.)

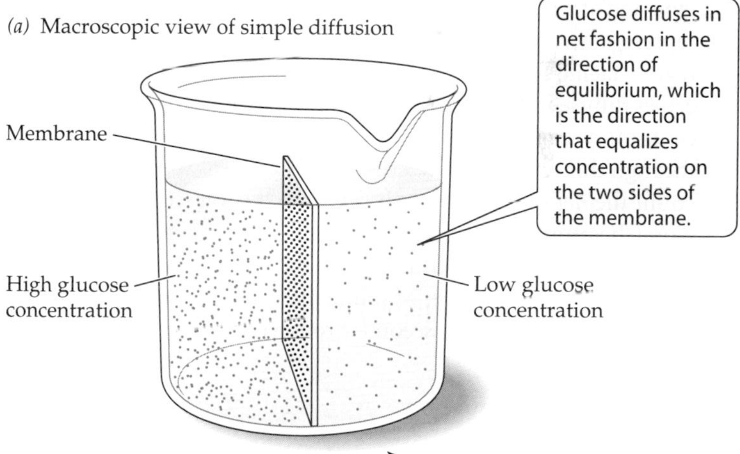

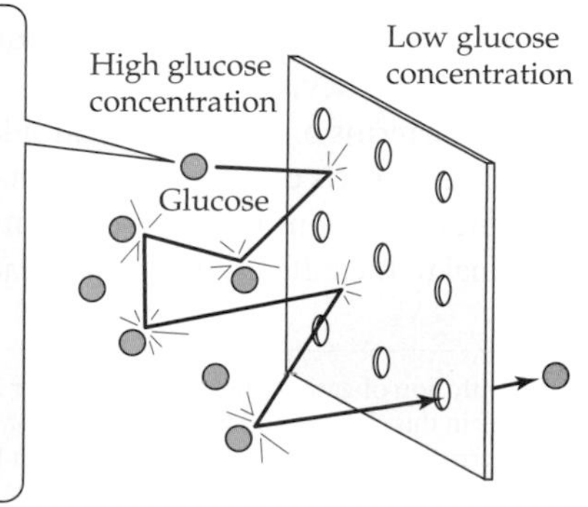

---

[2] We say much more about the second law of thermodynamics in Chapter 5.

[3] Appendix J (Basic Physics Terms) provides background on work and other physics concepts used in this chapter.

The mechanism of glucose transport that we have just described is **simple solute diffusion**. Such diffusion is *transport that arises from the molecular agitation that exists in all systems above absolute zero and from the simple statistical tendency for such agitation to carry more molecules out of regions of relatively high concentration than into such regions*. Note that simple diffusion could not possibly *increase* the difference in glucose concentration between the two sides of a membrane. The equilibrium state for the beaker in Figure 3.2 is a state of equal glucose concentration on both sides, and simple diffusion can change the glucose concentrations only *toward* equilibrium. Thus, simple diffusion is a *passive* transport mechanism.

Because molecular agitation is universal, simple diffusion is universal. Gases diffuse in air; water diffuses; even heat diffuses. As we now discuss the quantitative principles of the simple diffusion of solutes in solutions, it is worth noting that all simple-diffusion phenomena follow similar principles. Thus, later when we discuss the quantitative laws of gas, water, and heat diffusion,[4] you will notice similarities to the principles we now address. Finally, note that simple diffusion is often called just *diffusion*.

Concentration gradients give rise to the most elementary form of simple solute diffusion

Solutes diffuse between any two regions of a solution, whether or not a membrane is present. To understand the quantitative laws of solute diffusion, we start with diffusion in an open solution, which is the most general case. Consider an open solution in which a solute, such as glucose, is at a relatively high concentration (C_1) in one region and at a relatively low concentration (C_2) in another. Imagine a representative 1-cm^2 cross-sectional area in the low-concentration region, facing toward the high-concentration region. As solute molecules diffuse into the low-concentration region from the high-concentration region, we can count the numbers of molecules passing through this representative cross-sectional area. Specifically, we can count the *net* numbers that diffuse into the low-concentration region (the numbers diffusing in, minus the numbers diffusing out). Let J be the net rate of diffusion into the low-concentration region; that is, J is the net number of solute molecules passing into the low-concentration region per second through each unit of cross-sectional area. Then

$$J = D\frac{C_1 - C_2}{X} \tag{3.1}$$

where X is the distance separating the region of high concentration (C_1) from the region of low concentration (C_2), and D is a proportionality factor termed the **diffusion coefficient**. This formulation is often called the **Fick diffusion equation** after Adolf Fick (1829–1901), who devised it.

The terms on the right-hand side of the Fick equation reveal important aspects of diffusion. Note that the rate (J) at which solute molecules diffuse into the low-concentration region is directly proportional to the difference in concentration ($C_1 - C_2$). Note also that the rate increases as the distance separating the two concentrations (X) decreases: Diffusion is a notoriously slow process for transporting substances from one place to another in the macroscopic world, but when only a minute distance separates regions of differing concentration, diffusion can transport substances very rapidly. To illustrate (Table 3.1), when there is a concentration difference of a small solute between two regions of a water solution, half of all the solute molecules that must cross into the low-concentration region to create concentration equilibrium will do so in 32 years if the distance between regions is 1 meter, but half will cross in 100 nanoseconds if the distance is 10 nanometers, the approximate thickness of a cell membrane! The high rate of diffusion across minute distances helps explain why diffusion is a crucially important process in the lives of cells.

Looking again at the right-hand side of the Fick equation, the ratio $(C_1 - C_2)/X$ is called the **concentration gradient**. It expresses how much the concentration changes per unit of distance, and J, the diffusion rate, is proportional to it. Sometimes the term *concentration gradient* is used in a looser sense to refer just to the difference of concentration $(C_1 - C_2)$. An important aspect of the concentration effects on diffusion is that when one is considering the diffusion of a particular solute, the rate of diffusion is determined by the concentrations of *that* solute, not those of other solutes that might be present; each solute diffuses according to its own concentration gradient.

What determines the diffusion coefficient D? One important factor is the ease with which the solute of interest moves through the material separating the two concentrations; when diffusion is occurring through a cell membrane or an epithelium, this factor is termed the *permeability* of the membrane or epithelium to the solute (permeability is discussed later in this chapter). Another factor that determines D is temperature. As temperature rises, random molecular motions become more vigorous (see Chapter 8). Thus the rate of diffusion increases.

If a fluid current flows over the surfaces of an animal or cell, the flow can also affect the rate of diffusion. Suppose the concentration of a solute is 100 mM in the body fluids of an aquatic animal and 2 mM in the ambient water. The outward diffusion of the solute will tend to create a *boundary layer* of elevated solute concentration next to the animal's body surface (Figure 3.3). The presence of the boundary layer decreases the rate of diffusion. One way to understand this effect in terms of Equation 3.1 is to consider the effect of the boundary layer on X, which is the distance separating the two concentrations, 100 mM and 2 mM; without the boundary layer, these two concentrations would be separated just by the thickness of the outer membrane or epithelium of the animal, but with the boundary layer, the two concentrations are separated in addition by the thickness of the boundary layer itself, which increases X. Increasing the flow of water over an animal or cell tends to carry solute away from areas of accumulation at the animal or cell surface, thereby decreasing boundary-layer thickness, decreasing X, and increasing the rate of diffusive solute loss from the animal or cell.

Electrical gradients often influence the diffusion of charged solutes at membranes

Many solutes of biological importance—such as Na^+, Cl^-, and other inorganic ions—bear an electrical charge. The motion of charged

[4] Diffusion of gases is discussed in Chapter 20. Diffusion of liquid water is discussed later in this chapter, and that of water vapor, in Chapters 20 and 25. Diffusion of heat (thermal conduction) is covered in Chapter 8.

TABLE 3.1 The time required for diffusion through water to halve a concentration difference Values are calculated for small solutes such as O_2 or Na^+. For each distance between solutions, the time listed is the time that will be required for diffusion to transport half the solute molecules that must move to reach concentration equilibrium. It is assumed that no electrical effects exist, and thus only diffusion based on concentration effects is occurring.

Time required to halve a concentration difference by diffusion	Distance between solutions	A biological dimension that exemplifies the distance specified
100 nanoseconds	10 nanometers	Thickness of a cell membrane
100 milliseconds	10 micrometers	Radius of a small mammalian cell
17 minutes	1 millimeter	Half-thickness of a frog sartorius muscle
1.1 hours	2 millimeters	Half-thickness of a human eye lens
4.6 days	2 centimeters	Thickness of the human heart muscle
32 years	1 meter	Length of a long human nerve cell

Source: After Weiss 1996.

solutes, besides being affected by concentration gradients, is influenced by forces of *electrical attraction* or *repulsion*, forces that do not affect the movement of uncharged solutes such as glucose. As individual ions or other charged solutes move ceaselessly because of atomic-molecular agitation, their paths of motion are affected by the attraction of positive charges toward negative ones, and by the tendency of like charges (e.g., two positive ones) to repel.

To understand *where* such electrostatic effects need to be considered in the analysis of diffusion, a key principle is that solutions at large are electrically neutral. In **bulk solution**—the solution that is away from contact with a membrane—the concentrations of positive and negative charges are equal (**Figure 3.4**). Thus the net charge in any region of bulk solution is zero, and different regions do not differ in net charge. Accordingly, in bulk solution the charges on solutes do not affect their diffusion, and diffusion simply follows the *concentration*-based principles we have already discussed.

In contrast to bulk solution, electrical attraction and repulsion can play a large role in diffusion *along or across cell membranes or epithelia* because the lipid bilayers in cell membranes can maintain separation of oppositely charged ions by acting similarly to capacitors in electrical circuits.[5] Within a few *nanometers* of a cell membrane or epithelium, the solution on one side may have a net positive charge because positive ions outnumber negative ones; and conversely, the solution on the other side may have a net negative charge because negative ions outnumber positive ones (see Figure 3.4). Under these circumstances the diffusion of electrically charged solutes may be greatly affected by electrical attraction and repulsion.

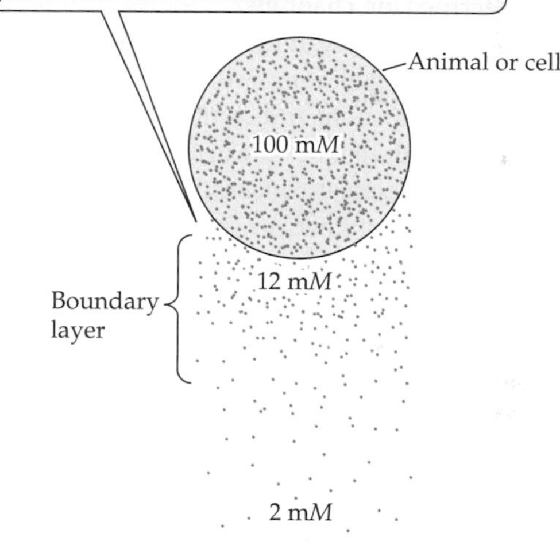

Figure 3.3 Diffusional concentration of a solute in a boundary layer next to an animal or cell In this example, the concentration of the solute is 100 m*M* inside the animal or cell, and 2 m*M* in the open environmental solution.

Biological aspects of diffusion across membranes: Some solutes dissolve in the membrane, others require channels

Principles we discussed at the start of Chapter 2 return to center stage when we consider the chemical mechanisms of the diffusion of solutes through cell membranes or epithelia. The interior of a cell membrane is hydrophobic because it consists principally of the hydrocarbon tails of phospholipid molecules composing the lipid bilayer (see Figure 2.1). The ease with which a solute can diffuse directly through the lipid interior of a cell membrane thus depends on the hydrophobic or hydrophilic nature of the solute.

Let's start by considering *lipid solutes*, such as steroid hormones and fatty acids, which are hydrophobic. Molecules of these solutes *dissolve* in the lipid interior of a cell membrane. In this dissolved state, they make their way from the side of the membrane where they are more concentrated to the side where they are less concentrated because of molecular agitation and the other principles of simple diffusion. *Molecular oxygen* (O_2), which is small and nonpolar, is also generally believed to make its way through cell membranes by simple diffusion through the lipid layers, without need of channels or other proteins, as are certain signaling molecules besides steroid hormones, such as *thyroid hormones* and *nitric oxide*.

[5] Appendix J (Basic Physics Terms) discusses capacitance and other physics concepts used in this chapter.

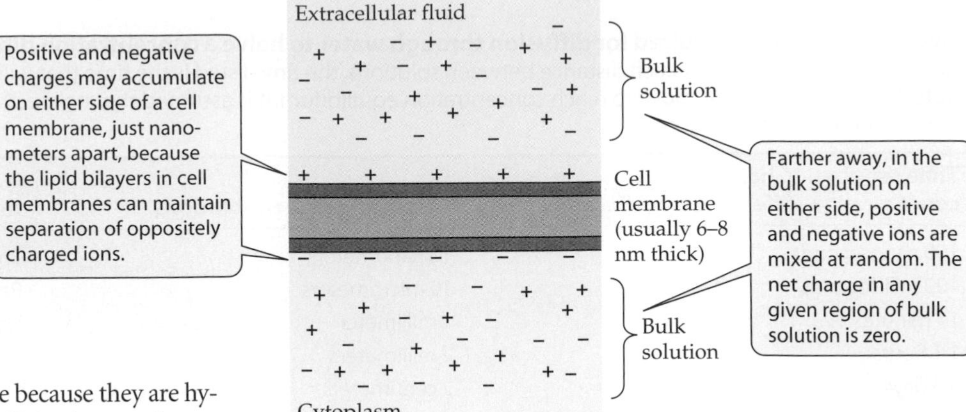

Figure 3.4 Ionic charge separation occurs only within nanometers of membranes Thus, only in the vicinity of membranes do conditions exist for the diffusion of charged solutes to be affected by electrical attraction and repulsion.

Inorganic ions present a very different picture because they are hydrophilic and therefore have very low solubilities in membrane lipids. Studies using experimental *all-phospholipid* membranes demonstrate that the rates of simple diffusion of ions directly through membrane lipids are exceedingly low. However, ions of physiological importance—such as Na^+, K^+, Ca^{2+} (calcium), and Cl^-—can sometimes move passively through *actual* cell membranes at very rapid rates. They are able to do this because their passive transport through cell membranes is mediated by integral membrane proteins termed **ion channels.**

The defining characteristic of ion channels is that they permit the *passive* transport of inorganic ions by diffusion through a membrane. This means that the directions and eventual equilibrium states of the ion movements through channels are determined not by the channels themselves, but by chemical-concentration and charge gradients across the membranes. Another key characteristic of channels is that the ions that pass through them do not bind to the channel proteins. Structurally, an ion channel consists of one or more protein molecules that extend across the full thickness of a cell membrane or intracellular membrane and that encircle a lipid-free central passageway through the membrane (see Figure 2.4). From this fact and the fact that ions do not bind to channels, one might think that channels are simple holes through the membrane. Actually, however, they are intricate protein structures that are *selective* in determining which ions can pass readily through them. Some ion channels, for instance, are specific for Na^+. They permit Na^+ to move through the cell membrane at a relatively high rate but do not permit rapid passage of other ions. Other channels are specific for K^+. Still others allow Na^+ and K^+ to pass with equal ease. Some channels allow several sorts of ions to pass, but even the least selective channels discriminate between anions and cations.[6]

Many ion channels are **gated channels**, meaning that they can "open" and "close" because the proteins of which they are composed are able to undergo conformational changes that cause their central passageways to increase or decrease the ease with which ions pass through. Four categories of gated ion channels, distinguished by the mechanisms of gating, are recognized (**Figure 3.5**). **Voltage-gated channels** open and close in response to changes in the voltage difference across a membrane; they are very important in the generation of nerve impulses. **Stretch-gated** (**tension-gated**) **channels** open or close in response to stretching or pulling forces that alter the physical tension on a membrane. **Phosphorylation-gated channels** open or close according to whether the channel proteins are phosphorylated; they are under the control of protein kinases (see page 53), which are often themselves controlled by second messengers (see page 62). **Ligand-gated channels**, discussed in Chapter 2 (see page 56), act both as *receptors* of extracellular signals and as ion channels. They are closed to passage of ions in the absence of a signal, but when their receptor sites bind to receptor-specific ligands (e.g., neurotransmitter molecules), they open, allowing the passage of selected ions.

The **permeability of a cell membrane to a solute** is defined to be the ease with which the solute can move through the membrane by diffusion. The concept of permeability applies both to solutes that dissolve in the membrane interior and to solutes that pass through channels. When Equation 3.1 is applied to solute diffusion through a membrane, the membrane permeability to the solute of interest helps determine D, the diffusion coefficient. For solutes like lipids and O_2 that dissolve in the lipid bilayer to pass through a membrane, the permeability of a cell membrane depends on the set of factors that affect solute motion through lipid, such as the molecular size of the solute. For inorganic ions, the permeability of a cell membrane depends on the number of channels per unit area of membrane and on the proportion of the channels that are open. Permeability to an ion can be modulated rapidly up and down by the opening and closing of the gated channels for the ion.

Cell membranes are often described as **selectively permeable** because they permit some solutes to pass through by diffusion with greater ease than others. One reason for selective permeability is the relative representation of different sorts of ion channels in a membrane. For example, if a membrane has a high density of Na^+ channels and a low density of Ca^{2+} channels, it is more permeable to Na^+ than to Ca^{2+}.

Diffusion of ions across cell membranes is determined by simultaneous concentration and electrical effects

The final aspect of the mechanism of simple diffusion that we need to consider is the manner in which concentration effects and electrical effects interact in affecting the diffusion of ions and other charged solutes. We have already seen that there is a *concentration gradient* across a cell membrane for each major solute. The diffusion

[6] *Anions* are negatively charged; *cations* are positively charged. The ion selectivity of channels is discussed in more detail in Chapter 11.

(a) Voltage-gated channel

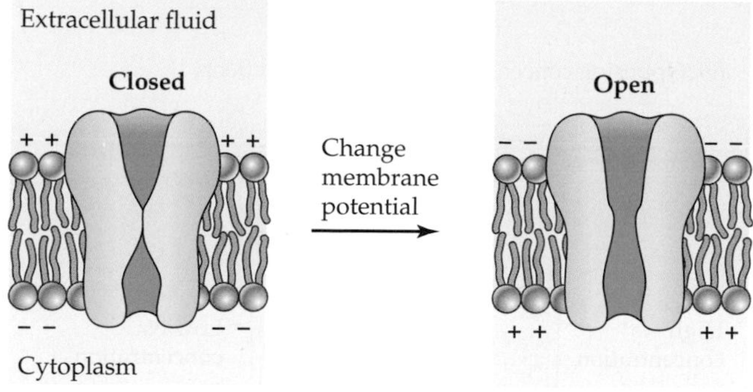

(b) Stretch- or tension-gated channel

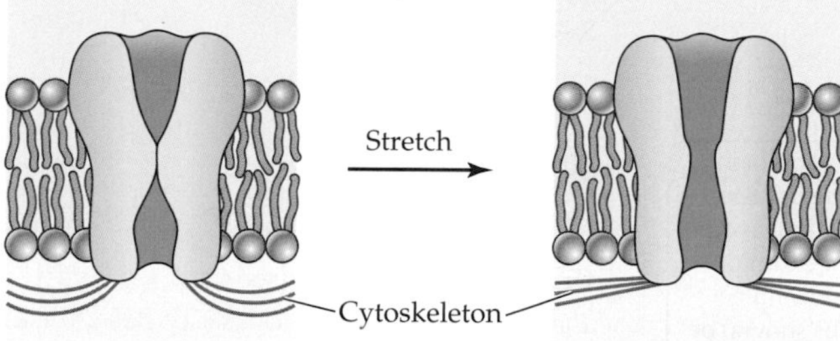

(c) Phosphorylation-gated channel

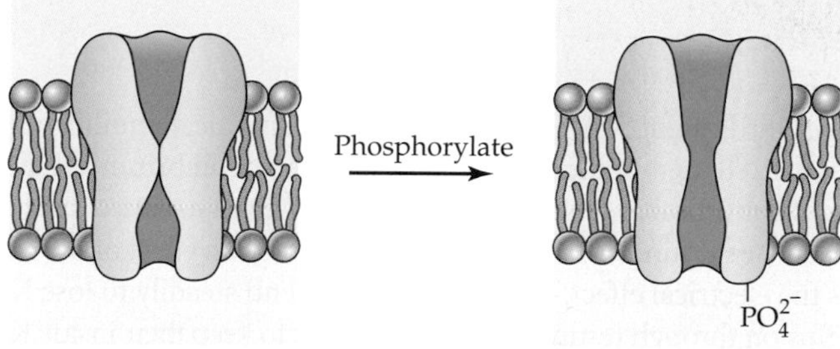

(d) Ligand-gated channel

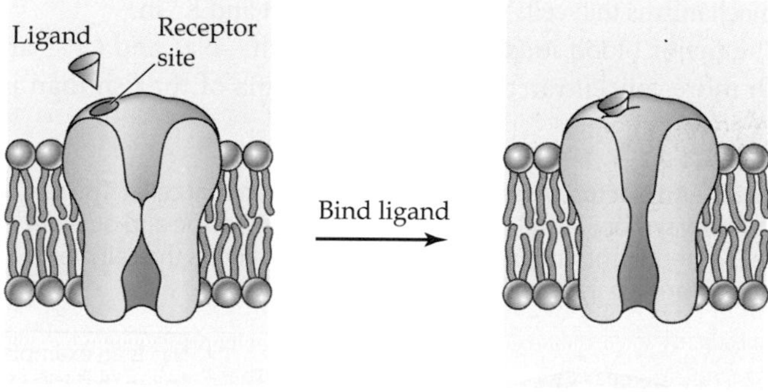

Figure 3.5 Gated ion channels Gated ion channels are classified into four functional categories based on which property must change for them to open and close: *(a)* transmembrane voltage, *(b)* stretch or tension, *(c)* phosphorylation, or *(d)* noncovalent binding with ligands at receptor sites.

of each solute is affected by its own concentration gradient but not the concentration gradients of other solutes. All cell membranes also exhibit charge separation. Thus there is always an **electrical gradient** across a cell membrane, defined to be the difference in voltage (electrical potential) between the two sides. The electrical gradient affects the diffusion of all charged solutes.

It is obvious that the diffusion of an ion or other charged solute must depend on *both* that solute's concentration gradient *and* the electrical gradient. A problem that soon arises in trying to *quantify* this dual effect is that concentration gradients and electrical gradients are expressed in different units: molarity and volts, respectively. Walther Nernst (1864–1941), an early winner of the Nobel Prize, goes down in history for solving this problem. Using his *Nernst equation*, which is discussed in Chapter 11, investigators can directly compare concentration and electrical gradients to determine which is stronger, and thus predict the diffusion of an ion.

Here we take a more informal approach to understanding the dual effects of concentration and electrical gradients on an ion's diffusion across a cell membrane. We term the influence of the concentration gradient across the membrane on the ion's diffusion the **concentration effect** on the ion, and we term the influence of the electrical gradient on the ion's diffusion the **electrical effect** on the ion.

Sometimes the electrical charge difference across a membrane pulls an ion in the same direction as the ion is moving because of its concentration gradient as in Figure 3.6*a*. In such cases the electrical and concentration effects reinforce each other, and the presence of the electrical gradient *accelerates* diffusion compared to the rate at which it would proceed if only the concentration gradient were present.

On the other hand, the electrical charge difference and the concentration gradient sometimes act *in opposite directions*, as in Figure 3.6*b*. If the electrical effect on Na^+ diffusion in Figure 3.6*b* is weaker than the concentration effect, Na^+ ions diffuse, as shown, in net fashion from left to right—the direction favored by the concentration gradient—but *more slowly* than if the opposing electrical gradient were absent. If the electrical effect in Figure 3.6*b* is stronger than the concentration effect, Na^+ ions diffuse in net fashion from right to left—the direction favored by the electrical gradient—even though that direction of diffusion is opposite to what the concentration gradient alone would cause. A large voltage difference can cause an ion or other charged solute to diffuse in a direction that *increases* its concentration gradient!

A charged solute is *at equilibrium* across a membrane when the concentration effect on its diffusion and the electrical effect are equal but opposite. Under such conditions the solute will not move in net fashion in either direction. Such an equilibrium is called an **electrochemical equilibrium**, stressing that both electrical and "chemical" (concentration) effects are involved. The concept of an electrochemical equilibrium is, in fact, an entirely general concept that can be applied whether a solute is charged or uncharged (for an uncharged solute, no electrical effect exists, and thus electrochemical equilibrium prevails when the concentration gradient is zero).

Diffusion often creates challenges for cells and animals

Because living creatures are inherently nonequilibrium systems and simple diffusion moves solutes toward equilibrium, diffusion often creates challenges that must be met by energy expenditure. This point is illustrated by two of our focal examples, the individual animal cell and the freshwater fish (see Figure 3.1*a* and *b*).

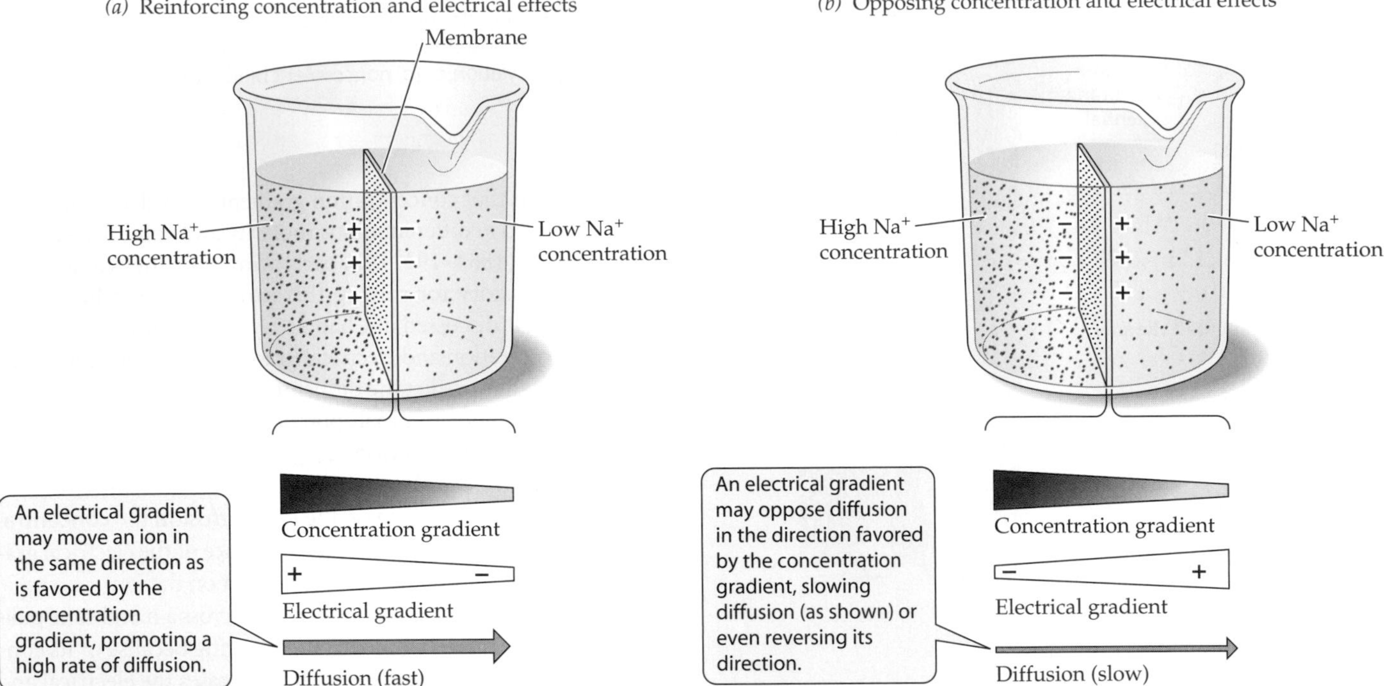

Figure 3.6 Ion diffusion depends on the dual effects of concentration and electrical gradients *(a)* Fast diffusion resulting from reinforcing concentration and electrical effects. *(b)* Slow diffusion resulting from opposing concentration and electrical effects. In both parts the different dot densities on the two sides of the membrane in the beaker symbolize the different Na^+ concentrations.

Considering the cell first, let's examine the concentration and electrical effects on each of the three principal ionic constituents: Cl^-, Na^+, and K^+ (Figure 3.7). A typical cell membrane is positively charged on the outside and negatively charged on the inside. The concentration and electrical effects on the diffusion of Cl^- across the cell membrane are therefore opposite to each other because Cl^- is more concentrated outside a cell than inside (favoring inward diffusion) but the inside of a cell membrane is negatively charged (repelling Cl^-). The magnitudes of the concentration and electrical effects on Cl^- are typically about equal, meaning that Cl^- is typically at or near electrochemical equilibrium across a cell membrane. Na^+ presents an extremely different picture, however.

The concentration and electrical effects on Na^+ diffusion are in the same direction; Na^+ is more concentrated on the outside of a cell than the inside, favoring inward diffusion, and the negative charges at the inside of the cell membrane also tend to draw Na^+ in by electrical attraction. Thus, *Na^+ is very far from electrochemical equilibrium across the cell membrane and has a great tendency to diffuse from the extracellular fluid into a cell.* Sets of so-called *resting channels*, which are open all or most of the time, permit Na^+ to leak across cell membranes, and cells therefore experience a steady inward diffusion of Na^+. This diffusion presents a challenge because it would do away with the Na^+ concentration gradient if it were unopposed. Thus, for a cell to maintain its normal Na^+ concentration gradient across its cell membrane, it must use some of its energy resources to do work to oppose the effects of Na^+ diffusion.

The third major ion, K^+, is not at electrochemical equilibrium across the cell membrane, but is not as far from equilibrium as Na^+. The concentration and electrical effects on K^+ tend to cancel rather than add (see Figure 3.7). However, the concentration effect on K^+ exceeds the electrical effect. Thus, animal cells tend steadily to lose K^+ by diffusion through resting channels. For cells to keep their inside K^+ concentration high, they must use energy to counteract this diffusion. When we address active transport later in this chapter, we will discuss the mechanisms that cells use to move Na^+ out and K^+ in.

The major blood ions of a freshwater fish—Na^+ and Cl^-—are much more concentrated in the blood plasma of the fish than in

Figure 3.7 An electrochemical view of a typical animal cell The size of the lettering symbolizes relative ion concentrations inside and outside the cell. A^- represents anionic proteins that cannot cross the cell membrane and thus are trapped inside the cell.

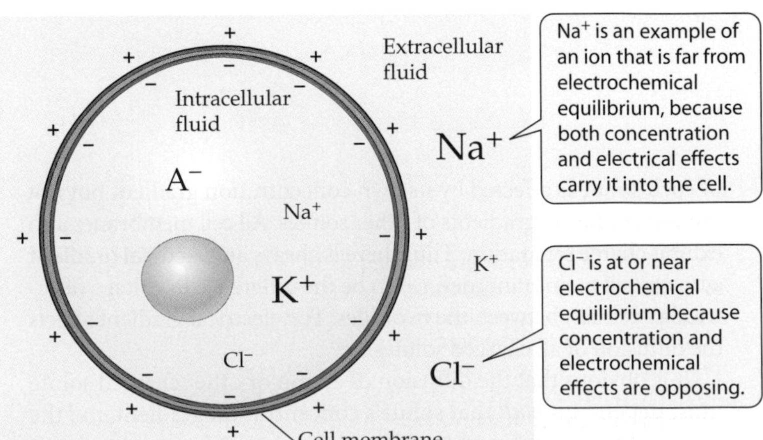

pond water, suggesting that the ions tend to diffuse outward across fish's gills (see Figure 3.1*b*). Now we recognize that electrical effects need to be taken into account as well as concentration effects. When we take both types of effects into consideration, however, we come to the same conclusion: For both Na^+ and Cl^-, the direction of simple diffusion is from the blood into the pond water. Thus the ceaseless and unstoppable random motions of ions present a challenge to a freshwater fish, causing the fish to lose these important ions steadily from its blood. A fish could theoretically block the losses by making its whole body impermeable to the ions. However, fish need to breathe, and gills that are permeable to O_2 are also permeable to Na^+ and Cl^-. Fish must replace the Na^+ and Cl^- they lose by diffusion. This task is accomplished by active transport and costs energy, as we will discuss soon.

Concentration gradients can create electrical gradients that alter concentration gradients

Thus far, we have discussed the diffusion of ions as if the process were purely the *consequence* of concentration and electrical gradients. Ion diffusion is actually more dynamic, however. Because cell membranes and epithelia are selectively permeable—excluding some solutes while allowing the passage of others—the concentration gradients of ions across cell membranes or epithelia can *create* electrical gradients that may then affect ion diffusion and ion concentration gradients.

To see how this process takes place, let's study an experiment in which the diffusion of K^+ along its concentration gradient produces a counteracting electrical gradient. In our laboratory setup (Figure 3.8*a*), a selectively permeable membrane that is *permeable only to* K^+ separates two solutions. At the start, the solutions on the two sides are equimolar, one composed of K^+ and A_1^- (an anionic solute) and the other composed of Na^+ and A_2^- (another anionic solute). After the apparatus is set up in this way, K^+ will start to diffuse from left to right across the membrane because of its concentration gradient. This diffusion *creates* a voltage difference across the membrane because K^+ is charged; specifically, the diffusion of K^+ causes more positive charges—K^+ ions—to accumulate along the right-hand surface of the membrane than the left-hand surface. As the K^+ diffusion continues, this voltage difference will steadily grow larger, and it will keep growing larger until the electrical effect on K^+ diffusion (repelling K^+ from the right) fully counteracts the concentration effect (favoring diffusion to the right). At that point, illustrated by the equilibrium diagram (Figure 3.8*b*), K^+ will be at electrochemical equilibrium, and no further net diffusion of K^+ will occur.

Reviewing this experiment, you will see that when a membrane is selectively permeable and bars some solutes from crossing, the diffusion of a permeating ion along its concentration gradient can *create* an electrical gradient. This sort of process produces the voltage difference, termed the *membrane potential*, that typically exists across the cell membrane of an animal cell (see Figure 3.7), as discussed further in Chapter 11.

When a membrane is permeable to more than one sort of ion, an electrical gradient produced by diffusion of any one ion will affect the diffusion of the others. A complex type of electrochemical equilibrium involving diffusion of multiple ions (and water) then tends to develop. The **Donnan equilibrium**, which you will see mentioned in advanced literature, is this type of equilibrium. Specifically, a Donnan equilibrium occurs across a membrane when a number of ions (anions and cations) can cross the membrane but there is a set of nonpermeating ions (ions that cannot cross the membrane) that are more abundant on one side of the membrane than the other. Animal cells tend toward Donnan equilibrium because they have high concentrations of nonpermeating anionic proteins and nucleic acids within them.

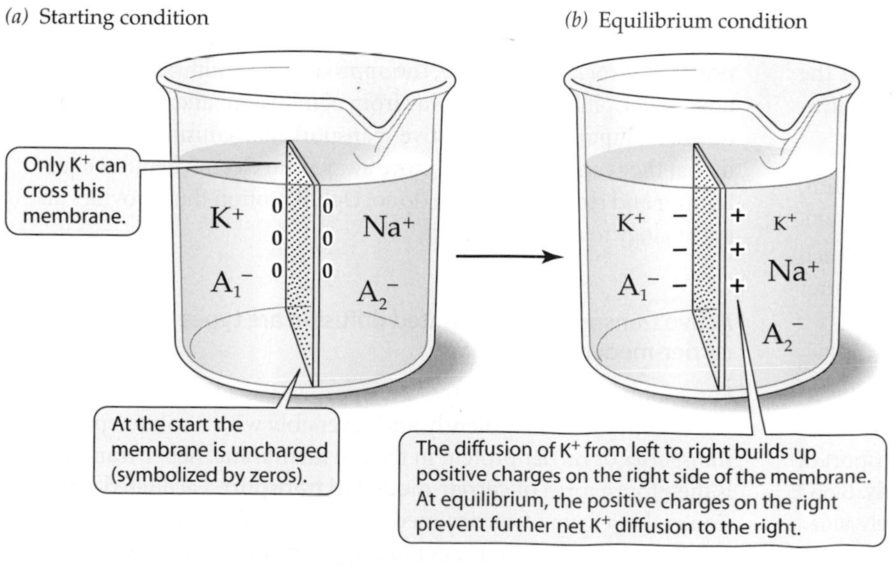

(*a*) Starting condition

Only K^+ can cross this membrane.

At the start the membrane is uncharged (symbolized by zeros).

(*b*) Equilibrium condition

The diffusion of K^+ from left to right builds up positive charges on the right side of the membrane. At equilibrium, the positive charges on the right prevent further net K^+ diffusion to the right.

Figure 3.8 The development of an equilibrium in which a voltage difference generated by diffusion exactly opposes the remaining concentration gradient Because K^+ is more concentrated on the left at the start (*a*), it diffuses to the right, causing accumulation of positive charges on the right side of the membrane. K^+ diffusion to the right continues—building up an ever greater voltage difference across the membrane—until eventually net K^+ diffusion stops (*b*) because the electrical effect on K^+ diffusion exactly opposes the remaining concentration effect. The size of the lettering symbolizes relative ion concentrations. A_1^- and A_2^- represent anionic solutes that cannot cross the membrane.

Summary: Passive Solute Transport by Simple Diffusion

- A mechanism of solute transport is *passive* if it transports only *toward* electrochemical equilibrium. For an uncharged solute, the electrochemical equilibrium is the same as concentration equilibrium. For a charged solute, the electrochemical equilibrium is achieved when the voltage difference exactly counterbalances the solute's concentration gradient across a cell membrane or epithelium.
- Simple diffusion is the most straightforward type of passive solute transport. The fundamental mechanism of simple diffusion is exemplified most clearly by the diffusion of an uncharged solute. Molecular agitation tends by simple statistics to carry more molecules of such a solute out of regions of high concentration than into such regions, thereby tending to produce equality of concentration everywhere. If a solute is charged and diffusing where a voltage difference exists, forces of electri-

cal attraction and repulsion affect the paths followed by molecules or ions during molecular agitation and thus contribute to diffusion.

- Voltage differences can occur across cell membranes or epithelia because the lipid layers in cell membranes act as capacitors, permitting positive and negative charges to be unequally distributed on the two sides. Bulk solutions are electrically neutral, however, and voltage differences are therefore not a factor in simple diffusion within bulk solutions.
- When lipid solutes such as steroid hormones and fatty acids cross cell membranes by simple diffusion, they typically dissolve in the lipid layers of the membranes. Ions, however, cannot cross membranes by this mechanism because of their poor solubility in lipids; thus their diffusion requires mediation by membrane proteins. Inorganic ions typically cross membranes by simple diffusion through ion channels. Channels are often gated—that is, opened and closed by factors such as ligand binding.
- The permeability of a membrane to a lipid solute depends on how readily the solute dissolves in and moves through the membrane lipid layer. The permeability of a membrane to an inorganic ion depends on the number of channels for the ion per unit of membrane area, and can be changed by the opening and closing of channels.

Passive Solute Transport by Facilitated Diffusion

Glucose, amino acids, and other *polar organic solutes* can move passively across cell membranes, yet they do not typically do so by simple diffusion, explaining why they were not discussed in the last section. These solutes cannot diffuse through the lipid interior of cell membranes because they are hydrophilic, and there are no channels for them. Passive transport of the polar organic solutes across cell membranes occurs typically by the noncovalent and reversible *binding* of the solutes to solute-specific *transporter* (*carrier*) proteins in the membranes, a process called **facilitated diffusion**. Facilitated diffusion has three defining properties:

1. It always occurs in the direction of electrochemical equilibrium. This is why it is a form of passive transport and is called diffusion.
2. Solutes transported in this way move across membranes much faster than they could if they did not associate with transporter proteins. This is why the transport is called *facilitated*.
3. The mechanism requires solutes to bind reversibly with binding sites on the transporter proteins. This is why it is not a type of simple diffusion.

An important example of facilitated diffusion is the transport of glucose from the blood into cells throughout an animal's body. Blood glucose typically crosses cell membranes to enter cells passively, aided by glucose transporters.[7] One of the most important effects of the hormone insulin in vertebrates is that it stimulates certain types of cells (notably muscle and adipose cells) to increase the numbers of transporter molecules in their cell membranes, thereby poising the cells to be able to take up glucose rapidly.

[7] Because cells continuously use glucose, the concentration gradient for glucose is *from* the blood *into* the cells.

Active Transport

Animals are able to transport many solutes across their cell membranes and epithelia in directions away from electrochemical equilibrium by using inputs of energy from metabolism. Mechanisms capable of such transport are called **active-transport mechanisms** or **pumps**. Active-transport mechanisms do not *necessarily* move solutes away from electrochemical equilibrium, but they are *capable* of doing so.

Active transport plays critical roles in all three of our focal examples. To see this, let's look first at the individual animal cell. We have seen that K^+ is more concentrated inside a cell than in the extracellular fluids bathing the cell, and Na^+ is less concentrated inside than out (see Figure 3.1*a*). We have also seen that diffusion tends to obliterate these concentration differences by carrying K^+ outward and Na^+ inward. Cells keep the inside K^+ concentration high, and the inside Na^+ concentration low, by using an active-transport mechanism that transports K^+ in and Na^+ out *at a cost*, using metabolic energy.

Active transport also plays critical roles in our other two focal examples—the freshwater fish and the vertebrate small intestine. In the case of the freshwater fish (see Figure 3.1*b*), Na^+ and Cl^- are so much more concentrated in the blood than in pond water that the two ions tend to diffuse out of the fish. A major mechanism by which a freshwater fish replaces lost ions is to carry out active transport of Na^+ and Cl^- from the dilute pond water into its concentrated blood. By means of active-transport mechanisms in their gills, many fish can pump Cl^- from freshwater, where it is as dilute as 0.02 m*M*, into their blood, where the concentration is over 100 m*M*. In the small intestines of vertebrates (see Figure 3.1*c*), active-transport mechanisms help transport sugars, such as glucose, and amino acids from digested food into the blood. Active transport also plays countless other, vital roles in the lives of animals.

An instructive alternative name for *active transport* is **uphill transport**. For a rock on a hillside, the approach to equilibrium is to roll downhill. Uphill motion is away from equilibrium and therefore requires an input of energy. Active-transport mechanisms are similar in that they can make solutes move away from electrochemical equilibrium and require energy to do so. Uphill motion thus provides an apt analogy to active transport.

Active transport and facilitated diffusion are types of carrier-mediated transport

For a solute to undergo active transport across a cell membrane, it must combine noncovalently and reversibly with a solute-specific *transporter* or *carrier* protein in the cell membrane. Active transport is therefore a form of **carrier-mediated transport**. Facilitated diffusion is also a form of carrier-mediated transport because, as we have noted already, it, too, requires binding of the transported solute to a transporter (carrier) protein. The two forms of carrier-mediated transport are distinguished by whether transport is driven by metabolic energy. In *facilitated diffusion*, no mechanism exists for metabolic energy to drive the process; thus transport is always toward electrochemical equilibrium. In *active transport*, metabolic energy is used to drive the process, so transport can be away from electrochemical equilibrium.

We have now reached a point where we can fully define **transporters (carriers)**, one of the five principal classes of membrane proteins (see Table 2.1). A transporter or carrier is a membrane protein that mediates *active* or *passive* transport across the membrane by binding noncovalently and reversibly with solute. *Carrier* is the older term and is the basis for the expression *carrier-mediated transport.* Physiologists once believed that the molecular mechanism of carrier-mediated transport in animals was a true carrier mechanism. The "carrier" proteins were hypothesized to link up with solutes on one side of a cell membrane and then carry the solutes through the membrane by moving physically from one side of the membrane to the other. As more has been learned about carrier proteins, this sort of mechanical model has become less plausible. Current models of carrier-mediated transport envisage proteins that do not move through a cell membrane but instead stay in one place while *intramolecular conformational changes* usher solutes through. The actual mechanisms remain unknown. The solutes that bind noncovalently and reversibly with a transporter are the **ligands** of the transporter.

Carrier-mediated transport, whether active or passive, exhibits saturation kinetics (see page 43). Saturation occurs because solute molecules must bind with transporter molecules. When the concentration of solute is high enough for all transporter molecules to be bonded with solute and engaged in transport all the time, the transport rate is maximal and cannot increase if more solute is made available.

Basic properties of active-transport mechanisms

Active-transport mechanisms in animal cells draw their energy directly or indirectly from ATP made by the catabolism of foodstuffs (see Chapter 6). Because of the requirement for ATP, active transport is sensitive to factors that affect a cell's ATP supply, such as poisons or O_2 deficiency. In some sorts of cells that are very involved with transport, such as certain kidney-tubule cells, active-transport mechanisms are the single biggest consumers of ATP and can use up to 40% of the ATP a cell makes.

A great diversity of active-transport mechanisms is known. Thus although any particular active-transport mechanism is specific for certain solutes, a great variety of solutes can be actively transported in one situation or another, including inorganic ions, amino acids, and sugars. A fascinating and highly consequential aspect of active-transport physiology is that active-transport mechanisms seem *not* to have evolved for two of the most important commodities in physiology: H_2O and O_2. Current evidence indicates that *H_2O and O_2 always move by passive transport.*

An active-transport mechanism that transports ions is potentially able to create a voltage difference across a membrane because the mechanism moves charges. Active ion-transport mechanisms do not *necessarily* generate voltage differences across membranes, however, because they often simultaneously transport two ions, and the transport of two ions may be electrically neutral. For an example, consider the "proton pump" responsible for acidifying stomach contents (Figure 3.9). This active-transport mechanism transports H^+ (protons or hydrogen ions) into the stomach cavity and K^+ in the opposite direction *in a 1:1 ratio.* Mechanisms of this sort, which do not generate an imbalance of charge across a membrane, are termed **electroneutral (nonelectrogenic)**. An active-transport mechanism is **electrogenic** if the actions of the mechanism create a charge imbalance across a membrane. An example of an electrogenic mechanism is provided by the famous and important **Na^+–K^+ pump**, which is found in the basolateral membranes of all epithelial cells and is ubiquitous in other sorts of animal cells as well. During each cycle of pumping, this active-transport mechanism transports three Na^+ ions out of a cell, but only two K^+ ions in (Figure 3.10*a*). It thus creates a charge difference across the cell membrane (outside positive).

At this point we can complete our discussion of the first of our three focal examples, the single animal cell (see Figure 3.1*a*). Earlier we saw that a cell must actively transport Na^+ outward and K^+ inward across its cell membrane to maintain differences of ion composition between the cell's intracellular fluid and the extracellular fluid. The Na^+–K^+ pump carries out this transport using ATP-bond energy. Figure 3.10*b* summarizes the patterns of active and passive ion transport between an animal cell and the extracellular fluids that bathe it.

Primary and secondary active transport differ in their cellular-molecular mechanisms

One of the best-known transporter proteins involved in active transport is **Na^+–K^+-ATPase**. This molecule *is* the Na^+–K^+ pump that we discussed just a moment ago. The molecule is called an *ATPase* be-

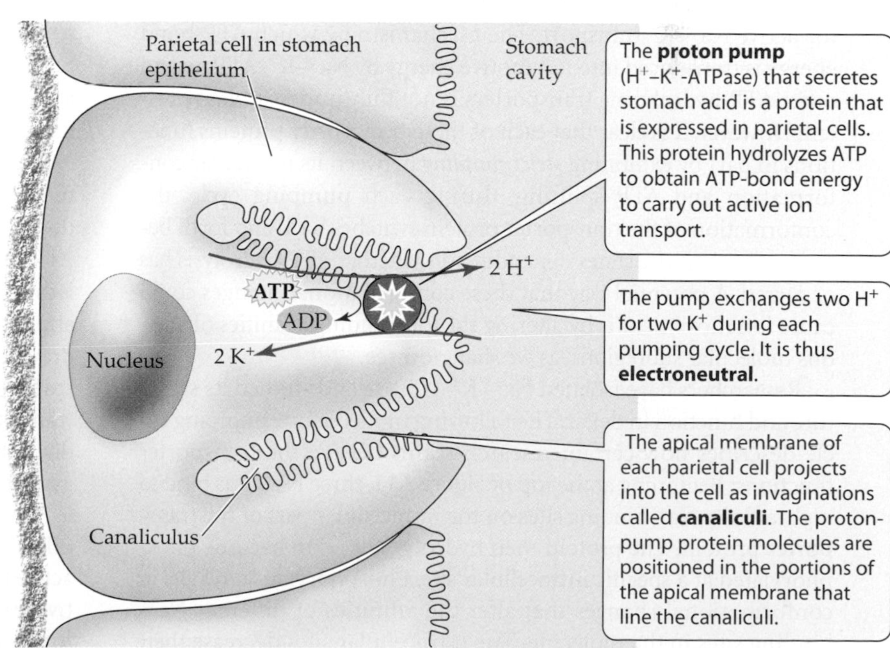

Figure 3.9 Electroneutral active transport is responsible for secretion of stomach acid in the vertebrate stomach lining This acid-secreting *parietal cell* is bounded by other types of cells in the epithelium lining the stomach. The singular of *canaliculi* is *canaliculus.*

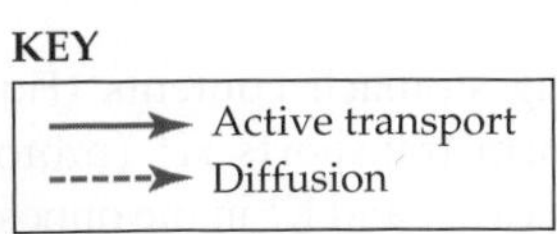

(a) The action of the Na^+–K^+ pump

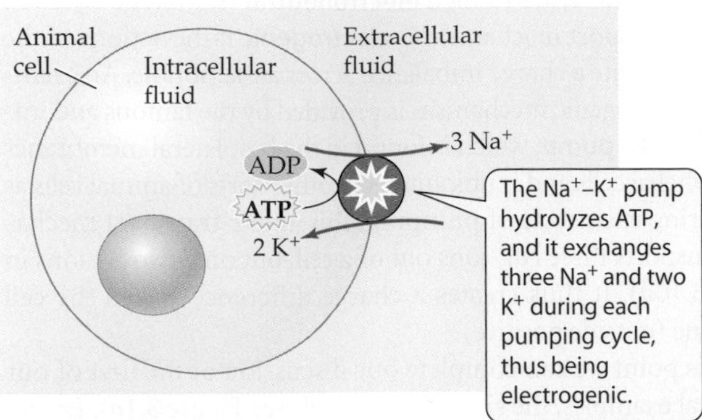

(b) An animal cell in summary

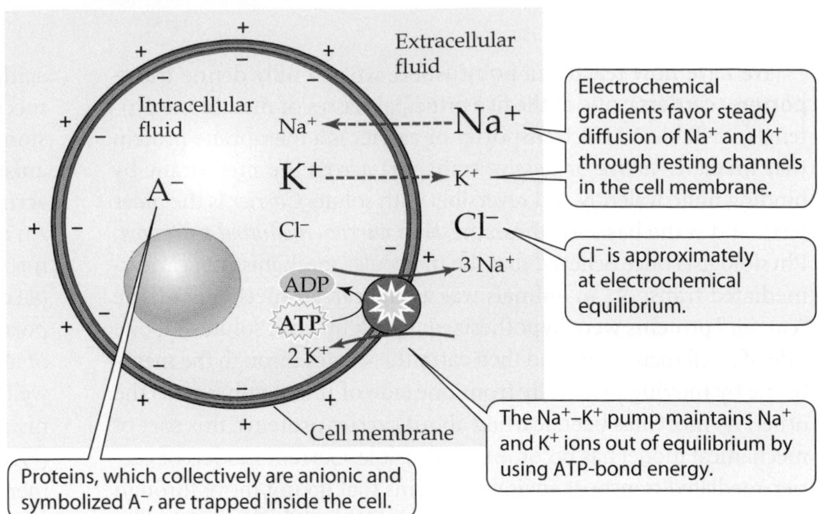

Figure 3.10 Summary of active and passive ion transport in a typical animal cell *(a)* The action of the Na^+–K^+ pump in the cell membrane. *(b)* Active and passive ion transport processes, and related phenomena, in a cell. As usual, the size of the lettering in the solutions symbolizes relative ion concentrations inside and outside the cell. The Na^+–K^+ pump contributes a portion of the charge difference across the cell membrane (outside positive) because of its electrogenic nature, but much of the charge difference originates from the sort of ion-diffusive processes exemplified in Figure 3.8 (discussed in depth in Chapter 11). The reason that Cl^- is approximately at electrochemical equilibrium is that its concentration gradient across the cell membrane is almost exactly opposed by the voltage difference across the membrane.

cause *it is an enzyme as well as a transporter*. It taps ATP-bond energy directly, catalyzing the hydrolysis of ATP to obtain the energy it uses for active Na^+–K^+ transport. The mechanism by which ATP-bond energy is transduced into ion motive energy by Na^+–K^+-ATPase and other ATP-hydrolyzing transporters is not fully understood at present. However, it is clear that each of these transporter proteins functions in part by exhibiting *strict coupling* between its molecular conformation and ATP splitting; during each pumping cycle, the conformation of the transporter protein switches back and forth between two or more states depending on whether ATP hydrolysis has occurred. A potential way that these conformational changes could cause ion movement is by altering the ion-binding affinities of various molecular subregions, as we shall now see.

Researchers have isolated Na^+–K^+-ATPase and studied its structure and function in detail. The following model of one pumping cycle describes how certain research teams believe the transporter functions: Beginning at the top of Figure 3.11, three Na^+ ions bind to high-affinity Na^+-binding sites on the intracellular part of the transporter protein. The protein then hydrolyzes ATP to become phosphorylated at a specific intracellular site. Phosphorylation results in conformational changes that alter the affinities of different Na^+-binding sites in the molecule. The intracellular sites decrease their Na^+ affinity while sites in other parts of the molecule develop higher Na^+ affinity. These changes draw Na^+ into the deep interior of the molecule. Some of the most recent research shows that while the Na^+ ions are being transported, they become temporarily "occluded" within the transporter, meaning that they cannot be accessed from either the extracellular fluid or the intracellular fluid. The conformational change induced by phosphorylation ultimately causes the Na^+ ions to move all the way through to the extracellular fluid. When the transporter undergoes the changes that transport Na^+ out of the cell, K^+-binding sites on the extracellular side of the transporter molecule simultaneously develop high K^+ affinity, as Figure 3.11 shows. After the transporter binds two K^+ ions, it is dephosphorylated, and its conformation changes again. This time the external binding sites reduce their K^+-binding affinity, and sites within the molecule increase their affinity for K^+. The K^+ ions are drawn into the deep interior of the molecule, where they also become temporarily occluded as they are transported, ultimately to be released into the intracellular fluid.

Besides Na^+–K^+-ATPase, other ATP-hydrolyzing transporter molecules are known. One is a **Ca^{2+}-ATPase** that is responsible for a critical Ca^{2+} pump in the sarcoplasmic reticulum of muscle (see Chapter 17). Another is **H^+–K^+-ATPase**, which is the engine of the "proton pump" that acidifies stomach contents (see Figure 3.9). Still another is a so-called *vesicular* or *v-type* **H^+-ATPase** that is increasingly recognized to be important in ion transport by gills and kidneys.

Two categories of active transport are presently recognized: primary and secondary. **Primary active transport** draws energy immediately from ATP. The transporter in primary active transport is an ATPase. Thus the active transport of Na^+ and K^+ by Na^+–K^+-ATPase is primary active transport, as is the transport of Ca^{2+} by Ca^{2+}-ATPase, and that of protons by H^+–K^+-ATPase. **Secondary active transport** draws energy, in an immediate sense, not from ATP, but from an electrochemical gradient of a solute, such as Na^+. In secondary active transport, although ATP is required to create the electrochemical gradient that drives transport, ATP is not the immediate source of energy tapped by the transporter protein.[8]

Glucose absorption in the small intestines of a hummingbird or other bird, or of a mammal, provides a classic example of secondary active transport. Thus, to understand the principles of secondary active transport and how energy is provided for it, we may now return to the focal example with which we opened this chapter. Specifically, let's consider how glucose is transported into a single intestinal ep-

[8] Primary active transport is sometimes called *direct* active transport, and secondary active transport is sometimes called *indirect*.

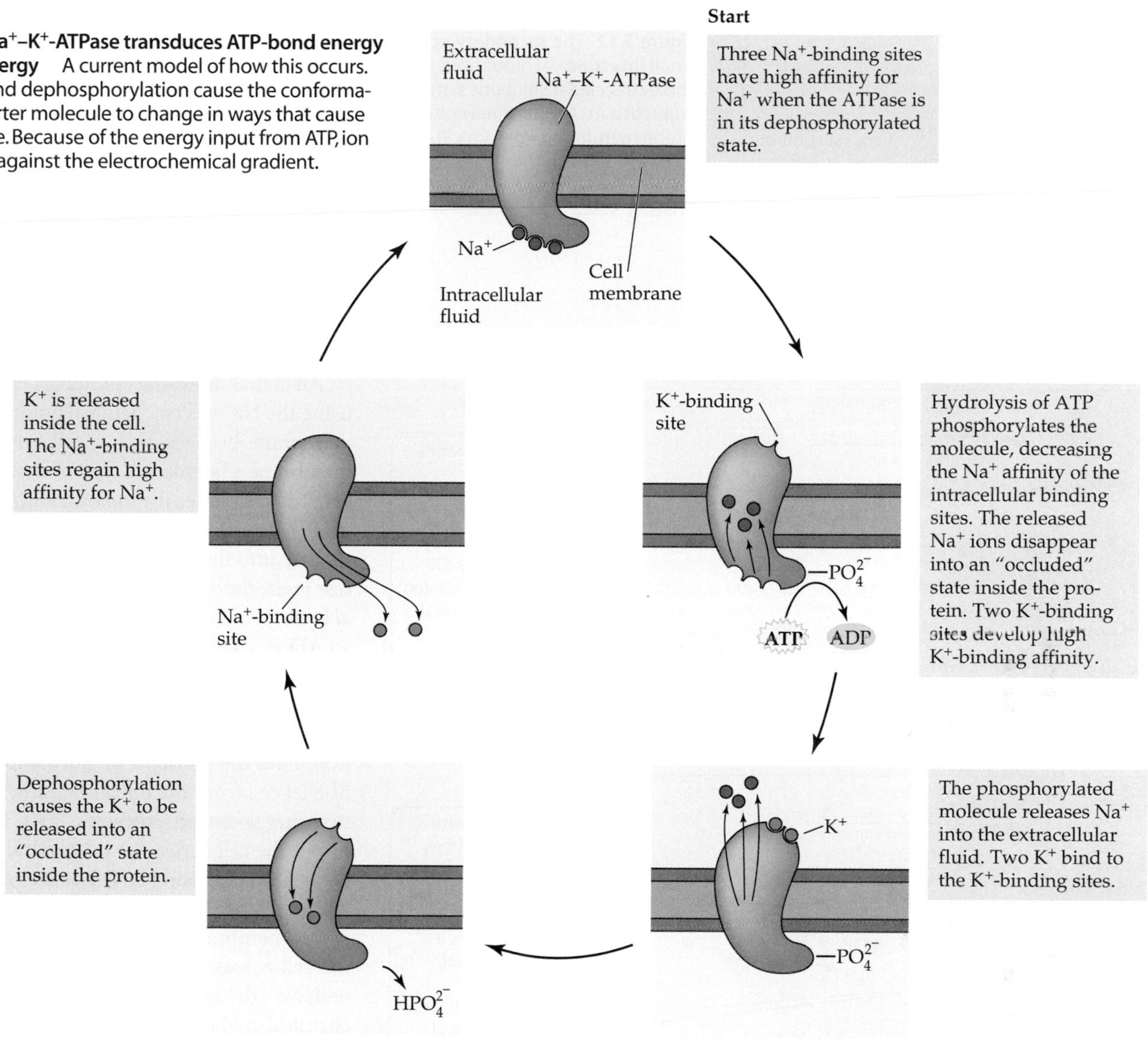

Figure 3.11 The Na^+–K^+-ATPase transduces ATP-bond energy into ion motive energy A current model of how this occurs. Phosphorylation and dephosphorylation cause the conformation of the transporter molecule to change in ways that cause Na^+ and K^+ to move. Because of the energy input from ATP, ion movement can be against the electrochemical gradient.

Start

Three Na^+-binding sites have high affinity for Na^+ when the ATPase is in its dephosphorylated state.

Extracellular fluid

Na^+–K^+-ATPase

Na^+

Cell membrane

Intracellular fluid

Hydrolysis of ATP phosphorylates the molecule, decreasing the Na^+ affinity of the intracellular binding sites. The released Na^+ ions disappear into an "occluded" state inside the protein. Two K^+-binding sites develop high K^+-binding affinity.

K^+-binding site

$—PO_4^{2-}$

ATP

ADP

The phosphorylated molecule releases Na^+ into the extracellular fluid. Two K^+ bind to the K^+-binding sites.

K^+

$—PO_4^{2-}$

Dephosphorylation causes the K^+ to be released into an "occluded" state inside the protein.

HPO_4^{2-}

K^+ is released inside the cell. The Na^+-binding sites regain high affinity for Na^+.

Na^+-binding site

ithelial cell from the intestinal lumen across the apical membrane of the cell (Figure 3.12). Our story must actually begin with the basolateral membrane of the cell. Recall that Na^+–K^+-ATPase (the Na^+–K^+ pump) is found in the basolateral membranes of all epithelial cells. The Na^+–K^+-ATPase in the basolateral membrane of an intestinal epithelial cell steadily transports Na^+ out of the cell, using ATP-bond energy. This activity of the Na^+–K^+-ATPase creates and maintains a Na^+ electrochemical gradient across the part of the cell membrane that faces into the gut lumen: the cell's *apical membrane*.[9] Because of the action of the ATPase, Na^+ is less concentrated on the inside of the apical membrane than the outside; furthermore, the inside of the apical membrane is negative relative to the outside (Figure 3.12*a*). From the perspective of energetics, the Na^+ electrochemical gradient across the apical membrane is an *energy store* that can be used to drive transport processes across the membrane.

[9] The electrochemical gradient is actually created across all parts of the cell membrane, but we focus here on the apical part because it is where glucose is transported into the cell from the intestinal lumen.

To understand how the Na^+ electrochemical gradient across the apical membrane functions as stored (potential) energy, imagine that there is a unique sort of pinwheel in the apical membrane: a "molecular pinwheel" that is turned by passing Na^+ ions (Figure 3.12*b*). Because of the electrochemical gradient for Na^+ across the apical membrane, Na^+ ions tend to diffuse inward. As Na^+ ions diffuse in this way, they spin the pinwheel, demonstrating that work can be done by the Na^+ electrochemical gradient, and thus the gradient can function as a source of energy. Now, the real point of this fanciful exercise comes to light when we ask, "What is the *ultimate* source of energy that turns the pinwheel?" The answer is ATP. *All the energy for ion transport in this system comes initially from ATP*, used by the Na^+–K^+-ATPase to transport Na^+ out of the cell across the basolateral membrane. A portion of the ATP-bond energy used by the Na^+–K^+ pump is stored as the potential energy of the apical Na^+ electrochemical gradient. This potential energy sets Na^+ in motion across the apical membrane, and this Na^+ movement can be used to do work. Although the work in our thought exercise (spinning of a pinwheel) is fanciful, we see that *work can be done in the apical mem-*

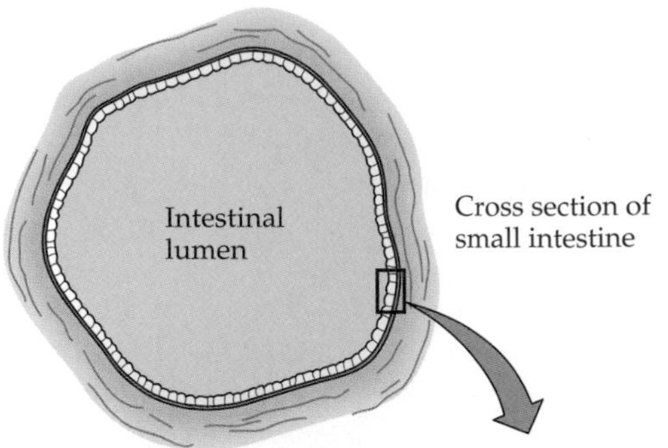

Figure 3.12 The secondary active transport of glucose into an epithelial cell of the vertebrate small intestine Throughout this figure, the size of the lettering used for solutes in the intracellular and extracellular solutions symbolizes relative solute concentrations inside and outside the cell. Glu = glucose. *(a)* An intestinal epithelial cell emphasizing the actions of the Na^+–K^+-ATPase in the basolateral membrane. *(b)* A fanciful cell with a Na^+-driven "molecular pinwheel" in the apical membrane. *(c)* A real cell with the Na^+–glucose cotransporter protein in the apical membrane. After glucose is inside the cell, it crosses the basolateral membrane into the blood by mechanisms discussed in Chapter 4 (see Figure 4.19).

(a) Na^+–K^+-ATPase in the basolateral membrane and the Na^+ electrochemical gradient it generates in the apical membrane

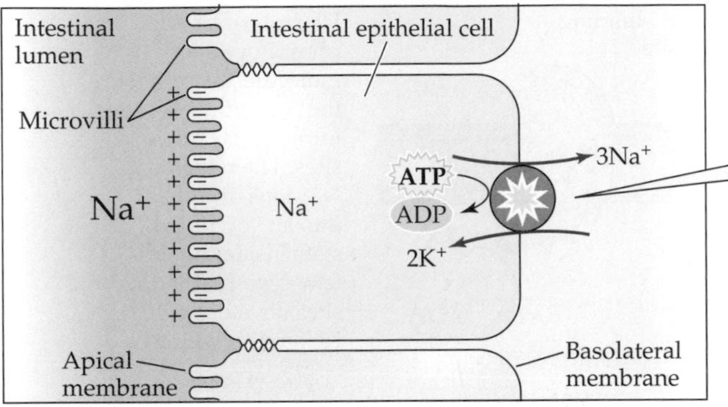

(b) Hypothetical Na^+-driven "pinwheel" in apical membrane

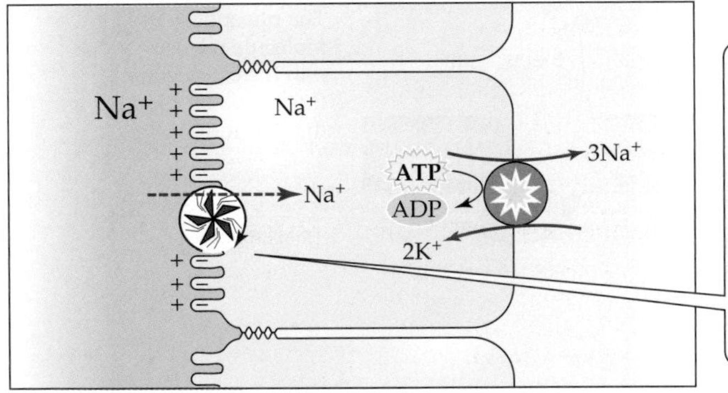

(c) Na^+–glucose cotransporter in apical membrane

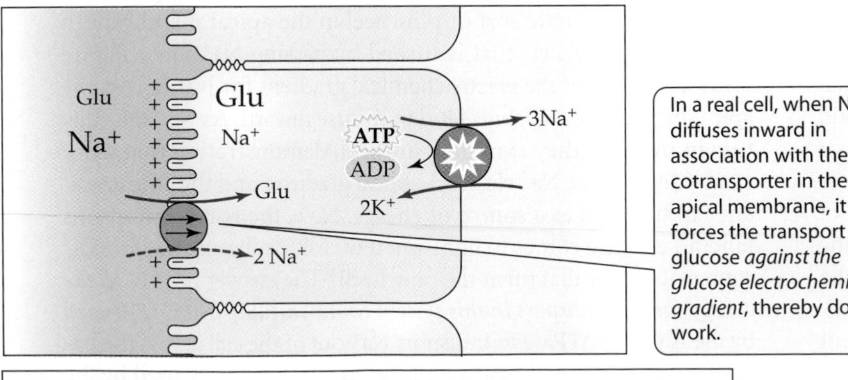

brane using energy from ATP hydrolyzed in the basolateral membrane.

An actual intestinal epithelial cell is able to do work using the Na^+ electrochemical gradient across its apical membrane because such a cell possesses in its apical membrane a fascinating type of dual transporter molecule that facilitates the *linked* transport of Na^+ and glucose (Figure 3.12*c*). This transporter protein can move glucose into the cell *even when the concentration of glucose inside the cell exceeds the concentration in the intestinal lumen*, but the transporter does not use ATP and is not an ATPase. The transporter molecule has the following essential attribute: For every two Na^+ ions it carries across the membrane, it *must*, because of its particular chemistry, carry a molecule of glucose in the same direction. Thus the transport of glucose into an intestinal epithelial cell from the intestinal lumen occurs by way of two energy-driven processes. First, Na^+ is transported out of the cell across the basolateral membrane by the Na^+–K^+ pump using ATP-bond energy, a process that maintains the Na^+ electrochemical gradient across the apical membrane. Second, glucose is transported into the cell across the apical membrane in linked fashion with Na^+, driven by the energy stored in the Na^+ electrochemical gradient across the apical membrane.

Most investigators today consider this type of transport of glucose to be a form of *active* transport because it meets three critical criteria: (1) The mechanism can move glucose away from glucose equilibrium, from a solution of low concentration into one of high concentration; (2) the glucose transport is carrier-mediated; and (3) the energy source for the uphill transport of glucose is metabolism. The active transport of glucose deserves, nonetheless, to be distinguished from forms of active transport that use ATP-bond energy directly. This is why *secondary active transport* (transport that draws energy from an electrochemical gradient of a solute) is distinguished from primary active transport (transport that draws energy from ATP).

Available evidence indicates that when *organic solutes* are *actively* transported by animal cells, secondary active transport is the usual mechanism. In the vertebrate intestine, for example, secondary-active-transport mechanisms are responsible for the active uptake of not only glucose but also certain other sugars, amino acids, and water-soluble vitamins (see Chapter 4). A variety of apical-membrane transporter proteins that move two solutes simultaneously, in linked fashion, are involved. When a transporter protein moves two solutes in coupled fashion in *one* direction, the transport is called **cotransport**, and the protein is a **cotransporter**. When a transporter moves two solutes in coupled fashion in *opposite* directions, the transport is **countertransport**, and the protein is a **countertransporter**.[10] In our focal example, the transporter for glucose is a cotransporter. All vertebrates have this cotransporter in the apical membranes of their intestinal epithelial cells. Hummingbirds are distinguished, however, by exhibiting extraordinary activity of this cotransporter, far higher than that observed in omnivorous vertebrates. Their exceptional glucose cotransporter activity seems almost surely to be an adaptation to the exceptional glucose yield of their nectar diet.

Active transport across an epithelium does not imply a specific transport mechanism

Scientists who study *epithelia* often know that active transport occurs across a particular epithelium without knowing much, if anything, about the specific molecular mechanisms that carry out the transport. There are, in fact, two different perspectives in the study of active transport across epithelia. The simpler *whole-epithelium* perspective—exemplified by Figure 3.13*a*—regards the epithelium as a "black box": The behavior of the epithelium itself is described without the underlying cell-membrane mechanisms being identified. The more-complex *cell-membrane* perspective—exemplified by the options in Figure 3.13*b*—seeks to identify the cell-membrane mechanisms of transport, including all the particular proteins that are involved (e.g., ATPases, cotransporters, and channels) in both the apical and basolateral membranes of the epithelial cells.

To make this more clear, suppose that a researcher carries out the following measurements on an epithelium: First the researcher measures the concentration of the solute of interest on the two sides of the epithelium, and if the solute is charged, the researcher measures the voltage difference across the epithelium. From these data, the electrochemical gradient across the epithelium is calculated. Next the re-

[10] Cotransporters and countertransporters are sometimes called *symporters* and *antiporters*, respectively.

Figure 3.13 Two perspectives on epithelial active transport Whole-epithelium *(a)* and cell-membrane *(b)* views of glucose transport across the epithelium of the vertebrate small intestine. Each panel in *(b)* depicts a potential mechanism by which transporter proteins and channel proteins in the apical and basolateral membranes could endow the whole epithelium with the property shown in *(a)*. In each diagram, proteins are labeled with the solutes they transport. Free-standing Glu symbols show relative glucose concentrations in the intracellular and extracellular fluids. Explanations of the four plausible cell-membrane mechanisms in *(b)*: *Upper left:* Glucose could cross the apical membrane by means of **primary active transport** mediated by an ATPase and powered with energy derived directly from ATP. The high glucose concentration thereby created within the cell would favor diffusion of glucose out of the cell, so transport across the basolateral membrane could occur by facilitated diffusion. *Upper right:* Glucose enters the cell across the apical membrane by **facilitated diffusion**. This can occur only if the cell concentration of glucose is low enough to be below the intestinal-lumen concentration, but the problem of then making glucose move out across the basolateral membrane could be solved by the presence of a glucose-transporting ATPase that uses ATP-bond energy to make glucose to move against its concentration gradient. *Lower left:* Glucose enters the cell across the apical membrane by **secondary active transport**, involving the Na^+–K^+-ATPase in the basolateral membrane and a Na^+–glucose cotransporter in the apical membrane (see Figure 3.12*c*). Then glucose could cross the basolateral membrane by simple diffusion through a glucose channel. *Lower right:* This is what actually happens. Glucose enters the epithelial cell across the apical membrane by **secondary active transport** and leaves across the basolateral membrane by **facilitated diffusion** mediated by a glucose-transporter protein (see Figure 4.19 for full detail).

(a) Whole-epithelium view of glucose transport

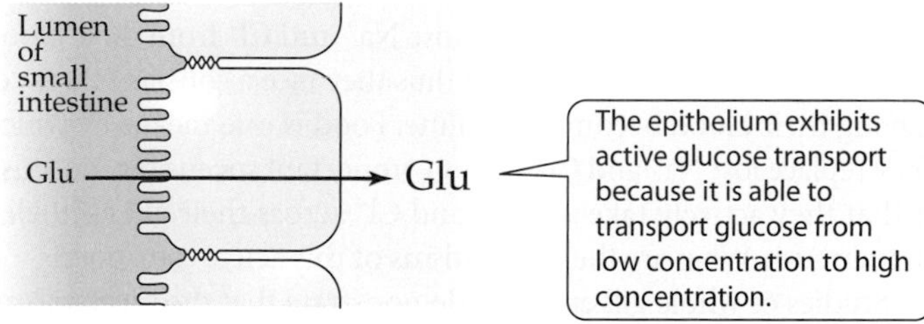

(b) Four plausible cell-membrane mechanisms of glucose transport

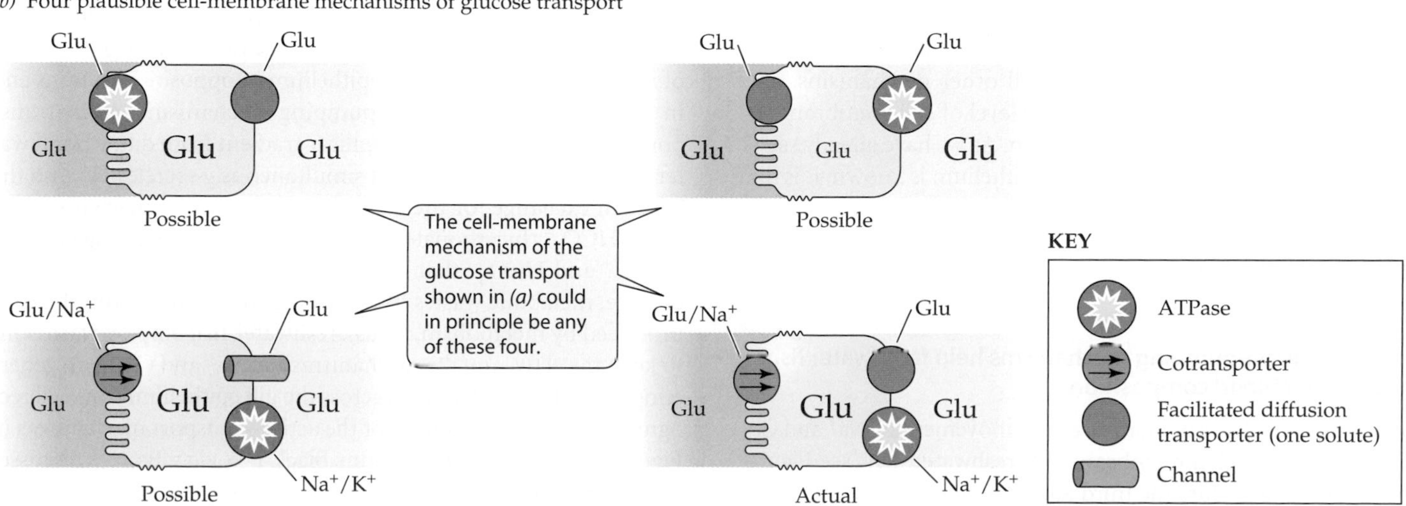

BOX 3.1 Cellular Mechanisms of Ion Pumping in Freshwater Fish Gills

What are the cell-membrane proteins and other cellular mechanisms that endow the gill epithelium of freshwater fish with its pumping abilities for Na^+ and Cl^-? Figure A presents two possible models for the pumping of Na^+ across the gill epithelium from pond water to the blood plasma of a fish. Before we discuss these models, it is important to recognize that for any postulated model to hold promise of being correct, it must propose ways for the epithelial cells to stay in steady state; for instance, a model that involved the endless buildup of potassium in a cell could not possibly be correct because no cell can survive that sort of nonsteady-state condition. Now let's take a look at each of the two models illustrated in Figure A in turn.

According to the model favored in the past (Figure A1), two principal membrane proteins are involved in Na^+ transport: the ubiquitous Na^+–K^+-ATPase in the basolateral membrane, and a countertransport protein for Na^+ and NH_4^+ (ammonium) or Na^+ and H^+ in the apical membrane. ATP-bond

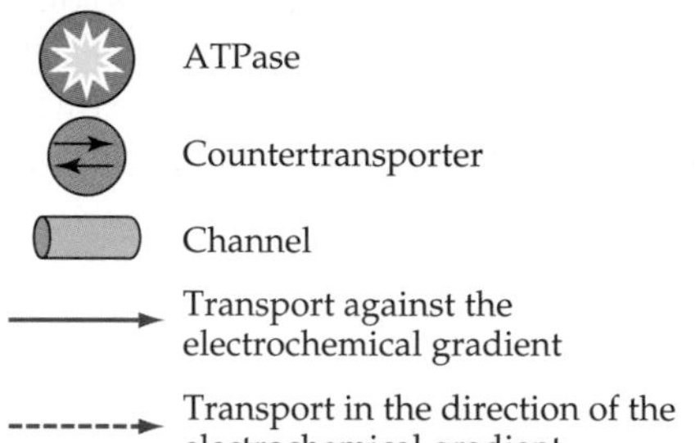

Figure A Two models of the cellular mechanism of active Na^+ uptake across the epithelial cells of freshwater fish gills

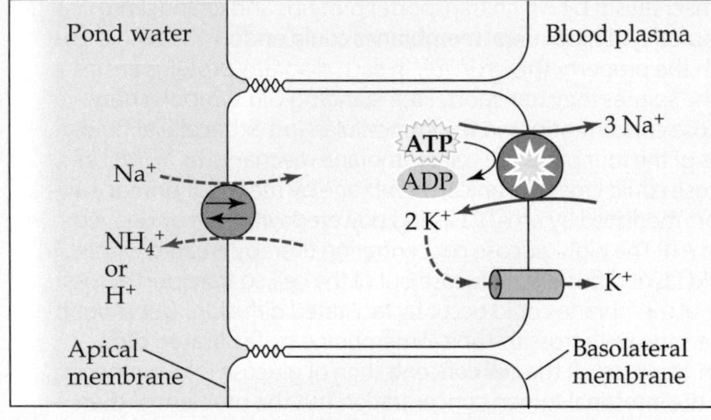

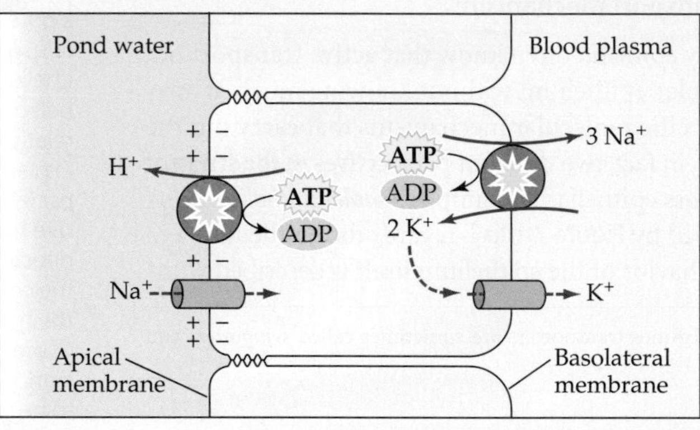

searcher studies whether the epithelium can move the solute of interest from one side to the other in a way that *increases* the electrochemical gradient across the epithelium. If so, it is common practice to say that "the epithelium actively transports the solute." By this standard, it is easy to show that the epithelium of the small intestine of a vertebrate actively transports glucose (see Figure 3.13*a*).

It is always the case, however, that many different cell-membrane mechanisms could permit an epithelium to function as it does. In the case of the intestinal epithelium, for example, *any* of the four mechanisms in Figure 3.13*b* would permit the epithelium to transport glucose as it does, and in fact, still other mechanisms, not shown, could also do so. Research at the level of cell membranes is required to identify the actual mechanism. As we have already seen, the actual mechanism in the intestinal epithelium is known; it is the one shown at the lower right of Figure 3.13*b*. Often, however, physiologists have not fully resolved the molecular mechanisms of epithelial transport.

Two epithelial ion-pumping mechanisms help freshwater fish maintain their blood composition

Now we can complete our analysis of the movement of Na^+ and Cl^- across the gill epithelial membranes of freshwater fish (see Figure 3.1*b*), bringing to closure the third and final of our focal examples. We have seen that freshwater fish lose Na^+ and Cl^- from their blood into pond water by diffusion, and thus they face a constant threat of having their blood become too dilute. Food is one means by which they replace lost Na^+ and Cl^-. A more important means for most fish is that they actively take up Na^+ and Cl^- across their gill epithelial membranes. What are the mechanisms of this active transport?

Studies of whole gill epithelia demonstrate that the mechanisms of active Na^+ uptake and active Cl^- uptake are different and independent (Figure 3.14). Each active-transport mechanism is electroneutral (or approximately so) because each exchanges two ions of identical charge across the epithelium in opposite directions and in a 1:1 ratio. When the Na^+-pumping mechanism actively transports Na^+ against its electrochemical gradient from dilute pond water into concentrated blood, it simultaneously secretes H^+ into the water in exchange for the Na^+. The Cl^--pumping mechanism secretes HCO_3^- (bicarbonate ion) into the water in exchange for Cl^-. The H^+ and HCO_3^- that are exchanged for Na^+ and Cl^- are, in essence, metabolic wastes because both are derived from the CO_2 produced by metabolism. Thus, freshwater fish employ waste ions to keep the active uptake mechanisms for Na^+ and Cl^- from generating large charge differences across the gill epithelium. You will recognize that our presentation of the active-transport mechanisms in Figure 3.14 is a whole-epithelium, black-box view. Box 3.1 discusses the cell-membrane view.

BOX 3.1 *(Continued)*

energy is invested in the process by the Na^+–K^+ pump, which lowers the intracellular Na^+ concentration and helps make the inside of the epithelial cell negative. Na^+ enters the cell across the apical membrane by carrier-mediated diffusion, following its electrochemical gradient; the Na^+ binds to the countertransporter during this process, and the countertransporter is postulated to move either H^+ or NH_4^+ outward in exchange, thereby maintaining charge balance. Once Na^+ is in the cell, it makes the next step to the blood by being pumped by the Na^+–K^+ pump. Potassium channels in the basolateral membrane allow the K^+ pumped into the cell by the Na^+–K^+ pump to return to the blood.

Recently the model just described has been discarded in favor of a new model (Figure A2), which postulates that ATP-bond energy is invested principally by a strongly electrogenic proton pump (an H^+-ATPase) in the apical membrane. By pumping H^+ out, this pump produces negative charge on the inside of the apical membrane. The negativity establishes an electrochemical gradient favoring the inward diffusion of Na^+, which enters by way of a channel. In this model, Na^+ and H^+ are exchanged in a 1:1 ratio because the motive force for Na^+ entry originates from H^+ exit. After Na^+ is in the cell, it continues its journey to the blood by being transported by the Na^+–K^+ pump, much as in the old model. The source of H^+ for both models is the reaction of the metabolic waste CO_2 with water (see Figure 3.14).

The contrast between the two models in Figure A emphasizes that the details of membrane physiology are hardly irrelevant to our concepts of animal performance. For many years, on the basis of the old model, physiologists believed that the nitrogenous waste NH_4^+ could be excreted by exchange for Na^+, but the new model indicates that this exchange does not occur.

The mechanism of the Cl^- pump is not well resolved, but Figure B shows a current model for the proteins and processes in the apical membrane. A countertransport protein in the apical membrane that carries out a strict 1:1 exchange of Cl^- and HCO_3^- is the best-understood element of the Cl^- transport mechanism. This protein was postulated for many years to be an ATPase, but that hypothesis has recently been discarded. Current models postulate, instead, that the energy for countertransport comes from an electrochemical gradient for HCO_3^- across the apical membrane. The gradient favors outward HCO_3^- diffusion, which drives the countertransport protein, bringing Cl^- into the cell against the Cl^- electrochemical gradient (a form of secondary active transport). In the model shown in Figure B, a proton pump driven by ATP-bond energy induces the steady formation of HCO_3^- inside the cell by pulling the CO_2/HCO_3^- reaction to the left by mass action. The outward pumping of protons may also lower the local HCO_3^- concentration on the outside of the apical membrane because extruded protons will react there with HCO_3^-. The steady formation of HCO_3^- in the cell and steady depletion of HCO_3^- outside the cell maintain the electrochemical gradient for outward HCO_3^- transport across the apical membrane.

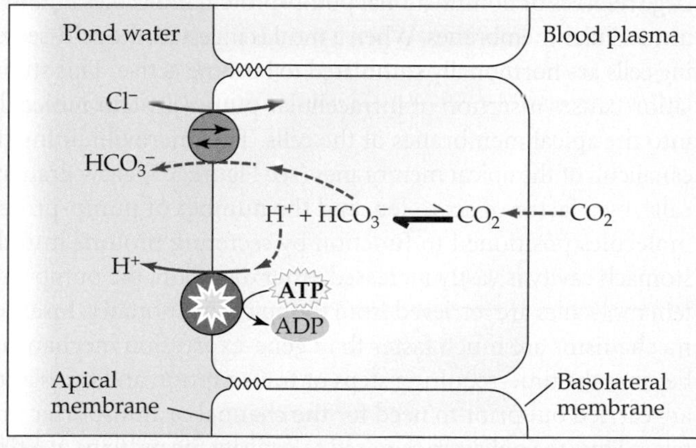

Figure B A model of active Cl^- uptake across the apical membranes of cells in the gill epithelium of freshwater fish Although active transport of protons out of the cell is depicted in the apical membrane, an alternative model is that it occurs in the basolateral membrane. (See Figure A for a key to symbols.)

Figure 3.14 A whole-epithelium view of active ion transport across the gill epithelium of a freshwater fish The Na^+ and Cl^- active transport mechanisms are different and independent. Carbon dioxide (CO_2) from metabolism reacts with body water to form bicarbonate ions (HCO_3^-) and protons (H^+). The latter are exchanged for Cl^- and Na^+ by the active-transport mechanisms. The pumping mechanisms exist within single epithelial cells.

Bicarbonate ions (HCO_3^-) and protons (H^+) are exchanged for Cl^- and Na^+, respectively, by the active-transport mechanisms.

Summary: Active Transport

- Active-transport mechanisms—also called uphill-transport mechanisms or pumps—are able to convert energy obtained from the catabolism of foodstuffs into solute motive energy and thus can transport solutes away from electrochemical equilibrium. Such mechanisms are known for many solutes, but not O_2 or H_2O.
- Solutes must bind noncovalently to a transporter protein for active transport to occur. Thus, active transport is a type of carrier-mediated transport. A second type is facilitated diffusion, which differs in that it cannot tap metabolic energy and is therefore strictly toward equilibrium.
- Active transport is primary if the transporter protein is an ATPase and thus draws energy directly from ATP bonds. The best-known transporter of this sort is Na^+–K^+-ATPase, which is found in the basolateral

membranes of all epithelia. Primary-active-transport mechanisms pump ions. Active transport is secondary if the immediate source of energy for transport is a solute electrochemical gradient, rather than ATP. Secondary-active-transport mechanisms depend on transporter proteins (cotransporters or countertransporters) that obligatorily transport two solutes simultaneously. Organic solutes are pumped by secondary-active-transport mechanisms.

- Active transport of ions is electrogenic if it produces a voltage difference but electroneutral if it does not.

Modulation of Channels and Transporters

Like enzyme proteins, channel and transporter proteins can be modulated in numerous different ways, providing many opportunities for the regulation of cell and tissue function. A detailed treatment of this topic would be at least as long and full of possibilities as our discussion of enzyme modulation in Chapter 2. Instead, the goal in this case will be simply to outline known options, many of which are analogous to the options available for modulating enzymes.

- *Modulation of gene expression.* The types and amounts of channel and transporter proteins that are synthesized by a cell are under control of gene expression. Thus, for example, differential gene expression produces different suites of channel and transporter proteins in different tissues during early development, helping to establish the mechanistic basis for tissue specialization. All cells of a mammal, for example, have the genes that code for the proton pump, but only certain cells—notably the acid-secreting cells of the stomach—vigorously express the genes. Gene expression is also modulated during the life of an animal in response to an individual's circumstances, often under hormonal control. The adrenal hormone aldosterone, for instance, increases Na^+ reabsorption in the kidney tubules, thus increasing retention of Na^+ in the body. It exerts this effect in part by increasing expression of the genes that code for Na^+–K^+-ATPase in kidney-tubule cells. With aldosterone present, more molecules of the ATPase are synthesized and inserted in the basolateral membranes of the cells. As is characteristic of modulation by gene expression, these changes cannot be implemented instantly but instead require many hours, at least, to be fully effected.
- *Multiple molecular forms.* The existence of multiple molecular forms of channel and transporter proteins is common. For example, just as the genome of an individual animal may code for the synthesis of two or more LDH (lactate dehydrogenase) enzyme proteins, it may code for the synthesis of two or more voltage-gated Ca^{2+} channels. One molecular form of voltage-gated Ca^{2+} channel is expressed in nerve cells of vertebrates, whereas a different form is expressed in certain muscle cells, for instance. Different species may also have different molecular forms of channels and transporters. Because different molecular forms can exhibit distinct transport and modulation characteristics, the presence of different forms in different tissues of a species or in different species provides opportunities for adaptation.
- *Noncovalent and covalent modulation.* The channel and transporter proteins resident in a membrane often are subject to modulation. One form of modulation is analogous to the allosteric modulation of enzymes in that it is mediated by noncovalent and reversible bonding with modulating agents. Ligand gating of channels provides an illustration (see Figure 3.5*d*). Covalent modulation is also possible. Both noncovalent and covalent modulation provide for very rapid adjustment of channel or transporter function.
- *Insertion-and-retrieval modulation.* Because channel and transporter proteins perform their functions only when in the cell membrane (or in a function-specific intracellular membrane), the *location* of the proteins in a cell provides a means of controlling their activity. Cells often have reserves of cell-membrane channel or transporter proteins held elsewhere than in the cell membrane. In such cases the protein molecules are *inserted* or *retrieved* from the cell membrane. An excellent example is provided by the proton pump that acidifies the stomach contents after a meal. Resting acid-secreting cells (see Figure 3.9) have relatively few molecules of H^+–K^+-ATPase in their apical cell membranes, but they have large reserves of nonfunctional pump-protein molecules bound to intracellular membranes. When a meal is ingested, the acid-secreting cells are hormonally stimulated to become active. This stimulation causes insertion of intracellular pump-protein molecules into the apical membranes of the cells. The microvilli lining the canaliculi of the apical membranes (see Figure 3.9) grow dramatically, by a factor of six to ten, and the number of pump-protein molecules positioned to function by secreting protons into the stomach cavity is vastly increased. After digestion, the pump-protein molecules are retrieved from the apical membranes. Insertion mechanisms are much faster than gene-expression mechanisms because the time-requiring steps of transcription and translation are carried out prior to need for the channel or transporter proteins. Thus the only step required to activate the proteins at a time of need is to insert them in the membranes where they are active, a process that can be completed in minutes.

Osmotic Pressure and Other Colligative Properties of Aqueous Solutions

Having discussed the mechanisms of *solute* transport, we now turn to the transport of the *solvent* in aqueous solutions, water. A logical starting point is to discuss the **colligative properties** of aqueous solutions because osmosis of water, discussed in the next section, is fundamentally a colligative phenomenon. The colligative properties of aqueous solutions are the properties that depend simply on the *number* of dissolved entities per unit of volume rather than the chemical nature of the dissolved entities.

The three colligative properties of greatest significance for the study of animal physiology are the osmotic pressure, freezing point, and water vapor pressure of a solution.[11] A useful way to define **osmotic pressure** is that it is the property of a solution that allows one to predict whether the solution will gain or lose water by osmosis

[11] A fourth colligative property of occasional interest is the **boiling point**, the lowest temperature that will cause boiling.

when it undergoes exchange with another solution; in other words, if you know the osmotic pressures of two solutions, you can predict the direction of osmosis between them. The **freezing point** of a solution is the highest temperature capable of inducing freezing. For a formal definition of **water vapor pressure**, we defer to Chapter 25, but here we can say that the water vapor pressure measures the tendency of a solution to evaporate.

The study of colligative properties requires a shift in view, compared to earlier sections of this chapter, because the colligative properties—unlike solute-transport mechanisms—do not depend on the particular chemical nature of solutes. In Figure 3.15, the two upper drawings are microscopic views of two complex solutions. The same two solutions are shown again in the two lower drawings, except that in the lower drawings, the chemical identity of each dissolved entity is not recognized; instead, each dissolved entity is represented just with a dot. For quantifying colligative properties, the view in the lower drawings is more straightforward because the colligative properties are independent or approximately independent of the chemical nature of dissolved entities. They depend instead on the simple matter of *how many dissolved entities* (of any kind) *are present per unit of volume of solution.* Whether a dissolved entity is a glucose molecule, a large protein molecule like albumin, or a Na^+ ion, each makes an approximately equal contribution to the quantitative magnitude of the colligative properties of a solution. Thus, the two solutions in Figure 3.15 closely resemble each other in their osmotic pressures, freezing points, and water vapor pressures, even though they are very different in their chemical mixes of solutes.

All the colligative properties of a solution change in magnitude as the concentration of dissolved entities changes. Raising the concentration of dissolved entities in a solution increases the osmotic pressure of the solution. It also lowers the solution's freezing point and lowers its water vapor pressure. Speaking exactly, the osmotic pressure is approximately proportional to the concentration of dissolved entities; doubling the concentration approximately doubles the osmotic pressure. The other colligative properties also exhibit simple proportionalities to concentration, but these proportionalities can be a little less obvious because of the way the properties are usually expressed. Considering the freezing point and water vapor pressure, when the concentration of dissolved entities in a solution is doubled, what approximately doubles is the difference (sign ignored) between the actual freezing point or the actual water vapor pressure of the solution and the freezing point or the water vapor pressure of pure water. To see this more clearly, consider the **freezing-point depression**, defined to be the difference (sign ignored) between the actual freezing point of a solution and the freezing point of pure water. The freezing-point depression of a solution is approximately proportional to the concentration of dissolved entities. For example, if the freezing-point depression is 0.4°C before the concentration of dissolved entities is doubled, it is about 0.8°C after. This would mean that the freezing point itself is lowered from –0.4°C to –0.8°C. The **water-vapor-pressure depression** is the difference (sign ignored) between a solution's water vapor pressure and the water vapor pressure of pure water under the same conditions. It is approximately proportional to the concentration of dissolved entities.

A useful corollary of the points already made is that the osmotic pressure, the freezing-point depression, and the water-vapor-pressure depression of any solution *are all proportional to each other.* This means that if you have measured any one of these properties in a solution, you can calculate the others, a principle that we will see shortly has considerable practical significance.

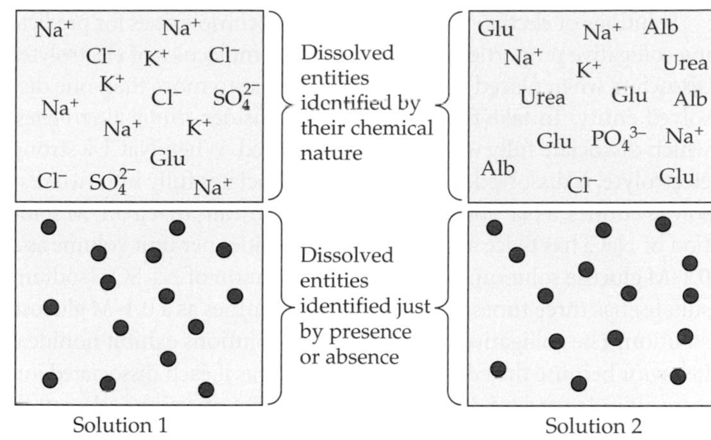

Figure 3.15 Magnified views of two solutions that are similar in colligative properties All the drawings depict identical volumes. The upper drawings of solutions 1 and 2 show the chemical nature of each dissolved entity in the solutions. The lower drawings of the same two solutions, however, show merely the numbers of dissolved entities within the volumes shown, disregarding chemical differences. Because the two solutions have identical numbers of dissolved entities per unit of volume, they are virtually the same in their osmotic pressures, freezing points, and other colligative properties. Alb = albumin; Glu = glucose.

Only dissolved materials affect the colligative properties of solutions. Suspended materials, such as clay particles in river water or red blood cells in blood plasma, do not affect the osmotic pressures or the other colligative properties of solutions, because they are not dissolved.

Solutions of *nonelectrolytes* that are equal in their molar chemical concentrations usually are the same in their osmotic pressures and other colligative properties. This is easy to understand when you consider that a mole of solute contains a fixed number of molecules (Avogadro's number—6×10^{23}) regardless of the type of solute; in addition, when nonelectrolytes go into solution, each individual molecule remains intact in the solvent, meaning that each constitutes a single dissolved entity. From these principles, if several solutions of nonelectrolytes have equal molarities, they are also equal in their numbers of dissolved entities per unit volume. A 0.1-*M* solution of glucose, for example, has the same number of dissolved entities per unit volume as a 0.1-*M* solution of urea. For the determination of colligative properties, it does not matter that the dissolved entities are glucose molecules in one solution and urea molecules in the other. Thus, both solutions are virtually identical in their osmotic pressures and other colligative properties. Similarly, the colligative properties of a 0.1-*M* solution of a large, nondissociating protein such as serum albumin (molecular mass ~66,000 daltons) are essentially the same as those of the equimolar solutions of the lower-mass solutes: glucose (180 daltons) and urea (60 daltons).[12]

[12] To be strictly accurate, solutions of various nonelectrolytes have the same ratio of dissolved entities to water molecules when they are of the same *molality*, not molarity. Thus, it is solutions of identical molality that exhibit identical colligative properties. Biologists usually measure and discuss molarities, and for most purposes this habit does not introduce consequential errors.

Solutions of *electrolytes* present additional complexities for predicting colligative properties because individual molecules of electrolytes dissociate when placed in solution, giving rise to more than one dissolved entity. To take the simplest case, consider *strong electrolytes*, which dissociate fully when they are dissolved. When NaCl, a strong electrolyte, is dissolved, each molecule dissociates fully into two dissolved entities, a Na^+ ion and a Cl^- ion. This means that a 0.1-*M* solution of NaCl has twice as many dissolved entities per unit volume as a 0.1-*M* glucose solution. Similarly, 0.1-*M* solution of Na_2SO_4 (sodium sulfate) has three times as many dissolved entities as a 0.1-*M* glucose solution. The colligative properties of salt solutions exhibit nonideal behavior because they do not behave exactly as if each dissociated ion were an independent dissolved entity. In most situations relevant to animal physiology, however, there is a relatively close approximation to ideal behavior. From the viewpoint of ideal behavior, for example, we would expect the 0.1-*M* solution of NaCl to have an osmotic pressure and freezing-point depression that are 2 times higher than those of a 0.1-*M* glucose solution. The actual osmotic pressure and freezing-point depression are about 1.9 times higher in the NaCl solution.

Physiologists usually express osmotic pressure in osmolar units

The most commonly used system of units for osmotic pressure in biology today is the **osmolarity** system. A **1-osmolar** (***Osm***)solution is defined to be one that behaves osmotically as if it has 1 Avogadro's number of independent dissolved entities per liter. Another way to say the same thing is that a solution has an osmotic pressure of 1 osmolar if it has the same osmotic pressure as a 1-*M* solution of ideal nonelectrolyte. Some solutions relevant to animal physiology, such as seawater and the bloods of many marine animals, are concentrated enough to be about 1 osmolar. Many solutions in animal physiology, however, are more dilute and typically are expressed in units of **milliosmolarity;** a **1-milliosmolar** (**m*Osm***) solution behaves osmotically as if it has 0.001 Avogadro's number of independent dissolved entities per liter.

Osmotic pressures can be measured in several ways

For day-to-day practical purposes in physiology and medicine today, people measure osmotic pressures by employing instruments that actually measure either the freezing-point depression or the water-vapor-pressure depression. These are called *freezing-point osmometers* and *vapor-pressure osmometers*. The instruments take advantage of the point, stressed earlier, that all the colligative properties of a solution are typically proportional to each other. Thus the osmotic pressure can be calculated once the freezing-point depression or the water-vapor-pressure depression has been measured.[13]

Physical chemists have devised methods of measuring osmotic pressure itself, rather than calculating it from other colligative properties. These methods are the gold standard of osmotic-pressure measurement, but usually they are not employed for day-to-day practical work by biologists because they are more complex to carry out than the methods already mentioned. Figure 3.16 illustrates the basic approach to the direct measurement of the osmotic pressure of a solution. The solution is placed within a piston device as shown in Figure 3.16*a*. The end of the piston consists of a **semipermeable membrane**, defined to be a membrane permeable only to water. The reason for using such a membrane is that it ensures the solution will not be corrupted by losing some of its solute during the measurement process. A semipermeable membrane is a laboratory device; no animal or plant membranes are semipermeable. Note in Figure 3.16*a* that whereas the solution of interest is on one side of the semipermeable membrane, *pure water* is on the other side. If the piston is simply allowed to move freely, water molecules will travel by osmosis from the side filled with pure water to the side filled with the solution, and the changes in volume of the two fluid compartments will push the piston to the right (Figure 3.16*b*).

This piston movement can be prevented by exertion of force on the piston as shown in Figure 3.16*c*. The force will raise the hydrostatic pressure in the solution to be higher than that in the pure water, and the elevated hydrostatic pressure in the solution will tend to force water molecules through pores of the membrane by pressure-

Figure 3.16 How to measure the osmotic pressure of a solution directly *(a)* A piston osmometer. The semipermeable membrane is permeable only to water. *(b)* A later state of the system if the piston is free to move. Water moves into the solution by osmosis. *(c)* The stable state of the system if the solution is subjected to increased hydrostatic pressure by a force applied to the piston and the difference of hydrostatic pressure across the membrane is equal to the osmotic pressure of the solution. The open sidearm on the pure-water compartment in *(c)* permits the hydrostatic pressure in that compartment to remain constant as the hydrostatic pressure in the solution compartment is increased.

(a) A piston device for direct measurement of osmotic pressure

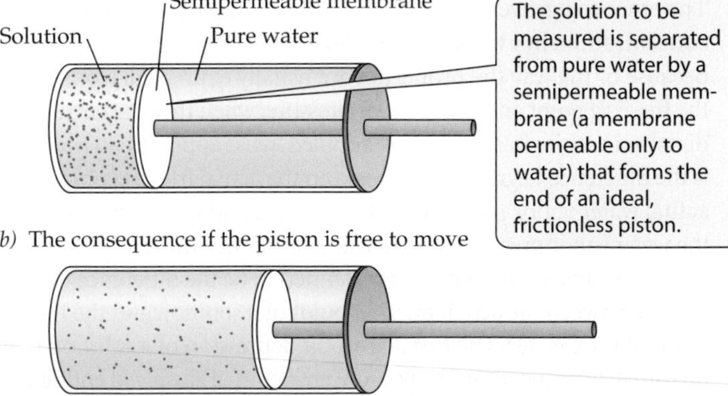

(c) The consequence if the solution is placed under pressure that exactly opposes osmosis

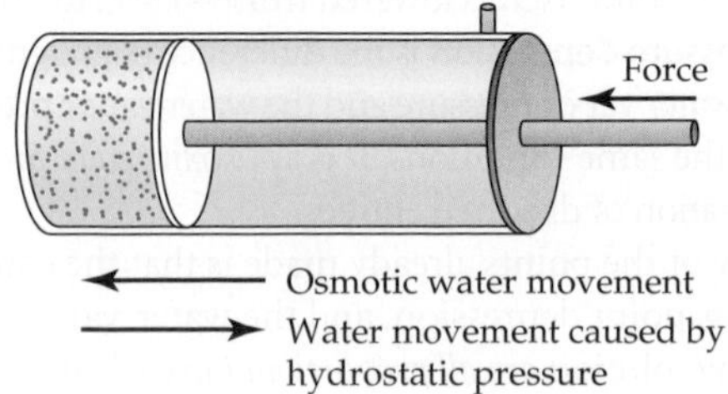

[13] Prior to about 1980, it was very common for biologists who measured freezing-point depression to leave their results in temperature units rather than converting them to osmolarity. Thus, in the literature of that era you will find osmotic pressures expressed in degrees Celsius! The equivalency is that a freezing-point depression of 1.86°C corresponds to 1 osmolar.

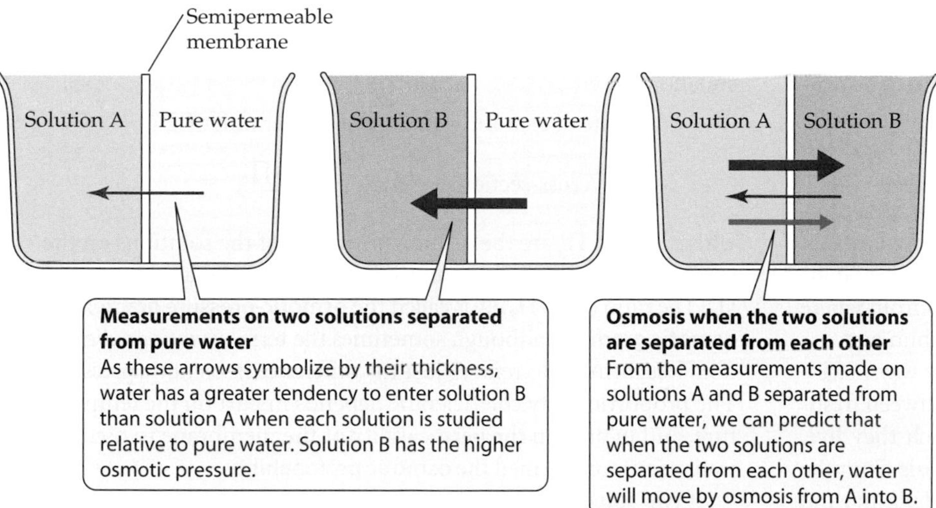

Figure 3.17 Predicting the direction of osmosis between two solutions from measurements made independently on each The red arrows in the two diagrams on the left symbolize the tendency for water to move by osmosis into each of two solutions, A and B, when the solutions are separated by a semipermeable membrane from pure water. The thickness of each arrow is proportional to the hydrostatic pressure difference that is required between the solution and the pure water to block any net transfer of water by osmosis; thus the thicknesses of the red arrows are proportional to the osmotic pressures of the two solutions. From this information, we can predict, as shown on the right, that when solutions A and B are separated from each other, water will move by osmosis from A into B. In the right-hand diagram, the red arrows are repeated from the diagrams on the left, and the blue arrow shows the difference between them.

driven streaming mass flow (termed **ultrafiltration**) from the solution into the pure water. For any given solution, there is a particular difference of hydrostatic pressure between the solution and pure water that will exactly prevent the piston from moving in either direction from its initial position, as in Figure 3.16*c*. This is the difference of hydrostatic pressure that forces water out of the solution at the same rate that osmosis tends to carry water into the solution. Thus this difference of hydrostatic pressure serves as a measure of the degree to which the solution tends to take on water by osmosis.

To a physical chemist, the specified difference of hydrostatic pressure *is* the osmotic pressure of the solution. Stated succinctly, *the osmotic pressure of a solution is the difference of hydrostatic pressure that must be created between the solution and pure water to prevent exactly any net osmotic movement of water when the solution and the pure water are separated by a semipermeable membrane.* Historically, osmotic pressures came to be called *pressures* precisely because they can be measured as hydrostatic pressures; and to this day, in physical chemistry they are usually expressed in pressure units, such as pascals or atmospheres.[14]

To see the logic of the physical chemist's approach, suppose you place two solutions, called A and B, in the piston device at different times so that you can measure each solution's osmotic pressure, and suppose that the hydrostatic pressure required to negate osmosis into B is greater than that required to negate osmosis into A. From these results, you would know that when the solutions are studied in relation to pure water, B has a greater tendency to take on water by osmosis than A, as shown in the two images to the left in Figure 3.17. Knowing this, you could then accurately predict that if the two solutions, A and B, are placed on opposite sides of a water-permeable membrane, as at the right in Figure 3.17, the direction of water transport by osmosis will be from A into B. Recall that one way to define osmotic pressure is that it is the property of a solution that allows one to predict whether the solution will gain or lose water by osmosis when it undergoes osmotic exchange with another solution. The physical chemist's approach permits measurement of this property and is in fact the most fundamental of all ways to measure osmotic pressure (again, the gold standard). The measurements of osmotic pressure that biologists obtain using freezing-point osmometers or vapor-pressure osmometers typically provide the same information by different means.[15]

Summary: Osmotic Pressure and Other Colligative Properties of Aqueous Solutions

- Colligative properties depend on the number of dissolved entities per unit of volume of solution rather than the chemical nature of the dissolved entities. The three colligative properties of greatest importance in biology are the osmotic pressure, freezing point, and water vapor pressure. Because the three are quantitatively related, any one can be calculated from the others.
- Osmotic pressures are generally expressed in osmolar units by biologists today. A 1-osmolar (*Osm*) solution behaves osmotically as if it has 1 Avogadro's number of independent dissolved entities per liter.
- Physical chemists measure osmotic pressures by using hydrostatic pressure, explaining why osmotic pressures are called *pressures*.

Osmosis

Osmosis is the passive transport of water across a membrane, which can be a cell membrane, an epithelium, or an artificial membrane. Sometimes osmosis is referred to as "diffusion of water." This is misleading as applied to biological membranes because strong evidence indicates that the rate of osmosis across some sorts of biological membranes is greater than can be explained by simple water diffusion. Because osmosis is a form of passive transport, it is strictly toward equilibrium, regardless of its rate or mechanistic details. Active water transport is not generally believed to occur.

[14] The equivalency to convert to osmolar units is that a 1-osmolar solution has an osmotic pressure in pressure units of 22.4 atmospheres (technically, at 0°C, because temperature matters to a small degree in the conversion).

[15] The physical chemist's approach is the *direct* method of measuring osmotic pressure. All other approaches are *indirect*. See Box 5.3 for a discussion of the fundamental distinctions between direct and indirect measurements.

When two solutions exchange water by osmosis, *water always moves from the one with the lower osmotic pressure into the one with the higher osmotic pressure.* This attribute of osmosis tends to be confusing at first because we are accustomed to thinking of things moving passively from high to low; heat moves from high temperature to low, glucose diffuses from high concentration to low, and so forth. Water undergoing osmosis, however, moves from low osmotic pressure to high. One way to remember this crucial point is to note that water itself is more abundant per unit volume where dissolved matter is less abundant. Thus, during osmosis water moves from where *it* is more abundant per unit volume to where *it* is less abundant.

Osmosis is important in the lives of animals in many ways. Striking examples are provided by the osmotic exchanges between freshwater animals and the environmental waters in which they live. Freshwater animals ceaselessly take on water by osmosis from the pond or stream waters they inhabit because their blood plasma and other body fluids have far higher concentrations of dissolved entities—higher osmotic pressures—than freshwater has (Figure 3.18). A 100-g goldfish (*Carassius auratus*) living in freshwater, for example, gains about 30 g of water per day by osmosis—a third of its body weight! Osmotic influxes like this tend to bloat freshwater fish and dilute their blood. Thus the fish must continually expend energy to rid themselves of excess water.

As you study transport, never forget that osmotic pressures have just one major claim to consideration: They govern the movements of *water* across membranes. It is easy to slip into a habit of thinking that osmotic pressures affect solute movements. They do not do so directly. When solutes diffuse, they are affected by gradients of their specific chemical concentrations. Only water follows the gradient of osmotic pressure.

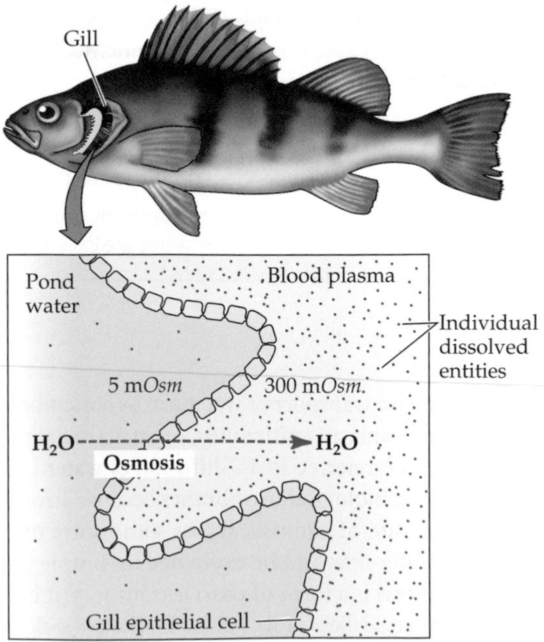

Figure 3.18 Osmotic uptake of water by a freshwater animal Because freshwater animals have osmotic pressures in their blood that greatly exceed the osmotic pressure of freshwater, they gain water by osmosis, especially across their gills or other highly permeable body surfaces. Each dot represents a single osmotically effective dissolved entity.

Quantification and terminology

The *rate* at which water crosses a membrane by osmosis follows the equation

$$\text{Rate of osmotic water transport per unit of cross-sectional area} = K\frac{\Pi_1 - \Pi_2}{X} \quad (3.2)$$

where Π_1 and Π_2 are the osmotic pressures of the solutions on the two sides of the membrane, and X is the distance separating Π_1 and Π_2. The ratio $(\Pi_1 - \Pi_2)/X$ is called the **osmotic-pressure gradient** or **osmotic gradient**, although sometimes the term *osmotic gradient* is used less formally to refer to just the difference in osmotic pressure. The proportionality coefficient K depends in part on the temperature, and in part on the permeability of the membrane to osmotic water transport, termed the **osmotic permeability.**

If two solutions, A and B, have the same osmotic pressure, they are said to be **isosmotic.** If solution A has a lower osmotic pressure than solution B, A is said to be **hyposmotic** to B, and B is said to be **hyperosmotic** to A.[16] Note that the terms *isosmotic, hyposmotic,* and *hyperosmotic* are *relative*; the two solutions being compared must be specified for the terms to have meaning. In this system of terminology, the *direction* of net water movement by osmosis is from the hyposmotic solution into the hyperosmotic one. Osmotic equilibrium is reached when two solutions become isosmotic.

Hydrostatic pressures develop from osmotic pressures only when two or more solutions interact

For most people, the most confusing aspect of osmosis and osmotic pressures is that osmotic pressures are called *pressures.* To avoid confusion, it is important to be clear that *a solution by itself does not exert a hydrostatic pressure because of its osmotic pressure.* In pressure units, a 1-*M* glucose solution has an osmotic pressure of about 22 atmospheres. Thus this simple sugar solution would explode glass bottles if it exerted a hydrostatic pressure equal to its osmotic pressure! The glucose solution can be stored in exactly the same sorts of containers as pure water precisely because the osmotic pressure of an isolated solution is not a hydrostatic pressure. Osmotic pressures are named *pressures* because of a historical accident—namely, that the earliest scientists to study them were physical chemists, who (as we have seen) used hydrostatic pressure as *a measurement device* to quantify osmotic pressures.

There are circumstances relevant to animal physiology when osmotic pressures generate hydrostatic pressures. These circumstances *always involve interaction between two or more solutions* across a membrane. If two solutions of differing osmotic pressure are separated by a water-permeable membrane and the solution of higher osmotic pressure is contained in some way that limits its freedom to expand, then osmosis will create an elevated hydrostatic pressure in

[16] The *tonicity* system of nomenclature, often inappropriately used, is not equivalent to the *osmoticity* system mentioned here. The tonicity system refers to effects on cellular volume. An **isotonic** solution is one into which cells of a specified kind can be placed without the volumes of the cells being affected. If cells are placed in a **hypotonic** solution, they swell because of an osmotic influx of water. In a **hypertonic** solution, they shrink because of osmotic water loss. In many cases, solutions isosmotic to cells are not isotonic in the long run, emphasizing the difference between the two systems of nomenclature.

that solution. The reason is simply that transport of water by osmosis adds matter (water molecules) to the solution with the higher osmotic pressure, and if that solution is prevented from freely expanding to accommodate the increased matter, it will be pressurized. This is why human red blood cells dropped into freshwater enlarge and burst. When a red blood cell first falls into freshwater, its cell membrane is intact and, like a rubber balloon, tends to resist expansion. Thus as osmosis carries water molecules into the cell from the surrounding freshwater, an elevated pressure develops inside the cell, as in a balloon being blown up. Eventually the pressure becomes so great that the cell membrane bursts.

Water may dissolve in membranes or pass through water channels during osmosis

Experiments with artificial all-phospholipid membranes show that water molecules can dissolve in phospholipid bilayers and move through them by the random molecular motions of simple diffusion. This result seems odd at first because H_2O is a polar molecule. Evidently H_2O is small enough (0.3 nm in diameter) for the dissolve-and-diffuse mechanism to operate.

Osmosis across many real cell membranes is thought to occur entirely by simple diffusion of water through the lipid bilayers. On the other hand, channel proteins called **water channels** provide avenues for water molecules to move by osmosis through the cell membranes of many sorts of cells without having to encounter the lipid bilayers. The study of water channels is an exploding area of research at present. Some cells and tissues known to have water channels in critical cell membranes are vertebrate red blood cells, vertebrate kidney-tubule epithelia, and the bladder and skin of amphibians. Cell membranes endowed with water channels are said to exhibit **channel-mediated water transport**. In the past decade, an important family of specific water-channel proteins named **aquaporins** has been described. Some water channels may prove to be fairly nonspecific, allowing some solutes as well as H_2O through. The aquaporins are in general remarkably specific to H_2O.

The rate of water *transport* through water channels exceeds the rate of water *diffusion* through the channels. For example, when human red blood cells, which possess water channels, are placed in a hyposmotic solution, they take on water by osmosis five to ten times faster than is accounted for by just water diffusion. Some sort of bulk flow (streaming) of water evidently occurs through water channels. To explain this, an old hypothesis is that a difference of osmotic pressure on opposite sides of a water channel generates a hydrostatic-pressure gradient *within* the channel that drives bulk water flow through. A newer concept is that water molecules make their way through water channels in strings held together by hydrogen bonds and that somehow this collective movement speeds transport. Whatever happens, osmosis through water channels is passive. It does not employ metabolic energy, and it is always in the direction of equalizing osmotic pressure on the two sides.

Water channels are subject to modulation like other membrane proteins. At least some aquaporins, for instance, have phosphorylation-modulation sites and are evidently modulated covalently by protein kinases. Good evidence also exists for modulation of the water permeability of cell membranes by *insertion* and *retrieval* of water-channel molecules. Insertion increases the number of water channels in a membrane and thus increases membrane water permeability. Retrieval exerts opposite effects. Some kidney-tubule epithelia, to give a specific example, undergo rapid changes in water permeability by the insertion and retrieval of aquaporins under control of the hormone vasopressin (antidiuretic hormone). Responses of this sort are important in the ability of people and other vertebrates to raise and lower their urine concentration (see Chapter 27).

Osmosis and solute physiology often interact

Because the osmotic pressures of solutions depend on solute concentrations, osmotic water movements are interrelated with solute physiology. The three concepts we now discuss demonstrate the often complex interrelations between water and solutes.

NONPERMEATING SOLUTES OFTEN CREATE PERSISTENT OSMOTIC-GRADIENT COMPONENTS ACROSS CELL MEMBRANES OR EPITHELIA If a cell membrane or epithelium is impermeable to a solute and the solute is more concentrated on one side than the other, the solute creates a persistent difference of osmotic pressure across the cell membrane or epithelium. The osmotic-gradient component attributable to the nonpermeating solute persists because the solute cannot diffuse to concentration equilibrium across the membrane or epithelium.

This principle is critically important for understanding water movement across the walls of blood capillaries, for example. Blood plasma typically contains relatively high concentrations of dissolved proteins that do not cross blood-capillary walls. These proteins create a persistent tendency for the blood inside the capillaries to take up water osmotically across the capillary walls from the tissue fluids surrounding the capillaries. On the other hand, the hydrostatic pressure of the blood (the *blood pressure*) tends to force water through the capillary walls from the blood inside to the tissue fluid outside. The net effect of these processes, as we will discuss in Chapter 23, is for blood to lose water to the tissue fluids as it flows through the capillaries, but this loss would be far greater than it is, were it not for the osmotic effect of blood-plasma proteins trapped inside the capillaries. The portion of the blood-plasma osmotic pressure that is caused by nonpermeating dissolved proteins is termed the **colloid osmotic pressure** or **oncotic pressure** of the blood.

PASSIVE SOLUTE TRANSPORT AND OSMOSIS INTERACT When permeating solutes diffuse across a cell membrane or epithelium because they are not at electrochemical equilibrium, their movement tends to alter the osmotic-pressure gradient across the membrane by removing osmotically effective solute from one side and transferring it to the other. On the other hand, osmosis of water across a cell membrane or epithelium tends to alter the electrochemical gradients of solutes because solutes tend to become concentrated on the side losing water and diluted on the opposite side. In addition to these interactive effects of solute diffusion and osmosis on osmotic and electrochemical gradients, the rates of passive solute transport and passive water transport frequently interact by direct mechanical, chemical, or electrochemical coupling of solute and water movements. When water is crossing a membrane in one direction, for example, solute may tend to move physically with the water in the same direction by a phenomenon (*solvent drag*) roughly analogous to bubbles being carried along by a water current.

ACTIVE SOLUTE TRANSPORT PROVIDES A MEANS TO CONTROL PASSIVE WATER TRANSPORT Animals are generally believed to lack active water transport. However, active solute transport can create a gradient of osmotic pressure, thereby setting up conditions favorable to osmosis in one direction or another. In this way, active solute transport provides a means by which animals can use metabolic energy to exert control over the direction of water transport. Suppose, for example, that it would be adaptive for an animal to force water to cross a membrane from left to right. The animal could employ an active solute pump to concentrate a solute on the right side of the membrane, so that the solution on the right would become hyperosmotic to that on the left. Then water would cross the membrane by osmosis in the adaptive direction. Indirectly, in this way, ATP-bond energy would be used to provide the motive force for water transport. This principle is believed to be used by animals in many situations; for example, many insects use the principle to concentrate their urine (see Chapter 27), and vertebrates are believed to use it to absorb water into their blood from the materials passing through their intestines.

Summary: Osmosis

- Osmosis is the passive transport of water across a membrane, such as a cell membrane or an epithelium. Osmosis always occurs from the solution of lower osmotic pressure (more abundant water) to the solution of higher osmotic pressure (less abundant water). Osmosis is always toward equilibrium, tending to bring the solutions on the two sides of a membrane to equal osmotic pressure.
- Two solutions are isosmotic if they have the same osmotic pressure. If two solutions have different osmotic pressures, they are described as hyposmotic and hyperosmotic to each other, the one with the lower osmotic pressure being the hyposmotic one.
- A single solution exerts no increase in hydrostatic pressure because of its osmotic pressure. However, hydrostatic pressures can be generated by osmosis between two solutions interacting across a membrane.
- In some cell membranes, the only mechanism by which osmosis occurs is that H_2O molecules dissolve in the membranes and move through by molecular agitation. Other cell membranes, such as the membranes of red blood cells and certain kidney tubules, however, have aquaporins or other water channels, providing a second path for osmosis to occur. The presence of aquaporins or other water channels speeds osmosis, and often the rate at which water passes through them is faster than simple diffusion can explain.

Looking Forward

The first aquaporin was described in 1992, and by now physiologists recognize the discovery as a revelation. The rapid osmosis that occurs across some cell membranes—osmosis that had already been documented in the 1950s to be far too fast to occur just by simple diffusion—is finally accounted for as a property of defined integral proteins. Diseases caused by aquaporin malfunction could only be mysterious while the very existence of aquaporins remained unknown. Since the discovery of aquaporins, researchers have established that certain types of hearing loss, cataract development, and kidney disease are caused by aquaporin malfunction. Looking forward to the long-term future, we can have no doubt that new discoveries in the molecular biology of solute and water transport will revolutionize understanding of animal physiology and disease. Looking forward to Chapter 4, we will soon see in greater detail how transport physiology plays pivotal roles in one of the most essential of animal functions, the acquisition of nutrients from the environment.

Study Questions

1. People often say things like, "A city has reached equilibrium size," or "A person has reached an equilibrium between needs and wants." Discuss whether these uses of *equilibrium* are compatible with the word's thermodynamic meaning.

2. What are the similarities and differences between the mathematical equation for the rate of simple solute diffusion and the equation for the rate of osmosis?

3. Consider three groups of solutes: (1) steroid hormones, fatty acids, and other lipids; (2) inorganic ions; and (3) polar organic solutes such as glucose or amino acids. What is the principal mechanism by which each group crosses cell membranes passively? Why do members of the first group cross in a fundamentally different way from those of the other two groups?

4. (a) Life-threatening diarrhea is a shockingly common problem in the developing world. People suffering from life-threatening diarrhea are often Na^+-depleted, and to save their lives, replacing Na^+ is essential. However, "raw" Na^+ in the intestines is not absorbed. Drinking a solution of NaCl does not, therefore, replenish body Na^+. In fact, drinking such a solution can actually worsen a person's situation by osmotically dehydrating the blood and other body fluids. Explain how the drinking of an NaCl solution could have this effect. (b) One of the greatest physiological discoveries of the twentieth century was the discovery that drinking a solution of mixed glucose and NaCl can promote restoration of the body's Na^+. With the glucose concentration high enough in the solution, glucose "drives" the glucose–Na^+ cotransporter in the apical membranes of intestinal epithelial cells, promoting Na^+ uptake by sick people. Explain the concept behind this manipulation of the cotransporter for therapeutic ends. The approach has saved millions of lives.

5. Explain why active transport of an ion shows saturation kinetics, whereas transport of an ion through a channel does not.

6. Whereas electrical currents are carried by electrons in copper wire, they are carried by ions in aqueous solutions. Explain how an active-transport mechanism can create a *current* across a membrane.

7. One way to produce freshwater from seawater is "reverse osmosis," in which high hydrostatic pressures are used to force water to move against its osmotic gradient, from seawater to freshwater, across a membrane. How would you calculate the minimal hydrostatic pressures required? Why might it be preferable to use salty water from a coastal bay diluted with river water rather than full-strength seawater as the water source?

8. Outline all the principal transport processes at work in the three focal examples shown in Figure 3.1.

9. When we discussed the microscopic mechanism of simple diffusion, we made the following point: After the concentrations of glucose on the two sides of a membrane have become equal, glucose molecules continue to move at random from left to right and from right to left; the numbers of glucose molecules going in the two directions are equal, however, ex-

plaining why the two concentrations stay equal once they have become equal. Taking advantage of the options provided by multiple isotopes of elements, how would you determine, in an actual physical system, whether the point we have made here is true?

10. The cell membranes of mammalian red blood cells are permeable to urea. If red blood cells are dropped into an isosmotic solution of urea, although they do not swell and burst as quickly as when they are dropped simply into water, they eventually swell and burst. Explain. Also discuss how you would design a solution into which red cells could be placed without ever swelling. (Hint: Think about whether urea will stay on the outside of the cells and the implications for osmotic pressures.)

Selected References

Alberts, B., A. Johnson, J. Lewis, M. Raff, K. Roberts, and P. Walter. 2002. *Molecular Biology of the Cell*, 4th ed. Garland, New York.

de Groot, B. L., and H. Grubmüller. 2001. Water permeation across biological membranes: Mechanism and dynamics of aquaporin-1 and GlpF. *Science* 294: 2353–2357. [See also the commentary on pp. 2304–2305 of the same issue of the journal.]

Evans, D. H. (ed.). 1998. *The Physiology of Fishes*, 2nd ed. CRC Press, Boca Raton, FL. [Includes chapters on ion transport and the role of transport in response to acid.]

Gerencser, G. A. (ed.). 1984. *Chloride Transport Coupling in Biological Membranes and Epithelia.* Elsevier, New York.

Greger, R. (ed.). 1988. *NaCl Transport in Epithelia.* Advances in Comparative and Environmental Physiology, vol. 1. Springer, New York.

Harvey, B., I. Lacoste, and J. Ehrenfeld. 1991. Common channels for water and protons at apical and basolateral cell membranes of frog skin and urinary bladder epithelia. *J. Gen. Physiol.* 97: 749–776.

Karasov, W. H., and J. M. Diamond. 1988. Interplay between physiology and ecology in digestion. *BioScience* 38: 602–611.

Kinne, R. K. H. (ed.). 1990. *Basic Principles in Transport.* Karger, Basel, Switzerland.

Kinne, R. K. H. (ed.). 1990. *Comparative Aspects of Sodium Cotransport Systems.* Karger, Basel, Switzerland.

Kirschner, L. B. 1991. Water and ions. In C. L. Prosser (ed.), *Environmental and Metabolic Animal Physiology*, pp. 13–107. Wiley-Liss, New York.

Kirschner, L. 1997. Extrarenal mechanisms in hydromineral and acid-base regulation of aquatic vertebrates. In W. H. Dantzler (ed.), *Comparative Physiology*, vol. 1 (Handbook of Physiology [Bethesda, MD], section 13), pp. 577–622. Oxford University Press, New York.

Lodish, H., A. Berk, S. L. Zipursky, P. Matsudaira, D. Baltimore, and J. Darnell. 2000. *Molecular Cell Biology*, 4th ed. W. H. Freeman, San Francisco.

Maloney, P. C., and T. H. Wilson. 1985. The evolution of ion pumps. *BioScience* 35: 43–48.

Perry, S. F. 1997. The chloride cell: Structure and function in the gills of freshwater fishes. *Annu. Rev. Physiol.* 59: 325–347.

Sasaki, S., K. Ishibashi, and F. Marumo. 1998. Aquaporin-2 and -3: Representatives of two subgroups of the aquaporin family colocalized in the kidney collecting duct. *Annu. Rev. Physiol.* 60: 199–220.

Scholander, P. F., and H. T. Hammel. 1976. *Osmosis and Tensile Solvent.* Springer, New York.

Sidell, B. D., and J. R. Hazel. 1987. Temperature affects the diffusion of small molecules through cytosol of fish muscle. *J. Exp. Biol.* 129: 191–203.

Somero, G. N., C. B. Osmond, and C. L. Bolis (eds.). 1992. *Water and Life.* Springer, New York.

Spring, K. R. 1998. Routes and mechanism of fluid transport by epithelia. *Annu. Rev. Physiol.* 60: 105–119.

Stein, W. D. 1990. *Channels, Carriers, and Pumps. An Introduction to Membrane Transport.* Academic Press, New York.

Toyoshima, C., H. Sasabe, and D. L. Stokes. 1993. Three-dimensional cryo-electron microscopy of the calcium ion pump in the sarcoplasmic reticulum membrane. *Nature* 362: 469–471.

Verkman, A. S. 1993. *Water Channels.* Landes, Austin, TX.

Weiss, T. F. 1996. *Cellular Biophysics.* MIT Press, Cambridge, MA.

PART II

Food, Energy, and Temperature

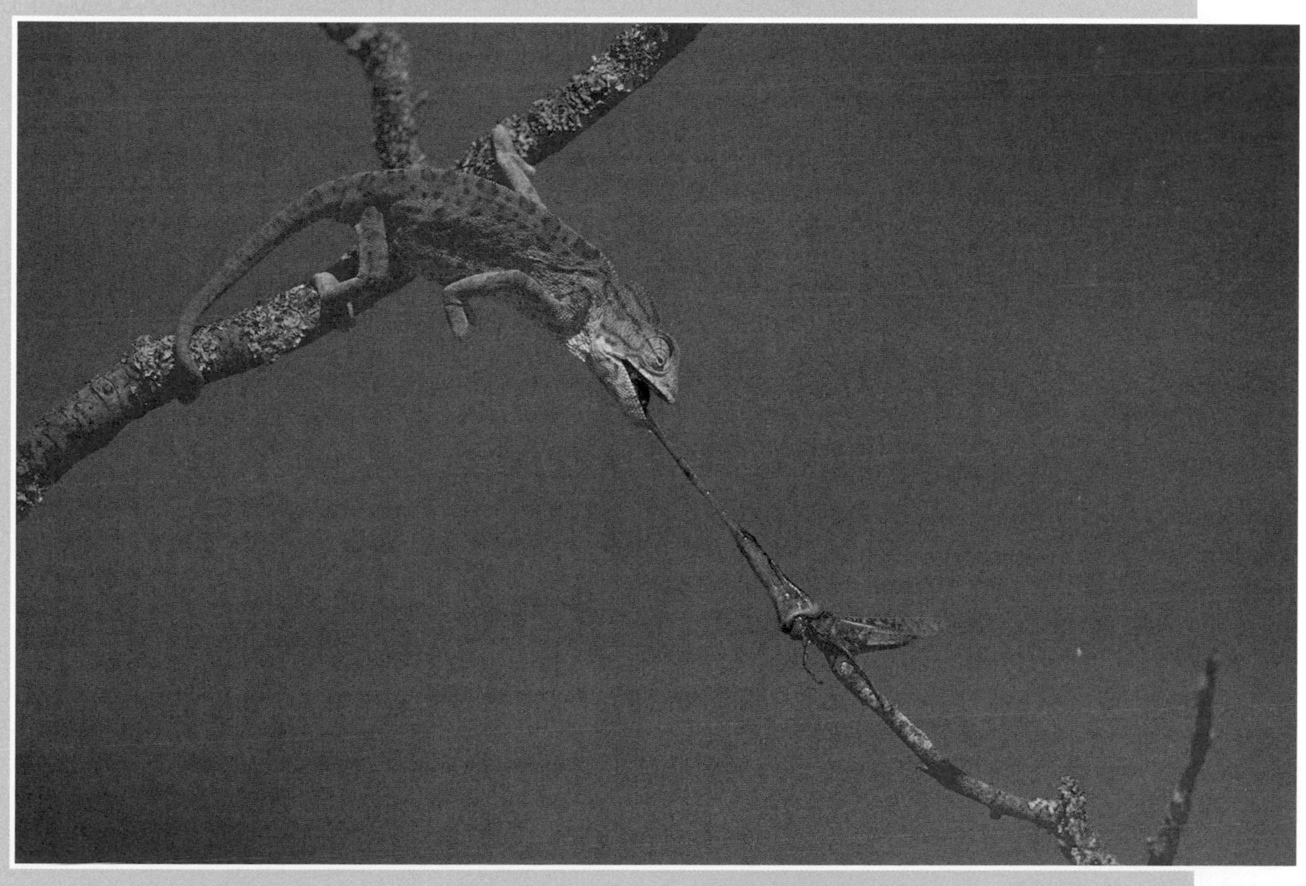

Part II Food, Energy, and Temperature

Previous page: Animals obtain the energy and chemical building blocks they need by eating other organisms. To succeed they must overcome the defenses of the organisms they eat.

4 Nutrition, Feeding, and Digestion

The migration of wildebeests and zebras in the Serengeti ecosystem has justly been called one of the wonders of the living world. At the start of each rainy season, a herd of a million or more migrates 200 km (over 100 miles) from the far north of the Serengeti—where rivers flow year-round—to the dry southeast. As the rains fall in the southeast, parched grasses spring to life, and dry streambeds fill with running water. The animals thrive there during the rains. Indeed, they give birth to most of their young there. When the rains end months later, however, the thin soils of the southeast dry out quickly, and the animals are forced by lack of water into a mass exodus that ultimately takes them back to their starting place in the north. Many die along the way, victims of crocodiles at river crossings or other hardships.

No one is certain why the wildebeests and zebras migrate. Why do they not stay year-round in the north, where water is available in all seasons, rather than migrating a great distance into an area—the southeast—that cannot possibly be a permanent home, because it is too dry for half the year? The most compelling modern hypothesis is that the journey to the southeast is a quest for mineral nutrients. The soils of the southeast, being of relatively recent volcanic origin, are especially rich in Ca^{2+} and other needed minerals. Because of the difference in soils between the north and the southeast, grasses in the southeast contain 1.5 times as much Ca^{2+} per gram of tissue weight as grasses in the north. Similarly, the grasses in the southeast are 1.2 to 1.6 times richer in magnesium, phosphorus, copper, and nitrogen. Lactating mothers and growing young require especially large amounts of Ca^{2+} and other minerals. According to the mineral-nutrition hypothesis, the reason the fabled herds migrate to the southeast, and calve there, is that the migration assures their young of adequate minerals for development.

Nutrition is the study of the chemical components of animal bodies and how animals are able to synthesize the chemical components of their bodies from the chemical materials they collect from their environments. Nutrition also includes the study of the energy available from foods. Nutrition, therefore, is a very broad topic, encompassing not only mineral nutrition but many additional aspects. An instructive starting point for the study of nutrition is to look at the chemical composition of a familiar example, the human body (Figure 4.1). Excluding water, the human body is composed principally of proteins and lipids. Minerals are third in overall importance, followed by two additional categories of organic molecules: nucleic acids and carbohydrates. From an atomic viewpoint,

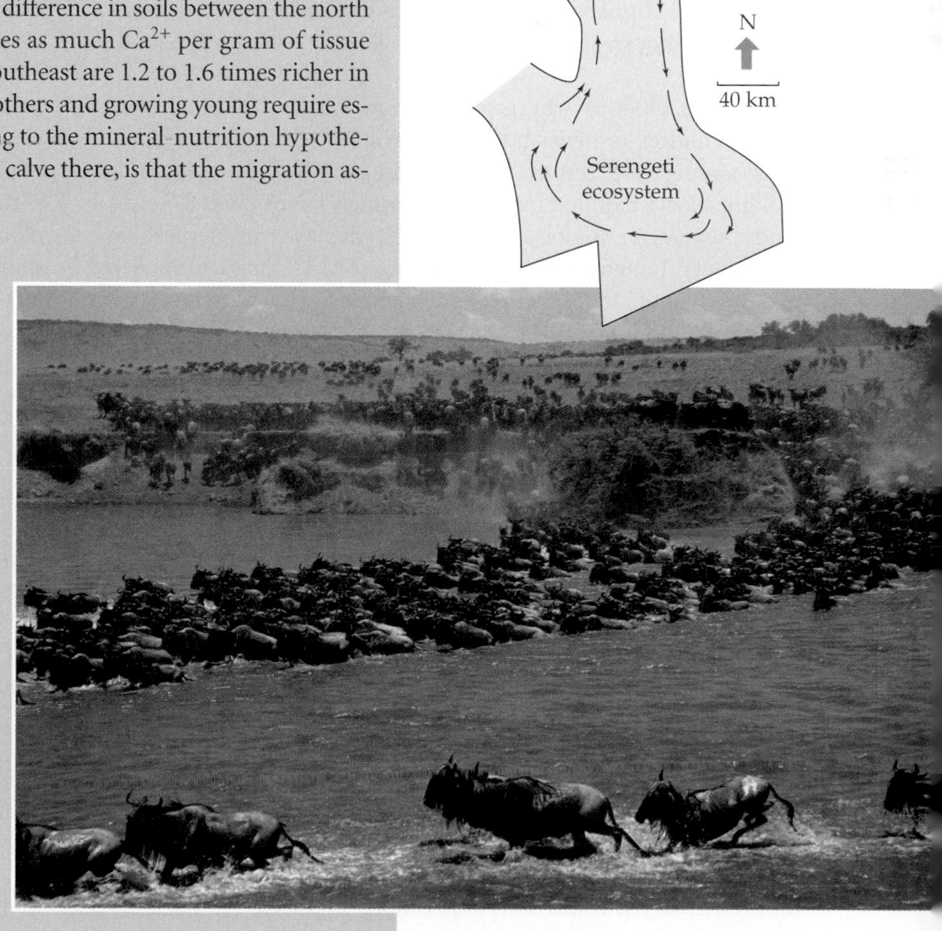

The great Serengeti migration: A quest for minerals Wildebeests and zebras, in herds sometimes numbering a million or more, travel from the far north of the Serengeti ecosystem to the southeast at the start of the rainy season and return to the north for the dry season each year. According to one prominent hypothesis, the trek to the southeast is a quest for minerals.

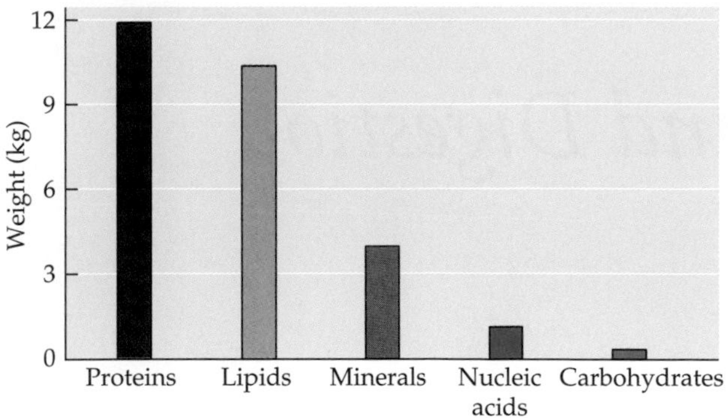

Figure 4.1 The composition of the adult human body The weights in kilograms of the five principal body components, other than water, in an average adult weighing a total of 70 kg. About 42 kg of the total body weight is water (not shown).

the most abundant types of atoms in the human body are carbon, hydrogen, and oxygen—the ubiquitous building blocks of organic compounds. Nitrogen, calcium, and phosphorus are additional abundant types of atoms—nitrogen because it appears in all proteins and nucleic acids, calcium because of its importance in the skeleton, and phosphorus because it occurs not only in the skeleton but also in nucleic acids and the phospholipids of all cell membranes. At least 18 other types of atoms are required to construct the human body. Humans and other animals must acquire all the atomic building blocks of their bodies—plus energy—from the foods they eat.

It seems obvious why growing animals must feed. Whether human children or newly born wildebeests or zebras, growing animals (among other things) must accumulate all the diverse atomic building blocks required for their increasingly large bodies from the foods they eat. A paradox arises however, when we consider feeding by adults: Although feeding is often dangerous, animals continue to feed throughout life. Danger arises during feeding for several reasons. The most straightforward is that animals often leave safe refuges to feed, as when a squirrel leaves its nest cavity to search in the open for nuts or seeds. If feeding can be dangerous, why do animals feed regularly throughout their adult lives? Why can't an adult animal simply make do with the chemical building blocks and energy it already has and thus opt for safety by not feeding at all? One of the principal reasons that animals must feed throughout life is the dynamic state of body constituents emphasized in Chapter 1 (see page 10). Individual cells and molecules in an animal's body age or become damaged, and ultimately they are discarded and replaced. During this turnover, some of the chemical building blocks of the old cells and molecules are lost from the body and must be replaced with new chemical building blocks from foods. New amino acids are needed to replace amino acids lost in the breakdown of old proteins, for example. New chemical-bond energy must also be acquired because after chemical-bond energy is used in metabolism, it cannot be reused (see Chapter 5). Feeding in the face of danger really is no paradox. Because animals are physiologically dynamic—not static—systems, feeding throughout life is essential regardless of its implications for safety.

The simple act of feeding does not necessarily meet an animal's nutritional needs, even if the food ingested contains the chemical building blocks and energy that the animal requires. This is true because the components of foods must be absorbed from the gut to be of use. The fate of cellulose in human foods emphasizes this point. Cellulose is too big a molecule to pass from the gut lumen into the bloodstream, and humans lack digestive enzymes that could break up cellulose into smaller molecular pieces. Thus, cellulose in our food simply passes into our feces, unutilized. For all animals, the acquisition of nutrients depends not only on feeding but also on having mechanisms of digestion and absorption to process the foods obtained. Thus, nutrition, feeding, digestion, and absorption form an interrelated set of needs and processes, each of which is essential in the lives of animals.

Nutrition

A useful way to understand animal nutrition is to consider the major components of the body—such as proteins or lipids—and ask several questions about each. Why is each sort of component needed? How do animals acquire each sort? What peculiar challenges are likely to be faced for an animal to obtain adequate amounts of each? These are the questions that dominate this section.

Proteins are "foremost"

Among the major components of the body, proteins deserve to be discussed first because they occupy a position that is both uniquely important and uniquely tenuous in animal nutrition. The very word *protein* is derived from the Greek for "foremost," referring to the preeminent position that proteins occupy in the structure and function of all animals. One way to see the importance of proteins is to look at their abundance in the bodies of animals. About half the organic matter in a mammal is protein (see Figure 4.1), and proteins are similarly abundant in other types of animals. A second way to see the importance of proteins is to consider the many essential roles they play. Enzyme proteins *speed and regulate biochemical reactions.* Muscle proteins are essential for *locomotion,* one of the most characteristic features of animals. Structural proteins, such as collagen in connective tissue and keratin in skin, *determine the structural properties of tissues.* Other proteins are so diverse in function as to defy simple classification. They include the receptor, channel, and transporter proteins in cell membranes; globular proteins dissolved in blood plasma (e.g., albumin); antibodies; proteins that function as hormones; oxygen-transport proteins such as hemoglobin; protein venoms; and the transparent crystallin proteins of eye lenses.

A set of 20 to 22 amino acids, called the **standard amino acids**, is required for protein synthesis in all organisms. The structures of five are shown in Figure 4.2*a*, and all the standard amino acids are listed (along with their usual abbreviations) in Appendix I. Besides the standard amino acids, over 200 other amino acids are known in organisms and play diverse roles, although they are not involved in protein synthesis. From the viewpoint of primary structure (see Box 2.1), **proteins** are *defined* to be strings of amino acids composed of many amino acid units. Relatively short strings are often distin-

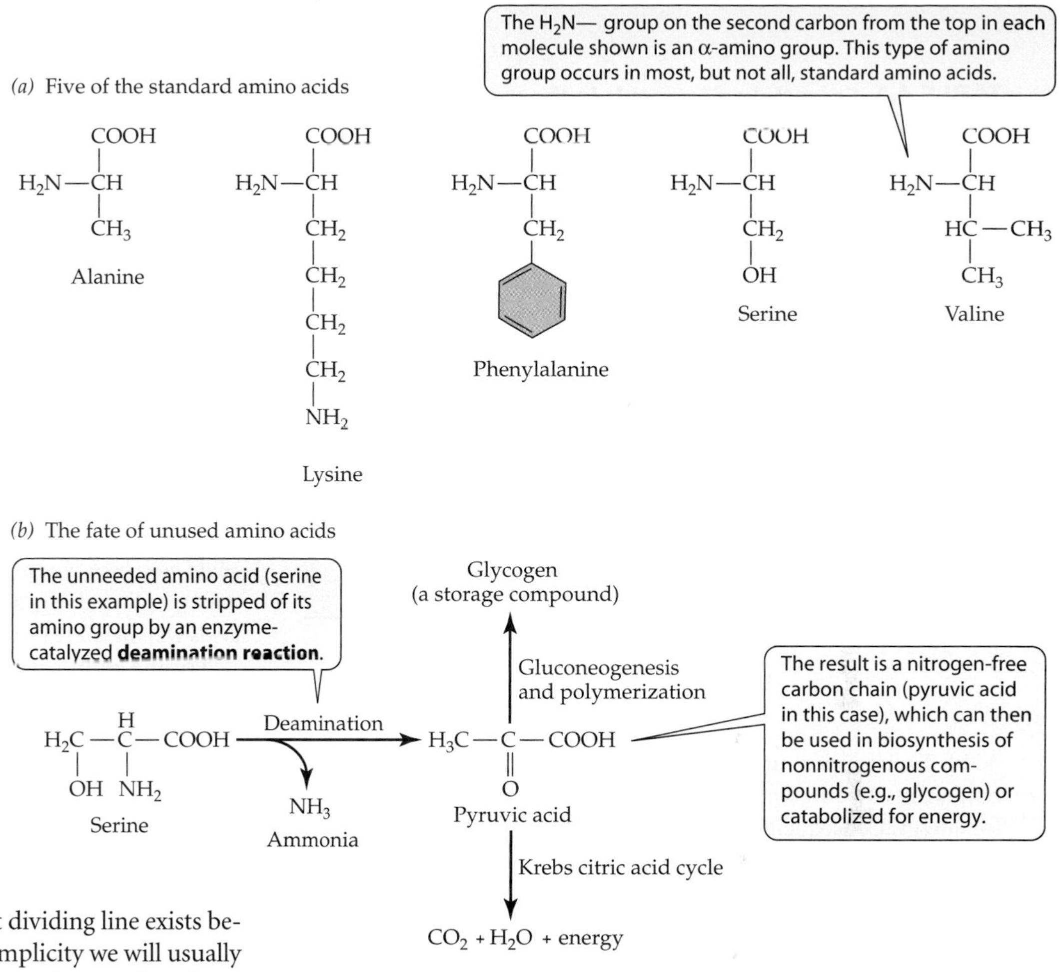

Figure 4.2 Amino acid chemistry *(a)* Structures of five of the standard amino acids required for protein synthesis. All amino acids contain nitrogen even If lacking an α-amino group. Amino acids with α-amino groups may contain additional nitrogen, as exemplified by lysine. Box 2.1 discusses the relation between amino acids and protein structure. *(b)* The fate of amino acids that are ingested but that an animal either does not need or is unable to use for protein synthesis. An alternative to the deamination reaction shown here is a **transamination reaction** that transfers the amino group from the unneeded amino acid to another carbon chain, possibly synthesizing another amino acid (only nonessential amino acids can be made in this way). Although transamination reactions can alter the mix of amino acids obtained from a meal, in the end all the amino acids that cannot be promptly used are deaminated, because amino acids are not typically stored for future use.

guished as **polypeptides**, although no strict dividing line exists between proteins and polypeptides, and for simplicity we will usually use the word *protein* to refer to both. **Dipeptides** consist of two amino acids; **tripeptides**, three. All amino acids contain nitrogen. Proteins are about 16% nitrogen by weight, in contrast to carbohydrates and lipids, which do not contain nitrogen.

A fact that colors every aspect of the nutritional biology of proteins is that nitrogen is a limiting element in many ecosystems, including many terrestrial ecosystems and more than half the oceans. Nitrogen limitation may at first seem strange because about 78% of the gas in the atmosphere is molecular nitrogen, N_2. However, animals, plants, and most microbes cannot use N_2 as a source of nitrogen for building proteins or other nitrogen-containing compounds. Instead, as their source of nitrogen they require the element in chemically combined, nongaseous ("fixed") forms. Plants and algae most commonly use nitrate (NO_3^-), ammonium (NH_4^+), and other inorganic nitrogen-containing ions or compounds as their nitrogen sources. The nitrogen available in these chemical forms in an ecosystem is often low enough that the growth and reproduction of plants or algae are held back by nitrogen insufficiency. Because herbivorous animals most commonly acquire nitrogen from proteins and other nitrogen-containing organic compounds in the tissues of plants and algae they eat, nitrogen limitation can be relayed up food chains. If the production of plants and algae in an ecosystem is nitrogen-limited, growth and reproduction in herbivore populations may be nitrogen-limited, and fewer carnivores may be able to live there than if herbivores were more successful.

Of all the nutritional vulnerabilities of animals, one of the greatest is that they are biochemically unable to synthesize about ten of the standard amino acids at rates sufficient to meet their requirements (even if nitrogen is freely available). The amino acids that cannot be adequately synthesized are called **essential amino acids** because there is an "essential" requirement that they be acquired *fully formed* from food or another *outside* source.[1] Experiments to determine the list of essential amino acids in a species are challenging. Thus, ambiguity often exists about the list even for relatively well-studied species. The set of essential amino acids also varies a bit from one animal group to another. Table 4.1 lists the ten amino acids that are essential for the best-studied animal model, the growing laboratory rat. As the table notes, eight of the ten are required in the diet of adult humans, and nine are essential in children. The ten essential amino acids of the growing rat are also essential in insects, fish, and birds, although some insects and birds also have additional amino acids that are essential.

Possibly the inability of animals to synthesize certain amino acids would not be a vulnerability if animals kept stores of amino acids or proteins in their bodies in the same way they store fat. If stores of "extra" amino acids were maintained, presumably there would always be some of each kind available. However, animals do not generally store amino acids for future use, either as free amino acids or as storage proteins. Instead, when an animal eats amino acids in excess of those it needs for the synthesis of functioning proteins at the time, it promptly strips the nitrogen-containing amino groups (—NH_2) from the carbon chains of the excess amino acids (Figure

[1] Essential amino acids are sometimes called *indispensable amino acids.*

TABLE 4.1 Essential amino acids in the most thoroughly studied model, the growing laboratory rat, and in people

Essential amino acids in the growing rat	Ways that people are similar and different
Arginine	Arginine not essential in humans
Histidine	Histidine essential for children, probably not for healthy adults
Isoleucine	All essential in children and adults
Leucine	
Lysine	
Methionine	
Phenylalanine	
Threonine	
Tryptophan	
Valine	

Sources: After Burton and Foster, 1988; Morris, 1991.

4.2*b*). The stripped nitrogen is excreted, and the nitrogen-free carbon chains that are formed are used for biosynthesis or catabolized for energy. Consequently, the amino acids in an animal's body at any one time are generally constituents of *functioning* proteins.

Another important fact about the amino acid physiology of animals concerns the source of nitrogen for synthesis of **nonessential amino acids**—that is, those amino acids that an animal is capable of synthesizing. A property of the biochemical pathways used by animals for such synthesis is that generally the nitrogen-containing amino groups are derived from other amino acids. Thus, for a nonessential amino acid to be synthesized, other amino acids must be available to serve as amino-group donors.

With these background points in mind, you will see that animals employ a "just in time" strategy for acquiring the raw materials they need for protein synthesis. In automobile-assembly plants, the standard practice throughout most of the twentieth century was to maintain large stores of all the necessary parts. Thus, if there were problems in receiving new shipments of certain parts, manufacturers could still make cars for a protracted period by drawing on the stores. The approach used nowadays in many assembly plants, however, is the "just in time" strategy. Large stores of parts for future use are not maintained. Instead, all parts are brought to the assembly plant just in time to be used. The "just in time" approach eliminates storage costs, but it makes car assembly more vulnerable to parts shortages; if shipments of *any* required parts are interrupted, assembly of cars halts. Animals use a "just in time" strategy in building their proteins. The amino acid building blocks that an animal needs for protein synthesis must arrive at the same time they are actually needed. What is more, the essential amino acids must arrive fully formed. The nonessential amino acids either must arrive preformed or must be synthesized on the spot; synthesis requires the ingestion of alternative amino acids that can serve as nitrogen donors for making the synthesized amino acids.

To illustrate the weighty consequences of this strategy for protein synthesis, consider an animal that eats too little lysine in a particular week but for which lysine is an essential amino acid. The animal will be unable to synthesize lysine-containing proteins in the full amounts needed during the week in question. If the next week again brings in too little lysine, the shortages of proteins will worsen. The animal's health will deteriorate because the functions of the unmade proteins are not carried out. Lysine deficiency, in fact, is a serious health problem for millions of people in impoverished parts of the world.

The failure of animals to store amino acids for future use leads to wastage of available essential amino acids when other essential amino acids are in too-short supply for protein synthesis. Returning to the animal we were just discussing, suppose, to illustrate, that a particular meal provides all the essential amino acids in abundance, except that lysine is in short supply. Needed proteins will go unmade because of the lysine shortage. In addition, as the meal is metabolized, the essential amino acids, besides lysine, *that are abundant in the meal and that could have been used to make needed proteins* will not be fully used and will be broken down, as discussed earlier. These other essential amino acids will therefore be wasted.

For the amino acids in a meal to be used optimally for protein synthesis, the meal typically must supply all essential amino acids in the amounts required. Mixing foods is a mechanism to reduce or eliminate amino acid wastage. To illustrate, consider an animal for which both lysine and methionine are essential. If food A contains little lysine but much methionine and food B contains lots of lysine but little methionine, eating the two together will allow the animal to utilize the amino acids from the foods more fully than if either food were eaten alone. Foods must be eaten nearly simultaneously to complement each other in this way. In the human diet, combinations like corn and beans, or milk and cereal, eaten approximately simultaneously, increase the utilization of amino acids for protein synthesis because the foods tend to make up for each other's shortcomings in essential amino acids.

Where do essential amino acids originate in ecosystems? Plants and algae are able to synthesize from scratch all 20 to 22 of the standard amino acids required for protein synthesis. For most animals, plants and algae are the principal ultimate sources of essential amino acids.[2] Herbivores eat the essential amino acids in their plant or algal foods, and then the molecules are relayed up food chains when carnivores eat the herbivores.

Lipids are required for all membranes and are the principal storage compounds of animals

Lipids are often about as abundant in the structure of animal bodies as proteins (see Figure 4.1). They are structurally very diverse and thus problematic to define. They are usually defined as organic molecules, composed principally of carbon and hydrogen, that are predominantly nonpolar and therefore hydrophobic. The chemical components that most lipids have in common are **fatty acids**, which are hydrocarbons built on a backbone consisting of a chain of carbon atoms (Figure 4.3*a*). One important chemical category of lipids is the **triacylglycerols** (**triglycerides**), also known as **fats** and **oils**; a triacylglycerol consists of three fatty acid molecules combined (es-

[2] Bacteria or other heterotrophic microbes may also act as ultimate sources, as discussed later in this chapter.

terified) with a molecule of glycerol (Figure 4.3*b* and *c*). Waxes, phospholipids (see Figure 2.2), and sterols (see Figure 14.1) are additional categories of lipids.

Fatty acids are chemically diverse, a factor that is important in the diversity of lipids. Over 50 different fatty acids have been isolated from organisms. The majority of the fatty acids in organisms have an even number of carbon atoms, between 8 and 24, in their carbon-chain backbone, and they are unbranched. Besides the number of carbon atoms, additional features that impart diversity to fatty acids—and help define their properties—are their degree of saturation and the positions of double bonds. A fatty acid is **saturated** if all the bonds between carbon atoms are single bonds, as exemplified by palmitic acid (see Figure 4.3*a*). A fatty acid is **unsaturated** if one or more of the bonds between carbon atoms are double. **Polyunsaturated** fatty acids are the subset of unsaturated fatty acids that contain multiple (two to six) double bonds, exemplified by linoleic acid (see Figure 4.3*a*).

(*a*) Two fatty acids

Palmitic acid (16.0ω0)

Linoleic acid (18.2ω6)

(*b*) Glycerol

(*c*) A triacylglycerol

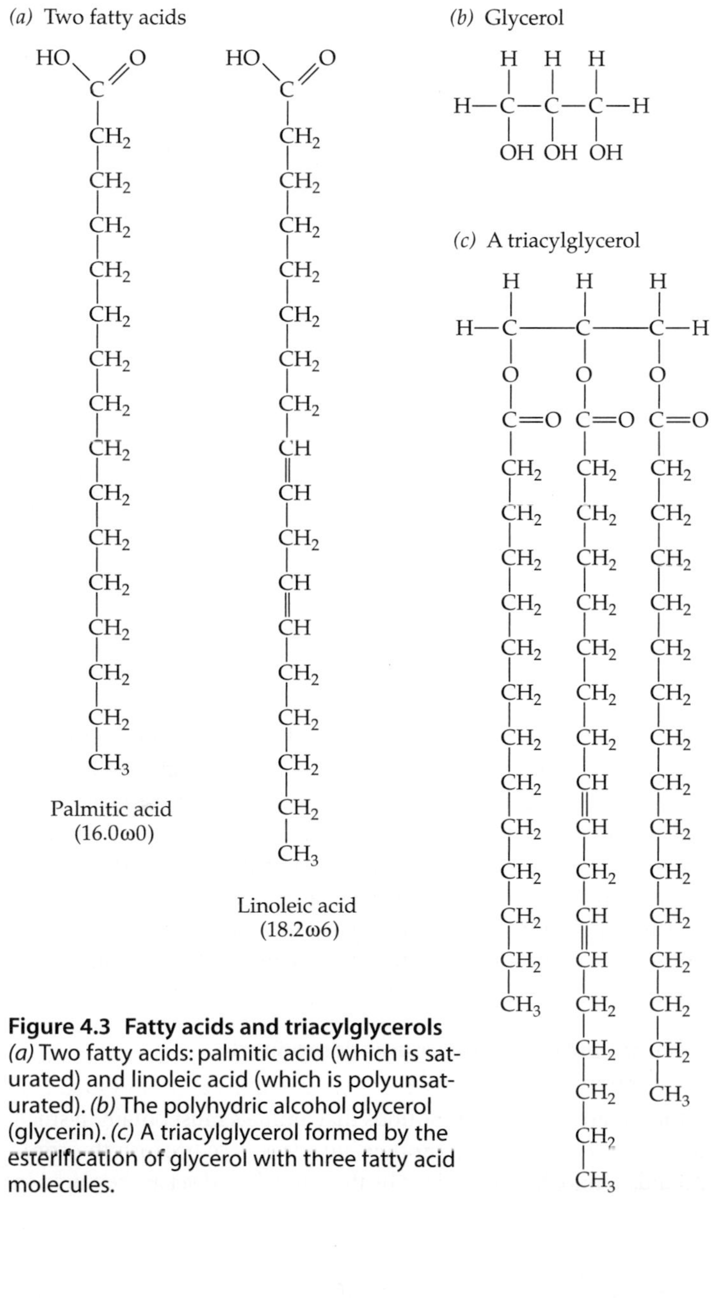

Figure 4.3 Fatty acids and triacylglycerols *(a)* Two fatty acids: palmitic acid (which is saturated) and linoleic acid (which is polyunsaturated). *(b)* The polyhydric alcohol glycerol (glycerin). *(c)* A triacylglycerol formed by the esterification of glycerol with three fatty acid molecules.

A common system for symbolizing fatty acids specifies three integer numbers, as in the example "18.2ω6." The first integer, preceding the decimal point, is the number of carbons in the fatty acid. The second, written between the decimal and the Greek letter omega (ω), is the number of double bonds. The third, which follows ω, is the position of the first double bond encountered when the molecule is scanned from its methyl (—CH_3) end. The example "18.2ω6" refers to linoleic acid (see Figure 4.3*a*), which has 18 carbons and 2 double bonds, with the first double bond at the sixth position from the methyl end. In common parlance, linoleic acid is an "omega-6" (ω6) fatty acid.

Lipids play several functional roles. For example, phospholipids and cholesterol are principal *components of cell and intracellular membranes* (see Chapter 2). Lipids are also *storage compounds* in both animals and plants. Lipids serve especially well as energy stores because they far exceed proteins and carbohydrates in their energy value per unit weight (Table 4.2). The high energy density of lipids means that the weight of material that an animal must carry around to store a given amount of chemical-bond energy is far lower if lipids are the storage compounds than if other compounds are used. A third functional role played by lipids is that in fully terrestrial animals such as mammals and insects, lipids in or on the integument *reduce the permeability of the integument to water*; they thus slow the evaporative loss of body water and permit the animals to live in the open air. Other roles of lipids are numerous. Triacylglycerols, for example, are key components of the sound-focusing structures in echolocating dolphins, and steroids play hormonal roles in both vertebrates and invertebrates.

Animals must synthesize a great variety of lipids to construct their cells and tissues and meet other requirements for lipids in their bodies. They are able to meet these biosynthetic requirements using a variety of organic carbon chains that they obtain in their diets. Indeed, carbon chains used to synthesize lipids may be derived from dietary carbohydrates and proteins, as well as dietary lipids (we are well aware that we can add body fat by eating sugar). This biochemical flexibility is one reason that lipid nutrition is less likely to be fraught with problems than protein nutrition. Another reason is the fact that animals maintain stores of lipids; contrary to the case with proteins,

TABLE 4.2 The energy values of mixed lipids, carbohydrates, and proteins in aerobic catabolism[a]

	Energy value in kilocalories (kcal) per gram	Energy value in kilojoules (kJ) per gram
Mixed lipids	9.3–9.5	39–40
Mixed carbohydrates	4.0–4.1	17
Hydrated glycogen	1.1	4.6
Mixed proteins	4.3–4.8	18–20

[a] Data are for mammals and are measured or estimated by bomb calorimetry (see Chapter 5). The values for proteins depend on the chemical form of nitrogen excreted; thus, moderately different values apply to groups of animals that differ from mammals in their nitrogenous excretory compounds (e.g., protein that yields 4.77 kcal/g in urea-producing animals such as mammals yields more, about 4.85 kcal/g, in ammonia producers but less, about 4.34 kcal/g, in uric acid producers). All values are for dry foodstuff material except those for hydrated glycogen, which apply to glycogen combined with its usual water of hydration.

lipids obtained in excess of needs at one time can be held over in the body for use at future times.

Problems can arise in lipid nutrition, however, because although animals have extensive capabilities to synthesize and structurally modify fatty acids, many types of animals—including mammals—lack the enzymes needed to create double bonds at the omega-3 and omega-6 positions. The omega-3 and omega-6 fatty acids therefore cannot be synthesized and must be obtained from foods or other *outside* sources, at least during critical life stages. The omega-3 and omega-6 fatty acids thus are **essential fatty acids.** Essential amino acids, you will recall, are specific compounds. The situation with essential fatty acids is usually a little different because what animals must obtain in their diets are fatty acid carbon chains with double bonds in particular positions, rather than specific compounds. If the required sequences of carbon atoms and double bonds are ingested, they can usually be employed to synthesize fatty acids other than the specific fatty acids actually eaten; a simple example is that relatively long fatty acids with double bonds at the omega-3 and omega-6 positions can meet the dietary requirement for shorter omega-3 and omega-6 fatty acids because the long ones can be shortened. The simplest fatty acid that will meet needs of the omega-6 sort is linoleic acid (see Figure 4.3*a*); the simplest that will meet needs of the omega-3 sort is α-linolenic acid. Exceptions to this general picture are known. Cats and certain fish, for example, have dietary fatty acid requirements that are more chemically specific than the requirements of most vertebrates. These requirements must be taken into account in the design of pet foods and foods for use in aquaculture.

Carbohydrates are low in abundance in many animals but highly abundant when they play structural roles

The simplest carbohydrates are **monosaccharides** such as glucose and fructose (Figure 4.4*a*). Other types of carbohydrates are composed of two or more monosaccharides bonded together. Those consisting of two monosaccharides are called **disaccharides**; examples are sucrose (glucose + fructose) (Figure 4.4*b*), lactose (glucose + galactose), and trehalose (glucose + glucose). **Polysaccharides** have over ten monosaccharides per molecule and often consist of hundreds or thousands of polymerized monosaccharide units.

Carbohydrates play three principal functional roles in animals and plants. First, large polysaccharides often *provide structural support and shape* to cells and other parts of organisms. The most important structural carbohydrate in animals is the polysaccharide *chitin*, which is the principal component of the exoskeletons of insects and other arthropods. Over half the dry weight of many arthropods consists of chitin. The structural polysaccharides of plants and algae, which include *cellulose* and *hemicelluloses*, are enormously abundant in ecosystems and thus are potentially major food sources for animals. Chitin and cellulose are the two most abundant organic compounds in the biosphere!

In their second principal role, carbohydrates function as *storage* compounds. Like the structural carbohydrates, the storage carbohydrates are polysaccharides. They are accumulated and broken down far more dynamically than structural polysaccharides, however. *Starch*, a form of polymerized glucose, is one of the principal storage carbohydrates of plants. *Glycogen*, also a form of polymerized glucose, is the principal storage carbohydrate of animals. The amount of glycogen stored by animals is typically much less than the amount of lipid stored (see Figure 4.1). Glycogen is probably stored in relatively small amounts because it is far inferior to lipid as an energy-storage compound. Molecules of glycogen are highly hydrated within the tissues of living animals, and because of the added weight of their water of hydration, they yield a low amount of energy per unit of total weight (see Table 4.2). Glycogen serves more as a storage form of *glucose* than as a simple energy store. Glucose is a required fuel for the central nervous system in vertebrates. Glucose is also the fuel required for the fastest form of ATP production in skeletal muscles. Stores of glycogen supply glucose to the brain and skeletal muscles in times of need.

Figure 4.4 Carbohydrate chemistry

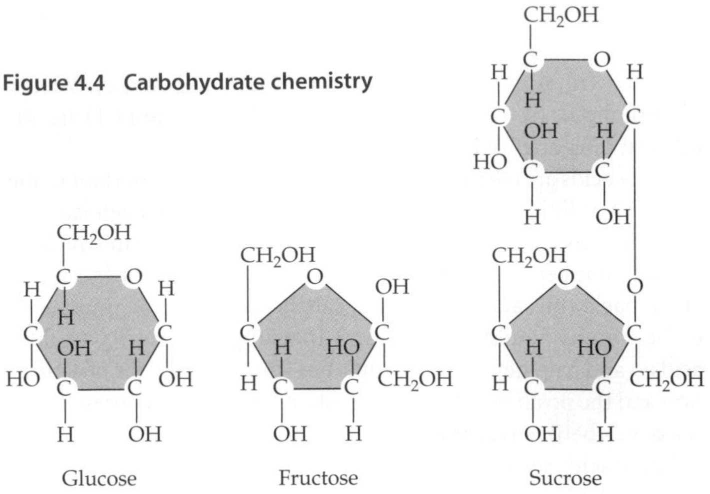

(*a*) Two monosaccharides (*b*) A disaccharide

The third principal functional role of carbohydrates is as *transport* compounds. The transport carbohydrates are small molecules, monosaccharides or disaccharides, that dissolve in the blood or other moving body fluids. Glucose is the principal transport carbohydrate ("blood sugar") of vertebrates and most other groups of animals, although many insects employ the disaccharide trehalose in this role. Lactose is the main carbohydrate in the milks of most mammals, serving a transport function from mother to young.

Animals typically have a great deal of latitude in the biochemical mechanisms by which they can synthesize the variety of carbohydrates they need to construct their cells and tissues. The carbon chains used to synthesize carbohydrates may be derived from dietary proteins and lipids, as well as dietary carbohydrates. Moreover, there are no essential carbohydrates; animals can synthesize all the carbohydrates they need. The one noteworthy "nutritional problem" that animals have with carbohydrates is that many animals are unable to digest some or all of the structural polysaccharides, meaning that they are potentially unable to utilize some of the most abundant organic materials on Earth.

Vitamins are essential organic compounds required in small amounts

Vitamins are organic compounds that animals must obtain in small quantities from food or another *outside* source because the compounds cannot be synthesized by the animals and yet are required in

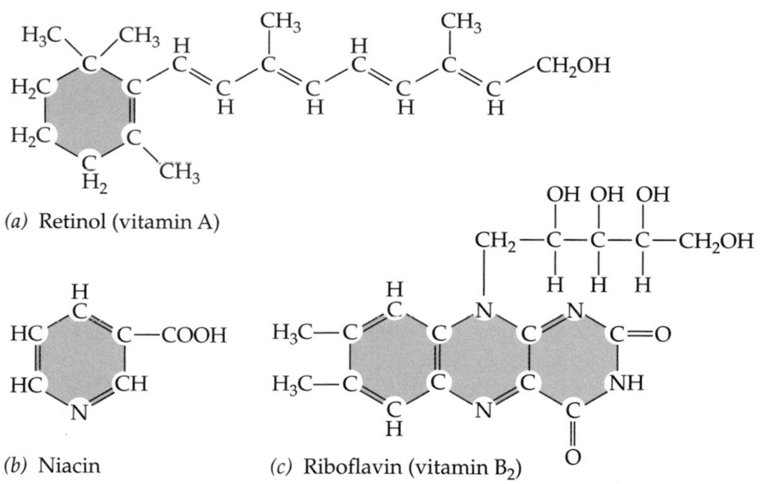

(a) Retinol (vitamin A)

(b) Niacin

(c) Riboflavin (vitamin B_2)

Figure 4.5 Vitamin structures

small amounts. Possibly this definition will seem strange because of its emphasis on small quantities. For historical reasons, however, compounds that are required to be ingested in relatively large amounts, such as essential amino acids, are not considered vitamins. Vitamins initially came to light when, in about 1880, scientists tried to maintain normal growth in mice by feeding them purified diets consisting only of carbohydrates, lipids, nutritionally complete proteins, and inorganic salts. The mice could not maintain normal growth. Decades of research then gradually revealed the need for small (often minute) amounts of additional organic compounds, which were called *vitamins*.

The need for vitamins in modern-day animals is often a consequence of what might be called "opportunism" during evolution. Vitamin A illustrates this point. The term *vitamin A* refers to *retinol* (Figure 4.5) and several closely related compounds. Animals cannot synthesize the complex retinol structure from small building blocks. Instead, they are dependent on plants or algae to make the structure. In plants and algae, the structure occurs as a subpart of *carotenoid pigments* such as *β-carotene*. An animal can obtain vitamin A by eating carotenoids and liberating the vitamin-A structure from these large molecules. The animal can alternatively obtain vitamin A by eating another animal that obtained the vitamin A by breaking down carotenoids. Either way, the source of the structure is plant or algal biosynthesis.

In animals, the vitamin-A structure is incorporated into crucial visual pigments called *rhodopsins*. To synthesize a rhodopsin, an animal converts vitamin A to its aldehyde *retinal*, which then is conjugated with a protein (*opsin*) synthesized by the animal to form the rhodopsin.[3] *Remarkably, the light-absorbing portion of the rhodopsin molecule is the part derived from vitamin A!* In plants and algae, the carotenoids serve as photon-collecting molecules for photosynthesis. Evidently, as vision evolved, animals that ingested the plant or algal photon-absorbing molecules in their foods "opportunistically" adopted the vitamin-A substructure from the molecules to be the photon-absorbing portion of their visual pigments. This strategy means that animals need not be able to synthesize the photon-absorbing substructure of their rhodopsins, but they are vulnerable to shortages if they are unable to get enough in their foods.

Because vitamins are consequences of opportunism during evolutionary history, it is not surprising to find that they are extremely diverse in their chemical structures (see Figure 4.5). They are also very diverse in their functions, meaning that there is no simple way to summarize what they do. The use made of most vitamins is that they are incorporated into coenzymes and other molecules as *key molecular subsystems*, necessary for the proper function of the larger molecules of which they are made parts.

Vitamins are usefully subdivided into a *water-soluble* set and *lipid-soluble* set. The water-soluble vitamins include the B vitamins and vitamin C (Table 4.3). The B vitamins are essential nutrients for virtually all animals because they are commonly incorporated into coenzymes required for reactions of fundamental importance. The lipid-soluble set of vitamins includes vitamins A, D, E, and K. All four are required by vertebrates. Invertebrates, on the other hand, may or may not require them. The physiological uses of the lipid-soluble vitamins (Table 4.4) are more specialized than those of the B vitamins, helping to explain why the need for them is not nearly so universal as that for B vitamins.

Elemental nutrition: Many minerals are essential nutrients

Many chemical elements—often referred to as **minerals** in the study of nutrition—are required by animals in addition to the carbon, hy-

TABLE 4.3 Water-soluble vitamins and their physiological roles Of the vitamins listed, all except ascorbic acid (vitamin C) are classified as B vitamins, although some are not usually referred to using "B" symbols.

Vitamin	Physiological role in animals
Thiamin (B_1)	Coenzyme for oxidative decarboxylation reactions
Riboflavin (B_2)	Coenzyme for oxidation–reduction reactions (required for synthesis of FAD, flavin adenine dinucleotide)
Niacin	Coenzyme for oxidation–reduction reactions (required for synthesis of NAD, nicotinamide adenine dinucleotide)
Pyridoxine (B_6)	Coenzyme for reactions of amino acid metabolism
Pantothenate	Coenzyme for acetyl-transfer reactions (required for synthesis of coenzyme A)
Folate	Coenzyme for single-carbon-transfer reactions (as in nucleic acid synthesis)
Cobalamin (B_{12})	Coenzyme for single-carbon-transfer reactions (as in nucleic acid synthesis)
Biotin	Coenzyme for carboxylation reactions
Ascorbic acid (C)	Protects cells against damage by reactive oxygen compounds (antioxidant). Plays many additional roles in reactions involving oxygen

[3] Rhodopsin is discussed further in Chapter 13 (see Figure 13.13).

TABLE 4.4 Lipid-soluble vitamins and their physiological roles Unlike the water-soluble vitamins, lipid-soluble vitamins are usually referred to by their vitamin designations rather than chemical names. The reason is that multiple related compounds can often meet the vitamin requirement for lipid-soluble vitamins.

Vitamin	Physiological role in animals
Vitamin A	Light-activated component of visual pigments. Also needed for normal bone growth, reproductive function (e.g., sperm production), cell membrane integrity, and other functions, but exact biochemical mechanisms are not always known. Vitamin A is a regulator of gene transcription and can cause deranged development of a fetus if ingested in artificially large amounts by the mother during pregnancy.
Vitamin D	Activator of pathways of calcium and phosphorus metabolism; acts by binding like a hormone to specific receptors
Vitamin E	Protects cells against damage by reactive oxygen compounds (antioxidant), preserving integrity of critical molecules, especially membrane phospholipids
Vitamin K	Coenzyme for production of blood-clotting factors

drogen, oxygen, and nitrogen that predominate in organic molecules. One reason a wide diversity of elements is essential is that many enzymes and other proteins have trace amounts of "unusual" atoms in their structures. In fact, about 40% of all proteins[4] are metal-containing *metalloproteins*. Collectively, the metal atoms in protein structures include iron (as in cytochrome pigments and hemoglobin), copper, molybdenum, zinc, manganese, vanadium, selenium, cobalt, and others. Humans and other mammals require all of the named metals for proper nutrition. In addition, phosphorus is required for the synthesis of phospholipids, nucleic acids, and bone; sodium, chlorine, and potassium are essential solutes in the body fluids; and iodine is needed for thyroid hormones. Other kinds of animals have lists of elemental needs similar to those of mammals, but animals vary in the relative quantities of elements they need. The groups of animals that do not build calcareous skeletons, for example, need less calcium than mammals do.

Mineral deficiencies are a potential problem for terrestrial animals in a great many regions of the world because of mineral-deficient soils or soil-chemistry processes that render minerals unavailable. Sodium and phosphorus deficiencies, for example, are very common threats to livestock. Iodine deficiencies provide another example; terrestrial regions so expansive that they are occupied by over a billion people have soils that provide too little iodine for naturally growing foods to meet human iodine needs. Iodine deficiency is prevented in wealthy nations by the use of dietary iodine supplements (e.g., iodized salt), but insufficient iodine ingestion is a major cause of mental retardation and brain damage in impoverished nations. As we noted earlier in discussing wildebeests and zebras, calcium, magnesium, and copper concentrations are too low in the soils of much of tropical Africa to meet readily the needs of lactating mothers and growing young.

[4] Including those of plants and microbes as well as animals.

Summary: Nutrition

- Essential organic nutrients are organic compounds that animals must obtain from other organisms because the animals are biochemically unable to synthesize them. About ten of the standard amino acids required for protein synthesis are essential in most sorts of animals. Omega-3 and omega-6 fatty acids are essential in many animals, and vitamins are essential.
- Proteins often present particular nutritional problems because nitrogen can be in short supply in ecosystems and essential amino acids may be unavailable in the amounts needed. Because animals employ a "just in time" strategy for protein synthesis (rather than storing amino acids for future use), the essential amino acids must be eaten together in the amounts they are needed if they are to be used. A shortage of any one essential amino acid can cause wastage of available supplies of others.
- Vitamins are very diverse in their chemical structures. What they have in common is not their chemical nature but the fact that they are organic compounds required in small (often minute) amounts. Most are employed as substructures in molecules of critical importance for animal function. The water-soluble vitamins, notably the B vitamins, are essential for most or all animals. The lipid-soluble vitamins, such as vitamins A and K, are more specialized in their functions and thus not as universally required.
- Minerals are also essential nutrients. Metal atoms occur in about 40% of proteins. Minerals are also required as constituents of body fluids and skeletons. Over 20 chemical elements are required for the construction of animal bodies.
- Structural carbohydrates such as chitin, cellulose, and hemicelluloses are the most abundant organic compounds on Earth, but many animals lack enzymes required to digest them and thus are unable (on their own) to tap those carbohydrates for nutritional value.

Feeding

Animals employ a stunning diversity of feeding mechanisms. Consider, for example, three animals that live in the oceans: orcas, blue whales, and reef-building corals. Orcas feed in a way that is very familiar to us. They have large teeth and strong jaw muscles, and they feed by attacking other animals that are modestly smaller than they are, such as fish and seals (**Figure 4.6**). Blue whales (see Figure 4.10) lack teeth and do not attack individual prey items. Instead, they are *suspension feeders*, meaning they ingest organisms suspended in the water that are minute by comparison to their own size. They capture their tiny prey by sieving them out of the water using closely spaced plates hanging from each side of the upper jaw. Reef-building corals (see Figure 4.12) are photosynthetic; they get much of their nutrition from algae with which they live in symbiotic associations. All three of the feeding modes exemplified by the orca, blue whale, and corals—attack of individual prey items, suspension feeding, and association with symbiotic microbes—are widespread in the animal kingdom.

(a)

(b)

Figure 4.6 Some species feed by targeting and subduing individual food items *(a)* Orcas (*Orcinus orca*), also known as killer whales, eat fish, sea otters, and seals by attacking them with sharp teeth. *(b)* Groups (pods) sometimes specialize behaviorally in feeding on certain sorts of prey. Some groups, for example, eat mostly fish, whereas others eat mostly marine mammals.

Many animals feed on organisms that are individually attacked and ingested

Many animals, both terrestrial and aquatic, are like orcas in that their feeding is directed at individually targeted food items. This feeding mode requires that the individual food organisms be located, identified, subdued, and ingested. In contrast to the simplicity of these basic imperatives, the array of actual mechanisms employed is wonderful in its diversity. Mammals that feed in this way grasp prey with teeth, and usually they use their teeth to reduce the prey to smaller pieces by tearing or chewing before swallowing. Fish often grasp prey with teeth, but they usually swallow their prey whole. Sea stars envelop prey in their evaginated stomachs, sometimes after overcoming defenses by mechanical forces developed with their tube feet. Sea urchins—another group of echinoderms—grasp, dig, or scrape food using a set of five teeth that close and open radially, driven by a unique and intricate array of skeletal elements and muscles. Fragile butterflies harvest the nectar of flowers with long and delicate proboscises. Peregrine falcons violently strike flying birds at high speeds in midair, then circle back to grasp their victims as the latter—stunned—fall toward ground in the grip of gravity.

Related animals that share a particular mechanism of feeding usually exhibit specialized variations in the mechanism. Bird bills provide a familiar example. Virtually all birds collect food items using their bills. The bills, however, exhibit a wide range of specializations. Some bird species have long, strong, sharp bills that they use as chisels to reach prey within tree trunks (Figure 4.7*a*); other species have stout bills that help them crush seeds (Figure 4.7*b*); still others have slender bills that they insert, stiletto-like, deep into mud to collect burrowing prey (Figure 4.7*c*).

The food-collection devices of grazing and browsing mammals—not just their teeth and lips, but also their sensory and prioritization mechanisms—provide another, although more subtle, vertebrate example of variations on a theme. Grazing can at first look as nonselective as lawn mowing. In fact, however, grazers and browsers in their natural habitats usually are selective about which plants or parts of plants they eat; indeed, species of grazers that are ostensibly similar sometimes feed on very different sets of the plant species available. The teeth, jaws, and lips of zebras, for example, enable them to feed more readily on tall grasses than wildebeests can. Wildebeests, in contrast, are more effective at ingesting short grasses in large quantities, in part because they have an exceptionally blunt muzzle and wide row of incisors. A third species that feeds with wildebeests and zebras on grasslands, Grant's gazelle, tends to choose herbs, sedges, and small woody plants more than the grasses themselves. Differences such as these are believed by most biologists to help account for the diversity of coexisting species.

Snails provide a dramatic invertebrate example of diversification in the use of a single basic feeding mechanism. The primitive feeding device of snails is a scraping organ, the **radular apparatus**. It consists of a ribbonlike band of connective tissue (the radula), studded with chitinous teeth, stretched over a cartilaginous rod (Figure 4.8*a*). Most snails use this apparatus to rasp organic matter off objects, such as algae off rocks. The cartilaginous rod is pressed against a surface, and muscles at the two ends of the band of connective tissue pull the band back and forth over the rod to produce rasping action. The radular teeth are elaborate (Figure 4.8*b*) and vary in shape from species to species. Possibly this variation in shape, like that of bird bills, suits different species to feed on different foods. In some species, the teeth are specialized by being hardened with iron or silica compounds, enabling the snails to rasp tissue from armored food sources.

The most exotic specializations of the radular feeding mechanism occur in species that attack and feed on living animals. One specialization for carnivory is the use of the radula as a *drill* to gain entry to

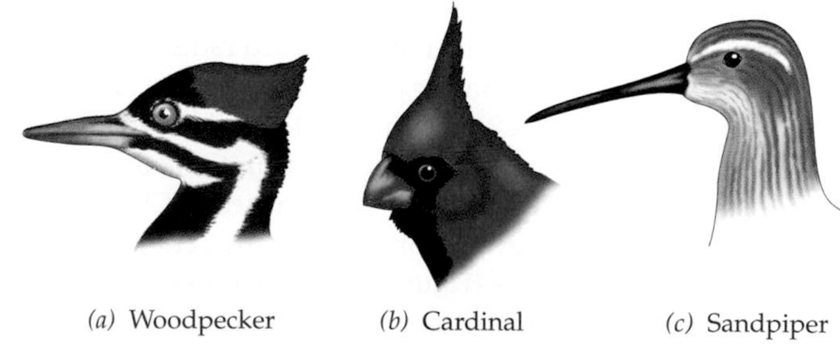

Figure 4.7 Specialization of a vertebrate feeding apparatus Bill specialization in a woodpecker, a cardinal, and a sandpiper (a type of shorebird).

(a) The mouth cavity of a snail showing the radular apparatus

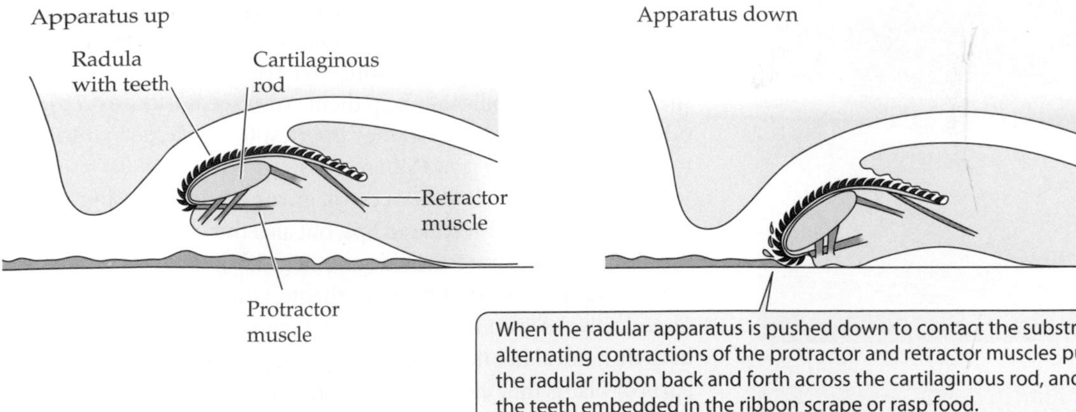

A herbivorous snail feeding by rasping algal growth off a rock.

(b) A radula viewed side-to-side, showing rows of teeth

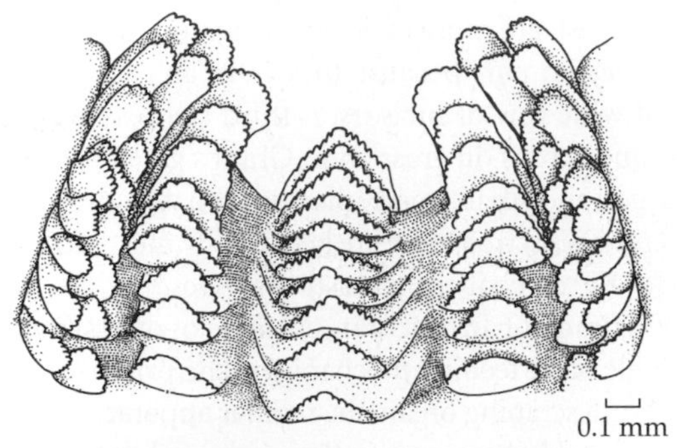

(c) The shells of two clams and a snail that were killed by carnivorous drill snails

(d) The harpoonlike tooth of a cone snail

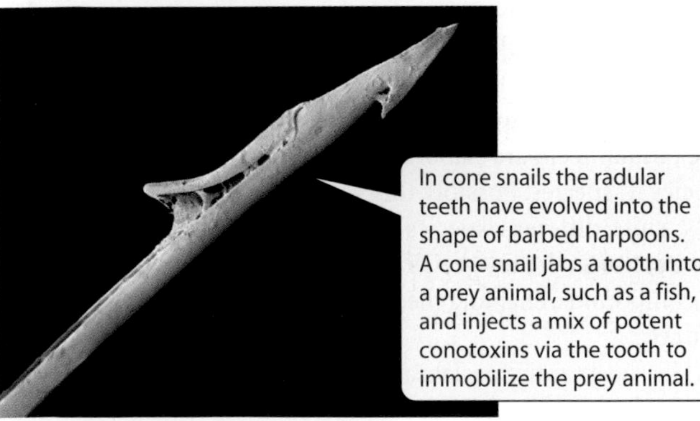

Figure 4.8 Specialization of an invertebrate feeding apparatus The primitive form of the radular apparatus in snails is a scraping or rasping organ used to scrape algae off rocks, rasp flesh off dead animals, and so forth. The radular apparatus becomes greatly specialized in carnivorous snails that prey on living animals. *(a)* Longitudinal sections of the primitive radular apparatus in the mouth cavity of a snail. *(b)* A magnified drawing of a radula of the primitive sort seen in *(a)*, viewed side-to-side, showing the intricate shapes and multiple rows of teeth; these details are species-specific. *(c)* The telltale signs of specialized radular feeding by carnivorous snails called drills. *(d)* The highly specialized radular tooth of a fish-eating cone snail. (*b* from Hyman 1967.)

shelled animal prey. Carnivorous snails with this specialization press their cartilaginous rod, with the tooth-studded radula at the end, against the shell of a clam or similar prey item. They then rasp a hole through the shell (Figure 4.8*c*) by turning the rod back and forth around its long axis. The most extreme specialization of the radula occurs in the carnivorous cone snails we discussed in Chapter 2. In these, the radular teeth have become detachable, elongated harpoons, complete with barbs (Figure 4.8*d*). Cone snails synthesize some of the most toxic venoms known (conotoxins), and they use their highly modified teeth to inject the venoms into fish or other prey (see Figure 2.24). That's quite a difference from ancestral snails using their ancestral radular teeth to scrape algae off rocks!

TOXIC COMPOUNDS Toxic compounds incorporated into venoms or employed in other ways are involved in the feeding biology of a wide diversity of animals that attack individual food items. Predators that employ poisons to attack prey include not just cone snails but also scorpions, spiders, many snakes, and many coelenterates, such as jellyfish. The poisons used by all these predators are proteins. The most common actions of the poisons, as we saw in Chapter 2, are to inflict structural damage on cell or basement membranes, or to interfere with nerve and muscle function by combining with receptor or channel proteins.

Another way that toxic compounds play roles in predation is that they are employed as defenses by potential prey. Bees, wasps, and some skates, for instance, have defensive venoms containing protein or polypeptide active agents. Far more common are toxic or repellent compounds within the tissues of sessile prey. Sessile organisms such as sponges or plants may seem at first to be easy meals—sitting in plain view, waiting to be eaten. Many, we now realize from advances in analytical chemistry, are far from defenseless, however, because they synthesize and deposit within their tissues chemicals that discourage attack. Besides sponges and plants, other groups sharing this property include seaweeds, soft corals, and tunicates. The compounds *within the tissues* of these organisms that deter consumption are called **secondary compounds, secondary metabolites,** or **allelochemicals.** Some are merely distasteful; others disturb the digestion or metabolism of herbivores or predators. The compounds are chemically very diverse, including alkaloids, terpenes, polymerized phenolics (e.g., tannins), steroids, nonprotein amino acids, polyethers, and quinones.

Some herbivores and predators have become specialists at defeating defenses based on secondary compounds. They may, for instance, have a specialized biochemistry that enables them to detoxify defensive chemicals, or they may be able to tolerate the chemicals or rapidly excrete them. Thus the battle to eat and avoid being eaten, on both sides, is often waged biochemically as well as mechanically.

Suspension feeding is common in aquatic animals

Suspension feeding, as noted earlier, is feeding on objects, suspended in water, that are very small by comparison to the feeding animal.[5] Clams, oysters, mussels, and scallops—which are often many *centimeters* long—typically feed on suspended particles that are 5 to 50 *micrometers* in size. They thus qualify as suspension feeders. Blue whales that are 20 to 30 *meters* long feed on shrimplike krill that are 2 to 3 *centimeters* long and on copepods that are even smaller. Blue whales thus also qualify as suspension feeders. A requirement for suspension feeders is to collect large numbers of food items because each food item is so small. Although suspension feeders often display some evidence of selectivity, they typically do not single out and attack food items individually. Instead, they collect food items in numbers.

In both freshwater and marine systems, suspension feeding is very common. A factor that helps explain this is that much of the primary production in aquatic systems is carried out by algal cells suspended individually in the water. These algal cells are then often eaten by tiny animals suspended in the water. Dead algal cells and animals constitute additional suspended food material. The food items of a suspension feeder may be mobile or not, algae or animals, living or nonliving.

Besides the abundance of suspended food items in the water, there are likely additional factors that help explain why suspension feeding has evolved as often as it has. One of the most important factors is probably that suspension feeding permits animals to feed lower on food chains, where the food energy available per unit time tends to be particularly great. Ecologists find that as food is passed through the steps of a food chain, only about 10% of the energy value of the food makes the transition through each step. Feeding lower on a food chain reduces the number of steps and thereby increases the energy available.

For an animal the size of a whale, one possibility for feeding is to eat fish just somewhat smaller than the whale. Such fish might live by eating other fish just somewhat smaller than they are, and so forth. The entire food chain leading to the whale might thus look like that in Figure 4.9*a*. Note that there are four steps in this food chain between the photosynthetic organisms (the algal cells) and the prey of the whale (the large fish). Suppose that the population of algal cells grows and reproduces sufficiently over a period of a month to produce 10,000 units of food energy. Using the figure of 10%, we can then estimate that the population of small crustaceans

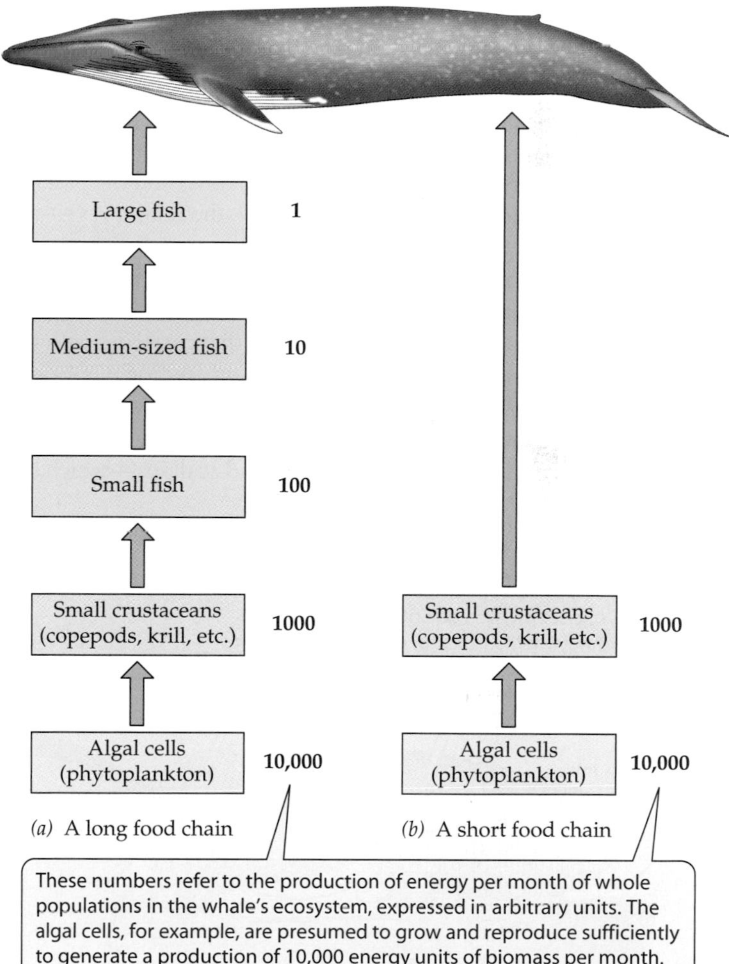

Figure 4.9 Short food chains deplete energy less than long food chains do In this thought exercise, we imagine a whale with two possible food chains. In *(a)*, the whale eats fish just a bit smaller than itself, and the food chain leading to the whale's prey is four steps long. In *(b)*, the whale eats animals that are minute by comparison to its own size, and the food chain leading to the whale's prey is one step long. The short food chain *(b)* depletes far less of the energy, prior to consumption by the whale, than the long food chain *(a)* does.

[5] Suspension feeding can in principle occur in air; in fact, some terrestrial animals, such as certain spiders, function as suspension feeders. However, little if any primary production occurs in the atmosphere, and potential food items for suspension feeders are sparse in air. Thus, suspension feeding is far more common in aquatic animals than terrestrial ones.

will grow and reproduce enough to produce 1000 units of food energy during the month (10% of 10,000). Next, the food energy produced by the population of small fish during the month will be 100 units. That produced by the population of medium-sized fish will be 10 units, and finally, that produced by the population of large fish will be 1 unit. This single unit would be the food available for a whale population at the top of the food chain. Thus a long food chain with many steps (see Figure 4.9*a*) depletes most of the energy value of algal production before the energy reaches a form that the whales eat. In contrast, whales can function as suspension feeders and eat organisms much smaller than they are, such as the small crustaceans (Figure 4.9*b*). In this case, there is just one step between the photosynthetic algal cells and the prey of the whales. The energy *available to the whales* in a month is thus 1000 units rather than 1. Shortening the food chain in this example increases the food available to the whales by a factor of 1000!

Many of the largest animals on Earth, and some of the most productive animal populations in the world, are suspension feeders. The large individual size and high productivity observed in suspension feeders are believed to be consequences of, and evidence for, the energy advantages of suspension feeding. Blue whales, the suspension feeders we have already highlighted, are by far the largest animals that have ever lived. The two largest species of fish alive today are both suspension feeders: the whale shark (12 tons) and the basking shark (5 tons). The largest rays alive today—the manta rays—are suspension feeders. About 30% of all the biomass of fish caught commercially in the world each year consists of herrings, sardines, menhadens, anchovies, shads, and other clupeid species. The clupeid fish are stunningly productive animals, a fact that correlates with the specialization of many species for suspension feeding. Crabeater seals, which are by far the most abundant and productive seals in the world, are not actually "crab eaters" but eat small crustaceans by suspension feeding. Clams, oysters, mussels, and scallops—which are suspension feeders—can be harvested in enormous numbers because their populations are very productive.

For an animal to have a shortened food chain and reap the energy advantages of suspension feeding, the *mechanistic* challenge that it must meet is that it must be able to gather food items far smaller than it is, and do so in numbers sufficient to meet its food needs. A blue whale feeding on krill, for example, must have a mechanism of collecting from the water great numbers of objects that are individually a thousandth of its size. A rough analogy would be for a person to live by collecting objects the size of individual tomato seeds from the air.

The mechanism employed by suspension-feeding whales is based on sets of keratin-rich plates, called **baleen plates**, that hang from the upper jaw on the two sides of the head, each plate oriented approximately perpendicular to the longitudinal axis of the whale's body (Figure 4.10). In an adult blue whale, there are hundreds of these plates on each side of the mouth cavity; most are a meter or more long, and they hang like pages of a book from front to rear. The inside (medial) edge of each plate frays into strands of keratin, and the frayed strands of adjacent plates become tangled with each other. In this way, the entire inner face of the baleen array on each side of the whale's mouth consists of a thick mat of tangled keratin fibers. A suspension-feeding whale typically feeds by taking a huge mass of water into its mouth cavity and forcing the water out laterally on each side through the baleen arrays. Krill, copepods, or tiny fish in the water cannot fit through the mats of fibers. Thus these small food items are *sieved out* in huge numbers as the water passes through the fibrous mats. The whale collects the sieved organisms off the inner faces of its baleen arrays with its tongue and then swallows.

When mechanical sieving is the mechanism an animal uses to remove suspended food items from the water, suspension feeding by the animal may be termed **filter feeding** because of its resemblance

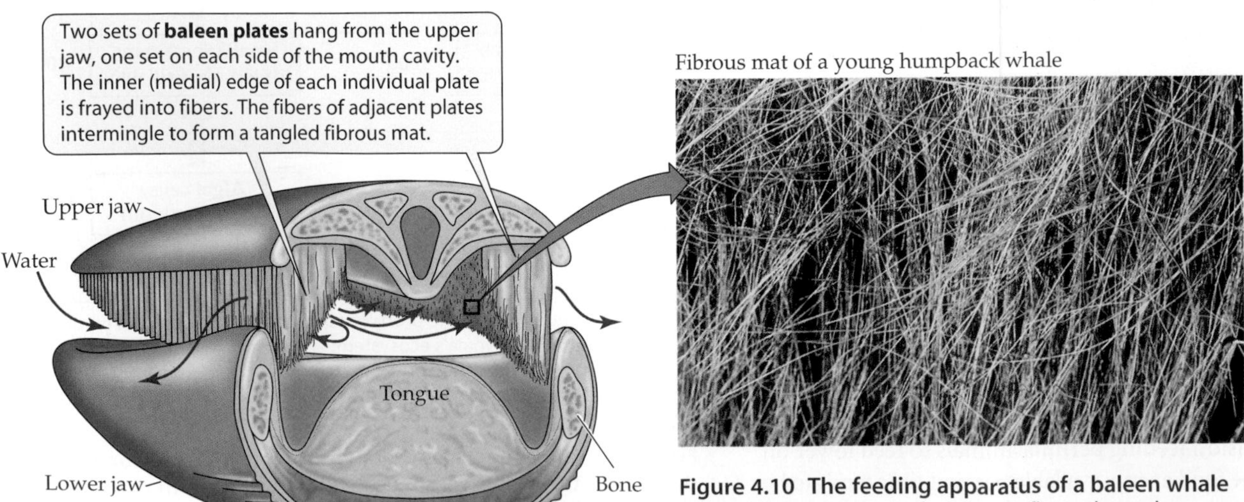

Figure 4.10 The feeding apparatus of a baleen whale In these toothless whales, water enters at the mouth, flows through an array of baleen plates on each side of the mouth cavity from the inside, and exits laterally. A thick fibrous mat on the inside face of each array of baleen plates forms a sieve. Small food items accumulate on the inside face of each fibrous mat because of sieving and are licked off by the tongue to be swallowed. (Drawing after Slijper 1979.)

Figure 4.11 Suspension-feeding mechanisms based on spines in vertebrates and invertebrates *(a)* A gill arch of a suspension-feeding clupeid fish (a shad), showing the elongated, closely spaced gill rakers used to collect tiny food items. The inset illustrates that the gill rakers of a nonsuspension-feeding fish have a very different morphology. Gill rakers are skeletal elements covered with ordinary epidermis; they are not involved in gas exchange. When fish breathe (see Figure 21.10), they pump a stream of water unidirectionally through their arrays of gill rakers and across their gills. In suspension-feeding species, the gill-raker arrays collect tiny food items from the water stream. Although sieving occurs, there is evidence in some species (e.g., menhaden) that mechanisms in addition to sieving are involved in the collection of food items by the gill rakers. *(b)* Arrays of closely spaced spines—termed *feeding setae*—used in suspension feeding by krill. Although the whole animal shown is a generalized krill, the feeding appendage is that of the Antarctic krill *Euphausia superba*. Although Antarctic krill are several centimeters long, they are able to feed (in part) directly on algal cells using their arrays of spines. Their populations are so productive that they form the principal food base for blue whales and crabeater seals.

(a) Gill rakers employed by suspension-feeding fish

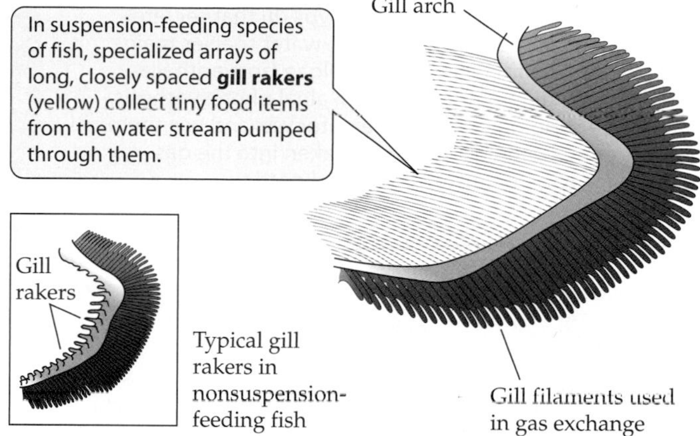

(b) Setae employed for suspension feeding by krill

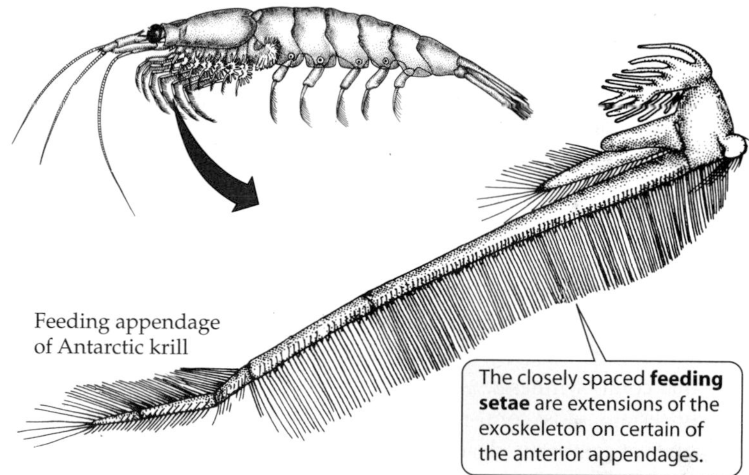

to the action of a simple filter. Whales that use baleen are filter feeders, for example. However, not all suspension feeders are filter feeders. Besides mechanical sieving, suspension-feeding devices sometimes employ electrostatic attractions, chemical attractions, targeted grasping by tiny appendages, and other processes to collect food items from the water. The term *filter feeding* applies only when mechanical sieving is the mechanism.

In suspension-feeding fish, arrays of specialized gill rakers are employed to collect food items. Although the gill rakers of most fish are relatively short and widely spaced, those of suspension-feeding fish are elongated and closely spaced (Figure 4.11*a*). The stream of water that fish force across their gills passes through the arrays of gill rakers, and small food items are sieved out. The mechanism seems to be no more than simple sieving—filter feeding—in most species of fish, but not all. Some invertebrate suspension feeders employ intricate nets woven of mucus to collect food items from the water. The nets may be internal (as in tunicates) or external (as in some stream insects). Other invertebrate suspension feeders, such as the bivalve molluscs, use cilia. Still others, such as krill, copepods, barnacles, and other sorts of small arthropods, collect food using elaborate arrays of rigid spines (setae) formed as extensions of the exoskeleton (Figure 4.11*b*).

Symbioses with microbes often play key roles in animal feeding and nutrition

Various types of animals maintain symbioses of nutritional importance with three different categories of microbes. One important distinction in categorizing the microbes is whether they are *heterotrophs* or *autotrophs*. As their sources of energy, heterotrophic microbes require organic compounds of external origin; they take in these organic compounds and break them down to release energy. Autotrophic microbes, in contrast, are able to synthesize organic molecules from inorganic precursors (e.g., CO_2 and NO_3^-) using nonorganic energy sources. Some autotrophs are *photosynthetic*; their source of energy for the synthesis of organic compounds is photon energy from the sun. Other autotrophs are *chemosynthetic* and obtain their energy from energy-releasing inorganic chemical reactions (oxidation reactions). In summary, therefore, the three categories of microbes with which animals maintain symbioses are **photosynthetic autotrophs**, **chemosynthetic autotrophs**, and **heterotrophs**.

SYMBIOSES WITH PHOTOSYNTHETIC AUTOTROPHS (PHOTOAUTOTROPHS)

Several sorts of aquatic animals obtain organic food molecules from internal populations of algae—photosynthetic autotrophs—with which they maintain symbiotic associations. After synthesizing organic molecules from inorganic precursors employing sunlight, the algae export some of the organic molecules to the tissues of their animal host, where the animal's cells use the organic molecules of algal origin as food molecules. All the animals characterized by these symbioses also have other modes of feeding. Nonetheless, the algal symbionts may be so important that their photosynthetic products are able to meet 100% of their host's energy needs.

The reef-building corals are the most famous of the animals that feed by obtaining inputs of organic compounds produced by endogenous algal symbionts (Figure 4.12). Over 500 species of warm-

Figure 4.12 The reef-building corals of warm waters need light because they are symbiotic with algae Reef-building corals *(a)* are colonies of polyps *(b)* that secrete skeletal material. The polyps of warm-water species maintain a symbiosis with dinoflagellate algae (zooxanthellae). In addition to gaining nutrition from algal photosynthesis, polyps have stinging cells with nematocysts and use them to capture small animals, which are taken into the gastrovascular cavity for absorption and digestion.

(*a*)

(*b*)

About a dozen polyps

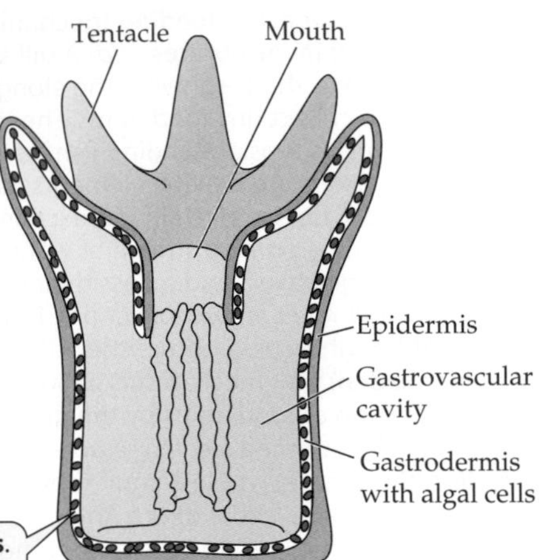

A single polyp in cross section

The algal populations live in the **gastrodermis**. Photosynthetic products from the algae pass directly to the animal cells in each polyp.

water, reef-building corals are known, and all have algal symbionts. Besides providing food energy to the coral animals, the algal symbionts metabolically facilitate formation of the calcareous skeleton that corals synthesize. Although the mechanism of this facilitation is uncertain, most coral-reef physiologists believe that healthy populations of symbiotic algae are essential for skeleton to be made rapidly enough for reefs to be formed and maintained. A topic of great current importance in coral-reef biology is the *stability of the symbiotic association.* Stresses can destabilize the association so that algal symbionts leave the coral polyps, an unhealthy condition that can lead to polyp death and reef disintegration. Corals that have lost their symbionts are called *bleached* because they no longer have chlorophyll or other algal pigments.

A survey of the animal kingdom reveals that in most cases of symbiotic associations between animals and algae, the algae belong to the taxonomic group known as *dinoflagellates* and are called **zooxanthellae** (pronounced as if the *x* were a *z*; singular *zooxanthella*).[6] The algal symbionts of reef-building corals are dinoflagellates. Similarly, the true giant clams (tridacnids), which live in coral-reef ecosystems, have dinoflagellate symbionts, as do some marine sponges, jellyfish, anemones, and nudibranchs. In freshwater animals that have algal symbionts—such as some of the freshwater hydras, sponges, and flatworms—the algae belong to taxonomic groups other than the dinoflagellates. During the reproduction of animals that have algal symbionts, each new animal organism must acquire the symbionts. Sometimes, as exemplified by many corals, mothers include algal cells in eggs, so the symbionts are present from conception. In other cases, exemplified by the giant clams, the young start without symbionts and become "infected" with them during early development.

All animals that depend on algal symbionts for nutrition must ensure that their symbionts receive adequate light for photosynthesis. This is a major reason why reef corals and giant clams occur only in clear or shallow waters. It also helps explain why any human activity that makes water turbid tends to destroy coral reefs; water turbidity can *starve* coral polyps!

SYMBIOSES WITH CHEMOSYNTHETIC AUTOTROPHS (CHEMOAUTOTROPHS)

Until the late 1970s, chemosynthetic autotrophs were not considered to be of importance as animal symbionts. That view changed abruptly with a highly eventful expedition to the deep sea. The lightless bottom of the deep sea is generally an energy-poor habitat, and with very little food available, the animals living there tend to be small and sparse. In 1977, a group of geologists exploring deep-sea tectonic spreading centers made the stunning discovery that crowded communities of large animals often live near such centers; the animals include clams as big as quart bottles and worms as big as baseball bats. The communities are now named **hydrothermal-vent communities** because they occur at places where warm water is rising out of underwater cracks (vents) in Earth's crust.

Chemoautotrophic **sulfur-oxidizing bacteria** are the key to the profusion of animal life in the hydrothermal-vent communities. These bacteria oxidize inorganic sulfide (S^{2-}) to obtain energy for the synthesis of organic molecules. The source of the S^{2-} in hydrothermal-vent communities is *local* reduction of sulfate (SO_4^{2-}) by the heat of Earth. Seawater is rich in SO_4^{2-}. At tectonic spreading centers, seawater that seeps into cracks in Earth's crust comes into contact with hot rock of Earth's core and is heated to temperatures as high as 350°C. Inorganic chemical reactions between the seawater and rock at the prevailing high temperatures and pressures chemically reduce all the sulfur in the heated water from SO_4^{2-} to S^{2-}. Heated water has a reduced density because of its elevated temperature. Thus, water heated in the cracks of Earth's crust rises back up into the open sea, carrying S^{2-} with it in the form of hydrogen sulfide (H_2S) (**Figure 4.13**). Chemoautotrophic bacteria in the vent

[6] *Zooxanthellae* are "golden-colored" algal symbionts. Most are dinoflagellates, but some belong to other taxa of algae, such as the diatoms or cryptophytes.

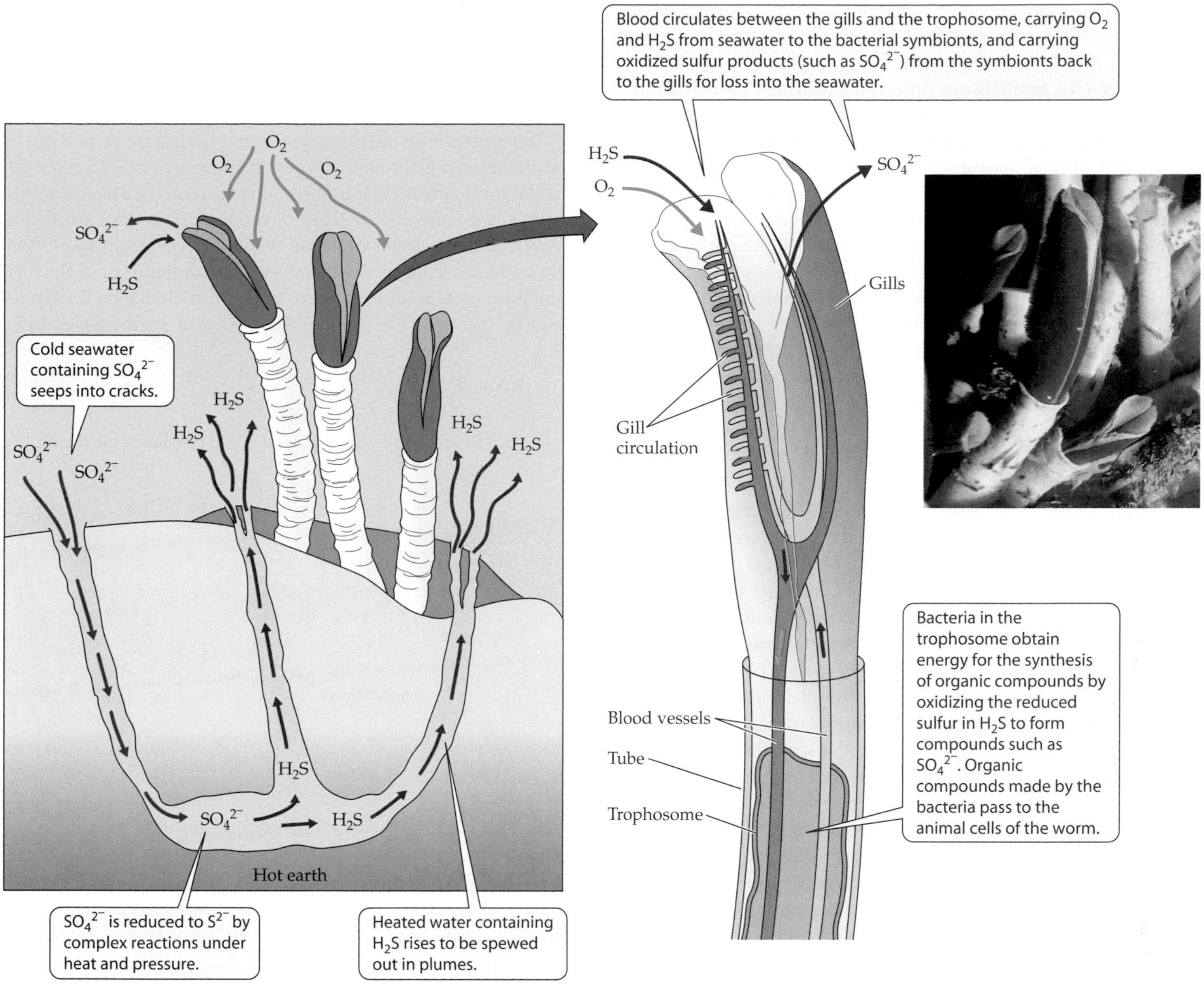

Figure 4.13 Hydrothermal-vent worms are symbiotic with chemoautotrophic bacteria The hydrothermal-vent worm *Riftia pachyptila* has no mouth, gut, or anus. Its food comes from sulfur-oxidizing chemoautotrophic bacteria that live in its trophosome. Organic molecules synthesized by the bacteria enter the animal tissues of the worm and meet the worm's nutritional needs. The worms must live near hydrothermal vents because the vents are their source of H_2S. (After Arp, Childress, and Fisher 1985.)

communities take up the H_2S, oxidize the S^{2-} to chemical forms such as SO_4^{2-}, and employ the energy released to synthesize organic molecules. Sulfur, you will note, acts as an *energy shuttle* in these communities. As seawater passes down into cracks in the seafloor and contacts the hot Earth, sulfur acquires energy by undergoing reduction. The sulfur later yields this energy when it is reoxidized by sulfur-oxidizing bacteria. In this way, the bacteria are able to draw on Earth's heat as an energy source for primary production.

Many animals in the hydrothermal-vent communities obtain organic nutrient molecules by maintaining symbiotic associations with sulfur-oxidizing bacteria. The most dramatic example is provided by the vestimentiferan worm *Riftia* (see Figure 4.13), a large worm (up to 1.5 m long) named after the undersea rifts (cracks in Earth's crust) near which it lives. *Riftia* has no mouth, no gastrointestinal tract, and no anus. It cannot ingest food! However, about a fifth of its body is filled with a tissue called the *trophosome*, where abundant populations of sulfur-oxidizing bacteria live. The blood hemoglobin of the worm transports not only O_2 but also H_2S. As the blood flows through the worm's gills, it picks up H_2S from the vent water; it then delivers it to the symbionts. The symbionts oxidize the S^{2-} and use the energy released to synthesize organic compounds, some of which enter the worm's tissues and meet the worm's food needs. Clams and snails in the hydrothermal-vent communities also have sulfur-oxidizing bacterial symbionts in their tissues (e.g., gills), although they remain able to ingest and digest foods. A key to the chemoautotrophic food base of the hydrothermal-vent communities is that the water of the deep seas is well oxygenated; O_2 is required for sulfur-oxidizing bacteria to oxidize H_2S. Animals with such bacteria as symbionts have now been discovered also in other regions of the ocean (e.g., petroleum seeps) where O_2 and H_2S occur together.

SYMBIOSES WITH HETEROTROPHIC MICROBES Although heterotrophic microbes resemble animals in requiring organic compounds from external sources, animals can potentially gain nutritional advantages from symbiotic associations with heterotrophs because such microbes often provide metabolic capabilities that the animals lack. When animals possess symbiotic heterotrophic microbes, the microbes typically occur as mixed communities of multiple microbial types, living in the lumen (the hollow central core) of the gut. In vertebrates and most other animals, the gut lumen is anaerobic (devoid of O_2). Thus the microbes are anaerobes. They are often called **fermenting microbes** to emphasize that they carry out fermentation. **Fermentation** refers to several sorts of enzyme-catalyzed reactions that occur without O_2, such as reactions that accomplish the anaerobic breakdown of compounds to liberate energy for metabolic use. Animals that maintain specialized symbiotic associations with fermenting microbes are often themselves described as **fermenters**.

VERTEBRATE FOREGUT FERMENTERS Although all vertebrates have populations of fermenting microbes in their guts, certain types of vertebrates are distinguished because they have gut chambers specialized to provide suitable habitat for microbial communities, and they have particularly large and thriving cultures of mixed microbes living in these chambers. The gut chambers that play this specialized role are typically enlarged and are places where the gut contents move through slowly. The specialized gut chambers also provide a suitable, nonacidic fluid environment for the microbial communities. In many cases the specialized gut chambers are part of the foregut, which is the part of the gut composed of the esophagus and stomach. Vertebrate species that have specialized foregut chambers housing communities of fermenting microbes are called **foregut fermenters**. The microbial communities commonly include bacteria, protists, yeasts, and fungi.

Ruminant mammals provide the most famous example of foregut fermenters. In these mammals, all of which are herbivores, the stomach has a complex structure, consisting of several compartments (Figure 4.14). The first and largest compartment, to which the esophagus connects, is the **rumen**, where communities of fermenting microbes thrive in a nonacidified setting. When a ruminant swallows grasses, leaves, or other plant parts, the microbes colonize the materials and ferment them. The acid-secreting compartment of the ruminant stomach follows the rumen; thus the contents of the rumen are subject to ordinary stomach digestion only after they are allowed to pass on from the rumen. Ruminants include sheep, antelopes, cattle, goats, camelids, moose, deer, giraffes, and buffaloes. Not all vertebrate foregut fermenters are ruminants. Others—all of which are herbivores like ruminants—include kangaroos, hippos, colobus monkeys, sloths, and a rain forest bird called the hoatzin.

The functions of nutritional importance that are carried out by the mixed microbial communities of the rumen or other foregut fermentation chamber fall into three broad categories. The first is *synthetic*. Rumen microbes are able to synthesize *B vitamins*, which vertebrate animals themselves cannot synthesize. The microbes of sheep and cattle, for instance, synthesize sufficient amounts of all the B vitamins to meet the animals' needs fully, meaning that no B vitamins need be ingested. The rumen microbes also synthesize *essential*

(*a*) An African buffalo showing the size and position of the rumen

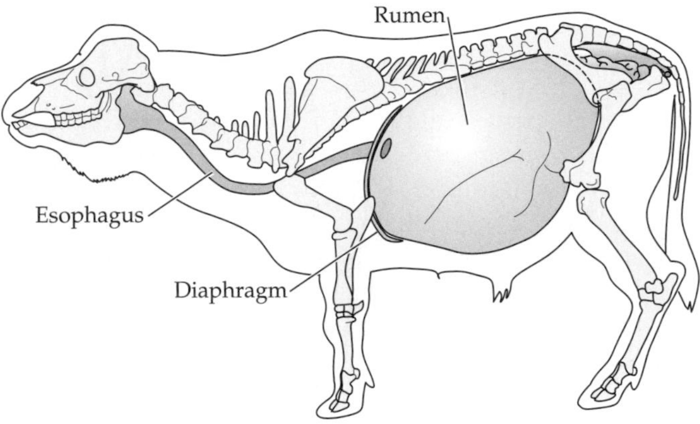

(*b*) The digestive tract of a domestic sheep

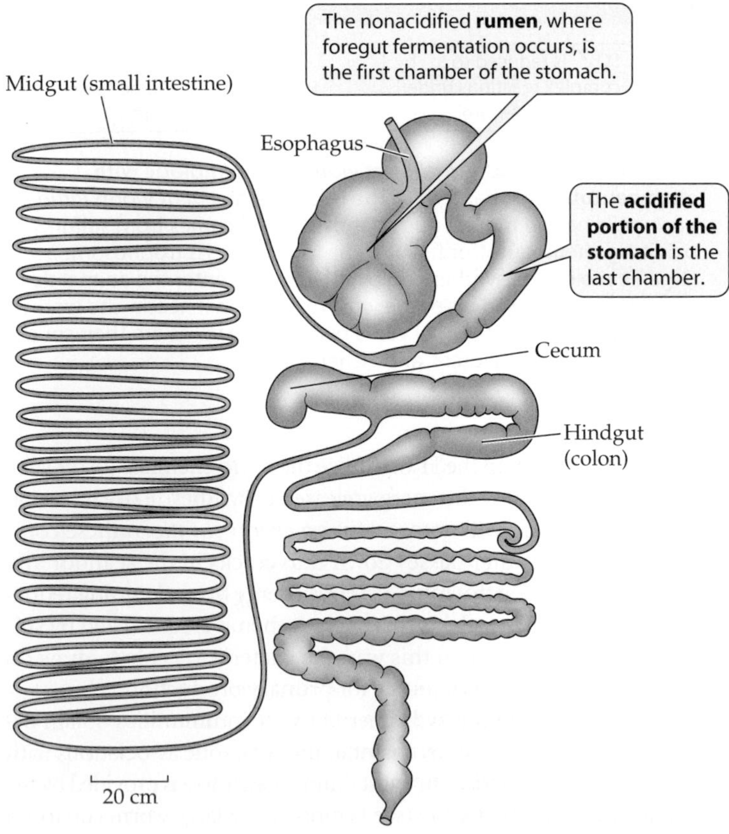

Figure 4.14 The digestive tract of ruminants *(a)* An African buffalo, showing the position and size of the rumen. *(b)* The digestive tract of a domestic sheep; the stomach of a ruminant consists of several chambers, the first of which is the rumen. To create this drawing, the digestive tract was removed in its entirety and laid out as shown. Ruminants periodically regurgitate fermenting material from the rumen into the mouth, break the material up by further grinding using their teeth, and reswallow it. (*a* after Hofmann 1989; *b* after Stevens 1977.)

amino acids. As the microbial cells grow and divide in an animal's rumen, they synthesize their own cellular proteins from nitrogenous and nonnitrogenous compounds in the animal's food. Some of the amino acids they synthesize to make their cellular proteins are essential amino acids for the animal. When the microbial cells later pass down the animal's digestive tract to the acid stomach and intestines, their protein constituents are digested by the animal's digestive enzymes, and the amino acid products of digestion are absorbed for use in the animal's metabolism. The essential amino acids obtained by the animal in this way reduce its requirement for essential amino acids in its ingested food.

The second major function carried out by the microbes of foregut fermenters is the *fermentative breakdown of compounds that the animal cannot digest*, notably cellulose and some of the other structural carbohydrates of plant cell walls, such as hemicelluloses. As we noted earlier, cellulose is one of the most abundant organic compounds in the biosphere and thus a major potential source of energy; yet vertebrates are unable to digest it, so —left to their own devices—they allow it to pass entirely into their feces. Some of the microbes in gut microbial communities produce a complex of enzymes, called **cellulase**, that breaks down cellulose into compounds that their animal host can absorb and metabolize. **Short-chain fatty acids** (**SCFA**s), sometimes called **volatile fatty acids**, are the principal useful end products of the microbial breakdown of plant structural carbohydrates. SCFAs are fatty acids, structurally similar to those in Figure 4.3a, that consist of relatively few carbon atoms; they include acetic acid (two carbons), propionic acid (three), and butyric acid (four). SCFAs are readily absorbed and metabolized by animals. Other products of the fermentation of structural carbohydrates include CO_2 and methane (CH_4), which in foregut fermenters are voided to the atmosphere by retrograde passage up the esophagus. Methane is a potent greenhouse gas, and as absurd as it may seem, the collective burps of proliferating populations of cattle around the world are a factor in the worrisome global heat budget.

The third noteworthy function of the microbial communities of foregut fermenters is actually related to the first. The microbial communities *permit waste nitrogen from animal metabolism to be recycled into new animal protein rather than being excreted*. To see how this occurs, consider first that mammals incorporate nitrogen from their metabolic breakdown of proteins mostly into a waste compound called *urea* (see page 750). We ourselves and many other species excrete the urea in our urine. In ruminants and other foregut fermenters, however, the urea can diffuse from the blood into the fermentation region of the gut, where certain of the microbes make ammonia (NH_3) from it. Other microbes are then able to use the NH_3 as a nitrogen source for the *synthesis of proteins*. When the microbial proteins are later digested, the animal obtains amino acids, which it uses to synthesize animal proteins. In this way, nitrogen—which is potentially a very valuable asset—is saved from excretion and recycled into the animal's tissues.

VERTEBRATE HINDGUT AND MIDGUT FERMENTERS Some herbivorous vertebrates, called **hindgut fermenters**, are specialized to support communities of mixed, fermenting microbes in their hindgut, the part of the gut corresponding to the mammalian large intestine. Many species of mammals, for example, are hindgut fermenters. These mammals have an enlarged cecum or enlarged colon—or both—where thriving microbial communities reside (Figure 4.15). The mammalian hindgut fermenters include rabbits, horses, zebras, rhinos, apes, elephants, and some rodents. Hindgut fermentation in the cecum or colon—or both—occurs also in many ground-dwelling herbivorous birds—ostriches, geese, grouses, and chickens—and some herbivorous lizards and turtles. Quite a few herbivorous fish are classed as **midgut fermenters** because part of the midgut (the gut between the foregut and hindgut) serves as the principal fermentation chamber. These fish include tilapias, carps, and catfishes—all noted for their usefulness in aquaculture.

The functions of microbial communities in hindgut fermenters differ from those in foregut fermenters because the hindgut is posterior to the stomach and the small intestines, meaning (among other things) that microbes from hindgut communities are not automatically digested. One function occurs in hindgut fermenters in much the same way as in foregut fermenters—namely, the microbial breakdown of cellulose and other structural carbohydrates to form short-chain fatty acids used in animal metabolism. SCFAs, in part because of their small molecular sizes and other unique properties (they are both water- and lipid-soluble), readily cross the gut wall to enter an animal's blood wherever they are produced in the gut. Thus the SCFAs produced by hindgut or midgut fermentation of structural carbohydrates are readily available to the animal.

On the other hand, the roles of microbes in supplying B vitamins, supplying essential amino acids, and recycling nitrogen are not fulfilled in the same way in hindgut fermenters as in foregut fermenters—if they occur at all. Let's first consider the B vitamins. To be absorbed, they require specialized absorption mechanisms, present mainly in the small intestines. The B vitamins synthesized by hindgut microbes do not pass through the small intestines and thus tend mostly to be lost in an animal's feces. Similarly, for an animal to gain from microbial synthesis of essential amino acids (or nitrogen recycling), the cellular proteins of the microbes must be digested. When microbial growth occurs in the cecum or colon, the microbes are eliminated in the feces, undigested. As odd as it may seem at first, a *common* solution to the problems just described is for hindgut fermenters to eat defecated material. Some eat ordinary feces; others (such as rabbits) eat special soft feces derived from the contents of the cecum. In either case, the animals ingest microbes and microbial products in defecated material just like other foods and thus can digest and absorb them.

INVERTEBRATE SYMBIOSES WITH HETEROTROPHIC MICROBES A variety of invertebrate animals also maintain critical symbiotic associations with heterotrophic microbes in their digestive tracts. The "lower" termites are the most famous example. Because termites are wood eaters, much of the nutritional value of their food lies in cellulose. Some invertebrates, unlike vertebrates, synthesize cellulose-digesting enzymes; certain insects, for example, are entirely self-sufficient in their capacity to digest and metabolize cellulose. More commonly, however, both insects and other invertebrates rely on microbial symbionts to aid cellulose breakdown. The lower termites maintain communities of flagellated, anaerobic protists in their hindguts. These protists ferment cellulose to acetic acid, which is used by the termites. In a similar vein, scarabid beetles—a very large and often ecologically important

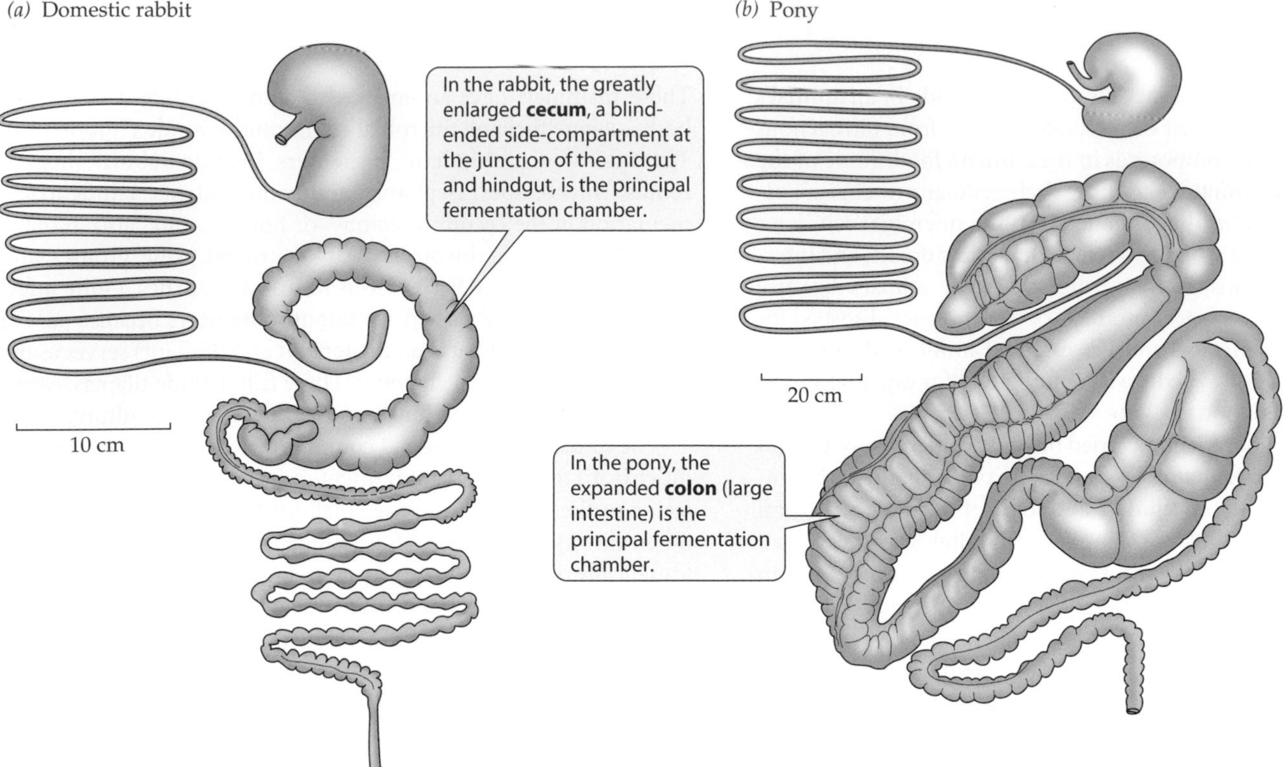

Figure 4.15 The digestive tracts of two hindgut fermenters *(a)* A rabbit. *(b)* A pony. Both species have simple, one-chambered stomachs. Their fermentation chambers are parts of the hindgut. See Figure 4.14 for labeling of other parts of the digestive tract and for information on the format of the drawings. Rabbits produce hard and soft feces. The hard feces are the familar "rabbit pellets." The soft feces, which are less familiar, consist of material from the cecum that is enclosed in a mucous membrane and that passes through the full length of the hindgut without being altered. Soft feces are eaten as they emerge from the anus. (After Stevens 1977.)

group of insects—depend on communities of gut bacteria to ferment cellulose to SCFAs. Microbial symbionts of invertebrates may also provide B vitamins, essential amino acids, and nitrogen recycling. For instance, one of the symbionts of shipworms—highly modified bivalve molluscs that eat wood—is a bacterium that not only degrades cellulose but also synthesizes protein from scratch by using atmospheric N_2 as a nitrogen source!

Some roles of microbial symbionts in invertebrates are novel by comparison to their roles in vertebrates. Sterol synthesis in arthropods provides a prominent example. Arthropods are peculiar in that, unlike most animals, they are unable to synthesize required precursors for making sterols. Sterols or sterol precursors are therefore *essential* nutrients for arthropods. Microbial symbionts commonly supply many or all of the sterol precursors in insects. Another distinctive role for symbionts in invertebrates occurs in those that feed on vertebrate blood—such as leeches, tsetse flies, ticks, and sucking lice. The bloodsucking animals seem to have a universal requirement for particular sorts of bacterial symbionts, which help them digest blood and may produce antibiotics that prevent the decay of blood during processing.[7]

[7] Similar symbionts are also required by vampire bats.

Summary: Feeding

- One mode of feeding is for animals to target and ingest individual food items, such as when orcas or cone snails catch fish. Toxic compounds, such as venoms and secondary compounds, often play roles as weapons or defenses in feeding of this sort.
- Suspension feeding is a second major mode of feeding. In this mode, an animal feeds on living or nonliving food items, suspended in water, that are individually tiny in comparison to the animal and that typically are collected in numbers rather than being individually targeted. Suspension feeding permits animals to feed lower on food chains and thus gain access to higher food productivity. Many of the largest and most productive animals on Earth are suspension feeders.
- Symbiotic associations with microbes are a third major mode of feeding in animals. Some animals maintain symbiotic associations with algae; warm-water reef corals, which are symbiotically associated with dinoflagellate zooxanthellae, are prominent examples. Other animals, notably species in hydrothermal-vent communities, are symbiotically associated with chemosynthetic autotrophs. Many herbivorous animals, both vertebrate and invertebrate, maintain symbiotic associations with heterotrophic, anaerobic fermenting microbes.
- Fermenting microbes most commonly provide three sorts of nutritional advantages to their animal hosts. They synthesize vitamins and essential amino acids; they break down structural polysaccharides such as cellulose so that the host can gain food value from them; and they permit waste nitrogen to be recycled for use in animal protein synthesis. Although all three functions are straightforward in foregut fermenters, some of the functions require consumption of feces by hindgut fermenters.

Digestion and Absorption

Of all the topics in physiology, digestion and absorption are often the first to be studied in detail in a student's education. Many students learn about the parts of the digestive tract and the major digestive enzymes on several occasions before studying physiology at the university level. These early treatments rightfully emphasize humans and other vertebrates. A problem can arise in the *comparative* study of animal digestion and absorption, however, because vertebrate anatomy and physiology do not provide adequate models for understanding the anatomy and physiology of many other animal groups. The comparative study of digestion and absorption, therefore, calls for a particular effort to be open-minded.

Digestion is defined to be the process of splitting up ingested food molecules into smaller chemical components that an animal is capable of distributing to the tissues of its body. For example, digestion splits up ingested proteins—which typically cannot be absorbed and distributed throughout the body—into amino acids that can be taken up from the digestive tract and distributed. **Extracellular digestion** is digestion in an extracellular body cavity, such as the lumen of the stomach or intestines. It is the principal mode of digestion in vertebrates, arthropods, and many other animals. In **intracellular digestion**, food particles are taken into specialized cells prior to digestion, and digestion occurs within the cells. Digestion is principally intracellular in sponges, coelenterates, flatworms, and some molluscs.

In human physiology, *absorption* is defined to be the transfer of products of digestion from the lumen of the gastrointestinal tract to the blood or lymph. Because digestion is almost entirely extracellular in humans, digestion and absorption occur sequentially: digestion first, absorption second. However, these organizing concepts are inadequate for the animal kingdom as a whole. When digestion is predominantly intracellular, *absorption* is defined to be the transfer of food particles from the gut lumen into the cells that digest the particles intracellularly, and only later are digestive products passed to the blood. *Absorption* of small molecules dissolved in seawater, such as amino acids and B vitamins, may even occur directly from the seawater into the bloods of marine animals, without being preceded by ingestion or digestion. Defined generally, **absorption**—also called **assimilation**—is the entry of molecules into the living tissues of an animal from outside those tissues.[8]

To understand digestion and absorption more thoroughly, our first step will be to examine the major features of the digestive-absorptive systems in some of the principal groups of animals. Continuing caution about terminology will be required during this study. Biologists have often tried to apply a single set of anatomical terms—typically drawn from human anatomy—to the anatomy of digestive systems throughout the animal kingdom. Thus, many very diverse digestive systems are described as consisting of the parts we ourselves have: esophagus, stomach, small intestine, pancreas, and so forth. This use of human terms can be extremely misleading because the *functional* properties of organs in invertebrates sometimes differ dramatically from those of like-named organs in humans. One must be very cautious about drawing *any* conclusions about function from names alone.

[8] The lumen of an animal's gut is continuous with the ambient air or water. Thus the gut lumen is outside the animal, and materials pass into the animal's living tissues when they leave the lumen.

Vertebrates, arthropods, and molluscs represent three important digestive-absorptive plans

Modern morphologists recognize four sequential segments of the tubular digestive tract in vertebrate animals: **headgut, foregut, midgut,** and **hindgut** (see Figures 4.14 and 4.15). The headgut of a vertebrate consists of the parts of the digestive tract in the head and neck, such as the lips, buccal cavity, tongue, and pharynx. The principal functions of the headgut are to capture and engulf food, and to prepare the food for digestion. The preparation of food by the headgut is quite minimal in some vertebrates, such as many predatory fish, which swallow food quickly and whole. In other vertebrates, preparation is extensive and involves chewing or grinding, addition of lubricating agents, and addition of digestive enzymes.

The foregut of a vertebrate consists of the parts of the digestive tract between the headgut and the intestines. Thus the foregut consists of the esophagus and stomach, and sometimes a crop (storage chamber) or gizzard (grinding chamber). The function of the esophagus is to move food from the headgut to the stomach. With few exceptions, the stomach of a vertebrate secretes acid (HCl) and pepsins, a set of protein-digesting enzymes. The vertebrate stomach stores ingested food, initiates protein digestion, and breaks up food by a combination of muscular, acid, and digestive-enzyme effects.[9]

The midgut and hindgut are the first and second segments of the intestines. In humans, they are distinguished by diameter: The midgut is of smaller diameter and thus termed the *small intestine*; the hindgut is of larger diameter and called the *large intestine*. The two segments are of similar diameter in many vertebrates; thus the distinction between "small" and "large" often does not apply. The midgut of a vertebrate is the principal site of digestion of proteins, carbohydrates, and lipids. It is also typically the principal site of absorption of the products of digestion of all three categories of foodstuffs, as well as vitamins, minerals, and water. The main functions of the hindgut are to complete the absorption of needed water and minerals from the gut contents prior to elimination and to store wastes between defecations. The removal of water and minerals by the hindgut may sound like a humble function, but it is vitally important. In humans, for example, about 7 L of water—along with Na^+, Cl^-, and other major ions—is secreted from the blood into the gut per day to facilitate processing of food. Most of this water and electrolytes must be reclaimed, as evidenced by the life-threatening consequences of diseases that interfere with reabsorption, such as cholera.

Besides the headgut, foregut, midgut, and hindgut, the vertebrate digestive system includes two additional parts that secrete important materials into the midgut: the **pancreas** and the **biliary system**. The pancreas secretes many digestive enzymes into the midgut. The biliary system, which is a part of the liver, secretes bile, which plays a crucial emulsifying role in the digestion of lipids.

[9] In this general overview, we cannot address the full range of vertebrate variation. Thus, for example, the special stomach features of ruminants—discussed already—are not repeated here. Nor are the special hindgut features of hindgut fermenters.

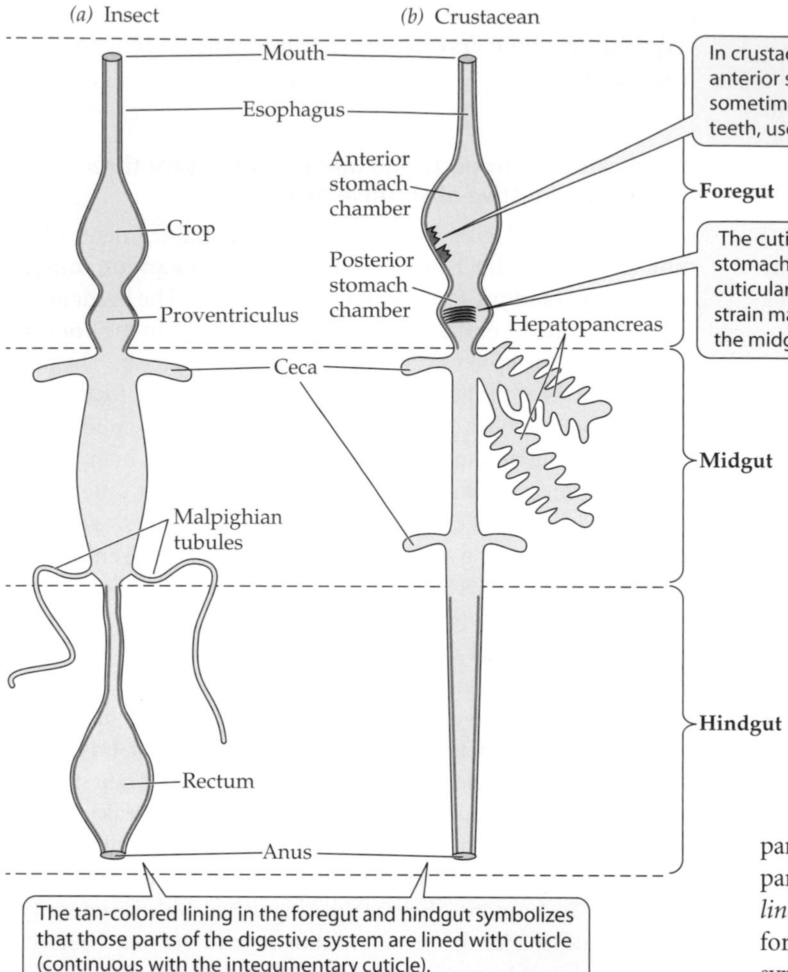

Figure 4.16 The digestive systems of two types of arthropods: insects and crustaceans *(a)* An insect. *(b)* A crustacean. Although the excretory system of crustaceans is separate from the digestive system, in insects the Malpighian tubules—which constitute the initial part of the excretory system—connect to the digestive system at the junction of the midgut and hindgut; thus the insect hindgut is actually part of both the digestive system and the excretory system.

Muscular action is the principal mechanism by which food and other materials are moved during passage through the digestive systems of vertebrates. The muscles of the gut wall, which are composed of mostly smooth muscle cells (see page 484), are arrayed in two primary layers: an outer layer of *longitudinal* muscles that *shorten* the gut when they contract, and an inner layer of *circular* muscles that *constrict* the gut when they contract. **Peristalsis** is one of the gut's principal modes of muscular activity. It is a highly coordinated pattern of contraction in which constriction of the gut at one point on its length initiates constriction at a neighboring point farther along the gut, producing a "wave" of constriction that moves progressively along the gut, propelling food material before it. **Segmentation** is a second important mode of muscular activity, in which constrictions of circular muscles appear and disappear in patterns that push the gut contents *back and forth*. The control of peristalsis and segmentation is intricate but depends principally on the *enteric nervous system* (see page 274), a network of nerve cells within the gut walls. **Gut motility** is a general term that refers to peristalsis, segmentation, or other muscular activity by the gut in vertebrates or invertebrates.

THE DIGESTIVE SYSTEM OF ARTHROPODS Arthropods deserve special attention in a survey of digestive and absorptive systems because they constitute the most species-rich and numerous phylum of animals (Figure 4.16). The two principal groups of arthropods are the insects and the crustaceans (including crabs, shrimp, crayfish, and lobsters). Digestion in insects and crustaceans is principally extracellular. Another similarity to vertebrates is that in both groups, food is moved through the digestive tract by muscular contraction.

Morphologists describe the digestive tract of an insect or crustacean as consisting of a **foregut**, **midgut**, and **hindgut**, although the terms are not intended to suggest homology or even functional analogy to the like-named parts of vertebrates. The foregut and hindgut of an arthropod are the parts of the digestive tract at the anterior and posterior ends that are *lined with a thin chitinous layer*. Like the outer integument, the foregut and hindgut have an ectodermal developmental origin and synthesize chitinous exoskeleton material, called *cuticle*. The midgut of an insect or crustacean is of different (endodermal) developmental origin and lacks a cuticular lining.

Insects (see Figure 4.16*a*) commonly have an enlarged storage chamber, termed a *crop*, in the foregut. Where the foregut meets the midgut in some species, the foregut also includes a muscular chamber, called a *proventriculus* or *gizzard*, which acts as a grinding organ. The midgut is the principal site of both digestion and absorption in an insect. Numerous digestive enzymes are secreted into the midgut, which is maintained at a near-neutral pH. The midgut typically bears anterior diverticula—called *ceca* (singular *cecum*)—that help with absorption and sometimes house microbial symbionts. The excretory tubules of insects, called *Malpighian tubules*, empty into the gut at the junction of the midgut and hindgut. Thus, urine joins other gut material at that point and travels through the hindgut. Because the hindgut plays a crucial role in adjusting the final composition and quantity of the urine, it is part of both the excretory system and the digestive system of an insect. Some nutrient absorption occurs in the hindgut.

The foregut of crustaceans (see Figure 4.16*b*) consists of a tubular esophagus followed by a saclike structure called the *stomach* (*proventriculus*). The stomach is often divided into two chambers. The anterior ("cardiac") stomach chamber is particularly muscular, and often part of it is specialized as a *gastric mill*, having its cuticular lining thrown into ridges with cutting edges. The gastric mill grinds food, frequently with the aid of digestive enzymes that arrive in the

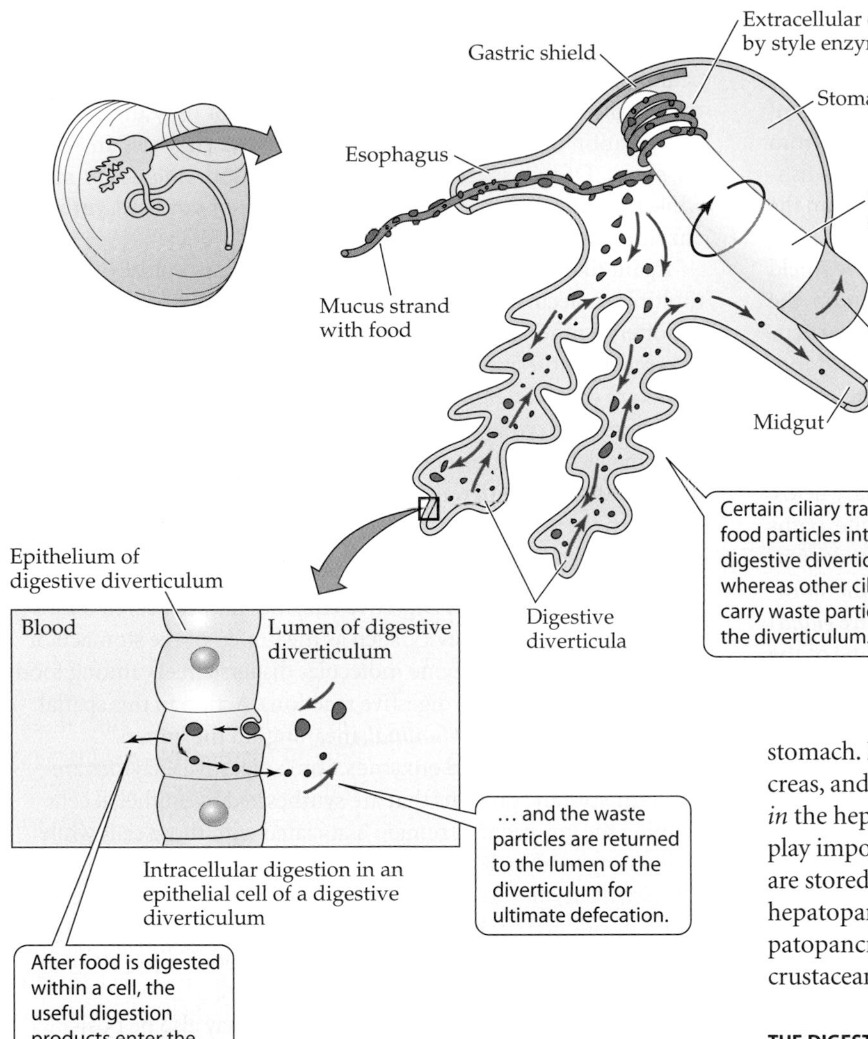

Figure 4.17 The stomach of a clam In bivalve molluscs, such as clams, the stomach and associated digestive diverticula are the most critical parts of the digestive system. In many species, digestion occurs primarily in epithelial cells of the digestive diverticula. Certain ciliary tracts bring food particles to those epithelial cells from the stomach; other ciliary tracts carry waste particles from the cells back to the stomach. The digestive tract inside the clam is presented diagrammatically, not realistically.

anterior stomach chamber by retrograde movement from the midgut. The second stomach chamber of a crustacean—the posterior ("pyloric") chamber—is smaller than the first. Its cuticular lining often bears bristlelike *setae* (singular *seta*), which act collectively as a sieve, keeping food particles from leaving the foregut until the gastric mill has reduced them to small size. Acid conditions are created in the stomach of a crustacean during digestion, but not nearly so acid as in the human stomach.[10]

The midgut of crustaceans varies in complexity from being a fairly simple tube in some species to bearing elaborate arrays of anterior and posterior diverticula (ceca) in others. The **hepatopancreas**, an organ of enormous importance in crustacean digestion, connects to the anterior midgut via ducts. The hepatopancreas is poorly named because its name greatly exaggerates any similarity it bears to the vertebrate liver and pancreas. The hepatopancreas consists of a branching array of blind-ended tubules lined with epithelium. One of its several functions is to secrete digestive enzymes that flow by way of its ducts into the midgut and then retrograde to the stomach. Food particles from the midgut also *enter* the hepatopancreas, and much extracellular digestion and absorption occur *within* the hepatopancreas. In addition, the cells of the hepatopancreas play important storage and sequestration roles; lipids and glycogen are stored in the hepatopancreas, and toxins may be sequestered in hepatopancreas cells. Current evidence indicates that the hepatopancreas is both the principal source of digestive enzymes in crustaceans and the principal site of nutrient absorption.

[10] The pH is 4 or higher in crustaceans, but less than 2 in humans.

THE DIGESTIVE SYSTEM OF BIVALVE MOLLUSCS The bivalve molluscs—which include clams, mussels, oysters, and scallops—provide an outstanding example of how very different from vertebrates some animals can be in their digestive and absorptive processes. The bivalves are also worth considering because many species are very productive and are of great ecological or commercial importance.

The gross morphology of the digestive tract in bivalves is not particularly remarkable. Bivalves are said to have an **esophagus**, **stomach**, **intestine**, **rectum**, and so forth. Striking differences from the vertebrates and arthropods come to light when details are examined, however. First, bivalve molluscs *depend strongly on ciliary action, rather than muscular contraction, to move food through their digestive tracts.* This is important in part because ciliary action is capable of *sorting* particles according to size and other features. A second major difference is that in many bivalves, *digestion is primarily intracellular.*

The key to understanding bivalve digestion is to understand the complex array of processes occurring in the stomach (Figure 4.17). One striking feature of the stomach is the presence of a long, slender gelatinous rod called the **crystalline style.** Many an oyster lover has been startled to stumble upon this unfamiliar structure while preparing fresh oysters for eating. The style has the consistency of gelatin and is as perfectly clear and colorless as ice. It is synthesized by the *style sac.* Because the protruding end of the style disintegrates during food processing, as we will see, new style material must be added regularly by the style sac. One end of the style remains sheathed by the style sac, and cilia within the sac turn the style on its long axis.

The style has several functions. One is to help pull strands of mucus, containing collected food particles, into the stomach from the mouth. It does this by winding the strands onto itself, like a fishing reel drawing in line. The end of the style that protrudes from the style sac pushes against a chitinous plate, the *gastric shield*, on part of the stomach wall. The turning of the style against the gastric shield helps break up food particles by mechanical action. In addition, the material substance of the style contains digestive enzymes, particularly amylases (starch-digesting enzymes), which are released into the stomach cavity as the turning of the style against the gastric shield causes style material to disintegrate. Once released, the enzymes carry out extracellular digestion.

The wall of a bivalve's stomach is lined with complex ciliary fields, which sort food particles and direct them along intricate paths through the diverse regions of the stomach. Ciliary action carries materials into and out of **digestive diverticula** connected to the stomach. These diverticula (often misleadingly called the *digestive gland*) are composed of branching, blind-ended tubes and form one of the most critical structures in the bivalve digestive-absorptive system. The digestive diverticula, in fact, are the principal sites of digestion and absorption in many species of bivalves. Food particles are carried into the diverticula by certain ciliary tracts. The food particles are then taken into cells lining the diverticula by phagocytosis and pinocytosis. The cells possess intracellular digestive enzymes, and much of digestion occurs intracellularly within them. The nutrient products of intracellular digestion are believed to be passed from the digestive cells of the diverticula directly into blood bathing the diverticula. Cells of the diverticula also play other roles, such as being major storage sites for lipids.

In some types of bivalves, the tubular midgut is also implicated in digestion and absorption. A bizarre and poorly understood aspect of the bivalve digestive tract is that the hindgut passes directly through the ventricle of the heart on its way to the anus!

Digestion is carried out by specific enzymes operating in three spatial contexts

In all animals, digestion is carried out for the most part by **hydrolytic enzymes**, which catalyze the breakup of large molecules into smaller parts by bond-splitting reactions with H_2O, termed *hydrolytic reactions*. In common with all other enzymes (see Chapter 2), each digestive enzyme is specific (more or less) in the type of chemical bond it can break. Because of the specificity of digestive enzymes, a species can digest only those ingested molecules for which it has enzymes that hydrolyze the specific types of bonds in the molecules. The enzyme *chitinase*, for example, is required for a species to digest chitin;[11] the enzyme initiates chitin breakdown by attacking key bonds in the large chitin polysaccharide to divide it into disaccharide (chitobiose) subunits. Whereas many species of rodents and bats synthesize chitinase, humans and rabbits do not. Thus, mammalian species differ in their ability to digest chitin. The rodents and bats that synthesize chitinase can digest it, but humans and rabbits cannot. An important corollary of the points already made is that the nutritional value of a food depends on not only the chemical composition of the food but also the digestive enzymes an animal possesses. Of two species eating the same food, one might starve while the other thrives, depending on whether or not they synthesize enzymes that can digest the food.

[11] Chitin (see page 98) is a structural polysaccharide of great importance in the exoskeletons of insects and other arthropods.

Again like other enzymes, digestive enzymes typically occur in multiple molecular forms. Different molecular forms of an enzyme are often synthesized by different species. In addition, if a digestive enzyme is synthesized by two tissues in one species, the molecular forms are often different in the tissues.

Digestive enzymes act in *three spatial contexts* in animals. An enzyme that acts in one context in one group of animals may act in another context in a different group. Thus the three spatial categories of enzyme action do not bear a one-to-one correspondence with chemical categories of enzymes.

1. **Intraluminal enzymes.** Digestive enzymes may be secreted into the lumen of a body cavity, such as the lumen of the stomach or midgut, where the enzyme molecules disperse freely among food materials and catalyze digestive reactions. Action in this spatial context is termed *intraluminal*, meaning "in the lumen."
2. **Membrane-associated enzymes.** Some digestive enzymes are cell-membrane proteins that are synthesized by epithelial cells lining the gut and that remain associated with those cells while digesting food molecules in the gut lumen. These *membrane-associated enzymes* are located in the apical membranes of the cells and positioned so that their catalytic sites are exposed to the gut lumen. Food molecules must approach the epithelium to be digested by the membrane-associated enzymes.
3. **Intracellular enzymes.** Digestive enzymes may also be positioned inside cells and act intracellularly. For food particles or molecules to be affected by these enzymes, the particles or molecules must be taken inside cells.

Whereas intraluminal enzymes and membrane-associated enzymes are responsible for extracellular digestion, intracellular enzymes are responsible for intracellular digestion. It is common for an animal to have digestive enzymes functioning in all three spatial contexts. However, the emphasis on the three classes varies from one group of animals to another, as we have already seen. In some clams, for example, the great majority of digestive reactions are carried out by intracellular enzymes. In vertebrates, by contrast, the majority of digestive reactions are extracellular, catalyzed by intraluminal or membrane-associated enzymes; only some of the final reactions of protein digestion occur intracellularly.

CARBOHYDRATE DIGESTION Disaccharides are sometimes ingested as foods, as when hummingbirds feed on sucrose-rich flower nectars, infant mammals ingest lactose in their mother's milk, or people eat candies rich in sucrose. Disaccharides are also produced in the digestive tract by the digestion of more-complex ingested carbohydrates. Enzymes that hydrolyze disaccharides into their component monosaccharides are called **disaccharidases**. They include *sucrase*—which breaks sucrose into two monosaccharides—*lactase*, *trehalase*, and others. In vertebrate animals, the disaccharidases are membrane-associated enzymes found in the apical membranes of the midgut epithelium.

In the digestion of a polysaccharide, typically two enzymes act in sequence. The first enzyme splits the polysaccharide into disaccharides or oligosaccharides (short chains of three or more monosaccharides). Then the second enzyme splits the products of the first into monosaccharides. As stressed several times already, the enzymes that carry out the initial breakup of *structural* polysaccharides are far from universally distributed in the animal kingdom. *Cellulase*, required for the initial breakup of cellulose, is not synthesized by any vertebrates, for example, meaning that vertebrates cannot tap cellulose for its food value unless they have symbiotic associations with microbes that ferment cellulose. *Chitinase*, required for the initial breakup of chitin, is synthesized by some species of vertebrates and invertebrates, but not others; thus, only some animals can obtain food value from the exoskeletons of insects and other arthropods.

Although the ability to digest cellulose or chitin is spotty, most animals are able to digest starch and glycogen, two of the principal *storage* polysaccharides. The first enzyme to act on starch or glycogen is **amylase**, which is found in most animals. Amylase hydrolyzes starch or glycogen to yield the disaccharides maltose and isomaltose, plus oligosaccharides such as maltotriose. Molecular forms of amylase are present in both the saliva and the pancreatic juice of mammals; amylase thus acts as an intraluminal enzyme in the mouth, foregut, and midgut. The disaccharides and oligosaccharides produced by amylase digestion are hydrolyzed to yield free glucose by *maltase*, a disaccharidase, and by other enzymes.

PROTEIN DIGESTION Protein digestion involves a much larger array of enzymes than carbohydrate (or lipid) digestion because the proteins include, by far, the greatest diversity of types of chemical bonds that must be hydrolyzed for digestion to occur. As with polysaccharides, the complete breakdown of a protein involves sequential action of more than one enzyme. Three or more enzymes are often required. The enzymes that digest proteins and polypeptides are categorized as **endopeptidases** and **exopeptidases.** Endopeptidases break bonds between amino acids (i.e., peptide bonds) *within* chains of amino acids; exopeptidases split off terminal amino acids from amino acid chains.

An interesting problem associated with protein digestion is that digestive enzymes targeted at proteins have the potential to attack an animal's own body substance. When protein-digesting enzymes function intraluminally, they are typically synthesized in inactive forms called **proenzymes** or **zymogens.** The enzymes are activated only after their arrival at the location where they are to act in food digestion.

Protein digestion in vertebrates usually begins in the stomach with the action of a set of intraluminal endopeptidases named *pepsins*, which require acid conditions to be enzymatically effective. Pepsins are secreted by stomach cells—and in some vertebrates by esophageal cells—as proenzymes called *pepsinogens*. Exposure to acid in the stomach lumen activates the pepsins by cleaving the pepsinogens.

When proteins and protein fragments arrive in the midgut of a vertebrate, they are subjected to further intraluminal digestion by enzymes synthesized by the pancreas. The midgut intraluminal enzymes are secreted by the pancreas as proenzymes, which are carried in the pancreatic juice to the midgut; the proenzymes are activated in the midgut by enzyme-catalyzed cleavage reactions that release the active peptidases. The pancreatic enzymes include endopeptidases, notably *trypsin*, *chymotrypsin*, *elastase*, and *collagenase* (not all are found in all species). The pancreatic enzymes also include exopeptidases known as *carboxypeptidases A* and *B*. Figure 4.18 shows how these enzymes act in concert to break up a protein.

After the intraluminal protein-digesting enzymes of the vertebrate stomach and midgut have had their effect, the resulting product is a mix of both free amino acids and short amino acid chains called *oligopeptides*. The next step in protein digestion is for the oligopeptides to be further hydrolyzed. They are digested by a great diversity of membrane-associated endo- and exopeptidases (about 20 in humans) located in the apical membranes of the midgut epithelium. Following the action of the membrane-associated peptidases, the amino acids from the original proteins are in the form of free amino acids and di- and tripeptides. These products are transported into the digestive-absorptive cells of the midgut epithelium (see Figure 2.5). There, protein digestion is completed by intracellular peptidases, which hydrolyze the di- and tripeptides. In the end, therefore, the products passed to the blood are mostly free amino acids.

LIPID DIGESTION Lipid digestion is both simpler and more complex than protein digestion. It is simpler because fewer types of chemical bonds need to be hydrolyzed, and thus fewer enzymes are required. It is more complex because although digestive enzymes are in the aqueous phase, lipids are insoluble in water. Successful lipid digestion depends not just on enzymes but also on nonenzymatic, emulsifying processes that break up lipids into small droplets having lots

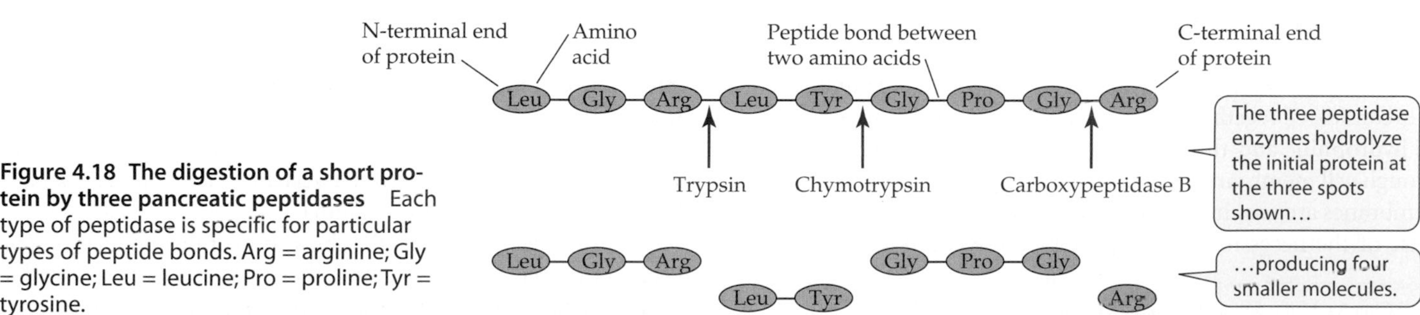

Figure 4.18 The digestion of a short protein by three pancreatic peptidases Each type of peptidase is specific for particular types of peptide bonds. Arg = arginine; Gly = glycine; Leu = leucine; Pro = proline; Tyr = tyrosine.

of surface area relative to their lipid content. The principal digestive enzymes of lipids are **lipases.** They act at lipid–water interfaces, and therefore their effectiveness depends on the surface area produced by emulsifying processes. Emulsification can be brought about by mechanical agitation or by chemical action.

The midgut is the principal site of lipid digestion in vertebrates. One reason for the vigor of lipid digestion in the midgut is the presence there of *pancreatic lipases*, which are secreted by the pancreas as active enzymes and act as intraluminal enzymes in the gut. A second important factor promoting effective lipid digestion in the midgut is that chemical emulsifying agents—namely, **bile salts** and other bile constituents—are secreted into the midgut by the biliary system of the liver. Bile salts are amphipathic molecules (see page 31) that act as detergents, breaking up lipids into droplets having a large collective surface area at which lipases can act. Interestingly, bile salts are not destroyed in the gut. Instead, they are reabsorbed downstream by active transport in terminal portions of the midgut and returned to the liver biliary system, which again secretes them upstream. Any given bile-salt molecule may in fact be recycled in this way several times during the digestion of a single meal.

The products of the digestion of fats and oils (triacylglycerols) by lipases are free fatty acids, 2-monoacylglycerols (glycerol esterified at the middle position with a single fatty acid), and glycerol (see Figure 4.3). Lipids such as phospholipids and cholesterol esters (containing ring structures) require additional enzymes, besides lipases, for digestion. These enzymes include pancreatic *phospholipases* and *esterases*.

Absorption occurs by different mechanisms for hydrophilic and hydrophobic molecules

Absorption can mean different things in different sorts of animals, as stressed earlier. Here we discuss just the type of absorption that occurs in vertebrates, arthropods, and other animals in which digestion is principally extracellular. Absorption in these animals consists primarily of the transport of chemically simple compounds—such as monosaccharides, free amino acids, and free fatty acids—across the epithelial cells lining the digestive tract, from the gut lumen into blood or lymph.

Three mechanisms of transport are involved in absorption: simple diffusion, facilitated diffusion, and active transport.[12] When nutrient molecules are actively transported, *secondary active transport* is the form of active transport involved in both vertebrates and invertebrates. Thus, in the active transport of nutrient molecules, the transporter proteins are cotransporters or countertransporters that use energy from metabolically generated electrochemical gradients of inorganic ions (e.g., Na^+) to transport nutrient molecules uphill (see page 76).

Monosaccharides, *amino acids*, and *water-soluble vitamins* are aptly considered together in the study of absorption because all three are hydrophilic and thus do not pass readily by simple diffusion through cell membranes (see page 74). Transporter proteins in cell membranes are required for these compounds to be absorbed. Because of the necessity of transporters, these compounds are absorbed in only certain parts of the gut—namely, the regions that synthesize the transporters.

The absorption of glucose in the vertebrate midgut is the best understood of all transporter-mediated absorptive systems. For glucose to be absorbed into the bloodstream across a midgut digestive-absorptive epithelial cell, the glucose must first enter the cell from the gut lumen across the apical, brush-border cell membrane, and then it must exit the cell into the blood across the basolateral cell membrane. The transport of glucose across both cell membranes is principally transporter-mediated (carrier-mediated). The best-known transporter in the apical membrane is a cotransporter named *sodium–glucose transporter 1* (*SGLT1*) that mediates the secondary active transport of glucose using energy from the Na^+ electrochemical gradient across the apical membrane. Figure 4.19 shows this transporter, and Figure 3.12 explains in detail how it mediates the active transport of glucose into an epithelial cell. Modern molecular techniques have established that the SGLT1 protein is very similar, and probably homologous, in most or all vertebrates. After glucose has entered a midgut epithelial cell by *active* transport, it leaves the cell to enter the blood across the cell's basolateral membrane by transporter-mediated *passive* transport—that is, by facilitated diffusion. The best-known transporter protein mediating this facilitated diffusion is GLUT2 (see Figure 4.19).

The absorption of other monosaccharides besides glucose by vertebrates exhibits similarities and differences relative to glucose absorption. Galactose—a monosaccharide released by the digestion of lactose ("milk sugar")—is transported into vertebrate midgut epithelial cells actively by SGLT1. But fructose—a monosaccharide product of sucrose digestion—enters the cells principally by facilitated diffusion mediated by an apical-membrane transporter, GLUT5 (see Figure 4.19). GLUT2 transports fructose and galactose, as well as glucose, out of the cells.

Amino acid absorption in the vertebrate midgut is complicated (partly because of the diversity of amino acids) and not thoroughly understood. Transporter proteins clearly play pivotal roles in amino acid absorption. Up to seven distinct amino acid transporters are found in the apical, brush-border membranes of digestive-absorptive epithelial cells. Each transporter is specialized to transport a distinct *set* of amino acids (e.g., the set of cationic amino acids) into the epithelial cells. Most of the transporters carry out secondary *active* transport. Some of these transporters draw energy from the Na^+ electrochemical gradient across the apical membranes of the cells, much as SGLT1 does. Others are dependent for energy on electrochemical gradients of other ions. Additional transporters are involved in moving dipeptides into epithelial cells and in moving amino acids out of the cells into the blood across the basolateral membranes.

Transporters are also known in the apical membranes of vertebrate midgut epithelial cells for many of the water-soluble vitamins. Active transporters driven by the Na^+ electrochemical gradient are known for some of the B vitamins, for example. Many primates (including humans), and some other vertebrates, also possess a Na^+-driven active transporter for vitamin C.

The principles that govern the absorption of the products of *lipid* digestion typically are different from those that govern the absorption of hydrophilic molecules. Fatty acids and monoacyglycerols (mono-

[12] See Chapter 3 for a detailed discussion of these transport mechanisms.

glycerides) produced by lipid digestion, because they are hydrophobic, are able to dissolve in the lipid-rich interior of cell membranes. Thus they can move through cell membranes readily by simple diffusion (see page 69). In fact, simple diffusion—unmediated by transporter proteins—often seems to account for much or all of the transport of fatty acids and monoacylglycerols into gut epithelial cells. Transporters for these compounds—some active—are known, however, and clearly they deserve more study.

To fully understand the absorption of fatty acids and monoacylglycerols, one must take into account that most are poorly soluble in the aqueous solutions inside and outside epithelial cells. Vertebrates have mechanisms to emulsify or solubilize them in both solutions. In the midgut lumen, as fatty acids and monoacylglycerols are produced by lipid digestion, they are dramatically solubilized by combining with bile salts to produce minute disc- or sphere-shaped molecular aggregates called *micelles*; a micelle is smaller than 10 nm and maintains its lipid constituents in solution by means of the emulsifying effects of the amphipathic bile salts. Lipid-soluble vitamins participate in micelle formation as well. The solubilization of fatty acids, monoacylglycerols, and lipid-soluble vitamins by micelle formation greatly aids the absorption of the molecules. Micelles themselves are not absorbed, however. Instead, the fatty acids and other molecules dissociate from micelles next to the apical membranes of gut epithelial cells, then move through the membranes by simple diffusion.

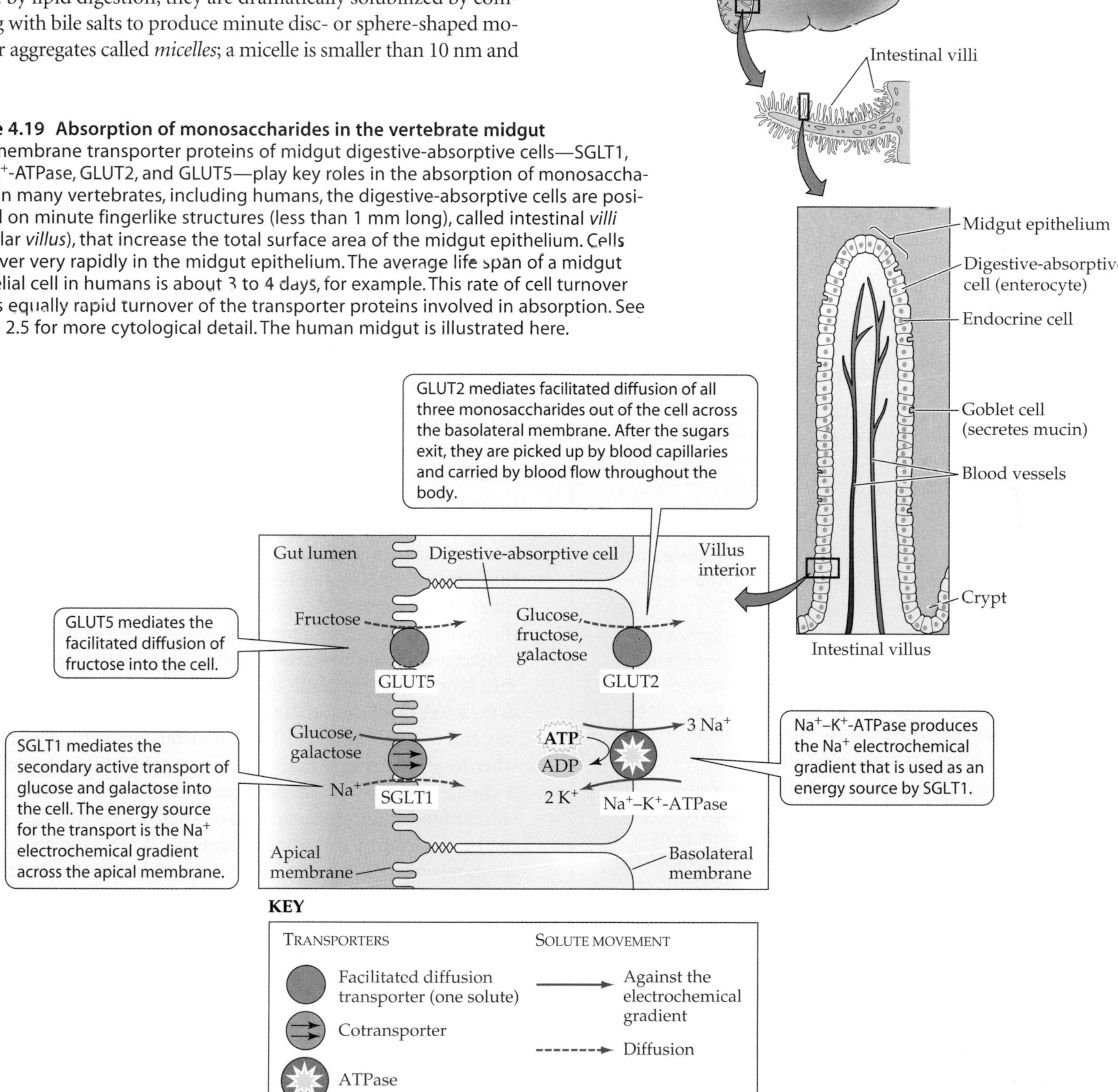

Figure 4.19 Absorption of monosaccharides in the vertebrate midgut Four membrane transporter proteins of midgut digestive-absorptive cells—SGLT1, Na^+–K^+-ATPase, GLUT2, and GLUT5—play key roles in the absorption of monosaccharides. In many vertebrates, including humans, the digestive-absorptive cells are positioned on minute fingerlike structures (less than 1 mm long), called intestinal *villi* (singular *villus*), that increase the total surface area of the midgut epithelium. Cells turn over very rapidly in the midgut epithelium. The average life span of a midgut epithelial cell in humans is about 3 to 4 days, for example. This rate of cell turnover means equally rapid turnover of the transporter proteins involved in absorption. See Figure 2.5 for more cytological detail. The human midgut is illustrated here.

After the products of lipid digestion are inside gut epithelial cells, enzyme-catalyzed metabolic pathways synthesize intracellular triacylglycerols, phospholipids, and cholesterol esters from them. These complex lipids are then combined inside the cells with proteins to form **lipoprotein molecular aggregates**, especially ones called **chylomicrons**. The formation of chylomicrons sets the stage for the lipids to be carried throughout the body by way of the lymphatic and circulatory systems, because the proteins in the chylomicrons are amphipathic and emulsify the lipids in the aqueous solution of the lymph or blood. The chylomicrons produced in a midgut epithelial cell are assembled into secretory vesicles by the cell's Golgi apparatus. The vesicles then fuse with the cell's basolateral membrane to release the chylomicrons into the midgut lymphatic circulation, which carries the chylomicrons into the bloodstream.

The absorption of *short-chain fatty acids* (*SCFAs*) produced by microbial fermentation is a special case because these particular fatty acids (e.g., acetic acid) are both water-soluble and lipid-soluble. The SCFAs do not require emulsification, because of their water solubility, and they tend to be absorbed wherever they are produced in the gut. The least-complicated model for their absorption is that they simply diffuse through cell membranes and cell interiors by being lipid- and water-soluble. Detailed studies show, however, that their absorption is sometimes transporter-mediated in ways that remain poorly understood.

Summary: Digestion and Absorption

- Digestion is the process of splitting food molecules into smaller parts that an animal can take into its living tissues and distribute throughout its body. Absorption (assimilation) is the process of taking organic compounds into the living tissues of an animal from the gut lumen or other places outside those tissues. In vertebrates, arthropods, and some other groups, digestion precedes absorption. However, absorption precedes digestion in certain other animals, such as certain bivalve molluscs.
- Digestion is carried out mostly by hydrolytic enzymes, each of which catalyzes the splitting of specific types of chemical bonds. The enzymes may be intraluminal, membrane-associated, or intracellular. Intracellular digestion is carried out by intracellular enzymes. Extracellular digestion is carried out by intraluminal and membrane-associated enzymes.
- The absorption of relatively simple hydrophilic compounds—such as monosaccharides, amino acids, and water-soluble vitamins—usually requires transporter proteins in the cell membranes involved and occurs by either facilitated diffusion or secondary active transport. The absorption of the hydrophobic fatty acids and monoacylglycerols produced by lipid digestion can occur to a large extent by simple diffusion across cell membranes. However, complexities arise in the absorption of these compounds because (being hydrophobic) they need to be emulsified in the aqueous solutions outside and inside cells while being absorbed. SCFAs are a special case because they are water- and lipid-soluble.
- In vertebrates, the midgut is typically the most important site of digestion and absorption, in part because pancreatic and biliary secretions enter the midgut and also because the apical membranes of midgut epithelial cells are richly populated by membrane-associated digestive enzyme proteins and by transporter proteins.
- An animal's digestive and absorptive capabilities are major determinants of the nutritional value of foods because ingested organic compounds can be utilized only to the extent that they can be digested and absorbed.

Responses to Eating

When you eat a meal, your gastrointestinal system responds to the arrival of food in intricate ways that enable the system to carry out its function of making the constituents of the food available to your body. These responses belong to the category of *acute responses* defined in Chapter 1 (Table 1.2). They occur within a short time after the arrival of food, and because of their timing, they are carried out by means of physiological mechanisms already existing before the meal was ingested. Cellular activities that are modified to process a meal include the secretion of digestive enzymes, secretion of acid, and contraction of muscles. Nerve, endocrine, and paracrine cells control and coordinate the responses.[13]

After chewing and secretion of saliva, the first step to consider is the transit of food from the mouth to the stomach. Each mouthful of food that we force by *voluntary* swallowing to the pharynx in the back of the mouth initiates an *involuntary* wave of peristaltic muscular contraction that passes from the pharynx along the length of the esophagus. Complexities arise from the presence of sphincters at the two ends of the esophagus. A **sphincter** is a circular muscle that can contract tightly and steadily (tonically) for long periods of time, thus preventing exchange between one segment of the gut and another. When a peristaltic wave initiated by the presence of food in the pharynx reaches the upper esophagus, the upper esophageal sphincter transiently relaxes, allowing the food to pass. The food is then propelled by peristalsis down the esophagus, and when it reaches the lower esophageal sphincter, the latter opens transiently to permit entry to the stomach. The sphincters are kept closed except when food is passing so as to contain ingested food in the stomach.

In the human stomach, the upper (initial) part serves as a reservoir for food; it is the principal place where a large quantity of food can be housed while awaiting processing. When food arrives from the esophagus, the muscles in the walls of this part of the stomach relax in response to autonomic commands from the parasympathetic nervous system (see page 271). The lumen then can expand to accommodate the food.

[13] As discussed in Chapter 10 (see Figure 10.2), *endocrine cells* secrete hormones into the blood, which carries them to often distant target cells where they exert their effects. *Paracrine cells* secrete communicatory chemicals into the fluids between cells, where the chemicals affect other neighboring cells.

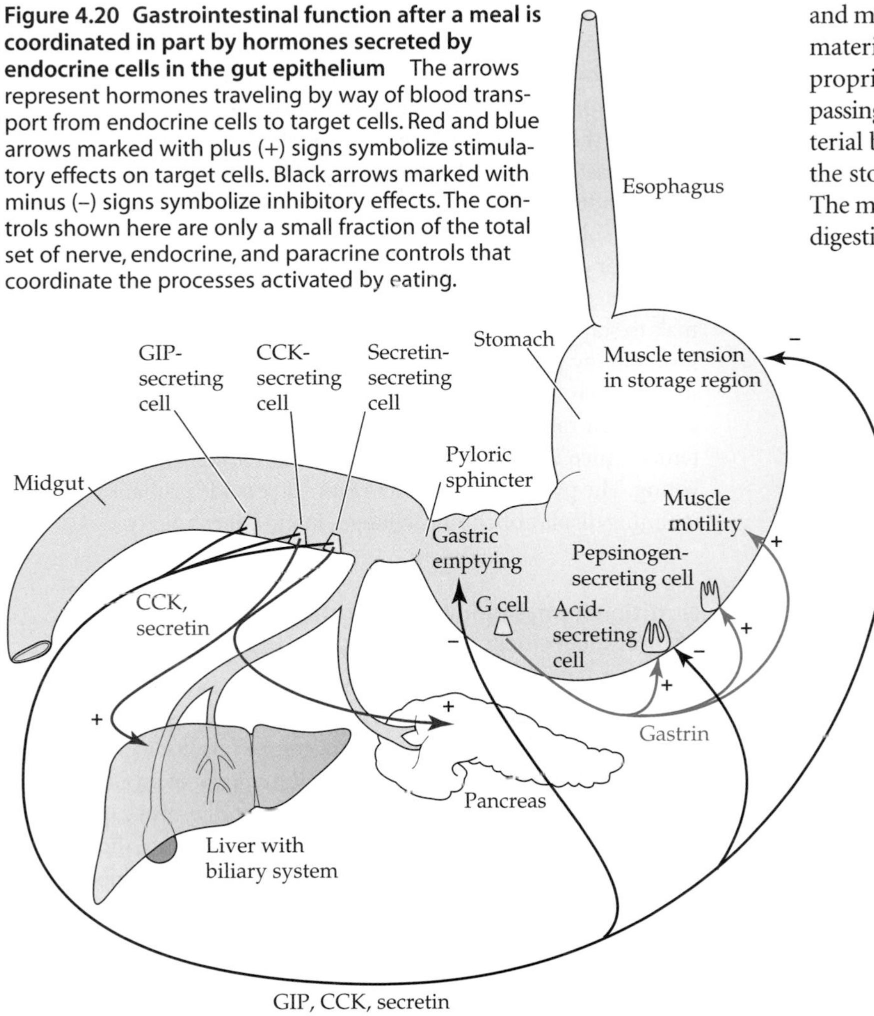

Figure 4.20 Gastrointestinal function after a meal is coordinated in part by hormones secreted by endocrine cells in the gut epithelium The arrows represent hormones traveling by way of blood transport from endocrine cells to target cells. Red and blue arrows marked with plus (+) signs symbolize stimulatory effects on target cells. Black arrows marked with minus (–) signs symbolize inhibitory effects. The controls shown here are only a small fraction of the total set of nerve, endocrine, and paracrine controls that coordinate the processes activated by eating.

The presence of food in the stomach exerts mechanical and chemical effects on endocrine cells, called *G cells*, located in the mucosal lining of the lower (final) part of the stomach. The G cells are stimulated to secrete into the blood a polypeptide hormone called **gastrin**. As Figure 4.20 shows, when gastrin reaches other stomach cells by way of blood circulation, it stimulates both the acid-secreting cells (see Figure 3.9) and the pepsinogen-secreting cells to carry out secretory activity, and it stimulates the muscle layers of the nonstorage part of the stomach to begin and sustain peristaltic contractions. This motility of the stomach physically mixes and breaks up food at the same time that acid and pepsins initiate digestion of the meal. Additional controls besides those exerted by G cells and gastrin are involved in modulating stomach function. For example, besides being stimulated by gastrin, the acid-secreting cells are stimulated in at least two additional ways. First, the smell and taste of food cause stimulatory signals to be passed from the brain to the acid-secreting cells by way of the parasympathetic nervous system. Second, molecules from the midgut (e.g., amino acids) reach the cells by way of the circulation and stimulate them.

The midgut (small intestine) can accommodate only a relatively small amount of material at a time. A sphincter between the stomach and midgut, called the *pyloric sphincter* (see Figure 4.20), meters material into the midgut by opening only enough to let an appropriate amount through. At the end of each peristaltic wave passing along the stomach toward the sphincter, most of the material being digested is refluxed back toward the main body of the stomach. Only a metered amount passes into the midgut. The midgut, as we have already seen, is the principal site of both digestion and absorption in the human gastrointestinal tract.

One of the most important principles of gut coordination is that endocrine cells of many sorts are found in the epithelium lining much of the gut, and the *types* of endocrine cells differ from one region of the gut to another, meaning that each region can send out distinctive hormonal messages. An illustration is provided by three classes of endocrine cells that occur in the epithelium of the upper midgut. These three classes of cells secrete three polypeptide hormones: **secretin**,[14] **cholecystokinin** (**CCK**), and **gastric inhibitory polypeptide** (**GIP**) (see Figure 4.20).

The endocrine cells in the upper midgut are stimulated to secrete their respective hormones by acidity and by exposure to nutrient molecules. Thus, when acidifed digesting food is metered into the upper midgut from the stomach, it causes release of the three hormones we have mentioned into the blood. Secretin and CCK synergistically stimulate the pancreas and the biliary system of the liver to secrete into the midgut the array of critical digestive agents we have already discussed, notably pancreatic digestive enzymes and bile. An additional critical pancreatic function stimulated by secretin is the secretion of bicarbonate (HCO_3^-), which neutralizes acid from the stomach. The midgut hormones also affect the stomach. All three inhibit gastric-acid secretion, gastric emptying, and muscle contraction in the storage part of the stomach, for example; in this way they promote retention of food in the stomach when acidity and nutrient concentrations in the midgut are high, signifying that the midgut—for the moment—already has adequate input from the stomach. Stomach function is also affected by nerve-mediated reflexes from the midgut; excessive acidity or accumulation of unabsorbed digestive products in the midgut, for instance, can inhibit stomach motility by this means.

Motility of the midgut is under control of a complex array of hormonal and neuronal influences, both stimulatory and inhibitory. Segmentation is the principal type of midgut motility. It pushes the midgut contents back and forth, thereby mixing food products, intraluminal digestive enzymes, and bile. Segmentation also ensures contact of all digesting material with the walls of the midgut, where the membrane-associated digestive enzymes and absorption transporters are found. Besides segmentation, the midgut exhibits progressive, peristaltic contractions that move material toward the hindgut.

[14] Secretin, discovered in about 1900, was the first hormone known to science.

Just as humans exhibit responses to the arrival of food, so also do all other animals. Recall, for instance, that in oysters, mussels, and clams, much of digestion occurs intracellularly in the digestive diverticula of the stomach (see Figure 4.17). Between meals, the principal cells lining the digestive diverticula become quite small. When food arrives, the cells *grow dramatically*, increasing in height as they take on food particles. They rapidly synthesize new cell membrane at this time to meet the membrane requirements of phagocytosis and pinocytosis. Certain cells, which lack flagella between meals, grow flagella when food is present, permitting them to circulate fluid and food particles within the diverticula by flagellar action. All these changes reverse after food has been processed.

Changes in Digestion and Absorption in Additional Time Frames

The digestive and absorptive physiology of animals is adjusted—often in adaptive ways—in all of the five time frames emphasized in Chapter 1 (see Table 1.2). In the preceding section we looked at *acute* responses displayed shortly after eating. Here we look at changes in digestion and absorption that occur in the other four principal time frames: chronic, clock-driven, developmental, and evolutionary.

The nutritional physiology of individuals responds chronically to changed conditions

Individual animals commonly restructure their digestive-absorptive systems when they are exposed to new sets of environmental conditions for extended periods of time. The sugar-transporter proteins in the midgut epithelial cells of a person (see Figure 4.19) provide an excellent first example. The numbers of molecules of SGLT1, GLUT2, and GLUT5 *per cell* are not static in an adult person. Instead, the numbers of these molecules per cell become relatively low if a person eats little carbohydrate for many days. Then, if the person changes his or her diet so that a lot of carbohydrate is ingested for several days, the numbers of transporter molecules per cell increase. Thus the capacity for carbohydrate absorption undergoes chronic changes. This sort of chronic response is hardly limited to people or to sugars. Many vertebrates, ranging from fish to mammals, are known from experiments to exhibit acclimation of midgut sugar transporters of the sort just described. Midgut amino acid transporters are also upregulated or downregulated in a variety of vertebrates when dietary protein is chronically increased or decreased.

Sustained changes in diet often also bring about chronic adjustments in the quantities of digestive enzymes produced. The intraluminal digestive enzymes secreted by the pancreas in vertebrates provide dramatic examples. In a variety of types of vertebrate animals, a sustained increase in the amount of carbohydrate in the diet results in chronically increased pancreatic secretion of amylase. Similarly, sustained increases in the amounts of dietary proteins or lipids lead to chronically heightened secretion of pancreatic peptidases or lipases. Laboratory rats provide the best-studied model of these chronic responses. In a rat, changes in pancreatic enzyme secretion start within 24 hours after a change in diet and require about a week to be fully expressed. The changes can be large: Enriching a rat's diet with a particular class of foodstuff may be followed by a fivefold or greater increase in pancreatic secretion of digestive enzymes targeted at the foodstuff. Thus an animal that has been eating certain sorts of foods for a week is in a far better position to digest those foods than one that has had little recent experience with them.

The gastrointestinal morphology of an animal is also subject to chronic adjustments. When a mouse or small bird is placed chronically at subfreezing air temperatures, its intestines grow. House wrens, for example, gradually increase their intestinal length by 20% when placed at –9°C after living at +24°C. For small birds or mammals to stay warm in cold environments, they must metabolize more food energy per day than when they are living in warm environments. Their midgut growth in cold environments helps them to increase their rates of food digestion and absorption. In a similar vein, female mice increase their midgut length when they are nursing young. The pythons described in Box 4.1 provide probably the most stunning display of chronic changes in gut morphology.

Nutritional physiology is under the control of biological clocks

Animals undergo many sorts of periodic changes in their nutritional physiology under the control of biological clocks. The activity of disaccharidases in the midgut of laboratory rats, for example, rises at night and falls during the day under the control of an endogenous biological clock. The rhythm ensures greatest digestive capability at the time the rats are most likely to ingest food, because the rats feed principally at night. On a different scale of time, some hibernating species of mammals, such as certain ground squirrels, increase their ingestion and absorption of food energy—and their storage of body fat—in the autumn of each year, even when they are kept under unvarying environmental conditions in a laboratory (Figure 4.21). *Circannual* ("approximately annual") *biological clocks* control these changes in nutritional physiology. In fact, such clocks were originally discovered by research on the nutritional physiology of ground squirrels.

Nutritional physiology undergoes programmed changes during development

The digestive systems of animals commonly undergo developmentally programmed changes between birth and adulthood. An excellent example is provided by the developmental timing of the capacity for lactose digestion in mammals. The disaccharide lactose is the principal carbohydrate in the milks of most species of mammals, including humans. Digestion of lactose requires lactase, a membrane-associated midgut enzyme. Newborn mammals synthesize large amounts of lactase. Adults, however, typically produce little or no lactase. The synthesis of lactase undergoes a programmed developmental change, diminishing as young animals mature beyond the suckling stages of their lives. Simultaneously, programmed increases occur in the production of other disaccharidases, such as sucrase and maltase. The substrates of these other disaccharidases appear in the diet only as young mammals start to eat solid foods.

Amphibians provide some of the most dramatic examples of programmed developmental changes in nutritional physiology:

BOX 4.1 Pythons: Extreme Examples of Feast and Famine

One of the most extraordinary stories in the annals of eating has unfolded with research on pythons. The snakes obtain food by waiting for suitable prey animals to wander near. With this strategy, many weeks may sometimes pass between meals. One adaptation of the pythons to their feast-and-famine type of life is that they are equipped to ingest large animals if the opportunity arises. The snakes can weigh more than a person and can easily eat animals that weigh 70% as much as they do. Thus, stories of goats, antelopes, and teenagers being consumed are true.

Unlike humans and many other animals that eat on a daily basis and maintain their digestive systems in a state of readiness between meals, pythons deconstruct their gastrointestinal systems between meals. Burmese pythons (*Python molurus*), for example, undergo extensive deconstruction of their digestive apparatus if they are deprived of food for a month. Then, when they obtain food, they rapidly reconstruct the apparatus. In the first 24 hours after feeding, these pythons *double* the mass of their midgut, largely by the growth of new gut epithelium (critical to digestion and absorption). Simultaneously, they undergo massive synthesis of midgut transporter proteins; the total numbers of molecules of both glucose transporters and amino acid transporters may increase by 20 times or more! In extreme cases, the metabolic rates of pythons increase 40-fold after eating. Part of the reason for their steep increase in energy use after eating is that they need to expend energy to rebuild their digestive systems to process the food.

Whereas the tadpoles of frogs and toads are herbivores or omnivores, the adults are carnivores. The gut of a frog or toad is rapidly shortened and restructured during metamorphosis.

Nutritional physiology undergoes evolutionary change

Related species often differ in their digestive physiology in ways that are well correlated with the types of foods they eat, providing comparative evidence that digestive physiology evolves in parallel with diet. The distributions of disaccharidases in relation to disaccharides in the diet provide well-documented illustrations. For example, sucrase tends to be synthesized in abundance just by species that consume sucrose-rich foods. Although sucrose is common in our kitchens in the form of "table sugar," it is actually not a common sugar in nature; flower nectars that are rich in sucrose are relatively distinctive foods. Species of birds, bats, and insects that specialize in feeding on sucrose-rich nectars synthesize abundant sucrase, but birds, bats, and insects that do not eat such nectars usually do not synthesize sucrase in particular abundance. The distributions of trehalase and dietary trehalose parallel each other in equally striking ways. The disaccharide trehalose has a very limited distribution in the biosphere. One of the few places it is found in abundance is the blood of insects. Mammal species that do not eat insects typically synthesize little or no trehalase. By contrast, insectivorous mammals, such as insectivorous bats, exhibit high trehalase activity.

The midgut glucose transporters of vertebrates are strikingly correlated with the amounts of starch in their diets. Starches from plants yield glucose when digested. Vertebrate species that are principally plant eaters (herbivores), from fish to mammals, tend to show higher activities of midgut glucose transporters than related carnivorous vertebrates. Such differences between species could result partly from acclimatization to different diets. To test this pos-

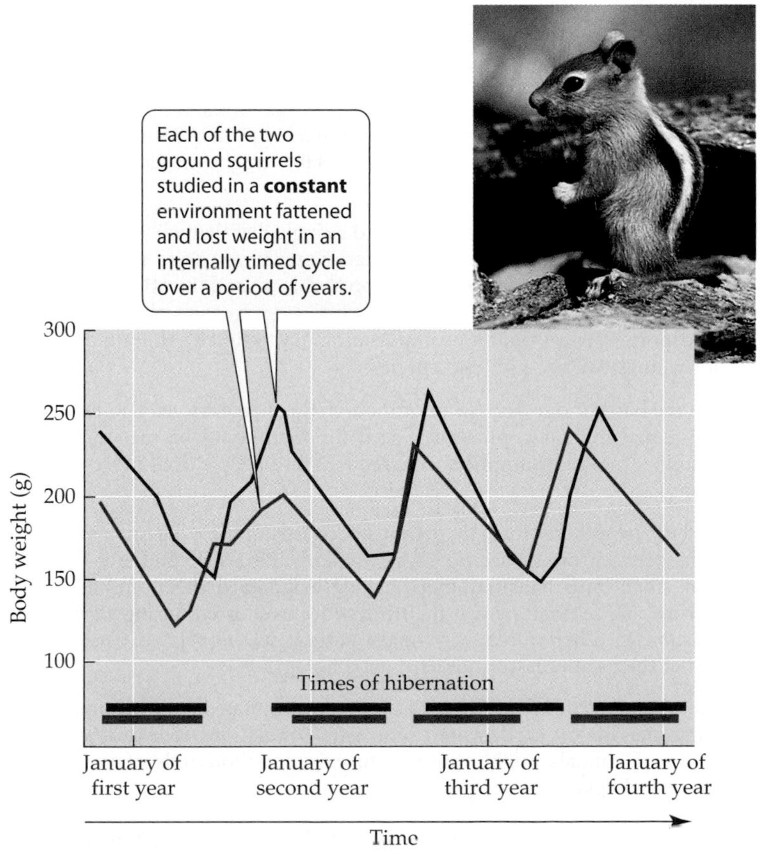

Figure 4.21 Endogenously timed annual rhythms in nutritional physiology Two golden-mantled ground squirrels (*Spermophilus lateralis*) were kept under *constant* environmental conditions in a laboratory for over 3 years, during which they received no information about the time of year. The plotted lines show how their body weights changed; each ground squirrel lost a great deal of weight when it hibernated and fattened between periods of hibernation. The bars near the bottom mark the time periods when each animal hibernated (color coded to match the plots of body weight). The changes in eating behavior, fat storage, and hibernation were under the control of endogenous timekeeping mechanisms that are called *circannual clocks* because they do not maintain exact 365-day rhythms when animals receive no information about time of year (*circa* means "about"; thus, *circannual* means "about a year"). Some other species of ground squirrels have been shown to maintain circannual rhythms in nutritional physiology for a *decade* under constant laboratory conditions. (After Pengelley and Asmundson 1969.)

TABLE 4.5 Percentages of adults who synthesize sufficient lactase to process a test dose of lactose without clinically significant difficulty in various human populations

Group	Percentage synthesizing lactase
Dutch	100
Danes	97
White Americans	76
Greeks	47
Mexican-Americans	44
Indigenous people of Australia	33
Italians	29
Afro-Americans	19
Members of the Ibo and Yoruba tribes (Nigeria)	11
Native Americans (Oklahoma)	5
Asians in the United States	0

Source: After Buller and Grand 1990.

sibility, a few experiments have succeeded in comparing related herbivores and carnivores living on a single diet in the laboratory. Under such conditions, the herbivores still exhibit greater glucose-transporter activities, indicating that evolved genetic differences exist between species of herbivores and carnivores.

A fascinating example of evolutionary change *within* a species is provided by the genetics of *adult lactase production* in people. The worldwide developmental "norm" is that humans follow the pattern typical of most mammals: Whereas babies synthesize abundant lactase, adults do not. Midgut lactase levels typically fall to 10% or less of newborn levels by the time people are 5 to 10 years old. A minority of the world's people, however, retain high lactase levels in adulthood. Adult lactase synthesis is under genetic control.[15] Human populations exhibit extreme diversity in the frequencies of the relevant genes and therefore in the percentages of adults who synthesize enough lactase to process a test dose of lactose without difficulty (Table 4.5). Nearly 100% of adults in northern European populations, such as the Dutch and Danes, synthesize lactase in abundance; in sharp contrast, only 20% or less of adults synthesize abundant lactase in many other populations, such as Africans and Asians.

Whether an ordinary meal that contains lactose will cause difficulty for a person depends on more than the person's genetic makeup. However, adults who exhibit genetically low lactase synthesis are at risk of having difficulty when they ingest lactose. If such people ingest cow's milk or milk products, the lactose from the foods may remain largely undigested (and thus unabsorbed) as it passes through their digestive systems. Thus, gut bacteria may attack the lactose as a nutrient source and produce pathological levels of bacterial metabolites. People afflicted by this and other consequences of inadequate lactose digestion are often called *lactose-intolerant*.

[15] Persistence of high lactase synthesis into adulthood, according to present evidence, seems to be inherited as an autosomal recessive genetic character.

The differences among various human populations in the frequencies of genes for lactase synthesis are, without doubt, results of past evolutionary processes (evolution is defined to be a change in gene frequencies). Some scientists argue that natural selection has been responsible for these differences; for instance, some postulate that the early advent of dairy farming in northern Europe, thousands of years ago, favored genes for abundant adult lactase synthesis in northern European populations because the genes were adaptive for people who ate lots of dairy products. However, no compelling evidence yet favors such natural-selection arguments. Thus, it remains possible that genetic drift or other processes of nonadaptive evolution caused the gene-frequency differences for lactase synthesis. Today, as disparate populations intermarry, gene frequencies are modified by the intermingling of low- and high-lactase gene pools. For this reason, if no other, we know that evolution is continuing in this significant aspect of human nutritional physiology.

Study Questions

1. Malnutrition of children is a profound global problem, and protein deficiencies are the most common form of childhood malnutrition. Why are people so vulnerable to protein deficiencies?
2. Explain how and why the absorption of hydrophobic and hydrophilic organic molecules differs.
3. List three major functions played by each class of foodstuffs: proteins, carbohydrates, and lipids.
4. Many of the most important vertebrate species in agriculture and aquaculture are foregut, midgut, or hindgut fermenters. Name an example of each, and discuss why species of this sort have advantages in agriculture and aquaculture.
5. Explain why wastage of essential amino acids can occur and how eating two foods simultaneously can prevent wastage.
6. Many species of hindgut fermenters, including horses, eat their feces to a greater extent when they are suffering from protein insufficiency than when they have plenty of protein. How might increased fecal consumption help?
7. It can be difficult or impossible to find a single diet that will serve the needs of both an herbivore and a carnivore. If you were trying to determine if two related species, one an herbivore and the other a carnivore, differed *genetically* in their digestive enzymes or absorption transporters, why would it be important to try to find a common diet for your test subjects of both species?
8. On the basis of our discussion of food-chain energetics, explain how the global human population's need for food could be met more readily if meat consumption were reduced and people lived more on plant foods.
9. Discuss hypotheses for why pythons deconstruct their digestive systems between meals (see Box 4.1). Similarly, discuss hypotheses for why small birds and mammals, having lengthened their intestines during cold seasons, permit their intestines to shorten during warm seasons. In other words, why might natural selection favor plastic, rather than static, gastrointestinal morphology?
10. Most people tend to think that if the sun ever stopped emitting light, all animal life on Earth would soon end. Populations of hydrothermal-vent animals might continue to exist for thousands of years, however. Explain.

Key References

Chivers, D. J., and P. Langer (eds.). 1994. *The Digestive System in Mammals: Food, Form and Function*. Cambridge University Press, New York. Interesting and informative papers on most topics in the modern study of mammalian feeding and digestion.

Howe, H. F., and L. C. Westley. 1988. *Ecological Relationships of Plants and Animals*. Oxford University Press, New York. A captivating and far-reaching discussion of relationships between plants and plant-eating animals in terrestrial ecosystems. Extensive treatment of plant secondary compounds and of principles of evolutionary ecology relevant to feeding.

Karasov, W. H., and I. D. Hume. 1997. Vertebrate gastrointestinal system. In W. H. Dantzler (ed.), *Comparative Physiology*, vol. 1 (Handbook of Physiology [Bethesda, MD], section 13), pp. 409–480. Oxford University Press, New York. A detailed, modern treatment of comparative digestion, fermentation, and absorption in vertebrates, emphasizing processes rather than morphology.

Morris, J. G. 1991. Nutrition. In C. L. Prosser (ed.), *Environmental and Metabolic Animal Physiology*. (*Comparative Animal Physiology*, 4th ed.), pp. 231–276. Wiley-Liss, New York. A detailed yet readable and concise survey of animal nutritional requirements.

Sterner, R. W., and J. J. Elser. 2002. *Ecological Stoichiometry: The Biology of Elements from Molecules to the Biosphere*. Princeton University Press, Princeton, NJ. A challenging book that synthesizes information on elemental nutrition in a groundbreaking, almost revolutionary, way. The chapter on biological chemistry is particularly relevant to this chapter.

Stevens, C. E., and I. D. Hume. 1995. *Comparative Physiology of the Vertebrate Digestive System*, 2nd ed. Cambridge University Press, New York. The best available book-length treatment of digestion and digestive systems throughout the vertebrates.

Wright, S. H., and G. A. Ahearn. 1997. Nutrient absorption in invertebrates. In W. H. Dantzler (ed.), *Comparative Physiology*, vol. 2 (Handbook of Physiology [Bethesda, MD], section 13), pp. 1137–1205. Oxford University Press, New York. A detailed, modern treatment of comparative nutrient absorption and digestion in invertebrates, emphasizing processes more than morphology.

Selected References

Baumann, P., N. A. Moran, and L. Baumann. 1997. The evolution and genetics of aphid endosymbionts. *BioScience* 47: 12–20.

Becerra, J. X. 1997. Insects on plants: Macroevolutionary chemical trends in host use. *Science* 276: 253–256.

Bels, V. L., M. Chardon, and P. Vandewalle (eds.). 1994. *Biomechanics of Feeding in Vertebrates*. Advances in Comparative and Environmental Physiology, vol. 18. Springer, New York.

Blaxter, K., and I. MacDonald (eds.). 1988. *Comparative Nutrition*. Libbey, London.

Burton, B. T., and W. R. Foster. 1988. *Human Nutrition*, 4th ed. McGraw-Hill, New York.

Chang, E. B., M. D. Sitrin, and D. D. Black. 1996. *Gastrointestinal, Hepatobiliary, and Nutritional Physiology*. Lippincott-Raven, Philadelphia.

Choat, J. H., and K. D. Clements. 1998. Vertebrate herbivores in marine and terrestrial environments: A nutritional ecology perspective. *Annu. Rev. Ecol. Syst.* 29: 375–403.

Coale, K. H., K. S. Johnson, S. E. Fitzwater, R. M. Gordon, S. Tanner, F. P. Chavez, L. Ferioli, C. Sakamoto, P. Rogers, F. Millero, P. Steinberg, P. Nightingale, D. Cooper, W. P. Cochlan, M. R. Landry, J. Constantinou, G. Rollwagen, A. Trasvina, and R. Kudela. 1996. A massive phytoplankton bloom induced by an ecosystem-scale iron fertilization experiment in the equatorial Pacific Ocean. *Nature* 383: 495–501.

Dissanayake, C. B., and R. Chandrajith. 1999. Medical geochemistry of tropical environments. *Earth-Sci. Rev.* 47: 219–258.

Douglas, A. E. 1995. The ecology of symbiotic micro-organisms. *Adv. Ecol. Res.* 26: 69–103.

Eisner, T., and J. Meinwald (eds.). 1995. *Chemical Ecology: The Chemistry of Biotic Interaction*. National Academy Press, Washington, DC.

Ferraris, R. P., and J. Diamond. 1997. Regulation of intestinal sugar transport. *Physiol. Rev.* 77: 257–302.

Finney, L. A., and T. V. O'Halloran. 2003. Transition metal speciation in the cell: Insights from the chemistry of metal ion receptors. *Science* 300: 931–936.

Foley, W. J., and C. McArthur. 1994. The effects and costs of allelochemicals for mammalian herbivores: An ecological perspective. In D. J. Chivers and P. Langer (eds.), *The Digestive System in Mammals: Food, Form and Function*, pp. 370–391. Cambridge University Press, New York.

Gee, J. H. R. 1991. Specialist aquatic feeding mechanisms. In R. S. K. Barnes and K. H. Mann (eds.), *Fundamentals of Aquatic Ecology*, pp. 186–209. Blackwell, London.

Grieshaber, M. K., and S. Völkel. 1998. Animal adaptations for tolerance and exploitation of poisonous sulfide. *Annu. Rev. Physiol.* 60: 33–53.

Hofmann, R. R. 1989. Evolutionary steps of ecophysiological adaptation and diversification of ruminants: A comparative view of their digestive system. *Oecologia* 78: 443–457.

Jackson, S., and J. M. Diamond. 1996. Metabolic and digestive responses to artificial selection in chickens. *Evolution* 50: 1638–1650.

Jørgensen, C. B. 1990. *Bivalve Filter Feeding*. Olsen & Olsen, Fredensborg, Denmark.

Karasov, W. H., and J. M. Diamond. 1985. Digestive adaptations for fueling the cost of endothermy. *Science* 228: 202–204.

Karasov, W. H., and J. M. Diamond. 1988. Interplay between physiology and ecology in digestion. *BioScience* 38: 602–611.

Kunze, W. A. A., and J. B. Furness. 1999. The enteric nervous system and regulation of intestinal motility. *Annu. Rev. Physiol.* 61: 117–142.

Martinez del Rio, C. 1994. Nutritional ecology of fruit-eating and flower-visiting birds and bats. In D. J. Chivers and P. Langer (eds.), *The Digestive System in Mammals: Food, Form and Function*, pp. 103–127. Cambridge University Press, New York.

McNaughton, S. J. 1988. Mineral nutrition and spatial concentrations of African ungulates. *Nature* 334: 343–345.

McNaughton, S. J. 1990. Mineral nutrition and seasonal movements of African migratory ungulates. *Nature* 345: 613–615.

Mellinger, J. (ed.). 1990. *Animal Nutrition and Transport Processes. 1. Nutrition in Wild and Domestic Animals*. Karger, Basel, Switzerland.

Mertz, W. 1981. The essential trace elements. *Science* 213: 1332–1338.

Michal, G. (ed.). 1999. *Biochemical Pathways. An Atlas of Biochemistry and Molecular Biology*. Wiley, New York.

Morton, B. 1983. Feeding and digestion in Bivalvia. In A. S. M. Saleuddin and K. M. Wilbur (eds.), *The Mollusca*, vol. 5, pp. 65–147. Academic Press, New York.

Motta, P. J., and C. D. Wilga. 2001. Advances in the study of feeding behaviors, mechanisms, and mechanics of sharks. *Environ. Biol. Fishes* 60: 131–156.

Nagy, K. A. 2001. Food requirements of wild animals: Predictive equations for free-living mammals, reptiles, and birds. *Nutr. Abstr. Rev., Ser. B* 71: 1R–12R.

Olivera, B. M., J. Rivier, C. Clark, C. A. Ramilo, G. P. Corpuz, F. C. Abogadie, E. E. Mena, S. R. Woodward, D. R. Hillyard, and L. J. Cruz. 1990. Diversity of *Conus* neuropeptides. *Science* 249: 257–263.

Pandian, T. J., and F. J. Vernberg (eds.). 1987. *Animal Energetics*, 2 vols. Academic Press, New York.

Paul, V. J. (ed.). 1992. *Ecological Roles of Marine Natural Products*. Comstock, Ithaca, NY.

Reilly, S. M., and G. V. Lauder. 1992. Morphology, behavior, and evolution: Comparative kinematics of aquatic feeding in salamanders. *Brain Behav. Evol.* 40: 182–196.

Saffo, M. B. 1992. Invertebrates in endosymbiotic associations. *Amer. Zool.* 32: 557–565.

Schultz, S. G. (ed.). 1991. *The Gastrointestinal System*, 4 vols. Handbook of Physiology (Bethesda, MD), section 6. American Physiological Society, Bethesda, MD.

Scornik, O. A. 1982. Protein synthesis and degradation during growth. In C. T. Jones (ed.), *The Biochemical Development of the Fetus and Neonate*, pp. 865–894. Elsevier Biomedical Press, New York.

Secor, S. M., and J. Diamond. 1997. Effects of meal size on postprandial responses in juvenile Burmese pythons (*Python molurus*). *Amer. J. Physiol.* 41: R902–R912.

Spallholz, J. E. 1989. *Nutrition: Chemistry and Biology*. Prentice Hall, Englewood Cliffs, NJ.

Tunnicliffe, V. 1992. Hydrothermal-vent communities of the deep sea. *Amer. Sci.* 80: 336–349.

Urich, K. 1994. *Comparative Animal Biochemistry*. Springer, New York.

Vonk, H. J., and J. R. H. Western. 1984. *Comparative Biochemistry and Physiology of Enzymatic Digestion*. Academic Press, New York.

Wainwright, P. C., and S. M. Reilly (eds.). 1994. *Ecological Morphology*. University of Chicago Press, Chicago.

Ward, J. E., L. P. Sanford, R. I. E. Newell, and B. A. MacDonald. 1998. A new explanation of particle capture in suspension-feeding bivalve molluscs. *Limnol. Oceanogr.* 43: 741–752.

Weindruch, R. 1996. Caloric restriction and aging. *Sci. Amer.* 274(1): 46–52.

Weiss, S. L., E. A. Lee, and J. Diamond. 1998. Evolutionary matches of enzyme and transporter capacities to dietary substrate loads in the intestinal brush border. *Proc. Natl. Acad. Sci. U.S.A.* 95: 2117–2121.

Wildish, D., and D. Kristmanson. 1997. *Benthic Suspension Feeders and Flow*. Cambridge University Press, Cambridge.

Woods, S. C., R. J. Seeley, D. Porte, Jr., and M. W. Schwartz. 1998. Signals that regulate food intake and energy homeostasis. *Science* 280: 1378–1382.

Wright, P. A., and P. M. Anderson (eds.). 2001 *Nitrogen Excretion*. Academic Press, New York.

See also References in Appendix A.

5 Energy Metabolism

In 1986 explorers set out to reach the North Pole by dogsled from Ellesmere Island at latitude 83° north. Their objective was to make the journey without any resupply along the way and thus to reenact the famous expedition by Admiral Robert Peary in 1909. Each of the modern explorers' sleds was huge—16 feet (5 m) long, and loaded with 1400 pounds (630 kg) of material. Of the two-thirds of a ton on each sled, most was food for people and dogs. Most of the weight of the food, moreover, was required to meet energy needs; if the only food materials that had needed to be hauled were vitamins, minerals, amino acids, and other sources of chemical building blocks for biosynthesis, the pile of food on each sled would have been much smaller. Two people's needs for energy were to be met by the supplies on each sled. The sled needed to be pulled and pushed over numerous ice ridges 20 to 60 feet high on the way to the pole. Even during progress over level ice fields, the way for the sled's runners needed often to be cleared by manpower. A team of sled dogs had to be maintained and fed to do most of the hauling. A trek to the North Pole by dogsled would be immeasurably easier if there were no need for food energy for people and dogs. The need for energy is equally consequential in the natural world. Animals regularly risk their lives to obtain energy, or die because they did not obtain enough.

Why do animals need energy? Even if the answer to that question seems obvious, why do animals need new inputs of energy on a regular basis? Why do people typically need new food energy every day, for instance? Why not take in an adequate amount of energy early in life and simply reuse it, over and over, for the rest of life? These are some of the key questions addressed in this chapter.

Other questions to be discussed are more practical. If you were planning an Arctic expedition, how would you use physiological principles to predict the energy needs of you and your companions? Knowing the energy needs of your team, how would you calculate the amount of food to pack on your sleds? The food on the sleds of the North Pole explorers was rich in fats and oils; the meat they carried, for instance, was a 50:50 mix of ground beef and lard (pure fat). Why did they emphasize lipids?

Energy metabolism, the subject of this chapter, is the sum of the processes by which energy is acquired, transformed, channeled into useful functions, and dissipated. It consists of two subsets of processes mentioned in Chapter 2: *catabolic processes* that break down organic molecules to release energy and *anabolic processes* that use energy to construct molecules.

The burden of food A trek to the North Pole from the nearest land requires hundreds of pounds of food to be hauled per explorer, to supply energy for the explorers and their dogs.

Why Animals Need Energy: The Second Law of Thermodynamics

Animals are *organized* or *ordered* systems. As we saw in discussing the dynamic state of body constituents in Chapter 1, the atoms composing an animal's body are routinely exchanged with atoms in the environment. The *organization* of atoms in the body persists, however, even as particular atoms come and go. As this observation suggests, organization is a far more fundamental feature of animals than their material construction (see page 10).

The **second law of thermodynamics**, one of the greatest achievements of intellectual history, provides fundamental insight into the nature of organized systems. The law applies to **isolated systems**. An isolated system is a part of the material universe that exchanges nothing—neither matter nor energy—with its surroundings. Animals, you will recognize, are not isolated systems. In fact, true isolated systems are difficult to create even in the highly controlled setting of a physics laboratory. Thus the concept of an isolated system is largely an abstraction. Nonetheless, we can gain insight into animal energetics by analyzing isolated systems. The second law of thermodynamics states that *if an isolated system undergoes internal change, the direction of the change is always toward greater disorder.*

A corollary of the second law is that order can be maintained or increased within a system only if the system is *not* isolated. If "energy" is permitted to enter a system from the outside, order may be maintained or increased within the system. **Energy** is defined in the field of mechanics to be the capacity to do mechanical work, measured as the product of force and distance. A broader definition that is often more useful for biologists is that *energy is the capacity to increase order.*

To illustrate the second law of thermodynamics, let's examine an isolated system that consists of a closed loop of copper pipe filled with water, with the water initially flowing around and around in the loop. We need not know how the water started moving; it is enough to know that the loop of pipe has water flowing in it and that the loop exchanges no energy or matter with its surroundings. Focus on the motions of atoms and molecules *on an atomic-molecular scale.* The distribution of atomic-molecular motions in the system we have described is highly nonrandom initially, because each water molecule is moving in an ordered way in its direction of travel around the loop of pipe (Figure 5.1*a*). In addition, all of the water molecules—and all of the copper atoms in the wall of the pipe—undergo ceaseless random motions on an atomic-molecular scale.[1]

Over time, the energy of *directional* motion of the water molecules is gradually transformed to energy of *random* motion (Figure 5.1*b*). The energy changes in this manner because when water molecules collide with copper atoms or other water molecules, some of their energy of directional motion is transferred in such a way as to increase the intensity of random motions of the atoms and molecules with which they collide. Consequently, as time passes, the rate of flow of water around the pipe diminishes. The energy of nonrandom flow is translated into random molecular agitation (heat), and the temperature of the system rises. Eventually, the flow of water ceases. At that point all the water molecules and copper atoms in the system display merely random motions, and the original order in the system (the directional motion of water molecules) is entirely degraded to disorder. This inevitable outcome represents the second law of thermodynamics in action.

The only way to keep the water flowing, and thus to maintain the original order in our water-filled loop of pipe, would be to convert the system into an **open system**, a system that is *not* isolated. If there were a pump in the loop of pipe, and if electricity were provided to the pump from outside, the initial nonrandom state could be sustained indefinitely. That is, an energy input to the system could cre-

[1] Random atomic-molecular motions are a universal property of matter that is at a temperature above absolute zero. In fact, the *temperature* of an object is a measure of the intensity of these motions in an object.

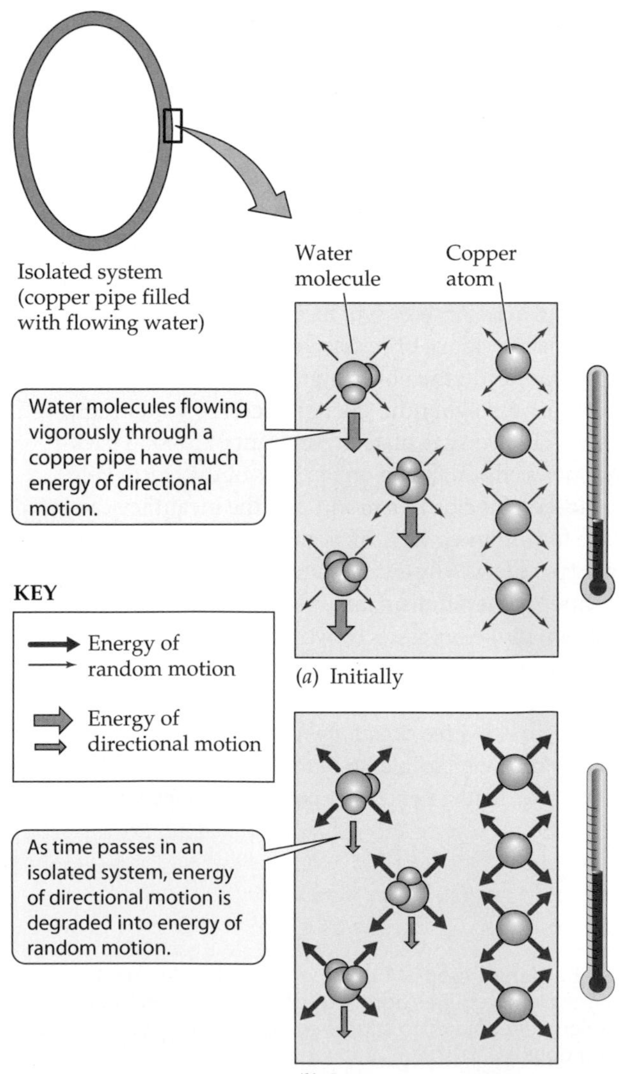

Figure 5.1 The second law of thermodynamics in action Energy of directional motion is converted to energy of random motion as time passes from an initial time *(a)* to a later time *(b)* in an isolated system. This transformation of energy increases the intensity of the random motions of both the water molecules and the copper atoms in the system, causing the system temperature to rise.

ate order in the system as rapidly as processes within the system tended to diminish order.

In terms of their thermodynamics, animals must function as open systems. Without an energy input, the blood coursing through an animal's circulatory system will slow to a halt, just as the water in the loop of pipe does when there is no energy supplied from outside. Without an energy input, vital molecules in an animal's tissues will become more disorganized, as by spontaneously breaking down; thus eventually they will lose their critical structural and functional properties. Without an energy input, positive and negative ions will distribute themselves randomly across an animal's cell membranes; this randomization of electrical charges, among other things, will make nerve impulses impossible. The second law of thermodynamics dictates that if an animal were required to function as an isolated system, all forms of order within its body would decay. This loss of order would eventually kill the animal because order is essential for life. *Animals require energy from the outside because energy is necessary to create and maintain their essential internal organization.*

Fundamentals of Animal Energetics

To understand animal energetics more thoroughly, the first step is to recognize that energy exists in different *forms*, and the various forms differ in their significance for animals. We will focus here on four forms of energy of particular importance: chemical energy, electrical energy, mechanical energy, and heat. **Chemical energy** (**chemical-bond energy**) is energy liberated or required when atoms are rearranged into new configurations. Animals obtain the energy they need to stay alive by reconfiguring atoms in food molecules, thereby liberating chemical energy. **Electrical energy** is energy that a system possesses by virtue of the separation of positive and negative electrical charges. All cell membranes possess electrical energy because there is charge separation across them, as we saw in Chapter 3. There are two forms of energy of motion (kinetic energy) that are important for animals. One, **mechanical energy**, is energy of *organized* motion in which many molecules move simultaneously in the same direction. The motion of a moving arm, or that of circulating blood, provides an example. **Heat**, often called **molecular kinetic energy**, is the energy of *random atomic-molecular* motion. Heat is the energy that matter possesses by virtue of the ceaseless, random motions of all the atoms and molecules of which it is composed.[2]

The forms of energy vary in their capacity for physiological work

Although, by definition, all forms of energy are capable of doing work in one context or another, all forms of energy are *not equally* capable of doing physiological work in animals. **Physiological work** is any process carried out by an animal that increases order. For example, an animal does physiological work when it synthesizes macromolecules such as proteins, generates electrical or chemical gradients by actively transporting solutes across cell membranes, or contracts muscles to move materials inside or outside its body (or set its whole body in motion).

How do the forms of energy that we have mentioned—chemical energy, electrical energy, mechanical energy, and heat—differ in their ability to do physiological work? Animals can use chemical energy (directly or indirectly) to do *all* forms of physiological work; hence, for animals, chemical energy is *totipotent* (*toti*, "all"; *potent*, "powerful"). Animals use electrical and mechanical energy to accomplish some forms of physiological work, but neither form of energy is totipotent. For example, animals use electrical energy to set ions in motion and mechanical energy to pump blood, but they cannot use either form of energy to synthesize proteins. Finally, *animals cannot use heat to do any form of physiological work.*

This last point is important. According to thermodynamics, a system can convert heat to work only if there is a temperature difference between one part of the system and another. Temperature is a measure of the intensity of random atomic-molecular motions. If the intensity of random motions differs from one place to another within a system, this difference represents a form of order, and the system can be used as a machine to convert heat to work; the high temperatures in the cylinders of an internal combustion engine, for example, permit heat from the burning of fuel to be converted into mechanical energy that propels a car. The purely random molecular motions that exist throughout a physical system at a uniform temperature cannot do work, however. Within cells—the relevant functional systems of organisms—temperature differences from place to place are very small and transient, when they exist at all. Hence, cells cannot in theory use heat to do physiological work, and biological experiments confirm that they cannot. Heat is hardly unimportant to animals; as discussed later in this chapter and in Chapter 8, inputs of heat influence animal metabolic rates and affect the abilities of macromolecules such as proteins to carry out their functions. However, heat has no importance as a source of energy for physiological *work* because heat cannot do work in organisms.

The forms of energy are placed into two categories based on their ability to do physiological work. **High-grade energy** can do physiological work; chemical, electrical, and mechanical energy are included in this category. **Low-grade energy**—heat—cannot do physiological work. When we say that animals **degrade** energy, we mean that they transform it from a high-grade form to heat.

Transformations of high-grade energy are always inefficient

When organisms transform energy from one high-grade form to another high-grade form, the transformation is always incomplete, and some energy is degraded to heat. The **efficiency** of an energy transformation is defined as follows:

$$\text{Efficiency of energy transformation} = \frac{\text{output of high-grade energy}}{\text{input of high-grade energy}} \tag{5.1}$$

The efficiency of energy transformation is typically much less than 1. For example, when a cell converts chemical-bond energy of

[2] The study of forms of energy is part of thermodynamics. Some specialists in thermodynamics emphasize energy transfer. For them, heat and mechanical work are not forms of energy, but rather *heating* and *working* are processes that *transfer* energy.

glucose into chemical-bond energy of adenosine triphosphate (ATP), at most only about 70% of the energy released from glucose is incorporated into bonds of ATP; the other 30% is lost as heat. When, in turn, a muscle cell uses the chemical-bond energy of ATP to contract, typically a maximum of only 25% to 30% of the energy liberated from the ATP appears as energy of muscular motion; again, the remainder is lost as heat. The contraction efficiency of muscles in fact depends on the type of work they are doing. If you plant one of your feet on a wall and use your leg muscles to push your foot steadily against the wall (isometric exercise), no motion occurs, and the muscles' efficiency in producing motion is zero. If you ride an exercise bike, however, the efficiency of your leg muscles in producing external motion might be as high as 25% to 30%. Even then, 70% to 75% of the energy liberated from ATP in the process would become heat inside your body.

You may be familiar with the Hollywood image of a jungle explorer caught in quicksand. The victim sinks deeper every time he moves. In certain ways an animal's use of food energy is analogous to this image. To make use of the chemical energy from a meal, an animal *must* transform the energy, usually in multiple steps. Each step, however, robs the energy of part of its value because energy transformations are always inefficient and degrade energy to heat. Thus, with each step an animal takes to use the chemical energy in its food, the resource shrinks, just as each move of the hapless jungle explorer lowers his body further into the quicksand.

Animals use energy to perform three major functions

It is helpful to visualize the energy used for physiological work as a resource that "flows" through an animal during its lifetime. Chemical energy enters the animal from the external environment (Figure 5.2). Then forms of energy derived from the chemical energy pass back into the external environment. This flow of energy begins as **ingested chemical energy** (also termed **ingested energy**), the energy present in the chemical bonds of an animal's food. Some of this ingested energy is not absorbed because following digestion it remains in chemical structures that the animal is unable to absorb. This portion of the energy never really enters the animal's tissues and is egested in the feces. The egested chemical-bond energy is known as the animal's **fecal chemical energy** (**fecal energy**). The remainder of the ingested energy is taken into the living tissues of the animal. The chemical-bond energy of the organic compounds that are absorbed or assimilated is known as the animal's **absorbed chemical energy** (**absorbed energy**) or **assimilated chemical energy.**[3] It is the energy available for the performance of physiological work by the animal's cells.

An animal uses its absorbed chemical energy to carry out three major tasks:

1. *Biosynthesis.* An animal synthesizes its body constituents, such as its proteins and lipids, using absorbed energy. In this process, called **biosynthesis**, some of the absorbed energy remains in chemical form because the products of biosynthesis are molecules with chemical-energy content. During *growth*, chemical energy accumulates in the body as biosynthesized products, which are used to assemble new cells and tissues (see Figure 5.2). Some of the chemical energy accumulated in body tissues through growth (e.g., the chemical energy of fat) may be used by an animal as food energy during times of hunger or starvation; ultimately, all of the energy accumulated in body tissues becomes food for predators or decay organisms when the animal dies. Besides contributing to tissues during growth, biosynthesis also produces *organic compounds that are exported from the body during an animal's life*, taking chemical-bond energy

[3] The term *digestible energy* is widely used as a synonym, but it is not used in this book because the absorbed chemical energy depends on not only digestive but also absorptive processes.

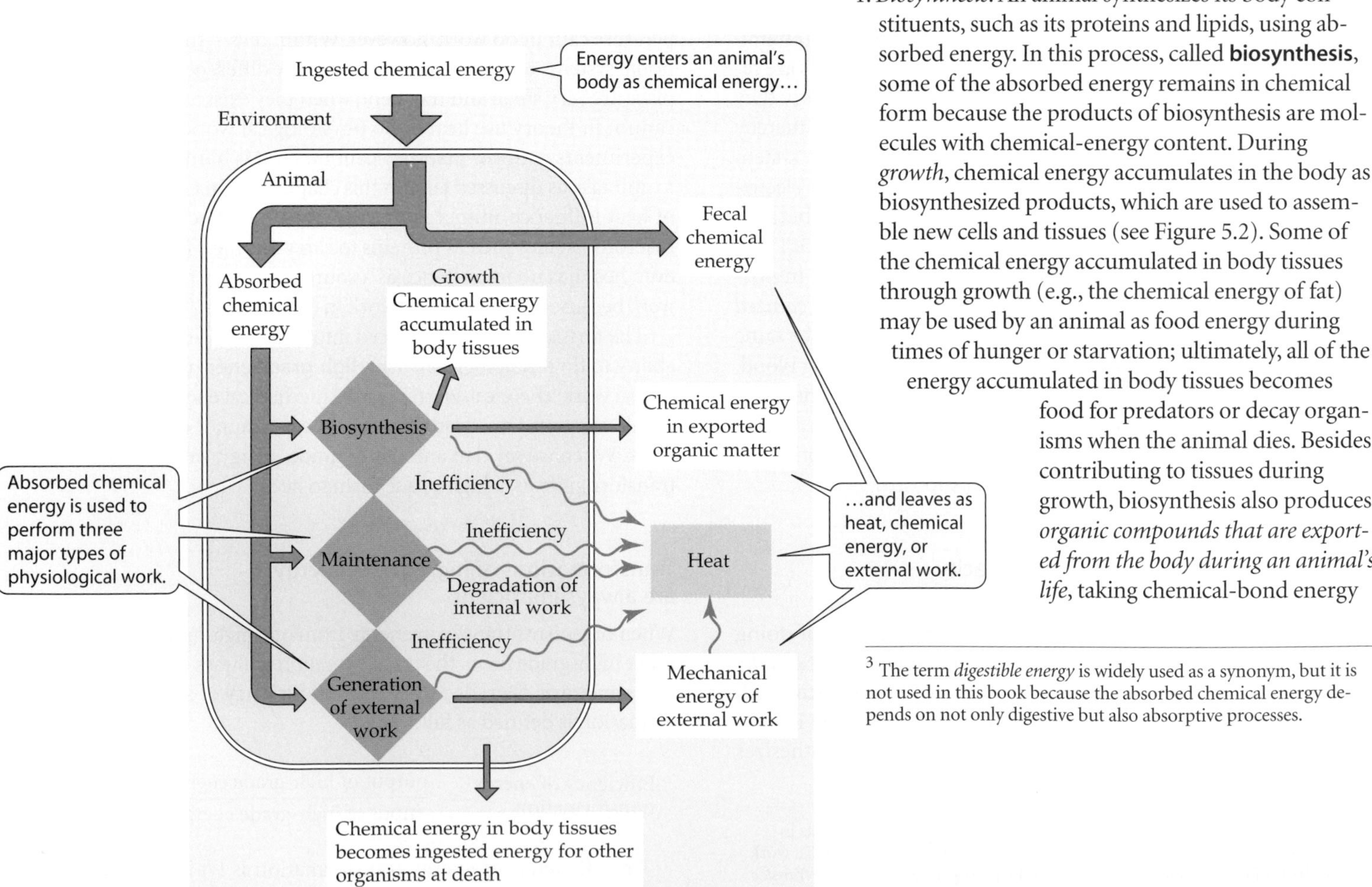

Figure 5.2 The uses of energy by an animal

with them. Gametes, milk, mucoid secretions, sloughed skin or hair, and shed exoskeletons are just a few of the organic products that animals synthesize and lose to the environment during their lives. Whether biosynthesis yields growth or exported organic products, this form of physiological work typically involves elaborate biochemical pathways requiring many steps. Heat is produced during biosynthesis because each step is inevitably inefficient.

2. *Maintenance.* An animal's **maintenance** functions are all the processes that maintain the integrity of its body; examples include circulation, respiration, nervous coordination, gut motility, and tissue repair. With only trivial exceptions, *the energy used for maintenance is degraded entirely to heat within the body.* To see why, consider the circulation of the blood. First, the chemical-bond energy of the absorbed food molecules that are used as fuel is converted into chemical-bond energy of ATP, and energy is lost as heat in the process because of the inefficiency of the conversion. Additional energy is lost as heat when the energy of ATP is used to drive contraction of the heart muscle. Finally, a small fraction of the chemical-bond energy originally obtained from food molecules appears as mechanical energy in the motion of the blood ejected from the heart. Even that mechanical energy is degraded to heat within the body, in overcoming the viscous resistances that oppose motion of the blood through the blood vessels. Mechanical work that takes place inside an animal's body is termed **internal work.** Several maintenance functions, such as the circulation of the blood, are kinds of internal work. The energy of internal work is entirely (or almost entirely) degraded to heat within the body.

3. *Generation of external work.* Animals perform **external work** when they apply mechanical forces to objects outside their bodies. A mouse running across a field and a bicyclist ascending a hill, for example, are performing external work using their leg muscles. Much of the absorbed chemical energy used to fuel external work is degraded to heat within the body (e.g., in using ATP to set muscles in motion). When external work is performed, however, some energy leaves the body as mechanical energy transmitted to the environment. The fate of that energy depends on whether it is stored. *Energy of external work is stored if it is converted into increased potential energy of position.* As a bicyclist ascends to the top of a hill, part of his energy of external work is stored as increased potential energy of position because his body and bike move to a higher position in Earth's gravitational field (Figure 5.3). When he later descends, this potential energy of position is converted into mechanical energy (the motion of his body and the bike downhill) and then to heat. On the other hand, if a mouse scurries a *horizontal* distance across a field—or a person bicycles along a horizontal road—no energy is stored in repositioning the body relative to gravity, and the mechanical energy transmitted to the environment is entirely, and quickly, degraded to heat in overcoming resistances to motion.

Figure 5.3 The energy driving upward locomotion is converted into heat and potential energy of position As this bicyclist goes uphill, much of his mechanical energy of external work becomes heat, but a fraction is stored as potential energy because he is propelling the mass of his body and bicycle higher in Earth's gravitational field.

Reviewing the flow of energy through an animal's body (see Figure 5.2), we can see that all uses of energy by animals generate heat. Thus, all living animals produce heat. Because frogs, fish, clams, and other poikilotherms are often cool to the touch, one can get the erroneous impression that they do not produce heat. However, such animals are cool not because they fail to produce heat, but because their rates of heat production are so low and their bodies are so slightly insulated that they are not warmed by the heat that they produce (see page 219). Animal heat, which is universal, has been studied for centuries, leading to fundamental discoveries about the nature of life (Box 5.1).

Another point to stress as we conclude our discussion of energy flow through animals is that the conversion of chemical-bond energy to heat is one-way: No animal or other living creature is able to convert heat back to chemical-bond energy or any other form of high-grade energy. Thus, energy is not recycled within individual animals or within the biosphere as a whole. This principle provides the answer to a key question we asked at the start of this chapter—namely, why animals need to obtain food energy regularly throughout their lives. When an animal ingests and uses totipotent chemical-bond energy, it converts much of it in a one-way, irreversible fashion to heat, which is useless for physiological work. Thus, as an animal uses the chemical-bond energy from a meal, it inevitably develops a need to eat again to acquire new chemical-bond energy. The biosphere as a whole requires a continuing input of high-grade photon energy from the sun for much the same reason. The photon energy captured in bonds of organic compounds by photosynthesis is converted to heat by plants as well as animals, meaning that new photon energy is required if organ-

BOX 5.1 Views on Animal Heat Production

Heat is an inevitable *by-product* of the use of high-grade, chemical-bond energy to create and maintain the vital organization of living organisms. Interestingly, from the time of Aristotle until the nineteenth century, the significance of heat was generally viewed quite oppositely. Far from being a by-product, heat was usually seen as a primary source of life, a vital force that endowed many parts of organisms with their living attributes. This "vital heat" was thought to differ from the heat of a fire. It was believed to originate exclusively in the heart, lungs, or blood and to suffuse the rest of the body. When William Harvey described the circulation of the blood in the early seventeenth century, one of the principal roles attributed to the newfound circulation was transport of "vital heat" from tissues where it was produced to other tissues, which it animated.

The old view of animal heat began to change at about the time of the American Revolution, when Antoine Lavoisier, in France, showed that the ratio of heat production to CO_2 production was about the same for a guinea pig as for burning charcoal. From this and other evidence, Lavoisier and the Englishman Adair Crawford argued that animal respiration is a slow form of combustion, and animal heat is the same as the heat produced by fire. Still, for several more decades, all animal heat was believed to originate in the lungs, and the lungs were thought to be the exclusive site of O_2 utilization. Not until 1837 did Heinrich Gustav Magnus show that the blood takes O_2 from the lungs to the rest of the body and returns CO_2. Evidence for the all-important concept that tissues throughout the body make heat came a decade later when Hermann von Helmholtz demonstrated that muscular contraction liberates heat. In 1872, Eduard Pflüger presented evidence that all tissues consume O_2.

The discovery that all tissues use O_2 and produce heat was one of several lines of thought and investigation that came together in the nineteenth century to spawn our modern understanding of animal energetics. Other important developments were the flowering of the science of thermodynamics (sparked by the Industrial Revolution) and profound changes in the understanding of energy. In the 1840s, Julius Robert von Mayer in Germany and James Joule in England developed the seminal concept that heat, motion, electricity, and so on are all forms of one thing: energy. Mayer, a physician, conceptualized clearly for the first time the nature of animal energy transformations described in this chapter.

ic compounds are to continue to be available. The heat that all organisms collectively make is radiated from Earth into outer space.

Summary: Fundamentals of Animal Energetics

- Forms of energy vary in their capacity to do physiological work. Chemical-bond energy is totipotent for animals. Electrical and mechanical energy can do certain types of physiological work but are not totipotent. Heat cannot perform physiological work of any kind.
- Animals use their absorbed chemical energy for three major functions: biosynthesis, maintenance, and generation of external work. Biosynthesis, which preserves some of the absorbed energy in the form of chemical energy, includes both growth and the synthesis of organic materials that are exported from the body during an individual's life.
- Some energy is degraded to heat (low-grade energy) whenever one high-grade form of energy is transformed to another. Energy transformations are always inefficient.
- Energy is not recycled, either in individual animals or in the biosphere as a whole.
- Animals take in chemical-bond energy and put out heat, chemical-bond energy, and external work.

Metabolic Rate: Meaning and Measurement

We have seen that an animal takes in chemical energy in its food, and in the process of living it releases chemical energy, heat, and external work. The energy converted to heat and external work is defined by physiologists to be **consumed**, because an animal cannot use that energy to do additional physiological work. An animal's **metabolic rate** is its *rate of energy consumption*: the rate at which it converts chemical energy to heat and external work. Energy is measured in *calories* or *joules*, and metabolic rates are expressed in *calories per unit time* or *watts* (Box 5.2). Heat is always the main component of the metabolic rate.

You will recall that at the start of this chapter, we wondered how it is possible to predict the food needs of people and sled dogs during a polar expedition. Their metabolic rates are the basis for making this prediction because once you know their daily energy consumption, you can estimate the total energy that the explorers and their dogs will need over the days of their trek. Speaking broadly, metabolic rates are significant for three reasons:

1. An animal's metabolic rate is a principal determinant of how much food it needs.
2. An animal's metabolic rate provides a quantitative measurement of the *total activity* of all its physiological mechanisms, because every energy-using process contributes heat. The metabolic rate, roughly speaking, represents an animal's *intensity* of living.
3. Ecologically, the metabolic rate measures the drain an animal places on the physiologically useful energy supplies of its ecosystem because the metabolic rate is the pace at which the animal degrades the chemical energy of organic compounds in the ecosystem.

Direct calorimetry: The metabolic rate of an animal can be measured directly

Physiologists sometimes measure metabolic rates directly using a **direct calorimeter**, a device that measures the rate at which heat leaves an animal's body (Box 5.3). Modern direct calorimeters are technically complex instruments. The basic operation of a direct calorimeter is illustrated nicely, however, by the relatively simple device that Antoine Lavoisier used in the first measurements of animal heat production (Figure 5.4). The heat leaving the test animal melted the sur-

BOX 5.2 Units of Measure for Energy and Metabolic Rates

The traditional unit of measure for energy is the **calorie** (**cal**), which is the amount of heat needed to raise the temperature of 1 g of water by 1°C. Although the calorie is defined as a quantity of heat, it can be used as a unit of measure for motion, electricity, or any other form of energy because the forms of energy bear strict equivalencies to each other. A **kilocalorie** (**kcal**) is 1000 cal. Sometimes the kilocalorie is written *Calorie*, with a capital *C*. (Popular books and periodicals sometimes write *calorie* with a lowercase *c* when they mean kilocalorie, an unfortunate practice that can create three orders of magnitude of confusion.) If energy is expressed in calories or kilocalories, then *rates* of energy exchange or transformation—such as metabolic rates—are expressed in calories or kilocalories *per unit time*.

The fundamental unit of measure for energy in the SI system of units is the **joule** (J), named in honor of James Joule. Appendix B (The Système International and Other Units of Measure) discusses the derivation of the joule from the SI base units. A **watt** (**W**), which is equivalent to 1 joule/second (J/s), is the fundamental SI unit for rates of energy exchange or transformation.

One calorie is equivalent to 4.186 J, a relation that permits the interconversion of units in the calorie and SI systems. For example, as you sit quietly reading this page, your body is likely producing heat at a rate near 23 cal/s, equivalent to 23 × 4.186 = 96 J/s, or 96 W. That is, you are producing heat about as rapidly as a 100-W light bulb.

rounding ice, and Lavoisier collected the meltwater over measured periods of time. By knowing the amount of heat required to melt each gram of ice, he could calculate the animal's rate of heat output, and thus its metabolic rate.

Sometimes, to carry out direct calorimetry, one must also measure external work, the second component of metabolic rate. If an animal under study is at rest, no external work is being performed; thus a measurement of heat production alone encompasses a resting animal's entire metabolic rate. When external work *is* being performed, the energy of the external work is often rapidly degraded to heat (as in the case discussed earlier of a mouse scurrying over a horizontal surface). In such cases, a measure of heat production includes the energy of external work, so heat production remains a sufficient determination of metabolic rate. If some energy of external work fails to be converted to heat because it is stored, however, that energy must be measured independently and added to heat production to quantify an animal's metabolic rate by direct calorimetry.

Indirect calorimetry: Animal metabolic rates are usually measured indirectly

Indirect calorimetry measures an animal's metabolic rate by means other than quantifying heat and work (see Box 5.3). Biologists today usually measure metabolic rates indirectly because the methods of

BOX 5.3 Direct Measurement versus Indirect Measurement

An investigator who wants to measure a particular trait of an animal or environment must decide whether to use a *direct* or *indirect* method of measurement. The distinction between direct and indirect methods is very important and yet can seem subtle at first. The two types of measurement give results that can be expressed in the same units; a metabolic rate, for example, can be written down in watts whether it is measured directly or indirectly. The distinction between direct and indirect methods is in the property that is *actually measured*.

Rigorously speaking, any physiological trait is *defined* by specific properties. A direct method of measurement records the *very same properties specified by the definition of a trait*. An animal's metabolic rate, for example, is defined as its rate of production of heat and external work. Thus a direct method of measuring metabolic rate actually quantifies heat and external work. Direct methods are the gold standards of measurement because their results relate unambiguously to the trait being studied.

On the other hand, an indirect method of measurement records *properties different from those specified by the definition of a trait*. The properties that are actually measured and quantified by an indirect method serve as "stand-ins" or "proxies" for the properties that define the trait. For instance, some indirect methods for measuring metabolic rate record O_2 consumption. Researchers use such methods because O_2 consumption is relatively easy to measure and often closely reflects an animal's rate of production of heat and external work. After measuring the amount of O_2 consumed per unit of time, researchers often convert the amount of O_2 to calories or joules, units of energy. It is important to understand that the act of converting the measurement results does not change the nature of the measurement method. A measure of metabolic rate obtained by recording O_2 consumption is an indirect measure regardless of the units in which it is ultimately expressed.

Why does the distinction between direct and indirect methods matter? A direct method, if carried out carefully with suitable instruments, *must* provide information of the sort desired because it measures exactly what the definition specifies. An indirect method, however, may introduce errors regardless of how carefully it is carried out, because it is measuring something different from what is stated by the definition. Why, then, are indirect methods ever used? As the example of O_2 consumption illustrates, indirect methods often have practical advantages; they may be cheaper or easier than direct methods. The accuracy of an indirect method must be judged by how well its results agree with a direct method. Indirect methods are usually less accurate than direct ones, at least under certain circumstances, so the choice of method involves a compromise. In the measurement of metabolic rate, the indirect methods commonly introduce uncertainties of 1% to 5% in the results, but they might reduce costs of time and money by tenfold, making them attractive.

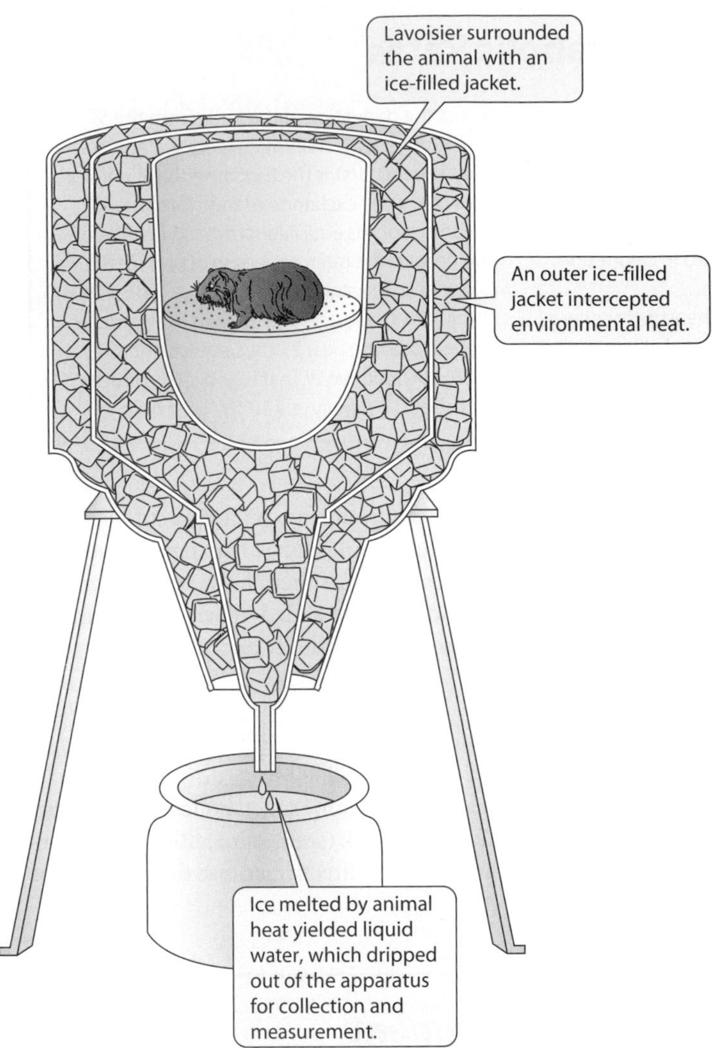

Figure 5.4 Lavoisier's direct calorimeter Heat from the general environment must be excluded from measurement of animal heat. In Lavoisier's device, heat entering from the air surrounding the calorimeter was intercepted by an outer ice-filled jacket. Modern direct calorimeters, although they measure heat in a different way and more precisely, still reflect the fundamental design considerations that Lavoisier introduced. (After Lavoisier 1862.)

indirect calorimetry are cheaper and easier than those of modern, sophisticated direct calorimetry. Here we consider two methods of indirect calorimetry: (1) measuring an animal's rate of respiratory gas exchange with its environment (termed *respirometry*) and (2) measuring the chemical-energy content of the organic matter that enters and leaves an animal's body (the *material-balance method*).[4]

INDIRECT CALORIMETRY BASED ON RESPIRATORY GAS EXCHANGE An animal's rate of oxygen consumption provides a convenient and readily measured estimate of its metabolic rate. To understand the use of O_2 consumption for this purpose, consider first an oxidation reaction occurring in a test tube. If a mole of glucose ($C_6H_{12}O_6$) is burned completely, chemical stoichiometry dictates that 6 moles of O_2 will be used, and 6 moles of CO_2 will be produced. The reaction will also release heat, known as *heat of combustion*. For glucose, the heat of combustion during complete oxidation is about 2820 kilojoules (kJ), or 673,000 calories (cal), per mole (mol). The following equation applies:

$$C_6H_{12}O_6 + 6\,O_2 \rightarrow 6\,CO_2 + 6\,H_2O + 2820\text{ kJ/mol} \quad (5.2)$$

Note that when glucose is oxidized, a fixed proportional relation prevails between the amount of heat produced and the amount of O_2 used: 2820 kJ of heat per 6 mol of O_2. Similarly, a proportional relation exists between heat production and CO_2 production: 2820 kJ per 6 mol of CO_2. Knowing these relations, if you oxidize an *unknown* quantity of glucose in a test tube and you measure only the amount of O_2 used or the amount of CO_2 produced, you can calculate the exact amount of heat produced.

When an animal metabolically oxidizes glucose (or any other chemical substance), if the chemical end products are the same in the animal as in a test tube, then the stoichiometric relations that prevail among heat production, O_2 consumption, and CO_2 production in the animal will be the same as those in the test tube. This important principle, established by Max Rubner and Wilbur Atwater in the 1890s, is true even though the *intermediate* steps of the metabolic oxidation reactions in an animal differ from the intermediate steps of test-tube oxidation. Because of this principle, if an animal were to oxidize glucose to CO_2 and H_2O and you were to measure either the animal's O_2 consumption or its CO_2 production in the process, you could calculate the animal's heat production, just as you could calculate heat production in the test tube discussed earlier. This is the rationale for estimating animal metabolic rates from rates of respiratory exchange of O_2 and CO_2.

Table 5.1 provides conversion factors for calculating the amount of heat generated when a milliliter of O_2 is consumed or a milliliter of CO_2 is produced.[5] To understand how to use the table, imagine that an animal consumes O_2 at a rate of 10 mL/minute, and suppose you know that the animal's cells are oxidizing only glucose (a carbohydrate). The animal's metabolic rate would then be 10 mL/minute × 21.1 J/mL = 211 J/minute.

As you can see from Table 5.1, problems can arise in the use of respiratory gas exchange to measure metabolic rates because the correct conversion factor for calculating heat production from O_2 consumption or CO_2 production is not a simple, fixed number. Instead, the conversion factor varies depending on the foodstuffs being oxidized. If animals oxidize only glucose, calculating their metabolic rates by measuring their rates of O_2 consumption is unambiguous, as we have already seen. However, animals oxidize a variety of foodstuffs, which yield different quantities of heat per unit volume of O_2 consumed or CO_2 produced (see Table 5.1). This consideration introduces the possibility of ambiguity or inaccuracy. Returning to our previous example, we saw that if an animal consumes O_2 at a rate of 10 mL/minute, its metabolic rate is 211 J/minute (10 mL/minute ×

[4] Additional methods of indirect calorimetry that are used for active or free-living animals are discussed in Chapter 7.

[5] It is a universal convention in metabolic studies to express gas volumes at standard conditions of temperature and pressure (STP), and that is the way they are expressed here. For more information, see Appendix D: Gases at Standard Temperature and Pressure.

TABLE 5.1 Ratios of heat production to O_2 consumption and CO_2 production during the aerobic catabolism of carbohydrates, lipids, and proteins Values given are for representative mixtures of each of the three foodstuffs. Gas volumes are at standard conditions of temperature and pressure (see Appendix D: Gases at Standard Temperature and Pressure).

Foodstuff	Heat produced per unit O_2 consumed (J/mL O_2)	Heat produced per unit CO_2 produced (J/mL CO_2)
Carbohydrates	21.1	21.1
Lipids	19.8	27.9
Proteins[a]	18.7	23.3

Source: After Brown and Brengelmann 1965.
[a] For proteins, values depend on the metabolic disposition of nitrogen; those tabulated apply to mammals and other animals in which urea is the dominant nitrogenous end product.

21.1 J/mL) if its cells are oxidizing carbohydrates. Its metabolic rate is only 198 J/minute, however, if its cells are oxidizing lipids (10 mL/minute × 19.8 J/mL). Hence an investigator cannot calculate metabolic rate (the rate of heat production) exactly from measurements of O_2 consumption or CO_2 production without knowing the exact mixture of foodstuffs that the study animal's cells are oxidizing.

An animal's recent diet may not provide accurate insight into the foodstuffs its cells are oxidizing, because animals store and interconvert foodstuffs. We can determine the foodstuffs that cells are oxidizing only by looking at indices of *cellular function*. We can obtain one useful index of cellular function by simultaneously measuring CO_2 production and O_2 consumption and taking their ratio:

$$\frac{\text{moles of } CO_2 \text{ produced per unit time}}{\text{moles of } O_2 \text{ consumed per unit time}}$$

This ratio is called the **respiratory exchange ratio (*R*)** when measured at the respiratory organs (e.g., lungs); it is called the **respiratory quotient (*RQ*)** when measured at the level of the cells.[6] Under many circumstances, *R* and *RQ* are equal, and we limit ourselves here to that straightforward situation. The *R* and *RQ* values provide a signature of the particular sorts of foodstuffs being oxidized by an animal's cells (Table 5.2). If an animal exhibits a value of *R* near 1.0, for example, its cells are likely oxidizing mostly carbohydrates. On the other hand, if an animal's *R* value is near 0.7, its cells are likely catabolizing mostly lipids. An *R* or *RQ* value near 1.0 or 0.7 strongly suggests which conversion factor from Table 5.1 should be used to calculate an animal's metabolic rate from its O_2 consumption. Specifically, an *R* value near 1.0 suggests use of the carbohydrate factor, 21.1 J/mL O_2, whereas an *R* value near 0.7 suggests use of the lipid factor, 19.8 J/mL O_2.

Unfortunately, *R* values that are not close to 1.0 or 0.7 are often difficult to interpret. For example, if an animal has an *R* value of 0.8, a researcher cannot simply conclude that its cells are catabolizing proteins, because the catabolism of a mixture of carbohydrates and lipids (or of all three foodstuffs) could also produce an *R* value of 0.8. The potential for ambiguity stems from the fact that although three unknowns exist (the proportions of the three foodstuffs being oxidized), the *R* or *RQ* value encompasses only two knowns (O_2 consumption and CO_2 production). The ambiguity can be resolved, but only by the measurement of additional indices of cellular function besides O_2 consumption and CO_2 production.

Investigators often sidestep the uncertainty concerning foodstuffs by using a "representative" conversion factor of 20.2 J/mL O_2 (4.8 cal/mL O_2) to calculate an animal's metabolic rate from its O_2 consumption. The representative factor approximates the heat produced by an animal that is assumed to be catabolizing a representative mixture of carbohydrates, lipids, and proteins. Using a representative factor of 20.2 J/mL O_2 to calculate metabolic rate does not do away with the uncertainty we have been discussing. To illustrate, suppose that an investigator uses the conversion factor of 20.2 J/mL O_2 but that the animal was actually oxidizing only carbohydrates. Because the true conversion factor (for carbohydrates) is 21.1 J/mL O_2, the investigator would underestimate the metabolic rate by 4.5% by using the approximate factor of 20.2 J/mL O_2. If the animal were oxidizing only proteins, the true conversion factor (according to Table 5.1) would be 18.7 J/mL O_2; thus the investigator would overestimate the metabolic rate by 8% by using the approximate factor. The use of the representative conversion factor is a "good news, bad news" situation. The bad news is that the metabolic rate can be misestimated by ±5% to 8% if one ignores the foodstuff question and simply multiplies the O_2 consumption by the representative conversion factor to calculate metabolic rate. The good news is that the use of this conversion factor is convenient, and for many purposes an error of ±5% to 8% may not be much of a worry.[7]

Among all the methods available to measure metabolic rates, the most common approach used today for routine metabolic rate studies is to measure the rate of O_2 consumption—nothing more—and "live with" the relatively small potential errors that are inherent in the method. Metabolic rates, in fact, are very often expressed simply

[6] To calculate *R* or *RQ*, one can divide the *volume* of CO_2 produced per unit time by the *volume* of O_2 consumed per unit time. This is true because all gases occupy essentially the same volume per mole under given conditions (that value is approximately 22.4 L per mole at standard conditions of temperature and pressure).

[7] If one measures only CO_2 production and calculates metabolic rate with a representative conversion factor, the potential error is much greater: ±15% to 20%. You can see the principal reason in Table 5.1; namely, the relationship between CO_2 production and heat production depends very strongly on foodstuff. Because of the large potential error, CO_2 production—measured by itself—is hardly ever used as an indirect measure of metabolic rate in laboratory studies.

TABLE 5.2 Respiratory exchange ratios (*R* values) during the aerobic catabolism of carbohydrates, lipids, and proteins Respiratory quotients (*RQ* values) are equal in steady state.

Foodstuff	Respiratory exchange ratio
Carbohydrates	1.0
Lipids	0.71
Proteins	0.83[a]

Source: After Kleiber 1975.
[a] The protein *R* value is for animals such as mammals in which urea is the dominant nitrogenous end product. Different values apply to animals that produce ammonia or uric acid as their principal nitrogenous end product.

as rates of O_2 consumption. Besides its technical ease of accomplishment (Box 5.4) and relatively small inherent uncertainty, a third advantage of using O_2 consumption to measure metabolic rate is that external work by an animal need never be measured independently when this method is used. Because the O_2 consumption of an animal is proportional to the ultimate yield of heat from the foodstuffs it catabolizes, the heat equivalent of any external work performed by the animal is included in the metabolic rate computed from its O_2 consumption. An important point to keep in mind about the O_2 consumption method, however, is that O_2 consumption reflects only *aerobic* catabolism. Thus, if part or all of an animal's catabolism is anaerobic, one must turn to alternative methods, such as direct calorimetry, to measure metabolic rate (see Chapter 6).

INDIRECT CALORIMETRY BASED ON MATERIAL BALANCE Besides methods based on respiratory gas exchange, the second most commonly used approach for measuring animal metabolic rates today is a type of indirect calorimetry in which investigators measure the chemical-energy content of *organic materials* entering and leaving an animal's body. This approach, which is widely used in agricul-

BOX 5.4 Respirometry

Respirometry is the process of measuring an animal's gas exchange with its environment. The devices used are called *respirometers*. For studies of metabolic rate, the most common type of respirometry is the measurement of an animal's rate of O_2 consumption. Two basic respirometry configurations are used: *closed* and *open*.

In **closed respirometry** (Figure A), air does not flow through the animal chamber during measurement. The principal practical challenge with closed respirometry is to prevent temperature-induced changes in gas volume from confounding the results. This problem is solved in the closed respirometer shown in Figure A by connecting two identical chambers to either side of a *manometer*, a device that measures gas-pressure differences (the U-shaped tube filled with water in Figure A). If the environmental temperature changes, the gas pressure within both chambers rises equally, exerting equal and opposite increases of pressure on the two sides of the manometer; in this way, the position of the water in the manometer is unaffected by changes of environmental temperature.

In the closed respirometer, as the animal uses O_2, the CO_2 absorbent in the animal chamber removes exhaled CO_2 from the air in the chamber. Thus the animal's O_2 consumption causes the total gas volume in the animal chamber to decrease relative to that in the other (animal-free) chamber; this decrease in the gas volume in the animal chamber causes the water in the manometer to shift, rising higher in the left arm than the right. At timed intervals, enough pure O_2 is injected from the syringe into the animal chamber to make the manometer return exactly to the initial, unshifted position shown in Figure A. The amount of O_2 injected at any one moment in this procedure must equal the amount of O_2 that the animal consumed during the interval of time preceding the injection. By measuring the O_2 injected and knowing the length of the preceding time interval, a researcher can calculate the animal's O_2 use per unit of time.

In **open respirometry** (Figure B), air flows through the animal chamber (or through a mask worn by the animal) during the measurement of O_2 consumption. The rate of airflow is measured as carefully as possible. A precision O_2 meter, typically using an electrochemical or paramagnetic cell for O_2 detection, measures the O_2 concentration of the flowing air just before the air enters the animal chamber and just afterward. The researcher calculates the animal's rate of O_2 consumption by taking account of the volume of air passing through the chamber per unit of time and the amount of O_2 extracted from each unit volume. Although open respirometry requires the use of far more costly equipment than closed respirometry does, it permits continuous, minute-by-minute (even instantaneous) monitoring of an animal's rate of O_2 consumption.

Figure B An open respirometer

Air enters
Carefully measured airflow
20%
21%
Air exits

Figure A A closed respirometer

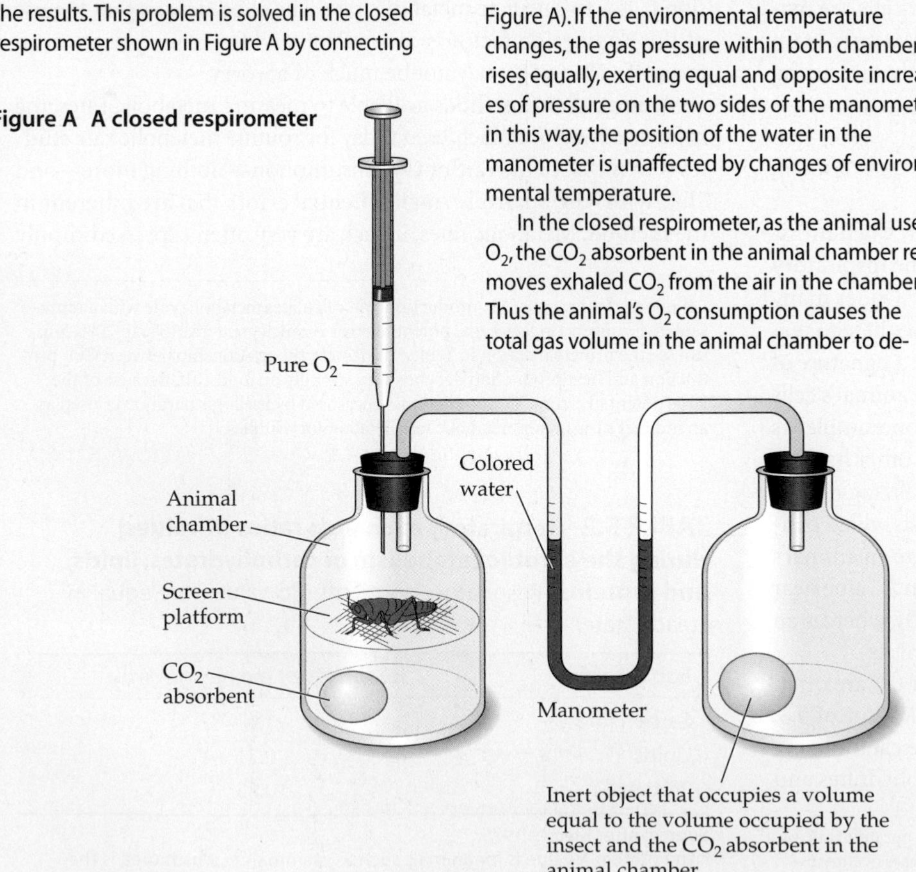

tural production research, is described as the study of **material balance.** To apply the method, one measures the chemical-energy content of all the food an animal eats over a period of time, as well as the chemical-energy content of the feces and urine eliminated over the same period of time.[8] Subtracting the energy content of the excreta from that of the food then gives an estimate of the animal's metabolic rate. The logic of the method is straightforward: Any energy that an animal ingests as chemical energy, but does not void as chemical energy, must be consumed.

Complications arise if the animal under study is increasing or decreasing its biomass. If, for example, an animal is growing and thus increasing the chemical-energy content of its body, some of the chemical energy ingested but not voided is nonetheless not consumed; an estimate of this quantity must enter the calculation of metabolic rate. Another type of complication is that chemical energy may enter or leave an animal's body in other ways than in food, feces, and urine. For instance, an animal could lose chemical energy by shedding feathers or secreting mucus. For the material-balance method to be applied, the chemical-energy content of *all* significant inputs and outputs of organic material must be measured.

To use the material-balance method, measurements of ingestion, egestion, and other relevant processes must extend over a substantial period of time—typically 24 hours or more—so that average, steady-state rates of input and output of chemical energy will be quantified. The metabolic rate calculated from the method is the animal's average rate over the entire study period. Thus the material-balance method is suited only for *long-term measurements of average metabolic rates*. To measure minute-to-minute variations in metabolic rate, the methods of choice are those based on respiratory gas exchange or direct calorimetry.

[8] An instrument known as a *bomb calorimeter* is used to measure the energy values of the organic materials. It does this by burning them in pressurized, pure O_2 and measuring the heat evolved.

Summary: Metabolic Rate

- An animal's metabolic rate is the rate at which it converts chemical energy into heat and external work.
- Metabolic rate is important because it helps determine the amount of food an animal needs, and therefore the food energy that the animal removes from its ecosystem. Metabolic rate also measures an animal's intensity of existence.
- An animal's rate of O_2 consumption is the most common measure of metabolic rate. Metabolic rates can also be measured by direct calorimetry or studies of material balance.

Factors That Affect Metabolic Rates

Now that we have discussed how metabolic rates are defined and measured, we can turn our attention to the experiences of animals and the processes within them that influence their metabolic rates. The two factors that typically exert the greatest effects on an animal's metabolic rate are the intensity of its physical activity (e.g., speed of running) and the temperature of its environment. Other factors that commonly influence animal metabolic rates include the ingestion of food, age, gender, time of day, body size, reproductive condition, hormonal state, psychological stress, and for aquatic animals, the salinity of the ambient water. Table 5.3 provides an overview of many of these factors and identifies where they are discussed in this book.

TABLE 5.3 Some factors that affect the metabolic rates of individual animals

Factor	Response of metabolic rate	Chapter(s) where discussed in this book
Exerting largest effects		
Physical activity level (e.g., running speed)	↑ with rising activity level	6, 7
Environmental temperature	*Mammals and other homeotherms:*	8
	Lowest in thermoneutral zone	
	↑ below thermoneutral zone	
	↑ above thermoneutral zone	
	Fish and other poikilotherms:	
	↑ with increasing temperature	
	↓ with decreasing temperature	
Exerting smaller effects		
Ingestion of a meal (particularly protein-rich)	↑ for several to many hours following ingestion	5
Body size	Weight-specific rate ↑ as size ↓	5
Age	Variable; in humans, weight-specific rate ↑ to puberty, then ↓	
Gender	Variable; in humans, ↑ in male	
Environmental O_2 level	Often ↓ as O_2 ↓ below a threshold, not affected above threshold	6, 21
Hormonal status	Variable; example: ↑ by excessive thyroid secretions in mammals	14
Time of day	Variable; in humans, ↑ in daytime	10
Salinity of water (aquatic animals)	Variable; in osmoregulating marine crabs, ↑ in dilute water	26

Ingestion of food causes metabolic rate to rise

Among the factors that affect metabolic rate, the ingestion of food deserves some extended consideration, even though it is not the most influential factor quantitatively, because it must be taken into account in almost all metabolic studies. Under many circumstances, if an animal has been fasting and then consumes food, its metabolic rate temporarily increases following the meal *even though all other conditions are kept constant*. This increase in metabolic rate caused by food ingestion is known as **specific dynamic action** (**SDA**), the **calorigenic effect of ingested food**, or the **heat increment of feeding** (Figure 5.5). We may often not notice this process in our day-to-day lives, but it is very apparent at certain times. Think back, for example, to a festive holiday dinner when everyone ate lots of high-protein food such as poultry or other meat. After such a meal, people may feel so warm that they remove sweaters and loosen neckties or scarves. The reason for the sense of excessive warmth is the SDA of the ingested protein. The occurrence of SDA means that a certain portion of the energy available from a meal is degraded to heat in processing the meal; only the remaining portion of the energy is available for subsequent physiological uses.

The *magnitude* of the SDA following a meal is the total *excess* metabolic heat production induced by the meal, integrated from the time metabolism first rises to the time that it falls back to the background level. Thus the blue area in Figure 5.5*a*, showing the integrated difference between the actual metabolic rate after a meal and the metabolic rate that would have prevailed without eating, represents the magnitude of the SDA. The magnitude of the SDA occurring after a meal of a particular type of food tends to be roughly proportional to the amount of food eaten, for given animals under given conditions (Figure 5.5*b*). Doubling the amount of food eaten tends approximately to double the SDA. Protein foods exhibit much higher SDAs, in proportion to the amount eaten, than lipids or carbohydrates. Although traditionally the SDA of a protein meal has been considered rather dogmatically to be equivalent to 25% to 30% of the total energy value of the meal, recent research indicates that the percentage (while high) varies with prevailing conditions.

The mechanism of SDA remains uncertain. Although digestive processes make a contribution, strong evidence exists that the SDA arises mainly *after the absorption* of digestive products from the gastrointestinal tract, as a consequence of cellular processing of the absorbed organic compounds. An important cause of the SDA associated with protein meals, for example, is believed to be the energy expenditure required to synthesize nitrogenous waste products (e.g., urea in mammals) to dispose of nitrogen from excess amino acids.

The SDA is a relatively short-term phenomenon, but sometimes an animal's diet induces a *semipermanent*, or *chronic*, change in its metabolic rate. This effect was discovered only about 20 years ago. Experiments revealed that if laboratory rats are enticed to eat unusually large amounts of food day after day (as by the addition of sweets to their food), some individuals do not fatten, because their metabolic rates chronically rise, turning the excess food energy into heat. The long-term increase in metabolic rate induced by persistent overeating was dubbed **diet-induced thermogenesis** (**DIT**). Research on DIT has been intense ever since its discovery because DIT is an anti-obesity process of potential human importance. The relation between DIT and SDA is confused at present. We say more about DIT in Chapters 6 and 8.

Basal Metabolic Rate and Standard Metabolic Rate

Physiologists, ecologists, and other biologists often wish to compare metabolic rates. A physician, for example, might want to know how the metabolic rate of a particular patient compares with the average metabolic rate of all people of similar age, because some diseases are distinguished by abnormal rates of energy consumption. An ecologist might want to compare the metabolic rates of two species in an

(*a*) The concept of SDA

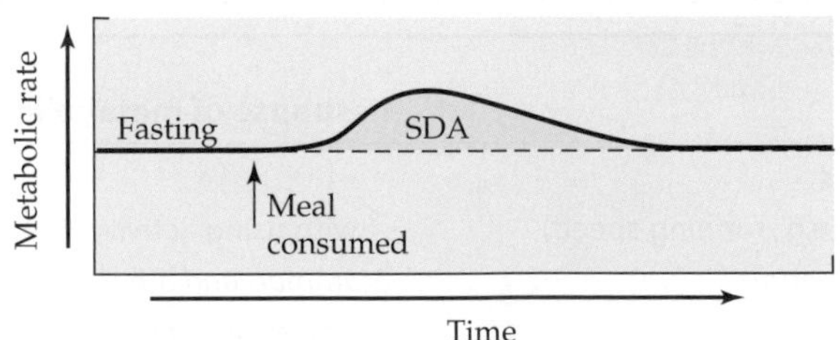

(*b*) SDA in flatfish

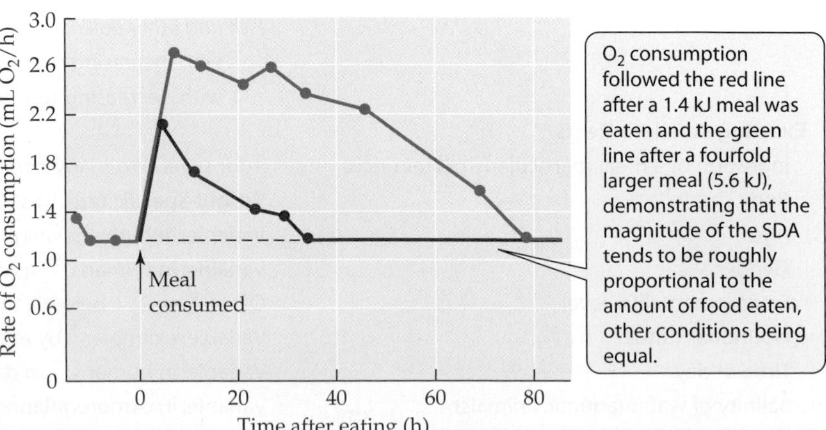

Figure 5.5 Specific dynamic action (*a*) A resting, fasting animal is fed at the time marked and then remains at rest. The SDA begins after a delay, which may be up to an hour. The solid line shows the animal's actual metabolic rate. The dashed line depicts what the rate would have been, had the meal not been eaten. The area shaded blue is the magnitude of the specific dynamic action (SDA). There is enormous variation in the timing of this process; the SDA might be over in a few hours in a mouse, in 12 hours in a cow, and in 1 to 3 days in a fish. (*b*) Actual data for predatory flatfish (*Pleuronectes*) fed two different-sized meals of fish meat. (*b* after Jobling 1993.)

attempt to learn which species is more likely to place high demands for food on an ecosystem.

In making comparisons, it is often important to standardize factors that could confound results. For example, physicians typically standardize food ingestion during diagnostic metabolic studies by having patients fast for at least 12 hours prior to measurements of their metabolic rates; otherwise, some patients would have their measured metabolic rates elevated by SDA, whereas others would not, creating confounding variation. Similarly, biologists who want to know if species have inherently different metabolic rates typically standardize physical activity, so that results are not confounded by having the individuals of one species walking around while those of the other rest during measurement. A number of standardized measures of metabolic rate have been defined to facilitate valid comparisons. The two most commonly used standardized measures are the *basal metabolic rate* and the *standard metabolic rate*.

The **basal metabolic rate** (**BMR**) is a standardized measure of metabolic rate that applies to *homeotherms*, animals that physiologically regulate their body temperatures, such as mammals and birds. For each homeothermic species, there is a range of environmental temperatures within which the metabolic rate is minimal. This range, called the *thermoneutral zone*, is illustrated in Figure 8.22. The basal metabolic rate of a homeotherm is the animal's metabolic rate while it is (1) in its thermoneutral zone, (2) fasting, and (3) resting.

The concept of **standard metabolic rate** (**SMR**) applies to *poikilotherms* (*ectotherms*), animals that allow their body temperatures to fluctuate freely with variations in environmental temperature, such as amphibians, molluscs, and most fish. The standard metabolic rate is the metabolic rate of a poikilothermic animal while it is (1) fasting and (2) resting. The SMR is specific for the prevailing body temperature; thus there are as many SMRs as there are different body temperatures.

Both of the standardized measures of metabolic rate mentioned here call for subjects to be fasting and resting. In studies of metabolic rate, the term **fasting** (or **postabsorptive**) has a specific meaning—namely, that a subject's last meal took place long enough ago for the SDA of the meal to be over. The term *resting* can have somewhat different meanings in different studies because inducing animals to rest is often not simple. Different levels of rest are formally recognized in some subdisciplines of physiology. Fish physiologists, for instance, often use the term **routine metabolic rate** to refer to metabolic rates of reasonably quiet fish exhibiting only small, spontaneous movements and reserve *standard metabolic rate* to describe fish that have been coaxed to a truly minimal level of activity.

The Relation between Metabolic Rate and Body Size

How does metabolic rate vary with body size within sets of phylogenetically related species? This simple question turns out to have an intricate and profoundly important answer. A comparison between meadow voles and white rhinos provides a revealing starting point for understanding the relation between metabolic rate and body size. Both of these species are mammals, and both eat similar foods, being "pure grazers" that eat little else besides grassland plants. They are very different in body size, however. An interesting way to gain insight into the effect of their different body sizes on their metabolic rates is to compare how much food they must eat to meet their metabolic needs. If we pile up all the grass that a vole and a rhino must eat in a week under similar measurement conditions, we find, not surprisingly, that the rhino requires more food than the vole. However, a week's pile of food for the vole is larger than the vole itself, whereas the pile for the rhino is much smaller than the rhino (Figure 5.6). This disparity indicates that *the energy needs of the species are not proportional to their respective body sizes.*

Resting metabolic rate is an allometric function of body weight in related species

To fully understand the relation between metabolic rate and body size, it is important to compare large numbers of related species, not just two. The BMRs of over 600 species of placental mammals have been measured. When all these BMRs are plotted against the body weights of the species, statistics can be used to draw the best-fitted line through the data (see Appendix E: Fitting Lines to Data). From this sort of analysis, physiologists have discovered that the pattern we have observed in comparing voles and rhinos is in fact a general pattern that applies across the full range of mammalian body sizes. Although the total BMR of species of placental mammals increases with body weight, it does not increase in proportion to weight. Instead, the total BMR increases less than proportionally with body

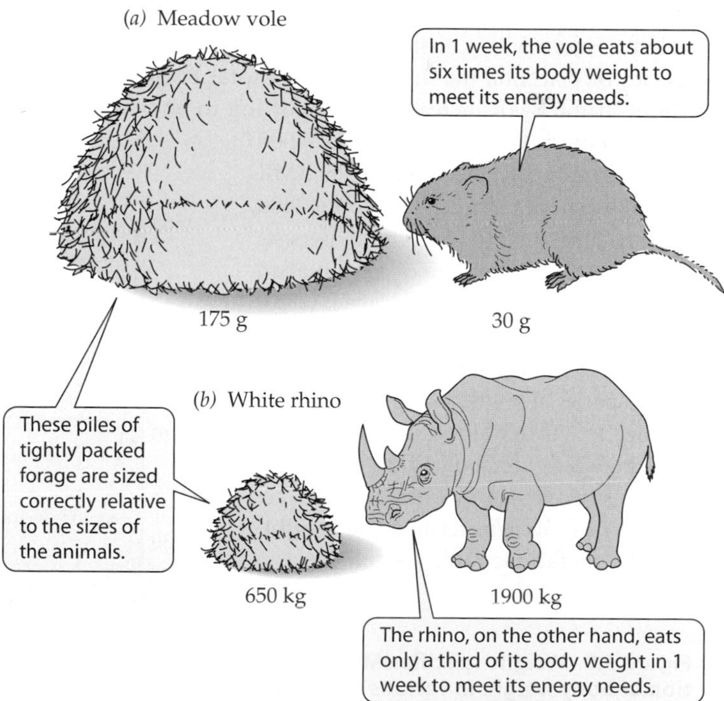

Figure 5.6 The effect of body size on weekly food requirements Both species—*(a)* the 30-g meadow vole (*Microtus pennsylvanicus*) and *(b)* the 1900-kg white rhino (*Ceratotherium simum*)—are grazers. (Calculated from Golley 1960 and Owen-Smith 1988, assuming 70% moisture content in the forage.)

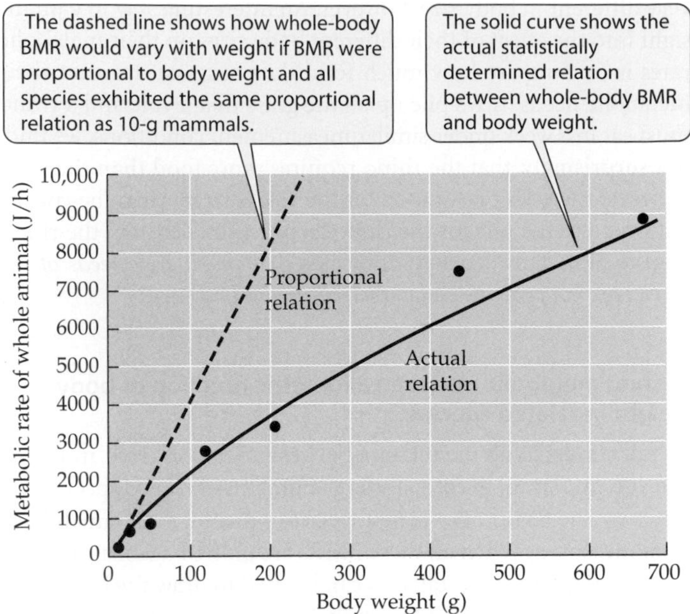

Figure 5.7 BMR as a function of body weight in various species of placental mammals The curve showing the actual relation is statistically fitted to data for all sizes of mammals, although this plot includes body weights up to only 700 g. The points are data for seven North American species (see Figure 5.8 for identifications), illustrating that although the statistical line runs through the data, individual species do not necessarily fall right on the line. (After Hayssen and Lacy 1985.)

weight (Figure 5.7). Consider, for example, that an average 10-g species of placental mammal exhibits a total BMR of about 400 J/hour. If the BMR increased in proportion to body size, a 100-g species would have a total BMR of 4000 J/hour. In actuality, the average total BMR of a 100-g species is much less, about 2200 J/hour. This quantitative trend persists through the entire range of mammalian weights. For instance, the average BMR of a 400-g species of placental mammal is only about 2.7 times higher than that of a 100-g species, not 4 times higher.

An alternative way to examine the relation between metabolic rate and body weight is to calculate the metabolic rate *per unit of body weight*, termed the **weight-specific metabolic rate**, and plot it as a function of body weight. Figure 5.8 presents the data points and curve from Figure 5.7 in this new way. This representation illustrates that the weight-specific BMR of mammals decreases as weight increases. Under basal conditions, a 670-g desert cottontail rabbit produces only about 40% as much metabolic heat per gram as a 21-g white-footed mouse. These trends become even more dramatic if we examine mammals still larger than those represented in Figure 5.8. A 70-kg human produces about 10% as much heat per gram as the mouse; and a 4000-kg elephant, about 5% as much. The basal energy expenditure of a gram of mammalian tissue is far lower if it is a gram of elephant than a gram of mouse!

Suppose that instead of mammals, we look at many species of fish of different body weights—or many species of crustaceans of different sizes. Do poikilothermic animals (ones having variable body temperatures) exhibit the same sorts of relations between metabolic rate and body size? Yes, for phylogenetically related poikilothermic species of different body sizes, the standard metabolic rate (SMR) at a particular body temperature usually varies with body size in the same basic pattern as seen in mammals. The total SMR increases with body weight, but it increases less than proportionally. Thus the weight-specific SMR decreases as body weight increases.

The patterns we are discussing are of such great generality that a single type of equation describes the relation between resting metabolic rate and body weight within most groups of related species. If M symbolizes the total (whole-animal) BMR or SMR, and W symbolizes body weight, then usually

$$M = aW^b \qquad (5.3)$$

where a and b are constants. The constants in the equation are determined statistically. To determine a and b for placental mammals, for example, researchers begin by tabulating the total BMR and the body weight for as many species as possible. The raw information used for the process, in other words, is a series of points on M, W co-

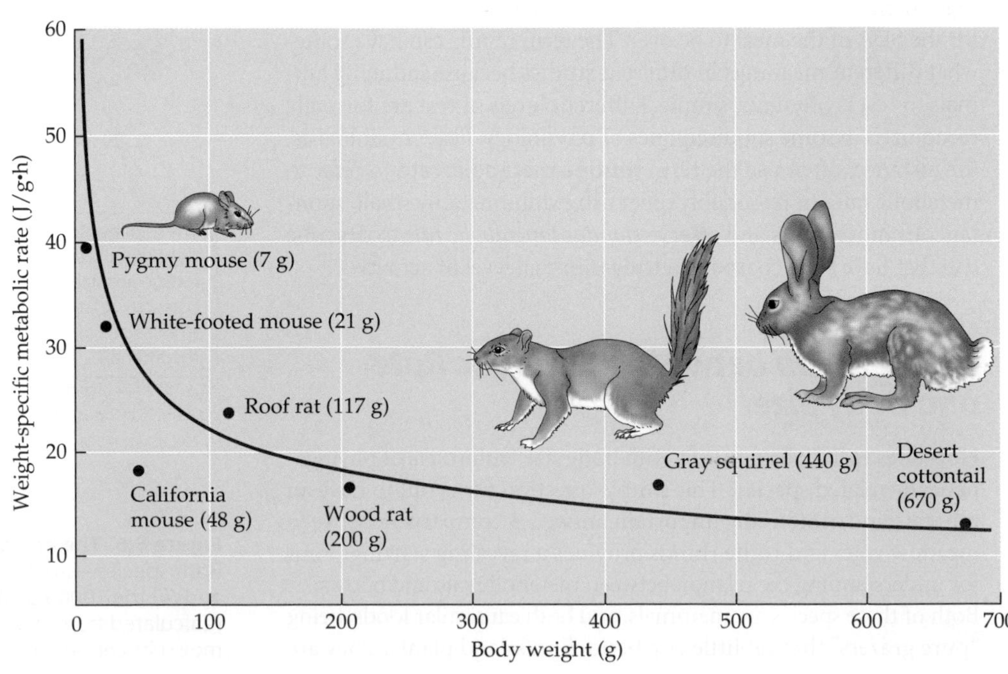

Figure 5.8 Weight-specific BMR as a function of body weight in various species of placental mammals The *x* axis (body weight) and animals are the same as in Figure 5.7, but here the BMRs (*y* axis) are expressed per gram of body weight. Small mammals have higher weight-specific BMRs than big ones. (After Hayssen and Lacy 1985.)

ordinates, such as those plotted on the graph in Figure 5.7. The researchers then use a statistical algorithm to fit an equation of the form $M = aW^b$ as closely as possible to the data points. Such a statistical procedure calculates the values of a and b that make the equation match the data points as closely as possible (see Appendix E: Fitting Lines to Data).

Note that the values of a and b depend not only on the animal group being studied but also on the particular data used. Thus if two researchers independently investigated mammals using data for different sets of species, they would obtain slightly different values of a and b.

If b equaled 1.0, Equation 5.3 would become $M = aW$, a proportional relation. However, b is almost always less than 1.0. If b is not 1.0, an equation like Equation 5.3 (having the general form $Y = aX^b$) is nonproportional and is known as an **allometric equation**.[9] Resting metabolic rate (BMR or SMR) is said to be an allometric function of body weight because the relation between metabolic rate and weight follows the form of the allometric equation.

Biologists have discovered that the exponent b in the allometric relation between metabolic rate and body weight exhibits an impressive consistency in its value from one phylogenetic group to another.[10] *The value of* b *for the resting metabolic rates of diverse groups of animals is usually about 0.7.* This is true of mammals, fish, crustaceans, and most other animal groups. Most commonly, b is between 0.65 and 0.75; it is nearly always between 0.6 and 0.9. Universal biological principles seem to be at work in determining b.

Unlike the value of b, the value of a is not at all consistent from one phylogenetic group to another. From Equation 5.3, you can see that $M = a$ when $W = 1$. Thus, a is the metabolic rate of a 1-gram animal (real or theoretical) in the phylogenetic group under consideration. Some phylogenetic groups (e.g., mammals) have intrinsically much higher metabolic rates than others (e.g., fish) and thus also have much higher values of a.

What is the mathematical relation between *weight-specific* resting metabolic rate and body weight? This relation is easily derived if both sides of Equation 5.3 are divided by W, yielding

$$M/W = aW^{(b-1)} \tag{5.4}$$

The expression M/W is the weight-specific BMR or SMR, and you can see that it is an allometric function of W. The value of a is the same as in Equation 5.3, but the exponent in Equation 5.4 is $(b - 1)$. Because b is usually about 0.7, the exponent here is usually about –0.3. The negative value of $(b - 1)$ signifies what we have already said; namely, weight-specific resting metabolic rate *decreases* with increasing body weight. Four examples of Equation 5.4, fitted to four different groups of vertebrates, are presented in Figure 5.9. Note the similarity of the exponents but the differences in a, signifying different metabolic intensities in the four types of animals.

A useful property of Equations 5.3 and 5.4 is that they are linear equations when plotted on log–log coordinates (see Appendix G: Exponential and Allometric Equations). For example, taking the logarithm of both sides of Equation 5.3 yields

$$\log M = \log a + b \log W \tag{5.5}$$

Note that the dependent variable in Equation 5.5 ($\log M$) equals a constant ($\log a$) plus the independent variable ($\log W$) multiplied by another constant (b). This means that Equation 5.5 describes a straight line. Thus a plot of $\log M$ as a function of $\log W$ is linear, and similarly, a plot of $\log M/W$ against $\log W$ is linear. Data relating metabolic rate to weight are nearly always graphed on a log–log plot, as exemplified by Figure 5.10. An advantage of logarithmic axes is that they can accommodate very wide ranges of values. Figure 5.10*a*, for instance, permits species of mammals ranging in weight from 77 g to almost 200,000 g to be analyzed together on one graph.

The metabolic rate of active animals is often also an allometric function of body weight

Sustained vigorous physical activity causes an animal's rate of aerobic metabolism to reach a maximum. How does the maximal rate of aerobic metabolism compare to the resting metabolic rate, and how does it vary with body weight within sets of phylogenetically related

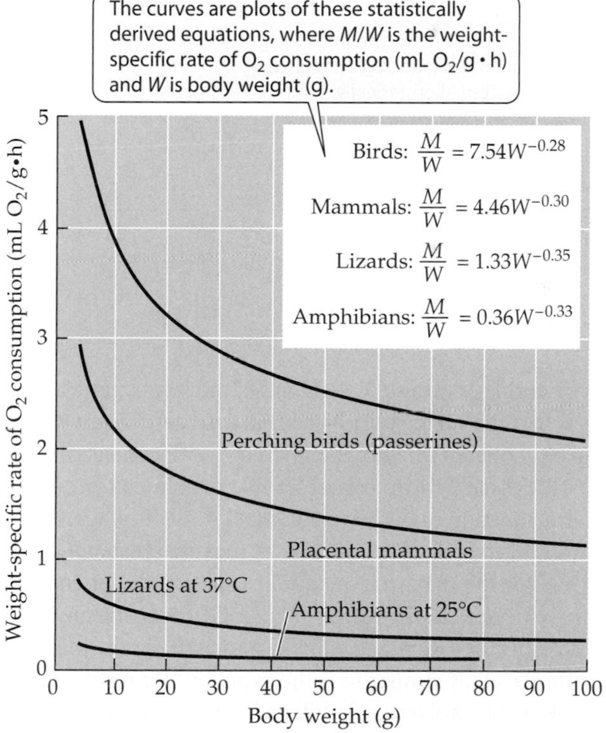

Figure 5.9 Weight-specific metabolic rate as a function of body weight in four groups of vertebrates The lines for birds and mammals show BMR. The line for lizards shows the SMR when the lizards are at the same body temperature as placental mammals, 37°C. The line for amphibians shows the SMR in temperate-zone anurans and salamanders at a body temperature of 25°C. The lines are based on the statistically derived equations shown, where M/W is weight-specific rate of O_2 consumption (mL O_2/g • h) and W is body weight (g). (Sources for equations: birds, Lasiewski and Dawson 1967; mammals, Hayssen and Lacy 1985; lizards, Templeton 1970; amphibians, Whitford 1973.)

[9] This form of equation is discussed in Appendix G: Exponential and Allometric Equations.

[10] The exponent b is dimensionless and thus has no units of measure.

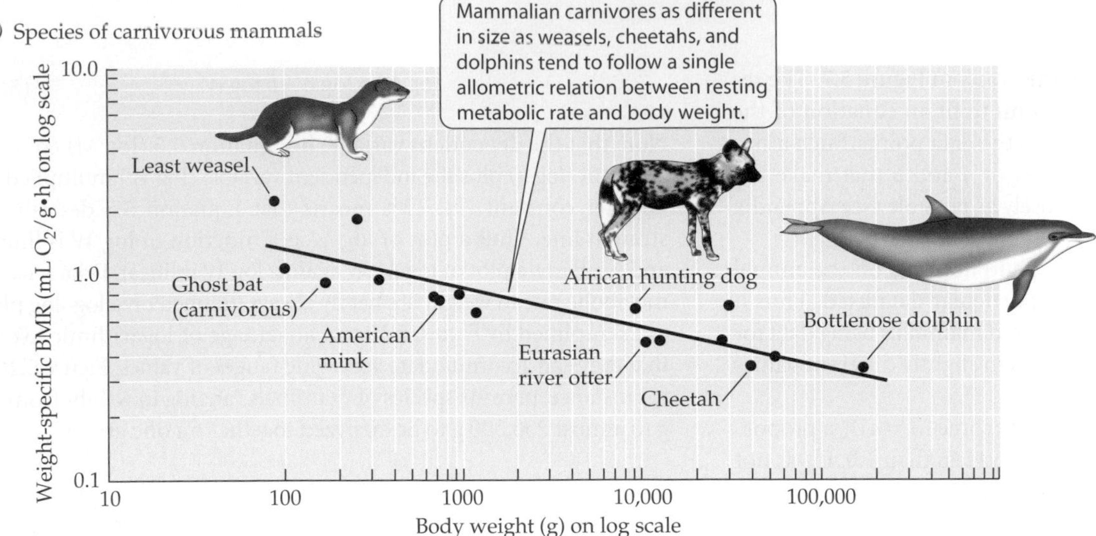

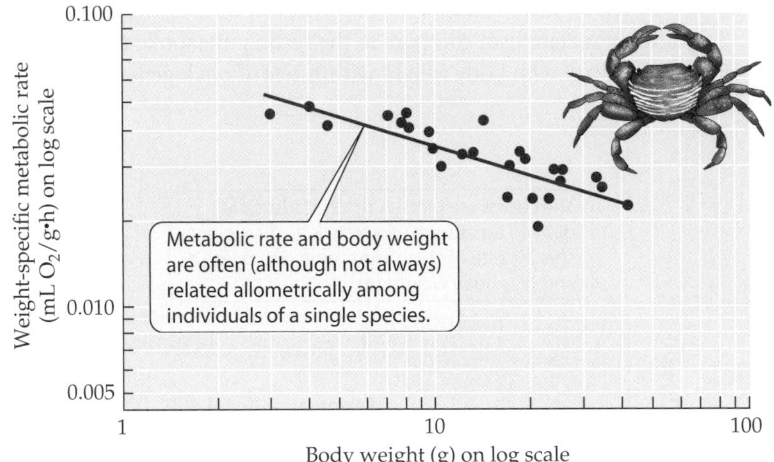

Figure 5.10 Metabolic rate and body weight are related linearly on log–log coordinates *(a)* Weight-specific BMR as a function of body weight for mammalian species that eat primarily vertebrate flesh, plotted on log–log coordinates. The points represent individual species; the line is statistically fitted to them. *(b)* A log–log plot of weight-specific metabolic rate as a function of body weight in a common Pacific shore crab (*Pachygrapsus crassipes*) at a body temperature of 16°C. Each point represents a particular individual. The line is statistically fitted to the points. See Appendix F (Logarithms) for the axis layouts. (*a* after McNab 1986; *b* after Roberts 1957.)

species? A useful rule of thumb in vertebrates and some groups of invertebrates is that the exercise-induced maximal aerobic metabolic rate tends to be about ten times higher than the resting metabolic rate (BMR or SMR). Usually the maximal aerobic metabolic rate (M_{max}) is an allometric function of body weight: $M_{max} = a'W^{b'}$, where a' and b' are the constants that apply in the case of maximal metabolic rate. The value of a' for the maximal metabolic rate tends to be roughly ten times the value of a for the resting metabolic rate (corresponding to the tenfold difference we noted). The exponent b' for the maximal metabolic rate is usually similar to b, the exponent for resting metabolic rate. However, b' and b clearly differ a bit in mammals and some other groups that can be analyzed in close detail using available data; in placental mammals, for example, whereas the exponent for resting metabolic rate is about 0.7, that for maximal metabolic rate is about 0.85. The difference in exponents is noteworthy because it has theoretical significance, as we will soon discuss.

What about the *average* metabolic rates of animals living in nature? Animals in nature sometimes rest and sometimes are active. Recognizing that both resting metabolic rate and maximal aerobic metabolic rate are typically allometric functions of body weight, we might expect that *average daily* metabolic rate is also an allometric function of weight within sets of phylogenetically related animals. It is.[11]

The metabolic rate–body weight relation has important physiological and ecological implications

The allometric relation between metabolic rate and body weight suggests that different-sized, but related, species are likely to differ in how well their cells and tissues are endowed with mitochondria and other components of the aerobic catabolic apparatus. Cellular properties indeed often vary allometrically with body size. In animals as diverse as mammals and fish, for example, the skeletal muscles of small species have more mitochondria per unit of tissue than those of large species. The density of mitochondria varies allometrically with body weight, paralleling the relation between weight-specific metabolic rate and weight.

Because the breathing and circulatory systems are responsible for delivering the O_2 consumed by the tissues, you might predict that

[11] Chapter 7 discusses methods used to measure the average daily metabolic rates of free-living animals.

TABLE 5.4 Resting heart rate, and heart size relative to body weight, in seven species of mammals

Species and average body weight	Resting heart rate (beats/min)[a]	Heart weight per unit body weight (g/kg)[b]
African elephant (4100 kg)	40	5.5
Horse (420 kg)	47	7.5
Human (69 kg)	70	5.2
Domestic dog (19 kg)	105	9.2
Domestic cat (3 kg)	179	4.1
Roof rat (0.34 kg)	340	2.9
Lab mouse (0.03 kg)	580	4.0

Source: After Seymour and Blaylock 2000.

[a] According to the source of these data, the statistical relation between resting heart rate (*RHR*) and body weight (*W*) in mammals is *RHR* (beats/min) = $227W^{0.23}$, where *W* is in kilograms.

[b] Although heart weight per unit body weight varies, it shows little or no consistent relation to body size.

key features of respiratory and circulatory physiology are also allometrically related to body size within sets of related species. If such predictions made millionaires, you would be one!

The resting heart rate in placental mammals, to cite one example, varies with body weight in almost exactly the same functional relation as weight-specific BMR, small species having far higher heart rates than large ones (Table 5.4). This pattern makes sense in view of the fact that statistically, all mammals have about the same size of heart in relation to their body size. Small mammals require more O_2 per gram of body weight than large ones, but their hearts are no bigger relative to their body size than the hearts of large mammals. Thus the hearts of small mammals must beat faster than those of large mammals for O_2 to be delivered at a greater rate per unit of body weight. Various species of mammals also have about the same size of lungs in relation to body size. Thus, by much the same logic, small mammals must breathe more rapidly than large ones. Whereas humans, for example, breathe about 12 times per minute when resting, mice breathe about 100 times per minute. The metabolism–weight relation pervades almost every aspect of an animal's physiology.

Numerous ecological and practical implications also arise from the allometric relation between metabolic rate and body weight. We have already seen in our initial comparison of voles and rhinos that within sets of phylogenetically related animals, small-bodied species typically require food at a greater rate per unit of body weight than large-bodied species (see Figure 5.6). Ecologically, the uninitiated might expect that the basal food requirements of 3500 mice, each weighing 20 g (total weight 70,000 g), would place no greater demands on a woodland ecosystem than a single 70,000-g deer would. Because the weight-specific BMR of a 20-g mouse is about eight times greater than that of a deer, however, the total basal food requirement of only about 440 mice is equivalent to that of a single deer.

When whole ecosystems are analyzed, ecologists sometimes observe that the allometric relation between metabolic rate and body weight has a significant structuring effect. Consider, for example, the woodland savannahs of eastern and southern Africa. These ecosystems are among the marvels of life on Earth, in part because they support such an extreme diversity of coexisting antelopes and other medium-sized to large mammalian herbivores (Figure 5.11). From aerial surveys of major national parks in Africa, we know the average numbers of many herbivore species per square kilometer. Multiplying numbers by body weights, we can calculate the average population biomass per square kilometer of each species. Population biomass per square kilometer turns out to be a regular function of body size; for example, whereas all the warthogs (a relatively small species) living per square kilometer together weigh about 95 kg, the zebras in a square kilometer collectively weigh 460 kg, and the elephants weigh 1250 kg (Table 5.5). Metabolic allometry, while not the only factor at work, helps explain this trend, because each kilogram of a large-bodied species requires less food than each kilogram of a small-bodied species.

The allometric relation between metabolic rate and body size also means that related small and large species process foreign chemicals differently. Small species, for instance, tend to be more prone to accumulating environmental pollutants. Because the small species eat food and breathe air at greater rates per unit of body weight than their larger counterparts, they tend to receive greater weight-specific doses of food-borne and airborne toxins such as pesticides; the toxins may then accumulate more readily to high concentrations in their tissues. On the other hand, the high weight-specific metabolic rates of small species also mean that small animals tend to catabolize or excrete some substances faster per unit weight than related larger animals. One practical consequence of this fact is that small species may require relatively high doses of a veterinary drug per unit weight to achieve and sustain the drug's intended effect.

Figure 5.11 Herbivores of different body sizes coexisting on an African grassland Species vary in population biomass in a way that depends in part on body size and metabolic allometry.

TABLE 5.5 Biomasses of populations of selected herbivores living in mixed communities in African national parks Species are listed in order of increasing individual size. These species were chosen for listing because they are statistically about average in population biomass for their body sizes.

Species	Average biomass of whole population per square kilometer (kg/km^2)	Average individual body weight (kg)
Oribi (*Ourebia ourebi*)	44	13
Gray duiker (*Sylvicapra grimmia*)	62	16
Gray rhebok (*Pelea capreolus*)	105	25
Warthog (*Phacochoerus aethiopicus*)	95	69
Waterbuck (*Kobus ellipsiprymnus*)	155	210
Greater kudu (*Tragelaphus strepsiceros*)	200	215
Plains zebra (*Equus burchelli*)	460	275
White rhino (*Ceratotherium simum*)	2400	1900
African elephant (*Loxodonta africana*)	1250	3900

Source: After Owen-Smith 1988.

The explanation for allometric metabolism–weight relations remains unknown

The fact that b, the allometric exponent, is often about 0.7 in widely diverse sorts of animals is profoundly intriguing. The exponent, in fact, is often about 0.7 in both microbes and plants! For a century, some of the greatest minds in biology have grappled with the questions of *why* metabolic rate and body weight are related allometrically and *why* the allometric exponent so consistently assumes the value it does. Great minds have been drawn to these questions because of a conviction that the allometries are manifestations of fundamental—and possibly universal—organizing principles of life. There is as yet, however, no consensus about how to explain the allometries.

Eighty years ago, the problem seemed solved. Physiologists thought they then understood the reasons for not only the allometric relation between metabolic rate and weight, but also the particular value of b. The theory offered at that time has been reinvented by every generation of biologists because it seems so "obvious." Thus an understanding of the theory's flaws remains important even today. At the time the theory first appeared in the early twentieth century, all the data on metabolism–weight relations were on mammals, and mammals therefore dominated thinking about the subject. During that period, Max Rubner articulated an explanatory theory that is still known as *Rubner's surface "law."*

Euclidean geometry provides the starting point for understanding this "law" that is not a law. Recall from your study of geometry that the surface area s of a sphere is proportional to the square of r, the sphere's radius: $s \propto r^2$. The volume v of a sphere, however, is proportional to the cube of the radius: $v \propto r^3$. From the rules of exponents, we can write $r^2 = (r^3)^{2/3}$. Thus, $s \propto (r^3)^{2/3}$; and substituting v for r^3, we get

$$s \propto v^{2/3} \tag{5.6}$$

In words, as spheres increase in size, their surface area increases only as the two-thirds power of their volume, signifying that the increase in surface area is less than proportional to that in volume. Thus, big spheres have less surface area per unit of volume (or weight) than little spheres. Similar relationships hold true for all sets of geometrically similar objects. Whether you consider cubes, cylinders, hearts, or whole animals, as the objects within a geometrically similar set become larger, the area of the outside surface is expected to increase approximately as the two-thirds power of volume, and the ratio of outside-surface area to volume declines.

Rubner's surface "law" stated that the BMR of a mammal is proportional to its body-surface area[12] and that the allometric relation between BMR and body weight is a corollary of this proportionality. Rubner's explanation of the allometric relation rested on four logical steps:

1. Mammals maintain high, relatively constant body temperatures (near 37°C) and thus tend to lose heat to the environment when studied at thermoneutral environmental temperatures.
2. Because heat is lost across the outer body surfaces, the rate of heat loss from a mammal is approximately proportional to the animal's body-surface area.
3. Small mammals have more surface area per unit of weight than large mammals and thus lose heat more rapidly per unit of weight.
4. Heat lost must be replaced metabolically for a mammal to stay warm. Thus, small mammals must produce heat at a greater rate per unit of weight than large ones.

The surface "law" as just outlined can hardly be faulted as a thought exercise. Why, then, do most physiologists today believe that it is not the correct mechanistic explanation of the allometric relation between BMR and body weight? The answer is that data contradict the "law" in two respects. First, although the surface "law" predicts an allometric exponent, b, equal to about 0.67 ($\frac{2}{3}$), most physiologists who have estimated values of b have concluded that b is statistically higher than 0.67 to a significant degree. Second, by now we realize, as emphasized in this chapter, that poikilothermic animals—such as crabs, fish, and frogs—display allometric relations between metabolic rate and body weight (see Figures 5.9 and 5.10). Rubner's "law" cannot possibly explain these relations in poikilotherms because the reasoning behind the "law" applies only to animals that warm their bodies to elevated, regulated temperatures using metabolic heat production. Since the surface "law" cannot explain the allometric relations in the vast majority of animals, it seems unlikely to be the correct explanation even for mammals.

Since the time in the mid-twentieth century when the surface "law" started to be rejected by most physiologists, a number of alternative hypotheses have been put forward. Most have attracted little enthusiasm. Just recently, however, a renewed flurry of interest in the

[12] A modern holdover of the early emphasis on body-surface area is that surface areas are employed to calculate certain sorts of critical variables in the contemporary practice of medicine. In breast cancer chemotherapy, for example, the doses of chemotherapeutic agents administered to a woman are calculated from her body-surface area.

allometric relations between metabolic rate and body size has developed because of the publication in prominent journals of two new theories explaining the relations.

One of the new theories is based on the mathematics of *fractal geometry*. According to this theory, the allometric relations observed between metabolic rate and body weight occur because key internal transport systems of animals, such as their blood vascular systems, branch according to fractal-geometric principles (Figure 5.12*a*). The internal transport of O_2 and metabolic fuels is critical for metabolism. This theory stresses that in fractally structured transport systems, *rates* of transport are constrained in distinctive ways by the geometry. The theory is based on complex computer models of in-

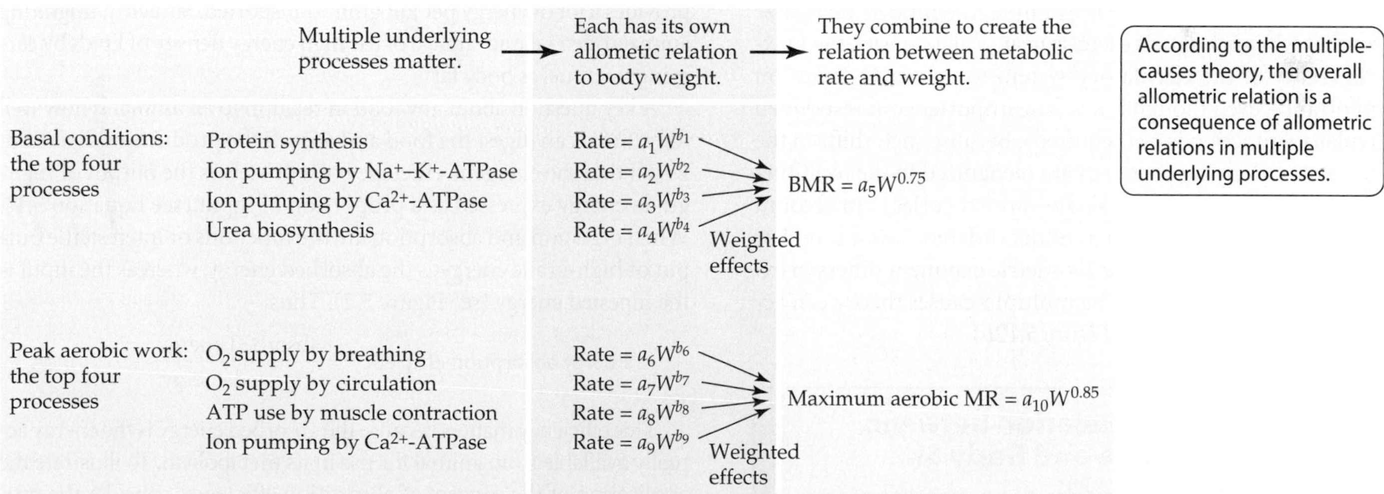

Figure 5.12 Two contemporary theories to explain the allometric relation between metabolic rate and body weight *(a)* The fractal theory argues that with the invention of fractal geometry by Benoit Mandelbrot in about 1980, it is now possible to understand properly the constraints faced by internal transport systems, which are argued to branch according to fractal-geometric principles. These constraints, it is theorized, help give rise to allometric relations between metabolic rate and body size. Fractal systems, as you can see in the example shown, are "self-similar" at multiple scales in that the patterns of branching of fine elements are miniatures of the patterns of branching of large elements. *(b)* The multiple-causes theory argues that the overall allometric relation between metabolic rate and body weight arises from the interplay of allometric relations in multiple processes that underlie an animal's metabolic rate. The theory is illustrated here with information on placental mammals. Because the key processes that underlie metabolism shift with an animal's metabolic intensity, the multiple-causes theory predicts that the allometric relation between BMR and body weight differs in *b* as well as *a* from the relation between maximum aerobic metabolic rate and body weight. (Left drawing in *a* from Vesalius's *De Humani Corporis Fabrica*, 1543, as reproduced in Saunders and O'Malley 1950; right drawing after Mandelbrot 1983. *b* after Darveau et al. 2002.)

ternal transport that examine how the constraints of fractal geometry interact with evolutionary selection pressures to maintain optimized transport capabilities as animals evolve different body sizes. This theory, championed by people like Geoffrey West and James Brown, predicts that an allometric exponent of 0.75 ($\frac{3}{4}$) will occur almost universally—not just in animals but also in plants—because of the fundamental fractal nature of transport systems.

The second major contemporary theory is different from all other theories ever put forth in that it is a *multiple-causes* theory. Rubner's theory and the fractal theory exemplify single-cause theories; each postulates that a single property accounts for allometric relations.[13] The multiple-causes theory, argued by people like Peter Hochachka and Raul Suarez, is based on modern hierarchical models of how biochemical processes in metabolism are controlled. According to these control models, the rate of any major process depends on an interaction of the rates of multiple underlying processes (Figure 5.12*b*). For example, the rate of resting metabolism depends on at least four underlying rates: (1) the rate at which ATP is used to replace proteins that break down, (2) the rate at which Na^+–K^+-ATPase in cells throughout the body uses ATP to pump ions, (3) the rate at which Ca^{2+}-ATPase uses ATP, and (4) the rate at which ATP is used to synthesize nitrogenous waste molecules. *Each* of these underlying rates exhibits its own allometric relation to body size. According to the multiple-causes theory, the rates of all the relevant underlying processes interact in a quantitatively defined way—some processes having greater effect than others—to determine the overall relation between metabolic rate and body size.

An important attribute of the control models used in the multiple-causes theory is that the models openly recognize and stress that changing circumstances often result in shifts in the relative importance of underlying processes. For example, the models recognize and stress that when mammals exercise near peak intensity, the maximum capability of the circulatory system to deliver O_2 takes on great importance, even though it has little importance at rest (when peak circulatory function is not required). Because such shifts in the importance of underlying processes are recognized in the multiple-causes theory, the theory (unlike single-cause theories) can account for shifts in the allometric exponent under different conditions. For example, earlier we noted that the allometric exponent differs in exercising and resting mammals. The multiple-causes theory can account for this phenomenon (see Figure 5.12*b*).

Summary: The Relation between Metabolic Rate and Body Size

- BMR, SMR, and other measures of resting metabolic rate are allometric functions of body weight within phylogenetically related groups of animals ($M = aW^b$, where b is usually about 0.7). Small-bodied species tend to have higher weight-specific metabolic rates than related large-bodied species, an effect so great that the weight-specific BMR is 20 times higher in mice than in elephants, for example.
- The maximal aerobic metabolic rate also tends to be an allometric function of body weight in sets of related species, but sometimes the allometric exponent for maximal metabolic rate differs from that for resting metabolic rate.
- The allometric relation between metabolic rate and weight exerts important effects on the organization and structure of both individual animals and ecosystems. Heart rates, breathing rates, mitochondrial densities, and dozens of other features of individual animals are allometric functions of body weight within sets of phylogenetically related species. In ecosystems, population biomasses and other features of community organization may vary allometrically with individual body size.
- Scientists are not agreed on the explanation for the allometric relations between metabolic rate and body weight. Rubner's surface "law," based on heat loss from homeothermic animals, does not provide a satisfactory explanation. The newest theories are based on fractal geometry and multiple causes.

[13] For Rubner, heat loss through surfaces was the single overriding property that gave rise to allometric relations. According to the fractal theory, allometric relations arise from the fractal branching of internal transport systems.

Energetics of Food and Growth

Food and growth are important topics in animal energetics, aptly discussed together because one animal's growth is another's food. A consequential attribute of foods as energy sources is that lipids are at least twice as high as proteins or carbohydrates in their energy density—that is, their energy value per unit weight (see Table 4.2). We asked at the start of this chapter why polar explorers carry lipid-rich foods, such as meat mixed with pure lard. If they are going to pull, push, and lift their food for many miles before they eat it, they should choose food that provides a lot of energy per kilogram transported. Similarly, migrating animals often take advantage of the high energy density of lipids by carrying their fuel as body fat.

A key question about any food in relation to an animal is how well the animal can digest the food and absorb the products of digestion. Recall that the energetic efficiency of a process is the output of high-grade energy expressed as a proportion of input (see Equation 5.1). When digestion and absorption are the functions of interest, the output of high-grade energy is the absorbed energy, whereas the input is the ingested energy (see Figure 5.2). Thus,

$$\text{Energy absorption efficiency} = \frac{\text{absorbed energy}}{\text{ingested energy}} \tag{5.7}$$

This efficiency matters because the absorbed energy is the energy actually available to an animal for use in its metabolism. To illustrate the application of the concept of absorption efficiency, consider the processing of ingested cellulose by people and ruminants. Because people cannot digest cellulose, they cannot absorb it, and their absorption efficiency for the compound is essentially 0%; if they eat only cellulose, they starve. Ruminants such as cows, in contrast, commonly achieve about 50% absorption efficiency for cellulose because their rumen microbes ferment cellulose into compounds that the animals can absorb; thus, ruminants are able to use about half of the energy available from cellulose in their own metabolism. This example illustrates how the physiology of digestion and absorption, discussed in Chapter 4, bears on the physiology of energy.

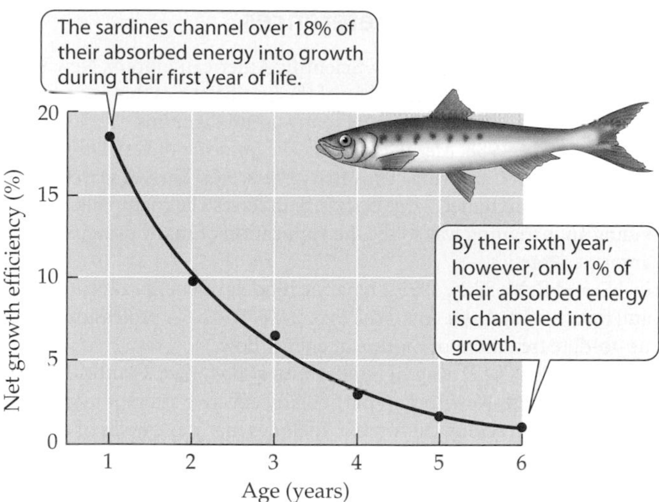

Figure 5.13 Net growth efficiency during each year of life in Pacific sardines (*Sardinops sagax*) When their populations are thriving, these fish are a major food source for seals, predatory fish, birds, and humans. (After Lasker 1970.)

Growing animals accumulate chemical-bond energy in their bodies (by adding tissue consisting of organic molecules). When growth is the energy input–output process of interest, the output of high-grade energy is the chemical-bond energy of added tissue, whereas the input is food energy. Two types of growth efficiency, termed **gross growth efficiency** and **net growth efficiency**, are defined on the basis of whether the food energy is expressed as the ingested energy or the absorbed energy:

$$\text{Gross growth efficiency} = \frac{\text{chemical-bond energy of tissue added in net fashion by growth}}{\text{ingested energy}} \tag{5.8}$$

$$\text{Net growth efficiency} = \frac{\text{chemical-bond energy of tissue added in net fashion by growth}}{\text{absorbed energy}} \tag{5.9}$$

The growth efficiency of animals (gross or net) typically declines with age (Figure 5.13). Besides being of interest in the analysis of energy flow in ecological communities, growth efficiencies are of great concern in agriculture and aquaculture because they affect the amount of product (such as meat) that is obtained in return for the investment in food to feed farmed or cultured animals. In the production of broiler chickens, for example, the birds are slaughtered at just 2 to 3 months of age because at that point they are large enough to be meaty but their growth efficiency—their growth in return for feed provided—is declining.[14]

Conclusion: Energy as the Common Currency of Life

Energy features in virtually every biological process and in many inanimate processes as well. It is a factor in animal growth, body maintenance, migration, photosynthesis, building construction, automobile operation, ecosystem degradation, and war.

When scientists attempt to analyze complex systems—from individual animals to whole ecosystems or even the whole biosphere—they inevitably come up with long lists of processes that they must take into account. The study of individual processes may be straightforward, but the step of integrating knowledge of numerous processes is usually not. One of the greatest challenges in the integrated study of complex systems is to find a common set of units of measure—a "common currency"—in which all the operative processes can be expressed so that they can be compared, added, or multiplied.

Energy is probably the single most promising common currency. In the study of an individual animal, for example, processes as diverse as growth, running, nerve conduction, blood circulation, tissue repair, and thermoregulation can all be expressed in units of energy. Thus their costs can be summed to estimate a total cost of life for the individual, and the impact of a whole population on an ecosystem can be determined by multiplication of the cost per individual by the number of individuals present. Few, if any, other properties come close to energy in their potential to serve as common currencies in this way.

POSTSCRIPT: *The Energy Cost of Mental Effort*

Not the least of the energy costs of analyzing complex systems is the cost of operating our brain. This cost has some most interesting and unexpected properties, however. From studies of tissue metabolic rates, we know that the human brain accounts for about 20% of resting metabolic rate; loosely put, a fifth of our food is for our brain when we are at rest. This cost resembles an "idling" cost; the energy is expended whether we subjectively feel we are doing hard mental labor or not. The prominent physiologist Francis Benedict (1870–1957) wanted to know how much, exactly, the brain's energy needs increase with mental "effort." So, of course, he recruited a group of college students to find out. He told the students on one occasion to sit for tens of minutes keeping their minds as blank as possible. Then he had them spend an equal amount of time working mental arithmetic problems at a fevered pace. Measures of their metabolic rates under the two conditions indicated that the increase in energy consumption caused by an hour of hard mental effort is slight, equivalent to the energy of half a peanut! Thus the brain's high costs are largely steady costs, and thinking hard is not a way to stay slim.

Study Questions

1. Assuming that 10 people plan to trek 500 miles to the North Pole, outline the steps you would take to calculate the amount of food they should pack, taking into account the number of sled dogs needed and the food needed for the dogs.
2. Suppose you use a tire pump to inflate a tire on a bicycle. The elevated pressure created in the tire represents a form of potential energy because the release of the pressure can do mechanical work (such as making a pinwheel turn). The potential energy in the tire is derived from chemical-bond energy in your food. Trace the energy from the time it enters your mouth at a meal until it ends up in the tire, identifying losses of energy as heat along the way.

[14] Feed accounts for 60% to 75% of a farmer's costs.

3. Define *absorbed energy* (*assimilated energy*). Then list the major categories of use of absorbed energy, and specify the fate of energy used in each category. Explain rigorously why heat is a fate of energy used in any and all categories.

4. Small animals tend to expire sooner than related large ones if forced to live on stored supplies. For instance, suppose you have a mouse and a dog that both start with body stores of fat equal to 20% of body weight. Explain why the mouse would be likely to die sooner if these animals could not find any food and thus had to live on their fat reserves. Which one would die sooner if they were trapped underwater and had only their stores of O_2 to live on while trying to escape?

5. Suppose that over a period of 4 hours, a dog was observed to consume 20 L of O_2 and produce 14 L of CO_2. Making use of Tables 5.1 and 5.2, estimate the dog's total heat production over the 4 hours. Explain why Table 5.2 is relevant to your calculation.

6. Poultry scientists are doing research on the design of diets that are nutritionally complete for chickens but minimize the SDA. These scientists believe that such diets would be particularly helpful to the poultry industry in southern states during the heat of summer. Why might this be true?

7. Before Mayer and Joule came along (see Box 5.1), people were well aware that if a person cranked a drill or the like, heat appeared. For instance, the drilling of the bores of cannons was legendary for the heat produced. However, heat per se was believed to be neither created nor destroyed, and thus no one thought that the motion associated with drilling *turned into* heat. Mayer and Joule go down in history in part because they demonstrated the real relation between motion and heat. Suppose that you were alive in the early nineteenth century and, like Mayer and Joule, you hypothesized that animal motion could turn into heat. Design an experiment that would provide a rigorous test of your hypothesis.

8. Suppose you are measuring the metabolic rate of a young, growing cow by using the material-balance method. What procedures could you use to take account of the cow's growth, so that you measure a correct metabolic rate?

9. Suppose you have measured the average rate of O_2 consumption of two groups of laboratory rats that are identical, except that one group was injected with a hormone that is being tested to see if it affects metabolic rate. If the hormone-treated group has a rate of O_2 consumption 5% higher than the other, there are physiological reasons why you cannot conclude that the hormone has changed the metabolic rate. Explain, referring to Table 5.1. According to the table, what might the hormone have done to change the rate of O_2 consumption without changing the metabolic rate?

10. Only eight species of existing land mammals grow to adult body weights over 1000 kg (1 megagram). All are herbivores that employ fermentative digestion. These "megaherbivores" are the two species of elephants, the four species of rhinos, the hippo, and the giraffe. What are the metabolic pros and cons of such large size? Can you suggest why no terrestrial carnivores achieve such large size?

11. If there are many species of herbivores in a grassland ecosystem, and if the species as populations are equally competitive in acquiring food, predict b in the following allometric equation: population biomass per square kilometer = aW^b, where W is individual body weight. Do the data in Table 5.5 follow your equation? What hypotheses are suggested by the comparison?

Key References

Atkins, P. W. 1984. *The Second Law.* Scientific American Library, New York. A serious book on the second law of thermodynamics that is accessible to general readers, containing many useful diagrams.

Owen-Smith, R. N. 1988. *Megaherbivores: The Influence of Very Large Body Size on Ecology.* Cambridge University Press, New York. A searching discussion of extremely large body size in terrestrial mammals, providing an intriguing way to see the application of many principles of animal energetics.

Perrin, N., and R. M. Sibly. 1993. Dynamic models of energy allocation and investment. *Annu. Rev. Ecol. Syst.* 24: 379–410. A professional and up-to-date treatment of models of energy flow.

Peters, R. H. 1983. *The Ecological Implications of Body Size.* Cambridge University Press, New York. A particularly effective attempt to discuss allometry in an ecological context. Includes many examples of allometric equations and their ecological applications.

Schmidt-Nielsen, K. 1984. *Scaling: Why Is Animal Size So Important?* Cambridge University Press, New York. A discussion of the importance of size by a master writer and physiologist. The author was a pioneer in the modern study of many branches of animal physiology, including scaling.

Selected References

Alexander, R. M. 1999. *Energy for Animal Life.* Oxford University Press, New York.

Beamish, F. W. H., and E. A. Trippel. 1990. Heat increment: A static or dynamic dimension in bioenergetic models? *Trans. Amer. Fish. Soc.* 119: 649–661.

Bishop, C. M. 1999. The maximum oxygen consumption and aerobic scope of birds and mammals: Getting to the heart of the matter. *Proc. R. Soc. London* [*Biol.*] 266: 2275–2281.

Bridges, C. R., and P. J. Butler (eds.). 1989. *Techniques in Comparative Respiratory Physiology: An Experimental Approach.* Cambridge University Press, New York.

Brody, S. 1945. *Bioenergetics and Growth.* Reinhold, New York. [Considered a classic.]

Brown, A. C., and G. Brengelmann. 1965. Energy metabolism. In T. C. Ruch and H. D. Patton (eds.), *Physiology and Biophysics*, 19th ed., pp. 1030–1049. Saunders, Philadelphia. [Another classic.]

Brown, C. R., and J. N. Cameron. 1991. The induction of specific dynamic action in channel catfish by infusion of essential amino acids. *Physiol. Zool.* 64: 276–297.

Brown, J. H., and G. B. West (eds.). 2000. *Scaling in Biology.* Oxford University Press, New York.

Calder, W. A., III. 1984. *Size, Function, and Life History.* Harvard University Press, Cambridge, MA.

Calder, W. A., III. 1987. Scaling energetics of homeothermic vertebrates: An operational allometry. *Annu. Rev. Physiol.* 49: 107–120.

Costa, D. P., and G. L. Kooyman. 1984. Contribution of specific dynamic action to heat balance and thermoregulation in the sea otter *Enhydra lutris. Physiol. Zool.* 57: 199–203.

Darveau, C.-A., R. K. Suarez, R. D. Andrews, and P. W. Hochachka. 2002. Allometric cascade as a unifying principle of body mass effects on metabolism. *Nature* 417: 166–170.

Dobson, G. P., and J. P. Headrick. 1995. Bioenergetic scaling: Metabolic design and body-size constraints in mammals. *Proc. Natl. Acad. Sci. U.S.A.* 92: 7317–7321.

Dyson, F. J. 1954. What is heat? *Sci. Amer.* 191(3): 58–63.

Giampietro, M., and D. Pimentel. 1991. Energy efficiency: Assessing the interaction between humans and their environment. *Ecol. Econom.* 4: 117–144.

Heglund, N. C., and G. A. Cavagna. 1985. Efficiency of vertebrate locomotory muscles. *J. Exp. Biol.* 115: 283–292.

Heusner, A. A. 1991. Size and power in mammals. *J. Exp. Biol.* 160: 25–54.

Houlihan, D. F., C. P. Waring, E. Mathers, and C. Gray. 1990. Protein synthesis and oxygen consumption of the shore crab *Carcinus maenas* after a meal. *Physiol. Zool.* 63: 735–756.

Hulbert, A. J., and P. L. Else. 2000. Mechanisms underlying the cost of living in animals. *Annu. Rev. Physiol.* 62: 207–235.

Jobling, M. 1993. Bioenergetics: Feed intake and energy partitioning. In J. C. Rankin and F. B. Jensen (eds.), *Fish Ecophysiology*, pp. 1–44. Chapman and Hall, London.

King, J. R., and M. E. Murphy. 1985. Periods of nutritional stress in the annual cycles of endotherms: Fact or fiction? *Amer. Zool.* 25: 955–964.

Kleiber, M. 1975. *The Fire of Life*, 2nd ed. Krieger, Huntington, NY. [Considered a classic.]

Kooijman, S. A. L. M. 1993. *Dynamic Energy Budgets in Biological Systems.* Cambridge University Press, New York.

LaBarbera, M. 1989. Analyzing body size as a factor in ecology and evolution. *Annu. Rev. Ecol. Syst.* 20: 97–117.

Levine, J. A., N. L. Eberhardt, and M. D. Jensen. 1999. Role of nonexercise activity thermogenesis in resistance to fat gain in humans. *Science* 283: 212–214.

Lindstedt, S. L., and W. A. Calder III. 1981. Body size, physiological time, and longevity in homeothermic animals. *Q. Rev. Biol.* 56: 1–16.

Lomax, E. 1979. Historical development of concepts of thermoregulation. In P. Lomax and E. Schönbaum (eds.), *Body Temperature*, pp. 1–23. Dekker, New York.

Lovegrove, B. G. 2000. The zoogeography of mammalian basal metabolic rate. *Amer. Nat.* 156: 201–219.

Mandelbrot, B. B. 1983. *The Fractal Geometry of Nature.* Freeman, San Francisco.

McCarthy, I. D., D. F. Houlihan, and C. G. Carter. 1994. Individual variation in protein turnover and growth efficiency in rainbow trout, *Oncorhynchus mykiss* (Walbaum). *Proc. R. Soc. London [Biol.]* 257: 141–147.

McLean, J. A., and G. Tobin. 1987. *Animal and Human Calorimetry.* Cambridge University Press, New York.

McNab, B. K. 1988. Complications inherent in scaling the basal rate of metabolism in mammals. *Q. Rev. Biol.* 63: 25–54.

McNab, B. K. 1992. A statistical analysis of mammalian rates of metabolism. *Funct. Ecol.* 6: 672–679.

Mendelsohn, E. 1964. *Heat and Life: The Development of the Theory of Animal Heat.* Harvard University Press, Cambridge, MA.

Miller, P. J. (ed.). 1996. *Miniature Vertebrates: The Implications of Small Body Size.* Symposia of the Zoological Society of London, no. 69. Clarendon, Oxford, UK.

Nagy, K. A. 1987. Field metabolic rate and food requirement scaling in mammals and birds. *Ecol. Monogr.* 57: 111–128.

Pandian, T. J., and F. J. Vernberg (eds.). 1987. *Animal Energetics*, 2 vols. Academic Press, New York.

Pennycuick, C. J. 1992. *Newton Rules Biology: A Physical Approach to Biological Problems.* Oxford University Press, New York.

Prusiner, S., and M. Poe. 1970. Thermodynamic considerations of mammalian heat production. In O. Lindberg (ed.), *Brown Adipose Tissue*, pp. 263–282. American Elsevier, New York.

Rolfe, D. F. S., and G. C. Brown. 1997. Cellular energy utilization and molecular origin of standard metabolic rate in mammals. *Physiol. Rev.* 77: 731–758.

Sinervo, B., and R. B. Huey. 1990. Allometric engineering: An experimental test of the causes of interpopulational differences in performance. *Science* 248: 1106–1109.

Webster, A. J. F. 1983. Energetics of maintenance and growth. In L. Girardier and M. J. Stock (eds.), *Mammalian Thermogenesis*, pp. 178–207. Chapman and Hall, London.

Weibel, E. R. 1991. Fractal geometry: A design principle for living organisms. *Amer. J. Physiol.* 261: L361–L369.

West, G. B., J. H. Brown, and B. J. Enquist. 1997. A general model for the origin of allometric scaling laws in biology. *Science* 276: 122–126.

West, G. B., J. H. Brown, and B. J. Enquist. 1999. The fourth dimension of life: Fractal geometry and allometric scaling of organisms. *Science* 284: 1677–1679.

White, C. R., and R. S. Seymour. 2003. Mammalian basal metabolic rate is proportional to body mass$^{2/3}$. *Proc. Natl. Acad. Sci. U.S.A.* 100: 4046–4049.

Wieser, W. 1984. A distinction must be made between the ontogeny and the phylogeny of metabolism in order to understand the mass exponent of energy metabolism. *Respir. Physiol.* 55: 1–9.

See also references in Appendix A.

6 *Aerobic and Anaerobic Forms of Metabolism*

A startled crayfish swims backward at jetlike speed by flipping its tail in a series of powerful contractions and relaxations. In this way, a crayfish discovered under a rock in a stream can move to another rock and disappear almost before its presence is appreciated. The tail muscles, which account for a quarter of a crayfish's weight in some species, are a powerful tool for escape and survival. To contract, however, the crayfish's muscles—like all other muscles—require adenosine triphosphate (ATP) as a source of energy. When a crayfish is startled, its first tail flip occurs instantly and is followed in short order by multiple additional flips. Each of these massive contractions requires a substantial amount of ATP. How is the ATP made available so promptly? The same question can be asked about the ATP that a human sprinter requires to run 100 m in 10 s.

Burst exercise is a general term that refers to sudden, intense exercise. Besides escaping crayfish and sprinting people, burst exercise is illustrated by salmon leaping waterfalls during their upstream journey, cheetahs racing toward antelopes, and scallops jetting away from danger by clapping their shells. In most sorts of animals, burst exercise cannot be continued for long periods. Crayfish escaping by flipping their tails become exhausted after 15 to 30 flips, just as human sprinters are exhausted or nearly exhausted at the ends of their races.

A second general form of physical activity is **sustained exercise**, defined to be exercise that can be continued at a steady rate for a long period of time. Jogging by people, migratory flight by birds, and steady cruising by fish or crayfish are examples. During sustained exercise, the locomotory muscles must be supplied with ATP minute after minute for long, uninterrupted periods of time. What are the mechanisms that supply ATP in this way?

Comparing sustained exercise and burst exercise, do the mechanisms that supply ATP during sustained exercise differ from the higher-intensity, shorter-duration mechanisms used during burst exercise? Do different types of animals differ in their ATP-producing mechanisms? Is exercise ever forced to stop because of limitations in ATP-generating mechanisms? These are some of the questions we discuss in this chapter.

Each cell must make its own ATP because ATP is not transported from one cell to another. Moreover, ATP is not stored by cells to any substantial extent. Generally, therefore, the *rate* at which a cell can undergo muscular contraction or carry out other forms of physiological work depends strictly on the rate at which that *very cell* is able to produce ATP.

The biochemistry of survival Crayfish can escape rapidly from danger by tail flipping, provided that their tail muscles can generate adenosine triphosphate (ATP) at a high enough rate for rapid, powerful contraction to occur.

The following two complementary reactions, taken together, serve as a crucial *energy shuttle* and *energy transduction* mechanism in cells:

$$\text{ADP} + \text{P}_i + \text{energy from foodstuff molecules} \rightarrow \text{ATP} \quad (6.1)$$

$$\text{ATP} \rightarrow \text{ADP} + \text{P}_i + \text{energy usable by cell processes} \quad (6.2)$$

A cell uses energy from carbohydrates, lipids, or other foodstuff molecules to drive Equation 6.1, causing ATP to be formed from adenosine diphosphate (ADP) and inorganic phosphate ions (P_i = HPO_4^{2-}); in this way, the energy from the bonds of the foodstuff molecules is moved into bonds of ATP and becomes poised for use in physiological work. Then the contractile proteins of muscle cells, ion pumps, or other energy-requiring systems split the ATP, as shown in Equation 6.2, releasing the energy they need. Cellular energy-demanding processes are unable to draw energy directly from the bonds of foodstuff molecules. Thus they are utterly dependent on the cellular processes that drive Equation 6.1 and can take place only as fast as those processes supply ATP.

Mechanisms of ATP Production and Their Implications

There are two major categories of catabolic, biochemical pathways by which animals release energy from foodstuffs to synthesize ATP and thus make energy available for the performance of physiological work. Some of the pathways, termed **aerobic**, require O_2; others, termed **anaerobic**, can function without O_2. In this section we will examine the mechanistic features of the aerobic and anaerobic catabolic pathways. In later sections we will look at the interplay between the aerobic and anaerobic modes of energy release when animals engage in burst or sustained exercise, and when they face environmental stresses such as O_2 deficiency. A compelling reason to start with a study of the mechanistic features of the pathways of ATP production is that the overall performance of animals during exercise and other situations often depends on the mechanistic peculiarities of the particular biochemical pathways by which energy is extracted for use from foodstuff molecules.

Aerobic catabolism consists of four major sets of reactions

Each cell in most animals possesses **aerobic catabolic pathways:** pathways that completely oxidize foodstuff molecules to CO_2 and H_2O using O_2, capturing in ATP bonds much of the chemical energy thus released. The aerobic pathways typically can oxidize *all* the major classes of foodstuffs. We emphasize here just the catabolism of carbohydrates. Furthermore, our aim is to provide an overview, not duplicate detailed treatments available in texts of cellular physiology or biochemistry. The principal aerobic catabolic pathway can be subdivided into four major sets of reactions: glycolysis, the Krebs citric acid cycle, the electron-transport chain, and oxidative phosphorylation.

GLYCOLYSIS **Glycolysis** is the series of enzymatically catalyzed reactions shown in Figure 6.1*a*, in which glucose (or glycogen) is converted to pyruvic acid. The enzymes and reactions of glycolysis occur in the cytosol. The first step in glycolysis is that glucose is phosphorylated at the *cost* of an ATP molecule to form glucose-6-phosphate. Glucose-6-phosphate is then converted to fructose-6-phosphate, and the latter is phosphorylated—also at the *cost* of an ATP molecule—to form fructose-1,6-diphosphate. The latter is cleaved to form two three-carbon molecules: dihydroxyacetone phosphate and glyceraldehyde-3-phosphate. These compounds are interconvertible, and when glucose is being catabolized for release of energy, the former is converted to the latter, yielding two molecules of glyceraldehyde-3-phosphate. The reactions subsequent to glyceraldehyde-3-phosphate in Figure 6.1*a* are all multiplied by 2 to emphasize that two molecules follow these pathways for each glucose molecule catabolized.

The next reaction *is the only oxidation reaction* in glycolysis and is particularly significant for understanding not only aerobic but also anaerobic catabolism. Glyceraldehyde-3-phosphate is oxidized, with the addition of inorganic phosphate (P_i), to a three-carbon diphosphate: 1,3-diphosphoglyceric acid. Although this reaction is an oxidation reaction, it does not in itself require O_2; instead, it occurs by the simultaneous reduction of one molecule of nicotinamide adenine dinucleotide (NAD) per molecule of glyceraldehyde-3-phosphate. NAD is a coenzyme synthesized from the vitamin niacin. When a molecule of glyceraldehyde-3-phosphate is oxidized, two hydrogen atoms are removed from it and transferred to NAD, forming $NADH_2$.[1]

An alternative way to think about a reaction of this sort is that electrons are transferred. NAD is the *immediate electron acceptor* for this reaction because it combines with the electrons (hydrogens) removed from glyceraldehyde-3-phosphate. The fate of the reduced NAD ($NADH_2$) is of critical importance, as we discuss later.

The 1,3-diphosphoglyceric acid formed from the oxidation of glyceraldehyde-3-phosphate is next converted to a monophosphate, 3-phosphoglyceric acid, with the formation of one ATP per molecule. The 3-phosphoglyceric acid is converted in two steps to phosphoenolpyruvic acid, and the latter reacts to form pyruvic acid, again with the formation of one ATP per molecule.

Three important consequences of glycolysis deserve note:

1. Each molecule of glucose is converted into two molecules of pyruvic acid.
2. Two molecules of NAD are reduced to $NADH_2$ per molecule of glucose catabolized.
3. Two molecules of ATP are used and four are formed for each glucose processed, providing a *net yield of two ATP molecules per glucose molecule.*

THE KREBS CITRIC ACID CYCLE During aerobic catabolism, the pyruvic acid formed by glycolysis is oxidized in the mitochondria by a cyclic series of enzymatically catalyzed reactions called the **Krebs citric acid cycle**. This set of reactions, named after Hans Krebs, who in 1937 was the first to envision its features, is diagrammed in Figure 6.1*b*.

[1] The reduction of NAD is symbolized "NAD + 2 H $\rightarrow$ $NADH_2$" or "NAD $\rightarrow$ $NADH_2$" in this book because these simplified expressions compactly emphasize the features of relevance for us. The true reaction is $NAD^+ + 2\ H \rightarrow NADH + H^+$.

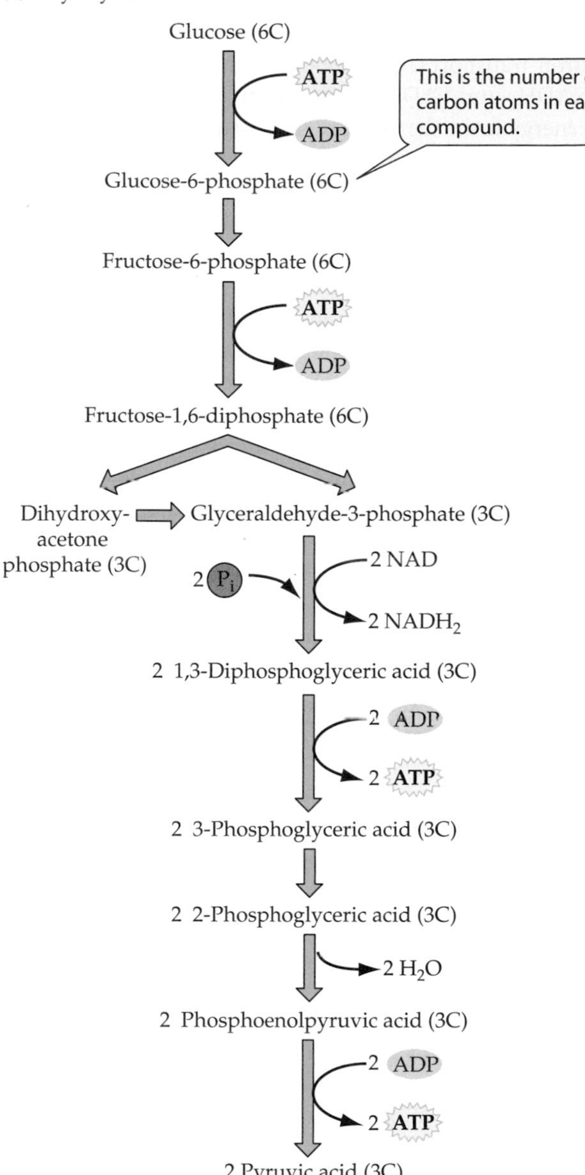

Figure 6.1 The major reactions of glycolysis and the Krebs citric acid cycle A molecule of H_2O enters the reactions at each asterisk (*) in *(b)*. The expression for reduction of NAD, NAD → $NADH_2$, is shorthand; the actual reaction is $NAD^+ + 2\ H \rightarrow NADH + H^+$. P_i = inorganic phosphate.

For simplicity the reactions involved in the oxidation of just one molecule of pyruvic acid are shown, although two molecules are in fact processed for each molecule of glucose.

Pyruvic acid enters the Krebs cycle through a complex set of reactions in which it is oxidatively decarboxylated, forming CO_2 and a two-carbon acetyl group that is combined with coenzyme A in the form of acetyl coenzyme A.[2] In the process, a molecule of NAD is reduced. Acetyl coenzyme A then reacts with oxaloacetate, with the end result that coenzyme A is released and the acetyl group is condensed with oxaloacetate (four-carbon) to form citrate (six-carbon). In the ensuing series of reactions, oxaloacetate is ultimately regenerated and then again can combine with acetyl coenzyme A. For our purposes, there is no need to review the reactions stepwise. It is more important to emphasize the overall outcomes of the reactions:

- *The six carbons of each glucose molecule catabolized emerge in the form of six molecules of CO_2 as the pyruvic acid molecules produced by glycolysis are processed by the Krebs cycle.* The CO_2 is formed by decarboxylation reactions. Such reactions occur at two points within the Krebs cycle: in the conversion of isocitrate to α-ketoglutarate and in the conversion of α-ketoglutarate to succinyl coenzyme A. These two decarboxylations, plus the one in the reaction of pyruvic acid to form acetyl coenzyme A, account for the formation of three molecules of CO_2 for every molecule of pyruvic acid processed (six molecules per glucose).
- *For each glucose molecule catabolized, the Krebs cycle produces eight molecules of $NADH_2$ and two molecules of $FADH_2$.* Oxidation–reduction reactions produce reduced coenzyme molecules. Such reactions occur at four points within the Krebs cycle. At three of these, NAD is reduced, forming $NADH_2$. At one (the oxidation of succinate to fumarate), another coenzyme, flavin adenine dinucleotide (FAD), is reduced, forming $FADH_2$.[3] Recall also that one $NADH_2$ is formed in the reaction of pyruvic acid to form acetyl coenzyme A. Considering all the oxidation–reduction reactions, the processing of each pyruvic acid molecule results in four $NADH_2$ and one $FADH_2$ (thus, eight $NADH_2$ and two $FADH_2$ per glucose).
- *Two molecules of ATP are produced in the Krebs cycle for each molecule of glucose catabolized.* Guanosine triphosphate (GTP) is formed from guanosine diphosphate (GDP) when succinyl

[2] The critically important coenzyme known as *coenzyme A* is synthesized from pantothenic acid, a B vitamin.

[3] Flavin adenine dinucleotide is synthesized from riboflavin (vitamin B_2).

coenzyme A reacts to form succinate in the Krebs cycle. GTP donates its terminal phosphate group to ADP, resulting in GDP and ATP. Thus, one molecule of ATP is generated for each molecule of pyruvic acid processed (two ATPs per glucose).

ELECTRON TRANSPORT, OXIDATIVE PHOSPHORYLATION, AND THE ROLE OF O_2 The final two of the four sets of reactions in the aerobic catabolic pathway are the *electron-transport chain* and *oxidative phosphorylation*. We discuss them together here because they are often tightly linked.

A paradox you may have noticed is that thus far, in discussing *aerobic* catabolism, we have not mentioned the involvement of O_2. The reason is that O_2 is in fact not a participant in any of the reactions of glycolysis or the Krebs cycle; and in a very narrow sense, all those reactions can proceed without O_2. Nonetheless, O_2 is essential. The reason it is essential lies in the disposition of the reduced coenzymes, $NADH_2$ and $FADH_2$.

$NADH_2$ and $FADH_2$ cannot serve as final resting places for electrons, because NAD and FAD are present in only limited quantities in a cell. Whenever one of the mainstream molecules in glycolysis or the Krebs cycle is oxidized, the electrons (or hydrogens) removed are transferred to NAD or FAD. Thus, as stressed earlier, NAD and FAD are the *immediate* electron acceptors in glycolysis and the Krebs cycle. However, if the $NADH_2$ and the $FADH_2$ thereby formed were simply allowed to accumulate, a cell would soon run out of NAD and FAD. Running out of NAD and FAD would bring glycolysis and the Krebs cycle to a halt because NAD and FAD are required. Thus, NAD and FAD cannot serve as *final* electron acceptors. The electron-transport chain regenerates NAD and FAD by removing electrons (hydrogens) from $NADH_2$ and $FADH_2$. Oxygen is required for the ordinary operation of the electron-transport chain, and in that way O_2 is necessary for glycolysis and the Krebs cycle to function as they do during aerobic catabolism.

The **electron-transport chain** is a series of four major protein complexes, plus other compounds, located in the inner membranes of mitochondria (Figure 6.2). A key property of the constituents of the electron-transport chain is that each is capable of undergoing reversible reduction and oxidation. The compounds function in a discrete order—that is, a chain. The chain takes electrons from $NADH_2$ and $FADH_2$ and passes them in sequence from one compound to the next in a series of reductions and oxidations (see Figure 6.2). Finally, the last compound in the electron-transport chain, known as complex IV or cytochrome oxidase, passes the electrons—along with "accompanying" protons (H^+ ions)—to oxygen, reducing the oxygen to water.[4] In this way, *O_2 acts as the final electron acceptor*. The net effect of the operation of the electron-transport chain is to take electrons from $NADH_2$ and $FADH_2$ and pass them to O_2.

The special role played by O_2 is important. The constituents of the electron-transport chain (e.g., cytochrome b-c_1 and cytochrome oxidase), just like NAD and FAD, are present in limited quantities in a cell and therefore cannot act as terminal electron acceptors. In contrast, O_2 is continuously supplied to a cell, and the product of its reduction, water, can be voided into the environment, *thereby carrying electrons out of the cell*. An adult person produces over 1.5 L of water per day in the process of voiding electrons from cells!

The electron-transport chain is not just a mechanism for the oxidation of $NADH_2$ and $FADH_2$. It is also pivotally involved in the transfer of energy from the bonds of foodstuff molecules to ATP. Molecular oxygen has a much higher affinity for electrons than the coenzymes NAD and FAD, and there is a large decline in free energy as the electrons originally taken from foodstuff molecules are passed through the electron-transport chain. A considerable portion of this energy is captured in bonds of ATP. The process of forming ATP from ADP using energy released by the transport of electrons through the electron-transport chain is called **oxidative phosphorylation.** Of the four protein complexes in the electron-transport chain, three (complexes I, III, and IV) participate in ATP synthesis (see *electron flux* in Figure 6.2). Each complex is poised to produce one ATP from one ADP per pair of electrons transported. Thus, in principle, for each pair of electrons passed through the full length of the electron-transport chain from $NADH_2$ to O_2, three molecules of ATP can be produced. Box 6.1 discusses the mechanism.

A common mode of expressing the yield of ATP by oxidative phosphorylation is as a **P/O ratio** (P = phosphate, O = oxygen), defined to be the number of ATP molecules formed per *atom* of oxygen reduced to water. For example, if a pair of electrons passes from $NADH_2$ to an atom of oxygen and three ATP molecules are formed, P/O = 3.

Although we have been implying that the processes of electron transport and oxidative phosphorylation are obligatorily linked, this is actually not the case. In fact, it is possible for electrons to pass through the electron-transport chain without any formation of ATP at all.

Linkage of electron transport with oxidative phosphorylation is called **coupling.** In principle, the P/O ratio for electrons originating from $NADH_2$ can vary between 3 and 0, depending on the degree of coupling: When electron transport and oxidative phosphorylation are *fully ("tightly") coupled*, the P/O ratio is, in principle, 3; but if the processes are *completely uncoupled*, the P/O ratio is 0 (zero). Intermediate states of coupling are possible. The tightly coupled state is by far the most usual. An interesting and important discovery from modern research using nuclear magnetic resonance is that when a fully coupled state prevails, the P/O ratio is not in fact 3, the theoretical maximum. Instead, the P/O ratio for electrons originating from $NADH_2$ is often about 2.6 to 2.8 in fully coupled cells. The reasons are only partly known (see Box 6.1).

Full uncoupling works against ATP production; in the fully uncoupled state, even though electron transport continues to convert $NADH_2$ and $FADH_2$ to NAD and FAD—thereby permitting continued oxidation of foodstuff molecules by glycolysis and the Krebs cycle—little or none of the energy released by electron transport is captured as ATP. The energy instead is immediately converted to heat (see Box 6.1). Uncoupling is therefore a specialized biochemical state that is ordinarily employed only when there is an advantage to converting food energy to heat rather than ATP. Uncoupling of oxidative phosphorylation is a major mechanism of increasing metabolic heat production in many mammals in cold seasons, as discussed in Chapter 8. It also sometimes serves as a mechanism of

[4] Two electrons and two protons combine with one oxygen atom. A molecule of oxygen (O_2) therefore can react with four electrons and four protons.

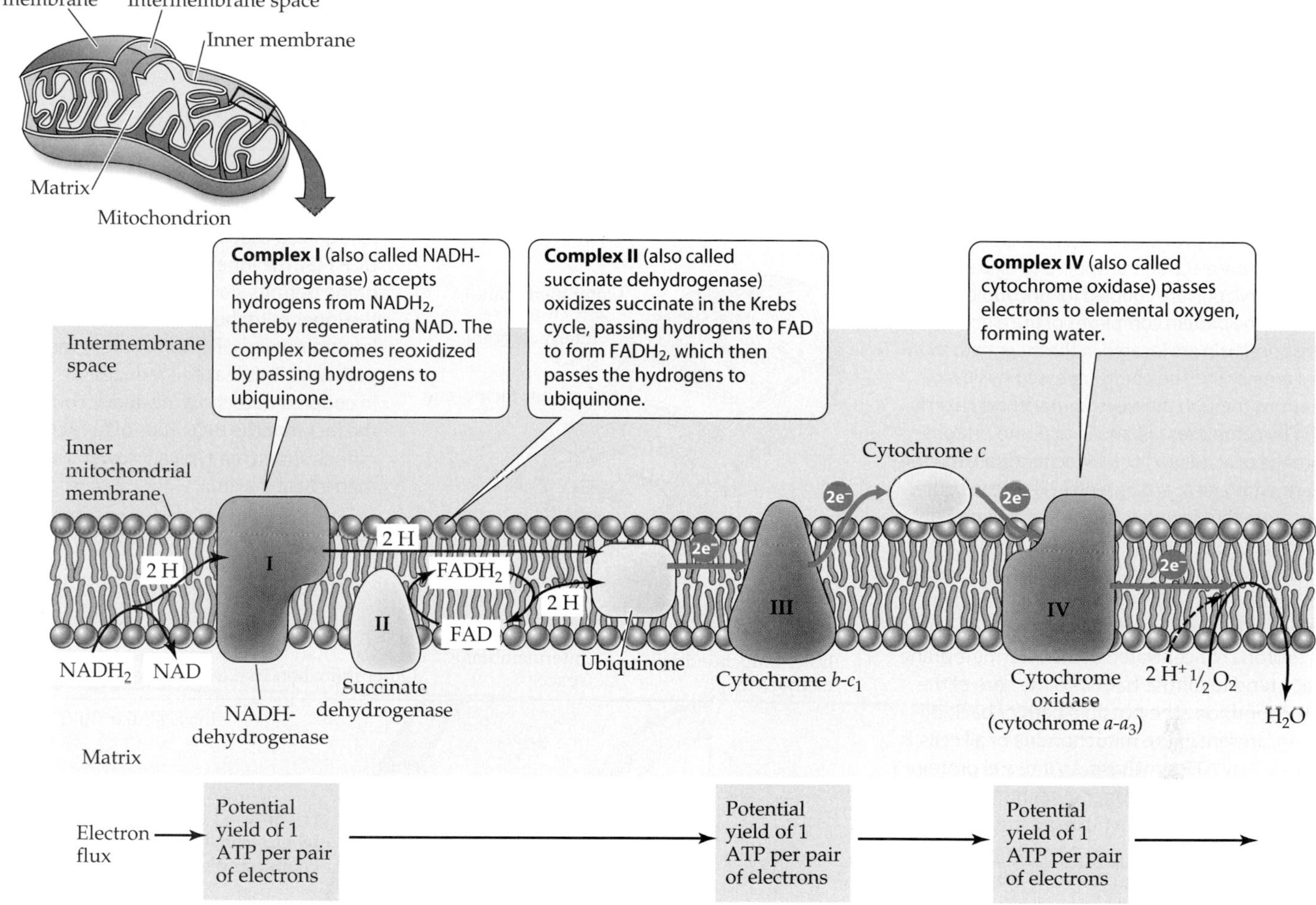

Figure 6.2 The electron-transport chain The electron-transport chain consists of four enzyme protein complexes, called complexes I through IV, embedded in the inner mitochondrial membrane, plus ubiquinone and cytochrome *c*. The three purple complexes (I, III, and IV) capture energy from the flux of electrons to synthesize ATP from ADP. The immediate mechanism by which these complexes capture energy is that they use it to transport protons across the inner mitochondrial membrane, creating a proton electrochemical gradient that then is used to synthesize ATP as described in Box 6.1. Ubiquinone, also called coenzyme Q, is a lipid-soluble compound that is mobile in the lipid core of the membrane. Ubiquinone that is reduced by receiving hydrogens from NADH-dehydrogenase or $FADH_2$ becomes reoxidized by passing electrons to complex III, also called cytochrome b-c_1. Complex III then passes the electrons to cytochrome *c* (a water-soluble compound), which in turn passes them to complex IV, also called cytochrome oxidase (or cytochrome a-a_3, COX, or cytochrome *c* oxidase). Cytochrome oxidase then passes the electrons to elemental oxygen, forming water. e^- = electron. (After Saraste 1999.)

body-weight regulation by "burning off" the energy of excess food molecules.[5]

THE YIELD OF ATP FROM AEROBIC CATABOLISM Now let's look at the total yield of ATP from the aerobic catabolism of glucose when coupling is tight. For each molecule of glucose that is catabolized, two molecules of $NADH_2$ are produced in glycolysis. Furthermore, eight $NADH_2$ molecules and two $FADH_2$ molecules are formed in the reactions subsequent to glycolysis, notably the Krebs cycle. When the ATP molecules produced from all these $NADH_2$ and $FADH_2$ molecules are added, the theoretical maximum yield of ATP by oxidative phosphorylation is 34 ATP molecules per glucose.[6] As we have noted already, even in tightly coupled cells, this theoretical yield is not fully achieved in reality. The round number 34 is the one usually cited as the net yield, however, and in future pages of this book we will use it for simplicity, without further qualification.

Besides the ATP molecules made by oxidative phosphorylation, additional ATP molecules are generated by several reactions in glycolysis and the Krebs cycle: the reaction of 1,3-diphosphoglyceric acid to form 3-phosphoglyceric acid, that of phosphoenolpyruvic acid to form pyruvic acid, and that of succinyl coenzyme A to form succinate (see Figure 6.1). These phosphorylations, in contrast to oxidative phosphorylations, are known collectively as **substrate-level phosphorylations** because they occur immediately in the reactions of substrates of glycolysis and the Krebs cycle. Altogether, six ATP

[5] Recall, for example, the rats mentioned in Chapter 5 that remain slim despite chronic overeating by undergoing diet-induced thermogenesis (DIT). Uncoupling is the principal mechanism of DIT in those animals.

[6] The assumption here is that P/O = 3 for $NADH_2$ oxidation and P/O = 2 for $FADH_2$ oxidation (see Figure 6.2).

BOX 6.1 The Biochemistry of Coupling and Uncoupling of Oxidative Phosphorylation: The Chemiosmotic Theory

The **chemiosmotic theory** is currently believed to provide a correct understanding of how electron transport and oxidative phosphorylation are coupled (or uncoupled). Recall that the protein complexes of the electron-transport chain are located in the inner mitochondrial membrane. The energy released by flow of electrons through the electron-transport chain is used by complexes I, III, and IV to pump protons from the core (matrix) of a mitochondrion into the intermembrane space, as seen in (1) in the figure. Thus, from the viewpoint of energy, the immediate result of electron transport is the formation of a proton electrochemical gradient (2), which constitutes an energy store, just as the ion gradient created in Figure 3.12 represents an energy store.

Protons concentrated in the intermembrane space tend to diffuse back into the core of the mitochondrion. One possible path of back-diffusion, present in the mitochondria of all cells, is provided by **ATP synthase**, an integral protein complex found in the inner mitochondrial membrane. As shown in (3) in the figure, ATP synthase captures energy released by the back-diffusion of protons and uses it to synthesize ATP from ADP. Thus, *ATP synthase provides the mechanism of coupling*, the mechanism that enables energy released by electron transport to be used to make ATP.

Besides ATP synthase, the mitochondria of *some*, specialized, cells have in the inner mitochondrial membrane a second type of protein, called **uncoupling protein 1** (**UCP1**), that provides a path for protons to diffuse from the intermembrane space to the inner core without ATP synthesis (4). Uncoupling protein is unable to synthesize ATP. Thus, when protons back-diffuse through it, the energy released by the diffusion appears as heat. In a tissue with UCP1, the degree of coupling (represented as P/O, the ratio of ATP molecules formed per atom of oxygen reduced) is controlled by adjustment of the extent to which protons back-diffuse by way of the two possible paths: ATP synthase and uncoupling protein. This control is mediated by a variety of agents, including hormones, fatty acids, and Ca^{2+} ions.

Although mitochondrial physiologists do not dispute that UCP1 permits protons to reenter the inner mitochondrial core without ATP synthesis, there are at least four different hypotheses regarding how exactly UCP1 functions to achieve this outcome. The concept that UCP1 provides a straightforward back-diffusion path for protons is the simplest of the models.

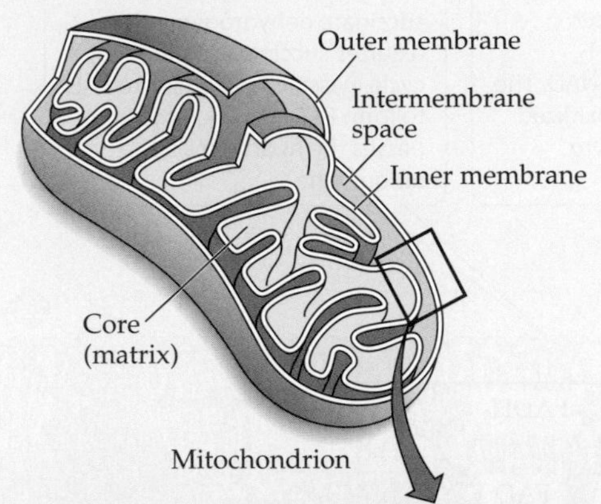

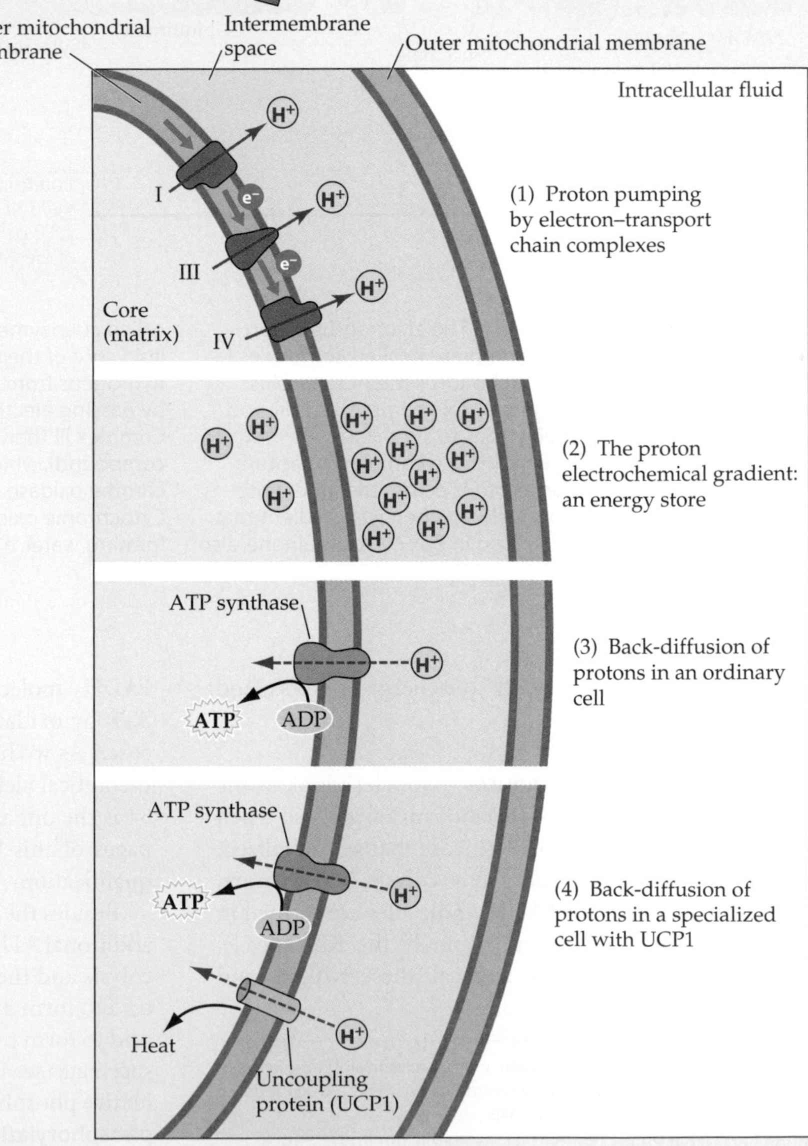

On the basis of recent research, it is now recognized that in mitochondria lacking UCP1 (3), the inner mitochondrial membrane permits slow leakage of protons into the inner mitochondrial core by way of unknown, nonspecific, non-protein-mediated pathways. This leakage reduces the ATP yield in coupled mitochondria and accounts for the fact that the P/O ratios of fully coupled mitochondria are typically lower than their theoretical maxima.

molecules are formed by substrate-level phosphorylation per molecule of glucose.

In total, 40 molecules of ATP are generated for each glucose molecule catabolized. However, two ATP molecules are consumed at the start of glycolysis (see Figure 6.1*a*). Thus the *net* yield is 38 molecules of ATP per glucose. The *energetic efficiency* of this ATP production—defined by Equation 5.1—is not altogether certain and can vary even in a tightly coupled cell, depending on metabolite concentrations and other conditions.[7] In glucose catabolism, assuming tight coupling, the efficiency is usually considered to be 60% to 70%. That is, 60% to 70% of the chemical energy released by oxidizing glucose to CO_2 and H_2O is captured as chemical-bond energy of ATP.

O_2 deficiency poses two biochemical challenges: Impaired ATP synthesis and possible redox imbalance

What are the biochemical implications if cells are denied O_2 or supplied with O_2 at an inadequate rate? To explore this question, let us for simplicity consider cells deprived entirely of O_2, even though for many animals this would be an unrealistically extreme state.

Without O_2, electrons entering the electron-transport chain cannot be discharged at the end by the reduction of oxygen. Thus the electron-transport chain becomes a dead end for electrons. The consequences for most animals are easily summarized. Soon after the supply of O_2 is cut off, all the cytochromes and other constituents of the electron-transport chain become fully reduced because electrons entering the chain accumulate. The electron-transport chain then can no longer accept further electrons, and two major consequences follow. First, oxidative phosphorylation cannot take place, eliminating a cell's ability to produce 34 out of the 38 ATP molecules that are ordinarily produced in net fashion by the aerobic catabolism of glucose. Second, the electron-transport chain can no longer serve as a mechanism for reoxidizing the reduced coenzyme molecules, $NADH_2$ and $FADH_2$, that are produced by glycolysis and the Krebs cycle. The cell's supply of NAD and FAD is thus threatened. A complete failure to regenerate NAD and FAD would mean that a cell could produce no ATP at all, because NAD is needed for glycolysis and both NAD and FAD are required for operation of the Krebs cycle.

The inability of the electron-transport chain to regenerate NAD and FAD from $NADH_2$ and $FADH_2$ in the absence of O_2 is usefully described as a problem of **reduction–oxidation balance** (**redox balance**). Redox balance is a key concept in the study of all cellular compounds that undergo alternating reduction and oxidation. By definition, a cell is in redox balance for such a compound if the cell possesses the means to remove electrons from the compound as fast as electrons are added to it.

Applying the concept specifically to NAD, a cell is in redox balance for NAD if it can convert $NADH_2$ back to NAD (a process that removes electrons) as fast as NAD is being converted to $NADH_2$ (a process that adds electrons). During aerobic catabolism, O_2 and the electron-transport chain together provide the means of maintaining redox balance for both NAD and FAD, thereby ensuring that NAD and FAD are steadily available to perform their essential roles in glycolysis and the Krebs cycle. *To make ATP without O_2, a cell must possess alternative mechanisms that will permit redox balance to be maintained while at least some ATP-generating reactions are sustained.*

Certain tissues possess anaerobic catabolic pathways that synthesize ATP

Only some tissues have evolved an ability to make ATP without O_2 at significant rates. The mammalian brain is a prime example of a tissue that lacks this ability. The brain succumbs rapidly when denied O_2 (as during cardiac arrest) because, lacking a significant anaerobic catabolic pathway, when it lacks O_2 it is unable to synthesize ATP at the rate it needs it. Many other vertebrate tissues possess an anaerobic catabolic pathway that can produce ATP at a substantial rate without O_2. These other tissues are therefore less rigidly dependent on an O_2 supply than the mammalian brain is.

Anaerobic glycolysis is the principal anaerobic catabolic pathway of vertebrates

Vertebrate skeletal muscle and many other vertebrate tissues are able to use the substrate-level phosphorylations of glycolysis to transfer energy from the bonds of glucose to ATP at substantial rates in the absence of O_2. For a tissue to have this ability, it must have a way of maintaining redox balance for NAD without O_2. Glycolysis requires NAD to serve as the immediate electron acceptor for the oxidation of glyceraldehyde-3-phosphate, and in using NAD in this way, it converts the NAD to $NADH_2$ (see Figure 6.1*a*). Under anaerobic conditions in skeletal muscles and other such tissues, the $NADH_2$ molecules are reoxidized to NAD by passing their electrons (hydrogens) to pyruvic acid, reducing the latter to lactic acid (Figure 6.3). Pyruvic acid, in essence, is used as the final electron acceptor by the anaerobic cell. The entire sequence of reactions that converts glucose to lactic acid is called **anaerobic glycolysis**.

The principal determinant of whether a cell can carry out anaerobic glycolysis at a substantial rate is its expression of the enzyme lactate dehydrogenase (LDH), the enzyme that reduces pyruvic acid. LDH occurs in multiple molecular forms, as discussed in detail in Chapter 2 (see page 46). *For a cell to carry out anaerobic glycolysis, it must synthesize significant amounts of one or more LDH forms that are catalytically effective in converting pyruvic acid to lactic acid under the intracellular conditions that prevail during O_2 deprivation.* In cells meeting these criteria, a true redox balance exists during anaerobic glycolysis. As a study of Figure 6.3 reveals, one molecule of pyruvic acid is produced for each molecule of NAD that is reduced to $NADH_2$. Thus the supply of pyruvic acid keeps exact pace with the need for it as an electron acceptor. A cell with appropriate amounts and types of LDH is thus able to convert each $NADH_2$ that it forms back to NAD without O_2. This means that from a strictly biochemical perspective, a steady production of ATP can be maintained.

Glucose and glycogen are the only foodstuff molecules that can serve as fuels for anaerobic glycolysis under most conditions. The amount of ATP synthesized in net fashion by anaerobic glycolysis is two molecules

[7] For example, ATP does not have a fixed energy value, as often implied by introductory biology texts. Instead, the "energy content" of ATP depends on how close, or far away, the ATP–ADP interconversion reaction is to chemical equilibrium, which in turn depends on relative concentrations of the reactants in a particular cell.

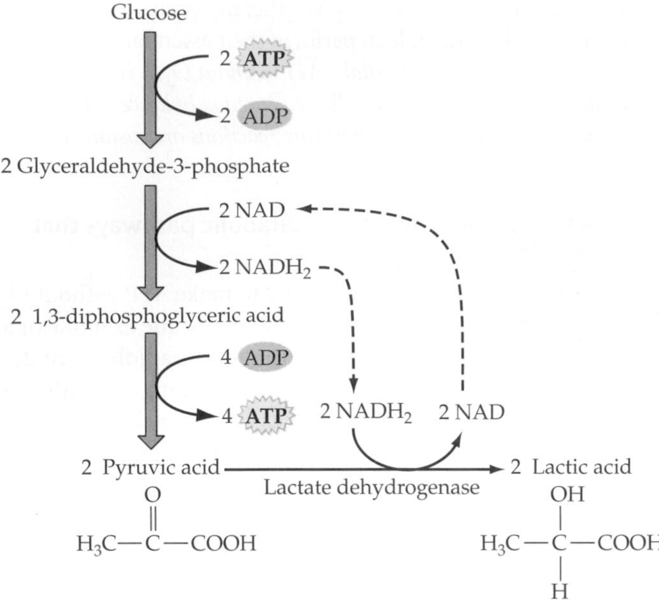

Figure 6.3 Anaerobic glycolysis The reduction of pyruvic acid is catalyzed by the enzyme lactate dehydrogenase (LDH) and permits redox balance to be maintained without O_2. The net yield of ATP is two ATP molecules per glucose molecule if the initial fuel is glucose itself, as shown here. If glycogen is used as fuel, there is a net yield of three ATPs per glucose unit catabolized.

of ATP per glucose molecule catabolized (see Figure 6.3). This yield of ATP is vastly lower than that from the aerobic catabolism of glucose, but as we will see, the anaerobic pathway can nonetheless be an impressively important source of ATP. The major reason for the comparatively low yield of ATP per glucose molecule during anaerobic glycolysis is that only a small fraction (about 7%) of the free energy available from glucose is released by the conversion of glucose to lactic acid. Lactic acid is itself an energy-rich molecule.

Anaerobic glycolysis is not the only mechanism of anaerobic catabolism in animals. However, we take up other anaerobic pathways later and now address the important matter of the disposition of end products.

What happens to catabolic end products?

For an animal to make use of any ATP-producing pathway, it must possess satisfactory means of disposing of the chemical end products generated. The principal products of *aerobic* catabolism, CO_2 and H_2O, are fully oxidized and not capable of being tapped for further energy. Animals typically dispose of CO_2 and H_2O by voiding them into the environment, as when we exhale CO_2 by breathing.

The end products of anaerobic catabolic pathways, in contrast to CO_2 and H_2O, are always organic compounds that are far from fully oxidized and possess considerable further potential to yield energy. The high energy value of these products places a premium on retaining them in the body for future use as energy sources; excreting lactic acid, for instance, would be equivalent to voiding over 90% of the energy value of catabolized carbohydrates. On the other hand, unlimited retention of the organic end products of anaerobic catabolism is usually not possible, because the end products typically exert harmful effects if allowed to accumulate to high concentrations. These two competing considerations seem to have been major operative factors in the evolution of animal strategies for disposing of anaerobic end products. A third factor of importance is time. An animal that makes only short-term use of anaerobic catabolic pathways is more likely to be able to retain all the end-product molecules without adverse effects than an animal that uses anaerobic catabolism for days or weeks on end.

THE DISPOSITION OF LACTIC ACID IN VERTEBRATES When vertebrates use anaerobic glycolysis, their universal strategy is to retain the lactic acid they produce in their bodies, and they ultimately get rid of the lactic acid *metabolically*. The metabolism of lactic acid requires O_2 and thus can take place only when and where O_2 becomes available. Lactic acid itself is a metabolic cul-de-sac (**Figure 6.4**). Thus, for lactic acid to be metabolized, it must first be converted back to pyruvic acid by a reversal of the very reaction that formed it. The conversion of lactic acid to pyruvic acid is an oxidation reaction, with NAD acting as the immediate electron acceptor. After conversion to pyruvic acid, the carbon chains of lactic acid are metabolized by one of two major paths, both of which require O_2 (see Figure 6.4):

1. One path is for the carbon chains to be fully oxidized by way of the Krebs cycle and electron-transport chain. This process *makes ATP* (36 ATP molecules per pair of lactic acid molecules). It requires O_2 because the electron-transport chain must be functional.

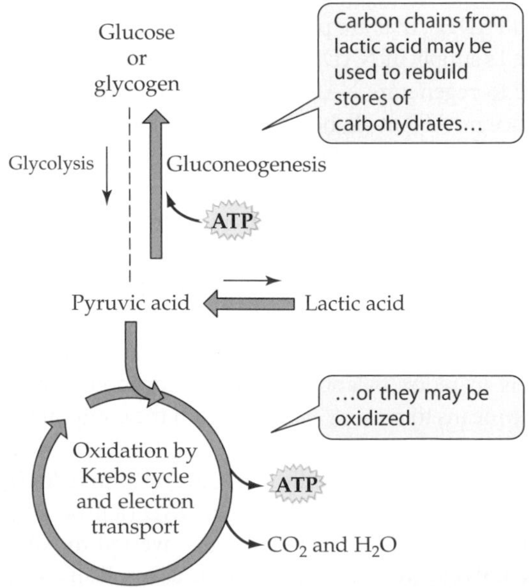

Figure 6.4 Major paths by which lactic acid is metabolized when O_2 is available Some carbon from lactic acid may also be incorporated into amino acids and proteins by way of Krebs-cycle intermediates.

2. The second path is for the carbon chains to be used to form glucose or glycogen, thereby replenishing a cell's carbohydrate stores. The conversion of lactic acid or pyruvic acid to glucose or glycogen is a form of **gluconeogenesis** ("new formation of glucose"). The process *uses ATP* (six ATP molecules per glucose molecule formed). It requires O_2 because the ATP must be made aerobically.

To determine the relative importance of the two paths followed by lactic acid, investigators inject animals with lactic acid containing the unusual carbon isotope ^{14}C. They then monitor the fractions of the ^{14}C that show up in CO_2 and carbohydrate. These fractions vary, even within single species, in ways that are still poorly understood. As a gross generalization, the poikilothermic groups of vertebrates—fish, amphibians, and reptiles—use carbon chains of lactic acid principally for gluconeogenesis, whereas mammals either emphasize oxidation of the carbon chains in the Krebs cycle or employ both oxidation and gluconeogenesis about equally. Lactic acid may be metabolized by the same tissues that produce it, or it may be carried by the blood to other tissues that metabolize it. Quite often, tissues that are well supplied with O_2 metabolize lactic acid made by other tissues; this can occur even while lactic acid production continues.

The functional roles of ATP-producing mechanisms depend on whether they operate in steady state or nonsteady state

One of the most consequential properties of any ATP-producing mechanism is whether or not it can operate in **steady state**. Ideally, a mechanism is in steady state if (1) it produces ATP as fast as ATP is used, (2) it uses raw materials (e.g., foodstuff molecules) no faster than they are replenished, and (3) its chemical by-products (besides ATP) are voided as fast as they are made. When a cell uses a **steady-state mechanism** of ATP production, even as ATP is made and used in the cell, the cell remains essentially constant in its levels of ATP and of the precursors and by-products of ATP production. Thus, at least in principle, a steady-state mechanism can go on and on indefinitely, free of any intrinsic limitations.

In contrast, a **nonsteady-state mechanism** of ATP production depletes supplies, accumulates products, or otherwise alters the conditions of its own operation at rates fast enough that the mechanism is *self*-terminating. For this reason, a nonsteady-state mechanism cannot persist for long by comparison to steady-state ones.

The concepts of steady-state and nonsteady-state mechanisms are relative, rather than absolute, but they provide useful organizing principles in the study of ATP synthesis. In vertebrates and many invertebrates, aerobic ATP production is considered typically to be a steady-state process, whereas anaerobic ATP production is usually categorized as a nonsteady-state process.

EXAMPLES FROM HUMAN BIOLOGY To understand the practical meaning of steady-state and nonsteady-state mechanisms better, let's examine some examples in our own lives. Routine daily living provides a good illustration of a circumstance in which ATP is produced by aerobic catabolism functioning in steady state. Consider what you are doing now, for instance. As you read this page, unless you have bizarre reading habits, you are in an aerobic steady state. You are making all your ATP by aerobic catabolism; your needs for ATP are fully met; you are taking in O_2 at the rate you use it; and you are voiding CO_2 and H_2O approximately as you produce them. The one way in which you may not be in steady state is in regard to the foodstuff molecules you are using as fuels. If a person is not eating while reading, fuel levels are obviously falling. However, the rate of decline during reading is very low relative to fuel stores, and intakes of fuels probably closely match uses when averaged over a day (including meals). Thus, even with respect to fuels, the departure from steady state is merely technical.

The capability of the aerobic catabolic pathways to operate in steady state is of great significance for meeting the ATP requirements of daily living, because the aerobic production of ATP can go on and on without self-limitations. Aerobic catabolism can meet our ATP needs for a lifetime!

A mile race is a good example of an activity that is fueled substantially by anaerobic ATP production functioning in nonsteady state. As we will see in detail later in this chapter, the maximum rate at which aerobic catabolism can make ATP is far less than the rate at which ATP is used in a competitive mile race. Thus, additional ATP must be provided by anaerobic glycolysis throughout such a race. Anaerobic glycolysis is a nonsteady-state mechanism of ATP synthesis during a mile race because the lactic acid that it produces accumulates in the body to such high levels that the acidification of tissues eventually makes continued running *physiologically impossible*; the process thus is self-limiting. The mile race contrasts with routine daily living in that it cannot go on indefinitely, and the key reason for the difference lies in the mechanisms of ATP production: nonsteady-state during the mile race, steady-state during daily living.

Phosphagens provide an additional mechanism of ATP production without O_2

Phosphagens are compounds that serve as temporary stores of high-energy phosphate bonds. They occur in the skeletal muscles of vertebrates and muscles of many invertebrates. *Creatine phosphate* (Figure 6.5*a*) is the phosphagen of vertebrate muscle and also occurs in some groups of invertebrates. The most widespread phosphagen of invertebrates (including crayfish) is *arginine phosphate* (see Figure 6.5*a*), and other phosphagens are known, especially among annelids. These compounds are synthesized by use of high-energy phosphate bonds taken from ATP, and later they can donate the phosphate bonds to ADP to form ATP. The reversible reactions for creatine phosphate and arginine phosphate, catalyzed by *phosphagen kinases* (*creatine kinase* and *arginine kinase*), are shown in Figure 6.5*b*.

Chemical mass action determines the direction of a phosphagen reaction.[8] Consider the reaction of creatine phosphate and creatine (see Figure 6.5*b*), for example. During times when a vertebrate muscle cell is at rest, when ATP supplies in the cell are relatively untaxed and ATP concentrations are high, the reaction is shifted to the left.

[8] See Chapter 2 (page 46) for a discussion of the meaning of mass action.

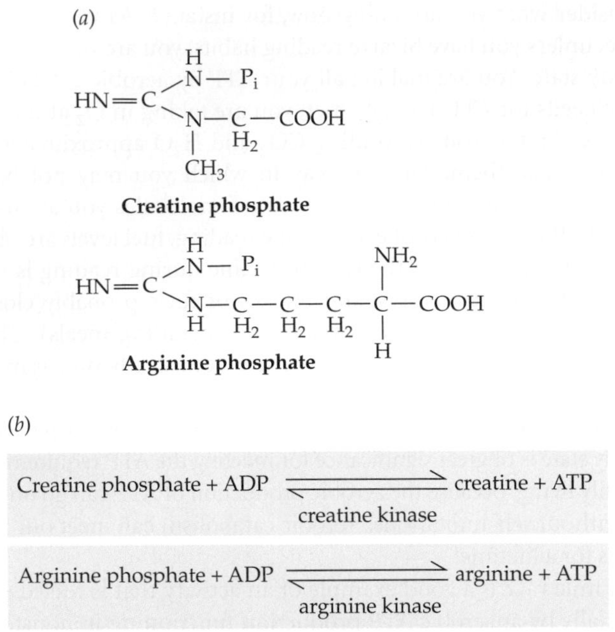

(*a*)

H
N— P_i
HN=C
N—C—COOH
| H_2
CH_3
Creatine phosphate

H NH₂
N— P_i |
HN=C
N—C—C—C—C—COOH
H H_2 H_2 H_2 |
H
Arginine phosphate

(*b*)

Creatine phosphate + ADP ⇌ creatine + ATP
creatine kinase

Arginine phosphate + ADP ⇌ arginine + ATP
arginine kinase

Figure 6.5 Two important phosphagens and the reactions they undergo *(a)* Creatine phosphate and arginine phosphate. (P_i = inorganic phosphate). *(b)* The reversible reactions by which these phosphagens are formed from ATP or can donate their phosphate groups to ADP to make ATP. The reactions are catalyzed by creatine kinase and arginine kinase.

Thus, most of the creatine in the resting cell becomes phosphorylated. The molar concentration of creatine phosphate in a vertebrate muscle cell might then be three to six times the concentration of ATP. Later, during times when ATP supplies are taxed and ATP concentrations fall, the reaction is shifted to the right, *forming ATP without a simultaneous need for O_2*. Each ADP molecule in a vertebrate muscle cell can be rephosphorylated three to six times by this mechanism, assuming the concentration ratios mentioned earlier. Anaerobic production of ATP in this way is a nonsteady-state mechanism because it exhausts the supply of creatine phosphate and causes buildup of the end product creatine.

Internal O_2 stores may be used to make ATP

For certain purposes, it is useful to recognize that aerobic catabolism can function in two different modes, steady-state and nonsteady-state, depending on its source of O_2. Up to now, whenever we have discussed aerobic catabolism, we have assumed that O_2 is taken into the body from the environment and distributed to cells as it is used. Aerobic catabolism is a steady-state mechanism under such conditions. Aerobic catabolism, however, can also draw its O_2 from preexisting stores of O_2 in a cell or tissue. Aerobic catabolism using O_2 stores is a nonsteady-state process because it depletes the O_2 stores on which it depends. Muscle cells often have enhanced O_2 stores because they contain *myoglobin*, a red O_2-binding pigment that is actually a type of hemoglobin (see Chapter 22). The myoglobin binds with O_2 when O_2 is abundant. Aerobic catabolism then can draw on these O_2 stores at other times.

Summary: Mechanisms of ATP Production and Their Implications

- Aerobic catabolism occurs in four steps: (1) glycolysis, (2) the Krebs citric acid cycle, (3) electron transport, and (4) oxidative phosphorylation. It can oxidize carbohydrates, lipids, or proteins to produce ATP. The nominal net yield of ATP in carbohydrate oxidation is 38 ATP molecules per glucose molecule.
- Anaerobic glycolysis is a redox-balanced process by which ATP can be made without O_2 in certain tissues. Whether a tissue can employ anaerobic glycolysis depends on the amount and type of its lactate dehydrogenase (LDH). Anaerobic glycolysis typically can use only carbohydrate fuel. It releases less than 10% of the energy of glucose. Thus it produces just two ATP molecules per glucose molecule, and the lactic acid produced is itself an energy-rich compound.
- Vertebrates retain the lactic acid that they produce. The metabolism of lactic acid requires O_2, either to oxidize the lactic acid via the Krebs citric acid cycle—thereby producing ATP—or to convert lactic acid to glycogen or glucose—a process that uses ATP.
- Phosphagens in muscle cells serve as temporary stores of high-energy phosphate bonds, which they can transfer to ADP to make ATP anaerobically.

Comparative Properties of Mechanisms of ATP Production

We have now identified four mechanisms of ATP production: (1) aerobic catabolism by use of O_2 acquired simultaneously from the environment, (2) anaerobic catabolism, (3) anaerobic ATP production by use of phosphagens, and (4) aerobic ATP production by use of internal O_2 stores. To compare the functional properties of these four, we now discuss several key questions. Table 6.1 summarizes much of what will be said.

Question 1: What is each mechanism's total possible ATP yield per episode of use?

If an animal uses a particular mechanism to make ATP and continues using it until no more ATP can be made, what is the total amount of ATP that can be produced? Because aerobic catabolism using environmental O_2 operates in steady state, it is capable of supplying an indefinite amount of ATP (a lifetime's amount). In contrast, the three mechanisms that operate in nonsteady state are limited in the amount of ATP they can produce during any one episode of use because of their mechanistic self-limitations. For example, the amount of ATP that can be made from phosphagens is limited by the quantity of phosphagens available; after the phosphagens originally present have been used, no more ATP can be made until (at some later time) the phosphagens are regenerated. Table 6.1 shows representative

TABLE 6.1 Comparative properties of mechanisms of ATP production in vertebrates, including numerical estimates for some of the properties in human beings

Mechanism of ATP production	Mode of operation (mandatory or assumed)	Total possible ATP yield per episode of use[a] (moles)	Rate of acceleration of ATP production at onset of use	Peak rate of ATP production[b] (μmol ATP/g · min)	Rate of return to full potential for ATP production after use
Aerobic catabolism using O_2 from environment	Steady state	Very large (~200 in a marathon, $> 4 \times 10^6$ in a lifetime)	Slow	Moderate (30 with glycogen fuel, 20 with fatty acid fuel)	—
Anaerobic glycolysis	Nonsteady state	Moderate (1.5)	Fast	High (60)	Slow
Phosphagen use	Nonsteady state	Small (0.4)	Fast	Very high (96–360)	Fast
Aerobic catabolism using O_2 preexisting in body	Nonsteady state	Small (0.2)	Fast	High	Fast

[a] Numerical estimates of total yields are computed from information in Åstrand and Rodahl 1986; a 75-kg person living 70 years is assumed.
[b] Peak rates of production are from Hochachka and Somero 2002.

yields of ATP in humans. Phosphagens and O_2 stores permit only relatively small quantities of ATP to be made per episode of use. The yield from anaerobic glycolysis is larger but modest.

Question 2: How rapidly can ATP production be accelerated?

In vertebrates and many other animals (although not insects), aerobic catabolism using environmental O_2 requires a relatively long time (minutes) to accelerate its rate of ATP production fully to a new high level. Acceleration is slow because this mechanism of ATP production is *not self-contained in cells*; the mechanism requires inputs of O_2, and the pulmonary and circulatory systems that transport O_2 from the environment to the cells increase their rate of O_2 delivery gradually, not instantly.

By contrast, the other three mechanisms can accelerate very rapidly because all *are self-contained in the cells*. At the start of a bout of exercise, for example, a vertebrate muscle cell contains not only all the enzymes of anaerobic glycolysis but also the glycogen fuel required. Nothing needs to be brought to the cell for anaerobic glycolysis to take place, and in fact a cell is capable of nearly instantly stepping up its glycolytic rate of ATP production to a high level. ATP production by use of phosphagens and O_2 stores can also accelerate rapidly because the phosphagens and O_2 stores are already present in cells when the mechanisms are called into play.

In both vertebrates and crustaceans such as crayfish, the first stages of burst exercise receive their ATP supply mainly from anaerobic glycolysis, phosphagen use, and use of O_2 stores, precisely because these are the three mechanisms that can rapidly accelerate (Box 6.2). As bizarre as it might be to imagine, vertebrates would be *incapable* of burst exercise if aerobic catabolism using O_2 from the environment were their only means of making ATP.

Question 3: What is each mechanism's peak rate of ATP production (peak power)?

After a mechanism has accelerated its ATP production to the fastest rate possible, how fast can ATP be made? As shown in Table 6.1, the peak rate at which anaerobic glycolysis can make ATP is much greater than the peak rate of ATP production by aerobic catabolism using environmental O_2, and the rate of ATP production by use of phosphagen is greater yet. Although the phosphagen mechanism cannot make a lot of ATP, it can make its contribution very rapidly; thus it can briefly support very intense exertion. Anaerobic glycolysis can make a modest amount of ATP at a high rate, and aerobic catabolism using environmental O_2 can make an indefinite amount at a relatively low rate.

Question 4: How rapidly can each mechanism be reinitialized?

Whenever ATP has been made by a nonsteady-state mechanism, cells are left in an altered state and must be returned to their original state before the mechanism can be used again to full effect. When internal O_2 stores have been used, the stores must later be recharged. When phosphagens have been used, they must be remade. When lactic acid has accumulated, it must be metabolized. In a word, the nonsteady-state mechanisms must be *reinitialized*.

The reinitialization of anaerobic glycolysis requires much more time than that of the phosphagen or O_2-store mechanisms because the length of time required to rid tissues of lactic acid is much greater than the time required to rebuild phosphagens or O_2 stores. In people, for example, a substantial accumulation of lactic acid may require 15 to 20 minutes for half dissipation and even 1 to 2 hours for full dissipation, whereas the half-time for reconstituting phosphagens and O_2 stores is just 30 seconds. Fish, frogs, lizards, and other poikilothermic vertebrates require even more time than mammals to dissipate lactic acid; 1 to 10 hours may be required for half dissipation.

Conclusion: All mechanisms have pros and cons

A review of Table 6.1 and the preceding discussion reveals that each mode of ATP production has pros and cons; none is superior in all respects. Consider, for example, the relative pros and cons of the two major sources of ATP in vertebrates: aerobic catabolism using environmental O_2 and anaerobic glycolysis.

BOX 6.2 Genetic Engineering as a Tool to Test Physiological Hypotheses

Experiments based on genetic engineering are increasingly being used to test physiological hypotheses. An example is provided by studies of the role of creatine phosphate. Because creatine phosphate in mammalian muscle can be mobilized extremely rapidly to make ATP at a high rate (see Table 6.1), physiologists have long hypothesized that the phosphagen serves as a principal source of ATP during the first seconds of burst exercise. ATP synthesis from creatine phosphate depends on the enzyme creatine kinase (CK) (see Figure 6.5*b*). If the hypothesized role of creatine phosphate in burst exercise is correct, lowering the levels of CK in muscle cells by genetic engineering should interfere with burst exercise. A failure of this prediction to hold true would be evidence against the hypothesis.

Mutant mice deficient in CK have been reared by genetic engineering. Their muscles exhibit subnormal ability to perform burst activity, and this performance deficit increases with the extent of their CK deficiency. Muscles from mice engineered to produce unusually *high* amounts of CK contract faster than normal muscles in the first moments of isometric twitches.

These experiments using genetically engineered mice support the hypothesized role of creatine phosphate in burst exercise. At the same time, detailed studies of the metabolic chemistry of muscles from CK-deficient mice provide a fascinating new revelation: *Cell* physiology undergoes compensatory adjustments to loss of CK function. Thus the contractile performance of CK-deficient muscle cells is probably not as impaired as it would be if CK deficiency could be produced independently of any other change.

Looking just at biochemical maps, one can get the impression that aerobic catabolism is far superior to anaerobic glycolysis. Aerobic catabolism fully releases the energy of food molecules (producing 38 ATPs per glucose), whereas anaerobic glycolysis unlocks only a small fraction of the energy value of food molecules (producing 2 ATPs per glucose). Moreover, anaerobic glycolysis typically is able to use only carbohydrate fuels, produces only a limited total amount of ATP in any one episode, and creates a product, lactic acid, that requires a long time to be cleared from the body.

Anaerobic glycolysis also has advantages in comparison with aerobic catabolism, however. Because it does not require O_2, it can provide ATP when O_2 is unavailable or supplement aerobic ATP production when O_2 is available. Furthermore, because anaerobic glycolysis can accelerate very rapidly and reach an exceptionally high rate of ATP production, it is well suited to the support of burst exercise.

Two Themes in Exercise Physiology: Fatigue and Muscle Fiber Types

Fatigue has many causes, most of which are poorly understood

Muscle fatigue is the lessening of a muscle's ability to generate peak forces and maintain power output. It is a critical physiological phenomenon—having life-and-death consequences in natural settings—that has defied full understanding. Fatigue clearly has multiple causes, depending on the type and duration of exercise. The fatigue associated with lactic acid accumulation is one of the best-understood types of fatigue. In people or other vertebrates undergoing intense exercise that involves sustained net lactate production, it is common to find that individuals are profoundly overcome with fatigue when lactic acid accumulates to a certain level. Virtual paralysis often sets in. Decades ago, the lactate ion was considered to be a specific "fatigue factor." Now, however, acid–base disturbances seem certain to be the actual cause of the fatigue associated with lactic acid accumulation. Cells and body fluids are made unusually acidic by a buildup of lactic acid, and this pH shift causes numerous disruptions of normal muscle function that interfere with muscle performance.

Other sorts of fatigue occur in other circumstances. One sort that is fairly well understood is the fatigue that develops after a long time during *sustained* exercise that is vigorous enough to require a rate of aerobic catabolism representing a substantial fraction of an individual's maximal rate of aerobic catabolism. When people engage in this sort of exercise, fatigue is associated with (perhaps caused by) inadequate muscle glucose. In the first tens of minutes of such exercise, glucose from muscle glycogen suffices to fuel the portion of muscle ATP production that is based on glucose oxidation. As time passes, however, muscle glycogen is depleted, and the muscles become dependent on glucose brought to them from other organs (notably the liver) by blood flow (Figure 6.6). Inadequacies in the rate of this glucose resupply are associated with fatigue. Another factor believed to contribute to fatigue in this sort of exercise is accumulation of critical ions (e.g., Ca^{2+}) in inappropriate cellular locations as a consequence of the ion fluxes involved in muscle activation and control (see p. 469). Overheating is another factor that can cause fatigue.

The muscle fibers in the muscles used for locomotion are heterogeneous in functional properties

The muscle tissue of an animal is typically heterogeneous in its metabolic and contractile properties. In vertebrates, for example, several different types of muscle fibers (muscle cells) occur in an individual. Two of the principal types in mammals are the **slow-oxidative** (**SO**) muscle fibers and the **fast-glycolytic** (**FG**) fibers. The SO fibers are called *slow* because they contract and develop tension relatively slowly, in contrast to the FG fibers, which are *fast* in contracting and developing tension. The SO fibers are called *oxidative*, whereas the FG fibers are called *glycolytic*, because of differences in their physiology of ATP production. The SO fibers are poised principally to make ATP by aerobic catabolism; they have high levels of key enzymes specific to aerobic catabolism, such as succinic dehydrogenase (a Krebs-cycle enzyme), and are well endowed with mitochondria. The FG fibers, by contrast, are poised

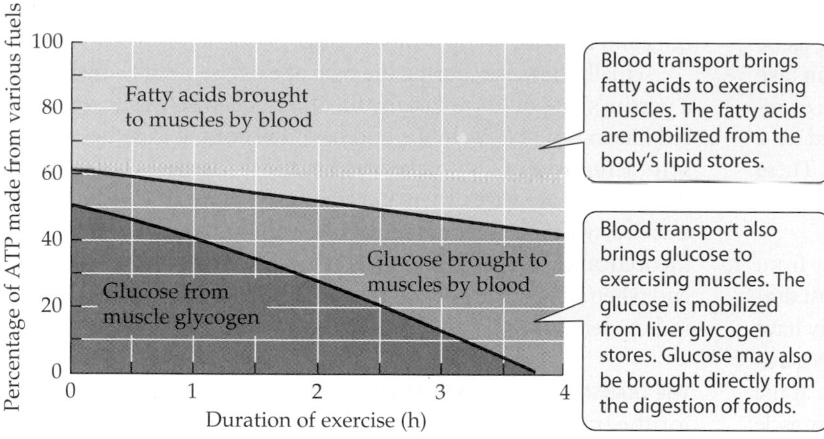

Figure 6.6 The fueling of intense but sustained muscular work in humans During 4 hours of exercise that requires about 80% of a person's maximal rate of aerobic catabolism, muscle glycogen is depleted, but glucose is still required. As time passes and muscle glycogen is used up, glucose must be brought to the muscles by blood transport from elsewhere in the body at an increasing rate. A failure of this circulatory resupply of glucose to keep up with muscle needs is believed to be an important cause of fatigue during this type of exercise. An important fact that helps explain why shortage of glucose can be a fatigue factor is that the human body contains only about 450 g (one pound) of glycogen even when glycogen stores are fully loaded. (After Coyle 1991.)

principally to make ATP anaerobically; they are high in enzymes of anaerobic glycolysis, such as lactate dehydrogenase, and are relatively sparse in mitochondria.

Correlated with the differences in the catabolic pathways they use, the SO and FG fibers differ in how readily they can take up and store O_2. The SO fibers, for example, are relatively rich in myoglobin, which not only helps store O_2 within the fibers but also aids diffusion of fresh O_2 into the fibers. FG fibers are low in myoglobin. Because of the difference in myoglobin, SO fibers are reddish in color, whereas FG fibers are whitish, explaining why they are sometimes called *red* and *white* fibers.

SO and FG fibers also differ in their power-generation and fatigue properties in ways that correlate with the features we have already discussed. The properties of SO fibers are predictable from the fact that they principally employ steady-state aerobic catabolism to make ATP: Although SO fibers have relatively low peak mechanical-power outputs, they are relatively resistant to fatigue and readily sustain work over long periods. FG fibers represent the opposite extreme. They can put out a lot of power, but they rely strongly on nonsteady-state mechanisms of ATP production, accumulate lactic acid, and fatigue quickly.

In mammals, major muscles are typically built of mixes of intermingled SO fibers, FG fibers, and other fiber types. Fish, on the other hand, often have large muscle masses composed principally of a single type of muscle fiber similar to the mammalian SO or FG fibers; thus whole blocks of muscle in fish are red or white and exhibit the performance properties of red and white fibers.

The Interplay of Aerobic and Anaerobic Catabolism during Exercise

When animals engage in exercise, their performance typically reflects the underlying mechanisms that they are using to produce the ATP required. Fish that are cruising about at relatively leisurely speeds, for example, do so with their red swimming muscles, employing steady-state aerobic catabolism to make ATP. Cruising therefore can be sustained for long periods. However, fish that engage in sudden, intense exertion—such as cod avoiding a trawling net or salmon leaping waterfalls on their upstream journey—use their white swimming muscles and anaerobic glycolysis to generate the high power they require.[9] They thus accumulate lactic acid, and if they must perform repeatedly in a short time, they are in danger of fatigue from acidification of their tissues.

Crustaceans similarly illustrate that exercise performance reflects the underlying mechanisms of ATP production. When crayfish, lobsters, and crabs walk or cruise about at modest, sustainable speeds, they produce ATP by steady-state aerobic catabolism. When a crayfish or lobster employs tail flipping to power itself rapidly away from danger, however, the tail muscles require ATP at a greater rate than aerobic catabolism can provide. Thus the muscles turn to mechanisms that can produce ATP at exceptionally high rates. In species of crayfish that have received detailed study, the tail muscles exhibit high levels of the "anaerobic ATP production" enzymes arginine kinase and lactate dehydrogenase, and low levels of aerobic catabolic enzymes. The primary mechanism of ATP production at the start of escape swimming by tail flipping is use of the phosphagen arginine phosphate. Then later, if tail flipping continues, anaerobic glycolysis is brought into ever greater play to meet the ATP demands of the tail muscles. Because both of these mechanisms are nonsteady-state and self-limiting, tail flipping cannot be sustained for long. Crayfish fatigue during tail flipping, and part of the reason is believed to be tissue acidification from lactic acid accumulation.

When lactic acid accumulates in a vertebrate or invertebrate, the slow rate at which it is metabolized itself affects an animal's performance. Each animal is believed to have a certain absolute maximum accumulation of lactic acid that it can tolerate. This means that if lactic acid accumulates in a particular bout of exercise, the animal is limited in the extent to which it can use anaerobic glycolysis in a subsequent bout of exercise until it has fully metabolized the lactic acid from the first bout. A human example is provided by competitive racers who compete in two types of high-speed, burst-type races in a single track meet. It is well established that performance in

[9] When we speak of red and white muscles in this context, we are referring to the distinction between the myoglobin-rich, SO-like muscles and the myoglobin-poor, FG-like muscles discussed in the previous section. This is an entirely different matter from the overall hue of a fish's muscle tissue, whether off-white in a cod or orange in a salmon.

the second race is impaired if lactic acid accumulated in the first race has not yet been fully metabolized. As noted earlier, clearing lactic acid from the body takes tens of minutes in mammals and can take hours in poikilotherms. For animals in nature, the slow rate of removal of lactic acid may well mean that an animal is impaired for a substantial time in its ability to pursue prey or escape danger. These important behavioral consequences are additional manifestations of the biochemistry of ATP production.

Some types of animals follow patterns that are different from those of vertebrates and crustaceans. Insects constitute the most dramatic and important example. Insect flight muscles typically have little or no ability to make ATP anaerobically and remain fully aerobic even when they suddenly increase their power output, as at the onset of flight. Correlated with the fact that virtually all flight-muscle work is aerobic, insect flight muscles contain very high levels of aerobic catabolic enzymes, and sometimes half of their tissue volume is occupied by mitochondria! The flight muscles of certain insects, in fact, are the most aerobically competent of all animal tissues as judged by the rate per gram at which they can synthesize ATP by aerobic catabolism. A characteristic of insects that helps explain their aerobic fueling of burst exercise is their tracheal breathing system, which provides O_2 directly to each flight-muscle cell by way of gas-filled tubes and which often is ventilated by the action of the flight muscles themselves (see page 570).

Other, less dramatic differences from the vertebrate and crustacean pattern are found in other animal groups. For example, when squids, octopuses, and scallops use anaerobic catabolism during burst swimming, their swimming muscles reduce pyruvic acid to *octopine* rather than lactic acid. The octopine-generating anaerobic pathway provides a high rate of ATP synthesis, as in vertebrates, but has different detailed biochemical implications.

Metabolic transitions occur at the start and end of vertebrate exercise

Multiple metabolic processes are typically involved in the provision of ATP for all forms of vertebrate exercise. To understand this attribute of vertebrate exercise fully, one must take into account the intensity of exercise. A useful way to do this is to reference the cost of exercise to an individual's maximal rate of aerobic catabolism.

A given individual in a particular state of training is capable of a certain maximal rate of O_2 consumption. Exercise that requires exactly this maximum is called **maximal exercise.** Exercise requiring less than the maximal rate of O_2 consumption is called **submaximal exercise.** As strange as it may initially sound, a third category of exercise in this system of reference is **supramaximal exercise**, exercise that requires more than an individual's maximal rate of O_2 consumption.

Consider a bout of submaximal exercise that starts and ends abruptly, requires about 80% of an individual's maximal rate of O_2 consumption, and lasts 30 minutes or so. Let's assume we are talking about a person who is running, although the principles we will develop apply to all or most vertebrates. Figure 6.7*a* shows how the person's rate of O_2 consumption by breathing would change during this bout of exercise if *all* ATP were made on a moment-by-moment basis by aerobic catabolism using environmental O_2. The person's rate of O_2 consumption would increase stepwise at the start of running and decrease stepwise at the end.

Actually, however, during the type of exercise we are considering, a person's rate of O_2 uptake by breathing changes as shown by the red line in Figure 6.7*b*. In the middle of the bout of exercise, the person's actual rate of O_2 uptake matches the theoretical O_2 demand of the exercise. However, there is a transition phase at the start of the exercise, when the person's actual rate of O_2 uptake is lower than the theoretical O_2 demand, and there is another transition phase at the end when the person's actual rate of O_2 uptake exceeds the theoretical O_2 demand.

THE TRANSITION PHASE AT THE START: THE OXYGEN DEFICIT The reason for the transition phase at the start of the exercise we are considering is the fact, already mentioned, that the respiratory and circulatory systems in vertebrates do not instantly increase the rate at which they deliver O_2 to the body. Instead, even if *exercise* starts *abruptly*, *O_2 delivery* to the tissues increases *gradually*. In people, 1 to 4 minutes of time is required for the pulmonary and circulatory systems to accelerate fully. While the pulmonary and circulatory systems are accelerating their actual rate of O_2 delivery at the start of a bout of exercise, the body's supply of O_2 from the environment (actual O_2 uptake) is less than its theoretical O_2 demand for the exercise. This difference is termed an **oxygen deficit** (see Figure 6.7*b*).

During the period of the oxygen deficit, the full ATP demand of exercise is *not* met by aerobic catabolism based on environmental O_2. How then is the ATP demand met? What mechanisms provide ATP in addition to aerobic catabolism based on environmental O_2? During the period of the oxygen deficit in the sort of exercise we are discussing, the answer is that ATP is contributed by anaerobic glycolysis, use of phosphagens, and use of O_2 stores.[10] *These three mechanisms, in fact, are essential for exercise to start in a stepwise way.* They make up for the slow acceleration of ATP production by aerobic catabolism based on environmental O_2, permitting the *overall* rate of ATP production to increase abruptly to a high level when exercise begins.

THE PAY-AS-YOU-GO PHASE During the exercise we are discussing or any other sort of submaximal exercise, the pulmonary and circulatory systems ultimately accelerate their rate of O_2 delivery sufficiently to meet the full O_2 demand of the exercise. At that point, the exercise is said to enter a **pay-as-you-go phase** (see Figure 6.7*b*) because thereafter its full O_2 cost is met on a moment-to-moment basis by use of O_2 taken up from the environment by breathing. With all ATP being made by steady-state aerobic catabolism in the pay-as-you-go phase, the exercise can in principle be sustained indefinitely.

As stressed earlier, ATP production by anaerobic glycolysis, use of phosphagens, and use of O_2 stores can self-terminate because all three of these processes represent nonsteady-state mechanisms of

[10] Anaerobic glycolysis, use of phosphagens, and use of O_2 stores are sometimes termed the *mechanisms of oxygen deficit*. In vertebrates, the three are subdivided into two categories. Anaerobic glycolysis is the **lactacid** mechanism of oxygen deficit because it produces lactic acid. Phosphagen use and the use of O_2 stores are the **alactacid** mechanisms of oxygen deficit.

(a) Theoretical rate of O_2 consumption assuming all ATP is made using O_2 from the atmosphere

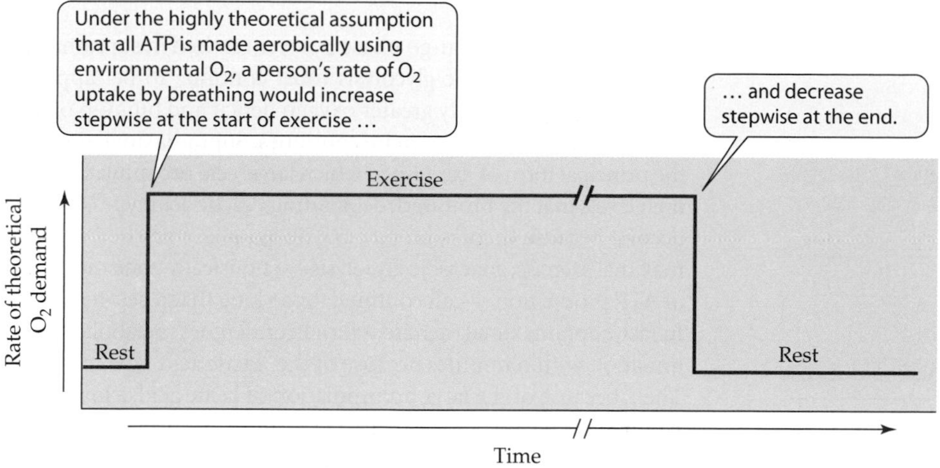

(b) Actual rate of O_2 consumption compared to theoretical

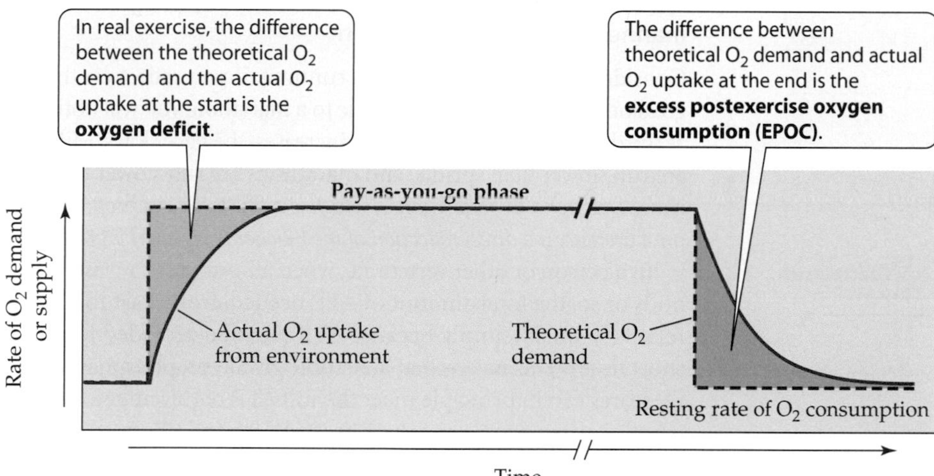

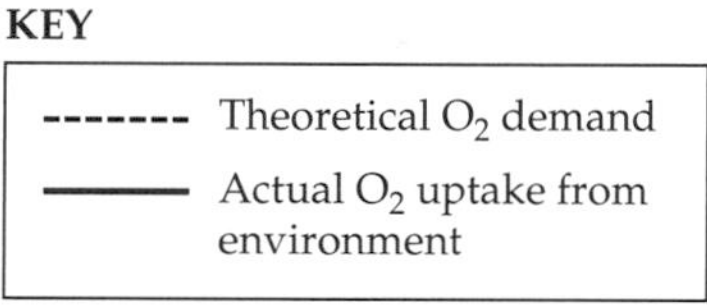

Figure 6.7 The concepts of oxygen deficit and excess postexercise oxygen consumption *(a)* The rate of O_2 demand of a person who is initially at rest, suddenly starts vigorous submaximal exercise, continues for 30 minutes or so, and then suddenly stops, assuming unrealistically that all ATP is produced aerobically at all times. *(b)* The actual rate of O_2 uptake from the environment (red line) of the person in *(a)* showing that there is an initial transition period during which the full ATP demand is not met by O_2 uptake, then a period when O_2 uptake matches the full O_2 demand, and finally a transition period when the person's actual rate of O_2 uptake exceeds the resting rate even though the person is at rest.

ATP production. During submaximal exercise, these nonsteady-state mechanisms—when employed in the early minutes —do not self-terminate because the need for them ends when the the pay-as-you-go phase begins. In brief, the initial oxygen-deficit phase in which nonsteady-state mechanisms of ATP production are used is *replaced* by the pay-as-you-go phase in which steady-state aerobic catabolism (using environmental O_2) is the sole mechanism of ATP production.

THE TRANSITION PHASE AT THE END: EXCESS POSTEXERCISE OXYGEN CONSUMPTION At the end of the exercise bout we are describing, the exercising person suddenly stops running, but thereafter his or her actual rate of O_2 uptake does not suddenly drop in stepwise fashion. Instead, it declines gradually, remaining above the resting rate of O_2 consumption for many minutes (see Figure 6.7*b*). This elevation of the actual rate of O_2 uptake above the resting rate of O_2 uptake—even though the person is behaviorally at rest—is termed **excess postexercise oxygen consumption (EPOC).** People describe it as "breathing hard" after exercise. An older term for the phenomenon, *oxygen debt*, has been discredited because it was based on an assumption that the cause of the EPOC is the need to metabolize lactic acid. Actually, we now know that in many cases among vertebrates, there is a nearly total lack of correspondence between the time course of the EPOC and the time course of lactic acid metabolism. In some instances, for example, lactic acid remains present at elevated levels long after the EPOC is over. The term *EPOC* is preferred over *oxygen debt* because it is functionally neutral and implies no particular mechanistic explanation. The reasons for the EPOC are not well understood at present.

THE TRANSITION PHASES VARY IN THEIR NATURE DEPENDING ON THE INTENSITY OF EXERCISE Different intensities of exercise in vertebrates produce different sorts of transition phases. If a vertebrate undertakes *light submaximal* exercise that requires less than 50% to 60% of the maximal rate of O_2 consumption, transition phases still occur at the start and end, as shown in Figure 6.8*a*. However, lactic acid does not accumulate at these exercise intensities. This means that the only processes required to "reinitialize" the body at the end of exercise are the replenishing of O_2 stores and phosphagen stores, both of which occur very rapidly (see Table 6.1). In *heavy submaximal* exercise requiring more than 50% to 60% of maximal O_2 consumption (Figure 6.8*b*), as we have already discussed, net accumulation of lactic acid occurs at the start, and thus the longer process of metabolizing lactic acid must occur at the end.

The most striking sorts of transition phases occur during *supramaximal* exercise (Figure 6.8*c*). Such exercise demands ATP at a greater rate than can *ever* be supplied by steady-state aerobic catabo-

(a) Light submaximal exercise

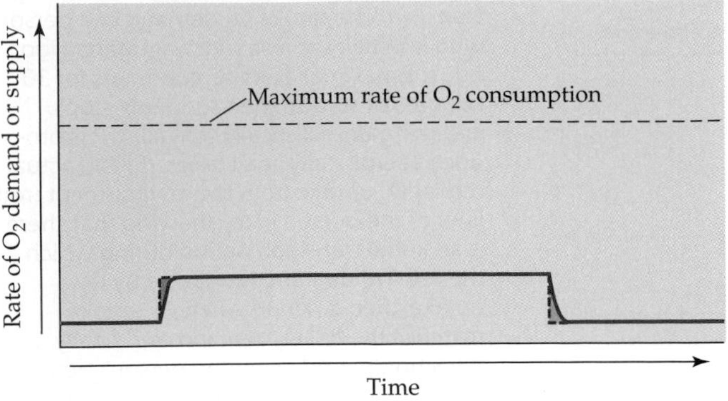

(b) Heavy submaximal exercise

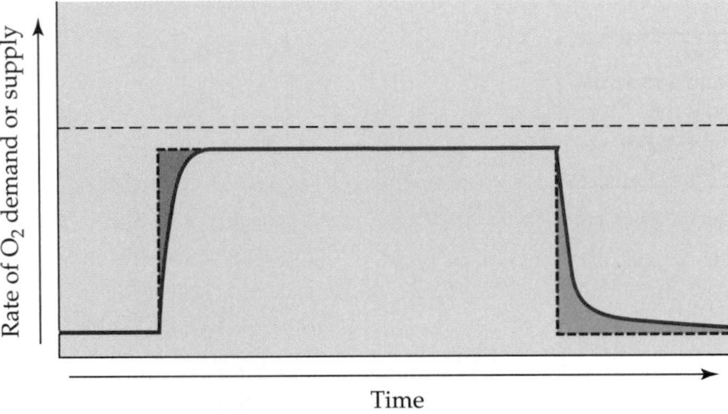

(c) Supramaximal exercise

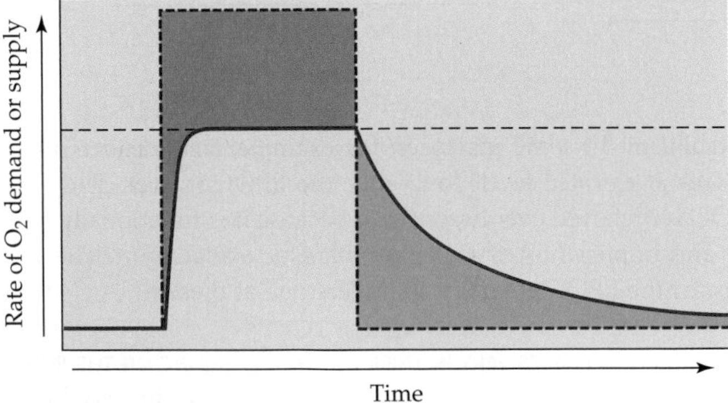

KEY

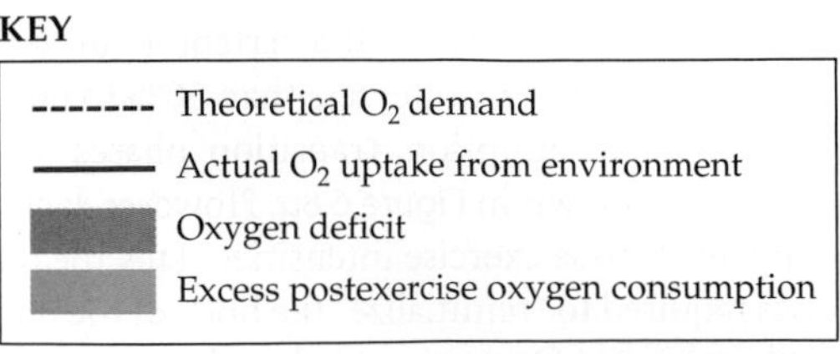

Figure 6.8 Stylized O_2 supply–demand diagrams for light submaximal, heavy submaximal, and supramaximal exercise The format is as in Figure 6.7. The patterns illustrated are those observed in mammals. Two important determinants of exercise performance are an individual's maximal rate of O_2 consumption and maximal oxygen deficit. Both are increased by repeated exercise, thereby increasing performance. A person, for example, might increase his or her maximal rate of O_2 consumption by 10% to 30% through appropriate training.

lism. Thus, a pay-as-you-go phase is never reached in supramaximal exercise, and anaerobic glycolysis must continue to be tapped for ATP—causing a steadily greater oxygen deficit and buildup of lactic acid—for as long as the exercise continues. Supramaximal exercise is the principal form of exertion in which lactic acid accumulates to such high levels that the profound, debilitating, "lactic acid type" of fatigue occurs. Because there is no pay-as-you-go phase reached in supramaximal exercise, anaerobic glycolysis—a nonsteady-state mechanism of ATP production —can continue for so long that it self-terminates. In fact, supramaximal exercise typically undergoes metabolic self-termination within minutes because of the "lactic acid type" of fatigue. Then, because of the large accumulation of lactic acid, a long time is required for the body to be cleared of lactic acid, during which an individual's capability for supramaximal exertion is impaired.

The ATP source for all-out exercise varies in a regular manner with exercise duration

Considering human competitive running, if you reflect on the progression from a sprint to a mile race to a marathon, you will notice that as the duration of all-out exertion increases, the pace slows. Mile races are run slower than sprints, and marathons are run slower than mile runs. *This trend in performance, which is a fairly general property of animal exercise, is a direct reflection of the biochemistry of ATP production.*

In a person or other vertebrate, when all-out exertion lasts 10 seconds or so, the total amount of ATP needed, from start to finish, is relatively small (simply because ATP must be provided for only a short time). This means that anaerobic glycolysis, phosphagens, and O_2 stores can in principle meet the full ATP requirement. In actual practice, when people run the 100-yard dash (or 100-m dash), these particular mechanisms of ATP production meet at least 90% of the ATP cost; some champions in the event hold their breath from start to finish! Anaerobic glycolysis and the use of phosphagens and O_2 stores—while unable to make a great quantity of ATP—are able to produce ATP at exceptionally high *rates* (see Table 6.1). Thus, when these mechanisms are sufficient to meet most of the cost of running, the pace of running can be very fast, as it is in the 100-yard dash.

In a mile race (about the same as a 1500-m race), the total ATP requirement is much greater than in a 100-yard dash. Even if anaerobic glycolysis, phosphagens, and O_2 stores are fully exploited to make as much ATP as they possibly can during a mile race, they cannot meet more than 25% to 50% of the total ATP need. The rest of the ATP must be made by steady-state aerobic catabolism using O_2 from the environment, a mechanism that cannot produce ATP as fast as the other mechanisms (see Table 6.1). Thus, taking into account the rates of all the processes that are required to contribute ATP during a mile race, the overall rate of ATP production during such a race is lower than that during a 100-yard dash. The pace of the mile must therefore be slower. The pace of a marathon must be slower yet because the total ATP requirement of a marathon is so great that only 2% to 3% of it can be met using anaerobic glycolysis, phosphagens, and O_2 stores. In the marathon, almost all the ATP is made by steady-state aerobic catabolism, limiting the pace to that which is permitted by that mechanism.

Figure 6.9 shows how the mechanisms of ATP production during all-out exertion vary with the duration of exertion in world-class competitive runs. The upper line shows how pace varies with dura-

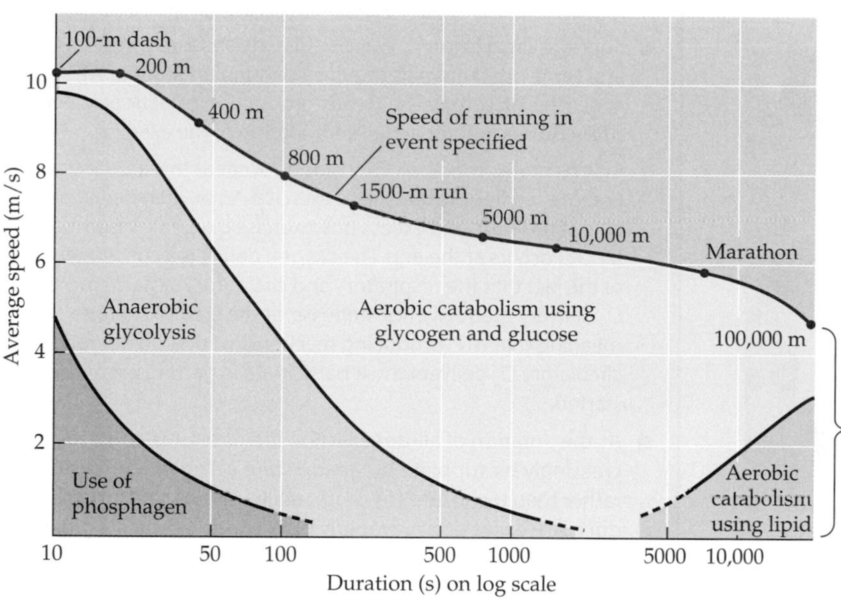

Figure 6.9 The mechanisms of meeting the ATP costs of world-class competitive running The upper line shows the average pace of races as a function of the length of time the pace is maintained, based on current world records for men in the specified competitive events. The space below the upper line shows the approximate fraction of ATP made by aerobic catabolism, anaerobic glycolysis, and use of creatine phosphate. For simplicity, synthesis of ATP by use of O_2 stores is not shown as a separate category. (Data on ATP after Newsholme et al. 1992.)

tion. The subdivision of the space below the upper line shows how ATP is made. With increasing duration, ATP production shifts from being principally anaerobic (based on phosphagens and glycolysis) to being chiefly aerobic. Moreover, at marathon and ultramarathon distances, aerobic catabolism shifts toward substantial use of lipid fuels, which permit only a lower rate of ATP synthesis than carbohydrate fuels (see Table 6.1). Similar trends apply to vertebrates living in the wild. Thus, ecologically relevant performance depends in a regular way on the biochemical mechanisms by which ATP is made. As the primary mechanism of ATP synthesis shifts from phosphagens to anaerobic glycolysis and then to aerobic catabolism based on carbohydrate and lipid fuels, the pace slows.

Related species and individuals within one species are sometimes poised very differently for use of aerobic and anaerobic catabolism

Related species sometimes have evolved very different emphasis on aerobic and anaerobic ATP production during intense exertion, and these differences can have important life-history consequences. Thus the biochemistry of ATP production is one of the ways that species become specialized to live as they do. The terrestrial amphibians we briefly noted in Chapter 2 (see Figure 2.10) provide a classic illustration.

Some species of terrestrial amphibians, such as the common leopard frog (*Rana pipiens*) and many other ranid and hylid frogs, have a biochemistry that emphasizes anaerobic glycolysis as the principal mechanism of ATP production during all-out exertion. If you chase a leopard frog, at first the frog flees by jumping away very rapidly, but within a few minutes it collapses in fatigue. Both the high speed of the initial jumping and the quick fatigue reflect the emphasis on anaerobic glycolysis to make ATP. Many of the bufonid toads, such as the western toad (*Bufo boreas*), exemplify the opposite extreme. The toads have a biochemistry that emphasizes aerobic production of ATP during all-out exertion. When chased, they do not flee as quickly as leopard frogs, but they can continue jumping away at a steady pace for a long time. Both their slow speed and their resistance to fatigue reflect the aerobic fueling of their exercise.

Lactate dehydrogenase represents a key respect in which the frogs and toads differ; the frogs synthesize greater amounts of lactate dehydrogenase than the toads. A parallel, and very dramatic, example is provided by species of fish. The supremely active skipjack tuna exhibits over 1000 times the activity[11] of lactate dehydrogenase in its white swimming muscles as sluggish fish species do.

At a different scale, individuals within a single species are often poised differently for use of aerobic and anaerobic catabolic pathways. A particularly intriguing example of this sort of variation is that people vary widely in the fiber composition of their muscles, as illustrated in Figure 6.10. The swimmer (Figure 6.10*a*) competes in sprints that require intense exertion for short periods of time. The cyclist (Figure 6.10*b*) competes in sustained cycling races. Small pieces of tissue (called *biopsy samples*) were removed from identical thigh muscles of the two men and subjected to a histochemical procedure that darkly stains just slow-oxidative (SO) muscle fibers. The images to the right in Figure 6.10 show that few of the muscle fibers of the sprinter are SO fibers, whereas most of the fibers of the cyclist are SO fibers. Most researchers believe that differences of the sort exhibited in the thigh muscles of the short-distance swimmer and the long-distance cyclist are relatively fixed. That is, people like the cyclist, it is thought, have principally SO fibers from birth. They discover through experience that their muscles, although not extremely powerful, are resistant to fatigue; and if such people are inclined toward athletic competition, they choose sports like long-distance cycling that are well served by slow-oxidative (SO) muscle fibers. In contrast, people like the sprinter, whose muscles are low in SO fibers and presumably high in fast-glycolytic (FG) fibers, gravitate toward sports in which high power—rather than long-term endurance—is a key to success.

[11] Enzyme activity is measured as the maximum rate at which substrate can be converted to product by a unit weight of tissue.

(a)

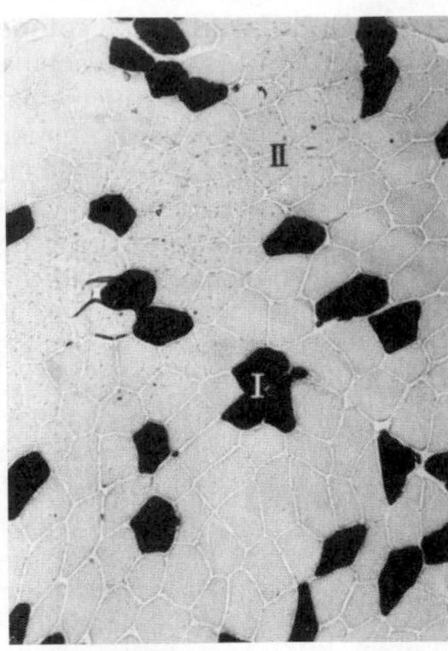

(b)

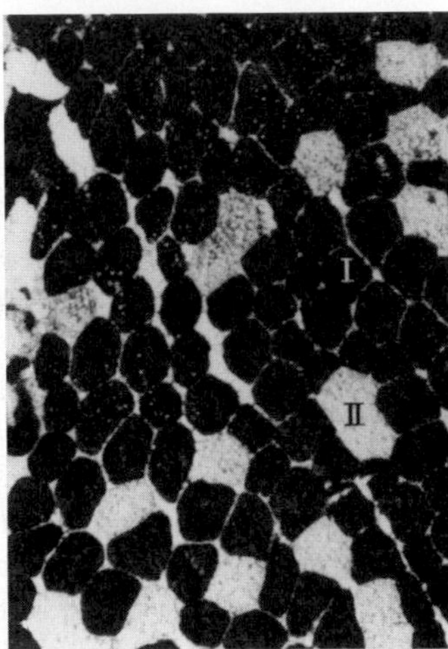

Figure 6.10 Two top athletes who differ in the fiber composition of their thigh muscles *(a)* The swimmer competes in 50-m sprints. *(b)* The cyclist competes in long-distance races. Shown to the right of each man is a microscopic section of his vastus lateralis, a thigh muscle, stained in a way that makes SO fibers dark. The sections are labeled with a different nomenclatural system from the one we use in this chapter: I = SO fibers; II = either FG fibers or other fast-contracting fibers. (Courtesy of Rudolf Billeter-Clark and Hans Hoppeler; from Billeter and Hoppeler 1992.)

Summary: The Interplay of Aerobic and Anaerobic Catabolism during Exercise

- Behavior and biochemistry are linked during physical activity because attributes of performance depend on how the ATP for muscular effort is synthesized.
- Submaximal forms of exercise can be supported entirely (except during transition phases) by aerobic catabolism using O_2 taken in from the environment by breathing. From the viewpoint of ATP supply and demand, submaximal forms of exercise can thus be sustained indefinitely.
- Supramaximal forms of exercise in vertebrates, crustaceans, and some other animals require a continuing input of ATP from anaerobic glycolysis. The steady production of lactic acid eventually causes metabolic self-termination of the exercise.
- In vertebrates, metabolic transitions occur at the start and the end of even light submaximal exercise. An oxygen deficit occurs at the start, and excess postexercise oxygen consumption (EPOC) occurs at the end. The oxygen deficit is a consequence of the fact that the respiratory and circulatory systems increase O_2 delivery gradually, not stepwise, at the start of exercise, meaning that ATP-producing mechanisms that do not require circulatory O_2 delivery must participate in getting exercise started.
- As the duration of all-out exertion increases, ATP must increasingly be supplied by steady-state aerobic catabolism, rather than by nonsteady-state mechanisms that can produce ATP exceptionally rapidly but cannot produce a great deal of it. Thus the pace of all-out exertion declines as duration increases.
- Closely related species, and even individuals within one species, often differ significantly in their emphasis on aerobic and anaerobic mechanisms of producing ATP for exercise. These metabolic differences help explain differences in exercise performance.

Responses to Impaired O_2 Influx from the Environment

Besides vigorous exercise, impaired O_2 influx from the environment is the second major reason that animals turn to anaerobic catabolic pathways to make ATP. Many animals experience reduced O_2 influx from their environments during parts of their lives. The situation can arise in two ways. First, O_2 influx may be reduced because the concentration of O_2 in the environment is low. Alternatively, animals may enter environments in which they cannot breathe, as is true of mammals—and many species of turtles—during diving. Under either set of conditions, it is common for at least some tissues to experience **hypoxia** or **anoxia**, defined respectively to be an especially low level of O_2 in tissues or an absence of O_2 from tissues.

All vertebrate animals are limited to the mechanisms of ATP production we have already discussed; the only known exceptions (as we will see) are two species of fish. When vertebrates turn to anaerobic catabolism during periods of reduced O_2 influx from the environment, they produce lactic acid, which they never excrete. In sharp contrast, invertebrates adapted to life without influx of O_2 rarely produce lactic acid. Instead, they usually employ anaerobic catabolic pathways that are far more elaborate than anaerobic glycolysis and that yield a variety of different products, which often are excreted.

Metabolic depression is an important theme in the study of animals living with reduced O_2 influx. Metabolic depression is a reduction in the ATP needs of an animal (or specific tissue) to levels below the needs ordinarily associated with rest in a way that does not present an immediate physiological threat to life. Although metabolic depression is not universal, it is increasingly known to be part of the

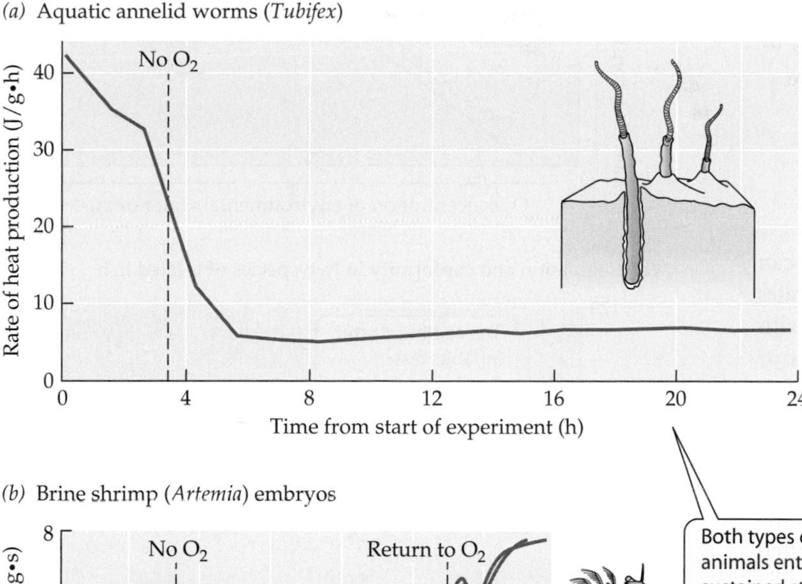

Figure 6.11 Metabolic depression in two invertebrates experiencing anoxia The animals studied were two types that are highly tolerant of anoxia: *(a)* adults of *Tubifex tubifex*, a species of small annelid worm commonly found in O_2-free bottom waters and muds of lakes and ponds, and *(b)* embryos of brine shrimps (*Artemia*). In each case the animals were living in water near equilibrium with the atmosphere at the start of the experiment. Then, at the time marked "no O_2," they were switched into O_2-free water. The metabolic rates of the animals were quantified directly, by measurements of heat production. Both types of animals responded to the lack of O_2 by undergoing a profound drop in their metabolic rates. The worms in *(a)* remained anoxic to the end of the experiment. The embryos in *(b)* were returned to oxygenated water 13 hours into the experiment. Two groups of embryos were studied in *(b)*, accounting for the two lines. The embryos were exposed to ammonia while in O_2-free water, as a way of raising their pH. Their response to ammonia supports the hypothesis that metabolic depression in the embryos is partly dependent on a low pH. (*a* after Famme and Knudsen 1984; *b* after Hand and Gnaiger 1988.)

suite of responses that animals show to reduced O_2 influx (Figure 6.11). All anaerobic catabolic pathways produce far less ATP per food molecule than aerobic catabolism. Thus, when animals turn from aerobic to anaerobic catabolic pathways as their means for *long-term* ATP production, they are in a biochemically diminished position to make ATP. One option is to reduce the rate at which they *need* ATP: metabolic depression.

Diving vertebrates exploit anaerobic glycolysis during protracted dives

When mammals, birds, and turtles dive, they usually limit their time underwater to a short enough period that they can meet their ATP needs by aerobic catabolism, using their O_2 stores, which often are exceptionally large. By diving in this pattern, they avoid accumulating lactic acid, and therefore they need not spend long periods at the surface between dives to metabolize lactic acid. Dives of exceptionally long duration, however, cause vertebrates to resort to anaerobic glycolysis. Mammals and birds, as we discuss in great detail in Chapter 24, metabolically subdivide their bodies during protracted dives. Their brains obligatorily require O_2. By their metabolic subdivision, they deliver certain of their O_2 stores preferentially to their brains, ensuring that brain catabolism remains aerobic, while simultaneously, large portions of their bodies run out of O_2 and become fully dependent on anaerobic glycolysis. Diving crocodilians, sea turtles, and lizards must also keep their brains aerobic. Certain of the freshwater and terrestrial turtles, however, are dramatically different in that they can tolerate total-body anoxia—full O_2 depletion of all tissues—during protracted dives.

An informative way to understand the implications of brain anoxia in diving turtles is to look briefly at the threat anoxia poses to vertebrate central nervous tissue. In mammals, catastrophe strikes when the brain is deprived of O_2 because the ATP requirement of mammalian brain cells per unit time far exceeds the rate at which the cells can make ATP by anaerobic means. Within seconds after O_2 influx to the mammalian brain is cut off, the concentration of ATP in brain cells starts falling precipitously. Soon, ATP-dependent ion pumps (e.g., Na^+–K^+-ATPase) are unable to pump ions across the cell membranes rapidly enough to maintain normal membrane polarization. The cell membranes thus depolarize, with numerous cataclysmic consequences; for instance, nerve impulses (action potentials) become impossible, and voltage-gated Ca^{2+} channels are inappropriately triggered to open, allowing Ca^{2+} to flood into the cells in quantities that inappropriately and disastrously trigger a variety of Ca^{2+}-mediated responses.

The turtles that tolerate brain anoxia employ metabolic depression of the brain as a key mechanism of maintaining the integrity of brain tissue during anoxia. Synaptic transmission between brain cells is suppressed in the absence of O_2, and ion-mediated bioelectrical activity of cells is reduced so that the brain becomes electrically relatively silent. This approach has a significant cost: The turtles cease to be behaviorally alert, and they become comatose. Because of the adjustments, however, the brain ion pumps have much-reduced work to do to maintain cell-membrane polarization. The ATP requirement of maintaining tissue integrity is lowered, anaerobic glycolysis is able to meet the ATP requirement, and brain ATP concentrations do not fall. With its entire body anoxic, a turtle accumulates lactic acid, which can reach extraordinary concentrations during prolonged anoxia. The shell and the bones of a turtle play key roles in preventing lethal acidification under these circumstances, by buffering the acid.

Animals faced with reduced O_2 availability in their usual environments may show conformity or regulation of aerobic ATP synthesis

When animals are living in the environments in which they can breathe, such as fish in water or mammals in air, how do their rates of aerobic ATP production change when they are confronted with changes in the concentration of O_2 in the environmental water or air? As the environmental O_2 level is lowered, the usual pattern is for an animal's rate of aerobic catabolism—and rate of O_2 consumption—to be unaffected over a certain range of O_2 levels (Figure 6.12*a*). This maintenance of a steady rate of aerobic catabolism (and aerobic ATP production) regardless of the level of O_2 in the environment is termed **oxygen regulation** (see Figure 1.5). It often involves active responses by the animals, such as increases in breathing rate as the O_2 level in the water or air decreases. Then, if the environmental O_2 level is lowered further, animals typically become unable to maintain an unaltered rate of aerobic catabolism, and their rates of O_2 consumption fall as the environmental O_2 level falls, a condition termed **oxygen conformity** (see Figure 6.12*a*). High altitudes are an intriguing situation in which these concepts apply to terrestrial species (Box 6.3).

When related species are compared under similar test conditions, their abilities for regulation and conformity often correlate with the types of habitats in which they live. These abilities, therefore, seem to have evolved in parallel with habitat selection. An illustration is provided by two species of related freshwater fish living in a single Alabama watershed (Figure 6.12*b*). One, the redline darter, occurs in fast-flowing streams where O_2 levels tend to be consistently high because turbulence promotes aeration. The other, the slack-water darter, is found in slow-moving streams where the O_2 concentration may be only a third as high as in fully aerated water. As the figure shows, the slack-water darter, which often experiences low-O_2 waters, exhibits a much broader range of oxygen regulation than the redline darter, which rarely must cope with low O_2 levels. Indeed, lowering the concentration of O_2 to 40% to 50% of the fully aerated level causes a sharp depression of O_2 uptake and deaths in redline darters while not affecting the O_2 uptake of slack-water darters at all.

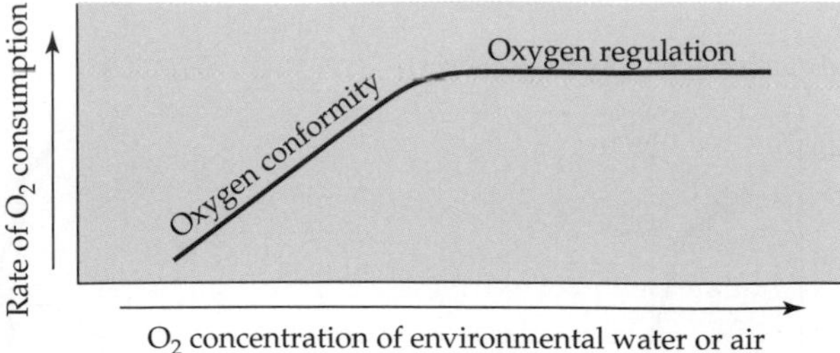

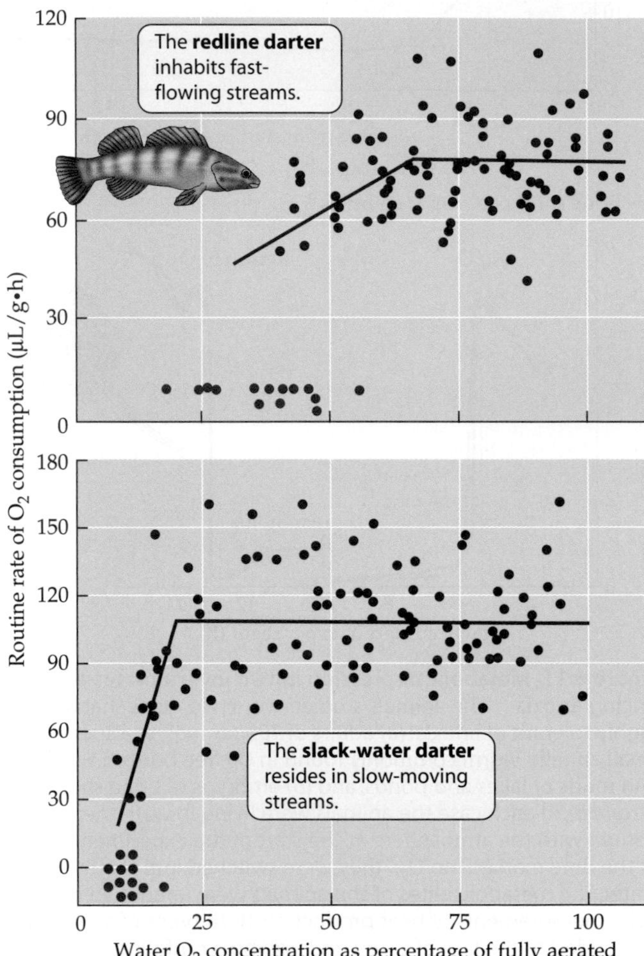

Figure 6.12 Oxygen regulation and conformity *(a)* The general concepts of oxygen regulation and conformity. *(b)* Rates of O_2 consumption of two species of fish in the genus *Etheostoma*—the redline darter (*E. rufilineatum*) and the slack-water darter (*E. boschungi*)—during routine activity at 20°C. Red circles mark ambient O_2 levels at which deaths occurred. On the *x* axis, 100 is the O_2 concentration of water that is at equilibrium with the atmosphere and therefore fully aerated; values higher than 100 were created in some studies by bubbling water with pure O_2. In the technical literature, O_2 levels are expressed as O_2 partial pressures, as explained in Chapter 20. (After Ultsch, Borschung, and Ross 1978.)

Aquatic anaerobes: Some aquatic animals are capable of protracted life without O_2

Water is far from being as rich a source of O_2 as air (see Table 20.1), and among the water-breathing aquatic animals are quite a few that can function as **anaerobes**, able to survive whole-body anoxia for long periods of time. Some of the best-studied examples among invertebrates are certain species of clams, mussels, and other bivalve molluscs. Ribbed mussels (*Geukensia demissa*), for example, live in the mud of salt marshes, where they can become buried; they are able to survive in an atmosphere of pure N_2 for 5 days. Certain marine clams that live in seas prone to O_2 depletion, such as the ocean quahog (*Arctica islandica*), can live for 1 to 2 months in O_2-free water.

Worms and other invertebrates that live at the bottoms of lakes or ponds are other animals that may experience severe and prolonged O_2 deprivation (see Figure 1.13), and some are among the most tolerant to anoxia of all free-living animals. Recently, for example, certain annelid worms in the genus *Tubifex*, which burrow in O_2-free sediments in pond or lake bottoms, have been shown not only to survive but to feed, grow, and *reproduce* while deprived of O_2 for 7 months!

Anaerobes are rare among aquatic vertebrates. Nonetheless, two species of related cyprinid fish are known to have extraordinary abilities to live without O_2. One of these is the common goldfish (*Carassius auratus*), which is reported to survive in O_2-free water for 11 to 24 hours at 20°C and 1 to 6 days at 10°C. No wonder goldfish survive the tender loving care of 5-year-olds! The other species, which is even more tolerant of anoxia than the goldfish, is the Crucian carp (*Carassius carassius*), a common inhabitant of European ponds, which can survive without O_2 for several months at temperatures below 10°C. The ability of Crucian carp to live without O_2 permits them to outcompete other fish in ponds that become O_2 depleted. Thus their physiology is a key to their ecological success. For both invertebrate and vertebrate anaerobes, metabolic depression (see Figure 6.11) is often one of the key strategies used to survive anoxia. It lowers ATP requirements and thus relaxes demands on the ATP-producing mechanisms that are available in the absence of O_2.

Mechanisms of invertebrate anaerobiosis

The most common principal products of the anaerobic biochemical pathways used for ATP synthesis by invertebrate aquatic anaerobes are acetic acid, succinic acid, propionic acid, and the amino acid alanine. These products obviously signal that the biochemical pathways of anaerobic catabolism in these animals differ from simple anaerobic glycolysis. The pathways are elaborate, typically involving some of the reactions of the Krebs cycle and additional other reactions besides glycolytic ones, even sometimes including truncated segments of the electron-transport chain. Importantly, some of these reactions generate ATP. Thus the pathways of anaerobic catabolism in inverte-

BOX 6.3 Human Peak O_2 Consumption and Physical Performance at High Altitudes

When people (and other mammals) are exposed to decreased atmospheric concentrations of O_2, they marshal vigorous physiological defenses, discussed in Box 21.1 and Box 22.3. Because of these responses, resting or moderately active people show a substantial degree of oxygen regulation as the ambient O_2 level falls with increasing altitude. Here, however, we consider the most demanding of circumstances: the capability of mountaineers for all-out physical effort at the highest altitudes on Earth.

The figure shows that when people are asked to work hard enough that they take in O_2 at their peak rate, their maximal rate of O_2 consumption becomes a smaller and smaller fraction of their rate at sea level as altitude increases. The cost of any particular form of exercise remains the same regardless of altitude, however. Thus a rate of climbing that is distinctly submaximal at low altitudes may become maximal, or even supramaximal, at high altitudes. At altitudes near the top of Mt. Everest (8848 m), the maximal rate of O_2 consumption is so low that even minimal rates of climbing require about the maximal possible rate of O_2 consumption.

Work that requires 100% of maximal O_2 consumption is always extremely taxing—barely possible. In the 1920s, E. F. Norton, climbing without supplemental O_2 at 8500 m on Mt. Everest, reported his attempts to take 20 consecutive uphill steps. He never could and did not reach the top. Starting in 1978, people have reached the peak of Mt. Everest without use of supplemental O_2. Reinhold Messner and Peter Habeler, the first to succeed, reported climbing so slowly near the top that, even though they felt they were working at their limits, they required an hour to cover the final 100 m! Physiological data gathered on climbers who have reached the peak of Mt. Everest without supplemental O_2 indicate that the feat is so close to the margin of what is possible that success or failure depends on whether the barometric pressure is relatively high or low on the day of the climb (O_2 is slightly more concentrated in the air on a day of high barometric pressure).

One of the intriguing facts about our planet is that Mt. Everest is nearly exactly as high as a mountain could be and still be climbed by people without supplemental O_2. If a mountain were 400 m higher than Everest, maximal human O_2 consumption at its top would be 25% lower than on the top of Everest, and probably not even the most elite mountaineer could reach the top breathing ambient air, regardless of the barometric pressure on the day of the climb.

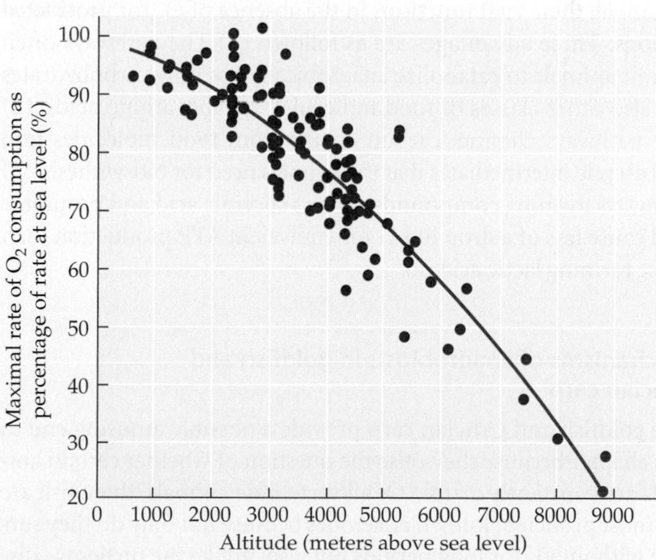

The maximal rates of O_2 consumption of human mountaineers at increasing altitudes Values are expressed as percentages of the values at sea level. (After Fulco, Rock, and Cymerman 1998.)

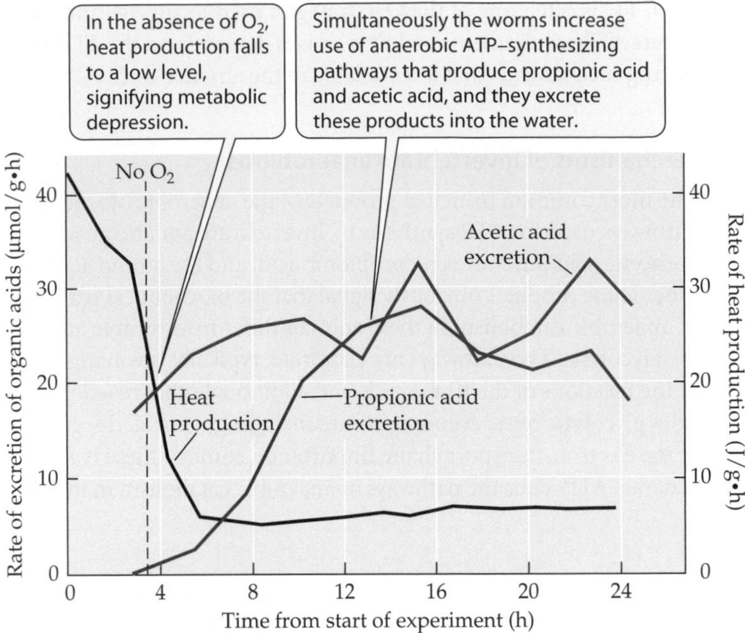

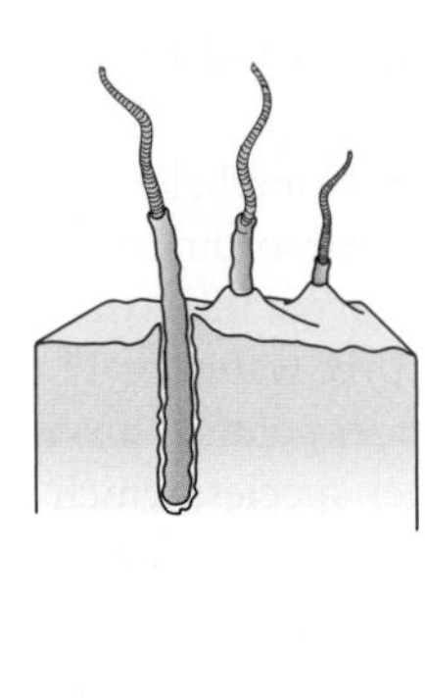

Figure 6.13 Excretion of anaerobic end products by *Tubifex* worms exposed to anoxia These data are from the same study as in Figure 6.11*a*. The principal chemical end products of anaerobic catabolism in this species of aquatic annelid (*Tubifex tubifex*) are acetic acid and propionic acid. The worms were switched from a high-O_2 environment into O_2-free water at the time marked "no O_2." They excrete the end products of anaerobic catabolism into the environmental water, thereby preventing end-product accumulation in their bodies. The worms sustain a low rate of anaerobic catabolism even when O_2 is available, accounting for the excretion of acetic acid prior to transfer into O_2-free water. (After Famme and Knudsen 1984.)

brate anaerobes typically yield more ATP per food molecule than anaerobic glycolysis does. A challenge that all the pathways must meet to be biochemically viable is that redox balance must be maintained without O_2. Invertebrate anaerobes often excrete their anaerobic end products or derivatives of them (Figure 6.13). Excretion wastes the energy value of the carbon compounds excreted, but it helps limit acidification of the body fluids and helps prevent self-limitation of the ATP-generating mechanisms.

Besides their relatively high ATP production per food molecule, the elaborate biochemical pathways of invertebrate anaerobes may have additional advantages over anaerobic glycolysis for animals that sustain all their vital functions in the absence of O_2 for protracted periods. Three advantages are as follows: (1) The pathways often permit animals to catabolize anaerobically not only carbohydrates but also other classes of food molecules, notably amino acids. (2) The pathways channel carbon chains from food molecules into Krebs-cycle intermediates that the animals need for biosynthesis. (3) Pathways forming compounds such as succinic acid and propionic acid cause less of a drop in pH for equivalent ATP production than those forming lactic acid do.

Mechanisms of anaerobiosis in goldfish and Crucian carp

The goldfish and Crucian carp provide a possibly amusing end to this chapter because they raise the question of whether certain animals are constantly drunk. Of all vertebrate animals, these fish are the most proficient known anaerobes because not only do they survive without O_2 for long periods but also, unlike the turtles we discussed earlier, they remain conscious and capable of responding behaviorally to their environments. The *swimming muscles* of these fish possess an unusual form of the enzyme alcohol dehydrogenase, a form that strongly favors the formation of ethanol under prevailing tissue conditions. All tissues, including the brain, synthesize ATP by anaerobic glycolysis and produce lactic acid. The swimming muscles convert lactic acid (some of it brought to the muscles from the other tissues) to ethanol and CO_2. The production of ethanol does not increase ATP yield. Rather, its principal advantage seems to be that it makes possible the *excretion* of the carbon chains produced by anaerobic glycolysis. Unlike lactic acid, ethanol is lost across the gills into the water that the fish inhabit. This excretion limits end-product accumulation, thereby helping to prevent self-limitation of the ATP-generating mechanisms.

The brains of goldfish and Crucian carp are believed to exhibit much less metabolic depression than those of turtles. The more-limited metabolic depression permits the fish to remain conscious and responsive even during anoxia. It also increases demands for anaerobic ATP synthesis by brain cells.

The fish do not get drunk! Their rate of excretion of ethanol is great enough to keep their tissue concentrations of ethanol below inebriating levels.

Summary: Responses to Impaired O_2 Influx from the Environment

- Animals adapted to living without O_2 frequently undergo metabolic depression when deprived of O_2. Metabolic depression can be so profound as to lower an animal's metabolic rate to less than 5% of the usual rate, thus greatly reducing the rate that ATP must be supplied by catabolic mechanisms.
- Invertebrate anaerobes deprived of O_2 produce ATP by means of a diversity of complex anaerobic catabolic pathways that generate end products such as acetic acid, succinic acid, and propionic acid. The invertebrates commonly excrete these organic products during anoxia as a way of avoiding end-product accumulation in their bodies.

- Virtually all vertebrates use simple anaerobic glycolysis to produce ATP in tissues deprived of O_2, and vertebrates invariably retain lactic acid in their bodies, setting the stage for potential metabolic self-limitation. Usually when vertebrates experience anoxia, it is strictly regional; whereas some tissues become anoxic, others—most notably the central nervous system—must retain an O_2 supply. Only a few vertebrates can tolerate total-body anoxia.
- Turtles capable of total-body anoxia employ anaerobic glycolysis to make ATP. A key part of their strategy for survival is a metabolic depression of the central nervous system that is so profound that they become comatose. Goldfish and Crucian carp undergoing total-body anoxia have the unusual ability to convert lactic acid to ethanol, which they can excrete, thereby preventing end-product accumulation in their bodies. They maintain enough brain activity during anoxia to remain aware of their environments.

Study Questions

1. One approach to conservation of fish populations is to release unwanted fish accidentally caught in trawling nets. Such fish often have very high concentrations of lactic acid in their bodies. Why do you think they have these high concentrations of lactic acid, and how might their survival after release be affected by the lactic acid levels?
2. Explain how the reactions that produce and use ATP serve, together, as an *energy shuttle* mechanism in cells.
3. Explain the concept of redox balance. What conditions must exist, for example, for cytochrome oxidase to be in redox balance?
4. How does the reduction of pyruvic acid create a state of redox balance in anaerobic glycolysis?
5. Outline the chemiosmotic theory of how oxidative phosphorylation is coupled with electron transport. How does uncoupling occur in tissues with uncoupling protein? Under what circumstances would uncoupling be disadvantageous, and under what circumstances might it be advantageous?
6. Using two or three carefully chosen examples, illustrate the point that during physical activity, behavior and biochemistry are intimately linked, such that an animal's exercise performance depends on the mechanisms that are making ATP for the exercise.
7. Assuming that an animal uses a catabolic pathway that produces organic products, such as lactic acid or propionic acid, compare the pros and cons of retaining or excreting the organic molecules.
8. Why is it important to distinguish *temporary* electron (hydrogen) acceptors in cells from *final* electron acceptors. What are the unique advantages of elemental oxygen as an electron acceptor?
9. Why does an oxygen deficit occur at the start of submaximal exercise in vertebrates? What are the mechanisms of ATP production during the oxygen deficit phase, and how is ATP made in the ensuing pay-as-you-go phase?
10. A single individual can differ from time to time in his or her maximal rate of O_2 consumption. For example, athletic training in people can raise the maximal rate of O_2 consumption by 10% to 30%, whereas going to high altitudes can lower it (see Box 6.3). Explain how these sorts of changes in the maximal rate of O_2 consumption can make a single type of exercise (such as jogging at 6 miles per hour) shift from being submaximal to supramaximal, or vice versa. What are the physiological implications of such shifts?
11. There has been a great deal of debate over whether the ratio of SO to FG fibers in the muscles of individual people or other animals is fixed genetically. Researchers have asked the questions: Can the ratio of fiber types be altered during an individual's lifetime by various sorts of training or other experiences, or is it invariant? Design experiments or other sorts of studies that would help answer this question.
12. The species of turtles that tolerate total-body anoxia store especially large amounts of glycogen in their brains, in comparison to other species of turtles. How might such glycogen stores be advantageous? (Consider whether or not liver glycogen would be "just as good.")

Key References

Alberts, B., A. Johnson, J. Lewis, M. Raff, K. Roberts, and P. Walter. 2002. *Molecular Biology of the Cell*, 4th ed. Garland, New York. A superior and time-proven presentation of all aspects of cellular physiology, including the catabolic mechanisms, emphasizing mammals. Excellent illustrations for understanding difficult concepts.

Åstrand, P.-O. and K. Rodahl. 1986. *Textbook of Work Physiology: Physiological Bases of Exercise*, 3rd ed. McGraw-Hill, New York. A classic treatment of the physiology of all forms of human work, from housework to Nordic skiing. New edition due out soon.

Hand, S. C., and I. Hardewig. 1996. Downregulation of cellular metabolism during environmental stress: Mechanisms and implications. *Annu. Rev. Physiol.* 58: 539–563. A compact, but fascinating and thought-provoking, review of the physiology of metabolic dormancy. See also S. C. Hand's 1991 paper listed in the Selected Readings for a great deal more on the natural history of metabolic dormancy.

Hochachka, P. W., and G. N. Somero. 2002. *Biochemical Adaptation*. Oxford University Press, New York. A definitive synthesis of the themes discussed in this chapter.

Rome, L. C., P. T. Loughna, and G. Goldspink. 1985. Temperature acclimation: Improved sustained swimming performance in carp at low temperatures. *Science* 228: 194–196. A report on an intriguing interplay between muscle types, pathways of ATP production, power generation for swimming, and temperature acclimation in a fish. A beautiful illustration of the interaction of physiology and environment at multiple levels of organization.

Shephard, R. J., and P.-O. Åstrand (eds.). 2000. *Endurance in Sport*, 2nd ed. Blackwell, Oxford. Excellent chapters by many different authorities on most of the major topics in human exercise performance. Chapters 22 and 23 on metabolism in muscle are especially relevant to our discussion here.

Selected References

Adamczewska, Z. M., and S. Morris. 2001. Metabolic status and respiratory physiology of *Gecarcoidea natalis*, the Christmas Island red crab, during the annual breeding migration. *Biol. Bull.* 200: 321–335.

Bennett, A. F. 1972. The effect of activity on oxygen consumption, oxygen debt, and heart rate in lizards *Varanus gouldii* and *Sauromalus hispidus*. *J. Comp. Physiol.* 79: 259–280.

Bennett, A. F. 1991. The evolution of activity capacity. *J. Exp. Biol.* 160: 1–23.

Billeter, R., and H. Hoppeler. 1992. Muscular basis of strength. In P. V. Komi (ed.), *Strength and Power in Sport*, pp. 39–63. Blackwell, Oxford.

Boutilier, R. G., and J. St-Pierre. 2000. Surviving hypoxia without really dying. *Comp. Biochem. Physiol. A* 126: 481–490.

Brooks, G. A., T. D. Fahey, T. P. White, and K. W. Baldwin. 2000. *Exercise Physiology: Human Bioenergetics and Its Applications*, 3rd ed. Mayfield Publishing, Mountain View, CA.

Chih, C. R., and W. R. Ellington. 1983. Energy metabolism during contractile activity and environmental hypoxia in the phasic adductor muscle of the bay scallop *Argopecten irradians concentricus*. *Physiol. Zool.* 56: 623–631.

Clanton, T. L., and P. F. Klawitter. 2001. Physiological and genomic consequences of intermittent hypoxia—Invited review: Adaptive responses of skeletal muscle to intermittent hypoxia: The known and the unknown. *J. Appl. Physiol.* 90: 2476–2487.

Coughlin, D. J., and L. C. Rome. 1996. The roles of pink and red muscle in powering steady swimming in scup, *Stenotomus chrysops*. *Amer. Zool.* 36: 666–677.

DeZwann, A., and V. Putzer. 1985. Metabolic adaptations of intertidal invertebrates to environmental hypoxia (a comparison of environmental anoxia to exercise anoxia). *Symp. Soc. Exp. Biol.* 39: 33–62.

Dickson, K. A. 1996. Locomotor muscle of high-performance fishes: What do comparisons of tunas with ectothermic sister taxa reveal? *Comp. Biochem. Physiol. A* 113: 39–49.

Dohm, M. R., J. P. Hayes, and T. Garland, Jr. 1996. Quantitative genetics of sprint running speed and swimming endurance in the laboratory mouse (*Mus domesticus*). *Evolution* 50: 1688–1701.

Ellington, W. R. 2001. Evolution and physiological roles of phosphagen systems. *Annu. Rev. Physiol.* 63: 289–325.

England, W. R., and J. Baldwin. 1983. Anaerobic energy metabolism in the tail musculature of the Australian yabby *Cherax destructor* (Crustacea, Decapoda, Parastacidae): Role of phosphagens and anaerobic glycolysis during escape behavior. *Physiol. Zool.* 56: 614–622.

Famme, P., and J. Knudsen. 1984. Total heat balance study of anaerobiosis in *Tubifex tubifex* (Müller). *J. Comp. Physiol. B* 154: 587–591.

Fitts, R. H. 1996. Cellular, molecular, and metabolic basis of muscle fatigue. In L. B. Rowell and J. T. Shepherd (eds.), *Exercise: Regulation and Integration of Multiple Systems* (Handbook of Physiology [Bethesda, MD], section 12), pp. 1151–1183. Oxford University Press, New York.

Gatten, R. E., Jr. 1985. The uses of anaerobiosis by amphibians and reptiles. *Amer. Zool.* 25: 945–954.

Gleeson, T. T. 1996. Post-exercise lactate metabolism: A comparative review of sites, pathways, and regulation. *Annu. Rev. Physiol.* 58: 565–581.

Hand, S. C. 1991. Metabolic dormancy in aquatic invertebrates. *Adv. Comp. Environ. Physiol.* 8: 1–50.

Henry, R. P., C. E. Booth, F. H. Lallier, and P. J. Walsh. 1994. Post-exercise lactate production and metabolism in three species of aquatic and terrestrial decapod crustaceans. *J. Exp. Biol.* 186: 215–234.

Herbert, C. V., and D. C. Jackson. 1985. Temperature effects on the responses to prolonged submergence in the turtle *Chrysemys picta bellii*. *Physiol. Zool.* 58: 655–681.

Hochachka, P. W. 1986. Defense strategies against hypoxia and hypothermia. *Science* 231: 234–241.

Hochachka, P. W. 1998. Mechanism and evolution of hypoxia-tolerance in humans. *J. Exp. Biol.* 201: 1243–1254.

Hochachka, P. W., and M. Guppy. 1987. *Metabolic Arrest and the Control of Biological Time*. Harvard University Press, Cambridge, MA.

Hochachka, P. W., P. L. Lutz, T. J. Sick, and M. Rosenthal. 1993. *Surviving Hypoxia: Mechanisms of Control and Adaptation*. CRC Press, Boca Raton, FL.

Hylland, P., S. Milton, M. Pek, G. E. Nilsson, and P. L. Lutz. 1997. Brain Na^+/K^+-ATPase activity in two anoxia tolerant vertebrates: Crucian carp and freshwater turtle. *Neurosci. Lett.* 235: 89–92.

Jackson, D. C. 2000. Living without oxygen: Lessons from the freshwater turtle. *Comp. Biochem. Physiol. A* 125: 299–315.

Jones, D. R. 1982. Anaerobic exercise in fish. *Can. J. Zool.* 60: 1131–1134.

Klingenberg, M., and K. S. Echtay. 2001. Uncoupling proteins: The issues from a biochemist point of view. *Biochim. Biophys. Acta Bioenergetics* 1504: 128–143. [The entire issue of this journal is devoted to papers on mitochondrial energetics and uncoupling proteins.]

Kooyman, G. L. 1989. *Diverse Divers*. Springer, New York.

Lehninger, A. L., D. L. Nelson, and M. M. Cox. 1993. *Principles of Biochemistry*, 2nd ed. Worth, New York.

Lutz, P. L., and G. E. Nilsson. 1997. Contrasting strategies for anoxic brain survival—Glycolysis up or down. *J. Exp. Biol.* 200: 411–419.

Lutz, P. L., and K. B. Storey. 1997. Adaptations to variations in oxygen tension by vertebrates and invertebrates. In W. H. Dantzler (ed.), *Comparative Physiology*, vol. 2 (Handbook of Physiology [Bethesda, MD], section 13), pp. 1479–1522. Oxford University Press, New York.

McArdle, W. D., F. I. Katch, and V. L. Katch. 2001. *Exercise Physiology: Energy, Nutrition, and Human Performance*, 5th ed. Williams & Wilkins, Baltimore, MD.

Nedergaard, J., V. Golozoubova, A. Matthias, A. Asadi, A. Jacobsson, and B. Cannon. 2001. UCP1: The only protein able to mediate adaptive nonshivering thermogenesis and metabolic inefficiency. *Biochim. Biophys. Acta Bioenergetics* 1504: 82–106. [The entire issue of this journal is devoted to papers on mitochondrial energetics and uncoupling proteins.]

Nilsson, G. E. 1995. Oxygen availability: Brain defence mechanisms. In P. W. Hochachka and T. P. Mommsen (eds.), *Environmental and Ecological Biochemistry* (Biochemistry and Molecular Biology of Fishes, vol. 5), pp. 19–44. Elsevier, New York.

Officer, C. B., R. B. Biggs, J. L. Taft, L. E. Cronin, M. A. Tyler, and W. R. Boynton. 1984. Chesapeake Bay anoxia: Origin, development, and significance. *Science* 223: 22–27.

Ott, M. E., N. Heisler, and G. R. Ultsch. 1980. A re-evaluation of the relationship between temperature and the critical oxygen tension in freshwater fishes. *Comp. Biochem. Physiol. A* 67: 337–340.

Pierce, V. A., and D. L. Crawford. 1997. Phylogenetic analysis of glycolytic enzyme expression. *Science* 276: 256–259.

Portner, H. O., D. M. Webber, R. K. O'Dor, and R. G. Boutilier. 1993. Metabolism and energetics in squid (*Illex illecebrosus, Loligo pealei*) during muscular fatigue and recovery. *Amer. J. Physiol.* 265: R157–R165.

Rolfe, D. F. S., and G. C. Brown. 1997. Cellular energy utilization and molecular origin of standard metabolic rate in mammals. *Physiol. Rev.* 77: 731–758.

Rothwell, N. J., and M. J. Stock. 1979. A role for brown adipose tissue in diet-induced thermogenesis. *Nature* 281: 31–35.

Rowell, L. B., and J. T. Shepherd (eds.). 1996. *Exercise: Regulation and Integration of Multiple Systems*. Handbook of Physiology (Bethesda, MD), section 12. Oxford University Press, New York.

Saraste, M. 1999. Oxidative phosphorylation at the *fin de siècle*. *Science* 283: 1488–1493.

Seymour, R. S. 1997. Plants that warm themselves. *Sci. Amer.* 273(3): 104–109.

Shick, J. M., J. Widdows, and E. Gnaiger. 1988. Calorimetric studies of behavior, metabolism and energetics of sessile intertidal animals. *Amer. Zool.* 28: 161–181.

Sinervo, B., and R. B. Huey. 1990. Allometric engineering: An experimental test of the causes of interpopulational differences in performance. *Science* 248: 1106–1109.

Somero, G. N., and J. J. Childress. 1980. A violation of the metabolism-size scaling paradigm: Activities of glycolytic enzymes in muscle increase in larger-size fish. *Physiol. Zool.* 53: 322–337.

Suarez, R. K. 1996. Upper limits to mass-specific metabolic rates. *Annu. Rev. Physiol.* 58: 583–605.

Taigen, T. L., and F. H. Pough. 1985. Metabolic correlates of anuran behavior. *Amer. Zool.* 25: 987–997.

Ultsch, G. R., H. Borschung, and M. J. Ross. 1978. Metabolism, critical oxygen tension, and habitat selection in darters (*Etheostoma*). *Ecology* 59: 99–107.

van Deursen, J., W. Ruitenbeek, A. Heerschap, P. Jap, H. ter Laak, and B. Wieringa. 1994. Creatine kinase (CK) in skeletal muscle energy metabolism: A study of mouse mutants with graded reduction in muscle CK expression. *Proc. Natl. Acad. Sci. U.S.A.* 91: 9091–9095.

West, J. B. 1996. Physiology of extreme altitude. In M. J. Fregly and C. M. Blatteis (eds.), *Environmental Physiology*, vol. 2 (Handbook of Physiology [Bethesda, MD], section 4), pp. 1307–1325. Oxford University Press, New York.

See also references in Appendix A.

7 The Energetics of Aerobic Activity

Terns and other seabirds often congregate in spectacular, boisterous colonies during their breeding seasons. In one colony in the Shetland Islands north of the Scottish mainland, for example, almost 200,000 adult seabirds of mixed species—including over 11,000 Arctic terns (*Sterna paradisaea*)—are known to nest and raise their young together each spring and summer. To meet their own energy needs and those of their growing chicks, the adults spend numerous hours per day flying over the surrounding ocean, searching for and capturing food, mostly small fish. This sustained feeding activity must be aerobic; that is, the ATP for the exertion must be produced by aerobic catabolism, using O_2 delivered to the exercising muscles from the atmosphere. Anaerobic mechanisms of ATP production could not possibly supply the ATP because only very limited total amounts of ATP can be made anaerobically (see Chapter 6). In addition to spending long periods of time flying over the ocean searching for food, Arctic terns are among the world's most fabled migratory species, traveling each year between Earth's two polar regions, a round-trip of over 35,000 km. Migratory travels of all animals must be aerobic because only aerobic catabolism can meet the long-sustained ATP needs of such exertion.

In this chapter we examine the energy costs of routine daily life and of sustained, aerobic physical activity in animals. The chapter is thus, in a sense, a continuation of Chapter 6. There, in discussing exercise, we emphasized the interplay of aerobic and anaerobic catabolism as sources of ATP. Here we focus on forms of exercise that are aerobically fueled; this focus is justified because although highly anaerobic forms of exertion (e.g., sprints) sometimes spell the difference between life and death, aerobic types of exercise predominate in the lives of most animals in regard to time spent and energy required. As a first look at the energy costs of sustained exercise, Table 7.1 presents the metabolic rates of people engaged in various activities. Many sustained activities increase a person's metabolic rate by a factor of two to ten over the resting rate. If long periods of time are spent in such activities, a person's overall energy expenditure is significantly affected.

The routine daily flights of these terns to go to and return from foraging areas over the ocean depend on aerobic catabolism to supply energy for muscular work The sustained activities of animals, such as the foraging flights of terns, elevate the energy costs of daily life.

TABLE 7.1 Representative metabolic rates of young adult humans of average build during sustained forms of exercise

Type of activity	Metabolic rate[a] (kJ/minute)
Lying down	6.3
Sitting	7.1
Standing	8.8
Walking at 2 miles per hour (mph)	12
Walking at 4 mph	21
Bicycling at 13 mph	32
Jogging at 7 mph	59
Crawl swimming at 2 mph	59
Running at 10 mph	84

Source: After Åstrand and Rodahl 1986.
[a] All forms of locomotion are assumed to be on level ground.

How Active Animals Are Studied

Studies of actively moving animals present challenging methodological problems, whether carried out in the laboratory or in the wild. One important question that is usually studied under laboratory conditions is the relation between the speed of locomotion and the metabolic rate. One can measure metabolic rate straightforwardly in a laboratory, using the techniques discussed in Chapter 5. A more problematic question is how to control an animal's speed of locomotion.

Fifty years ago, physiologists generally despaired over the problems of controlling the speeds of moving animals. By now, however, many species of running, flying, and swimming animals have been trained to exercise on or in devices that control their speeds. For running or walking animals, the device most commonly used is a motor-driven treadmill (Figure 7.1). The animal stands on a belt, which is driven round and round by a motor. For the animal to keep its position, it must run or walk at the same speed as the belt that is passing beneath its feet. The treadmill can be tilted at an angle relative to the horizontal to simulate uphill or downhill running. Animals as diverse as cockroaches, land crabs, turkeys, and cheetahs have been trained to run and walk on treadmills. Lobsters and crayfish have even been trained to use underwater treadmills. As the speed of a treadmill is varied, the subject's rate of O_2 consumption is measured, permitting investigators to describe the relation between the speed and the metabolic cost of locomotion.

Wind tunnels are used to control the speeds of flying animals. A bird, insect, or bat in a wind tunnel must fly into the forced air current produced by the tunnel at the same speed as the current if it is to maintain its position. For swimming animals such as fish, a device analogous to a wind tunnel—but filled with water instead of air—is commonly used to control speed.

For field studies of the energetics of actively moving terrestrial animals, one of the most important innovations is the **doubly labeled water method**, a technique that permits one to quantify the metabolic rate of an individual that is entirely free to engage in its normal behaviors in the wild. The method takes its name from the fact that the animal under study is injected with water labeled with unusual isotopes of both hydrogen and oxygen. One commonly used pair of isotopes consists of deuterium (a hydrogen isotope) and oxygen-18.[1] The method is then called the **$D_2{}^{18}O$ method**. We explain the method here with reference to these particular isotopes, but the rationale is the same regardless of the isotopes used.

For an animal to be studied by use of the $D_2{}^{18}O$ method, it is injected at the start of the study period with measured amounts of both $D_2{}^{16}O$ and $H_2{}^{18}O$, which mix with its body water. Later, at the end of the study period, a blood sample is taken to determine how rapidly the deuterium and oxygen-18 were eliminated from the animal's body during the study period. In the interval between the initial injection of the isotopes and the collection of the final blood sample, the subject is released and is free to live normally in its natural habitat. The method measures the animal's *average* rate of CO_2 production during the study period. The animal's average metabolic rate is then computed from its rate of CO_2 production.[2]

The reason the doubly labeled water method works is that the oxygen atoms in expired CO_2 are in isotopic equilibrium with the oxygen atoms in body water; this means that if the body H_2O consists of given proportions of $H_2{}^{16}O$ (ordinary water) and $H_2{}^{18}O$, expired CO_2 contains oxygen-16 and oxygen-18 in the same proportions.[3] After the concentration of $H_2{}^{18}O$ is elevated in an animal's body at the start of a study, the excess atoms of oxygen-18 are gradually lost by way of expired CO_2. The *rate* of loss of oxygen-18 thus depends on the subject's rate of CO_2 production. A problem is that atoms of oxygen-18 are lost from the body in the form of H_2O as well as CO_2 because H_2O that leaves the animal's body by evaporation, urination, or other mechanisms consists in part of $H_2{}^{18}O$.

[1] The ordinary isotope of oxygen is oxygen-16.

[2] As discussed in Chapter 5 (see Table 5.1), uncertainties are involved in calculating metabolic rate from the rate of CO_2 production and need to be kept in mind.

[3] The reason for the isotopic equilibrium between oxygen in H_2O and CO_2 is that atoms of oxygen are freely exchanged between molecules of H_2O and CO_2 during the reactions of the Krebs citric acid cycle (see Figure 6.1).

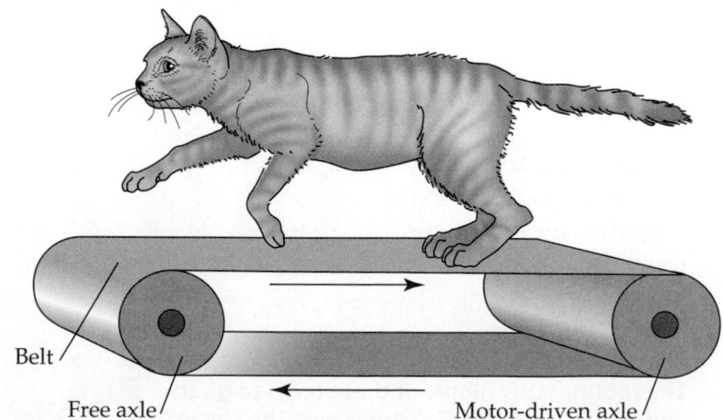

Figure 7.1 A treadmill provides a way to control the running or walking speed of an animal during study When the motor-driven axle turns, the belt moves as indicated by the arrows. After being trained, animals run at the same speed as the belt to keep their position relative to the surrounding room.

Thus, after an animal has been injected with oxygen-18, its *total* rate of elimination of the oxygen-18 is in fact a function of *both* its rate of CO_2 expiration *and* its rate of H_2O loss. The hydrogen isotope—deuterium—is injected in the doubly labeled water method to obtain an independent measure of the rate of H_2O loss; because CO_2 contains no hydrogen, after an animal has been injected with deuterium, its rate of loss of the unusual hydrogen isotope from its body depends just on its rate of H_2O elimination. Knowing the animal's rate of H_2O loss from the hydrogen-isotope data, one can calculate how much oxygen-18 is lost in H_2O during the study period. Knowing this value, one can then subdivide the total rate of oxygen-18 loss into two components: the rate of loss in H_2O and the rate of loss in CO_2. In this way, the rate of CO_2 expiration—and the metabolic rate—is calculated.

The most common use of the doubly labeled water method is to determine the average metabolic rate of an animal in the wild during the whole 24-hour day, termed the **average daily metabolic rate** (**ADMR**) or **field metabolic rate** (**FMR**). The animal might sleep for part of the day, stand guard for another part, and move about actively during still another part. The doubly labeled water method measures the metabolic cost of all these behavioral states lumped together.

Time–energy budgets represent an older, but still useful, approach for estimating the average daily metabolic rates of animals free in their natural habitats. To construct a time–energy budget, an investigator first categorizes all of an animal's behaviors into a few categories. For instance, the categories might be sleeping, standing guard, and actively moving. Using laboratory studies or other sources of information, the investigator then estimates the animal's rate of energy expenditure (metabolic rate) while it engages in each category of behavior. Finally, field observations are used to estimate the amount of time the animal spends in each sort of behavior during a day. To get the total energy cost of each behavior, the time spent in the behavior is multiplied by the rate of energy expenditure during the behavior. The energy costs of all behaviors are then summed to obtain the total daily energy expenditure.

An example of a time-energy budget is provided by the data on African penguins (*Spheniscus demersus*) in Table 7.2. These penguins live on southern African shores and forage for fish by underwater swimming. In the time-energy budget, three categories of behavior are recognized: maintenance on land (which includes costs of thermoregulation), swimming underwater, and resting at the water's surface while at sea. The values in column *C* show the estimated daily costs of the three behaviors separately. Summation of the daily costs of all three yields the total daily energy expenditure of a penguin, about 1900 kJ.

A third technique useful for studying the physiology of activity in the wild is **radiotelemetry**, which involves placing a radio transmitter on or in an individual and arranging for it to transmit physiological data by radio. Technological innovation is constantly expanding the types of data that can be monitored in this way and the types of biological situations in which the method can be applied. Just recently, telemetry of the rates of O_2 consumption of people has become possible (Figure 7.2). Unfortunately, the size of the telemeters used for this purpose and the fact that they require the subject to breathe from a mask mean that most animals cannot now be studied by direct telemetric measurement of their O_2 consumption. Radiotelemetry nonetheless has wide application in studies of the energetics of animals besides people because small, unobtrusive transmitters can be used to measure heart rates or behavioral activity levels in many sorts of animals. Metabolic rates sometimes can be estimated from heart-rate data gathered by radiotelemetry. Moreover, telemetrically gath-

TABLE 7.2 A time–energy budget for an adult African penguin (*Spheniscus demersus*) during the breeding season

Behavior	*A* Hours per day devoted to behavior[a]	*B* Hourly cost of behavior[b] (kJ/h)	*C* Daily cost of behavior: *A* × *B* (kJ)
Maintenance on land	19.5	49	956
Swimming underwater	2.0	358	716
Resting at water's surface	2.5	83	208
Total daily energy expenditure (sum of column *C*):			**1880 kJ/day**

Source: After Nagy, Siegfried, and Wilson 1984.
[a] Time devoted to each behavior was determined from field observations of African penguins.
[b] The hourly energetic cost of each behavior was estimated from comparative laboratory studies of multiple bird species.

Figure 7.2 Modern telemetry devices permit information on O_2 consumption, heart rate, and pulmonary ventilation to be radioed from a freely moving athlete The rate of O_2 consumption is measured by the principle of open-flow respirometry (see Box 5.4). Valves in the mask allow the athlete's breathing to cause a measured stream of air to flow through the mask. An O_2 probe measures how much O_2 is removed from each unit volume of the flowing air. A full accounting of the energetics of the particular exercise pictured would also require measurement of anaerobic catabolism.

ered data on the amounts of time devoted to various behaviors can be employed to estimate metabolic rates by use in time–energy budgets.

The Energy Costs of Defined Exercise

The major forms of animal locomotion typically exhibit characteristic relations between the metabolic cost per unit of time and the speed at which animals move. For fish swimming, metabolic rate typically increases in a *J-shaped* power function as speed increases (Figure 7.3).[4] This relation occurs because the drag that a fish must overcome to move through water increases approximately in proportion to the square of its speed of swimming.

When a mammal runs at sustained speeds, its metabolic rate usually increases as a *linear* function of speed, as is shown for six species in Figure 7.4. Other running or walking animals, such as running insects and terrestrial crabs, also usually exhibit linear relations between metabolic rate and speed.

For birds flying at sustained speeds by flapping their wings, aerodynamic theory predicts a *U-shaped* relation between metabolic rate and speed (Figure 7.5*a*). The actual existence of this sort of relation has been difficult to test in practice because of technical difficulties in measuring the metabolic rates of birds flying at low speeds in wind tunnels. Recently, however, investigators have conducted novel studies using analyses of wing movements and forces to estimate the rate of energy investment in flight at all possible flight speeds, ranging from 0 (hovering) upward. The results for magpies in Figure 7.5*b* (red line), obtained by this method, confirm in a rough sense that birds actually exhibit a U-shaped sort of relation between metabolic rate and flight speed, as aerodynamic theory predicts. The data for the four other species of birds in Figure 7.5*b* are from traditional studies of rates of O_2 consumption. Of the four species, only the budgerigars (parakeets) exhibit a U-shaped relation (yellow line), but the data for the other three are incomplete because of no information at low speeds.

For all three forms of locomotion that we have discussed—swimming by fish, running by terrestrial animals, and flying by birds—small-bodied species tend to require greater weight-specific metabolic rates to move at any particular speed than large-bodied species require. Note, for example (see Figure 7.4), that the weight-specific metabolic rate required to run at about 1 km/hour is far higher for a mouse than for a rat, and the rate required for a rat is higher than that for a dog. Similarly, small birds like budgerigars have higher weight-

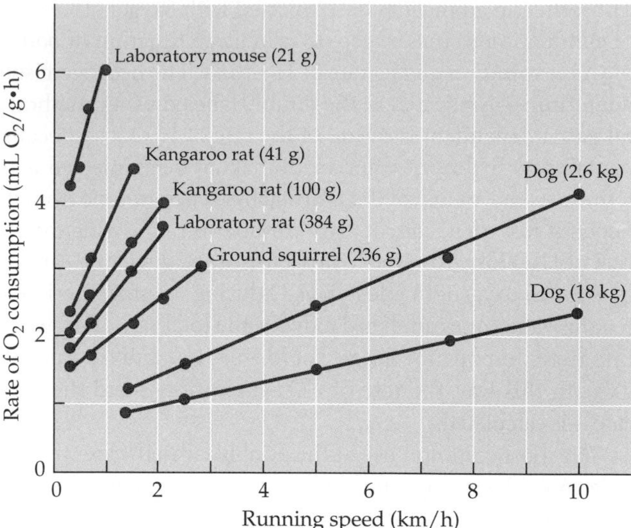

Figure 7.4 The rate of O_2 consumption as a function of running speed in six species of mammals of different body sizes studied on treadmills The species are laboratory mice (*Mus domesticus*, average weight 21 g), Merriam's kangaroo rats (*Dipodomys merriami*, 41 g), banner-tailed kangaroo rats (*D. spectabilis*, 100 g), laboratory rats (*Rattus norvegicus*, 384 g), round-tailed ground squirrels (*Spermophilus tereticaudus*, 236 g), and domestic dogs (mongrels weighing 2.6 kg and Walker foxhounds weighing 18 kg). (After Taylor, Schmidt-Nielsen, and Raab 1970.)

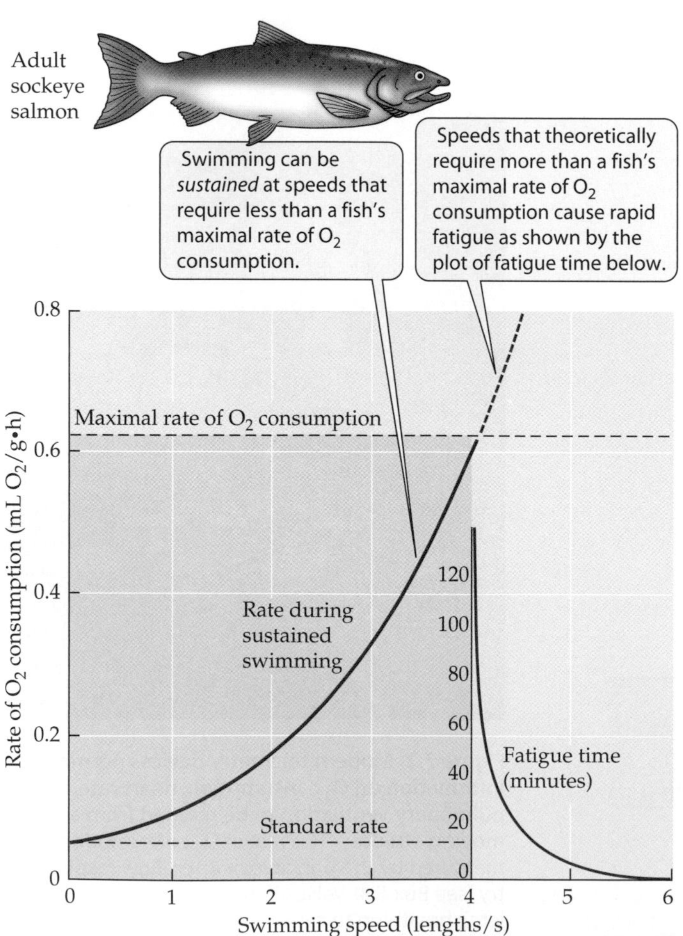

Figure 7.3 The rate of O_2 consumption as a function of swimming speed in yearling sockeye salmon (*Oncorhynchus nerka*) studied in a water tunnel The fish studied were about 18 cm long and weighed about 50 g. (After Brett 1964.)

[4] If M is metabolic rate and u is speed, the expected relation between M and u for swimming fish is $M = a + bu^x$, where a, b, and x are constants for a particular fish species. The characteristics of power functions like this are discussed in Appendix G: Exponential and Allometric Equations.

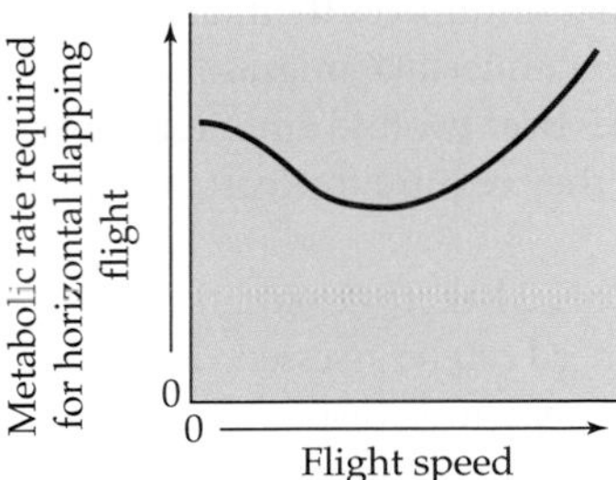

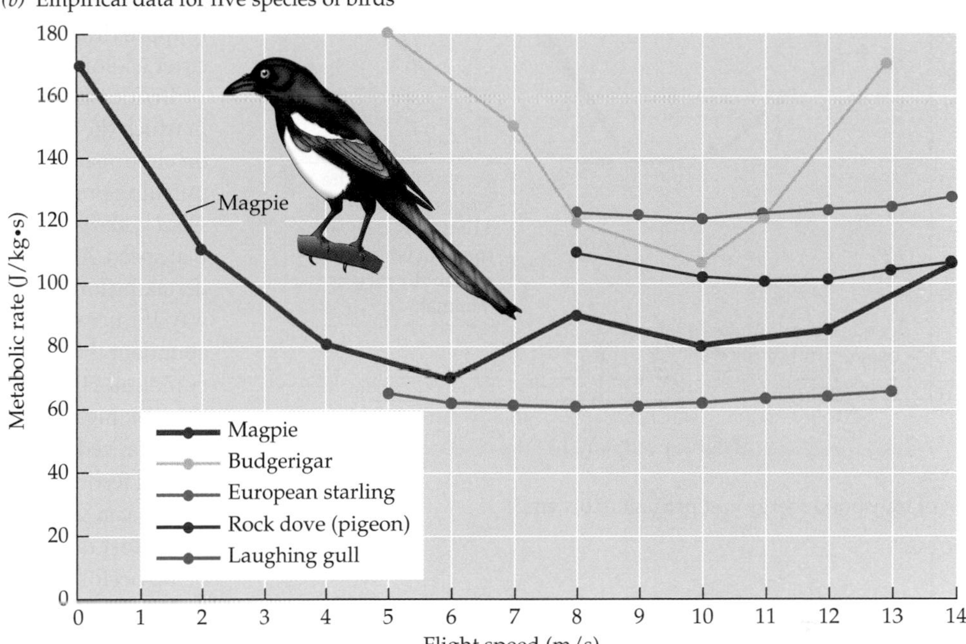

Figure 7.5 The metabolic rate of birds moving by flapping flight as a function of flight speed *(a)* The predicted relation between energy output and speed based on aerodynamic theory. *(b)* The empirically measured rate of output of metabolic energy as a function of flight speed in five species of birds studied in wind tunnels. The results for magpies (*Pica pica*) were obtained by analysis of data on wing movements and forces. The results for the other four species were gathered by the measurement of O_2 consumption and have been calculated to be in the same units as the magpie data. (*a* after Alerstam and Hedenström 1998; *b* after Dial et al. 1997.)

specific metabolic rates than large ones (e.g., gulls) when flying at the same speed (see Figure 7.5*b*). The trend is also evident in fish. A 150-g salmon that increases its weight-specific metabolic rate by the same amount as a 50-g salmon is able to swim at about the same number of body lengths per second—meaning that because it has a longer body, it can swim at a greater absolute speed than the 50-g fish for the same weight-specific metabolic effort.

The most advantageous speed depends on the function of exercise

We are well aware that we can judge the ideal speed of a vehicle in more than one way, depending on what we are attempting to accomplish by travel in the vehicle. For instance, we might want to get to our destination as fast as possible, or we might want to travel at a speed that promotes energy efficiency. Similarly, there are several ways to judge the most advantageous speed for an animal to run, fly, or swim, depending on the function that is being performed by the exercise.

The study of flapping flight by birds is particularly revealing for improving one's understanding of the different perspectives on speed that can be relevant in various situations. Figure 7.6*a* shows the weight-specific *energy cost per unit of time* (i.e., weight-specific metabolic rate) of budgerigars as a function of speed. Figure 7.6*b* presents the same data in a different way. It shows the weight-specific *energy cost per unit of distance traveled* as a function of speed. The following equation explains how the two plots are related:

$$\frac{\text{energy}}{\text{time}} \div \frac{\text{distance}}{\text{time}} = \frac{\text{energy}}{\text{distance}} \qquad (7.1)$$

The first expression on the left-hand side of Equation 7.1 is the metabolic rate; the second expression is the speed. Thus, to calculate the energy cost per unit of distance when an animal is traveling at a certain speed, one simply divides the metabolic rate at that speed by the speed itself. Each of the six data points in Figure 7.6*a* depicts the metabolic rate at a particular speed. Dividing each metabolic rate by the corresponding speed produces the six data points in Figure 7.6*b*. Thus *the two plots are merely different ways of looking at the same information.* The energy cost per unit of distance at a particular speed is called the animal's **cost of transport** at that speed.

To see the significance of the two ways of analyzing budgerigar flight costs in Figure 7.6, consider these two questions: First, if a budgerigar takes off with certain fuel reserves and does not eat during flight, at what speed should it fly to stay airborne for the longest time before it runs out of fuel? Alternatively, at what speed should it fly to cover the greatest possible distance before it runs out of fuel? Clearly, to stay aloft for the longest possible *time*, the bird should fly at the speed that minimizes its energy cost per unit of time—the speed at which the metabolic rate in Figure 7.6*a* is minimal. By contrast, to cover the greatest *distance*, the bird should fly at the speed that minimizes its energy cost per unit of distance—the speed at which the cost of transport in Figure 7.6*b* is minimal. The two speeds are different. Thus, on the basis of these theoretical considerations, the most adaptively advantageous speed depends on the function

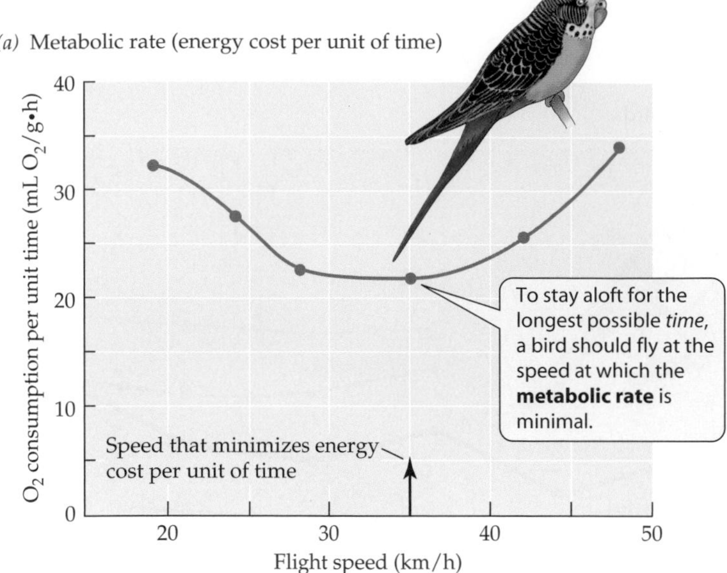

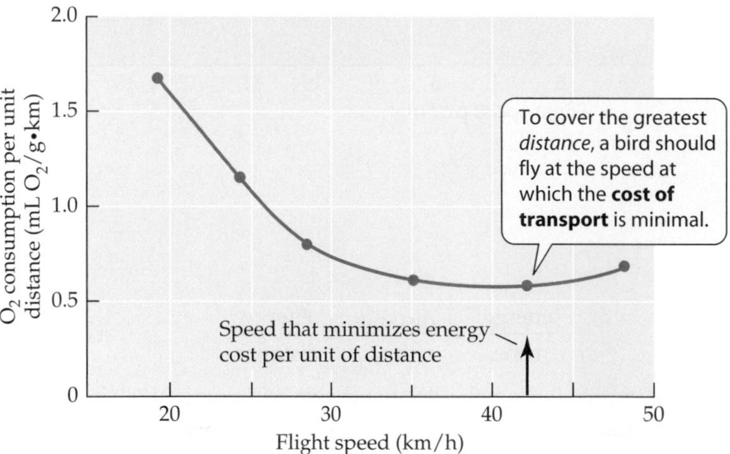

Figure 7.6 Two ways to view the energetics of flapping flight by budgerigars *(a)* Metabolic rate as a function of speed. *(b)* Cost of transport as a function of speed. The *metabolic rate* is far higher than minimal at the speed at minimizes the *cost of transport*. To minimize metabolic rate, the flight speed must be significantly lower than the speed that minimizes cost of transport. The birds (*Melopsittacus undulatus*) were studied in a wind tunnel, and measurements of their rates of O_2 consumption were made to assess their rates of energy expenditure. Both graphs express energy expenditure in weight-specific terms. Figure 7.5*b* presents the data in *(a)* in transformed units. (After Tucker 1968, 1969.)

being performed by flight. If the function of flight is best served by staying airborne as long as possible, the speed marked in Figure 7.6*a* would be most advantageous. On the other hand, if the function is to cover as much distance as possible—as when a bird migrates—the speed marked in Figure 7.6*b* would be most beneficial.

Data collected on the behavior of birds in the wild over the past 20 years strongly suggest that birds actually tend to follow these principles in selecting their flight speeds. When skylarks (*Alauda arvensis*), for example, engage in song flights that act as displays to attract mates, they fly at about the speed that minimizes their energy cost per unit of time in the air. When they undertake migratory flights, however, their average speed increases over twofold and more closely approximates the speed at which they require the least energy to cover a kilometer of distance.

Besides the perspectives already stressed, there are additional ways in which the adaptive advantage of speed can be assessed, depending on circumstances. When an animal runs a long distance to escape a pursuing predator, for example, its shear maximization of *sustained speed* is likely to be paramount, regardless of the efficiency of travel at that speed. Another perspective arises when we consider animals undertaking long-distance migrations during which they feed along the way. If migration brings an animal into habitats where food is more abundant than in the habitat it is leaving, a substantial body of theory indicates that—energetically speaking—the optimal speed of travel is a bit higher than the speed at which the animal's cost of transport is minimized; this is so because faster travel allows faster access to enhanced feeding possibilities.

You can see from this discussion that travel at a speed that minimizes cost of transport is relevant only when certain functions are being performed by exercise; it is not a universal standard of performance. Nonetheless, the study of travel at the minimal cost of transport (minimal energy cost per distance covered) has provided powerful insights into many types of animal performance—insights we now discuss.

The minimal cost of transport depends in regular ways on mode of locomotion and body size

Suppose that for each species of animal we study, we identify the *minimal* weight-specific cost of transport—that is, the minimal energy cost to move 1 kg of body weight 1 m—displayed by the species during horizontal locomotion, regardless of the speed at which the minimum occurs. Then we can plot the minimal cost of transport as a function of body weight, as in Figure 7.7. This sort of plot reveals one of the most striking patterns ever discovered in exercise physiology: Namely, species that engage in a particular primary mode of locomotion—running, flying, or swimming—tend to exhibit a regular relation between minimal cost of transport and body size. Furthermore, the relation for runners differs from that for fliers, and both relations differ from that exhibited by swimming fish. To interpret the lines showing these three relations in Figure 7.7, note that the graph employs log–log coordinates. Minimal cost of transport is an allometric function of body weight for animals that engage in each primary mode of locomotion, and for this reason the relations plot as straight lines when both axes are logarithmic.[5]

To appreciate the full significance of the regular relations shown in Figure 7.7, consider first the fliers—the species for which the primary mode of locomotion is flapping flight. These species are taxonomically very diverse—including insects, bats, and birds—yet they all fall approximately on a single line in terms of their relation between mini-

[5] For each mode of locomotion, minimal weight-specific cost of transport = aW^b, where W is body weight, a and b are constants, and b is typically about −0.2 to −0.4. See Chapter 5 (p. 139) and Appendix G for discussions of allometric functions and their shapes on various sorts of plots.

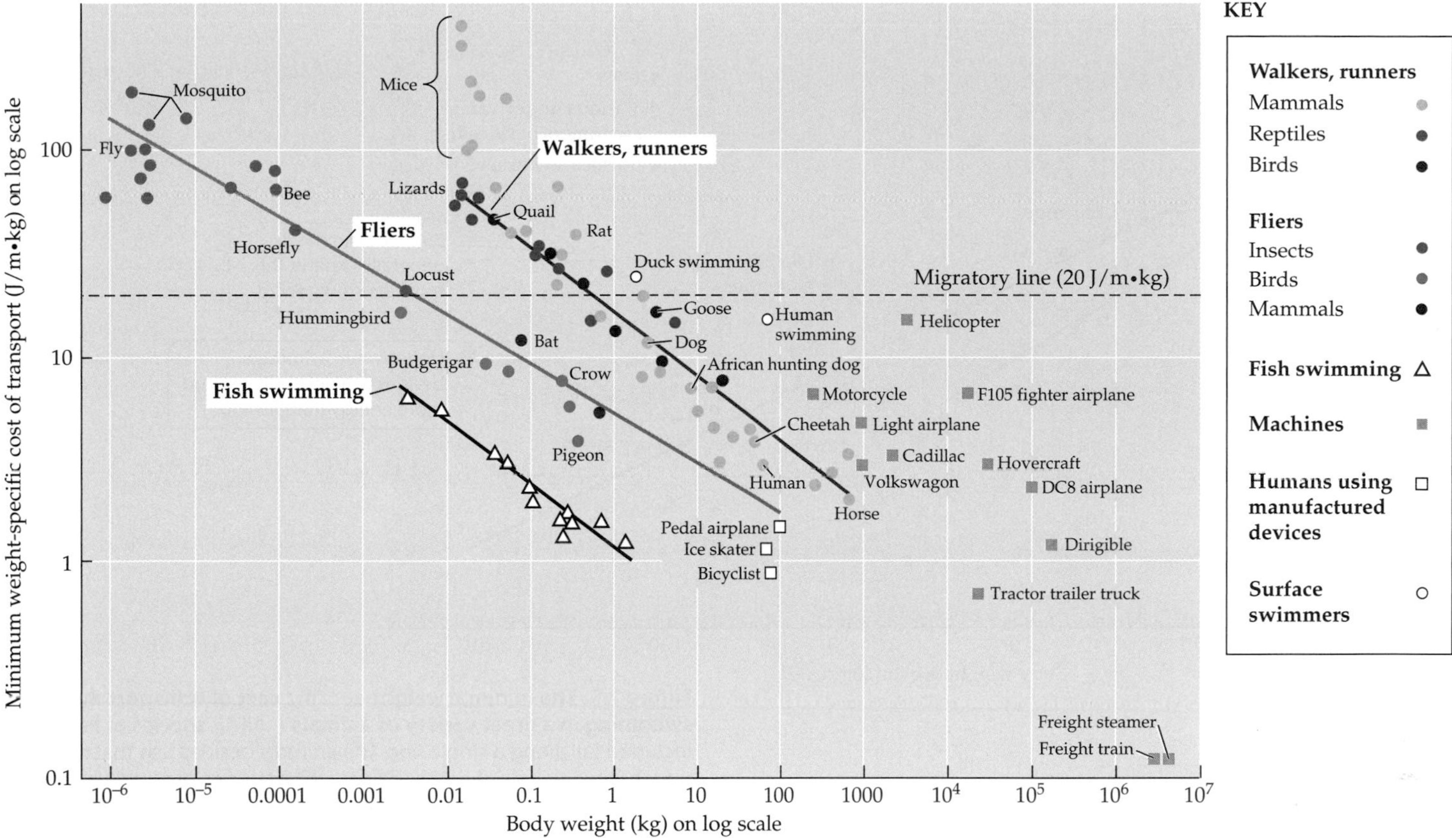

Figure 7.7 The minimal weight-specific cost of transport in relation to body weight for running, flying, and swimming animals and for machines Representative species and machines are identified at random. The three solid lines show the relation between cost of transport and body weight for animals that employ walking or running, flapping flight, or swimming as their *primary* modes of locomotion. Only fish are included in the swimming line. (After Tucker 1975.)

mal cost of transport and body weight. Similarly, the walkers and runners plotted in Figure 7.7—which consist of lizards as well as mammals—fall on a single line. In fact, data are now available for over 150 species of animals that move primarily by running—including running insects, semiterrestrial crabs, running birds (e.g., roadrunners and ostriches), centipedes, and millipedes, as well as lizards and mammals—and, with a few exceptions, all the species fall statistically along a single line. The line for swimming in Figure 7.7 is just for fish. Nonetheless, data are available by now for about 20 species of fish, some of which are very different from one another in their body forms and swimming styles, and all fall along one line. The overall picture that emerges from analyses such as the one shown in Figure 7.7 is remarkable: Among animals engaged in their primary form of locomotion, the minimal cost of transport displayed by a species of given body size typically *depends principally on the species' mode of locomotion rather than its phylogenetic position or the details of its locomotor mechanisms.*

For animals of a particular body size, running is the most costly way to cover distance, whereas swimming (as practiced by fish) is the least costly. The differences in cost among the three forms of locomotion are more substantial than they might appear from a simple visual inspection of Figure 7.7 because the logarithmic scale used for cost of transport tends to make differences look smaller than they are. For a 100-g animal, the cost of running a unit of distance is about 4 times higher than that of flying the same distance and about 14 times greater than the cost of swimming! The high cost of running compared to swimming may initially strike you as odd, because we know from personal experience that the energy cost of covering distance is greater when *we* swim than when *we* run. Remember, however, that the lines in Figure 7.7 represent animals moving by their *primary* forms of locomotion. For fish, swimming is the primary form of locomotion and an exceptionally economical way to cover distance.

Among animals that share a single primary mode of locomotion, large-bodied species cover distance at less weight-specific cost than small-bodied species do. This means, among other things, that if two animals of different size set off on travels with equal proportions of body fat (which they use as fuel), the larger one will be able to cover more distance before running out of fat. Differences like this can be important in understanding migration.

When we consider the various sorts of animals that undertake long migrations, we encounter a remarkable fact: Although certain small or medium-sized fish and flying animals (insects, bats, and birds) undertake such migrations, long migrations are rare among small or medium-sized running animals. Some *large* running animals, such as reindeer (caribou), are noted for their long migrations, but as Figure 7.7 shows, their cost of covering distance is likely to be similar to that of relatively small fish!

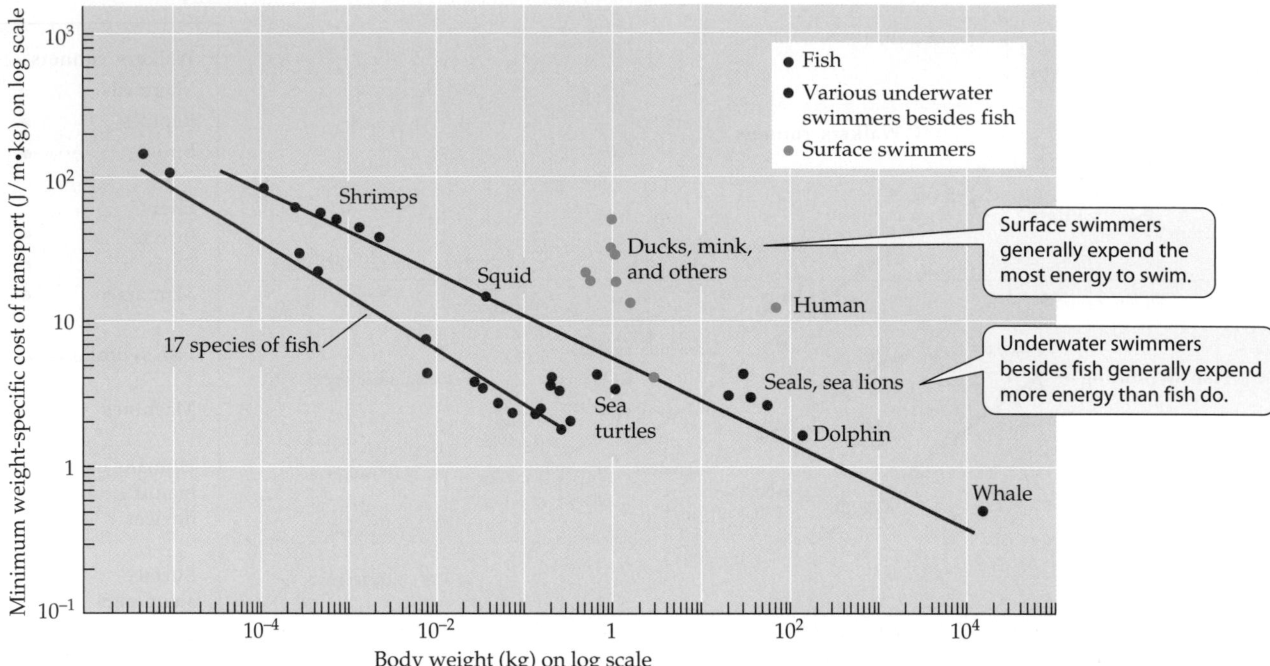

Figure 7.8 The minimal weight-specific cost of transport during swimming in a great variety of animals All 17 species of fish included fall along a single line. The animals besides fish that swim underwater include shrimps (*Palaemon*), a species of squid (*Loligo*), the green sea turtle (*Chelonia*), the harbor seal (*Phoca*), the California sea lion (*Zalophus*), the bottlenose dolphin (*Tursiops*), and the gray whale (*Eschrichtius*). These various species roughly fall on a single line. Animals swimming at the water's surface include humans, several species of ducks and geese, a mink (*Mustela*), a pengiun (*Eudyptula*), and the marine iguana (*Amblyrhynchus cristatus*); the iguana plots very close to the underwater-swimming sea turtles. (After Videler 1993.)

Vance Tucker pointed out years ago that if a line (the so-called *migratory line*) is drawn across Figure 7.7 at about 20 J/m•kg, most migratory species are found below the line, and species above the line are unlikely to be migrants. Although advances in the study of migration now permit a far more detailed analysis, the observation that most migratory species have costs of transport below a certain benchmark remains striking. Apparently, migration has a high chance of being favored by natural selection only if the cost of covering distance is relatively low. The body size that permits an adequately low cost of covering distance depends on the mode of locomotion.

Swimming has proved to be a particularly interesting mode of locomotion for study. As we have noted, even though different species of fish exhibit many different body shapes and employ a variety of propulsive strategies (some emphasize sinusoidal body movements, whereas others emphasize tail or fin propulsion), fish species usually fall along a single line in their relation between minimal cost of transport and body size. Other types of animals that travel by underwater swimming as their primary mode of locomotion—including shrimps, sea turtles, and marine mammals—are typically not as streamlined as fish and tend to exhibit higher costs of transport than fish (Figure 7.8). Costs of transport are in general even higher in humans and other primarily terrestrial animals—such as ducks—that swim on the water's surface rather than underwater (see Figure 7.8); this is true in part because swimming is not the primary mode of locomotion for the terrestrial animals and in part because swimming at the surface tends to be intrinsically more costly than underwater swimming.

For us humans, an interesting corollary of our high cost of swimming is that our own experience of the effort that must be expended to swim provides not the slightest insight into the effort required of a fish. In fact, if you want to get a sense of how strenuous it would be for a human-sized fish to cover distance by underwater swimming, climb on a bicycle. Bicycling ranks as one of the least costly of all animal-powered forms of locomotion, and the cost to cover distance by bicycling approaches the cost of swimming expected for fish of human size (see Figure 7.7).

Summary: The Energy Costs of Defined Exercise

- When a mammal or other animal runs, its metabolic rate typically increases linearly with its speed. Metabolic rate and speed are related by a J-shaped power function in swimming fish. For animals engaged in flapping flight, metabolic rate is expected to exhibit a U-shaped relation to speed, but this theoretical expectation is not always observed in real animals.
- Cost of transport is the energy cost of covering a unit of distance. The speed that minimizes the cost of transport is the speed that maximizes the distance that can be traveled with certain energy.

■ Running animals, animals flying by flapping flight, and swimming fish exhibit three distinctive and coherent allometric relationships between minimal cost of transport and body weight. For animals of any particular body size, running is the most expensive way to cover distance, flying is intermediate, and swimming by fish is the least expensive. Within any one locomotory group, the minimal weight-specific cost of transport decreases as body size increases.

The Maximal Rate of Oxygen Consumption

An animal's **maximal rate of oxygen consumption**, symbolized $\dot{V}_{O_2max}$ and sometimes called **aerobic capacity** or **maximal aerobic power**, is a key property for the study of aerobic activity.[6] One reason an animal's $\dot{V}_{O_2max}$ is of interest is that it determines the peak rate at which the animal can synthesize ATP by aerobic catabolism, and thus it determines how intensely the animal can exercise in a pay-as-you-go mode (see Figure 6.7*b*). Two species might exhibit the same relation between cost of transport and speed yet differ in $\dot{V}_{O_2max}$. In that case, although they would require the same amount of ATP to cover a kilometer *at any given speed*, the species with the higher $\dot{V}_{O_2max}$ would be able to cover the kilometer *faster* if need be because it would be able to make ATP more rapidly by aerobic catabolism and thus reach higher sustained speeds.

[6] In this chapter we assume that to measure $\dot{V}_{O_2max}$, investigators induce animals to exercise intensely, thereby raising their rates of O_2 consumption to peak levels. Another way to raise the rates of O_2 consumption of some animals, such as mammals, is to expose them to cold. Maximal rates of O_2 consumption induced by exercise and cold are often not the same because different organs participate in exercise and cold defense.

A second reason for interest in $\dot{V}_{O_2max}$ is more subtle and remarkable: In studies of humans and other vertebrates, the maximal rate of O_2 consumption induced by exercise provides a benchmark by which to judge the *strenuousness* of all *aerobic* physical activity. Specifically, the strenuousness to an individual of any particular form of aerobic exercise depends roughly on how high a proportion of the individual's $\dot{V}_{O_2max}$ is required. In average people, if an activity demands 35% or less of $\dot{V}_{O_2max}$, the activity can likely be continued for 8 to 10 hours in a day. As the percentage of $\dot{V}_{O_2max}$ required by activity increases, the activity becomes more strenuous; a type of activity that requires 75% of $\dot{V}_{O_2max}$, for instance, will probably be exhausting in 1 to 2 hours.

The use of $\dot{V}_{O_2max}$ as a benchmark is informative in a number of ways. One is the anylysis of human-powered flight (Box 7.1). A more practical application for most people is the analysis of the relation between exercise and ageing. As people age beyond young adulthood, their $\dot{V}_{O_2max}$ tends to decline; after age 30, the decline is about 9% per decade for sedentary people, although it is less than 5% per decade for people who stay active. Because of the decline in $\dot{V}_{O_2max}$, an activity that requires any particular *absolute* rate of O_2 consumption tends to demand an ever greater proportion of $\dot{V}_{O_2max}$ as people age. Thus, the activity becomes more strenuous. A form of exercise that demands 35% of $\dot{V}_{O_2max}$ in youth might require 75% in old age; the exercise would therefore shift from being sustainable for 8 to 10 hours in youth to being sustainable for only 1 to 2 hours in old age. This helps explain why jobs involving physical labor become more difficult for people to perform for a full workday as aging occurs. Work that requires O_2 consumption at 100% of $\dot{V}_{O_2max}$ is inevitably strenuous in the extreme; people can ordinarily continue it for only a few minutes. We saw in Box 6.3 that people exhibit a lower and lower $\dot{V}_{O_2max}$ as they go to higher altitudes. Although slow uphill walking requires just a small percentage of $\dot{V}_{O_2max}$ for relatively young people at low altitudes, it demands more or less 100% of $\dot{V}_{O_2max}$ at the highest altitudes on Earth. Thus, such walking shifts

BOX 7.1 Finding Power for Human-Powered Aircraft

The world was electrified in 1977 when a human-powered aircraft named *Gossamer Condor* first flew a mile. Two years later, Bryan Allen powered another such aircraft, the *Gossamer Albatross*, across the English Channel in a flight of 37 km (more than 20 miles) that required almost 3 hours. Then, in 1988, Kanellos Kanellopoulos achieved a still greater record by powering the aircraft *Daedalus* for a flight of 119 km requiring almost 4 hours.

Human-powered aircraft fly at altitudes of about 6 meters. At these heights, even a few *seconds* of loss of power are out of the question, because without continuous power the aircraft quickly land or crash. Thus the feats achieved by Bryan Allen, Kanellos Kanellopoulos, and the other pilots of human-powered aircraft require 3 to 4 hours of truly *uninterrupted* effort.

The physiological properties of human exercise discussed in this chapter dictate that the rate of O_2 consumption of a pilot must be no higher than 65% to 70% of his or her $\dot{V}_{O_2max}$ for there to be reasonable certainty of maintaining an uninterrupted effort for 3 to 4 hours. This biological fact set the engineering goals for the design of human-powered aircraft: An aircraft could not demand more power from its "power plant" than 65% to 70% of $\dot{V}_{O_2max}$. Engineering innovations in the 1960s and 1970s led to the invention of the *Gossamer Albatross* design, which required about 4.2 W of power for each kilogram of pilot body weight to fly. A person could be expected to require an actual rate of O_2 consumption of 50 to 55 mL O_2/kg•minute to provide this much power to the machine. Thus, *Gossamer Albatross* required a pilot with a $\dot{V}_{O_2max}$ of 74 to 81 mL O_2/kg•minute. The engineers knew that people with such high $\dot{V}_{O_2max}$ values, although unusual, actually exist (see Table 7.4). Bryan Allen provided the needed $\dot{V}_{O_2max}$ and the determination for the first human-powered 3-hour flight.

from being simple at low altitudes to being barely possible, or impossible, in high mountains (see Box 6.3).

The physiological causes of the limits on maximal O_2 consumption are hotly debated. Some physiologists argue that particular organ systems set the limits on $\dot{V}_{O_2max}$. For instance, some point to the circulatory system as being the "weak link" in mammals, arguing that all other organ systems could transport and use O_2 at a greater rate if it were not for more-restrictive limits on how fast the circulatory system can transport O_2. An alternative position is the hypothesis of **symmorphosis**, which states that all organ systems that serve a single function in an animal are interactively adapted to have approximately equal limits because it would make no sense for any one system to have evolved capabilities that could never be used because of more-restrictive limits in other systems. For biologists who subscribe to the concept of symmorphosis, the reason an animal cannot attain a rate of O_2 consumption higher than its $\dot{V}_{O_2max}$ is that multiple organ systems simultaneously reach their performance limits at $\dot{V}_{O_2max}$.

Aerobic scope for activity and **aerobic expansibility** are two concepts that are sometimes employed in the study of $\dot{V}_{O_2max}$. An animal's aerobic scope for activity at a particular temperature is usually defined to be the *difference* between its $\dot{V}_{O_2max}$ at that temperature and its resting rate of O_2 consumption at the same temperature. Its aerobic expansibility is the *ratio* of its $\dot{V}_{O_2max}$ over its resting rate of O_2 consumption. To illustrate, suppose that at a particular body temperature, a fish has a resting rate of O_2 consumption of 0.05 mL/g•hour and a $\dot{V}_{O_2max}$ of 0.30 mL/g•hour. Its aerobic scope for activity would be (0.30 – 0.05) = 0.25 mL/g•hour. Its aerobic expansibility would be 0.30/0.05 = 6.

$\dot{V}_{O_2max}$ differs among phyletic groups and often from species to species within a phyletic group

A rough but useful rule of thumb for vertebrates is that the $\dot{V}_{O_2max}$ induced by exercise is about ten times higher than the resting rate of O_2 consumption.[7] That is, the aerobic expansibility of vertebrates tends to be about 10. The consistency of the aerobic expansibility in vertebrates has some remarkable implications. For example, consider that the standard metabolic rates of fish, amphibians, and reptiles are typically no more than 1/10 to 1/4 as high as the basal metabolic rates of mammals and birds of similar body size (see page 219). Given that aerobic expansibility averages about 10 in all vertebrate groups, you can see that the *peak* rates of O_2 consumption of fish, amphibians, and reptiles are of the same order of magnitude as the *basal* (resting) rates of mammals and birds.

Table 7.3 illustrates this critically important relation with data on two pairs of vertebrates: (1) a salmon and a rat of similar size and (2) a monitor lizard and a guinea pig of similar size. Salmon and monitor lizards are among the most aerobically competent of all fish and reptiles. Yet their *peak* rates of O_2 consumption ($\dot{V}_{O_2max}$) resemble the *basal* rates of O_2 consumption in mammals, and $\dot{V}_{O_2max}$ in the mammals is far above $\dot{V}_{O_2max}$ in the salmon and lizards. Mammals and lizards of the same body weight typically resemble each other in the ATP cost to cover a unit of distance when they run at speeds that minimize their cost of transport (see Figure 7.7). Mammals, however, typically can make ATP aerobically far faster than lizards of the same size can. Thus, mammals can achieve far higher *sustained speeds* than lizards can. One of the principal hypotheses offered for the evolution of homeothermy is that it permitted faster sustained locomotion.

Within any one vertebrate phyletic group, $\dot{V}_{O_2max}$ per gram of body weight tends to be an allometric function of body weight, with small species having a higher $\dot{V}_{O_2max}$ per gram than large ones (see Chapter 5). The line in Figure 7.9 shows the average relation between $\dot{V}_{O_2max}$ and size in mammals. Because of variation among species of any given size, there is scatter around the line. In fact, some species have $\dot{V}_{O_2max}$ values that are very different from the average—or "expected"—values for their size. Among such species, the pronghorn, found in grasslands of the American West, is the most extreme example known (see Figure 7.9). Its $\dot{V}_{O_2max}$ value is over three times higher than the value that would be average for a mammal of its size, indicating that it has

[7] In the case of mammals and birds, the $\dot{V}_{O_2max}$ is about ten times the basal metabolic rate.

TABLE 7.3 Rates of aerobic catabolism during rest and during sustained exercise of peak intensity in two pairs of vertebrates: A fish and mammal of similar body weight, and a lizard and mammal of similar weight

			Rates of O_2 consumption (mL O_2/g•h)		
Species	Body weight (g)	Test temperature[a] (°C)	Basal or standard rate of O_2 consumption	$\dot{V}_{O_2max}$	Aerobic scope (mL O_2/g•h)
Pair compared:					
Rat (*Rattus*)	230	30	0.9	4.6	3.7
Salmon (*Oncorhynchus*)	230	15	0.05	0.49	0.44
Pair compared:					
Guinea pig (*Cavia*)	880	30	0.6	3.7	3.1
Monitor lizard (*Varanus*)	670	40	0.11	1.0	0.89

Source: From Hill and Wyse 1989.
[a] Test temperatures for the fish and lizard were those at which aerobic scope was maximal.

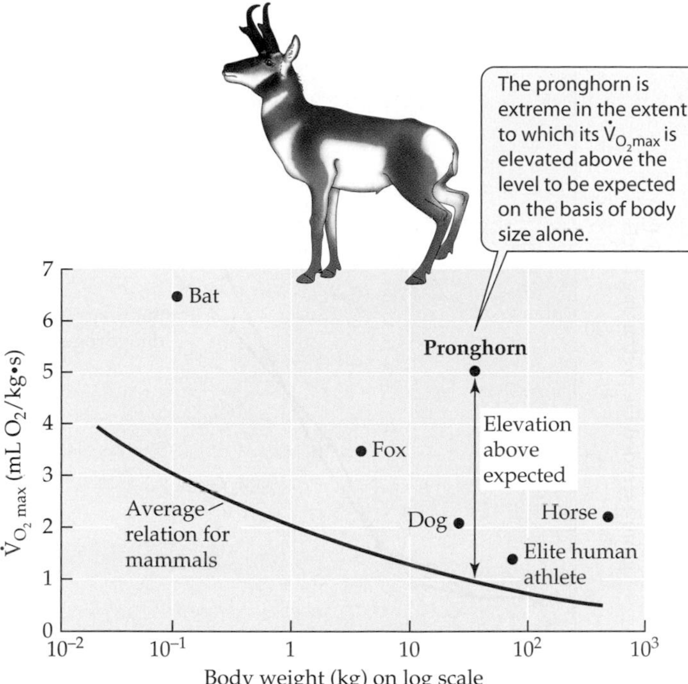

Figure 7.9 The pronghorn (*Antilocapra americana*) represents an extreme case of evolutionary specialization for high $\dot{V}_{O_2max}$ The line shows the average statistical relation between weight-specific $\dot{V}_{O_2max}$ and body weight in mammals; it is curved rather than straight in this plot because although the *x* axis is logarithmic, the *y* axis is not. Five nonhuman species that are strong athletic performers and that have $\dot{V}_{O_2max}$ values higher than average (for their body sizes) are plotted individually. The high $\dot{V}_{O_2max}$ of the pronghorn enables it to be the fastest sustained runner on Earth. (After Lindstedt et al. 1991.)

evolved a specialized ability to engage in sustained, aerobic exercise (see page 16). Pronghorns are grazers that are legendary for the speeds at which they flee danger. They are not quite as fast as cheetahs, but unlike cheetahs—which produce ATP anaerobically when running fast—pronghorns produce the ATP required for fast running *aerobically*. Cheetahs fatigue within a minute or two when running fast. Pronghorns, however, can run at speeds of at least 65 km/hour (about 40 miles/hour) on a sustained basis. They are the fastest sustained runners on Earth. They have exceptionally large lungs for their size and exceptional abilities to maintain high rates of blood circulation.

The highest known weight-specific values for $\dot{V}_{O_2max}$ occur in certain strong-flying insects among invertebrates and in certain small-bodied fliers—hummingbirds and bats—among vertebrates. In the flight muscles of such aerobic heroes, mitochondria often occupy 35% to 45% of tissue volume—very high values. With mitochondria packed this densely in muscle cells, a sort of competition arises between mitochondria and contractile elements for cell space; as mitochondria occupy more space, contractile elements have less. Animals seem never to go higher than about 45% mitochondria in working muscles, suggesting that the evolution of higher values would be pointless because the contractile apparatus would be too diminished to use the ATP that the mitochondria could make. The peak in mitochondrial packing helps set an ultimate peak on $\dot{V}_{O_2max}$ per gram of muscle.

$\dot{V}_{O_2max}$ varies among individuals within a species

Individuals of a species vary in $\dot{V}_{O_2max}$, as we saw in Figure 1.18. The phenomenon is readily documented in humans. For example, when groups of young men entering the military are tested for $\dot{V}_{O_2max}$, they typically exhibit a normal statistical distribution (bell curve) in $\dot{V}_{O_2max}$, with the low and high values being quite different (e.g., lows near 40 mL O_2/kg•minute and highs near 70 mL O_2/kg•minute for men drawn randomly from the general population). A significant portion of the variation in $\dot{V}_{O_2max}$ among individuals of a species often proves to be hereditary when tested. A recent, large study of people in the United States and Canada (the HERITAGE study) found, for example, that when the confounding effects of age, gender, and body composition are taken into account, about half of all individual variation in human $\dot{V}_{O_2max}$ can be attributed to genetic inheritance. In wild species, when individuals vary genetically in $\dot{V}_{O_2max}$, the capacity for aerobic exercise performance can potentially undergo evolution by natural selection (see page 26).

Recall that the strenuousness of any given form of sustained exercise depends on the fraction of $\dot{V}_{O_2max}$ that the exercise requires. In the context of individual variation in $\dot{V}_{O_2max}$, this principle helps explain why a single form of sustained exercise can be differently taxing to different individuals. Again using humans to exemplify, suppose that running at a particular speed requires an O_2 consumption of 30 mL O_2/kg•minute. Suppose also that two otherwise identical individuals have $\dot{V}_{O_2max}$ values of 40 and 70 mL O_2/kg•minute, respectively. For the former person the running would require 75% of $\dot{V}_{O_2max}$ and would be very taxing; for the latter, it would require just 43% and would be sustainable for hours.

Among successful human athletes, it is a consistent finding that individuals who excel in various sorts of competition tend to exhibit striking differences in $\dot{V}_{O_2max}$. As Table 7.4 shows, for example, men who compete at the world-class level in cross-country skiing or long-distance running have far higher average $\dot{V}_{O_2max}$ values than men who compete successfully in weight lifting or gymnastics. Based on careful analysis, the differences in Table 7.4—taken as a whole—cannot be attributed primarily to differences in training among men in the various events (except for the "general population"). The differences in $\dot{V}_{O_2max}$ among successful athletes in various events are believed to result to a substantial extent from the athletes' choosing to compete in events in which they have inherent abilities to succeed. A high $\dot{V}_{O_2max}$, meaning a high peak rate of aerobic ATP production, is an asset in long-sustained events, especially ones like cross-country skiing in which most major muscle masses are employed. Athletes who are endowed by heredity or by early developmental influences with a high $\dot{V}_{O_2max}$ choose sports in which aerobic ATP synthesis is particularly important. By contrast, those with low $\dot{V}_{O_2max}$ values choose sports like weight lifting, in which aerobic ATP production is less critical (they often have unusually high anaerobic competence).

Although we know almost nothing about nonhuman animals in this regard, it seems reasonable to assume that in them also, individuals with different aerobic competence elect different lifestyles. Thus, two individuals of a single species that differ in $\dot{V}_{O_2max}$ might differ in hunting strategies, for example.

TABLE 7.4 Average $\dot{V}_{O_2max}$ in male Swedish athletes who compete in various events at the world-class level

Event (or other category)	Average $\dot{V}_{O_2max}$ (mL O_2/kg•minute)
Highest values ever recorded	90–95
Cross-country skiing	84
Long-distance running	83
Canoeing	67
Ice hockey	63
Soccer	58
Weight lifting	53
Gymnastics	52
General population (young adult Swedish men)	44

Source: After Åstrand and Rodahl 1986.

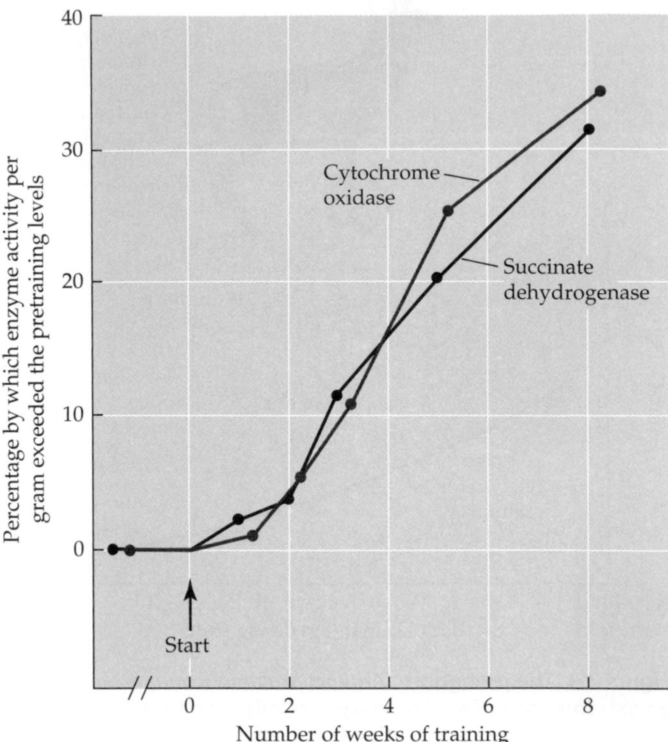

Figure 7.10 Mammalian skeletal muscles respond to endurance training by increases in Krebs-cycle and electron-transport enzymes The activities of succinate dehydrogenase (a Krebs-cycle enzyme) and cytochrome oxidase (an electron-transport enzyme) per gram of muscle tissue were measured in the thigh muscles of human subjects before and during 8 weeks of endurance training. The subjects were healthy men of college age who were previously not in endurance training programs. Their training for this study involved intense bicycle work (requiring about 80% of $\dot{V}_{O_2max}$) for 20 to 40 minutes per day during 4 days each week. (After Henriksson and Reitman 1977.)

$\dot{V}_{O_2max}$ responds to training

To gain insight into the effects of training on $\dot{V}_{O_2max}$, it is again appropriate to focus on people because a very large body of data is available on our species. When sedentary people participate in programs of repeated, prolonged exercise, they vary widely in how much their $\dot{V}_{O_2max}$ increases. Whereas some individuals show no change in $\dot{V}_{O_2max}$, others undergo considerable increases; heredity accounts for about half the variation in the response of $\dot{V}_{O_2max}$ to training (according to the HERITAGE study). The typical increase in $\dot{V}_{O_2max}$ is 10% to 30%.

What are the physiological mechanisms by which $\dot{V}_{O_2max}$ is increased? When people or experimental animals undergo training by repeated, prolonged exercise, one sort of response is that organ systems that deliver O_2 increase their potential delivery rates; the heart, for example, increases the volume of blood it can pump per beat and thus is able to circulate blood faster. In addition, skeletal muscles change in many ways that poise them better to make ATP aerobically. A consistent response to training in the skeletal muscles of people and experimental animals is an increase in the number of mitochondria and in the activities of enzymes of the Krebs cycle and electron-transport chain (Figure 7.10). Increases are also observed in the skeletal muscles in the density of blood capillaries, the activities of glucose transporters, and the activities of enzymes of fatty acid catabolism.

Summary: The Maximal Rate of Oxygen Consumption

- An animal's maximal rate of O_2 consumption ($\dot{V}_{O_2max}$) is significant for two principal reasons. First, it determines the maximal rate at which sustained, aerobic exercise can be performed. Second, it serves as a benchmark by which the strenuousness of submaximal aerobic work can be assessed. Sustained work becomes more strenuous—and more quickly fatiguing—for an individual as it demands a higher proportion of the individual's $\dot{V}_{O_2max}$.
- Major phyletic groups sometimes exhibit consistent differences in $\dot{V}_{O_2max}$. Among vertebrates—as a rough but important rule of thumb—$\dot{V}_{O_2max}$ in mammals and birds is about an order of magnitude higher than $\dot{V}_{O_2max}$ in fish, amphibians, and reptiles of similar body size, assuming that the latter groups are at body temperatures near mammalian and avian levels. Within a single phyletic group, $\dot{V}_{O_2max}$ typically tends to vary with body size in an allometric fashion, small-bodied species having higher $\dot{V}_{O_2max}$ per gram than large-bodied species have.
- Individuals of a species that are similar in age and gender typically vary considerably in $\dot{V}_{O_2max}$. Some of this variation can be attributed to differences in training. Usually a significant proportion can also be attributed to inheritance. Genetically based differences in $\dot{V}_{O_2max}$ among individuals of a species provide opportunities for natural selection to favor changes in the species' average $\dot{V}_{O_2max}$ over evolutionary time.

The Energetics of Routine and Extreme Daily Life

From studies using the doubly labeled water method, physiologists now know a great deal about the average daily metabolic rates (ADMRs) of terrestrial animals that are leading their routine lives in their natural habitats. Within sets of related animals, the ADMR per unit of body weight typically decreases allometrically with increasing body size. In birds, the *routine* ADMR in the wild is typically 2.5 to 3 times the basal metabolic rate (BMR). In rodents, the routine ADMR averages 3 to 4 times the BMR. People who lead moderately sedentary lives, such as students, scientists, and bankers, typically exhibit ADMRs that are 1.5 to 2 times higher than their BMRs.

Animals sometimes face *extreme* energy demands in their daily lives. This is often true of small mammals and birds during the cold of winter, for example, because of the high energy costs of thermoregulation. A question that has drawn the interest of physiologists in recent years is how high the ADMR can be. That is, what is an animal's maximum possible average metabolic rate that can be *sustained* day after day? Note that this is a very different question from asking how high the $\dot{V}_{O_2max}$ can be, because $\dot{V}_{O_2max}$ reflects a rate of metabolism that can be maintained for only tens of minutes, at most.

To gain insight into how high the human sustained metabolic rate can possibly be, investigators studied cyclists in the Tour de France, a long-distance bicycle race. In 1984 the race covered nearly 4000 km and lasted 22 days, during which the cyclists fought for the lead while going up and down 34 mountains. The athletes maintained stable body weights over the 3 weeks of the race by eating large amounts of food during their nighttime breaks. The metabolic rates of four cyclists were measured with the doubly labeled water method and averaged about 4.5 times higher than BMR. These data indicate that when well-conditioned people in a competitive situation expend the maximal effort that they can possibly maintain for periods of many days, their time-averaged metabolic rates are less than 5 times BMR (even though $\dot{V}_{O_2max}$ in such people is at least 10 times the BMR).

Table 7.5 presents a collection of data on the sustained metabolic rates of rodents and birds under circumstances that could be expected to call forth all-out effort, analogous to the Tour de France.

TABLE 7.5 Particularly high average daily metabolic rates measured in nonhuman mammals and birds

Animals and circumstances	Average daily metabolic rate as ratio of resting metabolic rate
Female laboratory mice nursing litters of 14 young	6.5
Females of three other species of mice nursing litters	3.7–6.7
Four species of mice living at –10°C	3.7–6.1
Nine species of birds rearing young	2.4–6.6

Source: After Hammond and Diamond 1997. See also Peterson, Nagy, and Diamond 1990.

Whereas some of the animals in the table were in cold environments, others were rearing young. As you can see, these estimates of maximal ADMR are in the range of 2.5 to 6.5 times the resting metabolic rate.

Clearly, if animals could achieve higher peak sustained metabolic rates than they do, they could be more physically active, raise more young, or survive in more-demanding environments. Thus, animals confront limits that are imposed by the rates at which they can use energy to their advantage in carrying out prolonged, ecologically critical activities. In view of the importance of such limits, we must wonder what *causes* them: Why are the peak sustained metabolic rates of mammals and birds not higher than 3 to 7 times their resting metabolic rates? That is a current question for research.

Ecological Energetics

Ecological energetics is the study of the dynamic interplay of energy supplies and needs in ecologically realistic settings. An example is provided by the study of colonies of terns and other seabirds in relation to fish populations upon which they prey. Using the doubly labeled water method or time–energy budgets, one can estimate the daily energy demand of individuals of each species and age-class of bird in a seabird colony. One can then calculate the energy demand of a whole colony by multiplying the energy demand of each particular species and age-class by the numbers of individuals of each type. Then the energy demand of the birds in the colony can be compared to the energy available from the fish populations on which the birds feed within the area of the ocean where they forage. Studies of this sort have revealed that seabird colonies sometimes consume ¼ or ⅓ of all the productivity of prey fish in their foraging areas. This sort of ecological energetic analysis has helped biologists better understand facets of seabird population dynamics. Some seabird colonies in the Pacific are devastated in El Niño years when the productivity of fish is reduced, and some colonies have been devastated by competition from human fishing. We can better understand such problems when we realize how high the energy needs of seabird colonies can be, relative to the energy available as food in their ecosystems.

A more elaborate illustration of the power of ecological energetic research is provided by Bernd Heinrich's analysis of costs and rewards in bumblebee foraging—an example of what Heinrich terms *bumblebee economics*. The starting point of his analysis is the recognition that in ecologically realistic settings, the acquisition of food has energy costs as well as energy rewards. By analyzing the interplay of the two, the study of ecological energetics can provide insight into the evolved and learned foraging behaviors of animals.

When bumblebees (*Bombus*) forage, they fly from one flower (or flower cluster) to another, landing on each one long enough to probe the flower with their tongues and collect available nectar. Two major *costs* of bumblebee foraging must be considered:

1. Flight is itself very costly. It can easily elevate the metabolic rate of a bumblebee to 20 to 100 times its resting rate. The cost of flight per unit of time is essentially independent of air temperature.
2. Temperature regulation for flight may also be costly. To fly, bumblebees require the temperature of their flight muscles to be

30°C or higher (see page 235). When the bees are flying, temperatures that high are maintained by the heat produced by the wing-flapping contractions of their flight muscles. When bees alight on flowers, however, they are at risk of quickly cooling to below the necessary flight temperature if the surrounding air is cool. If a bee were to cool in this way, it would be unable to take to the air again in search of other flowers. To keep their flight muscles warm while they are sitting on flowers, bees produce heat by a process analogous to human shivering.[8] The intensity and *energetic cost* of this shivering become greater as the air temperature decreases. Although shivering may be unnecessary at an air temperature of 25°C, shivering at 5°C may be so intense as to raise a stationary bee's metabolic rate to as high a level as prevails when the bee is flying.

Considering the costs of *both* flying and shivering, the *average* metabolic expenditure per unit of time required for a bee to forage tends to increase as the air becomes cooler. If the air is warm enough that no shivering is needed when the bee is sitting on flowers, the bee has a high metabolic rate when it is flying but a low rate when it is not. If the air is cold, the bee has a high metabolic rate all the time, whether flying or stationary.

Now let's turn to the energy *rewards* of foraging. The energy reward that can be obtained per unit of time from any particular species of flowering plant depends on (1) the volume of nectar obtained per flower, (2) the sugar concentration of the nectar, and (3) the number of flowers from which a bee can extract the nectar per unit of time. The third property depends on the spacing of the flowers and the difficulty of penetrating flowers to obtain their nectar.

Some species of plants yield sufficient sugar per flower that bumblebees can realize a *net* energy profit when foraging from them regardless of the air temperature. For example, the rhodora (*Rhododendron canadense*), a rhododendron with large flowers, typically yields sugar equivalent to about 1.7 J/flower. At 0°C, a large bee might expend energy at a time-averaged rate of 12.5 J/minute while foraging. Accordingly, the bee could break even energetically by taking the nectar from about 7 to 8 rhododendron flowers per minute. In fact, bees can tap almost 20 rhododendron flowers per minute. Thus, even at 0°C, bees foraging on rhododendron are able to meet their costs of foraging *plus* accumulate a surplus of nectar to contribute to the hive.

In contrast, some plants yield so little sugar per flower that they are profitable sources of nectar only when air temperatures are relatively high (and the bees' costs of foraging are thereby reduced). For example, bees typically visit blossoms of wild cherry (*Prunus*) only when the air is warm. The blossoms yield sugar equivalent to only about 0.21 J apiece. At 0°C, a bee would have to tap an unrealistically high number of cherry blossoms per unit of time—about 60 per minute—just to meet its costs of foraging.

The study of bumblebee foraging exemplifies how an accounting of energy costs and gains can reveal constraints on animals' daily lives. From our energy analysis of bees, we come to recognize that air temperature helps dictate the suite of flowers that bees can profitably exploit as sources of food. Thus, from an energy analysis we can understand why we see bumblebees foraging on wild cherry trees on warm days but not cold ones.

[8] The process in bees and other insects consists of high-frequency muscle contractions that produce heat (rather than mechanical work) as their primary product. The process is called *shivering* because of this resemblance to vertebrate shivering, but it is not homologous to shivering in mammals and birds.

Study Questions

1. How does the doubly labeled water method depend on the existence of isotopic equilibrium between the oxygen in H_2O and that in CO_2?
2. Think through your list of friends, and select one (theoretically) for study to determine his or her average daily metabolic rate. How would you carry out research to create a time–energy budget for your friend?
3. In your own words, explain why foraging on wild cherry flowers is beneficial for bumblebees in warm weather but not cold weather.
4. As noted in this chapter, the $\dot{V}_{O_2max}$ of people tends to decline after age 30 by about 9% per decade for sedentary individuals, but it declines less than 5% per decade for people who stay active. Consider the activities in Table 7.1. How would you expect sedentary and active people to differ in their capacities for each of those activities in old age? Explain.
5. For an animal engaging in sustained exercise, why is there not one single ideal speed?
6. List the possible reasons why two individuals of a certain species might differ in $\dot{V}_{O_2max}$.
7. Suppose that a bird's metabolic rate while flying at 30 km/hour is 8 kJ/hour. What is the bird's cost of transport when flying at 30 km/hour?
8. Looking at Figure 7.7, how would you say animals and machines compare in their efficiencies in covering distance?
9. African hunting dogs depend on sustained chases by groups of cooperating individuals to capture antelopes for food. If the members of two groups differ in their average $\dot{V}_{O_2max}$, how might the two groups differ in the strategies they use during hunting?
10. In mammals of all species, the peak rate of O_2 consumption of each mitochondrion is roughly the same. On the basis of patterns of how $\dot{V}_{O_2max}$ varies with body size in species of mammals, how would you expect the muscle cells of mammals of various body sizes to vary in how tightly they are packed with mitochondria? Explain your answer.
11. What is the hypothesis of symmorphosis? How might you evaluate or test the hypothesis?
12. Explain the concept that in high-performance muscle cells, mitochondria and contractile elements compete for space over scales of evolutionary time.

Key References

Alexander, R. M. (ed.). 1992. *Mechanics of Animal Locomotion.* Springer, New York. Quite a few uniquely valuable papers. The one by T. M. Casey on energetics, although becoming dated, covers an exceptional breadth of subject matter in a compact, readable way. The paper by M. B. Bennett provides a superb introduction to biomechanics and the physics of biomaterials, subjects outside the scope of the present chapter.

Åstrand, P.-O., and K. Rodahl. 1986. *Textbook of Work Physiology: Physiological Bases of Exercise*, 3rd ed. McGraw-Hill, New York. A definitive textbook treatment of work in the full range of human endeavor, from daily life to sport. New edition due out soon.

Bennett, A. F. 1991. The evolution of activity capacity. *J. Exp. Biol.* 160: 1–23. A searching and provocative discussion of when and how animals evolved their aerobic and anaerobic capabilities to produce ATP for exercise, emphasizing vertebrates.

Dickinson, M. H., C. T. Farley, R. J. Full, M. A. R. Koehl, R. Kram, and S. Lehman. 2000. How animals move: An integrative view. *Science* 288: 100–106. A short but broadly conceived introduction to the biomechanics and functional morphology of animal locomotion. Although the present chapter does not include these topics, they are intimately related to the energetic themes that the chapter stresses.

Heinrich, B. 1979. *Bumblebee Economics.* Harvard University Press, Cambridge, MA. One of the great essays on ecological energetics of the twentieth century. A rewarding and thoroughly enjoyable book.

Shephard, R. J., and P.-O. Åstrand (eds.). 2000. *Endurance in Sport*, 2nd ed. Blackwell, Oxford, UK. Numerous well-written chapters on all aspects of sustained athletic performance. For students interested in sport itself, the book includes chapters on most practical considerations and on many specific types of sport.

Selected References

Alerstam, T., and A. Hedenström. 1998. The development of bird migration theory. *J. Avian Biol.* 29: 343–369.

Alexander, R. M. 1984. Walking and running. *Amer. Sci.* 72: 348–354.

Alexander, R. M. 2003. *Principles of Animal Locomotion.* Princeton University Press, Princeton, NJ.

Autumn, K., D. Jindrich, D. DeNardo, and R. Mueller. 1999. Locomotor performance at low temperature and the evolution of nocturnality in geckos. *Evolution* 53: 580–599.

Brooks, G. A., T. D. Fahey, T. P. White, and K. W. Baldwin. 2000. *Exercise Physiology: Human Bioenergetics and Its Applications,* 3rd ed. Mayfield Publishing, Mountain View, CA.

Chai, P., and R. Dudley. 1995. Limits to vertebrate locomotor energetics suggested by hummingbirds hovering in heliox. *Nature* 377: 722–725.

Chassin, P. S., C. R. Taylor, N. C. Heglund, and H. J. Seeherman. 1976. Locomotion in lions: Energetic cost and maximum aerobic capacity. *Physiol. Zool.* 49: 1–10.

Davis, R. W., T. M. Williams, and G. L. Kooyman. 1985. Swimming metabolism of yearling and adult harbor seals *Phoca vitulina*. *Physiol. Zool.* 58: 590–596.

Dial, K. P., A. A. Biewener, B. W. Tobalske, and D. R. Warrick. 1997. Mechanical power output of bird flight. *Nature* 390: 67–70.

Dickinson, M. H., F.-O. Lehmann, and S. P. Sane. 1999. Wing rotation and the aerodynamic basis of insect flight. *Science* 284: 1954–1960.

Dohm, M. R., J. P. Hayes, and T. Garland. 1996. Quantitative genetics of sprint running speed and swimming endurance in laboratory house mice (*Mus domesticus*). *Evolution* 50: 1688–1701.

Ellington, C. P. 1991. Limitations on animal flight performance. *J. Exp. Biol.* 160: 71–91.

Fedak, M. A., and H. J. Seeherman. 1979. Reappraisal of energetics of locomotion shows identical cost in bipeds and quadrupeds including ostrich and horse. *Nature* 282: 713–716.

Full, R. J. 1991. The concepts of efficiency and economy in land locomotion. In R. W. Blake (ed.), *Efficiency and Economy in Animal Physiology*, pp. 97–131. Cambridge University Press, Cambridge.

Full, R. J. 1997. Invertebrate locomotor systems. In W. H. Dantzler (ed.), *Comparative Physiology*, vol. 2 (Handbook of Physiology [Bethesda, MD], section 13), pp. 853–930. Oxford University Press, New York.

Gans, C., A. S. Gaunt, and P. W. Webb. 1997. Vertebrate locomotion. In W. H. Dantzler (ed.), *Comparative Physiology*, vol. 1 (Handbook of Physiology [Bethesda, MD], section 13), pp. 55–213. Oxford University Press, New York.

Giampietro, M., and D. Pimentel. 1991. Energy efficiency: Assessing the interaction between humans and their environment. *Ecol. Econom.* 4: 117–144.

Hainsworth, F. R. 1981. Energy regulation in hummingbirds. *Amer. Sci.* 69: 420–428.

Hammond, K. A., and J. Diamond. 1997. Maximal sustained energy budgets in humans and animals. *Nature* 386: 457–462.

Harrison, J. F., and S. P. Roberts. 2000. Flight respiration and energetics. *Annu. Rev. Physiol.* 62: 179–205.

Jones, J. H., and S. L. Lindstedt. 1993. Limits to maximal performance. *Annu. Rev. Physiol.* 55: 547–569.

Komi, P. V. (ed.). 1992. *Strength and Power in Sport.* Blackwell, Oxford, UK.

McArdle, W. D., F. I. Katch, and V. L. Katch. 2001. *Exercise Physiology: Energy, Nutrition, and Human Performance*, 5th ed. Lippincott Williams & Wilkins, Philadelphia, PA.

Moreno, J., A. Carlson, and R. V. Alatalo. 1988. Winter energetics of coniferous forest tits Paridae in the north: The implications of body size. *Funct. Ecol.* 2: 163–170.

Mullen, R. K., and R. M. Chew. 1973. Estimating the energy metabolism of free-living *Perognathus formosus*. *Ecology* 54: 633–637.

Müller, U. K., E. J. Stamhuis, and J. J. Videler. 2000. Hydrodynamics of unsteady fish swimming and the effects of body size: Comparing the flow fields of fish larvae and adults. *J. Exp. Biol.* 203: 193–206.

Nagy, K. A., I. A. Girard, and T. K. Brown. 1999. Energetics of free-ranging mammals, reptiles, and birds. *Annu. Rev. Nutr.* 19: 247–277.

Norberg, U. M. 1990. *Vertebrate Flight.* Springer, New York.

O'Dor, R. K., and D. M. Webber. 1991. Invertebrate athletes: Trade-offs between transport efficiency and power density in cephalopod evolution. *J. Exp. Biol.* 160: 93–112.

Parker, K. L., C. T. Robbins, and T. A. Hanley. 1984. Energy expenditures for locomotion by mule deer and elk. *J. Wildl. Manage.* 48: 474–488.

Pedley, T. J. (ed.). 1977. *Scale Effects in Animal Locomotion*. Academic Press, New York.

Peterson, C. C., K. A. Nagy, and J. Diamond. 1990. Sustained metabolic scope. *Proc. Natl. Acad. Sci. U.S.A.* 87: 2324–2328.

Rothe, H.-J., and W. Nachtigall. 1987. Pigeon flight in a wind tunnel. *J. Comp. Physiol. B* 157: 91–98, 99–109.

Rowell, L. B., and J. T. Shepherd. 1996. *Exercise: Regulation and Integration of Multiple Systems.* (Handbook of Physiology [Bethesda, MD], section 12.) Oxford University Press, New York.

Schmidt-Nielsen, K. 1972. Locomotion: Energy cost of swimming, flying, and running. *Science* 177: 222–228.

Schreiber, E. A., and J. Burger (eds.). 2002. *Biology of Marine Birds.* CRC Press, Boca Raton, FL.

Secor, S. M., and J. Diamond. 1997. Determinants of the postfeeding metabolic response in Burmese pythons, *Python molurus*. *Physiol. Zool.* 70: 202–212.

Speakman, J. R. 1997. *Doubly Labelled Water: Theory and Practice.* Chapman and Hall, London.

Suarez, R. K. 1996. Upper limits to mass-specific metabolic rates. *Annu. Rev. Physiol.* 58: 583–605.

Taylor, C. R. 1985. Force development during sustained locomotion: A determinant of gait, speed and metabolic power. *J. Exp. Biol.* 115: 253–262.

Taylor, C. R. 1987. Structural and functional limits to oxidative metabolism: Insights from scaling. *Annu. Rev. Physiol.* 49: 135–146.

Taylor, C. R., S. L. Caldwell, and V. J. Rowntree. 1972. Running up and down hills: Some consequences of size. *Science* 178: 1096–1097.

Taylor, C. R., G. M. O. Maloiy, E. R. Weibel, V. A. Langman, J. M. Z. Kamau, H. J. Seeherman, and N. C. Heglund. 1980. Design of the mammalian respiratory system. III. Scaling maximal aerobic capacity to body mass: Wild and domestic mammals. *Respir. Physiol.* 44: 25–37.

Taylor, C. R., A. Shkolnik, R. Dmi'el, D. Baharav, and A. Borut. 1974. Running in cheetahs, gazelles, and goats: Energy cost and limb configuration. *Amer. J. Physiol.* 227: 848–850.

Townsend, C. R., and P. Calow (eds.). 1981. *Physiological Ecology: An Evolutionary Approach to Resource Use.* Sinauer, Sunderland, MA.

Tucker, V. A. 1975. The energetic cost of moving about. *Amer. Sci.* 63: 413–419.

Tucker, V. A. 2000. The deep fovea, sideways vision and spiral flight paths in raptors. *J. Exp. Biol.* 203: 3745–3754.

Videler, J. J. 1993. *Fish Swimming.* Chapman & Hall, London.

Wagner, P. D. 1996. Determinants of maximal oxygen transport and utilization. *Annu. Rev. Physiol.* 58: 21–50.

Wainwright, P. C., and S. M. Reilly (eds.). 1994. *Ecological Morphology: Integrative Organismal Biology.* University of Chicago Press, Chicago.

Weibel, E. R., C. R. Taylor, and L. Bolis. 1998. *Principles of Animal Design: The Optimization and Symmorphosis Debate.* Cambridge University Press, New York.

Weibel, E. R., C. R. Taylor, and H. Hoppeler. 1991. The concept of symmorphosis: A testable hypothesis of structure-function relationship. *Proc. Natl. Acad. Sci. U.S.A.* 88: 10357–10361.

Wickler, S. J., D. F. Hoyt, E. A. Cogger, and M. H. Hirschbein. 2000. Preferred speed and cost of transport: The effect of incline. *J. Exp. Biol.* 203: 2195–2200.

Woakes, A. J., and W. A. Foster (eds.). 1991. *The Comparative Physiology of Exercise.* Journal of Experimental Biology, vol. 160. Company of Biologists, Cambridge.

Wolf, T. J., P. Schmid-Hempel, C. P. Ellington, and R. D. Stevenson. 1989. Physiological correlates of foraging efforts in honey-bees: Oxygen consumption and nectar load. *Funct. Ecol.* 3: 417–424.

Yousef, M. K., M. E. D. Webster, and O. M. Yousef. 1989. Energy costs of walking in camels, *Camelus dromedarius. Physiol. Zool.* 62: 1080–1088.

See also references in Appendix A.

8 Thermal Relations

As this bumblebee flies from one flower cluster to another to collect nectar and pollen, temperature matters for the bee in two crucial ways. First, the temperature of the bumblebee's flight muscles determines how much power they can generate. The flight muscles must be at least as warm as about 35°C to produce enough power to keep the bee airborne; a bee with cooler flight muscles cannot fly. The second principal way in which temperature matters is that for a bumblebee to maintain its flight muscles at a high enough temperature to fly, the bee must expend food energy to generate heat to warm the muscles. In a warm environment, all the heat required may be produced simply as a by-product of flight. In a cool environment, however, as a bumblebee moves from flower cluster to flower cluster—stopping at each to feed—it must expend energy at an elevated rate even during the intervals when it is not flying, either to keep its flight muscles continually at a high enough temperature to fly or to rewarm them to flight temperature if it cools while feeding. Assuming that the flight muscles must be at 35°C for flight, they must be warmed to 10°C above air temperature if the air is at 25°C, but to 30°C above air temperature if the air is at 5°C. Thus, as the air becomes cooler, a bee must expend food energy at a higher and higher rate to generate heat to warm its flight muscles to flight temperature, meaning it must collect food at a higher and higher rate. Overall, the temperature of a bumblebee's flight muscles has a twofold significance: It determines how much food energy the bee must employ for heat production each day, and it determines how intensely the bee's muscles are able to perform their function of generating lift for flight. Tissue temperature has the same twofold significance in many other animals, including ourselves. Our rate of energy expenditure depends in part on how rapidly we must produce heat to keep our bodies warm, and the temperatures of our tissues help determine how our tissues perform.

Physiologists realize by now that there is great diversity in the types of thermal relations animals maintain with their environments; bumblebees and people exemplify only certain possibilities. To categorize the thermal relations of animals, one key concept is **endothermy**; if an animal's tissues are warmed by its metabolic production of heat, the animal is said to exhibit endothermy. A second key concept is **thermoregulation**, which refers to the maintenance of a relatively constant tissue temperature.[1] Classifying animals according to whether or not they exhibit endothermy and

For a foraging bumblebee, warming the thorax to a high temperature is a critical requirement The process adds to the bee's energy costs and food needs on cool days. However, the flight muscles in the thorax require high temperatures to produce sufficient power for flight. (The photograph is of *Bombus griseocollis*, courtesy of Bernd Heinrich.)

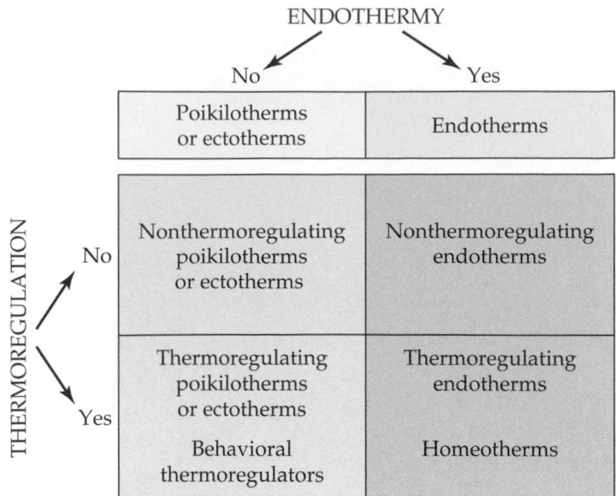

Figure 8.1 Animals fall into four categories of thermal relations based on whether they display endothermy and whether they display thermoregulation

whether or not they display thermoregulation gives us the matrix in Figure 8.1, which identifies four types of thermal relations that animals have with their environments.

Most animals are incapable of endothermy and thus fall on the left side of the matrix; although such animals produce heat metabolically,[2] they do not make heat fast enough or retain it well enough for their tissues to be warmed by their metabolic heat production. Nearly all species of fish, for example, are at essentially the same temperature as the water in which they swim. Animals of this sort are termed **ectotherms** because the thermal conditions outside their bodies determine their body temperatures (*ecto* = "outside"). They are also called **poikilotherms** because they have variable body temperatures (*poikilo* = "variable"); their body temperatures are high in warm environments but low in cool ones. A poikilotherm or ectotherm may or may not exhibit thermoregulation. When it displays thermoregulation and thus falls into the lower left category of our matrix, it does so by behavior; it keeps its tissues at a certain temperature by behaviorally choosing to occupy environments that produce that temperature in its body.

Animals that exhibit endothermy—that is, animals that warm their tissues by their production of metabolic heat—are termed **endotherms** and fall on the right side of the matrix in Figure 8.1. Although endotherms may or may not be thermoregulators, most in fact exhibit thermoregulation (placing them in the lower right category). Mammals and birds are outstanding examples of animals that exhibit both endothermy and thermoregulation. Other sorts of animals also do so. For example, many medium-sized to large insects, such as the bumblebees we have already discussed, exhibit both endothermy and thermoregulation in their flight muscles when flying. A **homeotherm** is an animal that thermoregulates by physiological means (rather than just by behavior). Mammals, birds, and insects like bumblebees are homeotherms. Under many circumstances, the principal way they thermoregulate is by adjusting how rapidly they produce and retain metabolic heat: They thermoregulate by modulating their endothermy!

Additional complexity arises in categorizing animal thermal relations because of the frequent occurrence of temporal and spatial variation. Speaking first of temporal variation, individuals may adopt different thermal relations to their environments at different times. In species of mammals that hibernate, for example, individuals are homeotherms during the seasons of the year when they are not hibernating, but often they exhibit neither endothermy nor thermoregulation when hibernating. Thermal relations may also exhibit spatial variation, differing from one region of an animal's body to another. The abdomens of active insects, for example, are typically neither endothermic nor thermoregulated even in individuals that exhibit endothermy and thermoregulation in their thoracic flight muscles; the abdomen of a foraging bumblebee, for instance, remains always at about the same temperature as the air, whether the bee has landed on flowers to feed or is flying from flower to flower. **Heterothermy** refers to a difference in thermal relations from one time to another or one body region to another within a single individual. Hibernating species of mammals exemplify **temporal heterothermy**. Flying bumblebees illustrate **regional** (i.e., spatial) **heterothermy**.

Temperature is always a major factor in the lives of individual animals, regardless of the particular thermal relations the animals exhibit. Whether animals are poikilotherms or homeotherms, for example, temperature is universally a principal determinant of their metabolic rates and therefore the rates at which they must acquire food. Moreover, tissue temperature is universally important, in poikilotherms as well as homeotherms, in helping to determine the functional properties of tissues and tissue constituents. Enzyme molecules, receptor molecules, and other protein molecules in tissues change their three-dimensional conformations as their temperatures change (see page 205). Tissue temperature thus affects whether protein molecules are in high-performance or low-performance conformations. Tissue temperature also affects the rates of biophysical processes (e.g., diffusion and osmosis), the rates of biochemical reactions, and the viscous physical states of cellular materials such as cell-membrane phospholipids.

The effects of temperature on the properties of whole ecosystems are just as notable as its effects on individual animals. Walking through a temperate woodland during the various seasons of the year provides a striking illustration. On a walk in summer, one witnesses vigorous photosynthesis by plants and sustained activity by mammals, birds, reptiles, insects, amphibians, and other animals. In the winter, however, plants and most animals become cold and quiescent; activity in the woodland becomes restricted largely to the mammals and birds that keep their tissues warm. We cannot always say with certainty whether temperature is the primary determinant of the seasonal changes that we observe in a particular species, because in a whole community of this sort, the responses of any one species may be ripple effects of impacts on others. The evolution of winter quiescence in many animals, for example, has surely been influenced by the effects of cold on plant growth, because the animals

[1] *Thermoregulation* is a specific type of *regulation* as defined in Chapter 1 (see Figure 1.5).

[2] Metabolic heat production is a universal feature of living organisms, as discussed in Chapter 5.

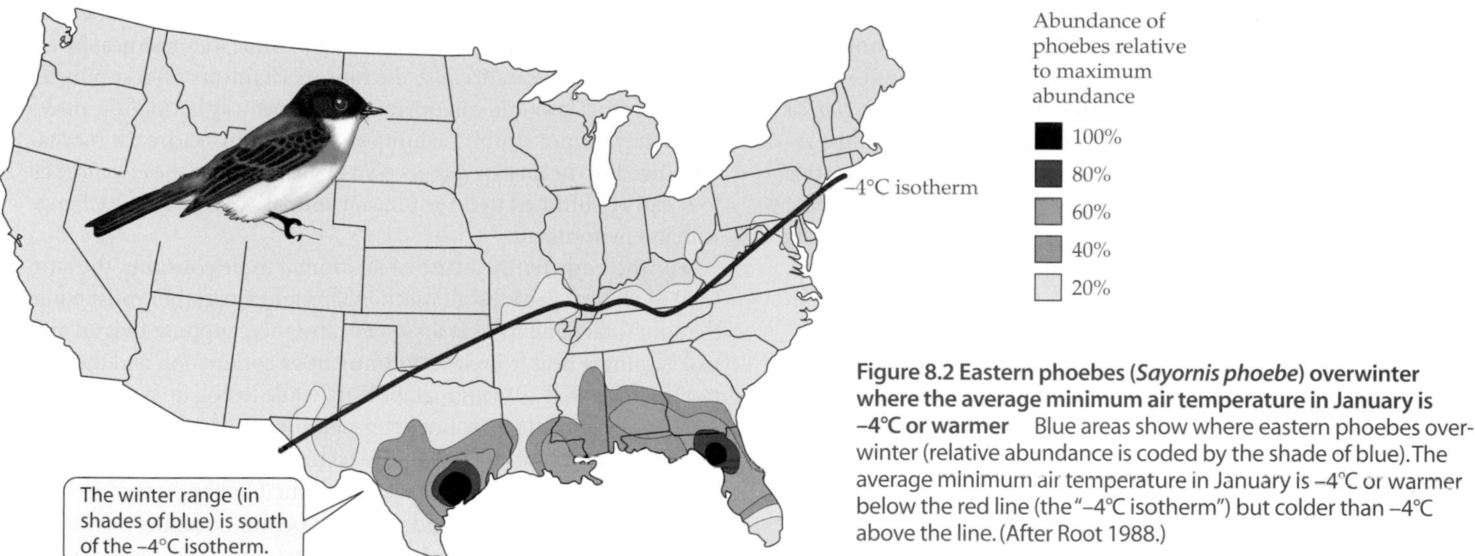

Figure 8.2 Eastern phoebes (*Sayornis phoebe*) overwinter where the average minimum air temperature in January is –4°C or warmer Blue areas show where eastern phoebes overwinter (relative abundance is coded by the shade of blue). The average minimum air temperature in January is –4°C or warmer below the red line (the "–4°C isotherm") but colder than –4°C above the line. (After Root 1988.)

face restricted food resources in winter. One cannot doubt, nonetheless, that much of the change in the animal life of a temperate woodland from summer to winter is a consequence of the seasonal change of temperature.

Biogeographers have built a strong case in recent years that temperature sometimes plays one of the single greatest roles of all environmental factors in determining the geographical ranges where species of animals occur. An example is provided by the winter range of eastern phoebes. In winter, when these birds occur only in the south-central and southeastern United States, the northern limit of their geographical range corresponds closely with a line that connects all the places where the average minimum air temperature is –4°C (**Figure 8.2**). Eastern phoebes in winter do not extend northward to a fixed latitude, mountain range, river, or other geographical limit. Instead, they apparently extend northward to a relatively fixed severity of winter cold stress. Where winter nights average warmer than about –4°C, these birds are to be found. Where winter nights average colder than –4°C, they do not occur. Of course, the data on eastern phoebes (see Figure 8.2) represent just a correlation. One cannot conclude from these data alone that the winter range limit of the phoebes is *caused* by winter temperature. However, many other bird species also exhibit northern limits to their winter ranges that correlate well with particular winter temperatures. Statistically, this evidence taken as a set suggests that winter temperature often plays a particularly important role in determining the winter range limits of species.

One reason that biologists today are acutely interested in the relation between temperature and species ranges is the threat of global warming. To the extent that temperature determines where species live, global warming will lead to changes in species ranges. On a warmed planet, for example, eastern phoebes can be expected to overwinter farther north than they now do. Shifts of this sort might bring species into new combinations, disrupting communities. The need for accurate predictions of global warming effects has become a major new reason for studying animal thermal relations.

Temperature and Heat

The distinction between temperature and heat is tricky, and it is important for understanding the thermal relations of animals. To elucidate the distinction, consider a simple inanimate system: two blocks of copper—one of which is ten times more massive than the other, and both of which have been sitting in a room at 20°C long enough that they are at temperature equilibrium with the room. If you measure the *temperature* of each block, you will find that it is 20°C, even though one block is small and the other is large. Suppose, on the other hand, that you remove and measure the *heat* from each block; suppose, for instance, that you place each block at absolute zero and measure the amount of heat liberated as the block temperature falls from 20°C to absolute zero. You will find that the large block yields ten times more heat than the small one. Thus, as the two blocks sit in the room at 20°C, their *temperatures* are the same and independent of the amount of matter in each block, but their contents of *heat* are different and directly proportional to the amount of matter in each block.

To understand in greater depth these contrasting attributes of temperature and heat, recall from Chapter 3 (see page 67) that the atoms and molecules within any substance undergo constant random motions on an atomic-molecular scale. The **temperature** of a substance is a measure of the *speed*—or *intensity*—of these incessant random motions.[3] In the two copper blocks sitting in the room at 20°C, the average speed of atoms during the random atomic-molecular motions is identical; thus, even though the blocks differ in size, they are the same in temperature. **Heat**, unlike temperature, is a form of energy; it is the energy that a substance possesses by virtue of the random motions of its atomic-molecular constituents (see page 127). The amount of heat in a piece of matter thus depends on the *number of atoms and molecules* in the piece, as well as the *speed* of

[3] Temperature, more specifically, is proportional to the product of molecular mass and the mean square speed of random molecular motions. The speeds of the motions are astounding. In a gas, molecules collide with each other, bounce apart, and then fly through free space until they collide with other molecules. At 20°C, the average speed during each period of free flight is about 500 m/s! The speed is lower at lower temperatures, and higher at higher temperatures.

each atom and molecule. A copper block with many copper atoms moving at a given average speed contains proportionally more heat energy than one with fewer atoms moving at the same speed.

A key property of temperature is that it dictates the *direction* of heat transfer. Heat moves by conduction or convection from a region of high temperature to one of low temperature. To refine this concept, suppose you have a large copper block at 20°C in contact with a tiny copper block at 30°C; although the large block contains more heat than the small one, heat will move from the small block into the large one because temperature, not energy content, dictates the direction of energy transfer. The net addition of heat to any object causes an increase in the temperature of the object. All in all, therefore, *temperature* and *heat* have intimate interactions:

- Heat moves by conduction or convection from high temperature to low.
- The transfer of heat raises the temperature of the object receiving heat and lowers the temperature of the object losing heat.
- In a simple physical system such as two solid objects in contact with each other, objects are at thermal equilibrium when their temperatures are the same because then heat does not tend to move in net fashion between them.

Heat Transfer between Animals and Their Environments

A living animal positioned in an environment, besides making heat internally because of its metabolism, exchanges heat with its surroundings by four distinct heat-transfer mechanisms: *conduction*, *convection*, *evaporation*, and *thermal radiation* (Figure 8.3). The animal may well gain heat by one mechanism of heat transfer while it simultaneously loses heat by another. A familiar illustration of this important point is that on a hot day in summer, people may simultaneously gain heat from the sun by thermal radiation while they lose heat by the evaporation of sweat. Because the four mechanisms of heat transfer follow distinct laws and can operate simultaneously in opposite directions, they cannot simply be lumped together. Instead, each mechanism needs to be analyzed in its own right, and then the effects of all four can be summed to determine an animal's overall heat exchange with its environment.

For the body temperature of an animal to be constant, the sum total of its heat gains by all mechanisms taken together must equal the sum total of all its heat losses. For instance, suppose that an animal is gaining heat from its environment by conduction and thermal radiation, as well as from metabolism, while losing heat by convection and evaporation. Its body temperature will be constant if and only if the sum of its heat gains by conduction, radiation, and metabolism per unit of time is exactly matched by the sum of its heat losses through convection and radiation per unit of time.

Figure 8.4 presents a simple model of an animal that will be useful as we discuss the individual mechanisms of heat exchange. The core of an animal's body is considered to be at a uniform **body temperature**, symbolized T_B. The temperature of the environment is called **ambient temperature, T_A**. The temperature of the body surface often differs from T_B and T_A and thus is distinguished as **surface temperature, T_S**. Separating the body core from the body surface are the outer layers of the body, where temperature gradually changes from T_B on the inside to T_S on the outside.

Conduction and convection: Convection is intrinsically faster

Conduction and convection are usefully discussed together because, in a sense, the two mechanisms of heat transfer define each other. What they have in common is that when heat moves through a material substance by either mechanism, the atoms and molecules of the substance *participate* in the transfer of heat. **Conduction** is the transfer of heat through a material substance that is macroscopically motionless. A familiar example of conduction is the transfer of heat through a block of copper. We know that if the temperature of one side of a copper block is raised, heat will move through the block and appear on the other side even though the copper undergoes no macroscopic motion. The way heat makes its way through such a macroscopically motionless substance is strictly by *atomic-molecular interactions*; if atoms (or molecules) on one side are especially agitated,

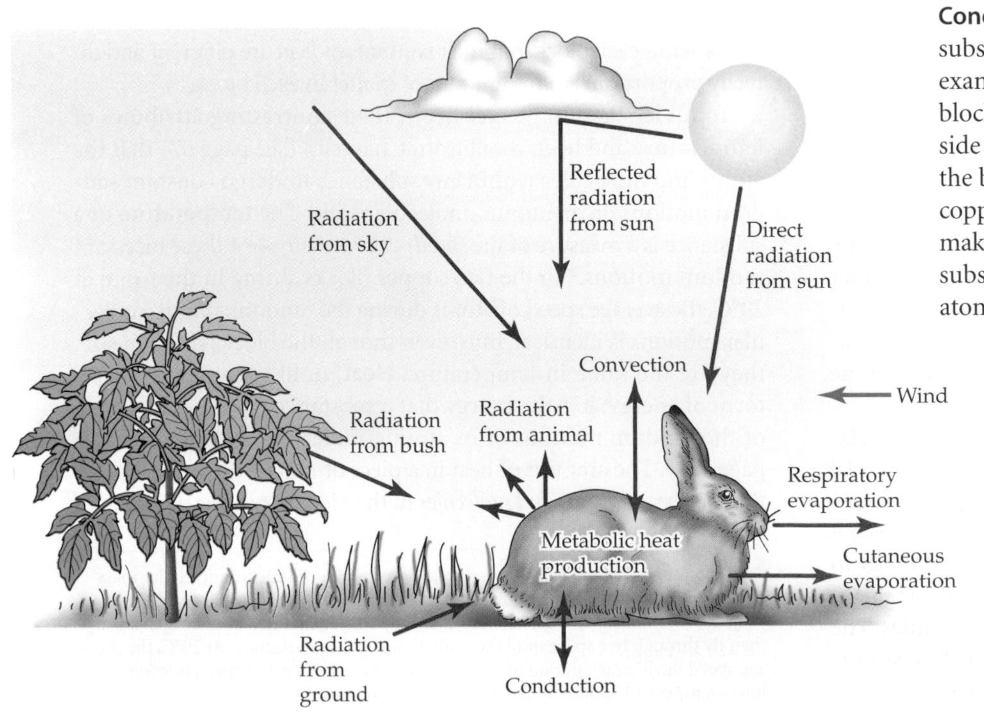

Figure 8.3 An animal exchanges heat with its environment by conduction, convection, evaporation, and thermal radiation The animal exchanges heat conductively with the ground and convectively with the wind. It loses heat by both respiratory and cutaneous evaporation. It receives thermal radiation from all objects in its surroundings and also emits thermal radiation toward all objects. Finally, it gains heat from its own metabolism. (The cutaneous evaporation from a rabbit is entirely of a nonsweating sort because rabbits lack sweat glands.)

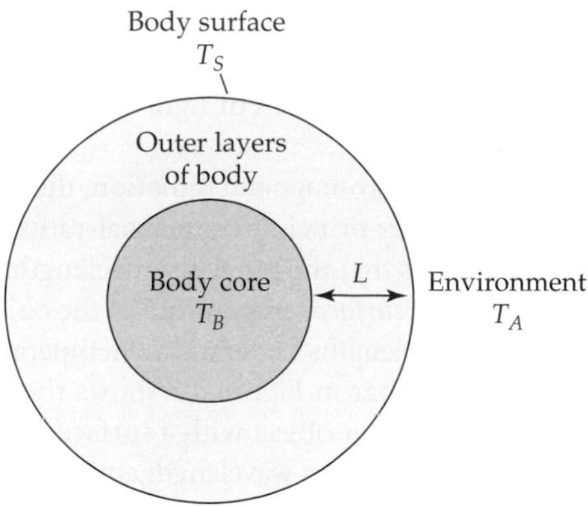

Figure 8.4 A model of an animal's body showing key temperatures The body core is at body temperature T_B, while the surrounding environment is at ambient temperature T_A. The temperature of the body surface is T_S. The outer layers of the body, separating the body core from the surface, have thickness L.

they increase the agitation of atoms farther into the substance by interatomic collisions, and by repetition of this process, successive layers of atoms relay the increased agitation through the whole thickness of the substance. Conduction mechanistically has much in common with simple solute diffusion (the movement of solute through a macroscopically motionless solution; see page 67), and conduction in fact is sometimes called *heat diffusion*.

Convection, in sharp contrast, is transfer of heat through a material substance *by means of macroscopic motion of the substance*. Fluid flow is required for convection. If a wind or water current is present, the macroscopic motion of matter carries heat from place to place. This transfer of heat is convection.

A critical difference between conduction and convection is that, for a given difference of temperature, heat transfer by convection is *much* faster than that by conduction. Consider, for example, a horizontal surface that is 10°C warmer than the surrounding air. If the air is moving at just 10 miles/hour (4.5 meters/second), convection will carry heat away from the surface about 70 times faster than if the air is perfectly still! The acceleration of heat transfer by fluid movement is familiar from everyday experience. We all know, for instance, that a wind greatly increases the thermal stress of a cold day.

THE LAWS OF CONDUCTION We can better understand conduction if we focus on a specific object, a sheet of material of thickness d. If the temperature on one side of the sheet is T_1, that on the other is T_2, and heat is moving through the sheet by conduction, then the rate of heat transfer H from one side to the other per unit of cross-sectional area is

$$H_{\text{conduction}} = k\frac{T_1 - T_2}{d} \qquad (8.1)$$

where k is a constant. The ratio $(T_1 - T_2)/d$ is called the **thermal gradient.** You can see from the equation that the rate of heat transfer by conduction through a sheet of material increases as the temperature difference between the two sides increases. In addition, the rate at which heat moves from one side of the sheet to the other decreases as the thickness of the sheet (d) increases. The coefficient k depends in part on the type of material through which conduction is occurring. Some biologically important materials, such as air, conduct heat poorly; they are said to exhibit low *thermal conductivity* and have low values of k. Other materials, such as water, exhibit higher thermal conductivity and higher k values (water's conductivity is about 20 times that of air).

Heat transfer through the fur of a furred mammal, or through a winter jacket worn by a person, is typically analyzed as a case of conduction because fur traps a layer of relatively motionless air around the body of a furred mammal, and a winter jacket envelops a person's body in a shell of relatively still air. The stillness of the air layer trapped by fur or a jacket is the key to the insulative value of the fur or jacket. To the extent that the air is motionless, heat must move through it by conduction; thus, heat moves much more slowly than if convection were at work. Indeed, from the viewpoint of physics, the benefit of fur or a jacket in a cold environment is that it favors an intrinsically slow mechanism of heat loss from the body, conduction, over an intrinsically more rapid one, convection. In Figure 8.4, the "outer layer" of the body might be taken to represent the fur or jacket. Equation 8.1 shows that increasing the thickness (d) of the motionless air layer trapped by the fur or jacket will tend to slow heat loss from an animal or person to a cold environment.

THE LAWS OF CONVECTION When air or water flows over an object, the rate of heat transfer by convection between the object and the moving fluid depends directly on the difference in temperature between the *surface* of the object and the fluid. Suppose, for instance, that the model animal in Figure 8.4 is exposed to a wind. Then the rate of convective heat transfer between the animal and the air per unit of surface area is calculated as follows:

$$H_{\text{convection}} = h_c(T_S - T_A) \qquad (8.2)$$

The animal will lose heat by convection if its surface temperature (T_S) exceeds the ambient air temperature (T_A); on the other hand, it will gain heat by convection if T_A is higher than T_S. The coefficient h_c, called the **convection coefficient**, depends on many factors, including the wind speed, the shapes of the body parts of the animal, and orientation to the wind. If the shape of a body part is approximately cylindrical (as is often true of the limbs or torso) and the wind is blowing perpendicularly to the cylinder's long axis, then

$$h_c \propto \frac{\sqrt{V}}{\sqrt{D}} \qquad (8.3)$$

where V is the wind speed and D is the diameter of the cylinder.

Equation 8.3 shows that the rate of heat transfer per unit of surface area by convection tends to increase with the square root of the wind speed. The rate of heat transfer per unit of surface area also tends to increase as the square root of the diameter of a cylindrically shaped body part is decreased; this physical law helps explain why body parts of small diameter (e.g., fingers) are particularly susceptible to being cooled in cold environments.

Evaporation: The change of water from liquid to gas carries much heat away

Evaporation of body water from the respiratory passages or skin of an animal takes heat away from the animal's body because water absorbs a substantial amount of heat whenever its physical state changes from a liquid to a gas. The amount of heat required to vaporize water, called the **latent heat of vaporization**, depends on the prevailing temperature and is 2385 to 2490 J (570–595 cal) per gram of H_2O at physiological temperatures. These are *large* values. Whereas heating a gram of liquid water from 0°C to 100°C requires 100 cal, changing a gram from a liquid to a gas requires 570 to 595 cal—almost six times as much. The enormous heat absorption that occurs when water vaporizes means that evaporation can be a highly effective cooling mechanism for an animal. The heat is absorbed from the body surface where the vaporization occurs, and it is carried away with the water vapor.[4]

Thermal radiation permits widely spaced objects to exchange heat

For terrestrial animals, including people, thermal-radiation heat transfer often ranks as one of the quantitatively dominant mechanisms of heat exchange with the environment, yet it tends to be the least understood of all the mechanisms. Although we are all familiar with radiant heating by the sun, such heating is only a special case of a sort of heat transfer that is in fact ubiquitous.

The first fact to recognize in the study of thermal-radiation heat transfer is that all objects *emit* electromagnetic radiation. That is, all objects are *original sources* of electromagnetic radiation. If you look at a wall, your eyes see electromagnetic radiation (light) coming from the wall, but that radiation is merely reflected; it originated from a light bulb or the sun and reflected off the wall to enter your eyes. As a completely separate matter, the wall also is the original source of additional electromagnetic radiation. The radiation *emitted* by the wall is at infrared wavelengths and thus invisible. It travels at the speed of light, essentially unimpeded by the intervening air, until it strikes a solid surface (such as your body), where it is absorbed. Simultaneously, your body emits electromagnetic radiation, some of which strikes the wall. In this way the wall and your body can exchange heat even though they are not touching and in fact may be far apart. Any two objects that are separated only by air undergo exchange of heat at the speed of light by thermal-radiation heat transfer.[5]

When objects emit electromagnetic radiation, they do so over a *range* of wavelengths. A key principle of thermal-radiation physics, illustrated in Figure 8.5, is that the range of wavelengths emitted by an object depends on the *surface* temperature of the object (T_S) and shifts toward shorter wavelengths as the surface temperature increases. The lowest thin black bar in Figure 8.5 shows the wavelengths emitted by an animal or other object with a surface temperature of about 30°C. Note that the shortest wavelength emitted by a surface at this temperature is between 3 and 4 µm; energy is also emitted over a broad range of longer wavelengths. All the emitted wavelengths are in the infrared range and thus invisible. The embers of a fire (middle thin black bar in the figure) emit at shorter wavelengths because they are hotter. They are in fact hot enough that the shortest wavelengths they emit are within the visible range. Because we *see* those wavelengths, we see the coals *glow*. The visible wavelengths emitted by the coals are limited to the red-orange end of the visible spectrum; thus the glow of the coals is red-orange. The sun is so hot that it emits electromagnetic energy (upper thin black bar in the figure) at all wavelengths of the visible spectrum and therefore glows with a nearly white light. The most important concept illustrated by Figure 8.5 is that the radiative emissions from organisms are of the same basic nature as those from a fire or the sun. The only reason we do not see organisms glow is that the wavelengths they emit are out of our visible range.

Another key principle of thermal-radiation physics is that the total *intensity* of radiation emitted by an object—summing the radiation emitted at all wavelengths—increases as surface temperature increases:

$$H_{\text{radiative emission}} = \varepsilon\sigma T_S^{\,4} \qquad (8.4)$$

In this equation, which is known as the **Stefan-Boltzmann equation**, H is the rate of emission per unit of surface area at all wavelengths combined, ε is a surface property called *emissivity* (*emittance*), σ is a constant called the Stefan-Boltzmann constant, and the surface temperature T_S must be expressed in absolute degrees (K).

[4] The opposite of evaporation, *condensation*, usually does not occur in biological systems; but when it does, heat is added to the site where water changes to its liquid form from its gaseous form. See Chapter 25 (page 672) for a detailed discussion of the physical laws of evaporation and condensation.

[5] Water, being far more opaque to infrared radiation than air, largely blocks this sort of heat transfer in aquatic environments.

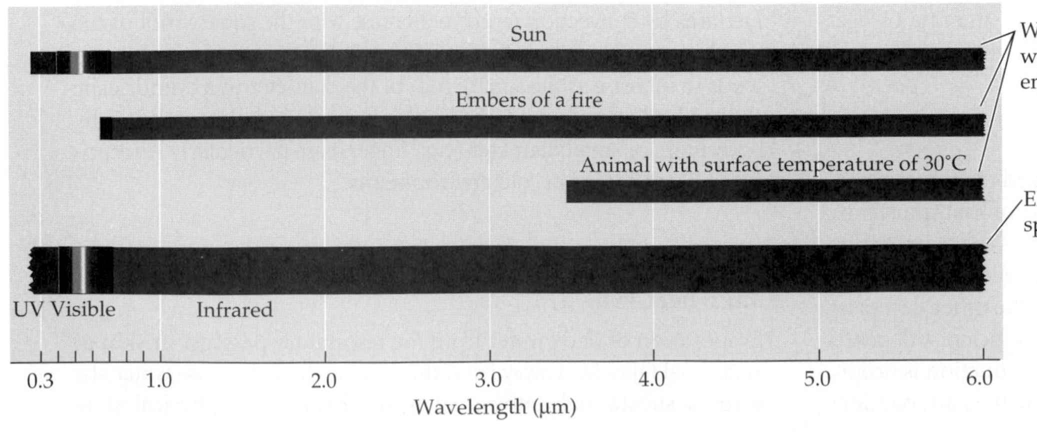

Figure 8.5 As objects reach higher surface temperatures, the ranges of wavelengths at which they emit thermal radiation extend to shorter wavelengths All three of the objects shown also emit energy at wavelengths longer than 6 µm (not shown).

A final key principle to recognize from radiation physics is that when electromagnetic radiation strikes an object, the radiant energy may be absorbed or reflected, or it may pass through. The fractions of the energy absorbed, reflected, and transmitted depend on the surface properties of the object and are wavelength-specific. The portion of the energy that is absorbed is converted into heat at the surface of the absorbing object, as illustrated in everyday experience by the fact that our skin is warmed by the radiant energy from a fire.

RADIANT EXCHANGES IN THE BIOSPHERE THAT DO NOT INVOLVE THE SUN In natural biological communities, the sun is usually the only object that is hot enough to emit energy at wavelengths shorter than 3 to 4 μm. The surface temperatures of animals, plants, rocks, and all other objects besides the sun are typically between –50°C and +50°C, and surfaces at such temperatures emit only wavelengths of 3 to 4 μm or longer (see Figure 8.5). Thus, if we exclude the sun from consideration, all radiant exchanges among objects in the biosphere are at such wavelengths: Various organisms and objects emit at 3 to 4 μm and longer, and the emitted radiation that they receive from other organisms and objects is at 3 to 4 μm and longer. This fact massively simplifies the analysis of radiant exchanges because although organisms and objects in the biosphere commonly differ from one another in surface temperature, all are essentially identical in their other radiative properties at wavelengths of 3 to 4 μm and longer. Specifically, all exhibit about the same value for ε in the Stefan-Boltzmann equation (Equation 8.4) at these wavelengths; and all are highly absorptive at these wavelengths, meaning that they absorb (rather than reflect or transmit) most energy that strikes them. Put loosely, organisms and objects in the biosphere do not differ in color at these wavelengths. If this idea sounds strange, recognize that the color you see with your eyes is a property at *visible* wavelengths of 0.4 to 0.72 μm. Whether the visible color of an organism or object is brown, green, or even white, the color at wavelengths of 3 to 4 μm and longer is, in all cases, nearly black.

Because all organisms and objects in the biosphere are virtually identical in ε and in their absorptive properties at wavelengths of 3 to 4 μm and longer, surface temperature (T_S) is the sole major determinant of radiative heat exchange when the sun is excluded from consideration. If two organisms or objects are exchanging heat radiatively, each can be imagined as emitting a beam of energy toward the other. Whereas the warmer of the two emits a relatively strong beam (see Equation 8.4), the cooler emits a relatively weak beam. Each absorbs most of the energy that it receives from the other. For these reasons, energy is passed in *net* fashion from the warmer object to the cooler one. Quantitatively, if the surface temperatures of the two objects (on the Kelvin scale) are T_1 and T_2, the rate of heat transfer between them is proportional to $(T_1^4 - T_2^4)$.

As examples, consider first a relatively cool lizard standing in the early nighttime hours near a rock that remains hot from the preceding day. The rock emits a relatively strong beam of energy toward the lizard, and the lizard absorbs most of this radiant energy; simultaneously, the lizard emits a weaker beam of energy toward the rock, and the rock absorbs that energy. The net effect is that the lizard is warmed by standing near the rock. A less familiar example is provided by a bird flying past cold trees during a winter night (Figure 8.6). The surface temperature of the bird (+15°C in Figure 8.6) is higher than that of the tree trunks (–10°C). In this case the beam of energy carrying heat away from the bird is more intense than the beam striking the bird from each tree, and the net effect of thermal-radiation heat transfer is to cause a loss of heat from the bird to the trees.

THE NIGHT SKY AS A RADIANT OBJECT The sky is one of the objects in the biosphere that deserves special note. Here we consider just the night sky; because the sun is absent at night, this section is a special case of the last section's discussion. In the atmosphere above us at night, each gas molecule, from just above Earth's surface to the limits of outer space, emits radiation as a function of its temperature. In this way, the surface of Earth is steadily showered with radiation emitted by the sky above. One way to express the intensity of this radiation is to pretend that the sky is a solid surface and ask what the temperature of this surface would have to be for it to emit at the intensity observed (assuming $\varepsilon = 1.0$). This temperature is called the **radiant temperature of the sky**. A characteristic of the radiant temperature of the clear night sky is that it is greatly below the simultaneous air temperature at ground level. For example, during a particular summer night in the Arizona desert when the air temperature near the ground was 30°C, the radiant temperature of the clear sky was simultaneously –3°C; that is, the sky on that warm night behaved like a subfreezing object. The low radiant temperature characteristic of the clear night sky explains how frosts can form on nights when the air temperature at ground level stays above freezing.

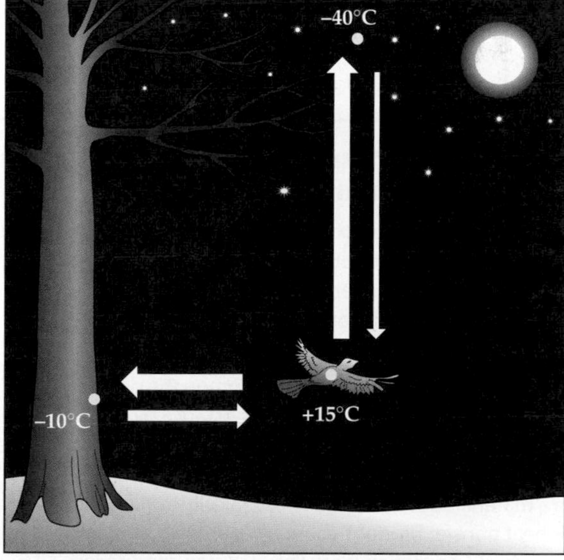

Figure 8.6 A bird loses heat in net fashion to tree trunks by thermal radiation as it flies past them on a cold winter night The bird also loses heat in net fashion to the night sky, as discussed later in the text. Over half of a bird's total heat loss may be by thermal-radiation heat transfer. The temperatures shown for the tree and bird are their surface temperatures; that shown for the sky is the radiant sky temperature expected on a night when the air temperature near the ground is –10°C.

When animals are exposed to the clear night sky, they emit a beam of radiation toward the sky. In return, they receive only a relatively weak beam of radiation from the sky (see Figure 8.6). Thus, animals tend to lose energy in net fashion to the clear night sky, which is often therefore said to act as a "radiant heat sink." The radiative loss of heat to the clear sky is often very considerable—so considerable that for animals already suffering cold stress, there can be substantial advantages to avoiding exposure to the clear sky. When small mammals burrow under the snow in winter, for example, one of the advantages they obtain is that they reduce their radiant heat losses at night by interposing the snow as a barrier to radiant exchange between themselves and the sky; however cold the snow may be, it is warmer than the clear night sky.

SOLAR RADIATION The sun is the one object in the biosphere that routinely emits radiation at wavelengths shorter than 3 to 4 μm (see Figure 8.5). Much of the solar radiant energy is at visible or near-visible wavelengths. The visible color of an object matters in relation to this energy because it affects the fraction of the energy that is absorbed. If an animal's body surfaces are essentially opaque (nontransparent), the analysis of the effects of the animal's visible color on the absorption of the visible and near-visible radiation from the sun is straightforward.[6] Dark surfaces absorb more of this solar radiation—and are heated more by it—than light ones. Black beetles, for instance, absorb the visible and near-visible wavelengths relatively well, whereas light-colored beetles tend more to reflect these wavelengths and absorb them relatively poorly. Animals that can change their skin color, such as many species of lizards, can increase and decrease the solar heating of their bodies by darkening and lightening, respectively.

[6] Additional complexities are introduced if the body surfaces of an animal are significantly transparent, as in the case of mammals and birds in which sunlight can pass through layers of their pelage or plumage.

Summary: Heat Transfer between Animals and Their Environments

- Besides making heat metabolically, animals exchange heat with their environments by conduction, convection, evaporation, and thermal radiation. An animal's body temperature depends on heat gains and losses; it is constant only if the sum total of gains equals the sum total of losses.
- Conduction and convection have in common the property that when heat moves through a material substance by either mechanism, the atoms and molecules of the substance participate in the transfer of heat. Conduction, also called thermal diffusion, occurs when a material substance is macroscopically motionless. Convection is heat transfer brought about by flow of a material substance (e.g., by wind). Convection is much faster than conduction.
- Evaporation is a potentially potent mechanism for heat transfer because the change of state of water from a liquid to a gas absorbs a great deal of heat. The heat is absorbed from the surface where evaporation occurs and is carried away with the water vapor.
- Thermal radiation heat transfer occurs by means of beams of radiant energy that all objects emit and that travel between objects at the speed of light. Because of thermal radiation, objects can exchange heat at a distance. Most thermal-radiation heat transfer in the biosphere occurs at invisible infrared wavelengths; because all objects are nearly black at such wavelengths, visible color plays little role, and the net transfer of heat is from the object with higher surface temperature to the one with lower surface temperature. Visible color, however, is a major factor in how well objects absorb the visible and near-visible wavelengths of solar radiation.

Poikilothermy (Ectothermy)

Poikilothermy is by far the most common type of thermal relation exhibited by animals. Amphibians, most fish, most reptiles, all aquatic invertebrates, and most terrestrial invertebrates are poikilotherms. The defining characteristic of poikilothermy is that the animal's body temperature is determined by equilibration with the thermal conditions of the environment and varies as environmental conditions vary. *Poikilothermy* and *ectothermy* are the same thing. The two terms simply emphasize different aspects of one phenomenon; whereas *poikilothermy* emphasizes the variability of body temperature, *ectothermy* emphasizes the fact that outside conditions determine the body temperature (see page 192). Poikilothermy manifests itself differently depending on whether the animal is aquatic or terrestrial. Aquatic poikilotherms typically have body temperatures that are essentially the same as water temperature. Terrestrial poikilotherms, however, do not necessarily have body temperatures that equal "air" temperature, because thermal-radiation heat transfer or evaporation on land can tend to draw the body temperature away from air temperature. For instance, if a frog or snail on land basks in the sun, its body temperature may be much higher than the air temperature. The animal nonetheless still meets the definition of poikilothermy or ectothermy, because its body temperature is simply a consequence of equilibration with the thermal conditions in its environment.

Poikilothermic or ectothermic animals are often called *cold-blooded* in nonscientific writing, in reference to their coolness to the touch under certain conditions. Many species, however, may have high body temperatures when in warm environments. For example, desert lizards and insects that are perfectly fine poikilotherms often have body temperatures that substantially exceed human body temperature! *Cold-blooded* is therefore not a suitable general term to describe poikilotherms or ectotherms.

Poikilotherms often exert behavioral control over their body temperatures

The natural environments of poikilotherms typically vary from place to place in thermal conditions. In a forest, for example, the temperature on the exposed forest floor might be higher than that under a log, and the temperature in a spot of sunlight might be higher yet. Poikilotherms in the wild can behaviorally choose where they position themselves and, in this way, control their body temperatures. One reason to stress this point near the start of our discussion of poikilothermy is that the situation of captive poikilotherms is so different and potentially misleading. In an aquarium or cage in a home or laboratory, the thermal conditions are often uniform, meaning that the body temperatures of captive poikilotherms are deterministically set by the environmental conditions. For example, if we have a fish in an aquarium and we set the water temperature to be 17°C, the body temperature of the fish is also set to be 17°C. From experiences like this, we can get the impression that poikilotherms are simply at the mercy of external conditions and have no options. The impression is really a misimpression arising from the artificial conditions, however. When the thermal environment is heterogeneous—as it nearly always is in nature—poikilotherms typically can and do exert at least some control over their body temperatures.

If a poikilotherm behaviorally maintains a relatively *constant* body temperature, it is said to exhibit **behavioral thermoregulation**. Sometimes behavioral thermoregulation is rather simple. In a lake or the ocean, for instance, various large water masses (such as those at the surface and at greater depth) often differ in temperature (see Figure 1.13). Fish that elect to stay in one water mass, rather than another, take on the temperature of the water they occupy and remain at that temperature for extended periods. The behavior of the fish is accordingly a simple form of behavioral thermoregulation.

In other cases, behavioral thermoregulation is far more complex and dynamic. Many lizards, for example, maintain relatively stable body temperatures during daylight hours (Figure 8.7), and they do so by complex, moment-to-moment behavioral exploitation of environmental opportunities for heating and cooling. A desert lizard, for instance, ordinarily emerges in the morning and basks in the sun until its body temperature rises to be within a "preferred" range that it maintains during its daily activity. Thereafter, the lizard keeps its body temperature within this range until nightfall by a variety of mechanisms. One common strategy is to shuttle back and forth between sun and shade; when its body temperature starts to drop too low, the lizard moves into sunlight, and then later, when its body temperature starts to rise too high, it enters a shady area. The lizard might also modify the amount of its body surface exposed to the direct rays of the sun by changing its posture and orientation to the sun. It might flatten itself against the substrate to lose or gain heat (depending on substrate temperature), and when the substrate has become very hot during midday, it might minimize contact by elevating its body off the ground or even climbing on bushes. By thus exploiting the numerous opportunities for heating and cooling in its thermally heterogeneous environment, a lizard may well maintain a body temperature that varies only modestly for long periods. The desert iguana, illustrated in Figure 1.11, for instance, typically maintains an average abdominal temperature of 38°C to 42°C during daylight hours, and it often keeps its temperature within 2°C to 3°C of the mean for hours on end. The species of lizard in Figure 8.7 is an equally accurate thermoregulator, although it has a lower preferred temperature.

Poikilotherms must be able to function over a range of body temperatures

A limitation of behavioral thermoregulation is that it is dependent on the thermal opportunities available in the environment, and thus it may be thwarted by changes of weather or other conditions outside an animal's control. A desert iguana, for example, may never reach a body temperature that is even close to its "preferred" level of 38°C to 42°C on a day that happens to be cloudy and cool. Similarly, a fish that would select a cool water mass if it could, cannot do so if all the water in its lake or pond is warm.

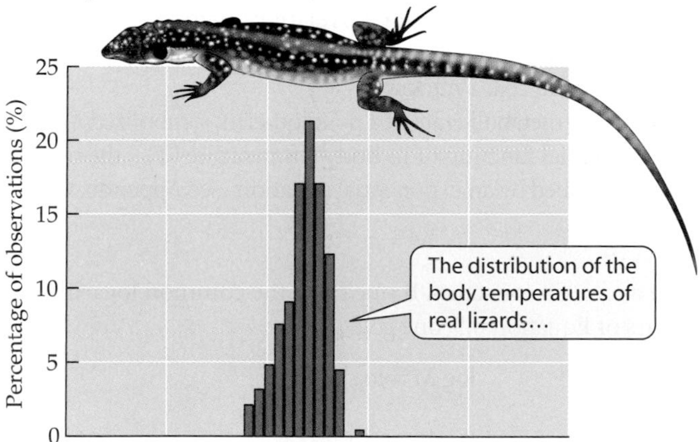

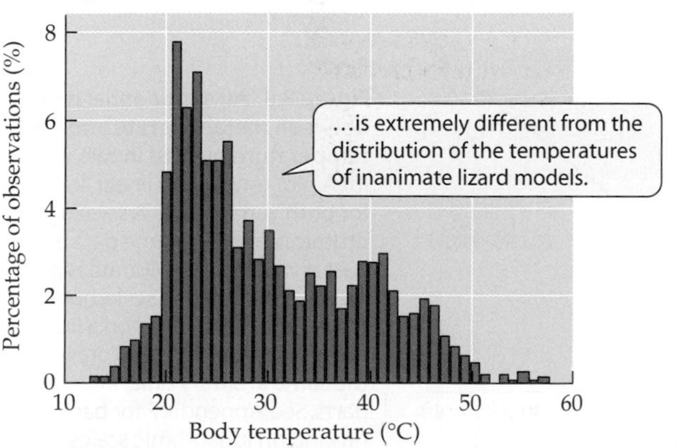

Figure 8.7 Behavioral thermoregulation revealed by comparison of living lizards with inanimate lizard models Multiple daytime measurements of the body temperatures of real lizards *(a)* and inanimate lizard models *(b)* were made. The lizards (*Podarcis hispanica*) were living freely on a Mediterranean island. The lizard models were placed as comprehensively as possible in all the various microhabitats available to real lizards during their daytime activities on the same island. The fact that the distribution of the body temperatures of the real lizards differs from that of lizard models shows that during their daytime activities, real lizards do not simply position themselves at random, but behave in ways that keep their body temperatures within a relatively narrow preferred range. Data on the *y* axis are the percentages of all observations in various 1°C intervals of temperature. Lizards and lizard models have identical temperatures when constrained to one location, side-by-side. Unambiguous demonstration of behavioral thermoregulation actually involves even more considerations than mentioned here, as discussed in the original research report. (After Bauwens, Hertz, and Castilla 1996.)

For these and other reasons, poikilotherms must typically be thermal generalists: They must be capable of functioning at a variety of different body temperatures. Species differ in how wide a range of body temperatures is acceptable. Some species, termed **eurythermal**, can function over quite wide ranges of body temperature; goldfish, for instance, maintain normal body orientation, feed, and swim about at body temperatures from 5°C to 30°C. Other poikilotherms, termed **stenothermal**, have comparatively narrow ranges of body temperatures over which they can function.

Poikilotherms respond physiologically to their environments in all three major time frames

The three major time frames of physiological response to the environment identified in Chapter 1 (see Table 1.2) provide a useful way to organize knowledge of the relations of poikilotherms to their thermal environments because poikilotherms respond to their environments in all three time frames. The next three sections discuss poikilotherms in the context of each of the three time frames. First, in the section coming up, we address the *acute responses* of poikilotherms to changes in their body temperatures. The acute responses are those that individual animals exhibit *promptly* after their body temperatures are altered. Next we address the *chronic responses* of poikilotherms, termed *acclimation* and *acclimatization*: What changes do individual animals undergo when they live in altered thermal environments, and therefore have altered body temperatures, for prolonged periods? Finally, in the third of the next three sections, we discuss *evolutionary changes*, the ways in which the physiology of poikilotherms may be modified by changes in the frequencies of genes when populations live in different environments over many generations.

Acute responses: Metabolic rate is an approximately exponential function of body temperature

When the body temperature of an individual poikilotherm is raised in a series of steps and its metabolic rate is measured promptly after each upward step, the usual pattern is that the resting metabolic rate increases approximately exponentially with the animal's body temperature (**Figure 8.8*a***). An *exponential* relation signifies that the metabolic rate increases by a particular *multiplicative factor* each time the body temperature is stepped up by a particular *additive increment* (see Appendix G: Exponential and Allometric Equations). For example, the metabolic rate might increase by a factor of 2 for each increment of 10°C in body temperature. Then, if the metabolic rate were 1 J/minute at 0°C, it would be 2 J/minute at 10°C, 4 J/minute at 20°C, and 8 J/minute at 30°C. The acute relation between metabolic rate and body temperature is usually, in fact, only *approximately* exponential. That is, the factor by which the metabolic rate increases for a given increment in temperature is usually not precisely constant from one temperature range to the next but might, for example, be 2.5 between 0°C and 10°C but only 1.8 between 20°C and 30°C.

The reason that the metabolic rate of a poikilotherm increases as its body temperature goes up relates back to the concept of activation energy discussed in Chapter 2 (see page 43). Each biochemical reaction involved in metabolism is characterized by a particular activation energy, a certain minimum energy level that a reacting molecule must attain in order to undergo the reaction (see Figure 2.13). As the temperature of a cell increases, all molecules in the cell tend to become more agitated and have higher energy levels. Svante Arrhenius demonstrated in the late nineteenth century that if one specifies any particular activation energy, the fraction of molecules that have that level of energy or more at any moment increases approximately exponentially as temperature increases. Reactions tend therefore to speed up approximately exponentially as cellular temperature rises. In this context, it is vital to recall that most metabolic reactions are enzyme catalyzed, and the enzymes determine the activation energies. Thus the detailed, quantitative relations between biochemical reaction rates and temperature in cells depend on the particular enzyme proteins that cells synthesize.

If the resting metabolic rate of a poikilotherm, symbolized M, were a true exponential function of its body temperature (T_B), the relation would be described by an exponential equation (see Appendix G):

$$M = a \cdot 10^{n \cdot T_B} \tag{8.5}$$

where a and n are constants. If one takes the common logarithm of both sides of Equation 8.5, one gets

$$\log M = \log a + n \cdot T_B \tag{8.6}$$

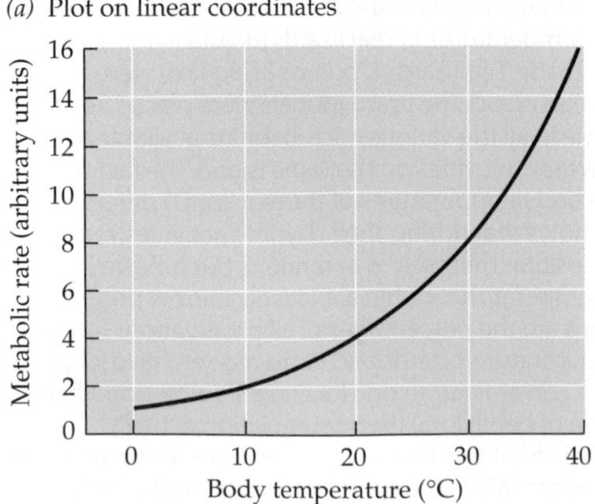

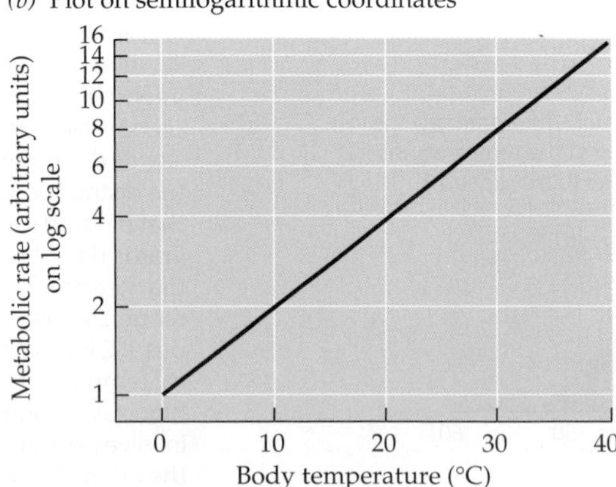

Figure 8.8 An exponential relation between metabolic rate and body temperature plotted in two ways *(a)* A plot employing linear scales for both variables. *(b)* A semilogarithmic plot of the same data as in part *a*; while metabolic rate is plotted on a logarithmic scale, body temperature is plotted on a linear scale. Metabolic rate is expressed in the same arbitrary units in both parts. See Appendix F for background on logarithmic scales.

According to this second equation, log M is a *linear* function of T_B (log a and n are constants).

We see that if M is an exponential function of T_B as in Equation 8.5, log M is a linear function of T_B (Equation 8.6). This result represents the basic reason that physiologists usually plot metabolism–temperature data for poikilotherms on semilogarithmic coordinates. The logarithm of the animal's metabolic rate is plotted on the y axis, and the animal's body temperature itself is plotted on the x axis. The curve of Figure 8.8*a* is replotted on semilogarithmic coordinates in Figure 8.8*b*, illustrating the "linearizing" effect of semilogarithmic coordinates. A similar comparison is provided in Figure 8.9 using data on actual animals. As we have emphasized, metabolic rate in fact is usually an *approximately* exponential function of body temperature, not a truly exponential one. Thus the semilogarithmic plot for actual animals is typically not precisely linear, as exemplified in Figure 8.9*b*.

One simple way to describe an exponential relation between metabolic rate (or any other rate) and temperature is to specify the multiplicative factor by which the rate increases when the body temperature is increased by a standardized increment of 10°C. This factor is called the **temperature coefficient, Q_{10}**:

$$Q_{10} = \frac{R_T}{R_{(T-10)}} \tag{8.7}$$

where R_T is the rate at any given body temperature T, and $R_{(T-10)}$ is the rate at a body temperature 10°C lower than T. To illustrate, if the resting metabolic rate of an animal is 2.2 J/minute at a body temperature of 25°C and 1.0 J/minute at 15°C, the Q_{10} is 2.2. As a rough rule of thumb, the Q_{10} for the metabolic rates of poikilotherms is usually between 2 and 3. If metabolic rate were a truly exponential function of body temperature, you could calculate the Q_{10} of an animal from data for *any* two body temperatures that are 10°C apart and always get the same value. Because metabolic rate is not a truly exponential function of temperature, however, the Q_{10} of an animal in fact varies with the particular range of body temperatures considered.[7]

As you might guess, the heart rates and breathing rates of poikilotherms typically increase as body temperature is raised, up to a point. Behavioral rates are also affected by body temperature; frogs, insects, and fish, for instance, often move more rapidly at high than at low body temperatures, again up to a point.

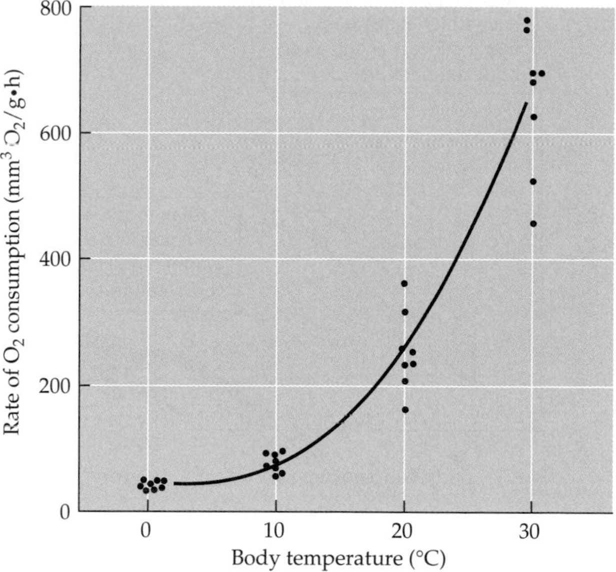

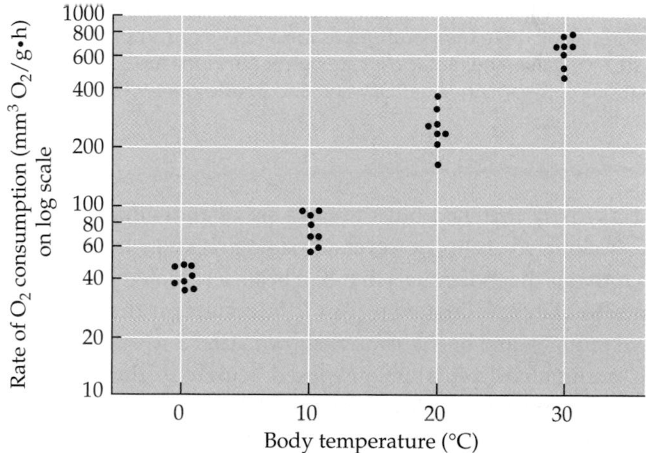

Figure 8.9 The relation between metabolic rate and body temperature in tiger moth caterpillars (family Arctiidae), plotted in two ways The metabolic rate was measured as the rate of O_2 consumption. *(a)* A plot employing linear scales for both variables. *(b)* A semilogarithmic plot. See Appendix F for background on logarithmic scales. (After Scholander et al. 1953.)

Chronic responses: Acclimation often blunts metabolic responses to temperature

When an individual poikilotherm is kept chronically at one body temperature for a number of weeks and then is kept chronically at a different body temperature for a number of weeks, the details of its acute metabolism–temperature relation usually change. Such a change is an example of *acclimation* (see page 14). Understanding this sort of acclimation and its implications can be tricky. The best way to gain clear insight is to discuss the actual procedures that are followed to study acclimation. To this end, let's discuss the acclimation study in Figure 8.10.

In the experiment represented by Figure 8.10, a group of lizards, named the "33°C-acclimated" group, was maintained for 5 weeks at 33°C. At the end of this chronic exposure to 33°C, the lizards were exposed acutely (i.e., briefly) to three different body temperatures—16°C, 28°C, and 33°C—and their resting metabolic rates were measured at each of the three. The line labeled "33°C-acclimated" shows the results. It represents the *acute* relation between resting metabolic rate and body temperature for lizards that were living *chronically* at 33°C during the weeks before the measurements were made.

Another group of lizards, named the "16°C-acclimated" group, was maintained for 5 weeks at 16°C. These 16°C-acclimated lizards

---

[7] An equation called the *van't Hoff equation* permits Q_{10} to be calculated for ranges of temperature that are less than 10°C wide. Thus, for detailed analysis the temperature scale need not be artificially divided into 10°C segments.

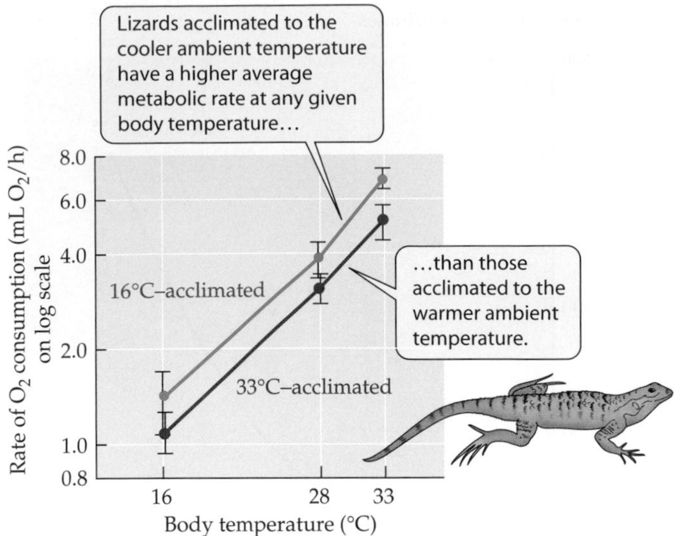

Figure 8.10 Acclimation of metabolic rate to temperature in a poikilotherm One group of fence lizards (*Sceloporus occidentalis*) was acclimated for 5 weeks to 33°C prior to testing. A second, otherwise identical group was acclimated to 16°C for 5 weeks prior to testing. After the 5 weeks, the animals in each group were placed briefly at body temperatures of 16°C, 28°C, and 33°C, and their standard metabolic rates were measured at all three temperatures as rates of O_2 consumption. The circles show the average metabolic rates; error bars indicate ±2 standard deviations of the mean. (After Dawson and Bartholomew 1956.)

were a closely matched but different set of individuals from the 33°C-acclimated group; however, physiologists know from other research that if the individuals that had been acclimated to 33°C were themselves later acclimated to 16°C, the results for the 16°C-acclimated group would be the same as shown. After 5 weeks at 16°C, the 16°C-acclimated lizards were exposed acutely to the same three study temperatures employed for the 33°C-acclimated group, and their metabolic rates were measured. The line labeled "16°C-acclimated" in Figure 8.10 shows the results and thus represents the acute relation between resting metabolic rate and body temperature for lizards that were living chronically at 16°C.

As Figure 8.10 shows, the acute metabolism–temperature relation is altered when lizards have been living chronically at 16°C rather than 33°C. Lizards acclimated to the cooler ambient temperature, 16°C, have a higher average metabolic rate at any given body temperature than those acclimated to the warmer ambient temperature, 33°C. Although this specific sort of change during temperature acclimation is not universal, it is the most common type of acclimation response in poikilotherms and has been observed in well over half the species studied.

What is the *significance* of this acclimation response? One way to understand the significance is provided by Figure 8.11*a*. As a thought exercise, imagine that we have some lizards that are at 33°C and have been living at that temperature for 5 weeks. The average metabolic rate of these lizards—that is, the metabolic rate of 33°C-acclimated lizards at 33°C—is marked *x* in Figure 8.11*a*. Imagine now that we suddenly lower the temperature of these lizards to 16°C and leave the lizards at 16°C for 5 weeks. The key question we need to address is, How will their average metabolic rate change from the moment their temperature is lowered? First, what will happen during the *first hour*? In other words, what is the *acute* (prompt) response of the lizards to the change of their temperature? As the animals cool from 33°C to 16°C during the first hour, their average metabolic rate will decline along the acute-response line for 33°C-acclimated animals, following the thin arrows from *x* to *y*. Immediately after the lizards have cooled fully to 16°C, their average metabolic rate will be *y*, the metabolic rate of 33°C-acclimated lizards at 16°C. Note that the drop of body temperature causes a profound fall in metabolic rate. Now the most critical question to understand is, *What will happen to the average metabolic rate of the lizards during the following 5 weeks at 16°C?* The answer is that it will *rise* from *y* to *z* because during those 5 weeks the lizards will become 16°C-acclimated animals; at the end of the 5 weeks, they will have the metabolic rate of 16°C-acclimated animals at 16°C (*z*). Acclimation in these lizards thus reduces—blunts—the effect of the change of their body temperature. Although cooling to 16°C initially lowers the lizards' average metabolic rate by a profound amount, the metabolic rate is lowered to a lesser extent after acclimation has occurred. Put another way, acclimation tends to return the metabolic rate toward its level prior to the drop in body temperature.

After a physiological rate has been raised or lowered by an abrupt change in body temperature, any subsequent, long-term tendency for it to return toward its original level even though the new temperature continues is called **compensation**. The rise from *y* to *z* in Figure 8.11*a* is an example of compensation. Compensation is **partial** if the rate returns only partially to its original level, as in Figure 8.11*a*. If a rate were to return all the way to its original level, as in Figure 8.11*b*, the compensation would be **complete**. When compensation occurs, it is nearly always partial.

An alternative way to understand the *significance* of the sort of acclimation response we have been discussing is presented in Figure 8.12. Fish of a particular species were acclimated to 10°C, 20°C, and 30°C by being kept at the three temperatures for several weeks. The 30°C-acclimated fish were then tested acutely at all three temperatures, resulting in the lowermost solid line in Figure 8.12. Similarly, the 20°C- and 10°C-acclimated fish were tested acutely at the three temperatures. Note that each of the three solid lines is an *acute-response line*: Each shows how the metabolic rate of an acclimation group varies when it is measured promptly after changes of body temperature. Now let's construct the *chronic-response line* for this species: The chronic-response line will show how metabolic rate varies with temperature when the fish are permitted to live at each temperature for weeks before their metabolic rate is measured. The three bold circles are the metabolic rates of the fish when living chronically at the three temperatures. For instance, the bold circle at the left is the metabolic rate at 10°C of fish that have been living at 10°C for several weeks (10°C-acclimated fish), and the bold circle at the right is the metabolic rate at 30°C of fish that have been living at 30°C. We obtain the chronic-response line by connecting the three bold circles. The chronic-response line has a shallower slope than *any* of the acute-response lines. In words, this means that *if the fish are allowed to acclimate to each temperature before their metabolic rate is measured, their metabolic rate is less affected by changes of body temperature than if they are shifted rapidly from one temperature to another.*

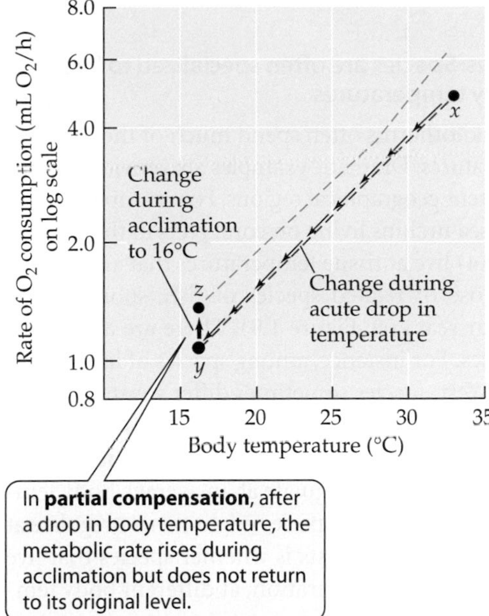

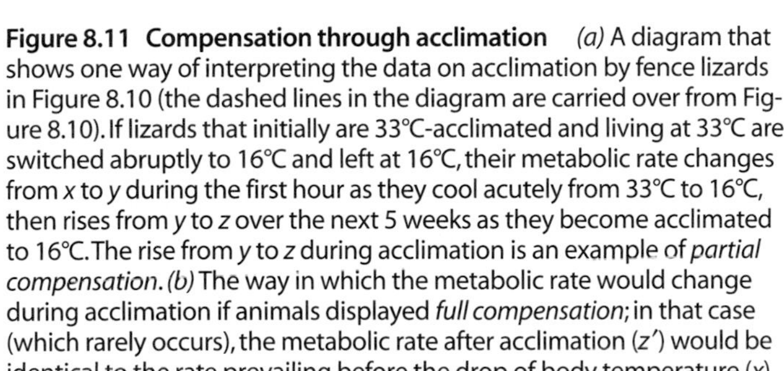

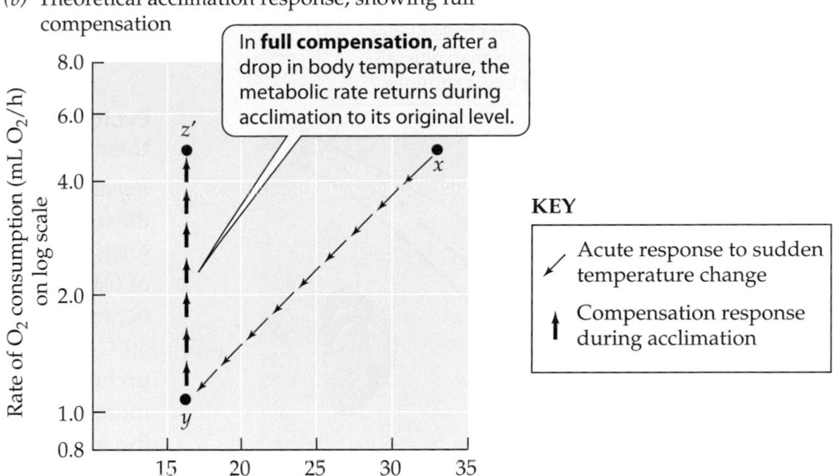

Figure 8.11 Compensation through acclimation *(a)* A diagram that shows one way of interpreting the data on acclimation by fence lizards in Figure 8.10 (the dashed lines in the diagram are carried over from Figure 8.10). If lizards that initially are 33°C-acclimated and living at 33°C are switched abruptly to 16°C and left at 16°C, their metabolic rate changes from *x* to *y* during the first hour as they cool acutely from 33°C to 16°C, then rises from *y* to *z* over the next 5 weeks as they become acclimated to 16°C. The rise from *y* to *z* during acclimation is an example of *partial compensation*. *(b)* The way in which the metabolic rate would change during acclimation if animals displayed *full compensation*; in that case (which rarely occurs), the metabolic rate after acclimation (*z′*) would be identical to the rate prevailing before the drop of body temperature (*x*).

In terms of the biochemistry and molecular biology of metabolism, many responses can occur during acclimation. The most common and crucial response is that cells modify their *amounts* of key, rate-limiting enzymes, notably enzymes of the Krebs cycle and the electron-transport chain. During acclimation to cold temperatures, greater amounts of these enzymes are synthesized.[8] The accumulation of the enzymes to higher concentrations in cells requires time; this is why the acclimation response is not observed immediately after a drop in temperature but requires a more extended length of time to be consummated. As the amounts of key, rate-limiting enzymes increase in cells, the presence of the increased enzymes tends to speed metabolic reactions, accounting for the compensation observed (see Figure 8.11*a*). During acclimation to warm temperatures, enzyme amounts are reduced. Thus, although a shift to a higher body temperature initially speeds an animal's metabolism dramatically, the metabolic rate tends to slow as acclimation occurs (another manifestation of compensation).

[8] Chapter 2 reviews the effects of enzyme concentration and the processes by which cells modify it.

These acclimation responses provide an outstanding example of how animals can modify their own biochemistry in potentially adaptive ways. In the study of biochemistry, it is sometimes easy to get the impression that cells are simply like miniature reaction vessels in which test-tube reactions take place. However, because most reactions must be catalyzed by enzymes to occur and the cells make the enzymes, cells in fact orchestrate their own biochemistry.

A classic example of *acclimatization* by poikilotherms living in their natural habitats is presented in Figure 8.13.[9] Mussels of a single species were collected at three latitudes along the West Coast of the United States. Each group was acutely exposed to three or four test temperatures, and the rate at which the animals pumped water

[9] The distinction between *acclimation* and *acclimatization* is discussed in Chapter 1 (see page 14).

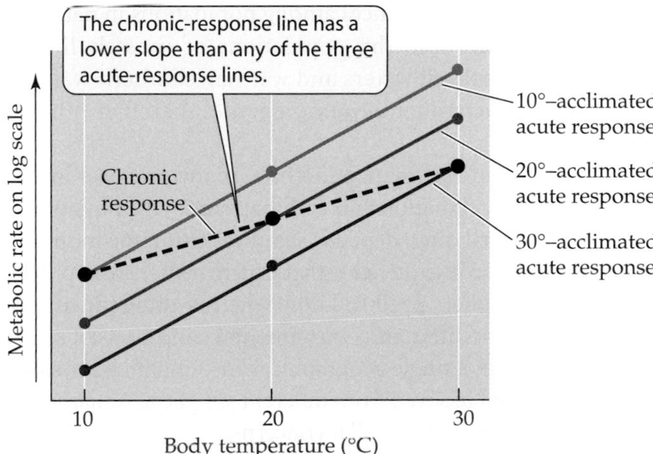

Figure 8.12 Because of acclimation, the chronic metabolism–temperature curve is relatively flat compared to the acute metabolism–temperature curves The three solid lines show the acute relations between metabolic rate and body temperature for hypothetical fish when 10°C-, 20°C-, and 30°C-acclimated. The dashed line shows the relation between metabolic rate and body temperature when the fish live chronically at each temperature.

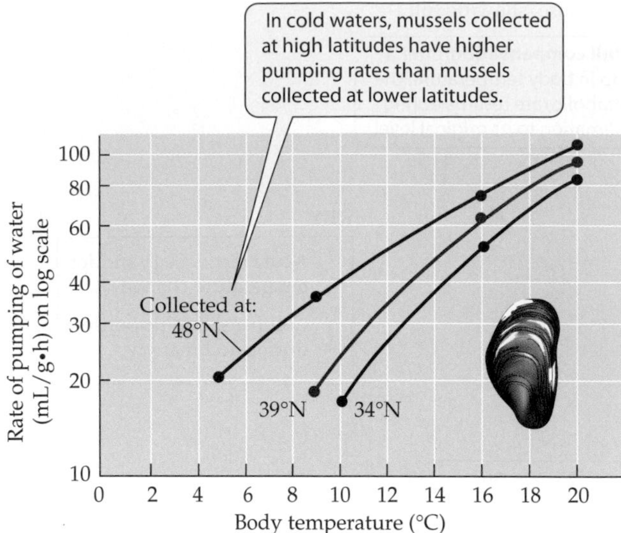

Figure 8.13 Acclimatization in mussels Mussels are extremely important members of intertidal and subtidal marine communities. They pump water through their bodies by ciliary action. Being suspension feeders (see Chapter 4, page 103), they acquire food as well as O_2 from the water they pump. Mussels of the most abundant West Coast species (*Mytilus californianus*) were collected from nature at the three latitudes indicated. Soon after collection, each group was exposed to three or four body temperatures, and its rate of pumping was measured. The three lines are acute-response lines for three groups of mussels acclimatized to different latitudes. (After Bullock 1955.)

across their gills was measured. Because the three groups are thought to have been genetically similar, the differences among their acute-response curves represent acclimatization. Mussels from relatively cold, high-latitude waters pump water across their gills faster at any given test temperature than mussels from warmer, lower-latitude waters do. The acclimatization observed thus compensates for temperature differences among populations in nature. Because of acclimatization, the populations of mussels living in relatively cold, northern waters and warmer, southern waters are more similar to each other in pumping rates than they otherwise would be.

Acclimation and acclimatization have many practical implications. Because of acclimation and acclimatization, the physiology of an individual animal often depends significantly on the individual's recent *history*. One consequence is that experimenters who purchase animals from suppliers need to know whether different suppliers keep animals under different environmental conditions; if so, even animals collected in a single geographical area might be physiologically different when received from different suppliers. Aquarium and aquaculture managers face similar concerns.

As the study of thermal acclimation and acclimatization in poikilotherms moves into the twenty-first century, a question receiving ever-increasing emphasis is whether the acclimation and acclimatization responses are always beneficial for the animals. Evolutionary physiologists interested in this issue stress that a response cannot be assumed to be beneficial just because it occurs (see page 24). The weight of evidence at present is that the responses observed are usually advantageous to the animals, but ever more-sophisticated insight can be expected in the decades ahead.

Evolutionary changes: Species are often specialized to live at their respective body temperatures

Related species of poikilotherms often spend much of their time at different body temperatures. Dramatic examples are provided by animals that live in different geographical regions. For example, species of fish, sponges, and sea urchins living on coral reefs in the tropical oceans (see Figure 1.14) live at tissue temperatures that are 25°C to 30°C higher than those of related species of fish, sponges, and urchins living in polar seas (see Figure 1.9). There are also many more-subtle differences. For instance, among species of lizards that live in the American West, species sometimes differ substantially in the behaviorally regulated "preferred" body temperatures they maintain during the daylight hours of each day. Whereas one species might employ behavior to thermoregulate at an average body temperature of 34°C, another coexisting species might thermoregulate at 40°C. A key question for physiologists is whether species that live and reproduce, generation after generation, at different body temperatures have evolved adaptations to their respective temperatures.

Some physiological differences between species living at different body temperatures are so dramatic that there can be no doubt about the existence of evolved, adaptive specializations. For example, many Antarctic species of fish thrive at temperatures near freezing and die of heat stress when warmed to above 4°C to 6°C. Meanwhile, many tropical species of fish thrive at tropical temperatures and die of cold stress if cooled to below 4°C to 6°C. One can hardly doubt that these Antarctic and tropical species have evolved adaptive specializations to their respective body temperatures.

Most differences between species are not so categorical and are more challenging to interpret. One problem in the study of evolutionary adaptation to temperature is that the vast majority of animal species cannot be bred in captivity, and therefore, individuals must be collected from nature for study. When biologists try to interpret data gathered on wild-caught adults, they must always worry that differences may exist between sets of animals of different species—not because the species differ genetically, but because the study animals underwent their early development under different conditions in their respective natural habitats. Another problem in the study of evolutionary adaptation is sometimes that species from thermally different environments are unable to live successfully at a single temperature; in such cases, biologists cannot do the "obvious" experiment of comparing species in a single laboratory environment. Because of these concerns and others, physiologists working on evolutionary differences among species confront particularly challenging methodological issues, and the development of improved methods is an active area of contemporary research.

LIZARDS WITH DIFFERENT PREFERRED BODY TEMPERATURES Of what advantage is thermoregulation? A plausible hypothesis is that when a species thermoregulates, its tissues and cells can improve their performance by becoming thermally specialized to function at the body temperatures maintained. In the complete absence of thermoregulation, tissues are equally likely to be at almost any temperature, and therefore specialization to function at particular temperatures might be disadvantageous. Without specialization, tissues might be "jacks-of-all-trades, masters of none." On the other hand, if thermoregulation occurs and tissue temperature is thereby maintained for substantial periods of time in a narrow range, a tissue might profit by becoming specialized (over evolutionary time) to function at temperatures in that range.

Species of lizards with different preferred body temperatures provide excellent material to test the hypothesis that tissues become specialized to the body temperatures maintained by thermoregulation. If the hypothesis is correct, species with relatively high preferred body temperatures are predicted to have tissues specialized to function at relatively high temperatures, whereas species with lower preferred temperatures are predicted to exhibit tissue specializations to lower temperatures.

Many tissue functions of lizards, when tested, seem in fact to be carried out best in various species when the species are at their respective preferred body temperatures. For example, in species that have preferred body temperatures near 40°C, testicular development at the onset of the breeding season is often most rapid and complete at such high temperatures; in other species that prefer body temperatures near 30°C, the testicles develop optimally near 30°C and are damaged by 40°C. The optimum body temperature for sprint running by various species of lizards is well correlated with the respective preferred body temperatures of the species (Figure 8.14). Hearing, digestion, and the response of the immune system to bacterial invasion are just some of the other processes known to take place optimally in at least certain sets of lizards when body temperatures are at preferred levels. There are exceptions to these patterns, and there are traits that seem in general not to be optimized at preferred temperatures. Nonetheless, the data on lizards indicate that thermoregulation and tissue thermal specialization have often evolved in tandem.

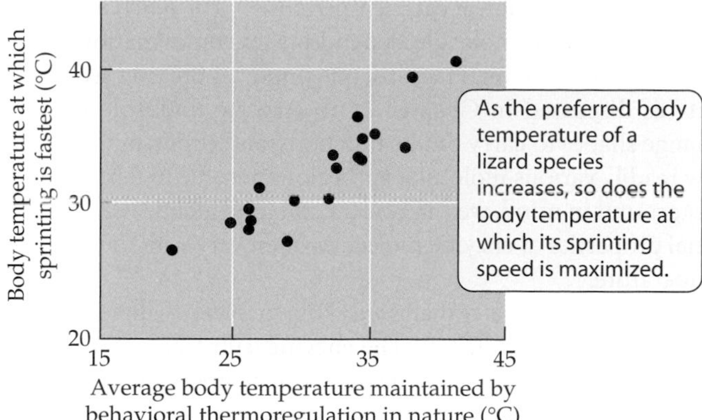

Figure 8.14 The body temperatures at which 19 species of iguanid lizards are able to sprint fastest correlate well with the behaviorally regulated preferred body temperatures of the species In each species of lizard, as the body temperature is raised, sprinting speed increases to a certain point, but then if the body temperature is raised further, sprinting speed starts to decline. The body temperature at which the sprinting speed is maximized is plotted on the *y* axis for each of the 19 species. Preferred body temperatures maintained by behavioral thermoregulation in nature are on the *x* axis. (After Huey and Kingsolver 1993.)

FISH AND INVERTEBRATES OF POLAR SEAS The oldest hypothesis on the tissue adaptations of aquatic poikilotherms is that the species of fish and invertebrates in polar seas maintain higher resting and average metabolic rates in cold waters than related temperate-zone or tropical species could maintain in the same waters. Today, most experts conclude that the hypothesis is correct, at least for certain groups of polar poikilotherms. This conclusion, however, follows 60 years of contentious debate, and the debate is likely to continue.

More-certain evidence for evolutionary specialization in polar poikilotherms comes from studies of isolated tissues. For instance, investigators have studied the rate of protein synthesis in isolated fish livers. At near-freezing tissue temperatures, the rate is much more rapid in livers taken from polar species than in those taken from temperate-zone species. Similarly, the skeletal muscles of polar fish are able to generate more mechanical power at polar temperatures than are the muscles of temperate-zone fish. Indeed, the muscles of the temperate-zone fish need to be as warm as 15°C to 25°C to generate the same power as the muscles of some Antarctic species can develop at 0°C. One reason that the skeletal muscles of polar species are particularly competent to generate power at low temperatures is that, at least in some cases, they are richer in mitochondria (and mitochondrial enzymes) than muscles of temperate-zone fish are. All these sorts of evidence point to the evolution in polar fish of distinctive physiological properties that permit them to function more vigorously at low body temperatures than unspecialized fish can.

One of the best-understood differences between polar and tropical species of fish is that they often synthesize different molecular forms of macromolecules such as proteins and lipids. We address this sort of evolutionary divergence in the next section.

Temperature and heat matter to animals because they affect the rates of processes and the functional states of molecules

One of the most important reasons to study poikilotherms is that they clarify the fundamental ways in which temperature and heat matter for the tissues of animals. Recall from Chapter 5 that heat energy cannot be used to do work by organisms. If heat cannot do work, then why does it even matter? Temperature and heat are profoundly important for animal tissues for two reasons. The first we have already discussed: The temperatures of tissues (which are determined by heat inputs and outputs) *affect the rates of tissue processes*—such as metabolic rates, rates of particular biochemical reactions, and rates of biophysical processes such as diffusion and osmosis. The second major way that temperature and heat are important for tissues is that the temperatures of tissues *affect the molecular conformations and, therefore, the functional states of molecules.*

The exact three-dimensional conformation of an enzyme molecule (or other protein molecule) depends on the prevailing temperature because three-dimensional conformation is stabilized by weak, noncovalent bonds, not strong, covalent bonds (see Box 2.1). When the temperature is modified, the various weak bonds in a molecule change in their relative strengths, and thus the molecule assumes a different conformation at each temperature.[10] The *functional properties* of an enzyme molecule depend on its conformation. This is probably true for several reasons; one important reason seems to be that (as discussed in Chapter 2) an enzyme molecule *must* flex (change shape) to carry out its functions, and conformation affects how readily various molecular subregions are able to flex. With this background in mind, we can understand in principle why the functional properties of enzyme molecules often vary with the prevailing temperature.

A functional property that is well known to vary with temperature is enzyme–substrate affinity. The enzyme–substrate affinities of all enzyme molecules change as the prevailing temperature is raised and lowered. Enzyme–substrate affinity is one of an enzyme molecule's most important functional properties because it determines how readily the molecule is able to form an enzyme–substrate complex (see page 44). Biochemists believe that a certain intermediate level of enzyme–substrate affinity is ordinarily ideal. Whereas too low an affinity can render an enzyme molecule incapable of forming complexes with substrate molecules, too high an affinity can make an enzyme molecule so prone to forming complexes with substrate that it becomes uncontrollable by regulatory processes. Figure 8.15*a* illustrates how the enzyme–substrate affinity of one particular enzyme molecule—lactate dehydrogenase (LDH) isolated from the muscles of a goby fish—varies with the prevailing temperature because of temperature-induced conformational changes in the protein.

Because the functional properties of enzymes depend on the prevailing temperature, any particular enzyme protein can be highly functional at certain tissue temperatures while being only marginally functional (or even nonfunctional) at other tissue temperatures. This fact raises the question of how animals having very different body temperatures can all have suitably functional enzymes.

An important part of the answer is that during evolution, species that have different body temperatures have often evolved different molecular forms of enzyme proteins. Not all species of vertebrates, for instance, have the same molecular form of LDH that the goby fish in Figure 8.15*a* has. If they did, species that ordinarily have low body temperatures would routinely have far higher enzyme–substrate affinities than ones that have high body temperatures. Instead, as Figure 8.15*b* shows, different species have evolved different molecular forms of LDH. The six species of poikilotherms shown in Figure 8.15*b*, some of which ordinarily live at very different body temperatures than others, have six different (although homologous) LDH proteins.[11] Although all six LDH proteins catalyze the same reaction, they differ in their detailed structures and functional properties, so the six exhibit different relations between enzyme–substrate affinity and temperature. The line for each species in Figure 8.15*b* is thickened and colored blue at the temperatures that correspond to the usual body temperatures of the species. For example, the line for the warm-water goby is thickened and colored blue at temperatures between 25°C and 40°C because that species of fish ordinarily has body temperatures of 25°C to 40°C, and the line for the Antarctic fish is thickened and colored blue at temperatures near –1°C because that species ordinarily has body temperatures near –1°C. All the blue, thickened segments are at about the same height on the *y* axis. Specifically, all fall within the span of the vertical blue bar. What this shows is that *all six species have about the same enzyme–substrate affinity when they are at their respective body temperatures.* The way they have achieved this remarkable condition, even though they live at body temperatures as much as 40°C apart, is by having evolved their different molecular forms (homologs) of the enzyme.

The conservation of enzyme–substrate affinity by the evolution of enzyme homologs that are adapted to function best at different temperatures is very common. One of the most striking and instructive examples is provided by the four species of barracudas in Figure 8.16. These four species, all closely related evolutionarily, behaviorally elect to live in waters that are just modestly different in temperature. For example, the waters occupied by *Sphyraena lucasana* average just 3°C to 4°C warmer than those occupied by *S. argentea*, and those occupied by *S. ensis* average just 3°C to 4°C warmer yet. Even these relatively small differences in habitat temperature (and body temperature) have led to the evolution of different molecular forms of the LDH protein. Consequently, the four species all have similar enzyme–substrate affinities when living at their respective temperatures.

At this point the second of the major ways in which temperature and heat matter for animals can be seen with full clarity. Particular enzyme molecules (and other sorts of protein molecules) are typically specialized to function best within certain temperature ranges. The protein molecules therefore *require* certain temperatures to function optimally. With few known exceptions, the tissues of the adults of any particular species are fixed in the homolog of each enzyme they synthesize; although a tissue may change the *amount* of the enzyme protein it synthesizes (as often occurs during acclimation or acclimatization), it cannot change the *type* of enzyme protein. Thus, individuals of a species of fish or other aquatic poikilotherm ordinary found in warm waters *require* warm tissue temperatures for their enzyme molecules to have ideal functional forms. Conversely, individuals of a cold-water species that has evolved different enzyme homologs *require* cold tissue temperatures for their particular types of enzyme molecules to have ideal functional forms. The same principles apply also to homeotherms; for instance, the LDH of pigs needs to be at about 37°C to have an appropriate enzyme–substrate affinity. Certain tissue temperatures, in brief, are crucial because the conformations and functional properties of proteins are not deterministically set by the chemical compositions of the proteins but depend as well on the prevailing temperature.

GLOBAL-WARMING FEARS Global warming threatens to subject many poikilotherms to changed tissue temperatures that will potentially force enzymes and other proteins to take on functional properties that are nonideal or even lethal. A key question in this regard is *how much* the tissue temperature must change for the change to have significant consequences. Data such as those on the barracudas (see Figure 8.16) suggest that the answer is sometimes, "not very much." The species of barracudas, which live in waters differing by 3°C to

[10] To be more exact, one current view is that a given molecule has multiple microconformations, which it "flickers" among from instant to instant. In this view, changes of temperature alter the probabilities that a molecule will be in its various microconformations. Thus, they alter the time-averaged conformation of the molecule. From this viewpoint, when we speak of *conformation* in the text, we mean *time-averaged conformation*.

[11] These are *interspecific enzyme homologs*. See page 47 in Chapter 2 for an extensive discussion of both LDH and the concepts of protein homology.

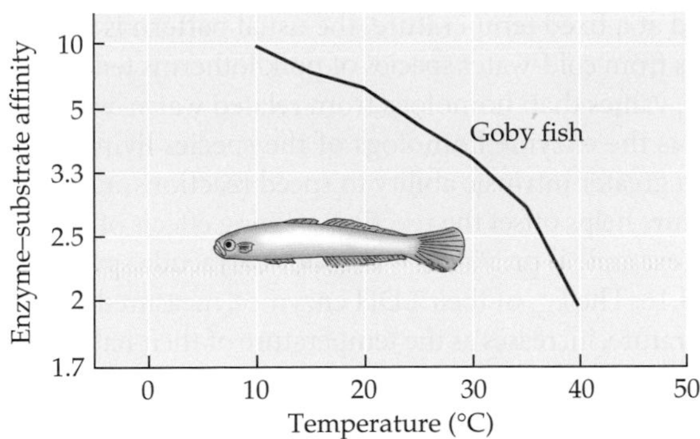

Figure 8.15 The affinity of the enzyme lactate dehydrogenase for substrate as a function of temperature This relation is shown in *(a)* for a goby fish (*Gillichthys mirabilis*), and in *(b)* for six species of poikilotherms—five fish and a desert lizard—that ordinarily live at different body temperatures. The blue, thickened portion of each line identifies the range of body temperatures ordinarily experienced by the species. Affinities associated with ordinary body temperatures all fall within the range marked by the vertical blue bar at left. The enzyme–substrate affinity shown in both plots is the affinity of muscle LDH (LDH-A_4) for pyruvic acid. Affinity is expressed as the inverse of the apparent Michaelis constant (m*M* pyruvate); see Chapter 2 (page 44) for background. The Antarctic fish are Antarctic notothenioids; the South American fish are also notothenioids; the barracuda is *Sphyraena idiastes*; the temperate-zone goby is *Gillichthys mirabilis*; the warm-water goby is *Gillichthys seta*; and the desert iguana is *Dipsosaurus dorsalis*. (After Hochachka and Somero 2002.)

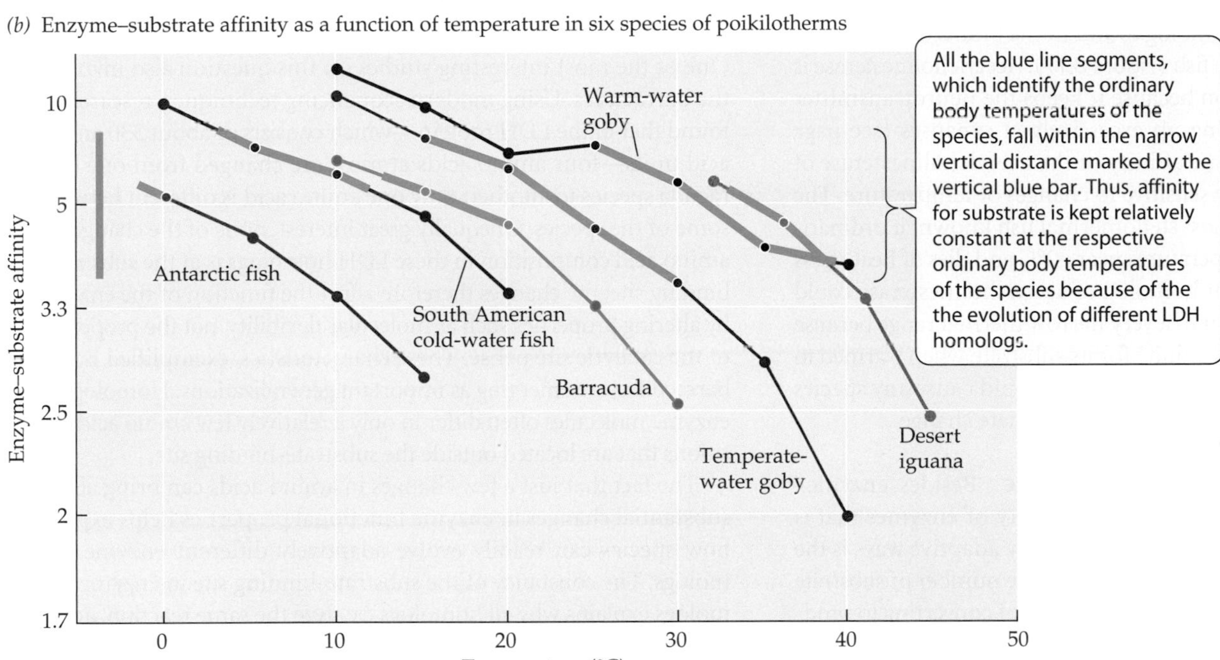

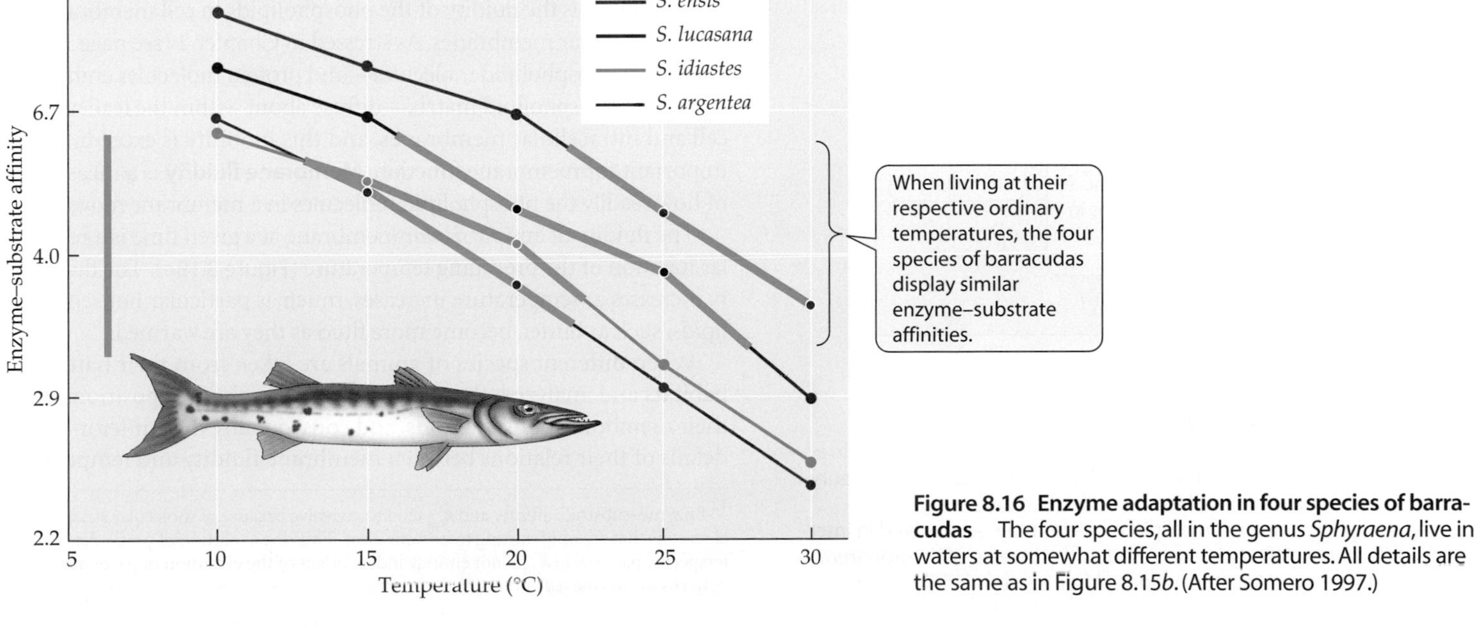

Figure 8.16 Enzyme adaptation in four species of barracudas The four species, all in the genus *Sphyraena*, live in waters at somewhat different temperatures. All details are the same as in Figure 8.15*b*. (After Somero 1997.)

4°C and have evolved different enzyme homologs, appear to be telling us that a 3°C to 4°C change in body temperature is sufficiently consequential for natural selection to favor the evolution of new molecular variants of key enzymes. A worry about human-induced global warming is that it may occur so rapidly that evolution will not immediately "keep up," and many poikilotherms (if they remain in their traditional geographical ranges) may be forced to function for years with nonoptimized molecular systems.

In situations when tissue temperatures are different from ideal, an important matter is the steepness of the relation between molecular functional properties and temperature. A classic example of a very steep relation is provided by the enzyme–substrate affinity of the acetylcholinesterase homolog found in the brains of *Pagothenia borchgrevinki*, an Antarctic fish (Figure 8.17). Acetylcholinesterase is essential for brain function because it keeps the neurotransmitter acetylcholine from building up excessively at synapses (see page 331). The enzyme–substrate affinity of the acetylcholinesterase of *Pagothenia* is exceptionally sensitive to changes of temperature. The species is also one of the most stenothermal fish known; it ordinarily lives its entire life at temperatures near –2°C and dies of heat stress at 4°C to 6°C. Its particular homolog of acetylcholinesterase could well be a factor in limiting it to a very narrow thermal range because the enzyme loses almost all affinity for its substrate when warmed to 5°C to 10°C. An enzyme form of this sort would cause any species possessing it to be unusually vulnerable to climate change.

MORE ON PROTEIN ADAPTATION TO TEMPERATURE Besides enzyme–substrate affinity, another functional property of enzymes that is known to vary among homologs in seemingly adaptive ways is the *catalytic rate constant*, k_{cat}. The k_{cat} measures the number of substrate molecules that an enzyme molecule is capable of converting to product per unit of time. If the k_{cat} of a particular type of enzyme, such as LDH, is measured at a fixed temperature, the usual pattern is that enzyme homologs from cold-water species of poikilotherms tend to exhibit higher k_{cat} values than homologs from related warm-water species show. Thus the enzyme homologs of the species living in cold waters have a greater intrinsic ability to speed reactions, an attribute that in nature helps offset the reaction-slowing effects of low temperatures. An example is provided by the four barracuda species shown in Figure 8.16. The k_{cat} of their LDH enzymes, measured at a fixed study temperature, increases as the temperature of their habitat decreases.[12]

An important question from the viewpoint of evolutionary biochemistry is how much the amino acid composition of a protein must change for the protein to take on new functional properties. One of the most interesting studies on this question also involved the barracudas. Using modern sequencing techniques, researchers found that in the LDH protein—which consists of about 330 amino acid units—four amino acids at most are changed from one barracuda species to another; only one amino acid is different between some of the species. Of equally great interest, none of the changes in amino acid composition in these LDH homologs is at the substrate-binding site; the changes therefore affect the function of the enzyme by altering properties such as molecular flexibility, not the properties of the catalytic site per se. These characteristics, exemplified by the barracudas, are emerging as important generalizations: Homologous enzyme molecules often differ in only a relatively few amino acid positions that are located outside the substrate-binding site.

The fact that just a few changes in amino acids can bring about substantial changes in enzyme functional properties helps explain how species can readily evolve adaptively different enzyme homologs. The constancy of the substrate-binding site in enzyme homologs explains why all homologs catalyze the same reaction, and it suggests that the ultimate act of helping covalent bonds to break and form during catalysis requires specific enzyme structures.

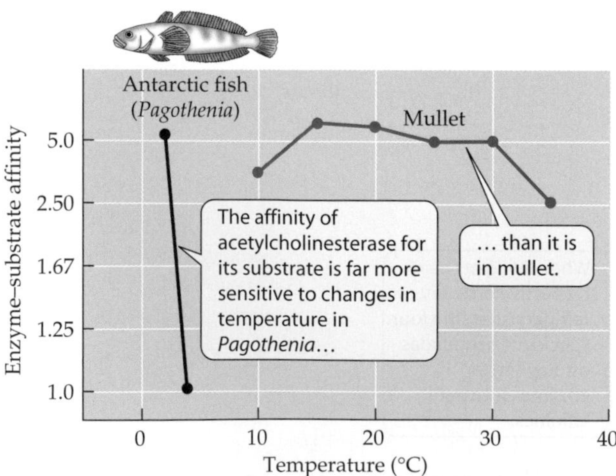

Figure 8.17 An enzyme very sensitive to temperature change A plot of the affinity of brain acetylcholinesterase for acetylcholine in a stenothermal, polar fish (*Pagothenia borchgrevinki*, pictured) and a eurythermal, warm-water species of fish, a mullet. Because acetylcholinesterase is a lipoprotein enzyme, lipid moieties may be involved in interspecific differences. Affinity is expressed as the inverse of the apparent Michaelis constant (m*M* acetylcholine). (After Somero 1997.)

LIPIDS As is true of proteins, the functional properties of lipids depend on the prevailing temperature as well as the chemical compositions of the molecules. One of the most important functional properties of lipids is the fluidity of the phospholipids in cell membranes and intracellular membranes. As stressed in Chapter 2 (see page 31), individual phospholipid molecules—and protein molecules embedded in the phospholipid matrix—diffuse about within the leaflets of cell and intracellular membranes, and this mobility is exceedingly important for membrane function. **Membrane fluidity** is a measure of how readily the phospholipid molecules in a membrane move.

The fluidity of any *particular* membrane at a given time is a regular function of the prevailing temperature (Figure 8.18*a*). The fluidity increases as temperature increases, much as particular household lipids, such as butter, become more fluid as they are warmed.

When different species of animals are taken from their natural habitats and analyzed, they typically differ in the compositions of their membrane phospholipids, and consequently they differ in the details of their relations between membrane fluidity and tempera-

[12] Enzyme–substrate affinity and k_{cat} tend to coevolve because of molecular structural reasons that are only starting to become clear. Thus the evolution of particular interspecific patterns in k_{cat} is not entirely independent of the evolution of particular patterns in enzyme–substrate affinity.

(a) The effect of temperature on membrane-lipid fluidity in a perch

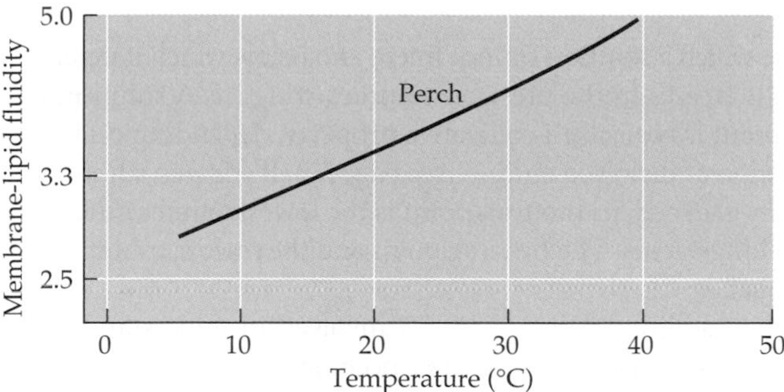

(b) The effect of temperature on membrane-lipid fluidity in six vertebrate species

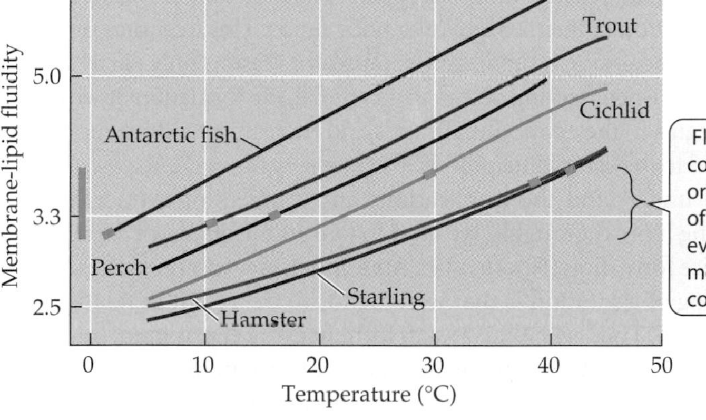

Figure 8.18 The fluidity of lipid-bilayer membranes from brain tissue as a function of temperature This relation is shown in *(a)* for a perch, and in *(b)* for six species of vertebrates—four fish that ordinarily live at different body temperatures, a mammal, and a bird. The light-blue, thickened portion of each line identifies the body temperatures ordinarily experienced by the species. Fluidities associated with ordinary body temperatures all fall within the range marked by the vertical blue bar at left. Fluidity is measured in terms of the mobility of a molecular probe, to which the units of measure refer. (After Somero 1997.)

ture. Figure 8.18*b* illustrates this pattern for six species of vertebrates. In the figure, the line for each species is thickened and colored blue at the temperatures that correspond to the usual body temperatures of the species. The blue, thickened line segments all fall within a narrow range on the *y* axis, signifying that all six species have about the same membrane fluidity when existing at their respective, normal body temperatures. The maintenance of a relatively constant membrane fluidity regardless of tissue temperature is called **homeoviscous adaptation** (*homeoviscous* = "steady viscosity"). Homeoviscous adaptation is possible because the chemical composition of membrane phospholipids is not fixed but instead can differ between species. If all animal species had the same membrane phospholipid composition—for instance, if all had the same composition as the perch in Figure 8.18*a*—the species with high body temperatures would have very fluid membranes, whereas those with low body temperatures would have stiff membranes. In reality, all have about the same membrane fluidity because species that have evolved to operate at different body temperatures have also evolved systematically different phospholipid compositions. As we saw in studying proteins, again this means that tissue temperature is important because it must be "matched" to the particular molecules present; *a tissue in which cell membranes are built of particular phospholipids will have the "correct" membrane fluidity only if its temperature is correct.*

The best-understood chemical basis for homeoviscous adaptation is modification of the number of double bonds in the fatty acid tails of the membrane phospholipids. Double bonds create bends in the fatty acid tails (see Figure 2.2*a*), and these bends interfere with close packing of the tails in a membrane. Thus, membrane fluidity tends to increase as the number of double bonds increases—that is, as the lipids become more chemically unsaturated. We saw in Chapter 2 (Figure 2.3) that among fish species native to different thermal environments, the degree of unsaturation of brain phospholipids increases as habitat temperature decreases: Whereas polar species have highly unsaturated lipids that, because of their chemical structure, remain reasonably fluid at polar temperatures, tropical species have much more saturated lipids that, because of their chemical structure, resist becoming too fluid at tropical temperatures.

Whereas individual adult animals usually cannot alter their protein homologs, typically they are able to alter the membrane phospholipids that they synthesize. The phospholipid composition of cell membranes and intracellular membranes is commonly restructured during acclimation and acclimatization in ways that promote homeoviscous adaptation. The restructuring of membrane phospholipids by an individual exposed to a chronically changed temperature typically requires many days or more. However, some fish in desert ponds undergo substantial phospholipid restructuring on a day–night cycle, thereby keeping membrane fluidity relatively constant even as the ponds heat up during the day and cool at night.

Poikilotherms at high temperatures: Heat-shock proteins help repair damage

If poikilotherms are briefly exposed to stressfully high but nonlethal tissue temperatures, they promptly synthesize proteins called **heat-shock proteins** (**HSPs**) or **stress proteins**. This response is now known to occur in all groups of animals and other cellular living organisms. The heat-shock proteins belong to several sets of protein families of characteristic molecular weights (especially 70 and 90 kilodaltons) and display highly conserved amino acid sequences, indicating that they are evolutionarily related in most or all animals. They are principally inducible proteins (see page 49); most are absent except during times when elevated temperature or another stress has elicited expression of the genes that encode them.

Functionally, the heat-shock proteins are types of **molecular chaperones**, proteins that use ATP energy to guide the folding of other proteins into correct three-dimensional configurations. High, nonlethal temperatures, according to the current working hypothesis, cause a variety of ordinary cellular proteins to become reversibly denatured, which means that their primary structures remain intact but their patterns of three-dimensional folding (tertiary structures) are disrupted (see Box 2.1). Many of these disrupted proteins, although not denatured in an inherently irreversible way, will not repair if left to themselves. Heat-shock proteins assist repair by preventing proteins in unfolded states from aggregating with each other and by promoting folding patterns that tend to restore disrupted proteins to their correct three-dimensional structures. Recovery from heat stress can be expensive, both because heat-shock proteins use ATP to carry out their repair functions and because irreparably damaged proteins must be resynthesized (at an ATP cost).

Rocky shores along the seacoast are known from recent research to be one of the ecological settings in which heat-shock proteins routinely play critical roles. Mussels, snails, and other attached or slow-moving animals living on the rocks can experience heat stress on clear, hot days when the tide goes out and they are exposed to the sun. During or soon after such events, these animals express heat-shock proteins. Species differ in how high their tissue temperatures must rise for such expression. Species positioned on the rocks where they ordinarily experience relatively low body temperatures express heat-shock proteins at lower temperatures than do species positioned where they routinely experience higher temperatures.

Poikilotherms threatened with freezing: They may survive by preventing freezing or tolerating it

If poikilotherms are exposed to temperatures even slightly colder than those necessary to freeze water, they themselves face a threat of freezing. Animal body fluids have lower freezing points than pure water because of their solute content (see page 83). However, they typically freeze at –0.1°C to –1.9°C (depending on the animal group) unless they are specially protected.

THE FREEZING PROCESS IN SOLUTIONS AND TISSUES To understand the threat of freezing and the possible strategies that animals might use to avoid freezing damage, the first step is to examine the freezing process. An important and seemingly strange point to mention at the outset is that when aqueous solutions are progressively cooled, they commonly remain unfrozen even when their temperatures have fallen below their freezing points, a phenomenon called **supercooling**. Supercooling is an intrinsically unstable state, and a supercooled solution can spontaneously freeze at any moment. Nonetheless, solutions in the supercooled state sometimes remain supercooled for great lengths of time.

An important determinant of a supercooled solution's likelihood of freezing is its extent of supercooling; freezing becomes more likely as the solution's temperature drops further below its freezing point. If the temperature of a supercooled solution is gradually lowered while the solution is not otherwise perturbed, a temperature is reached at which the likelihood of freezing becomes so great that the solution spontaneously freezes within a short time. This temperature is called the **supercooling point** of the solution.

Exposure to ice *induces* freezing in a supercooled solution. This has two important implications. First, an unfrozen but supercooled solution immediately freezes if it is seeded with even just a tiny ice crystal, regardless of its extent of supercooling. Second, if a solution, initially at 0°C, is gradually cooled in the presence of an ice crystal, the crystal will prevent supercooling.

An important application of this second point is that the cooling of a solution in the presence of an ice crystal permits determination of the solution's **freezing point**. The freezing point is the temperature above which a solution cannot freeze and below which it deterministically freezes in the presence of preexisting ice. A solution's freezing point is typically a colligative property, depending on the concentration of dissolved entities (see page 82). If a frozen solution is gradually warmed, its **melting point** is the lowest temperature at which melting occurs. The freezing point and the melting point are usually equal.

In tissues, additional complexity is involved in understanding freezing because the *location* of freezing becomes a factor. Under natural conditions, freezing within cells (*intracellular* freezing) almost always kills the cells in which it occurs. Intracellular freezing is thus fatal for animals unless they can survive without the cells affected. On the other hand, many animals are remarkably tolerant of widespread ice formation in their *extracellular* body fluids. This tolerance of extracellular freezing is significant because, for reasons only partly understood, when freezing occurs in an animal, the formation of ice often begins in the extracellular fluids and thereafter tends to remain limited to the extracellular fluids.

To understand the implications and dangers of extracellular freezing more thoroughly, we need to look at the *process* of extracellular ice formation (Figure 8.19). An important attribute of the slow freezing of a solution is that water tends to freeze out in relatively pure form. Thus, when ice crystals form in extracellular fluid, solutes tend to be excluded from the ice crystals and accumulate in the portion of the extracellular fluid that remains unfrozen, raising the total solute concentration of the unfrozen fluid (Figure 8.19*b*). The freezing point of the unfrozen fluid is lowered by the increase in its solute concentration. Thus the formation of ice in extracellular fluid after a stepwise drop in temperature is a *self-limiting* process: Water freezes out of the extracellular fluid only until the freezing point of the unfrozen fluid becomes low enough to equal the prevailing temperature, and then the ice crystals and concentrated fluid exist side-by-side at equilibrium.

The intracellular and extracellular fluids have similar osmotic pressures in an unfrozen animal, meaning that water has little or no tendency to enter or leave cells by osmosis. This benign state of affairs is disrupted by freezing in an animal's extracellular fluids. Im-

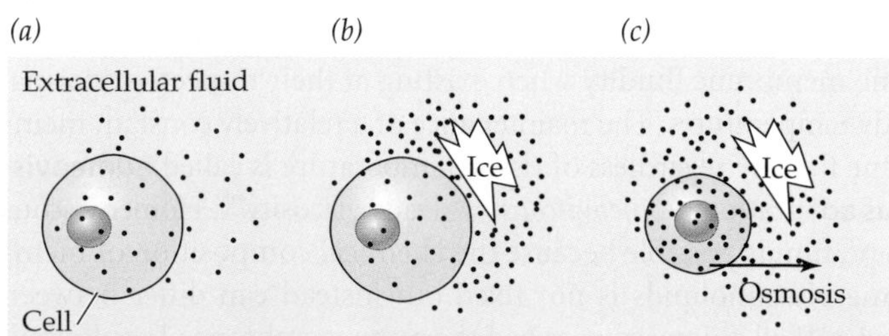

Figure 8.19 The process of extracellular freezing in a tissue Dots represent solute molecules. *(a)* The unfrozen condition; the intracellular fluid inside the cell and the surrounding extracellular fluid are the same in osmotic pressure. *(b)* Slow extracellular freezing produces ice that consists of virtually pure water; solutes excluded from the ice elevate the osmotic pressure of the unfrozen extracellular fluid. *(c)* Because of the difference in osmotic pressure created in part *b*, water leaves the cell by osmosis, shrinking the cell and raising the osmotic pressure of the intracellular fluid.

mediately after extracellular ice forms, the unfrozen extracellular fluids are osmotically more concentrated than the intracellular fluids (see Figure 8.19*b*). Thus the ice formation leads to the osmotic loss of water from cells (Figure 8.19*c*). This loss of intracellular water is itself self-limiting; it stops after the intracellular osmotic pressure has risen to equal the extracellular osmotic pressure. Within limits, the osmotic loss of water from cells is protective: By concentrating the intracellular fluids and thus lowering the intracellular freezing point, the loss of water from cells helps prevent intracellular freezing, which usually is fatal.

THE ADAPTIVE RESPONSES OF ANIMALS TO FREEZING CONDITIONS: INTRODUCTION Many poikilotherms behaviorally avoid environments where freezing conditions prevail. For example, many species of frogs, turtles, and crayfish move to the bottoms of lakes and ponds during winter. This location is a safe microhabitat because lakes and ponds do not normally freeze to the bottom.

Many poikilotherms, in contrast, are actually exposed to freezing conditions and must cope physiologically. The mechanisms by which they do so are classified into three types: (1) production of antifreeze compounds, (2) supercooling, and (3) tolerance of freezing. Antifreeze production and supercooling are mechanisms of *preventing* freezing. Usually, species that employ antifreezes and supercooling are **freezing-intolerant**; they die if they freeze and thus are absolutely dependent on successful prevention. On the other hand, some species are **freezing-tolerant**; they have evolved the ability to survive extensive freezing of extracellular body water and typically respond to freezing conditions by freezing. It remains largely a mystery why some species have evolved along lines of freezing intolerance, whereas others, sometimes closely related, have evolved toward freezing tolerance.

PRODUCTION OF ANTIFREEZE COMPOUNDS Many animals gain protection against freezing by synthesizing **antifreeze compounds**, defined to be dissolved substances that are added to the body fluids specifically to lower the freezing point of the body fluids. Two types of physiologically produced antifreezes are recognized.

1. *Colligative antifreezes.* Some antifreezes lower the freezing point of the body fluids strictly by colligative principles; they affect the freezing point by increasing the total concentration of solutes in the body fluids, not by virtue of their particular chemical properties. The most common of these colligative antifreezes are polyhydric alcohols, especially glycerol, sorbitol, and mannitol.
2. *Noncolligative antifreezes.* Some antifreezes lower the freezing point of the body fluids because of specialized chemical properties. Certain polypeptides and glycopeptides produced by a variety of insects and marine fish are the best-understood antifreezes of this sort. They are believed to act by binding (through weak bonds like hydrogen bonds) to nascent ice crystals in geometrically specific ways, thereby suppressing growth of ice by preventing water molecules from freely joining any crystals that start to form. The noncolligative antifreezes can be quite dilute and yet highly effective because they depress the freezing point hundreds of times more than can be accounted for by simple colligative principles. The noncolligative antifreezes, however, do not depress the melting point any more than colligative principles explain. Thus, solutions containing these antifreezes exhibit the unusual property—termed **thermal hysteresis**—that their freezing points are substantially lower than their melting points. The noncolligative antifreezes are usually called **thermal-hysteresis proteins** or **antifreeze proteins**.

Antifreezes are synthesized principally by certain species in two sets of animals: the marine teleost fish (the numerically dominant group of marine bony fish) and the insects. The marine teleost fish, in comparison to most other aquatic animals, face particular problems of freezing because their body fluids are osmotically more dilute than seawater (see page 692). Marine invertebrates generally have body fluids that are as concentrated as seawater; their freezing points thus match the freezing point of seawater, and they are not threatened with freezing unless the water in which they are living freezes. Freshwater animals of all kinds have body fluids that are more concentrated than freshwater; thus their freezing points are below the freezing point of freshwater, and they also do not freeze unless the water in which they live freezes. By contrast, marine teleost fish have blood and other body fluids that freeze at temperatures (–0.6°C to –1.1°C) higher than the freezing point of seawater (–1.9°C) because their body fluids are more dilute than seawater. The marine teleosts therefore can potentially freeze even when swimming in unfrozen seas!

A great many of the marine teleost species that live at polar and subpolar latitudes prevent freezing by synthesizing antifreeze proteins. These proteins are found in the blood, peritoneal fluid, pericardial fluid, and most other extracellular fluids of the fish. Although some polar species maintain high antifreeze-protein concentrations in all months, most species synthesize antifreeze proteins just in the cold seasons. The winter flounder (*Pleuronectes americanus*) is one of the best-known antifreeze-producing species. It has an unusually large number of copies of the gene for antifreeze-protein synthesis (about 100), and these genes are transcribed in an anticipatory way as winter approaches (Figure 8.20) under photoperiodic control.

Antifreezes are also found in the body fluids of many species of insects during winter. Colligative antifreezes such as glycerol, sorbitol, and mannitol are common and sometimes accumulate to impressive levels; in extreme cases, for example, 15% to 25% of an insect's overwintering body weight consists of glycerol and other polyhydric alcohols. Antifreeze proteins are also known in many insects. Insects, when devoid of antifreezes, have freezing points higher than –1°C. With high levels of antifreeze solutes, however, they may exhibit freezing points as low as –10°C or, in one known case, –19°C. Even freezing points this low, however, may be well above temperatures that insects encounter in many terrestrial environments. Prevention of freezing in insects is thus often a result of both antifreezes and supercooling working in concert. Both polyhydric alcohols and antifreeze proteins promote supercooling and may have other favorable effects.

SUPERCOOLING Supercooling is a perfectly ordinary, commonplace phenomenon in both the inanimate and animate worlds; animals do not *cause* themselves to supercool. However, animals can modify their probabilities of spontaneous freezing during supercooling.

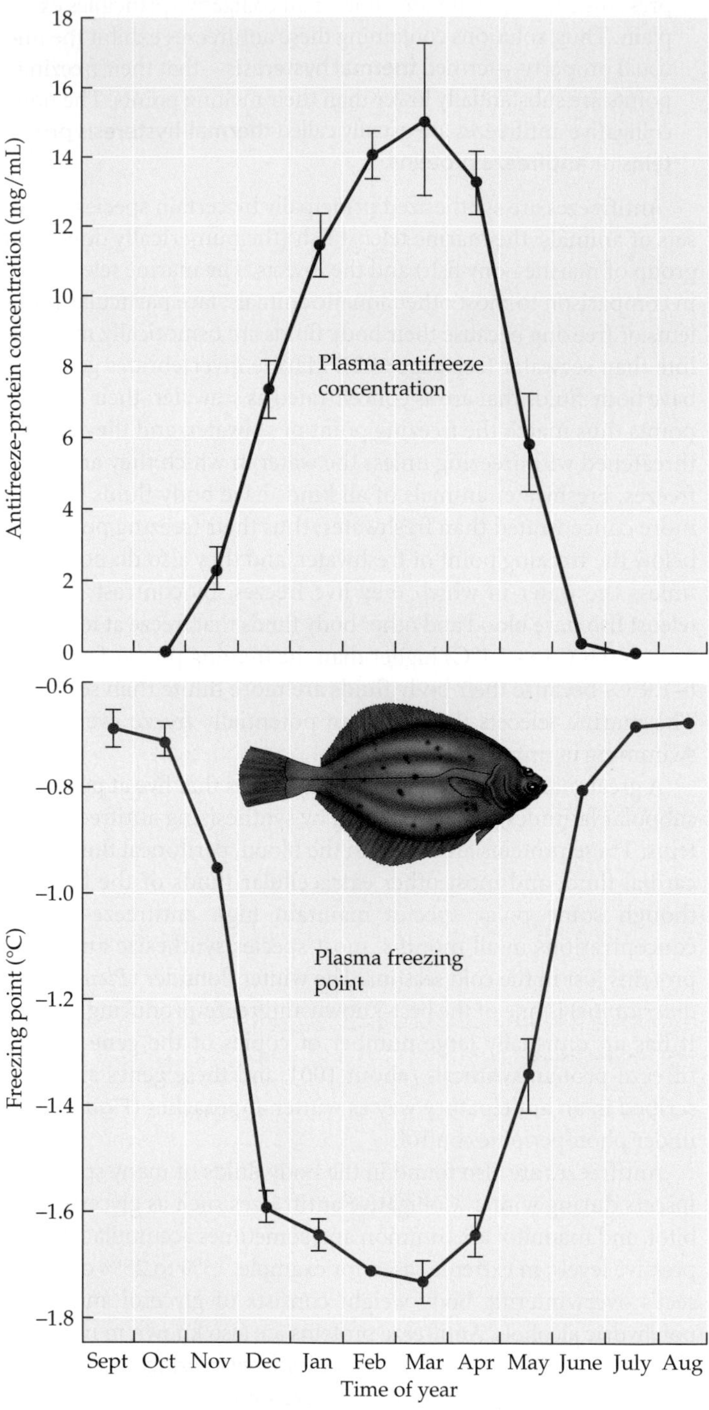

Figure 8.20 Seasonal changes in antifreeze protection in winter flounder (*Pleuronectes americanus*) The concentration of antifreeze protein in the blood plasma rises as winter approaches because of increased expression of the genes coding for antifreeze protein. The freezing point of the plasma is synchronously lowered and in winter is below the lowest winter temperatures, ensuring protection against freezing. The winter flounder is an important species in commercial fishing, noted for the fact that it spawns in frigid waters in late winter or early spring. (After Fletcher et al. 1998.)

Many animals, in fact, undergo adjustments whereby they enjoy low probabilities of spontaneous freezing even when they are supercooled to temperatures far below their freezing points. At one level of understanding, animals achieve these ends by altering the quality or quantity of **ice-nucleating agents** in their bodies. Ice-nucleating agents are dissolved or undissolved substances that act as foci for the initiation of freezing. An animal containing an abundance of ice-nucleating agents may freeze when its body temperature is barely below the freezing point of its body fluids. In contrast, an animal that has substantially cleansed its body of ice-nucleating agents may have a supercooling point that is far below its freezing point.

Extensive supercooling is the principal means by which the overwintering life stages of many freezing-intolerant species of insects survive, and it is among these insects that the greatest known capacities to supercool are found. An ability to supercool to 20°C to 25°C below the freezing point of the body fluids—and remain unfrozen for prolonged periods of time—is about average for such insects, and prolonged supercooling to 30°C to 35°C below the freezing point is not uncommon. At the extreme, there are now several known examples of insects that remain unfrozen at –50°C to –65°C by virtue of extensive supercooling combined with antifreeze depression of their freezing points. These species can overwinter in exposed microhabitats, such as plant stems, in some of the most severe climates on Earth.[13]

Less-extreme supercooling is employed by a variety of other types of animals to avoid freezing. For example, some species of *deep-water* marine teleost fish found in polar seas have been shown to have freezing points of about –1.0°C yet swim about unfrozen in waters that have a temperature of about –1.9°C. Supercooled fish in deep waters are unlikely to encounter floating ice crystals that would induce them to freeze. Shallow-water polar species, which commonly encounter floating ice crystals, have antifreezes!

TOLERANCE OF FREEZING An ability to survive extracellular freezing is far more widespread than was appreciated even 20 years ago. In the intertidal zone along ocean shores at high latitudes, sessile or slow-moving invertebrates clinging to rocks frequently experience freezing conditions when exposed to the air during winter low tides. Many of these animals—including certain mussels, barnacles, and snails—actually freeze and survive; some tolerate solidification of 60% to 80% of their body water as ice. Increasing numbers of insect species are also known to tolerate freezing of their blood; some survive temperatures below –50°C in their frozen state. One of the extreme examples is the insect called the woolly bear on Ellesmere Island in the Arctic; it overwinters, frozen, in exposed sites on top of the pack ice, tolerating a body temperature of –70°C! Certain amphibians that overwinter on land, such as spring peepers (*Hyla crucifer*) and wood frogs (*Rana sylvatica*), survive freezing at body temperatures of –2°C to –9°C (**Figure 8.21**).

A common property of freezing-tolerant animals is that they *limit the degree of supercooling* that can occur in their extracellular fluids. In this way they promote freezing in their extracellular fluids, where it is safe. To limit supercooling, some species expose themselves to ice

[13] From a practical viewpoint, supercooling disruption is a potential insect-control tool. Some bacteria and other microbes are known to act as highly effective ice-nucleating agents. Such microbes are being investigated as biological control agents against insect pests that depend on supercooling for winter survival.

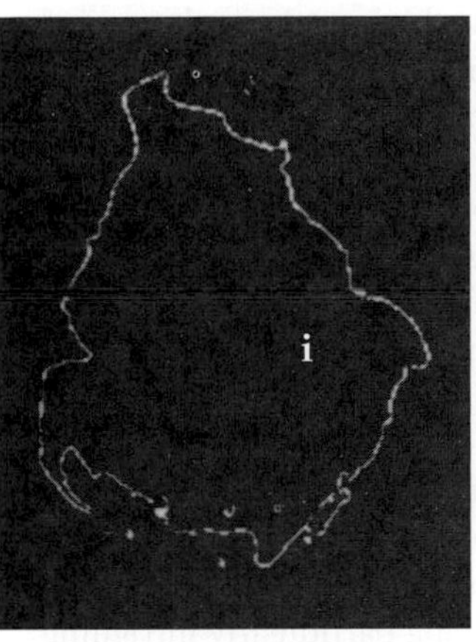

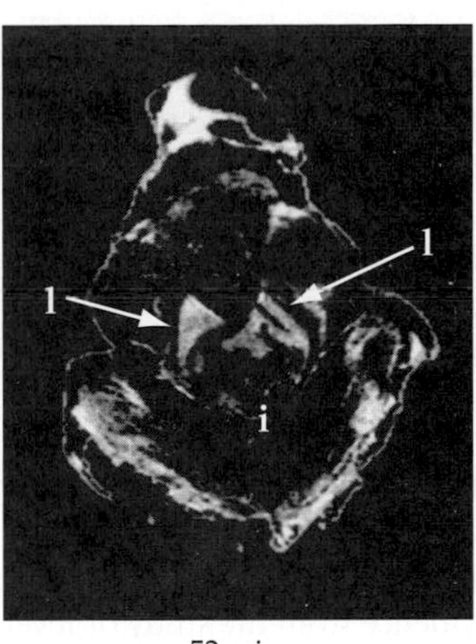

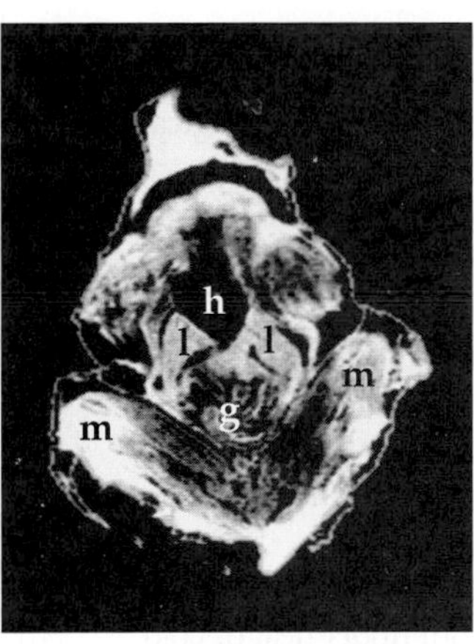

Figure 8.21 Frozen wood frogs (*Rana sylvatica*) thaw approximately synchronously throughout the body Magnetic resonance imaging (MRI) was used to detect the state of water within the body of a thawing frog: Ice is dark; liquid water is light. Images were taken at specified times after the frozen frog was placed at +4°C. Contrary to what might be expected, wood frogs do not thaw from outside to inside. Instead, deep and superficial regions of a frog's body thaw approximately simultaneously, probably because deep regions have lower melting points than superficial ones have. Synchronous thawing may ensure that blood flow to thawed tissues can start promptly. g = gut; h = heart; i = ice; l = liver; m = leg muscle. (From Rubinsky et al. 1994.)

and have body surfaces that permit external ice to induce freezing (inoculative freezing) of their extracellular fluids. Some synthesize ice-nucleating agents (e.g., proteins or lipoproteins) and add them to their extracellular fluids.

The ability of animals to *tolerate* freezing depends in part on the addition of particular organic solutes to their body fluids. Polyhydric alcohols (principally glycerol) are the primary organic solutes promoting tolerance of freezing in insects. Glucose is the solute of primary importance in most freezing-tolerant amphibians. Insects build up their blood concentrations of polyhydric alcohols in a preparatory manner over days or weeks prior to freezing. Amphibians, however, suddenly produce massive quantities of glucose from glycogen when ice formation starts in their bodies.

The organic solutes that promote tolerance of freezing in freezing-tolerant animals enter both the intracellular and the extracellular fluids of the animals, thereby increasing the amount of solute in both places. The increased solute in the extracellular fluids limits the amount of extracellular ice formation that occurs before the concentration of the unfrozen extracellular fluids rises high enough to prevent further freezing (see Figure 8.19*b*). The increased solute in the intracellular fluids limits the amount of water that must be lost from the cells for cells to come to osmotic equilibrium with freeze-concentrated extracellular fluids (see Figure 8.19*c*), thus limiting cell shrinkage. These are thought to be the principal ways by which the organic solutes aid the tolerance of freezing by poikilotherms.

Summary: Poikilothermy (Ectothermy)

- Poikilotherms, also called ectotherms, are animals in which body temperature (T_B) is determined by equilibration with external thermal conditions. They often thermoregulate. Their mechanism of thermoregulation is behavioral; a poikilotherm controls its T_B by positioning its body in environments that will bring T_B to the set-point level.
- The resting metabolic rate of a poikilotherm is usually an approximately exponential function of its T_B. The Q_{10} is typically 2 to 3. The metabolism–temperature curves of poikilotherms are often plotted on semilogarithmic coordinates because exponential functions are straight on such coordinates.
- When poikilotherms acclimate to cold or acclimatize to low-temperature environments in nature, their most common response is partial compensation. Partial compensation returns an animal's metabolic rate toward the level that prevailed prior to the change in environment, and thus it blunts the effect of environmental change. The most common known mechanism of partial compensation is for cells to change their concentrations of key, rate-limiting enzymes.
- Different species of poikilotherms that have long evolutionary histories of living at different body temperatures frequently display evolved physiological differences that suit them to function best at their respective body temperatures. Species of lizards sprint fastest at their respective preferred body temperatures, and polar species of fish function at higher rates in frigid waters than temperate-zone species can. The important mechanisms of adaptation to different body temperatures include molecular specialization: Species with evolutionary histories in different environments often synthesize different homologs of protein molecules and different suites of cell-membrane phospholipids. The evolution of *structurally* distinct proteins and phospholipids conserves *functional* properties of the molecules; because of evolved molecular specialization, species living in different thermal environments exhibit similar enzyme–substrate affinities and similar membrane-lipid fluidities.
- When exposed to heat stress, poikilotherms (and other animals) synthesize heat-shock proteins, molecular chaperones that (at an ATP cost) guide reversibly denatured proteins back into correct molecular conformations.
- When exposed to threat of freezing, some poikilotherms actually freeze and are freezing-tolerant; freezing must be limited to the extracellular body fluids, however. Other poikilotherms are freezing-intolerant and exploit one of three strategies—behavioral avoidance, antifreeze production, or supercooling—to avoid freezing. Antifreezes lower the freezing point. Stabilization of supercooling permits animals to remain unfrozen while at temperatures below their freezing points.

Homeothermy in Mammals and Birds

Homeothermy, the regulation of body temperature by physiological means, gives mammals and birds a great deal more independence from external thermal conditions than is observed in lizards, frogs, or other poikilotherms. On a cool, cloudy day, a lizard or other behaviorally thermoregulating poikilotherm may be unable to reach its preferred body temperature, because warming in such animals depends on a source of outside heat. A mammal or bird, however, produces its own heat for thermoregulation and thus can maintain its usual body temperature whether the environment is warm, moderately cold, or subfreezing.

Placental mammals typically maintain deep-body temperatures averaging about 37°C when they are at rest and not under heat or cold stress.[14] Birds maintain higher temperatures under similar conditions: about 39°C. One of the most remarkable attributes of mammals and birds is that the average body temperatures of thermally unstressed animals do not vary much with climate. One might expect, for instance, that species of mammals living in the Arctic would have lower average body temperatures than related species living in the tropics. Actually, however, differences of this sort are slight, if present at all.

Deep-body temperature is not absolutely constant. Daily cycles of body temperature occur; the body temperatures of mammals and birds are typically 1.5°C to 2.0°C higher during their active phases each day than during their resting phases. Moreover, in some species the body temperature is permitted to rise when individuals are under heat stress, or it is permitted to fall in winter.

Regardless of the variations that occur, the body temperatures of mammals and birds are among the most stable in the animal kingdom. Thus, one consequence of homeothermy is that cellular functions are able to be specialized to take place *especially reliably* at certain temperatures. On the other hand, as we will see, homeothermy has a very high energy cost and greatly increases the food requirements of mammals and birds in comparison to like-sized reptiles or fish.

Thermoregulation by a mammal or bird requires neurons (nerve cells) that sense the current body temperature and also requires thermoregulatory control centers in the brain that, by processing thermal sensory information, properly orchestrate the use of heat-producing and heat-voiding mechanisms in ways that stabilize the body temperature (Box 8.1). The detection of body temperature in a mammal or bird occurs in multiple parts of the body; thermosensitive neurons of importance are found in the skin, spinal cord, and brain, and sometimes also in specialized locations such as the scrotum. The principal control centers—which process all the multiple sensory inputs and command the thermoregulatory mechanisms—are located in the hypothalamus and the associated preoptic regions of the brain.

A behavioral thermoregulator like a fish or a lizard must also have thermoregulatory control centers that receive and process thermosensory information to orchestrate the processes of thermoregulation (see Box 8.1). Physiologists generally hypothesize that during the course of vertebrate evolution, there has been a thread of continuity in the control centers. According to this hypothesis, the control centers were already present in a rudimentary way when the only vertebrates were fish and the only thermoregulatory mechanisms to control were behaviors. Recognizing that some modern reptiles pant, change color to aid their thermoregulation, or employ other physiological mechanisms, physiologists usually conclude that the early reptilian ancestors of mammals and birds probably had some physiological thermoregulatory mechanisms that supplemented their dominant behavioral mechanisms. The control centers of those reptiles would therefore have had physiological as well as behavioral mechanisms to control. Then, as mammals and birds appeared, the control centers assumed control of predominantly physiological mechanisms. There is some evidence for this sort of scenario—with the evidence from the study of fever being partidularly intriguing (see Box 8.1). Researchers nonetheless still face daunting challenges to assess whether the hypothesized homologies in the control centers of various vertebrate groups are real.

A comparison of modern reptiles with mammals and birds suggests that the single most revolutionary step that occurred in the evolution of mammalian and avian homeothermy was the evolution of endothermy. Modern reptiles are essentially devoid of an ability to warm their bodies by metabolic heat production. Mammals and birds, in dramatic contrast, have an endogenous ability to stay warm in cold environments because of endothermy. With endothermy plus their physiological mechanisms of keeping cool in hot environments, mammals and birds are able to maintain relatively constant tissue temperatures over exceedingly wide ranges of environmental conditions. Biologists debate when this sort of sophisticated homeothermy first appeared in evolution. Certain authorities believe that it was already present in some dinosaurs.

Metabolic rate rises in cold and hot environments because of the costs of homeothermy

The resting metabolic rate of a mammal or bird typically varies with ambient temperature as shown in Figure 8.22. Within a certain range of ambient temperatures known as the **thermoneutral zone** (**TNZ**), the resting metabolic rate is independent of ambient temperature. The lowest ambient temperature in the TNZ is termed the **lower-critical temperature**; the highest is the **upper-critical temperature**. The lower-critical and upper-critical temperatures depend on the species, and they can also be affected by acclimation or acclimatization. An animal's **basal metabolic rate** is its metabolic rate when resting and fasting in its thermoneutral zone. The resting metabolic rate of a mammal or bird increases as the ambient temperature falls below the animal's lower-critical temperature or rises above its upper-critical temperature. These increases in metabolic rate in cold and warm environments arise from the animal's need to perform physiological work to keep its deep-body temperature constant when the ambient temperature is low or high.

The shape of the metabolism–temperature curve depends on fundamental heat-exchange principles

Before we study the physiological mechanisms used by mammals and birds to thermoregulate, it is important to analyze why the me-

[14] Marsupials, some of the primitive placental mammals, and especially monotremes have lower body temperatures; the platypus, for example, exhibits a deep-body temperature of 30°C to 33°C.

(a) The general relation

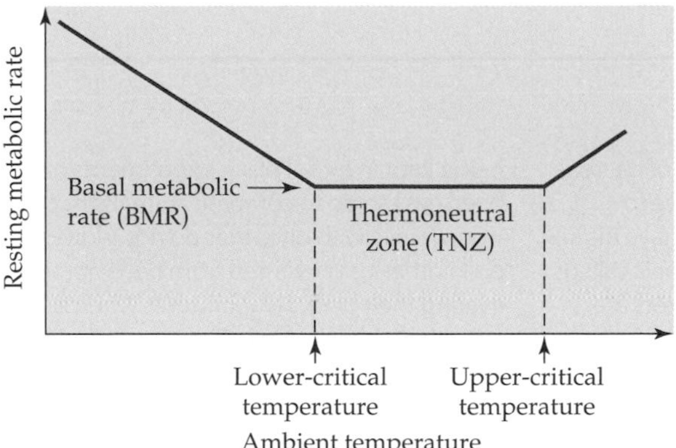

(b) An example

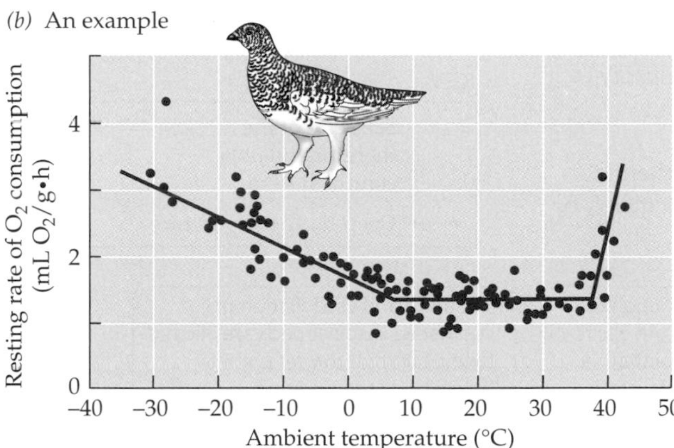

Figure 8.22 The relation between resting metabolic rate and ambient temperature in mammals and birds *(a)* The general relation and the terminology used to describe it. *(b)* An example, the metabolism–temperature relation of the white-tailed ptarmigan (*Lagopus leucurus*). (*b* after Johnson 1968.)

tabolism–temperature curve has the specific shape it does. A useful first step for this analysis is to recognize the concept of **dry heat transfer**, defined to be heat transfer that does not involve the evaporation (or condensation) of water. Dry heat transfer occurs by conduction, convection, and thermal radiation. As stressed earlier, these three mechanisms of heat transfer must be analyzed separately in heterogeneous thermal environments. However, in a uniform thermal environment such as a laboratory cage, where the radiant temperatures of all objects are similar to air temperature, these three mechanisms of heat transfer can be meaningfully lumped together because the rate of heat transfer by each of the three between an animal and its environment tends to increase as the difference in temperature between the animal's body and the environment ($T_B - T_A$) increases (see Figure 8.4).

In a uniform thermal environment, by lumping together the three mechanisms of dry heat transfer, we find that when factors other than temperature are held constant, then

$$\text{Rate of dry heat transfer} \propto T_B - T_A \qquad (8.8)$$

BOX 8.1 Thermoregulatory Control, Fever, and Behavioral Fever

Of all the physiological control systems, the system for thermoregulation is the one that usually seems the most straightforward conceptually. Virtually every introductory treatment of control theory in the physiological literature uses the thermoregulatory control system as its central example. This is undoubtedly true because analogies are readily drawn with engineered thermal control systems, which are common in our everyday lives.

The thermostat of a house with a furnace and air conditioner controls heat production by the furnace and heat removal by the air conditioner to maintain a stable air temperature in the house. From the point of view of control theory, the air temperature is the **controlled variable** (see Box 1.1), and the furnace and air conditioner are **effectors**, instruments that are capable of changing the controlled variable. The thermostat itself actually includes three separate elements that are essential for a control system:

1. A **sensor**, a device that can measure the controlled variable so that the control system knows its current level (the current air temperature).
2. A **set point** or **reference signal**. The set point is a type of information that remains constant in a control system even when the controlled variable goes up and down, and that tells the system the desired level of the controlled variable. We usually call the set point of a home thermostat its "setting." If, for example, we "set" the thermostat to 20°C, the device is able to retain that set-point information in an invariant form, so that the air temperature detected by the sensor can be compared to it. An important point to recognize is that a thermostat does not remember its set point by having inside it an object that is kept literally at the set-point temperature. Instead, the set point is *represented* in the thermostat by a physical system that is not a temperature but corresponds to a temperature (in some old-fashioned thermostats, the set point is represented by a screw that protrudes to a certain extent).
3. A **controller**, a mechanism that compares the set point with the current level of the controlled variable to decide whether the controlled variable is too high or low.

The control system in a house, considered as a whole, operates as a **negative feedback** system (see Box 1.1). It controls the effectors to bring the controlled variable back toward the set point. Thus if the air temperature goes below the set point, the furnace is commanded to add heat to the house, and if the air temperature goes above the set point, the air conditioner is commanded to remove heat.

By analogy, it is easy to describe the thermoregulatory control system of a lizard or mammal (or any other thermoregulator) in terms of the same basic concepts. The principal *effectors* in a lizard (a behavioral thermoregulator) are the skeletal muscles that move the limbs and control posture. Effectors in a mammal might include muscle cells that can produce heat by shivering, sweat glands that can promote evaporative cooling, hair-erector muscles that determine how fluffed the pelage is, and so forth. A lizard or mammal has multiple *sensors*: temperature-sensitive neurons that measure the current temperatures of the skin, spinal cord, and brain. These sensors send their temperature data to a *controller* in the brain that compares the current temperatures to a *set point* to decide what to do. If a lizard is too cool, the controller commands behaviors that move the lizard to a warmer

BOX 8.1 *(Continued)*

place. If a mammal's body temperature is too low, the controller might activate shivering, deactivate sweating, and command fluffing of the pelage. The exact nature of the controller and set point in an animal remain far from fully understood because they consist of many tiny neurons in the depths of the brain. As in the case of the home thermostat, however, we recognize that the set point is not literally a temperature in the brain but is *represented* in some way by neurons. No one knows for certain yet how the set point is represented.

If we disregard these uncertainties and simply use the terminology of control theory, the set point of a lizard or mammal can be adjusted to different "settings" at different times, just as the setting of a home thermostat can be adjusted. **Fever** provides an elegant example of resetting of the thermostat. In a classic fever, people go through alternating periods of shivering and sweating, a pattern that reflects the changes in set point. At the start of a bout of fever, as the figure shows, the set point is raised, and the controller issues commands that bring about shivering, restriction of cutaneous blood flow, and other processes that raise the body temperature to the new set point. Later, when the set point falls back to normal, commands are issued for sweating, flushing of the skin with blood, and other processes that bring the body temperature to the lowered set point. The rise of the set point during fever results from a series of events set in motion by the presence in the body of chemicals known as **pyrogens**, among which are lipopolysaccharides in bacterial cell walls.

One of the most remarkable discoveries in thermal biology in the twentieth century was that lizards, other reptiles, amphibians, and fish sometimes develop fevers. When these animals are infected with certain sorts of bacteria, their set points go up, and they move to warmer environments where they have higher body temperatures! This sort of fever is called **behavioral fever** because the effectors are skeletal muscles that modify behaviors. However, there are many reasons to believe that the behavioral fevers in poikilothermic vertebrates are homologous to physiological fever in mammals. Thus the presence of fever in poikilothermic vertebrates is an intriguing type of evidence that there is evolutionary continuity between the thermoregulatory control systems in poikilothermic and homeothermic vertebrates.

Because a lizard or other poikilotherm can be prevented from developing fever merely by being kept in a cool place, experiments have been easy to do to compare animals that develop fevers with ones that do not. Individual poikilotherms prevented from behaviorally elevating their body temperatures when infected with bacteria often suffer greater morbidity and mortality than do individuals that are allowed to develop fevers, suggesting that fever evolved as a defense against disease.

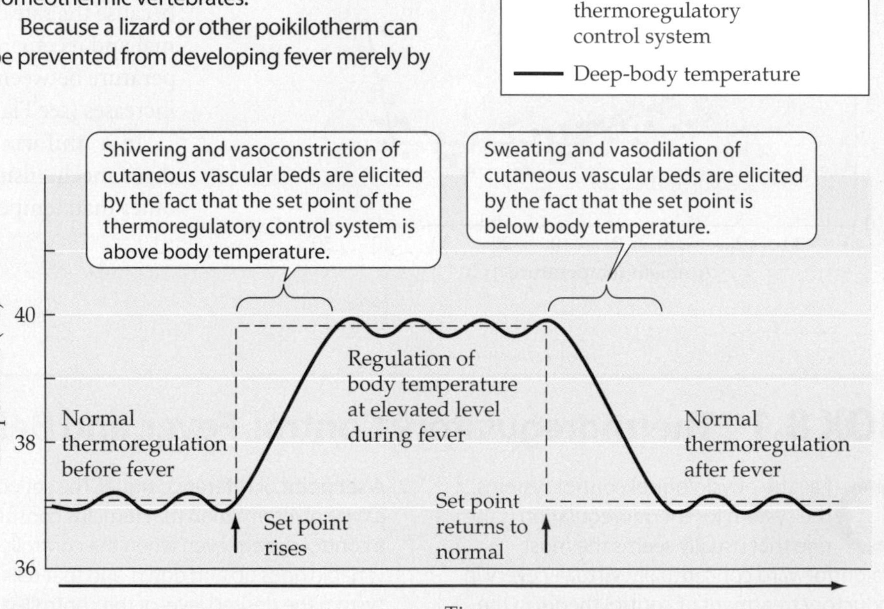

The relation between the set point of the thermoregulatory control system and body temperature during a bout of fever When the set point jumps up at the start and falls back down at the end, the thermoregulatory control system detects the mismatch between the set point and the body temperature and commands vigorous effector responses to correct the mismatch.

Heat moves out of an animal's body by dry heat transfer when T_B exceeds T_A; conversely, heat moves into the body when T_B is less than T_A. The rate of dry heat transfer is proportional to $(T_B - T_A)$ in either case, and thus $(T_B - T_A)$ can be thought of as the "driving force" for dry heat transfer.

To analyze the shape of the metabolism–temperature curve, Equation 8.8 *taken by itself* can be used at ambient temperatures that are in and below the thermoneutral zone (TNZ). At temperatures above the TNZ, evaporative heat transfer is too important to be ignored. At ambient temperatures within and below the TNZ, however, evaporative heat transfer is in general minor enough that it can be disregarded for purposes of a conceptual analysis, explaining why the equation for dry heat transfer (Equation 8.8) can be used. The body temperature of a mammal or bird is typically higher than the animal's upper-critical temperature. Thus, when the ambient temperature is within or below the TNZ, $(T_B - T_A)$ is positive and dry heat transfer carries heat out of the body.

If an animal is losing heat to its environment, the only way it can maintain a constant body temperature is to make heat *metabolically* at a rate that matches its rate of heat loss; symbolically, if M is the animal's metabolic rate, M must equal the animal's rate of heat loss. If we assume that Equation 8.8 describes a mammal's or bird's rate of heat loss at ambient temperatures within and below its TNZ, then at those ambient temperatures, $M \propto (T_B - T_A)$. We can rewrite this expression as an equation by introducing a proportionality coefficient (C):

$$M = C(T_B - T_A) \tag{8.9}$$

This equation, which is a key equation for analyzing a mammal's or bird's thermal relations, is the **linear heat-transfer equation**, also described at times as a version of **Newton's law of cooling** or **Fourier's law of heat flow**. The coefficient C, which is called the animal's **thermal conductance**, is a measure of *how readily* heat can move by dry heat transfer between the animal and its environment.

To see the significance of C, suppose that two placental mammals are in the same environment and therefore have the same driving force for dry heat loss ($T_B - T_A$), but one has a higher thermal conductance than the other. The one with the higher C will lose heat faster because heat can move out of its body more readily than it can move out of the body of the other. Therefore the one with the higher C will require a higher metabolic rate to stay warm.

An animal with a *high* C can be thought of as having a *low resistance* to dry heat loss; and conversely, an animal with a *low* C can be thought of as having a *high resistance* to dry heat less. Physiologists thus define an animal's **resistance to dry heat loss** as being the inverse of C: $1/C$. The resistance to dry heat loss is often called **insulation** (I). Thus $I = 1/C$. The linear heat-transfer equation can therefore also be written as follows:

$$M = \frac{1}{I}(T_B - T_A) \tag{8.10}$$

An important point to note about the concept of insulation (I) introduced here is that it is not simply a measure of the heat-retaining properties of the fur or feathers. Instead, insulation is a measure of the body's *overall* resistance to dry heat loss. For instance, because both posture and fur affect a mammal's resistance to dry heat loss, the value of I for a mammal depends on its posture as well as its fur (and also on additional factors).

THE THERMONEUTRAL ZONE Let's now use the concepts we have developed to understand why the metabolism–temperature curve of a mammal or bird has the shape it does in the thermoneutral zone. The defining property of the TNZ is that an animal's metabolic rate (M) remains constant at all the different ambient temperatures in the TNZ. This property probably seems impossible or paradoxical at first. After all, if T_A changes, then ($T_B - T_A$) changes, and Equation 8.10 suggests that M would have to change. The answer to the paradox is that in its TNZ, a mammal or bird modulates its insulation to a variety of different values. *Modulation of insulation against a background of constant metabolic heat production is in fact the principal means by which a mammal or bird thermoregulates in its thermoneutral zone.*

As the ambient temperature is lowered in the TNZ, a mammal or bird responds by increasing its insulation, I.[15] This increase in the animal's resistance to heat loss counterbalances the increase in the driving force for heat loss, ($T_B - T_A$), so that the animal's actual rate of heat loss remains constant (or nearly so). The animal's rate of metabolic heat production therefore can also remain constant. These points are mathematically apparent in Equation 8.10. In the TNZ, as T_A decreases and ($T_B - T_A$) therefore increases, I is increased in a precisely counterbalancing way so that the ratio $(T_B - T_A)/I$ in the equation remains constant. The metabolic rate of the animal, M, can therefore be constant.

The width of the TNZ varies enormously from species to species, depending in part on the extent to which various species are able to modulate their insulation. Small-bodied species tend to have narrower TNZs than large-bodied species. Species of mice, for instance, often have TNZs extending just from about 30°C to 35°C. At another extreme, Eskimo dogs have a TNZ extending from –25°C to +30°C—a range of 55°C!

TEMPERATURES BELOW THERMONEUTRALITY Unlike the case within the thermoneutral zone, the principal means by which a mammal or bird thermoregulates at ambient temperatures below thermoneutrality is modulation of its rate of metabolic heat production. Below the TNZ, as the environment becomes colder, a mammal or bird must raise its rate of metabolic heat production to higher and higher levels if it is to stay warm, very much in the way that a furnace in a house must increase the rate at which it produces heat as the air outside becomes colder.

To see the significance of the lower-critical temperature, consider an animal, initially in its TNZ, that is subjected to a steadily declining ambient temperature. As the T_A declines while remaining in the TNZ, the rate at which the animal loses heat to its environment stays constant because the animal increases its insulation, I, to an extent that fully counterbalances the increase in the driving force for dry heat loss ($T_B - T_A$). Insulation cannot be increased without limit, however. An animal's lower-critical temperature represents the T_A below which its insulatory adjustments become inadequate to counterbalance fully the increase in the driving force favoring heat loss. Thus, as the T_A falls below the lower-critical temperature, the rate at which the animal loses heat increases, and the animal must increase its rate of heat production to match the increased rate of heat loss.

The insulation of a mammal or bird sometimes becomes maximized at the lower-critical temperature. Cases like this are particularly straightforward to understand in terms of the linear heat-transfer equation. If an animal maximizes its insulation at the lower-critical temperature, then its value of I at ambient temperatures below the TNZ is a *constant* (equaling its maximum value of I). In addition, because $C = 1/I$, the animal's value of C below the TNZ is a *constant* (equaling its minimum value of C). Thus, for such an animal, T_B, I, and C in the linear heat-transfer equation are all constants below the TNZ. Therefore the equation—whether written as Equation 8.9 or as Equation 8.10—is a simple linear equation (accounting for its name) that has two variables: M and T_A. If we plot M as a function of T_A for this linear equation (using Equation 8.9), we obtain a straight line having two particular properties, illustrated in Figure 8.23*a*: First, the slope of the line is $-C$; second, the line intersects the x axis at the ambient temperature that is equal to T_B.

As a model of an animal's metabolism–temperature curve, the plot in Figure 8.23*a* is flawed, of course, because it ignores the fact that an animal's metabolic rate (M) does not truly fall below the basal level. Figure 8.23*b* is thus a more realistic model. By examining Figure 8.23*b*, you can see that the portion of an animal's metabolism–temperature curve below the TNZ is simply a plot of the linear heat-transfer equation.

[15] We discuss the mechanisms of increasing insulation on page 219. Examples of mechanisms are that a mammal might curl up into a compact posture and fluff its fur.

(a)

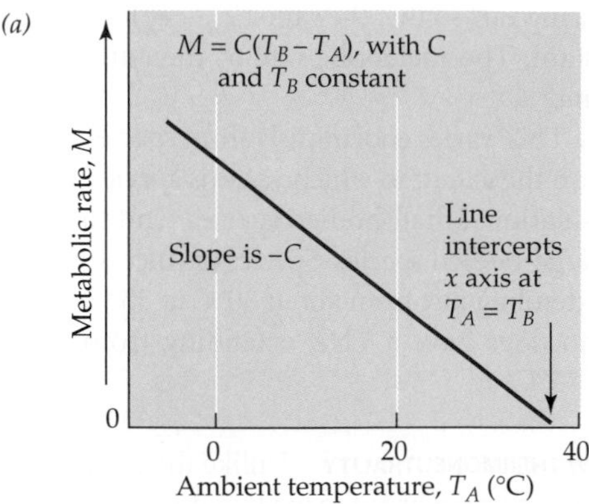

(b)

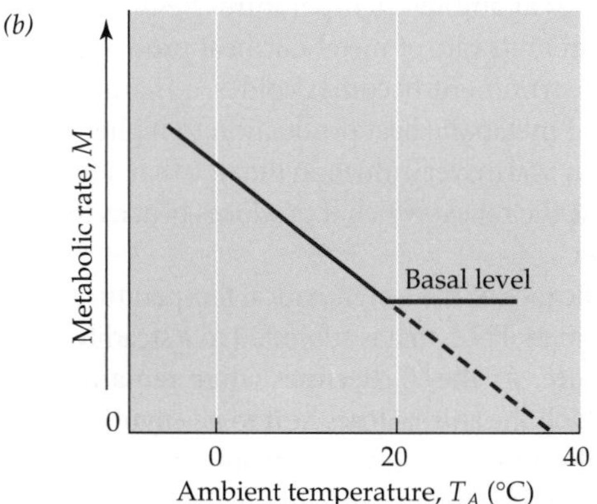

(c)

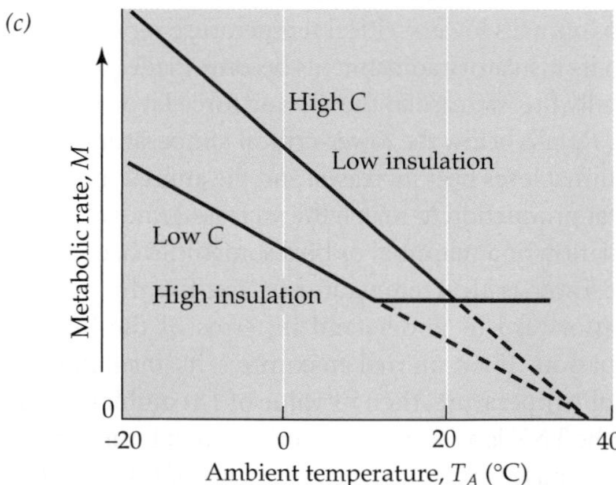

Figure 8.23 A model of the relation between metabolic rate and ambient temperature in and below the thermoneutral zone *(a)* A plot of the linear heat-transfer equation (the version in Equation 8.9) when C and T_B are constant ($T_B = 37$°C). *(b)* The plot from part *a*, with realism added in that the metabolic rate is not allowed to go below the basal level. *(c)* A comparison of two animals that are identical except that at temperatures below thermoneutrality, one maintains a high C (low insulation) relative to the other.

The fact that the slope of the metabolism–temperature curve below the TNZ is equal to $-C$ (for animals that maintain a constant C) provides a useful tool for the visual interpretation of metabolism–temperature curves. As shown in Figure 8.23*c*, if two otherwise similar animals differ in thermal conductance (C) below the TNZ, the relative slopes of their metabolism–temperature curves mirror their differences in C: The animal with the higher C (lower insulation) has a steeper slope than the one with the lower C (higher insulation). Using this principle, one can look at Figure 8.34*b* (see page 228), for example, and tell at a glance that the winter fox has lower conductance and higher insulation than the summer fox. Figure 8.23*c* also brings out the energy advantages of high insulation. Note that the animal with relatively high insulation (low C) has a relatively low metabolic rate at any given ambient temperature below the TNZ, just as a well-insulated house, to stay warm, requires a lower rate of heat production by its furnace than a poorly insulated one requires.[16]

TEMPERATURES ABOVE THERMONEUTRALITY Mammals and birds employ two principal processes to cope with ambient temperatures above their thermoneutral zones. First, most mammals and birds actively increase the rate at which water evaporates from certain of their body surfaces; such **active evaporative cooling** most often occurs by sweating, panting, or gular fluttering (Figure 8.24). Second, some mammals and most birds allow their body temperatures to rise to unusually high levels, a phenomenon called **hyperthermia**.

Both active evaporative cooling and hyperthermia can cause an animal's metabolic rate to rise at temperatures above thermoneutrality. Active evaporative cooling causes a rise in metabolic rate because physiological work must be done to enhance water evaporation (panting, for example, requires an increase in the rate of breathing). Hyperthermia can also cause a rise in metabolic rate because tissues tend to accelerate their metabolism when they are warmed; according to recent research, hyperthermia does not always induce metabolic acceleration in mammals and birds, but in some cases it does.

To appreciate more fully the processes at work above the TNZ, it is informative to take a dynamic approach by considering an animal that is initially within its TNZ and subjected to a steadily increasing ambient temperature. As the T_A rises, the driving force for dry heat loss ($T_B - T_A$) decreases, meaning that the animal faces a greater and greater challenge to get rid of its *basal* metabolic heat production. While the T_A remains in the TNZ, the animal responds to the rising T_A by decreasing its resistance to dry heat loss, its insulation. Consequently, even high in the TNZ, metabolic heat is carried away as fast as it is produced by a combination of dry heat transfer and *passive* evaporation. This handy state of affairs comes to an end when the T_A reaches the upper-critical temperature and goes higher. Near the upper-critical temperature, insulation either reaches its minimum or, at least, becomes incapable of sufficient further reduction to offset additional decreases in ($T_B - T_A$). Thus, as the T_A rises above the upper-critical temperature, the rate of dry heat loss tends to fall so low that the combination of dry heat loss and passive evaporation becomes inadequate to void metabolic heat. Both of the principal responses of mammals and birds—active evaporative cooling and hyperthermia—serve to promote heat loss so that animals are not overheated by their metabolic heat production. Hyperthermia does this because a rise in T_B increases the driving force for dry heat loss ($T_B - T_A$).

[16] Although the slopes of metabolism–temperature curves were used to calculate C quantitatively some years ago, better approaches to the calculation of C have been developed. Thus, the use of slopes today should be reserved for just qualitative, visual interpretation.

Figure 8.24 Gular fluttering is one means of actively increasing the rate of evaporative cooling During gular fluttering, which occurs just in birds, an animal holds its mouth open and vibrates the floor of the mouth, termed the gular area (arrow). In this way, airflow is increased across the moist, vascular mouth membranes, promoting a high rate of evaporation. The bird seen here is a young great egret (*Casmerodius albus*) on a hot day in Florida. In some species the gular area vibrates up and down at 800 to 1000 cycles/minute during gular fluttering.

If the T_A keeps rising and becomes so high that it exceeds T_B, heat stress becomes extraordinary, because when the T_A is above the T_B, dry heat transfer carries environmental heat *into* the body![17] Then active evaporative cooling must assume the entire burden of removing heat from the body.

From a quick glance at the metabolism–temperature curve above the TNZ, it may seem extremely paradoxical that a mammal or bird increases its metabolic rate—its rate of internal heat production—when it is under heat stress. To understand this paradox, it is important to recall the very large amount of heat carried away by the evaporation of each gram of water (see page 196). Although an animal must increase its metabolic rate to pant, gular flutter, or otherwise actively increase its rate of evaporation, the amount of heat carried away by the evaporation of each gram of water far exceeds the heat produced per gram by the physiological processes that accelerate evaporation.

Homeothermy is metabolically expensive

One of the most important attributes of homeothermy in mammals and birds is that it is metabolically expensive in comparison to vertebrate poikilothermy. Homeothermy in mammals and birds in fact provides an outstanding example of a point we stressed in Chapter 1; namely that when physiological *regulation* and *conformity* are compared, the greatest downside of regulation is that its energy costs are high.

To quantify the cost of homeothermy, physiologists have compared the metabolic rates of vertebrate homeotherms and poikilotherms at similar tissue temperatures. Specifically, they have compared the basal metabolic rates of mammals and birds with the resting metabolic rates of like-sized poikilotherms held at the same body temperatures as the mammals and birds. A typical experiment would be to obtain a 100-g placental mammal and place it in its thermoneutral zone, and simultaneously obtain a 100-g lizard and place it in a chamber at 37°C so that its body temperature matches that of the mammal. If both animals are at rest and fasting and you measure their metabolic rates, you will obtain the basal metabolic rate (BMR) of the mammal and the standard metabolic rate (SMR) of the lizard at mammalian body temperature. Typically what you will find is that the metabolic rate of the mammal is four to ten times higher than that of the lizard, even though the cells of the two animals are at one temperature and even though the mammal's metabolic rate under these conditions is its *minimal* rate! Many studies of this sort have been carried out on a variety of species, and they have confirmed repeatedly that the BMRs of mammals and birds are four to ten times the SMRs of poikilothermic vertebrates at mammalian or avian body temperatures (see Figure 5.9). *Metabolic intensity stepped up dramatically when vertebrates evolved homeothermy.* In fact, there is every reason to believe that the high metabolic intensities of mammals and birds *permit* them to be homeothermic.

If mammals, birds, and poikilothermic vertebrates studied as we have just described are transferred to cold ambient temperatures, the metabolic rates of the mammals and birds *rise*, whereas the metabolic rates of the poikilotherms *fall*. At cold ambient temperatures, therefore, the difference in metabolic intensity between homeotherms and poikilotherms is far greater than just four- to tenfold.

The high metabolic rates of mammals and birds mean that they must acquire much more food than poikilothermic vertebrates of equal body size. An intriguing illustration of the consequences is provided by desert communities with simplified vertebrate biotas dominated by mammals and lizards Population densities of the mammals are often far lower than those of the lizards, indicating that there is a trade-off between the food needs of individual animals and the numbers that can exist in an ecological community.

Insulation is modulated by adjustments of the pelage or plumage, blood flow, and posture

In this section and several that follow, we turn to the mechanisms that mammals and birds employ to thermoregulate physiologically. The first question we address is how mammals and birds modulate their resistance to dry heat transfer, their insulation. They employ several distinct physiological mechanisms, which, as we have seen, are the predominant mechanisms of thermoregulation within the thermoneutral zone.

One means of varying insulation is erection or compression of the hairs or feathers. Each hair or feather can be held upright or al-

[17] For the T_A to exceed T_B, it must rise substantially above the upper-critical temperature. When the T_A is just moderately above the upper-critical temperature, it is typically below the T_B.

lowed to lie flat against the skin by the contraction or relaxation of a tiny muscle at its base, under control of the sympathetic nervous system. These responses are termed **pilomotor responses** in mammals and **ptilomotor responses** in birds. As the ambient temperature falls within the TNZ, the hairs or feathers are erected. In this way the pelage or plumage is fluffed out and traps a thicker layer of stagnant air around the animal, increasing the resistance to heat transfer through the pelage or plumage (see Equation 8.1).

Another mechanism of modulating insulation is the use of **vasomotor responses** in blood vessels (see page 622), responses which alter the rate of blood flow to the skin and other superficial parts of the body. Constriction of arterioles supplying superficial vascular beds at cool ambient temperatures, brought about by stimulation by the sympathetic nervous system, results in retarded transport of heat to body surfaces by blood flow; vasodilation at warm ambient temperatures enhances blood transport of heat to body surfaces where the heat is readily lost.

Insulation may also be modified by **postural responses** that alter the amount of body surface area directly exposed to ambient conditions. At low ambient temperatures, for example, mammals often curl up, and some birds tuck their heads under their body feathers or squat so as to enclose their legs in their ventral plumage. Many birds hold their wings away from their bodies when ambient temperatures are high.

In addition to the insulative properties that can be modulated by an individual animal, there are also properties that affect insulation but are more or less fixed for any given individual. Outstanding among these is body size. At temperatures *below thermoneutrality*, small size tends to increase the weight-specific rate at which animals lose heat—and thus the weight-specific cost of thermoregulation—because relatively small animals have more body surface per unit of weight than large ones have.[18] Small size also tends to enhance heat loss because small animals cannot have as thick pelage or plumage as large ones can have. Whereas large mammals commonly have pelage that is at least 5 to 6 cm thick, mice could not conceivably have such thick pelage; a mouse with 5 to 6 cm of pelage would be trapped inside its own hair!

Heat production is increased below thermoneutrality by shivering and nonshivering thermogenesis

When a mammal or bird is below its lower-critical temperature, it must increase its rate of heat production as the ambient temperature falls. Although all metabolic processes produce heat as a by-product, mammals and birds have evolved mechanisms, termed **thermogenic mechanisms**, that are specialized to generate heat for thermoregulation. One of these, shivering, is universal in adult mammals and birds.

SHIVERING **Shivering** is unsynchronized contraction and relaxation of skeletal-muscle motor units in high-frequency rhythms, mediated by motor neurons (nerve cells) of the somatic nervous system. Skeletal muscles can basically contract in two patterns. When muscles are being employed in locomotion to move a limb, all the motor units in each muscle contract synchronously, and antagonistic muscles contract alternately. When the same muscles are employed in shivering, various motor units within each muscle contract more or less at random relative to each other, antagonistic muscles are activated simultaneously, and the muscles quiver. Either mode of contraction uses ATP and liberates heat. When a muscle shivers, the conversion of ATP-bond energy to heat becomes the primary function of contraction because no useful mechanical work is accomplished.

NONSHIVERING THERMOGENESIS The concept of **nonshivering thermogenesis (NST)** is most readily understood by taking a look at the classic studies on laboratory rats that originally led to the discovery of NST. If lab rats that have been living at warm temperatures are transferred to a room at 6°C, they shiver violently during their first days in the new environment. If one observes them visually over the next few weeks as they acclimate to 6°C, however, they appear gradually to stop shivering even though they continue to maintain elevated metabolic rates. This visual observation suggests that during acclimation to cold, the rats develop mechanisms of thermogenesis that do not involve shivering. To test if this is in fact the case, cold-acclimated rats can be injected with curare, a plant extract that blocks the contraction of skeletal muscle and therefore prevents shivering. Curare-injected, cold-acclimated rats continue to have elevated metabolic rates and thermoregulate, confirming that they have well-developed nonshivering thermogenic mechanisms. Whereas shivering is universal in mammals and birds, NST is not. NST is best known, and very common, in placental mammals. It has been reported in the young of a few species of birds (e.g., ducklings), but its occurrence in adult birds remains controversial.

Of all the possible sites of NST in placental mammals, the one that is best understood and dominant is **brown adipose tissue (BAT)**, also called **brown fat.**[19] This is a specialized type of adipose tissue—often reddish brown—that differs greatly in both structure and function from the commonplace white fat that is more familiar. Brown fat tends to occur in discrete masses, located in such parts of the body as the interscapular region, neck, axillae, and abdomen (Figure 8.25). Deposits of BAT receive a rich supply of blood vessels and are well innervated by the sympathetic nervous system. The cells of BAT are distinguished by great numbers of relatively large mitochondria. The rich, red blood supply of BAT and the abundant, yellow cytochrome pigments in its mitochondria impart to the tissue its distinctive color.

When the sympathetic nervous system releases norepinephrine in BAT, the tissue responds by greatly increasing the rate of oxidation of its stored lipids, resulting in a high rate of heat production. BAT is biochemically specialized to undergo uncoupling of oxidative phosphorylation from electron transport (see Box 6.1). Indeed, uncoupling of oxidative phosphorylation is the mechanism by which BAT is able to produce heat rapidly because uncoupling does two things: (1) It releases aerobic catabolism from ordinary controls, permitting unbridled rates of lipid oxidation; and (2) it causes the chemical-bond energy of oxidized lipid molecules to be released immediately as heat (rather than being stored in ATP). What gives BAT its special-

[18] Be certain that you do not extend this argument to the thermoneutral zone, where it probably does not apply (see page 142).

[19] Brown adipose tissue does not occur in birds. In young birds that show NST, skeletal muscles are apparently the NST site.

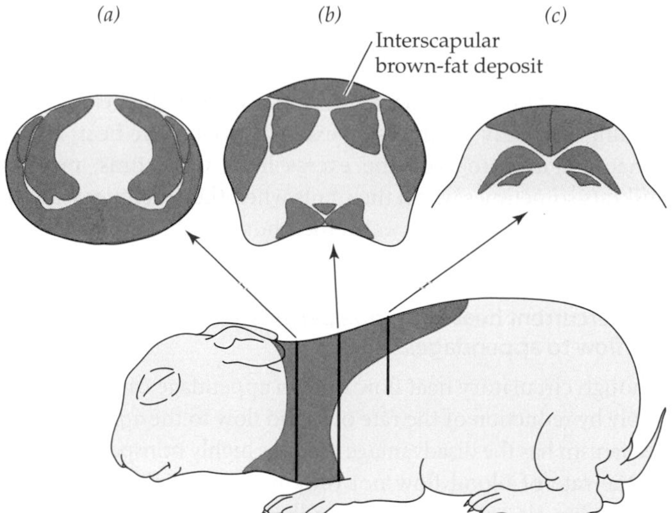

Figure 8.25 The deposits of brown adipose tissue in newborn rabbits *(a), (b),* and *(c)* are cross sections of the body at the positions indicated on the side view. Brown adipose tissue also occurs in discrete deposits in adult mammals that have the tissue. (After Dawkins and Hull 1964.)

ized ability to undergo uncoupling is the presence of a distinctive protein, **uncoupling protein 1** (**UCP1**; **thermogenin**), in the inner membranes of its mitochondria (see Box 6.1).

Norepinephrine released in BAT binds to β-adrenergic receptors (and other receptors) in the cell membranes of the BAT cells. These receptors are G protein–coupled receptors; as discussed in Chapter 2 (see Figure 2.26), the binding of norepinephrine to the receptors activates G proteins in the cell membranes and leads to the intracellular production of the second messenger cyclic AMP. Cyclic AMP then brings about the activation (by phosphorylation) of an intracellular lipase enzyme that rapidly hydrolyzes triacylglycerols stored in the cells to release free-fatty-acid fuels for mitochondrial oxidation. Simultaneously, by a mechanism that remains ambiguous, existing molecules of the uncoupling protein UCP1 are activated, and thus the mitochondria carry out the lipid oxidation in an uncoupled state. If norepinephrine stimulation continues for tens of minutes or longer, increased amounts of UCP1 are synthesized because β-adrenergic activation stimulates increased transcription of the gene that encodes UCP1.

Brown fat—like NST in general—is particularly prominent in three types of placental mammals: (1) cold-acclimated or winter-acclimatized adults (particularly in species of small to moderate body size), (2) hibernators, and (3) newborn individuals. During acclimation to cold or acclimatization to winter, BAT often markedly increases its potential to produce heat; deposits of the tissue grow, BAT cells increase their numbers of mitochondria, and the mitochondria become richer in uncoupling protein. BAT in newborns (including human babies) and hibernators is one of the major topics in Chapter 9.

Regional heterothermy: In cold environments, allowing some tissues to cool can have advantages

Appendages such as legs, tails, and ear pinnae present particular thermal challenges when mammals and birds are below thermoneutrality. The appendages are potentially major sites of heat loss because they have a great deal of surface area relative to their sizes, are often thinly covered with fur or feathers, and exhibit (because of their dimensions) intrinsically high rates of convective heat exchange (see Equation 8.3). If a mammal or bird in a cold environment were to keep its appendages at the same temperature as its body core, the appendages would contribute disproportionately, for their weight, to the animal's overall weight-specific metabolic cost of homeothermy.

A mammal or bird can limit heat losses across its appendages in cool environments by allowing the appendages to cool. The difference between the temperature of an appendage and the ambient temperature is the driving force for heat loss from the appendage. Allowing the appendage to cool toward ambient temperature reduces this driving force, in effect compensating for the appendage's relatively low resistance to heat loss. Cooling of the appendages, a type of regional heterothermy, is in fact very common. When the ambient temperature is low, the tissues of appendages—especially their distal parts—are often 10°C to 35°C cooler than tissues in the core parts of an animal's thorax, abdomen, and head (Figures 8.26 and 8.27).[20]

The usual mechanism by which appendages are allowed to cool is the curtailment of circulatory heat delivery to them. Many appendages (or parts of appendages) consist in large part of bone, tendon, cartilage, skin, and other tissues that metabolically are relatively inactive. Such appendages typically do not have sufficient endogenous heat production to keep themselves warm in cold environments. Their temperatures depend on how rapidly heat is brought to them from the thorax, abdomen, or head by the circulating blood, and curtailing circulatory heat delivery lets them cool. Heat delivery to an appendage may be curtailed simply by the re-

[20] Because regional heterothermy reduces the total metabolic cost of maintaining a given core body temperature, it effectively increases insulation (I) in the linear heat-transfer equation (Equation 8.10).

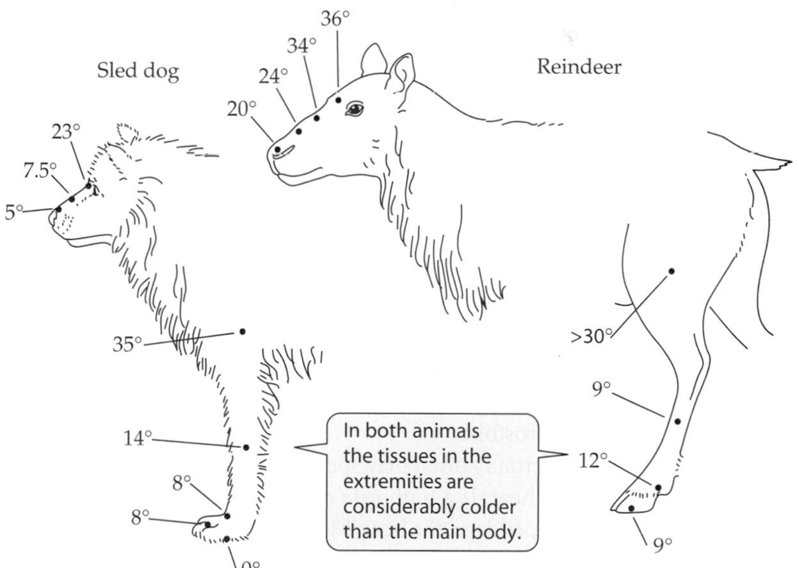

Figure 8.26 Regional heterothermy in Alaskan mammals The air temperature was –30°C when these data were gathered. The temperatures shown are subcutaneous temperatures (°C) at various spots on the body. Note that the footpads of the sled dog were at 0°C. Both animals had deep-body temperatures in the typical mammalian range: 37°C to 38°C. (After Irving and Krog 1955.)

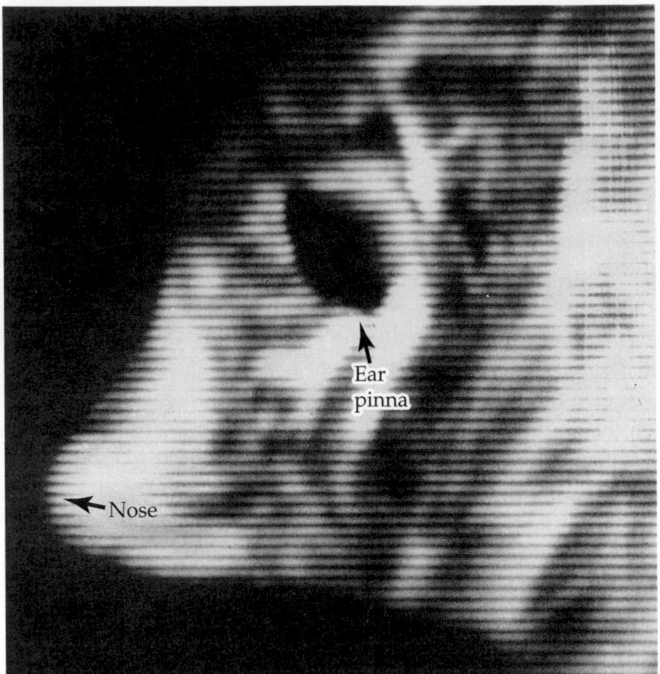

Figure 8.27 A thermal map of an opossum showing regional heterothermy in the pinna of the ear In this image, which was produced by infrared radiography, shades of gray represent temperatures on the animal's body surface. Surfaces that were close to ambient temperature are colored black; those that were much warmer than ambient are colored white; intermediate temperatures are represented by intermediate gray tones. The animal, a Virginia opossum (*Didelphis marsupialis*), was studied at an ambient temperature of 10°C. Note that the surface temperature of the ear pinna, a naked flap of skin, was virtually the same as ambient, whereas other parts of the body, notably the snout, were much warmer than ambient.

striction of blood flow to the appendage, but as we will see in the next section, more-elaborate mechanisms of restricting heat delivery are usually employed.

Species that have long evolutionary histories in frigid climates often display exquisite control over the extent of appendage cooling. For example, in a variety of Arctic canids—including foxes and wolves, as well as sled dogs—the tissues of the footpads are routinely allowed to cool to near 0°C in winter (see Figure 8.26), but even when the feet are in contact with much colder substrates (e.g., –30°C to –50°C), the footpads are not allowed to cool further. Thus the footpads are kept at the lowest temperature that does not subject them to a risk of frostbite.

Appendages actually often play special roles in the dissipation of excess metabolic heat. If a high rate of circulatory heat delivery is provided to an appendage, the heat is lost readily to the environment because of the ease of heat loss from appendages. Whereas animals curtail circulatory heat delivery to their appendages when heat conservation is advantageous, they often augment heat delivery to their appendages when they need to get rid of heat. In a cool environment, for example, when black-tailed jackrabbits are at rest, they limit blood flow to their huge ear pinnae—so much so that the pinnae become as cool as the air; but when the jackrabbits run, they increase blood flow and pinna temperature considerably (Figure 8.28). Running evidently produces an excess of metabolic heat, and the pinnae are used to void the excess heat. Opossums, rats, and muskrats sometimes warm their tails when they exercise; seals heat up their flippers; and goats warm their horns.

Countercurrent heat exchange permits selective restriction of heat flow to appendages

Although circulatory heat flow into an appendage can be curtailed simply by reduction of the rate of blood flow to the appendage, this mechanism has the disadvantage that it is highly nonspecific. Reducing the rate of blood flow not only limits heat flow into an appendage but also subjects the appendage to a reduced rate of oxygen supply, a reduced rate of waste removal, and other consequences. Heat flow to an appendage can be *selectively* curtailed by **countercurrent heat exchange**, a process that depends on a specialized morphological arrangement of the blood vessels carrying blood to and from the appendage.

To understand countercurrent heat exchange, let's examine the two different arrangements of the arteries and veins in a limb diagrammed in Figure 8.29. The arteries are located deep within the appendage. In Figure 8.29*a* the veins are superficial, but in Figure 8.29*b*

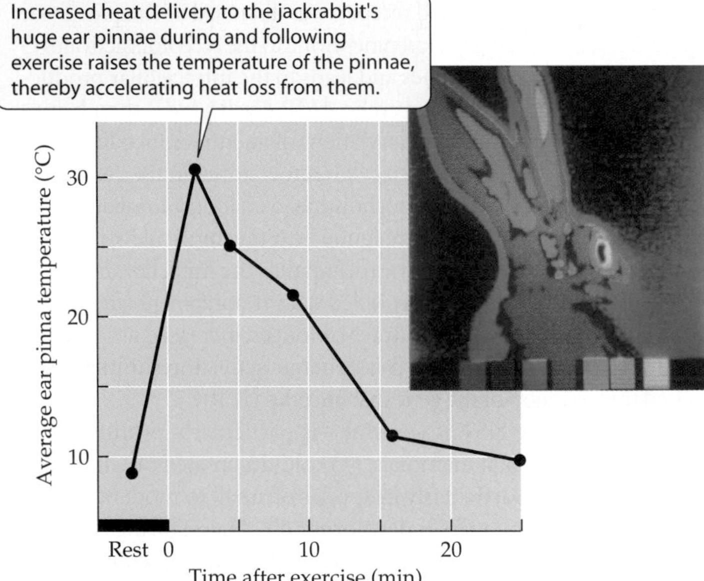

Figure 8.28 Heat loss across appendages is sometimes modulated in ways that aid thermoregulation The average surface temperature of the ear pinnae of this jackrabbit (*Lepus californicus*) was near ambient temperature (8°C) when the rabbit was resting but increased to over 30°C following a 6-minute bout of running. (Pinna temperature could not be measured during running itself. Data are presented for the period of rest before running and for various times after running stopped.) The inset is an infrared radiograph of the jackrabbit when it had an elevated ear-pinna temperature after exercise. In this presentation, temperature is color-coded. The color blocks at the bottom symbolize increasing temperatures from left to right. The environment fell into the range of temperatures coded by green. Part of the right ear pinna was warm enough to fall within the much higher temperature range coded by yellow. (After Hill, Christian, and Veghte 1980.)

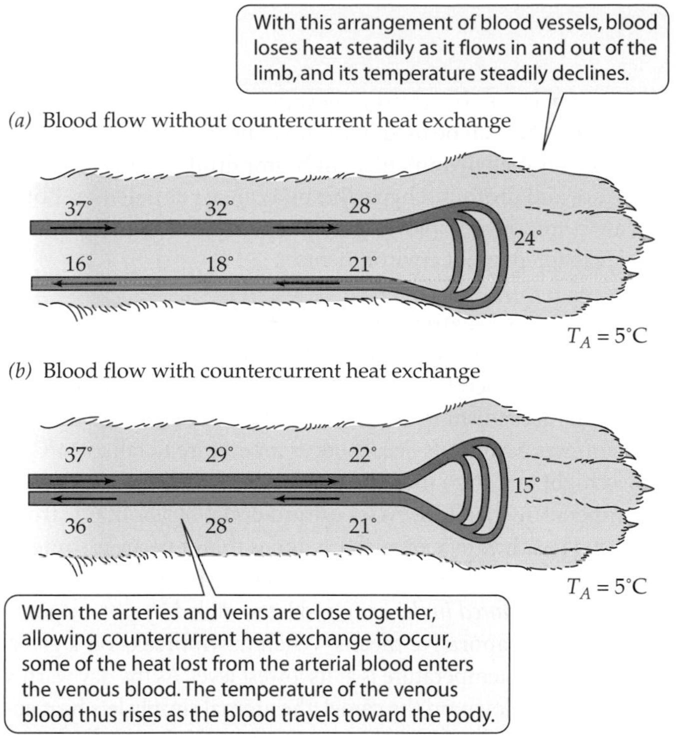

With this arrangement of blood vessels, blood loses heat steadily as it flows in and out of the limb, and its temperature steadily declines.

(a) Blood flow without countercurrent heat exchange

37° 32° 28°
24°
16° 18° 21°
$T_A = 5°C$

(b) Blood flow with countercurrent heat exchange

37° 29° 22°
15°
36° 28° 21°
$T_A = 5°C$

When the arteries and veins are close together, allowing countercurrent heat exchange to occur, some of the heat lost from the arterial blood enters the venous blood. The temperature of the venous blood thus rises as the blood travels toward the body.

Figure 8.29 Blood flow with and without countercurrent heat exchange Arrows show blood flow in both parts. All temperatures are in degrees Celsius (°C). *(a)* In this arrangement, which does not permit countercurrent heat exchange, the veins returning blood from the limb are just under the skin and separate from the arteries that carry blood into the limb. *(b)* In this case, countercurrent heat exchange can occur because the veins returning blood from the limb are closely juxtaposed to the arteries carrying blood into the limb. In part *b* the arterial blood is cooled more than in part *a* because of the close proximity of cool venous blood. Furthermore, in *b*, more heat is returned to the body than in part *a* because heat that enters the venous blood is carried back to the body rather than being lost to the environment.

the veins are closely juxtaposed to the arteries. The vascular arrangement in Figure 8.29*a* does nothing to conserve heat; as blood flows into the appendage through the arteries and then flows back through the veins, it loses heat all along the way, without any opportunity to regain it. In contrast, the vascular arrangement in Figure 8.29*b* promotes heat conservation because it encourages a transfer of heat from the arterial blood to the venous blood rather than simple heat loss from the blood; the venous blood flowing in close proximity to the arteries picks up heat lost from the arterial blood and carries it back to the body core, thereby helping to keep heat in the body. If the area of contact between the veins and arteries in Figure 8.29*b* is sufficiently extensive, blood may be little cooler when it reenters the body core than it was when it flowed into the appendage. The heat exchange in Figure 8.29*b* is *countercurrent heat exchange*. By definition, such heat exchange depends on the transfer of heat between two closely juxtaposed fluid streams flowing in opposite directions (*counter* = "opposite"; *current* = "flow").

A particularly useful way to conceive of the effect of countercurrent heat exchange in an appendage is to think of it as *short-circuiting the flow of heat into the appendage.* Figure 8.30 illustrates that in the presence of a suitable vascular arrangement, although *blood* flows all the way to the end of an appendage before returning to the body core, *heat* tends to flow only part of the length of the appendage before it short-circuits from the arteries to the veins and starts its return to the body core. This short-circuiting impedes the access of heat to the outer extremities of the appendage. The outer extremities are therefore cooler than they otherwise would be, limiting heat loss to the environment.

A vascular countercurrent exchange system short-circuits the flow of only those commodities that are able to pass through the walls of the blood vessels involved. Heat is short-circuited by the vascular systems we have been discussing precisely because heat can pass through the walls of arteries and veins. If oxygen, nutrients, or wastes could pass through the walls of arteries and veins, they, too, would be short-circuited. However, they cannot pass through such thick-walled vessels, and thus they travel with the blood all the way to the outer limits of an appendage and back. This is how selectivity is achieved, explaining how a vascular system can conserve heat while not affecting the flow of other commodities in and out of an appendage.

Vascular arrangements that meet the prerequisites for countercurrent heat exchange (close juxtaposition of arteries and veins) are commonly found in appendages that display regional heterothermy. Such vascular arrangements are known, for example, in the arms of people, the legs of many mammals and birds, the flippers and flukes (tail fins) of whales, the tails of numerous rodents, and the ear pinnae of rabbits and hares. Anatomically the vascular arrangements vary from relatively simple to highly complex. All that exists in some simple cases is ordinary veins and arteries touching each other; this is the case in the human arm. Greater complexity is found in the flippers and flukes of whales, where the major arteries are almost completely *surrounded* by venous channels; with this arrangement, heat leaving the arteries can hardly help but enter venous blood. Another type of complex arrangement found in some animals (e.g., armadillos and some storks) is that the main arteries and veins in a limb split up to form a great many fine vessels that intermingle. A complex network of tiny vessels like this is termed a **rete mirabile** ("wonderful net") or simply a **rete.**[21]

[21] The word *rete* is pronounced with both syllables rhyming with eat: "ree-tee." *Rete mirabile* and *rete* are general terms used to refer to intricately complex systems of small-diameter arterial and venous vessels wherever they occur. We will encounter many additional examples in this book.

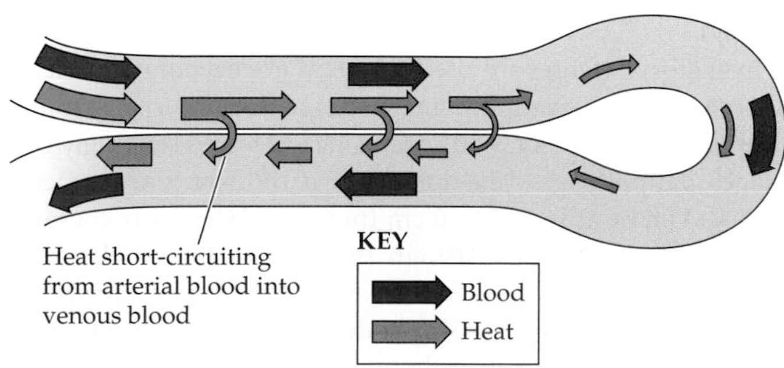

Figure 8.30 Countercurrent heat exchange short-circuits the flow of heat in an appendage In a vascular countercurrent exchanger, commodities that can pass through the walls of the blood vessels short-circuit from one fluid stream to the other. Other commodities travel with the blood. This illustration shows heat being short-circuited. The widths of the arrows symbolize the relative magnitudes of heat flow and blood flow from place to place.

A common way for countercurrent heat exchange to be controlled is for an appendage to have *two* sets of veins, only one of which is juxtaposed to the arteries. Countercurrent exchange can then be activated or deactivated through control of the set of veins in use. In the arm of a person, for example, one set of veins is deep in the arm and closely juxtaposed to the arteries, whereas a second set is just under the skin. Under control of the autonomic nervous system, the deep set of veins is used when there is a premium on heat conservation, but the superficial set is used when heat loss is advantageous. These controls explain why the superficial veins of our arms seem to disappear on cold days while bulging with blood on warm days.

Mammals and birds in hot environments: Their first lines of defense are often not evaporative

Sweating, panting, and other modes of actively increasing the rate of evaporative cooling are so easy to observe when they occur that they are often thought to be the principal or only means by which mammals and birds cope with high environmental or metabolic heat loads. Evaporation, however, has a potentially lethal price: It carries body water away. Although evaporative cooling may solve problems of temperature regulation, it may create problems of water balance. For many mammals and birds, especially species that have long evolutionary histories in hot, arid climates, active evaporative cooling is in fact a *last* line of defense against heat loading. Other defenses are marshaled preferentially, and only when these other defenses have done as much as they can is body water used actively to void heat. In this section we discuss the nonevaporative defenses. When these defenses are employed as the preferential or first-line defenses, they act as water-conservation mechanisms.

Behavioral defenses are one set of commonly employed nonevaporative defenses. Examples abound. Desert rodents, for instance, construct burrows, which they occupy during the day (see Figure 1.15), and most emerge on the desert surface only at night. By means of these behaviors, they evade the extremes of heat loading that could occur in deserts. Mammals and birds that are active during daylight hours often seek shade during the heat of the day, and they often restrict their activity at particularly hot times, thus minimizing their metabolic heat loads. Resting camels shift the positions of their bodies to present a minimum of surface area to the sun throughout hot days.

Insulatory defenses are also important nonevaporative defenses in some cases. For example, some species of large, diurnal mammals and birds native to hot, arid regions have evolved strikingly thick pelages and plumages. The dorsal pelage of dromedary camels in summer can be at least 5 to 6 cm thick, and when ostriches erect their plumage, it can be 10 cm thick. Such thick pelages and plumages probably evolved because in very hot environments, they can act as heat shields, increasing body insulation and thereby acting as barriers to heat influx from the environment. The outer surface of the dorsal pelage of camels and sheep has been measured to get as hot as 50°C to 80°C when exposed to solar radiation on hot days! The pelage shields the living tissues of the animals from these enormous heat loads, and much of the heat intercepted by the pelage disperses back into the environment by reradiation and convection, without reaching the living tissues.

Body temperature is a third nonevaporative attribute of mammals and birds that can be used in the first line of defense against the challenges of hot environments. High-amplitude cycling of body temperature and profound hyperthermia can act as defenses. Both, in fact, are commonly employed as water-conservation mechanisms by species adapted to hot environments.

CYCLING OF BODY TEMPERATURE Dromedary camels provide a classic and instructive example of how animals can employ high-amplitude cycling of body temperature as a nonevaporative defense and water-conservation mechanism in hot environments. A dehydrated dromedary in summer permits its deep-body temperature to fall to 34°C to 35°C overnight and then increase to over 40°C during each day. Its body temperature therefore cycles up and down by about 6°C from day to day. The advantage of such cycling is that it permits some of the heat that enters the body during the intensely hot part of each day to be temporarily *stored in the body* and later voided by nonevaporative rather than evaporative means. When dawn breaks on a given day, a camel's body temperature is at its lowest level. As the day warms and the sun beats down on the camel, the animal simply lets heat accumulate in its body, rather than sweating to void the heat, until its body temperature has risen by 6°C. Physiologists have measured that about 3.3 J (0.8 cal) is required to warm 1 g of camel flesh by 1°C. From this figure, one can calculate that a 400-kg camel will accumulate about 7920 kJ (1900 kcal) of heat in its body by allowing its body temperature to rise 6°C; to remove this amount of heat by evaporation would require over 3 L of water, but the camel simply stores the heat in its body. Later, after night falls and the environment becomes cooler, conditions become favorable for convection and radiation to carry heat out of the camel's body. At that point the camel is able to get rid of the heat stored during the day by nonevaporative means. Its body temperature falls overnight to its minimum, poising the animal to take full advantage of heat storage during the following day, thereby again saving several liters of water.

Many large-bodied species of mammals native to hot, arid regions are known to undergo high-amplitude cycles of body temperature on a daily basis as the dromedary camel does. Some small-bodied species take advantage of cycling by undergoing multiple cycles each day. Antelope ground squirrels, for example, forage during daylight hours in hot deserts by undertaking numerous forays from their burrows. They heat up during each foray and then cool nonevaporatively during each return underground.

HYPERTHERMIA Many mammals and birds employ controlled, profound hyperthermia as a principal nonevaporative and water-conserving mechanism of coping with hot environments. Because a rise in body temperature entails heat storage, the benefits of hyperthermia are to some extent the very ones we have just noted in discussing cycling. In addition, however, a high body temperature *in and of itself* holds advantages for water conservation. As mentioned already, under conditions when dry heat *loss* occurs, a high T_B promotes such nonevaporative heat loss by elevating the driving force ($T_B - T_A$) that favors it. A high T_B also aids water conservation under conditions when an animal has stored as much heat as it can and yet the environment is so hot that dry heat *gain* occurs. Under such conditions, evaporation of water must be used to get rid of all the heat that enters an animal's body. A high T_B impedes the heat gain from the environment by decreasing the driving force ($T_A - T_B$) that favors heat influx, and thus the high T_B reduces the rate at which body water must be evaporated to void the incoming heat.

Birds almost universally permit their body temperatures to rise to profoundly high levels when in hot environments; whereas resting birds typically have body temperatures near 39°C in the absence of heat stress, they commonly have body temperatures as high as 44°C to 45°C in hot environments. Among mammals, profound hyperthermia typically occurs only in species with long evolutionary histories in hot, arid climates, but among such species it is common. Certain antelopes native to the deserts and dry savannas of Africa provide the extreme examples. Two such species, the beisa oryx (*Oryx beisa*) and Grant's gazelle (*Gazella granti*), sometimes permit their rectal temperatures to reach 45.5°C to 47°C (114°F–116°F) without ill effect!

KEEPING A COOL BRAIN A large body of evidence indicates that the brain is kept cooler than the thorax and abdomen in quite a few species of mammals and birds when the animals are in warm or hot environments, especially during exercise. Although some recent research reports have argued that brain cooling is an artifact, the consensus view at present is that it is real. To cite an extreme example, when a Thomson's gazelle (*Gazella thomsonii*) runs vigorously in a warm environment, its brain is kept as much as 2.7°C cooler than its thorax. Camels and dogs are other species that exhibit brain cooling.

The advantage of brain cooling is believed to be that it permits an animal to take enhanced advantage of the benefits of high-amplitude body-temperature cycling and hyperthermia. The brain tolerates less of an elevation of temperature than most organs. Thus the bulk of an animal's body can cycle to a higher temperature, and become more hyperthermic, if the brain can be kept from being as hot as most of the body.

What is the mechanism of brain cooling? In many cases, the key process is cooling of arterial blood supplying the brain by countercurrent heat exchange (Figure 8.31). The arteries carrying blood toward the brain from the heart come into intimate contact with veins or venous blood draining the nasal passages and other upper respiratory passages. The site of this contact in many of the mammals involved is the cavernous sinus located at the base of the skull; there the arteries divide into a plexus of small vessels (the *carotid rete mirabile*) that is immersed in a lake of venous blood. As noted, the venous blood juxtaposed to the arteries is traveling back toward the heart from the upper respiratory passages. Blood in the upper respiratory passages is cooled by the inevitable evaporation of water from the walls of the respiratory passages into breathed air. As the cooled venous blood traveling back to the heart flows by the arteries, it cools the arterial blood traveling toward the brain.

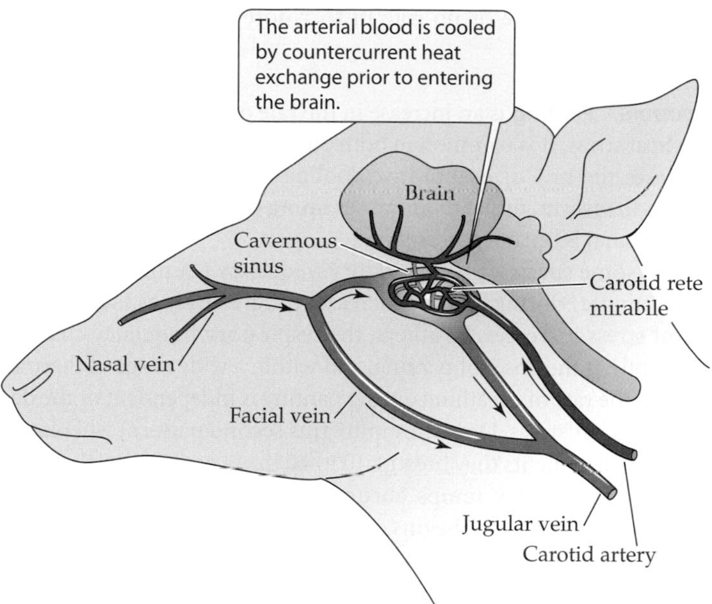

Figure 8.31 Structures hypothesized to be responsible for cooling the brain in sheep and other artiodactyls The carotid artery branches and anastomoses in the cavernous sinus, forming the carotid rete mirabile. Venous blood from the upper respiratory passages flows around the vessels of the carotid rete. (After Maloney and Mitchell 1997.)

Active evaporative cooling is the ultimate line of defense against overheating

Active facilitation of evaporation is the ultimate line of defense for mammals and birds faced with high environmental or metabolic (e.g., exercise-induced) heat loads. If heat is accumulating in the body to excessive levels and all the other means we have already discussed fail to stop the accumulation, active evaporative cooling becomes the only mechanism available to reestablish a balance between heat gain and heat loss. As stressed earlier, the loss of water during evaporative cooling can dehydrate an animal if replacement water is not readily available; this probably explains why species native to arid habitats employ other defenses against overheating before turning to evaporative cooling. Four major mechanisms of active evaporative cooling are known: sweating, panting, gular fluttering, and saliva spreading.

SWEATING During **sweating**, fluid is secreted by way of the sweat-gland ducts through the epidermis of the skin onto the skin surface. A slow rate of water loss occurs through the substance of the skin when an animal is not sweating.[22] Sweating increases the rate of cutaneous evaporation by a factor of 50 or more by wetting the outer surface of the skin. Sweat is not pure water but instead is a saline solution. Concentrations of Na^+ and Cl^- in sweat are lower than in the blood plasma, and during acclimation to hot conditions the salinity of sweat becomes reduced. Nonetheless, prolonged sweating can impose a significant drain on the body's pool of Na^+ and Cl^-. Secretion by the sweat glands is activated by the sympathetic nervous system.

A capability to sweat vigorously is found in a variety of mammals, including humans, horses, camels, and some kangaroos. Sweat production can be profuse. Humans working strenuously in the desert, for example, can attain sweating rates of 2 L/h! Many types of mammals, however, do not sweat. Rodents, rabbits, and hares lack integumentary sweat glands. Although dogs and pigs have sweat glands, the secretion rates of the glands are so low that sweating ap-

[22] This is termed *transpirational water loss* or *insensible* ("unperceived") *water loss.*

pears to play little or no role in thermoregulation. Birds do not sweat.

PANTING **Panting** is an increase in the rate of breathing in response to heat stress. It is common in both birds and mammals. Panting increases the rate of evaporative cooling because water evaporates from the warm, moist membranes lining the respiratory tract into the air that is breathed in and out.

In some species, the respiratory frequency (number of breaths per minute) during panting increases progressively as the extent of heat stress increases. In others, the respiratory frequency changes abruptly at the onset of panting, and within a wide range of thermal stress, the rate of breathing during panting is independent of the degree of heat stress. Dogs exemplify this second pattern; whereas in cool environments they breathe 10 to 40 times per minute, their respiratory frequency jumps abruptly to 200 or more breaths per minute when panting begins. Analysis indicates that animals with such a stepwise change in respiratory frequency often pant at the *resonant* frequency of their thoracic respiratory structures. Use of the resonant frequency means that less muscular work must be done to pant (and less heat is produced by the muscular work) because the thorax has an intrinsic tendency to "vibrate" between its inspiratory and expiratory positions at that frequency.

By comparison to sweating, panting holds certain advantages. One is that no salts are lost during panting, because evaporation occurs within the body and only pure water vapor leaves the body in the exhalant air. A second advantage of panting is that it *forcibly* drives air saturated with water vapor away from the evaporative surfaces.

Panting also has liabilities in comparison to sweating. Because of the muscular effort required for panting, evaporation of a given quantity of water is likely to require more energy—and entail more heat production—when panting is employed than when sweating is. Another potential liability of panting is that it can induce *respiratory alkalosis*, an elevation of the pH of the body fluids caused by excessive removal of carbon dioxide (see page 604). Ordinarily, when animals are not panting, ventilation of the respiratory-exchange membranes deep in their lungs (e.g., the alveolar membranes of mammals) is closely regulated so that the rate at which CO_2 is voided is equal to the rate of metabolic production of CO_2. During panting, the potential exists for breathing to carry CO_2 away faster than it is produced, because the rate of breathing is increased for thermoregulation rather than just being governed by metabolic needs. If CO_2 is carried away by breathing faster than it is produced by metabolism, the concentration of CO_2 in the blood will fall, causing the following reactions in the blood to shift to the left:

$$CO_2 + H_2O \rightleftharpoons H_2CO_3 \rightleftharpoons H^+ + HCO_3^- \qquad (8.11)$$

Consequently, the concentration of H^+ in the blood will fall, and the pH of the blood will rise. Such excessive alkalinity—*alkalosis*—can have major deleterious effects because many enzymes and cellular processes are acutely sensitive to pH. (In middle school, we probably all witnessed friends making themselves dizzy by deliberately breathing too rapidly.)

From extensive research, physiologists now know that little or no alkalosis develops during panting in many species of mammals and birds when the heat stress to which they are exposed is light to moderate. These animals avoid alkalosis by restricting the increased air movement during panting to just their upper airways,[23] where no exchange of CO_2 occurs between the air and blood (Figure 8.32); because the increased airflow is restricted to the upper airways, the respiratory-exchange membranes deep in the lungs receive about the same rate of airflow during panting as they usually do. By contrast, when heat stress becomes extreme, resting but panting animals often develop severe alkalosis. Some panting species have evolved superior tolerance to alkalosis.

[23] In birds, both the upper airways and air sacs may be involved.

(*a*) Shallow breaths superimposed on deep breaths (compound breathing)

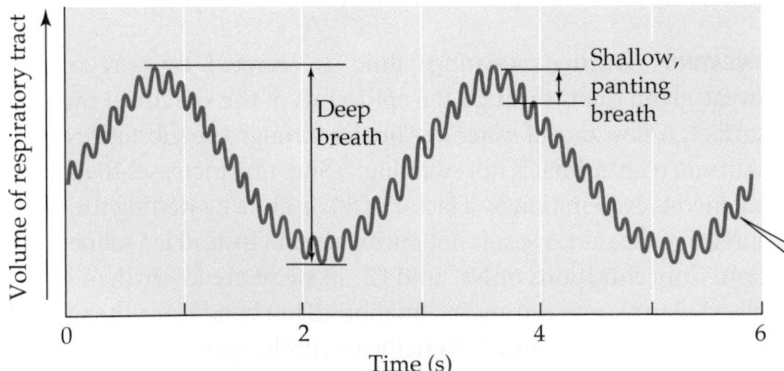

(*b*) Shallow breaths occasionally interrupted by deep breaths

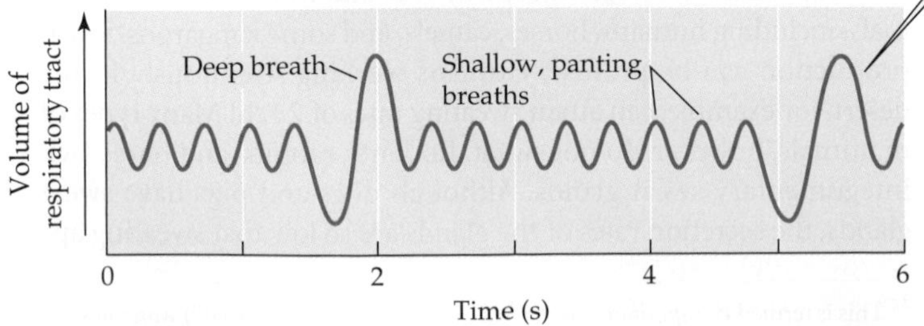

Figure 8.32 Two breathing patterns that limit hyperventilation of the respiratory-exchange membranes during panting Shallow, panting breaths move relatively small volumes of air in and out of the respiratory passages and are too shallow to carry air to the respiratory-exchange membranes deep in the lungs. Deep breaths (resembling ordinary breaths) move relatively large volumes of air in and out, and ventilate the respiratory-exchange membranes. In (*a*) shallow breaths are superimposed on deep breaths; in (*b*) periods of shallow breathing alternate with periods of deep breathing.

GULAR FLUTTERING Many birds (but not mammals) augment evaporative cooling by rapidly vibrating their gular area (the floor of the mouth cavity) while holding the mouth open, a process termed **gular fluttering** (see Figure 8.24). The process is driven by flexing of the hyoid apparatus and promotes evaporation by increasing the flow of air over the moist, highly vascular oral membranes. Gular fluttering usually occurs at a consistent frequency, which apparently matches the resonant frequency of the structures involved. Gular fluttering is commonly employed simultaneously with panting.

Gular fluttering shares certain positive attributes with panting: It creates a vigorous, forced flow of air across evaporative surfaces and does not entail salt losses. Unlike panting, gular fluttering cannot induce severe alkalosis, because it enhances only oral airflow, and CO_2 is not exchanged between air and blood across oral membranes. Gular fluttering involves the movement of structures that are less massive than those that must be moved in panting; thus it entails less muscular work—and less heat production—to achieve a given increment in evaporation.

SALIVA SPREADING Many rodents and marsupials spread saliva on their limbs, tail, chest, or other body surfaces when under heat stress. Spreading of saliva on furred regions of the body is a relatively inefficient use of body water for cooling because the evaporative surface created—on the outer surface of the fur—is insulated from the living tissues of the animal's body by the pelage. Nonetheless, saliva spreading is useful in certain circumstances. For many rodents, saliva spreading is the only means available to increase evaporative cooling, and the animals use it in emergency situations to save themselves from lethal hyperthermia while they seek a cool environment.

(a) Acclimatization of peak metabolic rate

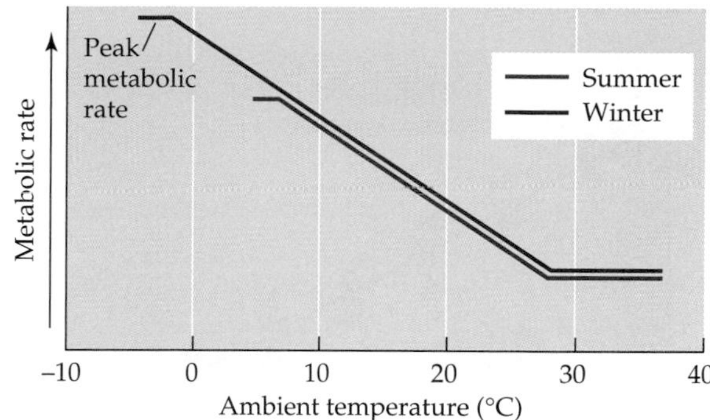

(b) Insulatory acclimatization

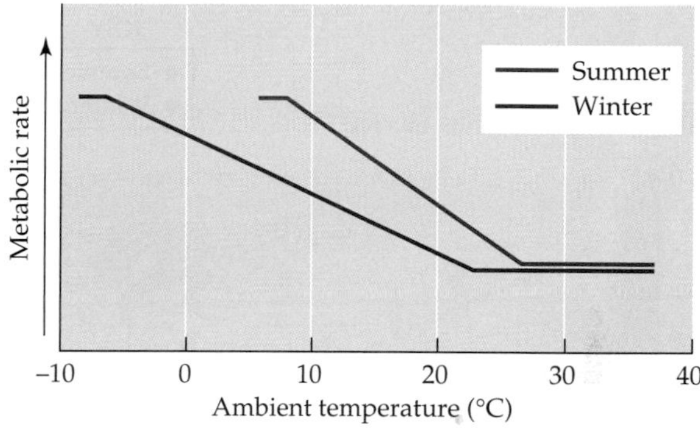

Figure 8.33 Two types of seasonal acclimatization *(a)* Acclimatization of peak metabolic rate without insulatory acclimatization. *(b)* Insulatory acclimatization without acclimatization of peak metabolic rate. The plateau of each curve at the left indicates where metabolic rate has peaked.

Mammals and birds acclimatize to winter and summer

When individual mammals and birds live chronically in cold or warm environments, they usually undergo long-term alterations in their thermoregulatory physiology. During acclimatization to winter, for example, a mammal or bird typically exhibits one or more of three sorts of chronic responses.[24] Because the change of seasons is complex, these responses are not necessarily triggered solely (or even primarily) by the drop in temperature as winter approaches but may be triggered by photoperiod (shortening days) or other seasonal cues.

One possible chronic response to winter is **acclimatization of peak metabolic rate.** When a mammal or bird exhibits this response, it increases the maximal rate at which it can produce heat by sustained, aerobic catabolism. If an animal displays just this sort of acclimatization, although the metabolic rate it requires to thermoregulate at any given ambient temperature remains unchanged, it can thermoregulate in colder environments than it could before (Figure 8.33*a*). The development in winter of enlarged brown adipose tissues in which cells are biochemically especially well poised for heat production is a common mechanism by which small and medium-sized mammals increase the rate at which they can produce heat and thus undergo acclimatization of peak metabolic rate.

A second possible chronic response to winter is **acclimatization of metabolic endurance**, meaning an increase in the length of time that a high rate of metabolic heat production can be maintained. Although current evidence indicates that this sort of acclimatization is common, little is known about its mechanisms.

The third major sort of chronic response that a mammal or bird might exhibit during winter is **insulatory acclimatization**, an increase in its maximal resistance to dry heat loss (maximal insulation). If this sort of acclimatization occurs, the metabolic rate required to thermoregulate at any particular ambient temperature below thermoneutrality is reduced, and even if an animal's peak metabolic rate remains unchanged, the animal is able to thermoregulate in colder environments than it could before (Figure 8.33*b*). The most obvious way for insulatory acclimatization to occur is for an animal to molt into a more protective pelage or plumage in winter, but other determinants of insulation (such as peripheral blood flow) can also change.

Of three chronic responses to winter we have described, two (or all three) can occur together. For example, an animal might exhibit both acclimatization of peak metabolic rate and insulatory acclimatization. These two responses acting in concert could produce great enhancements of an animal's thermoregulatory ability.

[24] Acclimatization to summer, to some extent, involves simply the same chronic responses in reverse.

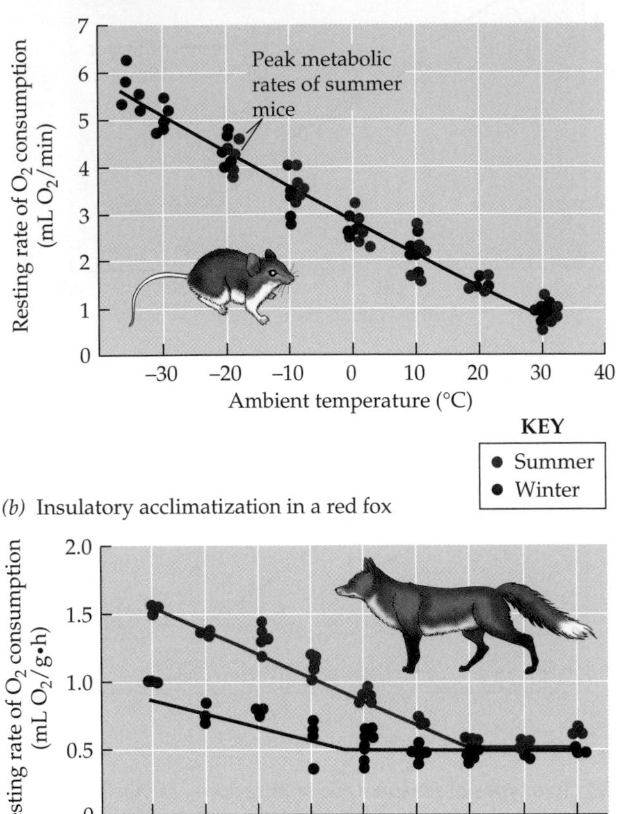

Figure 8.34 Seasonal acclimatization in two species of mammals *(a)* The deer mice (*Peromyscus maniculatus*) studied had the same insulation in winter and summer, but their peak metabolic rates rose in winter, meaning they could thermoregulate at lower ambient temperatures. *(b)* A single red fox (*Vulpes vulpes*) individual, studied in both seasons, had far greater insulation in winter than in summer. (After Hart 1957.)

Acclimatization of peak metabolic rate occurs in perhaps half of small birds and is the norm in small and medium-sized mammals. As for insulatory acclimatization, some small species of mammals and birds fail to exhibit it and thus undergo only metabolic forms of acclimatization (Figure 8.34*a*). Among the mammals that undergo insulatory acclimatization, medium-sized and large species tend to exhibit greater changes in insulation between summer and winter than small species. Red foxes (Figure 8.34*b*), collared lemmings, and varying hares in northern Alaska nicely illustrate the advantages of insulatory acclimatization. The air temperature there averages –30°C in winter and +5°C in summer. For the foxes, lemmings, and hares, the metabolic cost of thermoregulating at –30°C in winter is not much higher than the cost of thermoregulating at +5°C in summer, because of their winter increase in insulation.[25]

Mammals and birds commonly acclimatize to heat stress as well as cold stress, as seen in Chapter 1 (Figure 1.7). Among people, acclimatization to heat stress in fact occurs much more rapidly than that to cold stress. Partly for that reason, we tend to notice our own acclimatization to heat more than our acclimatization to cold.

Evolutionary changes: Species are often specialized to live in their respective climates

Abundant evidence suggests that the thermoregulatory physiology of mammals and birds has undergone evolutionary adaptation to different climates. One sort of evidence is provided in Figure 8.35, which is one of the classic sets of data in animal physiology. Species of mammals native to the Arctic and the tropics differ dramatically in their thermal relations; compared to tropical species, Arctic species have lower-critical temperatures and increase their metabolic rates proportionally less above basal levels at ambient temperatures below thermoneutrality. Direct studies of pelage insulation demonstrate that the Arctic species typically have thicker and better insulating pelages than similarly sized tropical species. As a consequence of all these differences, Arctic species are in a far better position to thermoregulate under Arctic conditions than tropical species are. Arctic foxes (*Alopex*) and reindeer (*Rangifer*) have lower-critical temperatures below –30°C! This means that just their basal rates of heat production are sufficient to maintain homeothermy until the air becomes exceedingly cold.

In hot climates, a major pattern that has emerged with ever-increasing clarity in recent decades is that species of both mammals and birds native to such climates often have lower basal metabolic rates than are observed in related species native to temperate or cold climates. The evolution of an exceptionally low BMR has probably been favored in hot climates because, with a low BMR, an animal has a particularly low internal heat load.

As mentioned earlier, body temperature is basically a *conserved* character; within any taxonomic group of mammals or birds, the core body temperature maintained in the absence of heat or cold stress tends to be the same in species from various climates. Adaptation of body temperature to different climates is clearly evident, however, in mammals exposed to heat stress; species native to hot climates typically tolerate greater degrees of hyperthermia than species native to temperate or cold climates do.

Mammals and birds sometimes escape the demands of homeothermy by hibernation, torpor, or related processes

Many species of mammals and birds allow their body temperatures to fall in a controlled manner under certain circumstances. **Controlled hypothermia** is the general term for this sort of phenomenon; **hypothermia** is the state of having an unusually low body temperature, and in the cases we are discussing, it is "controlled" because the animals orchestrate their entry into and exit from hypothermia rather than being forced.

The most well-known and profound forms of controlled hypothermia are **hibernation**, **estivation**, and **daily torpor**. These, by definition, are all states in which an animal *allows its body temperature to approximate ambient temperature within a species-specific range of*

[25] Acclimation of mammals or birds to cold in a laboratory sometimes has dramatically different effects than acclimatization to winter has, meaning that results of laboratory acclimation studies on these sorts of animals cannot confidently be extrapolated to acclimatization in nature. The distinction between acclimation and acclimatization is discussed in Chapter 1 (see page 14). One reason for the difference between acclimation to cold and acclimatization to winter in mammals and birds is that some of the winter responses occur only when there are changes in photoperiod, and usually just temperature is altered in laboratory studies.

KEY

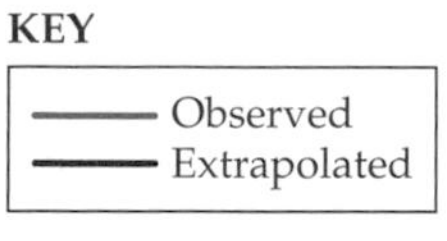

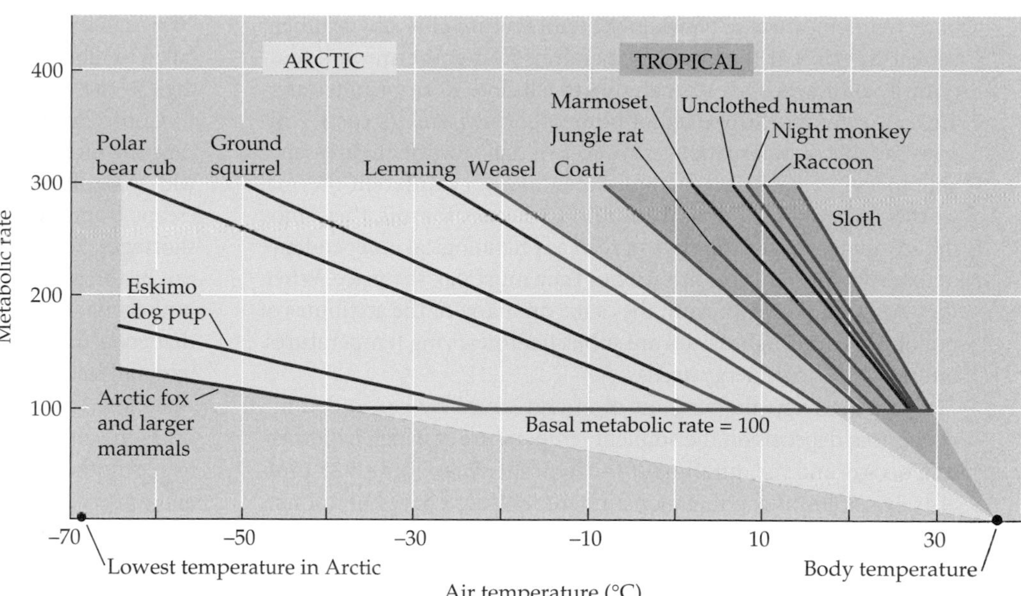

Figure 8.35 Mammalian physiological specialization to different climates Species found in the Arctic (Alaska) expend less energy to thermoregulate at cold ambient temperatures, and they can thermoregulate at lower temperatures, than species found in the tropics (Panama) can. In this presentation, each species' basal metabolic rate is set equal to 100, and metabolic rates outside the thermoneutral zone are expressed relative to basal; this convention facilitates comparison in certain ways but means that the slopes of the metabolism–temperature curves below thermoneutrality can be used in only a qualitative way to compare insulation. (After Scholander et al. 1950.)

ambient temperatures. Much evidence indicates that the three are different manifestations of a single physiological process. The three are distinguished by differences in their durations and seasons of occurrence. When an animal allows its body temperature to fall close to ambient temperature for periods of several days or longer during winter, the process is termed *hibernation*. When this occurs during summer, it is called *estivation*. When an animal permits its body temperature to fall close to ambient temperature for only part of each day (generally on many consecutive days), the process is termed *daily torpor* in any season. Figures 8.36 and 8.37 illustrate the sorts of changes in body temperature and metabolic rate that occur from the start to the finish of episodes of controlled hypothermia.

Figure 8.36 Changes in body temperature during hibernation A woodchuck (groundhog) (*Marmota monax*) was implanted with a small temperature transmitter that broadcast its body temperature continuously, and after it healed from the surgery, it was studied at an air temperature of 6°C. The record shows its body temperature during a 12-day episode of hibernation. (After Armitage, Woods, and Salsbury 2000.)

Hibernation, estivation, and daily torpor permit mammals and birds to escape the energy demands of homeothermy. As stressed earlier, homeothermy is energetically costly. A hamster, for example, needs to acquire and consume a great deal of food energy to keep its

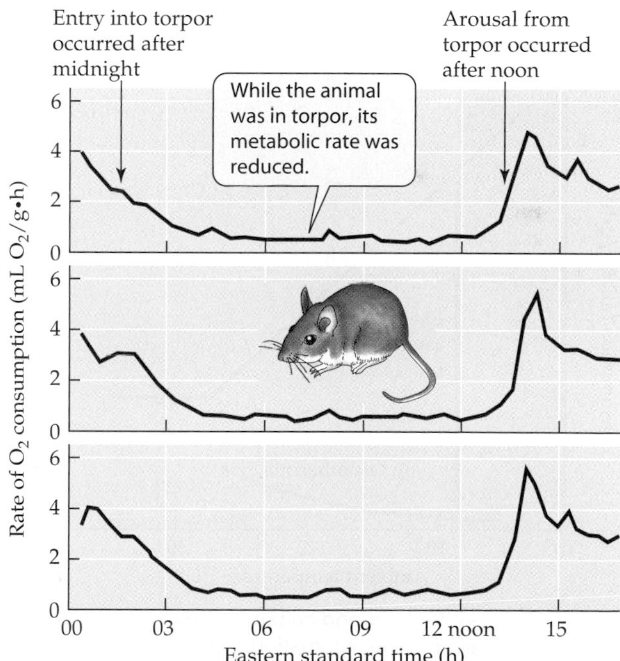

Figure 8.37 Changes in metabolic rate during daily torpor The three panels, from top to bottom, show the rate of O_2 consumption during three consecutive days of a white-footed mouse (*Peromyscus leucopus*) studied at an air temperature of 13°C to 15°C with an abundance of food. The *x* axis shows time of day on a 24-h scale (00 = midnight; 03 = 3:00 AM; 15 = 3:00 PM) The animal required a resting metabolic rate of about 3.0 mL O_2/g•h to be able to maintain high body temperatures. It underwent a prolonged episode of daily torpor on each day, as indicated by the drop in metabolic rate. Its body temperature during an episode of torpor similar to those shown was 16.8°C. (After Hill 1975.)

body temperature at 37°C when the temperature of its environment is near freezing. If the hamster abandons homeothermy and temporarily allows its body temperature to fall close to ambient temperature, it is temporarily freed of homeothermy's energy costs. Animals capable of hibernation, estivation, or daily torpor are in essence able to switch back and forth between two very different thermal worlds. When they function as ordinary homeotherms, they enjoy the advantages of homeothermy, such as physiological independence of external thermal conditions; but they must pay the price. When they suspend homeothermy, they take on many of the attributes of poikilotherms: Their tissues are subjected to varying temperatures, but they have low energy needs.

Quantitatively, the amount of energy saved by controlled hypothermia depends on the ambient temperature at which hypothermia occurs and the duration of the hypothermia. Figure 8.38 presents examples of the metabolism–temperature curves of animals when they are and are not in hypothermia, illustrating how animals can switch back and forth between homeothermic metabolism–temperature relations and poikilotherm-like relations. At any given ambient temperature, the difference between the two curves shows the degree to which animals can reduce their energy costs per unit of time by entering hypothermia; the amount of energy saved per unit of time becomes greater as the ambient temperature falls. If a hibernating animal remains in hibernation at low ambient temperatures for long periods of time, its total energy savings can be enormous. For example, free-living ground squirrels of at least two species, living in cold climates, have been measured to expend only 10% to 20% as much energy per month by hibernating as they would if they failed to hibernate, and they reap these monthly savings throughout their 7- to 8-month hibernating seasons.

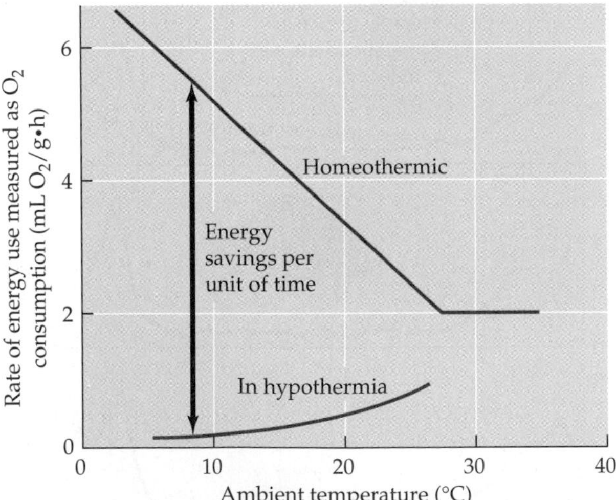

Figure 8.38 Energy savings depend on temperature In common with other species that undergo hibernation, estivation, or daily torpor, kangaroo mice (*Microdipodops pallidus*) alternate between two metabolism–temperature relations. The double-headed arrow shows how much a kangaroo mouse's rate of energy utilization is reduced when the animal is in hypothermia instead of being homeothermic. The amount of energy saved by being in hypothermia is greater at low ambient temperatures than higher ones because the metabolic cost of homeothermy is particularly high at low ambient temperatures, whereas the cost of hypothermia is particularly low at low ambient temperatures. At ambient temperatures of 5°C to 25°C, the body temperatures of these mice when in hypothermia are just 1°C to 3.5°C higher than ambient. (After Brown and Bartholomew 1969.)

Controlled hypothermia also permits mammals and birds to escape the high water demands of homeothermy. This point is not as widely significant as the escape from energy demands, because the escape from water demands matters just for animals that face water shortages. Sometimes, nonetheless, the escape from water demands can be the most important consequence of entering controlled hypothermia; this is especially true for animals in estivation or ones that enter daily torpor in hot, dry environments. As we will discuss in detail in Chapter 26, homeotherms have relatively high rates of water loss. One reason is simply that they must breathe rapidly to acquire the amounts of O_2 they need for their high rates of metabolism. Another is that the air they exhale tends to be relatively warm, and warm air holds more water vapor (which is exhaled with the air) than cooler air. Entry into controlled hypothermia reduces an animal's rate of water loss by reducing its breathing rate and the water vapor content of its exhaled air. With its rate of water loss reduced, the animal has a reduced need to acquire water, a possibly lifesaving advantage in arid habitats.

IN WHAT RESPECTS IS "CONTROLLED" HYPOTHERMIA CONTROLLED? Mammals and birds that display controlled hypothermia orchestrate their entry into and exit from hypothermia, and they exhibit control over their situation in other respects as well. The most dramatic evidence of the controlled nature of hibernation, estivation, and daily torpor is the fact that animals are able to arouse from these conditions. **Arousal** is the process of rewarming the body by metabolic heat production. The animals do not require outside warming to return to homeothermy. Instead, they are in control: They return to homeothermy on their own by employing intense shivering and, in mammals, nonshivering thermogenesis to warm their tissues. All episodes of controlled hypothermia *end* with arousal. In addition, hibernating animals universally undergo periodic, short arousals during the period of time they are hibernating; for instance, an animal that hibernates for 6 months might arouse for a few hours every 14 days or so. We discuss the possible functions of periodic arousals in Chapter 9.

A second type of control exhibited by animals in controlled hypothermia is *biochemical downregulation* of metabolism, a phenomenon that has recently been convincingly argued to occur in certain small and medium-sized mammalian hibernators. When these animals enter hibernation, their metabolism is depressed in a controlled manner by being downregulated biochemically, and while they are in hibernation, their metabolic rates are determined by the biochemical controls rather than by simple tissue-cooling effects ("Q_{10} effects"). Until about 15 years ago, the almost-universal view was that animals initiate their entry into hibernation by turning off thermoregulation. According to this older view, the sequence of events during entry into hypothermia is that thermoregulation is deactivated, body temperature therefore falls, and tissue metabolic rates decline because the tissues cool. The new view is that biochemical downregulation of tissue metabolism is the first step in the sequence, and the body temperature falls as a consequence of the reduced metabolic rate. Either scenario represents control by the animal. Future research will be required to sort out which view applies to which species.

A final, fascinating sort of control exhibited by animals in controlled hypothermia is the control displayed when their body temperatures start to fall too low. Each species that undergoes hibernation, estivation, or daily torpor has a species-specific range of body temperatures that it can tolerate, and for an animal to survive hypothermia, it must respond if its body temperature starts to go below the tolerable range. *Within* the tolerable range, animals typically let their body temperatures drift up and down as the ambient temperature rises and falls. For instance, if an animal can tolerate a T_B as low as 3°C and the T_A varies between 5°C and 15°C, the animal typically will allow its T_B to vary as the T_A varies (always being a bit higher than the T_A). What happens, however, if the ambient temperature falls below 3°C? Frequently (although not always), the animal exerts control in one of two life-preserving ways. One way the animal may respond is to arouse. More remarkably, it may start to *thermoregulate* at a reduced body temperature, its thermoregulatory control system functioning with a lowered set point. For example, an animal that must stay at a body temperature of at least 3°C to survive may keep its body temperature at 3°C even if the ambient temperature drops to –10°C or –20°C, increasing its metabolic rate as the ambient temperature falls so as to offset the increasing cooling effect of the air (see Figure 9.12).

DISTRIBUTION AND NATURAL HISTORY Hibernation is known to occur in at least six different orders of mammals. Species that hibernate include not only certain hamsters and ground squirrels but also dormice, jumping mice, marmots, woodchucks, some bats, some marsupials, and some monotremes. Because of its seasonal nature, hibernation is often preceded by long-term preparation. Hibernating mammals, for instance, typically store considerable quantities of body fat during the months before their entry into hibernation (see Figure 4.21). Hibernation is rare in birds; it may occur in just one species, the poorwill (*Phalaenoptilus*). We discuss mammalian hibernation at more length in Chapter 9.

Estivation is not nearly as well understood as hibernation, partly because it is not as easy to detect. It has been reported mostly in species of desert ground squirrels.

Daily torpor is widespread among both mammals and birds in both warm and cold situations. It occurs, for example, in numerous species of bats and rodents and in certain hummingbirds, swallows, swifts, and caprimulgid birds (e.g., nightjars and poorwills). Animals undergoing daily torpor are homeothermic for part of each day, and they feed during their homeothermic periods. When bats are undergoing daily torpor, they become hypothermic during daylight hours and forage at night; hummingbirds, in contrast, become torpid at night and feed in daylight. In some species, the proclivity to enter daily torpor is seasonally programmed. More generally, however, daily torpor seems to be employed in any season as an immediate response to hardship. Many species, for example, undergo daily torpor only when they are suffering food shortage; in some cases they increase the length of time they spend in torpor each day as food shortage becomes more severe.

CONTROLLED HYPOTHERMIA IN WHICH THE BODY TEMPERATURE REMAINS WELL ABOVE AMBIENT TEMPERATURE Over the last 20 years, there has been an ever-increasing realization that many species of birds undergo hypothermia without ever allowing their body temperatures to approximate ambient temperature. Such shallow hypothermia is not properly called hibernation, estivation, or daily torpor; it is presently nameless because physiologists have reached no consensus on a term for it. Black-capped chickadees (*Poecile atricapilla*) provide an excellent example. They sometimes allow their core body temperature to fall by roughly 7°C while sleeping overnight in freezing-cold winter weather. They then have body temperatures, 31°C to 34°C, that are distinctly hypothermic but nonetheless far above ambient temperature. Their hypothermia does not eliminate their need to expend metabolic energy to stay warm. However, because the hypothermia reduces the difference between body temperature and ambient temperature, the birds lose heat more slowly—and have lower costs for thermoregulation—than if they maintained higher body temperatures. Chickadees do not feed at night and are so small (11 g) that they may virtually exhaust their body fat in a night of frigid weather; their hypothermia helps them survive until they can feed the next day.

Bears of some species—and possibly raccoons, skunks, and other such mammals—also exhibit a form of shallow hypothermia, sometimes called *winter sleep* or *carnivorean lethargy*. Black bears in their winter dens, for instance, have body temperatures of 31°C to 33°C—about 5°C below the placental norm.

Summary: Homeothermy in Mammals and Birds

- Homeothermy—thermoregulation by physiological means—is energetically expensive.
- The principal way that a mammal or bird thermoregulates in its thermoneutral zone is that it varies its body insulation to offset changes in the driving force for dry heat loss ($T_B - T_A$). Insulation can be modulated by changes in posture, cutaneous blood flow, the thickness of the stagnant air layer trapped by the pelage or plumage, and regional heterothermy.
- Below thermoneutrality, variation in the rate of metabolic heat production (thermogenesis) is the principal mechanism of thermoregulation. The two most prominent mechanisms of increasing heat production are shivering—found in both mammals and birds—and nonshivering thermogenesis (NST)—found mainly in placental mammals. The principal site of NST in mammals is brown adipose tissue, which, by synthesizing uncoupling protein 1, is able to employ uncoupling of oxidative phosphorylation to achieve very high rates of lipid oxidation with immediate heat release.
- Regional heterothermy, often highly expressed at ambient temperatures below thermoneutrality, usually depends on countercurrent heat exchange. Close juxtaposition of arteries and veins short-circuits the flow of heat into appendages.
- Above thermoneutrality, species with long evolutionary histories in hot, dry environments typically use nonevaporative mechanisms as first lines of defense. Two common nonevaporative mechanisms are hyperthermia and cycling of body temperature. When active evaporative cooling occurs, the specific mechanisms employed to increase the rate of evaporation are sweating (only in certain mammals), panting (mammals and birds), gular fluttering (only birds), and saliva spreading (only certain mammals). Both hyperthermia

and the effort involved in active evaporative cooling can cause metabolic rate to rise above thermoneutrality.

- Acclimatization to changing seasons is the norm and may involve one or more of three mechanisms: acclimatization of peak metabolic rate, acclimatizaion of metabolic endurance, and insulatory acclimatization.
- Controlled hypothermia permits animals to evade the high energy costs and water costs of homeothermy. During hibernation, estivation, and daily torpor, T_B is allowed to fall close to T_A within a species-specific range of T_A. Forms of shallow hypothermia also occur.

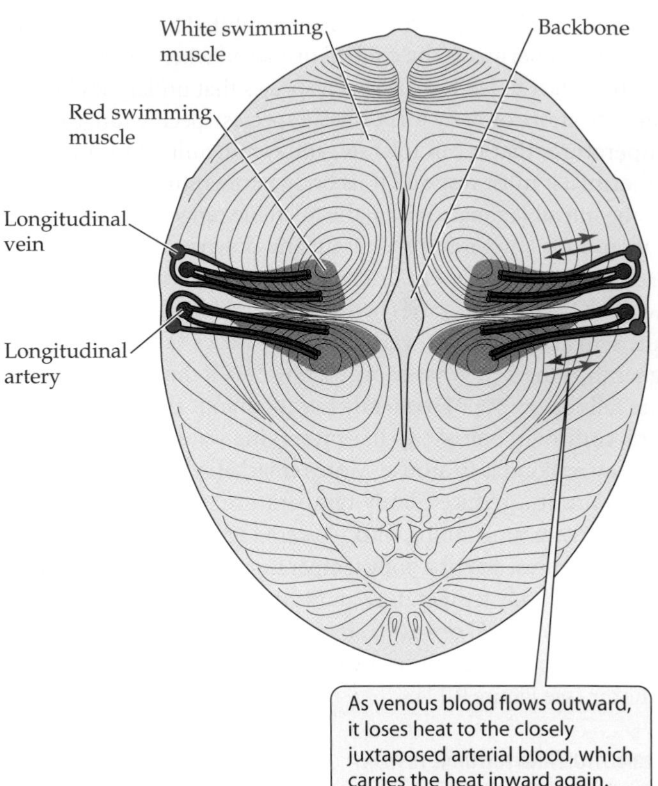

Figure 8.39 A cross section of a tuna showing the nature of the blood supply to the red swimming muscles The longitudinal arteries, carrying blood along the length of the body, give off small arteries that penetrate (toward the backbone) into the muscles. Small veins running in close juxtaposition to the small arteries return blood peripherally to the longitudinal veins, which lead back to the heart. Red vessels and arrows refer to arterial blood flow; blue vessels and arrows refer to venous flow.

Warm-Bodied Fish

The body temperatures of 99% of all species of fish closely approximate water temperature. However, in tunas, lamnid sharks, and billfishes, temperatures within *certain body regions* exceed water temperature, sometimes substantially. All the warm-bodied fish are large, streamlined, fast-swimming predators that lead wide-ranging lives and feed on such speedy prey as squid and herring. The lamnid sharks include the great white shark, mako, porbeagle, and salmon shark. The billfishes include the marlins and swordfish.

In tunas and lamnid sharks, the red (dark) swimming muscles are warmed above water temperature.[26] These muscles provide the power for steady swimming in these vigorously active animals, and the contractile activity of the muscles produces the heat that warms the muscles. A critical principle to recognize, however, is that *a high rate of heat production is never in itself adequate to elevate tissue temperature in gill-breathing animals*. If metabolic heat is carried freely to the gills by the circulation of the blood, the heat is lost so readily to the surrounding water across the gills that no significant elevation of body temperature can occur. Thus, for a region of the body to be warmed, transport of heat out of that body region by the circulation must be impeded. Not just in the red swimming muscles of tunas and lamnid sharks, but universally in warm-bodied fish, the mechanism of impeding heat loss is countercurrent heat exchange.

The vasculature of the red swimming muscles in tunas and lamnid sharks is diagrammed in Figure 8.39. Note that the red muscles are found deep in the body near the spinal column in these fish (a placement that is unusual compared to that in most other fish). The major longitudinal arteries and veins that carry blood along the length of the body, to and from the swimming muscles, run just under the skin on each side of the body (again, an unusual placement). Small arteries branch off from the longitudinal arteries and penetrate inward to the red muscles. In turn, blood is brought outward from the muscles in veins that discharge into the longitudinal veins leading back to the heart. The arteries carrying blood inward to the red swimming muscles and the veins carrying blood outward from those muscles are closely juxtaposed, forming countercurrent exchange networks. Figure 8.39 is highly simplified in the way it presents these networks. In fact, the arteries and veins going to and from the red muscles branch profusely, forming thick layers of vascular tissue in which huge numbers of minute arterial and venous vessels, each only about 0.1 mm in diameter, closely intermingle—a true rete mirabile (see page 223). Because of the countercurrent-exchange arrangement, much of the heat picked up by the venous blood in the red muscles is transferred to the ingoing arterial blood rather than being carried to the periphery of the body and the gills, where it would readily be lost to the water. Thus, heat produced by the red swimming muscles tends to be retained within them.

Bluefin tunas, which reach body weights of 700 kg and are the largest of all tunas, maintain fairly constant red-muscle temperatures over a wide range of water temperatures (Figure 8.40). In most other tunas, such as the yellowfin and skipjack tunas, red-muscle temperature is elevated over water temperature by a relatively constant amount regardless of the water temperature. Referring back to our scheme for classifying animal thermal relations (see Figure 8.1), all the tunas are endotherms, but species differ in whether they also thermoregulate. Whereas yellowfin and skipjack tunas are en-

[26] The roles of the red and white muscles in powering swimming in fish are discussed in Chapter 6 (see page 161).

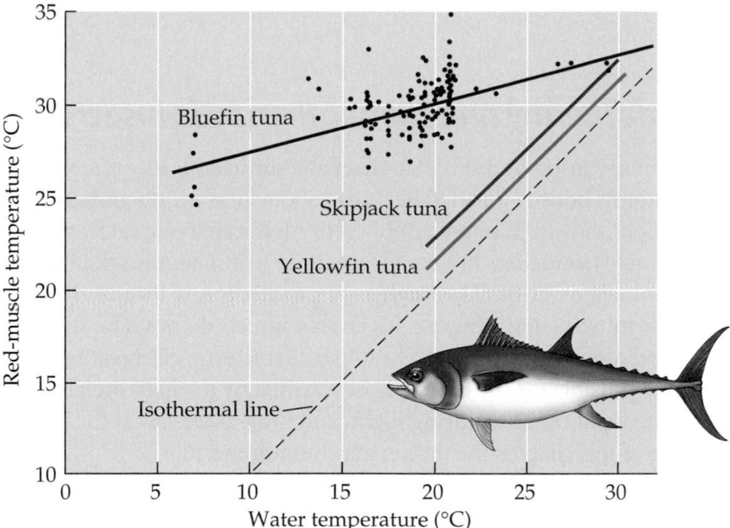

Figure 8.40 Red-muscle temperatures of tunas at various ambient water temperatures The upper line and data points are for wild bluefin tunas (*Thunnus thynnus*) captured in waters of various temperatures. The other two solid lines show the average relation between red-muscle temperature and water temperature in small, captive skipjack tunas (*Katsuwonus pelamis*) and yellowfin tunas (*Thunnus albacares*) swimming in an aquarium; larger, wild fish of these species are sometimes observed to exhibit greater temperature differentials between muscles and water (e.g., 5–10°C in skipjacks). The isothermal line shows how tissue temperature would vary with water temperature if there were no endothermy and tissue temperature simply equaled water temperature. The image is a bluefin tuna. (After Carey and Teal 1969; Dizon and Brill 1979.)

dotherms without being thermoregulators,[27] bluefin tunas are endothermic thermoregulators (homeotherms).

The warming of the red swimming muscles in tunas and lamnid sharks is generally thought to aid power development and locomotory performance, although exactly how is debated. Any aid to the performance of the swimming muscles would be significant for fish that are so dependent on high-intensity exertion for their livelihood.

The swimming muscles are not the only tissues kept warm in tunas and lamnid sharks. In certain species, the stomach and other viscera are warmed when food is being digested. The brain and eyes are also warmed in some species. Each warmed organ is served by arteries and veins that form a rete mirabile, which short-circuits the outflow of heat produced in the organ, thereby favoring heat accumulation in the organ.

The billfishes have been saved for separate discussion because of two ways in which they differ from the tunas and lamnid sharks. First, in the billfishes, only the brain and the retinas of the eyes are warmed. Second, the billfishes possess "heater tissues" specialized for exceptional heat output.[28]

[27] There is some evidence for active thermoregulation in these fish. For example, they decrease heat retention when they are highly active in warm water, thus preventing their activity from driving their muscle temperature too high.

[28] In tunas and lamnids, the rate of heat production in each warmed organ or tissue is simply the ordinary rate according to available evidence.

The heater tissues of billfishes are derived from portions of the extraocular eye muscles (the muscles on the outside of each eyeball that serve to turn the eyeball to look in various directions). These portions of the muscles have lost most of their contractile apparatus and are very rich in mitochondria. Current evidence suggests that they produce heat at a high rate by a "futile cycle" of Ca^{2+} pumping: ATP is used to transport Ca^{2+} actively from one intracellular compartment to another, and then the Ca^{2+} leaks back to where it started, where once again ATP is used to pump it; the principal net result is breakdown of a lot of ATP to release heat. The heat produced by the heater tissues is retained in the head by countercurrent vasculature and in that way warms the brain and retinas. If warming of the *brain* by specialized *eye* muscles sounds impossible, remember that in a fish, the eyes and eye muscles are far larger than the brain! Warming of the brain and the retinas is hypothesized to aid marlins, swordfish, and other billfishes in their pursuit of prey because the tissues are kept from becoming cold when the fish swim through cold water.

A principal goal of modern animal physiology is to understand the evolutionary history of physiological features, as discussed in Chapter 1. One of the best examples of a modern approach taken toward this task is provided by the research of Barbara Block and her colleagues on the family tree (the phylogeny) of the warm-bodied teleost fish and their close relatives.[29]

From study of DNA nucleotide sequences in the gene for mitochondrial cytochrome *b* in a variety of species, Block's team constructed the family tree shown in Figure 8.41. A big advantage of using molecular data to construct a family tree of this sort is that the proposed tree is based on information that is independent of the physiology or anatomy of interest. If you ignore for a moment the colors in Figure 8.41 and simply focus on the lines and their patterns of branching, those lines and branching patterns represent the family tree that Block's team constructed without any reference to the physiology of endothermy. Only *after* the family tree was completed were physiological or anatomical features—symbolized by the colors and letters—mapped onto it.

One of the physiological features mapped onto the tree is endothermy in the red swimming muscles. Specifically, all the fish that exhibit such endothermy are colored red. When endothermy is mapped in this way onto the independently derived family tree as it is in Figure 8.41, a coherent pattern emerges, strongly suggesting that red-muscle endothermy appeared in the evolutionary history of these fish at the spot marked A. Not just tunas, but also bonitos, exhibit some degree of morphological shift of their red swimming muscles toward the unusual location near the spinal column, and they are the only teleost fish that do. Thus, spot B is apparently where the shift in red-muscle position started. Evidently, the new red-muscle position evolved first, and then the tunas (but not the bonitos) capitalized on it to evolve red-muscle endothermy. Note also that the three groups with some sort of endothermy—the billfishes, tunas, and butterfly mackerel—do not form a coherent group; they are separated by lines of the family tree in which there is no endothermy. The family tree thus indicates that endothermy evolved separately three times in teleost fish (spots A, C, and D).

[29] Block's team did not study the sharks, which are not teleosts.

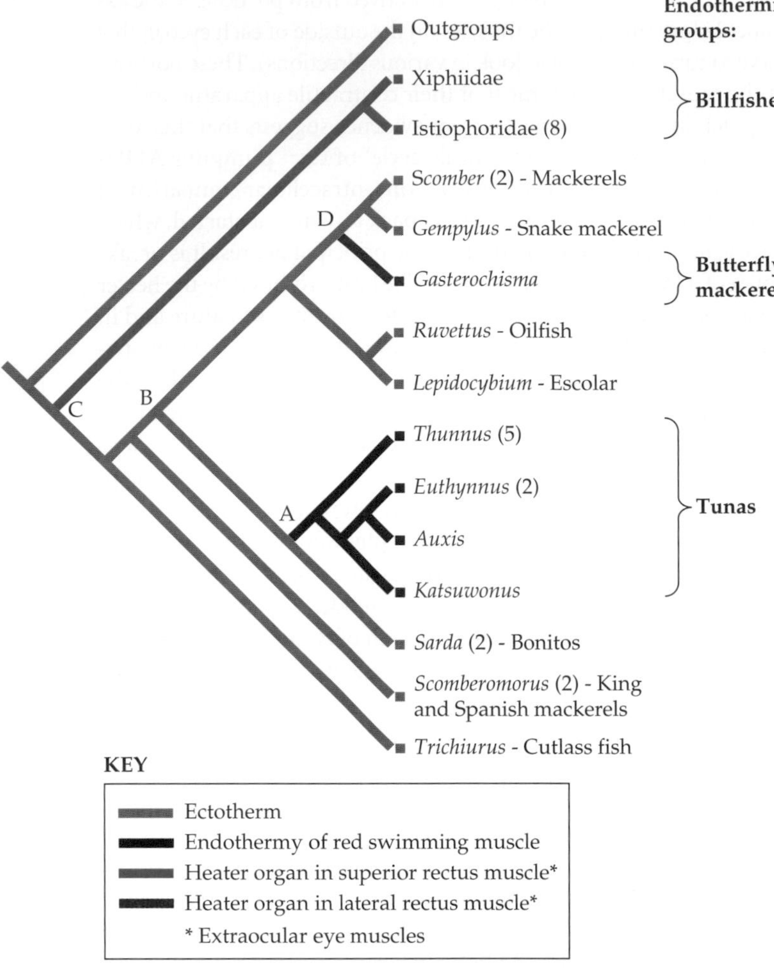

Figure 8.41 A family tree (phylogeny) of teleost fishes belonging to the suborder Scombroidei Physiological features (see key) are mapped onto the family tree, which was derived independently of physiological information. The colors of the little boxes on the right-hand side of the tree indicate the actual, known physiological nature of the various modern-day fish specified. The coloring of the lines of the family tree represents an interpretation of past history. Endothermy occurs only in the groups so identified at the right side of the diagram. Numbers are numbers of species studied if greater than one. "Outgroups" are other species of fish used to establish a base of comparison. Refer to the text for an explanation of the significance of the letters at branch points of the tree. (After Block and Finnerty 1994.)

Summary: Warm-Bodied Fish

- Tunas, lamnid sharks, and billfishes are distinguished from other fish by exhibiting endothermy in certain body regions. The tissues that are endothermic in tunas and lamnids are the red swimming muscles and—sometimes—the stomach, other viscera, brain, and retinas. In billfishes, only the brain and retinas are endothermic.
- A countercurrent vascular array that short-circuits outflow of heat from a tissue is required for the tissue to be endothermic in gill-breathing animals such as these fish.
- Ordinary metabolic heat production is the source of heat for endothermy in all cases except the billfishes, which have specialized "heater" tissues that are derived from extraocular eye muscles.

Endothermy and Homeothermy in Insects

A solitary insect at rest metabolizes at a sufficiently low rate that no part of its body is warmed by its metabolic heat production. Insects in flight, however, often exhibit very high metabolic rates; species that are strong fliers in fact release more heat per gram than active mammals or birds. This high heat production is localized in the flight muscles in the thorax. Given that insects do not have the profound problems of retaining heat that characterize gill-breathers, it is quite possible for the thorax to be warmed by the high metabolism of the flight muscles during flight, and thus, as we saw at the beginning of this chapter, the thorax may be endothermic.

Some insects that display thoracic endothermy during flight do not thermoregulate; examples are provided by certain species of small geometrid moths, which maintain a thoracic temperature that is about 6°C above air temperature regardless of what the air temperature is. Other sorts of insects physiologically thermoregulate during flight and thus exhibit thoracic homeothermy. The thermal relations of endothermic insects are particularly complex because they exhibit both temporal and spatial heterothermy. The insects exhibit endothermy only when they are active, not when they are resting. Moreover, even when they exhibit endothermy, they usually do so just in their thorax, not their abdomen.

Historically, sphinx moths were the first sorts of insects discovered to display thoracic physiological thermoregulation during flight, and to this day they are model examples of the phenomenon. Sphinx moths are particularly large insects; certain species weigh as much as several grams and thus are similar in weight to some of the smallest mammals and birds. Flying sphinx moths closely regulate their thoracic temperatures. Those in Figure 8.42, for example, maintained thoracic temperatures within a narrow range, 38°C to 43°C, over a wide range of air temperatures. Thermoregulation is not limited just to insects of such large body size. Worker bumblebees

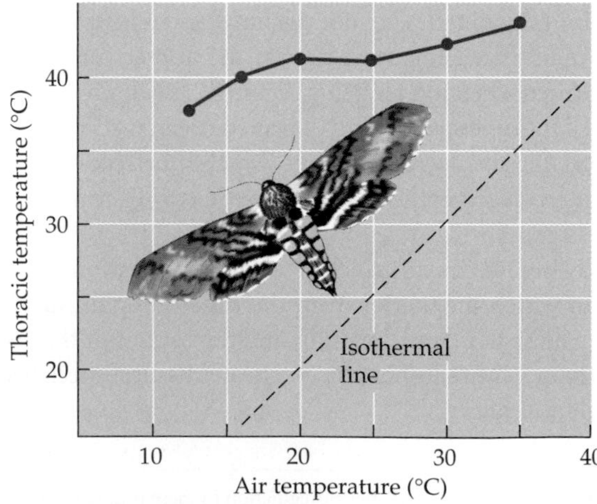

Figure 8.42 The average thoracic temperature of freely flying sphinx moths (*Manduca sexta*) as a function of air temperature The isothermal line shows how thoracic temperature would vary with air temperature if there were no endothermy or thermoregulation and the temperature of the thorax simply equaled air temperature. (After Heinrich 1971.)

(*Bombus vagans*), averaging 0.12 g in body weight, for instance, maintain thoracic temperatures near 32°C to 33°C whether the air temperature is 9°C or 24°C when they are foraging. The list of insects known today to exhibit thoracic homeothermy during flight includes not only sphinx moths and bumblebees, but also certain other lepidopterans and bees, some dragonflies, and some beetles.

Although endothermy and physiological thermoregulation occur principally during flight in insects, a few types of insects display the phenomena during solitary terrestrial activities. In nearly all such cases, the primary source of heat is the flight muscles, which instead of being used to fly, are activated to "shiver" (as discussed shortly). Dung beetles sometimes become markedly endothermic while working in dung piles and rolling their dung balls. Some crickets and katydids thermoregulate while they sing.

The insects that thermoregulate during flight require certain flight-muscle temperatures to fly

The flight muscles of an insect must be able to generate mechanical power at a certain minimal rate (which is species-specific) for the insect to be able to fly. Within a broad range of temperatures, the power output that flight muscles can attain increases as their temperature increases. Thus the temperature of an insect's flight muscles is potentially an important determinant of whether the insect can fly. Tiny insects such as fruit flies, mosquitoes, and midges have such high surface-to-volume ratios that the activity of their flight muscles cannot warm the thorax significantly. Correlated with their inability to be endothermic, the tiny insects commonly can fly over very broad ranges of thoracic temperatures, including, in some species, thoracic temperatures as low as 0°C to 5°C. An important facet of the flight physiology of these tiny, poikilothermic fliers is that they apparently require only a modest fraction of their maximal power output to stay aloft; thus, they can fly at relatively low thoracic temperatures, at which their power output is substantially submaximal. In sharp contrast, many medium-sized and large insects, including the species known to thermoregulate, require a near-maximal power output from their flight muscles to take off and remain airborne. They therefore require that their flight muscles be at high temperatures to fly. A sphinx moth (*Manduca sexta*), for example, cannot fly unless its thorax is at least as warm as 35°C to 38°C, and worker bumblebees (*Bombus vagans*) require about 29°C.

The need for high flight-muscle temperatures for flight raises the question of how resting insects are able to get warm enough to take off. Because insects typically cool to environmental temperature when they are fully at rest, an insect that requires a high flight-muscle temperature to fly will often be too cold to take off after it has been resting for a while. Diurnal species may be able to warm their flight muscles to flight temperature by basking in the sun. Most species, however, have an endogenous ability to warm their flight muscles to flight temperature, a phenomenon known as **physiological preflight warm-up**.

Physiological preflight warm-up is accomplished by contraction of the flight muscles in a nonflying mode, a process often called **shivering** (not homologous to vertebrate shivering). Several forms of shivering are known. In many types of insects, including moths and butterflies, what happens during shivering is that the muscles responsible for the upstroke and downstroke of the wings contract simultaneously (rather than alternately as they do in flight), thus working against each other. The wings merely vibrate during shivering, rather than flapping, but heat is evolved by the muscular contraction, warming the flight muscles. When a sphinx moth warms from a low temperature, its flight muscles shiver in this manner at an ever-higher intensity as its thoracic temperature increases to the flight level. Then suddenly the pattern of muscular contraction changes, the wings are driven through the flapping motions of flight, and the moth takes to the air.

Solitary insects employ diverse mechanisms of thermoregulation

Innovative investigators continue to progress in understanding the mechanisms that insects employ to thermoregulate, despite the obstacles of working on such small animals. Some species employ evaporative cooling in hot environments. Here we focus on the processes of keeping warm in cool environments.

As the ambient temperature drops, one mechanism of maintaining a constant thoracic temperature is for an insect to increase its rate of heat production, much as mammals and birds do below thermoneutrality. Many insects do this when they are *not flying*. Heat is made in these circumstances by shivering of the flight muscles, and because the muscles can engage in various intensities of shivering, they can modulate their rate of heat production to serve thermoregulatory needs. Honeybees and bumblebees, for example, often maintain high and stable body temperatures for long periods when working in the hive by increasing and decreasing their rates of shivering heat production as the air temperature falls and rises. An intriguing example is also provided by the brood incubation of queen bumblebees (Figure 8.43). A queen, which overwinters alone and thus is solitary when she rears her first brood in the spring, incubates her brood by keeping her abdomen at an elevated temperature and pressing it against the brood. Heat is brought to her abdomen from her thorax, where it is produced by her flight muscles. As the ambient air temperature falls, the queen thermoregulates by increasing her rate of heat production.

Modulation of shivering can also be used to thermoregulate during *intermittent flight*. Bumblebees do this, for instance. As a bumblebee, such as that pictured at the start of this chapter, flies from flower to flower during foraging, it can shiver or not shiver while it is clinging to each flower. More shivering of this sort occurs as the air temperature falls, and thus the bee's overall metabolic rate increases as air temperature decreases.

When insects *fly continuously*, their flight muscles are employed in flight movements all the time and cannot shiver. Investigators hypothesized years ago that under these circumstances, the rate of heat production by the flight muscles would be determined by the requirements of flight and not modulated to serve thermoregulation. Early, seminal experiments on sphinx moths supported the truth of this hypothesis, because when the moths flew at a certain speed, their metabolic rates proved to be essentially constant whether the air temperature was 15°C or 30°C. Until very recently, therefore, the prevailing paradigm has been that insects in continuous flight do not modulate their rates of heat production as a means of thermoregulating. How, then, do they thermoregulate?

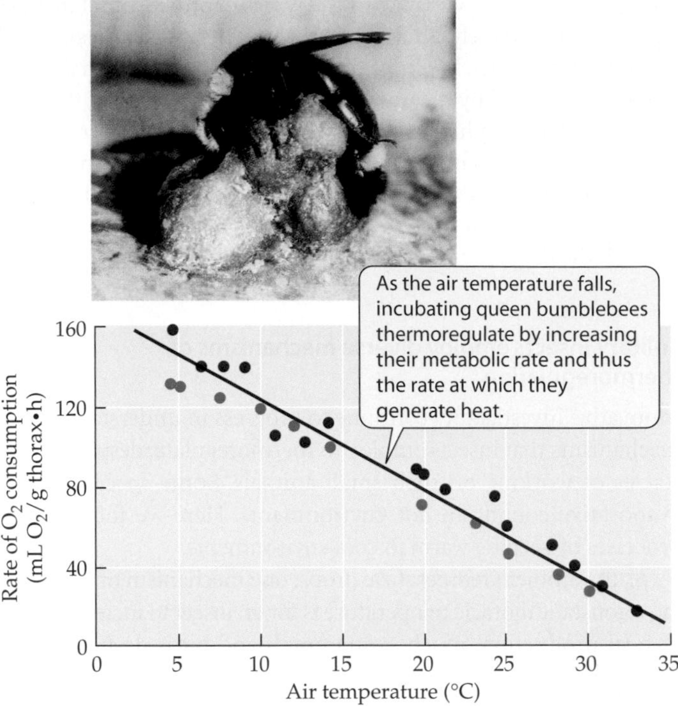

Figure 8.43 The rate of O_2 consumption by queen bumblebees as a function of air temperature when they are incubating their broods In the species studied, *Bombus vosnesenskii*, a queen incubates her brood by pressing her abdomen against it as seen in the photograph. In the plot, the two colors of symbols refer to two different individuals. (After Heinrich 1974; photograph courtesy of Bernd Heinrich.)

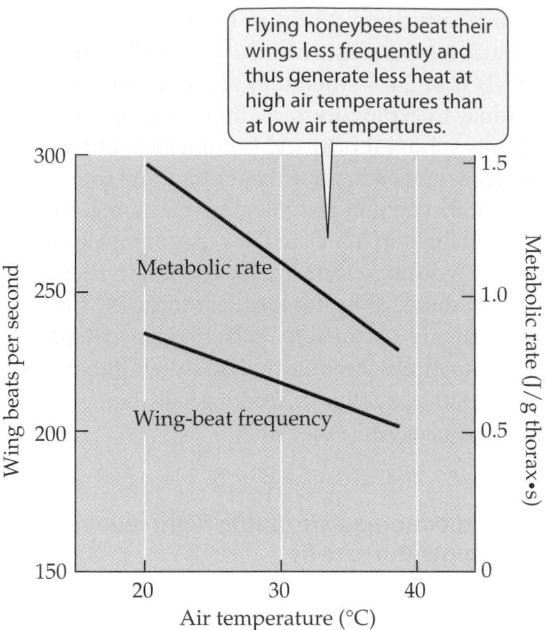

Figure 8.44 The effect of air temperature on wing-beat frequency and metabolic rate in flying honeybees (*Apis mellifera*) The lines are average-response lines, based on hundreds of individual measurements. (After Harrison et al. 1996.)

Studies on continuously flying sphinx moths, bumblebees, and some other insects reveal that their primary mechanism of thermoregulation is much akin to that used by mammals and birds in the thermoneutral zone; namely, they vary their *insulation*. A flying insect keeps its thorax at a steady temperature by modulating how readily heat can exit the thorax. This modulation is accomplished at least in part by control of the rate of blood flow between the thorax and abdomen. In continuously flying sphinx moths, when the air temperature is low the heart beats weakly and blood circulates only slowly between the thorax and abdomen; thus, heat produced by the flight muscles in the thorax tends to remain in the thorax, which retains the heat effectively because it is densely covered with furlike scales. As the air temperature is raised, the heart beats more vigorously and circulates blood to the abdomen more rapidly; in this way, heat is transported at an increased rate out of the thorax into the abdomen, where it can be relatively easily lost to the environment because the abdomen has a large and thinly "furred" surface.

Just in recent years, the old paradigm of thermoregulation during continuous flight—which essentially held that all insects resemble sphinx moths—has been challenged with new data. These data indicate that during continuous flight, some insects—notably honeybees and some other bees—modulate their rate of heat *production* by their flight muscles in ways that serve thermoregulation. When honeybees fly from place to place, they beat their wings more slowly—and slow their progress—as the air temperature rises (**Figure 8.44**). Even when they engage in hovering, an ostensibly invariant activity, they generate less heat at high air temperatures than at low ones, by unknown means. These discoveries indicate that the diversity of thermoregulatory mechanisms employed by solitary insects sometimes includes modulation of heat production during continuous flight.

Colonies of social bees and wasps often display sophisticated thermoregulation

Physiological regulation of colony temperature is widespread in colonies of social bees and wasps. Honeybees (*Apis mellifera*) provide the best-studied example. Thermoregulation by honeybee hives is so dramatic that it was recognized for almost two centuries before thermoregulation by solitary insects was first demonstrated.

A honeybee hive that is rearing a brood maintains the temperature of its brood combs within a narrow range, about 32°C to 36°C, even if the air temperature outside the hive falls to –30°C or rises to +50°C. When the air outside the hive is cold, worker bees cluster together within the hive and shiver. When the air outside becomes warm enough that the hive is threatened with metabolic overheating, workers disperse within the hive and fan with their wings in a cooperative pattern that moves fresh air from outside the hive across the brood combs. At very high outside air temperatures, workers also collect water and spread it within the nest, where it evaporates into the airstream produced by fanning. Honeybees provide an outstanding example of coevolution between thermal requirements and thermoregulation. Their broods of young require temperatures of about 32°C to 36°C for proper development. Thus the sophisticated thermoregulation of the hive by the workers is essential for a hive's reproductive success.

Summary: Endothermy and Homeothermy in Insects

- Many solitary insects, especially those of medium to large size, display thoracic endothermy or homeothermy during flight or certain other sorts of activity. Warming of the flight muscles increases their power output. Often in these insects, a certain minimum flight-muscle temperature is required for flight.
- When insects are not flying, activation of the flight muscles in a nonflight mode—termed *shivering*—is the mechanism they employ to warm the thorax. Shivering is used for preflight warm-up. In addition, modulation of shivering is used by nonflying insects, such as bees working in their hives, as a mechanism of thermoregulation.
- When insects are flying, one common and long-known mechanism of thermoregulation is modulation of thoracic insulation, brought about by raising and lowering circulatory transport of heat out of the thorax. Recently some insects have been found to thermoregulate in flight by varying their intensity of flight and thus the rate at which heat is generated by the work of flight.
- Colonies of social bees and wasps sometimes employ group efforts to maintain exquisitely stable hive temperatures.

Study Questions

1. As discussed in Chapter 1 (see page 10), Claude Bernard, a nineteenth-century French physiologist often considered the father of modern animal physiology, is still remembered today for his famous dictum: "Constancy of the internal environment is the condition for free life." Does the study of thermal relations lend support to his dictum? Explain.
2. There is currently a worldwide movement to create protected marine parks. If the parks have an Achilles heel, it is that they have fixed geographical positions, just in the way that Yellowstone National Park is at a fixed geographical location. Suppose that a certain endangered species of fish exists only in a marine park. If the ocean temperature rises in the park because of global warming, explain what physiological problems the species of fish might confront. How might the species face a brighter future if parks could have moveable boundaries rather than fixed ones?
3. Referring to Figure 8.10, suppose you have some lizards that are at 16°C and have been living at that temperature for 5 weeks. What is their resting metabolic rate? If the lizards are suddenly shifted to a room at 33°C, trace on the graph how their metabolic rate will change from the moment they are placed in the new room until 5 weeks have passed. According to the graph, will they exhibit compensation?
4. Discuss ways that the cryobiology ("freezing biology") of insects could be manipulated to control insect pests. One factor to consider is that certain bacteria and fungi act as highly effective ice nucleators.
5. Suppose a state government wants to know if fish along the state's coastline are experiencing heat stress. Using your knowledge of heat-shock proteins, design tests that the state could carry out.
6. Suppose you travel to a tropical place like Bermuda and watch the coastal poikilotherms, such as fish, crabs, and starfish, swim and crawl about in the warm waters. Suppose then that you travel to northern Maine and watch the related species of poikilotherms in the cold waters there. In the abstract, it would not be unreasonable to expect to see the animals in Maine moving about in slow motion compared to those in Bermuda. In fact, however, rates of locomotion are likely to look to your eye to be more similar than different in the two places. Design experiments to assess whether the Maine animals are especially able to be active in cold waters. If you find that they are, how might their high ability for activity in cold waters be explained? For each hypothesis you present, design an experiment to test the hypothesis.
7. During winter, when people are in a well-insulated house, they usually feel comfortable if the air temperature is near 22°C (72°F). If you have ever spent a night in a poorly insulated cabin in winter, however, you will recognize that paradoxically, when people are in poorly insulated buildings, they often feel chilly even when the air inside is heated to 22°C or higher. One important reason for the difference in how warm people feel in the two sorts of buildings is that even if a well-insulated and poorly insulated building are identical in the air temperature inside, they differ in thermal-radiation heat transfer. Specifically, a person standing in the two types of buildings experiences different heat exchange by thermal radiation in the two. Explain how thermal-radiation heat transfer accounts for the sense of chill in the poorly insulated building. (Hint: Think of the exterior walls of the two types of buildings, and specifically of the temperatures of the interior surfaces of those walls.)
8. What is homeoviscous adaptation? Although we discussed it in our study of poikilotherms, the phenomenon was actually first discovered about a century ago in studies of pigs in Sweden. Some pigs were dressed in blankets during winter while others were allowed to roam about stark naked. When their subcutaneous fat was analyzed, the two sets of pigs turned out to have laid down lipids of differing chemical composition. How could different lipids give the two groups similar lipid fluidities?[30]
9. Suppose you are trying to choose between two winter jackets. Suppose also that you have a heat-producing mannequin available for your use and you are able to adjust the mannequin's rate of heat production. According to Equation 8.10, insulation is equal to $(T_B - T_A)/M$ (this is in fact a general equation for insulation). How would you make a quantitative comparison of the insulation provided by the two jackets?
10. In the rete mirabile serving the red swimming muscles of tunas, some key enzymes of catabolism and other processes show gradients of concentration: They are more concentrated at the cold end of the rete, and less concentrated at the warm end. These variations parallel variations that are often seen in the thermal acclimation of poikilotherms, when enzyme concentrations rise during acclimation to cold and fall during acclimation to heat. What do you think could be some of the reasons for these spatial and temporal variations in enzyme concentration? Why not have the highest observed enzyme concentrations everywhere and all the time?

Key References

Heinrich, B. 1993. *The Hot-Blooded Insects: Strategies and Mechanisms of Thermoregulation.* Harvard University Press, Cambridge, MA. A searching and enlightening treatment (although becoming dated) of all aspects of insect thermal relations written by a pioneer in the field. This book does a wonderful job of putting the physiological information into the larger context of insect life histories and ecology.

Heldmaier, G., and T. Ruf. 1992. Body temperature and metabolic rate during natural hypothermia in endotherms. *J. Comp. Physiol. B* 162: 696–706; Ortmann, S., and G. Heldmaier. 2000. Regulation of body temperature and energy requirements of hibernating Alpine marmots

[30] To avoid introducing any confusion, it may be important to mention that the subcutaneous lipid deposits of mammals (which are known as *depot fats*) consist of ordinary triacylglycerols, not phospholipids like membrane lipids, but the basic concepts of homeoviscous adaptation remain the same.

(*Marmota marmota*). *Amer. J. Physiol.* 278: R698–R704. These companion papers grapple in an illuminating way with the question of what governs the metabolic rates of hibernators: their tissue temperatures or some sort of biochemical metabolic depression. The first paper takes a very broad approach. The second focuses on just one species and is simpler to follow.

Hochachka, P. W., and G. N. Somero. 2002. *Biochemical Adaptation: Mechanism and Process in Physiological Evolution.* Oxford University Press, New York. A definitive, stimulating treatment of the biochemistry and molecular biology of temperature, written by the two leaders in the field.

Huey, R. B., and J. G. Kingsolver. 1993. Evolution of resistance to high temperature in ectotherms. *Amer. Nat.* 142: S21–S46. This rewarding paper presents a broadened view of the acute responses of poikilotherms to changes in their body temperatures and discusses the evolution of acute-response patterns as studied by both retrospective and experimental approaches. The thoughts are germane to projecting global-warming scenarios.

Ortmann, S., and G. Heldmaier. 2000. *See Heldmaier and Ruf 1992 earlier in this list.*

Steudel, K., W. P. Porter, and D. Sher. 1994. The biophysics of Bergmann's rule: A comparison of the effects of pelage and body size variation on metabolic rate. *Can. J. Zool.* 72: 70–77. This interesting paper cannot help but excite curiosity over the long-raging debate as to whether Bergmann's rule rests on valid physiological reasoning. It clarifies some key issues and provides references to some of the earlier literature in the debate (although it does not go back as far as the initial contentious interchange between Laurence Irving and Ernst Mayr).

Selected References

Altringham, J. D., and B. A. Block. 1997. Why do tuna maintain elevated slow muscle temperatures? Power output of muscle isolated from endothermic and ectothermic fish. *J. Exp. Biol.* 200: 2617–2627.

Aschoff, J. 1981. Thermal conductance in mammals and birds: Its dependence on body size and circadian phase. *Comp. Biochem. Physiol. A* 69: 611–619.

Bakken, G. S., W. R. Santee, and D. J. Erskine. 1985. Operative and standard operative temperature: Tools for thermal energetics studies. *Amer. Zool.* 25: 933–943.

Bernal, D., K. A. Dickson, R. E. Shadwick, and J. B. Graham. 2001. Review: Analysis of the evolutionary convergence for high performance swimming in lamnid sharks and tunas. *Comp. Biochem. Physiol. A* 129: 695–726.

Bligh, J., and K. Voigt (eds.). 1990. *Thermoreception and Temperature Regulation.* Springer, New York.

Block, B. A., and J. R. Finnerty. 1994. Endothermy in fishes: A phylogenetic analysis of constraints, predispositions, and selection pressures. *Environ. Biol. Fishes* 40: 283–302.

Brück, K. 1986. Basic mechanisms in thermal long-term and short-term adaptation. *J. Thermal Biol.* 11: 73–77.

Burggren, W. W., and B. R. McMahon. 1981. Oxygen uptake during environmental temperature change in hermit crabs: Adaptation to subtidal, intertidal, and supratidal habitats. *Physiol. Zool.* 54: 325–333.

Campbell, G. S., and J. M. Norman. 1998. *An Introduction to Environmental Biophysics*, 2nd ed. Springer, New York.

Carey, F. G. 1982. A brain heater in the swordfish. *Science* 216: 1327–1329.

Chappell, M. A. 1980. Thermal energetics and thermoregulatory costs of small arctic mammals. *J. Mamm.* 61: 278–291.

Chen, L., A. L. DeVries, and C.-H. C. Cheng. 1997. Convergent evolution of antifreeze glycoproteins in Antarctic notothenoid fish and Arctic cod. *Proc. Natl. Acad. Sci. U.S.A.* 94: 3817–3822. (See also 94: 3811–3816.)

Christian, K. A., and B. W. Weavers. 1996. Thermoregulation of monitor lizards in Australia: An evaluation of methods in thermal biology. *Ecol. Monogr.* 66: 139–157.

Costanzo, J. P., R. E. Lee, Jr., and M. F. Wright. 1991. Glucose loading prevents freezing injury in rapidly cooled wood frogs. *Amer. J. Physiol.* 261: R1549–R1553.

Costanzo, J. P., J. D. Litzgus, J. B. Iverson, and R. E. Lee, Jr. 2000. Seasonal changes in physiology and development of cold hardiness in the hatchling painted turtle *Chrysemys picta*. *J. Exp. Biol.* 203: 3459–3470.

Dawson, W. R., R. L. Marsh, and M. E. Yacoe. 1983. Metabolic adjustments of small passerine birds for migration and cold. *Amer. J. Physiol.* 245: R755–R767.

Duman, J. G. 2001. Antifreeze and ice nucleator proteins in terrestrial arthropods. *Annu. Rev. Physiol.* 63: 327–357.

Feder, M. E., and G. E. Hofmann. 1999. Heat-shock proteins, molecular chaperones, and the stress response: Evolutionary and ecological physiology. *Annu. Rev. Physiol.* 61: 243–282.

Fletcher, G. L., C. L. Hew, and P. L. Davies. 2001. Antifreeze proteins of teleost fishes. *Annu. Rev. Physiol.* 63: 359–390.

Florant, G. L. 1998. Lipid metabolism in hibernators: The importance of essential fatty acids. *Amer. Zool.* 38: 331–340.

Fregly, M. J., and C. M. Blatteis. 1996. *Environmental Physiology*, vol. 1. (Handbook of Physiology [Bethesda, MD], section 4). Oxford University Press, New York. (Many articles on thermal relations by recognized authorities.)

Fudge, D. S., E. D. Stevens, and J. S. Ballantyne. 1997. Enzyme adaptation along a heterothermic tissue: The visceral retia mirabilia of the bluefin tuna. *Amer. J. Physiol.* 272: R1834–R1840.

Garland, T., Jr., and S. C. Adolph. 1991. Physiological differentiation of vertebrate populations. *Annu. Rev. Ecol. Syst.* 22: 193–228.

Gates, D. M. 1980. *Biophysical Ecology.* Springer, New York.

Geiser, F., and T. Ruf. 1995. Hibernation versus daily torpor in mammals and birds: Physiological variables and classification of torpor patterns. *Physiol. Zool.* 68: 935–966.

Goldman, K. J. 1997. Regulation of body temperature in the white shark, *Carcharodon carcharias*. *J. Comp. Physiol. B* 167: 423–429.

Graether, S. P., M. J. Kuiper, S. M. Gagné, V. K. Walker, Z. Jia, B. D. Sykes, and P. L. Davies. 2000. β-Helix structure and ice-binding properties of a hyperactive antifreeze protein from insects. *Nature* 406: 325–328.

Graves, J. E., and G. N. Somero. 1982. Electrophoretic and functional enzymic evolution in four species of eastern Pacific barracudas from different thermal environments. *Evolution* 36: 97–106.

Heinrich, B., and H. Esch. 1994. Thermoregulation in bees. *Amer. Sci.* 82: 164–170.

Heldmaier, G., and M. Klingenspor (eds.). 2000. *Life in the Cold.* Springer, New York.

Hill, R. W., and D. L. Beaver. 1982. Inertial thermostability and thermoregulation in broods of redwing blackbirds. *Physiol. Zool.* 55: 250–266.

Holland, L. Z., M. McFall-Ngai, and G. N. Somero. 1997. Evolution of lactate dehydrogenase-A homologs of barracuda fishes (genus *Sphyraena*) from different thermal environments: Differences in kinetic properties and thermal stability are due to amino acid substitutions outside the active site. *Biochemistry* 36: 3207–3215.

Irving, L. 1972. *Arctic Life of Birds and Mammals.* Springer, New York.

Jessen, C. 2001. *Temperature Regulation in Humans and Other Mammals.* Springer, New York.

Johnston, I. A., and A. F. Bennett (eds.). 1996. *Animals and Temperature: Phenotypic and Evolutionary Adaptation.* Cambridge University Press, New York.

Johnston, I. A., and P. Harrison. 1985. Contractile and metabolic characteristics of muscle fibers from Antarctic fish. *J. Exp. Biol.* 116: 223–236.

Karasov, W. H., and J. M. Diamond. 1985. Digestive adaptations for fueling the cost of endothermy. *Science* 228: 202–204.

Kluger, M. J. 1979. *Fever: Its Biology, Evolution, and Function*. Princeton University Press, Princeton, NJ.

Kluger, M. J. 1991. Fever: Role of pyrogens and cryogens. *Physiol. Rev.* 71: 93–127.

Lee, R. E., Jr., and J. P. Costanzo. 1998. Biological ice nucleation and ice distribution in cold-hardy ectothermic animals. *Annu. Rev. Physiol.* 60: 55–72.

Linton, T. K., I. J. Morgan, P. J. Walsh, and C. M. Wood. 1998. Chronic exposure of rainbow trout (*Oncorhynchus mykiss*) to simulated climate warming and sublethal ammonia: A year-long study of their appetite, growth, and metabolism. *Can. J. Fish. Aquat. Sci.* 55: 576–586.

Logue, J. A., A. L. DeVries, E. Fodor, and A. R. Cossins. 2000. Lipid compositional correlates of temperature-adaptive interspecific differences in membrane physical structure. *J. Exp. Biol.* 203: 2105–2115.

Loomis, S. H. 1995. Freezing tolerance of marine invertebrates. *Oceanogr. Mar. Biol. Annu. Rev.* 33: 337–350.

Lynch, C. B. 1992. Clinal variation in cold adaptation in *Mus domesticus*: Verification of predictions from laboratory populations. *Amer. Nat.* 139: 1219–1236.

Manis, M. L., and D. L. Claussen. 1986. Environmental and genetic influences on the thermal physiology of *Rana sylvatica*. *J. Thermal Biol.* 11: 31–36.

McKechnie, A. E., and B. G. Lovegrove. 2002. Avian facultative hypothermic responses: A review. *Condor* 104: 705–724.

Moerland, T. S. 1995. Temperature: Enzyme and organelle. In P. W. Hochachka and T. P. Mommsen (eds.), *Environmental and Ecological Biochemistry*, pp. 57–71. Elsevier, New York.

Pörtner, H. O., and R. C. Playle (eds.). 1998. *Cold Ocean Physiology*. Cambridge University Press, New York.

Pough, F. H., and R. M. Andrews. 1984. Individual and sibling-group variation in metabolism of lizards: The aerobic capacity model for the origin of endothermy. *Comp. Biochem. Physiol. A* 79: 415–419.

Powers, D. A., M. Smith, I. Gonzalez-Villasenor, L. DiMichele, D. Crawford, G. Bernardi, and T. Lauerman. 1993. A multidisciplinary approach to the selectionist/neutralist controversy using the model teleost, *Fundulus heteroclitus*. *Oxford Surv. Evol. Biol.* 9: 43–107.

Prinzinger, R., A. Pressmar, and E. Schleucher. 1991. Body temperature in birds. *Comp. Biochem. Physiol. A* 99: 499–506.

Roberts, S. P., and J. F. Harrison. 1998. Mechanisms of thermoregulation in flying bees. *Amer. Zool.* 38: 492–502.

Rolfe, D. F. S., and G. C. Brown. 1997. Cellular energy utilization and molecular origin of standard metabolic rate in mammals. *Physiol. Rev.* 77: 731–758.

Rommel, S. A., D. A. Pabst, and W. A. McLellan. 1998. Reproductive thermoregulation in marine mammals. *Amer. Sci.* 86: 440–448.

Root, T. 1988. Energy constraints on avian distributions and abundances. *Ecology* 69: 330–339.

Rubinsky, B., S. T. S. Wong, J.-S. Hong, J. Gilbert, M. Roos, and K. B. Storey. 1994. 1H magnetic resonance imaging of freezing and thawing in freeze-tolerant frogs. *Amer. J. Physiol.* 266: R1771–R1777.

Schulte, P. M., M. Gomez-Chiarri, and D. A. Powers. 1997. Structural and functional differences in the promoter and 5′ flanking region of *Ldh-B* within and between populations of the teleost *Fundulus heteroclitus*. *Genetics* 145: 759–769.

Seymour, R. S. 1997. Plants that warm themselves. *Sci. Amer.* 276(3): 104–109.

Somero, G. N. 1997. Temperature relationships: From molecules to biogeography. In W. H. Dantzler (ed.), *Comparative Physiology*, vol. 2 (Handbook of Physiology [Bethesda, MD], section 13), pp. 1391–1444. Oxford University Press, New York.

Somero, G. N. 2002. Thermal physiology and vertical zonation of intertidal animals: Optima, limits, and costs of living. *Integr. Comp. Biol.* 42: 780–789.

Somero, G. N., C. B. Osmond, and C. L. Bolis (eds.). 1992. *Water and Life*. Springer, New York.

Southwick, E. E., and G. Heldmaier. 1987. Temperature control in honey bee colonies. *BioScience* 37: 395–399.

Storey, K. B. 1997. Organic solutes in freezing tolerance. *Comp. Biochem. Physiol. A* 117: 319–326.

Storey, K. B., and J. M. Storey. 1996. Natural freezing survival in animals. *Annu. Rev. Ecol. Syst.* 27: 365–386.

Tanaka, H., Y. Takagi, and Y. Naito. 2000. Behavioural thermoregulation of chum salmon during homing migration in coastal waters. *J. Exp. Biol.* 203: 1825–1833.

Tieleman, B. I., and J. B. Williams. 1999. The role of hyperthermia in the water economy of desert birds. *Physiol. Biochem. Zool.* 72: 87–100.

Tomanek, L., and G. N. Somero. 1999. Evolutionary and acclimation-induced variation in the heat-shock responses of congeneric marine snails (genus *Tegula*) from different thermal habitats: Implications for limits of thermotolerance and biogeography. *J. Exp. Biol.* 202: 2925–2936.

Ultsch, G. R. 1989. Ecology and physiology of hibernation and overwintering in freshwater fishes, turtles and snakes. *Biol. Rev. (Cambridge)* 64: 435–516.

Vogt, F. D. 1986. Thermoregulation in bumblebee colonies. I. Thermoregulatory versus brood-maintenance behaviors during acute changes in ambient temperature. *Physiol. Zool.* 59: 55–59.

Walsberg, G. E. 1983. Coat color and solar heat gain in animals. *BioScience* 33: 88–91.

Weathers, W. W. 1981. Physiological thermoregulation in heat-stressed birds: Consequences of body size. *Physiol. Zool.* 54: 345–361.

Weathers, W. W., and K. A. Sullivan. 1993. Seasonal patterns of time and energy allocation by birds. *Physiol. Zool.* 66: 511–536.

Wolf, B. O., and G. E. Walsberg. 1996. Thermal effects of radiation and wind on a small bird and implications for microsite selection. *Ecology* 77: 2228–2236.

Yagil, R. 1985. *The Desert Camel: Comparative Physiological Adaptation*. Karger, New York.

See also references in Chapter 9 and Appendix A.

9 Food, Energy, and Temperature at Work: The Lives of Mammals in Frigid Places

Reindeer (*Rangifer tarandus*), which occur in Siberia and other far-northern regions, typically give birth in May, when the ground remains snow-covered and the air often cools to below 0°C overnight. At the moment of its birth, a reindeer calf experiences a drop in its environmental temperature from 37°C inside its mother's uterus to the prevailing air temperature. If the air temperature is –3°C, the calf's environmental temperature plummets by 40°C at birth. The air sometimes is much colder, and some calves experience a drop of 50°C to 60°C when they are born. Newborns are wet—covered with amniotic fluid—and a strong wind may blow.

Reindeer calves must thermoregulate on their own from the moment they are born, or die, because they do not huddle with each other or adults, and they have no nest to protect them. They are perhaps the most precocial (adultlike) of all the newborns of land mammals. They stand the moment they are born. By the second day they are able to run fast enough to outdistance a man. Because herds typically move from place to place incessantly to find food, the newborn calf has no luxury of resting to gather strength. It must keep up. Within a week, a reindeer calf can swim across broad rivers.

Biologists are often so preoccupied with adults that it can be easy to forget that a mammalian life consists of all the stages from conception to death. Each individual reindeer—and each individual of every other species—must survive first as a youngster if it is ever to have the chance of surviving (and reproducing) as an adult. Although much less is known about the physiology of young mammals than adults, there is every reason to think that the young stages are particularly important in the evolution of many species, because many individuals die when young, and natural selection may therefore be intense at that time.

We will return to the young of reindeer shortly. First, however, let's focus on adult reindeer and the environment where they live—matters that set the context for fully understanding the young.

A newborn reindeer calf must thermoregulate on its own Although mother reindeer feed their young, they do not keep them warm, and the Arctic environment where birth occurs may be very cold.

Food, Nutrition, Energy Metabolism, and Thermoregulation in the Lives of Adult Reindeer

Reindeer, as a species, are probably the most adapted to cold exposure of all inland mammals. An intriguing aspect of animal species that are extremely well suited physiologically to live in stressful habitats is that their very presence can permit other species to exist there as well. In deserts, the existence of rodents that have evolved extremely low requirements for ingested water permits many water-dependent predators to exist as well; the rodents provide the watery food the predators need. Similarly, in the Far North, the existence of reindeer permits the existence there of wolves and other predators—including indigenous human cultures. During preindustrial times (or in indigenous cultures living apart from modern comforts today), people could not possibly have survived on their own in the interior of the Far North. The existence of humankind there was made possible, all around the world, by herds of reindeer, which served as sources of food and of highly insulating pelts that people employed for necessary shelter and clothing.

Because of the intimate relations between reindeer and people, reindeer were domesticated to some degree in many regions. Today, therefore, the species consists of both wild populations and numerous domesticated strains. Caribou are considered by most mammalogists to be the same species.

Reindeer have many characteristics that help them prosper in the Far North. Although adult reindeer weigh one-third as much as adult moose (another northern species), the feet of reindeer are so unusually broad that the contact area of a reindeer with the ground is about the same as that of a moose. Thus the downward gravitational force on each unit area of a reindeer's feet is very low compared to that of a moose, explaining how reindeer can readily negotiate snow fields that stop moose.

Adult reindeer in their winter pelage have lower-critical temperatures of –30°C or lower in still air, and because the air temperature in the Arctic rarely goes below –50°C, reindeer in still air never have to increase their metabolic rates much above basal (Figure 9.1). As an amusing comparison, Laplanders and Norwegians equipped just with the insulation "nature gave them" (i.e., naked) have lower-critical temperatures of +24°C to +27°C; this is true even though the people are of about the same body size as the reindeer represented in Figure 9.1. Only by dressing in something like a reindeer pelt can a person have a lower-critical temperature that is compatible with Arctic life.

One reason for the low energy costs of adult reindeer in the frigid air of winter is their pelage. A dense underfur of fine hairs thickly covers their skin, and the longer "guard" hairs of the pelage, which protrude beyond the underfur and hang over its outer surface, are extensively vacuolated ("hollow"). In common with other large species of mammals, reindeer undergo dramatic seasonal molts. In winter a reindeer's fur is 3 to 4 cm thick over much of its body (Figure 9.2). All body surfaces of a reindeer, even the nose, have a hair covering.

Besides the pelage, another reason for the low energy costs of adult reindeer in frigid air is that, by employing regional heterothermy, they do not keep their legs and other exposed body parts as warm as their body core (see Figure 8.26). Reflecting on what we learned about lipid fluidity in our study of fish and other poikilotherms in Chapter 8, an interesting question about the heterothermic legs of reindeer is whether the fats in their legs are hard and stiff where the limbs are cold, much as butter is hard when cold.

Actually, it is a matter of ancient knowledge that fats from the outer extremities of reindeer legs—and also those from the hooves of cattle—are particularly fluid, compared to body-core fats; people have long used these fats (actually oils) to soften boot leather and give flexibility to leather bow strings in cold climates. Homeoviscous adaptation thus exists from one end of a heterothermic leg to the other: Whereas the fats in the upper leg of a reindeer or cow have chemical structures that give them a gel-like texture at 37°C and become hard at 0°C, the fats from the outer extremities have different chemical structures that give them a gel-like or oily texture near 0°C. In the marrow of the limb bones of reindeer, oleic acid—an unsatu-

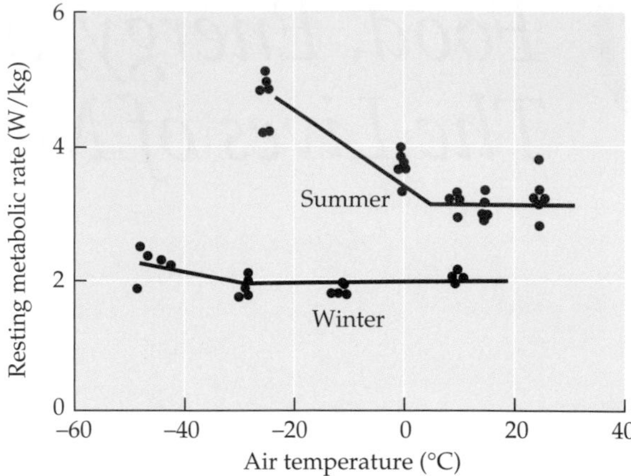

Figure 9.1 Resting metabolic rate as a function of air temperature down to –50°C in adult reindeer The same Norwegian animals, weighing about 70 kg, were studied in winter and summer. The air was still. Lines are drawn by eye to approximate the trends in the data. (After Nilssen, Sundsfjord, and Blix 1984.)

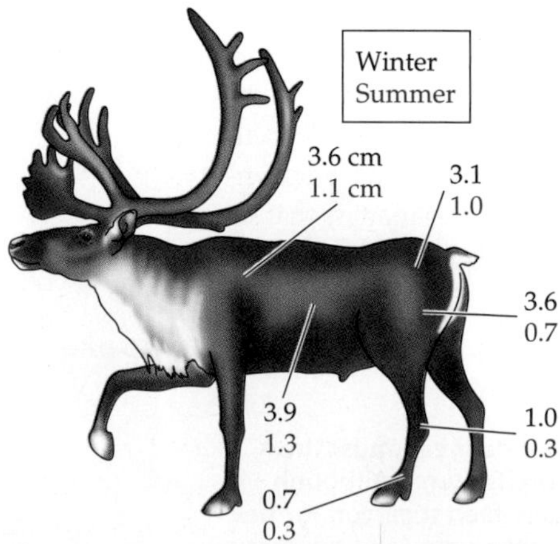

Figure 9.2 Fur thickness of adult reindeer in winter and summer The thickness was measured perpendicular to the skin surface and is expressed in centimeters. (After Johnsen et al. 1985.)

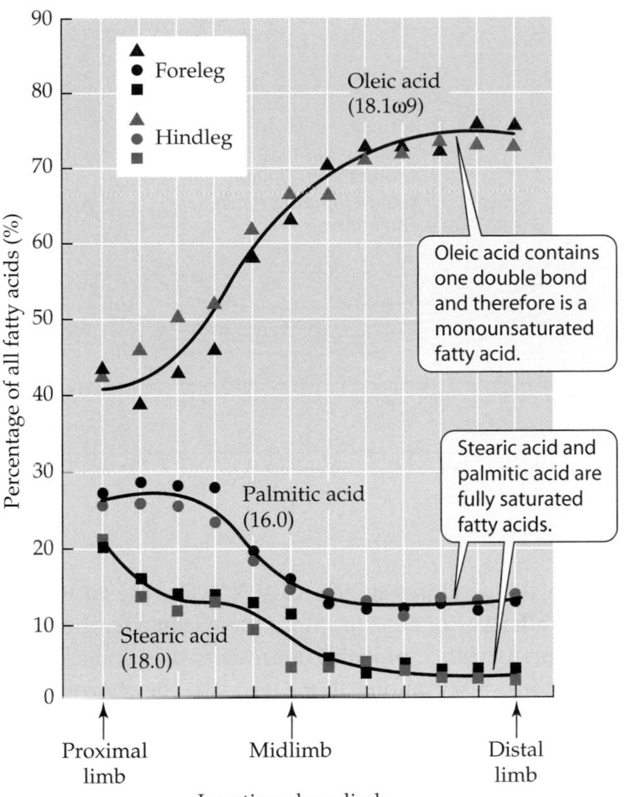

Figure 9.3 Fatty acid composition of bone marrow lipids in the legs of reindeer Marrow lipids were sampled at 12 locations from the proximal (upper) end of each limb to the distal (lower) end. The proximal locations were the proximal ends of the humerus and femur; the midlimb locations were the distal ends of the radius and tibia; the distal locations were the third phalanges. In the numbering system for the fatty acids (discussed in Chapter 4, see page 97), the number before the decimal is the number of carbon atoms; that following the decimal is the number of double bonds; and that following ω (omega) designates the position of a key sort of double bond (ω is not written when there are no double bonds). (After Meng, West, and Irving 1969.)

rated fatty acid—becomes a more important constituent of the lipids as one moves out along a leg, whereas stearic acid—a saturated fatty acid—becomes less important (Figure 9.3). In this way, the bone marrow lipids in the legs of reindeer are reminiscent of the brain phospholipids of fish from various climates (see Figure 2.3). The same basic trends exist even in the legs of many tropical mammals. Thus, although the trends are significant for reindeer, they do not seem to be specific adaptations to a truly frigid climate.

Food and nutrition represent great challenges in the environments where reindeer live. Like other deer, reindeer are ruminants. Thus, rumen processes, as well as digestive and absorptive processes, play pivotal roles in their physiology of food and nutrition. A key to the survival of reindeer in the Far North is that they eat a great diversity of plants (37 genera were found in the rumens of one herd). Moreover, they obtain an exceptional degree of nutritional benefit from lichens ("reindeer moss") and a variety of other species of far-northern plants that are not much eaten by other mammals.[1] The exact mechanisms by which reindeer are able to exploit their unusual foods remain poorly known. Nonetheless, studies of ingestion and egestion show, for example, that reindeer obtain twice the nutrient value from lichens as sheep or cows do.

The seasonal cycle of plant growth in the Far North is dramatic—not just because winters are cold, but also because of winter darkness (in some places that reindeer occupy, the sun does not rise for 3 months). In most places, the summer foods of reindeer are collectively high in protein and mineral nutrients, and they are high in the proportion of total carbohydrate that is in readily digestible forms rather than cellulose or hemicelluloses. By contrast, the lichens and senescent vascular plants that dominate the reindeer's winter diets are collectively the opposite: low in protein, low in minerals, low in highly digestible carbohydrates, and high in cellulose and hemicelluloses. Lichens are often the salvation of reindeer in winter *in terms of supplying energy* because they are abundant, and as we have already said, reindeer can tap a high proportion of their nutrient value. Lichens, however, typify the nutritional problems that reindeer face in winter, in that, being low in protein and low in minerals like Na^+, they are not even close to being nutritionally complete. Figure 9.4, which summarizes the composition of the foods available to reindeer throughout the year in Finland, is well worth close study, because it illustrates dramatically that animals in the wild—away from veterinarians and carefully manufactured feeds—often face substantial nutritional stresses.

Reindeer and the microbial symbionts in their rumens make adjustments as the seasons change. For instance, certain strains of reindeer fatten dramatically as winter approaches, thereby reducing their need for winter food. In addition, the community of fermenting microbes in the rumen changes in composition in ways that respond to the shifts in the types of foods eaten (Table 9.1); microbes that digest

[1] For simplicity of language, we will use the term *plants* in a loose, comprehensive sense to refer to all the photosynthetic organisms that reindeer eat, including lichens and mosses, as well as true plants.

TABLE 9.1 Responses of the rumen microbes of reindeer to seasonal changes in diet The mixed communities of fermenting microbes in rumen fluid were classified using standard microbiological methods. The winter microbial community was more able to ferment plant fiber, including particular structural carbohydrates such as cellulose, and less able to carry out hydrolysis of proteins, than the summer microbial community.

Food breakdown activity	Percentage of microbes that could carry out each activity (%)	
	Summer	Winter
Fiber digestion	31	74
Cellulose digestion	15	35
Hemicellulose (xylan) digestion	30	58
Proteolysis (protein hydrolysis)	51	28
Starch digestion	68	63

Source: After Orpin et al. 1985.

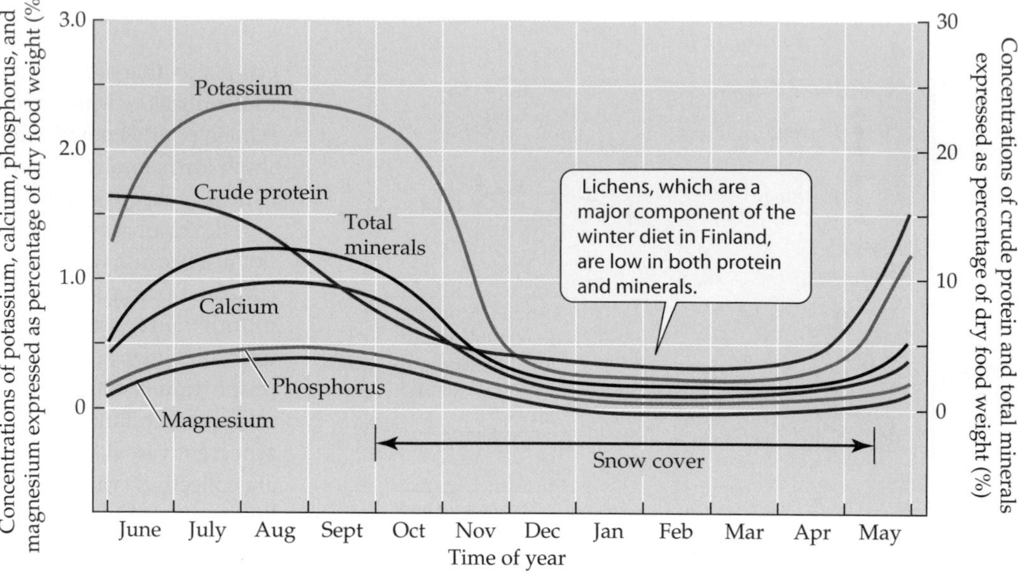

Figure 9.4 Seasonal changes in diet: the protein and mineral content of the foods eaten by Finnish reindeer Note that the scales on the *y* axes begin above the bottom of the graph. (After Nieminen 1980.)

woody, fibrous plant material, for example, increase in winter when reindeer ingest considerable amounts of such material. Despite such adjustments, by the time spring arrives after a long winter, reindeer have lost body weight, are often somewhat emaciated, and may exhibit blood mineral levels diagnostic of mineral deficiency. In some places, they are renowned for their "mineral appetites" in spring.

In addition to the physiological and anatomical features we have already discussed, the distinctive feeding behaviors and migratory behaviors of reindeer herds play key roles in their success in the Far North. Food is thinly distributed there, even for animals that feed on a diversity of plants. Thus, herds must range widely to obtain enough food. Reindeer herds are legendary for being incessantly on the move during daylight hours, covering large areas of ground every day, nibbling as they go. The extraordinary annual migrations of reindeer herds are additional behavioral adaptations, helping them to find not only food but also shelter from wind during winter. As winter approaches, the norm is for a herd to migrate 1000 km from its summer site to its winter site. Recent satellite-based tracking studies in Alaska and the Yukon have revealed that some herds make a round-trip of 5000 km per year.

The movements of the adults in a herd are part of reality for all calves born into the herd. Calves must join in the wide-ranging daily movements of their herd soon after birth, as we have already mentioned, and by autumn of their first year of life, they must be ready for the annual migration.

Newborn Reindeer

When reindeer are born, they already have a well-developed pelage, consisting of woolly, hollow hairs. Their fur provides substantial insulation as soon as the uterine fluids have evaporated away and the hairs are dry. From the moment of birth, reindeer also exhibit a typical homeothermic relation between their metabolic rate and the air temperature, as shown by the red line in Figure 9.5. Newborns are able to raise their rate of metabolic heat production to at least twice their resting rate. By virtue of the combined effects of their pelage insulation and this thermogenic ability, newborns are able to keep their body temperatures at 39°C to 40°C when the air is –20°C to –25°C (a difference of 60°C or more) for at least a few hours in still air. This performance probably represents the pinnacle of thermoregulatory ability among all the terrestrial newborns on Earth. That said, it remains true that reindeer are born into an environment that can be very harsh, and the thermoregulatory abilities of newborns are far inferior to those of adults. Many newborns die if they get wet, or if the wind blows briskly or the air temperature remains very low for a day or more.

Reindeer calves grow rapidly compared to other deer. Partly as a consequence of their increasing body size, their metabolism–temperature relation becomes noticeably more favorable with each passing week (see Figure 9.5). When calves are 2 weeks old, the weight-specific rate of heat production that they require to stay

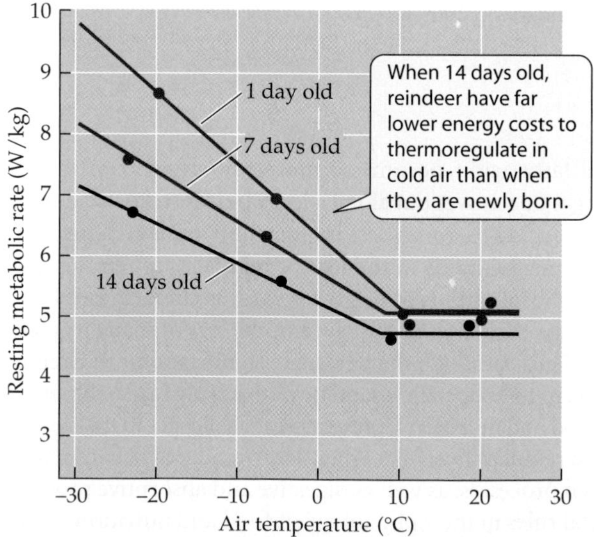

Figure 9.5 Resting metabolic rate as a function of air temperature in newborn and growing reindeer The air was still when these measurements were made. Dots are averages for the sets of animals studied at various conditions. (After Markussen, Rognmo, and Blix 1985.)

BOX 9.1 Knockout Mice Clarify the Function of Brown Fat

Very recently, molecular genetic tools have been used to produce laboratory mice that cannot synthesize the type of uncoupling protein, known as *uncoupling protein 1* (*UCP1*), that is involved in nonshivering thermogenesis (NST). The gene coding for UCP1 is inactivated in these mice.

Surprisingly, the knockout mice without UCP1 can be acclimated to live at 4°C, a temperature at which their metabolic rates are four times basal. To thermoregulate, they produce all their heat by shivering, according to present evidence. The discovery that the knockout mice depend on shivering provides compelling evidence that, at least in laboratory mice, brown fat is the only tissue that carries out NST; UCP1 occurs just in brown fat, and because the ablation of UCP1 eliminates NST, brown fat must be the sole tissue that can carry out NST. The discovery also provides evidence that molecular homologs of UCP1, such as UCP2 and UCP3, are not involved in NST. Such homologs occur in a variety of tissues. Because the targeted ablation of UCP1 eliminates NST, the other homologs are evidently not involved in NST.

warm at an air temperature of –20°C is already reduced to only 70% of that required at birth. For understanding the rapid growth of reindeer calves, it is undoubtedly significant that the milk produced by their mothers is about 20% lipid, compared to about 4% lipid in cow's milk. Reindeer milk is accordingly very energy dense (see Table 4.2), having about three times the energy value per liter as cow's milk has. It is also particularly rich in protein. The nutrient-rich milk of reindeer aids the rapid growth of the calves. In addition, calves may start eating vegetation within 2 to 3 days of being born, and by 2 weeks after birth, they eat considerable amounts of plants, while continuing to nurse. By late autumn of their first year, when they are 5 to 6 months old and fully stop nursing, they have reached 50% to 60% of their adult weight. This extent of growth in the first season of life far exceeds the average for other, related mammals and is believed to be important in enabling young reindeer to participate successfully in their herd's long migration to its wintering area.

A great deal of interest has focused on the mechanisms by which newborn reindeer and other young mammals increase their metabolic heat production for thermoregulation. Nonshivering thermogenesis by brown fat plays an extraordinary role in this regard. Brown fat is nearly always the principal thermogenic tissue in newborn placental mammals, and nonshivering thermogenesis is thus the principal mechanism of thermogenesis (Box 9.1). *Why* brown fat and nonshivering thermogenesis should be of paramount importance in newborns is largely an unresolved mystery, as we will discuss later.

The ways in which brown adipose tissue and nonshivering thermogenesis have been identified in reindeer calves illustrate well the steps that physiologists typically take to identify the tissue and process. The first study indicating that brown fat occurs in newborn reindeer involved microscopic examination of adipose tissues. The investigators observed that the fatty tissue between the scapulae (shoulder blades) of newborn calves is reddish brown and, suspecting it to be brown fat, prepared it for microscopy. The microscopic approach to identifying brown fat is based on the fact that the tissue usually differs from white fat in several structural ways. For example, brown fat is much richer in mitochondria, and its intracellular fat deposits are multilocular (meaning there are many small lipid droplets per cell) rather than unilocular (one droplet per cell) as in white fat. Microscopic studies performed in about 1980 indicated that most of the major deposits of adipose tissue in the body of a newborn reindeer are brown fat. Nonetheless, physiologists assume that function, not just morphology, must ultimately be directly studied for function to be understood. Thus, soon after microscopy had established the likely presence of brown fat in reindeer, investigators tested calves to determine their responses to injected norepinephrine.

As you will recall from Chapter 8, thermogenesis by brown fat is ordinarily activated by the sympathetic nervous system, secreting norepinephrine. On the basis of this fact, an animal's response to a norepinephrine injection has often been used as an informative (although relatively crude) test for brown-fat function. When newborn reindeer are injected subcutaneously with a standard dose of norepinephrine, they respond with a large increase in their rate of O_2 consumption (Figure 9.6). This response to norepinephrine, coupled with the microscopic observations discussed earlier, convinced physiologists that brown fat and nonshivering thermogenesis are important for heat production in newborn reindeer.

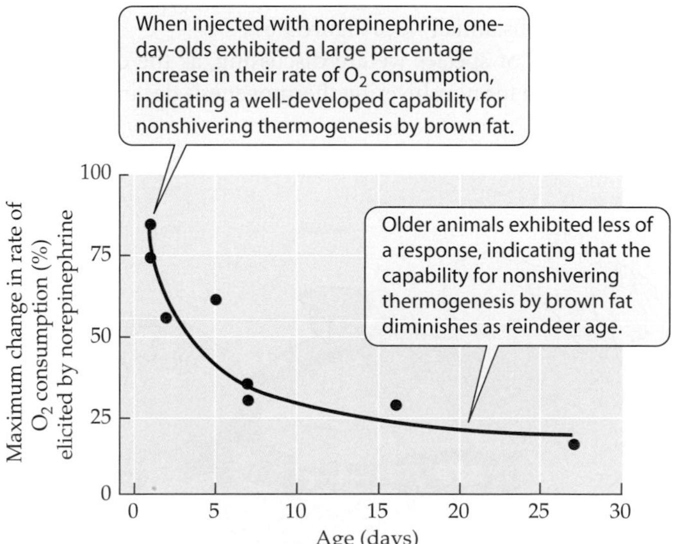

Figure 9.6 A test for brown-fat thermogenesis in newborn and growing reindeer Reindeer of various ages were injected subcutaneously with a standard dose of norepinephrine per kilogram of body weight. (After Soppela et al. 1986.)

Always, however, thoughtful scientists are wondering if their standards of evidence are adequate. Anyone who has spent months of his or her life gathering data becomes aware that there is a risk of being fooled. Maybe, one fears at times, the hard-won data do not say exactly what we imagined they say. Because of these concerns, people using the microscopic approach kept reassessing whether the microscopic criteria used to identify brown fat were infallible. By the last decade of the twentieth century, these investigators had concluded—on the basis of hundreds of studies of brown fat in various mammals—that in fact the traditional microscopic criteria are not 100% reliable. On unusual occasions, white fat can seem to be brown fat, or vice versa, when assessed according to the traits observed using traditional microscopy.

Thus a third test for brown fat was devised, employing a distinctive molecular marker. The mitochondrial protein known specifically as *uncoupling protein 1* (*UCP1*) is believed to occur exclusively in brown fat, not white fat or other tissues. Antibodies to UCP1 can be produced in a laboratory. Then, UCP1 in a newborn reindeer or other animal can be unambiguously identified by its antigen–antibody reaction with the antibodies. Studies have shown that all of the major adipose-tissue deposits in newborn reindeer react with UCP1 antibodies, demonstrating more convincingly than ever that the tissues are brown fat.

Reindeer newborns are not alone in being well endowed with brown fat. In fact, most placental mammals, including humans (Figure 9.7), have extensive masses of brown fat at birth. In large-bodied species, such as reindeer and humans, the brown fat of newborns typically declines rapidly with age. This decline is particularly rapid in ruminants. In reindeer, sheep, goats, and other species of ruminants that have been studied, all the brown fat present in newborns undergoes a genetically programmed transformation into white fat by 1 month of age; this is demonstrated by the fact that in 1-month-old animals, the gene for UCP1 is no longer expressed and no tissue reaction with UCP1-specific antibodies occurs. Simultaneously, the metabolic response of the young animals to a norepinephrine injection declines substantially (see Figure 9.6).

In the sorts of species we are discussing, as the capacity of a young mammal for nonshivering thermogenesis declines, shivering becomes the sole mechanism of increasing heat production for thermoregulation. This transition is quite obvious in people; whereas we rely principally on nonshivering thermogenesis when we are newly born, we become dependent principally on shivering thermogenesis in youth and adulthood. In reindeer, as already said, the transition occurs relatively rapidly. Shivering becomes their only substantial mechanism of increasing metabolic heat production for thermoregulation by the time they are 1 month old, as far as we know now. This is not to say that young reindeer shiver a lot. By the time they are a month old, the warm air temperatures of summer have started to prevail, and the need for any sort of thermoregulatory thermogenesis is reduced. When winter arrives, the young reindeer are 6 months old and—similar to adults—have lower-critical temperatures of about –30°C.

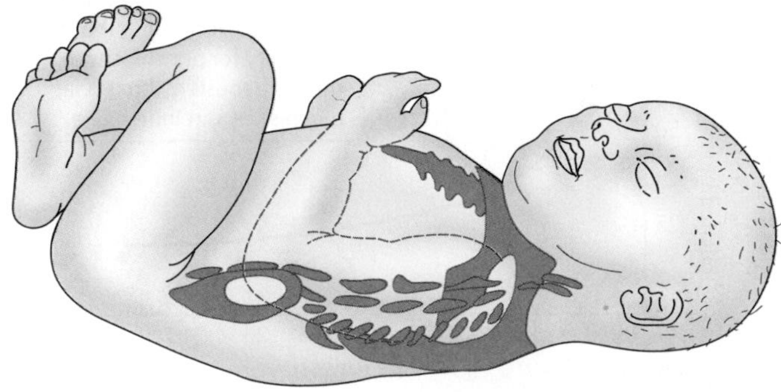

Figure 9.7 Brown adipose tissue in a human infant The tissue occurs in discrete masses in many parts of the body. (After Hull 1973.)

The primary role of brown fat in newborn ruminants, such as reindeer and sheep, seems to be to act as a thermogenic bridge during the immediate transition between life in the uterus and life outside the mother. For brown fat to play this role, it must develop extensively prior to birth, a fact that raises numerous interesting questions. For example, do fetuses develop more brown fat when the environment into which they will be born is cold rather than warm?

This question has been studied in sheep, in an indirect way, by shearing off the fur of some mothers during their final month of pregnancy and comparing their newborns with the newborns of unshorn mothers (in these experiments, shearing was used to mimic colder weather for the mothers by lowering their insulation). Lambs born to shorn mothers have more brown fat at birth than lambs born to unshorn mothers have. Moreover, whereas the newborns of shorn mothers never shiver when exposed to a low (14°C) air temperature, many of the newborns of unshorn mothers shiver. These results indicate that the intrauterine development of brown fat by fetuses is modified, depending on the cold stress experienced by their mothers, in ways that help ensure that nonshivering thermogenesis will be sufficient for thermoregulation in the newly born.

Another interesting question is how the large masses of brown fat in unborn, near-term young are regulated so that they do not catabolize large amounts of the young's foodstuffs. Logic suggests that, prior to birth, a young animal will profit most from the foodstuffs it has by using them to grow and mature; brown fat, if uncontrolled, could turn large quantities of foodstuffs into heat and chemical wastes. Experiments reveal that if near-term sheep fetuses are cooled inside the uteruses of their mothers (Figure 9.8), they do not activate their brown fat or engage in nonshivering thermogenesis. On the other hand, if blood flow from their placenta is blocked by pinching shut the umbilical cord, they rapidly exhibit substantial nonshivering thermogenesis. These results and others indicate that near-term sheep in the uterus receive from their placenta some sorts of signaling molecules (possibly prostaglandins) that inhibit their activation of their brown fat. These signaling molecules are cut off at birth by the severing of the umbilical cord, and the brown fat is then able to respond vigorously to the chilly outside environment by producing heat. A similar process may occur in reindeer. Future research will tell.

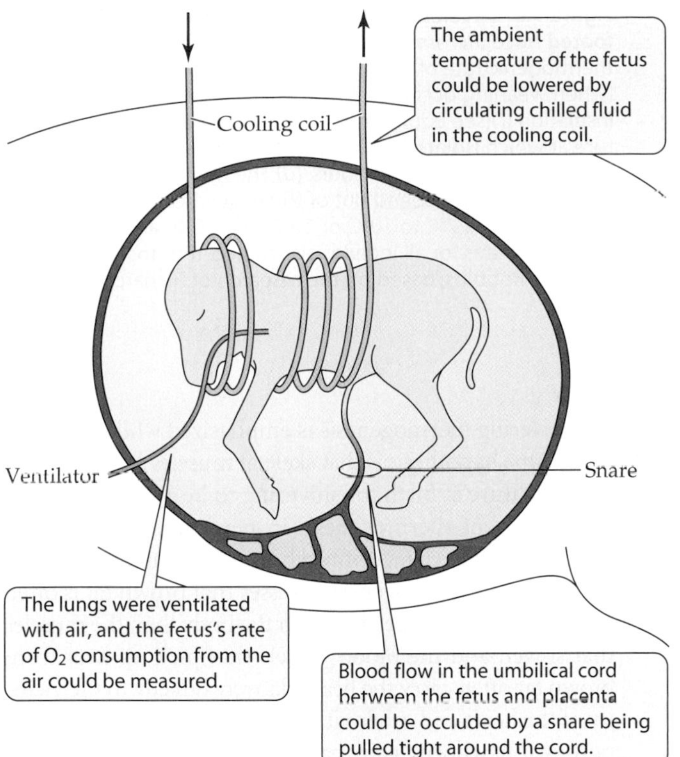

Figure 9.8 A near-term sheep fetus in the uterus prepared for study of "simulated birth" (From Power et al. 1987.)

Lifetime Patterns of Thermoregulation and Thermogenesis in Small Mammals

Mammals the size of mice and rats have only very modest capabilities to thermoregulate when they are first born. The white-footed mouse (*Peromyscus leucopus*), one of the most abundant native small mammals in North America, provides a typical example of the course of development in such animals.

White-footed mice occur in northern states like Michigan and Wisconsin, as well as into Canada. In these areas they give birth to their first litters of young each year in March and April, when the cold of late winter still prevails. Their litters consist of four to six littermates. The young are born without fur, and although they respond to cold exposure by increasing their metabolic rates in the first days after they are born, their peak rates of heat production at those ages are not high.

In the days right after birth, if all the young in a litter of white-footed mice huddle together within the nest that their mother constructs for them, they are able collectively to thermoregulate reasonably well for a few hours even when the air outside the nest is near freezing. Thus a newborn litter can stay relatively warm for a while when its mother is away foraging. However, if a single newborn white-footed mouse is removed from the nest and studied by itself, it cannot marshal a high enough metabolic rate to stay warm even when the air temperature is +25°C—a temperature higher than "room temperature" in American buildings. At an ambient temperature of 25°C, the body temperature of a solitary 2-day-old soon becomes 28°C to 29°C—not because the animal is in some sort of controlled hypothermia, but because its thermoregulatory abilities are overwhelmed. The young of lemmings and other small mammals characteristic of the Far North are not much different; to thermoregulate in their first days of life, they require a protective microhabitat, including the nest their mother provides and the siblings with which they huddle.

The length of the nestling period of white-footed mice is typical of that of most mice and rats: about 3 weeks. During that time, young mice must become physiologically capable of living independently. As young white-footed mice grow and mature during their nestling period, the peak rate at which they can make heat (their peak weight-specific metabolic rate) increases dramatically (**Figure 9.9*a***). Moreover, they develop fur, and their resistance to heat loss (their total body insulation) increases (**Figure 9.9*b***). Thus the mice become increasingly able to thermoregulate as isolated individuals, away from their nest and siblings (**Figure 9.9*c***). By 18 days of age, a lone youngster can thermoregulate for a number of hours even in freezing-cold air, and it capitalizes on this newfound "constancy of its internal environment" by making excursions outside its nest (**Figure 9.9*d***).

Regarding the *mechanisms* of thermogenesis in developing mice and rats, the evidence available indicates that shivering is not functional in most species for roughly the first week of postnatal life. Brown fat, in contrast, is present at birth. The brown fat is believed to be responsible for the very modest abilities of the young to increase their metabolic rates during their first days of nestling life. Moreover, the brown fat grows as the young grow; in fact, studies of laboratory rats have shown that the maturation of brown fat as a thermogenic tissue outpaces body growth for a while in these animals, so the brown fat is able to produce ever more heat per unit of body weight as time passes. In white-footed mice, the increase in peak weight-specific metabolic rate as nestlings age (see Figure 9.9*a*) is almost surely a consequence of the growth and maturation of brown fat.

Clearly the trajectory of brown-fat development in small mammals is extremely different from the trajectory in large mammals. In reindeer or sheep, brown fat is maximally or near-maximally developed at birth, and it starts to wane soon after birth, so it is gone by about a month of age. In small mammals like mice and rats, although brown fat is present at birth, it is far from fully developed then, and for much of the first month of life it develops an ever-greater thermogenic ability.

As a corollary of these differences in developmental trajectory, the adults of large and small mammals tend to exhibit consistent differences in brown fat. In the adults of most large-bodied species of mammals, brown fat is not present to any great extent; it is either entirely absent, or it is present in such ambiguous ways that its existence is a topic for debate among experts. By contrast, brown fat is obviously present in the adults of most species of placental mammals smaller than about 5 kg. It becomes particularly prominent in these adults when they are acclimated to cold or acclimatized to winter, as we saw in Chapter 8; and it is prominent in hibernators, as we will discuss later in this chapter.

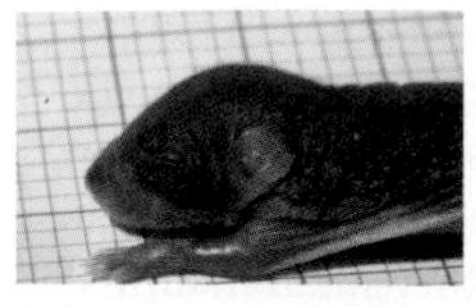

4 days

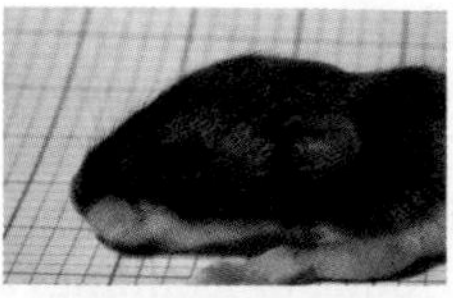

10 days

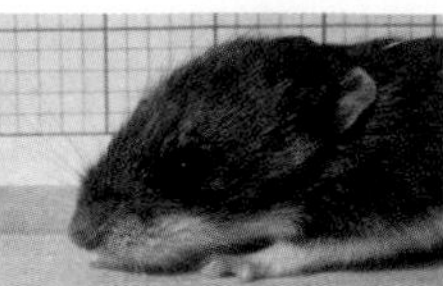

14 days

(*a*) Peak rate of metabolic heat production (cal/g•h)

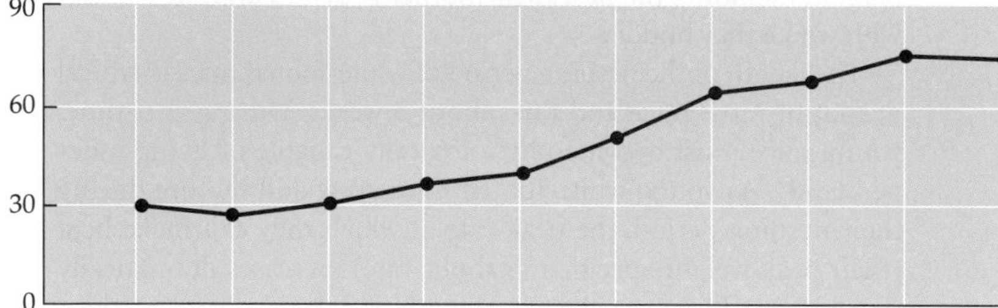

(*b*) Whole body-insulation below thermoneutrality (°C•g•h/cal)

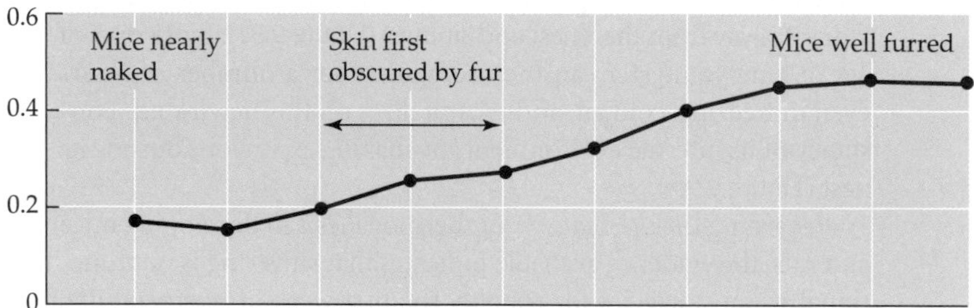

(*c*) Lowest ambient temperature at which isolated individuals can thermoregulate (°C)

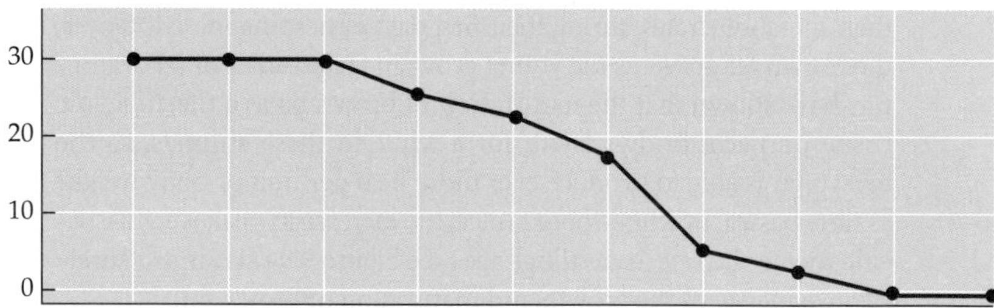

(*d*) Time per day spent out of nest (h)

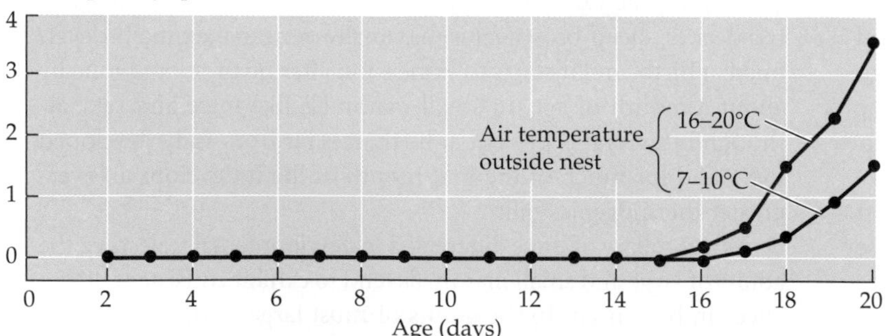

Figure 9.9 Development of thermoregulation in white-footed mice (*Peromyscus leucopus*) *(a)* Changes in the peak thermogenic rate of young between birth and 3 weeks of age (when weaning occurs). *(b)* The development of whole-body insulation (*I*)(see Equation 8.10). *(c)* The lowest air temperature at which isolated individuals of various ages can thermoregulate for 2.5 to 3.0 hours. *(d)* The number of hours per day that nestlings spend out of the nest when the air temperature outside is 7°C to 10°C, or 16°C to 20°C. In all parts, the data are means for all individuals studied. (Photos courtesy of Robert J. Robbins; based on the research of Richard W. Hill.)

Looking back over our discussion of the development of thermoregulation in large and small species, it is striking to observe that in *both* large and small placental mammals, brown fat and nonshivering thermogenesis are the favored means of producing heat for thermoregulation at birth and during the period immediately following birth. Conversely, shivering is not favored at those times. Only rudimentary hypotheses are presently available to explain why nonshivering thermogenesis is emphasized while shivering is not. One hypothesis is that skeletal muscles are inherently too immature at birth for shivering to be a viable primary mechanism of thermogenesis in newborns. Another hypothesis for the reliance of newborn placental mammals on nonshivering thermogenesis stresses that brown fat is often positioned, on average, deeper in the body than the muscles that shiver, and the blood vessels that pick up heat from brown fat often carry the heat relatively directly to the vicinity of the heart and other vital organs. According to this hypothesis, newborns gain particularly great benefits from using whatever heat they produce to warm their *body core*, and thus, natural selection has favored nonshivering thermogenesis in them because it is poised better than shivering to do that.

Neither of these hypotheses has been tested, and both make unverified assumptions. Thus the question of why newborn placental mammals so consistently use brown fat and nonshivering thermogenesis to produce heat remains a tantalizing challenge for future research. The fact that we humans rely on brown fat for thermogenesis when we are newly born adds interest to finding an answer.

The Effect of Body Size on Mammals' Lives in Cold Environments

A retrospective look at the topics we have discussed so far in this chapter makes clear that body size is a principal determinant of the options available to mammals for thermoregulation. Although both reindeer and white-footed mice are warm in the days following their birth, they are warm for different reasons. Newborn reindeer are so large that they have few options for using protective microhabitats; they cannot burrow underground, and on the tundra, they cannot readily find other refuges. Thus, newborn reindeer must *physiologically* confront the full harshness of their environment if they are to stay warm. Fortunately, large size—while limiting behavioral options—has physiological advantages; because of their size, newborn reindeer have a surface-to-volume ratio that is favorable for retaining heat in cold environments, and they can have a thick pelage. The interplay between behavior and physiology is almost opposite for small-bodied newborns like white-footed mice. They are smaller than the little fingers on our hands, and just as naked. Thus, physiologically, there is no chance that newborn mice could evolve mechanisms that would allow them to thermoregulate while fully exposed to a cold external environment. Being small, however, they can exist in highly protective microhabitats; their mothers can place them in secluded, benign spots like underground burrows and can readily ensconce them in an insulating nest. The reasons newborn white-footed mice are warm are principally behavioral: When their mother is present, they are warmed by her; and when she is absent, they benefit from the behavioral provisions she has made for them.

The same trade-offs between physiology and behavior are equally evident in the lives of adult mammals of different body sizes. During winter in Earth's frigid places, small nonhibernating mammals such as lemmings in the Arctic and pikas in the high mountains, are able to escape the biting cold and howling winds of the larger environment by living under the snow (see Figure 1.16) or in other protective hideaways. Their ability to escape in this way is fortunate because, physiologically, a mouse- or rat-sized adult could not survive full exposure to the cold of winter in such places. Large mammals, in contrast, are in a far more advantageous position in terms of their physiology of thermoregulation; the body size of an adult reindeer is one of the major reasons it can have a lower-critical temperature below –30°C. The physiological advantage of large size is itself fortunate, because large size limits behavioral options. If a large mammal, such as an adult reindeer in the Arctic or bighorn sheep in the high mountains, were to remain for the winter where it spends the summer, it would be unable to escape the severity of the cold season by burrowing under the snow.

The single greatest behavioral option for large-bodied species is migration. In fact, as we saw in Chapter 7 (see Figure 7.7), migration is energetically more feasible for large species than small ones. Often, therefore, with the approach of winter, as the small mammals in a place go underground or under the snow, the large ones get out. Bighorn sheep trek to the lowlands, and reindeer often migrate into more-forested areas where they can find windbreaks, as well as better winter feeding grounds.

Because body size is one of an animal's most consequential attributes, it is rewarding to ponder the pros and cons of different sizes. If natural selection starts to favor an increase or decrease in the body size of a species, what are all the implications, and the potential trade-offs, over the full span of an animal's life? We have touched on only some of the highpoints in our discussion here.

Hibernation as a Winter Strategy: New Directions and Discoveries

The hibernating species of placental mammals are noted for retaining brown fat as adults. In fact, the association of brown fat with hibernation is so dramatic that for many decades brown fat was named "the hibernation gland." Brown fat was first described by Conrad Gessner in 1551, in studies of adult European marmots, which are hibernators. For four centuries thereafter, however, no one knew that brown fat was a thermogenic tissue, either in adult hibernators or in newborns

Today we know that in adult hibernators, brown fat is the lead thermogenic tissue during arousal (emergence from hibernation). Intense thermogenesis by brown fat is activated at the very beginning of the arousal process. Because of the placement of brown fat in the body (some deposits actually surround major arteries), and because heat from brown fat tends to be carried by blood flow to the body core, the vital organs such as the heart, lungs, and brain are hypothesized to be particularly warmed by brown fat–generated heat. This focused warming of the vital organs, starting early in arousal, may be important in poising these organs to play their roles in the overall sequence of events by which homeothermy is gradually restored throughout the body.

Among the mammals that live in the most frigid places on Earth, hibernation is employed as a strategy to cope with the stresses of winter by a number of species. Our focus here will be on two of them, the Arctic ground squirrel (*Spermophilus parryii*) and the alpine marmot (*Marmota marmota*). The ground squirrel, which weighs 500 to 800 g, is found widely in Alaska, the Yukon, and neighboring parts of Canada. The marmot, which is larger and weighs about 4 to 5 kg, occurs high in the European Alps.

The occurrence of hibernation displays a curiosity-provoking relation to body size. Mammals the size of alpine marmots are the biggest that hibernate.[2] Larger species that occupy frigid environments, like reindeer, typically remain active and fully homeothermic through all seasons. One reason the larger species have not evolved hibernation is probably that the energy savings afforded by hibernation decline with body size (**Figure 9.10**). Among species of small and medium-sized mammals that weigh 5 kg or less, for reasons that remain largely obscure, hibernation is not at all universal. In the Arctic, for example, whereas ground squirrels hibernate, lemmings and flying squirrels do not. At our present level of understanding, we can conclude from this only that the two modes of dealing with the cold of winter—hibernating and staying active—must each have pros and cons for small mammals. Neither strategy is so obviously "better" that it has become universal.

As we discuss Arctic ground squirrels, alpine marmots, and other hibernators, it is worth reflecting on the fact that much of what we know about them today is a consequence of a technological revolution that has occurred in the past 30 years in which biologists, engineers, and computer scientists have teamed up to create ever-better technologies for the remote monitoring of physiological traits. At the beginning of this revolution, small radio transmitters were designed that could be implanted in animals and report their body temperatures to investigators by radio. Today's transmitters employed in such *radiotelemetry* are able to operate on their built-in battery power for many continuous months, and the data they transmit are recorded automatically by computers. A recent new advance is the introduction of *data loggers* to physiological research. These are basically dedicated microcomputers that have large amounts of memory and that log the temperatures they measure periodically into memory. When a data logger has been implanted in an animal, it stores the information it gathers internally over a long period of time, rather than reporting it by radio. Later the logger is surgically reclaimed, and its accumulated information is off-loaded.

[2] Bears do not hibernate, but undergo just a form of shallow hypothermia, as discussed in Chapter 8 (see page 231).

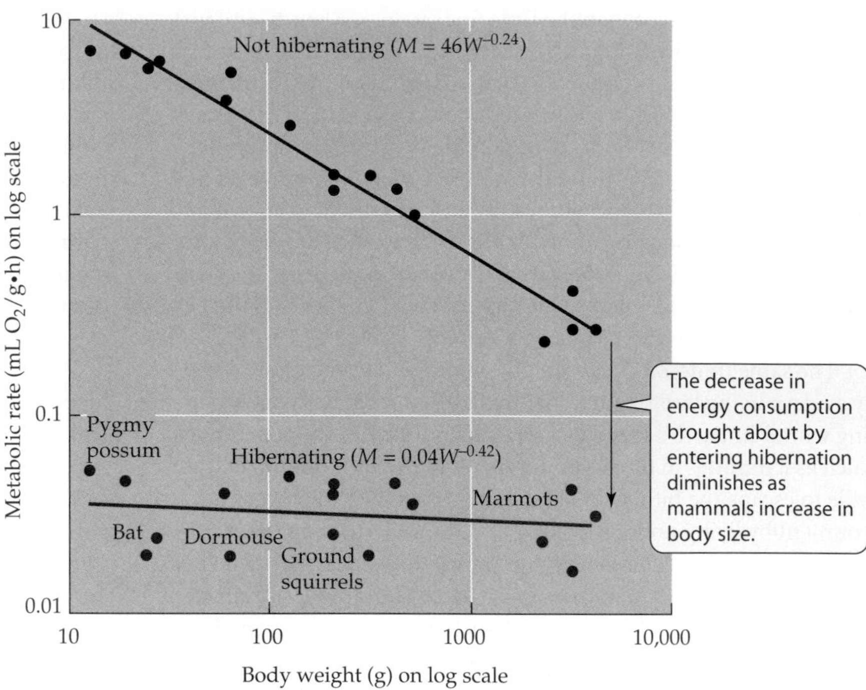

Figure 9.10 The allometry of energy savings by hibernators Using log-log coordinates, the black line shows the relation between resting metabolic rate and body weight for hibernating species when not hibernating; this metabolic rate decreases substantially as body weight increases. The red line shows the metabolic rate during hibernation as a function of body weight; this metabolic rate is nearly the same for animals of all sizes. The equations for the two lines are shown (M = metabolic rate; W = body weight). The ambient temperature in the studies shown averaged 6°C to 9°C. (After Heldmaier and Ruf 1992.)

With these advances in technology, records like that in Figure 9.11 have become possible, thereby revolutionizing our knowledge of the physiology of hibernation. The record in the figure shows the body temperature of a free-living, undisturbed Arctic ground squirrel for the full duration of its 8-month hibernation season.

Arctic ground squirrels supercool during hibernation and arouse periodically throughout their hibernation season

Although most hibernating species spend the winter in microhabitats where the temperature stays above freezing, Arctic ground squirrels cannot do so over much of the range where the species occurs. The reason is *permafrost*—soil that never melts throughout the year. At the Arctic Circle, permafrost starts at 0.8 to 1.0 m below the ground surface. The ground squirrels can burrow only in unfrozen soil, and only the soil above the permafrost thaws in the summer. Thus, however deep the animals might dig their burrows if they could, they cannot go deeper than about a meter, and at that depth the ambient temperature in the winter can drop to far below freezing, even sometimes to –25°C. The chamber or cavity where an individual hibernates is called its **hibernaculum** (plural *hibernacula*). Arctic ground squirrels probably represent the extreme in the ambient temperatures they experience in their hibernacula; no other hibernating species on Earth is known to cope with temperatures this low.

Automated long-term records of the body temperatures of hibernators have made clear that all hibernating species periodically undergo temporary arousals during their hibernation

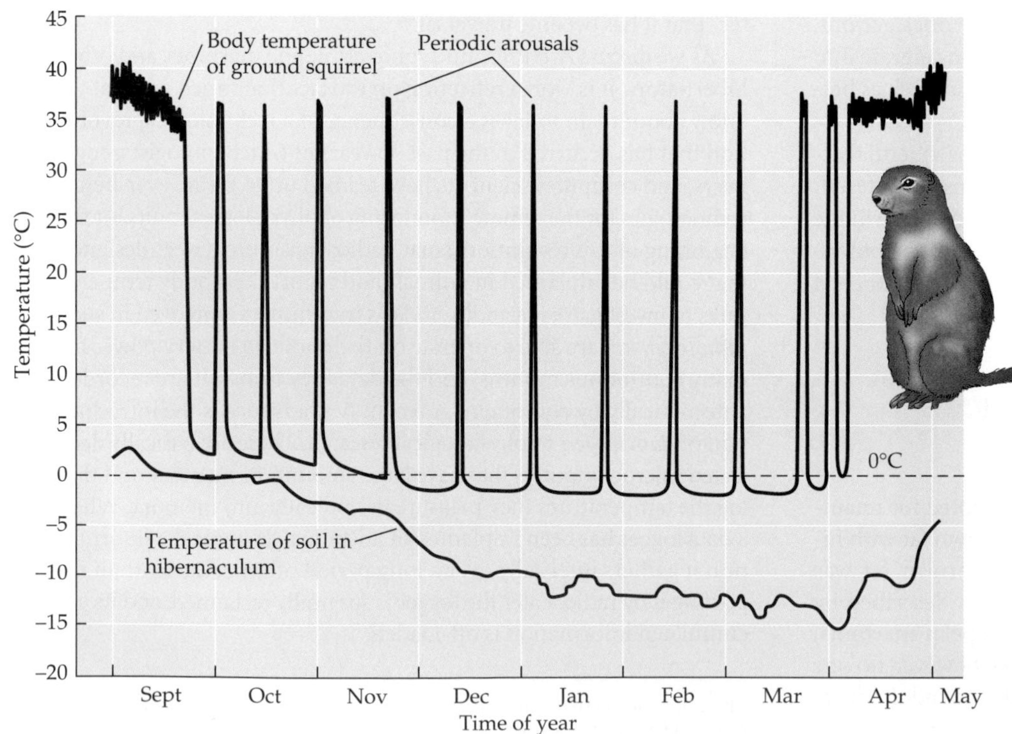

Figure 9.11 The body temperature of an Arctic ground squirrel during its hibernation season in Alaska The body temperature was recorded with a data logger. The soil temperature of the ground squirrel's hibernaculum is also shown. (From Boyer and Barnes 1999.)

season. The animals typically remain in their hibernacula during these arousals. Thus we can readily understand why early naturalists believed that hibernation simply started in the early winter and ended in the spring; hibernating animals disappear into their hibernacula for the entire period. Periodic arousals are universal, however. The Arctic ground squirrel in Figure 9.11 aroused 11 times between October and April before it aroused for the final time and emerged from hibernation. For this ground squirrel, each continuous period of hypothermia was 1 to 3 weeks long. Between these hibernation periods, the animal raised its body temperature to 36°C to 37°C for 12 to 24 hours during each arousal.

During its first two bouts of hibernation, the ground squirrel in Figure 9.11 allowed its body temperature to approach to within 1°C to 2°C of the soil temperature in its hibernaculum. This was possible because the soil temperature remained no lower than 0°C in the early winter. As winter progressed, however, the soil temperature fell to far below 0°C, presenting the ground squirrel with much the same challenge—a threat of freezing—as we addressed in Chapter 8 when discussing poikilotherms in winter (see page 210). One way that Arctic ground squirrels meet this challenge is by supercooling. The freezing point of their body fluids during winter remains the same as it is in summer: –0.6°C (the typical value for mammals). Yet, as exemplified in Figure 9.11, the ground squirrels sometimes cool during hibernation to body temperatures of –2°C to –3°C without freezing. Supercooling is rare in mammals, and possibly the ground squirrels have unusual adaptations to stabilize the supercooled state.

The principal way that Arctic ground squirrels cope with the threat of freezing during their hibernation bouts is by *thermoregulating* at body temperatures of –2°C to –3°C, thereby maintaining their tissues well above the soil temperature when the soil becomes profoundly cold. Illustrating this pattern, the ground squirrel in Figure 9.11 kept its body temperature 10°C above the soil temperature during its bouts of hibernation in midwinter. The endothermy that hibernating Arctic ground squirrels display at subfreezing ambient temperatures requires that they increase their rates of heat production, which they do by brown-fat thermogenesis (**Figure 9.12**). As noted in Chapter 8, it is quite common for mammals and birds that are in states of controlled hypothermia to prevent their body temperatures from falling lower than certain levels by using metabolic heat production. The increased energy expenditure lowers the energy savings of hibernation, however.

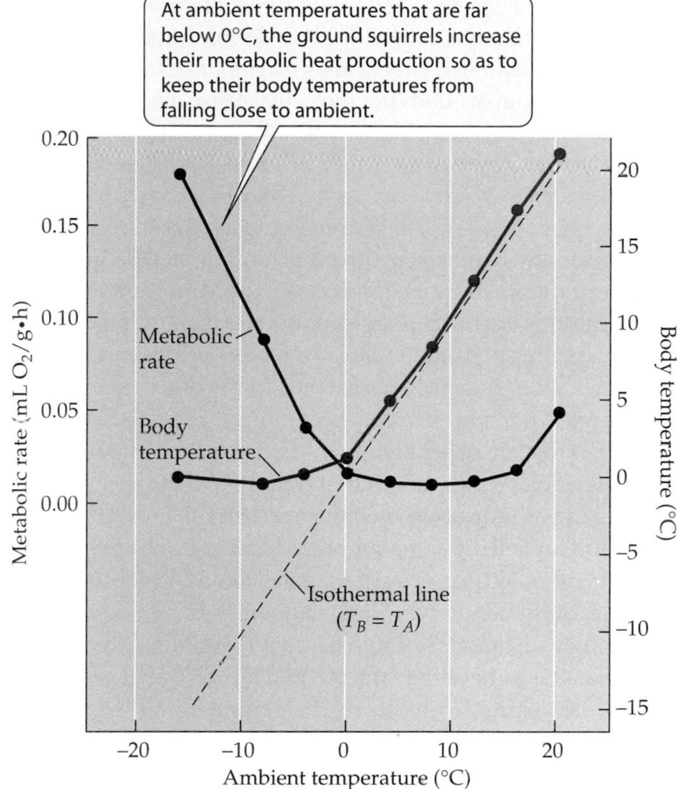

Figure 9.12 Relations among metabolic rate, body temperature, and ambient temperature in hibernating Arctic ground squirrels As the ambient temperature drops from 0°C to –20°C, the body temperature is held higher and higher above ambient because of ever-accelerating metabolic heat production. The isothermal line is a line of equality between body temperature (T_B) and ambient temperature (T_A). (After Barnes and Buck 2000.)

The composition of the lipids consumed before hibernation affects the dynamics of hibernation

As stressed in Chapter 4, the composition of the foods that animals eat is often as significant as the amount of food. A dramatic illustration of the interaction between food composition and physiology has emerged in recent years from studies of hibernators. Besides brown fat, hibernators often accumulate large stores of white fat with the approach of winter. The white fat is not a thermogenic tissue like brown fat. Instead, it is a storage tissue; lipids are deposited in the tissue as fattening occurs and later are mobilized from the tissue to meet metabolic needs (including the need to replenish lipids oxidized in brown fat). The lipids stored are triacylglycerols (see Figure 4.3*c*) and thus meet the chemist's definition of fats and oils. They are accumulated in droplets within specialized cells called *adipocytes*.

Fattening requires time; animals may start eating enough extra food to lay down fat months before hibernation begins. The actual stresses of winter are thus not suitable as cues for fattening. Instead, fattening must be set in motion by processes that occur well in advance of winter stresses. In some species, the primary controls of fattening are photoperiodic; the shortening of day length in autumn, for example, may serve as a cue for fattening to begin. In other species, the timing of fattening is endogenously programmed. Golden-mantled ground squirrels (*Spermophilus lateralis*), for instance, fatten on an approximately annual cycle even if they have no outside information on the time of year, as Figure 4.21 showed. Both their fattening and their hibernation itself are under control of a circannual biological clock (see Chapter 10).

The storage lipids of hibernators (and other mammals) reflect in their composition the suites of fatty acids present in the foods the animals eat. Each triacylglycerol molecule is built from three fatty acid molecules. The fatty acids fall into three chemical categories (see page 97). Saturated fatty acids, also termed *saturates*, contain no carbon–carbon double bonds. Monounsaturated fatty acids—*monounsaturates*—contain one carbon–carbon double bond per molecule. Finally, polyunsaturated fatty acids—*polyunsaturates*—contain two or more such double bonds per molecule. As we discussed in Chapter 4, mammals are incapable of synthesizing most polyunsaturates from scratch. Plants, however, often synthesize substantial amounts of polyunsaturates—fatty acids such as linoleic acid (see Figure 4.3*a*) and α-linolenic acid—which mammals can use directly or employ as substrates for the synthesis of other polyunsaturates. Because of this relationship, animals eating foods rich in polyunsaturates typically deposit fats that are richer in polyunsaturates than the fats deposited by animals eating polyunsaturate-poor foods are. This correlation between animals and their foods extends to the other classes of fatty acids as well. Individuals eating monounsaturate-rich diets tend to deposit fats that are richer in monounsaturates than the fats deposited by individuals eating monounsaturate-poor diets are.

Biochemists have postulated that storage fats must be in a fluid state to be capable of being mobilized and metabolized. If this is true, fats composed primarily of saturates could become useless during hibernation because the body temperatures of hibernators are low enough to cause saturate-rich fats to solidify (for a mental image, recall how hard the saturate-rich fats on beefsteaks become when at refrigerator temperature). Reasoning from these thoughts, researchers hypothesized that hibernators might hibernate in a more effective manner if they deposited fats rich in mono- and polyunsaturates during their fattening periods. Such fats tend to remain fluid at far colder temperatures than fats composed strictly of saturates do, as we have noted. Thus fats rich in mono- and polyunsaturates would be more likely to remain fluid within the body of an animal in hibernation.

A great deal of evidence now exists indicating that the lipid composition of the diet of hibernators affects the dynamics of their hibernation; relatively high levels of mono- and polyunsaturates, as predicted, improve the effectiveness of hibernation. Laboratory studies using defined diets provide one sort of data supporting this conclusion. For example, chipmunks in one study were fed three diets: a particularly polyunsaturate-rich diet, an intermediate diet, and a particularly saturate-rich diet. Those on the polyunsaturate-rich diet, compared to those on the saturate-rich diet, were more likely to hibernate, tolerated lower body temperatures, had lower metabolic rates, and had longer bouts of hibernation (meaning they aroused less frequently) (Table 9.2).

Another sort of study used to evaluate the predicted relation between diet and hibernation has been to look at correlations between white-fat composition and hibernation performance in free-living animals. A recent study of alpine marmots illustrates this approach. Because they eat different foods, individual free-living marmots differ in the percentage of polyunsaturates in their white fat when they start their hibernation season. Those that are particularly high in polyunsaturates tend to lose less weight during the winter and have lower body temperatures during hibernation than do those that are relatively low in polyunsaturates.

Whereas the studies carried out thus far demonstrate that the dynamics of hibernation are affected by diet, they do not in fact demonstrate that the fluidity of white fat is the *reason* for the relation. Possibly white-fat fluidity is what matters most. However, the fatty acid composition of the diet can also affect the composition of membrane phospholipids, and it may even affect the synthesis of prostaglandins or other hormonal substances. Thus, future research will be required to determine mechanistically *why* diet affects hibernation.

Although periodic arousals detract from the energy savings of hibernation, their function is unknown

Periodic arousals significantly reduce the energy savings of hibernation. To illustrate, we can use calculations done by Lawrence Wang for Richardson's ground squirrels (*Spermophilus richardsonii*) hibernating in Alberta, Canada. Suppose that these ground squirrels would use 100 units of energy over the course of their winter hibernation season if they did not hibernate. They actually use 12 units of energy. However, of the 12 units they use, 10 are expended for their periodic arousals. Thus, if they did not periodically arouse and simply remained in continuous hibernation, their expenditure of energy would be 2 units. Putting these numbers slightly differently, of all the energy these ground squirrels use during the hibernation season, over 80% is used for periodic arousals!

A qualitatively similar pattern is exhibited by other species. Alpine marmots, for example, reap great energy rewards by entering hibernation each winter (Figure 9.13). However, about two-thirds of their energy expenditure during their hibernation period is for arousals, and thus their winter energy expenditure could be only one-third as great if they did not arouse.

TABLE 9.2 Hibernation performance in chipmunks (*Eutamias amoenus*) fed three diets All values are means. All differences between the group on a polyunsaturate-rich diet and the group on a saturate-rich diet are statistically significant.

Diet	Percentage that hibernated (%)	Lowest body temperature that did not provoke arousal (°C)	Rate of O_2 consumption at an ambient temperature of 2°C (mL O_2/g•h)	Length of each continuous hibernation bout at an ambient temperature of 5°C (hours)
Polyunsaturate-rich	100	0.6	0.034	138
Intermediate	100	1.2	0.047	110
Saturate-rich	75	2.2	0.064	92

Source: After Geiser and Kenagy 1987.

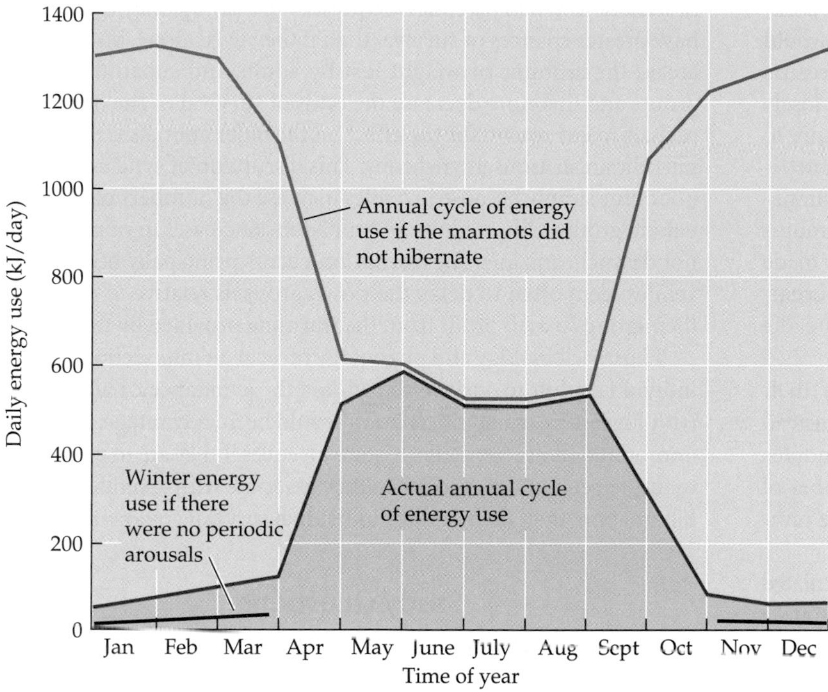

Figure 9.13 The annual cycle of energy use in alpine marmots (blue line) Also shown are the cycle of energy use that would exist if the marmots did not hibernate (green line), and the energy costs in winter if the marmots did not undergo periodic arousals during hibernation (black lines). (After Heldmaier 1993.)

Given that animals pay an energy price for periodic arousals, it seems that the arousals must have physiologically important functions. In the early days of research on this topic, a lively hypothesis for periodic arousals was that they allow hibernators to void wastes. Investigators assumed, for instance, that the nitrogenous waste product urea tends to accumulate during bouts of hibernation, and thus animals need to wake up to urinate. One piece of contrary evidence that came to light when researchers tested the hypothesis is that golden-mantled ground squirrels arouse at times that are unrelated to bladder volume. Then, studies of blood urea in the ground squirrels revealed that the urea concentration does not increase, but *declines*, during a bout of hibernation! Several other seemingly obvious reasons for periodic arousal have also proved to be far from obvious when studied. In the past decade, a new candidate has been put forward: It has been suggested that animals are unable to sleep when in hibernation and thus must wake up periodically in order to catch up on sleep. Some of the evidence for this hypothesis is that electroencephalographic patterns typical of sleep deprivation have been observed in some hibernators at the start of arousal periods, and the animals sleep a lot during their arousals.

Some years ago, J. S. Willis titled a detailed review of this topic "The Mystery of the Periodic Arousal." To this day, no single hypothesis has received strong enough support to be a consensus hypothesis explaining the function of periodic arousals, and in fact there may not be just a single function. Thus, although hibernation has been recognized for centuries and studied in detail for decades, the understanding of periodic arousals still represents an area of mystery, and it is potentially a frontier for major new insights into the relations among sleep, restorative processes, body temperature, and other attributes.

The intersection of sociobiology and physiology: Social hibernation may save energy

Most of the world's 14 species of marmots, unlike ground squirrels, hibernate in social groups and thus are known as **social hibernators**. Alpine marmots live throughout the year in social groups of adults and youngsters (youngsters do not quickly disperse from their natal area but instead remain there for up to several years). A typical social group of alpine marmots consists of a dominant male and female and an assemblage of related, younger animals of various ages. All these animals hibernate together, in sets that number up to 20 individuals.

Social hibernation in alpine marmots increases the probability of overwinter survival, particularly among the young-of-the-year, termed *juveniles*. One reason that survival is aided relates to the fact that these marmots do not allow their body temperatures to go below 3°C to 4°C in winter, yet temperatures in their hibernacula are often lower, meaning that hibernating animals must elevate their metabolic rates to keep their body temperatures from going too low. Animals huddled together in a large cluster are better insulated and require less of an increase in metabolic rate than those hibernating alone or in small groups.

One of the most remarkable revelations in the recent annals of the radiotelemetric study of hibernation is the discovery that adults and subadults in a hibernating group of alpine marmots usually undergo their periodic arousals in close synchrony (Figure 9.14). This

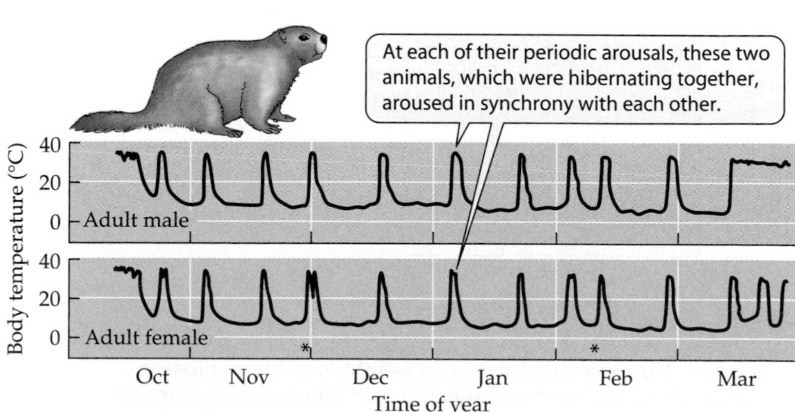

Figure 9.14 Marmots arouse synchronously The body temperatures of two adult marmots that were hibernating together. Outside disturbances may have affected arousal of the animals at the two times marked with asterisks (*). (After Arnold 1988.)

synchrony surely lowers their mutual energy costs to arouse. If a single adult were to arouse alone within a hibernating group, it would experience heightened energy costs by being in the group, because the cold tissues of the other animals in the group, pressed closely against it, would increase the heat production it would require to raise its own body temperature. What actually happens is diametrically opposite. Adults and subadults in a group arouse simultaneously, thereby lowering each other's energy costs to arouse by mutually warming each other. The simultaneity of this arousal is made particularly intriguing by the fact that all the animals are hibernating, and therefore seemingly comatose, when they initiate the process!

Juveniles in a hibernating group often do not initiate their arousals simultaneously with the adults and subadults, but instead lag behind. When an arousal is initiated by the adults and subadults, juveniles often let the older animals warm them for a number of hours (or even days). Only after receiving this benefit do the juveniles invest their own energy in the arousal process.

By statistical calculations, an *index of synchrony* can be calculated for arousals in groups of marmots. A high index for a hibernating group means that the individuals monitored with radiotelemeters in the group aroused highly synchronously. A low index means that the individuals aroused in poor synchrony. Based on data for 14 hibernating groups, Figure 9.15 shows the relation between loss of body weight and synchrony. As you can see, individuals in highly synchronized groups lost approximately 20% to 25% of their body weight during the winter hibernation season. Individuals in some poorly synchronized groups, however, lost about 40% to 45%—roughly twice as much. Because relatively great loss of body weight lowers animals' likelihood of survival, this analysis reveals that synchrony during periodic arousals has important survival consequences.

Further analysis of the data available also reveals that the presence of juveniles is very important in determining the dynamics of group hibernation. When juveniles are present in a group, they themselves have greater chances of survival than if they lived alone, but they increase the amount of weight lost by adults and subadults in the group, and therefore decrease the odds of survival of the older animals. A major reason for the effect on the older animals is that juveniles diminish arousal synchrony. This disruption of synchrony does not occur simply because juveniles increase the numbers of individuals; in groups of just adults and subadults, increases in numbers are not desynchronizing. The disruption occurs principally because juveniles seem often to delay their own arousals relative to others in their group so as to profit from the warming provided by the others.

Thus, sociobiology and physiology meet at an intersection. For an individual adult to survive the winter, the avoidance of all contact with juveniles during hibernation would be an advantage. Without immortality, however, adults require juveniles if they are to pass genes to future generations, and when they associate with juveniles during hibernation, their own odds of individual survival go down.

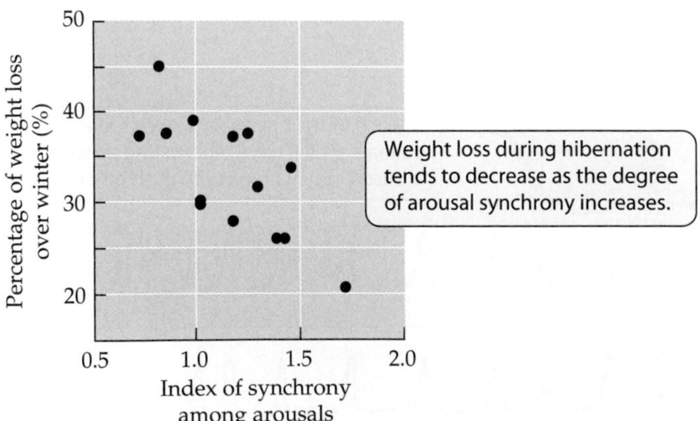

Figure 9.15 The correlation between weight loss during hibernation and the degree of arousal synchrony in hibernating groups of alpine marmots Each point refers to a distinct group of hibernating marmots and shows both the group's average weight loss during winter and the group's degree of arousal synchrony. High values of the index of synchrony signify high degrees of arousal synchronization among group members. (After Ruf and Arnold 2000.)

Study Questions

1. What are the relative advantages and disadvantages of large and small body size? In your explanation, try to go beyond the points mentioned in this chapter.
2. When the composition of the community of rumen microbes in a ruminant changes, there are several possible causes. What are some of them? Could you do experiments to decide what the actual causes are? Explain.
3. Long migrations such as those of reindeer are rare in animals that walk across the land, despite being common in flying birds. In what way are reindeer similar to birds in their energetics of covering distance? (Review Chapter 7 if you are stumped.)
4. List some possible hypotheses for the function of periodic arousal, and design experiments to test your hypotheses as rigorously as possible.

Key References

Boyer, B. B., and B. M. Barnes. 1999. Molecular and metabolic aspects of mammalian hibernation. *BioScience* 49: 713–724. A very accessible treatment of many of the current issues in hibernation research, including modification of gene expression.

Heldmaier, G., and M. Klingenspor (eds.). 2000. *Life in the Cold*. Springer, New York. This is the symposium volume for the 11th International Hibernation Symposium. As such, it provides up-to-date articles, as well as links to the wider literature, in most areas of hibernation research.

Nedergaard, J., and B. Cannon. 1990. Mammalian hibernation. *Philos. Trans. R. Soc. London* 326B: 669–686. Probably the most delightful, scientifically serious article ever written on hibernation.

Selected References

Brück, K. 1970. Nonshivering thermogenesis and brown adipose tissue in relation to age and their integration in the thermoregulatory system. In O. Lindberg (ed.), *Brown Adipose Tissue*, pp. 117–154. American Elsevier, New York. [This older reference is included because it is an English-language summary of a series of definitive studies, carried out by the author, mainly on development of thermoregulation in guinea pigs.]

Frerichs, K. U., C. B. Smith, M. Brenner, D. J. DeGracia, G. S. Krause, L. Marrone, T. E. Dever, and J. M. Hallenbeck. 1998. Suppression of protein synthesis in brain during hibernation involves inhibition of protein initiation and elongation. *Proc. Natl. Acad. Sci. U.S.A.* 95: 14511–14516.

Heldmaier, G., and T. Ruf. 1992. Body temperature and metabolic rate during natural hypothermia in endotherms. *J. Comp. Physiol. B* 162: 696–706.

Johnsen, H. K., A. Rognmo, K. J. Nilssen, and A. S. Blix. 1985. Seasonal changes in the relative importance of different avenues of heat loss in resting and running reindeer. *Acta Physiol. Scand.* 123: 73–79.

Markussen, K. A., A. Rognmo, and A. S. Blix. 1985. Some aspects of thermoregulation in newborn reindeer calves (*Rangifer tarandus tarandus*). *Acta Physiol. Scand.* 123: 215–220.

Nedergaard, J., E. Connolly, and B. Cannon. 1986. Brown adipose tissue in the mammalian neonate. In P. Trayhurn and D. G. Nicholls (eds.), *Brown Adipose Tissue*, pp. 152–213. Edward Arnold, London.

Nedergaard, J., V. Golozoubova, A. Matthias, I. Shabalina, K. I. Ohba, K. Ohlson, A. Jacobsson, and B. Cannon. 2001. Life without UCP1: Mitochondrial, cellular and organismal characteristics of the UCP1-ablated mice. *Biochem. Soc. Trans.* 29: 756–763.

Nilssen, K. J., J. A. Sundsfjord, and A. S. Blix. 1984. Regulation of metabolic rate in Svalbard and Norwegian reindeer. *Amer. J. Physiol.* 247: R837–R841.

Rothwell, N. J., and M. J. Stock. 1985. Biological distribution and significance of brown adipose tissue. *Comp. Biochem. Physiol. A* 82: 745–751.

Ruf, T., and W. Arnold. 2000. Mechanisms of social thermoregulation in hibernating alpine marmots (*Marmota marmota*). In G. Heldmaier and M. Klingenspor (eds.), *Life in the Cold*, pp. 81–94. Springer, New York.

Soppela, P., M. Nieminen, S. Saarela, J. S. Keith, J. N. Morrison, F. MacFarlane, and P. Trayhurn. 1991. Brown fat-specific mitochondrial uncoupling protein in adipose tissues of newborn reindeer. *Amer. J. Physiol.* 260: R1229–R1234.

Syroechkovskii, E. E. 1995. *Wild Reindeer.* Smithsonian Institution Libraries, Washington, DC.

Wang, L. C. H., and T. F. Lee. 1996. Torpor and hibernation in mammals: Metabolic, physiological, and biochemical adaptations. In M. J. Fregly and C. M. Blatteis (eds.), *Environmental Physiology*, vol. 1 (Handbook of Physiology [Bethesda, MD], section 4), pp. 507–532. Oxford University Press, New York.

Willis, J. S. 1982. The mystery of periodic arousal. In C. P. Lyman, J. S. Willis, A. Malan, and L. C. H. Wang (eds.), *Hibernation and Torpor in Mammals and Birds*, pp. 92–101. Academic Press, New York.

PART III

Integrating Systems

Part III Integrating Systems

Previous Page: Compound eyes have many facets called ommatidia. In the compound eyes of this shrimp, it is the dark ommatidia that are directed at the camera.

10 Neural and Endocrine Control, Nervous Systems, and Biological Clocks

The jumping lemur in this picture, like all animals, is a collection of cells. The activities of these cells must be controlled and coordinated in order for the lemur to function and behave as a lemur, rather than as just a mass of cells. This control extends to all parts of the lemur's body. In preparation for the exertion of the leap, the metabolic rate of the lemur's muscles may increase, as well as blood pressure in vessels to the limbs and head. Blood flow to the gut and other organs may decrease. Certainly sensory data must be collected and evaluated, and commands must be sent to control muscles. The lemur's nervous system and endocrine system work together to integrate all these activities, and to control all the effectors that bring about the changes.

Every animal must have an organization for control and coordination. Chapter 2 introduced controls within cells, mediated by cell-membrane receptors, second messengers, enzymes, transcription factors, and the like. Our focus here is control, coordination, and integration of activities among cells.

An **effector** is a tissue or organ that carries out functions as directed by the nervous system (or endocrine system). Control of effectors (such as muscles, glands, and target cells in other organ systems) provides for all output functions, such as movement, behavior, and physiological regulation. **Integration** summates and coordinates these controls so that they work together harmoniously. *Cellular integration* refers to processes within a cell; *integration* on the larger, organismal scale refers to the selective combination and processing of sensory, endocrine, and central nervous system information in order to promote the harmonious functioning of the whole organism within its environment. This integration of organismal functioning is achieved through the activities of the nervous and endocrine systems. The integrative functions of the nervous system and endocrine system ensure that an animal's responses are smooth and coordinated, rather than clashing or disjointed.

A **control system**, as first presented in Box 8.1, sets a level of a particular variable (temperature, blood pressure, muscle force, and so on) that is being controlled. Control systems are often, but not always, stabilizing. Physiological control is usually exerted via the nervous system and endocrine system. Just as some cells are specialized to produce movements, secrete acid, or carry oxygen, nerve cells and endocrine cells are specialized for coordination. Fifty years ago, these two cell types were be-

How does this lemur control and integrate the functions of all the cells in its body?

lieved to be distinctly different from one another, but more-recent studies show that there are many interactions and intergradations between neural control and endocrine control.

This chapter introduces how animals employ nervous systems and endocrine systems as physiological control systems. We describe general principles of both neural control and endocrine control, and then we examine the organization of nervous systems. Finally, to emphasize that physiological control is dynamic rather than static, we discuss biological clocks, mechanisms that permit animals to keep track of the passage of time without requiring external signals.

The Physiology of Control: Neurons and Endocrine Cells Compared

In an organism, physiological control systems regulate organ systems—digestive, respiratory, circulatory, and excretory—and govern the activities of muscles and glands to generate behavior. This physiological control is accomplished in two basic ways: via the nervous system and via the endocrine system.

Neurons transmit electrical signals to target cells

Because neurons are commonly likened to the wires in a telephone or computer network, most people have an intuitive understanding of what these cells do. A **neuron** is a cell that is specially adapted to generate an electrical signal, most often in the form of a brief, self-perpetuating impulse called an *action potential*. We discuss neurons in Chapter 11; here we introduce only the broad features of neural control. As Figure 10.1 reveals, a neuron has four parts—dendrites, cell body, axon, and presynaptic terminals—that generally correspond to its four functions—input, integration, conduction, and output—as a controller cell within an animal's body.

A neuron receives input, signals from other neurons or sensory cells, at specialized contact points called **synapses**. Usually, the synapses occur along branching processes known as **dendrites**, although synapses may occur on the cell body as well. Impulses arriving at synapses from *presynaptic* cells cause the release of chemical substances called **neurotransmitters**, which cross the synapse and exert specific physiological effects on the *postsynaptic* cell.

The **cell body** is commonly the site of signal integration and impulse generation. A single neuron may receive thousands of synaptic contacts from other neurons. The neurotransmitters released across some synapses excite the neuron; those released across other synapses inhibit it. From moment to moment, the cell membrane of the cell body combines the inhibitory and excitatory synaptic inputs, and if excitatory inputs surpass inhibitory inputs, the neuron may respond by generating an action potential.

The long slender **axon** is the conduction component of the neuron, serving to propagate action potentials between the axon hillock and the presynaptic terminals. The microscopic axons from individual neurons sometimes collect together in long macroscopically visible bundles that are called *tracts* in the central nervous system and *nerves* in the peripheral nervous system.[1]

[1] The central and peripheral nervous systems will be discussed in depth later in the chapter.

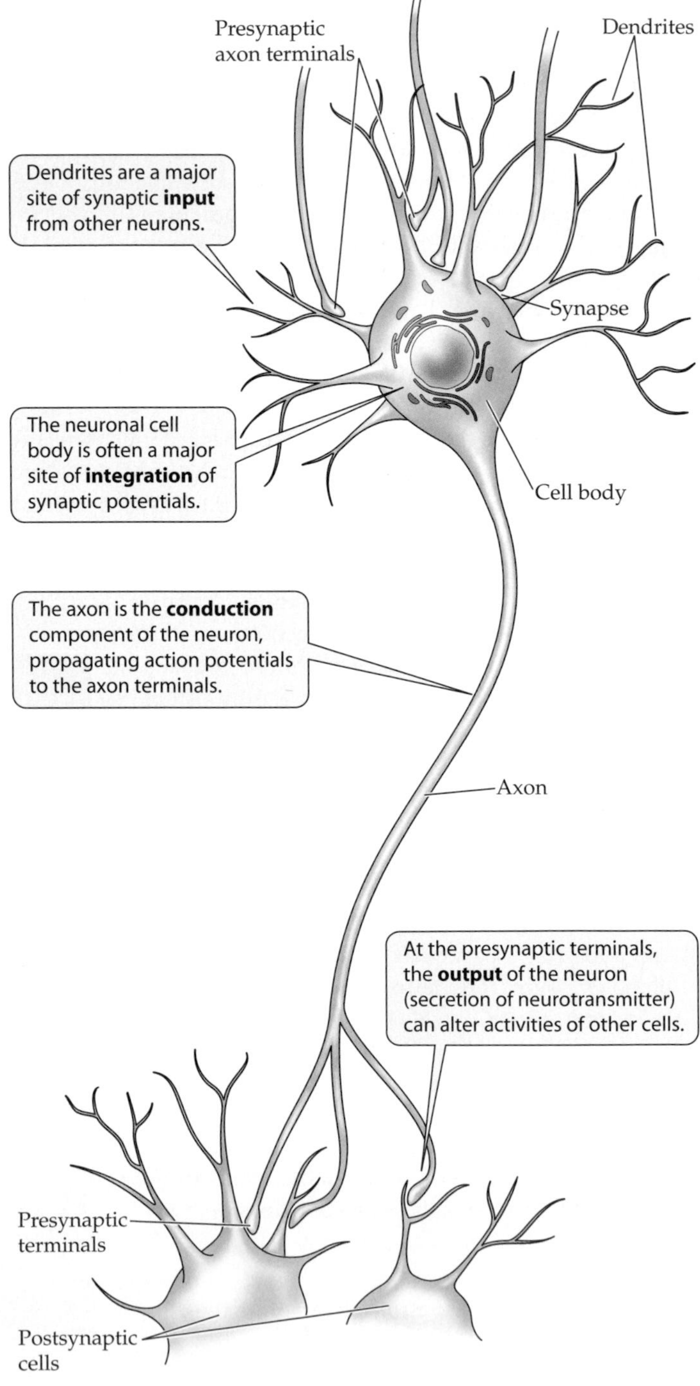

Figure 10.1 Neurons have four functional regions that typically correspond to their four major structural regions The descriptions in the figure provide a functional model of a neuron, showing typical functional properties that it mediates. The labels identify the structural parts of a neuron that are associated with these functions. The correlation between structures and functional properties is imperfect: Synaptic input often occurs at the cell body as well as the dendrites, for example, and some dendrites can generate action potentials. In contrast, some local neurons generate no action potentials at all, and thus lack a separate function of active conduction.

The axon usually divides into several **presynaptic terminals**, the place where neuronal output occurs. The presynaptic terminals form synapses with other neurons or other types of cells, such as muscle fibers. An action potential arriving at the presynaptic terminals triggers the release of molecules of *neurotransmitter* across the synapses that exert a specific physiological effect—excitatory or inhibitory—on the target cell. Neurons that form synaptic endings on a target cell are said to **innervate** the targeted cell.

The extended networks of neurons in an animal's body constitute its *nervous system*. As parts of a nervous system, neurons perform various roles. Some neurons perform sensory functions by initiating signals in response to physical or chemical stimuli. As we have just described, other neurons integrate signals arriving from other cells, generate nerve impulses of their own, and transmit these signals over distances that can be very long, at least on a cellular scale.[2] In nervous systems, neurons that relay sensory signals to integrative centers (such as the brain) are called **afferent neurons** (*afferent*, "to bring toward"). Other neurons, called **efferent neurons** (*efferent*, "to carry off") relay control signals ("orders") from the integrative centers to target cells that are under nervous control, such as muscle cells or secretory cells. Neurons that are entirely within the central nervous system are called **interneurons**.

Neural control has two essential features: It is fast and addressed. Neuronal signals are *fast*: They travel very rapidly and begin and end abruptly. A mammalian neuronal axon, for example, might conduct impulses along its length at 20 to 100 m/s, and it might be capable of transmitting 100 or more impulses in a second. The connections of neurons are said to be *addressed* because they provide highly discrete lines of communication (like a letter or a telephone call). A neuron normally must make synaptic contact with another cell to exert control, and it normally innervates only relatively few cells that are its potential targets. Nervous lines of communication, therefore, provide opportunities for fine control of other cells both *temporally* and *spatially*, sending fast, rapidly changing signals to some potential targets and not to others.

Endocrine cells broadcast hormones

Whereas the signals of the neurons in nervous systems are precisely targeted, the signals produced by the *endocrine system* are broadly distributed throughout the animal's body. **Endocrine cells** release signaling chemicals, called **hormones**, into the blood (or sometimes just into other extracellular fluids). These chemicals are carried throughout the body by the blood, bathing the tissues and organs at large. For a hormone to elicit a specific response from a cell, the cell must possess *receptor proteins* for that hormone. Thus, cells of only certain tissues or organs respond to a hormone and are called *target cells*. The responsiveness of target cells is under control of gene expression; that is, the tissues that respond are tissues that express the genes encoding the receptor proteins. Hormones can produce effects in target cells at low blood concentrations, approximately 10^{-5} to 10^{-12} *M*.

[2] When you wiggle your toes, the axons of neurons in the lumbar spinal cord (in the small of your back) conduct action potentials to excite muscle fibers near your ankle, a distance of nearly a meter. Most cells, in contrast, are approximately 10 μm in diameter.

Endocrine cells often are organized into organs called **endocrine glands**. Endocrine glands are *ductless glands*, because hormones are secreted into the blood flowing in capillaries that run through the gland among the endocrine cells. Endocrine glands differ in this respect from *exocrine glands*, such as the salivary glands, that empty their secretions into ducts through which the secretions flow to points of release. Examples of endocrine glands include the thyroid, adrenal, and pineal glands of vertebrates. Some endocrine cells are not organized into discrete glands, and instead are found scattered among other sorts of cells (e.g., the cells of the vertebrate stomach wall that secrete the hormone gastrin to control the secretion of digestive enzymes and acid in the stomach; see Chapter 4, Figure 4.20).

Physiologists recognize two basic types of endocrine cells: nonneural endocrine cells and neurosecretory cells. **Nonneural endocrine cells** are *epithelial* hormone-producing cells that do not resemble nerve cells (Figure 10.2*a*). Such endocrine cells may or may not be innervated by neurons. Nonneural endocrine cells are often controlled by hormones from other endocrine cells, whether or not they are also under direct neural control. **Neurosecretory cells** (Figure 10.2*b*) are specialized neurons that produce hormones and secrete them into the blood (rather than secreting a neurotransmitter at a synapse). Their secretions are known as **neurohormones**.

Other neurons make synaptic contacts with neurosecretory cells and control their release of neurohormones largely via these synapses. Typical neurons influence other cells synaptically (Figure 10.2*c*), but neurosecretory cells do not do so. Instead, their major signaling output is chemical secretion into the blood. *Neurosecretory cells constitute a direct interface between the endocrine and nervous types of control systems: They are endocrine cells immediately controlled by the nervous system.*

Some cells secrete "hormonelike" **paracrine** or **autocrine** substances, which are intergrades between hormones and neurotransmitters. A paracrine substance, for example, travels through the intercellular fluid to a distance of a few cell diameters in all directions from the secreting cell (Figure 10.2*d*), permitting the secreting cell to orchestrate the function of a highly localized patch of contiguous cells. An autrocrine substance acts on the very cell that releases it.

Endocrine control has two essential features: It is slow and broadcast. Individual hormonal signals are relatively *slow* because they operate on much longer timescales than individual nervous signals. Initiation of hormonal effects requires at least several seconds or minutes because a hormone, once released into the blood, must circulate to target tissues and diffuse to effective concentrations within the tissues before it can elicit responses. After a hormone has entered the blood, it may act on targets for a substantial amount of time before metabolic destruction and excretion decrease its concentration to ineffective levels. In the human bloodstream, for example, the hormones vasopressin, cortisol, and thyroxine display half-lives of about 15 minutes, 1 hour, and nearly 1 week, respectively. Thus a single release of hormone may have protracted effects on target tissues.

Unlike addressed neural control, endocrine control is said to be *broadcast*. Once a hormone is released into the blood, all cells in the body are potentially bathed by it. The specificity of hormone action depends on which cells have receptor molecules for the hormone. Many types of cells may respond to the hormone, perhaps with different types responding in different ways. Alternatively, a hormone

(a) Nonneural endocrine cell

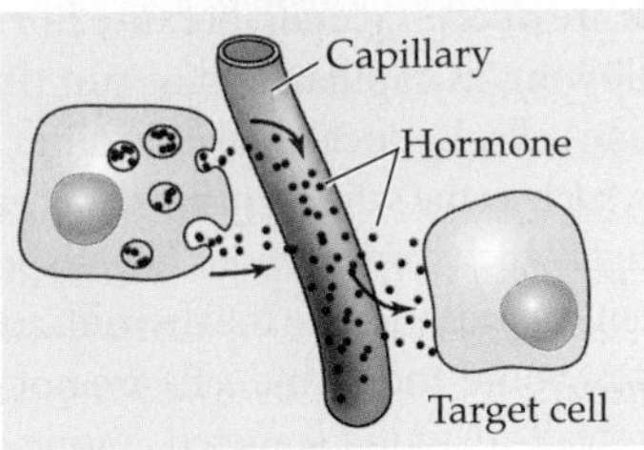

(b) Neurosecretory cell

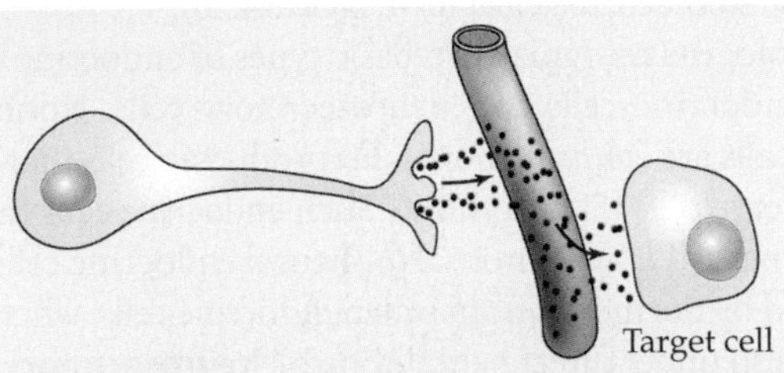

(c) Neuron

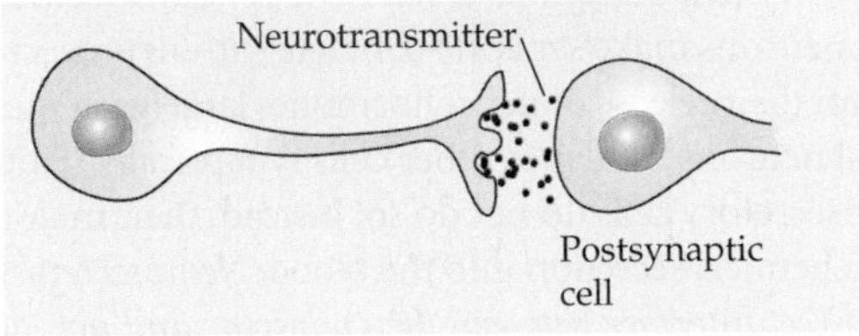

(d) Paracrine and autocrine signaling

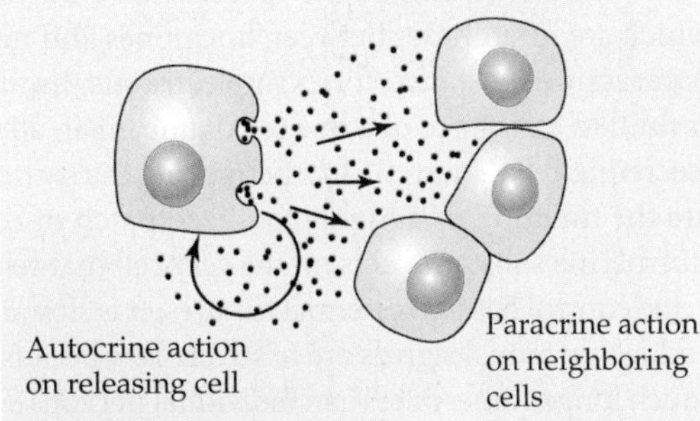

Figure 10.2 Both nonneural cells and some neurons secrete hormones *(a)* A nonneural endocrine cell secretes hormone molecules that enter a capillary and are carried throughout the bloodstream. *(b)* A neurosecretory cell is a neuron that secretes hormones into the blood in similar fashion. *(c)* An ordinary neuron (shown for comparison) secretes neurotransmitter molecules that act very locally on receptor molecules of the postsynaptic cell. *(d)* In paracrine and autocrine signaling, chemicals diffuse locally to activate receptors on neighboring cells (paracrine) or on the same cell (autocrine).

may affect only one type of target cell, because only those target cells have the kind of receptor to which the hormone attaches. Although in principle hormones may exert either limited or widespread effects, in practice they commonly affect at least a whole tissue, and often multiple tissues.

Nervous systems and endocrine systems tend to control different processes

Neural lines of communication are capable of much finer control—both temporal and spatial—than is possible for endocrine systems. Not surprisingly, the two systems tend to be used to control different functions in the body. *Whereas the nervous system controls predominantly the fine, rapid movements of discrete muscles and glands, the endocrine system typically controls more widespread, prolonged activities* such as metabolic changes.

Consider, for example, a task that requires the modulation of many tissues over a protracted period, such as the control of metabolism or growth. An animal's nervous system *could* carry out a coordination task of this sort. To do so, however, the nervous system would need to run thousands of discrete axons between integrating centers and controlled cells, and send trains of impulses through all these axons for as long as modulation is required. An endocrine gland can accomplish this task with greater economy than the nervous system could, by secreting a single long-lasting chemical into the blood. For this reason, growth, development, and reproductive cycles—processes that involve many tissues and occur on timescales of days, months, or years—are often under primarily hormonal control.

As a practical matter, most tissues in an animal's body are under dual control of the nervous and endocrine systems. Skeletal muscle illustrates the relationship of this dual-system control. A typical muscle contains thousands of muscle fibers and is innervated by hundreds of motor neurons. Each motor neuron innervates a separate set of muscle fibers, controlling the contraction of just these fibers. The nervous system can activate a few, many, or all of the motor neurons. In this way it graduates the contraction of the entire muscle by adjusting from moment to moment the number of active motor neurons and activated fibers. At the same time that the nervous system controls the muscle cells' contractile activity, the hormone insulin provides endocrine control of their metabolic activity. Insulin facilitates their uptake of glucose from the blood and their rate of glycogen synthesis. The spatial and temporal distinctions between the two types of control are evident in this example: The nervous system controls moment-to-moment, differential contractile actions of muscles, while the endocrine system provides long-term control of their metabolism.

Nervous and endocrine systems can exert control over each other, as well as over other targets. *Interaction between the nervous and endocrine systems occurs in both directions.* Nervous systems can affect the function of endocrine cells, as in innervated endocrine glands. Likewise, hormones can modulate nervous system function; for example, certain neurons in mammalian brains are responsive to sex steroid hormones.

Summary: The Physiology of Control: Neurons and Endocrine Cells Compared

- Control by a nervous system involves neurons that send axons to discrete postsynaptic cells. Neurons generate rapidly conducting action potentials to control the specific targets on which they end. They exert fast, specific control by releasing neurotransmitters at synapses.

- Endocrine cells release hormones into the bloodstream to mediate endocrine control. All body cells are potential targets of hormones, but not all need to have receptors to the hormones. Hormonal control is slower, longer lasting, and less specific than neural control.

The Organization and Evolution of Nervous Systems

Nervous system function depends on "wiring"—the anatomical organization by which neurons are connected into circuits. We will discuss nervous system organization in this chapter, in order to prepare for the following chapters on neurophysiology.

Nervous systems organize neurons into functional circuits

To set the stage, let's consider a simple example. Suppose you walk into the kitchen and surprise a cockroach. The cockroach jumps, exhibiting a *startle response* in which it turns away from the disturbance and prepares to run. This simple behavioral act is mediated by electrical signals within the cockroach's nervous system.

The cockroach's jump is a **reflex**, a simple behavioral response to a distinct stimulus. Air currents or airborne sound waves vibrate the filiform hairs (wind receptors) at the cockroach's posterior end (Figure 10.3*a*), providing the stimulus that evokes the reflex. This stimulus initiates a brief series of action potentials in sensory neurons located at the bases of the hairs. The action potentials travel along the conducting afferent processes (axons) of the sensory neurons toward the central nervous system (CNS), where the sensory neurons contact other neurons within the CNS. In the cockroach, the sensory axons make synaptic contacts with a few large *interneurons* (neurons that do not extend outside the CNS). These synapses are excitatory, so the barrage of action potentials from the sensory neurons excites the interneurons, which generate their own action potentials (Figure 10.3*b*).

The interneuron axons extend anteriorly in the ventral nerve cord (part of the CNS). They in turn make synaptic contact with efferent **motor neurons**, whose outgoing axons exit the CNS and innervate a muscle. The interneurons synaptically excite the motor neurons, which in turn excite the extensor muscles of the legs that produce the jump. At the same time, the interneurons inhibit motor neurons that excite the antagonist flexor muscles of the cockroach's legs.

As the barrage of action potentials in Figure 10.3 indicates, this startle response happens very quickly: It's less than 150 ms from

(a) Reflex arc for startle response

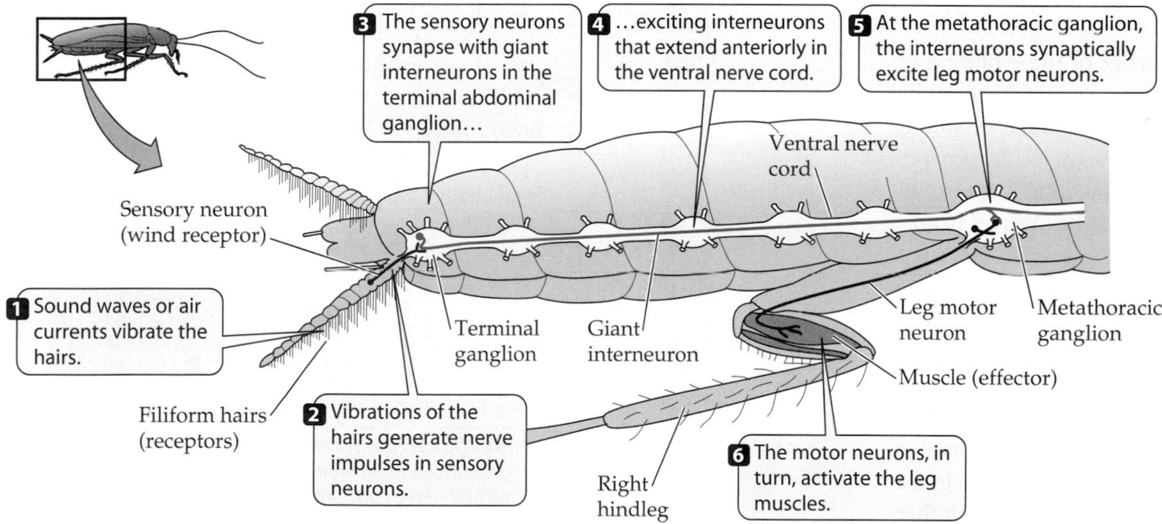

(b) Stimulus, nerve impulses, response

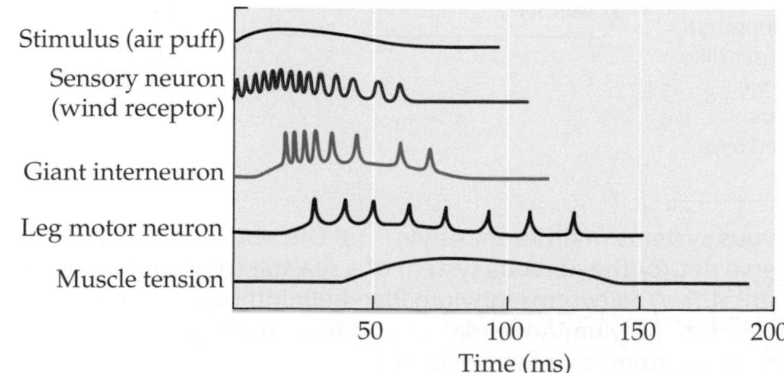

Figure 10.3 The neural circuit mediating the startle response in the cockroach *Periplaneta* *(a)* Hairlike wind receptors located on an abdominal cercus trigger this reflex. *(b)* Nerve and muscle cells in the reflex circuit respond to a controlled puff of air lasting 50 ms. The action potentials in successive neurons in the circuit lead to contraction (tension) in the muscle. (After Camhi 1984.)

stimulus to jump! This simple reflex in a cockroach illustrates the following functional features of nervous systems:

- Neurons are organized in circuits in such a way that they can elicit a coordinated, adaptive response of effectors.
- Sensory receptor cells (also excitable cells) transform environmental stimuli into electrical signals.
- Central interneurons integrate signals from sensory receptors and other signals arising within the animal, generating an integral pattern of impulses.
- Motor commands are sent out from the CNS to effectors.

The organization of neurons into functional nervous systems is what allows for the complexity of the neural control of animal behavior. We can define a **nervous system** as an organized constellation of cells (neurons and support cells) specialized for the repeated conduction of electrical signals within and between cells. These signals pass from sensory cells and neurons to other neurons and to effectors. Nervous systems integrate the signals of converging neurons, generate new signals, and modify the properties of neurons based on their interactions. Nearly all multicellular animals have nervous systems, of varying degrees of complexity.

Multicellular animals have evolved complex nervous systems

We have little direct knowledge of the evolution of nervous systems, which are rarely preserved in the fossil record. Theories of nervous system evolution are based on the anatomy and the molecular genetics of living groups, a risky proposition because all groups are evolved and none can be taken as primitive. Comparative studies of living animals show that *neurons* of nervous systems in metazoan (multicellular) animals are all quite similar. Neurons of all phyla have common molecular bases for their excitability and intercellular communication, with homologous voltage-gated channels and synaptic mechanisms, and similar morphology. The evolution of nervous systems, then, has involved principally changes in the complexity of *organization* of the neurons, rather than changes in the neurons themselves.

The simplest (and presumably most primitive) form of nervous system is termed a **nerve net**. In a nerve net, neurons are dispersed in a thin layer, seemingly at random (Figure 10.4*a*). Nerve nets are present in cnidarians such as jellyfish and *Hydra*, as well as in peripheral parts of the body in many more-complex invertebrate groups and in the intestines of vertebrates. The simplest nerve nets are *unpolarized*; that is, neurons make synaptic contacts where they cross, so that an excitation initiated in one neuron can spread in all directions along multiple, diffuse paths. Even in Cnidaria, considered the most primitive phylum having a nervous system, more elaborately organized elements may augment unpolarized nerve nets. Some jellyfish have a more directionally oriented polarized network of through-conducting neural pathways with primitive integrative centers, as well as the unpolarized net.

Views of nervous system evolution are complicated by examples of nonnervous coordinating systems. For example, many cnidarians possess apparently nonneural epithelial conducting systems, as well as nervous systems. The epithelial cells generate action potentials

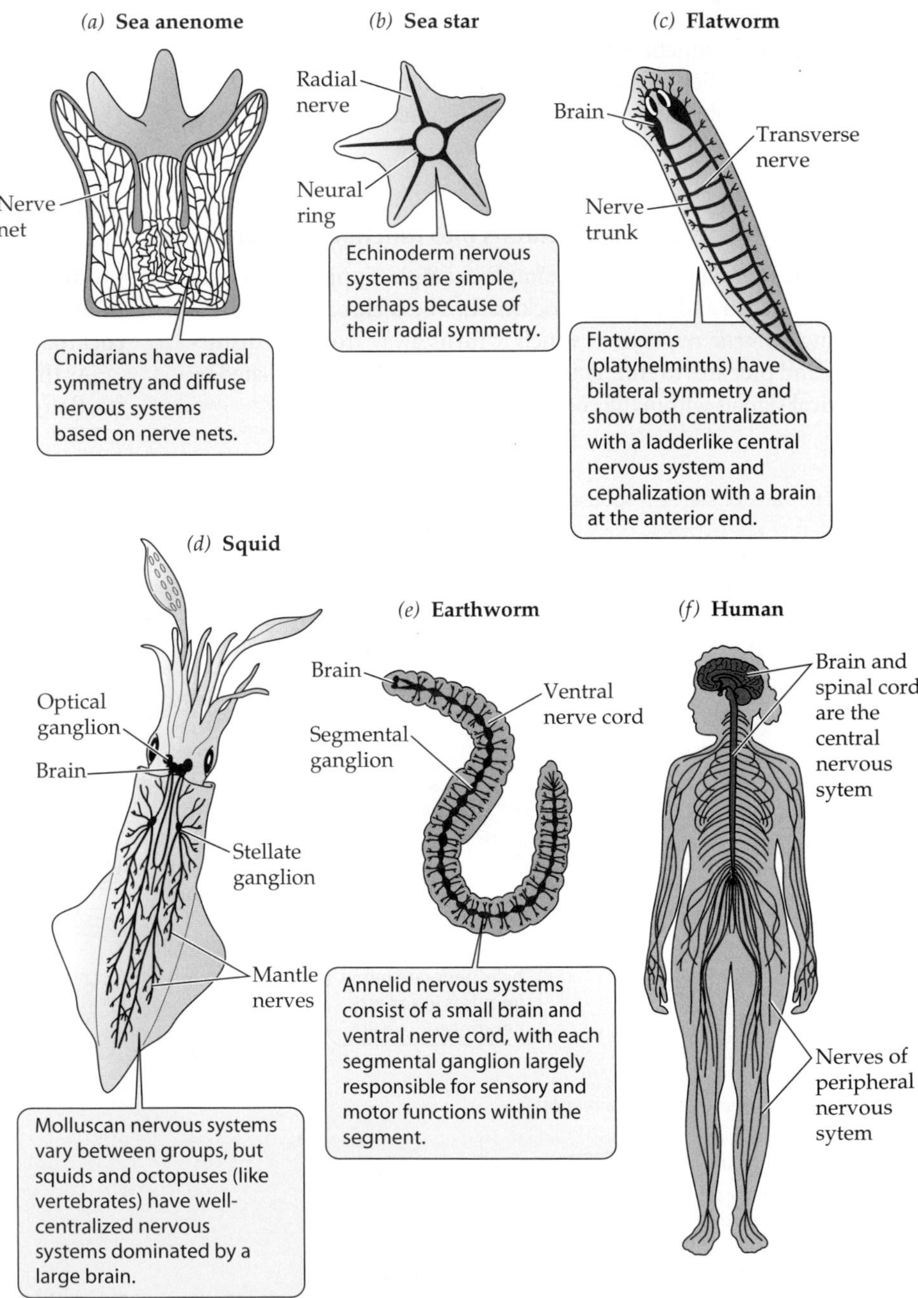

Figure 10.4 Nervous systems of different phyla *(a)* The sea anemone (phylum Cnidaria) has a nerve net. *(b)* The nervous system of a sea star (phylum Echinodermata) is radially symmetrical. *(c–f)* Flatworms (phylum Platyhelminthes) *(c)*, squids (phylum Mollusca) *(d)*, earthworms (phylum Annelida) *(e)*, and humans (phylum Chordata) *(f)* all display central nervous systems that feature brains.

that are propagated from cell to cell through low-resistance electrical junctions. Thus, nervous systems could have evolved after preexisting epithelial conduction networks that also coordinated the activity of arrays of muscle cell effectors.

Cnidarian nervous systems appear to be not very centralized, with fibers running in all directions and little apparent organization into central integrating areas (see Figure 10.4*a*). Cnidarians have *radial symmetry*, a body form with no front or back and with apparently limited potential for the evolution of nervous system centralization. Echinoderms, considered to be evolutionarily more advanced but to have secondarily evolved radial symmetry, also have relatively simple and uncentralized nervous systems (Figure 10.4*b*). In contrast, all groups with *bilateral symmetry* (Figure 10.4*c–f*) show evolutionary trends of increasing centralization and complexity of nervous system organization.

NERVOUS SYSTEMS OF ANIMALS WITH BILATERAL SYMMETRY EXHIBIT CENTRALIZATION AND CEPHALIZATION Two major trends characterize the evolution of nervous systems in the bilaterally symmetrical metazoan phyla: centralization and cephalization. **Centralization** of nervous systems refers to a structural organization in which integrating neurons are collected into central integrating areas rather than being randomly dispersed. **Cephalization** is the concentration of nervous structures and functions in the head. Both trends can be seen even in flatworms, which belong to the phylum Platyhelminthes, considered the most ancient phylum to have bilateral symmetry (see Figure 10.4*c*). Apparently the presence of a distinct anterior end and the development of a preferred direction of locomotion in bilateral animals have been important in the evolution of centralized, cephalized nervous systems.

In a **central nervous system** (**CNS**), most neuronal cell bodies and synaptic contacts are located in the nerve cords, and neural control pathways become increasingly routed through the CNS. Flatworms and organisms of more-complex bilaterally symmetrical phyla exhibit longitudinal *nerve cords* that constitute a distinct central nervous system. Motor neurons extend out from the CNS to effectors, and sensory neurons extend from the periphery of the body into the CNS. Increasing numbers of *interneurons*—neurons that are neither sensory nor motor and are confined to the CNS—make their appearance and enhance the integrative capabilities of the nervous system.

The **peripheral nervous system** (**PNS**)—consisting of all the processes of sensory and motor neurons that extend outside the CNS—also becomes increasingly consolidated. Instead of a random meshwork of processes running in all directions in an unpolarized nerve net, the peripheral sensory and motor processes condense into *nerves*, discrete bundles of nerve axons running between the CNS and the periphery (see Figure 10.4*c–f*).

Cephalization, the other general evolutionary trend in nervous system organization, involves varying degrees of anterior concentration of nervous system organization. In the most primitive of centralized nervous systems, each region of the CNS largely controls just its own zone or segment of the body (see Figure 10.4*c,e*); indeed, elements of such segmental or regional organization persist in all higher phyla, including vertebrates. In most bilaterally symmetrical animals, however, the most anterior part of the CNS exerts a considerable degree of domination and control over other regions. This anterior part, typically larger and containing more neurons than other parts, is called the *brain* (see Figure 10.4*c–f*). A **brain** is a general term for an anterior enlargement of the CNS.

Cephalization is thought to have been an evolutionary adaptation resulting from the tendency of bilaterally symmetrical animals to move forward, so that information about newly encountered parts of the environment impinges first on the front of the animal. As a correlate of forward motion, most groups with bilateral symmetry have evolved anterior placement of many of their major sense organs, and the brain, which receives environmental information via these sense organs, has become dominant over the rest of the CNS. Among vertebrates, a particular manifestation of cephalization is that the relative size of the forebrain (anterior part of the brain) increases successively in reptiles, birds, nonprimate mammals, and primates. Along with this development, functions formerly controlled by the spinal cord or brainstem come increasingly under forebrain control.

ARTHROPOD CENTRAL NERVOUS SYSTEMS ARE ORGANIZED AS CHAINS OF SEGMENTAL GANGLIA Central nervous systems exhibit two different major forms of organization of their central nervous systems: *Ganglionic* central nervous systems characteristic of protostomes, and *columnar* nervous systems characteristic of vertebrates and other deuterostomes. To see the features of ganglionic nervous system organization, we focus here on arthropods. Aspects of the organization of a ganglionic nervous system are present in arthropods, annelids, and molluscs.

In arthropods, the central nervous system consists of a chain of segmental **ganglia** (singular *ganglion*) (swellings containing discrete aggregations of nerve cell bodies and processes), linked by paired bundles of axons called **connectives** (Figure 10.5). The cockroach we startled earlier (see Figure 10.3) offers a fine example of the nervous system of an arthropod. The central nervous system of a cockroach consists of an anterior *brain* and a *ventral nerve cord* that is linked to the brain by connectives encircling the esophagus. The ventral nerve cord is a chain of ganglia linked by connectives—one ganglion for each thoracic and abdominal body segment. (Some arthropods show secondary fusion of some of these segmental ganglia.)

Each ganglion consists of an outer *rind* and an inner *core*. The rind consists mostly of cell bodies of neurons and is devoid of axons and synapses. Indeed, nearly all neuronal cell bodies of arthropods are confined to the outer rinds of the central ganglia, the major exceptions being cell bodies of sensory neurons, many of which are located in the peripheral nervous system. The inner core of each ganglion contains two regions: a region of synaptic contacts between axons and dendrites that is termed the **neuropile**, and a region of **tracts** (bundles) of axonal processes within the ganglion.

In arthropod or other ganglionic nervous systems, there are four terms for a bundle of nerve axons, depending on where the bundle is located. In the peripheral nervous system a bundle of axons is a **nerve**, between ganglia in the CNS it is a **connective**, within a ganglion it is a **tract**, and between right and left sides of a bilaterally symmetrical ganglion it is a **commissure**. The terms *nerve*, *tract*, and *commissure* have the same meanings for vertebrate nervous systems, but vertebrate nervous systems do not have connectives.

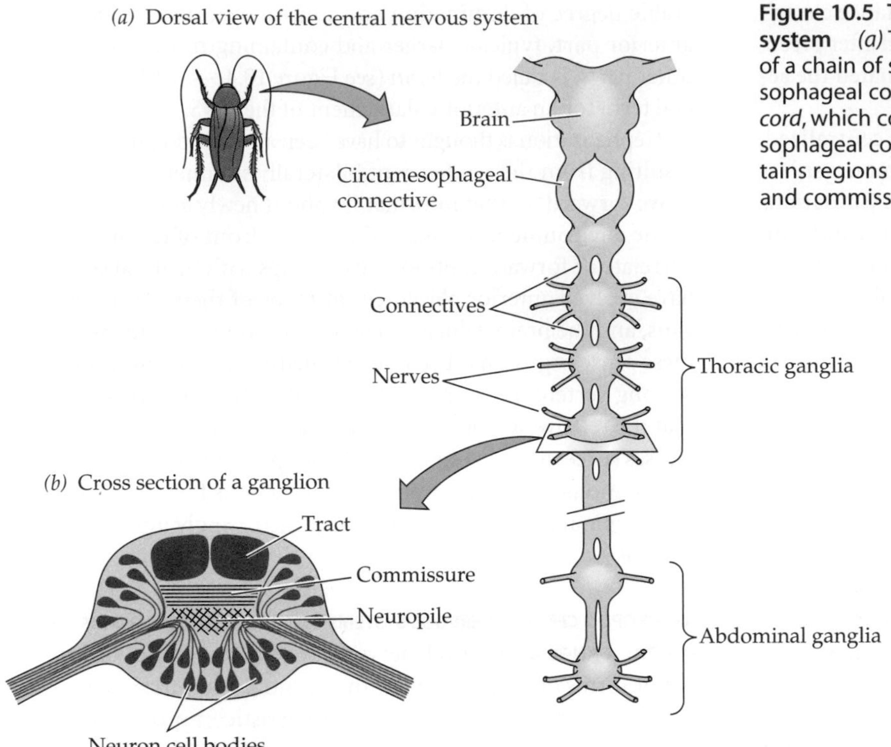

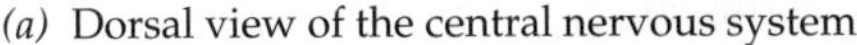

Figure 10.5 The organization of an arthropod central nervous system *(a)* The CNS, which is shown here in a dorsal view, consists of a chain of segmental ganglia linked by connectives. The circumesophageal connectives link the anterior *brain* to the *ventral nerve cord*, which consists of the linked ganglia posterior to the circumesophageal connectives. *(b)* A ganglion, shown in cross section, contains regions of cell bodies, of synaptic neuropile, and of axons (tracts and commissures).

THE VERTEBRATE CENTRAL NERVOUS SYSTEM IS A CONTINUOUS COLUMN Vertebrate central nervous systems, in contrast to those of arthropods, consist of a continuous column of neural tissue, with cell bodies and synaptic areas intermingled. The central nervous system of vertebrates consists of a brain and a spinal cord (Figure 10.6). It differs from the ganglionic central nervous systems of arthropods (and other protostomes) in several respects. The vertebrate central nervous system is dorsal and hollow, and it develops from a neural tube that invaginates from the dorsal surface of the embryo. The nerve cords of arthropods, in contrast, are ventral and solid, do not arise by invagination, and have connectives that typically remain paired. The vertebrate CNS, reflecting its origin as a continuous tube, is not clearly divided into ganglia and connectives, as is the arthropod CNS (compare Figure 10.6*a* with Figure 10.5*a*).

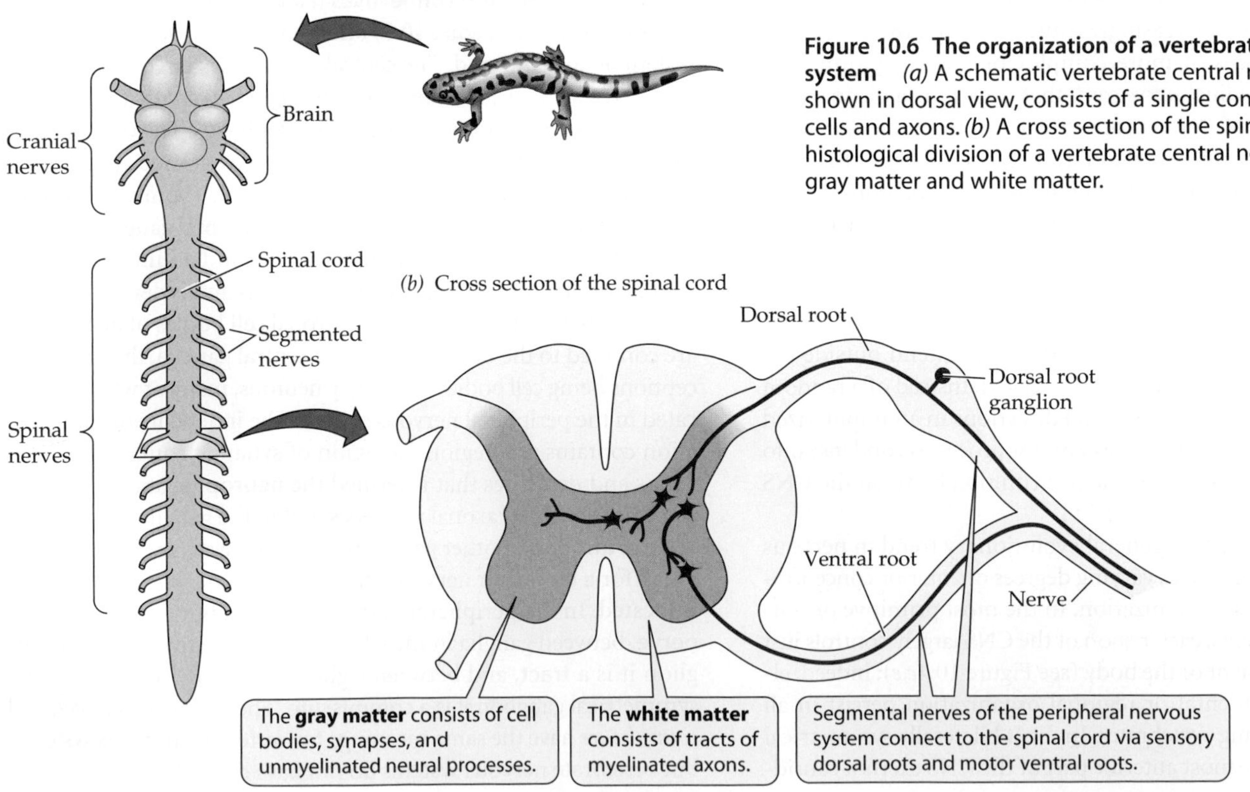

Figure 10.6 The organization of a vertebrate central nervous system *(a)* A schematic vertebrate central nervous system, shown in dorsal view, consists of a single continuous column of cells and axons. *(b)* A cross section of the spinal cord shows the histological division of a vertebrate central nervous system into gray matter and white matter.

Summary: The Organization and Evolution of Nervous Systems

- Animals have evolved nervous systems with varying degrees of centralization and complexity. It is tempting to consider simple nervous systems (such as the nerve nets of Cnidaria) to be primitive, but the major stages in nervous system evolution are unknown.
- Most phyla of animals have bilateral symmetry and have evolved central nervous systems that centralize control functions. Afferent neurons convey sensory information into the central nervous system, and efferent neurons convey outward motor commands to effectors. Central nervous systems usually have some degree of cephalization (concentration of neural structures into a clear anterior brain).
- Arthropods have a ganglionic nervous system, one major form of nervous system organization. The arthropod CNS is a ventral ladderlike chain of segmental paired ganglia joined by connectives. A vertebrate central nervous system, in contrast, is a continuous column of cells and axons.

The Vertebrate Nervous System: A Guide to the General Organizational Features of Nervous Systems

The nervous systems of most animals tend to share common organizational features of general structure. Here we discuss these organizational features using vertebrate nervous systems as examples. Keep in mind, however, that many of these organizational features apply to nervous systems in general. The vertebrate nervous system is organized into different regions that are discrete in gross structure, although neurons and their functions may cross these boundaries.

Nervous systems have central and peripheral divisions

For a vertebrate, the **central nervous system** (**CNS**) consists of the brain and spinal cord, and the **peripheral nervous system** (**PNS**) consists of the afferent and efferent nerves that connect the CNS to various parts of the body (Figure 10.7). The vertebrate PNS also includes *peripheral ganglia*, which are collections of neuronal cell bodies associated with peripheral nerves. (These should not be confused with the *central* ganglia of arthropod nervous systems.) As in other animals, the vertebrate PNS conveys sensory input to the CNS, and it conveys motor output (to control muscles and other effectors) from the CNS to the periphery. Effector functions include contraction, secretion, emission of light or heat, and other actions.

The central nervous system controls physiology and behavior

The vertebrate CNS demonstrates the two general principles of organization of complex nervous systems: centralization and cephalization. The nerves connected to the brain are called **cranial nerves**; those attached to the spinal cord are called **spinal nerves.** The spinal cord receives sensory input through the dorsal roots (see Figure 10.6*b*) and sends motor output to the periphery via ventral roots. These roots collect into segmental nerves (described shortly) as parts of the PNS. The brain receives sensory input from cranial nerves, and also via ascending pathways from the spinal cord. Likewise, although brain neurons directly control effectors of the head, much of the brain's output serves to control or modulate the spinal cord.

Histologically, the vertebrate CNS consists of two types of tissue: gray matter and white matter. **Gray matter** is composed of intermingled neuronal cell bodies, processes, and synaptic contacts. **White matter**, on the other hand, consists entirely of tracts of myelinated axons; it is the myelin that imparts a distinctive white appearance to the tissue. In the spinal cord the white matter is external, and the gray matter is internal (see Figure 10.6*b*). This arrangement does not hold in the brain, where gray matter is often external to white matter and the arrangement is considerably more complicated. For example, brains have a more layered gross organization: Often the outer surface layer is called a **cortex**, and the inner layer is a **medulla.**

Functionally, all neural activity is funneled into the CNS via afferent sensory neurons and sent out from the CNS by efferent motor neurons to effectors (see Figure 10.7). All significant integration and processing of neural activity occurs in the CNS; there are no sensory-to-motor synapses in the PNS and therefore no peripheral integration. Two types of circuits characterize the functional organiza-

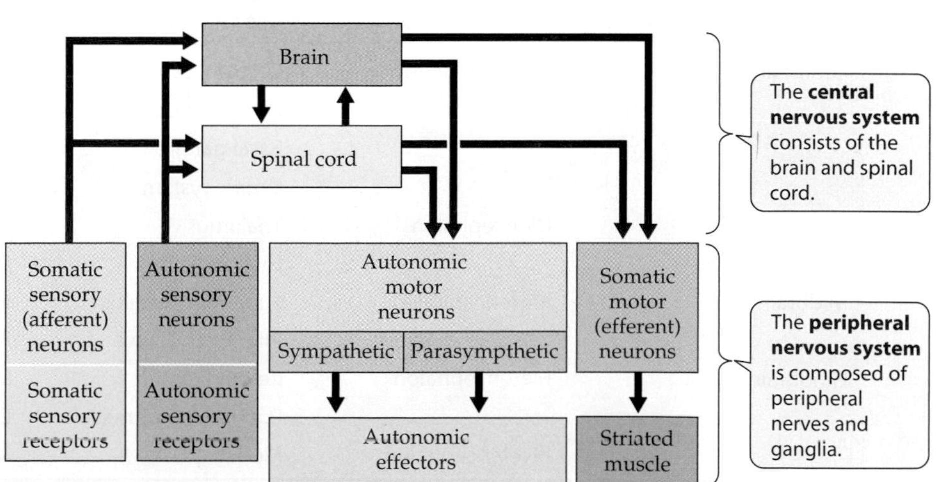

Figure 10.7 Divisions of the vertebrate nervous system are interconnected The most basic distinction is between the central nervous system (CNS) and the peripheral nervous system (PNS). The PNS has sensory and motor divisions. The somatic nervous system includes somatic receptors and afferent sensory neurons (these might be parts of the same sensory cell, or different cells), and efferent motor neurons controlling striated skeletal muscle. The autonomic nervous system includes autonomic sensory neurons and efferent neurons controlling internal autonomic effectors. (The enteric division of the autonomic nervous system is not shown.)

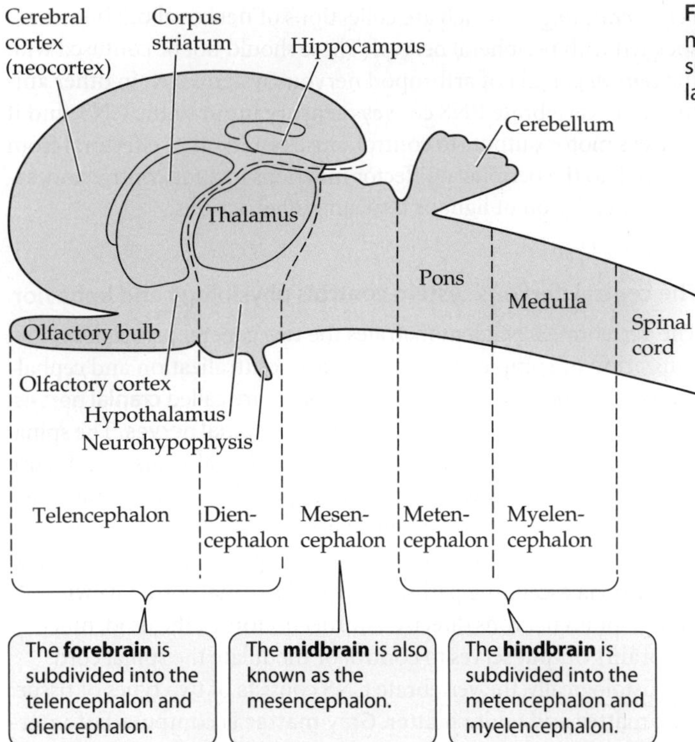

Figure 10.8 A schematic of vertebrate brain structure The brain has three major parts—forebrain, midbrain, and hindbrain—that are divided into five subdivisions, as shown. Some major structures of these five brain regions are labeled. In this side view, the anterior end is to the left.

tion of the spinal cord: local and ascending/descending. In local segmental circuits (such as simple spinal reflexes; see Chapter 18) sensory and local interneurons control motor output of the same segment. This local control is thought to be primitive. In addition to this local control, sensory input is relayed to the brain and contributes to higher integration and descending control of the spinal cord from the brain.

All vertebrate brains share a common structural organization, exhibiting three major regions: **forebrain**, **midbrain**, and **hindbrain** (Figure 10.8). Five subdivisions of these regions are recognized: **telencephalon** and **diencephalon** (forebrain), **mesencephalon** (midbrain), and **metencephalon** and **myelencephalon** (hindbrain). Each of these divisions contains many tracts of nerve axons and many clusters of cell bodies termed **nuclei** (singular *nucleus*).[3] Two major outgrowths of the dorsal surface of the brain become increasingly prominent in higher vertebrates: the **cerebellar cortex** of the metencephalon and the **cerebral cortex** of the telencephalon. The cerebral cortex includes the olfactory cortex, hippocampus, and (in mammals) the neocortex. Table 10.1 lists some brain structures and their functions. We'll discuss several of these brain regions in Chapters 13 and 18.

The CNS, along with the endocrine system, exerts control over the functions of an animal's organ systems and over its behavior: all movements and externally observable activity. We will take the simplifying view that behavior is the province of the skeletal muscles, controlled via the somatic nervous system. In contrast, other effectors, controlled via the autonomic nervous system, exert physiological control of internal organ systems. This distinction is not completely true, because (for example) reproductive behavior includes glandular secretion and smooth muscle contraction, while breathing (lung ventilation) is a physiological function controlled by skeletal muscles of the diaphragm and ribs. Still, the distinction between control of behavior through skeletal muscle and control of physiology through other effectors, although imprecise, is a useful generalization.

In general, CNS control may be local (involving only a restricted part of the CNS, such as a segment of the spinal cord) or more widespread and hierarchical. As an example of the latter case, segmental spinal motor neurons control muscles of the diaphragm to generate

[3] In vertebrates, clusters of neuronal cell bodies are usually called *nuclei* in the CNS and *ganglia* in the PNS. The basal ganglia of the brain constitute an exception to this terminology; they are actually central nuclei.

TABLE 10.1 Some major areas of the mammalian brain and their general functions. Most functions involve several brain areas, and many brain areas and functions are omitted.

Major brain division	Brain subdivision	Area	Major functions
Forebrain	Telencephalon	Cerebral cortex	Higher sensory, motor, and integrative functions
		Hippocampus	Learning and memory
		Basal ganglia	Motor control
		Limbic system	Emotions
	Diencephalon	Thalamus	Major sensory relay
		Hypothalamus	Homeostatic and endocrine regulation; circadian clock
Midbrain	Mesencephalon	Superior colliculus	Visual integration
		Inferior colliculus	Auditory integration
Hindbrain	Metencephalon	Cerebellum	Motor coordination
		Pontine motor nuclei	Descending motor control
	Myelencephalon	Medulla	Autonomic and respiratory control

breathing movements in a mammal. The motor neurons receive rhythmic excitatory commands from brainstem neurons that generate a respiratory rhythm. Sensors of blood CO_2 and O_2 levels speed up or slow down the respiratory rhythm at the level of the brainstem, but conscious controls from the forebrain also modify the respiratory pattern—for example, in vocalization.

Four principles of functional organization apply to all mammalian and most vertebrate brains

PRINCIPLE 1: LOCALIZATION OF FUNCTION Neurons in different anatomical regions of the brain play different functional roles. That is, you can point to a part of the brain and reasonably ask what functional activities occur there.

This point has been controversial in the past. In the early nineteenth century, Franz Joseph Gall advanced a theory that personality traits were located in different bumps on the convoluted surface of the brain. He believed that these "brain bumps" could be felt on the skull and could be used as predictors of personality (Figure 10.9*a*). The resulting pseudoscience of *phrenology* was transiently popular, and the backlash against it made scientists skeptical of theories advancing localization of function for over a century.

Many studies amply confirm, however, that specific brain areas have specialized functions. Figure 10.9*b*, for example, shows some major areas of the cerebral cortex that are involved in sensory processing and in speech. Several kinds of studies demonstrate localization of function: stimulation studies, lesion studies that destroy or isolate brain tissue in a region, strokes (that interfere with blood circulation to a particular part of the brain, resulting in death of neurons and disruption of a particular sensory or motor process), and functional imaging studies.

Modern imaging studies are particularly important in clarifying this issue. Imaging methods such as functional magnetic resonance imaging (fMRI) (Figure 10.10) and positron emission tomography (PET) show localized increased metabolic activity that results from increases in neuronal electrical activity. These areas of increased activity show where particular functions are localized. Note that localization of function does *not* mean that a particular brain area is involved in only one function, or that a function involves only one discrete brain area. A function is often apt to involve a *circuit* in the brain (a network of synaptically interconnected and interacting neurons, which may be widespread) rather than a discrete *center*.

PRINCIPLE 2: SIZE MATTERS In general, the more neurons that are present in a brain area, the more complex the integration that occurs there. Thus, mammalian brains, with tens of billions of neurons, perform more-complex integration than insect brains with tens of thousands of neurons do. By the same token, the massive enlargement of the cerebral cortex in the course of mammalian evolution (as a result of the developmental proliferation of greatly increased numbers of neurons) attests to major increases in processing capability of mammalian and especially primate brains. Primate brains are not "more evolved" than the brains of fish and amphibians, because fish and amphibians have had as long a time to evolve and can be considered to be as well adapted to their lives as primates are. Nevertheless, the great expansion of cerebral cortex in primates has allowed types of neural function (e.g., language, culture) not found in brains with fewer neurons.

Figure 10.9 Localization of function in the mammalian brain: old and current views *(a)* Nineteenth-century phrenologists wrongly believed that personality traits were very localized and could be "read" from skull bumps. *(b)* A modern view of localization of function emphasizing areas involved in spoken language. This surface view of the human left cerebral hemisphere shows primary sensory areas (visual cortex, primary auditory cortex, somatosensory cortex), motor areas, and areas involved in language.

(a) Phrenology

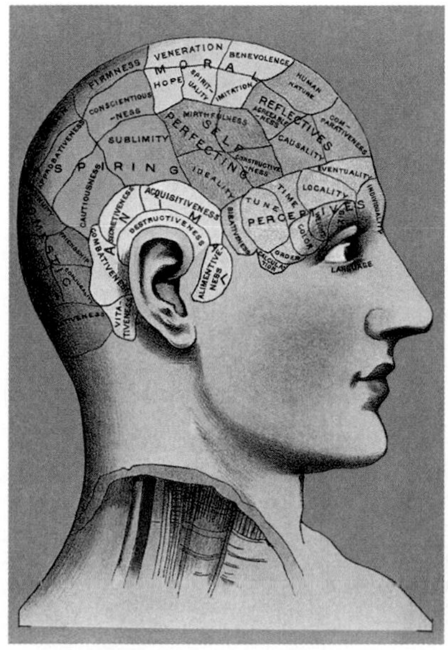

(b) Brain areas in speech

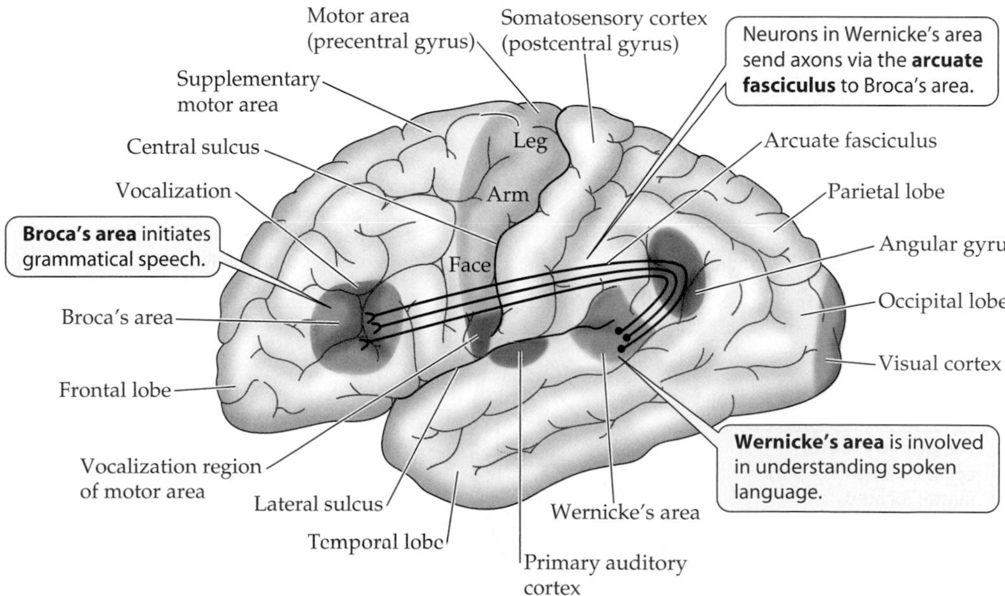

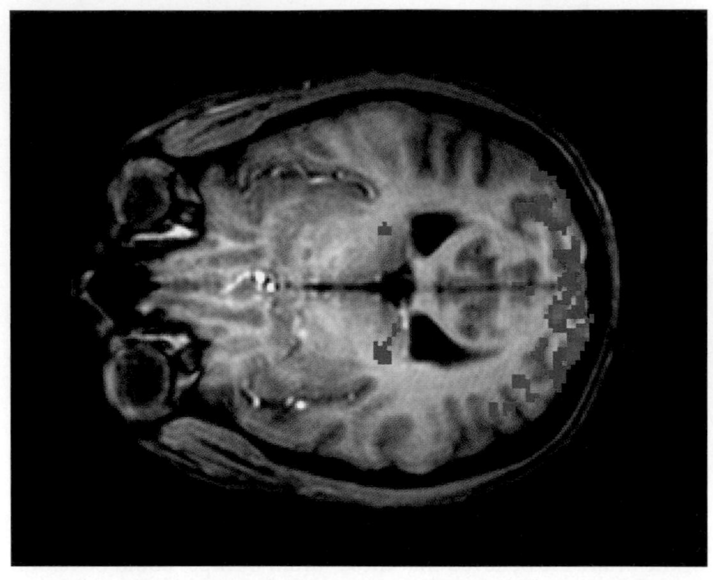

Figure 10.10 Functional neuroimaging demonstrates localization of function in human cerebral cortex Functional magnetic resonance imaging (fMRI) shows increased activity of particular brain areas. This fMRI sectional image of a brain (seen from above; left is anterior) shows increased neural activity during a visual stimulus. The neural activity (measured as increased blood flow) increases in the lateral geniculate nucleus of the hypothalamus and the primary visual cortex (posterior). (See Figure 13.19 for a diagrammatic view of this pathway.) (From Chen et al. 1999.)

PRINCIPLE 3: MAPS IN THE BRAIN The brain maintains information about the body's anatomical organization in terms of topographic representations, or **maps**. The brain contains many mapped representations that record and recall the parameter of *where* a stimulus occurs or an effector is to be controlled. For example, the sensory surface of the body is mapped onto the primary somatosensory area of the cerebral cortex (Figure 10.11*a*) to form a **somatotopic map** (a map of the body projected to a brain area). An analogous somatotopic map of motor control exists on the surface of the primary motor cortex (Figure 10.11*b*); any point stimulated will elicit movement of the corresponding part of the body.

There are additional sensory and motor maps in other cortical areas, as well as in many noncerebral structures. The visual system, for example, exhibits a point-to-point correspondence between areas in the visual field in the outside world, and areas in the retina, which we'll discuss further in Chapter 13. This geometric orientation is preserved in visual pathways, so that the retinal map of the world is projected to each of the major visual centers in the brain. On the other hand, maps in the brain are by no means universal; many areas of the mammalian brain lack topographic organization.

PRINCIPLE 4: PLASTICITY OF NEURAL CIRCUITS We tend to think of the synaptic connections between neurons as "hardwired"—rather fixed and unchanging—but there is abundant evidence that this is not the case. Instead, synapses are *plastic*: They change with development, maturation, and experience. Noting a difference between short-term memory and long-term memory will clarify this point: *Short-term memory*, memory of events of the last minutes to hours, can be dis-

Figure 10.11 Maps in the human brain In what is called a somatotopic map, parts of the body are mapped onto the cerebral cortex in a way that preserves their anatomical position on the body. *(a)* Neurons in the primary motor cortex control movements of parts of the body, and are arranged in a somatotopic map. *(b)* Sensory receptor locations are mapped somatotopically onto the somatosensory cortex. A drawing of the body projected onto the brain is called a *homunculus*, meaning "little person."

(a) Motor homunculus

Hip
Trunk
Shoulder
Elbow
Wrist
Hand
Little finger
Ring finger
Middle finger
Index finger
Thumb
Knee
Ankle
Toes
Neck
Brow
Eyelid and eyeball
Face
Lips
Jaw
Tongue
Swallowing
Mastication
Salivation
Vocalization

Some areas of the body (face, hands) have disproportionately large representations.

The **primary motor cortex** is a band of gray matter located just anterior to the central sulcus (see inset).

(b) Sensory homunculus

Leg
Hip
Trunk
Neck
Head
Shoulders
Arm
Elbow
Forearm
Wrist
Hand
Little finger
Ring finger
Middle finger
Index finger
Thumb
Eye
Nose
Face
Foot
Toe
Genitals
Upper lip
Lower lip
Teeth, gums, jaw
Tongue
Pharynx
Intra-abdominal

Sensory receptors on the body surface project to the **somatosensory cortex**, a region of the cerebral cortex located just posterior to the central sulcus (see inset).

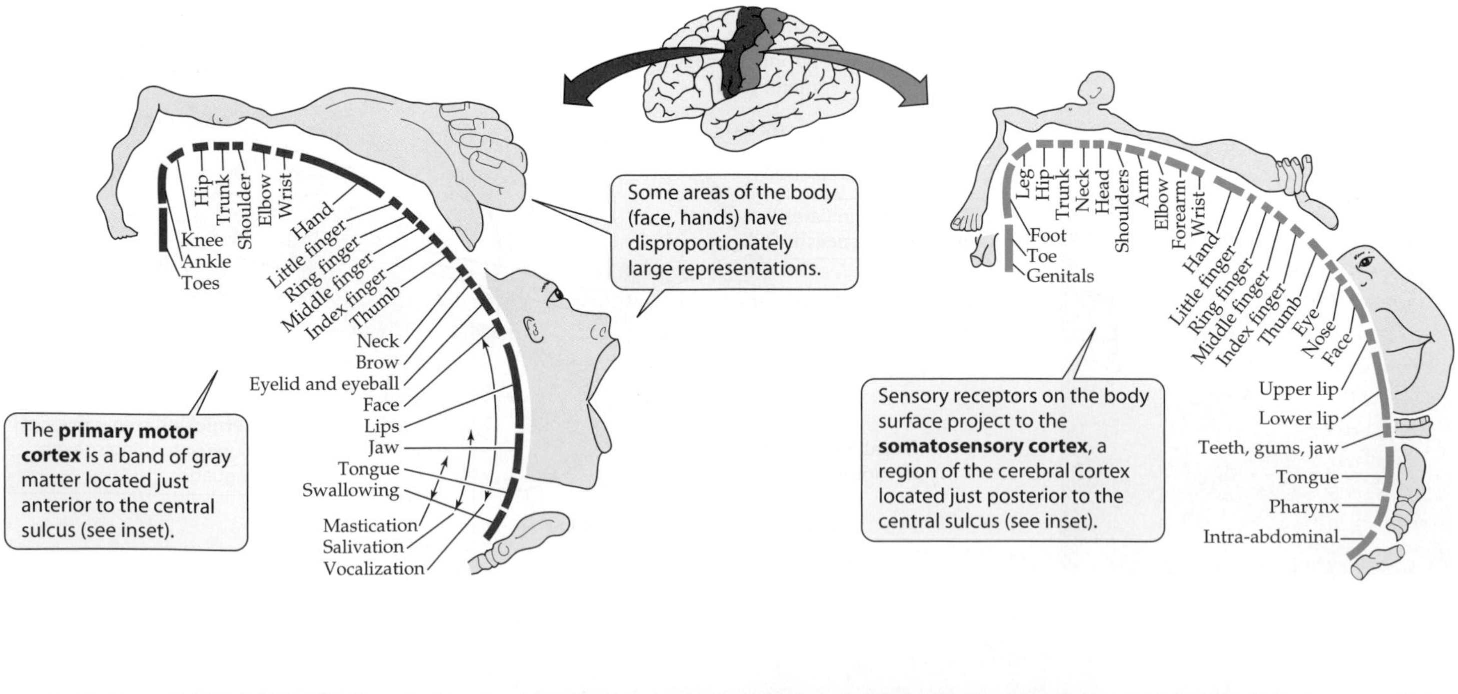

rupted by a concussion or by electroconvulsive shock. *Long-term memories*, in contrast, are not lost after the concussion, because they are thought to have been stored more permanently as changes in the "wiring"—changes in the strengths of synaptic interconnections of neurons. This storage of long-term memory may be an example of how neural connections are plastic, or subject to change in strength and effectiveness.

We know also that synaptic connections are plastic during development of the nervous system; they are made and broken as an animal's brain matures, and they are subject to competition between neuron endings for synaptic sites on a postsynaptic target cell. Scientists believe that plasticity in development and the plasticity associated with learning and memory may have similar mechanisms; we explore this topic further in Chapter 12.

The peripheral nervous system has somatic and autonomic divisions that control different parts of the body

Physiologists recognize two great divisions of the vertebrate peripheral nervous system: the *somatic* and the *autonomic* nervous systems (see Figure 10.7). The **somatic nervous system** is the part of the peripheral nervous system that controls the skeletal (striated) muscles; skeletal muscles are thus called **somatic effectors.** The autonomic nervous system, on the other hand, is the part that controls **autonomic effectors** (or *internal effectors*), defined to include all neuron-controlled effectors other than the striated muscles, such as cardiac muscle, smooth (nonstriated) muscles, and glands.

Autonomic effectors exert most of their effects on visceral internal organs, internally and invisibly. The somatic nervous system controls most observable behavior and it is the part of the nervous system with which we are most familiar. It includes somatic sensory receptors (touch, hearing, vision, taste, olfaction, and so forth) and controls the muscles of movements, breathing, and speech. We will consider aspects of the somatic nervous system in subsequent chapters; here we will consider only its overall organization.

SOMATIC AND AUTONOMIC ORGANIZATION As we have already noted, the peripheral nervous system consists of cranial and segmental spinal nerves that convey sensory information in from different regions of the body, and convey motor signals out to control effectors. The cell bodies of spinal sensory neurons are located in **dorsal root ganglia**, enlargements of the dorsal roots outside the spinal cord (see Figure 10.6*b*). The ventral roots contain axons of both somatic motor neurons that innervate skeletal muscles and autonomic neurons that innervate autonomic ganglia (discussed shortly).

The cranial nerves are more specialized in function. Mammals have 12 pairs of cranial nerves. Some are associated with the major sense organs of the head (I, olfactory; II, optic; VIII, auditory); others have motor or mixed functions. The vagus nerve (X) is a major component of the autonomic nervous system, and it also innervates the larynx and other somatic components.

The **autonomic nervous system** (**ANS**) (see Figure 10.7) is usually defined as if it were a motor system—the division of the peripheral nervous system that controls the autonomic effectors. However, the autonomic nervous system also includes sensory neurons that convey afferent signals from internal organs to the CNS. The autonomic effectors that are controlled at least partly by the vertebrate autonomic nervous system include the following:

- Smooth muscles throughout the body, such as those in the gut wall, blood vessels, eyes (iris muscles), urinary bladder, hair follicles, spleen, airways of the lungs, and penis
- Many exocrine glands, such as sweat glands, tear glands, and the exocrine pancreas
- A few endocrine glands, notably the adrenal medullary glands (chromaffin tissue) that secrete epinephrine (adrenaline)
- Acid-secreting cells of the stomach
- The pacemaker region and other parts of the heart
- The brown adipose tissue of mammals (a heat-producing tissue)
- The swim bladders and integumentary chromatophores (color-change cells) of fish

Many invertebrate animals have nervous system divisions that control visceral functions. These divisions are sometimes described as *autonomic* by analogy to vertebrate autonomic nervous systems.

THE THREE DIVISIONS OF THE AUTONOMIC NERVOUS SYSTEM The three commonly recognized divisions of the vertebrate autonomic nervous system—*sympathetic*, *parasympathetic*, and *enteric*—were first described by John Langley (1852–1925). Langley based his classification on anatomy, not function. Thus, his divisions are not always distinctly different in function. Today, his scheme is considered valid for mammals. It is often used for nonmammalian vertebrates also, although its application to these other groups remains debatable.

The mammalian **sympathetic division** and **parasympathetic division** functionally link the central nervous system (CNS) and the autonomic effectors. Both systems are distinguished by their morphology and function. The enteric division has some connection with the CNS but is largely contained in the gut.

ANATOMY OF THE SYMPATHETIC AND PARASYMPATHETIC DIVISIONS Autonomic motor pathways are characterized by a *peripheral synapse*—that is, a synapse interposed between the CNS and an effector tissue (Figure 10.12*a*). This peripheral synapse is a characteristic of the sympathetic and parasympathetic divisions of all vertebrates, so autonomic motor signals in both divisions traverse a two-neuron relay to reach their effectors from the CNS. (In contrast, in the somatic nervous system, motor signals travel along single neurons after leaving the CNS, without a peripheral synapse; Figure 10.12*b*.)

The peripheral synapses between the first and second neurons in the sympathetic and parasympathetic divisions are located within clusters of neuronal cell bodies called **autonomic ganglia** (singular *ganglion*). The neurons that extend from the CNS to the ganglia are termed **preganglionic** neurons, and those extending from the ganglia to the effectors are termed **postganglionic** neurons. The mammalian sympathetic and parasympathetic divisions differ in the positions of the ganglia. In the parasympathetic division, ganglia are located mostly at or near the effectors, so preganglionic parasympathetic neurons are long and postganglionic parasympathetic neurons are short (see Figure 10.12*a*). In contrast, sympathetic ganglia are located mostly near the spinal cord, so preganglionic sympathetic neurons are short, and postganglionic sympathetic neurons are long.

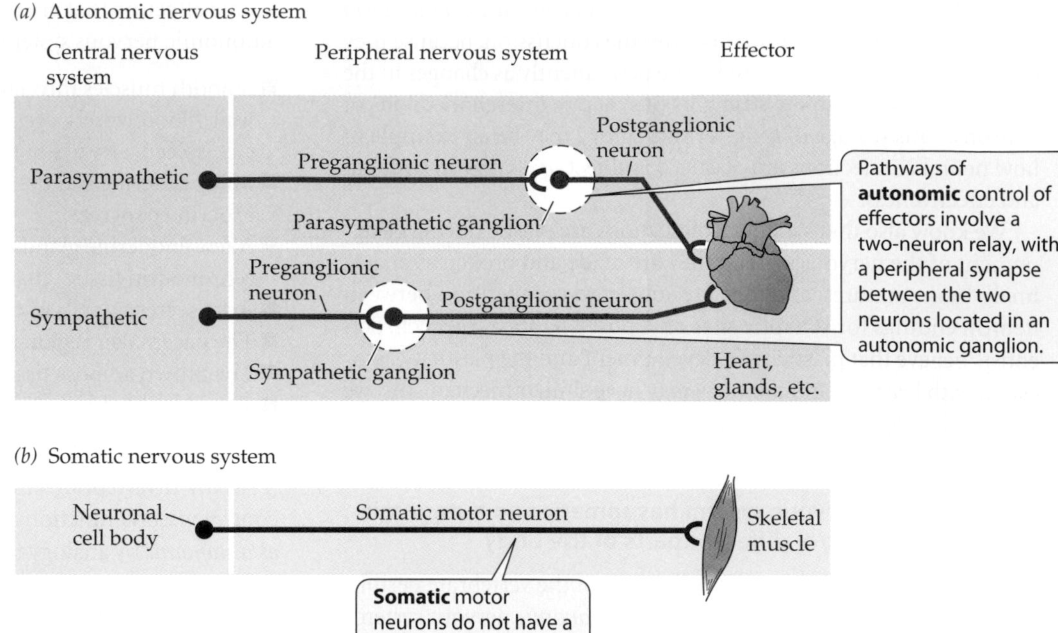

Figure 10.12 The organization of the mammalian autonomic and somatic nervous systems, including examples of effector organs *(a)* The heart is under autonomic control. Sympathetic ganglia are typically close to the spinal cord; parasympathetic ganglia are typically near target organs. Preganglionic autonomic neurons secrete the neurotransmitter acetylcholine (ACh); postganglionic sympathetic neurons secrete norepinephrine (NE), and postganglionic parasympathetic neurons secrete ACh. *(b)* Skeletal muscle is under somatic control. The enteric nervous system, not shown, has some connection to the CNS but is largely restricted to the gut.

The *nerves* of the parasympathetic and sympathetic divisions are associated with different regions of the CNS in mammals (Figure 10.13). Parasympathetic preganglionic neurons exit the CNS from two regions: cranial and sacral. For this reason, the parasympathetic division is sometimes called the *craniosacral* division. The cranial group of nerves consists of four of the pairs of cranial nerves: the oculomotor, facial, glossopharyngeal, and vagus nerves. The sacral group of nerves emerges from the posterior part of the spine.

In the sympathetic division the preganglionic neurons emerge in nerves of the thoracic and lumbar regions of the spine (see Figure 10.13). For this reason, the sympathetic division is also called the *thoracolumbar* division. Most of the nerves terminate in sympathetic ganglia immediately lateral to the spine. These **paravertebral ganglia** occur segmentally (at regularly repeating intervals) along the length of the spine and are interconnected by longitudinal nerve connectives, forming a **sympathetic chain** on each side of the vertebral column. Some preganglionic sympathetic neurons terminate in ganglia more distant from the spine, as in the celiac (solar) plexus. Some directly innervate the medullary tissue of the adrenal glands (see Chapter 14).

FUNCTIONAL PROPERTIES OF THE SYMPATHETIC AND PARASYMPATHETIC DIVISIONS The postganglionic neurons of the parasympathetic and sympathetic divisions release different chemical neurotransmitter substances at their synapses with effector cells. Parasympathetic postganglionic neurons typically release acetylcholine and thus are termed **cholinergic**. Most sympathetic postganglionic neurons release mainly catecholamines—chiefly norepinephrine (noradrenaline) in mammals—and thus are called **adrenergic** (or noradrenergic).

Autonomic effectors may be innervated by one or both ANS divisions. In mammals, for example, most smooth muscles of blood vessels and the piloerector (hair-erecting) muscles of the hair follicles receive only sympathetic innervation. Many effectors, in contrast, receive both sympathetic and parasympathetic innervation (see Figure 10.13). The responses elicited by the two divisions in such cases are usually antagonistic. For instance, the pacemaker region of the heart (cells that initiate the heartbeat) is innervated by both divisions. Parasympathetic impulses decrease the heart rate; sympathetic impulses increase it.

The overall roles of the parasympathetic and sympathetic divisions reflect the tendency of the two divisions to act antagonistically and demonstrate the integrative actions of autonomic control (Table 10.2). The parasympathetic division (especially its cranial

TABLE 10.2 Major actions of sympathetic and parasympathetic divisions in vertebrates

Process	Parasympathetic effect	Sympathetic effect
Digestion: gastrointestinal secretion and motility	Stimulates	Inhibits
Heartbeat	Slows	Increases rate and force
Blood vessels	Usually dilates	Constricts vessels to kidneys and gut; dilates vessels to skeletal muscles
Blood pressure	Decreases	Increases
Lung passages	Constricts	Dilates
Secretion of epinephrine and norepinephrine by adrenal medullary glands	—	Stimulates

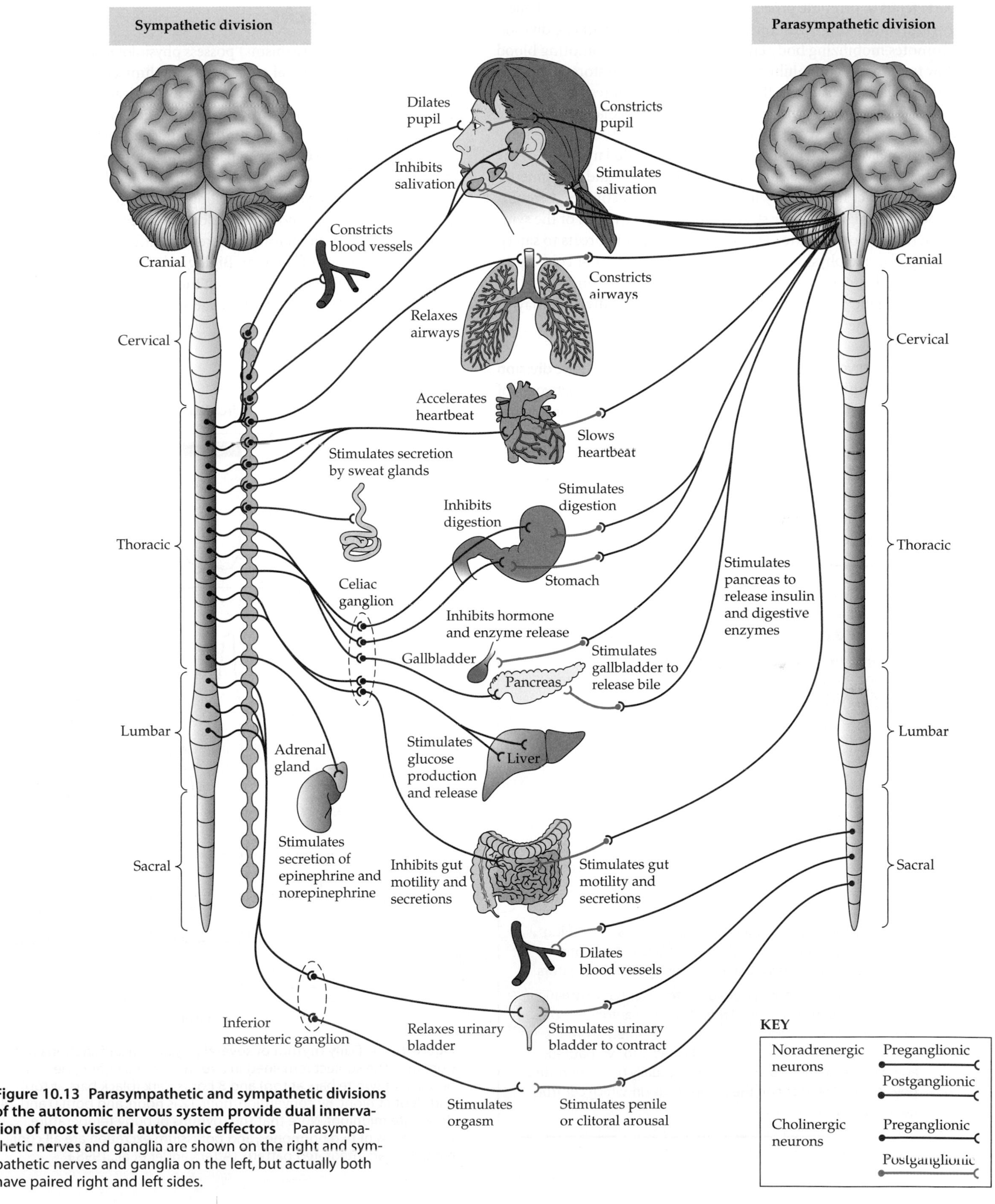

Figure 10.13 Parasympathetic and sympathetic divisions of the autonomic nervous system provide dual innervation of most visceral autonomic effectors Parasympathetic nerves and ganglia are shown on the right and sympathetic nerves and ganglia on the left, but actually both have paired right and left sides.

part) tends to promote processes that restore body reserves of energy (e.g., stimulating digestion). In contrast, the sympathetic division promotes mobilizing body energy reserves (e.g., promoting blood flow to muscles) and inhibits some processes that restore reserves; it is particularly activated in the face of stress, and it readies the body to meet stress. Direct sympathetic innervation mediates most of the effects listed for the sympathetic division in Table 10.2, but the sympathetic nervous system also stimulates the medullary tissue of the adrenal glands to secrete epinephrine (adrenaline) and norepinephrine, which act hormonally to complement its activities. Circumstances that might provoke strong, concerted activation of the sympathetic division include vigorous exertion, pain, threats to safety, and exposure to physical extremes such as severe heat or cold. Aptly, the sympathetic division has been said to prepare an individual for "fight or flight"; in contrast, the parasympathetic division may be said to promote "rest and digest" functions.

THE ENTERIC NERVOUS SYSTEM CONTROLS THE GUT The **enteric division** of the autonomic nervous system consists of elaborate networks of neurons located entirely within the walls of the gut. The enteric division controls peristalsis, segmentation, and other patterns of contraction of the smooth muscles of the gut wall that serve to move digested materials through the gut. The enteric nervous system's function is largely autonomous of CNS control, although the CNS can modulate enteric neurons and synapses via sympathetic and parasympathetic nerves.

Summary: The Vertebrate Nervous System: A Guide to the General Organizational Features of Nervous Systems

- The central nervous system of vertebrates consists of the brain and spinal cord. Cranial and spinal nerves emanate from the CNS to form the peripheral nervous system. The brain is divided into a forebrain, midbrain, and hindbrain, and the forebrain is enlarged in birds and especially in mammals.
- Vertebrate brain functions are somewhat localized. On the other hand, brain functions are also somewhat distributed, involving circuits rather than centers.
- Many vertebrate brain regions preserve the orderly spatial arrangements of the corresponding external world, for example, as somatotopic maps of body sensory input and motor output.
- Brains change with development, experience, and learning and memory. Understanding the structural and synaptic bases of these changes is a major challenge to investigators.
- The peripheral nervous system of vertebrates has a somatic division that controls skeletal muscle and an autonomic division that controls effectors associated with internal organs. The autonomic nervous system is divided into sympathetic and parasympathetic divisions, which usually have opposite physiological effects, and the enteric division, which controls gut contraction.

Biological Clocks

Animals (and other organisms) possess physiological timing mechanisms termed **biological clocks** that rhythmically modulate the functioning of cells, tissues, and organs. Biological clocks endow an animal with an intrinsic **temporal organization**, a timed pattern of change in physiology or behavior that is independent from a change in environment (see Chapter 1, pages 14–15, for discussion of temporal frameworks in animal physiology). Biological clocks are functions of the nervous system, controlling physiological and behavioral processes via nervous and neuroendocrine output. An animal characteristically is physiologically different at different times of day, or in different seasons of the year. Biological clocks orchestrate these changes, controlling and integrating the alteration of physiological states. Figure 10.14 shows the daily variation in several physiological functions in a human.

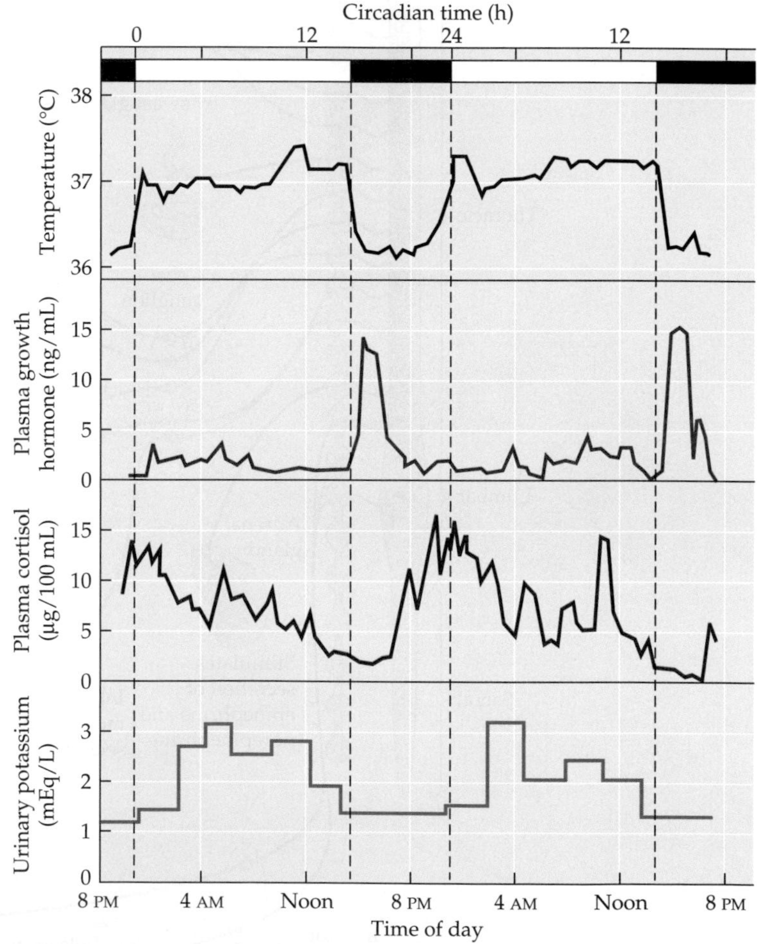

Figure 10.14 Daily rhythm of several physiological functions in a human The subject remained in a room under a light cycle of 16 hours light (white bars at top) and 8 hours dark (black bars at top) and slept at night. By convention, hour 12 is noon, and both 0 and 24 designate midnight. Note that the rhythms of body temperature, blood levels of two hormones, and potassium excretion are all coordinated in an integrated pattern. (After Vander et al., 1998.)

TABLE 10.3 Some processes that show circadian rhythmicity in animals and other eukaryotes

Locomotor activity in many vertebrates and invertebrates
Sleep–wake cycles in many animals
Metabolic rate in many animals
Variations of body temperature (including torpor) in birds and mammals
Urine output and drinking in mammals
Adrenocortical hormone secretion and epidermal mitosis in mammals
Integumentary color change in fish and crabs
Oviposition, mating, and emergence of adults from pupae in insects
Female pheromone release and male pheromone sensitivity in insects
Mating in *Paramecium*
Bioluminescence and photosynthetic capacity in dinoflagellate algae

Organisms have endogenous rhythms

Activities of organisms often change regularly over time. A **rhythm** is a regular, cyclical variation in function. Rhythms that continue in the absence of environmental information about time are termed **endogenous rhythms.** The first organisms in which endogenous rhythms were demonstrated were certain plants that raise their leaves during some times of day and lower them at others.[4] The plants in question are said to exhibit a *daily rhythm* of leaf movements.

In 1729 the Frenchman M. de Mairan reported that some plants continue to raise and lower their leaves in approximately a daily rhythm even when they are kept in constant darkness and at constant temperature. That is, their rhythm of leaf movements continues in more or less a daily pattern even when the plants are denied environmental sources of information about the time of day. Experiments of a similar nature have since been performed on many plant and animal systems, and daily rhythms in many types of function in many types of organisms have been shown to persist in constant laboratory environments.

An endogenous rhythm that has a period of about a day is termed a **circadian rhythm** (*circa*, "about"; *dies*, "a day"). Circadian rhythms that persist in the absence of environmental information about the time of day appear to occur in all eukaryotes (and some prokaryotes). Thus (although not all daily rhythms prove to be endogenous when tested) a capacity for endogenous rhythmicity is believed to be an ancient feature of life. Table 10.3 lists some examples of the known endogenous circadian rhythms.

Circadian rhythms under normal circumstances are tightly coupled to environmental cues such as a daily light–dark cycle. Figure 10.15*a* shows the behavioral activity and metabolic rate of a chaffinch (*Fringilla* sp.) initially kept on a normal light–dark cycle. In diurnal species such as the chaffinch, locomotor activity and many other physiological variables increase during the day, usually starting at or shortly before dawn. Two rhythms are said to be **in phase** if they occur synchronously, so the chaffinch's activity is in phase with the light cycle.

To show that a daily rhythm is an endogenous circadian rhythm, an experimenter must remove the environmental timing information, in this case the light–dark cycle. In the presence of constant, dim light (Figure 10.15*b*), the activity rhythm persists but fails to remain synchronized with the time dawn would have come. Instead, it is a little shorter than a day, with a period of about 23 hours. The **period** of a rhythm is the amount of time between a particular part of the rhythm in one cycle (one day) and that same part in the next

[4] In our discussion of rhythms and biological clocks, we will consistently use the words *day* and *daily* to refer to the 24-hour day, not just to hours of daylight.

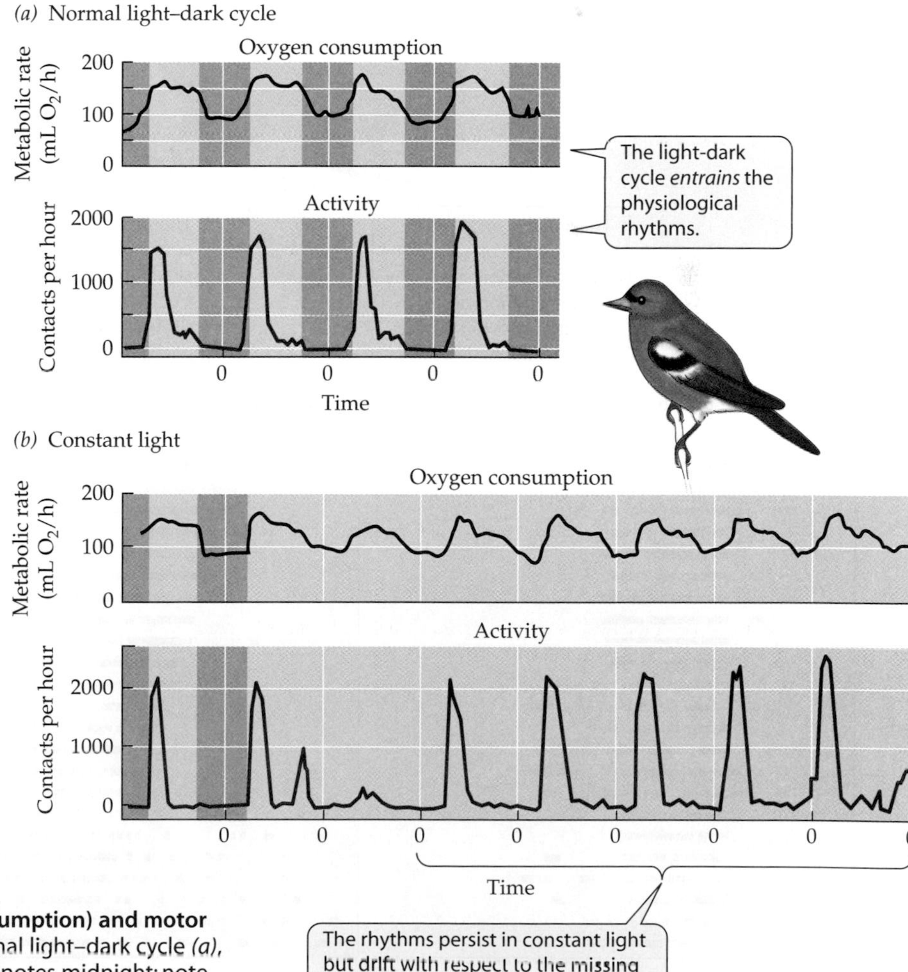

Figure 10.15 Circadian rhythm of metabolic rate (O_2 consumption) and motor activity for a chaffinch The bird was initially under a normal light–dark cycle *(a)*, but then was kept in constant light *(b)*. Zero on the *x* axis denotes midnight; note that the rhythms start earlier in each successive 24-hour period.

cycle. Typically the period is measured as the time between the start of one day's episode of activity and the start of the next day's. The 23-hour period determined in this way (see Figure 10.15*b*) is then the period of the chaffinch's activity rhythm at that level of constant illumination. When environmental cues are absent, the biological rhythm that persists is said to **free-run** or to be a **free-running rhythm.** Free-running circadian rhythms have periods that are close to but not exactly equal to 24 hours.

The difference between a free-running rhythm and one that is synchronized to environmental cues is easier to see if the periods of activity on successive days are stacked one below the last. Figure 10.16 shows the activity rhythms of two nocturnal flying squirrels (*Glaucomys volans*): one in a normal light–dark cycle and the other in constant darkness. The light–dark cycle (when present) synchronizes the activity rhythm, bringing it into phase, so that the onsets of activity periods are lined up at the same time each day. In the absence of environmental information about the time of day, the squirrel's endogenous rhythm of activity persists, but because the period of the endogenous circadian rhythm is not precisely 24 hours, the free-running rhythm drifts to occur later and later each day.

The process by which a biological rhythm is brought into phase with an environmental rhythm is called **entrainment.** During the process, the biological rhythm is said to become **entrained** by environmental cues, as illustrated in Figure 10.16*a*. An environmental cue that is capable of entraining (setting the phase) of a biological rhythm is called a **phasing factor** or **zeitgeber** (a term adopted from German and meaning "time-giver"). In nature, the onset of darkness cues the activity of nocturnal flying squirrels—not so much directly as indirectly—by resetting the controller of the circadian rhythm. The squirrels do not wait in total ignorance each day to see when darkness will arrive. Rather, they have an endogenous sense of the time of day, and the onset of darkness simply serves as a cue that maintains a *precise* 24-hour rhythm in a system that, in itself, would maintain an *approximate* 24-hour rhythm.

Several types of environmental stimuli serve as phasing factors for circadian rhythms. Daily cycles of light intensity entrain the great majority of rhythms. In addition, rhythms can be entrained by cycles of temperature, sound, food availability, social interaction, or other parameters.

Biological clocks generate endogenous rhythms

A *biological clock* is a physiological mechanism that times an endogenous rhythm. This statement does not explain the mechanism of how the clock works; it merely presents the logical necessity of the existence of such a mechanism. Most (but not all) biological clocks are located in the animal's nervous system, as might be expected for their control functions. In the examples of the preceding section, when we speak of entraining a circadian rhythm to a light–dark cycle, it is really the biological clock that is entrained. The endogenous rhythm is the *output* of the clock, controlling effectors that allow investigators to see what the clock is doing.

Often a biological clock is localized in a discrete region of the nervous system. For example, the biological clock controlling circadian rhythms in vertebrates is located in the suprachiasmatic nucleus of the brain (Figure 10.17). In insects and molluscs, the eyes—or structures closely

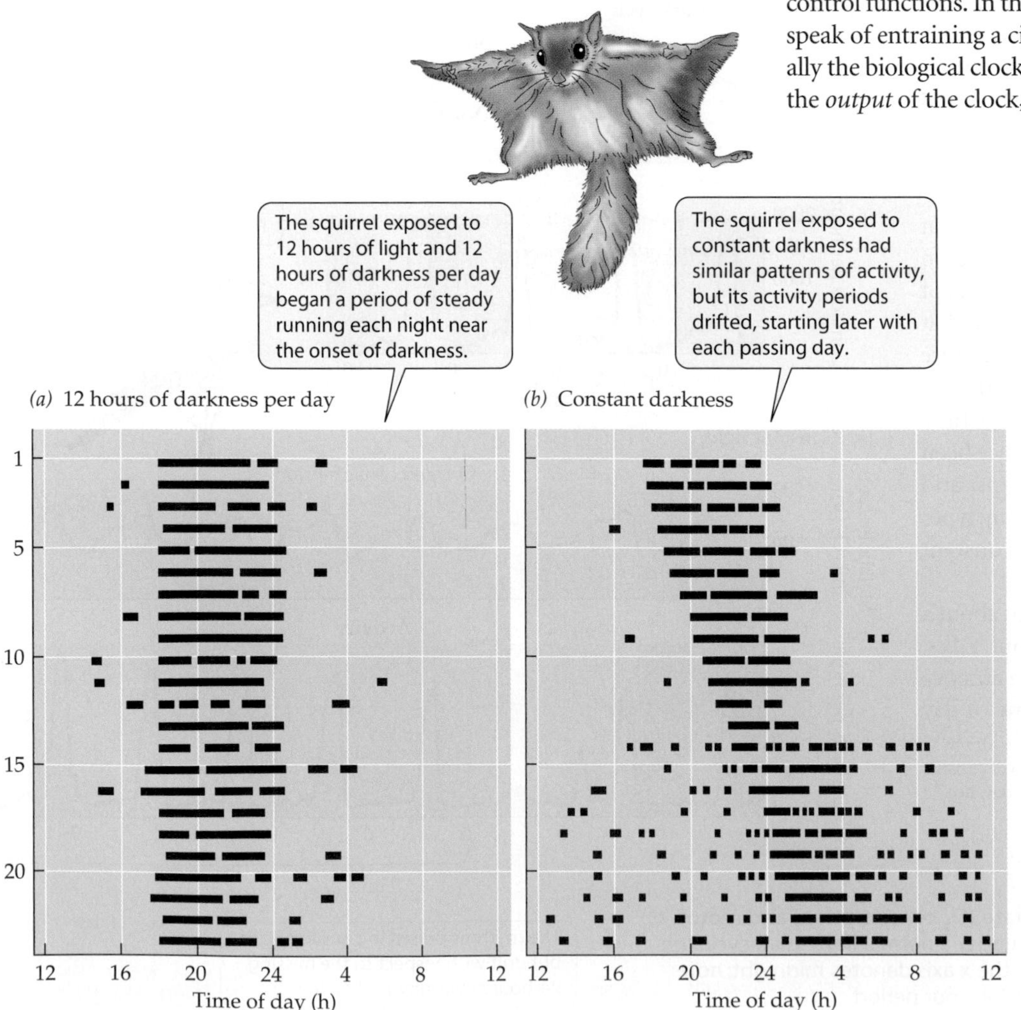

Figure 10.16 Activity rhythms of two nocturnal flying squirrels (*Glaucomys volans*) Activity was recorded on a running wheel over a period of 23 days at 20°C. Each horizontal line represents a 24-hour day. Times on the *x* axis are expressed in the 24-hour format; thus, hour 12 is noon and hour 24 is midnight. Turning of the running wheel activated a pen to record a short vertical line for each rotation of the wheel; these vertical blips are usually so close together that they are fused and give the appearance of a heavy, continuous horizontal bar during periods of steady running. The activity pattern of the squirrel in *(a)* was entrained by the environmental light–dark cycle. The free-running rhythm of the squirrel in *(b)* was about 21 minutes longer than 24 hours. (After Campbell 2002.)

(a) Location of the SCN

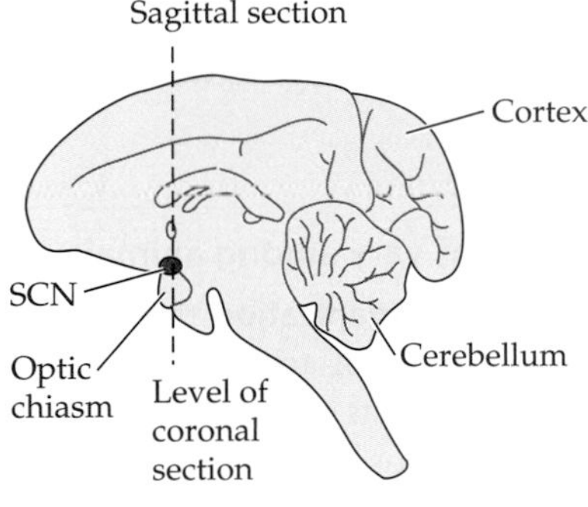

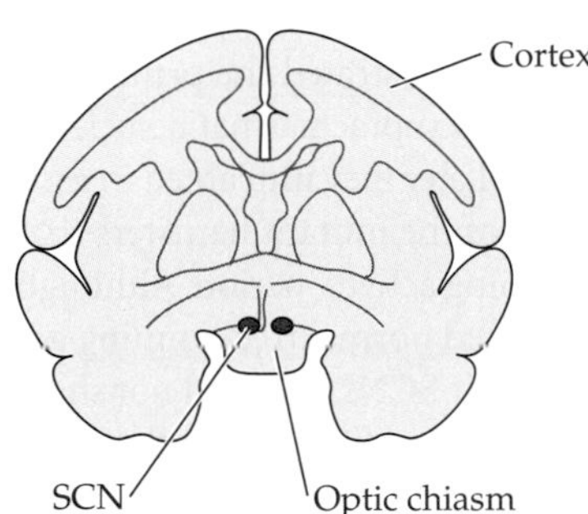

(b) Loss of free-running rhythms upon destruction of the SCN

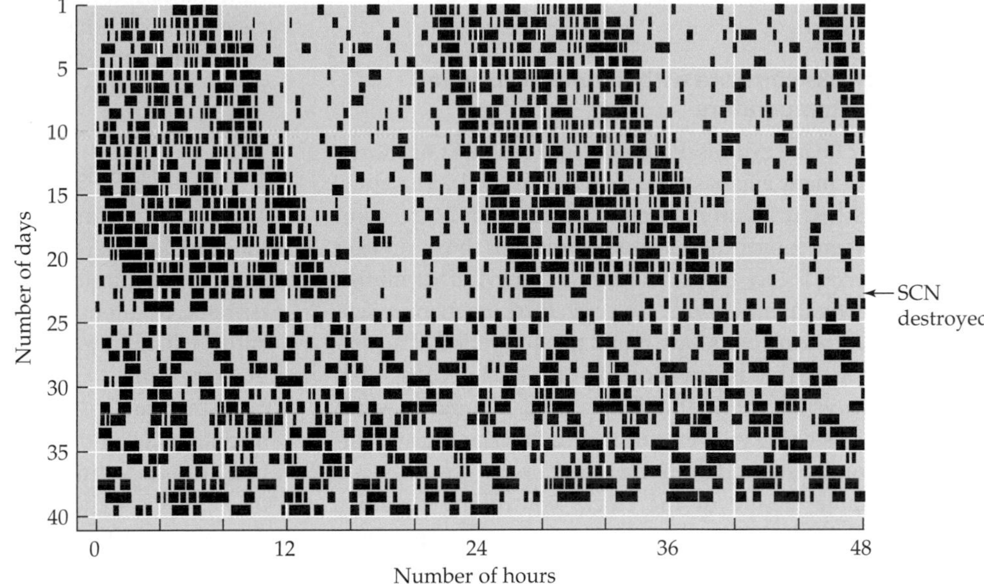

Figure 10.17 The suprachiasmatic nucleus in the hypothalamus of the brain is the major circadian clock of mammals *(a)* The location of the SCN in the ventral hypothalamus, above the optic chiasm (the crossing of the optic nerves). The sagittal section shows a side view near the midline (anterior to the right). *(b)* A free-running circadian activity is lost following destruction of a mammal's SCN.

associated with the eyes—often act as the principal circadian control centers or pacemakers. If the optic lobes of the brain are transplanted from one cockroach to another, for example, the recipient takes on the rhythms of the donor.

Control by biological clocks has adaptive advantages

The major adaptive advantage of biological clocks is that they are predictive: They enable the animal to anticipate the effects on the body of regular environmental changes. Biological clocks exert *feed-forward control* over effectors, in contrast to homeostatic feedback control (see Box 8.1). Feed-forward control permits anticipation of and preparation for upcoming events, rather than correcting for changes after they happen. An animal that is strictly dependent on external cues must wait until the cues appear before it knows to act. An animal with an internal clock, however, can anticipate when action will be necessary and can prepare, even hours in advance.

Clocks also permit timing of processes during periods of the 24-hour day when environmental cues about time are vague or unreliable. For instance, consider a nocturnal animal that emerges from its nest each evening at dusk, spends the night in feeding and other activities, and retires to its nest each dawn. If it has a circadian clock, it can entrain the clock to the daily environmental cycle defined by dawn and dusk, cues that are obvious and reliable. Then the clock can time the animal's activities accurately throughout the dark of night, when environmental sources of time information are less obvious and less reliable.

In addition to having predictive value, biological clocks may control a whole suite of activities. The same biological clock in a mammal may control circadian rhythms of oxygen consumption, body temperature, lung function, and concentrations of many hormones, as well as locomotor activity and the sleep–wake cycle (see Figure 10.14).

One way to assess the adaptive advantages of biological clocks is to disrupt their normal function. Such disruption of circadian timing of the body can have detrimental effects. The malaise known as *jet lag*, for instance, results when a person suddenly experiences a shift of many hours in the phase of the environmental cycle. This shift disrupts the normal phase relation between internal biological rhythms and the environmental cycle, and the physiological processes regulated by the clock become dis-integrated until the clock reentrains to the shifted phasing factors.

Circadian clocks also enable animals to measure or detect changes in **photoperiod**, the number of hours of daylight in a 24-hour day. Many animals depend on changes in photoperiod over the course of the year, for timing annual events in their life cycles. For example, the long photoperiods of spring may be used as a cue for reproduction or migration. Sometimes circadian clocks make animals especially light-sensitive at particular hours of the 24-hour day. They then "know" when the photoperiod in their environment is long because the interval of daylight in a day extends sufficiently to overlap their especially light-sensitive hours.

Finally, circadian clocks enable some animals to sense direction of orientation or movement. Consider the fact that, if you see the sun on the horizon, you know the direction is west only if you know the time is afternoon. Similarly, certain animals can determine compass directions from the position of the sun, but only if they know the time of day. In these animals, circadian clocks provide the time-of-day information necessary to use the sun as a compass. Homing pigeons, for example, will orient at wrong directions to the sun if their circadian clocks have been abnormally shifted (see Chapter 16, page 452).

Endogenous clocks correlate with natural history, and compensate for temperature

Free-running circadian rhythms of animals may be longer or shorter than 24 hours. For many animals the period is correlated with natural history. Whereas nocturnal animals often have periods of free-running rhythm that are longer than 24 hours (and that thus drift later and later each day, as Figure 10.16*b* illustrates), many diurnal animals have periods shorter than 24 hours. However, there are many exceptions. Moreover, experimenters have found that the period of an animal's free-running rhythm is also affected by the level of constant illumination under which the rhythm is measured. The way in which light pulses or light–dark cycles entrain a circadian rhythm also differs somewhat for diurnal and for nocturnal animals.

One remarkable feature of the clocks controlling circadian rhythms is that their timing is relatively insensitive to temperature. Although some clocks can be entrained by temperature changes, the free-running period of the clock itself does not speed up or slow down much with changes in cellular temperature. As seen in Chapter 8 (page 200–202), the rates of most metabolic processes are quite sensitive to body temperature; heart rate, breathing rate, and metabolic rate, for example, are likely to double or triple if the body temperature of an animal is raised by 10°C. In sharp contrast, the frequencies of free-running circadian rhythms typically increase by less than 5% when body temperature is elevated by 10°C.

A biological clock would obviously be of little use if it were highly sensitive to temperature; imagine the chaos if our wristwatches were to double their rate when warmed by 10°C! The low thermal sensitivity of biological clocks is therefore adaptive. Given, however, that the primary timing mechanisms of these clocks operate on a cellular level, how do the clocks manage to be so immune to the thermal effects that so strongly influence most metabolic processes? This question awaits further investigation.

Clock mechanisms involve rhythms of gene expression

How biological clocks *work* has remained one of the great mysteries of physiology until recently, when new molecular studies of genetic mutants provided revolutionary insights. Investigators have identified mutations that modify or disrupt clock function in fruit flies (*Drosophila*), in hamsters, and in several other model organisms. For example, a mutation in golden hamsters (*Mesocricetus*) causes the activity rhythm of the animals to exhibit an exceptionally short (20-hour) free-running period. By determining the biochemical consequences of such mutations, investigators have identified many of the key components of clock mechanisms.

The timekeeping mechanism in a cell is often called a **circadian oscillator**. Circadian oscillators typically depend on a rhythmic alternation between enhanced and inhibited expression of key clock genes (such as a gene named *period* in fruit flies and one named *Clock* in mice). Enhanced expression of a clock gene results in increased transcription and translation, leading to increased levels of the protein coded by the gene. This protein is then postulated to suppress, directly or indirectly, the expression of the gene by interfering with transcription factors that promote its expression. Such an oscillator can cycle back and forth between two states of gene expression in much the same way that a pendulum swings between two extremes of position.

The loci of biological clock functions vary among animals

Multicellular animals exhibit **circadian organization** throughout their bodies: Many tissues may be capable of acting as circadian clocks, but typically one region (or more) acts as a *master clock* that entrains, or imposes its rhythm on, other tissues. The entrainment ensures that arrays of tissues and organs ordinarily exhibit synchronous rhythms.

In mammals, the master circadian clock is the **suprachiasmatic nucleus (SCN)** of the hypothalamic region of the diencephalon (Figure 10.17*a*). A stunning experiment demonstrated the primacy of the SCN. Researchers destroyed the paired suprachiasmatic nuclei in a group of genetically normal hamsters. Later they implanted in each hamster paired SCNs taken from one of the mutant hamsters that exhibits an unusual, 20-hour free-running activity period. Although the genetically normal hamsters exhibited normal free-running activity rhythms before destruction of their SCNs, they did not show circadian rhythms of activity when they lacked SCNs. After the hamsters received replacement SCNs, they once again exhibited circadian rhythms, but the free-running period of the rhythms was the unusual, short period characteristic of the mutant donor hamsters.

In a mammal, most circadian rhythms are obliterated if the SCN is inactivated (Figure 10.17*b*). Thus the SCN appears to be the principal seat of rhythmic control for rhythms in the body. Studies of SCNs removed from the body show that many of the individual neurons within the SCN are independently rhythmic. Communication between the SCN and the rest of the body remains inadequately understood, although it is clear that the SCN communicates both by nerve impulses and by the secretion of hormones or other diffusible substances. Neural connections from the eyes in mammals provide information to the SCN about the daily light–dark cycle in the environment.

The SCN is not the only anatomical location of circadian control in mammals, although it is the principal control center and the best understood. The retinas of the eyes are also endogenously rhythmic; however, the retinas do not seem to exert substantial direct control over other tissues. Sometimes certain circadian rhythms besides the retinal rhythms persist in mammals after SCN inactivation, pointing to additional clocks.

One important output of the SCN clock controls the pineal gland. The **pineal gland** is a small, unpaired gland that forms embryologically as an evagination of the roof of the brain and is found in virtually all vertebrates. Its principal hormonal secretion is **melatonin**, a compound synthesized from the amino acid tryptophan. In mammals, according to current evidence, the pineal is not independently rhythmic. Nevertheless, it secretes melatonin in a circadian rhythm because of circadian control from the SCN. Pineal melatonin is secreted at night in mammals and virtually all other vertebrates. Thus, melatonin is sometimes called the *darkness hormone*. Pineal melatonin is of great importance in controlling many seasonal rhythms, such as reproduction.

The pineal physiology of nonmammalian vertebrates often differs from that of mammals in two important ways. First, the nonmammalian pineal gland is often endogenously rhythmic and thus can act as a primary circadian control center. Second, the pineal is often light-sensitive and acts as a "third eye," providing extraocular information on the environmental day–night cycle (light may reach the pineal through the skull). Interaction between the SCN and the pineal in the control of circadian rhythms in nonmammalian vertebrates is complex, diverse, and not well understood. The retinas of the eyes in nonmammalian vertebrates are endogenously rhythmic, but as in mammals, the retinas seem not to serve as circadian control centers for the rest of the body.

Circannual and circatidal clocks: Some endogenous clocks time annual or tidal rhythms

Daily rhythms have been the most studied of all biological rhythms, and so we have emphasized them in our discussion of biological clocks. However, animals exhibit rhythmic physiological and behavioral variations that operate on other timescales as well. Annual rhythms of reproduction, migration, fat accumulation, dormancy, and so on are well-known examples. Animals living along the seashore often display rhythms synchronized with the tides, which usually rise and fall every 12.4 hours (half a lunar day). For instance, fiddler crabs that scavenge for food on the sand or mud exposed by low tide become rhythmically more active at each time of low tide.

Some, but not all, annual and tidal rhythms are endogenous: They persist even when animals are placed in a laboratory environment where they are denied environmental information about the time of year or the time of the tidal cycle. Typically the periods of the free-running endogenous rhythms are only *approximately* a year or a tidal cycle in length. Thus the endogenous rhythms are termed **circannual** or **circatidal** (*circa*, "about").

Under natural conditions, of course, certain environmental parameters vary in phase with the annual or tidal cycle. The endogenous circannual and circatidal rhythms of animals become entrained, so in nature the biological rhythms are kept in phase with the actual seasons and tides. For instance, the annual cycle of photoperiod length (long days in summer, short days in winter) is the phasing factor for certain circannual rhythms; and features of ebbing and flowing tidal water, such as mechanical agitation, serve as phasing factors for some of the circatidal rhythms. It is not clear whether the endogenous timing mechanisms for circannual and circatidal rhythms depend on circadian oscillators. Arguments have been presented on both sides. As yet, the nature of these timing mechanisms remains unresolved.

Interval or "hourglass" timers can time shorter intervals

In addition to circadian oscillators—which rhythmically cycle—animals appear to possess physiological timing mechanisms that permit timing of *parts* of days by functioning like stopwatches or hourglasses. These noncyclic timers are called **interval timers** or **hourglass timers**. Once activated on a given day, they measure the passage of time; but like stopwatches, they are noncyclic and must be restarted to operate again. Male pigeons, for example, seem to use an interval timer to determine how long they incubate eggs in a particular stint; they stay on the eggs for a relatively fixed length of time after they start, regardless of the time of day when they start. Recent research locates the neurophysiological sites of the interval timers of birds and mammals in different parts of the brain from the circadian oscillators.

Summary: Biological Clocks

- A circadian rhythm has a period of about a day. It is an example of an endogenous rhythm, one that does not require sensory information for timing.
- A circadian rhythm of an animal will drift or free-run in constant darkness, when there are no sensory timing cues. A light–dark cycle entrains the circadian rhythm to exactly 24 hours.
- A biological clock is the physiological basis of an animal's ability to time an endogenous rhythm. Biological clocks exert rhythmically changing control, modulating the outputs of the nervous and endocrine systems to prepare an animal for daily changes and seasonal changes. In mammals, the suprachiasmatic nucleus (SCN) of the brain is the principal biological clock for circadian rhythms.
- Animals may possess other timing functions that time periods much shorter (such as circatidal) or longer (such as circannual) than those of circadian rhythms.

Study Questions

1. How does the physiological control exerted by the nervous system and endocrine system relate to the concept of homeostasis?
2. Contrast neural control and endocrine control of effectors. What are their advantages and disadvantages? For what sorts of control functions are they best suited?
3. How, in general, is the vertebrate autonomic nervous system organized? Is it redundant to have separate sympathetic and parasympathetic control of many organs?
4. The sympathetic and parasympathetic divisions of the autonomic nervous system employ the same neurotransmitter (acetylcholine) for preganglionic neurons, but different neurotransmitters for postganglionic neurons (norepinephrine for sympathetic, and acetylcholine for parasympathetic) (see Figure 10.12). Would things work if the situation were reversed—that is, if the preganglionic neurotransmitters were different and the postganglionic neurotransmitters were the same? Why or why not?
5. What adaptive advantages might centralization and cephalization offer in the evolution of nervous system organization?
6. Mammals have more-cephalized brains than fish and amphibians, particularly in terms of expansions of the cerebral cortex. Does this increased cephalization make mammals more advanced and fish and amphibians more primitive? Defend and/or criticize this viewpoint.
7. The paired suprachiasmatic nuclei (SCNs) usually function as the master circadian clock in mammals. In some circumstances the circadian rhythms of animals may become split, with some effectors following one free-running rhythm and some following another. Give two hypotheses of how such a split might happen—one compatible with an SCN always being a master clock and one not.
8. One of the first genes determined to control circadian clock timing is the *per* gene in *Drosophila*. Mutants of this gene have shorter or longer circadian free-running rhythms, as well as shorter or longer

periods of a much faster rhythmic courtship song. What does this observation suggest about the relationship between circadian rhythms and shorter, hourglass-timing rhythms?

9. For a mammal such as the nocturnal flying squirrel shown in Figure 10.16, short pulses of light can shift the phase of a circadian rhythm that is free-running in darkness. A light pulse phase-shifts the rhythm differently if it comes at different phases of the free-running rhythm: Generally light pulses near the beginning of subjective night cause phase delays (negative numbers), and light pulses near the end of subjective night cause phase advances (positive numbers). A phase-response diagram quantifies this effect. Review the phase-response curve below and describe the results that it shows. Describe also how light pulses at exactly 24-hour intervals would entrain the animal's circadian rhythm.

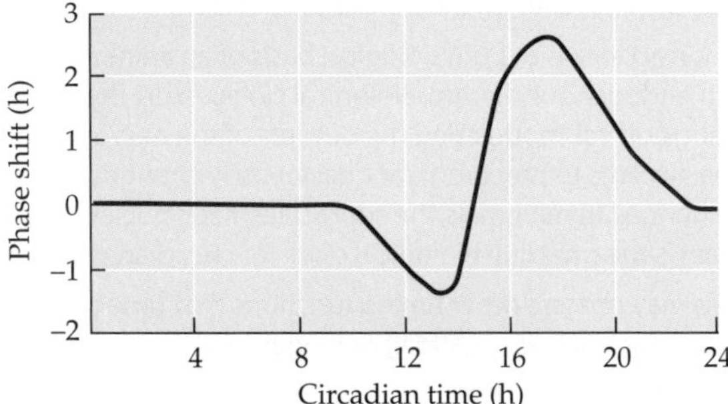

Key References

Camhi, J. 1980. The escape system of the cockroach. *Sci. Amer.* 243(6): 158–172. Readable analysis of the cockroach neural circuit described at the beginning of this chapter.

Gershon, M. D. 1998. *The Second Brain.* HarperCollins, New York. An engaging popular book on the enteric nervous system and control of the gut. Good treatment of paths of discovery of neurotransmitters and autonomic control.

Nauta, W. J. H., and M. Feirtag. 1979. The organization of the brain. *Sci. Amer.* 241(3): 88–111. An unusually straightforward and engaging introduction to the anatomy of vertebrate brains.

Selected References

Arbas, E. A., R. B. Levine, and N. J. Strausfield. 1997. Invertebrate nervous systems. In W. H. Dantzler (ed.), *Comparative Physiology*, vol. 2 (Handbook of Physiology [Bethesda, MD], section 13), pp. 751–852. Oxford University Press, New York.

Arbas, E. A., I. A. Meinertzhagen, and S. R. Shaw. 1991. Evolution in nervous systems. *Annu. Rev. Neurosci.* 14: 9–38.

Arendt, J. 1995. *Melatonin and the Mammalian Pineal Gland.* Chapman & Hall, New York.

Barinaga, M. 2002. How the brain's clock gets daily enlightenment. *Science* 295: 955–957.

Bern, H. A. 1990. The "new" endocrinology: Its scope and impact. *Amer. Zool.* 30: 877–885.

Brady, J. (ed.). 1982. *Biological Timekeeping.* Cambridge University Press, New York.

Darlington, T. K., K. Wager-Smith, M. F. Ceriani, D. Staknis, N. Gekakis, T. D. L. Steeves, C. J. Weitz, J. S. Takahashi, and S. A. Kay. 1998. Closing the circadian loop: CLOCK-induced transcription of its own inhibitors *per* and *tim*. *Science* 280: 1599–1603.

Donald, J. A. 1998. Autonomic nervous system. In Evans, D. H. (ed.), *The Physiology of Fishes*, 2nd ed., pp. 407–439. CRC Press, Boca Raton, FL.

Dunlap, J. C., J. J. Loros, and P. J. DeCoursey (eds.) 2004. *Chronobiology: Biological Timekeeping.* 407–439. Sinauer, Sunderland, MA.

Farner, D. S. 1985. Annual rhythms. *Annu. Rev. Physiol.* 47: 65–82.

Gekakis, N., D. Staknis, H. B. Nguyen, F. C. Davis, L. D. Wilsbacher, D. P. King, J. S. Takahashi, and C. J. Weitz. 1998. Role of the CLOCK protein in the mammalian circadian mechanism. *Science* 280: 1564–1569.

Hastings, J. W., B. Rusak, and Z. Boulos. 1991. Circadian rhythms: The physiology of biological timing. In C. L. Prosser (ed.), *Neural and Integrative Animal Physiology* (Comparative Animal Physiology, 4th ed.), pp. 435–546. Wiley-Liss, New York.

Jansen, A. S. P., X. V. Nguyen, V. Karpitskiy, T. C. Mettenleiner, and A. D. Loewy. 1995. Central command neurons of the sympathetic nervous system: Basis of the fight-or-flight response. *Science* 270: 644–646.

Menaker, M. 2003. Circadian photoreception. *Science* 299: 213–214.

Moore, R. Y. 1995. Organization of the mammalian circadian system. In D. J. Chadwick and K. Ackrill (eds.), *Circadian Clocks and Their Adjustment* (Ciba Foundation Symposium, 183), pp. 88–106. Wiley, Chichester, UK.

Moore-Ede, M. C., F. M. Sulzman, and C. A. Fuller. 1982. *The Clocks That Time Us.* Harvard University Press, Cambridge, MA.

Nilsson, S. 1983. *Autonomic Nerve Function in the Vertebrates.* Springer, New York.

Nilsson, S., and S. Holmgren (eds.). 1994. *Comparative Physiology and Evolution of the Autonomic Nervous System.* Harwood, Langhorne, PA.

Northrop, R. B. 2000. *Endogenous and Exogenous Regulation and Control of Physiological Systems.* Chapman & Hall/CRC, Boca Raton, FL.

Palmer, J. D. 1995. *The Biological Rhythms and Clocks of Intertidal Animals.* Oxford University Press, New York.

Pittendrigh, C. 1961. On temporal organization in living systems. *Harvey Lect.* 56: 93–125.

Pittendrigh, C. 1993. Temporal organization: Reflections of a Darwinian clock-watcher. *Annu. Rev. Physiol.* 55: 16–54.

Plautz, J. D., M. Kaneko, J. C. Hall, and S. A. Kay. 1997. Independent photoreceptive circadian clocks throughout *Drosophila. Science* 278: 1632–1635.

Ralph, M. R., and M. W. Hurd 1995. Circadian pacemakers in vertebrates. In D. J. Chadwick and K. Ackrill (eds.), *Circadian Clocks and Their Adjustment* (Ciba Foundation Symposium, 183), pp. 67–87. Wiley, Chichester, UK.

Ralph, M. R., R. G. Foster, F. C. Davis, and M. Menaker. 1990. Transplanted suprachiasmatic nucleus determines circadian period. *Science* 247: 975–978.

Reppert, S. M., and D. R. Weaver 2001. Molecular analysis of mammalian circadian rhythms. *Annu. Rev. Physiol.* 63: 647–676.

Ritzmann, R. E. 1993. The neural organization of cockroach escape and its role in context-dependent orientation. In R. D. Beer, R. E. Ritzmann, and T. McKenna (eds.), *Biological Neural Networks in Invertebrate Neuroethology and Robotics*, pp. 113–138. Academic Press, Boston.

Swanson, L. W. 2003. *Brain Architecture: Understanding the Basic Plan.* Oxford University Press, New York.

Takahashi, J. S., F. W. Turek, and R. Y. Moore (eds.). 2001. *Circadian Clocks.* Handbook of Behavioral Neurobiology, vol. 12. Kluwer Academic/Plenum, New York.

Toates, F. M. 1975. *Control Theory in Biology and Experimental Psychology.* Hutchinson Educational, London.

Turek, F. W., and O. Van Reeth 1996. Circadian rhythms. In M. J. Fregley and C. M. Blatteis (eds.), *Environmental Physiology*, vol. 2 (Handbook of Physiology [Bethesda, MD], section 4), pp. 1329–1359. Oxford University Press, New York.

Ulinski, P. S. 1997. Vertebrate nervous system. In W. H. Dantzler (ed.), *Comparative Physiology*, vol. 1 (Handbook of Physiology [Bethesda, MD], section 13), pp. 17–53. Oxford University Press, New York.

Underwood, H. A., G. T. Wassmer, and T. L. Page. 1997. Daily and seasonal rhythms. In W. H. Dantzler (ed.), *Comparative Physiology*, vol. 2 (Handbook of Physiology [Bethesda, MD], section 13), pp. 1653–1763. Oxford University Press, New York.

Yates, F. E. 1982. Outline of a physical theory of physiological systems. *Can. J. Physiol. Pharmacol.* 60: 217–248.

11 Neurons

Just as physiology underlies animal behavior, the behavior of animals serves important functions in their physiology. Feeding, escape, reproduction, and other behaviors are controlled by the fast, addressed signaling mechanisms of the nervous system. These mechanisms are a property of neurons, whose long cablelike axons convey electrical signals rapidly and faithfully between distant parts of the body. The neuronal activities underlying behavior have been studied extensively in squids. Locomotion in a squid, whether for capturing a meal or to avoid becoming one, depends on jet propulsion; the contraction of muscles in the squid's outer mantle expels seawater through a moveable siphon, propelling the animal in the opposite direction. Sensory neurons such as those in the eyes encode information about the squid's environment and convey signals to the brain, where they are integrated into a decision to attack or retreat. The brain then sends commands to the mantle muscles, in part through a set of giant axons.

As you will discover in this chapter, squid giant axons have played an important role in our understanding of neuronal functions. The diameter of these giant axons can be as large as 1 mm (1000 μm), and for over half a century investigators have taken advantage of this prodigious cellular size to perform noteworthy experiments that have revealed the mechanisms of neuronal signaling. Sir Alan Hodgkin (1914–1998), who received the Nobel Prize in 1963 for his work on squid axons, recalled that a colleague had remarked (not, he thought, with the greatest tact) that it was the squid that really ought to be awarded the prize!

This chapter describes the electrical basis of neuronal function—the ability of neurons to generate electrical signals and propagate them over relatively large distances. The cellular mechanisms of neuronal signaling are similar in all animals, whether we examine neurons of squid, cockroaches, jellyfish, or humans.

Squid axons are important to physiologists, and to the squid

The Cellular Organization of Nervous Systems

Nervous systems are composed primarily of neural tissue, which in turn is composed of discrete cells: neurons (nerve cells) and glial cells, as well as connective tissue cells and cells of the circulatory system. The cellular organization of nervous systems is a corollary of the **cell theory**, which states that organisms are composed of cells, that these cells are the structural and functional units of organization of the organism, and that all cells come from preexisting cells as a result of cell division. Matthias Schleiden (1804–1881) and Theodor Schwann (1810–1882) formulated the cell theory in 1839.

The cell theory gained widespread and rather rapid acceptance—except as applied to nervous systems. Instead, the dominant view of the organization of nervous systems in the latter half of the nineteenth century was the reticular theory, most strongly argued by J. Gerlach (1820–1896) and Camillo Golgi (1843–1926). The **reticular theory** held that nervous systems were composed of complex, continuous meshworks of cells and processes in protoplasmic continuity with each other (i.e., the cells ran together without any boundaries).

The reticular theory was supplanted only gradually, over the first third of the twentieth century, by an outgrowth of the cell theory known as the **neuron doctrine**, which states that neurons are anatomically distinct and are the structural, functional, and developmental units of organization of nervous systems. Santiago Ramón y Cajal (1852–1934), the main champion of the neuron doctrine, used special staining techniques to demonstrate convincingly that neurons are contiguous (in contact with each other) but are not continuous (connected without interruption). The debate on contiguity versus continuity persisted until the 1950s, when electron microscopy permitted resolution of cell membranes and rigorously demonstrated the discontinuity of neurons in contact.

Neurons are structurally adapted to transmit action potentials

It has been said that a neuron is a cell that is trying to be a telephone cable. This statement combines absurdity with the germ of a useful concept. **Neurons** are specialized for generating electrical impulses and transmitting these impulses from place to place within the body, sometimes over considerable distances. A neuron consists of a **cell body**, or **soma** (plural *somata*) (also called the *perikaryon*), which is the region that contains the nucleus, and one or more *processes* arising from it (Figure 11.1).

The cytology of a neuronal soma is broadly similar to that of nonneuronal cells. It contains a nucleus and most of the organelles and cytoskeletal elements familiar to cytologists: mitochondria, Golgi apparatus, smooth endoplasmic reticulum (ER), rough ER, microtubules, neurofilaments, and actin microfilaments. Neurons are very active in protein synthesis and thus have extensive, well-de-

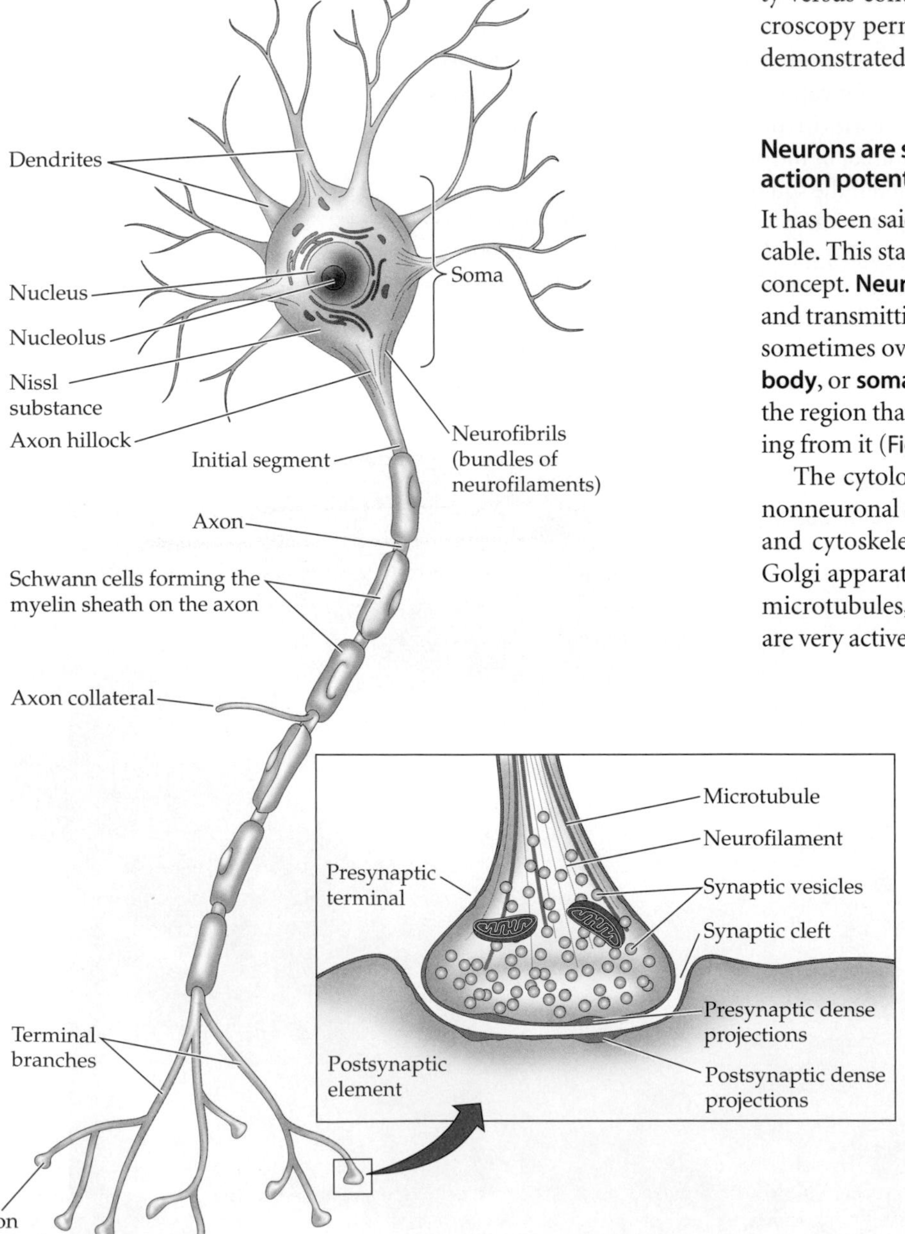

Figure 11.1 The structure of neurons Every neuron has a cell body (soma or perikaryon) and processes usually classified as axons and dendrites. The inset shows the structure of the very end of the axon, the axon terminal. The soma contains organelles, including rough endoplasmic reticulum (ER), Golgi apparatus, and mitochondria. Stained aggregates of rough ER appear in light microscopy as Nissl substance. Cytoskeletal elements—microtubules and neurofilaments (see the inset)—are present in the soma, dendrites, and axon. The axon of this neuron is myelinated, with periodic thickenings of myelin insulation around its axons. (The importance of the myelin sheath to the conduction of nerve impulses is discussed later in this chapter; see page 307.) The axon ends in terminals, where synaptic vesicles (see the inset) store molecules of neurotransmitter for synaptic transmission.

veloped rough ER, aggregates of which can be stained to appear in light microscopy as *Nissl substance.*

Neurons can be classified according to the number of processes emanating from the soma. Neurons may be unipolar (having one process), bipolar (two processes), or multipolar (three or more processes). Despite this seemingly simple classification scheme, neuronal processes exhibit a bewildering geometric variety and complexity. Early anatomists attempted to bring order to this variety by classifying processes as *axons* and *dendrites.* Their classifications were usually based on vertebrate spinal motor neurons (see Figure 11.1) and are useful for cells resembling spinal motor neurons in form. Definitions of dendrites and axons, however, are based on a mixture of functional and morphological criteria that do not always coincide in a single neuron. Functionally, a **dendrite** is considered to be a *receptive element* of a neuron that conveys information toward the soma (see Figure 10.1). An **axon**, on the other hand, is the *output element* of a neuron, carrying information away from the cell body to other cells. This functional classification applies to most, but not all, neurons.

The dendrites of spinal motor neurons are relatively short and branch repeatedly (*dendrite* is Greek for "branch"). Dendrites of most neurons have continuously varying diameters and lack myelin sheaths (which will be discussed shortly). In general, the broader dendritic trunks resemble the soma in fine structure; they contain rough ER, mitochondria, microtubules, neurofilaments, and an occasional Golgi apparatus (see Figure 11.1). Thinner dendritic branches may lack Golgi apparatus and rough ER. The dendrites of many vertebrate neurons bear numerous short, thin protrusions termed *dendritic spines* that, when present, are important sites of synaptic input.

The axon of a neuron is classically single and long, with a relatively constant diameter and few collateral branches. The larger vertebrate axons are surrounded by **myelin sheaths**—multiple wrappings of insulating glial cell membranes (see the next section) that increase the speed of impulse transmission. Not all axons are myelinated; the smaller axons of vertebrate neurons and nearly all invertebrate axons lack myelin and are termed *unmyelinated.* At the fine structural level, axons contain microtubules, neurofilaments, elongated mitochondria, and sparse smooth endoplasmic reticulum (see Figure 11.1). Axons generally lack rough ER and Golgi apparatuses. Functionally, the axon is usually the portion of the neuron that supports action potentials, which *propagate* or conduct along the axon without decrement, carrying information away from the cell body to the axon terminals.

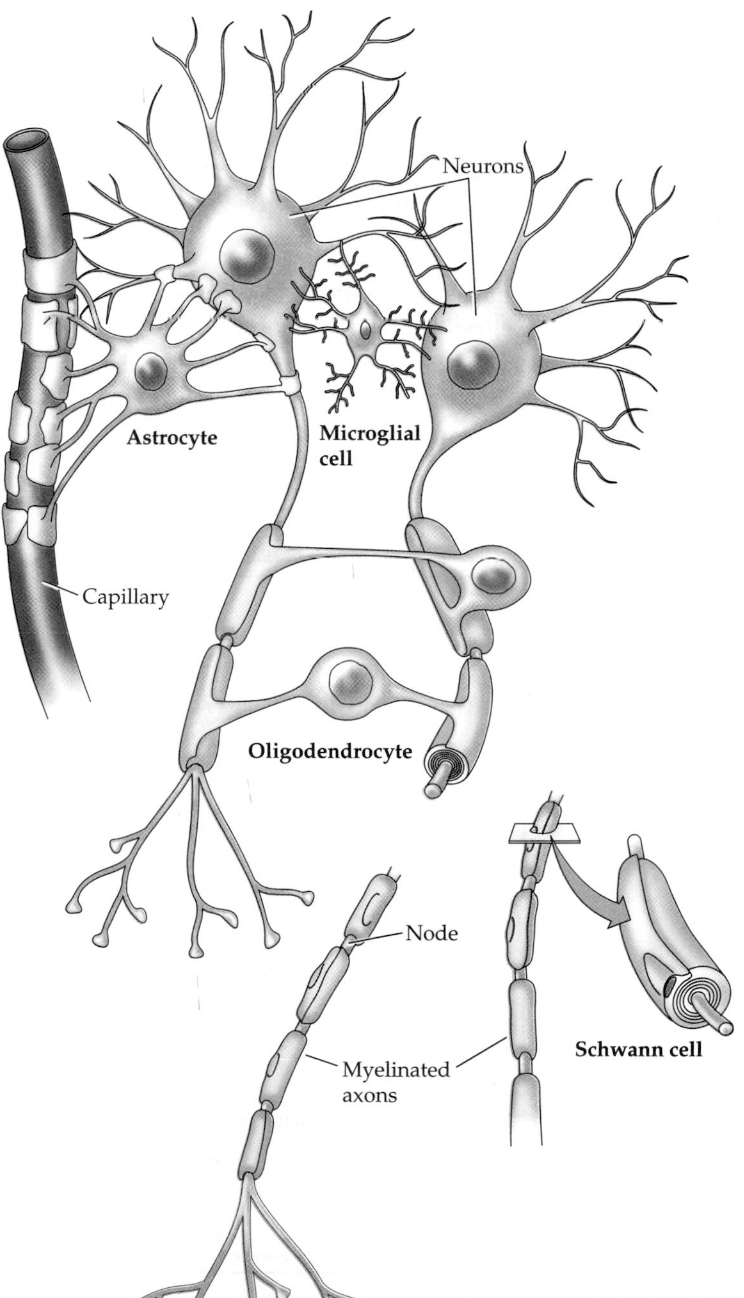

Glial cells support neurons physically and metabolically

Cells that are referred to collectively as **glia** or **neuroglia** ("nerve glue") surround the neurons (Figure 11.2). Rudolf Virchow (1821–1902) discovered and named the neuroglial cells in 1846 and thought that their primary function was to bind the neurons together and maintain the form and structural organization of the nervous system. The ratio of glia to neurons increases with increasing evolutionary complexity, from brains of fish to mammals. Glial cells are estimated to make up half the volume of the mammalian brain and to outnumber neurons by ten to one.

Figure 11.2 Glial cells There are four types of glial cells. Schwann cells ensheathe axons (myelinated and unmyelinated) in the peripheral nervous system. Oligodendrocytes ensheathe axons in the CNS. Astrocytes are metabolic support cells in the vertebrate central nervous system. Microglial cells are phagocytes related to cells of the immune system.

These measures suggest that glial cells are important in nervous system function, perhaps in ways that are not yet fully understood.

Different types of glial cells play diverse functional roles in nervous systems. In vertebrates, ensheathing glial cells called **Schwann cells** (in the peripheral nervous system, or PNS) and **oligodendrocytes** (in the central nervous system, or CNS) envelop the axons of neurons (see Figure 11.2). The glial sheath can be a simple encircling of an *unmyelinated* axon or a group of axons, or a *myelin sheath* consisting of multiple concentrically wrapped layers of glial membrane that insulate the axon and increase the velocity of nerve-impulse propagation. Other glial cells called **astrocytes** line the outside surfaces of capillaries in the vertebrate CNS and act as metabolic intermediaries between the capillaries and neurons. Astrocytes take up neurotransmitters from extracellular space, and help supply metabolic substrates to neurons. They also regulate extracellular ion concentrations and play important roles in nervous system development. *Microglial cells* mediate immune responses in neural tissue and may act as phagocytes, consuming pathogens and cell debris in brain injury.

> **Summary: The Cellular Organization of Nervous Systems**
>
> - Neurons are the principal cells of nervous systems. They have long processes (dendrites and axons) that are specialized to receive signals from other neurons (dendrites) and to generate and propagate action potentials (axons).
> - Glial cells are the support cells of the nervous system. Schwann cells (in the PNS) and oligodendrocytes (in the CNS) form sheaths around neuronal axons, including insulating myelin sheaths around myelinated axons. Astrocytes surround capillaries and act as metabolic intermediaries between neurons and their circulatory supply. Microglial cells serve immune and scavenging functions.

The Ionic Basis of Membrane Potentials

What are the properties of the electrical signals of neurons, and how are these signals generated? Let's begin with a brief review of basic electrical concepts. Protons and electrons have *electrical charge*, and **ions** are atoms or molecules that bear a net charge because they have unequal numbers of protons and electrons. The net movement of charges constitutes an electric **current** (I), which is analogous to the hydraulic current of fluids flowing in a system of pipes. The separation of positive and negative electrical charges constitutes a **voltage**, or electrical *potential difference* (V). This potential difference can do work when charges are allowed to flow as current. Voltage is analogous to a height difference or head of pressure in a hydraulic system, allowing water to flow downhill.

Figure 11.3 shows a simple electric circuit, that of a flashlight. A battery provides voltage; closing the switch allows current to flow through the electrical circuit. The electric current in the flashlight is the flow of free electrons along metal wires. Current flows through the lightbulb filament, which acts as **resistance** (R) that limits the current flow. Consequently, the filament heats and glows, emitting light.

Electrical circuits in cells are similar to the circuit in a flashlight, but they differ in some important ways. In cells, both the inside and outside media are *aqueous* solutions, in which the electrical charges are ions rather than free electrons. Furthermore, all currents in cells are carried by ions, and any voltage or potential difference results from local imbalances of ion charges. Recall from Chapter 3 (see Figure 3.4) that fluids farther than a few nanometers from a membrane are electrically neutral, with equal numbers of positive and negative charges.

Because of this *charge neutrality of bulk solutions*, the only portion of a cell that *directly* determines its electrical properties is its outer-limiting cell membrane. Any electrical activity of a nerve cell is a property of the cell membrane, and the electrical potentials observed are called *transmembrane* potentials. The only immediately important attribute of the rest of the cell is the concentration of ions in solution in the intracellular fluid.

Cell membranes have passive electrical properties: Resistance and capacitance

All cells respond to electric currents, but not all cells generate action potentials (nerve impulses). The universal responses are *passive* re-

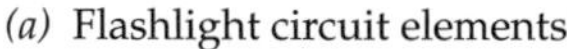

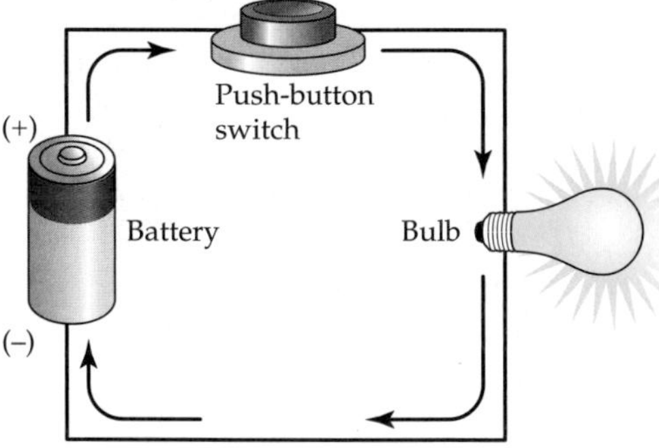

(*b*) Circuit diagram for the flashlight

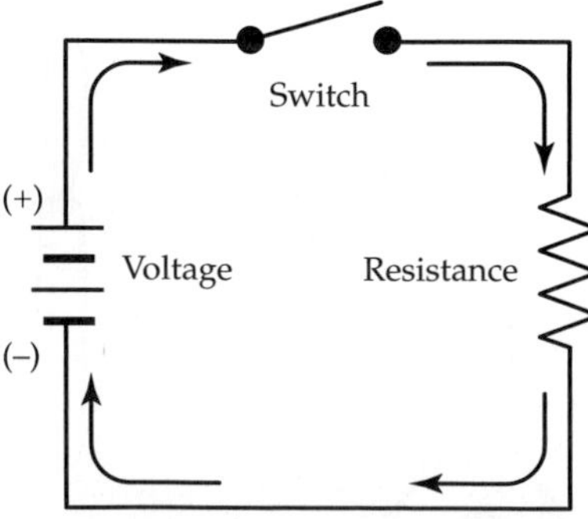

Figure 11.3 The simple electrical circuit of a flashlight (*a*) Elements of the electrical circuit. (*b*) Circuit diagram. When the switch is closed, electric current flows through the resistance of the lightbulb filament, causing it to glow. Current must flow through the complete circuit. (Note that, by convention, current always flows from plus to minus.)

sponses (meaning that the cell's electrical properties do not change), but action potentials are active responses in which the properties do change. A cell's passive responses depend on the **passive electrical properties** of the membrane, principally its resistance and capacitance. A membrane exhibits *resistance* (measured in ohms, Ω) resulting from the fact that ions must flow through restrictive ion channels because the membrane's lipid bilayer is impermeable to ions. A membrane exhibits **capacitance** (measured in farads, F) because of the insulating properties of the bilayer. A capacitor has two conducting plates separated by an insulating layer; the conducting fluids on either side of the membrane act as plates, and the lipid bilayer separates and stores oppositely charged ions.

The resistance and capacitance of a cell's membrane depend on membrane area; specific membrane resistance and capacitance are measured per unit of area (e.g., $R_m = 100\ \Omega\cdot\text{cm}^2$; $C_m = 1\ \mu\text{F/cm}^2$).[1] Whereas the specific membrane capacitance does not change, resistance may or may not change (depending on the behavior of specific populations of ion channels). *When we speak of a cell's passive electrical properties, we mean those conditions in which membrane resistance does not change.* A cell's passive electrical properties govern how voltages change over space and time along neuronal axons. Passive electrical properties do not explain the generation of action potentials (in which resistances change), but they are important for understanding how neurons generate and propagate action potentials.

We can use a squid giant axon to demonstrate the passive electrical properties of cells. The largest axons of a common squid may be 2 cm long and 700 to 1000 μm in diameter. Because these axons are so large, it is relatively easy to cut out a length of the axon, ligate the ends, and penetrate the isolated axon with a microelectrode (Figure 11.4). The microelectrode consists of a glass capillary that has been heated and pulled to a fine tip (<1 μm in diameter) so that it can penetrate the cell membrane without causing damage. The capillary is filled with a solution of strong electrolyte, such as 3 *M* KCl, to minimize its electrical resistance.

When the tip of the microelectrode is outside the axon (Figure 11.4*a*), a voltmeter records no potential difference (voltage) between the recording microelectrode and a neutral (reference) electrode suspended in the surrounding saline bath. Both electrodes are electrically neutral because of the charge neutrality of bulk solutions.

When the recording microelectrode is advanced just past the axon membrane into the cytoplasm (Figure 11.4*b*), the voltmeter records a potential difference (Figure 11.4*c*). This potential difference across the axon membrane is the **resting membrane potential** (V_m). For all known cells, the polarity of the resting membrane potential is *inside-negative*; that is, the inner membrane surface is negative with respect to the outer membrane surface.[2] This example also reveals the axon's **membrane resistance** (R_m) to current flow, because without such resistance, ions would freely diffuse across the membrane, and it would not be able to maintain a potential difference.

[1] The resistance of a whole cell's membrane is usually called its *input resistance*.

[2] By convention, the membrane potential is given as the inside value, with the outside considered zero.

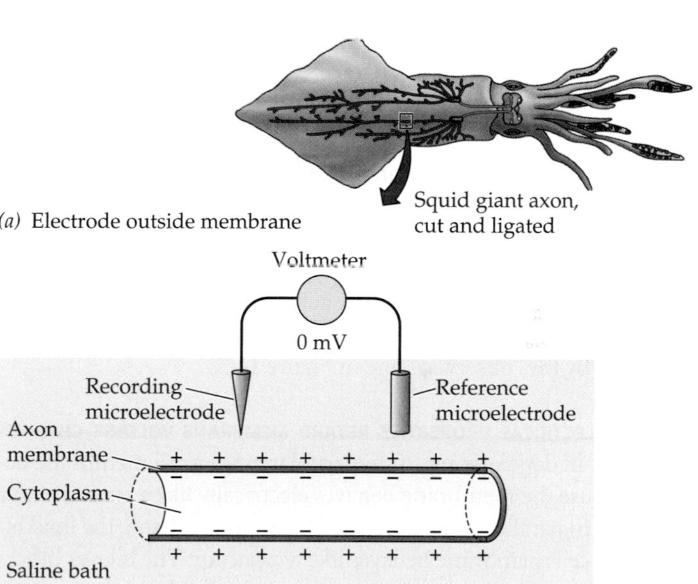

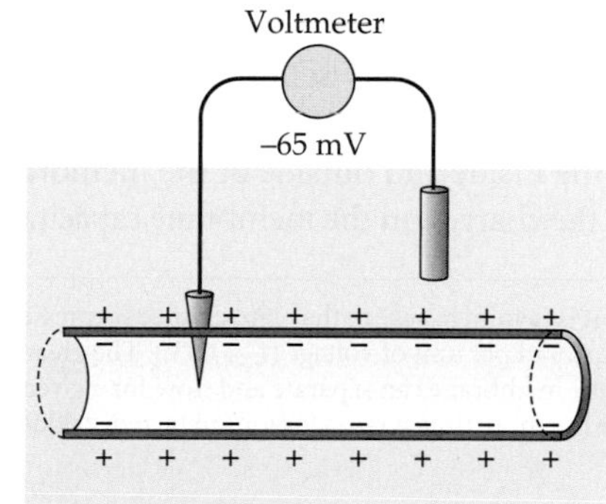

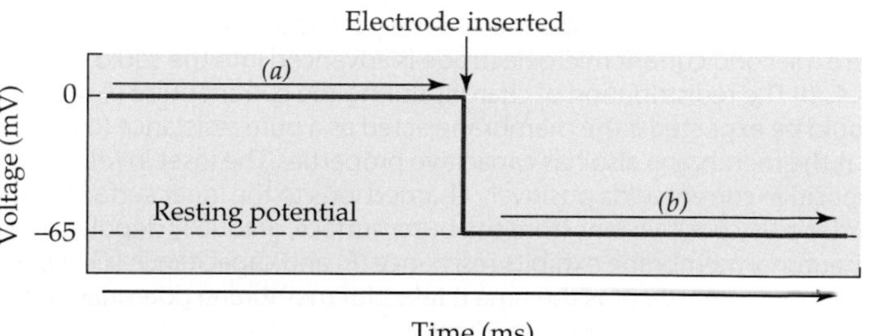

Figure 11.4 Recording the resting membrane potential of a squid giant axon *(a)* A voltmeter measures the potential difference between a glass capillary microelectrode (the recording electrode) and a reference electrode in the saline bath around the axon. When the microelectrode is outside the axon, there is no potential difference between the two electrodes. *(b)* The recording microelectrode has been advanced through the axon membrane, and the resting membrane potential (V_m) is recorded. *(c)* The output of the voltmeter, recorded on a chart writer or an oscilloscope, demonstrates that the resting membrane potential is inside-negative, a condition true for all cells. (By convention, negative is down for intracellular recording.)

What would happen to the resting membrane potential if we inserted a second microelectrode into the squid axon and generated a pulse of electric current between it and an extracellular electrode (Figure 11.5)? The current pulse would either depolarize or hyperpolarize the membrane, depending on the direction of the current. **Depolarization** is a *decrease* in the absolute value of the membrane potential toward zero (becoming less negative inside the cell). **Hyperpolarization** is an *increase* in the absolute value of the membrane potential away from zero (becoming more negative inside the cell).

In the example of Figure 11.5, we apply a current that flows outward across the membrane to cause a depolarization of the membrane potential, perhaps from –65 to –60 mV. (Remember that we are not considering action potentials yet, so assume that this depolarization is too small to trigger an action potential.) According to Ohm's law, the current should change the membrane potential by an amount proportional to the resistance to current flow:

$$\Delta V = IR \tag{11.1}$$

where ΔV is the change in potential (termed a **graded potential**), I is the current (in amperes), and R is the resistance (in ohms). If the membrane exhibited only resistance, the change in membrane potential would occur instantaneously, as shown by the "theoretical" line in Figure 11.5*b*. However, the actual change in membrane potential occurs more gradually, reaching a plateau after a short delay, as shown by the "observed" line in Figure 11.5*b*.

PASSIVE ELECTRICAL PROPERTIES RETARD MEMBRANE VOLTAGE CHANGES

The delay in depolarizing (or hyperpolarizing) a cell membrane occurs because the membrane behaves electrically like a resistor and a capacitor in parallel (see Figure 11.5*c*). On the one hand, the lipid bilayer of a cell membrane behaves like a *capacitor*: The bilayer blocks the exchange of ions between the extracellular fluid and the intracellular fluid, and its insulative properties enable oppositely charged ions to accumulate along the inner and outer surfaces of the membrane. On the other hand, the membrane-spanning ion channels behave like *resistors*: They allow ions to flow across the membrane at a rate governed by the structure of the channels and the potential difference between the inside and outside of the membrane. Current first redistributes the charges on the membrane capacitance (capaci-

[3] To understand this, it is helpful to realize that **capacitance** (expressed in farads) is the amount of charge stored per unit of voltage ($C = Q/V$). The greater the capacitance, the more ions the membrane can separate and store for a given potential difference, and therefore the more time it takes for ions to be redistributed in response to a pulse in current.

(a) A current pulse changes membrane potential

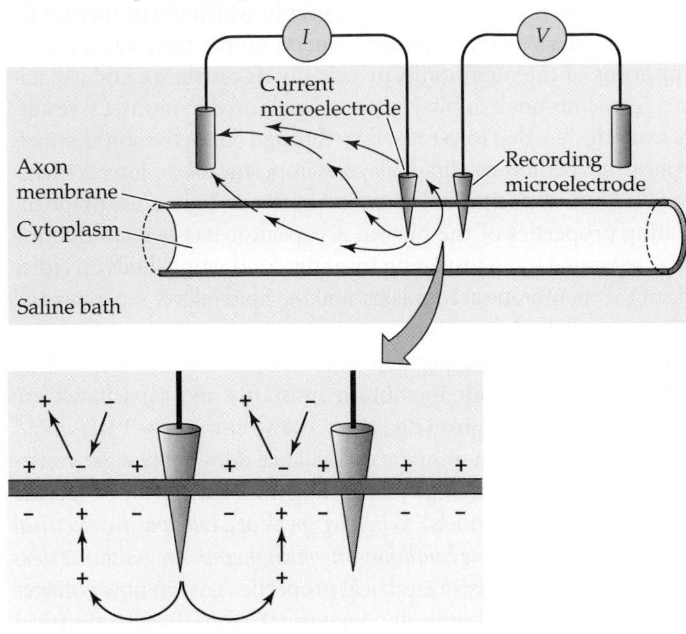

(b) Time course of voltage change

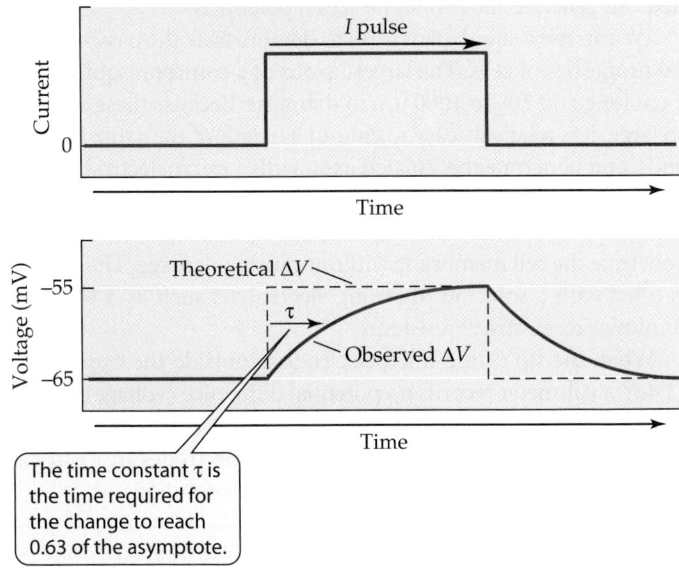

(c) Membrane resistance and capacitance

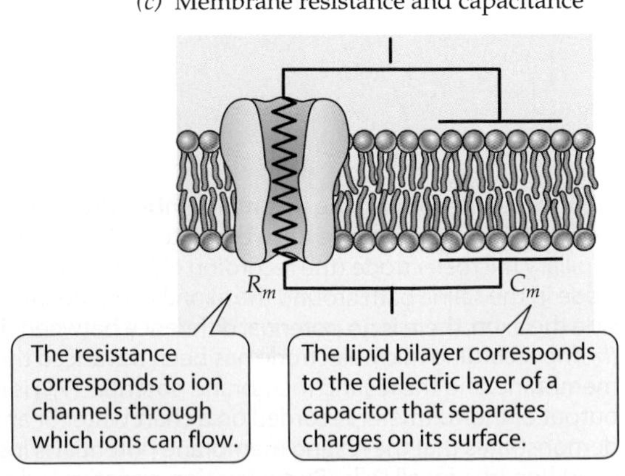

Figure 11.5 Changes in the membrane potential: the membrane time constant *(a)* Here a second current microelectrode is advanced into the squid axon shown in Figure 11.4. *(b)* The redistribution of charges during the current pulse (*I*) occurs more slowly than would be expected if the membrane acted as a pure resistance (dashed line), indicating that the membrane also has capacitive properties. The inset in *(a)* shows why: An outward capacitive current adds positively charged ions to the inner surface of the membrane and carries away positive ions from the outer surface, gradually depolarizing the membrane. *(c)* A neuronal membrane exhibits resistance (*R*) and capacitance (*C*) in parallel. The membrane *time constant* (τ) is the time it takes for membrane potential to reach 0.63 of its final value. In many neurons τ is 2 to 20 ms.

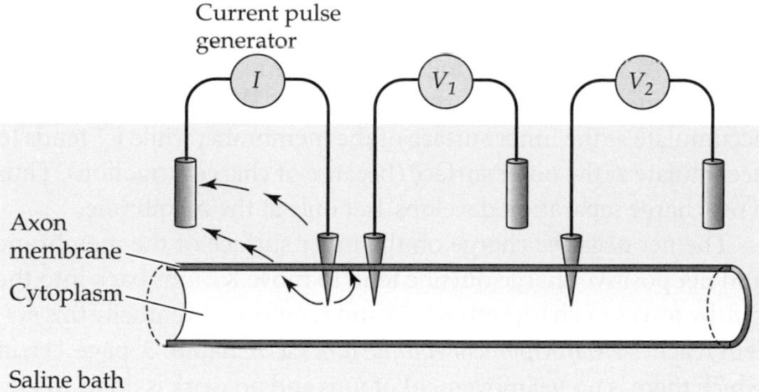

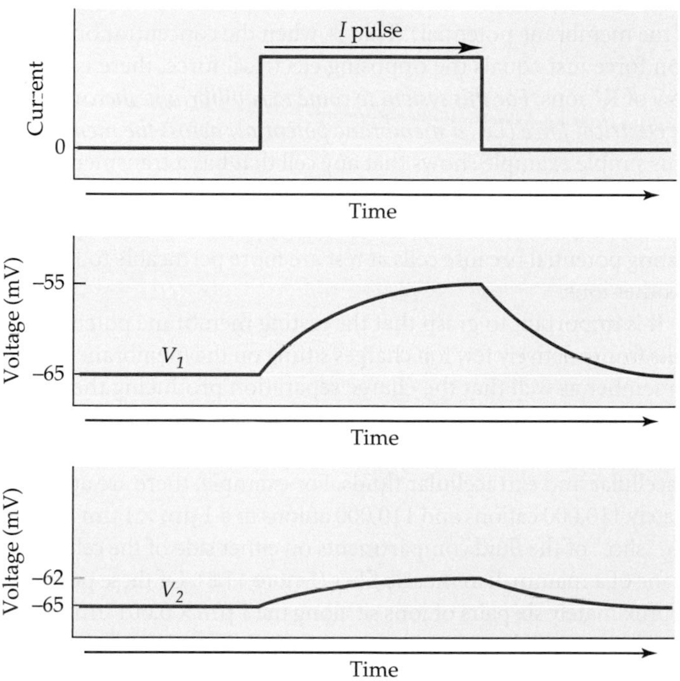

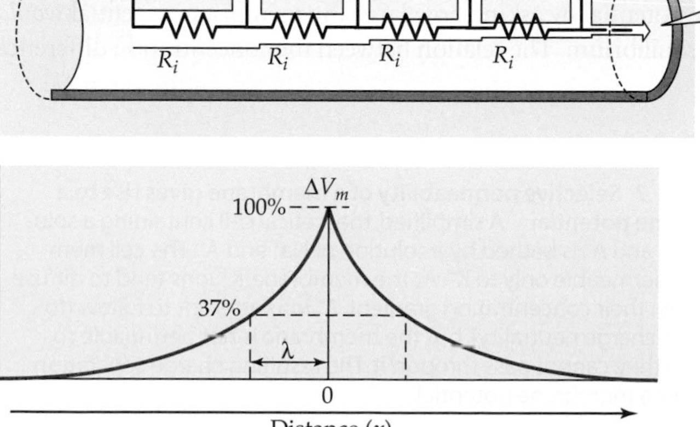

Figure 11.6 Graded potentials decrease exponentially with distance The amplitude of a voltage change decreases with distance along the axon. *(a)* A third electrode (V_2) is added to the setup diagrammed in Figure 11.5 and is used to measure the potential change at some distance from the source of current (*I*). *(b)* The voltage change measured at V_2 is smaller than that at V_1, which is closer to the source of current. This decremental spread of graded potentials is referred to as *electrotonic* conduction. *(c)* The decrement in voltage change is exponential with distance from *I*. Arrows show the local paths of current flow that depolarize the membrane. (Capacitances are ignored.) The membrane *length constant* (λ) describes the exponential decay of a change in voltage (ΔV) with distance.

tive current) and then flows through the membrane resistance (resistive or ionic current). This redistribution of charges slows (retards) the change in voltage on the membrane, by a factor that increases if the resistance or capacitance is increased.[3]

The exponential time course of the voltage change shown in Figure 11.5 is described by the **time constant**, τ (tau), the time it takes the voltage change to reach 63% of its final value. The time constant of a cell depends on the resistance and capacitance of its membrane:

$$\tau = RC \tag{11.2}$$

where R is the cell's resistance (input resistance) and C is capacitance. For many cells τ is in the range of 2 to 20 ms.

PASSIVE ELECTRICAL PROPERTIES LIMIT THE SPREAD OF GRADED POTENTIALS

How does a change in voltage spread over *distance* in a membrane? (Remember that we are considering only the *passive* electrical properties of a membrane.) Suppose we insert another electrode into the squid axon, this one farther away from the other two (Figure 11.6*a*). The V_2 electrode will record a smaller voltage displacement in response to a current pulse than the nearer V_1 electrode (Figure 11.6*b*). The voltage change (ΔV) will decrease exponentially with distance from the source producing it, a property called **passive spread** (**decremental spread**) or **electrotonic conduction.** The steepness of this decrease with distance is described by the membrane **length constant, λ** (lambda), which represents the distance at which the decaying voltage change (ΔV_m) is 37% of its value at the origin (Figure 11.6*c*).

The reason for this decrease with distance is that as current flows along the inside of the axon, some of it leaks out through ion channels. For simplicity we ignore the slowing effects of membrane capacitance, and we lump the resistive pathways as R_m (the resistance to current flow out across the membrane ion channels in a segment) and R_i (the resistance to current continuing down the axon to the next segment). An axon (or dendrite) with a high R_m value and a low R_i value will have a large λ value. We will consider the membrane length constant again later, when we discuss action-potential propagation.

The electrical properties of membrane resistance (R_m), membrane capacitance (C_m), and resting membrane potential (V_m), and the related time constant (τ) and length constant (λ), adequately describe the passive electrical properties of a neuron or any other cell. Because the passive electrical properties of axons are similar to those of underwater telephone cables, these passive electrical properties are often called *cable properties*. For the same reason, the equations describing the length constant and time constant of neuronal membranes are called *cable equations*.

Resting membrane potentials depend on selective permeability to ions: The Nernst equation

Our model of the passive electrical properties of cells describes the inside-negative resting membrane potential maintained by all living cells. A cell membrane's permeability to different kinds of ions is the mechanism that establishes and maintains this voltage. Dissolved ions have charges and attract polar water molecules around them. The charged ions cannot mix with the nonpolar tails of the lipid molecules in the center of the membrane bilayer, so the ions cannot pass through the bilayer (see Chapter 3). Instead, ions must pass (if they pass at all) by way of the protein ion channels that span the bilayer. There are many kinds of ion channels, each kind *selectively permeable* to specific ions. Moreover, some ion channels can open and close, which means that *membrane permeability to specific ions is a controlled condition.*

How does a membrane's selective permeability to ions produce a membrane potential? Consider a simplified cell (Figure 11.7) that contains a solution of potassium ions (K^+) and nonpermeating anions (represented by A^-), such as charged proteins. (The identity of A^- is unimportant; all we're concerned with is its charge.) The cell is in a bath of two nonpermeating ions (Na^+ and A^-); we stipulate that the cell membrane is permeable only to K^+. When the cell is placed in the bath, K^+ tends to diffuse out of the cell, down its concentration gradient. Because charge neutrality is always maintained in bulk solution, A^- would tend to follow the K^+ across the membrane; however, the membrane is not permeable to A^-. Therefore, A^- tends to accumulate at the inner surface of the membrane, while K^+ tends to accumulate at the outer surface (because of charge attraction). Thus a net charge separation develops, but only at the membrane.

The net negative charge on the inner surface of the membrane and net positive charge outside tend to move K^+ ions back into the cell, by forces of charge attraction and repulsion. Eventually this system reaches *electrochemical equilibrium* (see Chapter 3, page 71), in which there is no *net* movement of ions and no work is done. In Figure 11.7, equilibrium is reached when the tendency for K^+ ions to diffuse out of the cell (down the concentration gradient) is exactly balanced by their tendency to move in (down the electrical gradient of the membrane potential). That is, when the concentration-diffusion force just equals the opposing electrical force, there is no *net* flow of K^+ ions. *For this system to come to equilibrium, there must be an electrical force (i.e., a membrane potential) across the membrane.* This simple example shows that any cell that has a transmembrane concentration difference of a permeating ion tends to generate a membrane potential. It is a good starting point for visualizing a cell's resting potential because cells at rest are more permeable to K^+ than to other ions.

It is important to grasp that the resting membrane potential results from relatively few ion charges sitting on the membrane, and to remember as well that the charge separation producing the membrane potential is an extraordinarily local phenomenon. Charge neutrality always prevails in the bulk solutions that make up the intracellular and extracellular fluids. For example, there are approximately 110,000 cations and 110,000 anions in a 1 μm × 1 μm × 0.001 μm "slice" of the fluid compartments on either side of the cell membrane of a mammalian muscle fiber (Figure 11.8). Of these portions, approximately six pairs of ions sit along the 1 μm × 0.001 μm area of membrane, and these six pairs are responsible for the charge imbalance that amounts to a robust –90-mV resting membrane potential! Thus the movement of only a few ions in a region can establish (or change) a membrane potential without disrupting the overall charge neutrality of the intracellular and extracellular fluids.

Although living cells are not at electrochemical equilibrium, we can describe the contribution of a permeating ion species to membrane potential by asking how large the membrane potential *would be* at equilibrium. The relation between the concentration difference

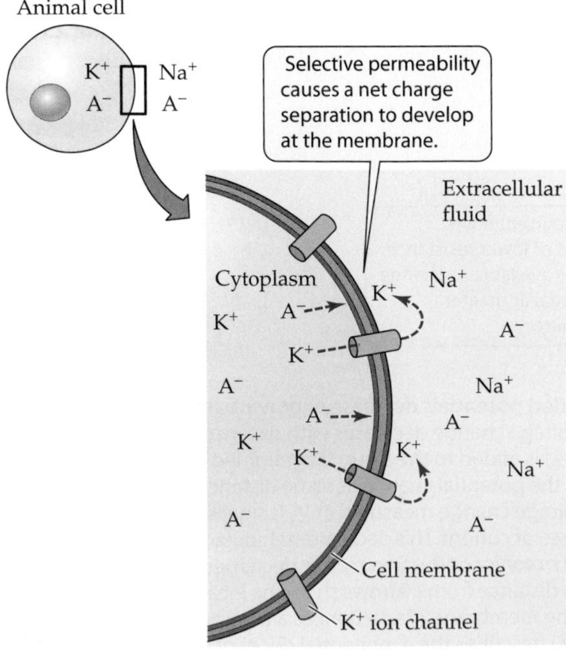

Figure 11.7 Selective permeability of a membrane gives rise to a membrane potential A simplified, theoretical cell containing a solution of K^+ and A^- is bathed by a solution of Na^+ and A^-. The cell membrane is permeable only to K^+. At the membrane, K^+ ions tend to diffuse out, down their concentration gradient. A^- ions attempt to follow (to maintain charge neutrality), but the membrane is not permeable to them, so they cannot pass through it. The resulting charge separation produces a membrane potential.

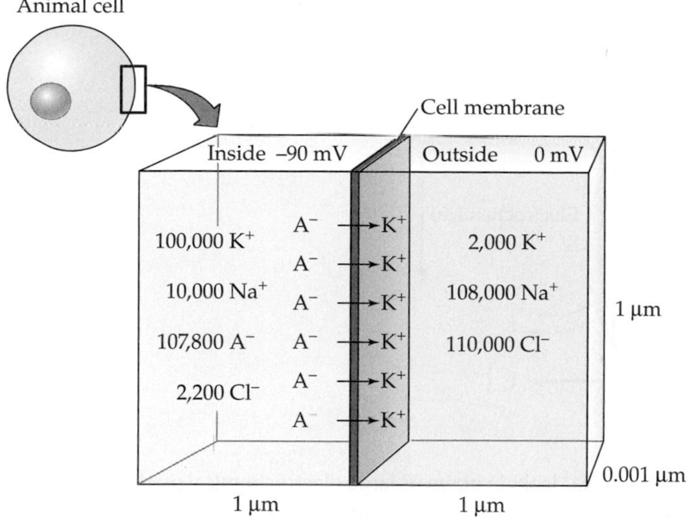

Figure 11.8 The membrane potential results from relatively few charges sitting on the membrane A small patch of membrane (1 µm × 0.001 µm in area) from a mammalian muscle fiber, with a small volume (1 µm × 1 µm × 0.001 µm) of adjacent cytoplasm and extracellular fluid. As in Figure 11.7, assume that the membrane is most permeable to K^+. Of the 110,000 cations and 110,000 anions in each fluid compartment, only *six* pairs of ions need to sit on the membrane and charge its capacitance to produce a membrane potential of –90 mV. (After Schmidt 1985.)

of a permeating ion across a membrane and the membrane potential at equilibrium is given by the **Nernst equation**:

$$E = \frac{RT}{zF} \ln \frac{C_{out}}{C_{in}} \qquad (11.3)$$

in which E is the membrane potential (E stands for *electromotive force*, an older term for voltage), R is the gas constant, T is absolute temperature, z is the valence of the ion species (charge for the kind of ion), F is Faraday's constant (charge per mole of ions), and C_{out} and C_{in} are the ion concentrations on the two sides of the membrane. Notice that *the larger the concentration difference across the membrane, the larger the membrane potential* at which the ion species is in equilibrium. The reason for this relation is that increasing the concentration difference increases the concentration gradient on the ion species and thus increases the electrical force necessary to oppose it.

We can simplify the Nernst equation by calculating R/F, converting to $\log_{10}$, and considering an ion of a given valence at a given temperature. For K^+, a monovalent cation, at 18°C (British room temperature),

$$E(\text{in mV}) = 58 \log_{10} \frac{C_{out}}{C_{in}} \qquad (11.4)$$

For a mammal at 37°C,

$$E(\text{in mV}) = 61 \log_{10} \frac{C_{out}}{C_{in}} \qquad (11.5)$$

Thus, for our simplified cell permeable only to K^+, if the internal K^+ concentration is, say, 100 mM and the external concentration is 10 mM, at 18°C, $E = 58 \log_{10} (0.1) = -58$ mV. (By convention, the minus sign means that the inside of the membrane is negative relative to the outside.)

The value of –58 mV is the **equilibrium potential** for potassium (E_K) in our system as we have defined it—that is, the value of the membrane potential at which K^+ ions are at electrochemical equilibrium and the internal K^+ concentration is 10 times the external concentration. In other words, the electrical force holding K^+ inside the cell is just balanced by the chemical (concentration) force for K^+ diffusion out of the cell.

The Nernst equation relates membrane potential to the concentration ratio of only one ion species. For this and other reasons, the generation of membrane potentials in real cells is considerably more complex than in this simplified model.

Ion concentration differences result from active ion transport and from passive diffusion

All cells maintain higher concentrations of potassium and lower concentrations of sodium and chloride in the intracellular fluids than are present in the surrounding extracellular fluid (Figure 11.9*a*). The concentrations of these ions differ from organism to organism, as shown in Table 11.1. Despite quantitative differences, the *concentration ratios* of ions in all cells are similar to those represented in the table. The difference in ion concentrations between the intracellular and extracellular fluids results from a combination of two processes: (1) the active transport of some ions, and (2) the passive distribution of other ions.

Examine Figure 11.9*a* and consider the Nernst equation. It is impossible for both Na^+ and K^+ ions to be in passive equilibrium, because the ratios of their concentrations differ. Whereas K^+ would require an inside-negative membrane potential to be in equilibrium, Na^+ would require an inside-positive potential (to counteract inward Na^+ diffusion). In fact, *neither* ion species is in passive equilibrium in cells, because both ions are maintained at nonequilibrium levels by *pumps*—that is, by active ion transport that requires the input of energy from hydrolysis of ATP.

The most important pump is the *Na^+–K^+-ATPase pump* (see Chapter 3, page 75), which actively transports Na^+ out of the cell and K^+ into it. For most cells, Na^+ ions slowly leak into the cell and K^+ ions slowly leak out. (The channels through which these leaks occur are normally open and are called *leakage channels*.) The Na^+–K^+ ATPase pump counteracts these leaks, using ATP energy to pump Na^+

TABLE 11.1 Concentrations of major ions in intracellular fluid (cytoplasm) and extracellular fluid

	Squid axon			Mammalian muscle		
Ion type	**Out (m*M*)**	**In (m*M*)**	**Out/In**	**Out (m*M*)**	**In (m*M*)**	**Out/In**
Na^+	440	50	8.8	145	12	12.1
K^+	20	400	0.05	4	155	0.03
Cl^-	560	60	9.3	120	3.8	31.6
A (organic anions)	—	270	—	—	—	—

(*a*) Ion concentrations

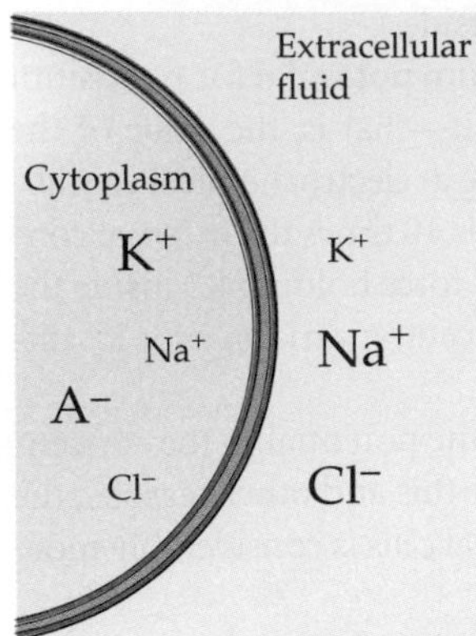

(*b*) Pump maintains Na+, K+ concentrations

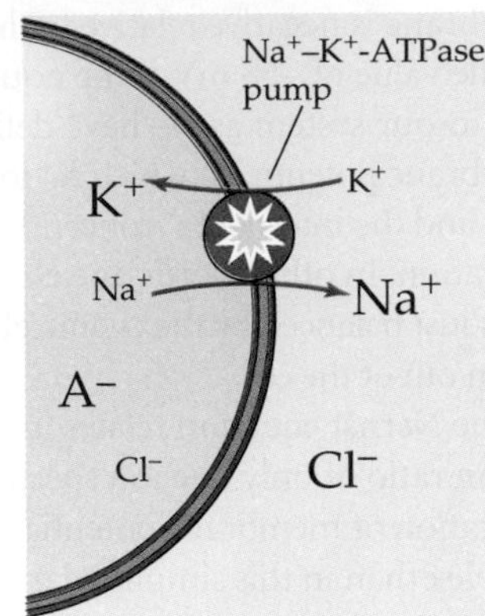

(*c*) Steady state

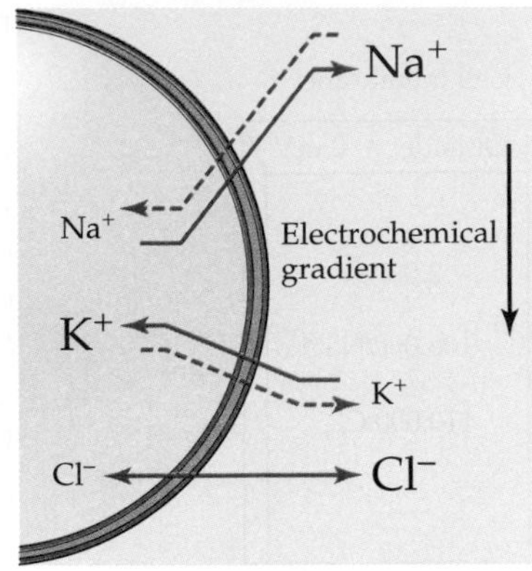

Figure 11.9 Ion pumps help maintain the concentration of major ions in intracellular and extracellular fluids *(a)* All cells maintain low intracellular concentrations of Na^+ and Cl^- and high concentrations of K^+ and nonpermeating anions (A^-) within intracellular fluids, relative to extracellular fluid. (Symbol sizes represent relative concentrations.) *(b)* An active sodium–potassium exchange pump transports Na^+ out and K^+ in, counteracting the tendency of Na^+ to diffuse in and K^+ to diffuse out. *(c)* Here the Na^+ and K^+ concentrations are maintained in a steady state across the membrane. Na^+ leaks in down its large electrochemical gradient, slowly because its permeability is low. K^+ leaks out slowly, its higher permeability being offset by a smaller electrochemical gradient (less steeply downhill). These slow passive leaks (dashed arrows) are counteracted by active transport by the Na^+–K^+ exchange pump (solid arrows). Chloride, in contrast, can be at passive equilibrium in some cells; unequal Cl^- concentrations result from nonpermeating anions (A^-) in the cytoplasm.

out as fast as it leaks in, and to pump K^+ in as fast as it leaks out (Figure 11.9*b*). The function of the Na^+–K^+-ATPase pump is analogous to a bilge pump in a boat, bailing water out as fast as it leaks in. Another good analogy for the pump is a battery charger, which can work in the background to prevent the "batteries" of Na^+ and K^+ concentration distributions from running down. For known Na^+–K^+ exchange pumps, the ratio of Na^+ and K^+ pumped is 3:2. (We will consider a secondary consequence of the 3:2 ratio shortly.)

On the other hand, a permeating ion species may have very different concentrations inside and outside a cell *without* being pumped. Consider the Cl^- ions in Figure 11.9*a*. Because cells have large intracellular concentrations of nonpermeating anions, permeating ions such as Cl^- must become distributed unequally across the membrane. The A^- ions that limit the tendency of K^+ ions to diffuse out of the cell (see Figure 11.7) also limit the tendency of Cl^- ions to diffuse in. As described by the Nernst equation, the membrane potential counteracts the tendency of Cl^- ions to diffuse from their high extracellular concentration to the lower intracellular concentration (Figure 11.9*c*). In many cells, Cl^- ions are in passive equilibrium despite strikingly unequal concentrations inside and outside the cell. This equilibrium, sometimes called a *Donnan equilibrium* or *Gibbs–Donnan equilibrium*, explains how nonpermeating anions inside the cell can lead to unequal concentrations across the membrane of permeating ions such as Cl^-.

Donnan-type phenomena are one way in which inequalities of ions such as Cl^- are maintained in living cells. The nonpermeating anions (A^-) in cells are a mixture of relatively small organic anions such as amino acids, along with proteins and other large molecules that bear net negative charges. Chloride ions appear to be passively distributed across the membranes of many cells, even though the ratio of outside to inside Cl^- concentrations may be quite large, as it is, for example, in mammalian muscle (see Table 11.1). Neurons of the vertebrate central nervous system, however, actively transport Cl^- out, so the ratio of $[Cl^-]_{out}$ to $[Cl^-]_{in}$ is larger than predicted by Donnan equilibrium, and E_{Cl} is more negative.

Now we are in a position to explain ion concentration distributions of living cells. Neither Na^+ nor K^+ is at equilibrium in Figure 11.9, so passive diffusion alone produces net movement of Na^+ and K^+. (This disequilibrium results from the pump having changed Na^+ and K^+ concentrations from their equilibrium levels.) Na^+ especially is far out of equilibrium because both the concentration-diffusion gradient and the electrical gradient of the inside-negative membrane potential drive Na^+ inward. Because the membrane is only slightly permeable to Na^+ at rest, Na^+ enters only slowly and is pumped out as fast as it diffuses in. K^+ is closer to equilibrium but not at it. The cell loses K^+ passively at a slow rate because although permeability to K^+ is large, the driving force for K^+ flow ($V_m - E_K$) is small. The slow passive loss of K^+ is also counteracted by the sodium–potassium exchange pump. Thus the cation concentrations are maintained in a *steady state* in which passive-transport leaks are counteracted by active-transport pumps. Metabolic energy is required to maintain concentrations different from the concentrations at equilibrium.

Figure 11.9*c* summarizes the roles of active and passive transport in maintaining the steady-state concentrations of Na^+ and K^+ ions in intracellular and extracellular fluids: Passive leaks of Na^+ into the cell and of K^+ out of the cell are counteracted by active ion transport. Chloride ions are passively distributed in this cell. The balance of active and passive transport that control concentrations of different ions is important because, as we will see, the steady-state concentrations of K^+, Na^+, and Cl^- ions all contribute to membrane potentials in animal cells.

Membrane potentials depend on the permeabilities to and concentration gradients of several ion species: The Goldman equation

The ion concentrations of living cells are in a steady state in which the ions are unequally distributed across the cell membrane and many are out of equilibrium. *The resting membrane potential is largely determined by K^+ concentrations because the cell membrane is more permeable to K^+ than to other ions.* If the membrane were permeable only to K^+, then the membrane potential would be exactly equal to the K^+ equilibrium potential (i.e., $V_m = E_K$), as predicted by the Nernst equation employing the K^+ concentrations across the membrane. Because the membrane is somewhat permeable to other ions, however, they also contribute to the membrane potential.

The contribution of each ion is weighted by its ability to permeate the membrane, the more-permeating ions having more effect. The value of the membrane potential (V_m) produced by the contributions of several permeating ion species can be determined by the **Goldman equation**:

$$V_m = \frac{RT}{F} \ln \frac{P_K[K^+]_o + P_{Na}[Na^+]_o + P_{Cl}[Cl^-]_i}{P_K[K^+]_i + P_{Na}[Na^+]_i + P_{Cl}[Cl^-]_o} \quad (11.6)$$

in which P_K, P_{Na}, and P_{Cl} are relative permeability values for potassium, sodium, and chloride ions, respectively. (The chloride term in the equation is inverted to reflect its negative charge.) In principle, it is necessary to add a term in the Goldman equation for every permeating ion species, but in practice it is necessary to include terms only for Na^+, K^+, and Cl^-. The contributions of other ion species can be neglected, by reason of either low permeability of the membrane to those ions (e.g., HCO_3^-) or low concentration of those ions[4] (e.g., $[H^+] = 10^{-7}\ M$).

In fact, for some purposes it is useful to consider sodium and potassium only, ignoring chloride. In such a simplification of the Goldman equation (**Figure 11.10**), we can view the membrane potential as a result of the membrane's relative permeabilities to sodium and potassium ions, visualized with a sliding voltage scale, rather like a thermometer but in units of voltage. Consider a squid axon with the following concentrations:

$$[K^+]_{out} = 20\ mM \qquad [Na^+]_{out} = 440\ mM$$
$$[K^+]_{in} = 400\ mM \qquad [Na^+]_{in} = 44\ mM$$
$$[K^+]_{out} = 0.05 \qquad [Na^+]_{out} : [Na^+]_{in} = 10$$
$$E_{Na} = 58 \log\ 0.05 = -75\ mV\ \text{(inside-negative)}$$
$$E_{Na} = 58 \log\ 10 = +58\ mV\ \text{(inside-positive)}$$

[4] Strictly speaking, the ion concentration values in the Nernst equation and the Goldman equation should be those for *free* (unbound) ions rather than total concentrations. The **activity** of an ion is the concentration of the ion in its dissociated, freely diffusible form. Monovalent ions dissociate relatively completely in cytoplasm and extracellular fluids, so not much correction is needed for ion activities. Divalent ions such as Ca^{2+}, however, are predominantly bound in cytoplasm, so corrections for their activity are important.

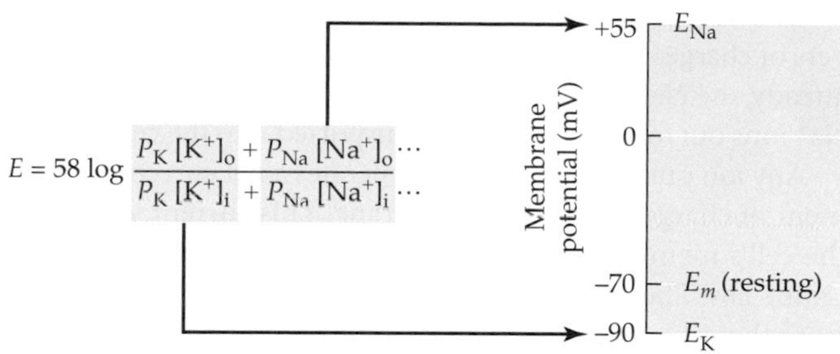

Figure 11.10 The Goldman equation and the "voltage thermometer" A simplified Goldman equation describes membrane potential in terms of relative permeabilities (*P*) of the membrane to K^+ and Na^+. The voltage scale graphs the membrane potential determined by these permeabilities. For the resting membrane, $P_K >> P_{Na}$, so E_m is close to E_K. If P_{Na} increases to become greater than P_K, E_m will approach E_{Na}. Each arrow relates the dominant term in the equation to the value of E_m toward which it drives the membrane.

If the cell were permeable only to K^+, then V_m would equal E_K, or –75 mV; and if it were permeable only to Na^+, then V_m would equal E_{Na}, or +58 mV. The actual membrane potential can be anywhere between these values and is governed by the ratio of membrane permeabilities to Na^+ and K^+. The "voltage thermometer" shows that when permeability to potassium is much greater than permeability to sodium, the membrane potential approaches E_K (see Figure 11.10):

$$\text{If } P_K = 10 \times P_{Na},$$

$$\text{then } V_m = 58 \log_{10} \frac{10[20]+1[440]}{10[400]+1[44]} = 58 \log(644/4044) = -46.3\text{mV}$$

In contrast, when permeability to sodium is much higher than peremeability to potassium, the membrane potential approaches E_{Na}. This visualization of membrane potential in terms of the Goldman equation will be important for our consideration of action potentials later in this chapter.

Electrogenic pumps also have a small direct effect on V_m

Our explanation to this point about the generation of membrane potentials has been termed the ionic hypothesis. The **ionic hypothesis** argues that the concentrations of ions inside and outside a cell are maintained in a *steady state* by a mixture of active-transport processes (ATPase pumps) and passive-transport processes (diffusion and Donnan effects). The ionic hypothesis further asserts that the concentrations of ions inside and outside the cell, and the permeability of the cell membrane to these ions, determine the resting membrane potential (V_m) as described by the Goldman equation. The ionic hypothesis is substantially accurate, and provides a useful description of the factors giving rise to membrane potentials in living cells. A more complete explanation of the causes of membrane potentials, however, must include the fact that some ion pumps are electrogenic.

There are two kinds of active ion-transport mechanisms: electroneutral pumps and electrogenic pumps. An **electroneutral pump** transports equal quantities of charge inward and outward across a membrane and thus changes ion concentrations without generating an electric current. An **electrogenic pump** transports unequal numbers of charges inward and outward across the membrane. As noted already, the Na^+–K^+ exchange pump has a 3:2 ratio, transporting 3 Na^+ ions out for each 2 K^+ ions transported into the cell.

Any ion pump that is not 1:1 generates a net current (net movement of charge) across the membrane. This current, acting across the cell's membrane resistance, directly generates a potential, via Ohm's law. The potential resulting from pump current changes V_m from the value predicted by the Goldman equation. Thus an electrogenic pump has two functional properties: It changes concentrations to offset passive leaks (its major function), and it alters V_m directly via the pump current (a smaller secondary function).

The 3:2 sodium–potassium pump generates an outward ionic current (outward movement of positive charge) that hyperpolarizes the cell to a level more inside-negative than is predicted by the Goldman equation. Because sodium–potassium exchange pumps can be selectively poisoned with toxins such as *ouabain*, their electrogenic contribution to resting membrane potentials can be measured as the initial change in V_m before concentrations change. In many neurons the direct contribution of an electrogenic pump accounts for only a few millivolts of the resting membrane potential, although electrogenicity can make a larger contribution in some invertebrate neurons.

Summary: The Ionic Basis of Membrane Potentials

- Cell membranes have properties of electrical resistance and capacitance, which allow them to maintain a voltage (membrane potential) and regulate current flow across the membrane. Cells have inside-negative resting membrane potentials. The passive electrical properties of membranes determine how membrane potentials change with time (the time constant, τ) and with distance (the length constant, λ).
- Membrane potentials depend on selective permeability to ions. Any ion species to which the membrane is permeable will tend to drive the membrane potential toward the equilibrium potential for that ion. The Nernst equation calculates the equilibrium potential of a single ion species in terms of its concentrations on both sides of the membrane.
- All cells have higher concentrations of K^+ inside than outside, higher concentrations of Na^+ outside than inside, and higher concentrations of Cl^- outside than inside. Ion concentrations inside and outside cells are maintained by active ion pumps, as well as by passive Donnan-equilibrium effects.
- Membrane potentials depend on the permeabilities to and concentration gradients of several ion species: The resting membrane is dominated by permeability to K^+, so the resting membrane potential is near E_K. The Goldman equation describes how changing the membrane permeability of an ion species changes the membrane potential.
- In addition to their major role of maintaining the nonequilibrium concentrations of ions, electrogenic ion pumps generate a current that makes a small, direct contribution to V_m. In addition, only those ions that are freely diffusible contribute to V_m, so corrections for bound ions may be necessary.

The Action Potential

Excitable cells such as neurons, muscle fibers, and a few others have the ability to generate electrical signals. The hallmark electrical signal of an excitable cell is the *action potential.* Action potentials (which in neurons may also be called *nerve impulses*) are one of the most important kinds of electrical signals underlying the integrative activity of nervous systems. Some kinds of neurons do not generate action potentials, however, so the association of neurons with action potentials is not universal.

Action potentials are voltage-dependent, all-or-none electrical signals

Action potentials result from *voltage-dependent* changes in membrane permeabilities to ions because the ion channels that produce action potentials are *voltage-gated.* An action potential is initiated by a change in the resting membrane potential, specifically by a depolarization sufficiently strong to open the voltage-gated channels. The voltage dependence of ion permeabilities is a critical feature of action potentials, and it makes action potentials fundamentally different from resting potentials or from graded potentials.

Action potentials have characteristic features. An **action potential** is a momentary reversal of membrane potential from about –65 mV (inside-negative) to about +40 mV (inside-positive)—a voltage change of about 100 mV, lasting about 1 ms, followed by restoration of the original membrane potential (Figure 11.11*a*). The action potential is triggered by any depolarization of the membrane that reaches a critical value of depolarization, the **voltage threshold.** After the suprathreshold (above threshold) depolarization, the action potential has a rapid *rising phase* that reaches a peak more positive than zero potential (*overshoot*) followed by a rapid repolarization (the *falling phase*). In the squid axon and in many other neurons, the action potential is followed by an *undershoot*, a transient after-hyperpolarization lasting a few milliseconds.

To illustrate the voltage-dependent properties of action potentials, let's perform a hypothetical experiment using a squid giant axon (Figure 11.11*b*). As we did in Figure 11.6, we penetrate the axon with two glass capillary microelectrodes—one to apply current pulses and one to record voltage. The first three inward-flowing current pulses hyperpolarize the membrane in the vicinity of V_1; the amount of hyperpolarization is proportional to the strength of each current pulse. This relation follows from Ohm's law and (ignoring the time constant) indicates that with hyperpolarization the membrane resistance does not change. Thus, hyperpolarization can't induce action potentials, because it doesn't change the permeabilities of membrane ion channels. Weak, outward-flowing current pulses in the opposite direction (pulses 4 and 5 in Figure 11.11*c*) elicit small depolarizations that approximately mirror the preceding hyperpolarizations, again indicating no significant change in membrane resistance.

Stronger depolarizing currents (pulses 6 to 8 in Figure 11.11*c*) that exceed the voltage threshold produce action potentials. However, a stronger depolarizing current (beyond threshold) does not produce a larger action potential (compare the responses to pulses 6 and 7). Instead, action potentials are **all-or-none** phenomena; that is, a

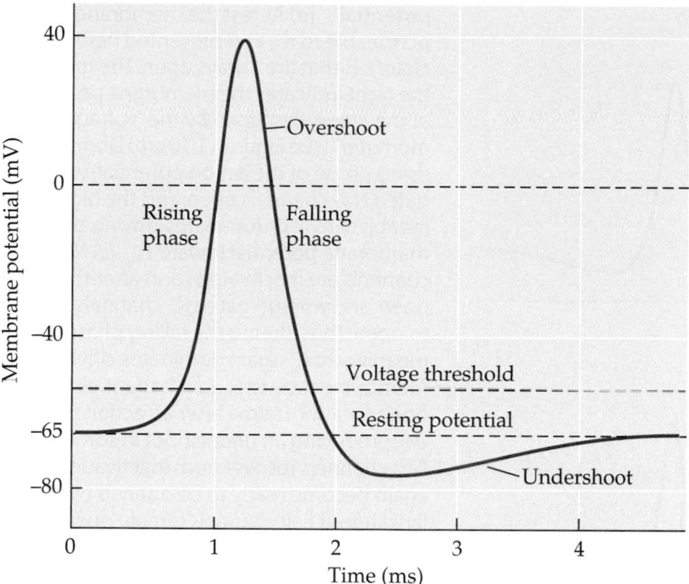

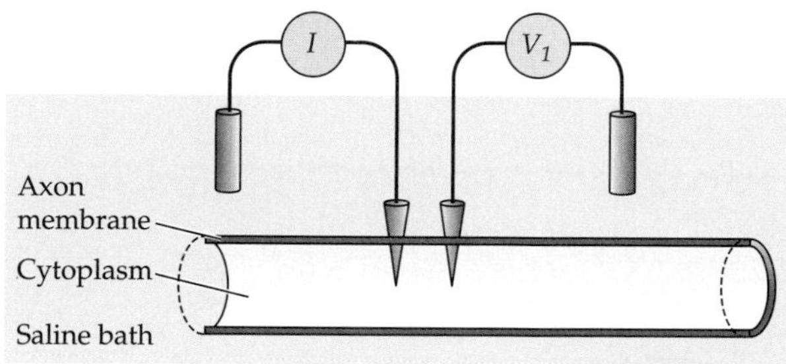

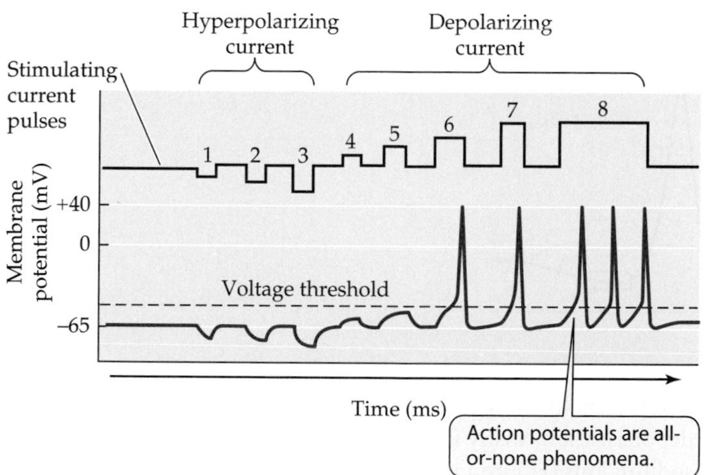

Figure 11.11 General features of action potentials *(a)* An action potential is a brief voltage change characterized by a rising phase that overshoots zero and a falling phase (repolarization) that may be followed by an after-hyperpolarization, or undershoot. *(b)* Recording action potentials in a squid giant axon, using a stimulating electrode (*I*) and two recording electrodes (V_1 and V_2). *(c)* Responses of the axon to stimulating current pulses. Current pulses flowing inward across the membrane capacitance (arrows) hyperpolarize the membrane (pulses 1–3). Small pulses of outward current (4, 5) elicit depolarizations that are nearly the mirror image of the corresponding hyperpolarizations; but as a critical *voltage threshold* is reached (6), the depolarization triggers an action potential. A larger pulse of depolarizing current (7) produces an identical action potential, and prolonged suprathreshold current (8) elicits a train of action potentials.

depolarization below threshold elicits no impulse, but all suprathreshold depolarizations produce complete impulses substantially alike in amplitude and duration.

Immediately following an action potential, another action potential cannot be generated for at least 1 ms (the **absolute refractory period**) and is harder to generate for a few milliseconds longer (the **relative refractory period**). The refractory periods show that the voltage threshold can change, rather than being a fixed value. Because of the all-or-none property of the action potential and the succeeding refractory period, impulses cannot summate. Instead, a prolonged suprathreshold depolarizing current (pulse 8 in Figure 11.11*c*) can elicit a *train* of discrete action potentials. For many neurons, the frequency of impulses in a train increases with increasing strength of depolarizing current (within limits).

An action potential, once initiated, *propagates* along the axon without a decrease in amplitude and at a constant velocity that depends on the diameter of the axon (among other factors). If in Figure 11.11*b* and *c* a remote electrode measured voltage at the end of the axon (not shown), it would record each action potential that the local electrode (V_1) records, with no decrease in amplitude. Each impulse recorded remotely follows the impulse at V_1 by a short latency that represents the time required for the impulse to propagate along the axon between the two electrodes. The distant V_2 electrode would not record the subthreshold depolarizations and hyperpolarizations, because they are not propagated; instead they spread decrementally and so are weakened before reaching V_2 (compare Figure 11.6).

In summary, action potentials are all-or-none electrical signals in excitable cells that propagate rapidly and without degradation over long distances. This ability to send signals over long distances rapidly and without distortion was presumably an important factor allowing the evolution of large animals whose complex physiology and behavior require extensive neural coordination.

Action potentials result from changes in membrane permeabilities to ions

The permeability terms in the Goldman equation (Equation 11.6) show that any factor that changes the permeability of the membrane to one or more ion species will change the value of the membrane potential. *An action potential results from intense, localized increases in permeabilities to specific ions*, increases that are both voltage- and time-dependent. What's more, the permeability increases are selective for specific ions: first sodium and then potassium.

PERMEABILITIES AND ION CHANNELS Let's follow the rise and fall of one action potential to see when and how these changes in the mem-

(a) Resting membrane potential

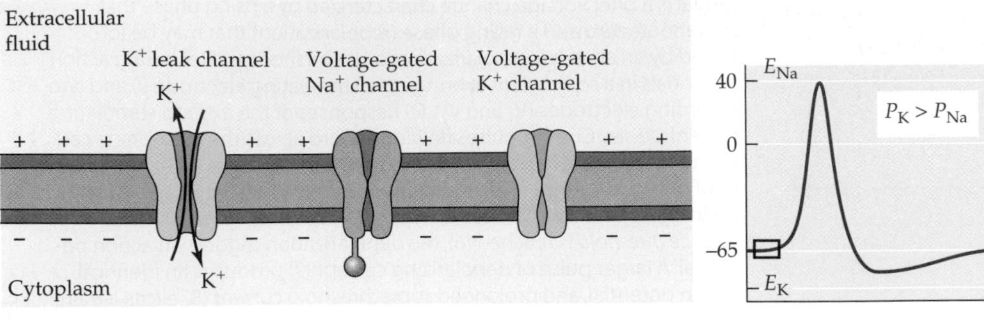

(b) Rising phase

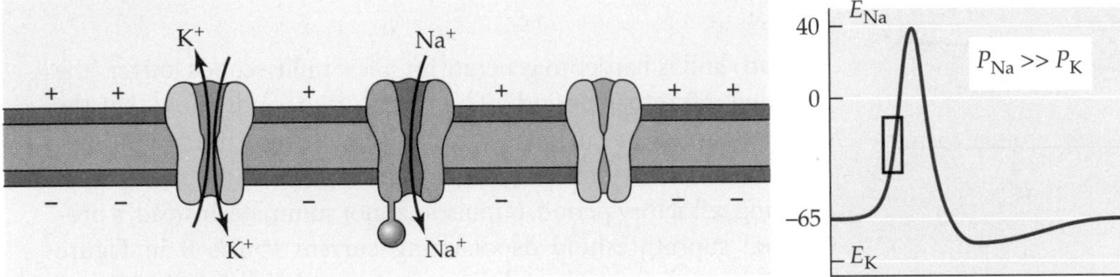

(c) Falling phase

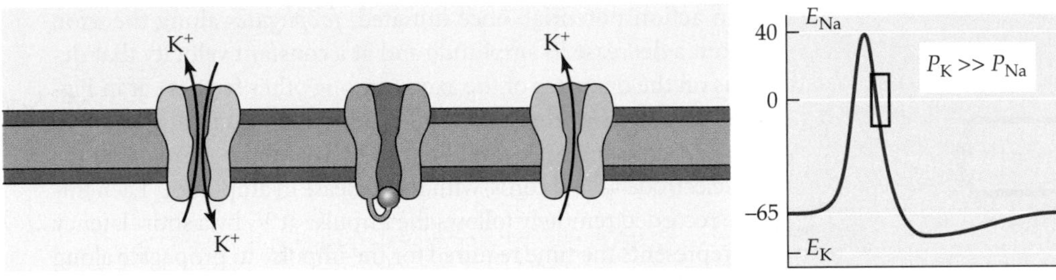

(d) Recovery

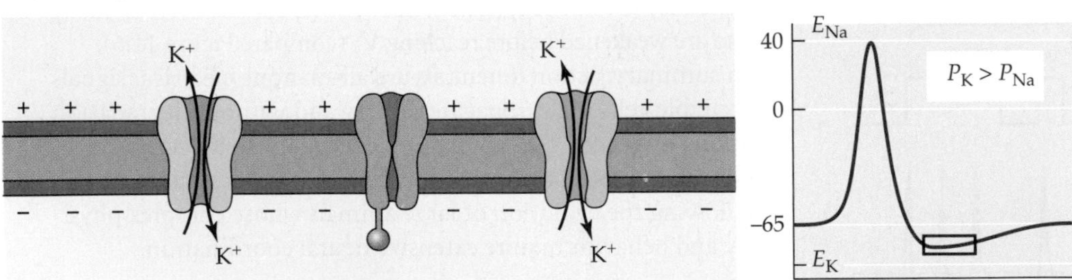

Figure 11.12 Membrane permeability changes that produce an action potential *(a)* At rest the membrane is most permeable to K^+, as represented here by leak channels that are always open. The graph on the right indicates the membrane potential at this stage, described by the "voltage thermometer" (see Figure 11.10). *(b)* During the rising phase of the action potential, voltage-gated Na^+ channels open, and the high permeability to Na^+ dominates, driving the membrane potential toward E_{Na}. *(c)* Na^+ channels are inactivated soon after they open, and voltage-gated K^+ channels begin to open. Thus, during the falling phase permeability to K^+ again dominates, driving the membrane toward E_K. *(d)* K^+ channels remain open for a short time after an action potential, producing an undershoot in some cells. Na^+ channels recover from inactivation and again become ready to be opened by depolarization. (Leak channels remain open throughout, but their effects are swamped by the more numerous voltage-gated channels.) (After Bear et al. 2001.)

brane's permeability to sodium and potassium ions occur. At the resting membrane potential of –65 mV, the membrane is most permeable to K^+ ions (Figure 11.12*a*). Neurons contain some K^+ channels that are normally open and are not voltage-gated. These *leakage channels* allow K^+ to diffuse across the membrane following the electrochemical gradient. The K^+ leak channels remain open throughout an action potential, but the more numerous voltage-gated channels swamp their effects.

The rising phase of the action potential (depolarization and polarity reversal) begins when a stimulus depolarizes the membrane past threshold. Voltage-gated Na^+ channels open in response to the depolarization, vastly increasing the membrane's permeability to sodium ions (Figure 11.12*b*). Because of the much higher concentration of Na^+ outside the cell, Na^+ rushes in, driving the membrane potential toward E_{Na} (which is inside-positive). The inward-rushing sodium current is the cause of depolarization and polarity reversal at the rising phase of the action potential. Just as a dominant permeability to K^+ at rest makes the resting membrane potential inside-negative, the inflow of Na^+ during the rising phase of the action potential makes the membrane momentarily inside-positive.

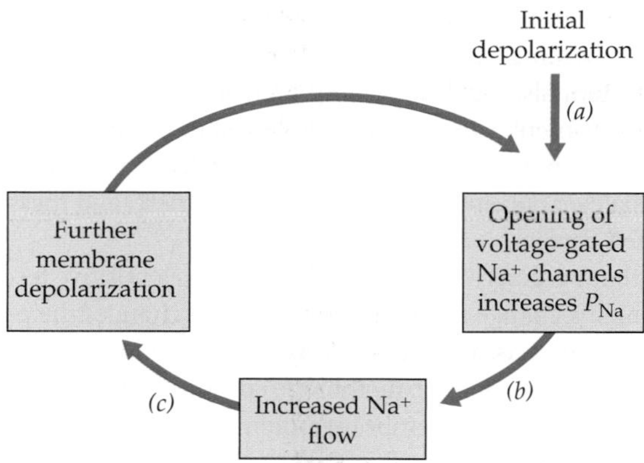

Figure 11.13 The Hodgkin cycle produces the rising phase of the action potential The critical feature of the cycle is that permeability to Na^+ is *voltage-dependent.* *(a)* Any depolarization increases P_{Na} by opening voltage-gated Na^+ channels. *(b)* The increased permeability to Na^+ allows inflow of Na^+ down its electrochemical gradient, which further depolarizes the membrane *(c)*. The cycle intensifies as each depolarization step opens additional Na^+ channels.

The falling phase of the action potential results from two changes in the membrane's permeability to ions (Figure 11.12*c*). First, the opening of the voltage-gated sodium channels is rapidly terminated by a process called **Na^+ channel inactivation**, which abruptly decreases permeability to Na^+. Second, after a slight delay, voltage-gated potassium channels open, greatly increasing permeability to K^+. Potassium ions flow out and drive the membrane toward E_K.

At the conclusion of an action potential, the membrane remains highly permeable to K^+ for a brief time (Figure 11.12*d*). Voltage-gated potassium channels remain open for a few milliseconds, producing a characteristic undershoot (after-hyperpolarization) in many neurons. The voltage-gated sodium channels recover from inactivation and again become ready to be opened by depolarization.

In summary, the action potential results from three overlapping permeability changes:

1. Increased permeability to Na^+, caused by the rapid opening of voltage-gated Na^+ channels
2. Decreased permeability to Na^+, caused by the inactivation of Na^+ channels
3. Increased permeability to K^+, caused by the slower opening of voltage-gated K^+ channels

All three permeability changes are initiated by depolarization of the membrane, and thus are characterized as *voltage-dependent* permeability changes. The voltage dependence of neuronal membrane permeabilities permits action potentials and gives the action potentials their unique all-or-none property.

THE HODGKIN CYCLE EXPLAINS THE RISING PHASE OF THE ACTION POTENTIAL

To see how voltage dependence makes an action potential all-or-none, let's examine the increase in permeability to sodium that underlies the rising phase of the action potential. We have discussed how increased permeability to and inflow of Na^+ depolarizes the membrane. *The critical feature of action-potential generation is that the permeability to Na^+ that produces depolarization itself depends on depolarization.* The **Hodgkin cycle** describes the effects of depolarizing an excitable membrane in which the permeability to sodium (P_{Na}) is voltage-dependent. (The cycle is named after Sir Alan Hodgkin, who was a corecipient of the Nobel Prize for his work clarifying the ionic mechanism of action potentials.) The cycle (Figure 11.13) consists of three processes that feed back on each other in a cyclic manner.

The Hodgkin cycle describes a positive feedback loop that starts with depolarization: Changing V_m changes P_{Na}, and (as predicted by the Goldman equation) changing P_{Na} changes V_m. At rest, the membrane is 20 to 50 times as permeable to K^+ as to Na^+, so the resting V_m is near E_K. Subthreshold depolarizations open some voltage-gated Na^+ channels, but not enough to overcome the effects of the higher resting permeability to K^+. At threshold, the current carried by Na^+ inflow just equals the K^+ current, and at any depolarization above threshold the Hodgkin cycle "wins." The regenerative increase in P_{Na} in the Hodgkin cycle makes the membrane transiently much more permeable to Na^+ than to K^+, so V_m approaches E_{Na} (+40 to +55 mV inside-positive).

The Hodgkin cycle explains only the rising phase of the action potential, since if the cycle alone were operating, the membrane potential would remain near E_{Na} indefinitely. Instead, the polarity reversal lasts only about a millisecond because the sodium channels become inactivated and voltage-gated potassium channels open, causing the membrane to repolarize rapidly.

SINGLE-CHANNEL CURRENT RECORDING FROM ION CHANNELS The changes in membrane permeability that cause action potentials can be visualized as the actions at individual ion channels. (They can also be seen as whole-cell ionic currents, which are discussed in the next section.) Evidence at the level of single ion channels comes from **single-channel current recording**, also termed **patch-clamp recording**.

In this procedure, a patch of membrane containing a single Na^+ ion channel is sealed by suction onto the smoothed tip of a fine glass micropipette electrode and then detached from the cell (Figure 11.14*a*), so that any current must flow through a channel in the isolated patch. The electrode records the opening and closing of the membrane ion channel by recording the ionic current that flows through the single channel when it is open. In response to a depolarization (Figure 11.14*b*), the channel opens, allowing Na^+ ions to produce an inward-flowing current, and then closes again in about a millisecond. By providing data about the opening and closing of single channels, patch-clamp recording allows direct visualization of the permeability changes underlying action potentials.

To show a more complete picture of the ion currents flowing in and out of a neuron during an action potential, let's consider three responses of voltage-gated Na^+ channels and three responses of voltage-gated K^+ channels to depolarization that is similar to the depolarization in an action potential (see Figure 11.15). Recordings of the voltage-gated Na^+ channels and voltage-gated K^+ channels reveal

(a) Extraction of Na^+ channel

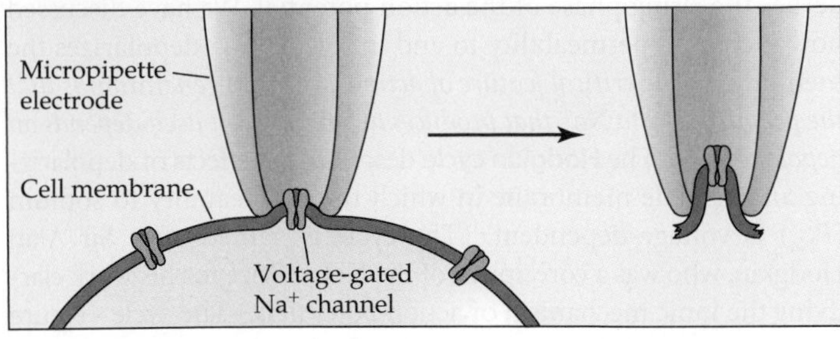

(b) Current flowing through Na^+ channel

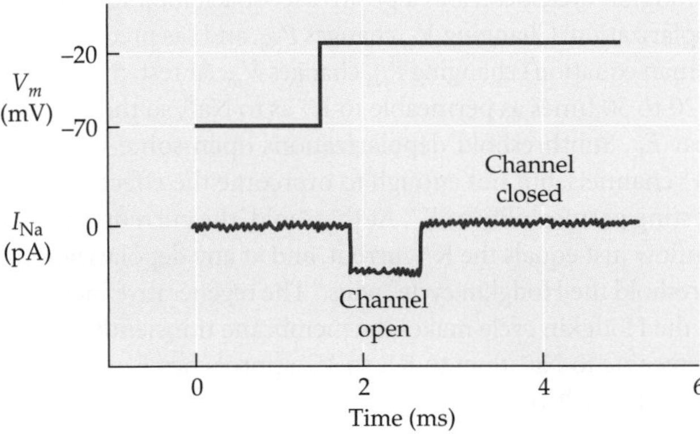

Figure 11.14 Patch-clamp recording of single-channel currents *(a)* A fine, fire-polished glass electrode is fused to the membrane with suction, making what is known as a *gigaohm seal*, and the patch of membrane is pulled away from the cell. The electrode will then record current flowing through the channel when it opens. *(b)* A voltage-gated Na^+ channel is closed at resting potential (–70 mV), and no current flows through it. When the membrane patch is depolarized, the channel opens transiently, allowing an inward current carried by Na^+ ions. (Note that inward is toward the cytoplasmic side, not relative to the pipette, and that ion concentrations of the solutions in the bath and inside the electrode on either side of the patch are similar to the respective concentrations inside and outside the cell before detachment.) pA = picoampere (a measure of electrical current).

conspicuous differences in latency and action of each type of channel. The channels are normally closed at resting potential, and depolarization increases the *probability* that they will open.

The voltage-gated Na^+ channels have a short latency and open first but rapidly become inactivated (Figure 11.15*a*). The K^+ channels open with a slightly longer latency, do not become inactivated, and tend to stay open until the depolarization ends (Figure 11.15*b*). The six individual single-channel currents illustrate the three effects of depolarization on Na^+ and K^+ channels: (1) Na^+ channels open first in response to the depolarization, then (2) they are inactivated during depolarization, and (3) K^+ channels open slightly later than the Na^+ channels but do not become inactivated.

VOLTAGE-CLAMP EXPERIMENTS SHOW WHOLE-CELL IONIC CURRENTS Before the development of single-channel current recording, researchers used a whole-cell current-measuring technique called a *voltage clamp* in experiments to study action-potential generation, and these experiments became a cornerstone of the physiological investigation of action potentials. A **voltage clamp** is an electronic device that allows the experimenter to measure whole-cell ionic currents, by setting membrane potential very rapidly to a predetermined value, delivering whatever current is necessary to keep it there, and measuring the imposed current.

Recall that we described the Hodgkin cycle as a positive feedback loop in which a change in membrane potential changes the permeability to sodium ions, and vice versa. As the Hodgkin cycle exemplifies, any ion flow through the membrane constitutes an ionic current that tends to change the membrane potential. Clamping the membrane potential uncouples the feedback loop of the Hodgkin cycle. To keep the potential constant, the clamp circuit must generate an opposing (bucking) current that is exactly opposite to the net ionic current (current carried by ion flows through ion channels). *By measuring the bucking current, the experimenter has an accurate measure of the amplitude and time course of the net ionic current, because the two must be equal and opposite to each other.* Hence a voltage clamp uncouples the feedback loop of the Hodgkin cycle (at point *c* in Figure 11.13) so that ionic currents resulting from permeability changes are prevented from changing the membrane potential.

In 1952, Alan Hodgkin and Andrew Huxley published a series of landmark papers in which they used voltage-clamped squid axons to demonstrate and quantify the voltage-dependent permeability changes underlying the action potential. Figure 11.16 shows the most fundamental result of such a voltage-clamp experiment. When the membrane potential is clamped to a hyperpolarized value (see Figure 11.16*a*), the current-measuring circuit shows only a brief blip of capacitive current required to set the membrane potential to a new level (changing the charge stored by the membrane capacitance). The capacitive transient is not ionic current (i.e., not current flowing through ion channels), but just a shift in accumulated charges on either side of the membrane. Following the capacitive transient there is only a slight leakage current resulting from holding the membrane at a hyperpolarized level (too small to appear within the scale of the current record shown in Figure 11.16*a*). Hyperpolarization thus leads to no significant flow of ionic current because it doesn't increase permeability to any ions.

In contrast, clamping the membrane potential to a value more *depolarized* than the resting potential produces quite different effects (see Figure 11.16*b*). Following the initial capacitive transient, bucking current is required to hold the membrane at the set value. The bucking current (which is not shown) flows first outward and then inward. Because the bucking current is equal and opposite to the net ionic current, this pattern shows that there is an *early inward ionic current* that is reversed in 1 to 2 ms to a *later outward ionic current.* Depolarization of the membrane thus induces permeability changes that (if the currents are carried by cations) result in first an inward movement of cations and then an outward movement of cations. If the membrane were not clamped, these ionic currents would produce first a depolarization and then a repolarization of the membrane, as in an action potential.

Hodgkin and Huxley proposed that the early inward ionic current (which generates the rising phase of the action potential in unclamped axons) is an influx of Na^+ ions. How could this prediction

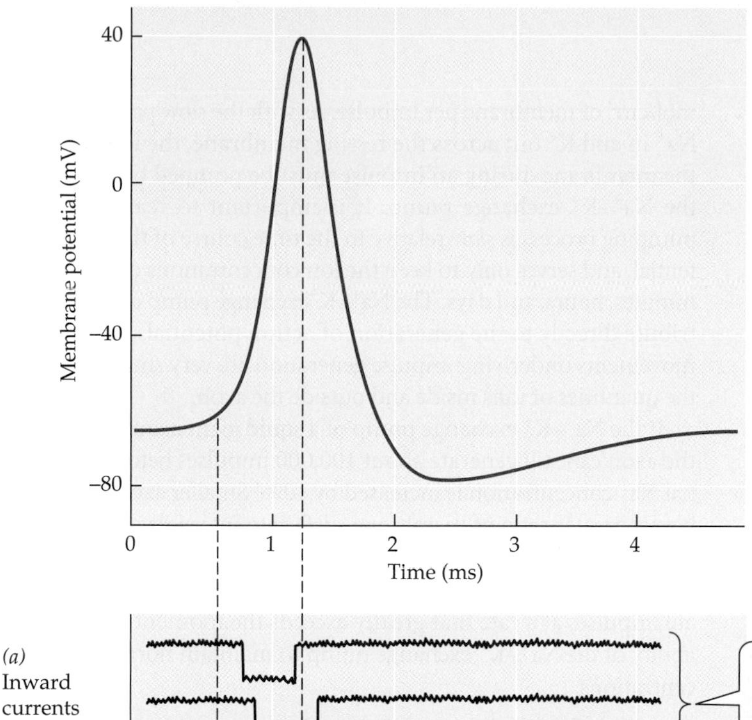

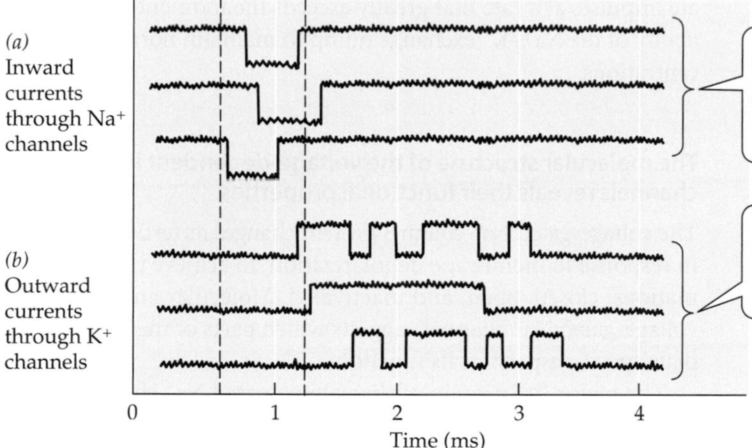

Figure 11.15 Patch-clamp recording of single-channel currents underlying an action potential These diagrams illustrate hypothetical patch-clamp recordings of inward currents through three representative voltage-gated Na^+ channels *(a)*, and outward currents through three representative voltage-gated K^+ channels *(b)* of the hundreds that produce the action potential. (These could be successive responses of the same voltage-gated Na^+ channel, or responses of different voltage-gated Na^+ channels; similarly, the voltage-gated K^+ channels could be the same or different.) Note that the voltage-gated Na^+ channels open in a narrow time window that corresponds to the rising phase of the action potential. The extended permeability to K^+ can lead to an after-hyperpolarization of the membrane. (See Figure 11.11*a* for a description of the different phases of the action potential diagrammed at the top of this illustration.) (After Bear et al. 2001.)

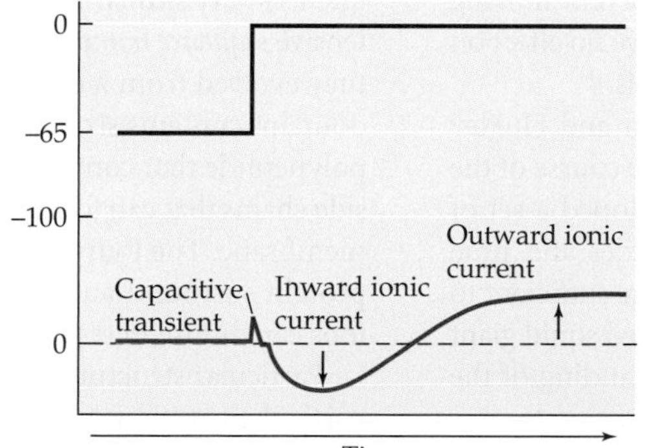

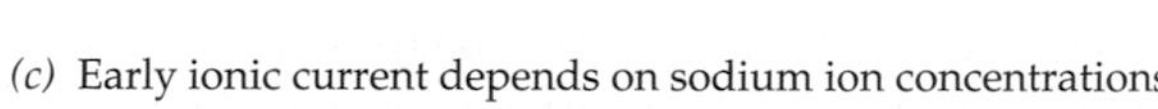

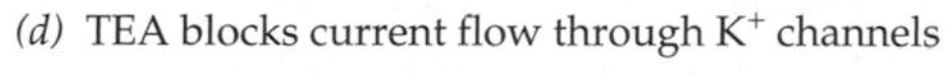

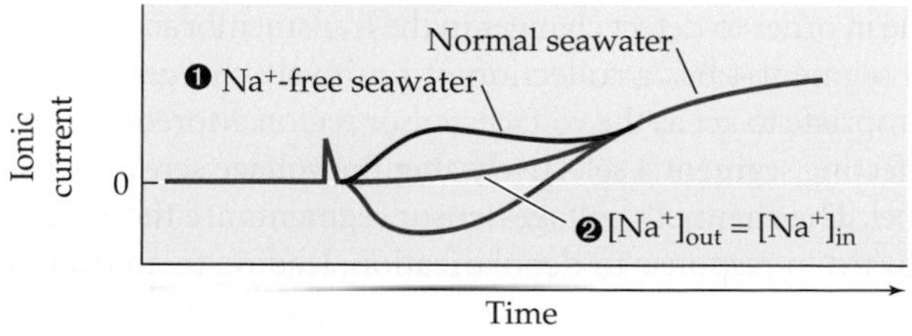

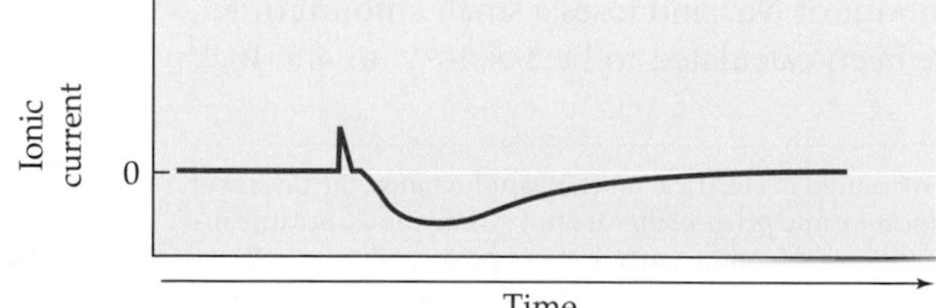

Figure 11.16 A voltage-clamp experiment reveals ionic currents during the action potential *(a)* The membrane potential is clamped at a hyperpolarized level (–100 mV) relative to the resting potential (–65 mV). After a brief capacitive transient, this hyperpolarization results in no significant ionic current. *(b)* The membrane is clamped at a depolarized level (0 mV). Depolarization induces an early *inward ionic current*, followed by a later *outward ionic current* that persists as long as the depolarization is maintained. *(c)* A voltage-clamp demonstration that the early inward current is carried by sodium ions. ❶ Bathing an axon in Na^+-free seawater (Na^+ replaced by choline) results in reversal of the early inward current to an outward current. ❷ When the outside Na^+ concentration ($[Na^+]_{out}$) is made equal to the inside Na^+ concentration ($[Na^+]_{in}$), a voltage clamp to zero elicits no early current. The late outward current is unaffected by changes in Na^+ concentration because it is not carried by Na^+ ions. *(d)* Tetraethylammonium (TEA) ions block K^+ channels, leaving only the early inward (Na^+) current.

be tested? Hodgkin and Huxley replaced the Na^+ in the seawater with which they bathed the axon with choline, an nonpermeating cation. In the absence of extracellular Na^+, the early inward current was replaced by an early outward current (Figure 11.16*c*). That is, depolarization induced an increase in permeability to Na^+, which, in the absence of extracellular Na^+, resulted in Na^+ diffusion *outward* down its concentration gradient.

This interpretation predicts that if the Na^+ concentration is equal on both sides of the membrane, there will be no Na^+ concentration gradient and no early Na^+ current in either direction. Hodgkin and Huxley replaced about 90% of the extracellular Na^+ with nonpermeating ions so that $[Na^+]_{in} = [Na^+]_{out}$. When the membrane was clamped to 0 mV (so that there was no voltage gradient), there was no early current (see Figure 11.16*c*).

Further evidence that the early inward current is carried by Na^+ was provided by experiments in which a squid axon in normal artificial seawater was clamped to the sodium equilibrium potential (E_{Na} = +50 mV). There was no resultant early current because there was no driving force on Na^+ ions at E_{Na}. Clamping the membrane at a level beyond E_{Na} (more inside-positive than E_{Na}) resulted in an early outward current, representing Na^+ efflux toward E_{Na}. These experiments demonstrate that Na^+ ions carry the early inward current during a voltage clamp, but not the later outward current, which is unchanged by changing Na^+ concentrations. Other experiments demonstrate that the late current is a K^+ efflux.

Pharmacological agents used in conjunction with a voltage clamp confirm that sodium and potassium currents flow through separate ion channels. Certain drugs can selectively block Na^+ and K^+ channels when they are applied to the membrane. For example, tetrodotoxin (TTX), an extremely poisonous substance found in puffer fish, selectively blocks voltage-dependent Na^+ channels. If a squid axon is bathed in seawater containing TTX and is voltage-clamped to a depolarized level such as 0 mV, the early inward Na^+ current is blocked. The delayed outward (K^+) current, however, is completely unaffected. On the other hand, tetraethylammonium (TEA) ions selectively block the delayed outward current flowing through K^+ channels (see Figure 11.16*d*). TEA ions have no effect on the early inward current flowing through Na^+ channels.

From their voltage-clamp experiments, Hodgkin and Huxley were able to quantify the voltage dependence and time course of the changes in permeability to Na^+ and K^+[5] They developed a set of equations by which they showed that the magnitudes and time courses of these three voltage-dependent processes are sufficient to describe the behavior of action potentials in unclamped squid giant axons. These studies remain critical for our understanding of the physiology of excitable membranes.

ION MOVEMENTS IN ACTION POTENTIALS DO NOT SIGNIFICANTLY CHANGE BULK ION CONCENTRATIONS In the generation of an action potential, a neuron gains a small amount of Na^+ and loses a small amount of K^+. These amounts have been calculated to be 3×10^{-12} to 4×10^{-12} mol/cm^2 of membrane per impulse. As with the slow passive leaks of Na^+ in and K^+ out across the resting membrane, the ions crossing the membrane during an impulse must be pumped back again by the Na^+–K^+ exchange pump. It is important to realize that the pumping process is *slow* relative to the time course of the action potential, and serves only to keep the ion concentrations constant over minutes, hours, and days. The Na^+–K^+exchange pump does not contribute directly to the generation of action potentials, and the ion movements underlying impulse generation are very small relative to the quantities of ions inside and outside the axon.

If the Na^+–K^+ exchange pump of a squid giant axon is poisoned, the axon can still generate about 100,000 impulses before the internal Na^+ concentration is increased by 10%! Smaller axons, however, have a greater ratio of membrane surface to internal volume, so the concentration changes produced by impulses are greater. Therefore, the smallest axons (0.1 μm in diameter) presumably cannot generate impulses at a rate that greatly exceeds the moment-to-moment ability of the Na^+–K^+ exchange pump to maintain normal ion concentrations.

[5] Permeabilities are often measured in electrical units of **conductance**, the inverse of resistance ($g = 1/R$). *Conductance* and *permeability* are not synonymous, because increasing ion concentrations increases conductance, but not permeability.

The molecular structure of the voltage-dependent ion channels reveals their functional properties

The voltage-gated Na^+ channel protein changes its tertiary structure in response to membrane depolarization, to achieve three conformations: closed, open, and inactivated. Molecular analysis of the voltage-gated Na^+ channel suggests which parts of the protein mediate specific aspects of its function.

The major (α) protein of the voltage-gated Na^+ channel forms the channel itself. It consists of a single polypeptide chain (Figure 11.17). (There are two smaller, variable β peptide subunits that interact with and modify the single α protein. However, the α protein alone is sufficient to produce voltage-gated Na^+ currents, and we will consider it only.) The Na^+ channel α polypeptide chain has four *domains* or regions, each domain consisting of amino acid sequences very similar to the other three. Thus there is said to be extensive *sequence homology* among the four domains, suggesting that they evolved from a common ancestral peptide. Each of the four domains contains six *membrane-spanning segments*, regions of the polypeptide that contain predominantly hydrophobic amino acid side chains that can form α-helices and cross the lipid bilayer of the membrane. The four domains of the voltage-gated Na^+ channel α protein surround an aqueous channel pore, through which Na^+ ions can diffuse in response to depolarization (see Figure 11.17*b*).

Particular structural regions of the channel protein impart to it particular functional properties. For example, the region of the protein that responds to voltage is membrane-spanning segment 4 of each domain (see Figure 11.17*a* and *b*). The voltage-sensor region of the channel must be charged, but it must also be in or close to the membrane in order to detect changes in the transmembrane electric field, and segment 4 has a collection of positively charged amino acids appropriate to act as the voltage-sensor region. Moreover, mutations affecting segment 4 selectively alter the voltage sensitivity of the channel. The channel's voltage-sensor segments are thought to pivot outward in response to depolarization, leading to an overall

(a) Topology of voltage-gated Na^+ channels

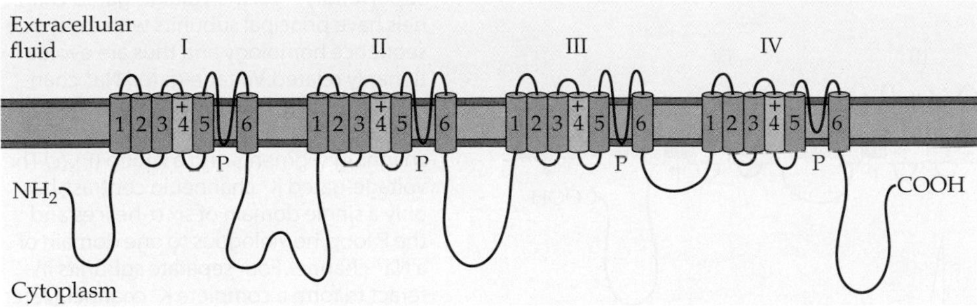

(b) Surface view of a Na^+ channel

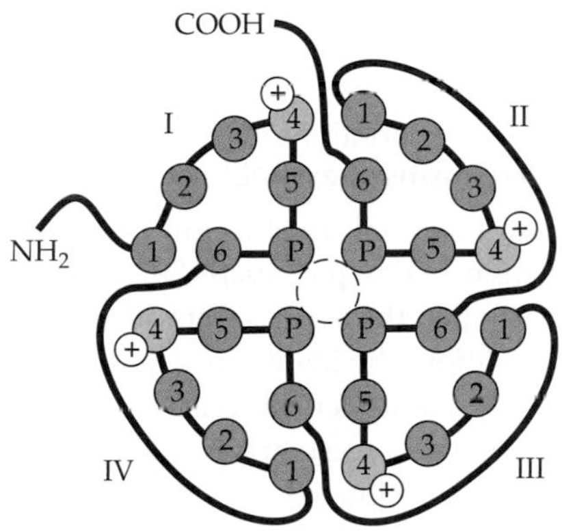

(c) Voltage-dependent conformational change

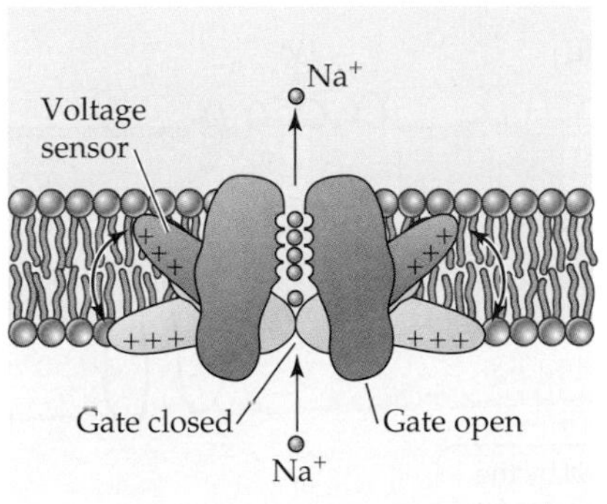

Figure 11.17 The molecular structure of voltage-gated Na^+ channels *(a)* The predicted structure of the principal (α) subunit of the voltage-gated Na^+ channel. This subunit is the pore-forming channel itself and is a single polypeptide chain, with four homologous domains labeled I, II, III, and IV. Each domain has six membrane-spanning α-helical segments (labeled 1 through 6). Segment 4 of each domain (the fourth membrane-spanning α-helical segment) is thought to be the voltage sensor. *(b)* A hypothetical surface view of a Na^+ channel. The P loops between segments 5 and 6 of each domain are thought to line the pore and determine ion selectivity. Other segments (not shown) mediate channel inactivation, intracellular regulation, and association with other channel subunits. *(c)* A hypothetical three-dimensional structure of the channel, showing closed and open conformations. The four domains surround a central pore. The segment-4 voltage sensors are thought to rotate and slide outward in response to depolarization, leading to channel opening.

conformational change in the channel from closed to open (see Figure 11.17*c*).

Another structural correlate of a critical channel function is the P loop connecting segments 5 and 6 of each domain. This loop lines the pore of the ion channel (see Figure 11.17*b*) and helps mediate ion *selectivity*. Mutations in the P-loop region alter ion selectivity in ways consistent with this idea. Finally, the cytoplasmic loop between domains III and IV appears to mediate inactivation of the Na^+ channel; it is thought to act as a "ball on a string" that can block the (open) channel from the cytoplasmic side.

Other voltage-gated channels are structurally similar to the voltage-gated Na^+ channel (Figure 11.18). Channels showing such similarity include the K^+ channels that repolarize the membrane in an action potential, as well as Ca^{2+} channels involved in neurotransmitter release. (Like the Na^+ channels, these channels have modulatory polypeptides, but only the principal α protein of each is considered here.) Na^+, Ca^{2+}, and K^+ ion channels are similar in overall structure and have extensive homology in amino acid sequence. The sequence homology is greatest in certain regions (conserved regions), such as the voltage-sensor region of membrane-spanning segment 4. Their homology suggests that the various ion channels are evolutionarily related, and therefore they are referred to as the **voltage-gated channel superfamily** of membrane proteins.

The Na^+ and Ca^{2+} channels are most similar in structure (see Figure 11.18*a* and *b*). The K^+ channels (see Figure 11.18*c*) consist of four identical subunits that resemble one of the four domains of a Na^+ channel. Evidently four K^+ channel proteins interact as subunits to form a channel, aligning like the four domains of a Na^+ channel protein around a central pore. There are several subtypes of each kind of ion channel, with an especially large number of K^+ channel subtypes. The evolution of the voltage-gated channel superfamily is discussed in Box 11.1.

What is the structural basis of the ion selectivity of the channels? The ions Na^+ and K^+ are chemically similar monovalent cations (of the same electromotive series of the periodic table). How can one kind of channel be selectively permeable to Na^+ and another be selectively permeable to K^+? Actually, neither the Na^+ channel nor the K^+ channel is absolutely selective; both channels are slightly permeable to the "wrong" ion and are also permeable to other ions that are not normally present in organisms. By studying the relative permeabilities of channels to different ions, one can infer something about the structural elements of the channels that determine their ion-selective properties.

Recall that ions in solution are normally hydrated—that is, surrounded by a shell of water molecules. The water molecules around a cation have their electronegative oxygen atoms facing inward toward the ion, held to it by charge attraction. A "naked" (unhydrated) Na^+ ion is smaller than a naked K^+ ion, but it can attract a larger shell of water molecules. Evidence indicates that the voltage-dependent Na^+ channel has a narrow selective region with a minimum

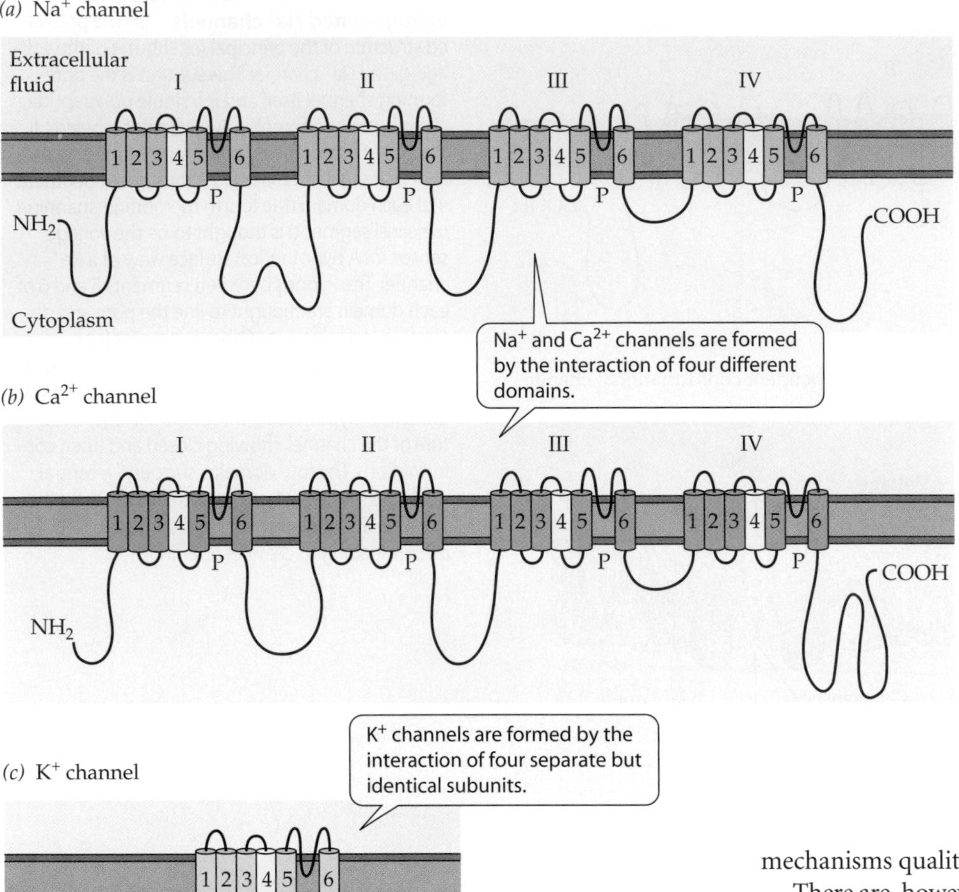

Figure 11.18 The voltage-gated channel superfamily All the voltage-gated channels have principal subunits with extensive sequence homology and thus are evolutionarily related. Voltage-gated Na^+ channels *(a)* and Ca^{2+} channels *(b)* have four domains, each with six α-helical membrane-spanning segments and a P loop (P). *(c)* The voltage-gated K^+ channel, in contrast, has only a single domain of six α-helices and the P loop, homologous to one domain of a Na^+ channel. Four separate subunits interact to form a complete K^+ channel. (After Kandel et al. 1995.)

pore size of 0.3 × 0.5 nm. Hydrated ions cannot pass through the pore, but it is lined by oxygen atoms that compete with the water molecules of the hydrated Na^+ ion, so all but one of the water molecules are left behind when the ion diffuses through the channel. A K^+ ion with one water molecule of hydration is larger, so a singly-hydrated K^+ is considerably less likely to pass through the Na^+ channel.

Permeation through K^+ channels is more completely understood but more complicated (Figure 11.19). Recent X-ray diffraction studies have clarified the three-dimensional structure of K^+ channels and show that K^+ ions completely lose the water of hydration in passing through the narrow pore of the K^+ channel's selectivity filter. There are four K^+ ion sites in the selectivity filter, but typically only alternate sites are occupied. Entry of a K^+ ion from the cytoplasm into the inner chamber of the channel displaces other ions by charge repulsion (see Figure 11.19*c*), "knocking" ions to the next site. Although the basis of Na^+ exclusion is not completely clear, these studies are elucidating the molecular basis of an ion channel's ability to be both highly permeable and highly selective.

There are variations in the ionic mechanisms of excitable cells

How universal are the ionic mechanisms of action potentials? Studies have shown that the basic aspects of impulse generation elucidated in squid axons apply to most excitable cells. The action potentials of vertebrate and invertebrate unmyelinated axons, amphibian myelinated axons, and vertebrate skeletal twitch muscle fibers have ionic mechanisms qualitatively similar to those of squid axons.

There are, however, variations on the theme: Some neurons may lack some voltage-gated channels or may possess additional channels. Studies reveal a wide range of ion channels in excitable membranes, each identified by its unique current signature (Table 11.2). The variations on action potential–generating mechanisms may endow the cell with special features, such as the ability to generate spontaneous action potentials or bursts of action potentials. Here we consider two examples of variations on the usual action-potential theme.

GRADED POTENTIALS IN NONSPIKING NEURONS Not all nerve cells generate action potentials. Researchers report increasingly numerous examples of **nonspiking neurons** (neurons that do not generate the sharp "spikes" of action potentials). Nonspiking neurons produce only *graded* membrane-potential changes in response to a stimulus or synaptic input because they substantially lack voltage-gated sodium channels.

Figure 11.20 compares signal transmission in spiking and nonspiking neurons. The input of both neurons (sensory stimulation or synaptic input) is graded in amplitude, and so is their output (neurotransmitter release). Spiking neurons encode the graded input signals into trains of action potentials for long-range transmission; the action potentials are recoded at the terminal to control graded release of neurotransmitter. Nonspiking neurons, in contrast, are typically compact cells with short axons or no axons, so a graded potential change at one part of the cell can spread passively (electrotonically) to the terminal without major decrement.

The input and output of spiking and nonspiking neurons are the same, but the short-axon nonspiking neuron does not require spike

BOX 11.1 The Evolution of Voltage-Gated Channels

The fact that voltage-gated channels have striking sequence homology and overall structural similarity shows that they are clearly evolutionarily related. The sequence of their evolutionary divergence is less certain, but the following sequence is likely (see the figure).

K^+ channels are thought to have evolved first. Primitive ion channels may have resembled bacterial K^+ channels, which consist of four subunits, each with two transmembrane segments (2TM) (1). Similar 2TM K^+ channels occur widely, from bacteria to mammals, and are not voltage-gated.

K^+ channels with six transmembrane segments (6TM) (2) are activated by depolarization and are present in animals, plants, fungi, and protists. The 2TM K^+ subunit corresponds to segments 5 and 6 and the ion-selective P loop of the 6TM protein. The smaller K^+ channel protein may have combined with another protein that contained the segment-4 gating region to make the channel voltage-dependent.

Ca^{2+} and Na^+ channels are thought to have evolved from the 6TM K^+ channel protein by gene duplications and successive mutations, leading to the 4×6 TM structure of these channels (3). Other channel proteins may have evolved from 6TM K^+ subunits as well.

Some K^+ channels can be inactivated, and their subunits have an extra portion at the NH_2 terminal that acts as a "ball on a string" to close the channel's inner end (4). Other K^+ channels are activated by intracellular Ca^{2+} ions and have a Ca^{2+}-binding site near the COOH terminal (5). Cyclic nucleotide–activated cation channels have a binding site for cyclic nucleotides (6), and mutations at the P loop make the channels unselective among cations.

Increasing knowledge of the structure of channel proteins will likely permit further understanding of the proteins' evolution.

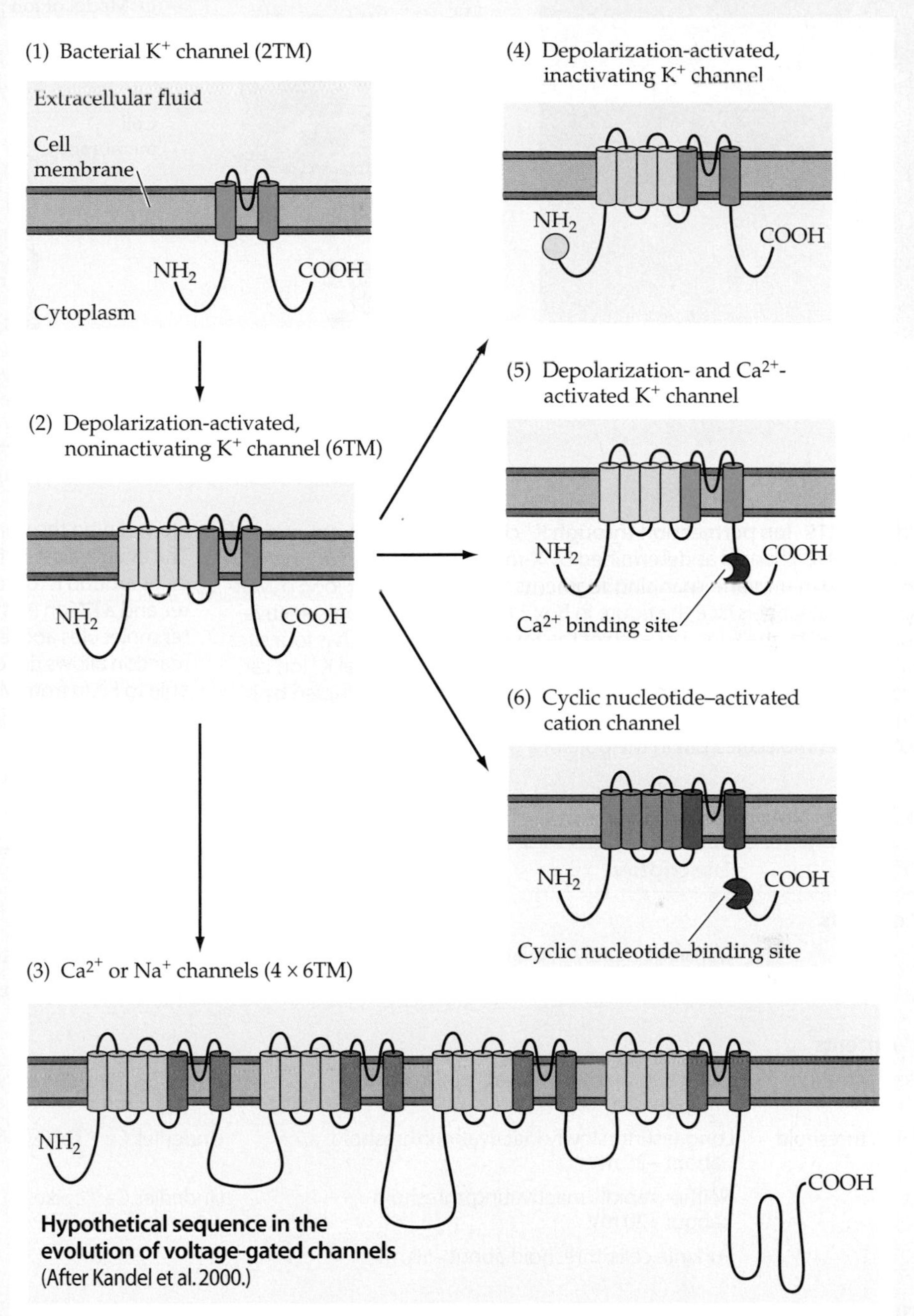

Hypothetical sequence in the evolution of voltage-gated channels (After Kandel et al. 2000.)

encoding to carry the signal over large distances. Examples of nonspiking neurons include the photoreceptors, bipolar cells, and horizontal cells of the vertebrate retina (see Chapter 13), granule cells of the olfactory bulb, and many arthropod interneurons.

PACEMAKER POTENTIALS OF SPONTANEOUSLY ACTIVE CELLS Many neurons are spontaneously active, generating action potentials at rather regular intervals without an external source of depolarization. The somata of some molluscan neurons, for example, generate action potentials in regular trains, or even in repetitive bursts, in the absence of synaptic input. Vertebrate cardiac muscle fibers and some other excitable cells are also spontaneously active.

The membrane potential of a spontaneously active cell, instead of maintaining a fixed resting value, undergoes a continuous ramp of

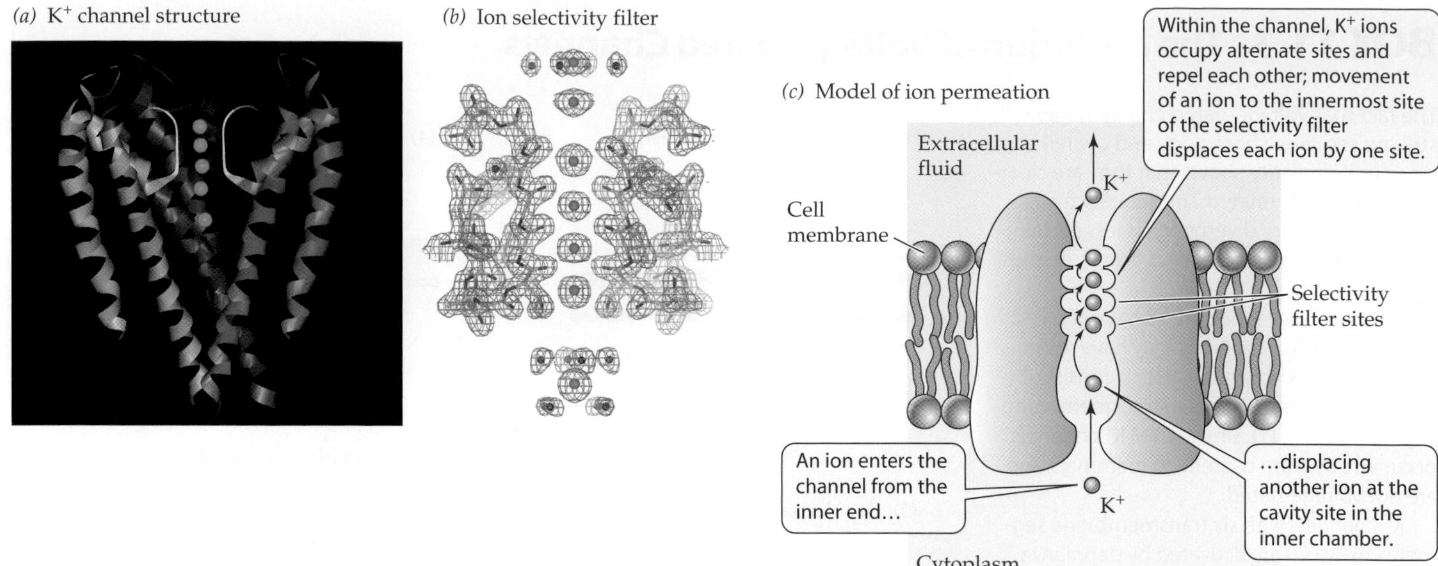

Figure 11.19 Ion permeation through K^+ channels *(a)* The structure of a bacterial K^+ channel as determined by X-ray diffraction. The subunits correspond to membrane-spanning segments 5 and 6 and the P loop of voltage-gated channels (see the figure in Box 11.1). The narrow ion-selective pore is lined by the P loop of each of the four subunits, and it has four sites that can be occupied by K^+ ions (green spheres). An additional K^+ ion can occupy an inner cavity below the selectivity filter. *(b)* Sites occupied by K^+ ions in and near the selectivity filter. K^+ ions are normally surrounded by polar water molecules, but in the pore of a potassium channel, oxygen atoms lining the pore compete with water molecules to attract the cation. The image shows a K^+ ion in the inner chamber with eight water molecules around it (bottom), four unhydrated K^+ ion sites at the selectivity filter, and a K^+ ion at the outer face of the pore, with a partial shell of four water molecules above it (top). *(c)* A model of K^+ ion permeation. This chain reaction allows the channel to be both highly selective and highly permeable to K^+. (*a* from Morais-Cabral, Zhou, and MacKinnon 2001; *b* from Zhou et al. 2001; *c* after Miller 2001.)

TABLE 11.2 Neuronal ionic currents

Current	Description	Function
Na^+ currents		
$I_{Na,t}$	Transient; rapidly activating and inactivating	Mediates action-potential rising phase
$I_{Na,p}$	Persistent: noninactivating	Enhances depolarization; sustained firing
Ca^{2+} currents		
I_T, low threshold	Transient; rapidly inactivating; threshold negative to –65 mV	Underlies rhythmic burst firing
I_L, high threshold	Long-lasting; slowly inactivating; threshold about –20 mV	Underlies Ca^{2+} spikes of dendrites; involved in synaptic transmitter release
I_N	Neither; rapidly inactivating; threshold about –20 mV	Underlies Ca^{2+} spikes of dendrites; involved in synaptic transmitter release
I_P	Purkinje cells; threshold about –50 mV	
K^+ currents		
I_K	Activated by strong depolarization	Mediates action-potential repolarization
I_C	Activated by increase in $[Ca^{2+}]_i$	Mediates action-potential repolarization and interspike interval
I_{AHP}	Slow after-hyperpolarization; sensitive to increase in $[Ca^{2+}]_i$	Slow adaptation of action-potential discharge; block by neuromodulators enhances excitability
I_A	Transient; inactivating	Delays onset of firing; lengthens interspike interval
I_M	Muscarine-sensitive; activated by depolarization; noninactivating	Contributes to spike frequency adaptation; block by neuromodulators enhances excitability
I_h	Depolarizing (mixed cation) current that is activated by hyperpolarization	Contributes to rhythmic spiking and rhythmic burst activity
$I_{K,leak}$	Contributes to neuronal resting potential	Block by neuromodulators changes E_m

Source: From McCormick 1999.

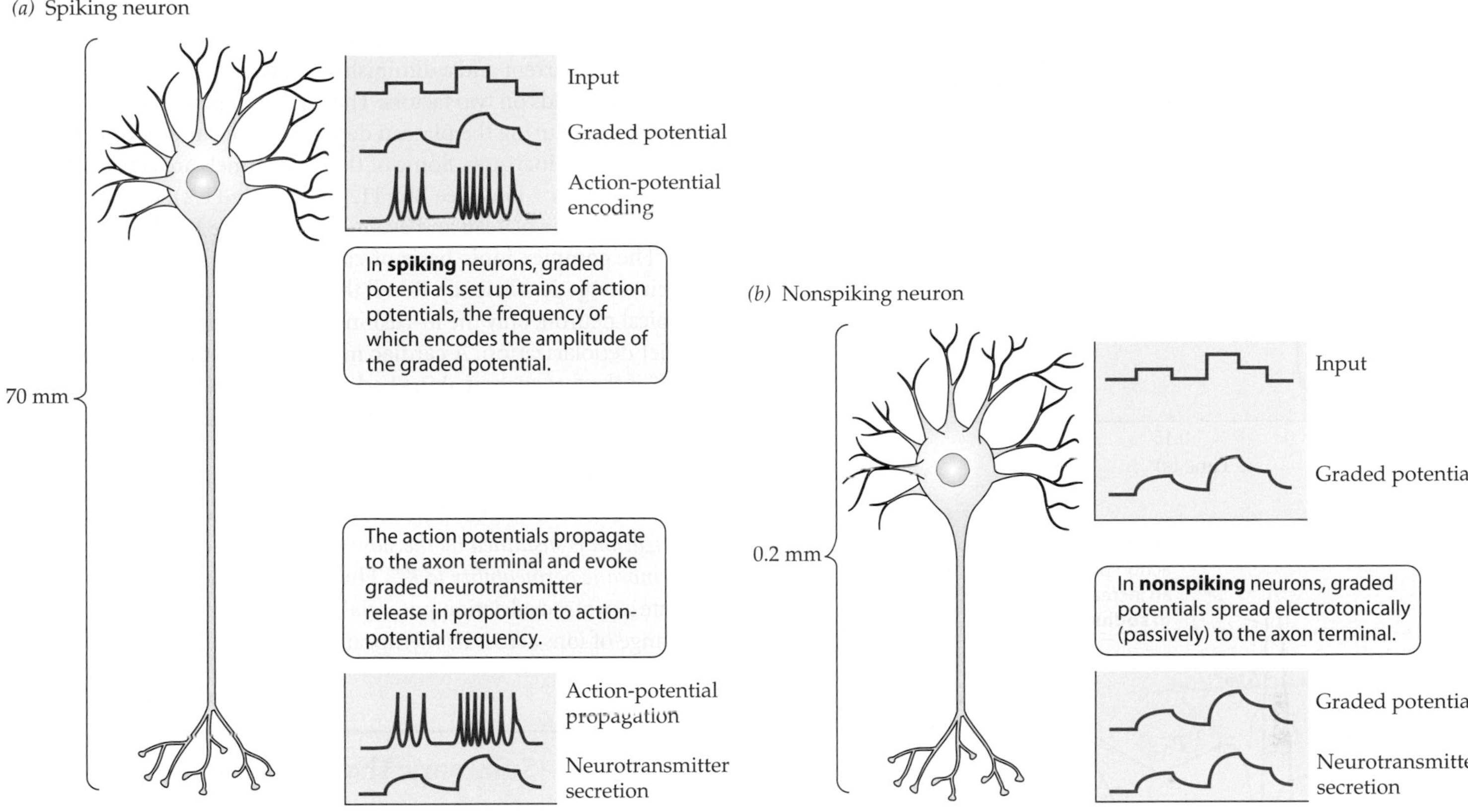

Figure 11.20 Nonspiking neurons do not generate action potentials Information transmission in an ordinary, spiking neuron *(a)* is contrasted with that in a nonspiking neuron *(b)*. *(a)* Input to the spiking neuron is graded in amplitude and evokes graded potentials in the dendrites and soma of the neuron. (The input could be an external stimulus for a sensory neuron, or synaptic input for a nonsensory neuron.) *(b)* Nonspiking neurons do not encode graded potentials into action potentials; instead the graded potentials spread electrotonically to the axon terminal. If the axon is short relative to its membrane length constant, the amplitude of the graded potential will be sufficient to evoke graded neurotransmitter release. Action-potential generation, then, may be an adaptation for signal transmission in long-axon neurons that is unnecessary in short-axon neurons.

depolarization between action potentials, until it reaches threshold for the generation of the next action potential. The repolarizing phase of an action potential restores the membrane to a relatively hyperpolarized level, from which the next ramp of depolarization begins. These ramp depolarizations are termed **pacemaker potentials** because they determine the rate of impulse generation by the cell. For example, in a cardiac muscle cell in the pacemaker region of a vertebrate heart, the greater the rate of depolarization during the ramp phase, the sooner the cell reaches threshold for the next action potential and, thus, the faster the heart rate. For vertebrate cardiac muscle fibers, norepinephrine increases the rate of depolarization during the ramp phase, while acetylcholine decreases it. (See Chapter 23 for discussion of heart rate control.)

The ionic basis of pacemaker potentials can be complex and may vary somewhat among cells. Spontaneously active cardiac muscle fibers have a slow inward cationic current (termed I_h) that is activated by hyperpolarization rather than by depolarization. This current is carried by both Na^+ and K^+ because the channels that open to produce it are not very selective among cations. I_h is activated at potentials more negative than –40 mV. The unique features of this current—the facts that it is inward and activated by hyperpolarization—give it pacemaker properties. I_h produces a ramp depolarization that triggers an action potential and then is inactivated at –40 mV. After the action potential, the repolarizing potassium current (I_K, flowing through voltage-gated K^+ channels) restores the membrane to a value near E_K. The hyperpolarization turns off I_K and turns on I_h for the next ramp depolarization.

CARDIAC MUSCLE ACTION POTENTIALS The action potentials of heart muscle fibers demonstrate a significant departure from the mechanisms of action-potential generation seen in squid axons. In contrast to most action potentials, which last about 0.4 to 3 ms, vertebrate cardiac muscle fibers have action potentials with typical durations of 100 to 500 ms (Figure 11.21*a*). The long duration of cardiac muscle action potentials is functionally important because action-potential depolarization is the necessary stimulus for myocardial contraction, and because cardiac muscle must contract for about 100 ms to pump blood effectively. A cardiac muscle fiber action potential has a

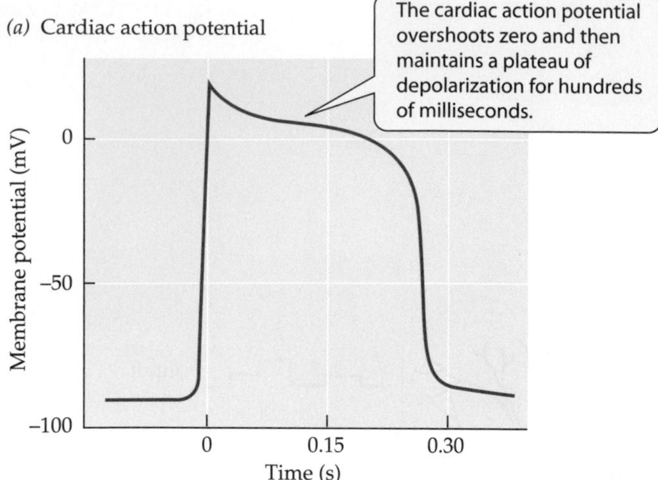

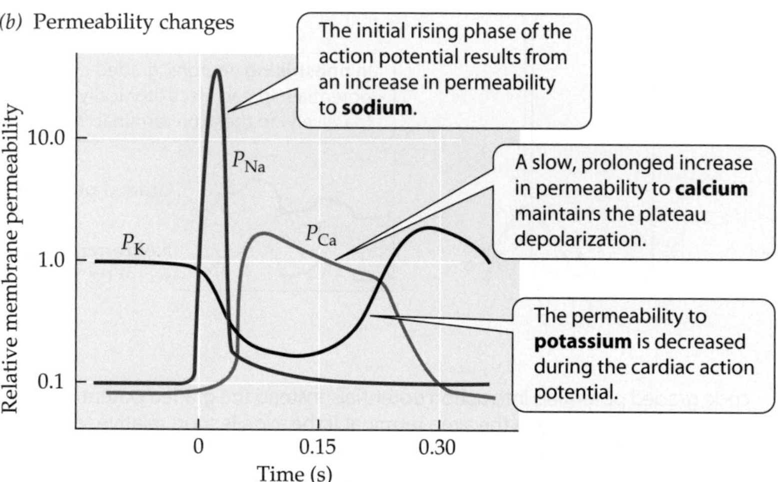

Figure 11.21 The cardiac muscle fiber action potential *(a)* An action potential in a vertebrate cardiac muscle fiber has a long duration. *(b)* Permeability changes underlie the cardiac action potential. Eventually, increased [Ca^{2+}] opens Ca^{2+}-activated K^+ channels, leading to repolarization.

rapid upstroke and a rapid initial recovery to near 0 mV, but it remains depolarized near zero for many milliseconds. This prolonged depolarization, the **plateau** of the action potential, gradually decreases and is followed by a relatively slow repolarization.

Two separate inward currents underlie the cardiac action potential. The first is a fast current resulting from increased permeability to Na^+ (Figure 11.21*b*), very similar to the squid axon's permeability to Na^+. The fast Na^+ current produces the rapid upstroke of the cardiac action potential and is inactivated within a few milliseconds. A second, slow inward current results mainly from increased permeability to Ca^{2+} (P_{Ca}) and helps produce the plateau. The Ca^{2+} channels take at least 20 ms to open, and their slow inward current is much weaker than the fast Na^+ current.

The other factor sustaining the plateau is a *decrease* during the plateau in permeability to K^+ from the resting level (in contrast to the increased permeability to K^+ that occurs in axons). Thus the plateau represents a balance between two small currents, a slow inward Ca^{2+} current and a diminishing outward K^+ current. Repolarization depends on two factors: The Ca^{2+} channels gradually become inactivated during the plateau depolarization, and permeability to K^+ gradually increases. Some of the K^+ channels are activated by intracellular Ca^{2+} ions (see Box 11.1 and Table 11.2), which accumulate during the plateau and eventually open the K^+ channels.

The complex ionic basis of cardiac action potentials increases their energetic efficiency. Recall that in a 1-ms action potential of a typical neuron, only the in-rushing fast sodium current sustains the brief depolarization. A cardiac muscle cell would be flooded with Na^+ if the fast current alone had to sustain a 100-ms action potential, and metabolically expensive ion-exchange pumps would be needed to clear the Na^+ ions from the cytoplasm. During a cardiac action potential, however, the duration of the fast sodium current is about the same as that of a standard action potential. Instead, *depolarization is sustained by the slow Ca^{2+} current and by the decreased membrane permeability to K^+*. Thus a cardiac muscle fiber can generate a protracted action potential with only a relatively modest exchange of ions across its membrane.

Summary: The Action Potential

- An action potential is a voltage change—a brief, transient reversal of membrane potential from inside-negative to inside-positive. Action potentials are all-or-none responses to any depolarization beyond a voltage threshold and are each followed by a brief refractory period.
- Action potentials result from voltage-dependent changes in membrane permeability to ions. Depolarization first opens voltage-gated Na^+ channels, allowing Na^+ ions to flow in and further depolarize the membrane toward E_{Na}. The voltage-gated Na^+ channels rapidly become inactivated to terminate the rising phase of the action potential; then voltage-gated K^+ channels open to repolarize the membrane.
- The effects of depolarization on membrane permeability to ions can be studied at the level of single channels by patch clamp, and at the whole-cell level by voltage clamp.
- Ongoing investigations are clarifying the molecular structures of voltage-gated channels. The principal protein subunit of a K^+ channel is a single chain with six transmembrane regions; a K^+ channel consists of four of these protein subunits around a central pore. Na^+ and Ca^{2+} channels consist of a single polypeptide chain with four similar domains; each domain corresponds to one of the four subunits of the K^+ channel. Functional attributes of the channels can be localized to particular regions of the proteins.
- Nonspiking neurons do not generate action potentials, and the ionic mechanisms of action potentials in excitable cells can vary. Calcium ions can make substantial contributions to action potentials in cardiac muscle cells and in some neurons. Other varieties of voltage-gated channels modify the excitable properties of neurons.

Propagation of Action Potentials

Consider that in large animals, single axons—such as those that control wiggling of your toes—can be at least a meter long. Now recall that an electrotonic voltage change at one point on a membrane decreases exponentially with distance. Indeed, with passive or electrotonic spread, the amplitude of a voltage change typically decreases to a third of its starting value in a fraction of a millimeter! Because of this decrement, passive electrotonic spread of a voltage change cannot serve as an electrical signal over long distances. Instead, there must be a mechanism to amplify or refresh the electrical signal.

We have seen that an action potential is all-or-none because the voltage-dependent, regenerative permeability increases bring the membrane potential toward a limiting value, which is the sodium equilibrium potential (E_{Na}). *Nondecremental propagation of the action potential is possible because the action potential at one location on an axon can itself initiate an action potential at a neighboring location, and the induced action potential will have the same all-or-none amplitude as the original action potential.* By repeating this process, a signal can travel a meter along an axon without any decrease in amplitude.

Local circuits of current propagate an action potential

An action potential at one locus on an axon depolarizes an adjacent locus by setting up local circuits of current flow (Figure 11.22). A complete local circuit is composed of a series of ionic and capacitive currents, which can be broken down into four components as shown in Figure 11.22*b*:

1. At the locus of the action potential, an ionic current begins with the inflow of sodium ions through open Na^+ channels into the intracellular fluid.
2. Ions flow in intracellular fluid, carrying the current to more-distant parts of the membrane (see Figure 11.6).

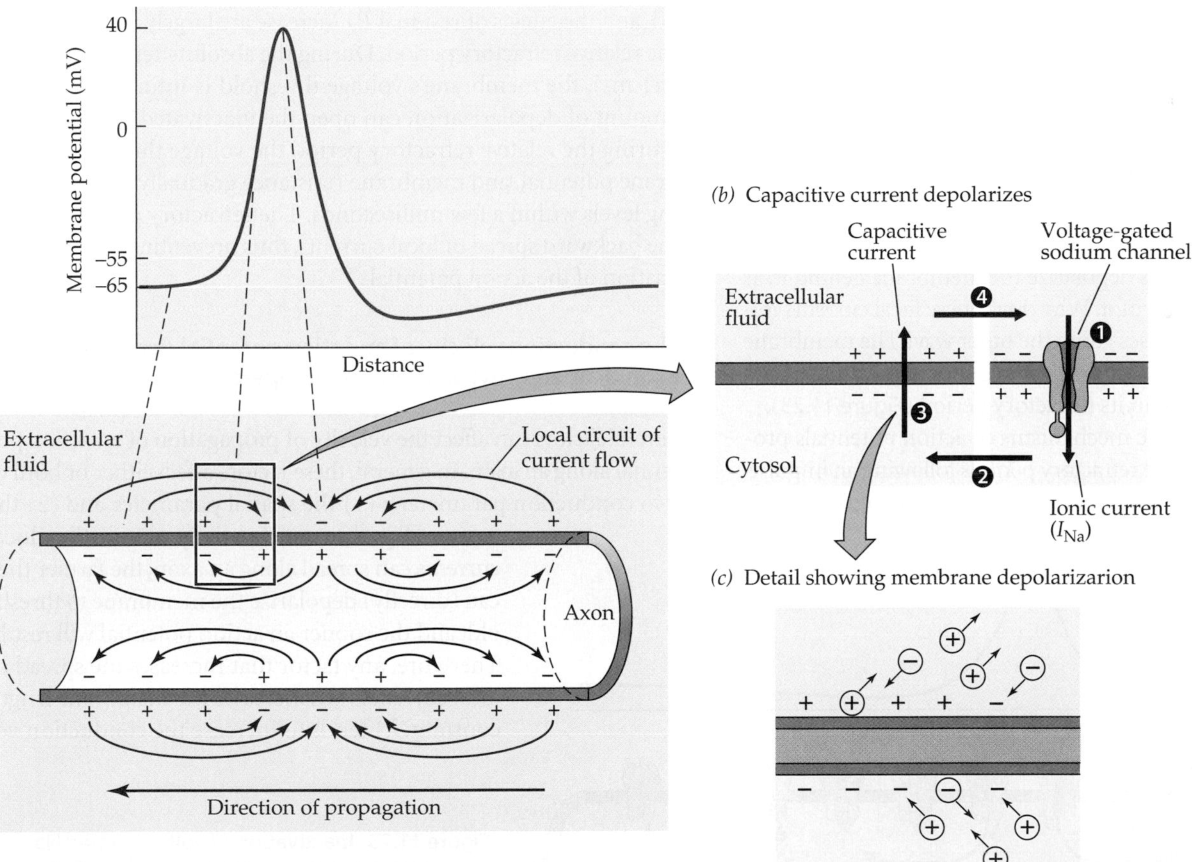

Figure 11.22 Propagation of an action potential An action potential is diagrammed here at an instant in time, while propagating from right to left. *(a)* The action potential induces local circuits of current flow along the axon, ahead of the advancing action potential and behind it. *(b)* The local circuit ahead of the action potential can be divided into four components as shown. These components are described in the text. *(c)* This diagram illustrates how the capacitive current on the left in *(b)* depolarizes the membrane ahead of the advancing action potential, by adding cations and removing anions at the inner side of the membrane, and adding anions and removing cations on the extracellular side. These changes in the distribution of charges on the membrane depolarize it, without any ions crossing the membrane. The depolarization resulting from the local currents opens voltage-gated sodium channels, leading (via the Hodgkin cycle) to an action potential at the new location.

3. At the membrane the ion movements change the distribution of charges on the membrane capacitance (Figure 11.22*c*): Cations accumulate along the membrane interior, displacing negative charges and repelling an equivalent number of cations from the membrane exterior. Although ions do not physically cross the membrane at this point, the movements of ions onto and off of the membrane surface constitute a capacitive current.
4. An (extracellular) ionic current completes the local circuit as cations move toward the locus of the action potential and anions move away.

During an action potential, local circuits of current such as the one described here spread the depolarization passively (electrotonically) along the surface of the membrane. *The action potential propagates to an adjacent portion of the axon because the capacitive depolarization produced by the local current lowers the membrane potential to threshold.* When this happens, the Hodgkin cycle takes over; the region of the membrane that reaches threshold undergoes a regenerative increase in permeability to Na^+ and generates its own action potential. In this way the nerve impulse passes along the entire length of the axon without any decrease in amplitude.

Membrane refractory periods prevent bidirectional propagation

An axon can conduct impulses equally well in either direction. For example, an action potential that is triggered via an electrode placed in the middle of an axon will be propagated in both directions from that point. Normally, however, impulses start at or near one end of an axon and travel along the axon in one direction. As an impulse is propagated, its local currents depolarize the membrane behind it, as well as the membrane ahead of it. Why don't these local currents initiate reverse-traveling impulses going the other way? The membrane behind a traveling impulse is not reexcited by the local currents because the membrane is still in its refractory period (Figure 11.23).

Three aspects of the ionic mechanisms of action potentials produce the absolute and relative refractory periods following an impulse, and thereby prevent reexcitation and bidirectional propagation:

1. The inactivation of sodium channels (which turns off the voltage-dependent increase in permeability to Na^+) lasts for at least a millisecond after an impulse passes a region of the membrane. Na^+ channel inactivation prevents the channels from entering the Hodgkin cycle until the action potential is far enough away to minimize local depolarization. This is the primary basis of the absolute refractory period.
2. The increased permeability to potassium (the slowest of the three voltage-dependent processes in onset) does not decrease to resting levels until after repolarization. The lingering P_K increase after an impulse may hyperpolarize the membrane toward E_K for a few milliseconds after the impulse. Thus (for those neurons whose action potentials have an undershoot) a region of membrane that has just generated an impulse is hyperpolarized *away* from its voltage threshold for new impulse generation.
3. The increase in P_K also renders a membrane refractory because it represents a decreased membrane resistance. The decreased resistance means that by Ohm's law ($V = IR$), local currents will cause a smaller voltage change, so more current is needed to depolarize the membrane to threshold.

Sodium-channel inactivation causes the absolute refractory period, and the effects of residual P_K increase are largely responsible for the relative refractory period. During the absolute refractory period (>1 ms), the membrane's voltage threshold is infinite because no amount of depolarization can open the inactivated Na^+ channels. During the relative refractory period the voltage threshold, membrane potential, and membrane resistance gradually return to resting levels within a few milliseconds. The refractory periods outlast the backward spread of local currents, thus preventing reverse propagation of the action potential.

The conduction velocity of an action potential depends on axon diameter, myelination, and temperature

Several factors can affect the velocity of propagation of an action potential along an axon. In general, these factors affect either or both of two conduction parameters: (1) the spatial parameter and (2) the temporal parameter. Spatially, the farther local currents can spread along an axon, the farther they can (directly) depolarize the membrane to threshold, and the sooner an action potential will result. Therefore, any factor that increases the spread of local currents (i.e., increases the membrane length constant, λ) tends to increase the conduction ve-

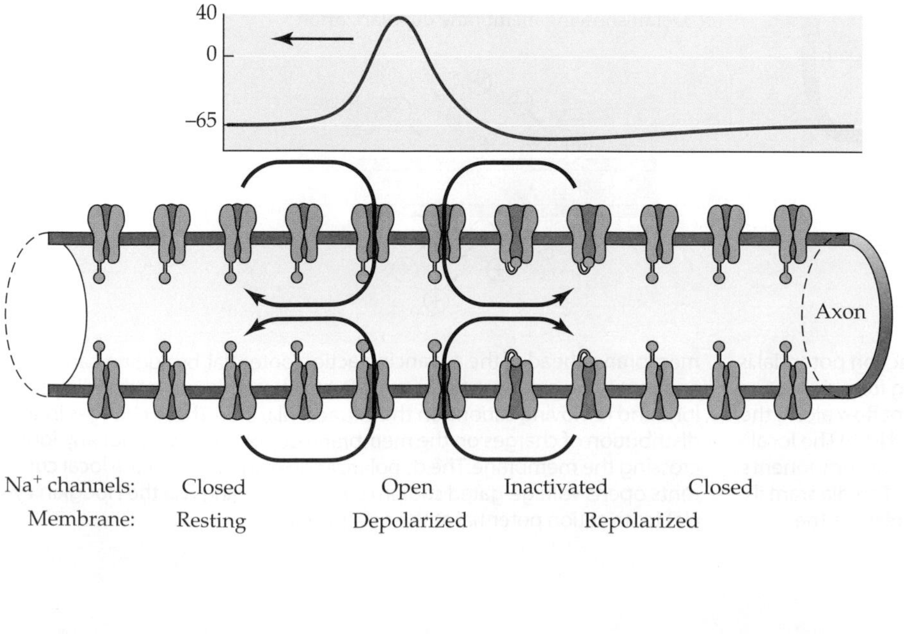

40
0
–65
Axon
Na^+ channels: Closed Open Inactivated Closed
Membrane: Resting Depolarized Repolarized

Figure 11.23 Inactivation of voltage-gated Na^+ channels prevents reverse propagation of an action potential An action potential is shown propagating from right to left; local currents depolarize the axon membrane ahead of the advancing action potential and behind it. The axon membrane behind the advancing action potential is refractory because its sodium channels are still inactivated. This refractory period prevents self-reexcitation by the trailing local currents.

locity of an action potential. Both large axon diameter and myelination increase the spatial spread of local currents.

With respect to the temporal parameter, the less time it takes the membrane to reach threshold, the faster the conduction velocity. Intrinsic membrane properties such as differences in the density of sodium channels may have minor effects on conduction velocity by influencing the spatial and temporal parameters. The three major evolutionary variables that influence conduction velocities are axon diameter, myelination, and temperature.

AXON DIAMETER AND CONDUCTION VELOCITY Large-diameter axons tend to conduct action potentials more rapidly than small-diameter axons. Many animal groups have evolved rapidly conducting neuronal giant axons (Box 11.2). Conduction velocity increases with axon diameter because larger-diameter axons have longer length constants, and thus farther spread of local currents. The length constant depends principally on two types of resistance (see Figure 11.6*c*): the resistance across the membrane (R_m) and the axoplasmic resistance (R_i) to current flow along the length of the axon. (The external longitudinal resistance is usually small and is ignored.) A slightly simplified equation for the length constant (λ) is

$$\lambda = K\left[\frac{R_m}{R_i}\right]^{1/2} \quad (11.7)$$

where K is a constant.

The membrane surface area increases proportionally with increasing axon diameter, which lowers R_m by adding resistances in parallel. On the other hand, R_i decreases in proportion to an increase in cross-sectional area of the axoplasm—that is, in proportion to the square of the diameter. The net effect is that the ratio R_m/R_i increases linearly with increasing diameter. If other factors are equal, the length constant and the conduction velocity should increase with the square root of the diameter.

Figure 11.24 shows that conduction velocity increases with increasing axon diameters. For some unmyelinated axons, this increase follows the square-root relationship, but many myelinated axons have a more nearly linear relationship of velocity to diameter. That the relation of velocity and diameter differs for different kinds of axons implies that other factors, including intrinsic membrane differences, are also involved.

MYELINATION INCREASES CONDUCTION VELOCITY Myelinated axons of vertebrates represent a tremendous evolutionary advance because they allow very high conduction velocities with relatively small axon diameters. A **myelinated axon** (Figure 11.25) is wrapped with 200 or more concentric layers of glial membrane (the membrane of Schwann cells in peripheral nervous systems and of oligodendrocytes in central nervous systems; see Figure 11.2). The glial cytoplasm is extruded from between the glial membrane layers so that the whole wrapping serves as an insulating layer. This multiply-wrapped insulating layer, termed **myelin**, stops at intervals of a millimeter or so along the length of the axon. The gaps at which the glial wrappings are absent (Nodes A, B, and C in Figure 11.25) are called the **nodes of Ranvier**.

Myelin electrically insulates the major part of the axon (the regions between nodes or **internodes**) nearly completely, leaving only the nodes of Ranvier as loci of ion flow across the axon membrane (see Figure 11.25). *In myelinated axons, action potentials occur only at the nodes of Ranvier*, in contrast to the continuous sweep of action potentials over an unmyelinated axon. Myelinated axons are therefore said to exhibit **saltatory conduction**, in which the action potential jumps (saltates) from node to node without active propagation in the internode.

The principal effect of myelin is to increase the membrane resistance of myelinated axon regions by 1000- to 10,000-fold over the resistance at the nodes of Ranvier. When a node of Ranvier undergoes an action potential, the local currents cannot leak out through the high membrane resistance of the adjacent internode, but instead must flow farther to the next node of Ranvier (see Figure 11.25). Thus, myelination greatly increases the spatial spread of local currents (i.e., the axon length constant) by this resistance effect.

A second, equally important function of myelin is to decrease membrane capacitance. If myelin only increased membrane resistance (without decreasing capacitance),

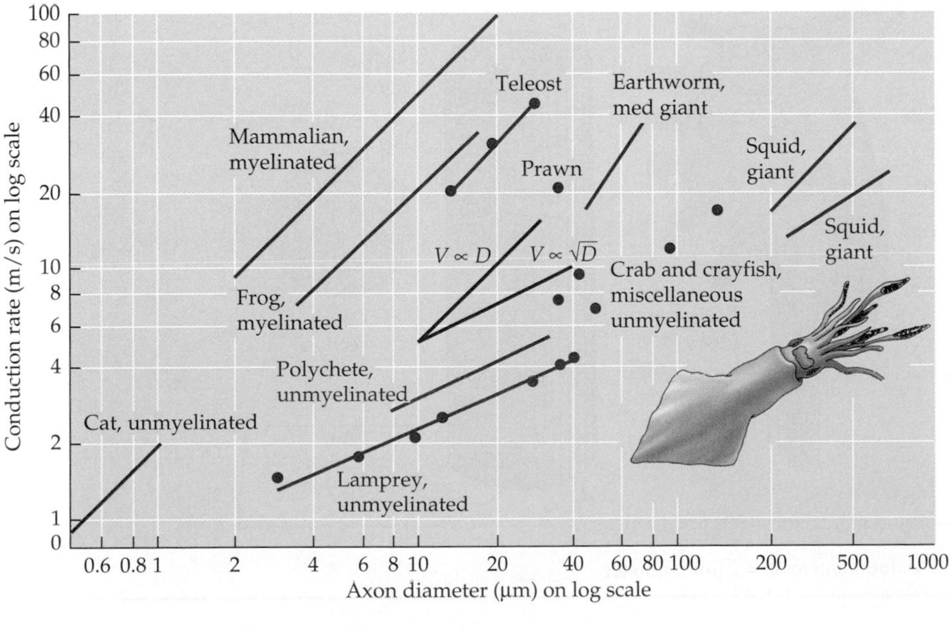

Figure 11.24 The velocity of nerve-impulse conduction increases with increasing axon diameter in both myelinated and unmyelinated axons Points not connected by lines are axons of different types. The two lines forming a V at the center show how velocity (V) and diameter (D) are related according to a simple proportional model and a square-root model. (After Bullock and Horridge 1965.)

BOX 11.2 Giant Axons

Because an increase in axon diameter increases the conduction velocity of an action potential and because animals often face circumstances in which a rapid response is advan- tageous for survival, **giant axons** have evolved in several animal groups. No particular diameter qualifies an axon as *giant*. Rather, the term is relative: A giant axon is of exceptional diameter in comparison to other axons in the same animal. Some axons are truly giant in cellular dimensions, such as the third-order giant axons in the squid, which may be 1 mm (1000 μm) in diameter. At the other extreme, the giant axons in the fruit fly *Drosophila* are only about 4 μm in diameter, but they are still an order of magnitude larger than other nearby axons.

Giant axons usually mediate behaviors for which speed and short latency of response are paramount, such as escape movements. The role of giant axons in squid locomotion provides an instructive example. A squid moves by jet propulsion. The animal contracts the muscles of its outer mantle to expel a jet of water from the mantle cavity through a moveable siphon. The giant axons serve to ensure a rapid and simultaneous contraction of all the mantle muscles—a necessary condition for fast, effective locomotion.

The squid actually possesses three sets of giant neurons, arranged in series(see the figure). In the brain are two partially fused *first-order* giant neurons (1), either of which can initiate activity of the entire propulsion system. Activation of a first-order giant neuron excites *second-order* giant neurons (2), which extend from the brain to paired stellate ganglia at the anterolateral margins of the mantle. Axons of several *third-order* giant neurons radiate from each stellate ganglion to the mantle muscles (3), and they are the motor axons that cause these muscles to contract.

The muscles at the posterior end of the elongate mantle are much farther from the stellate ganglion than are the anterior muscles. As an adaptation ensuring simultaneous contraction of these widespread muscles, the radiating axons of third-order giant neurons differ greatly in size. The largest and most rapidly conducting third-order giant axons extend to the most distant, posterior portion of the mantle. It is this arrangement of the giant axons that enables the signal for muscle contraction to arrive simultaneously at all parts of the mantle, as rapidly as possible.

Animals have followed several different evolutionary paths to achieve axons of large diameter. Giant axons can be *unicellular* (with one soma) or *multicellular* (with several somata). Unicellular giant axons are found in cockroaches, some annelids (such as *Protula*), and the medial giant axons of crayfish. Although most giant axons occur in invertebrates, the Mauthner neurons of fish and amphibians are examples of unicellular, myelinated vertebrate giant axons.

Multicellular giant axons may be syncytial or segmented. A *syncytium* (plural *syncytia*) results from a breakdown in plasma membrane boundaries between cells, forming large multinucleated "cells" of multicellular origin. Squid third-order giant axons develop by syncytial fusion of processes of 300 to 500 cells that retain discrete somata. In contrast, crayfish lateral giant axons and the giant axons of many annelids are made up of segmentally arranged cells that form low-resistance, end-to-end junctions. These junctions act as electrical synapses that allow direct electrical transmission of action potentials from cell to cell (see Figure 12.1), so the segmented axon *functions* as if it were one cell.

Clearly giant axons have evolved repeatedly, in diverse forms and developmental origins. This evolutionary convergence indicates a strong adaptive advantage of rapid impulse conduction. Giant axons such as the squid's are a selective *dis*advantage only in the neighborhood of Woods Hole, Massachusetts, and a few other locations of marine laboratories, where axonologists exert a negative selective pressure on the creatures that grow big axons for physiological experiments!

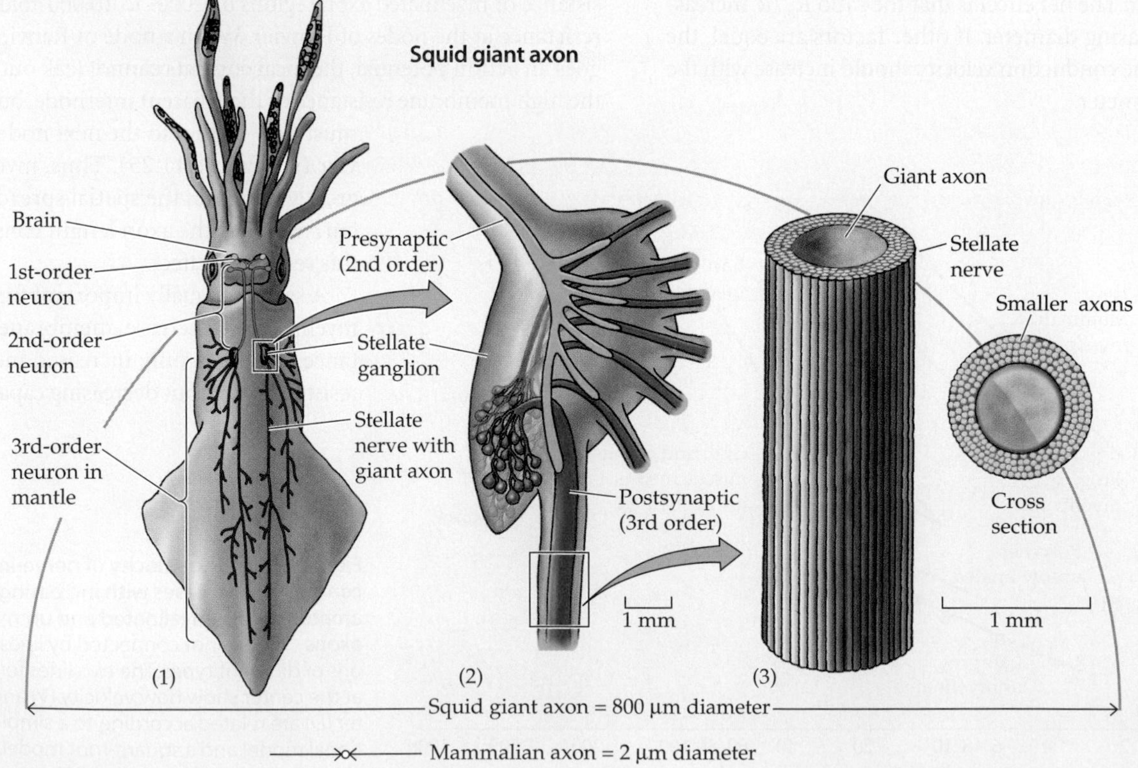

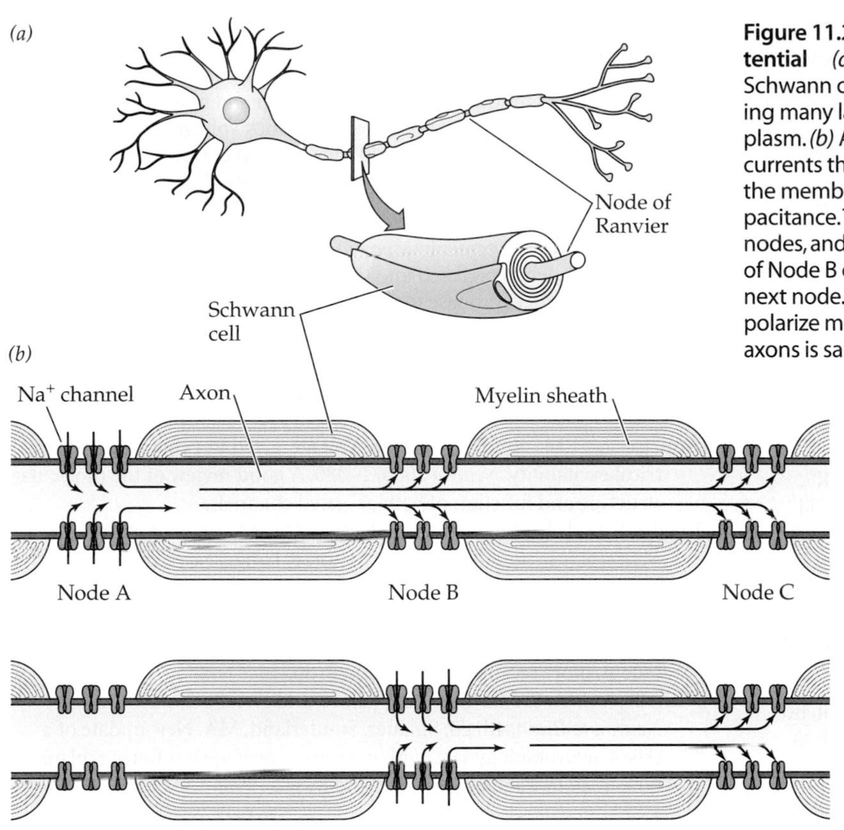

Figure 11.25 Myelinated axons speed the propagation of an action potential *(a)* Each segment of an axon in the PNS is myelinated by a Schwann cell. The Schwann cell wraps around the axon segment, providing many layers of Schwann cell membrane without intervening cytoplasm. *(b)* An action potential at one node of Ranvier (Node A) sets up local currents that flow along the axon, as in Figure 11.22. Currents cannot cross the membrane in the internode, which has very high resistance and low capacitance. Therefore, transmembrane current flow is restricted to the nodes, and the membrane length constant is much longer. Depolarization of Node B opens Na^+ channels, propagating the action potential to the next node. Note that some local current flows past the nearest node to depolarize more distant nodes. Action-potential propagation in myelinated axons is saltatory, jumping from node to node.

its effect on the axon length constant would be largely offset by an increase in the membrane time constant, τ. (Recall that the time constant is equal to the product of membrane resistance and membrane capacitance, R_mC_m.) An increase in the time constant would tend to *slow* conduction velocity because it would take more time for a current to depolarize a patch of membrane to threshold. Capacitance is inversely proportional to the distance separating the charges on the "plates" of a capacitor, which in this case is the distance between the axoplasm and the extracellular fluid. Myelin increases this distance in proportion to the number of glial membrane wrappings, so that capacitance is decreased about 1000-fold. Thus the increase in R_m of myelinated regions is offset by a decrease in C_m, and the membrane time constant is nearly unaffected.

Myelination, then, greatly increases conduction velocity by increasing the axon length constant without increasing the time constant. Currents from an action potential at one node must travel to the next node before crossing the membrane, and the currents are not slowed by having to displace much charge in the intervening myelinated internode.

Myelinated axons permit vertebrates to have neural coordinating and control systems with small-diameter axons that nonetheless conduct rapidly. A frog myelinated axon 12 μm in diameter has a conduction velocity of 25 m/s at 20°C. An unmyelinated squid giant axon must be about 500 μm in diameter to achieve the same 25-m/s velocity at 20°C! Thus, myelination allows the same velocity to be achieved with a 40-fold reduction in diameter and a 1600-fold reduction in axon cross-sectional area and volume. *With a reduction in axon dimensions, many more axons can be incorporated in a nervous system of a given size.*

Although myelin is usually considered to have evolved exclusively in vertebrates, some crustaceans have axons with analogous sheaths of glial wrappings that exhibit increased conduction velocities similar to those for vertebrate compact myelin. The 30-μm myelinated axons of a shrimp conduct at 20 m/s at 17°C, a velocity comparable to that of a 350-μm squid axon. Other myelinated shrimp axons 100 to 120 μm in diameter conduct at velocities exceeding 90 m/s at 20°C, rivaling the fastest mammalian myelinated axons. In contrast, unmyelinated lobster axons of the same 100- to 120-μm diameter have conduction velocities of only 8 m/s. Vertebrates, however, remain the only group with substantial numbers of myelinated neurons. Crustacean myelinated neurons are rare, specialized adaptations in high-velocity escape systems, but myelinated neurons are ubiquitous features of vertebrate nervous systems.

TEMPERATURE The gating of the voltage-dependent ion channels, and thus the time course of membrane depolarization to threshold, is temperature-dependent. Consequently, an increase of 10°C tends to nearly double conduction velocity (the value of Q_{10} is approximately 1.8; see Chapter 8, page 201) in both myelinated and unmyelinated axons. A frog myelinated axon 12 to 14 μm in diameter conducts at 25 m/s at 20°C, but a 3.5- to 4-μm cat myelinated axon conducts at the same 25 m/s at 37°C. *Thus the evolution of homeothermy in birds and mammals—along with myelination—has allowed further axon miniaturization and higher conduction velocities.*

Finally, the propagation of action potentials merely ensures that a neuron's electrical signals reach the end of the axon undiminished. For a neuron's signals to affect other cells, another process—synaptic transmission (the topic of Chapter 12)—is required.

Summary: The Propagation of Action Potentials

- Action potentials propagate because their underlying permeabilities to ions are voltage-dependent. Local circuits of current flow spread the depolarization along the axon, depolarizing a new region to threshold. Behind an advancing action potential, Na^+ channels remain inactivated long enough to prevent reexcitation by the local currents.

- The conduction velocity of an action potential depends on axon diameter, myelination, and temperature. Larger-diameter axons have higher conduction velocities because their length constants are longer, so local currents spread farther along the axon. Myelin greatly increases conduction velocity by increasing R_m (increasing the length constant) while decreasing C_m (preventing an increase in the time constant). Increasing temperature speeds the gating of channels so that the membrane responds faster to the local currents.

Study Questions

1. Suppose that the cell shown in Figure 11.9*a* is permeable to Na^+, K^+, and Cl^- but not to A^-. In the absence of ion pumps, Na^+ and Cl^- will diffuse in, the cell will become out of osmotic balance, and water will enter. Explain why a cell cannot be in thermodynamic equilibrium and also be in osmotic balance, unless there is an nonpermeating ion in the extracellular fluid (as would be the case if Na^+ were nonpermeating). What does this suggest about the evolutionary origin of the Na^+–K^+ exchange pump?

2. Unmyelinated axons conduct action potentials without decrement, but when myelinated axons lose myelin in demyelinating diseases such as multiple sclerosis, conduction of action potentials is blocked. Why?

3. Using Figure 11.24, compare and explain the difference (a) between the velocity of action-potential propagation in lamprey axons and frog myelinated axons of equal diameter, (b) between squid axons and frog myelinated axons of equal velocity, and (c) between frog and mammalian myelinated axons.

4. The ion flows across neuronal membranes at rest and during an action potential do not significantly change bulk ion concentrations, except for that of Ca^{2+} ions. Resting Ca^{2+} ion concentrations in cells are usually about 10^{-7} *M*, and Ca^{2+} ions exert physiological effects at concentrations of perhaps 10^{-5} *M*. Explain why relative changes of intracellular $[Ca^{2+}]$ are much greater than for, say $[Na^+]$ (12–50 m*M*).

5. For the ion concentrations in Table 11.1, calculate the equilibrium potentials for each ion species in squid axons and in mammalian muscle fibers.

6. Using the values in question 5, determine what ions are in passive equilibrium at a membrane potential of –60 mV in the squid at 18°C, and at –91 mV in mammalian muscle fibers at 37°C.

7. Suppose you voltage-clamp a squid axon from a resting membrane potential of –60 mV to a clamped value of +55 mV. Describe the early ionic current (say, at 0.5 ms after clamping). Use ionic concentrations from Table 11.1.

8. Suppose a squid axon at rest is 20 times as permeable to Na^+ as to K^+. Using the simplified version of the Goldman equation in Figure 11.10, calculate the resting membrane potential. If during the rising phase of an action potential the permeability to Na^+ increases to 100 times the permeability to K^+, at what value will the action potential peak?

9. Suppose you stimulate an axon so that you generate an action potential at both ends at the same instant. Describe the propagation of these action potentials. What happens when they meet?

10. With increased neuronal electrical activity in a brain area, the rates of glucose uptake, O_2 consumption, and blood flow increase. (These changes are the basis of activity imaging such as fMRI, which is illustrated in Figure 10.10). Why does neuronal activity increase local metabolic rate?

11. Toxins such as tetrodotoxin (from puffer fish and newts) and saxitoxin (from red-tide dinoflagellates) block voltage-gated Na^+ channels selectively. What effect would they have on currents in a voltage clamp to 0 mV? How do you suppose they evolved?

12. Part of the evidence for a "ball-and-string" model of Na^+ channel inactivation is that the proteolytic enzyme pronase can selectively remove inactivation when perfused into the axoplasm of a squid axon. Where would you expect pronase to cleave the Na^+ channel protein?

Key References

Armstrong, C. M., and B. Hille 1998. Voltage-gated ion channels and electrical excitability. *Neuron* 20: 371–380. A good review of the molecular structure and function of voltage-gated channels.

Kandel, E. R., J. A. Schwartz, and T. M Jessell (eds.). 1995. *Essentials of Neural Science and Behavior.* Appleton & Lange, Norwalk, CT. The clearest text treatment of neurophysiology, now somewhat dated.

Kandel, E. R., J. H. Schwartz, and T. M. Jessell (eds.). 2000. *Principles of Neural Science and Behavior*, 4th ed. McGraw-Hill, New York. An excellent, more recent, but heavier neuroscience text.

Nicholls, J. G., A. R. Martin, B. G. Wallace, and P. A. Fuchs. 2001. *From Neuron to Brain*, 4th ed. Sinauer, Sunderland, MA. New update of a classic neurobiology text, with narrower coverage but fuller explanations than in Kandel, Schwartz, and Jessell 1995.

Selected References

Aidley, D. J. 1998. *The Physiology of Excitable Cells*, 4th ed. Cambridge University Press, New York.

Arbas, E. A., R. B. Levine, and N. J. Strausfield. 1997. Invertebrate nervous systems. In W. H. Dantzler (ed.), *Comparative Physiology*, vol. 2 (Handbook of Physiology [Bethesda, MD], section 13), pp. 751–852. Oxford University Press, New York.

Arbas, E. A., I. A. Meinertzhagen, and S. R. Shaw. 1991. Evolution in nervous systems. *Annu. Rev. Neurosci.* 14: 9–38.

Catterall, W. A. 2000. From ionic currents to molecular mechanisms: The structure and function of voltage-gated sodium channels. *Neuron* 26: 13–25.

Catterall, W. A. 2001. A one-domain voltage-gated sodium channel from bacteria. *Science* 294: 2306–2308.

Dorsett, D. A. 1980. Design and function of giant fiber systems. *Trends Neurosci.* 3: 205–208.

Hess, P. 1990. Calcium channels in vertebrate cells. *Annu. Rev. Neurosci.* 13: 337–356.

Hille, B. 2001. *Ionic Channels of Excitable Membranes*, 3rd ed. Sinauer, Sunderland, MA.

Hille, B., and W. A. Catterall. 1999. Electrical excitability and ion channels. In G. Siegel, B. Agranoff, R. W. Albers, S. K. Fisher, and M. D. Uhler (eds.), *Basic Neurochemistry: Molecular, Cellular, and Medical Aspects*, 6th ed., pp. 119–137. Lippincott Williams & Wilkins, Philadelphia.

Hodgkin, A. L. 1964. *The Conduction of the Nervous Impulse.* Liverpool University Press, Liverpool, UK.

Hoffmann, F., M. Biel, and V. Flockerzi. 1994. Molecular basis for Ca^{2+} channel diversity. *Annu. Rev. Neurosci.* 17: 399–418.

Jan, L. Y., and Y. N. Jan. 1997. Cloned potassium channels from eukaryotes and prokaryotes. *Annu. Rev. Neurosci.* 20: 90–123.

Johnston, D., and S. M.-S. Wu. 1995. *Foundations of Cellular Neurophysiology.* MIT Press, Cambridge, MA.

Katz, B. 1966. *Nerve, Muscle, and Synapse*. McGraw-Hill, New York.

Laming, P. R., E. Syková, A. Reichenbach, G. I. Hatton, and H. Bauer (eds.). 1998. *Glial Cells: Their Roles in Behaviour*. Cambridge University Press, Cambridge.

Morais-Cabral, J. H., Y. Zhou, and R. MacKinnon. 2001. Energetic optimization of ion conduction rate by the K^+ selectivity filter. *Nature* 414: 37–42.

Neher, E., and B. Sakmann. 1992. The patch clamp technique. *Sci. Amer.* 266(3): 28–35.

Shepherd, G. M. 1994. *Neurobiology*, 3rd ed. Oxford University Press, New York.

Travis, J. 1994. Glia: The brain's other cells. *Science* 266: 970–972.

Vallee, R. B., and G. S. Bloom. 1991. Mechanisms of fast and slow axonal transport. *Annu. Rev. Neurosci.* 14: 59–92.

Waxman, S. G. 1983. Action potential propagation and conduction velocity. *Trends Neurosci.* 6: 157–161.

12 Synapses

The mouse shown here appears ready to jump off an object. This is no ordinary mouse; it is the celebrated *Doogie* mouse, smarter than average because it has been genetically engineered to overexpress a protein subunit of a neurotransmitter receptor molecule. This change augments a synaptic function that enhances learning and memory, producing a strain of mice with unusual abilities.

A **synapse**, as we have noted, is a specialized site of contact of a neuron with another neuron or with an effector. It is the locus where one cell (the *presynaptic* neuron) influences the function of another cell (the *postsynaptic* neuron or effector). Neurons are discontinuous with each other; even at synapses, they are typically separated by a 20- to 30-nm space called the synaptic cleft. Essentially all influences that neurons have on each other and on effectors are exerted at synapses.

The *Doogie* mouse introduces two sorts of synaptic function that illustrate the range of ways in which a synapse can work: *synaptic transmission* and *synaptic plasticity*. In **synaptic transmission**, a signal—usually an action potential—has an effect on a postsynaptic cell. The effect is rapid and transient, and it can be excitatory or inhibitory. This transmission of a signal across the synapse is the simplest kind of synaptic action. We introduce this fast-transmission function of synapses by examining a neuromuscular junction: the synapse between a neuron and a muscle fiber. Neuromuscular synaptic transmission causes muscle fibers to contract, as it does when the mouse jumps. The second sort of synaptic function is **synaptic plasticity**: the ability to change the functional properties of synapses. Synaptic actions can change the synapse itself to make it stronger or weaker, and they can also produce long-lasting changes in the postsynaptic cell. This modulation of synaptic processes is the basis of much functional change in the nervous system, both in development and in the changes involved in learning and memory.

Synaptic transmission, which carries signals from cell to cell, is a process distinct from the axonal transmission of action potentials within a cell. Synaptic transmission usually works chemically, but electrical transmission is possible in some cases. In chemical synaptic transmission, an action potential induces the release of a burst of neurotransmitter molecules, which diffuse across a gap (the synaptic cleft) between two excitable cells. At the postsynaptic side of the

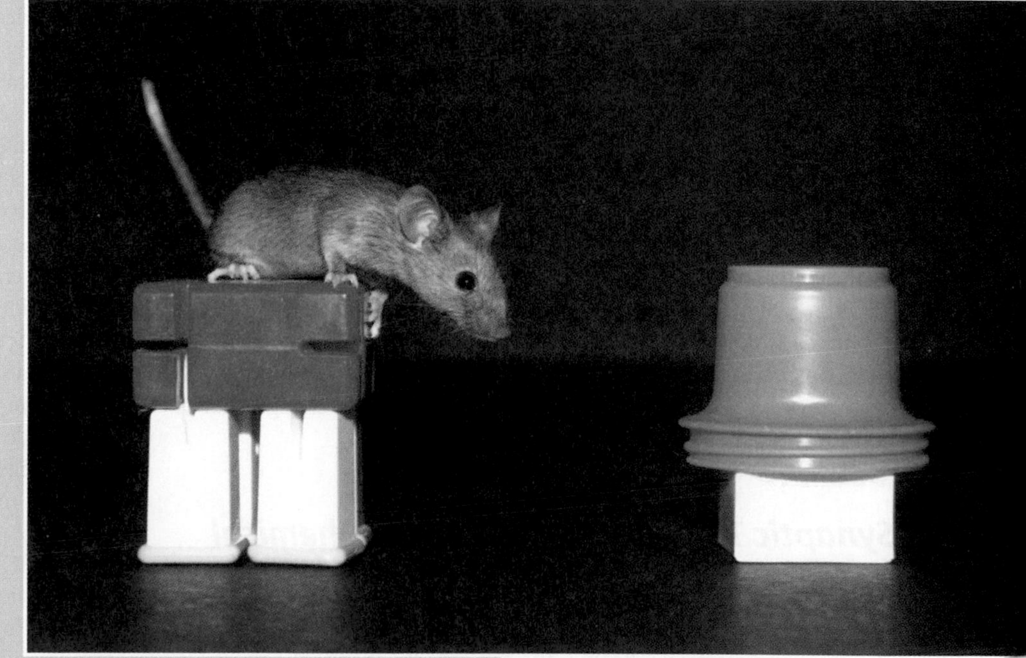

The *Doogie* mouse This genetically engineered mouse is synaptically smarter than the average mouse.

TABLE 12.1 Kinds of synapses

Characteristic	Chemical synapse: Ionotropic	Chemical synapse: Metabotropic	Electrical synapse
Mechanism and time course	Fast, ionotropic	Slow, metabotropic	Instantaneous current flow
Function	Signal transmission	Neuronal modulation	Electrical transmission
Effect	Excitation (fast EPSP), inhibition (fast IPSP)[a]	Excitation (slow EPSP), inhibition (slow IPSP), other (cytoplasmic and genetic)[a]	Electrical coupling

[a] EPSP = excitatory postsynaptic potential; IPSP = inhibitory postsynaptic potential.

synapse, the wave of neurotransmitter molecules elicits a synaptic potential (or postsynaptic potential), an electrical event that may excite or inhibit the postsynaptic cell. Synaptic transmission is directional, logical (excitatory or inhibitory), and changeable because its strength can be modified by experience.

You often hear the analogy that the brain is like a computer. Like all analogies, this one is imperfect, but useful here. Inasmuch as the brain is like a computer, the neuronal axons (which transmit action potentials long distances without degradation) are analogous to the wires in the computer, and the synapses are analogous to the logical junctions by which signals in one element affect others. It is the connections and properties of the logical junctions that determine the performance—and improvement in performance—of the brain computer, more than the properties of the wires.

Synapses have a diversity of actions, as befits their functional importance. A synapse can work chemically or electrically. It can excite the postsynaptic cell or inhibit it. A synaptic potential can result from a permeability increase or a permeability decrease. Synaptic action can be fast or slow, and it can *mediate* transmission or *modulate* the plastic properties of the synapse itself. Synapses can act directly on the membrane potential of the postsynaptic cell, or they can change the postsynaptic cell's metabolism. Some of these dichotomies go together, but some do not. In this chapter, we first consider electrical transmission and then the more widespread fast chemical transmission. Later in the chapter we examine the presynaptic and postsynaptic mechanisms of chemical synapses and study the distinction between *ionotropic* synaptic action (which is fast and changes ion permeability and thus membrane potential), and *metabotropic* synaptic action (which is slow and changes the metabolism of the postsynaptic cell). Table 12.1 classifies the different types of synapses. Only after considering these kinds of synaptic mechanisms can we try to explain (with the help of the *Doogie* mouse) the synaptic changes that may form much of the basis of learning and memory.

Synaptic Transmission Is Usually Chemical but Can Be Electrical

At the most fundamental level, there are two kinds of synapses: chemical and electrical. Both kinds can rapidly change the membrane potential of a postsynaptic cell. Electrical and chemical synapses play different functional roles, so both kinds are adaptive in nervous systems. We will discuss electrical transmission first, setting the stage for treatment of the more common and more complex chemical transmission process.

Fast synaptic transmission works in millisecond time: An action potential in a presynaptic neuron leads to a rapid postsynaptic voltage change, with a typical delay of less than a millisecond. The mechanism of this fast synaptic action was a subject of much debate in the first half of the twentieth century. One group (the "sparks") argued that synaptic transmission was by direct electrical means. The other group (the "soups") postulated that transmission was by a chemical mechanism. It is now clear that most synaptic transmission is chemical; depolarization of the presynaptic terminal causes it to release a chemical transmitter, which diffuses across the *synaptic cleft* to affect the postsynaptic cell (Figure 12.1*a*). Direct electrical transmission also occurs, but it is less common.

Electrical synapses transmit signals instantaneously

In an **electrical synapse**, electric currents from one cell flow directly into the next cell, changing its membrane potential (Figure 12.1*b* and *c*). Because this current flow is instantaneous, an electrical synapse has essentially no delay. Moreover, such current flow can usually be in either direction, so electrical synapses are often not polarized. Because of their instantaneous transmission of signals and their synchronizing ability, electrical synapses are found in nervous systems where speed is most important, and where synchronous activity of several cells is an advantage.

The major structural specialization for electrical transmission is the **gap junction**. A gap junction (as introduced in Chapter 2; see page 39) is a specialized locus where protein channels bridge the gap between two cells, directly connecting their cytoplasm. Gap junctions provide a low-resistance path for current flow, electrically coupling the cells that they join. Thus, any electrical change in one cell is recorded in the other, with some weakening but with negligible delay.

Figure 12.1*c* shows the typical effects of this electrical coupling. Depolarization or hyperpolarization of one cell produces a weaker corresponding change in the other cell. The voltage change is always accompanied by some decrease, but the amount of the decrease varies. Thus, electrical synapses differ in strength: The larger the gap-

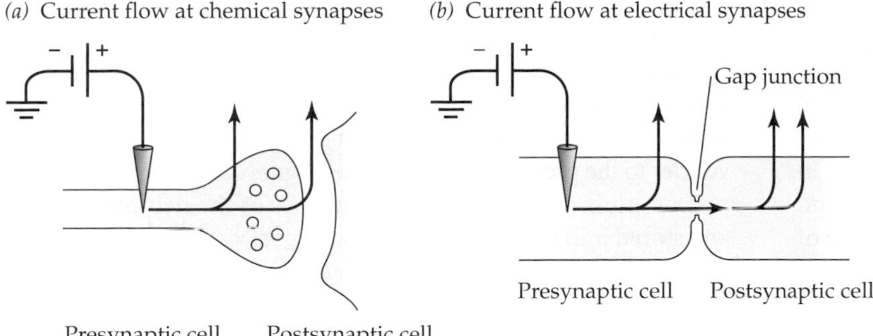

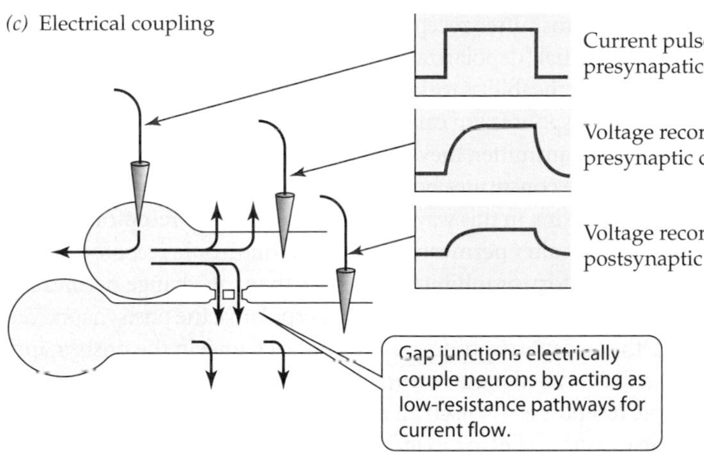

Figure 12.1 Electrical and chemical synapses Both kinds of synapses play roles in neural communication, but they work in fundamentally different ways. *(a)* In chemical synapses the currents escape between neurons and do not enter the postsynaptic neuron. Instead, released molecules of neurotransmitter carry the signal across the synaptic cleft. *(b)* Electrical synapses have low-resistance pathways that allow currents to flow directly between neurons. *(c)* Current flow through the low-resistance pathways of gap junctions electrically couples the neurons. If current is injected into one neuron, the depolarization (or hyperpolarization) spreads to another electrically coupled neuron, and the depolarization of one cell produces a smaller depolarization of the adjacent cell. (After Kandel et al. 2000.)

junction area, the lower the resistance between cells and the stronger the coupling.

Electrical synapses can act as short-latency synaptic relays, in which each presynaptic action potential triggers a postsynaptic action potential. In other cases, with weaker electrical coupling or high impulse thresholds, a presynaptic action potential may produce only subthreshold depolarization of the postsynaptic cell. Most electrical synapses are bidirectional, transmitting voltage changes roughly symmetrically in both directions. However, in the electrical synapse between the lateral giant axon and the giant motor neuron of crayfish, the electrical synapse is rectifying; that is, it allows current flow preferentially in one direction. Thus an action potential in the lateral giant axon excites the giant motor neuron, but excitation of the motor neuron by other pathways cannot "backfire" to the lateral giant axon.

The structure of gap junctions has been examined by electron microscopy and X-ray diffraction studies. Gap junctions narrow the space between adjacent cells, holding the pre- and postsynaptic membranes only about 3 nm apart (instead of the 20 to 30 nm that usually separates neighboring cells). In the region of close membrane apposition, the narrow gap separating the membranes is bridged by a regular array of channel structures termed **connexons** (Figure 12.2).

Each connexon is composed of hexamers of the protein *connexin* surrounding a 2-nm pore. Connexons are hemichannels; each connexon of one cell pairs with a connexon of the adjacent cell to form a channel that connects the cytoplasm of the cells. The channels of all the connexons are the low-resistance pathways that allow electric current to flow between the cells. They are large enough to allow the passage of most ions, as well as dye and tracer molecules smaller than about 1000 daltons. Connexon channels are thought to be open normally, but they can close in response to electrical or chemical changes in the cells.

Figure 12.2 The molecular structure of gap junctions A gap junction is a localized patch of close membrane apposition where protein hexamers called connexons provide channels that electrically connect the cytoplasm of the cells.

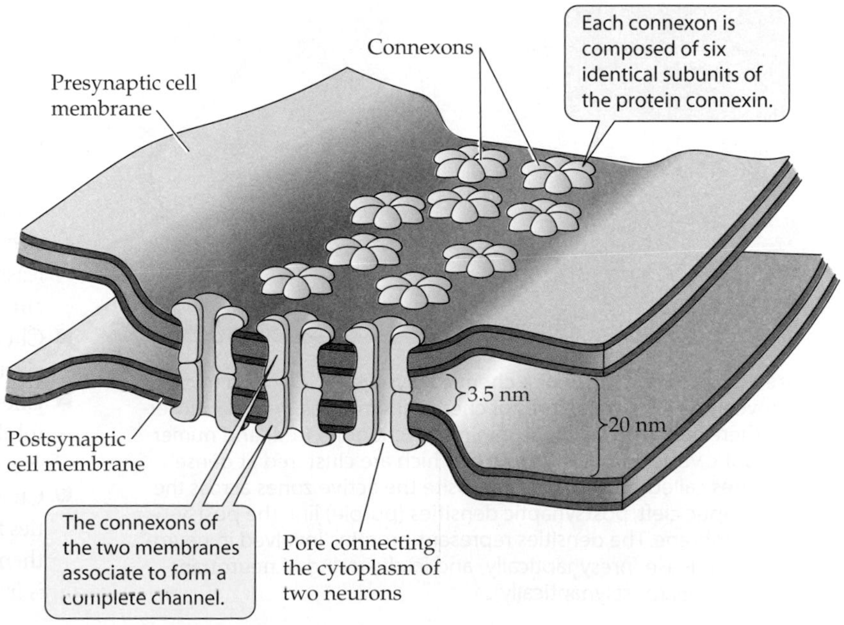

Where in nervous systems has natural selection favored the evolution of electrical synapses? In most cases it is where speed or synchronization is especially paramount. Electrical synapses can act as fast intercellular relays, in which each presynaptic action potential instantaneously triggers a postsynaptic action potential. Such relay electrical synapses are found in escape systems—for example, in crayfish and fish. In such fast escape responses, every millisecond saved by an electrical synapse can produce a selective advantage of not being eaten. Electrical synapses are also advantageous for groups of neurons that normally fire synchronously, such as the neurons controlling electric-organ discharge in fish.

Chemical synapses can condition and amplify signals

Unlike electrical synapses, **chemical synapses** have a discontinuity between the cells because the 20- to 30-nm synaptic cleft of a chemical synapse is a barrier to direct electrical communication (see Figure 12.1*a*). The presynaptic electrical signal is first transduced into a chemical signal, the release of neurotransmitter molecules from the presynaptic terminals. The molecules of neurotransmitter rapidly diffuse to the postsynaptic membrane, where they bind to receptor molecules that are specialized to generate an electrical or chemical change in response to the neurotransmitter binding.

Chemical synapses have a distinctive structure at the electron microscopic level (Figure 12.3). The axon terminal of the presynaptic neuron contains neurotransmitter molecules stored in **synaptic vesicles**, several thousand transmitter molecules per vesicle. At the synaptic cleft both the presynaptic and postsynaptic membranes appear denser and thicker than elsewhere because of local aggregations of proteins at these membranes. Dark tufts of electron-dense material on the cytoplasmic side of the presynaptic membrane mark **active zones** at which synaptic vesicles release their neurotransmitter.

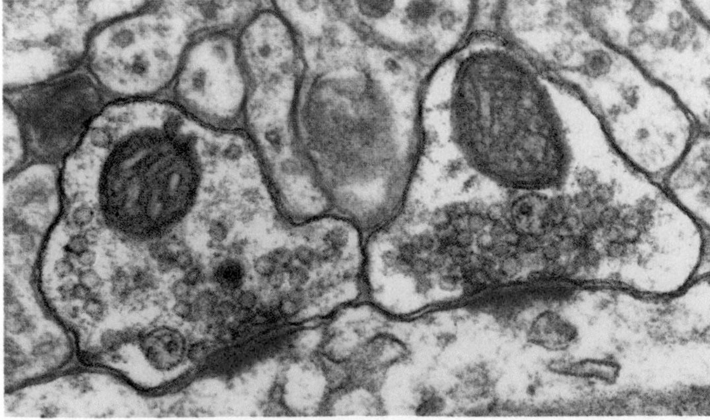

Figure 12.3 The structure of a chemical synapse Electron microscopy reveals distinctive features of chemical synapses. Presynaptic terminals (here colorized yellow) contain mitochondria (red) and numerous synaptic vesicles (green), some of which are clustered at dense release sites called active zones. Opposite the active zones across the (pink) synaptic cleft, postsynaptic densities (purple) line the postsynaptic membrane. The densities represent proteins involved in neurotransmitter release (presynaptically) and in clustering of neurotransmitter receptors (postsynaptically).

Other dense aggregates at the postsynaptic membrane are called **postsynaptic densities.**

A presynaptic neuron releases neurotransmitter molecules in response to an arriving action potential (Figure 12.4). Neurotransmitter is released by calcium-dependent exocytosis—fusion of synaptic vesicles to the presynaptic membrane (Figure 12.4*b*). The presynaptic neuron must have synthesized quantities of the neurotransmitter and stored it in the synaptic vesicles. The released neurotransmitter molecules bind to receptor proteins embedded in the postsynaptic membrane. Neurotransmitter receptors are transmembrane proteins that are effectors for change in the postsynaptic cell, usually producing a change in postsynaptic membrane potential.

Neurotransmitter receptors can produce fast changes in membrane potential (depolarization or hyperpolarization) by directly increasing permeability to ions (Figure 12.4*c*). Receptors mediating these fast responses are called *ligand-gated ion channels*; when they bind their transmitter, they open to allow ion flow. In this case, a single molecule constitutes both the receptor and the ion channel. Receptors working in this way are termed **ionotropic** receptors because they directly alter permeability to ions. Ionotropic receptors can be either excitatory or inhibitory, because they can change permeabilities to ions to either depolarize or hyperpolarize the postsynaptic cell.

Other receptors produce a metabolic change in the postsynaptic cell and are called **metabotropic** receptors (Figure 12.4*c*). A metabotropic receptor is not itself an ion channel; instead it activates an enzyme (an *intracellular effector*) to produce a *second messenger*—a cytoplasmic molecule such as cyclic AMP. Metabotropic receptors may change permeabilities to ions (and voltage) *indirectly*, by having their metabolic effects ultimately open or close ion channels that are different molecules from the receptor. Metabotropic receptors often have slower, long-lasting modulatory effects on synaptic processes.

Transmission at chemical synapses is necessarily slower than transmission at electrical synapses because the steps of transmitter release and receptor action take more time than the instantaneous spread of electric current. Fast chemical synapses typically have a *synaptic delay* of 0.3 to 3 ms, depending on species and temperature. Even though these delays are short, there must be countervailing adaptive advantages of chemical synapses that explain their prevalence. Several advantages can be recognized:

- Chemical synapses can amplify current flow. A presynaptic action potential can lead to the release of a few or many synaptic vesicles. Each vesicle contains a few thousand molecules of transmitter. Depending on the number of active zones and the size of the presynaptic terminal, the amount of transmitter released can open many channels and amplify the postsynaptic current.
- Chemical synapses can be either excitatory or inhibitory, unlike electrical synapses, which are nearly always excitatory.
- Chemical synapses are one-way; a presynaptic neuron excites or inhibits a postsynaptic cell, but not vice versa. Most electrical synapses are two-way.
- Chemical synapses are much more modifiable in their properties than electrical synapses are. Use and circumstance can make them stronger, and disuse can make them weaker. This plasticity is important for nervous system development and for learning.

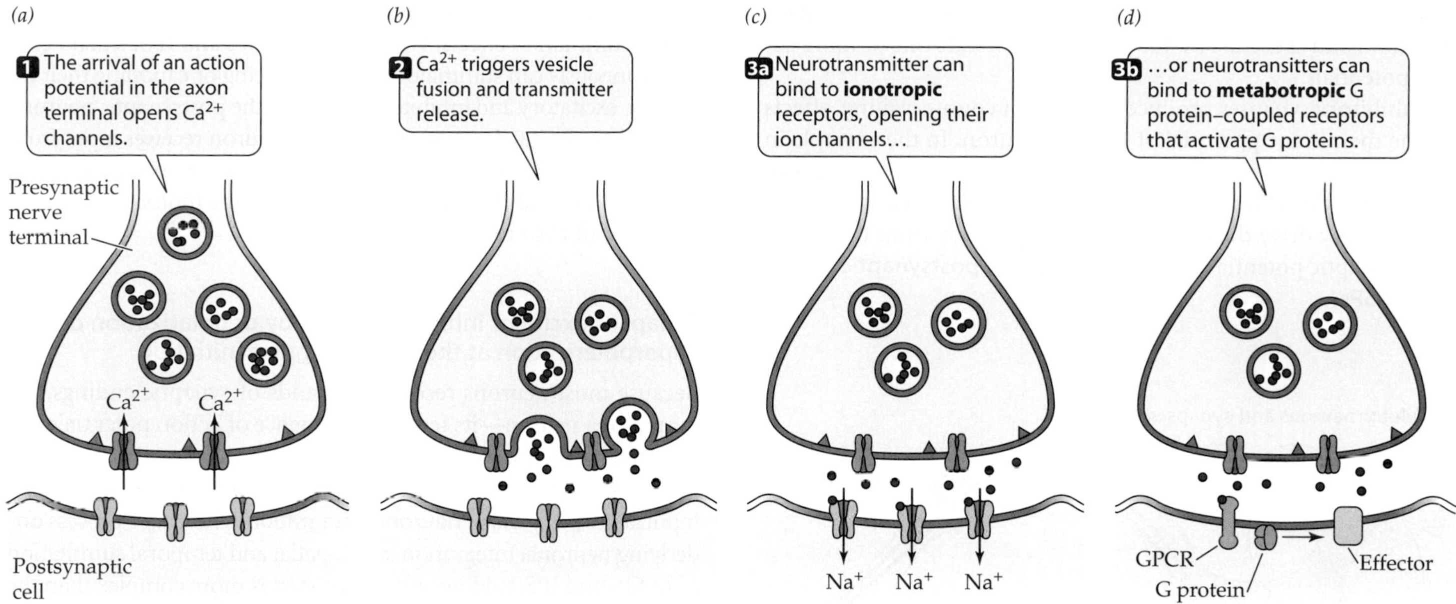

Figure 12.4 The function of a chemical synapse *(a)* A presynaptic action potential depolarizes the axon terminal and opens voltage-gated Ca^{2+} channels located near active zones. *(b)* Ca^{2+} ions enter the terminal and induce vesicles at the active zones to fuse with the presynaptic membrane and release neurotransmitter by exocytosis. Neurotransmitter molecules diffuse across the synaptic cleft and bind to postsynaptic receptor proteins. *(c)* Ionotropic neurotransmitter receptors open ion channels, creating an ionic current that changes the postsynaptic cell's membrane potential. In this example, sodium influx would cause depolarization. *(d)* Metabotropic receptors activate a metabolic cascade in the postsynaptic cell. Metabotropic receptors are G protein–coupled receptors (GPCRs) that activate G proteins to produce a second messenger such as cyclic AMP (cAMP). (*a–c* after Kandel et al. 2000.)

Summary: Synaptic Transmission Is Usually Chemical but Can Be Electrical

- Most synapses are chemical; some are electrical. Electrical synapses are very fast and usually are bidirectional. Gap junctions are the anatomical basis of electrical synapses; they contain connexons that allow current to flow directly between the cells, electrically coupling them.
- Chemical synapses are unidirectional, with a presynaptic neuron that releases neurotransmitter when stimulated, and a postsynaptic neuron (or effector) that bears receptor molecules to which the neurotransmitter binds.
- Neurotransmitter receptors may directly open their own ion channels, or they may act indirectly through a metabolic cascade to open, close, or change ion channels that are separate molecules.
- Electrical synapses mediate fast, synchronizing actions of neurons. Chemical synapses integrate neuronal functions, by fast (ionotropic) excitation and inhibition, or by slow (metabotropic) modulation of neuronal and synaptic properties.

Synaptic Potentials Control Neuronal Excitability

The primary function of synaptic transmission is to control the excitability of the postsynaptic cell. When a burst of neurotransmitter diffuses across a synapse, it generates a **synaptic potential**—a transitory, graded change in the resting membrane potential—in the postsynaptic cell. A synaptic potential that tends to *depolarize* the cell membrane is excitatory, and one that tends to *hyperpolarize* the cell membrane is inhibitory. **Excitation** is an increase in the probability that a cell will generate an impulse (an action potential), or if the cell is already generating impulses, excitation causes an increase in the impulse frequency. **Inhibition**, on the other hand, is a decrease in the probability of impulse generation or a decrease in impulse frequency.

Synapses onto a spinal motor neuron exemplify functions of fast synaptic potentials

We can witness the excitatory and inhibitory effects of synaptic potentials by stimulating the peripheral nerves containing axons that synapse on a cat motor neuron, while recording the changes in the motor neuron's membrane potential (Figure 12.5). In Figure 12.5*b*, stimulus A or B produces graded depolarizations in the motor neuron that have a brief rising phase and an exponential decay over a time course of 10 to 20 ms. These depolarizations are **excitatory postsynaptic potentials** (**EPSPs**). An excitatory synapse usually produces a very small EPSP, one that depolarizes the membrane usually no more than 0.5 mV.

EPSPs can be combined in two ways. If peripheral nerve A is stimulated rapidly, the resultant EPSPs combine in a process called **temporal summation**. Simultaneously occurring EPSPs produced by different nerves (stimulus A + B in Figure 12.5*b*) also combine, in a process called **spatial summation**. With sufficient presynaptic stimulation, EPSPs can summate temporally and spatially to the voltage threshold and cause the motor neuron to generate one or more action potentials.

Inhibitory synapses produce similar, but countervailing, effects on the membrane potential of a motor neuron. In the example in Figure 12.5c, stimulating peripheral nerve D or E evokes synaptic potentials that briefly hyperpolarize the motor-neuron cell membrane. Because they drive the membrane potential away from threshold, these synaptic potentials are called **inhibitory postsynaptic potentials** (**IPSPs**).

As Figure 12.5*c* demonstrates, IPSPs also produce temporal and spatial summation effects. What's more, EPSPs and IPSPs that occur simultaneously can summate spatially, reducing or canceling their respective excitatory and inhibitory effects on the postsynaptic neuron's membrane potential. A typical cat motor neuron receives input from about 10,000 synaptic terminals, and the moment-to-moment balance of EPSPs and IPSPs determines whether the motor neuron generates impulses or remains quiescent.

Synapses excite or inhibit a neuron by depolarization or hyperpolarization at the site of impulse initiation

Because most neurons receive thousands of synaptic endings, the output of a neuron—its temporal sequence of action potentials—is a complex function of its synaptic input. Thus a neuron's output is not the same as its input, but is instead an integral function of that input, a property called **neuronal integration**. The major process underlying neuronal integration is the spatial and temporal summation of EPSPs and IPSPs. However, this process is more complex than the simple algebraic summation of postsynaptic potentials. Neuronal integration also is a function of the spatial relationships of excitatory and inhibitory synapses to one another, as well as their relative proximity to the site of impulse initiation. For spinal motor neurons this site is the axon hillock.

Because postsynaptic potentials are *graded potentials*, their spread is governed by the cable properties of the postsynaptic cell membrane (see Chapter 11, page 287). Consequently, the functional effect of a synapse depends in part on where on a neuron it is. A synapse on the soma or cell body (called an **axosomatic** synapse) is only a short electrotonic distance from the axon hillock; that is, the amplitudes of postsynaptic potentials are decreased a small amount in their passive spread from the soma to the impulse initiation site.

A synapse onto a dendrite (**axodendritic** synapse), on the other hand, may be as much as 200 μm distant from the axon hillock. The electrotonic length of motor-neuron dendrites is estimated to be 1 to 2 λ (one or two times the membrane length constant), so a synaptic potential at the dendrite tip is decreased to 14% to 37% of its initial amplitude in its spread to the axon hillock. Although other factors may partially compensate for this decrease with distance, synapses that are closer to the

(*a*) Motor neurons and synapses

Spinal cord

Stimulus A

Stimulus B

Stimulus C

Stimulus D

Stimulus E

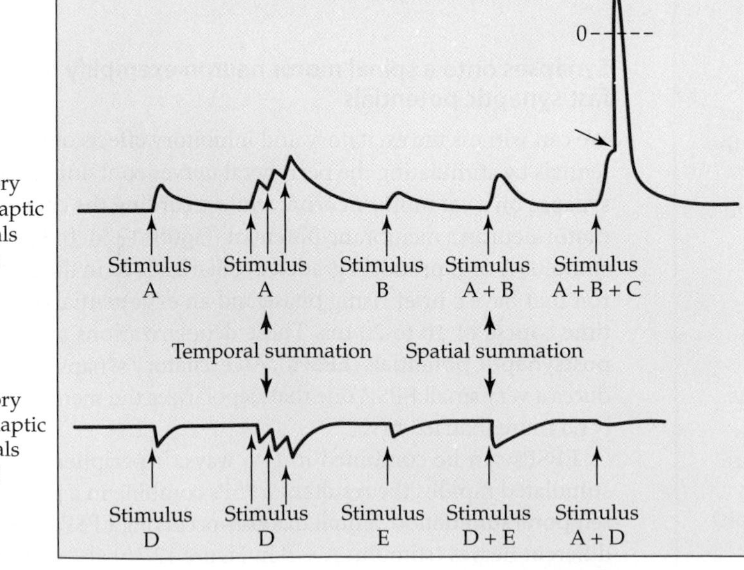

Figure 12.5 Excitatory and inhibitory postsynaptic potentials *(a)* Clusters of synapses onto a spinal motor neuron. Yellow synapses are excitatory; red synapses are inhibitory. *(b)* Stimulation of excitatory presynaptic neurons elicits excitatory postsynaptic potentials (EPSPs), graded depolarizations that depolarize V_m toward threshold. *(c)* Inhibitory postsynaptic potentials (IPSPs) are hyperpolarizing, moving V_m away from threshold. Both EPSPs and IPSPs can summate with other postsynaptic potentials of the same pathway (temporal summation) or a different pathway (spatial summation).

axon hillock have more effect on the output of the postsynaptic cell than synapses on the distal ends of dendrites have. Thus the summation of synaptic input is weighted by the electrotonic distance of the synapses from the axon hillock.

Summary: Synaptic Potentials Control Neuronal Excitability

- Most synapses in nervous systems are chemical synapses that mediate fast excitation and inhibition. Neurotransmitters act at receptors to open ion channels, to depolarize (EPSP) or hyperpolarize (IPSP) the postsynaptic neuron.
- EPSPs and IPSPs summate, so the membrane potential of the postsynaptic neuron is a moment-to-moment integral of synaptic input.
- Postsynaptic potentials are graded and spread passively to the axon hillock (the site of action-potential initiation). Therefore, more-distant synapses may have smaller effects on the neuron's output.

Fast Chemical Synaptic Actions Depend on Increases in Permeability to Ions

Next we turn to the mechanisms of action of fast chemical synapses, the sort that gave rise to the synaptic potentials in the preceding section. Such synapses are called *fast* because they produce postsynaptic potentials within a few milliseconds of a presynaptic action potential. Fast synaptic action is the conceptual model that has dominated scientific thinking about how nervous systems work. In fact, until about 25 years ago, fast synaptic transmission was the only kind known, or even suspected. All synapses were thought to work by producing fast EPSPs and IPSPs such as those shown in Figure 12.5. Although investigators know now that synapses can also work in other ways, fast chemical transmission is still considered the "workhorse" of synaptic mechanisms, the basis of most neuronal interaction in nervous systems.

To show the *function* of fast chemical synapses, in the previous section we considered synapses onto motor neurons in the central nervous system (CNS). Much of what we know about synaptic *mechanisms*, however, has been learned through the study of *peripheral* synapses. For technical reasons it is hard to study synaptic mechanisms in the CNS. Synapses in the vertebrate CNS are buried inside a large mass of tissue. They cannot be seen through a microscope, nor can their immediate environment be readily changed. Moreover, a single vertebrate CNS neuron may have tens of thousands of synapses from other neurons ending on it, each neuron producing its own synaptic effects. Hence, researchers have turned to simpler *model systems*, in which detailed investigations of synaptic transmission are easier, and have later applied their findings to the more complex central nervous system. This strategy is analogous to studying squid giant axons to clarify the mechanisms by which action potentials are generated.

The vertebrate skeletal neuromuscular junction (also called the *motor end-plate*) is the model system for chemical synaptic transmission from which much of our basic knowledge of synaptic physiology is derived (Figure 12.6). For most vertebrate skeletal muscles, each muscle fiber is innervated by only one motor neuron. (Hundreds of motor neurons may innervate the whole muscle, but only one innervates a given fiber.) The neuromuscular junction functions as a *relay* synapse: Each action potential in the motor neuron evokes a large EPSP in the muscle fiber, which reaches threshold and elicits an action potential in the muscle fiber. The muscle fiber action potential propagates to all parts of the muscle fiber, depolarizing its membrane and triggering its contraction (muscle contraction will be discussed in Chapter 17).

The vertebrate neuromuscular junction has several advantages as a model synapse. In addition to its anatomical simplicity and large synaptic response, the neuromuscular junction is accessible and microscopically visible at the muscle surface. The researcher can impale a muscle fiber with an electrode under visual control and can readily change the fluid bathing the junction.

Chemical synapses work by releasing and responding to neurotransmitters

The sequence of events of synaptic transmission at the vertebrate skeletal neuromuscular junction (Figure 12.7) is summarized here. Many of these steps and their sequence are characteristic of most synapses.

1. An approaching action potential in the motor neuron's axon depolarizes the presynaptic terminal, opening voltage-gated Ca^{2+} channels.
2. Ca^{2+} ions enter the terminal, causing synaptic vesicles to fuse with the presynaptic membrane and release neurotransmitter into the synaptic cleft. At the neuromuscular junction the neurotransmitter is acetylcholine (ACh).
3. Neurotransmitter molecules diffuse across the synaptic cleft to the postsynaptic membrane. Because the diffusion path is so short (50 nm at the neuromuscular junction, 20 to 30 nm at CNS synapses), neurotransmitter molecules reach the postsynaptic membrane in a diffusion time of only 20 to 50 μs.
4. Neurotransmitter molecules bind to postsynaptic neurotransmitter receptor molecules. These are ligand-gated channels that are opened by the binding of transmitter (the ligand). At the neuromuscular junction, ACh receptor molecules are located at the junctional folds of the muscle fiber membrane. The binding of ACh by ACh receptors opens ion channels to produce an *excitatory postsynaptic potential.*
5. The excitatory postsynaptic potential at the neuromuscular junction depolarizes the muscle fiber membrane to its threshold and initiates a muscle fiber action potential (which is generated by voltage-gated channels).
6. The action potential propagates to the ends of the fiber, depolarizing the entire membrane and thereby initiating contraction of the fiber (see Chapter 17).

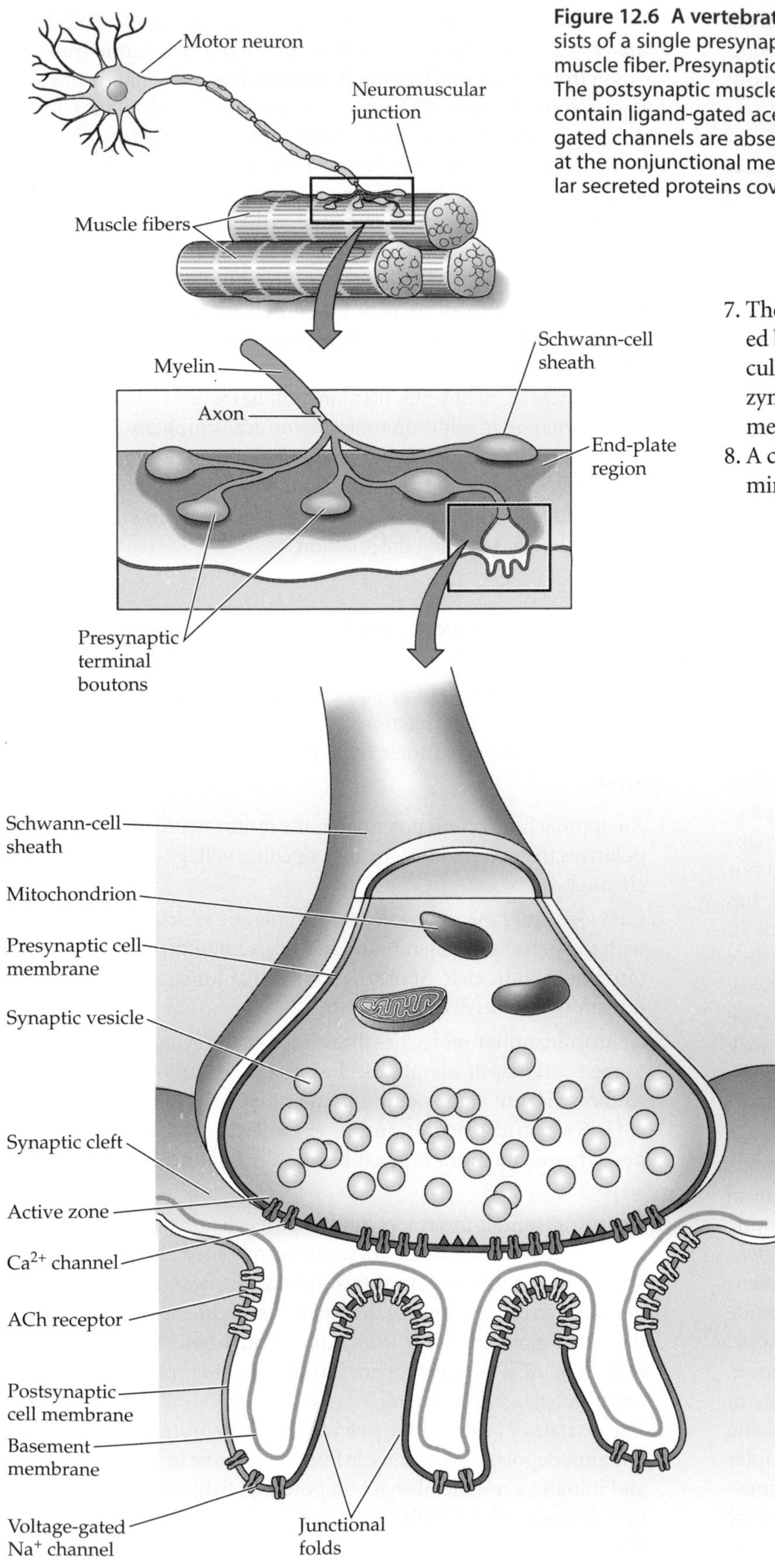

Figure 12.6 A vertebrate neuromuscular junction The neuromuscular junction consists of a single presynaptic motor neuron contacting a single postsynaptic skeletal-muscle fiber. Presynaptic vesicles fill part of the terminal and cluster near active zones. The postsynaptic muscle-fiber membrane is deeply infolded, and these junctional folds contain ligand-gated acetylcholine receptors at the upper parts of the fold. Voltage-gated channels are absent at the tops of the folds but are present deep in the folds and at the nonjunctional membrane. A basement membrane (or basal lamina) of extracellular secreted proteins covers the muscle fiber. (After Kandel et al. 2000.)

7. The action of the neurotransmitter at most synapses is terminated by enzymatic degradation or by reuptake. At the neuromuscular junction, ACh is destroyed by acetylcholinesterase, an enzyme located within the synaptic cleft and at the postsynaptic membrane.
8. A choline transporter retrieves choline into the presynaptic terminal for ACh resynthesis.

Postsynaptic potentials result from permeability changes that are neurotransmitter-dependent and voltage-independent

In Chapter 11 we learned that the permeability changes that produce an action potential are voltage-dependent, resulting from depolarization of the membrane. In contrast, the permeability changes that produce a synaptic potential depend on neurotransmitter and not on voltage. The molecules that control permeability changes (at fast synapses) are ligand-gated channels (see Chapter 2, page 57, for review). Ligand-gated channels open as a result of binding neurotransmitter, not in response to depolarization.

AN EPSP RESULTS FROM A SIMULTANEOUS INCREASE IN THE POSTSYNAPTIC MEMBRANE'S PERMEABILITY TO Na^+ AND K^+ An EPSP results from the flow of both Na^+ and K^+ ions. Even though Na^+ enters and K^+ leaves, the response is still an excitatory depolarization. Suppose that neurotransmitter molecules opened ion channels that allowed only Na^+ ions to flow into the cell. If this were so, then the EPSP would always displace V_m toward the sodium equilibrium potential, E_{Na} (see Chapter 11). As Figure 12.8 shows, however, this is not the case. If V_m is changed by current being passed through a second microelectrode in a muscle fiber, then EPSPs always displace the membrane toward a value near zero. This value (between 0 and –15 mV in different cells) is the **reversal potential** (E_{EPSP}), the value of V_m beyond which an EPSP reverses its polarity.

Why does the opening of the ligand-gated channels to produce an EPSP drive V_m toward zero? During an EPSP, P_{Na} and P_K increase *simultaneously*, rather than sequentially as in generation of the action potential. Therefore, the ionic currents resulting from P_{Na} and P_K drive the membrane toward a value (E_{EPSP}) intermediate between E_{Na} and E_K. (E_{EPSP} is slightly closer to E_{Na} than to E_K because P_{Na} is slightly greater than P_K, but we can simplify and consider the permeabilities equal.)

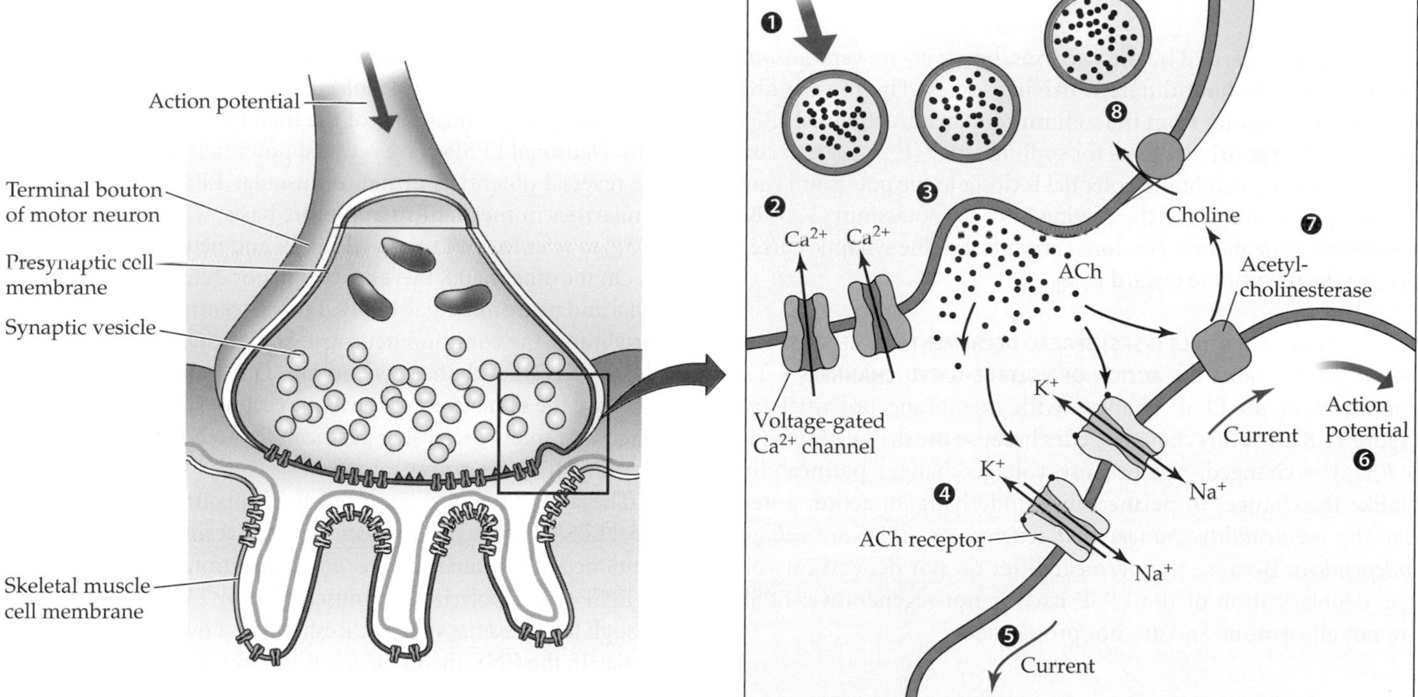

Figure 12.7 Summary of events in chemical synaptic transmission at the vertebrate neuromuscular junction ❶ An action potential depolarizes the axon terminal of the motor neuron. ❷ The depolarization opens voltage-gated Ca^{2+} channels. ❸ Depolarization of the terminal triggers vesicle exocytosis at an active zone, releasing acetylcholine (ACh). ❹ ACh diffuses rapidly across the synaptic cleft and binds to acetylcholine receptors at the postsynaptic membrane. ❺ The receptor channel opens to allow Na^+ and K^+ ion flow, producing a depolarizing excitatory postsynaptic potential (EPSP). The EPSP spreads to depolarize nearby regions to threshold and triggers a muscle-fiber action potential. ❻ The action potential propagates to all parts of the muscle fiber, eliciting contraction. ❼ Acetylcholinesterase hydrolyzes the acetylcholine into acetate and choline. ❽ Choline is actively transported back into the motor axon terminal to be resynthesized into acetylcholine.

Figure 12.8 Postsynaptic current at the neuromuscular junction drives the muscle-fiber membrane toward a reversal potential (E_{EPSP}) *(a)* A current-passing electrode is used to change the membrane potential prior to nerve stimulation. *(b)* Nerve stimulation and resultant neurotransmitter binding to postsynaptic receptors induces an ionic current that displaces the membrane potential toward E_{EPSP} from any set value. Therefore E_{EPSP} is called the *reversal potential* of the EPSP.

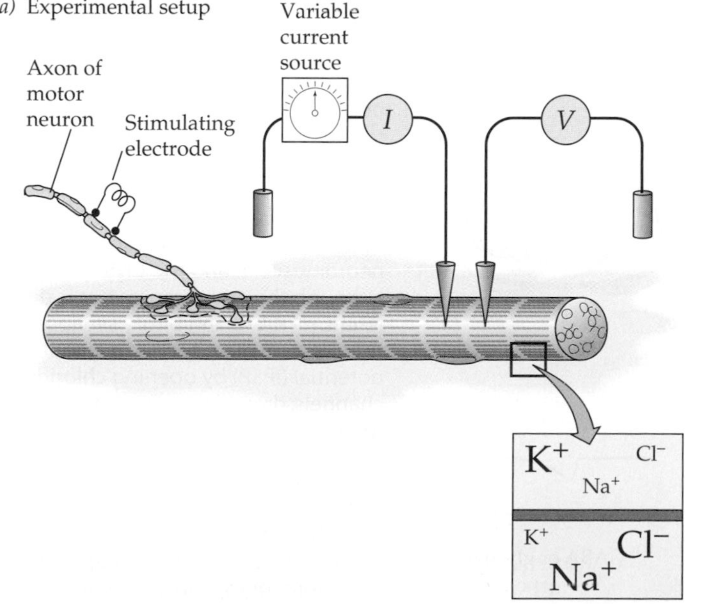

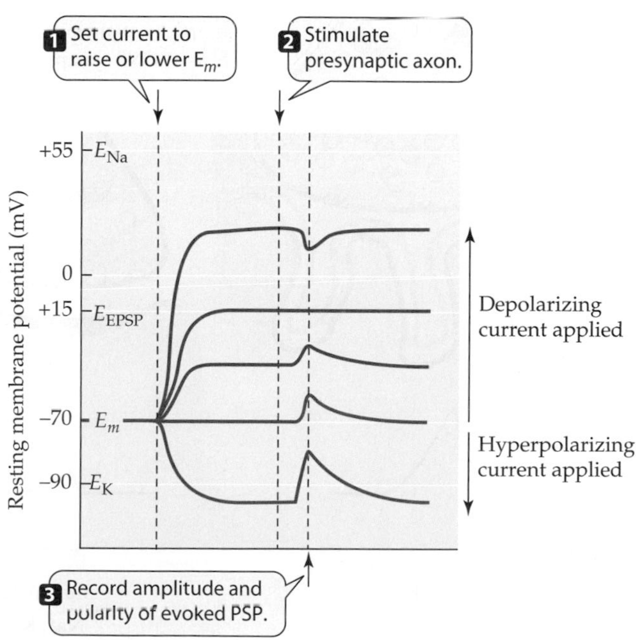

The *currents* carried by the two ions, however, are very *unequal* because the ions have different driving forces. The muscle fiber membrane is very far from the sodium equilibrium potential (E_{Na}), so there is a large driving force for sodium entry ($E_m - E_{Na}$). In contrast, the resting membrane potential is closer to the potassium equilibrium potential (E_K), so the driving force for potassium ($V_m - E_K$) is smaller. Consequently, Na^+ ions carry most of the synaptic current driving the membrane toward E_{EPSP}.

POSTSYNAPTIC POTENTIALS (PSPS) SPREAD DECREMENTALLY BECAUSE THEY DO NOT RESULT FROM THE ACTION OF VOLTAGE-GATED CHANNELS The amplitude of an EPSP changes with membrane potential (see Figure 12.8), but this change occurs because the driving force ($V_m - E_{EPSP}$) is changed, not because voltage changes permeability. Unlike the changes in permeability underlying an action potential, the *permeability changes that generate an EPSP are voltage-independent.* Because the permeabilities do not depend on voltage, depolarization of the EPSP itself is not regenerative; EPSPs are not all-or-none and are not propagated.

Neuronal EPSPs resemble neuromuscular EPSPs but are smaller

Fast EPSPs between neurons in the CNS are generally similar to those at the neuromuscular junction, but they also have some differences (Figure 12.9). They are similar in waveform, with a fast rise of 1 to 2 ms and an exponential return typically lasting 10 to 20 ms. The underlying ionic mechanisms of these PSPs are also similar. At both synapses, neurotransmitter molecules bind to postsynaptic receptors to increase the permeability of the membrane to both Na^+ and K^+ ions. Neuronal EPSPs have reversal potentials near zero, similar to the reversal potentials of neuromuscular EPSPs. Because of their similarities in mechanism and ionic basis, we use the same term, *EPSP*, to refer to both neuronal EPSPs and neuromuscular EPSPs.[1]

On the other hand, there are two major differences between neuronal and neuromuscular EPSPs. First, in central nervous systems of vertebrates, the common neurotransmitter mediating fast EPSPs is *glutamate*, rather than acetylcholine. Thus, some glutamate receptors have the same effect as the ACh receptors at the neuromuscular junction, rapidly increasing permeability to Na^+ and K^+ ions when they bind their neurotransmitter.

The second major difference between neuromuscular and neuronal EPSPs is their size. At a neuromuscular junction, a single presynaptic action potential liberates enough neurotransmitter to produce an EPSP that depolarizes the muscle fiber by 25 to 60 mV, more than enough to exceed the voltage threshold for a muscle-fiber action potential. In the CNS, the EPSP resulting from a single presynaptic action potential is typically a fraction of a millivolt in amplitude. The major cause of this difference is that a neuromuscular junction is large and contains many active zones so the axon terminal releases many vesicles of neurotransmitter per action potential. In contrast, at synapses in the CNS, each presynaptic action potential leads to the release of only one or a few vesicles. The smaller the quantity of neurotransmitter released, the fewer the receptors activated and the smaller the synaptic current generated.

The difference in amount of neurotransmitter release at neuromuscular and CNS neuronal synapses is an adaptation that largely determines the functional properties of the synapses. Recall that the ver-

[1] Neuromuscular EPSPs are sometimes called *end-plate potentials* (*EPPs*), named after the motor end-plate.

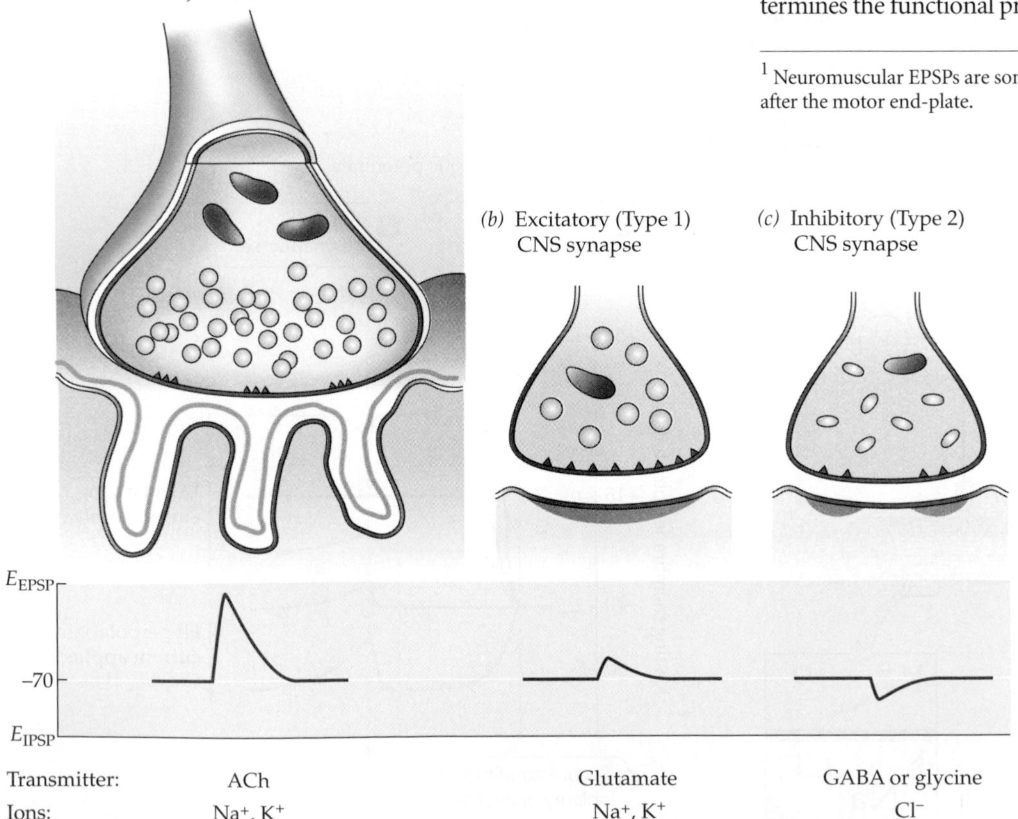

Figure 12.9 A neuromuscular synapse, a CNS excitatory synapse, and a CNS inhibitory synapse An excitatory synapse in the central nervous system *(b)* is similar in mechanism to a neuromuscular synapse *(a)*, but the EPSP is smaller because the presynaptic neuron terminal is smaller and releases fewer vesicles. The ionic basis of the EPSP is the same in both the neuromuscular and the CNS synapses, although the neurotransmitter is different. *(c)* A CNS inhibitory synapse produces an inhibitory postsynaptic potential (IPSP) by opening chloride channels, driving the postsynaptic membrane toward the chloride equilibrium potential (E_{Cl}). Thus, E_{Cl} is the reversal potential for the IPSP, or E_{IPSP}. Note that CNS excitatory synapses often have a different appearance (Type 1) in electron micrographs from CNS inhibitory synapses (Type 2).

tebrate skeletal neuromuscular junction is a relay synapse, at which each presynaptic action potential produces a postsynaptic action potential. In contrast, CNS synapses are usually integrating synapses, in which each of many presynaptic neurons has only a rather small effect on the postsynaptic neuron, and excitatory input from 20 to 50 neurons may have to be summated to depolarize the integrating neuron past its voltage threshold (see Figure 12.5).

Fast IPSPs can result from an increase in permeability to chloride

What about IPSPs? IPSPs are the major mechanism of synaptic inhibition in central nervous systems. The waveform of a fast IPSP resembles that of an EPSP but is typically hyperpolarizing rather than depolarizing. Most fast IPSPs result from an increase in permeability to Cl^- ions and are mediated by one of two neurotransmitters: gamma-aminobutyric acid (GABA) or glycine. We will consider IPSPs mediated by GABA as our example because most inhibitory synapses in mammalian brains use GABA as their neurotransmitter (see Figure 12.9*c*).

The common mechanism for GABA-mediated synaptic inhibition is as follows: As with other synapses, an action potential in the presynaptic, GABA-containing neuron causes vesicle fusion and GABA release. GABA diffuses to the postsynaptic membrane and binds to $GABA_A$ receptors that open to allow selective permeability to Cl^- ions. In most mammalian neurons, Cl^- is pumped out of the cell, so E_{Cl} is at a hyperpolarized value relative to E_m. Then when the permeability to Cl^- increases, Cl^- diffuses in, down its concentration gradient, hyperpolarizing the cell by driving E_m toward E_{Cl} (see Figure 12.9*c*). In other neurons, Cl^- is not pumped, so E_{Cl} and E_m are equal. Nevertheless, inhibitory synapses that increase permeability to Cl^- are still effective in inhibition because they "lock" the membrane potential at a value more hyperpolarized than the threshold voltage.

Excitatory and inhibitory synapses in the mammalian central nervous system have different characteristic appearances in electron micrographs (as shown schematically in Figure 12.9). Type 1 synapses are usually excitatory and have a relatively wide synaptic cleft with a relatively large area of prominent density under the postsynaptic membrane. Synaptic vesicles are round. Type 2 synapses are usually inhibitory and have a narrower cleft with smaller areas of membrane density that are more symmetrically distributed at the pre- and postsynaptic membranes. The vesicles appear flattened after aldehyde fixation for electron microscopy. This classification can be valuable in identifying which synapses in a micrograph are inhibitory and which are excitatory. The classification is not completely reliable, however, because the correlation of structural type with function is imperfect, and not all synapses fit the two types.

Overall, then, we see that at fast chemical synapses, neurotransmitters bind receptors and increase permeability to ions, leading to a PSP. Whether the PSP is excitatory or inhibitory depends on what kinds of ions flow through the ion channels when the channels open. Different neurotransmitters act at different receptors; they may produce different effects (e.g., glutamate → EPSP, GABA → IPSP) or similar effects (e.g., glutamate → EPSP, ACh → EPSP).

Note that in addition, a single neurotransmitter substance can have several different functions because it can act on different kinds of transmitter receptors. For example, acetylcholine produces fast EPSPs at neuromuscular junctions (by acting on one kind of acetylcholine receptor), but ACh produces IPSPs to inhibit the heart, by acting at a different kind of ACh receptor. We will consider this important principle in more depth later.

Summary: Fast Chemical Synaptic Actions Depend on Increases in Permeability to Ions

- At the vertebrate skeletal neuromuscular junction the neurotransmitter is acetylcholine. When stimulated, the presynaptic axon terminal releases acetycholine, which diffuses to postsynaptic receptors.
- Acetylcholine binding to its receptors opens ion channels to increase permeability to both Na^+ and K^+ ions. The resulting Na^+ and K^+ currents drive the membrane toward a value, E_{EPSP}, that is more depolarized than the threshold of the muscle fiber. Normally the amplitude of the EPSP at the junction is sufficient to exceed threshold and triggers a muscle-fiber action potential.
- The EPSP itself is a nonregenerative, nonpropagated local response because the neurotransmitter-dependent permeability changes are not voltage-dependent.
- Fast excitatory synapses in central nervous systems work by mechanisms similar to those at neuromuscular junctions. Neurotransmitter-gated ion channels increase membrane permeability to Na^+ and K^+ ions to produce depolarizing EPSPs.
- Neuronal EPSPs are much smaller than neuromuscular EPSPs because at neural synapses the postsynaptic membrane encompasses a small area that has a small number of receptor molecules, and the presynaptic axon releases less neurotransmitter, activating fewer postsynaptic receptors.
- Fast synaptic inhibition results from the opening of ion channels to increase permeability to chloride. E_{Cl} is commonly at a hyperpolarized value relative to the resting potential, leading to a hyperpolarizing IPSP.
- CNS excitatory and inhibitory synapses often have characteristic differences in their appearance in electron micrographs of the vertebrate CNS.

Presynaptic Neurons Release Neurotransmitter Molecules in Quantal Packets

Neurotransmitter molecules are synthesized and stored in the presynaptic terminal and are released by presynaptic impulses. Because acetylcholine is the neurotransmitter at the vertebrate neuromuscular junction, we will consider primarily cholinergic (acetylcholine-mediated) transmission here. However, mechanisms of neurotransmitter release are thought to be similar for all neurotransmitters, so much of this information is general.

Acetylcholine is synthesized and stored in the presynaptic terminal

Acetylcholine (ACh) is synthesized from choline and acetyl coenzyme A in the cytoplasm of the presynaptic terminal (Figure 12.10).

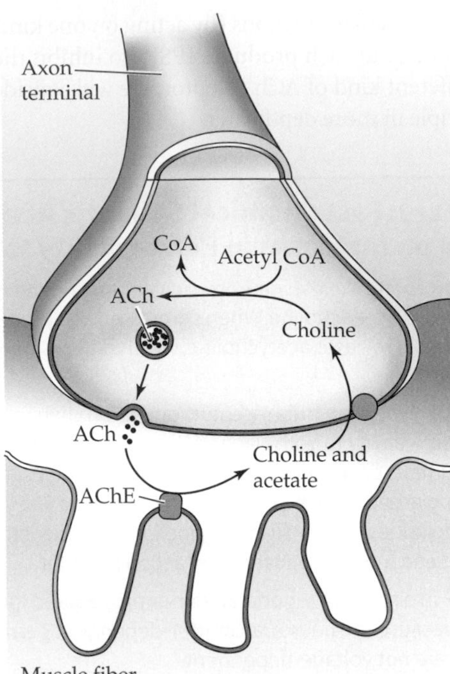

Figure 12.10 Acetylcholine synthesis and breakdown Acetylcholine is synthesized in the cytoplasm of the presynaptic terminal and then taken up into synaptic vesicles. Many vesicles are stored away from active-zone release sites; vesicles move from the storage compartment to release sites and dock to the presynaptic membrane at active zones. Acetylcholine is released by exocytosis of vesicles; the enzyme acetylcholinesterase (AChE) in the synaptic cleft terminates its action, hydrolyzing it to choline and acetate. A high-affinity transporter reabsorbs choline into the presynaptic terminal.

The reaction is catalyzed by the specific enzyme choline acetyltransferase. After synthesis in the cytoplasm, ACh is taken up into vesicles by a specific vesicular transporter molecule. Like all other proteins, choline acetyltransferase is synthesized in the cell body and must be transported the length of the axon. The availability of choline is the limiting factor in the rate of ACh synthesis. Choline is supplied via the blood and by choline uptake transporters at the presynaptic membrane, but as the available choline within the terminal is converted to ACh, synthesis slows and eventually ceases.

Acetylcholine in an axon terminal is stored in different compartments. Some vesicles are "docked" at release sites and constitute the rapidly releasable compartment of ACh. Other vesicles are anchored to the cytoskeleton distant from release sites in the interior of the terminal; these constitute the storage compartment of ACh.

Neurotransmitter release is voltage- and Ca^{2+}-dependent

Neurotransmitter release requires both presynaptic depolarization and Ca^{2+} ions. The normal regulator for neurotransmitter release is the depolarization of the presynaptic terminal by an action potential. Experiments indicate that the amount of neurotransmitter release increases with increasing depolarization over the physiological range. The release of ACh into the synaptic cleft depends on an influx of Ca^{2+} ions into the presynaptic terminal. If a neuromuscular junction is bathed in a saline solution that is free of Ca^{2+} ions, then a depolarization of the presynaptic membrane will not elicit an EPSP, because exocytosis does not occur.

All neurotransmitter release appears to be Ca^{2+}-dependent. Depolarization of the presynaptic membrane opens voltage-gated Ca^{2+} channels, allowing Ca^{2+} to enter and trigger neurotransmitter release. Because the intracellular concentration of Ca^{2+} is much lower than the extracellular concentration, there is a strong inward driving force on Ca^{2+} ions; Ca^{2+} entry locally increases the Ca^{2+} concentration near the channels. In preparations of the squid giant synapse (between second-order and third-order giant axons in the stellate ganglion; see Box 11.2), the presynaptic terminal is large enough to impale with a micropipette. Experimenters have injected the presynaptic terminal with aequorin, a protein that fluoresces in the presence of Ca^{2+}. Depolarization of the presynaptic terminal induces aequorin fluorescence, experimentally confirming that presynaptic depolarization leads to local Ca^{2+} entry. In fact, the only role of presynaptic depolarization is to trigger Ca^{2+} entry. In the squid giant synapse, investigators can inject Ca^{2+} directly into the presynaptic terminal with a microelectrode. Pulsed Ca^{2+} injection triggers postsynaptic EPSPs, indicating that Ca^{2+} entry is sufficient to release neurotransmitter.

Neurotransmitter release is quantal and vesicular

Acetylcholine is not released from the presynaptic membrane of a neuromuscular junction a molecule at a time; rather, it is released in multimolecular "packets" called **quanta** (singular *quantum*), units of about 5000 molecules each. Experimental evidence has demonstrated that each quantum is the equivalent of the contents of one synaptic vesicle.

The original evidence for quantal release comes from the vertebrate neuromuscular junction. If we impale a muscle fiber near the neuromuscular junction with an intracellular microelectrode (Figure 12.11), we can record a series of small depolarizations in the absence of any stimulation. These depolarizations have the same shape as a neuromuscular EPSP but are about $^1/_{50}$ the amplitude; thus, they are termed spontaneous **miniature EPSPs** (**mEPSPs**).

Each mEPSP is the postsynaptic response to a quantum. Miniature EPSPs typically have amplitudes of about 0.4 mV and occur nearly randomly over time. They do not represent responses to individual molecules of acetylcholine, because several thousand ACh molecules are required to produce a depolarization of 0.4 mV. The presence of spontaneous mEPSPs indicates that in the absence of presynaptic stimulation, there is a spontaneous, low-frequency release of ACh quanta—that is, 5000 molecules at a time.

Can we show that the release of ACh evoked by presynaptic depolarization is also quantal? The neuromuscular EPSP evoked by a presynaptic impulse has an amplitude of 20 to 40 mV and would require the nearly simultaneous discharge of 100 to 300 quanta. This number is too large to determine whether the release is quantal or not, since the difference between the response to, say, 200 and 201 quanta is not detectable.

(a) Experimental setup

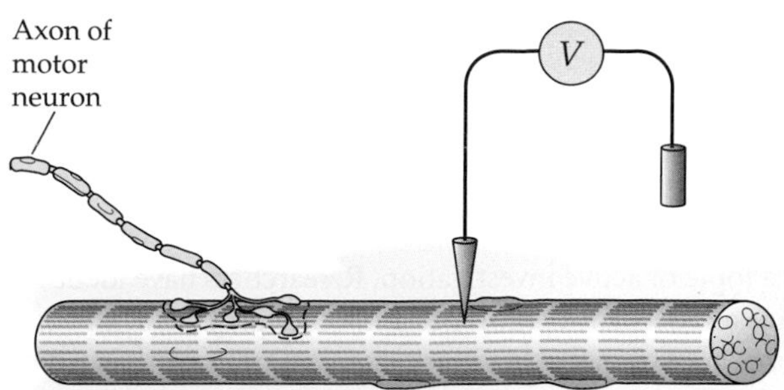

(b) Spontaneous miniature EPSP

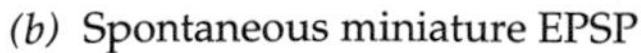

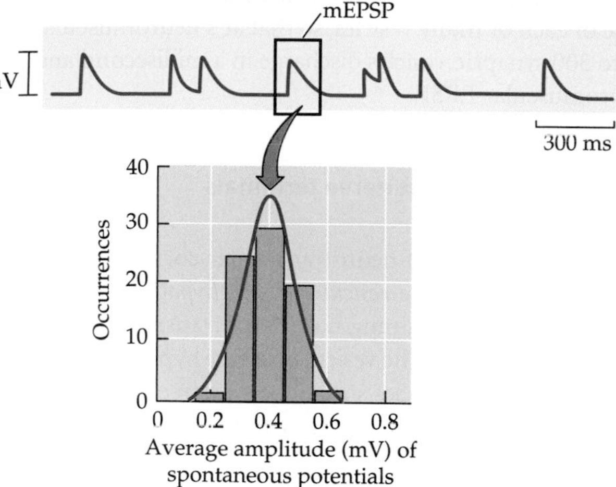

Figure 12.11 Spontaneous miniature excitatory postsynaptic potentials (spontaneous mEPSPs) *(a)* An electrode is positioned close to a neuromuscular synapse for recording in the absence of any stimulation. *(b)* The recording contains small (miniature) EPSPs at an average frequency of 1 to 2 per second. The average amplitude of spontaneous miniature EPSPs is 0.4 mV (histogram). Spontaneous mEPSPs show that the nerve terminal has a background low level of transmitter release in the absence of stimulation. The transmitter is released in multimolecular quantal packets, each quantum eliciting a miniature EPSP at the postsynaptic membrane.

Researchers solve this problem by *decreasing the number of quanta* released with an applied stimulus, so that only 1 or 2 quanta are released at a time into the synaptic cleft, rather than hundreds. To achieve this aim, they lower the Ca^{2+} concentration in the bath surrounding the muscle fiber, and they raise the bath concentration of Mg^{2+}, an ion that competitively inhibits the action of the Ca^{2+}. Are the resultant *evoked miniature EPSPs* quantal? As Figure 12.12 shows, evoked mEPSPs fall into amplitude classes that are multiples of the amplitudes of spontaneous mEPSPs. That is, each presynaptic stimulus evokes the release of 0, 1, 2, or 3 quanta (but never 1.5 quanta).

The experiment in Figure 12.12 illustrates that neurotransmitter release evoked by a presynaptic impulse is quantal, at least under conditions of low quantal content. Similar results occur with other types of synapses, and in some of these cases quantal release can be demonstrated without manipulation of extracellular ion concentrations. It is thought, then, that quantal release is the general rule for all chemical synapses.

Research has shown that cholinergic vesicles contain 10^3 to 10^4 molecules of acetylcholine, and that neurotransmitter is released in quanta of 10^3 to 10^4 molecules; thus the conclusion seems inescapable that a quantum corresponds to a synaptic vesicle. The synaptic vesicles fuse with the presynaptic membrane and discharge their content of neurotransmitter by exocytosis. The fusion of vesicles with the presynaptic membrane is sporadic in the resting termi-

Figure 12.12 Evoked miniature EPSPs (evoked mEPSPs) *(a)* A stimulating electrode evokes presynaptic action potentials in the motor axon, but because of the low Ca^{2+}, high Mg^{2+} bath, each action potential releases only a few quanta of neurotransmitter. *(b)* The amplitudes of the resultant EPSPs clustered around the size classes shown in the graph, indicating that the amplitude of each response resulted from the release of 0, 1, 2, or 3 quanta. (There were no recorded responses of 0.2 mV, for example, which would have been half a quantum.) Therefore, stimulus-evoked release of neurotransmitter, like spontaneous release, is quantal.

(a) Experimental setup

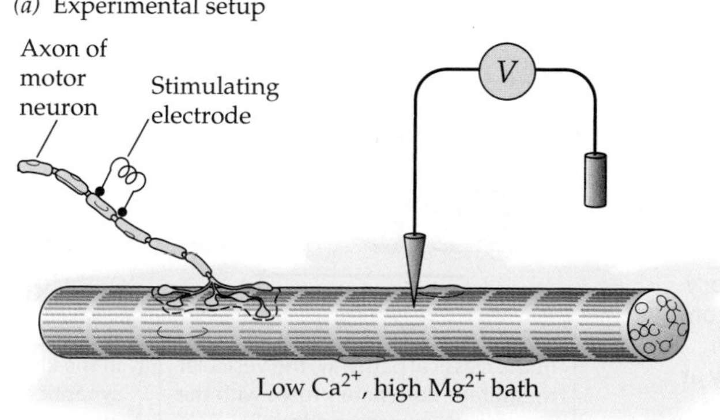

(b)

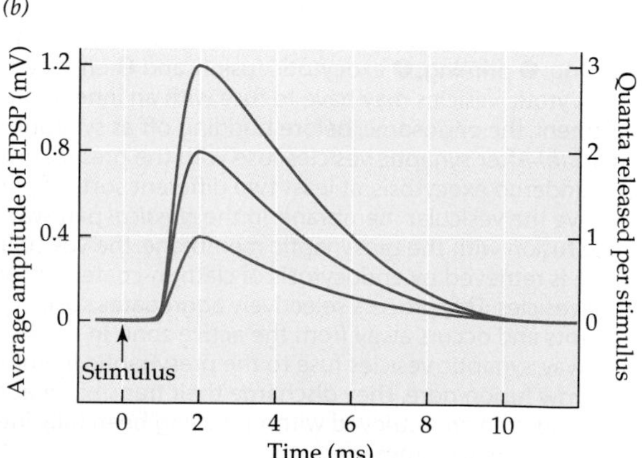

nal, producing spontaneous mEPSPs. Depolarization of the terminal by the presynaptic action potential greatly increases the *probability* of release of each of many vesicles, so that at a neuromuscular junction 150 to 300 synaptic vesicles discharge in a millisecond and produce a neuromuscular EPSP.

Synaptic vesicles are cycled at nerve terminals in distinct steps

The idea that quantal release of neurotransmitter corresponds to vesicular exocytosis is termed the *vesicular release hypothesis*, which is now widely accepted as explaining how neurotransmitter is released at all chemical synapses. The vesicular release hypothesis suggests that fusing exocytotic vesicles should be seen in electron micrographs, and that vesicular membranes must be recycled (to prevent vesicle depletion and expansion of the postsynaptic membrane with release). Exocytosis occurs rapidly, and it is difficult to show it directly in routinely fixed electron micrographs. On the other hand, electron micrographs of neuromuscular junctions stimulated at modest rates can show a temporary depletion of synaptic vesicles, but within minutes the vesicular membranes are pinched off and returned to the inside of the terminal by endocytosis.

The recycling of vesicular membranes allows synaptic vesicles to re-form and prevents the plasma membrane from expanding by the addition of vesicular membrane from exocytotic fusion. Figure 12.13 summarizes the processes of recycling vesicular membranes. Synaptic vesicles first are targeted to presynaptic release sites, then dock at release sites until Ca^{2+} ions trigger fusion of the vesicular and presynaptic membranes.

There appear to be at least two modes of fusion and subsequent retrieval of vesicular membranes: classical and kiss-and-run. In *classical* exocytosis the vesicular membrane merges with the terminal membrane and later is retrieved by endocytosis, a pinching off from the terminal membrane. Classical endocytosis selectively retrieves vesicular membranes with the aid of two proteins: clathrin and dynamin. In *kiss-and-run* fusion, the docked vesicle opens a fusion pore to release transmitter into the synaptic cleft, without completely becoming integrated into the terminal membrane; the vesicle is then reinternalized without requiring clathrin-mediated pinching off. The kiss-and-run pathway is faster and may predominate at lower rates of neurotransmitter release, while the classical exocytotic–endocytotic pathway is slower and may predominate at higher rates of release.

Several proteins play roles in vesicular release and recycling

The molecular mechanism of the exocytotic release of neurotransmitters is a topic of active investigation. Researchers have identified the proteins of vesicular membranes and other proteins with which they interact. Interestingly, these proteins are similar to those of exocytotic release in other cells, including yeast cells. These similarities suggest that a common mechanism of exocytotic secretion evolved in early eukaryotic cells and, although neurotransmitter release is much faster than other exocytotic secretion, the basic mechanism has been largely conserved.

Various proteins play roles in the several stages of vesicular release and recycling. The first stage is vesicular *mobilization*. At any time, many of the vesicles in a presynaptic terminal are located away from active zones in the *storage compartment*. Before these vesicles can release their contents, they must be mobilized to move up to release

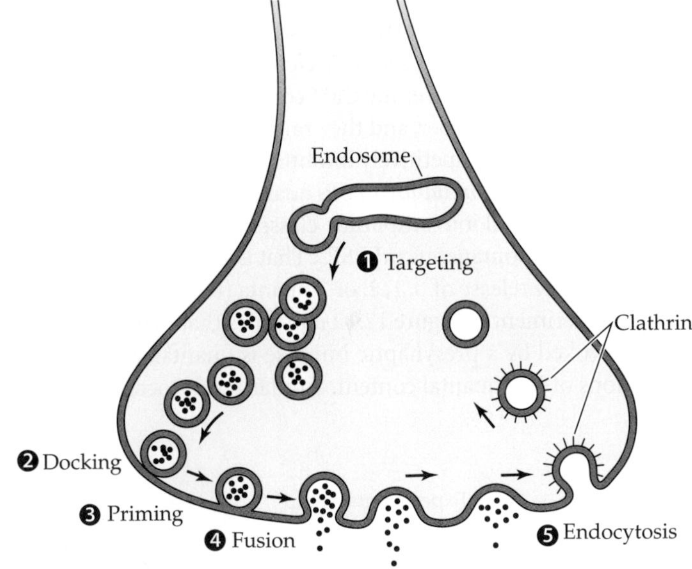

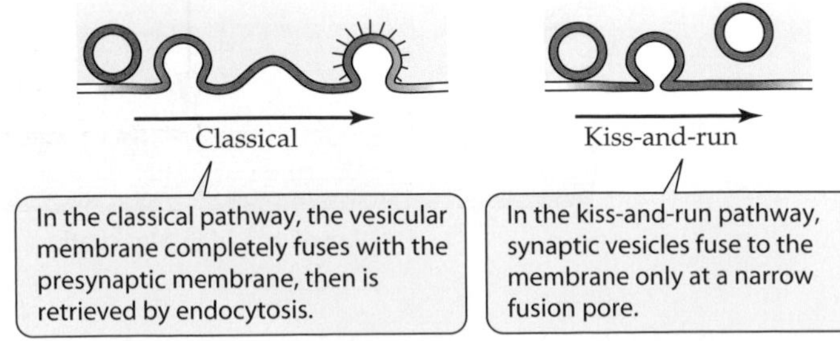

Figure 12.13 Synaptic vesicle recycling at the neuromuscular junction *(a)* Synaptic vesicles cycle though several steps: ❶ targeting; ❷ docking; ❸ priming; ❹ exocytotic fusion; and ❺ endocytosis. Endocytotic vesicles may have to fuse with an inner membrane compartment, the endosome, before budding off as synaptic vesicles again. *(b)* After synaptic vesicles fuse with the presynaptic membrane and undergo exocytosis, at least two different sorts of pathways can retrieve the vesicular membrane. In the *classical* pathway, after complete fusion with the presynaptic membrane, the vesicular membrane is retrieved by endocytosis of clathrin-coated pits into endocytotic vesicles. This process selectively aggregates specific vesicle components and occurs away from the active zone. In the *kiss-and-run* pathway, synaptic vesicles fuse to the presynaptic membrane only at a narrow fusion pore. They discharge their transmitter at the fusion pore and then are retrieved without having been fully integrated into the presynaptic membrane.

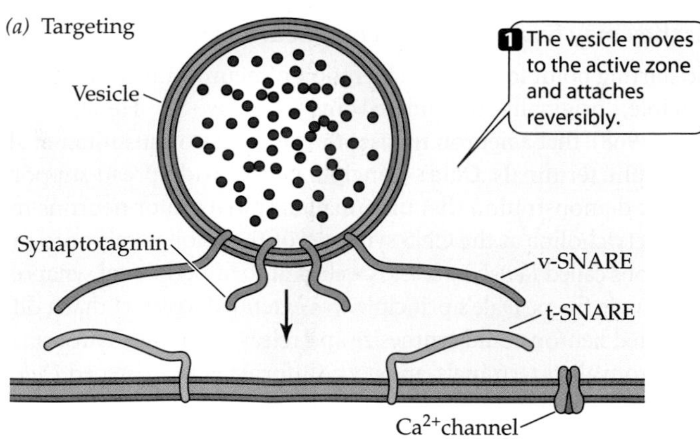

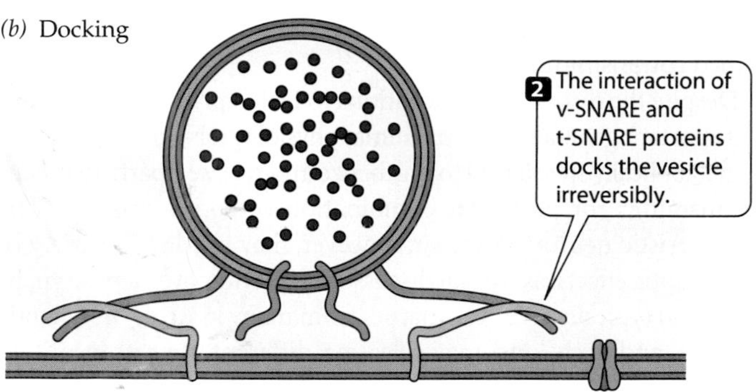

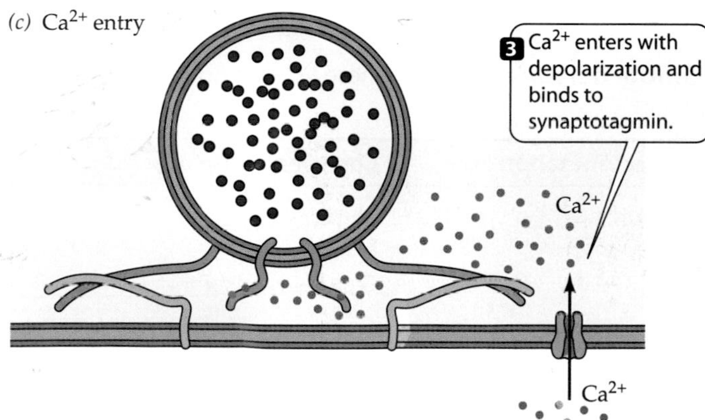

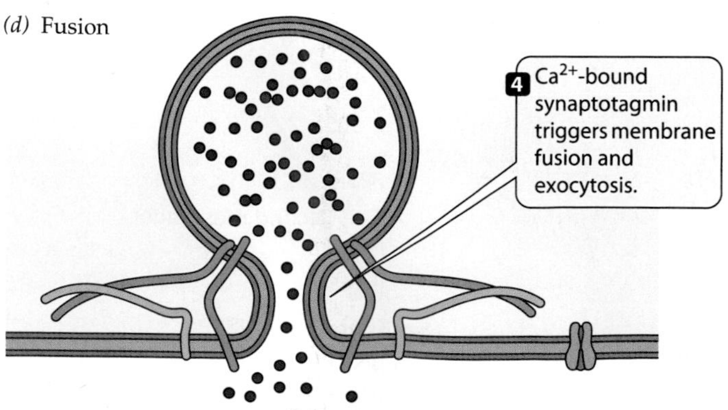

Figure 12.14 Vesicular docking and fusion release neurotransmitters Several proteins function in transmitter release from synaptic vesicles. Proteins called v-SNARES are associated with the vesicular membrane; and others, t-SNARES, are attached to the terminal membrane. ❶ In *targeting*, vesicles move to active zones, where they attach reversibly. ❷ *Docking* is mediated by the formation of a SNARE complex, the v-SNARES and t-SNARES interacting to hold the vesicle at the release site. ❸ *Ca^{2+} entry* through voltage-gated Ca^{2+} channels triggers ❹ *fusion* by the binding of Ca^{2+} to the vesicular protein synaptotagmin. The Ca^{2+}–synaptotagmin complex changes conformation to fuse the vesicular and terminal membranes.

sites at the active zones. In mobilization, the protein *synapsin* detaches the vesicle from the cytoskeleton, which permits the vesicle to migrate to an active zone. A vesicle may *attach* reversibly to the terminal membrane, followed by *docking* (which is irreversible).

Docking involves the interaction of proteins called *SNAREs* (Figure 12.14); vesicular v-SNAREs intertwine with terminal-membrane t-SNAREs to hold the docked vesicle at the active zone. (A major v-SNARE is called *vesicle-associated membrane protein* [*VAMP*]. It connects with the t-SNAREs *syntaxin* and *SNAP-25*. Cytoplasmic proteins called *NSF* and *SNAPs* may disassemble the SNARE complex after fusion.)

In response to the entry of Ca^{2+} through voltage-gated Ca^{2+} channels, a docked vesicle may form a fusion pore, which may lead to complete exocytotic fusion of the vesicle with the presynaptic membrane. The Ca^{2+} sensor for fusion is probably the vesicle protein *synaptotagmin*. Another protein, *rab3*, is active in vesicular mobilization and recycling, and may inhibit excessive fusion and release. The protein *dynamin* promotes reuptake of vesicles, and vesicular endocytosis is associated with the protein *clathrin*.

Summary: Presynaptic Neurons Release Neurotransmitter Molecules in Quantal Packets

- Small-molecule neurotransmitters are synthesized predominantly at axon terminals and are transported into synaptic vesicles.
- Neurotransmitters are released by presynaptic depolarization, which opens voltage-gated Ca^{2+} channels at active zones. Calcium ions trigger neurotransmitter release.
- Neurotransmitter is released in quantal packets, several thousand molecules at a time. Each quantum corresponds to a synaptic vesicle.
- Synaptic vesicles fuse with the presynaptic membrane to release their transmitter contents by exocytosis. Vesicular membranes are retrieved, refilled with transmitter, and recycled.
- Specific proteins associated with synaptic vesicles play different roles in vesicular targeting, docking, fusion, and retrieval.

Neurotransmitters Are of Two General Kinds

Several dozen chemical compounds have been identified as possible neurotransmitters, and the list of such compounds continues to increase. There are two major kinds of neurotransmitters: *small-molecule neurotransmitters* and *neuroactive peptides* (Table 12.2). Some neurotransmitters are very well known; others are in the process of discovery.

Our knowledge of different neurotransmitter systems bears little relation to their physiological prevalence. *Cholinergic* synapses (defined previously as synapses in which the neurotransmitter is acetylecholine) are best known because the neuromuscular junction is an easily studied model. Synapses in which the neurotransmitter is norepinephrine, or noradrenaline (termed *noradrenergic* or *adrenergic* synapses), are the next best known, because there are peripheral noradrenergic synapses in the sympathetic nervous system. It is estimated, however, that fewer than 10% of the synapses in the vertebrate CNS are cholinergic and fewer than 1% employ catecholamines such as norepinephrine. In many CNS synapses, the identity of the neurotransmitter is not known, but amino acids appear to be the most abundant and widespread neurotransmitters. This section discusses selected aspects of neurotransmitter function; it is beyond our scope to survey the subject comprehensively. We refer the reader to references at the end of the chapter that provide more extensive information.

Neurons have a characteristic neurotransmitter, but they may have more than one neurotransmitter

The starting point for our consideration of neurotransmitters is the principle, originally formulated in 1933 by Sir Henry Dale (1875–1968), that a neuron releases the same neurotransmitter at all its synaptic terminals. **Dale's principle** gained widespread support after the demonstration that mammalian spinal motor neurons release acetylcholine at the CNS synapses of their collaterals onto interneurons called *Renshaw cells*, as well as at their peripheral synapses with muscle fibers. Dale's principle was extended to assert that a differentiated neuron could synthesize and release only *one* neurotransmitter from all its terminals, and it was informally rechristened *Dale's law*. This restatement of Dale's principle is known to be widely untrue because an increasing number of cases have been found in which a neuron appears to contain and release more than one kind of neurotransmitter. Multiple neurotransmitters released by single neurons are termed **cotransmitters**.

Despite the discovery of exceptions to Dale's principle, it remains an important generalization in animal physiology. Neurons appear to be metabolically specialized to synthesize and release a particular neurotransmitter (or particular combination of cotransmitters). The characteristic neurotransmitter, however, may produce a variety of postsynaptic effects, as we will discuss shortly. Moreover, a postsynaptic neuron typically receives synapses from many kinds of presynaptic neurons, and each kind may release a different neurotransmitter.

TABLE 12.2 Some neurotransmitters and receptors of vertebrate central nervous systems These lists are not exhaustive; there are more transmitters, and more receptors for each transmitter.

Neurotransmitter	Receptor class	Direct/ionotropic	Indirect/metabotropic	Common mode of action
Amines				
Acetylcholine	Nicotinic	X		EPSP
	Muscarinic M_1–M_5		X	G protein → IPSP
Dopamine	d_1		X	
	d_2		X	
	d_3		X	
Norepinephrine	$\alpha_{1,2,3}$		X	
	$\beta_{1,2}$		X	
Serotonin	$5HT_1$		X	
	$5HT_2$		X	
	$5HT_3$	X		
Amino acids				
Glutamate	AMPA	X		EPSP
	NMDA	X		Ca^{2+} second messenger
	Metabotropic		X	DAG/IP_3
GABA	$GABA_A$	X		IPSP
	$GABA_B$		X	G protein → IPSP
Glycine		X		IPSP
Peptides			X	G protein–coupled (some tyrosine kinase)

Thus, although it is relatively sound to characterize a neuron in terms of the neurotransmitter it releases (e.g., as a cholinergic neuron), it is usually unsound to characterize it as responding to a particular neurotransmitter (e.g., as a cholinoceptive neuron), because any neuron may respond to several different neurotransmitters.

An agent is identified as a neurotransmitter if it meets five criteria

There is general agreement on the experimental criteria for identifying the neurotransmitter at a synapse:

1. The candidate neurotransmitter must be *present in the presynaptic terminal*, along with its synthetic machinery (enzymes, precursors).
2. The candidate neurotransmitter is *released upon presynaptic stimulation*, in amounts sufficient to exert a postsynaptic action.
3. When the candidate neurotransmitter is added to extracellular fluid in moderate concentrations, it should *mimic the effects of presynaptic stimulation*. For example, it should induce the same changes in permeability to ions as the synaptic action does.
4. A *mechanism for removal* of the candidate neurotransmitter should exist. This removal mechanism can be by enzymatic inactivation or by reuptake into cells. (We should point out, however, that many peptide neurotransmitters may have no specific removal mechanism.)
5. The *effects of drugs* on transmission at the synapse must be similar to and consistent with their effect on experimental application of the candidate neurotransmitter. For example, curare is an ACh receptor antagonist that blocks neuromuscular transmission, and it also blocks the effect of ACh experimentally applied through a micropipette.

Note that these criteria were developed at a time when only fast, direct synaptic transmission was known. Some criteria are more essential than others, and some may be modified as our views of synaptic processes expand. (As an example of these complications, gases such as nitrous oxide [NO] are thought to act at some synapses as retrograde messengers—molecules conveying signals from postsynaptic to presynaptic neurons. These are discussed in the last section of the chapter. Some physiologists would classify retrograde messengers as neurotransmitters, but we elect not to.)

It is difficult to demonstrate all of these criteria at a particular synapse, especially in the central nervous system. For example, the demonstration of release of a candidate neurotransmitter requires that it be collected in sufficient quantity to be assayed, and further requires demonstration that the released neurotransmitter came from the presynaptic terminal. Such demonstration is rarely possible in the brain, where neurotransmitter identification is often inferred solely from chemical evidence of neurotransmitter presence. Rigorous tests employing the full spectrum of criteria are important where possible, to prevent uncritical acceptance of every synaptically active agent as a presumed neurotransmitter. Because these criteria are so difficult to satisfy experimentally, we have a long list of *possible* and *probable* neurotransmitters and a shorter list of cases in which the neurotransmitter at a particular synapse is convincingly demonstrated.

Vertebrate neurotransmitters have several general modes of action

The principal synaptic neurotransmitters of vertebrates are summarized in Table 12.2. The list could be much longer; for example, at least 50 neuroactive peptides may act as neurotransmitters. It is difficult as well as simplistic to generalize about neurotransmitter functions because all neurotransmitters work in different ways and mediate different functional effects at different synapses. Nevertheless, here are some very broad generalizations about functional classes of neurotransmitters and synapses in the central nervous systems of vertebrates:

- Most synapses (numerically) in the CNS use amino acid neurotransmitters. Most fast EPSPs result from glutamate; most fast IPSPs, from GABA or glycine.
- Biogenic amines (acetylcholine, norepinephrine, dopamine, serotonin) are found in relatively few neurons, but these neurons have widely projecting endings. Many receptors for these neurotransmitters have slow actions that modulate neuronal activities, rather than mediating fast excitation or inhibition.
- Peptides are present in substantial minorities of CNS neurons. A neuroactive peptide may be co-released with one or more small-molecule neurotransmitters and may function as a cotransmitter with slow synaptic effects. Box 12.1 describes the roles of peptide neurotransmitters in pain perception.

Rather than surveying the metabolism and action of specific neurotransmitters, let's discuss a few neurotransmitter-related concepts that are important for understanding general synaptic functions.

MULTIPLE RECEPTORS Many neurotransmitters can mediate different postsynaptic actions at different effectors. For example, acetylcholine excites skeletal muscle via EPSPs but inhibits vertebrate heart muscle via hyperpolarizing IPSPs. These effects involve different permeability changes in the different postsynaptic cells (P_{Na} and P_K in skeletal muscle; P_K in heart muscle). Completely different postsynaptic receptors mediate these different effects: one a ligand-gated channel and one a G protein–coupled receptor. Actually the two different kinds of acetylcholine receptors were characterized many years ago by their pharmacology (i.e., the effects of drugs on them). The ACh receptor of skeletal muscle is stimulated by nicotine and hence is termed **nicotinic**. As noted previously, a nicotinic ACh receptor is blocked by curare. The ACh receptor of heart muscle is stimulated by muscarine and hence is termed **muscarinic**; it is blocked by atropine. Muscarinic ACh receptors are more numerous than nicotinic receptors within the vertebrate CNS.

Most neurotransmitters affect more than one kind of neurotransmitter receptor, and these may be of different classes (e.g., ligand-gated channels and G protein–coupled receptors). The different receptor classes that respond to a particular neurotransmitter were originally characterized by pharmacology, as with nicotinic and muscarinic acetylcholine receptors. More recently, researchers have cloned the genes for neurotransmitter receptors, enabling them to associate dozens of receptors with a particular neurotransmitter. Some well-studied neurotransmitter receptors are listed in Table 12.2.

BOX 12.1 Neuropeptides and Pain: The Nervous System Produces Natural Opiates

Morphine, heroin, and other opiate drugs have *analgesic* effects; that is, they decrease the strength and aversive quality of pain sensations. Opiate drugs are thus presumed to inhibit pathways in the CNS that convey and integrate sensory information about pain. During the 1970s, several groups of investigators demonstrated that the opiates bind stereospecifically (i.e., to one geometrical isomer but not the other) to anatomically discrete sites in the mammalian brain and that this specific binding can be blocked competitively by opiate antagonists such as the drug naloxone. Because stereospecific binding and competetive antagonism are characteristics of how ligands bind to receptor molecules, these results demonstrated that specific opiate receptor molecules exist in the brain. The finding of specific opiate receptors suggested the existence of endogenous opiates that bind to them because the receptors were unlikely to have evolved to bind derivatives of opium poppies. Thus the characterization of opiate receptors led to a search for endogenous opiates.

Several peptides have been isolated that bind to opiate receptors, and their binding can be antagonized by naloxone. These peptides include methionine-enkephalin and leucine-enkephalin (*met-* and *leu-enkephalin*)—each of which contains 5 amino acids—and β-endorphin, which contains 31 amino acids. The precursor from which enkephalins are derived differs from the precursor of β-endorphin, and the enkephalins are localized in neurons of many brain areas. Enkephalins exert relatively transient effects and are thought to function like neurotransmitters. β-Endorphin is largely limited to the pituitary and to hypothalamic neurons; it can exert a rather long-lasting analgesic effect and may act like a hormone. Further studies have revealed other endogenous opiates and have demonstrated the existence of several kinds of opiate receptors.

Another peptide, *substance P*, is involved in transmitting the pain signals that are inhibited by the endogenous opiates. Both the afferent sensory neurons and many of the CNS neurons that convey information about pain sensation contain substance P. The endogenous opiates act to regulate the activity of these substance P–containing pain neurons. Enkephalin-producing neurons inhibit the pain neurons, in part by presynaptic inhibition.

Presynaptic inhibition is a specific inhibitory interaction in which one axon terminal ends on another axon terminal (a configuraton called an *axo-axonal synapse*), and causes a *decrease* in the amount of neurotransmitter that the second ending releases per action potential. In this case, activation of enkephalin neurons decreases the release of substance P from pain sensory endings. This interaction might explain how acupuncture and electrostimulation treatments (which activate other nonpain sensory neurons to excite the enkephalin neurons) can inhibit pain sensations.

Study of the functional roles of peptides in nervous systems is one of the most rapidly expanding areas in physiology. We may find that a particular peptide such as an enkephalin acts in one place as a classical neurotransmitter, in another place as a hormone, and somewhere else as a paracrine agent (defined in Chapter 10, page 261) that alters the synaptic response to neurotransmitters.

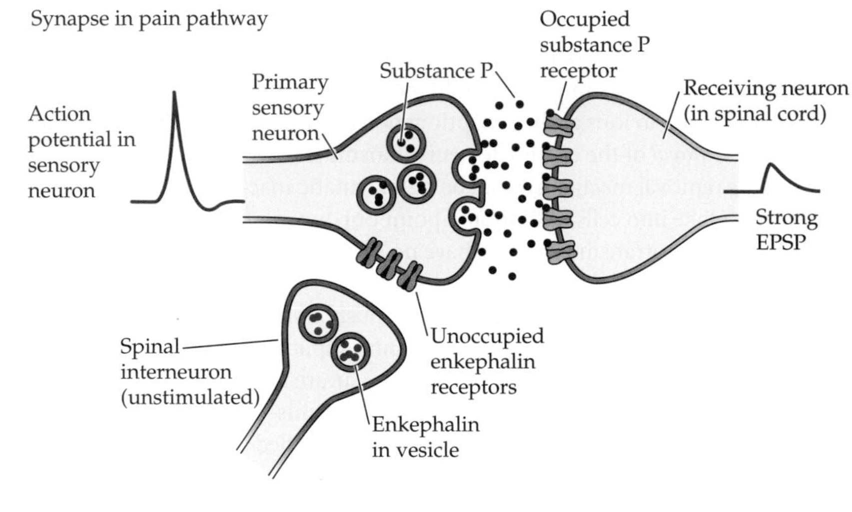

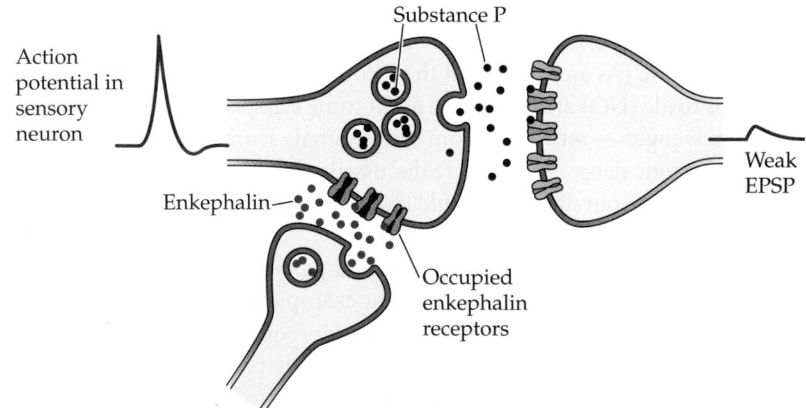

One way in which peptide transmitters may regulate the transmission of pain information from the peripheral pain receptors to the brain Pain sensory neurons release substance P at their synapses in the dorsal horn of the spinal cord. Interneurons containing the peptide transmitter enkephalin make synapses onto the axon terminals of the pain sensory neurons. Enkephalin released from the interneurons inhibits the release of substance P by presynaptic inhibition, so the receiving neuron in the spinal cord receives less excitatory stimulation and hence sends fewer pain-related impulses to the brain. Opiate drugs such as morphine presumably bind to enkephalin receptors, mimicking the pain-suppressing effects of the enkephalin system.

TERMINATION OF NEUROTRANSMITTER ACTION: ENZYMES AND REUPTAKE Neurotransmitters are generally active for only a short duration. Neurotransmitter release, diffusion, and receptor binding occur within a few milliseconds. For normal synaptic function, neurotransmitter molecules must be cleared from the synaptic cleft. The temporal and spatial effects of neurotransmitter action are limited in two ways: by *enzymatic destruction* of the neurotransmitter molecules, and by *reuptake* (using active transport to retrieve the neurotransmitter or its products).

At the neuromuscular junction, ACh is enzymatically digested by the enzyme acetylcholinesterase (AChE), located within the synaptic cleft (see Figure 12.10). AChE acts very rapidly; the entire sequence of ACh release, diffusion, binding to ACh receptors, and digestion requires about 5 ms. Some ACh molecules are destroyed even before they can bind to receptors. The products of ACh breakdown are choline and acetate. Choline is transported back into the presynaptic terminal by a specific high-affinity transporter in the terminal membrane. Thus the action of AChE both terminates the postsynaptic effects of ACh and (via the transporter) provides choline, the rate-limiting substrate for resynthesis of ACh in the presynaptic terminal.

The termination of neurotransmitter action by a localized enzyme is not a feature of all chemical synapses. For many neurotransmitters (catecholamines, amino acids), the termination and reuptake processes are combined. For example, the neurotransmitter norepinephrine (noradrenaline), is itself actively transported back into the presynaptic cell by a high-affinity transporter. Although enzymes that catabolize norepinephrine are present, they act more slowly than the reuptake system. Thus the synaptic action of norepinephrine is terminated by reuptake of the neurotransmitter rather than by enzymatic destruction. Glial cells can also actively take up neurotransmitters.

The "recycling" of neurotransmitter molecules—with or without enzymatic conversion—is a process distinct from the recycling of vesicular membranes discussed earlier. Whereas vesicular recycling involves endocytotic pinching off of *organelles* (multimolecular pieces of vesicular membrane), in neurotransmitter recycling a transporter actively transports *molecules* of neurotransmitter or metabolite across the membrane against a concentration gradient. The two processes occur independently of each other.

SYNTHESIS OF PEPTIDE AND SMALL-MOLECULE NEUROTRANSMITTERS Unlike small-molecule neurotransmitters, which are synthesized in the axon terminals, peptide neurotransmitters are synthesized in the cell body of a neuron and must be transported down the axon for release. All peptides are synthesized in the neuron cell body at ribosomal sites of the rough endoplasmic reticulum; the neuronal axon lacks ribosomes and performs no protein synthesis. Peptide neurotransmitters are chains of amino acids, typically 3 to 55 amino acids long.

Peptide neurotransmitters are synthesized as part of a larger precursor polypeptide, called a *propeptide*. A propeptide typically contains several copies of the peptide neurotransmitter within its amino acid sequence. After synthesis, the propeptide molecules are packed into secretory vesicles and transported down the axon; in the vesicle the propeptide is cleaved enzymatically into smaller pieces that may include several copies of the active peptide.

Peptide neurotransmitters cannot be retrieved once they have been released; eventually they are digested by nonspecific extracellular peptidases. Peptide neurotransmitters may become depleted as a result of the "long supply chain" for their synthesis, as well as the absence of their recycling. In contrast, for small-molecule neurotransmitters, resupply can usually keep up with release because the neurotransmitter can be locally retrieved or rapidly resynthesized in the axon terminal.

Neurotransmitter systems have been conserved in evolution

The best-known neurotransmitters that act at vertebrate synapses are also present in the nervous systems of the major invertebrate groups. For example, evidence (of variable quality) supports neurotransmitter roles for acetylcholine, GABA, glutamate, dopamine, and serotonin (also called *5-hydroxytryptamine*, or *5-HT*) among coelenterates, nematodes, annelids, arthropods, and molluscs. Genomic studies reinforce the evolutionary similarities of neurotransmitter systems in different phyla; for example, the genome of the nematode worm *Caenorhabditis* has been completely sequenced, so its entire collection of neurotransmitter enzymes and receptors is known. It has nicotinic ACh receptors, GABA receptors, excitatory glutamate receptors, and many G protein–coupled receptors related to their vertebrate counterparts.

The same neurotransmitters, however, are employed in different roles in different phyla. For example, most evidence indicates that in arthropods, glutamate is the major excitatory neuromuscular transmitter and acetylcholine is the major sensory neurotransmitter; their roles are reversed in the vertebrates (**Figure 12.15**).

These observations suggest that the evolution of neurotransmitter systems has been conservative; the same or related neurotransmitters and receptors are conserved across many phyla. Even the peptide neurotransmitters, which might be expected to show greater differences between phyla than the small-molecule amino acids and amines, are organized in protein families that indicate their evolutionary relationship across phyla.

Summary: Neurotransmitter Are of Two General Kinds

- Neurotransmitters can be small molecules or peptides. Perhaps a dozen small-molecule neurotransmitters and several dozen peptide neurotransmitters have been identified.
- A neuron can be identified by its characteristic neurotransmitter, but a single neuron may produce and release more than one neurotransmitter.
- For any neurotransmitter there are several receptors. Different kinds of receptors for a transmitter may coexist in the same organism and the same neuron.
- Most fast synapses in central nervous systems employ glutamate for EPSPs and GABA or glycine for IPSPs.
- Many receptors for small-molecule neurotransmitters, and probably for all peptides, are metabotropic and mediate slow synaptic potentials and modulatory responses.

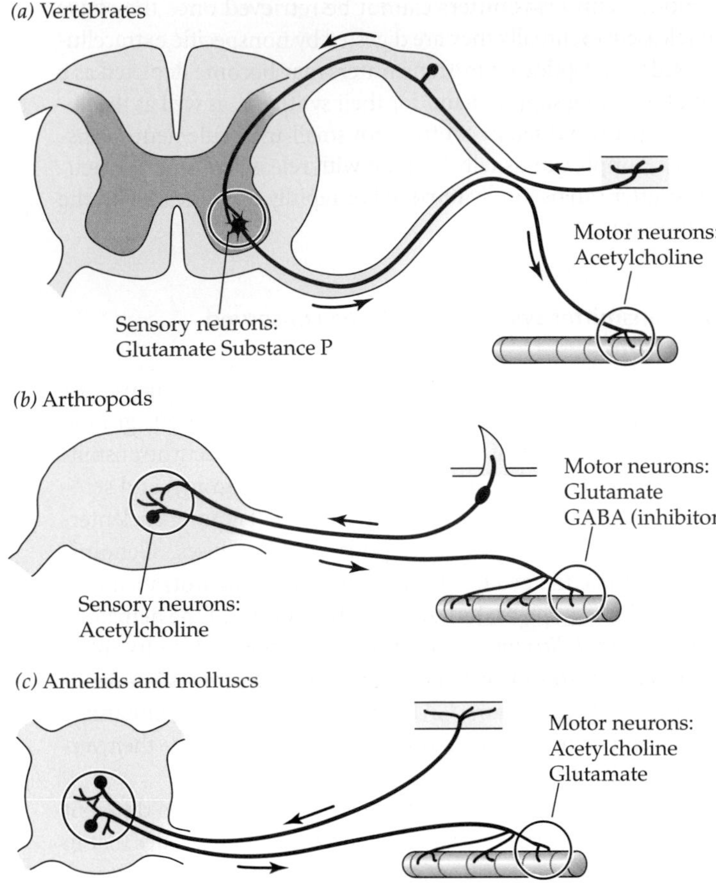

Figure 12.15 Major transmitters of peripheral neurons in vertebrates, arthropods, annelids, and molluscs Neurotransmitters are widely distributed among phyla, but the same neurotransmitters play different functional roles in the different phyla. For example, in vertebrates *(a)* glutamate is a major sensory neurotransmitter and acetylcholine is the major motor neurotransmitter, the reverse of the situation in arthropods *(b)*. *(c)* In annelids and mollusks, acetylcholine and glutamate can serve as neuromuscular neurotransmitters; sensory neurotransmitters are often not known.

- Peptides are synthesized in the neuronal cell body and transported down the axon packed in vesicles, unlike small-molecule transmitters, which are synthesized locally in axon terminals.
- The synaptic action of small-molecule neurotransmitters is terminated by reuptake or by enzymatic destruction.

Postsynaptic Receptors for Fast Ionotropic Actions: Ligand-Gated Channels

As we noted earlier, physiologists classify the postsynaptic neurotransmitter receptors into two broad categories, depending on their mechanism of action. **Ionotropic receptors** produce their effects *directly*; the neurotransmitter binds to the active site of the receptor protein, which is a *ligand-gated channel* (see Figure 2.23*a*)[2]. The receptor–channel opens in response to the binding to allow ions to pass into and out of the postsynaptic cell. **Metabotropic receptors** are not channels; instead they produce their effects *indirectly*. Neurotransmitter molecules bind to and activate a metabotropic receptor, which in turn initiates a metabolic cascade that may eventually modulate an ion channel in the postsynaptic membrane. Because an ionotropic receptor is a single macromolecular unit, it works rapidly and produces fast PSPs. Because a metabotropic receptor produces a cascade of reactions of separate proteins, its effects are slower and longer-lasting. Table 12.3 compares some of the key structural and functional differences between ionotropic and metabotropic receptors.

ACh receptors are ligand-gated channels that function as ionotropic receptors

The best-known examples of ionotropic receptors are the nicotinic acetylcholine (ACh) receptors that produce EPSPs at the vertebrate neuromuscular junction. Molecular biologists first isolated nicotinic ACh receptors in the electric organs of marine fish such as skates and rays (see Chapter 17), which consist almost entirely of modified neuromuscular junctions with dense concentrations of ACh receptors. These nicotinic ACh receptors are strikingly similar to those of other vertebrate neuromuscular junctions. A nicotinic ACh receptor is a glycoprotein composed of five homologous subunits designated α, β, δ, and γ, with α occurring twice (Figure 12.16*a*). Each subunit has four helical, membrane-spanning hydrophobic segments: M1 through M4 (Figure 12.16*b* and *c*). The polar M2 domains face the interior of the protein to form the central ion channel. The different subunits of the ACh receptor share considerable sequence homology and presumably evolved from a common ancestor.

Each of the two α subunits has an ACh-binding site on its extracellular side (Figure 12.16*d* and *e*). When the receptor binds two ACh molecules, the five subunits change their conformations to open a central channel large enough for Na^+ and K^+ ions to pass through it. The channel binds ACh and remains open only for a short time (about 1 ms). In fact, the channel may flicker open and closed during this brief interval. With prolonged exposure to ACh (e.g., if acetylcholinesterase is inhibited) the ACh receptor enters a third, *desensitized* state in which ACh is bound but the channel is closed. Desensitized receptors remain desensitized until the ACh molecules dissociate from the receptor.

As we have seen (page 320), during an EPSP the permeabilities of the postsynaptic membrane to Na^+ and to K^+ are nearly equal. However, because the resting membrane potential is much farther from E_{Na} than from E_K, the inward Na^+ current is much larger than the outward K^+ current. Thus, when a nicotinic ACh receptor channel opens, there is a net *inward* current flow (largely carried by Na^+ ions) into the postsynaptic cell.

The **synaptic current** is the sum of the net ionic currents though all of the activated ACh receptor; the synaptic current is the source of the EPSP. The current flowing through a single ACh receptor, and the additive nature of the currents flowing through more than one channel, can be demonstrated with the *patch-clamp technique* (de-

[2] To emphasize that the same molecule is a receptor and a channel, we can also call it a *receptor-channel*.

scribed in Chapter 11, on page 295) to record single-channel currents (**Figure 12.17**). By slowly increasing the ACh concentration in the pipette, we control the probability that a channel will open. At a low ACh concentration, only one channel opens at a time (top diagram in Figure 12.17*b*). Increasing the ACh concentration slightly causes each channel to open more frequently (center recording in Figure 12.17*b*), and increasing the ACh concentration still further causes both channels to open simultaneously for intermittent periods (bottom recording in Figure 12.17*b*).

To summarize, the patch-clamp experiment demonstrates four characteristics of nicotinic ACh receptors and the ionic currents that they produce once they are activated:

1. The opening of an ACh receptor channel is an all-or-none phenomenon.
2. The probability that a channel will open depends on the concentration of ACh at the receptor.
3. The net ionic current through the open channel provides that channel's contribution to a synaptic potential.
4. The currents through all open channels can be summated and constitute the synaptic current.

The properties of ACh receptors as ligand-gated channels can be compared to those of voltage-gated channels that underlie the action potential. The major difference is in the control of permeability changes of the molecules. In the voltage-gated sodium channel, permeability depends directly on the membrane potential (see Figures 11.13 and 11.14). Permeability of the ACh receptor is essentially voltage-independent and depends instead on the binding of the neurotransmitter ACh. As noted previously, this difference in the control of permeability to ions of the two molecules results in the difference in properties of the potentials they produce: Action potentials are all-or-none and propagated; synaptic potentials are graded in amplitude and spread decrementally.

TABLE 12.3 Ionotropic and metabotropic receptors: Structural, functional, and mechanistic differences

Characteristic	Ionotropic receptors	Metabotropic receptors
Receptor molecule	Ligand-gated channel receptor	G protein–coupled receptor
Molecular structure	Five subunits around an ion channel	Protein with seven trans-membrane segments; no channel
Molecular action	Open ion channel	Activate G protein; metabolic cascade
Second messenger	No	Yes (usually)
Gating of ion channels	Direct	Indirect (or none)
Type of synaptic effect	Fast EPSP or IPSP	Slow PSPs; modulatory changes (In channel properties, cell metabolism, or gene expression)

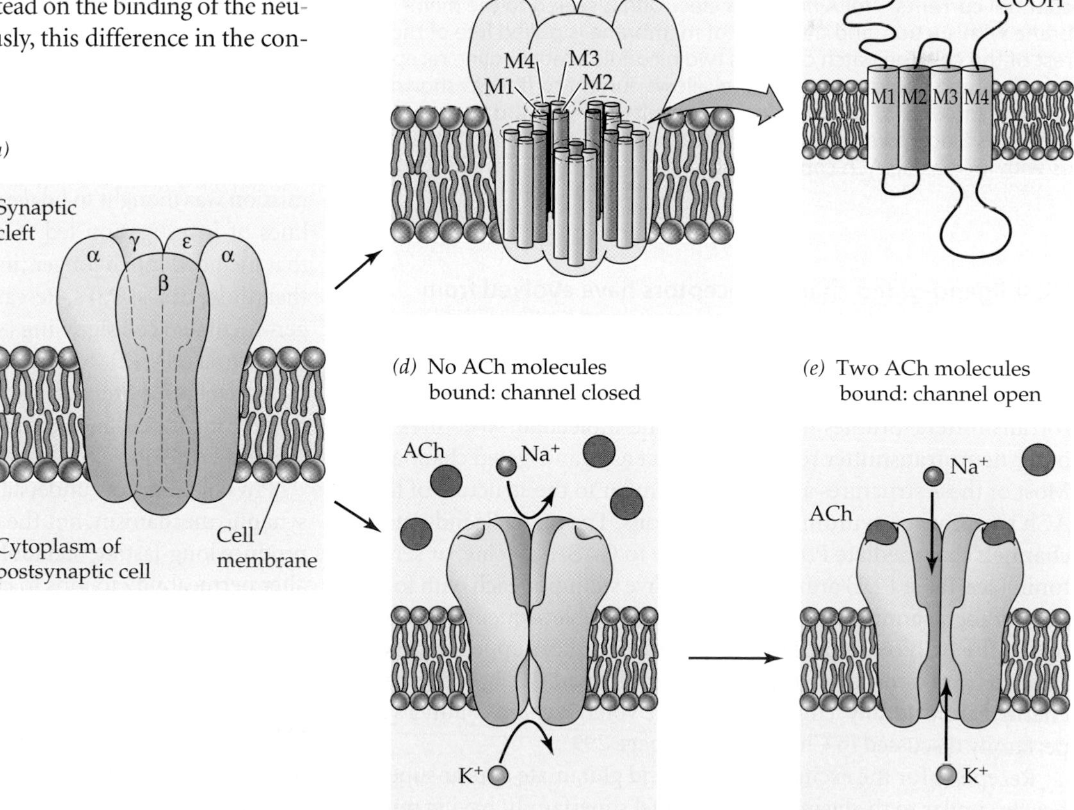

Figure 12.16 The molecular structure and function of a ligand-gated channel, the nicotinic acetylcholine receptor *(a)* The receptor has five protein subunits; the two α subunits each contain an ACh-binding site. *(b)* The five subunits surround a central ion channel, and each subunit has a similar structure, with four membrane-spanning segments (M1–M4). The M2 segment lines the pore. *(c)* Each subunit is a single polypeptide chain. *(d)* When ACh is not bound to the receptor, its inner channel, which narrows in the region of the lipid bilayer, is closed. *(e)* The binding of an ACh molecule to each of the two α subunit–binding sites leads to a conformation change that opens the channel to a diameter of 0.6 nm, allowing Na^+ and K^+ ions to flow through.

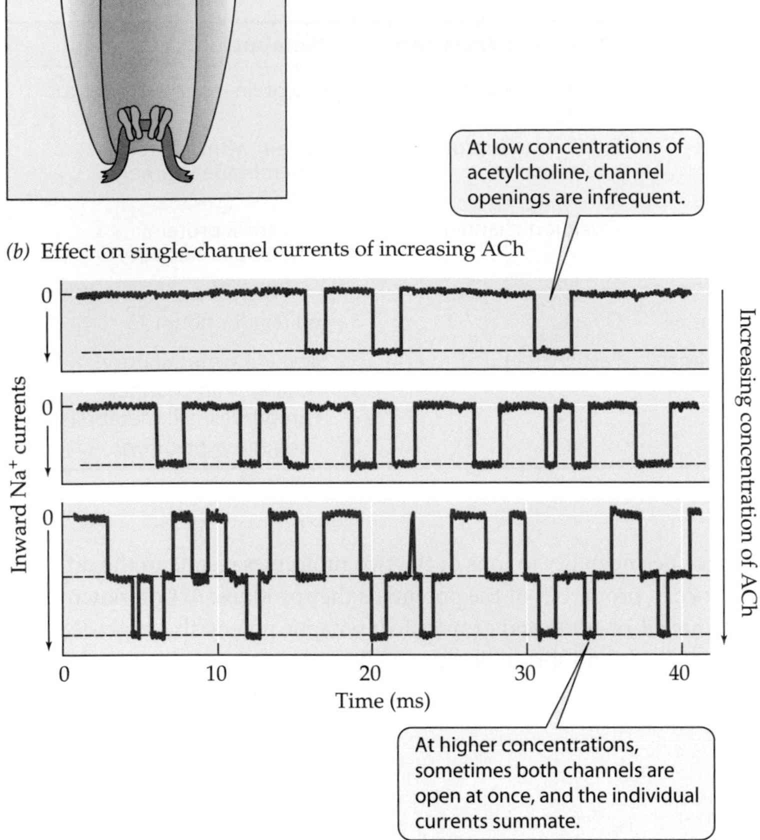

Figure 12.17 Patch-clamp recordings of acetylcholine receptor–channel currents *(a)* A fine glass electrode is sealed to the membrane with suction, and the patch of membrane is pulled free of the rest of the cell. The patch contains two nicotinic acetylcholine receptors. Opening of a receptor channel allows current to flow, as shown in the records in *(b)*. *(b)* Each opening results in an inward current, shown as a downward deflection. The frequency of opening increases with increasing ACh concentration.

Most ligand-gated channel receptors have evolved from a common ancestor

We have discussed nicotinic acetylcholine receptors in some detail, as examples of receptor proteins mediating direct, fast action of neurotransmitters. Studies have clarified the molecular structures of other neurotransmitter receptors that act as ligand-gated channels. Most of these structures are strikingly similar to the structure of the ACh receptors of neuromuscular junctions. The many ligand-gated channels that mediate PSPs in response to GABA, glycine, or serotonin (see Table 12.2) are composed of five subunits, each with four membrane-spanning segments, with considerable sequence homology to the ACh receptor. Therefore, these receptors appear to have evolved from a common ancestor and are termed the **ligand-gated channel superfamily**, comparable to the voltage-gated channel superfamily discussed in Chapter 11 (see page 299).

Receptors for the excitatory amino acid glutamate appear superficially similar to the ligand-gated channel superfamily, having multiple subunits of greater molecular weight and three or possibly four membrane-spanning segments. However, the glutamate receptors share little sequence homology with members of the ligand-gated channel superfamily, and probably evolved independently. Functions of glutamate receptors are discussed in the section titled "Synaptic Plasticity" (see page 341).

Summary: Postsynaptic Receptors for Fast Ionotropic Actions

- The receptors that produce fast postsynaptic potentials are ligand-gated channels. They are receptor–channels because the same molecule is both the receptor and the ion channel.
- The nicotinic acetylcholine receptor of the neuromuscular junction is the model ligand-gated channel. It contains five homologous subunits that surround a central ion channel that opens to allow Na^+ and K^+ ions to flow across the membrane.
- A ligand-gated channel opens briefly in response to binding two molecules of neurotransmitter, contributing to the synaptic current that produces a PSP.
- Ligand-gated channel receptors for GABA, glycine, and serotonin are evolutionarily related to nicotinic ACh receptors. Other ligand-gated channel receptors for glutamate appear to be a separate, evolutionarily unrelated family.

Postsynaptic Receptors for Slow, Metabotropic Actions: G Protein–Coupled Receptors

Our discussion of synaptic action so far has largely considered only fast synapses such as vertebrate neuromuscular junctions and synapses onto spinal motor neurons. In these synapses, neurotransmitter produces an EPSP or IPSP by means of a brief increase in permeabilities to ions. Until the 1980s such fast, direct synaptic transmission was thought to be *the* mechanism of synaptic action. Several lines of investigation led to a broader view of synaptic function that included much longer time courses and wider synaptic actions than those of fast PSPs. Researchers discovered (1) second messenger–mediated cell signaling (see Chapter 2; page 59) in nonneural and neural cells, (2) slower synaptic actions such as permeability-decrease synaptic potentials (to be discussed shortly), and (3) peptide neurotransmitters and cotransmitters that did not seem to produce classical fast PSPs.

Physiologists now understand that fast PSPs represent *one* kind of synaptic mechanism, not the only kind, and that synapses can also produce long-lasting metabolic effects. Metabotropic receptors can alter permeability to ions to change membrane potential indirectly (termed *indirect gating* of ion channels) but can also induce other metabolic changes that don't gate ion channels at all.

Metabotropic receptors act via second messengers

Metabotropic receptors typically act to increase the concentration of an intracellular second messenger. Recall from Chapter 2 that a **sec-**

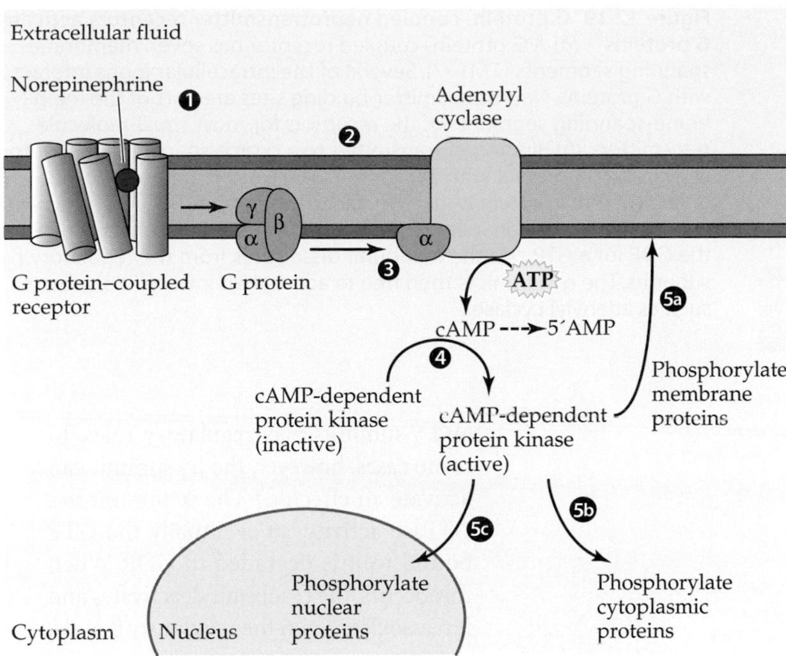

Figure 12.18 Metabotropic receptors: cyclic AMP as a second messenger Some neurotransmitters act via G protein–coupled receptors to alter cellular metabolism, rather than via ligand-gated channels. In this example, ❶ norepinephrine binds to a G protein-coupled receptor to activate a G protein, which in turn activates the enzyme adenylyl cyclase ❷. ❸ Adenylyl cyclase catalyzes the conversion of ATP to the second messenger cyclic AMP. ❹ Cyclic AMP activates a protein kinase, which phosphorylates proteins to activate or inactivate them. ❺(*a*) Cyclic AMP–dependent protein kinase can phosphorylate membrane proteins, which can indirectly gate them open or closed. The kinase can also phosphorylate cytoplasmic proteins (such as metabolic enzymes) ❺(*b*) and nuclear proteins that regulate gene expression ❺(*c*). Thus, neurotransmitter binding can have widespread, long-lasting effects on postsynaptic cells.

ond messenger is an intracellular signaling molecule that carries the signal to the interior of the cell, altering some activity of the cell in response to activation of a surface membrane receptor. The best-known second messenger is the cyclic nucleotide 3′,5′-cyclic adenosine monophosphate (**cyclic AMP** or **cAMP**).

Let's consider a specific example of how a neurotransmitter receptor acts via a second messenger to influence cellular metabolism (Figure 12.18). Consider a synapse at which the presynaptic neuron releases the neurotransmitter norepinephrine (noradrenaline). Norepinephrine acts on a postsynaptic metabotropic receptor that is not an ion channel. The receptor is a **G protein–coupled receptor** because when it is activated by binding neurotransmitter, it activates a membrane protein called a **G protein**. Both the G protein–coupled receptor and the G protein can diffuse laterally in the fluid mosaic membrane of the postsynaptic neuron, so when the G protein–coupled receptor is activated, it can bump into the G protein and activate it. Once it has been activated by the receptor, the G protein can activate another class of membrane protein termed an **intracellular effector**. In the case illustrated in Figure 12.18, the intracellular effector is an enzyme, adenylyl cyclase. Adenylyl cyclase (once activated by the active G protein) catalyzes the cytoplasmic conversion of ATP to cAMP.

Intracellular second messengers such as cAMP can exert widespread metabolic effects, by activating a **protein kinase**, an enzyme that phosphorylates proteins. An increase in intracellular cAMP concentration leads to increased activation of cAMP-dependent protein kinase (Figure 12.18). (Cyclic-AMP action is eventually terminated by the enzyme phosphodiesterase, which converts the cAMP molecules to 5′-AMP.) Activated cAMP-dependent protein kinase phosphorylates proteins, altering their structure and activity. Thus, stimulation of a neurotransmitter-dependent adenylyl cyclase leads to protein phosphorylation.

In postsynaptic neurons the proteins phosphorylated often include membrane ion channels, altering their gating to change the membrane's permeability to ions and thereby change membrane potential. This indirect gating of ion channels is one way in which metabotropic receptors function, but it is not the only way! Cyclic AMP–dependent protein kinase can also phosphorylate cytoplasmic proteins involved in control of cellular metabolism, and even nuclear proteins that regulate gene expression (Figure 12.19). Uncovering the fact that synaptic actions can control cellular metabolism and gene expression is one of the most important advances in our understanding of synaptic physiology in the last 25 years.

Cell signaling initiated by G protein–coupled receptors underlies the action of many other controls of cellular activities, as discussed in Chapters 2 and 14. G protein–coupled receptors constitute the major class of metabotropic receptor in synapses, so we will focus on their actions and second-messenger systems in greater detail.

The structure of G protein–coupled receptors

All G protein–coupled receptors have similar overall protein structures and have considerable sequence homology with each other. Thus, G protein–coupled receptors constitute an evolutionary superfamily, like the superfamilies of voltage-gated channels and ligand-gated channels. Because G protein–coupled receptors have seven transmembrane segments (Figure 12.19), they are sometimes termed the *7-TM superfamily*. For many 7-TM neurotransmitter receptors, ligands bind near the extracellular ends of some transmembrane domains, while the cytoplasmic loop between TM5 and TM6 and the cytoplasmic domain adjacent to TM7 mediate G-protein binding.

Many receptor molecules that are not activated by neurotransmitters also belong to the G protein–coupled receptor superfamily. For example, the sensory-receptor molecules rhodopsin and vertebrate olfactory receptor proteins (see Chapter 13) act via G proteins and share extensive sequence homology with neurotransmitter receptors. Receptors for peptide hormones (see Chapter 14) are also members of the G protein–coupled receptor superfamily. Clearly

(a) G protein–coupled receptor structure

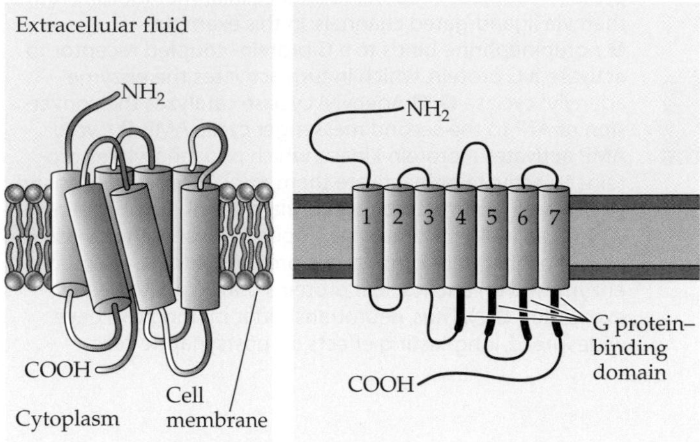

(b) G protein activation

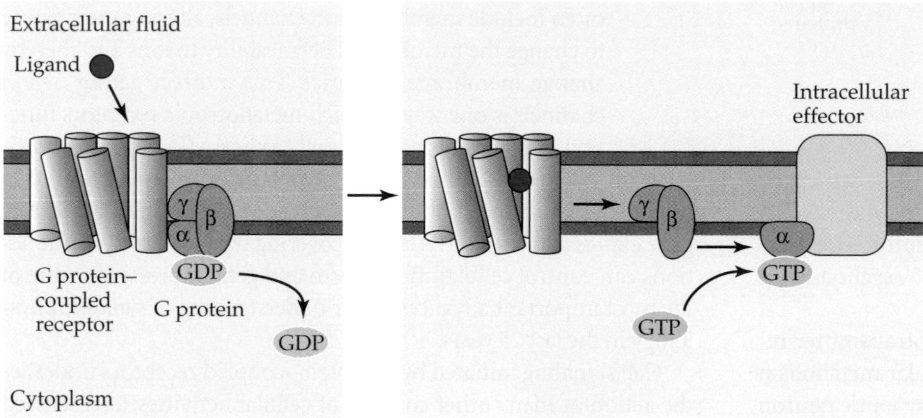

Figure 12.19 G protein–coupled neurotransmitter receptors activate G proteins *(a)* A G protein–coupled receptor has seven membrane-spanning segments (TM1–7). Several of the intracellular loops interact with G proteins. Neurotransmitter-binding sites are part of the membrane-spanning segments of the receptors for most small-molecule transmitters. *(b)* Binding of transmitter to a G protein–coupled receptor activates it to interact with a G protein. The G protein has three subunits; when the protein is inactive, the α subunit binds GDP. The G protein is activated when it encounters an activated receptor; it exchanges the GDP for a GTP, and the α subunit dissociates from the regulatory βγ subunits. The α subunit is then free to activate an intracellular effector such as adenylyl cyclase.

this mechanism of cellular response to extracellular signals arose early in evolutionary history and has been exploited for many functions.

G proteins act via intracellular effectors

G proteins are so named because they bind guanosine nucleotides, which regulate their activity. The G proteins activated by collision with G protein–coupled receptors are bound to the inner leaflet of the plasma membrane. They consist of α, β, and γ subunits (Figure 12.19). In their inactive form, the G-protein subunits are bound together, and the α subunit is bound to a molecule of guanosine diphosphate (GDP). When activated by a G protein–coupled receptor, the α subunit dissociates from GDP and acquires a molecule of guanosine triphosphate (GTP); the α subunit also dissociates from the β and γ subunits (which remain joined to each other). The activated α subunit diffuses laterally along the inner leaflet of the plasma membrane until it encounters an intracellular effector such as adenylyl cyclase. (Usually the dissociated α subunit is the active form of the G protein; the β and γ subunits play regulatory roles. In some cases, however, the βγ subunit can activate an effector.) The α subunit has GTPase activity, so eventually the GTP bound to it is degraded to GDP. When this occurs, the α subunit deactivates and it reassociates with the regulatory β and γ subunits.

Some G proteins can activate ion channels directly without employing a second messenger. This direct G-protein action is present in cholinergic synapses on cardiac muscle tissue, which, unlike skeletal muscle, is inhibited by acetylcholine. The inhibitory action of ACh is mediated by muscarinic ACh receptors, which are G protein–coupled receptors. In mammalian cardiac muscle fibers, the activated G protein binds directly to potassium channels, opening them to produce IPSPs (**Figure 12.20**). This mechanism demonstrates three significant features of synaptic function:

1. A particular neurotransmitter can mediate very different effects by activating different kinds of receptors.
2. G proteins can act on channels directly without an intervening second messenger.
3. Channels can be gated by signals other than voltage or direct neurotransmitter binding.

G protein–coupled receptors can act via second-messenger systems other than cyclic AMP. The number of known systems is small, although different kinds of cell-signaling mechanisms are still being discovered. One other second messenger is *cyclic GMP* (cyclic guanosine monophosphate), the production of which is similar to that of its adenosine analog cAMP. That is, some neurotransmitter receptors act via a G protein to activate a guanylyl cyclase to produce cyclic GMP, which activates a cGMP-dependent protein kinase, leading to protein phosphorylation. In addition, Ca^{2+} ions can themselves act as second messengers. For example, Ca^{2+} ions, entering the cytoplasm through voltage-gated Ca^{2+} channels or from other sources, can bind to the cytoplasmic protein **calmodulin**, activating it. Activated calmodulin can then bind

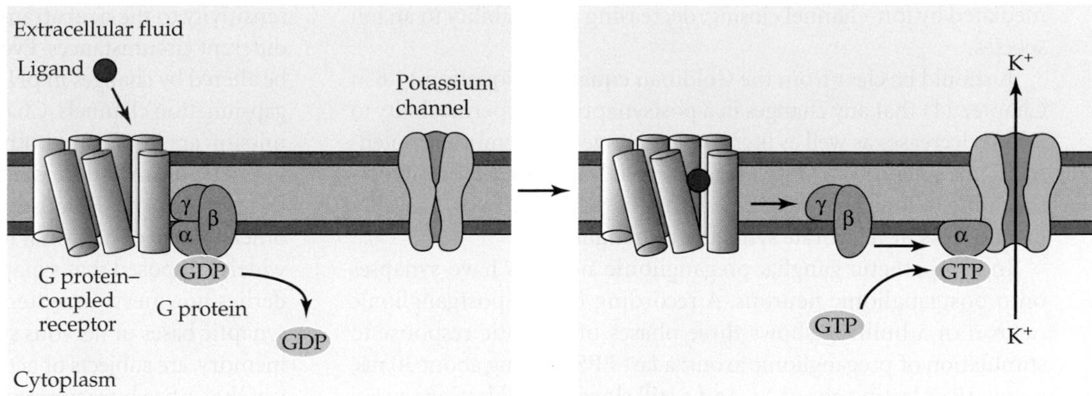

Figure 12.20 G proteins can themselves activate ion channels, without a second messenger In this example of acetylcholine-mediated inhibition of vertebrate heart muscle fibers, a muscarinic acetylcholine receptor (a G protein–coupled receptor) activates a G protein, which can itself gate a potassium (K^+) channel open without involving a second messenger.

and activate another kinase termed *calcium/calmodulin-dependent protein kinase*, or *CaM kinase*. Finally, products of phospholipid metabolism can mediate a more complex set of second-messenger effects.

Most lipids in membranes are phospholipids (see Chapter 2, page 31), in which one of the fatty acids in a neutral fat is replaced by a phosphate to which is bound a small organic residue. One common membrane phospholipid is phosphatidylinositol 4,5-bisphosphate (PIP_2). Neurotransmitters can act via G protein–coupled receptors to lead to the production of second messengers from PIP_2 (Figure 12.21). For example, norepinephrine can activate a G protein–coupled receptor, activating a G protein so that its α subunit activates a membrane-bound enzyme, *phospholipase C*. Phospholipase C catalyzes the hydrolysis of PIP_2 into **inositol trisphosphate** (**IP_3**) and **diacylglycerol** (**DAG**).

Both IP_3 and DAG act as second messengers, via different paths that reflect their structures. IP_3, which had formed the polar head of the phospholipid, is a polar molecule that freely diffuses through the cytoplasm. Its major action is to release Ca^{2+} ions from intracellular stores such as the endoplasmic reticulum. DAG, in contrast, is nonpolar and is constrained to the membrane lipid bilayer. It diffuses through the bilayer to activate *protein kinase C*. Thus, DAG as a second messenger leads to protein phosphorylation, as does cAMP. The Ca^{2+} released by action of IP_3 can enhance the activation of protein kinase C, as well as activating calmodulin and other Ca^{2+}-sensitive proteins.

The ability of neurotransmitters to stimulate the production of second messengers is important, in part because it provides one possible mechanism whereby synaptic transmission can mediate relatively slow and long-lasting effects. The direct action of neurotransmitters on ion channels in membranes has a time course of milliseconds. The time course of effects mediated by second messengers may be as short as tens of milliseconds (e.g., in slow PSPs mediated by decreased permeabilities to ions, discussed later in this section). At the other extreme, second-messenger actions may underlie synaptic changes involved in learning and memory, with a time course of days or years, as we will see.

Permeability-decrease synapses involve G protein–coupled receptors

G protein–coupled receptors mediate a variety of synaptic effects, as we have indicated. They may lead to indirect opening of ion channels, but they can also produce effects that close ion channels or act on postsynaptic targets that are not ion channels. Here we consider just one of these alternative possibilities: synaptic potentials that are

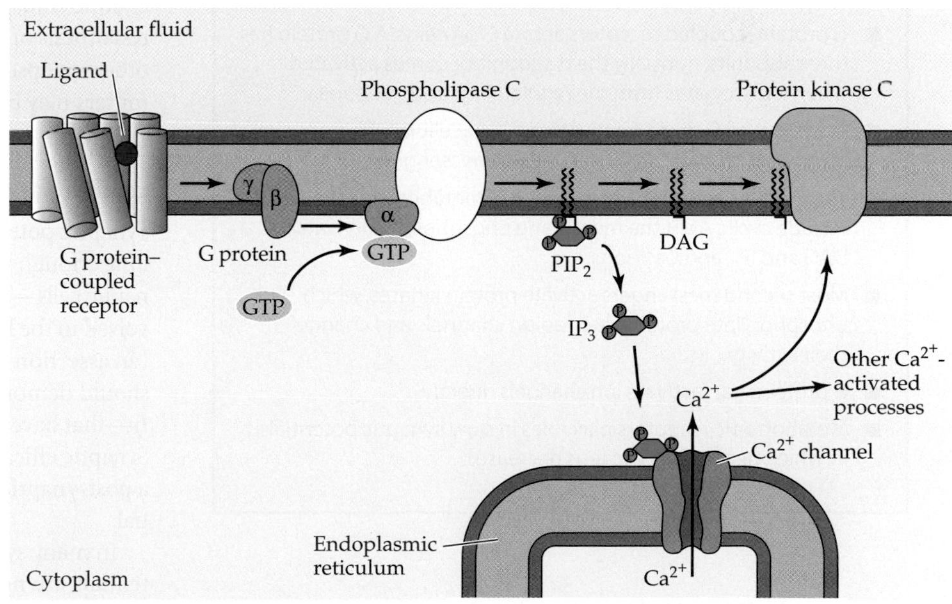

Figure 12.21 Diacylglycerol and inositol trisphosphate are other second messengers Another pathway of cellular regulation via G protein–coupled receptors is activation (via a G protein) of the enzyme phospholipase C, which cleaves the membrane phospholipid phosphatidylinositol bisphosphate (PIP_2) into two second messengers: diacylglycerol (DAG) and inositol trisphosphate (IP_3). DAG activates a protein kinase (kinase C), and IP_3 mobilizes Ca^{2+} ions from intracellular stores. Both pathways play important regulatory roles in cell metabolism.

mediated by ion-channel closing, decreasing permeability to an ion species.

It should be clear from the Goldman equation (Equation 11.6 in Chapter 11) that any changes in a postsynaptic cell's permeability to ions—decreases as well as increases—change the membrane potential. Thus a neurotransmitter receptor that leads to a *decrease* in permeability to ions also produces a synaptic potential. Such a mechanism occurs in vertebrate sympathetic ganglia.

In sympathetic ganglia, preganglionic neurons have synapses onto postganglionic neurons. A recording from a postganglionic neuron of a bullfrog shows three phases of synaptic response to stimulation of preganglionic axons: a fast EPSP lasting about 30 ms, a slow IPSP lasting about 1 s, and a still slower EPSP lasting several seconds. The fast EPSP is mediated by increased permeability to ions (like other fast EPSPs), but the slow IPSP and the slow EPSP are mediated by permeability *decreases*. As you might expect, the slow IPSP results from a decrease in the resting membrane permeability to Na^+, moving the membrane potential *away* from E_{Na}; in contrast, the slow EPSP results from a decrease in permeability to K^+, moving the membrane potential away from E_K. The permeability decrease results from a G protein–coupled receptor acting indirectly to gate the closing of channels.

Summary: Postsynaptic Receptors for Slow Metabotropic Actions

- Many neurotransmitter receptors act via second messengers, triggering metabolic cascades in postsynaptic neurons. These metabotropic receptor effects are often slow and long-lasting.
- G protein–coupled receptors are the major receptors of metabotropic synapses. All G protein–coupled receptors have seven membrane-spanning segments, and all are evolutionarily related.
- G protein–coupled receptors act via G proteins. A G protein has three subunits; normally the α subunit becomes activated when it dissociates from the regulatory β and γ subunits.
- An activated G protein activates an intracellular effector, usually to produce an intercellular second messenger.
- Second messengers of importance in metabotropic synapses include cyclic AMP, the membrane phospholipid derivatives DAG and IP_3, and Ca^{2+} ions.
- Most second messengers activate protein kinases, which phosphorylate proteins such as ion channels and change their activity.
- G proteins can activate ion channels directly.
- Metabotropic receptors play roles in slow synaptic potentials in which permeability to ions decreases.

Synaptic Plasticity: Synapses Change Properties with Time and Activity

The parameters of each step of synaptic transmission may change quantitatively over time. Presynaptically, neurotransmitter can be synthesized, stored, and released at different rates, and postsynaptic sensitivity to the neurotransmitter may be increased or decreased in different circumstances. Even electrically mediated transmission can be altered by changes in pH and in Ca^{2+} concentration that can close gap-junction channels. Changes in the parameters of synaptic transmission are important both for homeostatic regulation of transmitter metabolism and for **synaptic plasticity**—changes in synaptic efficacy over time. Because synaptic functions are more labile than other aspects of neuronal function, such as axonal conduction, it is widely supposed that synaptic plasticity is the mechanism that underlies how nervous system function changes over time. Thus the synaptic bases of nervous system development, and of learning and memory, are subjects of active current investigation. We cannot survey the entire burgeoning field, but we will present examples that demonstrate some of the major issues.

Neurotransmitter metabolism is regulated homeostatically

The metabolism of neurotransmitters must be regulated just as other aspects of metabolism are regulated. Consider the following example: A stimulated superior cervical ganglion releases 10% of its acetylcholine content per minute, without diminishing the total ACh content of the ganglion. The rate of stimulated release is 50 times greater than the resting release rate. Clearly, the rate of synthesis of ACh following stimulated release must also increase 50-fold if there is no depletion of ACh.

Although not all the mechanisms controlling the increased synthesis of ACh are known, the increased availability of free choline plays a major role, since choline is normally the rate-limiting substrate in ACh synthesis. With increased hydrolysis of ACh by acetylcholinesterase in the synaptic cleft, more choline is taken up by the choline transporter at the presynaptic terminal and is available for resynthesis of ACh by choline acetyltransferase (see Figure 12.10). At other synapses, controls of the synthesis and action of neurotransmitters may be more complex than in this simple example.

Learning and memory may be based on synaptic plasticity

Synaptic potentials have time courses of milliseconds to seconds—long enough to have a transient effect on the excitability of postsynaptic cells—but they are ephemeral nevertheless. If synapses are involved in the long-term behavioral changes of learning and memory (an assertion for which there is increasing evidence), then neurons should demonstrate changes in synaptic efficacy—*synaptic plasticity*—that have a suitably long time course of minutes, days, or weeks. Synaptic efficacy is usually measured as a change in the amplitude of a postsynaptic potential in response to a presynaptic action potential.

In many synapses, the amplitudes of individual postsynaptic potentials are not constant over time. **Synaptic facilitation** is an increase in amplitude of postsynaptic potentials in response to successive presynaptic impulses (**Figure 12.22*a***). A decrease in amplitude of postsynaptic potentials with successive presynaptic impulses is termed **synaptic antifacilitation** or *synaptic depression* (**Figure 12.22*b***). *Both synaptic facilitation and antifacilitation result from changes in the amount of neurotransmitter liberated per presy-*

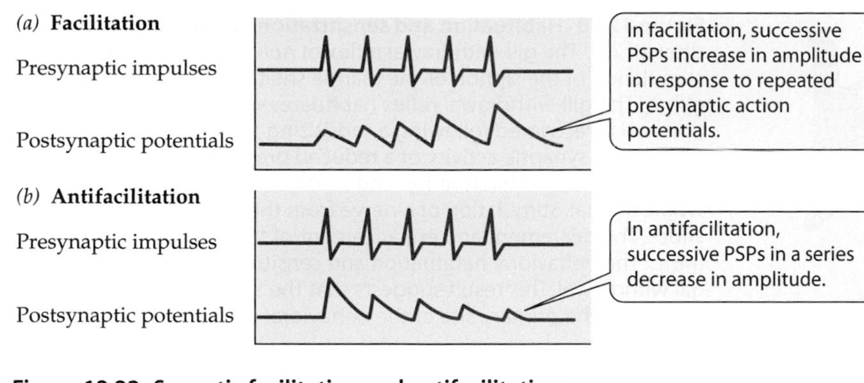

Figure 12.22 Synaptic facilitation and antifacilitation

naptic impulse. These changes are known to be calcium-dependent, but their mechanisms are otherwise incompletely understood.

Facilitation of synaptic transmission is often especially pronounced after *tetanic* stimulation of presynaptic neurons—that is, stimulation by trains of stimuli at a rate of about 10 to 100 per second for several seconds. The response to a single presynaptic impulse may be elevated severalfold after tetanic stimulation, and although the effect diminishes over time, it may persist for hours. This extended enhancement of synaptic response is termed **posttetanic potentiation.** Posttetanic potentiation indicates that synaptic efficacy can change with use, and these changes can be long-lasting. Particularly long-term potentiation changes have been reported in the hippocampus of the vertebrate brain, a region implicated in learning and memory functions. We will discuss these changes after considering the anatomically simpler example of the marine mollusc *Aplysia.*

HABITUATION AND SENSITIZATION IN *APLYSIA* Habituation and sensitization are two simple forms of behavior that occur in nearly all kinds of animals. **Habituation** is defined as the decrease in intensity of a reflex response to a stimulus when the stimulus is presented repeatedly. **Sensitization** is the prolonged enhancement of a reflex response to a stimulus, which results from the presentation of a second stimulus that is novel or noxious. Habituation and sensitization are considered simple forms of learning—that is, of modification of behavior with experience.

Reflexive gill withdrawal in *Aplysia* is a behavioral response that is subject to habituation and sensitization and is amenable to study of its synaptic basis. The gill of *Aplysia* withdraws in response to mechanical stimulation of the animal's siphon or mantle shelf (Figure 12.23*a*). The amplitude of gill withdrawal decreases with repeated low-frequency stimulation; that is, the response habituates. After a shock to the head, the response to siphon stimulation is again large; that is, it is *sensitized* by the head shock (Figure 12.23*b*).

Eric Kandel, a Nobel Prize–winning American neurobiologist, together with his colleagues, mapped the neural circuit of the gill-withdrawal reflex and determined the synaptic locus of the habituation and sensitization. Habituation of the gill-withdrawal response results from a waning of synaptic excitation of gill motor neurons by sensory neurons, and the time course of the decrease in sensory-to-motor EPSPs closely parallels the time course of behavioral habituation (Figure 12.23*c*). This decrease of EPSPs results not from any postsynaptic change, but rather from a decrease in the number of quanta of neurotransmitter released by the sensory nerve endings. Thus the synaptic basis of habituation in *Aplysia* is *antifacilitation* of the sensory synaptic terminals.

Sensitization of the gill-withdrawal response by head shock also occurs at the sensory-to-motor synapses (see Figure 12.23*c*). In contrast to habituation, a sensitizing stimulus increases the amount of neurotransmitter released per impulse at the sensory-neuron terminal. This facilitation apparently results from activation of synaptic endings of sensitizing interneurons that end on the sensory terminals in axo-axonal synapses. Thus, presynaptic *facilitation* is the synaptic basis of behavioral sensitization.

How is the amount of neurotransmitter release diminished during habituation and increased by sensitization? Evidence indicates that the calcium current entering the presynaptic terminal during an impulse is depressed during habituation. This finding suggests that there is a progressive, long-lasting inactivation of Ca^{2+} channels with habituation, allowing less Ca^{2+} to enter and to trigger neurotransmitter release. The presynaptic facilitation underlying sensitization is caused by an increased Ca^{2+} influx.

Some facilitating interneurons release 5-HT (serotonin), which acts to increase the amount of cAMP in the sensory terminals (Figure 12.24). The cAMP acts via a cAMP-dependent protein kinase to phosphorylate K^+ channels in the terminal, and thereby to decrease the K^+ current that normally terminates the action potential. This K^+ inactivation prolongs the action potential, leading to an increase in the Ca^{2+} influx and in resultant neurotransmitter release. Cyclic AMP–dependent protein kinase can also phosphorylate other proteins (e.g., to increase the mobilization of stored neurotransmitter to the release sites), and serotonin can also act at the synapse through other second messengers, such as diacylglycerol (DAG).

The studies on *Aplysia* outlined here have determined the anatomical location of two forms of behavioral plasticity in specific, identifiable synapses and have made considerable progress in defining the synaptic mechanisms producing these changes. Short-term habituation in *Aplysia* lasts about an hour, but with repeated training sessions habituation may persist for over 3 weeks. This long-term habituation depends on changes in the same sensory-to-motor synapses, but it requires new protein synthesis. Long-term habituation is driven by persistent serotonin-dependent elevation of cAMP, leading to the movement of cAMP-dependent protein kinase to the nucleus, where the kinase activates gene transcription via cAMP-dependent transcriptional regulatory proteins such as CREB (cAMP response element–binding protein).

CLASSICAL CONDITIONING IN *APLYSIA* Habituation and sensitization are examples of nonassociative forms of simple learning. *Aplysia* are also capable of **classical conditioning**, which is an example of associative learning. It involves sensitization mechanisms that make synaptic potentiation specific for the association of two stimuli. In classical conditioning an *unconditioned stimulus* (the sight of food in the case of Pavlov's dogs) reflexively elicits an

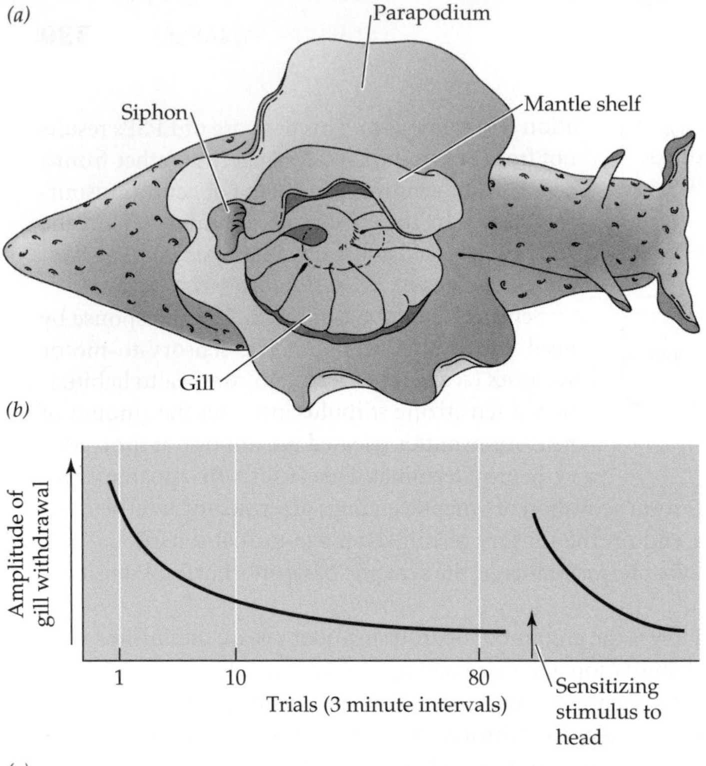

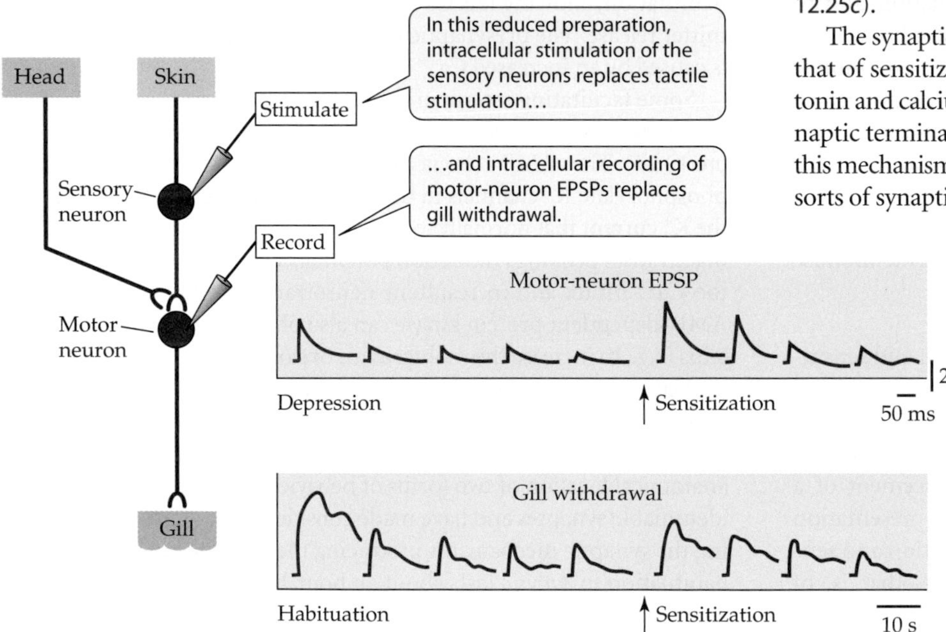

Figure 12.23 Habituation and sensitization in *Aplysia* gill withdrawal *(a)* The gill-withdrawal reflex of *Aplysia* is a response to stimulation of the siphon or the mantle shelf. The animal retracts the gill. *(b)* The gill-withdrawal reflex habituates with repeated stimulation and is recovered following a sensitizing stimulus to the head. *(c)* Changes in synaptic activity of a reduced preparation (diagrammed) are comparable to habituation and sensitization responses of the whole animal. Stimulation of a nerve from the head replaces head shock. The decrement and enhancement of the motor-neuron EPSPs mirror the behavioral habituation and sensitization, respectively, of gill withdrawal. This result suggests that the sensory-to-motor neuron synapse is the primary site of the behavioral plasticity.

unconditioned response (salivation). A *conditioning stimulus* (ringing a bell), which initially does not trigger a response, will come to evoke the response (*conditioned response*) only if it is repeatedly paired with the unconditioned stimulus. In Pavlov's example, then, the conditioned dogs salivate when the bell rings, having formed an association between the conditioning stimulus (the bell) and the unconditioned stimulus (the sight of food).

Kandel and his coworkers developed a classically conditioned form of gill withdrawal in *Aplysia*. In Kandel's experiment, an electric shock—the unconditioned stimulus (US)—applied to the mollusc's tail triggered gill withdrawal. Weak tactile stimulation of the mantle—the conditioning stimulus (CS)—was insufficient in itself to trigger gill withdrawal. Repeated unpaired presentation of both stimuli led to little response to the CS alone (Figure 12.25*a*). After repeated pairing of the two stimuli (Figure 12.25*b*), however, eventually the CS alone was sufficient to evoke gill withdrawal (Figure 12.25*c*).

The synaptic basis of classical conditioning is more complex than that of sensitization, but it depends on a similar mechanism: Serotonin and calcium ions potentiate cyclic-AMP production in a presynaptic terminal to enhance neurotransmitter release. The details of this mechanism are complicated, but the major point is that the same sorts of synaptic mechanisms that underlie sensitization and habituation can also be employed to explain classical conditioning.

LONG-TERM POTENTIATION IN THE HIPPOCAMPUS Synaptic changes in mammalian brains that are long-lasting have been explored extensively during the last 30 years, most notably in the hippocampus. The hippocampus is a deep forebrain structure that is strongly implicated in spatial learning (such as maze learning) and in memory formation. The circuitry of the hippocampus is well characterized, and because it is largely restricted to two dimensions, it can be studied in hippocampal slices that allow stable, long-term recordings from visible neurons (Figure 12.26*a*). Circuits in the hippocampus (and elsewhere in the brain) undergo prolonged changes when strongly stimulated, and some of these changes are associative; that is, they are specific to the pathways that are stimulated together.

We will examine associative long-term potentiation in a particular region (CA1) of the hippocampus. **Long-term potentiation (LTP)** is a long-lasting enhancement of synaptic transmission following intense stimulation (Figure 12.26*b*); it resembles posttetanic potentiation but is more prolonged. Associative LTP is specific to

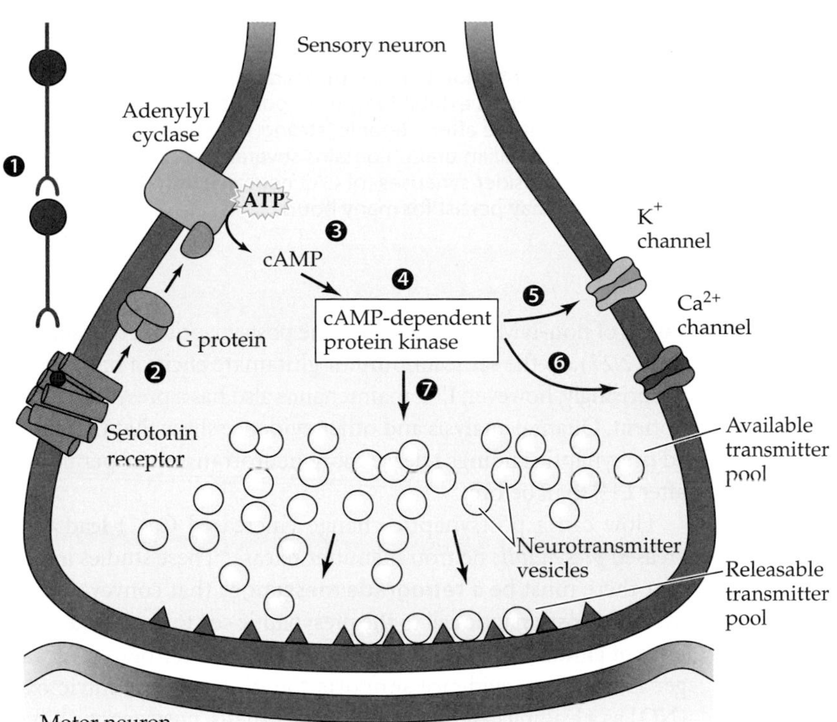

Figure 12.24 A model of Aplysia sensitization Transmitter release in the sensory terminals decreases with habituation and increases with sensitization. With habituation, fewer Ca^{2+} channels open in response to action potentials in the sensory terminal. The resulting decrease in Ca^{2+} influx decreases transmitter release. ❶ Sensitization is produced by cells thought to be serotonergic. ❷ Serotonin acts to activate an adenylyl cyclase in the terminals, which ❸ stimulates the synthesis of cAMP. ❹ cAMP in turn acts on a cAMP-dependent protein kinase to phosphorylate proteins, including ❺ a K^+ channel; this leads to a decrease in the repolarizing K^+ current and a broadening of the action potential. The increase in the duration of the action potential increases the time during which ❻ Ca^{2+} channels can open, leading to a greater influx of Ca^{2+} and increased release. The kinase also acts on Ca^{2+} channels an on ❼ mobilization of neurotransmitter vesicles. Other serotonin effects are not shown. (After Kandel et al. 1995.)

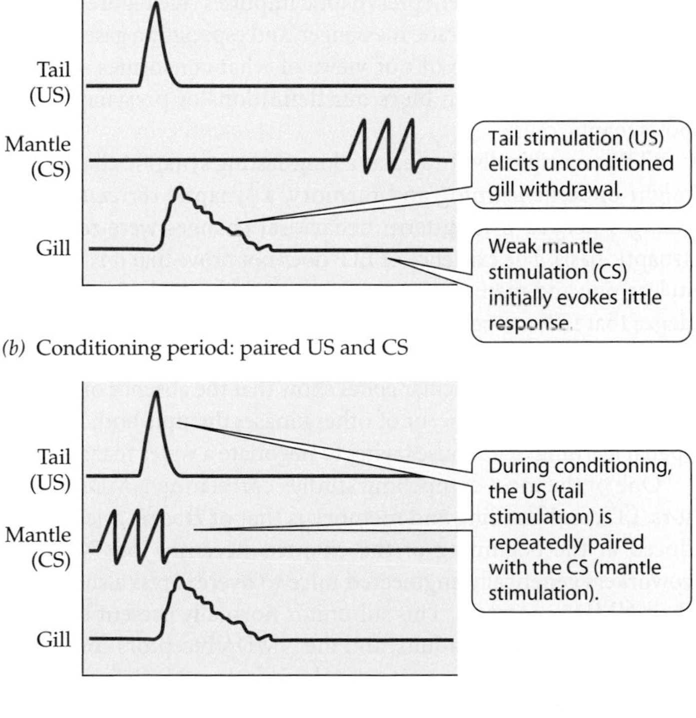

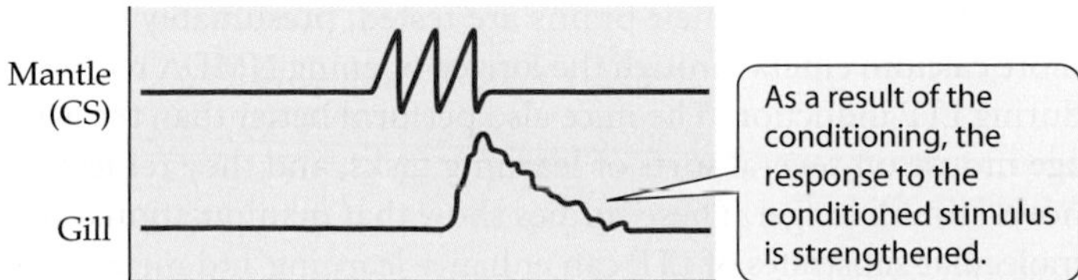

Figure 12.25 Classical conditioning in *Aplysia* *(a)* Circuit of conditioned gill withdrawal. Tail stimulation (US) elicits unconditioned gill withdrawal. Weak mantle stimulation (CS) initially evokes little response, but with repeated pairing with the US (tail stimulation) *(b)*, the response is strengthened *(c)*.

the activated pathway and requires the cooperative interaction of many presynaptic neurons to depolarize the postsynaptic CA1 cell adequately.

A critical feature of LTP is that any synapse that is active while the postsynaptic cell is strongly depolarized will be potentiated, for a period that can last hours in hippocampal slices and weeks in intact animals. As first pointed out by Donald Hebb, learning could depend on "successful" synapses getting stronger, with a "successful" synapse meaning one in which the presynaptic and postsynaptic neurons were active or depolarized at the same time: "Neurons that fire together wire together." A synapse that undergoes a long-term change in strength as a result of coincident activation of the pre- and postsynaptic neurons is called a *Hebbian synapse*. The synapses that undergo LTP in the CA1 region of the hippocampus are Hebbian because potentiation depends on the associated activity of presynaptic and postsynaptic cells.

Establishment of hippocampal LTP depends on two related glutamate receptors: NMDA receptors (so named because they also are activated by the drug *N*-methyl-D-aspartate) and non-NMDA receptors. Non-NMDA receptors produce fast EPSPs. NMDA receptors also produce EPSPs, but they have an important functional property: They work only when the postsynaptic cell is depolarized. As Figure 12.27*a* shows, an NMDA receptor is activated by glutamate, but at resting membrane potential the ion channel is blocked by a bound Mg^{2+} ion, so few or no ions flow through it. The small

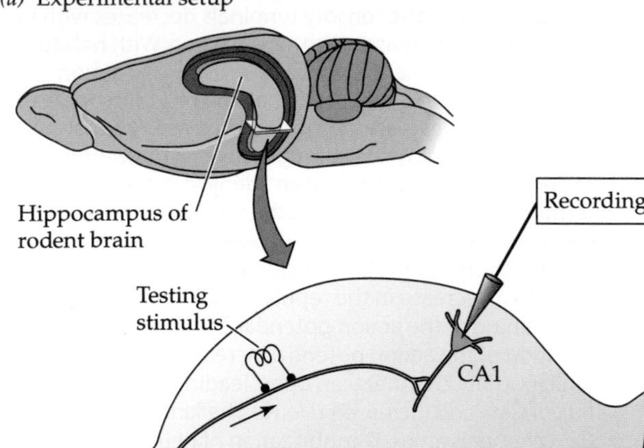

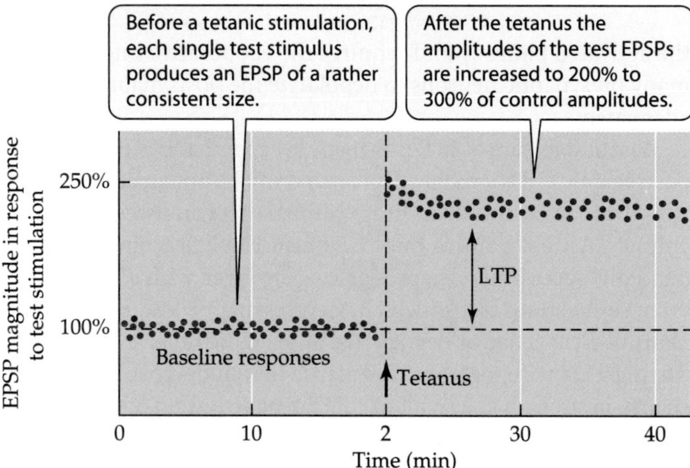

Figure 12.26 Long-term potentiation in the hippocampus Synapses in the hippocampus and in other brain areas exhibit long-term potentiation (LTP), a long-lasting increase in EPSP amplitude after a tetanic (strong, repeated train) stimulus. *(a)* The hippocampus of mammalian brains contains several synaptic pathways and types of neurons. Here we consider synapses (of CA3 neurons) onto CA1 neurons. *(b)* Long-term potentiation may persist for many hours.

EPSP produced under these conditions depends on the non-NMDA receptors. In contrast, when the postsynaptic cell is substantially depolarized, the bound Mg^{2+} ion is released; with glutamate binding to NMDA receptors, a considerable number of Na^+ and Ca^{2+} ions can enter the cell through the unblocked channels.

The Ca^{2+} ions entering the postsynaptic cell act as second messengers, activating Ca^{2+}-dependent signaling molecules such as Ca^{2+}/calmodulin-dependent kinase and kinase C (Figure 12.27*b*). Thus two features of NMDA receptors are important for the establishment of LTP: (1) Their full activation requires a Hebbian simultaneous depolarization of the presynaptic and postsynaptic cells, and (2) they have indirect, long-lasting second-messenger properties.

Because the establishment of LTP depends on postsynaptic activation of NMDA receptors, we would expect the maintenance of LTP also to be a postsynaptic phenomenon. This is partly true; the activation of kinases in the postsynaptic cell leads to enhanced activation of non-NMDA receptors in the postsynaptic membrane (Figure 12.27), so the same amount of glutamate elicits a bigger EPSP. Surprisingly, however, LTP maintenance also has a presynaptic component. Quantal analysis and other evidence show that the activated presynaptic endings release more neurotransmitter per impulse after LTP than before.

How can a postsynaptic change (increased Ca^{2+}) lead to increased *presynaptic* neurotransmitter release? These studies indicate that there must be a **retrograde messenger** that conveys a signal from the postsynaptic cell to the presynaptic cell to potentiate the release of glutamate. Although the identity of this retrograde messenger is not clear, evidence supports the diffusible gas nitric oxide (NO) as a leading candidate. In this hypothesis, postsynaptic kinases would activate the enzyme NO synthase, which catalyzes NO synthesis from arginine. NO is freely diffusible through membranes to the immediately adjacent cells and may activate guanylyl cyclase in the presynaptic terminal to increase the amount of neurotransmitter released by subsequent presynaptic impulses (see Figure 12.27*b*). The concept of a retrograde messenger, and especially a gaseous one, may require expansion of our views of what constitutes a neurotransmitter, and it even blurs our definitions of presynaptic and postsynaptic cells!

LTP was originally studied as a long-lasting synaptic change that might underlie learning and memory, a synaptic *correlate* of the changes needed if long-term behavioral changes were to have a synaptic basis. The existence of LTP does not prove that it is involved in learning and memory—only that it could be. Is there direct evidence that LTP is a necessary component of learning? The answer is yes, at least for spatial learning in mice. Studies that knock out (block the expression of) particular genes show that the absence of NMDA receptors, of CaM kinase, or of other kinases disrupts both LTP and spatial learning of a mouse trying to negotiate a water maze.

One of the most compelling studies correlating NMDA receptors, LTP, and learning and memory is that of *Doogie* mice, introduced at the beginning of the chapter. Recently Joe Tsien and coworkers genetically engineered mice to overexpress a subunit of their NMDA receptors. This subunit is normally present in developing mice but not in adults, and the NMDA receptors containing the juvenile subunit remain open longer than NMDA receptors with only adult subunit types. Mice of the *Doogie* strain (with NMDA receptors with the juvenile subunit) have more potent hippocampal LTP when their brains are tested, presumably because more calcium enters through the longer-opening NMDA receptors during LTP induction. The mice also perform better than the average mouse on several sorts of learning tasks, and they remember novel stimuli longer. These studies show that manipulation of the molecular substrates of LTP can enhance learning and memory, as

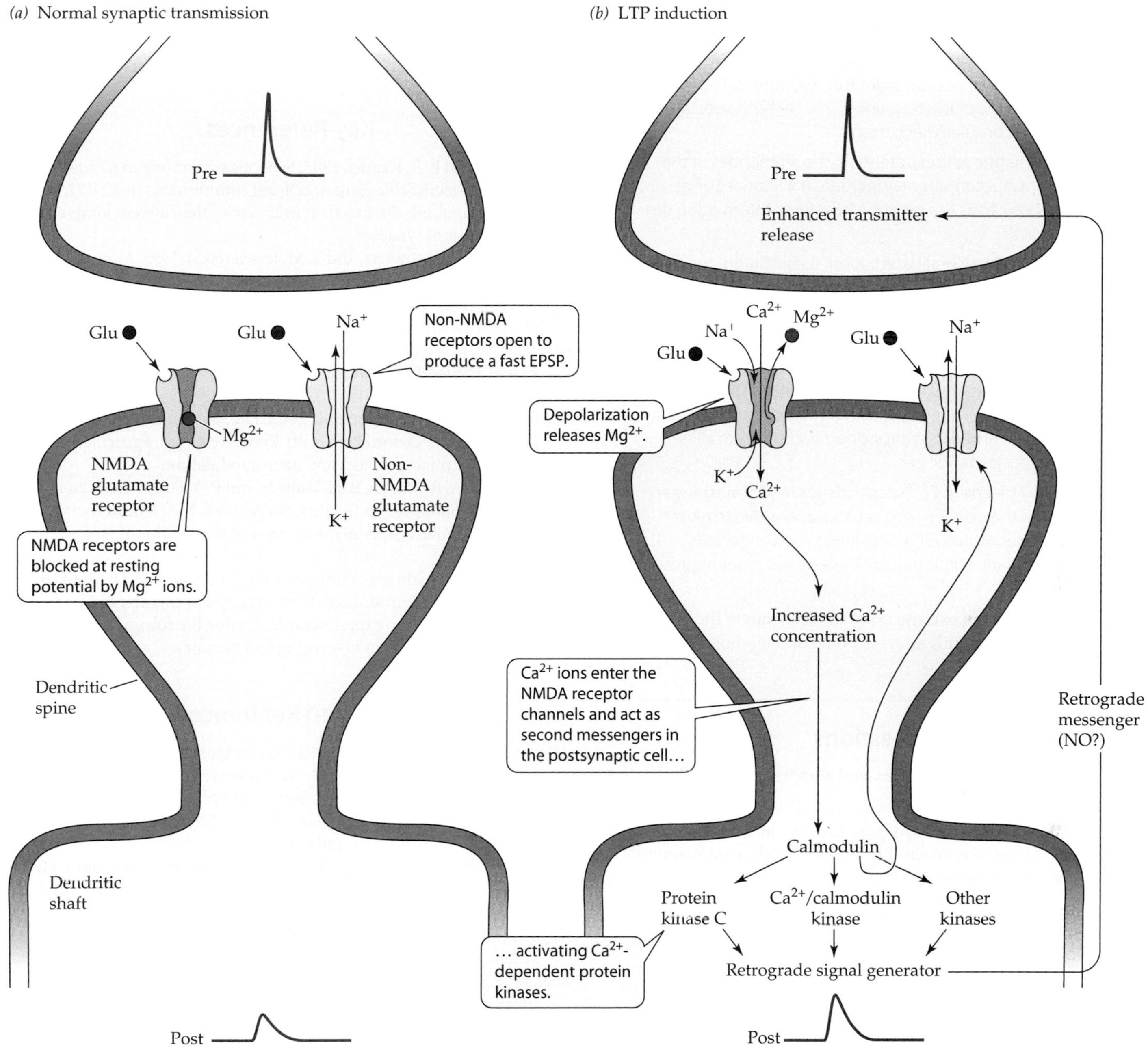

Figure 12.27 Induction and maintenance of LTP in the hippocampus Induction of long-term potentiation (LTP) is postsynaptic. *(a)* At the unpotentiated synapse, glutamate neurotransmitter is released and acts at different kinds of glutamate receptors, shown here as NMDA receptors and non-NMDA receptors. *(b)* Tetanic stimulation depolarizes the postsynaptic cell substantially, releasing the Mg^{2+} block and allowing glutamate to activate the NMDA receptors. The resultant phosphorylations enhance the postsynaptic response (e.g., via phosphorylating non-NMDA glutamate receptors) and also lead to increased glutamate releases from presynaptic neurons, via a retrograde messenger. (After Kandel et al. 1995.)

well as disrupt it. Thus the synaptic changes underlying LTP *are* causally related to at least some learning, although more than one metabolic pathway seems implicated in the learning-related synaptic changes.

The example of *Doogie* mice puts, if not a human face, at least a mouse face on the range of mechanisms and functions of synaptic physiology. Synapses transmit signals, electrically or (more commonly) by chemical transmission. Synapses also modulate properties—of synapses, of neurons, perhaps of more. Synaptic modulation clearly appears to underlie brain changes in learning and memory—illustrating that there is more to synaptic physiology than millisecond changes in membrane potential.

Summary: Synaptic Plasticity

- Neuronal stimulation that increases the rate of neurotransmitter release also increases rates of neurotransmitter resynthesis. The homeostatic mechanisms of this regulation involve both substrate availability and more-complex mechanisms.
- With a train of presynaptic action potentials, the amplitudes of the resultant postsynaptic potentials may increase (facilitation) or decrease (antifacilitation). Thus the synaptic transfer of information depends on its history.
- The synaptic bases of behavioral habituation, sensitization, and classical conditioning in *Aplysia* depend on second messenger–mediated control of the amount of neurotransmitter released at CNS synapses.
- Hippocampal long-term potentiation (LTP) is a long-lasting change in synaptic properties related to learning and memory. The induction of LTP depends on NMDA receptors that respond to both glutamate neurotransmitter and postsynaptic depolarization, to allow Ca^{2+} entry into the postsynaptic cell.
- LTP is maintained by means of Ca^{2+}-dependent second-messenger pathways that make the postsynaptic cell more sensitive to glutamate neurotransmitter. Moreover, a retrograde messenger acts presynaptically to enhance glutamate release by subsequent presynaptic stimuli.
- Studies that manipulate the expression of critical genes in the LTP metabolic pathway significantly affect learning and memory in mice.

Study Questions

1. What are the functional advantages and disadvantages of electrical synapses?
2. What adaptive advantage do synapses provide in nervous systems (i.e., why isn't the nervous system just a mesh of cells cytoplasmically connected through gap junctions)?
3. Why is it difficult to show vesicular release of neurotransmitter molecules by electron microscopy?
4. In his book *Ionic Channels of Excitable Membranes*, Bertil Hille characterized the importance of calcium ions: "Calcium channels … serve as the only link to transduce depolarization into all the nonelectrical activities controlled by excitation. Without Ca^{2+} channels our nervous system would have no outputs." Discuss this statement with reference to synaptic function.
5. What are the criteria for identification of the neurotransmitter at a particular synapse? Which two criteria are *sufficient* to conclude that a candidate is the neurotransmitter?
6. Why (in evolutionary terms) do you think there are so many kinds of neurotransmitters?
7. Why are there multiple receptor subtypes for each neurotransmitter? Why not a few receptors and lots of neurotransmitters acting on each receptor?
8. Discuss the functional differences between ionotropic and metabotropic receptors.
9. Would you classify NMDA receptors as ionotropic or metabotropic? Why?
10. Do you think there is one cellular mechanism of learning and memory, or several? Cite studies in *Aplysia* and in the mammalian hippocampus to support your answer.

Key References

Jessell, T. M., and E. R. Kandel. 1993. Synaptic transmission: A bidirectional and self-modifiable form of cell-cell communication. *Cell 72/Neuron* 10 (Suppl.): 1–30. Excellent overview of the different kinds and mechanisms of synapses.

Kandel, E. R., J. H. Schwartz, and T. M. Jessell (eds.). 1995. *Essentials of Neural Science and Behavior.* Appleton & Lange, Norwalk, CT. Excellent source on synaptic physiology, similar to *Principles* (below) but now somewhat out of date.

Kandel, E. R., J. H. Schwartz, and T. M. Jessell (eds.). 2000. *Principles of Neural Science*, 4th ed. McGraw-Hill, New York. The best reference on all things synaptic.

Levitan, I. B., and L. K. Kaczmarek. 2002. *The Neuron: Cell and Molecular Biology*, 3rd ed. Oxford University Press, New York. Particularly good on metabotropic receptors and neuromodulation.

Nicholls, J. G., A. R. Martin, B. G. Wallace, and P. A. Fuchs. 2001. *From Neuron to Brain*, 4th ed. Sinauer, Sunderland, MA. One of the clearest and most experiment-driven text discussions of synaptic physiology.

Tsien, J. Z. 2000. Building a brainier mouse. *Sci. Amer.* 282(4): 62–68. A straightforward introduction to the genetically engineered *Doogie* mouse strain, and to experiments indicating the roles of synaptic long-term potentiation in learning and memory.

Selected References

Abrams, T. W., and E. R. Kandel. 1988. Is contiguity detection in classical conditioning a system or a cellular property? Learning in *Aplysia* suggests a possible molecular site. *Trends Neurosci.* 11: 128–135.

Aidley, D., and P. Stanfield. 1996. *Ion Channels: Molecules in Action.* Cambridge University Press, Cambridge.

Akil, H., S. J. Watson, E. Young, M. E. Lewis, H. Khachaturian, and J. M. Walker, 1984. Endogenous opioids: Biology and function. *Annu. Rev. Neurosci.* 7: 223–255.

Augustine, G. J., M. P. Charlton, and S. J. Smith. 1987. Calcium action in synaptic transmitter release. *Annu. Rev. Neurosci.* 10: 633–693.

Bear, M. F., B. W. Connors, and M. A. Paradiso. 2001. *Neuroscience: Exploring the Brain*, 2nd ed. Lippincott Williams & Wilkins, Baltimore, MD.

Colquin, D., and B. Sakmann. 1998. From muscle endplate to brain synapses: A short history of synapses and agonist-activated channels. *Neuron* 20: 381–387.

Cooper, J. R., F. Bloom, and R. E. Roth. 2002. *The Biochemical Basis of Neuropharmacology*, 8th ed. Oxford University Press, New York.

Cotman, C. W., D. T. Monaghan, and A. H. Ganong. 1988. Excitatory amino acid neurotransmission: NMDA receptors and Hebb-type synaptic plasticity. *Annu. Rev. Neurosci.* 11: 61–80.

Eccles, J. C. 1982. The synapse: From electrical to chemical transmission. *Annu. Rev. Neurosci.* 5: 325–339.

Fatt, P., and B. Katz. 1951. An analysis of the end-plate potential recorded with an intracellular microelectrode. *J. Physiol.* 115: 320–370.

Feldman, R. S., J. S. Meyer, and L. F. Quenzer. 1997. *Principles of Neuropsychopharmacology*. Sinauer, Sunderland, MA.

Geppert, M., and T. C. Südhof. 1998. RAB3 and synaptotagmin: The yin and yang of synaptic membrane fusion. *Annu. Rev. Neurosci.* 21: 75–95.

Hammond, C. 1996. *Cellular and Molecular Neurobiology*. Academic Press, San Diego, CA.

Hawkins, R. D., E. R. Kandel, and S. A. Siegelbaum. 1993. Learning to modulate transmitter release: Themes and variations in synaptic plasticity. *Annu. Rev. Neurosci.* 16: 625–665.

Heuser, J. E., and T. S. Reese. 1979. Synaptic-vesicle exocytosis captured by quick-freezing. In F. O. Schmitt and F. G. Worden (eds.), *The Neurosciences: Fourth Study Program*, pp. 573–600. MIT Press, Cambridge, MA.

Hille, B. 2001. *Ion Channels of Excitable Membranes*, 3rd ed. Sinauer, Sunderland, MA.

Holtz, R. W., and S. K. Fisher 1999. Synaptic transmission and cellular signaling: An overview. In G. Siegel, B. Agranoff, R. W. Albers, S. K. Fisher, and M. D. Uhler (eds.), *Basic Neurochemistry*, 6th ed. Lippincott Williams & Wilkins, Philadelphia.

Katz, B. 1966. *Nerve, Muscle, and Synapse*. McGraw-Hill, New York.

Kuno, M. 1995. *The Synapse: Function, Plasticity, and Neurotrophism*. Oxford University Press, Oxford, UK.

Llinas, R. 1982. Calcium in synaptic transmission. *Sci. Amer.* 247(4): 56–65.

Lundberg, J. M., and T. Hokfelt. 1983. Coexistence of peptides and classical transmitters. *Trends Neurosci.* 6: 325–333.

Marder, E. 1998. From biophysics to models of network function. *Annu. Rev. Neurosci.* 21: 25–45.

Matthews, G. 1996. Neurotransmitter release. *Annu. Rev. Neurosci.* 19: 219–233.

McGeer, P. L., J. C. Eccles, and E. G. McGeer. 1987. *Molecular Neurobiology of the Mammalian Brain*, 2nd ed. Plenum, New York.

Powis, D. A., and S. J. Bunn (eds.). 1995. *Neurotransmitter Release and Its Modulation*. Cambridge University Press, Cambridge.

Shepherd, G. M. 1994. *Neurobiology*, 3rd ed. Oxford University Press, New York.

Shepherd, G. M., and S. D. Erulkar. 1997. Centenary of the synapse: From Sherrington to the molecular biology of the synapse and beyond. *Trends Neurosci.* 20: 385–392.

Südhof, T. C. 1995. The synaptic vesicle cycle: A cascade of protein-protein interactions. *Nature* 375: 645–653.

Zigmond, M. J., F. E. Bloom, S. C. Landis, J. L. Roberts, and L. R. Squire (eds.). 1999. *Fundamental Neuroscience*. Academic Press, San Diego, CA.

Zimmermann, H. 1993. *Synaptic Transmission: Cellular and Molecular Basis*. Georg Thieme, Stuttgart, Germany.

13 Sensory Processes

The antennae of moths—especially male moths—contain large numbers of chemosensory receptor neurons. For example, an antenna of a male sphinx moth (*Manduca sexta*) has about 260,000 chemoreceptor neurons, a substantial fraction of the neurons in the animal's nervous system. Many of these receptor neurons respond specifically to a pheromone, a signaling chemical produced by the female of the species (pheromones are discussed in Chapter 14). Male moths can detect the pheromone at a distance as great as a mile or two, and they respond by flying upwind in search of the female.

Sensory receptor cells such as the moth's chemoreceptor neurons provide **information** to the animal about its environment. Different sensory cells respond to different aspects of the environment (chemicals, light, sound, and so on) and they vary greatly in sensitivity and specificity. The pheromone chemoreceptor cells of the sphinx moth are near the extreme of sensitivity and specificity. All cells are somewhat responsive to aspects of their environment and thus subserve some functions that can be considered sensory. For example, protists respond to light, and bacteria respond to chemical gradients. Cellular responses of this kind presumably preceded the evolution of specialized sensory neurons. With the evolution of multicellularity and the attendant specialization of different cell types, it is likely that specialized sensory cells evolved in conjunction with neural coordinating systems. These sensory cells, called **receptor cells**, became specialized at providing information about the external environment, and also information about the internal environment. Sensory information is important, as indicated by the number of neurons devoted to it, and by the energetic cost of acquiring it. (The visual system of a blowfly, for example, consumes an estimated 13% of the energy of the animal's resting metabolism!)

A sensory receptor cell is a cell that is specialized to transform stimulus energy into a neural signal. Some receptor cells are neurons; others are nonneural cells that synaptically excite sensory neurons. The stimulus energy that excites different receptor cells may be chemical, mechanical, or electromagnetic. Because most neurons respond to synaptic transmitter molecules, we need to refine our definition of a sensory receptor as an excitable cell normally activated by stimuli other than synaptic activity. If a **stimulus** is a form of external energy (external to the cell) to which a sensory receptor can respond, then we have circular definitions in which sensory receptors are defined in terms of stimuli and vice versa. The interconnecting definitions of sensory receptors and stimuli point out that a receptor cell is uniquely adapted to detect a particular kind of stimulus.

A male *Antheraea polyphemus* moth A single pinnate antenna of a male moth may have as many as 260,000 chemosensory neurons, many of which are specialized to detect a chemical signal (pheromone) released by a female moth.

Organization of Sensory Systems

In the previous section we said that a sensory receptor (or simply **receptor**) is a *cell* specialized to convert stimulus energy into a neural signal. This use of the term *receptor* differs from that in Chapters 2, 10, and 12, where the term *receptor* referred to a neurotransmitter receptor or a hormone receptor: a *molecule* specialized to interact with molecules of synaptic transmitter or hormone, respectively. In this chapter the term *receptor* refers to a sensory cell, unless noted otherwise.

Some kinds of sensory receptor cells are neurons, with afferent axons that extend to the CNS. Other kinds of sensory receptors are epithelial cells, which lack axons but instead synaptically excite the axons of sensory neurons. Sensory receptors are commonly clustered together in **sense organs**, anatomical structures that are specialized for the reception of a particular kind of stimulus energy. Usually a sense organ contains many similar receptors, as well as several kinds of nonneural tissues. For example, the vertebrate eye is a sense organ that contains photoreceptor cells, as well as nonneural tissues such as those that make up the cornea and iris. We can also speak of **sensory systems** such as the vertebrate visual system, which includes the eyes and the central areas in the brain that are primarily concerned with processing visual information.

Receptors are classified by sensory modality, location, or form of stimulus energy

The most familiar way of classifying receptors is by **sensory modality**, a classification based on subjective human experience. Aristotle distinguished five primary senses: vision, hearing, touch, smell, and taste. Each of these sensory modalities may contain subdivisions termed **sensory qualities**, such as blue versus yellow light or salty versus sweet taste. Clearly a classification based on the old "five senses" is inadequate even for humans. There are additional sensory modalities of which we are aware (e.g., balance, temperature), as well as many others (e.g., muscle length and tension, blood oxygen partial pressure) that do not normally enter our consciousness. Early classifications of sensory modalities, moreover, are anthropocentric (human-centered) and thus are inadequate to account for the range of receptors in different kinds of animals. Many animals possess receptors sensitive to modalities and qualities of stimuli not sensed by humans, including electric and magnetic fields and ultraviolet radiation.

Receptors can also be classified according to *location*—that is, location of the source of the stimulus energy relative to the body. **Exteroceptors** are sensory cells that respond to stimuli outside the body; **interoceptors** respond to internal stimuli. Exteroceptors can be subdivided into *distance receptors*, for which the stimulus source is at a distance from the body (vision, hearing, and olfaction), and *contact receptors*, for which the stimulus contacts the body, as in touch and taste.

Probably the most useful classification of sensory receptors is in terms of the *form of effective stimulus energy* at the receptor surface (Table 13.1). Photoreceptors, electroreceptors, and magnetoreceptors all respond to different forms of electromagnetic energy; auditory receptors, touch receptors, and receptors for balance are all excited by mechanical stimuli. A practical limitation of this classification is that *any* form of stimulus energy may excite a particular receptor if there is enough of it. This observation leads to one of the most important generalizations in sensory physiology: *The sensory modality or quality of sensation associated with a stimulus depends solely on* which *receptors are stimulated, rather than on* how *they are stimulated.* For example, any stimulus that excites photoreceptors is perceived as light, whether the stimulus is actually light, a poke in the eye ("seeing stars"), or electrical stimulation of the optic nerve.

The principle that explains why any stimulation of a particular receptor is perceived as the same modality is called the principle of **labeled lines**. Let's consider the question of sensory quality from the viewpoint of a central nervous system. Stimuli such as light, sound, and touch do not act on the brain; instead they act on sensory receptors in the periphery. The brain receives *coded information* about stimuli, rather than the stimuli themselves. With rare exceptions, sensory information is coded as trains of action potentials. All action potentials are rather similar in form, so a brain cannot tell by the form of an action potential whether the action potential was initiated by light, sound, or touch. Different populations of receptors code different kinds of stimuli into action potentials, and the central nervous system must *decode* the action potentials into information about stimulus quality. The central nervous system performs this decoding by the *principle of labeled lines* (Figure 13.1), which states that any action potentials of a particular sensory axon are interpreted as a particular stimulus quality. Thus, any activity in central axonal projections from photoreceptors is interpreted as light, and any activity in central auditory projections is interpreted as sound. As we will see, the principle of labeled lines is an important component of sensory coding.

Because every action potential that is initiated by a sensory receptor are decoded as a particular stimulus modality, receptors normally have a high degree of *specificity*; that is, they are normally excited by

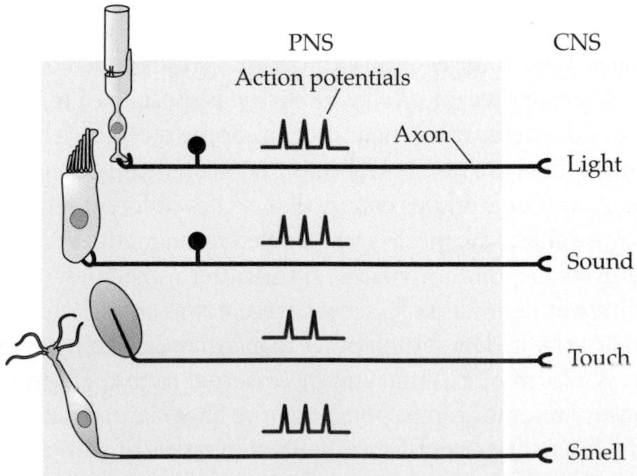

Figure 13.1 The principle of labeled lines in sensory systems Receptors sensitive to different kinds of stimuli send similar kinds of signals (action potentials) to the central nervous system (CNS). In the CNS, the signals are interpreted according to which lines (axons) convey the signals. (Note that some sensory receptor cells are neurons with axons that extend to the CNS; other sensory cells have no axons but instead synaptically excite sensory neurons.) PNS = peripheral nervous system.

TABLE 13.1 Classification of sensory receptors, based primarily on the kind of stimulus energy that excites them

Stimulus energy	Receptor modality	Stimulus perceived	Comments
Electromagnetic energy	Photoreceptors	Visible light	Some animals detect the polarity of visible and ultraviolet light
		Ultraviolet light	
		Infrared radiation?	Infrared detection may be thermal
	Electroreceptors	Electrical field or charge movement	
	Magnetoreceptors	Earth's magnetic field	Animals may sense polarity or angle of inclination of magnetic lines of force (see Chapter 16)
Thermal energy	Thermoreceptors	Hot Cold	"Infrared receptors" of rattlesnakes are actually sensitive heat receptors
Chemical energy	Chemoreceptors	Olfactory stimuli (distance chemoreceptors)	Chemical source is distant
		Taste (contact chemoreceptors)	Chemical source is nearby
		Internal chemoreceptors	O_2, CO_2, H^+, etc.
Mechanical energy	Mechanoreceptors	Touch, pressure	Skin or body surface
		Muscle length	Proprioceptors
		Muscle tension	Proprioceptors
		Joint position and movement	Proprioceptors
		Sound (auditory stimuli)	
		Balance and acceleration	Receptors detect gravitational or inertial forces
	Osmoreceptors	Osmotic pressure	Receptors detect mechanical stresses of osmotic swelling, etc.

only one form of stimulus energy. The term for this "best" form of stimulus energy is the **adequate stimulus** of a receptor. The adequate stimulus of a receptor is the *form* of stimulus energy to which the receptor is most sensitive, or the form to which it normally responds. This last phrase acknowledges that a few types of receptors (e.g., some electroreceptors) are very sensitive to more than one form of stimulus energy, although they are normally stimulated by only one form.

The specificity of receptors for one stimulus modality is achieved in two ways. First, sensory receptors usually are much more sensitive to one form of stimulus energy than to other forms. Second, the sense organs that house receptors act as filters; they diminish or block forms of stimulus energy other than the kind of energy for which the receptor is specialized. This **peripheral filtering** is an important function of the nonneural components of a sense organ. For example, photoreceptor cells in the retina of a vertebrate eye have some sensitivity to stimuli other than light, but they are usually not exposed to mechanical and thermal stimulation.

Sensory reception comprises a series of discrete operations

A sense organ mediates a series of discrete operations that convert stimulus energy into sensory signals, which are usually trains of action potentials in particular sensory axons (Figure 13.2). The first operation, peripheral filtration, is typically performed by nonneural tissues such as the iris of an eye. Peripheral filtering mechanisms not only block forms of stimulus energy other than the adequate stimulus, but also can control the amount of energy of the adequate-stimulus that reaches the receptors. For example, efferent impulses from the CNS regulate constriction of the iris of an eye during intervals of bright sunlight, thus limiting the amount of light that reaches the eye's photoreceptors.

The second, and most critical, operation in sensory reception consists of converting the energy of an adequate stimulus into an electrical signal called a *receptor potential.* The overall conversion process, called **sensory transduction**, occurs in the receptor cell and consists of four steps:

1. **Absorption.** In order for a stimulus to cause a change in a receptor, the stimulus energy must first be absorbed. A perfectly transparent photoreceptor, for example, could not function because it would absorb no light.
2. **Transduction.** The receptor converts absorbed stimulus energy into an electrical event, the *receptor potential.* The responses of all known receptors involve the production of electric currents and potentials.
3. **Amplification.** The energy that a receptor releases—the energy that generates action potentials—may be much greater than the energy provided by the stimulus. Thus, by releasing stored potential energy in a receptor, a very weak stimulus energy may trigger a strong receptor potential. In highly sensitive receptors (e.g., human photoreceptors), the amount of stimulus energy needed to elicit a sensory response is several orders of magnitude smaller than the energy of a receptor potential or of an action potential that it generates.
4. **Integration** and **spreading.** In most receptors, the site of sensory transduction is separated from the site that initiates action potentials; thus the receptor potential must spread from its point of origin to the impulse initiation site. Moreover, receptor potentials generated at spatially distinct sites may summate (integrate) just as distinct synaptic potentials can summate (see Figure 12.5).

Finally, in an operation called **coding and transmission**, most receptors convert their receptor potentials into a train of action po-

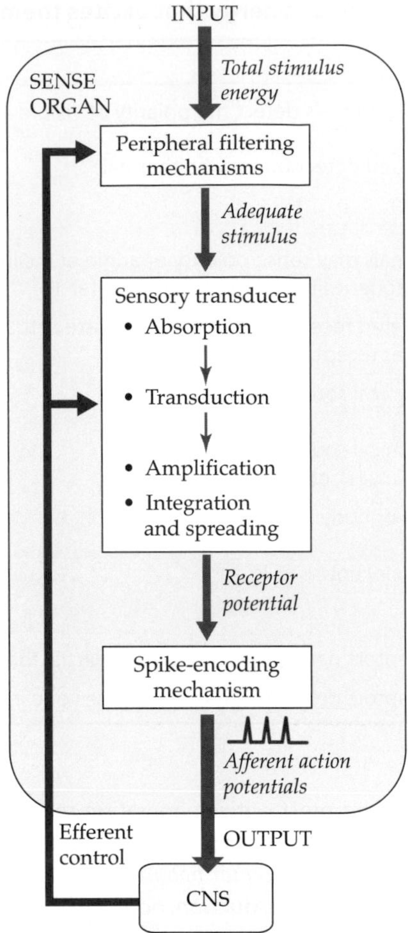

Figure 13.2 Functional processes in a generalized sensory system Sensory processes are shown here as a sequence of operations, in which the functional operators are shown in boxes and the output of each box is the input of the next box. Peripheral filtration occurs outside the sensory receptor cell; the functions in the sensory transducer box occur within the receptor cell. Action potentials may be encoded in the sensory cell, or in a synaptically excited neuron.

tentials that provide information about the strength and temporal properties of the stimulus. Afferent axons (whether or not they are part of the receptor cell) propagate the action potentials to the central nervous system.

Summary: Organization of Sensory Systems

- Sensory receptor cells respond to stimulation by a form of energy. Most sensory receptors are specialized to respond to one form of stimulus energy.
- Sense organs contain clusters of similar receptor cells and often filter stimulus energy so that only one sort of energy (the adequate stimulus) excites the receptors.
- Receptor cells transduce stimulus energy into an electrical response, usually a depolarizing receptor potential.
- The receptor potential in a sensory neuron spreads to a site of initiation of action potentials, and if the depolarization is large enough, it triggers action potentials that propagate to the central nervous system. Some sensory receptors are nonneural cells that synaptically excite sensory neurons to generate action potentials.

Receptor Functions and Their Control

The central operation of sensory processing in the preceding model is *transduction*, the process whereby a receptor converts the energy of its adequate stimulus into an electrical potential. To examine how transduction works, we will use as our model a widely studied example of a receptor found in crayfish.

Crayfish have pairs of *stretch receptors* located on the dorsal side of the abdomen—one pair on either side of the abdominal midline (Figure 13.3*a*). Each stretch receptor consists of a single sensory neuron that is associated with its own specialized muscle fiber, which spans an abdominal joint (Figure 13.3*b*). When the abdomen flexes ventrally, the receptor muscle stretches, distorting the sensory neuron's dendrites that ramify in the central, noncontractile region of the muscle fiber. Each stretch receptor neuron, with its muscle fiber and associated excitatory and inhibitory neurons (discussed later; see page 354), constitutes a complete miniature sense organ called a **muscle receptor organ** (MRO). A crayfish MRO is an important experimental model because an investigator can readily impale the large (up to 100 μm in diameter) stretch receptor sensory neuron with a recording electrode, and can record from the corresponding identified neuron in many individual animals.

When the receptor muscle is stretched, the membrane of its associated sensory neuron depolarizes from its resting membrane potential (E_m). This depolarization is graded in amplitude, increasing with increasing amounts of stretch, and it is sustained for as long as muscle stretch is maintained. The graded depolarization of a receptor in response to stimulation is a **receptor potential**. Like synaptic potentials (see Chapter 12), receptor potentials spread passively to the site where action potentials are generated. If the receptor muscle is stretched enough for the neuron's receptor potential to exceed threshold, the neuron generates action potentials (Figure 13.3*c*).[1] The greater the amount of stretch, the larger the amplitude of the receptor potential and the greater the frequency of the resulting action potentials. Accordingly, the crayfish stretch receptor demonstrates two important generalizations about receptor function:

1. An adequate stimulus elicits a graded receptor potential, the amplitude of which is a function of stimulus intensity.
2. The frequency of resultant action potentials in a receptor is a coded representation of the intensity of the adequate stimulus.

Although there are exceptions to these principles, they apply to the majority of receptors. The relationship of stimulus intensity to amplitude of the receptor potential and to impulse frequency differs

[1] Because receptor potentials in sensory neurons lead to the generation of action potentials, they are sometimes called *generator potentials*.

(a) Location of stretch receptors

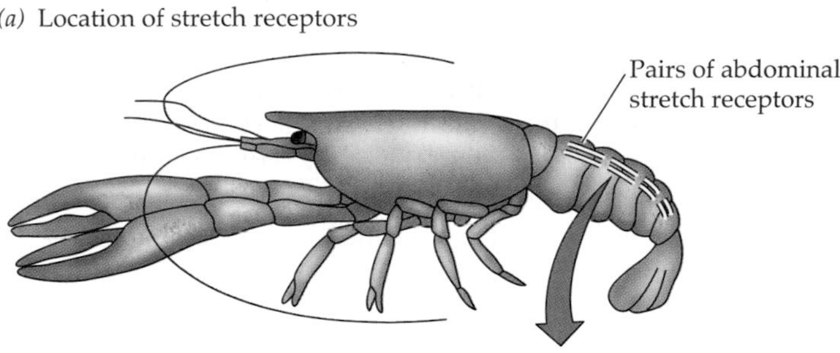

(b) A muscle receptor organ (MRO)

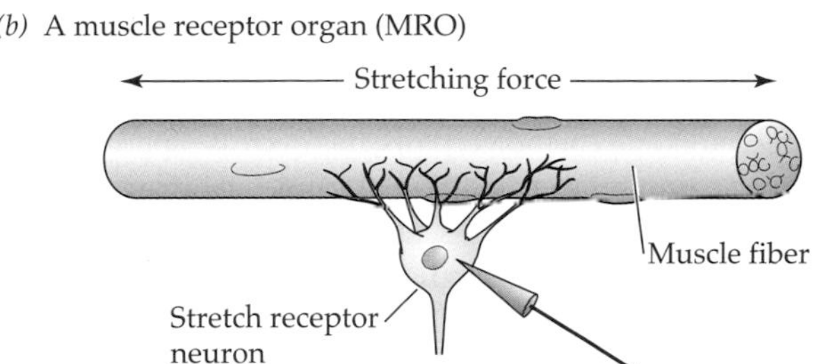

Figure 13.3 Crayfish stretch receptors exemplify sensory processes *(a)* Two receptors on each side of each abdominal segment span the joint. *(b)* Each stretch-receptor organ consists of a single sensory neuron and an associated specialized receptor muscle fiber, in which the sensory dendrites are embedded. *(c)* In excitation of a crayfish stretch receptor, stretch of the receptor muscle elicits a graded, depolarized *receptor potential*, the amplitude of which is proportional to the stimulus intensity. The receptor potential induces a train of action potentials, the frequency of which is a code of stimulus intensity.

(c) Excitation of a stretch receptor

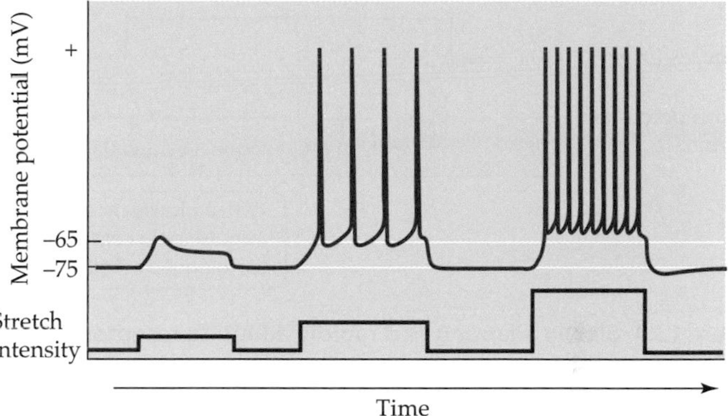

for different receptors. However, the frequency of impulses in a receptor is a code of stimulus intensity in most receptors.

Receptor potentials result from the influx of Na^+ ions

The receptor potential of a crayfish stretch receptor (and of most other receptors) results from an inward ionic current carried primarily by sodium (Na^+) ions. Presumably muscle stretch distorts the membranes of embedded dendritic endings, opening membrane ion channels that are linked to cytoskeletal and extracellular proteins. The resulting receptor potential passively spreads from the dendrites to the soma and to the initial segment of the axon where the receptor potential depolarization opens voltage-dependent sodium channels, generating action potentials.

The channels responsible for the receptor potential are quite different from the voltage-dependent sodium channels that generate action potentials. For mechanoreceptors, the receptor potential depends on stretch-sensitive ion channels. The channels that produce receptor potentials are not voltage-dependent and are unaffected by tetrodotoxin (TTX), which blocks most voltage-dependent sodium channels (see Chapter 11, page 298). The channels that generate receptor potentials are much less selective in their permeability to ions than are voltage-gated channels and will allow large ions such as tris and arginine to pass through them. The current producing a receptor potential has a *reversal potential*, the potential at which there is no net ionic flow through open channels (see Chapter 12, page 320). In the stretch receptor the value of the reversal potential is about +15 mV (inside-positive). This value is below (closer to zero than) the estimated sodium equilibrium potential, a difference reflecting the contribution to the receptor potential of other ions in addition to sodium ions.

Receptors adapt to sustained stimulation

You may have noticed that your perception of a steady stimulus, such as the smell of food cooking or the hum of a refrigerator tends to fade with time. Part of this decreased awareness is a higher brain function, but a large part of it results from **receptor adaptation**, a decrease with time in the response of a receptor to a steadily maintained stimulus. Receptors can adapt slowly or rapidly (Figure 13.4).

Slowly adapting receptors (often called **tonic** receptors) (see Figure 13.4*a*) tend to discharge a train of action potentials throughout the duration of the stimulus, usually with some decrease or adaptation of impulse frequency. The frequency of the impulses encodes information about the intensity of the stimulus, while the duration of the impulse train encodes information about stimulus duration. **Rapidly adapting receptors** (often called **phasic** receptors) (see Figure 13.4*b*), on the other hand, generate a short burst of action potentials at the onset of a stimulus but adapt rapidly and completely, ceasing impulse discharge during a prolonged stimulus. The initial frequency of the impulses encodes information about the rate of onset of the stimulus; the faster the rate of onset, the greater the frequency of impulses.

In the crayfish example discussed earlier, each pair of stretch receptors consists of two types: one slowly adapting (tonic) receptor neuron, and one rapidly adapting (phasic) receptor neuron. Thus they convey different information to the CNS: The slowly adapting receptor encodes abdominal *position* (a higher impulse frequency signals a more flexed position), and the rapidly adapting receptor encodes abdominal *movement* (a higher impulse frequency signals a faster rate of flexion).

A well-known example of a very rapidly adapting, phasic receptor is the mammalian Pacinian corpuscle (Figure 13.5*a*). The Pacinian corpuscle is an encapsulated nerve ending found subcutaneously and in deep dermal layers of the skin, as well as in mesenteries and other internal tissues. The sensory ending is unmyelinated and is encapsu-

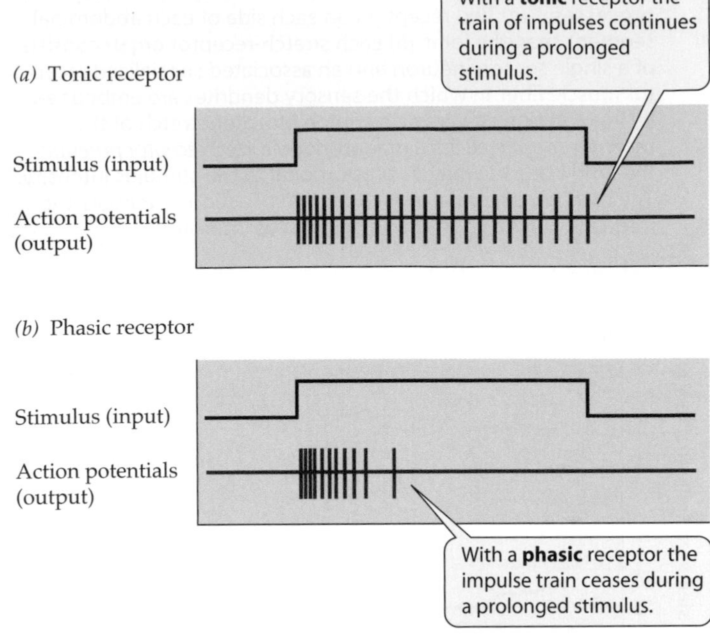

Figure 13.4 Slowly adapting and rapidly adapting receptors *(a)* Slowly adapting (tonic) receptors have a slow and incomplete decrease in impulse frequency. *(b)* Rapidly adapting (phasic) receptors have a rapid, complete decrease in impulse frequency. Phasic receptors thus convey information about *change* in stimulus intensity.

lated by many thin, concentric layers of epithelial cells. These multiple lamellae make up most of the volume of the Pacinian corpuscle and are collectively called the *onion skin* because of the resemblance of the corpuscle to a tiny onion. Pacinian corpuscles are mechanoreceptors, responding to mechanical deformation of the outer onion skin. They adapt extremely rapidly, normally generating only one impulse per stimulus onset, no matter how strong or prolonged the stimulus is. As Figure 13.5*b* shows, Pacinian corpuscles respond as vigorously to the sudden cessation of a prolonged stimulus as they do to its onset.

Because the Pacinian corpuscle is such a rapidly adapting receptor, we will use it to examine adaptation mechanisms. A review of our model sense organ (see Figure 13.2) suggests that receptor adaptation may occur at three sites: (1) the peripheral filtering mechanism, (2) the sensory transducer, and (3) the spike-encoding mechanism. Adaptation at either of the first two stages would decrease the amplitude of the receptor potential with time. Adaptation at the spike-encoding stage would decrease the impulse frequency even with a constant receptor potential.

For most receptors, adaptation occurs at more than one of these sites, the relative importance of the sites differing for different receptors. In the Pacinian corpuscle the most important site of adaptation is the peripheral filtering mechanism—namely, the onion skin surrounding the transducer (membrane of the nerve ending). In an intact corpuscle (Figure 13.6*a*), the amplitude of the receptor potential decreases to zero within 6 ms; therefore, adaptation occurs at or prior to the transduction step.

If most of the onion skin is dissected away, leaving a largely decapsulated corpuscle, the receptor potential adapts much more slowly (Figure 13.6*b*). This difference in response time shows that a major function of the onion skin is to act as a *frequency filter*, allowing sudden changes in mechanical displacement to pass through to the receptor membrane while filtering out stationary forces and slow changes. Apparently as a result of the elasticity of the lamellar membranes and the viscosity of fluid between lamellar layers, sustained pressure is dissipated in the lamellae while rapid changes are transmitted to the core with little loss. The pressure transmitted to the core (the adequate stimulus) has been calculated to be proportional to the *velocity* of displacement at the surface of the corpuscle.

Receptor adaptation, however, does not always depend on peripheral filtering mechanisms. For most receptors, the amplitude of the receptor potential decreases to a plateau during sustained stimulation. In photoreceptors and chemoreceptors, this decrease appears to be part of the transduction process itself, for example involving changes in the function of membrane channels, rather than the result of peripheral filtration. Finally, receptor adaptation may occur at the impulse-encoding (spike-encoding) stage (see Figure 13.2).

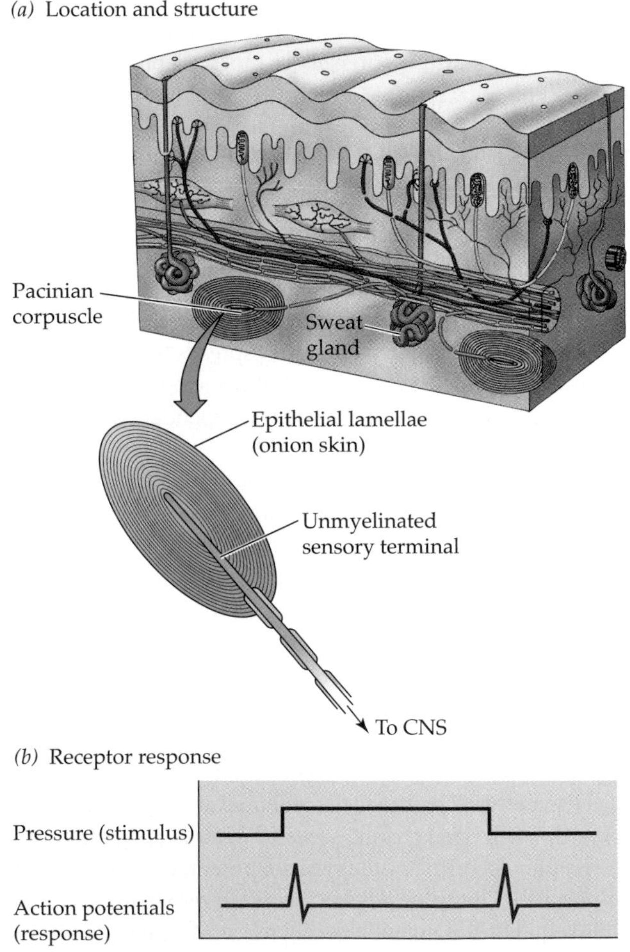

Figure 13.5 The Pacinian corpuscle is a rapidly adapting mechanoreceptor *(a)* Many concentric lamellar layers of connective tissue wrap the unmyelinated sensory terminal of the mechanoreceptor neuron. *(b)* The response of a Pacinian corpuscle is a single action potential at the onset and at the cessation of a mechanical stimulus.

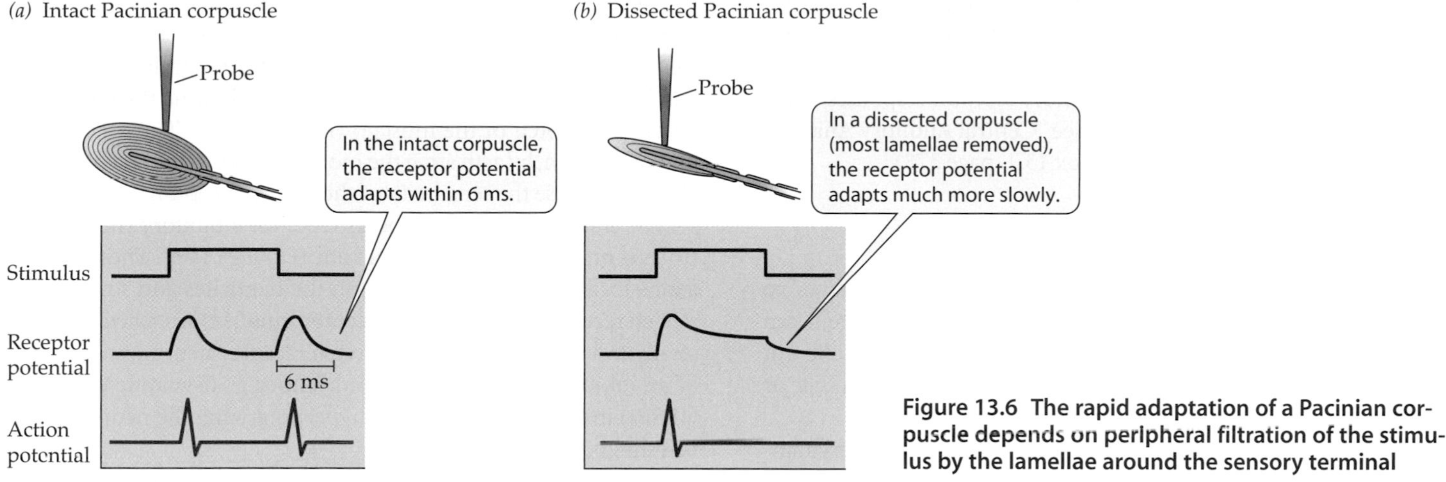

Figure 13.6 The rapid adaptation of a Pacinian corpuscle depends on peripheral filtration of the stimulus by the lamellae around the sensory terminal

In rapidly adapting and slowly adapting stretch receptors of crayfish, the receptor potentials of the two neurons adapt to nearly the same degree, but the rapidly adapting receptor ceases generation of action potentials in a few seconds, while the slowly adapting receptor continues to generate action potentials for several minutes. The major difference is that in the rapidly adapting receptor neuron, action potentials cease during sustained depolarization because the Na^+ channels remain inactivated to a greater extent than in the slowly adapting receptor.

Receptors encode information about modality, intensity, location, and timing of a stimulus

Recall that the central nervous system obtains information about the body's external and internal environments only from coded information carried to it by action potentials in sensory neurons. How is that information encoded? At least four aspects of a stimulus are encoded in receptor activity: stimulus modality (what is it?), intensity (how much?), temporality (when? how changing?), and location (where?).

As we have discussed, stimulus **modality** is coded by the principle of labeled lines. Anything that excites photoreceptors is interpreted as light, and anything that excites auditory receptors is interpreted as sound. Within a modality, different qualities of a stimulus (red versus green light, high-pitched versus low-pitched sound, salty versus sweet taste) are also encoded in part by labeled lines. In a familiar example, different classes of retinal cone photoreceptors are sensitive to different wavelengths of light and provide sensations of different colors. But as we have also mentioned, most receptors are rather broadly sensitive ("broadly tuned"), responding to more than one wavelength of light, or frequency of sound, or type of chemical odorant. For example, yellow, orange, and chartreuse light (of different wavelengths) will excite both red-sensitive and green-sensitive cones, but in different ratios.

Because receptors are not completely specific for stimulus quality, no one receptor serves as a pure labeled line. Therefore the CNS can get unambiguous information about stimulus quality only by comparing activity across several lines of incoming sensory information. This comparison-based information is usually termed an *across-fiber pattern* of sensory activity. As we will show for color vision and chemoreception, coding of stimulus quality usually involves combinations of labeled lines and across-fiber patterns of sensory activity.

Coding of stimulus **intensity** also has several components. Because the frequency of action potentials in a receptor neuron increases with increasing stimulus intensity, action-potential frequency is a code for stimulus intensity (see Figure 13.3). This relationship between frequency and intensity is modified by receptor adaptation, however, so *stimulus intensity* might mean the absolute magnitude of a stimulus for a slowly adapting receptor, but the rate of change of a stimulus for a rapidly adapting receptor.

Coding of stimulus intensity can also involve *recruitment* of additional receptors. Mechanoreceptors in your skin, for example, have different thresholds of sensitivity, some responding to light touch and some only to stronger mechanical stimulation. Thus a stronger stimulus will recruit additional receptors to respond, as well as increasing the impulse frequency in receptors already responding. Recruitment compensates for receptor saturation (the point at which an increase in stimulus intensity no longer causes an increase in action-potential frequency), and it extends the range of stimulus intensities that a population of receptors can code.

The **location** of a stimulus is largely coded via labeled lines. Tactile stimulation of different parts of the body surface excites different receptors, and the axons of these receptors project to different loci in the brain, so information about the anatomical position of the stimulus is preserved. The orderly projection of these stimulus positions is termed a *somatotopic map* of the body in the brain (see Figure 10.11).

Finally, the **temporality** of a receptor's impulse train (i.e., its occurrence in time) can encode attributes of a stimulus that differ depending on the type of receptor. Obviously the timing of an impulse train tells *when* a stimulus occurs. But as we have seen, the timing of an impulse train has different meanings for slowly adapting receptors and rapidly adapting receptors. For example, a short burst of action potentials from a slowly adapting receptor simply means that the stimulus lasted an equally short time (with the frequency of impulses encoding stimulus intensity from start to finish). A similarly short burst of action potentials from a rapidly adapting receptor, however, means that the stimulus had a sharp onset, but it tells the CNS nothing about the duration of the stimulus. Consequently, the

CNS must keep track of slowly adapting and rapidly adapting receptors on the basis of labeled lines. Temporal aspects of receptor responses may also encode other aspects of stimuli. For example, animals use slight time differences in auditory receptor responses of the two ears to localize a sound source (see "Central Auditory Analysis: Sound Localization," page 376 and Box 13.1, page 378).

Efferent control can adjust receptor sensitivity

Sense organs and their receptors are not merely passive "windows" to an animal's environment. Instead, the central nervous system can modulate the sensory information it receives, a process called **efferent control** of receptors and sense organs. Efferent control may excite or inhibit the response of a sensory receptor to a stimulus.

The crayfish stretch receptor offers a simple example of excitatory efferent control. As we have discussed, the dendrites of a crayfish stretch receptor neuron are embedded in the central, noncontractile region of a specialized muscle fiber (Figure 13.7*a*). The outer, contractile regions of the muscle fiber are innervated by axon branches of a motor neuron. (The same motor neuron also innervates adjacent muscles that extend the segments of abdomen.) Stimulating the excitatory motor neuron's axon causes the ends of the receptor muscle to contract, stretching the neuronal dendrites of the receptor cell and increasing its rate of impulse generation (beyond the rate caused by passive stretch of the muscle) (Figure 13.7*b*). Thus the central nervous system, by activating the motor neuron to the receptor muscle, can increase the sensitivity of the stretch receptor.

The crayfish stretch receptor also receives inhibitory efferent control, via one to three inhibitory axons (Figure 13.8*a*). These efferent axons form inhibitory synapses on the dendrites and soma of the stretch receptor neuron. If the receptor muscle is stretched to elicit a steady train of impulses in the stretch receptor neuron, stimulation of an inhibitory axon produces inhibitory postsynaptic potentials (IPSPs) in the receptor neuron, hyperpolarizing the neuron below threshold (Figure 13.8*b*).

The functional significance of efferent control of receptors is that it allows a central nervous system to set the level of activity of receptors, increasing their sensitivity by excitatory efferent control or decreasing sensitivity by inhibitory control. This level-setting can increase the range of stimulus intensities over which the receptor works, increasing sensitivity at low stimulus intensities and decreasing it at high intensities to prevent saturation of the receptor re-

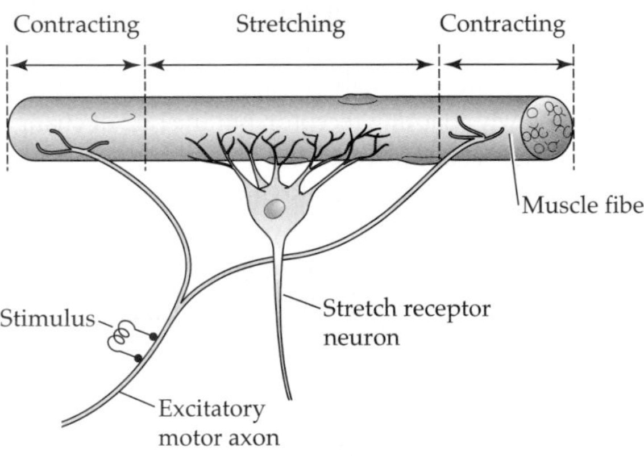

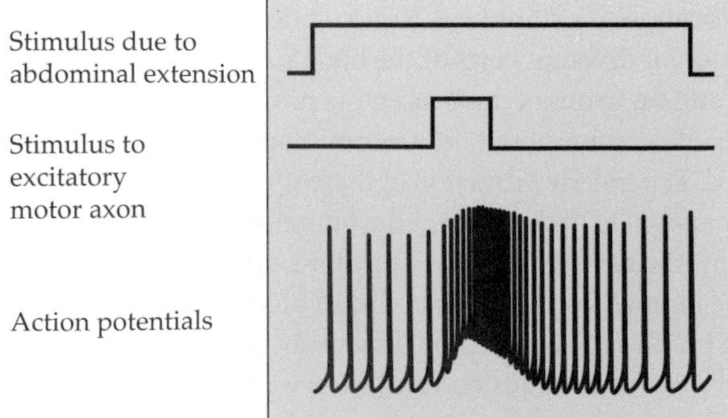

Figure 13.7 Sensory receptors are subject to excitatory efferent control from the CNS A crayfish stretch receptor responds to active contraction (as well as passive stretch) of its receptor muscle fiber. *(a)* Stimulation of the excitatory motor neuron contracts the ends of the muscle fiber, stretching the noncontractile central portion in which the dendrites of the receptor cell are embedded. *(b)* Tension resulting from motor neuron stimulation adds to the excitatory effects of ongoing passive stretch.

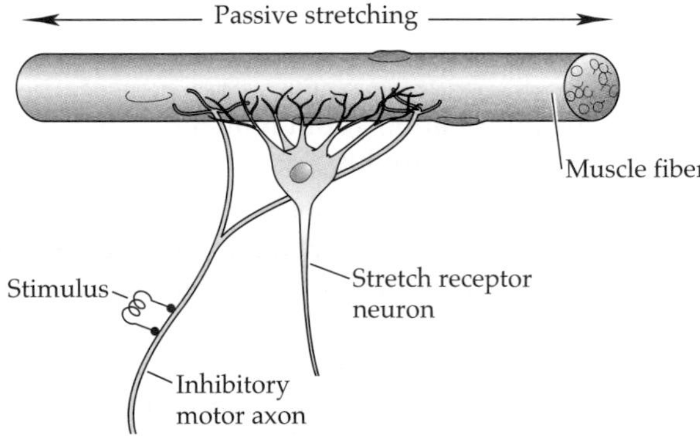

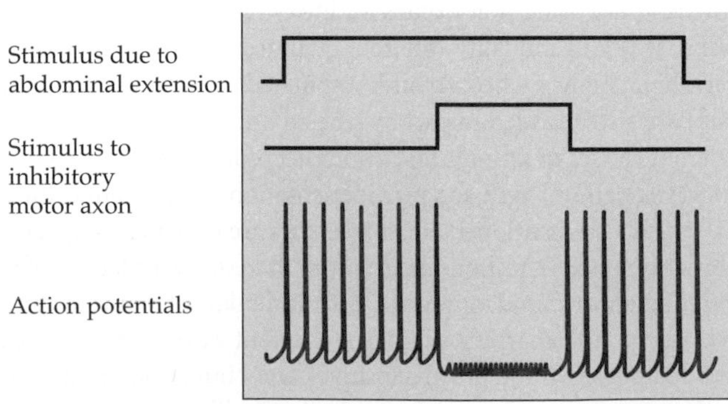

Figure 13.8 Receptors may also receive inhibitory efferent control *(a)* A crayfish stretch receptor receives inhibitory innervation that, when stimulated, hyperpolarizes the receptor membrane *(b)*, counteracting the depolarizing receptor potential and interrupting the train of action potentials resulting from passive stretch of the receptor muscle.

sponse. Efferent control can also alter receptor sensitivity during the animal's own active movement. Efferent control requires that the CNS monitor its own efferent output as well as sensory input, in order to interpret sensory information.

> **Summary: Receptor Functions and Their Control**
>
> - In sensory model systems (crayfish stretch receptor, Pacinian corpuscle), stretch opens ion channels to let Na^+ ions enter, producing a depolarizing receptor potential.
> - Receptors adapt to sustained stimulation, producing a response that diminishes over time. Slowly adapting (tonic) receptors signal the amount of a stimulus, and rapidly adapting (phasic) receptors signal changes in stimulus intensity. Receptor adaptation can occur at three sites: (1) the peripheral filtration stage, (2) transduction, or (3) the generation of action potentials.
> - Sensory receptors encode different aspects of a stimulus: type, strength, location, and time.
> - The sensitivity of many sensory receptors can be controlled by the central nervous system, a process called efferent control.

Photoreception

Humans depend on vision extensively. Perhaps for this reason, the vertebrate visual system is in many respects the best-known sensory system, and it serves as a model for studies of sensory processes. We therefore explore both photoreception and visual information processing in greater depth than the other sensory systems.

Photoreceptors and eyes have distinct evolutionary relationships

Animal photoreceptors contain photopigments—molecules that absorb and transduce light into an electrical signal—and most animal photoreceptors have an intracellular structure that increases the membrane surface area that bears the photopigments. In **ciliary** photoreceptors, modified cilia contain the phototpigment.[2] **Rhabdomeric** photoreceptors (*rhabdo*, "rod") have collections of microvilli that increase the membrane surface area. They are found in the protostome phyla (Annelida, Arthropoda, Mollusca, and so on) and are the most elaborately developed in arthropods.

As Figure 13.9 shows, photoreceptors appear to have evolved along either ciliary or rhabdomeric lines. There are exceptions to this generalization, however, since both categories of photoreceptors have been found in the same phylum, and even in the same species. The retinas of the scallop *Pecten*, for example, contain a proximal layer of rhabdomeric photoreceptors and a distal layer of ciliary photoreceptors. Such exceptions suggest that any phylogenetic generalization about lines of photoreceptor evolution is probably too simplistic.

[2]In fact, many kinds of sensory receptors have structures that are related to cilia, including photoreceptors of several phyla and mechanoreceptors and chemoreceptors of vertebrates and invertebrates.

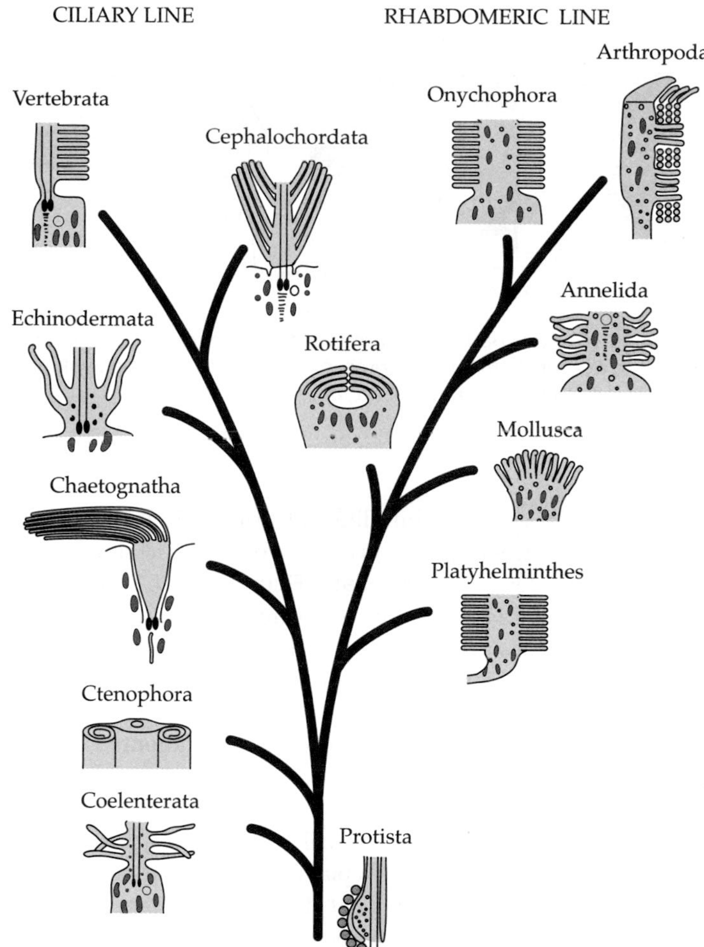

Figure 13.9 A simplified evolutionary relation of photoreceptive structures Two major lines of photoreceptor evolution have been proposed, one employing modified cilia (in deuterostomes) and the other a rhabdomeric arrangement of microvilli (in protostomes). Some protists such as *Euglena* have a photoreceptive structure (the paraflagellar body) associated with the flagellum. (After Eakin 1968.)

The evolution of eyes is as speculative an issue as the evolution of photoreceptors themselves. *The major selective advantage of localizing photoreceptors into eyes is to allow directionality of photoreceptor stimulation.* Photoreceptors bounded by screening pigment on one side respond to light from only one direction, permitting an animal to orient to its environment. A good starting point in considering the evolution of visual systems is the photosynthetic protist *Euglena*. The photoreceptive organelle, the eyespot of a *Euglena*, is a swelling on the side of the base of the flagellum, termed a *paraflagellar body*. The association of this apparently primitive photoreceptive organelle with a flagellum might be the evolutionary precursor of the ciliary line of photoreceptor cells. The paraflagellar body is shielded from light by screening pigment on one side. As a *Euglena* swims through the water, it rotates in a twisting motion so that light from a directional source alternately illuminates the paraflagellar body or is shaded by the screening pigment. The eyespot then sequentially samples light from different directions.

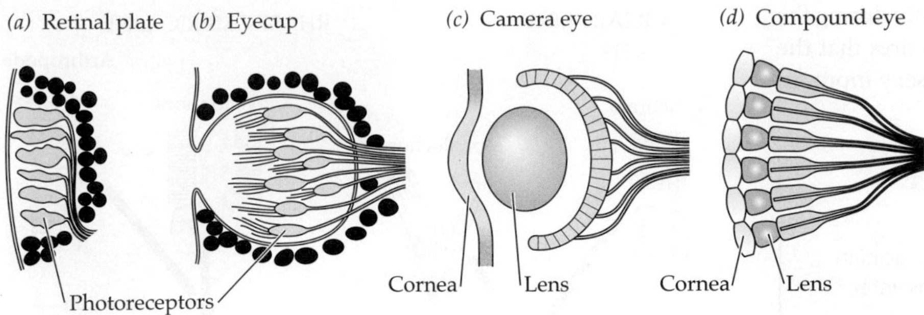

Figure 13.10 Four patterns of organization of eyes The simpler structures—retinal plate and eyecup *(a, b)*—may represent earlier stages in the evolution of the more-complex camera and compound eyes *(c, d)*.

Multicellular eyes have four different structural patterns, which may represent stages or pathways in eye evolution. The simplest eye may have consisted of a **retinal plate** (**Figure 13.10*a***), a flat layer of photoreceptors on a base of screening pigment. As with *Euglena*, the screening pigment shields the receptors from light on one side; thus, this simple eye can differentiate only the direction of light. Retinal plates are found in all major phyla.

An elaboration in eye organization is the **pigmented eyecup** (**Figure 13.10*b***). As with the retinal plate, an encircling pigment shield enhances the directional selectivity of the photoreceptors. If the opening in the pigment cup is small enough, this type of eye can form an image in the manner of a pinhole camera. The pinhole limits light entry, however, so a pigmented eyecup is not very photosensitive. A lens is needed in order to increase the light-gathering power of an eye, and this feature distinguishes the next level of morphological complexity of eyes.

The lens-bearing **camera eye** (**Figure 13.10*c***) is the characteristic visual organ of vertebrates, but it has evolved independently in many phyla. A parallel evolutionary trend is the **compound eye** (**Figure 13.10*d***), which consists of many optically independent facets called *ommatidia* (singular *ommatidium*). Each ommatidium has its own lens and pigment-shielded photoreceptors, and it focuses light from a small nonoverlapping piece of the visual field onto its photoreceptor membranes. Together, the ommatidia of a compound eye produce a noninverted *mosaic image* of the visual field.

Although compound eyes are very different in organization from the camera eyes of vertebrates, their developmental differentiation is under similar control of related genes (*pax* genes). This finding suggests (but does not demonstrate) that compound and camera eyes are more closely related evolutionarily than their structures would suggest.

The vertebrate eye focuses light onto retinal rods and cones

A vertebrate eye, as we have noted, is a camera eye. The lens system focuses an inverted image of the visual field on the **retina**, the photoreceptor-containing layer at the back of the eye that is analogous to the film in a camera (**Figure 13.11*a***). For terrestrial vertebrates the greatest amount of refraction occurs at the interface between the air and the cornea because the difference in the refractive index (which governs the bending of light rays) is greater at that interface than at the interface between the lens and the watery fluid within the eye. The iris diaphragm constricts the opening of the pupil to limit light entry and to enhance resolution to some extent, like the lens diaphragm of a camera.

Terrestrial vertebrates change the shape of the lens to focus the eye. In contrast, in aquatic vertebrates the cornea does little refraction, because its refractive index is similar to water. Instead, a thick, nearly spherical lens accomplishes most refraction; the refractive index of the lens is greater at its center than at its edges, minimizing distortion. Fish cannot focus their spherical lenses by shape change; instead the lens is displaced anteriorly like a camera lens.

The retina of the vertebrate eye is a developmental outgrowth of the brain. It contains rod and cone photoreceptors, as well as a network of neurons—horizontal cells, bipolar cells, amacrine cells, and ganglion cells—that perform the first stages of visual integration (**Figure 13.11*b***). At the back of the retina is a pigmented epithelium that absorbs light not captured by the photoreceptors. The retina is said to be inverted, with the photoreceptors in the outermost layer, farthest away from incoming light. This inverted structure is a consequence of the way in which the retina develops in the embryo, as the outer layer of a two-layered optic cup.

Although light must pass through all the retinal layers to reach the outer segments of the rods and cones, the retinal layers do not degrade the image greatly, because they are quite transparent. Some light scattering, however, does occur, and many retinas have a central high-acuity region in which the intervening cell layers and blood vessels are displaced to the side. In humans this region is the **fovea**, a depression 1.5 mm in diameter (5° of visual angle). The central 1° of the fovea contains tightly packed cones to the exclusion of other neurons. Rod photoreceptors are absent in this central part of the human fovea but outnumber cones elsewhere in the retina. Primates and some birds have well-developed foveas, and many other vertebrates have a less elaborate and broader area of relatively high acuity called the *area centralis*.

Another consequence of the inverted retina is that the axons of retinal ganglion cells, which form the optic nerve, come off the inner side of the retina, facing the lens. The axons exit through the retina at the *optic disc*, producing a blind spot in the visual field. We are normally unaware of this blind spot because it falls in the binocular field, and the other eye supplies the missing information. Moreover, we depend on the fovea for much of our high-acuity vision, and we make unconscious rapid eye movements, further decreasing any visual deficit resulting from the blind spot.

Retinal photoreceptors are classified as **rods** and **cones**. In mammals, cones are used in color vision and for high-acuity vision in the fovea; the more sensitive rods are used in dim light and do not contribute to color vision. Nocturnal animals tend to have retinas in which most or all receptors are rods, while cones predominate in

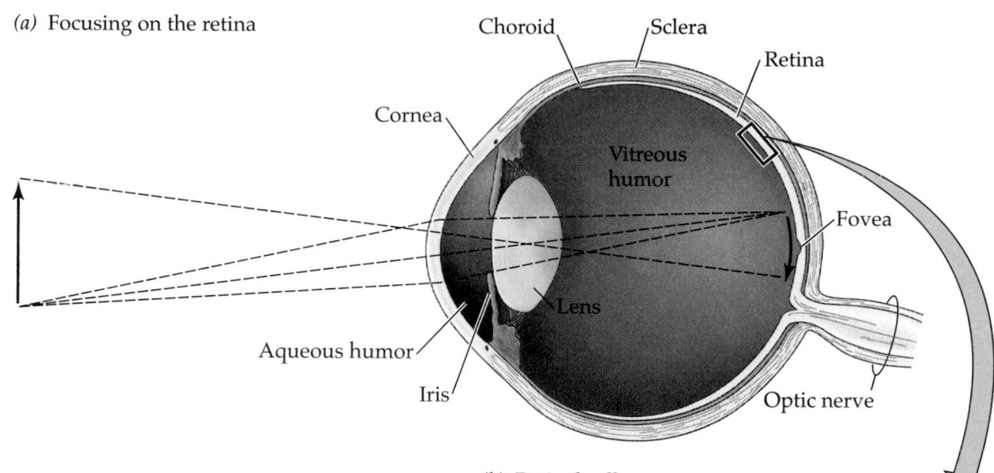

Figure 13.11 Structure of the mammalian eye and retina *(a)* The cornea and the lens focus an inverted image on the retina at the back of the eye. *(b)* The photoreceptors (rods and cones) are in an outer nuclear layer at the back of the retina. They are connected via bipolar cells to ganglion cells, in a "straight-through" pathway. Horizontal cells form a lateral pathway in the outer plexiform layer, and amacrine cells form a lateral pathway in the inner plexiform layer. The ganglion cells are the output of the retina, their axons forming the optic nerve. (After Dowling 1979 in Kandel, Schwartz, and Jessell 1995.)

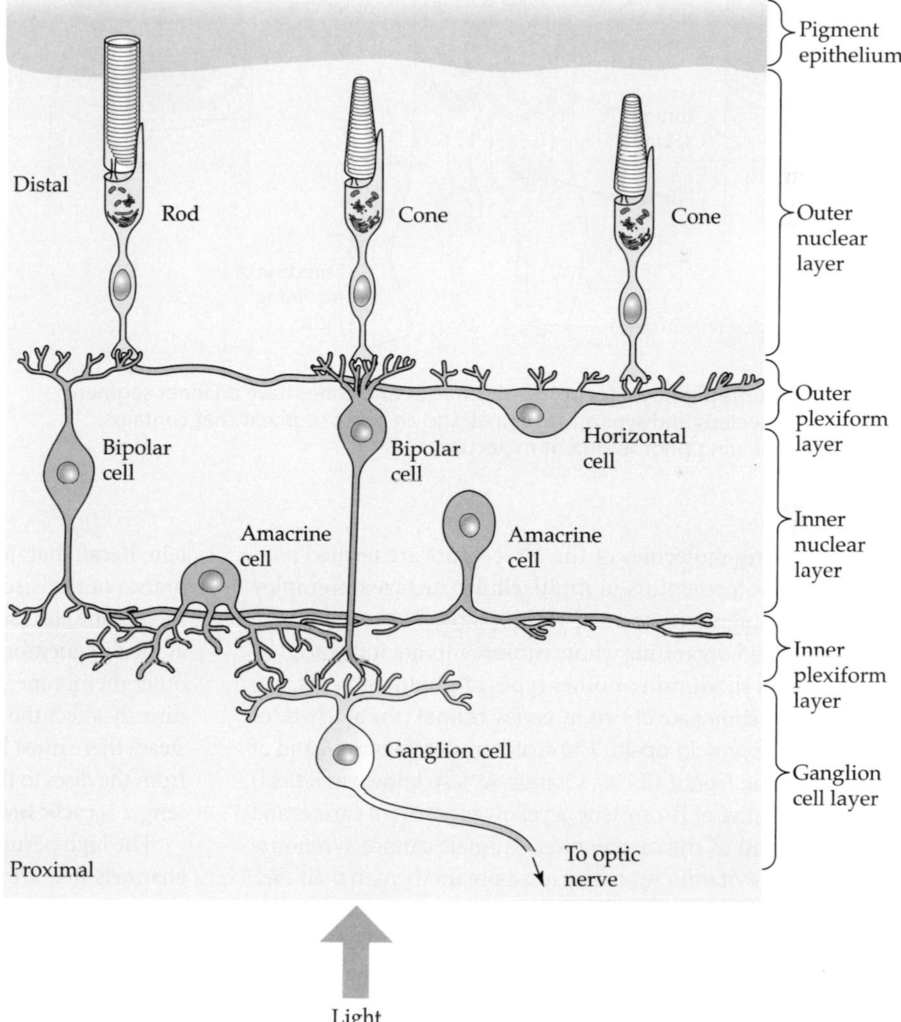

retinas of strongly diurnal animals. Because rods are larger and less diverse than cones, they have been studied more extensively.

Retinal rods and cones transduce light into a hyperpolarizing receptor potential

Retinal photoreceptors have a characteristic structure (Figure 13.12). Rods and cones consist of an **outer segment** containing the photosensitive membranes, and an **inner segment** containing the nucleus, mitochondria and other cell organelles, and the synaptic terminal. The inner and outer segments are connected by a short ciliary stalk, the outer segment being derived from a modified cilium. The photoreceptor outer segments contain many flattened *lamellae* of membranes derived from the plasma membrane. In the cones of mammals and of some other vertebrates, these lamellae retain continuity with the outer membrane, so the lumen of each lamella is continuous with extracellular space. In rods, in contrast, the lamellae become separated from the outer membrane and form internalized flattened **discs** (see Figure 13.12*a*). Several hundred to a few thousand discs, stacked like pancakes, fill the rod outer segment. The membranes of rod discs and cone invaginated lamellae contain the photopigment *rhodopsin*; thus they are the sites of the primary photochemical events that trigger photoreceptor excitation.

How is light phototransduced into the activity of rod and cone photoreceptors? This process has three stages: First, light activates rhodopsin photopigment; second, activated rhodopsin decreases the concentration of cyclic GMP in the photoreceptor cytoplasm; and third, the decrease in cyclic GMP closes ion channels, decreasing Na^+ influx and leading to hyperpolarization of the photoreceptor.

LIGHT ACTIVATES THE PHOTOPIGMENT RHODOPSIN Light must be absorbed as a necessary first step in the excitation of photoreceptors.

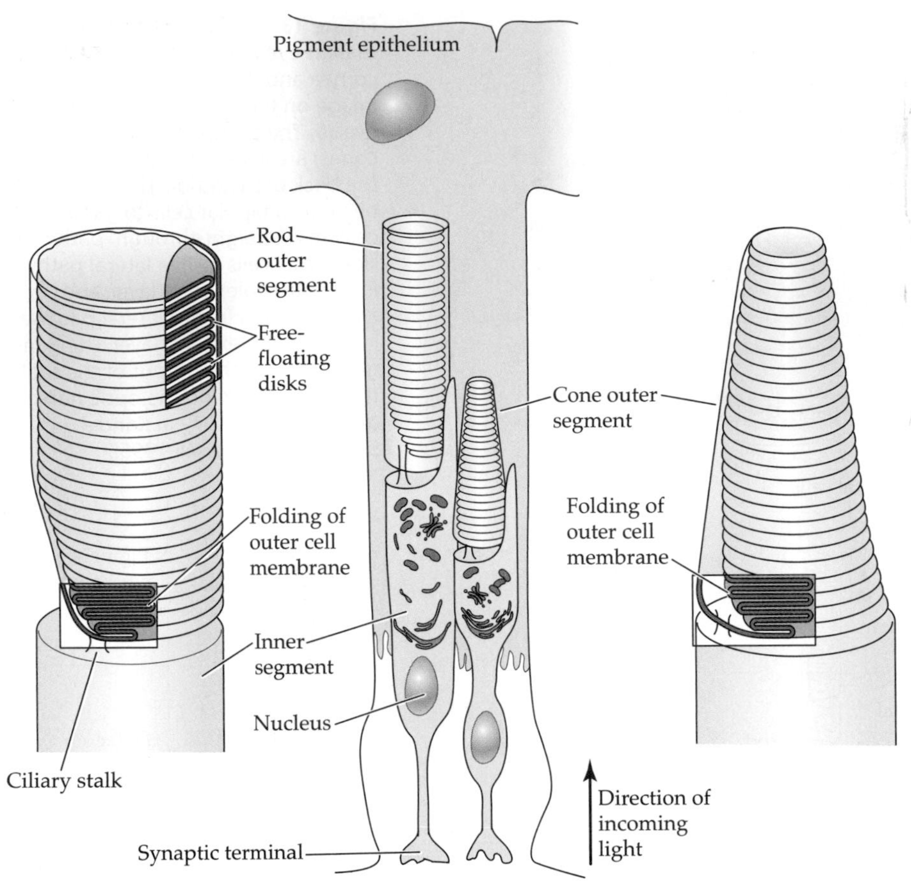

Figure 13.12 Vertebrate photoreceptors Both rods and cones have an inner segment that contains the nucleus and synaptic terminal, and an outer segment that contains ordered lamellae bearing photopigment molecules.

The light-absorbing molecules of these receptors are termed *photopigments*. All photoreceptors in multicellular metazoans employ carotenoid photopigments termed **rhodopsins**. Photoresponsive *protists* (single-celled organisms whose movements are influenced by light) may contain rhodopsin or other types of photopigments.

Rhodopsin is a conjugate of two moieties: **retinal** (the aldehyde of vitamin A) and the protein **opsin**. The structures of vitamin A and of retinal[3] are shown in Figure 13.13*a*. Vitamin A (also known as *retinol*) is an alcohol derivative of β-carotene, a yellow pigment of carrots and a parent compound of the carotenoids. Animals cannot synthesize carotenoids such as vitamin A, so they must obtain them in their diet. Retinal is conjugated to the protein opsin by a linkage of the aldehyde end to a lysine residue of opsin. Many rhodopsins with different light-absorption spectra have been characterized, and all of them contain the same carotenoids. Thus the differences among rhodopsins result from differences in the opsin moiety, and possibly in the linkage between retinal and opsin.

Opsins have a molecular weight of about 40,000 daltons and contain 2 short oligosaccharide chains and 30 or more attached phospholipids that help embed the protein in the photoreceptor membrane. Rhodopsin is a member of the G protein–coupled receptor superfamily (see Chapter 12, page 335), acting via the G protein transducin and sharing extensive sequence homology with the neurotransmitter receptors and hormone receptors in that superfamily. Figure 13.13*b* shows the three-dimensional structure of rhodopsin, with the seven transmembrane regions that are characteristic of a G protein–coupled receptor.

Light activates rhodopsin by isomerizing retinal, which leads to conformational changes in the whole molecule. In unstimulated photoreceptors the retinal in rhodopsin is in the 11-*cis* conformation, and the action of light is to photoisomerize this bent form to the straight, all-*trans* conformation (see Figure 13.13*a*). Because retinal is intimately associated with opsin, the photoconversion to all-*trans* retinal is followed by a series of spontaneous changes in conformation of opsin, producing activated rhodopsin, also called *metarhodopsin II*. This action of light is very rapid, producing activated rhodopsin in about 1 ms.

ACTIVATION OF RHODOPSIN DECREASES CONCENTRATIONS OF CYCLIC GMP IN PHOTORECEPTOR OUTER SEGMENTS Activated rhodopsin acts as a good G protein–coupled receptor, by activating a G protein. Recall that in rods, rhodopsin is localized at the internal membranes of the discs. Because the disc membranes and the outer cell membrane are discontinuous, the light-induced changes in rhodopsin at the disc membranes must somehow affect Na^+ permeability of the outer membrane. An electrical change in the disc membrane will not directly affect the surface membrane, because they are separate. Instead, there must be an *intracellular messenger* that conveys a change from the discs to the outer surface membrane. This intracellular messenger is **cyclic GMP** (**cyclic guanosine monophosphate**).

The high permeability to Na^+ of rods in the dark depends on Na^+ channels that are gated by cyclic GMP (Figure 13.14). The cGMP-gated channels are not voltage-gated, although they are structurally and evolutionarily related to voltage-gated channels. Instead they open when they bind cGMP on the cytoplasmic side. Light-induced change in rhodopsin activates a series of reactions at the disc membrane that result in an enzymatic degradation of cyclic GMP to 5′-GMP. Figure 13.14 shows this process: Activated rhodopsin (metarhodopsin II) activates the G protein **transducin**, which is closely related to the G proteins that mediate metabotropic synaptic actions (see Figure 12.19). The activated G protein activates **cGMP phosphodiesterase**, an enzyme in the disc membrane that hydrolyzes cGMP to 5′-GMP. Thus the second-messenger response to light in a rod outer segment is a *decrease* in cytoplasmic concentration of cGMP.

[3] Many freshwater fishes and freshwater stages of amphibians have a slightly different form of vitamin A and of retinal, respectively termed *vitamin* A_2 (3-dehydroretinol) and *retinal*$_2$ (3-dehydroretinal). These forms have two double bonds in the ring. We will consider only retinal$_1$, which has one double bond in its ring portion. Retinal$_1$ is the chromophore (light-absorbing part) of most rhodopsins, including those of mammals, birds, most marine fishes, and all invertebrate metazoans.

(a) Retinal and vitamin A

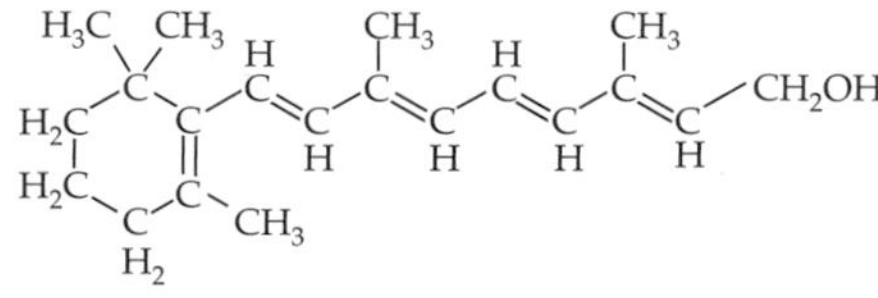

Complete structure of vitamin A (all-*trans*)

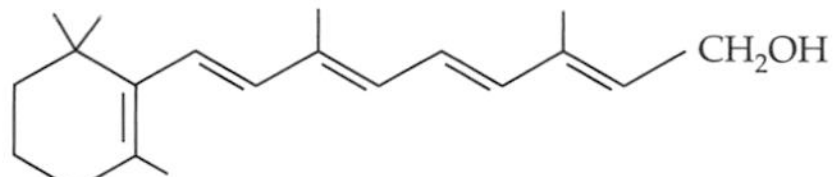

Condensed structure of vitamin A (all-*trans*)

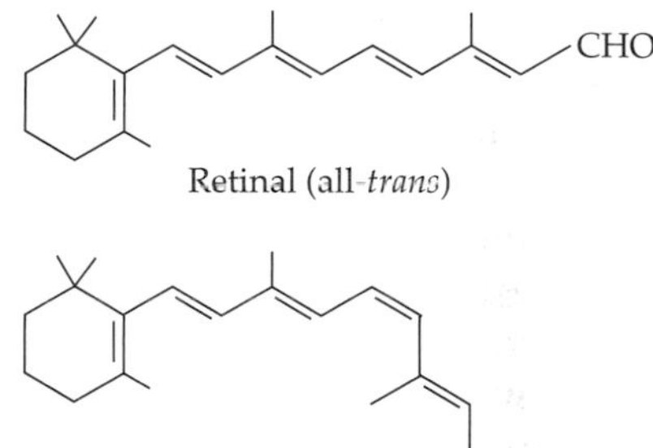

(b) Opsin

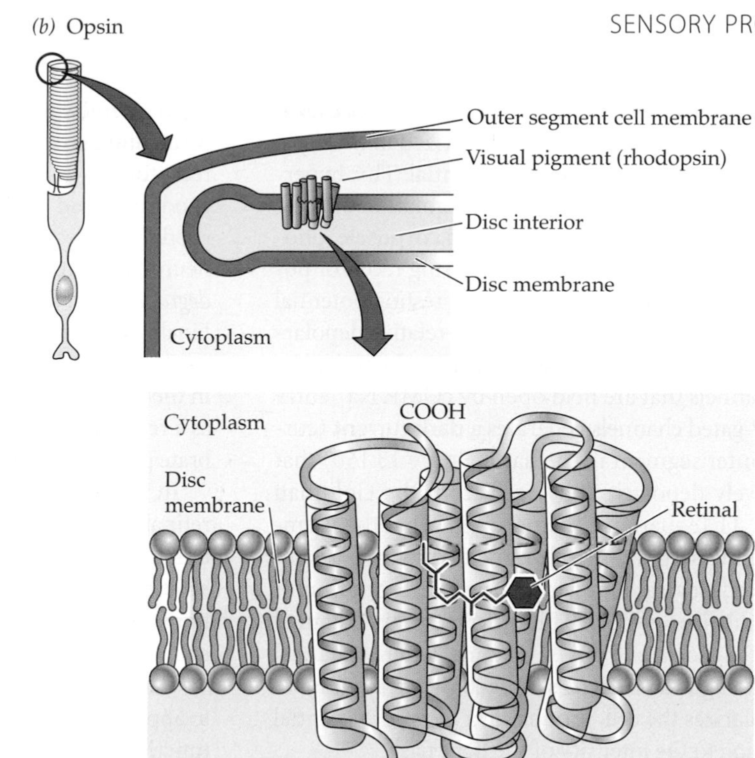

Figure 13.13 Rhodopsin is a photopigment composed of two parts: retinal and opsin *(a)* Chemical structures of vitamin A and of retinal. Vitamin A is shown both as a complete structure (top) and as a skeleton structure (middle). Vitamin A is converted to retinal, which has two isomers (11-*cis* and all-*trans*). *(b)* Three-dimensional structure of the protein (opsin) portion of vertebrate rhodopsin. Seven α-helical regions of the protein span the membrane; retinal is attached to an amino acid residue within the seventh membrane-spanning region.

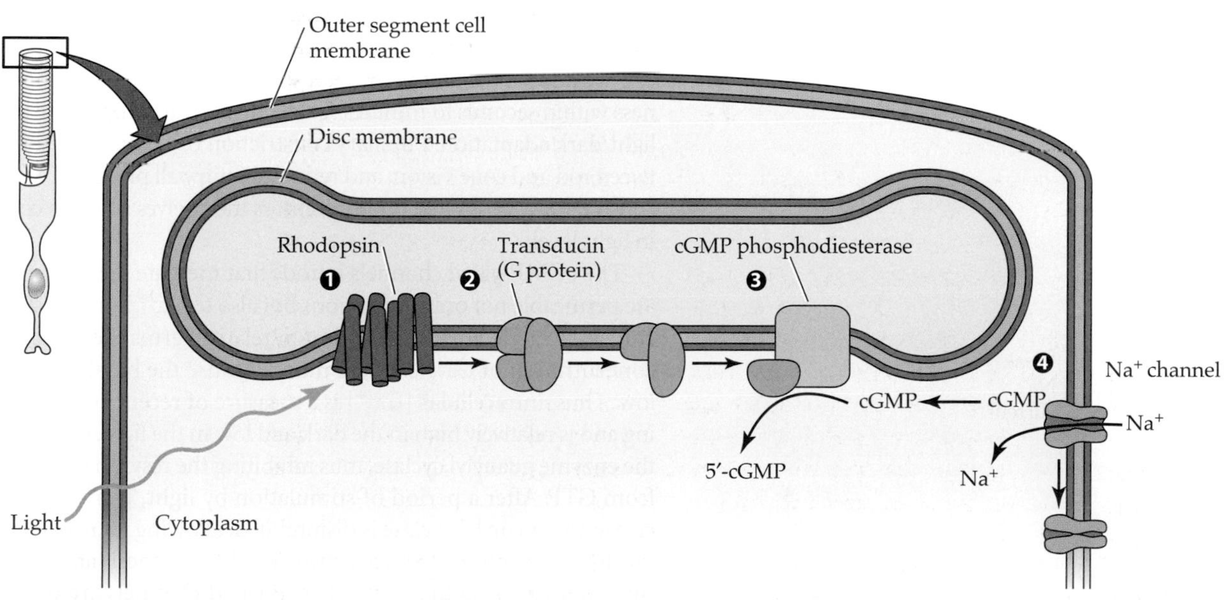

Figure 13.14 Phototransduction closes cation channels in the outer segment of the photoreceptor membrane In the dark, the cation channels are kept open by intracellular cGMP and conduct an inward current, carried largely by Na^+. When light strikes the photoreceptor, these channels are closed by a G protein–coupled mechanism. ❶ Rhodopsin molecules in the disc membrane absorb light and are activated. ❷ The activated rhodopsin stimulates a G protein (transducin in rods), which in turn activates cGMP phosphodiesterase. ❸ The phosphodiesterase catalyzes the breakdown of cGMP to 5′-GMP. ❹ As the cGMP concentration decreases, cGMP detaches from the cation channels, which close.

LIGHT HYPERPOLARIZES A RETINAL PHOTORECEPTOR BY CLOSING CYCLIC GMP-GATED CHANNELS A retinal rod is relatively depolarized in the dark, and light induces a *hyperpolarizing* receptor potential. This hyperpolarization is unlike the responses to sensory stimulation of most receptors (e.g., crayfish stretch receptor, Pacinian corpuscle, photoreceptors of invertebrates), which have depolarizing receptor potentials. Rods (and cones) have an unusually small resting potential in the dark (about –40 mV) (Figure 13.15*a*). This relative depolarization in the dark results from Na^+ ions flowing into the rod outer segment through channels that are held open by cGMP. Na^+ entry through these cGMP-gated channels generates a **dark current** (current flowing in the outer segment in the dark; Figure 13.15*b*) that keeps the rod relatively depolarized. (Recall from the Goldman equation in Chapter 11 [equation 11.6, page 291] that the membrane potential depends on the ratio of permeabilities to Na^+ and K^+.) The light-induced decrease in cytoplasmic concentration of cGMP shifts the equilibrium of cGMP binding to the Na^+ channel protein so that cGMP dissociates from the channel and the channel closes. The response to light thus decreases the dark current (Figure 13.15*c*) and hyperpolarizes the cell, producing a receptor potential that is graded according to the intensity of the light flash.

(a) Rod receptor potential

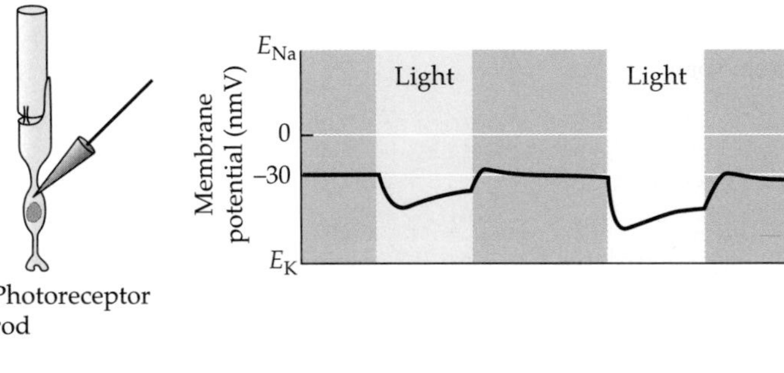

(b) Dark *(c)* Light

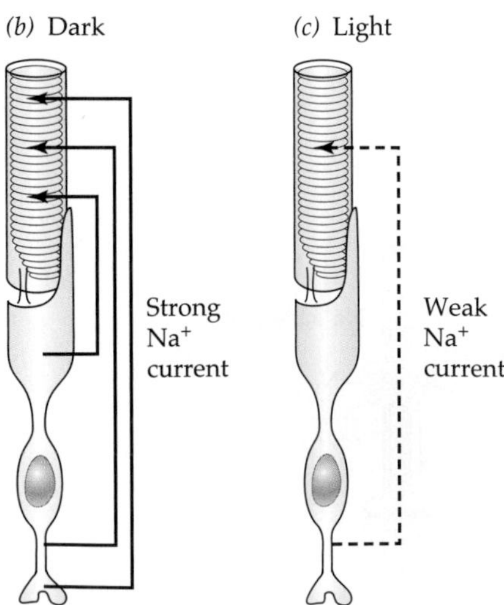

Figure 13.15 Light hyperpolarizes vertebrate photoreceptors *(a)* The rod receptor potential is a graded hyperpolarization; the brighter the light, the greater the hyperpolarization. *(b)* A *dark current* enters the rod outer segment in the dark, carried largely by Na^+ ions. *(c)* Light acts to decrease the dark current by closing cGMP-gated Na^+ channels.

In several ways the responses of vertebrate photoreceptors to light stimulation are not typical of other sensory receptors. Many sensory receptors do not employ G protein–coupled receptor molecules like rhodopsin and do not act through a metabolic cascade. Moreover, rhodopsin's second-messenger action is unlike that of metabotropic neurotransmitter receptors because activated rhodopsin regulates *degradation* of a second messenger (cGMP) rather than its synthesis. Finally we have the paradox of hyperpolarizing activation: An illuminated rod acts like a "resting" neuron (with E_m close to E_K), and a rod in the dark is like an "excited" neuron or other receptor, depolarized as a result of a high permeability to Na^+. (In contrast, most invertebrate photoreceptors have "typical" depolarizing receptor potentials).

In terms of consistency of response with other cells, a vertebrate retinal rod or cone is depolarized by dark (corresponding to excitation in other cells) and hyperpolarized by light (corresponding to inhibition in other cells). This semantic trap indicates that in using terms such as *excitation* of photoreceptors, we must be careful not to equate excitation with depolarization.

Calculations show that each photon absorbed by rhodopsin leads to approximately 10^6 Na^+ ions *not* entering a rod outer segment. One functional attribute of an intracellular messenger is to provide the necessary amplification for this process. Suppose that the activation of one rhodopsin molecule leads to activation of 500 phosphodiesterase molecules, causing the enzymatic degradation of 10^5 cGMP molecules. The decreased cGMP concentration could close 250 Na^+ channels, blocking entry of 10^6 Na^+ ions for about a second, providing the necessary amplification. This amplification of response to a single photon can hyperpolarize the rod by 1 mV.

LIGHT/DARK ADAPTATION Photoreceptor sensitivity also changes as a component of **light/dark adaptation**. When we step out from a dimly lit interior to the sunny outdoors, the bright light is dazzling if not painful at first, but we soon adapt to the new light level. Likewise, when stepping into a dim room from the bright outdoors, initially we can see little in the gloom, but our visual system adapts to the darkness within seconds to minutes. There are several components of this light/dark adaptation: Pupillary constriction or dilation, the shift between rod and cone vision, and retinal circuitry all play roles. Here we consider how vertebrate photoreceptors themselves adapt to changes in light level.

The cGMP-gated channels in rods that mediate the dark current are permeable not only to Na^+ ions but also to Ca^{2+} ions. Most of the current is carried by Na^+ ions, but in relative terms the cytoplasmic concentration of Ca^{2+} changes more because the basal $[Ca^{2+}]$ is so low. Thus, intracellular $[Ca^{2+}]$ is a measure of recent channel opening and is relatively high in the dark and low in the light. Ca^{2+} inhibits the enzyme guanylyl cyclase, thus inhibiting the resynthesis of cGMP from GTP. After a period of stimulation by light, $[Ca^{2+}]$ levels decrease and guanylyl cyclase is disinhibited, allowing increased cGMP synthesis to restore a balance between cGMP synthesis and degradation, and ensuring that some cGMP-gated channels are open and ready to be closed by light.

TERMINATION OF THE RESPONSE TO LIGHT AND REGENERATION OF RHODOPSIN Both rhodopsin and the G protein transducin must be inactivated after they are activated by light. Like other G proteins, transducin becomes activated when it unbinds GDP and binds GTP. Transducin (also like other G proteins) has GTPase activity, converting its bound GTP back to GDP and terminating its activation. Activated rhodopsin is a target for phosphorylation by a specific protein kinase. Phosphorylated rhodopsin then interacts with the regulatory protein *arrestin*, terminating its activation. Eventually the rhodopsin splits into all-*trans* retinal and opsin, a process termed *bleaching*. Bleaching proceeds spontaneously but very slowly over tens of seconds.

After the activation and inactivation of rhodopsin, its products must be regenerated to the original state, with retinal in the 11-*cis* configuration. Regeneration can occur in either of two ways: enzymatically or photochemically. For vertebrate receptors, most regeneration is by enzymatic means. After retinal and opsin split, the all-*trans* retinal is reconverted to 11-*cis* retinal by an isomerase. This process, however, is neither direct nor simple. Much of the enzymatic regeneration of vertebrate rhodopsin occurs not in the photoreceptors, but in the adjacent pigment epithelium.

A typical pathway for regeneration is as follows: (1) All-*trans* retinal is converted to all-*trans* retinol (vitamin A) by retinal oxidoreductase in the receptor cell. (2) Retinol migrates to the pigment epithelium (the pigment of which is melanin, not rhodopsin). (3) Vitamin A (retinol) is converted from all-*trans* to 11-*cis* in the pigment epithelium, by an isomerase. (4) Vitamin A (11-*cis*) migrates back to the photoreceptors and is converted to 11-*cis* retinal by retinal oxidoreductase. (5) The 11-*cis* retinal recombines with opsin, which remains stationary as an intrinsic protein of the photoreceptor membranes. It is not clear what functional advantage results from the major role of the pigment epithelium in the regeneration of vertebrate rhodopsin.

Rhodopsin also can be photoregenerated directly from its products, rather than enzymatically regenerated. Metarhodopsin can absorb a light photon and reconvert directly to 11-*cis* rhodopsin without requiring metabolic energy. Such direct photoregeneration is a minor pathway for vertebrate rhodopsin, accounting for perhaps 6% of all regeneration under normal light conditions. In contrast, in well-studied invertebrate examples (e.g., the fly *Ascalaphus* and the octopus *Eledone*), most regeneration of rhodopsin is by direct photoregeneration of activation products similar to vertebrate metarhodopsins.

Summary: Photoreception

- The vertebrate eye is a camera eye that focuses light onto retinal rod and cone photoreceptors. Rods and cones are unusual in that light produces a hyperpolarizing receptor potential.
- The photopigment rhodopsin, contained in membranes of photoreceptor outer segments, absorbs light. Activated rhodopsin acts via a G protein to decrease the concentration of cGMP in the cytoplasm, leading to closing of cGMP-gated Na^+ channels that keep the photoreceptor depolarized in the dark. Closure of these channels hyperpolarizes the photoreceptors.
- Rhodopsin is regenerated to 11-*cis* rhodopsin after activation. In vertebrates, most regeneration is a slow enzymatic process, part of which occurs outside the photoreceptors in the adjacent pigment epithelium.

Visual Sensory Processing

Photoreceptors respond to light, but the vertebrate visual system responds to patterns: changes in light level and color over space and time. This conversion from sensitivity to light to sensitivity to pattern occurs partly in the retina and partly in higher visual processing areas of the brain.

Retinal neurons respond to patterns of light and dark

The aspects of visual stimuli that are most important for the behavior of animals (including humans) are *patterns* of light, dark, and color, rather than the overall light level. In behavioral terms, the significant features of the visual world are spatial patterns of visual stimuli that represent objects in the world, and temporal patterns that indicate movements. To a frog, for example, a small dark area in the visual field may have great behavioral importance (a fly for lunch)—particularly if it moves relative to the rest of the visual field. Changes in overall illumination, in contrast, may be less important, merely indicating, for example, that a cloud has passed in front of the sun.

The photoreceptors themselves respond only to light level at one point in space. Therefore we would expect that visual systems might integrate signals from receptors in ways that abstract the behaviorally significant spatial patterns and movements of stimuli. This integration has been studied most extensively in the vertebrate visual system, where information about visual pattern and movement is abstracted first in the retina and then in central visual areas. In this section we consider how neural circuits in the retina perform the early stages of this visual integration.

The retina contains the rod and cone photoreceptors and four kinds of integrating neurons (see Figure 13.11*b*): horizontal cells, bipolar cells, amacrine cells, and ganglion cells. **Bipolar cells** receive input from photoreceptors at the *outer plexiform layer* in the back of the retina, and they synapse on amacrine and ganglion cells in the *inner plexiform layer*. **Horizontal cells** extend tangentially, connecting different regions of the outer plexiform layer. **Amacrine cells** mediate similar tangential interconnections in the inner plexiform layer. Retinal **ganglion cells** are the output of the retina; their axons form the optic nerve that extends to the brain.

We begin our examination of retinal integration by looking at the response properties of the retinal ganglion cells that provide the output of the retina. Steven Kuffler (1913–1980) and others recorded responses of mammalian retinal ganglion cells in the 1950s. In contrast to photoreceptors, ganglion cells respond to stimulation over a relatively large visual area. This brings us to an important concept—that of the **receptive field** of a sensory neuron. For the vertebrate visual system, the receptive field of a neuron is defined as the area of the retina from which the impulse activity of that neuron can be influenced by light. For a retinal rod or cone the receptive field essentially corresponds to the retinal area occupied by the receptor itself. For visual interneurons such as retinal ganglion cells, in contrast, the receptive field is typically much larger, embracing an area containing many receptors.

Let's examine the receptive fields of two ganglion cells in the retina of a cat. We can record from the ganglion cells with an extracellular microelectrode made of tungsten or stainless-steel wire, etched to a fine tip and insulated with resin except for a few microns at the tip. Such a microelectrode records impulses of neurons in the immediate vicinity. We can map the receptive fields by shining small spots of light on a screen in front of the eye.

Figure 13.16 shows the responses of two retinal ganglion cells to different kinds of light stimulation. The cells have receptive fields divided into two areas: a *center* and a *surround*. The first cell, termed an **on-center cell**, increases its rate of impulse discharge when the center of its receptive field is illuminated by a spot of light (see Figure 13.16*a*). The same spot of light, however, suppresses activity when it is presented in the larger surrounding part of the receptive field. The on-center cell is maximally stimulated when the entire center of its receptive field, but none of its surround, is illuminated. The cell's activity is maximally inhibited or suppressed when the surround, but not the center, is illuminated by an annulus (ring) of light. Such suppression is followed by an increased discharge when the light is turned off (an "off response"). Diffuse light stimulation, covering the entire receptive field of the ganglion cell, has little effect on the cell's activity because the excitatory effect of light at the center and the inhibitory effect of light in the surround are antagonistic, canceling each other out.

The second cell, termed an **off-center cell**, also has a receptive field with a concentric, antagonistic center and surround (see Figure 13.16*b*). The off-center cell, however, is inhibited by light in its center and excited by light in its surround. Its receptive field is thus the converse of the field of the on-center cell.

The two retinal ganglion cells just described are fairly typical of ganglion cells in cat and monkey retinas, which contain roughly equal numbers of the two types. (There are other classifying features of ganglion cell organization, which will be discussed later; see page 366.) Furthermore, other vertebrates, such as rabbits and frogs, have ganglion cells with properties that are more complex than those of cats. Nevertheless, the preceding description of response properties of on-center and off-center ganglion cells is a sufficient starting place for analysis of retinal integration.

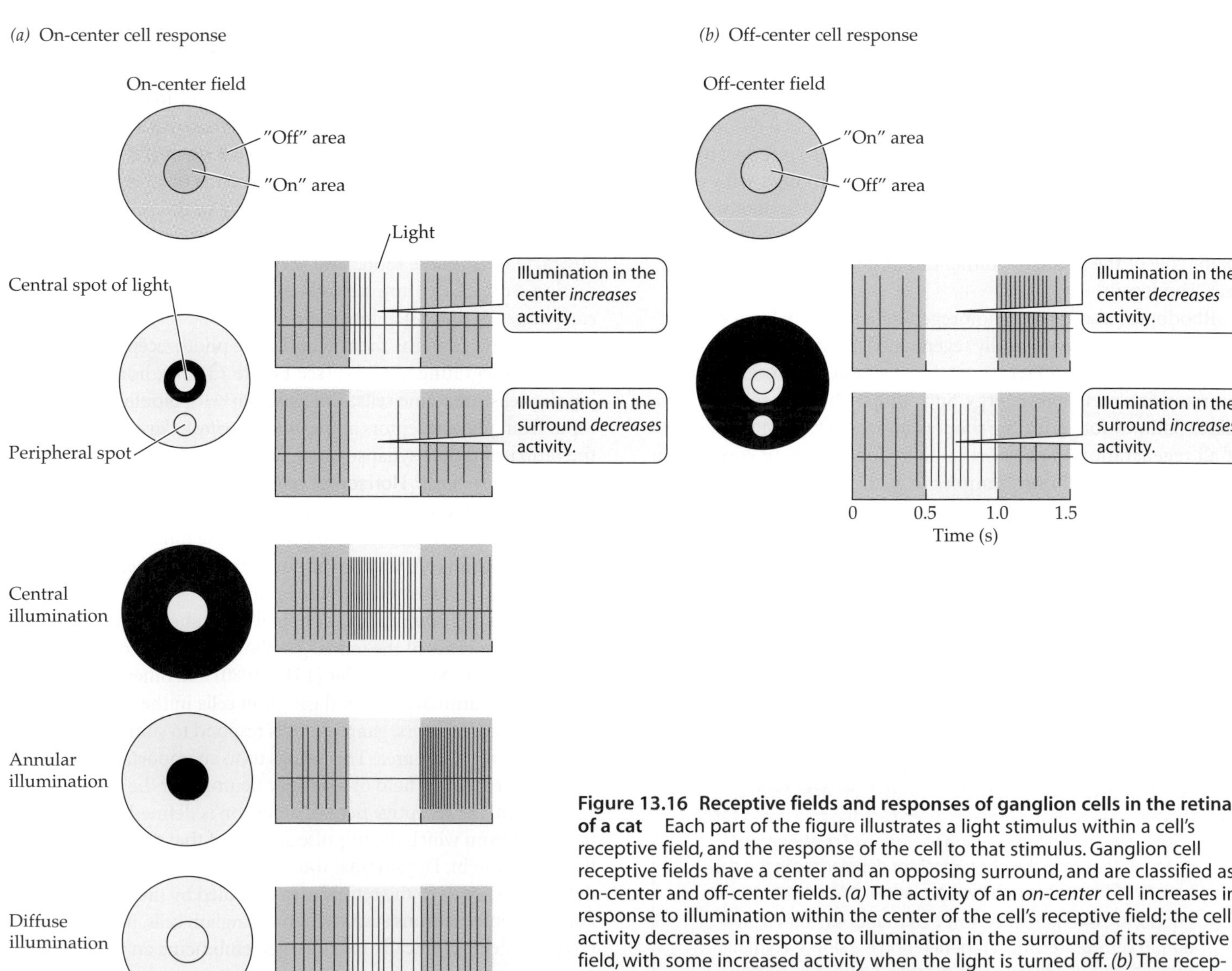

Figure 13.16 Receptive fields and responses of ganglion cells in the retina of a cat Each part of the figure illustrates a light stimulus within a cell's receptive field, and the response of the cell to that stimulus. Ganglion cell receptive fields have a center and an opposing surround, and are classified as on-center and off-center fields. *(a)* The activity of an *on-center* cell increases in response to illumination within the center of the cell's receptive field; the cell's activity decreases in response to illumination in the surround of its receptive field, with some increased activity when the light is turned off. *(b)* The receptive field and the activity of an *off-center* cell are analogous but reversed; light in the center of the receptive field decreases activity of the cell, and light in the surround increases activity. (After Kuffler, Nicholls, and Martin 1984.)

In order to see how the response properties and receptive fields of ganglion cells are derived from neural circuits in the retina, next we need to examine the synaptic connections and properties of the other retinal neurons. Figure 13.17 shows the basic synaptic organization of the vertebrate retina. Rods and cones (the photoreceptor cells) synapse on bipolar cells and horizontal cells. Horizontal cells synapse onto rods and cones, and through them, influence bipolar cells. Bipolar cells synapse on amacrine cells and ganglion cells. Amacrine cells (see Figure 13.11) synapse on ganglion cells. (In some retinas there are additional synapses—e.g., between photoreceptor cells. These will not be considered here.)

We can distinguish between two sorts of retinal pathways: **straight-through pathways** that project radially through the retina at right angles to its surface (photoreceptors → bipolar cells → ganglion cells) (see Figure 13.17*a*); and **lateral pathways** that extend along the retinal sheet, via horizontal cells and amacrine cells (see Figure 13.17*b*). The straight-through pathways give rise to the properties of the *center* of a ganglion cell's receptive field, and the lateral pathways give rise to the properties of the antagonistic *surround.*

To show how these two antagonistic pathways produce the receptive fields of retinal ganglion cells, investigators have recorded responses of retinal neurons to patterned light stimulation, usually in fish, amphibians, and turtles that have relatively large retinal neurons. Figure 13.18*a* illustrates how the straight-through pathway makes up the center of a ganglion cell's receptive field. Light in the center of the receptive field excites cones (or rods), hyperpolarizing them. A cone synapses onto two kinds of retinal bipolar cells: *H-type* bipolar cells in which light in the center of the receptive field hyperpolarizes the cell, and *D-type* bipolar cells in which light in the center of the receptive field depolarizes the cell.

The light stimulus in Figure 13.18*a* thus leads to hyperpolarization of an H-type bipolar cell, which synapses onto an off-center ganglion cell and hyperpolarizes it. Light in the center hyperpolarizes cones to depolarize a D-type bipolar cell, which synapses onto an on-center ganglion cell and depolarizes it. (The ganglion cells generate action potentials when they depolarize; however, rods and cones, bipolar cells, and horizontal cells never generate action potentials.)

Light in the surround of a ganglion cell's receptive field antagonizes these effects, in part by activating horizontal cells of the lateral pathway (Figure 13.18*b*). Stimulated cones in the surround area synapse onto horizontal cells, the lateral processes of which synapse onto cone terminals to oppose the effects of light on the cones (i.e., horizontal cells depolarize cone terminals). Thus, bipolar cells themselves have receptive fields with an antagonistic center and surround. Light in the *surround* of an H-type bipolar cell depolarizes it (leading to depolarization of an off-center ganglion cell), and light in the *surround* of a D-type bipolar cell hyperpolarizes it (leading to hyperpolarization of an off-center ganglion cell).

Horizontal cells do not mediate *all* of the lateral inhibition that produces the surround of a ganglion cell's receptive field. Amacrine cells provide a

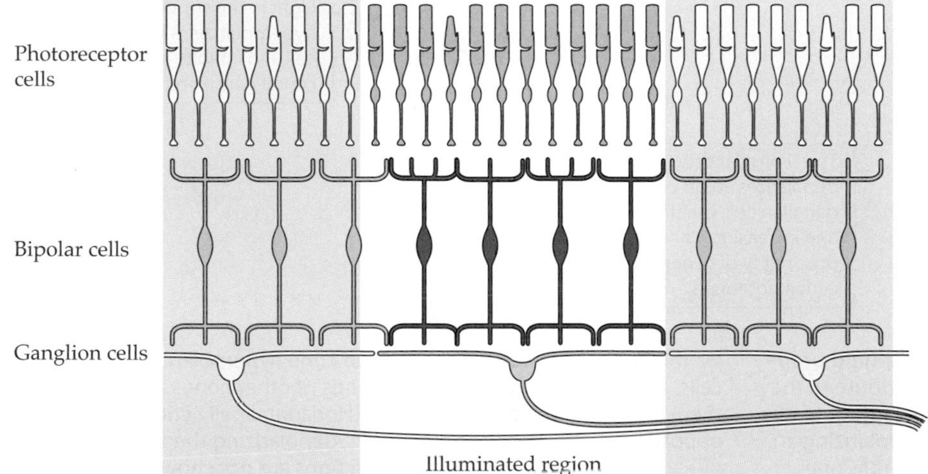

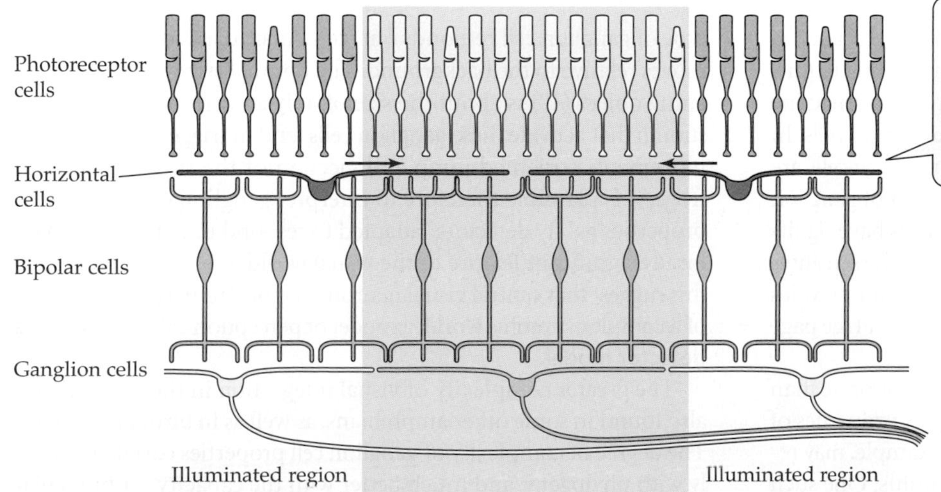

Figure 13.17 The synaptic connections of the retina produce the center–surround, concentric receptive field of retinal ganglion cells *(a)* The radial, straight-through pathways from receptors to bipolar cells to the dendritic tree of a retinal ganglion cell produce the center of the ganglion cell's receptive field. *(b)* Lateral pathways such as the horizontal cell pathways shown produce the surround of the ganglion cell's receptive field. (After Masland 1986.)

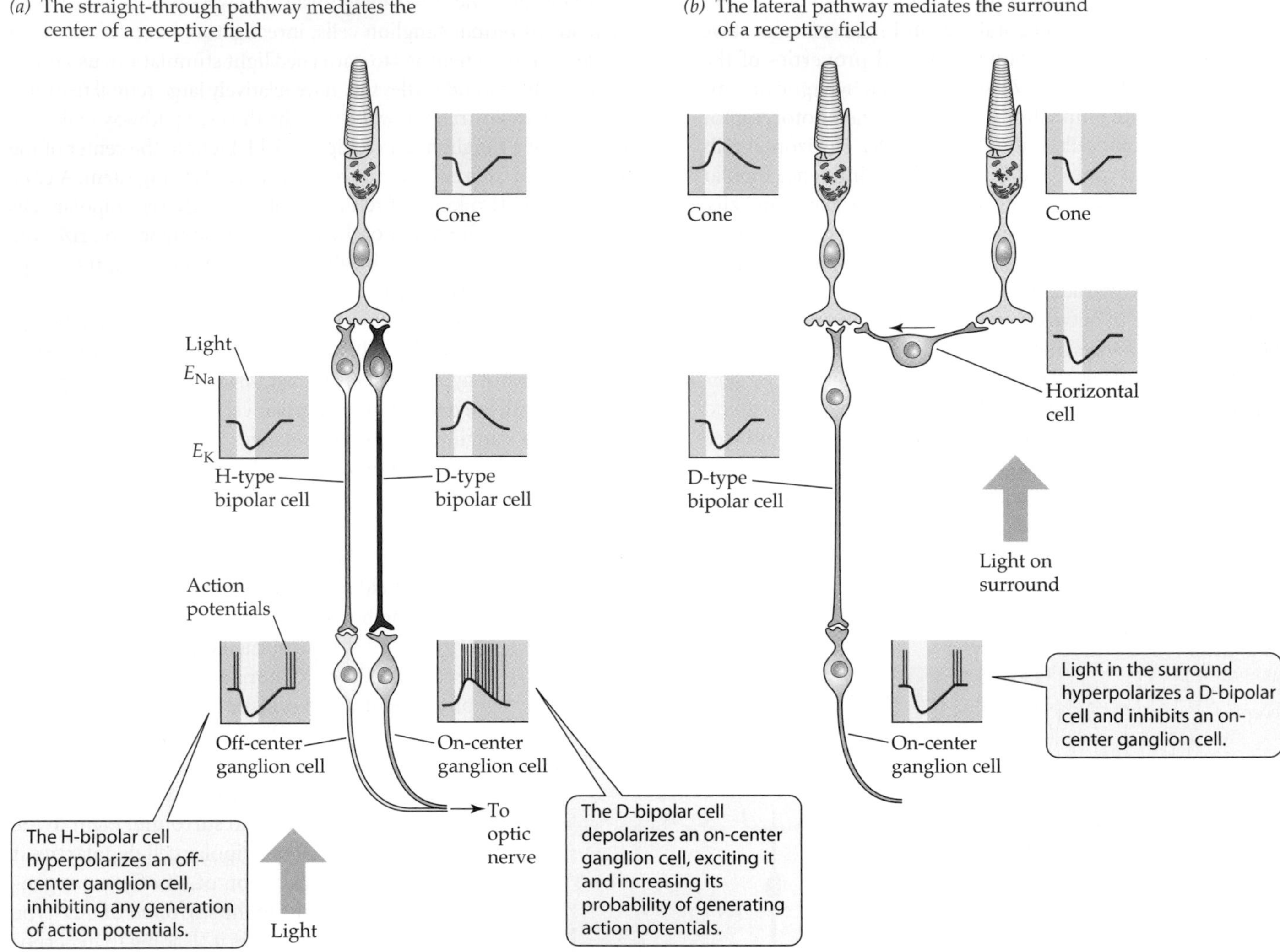

Figure 13.18 Retinal pathways determine the receptive-field properties of retinal ganglion cells (a) Cones and rods contribute to the *center* of a ganglion cell's receptive field via the straight-through pathway. A cone that is excited by light hyperpolarizes, hyperpolarizing an H-type bipolar cell and depolarizing a D-type bipolar cell. *(b)* Cones and rods contribute to the *surround* of a ganglion cell's receptive field via the lateral pathway. An illuminated cone hyperpolarizes horizontal cells, which end on the synaptic endings of other cones and depolarize them (opposing the effects of light). (Horizontal-cell actions also oppose light effects on H-bipolar cells, depolarizing them and exciting off-center ganglion cells; these interactions are not shown.)

second layer of lateral inhibitory interaction in the retina, enhancing the center–surround antagonism and in some cases imparting to ganglion cells a sensitivity to moving light stimuli. The roles of amacrine cells are more diverse and complex than those of horizontal cells. In mammals, horizontal cells are less important and amacrine cells are more important for determining the surround properties of ganglion cell receptive fields than for some lower vertebrates. Birds have significant efferent control of the retina, so the brain modifies the sensitivities and synaptic interactions of retinal cells. Additional complexities and differences among retinas are involved in color vision (see page 367).

Finally, in some species, retinal integration is more elaborate than we have indicated, with correspondingly more complex properties of retinal ganglion cells. The ganglion cells of frogs, for example, may respond only to quite specific features of a visual stimulus. One such type of ganglion cell responds only to a small, dark convex edge that moves relative to the background. The optimal size of a dark stimulating object (1°) is about the size of a fly at striking distance, and stimuli that activate these ganglion cells tend to trigger feeding-strike movements (oriented jump and ejection of the sticky tongue) of frogs. It is plausible therefore to interpret ganglion cells with these properties as "fly detectors" adapted to respond to a specific, behaviorally significant feature of the visual world. Observations such as this suggest that central visual neurons respond to particular features of a complex stimulus world, a model of perception called the **feature detector model.**

The greater complexity of visual integration in the frog retina is also found in some other amphibians, as well as in birds and rabbits. The degree of complexity of ganglion cell properties correlates poorly with phylogeny, and much better with the capacity for binocular vision. Species with laterally placed eyes and limited binocular overlap of the fields of the two eyes (e.g., frogs, rabbits) tend to have ganglion cells with more complex properties. Species with frontally placed eyes and extensive binocular overlap of fields (e.g., cats, primates) tend to have ganglion cells with simpler properties, as in Figure 13.16. Presumably it is adaptive for strongly binocular species to delay complex visual integration, performing it in areas (e.g., visual cortex) that receive binocular input, rather than in the (monocular) retina.

The vertebrate brain integrates visual information through parallel pathways

In the preceding discussion of the retina we have seen that the activities of ganglion cells (the retinal output) convey information about visual *pattern* rather than overall level of illumination. Activity in an on-center ganglion cell signals that a region of visual space (corresponding to the center of its receptive field) is *brighter* than the surround. Corresponding activity in an off-centered ganglion cell indicates that the center of its receptive field is *darker* than the surround. This responsiveness to stimulus pattern or contrast is continued in central visual projections in the brain. We will examine central integration of visual information as an example from which we can derive some general principles of the way in which a brain processes sensory information.

Visual information is conveyed over several different central pathways in the vertebrate brain. Vision is an important sensory component of many kinds of animal behavior, and different aspects or features of the visual world are important for different kinds of behavior. Thus it is perhaps not surprising that different pathways convey information about different aspects of a complex visual stimulus, such as color, fine details of form, and stimulus movements that elicit responsive eye movement.

In fish and amphibians the major visual projection of the optic nerve is to the optic tectum of the midbrain. The optic nerves cross the brain midline at the optic chiasm and connect to neurons in the contralateral (opposite) tectum. In mammals the region homologous to the optic tectum is the superior colliculus. Visual projections to the superior colliculus are important in visually guided movements such as those in prey capture and predator avoidance.

The major visual projection of mammals is the *geniculostriate system*. The axons of retinal ganglion cells that form the optic nerve synapse in a region of the thalamus termed the **lateral geniculate nucleus** (**LGN**). Neurons of the LGN project to the **primary visual cortex** at the posterior end of the cerebrum (Figure 13.19). Unlike the optic projections of lower vertebrates, the projections of most mammals are only partially crossed at the optic chiasm. In cats and primates the projections of the nasal (inner) half of the retina cross to the contralateral side, while those of the temporal (outer) half go to the ipsilateral (same side) LGN. This mixing of input from the two eyes allows mammals with forward-facing eyes to merge binocular input for depth perception at the visual cortex.

David Hubel and Torsten Wiesel contributed greatly to our understanding of the central processing of visual information with a series of studies of the response properties of visual neurons in cats and monkeys. They recorded from visual cells of anaesthetized animals while projecting patterned stimuli (light and dark bars, edges, and spots) on a screen in front of the animal. They found that LGN neurons had response properties similar to those of retinal ganglion cells: They responded to stimuli to only one eye and had concentric receptive fields with an antagonistic center and surround, either on-centered or off-centered. (The lateral geniculate nucleus is a layered structure; each layer receives input from either the contralateral eye or the ipsilateral eye, but the whole nucleus receives input from both eyes.)

The receptive fields of neurons in the visual cortex are quite different from those of lower levels of the visual pathway. We can differentiate two major kinds of neurons in the visual cortex: *simple cells* and *complex cells*. Both types are binocular: They can respond to visual stimuli presented to either eye, although one eye may predominate. Both kinds of cells have a preferred *axis of orientation* of visual stimuli, a term best explained if we describe the receptive fields of a few simple cells in the visual cortex of a cat.

First recall that a receptive field is the area on the retina in which light stimuli influence the activity of a particular neuron. **Simple cells** are characterized in terms of their receptive field properties. Figure 13.20 shows the receptive fields of three simple cells, as they might be mapped with small spots of light. Plus signs (+) denote areas in which a light spot excites a simple cell, increasing the frequency of its discharge of action potentials. Minus signs (–) denote areas in which the frequency of action potentials decreases when the light spot is on, and increases when the spot is turned off ("off response").

Each of the receptive fields in Figure 13.20 has a line drawn through it that defines its axis of orientation. To be most effective, a

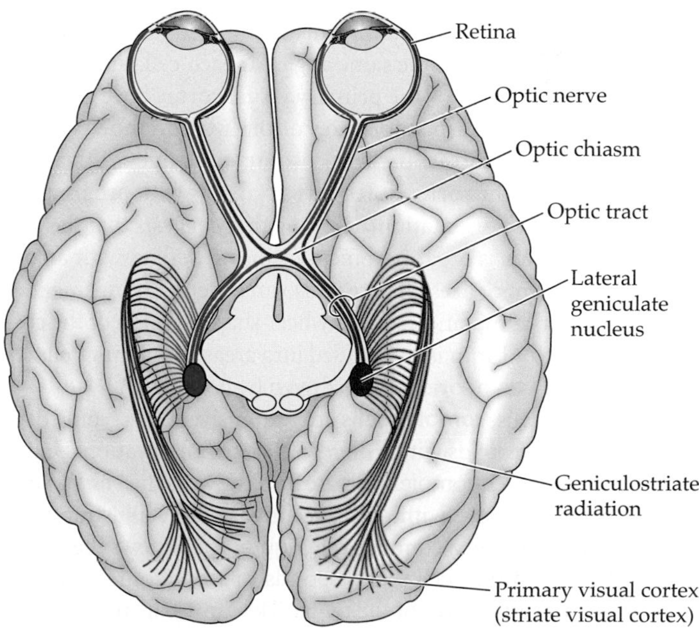

Figure 13.19 Central visual projections of a mammal As this cutaway view shows, only part of the optic tract crosses the midline at the optic chiasm. Therefore, a stimulus in the *left* visual field (the left half of the visible world) projects to the *right* lateral geniculate nucleus (LGN) via both eyes. Conversely, a stimulus in the *right* visual field projects to the *left* LGN and visual cortex.

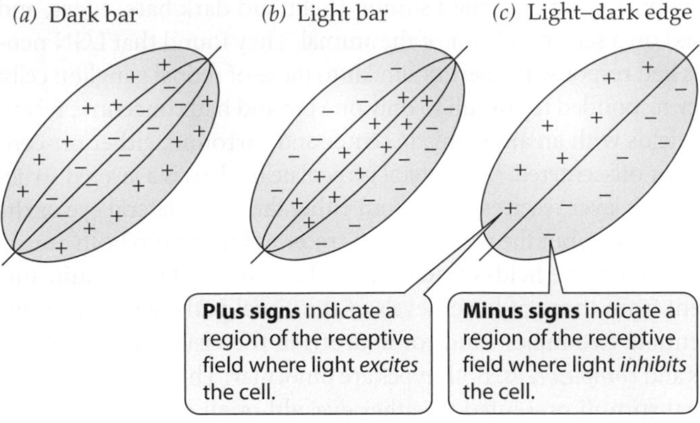

Figure 13.20 Receptive fields of simple cells in the visual cortex of a cat Each receptive field is an area of the retina that affects the activity of a single cortical cell. All three cells have the same axis of stimulus orientation; other cell fields have different axes of orientation. Cortical simple cells typically respond best to a dark bar *(a)*, a light bar *(b)*, or a light–dark edge *(c)*.

stimulus must be aligned with this axis of orientation. The optimal stimuli for these three cells are as follows: a dark bar in the central band of the receptive field; a light bar in the central band of the field; and an edge on the center line that is light on the upper left and dark on the lower right. In all cases the stimulus pattern must be correctly aligned and oriented. In Figure 13.20*a*, for example, a vertical bar will be a weak or ineffective stimulus because it falls on both excitatory and inhibitory areas. For the same reason, visual cortical cells are quite insensitive to changes in overall illumination.

The three receptive fields shown in Figure 13.20 have the same axis of orientation, which is to say that the best edge or bar stimulus in the visual field is at the same angle for each cell. In fact, all of the cells in a small area of the primary visual cortex have a similar axis of orientation. All possible axes of orientation, however, are represented in the cortex. If the recording electrode is advanced sideways through the cortex, it records from cells with progressively changing axes of orientation, a finding that suggests that the primary visual cortex is organized in columns or wedges with an orderly arrangement of axes. (There is also an orderly arrangement of bands of ocular dominance, so a pinwheel-shaped region of the cortex about 1 mm × 1 mm is subdivided into areas containing cells responding to all axes of orientation, driven by both eyes.)

Like simple cells, **complex cells** are responsive to bars and edges and have receptive fields with a preferred axis of orientation. Unlike simple cells, however, complex cells are insensitive to the *position* of a stimulatory bar or edge within the receptive field. As illustration of this defining feature of a complex cell, Figure 13.21 shows the receptive field of a complex cell that responds to a horizontal bar anywhere within a certain area of the retina. The same bar stimulus presented at an angle different from the preferred axis of orientation is much less stimulatory. No part of the receptive field of a complex cell can be defined as excitatory or inhibitory; instead the stimulus *pattern* (a dark horizontal bar) is excitatory. Complex cells then have strict requirements about stimulus form and orientation, but less-strict requirements about stimulus position within the field.

It is not completely clear how the receptive field properties of cortical simple and complex cells result from their synaptic input from other cells, such as LGN neurons. Early studies suggested a *hierarchical* organization in which many similar geniculate cells, the receptive field centers of which were aligned in a row, converged onto a cortical cell. Complex cells were envisioned to receive convergent synaptic input from several simple cells, and in turn to synapse upon still higher-order cells. Aspects of this hierarchical organization are no doubt correct, but complex cells have been shown to receive some synaptic input directly from geniculate axons.

These and other studies suggest a *parallel* organization of central visual projections. We mentioned in passing in the last section that there are three types of retinal ganglion cells that differ in size and in response properties. In the cat, these have been termed *W*-, *X*-, and *Y*-cells. X-cells are medium-sized and more responsive to fine detail of stationary visual patterns; Y-cells are large, more rapidly conducting, and more responsive to stimulus changes and movements. (W-cells are small and have complex and varied receptive-field properties that do not concern us here.) Primates have a similar division of pathways: *magnocellular* (large, Y-type), *parvocellular* (smaller, X-type), and *koniocellular* (very small and miscellaneous). The separation of X- and Y-type pathways is preserved in the lateral geniculate nucleus, and studies indicate that simple cells receive primarily X-type input, whereas Y-type input predominates for complex cells. Central visual pathways thus have elements of both parallel and hierarchical organization, the details of which are complicated and somewhat controversial.

The ways in which CNS neurons function in visual perception are also not clear. Simple and complex cells can be envisioned as *feature detectors*, responsive to line segments and contrast edges at particular orientations. We might see the world as a sort of line drawing composed of the activities of many such feature detectors. The response properties of known cortical cells, however, are also consistent with other theories of visual perception. Thus, at this point there remains a gulf between our understanding of the physiology of visual cells and our questions about mechanisms of visual perception.

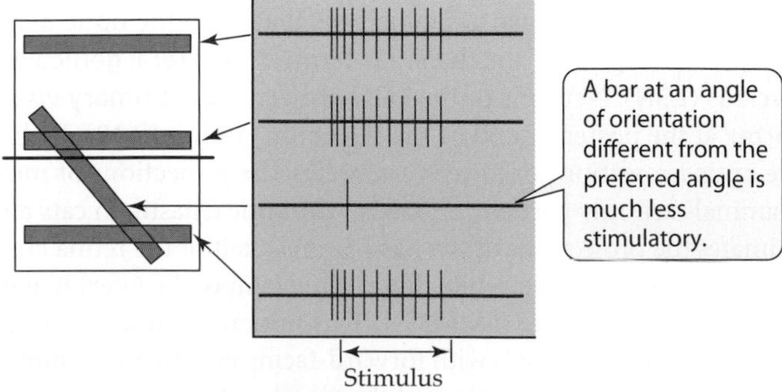

Figure 13.21 The receptive field of a complex cell in the visual cortex Complex cells respond best to a bar (or edge) at the correct angle of orientation, which in this case is horizontal. Unlike simple cells, however, the stimulatory bar may be anywhere within the receptive field.

COLOR VISION The ability to distinguish colors depends on differential sensitivities of photopigments to different wavelengths of light. Although many animals are color-blind, many other animals with well-developed diurnal visual systems have evolved color vision. Examples include several orders of insects, teleost fish, frogs, turtles, lizards, birds, and primates.

Theories of color vision are strongly based on human perceptual studies. In 1801, Thomas Young (1773–1829) proposed that human color perception was based on separate receptor classes sensitive to red, green, and blue light. This theory (termed the *trichromaticity theory*) was supported by perceptual observations that any color could be duplicated by a mixture of three primary colors.

The physiological basis of the trichromaticity theory has been clarified in the last 35 years. Humans and other primates have three populations of cone photoreceptors, which are sensitive to different wavelengths of light. Figure 13.22 shows the spectral characteristics of the three classes of primate cones, which we can determine by measuring, at different wavelengths, either light absorption (absorption spectrum) or effectiveness of stimulation (action spectrum). The three types of cones are termed *red cones*, *green cones*, and *blue cones*, loosely approximating the color of light to which they are most sensitive. The spectral sensitivities of the three types of cones are rather broad and overlapping, so the perception of color must be based on the ratio of excitation of different cone populations. For example, we might expect perception of long-wavelength red light to depend on an analyzer that is excited by red cones but inhibited by green cones. Just this sort of integration occurs in the retina.

Two kinds of *color opponent* processes occur in the mammalian retina: red–green opponency and blue–yellow opponency. These opponent processes explain why we can perceive a color as bluish green or as reddish yellow (orange), but we cannot perceive a color as reddish green or as bluish yellow. Many of the ganglion cells of a primate retina that are stimulated by cones have color-opponent properties. (Cones and rods stimulate different populations of cells in the retina.) Red–green opponent ganglion cells, for example, have concentric antagonistic receptive fields but are also color-opponent. One cell may be inhibited by red light in its receptive-field center and excited by green light in its surround, another may be excited by red light in the center and inhibited by green in the surround, and so on. Other classes of ganglion cells may be excited by blue light and inhibited by yellow (a sum of red and green cone input). Such cells lack center–surround antagonism.

As we might expect, central processing of color information becomes complicated at the level of the visual cortex. Color information appears to be integrated by clusters of cells in separate districts in the cortex, apart from other cells that are not color-selective but instead process information about brightness contrasts. This apparent segregation of color channels and so-called achromatic channels is another example of parallel organization in the mammalian visual system.

Mechanisms for color vision in animals other than mammals are always based on several populations of receptor cells with different spectral sensitivities, although these receptors are not always cones. Fish such as carp have three populations of cones analogous to those of primates, with absorption maxima of 455 nm, 530 nm, and 625 nm, respectively. In contrast, frogs have two or more classes of rod photopigments; their color vision involves input from both rods and cones. Although evidence for color vision is lacking in most invertebrates, insects have well-developed mechanisms for color vision, including receptors sensitive to ultraviolet radiation that is invisible to vertebrates. The honeybee, for example, has photoreceptors with spectral maxima of 350 nm, 450 nm, and 550 nm, respectively, and readily distinguishes both ultraviolet radiation and colors of visible light.

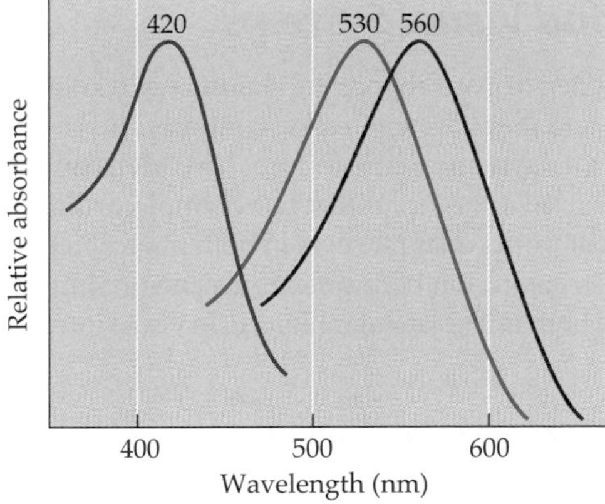

Figure 13.22 Spectral sensitivities of human retinal cones Spectral sensitivities were determined by measurement of the absorption spectra of single cones. The three types of cones have photopigments with different absorption maxima: 420 nm (blue cones), 530 nm (green cones), and 560 nm (red cones).

Summary: Visual Sensory Processing

- Neural circuits of the vertebrate retina integrate the responses of retinal photoreceptors to excite and inhibit retinal ganglion cells. Ganglion cell receptive fields may be excited or inhibited by light at the center of the field, while light in the surround antagonizes the effect of light in the center.
- Straight-through pathways (photoreceptor→bipolar cell→ganglion cell) produce the center (on- or off-center) of a ganglion cell's receptive field. Lateral pathways through horizontal cells and amacrine cells produce the antagonistic surround.
- Axons of ganglion cells make up the optic nerve, relaying visual information to several brain areas. The geniculostriate pathway projects to the lateral geniculate nucleus (LGN) and from there to primary visual cortex.
- Simple and complex cells in the primary visual cortex respond to light or dark bars or edges oriented at particular angles.
- Parallel pathways in the visual cortex convey information about different aspects of a visual stimulus, such as details of visual form, movement, color, and binocular determination of object distance.
- Color vision depends on the ratio of activation of three classes of cone photoreceptors sensitive to different wavelengths of light. Retinal circuitry integrates color contrasts based on red–green and blue–yellow opponencies.

Arthropod Visual Systems

The visual systems of arthropods are almost as well known as those of vertebrates, and they have interesting contrasts and similarities with vertebrate visual systems. Some features have been conserved; others may have evolved quite separately. For example, arthropod eyes are quite different from vertebrate eyes in both photoreceptor structure and optical structure, but both systems depend on similar rhodopsin changes and both utilize lateral inhibition in visual integration.

Arthropod eyes are compound eyes

Compound eyes are usually associated with the phylum Arthropoda,[4] although they also occur in a few annelids and molluscs and are absent in some crustaceans. As noted earlier (see page 356), each ommatidium has its own lens system and cluster of **retinular** cells (photoreceptors) (Figure 13.23*a*). The eight or more retinular cells of an ommatidium are arranged in a circle, like sections of an orange but more elongate. The sides of the ommatidium are lined with pigment cells, which are capable of shielding against the leakage of light between ommatidia. The light-sensitive photopigment of a retinular cell is localized in the membranes of microvilli.

Typically, all the microvilli of a retinular cell are arranged along one edge of the retinular cell, in an array termed a **rhabdomere**. In most eyes the rhabdomeres of all eight or more retinular cells are tightly packed along the central axis of an ommatidium (see Figure 13.23*a*). The resulting elongated zone of photosensitive microvilli is termed the **rhabdom** of the ommatidium. The rhabdom thus corresponds to the aggregated rhabdomeres of all the retinular cells of the ommatidium. The eyes of dipterans (flies) and hemipterans (true bugs) are unusual in that the rhabdomeres of an ommatidium remain separate rather than being joined into a single central rhabdom.

[4]Many arthropods also possess simple camera eyes called *ocelli* (singular *ocellus*).

In terms of how the eye forms images, there are two types of compound eyes: apposition eyes and superposition eyes. In **apposition eyes**, each ommatidium is optically isolated from other ommatidia and receives light from a narrow region of the visual field (Figure 13.23*b*). Light from outside this narrow *acceptance angle* is absorbed by lateral pigment cells and does not reach the rhabdom. In apposition eyes the distal end of the rhabdom is directly adjacent to the inner end of the crystalline lens.

In **superposition eyes**, in contrast, the rhabdoms are separated from the crystalline lenses by a wide optically clear zone. Because light can pass between ommatidia in this zone, it can reach a given rhabdom from many ommatidia. The lenses of superposition eyes are thick cylinders that refract eccentric light beams back at the same angle at which they enter the lens. Figure 13.23*c* shows the resultant convergence of light onto a rhabdom from many ommatidial lenses. Crayfish and lobsters achieve the same effect by prismatic reflection of light by the sides of squared crystalline lenses.

What is the functional significance of a superposition eye? It certainly does not allow visual acuity that is as great as that of an apposition eye (which in turn lacks the acuity of a camera eye such as that of a vertebrate). The adaptive value of a superposition eye is its great increase in light-gathering power. Light striking much of the eye surface can be brought to focus on a rhabdom. In contrast, in an apposition eye only the light falling on a single ommatidial lens is focused on the rhabdom. Superposition eyes therefore provide increased sensitivity to dim light (at some cost in resolving power), rather like opening the aperture of a camera lens.

Most arthropods that have superposition eyes are nocturnal (e.g., fireflies and moths), and most of those that have apposition eyes are diurnal (e.g., brachyuran crabs, diurnal insects). In superposition eyes,

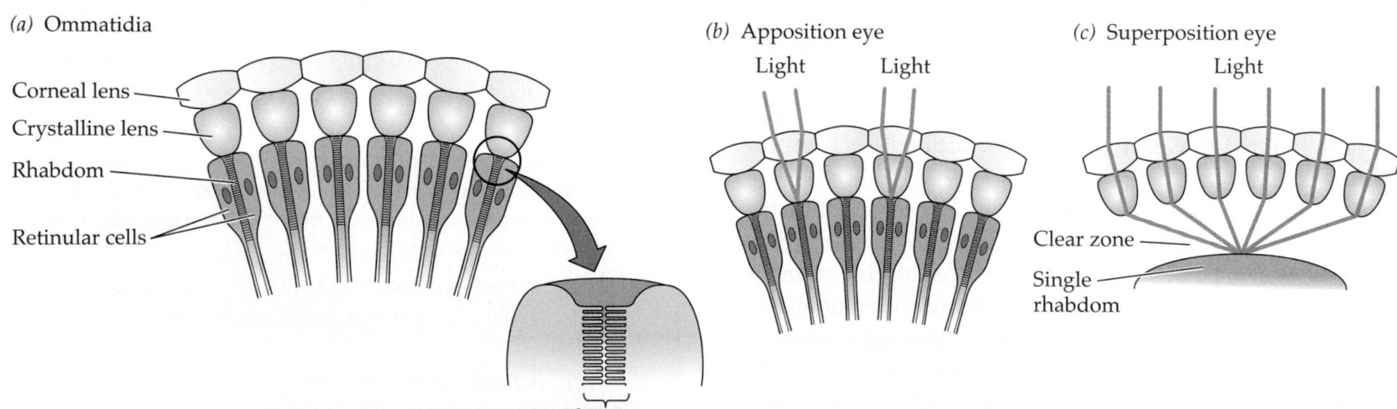

Figure 13.23 Arthropod compound eyes *(a)* A compound eye is made up of many *ommatidia*, or faceted optical elements. Each ommatidium has a corneal lens and a crystalline lens, which focus light from a narrow acceptance angle onto photosensitive microvilli of sensory retinular cells. The closely packed microvilli are collectively termed the *rhabdom*. *(b, c)* Compound eyes are of two types. In an apposition eye *(b)*, each ommatidium focuses light in its optical axis onto the rhabdom. Light not in the optical axis misses the rhabdom and is absorbed by screening pigment. In a dark-adapted superposition eye *(c)*, parallel light rays are focused by many ommatidial lenses onto a single rhabdom, increasing the light-gathering power. When the same eye is light-adapted (not shown), migrating screening pigment prevents light passage between ommatidia. Therefore, light-adapted superposition eyes function similarly to apposition eyes.

the degree of light passage between ommatidia is controlled by light-induced pigment migration. In the dark-adapted eye (e.g., at night), pigment is withdrawn from the clear zone (see Figure 13.23*c*), allowing light passage and image superposition. In the light-adapted state, pigment migrates between the ommatidia into the clear zone and blocks light passage, so image formation becomes similar to that of an apposition eye. Thus, superposition eyes function as such only when dark-adapted, under conditions where their increased light-gathering power is important.

Arthropod photoreceptors are depolarized by light

Arthropod and vertebrate photoreceptors have interesting similarities and differences in their transduction mechanisms. Both use similar rhodopsins to absorb light, and both employ G protein–coupled second-messenger cascades to amplify the effects of photoisomerization. The second-messenger mechanisms must differ, however, because arthropod photoreceptors increase permeability to generate a depolarizing receptor potential, and vertebrate photoreceptors decrease permeability to generate a hyperpolarizing receptor potential (see Figure 13.15). The receptor potential of arthropod photoreceptors results primarily from a permeability increase to Na^+ ions, but also to Ca^{2+} ions. The mechanism by which activated rhodopsin leads to the depolarizing receptor potential in arthropods is not completely clear. Genetic studies in *Drosophila* implicate IP_3 as the second messenger. Rhodopsin, via a G protein, appears to activate a phospholipase to form IP_3 (inositol trisphosphate) and DAG (diacylglycerol) from PIP_2 (phosphatidylinositol bisphosphate), as in some indirect synaptic actions (see Figure 12.21). IP_3 is thought to release Ca^{2+} then from intracellular stores, and the increased cytoplasmic $[Ca^{2+}]$ somehow opens cation channels to produce the receptor potential.

Mechanoreception

At the beginning of this chapter we considered the mechanisms of action of two mechanoreceptors: the crustacean stretch receptor and the mammalian Pacinian corpuscle. In this section we take a more functional view of mechanoreceptors. We examine selected examples of the ways in which mechanoreceptors are used to provide different sorts of functionally significant information to the animal. The examples we present are proprioceptors, receptors for equilibrium, and receptors for hearing.

Proprioceptors monitor the spatial relationship of the body

At the beginning of the chapter we drew the distinction between exteroceptors (responding to stimuli outside the body) and interoceptors (responding to internal stimuli). This distinction applies to many stimulus modalities; for example, mammals contain internal carotid body chemoreceptors sensitive to the partial pressure of oxygen in the blood, as well as external chemoreceptors. Nowhere else, however, is the distinction as useful as for mechanoreceptors.

All animals have exteroceptive mechanoreceptors sensitive to touch, pressure, water flow, or other surface stimulation. Probably all animals also have interoceptive mechanoreceptors that monitor movements, position, stresses and tensions of parts of the body. Such internal mechanoreceptors are termed **proprioceptors** (from Latin the *proprius*, which means "one's own"). Strictly speaking, proprioceptors are mechanoreceptors associated with the musculoskeletal system. They provide most of the information about muscle contraction, position, and movement of parts of the body, although other receptors, such as skin mechanoreceptors, may make a secondary contribution.

Proprioceptors are particularly important (and well studied) in animals with articulated joints and rigid skeletons—that is, arthropods and vertebrates. There has been considerable parallel evolution of proprioception in arthropods and vertebrates, sense organs quite different in structure having evolved independently in different groups to provide the same kinds of sensory information. For example, tetrapod vertebrates sense position of limbs with muscle receptors and joint receptors, and arthropods may use internal elastic sensory strands termed *chordatonal organs*, campaniform (bell-shaped) sensilla that detect cuticular strain, or arrays of sensory hairs or bristles (termed *hair plates*) that are stimulated by contact with the adjacent joint segment.

The best-known proprioceptive organ is the vertebrate **muscle spindle** organ that monitors the length of a skeletal muscle. Most skeletal muscles of vertebrates contain small groups of such organs embedded among the contractile muscle fibers, which are called **extrafusal fibers.** One component of each organ is a specialized muscle fiber termed an **intrafusal fiber.** The ends of the fiber are contractile, but the central sensory portion is noncontractile (Figure 13.24*a*). When an intrafusal fiber is stretched by elongation of the muscle, it activates sensory neurons, the endings of which are associated with the central sensory portion of the fiber. Figure 13.24*b* diagrams the response of a muscle-spindle sensory axon to stretch imposed on the muscle. Muscle spindles mediate the familiar stretch reflex (knee-jerk reflex), which is discussed in Chapter 18 (see page 492).

The functional organization of a muscle spindle is similar in several important respects to that of the crustacean abdominal stretch receptor organ (muscle receptor organ, MRO) discussed early in this chapter (see page 350, 354). The three most important similarities are as follows:

1. Both are proprioceptive organs arranged *in parallel* to the force-producing muscle fibers. Like the receptor muscle of the MRO, intrafusal fibers of muscle spindles are too few and weak to generate appreciable tension themselves. Because they are located in parallel to the tension-producing extrafusal fibers, muscle spindles do not sense muscle tension. Instead, their activity is related to *muscle length.* Contraction of extrafusal muscle fibers tends to decrease spindle sensory activity, by shortening the muscle.
2. In both sense organs there is a distinction between *rapidly adapting* and *slowly adapting* response properties. For the crustacean MRO, separate rapidly adapting and slowly adapting receptor neurons differ markedly in the rate of adaptation of their response. Mammalian muscle spindles also have two kinds of sensory neurons, termed *primary* and *secondary endings.* The smaller secondary endings have slowly adapting responses that provide information about muscle length but are rather insensitive to movement. The larger primary endings signal muscle

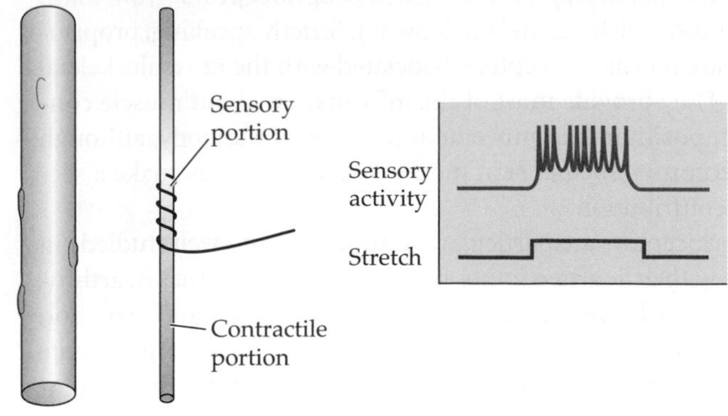

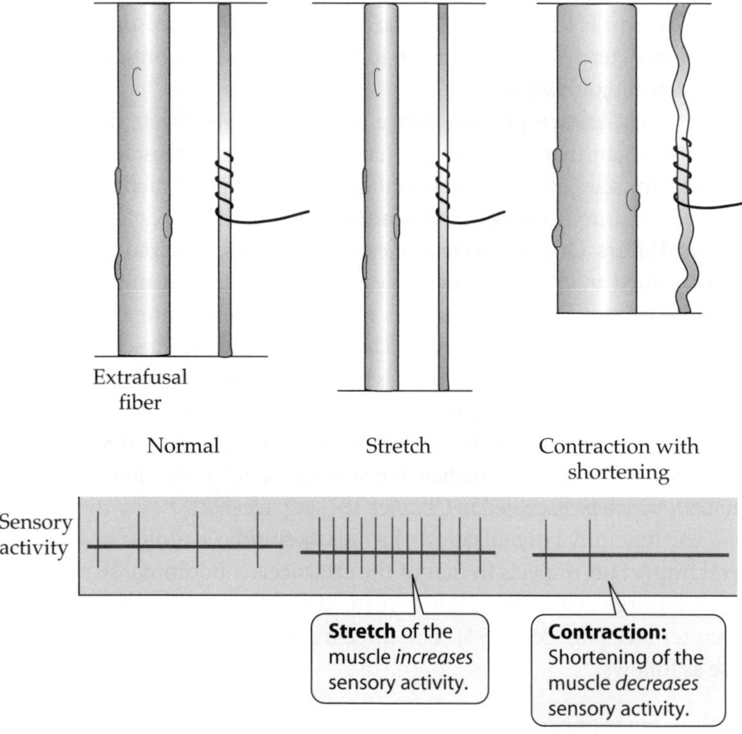

Figure 13.24 Vertebrate muscle-spindle stretch receptors
(a) Skeletal muscles contain muscle spindles, stretch receptors located in the central region of specialized intrafusal muscle fibers. Each muscle contains relatively few intrafusal fibers, as well as many tension-producing extrafusal fibers. Stretch of the intrafusal fiber produces a depolarizing receptor potential and a train of action potentials in the afferent sensory neuron of the muscle spindle. The cell bodies of the sensory neurons are in the dorsal root ganglia.
(b) Muscle spindles are sensitive to changes in muscle *length*.

length but also have a rapidly adapting component of their response that is particularly sensitive to *changes* in length. Their activity thus provides information about both muscle length (position) and rate of change in length (velocity of movement).

3. Both the crustacean MRO and the vertebrate muscle spindle are under efferent control of the central nervous system. The receptor muscle of the crustacean stretch receptor contracts when its motor neuron is stimulated, stretching the dendrites of the receptor cell and evoking a train of impulses (see Figure 13.7). Similarly, the intrafusal fibers of muscle spindles receive motor innervation; in birds and mammals, a separate class of small motor neurons termed *gamma* (γ) *motor neurons* supplies this innervation.

There are also several differences between muscle spindles and crustacean stretch receptors, although they are not as important as the similarities already listed: (1) Each muscle of a vertebrate typically has many muscle spindles, organized with considerable anatomical complexity. In contrast, each abdominal segment of a crayfish has a single pair of stretch receptor neurons on each side. (2) Crustacean stretch receptor neurons are fairly easy targets for intracellular recording, a technique that has not yet been possible for the fine sensory endings of muscle spindles. For these two reasons, we used the crustacean stretch receptor as the example to illustrate sensory responses at the beginning of the chapter. (3) Crustacean stretch receptors receive efferent inhibitory control (see Figure 13.8). In contrast, there is no inhibition of vertebrate muscle spindles.

The functional roles of muscle spindles and crustacean stretch receptors, along with their respective forms of efferent control, are treated in Chapter 18. Vertebrates have several other sorts of proprioceptors that monitor aspects other than muscle length. Two examples are *Golgi tendon organs*, which are embedded in tendons in series with working muscle fibers and monitor muscle tension, and *joint receptors* that are embedded in connective tissue at a joint and respond over a certain range of joint positions or movements.

A final note on proprioceptors is that in several cases in crustaceans, proprioceptors send information to the central nervous system by passive spread of a receptor potential, without the generation of nerve impulses. In these cases (thoracocoxal receptors at the bases of the legs of crabs, tail-fan receptors, gill-bailer receptors, and swimmeret receptors), the receptive endings are relatively close to the CNS and have large-diameter axons that increase the membrane length constant.

Equilibrium receptors detect gravity and acceleration to maintain balance

All animals tend to maintain an equilibrium of their body position with respect to gravity. This orientation to gravity may involve several sources of sensory information, including vision, proprioceptive and tactile receptors, and specialized organs of balance or equilibrium. Equilibrium receptors are thought to have been important in the early stages of the evolution of nervous systems. Sense organs for equilibrium reception appear to be particularly important in aquatic swimmers of many phyla, such as pelagic crustaceans and fish. These animals must orient in a three-dimensional world in which they are nearly neutrally buoyant and thus lack information about gravity from other sensory sources.

The most common and widespread organ of equilibrium sense is the **statocyst**, some version of which is present in members of all metazoan phyla. Many groups, including insects, however, lack statocysts. The essential feature of a statocyst is that a dense, mineral

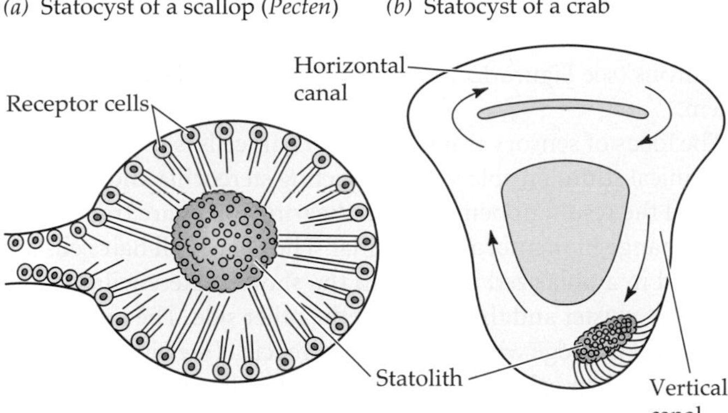

Figure 13.25 Invertebrate statocysts have different degrees of complexity *(a)* The statocyst of *Pecten* (a mollusc) appears as a relatively simple gravity detector. *(b)* The statocyst of a crab has canals that may detect angular acceleration. Arrows show fluid movement through canals.

structure termed a **statolith** contacts and stimulates mechanosensory cells that surround it (Figure 13.25). The statolith, which is usually composed of sand or of a calcareous concretion, has a greater specific gravity than the surrounding fluid. As a result, gravitational force presses the statolith against the underlying receptors. Statocyst receptor cells have hairlike projections often derived from cilia, which transduce bending or shearing forces into a depolarizing generator potential.

Statocysts provide unambiguous information about gravity only if the animal is stationary. Whenever the animal moves or is moved, the statolith will be pressed against receptors by a force proportional to acceleration. A simple statocyst of the sort shown in Figure 13.25*a* provides information about both gravity and acceleration, but these two kinds of information are mixed. One way to resolve this ambiguity in the response of a statocyst is to have a compound organ in which one part detects only acceleration. The statocyst of lobsters and crabs, located at the base of the antennule, is one such compound organ (Figure 13.25*b*). The statocyst walls are convoluted to form a horizontal canal and a vertical canal, within which fluid moves when the animal is rotated. The structure of these canals is analogous to the semicircular canals of the vertebrate vestibular organ (see the next section). Hairlike sensory cells (termed *thread hairs* and *free hook hairs*) are displaced by the moving fluid and signal rotational acceleration. The statolith and its associated receptor cells are located at the bottom of the statocyst. Statolith receptors do not adapt to constant stimulation, unlike thread hairs, which adapt with constant displacement and thus signal acceleration.

VERTEBRATE VESTIBULAR ORGANS Vertebrate organs of equilibrium are part of a sensory system termed the *acoustico-lateralis system*. The sensory cells within the acoustico-lateralis system are **hair cells** derived from the epithelium. Vertebrate hair cells are distinct from arthropod sensory bristles ("sense hairs") and of course are also different from mammalian hair. A vertebrate hair cell has a tuft of stereocilia at its apical end, resembling microscopic hairs. Bending of the stereocilia is transduced into a receptor potential. The acoustico-lateralis system includes the vestibular organs of equilibrium, the lateral line system of surface receptors in fish and amphibians, and the mammalian cochlea, an auditory organ that we will discuss shortly. The vestibular organs and cochlea together are termed the *membranous labyrinth*, which develops from the anterior end of the lateral line system. In different parts of the acoustico-lateralis system, hair cells are stimulated by mechanical stimuli derived from different sources: gravity, acceleration, water currents, and sound waves.

The basic structure of a vestibular organ consists of two parts: an **otolith organ** that functions as a gravity detector, and **semicircular canals** that detect angular acceleration. The otolith organ is divided into two bony chambers: the saccule and the utricle. The utricle connects to the semicircular canals. Sensory hair cells are located in discrete patches in the saccule, utricle, and enlargements at the ends of the semicircular canals termed *ampullae* (singular *ampulla*) (Figure 13.26). Groups of

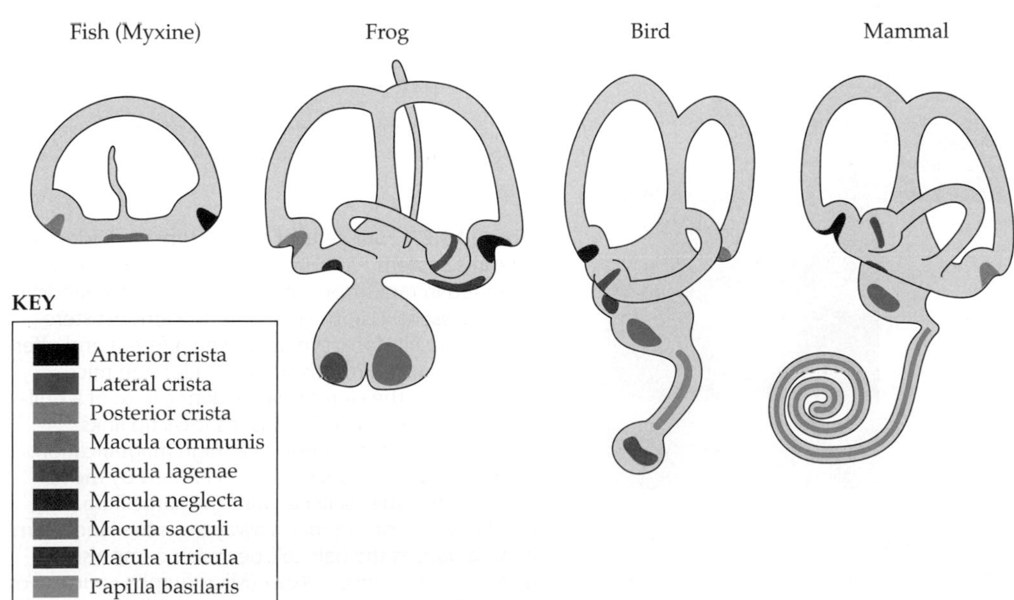

Figure 13.26 Evolution of the vertebrate labyrinth, as suggested by the labyrinths of living vertebrates A mammalian labyrinth contains a saccule and a utricle (which are responsive to gravity and linear acceleration), semicircular canals (responsive to angular acceleration), and a cochlea, the sound-sensitive hearing organ. Colored areas contain different populations of sensory hair cells.

hair cells of the saccule and utricle, termed *maculae* (singular *macula*), are attached to overlying otoliths by a gelatinous matrix. The sensory hairs of the ampullar receptors (*cristae*, singular *crista*) are embedded in a gelatinous flap termed a **cupula** (plural *cupulae*).

As the head is accelerated in one of three rotational planes, the endolymph fluid in one of the three semicircular planes lags behind the motion by static inertia, like water in a rotated glass. The movement of endolymph fluid relative to the cupula displaces the cupula, stimulating the associated hair cells.

TRANSDUCTION IN VERTEBRATE HAIR CELLS The hair cells of the vertebrate acoustico-lateralis system are all thought to work in basically the same way, their different sensory functions depending on whether sound, gravity, acceleration, or water current leads to their mechanical stimulation. We discuss transduction in hair cells here because it was first clarified for vestibular hair cells, but the mechanism is equally applicable to auditory hair cells. Hair cells are so named because they bear projecting cilia (looking like microscopic hairs) which impart to the cells exquisite sensitivity to minute mechanical stimuli (Figure 13.27).

For each hair cell there is a single **kinocilium** (plural *kinocilia*) and 20 or more **stereocilia** (singular *stereocilium*). The kinocilium is a true cilium, with the 9+2 arrangement of microtubules characteristic of motile cilia. Stereocilia are not true cilia; rather they are microvilli that contain actin microfilaments, and they are developmentally unrelated to the kinocilium. Unlike invertebrate mechanoreceptors, hair cells lack axons; instead they synaptically excite axons of sensory neurons (see Figure 13.27*a*) that project to the central nervous system.

The locus of sensory transduction in hair cells is the stereocilia. Mechanical stimuli displace the projecting stereocilia and kinocilium, and the resultant bending and shearing forces are transduced into a change in membrane potential. The cilia of all hair cells are arranged in a polarized array, with the shortest stereocilia at one side of the cluster and the longest at the other side. The kinocilium is located at the edge with the longest stereocilia (see Figure 13.27*a*), but auditory hair cells lack kinocilia, and the kinocilium is not necessary for mechanotransduction in other hair cells.

Displacement of the stereocilia in opposite directions produces opposite voltage effects (see Figure 13.27*c*). Displacement from the stereocilia toward the kinocilium depolarizes the hair cell and increases action-potential frequency in the sensory axon with which it synapses. Displacement from the kinocilium toward the stereocilia hyperpolarizes the cell and decreases sensory neuron activity.

Stereocilia are joined to their neighbors by filamentous *tip links* (see Figure 13.27*b*), and treatments that break the tip links disrupt transduction. Displacement of the stereocilia in the positive direction stretches the tip links and opens cation channels near the outer ends of the stereocilia, depolarizing the cell. Channels open within microseconds of displacement, and the receptor potentials of audi-

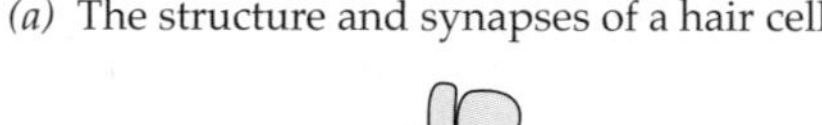

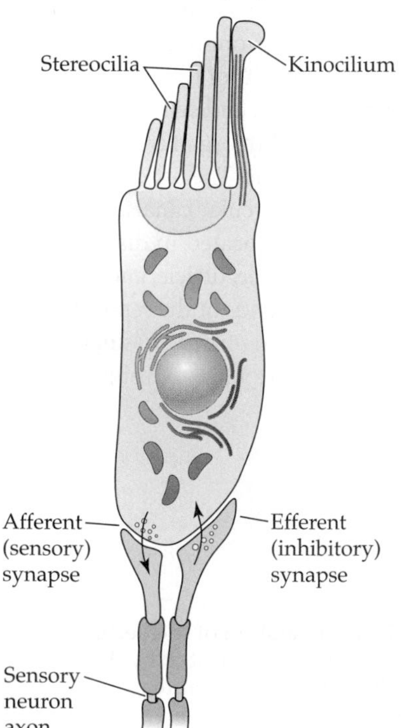

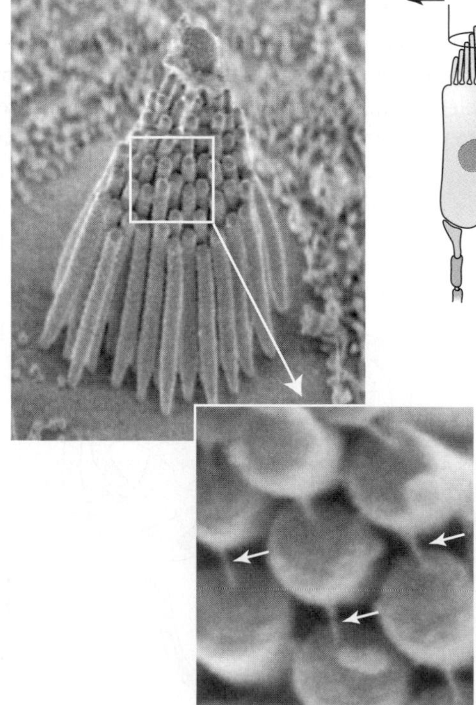

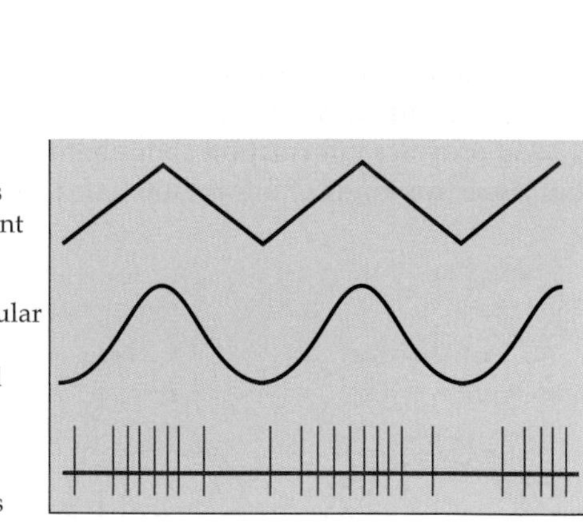

Figure 13.27 Hair cells of the vertebrate acoustico-lateralis system *(a)* Hair cells contain stereocilia arranged in ranks from shortest to longest. A kinocilium (if present) is at the end with the longest stereocilia. Hair cells form synapses with afferent and efferent nerve axons. *(b)* This scanning electron micrograph shows the stereocilia and kinocilium of a bullfrog saccular hair cell. The inset shows tip links between adjacent stereocilia at high magnification. *(c)* Hair cells are depolarized and excited by movements of the stereocilia toward the kinocilium. Displacement of stereocilia away from the kinocilium hyperpolarizes the hair cell, decreasing impulse frequency in the sensory axon. (Micrographs courtesy of Peter Gillespie; from Strassmaier and Gillespie 2002.)

tory hair cells can follow sound frequencies of several kilohertz. Because of this extraordinarily fast response, investigators believe that the forces exerted by attached tip links directly gate the channel opening. The channels themselves are nonselective cation channels, and the depolarizing receptor potential results largely from K^+ entry through the open channels.[5]

Auditory receptors transduce sound into electrical signals

One important use of mechanoreceptors, especially by vertebrates and insects, is for sound reception. Such receptors are termed **auditory receptors** or *phonoreceptors*, and the sense of hearing is termed **audition**. Sound consists of waves of compression of air or water, which propagate away from a vibrating source. Before considering the reception of sound, we need to consider briefly the physical characteristics of sound itself.

Suppose we have a loudspeaker, the cone or membrane of which is vibrating in and out. When the membrane pushes out, it compresses air molecules, momentarily increasing the air pressure. When the membrane vibrates back into the speaker, there is a momentary rarefaction of air molecules (they occupy more volume) and the pressure decreases. Therefore the vibrating speaker cone causes a repeating pressure wave in the air, the frequency of which is identical to the frequency of movement of the speaker cone. This wave of sound pressure propagates away from the speaker cone at the speed of sound. Because it is propagated, a sound-pressure wave has a wavelength as well as a frequency.

VERTEBRATE AUDITORY RECEPTORS ARE EXCITED BY SOUNDS OF DIFFERENT FREQUENCIES The auditory organs of all vertebrates are derived from largely homologous structures of the acoustico-lateralis system. The auditory receptors are hair cells similar to those of the related vestibular system and lateral line organs. In fact, some lateral line receptors of fish may be sensitive to the near-field displacement component of sound.

Early vertebrates evolved in an aquatic environment, an environment that presents a problem for sensitive sound detection. The body of a fish is largely aqueous, and therefore quite transparent to waterborne sound-pressure waves. Most fish have limited auditory sensitivity, achieved by their otolith organs (saccule and lagena). Because the otoliths are more dense than the surrounding tissues, they are displaced relative to the tissues by sound-pressure waves, exciting the attached hair cells.

In many fish the swim bladder functions as a pressure-to-displacement transformer of sound waves, increasing auditory sensitivity. Some fishes have evolved bony conduction links from the swim bladder to the saccule and lagena of the inner ear, an adaptation providing further increases in high-frequency responsiveness and overall sensitivity. Anuran amphibians, reptiles, and birds have an external tympanum, or eardrum, connected to the inner ear by a single bony ossicle,

[5] The dependence of the receptor potential of hair cells on K^+ is confusing: The scala media (outside auditory hair cells) contains a high concentration of K^+, unlike other extracellular fluids, and has a *positive* potential. There is little concentration gradient for K^+, but a large voltage gradient (inside-negative). Opening the mechanically gated cation channels allows K^+ (the major cation) to enter, carrying positive charge into the cell to depolarize it.

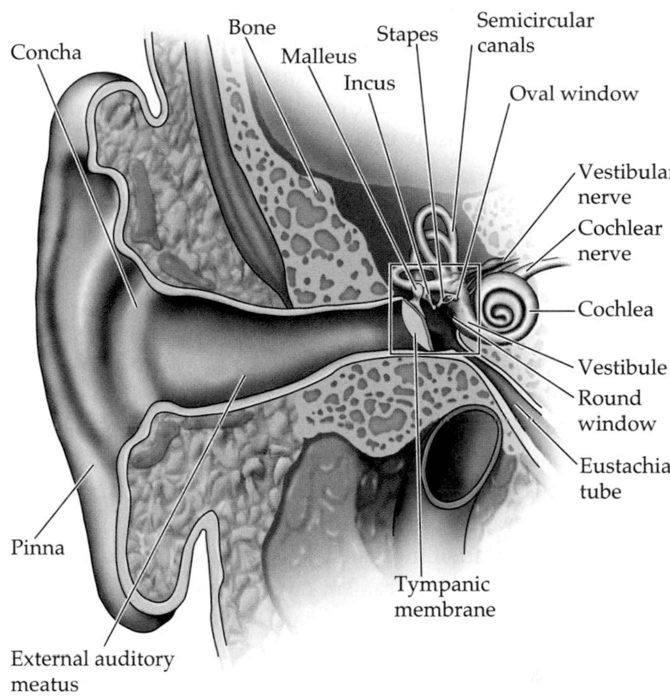

(b) A diagram with the cochlea unwound (side view)

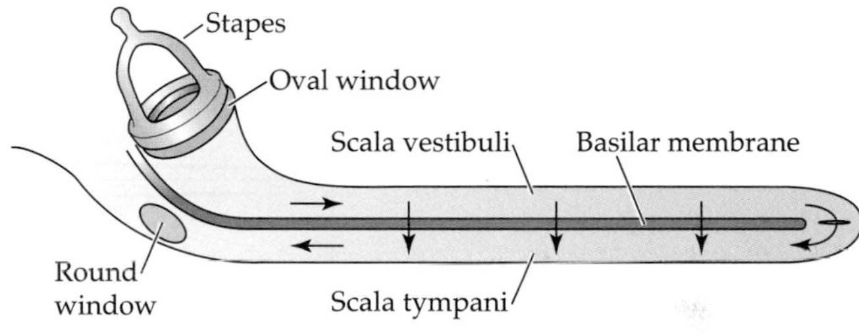

(c) The basilar membrane unwound (top view)

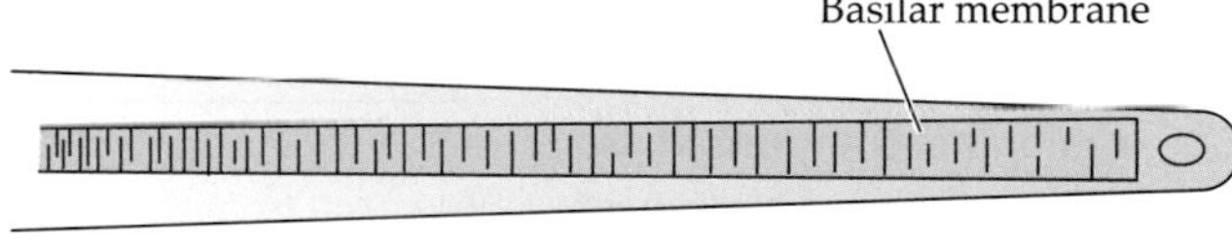

Figure 13.28 Anatomy of the human ear *(a)* An overview of the ear. *(b)* A diagrammatic representation of how the ear would appear if the inner-ear cochlea were unwound. *(c)* A view of the surface of the basilar membrane (unwound). The basilar membrane is narrower and stiffer at its basal end than at its apical end.

the columella. The function of this ossicle linkage is similar to that in mammals (see the next paragraph). The lagena cavity is progressively elongated in reptiles and birds, forming a tube homologous to the mammalian cochlea and containing a basilar membrane.

The structure of the mammalian ear is shown in Figure 13.28. The ear consists of three parts: an **external ear** distal to the eardrum, an air-filled **middle ear**, and a liquid-filled **inner ear**, which consists, in part, of the **cochlea**. Sound-pressure waves vibrate the eardrum (tympanic membrane), and this vibration is transmitted to the membranous *oval window* of the inner ear by three middle-ear ossicles: the *malleus*, *incus*, and *stapes*. Two middle-ear muscles, the tensor tympani and the stapedius, can contract to damp the movements of the ossicles, protecting the auditory membranes from damage by loud sounds. The eustachian tube connects the middle ear with the pharynx, equalizing pressure in the middle ear with environmental pressure.

The cochlea of the mammalian inner ear is a coiled tube that develops as an outpouching of the vestibular labyrinth (see Figure 13.26). The structure of the cochlea is more easily understood if we consider it to be uncoiled, as in Figure 13.28*b*. The cochlea is longitudinally divided into two major compartments: the scala vestibuli and the scala tympani (there is also a third compartment, which we will consider shortly). These longitudinal compartments are separated by the **basilar membrane**, a multicellular structure that bears the auditory hair cells. Sound-pressure waves are transformed into vibrations of a series of structures: eardrum, ossicles, oval window, basilar membrane, and round window (see Figure 13.28*b*).

This long succession of mechanical displacements points out the anatomical complexity of the ear. We can then ask, Why is the ear so complex? Of what functional significance are the various stages of transmission of mechanical displacement from the outside to the hair cells? We will examine two aspects of auditory function that will help make sense (no pun intended) of the anatomical complexity of the ear. These functional aspects are impedance matching and analysis of sound frequency.

The major function of the middle-ear ossicles is the efficient transfer of sound energy from air to the liquid of the inner ear. Airborne sound striking a liquid surface is almost all reflected; only about 1/30 of the sound energy is transferred to the liquid. (If you stand on the shore and talk to a fish, it cannot hear you very well.) The energy transfer is poor because liquids cannot be compressed, so they have a low volume of movement in response to sound pressure. Therefore there must be a corresponding increase in pressure for significant transfer of energy from the eardrum (vibrating in air) to the liquid medium of the inner ear. The middle-ear ossicles apply forces from a relatively large area (the eardrum) onto a much smaller area (the foot plate of the stapes, which covers the oval window). The concentration of force on a small area provides the necessary increase in pressure, allowing efficient transfer of sound energy.

Our second functional question is, How can auditory receptors respond differently to different frequencies of sound? Next we will consider the way in which movement of the basilar membrane stimulates the auditory hair cells located on it, but first let's examine the effects of sound waves on the basilar membrane itself. The basilar membrane varies in width and thickness along its length. It is narrow and rigid at the base of the cochlea near the oval window, and wider and more flexible at the apex, farthest from the oval window (see Figure 13.28*c*). (A rigid bony shelf spans the rest of the width of the cochlea.)

In the nineteenth century, Hermann von Helmholtz (1821–1894) formulated a resonance theory of cochlear mechanics. He thought that each portion of the basilar membrane was under tension, "tuned" like strings in a piano. Sound of a particular frequency would then induce resonant vibration of just the one part of the basilar membrane that was "tuned" to that frequency. Georg von Békésy (1899–1972) later tested and modified this theory. Using cochleas of cadavers, he found that the basilar membrane was not under tension and thus lacked strict resonant properties. Moreover, he found that sound induced a *traveling wave* that covered much of the length of the cochlea (Figure 13.29). Low-frequency sounds vibrated the entire length of the basilar membrane, with a maximum amplitude near the apex. High-frequency sounds vibrated only the region near the base, and for every frequency there was a different place of maximum amplitude of the traveling wave along the length of the basilar membrane.

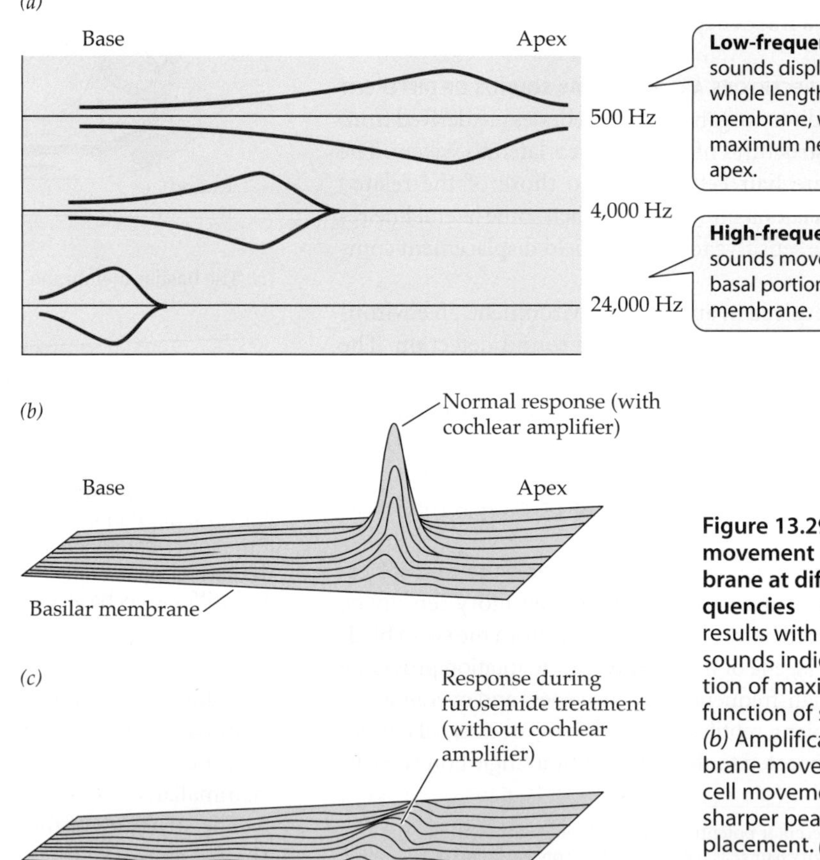

Figure 13.29 Amplitude of movement of the basilar membrane at different sound frequencies *(a)* Von Békésy's results with cadavers and loud sounds indicated that the position of maximal movement is a function of sound frequency. *(b)* Amplification of basilar-membrane movement by outer-hair-cell movements produces a sharper peak of membrane displacement. *(c)* Furosemide blocks hair-cell transduction, decreasing the amplitude of basilar-membrane displacement by disrupting the cochlear amplifier.

This observation forms the basis of the **place theory** of sound-frequency discrimination: The pitch or frequency of a sound is encoded by the *place* of maximum vibration (and thus maximal hair-cell stimulation) over the length of the basilar membrane. The place theory is somewhat similar to Helmholtz's resonance theory, in that both argue that the cochlea is *tonotopically* organized so that different places along the length of the basilar membrane respond to different frequencies. Von Békésy's place theory, however, depends on traveling waves rather than standing waves and postulates a broader responsive region to a single tone than Helmholtz envisaged.

Modern experiments have used emission of gamma rays to measure basilar-membrane movements for lower sound intensities in living cochleas. These experiments have somewhat confirmed von Békésy's observations, but they show that the basilar-membrane movements are more sharply localized than his measurements indicated.

How do movements of the basilar membrane excite the overlying auditory hair cells? First let's examine the structure of the part of the cochlea that contains the hair cells, a region termed the **organ of Corti**. Figure 13.30*a* shows a cross section of the cochlea, divided by the horizontal basilar membrane. The hair cells and various accessory structures of the organ of Corti sit on the basilar membrane and vibrate up and down with it. Stereocilia of the hair cells project into a separate fluid compartment of the cochlea, the *scala media*. The scala media is separated from the *scala vestibuli* by *Reissner's membrane*, a thin layer of cells that is transparent to sound pressure but separates the endolymph fluid of the scala media from the perilymph of the scala vestibuli and scala tympani.

The hair cells of the organ of Corti are of two kinds: Typically there are three rows of **outer hair cells** and a single row of **inner hair cells** (Figure 13.30*b*). Auditory hair cells lose their kinocilium during development and thus have only stereocilia. In other respects they are similar to the vestibular hair cells described earlier. The hair cells are covered by a flap of tissue termed the *tectorial membrane*. The stereocilia of the outer hair cells contact the tectorial membrane, but stereocilia of the inner hair cells are coupled to the membrane only by intervening viscous fluid. Displacement of the basilar membrane is thought to generate shearing forces of the tectorial membrane on the stereocilia.

The transduction mechanism of auditory hair cells is similar to that of vestibular hair cells: Displacement in one direction depolarizes the hair cell, and displacement in the opposite direction hyperpolarizes it. Auditory hair cells make synaptic contact with afferent nerve fibers of the auditory (cochlear) nerve and also receive efferent synapses. Most (80%–95%) of the afferent neurons synapse with the *inner* hair cells, which in humans represent only about 20% of the approximately 24,000 hair cells. Thus the inner hair cells are the major source of auditory input to the brain, and the more numerous outer hair cells may have other functions.

Intracellular responses of auditory hair cells have been recorded—a feat made difficult by the small size and inaccessibility of the cells. Both inner and outer hair cells depolarize in response to tones, the amount of depolarization depending on sound intensity and frequency. The response has both a steady DC (direct current) component and an AC (alternating current) component that mirrors the waveform of the sound. Hair cells respond over a narrower range of frequencies than the visible movements of the underlying basilar

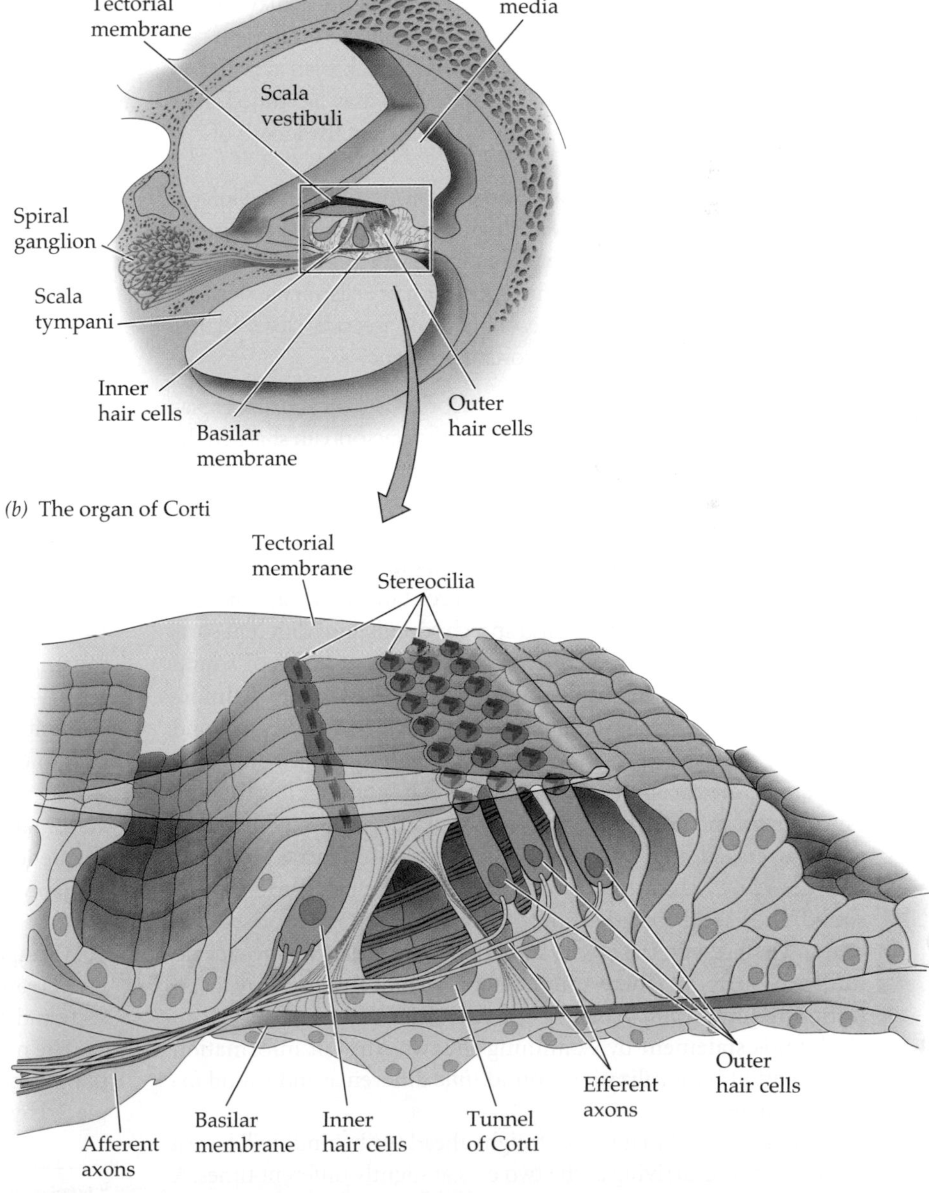

Figure 13.30 The structure of the mammalian cochlea *(a)* A cross section through one turn of the cochlea. *(b)* Detail of the organ of Corti.

membrane would suggest. There appear to be several components of frequency filtration that sharpen the tuning of hair cell responses:

- The basilar-membrane movements themselves are more sharply tuned than von Békésy's observations indicated. He used microscopic observation of responses to very loud sounds in cadavers. Responses to moderate sounds in living cochleas are more localized. In part this localization results from *amplification of basilar-membrane movement by movements* of outer hair cells in responses to sound. Isolated outer hair cells change length by as much as 4% in response to changes in their membrane potential. These movements are extremely fast and can follow frequencies of several kilohertz. Movements of outer hair cells in response to sound-induced receptor potentials probably amplify the local movement of the basilar membrane (see Figure 13.29*b*) and thus amplify the forces acting on nearby inner hair cells.
- Hair cells (at least in nonmammalian vertebrates) may have mechanical resonance properties. Hair cells of reptiles achieve narrow frequency responses without an overlying tectorial membrane, as do hair cells of amphibians, which lack even a basilar membrane. Thus, some hair cells themselves may be "tuned" to particular frequencies, perhaps by differences in the length of their stereocilia.
- Hair cells may have electrical resonance properties. Intracellular recordings of hair cells show a "ringing" oscillation in response to sudden displacements, at the frequency to which the cell is most sensitive. Evidence suggests that cochlear hair cells vary systematically in the frequency of this resonant response, cells closer to the cochlear base having higher resonant frequencies.

Thus, although von Békésy's place theory is a good starting explanation of auditory frequency selectivity, other factors can sharpen the tuning of hair cell responses. These "secondary filters" are subjects of much current attention in auditory physiology.

CENTRAL AUDITORY ANALYSIS: SOUND LOCALIZATION Vertebrates analyze sound in various ways. Your own brain contains pathways that are most sensitive to tones, clicks, and speech sounds, and different sorts of information are extracted from these different kinds of sound stimuli. We will consider only one aspect of central auditory information processing: How do animals localize the source of a sound stimulus?

Humans are capable of reasonably accurate auditory localization, especially in the horizontal (left–right) plane, but our abilities are greatly overshadowed by those of other vertebrates, such as owls and bats. Let's first examine the basic information we can use to localize sound; then we'll consider how this information is processed in a brain. We start with the observation that a single ear can provide no information about where a sound comes from; any auditory localization requires *comparison* of the responses of two ears. We can clarify this statement by examining the two sorts of information used in auditory localization: sound *time* difference and sound *intensity* difference.

First, any sound that is not straight ahead (or behind) will have a **time difference**, arriving at the two ears at slightly different times. A sound that is offset to the left will reach the left ear first, with a difference in time of arrival that increases with increasing offset. Second, an offset sound will have an **intensity difference**: It will be louder in the ear facing the sound source.

The intensity difference results from the fact that sounds do not go around corners well, so the far ear is said to be in a *sound shadow*, shielded from the sound by the head. Sound shadowing is a function of frequency because the head is an effective barrier to high-frequency sound (with wavelengths smaller than the head) but little barrier to low-frequency sound (with wavelengths bigger than the head). Because high-frequency sound has more sound shadowing, the difference in sound intensity between the ears is significant only for high-frequency sounds. For humans, both time differences and intensity differences provide information about right–left location of a sound source, and we have limited ability to localize the vertical origin of sounds.

Owls, on the other hand, have excellent abilities to localize sounds both horizontally (left–right) and vertically (up–down), and using this ability, they can catch mice hidden under leaf litter. Like other vertebrates, a barn owl uses time differences to determine left–right origin of sounds. Both owls and echolocating bats (**Box 13.1**) can use time differences as small as 10 *microseconds* in auditory localization! The owl's ability to localize the *vertical* component of a sound depends on a structural asymmetry of the two ears: The right ear points more upward; the left ear, more downward. Therefore, at higher sound frequencies (with more sound shadowing), a sound from above will stimulate the right ear more than the left, so the intensity difference between ears will code the vertical component of a sound source.

How does the owl's brain put together the information about the horizontal axis (coded by time differences) and the vertical axis (coded by intensity differences) of a sound source? Neurons in the inferior colliculus of the owl midbrain[6] map an orderly representation of the auditory space around the owl's head. Each neuron is said to be *space-specific*, responding to sound only from a particular direction (**Figure 13.31**). Nearby cells respond to sound from nearby areas, so the outside world of sound sources is mapped in two dimensions onto the inferior colliculus. This map is computed from two parallel pathways in the brainstem and midbrain—one processing time-difference (left–right) information, and one processing intensity-difference (up–down) information. Presumably it is this computed auditory map that allows an owl to localize mice in darkness to within 1° of the sound source.

INSECT AUDITORY RECEPTORS Many kinds of animals can detect low-frequency vibrations in air, in water, or within the substrate, presumably using sensory hairs or other undifferentiated mechanoreceptors. For example, cockroach cercal wind receptors (see Figure 10.3) respond to loud sounds at frequencies below 1 kHz. Specialized organs for sound detection, however, are largely confined to insects and vertebrates.

The most common form of auditory organs in insects is a **tympanal organ**, in which a thin cuticular tympanum (eardrum) is displaced by sound. Mechanosensory cells are attached to the tympa-

[6] *Inferior colliculus* is actually the term for the homologous auditory area in the mammalian midbrain. This area in the avian brain is termed the *mesencephalicus lateralis dorsalis* (*MLD*).

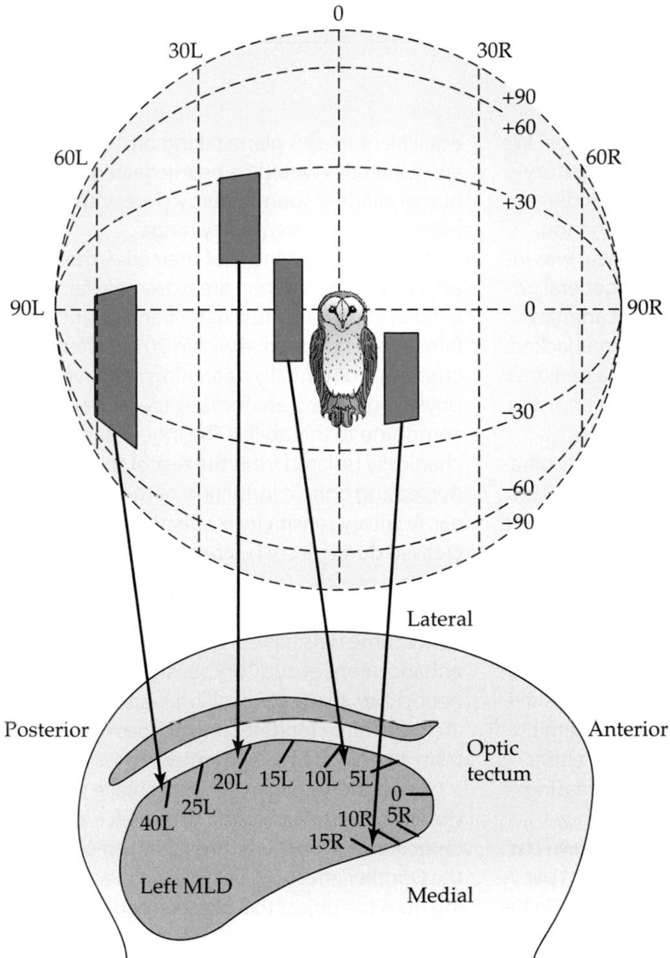

Figure 13.31 A map of auditory space in the brain of a barn owl Individual neurons in the midbrain auditory area (MLD) respond to sound stimuli from discrete locations in space around the owl's head, so the area of MLD maps the location of sound sources in auditory space. Rectangles represent the areas of auditory space from which sound sources evoke responses of individual space-specific neurons. Sounds to the left (L) or right (R) excite neurons in different positions along the MLD, as shown by the lines connecting the rectangles to recording electrodes. Only the horizontal component of the MLD map is shown; 10L means 10° to the left of center, 5R is 5° to the right, and so on. (After Knudson and Konishi 1978.)

num and are stimulated by its movement. Tympanal organs are air-filled cavities derived from some aspect of insects' tracheal systems. (The functional importance of an air-filled chamber is the same as for vertebrates, as described earlier.) Tympanal organs may occur at any of several locations in the body, including the thorax (e.g., in noctuid moths), the abdomen (locusts, cicadas), the legs (crickets, katydids), or the labial palps (sphyngid moths). This diversity, as well as the fact that nontympanal organs are also used for sound reception, shows that auditory organs have evolved repeatedly in different insect groups.

One of the simplest and best-studied tympanal organs is that of noctuid moths. Each of the paired thoracic organs contains only two neurons that respond to sound. The frequency sensitivity ranges from 3 to 150 kHz, with maximum sensitivity at 50 to 70 kHz. Most of this range is *ultrasonic*—above the frequency range of sound audible to humans (20 Hz–20 kHz). Moth auditory organs are most sensitive to the ultrasonic cries of echolocating bats (see Box 13.1). The two auditory cells (A_1 and A_2) respond similarly to ultrasonic pulses, but the threshold of A_1 is 20 decibels lower than the threshold of A_2. The cells convey no information about sound frequency; moths and most insects cannot discriminate frequencies of stimulating sound. Sound intensity is coded by the impulse frequency in each receptor, by a shorter response latency with increasing intensity, and by recruitment of the high-threshold A_2 cell.

The behavioral significance of auditory responses in moths has been well studied (as have central neuronal responses not considered here). Moth tympanal organs provide good directional information about a source of ultrasound, in part because the wavelength of the high-frequency sound is short relative to the size of the moth's body. If a bat emits ultrasonic pulses to the left of a moth, the response of the left "ear" (tympanal organ) will be greater, the moth's body creating an effective sound shadow for short-wavelength ultrasound. The left A_1 cell will respond with a shorter latency and a higher frequency than the right A_1 cell. As a result of this asymmetry, the moth will turn away from the sound source and fly away.

This response is effective for predator avoidance if the bat is distant, but not if the bat is close enough to detect the moth (because bats are stronger fliers). A nearby bat emits an ultrasonic cry loud enough to stimulate the A_2 cell, which triggers a very different response from the moth. Instead of turning away, the moth flies erratically or dives to the ground—responses presumably more adaptive than an attempt to "outrun" the bat once detected.

Most insect tympanal organs are somewhat similar to those of moths, in that they are sensitive detectors and encoders of sound intensity over a certain range of frequencies but are poor at detecting frequency differences. Some insects, however, have a limited ability to discriminate sound frequency, and some have specialized tympanal organs with which they can localize even relatively low-frequency sounds.

Summary: Mechanoreception

- Mechanoreceptors have many behavioral uses. In addition to surface mechanoreceptors that convey information about touch and pressure, mechanoreceptors can serve as proprioceptors that monitor body and limb position or muscle length and force; equilibrium receptors that provide information about acceleration, gravity and balance; and sound-detecting auditory receptors.
- Hair cells are sensitive and versatile vertebrate mechanoreceptors that transduce displacement of stereocilia into a receptor potential. They are the major receptors of vertebrate hearing and equilibrium sense.
- The structure of the vertebrate ear effectively conveys sound-pressure waves into the inner ear. Sounds of different frequencies stimulate hair cells at different locations along the length of the basilar membrane.
- Central auditory pathways of vertebrates sort coded information about sounds to discriminate different sound frequencies and locations. The auditory systems of insects, although less complex, can nonetheless provide them with behaviorally important information.

BOX 13.1 Echolocation

Many kinds of bats have poor vision and yet fly well at night, avoiding obstacles and catching insects at rates as high as two per second. They orient by emitting ultrasonic pulses (i.e., sound at frequencies too high to be audible to humans) and detecting echoes reflected by objects around them.

The Italian naturalist Lazzaro Spallanzani (1729–1799) provided the first evidence that "bats see with their ears" in 1793. Spallanzani placed hoods over the heads of bats and found that they were disoriented, flying into walls. When the bats were blinded, however, they flew with normal orientation and were still able to catch insects. In an ingenious set of experiments, Spallanzani plugged the ears of bats with either hollow brass tubes or tubes filled with wax. The bats with wax-filled tubes were disoriented, but those with hollow tubes (through which sound could pass) flew normally. Spallanzani correctly concluded that the ears were necessary for oriented flight in the dark, but the eyes were not.

Although Spallanzani's results were confirmed and extended by Charles Jurine (1751–1819) in 1794, they were disbelieved and became buried in obscurity. Why were Spallanzani's elegant and surprisingly "modern" experiments given so little credence? We must realize that in the eighteenth century the wave nature of sound was not understood, and thus the concept of sound too high in frequency for humans to hear was inconceivable. In fact, there was no general understanding of the possibility that animals had sensory capabilities that humans lacked.

"Spallanzani's bat problem" remained unexplained until 1938, when Donald Griffin, using newly developed ultrasound detectors, showed that bats emitted high-intensity ultrasound in the frequency range of 30 to 100 kHz. Griffin subsequently independently confirmed Spallanzani's experiments and demonstrated that bats orient by detecting the echoes of their ultrasonic cries.

Even with our present, relatively good understanding of the phenomenon, bat echolocation is a mind-boggling display of sensitivity and precision. Most species of bats emit ultrasonic cries, typically as pulses of either constant frequency (CF bats) or decreasing frequency within each pulse (frequency-modulating, or FM, bats). The little brown bat *Myotis lucifugus* is an example of an FM bat. A cruising *Myotis* emits pulses that sweep in frequency from 80 to 40 kHz. The sound energy of these pulses is an enormous 120 decibels, equivalent to a jet plane taking off 100 m away. Bat cries would be nearly deafening to humans if their sound energy were within the human auditory frequency range.

Despite the intensity of their cries, the echoes that bats detect are extremely faint. Echolocating bats must detect and orient to faint echoes that arrive within 20 ms after emitting a potentially deafening cry. Several physiological and anatomical specializations contribute to this ability. The inner ear is mechanically isolated from the rest of the skull, decreasing bone conduction of the cry to the ear. Auditory sensitivity is effectively decreased during a cry by contraction of tensor tympani muscles (see page 375), and the recovery of sensitivity after the cry is extremely rapid. Some bats have a selective short-term enhancement of auditory sensitivity during a period 2 to 20 ms after calling—an appropriate time frame for detecting echoes returning from an object 34 to 340 cm away.

Constant-frequency (CF) bats make use of the **Doppler shift** of sound to enhance echo detection. If the bat is approaching an object, the Doppler effect will shift the echoes returning from the object to a higher frequency—the same phenomenon that causes the familiar increase in sound frequency when a train or siren is approaching and then the decrease in frequency when the source passes and begins to recede. The ears of CF bats are sharply tuned to a frequency several kilohertz higher than their cries. Thus their auditory sensitivity is much greater to Doppler-shifted echoes (from an object being approached) than to the cry itself.

How does a bat's brain process the complex audi-

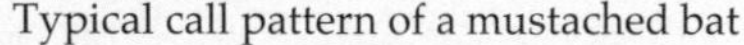
Typical call pattern of a mustached bat

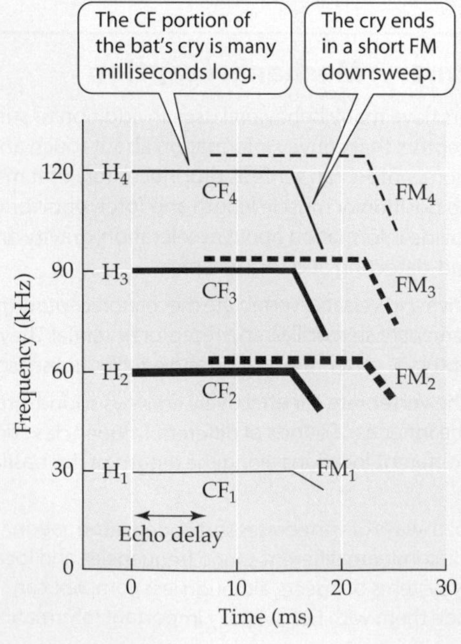

Echolocation in a mustached bat The cry of a mustached bat has both constant-frequency (CF) and frequency-modulated (FM) portions. The graph is a sonogram of sound frequency as a function of time. Solid lines represent the cry; the width of the line indicates loudness. The fundamental frequency (CF_1) is weak, and the bat hears it by bone conduction. Several overtones (CF_2, CF_3, CF_4) are louder. Echoes (dashed lines) return with a time delay (indicating distance) and with a Doppler-shifted frequency (indicating the bat's approach velocity). A large area of the bat's cerebral cortex is devoted to processing auditory information for echolocation. Some areas process the time delay of the FM sweep to determine target distance, and other areas process the degree of Doppler shift to determine relative velocity.

tory information to catch prey or avoid objects? Neural auditory processing of echolocation signals has received considerable study in the mustached bat (*Pteronotus*). The mustached bat is a CF/FM bat: Its call has a relatively long CF tone with a short FM downsweep at the end (see figure). The fundamental frequency of the tone is 30 kHz, with harmonics at 60, 90, and 120 kHz. The Doppler shift of the CF tone conveys information about approach velocity to the target, the delay of the echo's FM sweep gives information about the target's distance, and the strength of the echo indicates relative target size. A considerable area of the bat's cerebral cortex is devoted to processing different aspects of this coded information.

Bats, of course, must not only detect echoes, but must also determine their direction with great accuracy. Several auditory specializations contribute to this localization. For example, contralateral inhibition is well developed in bat auditory centers. Many central auditory neurons are excited by sound stimulation of one ear and inhibited by sound stimulation of the other ear. Behavioral experiments indicate that these neurons must be sensitive to differences in time of stimulation of the two ears as small as 0.01 ms. Several other auditory mechanisms contributing to echolocation have been explored. No matter how much we learn about the mechanisms of bat echolocation, however, the performance itself remains amazing. Many bats can echolocate insects as small as mosquitoes and fruit flies, and catch them at rates of two captures per second!

Echolocation is not confined to bats. Central American oilbirds orient through caves by echolocation, and dolphins employ sound pulses for echolocation, as well as for communication.

Chemoreception

The behavior of humans depends less on the chemical senses than does the behavior of many other kinds of animals. Consequently it is easy for us to underestimate the importance of chemoreceptors as a major source of sensory input for a variety of animals. Popular accounts of bloodhounds tracking escaped convicts, moths locating mates, and salmon returning to their "home" rivers—all by means of chemoreceptors—seem fantastic to us. Our lesser reliance than other animals on chemical senses in part may result from our lesser sensitivity to chemical stimuli, and in part it may reflect the common observation that our behavioral use of chemoreceptors is only partly conscious.

Sensory chemoreception is only one kind of chemical sensitivity of cells. Many animal cells have evolved sensitivities to hormones, synaptic transmitters, and growth factors. Moreover, nonmetazoan organisms such as bacteria and protozoans exhibit responses to chemicals in their environment. Such cells and organisms may serve as important model systems for studying chemosensory mechanisms, but we will not consider them, because they cannot be classified as differentiated chemoreceptor cells.

If we exclude these nonsensory cells, we can divide chemoreceptors into four loose categories:

1. **General chemical sensors** are relatively insensitive, nondiscriminating receptors, which, when stimulated, elicit protective responses of the organism. For example, stimulation of the skin of a frog with salt or acid solutions activates general chemical sensory endings that elicit avoidance or wiping movements of the legs.
2. **Internal chemoreceptors** respond to chemical stimuli within the body. Examples include blood glucose receptors, internal chemoreceptors of the digestive tract, and the carotid body chemoreceptors that respond to blood oxygen concentration (see Chapter 21, page 560).
3. **Contact chemoreceptors** have relatively high thresholds and respond to dissolved chemicals from a nearby source. Contact chemoreceptors usually play a role in feeding behavior, as in the familiar example of vertebrate taste receptors.
4. **Distance chemoreceptors**, or **olfactory receptors**, are more sensitive than contact chemoreceptors and are adapted to respond to external chemicals from a distant source.

The distinction between contact and distance chemoreception is relatively clear for animals that live in air. In these cases the stimuli of olfactory (distance) receptors are airborne chemicals, and the stimuli of taste or gustatory contact chemoreceptors are dissolved in liquid that contacts the receptive surface. Because stimuli of distance receptors also must dissolve in liquid at the receptive surface, however, this distinction is not as clear as we would like. For animals that live in water the distinction is even less clear, because all chemical stimuli are dissolved in the aquatic environment, whether their source is distant or nearby. Nevertheless the distinction remains useful, because many aquatic animals respond quite differently to dilute or distant stimuli than to more concentrated, proximate stimuli. Lobsters, for example, orient and search in response to chemicals at low concentrations (and presumably distant sources) that stimulate distance chemoreceptors on the antennules. High concentrations of chemicals may trigger feeding movements by stimulating "contact" chemoreceptors on the mouthparts. We will consider contact and distance chemoreceptors of insects and of terrestrial vertebrates, examples in which the distinction is relatively clear.

Insect contact chemoreceptors are localized in sensilla

The contact chemoreceptors of flies are among the most fully investigated sensory cells of any animal. Insect surface receptors are organized in **sensilla** (singular *sensillum*), specialized cuticular structures such as sensory hairs, pegs, plates, and pits. In flies such as the blowfly, *Phormia*, contact chemoreceptors are localized in sensory hairs of the *tarsus* (the terminal segment of the leg) and of the *labellum* (the tip of the extensible proboscis used for feeding). Stimulation of even a single tarsal sensory hair with sugar solution elicits extension of the proboscis. If increasing concentrations of salt or of quinine are added to the sugar solution, proboscis extension is increasingly inhibited. Stimulation of a labellar hair with sugar solution elicits drinking behavior, in which the proboscis works as a suction pump. As with stimulation of a tarsal hair, the addition of salt or quinine to the solution stimulating the labellar hairs inhibits the behavioral effects of sugar stimulation. Blowfly contact chemoreceptors are thus appealing to study, because they mediate clear behavioral choices and because it is easy to record the electrical activity of the relevant sensory neurons.

The contact chemosensory hairs ("taste hairs") of flies contain five sensory neurons. One neuron is a mechanoreceptor, the dendrite of which ends at the base of the hair. The other neurons are chemosensory and have single dendrites extending through the lumen of the hair to a pore at the hair tip. The dendrites are modified cilia, another example of sensory function of a ciliary derivative. Numerous investigators, most notably Vincent Dethier (1915–1993), have explored the chemical responsiveness of the sensory neurons. Chemical stimuli are applied through a small, fluid-filled pipette placed over the tip of a hair, and the resultant train of impulses is recorded extracellularly via an electrode in the stimulating pipette or by a second electrode at a crack made in the sidewall of the hair.

Figure 13.32 shows the responses of neurons in a tarsal taste hair to different stimuli. Different sensory cells in a hair can be identified by differences in size and waveform of their extracellularly recorded impulses. One of the sensory cells responds best to water, the response diminishing with increasing concentration of dissolved substances. Activity of this cell (termed the *water cell*) elicits proboscis extension and subsequent drinking if the fly is thirsty, but it does not induce proboscis extension in a water-sated fly.

A second cell (termed the *sugar cell*) responds most strongly to solutions of sugars. The frequency of spikes in the sugar cell increases with increasing concentration of sugar and thus encodes information about sugar concentration. Any stimulus that elicits proboscis extension in a water-sated fly activates the sugar cell above a certain minimum spike frequency.

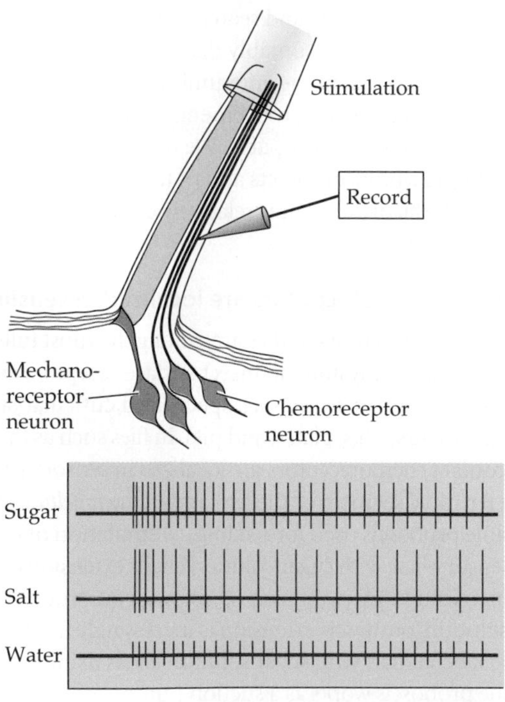

Figure 13.32 Responses of a blowfly contact chemoreceptor sensillum to different stimulatory solutions One of the three chemosensory neurons (sugar, salt, and water cells) is most responsive to one of the three kinds of stimulation shown. A mechanoreceptor neuron is also present.

A different cell (the *salt cell*) responds preferentially to a range of salts, particularly to monovalent cations. Salt cell activity is correlated with inhibition of proboscis extension, a finding that suggests that the CNS of a water-sated fly controls extension of the proboscis by assessing the logical function (sugar cells activity minus salt cell activity). Not all feeding deterrents activate the salt cell, however. Some may inhibit sugar cell activity directly, and some anions may stimulate the less fully characterized fifth cell of taste hairs (not shown). In tarsal hairs of another dipteran, the apple maggot fly *Rhagoletis*, a cell that may correspond to the blowfly fifth cell responds to an oviposition-deterrent pheromone. This and other examples suggest that some insect contact chemoreceptor neurons convey information about specific pheromones or food plant substances, rather than simply signaling nonspecific categories such as sugars and salts.

Insect contact chemoreceptors demonstrate several important features of sensory coding. The impulse frequency of a receptor response codes stimulus magnitude (chemical concentration). Each cell has a characteristic spectrum of sensitivity and acts (to some degree) as a *labeled line* for a stimulus quality (e.g., sugar) in the same way that different visual cones are labeled lines signaling colors. Chemoreceptor response spectra can be broad or narrow, and a stimulus may excite several cells, the activities of which must be integrated centrally. Most chemosensory systems are thought to code stimulus quality and magnitude in similar ways to this, but in other systems coding is more complicated and less clear, as we will see.

Insect olfactory receptors sense pheromones and other chemicals

The distance chemoreceptors of insects can be termed *olfactory receptors* because they respond to airborne odorants at low concentrations. The distinction between olfaction and taste is equally valid for insects and for terrestrial vertebrates, although the receptors of the two groups are not homologous. Insect olfactory receptors are localized in a variety of sensilla and are concentrated on the antennae. The internal structures of olfactory sensilla are similar to those of taste hairs, although olfactory sensilla may contain up to 50 or more sensory neurons. The ciliary dendrites of olfactory neurons may branch within the sensillum to end at or under pores in the cuticular wall of a peg. Odorant molecules are thought to adsorb to the hair surface, diffuse through the pores, and bind to sites on the dendrite membrane to stimulate a receptor cell.

Olfactory sensory processes have primary roles in many aspects of insect behavior, especially among the social insects. One important kind of use of olfactory receptors is in response to pheromones. As discussed in Chapter 14, a *pheromone* is a metabolically produced chemical that is released to the outside world by one individual and that elicits specific behavioral or systemic responses when detected by other individuals of the same species. Various insects have sex-attractant (and sex-deterrent) pheromones, oviposition-deterrent pheromones, trail-marking pheromones, alarm pheromones, and colony-recognition pheromones.

The best-studied examples of olfaction of pheromones concern sex attractants of moths. Female moths release pheromones that, when detected by males, induce the males to fly upwind to find the females. Males are extraordinarily sensitive to the female attractant, and they can be attracted over distances of miles. A female gypsy moth produces 1 mg of disparlure, its sex attractant. This amount is sufficient to attract a billion males if efficiently distributed.

The receptors for sex pheromones and other odors have been studied in many insects. We will discuss the silkworm moth, *Bombyx mori*, the first insect species for which a sex-attractant pheromone was characterized. *Bombyx* males have large pinnate or comblike antennae with up to 50,000 hair sensilla responsive to the sex-attractant pheromone; females have simple antennae that lack hair sensilla (although they have other types of olfactory sensilla, as we will see). Because of the elaborate branching and the density of pheromone-sensitive olfactory hairs, the antennae of *Bombyx* males are estimated to catch up to one-third of the odor molecules in the air passing through them.

The pheromone receptor cells are exquisitely sensitive and very specific in their responses. The major sex-attractant pheromone in *Bombyx* is bombykol, a 16-carbon unsaturated alcohol. The hair sensilla contain a sensory neuron specifically sensitive to bombykol. Electrophysiological recordings indicate that binding of a single molecule of bombykol elicits sufficient depolarization to generate an impulse. The threshold for the behavioral response of a male moth is reached with activation of about 200 receptor cells per second, which occurs at a phenomenally low concentration of 1000 molecules per square centimeter of air! Females, in contrast, completely lack receptors sensitive to bombykol.

Bombyx females release small amounts of bombykal (the corresponding aldehyde), along with bombykol. Bombykal activates separate receptor cells in males from those that respond to bombykol, a finding that demonstrates the specificity of response of pheromone receptors. Surprisingly, activation of bombykal receptors centrally inhibits the moths' behavioral response to bombykol. What is the adaptive value of such an inhibitory response to a component of the female's emission? When sex-attractant pheromones were first discovered, it was supposed that each species would employ a single, species-specific kind of molecule. The reality is more complicated, however. Females of many species release a mixture of two or more compounds, and the same compound may be released in more than one species. Bombykal is a putative sex attractant in another moth species (the sphinx moth, *Manduca sexta*). Thus the inhibitory response to bombykal in *Bombyx* may have evolved to suppress the attraction of *Bombyx* males to females of other species.

Not all olfactory receptors of insects, of course, respond to pheromones. Moths of both sexes possess olfactory sensilla sensitive to other odorants, and honeybees can behaviorally discriminate hundreds of kinds of odorants. Although the response spectrum of each species is different, a few generalities emerge from a wide range of studies of insect olfaction. The most important generalization is that olfactory receptor cells can be either *odor specialists*, responding to a narrow spectrum of kinds of chemicals, or *odor generalists*, responding to a broad spectrum.

The bombykol and bombykal receptors we have considered here are good examples of odor-specialist cells. In addition to such pheromone specialist cells, food odor specialist cells have also been described. Odor-generalist cells typically respond to a broad spectrum of plant and food odorants, different receptors having different relative sensitivities. For example, two generalist receptors of a moth may respond to geraniol, citral, and several other odorants, but one may be most sensitive to geraniol and the other most sensitive to citral.

Odor-specialist and odor-generalist receptors clearly provide different kinds of information to the central nervous system. Odor specialists act as *labeled lines*; that is, any activity in, say, a bombykol receptor neuron can be reliably interpreted as indicating the presence of bombykol, because the receptor is largely insensitive to anything else. Odor-generalist neurons, in contrast, are not specifically labeled lines, because any of various odorants can elicit their activity.

In order for the central nervous system to correctly interpret sensory activity as the presence of, say, geraniol, it must compare activities of many receptors rather than just monitoring a specifically labeled line. Generalist receptors are said to convey information about stimulus quality (e.g., the kind of odorant) by the *across-fiber pattern* of sensory activity. The distinction between labeled lines and the across-fiber pattern is of general importance in sensory coding, but it is a difference of degree. In color vision a red-sensitive cone can be viewed as a labeled line, but it has a rather broad response spectrum. A particular ratio of activation of red-, green-, and blue-sensitive cones (an across-fiber pattern) is interpreted as orange light. Across-fiber patterning is especially important in chemoreception, in which the range of stimulus qualities (e.g., kinds of odorants) is too great to allow narrow-spectrum specialist receptors and labeled lines for each kind. We will see other examples of this principle in chemosensory systems of vertebrates.

Vertebrate taste involves four or five taste qualities

When we speak of the taste of food, most of the complex perception of flavor to which we refer is actually mediated by olfactory processes, rather than by taste receptors. We have little awareness of the distinctness of olfactory and taste sensory systems. It is probably for this reason that even the central organizing principles of taste physiology are products of the twentieth century. The most fundamental of these organizing principles, derived from psychophysical studies, is that there are *four or five basic taste qualities: sweet, sour, salty, bitter, and* probably *umami* (the savory taste of monosodium glutamate). All true taste perception (as opposed to olfactory perception of flavors) in humans is thought to involve these four or five taste qualities, singly or in combination. The sensory physiology of taste in other vertebrates is thought to be broadly similar to that in humans.

Vertebrate taste receptor cells are grouped together in **taste buds**. In higher vertebrates, taste buds are confined to the tongue and back of the mouth, but some fish have taste buds over much of the body surface. The structure of a typical taste bud is shown in Figure 13.33*a*. There are at least three types of cells in a taste bud, all derived from epithelial cells. The *taste receptor* cells surround a shallow lumen or taste pore and have extensions into the pore that are presumably the sites of sensory transduction. *Supporting cells* are interspersed between the taste cells and appear to secrete fluid into the lumen. *Basal cells*, the third type, are derived from the epithelium adjacent to the taste bud and are precursors of the taste cells.

Studies with radioactive tracers have shown that individual taste receptor cells have lifetimes of only about 5 to 10 days, and they are constantly being replaced by new receptor cells derived from the basal cells. Because the receptor cells are epithelial rather than neu-

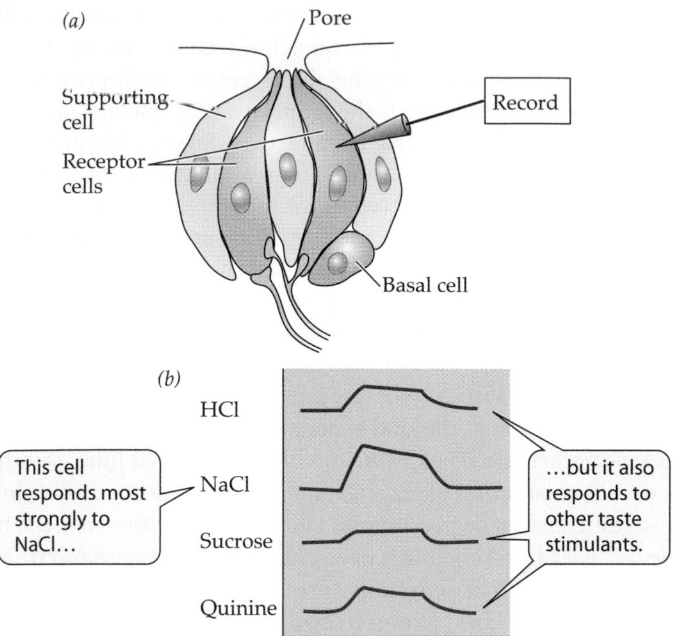

Figure 13.33 Mammalian taste *(a)* The structure of a mammalian taste bud, showing receptor cells, supporting cells, and a basal cell. *(b)* Intracellular recordings from a single receptor cell.

ral, they have no axons. Instead the receptors make synaptic contact with sensory neurons in the taste bud, in the same way that hair cells of the acoustico-lateralis system synapse with sensory neurons. As a result of the high rate of turnover of taste receptor cells, the synapses between these cells and their sensory neurons must be continuously broken and re-formed.

Taste buds of the human tongue have different distributions for different taste qualities, as determined by tests with small drops of solutions confined to localized regions of the tongue. The front of the tongue is most sensitive to sweet substances, with successively posterior zones of greatest sensitivity to salty-, sour-, and bitter-tasting substances. From these and other psychophysical studies, we would expect individual taste receptor cells to respond specifically to stimuli that evoke only one of the four perceptual taste qualities. This expectation, however, is not completely met.

Recordings from individual taste receptor cells and sensory axons show that each receptor responds preferentially to stimuli of one of the taste qualities, but also responds to other kinds of taste stimuli at a lower sensitivity. Figure 13.33*b* shows the responses of a single receptor cell, recorded intracellularly. The cell responds to stimulation with a graded depolarizing receptor potential, which will excite sensory axons that synapse with the cell. The cell responds most strongly to salt but also depolarizes in response to other stimuli. We can characterize the cell as a "salt-best" cell, but any information it provides to the central nervous system is ambiguous.

This finding is typical of most studies of vertebrate taste receptors. Most receptor cells (and sensory axons) respond preferentially to stimuli of one of the taste qualities, while also showing some response to the others. Therefore the coding of taste quality must be by the across-fiber pattern of incoming signals; the brain must integrate the *relative* activities of salt-best cells, sugar-best cells, and so forth in order to assess quality.

Taste receptors have much higher thresholds than do olfactory receptors. In humans, NaCl solutions must be at least 0.05 *M* to taste salty, and sucrose must be at least 0.17 *M* to taste sweet. Thresholds are lower for acids (0.001 *M*) and much lower for quinine (3×10^{-5} *M*). These values are still much higher than for olfaction, although some fish have taste buds on the skin surface that respond to amino acids with very low thresholds (10^{-8} to 10^{-10} *M*) that are comparable to their olfactory thresholds.

TASTE TRANSDUCTION MECHANISMS ARE DIVERSE Mechanisms of transduction in taste receptor cells differ for different taste modalities, and perhaps between species. Figure 13.34 shows some of the better-characterized mechanisms. The mechanism for salt reception is perhaps the simplest (see Figure 13.34*a*). Salt taste depends largely on $[Na^+]$. Na^+ ions enter through non-voltage-gated Na^+ channels to depolarize the taste cell. These channels, unlike voltage-dependent Na^+ channels, are blocked by the drug amiloride. Sour taste depends on acidity or $[H^+]$, and its transduction can involve H^+ ions entering through cation channels as with transduction in salt taste. In other cases, H^+ ions block K^+ channels (see Figure 13.34*b*), depolarizing the cell with a resistance increase.

Sweet taste, in contrast, is mediated by a G protein–coupled receptor molecule (see Figure 13.34*c*). When the receptor molecule binds sugar or another sweet-tasting compound, it acts via a G protein to activate adenylyl cyclase, increasing cAMP levels in the cell. This mechanism is strikingly similar to mechanisms of indirect action of synaptic transmitters (see Figure 12.18). Other sweet-tasting compounds, such as the amino acid arginine, appear to act by binding ligand-gated channels that are nonselective for cations. Bitter-taste transduction is less clear and may have several mechanisms. In the best-characterized example (see Figure 13.34*e*), a G protein–coupled receptor activates IP_3 metabolism (compare Figure 12.21), leading to Ca^{2+} release from intracellular stores.

Characterization of taste receptor mechanisms may be aided by the discovery of specific ligands for the receptor sites. For example, gymnemic acid specifically blocks sweetness, while miraculin (a glycoprotein of the "miracle fruit" *Synsepalum*) makes substances that are normally sour taste sweet, without affecting responses to salty, bitter, or sweet substances.

Vertebrate olfaction includes over a thousand receptor types

Vertebrates can employ olfaction not only in feeding behavior, but also for orientation and for social behavior. Vertebrate pheromones, although not as well characterized as those of insects, are implicated in some species in reproductive and maternal behavior, territoriality, and dominance. Olfactory systems are characterized by great sensitivity and by large numbers of qualitatively distinguishable odors. It is difficult to determine just how many kinds of odors humans and other animals can distinguish, but most estimates are in the hundreds.

Any conversation with a perfumer or a wine connoisseur will convince you of both the subtle discrimination of human olfaction and the difficulty in talking about it. The sensitivity of human olfaction is

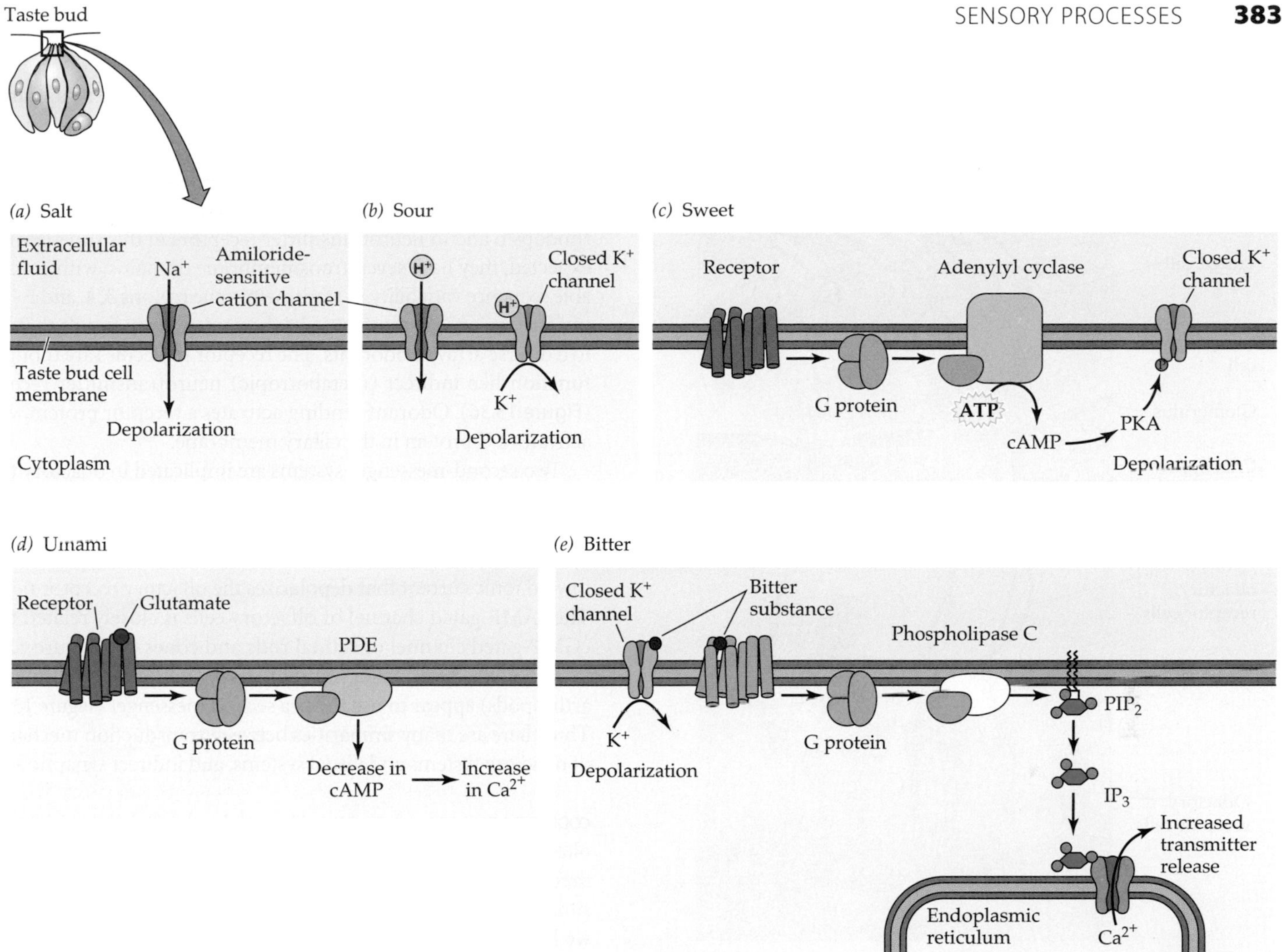

Figure 13.34 Taste-transduction mechanisms differ for different taste qualities All transduction mechanisms except the IP_3 action in *(e)* lead to *depolarization*, which spreads to the basal end of the cell and opens voltage-gated Ca^{2+} channels to allow Ca^{2+} entry and transmitter release. *(a)* For salt taste, sodium ions enter a taste bud cell through amiloride-sensitive cation channels, directly depolarizing the cell. *(b)* In sour taste, either H^+ ions enter the cell through amiloride-sensitive cation channels, or they close K^+ channels to produce depolarization. *(c)* Sweet taste is most commonly mediated by the binding of sugars to a G protein–coupled receptor, which acts via a G protein to activate adenylyl cyclase and produce cyclic AMP. Cyclic AMP then activates protein kinase A (PKA) to close a K^+ channel (by phosphorylating it), producing depolarization. *(d)* The amino acid glutamate (monosodium glutamate, MSG) stimulates the taste quality umami (a savory or meaty quality). Glutamate binds to a G protein–coupled receptor (related to synaptic metabotropic glutamate receptors) to activate a phosphodiesterase (PDE) and decrease the concentration of cAMP. The decrease in cAMP leads to an increase in intracellular Ca^{2+} concentration. *(e)* Bitter taste mechanisms can involve a G protein–coupled receptor for bitter substances that acts via a G protein and phospholipase C to produce IP_3. IP_3 liberates Ca^{2+} ions from intracellular stores, eliciting transmitter release without requiring depolarization. Other bitter substances bind to K^+ channels and close them to depolarize the cell.

quite impressive, but not nearly as impressive as that of some other vertebrates and invertebrates. For example, the human behavioral threshold for ethyl mercaptan in inhaled air is 7×10^{-13} *M*, or about 10^8 molecules per milliliter. For butyric acid, thresholds in molecules per milliliter are as follows: human, 7×10^9; *Triturus* (amphibian), 4×10^9; dog, 9×10^3; bee, 1.1×10^3. The physiological bases of this extraordinary sensitivity and of the discrimination of so many odor qualities have only recently begun to be clarified.

The anatomy of the olfactory system is rather similar for all vertebrates. The olfactory receptive surface, termed the *olfactory epithelium*, lines part of the internal nasal cavity. The area of the nasal mucosa that constitutes the olfactory epithelium varies greatly among species, being only 2 to 4 cm^2 in humans, but 18 cm^2 in dogs and 21 cm^2 in cats. Humans have an estimated 10^7 olfactory receptor neurons, and dogs have 2×10^8. In contrast, the mammalian auditory nerve contains only 3×10^4 axons, and the optic nerve, only 10^5 axons.

Unlike epithelial taste receptors, olfactory receptors are neurons. Each olfactory receptor is a bipolar sensory neuron with a cell body in the olfactory epithelium (Figure 13.35). A single, narrow dendrite extends from the cell body to the mucus-covered epithelial surface and ends in a knob that projects into the overlying layer of mucus (secreted by supporting cells and by underlying glands). From this knob project several olfactory cilia (typically 8 to 20 in mammals)

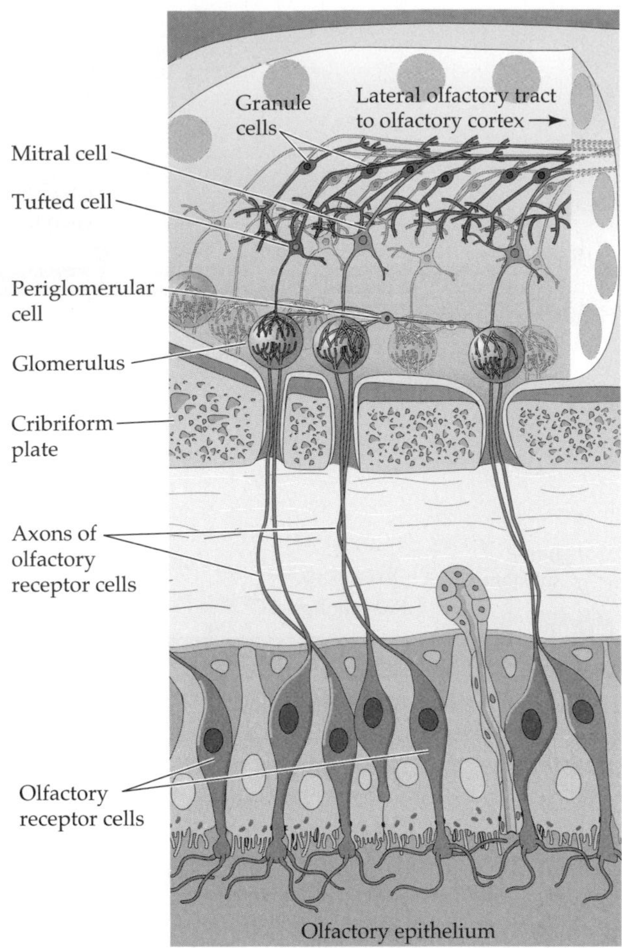

Figure 13.35 Vertebrate olfactory receptors Olfactory receptor cells are small bipolar neurons, the sensory cilia of which extend into the mucus layer of the nasal cavity. Their axons perforate the bone of the cribiform plate to end in glomeruli of the olfactory bulb of the brain. Mitral cells and granule cells integrate olfactory information, and the mitral cell axons carry the information to the olfactory cortex.

that extend and intermesh in the mucus layer. The membranes of these cilia are the sites of receptor transduction.

The receptor neurons have fine, unmyelinated axons that extend a short distance to the olfactory bulb of the forebrain, where they synapse with second-order cells. The receptor axons are typically only 0.2 μm in diameter, among the smallest axons in the nervous system. Olfactory neurons undergo continuous turnover throughout adult life, and they were the first mammalian neurons known to do so. The receptor neurons differentiate from basal cells in the epithelium, develop dendrites and axons, and have life spans of about 60 days before they degenerate.

The excitation of olfactory receptors is becoming better understood. Inhaled air (or water in aquatic vertebrates) passes across the olfactory epithelium, and odorant molecules are absorbed into the mucus layer. Transduction occurs at the surface membranes of olfactory cilia in the mucus layer, and most commonly involves cyclic AMP. Vertebrates contain a large family of genes that encode membrane proteins thought to act as odorant receptor molecules. As many as a thousand such genes may be expressed in a mammal (fewer in lower vertebrates), making the family of putative odorant receptor proteins one of the largest families in the genome.

The odorant receptor proteins are members of the G protein–coupled receptor superfamily, and they are similar in structure to rhodopsin and to neurotransmitter receptors in this superfamily. As expected, they have seven transmembrane domains, with considerable sequence variability in transmembrane regions 3, 4, and 5. These variable regions are thought to be the part of the molecule that binds to a diverse array of odorants. The receptor molecules are thought to function like indirect (metabotropic) neurotransmitter receptors (Figure 13.36). Odorant binding activates a receptor protein, which activates a G protein in the ciliary membrane.

Two second-messenger systems are implicated in olfactory transduction. In some cases the Gα subunit activates adenylyl cyclase, increasing the production of cAMP (see Figure 13.36*a*). Cyclic AMP binds to and opens a cation channel in the membrane, producing an inward ionic current that depolarizes the olfactory receptor neuron. The cAMP-gated channel of olfactory cells is closely related to the cGMP-gated channel of retinal rods and cones (see Figure 13.14). Other olfactory transduction pathways (both of vertebrates and arthropods) appear to use IP_3 as a second messenger (Figure 13.36*b*). Thus there are many similarities between transduction mechanisms of olfactory systems and visual systems, and indirect synaptic action.

CODING OF OLFACTORY QUALITY The psychophysical view of vertebrate olfaction is that vertebrates, including humans, can distinguish hundreds of kinds of odors and can make fine discriminations among similar odor mixtures, all at high levels of sensitivity. Until recently we had little idea how such discriminations could occur. The discovery of the large family of odorant receptor genes, each coding for a G protein–coupled receptor protein, has provided a good conceptual and technical basis for clarifying olfactory information processing. Certainly, with perhaps 1000 genes coding for odorant receptor proteins, there is substantial basis for discriminating many olfactory qualities.

Evidence suggests that each olfactory receptor cell expresses only one or a few of these genes. (Each receptor protein, however, may respond to several kinds of odorants.) The receptor cells that express a particular receptor protein may be widely distributed over the olfactory epithelium, but their axons all appear to converge on one or a few glomeruli at reproducible locations in the olfactory bulb. Thus the olfactory quality of an odorant is converted to a spatial pattern at the olfactory bulb, and lateral inhibitory interactions in the bulb may sharpen and refine quality discriminations. Many aspects of olfactory information processing appear to be similar for vertebrates and insects.

Many vertebrates have an accessory olfactory structure termed the **vomeronasal organ**, which projects via separate pathways and olfactory centers in the brain. Receptor cells of the vomeronasal organ are relatively narrowly tuned, responding to fewer compounds than the odor-generalist cells of the main olfactory epithelium do. Vomeronasal receptor cells express genes for a separate family of G protein–coupled receptors. The vomeronasal organ mediates many behavioral responses to pheromones (see Chapter 14); however, some of its responses are nonpheromonal, and some responses to pheromones are mediated by the main olfactory system.

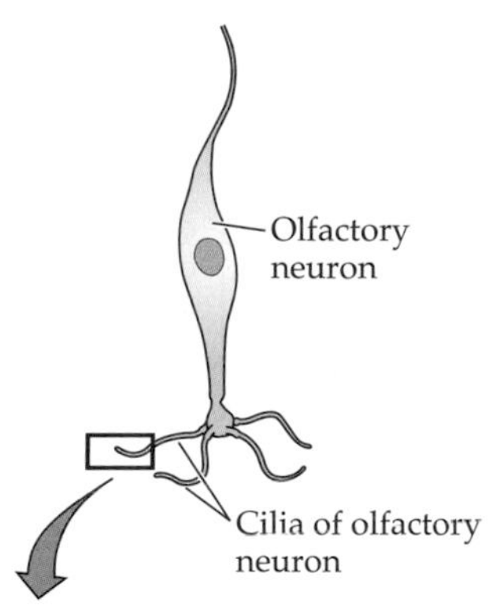

Figure 13.36 Olfactory transduction mechanisms in cilia membranes of olfactory neurons *(a)* Many odorants act to increase cyclic AMP. The odorant binds to an odorant receptor on the ciliary membrane; the receptor activates a G protein to activate adenylyl cyclase, producing cAMP. Cyclic AMP binds to and opens a cation channel, allowing entry of Na^+ and Ca^{2+} ions to depolarize the cell. Ca^{2+} binds to Ca^{2+}-activated Cl^- channels, augmenting the depolarization. *(b)* Some olfactory responses increase IP_3. This mechanism also starts with odorant binding to a G protein–coupled receptor, but in this case the G protein activates phospholipase C, forming IP_3 from PIP_2 (see Figure 12.21). IP_3 binds to and opens a calcium channel, letting Ca^{2+} enter to depolarize the cell. As in *(a)*, Ca^{2+}-activated Cl^- channels augment the depolarization.

(a) Increase in cAMP

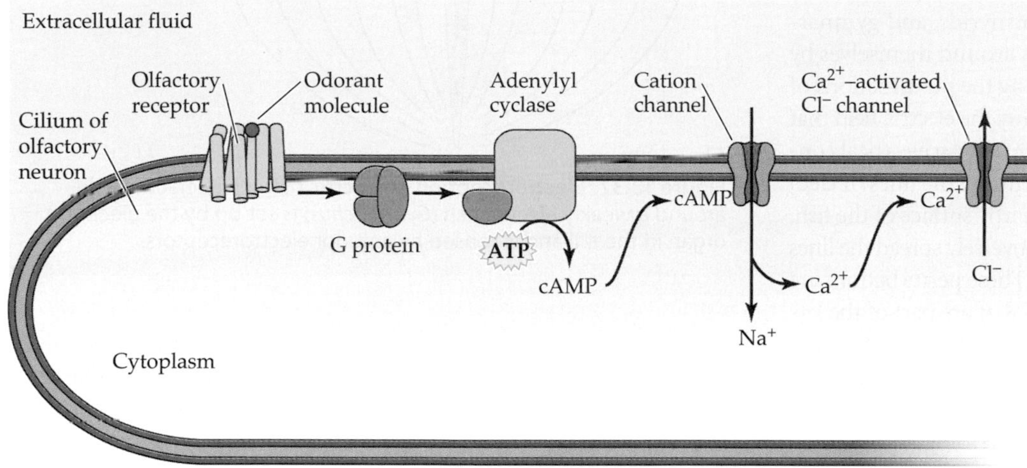

(b) Increase in IP_3

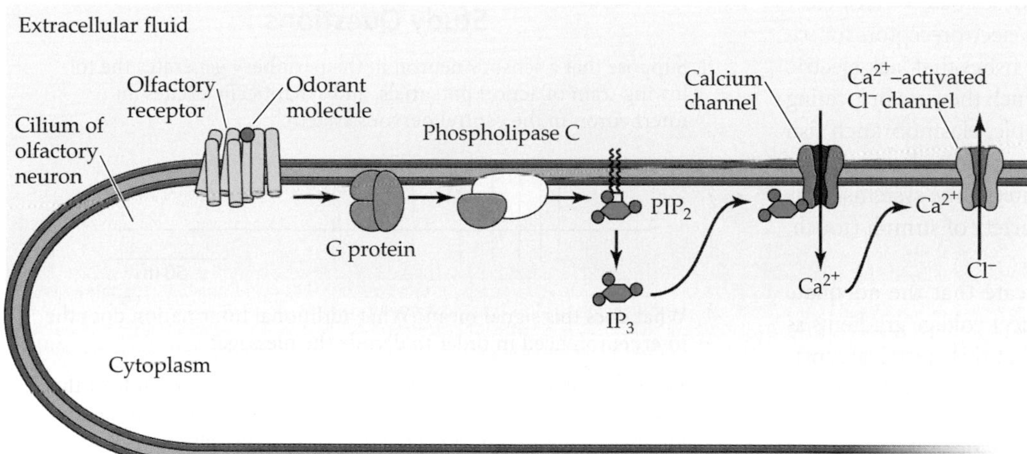

Summary: Chemoreception

- Most animals possess two sorts of external chemoreceptors: contact or taste chemoreceptors that respond to near-field chemicals at rather high concentrations, and olfactory or distance chemoreceptors that respond to low concentrations of chemicals from sources over a larger area. There are also internal chemoreceptors involved in homeostatic regulation.
- Taste chemoreceptors of mammals monitor several taste qualities: sweet, sour, salty, bitter, and perhaps umami. Insects have taste sensilla that provide analogous information.
- Olfactory chemoreceptors of vertebrates are neurons with cilia that contain receptor molecules. Each neuron expresses the gene for one of these membrane receptor proteins, and the neurons that express that same protein synapse in the same glomerulus of the olfactory bulb. Insect olfactory neurons have broadly similar receptor proteins and connection patterns.
- Transduction mechanisms of chemoreceptors are somewhat diverse. Olfactory receptor proteins are G protein–coupled receptors, which produce a second messenger (cAMP or, in some cases, IP_3). Taste sensory transduction may involve G protein–coupled receptors (sweet, bitter) or direct gating of ion channels (salty, sour).

Electroreception

Several kinds of fish possess electric organs (see Box 17.1). The strong electric organs of electric eels, rays, and catfish produce a powerful discharge used to stun prey and potential predators. Other less-familiar groups of fish have weak electric organs that, together with electroreceptors on the body surface, are used for orientation in the murky tropical waters that these fish inhabit. The mode of action of electro-orientation in weakly electric fish was first suggested by Hans Lissmann's behavioral studies in 1958 and was later confirmed physiologically.

Three groups of teleosts (gymnotids, mormyrids, and gymnarchids) can orient by producing electric fields around themselves by discharges of their electric organs, and detecting the perturbations of these fields by nearby objects. Figure 13.37 shows the electric field that is set up by electric-organ discharge in *Gymnarchus*. An electrical conductor such as a metal ball or plate will concentrate the lines of electric force, increasing the field density at the nearby surface of the fish. Poor conductors such as glass or plastic will conversely spread the lines of force and decrease the electric field density. These perturbations are sensed by electroreceptors in tuberous organs that are part of the lateral line system on the sides of the fish. Although this system would appear relatively insensitive, the fish can use it to make impressive discriminations. For example, a *Gymnarchus* was trained to discriminate between two porous clay flowerpots—one containing a glass rod 2 mm wide, and the other containing a glass rod 0.8 mm wide!

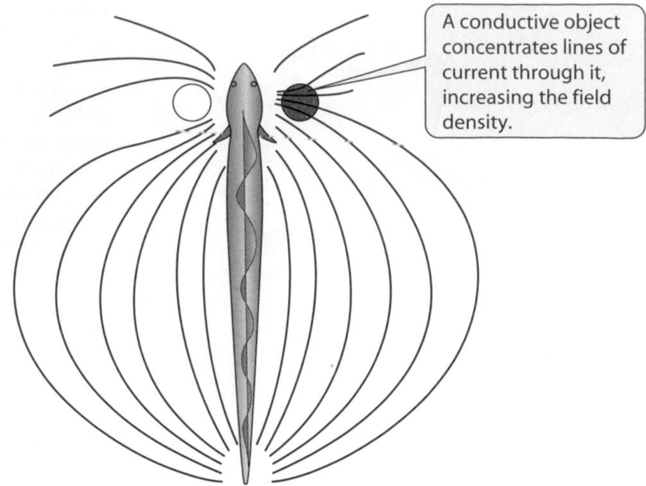

Figure 13.37 Electroreception in electric fish The electric field around a weakly electric fish (*Gymnarchus*) is set up by the electric organ in the tail and is sensed by anterior electroreceptors.

Electroreceptors and electric organs of fish are also used in other ways than the electric field orientation just described. Weakly electric fish also employ their electric organs and electroreceptors for social communication. Moreover, some other fishes that lack electric organs nevertheless have electroreceptors, which they use in locating prey and perhaps in other ways. For example, elasmobranch fish such as sharks and skates possess sense organs termed *ampullae of Lorenzini*, which are associated with their lateral line systems. Ampullae of Lorenzini are sensitive to a wide variety of stimuli (touch, temperature, salinity, and electrical stimuli).

Behavioral and physiological tests indicate that the adequate stimulus is electrical, and that sharks can detect voltage gradients as small as 0.005 V/cm. Calculations suggest that this electrical sensitivity would allow a shark to detect a camouflaged or buried flounder at 30 cm by sensing the electrical activity of its respiratory muscles. Sharks will attack buried electrodes rather than ground flounder extracts, a finding that suggests that electroreception is used for prey detection. Other calculations suggest that a shark swimming east or west through Earth's magnetic field will generate an induced current strong enough to stimulate the ampullary electroreceptors, but it is not clear whether sharks use this potential information for directional navigation.

Electroreceptors are just one of a variety of sensory receptors that seem foreign to humans, because they impart to some animals an environmental sensitivity that humans lack. In Chapter 16 we consider other sensory receptors that play roles in animal orientation and navigation, such as magnetoreceptors and sensory detectors of polarized ultraviolet light. Such examples remind us of the diversity of sensory processes, and the varied behavioral roles they play. Investigators of animal behavior use the word *Umwelt* to mean the sensory or perceptual world of an animal. The term suggests that an understanding of an animal's sensory world—the kinds of sensory information it receives and uses—is critical to understanding its behavioral ecology.

Study Questions

1. Suppose that a sensory neuron in the periphery generates the following train of action potentials, and synaptically excites an interneuron in the central nervous system:

What does this signal mean? What additional information does the interneuron need in order to decode the message?

2. Suppose that in a particular sensory receptor, the amplitude of the receptor potential increases linearly with the log of the stimulus intensity. The receptor potential depolarizes the cell to produce a train of action potentials, and the frequency of the action potentials increases linearly with increasing receptor-potential depolarization (above threshold). There is a maximum frequency of action potentials, beyond which additional increase in stimulus intensity does not further increase action-potential frequency. Graph these results as the amount of response as a function of stimulus intensity (or log intensity). If these results are typical of many receptors, what do they tell you about how sensory receptors encode stimulus intensity?
3. What is adaptation of a sensory receptor? At what stage(s) does it occur? What are its functions?
4. Where is rhodopsin localized in a vertebrate retinal rod? Where is the receptor potential generated? How does the transduction mechanism of the rod connect these two sites? How does it increase light sensitivity?
5. Diagram and describe how lateral inhibition via horizontal cells produces the surround of a retinal ganglion cell's receptive field.

6. Contrast the receptive-field properties of a simple cell and a complex cell in a mammal's visual cortex. How would you characterize these neurons as feature detectors?

7. Insects and vertebrates have analogous proprioceptors monitoring muscle length and tension, joint positions, limb forces, and so on. Insects, however, do not have statocysts to monitor the direction of gravitational force. How could an insect detect gravity when at rest? What could be the possible adaptive advantage of the *absence* of statocysts for a flying insect?

8. Vertebrates use acoustico-lateralis hair cells to detect water currents in lateral line organs of fish, gravity and acceleration in the labyrinth, and sound in the cochlea. How can similar receptor cells mediate all these responses?

9. Compare the mechanisms and accuracy of auditory localization of a sound source in a human, an owl, and a bat.

10. Mechanisms of chemoreceptor transduction are of two broad kinds, which parallel the two major mechanisms of synaptic transmission. Describe these, and suggest an evolutionary speculation for the similarities.

Key References

Carew, T. J. 2000. *Behavioral Neurobiology: The Cellular Organization of Natural Behavior*. Sinauer, Sunderland, MA. An excellent, engaging text on neuroethology, the interface of neurobiology and analysis of natural animal behavior.

Fain, G. L. 2003. *Sensory Transduction*. Sinauer, Sunderland, MA. By far the best and most current book on how sensory receptors work.

Hubel, D. H. 1988. *Eye, Brain, and Vision*. Scientific American Library, New York. A readable book on the mammalian visual system, emphasizing Hubel and Wiesel's characterizations of central visual neurons.

Selected References

Ashmore, J. F., and P. J. Kolston. 1994. Hair cell based amplification in the cochlea. *Curr. Opin. Neurobiol.* 4: 503–508.

Axel, R. 1995. The molecular logic of smell. *Sci. Amer.* 273(4): 154–159.

Bargmann, C. I., and J. M. Kaplan. 1998. Signal transduction in the *Caenorhabditis elegans* nervous system. *Annu. Rev. Neurosci.* 21: 279–308.

Barth, F. G., and A. Schmid (eds.). 2001. *Ecology of Sensing*. Springer, Berlin.

Bell, J., S. Bolinowski, and M. H. Holmes. 1994. The structure and function of Pacinian corpuscles: A review. *Prog. Neurobiol.* 42: 79–128.

Bradbury, J. W., and S. L. Vehrencamp. 1998. *Principles of Animal Communication*. Sinauer, Sunderland, MA.

Buck, L. B. 1996. Information coding in the vertebrate olfactory system. *Annu. Rev. Neurosci.* 19: 517–544.

Catania, K. C. 2002. The nose takes a starring role. *Sci. Amer.* 287(1): 54–59.

Corey, D. P. 2001. Transduction and adaptation in vertebrate hair cells. In C. I. Berlin and R. P. Bobbin (eds.), *Hair Cells: Micromechanics and Hearing*, pp. 1–25. Singular Thompson Learning, San Diego, CA.

Corey, D. P., and S. D. Roper (eds.). 1992. *Sensory Transduction: Society of General Physiologists, 45th Annual Symposium, Marine Biological Laboratory, Woods Hole, Massachusetts, 5–8 Sept. 1991.* Rockefeller University Press, New York.

Corwin, J. T., and M. E. Warchol. 1991. Auditory hair cells: Structure, function, development, and regeneration. *Annu. Rev. Neurosci.* 14: 301–333.

Delcomym, F. 1998. *Foundations of Neurobiology.* Freeman, New York.

Devor, M. 1996. Pain mechanisms. *Neuroscientist* 2: 233–244.

Doty, R. L. (ed.). 1995. *Handbook of Olfaction and Gustation.* Dekker, New York.

Døving, K. B., and D. Trotier. 1998. Structure and function of the vomeronasal organ. *J. Exp. Biol.* 201: 2913–2925.

Dowling, J. 1987. *The Retina: An Approachable Part of the Brain.* Belknap Press, Cambridge, MA.

Dulac, C. 1997. How does the brain smell? *Neuron* 19: 477–480.

Dulac, C. 2000. Sensory coding of pheromone signals in mammals. *Curr. Opin. Neurobiol.* 10: 511–518.

Dusenbery, D. B. 1992. *Sensory Ecology: How Organisms Acquire and Respond to Information.* Freeman, New York.

Farbman, A. I. 1992. *Cell Biology of Olfaction.* Cambridge University Press, Cambridge.

Fernald, R. D. 2000. Evolution of eyes. *Curr. Opin. Neurobiol.* 10: 444–450.

Finger, T. E., W. L. Silver, and D. Restrepo (eds.). 2000. *The Neurobiology of Taste and Smell*, 2nd ed. Wiley-Liss, New York.

Frisby, J. P. 1980. *Seeing: Illusion, Brain and Mind.* Oxford University Press, Oxford, UK.

García-Añoveros, J., and D. P. Corey. 1997. The molecules of mechanosensation. *Annu. Rev. Neurosci.* 20: 567–594.

Gilbertson, T. A., S. Damak, and R. F. Margolskee. 2000. The molecular physiology of taste transduction. *Curr. Opin. Neurobiol.* 10: 519–527.

Halliday, T. 1998. *The Senses and Communication.* Springer, New York.

Hansson, B. S. (ed.). 1999. *Insect Olfaction.* Springer, Berlin.

Hildebrand, J. G., and G. M. Shepherd. 1997. Mechanisms of olfactory discrimination: Converging evidence for common principles across phyla. *Annu. Rev. Neurosci.* 20: 595–631.

Hudspeth, A. J. 1989. How the ear's works work. *Nature* 341: 397–404.

Hughes, H. C. 1999. *Sensory Exotica: A World beyond Human Experience.* MIT Press, Cambridge, MA.

Kinnamon, S. C., and R. F. Margolskee. 1996. Mechanisms of taste transduction. *Curr. Opin. Neurobiol.* 6: 506–513.

Knudsen, E. 1981. The hearing of the barn owl. *Sci. Amer.* 245(6): 113–125.

Knudsen, E. I, and Konishi, M. 1978. Center-surround organization of auditory receptive fields in the owl. *Science* 202: 778–780.

Kolb, H. 2003. How the retina works. *Amer. Sci.* 91: 28–35.

Kruger, L. 1996. *Pain and Touch.* Academic Press, San Diego, CA.

Land, M., and R. Fernald. 1992. The evolution of eyes. *Annu. Rev. Neurosci.* 15: 1–29.

Lindemann, B. 2001. Receptors and transduction in taste. *Nature* 413: 219–225.

Lindemann, B. 1996. Taste reception. *Physiol. Rev.* 76: 719–766.

Lohmann, K. J., and S. Johnsen. 2000. The neurobiology of magnetoreception in vertebrate animals. *Trends Neurosci.* 23: 153–159.

Martin, K. 1994. A brief history of the "feature detector." *Cerebral Cortex* 4: 1–7.

McIlwain, T. T. 1996. *An Introduction to the Biology of Vision.* Cambridge University Press, New York.

McLaughlin, S., and R. F. Margolskee. 1994. The sense of taste. *Amer. Sci.* 82: 538–545.

Mombaerts, P. 1999. Molecular biology of odorant receptors in vertebrates. *Annu. Rev. Neurosci.* 22: 487–509.

Montmayeur, J.-P., and H. Matsunami. 2002. Receptors for bitter and sweet taste. *Curr. Opin. Neurobiol.* 12: 366–371.

Nobili, R., F. Mammano, and J. Ashmore. 1998. How well do we understand the cochlea? *Trends Neurosci.* 21: 159–167.

Oliver, G., and P. Gruss. 1997. Current views on eye development. *Trends Neurosci.* 20: 415–421.

Palmer, S. E. 1999. *Vision Science: Photons to Phenomenology.* MIT Press, Cambridge, MA.

Pickles, J. O. 1993. Early events in auditory processing. *Curr. Opin. Neurobiol.* 3: 558–562.

Reed, R. R. 1992. Signaling pathways in odorant detection. *Neuron* 8: 205–209.

Ressler, K. J., S. L. Sullivan, and L. B. Buck. 1994. A molecular dissection of spatial patterning in the olfactory system. *Curr. Opin. Neurobiol.* 4: 588–596.

Rodieck, R. W. 1998. *The First Steps in Seeing*. Sinauer, Sunderland, MA.

Schild, D., and D. Restrepo. 1998. Transduction mechanisms in vertebrate olfactory receptor cells. *Physiol. Rev.* 78: 429–466.

Shepherd, G. M. 1994. Discrimination of molecular signals by the olfactory receptor neuron. *Neuron* 13: 771–790.

Simmons, P. J., and D. Young. 1999. *Nerve Cells and Animal Behavior*, 2nd ed. Cambridge University Press, Cambridge.

Smith, D. V., and R. F. Margolskee. 2001. Making sense of taste. *Sci. Amer.* 284(3): 32–39.

Walker, R. G., A. T. Willingham, and C. S. Zuker. 2000. A *Drosophila* mechanosensory transduction channel. *Science* 287: 2229–2234.

Wolken, J. J. 1995. *Light Detectors, Photoreceptors, and Imaging Systems in Nature*. Oxford University Press, New York.

Zeki, S. 1993. *A Vision of the Brain*. Blackwell, London.

14 Endocrine and Neuroendocrine Physiology

Marine iguanas are lizards that live in the Galápagos archipelago. Although they are excellent swimmers, they do not venture far from the shores of their home volcanic islands. They are herbivores, and obtain their food by grazing on algae that grow in the intertidal zone. They can digest the algae only with the aid of microorganisms in their hindguts. Adult marine iguanas have few predators, except hawks that take females during the nesting season. Because of their algal diet and aquatic foraging behavior, marine iguanas don't appear to compete for food resources with other vertebrate animals, or with one another. However, they are exposed to natural and human-induced environmental stressors.

For the past several years, researchers have regularly taken blood samples from Galápagos marine iguanas to measure their levels of *corticosterone*, the main stress hormone in reptiles. As we'll see later in this chapter, corticosterone plays an important role in regulating carbohydrate, lipid, and protein metabolism in order to support survival in the face of a threat.

In 2001, an oil tanker ran aground near one of the Galápagos islands and spilled oil into the waters. Serendipitously, the researchers had taken blood samples from marine iguanas just 3 days prior to the oil spill on an island that was exposed to low-level contamination (32 km west of the spill). When they took samples again 10 days later, they found that corticosterone levels had tripled from those seen before the tanker ran aground. This finding strongly suggested that the spill had placed a stress on the population of exposed animals. By 11 months after the spill, more than 60% of the exposed animals had died. Marine iguanas on islands in the archipelago not exposed to oil contamination showed no increase in corticosterone levels and no increase in mortality. The researchers suggested that the animals exposed to the spill died because the microorganisms in their hindguts were killed by oil contamination of the algae they consumed. Without the microorganisms, the iguanas were unable to digest the algae, and they starved. This cautionary example illustrates both the sensitivity of wild animals to human perturbations of the environment and the interrelatedness of multiple systems required for survival.

In this chapter, we will examine the principles of hormonal regulation by focusing on a few well-studied examples. Our goal is to give you the tools to apply these principles in understanding the functions of hormones discussed in future chapters. As we explore endocrine physiology, we'll see that hormones play essential and integral roles in maintaining homeostasis. Because hormones influence a broad range of physiological processes, essentially every cell in an an-

The marine iguana (*Amblyrhynchus cristatus*)

imal participates in some endocrine function, and nearly every hormone participates in more than one physiological process.

Introduction to Endocrine Principles

In Chapter 10, we compared rapid, "addressed" neural control of physiological processes with slow, "broadcast" endocrine control. Chemical signals are used to achieve both types of control. However, in neural control, neurotransmitters released from axon terminals of neurons diffuse short distances to bind to receptor molecules on postsynaptic cells. Their short travel time to discrete sites ensures rapid, pinpointed control. Enzymes or reuptake mechanisms rapidly inactivate most neurotransmitters, so the neural signal is quickly terminated.

By contrast, hormones secreted from endocrine or neuroendocrine cells travel in the blood to distant target cells where they exert their effects. Whereas neurotransmitter molecules from a neuron typically reach a single postsynaptic cell, hormone molecules carried through the bloodstream can influence large populations of target cells, as long as the target cells express receptor molecules for the hormone. Thus transport of hormones over long distances permits widespread responses. These responses are initiated slowly, relative to responses to neural signals, because hormones require travel time to reach target cells. Furthermore, some hormones control gene expression and the synthesis of proteins by target cells, so the responses they initiate are exhibited only after a delay, when protein synthesis is accomplished. These responses may last as long as hours or days.

Processes controlled by endocrine systems in both vertebrate and invertebrate animals include water balance, metabolism, coping with a hostile environment, reproduction, and growth and development. Although researchers have accumulated a great deal of detailed knowledge regarding the functions and roles of hormones in vertebrate animals, our understanding of the endocrine systems of many invertebrate groups is still highly incomplete. Nevertheless, physiologists have found that basic principles of endocrine function—our focus in this chapter—apply to both vertebrate and invertebrate animals.

Defined specifically, a **hormone** is a metabolically produced chemical substance released by nonneural endocrine cells or by neurons; it exerts regulatory influences on the function of other distant cells reached via the blood; and it is effective at very low concentrations (as little as 10^{-12} *M*). The secretory cells that produce hormones secrete them into the surrounding extracellular fluid, from which they diffuse into capillaries.[1] The secretory cells may be organized into discrete organs termed **endocrine glands** (also called *ductless glands* because they lack outflow ducts), or they may be isolated cells distributed among cells of other tissues. Table 14.1 lists the major hormones and glands in mammals and describes the vast repertoire of their major functions.

Some substances are unambiguously hormones, such as thyroid hormones, secreted by the thyroid gland, and gastrin, secreted by G cells in the gastric mucosa of the lower part of the mammalian stomach (see Chapter 4). Many substances carried in the blood—CO_2, for example—are clearly not hormones. CO_2 is produced by metabolism and signals the respiratory centers of some animals to increase their breathing. However, specialized secretory cells do not release it, and it is found continuously in the blood at relatively high concentrations. Not all compounds, however, are so easily categorized. Many chemical signals released by certain cells affect the function of other nearby cells located in the same organ or tissue, but do not enter the circulatory system. As described in Chapter 10, **autocrine** and **paracrine** substances act in many ways like hormones, but are usually categorized separately. Furthermore, the same compound may be used both as a hormone and as another type of chemical signal in the same organism. In mammals, for example, cholecystokinin (CCK) is not only a hormone secreted by cells in the intestine, but also functions as a neurotransmitter or neuromodulator in the central nervous system. Recent studies have shown that **intracrines** are another example of signaling molecules that have different roles in different contexts. Intracrines are peptide growth factors or hormones that function *within* cells in addition to performing traditional hormonal, paracrine, or autocrine functions. Intracrines are either retained within the cell that synthesized them or internalized from the extracellular space.

Hormones bind to receptor molecules expressed by target cells

Although a hormone circulates past many cells, it interacts only with certain cells, called **target cells**, that respond to it. A target cell expresses specific *receptor molecules* that are recognized by the hormone. Consider thyroid hormones for example. These hormones, secreted by the thyroid gland, exert a wide range of metabolic, structural, and developmental effects on many different tissues (see Table 14.1). They have such widespread effects because many different cells of the body possess receptor molecules that recognize thyroid hormones. Typical target cells express thousands of receptor molecules for a particular hormone. In addition, many target cells express separate populations of different types of receptor molecules, so that they are capable of responding to more than one hormone. The *sensitivity* of a target cell to a particular hormone depends on the number of functional receptor molecules the target cell expresses for that hormone. The sensitivity of a target cell to a particular hormone can change under different conditions because the number of receptor molecules that recognize that hormone can increase (by upregulation) or decrease (by downregulation). These variations in the types and numbers of receptor molecules expressed by target cells contribute to the immense versatility of hormonal regulation in animals.

Concentrations of hormones in the blood vary

For hormones to serve as physiological regulators, their rates of synthesis and secretion must be controlled. Often, neurons or other hormones control these processes. Most endocrine cells synthesize and release some hormone all the time, but the rate of release is variable, depending on mechanisms of control. In general, the higher the rate at which a hormone is secreted, the higher its concentration in the

[1] In animals with open circulatory systems, the blood and extracellular fluid blend to form hemolymph (see Chapter 23). In these animals, hormones are released directly into, and circulated in, the hemolymph.

TABLE 14.1 Summary of major blood-borne hormones in mammals

Endocrine tissue	Hormone	Class of molecule	Main functions
Adrenal cortex	Aldosterone (mineralocorticoid)	All steroids	Stimulates Na reabsorption and K secretion in kidney
	Androgens		Act on bone to cause growth spurt at puberty; increase sex drive in females by action on brain
	Glucocorticoids		Part of stress response; affect metabolism of many tissues to increase blood glucose and cause protein and fat catabolism
Adrenal medulla	Epinephrine and norepinephrine	Catecholamines	Part of stress response; influence cardiovascular function and organic metabolism of many tissues
Anterior pituitary gland (adenohypophysis)	Prolactin (PRL)	All peptides	Promotes development of mammary gland during pregnancy; stimulates milk synthesis and secretion during lactation; promotes caring for young by both males and females in many species of fish, birds, and mammals
	Growth hormone (GH, somatotropin)		Stimulates growth and metabolism of bone and soft tissues; promotes protein synthesis, glucose conservation, and fat mobilization
			Stimulates secretion of insulin-like growth factors (IGFs, also called somatomedins) by liver
	Melanocyte-stimulating hormone (MSH)		In mammals, also produced in hypothalamus; reduces appetite, suppresses immune system; does not determine skin coloration; (causes skin darkening in amphibians, reptiles, and fish by stimulating dispersal of granules containing the pigment melanin)
	Adrenocorticotropic hormone (ACTH, corticotropin)		Stimulates glucocorticoid secretion by adrenal cortex and supports adrenal cortical tissue (tropic action)
	Thyroid-stimulating hormone (TSH, thyrotropin)		Stimulates hormone synthesis and secretion by thyroid gland; is tropic
	Follicle-stimulating hormone (FSH)		Stimulates sperm production in the testis, follicular growth in the ovary, and sex hormone production in male and female gonads; is tropic
	Luteinizing hormone (LH)		Stimulates sex hormone production in male and female gonads, ovulation, and development of corpus luteum; is tropic
	γ-lipotropin and β-endorphin		Possibly decreased perception of pain
Fat (adipose) tissue	Leptin	Peptide	Targets hypothalamus and other tissues to influence food intake (promotes weight loss), metabolic rate, and reproductive functions
Gastrointestinal (GI) tract	Stomach: gastrin; ghrelin Small intestine: secretin, cholecystokinin, glucose-dependent insulinotropic peptide (GIP); motilin	Peptides	Aid in digestion and absorption of nutrients by various actions on GI tract, liver, pancreas, and gallbladder
Gonad: ovary and testis	Estrogens and progesterone	Steroids	Maturation of follicle; ovulation; secondary sex characteristics; prepare uterus for pregnancy by actions during estrous and menstrual cycles
	Inhibin	Peptide	Inhibits secretion of FSH from anterior pituitary in males and females
	Activins	Peptide	Stimulate secretion of FSH; promote spermatogenesis and follicle development
	Relaxin	Peptide	May make cervix and pelvic ligaments more pliable
	Müllerian inhibiting substance	Peptide	Functions in male fetal development
	Testosterone (the major androgen)	Steroid	Sperm production and secondary sexual characteristics; sex drive in males
Heart: Cells in atria	Atrial natriuretic peptide (ANP)	Peptide	Promotes Na and water excretion in kidney

TABLE 14.1 ***(continued)***

Endocrine tissue	Hormone	Class of molecule	Main functions
Hypothalamus	Release- and release-inhibiting hormones	All peptides (except DA)	Stimulate or inhibit secretion of anterior pituitary hormones; releasing hormones are tropic
	Corticotropin-releasing hormone (CRH)		
	Thyrotropin-releasing hormone (TRH)		
	Growth hormone–releasing hormone (GHRH)		
	Somatostatin (SS, GHIH)		
	Gonadotropin-releasing hormone (GnRH)		
	Dopamine (DA, prolactin-inhibiting hormone)		
	Melanocyte-stimulating hormone–inhibiting hormone (MSH-IH)		
Kidney	Renin	Peptide	Cleaves angiotensinogen to form angiotensin I
	Erythropoietin (EPO)	Peptide	Stimulates red blood cell production in bone marrow
	Calcitriol (active form of vitamin D)	Steroid	Increases body calcium
Liver	Angiotensinogen	Peptide	Converted in blood to angiotensin II, which stimulates aldosterone secretion, promotes thirst, causes vasoconstriction, and secretion of vasopressin
	Insulin-like growth factors (also called somatomedins)	Peptides	Cell division and growth of many tissues (IGFs are also secreted by other tissues, such as muscle)
Pancreas (endocrine cells)	Insulin (B cells)	Peptides	Promotes uptake and storage of nutrients by most cells
	Glucagon (A cells)		Maintains blood levels of nutrients after a meal and during stress
	Somatostatin (D cells)		Inhibits digestion and absorption of nutrients by the gastrointestinal tract
Parathyroid gland	Parathyroid hormone	Peptide	Increases Ca and decreases plasma phosphate levels in the blood by action on kidney and bone; stimulates vitamin D activation by action at kidney, bone, and intestine
Pineal gland	Melatonin	Amine	Controls circadian rhythms; seasonal breeding, migration, hibernation; possibly sexual maturity; causes skin lightening in amphibians and lampreys
Placenta of pregnant female mammal	Estrogens and progesterone	Steroids	Support pregnancy and fetal and maternal development
	Chorionic gonadotropin	Peptide	Extends functional life of corpus luteum in ovary (horses, primates)
	Relaxin	Peptide	May make cervix and pelvic ligaments more pliable
	Placental lactogen	Peptide	May support mammary gland development during pregnancy; may change maternal metabolism to support fetus
Posterior pituitary gland (neurohypophysis)	Vasopressins (ADH)	Peptides (produced in hypothalamus and transported to posterior pituitary)	Water reabsorption at kidney; vasoconstriction
	Oxytocin		Stimulates uterine contraction during parturition and milk ejection from mammary glands during suckling
Skin	Vitamin D	Steroid	Increases body calcium
Thymus gland	Thymosin, thymopoietin	Peptides	Stimulate T-lymphocyte development and proliferation
Thyroid gland	Tetraiodothyronine (T4) and Triiodothyronine (T3) (thyroid hormones)	Iodinated amines	Increase metabolism of many tissues; necessary for normal growth and development of the nervous system
	Calcitonin	Peptide	Decreases blood Ca levels in some animals by action on bone (secreted by C cells)

Sources: After Bentley1998; Henderson 2000; Schmidt-Nielsen 1997; Sherwood 2004; Silverthorn 2004; and Widmaier, Raff, and Strang 2004.

TABLE 14.2 Peptide, steroid, and amine hormones

Property	Peptides	Steroids	Amine hormones: Catecholamines	Amine hormones: Thyroid hormones	Amine hormones: Melatonin
Site of secretion	Most sites in Table 14.1, except adrenal cortex and medulla, thyroid gland, and pineal gland	Adrenal cortex, gonads, and placenta	Adrenal medulla	Thyroid gland	Pineal gland
Structure	Chains of amino acids	Derived from cholesterol	Derived from tyrosine	Derived from tyrosine and iodine	Derived from tryptophan
Solubility	Water-soluble	Lipid-soluble	Water-soluble	Lipid-soluble	Water-soluble
Synthesis and storage	Synthesized at rough ER, processed in Golgi apparatus; stored in vesicles in advance of use	Synthesized on demand in intracellular compartments; not stored	Synthesized in the cytoplasm and stored in vesicles ("chromaffin granules")	Made prior to use and stored in a colloid island within the gland	Synthesized in the cytoplasm and stored in vesicles
Secretion	Exocytosis	Simple diffusion through cell membrane	Exocytosis	Simple diffusion through cell membrane	Exocytosis
Transport	Dissolved in plasma; some bound to carrier proteins	Bound to carrier proteins	Dissolved in plasma	Bound to carrier proteins	Dissolved in plasma
Half-life	Minutes	Hours	Seconds to minutes	Days	Minutes
Location of receptor molecules	Surface of target cell membrane	Cytoplasm or nucleus (some steroids bind to cell-surface receptors)	Surface of target cell membrane	Nucleus	Surface of target cell membrane
Action at target cell	Activate second-messenger systems or alter membrane channels	Alter gene expression; activated genes initiate transcription and translation	Activate second-messenger systems	Alter gene expression; activated genes initiate transcription and translation	Activate second-messenger systems
Response of target cell	Change activity of preexisting proteins, some of which may induce new protein synthesis	Synthesize new proteins; some may change activity of preexisting proteins	Change activity of preexisting proteins	Synthesize new proteins	Change activity of preexisting proteins

Sources: After Sherwood 2004; Silverthorn 2004; and Widmaier, Raff, and Strang 2004.

blood, and the greater its effect on target cells. Because hormone molecules secreted into the blood are enzymatically degraded at their targets or by organs (such as the liver and kidneys in vertebrates), they do not circulate indefinitely. The blood concentration of a hormone represents a balance between the *rate of addition* of hormone to the blood (by secretion) and the *rate of removal* of hormone from the blood (by metabolic destruction and excretion). Hormone concentration depends primarily on the rate of addition to the blood because the rate of removal is relatively constant. A hormone's **half-life**—the time required to reduce the concentration by one-half—indicates its rate of removal from the blood and thus the duration of its activity.

Some hormones may be converted to a more active form after secretion by a process termed **peripheral activation**. For example, thyroid hormone is secreted mainly as a four-iodine compound also known as tetraiodothyronine, or T_4. After T_4 is secreted, target and other tissues enzymatically remove one iodine to form triiodothyronine, or T_3, which is more physiologically active than T_4.

Most hormones fall into three chemical classes

Table 14.2 summarizes the characteristics of the following three chemical classes of hormones:

1. **Steroid hormones** are synthesized from cholesterol (Figure 14.1). In vertebrate animals, the gonads and the adrenal cortex secrete steroid hormones, as do the placentas of pregnant mammals. The molting hormones of arthropods (e.g., ecdysone) are also steroids. Steroid hormones are lipid-soluble, so they can pass through cell membranes to reach receptor molecules, located inside their target cells.
2. **Peptide** and **protein hormones** are structured from chains of amino acids (Figure 14.2). In vertebrates, they include antidiuretic hormones, insulin, and growth hormone. Examples of peptide and protein hormones in invertebrates include the gamete-shedding hormone of sea stars and the diuretic hormones of insects. Peptide and protein hormones vary enormously in

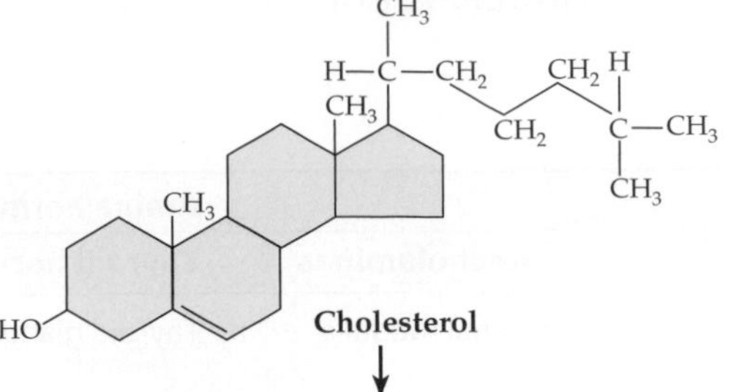

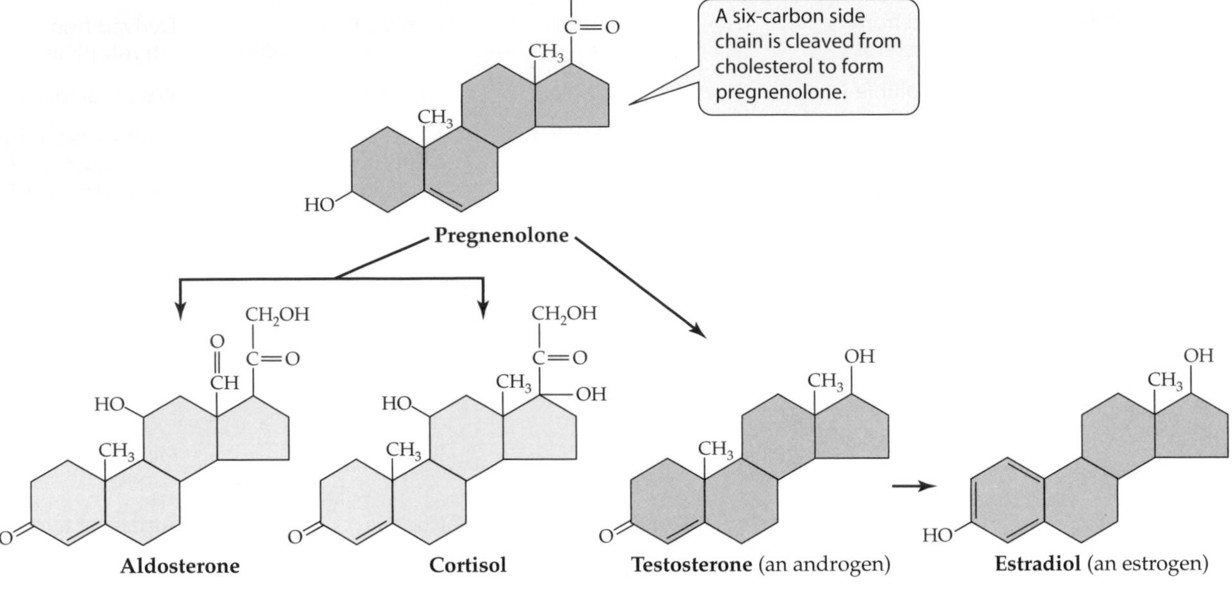

Figure 14.1 Steroid hormones are derived from cholesterol Steroidogenesis begins with the formation of pregnenolone. Different steroid-secreting cells possess different enzymes that modify pregnenolone to produce specific steroid hormones.

molecular size, from tripeptides (consisting of just 3 amino acid residues, such as thyrotropin-releasing hormone) to proteins containing nearly 200 amino acids (such as growth hormone). Often hormones consisting of assemblages of amino acids are simply called "peptide hormones" (blurring the size distinction), and we will usually follow that practice. Peptide hormones are soluble in aqueous solutions.

3. **Amine hormones** are modified amino acids (Figure 14.3). **Melatonin**, secreted by the vertebrate pineal gland (see Chapter 10), is derived from tryptophan, whereas the **catecholamines** and **iodothyronines** are derived from tyrosine. *Catecholamines* are found widely as synaptic transmitter substances in both invertebrates and vertebrates. However, three catecholamines also serve as hormones in vertebrates: epinephrine (also called adrenaline), norepinephrine (noradrenaline), and dopamine. *Iodothyronines*, the thyroid hormones, are found only in vertebrates. They are synthesized by the thyroid gland and have the unique property of being rich in iodine. Whereas melatonin and the catecholamines are soluble in water, the iodothyronines are soluble in lipids.

In the past, it was thought that most water-soluble hormones were transported simply dissolved in the blood, and that only lipid-soluble steroids and iodothyronines were transported bound to *carrier proteins*. However, it now appears that many peptide hormones are also transported bound to specific carrier-protein molecules. Water-soluble carrier proteins bind to hormone molecules reversibly and noncovalently, with varying degrees of affinity. Free and bound forms of the hormone are in equilibrium in the blood. Only the free form of the hormone is physiologically active. When some of the free hormone molecules leave a capillary to bind to receptors of a target tissue, some bound hormone molecules unbind from the carrier protein in the capillary, and the equilibrium in the blood is

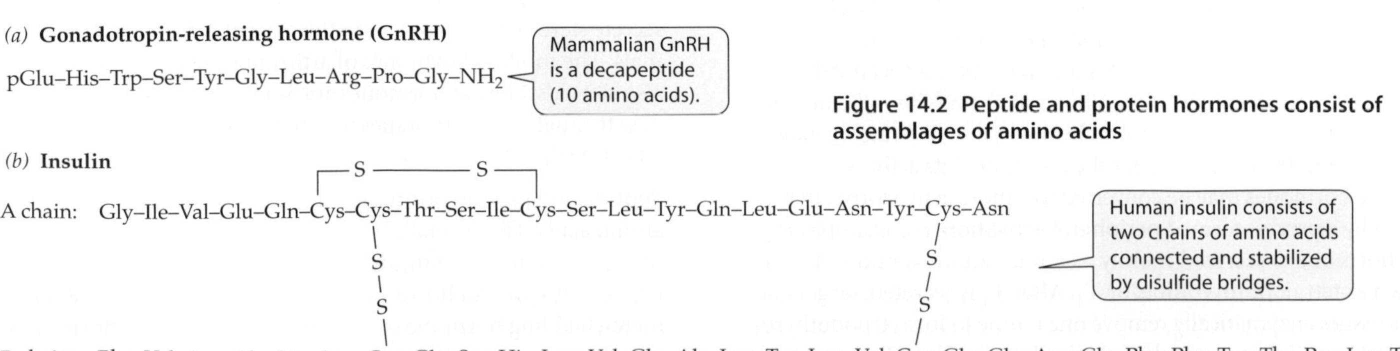

Figure 14.2 Peptide and protein hormones consist of assemblages of amino acids

Tyrosine

Catecholamines

Dopamine

Norepinephrine

Epinephrine

Thyroid hormones

Thyroxine (tetraiodothyronine, T_4)

Triiodothyronine (T_3)

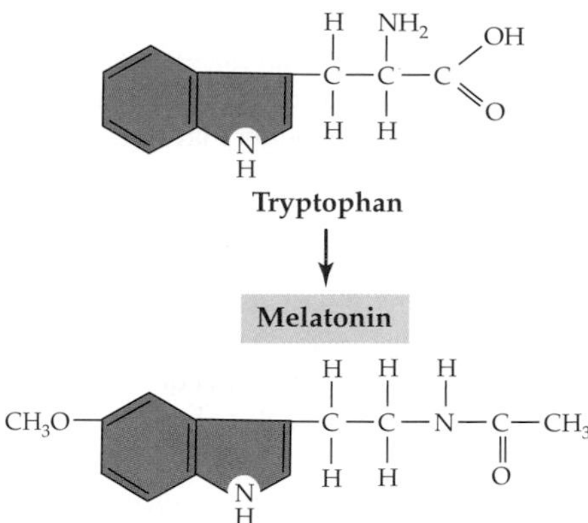

Figure 14.3 Amine hormones are derived from amino acids Specific biochemical pathways convert tyrosine into catecholamines or iodothyronines. Tryptophan is modified biochemically to produce melatonin.

maintained. Thus carrier proteins provide a reservoir of bound hormone that can be drawn upon. Carrier proteins also protect hormones from rapid inactivation and excretion and thus extend their half-life. In humans, the half-life of the steroid stress hormone cortisol is 60 to 90 minutes, and that of thyroxine is several days. In contrast, peptide hormones have half-lives in the range of a few minutes To maintain a steady level of a peptide hormone in the blood for a period longer than a few minutes, the endocrine cells must continue to secrete it. The water-soluble catecholamine epinephrine has a half-life of no more than a minute or two!

Hormone molecules exert their effects by producing biochemical changes in target cells

To initiate changes in target cells, hormones first bind to specific receptor molecules. Some hormones bind to intracellular receptors and others to cell-surface receptors. Three types of receptor molecules are important in mediating hormone actions: intracellular receptors, G protein–coupled membrane receptors, and enzyme-linked membrane receptors (see Chapter 2).

Lipid-soluble hormones (steroids and iodothyronines) bind to intracellular receptors. Because they are lipid-soluble, these hormones readily enter target cells by diffusing through the lipid bilayer of the cell membrane (see Figure 2.23). Their receptors are located either in the cytoplasm or in the nucleus. When the hormone molecule binds to the receptor molecule, it forms a hormone–receptor complex that acts as a transcription factor that interacts with the target cell's DNA to alter gene expression. By turning processes of transcription and translation on or off, the hormone directly influences the synthesis of proteins by the target cell. New proteins, which can be enzymes or structural proteins, carry out the target cell's physiological response. Because the production of new proteins requires time, there is a delay (ranging from many minutes to hours) between the hormone's binding to the intracellular receptor and the target cell's response. Once the proteins are synthesized, the response lasts until they are degraded. In addition, recent studies have discovered that certain target cells of some steroid hormones, such as aldosterone and estrogen, express *cell-surface* receptors for those hormones. When steroid hormones bind to receptors on the cell membrane, they do not exert their effects by genomic means, but instead function like water-soluble hormones.

The water-soluble peptide and catecholamine hormones do not diffuse into cells. Their receptor molecules are located in the cell membrane and have external hormone-binding sites. These receptors typically mediate hormone actions by altering membrane permeability or by activating an intracellular second-messenger system by way of a G protein (see Figures 2.25 and 2.26). The second messengers activated by hormones include cAMP, cGMP, Ca^{2+} ions, and phosphoinositides. A few peptide hormones, such as insulin and growth hormone, bind to enzyme-linked membrane receptors. Peptide hormones exert their effects primarily by changing the activities of existing proteins, although some may also alter gene activities. Target cells can change their physiology quickly by using preexisting proteins, so their responses to these hormones can be measured within minutes.

Summary: Introduction to Endocrine Principles

- Hormone molecules are synthesized, stored, and released by endocrine cells, travel through the circulating blood, exert their effects on target tissues, and are metabolically destroyed or excreted from the body.
- The magnitude of a hormone's effect depends on its concentration in the blood, which depends on a balance between the rate of synthesis and the rate of degradation or excretion. The rate of synthesis and secretion of a particular hormone is often governed by another hormone; some endocrine cells also receive neural input.

- Hormones are usually categorized into three main classes: steroids, peptides and proteins, and amines (melatonin, catecholamines, and iodothyronines). The same chemical messenger may function as a hormone in one context and as a neurotransmitter or neuromodulator in another.
- Hormones are carried in the blood in very low concentrations. Their half-lives vary depending on their chemical class, ranging from seconds to hours or days. Carrier proteins in the blood transport many hormones. Free and bound hormone molecules are in equilibrium in the blood. Only free hormone molecules are able to bind to receptor molecules in or on target cells.
- Both lipid-soluble and water-soluble hormones initiate biochemical changes in their target cells by binding to receptor molecules. Measurable responses to water-soluble hormones occur with a shorter delay than do responses to lipid-soluble hormones.

Synthesis, Storage, and Release of Hormones

In this section, we compare the cellular mechanisms of synthesis, storage, and secretion of two chemical classes of hormones: peptides and steroids. We use insulin as our primary example of a peptide hormone. **Insulin** is the vertebrate hormone that lowers the concentration of blood glucose by stimulating cells to take up glucose (as well as amino acids) from the blood. It also promotes the incorporation of food molecules into storage compounds. Insulin is synthesized in endocrine cells in the pancreas. Pancreatic endocrine cells are clustered in the **islets of Langerhans**, which are embedded in the pancreatic exocrine tissue that secretes digestive enzymes. The islet cells that produce and secrete insulin are called B or β (beta) cells.

Peptide hormones are synthesized at ribosomes, stored in vesicles, and secreted on demand

The insulin molecule consists of two peptide chains, designated A and B, connected by disulfide bonds (see Figure 14.2*a*). Although there is some variation among species in the amino acid sequences of the two chains, the general structure of the molecule, the amino acid sequences of certain regions, and the positions of the disulfide bonds are all highly conserved. The information determining the amino acid sequences of the chains is encoded in the DNA of the cell nucleus. The DNA nucleotide sequence of the insulin gene is *transcribed* into the nucleotide sequence of messenger RNA, which leaves the nucleus and interacts with ribosomes of the rough (granular) endoplasmic reticulum (ER). The nucleotide sequence of messenger RNA is then *translated* to determine the amino acid sequence of the polypeptide that is synthesized at the ribosomes.

The two peptide chains of insulin are synthesized as part of a single large precursor molecule, **preproinsulin** (Figure 14.4). The inactive preproinsulin molecule is then altered through *posttranslational processing* to produce the final product. The preproinsulin polypeptide chain consists of four regions, termed P, B, C, and A. Regions A and B become the A and B chains of the insulin molecule; C is the segment connecting these chains and will be cleaved away. The P segment functions as a leader or signal sequence to facilitate passage of the polypeptide chain through the membrane of the ER into its lumen; it is enzymatically cleaved from the polypeptide even before synthesis is completed. Without the P segment, the polypeptide is called **proinsulin**. Proinsulin contains sulfhydryl (—SH) groups, which are reduced to the three disulfide (—S—S—) bridges that fold the molecule. After the disulfide bridges form, the C segment is removed by enzymatic action, leaving the A and B chains linked by the disulfide bonds. These stages of posttranslational processing occur as the proinsulin molecules travel first to the Golgi apparatus in shuttle vesicles and then in secretory vesicles that bud off from the Golgi apparatus. The C fragments are retained in the secretory vesicles and secreted along with the mature insulin molecules. The C fragment is not physiologically active and is metabolized in the blood. It is clinically relevant, however, because clinicians measure its blood concentration in diabetics to monitor insulin secretion by the B cells.

The general sequence of synthesis of a preprohormone and posttranslational processing to produce a mature hormone is thought to be similar for all peptide hormones, and in fact for all secreted peptides. Another example of a preprohormone is pro-opiomelanocortin (POMC). Whereas preproinsulin is produced and processed by a single type of cell, the pancreatic B cell, POMC is synthesized by different types of cells, including cells in the anterior pituitary gland, brain, placenta, and skin. These different types of cells possess different enzymes to carry out posttranslational processing and so produce different end products for secretion. Some types of cells co-secrete more than one end product. For example, adrenocorticotropic hormone (ACTH) and melanocyte-stimulating hormone (MSH) are produced by different types of cells in the anterior pituitary by enzymatically cleaving POMC at different sites. The enzymes of the ACTH-secreting cells cleave POMC in such a way that another hormone, β-endorphin, is also produced and cosecreted with ACTH. β-endorphin plays an important role in pain control. Cells of the hypothalamus also process POMC to produce significant amounts of β-endorphin.

Insulin and other secreted peptides are stored in vesicles within cells that synthesize them. They are secreted by calcium-dependent exocytosis. When blood concentrations of glucose increase, B cells depolarize, causing voltage-gated Ca^{2+} channels to open and allow an influx of Ca^{2+} ions. The rate of insulin secretion is modulated by several factors, the most important of which are the levels of glucose and certain amino acids in the blood. As cells in the body respond to insulin's signal to take up glucose, the blood glucose concentration falls, and the B cells are no longer stimulated to secrete insulin. By this *negative feedback* mechanism, the blood glucose concentration is kept at a relatively constant level. Other factors also stimulate insulin secretion; these factors include parasympathetic stimulation of the islets of Langerhans and gastrointestinal hormones secreted by the digestive tract in the presence of food. Both of these stimuli act before blood glucose levels actually increase, and therefore function as *feed-forward* mechanisms. Sympathetic stimulation of the islets of Langerhans inhibits secretion of insulin. Without insulin promoting its uptake, glucose remains available in the blood to provide a source of energy for the sympathetic "fight-or-flight" response.

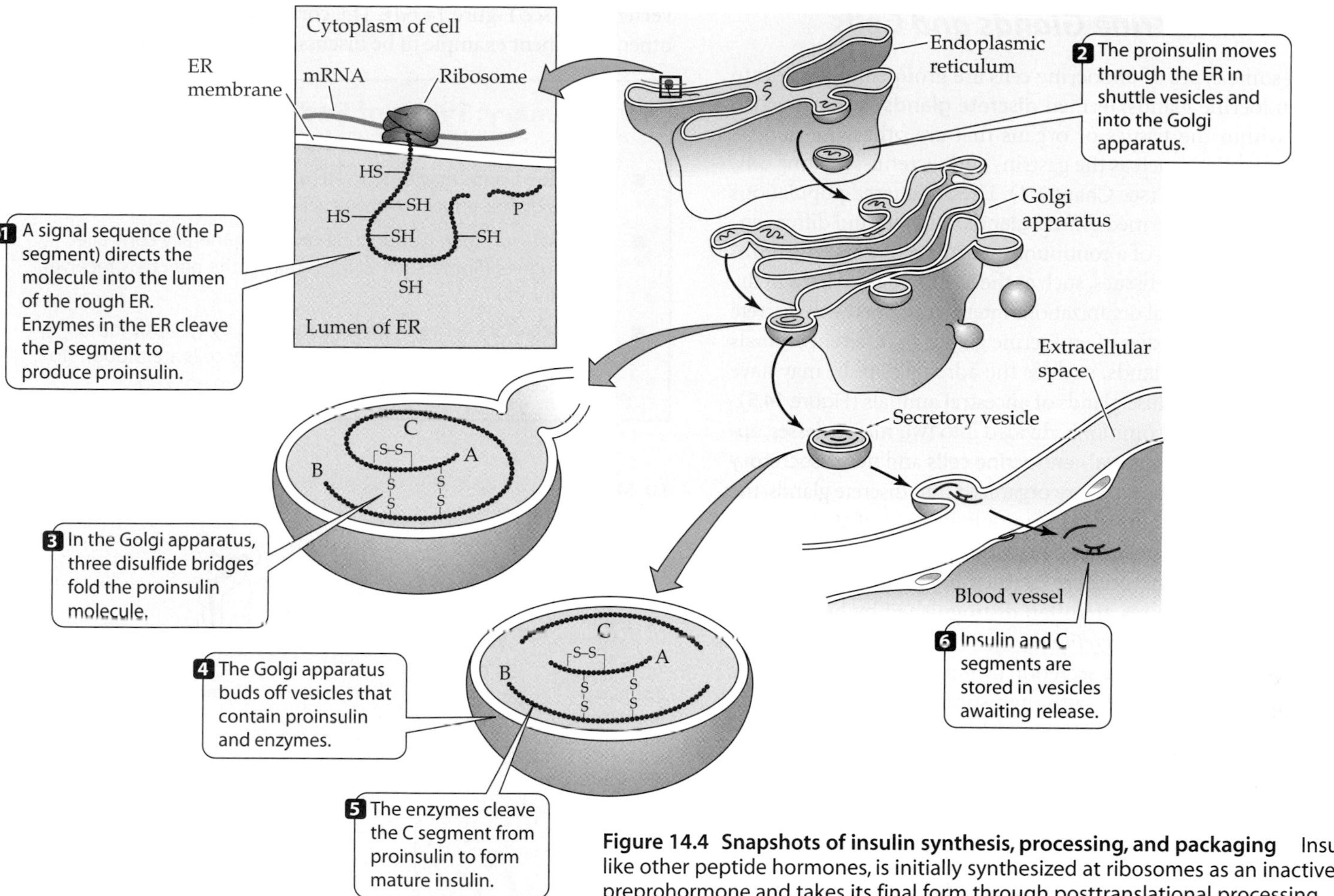

Figure 14.4 Snapshots of insulin synthesis, processing, and packaging Insulin, like other peptide hormones, is initially synthesized at ribosomes as an inactive preprohormone and takes its final form through posttranslational processing.

Steroid hormones are synthesized from cholesterol, are not stored, and are secreted by diffusion

In vertebrates, some cholesterol is obtained from animal fats in the diet. *Steroidogenic* (steroid-producing) endocrine cells and liver cells also synthesize cholesterol. Because cholesterol is not water-soluble, it circulates in the blood bound to lipoprotein carrier complexes. *Low-density lipoprotein* (*LDL*) complexes transport cholesterol *to* cells throughout the body, whereas *high-density lipoprotein* (*HDL*) complexes carry cholesterol *away from* cells to the liver. Steroidogenic cells express LDL receptor molecules on their cell membranes. When an LDL complex binds to the membrane receptor, the cell engulfs the complex by endocytosis and thereby obtains cholesterol.

In vertebrate animals, endocrine cells of the gonads and the adrenal cortex (and the placenta of pregnant mammals) produce steroid hormones. Once these cells synthesize or take in the cholesterol, enzymes cleave the six-carbon side chain to form pregnenolone (see Figure 14.1), which then enters one of several possible biochemical pathways. Different types of steroidogenic cells have different sets of enzymes, so that each produces a different major end product. The enzymes necessary for steroid synthesis are contained within intracellular compartments: Those involved in converting cholesterol to pregnenolone are located in the mitochondria, and most of the others are in the smooth ER. Thus pregnenolone molecules must travel from the mitochondria to the smooth ER for further enzymatic conversions.

Unlike peptide hormones, steroid hormones are not stored in the secreting cell. Instead, they are made from precursors stored in lipid droplets when the cell is stimulated, and immediately secreted. Secretion is accomplished by diffusion through the cell membrane. Thus, whereas the blood concentration of a peptide hormone is determined by the rate of release of preexisting stored hormone, the blood concentration of a steroid hormone is determined by the rate of synthesis—and immediate release—of the hormone.

Summary: Synthesis, Storage, and Release of Hormones

- Peptide hormones are synthesized by transcription of DNA, translation, and posttranslational processing. They are stored in vesicles and secreted on demand.
- Steroid hormones are synthesized from cholesterol. Different types of steroidogenic cells use different biochemical pathways and sets of enzymes to produce different steroid hormones. Steroid hormones are not stored, but are synthesized on demand and secreted by diffusion through the cell membrane.

Types of Endocrine Glands and Cells

Although some types of endocrine cells are grouped together into distinct endocrine glands, termed **discrete glands,** others may be scattered within the tissues of organs that are otherwise nonendocrine in function, such as the gastrin- and secretin-secreting cells of the mammalian gut (see Chapter 4). These scattered populations of endocrine cells are termed **diffuse glands.** Discrete and diffuse endocrine glands are ends of a continuous spectrum of tissue organization. Several endocrine tissues, such as the islets of Langerhans in the pancreas, have structural organizations intermediate between discrete and diffuse glands. Studies of endocrine glands in different animals suggest that discrete glands, such as the adrenal glands, may have evolved from more diffuse glands of ancestral animals (Figure 14.5).

Endocrine cells are commonly divided into two major classes, **epithelial** (also called **nonneural**) **endocrine cells** and **neurosecretory cells.** When cells of these types are organized into discrete glands, the glands are described by the same terminology: **epithelial** (**nonneural**) and **neurosecretory glands.** The endocrine secretions of either type of cell or gland are properly termed hormones, but those of neurosecretory structures are often distinguished by being called **neurohormones** or *neurosecretions.*

The signals that stimulate secretion by nonneural endocrine cells are usually other hormones, although some, such as the islets of Langerhans in the vertebrate pancreas, also receive neural input. By contrast, neurosecretory endocrine cells are always signaled to secrete their neurohormones by synaptic input from typical neurons. Thus neurosecretory cells interface directly with the nervous system. Both neurons and neuroscecretory cells typically generate action potentials and release their products by exocytosis. However, neurons release neurotransmitter at a synaptic gap. Neurosecretory cells release neurohormones into the blood. Through this process, they transduce neural signals into endocrine signals. The fundamental similarity between neurons and neurosecretory cells suggests evolutionary continuity between the neural and the endocrine control systems. We do not know, however, whether neurosecretory cells evolved from neurons, or vice versa, or whether both types of cells have a common ancestry.

The cell bodies of neurosecretory cells are located within the central nervous system (CNS), but their axons extend outside the CNS. Neurohormones are synthesized in the cell bodies, transported down the axons, and released at the ends of the axons. The axon terminals are often contained within a **neurohemal organ,** which is an anatomically distinct site for the release of neurohormones. It consists of one or more clusters of axon terminals and a rich supply of blood vessels or other circulatory specializations. Neurohemal organs occur in both vertebrates and invertebrates. The pars nervosa of the posterior pituitary gland is a prominent neurohemal organ in vertebrates (see Figure 14.6*a*). The corpus allatum of insects is another prominent example to be discussed later in this chapter.

Summary: Types of Endocrine Glands and Cells

- Endocrine glands may be discrete, diffuse, or intermediate. Many discrete glands appear to have evolved from diffuse glands.
- Epithelial (nonneural) endocrine cells are generally controlled by hormones. (Some, such as the B cells of the pancreas, receive neural input.)
- Neurosecretory cells are always controlled by synaptic input from neurons. Neurons and neurosecretory cells are thought to be related evolutionarily, but their origins are not known.

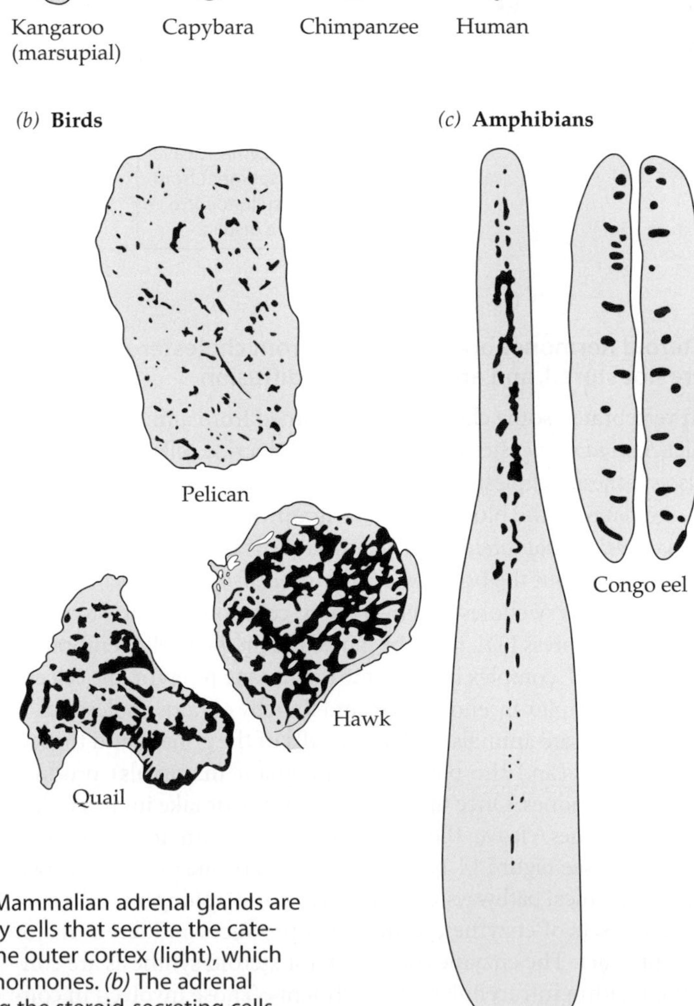

Figure 14.5 Discrete glands probably evolved from diffuse glands *(a)* Mammalian adrenal glands are discrete glands. The central medulla (dark), which consists of neurosecretory cells that secrete the catecholamines epinephrine and norepinephrine, is easily distinguished from the outer cortex (light), which consists of nonneural endocrine cells that secrete several different steroid hormones. *(b)* The adrenal glands of birds have catecholamine-secreting cells (dark) distributed among the steroid-secreting cells. *(c)* The diffuse adrenal tissues of amphibians consist of catecholamine-secreting and steroid-secreting cells that form patches (dark) on the kidney (light). (After Hartman and Brownell 1949.)

Control of Endocrine Systems: The Vertebrate Pituitary Gland

In this section, we will use the vertebrate pituitary gland as an example as we explore several different mechanisms of secretory control. It is important to keep in mind that the principles of control described in this context also apply to other endocrine tissues in both vertebrate and invertebrate animals. The pituitary gland lies immediately below the hypothalamus and consists of two parts: the **adenohypophysis**, commonly called the **anterior pituitary**, and the **neurohypophysis**, commonly called the **posterior pituitary**.

The posterior pituitary illustrates neural control of neurosecretory cells

The posterior pituitary (neurohypophysis) is a downgrowth of the brain (Figure 14.6*a*). It consists of three segments: (1) the **median eminence**, which forms part of the floor of the hypothalamus, (2) the **pars nervosa**, which means "nervous part" (also called the neural lobe or posterior lobe), and (3) an interconnecting segment, the **infundibular stalk**.

In most mammals, two peptide hormones are released into the blood in the pars nervosa: vasopressin and oxytocin. **Vasopressin**, also called **antidiuretic hormone** (**ADH**), limits the production of urine and also stimulates constriction of arterioles. **Oxytocin** causes contraction of the uterus during birth and ejection of milk by the mammary glands during suckling. Endocrinologists originally believed that the hormones of the pars nervosa were synthesized there. However, starting in the 1930s, research revealed that these hormones are actually synthesized by neurosecretory cells that have their cell bodies within the hypothalamus. In mammals (see Figure 14.6*a*), two paired clusters of cell bodies in the hypothalamus, the **paraventricular nuclei** and **supraoptic nuclei**, are the principal sites of production of these two peptides. Axons from the cell bodies of these nuclei extend in tracts along the infundibular stalk to the pars nervosa. When the neurosecretory cells generate action potentials, their axon terminals release hormones into the blood.

The hypothalamus–posterior pituitary connection illustrates one form of control of endocrine function: *neural control of neurosecretory cells*. The neurosecretory cells that produce and secrete vasopressin and oxytocin receive and integrate synaptic input from a host of typical neurons. Vasopressin cells, for example, receive input about blood volume and the osmotic concentration of body fluids. When they receive signals reporting high osmotic concentration and/or low blood volume, they secrete vasopressin, which triggers processes involved in retaining water. Likewise, oxytocin cells respond to signals from the mammary glands when suckling occurs or from the cervix of the uterus during labor and birth (see Chapter 15).

Vasopressin and oxytocin, which are very similar in their amino acid sequence (Table 14.3), are considered members of a **family** of hormones that are probably descended from a single ancestral peptide. They are both nonapeptides (have nine amino acids). In most mammals, they differ at only two amino acid sites, yet these structural differences underlie profound differences in function. In all vertebrates except cyclostomes (jawless fish such as the lamprey), vasopressin and oxytocin have separate and distinct functions. As shown in Table 14.3, vasopressins and oxytocins in different vertebrate species show additional variations in amino acid composition. There are many other examples of hormone families. For example, insulin is considered the prototype of the insulin family, which includes hormones with structures and genes related to those of insulin. Other members of the insulin family include relaxin, secreted by the ovary and thought to aid in the birth of viviparous vertebrates, and insulin-like growth factors (IGFs), secreted by the liver and other tissues such as muscle. Investigators analyze both chemical and genetic information about the peptides within families of hormones to learn about their common ancestral origins.

The anterior pituitary illustrates neurosecretory control of endocrine cells

The anterior pituitary (adenohypophysis) is subdivided into the *pars distalis*, *pars intermedia*, and *pars tuberalis* (see Figure 14.6*a*). The exact positions and relative sizes of these parts vary greatly from one animal group to another, and in some groups not all parts are present. All the hormones of the anterior pituitary are synthesized and secreted by endocrine cells within its tissues. Different specific populations of cells secrete different hormones. All anterior pituitary hormones are polypeptides, proteins, or glycoproteins (proteins with covalently bound carbohydrate chains).

TABLE 14.3 Vasopressins and oxytocins found in vertebrate animals

Common name	Found in	Amino acid site 1	2	3	4	5	6	7	8	9
Water conservation										
Arginine vasopressin (AVP) (Antidiuretic hormone, ADH)	Most mammals	Cys	Tyr	<u>Phe</u>	Gln	Asn	Cys	Pro	<u>Arg</u>	Gly(NH_2)
Lysine-vasopressin (LVP)	Pig, peccary, hippopotamus	Cys	Tyr	Phe	Gln	Asn	Cys	Pro	<u>Lys</u>	Gly(NH_2)
Arginine vasotocin (AVT)	Nonmammalian vertebrates	Cys	Tyr	<u>Ile</u>	Gln	Asn	Cys	Pro	<u>Arg</u>	Gly(NH_2)
Uterine contraction and milk ejection										
Oxytocin	Most mammals	Cys	Tyr	<u>Ile</u>	Gln	Asn	Cys	Pro	<u>Leu</u>	Gly(NH_2)
Mesotocin	Birds, reptiles, amphibians, lungfish	Cys	Tyr	Ile	Gln	Asn	Cys	Pro	Ile	Gly(NH_2)

Note: Underlines indicate differences in amino acids among the vasopressins and oxytocin.
Source: After Bentley 1998.

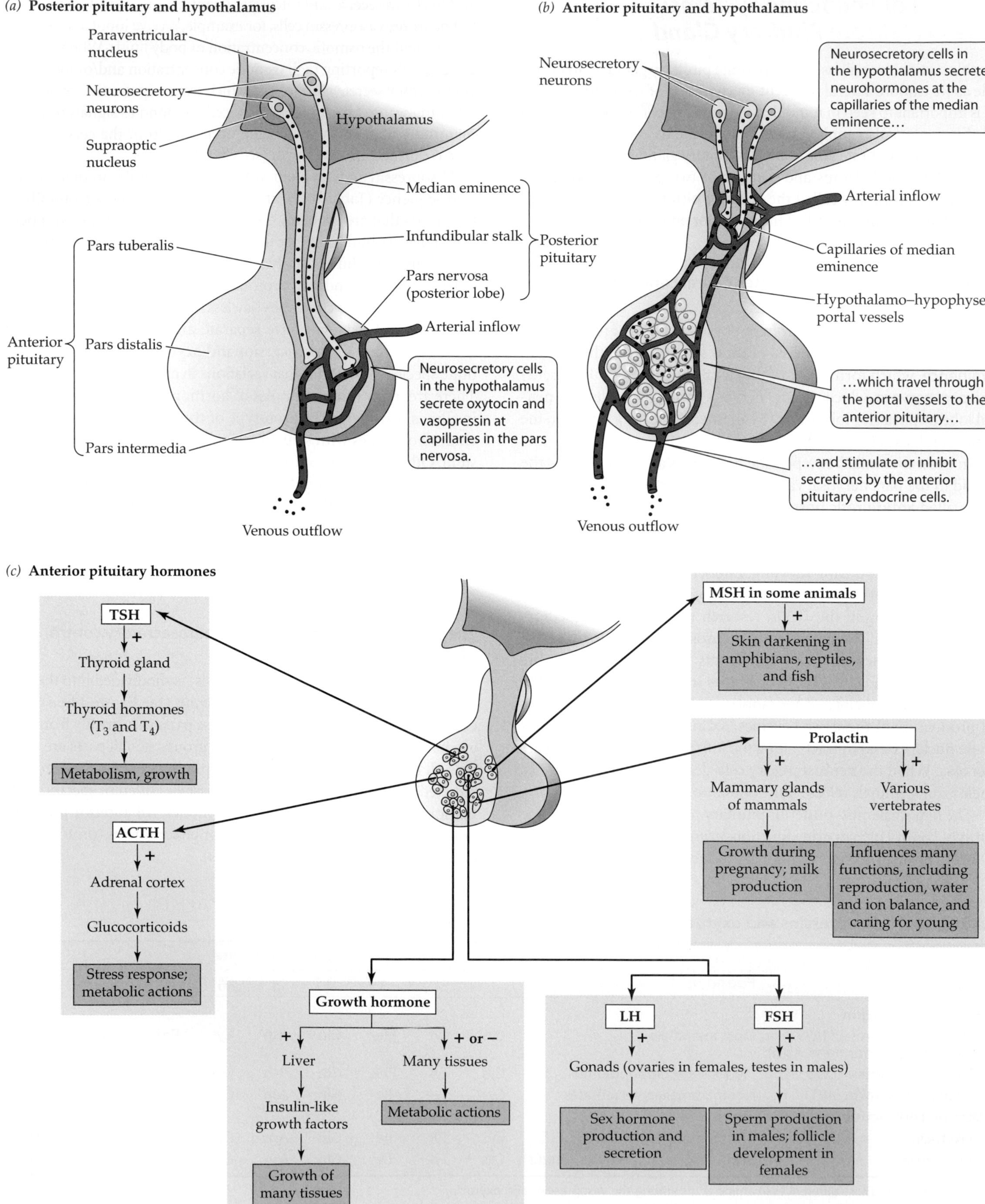

(a) Posterior pituitary and hypothalamus
Paraventricular nucleus
Neurosecretory neurons
Supraoptic nucleus
Hypothalamus
Median eminence
Infundibular stalk
Posterior pituitary
Pars tuberalis
Pars nervosa (posterior lobe)
Arterial inflow
Anterior pituitary
Pars distalis
Neurosecretory cells in the hypothalamus secrete oxytocin and vasopressin at capillaries in the pars nervosa.
Pars intermedia
Venous outflow
(b) Anterior pituitary and hypothalamus
Neurosecretory neurons
Neurosecretory cells in the hypothalamus secrete neurohormones at the capillaries of the median eminence…
Arterial inflow
Capillaries of median eminence
Hypothalamo–hypophyseal portal vessels
…which travel through the portal vessels to the anterior pituitary…
…and stimulate or inhibit secretions by the anterior pituitary endocrine cells.
Venous outflow
(c) Anterior pituitary hormones
TSH
+
Thyroid gland
Thyroid hormones (T_3 and T_4)
Metabolism, growth
ACTH
+
Adrenal cortex
Glucocorticoids
Stress response; metabolic actions
MSH in some animals
+
Skin darkening in amphibians, reptiles, and fish
Prolactin
+
Mammary glands of mammals
Growth during pregnancy; milk production
+
Various vertebrates
Influences many functions, including reproduction, water and ion balance, and caring for young
Growth hormone
+
Liver
Insulin-like growth factors
Growth of many tissues
+ or −
Many tissues
Metabolic actions
LH
FSH
+
Gonads (ovaries in females, testes in males)
Sex hormone production and secretion
Sperm production in males; follicle development in females

◀ **Figure 14.6 The vertebrate pituitary gland has two parts** *(a)* The pars nervosa of the posterior pituitary contains terminals of axons that extend along the infundibular stalk from cell bodies of neurosecretory cells located in the hypothalamus. *(b)* The anterior pituitary consists of nonneural endocrine cells. Secretion of anterior pituitary hormones is controlled by hypothalamic neurohormones, which are transported in the hypothalamo–hypophyseal portal system. *(c)* Separate populations of cells in the anterior pituitary secrete different hormones. These diagrams are based on the structure of the mammalian pituitary gland.

Anterior pituitary hormones are categorized into two main groups according to their target tissues. The hormones of one group exert their principal effects on nonendocrine tissues. **Growth hormone** (**GH**), for example, influences growth and nutrient metabolism in tissues such as bone and muscle. Other hormones of this group are **prolactin** and **melanocyte-stimulating hormone** (**MSH**). The second group includes hormones that control other endocrine glands. By convention, hormones that influence the functions of other endocrine glands have the suffix *-tropic* in their names, or are called *tropins*. **Thyroid-stimulating hormone** (**TSH**), for example, is also called *thyrotropin*. This anterior pituitary hormone supports and maintains the tissues of the thyroid gland and stimulates it to secrete thyroid hormones. If a target gland is deprived of input from its tropic hormone, the gland not only stops secreting hormone, but also shrivels in size. **Adrenocorticotropic hormone** (**ACTH**), **luteinizing hormone** (**LH**), and **follicle-stimulating hormone** (**FSH**) are also tropic hormones produced by the anterior pituitary. The functions of the anterior pituitary hormones are listed in Table 14.1 and Figure 14.6*c*.

What kinds of signals control the secretions of the anterior pituitary gland? The answer is, in large part, neurohormones secreted by the hypothalamus. The anterior pituitary and the hypothalamus are connected by a specific, dedicated vascular pathway (Figure 14.6*b*). Capillaries of the median eminence coalesce into *portal vessels* that travel the short distance along the infundibular stalk to the anterior pituitary and there branch to form capillary beds around the endocrine cells of the anterior pituitary. (*Portal vessels* are any vessels that carry blood from one capillary bed to another.) This whole system is termed the **hypothalamo–hypophyseal portal system.** The median eminence is a neurohemal organ in which the axons of several populations of hypothalamic neurosecretory cells terminate. These cells produce neurohormones that control the secretions of specific groups of anterior pituitary cells. Some hypothalamic neurohormones stimulate secretion of hormones and are called *releasing hormones* (RHs). Others, called *release-inhibiting hormones* (RIHs), inhibit secretion of anterior pituitary hormones. Each neurohormone is specific in its actions (see Table 14.1). Like the anterior pituitary tropic hormones, the RHs are also tropic. They both stimulate secretion by anterior pituitary endocrine cells and maintain their vigor.

The hypothalamic neurosecretory cells secrete their RHs and RIHs into the capillaries of the median eminence, and the neurohormones are carried in the portal system to the anterior pituitary, where they exert their effects on their specific target cells (Figure 14.6*c*). Thus, for example, *thyrotropin-releasing hormone* (TRH) from the hypothalamus stimulates the secretion of TSH in the anterior pituitary, and TSH travels through the general circulation to stimulate release of thyroid hormones from the thyroid gland. The portal system provides a short, direct path for neurohormones to travel quickly, without dilution, from the hypothalamus to the anterior pituitary.

We can think of the hypothalamo–hypophyseal portal system as providing an interface between the brain and much of the endocrine system. The posterior pituitary is actually an extension of the brain. The anterior pituitary is under control of the brain and regulates the functions of many other tissues, including several endocrine glands. The vertebrate pituitary gland illustrates how the sophisticated integrative capabilities of the CNS influence endocrine function. The CNS receives and integrates input from a multitude of sensory receptors that monitor conditions both within the animal and in its outside environment. Neurons provide synaptic input to neurosecretory cells, which, in turn, influence the physiology of other cells, including endocrine cells. (In some other glands, such as the islets of Langerhans, we know that neurons that are not part of the hypothalamo–hypophyseal system influence endocrine cells directly.)

Neural control of the endocrine system deserves emphasis, but endocrine control of the nervous system is also important. Overall, physiological control is performed by the two systems interacting mutually, not by one unilaterally controlling the other. Many instances of endocrine control of the nervous system are known; examples include the negative feedback effects of secreted hormones on neurosecretory cells (discussed in the next section) and the effects of hormones that direct stereotyped muscle contractions during metamorphosis in insects (described later in this chapter).

Hormones and neural input modulate endocrine control pathways

When the secretions of one endocrine gland act on another in a sequence, endocrinologists speak of the system as an **axis**. For example, a three-part axis (the hypothalamus–anterior pituitary–thyroid axis) mediates secretion of thyroid hormones. This axis represents a hormonal sequence from TRH to TSH to thyroid hormones. An example of a two-part axis is from growth hormone–releasing hormone secreted by the hypothalamus to growth hormone (GH) secreted by the anterior pituitary. GH does not stimulate secretion of another hormone. Similar axes are also found in invertebrate animals. In this section, we use the three-part hypothalamus–pituitary–adrenal cortex (HPA) axis to illustrate how rates of endocrine secretion can be *modulated*, or changed, by hormonal and neural influences.

The adrenal gland secretes several hormones (Figure 14.7). Here we limit our consideration to the **glucocorticoids** (cortisone, cortisol, corticosterone), a class of steroid hormones secreted in response to stress. Glucocorticoids are so named because (among other functions), they promote an increase in the blood concentration of glucose ("gluco") and because they are secreted by the adrenal cortex ("cortico"). The main glucocorticoid produced in humans is **cortisol**, whereas that in rats and marine iguanas is **corticosterone**. Glucocorticoid secretion increases under stressful conditions. Stressors experienced by animals include being wounded, being exposed to thermal extremes (birds and mammals) or other hostile environmental conditions (such as an oil spill), being forced to exercise vig-

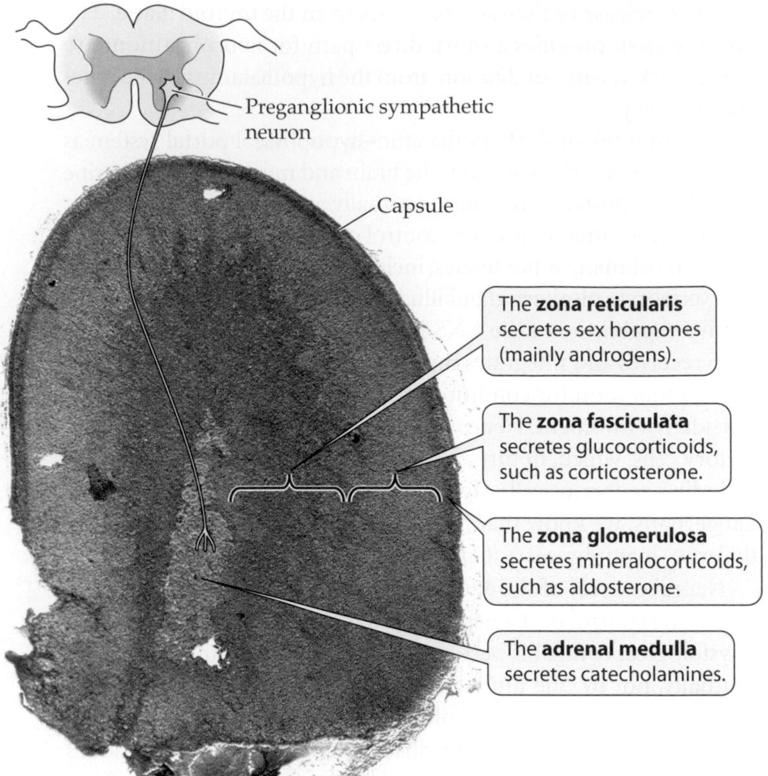

Figure 14.7 The adrenal gland consists of an inner medulla and an outer cortex The different portions characteristic of mammalian adrenal glands are visible in this section of a mouse adrenal gland. The medulla is homologous to a sympathetic ganglion, except that the postganglionic cells do not possess axons. The cortex, which surrounds the medulla, has three distinct layers, each specialized to secrete a different category of steroid hormones. (Courtesy of Judith Wopereis, Smith College.)

orously, and experiencing troublesome social conditions or high levels of emotion. Under stressful conditions, brain neural activity causes a rapid increase in hypothalamic secretion of **corticotropin-releasing hormone** (**CRH**) into the capillaries of the median eminence. CRH is carried in the portal system to the anterior pituitary, where it stimulates adrenocorticotropic cells to secrete the hormone ACTH into capillaries leading to the general circulation. ACTH is carried to the adrenal cortex, where it stimulates glucocorticoid secretion (Figure 14.8). Elevated levels of glucocorticoids influence several physiological processes during the stress response (described later in this chapter).

Any stage in this control pathway can be modulated. *Negative feedback* (see Figure 14.8) is the most widespread type of **hormonal modulation**. In negative feedback, a hormone controlled by a particular pathway causes changes that tend to suppress its own secretion. In the HPA axis, high glucocorticoid levels tend to suppress secretion of CRH by the hypothalamus and also reduce the responsiveness of the ACTH cells in the anterior pituitary to CRH. As a result, the pituitary secretes less ACTH, and the adrenal cortex receives less of a stimulus to secrete glucocorticoids. Mechanisms of negative feedback do not reduce hormone secretion to zero, but instead serve to *stabilize* blood concentrations of hormones. Occasionally, hormonal modulation involves *positive feedback*, such as that resulting in the explosive increase in oxytocin secretion during the process of birth (see Chapter 15).

In addition to feedback mechanisms, other types of hormonal modulation can affect endocrine control pathways. For example, hormones that are ancillary parts of a pathway can alter a target gland's response to a particular hormone. In the HPA pathway, vasopressin (VP) acts together with CRH to increase the secretion of ACTH from the anterior pituitary. Not all VP-secreting neurosecretory cells in the hypothalamus extend their axons to the posterior pituitary. Some terminate their axons on the capillary bed of the median eminence. When released, VP circulates through the portal vessels to the anterior pituitary. By itself, VP has little effect on the ACTH-secreting cells. However, when these cells receive signals from both VP and CRH, their secretion is greater than it would have been under the influence of CRH alone. This sort of effect, in which one hormone can amplify the effect of another, is called **synergism**.

A hormone can influence the effects of another hormone at the same target tissue in three different ways: by synergism (producing an enhanced response such as we have seen with VP and CRH), **permissiveness** (in which the presence of one hormone is *required* for

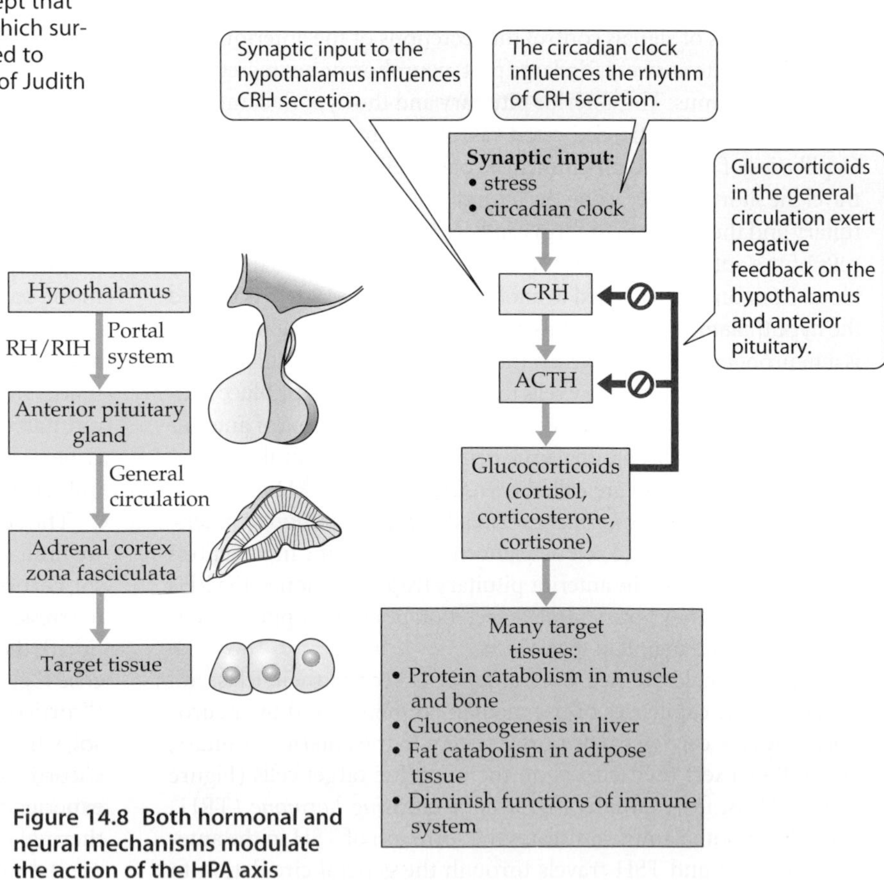

Figure 14.8 Both hormonal and neural mechanisms modulate the action of the HPA axis

the other to exert an effect), or **antagonism** (in which one hormone opposes the action of another). One example of permissiveness is that of cortisol permitting the catecholamines epinephrine and norepinephrine to cause constriction (narrowing of the diameter) of blood vessels, a function necessary to maintain normal blood pressure. Because it *must* be present for vasoconstriction to occur, basal levels of this glucocorticoid are necessary for homeostatic regulation of blood pressure. Without treatment, humans with a condition called "adrenal insufficiency" (who are unable to secrete hormones from the adrenal cortex) are at risk of death if they experience a stress, such as a hemorrhage, that requires systemic vasoconstriction.

An example of antagonism is the interaction between insulin and glucagon. We know that insulin secreted from B cells in the islets of Langerhans in the pancreas promotes uptake of glucose from the blood by many different tissues. **Glucagon** is a hormone secreted by *A cells* in the islets, and it functions to oppose the action of insulin: It stimulates the release of glucose and fatty acids into the blood. The balanced actions of these two hormones help to maintain stable levels of glucose in the blood.[2] In situations, such as stress, in which higher blood concentrations of glucose are required to respond to a crisis, glucagon secretion increases and insulin secretion decreases. Figure 14.9 shows blood glucose levels in dogs given insulin alone or in combination with glucagon and epinephrine. Epinephrine and glucagon are both antagonists of insulin, and they work synergistically to oppose insulin's action.

[2]Antagonism, in the context of endocrinology, means that two hormones have opposing actions, but it does not indicate their mechanisms of action. For example, opposing hormones may trigger different biochemical pathways in the target cell, or they may induce downregulation of receptors for the opposing hormone in the target cell.

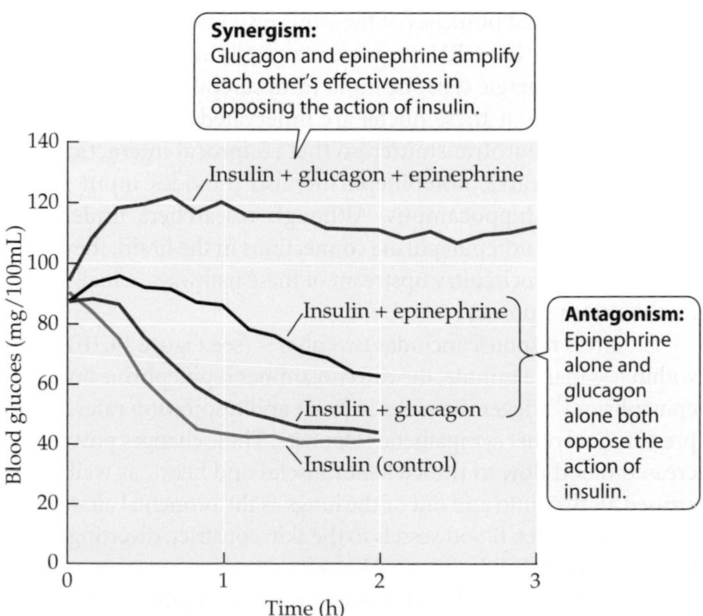

Figure 14.9 Interactions of insulin, glucagon, and epinephrine Blood glucose levels were measured in 29 dogs infused with insulin alone or in combination with glucagon and epinephrine. Glucagon and epinephrine both prevent insulin from stimulating cells to take up glucose from the blood. (After Sacca et al. 1979.)

Neural modulation also affects endocrine control pathways. In the HPA axis, neurons reporting increases or decreases in stress provide synaptic input to CRH neurosecretory cells in the hypothalamus (see Figure 14.8). Furthermore, neuronal biological clocks can influence hormone secretion. This is the case in the HPA axis: Blood levels of cortisol regularly rise and fall in a daily rhythm, being highest in the early morning and lowest during the evening. These cyclic changes in secretion are driven by a circadian clock in the brain that sends input to the hypothalamic CRH neurosecretory cells. Changes in lighting and feeding schedules can perturb these diurnal changes in glucocorticoid secretion, and stress can disrupt the diurnal rhythm.

Finally, many hypothalamic and pituitary hormones are secreted *in pulses*, so that there are intermittent brief periods of high hormone concentration in the blood. (When a hormone is continuously present, target cells typically downregulate their receptor molecules for that hormone; it is thought that one advantage of pulsatile release is that it prevents this *desensitization* of target cells.) Synaptic input to the neurosecretory cells governs the pulsatile release of hypothalamic hormones, which in turn causes pulsatile output of pituitary hormones. The amplitude of a pulse of hormone (i.e., the amount of hormone released during a pulse) can vary. For example, the CRH neurons secrete pulses of hormone two or three times per hour. The amplitude of each pulse increases in the early morning according to input from the circadian clock, and it also increases when neural activity signals the presence of one or more stressors. Ultimately, the *combination* of neural modulation and hormonal modulation determines the minute-to-minute blood levels of hormones.

Summary: Control of Endocrine Systems: The Vertebrate Pituitary Gland

- The vertebrate pituitary gland consists of the adenohypophysis (anterior pituitary) and the neurohypophysis (posterior pituitary). Posterior pituitary hormones (vasopressin and oxytocin) are secreted in response to neural activity. Secretions of anterior pituitary hormones are controlled by releasing hormones (RHs) and release-inhibiting hormones (RIHs) from the hypothalamus, which are transported to the anterior pituitary through the hypothalamo–hypophyseal portal system.
- The rate of hormone secretion is influenced by a combination of hormonal modulation (such as feedback mechanisms, synergism, permissiveness, and antagonism) and neural modulation (such as sensory input and clock mechanisms). Neural activity also governs the pulsatile release of hormones.

The Mammalian Stress Response

In this section we use the mammalian stress response (Figure 14.10) to illustrate and integrate several of the principles of endocrinology discussed in previous sections. The **stress response** is an adaptation that allows an animal to respond immediately in a generalized way to a threatening situation. During the stress response, feeding and reproduction are curtailed, heart and breathing rates increase, cognition and alertness are sharpened, metabolic processes release stored energy, and oxygen and nutrients are directed to the CNS and to

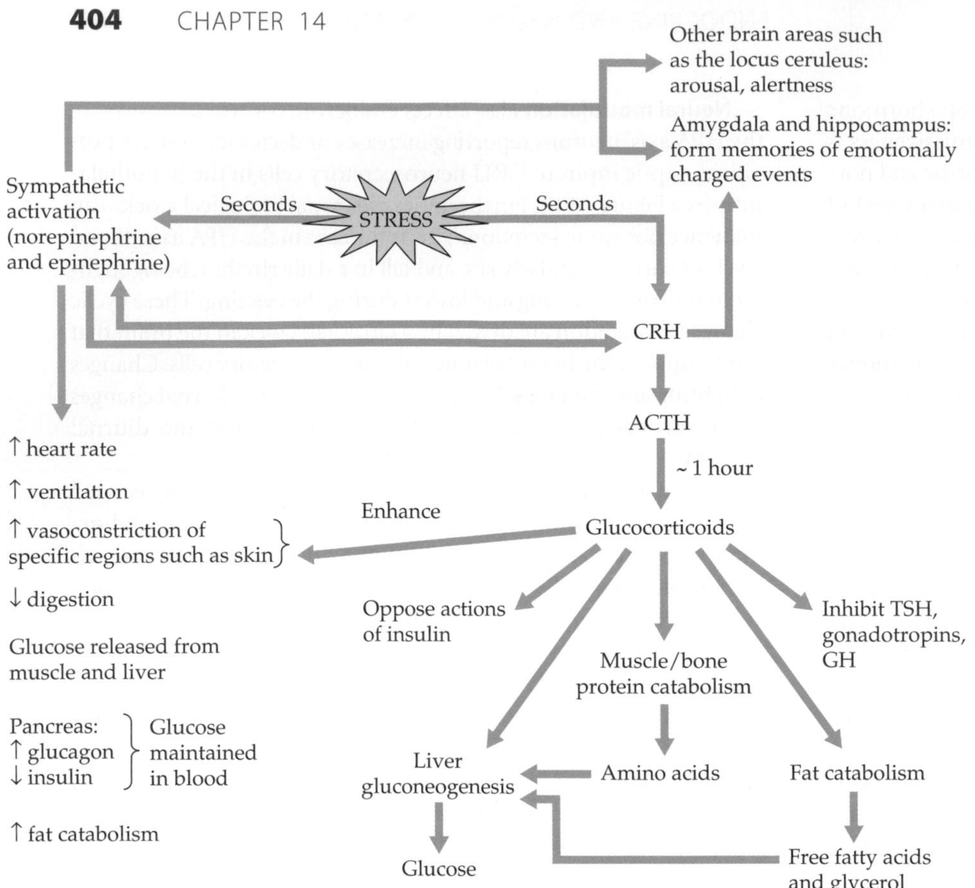

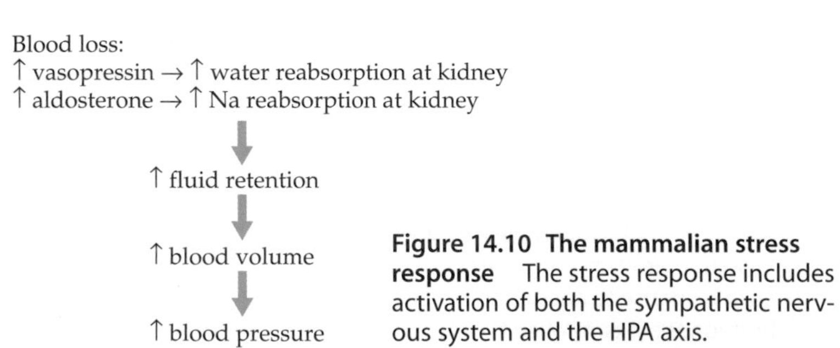

Figure 14.10 The mammalian stress response The stress response includes activation of both the sympathetic nervous system and the HPA axis.

those sites in the body that are stressed the most. All of these functions are aimed at ensuring survival in an acute crisis. Different stressors differentially turn on different components of the stress response; however, any one stressor that is potent enough will turn on a generalized *stress syndrome*.

In humans, and perhaps other animals as well, stressors over which an individual perceives a sense of control can be rewarding. For example, seeking novel situations generates stress, but also facilitates intellectual and emotional growth. Interestingly, feeding and sexual activity—both essential behaviors for biological success—also stimulate the stress system. When the stressor is no longer present, feedback mechanisms ensure that the stress response is turned off. In the classic example of a zebra chased by a lion, the threat is clear, the physiological response is swift, and the episode has a finite end. However, if a stressor (physical or emotional) persists for long periods, physiological responses that are adaptive in the short term become damaging in the long term.

The autonomic nervous system and HPA axis coordinate the stress response to an acute threat

The rat detects the cat and runs. Within seconds of the threat, the rat's sympathetic nervous system releases catecholamines (epinephrine and norepinephrine) from sympathetic nerve terminals and the adrenal medulla, and hypothalamic neurosecretory cells release CRH into the hypothalamo–hypophyseal portal system. A few seconds later, the anterior pituitary secretes ACTH. Thus two output systems, the sympathetic "fight-or-flight" system and the HPA axis, together mount the response to a stressor. Their functions are not independent, but intermingled. For example, in addition to its role as a neurohormone that stimulates ACTH secretion, CRH also acts as a neurotransmitter in other areas of the brain, where it stimulates the sympathetic nervous system. Researchers uncovered this additional role of CRH by injecting it into the brain ventricles of dogs and rats whose pituitary glands had been removed. These animals secreted no ACTH, because the ACTH-secreting cells were gone. However, injected CRH caused increases in blood concentrations of catecholamines and associated increases in blood pressure and heart rate. Experiments like these reveal that one of CRH's functions is to link the sympathetic and adrenocortical branches of the stress response. CRH also acts as a neurotransmitter or neuromodulator in the amygdala and hippocampus (which function together to form memories of emotionally charged events).

The two output branches of the stress response are also linked by norepinephrine. The CRH neurosecretory cells in the hypothalamus receive noradrenergic synaptic input from several different nuclei of the brain. Some of these nuclei are innervated by neurons using CRH as their neurotransmitter, so that reciprocal interactions are possible. Like CRH, norepinephrine also provides input to the amygdala and hippocampus. Although researchers understand many CRH and norepinephrine connections in the brain, they don't know what neurocircuitry upstream of these pathways actually turns on the stress response.

The stress response includes two phases (see Figure 14.10). First, within less than a minute, the catecholamines (epinephrine and norepinephrine)[3] trigger increases in heart and respiration rates, blood pressure, and other sympathetic responses. These changes provide increased blood flow to the skeletal muscles and heart, as well as increased air flow into and out of the lungs as the bronchial airways increase in diameter. Blood vessels to the skin constrict, diverting blood from sites of possible injury. Digestive functions are suppressed. Arousal of the CNS and alertness are promoted. Epinephrine stimulates the release of glucose into the blood by triggering the breakdown

[3] Catecholamines act quickly because they are stored in vesicles ready for release, and because, as we saw earlier in this chapter, their target cells respond through second-messenger systems that utilize preexisting proteins.

of glycogen stored in muscles and the liver, and it also stimulates the release of fatty acids from lipid stores. Epinephrine in the blood and norepinephrine from sympathetic nerve terminals both inhibit insulin secretion and stimulate glucagon secretion from the islets of Langerhans. Ordinarily, increased glucose in the blood would stimulate insulin secretion, which would promote the uptake of glucose from the blood by all tissues except brain and exercising skeletal muscle. By inhibiting insulin secretion and stimulating glucagon secretion, the catecholamines ensure plentiful levels of glucose in the blood to fuel physical exertion and maintain brain function.

Additional synergies occur during the first phase of the stress response. For example, epinephrine (in addition to CRH) appears to stimulate the secretion of ACTH, and ACTH may do more than stimulate glucocorticoid secretion. ACTH is known to facilitate learning, and it may contribute to an animal's preparedness in responding to a similar stressor in the future. Finally, ACTH is produced by cleaving of a preprohormone, pro-opiomelanocortin (POMC), and can be co-secreted along with other fragments of POMC, including β-endorphin. The latter substance (also produced by POMC cells in the hypothalamus) is an endogenous opiate and may contribute to *analgesia*; that is, it may decrease the animal's perception of pain.

In the second phase of the stress response (see Figure 14.10), glucocorticoids are secreted by the adrenal cortex. The effects of glucocorticoids on target tissues can be detected about an hour after the stress response is initiated.[4] Glucocorticoids reinforce the actions of the sympathetic nervous system and have additional metabolic effects that facilitate the release of usable sources of energy into the bloodstream. They stimulate the catabolism of protein in muscle and (at high doses) bone, and they stimulate the liver to use the released amino acids to produce glucose in a process called *gluconeogenesis*. The liver cells release this newly formed glucose into the blood. Like epinephrine and norepinephrine, glucocorticoids oppose the action of insulin and ensure fuel availability. Glucocorticoids also stimulate catabolism of fats so that fatty acids can be utilized as an alternative energy source by all tissues except the brain (which uses only glucose). The metabolic actions of glucocorticoids, coordinated with those of the catecholamines, ensure glucose availability to the brain in the face of required physical exertion and possible enforced fasting (for example, while hiding from a predator or recovering from a wound). The amino acids released by protein catabolism are also available for tissue repair.

In addition to their important metabolic effects, glucocorticoids increase their permissive effect on vasoconstriction stimulated by the catecholamines, as we saw earlier. They also inhibit the secretion of gonadotropins (FSH and LH), thyrotropin (TSH), and growth hormone (GH) from the anterior pituitary. Assuming the chase is short, and the rat pops into its burrow before the cat seizes it, the inhibitory effects of glucocorticoids on reproduction and growth are minimal. Safe in its burrow, the rat's sympathetic neural responses diminish, and the glucocorticoid molecules in its general circulation feed back negatively on the CRH and ACTH cells of the HPA axis. Thus the glucocorticoids themselves modulate the stress response. With decreased ACTH in the circulation, the adrenal cortex secretes decreased amounts of glucocorticoids, and concentrations in the blood return to basal levels.

If an animal experiences a severe wound and loses blood, several mechanisms are called into play to correct the loss of blood volume. Blood volume directly affects blood pressure, and blood pressure is necessary to ensure adequate delivery of oxygen and nutrients to the brain and other essential organs. Catecholamines released by the already-activated sympathetic nervous system stimulate the heart and blood vessels to maintain blood pressure. Vasopressin released from the posterior pituitary stimulates the nephrons in the kidney to retain water, and aldosterone secreted by cells in the zona glomerulosa of the adrenal cortex act at the nephrons to retain sodium, which exerts an osmotic holding effect on water. The combined actions of these hormones are aimed at conserving fluid volume, which translates to adequate blood volume and blood pressure.

The HPA axis modulates the immune system

While glucocorticoids modulate the stress response by negative feedback on the hypothalamic CRH cells, they also regulate functions of the immune system (Figure 14.11). The *immune system* works to prevent the invasion of foreign pathogens and to search out and destroy those that sneak through natural barriers. It neutralizes toxins and disposes of dead, damaged, or abnormal cells. The immune response often causes inflammation in response to infection or a wound. Glucocorticoids have anti-inflammatory effects and thus keep the immune system from overreacting and damaging healthy cells and tissues. Recently, researchers have found that a web of chemical pathways allows communication among the nervous, endocrine, and immune systems. These three systems interact continuously to maintain homeostasis as an animal navigates its daily life.

When certain cells of the immune system detect bacterial or viral pathogens or tumor cells, they release **cytokines**, which are chemical messengers (not antibodies). Like other chemical messengers, different cytokines bind with specific receptor molecules on target cells to communicate specific excitatory or inhibitory messages. In the stress response, certain cytokines travel in the blood to the hypothalamus, where they stimulate CRH neurosecretory cells. This chemical connection directly informs the CNS that the animal has detected an invading stressor. Using cytokines to turn on the stress response accomplishes two goals. First, the physiological responses of the HPA axis, such as the mobilization of energy stores, help the animal fight infection. Second, the glucocorticoids generated by the response inhibit the production of agents that cause inflammation (such as prostaglandins). By muting inflammation, they modulate (keep in check) the immune response.

Studies on experimental animals have indicated, on a whole-organism scale, a relationship between the stress response and inflammatory disease. For example, if the HPA axis of normal rats is disrupted by removing their pituitary glands, they become susceptible to inflammation. Genetic strains of rats have also shown this association. The Lewis strain of rat, for example, has an impaired HPA axis and secretes little CRH in response to stress; this strain of rats is highly susceptible to inflammatory and autoimmune disease. Inject-

[4] The effects of glucocorticoids take longer to appear because these steroid molecules must be produced from precursors prior to secretion. Once secreted, they exert genomic effects in their target cells, which then synthesize new proteins.

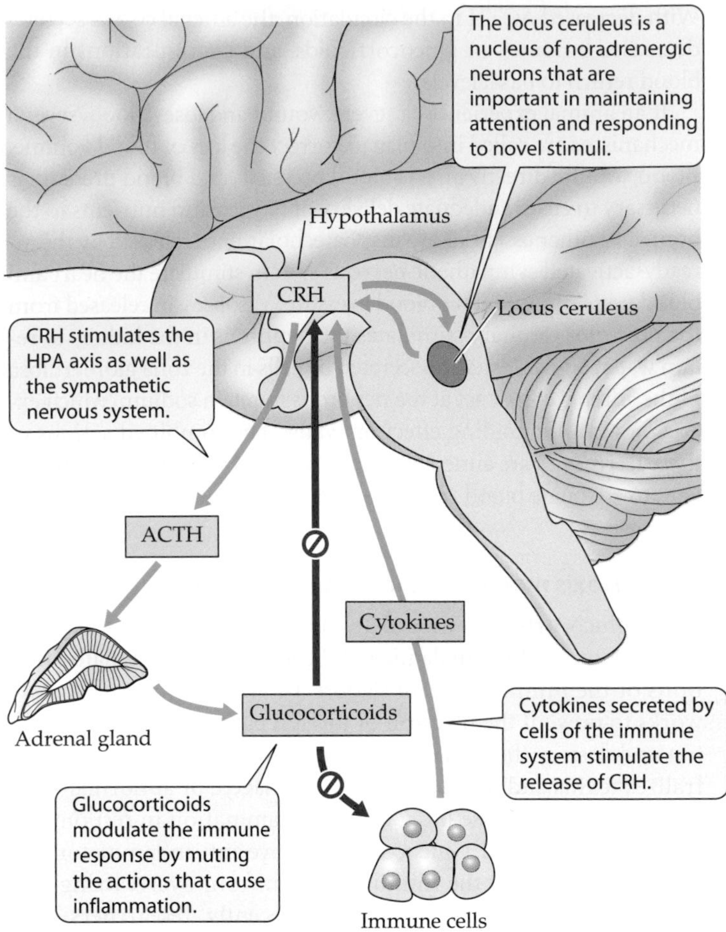

Figure 14.11 The CNS and the immune system interact during the stress response

ing small amounts of glucocorticoids into these animals improves their resistance to inflammation. Conversely, the Fischer strain of rat has an HPA axis that responds excessively to stress, and it is highly resistant to inflammatory disease, presumably because it secretes high levels of glucocorticoids.

The intricacies of communication among the nervous, endocrine, and immune systems are not limited to cytokines interacting with the CNS. Some cells of the immune system can also stimulate glucocorticoid secretion without involving the CNS. In the presence of pathogens, these cells synthesize the hormone ACTH and secrete it into the blood. This ACTH acts at the adrenal cortex in the same way it would have had it been secreted by cells in the anterior pituitary! This single example provides only a glimpse of the interrelatedness of the three systems. We have seen that cells of the immune system secrete hormones, that cells in the nervous system have receptors for cytokine signals produced by immune cells, and that hormones affect the functions of both the nervous system and the immune system. The field of *neuroimmunomodulation* is ripe for future investigations, which have the potential to reveal the mechanisms underlying "mind–body interactions."

Chronic stress causes deleterious effects

Acute stressors are of short duration, and the resulting stress responses are also brief. All of the components of the stress response, as we have seen, contribute positively to the survival of a rat getting away unscathed from a cat, or a gazelle narrowly escaping with a bleeding gash inflicted by the jaws of a hyena. However, the stress response can be maladaptive when it is induced in animals exposed to stressors for long periods, or exposed repeatedly to "psychological" stressors.

Researchers and clinicians suspect that continuously constricted blood vessels and retention of salt and fluid when no blood is lost contribute to hypertension (high blood pressure) and other cardiovascular maladies. Prolonged exposure to glucocorticoids also gives rise to pathological conditions. The catabolic actions of glucocorticoids cause muscle wasting and bone thinning. Their suppression of immune-system functions causes susceptibility to infections and disease. Chronic activation of the HPA axis suppresses reproductive functions, a phenomenon seen in highly trained athletes (both male and female), ballet dancers, starving animals, and persons with anorexia nervosa. High levels of glucocorticoids have also been implicated in causing atrophy of dendrites of neurons in the hippocampus, and even shrinkage of the hippocampus itself. Thus continuous or repeated bouts of stress can potentially damage a part of the brain that is especially important in forming memories of emotionally charged events. Because of their far-reaching effects, stress and anxiety are subjects of intense ongoing research.

Plasma glucocorticoid concentrations show seasonal variations

Much of our understanding of the stress response has been gained from experiments using laboratory animals. However, increasing numbers of investigators are now paying attention to the roles of the stress response, and glucocorticoids in particular, in wild animals in their natural environments. Although we know that blood concentrations of glucocorticoids vary according to a circadian rhythm, studies reveal that many amphibians, birds, mammals, and reptiles also show seasonal changes in glucocorticoid concentrations. In many (but not all) of these animals, glucocorticoids are highest during the breeding season. Three main hypotheses have been proposed to account for seasonal variations in glucocorticoids:

1. The *energy mobilization hypothesis* predicts that glucocorticoids should be secreted maximally during periods of the year when energy demands are highest.
2. The *behavioral hypothesis* is based on the idea that animals express different behaviors over the course of a year, and that different behaviors require different levels of glucocorticoids to bring about the physiological effects they produce.
3. The *preparative hypothesis* suggests that glucocorticoids prime the vascular, immune, cognitive, and metabolic systems in anticipation of stressors likely to occur during specific seasons, such as stressors related to breeding, disease, predation, or severe weather.

These hypotheses provide a framework for further studies of the roles of glucocorticoids in different species and the effects of different stressors on animals in their natural environments.

Summary: The Mammalian Stress Response

- The stress response is a generalized constellation of physiological changes aimed at ensuring survival when an animal is exposed to real or perceived hostile conditions.
- The major physiological actions in the stress response include mobilizing stored energy and inhibiting energy storage, enhancing cardiovascular and respiratory functions, increasing alertness and cognition, inhibiting feeding, digestion, and reproduction, and modulating immune function. If an animal experiences loss of blood, hormones are secreted that promote the retention of water and solutes.
- The same effects of the stress response that are essential for survival of an animal exposed to an acute stressor can be deleterious during periods of prolonged stress.
- Wild animals in their natural environments experience seasonal variations in blood concentrations of glucocorticoids.

Endocrine Control of Nutrient Metabolism in Mammals

The stress response employs several hormones to mobilize fuels and maintain blood pressure and fluid volume. In this section and the next, we examine the nature of these hormones in greater detail. In this section we consider the hormones that control the availability of nutrients to cells of the body; in the following section, we will look at those that maintain salt and water balance.

Animals acquire nutrients—carbohydrates, lipids, and proteins—when they eat, but many animals don't eat continuously. Still, their cells need nutrients all the time. Moreover, their cells may require the three major classes of nutrients in very different proportions than are found in digested foods. Thus the body must not only acquire nutrients, but must also manage their storage, mobilization, and molecular interconversions to provide continuous and appropriate nutrients to all tissues. Table 14.4 lists several hormones that influence nutrient metabolism. Two main hormones, insulin and glucagon, play major roles in nutrient metabolism, and they are supported by additional hormones.

TABLE 14.4 Hormones involved in mammalian nutrient metabolism

Hormone (source)	Main stimulus for secretion	Major actions
Insulin (B cells of the endocrine pancreas)	High blood glucose; high blood amino acids; gastrointestinal hormones; parasympathetic stimulation	Increases glucose uptake from the blood by noncontracting muscle cells and fat cells (hypoglycemic effect); promotes formation of glycogen (glycogenesis) from glucose in muscle and liver; stimulates fat cells to synthesize triglycerides from glucose; promotes the use of glucose in ATP production; increases uptake of free fatty acids and triglyceride synthesis by fat cells; inhibits breakdown of triglycerides; increases uptake of amino acids by muscle and liver, promotes protein synthesis, and inhibits protein degradation; inhibits gluconeogenesis
Glucagon (A cells of the endocrine pancreas)	Low blood glucose; high blood amino acids; sympathetic stimulation	Increases blood glucose levels by stimulating glycogenolysis and gluconeogenesis in liver cells (hyperglycemic effect); promotes breakdown of triglycerides to increase blood levels of free fatty acids and glycerol; inhibits triglyceride synthesis
Epinephrine (adrenal medulla)	Sympathetic stimulation during stress and exercise	Increases blood glucose levels by promoting glycogenolysis in both liver and muscle cells; often promotes triglyceride degradation to increase blood levels of free fatty acids and glycerol; inhibits insulin release and antagonizes glucose uptake by cells; stimulates glucagon and ACTH secretion
Glucocorticoids (adrenal cortex)	Stress	Enhance the actions of glucagon and catecholamines; increase blood glucose levels by antagonizing cellular glucose uptake and promoting glycogenolysis and gluconeogenesis; promote degradation of triglycerides to form free fatty acids and glycerol; promote degradation of proteins to form free amino acids
Thyroid hormones (thyroid gland)	Secreted continuously; increased by TRH and TSH, which are influenced by exposure to cold in human newborns (but not adults) and other animals	Promote oxidation of nutrients (raise metabolic rate); enhance effects of catecholamines on metabolism; stimulate growth hormone secretion and enhance the effects of growth hormone on synthesis of new proteins
Growth hormone (anterior pituitary gland)	Secreted continuously with a circadian rhythm; increased by GHRH (influenced by stress, exercise, hypoglycemia); inhibited by somatostatin	Promotes protein synthesis and growth; increases uptake of amino acids by liver and muscle; enhances breakdown of triglycerides to increase blood levels of free fatty acids; antagonizes glucose uptake by muscles
Androgens (gonads and adrenal cortex)	Secreted continuously from puberty onward	Promote protein synthesis and growth of muscle

Insulin regulates short-term changes in nutrient availability

Many mammals go through bouts of feeding separated by several hours of not feeding. To prevent alternations of feast and famine at the cellular level, mechanisms are set in motion that favor storage of nutrient molecules immediately after a meal and mobilization of nutrients from storage depots as the hours pass until the next meal. Insulin is the most important hormone involved in managing short-term fluctuations of nutrient availability.

During digestion, several factors, including rising concentrations of glucose and amino acids in the blood, gastrointestinal hormones, and parasympathetic activity, stimulate the B cells in the pancreatic islets of Langerhans to increase their secretion of insulin. Insulin is the dominant hormone in the blood during the fed state. It favors the *storage* of all three major classes of nutrients. It promotes the uptake of glucose, fatty acids, and amino acids from the blood into tissues like muscle and fat. In the case of glucose, for example, insulin binds to its receptor molecule on a target cell and triggers a series of signal transduction events that lead to the incorporation of specific glucose transporter molecules into the membrane of the target cell. The glucose transporter molecules allow the target cell to take up glucose by facilitated diffusion. Certain tissues of the body do not depend on insulin to stimulate the uptake of glucose. The brain and liver both have glucose transporter molecules that permit them to take up glucose continuously. Interestingly, exercising skeletal muscles do not depend on insulin for glucose uptake either. Resting skeletal muscles require insulin to promote incorporation of glucose transporter molecules into their cell membranes. However, the contractions of exercising muscles are sufficient stimuli for the incorporation of glucose transporter molecules.

Because insulin causes a decrease in blood glucose levels, it is said to exert a *hypoglycemic effect*. Insulin also sets in motion events that stimulate enzymes that synthesize nutrient storage molecules: glycogen from glucose, triglycerides (lipids) from fatty acids or glucose, and proteins from amino acids. At the same time insulin promotes the formation of glycogen, triglycerides, and proteins, it inhibits the enzymes that break down these large molecules.

Insulin secretion decreases as digestion comes to an end, and the ebbing of insulin levels in the blood is often the only endocrine change necessary for a shift to net *mobilization* of nutrients from stores. As insulin declines, breakdown of stored glycogen and lipids begins, and glucose and fatty acids are released into the blood. The overall pattern of insulin secretion—high levels in the fed state and low levels in the unfed state—provides negative feedback that tends to *stabilize blood concentrations of nutrients*.

Figure 14.12*a* shows the average rise and decline in plasma levels of insulin for several people after a high carbohydrate meal. With increased glucose in the blood, insulin secretion increases; as blood glucose levels decline, so do blood levels of insulin. The spike in blood glucose following the meal shows that blood nutrient concentrations are not completely stable. However, concentrations remain far more stable than they would without the negative feedback mediated by insulin. No other hormone within the body can lower blood glucose levels. This point is made dramatically clear by people with *diabetes mellitus*, who secrete abnormally low amounts of insulin or suffer from diminished tissue responsiveness to insulin. After a high-carbohydrate meal, untreated diabetics experience far higher blood glucose concentrations than nondiabetics. In fact, the blood glucose levels of diabetics become so high that their kidneys are unable to recover all the glucose filtered from the blood in the process of urine formation (see Chapter 27), and glucose is excreted in their urine and wasted. Chronic high levels of glucose cause damage to the eyes, kidneys, blood vessels, and nervous system.

Glucagon works together with insulin to ensure stable levels of glucose in the blood

Glucagon is a peptide hormone secreted by the A cells of the pancreatic islets. The main stimuli for its release are low levels of glucose in the blood, sympathetic stimulation of the A cells, and high levels of amino acids in the blood. Its main effect is to increase the production of glucose and its release into the blood. Because it causes blood glucose levels to rise, glucagon is said to exert a *hyperglycemic effect*, just the opposite of insulin's hypoglycemic effect. Glucagon stimulates cells in the liver to break down glycogen in a process called *glycogenolysis* and to release the resulting glucose into the blood. Glucagon also exerts effects opposite to those of insulin with regard to fats. It inhibits triglyceride (lipid) synthesis and stimulates adipose (fat) cells to break down triglycerides into fatty acids and glycerol and release these products into the blood. In addition, glucagon stimulates *gluconeogenesis* in liver cells. In gluconeogenesis, new glucose molecules are formed from noncarbohydrate molecules, mainly amino acids and glycerol. Proteins and fats are both mobilized from body tissues at low insulin levels. Thus amino acids from protein breakdown and glycerol from fat breakdown become available for gluconeogenesis in liver cells.

As blood glucose levels rise, glucagon secretion tends to decrease by negative feedback. Thus, from meal to meal, both insulin and glucagon contribute to stable blood glucose levels. Under conditions of stress (and exercise), however, sympathetic stimulation causes secretion of epinephrine from the adrenal medulla as well as increased synaptic input to the A cells. Epinephrine has the dual effect of stimulating the A cells to secrete glucagon and inhibiting the B cells from secreting insulin. This arrangement ensures increased glucose availability without hindrance from insulin.

Glucagon is secreted when blood levels of glucose and fatty acids are low, a condition typical of the unfed state. However, the rate of glucagon secretion is not increased by low levels of amino acids. Instead, glucagon secretion increases when blood levels of amino acids are *high*. Thus, although glucagon is the dominant hormone during the unfed state, it is often secreted during the fed state, depending on the nutrient composition of a meal. When a high-carbohydrate meal is consumed by healthy human subjects, blood levels of glucose rise, insulin secretion increases, and glucagon secretion decreases (Figure 14.12*a*). Under these conditions, the low levels of glucagon reinforce the actions of insulin. After a high-protein meal, however, *both* insulin and glucagon rise (Figure 14.12*b*). The rise in insulin promotes the incorporation of absorbed amino acids into body proteins. The rise in glucagon under these circumstances has an adaptive advantage because a high-protein meal in itself supplies little glucose, yet the brain requires glucose. Increased glucagon ensures an output of glucose from liver glycogen stores even in the face of high insulin levels.

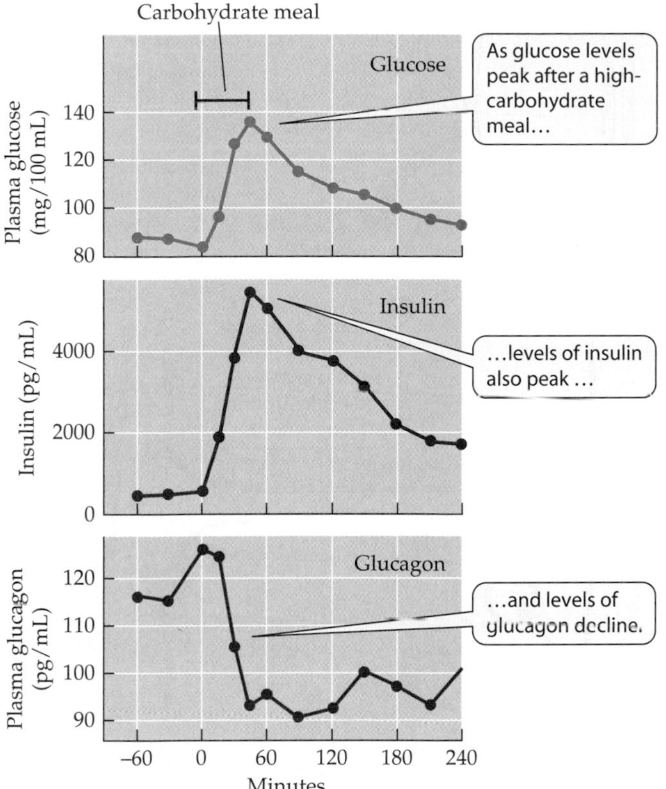

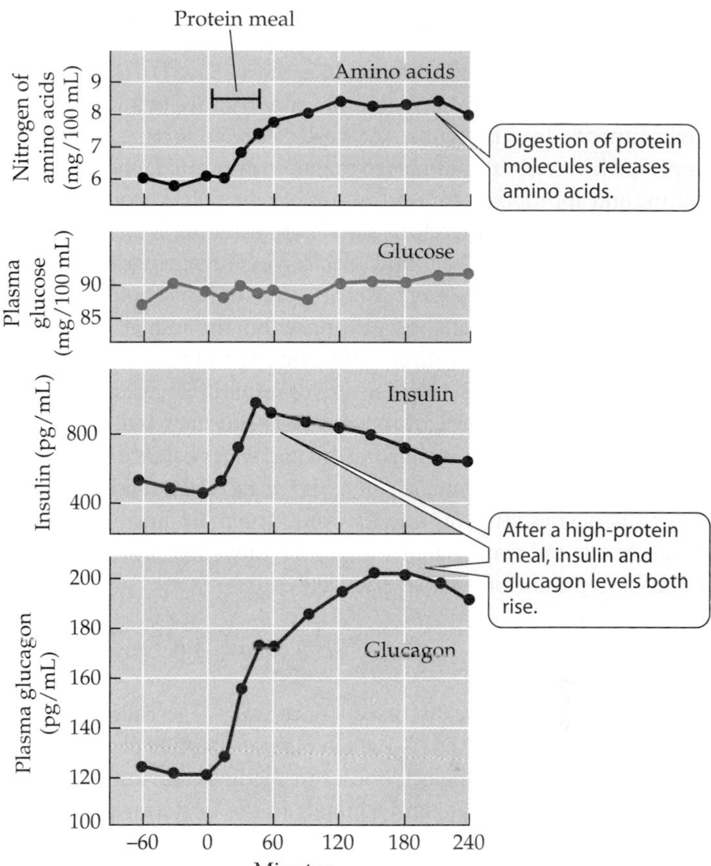

Figure 14.12 Hormone and nutrient levels in the blood of healthy human subjects before and after a meal *(a)* Levels for 11 subjects who consumed a high-carbohydrate meal of white bread and boiled spaghetti, corn, rice, and potatoes. *(b)* Levels for 14 subjects who consumed a high-protein meal of boiled lean beef. (After Müller et al. 1970.)

Other hormones contribute to the regulation of nutrient metabolism

Each of the other hormones listed in Table 14.4 exerts a unique regulatory role, but also influences the effects of other hormones. For example, both growth hormone and glucocorticoids act *synergistically* with epinephrine to enhance its effect on lipid breakdown. Other hormones play an essential *permissive* role in nutrient metabolism. Background levels of glucocorticoids, for example, are essential for preventing levels of blood glucose from plummeting during fasting and other stresses. Not only do glucocorticoids stimulate glucose formation, but they are also required for glucagon and epinephrine to exert their effects.

Several hormones play key roles in the growth of young animals. Growth hormone, thyroid hormones, and androgens are important because they promote the formation of proteins. For example, the presence of testicular androgens—principally testosterone—during puberty brings about the greater muscular development that occurs in boys as compared to girls. These hormones also work together synergistically to enhance each other's effects. Thyroxine, for example, enhances the effects of growth hormone in young animals.

Hormones also alter nutrient metabolism during exercise and fasting. Both circumstances require mobilization of metabolic fuels from stores. During both, insulin secretion declines. Without insulin's effects, glycogen breakdown, lipid breakdown, protein breakdown, and gluconeogenesis all occur at heightened rates. Glucagon secretion rises during both exercise and fasting and stimulates glycogenolysis and gluconeogenesis in liver cells. Other hormones may also play a role, but exercise and fasting do not necessarily elicit the same hormonal responses. For example, epinephrine secretion increases during exercise, but does not consistently increase during fasting. Furthermore, prolonged fasting, but not exercise, often leads to a decline in thyroid hormone levels; this decline serves to lower metabolic demands and conserve fuels.

Summary: Endocrine Control of Nutrient Metabolism in Mammals

- Insulin is secreted when nutrient molecules are abundant in the blood. It exerts a hypoglycemic effect by promoting uptake and storage of nutrients and inhibiting degradation of glycogen, lipids, and proteins. In the absence of insulin, nutrient molecules are mobilized to enter the blood from storage sites.
- Glucagon is secreted when blood glucose levels are low. It exerts a hyperglycemic effect by stimulating the breakdown of glycogen (glycogenolysis), the breakdown of triglyceride molecules, and the formation of glucose from noncarbohydrate sources (gluconeogenesis).
- Growth hormone, glucocorticoids, epinephrine, thyroid hormones, and androgens typically play permissive and synergistic roles in nutrient metabolism.

Endocrine Control of Salt and Water Balance in Vertebrates

We devote the last part of this book (Chapters 25–28) to the strategies animals use to maintain salt and water balance in a variety of environments. In this section, we explore the endocrine controls employed by vertebrate animals to maintain appropriate body fluid volume and appropriate concentrations of salts. Salts dissociate into ions in solution, and therefore are also referred to as *electrolytes*. (Physiologists often use "salt" and "electrolyte" interchangeably.) Because dissolved particles exert osmotic pressure that holds water, they have a direct effect on fluid volume within the animal. Furthermore, fluid volume has a direct effect on arterial blood pressure, which is essential for ensuring delivery of respiratory gases and nutrients to cells and removal of metabolic by-products (see Chapter 23). Thus the hormones involved in salt and water balance have far-reaching effects. The dominant salt in the extracellular fluid is NaCl, which dissociates into Na^+ and Cl^- ions. The main hormones involved in salt and water balance affect the movements of Na^+ and water. Cl^- typically follows Na^+ passively.

Antidiuretic hormones conserve water

Earlier in this chapter, we discussed vasopressin—also called antidiuretic hormone (ADH)—which is a nonapeptide produced by neuroendocrine cell bodies in the hypothalamus and released from their axon terminals in the posterior pituitary gland. This hormone acts to conserve water by preventing the production of a large volume of urine. All major classes of vertebrate animals produce hormones with antidiuretic action. Most mammals use *arginine vasopressin* (AVP), but some pigs and their relatives use *lysine-vasopressin* (LVP), and nonmammalian vertebrates use the closely related *arginine vasotocin* (AVT) (see Table 14.3). In all vertebrates, the target tissue of these hormones is the nephron of the kidney. Antidiuretic hormones stimulate the reabsorption of water from the lumen of the nephron. This means that, instead of being excreted in the urine, water is returned to the extracellular fluid.

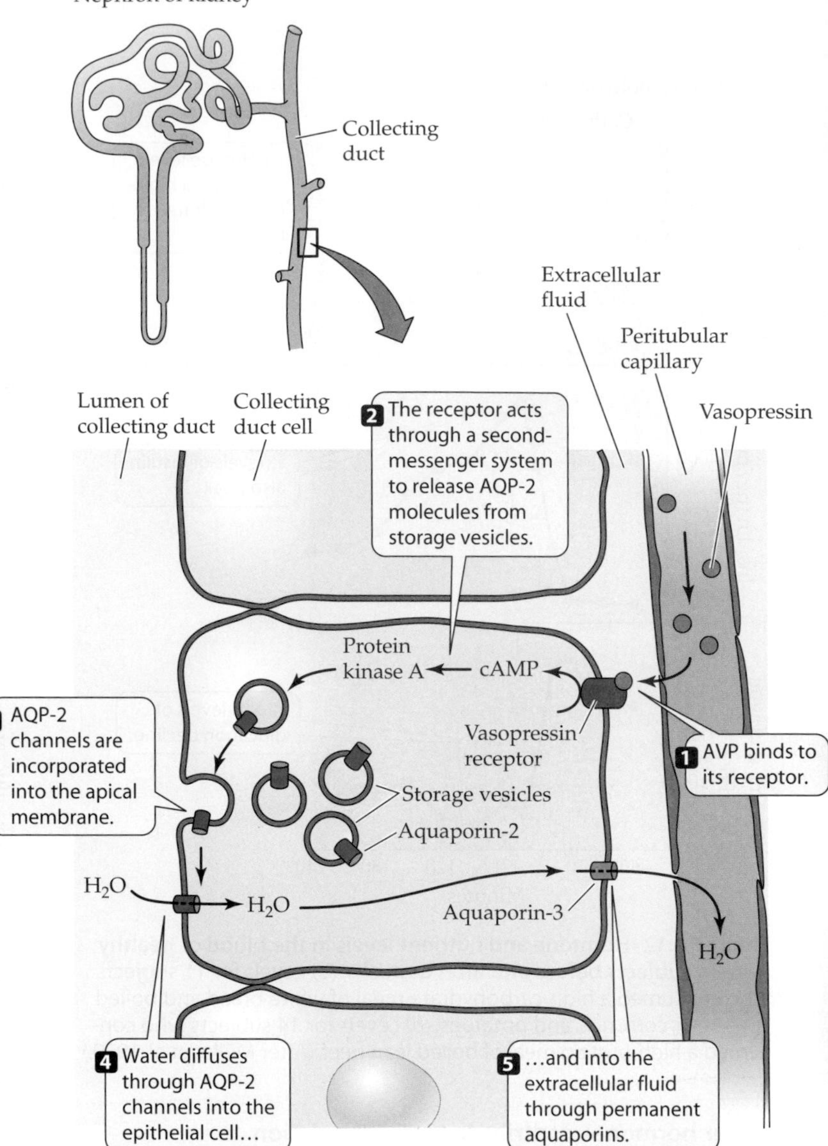

Figure 14.13 The action of an antidiuretic hormone Arginine vasopressin (AVP) stimulates the incorporation of AQP-2 into the apical membranes of epithelial cells in the collecting duct of the nephron, resulting in the return of water to the extracellular fluid.

The action of arginine vasopressin (AVP) has been studied extensively in mammals. Its effect is to stimulate the incorporation of specific **aquaporin** (AQP, water channel) molecules (see Chapter 3) into the membranes of epithelial cells in the collecting duct of the nephron (Figure 14.13). Different types of aquaporins exist permanently in various regions of the nephron. However, AQP-2 molecules are present in the apical membranes (those facing the lumen) of the cells of the collecting duct only when vasopressin is present. The epithelial cells of the tubules of the nephron are connected by *tight junctions*, which prevent movement of substances, including water, between cells. Thus water in the lumen of the tubule is destined for excretion unless it can pass through the epithelial cells back into the interstitial fluid and plasma. Receptor molecules for AVP are located on the basal side of the cells. When AVP is secreted from the posterior pituitary, it travels in the general blood circulation to the kidneys and binds to these receptors. Through second-messenger systems, AVP stimulates the movement of AQP-2 molecules from intracellular storage vesicles to the apical membrane facing the lumen. Experiments show that when exposed to AVP, the epithelial cells begin to increase their permeability to water within a minute, and reach peak permeability in about 40 minutes. Water moves out of the lumen by diffusion along its concentration gradient. It passes into the epithelial cell and out the basal end of the cell through a different type of AQP channel (aquaporin-3) that is always present and open.

When the extracellular fluid has a high osmotic concentration or the extracellular fluid volume is low, neurons in the CNS stimulate the AVP neuroendocrine cells in the hypothalamus to secrete AVP. In the presence of AVP, AQP-2 channels are incorporated into the apical membranes of the epithelial cells, allowing reabsorption of water. When the extracellular fluid has a low osmotic concentration or the extracellular fluid volume is large, the AVP neuroendocrine cells do not secrete AVP. In the absence of AVP, the AQP-2 channels are taken back into the cells' cytoplasm, and none (or very few) are present in the apical membranes. Therefore, water is not reabsorbed—no matter what the diffusion gradient. It will be excreted because it cannot escape the lumen.

The renin–angiotensin–aldosterone system conserves sodium

Aldosterone is a steroid hormone (see Figure 14.1) produced and secreted by the zona glomerulosa of the adrenal cortex. It is called a *mineralocorticoid* because it contributes to the balance of minerals—mainly sodium and potassium—in the body. The main target tissues of aldosterone are epithelial cells located in the late distal tubule and early collecting duct of the nephron (Figure 14.14). The effect of aldosterone is to increase the ability of these cells to reabsorb Na^+ (take it out of the lumen and return it to the interstitial fluid and plasma).[5] In addition, aldosterone stimulates the release of K^+ into the lumen (for excretion). When aldosterone is secreted, it enters its target cells, binds to cytoplasmic receptors within the cells, and stimulates them to synthesize new Na^+–K^+-ATPase pumps and Na^+ and K^+ channels for incorporation into the cell membrane. The pumps are incorporated into the basal membrane and the channels into the apical membrane of the epithelial cells.

The Na^+–K^+-ATPase pumps actively transport K^+ into the epithelial cell and Na^+ out of the cell. The pumps set up concentration gradients so that K^+ diffuses through its channels from a high concentration in the cytoplasm to a low concentration in the lumen of the tubule (to be excreted). Na^+ diffuses along its concentration gradient in the opposite direction. It passes through Na^+ channels into the cell (where the Na^+ concentration is low), and is then pumped from the basal side of the cell into the extracellular fluid (to be retained in the body).

Aldosterone is secreted under conditions of low arterial blood pressure. One way to correct low blood pressure is to increase the extracellular fluid volume (or preserve it, in the case of blood loss). Na^+ (and Cl^-) within the extracellular fluid exert an osmotic "hold" on water. This effect, combined with increased reabsorption of water, increases the volume of the extracellular fluid and therefore increases arterial blood pressure.

An intricate sequence of events and substances leads to the secretion of aldosterone and the correction of low blood pressure. When low blood pressure is detected, specialized cells in the kidney secrete a substance called **renin**. These cells, called **juxtaglomerular cells**, are modified smooth muscle cells located in the arterioles associated with the glomerulus of the nephron (see Figure 14.14). The juxtaglomerular cells are sensitive to stretch. When they are stretched *less* (as in the case of low blood pressure), they increase their secretion of renin. The cells are also innervated by sympathetic nerve fibers and receive sympathetic stimulation to secrete renin when blood pressure is low. Finally, renin secretion is regulated by paracrines from the *macula densa*, a group of specialized cells that monitor fluid and solute flow through the distal tubule. Birds, mammals, reptiles, amphibians, and bony and cartilaginous fish all have juxtaglomerular cells that secrete renin. Its actions have been studied most thoroughly in mammals.

[5]Aldosterone also has other target tissues, including the urinary bladder (amphibians, reptiles, and mammals), sweat glands (mammals), salt glands (birds and reptiles), salivary glands (mammals), and intestine (mammals, birds, and amphibians). At all of these target tissues, its effect is to retain Na^+ in the body.

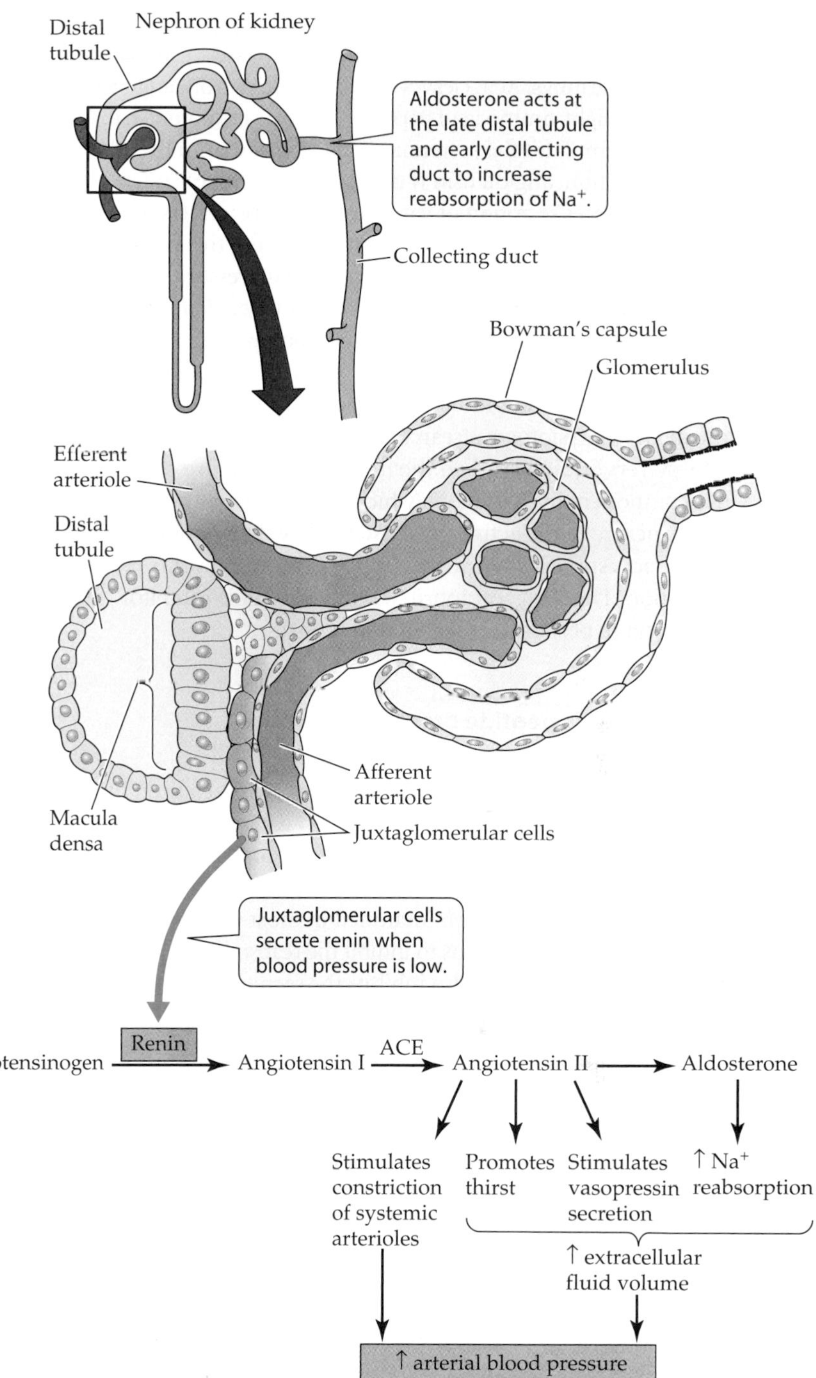

Figure 14.14 The renin–angiotensin–aldosterone system corrects low extracellular fluid volume and low blood pressure

Renin is the first substance in a chain that leads to the secretion of aldosterone (see Figure 14.14). Renin is secreted into the blood and, in mammals, has a half-life of about 20 minutes. It interacts enzymatically with a large protein called **angiotensinogen**, which is produced by the liver and is continuously present in the blood. Renin cleaves a peptide bond of angiotensinogen to produce a ten-amino acid peptide called **angiotensin I**. Angiotensin I does not have a physiological effect. Circulating in the blood, angiotensin I encounters a membrane-bound enzyme, called **angiotensin converting enzyme** (**ACE**), which is produced by the endothelial cells of the blood vessels, especially the capillaries in the lungs. ACE cleaves two amino acids from angiotensin I to make the eight-amino acid peptide hormone **angiotensin II**. Angiotensin II stimulates the secretion of aldosterone from the adrenal cortex. In addition, angiotensin II causes constriction of systemic arterioles, stimulates secretion of vasopressin, and promotes thirst and drinking. All of these actions contribute to raising the extracellular fluid volume and therefore raising arterial blood pressure.

Interestingly, the toxin of the Brazilian pit viper *Bothrops jararaca* kills by causing a catastrophic drop in blood pressure. A component in the toxin specifically blocks the action of ACE, and therefore halts production of angiotensin II and secretion of aldosterone. Knowing this physiology, researchers at in the pharmaceutical firm Bristol-Myers Squibb studied the molecular structure of the most active component in the venom, which gave them an understanding of the structure of the active site of ACE. They then designed a drug that interacts specifically with the active site and prevents its enzymatic action. This drug, captopril, was the first "ACE-inhibitor" drug used to treat hypertension (high blood pressure).

Atrial natriuretic peptide promotes excretion of sodium and water

Atrial natriuretic peptide (**ANP**) is produced by specialized muscle cells in the atria of the heart and is secreted when the heart muscle cells are stretched. High arterial blood pressure and large amounts of sodium, which lead to expanded extracellular fluid volume, stimulate ANP secretion. ANP has been identified in most vertebrates. It acts on the distal parts of nephrons to inhibit the reabsorption of Na^+ and increase its excretion. ANP also inhibits the secretion of vasopressin, renin, and aldosterone, thus counteracting the hormones that conserve water and Na^+. Finally, ANP increases the rate of blood filtration in the nephrons, which causes more fluid to be excreted. Thus, by its many effects, ANP produces a reduction in the extracellular fluid volume and corrects conditions of high arterial blood pressure. The mechanisms by which ANP exerts these actions are not yet understood and are being actively investigated. ANP has also been found to stimulate the excretion of Na^+ from the gills of bony fish.

Summary: Endocrine Control of Salt and Water Balance in Vertebrates

- Hormones continuously regulate the balance of salt and water in vertebrate animals.
- Vasopressins are peptide neurohormones that stimulate the conservation of water.
- Aldosterone is a steroid hormone that stimulates the conservation of Na^+. It is part of the renin–angiotensin–aldosterone system that is set in motion under conditions of low arterial blood pressure.
- Atrial natriuretic peptide exerts many different actions, all of which stimulate the excretion of Na^+ and water.

Hormones and Other Chemical Signals

In the examples so far described in this chapter, we have seen the far-reaching and complex interrelationships between the endocrine system and other systems within the organism. In future chapters, we will see comparable examples of hormonal regulation woven into

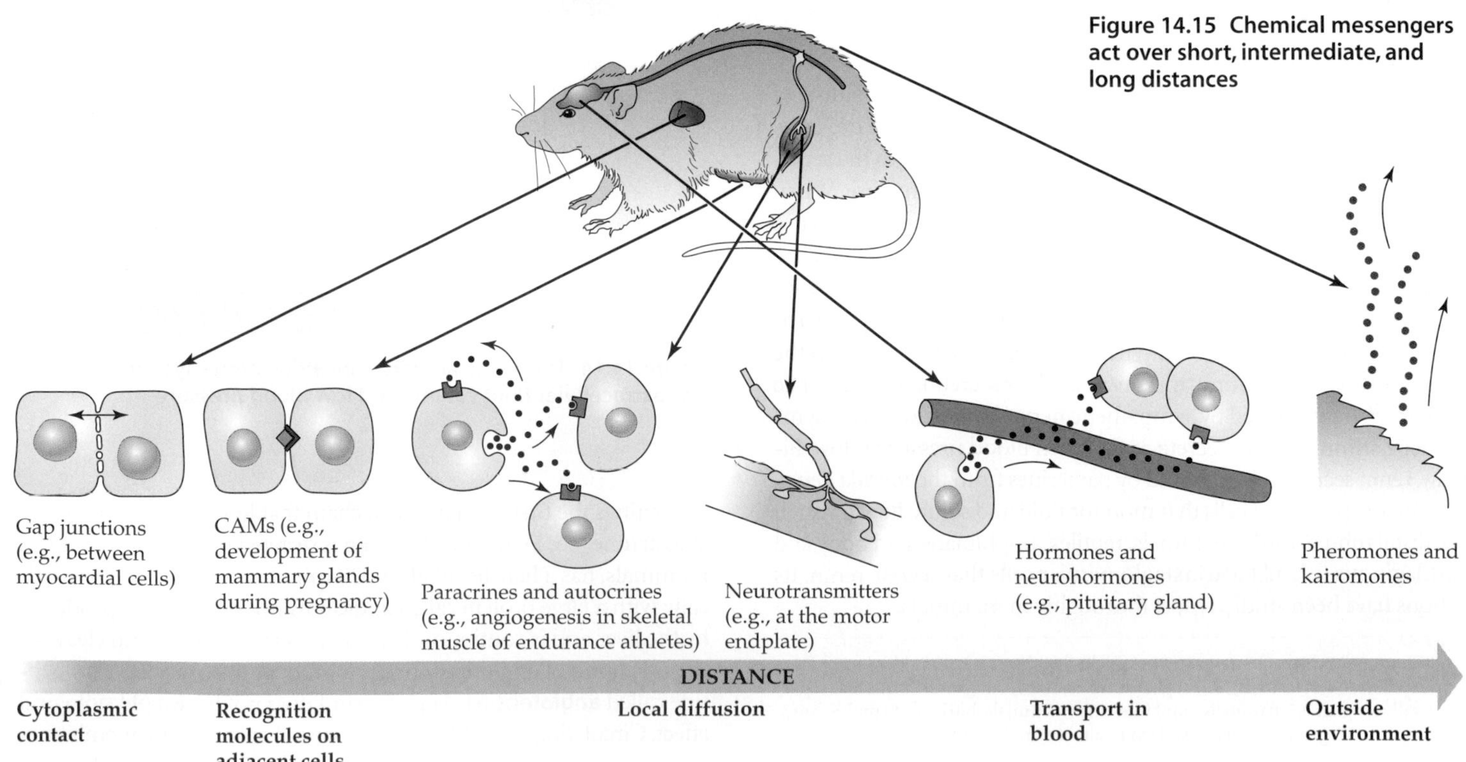

Figure 14.15 Chemical messengers act over short, intermediate, and long distances

the tapestry of overall homeostasis. Five prevailing patterns of hormonal control will appear repeatedly:

1. The control of any single system will very likely involve more than one hormone. For example, fluid balance requires not only hormones that control water release or conservation, but also other hormones that control solute concentrations in body fluids.
2. A hormone that affects the functions of one system will probably affect other systems as well. For example, a hormone that controls conservation or release of water will directly affect blood pressure.
3. Hormones may interact with each other synergistically, permissively, or antagonistically.
4. Most endocrine controls are inextricably associated with neural controls. We also know that the immune system interacts with the endocrine and nervous systems in some arenas; future research will probably reveal that all three systems are more intimately related than we now understand.
5. Many molecules that function as hormones in one context function as different types of chemical signals in different contexts. Vasopressin and oxytocin, for example, serve as neurotransmitters—not as hormones—in the CNS. To illustrate their neurotransmitter function, we describe in Box 14.1 the influence of oxytocin and vasopressin on male–female bonds in certain voles.

This list of five patterns of hormonal control gives an overview, but it is possible to make further generalizations about the complexity of endocrine function. Table 14.5 surveys several specific ideas to keep in mind when considering endocrine function.

Hormones and neurohormones fall along a "distance continuum" of chemical signals

In this and preceding chapters, we have seen that cells communicate with one another by a variety of mechanisms. Based on the distances involved, mechanisms of communication between cells can be broadly grouped into six main categories (Figure 14.15):

1. *Gap junctions* (see Chapter 12) are formed by connexon protein channels between adjacent cells. When these channels are open, they allow ions and other small molecules to diffuse directly from one cell to the next.

BOX 14.1 Can Mating Cause True Commitment?

The peptides arginine vasopressin (AVP) and oxytocin serve as hormones when released into the blood from the posterior pituitary gland. These same peptides serve as neurotransmitters when released by neurons in the CNS. Although many of their functions in the CNS are not yet understood, elegant studies of two different species of rodents, the prairie vole (*Microtus ochrogaster*) and the montane vole (*Microtus montanus*), have shown that AVP and oxytocin participate in the control of behaviors related to monogamy and social attachment.

The prairie vole is one of very few species of mammals that are monogamous. These animals form pair bonds in the process of mating. Over a 24-hour period, a pair will engage in 15 to 30 bouts of copulation, after which they undergo a transition in behavior. They show preference for each other's company, the male develops aggressive behaviors toward other males (which he didn't show before mating), and both parents care for their young even several weeks after weaning. By contrast, montane voles do not form pair bonds after mating, they breed promiscuously, and the males do not help to care for the young; in fact, even the females abandon their young 2 or 3 weeks after birth. These predictable behaviors can be demonstrated in the laboratory, and investigators have studied them using a variety of behavioral, physiological, and anatomical experimental paradigms.

Groundbreaking experiments showed that if oxytocin (but not AVP) is injected into the cerebral ventricles of the brain of an unmated female prairie vole, she will form a pair bond with a male without mating at all. Conversely, if an antagonist to oxytocin is administered by injection into the cerebral ventricles of a female before mating, she will not form a pair bond with the male, even though mating went ahead normally. Interestingly, oxytocin does not affect males in the same way. Instead, injection of AVP into the cerebral ventricles of an unmated male prairie vole elicits mate preference and aggressive behavior toward other males. The development of these behaviors is blocked if an antagonist to AVP is injected into the male prairie vole prior to mating.

In further experiments, researchers used labeling techniques to identify the distributions of postsynaptic receptors for oxytocin and AVP in the brains of the two types of voles. The monogamous prairie voles showed high densities of oxytocin receptors in the nucleus accumbens and high densities of AVP receptors in the ventral pallidum. The promiscuous montane voles did not show receptors in either of these areas, although they showed labeled receptors for AVP and oxytocin in other regions of the brain. The nucleus accumbens and the ventral pallidum are parts of the brain's reward circuitry, which is associated with reinforcement and conditioning (the same parts of the brain thought to be involved in cocaine or nicotine addiction). A current hypothesis to explain why prairie voles are monogamous is that mating stimulates the release of oxytocin and vasopressin. These peptides, in turn, activate the reward circuits, which reinforce the formation of pair bonds. Montane voles lack receptors for oxytocin and AVP in these brain areas, so they receive no reinforcement for pair-bond formation.

A monogamous mammal The prairie vole (*Microtus ochrogaster*) forms a pair bond with its mate.

TABLE 14.5 Generalizations about the complexity of endocrine and neuroendocrine function

Feature	Example
Some endocrine organs are specialized for a single function: to secrete hormones.	The exclusive function of the thyroid gland is to secrete hormones. The exclusive function of the anterior pituitary gland is to secrete hormones.
Some organs secrete hormones and also perform other functions.	The testes secrete testosterone and also produce sperm cells. The pancreas secretes hormones from specific endocrine cells (insulin from A, glucagon from B, and somatostatin from D cells) and also secretes digestive enzymes from exocrine structures.
A single endocrine organ may secrete more than one hormone, and different control mechanisms may control their secretion.	The anterior pituitary gland consists of discrete populations of endocrine cells, each of which secretes a specific hormone. Secretion is controlled by hypothalamic RHs and RIHs. The adrenal cortex secretes glucocorticoids and mineralocorticoids under the control of hormones. The adrenal medulla secretes epinephrine and norepinephrine in response to synaptic input from preganglionic fibers of the sympathetic nervous system.
More than one tissue may secrete the same hormone.	Both the pancreas and the hypothalamus secrete the hormone somatostatin (also called GHIH at the hypothalamus). Both the anterior pituitary and certain immune system cells secrete ACTH.
A single hormone may affect the functions of more than one target cell, and the actions of the target cells may produce different outcomes.	Glucocorticoids stimulate protein degradation in muscle, triglyceride degradation in adipose tissue, and gluconeogenesis in the liver. Oxytocin causes contraction of the uterine smooth muscle during birth and milk ejection from the mammary gland during suckling.
Secretion rates of hormones may vary over time in a cyclic pattern.	Glucocorticoid secretion varies on a diurnal cycle (high in the early morning and low in the evening). Secretion of estrogen and progesterone varies over the estrous or menstrual cycles of female mammals.
A single target cell may express receptor molecules for more than one hormone, and therefore may respond to more than one hormone.	Liver cells respond to insulin by taking up glucose and converting it into glycogen for storage (glycogenesis). The same liver cells respond to glucagon by breaking down glycogen (glycogenolysis), diminishing glycogen synthesis, and producing glucose from noncarbohydrate molecules (gluconeogenesis).
A single endocrine or neuroendocrine cell may secrete more than one hormone.	Because of the particular enzymes it uses in posttranslational processing, a POMC-expressing endocrine cell of the anterior pituitary secretes ACTH, β-endorphin, and other fragments. Some cells in the hypothalamus co-secrete AVP and CRH.
The same molecule may be a hormone in some contexts or a different type of chemical signal in others.	CRH is a neurohormone released into the hypothalamo–hypophyseal portal system and a neurotransmitter at other sites in the CNS. Norepinephrine is a hormone when released into the blood from the adrenal medulla and a neurotransmitter when released from postganglionic fibers of the sympathetic nervous system. Although the chemical nature of the signaling molecule remains the same, the nature of the response is determined by the mechanisms set in motion upon binding to the receptor molecule.
Hormones may interact with each other synergistically, permissively, or antagonistically.	Glucagon, epinephrine, and glucocorticoids interact synergistically to increase glucose levels in the blood. Insulin and glucagon exert antagonistic effects that influence glucose concentration in the blood; insulin decreases glucose in the blood whereas glucagon increases it. Glucocorticoids are necessary to permit the catecholamines to cause contraction of smooth muscle cells in the walls of small blood vessels.

2. *Cell adhesion molecules* (*CAMs*) on the external surface of cell membranes play important roles in signaling between adjacent cells involved in embryonic development, wound repair, and cellular growth and differentiation.
3. *Neurotransmitters* (see Chapter 12) are released by presynaptic neurons in response to electrical signals. They diffuse across a narrow synaptic gap to interact with receptor molecules on a postsynaptic cell, which may be a neuron, muscle cell, or endocrine cell.
4. *Paracrines* and *autocrines* diffuse relatively short distances to influence cells in the local environment—including themselves, in the case of autocrines.
5. *Hormones* and *neurohormones* are specialized for long-distance communication within the animal. (From our study of the stress response, we know that *cytokines* also communicate across long distances; however, they function locally as well.)
6. Finally, chemical signals are used outside the animal. Animals of the same species communicate with **pheromones**, whereas animals detect **kairomones** to obtain chemical information about members of a different species.

Having previously considered most of these classes of intercellular communication, we will now briefly turn our attention to the local chemical messengers—paracrines and autocrines—and the external chemical messengers—pheromones and kairomones.

Paracrines and autocrines are local chemical signals distributed by diffusion

Three main categories of locally acting chemical messengers are *neuromodulators*, *cytokines*, and *eicosanoids*. *Neuromodulators* are released from synaptic nerve terminals along with neurotransmitters and—like transmitters—bind to specific receptor sites on postsynaptic cells. They typically exert their effects more slowly and for a longer duration than neurotransmitters. One way in which neuromodulators modify the postsynaptic response to a neurotransmitter is to change the conductances of specific membrane channels. Like paracrines in general, neuromodulators can diffuse from their site of release and influence the responsiveness of not only their immediate postsynaptic cell, but also nearby cells that have appropriate receptor sites. Because of their long-term influences and their ability to affect a population of neighboring cells, neuromodulators have the potential to modify the functions of entire neuronal circuits.

Cytokines are peptides or proteins secreted by many different types of cells. They are made on demand when required to play a regulatory role. Acting locally as paracrines, cytokines control cell development and differentiation as well as the immune response. Cytokines involved in the immune response may also travel long distances in the blood, just like hormones. Because of their medical importance, our greatest knowledge of cytokines comes from studies of mammals. A particularly well-known large group of cytokines is the *interleukins*. Interleukin 2, for example, is a cytokine that enhances the activity of cytotoxic (killer) T cells in the immune response. Different cytokines control the process of *angiogenesis* (the development of new blood vessels). Several cytokines stimulate angiogenesis by acting as growth factors (such as vascular endothelial growth factor, VEGF), whereas others (such as endostatin) inhibit angiogenesis. Angiogenesis is important in the growing animal and also in wound healing, in rebuilding the uterine lining following menstruation, and in increasing the vascular supply to cardiac and skeletal muscle during endurance training by athletes (see Chapter 19).

Eicosanoids are lipid-derived paracrine molecules involved in inflammation and allergic responses. All eicosanoids are produced from arachidonic acid, a fatty acid constituent of the phospholipids in cell membranes. There are two main classes of eicosanoids: leukotrienes and prostanoids. Leukotrienes are secreted by certain white blood cells and cause constriction of the airways in the lungs, making it difficult to breathe during asthma attacks and the life-threatening allergic reaction called anaphylaxis. The prostanoids include thromboxanes and prostaglandins. Thromboxanes cause constriction of blood vessels and aggregation of platelets, two important processes necessary to close off a damaged vessel. The prostaglandins are found in many different tissues and are involved in pain, inflammation, fever, and numerous other functions.

These few examples of local chemical messengers show only a small sample of the exquisite variety of intercellular signaling mechanisms found in the animal kingdom. Animals use diverse classes of molecules as chemical messengers, they employ enzymes to modify the specificities of signaling molecules, and they use the same molecule in different contexts as a different type of chemical messenger. An interesting example is the gas nitric oxide (NO), which, as we have already learned, functions as a neurotransmitter and possibly as a neuromodulator (see Chapter 12). Although not included in the three main categories of paracrine signals, NO nevertheless functions as a paracrine to regulate the diameter of small blood vessels. It is synthesized by the endothelial cells that form the inner lining of blood vessels and diffuses into the adjacent smooth muscle cells in the wall of the vessel. It causes the smooth muscle cells to relax and the vessel to dilate.

Pheromones and kairomones are used as chemical signals between animals

Pheromones are chemical signals between members of the same species. They are produced within the animal and then released into the environment. In many animals, they convey information that signals social status (for example, gender or dominance) and sexual readiness, and they elicit behaviors that are typically stereotyped and not modified by experience. Pursuing a potential mate, copulatory behavior, and aggressive behavior to protect a territory are often set in motion by pheromones. Physiological functions, such as the onset of puberty and estrous cycling, are also influenced by pheromones.

Many vertebrate animals detect pheromones with a sensory structure called the *vomeronasal organ* (VNO) (see Chapter 13). In several amphibians and reptiles, the VNO is the major olfactory organ—that is, the VNO detects both general odors and pheromones. For example, the Komodo dragon specializes in eating carrion. This large lizard follows the scent of a recently dead animal by repeatedly sampling the air with its tongue and then touching the tongue to the roof of its mouth, where there are openings to the VNO. Many mammals, such as rodents, use the VNO to detect pheromones exclusively. Early experiments showed that surgically ablating the VNO resulted in impaired behaviors: Males would neither defend their territories nor mount females; females would not adopt the lordosis position for mating, and they seldom became pregnant. Invertebrate organisms also utilize pheromones, which they detect with a variety of chemosensory structures. The sensory apparatus used by male silkworm moths to detect *bombykol*, the powerful sex attractant pheromone released by females, is described in Chapter 13.

The molecular structures of most pheromones are not known, but those that have been identified vary widely. Bombykol is a fat-derivative alcohol; aphrodisin, the pheromone in the vaginal secretions of female hamsters, is a protein. The pheromone in the saliva of boars is a steroid; if a female pig detects this pheromone, she will stand motionless in the lordosis position.

Whereas pheromones released into the environment convey information between members of the *same* species, other categories of chemical signals released by animals give information (unintentionally) to members of *different* species. *Kairomones* are chemicals produced by animals of one species that are released into the environment—not for the purpose of communicating, but in the process of accomplishing some other function. Members of a second detecting species exploit these kairomones in a way that benefits them, but may be a detriment to the members of the releasing species. Kairomones elicit both behavioral and physiological responses. For example, one species may release a kairomone that reveals its location to a potential predator and elicits foraging behavior in the predator. Or a predator my release a kairomone that informs potential prey of its presence. Such an enemy-avoidance kairomone allows the prey to respond defensively. For example, water fleas of the genus *Daphnia* detect kairomones released by predatory fly larvae in the surrounding water; they respond by developing spines and "helmets" that may protect them from the predators. As researchers learn more about the physiological processes triggered by intraspecific and interspecific informational chemicals, we can look forward to enlarging our understanding of physiology in the context of ensuring survival and reproductive success in the natural world.

Summary: Hormones and Other Chemical Signals

- Chemical signals fall along a "distance spectrum" ranging from molecules that signal between individual cells and over short distances, to hormones and neurohormones that travel long distances in the blood, to chemical signals released by animals into the environment (pheromones and kairomones).
- Locally acting paracrines and autocrines include neuromodulators, cytokines, and eicosanoids. In addition, the gas NO functions as a paracrine.
- Pheromones are chemical signals released into the environment as cues to animals of the same species. Kairomones are chemicals released into the environment by an animal of one species that are exploited by an animal of a different species to its benefit, but not to the benefit of the releaser.

Insect Metamorphosis

In this section, we explore the hormones that control metamorphosis in insects. It is important to keep in mind that hormones play as varied and extensive roles in invertebrate animals as they do in vertebrates. Color changes, spawning, metamorphosis, water balance, nutrient metabolism, growth, maturation, and death are just a few of the functions controlled by hormones in invertebrates. We have chosen insect metamorphosis as our example because it serves to demonstrate that the same principles of endocrine function found in vertebrate animals also apply to invertebrates. In addition, because insects have great agricultural, medical, and forensic importance, their natural history and physiology have been studied extensively (Box 14.2).

Insects have separate sexes. They engage in courtship behavior that leads to sperm being deposited and stored in the female reproductive tract. Fertilization does not take place at the time of mating. It occurs later, usually just before the female lays her eggs. After various periods of time, depending on the species, the eggs hatch. From hatching onward, insects undergo a complex developmental journey to reach sexual maturity. The life cycles of most insects involve changes in the form of the animal at different stages. In other words, insects go through **metamorphosis** (*meta*, "change"; *morph*, "structure or form"). Although insects and vertebrate animals are evolutionarily divergent, we will find that their endocrine systems have independently evolved many similar features.

Insect metamorphosis may be gradual or dramatic

There are two main types of metamorphosis (Figure 14.16). Insects such as bugs, grasshoppers, and cockroaches go through gradual

BOX 14.2 Insects in Forensics and Medicine

Certain insects, such as the blowflies, feed and lay their eggs only on *necrotic* (dead) tissue. This characteristic has been employed by forensic scientists as one method to determine the time of death of bodies found under suspicious circumstances. Studies of many different insects have revealed a progression of species that occupy a corpse. The blowflies (bluebottles and greenbottles) arrive first and lay their eggs in moist regions. The eggs hatch, and maggots (larvae) develop and feed on the tissues. Flesh flies soon are drawn by olfactory cues to the decomposing flesh, and they too lay eggs that develop into maggots. Maggots of blowflies and flesh flies go through three larval instars. Depending on how long the cadaver is left to decompose, the maggots may pupate and metamorphose into adults. The presence of pupal cases gives clues to investigators about time elapsed since death.

As the maggots feed, beetles arrive to feed on both the corpse and the maggots. Wasps and other flies follow in progression. After several months, mites, along with skin and hide beetles, and finally clothes moths, explore the dried remains. Forensic entomologists collect pupal cases, excrement, and other evidence of the animals that roved the cadaver. They take into account environmental conditions (such as temperature and humidity), the species of insects on a corpse, and the details of the life cycle of each and "work backward" in time to estimate when death occurred. Insects can also give clues about the site of death. Because some insects have a very narrow range of habitats, a species of insect found on a corpse out of the insect's range indicates that the body was moved.

The larvae of some of the same insects that are attracted to corpses also feed on the necrotic tissues of wounds of living people. Indeed, during the American Civil War, surgeons recorded that wounds of soldiers with maggots on them were cleaner and healed more quickly than wounds not visited by maggots. These days, physicians usually clean wounds by *surgical debridement*, using instruments to scrape away dead tissue so that living tissue can grow in the healing process. However, they also use maggots of blowflies to debride wounds that are especially resistant to healing. Once or twice a week, blowfly larvae are applied to the surface of wounds at a density of five to eight larvae per square centimeter. The larvae are voracious feeders, but they do not eat living tissue. They seek out every bit of necrotic tissue in the wound, so they succeed in doing a far better job of cleaning than surgical instruments can do. Larvae applied to wounds must be sterile. The eggs laid by adult blowflies are soaked in solutions of Lysol or Clorox, and then kept in sterile culture dishes until they hatch into sterile larvae.

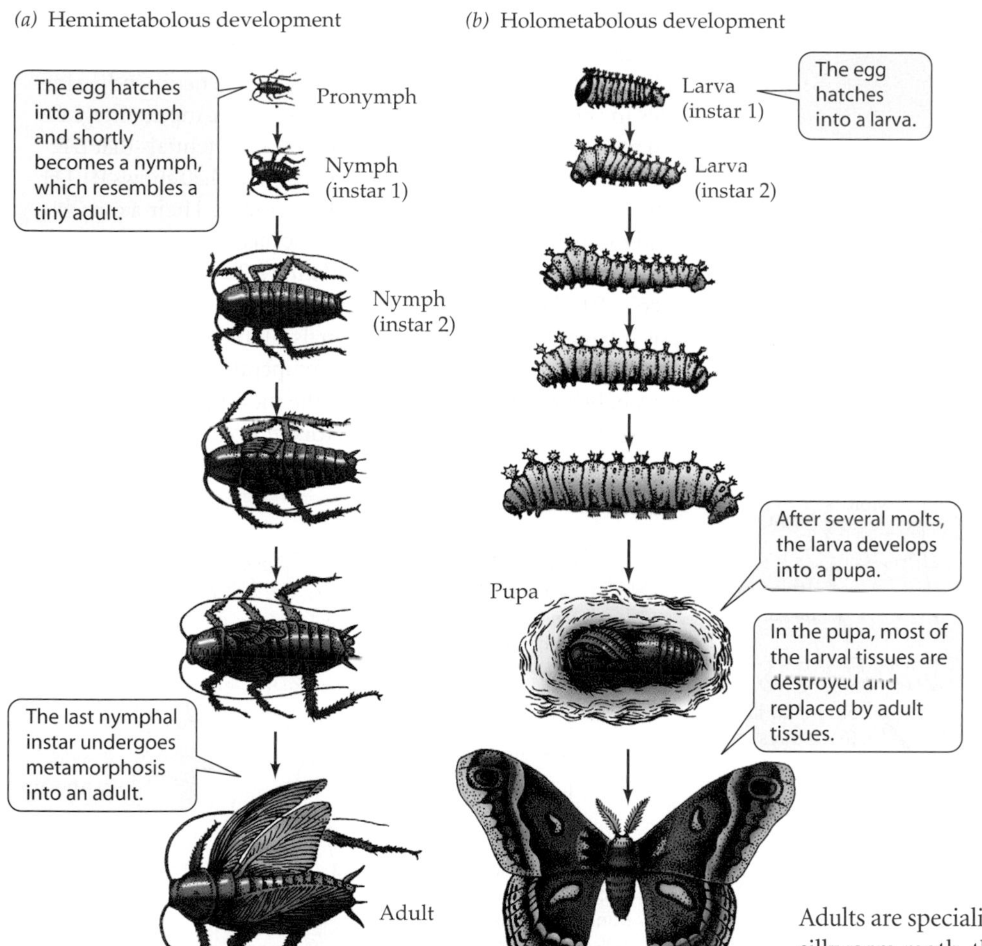

Figure 14.16 Two types of metamorphosis Most insects go through either *(a)* hemimetabolous or *(b)* holometabolous development.

metamorphosis and are referred to as **hemimetabolous** (*hemi*, "partial or gradual"; *metabolous*, "change"). In hemimetabolous insects, the immature (juvenile) forms, called *nymphs*, resemble the adult, except that they are smaller and have only immature wings and genital structures. Animals such as flies, beetles, butterflies, and moths go through dramatic metamorphosis and are referred to as **holometabolous** (*holo*, "complete"). In holometabolous insects, the *larva* becomes a *pupa* and then metamorphoses into an adult. The larva and pupa look completely different from the adult.

In hemimetabolous development, the egg hatches into a nymph, which goes through several **molts**, or **ecdyses** (Figure 14.16*a*). With each molt, epidermal cells underlying the cuticle (exoskeleton) synthesize a new cuticle, the old cuticle is shed (a process called *ecdysis*), and the new cuticle is expanded while it is still soft and pliable. To expand it, the animal takes air into the foregut and "puffs itself up." The swallowed air applies pressure on the hemolymph (blood) and forces it into narrow lanes. The increased pressure inside the body helps to fill out or unfurl external structures. Once expanded, the new cuticle hardens. The larger cuticle provides room for internal structures to grow before the next molt. The periods between molts are called **instars**. Hemimetabolous nymphs go through four to eight instars; each species has a characteristic number. The last nymphal instar undergoes metamorphosis into the adult. Adults have complete wings and are sexually mature. Adults do not grow or undergo additional molts.

In holometabolous development, the egg hatches into a *larva* (Figure 14.16*b*). Depending on the species, larvae are referred to as grubs, caterpillars, or maggots. Like hemimetabolous nymphs, holometabolous larvae go through several molts and expand the new cuticle. With each molt, the animal increases in size. The larvae of holometabolous insects are the forms that usually cause crop damage. Gypsy moth larvae, for example, have caused waves of severe defoliation of trees in the northeastern United States, and the larvae of the European corn borer (which feeds on stalks and leaves of young corn) cause damage resulting in the loss of several hundred million dollars annually in the United States. After several molts, holometabolous larvae enter a stage called the **pupa**, in which most of the larval tissues are destroyed and replaced by adult tissues. The pupa has a much thicker cuticle than the larva or adult. The pupa then metamorphoses into an adult. Adults are specialized for reproduction. In some species, such as the silkworm moth, the adult may not even feed. The holometabolous life cycle of the silkworm *Bombyx mori* is shown in Figure 14.17.

Hormones and neurohormones control insect metamorphosis

Three main hormones control insect metamorphosis: **prothoracicotropic hormone** (**PTTH**, a small protein), **ecdysone** (a steroid), and **juvenile hormone** (**JH**, a *terpene*, which is a fatty acid derivative). The chemical structures of JH and 20-hydroxyecdysone—the active form of ecdysone—are shown in Figure 14.18. Although the specific mechanisms of metamorphosis differ from species to species, investigators have shown that these three hormones typically function together. Several additional hormones play important roles in ecdysis (molting) and metamorphosis. **Bursicon** is a protein secreted by neurosecretory cells in the CNS. It stimulates the darkening (tanning) and hardening (by cross-linking of protein molecules) of the newly produced cuticle. **Eclosion hormone** (**EH**), **pre-ecdysis triggering hormone** (**PETH**), and **ecdysis triggering hormone** (**ETH**) are important in initiating the signals that control stereotyped movements during ecdysis. All of these hormones are peptides. EH is secreted by neuroendocrine cells in the brain. PETH and ETH are secreted by **Inka cells** located on the *tracheae*, the branched tubes of the respiratory system of insects that extend from openings in the body wall (see Chapter 21). Figure 14.19 illustrates the main neural and endocrine structures involved in insect metamorphosis, and Table 14.6 summarizes the endocrine and neuroendocrine secretions.

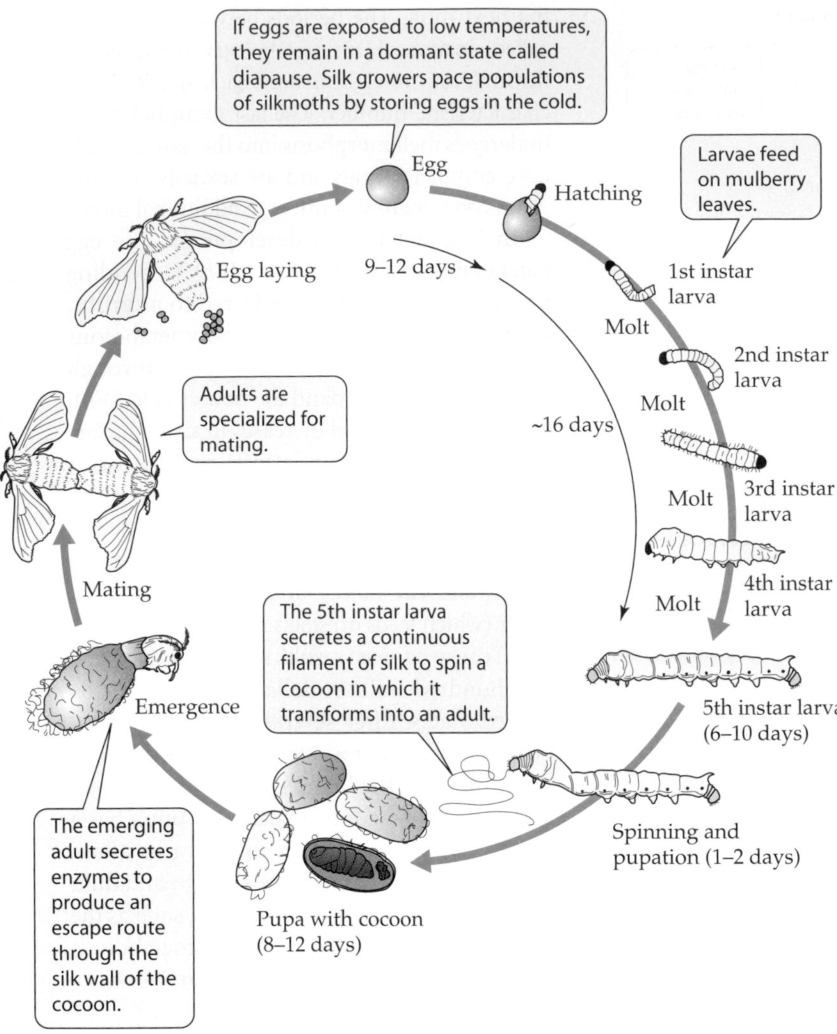

Figure 14.17 The silkworm *Bombyx mori* goes through holometabolous development

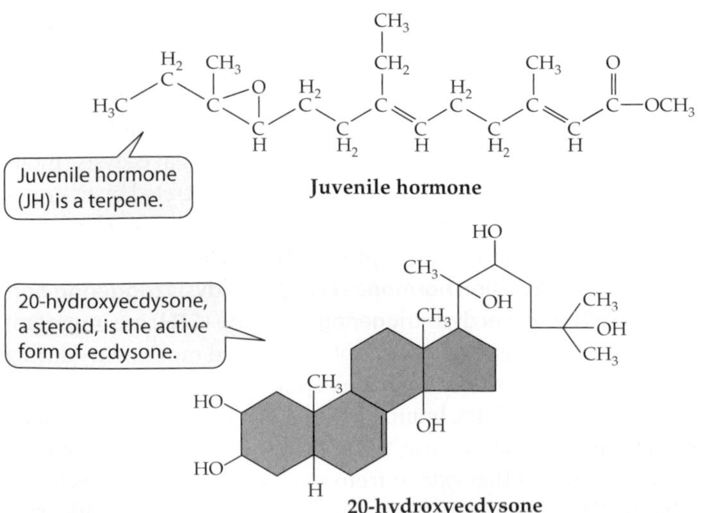

Figure 14.18 Juvenile hormone (JH) and 20-hydroxyecdysone are both lipid-soluble hormones

In both hemimetabolous and holometabolous insects, the molting process begins in the *brain*, where neuroendocrine cells produce the neurohormone PTTH. Like neuroendocrine cells in vertebrates, these cells receive synaptic input, integrate synaptic information, and generate action potentials that trigger hormone secretion by exocytosis. Figure 14.20 shows PTTH neuroendocrine cells in the brain of a moth. Their axons extend to a pair of structures closely associated with the brain, the **corpora allata** (singular *corpus allatum*). The axon terminals secrete PTTH from these structures into the hemolymph. Ecdysis is triggered by a variety of signals, all mediated ultimately by the nervous system. Day length, temperature, crowding, and certain behaviors are known to set the molting process in motion. Neurons that detect these factors send excitatory synaptic signals to the PTTH cells in the brain to stimulate secretion.

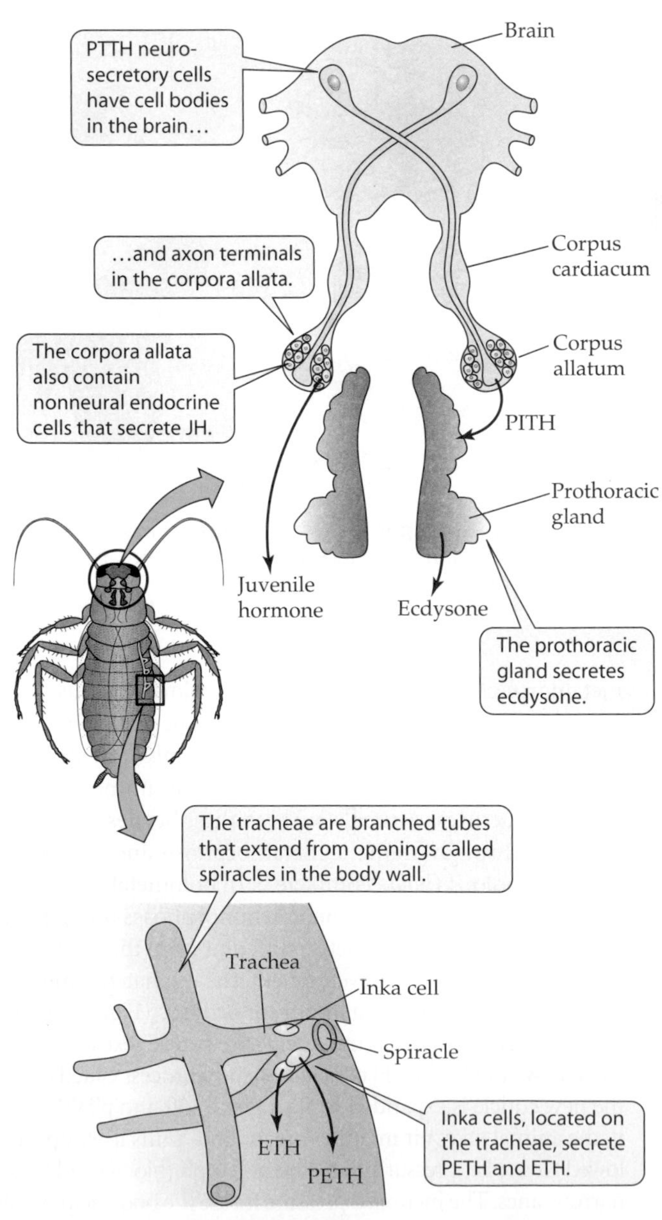

Figure 14.19 Endocrine and neuroendocrine structures involved in the control of insect metamorphosis

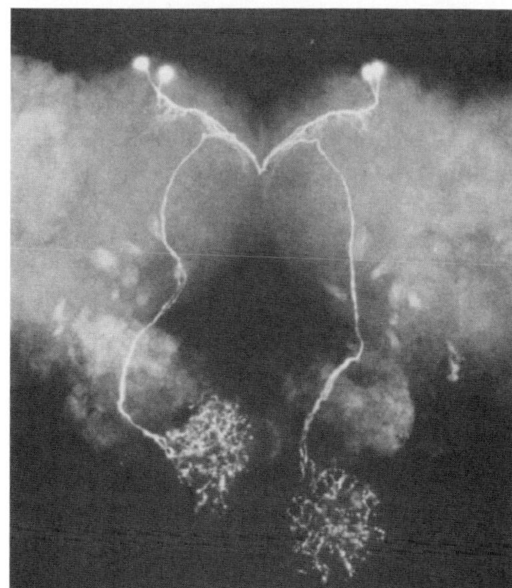

Figure 14.20 PTTH-secreting neuroendocrine cells Four PTTH-secreting cells in the brain of the moth *Manduca sexta* were stained with an immunofluorescent dye. The axons cross at the midline and extend to the corpora allata on the opposite side, where they form highly branched terminals. The width of the brain is about 1.5 mm. (Courtesy of Walter Bollenbacher; from O'Brien et al. 1988.)

The PTTH is carried in the hemolymph to the thorax, where it stimulates the **prothoracic glands** to secrete ecdysone. Ecdysone is a prohormone that is converted by peripheral activation to **20-hydroxyecdysone** by several different tissues in the body. The major target tissue of 20-hydroxyecdysone is the epidermis, which lies just below the cuticle. 20-hyroxyecdysone, also called β-*ecdysone* or *molting hormone*, stimulates the epidermis to secrete enzymes that digest away the old cuticle and synthesize a new one.

Juvenile hormone (JH) maintains juvenile (immature) characteristics in the developing animal. JH is released from nonneural endocrine cells in the corpora allata. Thus each corpus allatum is *both* a site for neurosecretory release of PTTH (a neurohemal organ) *and* a nonneural endocrine gland that secretes JH. When 20-hydroxyecdysone acts on the epidermis *and* the hemolymph level of JH is high, the insect will molt into another, larger juvenile form—a larger larva (in holometabolous insects) or nymph (in hemimetabolous insects). The corpora allata become inactive during the last larval or nymphal instar. Therefore, when ecdysone is secreted to initiate the next molt, there is little or no JH present. When 20-hydroxyecdysone is secreted in the presence of very low hemolymph levels of JH, the epidermis will produce adult or pupal structures. The nymph of a hemimetabolous insect will develop into an adult, and the larva of a holometabolous insect will develop into a pupa. In holometabolous insects, ecdysone is secreted again (and converted to 20-hydroxyecdysone) at the end of pupation. Because no JH is present, 20-hydroxyecdysone triggers metamorphosis into the adult form.

TABLE 14.6 Major hormones and neurohormones that control insect metamorphosis

Hormone	Type of molecule	Type of signal	Site of secretion	Major target tissue	Action
Prothoracicotropic hormone (PTTH)	Protein (~5000 molecular weight)	Neuroendocrine	Brain, with axon terminals extending to corpora allata	Prothoracic glands	Initiates molting (ecdysis) by stimulating release of ecdysone from prothoracic glands
Ecdysone (molting hormone)	Steroid	Endocrine	Prothoracic glands in larva/nymph; ovary in adult	Epidermis in larva/nymph; fat body in adult	When activated to 20-hydroxyecdysone, promotes cellular mechanisms to digest old cuticle and synthesize new one; stimulates production of yolk proteins in adult
Juvenile hormone (JH)	Terpene (fatty-acid derivative)	Endocrine	Corpora allata	Epidermis in larva/nymph; ovary in adult	Opposes formation of adult structures and promotes formation of larval/nymphal structures; functions as a gonadotropin in the adult
Eclosion hormone (EH)	Peptide	Neuroendocrine	Brain	Inka cells, possibly others	Promotes PETH and ETH secretion from Inka cells
Pre-ecdysis triggering hormone (PETH)	Peptide	Endocrine	Inka cells of tracheae	Neuronal circuits in brain	Coordinates motor programs to prepare for shedding the cuticle
Ecdysis triggering hormone (ETH)	Peptide	Endocrine	Inka cells of tracheae	Neuronal circuits in brain	Coordinates final motor programs for escaping from old cuticle
Bursicon	Large protein (~35,000 molecular weight)	Neuroendocrine	Brain and nerve cord	Cuticle and epidermis	Tans and hardens new cuticle

Sources: After Randall, Burggren, and French 2002; and Žitňan et al. 2003.

How do 20-hydroxyecdysone and JH exert their effects on target cells? Both hormones are soluble in lipids. Both pass through the cell membrane and bind with intracellular receptors in their target cells. The hormone–receptor complexes activate or inhibit the transcription of specific genes. Insects have *larval genes*, *pupal genes* (in holometabolous forms), and *adult genes*. As the animal goes through its life cycle, appropriate combinations of 20-hydroxyecdysone and juvenile hormone control expression of these genes. The proteins encoded by specific genes influence the functions of the target cells.

In the laboratory, it is possible to remove the corpora allata surgically and thus remove the source of JH. When this procedure is carried out on early instars, instead of molting into another juvenile form, the experimental animals metamorphose into tiny, sterile adults. Or the converse experiment can be done: Last-instar larvae or nymphs can be treated with additional JH, either by implanting donor corpora allata from earlier-stage instars or by applying JH to the animal's surface. These experimental animals do not metamorphose into adults, but instead molt into giant extra-instar larvae or nymphs. Silk growers have used this phenomenon to their advantage. The larvae of the silkworm (*Bombyx mori*) spin silk filaments to make their cocoons. Spraying late-instar larvae with an analog of JH prevents them from pupating, so they molt instead into extra-instar larvae. When these larger larvae pupate, they spin more silk to make larger cocoons. Using this technique, silk growers can obtain a 10% to 20% increase in usable silk filaments.

EH, PETH, and ETH are also essential for the progress of an insect through its life cycle. Every time the insect molts, it must break out of its old cuticle. Escaping the last nymphal cuticle (in hemimetabolous insects) or the pupal case (in holometabolous insects) is critical if the new adults are to survive and reproduce; indeed, all previous stages in the life cycle are aimed at this goal. Insects employ motor behaviors to shed their old cuticles. Holometabolous larvae, for example, generate very precise muscular movements. Before actually shedding the old cuticle, which has been previously digested away from the epidermis under the direction of 20-hydroxyecdysone, the animal performs *pre-ecdysis behavior* lasting about an hour. In pre-ecdysis, contractions of the body-wall musculature begin on the dorsal side of each segment and proceed toward the ventral side. *Ecdysis behavior*, in which the old cuticle is actually shed, takes about 10 minutes. It involves peristaltic waves of contractions starting from the posterior end and moving anteriorly. Similar behavioral sequences are required for ecdysis at the end of each instar in both holometabolous and hemimetabolous insects.

How are the skeletal muscles of the body wall signaled to contract? Current data indicate that the Inka cells located on tracheae secrete PETH and ETH. These peptides stimulate neuronal circuits in the central nervous system that control motor programs for coordinated muscle contractions. PETH and ETH are released prior to every molt in both holometabolous and hemimetabolous insects. How are the Inka cells signaled to secrete PETH and ETH? Eclosion hormone (EH) is known to promote Inka cell secretions, and additional endocrine and neuroendocrine substances also appear to be part of a sequence of signals to these cells. Thus neural activity and neuroendocrine and endocrine functions are all essential for coordinating the physiological processes leading to ecdysis.

After metamorphosis, the adult is poised for reproduction. For reproduction to succeed, the corpora allata, which became inactive and stopped secreting JH before the final molt, must begin to secrete JH again in the adult. JH functions as a gonadotropin in adults to support the production of fertile eggs and sperm. In addition, it stimulates the production of pheromones required for mating. Ecdysone is also produced in the adult—not by the prothoracic glands, but by the ovaries. JH, functioning now as a gonadotropin, causes its secretion. Ecdysone, converted to 20-hydroxyecdysone, stimulates the fat body, an organ that performs functions similar to those of the vertebrate liver, to produce yolk proteins, which are transported to the ovary for incorporation into eggs. With adults now capable of producing gametes and releasing chemical signals that attract mates, the insect life cycle is ready to be repeated.

Summary: Insect Metamorphosis

- Insect metamorphosis illustrates the convergent evolution of endocrine and neuroendocrine functions between vertebrate and invertebrate animals.
- Insects change form in the course of their life cycles. Hemimetabolous insects go through gradual metamorphosis, and holometabolous insects go through complete metamorphosis.
- Environmental and behavioral signals mediated by the nervous system initiate molting by providing synaptic input to the PTTH neuroendocrine cells in the brain. These cells secrete PTTH, which stimulates secretion of ecdysone from the prothoracic glands. Ecdysone is converted to 20-hydroxyecdysone by peripheral activation.
- 20-Hydroxyecdysone stimulates the epidermis to secrete enzymes required for the molting process. At each molt, the epidermis lays down a new cuticle beneath the old one.
- Under the control of PETH and ETH, the insect performs stereotyped pre-ecdysis and ecdysis movements in order to shed the old cuticle.
- JH, secreted by nonneural endocrine cells in the corpora allata, prevents metamorphosis into the adult form. The relative amounts of JH and 20-hydroxyecdysone in the hemolymph determine whether the epidermis will produce juvenile, pupal, or adult structures.
- In adults, JH functions as a gonadotropin, stimulates the production of sex-attractant pheromones, and stimulates the secretion of ecdysone, which promotes incorporation of yolk into eggs.

Study Questions

1. Explain why the effects of steroid hormones are seen after a longer delay than the effects of peptide hormones.
2. Construct a table to compare the hormones involved in water and salt balance. Include the site of secretion and molecular structure of each hormone, the principal stimuli that cause its secretion, its effect on target cells, and other information you consider important.

3. Insulin secretion is essential in metabolism. What type of molecule is insulin? What structure secretes it? What factors control its secretion? What is/are its target tissue(s)? What effect(s) does it exert?
4. Compare and contrast the functions of insulin and glucagon.
5. Compare and contrast the structure and functions of hypothalamic neurosecretory neurons that send their axons to the posterior pituitary gland with those of hypothalamic neurosecretory neurons that have axons terminating on the capillary network of the median eminence.
6. List the similarities and differences between prothoracicotropic hormone (PTTH) and ecdysone. Consider the structural and functional characteristics of both molecules.
7. In patients receiving cortisone drug therapy, adrenocorticotropic hormone (ACTH) secretion decreases, and the adrenal glands shrink in size. Using your knowledge of the HPA axis, feedback mechanisms, and the tropic functions of hormones, explain why patients experience these changes.
8. Using your knowledge of the stress response, explain why the physiological changes of the stress response are important to survival in the short term and deleterious when they persist over long periods of time.
9. Define/describe negative feedback in endocrine controls and explain its physiological significance.
10. Discuss antagonism, permissiveness, and synergism in the context of hormone interactions.

Key References

Bentley, P. J. 1998. *Comparative Vertebrate Endocrinology*, 3rd edition. Cambridge University Press. A detailed text replete with examples illustrating the physiology of hormones and their evolution in vertebrate animals.

Carrasco, G. A., and L. D. Van de Kar. 2003. Neuroendocrine pharmacology of stress. *Eur. J. Pharmacol.* 463: 235–272. Review of the integration of hormones, neural functions, and behaviors involved in the mammalian response to psychological and physical stressors.

Henderson, I. W. 2000. Endocrine system in vertebrates. In *Nature Encyclopedia of Life Sciences*. Nature Publishing Group, London. http//www.els.net/[doi:10.1038/npg.els.0001845]. An overview of several contemporary concepts of endocrine controls in vertebrate animals.

Nielsen, S., T-H Kwon, J. Frøkiær, and M. A. Knepper. 2000. Key roles of renal aquaporins in water balance and water-balance disorders. *News Physiol. Sci.* 15: 136–143. Overview of water channels in the mammalian nephron.

Raffin-Sanson, M. L., Y. de Keyzer, and X. Bertagna. 2003. Proopiomelanocortin, a polypeptide precursor with multiple functions: From physiology to pathological conditions. *Eur. J. Endocrinol.* 149: 79–90. An overview of the multiple physiological effects that can arise from a single prohormone.

Sapolsky, R. M., L. M. Romero, and L. U. Munck. 2000. How do glucocorticoids influence stress responses? Integrating permissive, suppressive, stimulatory, and preparative actions. *Endocrine Rev.* 21: 55–89. An interpretation of the multiple roles glucocorticoids play in the stress response.

Sternberg, E. M., and P. W. Gold. 2002. The mind-body interaction in disease. *Sci. Amer.* (special edition) 12(1): 82–89. An overview of the deleterious effects of long-term stress in mammals.

Wikelski, M., V. Wong, B. Chevalier, N. Rattenborg, and H. L. Snell. 2002. Marine iguanas die from trace oil pollution. *Nature* 417: 607–608. An illustration of stress imposed on a wild population by human perturbations of the environment.

Winslow, J. T., N. Hastings, C. S. Carter, C. R. Harbaugh, and T. R. Insel. 1993. A role for central vasopressin in pair bonding in monogamous prairie voles. *Nature* 365: 545–548. A classic paper showing how arginine vasopressin acting as a neurotransmitter in the brain affects behavior.

Žitňan, D., T. G. Kingan, J. L. Hermesman, and M. E. Adams. 1996. Identification of ecdysis-triggering hormone from an epitracheal endocrine system. *Science* 271: 88–91. First report of a previously unknown hormone linking endocrine and neural roles in metamorphosis.

Selected References

Baker, J. D., and J. W. Truman. 2002. Mutations in the *Drosophila* glycoprotein hormone receptor, *rickets*, eliminate neuropeptide-induced tanning and selectively block a stereotyped behavioral program. *J. Exp. Biol.* 205: 2555–2565.

Bradshaw, S. D., and F. J. Bradshaw. 2002. Arginine vasotocin: Site and mode of action in the reptilian kidney. *Gen. Comp. Endocrinol.* 126: 7–13.

Byrd, J. H., and J. L. Castner (eds.). 2001. *Forensic Entomology: The Utility of Arthropods in Legal Investigations*. CRC Press, Boca Raton, FL.

Carr, J.A. and C.H. Summers. 2002. Is stress more than a disease? A comparative look at the adaptiveness of stress. Integ. and Comp. Biol. 42: 505-507. [Introduction to a series of 12 papers in this volume presented in January 2001 at the symposium *Stress – Is It More Than a Disease? A Comparative Look at Stress and Adaptation*].

Chrousos, G. P. 1998. Stressors, stress, and neuroendocrine integration of the adaptive response. *Ann. N.Y. Acad. Sci.* 851: 311–335.

Chrousos, G. P. 1998. Stressors, stress, and neuroendocrine integration of the adaptive response. *Ann. N.Y. Acad. Sci.* 851: 311–335.

Colagiuri, S., and J. B. Miller. 2002. The 'carnivore' connection—evolutionary aspects of insulin resistance. *Eur. J. Clin. Nutr.* 56: 30–35.

Cone, R.D. (ed.) 2003. The Melanocortin System. Ann. N.Y. Acad. Sci. 994. [Papers presented in August 2002 at the Fifth International Melanocortin Meeting].

Denver, R. J. 1997. Environmental stress as a developmental cue: Corticotropin-releasing hormone is a proximate mediator of adaptive phenotypic plasticity in amphibian metamorphosis. *Hormones Behav.* 31: 169–179.

Døving, K. B., and D. Trotier. 1998. Structure and function of the vomeronasal organ. *J. Exp. Biol.* 201: 2913–2925.

Freeman, M. E., B. Kanyicska, A. Lerant, and G. Nagy. 2000. Prolactin: Structure, function, and regulation of secretion. *Physiol. Rev.* 80: 1523–1631.

Funk, C. D. 2001. Prostaglandins and leukotrienes: Advances in eicosanoid biology. *Science* 294: 1871–1875.

Gilbert, S. F. 2001. Ecological developmental biology: Developmental biology meets the real world. *Dev. Biol.* 233: 1–12.

Goffin, V., N. Binart, P. Touraine, and P. A. Kelly. 2002. Prolactin: The new biology of an old hormone. *Annu. Rev. Physiol.* 64: 47–67.

Goodman, H. M. 2001. Endocrinology concepts for medical students. *Adv. Physiol. Educ.* 25: 213–224.

Hadley, M. E., and C. Haskell-Luevano. 1999. The proopiomelanocortin system. *Ann. N.Y. Acad. Sci.* 885: 1–21.

Hartman, F.A. and K.A. Brownell. 1949. *The Adrenal Gland*. Lea & Febiger. Philadelphia.

Heymann, J. B., and A. Engel. 1999. Aquaporins: Phylogeny, structure, and physiology of water channels. *News Physiol. Sci.* 14: 187–193.

Højlund, K., M. Wildner-Christensen, O. Eshøj, C. Skjærbæk, J. J. Holst, O. Koldkjær, D. Møller Jensen, and H. Beck-Nielsen. 2001. Reference intervals for glucose, β-cell polypeptides, and counterregulatory factors during prolonged fasting. *Amer. J. Physiol.* 280: E50–E58.

Insel, T. R., and L. J. Young. 2001. The neurobiology of attachment. *Nature Rev. Neurosci.* 2: 129–136.

Kostron, B., K. Marquardt, U. Kaltenhauser, and H. W. Honegger. 1995. Bursicon, the cuticle sclerotizing hormone: Comparison of its molecular mass in different insects. *J. Insect Physiol.* 41: 1045–1053.

Levin, E. R., D. G. Gardner, and W. K. Samson. 1998. Natriuretic peptides. *N. Engl. J. Med.* 339: 321–328.

McEwen, B. S. 2002. *The End of Stress As We Know It.* Joseph Henry Press, Washington, DC.

Melo, L. G., S. C. Pang, and U. Ackermann. 2000. Atrial natriuretic peptide: Regulator of chronic arterial blood pressure. *News Physiol. Sci.* 15: 143–149.

O'Brien, M. A., E. J. Katahira, T. R. Flanagan, L. W. Arnold, G. Haughton, and W. E. Bollenbacher. 1988. A monoclonal antibody to the insect prothoracicotropic hormone. *J. Neurosci.* 8: 3247–3257.

Pryce, C. R., D. Rüedi-Bettschen, A. C. Dettling, and J. Feldon. 2002. Early life stress: Long-term physiological impact in rodents and primates. *News Physiol. Sci.* 17: 150–155.

Re, R. N. 2003. The intracrine hypothesis and intracellular peptide hormone action. *BioEssays* 25: 401–409.

Reinhardt, H. W., and E. Seeliger. 2000. Toward an integrative concept of control of total body sodium. *News Physiol. Sci.* 15: 319–325.

Romero, L. M. 2002. Seasonal changes in plasma glucocorticoid concentrations in free-living vertebrates. *Gen. Comp. Endocrinol.* 128: 1–24.

Romero, L. M., and M. Wikelski. 2001. Corticosterone levels predict survival probabilities of Galapagos marine iguanas during El Niño events. *Proc. Natl. Acad. Sci. U.S.A.* 98(13): 7366–7370.

Ruther, J., T. Meiners, and J. L. M. Steidle. 2002. Rich in phenomena—lacking in terms: A classification of kairomones. *Chemoecology* 12: 161–167.

Sawyer, T. K., V. J. Hruby, M. E. Hadley, and M. H. Engel. 1983. α–Melanocyte stimulating hormone: Chemical nature and mechanism of action. *Amer. Zool.* 23: 529–540.

Schradin, C., and G. Anzenberger. 1999. Prolactin, the hormone of paternity. *News Physiol. Sci.* 14: 223–231.

Sehnal, F., I. Hansen, and K. Scheller. 2002. The cDNA-structure of the prothoracicotropic hormone (PTTH) of the silkmoth *Hyalophora cecropia*. *Insect Biochem. Mol. Biol.* 32: 233–237.

Sherman, R. A., M. J. R. Hall, and S. Thomas. 2000. Medicinal maggots: An ancient remedy for some contemporary afflictions. *Annu. Rev. Entomol.* 45: 55–81.

Simoncini, T., and A. R. Genazzani. 2003. Non-genomic actions of sex steroid hormones. *Eur. J. Endocrinol.* 148: 281–292.

Takei, Y., and S. Hirose. 2002. The natriuretic peptide system in eels: A key endocrine system for euryhalinity? *Amer. J. Physiol.* 282: R940–R951.

Tsigos, C., and G. P. Chrousos. 2002. Hypothalamic–pituitary–adrenal axis, neuroendocrine factors and stress. *J. Psychosomatic Res.* 53: 865–871.

Vleminckx, K., and R. Kemler. 1999. Cadherins and tissue formation: Integrating adhesion and signaling. *BioEssays* 21: 211–220.

Watson, R.D., E. Spaziani, and W.E. Bollenbacher. 1989. Regulation of ecdysone biosynthesis in insects and crustaceans: a comparison. In Koolman, J. (ed.) *Ecdysone: From Chemistry to Mode of Action*, pp. 188–203. Thieme. Stuttgart.

Wilkins, M. R., J. Redondo, and L. A. Brown. 1997. The natriuretic-peptide family. *Lancet* 349: 1307–1310.

Young, L. J., M. M. Lim, B.Gingrich, and T. R. Insel. 2001. Cellular mechanisms of social attachment. *Hormones Behav.* 40: 133–138.

Žitňan, D., I. Žitňanová, I. Spalovská, P. Takáč, Y. Park, and M. E. Adams. 2003. Conservation of ecdysis-triggering hormone signaling in insects. *J. Exp. Biol.* 206: 1275–1289.

15 Reproduction

"Success," measured in biological terms, is accomplished by reproducing—passing one's genetic information on to offspring and ensuring the survival of as many of those offspring as possible until they are ready to reproduce in turn. Different animals use an astonishing variety of behavioral, morphological, and physiological strategies to achieve reproductive success. These strategies interact with environmental factors to influence the timing and outcome of reproductive events.

Horses, for example, breed only at certain times of the year. To succeed in mating, males and females must coordinate their behaviors, and they employ many neural, neuroendocrine, and endocrine functions to do so. When days begin to lengthen, mares go into *heat*—that is, they become sexually receptive. Sensory receptors in the mare detect light, and integrating mechanisms in her brain identify changes in day length. These changes set in motion hormonal changes that lead to ovulation. The female goes into heat at the time she ovulates. When she is in heat, she releases pheromones that attract males. As described in Chapter 14, *pheromones* are chemical signals that convey information between members of the same species about sexual readiness and social status. They trigger specific stereotyped behaviors or physiological changes that are important in mating and aggression.

In many vertebrate animals, the vomeronasal organ detects pheromones. With the least hint of female pheromones in the environment, a stallion raises his head, wrinkles his nose, curls his upper lip, and may also neigh. This *flehmen* (flared face) *behavior* draws airborne chemical signals into his vomeronasal organ. Should the vomeronasal organ detect pheromones and signal to his brain that a sexually receptive mare is in the vicinity, the stallion is attracted to her. Thus begins only one example of the amazing diversity of reproductive behaviors found (and yet to be found) in the animal kingdom. In this chapter, we first present an overview of reproduction and then focus on the physiology of reproduction in mammals.

Sexual and Asexual Reproduction

Although reproduction can be accomplished either sexually or asexually, most (but not quite all) animals use sexual reproduction at some point in their life cycle. In **sexual reproduction**, male and female **gametes**, or *germ cells* (eggs and sperm), are

Coordinating reproductive behavior A stallion uses flehmen behavior to draw airborne molecules into his vomeronasal organ. If the vomeronasal organ detects pheromones from a female in heat, the stallion's brain prompts him to pursue her.

produced by **gonads** (ovaries and testes). Unlike all other cells of the body, which include homologous pairs of chromosomes, gametes are produced by *meiosis*, so that they carry only one member of each pair of chromosomes (they are *haploid*) (see Appendix H, Mitosis and Meiosis). When two gametes join during fertilization to form a **zygote**, this new entity again has homologous pairs of chromosomes (it is *diploid*). Because each parent contributed one chromosome of each pair, the zygote has a unique genetic makeup. In addition, during meiosis, the members of a homologous pair of chromosomes may exchange segments with each other, so that genes are swapped from one to the other. Thus the process of meiosis allows not only the combination of different parental genotypes, but also the recombination of genes. Both of these factors contribute *genetic diversity* to a population of animals. The zygote and subsequent generations of cells divide by *mitosis*, maintaining the diploid number of chromosomes.

Gametes are either small, motile male **spermatozoa** (sperm) produced by **testes** (singular *testis*) or large, nonmotile female **ova** (eggs; singular *ovum*) produced by **ovaries.** In many species, separate male individuals produce spermatozoa, and separate female individuals produce ova. However, in some species, a single individual possesses both testes and ovaries and produces both types of gametes. These animals are called **hermaphrodites.** In *simultaneous hermaphrodites*, the two types of gonads are active at the same time, so the animal is technically able to fertilize itself. However, most simultaneous hermaphrodites (earthworms are an example) *copulate* (mate) to exchange gametes. In *sequential hermaphrodites*, the gonads become active in sequence. Certain reef fish, such as the clownfish in Figure 15.1*a*, develop first into males and then change into females, or vice versa, depending on the species and environmental circumstances.

Asexual reproduction does not involve the mixing of genes from different individuals. For example, an animal can reproduce asexually by budding off pieces of itself that develop into complete new individuals. Although this type of reproduction does not add genetic diversity to a population of animals, it can contribute to a rapid increase in the size of the population, giving it the advantage of outnumbering potential competitors. Because the genetic makeup of an asexually produced individual is identical to that of its parent, it is referred to as a **clone**.

Relatively few animals reproduce solely by asexual means. Many have life cycles that include sexual and asexual phases. Figure 15.2 shows the life cycle of the marine jellyfish *Aurelia*, which alternates sexual and asexual reproduction. In the *sexual* phase, umbrella-shaped free-swimming males and females produce gametes that are released into the water for fertilization. The resulting zygotes develop into **larvae**. Instead of developing into jellyfish, however, the larvae attach themselves to a rocky spot on the ocean floor. Each attached individual reproduces *asexually* by producing many new free-swimming individuals that are genetically identical. These new individuals mature into adult males or females that produce sperm or eggs for sexual reproduction, and the cycle is repeated.

Parthenogenesis (*partheno-*, "virgin"; *genesis*, "origin") is a type of asexual reproduction in which females produce eggs that do not require a male complement of chromosomes. A few invertebrate

Figure 15.1 Reproductive mechanisms are exceptionally diverse
(a) This clownfish is just one of over 50 species of reef fish known to change from male to female, or vice versa. Hormonal changes triggered by factors in the environment make these fish alternately capable of producing eggs or sperm. *(b)* This *Philodina gregaria* is a member of the strictly parthenogenetic class Bdelloidea of the phylum Rotifera. Molecular biologists estimate that her ancestors stopped reproducing sexually 80 to 100 million years ago. This animal is about half a millimeter in length. *(c)* Aphids alternate between sexual and asexual (parthenogenetic) reproduction. This image shows a female (right) giving birth. Commonly called "plant lice," aphids are agricultural pests and also transmit plant diseases. Females are typically about 1 mm in length. *(d)* The whiptail lizard *Cnemidophorus tesselatus* is an example of a strictly parthenogenetic vertebrate.

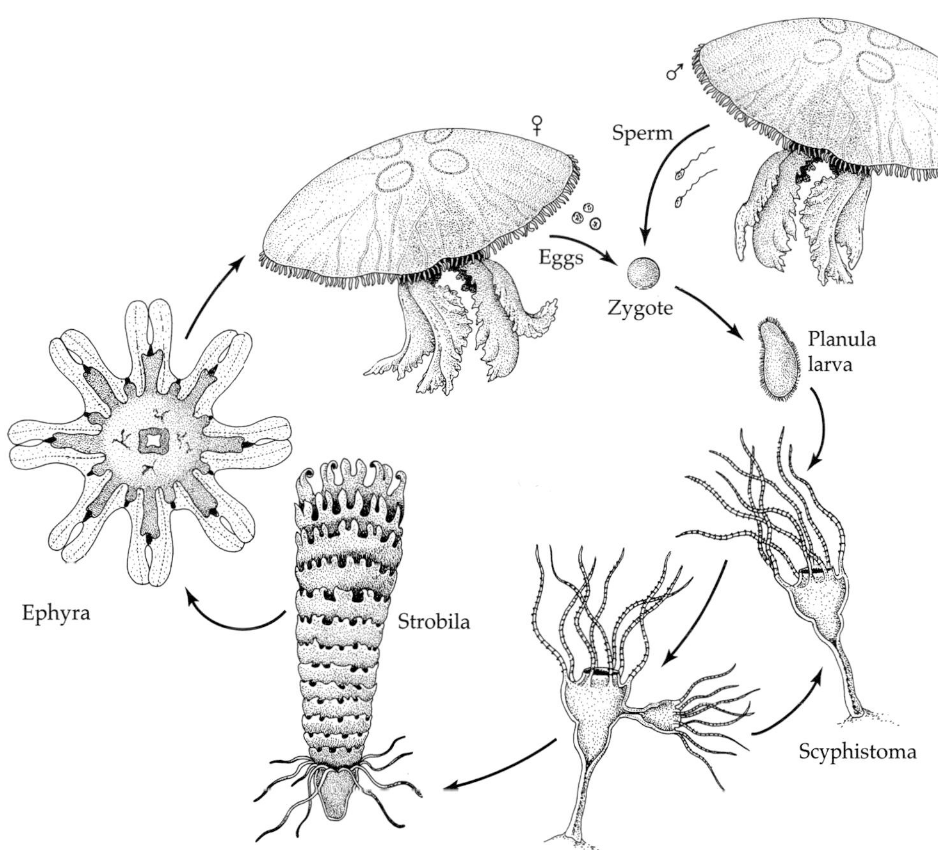

Figure 15.2 Life cycle of the jellyfish *Aurelia aurita* Male and female adults produce eggs and sperm. Fertilization occurs externally in the ocean waters. The resulting zygote develops into a nonfeeding larva called the *planula*. The planula attaches itself to the substrate and grows into a *scyphistoma*, which reproduces asexually by two means. In a process called *strobilation*, it divides transversely into many modular segments to form a strobila. Each segment breaks off and develops into a swimming form called an *ephyra*, which grows into an adult jellyfish. The scyphistoma can also bud to form more scyphistomae, which settle nearby and go through strobilation.

species use parthenogenesis as their only means of reproducing. For example, researchers have assiduously studied one class of rotifers, the Bdelloidea, which includes more than 350 different species (Figure 15.1*b*). They have never found a single male. The bdelloids produce eggs by mitosis. Mutations (permanent copying errors of the DNA) are the only means of introducing diversity into bdelloid genomes. Using molecular evidence, evolutionary biologists estimate that the bdelloid rotifers have been strictly parthenogenetic for the past 85 million years.

More commonly, invertebrate species, such as some flatworms, insects, and crustaceans, combine parthenogenesis with episodes of sexual reproduction. For example, certain aphids (Figure 15.1*c*) go through several parthenogenetic all-female generations in spring and summer. In autumn (probably in response to an environmental signal), the females lay eggs that hatch into a mixture of males and females. The two sexes mate and produce zygotes that remain dormant over the winter. In the following spring, each zygote gives rise by parthenogenesis to an explosion of genetically identical female clones. This life cycle is a tantalizing example of how environmental factors appear to influence specific reproductive outcomes, presumably by affecting patterns of gene expression.

Some vertebrate organisms, such as certain species of lizards and snakes, reproduce by parthenogenesis. The whiptail lizard *Cnemidophorus tesselatus* (Figure 15.1*d*) is thought to have originated no more than 100,000 years ago as a hybrid between two sexually reproducing species of the same genus. Whiptail lizards live in habitats that can be easily disrupted, such as river floodplains. Investigators note that, should most of the lizards in a floodplain perish in a flood, this parthenogenetic species—all females potentially able to produce offspring without having to locate a mate—would be able to repopulate the region more quickly than sexually reproducing species.

Although many animals combine asexual and sexual modes of reproduction, very few use asexual modes only. This fact is puzzling, because asexual reproduction has several advantages. It eliminates the metabolic costs of courtship and the hazards of exposure to predators while seeking a mate. It eliminates the problem of "wasted males" that fail to reproduce, but use resources that could have been used by reproductive females. It allows for rapid increases in population size, which means that abundant but transient food sources can be taken advantage of, or new niches in disturbed habitats filled quickly. Similarly, if predators take some members of a population, surviving individuals will nevertheless pass on the clonal genetic makeup to future generations. Given these clear advantages of asexual reproduction, why do most species that use only asexual reproduction go extinct over relatively short time spans? In other words, why does sexual reproduction confer longevity on a species?

Many theories have been proposed to explain the advantage of sexual reproduction—that is, the production of genetically diverse offspring. Some propose that reshuffling the genes in each new generation ensures that harmful mutations don't accumulate over many generations. Animals that reproduce asexually may go extinct because they have no such mechanism for avoiding the accumulation of too many harmful mutations. The hundreds of species of bdelloid rotifers, which appear to have survived without sexual reproduction for many tens of millions of years, provide a contrary example. Are mutation rates especially low in these asexually reproducing species? Or are mutations more neutral than harmful, or more easily repaired? These questions are currently being actively investigated.

Another theory to explain the importance of sexual reproduction is that genetic mixing with each generation confers the ability to maintain defenses against microbial pathogens (such as viruses and bacteria), and protozoan parasites. By producing successive generations of genetically diverse individuals, a sexually reproducing species presents pathogens and parasites with new defenses in each new generation. The invading organisms must evolve to keep up with continuing changes in their targeted hosts. In contrast, asexually reproducing animals have the same genetic makeup generation after generation, allowing invaders to evolve effective means to overcome their defenses. Why, then, have the bdelloid rotifers escaped reproductive oblivion? One proposed explanation is that adult rotifers can survive extreme desiccation. They shrink into dormant forms that can survive for several years. In this dormant, desiccated state, they are dispersed through the air to new sites, where they rehydrate, come out of dormancy, and establish new populations of cloned individuals. Dispersal allows them to escape pathogens or parasites adapted to their defenses at their old site. At their new site, until potential pathogens or parasites evolve strategies to overcome their defenses, the bdelloids can flourish. Studying such rare examples of asexual animals that are long-lived on an evolutionary timescale may provide answers to one of the major unanswered questions in biology: why species of sexually reproducing animals are more successful in surviving over long evolutionary periods.

Summary: Sexual and Asexual Reproduction

- Success of a species requires reproductive mechanisms that pass on genetic information from generation to generation and ensure survival of offspring until they are sexually mature.
- Most animals reproduce sexually, either exclusively or at some time during their life cycle.
- In sexual reproduction, haploid gametes produced by meiosis come together through fertilization to form the zygote. Meiosis contributes to genetic diversity.
- Sexual reproduction results in the mixing of genes from different individuals in each generation. Asexual reproduction results in offspring that are clones with the same genetic makeup as their parent.
- Very few species of animals that reproduce solely by asexual means are successful over long evolutionary periods. Current hypotheses proposed to account for their rapid extinction include the ideas that asexually reproducing animals accumulate excessive numbers of harmful mutations, and that the genetic homogeneity of successive generations makes them vulnerable to pathogens or parasites.

Mammalian Reproduction

All species of mammals have separate sexes. Males produce sperm in testes, and females produce eggs in ovaries. In all mammalian species, females produce milk and care for their young. Mammals are divided into three main groups based on their mode of reproduction. The **monotremes**, such as the echidna and duck-billed platypus, lay eggs. **Marsupials**, such as kangaroos and opossums, give birth to very immature young and typically house them in pouches, where they continue developing. **Eutherians**, such as mice and whales, form well-developed **placentas** that attach the developing young to the inner lining of the uterus. The reproductive system of mammals develops in close association with the urinary system, and the two are often referred to together as the **urogenital system.** Figure 15.3 shows the generalized structures and introduces the main components of the male and female urogenital systems of eutherian mammals.

Gametes are produced in ovaries and testes

Eggs and sperm (gametes) are formed by meiosis in a process called **gametogenesis.** Gametogenesis that produces sperm is called **spermatogenesis**, and gametogenesis of eggs is called **oogenesis.** Ga-

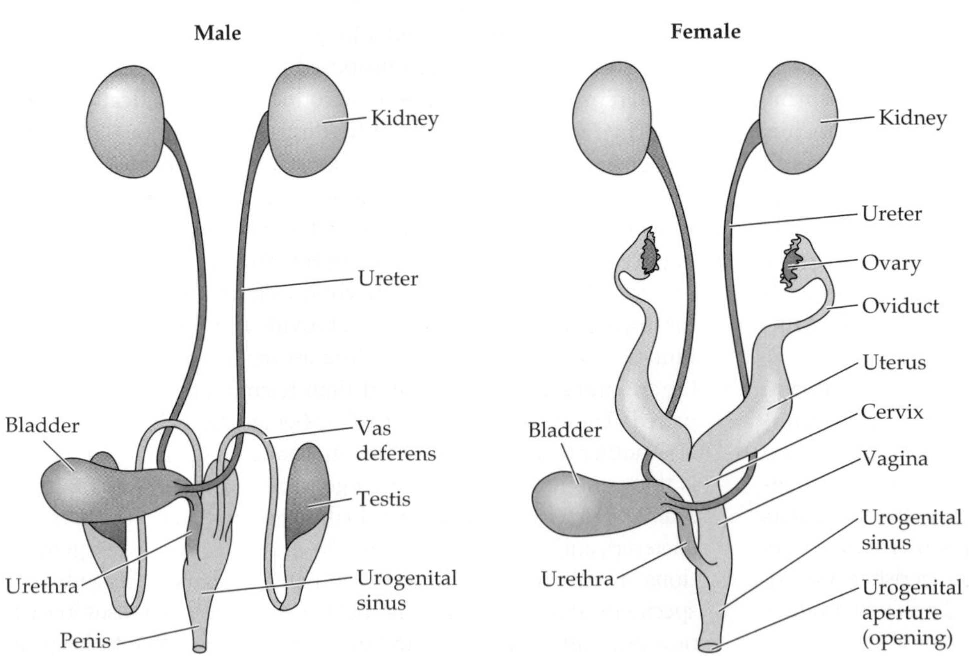

Figure 15.3 Generalized structure of the mammalian urogenital system In both males and females, the **ureters** descend from the kidneys to empty into the base of the bladder. In males, the **vasa deferentia** pass from the testes, loop around the ureters, and enter the **urogenital sinus** just posterior to the bladder. The general term *urogenital sinus* refers to a cavity that receives both urinary and reproductive products. In males, for example, both urine and semen exit the body through the same structure. In females, the **oviducts** (also called *fallopian tubes*) come into close association with the ovaries and extend to the **uterus.** The uterus narrows to a neck region, the **cervix**, which opens into the **vagina.** In most female mammals, the urethra and vagina both empty into the urogenital sinus, and both urine and ova leave the body from the same opening, the **urogenital aperture.** Only primates and some rodents have separate openings for the vagina and urethra.

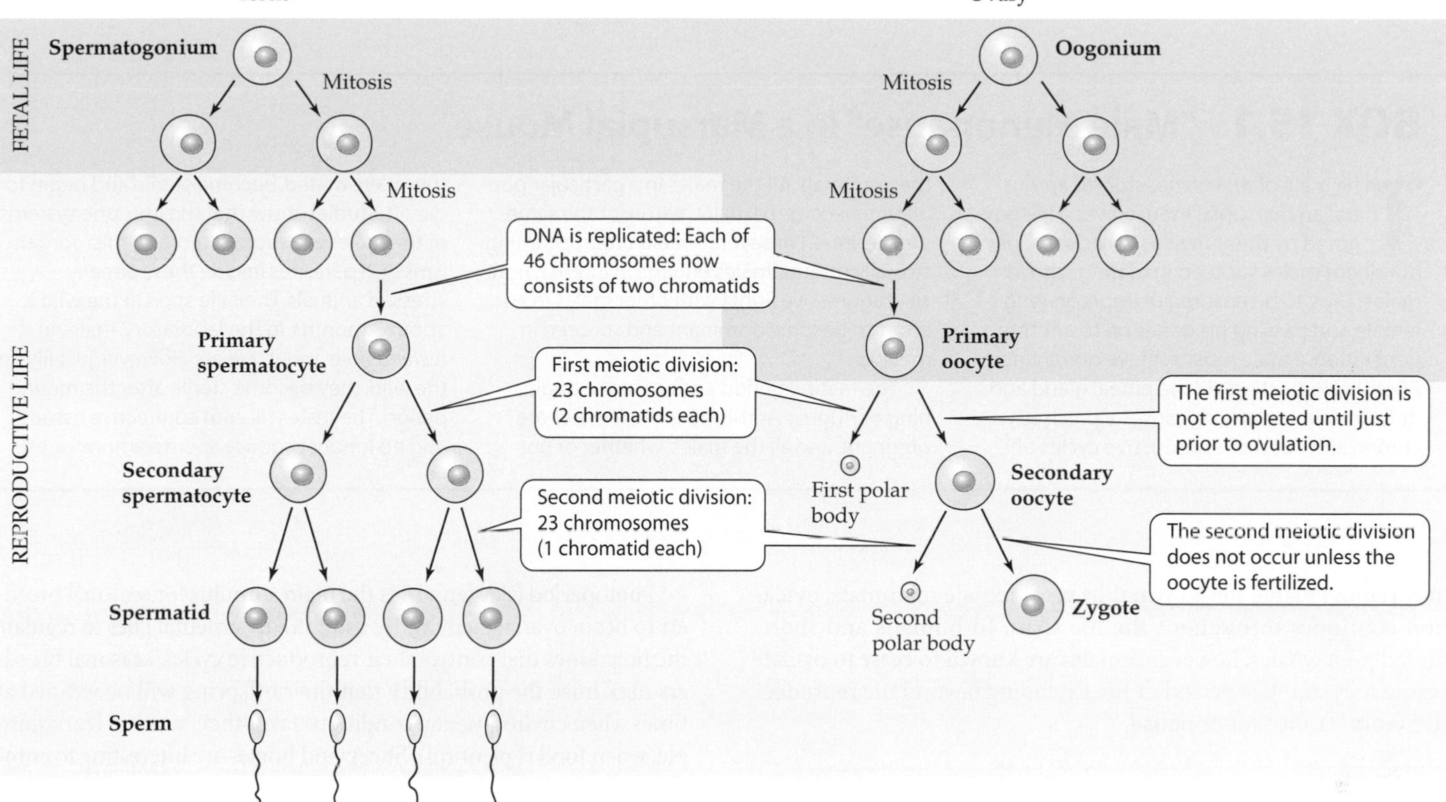

Figure 15.4 Gametogenesis In both males and females, gametogenesis starts with spermatogonia and oogonia dividing by mitosis during fetal life. In females, oogonia stop dividing by mitosis during fetal life and begin the first stage of meiosis, at which point their development is arrested. In males, spermatogonia begin dividing by mitosis again at puberty, and they provide a continuous supply of diploid cells that enter into meiosis throughout a male's reproductive life. Each primary spermatocyte formed will proceed through meiosis to yield four sperm cells. In females, each primary oocyte will yield only one secondary oocyte that may or may not be fertilized. The polar bodies formed by these meiotic divisions degenerate. In this figure, we use as an example the chromosome number characteristic of humans.

metes must have half the number of chromosomes that other cells in the animal (*somatic cells*) have so that the zygote formed by fertilization will have the correct number of chromosomes for the species. Humans have 23 pairs of homologous chromosomes, giving a total of 46 individual chromosomes. The total number of chromosomes is described as diploid, or $2n$ (twice the number of pairs). Horses, for example, have 32 pairs of chromosomes, giving a diploid number of 64. Sperm and eggs are haploid, or $1n$; each gamete has one member of each pair of homologous chromosomes. The number of chromosomes is halved through the process of meiosis. In essence, meiosis involves two cell divisions with only one replication of DNA. We will see however, that the timing of meiotic events in spermatogenesis differs markedly from that in oogenesis.

In both males and females, gametogenesis begins with a bout of mitosis in fetal life (Figure 15.4). In the developing male testes, diploid germ cells called **spermatogonia** divide by mitosis to increase in number (proliferate). This process stops at birth and starts up again at puberty. From then on, the diploid spermatogonia undergo meiosis to produce a continuous supply of sperm. At the beginning of the first meiotic division, the DNA of each chromosome replicates. Once the chromosomes of a spermatogonium are replicated, the cell is referred to as a **primary spermatocyte**. This cell then undergoes the first meiotic division to produce two **secondary spermatocytes**, each of which goes through the second meiotic division to form a total of four **spermatids**. The spermatids develop into mature sperm. In humans, sperm are produced throughout life, with only some diminishment with age. Not all male mammals are capable of reproducing throughout life, however. For example, *Antechinus stuartii*, an Australian marsupial mouse, ceases spermatogenesis immediately after the mating period (Box 15.1).

In the ovaries of female fetuses, diploid germ cells called **oogonia** divide by mitosis to produce more oogonia. Once this population has formed, no further mitosis of germ cells occurs. Recent studies have suggested that, in mice, the oogonia continue to divide after birth and into adulthood. It will be interesting to find if oogonia serve as stem cells after birth in other mammals. Knowledge of their existence and function could have significant clinical applications in regulating fertility. The chromosomes in the oogonia replicate in the first stage of meiosis, after which the cells are called **primary oocytes.** These cells will remain as primary oocytes, without any further divisions, until the female reaches reproductive maturity at puberty. Thus females produce all the oocytes they will ever have during fetal life. Once the female reaches sexual maturity, the oocytes, one or a few at a time, will be released— a process called **ovulation**. An individual oocyte completes the first meiotic division at the time of ovulation, and it will complete the second meiotic division only if fertilization occurs. Thus, in long-lived mammals, some primary oocytes remain in an arrested state for several years after fetal life. Oocytes of elephants and baleen whales, for example,

BOX 15.1 "Male Menopause" in a Marsupial Mouse

The male of *Antechinus stuartii*, an Australian marsupial mouse, lives a life governed by pressures to reproduce. Only dominant males succeed in mating with females. Thus, to be ensured of impregnating a female and passing his genes on to another generation, a male must achieve dominance. Photoperiod, acting via the pineal gland and its hormone, melatonin, causes a precise synchronization of the reproductive cycles of these animals. All the males in a particular population become sexually mature at the same time in June. For several weeks after becoming sexually mature, males engage in intense, frenzied, aggressive fights with other males in an effort to become dominant and succeed in mating.

The mating period ends near the beginning of August. At this time, the females are pregnant, and all the males, whether or not they have mated, become sterile and begin to die off. Studies show that the immune systems of the males are suppressed, and microorganisms and parasites invade these already stressed animals. Their life span in the wild is about 3 months. In the laboratory, male *Antechinus* live about 2 years. However, just like in the wild, they become sterile after the mating period. The testes fill with connective tissue and no longer produce sperm or hormones.

may remain viable for 60 years! In most female mammals, ovulation continues throughout the life span. In humans and short-finned pilot whales, however, females are known to cease to ovulate before they die. The period of life extending beyond the reproductive years is called **menopause**.

Oocytes mature periodically in menstrual cycles or estrous cycles

Most female mammals *ovulate* (release mature oocytes from the ovary) periodically. Only primates such as humans, gorillas, and chimpanzees, however, *menstruate*—that is, shed the uterine lining as a blood-tinged discharge from the vagina. Because menstruation occurs at intervals, these animals are said to have **menstrual cycles**. The majority of other female mammals ovulate periodically, but do not menstruate. Instead, their cycles are noticeable because they go into **estrus** (heat) at the time of ovulation. During estrus, they use behaviors and other signals—such as pheromones released by mares, described at the beginning of this chapter—to indicate that they are sexually receptive, thereby increasing the probability of mating near the time when a mature oocyte is released from the ovary. Their cycles are referred to as **estrous cycles**. Both kinds of cycles are triggered by sequences of hormonal changes. Some nonprimate mammals, such as dogs, experience vaginal bleeding; however, this bleeding is not menstruation. It occurs when the animal is in heat and the genital organs become engorged with blood.

Estrous cycles are defined as regularly repeated periods of sexual receptivity (estrus) separated by intervals characteristic of each species. Some animals, such as the red fox and other wild carnivores, have only one estrous cycle a year; they are said to be *monestrous*. Animals that have more than one estrous cycle a year are *polyestrous*. Mice and rats are examples of polyestrous animals that (like humans) can breed all year long. Mice have estrous cycles 4 or 5 days in length, and the gestation period (pregnancy) lasts 3 weeks. Other polyestrous animals, such as sheep and horses, go through several estrous cycles during a limited time of the year and are *anestrous* (without estrous cycles) during the remainder of the year. These animals are *seasonal breeders*. For example, the mare has estrous cycles of about 21 days in length during the breeding season, and she spends about 5 days in heat at the time of ovulation.

Photoperiod (day length) is the main stimulus for seasonal breeders to begin ovarian activity. By using environmental cues to regulate the hormones that control their reproductive cycles, seasonal breeders maximize the probability that their offspring will be weaned at times when environmental conditions favor their survival (for example, when food is plentiful). Sheep and horses are interesting to compare in this regard. To achieve the same end, these animals living at the same northern temperate latitude respond to day length in opposite ways. Although sheep begin cycling in late June and July (as days begin to shorten), horses begin ovarian activity in late winter or early spring (as days begin to lengthen). The gestation period of sheep is about 5 months, whereas that of horses is about 11 months. In both animals, breeding and length of gestation are coordinated to ensure birth and weaning at a time of year when the young can thrive.

Ovulation in mammals occurs in response to a surge of luteinizing hormone (LH) released from the anterior pituitary gland. (We will discuss LH in more detail in the next section of this chapter.) Animals are called *spontaneous ovulators* when the LH surge is produced by a series of hormonal events over the course of the estrous or menstrual cycle. In some animals, including domestic cats, mink, rabbits, and dromedary camels, the LH surge occurs only when the females mate. These animals are called *induced ovulators*. Physical stimulation of the cervix sets in motion the neurohormonal signals that trigger the LH surge, which in turn causes ovulation. The domestic cat, for example, is a polyestrous seasonal breeder. A female cat has repeating estrous cycles of about 16 days in length during the breeding season (January to October in northern temperate zones). Development of oocytes proceeds in the ovary with each cycle; however, ovulation occurs only with mating. The precise coordination of timing between ovulation and mating greatly increases the likelihood of fertilization.

Summary: Mammalian Reproduction

- Monotremes, marsupials, and eutherians all reproduce sexually and produce eggs or sperm that have a haploid complement of chromosomes.
- Reproductive males produce sperm continuously. Reproductive females periodically produce mature oocytes and ovulate in the course of estrous or menstrual cycles.

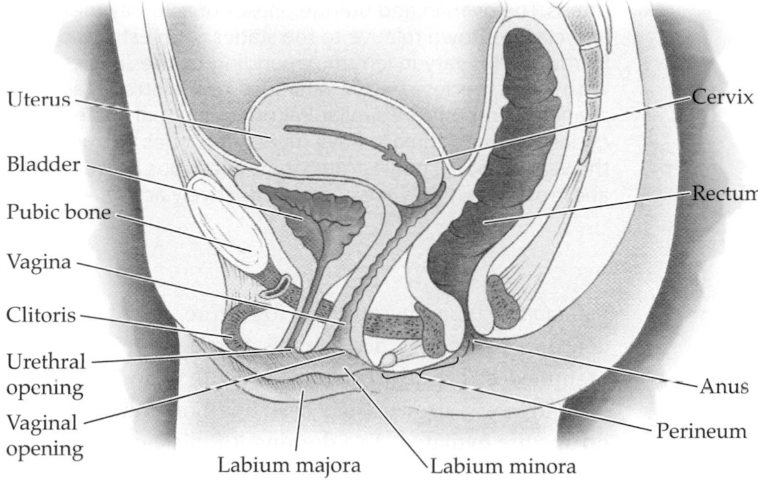

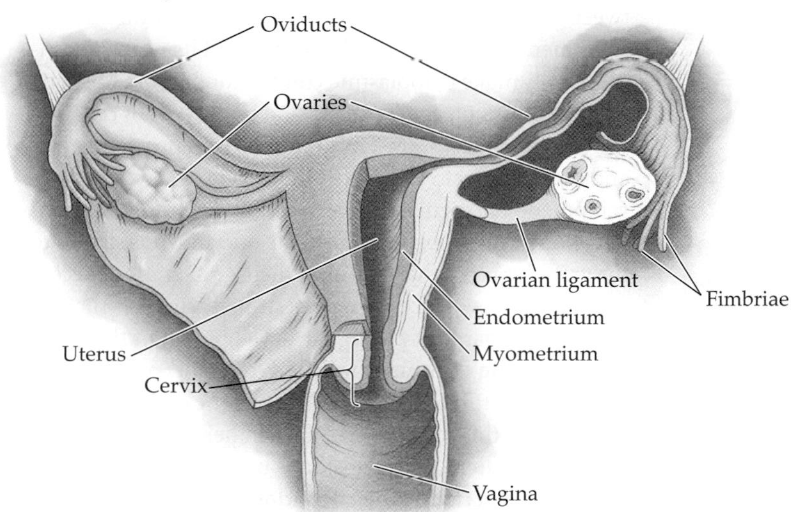

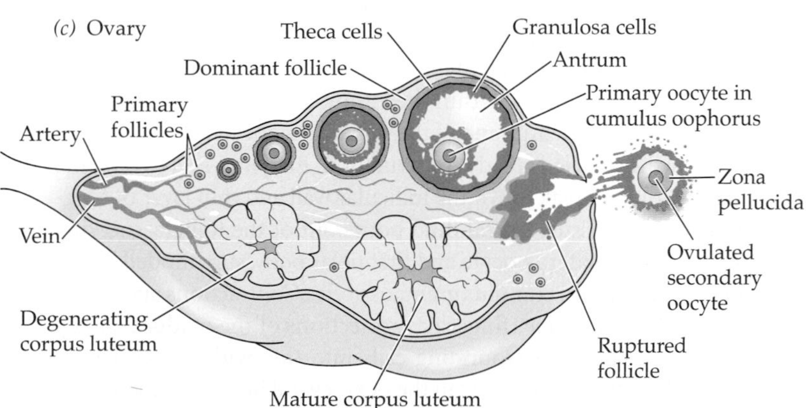

Figure 15.5 The human female reproductive system The female reproductive system includes *(a)* the external genitalia and *(a, b)* the internal organs. The external genitalia are collectively called the *vulva* and consist of the *labia majora*, *labia minora*, and *clitoris*. *(c)* The ovary is about 2 to 4 cm long. At any one time, it contains follicles and corpora lutea in various stages of development.

Hormonal Control of Female Reproduction

Both estrous and menstrual cycles are controlled by neurohormones released from the brain and hormones released from the anterior pituitary gland and ovaries. We will use the human menstrual cycle as our common thread in this section, while also noting its similarity to estrous cycles. Figure 15.5 shows the anatomy of the human female reproductive system. As in all reproductively mature female mammals, the ovaries release eggs periodically, and the inner lining of the uterus prepares for pregnancy on a cyclic basis. Unless they are interrupted by a pregnancy, human females typically have menstrual cycles throughout the entire year.

To understand menstrual and estrous cycles, we will need to follow events occurring over time at four different sites in the body: the hypothalamus, anterior pituitary gland, ovaries, and uterus. Figure 15.6 tracks these events throughout the human menstrual cycle. Prior to ovulation, the oocyte matures within a **follicle** in the ovary. In menstrual cycles, the glandular epithelial lining of the uterus, the **endometrium**, first sloughs off in menstruation and then regrows. In estrous cycles, the endometrium is not sloughed off but simply resorbed. Ovulation occurs in the middle of the menstrual cycle or near the end of estrus in the estrous cycle. After ovulation, the remaining cells of the follicle reconstitute to form an endocrine structure called a **corpus luteum**, and the endometrium becomes ready for possible implantation of an embryo. If fertilization and implantation do not occur, the corpus luteum stops functioning, and the endometrium is shed or resorbed. In the menstrual cycle, these changes of the uterus and ovaries are described as *phases*. Table 15.1 and Figure 15.6 show that the **follicular phase** of the ovary overlaps with the **menstrual** and **proliferative phases** of the uterus. The **secretory phase** of the uterus overlaps with the **luteal phase** of the ovary. For convenience, we will assume a 28-day cycle; however, normal human menstrual cycles range from 21 to 42 days in length, and normal menstrual periods last 2 to 8 days.

Although mammalian estrous cycles are of varying lengths and occur at varying intervals, ovarian and uterine events (except for menstruation) are similar to those of the menstrual cycle. The estrous cycle is divided into stages that are based on behavior (sexual receptivity during estrus) and changes in the ovary (see Figure 15.6). The period of follicular development that precedes estrus is called **proestrus**. Two stages follow estrus and ovulation: **metestrus**, during which the corpus luteum develops, and **diestrus**, during which the corpus luteum functions and then degenerates. Whereas the first day of the menstrual cycle is the first day of menstruation, the first day of the estrous cycle is the day of ovulation.

Dynamic cellular changes occur in the ovary over the course of a cycle

Figure 15.5*c* is a diagram of a section through the ovary. At any one time, the ovary has follicles—each containing a single primary oocyte—in many different stages of development. The

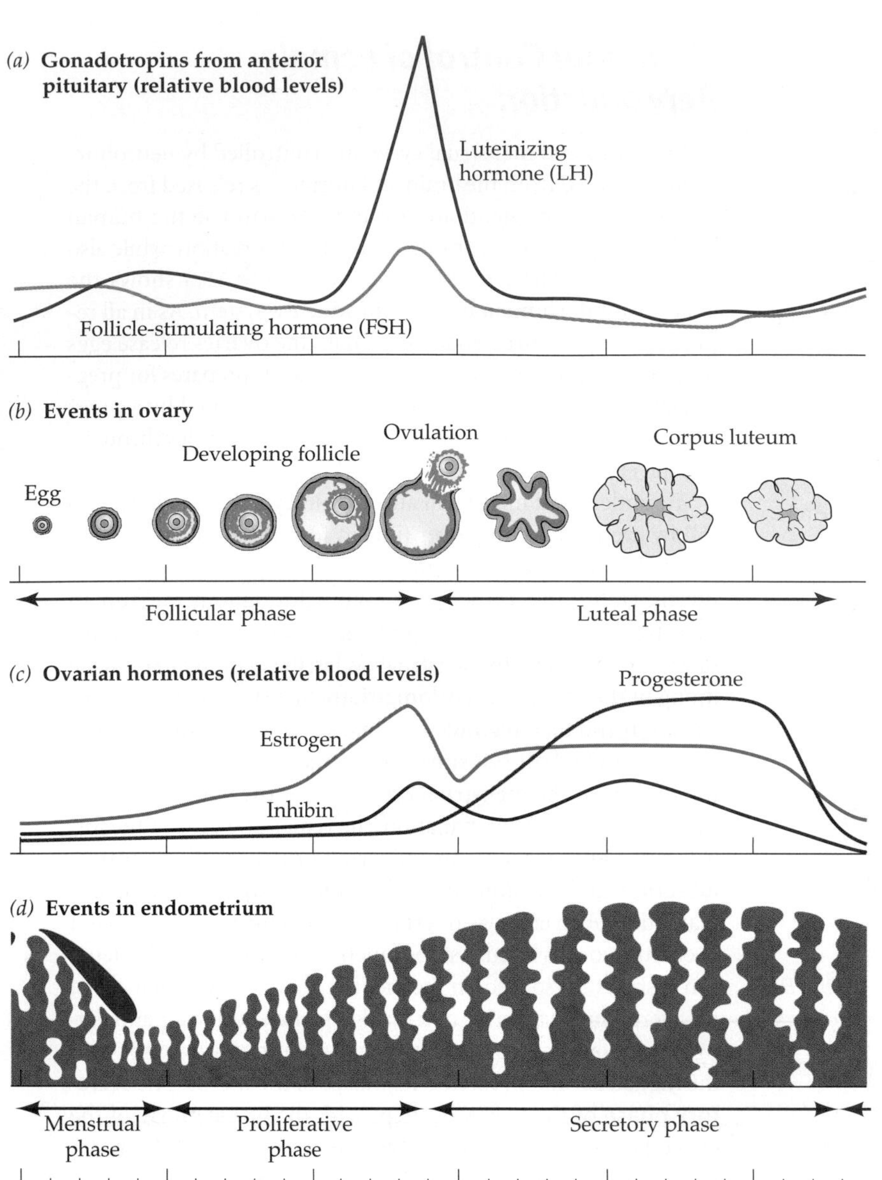

Figure 15.6 Events of menstrual and estrous cycles The blood levels of anterior pituitary and ovarian hormones vary in relation to cellular changes in the ovaries and uterus. The ovarian and uterine phases of a 28-day menstrual cycle are shown relative to the stages of an estrous cycle. Estrous cycles vary in length depending on the species. With some species-specific variations, gonadotropin and ovarian secretions are similar in both menstrual and estrous cycles. Whereas humans have menstrual cycles throughout the year, seasonal polyestrous or monoestrous breeders are anestrous when they are not reproductively active.

majority are **primary follicles**, which consist of a primary oocyte surrounded by a single layer of somatic **granulosa cells**. Throughout reproductive life, follicles from this finite population are recruited to develop into later stages. The development of follicles is called *folliculogenesis*. In successive stages of folliculogenesis, the primary oocyte increases in size and secretes the **zona pellucida** (an extracellular layer of glycoproteins). The granulosa cells increase in number and form multiple layers. Connective tissue around the follicle differentiates into a layer of **theca cells**, and a fluid-filled cavity called the **antrum** develops.

At the beginning of each follicular phase, several follicles begin to mature in both ovaries. In humans, after about a week, a single *dominant follicle* typically continues to mature, and the remainder of this cohort degenerates through *atresia* (programmed cell death).[1] In mammals that normally give birth to litters, atresia also occurs, but more follicles develop and are ovulated. As a dominant follicle develops, the oocyte, surrounded by granulosa cells, projects into the antrum as a structure called the **cumulus oophorus**.

During folliculogenesis, two-way communication between the oocyte and the granulosa cells coordinates development of the follicle. The oocyte and its nearest granulosa cells make cytoplasmic connections across the zona pellucida through *gap junctions*, and both types of cells also secrete *paracrine factors*. The granulosa cells provide signals and nutrients that (1) support the oocyte's growth and development, (2) regulate its meiosis, and (3) influence transcription of its genome. In turn, the oocyte sends signals to the granulosa cells that stimulate their proliferation. Recent experimental evidence from mice strongly suggests that the oocyte is in charge of pacing the development of the entire follicle. As the dominant follicle grows and matures, the primary oocyte completes the first meiotic division (which began when this female was a fetus in her mother's uterus, as we saw earlier). This division is unequal, however. One of the daughter cells, the **secondary oocyte**, retains one member of each pair of homologous chromosomes and essentially all of the cytoplasm from the primary oocyte (see Figure 15.4). The other cell, the **first polar body**, receives only the complementary chromosomes. The first polar body remains within the cumulus oophorus, just outside the cell membrane of the secondary oocyte.

Now the follicle is ready for ovulation. Enzymes secreted by the mature follicle break down the follicular wall and thin layer of overlying ovarian tissue. Antral fluid pours out of the follicle, carrying with it the secondary oocyte and first polar body, surrounded by the zona pellucida and cumulus granulosa cells. Contractions and ciliary action of the *fimbriae* (fingerlike projections of the oviduct) sweep the oocyte and its surrounding cells into the oviduct (see Figure 15.11). If fertilization occurs in the next several hours, the secondary oocyte will undergo the second meiotic division, and the fertilized oocyte will begin to develop. If not, all the cells will degenerate and die. The follicle cells remaining in the ovary reorganize into the cor-

[1]A human female has 200,000 to 400,000 primary oocytes in her ovaries at the time of puberty. Most of these are lost through cyclic bouts of atresia, and only about 400 will mature and be ovulated.

TABLE 15.1 Concurrent events in the uterus and ovary during the menstrual cycle

Days	Uterine phases	Ovarian phases
1–14	Menstrual: Days 1–5 Surface layers of the endometrium are sloughed off in menstruation. Proliferative: Days 6–14 The endometrium grows from about 1 mm in thickness to 3–5 mm and develops capillaries and glands. The smooth muscle wall of the uterus (myometrium) also grows.	Follicular: Several follicles, each containing a primary oocyte, begin to mature and secrete estrogen and inhibin. All of this cohort except a single dominant follicle will degenerate. This phase can vary in length from 10 to 20 days. Variations during follicular development contribute to variations in the length of the entire cycle.
14		Ovulation: The oocyte and the surrounding granulosa cells are released from the surface of the ovary. Occasionally more than one oocyte is ovulated. If fertilized, these offspring are fraternal (not identical) twins or triplets, because they have different sets of genes.
15–28	Secretory: Endometrial glands coil and secrete glycogen and enzymes. Blood vessels increase in number. If fertilization occurs, implantation begins about 7 days later, in the middle of the secretory phase.	Luteal: Follicle cells remaining after ovulation reorganize into the corpus luteum, which secretes estrogen, progesterone, and inhibin. If fertilization does not occur, the corpus luteum functions about 10–12 days and then degenerates. If fertilization does occur, the corpus luteum continues to function. This phase is not as variable as the follicular phase, and lasts about 2 weeks.

pus luteum. If fertilization does not occur, the corpus luteum secretes hormones for a finite time and then degenerates. If fertilization does occur, the corpus luteum continues to secrete hormones that are essential for establishing and maintaining pregnancy.

Hormones influence development of the follicle

Hormones secreted by the hypothalamus and anterior pituitary gland are required for development of the follicle (see Figure 15.6). In addition, hormones secreted by the follicle feed back to the hypothalamus and anterior pituitary, and also govern changes in the uterus. **Luteinizing hormone (LH)** and **follicle-stimulating hormone (FSH)**, secreted by the anterior pituitary, both peak at the time of ovulation and remain at lower levels during the rest of the cycle. These hormones are secreted in response to **gonadotropin-releasing hormone (GnRH)** secreted by neuroendocrine cells in the hypothalamus. These hypothalamic cells generate bursts of action potentials approximately every 1.5 hours (with some variation depending on the phase of the cycle) to cause pulsatile release of GnRH. In addition to feedback from follicular hormones, the GnRH neurons are influenced by many factors, including stress and nutrition. In most seasonal breeders, photoperiod influences the episodic activity of the GnRH neuroendocrine cells and, through this mechanism, affects the female's reproductive status.

GnRH travels through the hypothalamo–hypophyseal portal system to the anterior pituitary gland, where it stimulates the secretion of FSH and LH. These two gonadotropins travel in the general blood circulation to the ovaries. The target tissue of LH is the layer of theca cells enclosing the oocyte and granulosa cells; FSH acts on the granulosa cells. The coordinated functions of the two hormones and the two target tissues produce the hormone **estrogen** (Figure 15.7). LH binds to receptors on the cell membranes of the theca cells and stim-

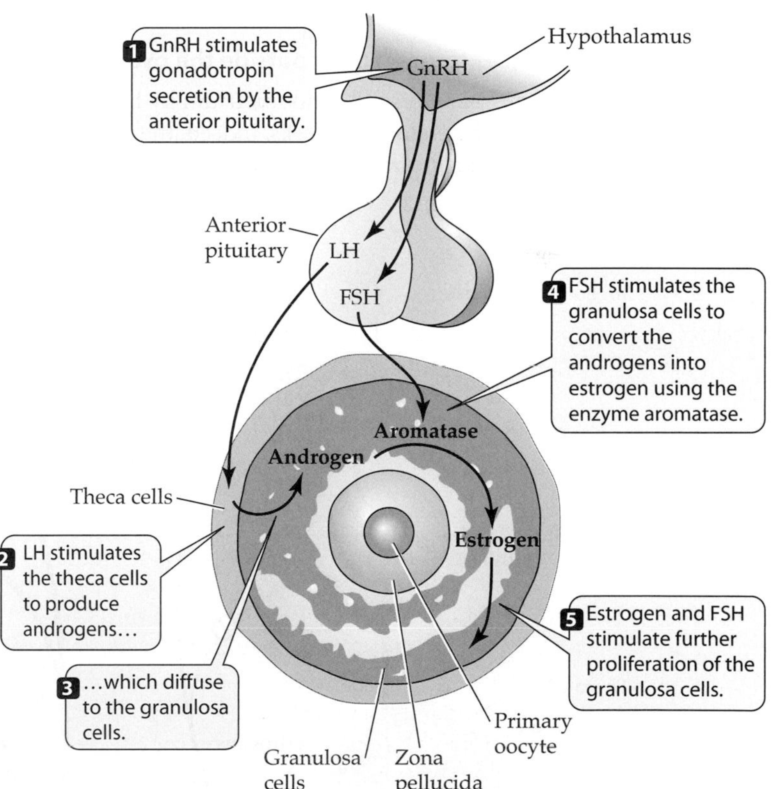

Figure 15.7 Hormonal control of estrogen secretion by the ovary Estrogen synthesis in the ovary requires both LH and FSH. During the follicular phase, estrogen and FSH stimulate the proliferation of granulosa cells, leading to increased secretion of estrogen. The main type of estrogen produced in the ovary is estradiol. However, we will use the generic term, estrogen, in this discussion.

ulates the cells to produce androgens (steroid hormones). The androgens diffuse through the theca cell membranes, through interstitial fluid between cells, and into the nearby granulosa cells. FSH binding to receptors on the cell membranes of the granulosa cells stimulates the action of the enzyme **aromatase**, which converts the androgens into estrogen. The reason for this division of labor is that theca cells, but not granulosa cells, possess the enzymatic machinery to synthesize androgens, the precursors of estrogen (see Figure 14.1).

The blood levels of estrogen increase slowly during the early stages of follicular development then rise dramatically prior to ovulation (see Figure 15.6). Estrogen itself plays a role in this increase by acting locally as a paracrine/autocrine agent within the ovary. In combination with FSH and growth factors, estrogen stimulates the proliferation of granulosa cells. This is a classic example of *positive feedback*, in which the production of estrogen by the granulosa cells stimulates mitotic multiplication of the granulosa cells themselves. As the cells increase in number, they produce increasing amounts of estrogen, which supports their continued proliferation (see Figure 15.7).

During the early part of follicular development, the granulosa cells of the developing follicles do not possess receptors for LH. Later, however, the granulosa cells of dominant follicles produce LH receptors and incorporate them into their cell membranes. LH plays a major role in triggering ovulation, as we will see shortly, and the LH receptors permit the granulosa cells to respond to it.

Estrogen influences target tissues outside the ovary

Estrogen stimulates growth of the endometrium by binding with specific receptors in the endometrial cells. Indeed, the rise in blood levels of estrogen during the late follicular development in the ovary corresponds to the proliferative phase of the uterus (see Figure 15.6). Furthermore, estrogen stimulates the endometrial cells to produce receptor molecules for another hormone, **progesterone**, in preparation for events that occur after ovulation.

Estrogen also feeds back to the anterior pituitary gland and probably to the hypothalamus. Estrogen's effects on these structures depend on its concentration in the blood. During early follicular development, when its blood level is low, estrogen exerts a *negative feedback* effect that keeps blood levels of LH and FSH low (Figure 15.8*a*). During late follicular development, however, estrogen (now at a high concentration in the blood) enhances the sensitivity of the gonadotropin (LH and FSH)-secreting cells in the anterior pituitary to GnRH (Figure 15.8*b*). It does so by stimulating them to incorporate increased numbers of GnRH receptors into their membranes. It probably also stimulates increased secretion of GnRH from the hypothalamus. Made exquisitely sensitive to GnRH, and very possibly receiving more frequent GnRH signals, the gonadotropin-secreting cells put out large amounts of LH. The peak of LH secretion, called the **LH surge**, plays an important role in ovulation, as we will see shortly. Just before ovulation, the granulosa cells increase their se-

Figure 15.8 Hormones of the hypothalamus, anterior pituitary, and ovary *(a)* Low levels of circulating estrogen exert negative feedback on the anterior pituitary and probably the hypothalamus, keeping LH and FSH secretions at relatively low levels. *(b)* High blood levels of estrogen exert positive feedback. Estrogen stimulates the gonadotropin-secreting cells of the anterior pituitary to increase their receptors for GnRH. Estrogen may also cause the hypothalamic GnRH cells to generate more bursts of action potentials. These effects lead to the LH surge and thus to ovulation. Inhibin from the granulosa cells also inhibits FSH secretion. *(c)* As long as the corpus luteum secretes estrogen, progesterone, and inhibin, these hormones ensure only low levels of LH and FSH secretion.

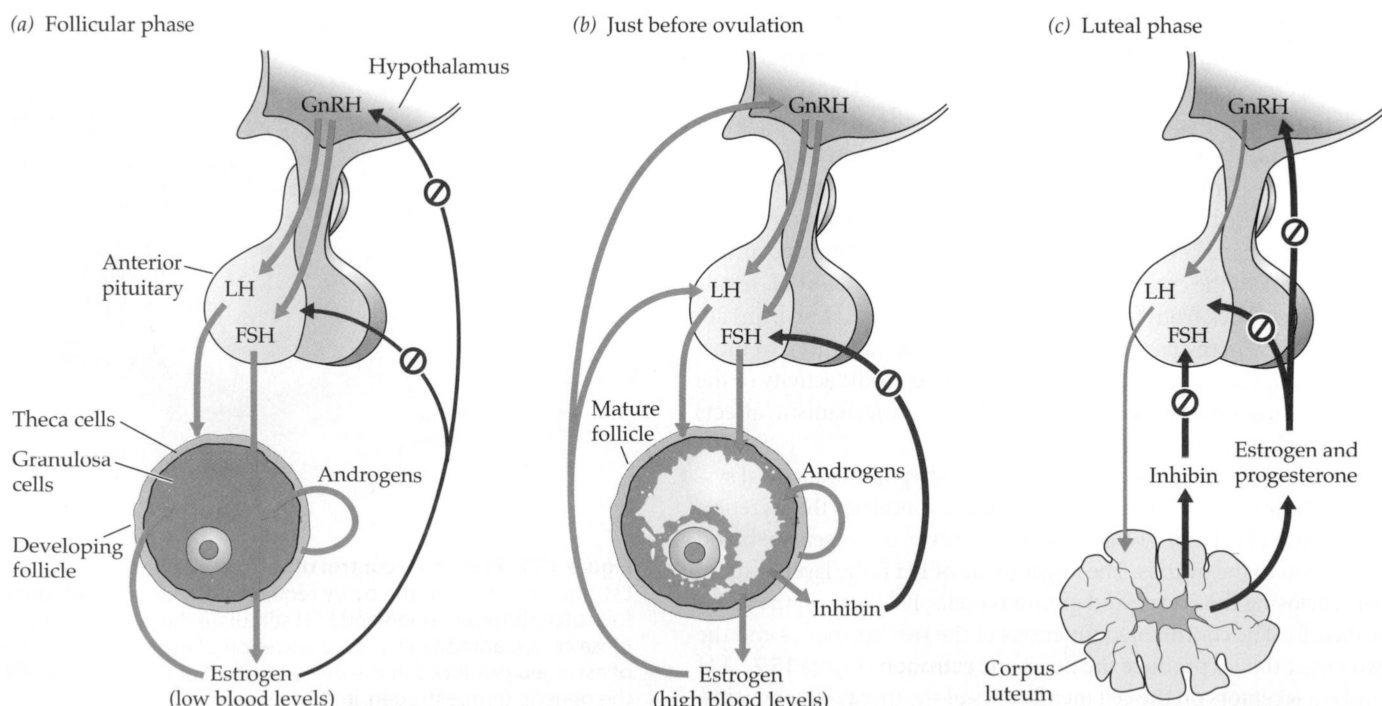

cretion of **inhibin**, a hormone that inhibits FSH secretion from the anterior pituitary gonadotropin cells. The granulosa cells also begin to secrete small amounts of progesterone. Table 15.2 summarizes the cells and hormones that are active during follicular development.

Ovulation is controlled by LH

The surge of LH from the anterior pituitary gland exerts multiple effects on the granulosa cells of the dominant follicle and brings about dramatic events within a period of hours. Under the control of LH, the granulosa cells secrete chemical mediators that cause the oocyte to complete the first meiotic division (see Figure 15.4). LH also causes the granulosa cells to begin secreting progesterone, decrease their secretion of estrogen, and release enzymes and prostaglandins that lead to breakdown of the follicular and ovarian membranes. As the membranes rupture, antral fluid, the oocyte, the first polar body, and the surrounding granulosa cells are released from the surface of the ovary. Finally, LH initiates morphological and biochemical changes in the remaining granulosa cells and theca cells, which transform into the corpus luteum. These *luteal cells* secrete three hormones: estrogen, progesterone, and inhibin.

The corpus luteum is essential for establishing and maintaining pregnancy

The LH surge is over, and the corpus luteum begins to function. The anterior pituitary secretes a low level of LH, which maintains the corpus luteum. Blood levels of progesterone, estrogen, and inhibin, secreted by the corpus luteum, steadily increase and peak in association with maximum endometrial thickness (see Figure 15.6). Progesterone is the major hormone secreted by the corpus luteum after ovulation. Estrogen is also secreted, but it never reaches blood levels as high as those prior to ovulation. Acting in concert, estrogen and progesterone exert *negative feedback* on the anterior pituitary (and

TABLE 15.2 Endocrine and neuroendocrine cells and the major secretions involved in female reproduction

Cells	Hormone	Action
Follicular development		
Hypothalamic GnRH cells	GnRH	Stimulates secretion of FSH and LH from anterior pituitary cells
Anterior pituitary gonadotropin cells	LH	Stimulates theca cells to secrete androgens during the early follicular phase; the LH surge triggers ovulation
	FSH	Stimulates aromatase action in granulosa cells to convert androgens to estrogen
Theca cells of ovarian follicle	Androgens	Diffuse from theca cells to granulosa cells for conversion to estrogen
Granulosa cells of ovarian follicle	Estrogen	Combined with FSH, stimulates proliferation of granulosa cells
		At low concentrations, has a negative feedback effect on the anterior pituitary (and possibly the hypothalamus), keeping FSH and LH secretions low
		At high concentrations, has a positive feedback effect and promotes LH surge
		Promotes growth of endometrium and development of endometrial progesterone receptors
	Inhibin	Inhibits FSH secretion
Luteal phase		
Corpus luteum cells	Progesterone (dominant)	Causes endometrium to become secretory; promotes relaxation of uterine and oviduct smooth muscles
	Estrogen	Combined with progesterone, reduces secretion of FSH and LH from the anterior pituitary and thereby inhibits folliculogenesis in primates and slows folliculogenesis in other mammals
	Inhibin	Inhibits FSH secretion
Pregnancy		
Placental cells	Chorionic gonadotropin	Secreted by embryonic placental cells in horses and primates. Rescues the corpus luteum and ensures its maintained function
	Progesterone	Opposes stimulatory effect of estrogen on uterine smooth muscle until late in pregnancy
		Stimulates secretion of prolactin from the anterior pituitary
		Combined with estrogen and prolactin, promotes growth of mammary glands
	Estrogen	Combined with progesterone and prolactin, promotes growth and development of mammary glands
		Combined with progesterone, prevents milk secretion by mammary glands
		Prepares the uterine smooth muscle for parturition by promoting production of oxytocin receptors and synthesis of connexins that form gap junctions
		Stimulates enzymatic breakdown of cervical collagen fibers
	Lactogen (chorionic somato mammotropin)	Alters maternal glucose and fatty acid metabolism to shunt glucose and fatty acids to the fetus; may contribute to lactational development

probably the hypothalamus) to keep GnRH, LH, and FSH secretions low (**Figure 15.8c**). Inhibin also suppresses FSH secretion. Decreased levels of gonadotropins in primates inhibit follicular development in the ovary; in other mammals, folliculogenesis can occur in all phases of the cycle, but is reduced during the luteal phase.

Fertilization and implantation are possible while the corpus luteum is active. Progesterone readies the uterus and oviducts for these events. The endometrium of the uterus is "primed" to respond to progesterone because it developed progesterone receptors earlier in response to estrogen. In progesterone's presence, the exocrine endometrial glands secrete glycogen and enzymes, and additional blood vessels develop. The thickened and secreting glandular endometrium is poised to support implantation and nourishment of an **embryo** if fertilization occurs.[2] If fertilization does not occur, the endometrium is sloughed off in mammals that menstruate or resorbed in mammals with estrous cycles. Progesterone also inhibits contractions of the smooth muscle cells of the walls of the uterus (the **myometrium**) and of the oviduct by opposing the stimulatory effect of estrogen. Keeping these muscles relaxed will provide a stable environment for successful implantation. Table 15.2 summarizes the cells and hormones that are active during the part of the cycle dominated by the corpus luteum.

The cells of the corpus luteum have a finite life span. If pregnancy occurs, the embryo of many mammals provides a hormonal or paracrine signal that rescues the corpus luteum, as we will see below. If pregnancy does not occur, the luteal cells stop secreting hormones and degenerate. Without a functional corpus luteum, progesterone, estrogen, and inhibin levels decrease. In humans, the hypothalamus and anterior pituitary gland are then freed of negative feedback, so that gonadotropin secretion begins again and initiates another round of follicular development in the ovary. In monestrous and seasonal polyestrous breeders, environmental signals integrated by the nervous system influence the secretion of GnRH and thereby determine when the next cycle of follicular development will occur.

In mammals that menstruate, the endometrium responds to the loss of luteal hormones by secreting prostaglandins, which initiate deterioration of its superficial layers. These chemical mediators cause constriction of the uterine blood vessels and contraction of the myometrium. Closing off the blood supply prevents oxygen and nutrients from reaching the cells in the endometrium, and they begin to die. The superficial layers will be lost in menstrual flow. The remaining base of the endometrium will proliferate in the next cycle. Following their initial constriction, the uterine blood vessels dilate. Increased blood flow causes the walls of weakened capillaries to break, and bleeding occurs. Menstrual flow is thus a mixture of sloughed-off endometrial tissues and blood.

Why do all mammals cyclically change the endometrium? Why do humans and a few other primates menstruate?

In most species, the deteriorated endometrial tissue is simply resorbed. However, some primates, such as humans and chimpanzees, produce too much blood and other tissue for complete resorption, and these products are lost in menstruation. The "overflow" can be explained at least in part by the fact that these primates have a larger uterus, relative to female body size, than do other mammals.

A more fundamental question is why all mammals go through a cyclic pattern of building up and tearing down the endometrium. Why not maintain it in a steady state ready for possible implantation? A possible explanation is that it is metabolically very costly to maintain a secretory endometrium. The body temperature of a human female increases after ovulation and is maintained at about 0.3°C higher during the luteal phase than during the follicular phase. Investigators have proposed that this increased body temperature reflects the higher metabolic rate required to maintain the endometrium in a secretory (metabolically active) state. Indeed, in vitro measurements of oxygen consumption of human endometrial tissue have shown that luteal-phase tissue consumes more oxygen than follicular-phase tissue. If it is metabolically "cheaper" to break down the endometrium (and in some cases lose blood and tissue entirely) than to maintain a constantly secretory endometrium, then cyclically shedding the endometrium would allow an animal to economize on the overall metabolic costs of reproduction. A female mammal would require less food and expend less energy foraging (and evading predators while doing so) during the nonsecretory phase of every estrous or menstrual cycle. Future experimental studies may yield more evidence to support the idea that cyclic changes in the endometrium reduce metabolic costs.

Summary: Hormonal Control of Female Reproduction

- The ovary produces oocytes and secretes hormones. Each estrous or menstrual cycle has three main ovarian phases: development of follicles, ovulation, and function of the corpus luteum. The uterine endometrium grows thicker prior to ovulation and becomes secretory after ovulation.
- Released in response to GnRH, the gonadotropins LH and FSH stimulate granulosa cells to secrete estrogen. Estrogen acts both as a paracrine/autocrine factor that stimulates proliferation of granulosa cells, and as a blood-borne hormone that stimulates growth of the endometrium and feeds back on the anterior pituitary and hypothalamus.
- A surge in secretion of LH controls ovulation.
- After ovulation, the follicular cells reorganize into the corpus luteum, which secretes progesterone, estrogen, and inhibin. These hormones inhibit or decrease folliculogenesis in the ovary by reducing secretion of gonadotropins from the anterior pituitary. Progesterone supports the secretory state of the endometrium and inhibits contraction of the smooth muscle of the myometrium and oviducts.
- The corpus luteum continues to function if fertilization takes place. If fertilization does not occur, the corpus luteum degenerates, and endometrial tissue is resorbed or given off as menstrual flow.

[2] A developing mammal is called an *embryo* once it implants in the endometrium. A human is an embryo from implantation to the end of the eighth week of pregnancy, and a *fetus* from the ninth week until birth.

Hormonal Control of Male Reproduction

We will use the human male to illustrate the basic anatomical and physiological features of male reproduction in mammals. Figure 15.9*a* shows the anatomy of the human male reproductive system, which has the same pattern of organization as that of other eutherian mammals (see Figure 15.3). The penis contains the **urethra**, which serves two urogenital functions (though not at the same time), allowing the passage of urine during micturition (urination) and of semen during ejaculation. The shaft of the penis is almost entirely filled by three compartments that can be expanded with blood during erection. The males of several species, such as bats and carnivores (but not humans), possess a **baculum** of cartilage or bone that extends the length of the shaft and provides stiffness to the penis for copulation. The paired testes are contained in the **scrotum**, a double sac that suspends the testes outside the body cavity. This anatomy is necessary because normal sperm production requires a temperature about 2°C lower than the internal body temperature. Marine mammals such as dolphins and seals must also maintain their testes at a

(*a*) External genitalia

Urinary bladder
Rectum
Pubic bone
Seminal vesicle
Prostate gland
Penis
Urethra
Vas deferens
Epididymis
Testis

The penis contains three compartments that fill with blood to cause erection.

The prostate gland and the seminal vesicles secrete most of the fluid released during ejaculation. Semen consists of this fluid plus sperm cells.

Figure 15.9 The human male reproductive system *(a)* The male reproductive system consists of external genitalia (penis and scrotum), testes, ducts, and accessory glands. *(b)* The testis is divided into 200–300 compartments containing coiled seminiferous tubules that are continuous with the epididymis and vas deferens. Each seminiferous tubule consists of a single layer of Sertoli cells. Connective tissue surrounding the tubules contains Leydig cells. *(c)* A mature sperm cell possesses a nucleus with a haploid set of chromosomes, a flagellum for motility, mitochondria that produce ATP, and an acrosome, which contains enzymes released during fertilization.

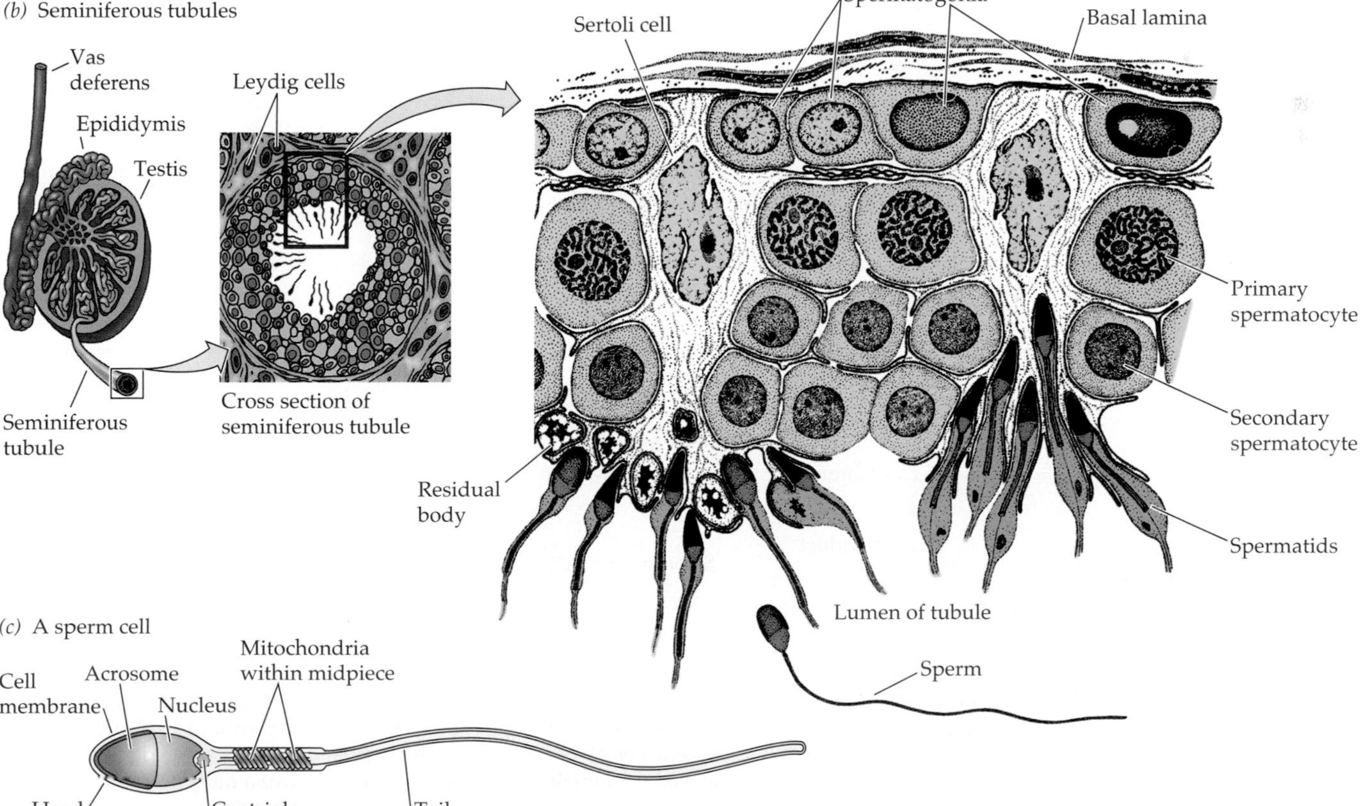

BOX 15.2 Sex Determination in Mammals

The majority of animals on Earth use sexual reproduction, at least at some time in their life cycle, and most sexually reproducing species have two different sexes, males and females. Animals use a variety of mechanisms to determine the sex of an individual. In certain reptiles, the temperature to which eggs are exposed during development determines the sex of hatchlings. Several different fish are known to change sex depending on social factors; for example, a female adult may become a male if there are no male fish in the local environment. In the insect family Hymenoptera, which includes bees, wasps, and ants, males develop from unfertilized (haploid) eggs and females from fertilized (diploid) eggs.

In most mammals, genes determine the sex, and set the stage for the following sequence:

- *Chromosomal sex* (XX female or XY male),
- *Gonadal sex* (whether an animal possesses testes or ovaries),
- *Hormonal sex* (whether the dominant hormones are androgens or estrogens), and
- *Phenotypic sex* (based on external genitalia and body type).

A mammalian embryo starts out in development with a pair of **indifferent gonads**, which have the potential to differentiate into either testes or ovaries. Each indifferent gonad is associated with two ducts, the **Müllerian duct** and the **Wolffian duct** (Figure A). If the gonads differentiate into testes, the Müllerian ducts will degenerate while the Wolffian ducts develop further. If the gonads develop into ovaries, the Müllerian ducts will continue to develop, and the Wolffian ducts will degenerate. The external genital structures of the developing embryo are also indifferent (Figure B). If the gonads develop into testes, the **genital tubercle**, **urethral folds** and **groove**, and **labioscrotal swellings** will develop into male structures. If the gonads develop into ovaries, these structures will become female external

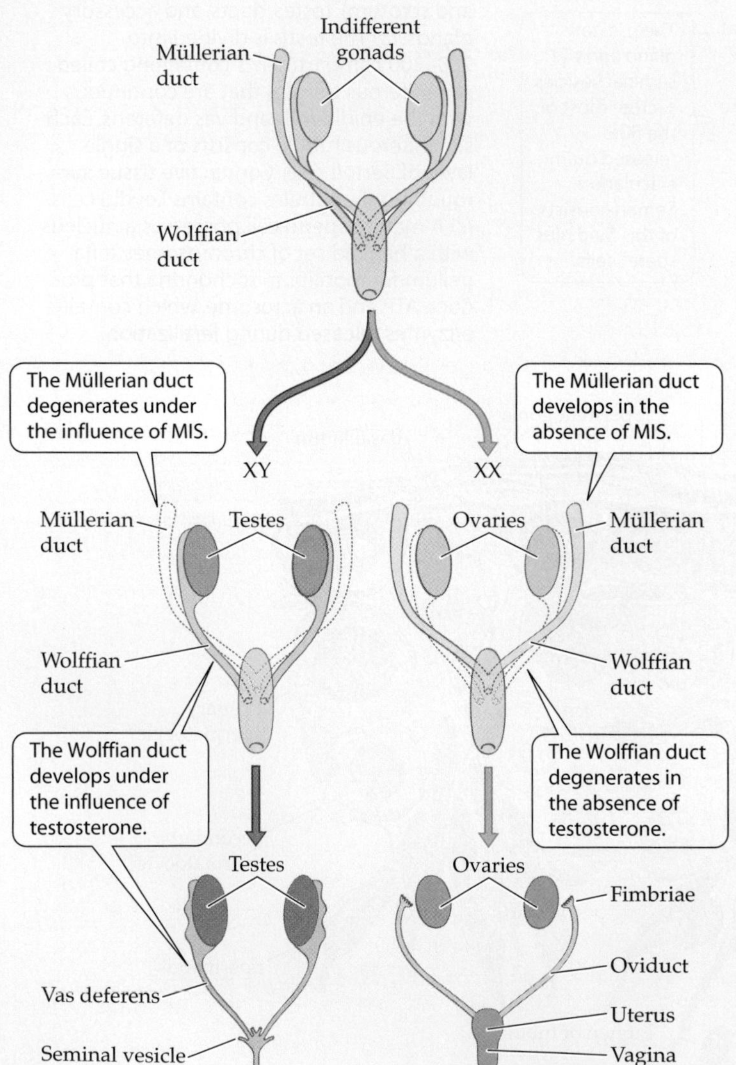

Figure A Differentiation of the internal urogenital structures The *SRY* gene determines whether bipotential structures develop male or female characteristics.

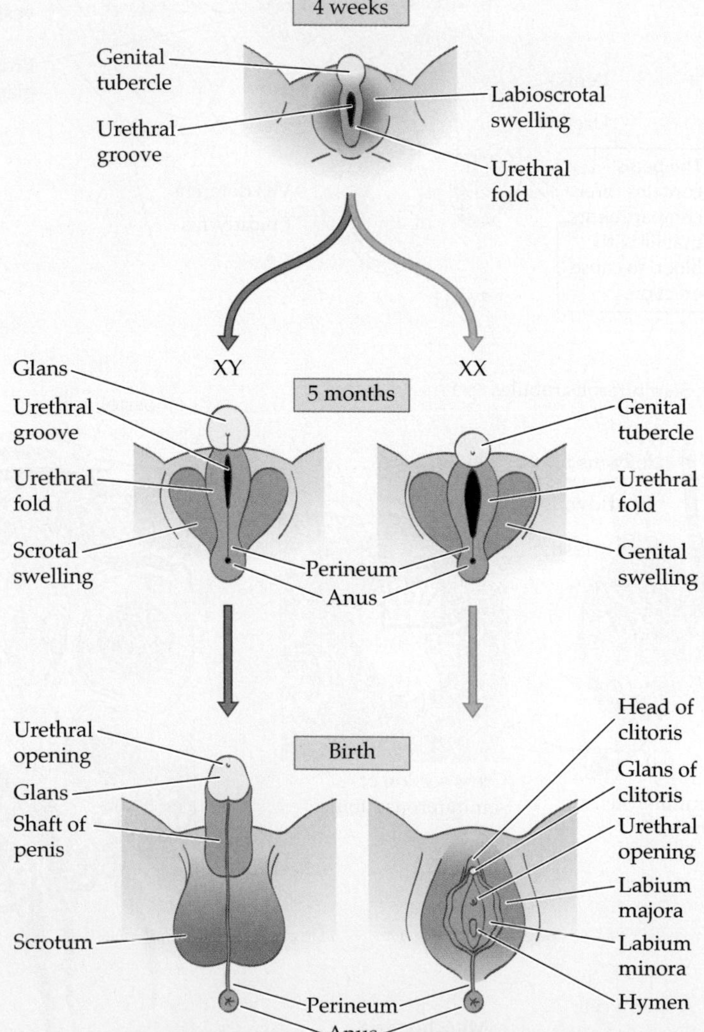

Figure B Differentiation of the external genitalia Undifferentiated external genitalia in the human begin to differentiate at about the seventh week of gestation. The external genitalia can be distinguished as male or female in the twelfth week of human development.

BOX 15.2 *(Continued)*

genitalia. These patterns of development have been known for many decades. In addition, it has long been known that mammals inheriting one Y chromosome (from the male parent) and one X chromosome (from the female parent) become males, whereas those inheriting two X chromosomes become females. However, not until the 1990s did investigators identify a gene on the Y chromosome of mice and humans, the ***SRY* gene**, which determines whether the undifferentiated gonad will become a testis or an ovary. (The *SRY* gene is named for its location in the Sex-determining *R*egion of the *Y* chromosome.)

Expression of the *SRY* gene sets in motion processes that result in the development of testes and other male urogenital structures. Females lack the *SRY* gene. In the absence of signals that initiate testicular development, the indifferent gonads differentiate into ovaries. How does the *SRY* gene induce the development of testes? In humans, around the seventh week of embryonic development, cells of the indifferent gonad in a genetically male embryo begin to express the *SRY* gene, which codes for a protein called *testis-determining factor* (TDF). TDF interacts with the DNA of the indifferent gonadal cells to activate additional genes that direct cellular differentiation into testicular Sertoli and Leydig cells. The functioning of these cells leads to differentiation of the other male structures. The Sertoli cells secrete **Müllerian inhibitory substance** (**MIS**), which causes degeneration of the Müllerian ducts. The Leydig cells secrete testosterone, which promotes development of the Wolffian ducts into the epididymi, vasa deferentia, and seminal vesicles (see Figure A). Testosterone and its metabolite dihydrotestosterone (DHT) also stimulate development of the prostate gland, penis, and scrotum (see Figure B).

In the female, in the absence of *SRY*-initiated gene products, the indifferent gonad develops into ovarian tissues. MIS is not secreted, so the Müllerian ducts develop into oviducts and uterus (see Figure A). Without testosterone, the Wolffian ducts degenerate, and the external genitalia develop into the clitoris, labia minora, openings of the vagina and urethra, and labia majora (see Figure B). In other words, the embryonic urogenital structures become female *unless SRY* and a cascade of following genes are expressed to specify male development. Differentiation of the female primary sexual structures begins around the eleventh week in human development. Estrogen does not appear to be required for initial sex determination of the female mammal, although estrogen is crucial at puberty and beyond.

We know that "hormonal sex" influences not only the development of urogenital structures, but also the development of structures in the brain—"brain sex," which in turn influences "behavioral sex." These broader hormonal influences are topics of active investigation.

---

cooler temperature; however, they house their testes within the body cavity. To keep them from overheating, parts of the circulatory system surrounding the testes carry blood that has been cooled by flowing through vessels close to the body surface.

Sperm are produced in the coiled **seminiferous tubules** that make up the bulk of the testicular tissue (Figure 15.9*b*). The seminiferous tubules merge to join the **epididymis** (plural *epididymi*), a coiled tube that forms a cap outside the capsule of the testis. The epididymis becomes continuous with the **vas deferens** (plural *vasa deferentia*; also called the *ductus deferens*), which empties into the urethra. In cross section, one can see that each seminiferous tubule consists of a single layer of **Sertoli cells** (also called *sustentacular cells*). Progressions of spermatocytes in various stages of spermatogenesis lie between the Sertoli cells, from spermatogonia on the periphery to spermatids at the lumen. The Sertoli cells support and regulate spermatogenesis. A basal lamina surrounds each seminiferous tubule and separates the Sertoli cells from the connective tissue that fills in the spaces between the tubules. **Leydig cells** (also called *interstitial cells*) located in the connective tissue secrete **testosterone** (an androgen). The Leydig cells also secrete testosterone during fetal life (Box 15.2), when this hormone influences sex differentiation and development. They stop secreting testosterone at birth and begin again at puberty.

Spermatogenesis is controlled by gonadotropins and also requires testosterone. Just as in oogenesis, GnRH is released from the hypothalamus (in a pulsatile fashion approximately every 1.5 hours) and FSH and LH are secreted from the anterior pituitary gland. The Leydig cells are the target tissue of LH. LH binds to receptors in the cell membranes of the Leydig cells and stimulates them to produce and secrete testosterone. Testosterone released into the general circulation provides negative feedback to the anterior pituitary and the hypothalamus. The Sertoli cells are the target tissue of FSH. Testosterone from the Leydig cells also influences the function of the Sertoli cells. FSH and testosterone together stimulate the Sertoli cells to secrete paracrine substances that promote proliferation of the spermatogonia and differentiation of the sperm. The Sertoli cells secrete several additional substances, including products that nourish the developing sperm; inhibin, which inhibits FSH secretion from the anterior pituitary; and fluid that fills the lumens of the seminiferous tubules. This fluid contains a protein called **androgen-binding protein** that binds testosterone and keeps it at a high concentration in the lumens. Table 15.3 summarizes the cells and hormones that are active in male reproduction.

As we've mentioned, testosterone and inhibin both exert negative feedback effects on the anterior pituitary and hypothalamus, keeping FSH and LH secretions relatively low and steady from day to day. This pattern is in distinct contrast to the variations of ovarian, hypothalamic, and anterior pituitary secretions that take place over the female estrous or menstrual cycle. In some mammalian species, however, testicular function does vary. For example, the testes of the Siberian hamster regress when days become short, and spermatogenesis is suspended until photoperiodic stimuli again lead to adequate secretion of GnRH and the gonadotropins.

The constant levels of hormones in most mammalian males permit continuous production of sperm. The sperm form in the walls of the seminiferous tubules. Mature sperm have a flagellar tail, an enzyme-filled organelle called the *acrosome*, and mitochondria stored in the *midpiece* (Figure 15.9*c*). Not yet motile, newly formed sperm enter the lumens of the seminiferous tubules. The fluid continuously secreted by the Sertoli cells moves the sperm along the tubules, out of the testis proper, and into the epididymis, where they undergo further maturation. Fluid is reabsorbed from the lumen of

TABLE 15.3 Endocrine and neuroendocrine cells and secretions involved in male reproduction

Cells	Hormone	Action
Hypothalamic GnRH cells	GnRH	Stimulates secretion of FSH and LH from anterior pituitary cells
Anterior pituitary cells	LH	Stimulates Leydig cells to secrete testosterone
	FSH	In combination with testosterone, stimulates Sertoli cells to support and regulate spermatogenesis
Leydig cells	Testosterone	Required for mitosis and meiosis of spermatogenesis. In combination with FSH, stimulates Sertoli cells to support and regulate spermatogenesis
		Exerts negative feedback on the anterior pituitary and the hypothalamus
Sertoli cells	Inhibin	Inhibits FSH secretion
		(Sertoli cells also secrete *nonhormonal* products, including substances that nourish the differentiating sperm, extracellular fluid that fills the seminiferous tubules and includes androgen-binding protein, and paracrine agents.)

the epididymal tubules, so that the sperm become highly concentrated. They are stored, ready for ejaculation, in the last part of the epididymis and in the vas deferens. Just prior to ejaculation, accessory glands (see Figure 15.9) secrete seminal fluids that carry the sperm out of the male reproductive tract. **Semen** consists of the fluids and sperm.

Summary: Hormonal Control of Male Reproduction

- The testes produce gametes and secrete hormones. Spermatogenesis occurs continuously in the seminiferous tubules, controlled by continuous secretion of GnRH, LH, and FSH.
- GnRH from neuroendocrine cells in the hypothalamus stimulates the secretion of LH and FSH from the anterior pituitary gland. LH targets the Leydig cells located in the connective tissue between the seminiferous tubules. FSH targets the Sertoli cells that make up the walls of the seminiferous tubules.
- Leydig cells secrete testosterone. Sertoli cells secrete inhibin, substances that nourish the developing sperm, the fluid that fills the lumens of the seminiferous tubules and contains androgen-binding protein, and paracrine substances.
- Both FSH and testosterone are required to regulate and promote spermatogenesis.

Fertilization, Pregnancy, and Birth in Eutherian Mammals

If mating occurs at a time when fertilization is possible, sperm cells travel through the female reproductive tract to meet the secondary oocyte and its surrounding cells in the oviduct. Fertilization and the beginning of development take place in the oviduct. We begin this section by exploring the events that occur during fertilization and subsequent implantation of an embryo in the uterine endometrium. Once implantation takes place, pregnancy is established, and developmental processes give rise to a new individual. In eutherian mammals, the placenta, a structure formed jointly by the embryo and the mother, supports growth and development until birth. After varying periods of gestation, depending on the species, birth (parturition) takes place. We conclude this section with a discussion of the physiological events that occur during birth. Much of our knowledge of the physiology of pregnancy and birth is based on studies of laboratory and domestic mammals.

Fertilization is the union of sperm and oocyte

Although millions of sperm enter the female reproductive tract, very few actually arrive at the oocyte, and only one fuses with it to form the zygote. While sperm are in the female reproductive tract, they go through a final process of maturation called **capacitation**. Flagellar movements of the tail are modified in capacitation to make the sperm capable of rapid forward swimming. In addition, changes occur in the cell membrane of the sperm head that are necessary for the sperm to reach the oocyte and ultimately fuse with its cell membrane. Several avenues of ongoing research suggest that the oocyte facilitates fertilization by releasing a chemical attractant, and that sperm cells possess chemoreceptors for that substance on their cell membranes.

Following ovulation, the oocyte is retained in the upper third of the oviduct, near the ovary. Travel toward the uterus is prevented because the oviduct is constricted. The high blood levels of estrogen at this stage of the cycle cause the smooth muscles of the oviduct wall to contract and decrease the diameter of its lumen. The lumen is too narrow for the oocyte and its surrounding cells to move toward the uterus, but wide enough for sperm cells to pass through. The oocyte is enclosed in the zona pellucida and surrounded by cumulus granulosa cells (Figure 15.10*a* and *b*). The sperm must traverse these structures before reaching the cell membrane of the oocyte. The head of the sperm has enzymes on its outer surface that break down the extracellular matrix between the cumulus granulosa cells. This enzymatic activity, combined with rapid movements of the flagellum, help it penetrate the layer of cells. To breach the zona pellucida, the sperm releases enzymes by exocytosis from the acrosome in its head. However, exocytosis of these enzymes, called the **acrosomal reaction**, is triggered only when the sperm head binds to specific protein molecules on the zona pellucida. The binding of sperm to the

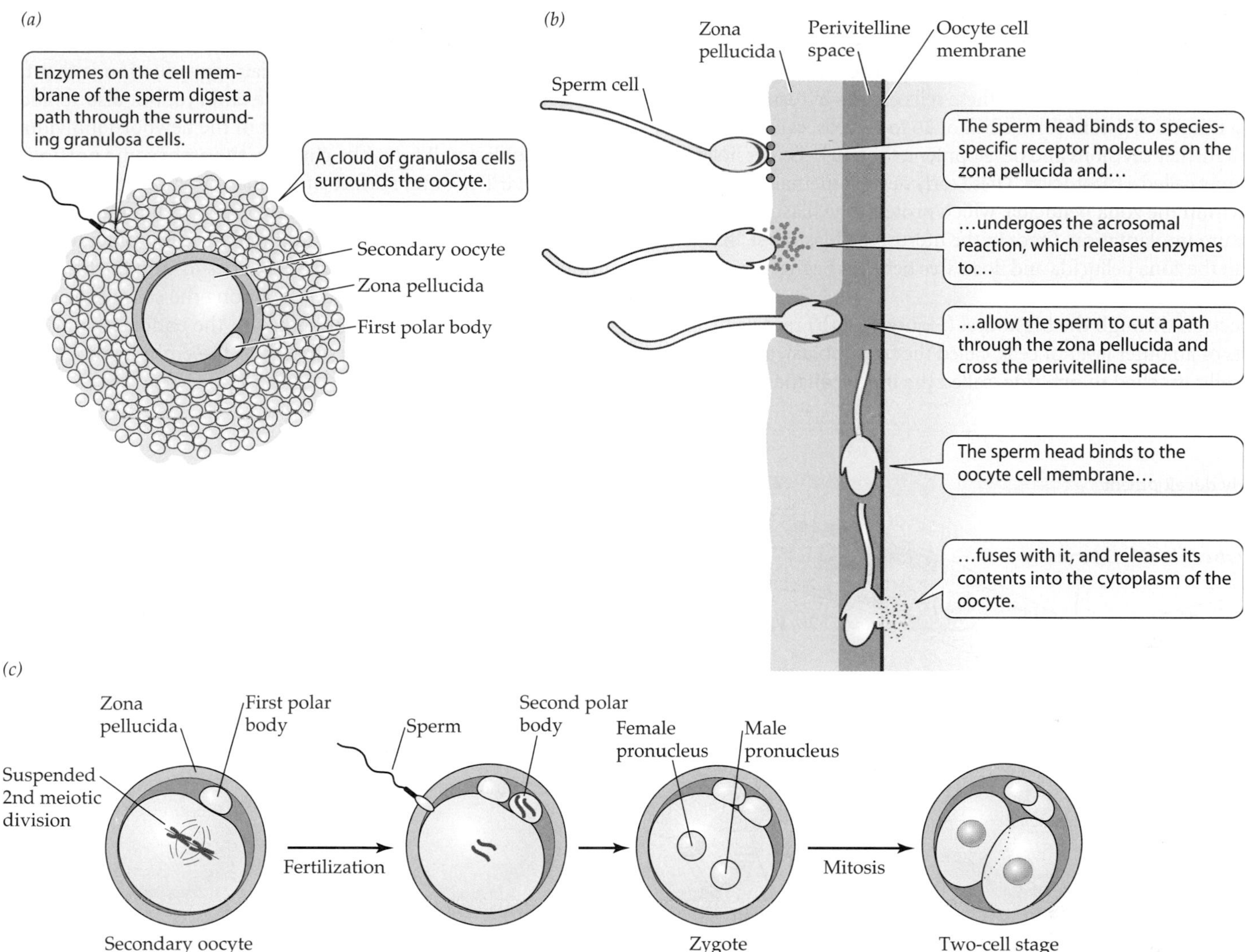

Figure 15.10 Fertilization To achieve fertilization, the sperm must *(a)* penetrate the layer of cumulus granulosa cells and *(b)* cut a channel through the zona pellucida surrounding the oocyte. *(c)* Fertilization allows the oocyte to complete meiosis, extrude the second polar body, and begin mitosis. (*b* after Primakoff and Myles 2002.)

zona pellucida is species-specific. In other words, sperm from a guinea pig, for example, cannot bind to sites on the zona pellucida surrounding the oocyte of a mouse or a human. After initiating the acrosomal reaction, the sperm head detaches from the binding site and (aided by the enzymes and the flagellar action of its tail) cuts a channel through the zona pellucida. It enters the *perivitelline space*, the narrow extracellular space between the zona pellucida and the oocyte, and adheres to the cell membrane of the oocyte. Only a few sperm get this far. The cell membrane of just one of these sperm fuses with the cell membrane of the oocyte, and the cytoplasms of the two gametes become continuous. This event is **fertilization**. The sperm tail does not enter the fertilized oocyte.

Two important processes quickly follow fertilization: completion of meiosis and prevention of more than one sperm from fertilizing the oocyte. In the **cortical reaction**, organelles in the oocyte's peripheral cytoplasm, called **cortical granules**, release substances into the extracellular space that modify the cell membrane as well as the zona pellucida. These alterations prevent adhesion or fusion of more than one sperm, achieving a **block to polyspermy**. Blocking the entry of more than one sperm ensures that only one haploid set of paternal chromosomes is admitted to the oocyte. Meanwhile, the oocyte completes the second meiotic division, forms the second polar body, and extrudes it from the cell (Figure 15.10*c*). Therefore, each gamete contributes a single haploid set of chromosomes. The two haploid sets of chromosomes join to make the zygote's diploid complement of chromosomes. At this stage, *conception* has occurred. However, the mother is not yet pregnant. Pregnancy is established when implantation takes place.

Development begins in the oviduct, and implantation establishes pregnancy

The zygote slowly moves down the oviduct. Its travel becomes possible because the rising levels of progesterone secreted by the corpus luteum oppose the estrogen that (by itself) stimulated the smooth muscles of the oviduct to contract; therefore, the muscles relax, and the lumen of the oviduct opens. Mitotic cell divisions take place during the journey to the uterus (Figure 15.11*a*). The zygote first divides by mitosis into two cells. Each of these cells divides in turn, and continuing mitotic divisions form a ball of 16 to 32 cells, called a *morula*. With further divisions and development, the ball becomes hollow, and is now called a **blastocyst**. These early developmental stages take place within the zona pellucida, which protects the blastocyst from premature implantation. Once in the uterus, the blastocyst "hatches" from the zona pellucida and floats free before it begins to implant in the endometrium.

A section through the blastocyst (Figure 15.11*b*) reveals that it consists of an outer layer of cells, called the **trophoblast**, and a cluster of cells attached to one side, called the **inner cell mass**. Its hollow interior, called the *blastocoel*, is filled with fluid. The inner cell mass will give rise to the embryo and some extraembryonic tissues. A fundamental principle of mammalian reproduction is that the fertilized egg gives rise not only to the embryo, but also to additional tissues that will not become part of the newborn individual. The trophoblast will form the **chorion**, the embryonic portion of the placenta. The outer surface of the blastocyst on the side of the inner cell mass is specialized to make contact with the endometrium and begin implantation. The trophoblast cells proliferate by dividing mitotically, and they secrete enzymes to help the blastocyst bury itself, that is become *implanted*, among the cells of the nutrient-rich endometrium. For the moment, the endometrium itself provides sufficient metabolic support to the embryo and its surrounding membranes.

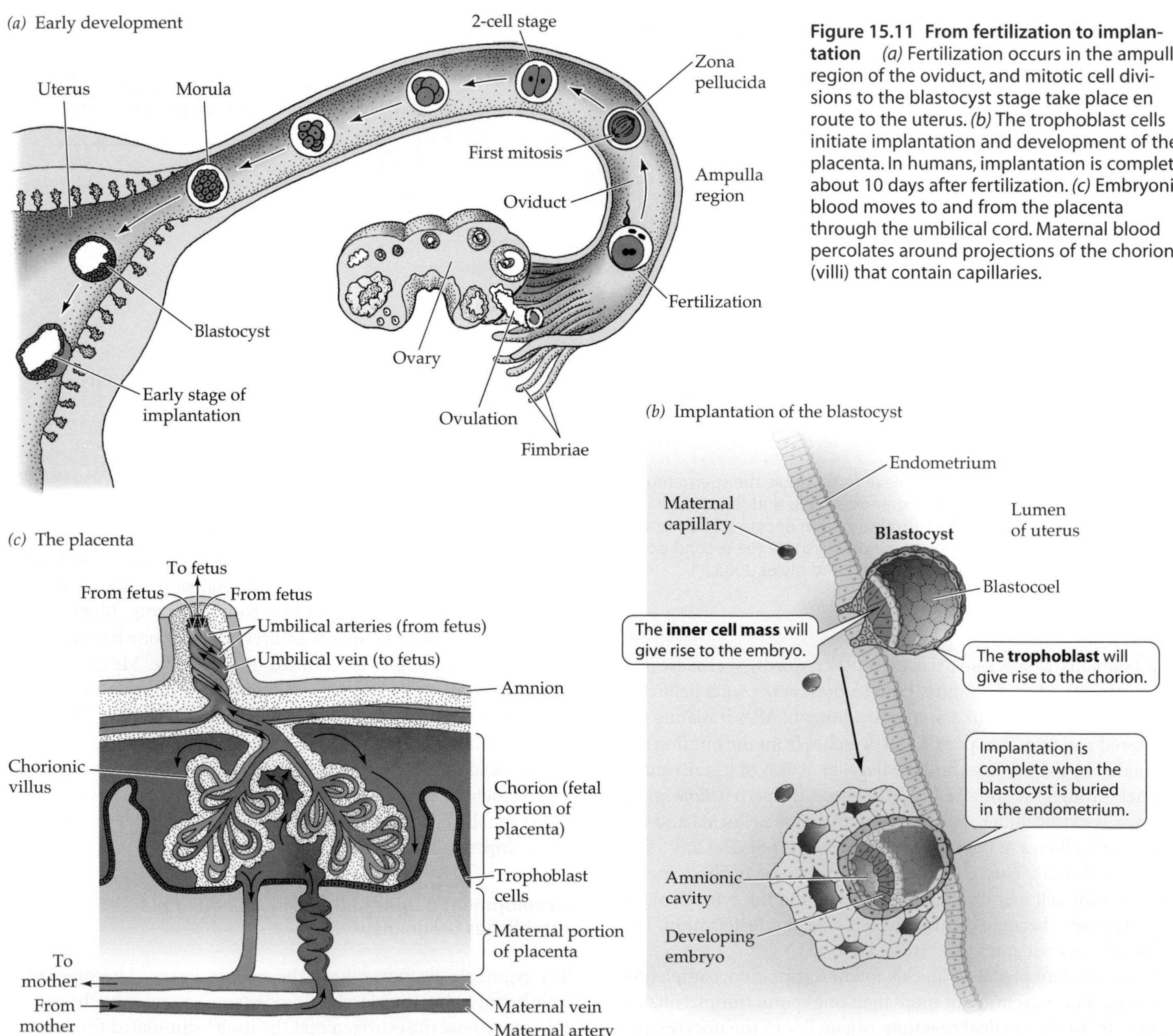

Figure 15.11 From fertilization to implantation *(a)* Fertilization occurs in the ampulla region of the oviduct, and mitotic cell divisions to the blastocyst stage take place en route to the uterus. *(b)* The trophoblast cells initiate implantation and development of the placenta. In humans, implantation is complete about 10 days after fertilization. *(c)* Embryonic blood moves to and from the placenta through the umbilical cord. Maternal blood percolates around projections of the chorion (villi) that contain capillaries.

To keep the endometrium in a secretory state during early pregnancy, the corpus luteum continues to secrete high levels of progesterone (and estrogen) until the placenta has developed enough to secrete adequate amounts of these hormones. Thus the corpus luteum must survive beyond the time when it would typically degenerate if pregnancy did not occur. In humans, the shift from dependence on hormones from the corpus luteum to dependence on the placenta happens after about 70 days (pregnancy lasts about 266 days). By contrast, the corpus luteum of the cow (which starts to degenerate after 16 days if pregnancy is not established) must function for 200 days of the 280-day long pregnancy. Different animals use different mechanisms to prolong the life of the corpus luteum. For example, in sheep and cows, the trophoblast cells of the embryo secrete a paracrine substance that prevents uterine cells from producing circulating prostaglandins that cause the corpus luteum to degenerate. In rodents, the hormone **prolactin**, released by the anterior pituitary gland, provides tropic support to the corpus luteum. In primates and horses, the chorion secretes the hormone **chorionic gonadotropin** (**CG**) (Figure 15.12), which extends the functional life of the corpus luteum.[3] Although CG functions similarly in both primates and horses, several lines of evidence indicate that these animals evolved CG independently.

The embryonic trophoblast and the maternal endometrium form the placenta

Within a hormonal milieu that supports continued development of the blastocyst, momentous changes take place in both the trophoblast and the inner cell mass. The trophoblast interacts with maternal uterine structures to form the placenta, which grows in tandem with the growing embryo/fetus throughout pregnancy. Within the placenta, the maternal blood comes into close association with the fetal blood, separated by only one or a few membranes. Propelled by the mother's heart, maternal blood continually moves through the placenta to provide O_2 and nutrients and remove CO_2 and metabolic by-products.[4] The fetus is connected to the placenta by the *umbilical cord.* Blood vessels form within the umbilical cord (Figure 15.11c), and the heart of the embryo/fetus pumps its blood through the placenta. The heart is the first organ to function in mammalian embryos. In humans, the heart starts to beat 22 days after fertilization.

As the placenta grows, it secretes hormones, including progesterone and estrogen (see Figure 15.12). Progesterone is essential for maintaining the endometrium and inhibiting contractions of the myometrium. Progesterone also contributes to preparing the mammary glands for lactation. Estrogen stimulates growth of the myometrium and also contributes to mammary development. The placental cells use the enzyme aromatase to convert androgens synthesized

[3]Because CG circulates throughout the body, including the kidneys, it gets excreted in the urine. Pregnancy tests assay urine samples for the presence of CG. If the assay is positive for CG, it indicates that an embryo has implanted itself in the endometrium.

[4]Substances such as drugs and alcohol in the maternal blood can also cross the placenta, and are known to cause physical and behavioral birth defects by disrupting development of the embryo/fetus.

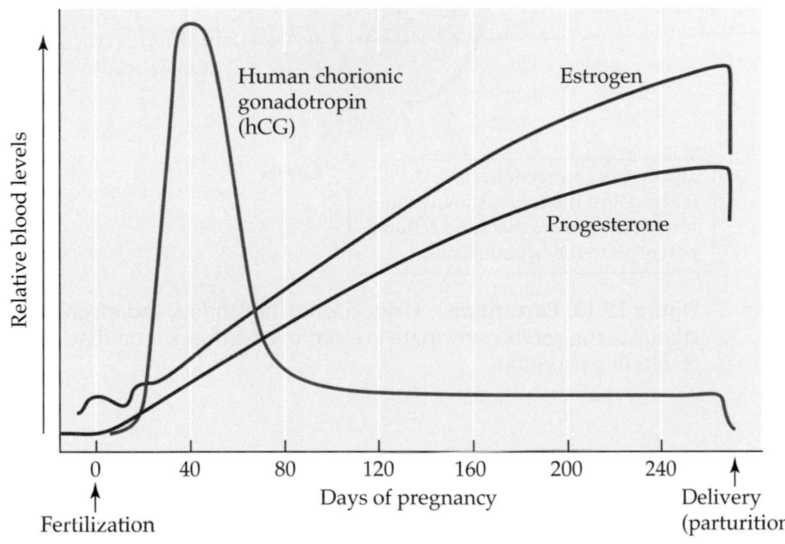

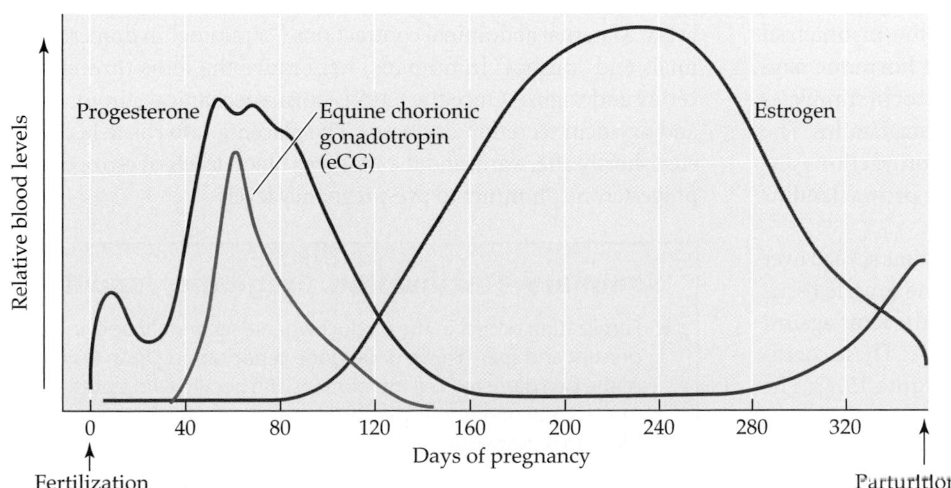

Figure 15.12 Hormonal levels during pregnancy During pregnancy in both *(a)* humans and *(b)* horses, chorionic gonadotropin sustains the life of the corpus luteum. As the placenta develops, it secretes estrogen and progesterone. *(b* after Allen 2001.)

in other tissues into estrogen. Although estrogen and progesterone are the dominant hormones produced by the placenta, this organ also secretes additional hormones that support physiological functions during pregnancy and prepare the mother's body for birth and lactation.

As the placenta is being established, the inner cell mass also changes. It will develop into the embryo itself as well as extraembryonic membranes. One extraembryonic membrane is the **amnion** (see Figure 15.11*b*). This membrane forms the *amnionic sac*, which is filled with *amnionic fluid* that has a composition similar to that of the embryo's extracellular fluid. Floating in the amnionic sac, the embryo/fetus is cushioned from possible physical injury and insulated from temperature extremes. Shortly after implantation, the embryo establishes three primary germ layers, **ectoderm**, **endoderm**, and **mesoderm**, which will participate in **organogenesis**: the formation of specific organs. In general, ectoderm gives rise to outer structures, such as the epidermis of the skin and the nervous system. Endoderm gives rise to inner structures, including the liver, pancreas, and the epithelial linings of the digestive tract and respiratory tract. Mesoderm forms structures in between, including nearly all of the skeletal muscle and circulatory, excretory, and reproductive systems.

Parturition requires cellular changes and orchestrated neural and hormonal signals

Uterine contractions are necessary to accomplish **parturition** (birth). The actual factors that signal when parturition should begin are not well understood. Researchers believe that one important factor is secretion of corticotropin-releasing hormone (CRH) in late pregnancy by the fetal portion of the placenta. You will recall from Chapter 14 that CRH is also secreted by the hypothalamus. Although the factor(s) that initiate parturition remain to be clarified, we know that several physiological changes occur in preparation for birth. Estrogen, present at ever-increasing levels (see Figure 15.12), prepares the myometrium for contractile activity in two important ways. First, estrogen stimulates the myometrial smooth muscle cells to synthesize *connexins*, the protein molecules that join together to make up *gap junctions* (see Chapter 12). The gap junctions allow electrical activity generated in the smooth muscle to spread from cell to cell, so that the entire myometrium produces highly coordinated contractions. Second, estrogen stimulates the myometrial smooth muscle cells to produce receptors for the hormone **oxytocin**, secreted by the posterior pituitary gland. Oxytocin stimulates the smooth muscle cells to produce and secrete prostaglandins. The smooth muscle cells are capable of generating contractions autonomously, and, in addition, both oxytocin and prostaglandins strongly stimulate contractile activity.

When parturition is imminent, forceful contractions spread over the body of the uterus toward the *cervix*, the uterine muscle tissue that opens into the vagina. The contractions force the fetus against the cervix, which stimulates mechanoreceptors there. These conditions set up the positive feedback loop shown in Figure 15.13: The mechanoreceptors in the cervix send action potentials to the hypothalamus, where they stimulate neurosecretory cells to secrete oxytocin. Having recently incorporated oxytocin receptors into their cell membranes, the myometrial smooth muscle cells respond with contractions. With increased contractions, the fetus stimulates the cervical mechanoreceptors even more, and more oxytocin is secreted. As myometrial contractions increase, the cervix dilates (widens) to permit passage of the fetus. These events reach a crescendo with delivery. Maternal abdominal contractions ("straining" in domestic animals and "pushes" in humans) help move the fetus through the cervix and vagina. Once the fetus is born, mechanical stimuli cease, and oxytocin secretion decreases. The placenta (afterbirth) is delivered shortly afterward, and the mother's blood levels of estrogen and progesterone plummet to pre-pregnancy levels.

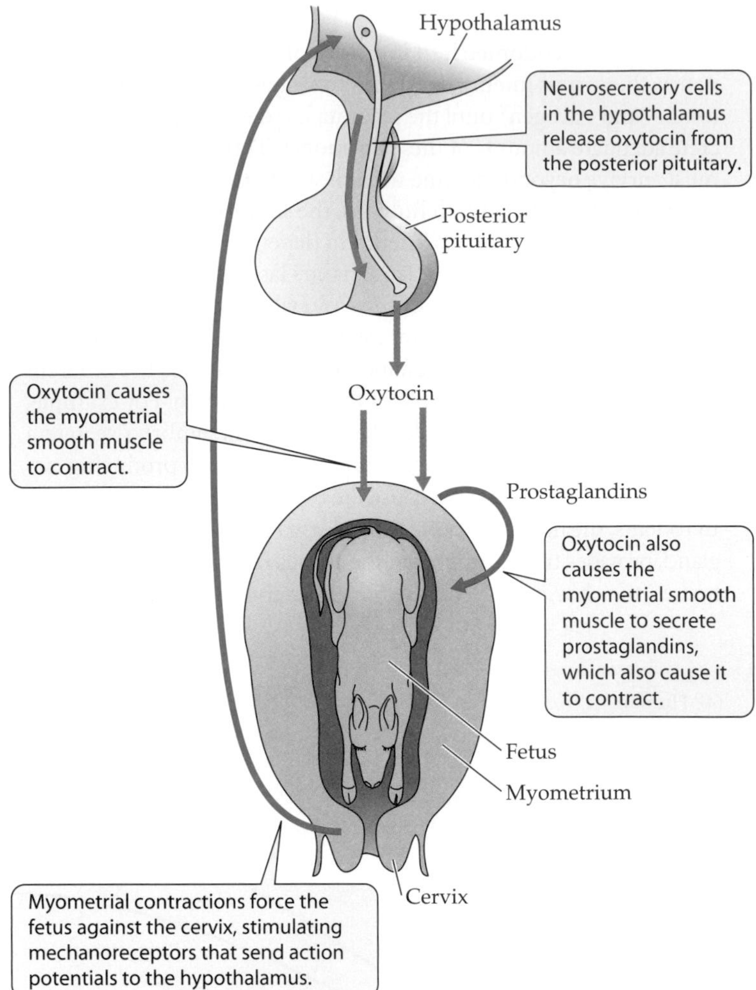

Figure 15.13 Parturition Oxytocin, prostaglandins, and mechanical stimuli at the cervix participate in a positive feedback loop that results in parturition.

Summary: Fertilization, Pregnancy, and Birth

- Fertilization occurs in the oviduct when a secondary oocyte is present and sperm have undergone capacitation. Upon fusion of the two gametes' cell membranes, the oocyte completes meiosis and undergoes the cortical reaction, which achieves a block to polyspermy.

- The zygote divides mitotically. Successive divisions produce a blastocyst, which implants in the endometrium. The corpus luteum continues to function if implantation takes place.
- Trophoblast cells of the blastocyst form the chorion, the embryonic part of the placenta. The inner cell mass of the blastocyst develops into the embryo and extraembryonic membranes. The embryo establishes three primary germ layers that develop into organs.
- The embryonic trophoblast and maternal endometrium form the placenta, which supports the growing embryo/fetus and secretes hormones.
- Birth is accomplished by coordinated contractions of the uterine myometrium and facilitated by a positive feedback loop that causes contractions to become more and more powerful.

Lactation

All female mammals have mammary glands that provide milk to their newborn young. The process of secreting and providing milk to the young is referred to as **lactation.** Milk is an essential source of food until the time of weaning for most newborns. Shortly after birth, before the mammary glands in many species begin to secrete milk, they secrete *colostrums*, which contain antibodies and vitamins.

Despite wide variations in size and placement,[5] all mammary glands have the same basic structure (Figure 15.14*a*). They consist of hollow glandular structures called **alveoli** connected to milk **ducts.** Each spherical alveolus is made up of a single layer of specialized secretory epithelial cells surrounding a central lumen. The lumen is continuous with a small duct, which feeds into a branched system of ducts draining toward the surface of the body. In some species of mammals, such as the cow, the ducts may open into *cisterns* or *sinuses* that allow storage of milk within the gland.

In a lactating female, the **alveolar epithelial cells** synthesize milk and secrete it into the lumen of the alveolus (Figure 15.14*b*). Contractile **myoepithelial cells** lie over the basal (outside) surface of the alveolar epithelial cells. When they contract, they squeeze the spherical alveolus and force milk from the lumen into the ducts. A network of capillaries surrounds each alveolus. The capillaries provide not only the components required for manufacturing milk, but also hormones that control the synthesis and release of milk.

Although all mammals have mammary glands that secrete milk, only marsupials and eutherians have **teats,** or nipples, from which the young can suckle. The monotremes have mammary glands, but their milk seeps through pores in the skin and is licked up by the newborn. Because marsupials do not form true placentas, their embryos develop in the uterus for short gestation periods. At a very early stage of development, the embryo migrates from the uterus to a teat, where it clamps on to obtain nourishment and continue development. The teats of marsupials are usually enclosed in a pouch. Kangaroos and wallabies are able to nurse two young at different stages of development simultaneously. A juvenile joey that has left the pouch comes back occasionally to drink from one teat, and a recently born joey stays firmly attached to a different teat and continues its embryonic development.

[5]For example, humans have two mammary glands on the ventral thorax, cows have four on the lower abdomen, and sows have 5 to 7 pairs lined up along the ventral thorax and abdomen.

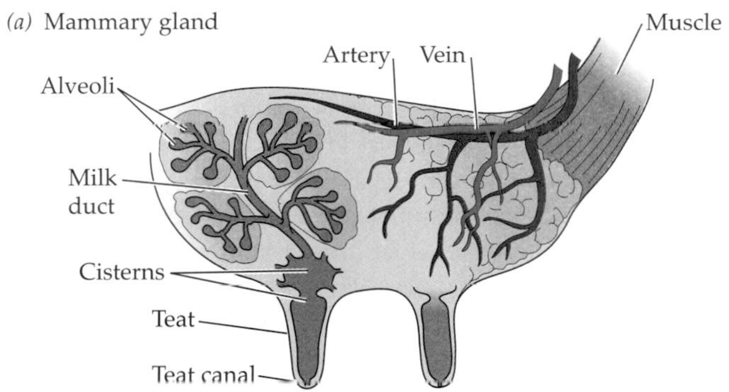

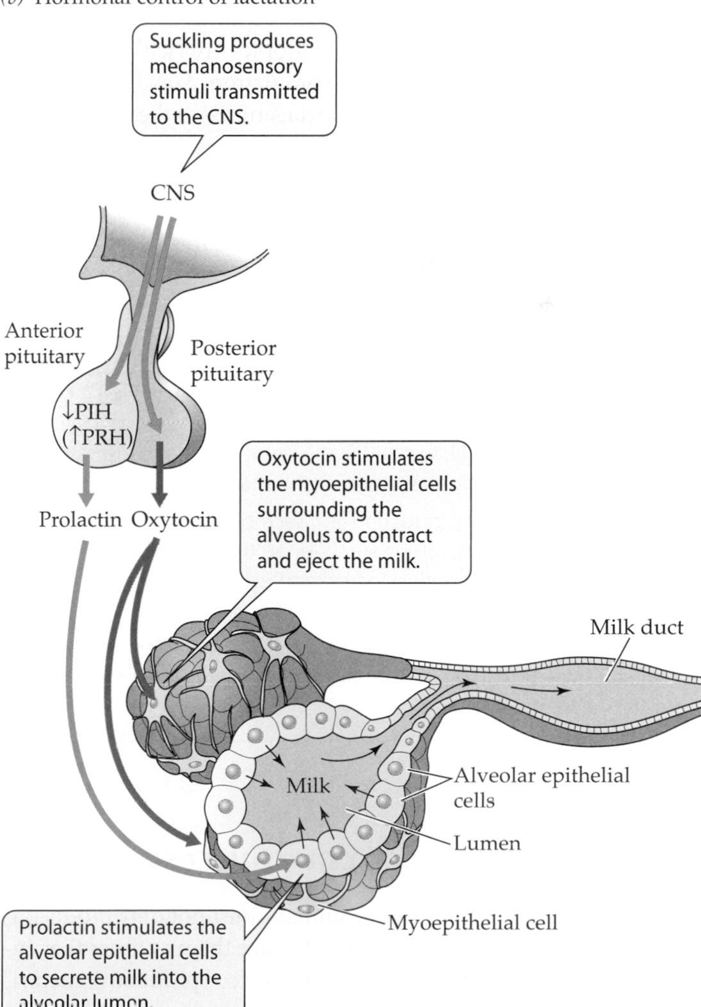

Figure 15.14 The mechanisms of lactation *(a)* The alveoli and milk ducts of the cow udder are typical of other mammary glands. *(b)* Suckling stimulates the secretion of prolactin and oxytocin.

During pregnancy, under the influence of estrogen, progesterone, and prolactin, the mammary glands become capable of producing milk and making it available to the newborn. Two main hormones control lactation (see Figure 15.14*b*). Prolactin from the anterior pituitary gland stimulates *milk production and secretion* by the alveolar epithelial cells, and oxytocin from the posterior pituitary gland stimulates *milk ejection*, also called "milk let-down."

In males and nonlactating females, prolactin secretion is maintained at low levels. A prolactin-inhibiting hormone (PIH), which is known to be **dopamine**, inhibits secretion of prolactin. Dopamine is secreted from neuroendocrine cells in the hypothalamus and carried to the anterior pituitary gland in the hypothalamo–hypophyseal portal system. During late pregnancy and lactation, dopamine secretion *decreases* and thereby permits secretion of prolactin. A prolactin-releasing hormone (PRH) may also influence the secretion of prolactin, but its identity is not yet known. Secreted prolactin is transported in the blood from the anterior pituitary gland to the mammary tissue, where it stimulates the alveolar cells to produce milk and secrete it into the alveolar lumen. However, the offspring cannot obtain milk by simply exerting negative pressure at the nipple. For milk to be ejected, the myoepithelial cells surrounding the alveoli must contract and force the milk out of the alveoli into the ducts leading to the nipple. Oxytocin from the posterior pituitary gland stimulates this process.

Suckling by the infant is a powerful stimulus for the secretion of both prolactin and oxytocin. The nipples are richly innervated and very sensitive to mechanical stimulation. Suckling provides mechanosensory information that is transmitted to the central nervous system, where it is integrated to determine the rate of secretion of prolactin and oxytocin. These neural signals affect the release of dopamine (and possibly a PRH) from the hypothalamus to increase prolactin secretion, and they stimulate hypothalamic neuroendocrine cells to secrete oxytocin. Thus, through a neuroendocrine pathway, the suckling infant ensures its own nourishment by stimulating both milk production and milk ejection with each bout of nursing.

Summary: Lactation

- All mammals produce milk to nourish their newborn offspring. Milk is produced by alveolar epithelial cells in the mammary gland.
- Prolactin stimulates milk production and secretion by alveolar epithelial cells. Suckling generates neural signals that decrease the release of dopamine (and possibly increase a PRH) to produce a surge in prolactin secretion.
- Oxytocin, also secreted in response to suckling, stimulates the contraction of myoepithelial cells that cause milk ejection from the mammary alveoli.

Maximizing Reproductive Success

To enhance reproductive success, animals must not only produce as many offspring as possible, but also produce them at times that ensure the greatest likelihood of their survival to reproductive maturity. These goals are necessarily constrained by environmental and developmental factors. For example, mammals living at northern latitudes, or in tropical regions that have seasons of heavy rainfall, must time the birth of their young so that they start to become independent during the least harsh times of the year. Large mammals that require long pregnancies must pace their breeding to accommodate time for development and suckling of their young. Elephants, for example, have gestation periods of 22 months and nurse their young for 3 years; the time between births of newborns ranges from 4 to 9 years. Hormonal mechanisms are required in all cases to control development of follicles within the ovary and prepare the endometrial lining of the uterus for implantation. But many other factors play important roles in maximizing reproductive success. We consider just a few examples here.

In many mammalian species, the time required for embryonic/fetal development is shorter than the interval between the best time to mate and the best time to give birth. These species have evolved ways to delay birth until a time that is optimal for survival of the young. Three main mechanisms are used: *sperm storage*, *delayed embryonic development*, and *delayed implantation* (also called *embryonic diapause*). Several species of bats in northern temperate zones store sperm or slow development of the embryo after it has implanted in the endometrium. Some female bats store sperm within their uteri for up to 6 months. By increasing the interval between mating and fertilization, they delay the time of birth. Slowing embryonic development also delays birth and, in bats, is thought to be a response to low temperatures.

Mammals as diverse as white-footed mice (*Peromyscus leucopus*), crabeater seals (*Lobodon carcinophagus*), and tammar wallabies (*Macropus eugenii*), all delay implantation of the blastocyst to delay birth. In small polyestrous animals like mice, delaying implantation (typically for 4 to 10 days) increases the spacing between litters and presumably ensures adequate maternal care of successive litters. In the monestrous crabeater seal, both mating and birthing are most successful in the spring. However, gestation requires only 7 months. Keeping the blastocyst in a state of diapause for 5 months after conception allows the seal to mate and give birth at the optimal times.

The marsupial tammar wallaby maintains young at two different stages. One embryo develops in the pouch, while a second stays in the uterus for 11 months in an arrested blastocyst stage (embryonic diapause). Gestation in these animals lasts only 26 or 27 days, but it is followed by a prolonged period in the pouch of lactation and further development. At its birth, a newly born embryonic joey crawls from the vagina into the mother's pouch and clamps onto a teat, where it suckles and continues to develop over several months. A few hours after giving birth, the mother goes through a *postpartum estrus*. She mates; ovulation and fertilization take place about 48 hours later. The zygote develops into a 100-cell blastocyst in about 7 days. Development is arrested at this stage, however, and the blastocyst will remain in diapause for 11 months. Diapause is maintained by lactational and seasonal factors. In the first several months, the joey currently in the pouch, suckling on a teat, provides mechanosensory stimulation that causes secretion of prolactin from the anterior pituitary gland. Prolactin both stimulates milk production and inhibits the corpus luteum. Without progesterone from the corpus luteum, the blastocyst does not develop. As lactational inhibition decreases, photoperiodic information (by way of the suprachiasmatic nucleus and pineal secretion of melatonin; see Chapter 10) continues to inhibit the corpus luteum, so that the blastocyst remains in diapause. When days begin to shorten, this change in photoperiod leads to reactivation of the blastocyst. Thus a developing tammar wallaby spends a year in the uterus—11 months as a blastocyst in diapause and about 1 month in gestation. Following birth, it moves to the pouch where it continues to grow and develop.

We have seen that the cycles of many mammals allow them to bear young either annually or at multiple times during a single year. However, large, long-lived mammals are known to nurse their young for long periods and space the births of their offspring over several years. Ovulation is often suppressed during lactation in response to high levels of prolactin. In human populations that do not use contraceptives, breast-feeding appears to play an important role in determining spacing between births. Studies of prolactin levels in both !Kung and American mothers show that frequent nursing episodes maintain high blood levels of prolactin, which are associated with *anovulation* (absence of ovulation) and *amenorrhea* (absence of menstruation). Finally, we should note that nutrition and the general health of any animal are important contributors to reproductive success. As we saw in Chapter 14, long-term stressors (such as malnutrition) halt the secretion of gonadotropins, and therefore the development of follicles.

Summary: Maximizing Reproductive Success

- Reproductive success involves not only producing as many offspring as possible, but also ensuring that they will survive and reach reproductive maturity.
- To be able to mate and give birth at optimal times, some mammalian species delay birth by processes that include sperm storage, slowing embryonic development, and delaying implantation (embryonic diapause).
- Lactation in some long-lived mammals inhibits folliculogenesis and results in the spacing of births over long intervals.

Study Questions

1. Estimate the age of a primary oocyte within a dominant follicle in the ovary of a woman on her eighteenth birthday.
2. Compare and contrast estrous cycles and menstrual cycles.
3. Only about a thousand giant pandas survive in the wild in China's bamboo forests, and about a hundred are in zoos worldwide. To prevent their extinction, breeding programs have been established in a few zoos. A female giant panda is fertile only once each year. When zoo scientists analyze her urine to determine the time of ovulation, what hormone(s) do they look for? Explain your reasoning.
4. Using your knowledge from Chapters 2 and 14, propose the cellular mechanisms by which estrogen stimulates cells of the endometrium to produce receptors for progesterone. Where does estrogen act? What cellular processes does it influence?
5. Predict the activity of GnRH-secreting neuroendocrine cells in the hypothalamus of a mare during the month of December.
6. Define and describe the strategies used by mammals to mate and give birth at optimal times of the year while also accommodating the time required for gestation.
7. Knowing what you do about the feedback effects of testosterone on the anterior pituitary gland, explain the reason that male athletes taking anabolic steroids become temporarily sterile.
8. Prolactin and oxytocin are both involved in lactation. Define, describe, and explain the functional significance of each of these hormones.
9. List in correct sequence the events that occur during the process of fertilization in mammals.
10. Define and describe sexual reproduction and asexual reproduction. Consider the advantages and disadvantages of each of these methods of reproducing.

Key References

Allen, W. R. 2001. Fetomaternal interactions and influences during equine pregnancy. *Reproduction* 121: 513–527. A review of pregnancy in mares, with special consideration of features unique to horses.

Barnes, R. S. K., P. Calow, P. J. W. Olive, D. W. Golding, and J. I. Spicer. 2001. *The Invertebrates: A Synthesis*, 3rd edition. Blackwell Science Ltd., Oxford. A combination of physiology and systematic descriptions of invertebrates.

Dukes, H. H., M. J. Swenson (ed.), and W. O. Reece (ed.). 1993. *Dukes' Physiology of Domestic Animals*, 11th edition. Comstock Publishing Associates, Ithaca, NY. A classic reference for functions of domestic animals, including reproductive processes.

Eppig, J. J., K. Wigglesworth, and F. L. Pendola. 2002. The mammalian oocyte orchestrates the rate of ovarian follicular development. *Proc. Natl. Acad. Sci. U.S.A.* 99: 2890–2894. A study of the nature of communication between the oocyte and granulosa cells in the ovarian follicle.

Gilbert, S.F. 2003. *Developmental Biology*, 7th ed. Sinauer, Sunderland, MA. A text for undergraduates that includes concepts in animal development, experimental techniques, and application of developmental biology in medical, ecological, and evolutionary contexts.

Judson, O. 2002. *Dr. Tatiana's Sex Advice To All Creation*. Metropolitan Books, Henry Holt and Company, New York. An amusing popular text accompanied by academic endnotes and extensive bibliography.

Lombardi, J. 1998. *Comparative Vertebrate Reproduction*. Kluwer Academic Publishers, Boston. Based on a course in reproductive biology given by the author at the University of North Carolina at Greensboro.

Niswender, G. D., J. L. Juengel, P. J. Silva, M. K. Rollyson, and E. W. McIntush. 2000. Mechanisms controlling the function and life span of the corpus luteum. *Physiol. Rev.* 80: 1–29. Review of mechanisms that control the corpus luteum in different animals.

Renfree, M. B., and G. Shaw. 2000. Diapause. *Annu. Rev. Physiol.* 62 : 353–375. A review of delayed implantation [diapause], a process used by many orders of mammals to time the birth of offspring to match environmental conditions that ensure survival of the young.

Suter, K. J., W. J. Song, T. L. Sampson, J.-P. Wuarin, J. T. Saunders, F. E. Dudek, and S. M. Moenter. 2000. Genetic targeting of green fluorescent protein to gonadotropin-releasing hormone neurons: Characterization of whole-cell electrophysiological properties and morphology. *Endocrinology* 141: 412–419. A method for marking and studying hypothalamic GnRH neuroendocrine cells, which are central to reproduction.

Selected References

Bahat, A., I. Tur-Kaspa, A. Gakamsky, L. C. Giojalas, H. Breitbart, and M. Eisenbach. 2003. Thermotaxis of mammalian sperm cells: A potential navigation mechanism in the female genital tract. *Nature Med.* 9: 149–150.

Bell, G. 1982. *The Masterpiece of Nature: The Evolution and Genetics of Sexuality*. University of California Press, Berkeley.

Berenbaum, S. A. 1999. Effects of early androgens on sex-typed activities and interests in adolescents with congenital adrenal hyperplasia. *Hormones Behav.* 35: 102–110.

Cooke, B. M., G. Tabibnia, and S. M. Breedlove. 1999. A brain sexual dimorphism controlled by adult circulating androgens. *Proc. Natl. Acad. Sci. U.S.A.* 96: 7538–7540.

Dawley, R. M. 1989. An introduction to unisexual vertebrates. In R. Dawley and J. P. Bogart (eds.), *Evolution and Ecology of Unisexual Vertebrates*, p. 1-18. New York State Museum, Albany.

Evans, J. P., and H. M. Florman. 2002. The state of the union: The cell biology of fertilization. *Nature Cell Biol.* 4 Suppl.: S57–S63.

Evans, N. 2002. Sex Begins in the Womb. Proceedings: Understanding the Biology of Sex Differences, Scientific Advisory Meeting in Palo Alto, CA, March 1, 2002. Society for Women's Health Research.

Frisch, R. E. 2002. *Female Fertility and the Body Fat Connection*. University of Chicago Press, Chicago.

Gilbert, S. F. 2001. Ecological developmental biology: Developmental biology meets the real world. *Dev. Biol.* 233: 1–12.

Goldstein, I., and the Working Group for the Study of Central Mechanisms in Erectile Dysfunction. 2000. Male sexual circuitry. *Sci. Amer.* 283: 70–75.

Hamilton, W. D. 1980. Sex versus non-sex versus parasite. *Oikos* 35: 282–290.

Herbison, A. E. 1998. Multimodal influence of estrogen upon gonadotropin-releasing hormone neurons. *Endocrine Rev.* 19: 302–330.

Hunt, P. A., and T. J. Hassold. 2002. Sex matters in meiosis. *Science* 296: 2181–2183.

Ikonomidou, C., P. Bittigau, M. J. Ishimaru, D. F. Wozniak, C. Koch, K. Genz, M. T. Price, V. Stefovska, F. Hörster, T.Tenkova, K. Dikranian, and J. W. Olney. 2000. Ethanol-induced apoptotic neurodegeneration and fetal alcohol syndrome. *Science* 287: 1056–1060.

Johnson, J., J. Canning, T. Kaneko, J. K. Pru, and J. L. Tilly. 2004. Germline stem cells and follicular renewal in the postnatal mammalian ovary. *Nature* 428: 145–150.

Judson, O. P., and B. B. Normark. 1996. Ancient asexual scandals. *Trends Ecol. Evol.* 11: 41–46.

LeVay, S. 1991. A difference in hypothalamic structure between heterosexual and homosexual men. *Science* 253: 1034–1037.

Loch-Caruso, R. K., K. A. Criswell, C. M. Grindatti, and K. A. Brant. 2003. Sustained inhibition of rat myometrial gap junctions and contractions by lindane. *Reprod. Biol. Endocrinol.* 1(1): 62. http://www.RBEj.com/content/1/1/62

Maston, G. A., and M. Ruvolo. 2002. Chorionic gonadotropin has a recent origin within primates and an evolutionary history of selection. *Mol. Biol. Evol.* 19: 320–335.

Matzuk, M. M., K. H. Burns, M. M. Viveiros, and J. J. Eppig. 2002. Intercellular communication in the mammalian ovary: Oocytes carry the conversation. *Science* 296: 2178–2180.

McCracken, J. A., E. E. Custer, and J. C. Lamsa. 1999. Luteolysis: A neuroendocrine-mediated event. *Physiol. Rev.* 79: 263–323.

Paria, B. C., J. Reese, S. K. Das, and S. K. Dey. 2002. Deciphering the cross-talk of implantation: Advances and challenges. *Science* 296 : 2185–2188.

Primakoff, P., and D. G. Myles. 2002. Penetration, adhesion, and fusion in mammalian sperm–egg interaction. *Science* 296: 2183–2185.

Ren, D., B. Navarro, G. Perez, A. C. Jackson, S. Hsu, Q. Shi, J. L. Tilly, and D. E. Clapham. 2001. A sperm ion channel required for sperm motility and male fertility. *Nature* 413: 603–609.

Ricci, C. N. 1987. Ecology of bdelloids: How to be successful. *Hydrobiologia* 147: 117–127.

Rommel, S. A., D. A. Pabst, and W. A. McLellan. 1998. Reproductive thermoregulation in marine mammals. *Amer. Sci.* 86: 440–448.

Savage, D. D., M. Becher, A. J. de la Torre, and R. J. Sutherland. 2002. Dose-dependent effects of prenatal ethanol exposure on synaptic plasticity and learning in mature offspring. *Alcohol Clin. Exp. Res.* 26: 1752–1758.

Scholtz, G., A. Braband, L. Tolley, A. Reimann, B. Mittmann, C. Lukhaup, F. Steuerwald, and G. Vogt. 2002. Parthenogenesis in an outsider crayfish. *Nature* 421: 806.

Shaw, G., J. L. Harry, D. J. Whitworth, and M. B. Renfree. 1997. Sexual determination and differentiation in the marsupial *Macropus eugenii*. In N. Saunders and L. Hinds (eds.), *Marsupial Biology: Recent Research, New Perspectives*, pp. 132–141. University of New South Wales Press, Sydney.

Short, R. V. 1998. Difference between a testis and an ovary. *J. Exp. Zool.* 281: 359–361.

Spehr, M., G. Gisselmann, A. Poplawski, J. A. Riffell, C. H. Wetzel, R. K. Zimmer, and H. Hatt. 2003. Identification of a testicular odorant receptor mediating human sperm chemotaxis. *Science* 299: 2054–2058.

Stern, J. M., M. Konner, T. N. Herman, and S. Reichlin. 1986. Nursing behaviour, prolactin and postpartum amenorrhoea during prolonged lactation in American and !Kung mothers. *Clin. Endocrinol.* 25: 247–258.

Strassmann, B. I. 1996. The evolution of endometrial cycles and menstruation. *Q. Rev. Biol.* 71: 181–220.

Swaab, D. F., and M. A. Hofman. 1995. Sexual differentiation of the human hypothalamus in relation to gender and sexual orientation. *Trends Neurosci.* 18: 264–270.

Tooby, J. 1982. Pathogens, polymorphism, and the evolution of sex. *J. Theor. Biol.* 97: 557–576.

Tyndale-Biscoe, H., and M. Renfree. 1987. *Reproductive Physiology of Marsupials*. Cambridge University Press, Cambridge.

Vandenbergh, J. G. 2003. Prenatal hormone exposure and sexual variation. *Amer. Sci.* 91: 218–225.

Wassarman, P. M., L. Jovine, and E. S. Litscher. 2001. A profile of fertilization in mammals. *Nature Cell Biology* 3: E59–E64.

Welch, D. B. M., and M. S. Meselson. 2001. Rates of nucleotide substitution in sexual and anciently asexual rotifers. *Proc. Natl. Acad. Sci. U.S.A.* 98: 6720–6724.

Western, P. S., and A. H. Sinclair. 2001. Sex, genes, and heat: Triggers of diversity. *J. Exp. Zool.* 290: 624–631.

Wilson, J. D., F. W. George, and M. B. Renfree. 1995. The endocrine role in mammalian sexual differentiation. Rec. Prog. Horm. Res. 50: 349–364.

16 Integrating Systems at Work: Animal Navigation

These newly hatched marine turtles crawling toward the sea are beginning an astounding migration during which they will spend up to three decades in the Atlantic Ocean before returning to their birth site to breed. In the Southern Hemisphere, green turtles (*Chelonia midas*) feed off the coast of Brazil and then migrate eastward across more than 2200 km of open ocean to nest at Ascension Island, an isolated speck of land in the middle of the South Atlantic. In the Northern Hemisphere, loggerhead turtles (*Caretta caretta*) hatch on Florida beaches, swim out to the Gulf Stream, and spend years circulating around the North Atlantic before returning to U.S. waters.

Because it may take an individual sea turtle 30 years to reach sexual maturity, researchers have been uncertain whether the turtles were returning to their *natal* beaches. Accumulated evidence suggests that they do. For example, no sea turtle tagged at Ascension Island has ever been found to nest elsewhere. Moreover, analysis of mitochondrial DNA indicates that green turtles nesting at different sites are genetically distinct, a fact suggesting that gene exchange between different colonies is relatively rare. The reproductive success of sea turtles is based on their finely tuned sense of place. As breeding season arrives, sea turtles move unerringly toward a precise destination, irrespective of their individual positions at sea.

Sea turtles are masters of animal navigation. As physiologists use the term, **navigation** is the act of moving on a particular course, or toward a specific destination, by using sensory cues to determine direction and position. As sea turtles demonstrate, navigational abilities facilitate specialized locomotor behaviors such as migration and homing. **Migration** is the periodic movement of an animal from one region to another. Migratory periods may be prompted by seasonal changes, as is often the case with migratory birds. A migratory cycle, however, may extend over an animal's lifetime, as is true of Pacific salmon, which return from the sea to their natal lakes to spawn and die. Whereas migration may involve movements on a global scale, homing is often a more localized behavior. **Homing** is the ability of an animal to find its way repeatedly to a *specific point*, most often its nesting or dwelling place, although some animals, such as honeybees (*Apis mellifera*), use their homing abilities to pinpoint food sources too.

The dividing line between migration and homing is not always clear, and some navigating animals appear to engage both abilities selectively. Homing performance, however, usually declines as the distance from home or a

Newly hatched sea turtles begin a lifetime of navigation

target increases, and an animal's homing abilities may be disrupted completely if it is passively displaced (e.g., blown off course by a storm, or deliberately moved by a researcher) from its path of travel.

It's not surprising that animal navigation has been an object of scientific interest to researchers. Lacking maps, compasses, and other human cultural accoutrements, how do navigating animals gather and interpret cues about their position on the planet? Do they possess physiological (sensory) mechanisms unavailable to humans? Is the ability to navigate mainly a result of learning and memory of past experience? And what is the evolutionary importance for the animal of the ability to navigate?

The Adaptive Significance of Animal Navigation

Physiologists presume that navigational abilities evolved because they represent an adaptive advantage to the animals that possess these skills. Before examining how animals navigate, let's consider *why* it may be advantageous for them to do so.

Navigational abilities promote reproductive success

Most (but hardly all) animals reproduce at a time and place that presents the lowest degree of environmental stress to the parents and their offspring. For example, long-distance migrants such as humpback whales (*Megaptera novaeangliae*) spend the summer feeding in the food-rich polar seas of the Northern Hemisphere. In the winter, however, they migrate to more tropical waters to reproduce, often traveling as much as 3200 km. Surprisingly, humpback whales apparently do not eat during the winter breeding season; only one out of 2000 caught in one study had food in the stomach. The winter migration is *not* to areas of high food productivity, but rather to areas of warm water that engender less thermoregulatory stress, especially for the young.

Many animals exhibit *natal philopatry*, the tendency either to stay in the area of their origin or to return to it to reproduce. In many species of migrating birds, such as the wood thrush (*Hylocichla mustelina*), males return every spring to reestablish the same territories each year. In this case, learned familiarity may contribute to the advantage of philopatry, but not in another case. Anadromous fish (i.e., fish that ascend rivers from the sea to breed), such as Pacific salmon, provide the most dramatic example of how animal navigation skills are employed to accomplish this reproductive aim (Figure 16.1).

(a)

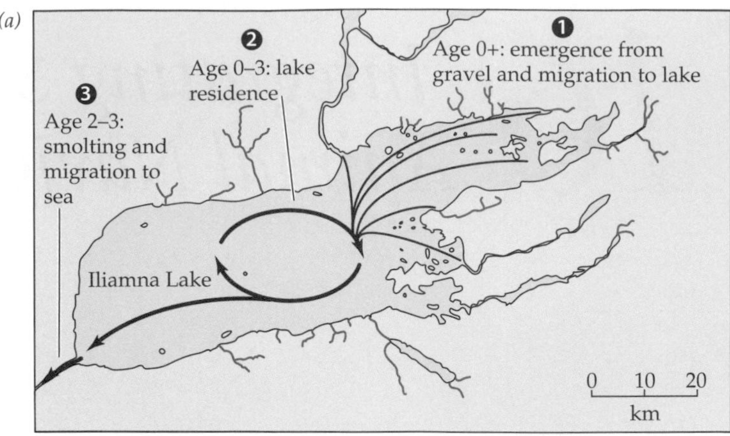

(b)

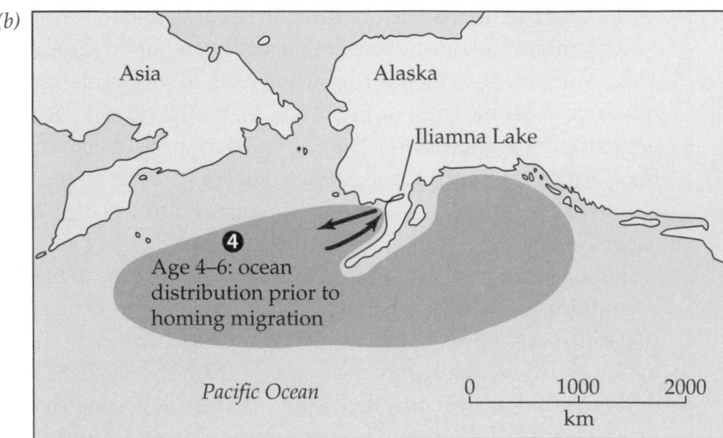

(c)

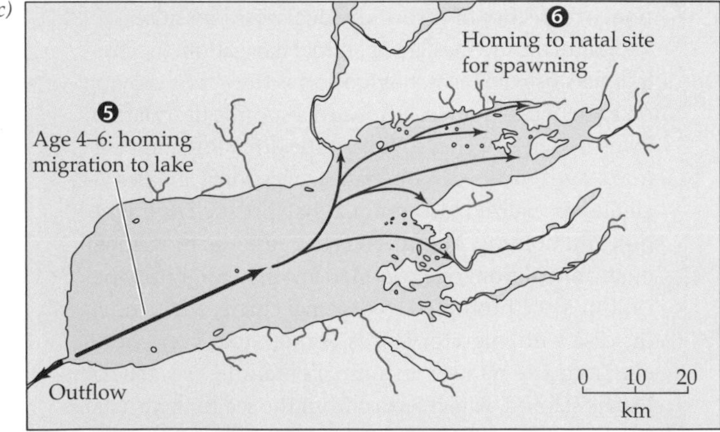

(d)

Figure 16.1 Salmon life-cycle migration and homing The life cycle of sockeye salmon (*Oncorhyncus nerka*) from Iliamna Lake in Alaska. *(a)* Newly hatched salmon emerge from gravel of streams and beaches and spend up to 3 years in the lake before transforming to smolts that migrate to the sea. *(b)* Ocean distribution of maturing salmon prior to their homing migration back to Iliamna Lake. *(c)* Homing migration back to the lake and to their natal site for spawning. *(d)* A male in the reproductive phase (left) has red sides, a deep body, and elongated, hooked jaws. (*a–c* from Dittman and Quinn 1996.)

Several species of salmon (*Oncorhynchus*) hatch in freshwater streams and lakeshores in the North Pacific rim. After spending between 1 and 3 years in downstream waters and lakes as juveniles, they metamorphose into smolts and migrate downstream to the Pacific. The maturing salmon spend a few years in the Alaskan Gyre, a large circulating current that offers plentiful foraging opportunities. The adult salmon of North American species share the gyre with those of Asiatic species. Although the various species feed together, they do not breed in the gyre. Spawning occurs only after adult salmon migrate through the open ocean to their natal river and then, using the imprinted scent of their natal lake or stream, home accurately to it. Here they reproduce once and die.

Presuming that natal philopatry evolved by natural selection, what is its adaptive advantage that has led to such impressive navigation ability in salmon? Probably the most important advantage is that the ability to home to the natal stream returns locally adapted individuals to appropriate environments. Salmon populations are often strongly adapted to local breeding environments, and this adaptation reduces the fitness of strays that disperse instead of returning to the natal site.

On the other hand, *perfect* site fidelity would be expected to be disadvantageous because conditions change from year to year at a breeding habitat, so there should be some dispersal to adjust to these temporal changes. Salmon populations appear to maximize reproductive fitness by natural selection favoring a balance of a high percentage of return to natal sites and a low percentage of dispersal. By using navigational skills to feed globally but spawn parochially, salmon thus maintain distinct populations that are well adapted to their breeding sites.

Navigational abilities facilitate food acquisition

The examples cited of sea turtles, humpback whales, and Pacific salmon all suggest a common conclusion: The best place to feed isn't always the best place to breed. Using their navigational abilities, migratory animals can position themselves favorably at food sources, accumulating chemical energy until they reach breeding readiness. By spending their summers feeding in polar seas, humpback whales take advantage of mineral-rich upwellings from the ocean floor that support an abundant food chain, as well as long summer days that maximize the time these animals can spend foraging.

Animal navigators also employ homing skills to acquire food. For example, several kinds of birds store or cache seeds, hiding them and then retrieving them days or months later. Clark's nutcrackers (*Nucifraga columbiana*) and pinyon jays (*Gymnorhinus cyanocephalus*) dig holes in sandy ground and bury seeds. An individual of either species is able to locate thousands of cache sites and also remember which sites it has already visited and emptied. A single Clark's nutcracker may hide 33,000 seeds in a season at approximately 6600 locations, and it can remember these locations and retrieve the seeds with a degree of success that is nothing less than astonishing.

In another example of using homing skills to acquire food, foraging honeybees may take hours or days to first locate a new source of nectar, but then new workers will appear at the food source within minutes of the first bee's return to the hive. Karl von Frisch (1886–1982) demonstrated that a returning forager performs a "dance" that conveys information to other bees about the direction and distance of the food source (Figure 16.2). The dance consists of looping turns alternating with a straight-line portion in which the signaling bee waggles her abdomen. Typically the dance is performed within the hive, on the vertical surface of a honeycomb, out of view of the sun. The angle between the sun and the food source is translated into an angle between vertical and the waggle component of the dance, so that other workers leaving the hive know the correct direction of the food source. Food location by honeybees certainly requires navigation, but it is somewhat atypical of homing behavior because it involves social cooperation and communication among individuals.

Migrating animals need navigation

For migrating animals, the adaptive value of navigation is difficult to separate from the adaptive value of the migration itself. If migratory movements have evolved by natural selection, they must convey

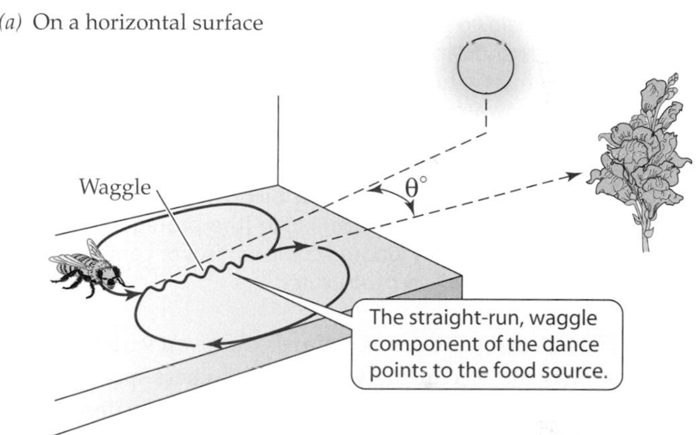

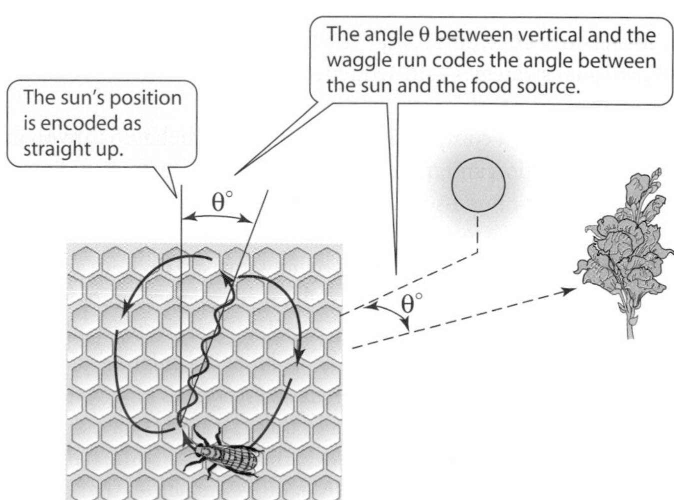

Figure 16.2 The waggle dance of honeybees conveys information about the direction and distance of a food source A returning forager that has discovered a nectar source dances either outside the hive on a horizontal surface *(a)* or inside the hive on a vertical surface *(b)*. Distance is coded by the duration of the waggle run.

adaptive advantages that outweigh their energetic and informational costs. Sometimes the adaptive advantages of migration are obvious; for example, birds that feed on flying insects in temperate summers must go south in the winter (when the insects at temperate latitudes are dormant), unless they hibernate or find a different winter food. Navigation is clearly important in migration, and not only for knowing which way is south. Most seasonal migrants do not just move south in the fall and north in the spring; instead they travel along defined routes to restricted destinations (more restricted for some than others). Therefore, detailed navigation ability of migrants presumably imparts its own selective advantage, one that is separable (at least in principle) from the migratory journey itself.

Navigational Strategies

In the novel *Something about a Soldier*, Mark Harris describes army inductees in World War II riding a train through New Jersey to the unknown destination of their next training camp:

> The men … engaged in quiet conversation … devoting themselves, in the main, to speculation concerning their destination. In this they were handicapped, lacking, as they did, the near presence of streets, avenues, schoolhouses, saloons, subway kiosks marked Uptown and Downtown, and other familiar landmarks by which they ordinarily brought themselves into relationship with the compass. They were now moving (they argued) either south or north or west. They dismissed only the possibility that they were traveling east, into the Atlantic Ocean. Jacob [the protagonist of the novel] observed that they ignored compromises or combinations, and he thought, *the fallacy of absolutes.*

Like Harris's army inductees, animals can use a variety of sensory cues to navigate. Accordingly, when researchers speculate about how animals navigate, and test their hypotheses, they need to avoid the *fallacy of absolutes*. A trip can involve several sequences of direction, and navigating its course can likewise involve combinations of mechanisms. While it is navigating, an animal may use several different sensory cues, and it may use them to implement a variety of navigational strategies—in sequence, in a hierarchy, or in simultaneous combinations.

Navigation is a complex instance of animal behavior, involving sensory integration and motor control, as well as learning and memory. Hence, it is often useful to approach this topic by studying the repertoire of behavioral mechanisms that animals exhibit as they navigate rather than to dive immediately into a search for underlying physiological mechanisms or cellular functions. Biologists who study animal behavior recognize five behavioral strategies that animals use to navigate (Table 16.1): *Trail-following, piloting, path integration, compass navigation, and map-and-compass navigation.*

It is noteworthy that the names of these strategies have analogies with human navigational techniques. But because some navigating species have sensory abilities and integrative capacities that humans lack, these named strategies may not actually encompass *all* the navigational techniques that animals employ. Moreover, terms like *compass* and *map* are metaphors when applied to animals; we cannot know how animals actually experience the compasses and maps that their behavior suggests they possess.

TABLE 16.1 A classification of strategies in animal navigation

Strategy	Description and source of information
Trail following, route learning	Continuous cues about trail, landmarks
Piloting	Landmarks (discontinuous)
Path integration	Integrated direction and distance information
Compass navigation	Celestial or magnetic cues, learned or genetic
Map-and-compass navigation	Compass cues, landmark- or gradient-based map

Source: After Papi 1992.

Trail following is the most rudimentary form of animal navigation

Most people are familiar with following a trail through a forest or across a field in which each spot of trampled earth follows in a continuous series from start to finish. Similarly, **trail following** is a navigational strategy that is accomplished by detection of an interconnected series of local sensory cues (i.e., cues that are immediately proximate to an animal). Although humans are used to thinking about trails in visual terms, for many navigating species of arthropods, such as ants, a trail may consist solely of olfactory cues.

Trail-laying ants have an abdominal scent gland that is used to lay a *trail pheromone*, a chemical signal that marks a path to a food source. Each ant senses the olfactory profile of its path of travel and, in turn, deposits olfactory markers that it and other members of the colony can use to travel between a nest and the food source. (As we will discuss shortly, other kinds of ants employ different navigational strategies.)

Piloting animals follow a discontinuous series of learned cues

A person who knows that a hot-dog stand is located at the base of a tower, and who heads toward the tower to find the stand, is employing piloting. Using this navigational strategy, **piloting** animals follow a discontinuous series of learned landmarks to determine where they are in their path of travel with respect to their destination. (A **landmark** is a discrete, sensed marker of position, usually but not necessarily visual.) The ability to use learned landmarks to navigate means that sensory cues need not be connected in a relatively unbroken series, as is the case with trail following. Moreover, because piloting is learned behavior, most animals that employ this strategy improve their navigational performance with age and experience. Although landmark-based learning such as piloting is considered relatively simple behavior, we will see that it is difficult to distinguish from more behaviorally sophisticated processes, such as map-and-compass navigation, a topic we will cover later (see page 457).

As our example of the hot-dog stand suggests, animals often use piloting as a strategy to find their way to a specific destination, a fact that Nikolaas Tinbergen (1907–1988) proved in a classic field experiment with digger wasps (*Philanthus triangulum*). Tinbergen placed a ring of pinecones around the entrance of a wasp's burrow. When the wasp emerged, she flew around the entrance for about 6 seconds before disappearing to hunt for food. Tinbergen established experimentally that the wasp used the pinecones as a local landmark for navigation by shifting the pinecone ring about 30 cm from the entrance while the wasp was away. On returning, the wasp flew to the center of the pinecone ring and was unable to find the nest. If the pinecone ring was moved farther than 1 or 2 m from the nest, the wasp even failed to locate the ring (and the nest). This latter observation suggests that, for the digger wasp, it is primarily the final moments in navigating to a destination that depend on piloting; the wasp uses other sensory cues to return to the correct neighborhood.

Birds that cache seeds also appear to use local landmarks as a primary mechanism in relocating their food. In one experiment, Clark's nutcrackers in an aviary hid pine seeds in an open arena strewn with landmark objects. After the birds had cached the seeds, the array of objects on one side of the arena was displaced by 20 cm, while the array on the other side was untouched. When the nutcrackers returned to search for their caches, they probed the ground at the "correct" places *with respect to the moved objects*, missing the caches by 20 cm on the manipulated side of the arena and retrieving them accurately on the undisturbed side. In the middle of the arena, the birds probed about 10 cm away from the caches, indicating that they were using information from both displaced and undisplaced landmarks.

Piloting is the navigational strategy employed in many instances of localized homing behavior and probably depends on a variety of familiar sensory cues, such as visual landmarks or odor profiles. Bees and other insects use landmarks to compose visual "snapshots" that are stored in memory in learning familiar paths. Like the digger wasps described previously, animals are most likely to pilot by the use of landmarks near home. For example, homing pigeons fitted with translucent contact lenses that obscure visual landmarks can still return to their home *area* (because they are using a different navigational strategy), but once in the area, they have difficulty finding their pigeon loft.

Path integration is a form of dead reckoning

As with piloting, animals that exhibit **path integration** behavior use this navigational strategy to accomplish homing, but unlike animals engaged in piloting, they do *not* refer to landmarks. In path integration, the navigating animal somehow keeps a running tally of its past directions and distances traveled, and it integrates this information in a manner that allows it to set a direct route back to its starting point at any time. Desert ants (*Cataglyphis fortis*) provide an excellent example of path integration behavior: They forage along tortuous, novel paths but can return from any point directly back to their hidden nesting site once they have located food (Figure 16.3).

Path integration in desert ants is equivalent to the method of dead reckoning often used by early mariners. *Dead reckoning* was a way of determining a ship's position at sea without using landmarks or star positions. The sailors kept a record (the "logbook") of the ship's direction and speed, and accumulated this information into an aggregate direction and distance. Mariners used the position of the sun to ascertain their direction, and they assessed speed by throwing overboard a piece of wood (the "log") attached to a long, knotted line that played out as the ship sailed. The number of knots that played out over a period of time gave the speed and was entered in the logbook.

Ants navigate in a featureless desert rather than a featureless ocean, but their strategy is similar to sailors' technique of dead reckoning. Ants gather directional information by using the sun's position (or polarized light) as a compass. Researchers, however, do not know how ants measure distance; perhaps they integrate proprioceptive information while walking, rather like counting steps for humans.

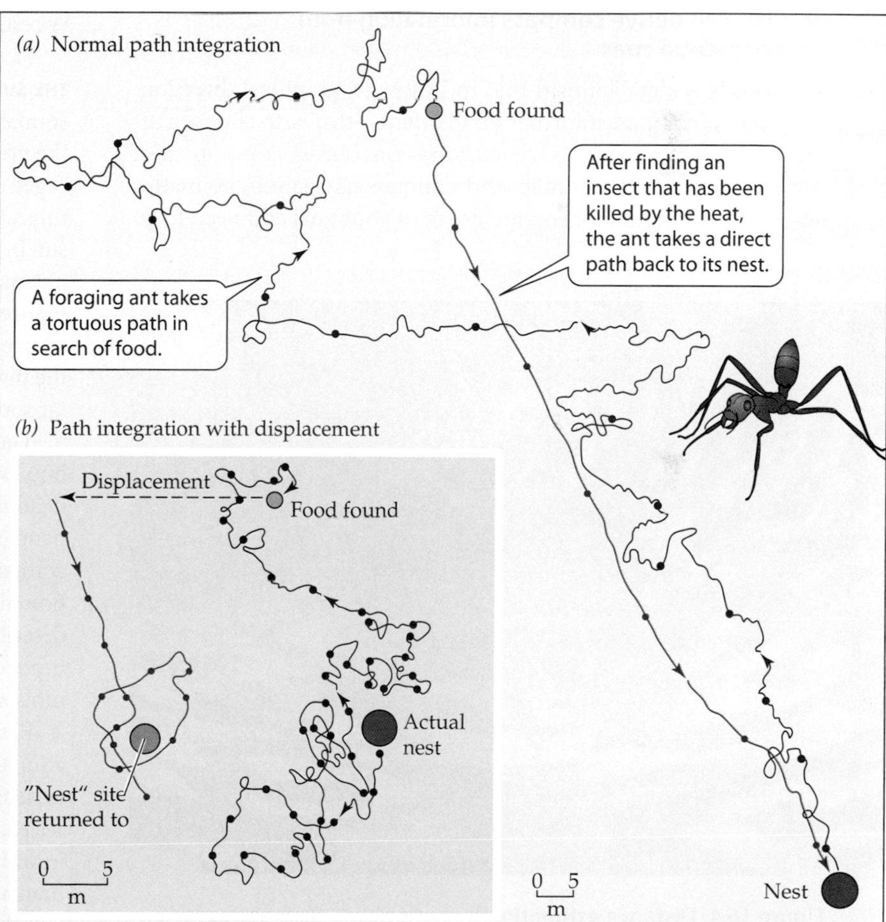

Figure 16.3 Path integration in desert ants *(a)* After finding food via a circuitous path, an ant takes a direct path back to its nest that does not depend on landmarks, but rather on path integration—adding all the direction and distance vectors in the tortuous outward path—and using a sun compass or polarized-sky-light compass to determine the direction back. *(b)* If an ant is artificially displaced when it has found food, it returns to where home would have been without the displacement, using external compass cues and ignoring local landmarks (except at the presumed nest location). Points mark positions at 1-minute intervals. (After Collett and Zeil in Healy 1998.)

Animals use various methods to estimate distance traveled. A honeybee, for example, estimates the distance between its hive and a food source on the basis of **optic flow**, an impression of the speed at which objects sweep through the bee's visual field as it flies. An object's apparent distance determines its contribution to optic flow. Nearby objects sweep by more quickly than distant ones, so they increase optic flow and heighten the sense of distance traveled.

To confirm that honeybees use optic flow, researchers manipulated the apparent distance of objects along their flight path. When bees were made to fly through a narrow tunnel with finely patterned walls (Figure 16.4), their optic flow increased and they overestimated their travel distances. Moreover, they communicated their overestimate to other bees in their waggle dance (see Figure 16.2), so that other bees (not flying through the tunnel) searched for the food source at too great a distance from the hive. This trick does not work on ants, suggesting that ants use methods other than optic flow to measure distance.

Animals can derive compass information from environmental cues

A **compass** is a mechanism that indicates geographical direction. Animals use compass information to **orient**—that is, to choose a direction in which to head. Orientation is a necessary component of compass navigation and map-and-compass navigation; no matter how you determine that you are north of your goal and need to go south, you need a mechanism to tell you which way is south. Animals derive compass information for navigation from a variety of environmental cues, such as the position of the sun and stars, the quality of polarized light, or the lines of force in Earth's magnetic field. It is useful to remember that we are using *compass* here as a metaphorical designation; investigators can only infer the presence of a compass on the basis of an animal's behavior.

Figure 16.4 Distance estimation by optic flow Bees assess the distance they travel by the flow of visual images past them as they move. The closer they are to objects they pass, the greater the objects' apparent motion is, and the greater the bees' estimation of distance. Bees that must fly through a narrow tunnel with densely patterned walls (shown in this composite photograph) consistently overestimate the distance they have traveled, because of the greater optic flow. Moreover, using a waggle dance, they signal their overestimate to other bees in the hive, so that the other bees (not flying in tunnels), although they fly in the correct direction, go too far in search of a food source. (From Srinivasan et al. 2000.)

To study compass-based navigational strategies, researchers usually manipulate the sensory basis of a suspected animal compass while they measure the animal's **initial orientation**, the direction in which the animal orients its body and movements at the start of an actual or intended journey. In some cases this measurement might consist of observing the flight of a bird to vanishing point—the point at which it moves beyond the observer's viewing horizon. In the case of migratory birds, measuring initial orientation may entail quantifying the direction of *migratory restlessness*.[1] Birds in migratory phase will orient in a cage, facing and hopping in the direction of intended migration. Researchers can alter the suspected compass cues and then observe whether the orientation direction of migratory restlessness changes.

THE SUN COMPASS Many animals use the position of the sun as a source of compass information. In an early example demonstrating the presence of a sun compass, Gustav Kramer (b. 1922) showed that caged songbirds in migratory phase changed the orientation of their migratory restlessness when he altered the apparent position of the sun by using mirrors. The other original demonstration of sun-compass navigation was for foraging bees. The bee's waggle dance (see Figure 16.2) conveys information about the direction of a nectar source as an angle relative to the sun's position. Other worker bees in the hive use this sun-compass information to fly directly to the nectar source.

The sun moves across the sky from east to west at about 15° per hour, so animals must know the time of day to determine a compass heading from the sun's position. (Details of the sun's path also depend on latitude and season.) An animal that navigates using a sun compass must integrate solar position and circadian time information in order to orient its path of travel. The caged birds and the bees described earlier could maintain their sense of direction at different times of the day using their circadian clocks to account for the sun's movement.

Researchers can disrupt a suspected sun compass by resetting the animal's circadian clock, usually by imposing an artificial light–dark cycle that is different from natural sunrise and sunset (see Chapter 10, page 277). If an animal experiences a shifted light–dark cycle, one in which the lights come on at noon and go off at midnight (rather than at 6:00 AM and 6:00 PM, respectively), the new light–dark cycle will entrain the animal's circadian clock, causing the animal to misinterpret the sun's position.

Consider the following example: Guided by its normal circadian clock, a pigeon flying south at 9:00 AM will orient 45° to the right of the sun, which will be in the southeast (Figure 16.5*a*). If the pigeon flies south at 3:00 PM, it will orient 45° to the left of the sun, which will be in the southwest. A pigeon whose circadian clock is set ahead

[1] Migratory restlessness is sometimes known by the German term *Zugunruhe*.

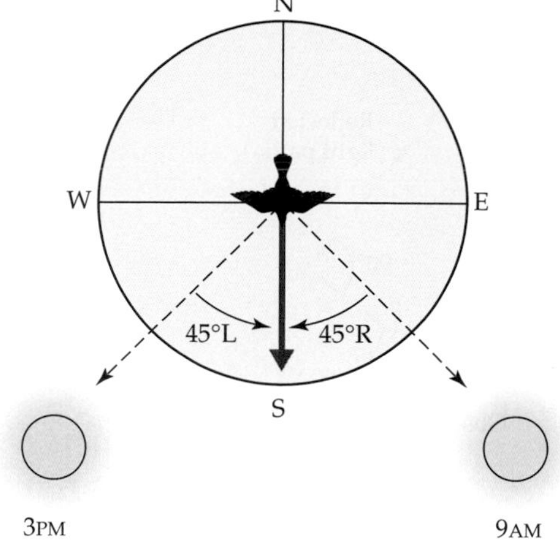

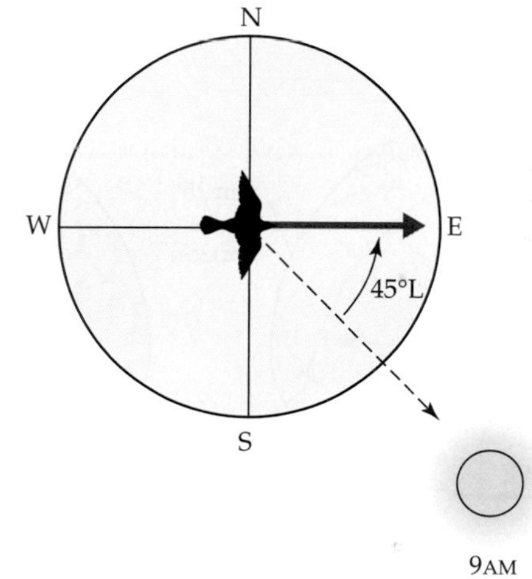

Figure 16.5 Homing pigeons use a sun compass on sunny days The direction in which individual pigeons vanished over the horizon from a release point north of home. *(a)* Control pigeons oriented in the homeward direction (south) when released at any time of day. They used the sun's position and their internal circadian clocks to determine which direction was south. *(b)* Pigeons whose circadian clocks had been shifted 6 hours ahead misinterpreted the sun's position and departed approximately 90° to the left of the homeward direction. If released at 9:00 AM, they thought it was 3:00 PM and departed 45° to the left of the sun's position (appropriate for 3:00 PM).

6 hours, however, will make predictable errors in sun-compass navigation (Figure 16.5*b*). If the pigeon thinks it is 3:00 PM when it is really 9:00 AM, the bird will fly due east, 90° from its intended southerly route because it will fly 45° to the left of the sun. Experiments like this one show that homing pigeons and most diurnal migrating birds use a sun compass as their primary mechanism of navigation.

THE POLARIZED-LIGHT COMPASS In addition to using a sun compass to navigate, many insects and birds appear to use polarized light as a source of compass information. This means of navigating is helpful on partly cloudy days, after sunset, or whenever the position of the sun is obscured from view.

Atmospheric particles polarize sky light. Recall from physics that white light from the sun is unpolarized; that is, it is a bundle of electromagnetic waves, each with an electrical vector (e-vector) that vibrates at right angles to the line of propagation of the light ray. In unpolarized light the e-vectors of different waves are at an infinite number of planes (Figure 16.6*a*). **Polarized light**, however, vibrates in only one plane with respect to its line of propagation. A fraction of the energy in a ray of sunlight will become polarized if it is reflected from a particle. Light reflecting from a surface at 90° from the incoming light ray is fully polarized; light reflecting at other angles is incompletely polarized.

Earth's atmosphere contains an abundance of suitably reflective particles—dust, water droplets, and ice crystals—that can polarize sunlight and provide the photic information that is the basis of an animal's polarized-light compass. Sky light in the vicinity of the sun reaches an observer's eye in a direct path, so it is unpolarized (see Figure 16.6*a*). Light reaching the eye from other parts of the sky, on the other hand, is reflected from atmospheric particles. As the reflection angle (from the sun, to the particles, to the eye) increases up to 90°, the proportion of light reaching the eye becomes increasingly polarized. Sky light, then, is maximally polarized in a band that is 90° away from the sun's position (Figure 16.6*b*).

Human eyes cannot detect useful differences in the polarization of sky light, but the eyes of many navigating insects (and birds) can. A bee, for example, has eyes that are sensitive to polarized ultraviolet light. As long as a patch of blue sky remains in view, a bee can detect the plane of light polarization and use this information to infer the position of the sun. Researchers cannot say with certainty *how* an animal experiences polarized light, but to extract compass information, it must be able to detect the angle of polarization, as well as gradients in the degree of light polarization (see Figure 16.6*b*).

The effect of light polarization detected by the bee may not be altogether different from what people experience when they wear polarized sunglasses. The polarizing filter in the lenses absorbs the horizontally polarized components of sunlight and passes the vertically polarized components. By tilting the head, a wearer of polarized glasses will observe that the sky appears to be noticeably darker in a band that is at right angles to the sun. Light reaching the eye from this part of the sky is maximally polarized, and the polarizing filter passes only the fraction of the light that is vertically polarized. The degree of darkness and the tilt angle of the wearer's head provide sensory cues from which the sun's position can be inferred.

How do the photoreceptor cells of insects respond to polarized light? First, an individual rhodopsin molecule must be differentially sensitive to the plane of light polarization. A rhodopsin molecule maximally absorbs light when the e-vector of the light is parallel to the long axis of the retinal chromophore (see Figure 13.13). Second, the many rhodopsin molecules in a photoreceptor must be aligned similarly, so that they all absorb light at the same plane of polarization.

In arthropod eyes, rhodopsin is localized in rhabdomeres, collections of microvilli arrayed at right angles to the long axis of the photoreceptor (retinular) cell (see Figure 13.23). The rhodopsin molecules are preferentially oriented so that the 11-*cis* retinal chromophore of each is parallel to the long axis of the microvillus. Therefore, light with its e-vector parallel to the long axis of the microvilli is

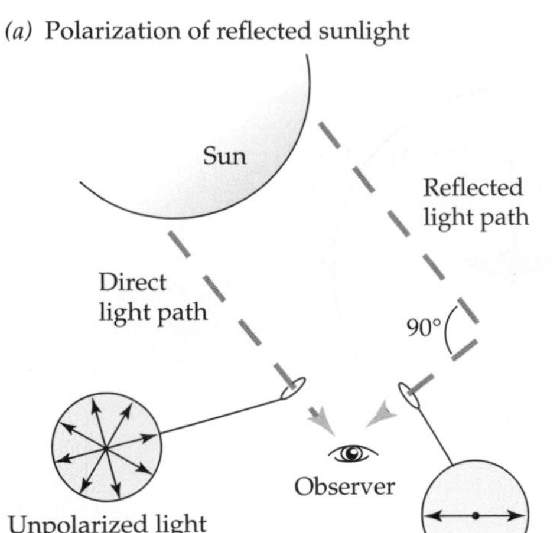

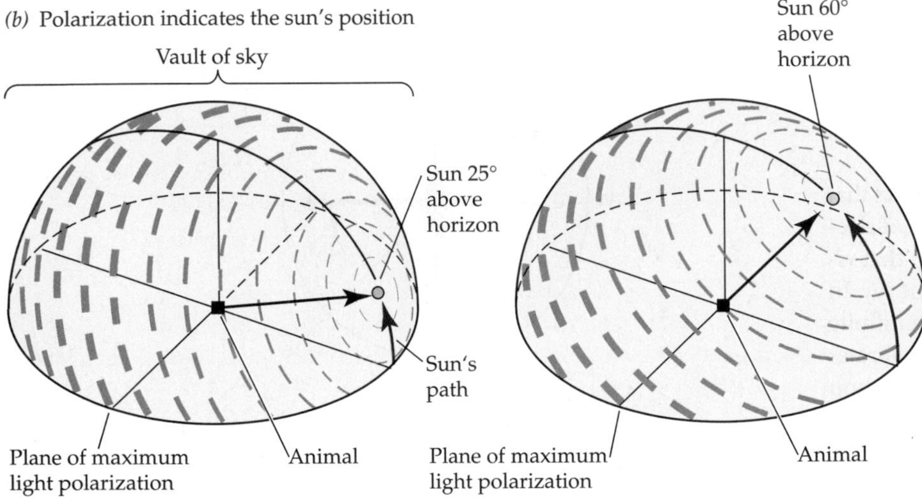

Figure 16.6 Polarization of sky light can aid in determining the sun's position *(a)* The blue sky results from reflected scatter of blue and ultraviolet sunlight by particles in the atmosphere. Sunlight is unpolarized; its electrical vector (e-vector) is at right angles to the direction of propagation of the light wave, but it can be at any direction. The insets show end views looking into the light; for unpolarized light, arrows show e-vectors at all orientations. In contrast, the reflected light is polarized, with its e-vector in only one direction (here shown as horizontal in the end view). *(b)* The pattern of polarized light at two solar positions: 25° (left) and 60° (right) above the horizon. The plane of polarization is at right angles to the plane of light scattering, and the degree of polarization (size of bars) is strongest at 90° from the sun. (*b* from Wehner 1997.)

preferentially absorbed. Different photoreceptor cells have their microvillar arrays aligned at right angles to each other and thus are differentially responsive to light at different planes of polarization.

THE STAR COMPASS Many species of birds that are normally diurnal migrate at night, perhaps to escape predation. They cannot effectively use the sun as a compass (although some nocturnally migrating birds do use the direction of the setting sun and sky-light polarization for initial nighttime orientation). The moon is an unreliable source of nocturnal directional information; it is visible at night only half the time, and its phases (resulting from a different periodicity than that of the sun) make lunar navigation a complex problem. The stars of the night sky provide more-reliable information and are used by nocturnal migrants as a **star compass**. Earth's rotation makes the stars appear to sweep across the sky like the sun, but the region around Polaris (the North Star) provides a stable reference point for north (in the Northern Hemisphere).

Experiments by the behaviorist Stephen Emlen convincingly demonstrated that some nocturnal migrants use star patterns in the night sky as a compass to determine their heading. Emlen raised indigo buntings (*Passerina cyanea*) so that they could see the night sky but not local landmarks. When the birds reached spring migratory phase, he measured their pattern of migratory restlessness. Caged buntings that could see the night sky exhibited the expected north-oriented pattern of restlessness, as did buntings that could see the identical Northern Hemispheric constellations projected in a planetarium (Figure 16.7*a*). When Emlen projected the same constellations so that they faced south in the planetarium, the buntings reversed their pattern of migratory restlessness by 180° to the south as well (Figure 16.7*b*).

To orient successfully, young birds must learn the elements of the night sky during a critical period prior to their first migration. Birds in the Northern Hemisphere learn that the northern sky rotates around Polaris (the North Star), and they learn star patterns within 35° of Polaris. Emlen raised indigo buntings so that they never saw the real night sky, but only artificial night skies in a planetarium. One group saw the normal pattern of stellar rotation around Polaris. The other group saw the normal star pattern, but instead of rotating around Polaris, the constellations rotated around Betelgeuse, a bright star in the constellation Orion. When the birds came into fall migratory condition, Emlen tested their orientation in a planetarium, under a stationary sky. The control birds oriented away from Polaris, that is, away from their stellar north. The experimental birds oriented away from Betelgeuse, indicating that they had learned a star map in which the stationary star (Betelgeuse) was considered north! Apparently rotation of the star pattern is necessary for learning the location of stellar north, but it is not necessary for using the pattern once that pattern has been learned.

MAGNETIC COMPASSES Earth acts like a gigantic bar magnet; it has north and south poles that are connected by magnetic lines of force (Figure 16.8*a*). Earth's magnetic field, in principle, provides a reliable source of directional information: The magnetic poles are close to the geographical poles, and they never change during an animal's lifetime. Because humans are quite unaware of magnetic fields, it is surprising to find that many species of animals can detect Earth's magnetic field and can use it to navigate.

Animals can sense the polarity and dip angle of Earth's magnetic field. Two qualities of Earth's magnetic field provide the navigational cues that animals use as a **magnetic compass**. Magnetic lines of

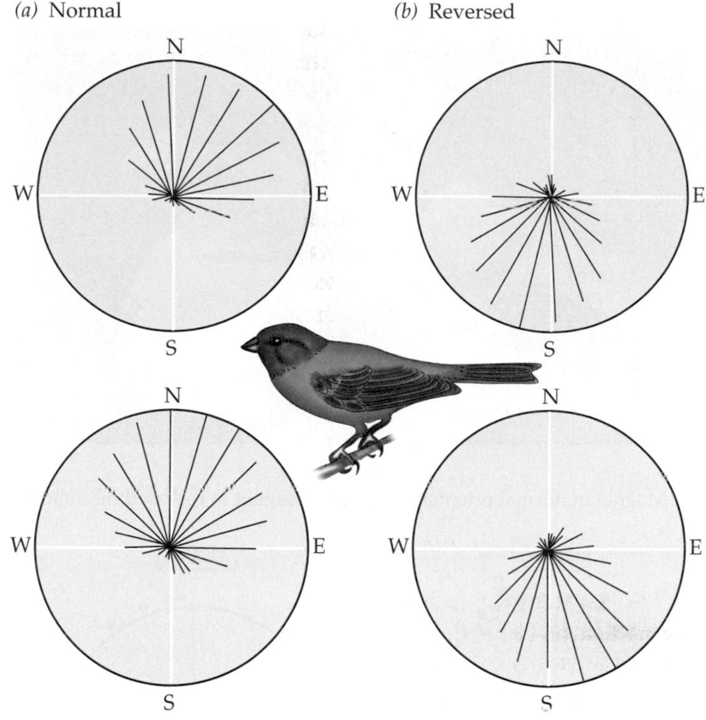

Figure 16.7 Planetarium experiments demonstrate that nocturnally migrating birds use star patterns for orientation Indigo buntings were raised so that they could see the night sky but not local landmarks. The orientation preferences of their migratory restlessness were then tested in a planetarium with either normal star patterns *(a)* or star patterns reversed in direction *(b)*. The results for two buntings shown here reveal that the star pattern is the dominant determinant of orientation direction.

force have a *polar component*: their north and south polarity. They also have an *inclination component*, the "dip angle" of the lines of force relative to Earth's surface (Figure 16.8*b*). Magnetic lines of force exit Earth nearly vertically at the south magnetic pole, and their angle of inclination steadily decreases at lower latitudes until they are parallel to Earth's surface at the magnetic equator. From there, their angle of inclination steadily increases until the lines of force are again nearly vertical at the north magnetic pole.

Accordingly, animals may use either a *magnetic polarity compass* (one sensing directional polarity) or an *inclination compass* (one sensing the magnetic inclination angle) to navigate. Experimental evidence shows that arthropods such as bees sense magnetic polarity, while birds and reptiles sense the inclination angle. Using an inclination compass, the direction "downward into the Earth" is always poleward, as close inspection of Figure 16.8*b* reveals. Thus the downward inclination is toward magnetic north in the Northern Hemisphere but toward magnetic south in the Southern Hemisphere. At the magnetic equator the lines of force are parallel to Earth's surface, and an animal's inclination compass cannot discriminate between north and south. For this reason, birds that obtain navigational cues from an inclination compass may become disoriented as they migrate past the magnetic equator if they don't have a backup source of navigational information.

Researchers infer the presence of a magnetic compass from animal behavior. To detect an animal's use of a magnetic compass, a researcher may experimentally alter the magnetic sensory cues that the animal receives. One way to do this is to place the test animal in a box made of a type of metal (mu metal) that shields it from Earth's magnetic field. Alternatively, researchers may alter the magnetic field in the animal's vicinity by using permanent magnets or by placing tiny Helmholtz coils around the suspected locations of the animal's

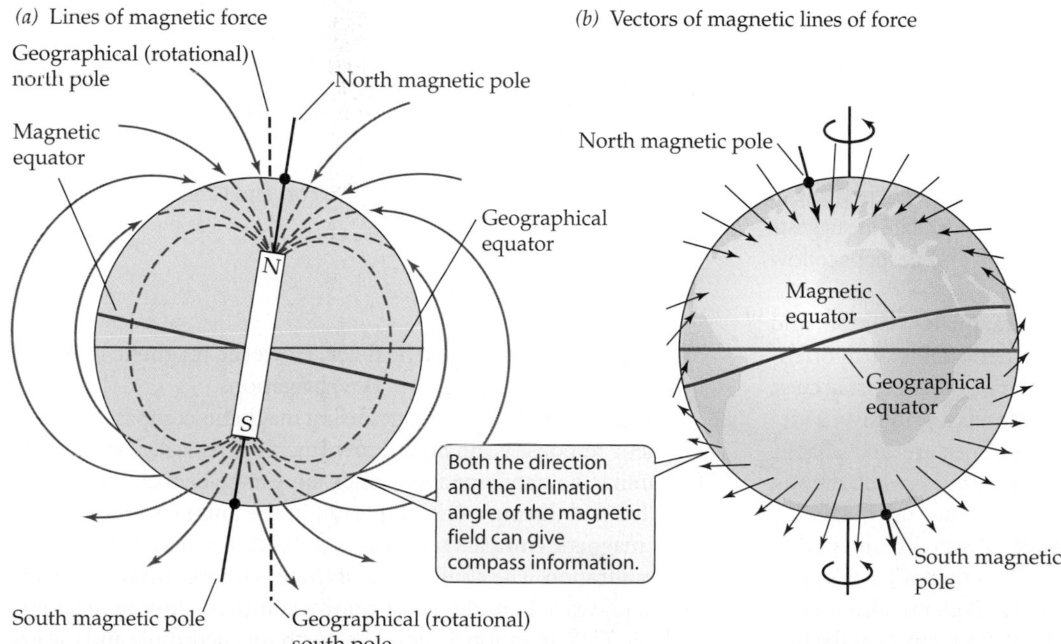

Figure 16.8 Earth's magnetic field can provide compass information *(a)* Lines of magnetic force leave the south magnetic pole vertically, curve around Earth's surface and reenter vertically over the north magnetic pole. *(b)* Vectors of magnetic lines of force at different points on Earth's surface. Arthropods and some vertebrates use a directional compass (arrowheads point toward magnetic north), and many vertebrates use an inclination compass (into the ground = toward the nearer pole). (*a* after Goodenough et al.; *b* after Wiltschko and Wiltschko 1996.)

magnetic sensory apparatus (Figure 16.9*a*). A Helmholtz coil consists of a pair of ring-shaped electromagnets. When an electrical current is passed through the coils, a magnetic field is established in the space between the coils that can be used to nullify or reverse the effect of Earth's magnetic field. By placing a Helmholtz coil to either side of a homing pigeon's head, for example, researchers can control its perception of magnetic fields.

The ability of animals to sense magnetic fields may be a primitive, relatively widespread mechanism, since it is found in some bacteria as well as many animals. Magnetic orientation is often used as a backup for other forms of compass navigation, as exemplified in homing pigeons. Recall from our discussion earlier that pigeons whose circadian clocks had been shifted forward 6 hours flew in the wrong direction when they attempted to navigate south by following their sun compass (see Figure 16.5). However, the clock-shifted birds had no trouble navigating in the desired direction on *cloudy* days, which suggests that they have another navigational mechanism that is not based on solar position.

Evidence demonstrates that this second navigational system in homing pigeons is a magnetic compass. Pigeons with magnets attached to their backs or heads home normally on sunny days but are disoriented on overcast days; control pigeons with brass bars attached to their backs instead of magnets home normally on both sunny and overcast days. What's more, homing pigeons wearing Helmholtz coils (see Figure 16.9*a*) experience similar disruptions of their homing abilities when released on cloudy days. When the inclination angle of the induced field between the coils matches that of Earth's magnetic field, the pigeons orient normally (Figure 16.9*b*). But when the inclination angle of the induced field is reversed, the pigeons fly in the opposite direction from their intended course home (Figure 16.9*c*).

Another line of evidence for animals' use of a magnetic compass comes from studying the effects produced by natural magnetic anomalies (distortions of Earth's magnetic field by geological peculiarities) on a navigating animal's path of travel. On overcast days, pigeons homing in a northeasterly direction toward Boston sometimes become disoriented at a major magnetic anomaly called Iron Mountain, in Rhode Island. The mountain's metallic composition bends the magnetic-field lines in its vicinity, sending the hapless birds off course. In the ocean, magnetic-field lines cluster together to form high-intensity ridges or spread apart to form low-intensity troughs that can be mapped by satellite. Whales and dolphins are more likely to swim ashore and become stranded at locations where troughs in magnetic-field intensity meet the shore. This correlation suggests that whales migrate using a magnetic compass and perhaps follow the path of magnetic troughs.

Other organisms, including salamanders, turtles, salmon, and bacteria, have been shown to orient to magnetic fields. Such orientation is commonly weaker than that to solar or other celestial cues, supporting the idea that magnetic orientation is a primitive, widespread mechanism, often used when other cues are unavailable. Monarch butterflies, for example, use a sun compass when the sun is available, but in the absence of celestial cues, they use magnetic-compass information to orient and navigate. Monarchs (and underwing moths) reverse orientation when the direction of Earth's magnetic field is reversed with Helmholtz coils. Bees are also able to detect magnetic fields, and the orientation of their dances at the hive can be influenced by magnetic fields. However, magnetic information plays little role in ordinary bee navigation.

(a) The experimental setup

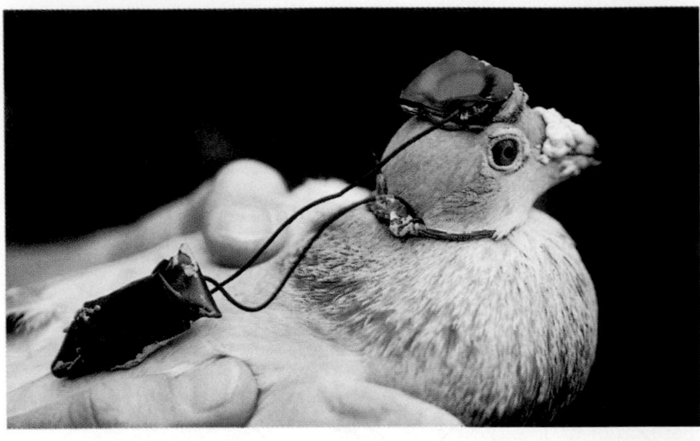

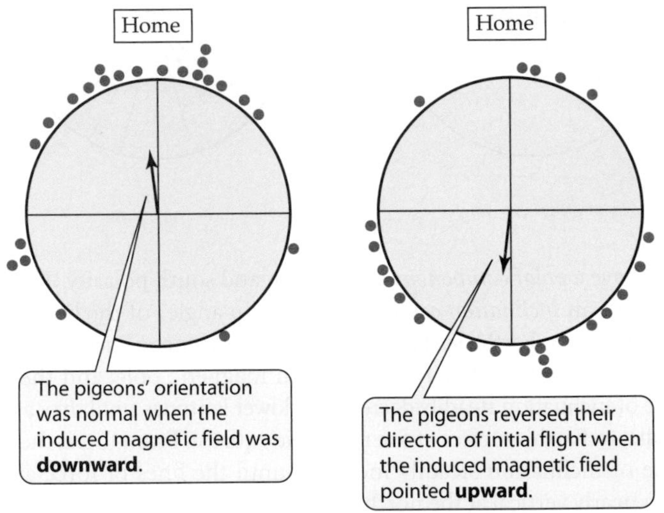

Figure 16.9 Changing the magnetic field changes the orientation of released pigeons *(a)* A small Helmholtz coil is shown attached to a pigeon's head, with a power pack on the pigeon's back. Reversing the direction of electrical current flow through the coil reverses the direction of the magnetic field. *(b, c)* Pigeons with Helmholtz coils were released south of home on overcast days. They interpreted the direction in which magnetic lines dip into Earth as north. Each dot outside the circle represents the direction in which a released pigeon vanished over the horizon. The arrow at the center is the mean vanishing bearing for the group. (After Walcott and Green 1974.)

For any animal that navigates using magnetic-compass cues, the magnetic response is innate and presumably genetically determined. But unlike the star compass—which appears to be fixed once it has been learned during a critical period in the animal's life—an animal's magnetic compass seems to be subject to at least some degree of recalibration. The ability to recalibrate is important because migrants to Arctic latitudes must resolve conflicts between magnetic and celestial navigational cues. The north magnetic pole and the geographical North Pole are not identical. Magnetic-field information points to a different location than celestial-compass information does, and the disparity between the two increases at higher and higher latitudes. For this reason, migrating Arctic birds must pause in their journey at intervals in order to allow their magnetic compasses to realign with celestial-compass cues such as polarized sky light. If prevented from pausing long enough to recalibrate, the birds may become disoriented.

What are the physiological mechanisms by which animals detect magnetic fields? There are three hypotheses. First, photochemical reactions such as photoexcitation of rhodopsin (see Chapter 13, page 358), could produce an excited chemical state decaying at a rate that depends on the strengths of magnetic fields. There is no direct evidence for this model, but it would explain findings that bird and salamander magnetoreception can be light-dependent.

The second hypothesis is that crystals of the ferromagnetic mineral *magnetite* in animals could serve as tiny magnets that would tend to orient like a compass needle. Such crystals have been found in bacteria, honeybees, salmon, pigeons, and elsewhere. These internal magnets may activate magnetoreceptor neurons in some way that remains to be clarified.

A third proposed mechanism of magnetoreception is electromagnetic induction, which may occur in elasmobranchs. As noted in Chapter 13, sharks, skates, and rays have lateral line receptors called *ampullae of Lorenzini*, which appear to function as exquisitely sensitive electroreceptors. Calculations suggest that a shark swimming east or west across magnetic lines of force will induce a tiny current that the ampullae of Lorenzini are sensitive enough to detect. Despite the theoretical calculations, however, there is no direct evidence that elasmobranchs actually use this mechanism to orient to geomagnetic fields.

Some animals appear to possess a map sense

Most animals probably do not use simple compass navigation in migration or homing without also using landmark or map information. In cases of **map-and-compass navigation**, the animal possesses in its brain some sort of representation of its position, and the position of its goal. This is the *map*, and the animal's *compass* provides bearings relative to the map.

For most animals that are thought to have a map sense, we do not know the map's basis, although it might be based on various sensory cues. A map could be based on geomagnetic information, since both the vertical angle and the strength of magnetic fields increase near the magnetic poles. The disorientation of homing pigeons and migrating whales in the vicinity of magnetic anomalies suggests that magnetic information is an important basis of their maps. Solar cues could also provide map information; the height of the sun above the horizon varies with the latitude, and the times of sunrise and sunset vary with longitude. For some migrating creatures, infrasounds (very-low-frequency sound waves—for example, from waves crashing on a distant shore) may serve as the basis of the map. Even olfactory cues are suspected of forming the basis of the map sense, as appears to be the case with pigeons. Whatever its basis, constructing and using a map represents a sophisticated instance of animal learning.

DISPLACEMENT EXPERIMENTS TEST FOR A MAP SENSE Displacement experiments show the difference between path integration and compass navigation, on the one hand, and complex map-and-compass navigation on the other. If a desert ant is experimentally moved while it is feeding, its return path makes no correction for this passive displacement (see Figure 16.3). Therefore, the ant has no map sense that allows it to detect its displacement relative to its goal.

Birds, in contrast, are more likely to correct for a passive displacement, whether experimental or by natural means such as a storm. If a bird migrating southwest purely by compass navigation became displaced, say, 100 km to the southeast, it would continue to go southwest. If the bird were using map-and-compass navigation, it would realize it had been displaced (apparently without using local, familiar landmarks) and would correct its path to the west or northwest.

Figure 16.10 illustrates the distinction between path integration and compass navigation. European starlings migrate from Baltic breeding grounds to wintering grounds in France and England. Mi-

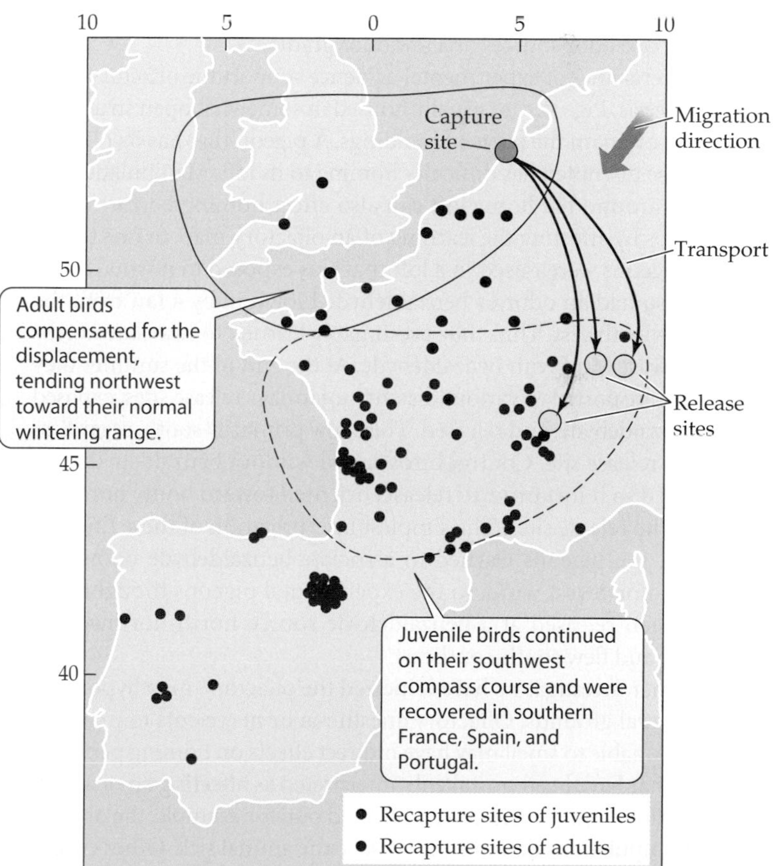

Figure 16.10 Migrating adult and juvenile starlings respond differently to displacement Starlings were captured in the Netherlands while in their autumn migration southwest from northeastern Europe, and they were transported to release sites in Switzerland. The normal wintering range is bounded by the solid line; a dashed line shows a similar area displaced to the same extent as the displacement of the released birds. The adults' compensation for displacement suggests the use of map-and-compass navigation. (After Perdeck 1958.)

grants were captured in the Netherlands, transported to Switzerland, and released. The juvenile starlings continued southwest (by compass navigation) and ended up in Spain, but the older, experienced birds that had developed a map sense could detect the displacement and reorient to a new northwestern path. Map-and-compass navigation is sometimes called *true navigation*, an unfortunate term because it (wrongly) suggests that other methods of an animal's finding its way are not really navigation. Because we find that compass orientation is relatively complex, we may expect the basis of the less-understood map sense to be complex as well.

PIGEONS MAY HAVE AN OLFACTORY MAP Some of the clearest studies of map sense have focused on pigeons, and the dominant hypothesis of map sense in pigeons is the *olfactory-map hypothesis*. Displaced pigeons can determine their position relative to home without using visual or proprioceptive information from the outward journey to the release site. Therefore, investigators conclude that the pigeons must have some sort of map sense that, with the compass information previously discussed, allows them to return from over 700 km away. According to the olfactory-map hypothesis, the pigeon's map sense is based on an odor profile somewhat analogous to the one used by homing salmon. The odor profile could consist of a mosaic of discrete odor sources or a few odor gradients.

Several lines of experimental evidence support the olfactory-map hypothesis. Pigeons are usually housed in somewhat open structures (lofts) often on the roofs of buildings. A pigeon that has its olfactory sense disrupted has difficulty homing to its loft. Manipulations of odors around the home loft can also affect homing performance, perhaps by affecting the learning of an olfactory map. In one example, pigeons were raised in a loft that was exposed to normal odors but also had an odor of benzaldehyde blown in by a fan from the north-northwest. Thus they presumably learned to associate north-northwest wind with benzaldehyde. At the end of the summer they were transported to various distant, unfamiliar release sites, exposed to benzaldehyde, and released. They flew primarily south, regardless of the release site. Control birds raised without benzaldehyde, but exposed to it just prior to release, oriented toward home normally from the release sites. The simplest interpretation of these findings is that the pigeons learned to associate benzaldehyde odor with north-northwest wind, so the experimental pigeons thought they had been released at a benzaldehyde source north-northwest of home, and flew south-southeast.

Other investigators have criticized the olfactory-map hypothesis on several grounds. Olfactory anesthesia or nerve cuts to make pigeons unable to smell may have indirect effects on homing performance that have been mistakenly interpreted as affecting an olfactory map. Some indirect effects can be ruled out; for example, the olfactory manipulations do not simply make the animal sick. Other effects, such as a possible indirect disruption of the pigeons' use of magnetic signals, are harder to rule out. In addition, some experiments have been hard to replicate, perhaps because investigators in different countries raise their birds in different styles of lofts that may affect olfactory-map learning. Moreover, critics argue that it is unlikely that odor cues/trails would be stable and strong enough at distances of hundreds of kilometers. Finally, critics argue that pigeons do not appear to zigzag like moths or lobsters orienting to odor plumes. These criticisms themselves have little evidence to support them, and we can consider the olfactory-map hypothesis to be somewhat supported by evidence but not proven.

Sea turtles exemplify the degree of our understanding of navigation

Sea turtles illustrate what we know and don't know about the behavioral mechanisms of navigational control. Turtles use different cues in navigation at different stages in their life cycle, and some of these cues are much easier to analyze than others. New hatchlings starting off the beach appear to engage in a sequence of orienting mechanisms. Loggerhead sea turtles hatching on Florida beaches initially orient toward a lighter sky, which will be toward water (because the water reflects more sunlight or moonlight and makes the sky lighter). This light orientation and a tendency to move down a slope take the hatchlings to the water, where they swim out continuously for 24 hours. The primary orientation cue for this frenzied outward swim is the waves; the turtles head into the waves, which come (with rare exceptions) from the open ocean. (During the exceptions, young turtles swim into the waves even if that orientation takes them back to shore.)

The young turtles can also detect Earth's magnetic field, and magnetic-compass orientation may replace wave orientation after a number of hours. This navigational sequence is thought to take the turtles out to the Gulf Stream, one of the currents flowing around the North Atlantic Subtropical Gyre (**Figure 16.11**). Adult turtles must use map-and-compass navigation to return to their natal beach to nest (in contrast to juveniles that may only have to reach the gyre). A few tracking studies indicate that the adults navigate by direct, straight-line routes. Recent evidence indicates that even the younger turtles can use magnetic information as a map component that helps keep them in the gyre as they mature.

Experimental replication of the parameters of the magnetic field at different points in their migration route leads to different swimming orientations that match the turtles' swimming direction at that location (see Figure 16.11). For example, caged turtles in an artificial magnetic field that replicates the natural magnetic field off the coast of Spain swim southward, a direction that keeps them in the North Atlantic gyre rather than washing into northern waters off Scandinavia. Similar experiments with salamanders and thrush nightingales also show that experimental replication of the magnetic-field vectors and strengths of a particular location can alter orientation behavior in predictable ways. These recent studies suggest that animals use magnetic information as a map, a finding that may begin to clear up the murkiest area in complex navigation.

Innate and Learned Components of Navigation

We have indicated that many elements of navigation, such as using landmarks, imprinting to a home stream, and constructing a map, are largely learned behaviors. Here we discuss the roles of innate ("wired-in") behavior and of learning in navigation, and we consider how vertebrates learn their spatial surroundings.

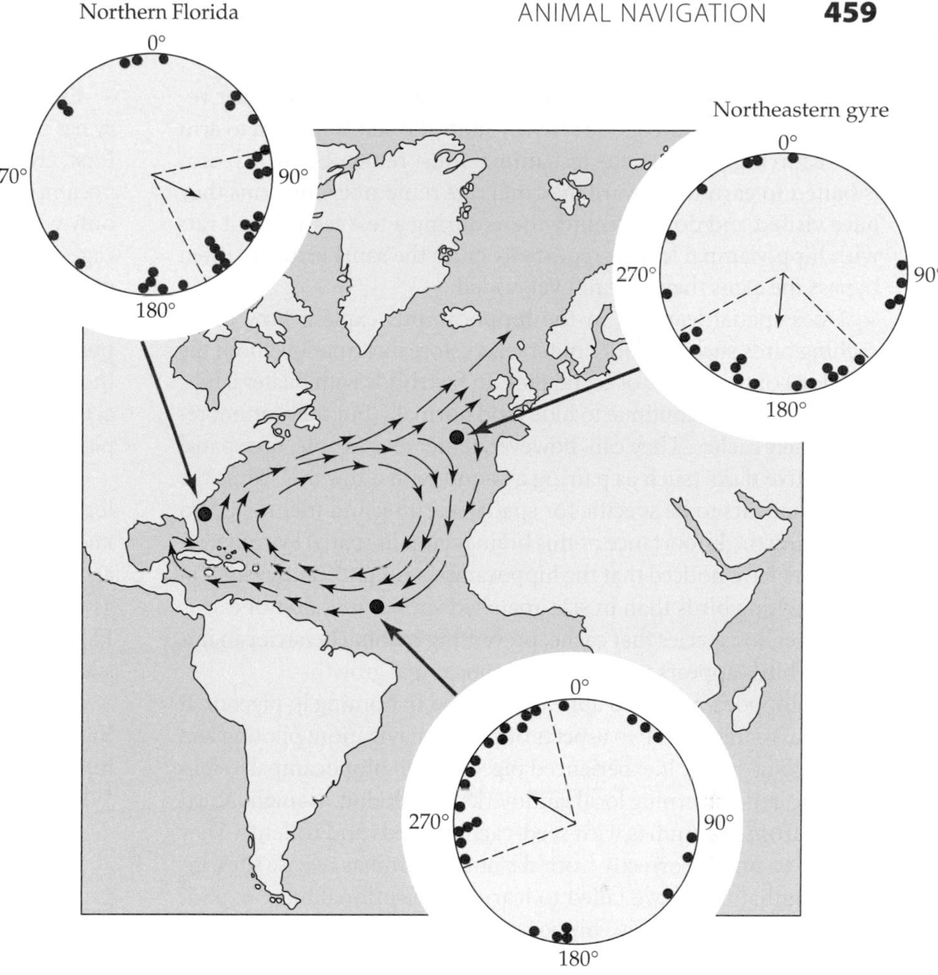

Figure 16.11 Hatchling turtles orient to artificial magnetic fields in ways that suggest a magnetic map Newly hatched loggerhead turtles were placed in artificial magnetic fields characteristic of three locations (red dots on map) along their prospective migratory route. Arrows on the map show major currents of the North Atlantic Subtropical Gyre, around which the turtles swim as they mature. Each dot in the orientation diagrams represents the mean angle of oriented swimming of one hatchling. The arrow in the center of an orientation diagram indicates the mean angle of orientation of each group; arrow length shows the statistical degree of orientation of the group. All three groups showed statistically significant orientation to the artificial magnetic fields. The angle of orientation differed greatly for the three groups tested with artificial magnetic fields corresponding to different locations. Moreover, the orientation at each simulated location was appropriate to keep turtles within the gyre, rather than (for example) straying into the fatally cold water of the North Atlantic and North Sea. Dashed lines represent the 95% confidence interval for the mean angle; data are plotted relative to magnetic north. (From Lohman et al. 2001.)

Some forms of navigation have strong innate aspects

Two examples demonstrate innate components of navigation that do not depend on prior learning. Monarch butterflies (*Danaus plexippus*) summer throughout the United States north of the Gulf states. Those east of the Rocky Mountains migrate south every fall to overwintering sites in forested mountains of central Mexico. In the spring the same individuals migrate north to the U.S Gulf Coast states, where females lay eggs on milkweed and die. A new generation continues the migration to summering grounds in the northern states and produces two or more short-lived generations in a summer. The autumn migrants that return to the same restricted overwintering sites in the following year are thus descendants, three to five generations removed, of ancestors that occupied the sites the previous winter! The navigational performance of the monarch butterfly, therefore, must have a strong innate, genetic component; the butterflies cannot have learned the overwintering location from previous experience.

Some birds also have innate, apparently genetic components of their migratory navigation. European blackcaps (*Sylvia atricapilla*), for example, migrate from Europe to Africa around the Mediterranean Sea—those in western Europe going southwest, via Gibralter and those in eastern Europe going southeast via Syria. When birds from the two sides of the divide are raised in isolation, they orient in cages in the correct direction for their area of origin, despite tbe lack of opportunity to learn their way. Cross-breeding experiments further demonstrate that the difference in orientation between eastern and western strains is genetic.

The hippocampus is a critical brain area for vertebrate spatial learning and memory

As we have discussed, an animal's ability to navigate often involves learning, such as the recognition of landmarks by some homing animals, or the imprinting of home stream odors by salmon. Researchers suspect that learning is a necessary component of many forms of animal navigation, basing their speculation on the observation that older and more experienced animals are usually better navigators than are juveniles of the same species. In most cases, we do not know exactly what animals learn, let alone how they learn the particular qualities or features that improve their navigational performance. What is understood, however, is that spatial learning is crucial for many sorts of navigation, and for vertebrates a major locus of spatial learning is the *hippocampus* of the brain.

By surgically creating hippocampal lesions in test animals and monitoring their performance in mazes, researchers have isolated the regions of the hippocampus that figure in spatial learning. One standard test of spatial learning in laboratory rodents is the *Morris water maze*, which consists of a hidden, submerged escape platform in a pool of turbid water. The test animal must learn the location of the platform from local cues (landmarks) that are situated outside the pool. Mice or rats with hippocampal lesions do not remember the location of the platform, and they continue to swim about randomly, trial after trial. Another means of testing spatial learning is the *radial-arm maze*, which consists of several blind corridors that extend symmetrically from a central chamber. The researcher repeatedly baits the end of one arm (or shifts the bait from arm to arm in a recurring pattern); the test animal must remember which arm is baited to earn the reward. Normal rats remember the arms they have visited and do not reenter them during a test session, but rats with hippocampal lesions repeatedly enter the same arm and often bypass the arms they have not yet visited.

Does spatial learning in the hippocampus explain how food-caching birds such as Clark's nutcrackers store information about the locations of thousands of cache sites? In fact, birds with bilateral hippocampal lesions continue to hide food normally, but they cannot recover their caches. They can, however, remember simple, nonspatial associative tasks (such as pairing a sound with a reward). Thus the deficit appears to be specific for spatial learning and memory. Also suggesting the importance of this brain region in spatial learning, researchers have noticed that the hippocampus is significantly larger in food-caching birds than in size-matched species that do not cache. Moreover, for species that cache, preventing caching behavior in immature birds appears to diminish hippocampal growth.

The hippocampus also appears involved in homing in pigeons. It is critical for *learning* two aspects of pigeon navigation: piloting and navigational maps. Inexperienced pigeons with hippocampal lesions are impaired in learning local landmarks for piloting, as might be expected from the studies with seed-caching birds and rodents. They also fail to orient correctly from distant, unfamiliar release sites, indicating that they have failed to learn a navigational map. Experienced pigeons that receive hippocampal lesions *after* learning a navigational map orient correctly from distant release sites, but they still have difficulty with local, landmark-based navigation. Therefore, the hippocampus appears necessary throughout life for landmark-based navigation but not for the use of a navigational map once that map has been learned. Interestingly, the left hippocampus appears necessary for map learning, but not the right.

Exactly *how* the hippocampus stores and retrieves spatial memory has not been determined, but two interesting findings are relevant. First, chronic electrical recordings from single neurons in the hippocampus of rats show that some neurons generate action potentials only when a rat is in a particular part of a familiar area (such as its cage or a familiar maze). These hippocampal neurons are termed **place cells** because their activity encodes the spatial position of the animal (**Figure 16.12**). The existence of place cells in the hippocampus suggests that this brain area creates some sort of a spatial map of the environment in the brain. The spatial representations of place cell activity are dynamic; as a rat is exposed to new environments, some place cells may change their specification to incorporate new areas.

The second finding is that the hippocampus is a principal site of long-term potentiation (LTP) (see Chapter 12, page 340). Gene knockouts that disrupt LTP in mice correspondingly affect place cell activity (see Figure 16.12*b*) and disrupt spatial learning in tests such as the Morris water maze. Overexpressions of genes that enhance LTP improve learning and memory of familiar objects, spatial tasks such as the Morris water maze, and other memory tasks.

Because of the importance of the hippocampus in spatial learning and memory, some investigators have suggested that the hippocampus somehow constructs or serves as a mental map. Whether or not this is the case, our increasing ability to relate hip-

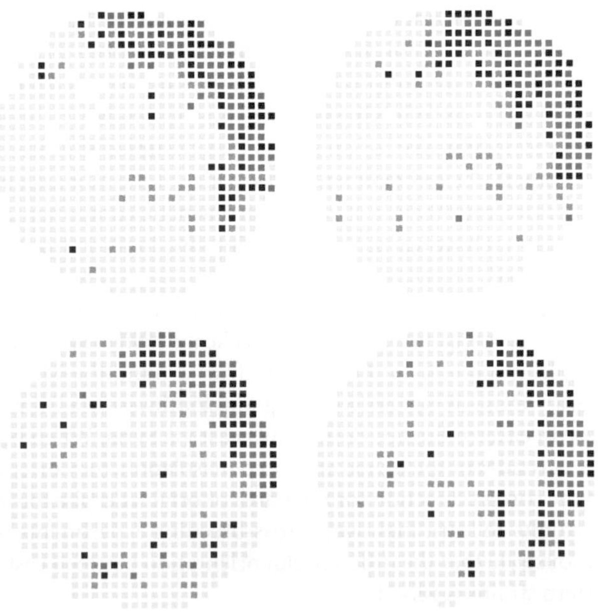

Figure 16.12 Place cells suggest how the hippocampus of the brain plays a role in spatial learning *(a)* A chronically implanted electrode records activity of an individual hippocampal pyramidal neuron while a mouse is moving around an arena. A TV camera records the mouse's position and correlates it with the neuronal activity. *(b)* An individual hippocampal neuron is active when the mouse is in a particular part of the arena. Data from four separate introductions of the mouse into the arena are presented as recorded in *(a)*. The neuron's activity is well correlated with one margin of the arena. Such correlations are learned during exploration of initially unfamiliar environments. (After Kandel, Schwartz, and Jessel 2000.)

pocampal mechanisms to map functions should provide an important link between brain processes and the complex spatial behaviors of navigation.

Recent studies like those of the hippocampus are starting to build a physiological basis of navigation, so the future holds promise of understanding the navigational basis of migration, homing, and other spatial behaviors in terms of their underlying physiology.

Study Questions

1. Explain the effect on sun-compass orientation of a circadian-clock shift to a light schedule that is 4 hours later (e.g., from lights-on at 6:00 AM to lights-on at 10:00 AM).
2. List the advantages and disadvantages to an animal of using information from magnetic, sun, star, and polarized-light compasses.
3. Experienced homing pigeons were thought not to need any information about the outward journey to a release site in order to home successfully. Much evidence for this conclusion came from depriving the birds of visual, magnetic, and vestibular information that could be used for path integration. What other sorts of studies would you want performed before accepting this conclusion?

Key References

Lohmann, K. J., S. D. Cain, S. A. Dodge, and C. M. F. Lohmann. 2001. Regional magnetic fields as navigational markers for sea turtles. *Science* 294: 364–366. A study demonstrating that animals can use magnetic information to serve as a map sense, as well as for compass orientation.

Wehner, R., M. Lehrer, and W. R. Harvey (eds.). 1996. Navigation. *J. Exp. Biol.* 199: 1–261. This multiauthor review volume contains interesting and readable reviews of most of the current issues in animal navigation. It is the best single starting point for further reading on the various questions raised in this chapter.

Selected References

Berthold, P. 1996. *Control of Bird Migration.* Chapman & Hall, London.

Collett, T. S., and M. Collett. 2000. Path integration in insects. *Curr. Opin. Neurobiol.* 10: 757–762.

Dingle, H. 1996. *Migration: Biology of Life on the Move.* Oxford University Press, New York.

Emlen, S. T. 1970. Celestial rotation: Its importance in the development of migratory orientation. *Science* 170: 1198–1201.

Emlen, S. T. 1975. The stellar orientation system of a migratory bird. *Sci. Amer.* 233(2): 102–111.

Esch, H. E., S. Zhang, M. V. Srinivasan, and J. Tautz. 2001. Honeybee dances communicate distances measured by optic flow. *Nature* 411: 581–583.

Etheredge, J. A., S. M. Perez, O. R. Taylor, and R. Jander. 1999. Monarch butterflies (*Danaus plexippus* L.) use a magnetic compass for navigation. *Proc. Natl. Acad. Sci. U.S.A.* 96: 1345–1346.

Giurfa, M., and E. A. Capaldi. 1999. Vectors, routes, and maps: New discoveries about navigation in insects. *Trends Neurosci.* 22: 237–242.

Gould, J. L. 1998. Sensory bases of navigation. *Curr. Biol.* 8: R731–R738.

Harris, M. 1957. *Something about a Soldier.* Macmillan, New York.

Healy, S. (ed.). 1998. *Spatial Representation in Animals.* Oxford University Press, Oxford, UK.

Keeton, W. T. 1971. Magnets interfere with pigeon homing. *Proc. Natl. Acad. Sci. U.S.A.* 68: 102–106.

Keeton, W. T. 1974. The mystery of pigeon homing. *Sci. Amer.* 231(6):96–107. (Reprinted in Eisner, T., and E. O. Wilson. 1975. *Animal Behavior: Readings from Scientific American.* Freeman, San Francisco.)

Labhart, T., and E. P. Meyer. 2002. Neural mechanisms in insect navigation: Polarization compass and odometer. *Curr. Opin. Neurobiol.* 12: 707–714.

Lehrer, M. (ed.). 1997. *Orientation and Communication in Arthropods.* Birkhäuser, Basel, Switzerland.

Lohmann, K. J. 1992. How sea turtles navigate. *Sci Amer.* 266(1): 100–106.

Lohmann, K. J., and S. Johnsen. 2000. The neurobiology of magnetoreception in vertebrate animals. *Trends Neurosci.* 23: 153–159.

Papi, F. (ed.). 1992. *Animal Homing.* Chapman & Hall, London.

Pennisi, E. 2003. Monarchs check clock to chart migration route. *Science* 300: 1216–1217.

von Frisch, K. 1967. *The Dance Language and Orientation of Bees.* Harvard University Press, Cambridge, MA.

Wehner, R. 2001. Bird navigation–computing orthodromes. *Science* 291: 264–265.

Wiltschko, R., and W. Wiltschko. 1995. *Magnetic Orientation in Animals.* Zoophysiology, vol. 33. Springer, Berlin.

PART IV

Muscle and Movement

Part IV Muscle and Movement

Previous Page: The grace and speed of this cheetah's movements are accomplished by precise control of its skeletal muscles.

17 *Muscle*

The intricate aerial ballet of this green lacewing is choreographed by flight muscles. Muscles are specialized for movement, and insects, like all other animals, use muscles to generate movements that accomplish physiological activities or behaviors. Hearts beat, gastrointestinal tracts mix and move food, insects follow pheromone trails in pursuit of mates, clams snap shut, and elephants use their trunks to touch and maneuver the bones of their deceased relatives. In this chapter we explore the physiology of muscles that underlie such activities.

Not all movements are generated by muscles, however. Chromosomes move during cell division, flagella and cilia make whiplike or wavelike motions, single cells such as amoebae or white blood cells travel from one place to another, and plants slowly bend toward sources of light. Organized motion is essential for survival, and natural selection continuously hones and improves mechanisms of movement. All organized movements, whether produced by muscles or not, depend on the generation of forces that require the conversion of chemical energy to mechanical energy. Muscle cells use the *molecular motor* **myosin** to convert chemical energy of ATP into mechanical energy of movement. Myosin is a large protein that interacts with another protein, **actin**, to generate force. Myosin and actin are referred to as *contractile proteins*.

All phyla have two categories of muscle cells: **striated** and **smooth** (or unstriated). Striated muscle cells have alternating transverse light and dark bands, giving them a striped appearance. The pattern of bands reflects the organization of the contractile proteins myosin and actin into regularly repeating units called *sarcomeres*. Smooth (unstriated) muscle cells also possess actin and myosin, but these proteins are not organized into sarcomeres.

In vertebrate animals, striated muscles make up **skeletal** (attached to bones) and **cardiac** (heart) muscles. Smooth muscles of vertebrates are found in hollow or tubular organs such as the intestine, uterus, and blood vessels. Invertebrate animals also have striated and smooth muscles, but they aren't always found in the same distribution as in vertebrates. In arthropods, for example, the skeletal (attached to the exoskeleton) and cardiac muscles are both striated, but so are muscles of the alimentary (digestive) tract.

In this chapter we consider muscles in light of the major themes of this book: mechanism and adaptation. We examine first the physiological and biochemical mechanisms that underlie muscle contraction, and then the adaptations of certain muscles specialized to per-

A green lacewing (*Chrysopa* sp.) in flight

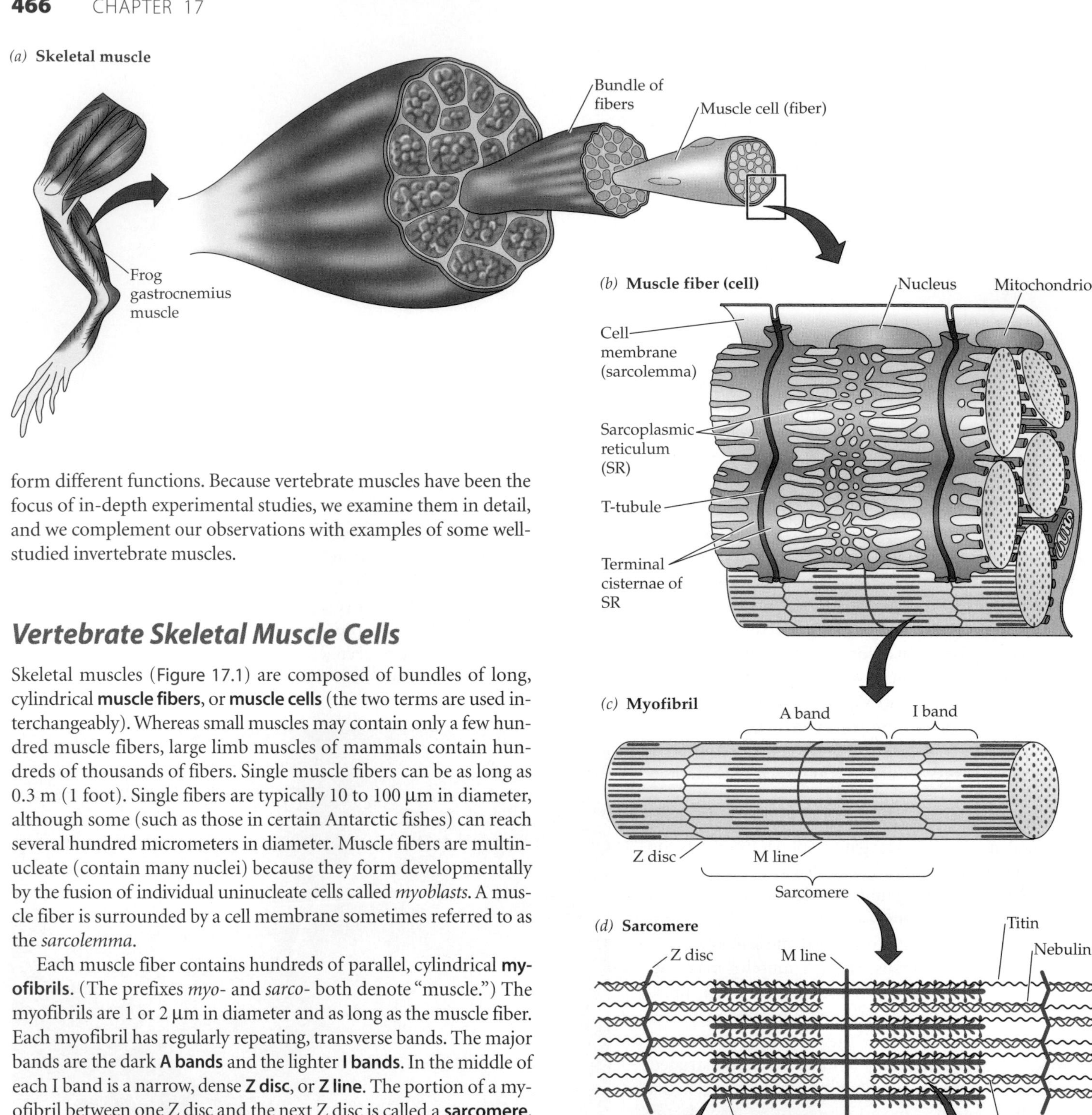

form different functions. Because vertebrate muscles have been the focus of in-depth experimental studies, we examine them in detail, and we complement our observations with examples of some well-studied invertebrate muscles.

Vertebrate Skeletal Muscle Cells

Skeletal muscles (Figure 17.1) are composed of bundles of long, cylindrical **muscle fibers**, or **muscle cells** (the two terms are used interchangeably). Whereas small muscles may contain only a few hundred muscle fibers, large limb muscles of mammals contain hundreds of thousands of fibers. Single muscle fibers can be as long as 0.3 m (1 foot). Single fibers are typically 10 to 100 μm in diameter, although some (such as those in certain Antarctic fishes) can reach several hundred micrometers in diameter. Muscle fibers are multinucleate (contain many nuclei) because they form developmentally by the fusion of individual uninucleate cells called *myoblasts*. A muscle fiber is surrounded by a cell membrane sometimes referred to as the *sarcolemma*.

Each muscle fiber contains hundreds of parallel, cylindrical **myofibrils.** (The prefixes *myo-* and *sarco-* both denote "muscle.") The myofibrils are 1 or 2 μm in diameter and as long as the muscle fiber. Each myofibril has regularly repeating, transverse bands. The major bands are the dark **A bands** and the lighter **I bands**. In the middle of each I band is a narrow, dense **Z disc**, or **Z line**. The portion of a myofibril between one Z disc and the next Z disc is called a **sarcomere.** Thus, one myofibril consists of a longitudinal series of repeating sarcomeres. The Z discs of adjacent myofibrils are lined up in register with each other, so the pattern of alternating A bands and I bands appears continuous for all the myofibrils of a muscle fiber. This alignment of banding within a muscle fiber gives the fiber its striated appearance. The striations are visible by light microscopy.

Figure 17.1 The organization of skeletal muscles The physiology of muscles can be explained in large part by their anatomical organization, shown here in successively more detailed views.

Higher magnification electron micrographs show that the myofibrils contain two kinds of **myofilaments.** The **thick filaments** are composed of the protein *myosin* and are confined to the A band of each sarcomere. A single thick filament consists of 200 to 300 myosin molecules. The **thin filaments** are composed of *actin*. A single thin filament consists of two chains of globular actin molecules wrapped around each other in a loose helix. Thin filaments are anchored to proteins in the Z discs. They extend from the Z disc partway into the A band of each flanking sarcomere, where they interdigitate with thick filaments. The central region of the A band, which contains only thick filaments and appears lighter than the rest of the A band, is called the **H zone.** A narrow dense region called the **M line** bisects the H zone.

In the M line, the thick filaments of the myofibril are webbed together with accessory proteins to maintain their regular spacing. The Z disc and M line ensure that neither the thick filaments nor the thin filaments float free. Cross sections of a myofibril show the relationship of thick and thin filaments in a sarcomere (Figure 17.2). A cross section through the I band shows only thin filaments. A section through the part of the A band in which the thick and thin filaments overlap shows each thick filament surrounded by six thin filaments. A section through the H zone shows only thick filaments.

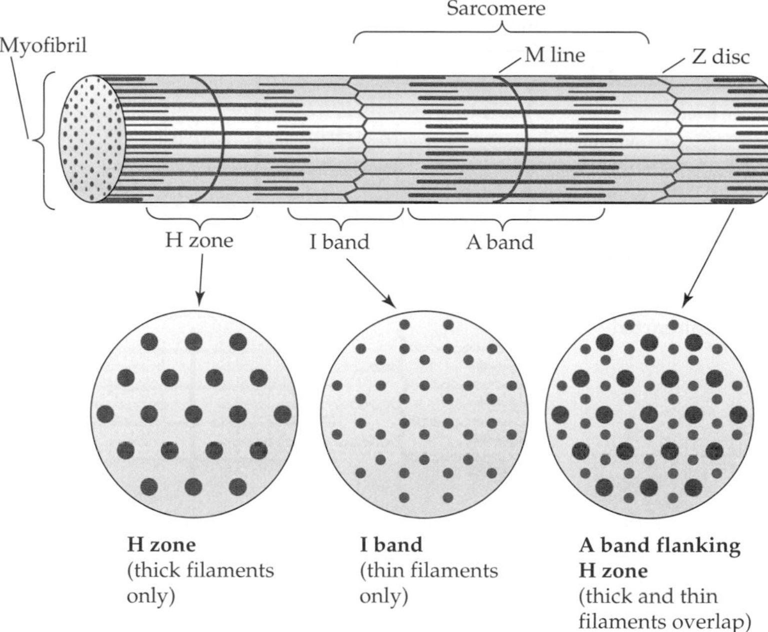

Figure 17.2 The arrangement of thick (myosin) and thin (actin) myofilaments in a sarcomere A longitudinal view of a cylindrical myofibril shows the parallel arrangement of the thick myosin myofilaments and the thin actin myofilaments. Cross sections illustrate the regions of overlap of the thick and thin myofilaments. Both the M line and the Z disc contain accessory proteins that anchor the thick and thin filaments. In vertebrate animals, the thick filaments are about 1.6 µm long and 12 to 15 nm across. The thin filaments are about 1.0 µm long and 7 to 8 nm across.

The myosin molecules have projections on them called heads or **cross-bridges** (see Figure 17.1*d, f*). When the muscle cell is stimulated to contract, the myosin cross-bridges interact transiently with the overlapping actin thin filaments. The interactions of the myosin cross-bridges with actin molecules generate the forces for muscle contraction. Muscle fibers contain specialized proteins in addition to actin and myosin. **Titin** and **nebulin** (Figure 17.1*d*) are structural proteins that help align the contractile proteins actin and myosin. Titin has elastic properties, and a single molecule spans the distance between the Z disc and the M line.[1] Titin confers the ability to spring back after the muscle fiber is stretched. *Nebulin* is inelastic; it runs the length of a thin filament and stabilizes it. **Troponin** and **tropomyosin** are protein molecules associated with the actin chains of the thin filaments (see Figure 17.1*e*). They regulate the process of contraction by controlling whether or not the myosin cross-bridges can interact with the thin filaments.

When a muscle fiber contracts, the thick and thin filaments do not shorten, but instead slide by one another. Investigators in the 1950s noted that the A band, which is the length of the thick filaments, does not shorten during contraction. Furthermore, the distance between the margins of the two H zones on either side of a Z disc (the length of the thin filaments) stays the same, no matter what the length of the adjacent sarcomeres. When a sarcomere changes length, the regions that also change length are the H zone and the I band. Two independent teams—H. E. Huxley and Hanson, and A. F. Huxley and Niedergerke—made these observations in 1954 and formulated the *sliding-filament theory* of muscle contraction, which has since been amply confirmed. It states that the force of contraction is generated by the cross-bridges of the thick filaments attaching to the thin filaments and actively pulling them toward the center of the sarcomere (Figure 17.3).

Thick and thin filaments are polarized polymers of individual protein molecules

Individual myosin molecules are large proteins of about 500 kilodaltons (kDa), each consisting of two globular heads joined to a long rod, or tail. The heads are the cross-bridges, and the tail contributes to the backbone of the thick filament (Figure 17.4). During polymerization the myosin molecules orient themselves with their tails pointing toward the center of the thick filament and their heads toward the ends. As a result, the two halves of the thick filament become mirror images of each other with a short bare zone of only tails in the middle of the filament. The cross-bridges on either side of the bare zone point in opposite directions.

Each actin molecule is a globular protein (42 kDa) called *G-actin*. G-actin monomers form chains of *F-actin* (filamentous actin). The two chains of F-actin wind around each other in a double helix (see Figure 17.1*e*). Like the myosin molecules in thick filaments, G-actin molecules in thin filaments are arranged so that those on one side of the Z disc have one orientation, and those on the other side have the opposite orientation. The consequence of the polarized organization of the thick and thin filaments is that the cross-bridges in contact

[1] Titin is the largest known protein. Composed of nearly 27,000 amino acids, it has a molecular weight of 3 million daltons. Whereas the thick and thin myofilaments of similar length are polymers made up of hundreds of myosin or actin molecules, a *single* molecule of titin extends from the Z disc to the center of the sarcomere!

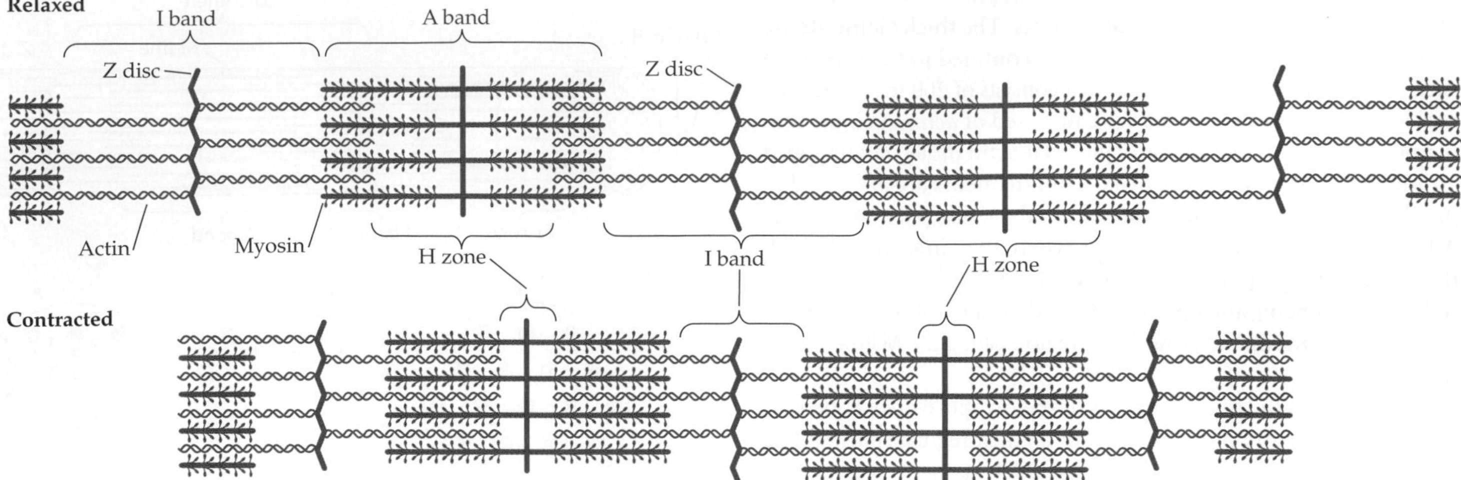

Figure 17.3 Muscle contraction produced by sliding filaments During contraction, the thick and thin myofilaments slide by one another. The polarized organization of the molecules in both types of myofilaments ensures that the thin filaments are drawn toward the center of each sarcomere. The myofilaments do not shorten, but the I band and H band of each sarcomere do. Because the sarcomeres are arranged in series in a myofibril, the entire myofibril shortens.

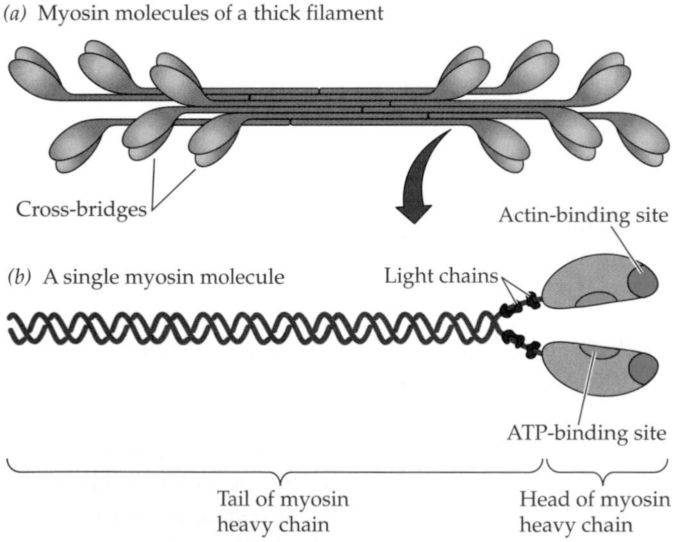

(a) Myosin molecules of a thick filament

Cross-bridges

Actin-binding site

(b) A single myosin molecule

Light chains

ATP-binding site

Tail of myosin heavy chain

Head of myosin heavy chain

Figure 17.4 Myosin molecules form the thick filament *(a)* About 200 to 400 myosin molecules polymerize to form a polarized thick filament. *(b)* Each myosin molecule contains two *heavy chains* of amino acids. The tail of the molecule consists of the two chains coiled around each other. The amino-terminal end of each heavy chain forms one of the heads. The head region has a surface for binding actin and a different site for binding and hydrolyzing ATP. A link region of the heavy chain connects the head (which projects from the thick filament) to the tail (which contributes to the backbone of the thick filament). In addition, the myosin molecule includes two smaller *light chains* associated with each head. Thus, each complete myosin molecule contains six polypeptide chains: two heavy and four light. The molecular composition of the heavy and light chains varies in different types of muscles. The different myosin *isoforms* of myosin heavy chains and light chains confer variations of functional properties, such as the rate at which the myosin ATPase hydrolyzes ATP. (After Sherwood 2004.)

with the thin filament can act like oars to pull the thin filaments toward the center of the sarcomere.

Muscles require ATP to contract

Myosin heads cyclically attach to actin molecules and then swivel to pull the actin and myosin filaments past each other. Each myosin head has two binding sites: one for actin and the other for ATP. The binding site for ATP is an ATPase with enzymatic activity that splits inorganic phosphate from the ATP molecule and releases energy. The energy is used to power cross-bridge action.

The cycle of molecular interactions underlying contraction is shown in Figure 17.5. Let's start with the myosin head bound to actin but not binding ATP (step 1 in Figure 17.5). This is the *rigor* conformation, as in *rigor mortis*, in which muscles of a dead person (or other animal) are rigidly fixed in place because of the absence of ATP in death. ATP is required for myosin to *unbind* from actin (step 2). In life the rigor stage of each cross-bridge cycle is brief because the globular myosin head readily binds ATP. It is important to understand that the detachment of myosin from actin requires the *binding* of ATP to change the conformation of myosin's actin-binding site, but not energy derived from the ATP.

Once released from actin, the myosin head hydrolyzes the ATP to ADP and inorganic phosphate (P_i) (step 3). A change in angle of the myosin head (termed *cocking*) accompanies hydrolysis, but the head does not release the ADP and P_i. The energy released by hydrolysis of ATP is stored in the myosin–ADP–P_i complex. The complex then binds actin (step 4), forming an actin–myosin–ADP–P_i complex. Initially actin binding is weak, but it triggers P_i release, tighter binding, and the power stroke (step 5). The myosin head swivels, pulling the attached actin toward the middle of the myosin filament. At the end of the power stroke, the ADP is released and the myosin remains tightly bound to the actin (step 6). A new molecule of ATP then binds to the myosin head, triggering its release from actin.

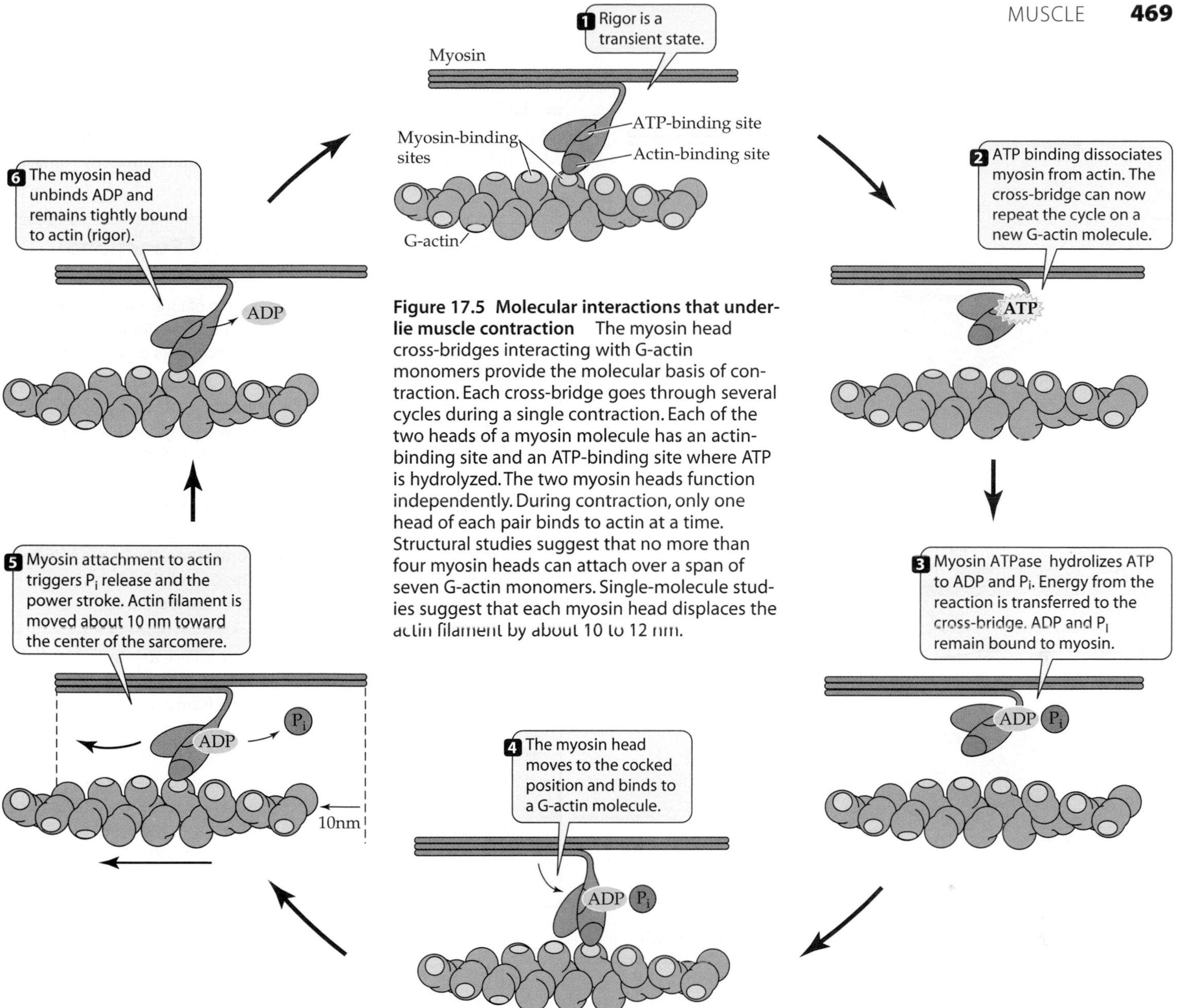

Figure 17.5 Molecular interactions that underlie muscle contraction The myosin head cross-bridges interacting with G-actin monomers provide the molecular basis of contraction. Each cross-bridge goes through several cycles during a single contraction. Each of the two heads of a myosin molecule has an actin-binding site and an ATP-binding site where ATP is hydrolyzed. The two myosin heads function independently. During contraction, only one head of each pair binds to actin at a time. Structural studies suggest that no more than four myosin heads can attach over a span of seven G-actin monomers. Single-molecule studies suggest that each myosin head displaces the actin filament by about 10 to 12 nm.

With each cycle, one ATP is consumed, and the myosin molecule moves the actin filament a short distance. During a single contractile event (produced when a muscle fiber generates an action potential), each cross-bridge repeats several binding/unbinding cycles. The cross-bridges work independently and asynchronously, so at any instant during a contraction a fraction of the cross-bridges is bound to actin while the rest are in other phases of the cycle. The summed effect of all the cross-bridge cycles is to pull the thin filaments toward the middle of the sarcomere. In living, relaxed muscle, every cross-bridge has stored energy and bound ADP and P_i, but it is temporarily unable to bind actin.

Calcium and the regulatory proteins tropomyosin and troponin control contractions

In a resting muscle, each myosin head has detached from actin, hydrolyzed the ATP, and stored the energy obtained from hydrolysis. It is "primed" for another cycle. However, the regulatory proteins **tropomyosin** (**TM**) and **troponin** (**TN**) prevent contraction by inhibiting the myosin head from binding to actin. TM is a long protein that is coiled along the groove between the two actin chains of the thin filament (Figure 17.6). A single TM molecule extends the length of seven globular actin molecules. Each TM molecule is associated with one TN molecule.

TN is a golf club–shaped complex of three subunits. The "handle" is troponin T (TN-T), which binds to tropomyosin. The "club" includes troponin I (TN-I), which binds to actin, and troponin C (TN-C), which binds Ca^{2+} ions. In the resting state (see Figure 17.6*a*), the TM molecule lies over the myosin-binding sites of the adjacent actin molecules and prevents myosin cross-bridges from binding to actin. For contraction to occur, TM's inhibition of cross-bridge binding is counteracted by Ca^{2+} binding to TN.

The physiological regulator of muscle contraction is calcium. When a Ca^{2+} ion binds to TN-C, it triggers conformational changes in TN, *(a)* No Ca^{2+} ions present (relaxed)

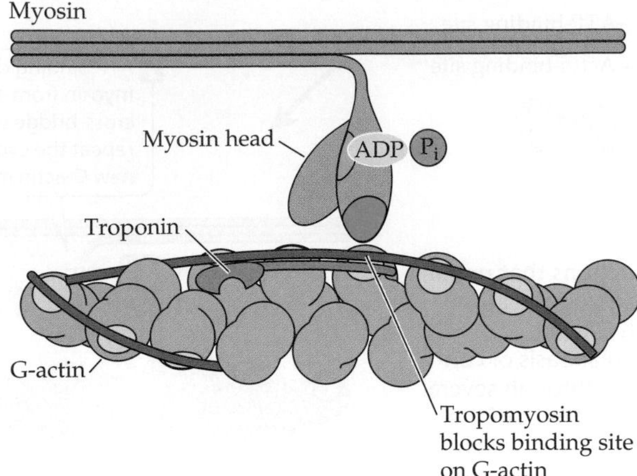

(b) Ca^{2+} ions released from the sarcoplasmic reticulum

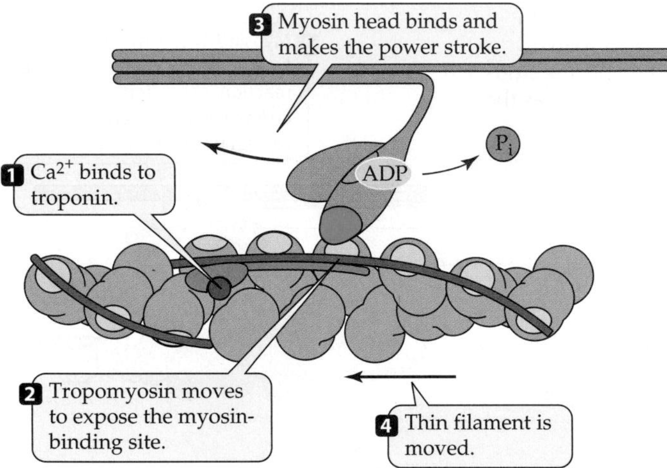

Figure 17.6 The regulation of contraction In combination with Ca^{2+} ions, troponin (TN) and tropomyosin (TM) regulate the interaction of myosin heads with actin molecules. Coiled TM molecules lie end to end (with a bit of overlap) in the groove of the actin spiral. Each TM molecule is the length of seven G-actin monomers. One TN molecule associates with one TM molecule. The TN molecule has three subunits: TN-I, TN-T, and TN-C. *(a)* When Ca^{2+} ions are *not* present in the cytoplasm, the subunit TN-I binds to two adjacent actin monomers, and the TN-T binds to the tropomyosin molecule. These connections hold TM in a position that covers the myosin-binding sites on actin and inhibits cross-bridge action. *(b)* The TN-C subunit binds to Ca^{2+} ions when they are released from the sarcoplasmic reticulum. The binding of Ca^{2+} ions to TN-C causes conformational changes that detach TN-I from actin and allow TM to roll over the actin surface. The changed position of TM, as well as allosteric changes, permits cross-bridge action.

TM, and probably actin that permit myosin cross-bridges to interact with actin (see Figure 17.6*b*). The changes that occur when Ca^{2+} binds with TN appear to involve both removal of TM's steric blocking of the myosin-binding sites on actin and also more subtle allosteric interactions among the proteins. Once interaction between actin and myosin is possible, the primed myosin cross-bridges are permitted to go through cross-bridge cycles until the Ca^{2+} is removed. The muscle will therefore contract only when Ca^{2+} ions are available to bind TN. In relaxed skeletal muscle fibers, the intracellular concentration of calcium is extremely low—less than 1×10^{-7} *M*, which is below the concentration that will induce (by mass action) calcium association with troponin.

Summary: Structural and Functional Organization of Skeletal Muscles

- Each whole muscle consists of a bundle of longitudinally arrayed muscle fibers that, in turn, consist of myofibrils made up of thick (myosin) and thin (actin) myofilaments organized into sarcomeres.
- A single myofibril consists of a series of sarcomeres aligned such that the muscle fiber, when viewed longitudinally, appears striped (striated). Titin and nebulin help to stabilize the thick and thin filaments in the sarcomeres.
- The contractile proteins actin and myosin polymerize in a polarized fashion to form the thin and thick filaments. When the filaments slide by each other during contraction, the heads of individual myosin molecules bind to sites on individual actin molecules and draw the thin filaments toward the center of each sarcomere.
- Each myosin head also functions as an ATPase to provide the energy required to power cross-bridge motion. In relaxed muscle, each "primed" cross-bridge is detached from actin and contains ADP, P_i, and stored energy obtained from the hydrolysis of ATP.
- The regulatory proteins troponin (TN) and tropomyosin (TM), located on the thin filament, inhibit myosin cross-bridges from interacting with actin, except when Ca^{2+} is present. When Ca^{2+} binds to TN-C, it triggers conformational changes that allow myosin cross-bridges to interact with myosin-binding sites on actin molecules.

Excitation–Contraction Coupling

Neural excitation triggers skeletal muscle contraction. Each skeletal muscle fiber is innervated by a motor neuron at the motor end-plate (see Figures 17.7 and 12.6). An action potential conducted to the axon terminal of the motor neuron releases acetylcholine, which binds to postsynaptic ACh receptors in the end-plate and causes permeability changes that result in depolarization of the muscle fiber membrane (sarcolemma). Depolarization of the muscle fiber is referred to as *excitation*. This excitation leads to rapid activation of the contractile machinery of the muscle fiber. The relationship between depolarization and contraction is called **excitation–contraction coupling.**

Excitation and contraction are coupled by two separate but intimately associated membrane systems. The first of these is a system of tubules that is continuous with the sarcolemma: the **transverse tubules**, or **t-tubules** (Figure 17.7). A single t-tubule dips into the muscle fiber at an angle perpendicular to the sarcolemma, transverse to the long axis of the muscle fiber. The t-tubule invaginations occur at regular intervals along the length of the sarcolemma. The position of invagination varies between phyletic groups, usually at the level of the Z discs (e.g., amphibian muscle) or at the junction of the A and

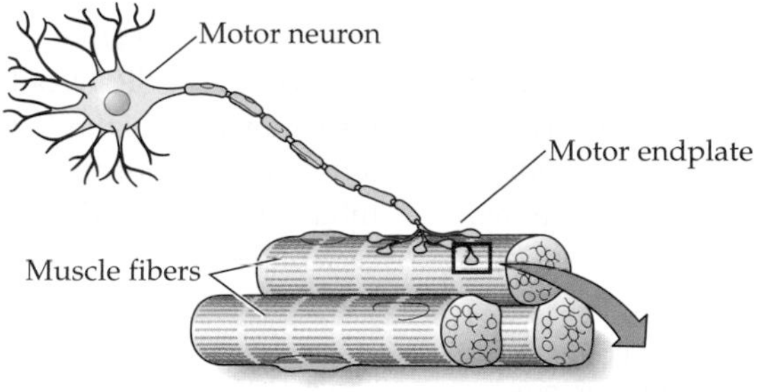

Figure 17.7 Excitation–contraction coupling Excitation–contraction coupling is accomplished by the interactions of components in two intimately associated membrane systems: the transverse tubular system and the sarcoplasmic reticulum. (After Silverthorn 2004.)

I bands (e.g., mammalian and reptilian muscles). Because the t-tubule membrane is a continuation of the outer sarcolemma, the tubule lumen is continuous with extracellular space. When the sarcolemma is depolarized, the t-tubules conduct this excitation deep into the interior of the muscle fiber. The t-tubules come into close association with the second membrane system required for excitation–contraction coupling, the sarcoplasmic reticulum (SR).

The SR is a network of branching tubules contained entirely within the muscle fiber.[2] Each myofibril is enveloped in SR. The SR membrane has Ca^{2+}-ATPase active-transport pumps (see Chapter 3,) that maintain a low concentration of Ca^{2+} ions in the cytoplasm and a high concentration of Ca^{2+} ions within the SR. The SR membrane also has calcium channels that open during excitation to allow Ca^{2+} ions to diffuse out of the SR into the cytoplasm. The Ca^{2+} ions bind to TN-C and trigger the conformational changes that permit cross-bridge interaction. The SR calcium channels open in response to depolarization conducted interiorly along the t-tubules.

The SR between two t-tubules is called an SR *compartment*. Each compartment of the SR forms a sleeve of branching tubules around each myofibril (see Figure 17.1*b*). Enlarged sacs called *terminal cisternae* (singular *cisterna*) lie next to the t-tubules. In resting muscle, Ca^{2+} is largely confined to the terminal cisternae of the SR. Once an action potential conducted along the t-tubule triggers the SR calcium channels to open, Ca^{2+} ions rapidly diffuse the short distance to the adjacent myofilaments, bind to troponin, and initiate the processes that will allow myosin cross-bridges to interact with actin. When vertebrate muscle fibers are stimulated to contract, sufficient Ca^{2+} diffuses to the myofibrils that every TN–TM complex moves to allow all cross-bridges to function.

How does depolarization of the t-tubule membrane produce Ca^{2+} release from the separate membrane system of the sarcoplasmic reticulum? In skeletal muscle, the two membrane systems are linked by two kinds of membrane proteins (Figure 17.7): the *ryanodine receptors* of the SR and the *dihydropyridine receptors* of the t-tubules. Both of these proteins are actually calcium channels, but they were both named for the drugs that bind to them specifically.

The ryanodine receptor (RyR) is located in the SR membrane facing a t-tubule. When open, it lets Ca^{2+} diffuse out of

[2] The sarcoplasmic reticulum is homologous to the smooth endoplasmic reticulum of other cells.

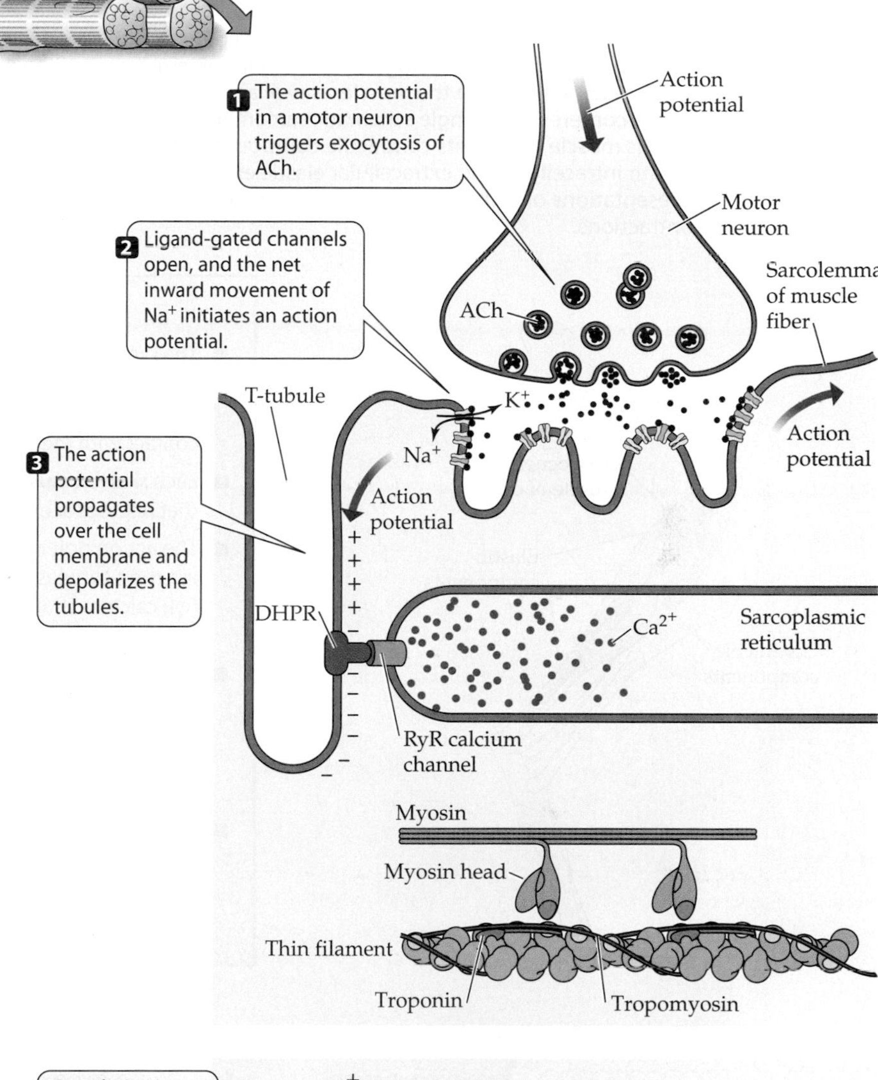

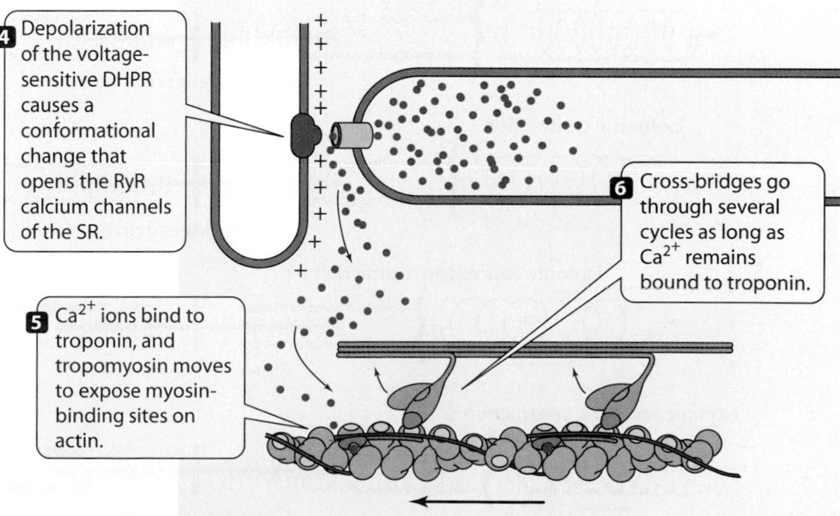

the SR into the cytoplasm. The dihydropyridine receptor (DHPR) is a voltage-sensitive calcium channel. However, in skeletal muscle, the DHPRs do not appear to permit much calcium flux from the extracellular fluid into the cytoplasm. Still, their sensitivity to voltage changes plays an important role in excitation–contraction coupling. About half of the DHPRs in the t-tubules form direct contacts with the RyRs of the SR. Depolarization of the t-tubule changes the conformation of the DHPR, which alters conformation of the attached RyR in the SR, opening it to release Ca^{2+} into the cytoplasm. The t-tubular DHPR and the SR Ca^{2+} channel (RyR) interact directly with each other.[3]

[3] Cardiac muscle also has both DHPR and RyR proteins (in different isoforms), but their interaction is indirect. The DHPR in cardiac muscle functions as a Ca^{2+} channel. It opens in response to depolarization and lets in extracellular Ca^{2+} from the t-tubular lumen. This Ca^{2+} from outside the cell opens the RyR calcium channel of the SR by a process called *Ca^{2+}-induced Ca^{2+} release*. The indirect coupling seen in cardiac muscle is probably more primitive and was replaced in skeletal muscle by evolution of the faster direct coupling of the two proteins.

Figure 17.8 Interaction between contractile and elastic components During contraction, the cross-bridge action within sarcomeres pulls on elastic elements within the muscle, as well as on immediately adjacent sarcomeres. *(a)* A single schematic sarcomere in a dog gastrocnemius muscle represents contractile components that are associated with intracellular and extracellular elastic elements. *(b)* Schematic representations of contractile and elastic elements at rest and during contractions.

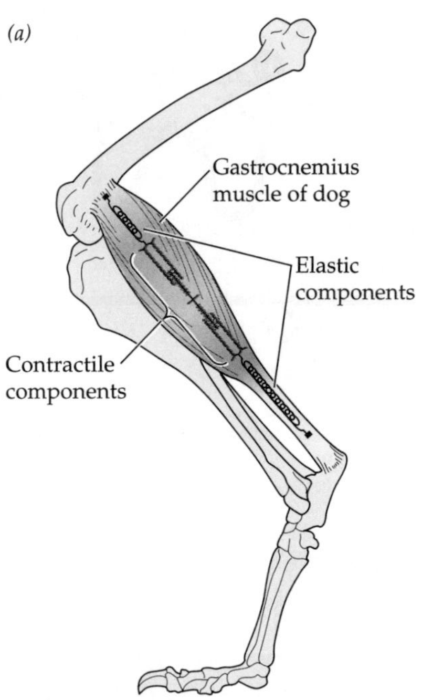

Summary: Excitation–Contraction Coupling

- The sarcoplasmic reticulum (SR) sequesters Ca^{2+} ions to keep the cytoplasmic concentration of Ca^{2+} low. The terminal cisternae of the SR possess RyR calcium channels. Transverse tubules include voltage-sensitive DHPRs that come into intimate contact with the RyRs of the SR.
- Each skeletal muscle contraction is initiated by an action potential in a motor neuron that releases acetylcholine, which in turn gives rise to muscle fiber action potential.
- The action potential propagates over the cell membrane of the muscle fiber, including the t-tubules, and depolarizes the DHPRs in the t-tubules. The DHPRs cause the RyR calcium channels to open and allow Ca^{2+} ions to diffuse out of the terminal cisternae of the SR into the cytoplasm.
- Ca^{2+} ions bind to TN and cause conformational changes of TN and TM that expose the myosin-binding sites of adjacent actin molecules. Previously primed myosin heads bind to the actin sites. Repeated cross-bridge cycles continue as long as sufficient Ca^{2+} is present. The cross-bridges move the thick and thin filaments relative to each other, pulling the thin filaments toward the center of the sarcomere.
- Once the muscle fiber action potential is over, the RyR channels close. The Ca^{2+}-ATPase pumps of the SR sequester Ca^{2+} back into the SR. As the Ca^{2+} concentration in the cytoplasm decreases, Ca^{2+} dissociates from TN, and the TN–TM complex again prevents contraction. The muscle relaxes.

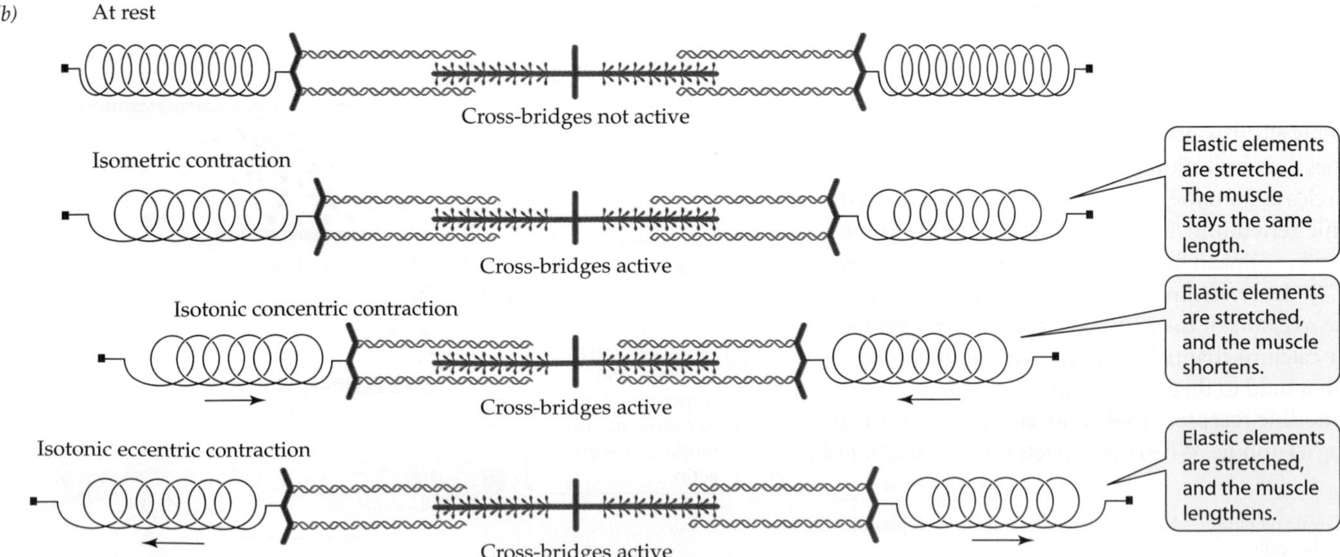

Whole Skeletal Muscles

Many skeletal muscles in vertebrates work in **antagonistic pairs** arranged around joints. When one muscle shortens, its antagonist lengthens; and vice versa. The quadriceps muscles on the front of the thigh and the hamstring muscles on the back of the thigh are a good example of an antagonistic pair of muscles. The hamstring muscles shorten to bend the knee joint. The quadriceps muscles shorten to straighten the knee joint. Often muscles work in combination with connective tissues. For example, grazing animals like camels or cows use muscles to pull their heads down to feed. Lowering the head stretches a ligament that attaches to the back of the head at one end and to the vertebral column at the other end. The stretched ligament stores energy like that in a stretched spring. This energy is expended as the ligament springs back to its original length, helping the muscles that lift the animal's head.

Muscle contraction is the force generated by a muscle during cross-bridge activity

Although the term *contraction* suggests that the muscle shortens during cross-bridge activity, this is not always the case. For example, you can "tighten up" your biceps without allowing your elbow joint to flex. Even though cross-bridge cycling occurs, the bones don't move, and the whole muscle stays the same length. This type of contraction is called **isometric** ("same length") **contraction**. The sarcomeres within the muscles shorten slightly during isometric contraction (the biceps "bulges") because they pull on elastic elements within the muscle. Elastic structures include the connective tissue surrounding individual muscle fibers, which continue into tendons, as well as components of the myofibrils such as titin and the cross-bridge links themselves. Figure 17.8 illustrates the relationship of contractile and elastic components in a muscle.

Whole muscles can indeed shorten. For example, when your hand brings a heavy book toward your face, the biceps muscle shortens to decrease the angle at your elbow. This type of contraction, in which the whole muscle shortens, is called **concentric contraction** because the muscular action brings the hand closer to the center of the body. Cross-bridges can also be active when the muscle is lengthening. For example, if you hold a 10-pound weight in your hand with your elbow bent, and slowly extend your arm, the sarcomeres of your biceps are lengthening at the same time that the cross-bridges are generating tension. Similarly, when you go through the motion of sitting down, or hike down an incline, the quadriceps muscles on the top of your thighs are actively contracting, but the muscles are actually longer than they are when the knee is not bent. The contractile activity in these cases is resisting stretch imposed by an external force. These contractions are called **lengthening**, or **eccentric**, **contractions.** Lengthening contractions are thought to produce minor damage to muscle fibers that lead to delayed soreness following exercise.

Concentric and eccentric contractions are both examples of **isotonic** ("same tension") **contractions** (Figure 17.8*b*), which we explain in the next sections. Most muscle activity involves dynamic combinations of both isometric and isotonic contractions. Physiologists separate these types of contractions experimentally in order to study particular properties of muscles.

A muscle exerts its force on a **load**. For example, when you lift an object with your hand, the load on which the biceps muscle exerts force includes the mass of the lower arm plus the mass of the object. The force of the muscle is opposed by the force of the load. The force exerted by a muscle on the load is called muscle **tension.**[4] If the tension (force) developed by a muscle is greater than the force exerted on it by a load, the muscle will change length. This is an isotonic contraction.

Once the muscle begins to change length, the tension it produces is constant and equal to the force of the load. Experimenters recording *isotonic* contractions measure changes in *length* of the muscle. If the force exerted by the load is greater than the muscle tension (e.g., an extremely heavy weight), the muscle will produce an isometric contraction. Experimenters recording *isometric* contractions measure the *tension* developed by the muscle. They usually attach the muscle to a very stiff force transducer that measures tension (force) while permitting only minuscule changes in length.

A twitch is the mechanical response of a muscle to a single action potential

Figure 17.9 shows the twitch response of the same mammalian muscle recorded under isometric and isotonic conditions. Both twitches have three phases: a latent period, a contraction phase, and a relaxation phase. The isometric twitch has a brief latent period before any contractile tension is recorded (see Figure 17.9*a*); this latent period largely reflects the time required for excitation–contraction coupling to occur. The isotonic twitch has a longer latent period (see Figure 17.9*b*). Before the muscle can lift the load and shorten, excitation–contraction coupling must occur, and the cross-bridges must develop enough tension isometrically to overcome the force exerted by the load. If the load were greater, the latency would be longer because additional time would be required to develop tension to equal the heavier load.

The velocity of shortening decreases as the load increases

Isotonic recordings are ideal for revealing that the load directly influences the velocity (speed) at which a muscle shortens. You know from experience that you can lift a pencil faster than an unabridged dictionary, and you can't lift your car at all. Figure 17.10 shows that the velocity of shortening is greatest when the muscle lifts no load, decreases progressively with increasing loads, and becomes zero when the load cannot be lifted—that is, when the muscle cannot produce a great enough isometric tension to overcome the load and shorten.

This **load–velocity relationship** is a fundamental property of muscle. Current models suggest that greater loads somehow decrease the rate at which the myosin heads detach from actin, and therefore slow the speed of shortening. This relationship is also referred to as the **force–velocity relationship** because the load exerts a force on the muscle, and—in isotonic contractions—the tension (force) produced by the muscle equals the force of the load.

---

[4] Physiologists often use the terms *muscle tension* and *muscle force* interchangeably. We know that cross-bridge action at the level of the sarcomere underlies the action produced by a whole muscle. The tension generated by a muscle fiber is directly proportional to the number of attached cross-bridges between the thick and thin filaments.

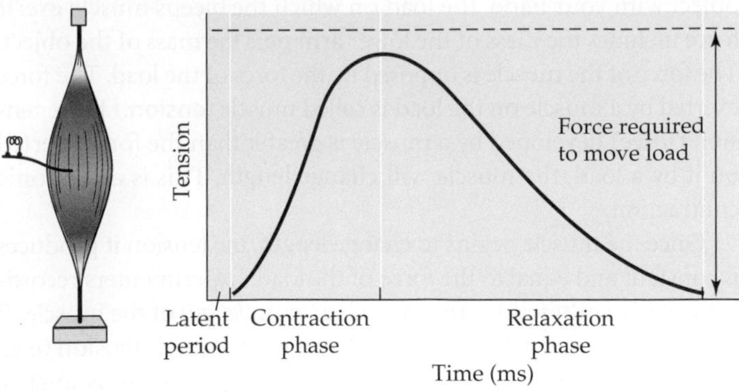

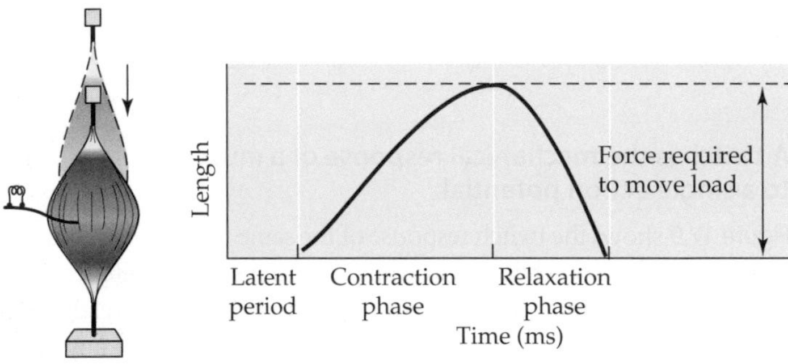

Figure 17.9 Recording isometric and isotonic contractions Isometric contractions are measured as changes in tension produced by the muscle; isotonic contractions are measured as changes in length of the muscle. *(a)* In the isometric experimental arrangement, the muscle contracts but is not allowed to move a load. *(b)* In the isotonic experimental arrangement, the muscle shortens and moves the load, once its contractile activity generates tension that equals the force of the load.

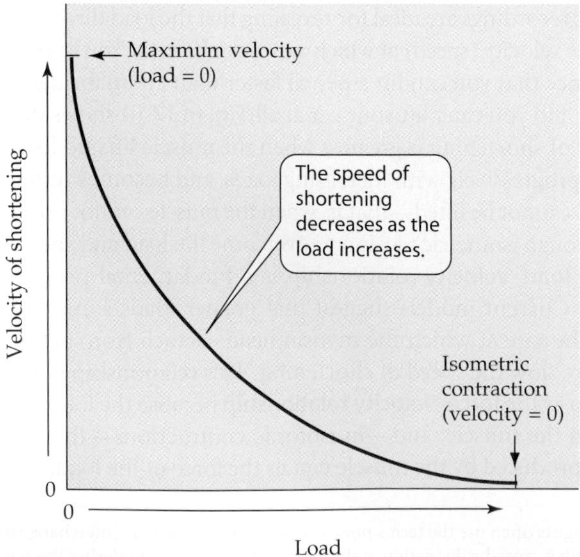

Figure 17.10 The load–velocity relationship of skeletal muscle This relationship is also referred to as the force–velocity relationship.

The frequency of action potentials determines the tension developed by a muscle

When a muscle is stimulated more than once within a brief period of time, the successive twitches produced add to each other, so the overall response is greater than the twitch response to a single stimulus. Such addition is called **summation.** Summation can be recorded either isometrically or isotonically. Figure 17.11 shows isometric records of summation. The electrical events triggering the contractions (the action potentials in the nerve axon and muscle fibers) are all-or-none and do not sum. However, because the action potentials last only 1 or 2 ms, and a muscle twitch lasts many milliseconds, the muscle can generate a second (or even multiple) action potential before the end of the twitch produced by the first action potential.

The amplitude of the summed contractions depends on the interval of time between stimuli. Low frequencies of stimulation with relatively long intervals between stimuli produce contractions that sum but are not fused. Higher-frequency stimulation produces a smoothly fused contraction called a **tetanus.** In mammalian muscle, the amplitude of the tetanus is usually three or four times the amplitude of a single twitch. In amphibian muscle, the tetanic response can exceed ten times the amplitude of a single twitch.

Sustained calcium in the cytoplasm permits summation and tetanus

As we noted earlier, a sufficient number of Ca^{2+} ions is released from the SR to allow every TN–TM complex to move away from the myosin-binding sites on the actin thin filaments. Thus, every cross-bridge in every stimulated muscle fiber is capable of interacting with actin and pulling the thin filaments toward the center of the sarcomere. If all cross-bridges are fully engaged, how is it possible to produce a tetanic force several times greater than the response to a single stimulus?

The answer is that the contractile apparatus requires *time* to pull on the various elastic components of the muscle. The connective tissue associated with the muscle fibers and the elastic components of the myofibrillar apparatus lie in series with each other, and they have elastic properties. The elastic structures are referred to as *series elastic elements*. For maximal tension to be recorded at the ends of the muscle fiber, the elastic elements must be stretched taut.

The Ca^{2+} released into the cytoplasm[5] by a single action potential is pumped back into the SR before the cross-bridges can stretch out the elastic elements fully. Successive action potentials, however, open the RyR channels frequently enough that the cytoplasmic concentration of Ca^{2+} keeps the actin-binding sites for myosin exposed, and cross-bridges can cycle repeatedly until the full contractile potential of the muscle fiber is realized.

The amount of tension developed by a muscle depends on the length of the muscle at the time it is stimulated

Whole skeletal muscles, because of their attachments to bones (or to exoskeleton in invertebrate animals), do not change greatly in length.

[5] The condition of the muscle fiber during the time Ca^{2+} ions are available to permit cross-bridge action is often referred to as the *active state*.

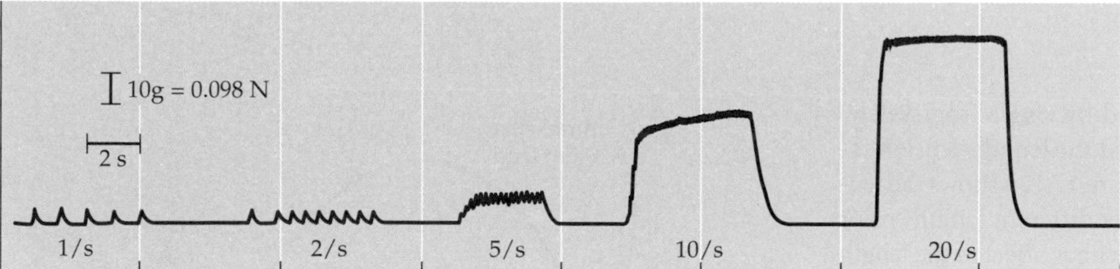

Figure 17.11 Summation and tetanus Increasing the frequency of stimulation produces summation of twitches up to a maximum contractile response called *fused tetanus*. In this example short trains of stimuli were applied to the sciatic nerve that innervates the gastrocnemius muscle of a frog. The muscle was allowed to rest briefly between trains of applied stimuli. No summation occurred when stimuli were applied at a rate of 1/s. Fused tetanus occurred at 20 stimuli/s. Experimenters use known weights to calibrate the recording apparatus. Because weight is the magnitude of the force of gravity on an object, it is expressed as the product of the mass of the object (kg) times the strength of the gravitational field (9.8 N/kg). Therefore, tension produced by the muscle is accurately expressed in units of Newtons (N). (Published values of muscle tension are often expressed in units of grams instead of Newtons.)

Nevertheless, muscles develop the most tension if they start contracting at an ideal initial length. Isometric recordings from isolated whole muscles illustrate this idea. Figure 17.12*a* shows the tension produced by a muscle when it was set at several different lengths prior to stimulation. Maximal tension was achieved when the muscle was set at lengths near its normal relaxed length in the animal. When the muscle was set at shorter lengths or stretched to longer lengths, the development of tension dropped off.

This **length–tension relationship** is entirely explained by the sliding-filament model of muscle contraction. Elegant

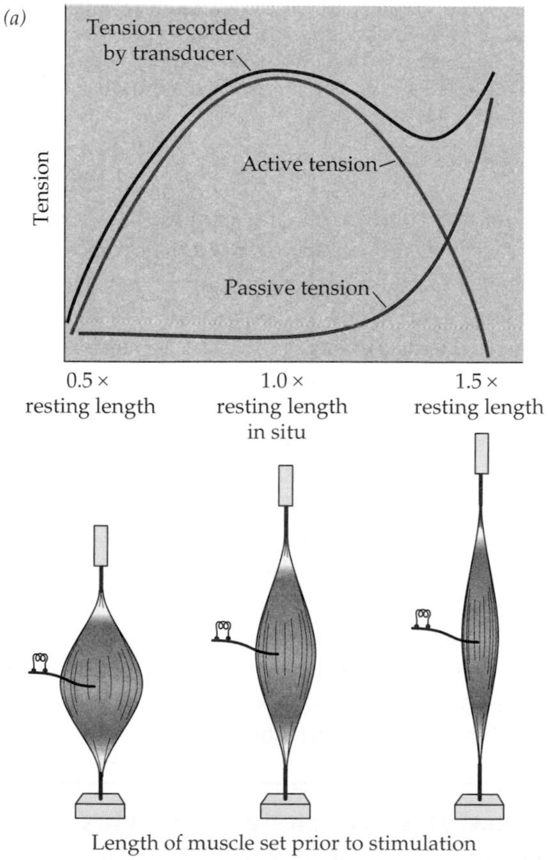

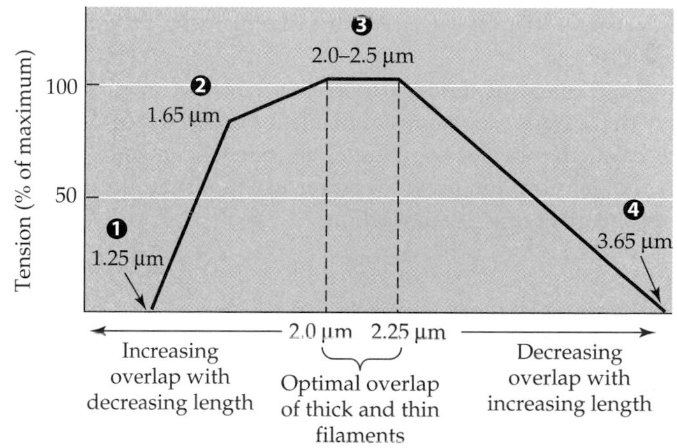

(a)

Tension recorded by transducer

Active tension

Passive tension

Tension

0.5 × resting length

1.0 × resting length in situ

1.5 × resting length

Length of muscle set prior to stimulation

(b)

Tension (% of maximum)

100

50

❶ 1.25 µm

❷ 1.65 µm

❸ 2.0–2.5 µm

❹ 3.65 µm

2.0 µm 2.25 µm

Increasing overlap with decreasing length

Optimal overlap of thick and thin filaments

Decreasing overlap with increasing length

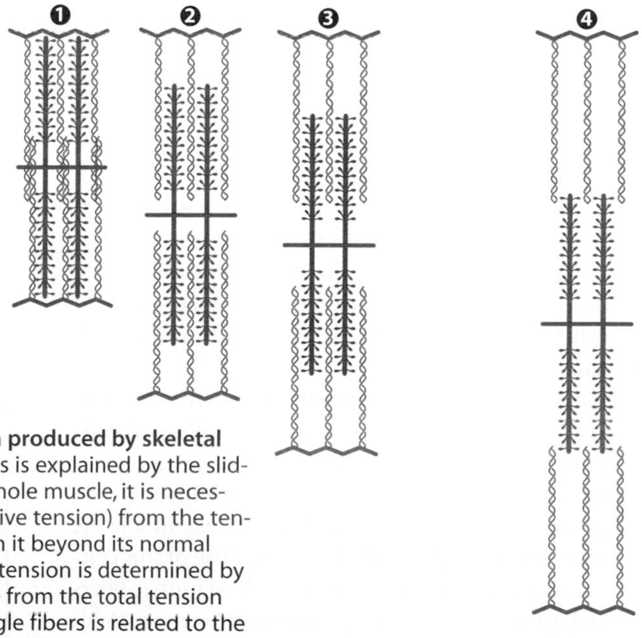

❶ ❷ ❸ ❹

Figure 17.12 The relationship between the length and the tension produced by skeletal muscle The length–tension relationship observed in whole muscles is explained by the sliding-filament theory. *(a)* When isometric tension is recorded from a whole muscle, it is necessary to distinguish the tension developed by the sarcomeres (the active tension) from the tension produced by the experimenter's pulling on the muscle to stretch it beyond its normal length (passive tension). The transducer records total tension. Active tension is determined by subtraction of the tension produced by passive stretch of the muscle from the total tension recorded during contraction. *(b)* Isometric tension recorded from single fibers is related to the sarcomere length set prior to stimulation. (*b* from Gordon, Huxley, and Julian 1966).

isometric recordings of tension developed by *single* frog skeletal muscle fibers unambiguously demonstrated the length–tension relationship at the level of the sarcomeres. Figure 17.12*b* shows the tension developed by single muscle fibers set at different lengths prior to stimulation. The set length of the muscle fiber affected the length of the sarcomeres within it and therefore the degree of overlap of the thick and thin filaments within each sarcomere. The experimenters plotted the amount of tension developed upon stimulation as a function of sarcomere length. Maximum tensions were recorded when the sarcomere lengths were set near those found in the intact animal.

The diagrams in Figure 17.12*b* illustrate that, at the lengths that yielded maximum tension, the overlap of thick and thin filaments permits optimal cross-bridge binding with actin. Stretching or compressing the sarcomeres led to less tension developed in response to stimulation. Sarcomeres set at longer-than-ideal lengths have less overlap of thick and thin filaments and therefore fewer available sites for myosin cross-bridges to bind. At sarcomere lengths that are shorter than ideal, the thin filaments overlap the H zone, where there are no myosin cross-bridges at all. At even shorter lengths, the thin filaments overlap each other, probably interfering with myosin cross-bridge action. Finally the thin filaments push up against the Z disc.

The striking agreement of the length–tension curve of single muscle fibers with the observed regions of filament overlap strongly implies that each cross-bridge contributes an independent and equal increment of tension, and provides powerful support for the sliding-filament theory of muscle contraction.

In general, the amount of work a muscle can do depends on its volume

Work performed by a muscle can best be understood by the use of isotonic recording. **Work** is the product of force produced by the muscle (i.e., tension) and the distance that the muscle shortens. Figure 17.13 shows the distance shortened by a muscle when it was given increasingly greater loads to lift. The muscle shortened the greatest distance when it lifted no load. It didn't shorten at all when the force of the load exceeded the maximum tension (force) it could develop. When the muscle lifted no load, although it shortened, it performed little work because it exerted negligible force (tension). When the muscle attempted to lift a very heavy load, it exerted isometric force (tension) but performed no work because it did not move the load. At intermediate loads, the muscle did increasing amounts of work, up to about 40% of the maximum load, and then it did progressively decreasing amounts of work while lifting loads of increasing mass.

The *force* exerted by a muscle is proportional to its cross-sectional area. In all muscles examined that use actin and myosin as contractile proteins, the diameters of the thick and thin filaments are essentially the same. This means that a cross section through the contractile components of any muscle would reveal essentially the same number of cross-bridges per unit of area. Because of this constant number of cross-bridges, most vertebrate skeletal muscle fibers (and many invertebrate muscle fibers) exert about the same amount of force per unit of area.

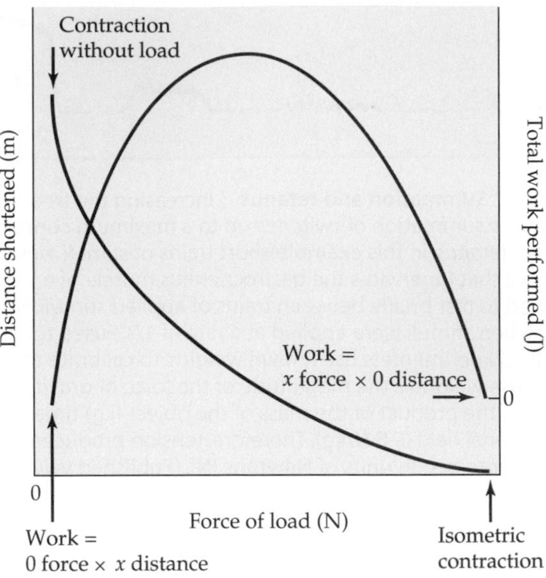

Figure 17.13 Work of contraction Isotonic recordings show that the muscle shortens the greatest distance when it lifts no load, and shortens progressively shorter distances with increasing loads. Multiplying the tension developed (equal to the force of the load, in Newtons, N) by the distance shortened for each load (in meters, m) gives a curve that represents work performed by the muscle (in Joules, J). (After Schmidt-Nielson 1997.)

The *length* of a muscle fiber does not contribute to the force it generates. However, the length is important in determining how much work the muscle can do. The sarcomeres in most vertebrates are about 2.5 μm long. Thus if each sarcomere contracts, for example, 10% of its length during a twitch, a muscle that has myofibrils consisting of 100 sarcomeres in series will shorten 25 μm. A muscle that had myofibrils consisting of 300 sarcomeres in series would shorten three times that distance. Assuming that both muscles had the same cross-sectional area, they would both exert the same force. However, the longer muscle would perform more work because work is the product of force times the distance shortened. Thus, muscles that have a greater volume of contractile machinery are generally capable of doing greater work.

Interestingly, the muscles of some animals have been so drastically modified that they possess hardly any contractile machinery at all. The electric organs of some fishes (Box 17.1) do not do contractile work. Instead, they produce electric shocks.

Summary: Whole Skeletal Muscles

- Cross-bridge activity within individual muscle fibers accounts for the tension (force) generated by a muscle. Force exerted by a muscle is proportional to the cross-sectional area of its contractile elements.
- The tension generated by a whole muscle is directly related to the number of actively contracting muscle fibers.
- The amount of tension developed by each contracting fiber in a muscle is determined by the frequency of action potentials

BOX 17.1 Electric Fish Exploit Modified Skeletal Muscles to Generate Electric Shocks

In addition to using skeletal muscles for locomotion, electric fish have modified them into **electric organs** used for stunning prey, exploring the environment, and even communicating. Most electric organs are made of flattened skeletal muscle cells that have few if any sarcomeres and do not contract. These cells, called *electrocytes*, are stacked into columns. A single column may consist of thousands of electrocytes, and an electric organ—depending on the fish—can contain 50 to 1000 columns.

Although the electrocytes don't contract, each responds to a neural signal by changing its membrane potential by some tens of millivolts. Because all of the electrocytes in a column respond to neural input simultaneously, their summed responses can produce large whole-animal voltage changes. The electric eel *Electrophorus electricus* from freshwater rivers of South America (see the figure) can produce electrical potentials of up to 600 V!

Some strongly electric fish were known in ancient times. Indeed, Aristotle described how the Mediterranean torpedo, which can deliver shocks of 50–60 V and several amperes of current, hides in wait to stun and then consume its prey. Francesco Redi (in 1671) and Stephano Lorenzini (in 1678) dissected torpedoes and concluded that the electric organ was derived from muscle. Nearly 300 years later, in the 1950s, fish with electric organs that generate only weak electric pulses were discovered in Africa and South America. Combined observations of electric organs in many unrelated fish indicate that electric organs have evolved independently at least six different times.

In different animals, electric organs are derived from different muscle lineages, including tail muscles, muscles of the body axis, and oculomotor muscles (muscles that move the eyes). The electrocytes, like all other vertebrate skeletal muscle cells, are innervated by presynaptic cholinergic motor neurons, possess postsynaptic nicotinic ACh receptors, and use acetylcholinesterase (AChE) to hydrolyze acetylcholine released upon stimulation. In addition to possessing electric organs that generate pulses of bioelectricity, most electric fishes also have electroreceptors (see Chapter 13) that give them the ability to detect electric fields from external sources. The strongly electric species are believed to use their electroreceptive capabilities to detect prey and potential predators. They use their electric organs to stun the prey and to ward off predators.

The weakly electric species use their electric organs for electrolocation and communication. Many weakly electric fish live in murky waters and are nocturnal. Electrolocation allows them to explore the environment without depending on vision. The electric organ generates constant discharges of a particular frequency that produce weak electric currents flowing out of the animal into the surrounding water. Nearby objects, which have different electrical properties from those of the water, alter the pattern of currents. These alterations are detected by electroreceptor organs in the skin of the animal, providing an "electric image."

For communication, weakly electric fish use discharges of their electric organs to indicate the sending fish's species, gender, and even individual identity. These signals are detected by the receiving fish's electroreceptor organs. Ongoing studies have revealed that the communication signals produced by electric organs can be modified by social interactions and environmental conditions.

In the process of learning how electric fish use electric organs in their native habitats, investigators have also made discoveries about convergent evolution, animal behavior, and cellular differentiation. Furthermore, because electric organs yield abundant amounts of nearly pure excitable membranes, they provide tissues for ever-more-sophisticated studies of channels and membrane receptors. Finally, the large quantities of acetylcholinesterase in many electric organs provide a plentiful source of enzyme for studies of anticholinesterases, which are of interest to both environmentalists (who find synthetic anticholinesterases in toxic wastes) and investigators seeking ways to prolong ACh signals at specific synapses.

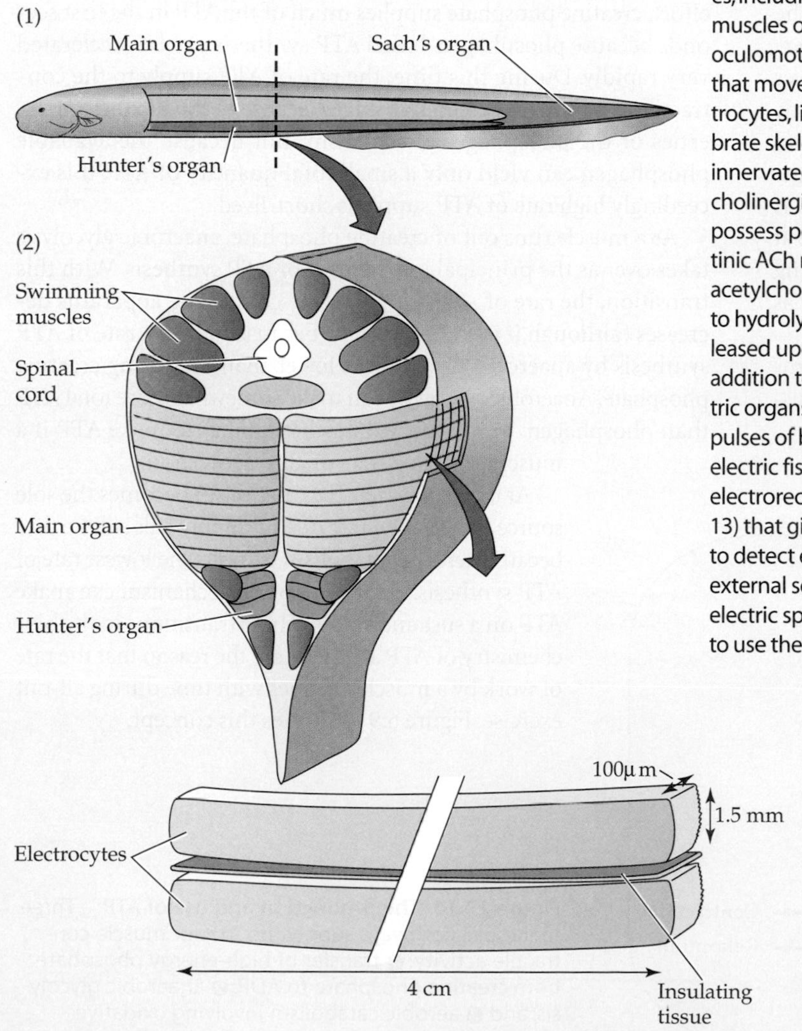

The electric eel *Electrophorus electricus* possesses both strong and weak electric organs (1) The strongly electric main organ and weakly electric Hunter's and Sach's organs form the bulk of the posterior mass of the animal. (2) A cross section reveals the electrocytes of several columns arranged on either side of the midline of the animal. Each column, which is separated by insulating tissue from adjacent columns, extends the length of the electric organ parallel with the long axis of the fish. Each electrocyte is shaped like a ribbon 4 cm long, 1.5 mm wide, and 100 μm thick. The posterior surface of each electrocyte receives a synapse (not shown) from a motor neuron extending from the spinal cord. (After Gotter, Kaetzel, and Dedman 1998.)

from its motor neuron (to produce summation of twitches and tetanus) and the length of the muscle fiber at the time it is stimulated (the length–tension relationship).

- The speed with which a muscle shortens decreases as the load it lifts increases (the load–velocity relationship).
- Work performed by a muscle is the product of tension (force) produced by the muscle and the distance it shortens.

Muscle Energetics

The muscle that does the work described in the previous section consumes energy to do so. In this section we examine the sources of energy available to muscle fibers and the ways in which energy is used by different types of muscle fibers.

ATP is the immediate source of energy for powering muscle contraction

ATP performs at least three functions in the contraction–relaxation cycle: (1) ATP binding to the cross-bridge is necessary for detachment of myosin from actin. (2) Hydrolysis of ATP activates the myosin cross-bridge in preparation for binding to actin and undergoing a power stroke. (3) Energy from the hydrolysis of ATP drives the calcium pump that transports Ca^{2+} ions into the sarcoplasmic reticulum. However, muscle contains only enough ATP (2–4 m*M*) to sustain contraction for a few seconds. Thus, nearly all forms of muscular work depend on regeneration of ATP while the muscle is working. The rate of muscular work strictly depends on the rate at which ATP is provided to the contractile apparatus, and a working muscle can be viewed as being in a race to resupply its ATP as fast as the ATP is used.

In broad outline, vertebrate muscle fibers possess three biochemical mechanisms that produce ATP (see Chapter 6 for a detailed discussion of ATP resupply):

1. *Use of the phosphagen creatine phosphate.* Phosphagens temporarily store high-energy phosphate bonds. The high-energy phosphate of creatine phosphate can be donated to ADP to produce ATP, as shown in Figure 6.5. Creatine phosphate is produced in resting muscle from creatine and ATP. The formation of ATP from creatine phosphate is driven by simple mass action.
2. *Anaerobic glycolysis.* This form of catabolism requires no oxygen (see Figure 6.3). It must have glucose or glycogen as fuel. In addition to ATP, it produces *lactic acid*, which in vertebrates is always retained in the body and disposed of metabolically.
3. *Aerobic catabolism.* This form of catabolism can use all three major classes of foodstuff as fuel (see Figures 6.1 and 6.2). It produces ATP principally by *oxidative phosphorylation.* Its other major products are CO_2 and H_2O.

Figure 17.14 summarizes major elements in the production and use of ATP in a vertebrate muscle fiber. As emphasized in Chapter 6, the three mechanisms of ATP production differ greatly in how fast they can make ATP when operating at peak output, how much ATP they can make, and how quickly they can accelerate their rate of ATP production. Table 17.1 summarizes attributes that were presented in greater detail in Table 6.1.

If a resting muscle is called upon suddenly to engage in all-out effort, creatine phosphate supplies much of the ATP in the first seconds because phosphagen-based ATP synthesis can be accelerated very rapidly. During this time, the rate of ATP supply to the contractile apparatus is exceedingly high because of the intrinsic properties of the phosphagen mechanism. But because the available phosphagen can yield only a small total quantity of ATP, this exceedingly high rate of ATP supply is short-lived.

As a muscle runs out of creatine phosphate, anaerobic glycolysis takes over as the principal mechanism of ATP synthesis. With this transition, the rate of ATP supply to the contractile apparatus decreases (although it is still very high) because the peak rate of ATP synthesis by anaerobic glycolysis is lower than that using creatine phosphate. Anaerobic glycolysis can make somewhat more total ATP than phosphagen, but it, too, exhausts its ability to make ATP if a muscle stays in a state of all-out exertion.

At that point, aerobic catabolism becomes the sole source of ATP. The rate of ATP supply falls still further because aerobic catabolism exhibits the lowest rate of ATP synthesis. But the aerobic mechanism can make ATP on a sustained basis. These transitions in the biochemistry of ATP synthesis are the reason that the rate of work by a muscle declines with time during all-out exercise. Figure 6.9 illustrates this concept.

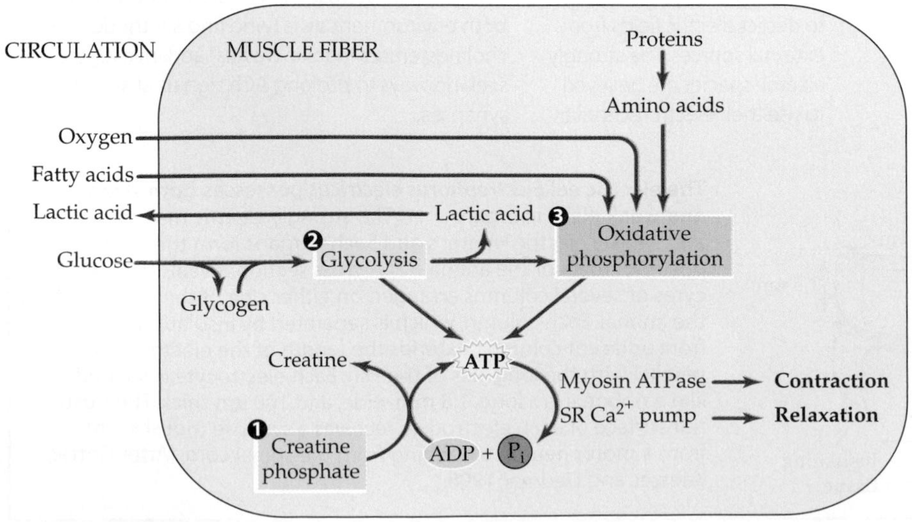

Figure 17.14 The production and use of ATP Three metabolic pathways supply the ATP for muscle contractile activity: ❶ transfer of high-energy phosphate from creatine phosphate to ADP; ❷ anaerobic glycolysis; and ❸ aerobic catabolism involving oxidative phosphorylation.

TABLE 17.1 Characteristics of the three principal mechanisms of ATP production in vertebrate muscle

	Aerobic catabolism	Anaerobic glycolysis	Use of phosphagen
Peak rate of ATP synthesis	Low	High	Very high
Maximum yield of ATP in one episode of use	Very high (indefinite)	Low	Very low
Rate of acceleration	Low	High	High

Note: See Table 6.1 for more detail.

Vertebrate muscle fibers are classified into different types

As discussed in Chapter 6, muscles of an animal are typically made up of a heterogeneous population of muscle fibers. By far the most common types of muscle fibers are **twitch fibers**. These fibers generate action potentials, and each action potential gives rise to a muscle twitch. Twitch fibers vary in the biochemical and metabolic features that give them different contractile abilities, as we discuss shortly. **Tonic muscle fibers** are relatively rare. They are found mainly in postural muscles of lower vertebrates.[6] Tonic muscle fibers do not generate action potentials, and they contract more slowly than the slowest twitch fibers. Their slow cross-bridge cycling permits the maintenance of long-lasting contractions with low energetic costs.

Twitch fibers are generally classified into three main categories: *slow oxidative* (*SO*), *fast oxidative* (*FO*), and *fast glycolytic* (*FG*) *fibers*. Twitch fibers are classified on the basis of differences in isoforms of the myosin ATPase and metabolic features of the cells. In mammals, the myosin heads in FG and FO fibers split about 600 ATP molecules per second, whereas those in SO fibers split ATP about half as rapidly. Because the rate of ATP hydrolysis governs the rate of cross-bridge cycling, higher ATPase activity allows faster contraction.

It is important to remember that the amount of tension developed *per cross-bridge cycle* is the same in both fast and slow types of muscle, but the number of cycles accomplished per unit of time differs. Earlier we saw that the velocity of contraction of a muscle fiber depends on the load being moved (the *load–velocity relationship*, or *force–velocity relationship*). This relationship applies to all muscle fibers but varies depending on the type. Thus the velocity of contraction depends on *both* the type of muscle fiber and the load against which it exerts force.

Recall from Chapter 6 (see page 160) that the SO fibers are mitochondria-rich and poised to make ATP principally by aerobic catabolism; they have small diameters and are rich in myoglobin (an intracellular hemoglobin that facilitates O_2 transport), red (because of the myoglobin), well supplied with blood capillaries, and slow to fatigue. The FG fibers have large diameters, are invested with fewer capillaries, have little myoglobin, and are *white*. They have few mitochondria and make ATP mainly by anaerobic glycolysis; they are rich in glycogen (the principal fuel of anaerobic glycolysis), quickly accumulate lactic acid, and fatigue rapidly. The FO fibers are intermediate. Although their isoform of myosin is different from that of FG fibers, the activity of their myosin ATPase is high, and they are therefore capable of rapid tension development. Unlike FG fibers, however, FO fibers are relatively rich in mitochondria, and because they make ATP aerobically, they are relatively resistant to fatigue.

Figure 17.15 shows that different types of fibers are often intermingled within a muscle. Table 17.2 provides a rough comparative guide to the three fiber types. The distinctions, while useful, should not be viewed too rigidly, however, because the fibers vary considerably in other characteristics. For example, the fibers of a given type may differ from each other because they have different isoforms of troponin, tropomyosin, or other proteins. Furthermore, varying conditions of use can cause one type of fiber to be converted into another.

Different fiber types are specifically adapted to subserve different functions that give muscles a broad repertoire of contractile abilities. SO fibers (and tonic fibers when present) do not generate much tension, but they operate efficiently and without fatigue. They are adapted for isometric postural functions and for small, slow movements. The myosin isoform in SO fibers has low ATPase activity and can therefore produce tension economically. FO fibers generate more tension and faster contractions, yet they are fatigue-resistant. They are adapted for repeated movements such as locomotion. FG fibers generate rapid contractions and large increments of tension, but they lack endurance. They are used for occasional, forceful, fast movements such as leaps or bursts of speed in escape or prey capture.

The ankle extensors in the cat hindlimb illustrate the functional roles of different fiber types. Three muscles—the soleus, medial gastrocnemius, and lateral gastrocnemius—comprise the ankle extensors. They all insert on the Achilles tendon at the heel. The soleus contracts slowly and consists entirely of SO fibers. It is most active in postural standing. The medial and lateral gastrocnemii are faster

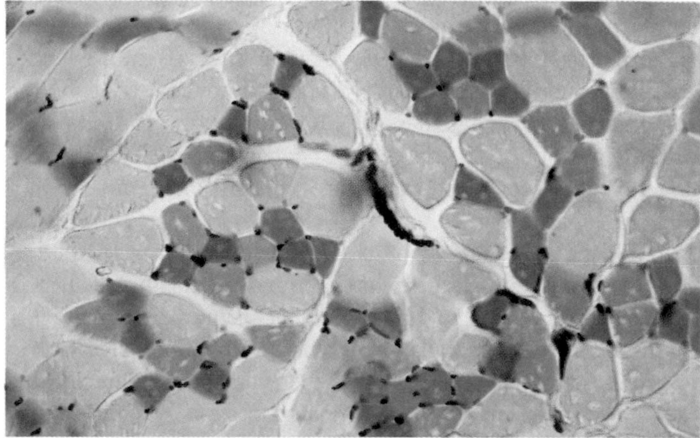

Figure 17.15 Whole muscles typically consist of mixtures of different types of fibers In this transverse section, small-diameter oxidative fibers stain dark because of their abundant mitochondria, and they are surrounded by many capillaries. Large-diameter glycolytic fibers have fewer mitochondria and fewer immediately adjacent capillaries.

[6] In mammals, tonic fibers occur only in extraocular muscles and as intrafusal fibers of muscle spindles.

TABLE 17.2 Characteristics of mammalian skeletal twitch muscle fibers

	Slow-oxidative	Fast-oxidative	Fast-glycolytic
Myosin ATPase activity	Slow	Fast	Fast
Speed to reach peak tension	Slow	Intermediate to fast	Fast
Duration of twitches	Long	Short	Short
Rate of Ca^{2+} uptake by sarcoplasmic reticulum	Intermediate	High	High
Resistance to fatigue	High	Intermediate	Low
Number of mitochondria	Many	Many	Few
Myoglobin content	High	High	Low
Color	Red	Red	White
Diameter of fiber	Small	Intermediate	Large
Number of surrounding capillaries	Many	Many	Few
Enzymes for glycolysis	Low	Intermediate	High
Ability to produce ATP using oxidative phosphorylation	High	High	Low
Force developed per cross-sectional area of entire fiber	Low	Intermediate	High
Function in animal	Posture	Standing, walking, rapid repetitive movements	Jumping, bursts of high-speed locomotion
Frequency of use by animal	High	Intermediate to high	Low

Sources: After Silverthorn 2004; Randall et al. 2002; and Sherwood 2004.
Note: The names of different types of skeletal muscle fibers vary in the literature. Slow-oxidative fibers are also called Type I; and fast-oxidative fibers, Type IIa. The fast-glycolytic fibers are Type IIb in small mammals and Type IIx in humans.

muscles of mixed fiber composition. For example, the medial gastrocnemius contains approximately 45% FG, 25% FO, and 25% SO fibers. (The remaining 5% are intermediate in their properties between FG and FO fibers.)

Because the FG fibers have relatively greater diameters (see Table 17.2), the 45% that they contribute to the muscle fibers contributes 75% of the maximal total tension of the medial gastrocnemius. However, walking and most running use only about 25% of the maximal tension of the medial gastrocnemius. This is the amount of tension produced by the FO and SO fibers without any contribution from the FG fibers. Thus these fatigue-resistant fibers are sufficient for most locomotion. The large force contributed by FG fibers is believed to be reserved for short bursts of contraction required in motions such as jumping.

In fish, the trunk muscles of the body are divided into separate regions of red slow muscle and white fast muscle. The muscle fibers in the two regions bear many histological, biochemical, and physiological similarities to mammalian fiber types. SO-like fibers are found in the red muscle, and FG-like fibers in the white muscle. The slow red muscle is less than 10% of the total trunk muscle in most fish species, and it never exceeds 25%. Yet only the slow red muscle is used at all speeds of steady cruising. The white fast muscle that constitutes the great bulk of the muscle mass is used only for bursts of high-speed swimming, and it fatigues very rapidly. The shear size of the white muscle is a testimony to the extreme importance of being able to accelerate rapidly when necessary to capture food or escape a predator.

Several animals possess exceptional muscles that are adapted for very rapid contractions. Box 17.2 describes the special organization of insect flight muscles that produce exceptionally rapid wing beats. Certain vertebrates possess rapidly contracting muscles that consist of fibers that are oxidative and fatigue-resistant. Hummingbird flight muscle, for example, can contract and relax at frequencies approaching 80 times/s (hertz, Hz), so the contraction–relaxation cycle is completed in less than 15 ms. Sound-producing muscles of insects, fish, birds, and bats can be even faster.[7] In all these cases there are extreme adaptations for rapid generation of tension, and also rapid relaxation.

Experimenters have shown that three main factors contribute to increased speeds of contraction: (1) myosin isoforms capable of rapid cross-bridge cycling, (2) troponin isoforms that have a low affinity for Ca^{2+} so that Ca^{2+} unbinds rapidly, and (3) increased density of Ca^{2+}-ATPase pumps in the SR for rapid relaxation. Large amounts of ATP are required to support rapid cross-bridge cycling and pump functions, and not surprisingly, these muscles require a well-developed SR, many mitochondria, and a rich supply of capillaries to deliver O_2 and nutrients.

The benefit of rapid contraction brings with it a cost of limited ability to generate tension. The reason is that space in cells is limited. In most muscles used for locomotion, about 90% of the space is filled by myofibrils; mitochondria, glycogen, and SR fill the remaining 10% of space. Consider the tail-shaker muscle of the rattlesnake, which can produce contractions up to 90 Hz (at optimum temperatures). Rattlesnakes make themselves conspicuous by rattling their tails continuously, sometimes for hours. The shaker-muscle fibers have high metabolic demands and require reserves of fuel, abundant mitochondria, as well as extensive SR.

The space required for these "supporting" components necessarily limits the space available for contractile proteins (the tension-generating components). Indeed, in rattlesnake shaker muscle fibers, only about 30% of the space is occupied by myofibrils. The remaining space is filled by SR (26%), glycogen (17%), and mitochondria (26%). The diminished contractile machinery results in less ability to generate tension. These muscle fibers illustrate a general point: that space in cells can be at a premium, and thus trade-offs may be required among various cell compartments.

[7] The sound-producing muscles of the male toadfish swim bladder are the fastest known vertebrate muscles, contracting at frequencies of up to 200 Hz. In insects, the sound-producing muscle of the shrill-chirping male cicada can contract and relax at a frequency of 550 Hz!

BOX 17.2 Insect Flight

Humans have long admired and envied the ability of other animals to fly. Insects especially—from lazily looping butterflies to dive-bombing mosquitoes—have captivated our attention. Physiologists studying the flight muscles of insects have begun to reveal the biological mechanisms that underlie insects' aerial feats. Insect flight muscles possess the familiar features of striated skeletal muscle fibers found in other animals. The myofibrils are organized into sarcomeres; t-tubules and sarcoplasmic reticulum are present; and Ca^{2+} ions bind to the TN–TM complex to permit cross-bridge cycling that produces tension.

Some insects, such as the dragonfly or locust, use **synchronous flight muscles** (1), so named because each muscle contraction is synchronized with the action potential that initiated it, as is the case in vertebrate skeletal muscle. Synchronous flight muscles are arranged vertically to the long axis of the animal, with one end attached to the wings and the other to the floor of the thorax. Contraction of the medial elevator muscles pulls the wings up, and contraction of the lateral depressor muscles pulls the wings down.

Several species of insects, (including flies, wasps, bees, and beetles), have evolved muscles that are unique to insects—**asynchronous flight muscles** (2)—that are capable of contracting at much faster frequencies than synchronous flight muscles or any other known muscles can. In asynchronous muscles, individual contractions are not synchronized with individual nerve action potentials. Instead they typically produce many oscillating contractions for each action potential. Each contraction produces a wing beat. Wingbeat frequencies of more than 1000 beats/s have been recorded in a midge! In addition, unlike fast-contracting muscles in other animals, such as the rattlesnake shaker muscle, asynchronous flight muscles also develop large amounts of tension with economy of ATP. These attributes give insects stunning agility to accomplish behaviors such as evading predators and seeking out and competing for mates.

Asynchronous muscles are not attached directly to the wings. Instead, vertical and longitudinal muscles attach to the walls of the thorax (see part *2* of the figure). These opposing pairs of muscles make use of the elastic properties of the thorax. When the vertical elevator muscles contract, they pull down on the roof of the thorax and deform its sides, causing the wings to move up through a hinge arrangement. When the longitudinal depressor muscles contract, the roof of the thorax bulges up and has the effect of moving the wings down.

The most striking functional feature of asynchronous flight muscle is that a single nerve action potential (which causes the release of Ca^{2+} from the SR) initiates a series of subsequent contractions that are each triggered by stretch. When the longitudinal muscles contract to elevate the roof of the thorax, they stretch the vertical muscles. The imposed stretch activates the contractile machinery of the vertical muscles. When the vertical muscles contract, they pull down on the roof of the thorax, and in turn stretch the longitudinal muscles. The release of elastic strain stored in the thoracic wall by contraction of one muscle aids in the stretching of the opposing muscle. Because the two sets of muscles are out of phase, contraction of one induces contraction of the other, and vice versa.

The frequency of the alternating contraction of vertical and longitudinal muscles depends on the resonant mechanical properties

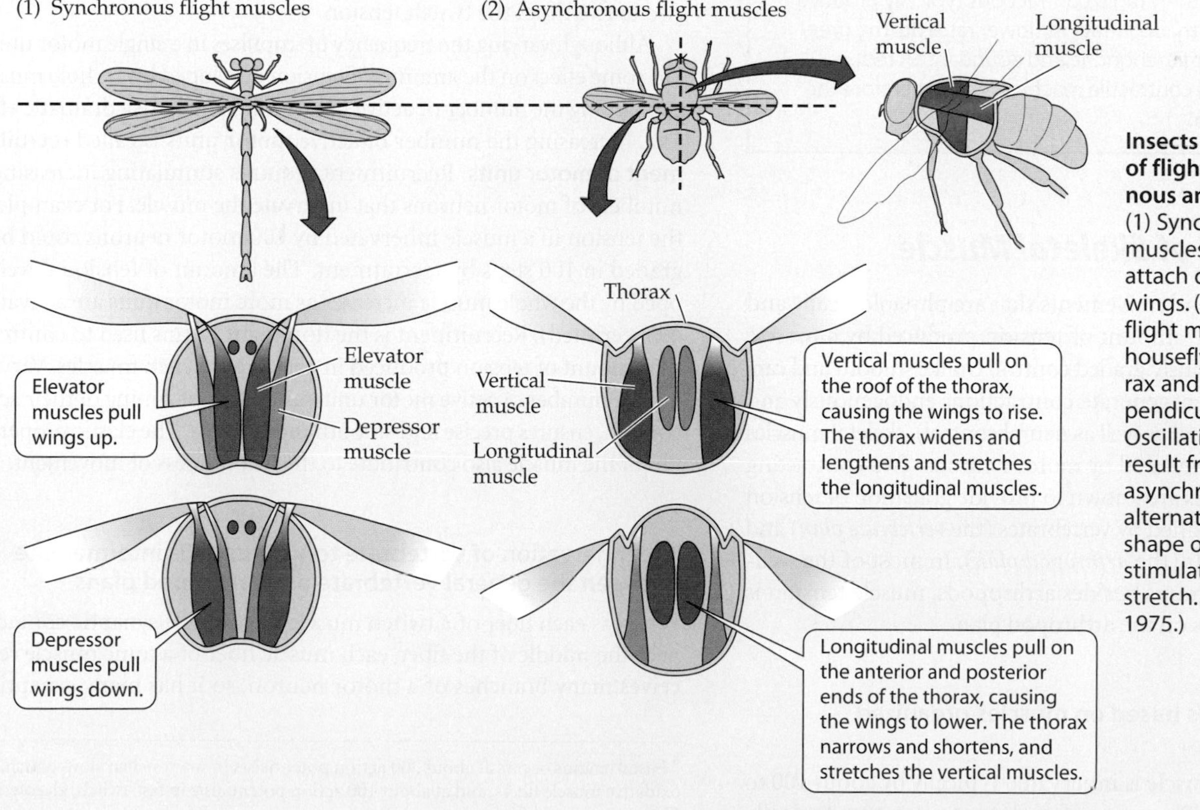

Insects exhibit two types of flight muscles: synchronous and asynchronous (1) Synchronous flight muscles of the dragonfly attach directly to the wings. (2) Asynchronous flight muscles of the housefly attach to the thorax and are arranged perpendicular to each other. Oscillating wing beats result from the two sets of asynchronous muscles alternately changing the shape of the thorax and stimulating each other by stretch. (After Pringle 1975.)

BOX 17.2 *(Continued)*

of the thorax and wings, not on the frequency of nerve action potentials. If the wings are clipped, the flight frequency increases because of decreased wing inertia and air resistance, but the frequency of initiating nerve action potentials is unchanged. As long as there is sufficient Ca^{2+} in the cytoplasm, the cross-bridges will function. Over time, the Ca^{2+} is pumped back into the SR, and contractions will cease until another action potential is generated by the motor neuron.

This property of many contraction–relaxation cycles happening in the continuous presence of Ca^{2+} ions in the cytoplasm is distinctly different from all other known skeletal muscle fibers. In all other fibers, Ca^{2+} ions are released from the SR and taken back up with each contraction–relaxation cycle. This difference provides an explanation of how asynchronous flight muscles can produce impressive amounts of tension with metabolic efficiency. The SR in asynchronous muscle occupies only about 3% of the volume of the muscle cells, simply because the muscle fiber does not pump Ca^{2+} ions into the SR with each contraction–relaxation cycle. Furthermore, because less ATP is needed to power the SR Ca^{2+}-ATPase pumps, fewer mitochondria are required to support the total metabolism of the cell, which is nevertheless significant because each cross-bridge cycle requires a molecule of ATP. The space that would have been occupied by both mitochondria and SR can instead be occupied by tension-generating myofibrils.

Thus, by evolving asynchronous muscles, insects saved space and energy. Because asynchronous muscles are found in many different insect groups, it is thought that asynchronous muscles evolved as many as seven to ten times. This example of convergent evolution suggests that asynchronous muscles have contributed to the impressive success that insects have achieved in terrestrial environments.

Summary: Muscle Energetics

- Contractile activity requires the hydrolysis of ATP to provide energy for cross-bridge power strokes and to support the Ca^{2+}-ATPase pumps of the sarcoplasmic reticulum.
- ATP is produced by three principal means: (1) transfer of the high-energy phosphate from creatine phosphate to ADP, (2) glycolysis, and (3) oxidative phosphorylation.
- Vertebrate muscle fibers are classified into different types on the basis of their biochemical and histological features, and each type is adapted to subserve different functions. Muscles usually contain a mixture of different fiber types.
- Muscles adapted for extremely rapid contractions typically produce less tension than muscles that contract at slower rates do. The presence of large numbers of mitochondria and abundant SR reduces the cross-sectional area of contractile machinery, and therefore the ability to generate tension.

Neural Control of Skeletal Muscle

To produce smooth, fluid movements that are physiologically and behaviorally useful, the amount of tension produced by a muscle must be under continuously graded control. Unlike smooth and cardiac muscles (which may generate contractions endogenously and may respond to hormonal as well as neural control), skeletal muscles contract only when stimulated by motor neurons. Two contrasting evolutionary approaches are known to provide gradation of tension in a muscle, one exemplified by vertebrates (the *vertebrate plan*) and the other by arthropods (the *arthropod plan*). In most of the well-studied invertebrate groups besides arthropods, muscle tension is controlled by variations on the arthropod plan.

The vertebrate plan is based on muscles organized into motor units

A vertebrate skeletal muscle is innervated typically by about 100 to 1000 motor neurons. The axon of each motor neuron typically branches to innervate multiple muscle fibers, and each muscle fiber receives synaptic contact by only one motor neuron. A motor neuron and all the muscle fibers it innervates are collectively termed a **motor unit** (Figure 17.16). When the motor neuron generates an action potential, all the muscle fibers in the motor unit generate action potentials and contract to produce a twitch. Trains of action potentials of increasing frequencies can produce summation of twitches up to fused tetanic contraction.[8] Thus the amount of tension produced by a single motor unit can be varied by variation in the frequency of action potentials generated by the motor neuron. In many vertebrate muscles, however, the amount of tetanic tension is only two to five times the twitch tension.

Although varying the frequency of impulses in a single motor unit has some effect on the amount of tension developed by a whole muscle, varying the number of active motor units can have a dramatic effect. Increasing the number of active motor units is called **recruitment of motor units**. Recruitment requires stimulating increasing numbers of motor neurons that innervate the muscle. For example, the tension in a muscle innervated by 100 motor neurons could be graded in 100 steps by recruitment. The amount of tension developed by the whole muscle increases as more motor units are activated (recruited). Recruitment is the dominant means used to control the amount of tension produced in vertebrate twitch muscles. Varying the number of active motor units, as well as the timing of their activation, ensures precise and smooth movements. The elastic properties of the muscle also contribute to the smoothness of movement.

The innervation of vertebrate tonic muscle is intermediate between the general vertebrate and arthropod plans

Whereas each fiber of a twitch muscle has a single synaptic contact near the middle of the fiber, each muscle fiber of a tonic muscle receives many branches of a motor neuron, so it has many synaptic

[8] Fused tetanus occurs at about 300 action potentials/s in mammalian slow-twitch, oxidative muscle fibers, and at about 100 action potentials/s in fast-twitch, glycolytic fibers.

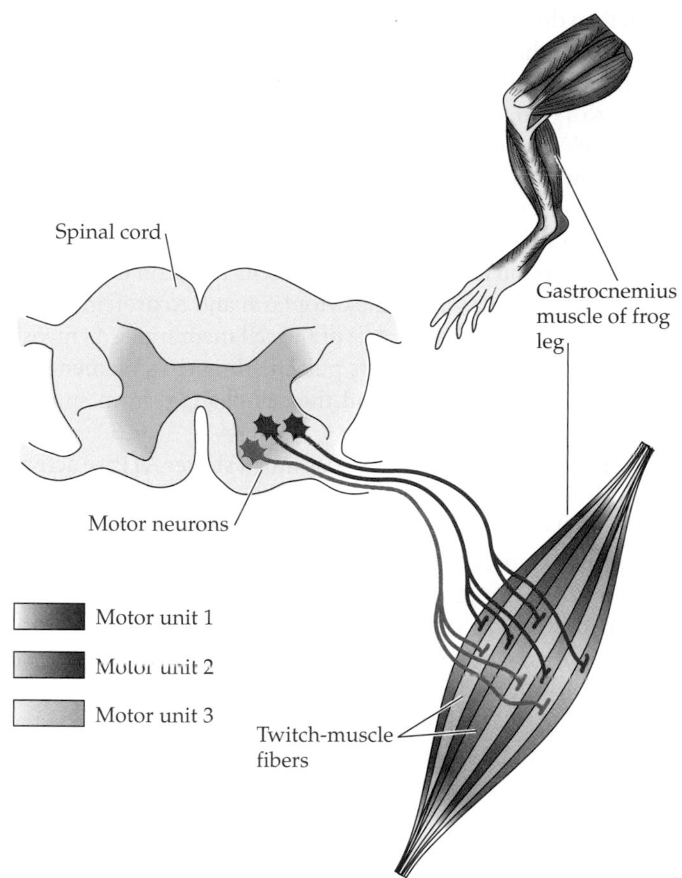

Figure 17.16 Vertebrate skeletal muscles consist of many different, independent motor units An action potential in the motor neuron of one motor unit stimulates an action potential and contraction in all of the muscle fibers it innervates. Varying the number of active motor units varies the amount of tension produced by the whole muscle.

contacts distributed over its length. This pattern, shown in Figure 17.17*a*, is termed **multiterminal innervation.** An action potential generated by a motor neuron results in an excitatory postsynaptic potential (EPSP) at each of the distributed junctions. The muscle fiber has little or no ability to generate action potentials, and the amount of tension generated depends directly on the amount of depolarization produced by the EPSPs.

The arthropod plan is based on multiterminal innervation of each muscle fiber by more than one neuron

Although the fibers of arthropod skeletal muscles share many features of vertebrate skeletal muscle, including the organization of thick and thin filaments into sarcomeres and excitation–contraction coupling by way of t-tubules and SR, they show interesting differences in their pattern of innervation. A typical arthropod muscle is innervated by one to ten motor neurons, in contrast to the hundreds or thousands of motor neurons that innervate a vertebrate muscle. Most individual arthropod muscle *fibers* are innervated by more than one motor neuron, a pattern termed **polyneuronal innervation** (Figure 17.17*b*).

As in tonic muscle, each neuron in arthropod skeletal muscle branches to provide multiterminal innervation. Arthropod muscle fibers typically do not generate all-or-none action potentials. (Insect flight muscles, which do generate action potentials, are an exception.) Because arthropod muscle fibers are polyneuronally innervated, the motor units of arthropods overlap; each muscle fiber is part of several motor units. Thus, arthropods have only a few overlapping motor units per muscle, whereas vertebrates have many, nonoverlapping motor units per muscle.

Some arthropod muscles are innervated by both excitatory and inhibitory motor neurons. This feature—distinctly different from vertebrate muscles, which are innervated solely by excitatory neurons—allows *peripheral inhibition*. The algebraic summation of

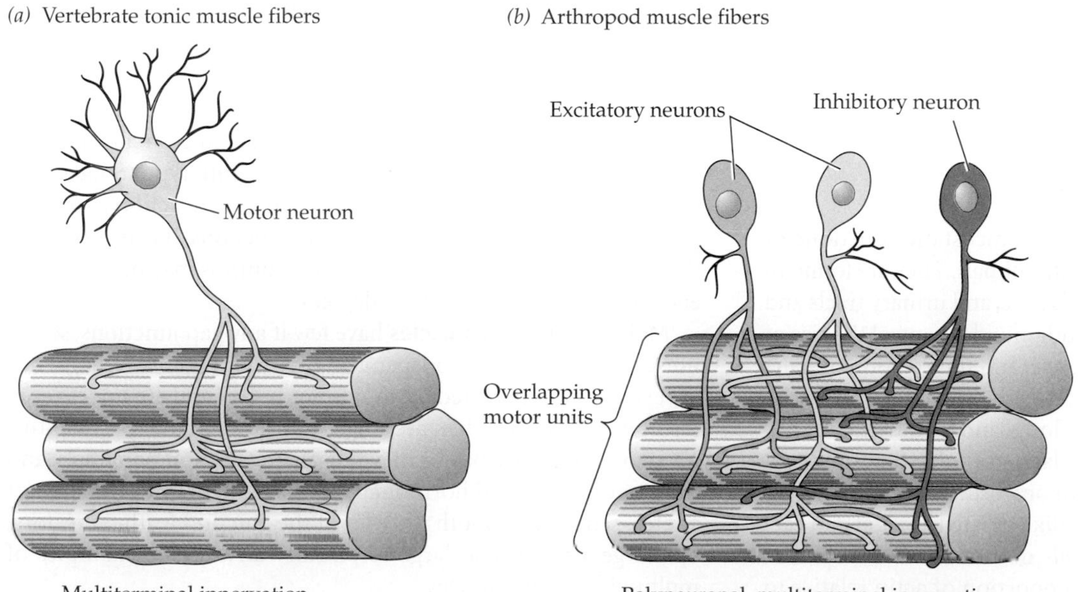

Figure 17.17 Innervation patterns of vertebrate tonic muscle fibers and arthropod muscle fibers *(a)* Each vertebrate tonic muscle fiber is innervated by a single axon that branches to make many synaptic contacts along the length of the fiber. Tonic fibers do not produce action potentials. *(b)* Muscles innervated according to the arthropod plan have overlapping motor units, multiterminal and polyneuronal innervation, and often peripheral inhibition. These muscle fibers often do not produce action potentials.

graded IPSPs and EPSPs *in the muscle fiber* determines the amount of tension developed. The greater the depolarization, the greater the amount of Ca^{2+} released from the SR, and the greater the tension developed. Thus the dominant mechanism for controlling tension in arthropod muscles is *controlling the degree of depolarization of muscle fibers*, which in turn depends on the frequency of action potentials in the excitatory and inhibitory motor neurons.

Arthropod muscle fibers have a range of speeds of contraction, but unlike vertebrate fibers, the velocity of contraction of arthropod muscle fibers is associated with different sarcomere lengths: Short-sarcomere fibers contract quickly, and long-sarcomere fibers contract slowly. Most arthropod muscles contain a variety of fibers with different sarcomere lengths and contraction speeds. Some muscles, however, are composed of all long-sarcomere slow fibers or all short-sarcomere fast fibers. One example of such a division is the group of abdominal flexor and extensor muscles of crayfish and lobsters. The slow flexor and extensor muscles each receive up to five excitatory motor neurons and one inhibitory neuron. Many of the fast muscles receive three excitatory axons and one inhibitory axon.

Summary: Neural Control of Skeletal Muscle

- The neuromuscular organization of vertebrates is characterized by many nonoverlapping motor units, each controlled by a single motor neuron. Each muscle fiber within a motor unit generates an action potential that spreads rapidly over the entire cell membrane and triggers the contractile response.
- The neuromuscular organization of arthropods is characterized by few motor neurons, overlapping motor units, and in some cases, peripheral inhibition. Each muscle fiber is typically innervated by more than one motor neuron, and each neuron makes multiple synaptic contacts on the fiber. The postsynaptic potentials produced at several points along the length of the fiber provide graded electric signals that trigger the contractile machinery and control the degree of tension developed.
- Vertebrate tonic fibers usually don't generate action potentials. Each fiber is typically innervated by a single motor neuron that makes multiple synaptic contacts along its length.

Vertebrate Smooth Muscle

Smooth muscles are important in the homeostatic functions of many different systems within vertebrate animals. They are found in the gastrointestinal, respiratory, reproductive, and urinary tracts and in the blood vessels. In addition, smooth muscles control the size of the pupil and shape of the lens in the eye, and the erection of hairs or feathers (see Chapter 8, page 219). Smooth muscle cells are small, spindle-shaped, typically 2 to 10 μm long, and uninucleate. They have a very rudimentary SR, and they lack t-tubules. Smooth muscle cells contain myosin, actin, and tropomyosin, but these proteins are not organized into sarcomeres. Unlike striated skeletal and cardiac muscle cells, smooth muscle cells do not contain troponin. Smooth muscle cells have a greater proportion of actin relative to myosin than do striated muscles—a difference that is reflected in the larger ratio of thin to thick filaments in smooth muscles (about 12 to 15 thin filaments per thick filament) relative to striated muscles (about 2 to 4 thin filaments per thick filament).

Although smooth muscle cells lack sarcomeres, electron micrographs show cross-bridges between thick and thin filaments, and contraction is accomplished by the sliding of the filaments past one another. The actin and myosin filaments are organized into diagonal bundles around the periphery of the cell. The actin filaments attach to **dense bodies** within the cytoplasm and to protein *attachment plaques* on the inner surface of the cell membrane. As myosin molecules of the thick filaments pull the thin actin filaments together, the bundles shorten, and the cell changes from spindle-shaped to globular.

The myosin in smooth muscle has much slower ATPase activity than the myosin isoforms found in different types of skeletal and cardiac muscle do. Thus, cross-bridge power strokes occur at a lower rate in smooth muscle. The contractions produced by smooth muscles last much longer than those of skeletal and cardiac muscles, and many smooth muscles can maintain contractions for long periods of time with only small expenditures of energy. The smooth muscle of the esophageal sphincter that guards the opening of the stomach is a good example. Except when food is swallowed, this muscle stays contracted continuously and prevents stomach acid and enzymes from entering the esophagus.

Smooth muscles perform a multitude of functions in the maintenance of homeostasis, and not surprisingly, they show considerable variety in their structural organization, functional properties, and control. Whereas skeletal muscle fibers are excited only via action potentials in the motor neurons innervating them, and cardiac muscle is activated by electrotonic spread of depolarization from spontaneously active pacemaker regions (which may be modulated by neural or hormonal signals), different smooth muscles can be excited spontaneously, or by neurotransmitters from the autonomic nervous system, hormones, paracrines, or even stretch.

One useful classification scheme differentiates vertebrate smooth muscle into two main types: *single-unit* and *multiunit* smooth muscles (Figure 17.18). In **single-unit smooth muscle**, the muscle cells are tightly electrically coupled by numerous gap junctions (see Figure 17.18*a*). Because of this coupling, groups of muscle cells are depolarized and contract together, functioning as a single unit. The smooth muscles of the gastrointestinal tract and small-diameter blood vessels are examples of the single-unit type. Single-unit smooth muscle is often spontaneously active, with action potentials propagating from cell to cell via the gap junctions. It can also be activated by stretch. Neural and hormonal controls may modulate the endogenous activity to varying degrees.

Multiunit smooth muscles have few if any gap junctions, so the muscle cells function as independent units (see Figure 17.18*b*). They are extensively innervated by autonomic nerves and are under more direct neural control than are single-unit smooth muscles. Multiunit smooth muscles may or may not generate action potentials, and they may be activated hormonally as well as neurally. They are not stretch-sensitive. Smooth muscles of the hair and feather erectors, eye, large arteries, and large respiratory airways are examples of multiunit smooth muscles.

(a) Single-unit smooth muscle

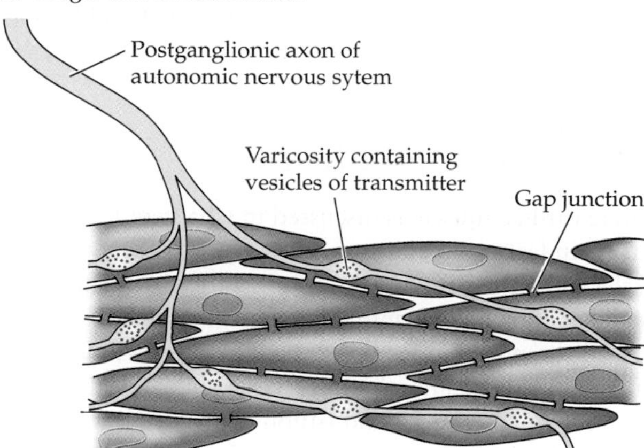

(b) Multiunit smooth muscle

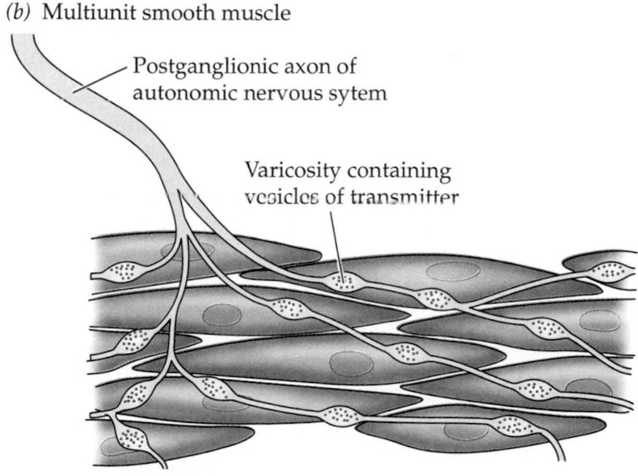

Figure 17.18 Smooth muscle Both multiunit and single-unit smooth muscle fibers receive innervation from the autonomic nervous system. *(a)* Single-unit smooth muscle cells are connected by gap junctions, so electrical signals can pass directly from cell to cell. *(b)* The individual muscle cells of multiunit smooth muscles are excited and contract independently.

The smooth muscle of the mammalian uterus changes between multiunit and single-unit depending on circulating levels of reproductive steroid hormones. For example, during late pregnancy the uterine smooth muscle cells express many gap junctions that electrically couple adjacent cells. Thus the uterus is able to function as a single-unit smooth muscle to produce coordinated contractions during the birthing process.

As in skeletal and cardiac muscle, the contraction of smooth muscle depends on the intracellular concentration of Ca^{2+} ions. In smooth muscle, excitation–contraction coupling is accomplished by some Ca^{2+} ions released from the SR (which consists of small membrane compartments in close association with the cell membrane) and some Ca^{2+} ions entering the cell directly across the cell membrane. An elaborate system of t-tubules and SR like that in skeletal muscle is not required, because the diameter of an entire smooth muscle cell is not much greater than the diameter of a single myofibril in a skeletal muscle fiber. Thus, Ca^{2+} ions entering the cell across the cell membrane diffuse only short distances in order to exert their effect. Depolarization of the surface membrane makes it permeable to Ca^{2+} ions, allowing Ca^{2+} entry down its concentration gradient. Both the cell membrane and the SR contain Ca^{2+} pumps that maintain a very low resting internal Ca^{2+} concentration. Thus, depolarization allows Ca^{2+} entry to trigger contractions, and contractions are terminated by Ca^{2+} being pumped back out.

Smooth muscles lack troponin, and contraction is not regulated by proteins on the actin filament as it is in skeletal and cardiac muscle. Instead, Ca^{2+} activates smooth muscle predominantly by *myosin-linked regulation.*[9] Ca^{2+} ions regulate the phosphorylation of myosin light chains by combining with the Ca^{2+}-binding protein *calmodulin*. The Ca^{2+}–calmodulin complex activates the enzyme *myosin light-chain kinase*, which phosphorylates the myosin light chains. As long as the myosin light chains are phosphorylated, repeated cross-bridge cycling takes place.

Relaxation is accomplished by the pumping of Ca^{2+} ions from the cytoplasm into the SR or out of the cell. As free Ca^{2+} ions are removed, Ca^{2+} ions unbind from calmodulin, and the myosin light-chain kinase becomes inactive. Another enzyme, *myosin light-chain phosphatase*, dephosphorylates the light chains. All of these processes are time-consuming and contribute to the long duration of smooth muscle contractions. Interestingly, in smooth muscles that produce especially prolonged contractions, the cross-bridges remain attached to actin in a **latch state** long after Ca^{2+} ions are gone. In this condition, ATP replaces the bound ADP extremely slowly, so the majority of cross-bridges "latched" to actin maintain tension without using ATP. The latch state can be terminated by phosphorylation of the myosin light chain. The detailed mechanisms responsible for maintaining the energy-saving latch state, and terminating it, are not yet known. It is possible, for example, that the relative activities of myosin light-chain kinase and myosin light-chain phosphatase are involved, or that proteins of the thin filament play a role.

Myosin-linked regulation of contraction (instead of troponin–tropomyosin–actin-linked regulation) also occurs in the muscles of molluscs and several other invertebrate groups. The muscles that hold the shells of bivalve molluscs (such as scallops) shut are known to remain contracted for hours or even days with very little O_2 consumption. Investigators have suggested that this "catch" phenomenon found in molluscan muscles is similar to "latch" in vertebrate smooth muscle. Thus an intermediate state of actin–myosin–ADP could also account for the economical maintenance of tension in molluscan muscle. Further experiments are needed to confirm this hypothesis.

Some smooth muscle cells undergo slow-wave changes in membrane potential in the absence of external stimulation. Slow waves recorded in these muscle cells may trigger action potentials if they exceed the voltage threshold of the fiber. In smooth muscle the inward current of the action potential is a Ca^{2+} influx, so action potentials lead to a direct increase in intracellular Ca^{2+} concentration. However, because action potentials are not required to open voltage-gated Ca^{2+} channels, even subthreshold depolarizations will allow an influx of Ca^{2+} ions that may produce measurable tension in the muscle. The amount of tension produced is directly related to the amount of Ca^{2+} available.

[9] Additional mechanisms, including those that involve proteins on the thin filament, also play a role in regulating contraction in smooth muscle.

A few smooth muscles lack innervation, but most are innervated by the autonomic nervous system (see Chapter 10). Postganglionic axons branch and ramify among the muscle cells (see Figure 17.18). The autonomic axons have repeated swellings, or varicosities, near their terminations, giving them a beaded appearance. Nerve activity releases transmitter molecules from the varicosities, and the transmitter molecules diffuse over the surface of the muscle cells until they encounter receptor molecules.

Unlike skeletal muscle fibers, smooth muscle cells lack distinct postsynaptic regions such as end-plates. Two examples illustrate both the power and the versatility of the autonomic nervous system and smooth muscles working in concert. In the first case, the smooth muscle of the urinary bladder is innervated by both divisions of the autonomic nervous system. The parasympathetic transmitter acetylcholine stimulates the smooth muscle cells to contract in the process of voiding urine, and the sympathetic transmitter norepinephrine inhibits the smooth muscle cells from contracting so that urine is retained.

In contrast, the smooth muscles in most blood vessels are innervated only by sympathetic postganglionic axons, and norepinephrine stimulates these cells to contract. Norepinephrine produces opposite responses—relaxation of bladder smooth muscles cells and contraction of blood vessel smooth muscle cells—because the two types of smooth muscle cells express different postsynaptic receptors that initiate different cellular responses

Vertebrate Cardiac Muscle

Vertebrate cardiac muscle is discussed in Chapters 11 and 23. We note its main features here to provide a comparison with smooth and skeletal muscle. Cardiac muscle is classified as striated because its myofibrils are organized into sarcomeres (Figure 17.19). The cells are typically branched instead of straight like skeletal muscle fibers or spindle-shaped like smooth muscle cells. They are usually uninucleate. In mammals, the SR and t-tubules are well developed, but they are variable in other vertebrate animals.

Cardiac muscle fibers have functional properties that contribute to their effectiveness in pumping blood. First, they are characterized by the presence of **intercalated discs** between adjacent cells. Intercalated discs include gap junctions and regions of tight mechanical contact called *desmosomes*. The desmosomes provide mechanical strength so that the force of contraction generated by one cell can be transmitted to the next. The electrical coupling at gap junctions ensures that all cells connected by gap junctions contract (beat) nearly synchronously.

TABLE 17.3 Characteristics of the three major types of muscles in vertebrates

	Skeletal	Multiunit smooth	Single-unit smooth	Cardiac
Structure	Large, cylindrical, multinucleate fibers	Small, spindle-shaped, uninucleate cells	Small, spindle-shaped, uninucleate cells	Branched uninucleate fibers, shorter than skeletal muscle fibers
Visible striations	Yes	No	No	Yes
Mechanism of contraction	Thick myosin and thin actin filaments slide by each other	Thick myosin and thin actin filaments slide by each other	Thick myosin and thin actin filaments slide by each other	Thick myosin and thin actin filaments slide by each other
Cross-bridge action regulated by Ca^{2+} ions	Yes	Yes	Yes	Yes
Innervation	Somatic nervous system initiates contractions	Autonomic nervous system initiates contractions	Autonomic nervous system modulates contractions	Autonomic nervous system modulates contractions
Spontaneous production of action potentials by pacemakers	No	No	Yes	Yes
Hormones influence function	No	Yes	Yes	Yes
Gap junctions present	No	No (few)	Yes	Yes
Transverse tubules	Yes	No	No	Yes
Sarcoplasmic reticulum	Abundant	Sparse	Sparse	Moderate
Source of Ca^{2+} ions for regulation	Sarcoplasmic reticulum	Extracellular fluid and sarcoplasmic reticulum	Extracellular fluid and sarcoplasmic reticulum	Extracellular fluid and sarcoplasmic reticulum
Troponin and tropomyosin	Both present	Tropomyosin only	Tropomyosin only	Both present
Ca regulation	Ca and troponin; tropomyosin–troponin complex moves to expose myosin-binding sites on actin	Ca and calmodulin; phosphorylation of myosin light chains	Ca and calmodulin; phosphorylation of myosin light chains	Ca and troponin; tropomyosin–troponin complex moves to expose myosin-binding sites on actin
Speed of contraction (reflecting myosin ATPase activity)	Varies from fast to slow depending on fiber type	Very slow	Very slow	Slow

Sources: After Silverthorn 2004; Randall et al. 2002; and Sherwood 2004.

Figure 17.19 Cardiac muscle Striated cardiac muscle fibers are connected by intercalated discs (arrows) that contain electrical gap junctions and mechanical connections called desmosomes.

A second property of cardiac muscle cells is that they are capable of generating endogenous action potentials at periodic intervals. Typically, specialized pacemaker cells with the fastest endogenous rate lead the contractile activity of the rest of the heart.

Finally, the action potentials of vertebrate cardiac fibers are very long, lasting typically 100 to 500 ms (see Figure 11.21). Their long duration ensures a prolonged contraction rather than a brief twitch. Indeed, the action potential lasts as long as the contraction. Because the cardiac cells are refractory during the prolonged action potentials, contractions cannot sum; thus the coordinated pumping of blood is ensured.

Table 17.3 summarizes the properties of vertebrate skeletal, smooth, and cardiac muscles.

Study Questions

1. Knowing the dimensions of a vertebrate skeletal muscle and the relationship between the SR and the myofilaments, estimate the approximate distance that a single calcium ion would travel from a terminal cisterna of the SR to a TN-binding site.
2. Experimenters can separate F-actin thin myofilaments from myosin thick myofilaments. First they homogenize muscle cells in a blender (to break cell membranes); then they place the homogenate in a Ca^{2+}-free "relaxing solution" that contains ATP. Explain why ATP must be present and Ca^{2+} ions must not be present in order to isolate thick and thin myofilaments from each other.
3. List and describe the events that take place (and the structures involved) between excitation of the cell membrane by an action potential and the initiation of cross-bridge action.
4. Combining your knowledge of rates of diffusion with your knowledge of muscle physiology, explain why it is advantageous for oxidative muscle fibers (which depend on aerobic metabolism to generate ATP) to have smaller diameters than glycolytic muscle fibers have.
5. What is the difference between a single cross-bridge power stroke and a single twitch of a skeletal muscle fiber?
6. If all cross-bridges are activated when a single action potential triggers Ca^{2+} release from the SR, why is the amount of tension produced by a train of action potentials greater than the amount of tension of a single twitch?
7. If muscle fibers innervated by the arthropod plan typically don't make action potentials, how are their contractile elements activated in a rapid and coordinated fashion?
8. Describe the organization of a motor unit in vertebrate skeletal muscle, and explain how recruitment of motor units influences the amount of tension produced by a whole muscle.
9. Two muscles have the same diameter, but one is twice as long as the other. Which muscle produces more work? Explain your answer.

Key References

Clark, K. A., A. S. McElhinny, M. C. Beckerle, and C. C. Gregorio. 2002. Striated muscle cytoarchitecture: An intricate web of form and function. *Annu. Rev. Cell Dev. Biol.* 18: 637–706. A description of many cytoskeletal protein molecules associated with the sarcomeres, and their contributions to contractile function.

Franzini-Armstrong, C., and F. Protasi. 1997. Ryanodine receptors of striated muscles: A complex channel capable of multiple interactions. *Physiol. Rev.* 77: 699–729. A review of the functions of the calcium channels of the sarcoplasmic reticulum.

Gordon, A. M., E. Homsher, and M. Regnier. 2000. Regulation of contraction in striated muscle. *Physiol. Rev.* 80: 853–924. A description of studies of the structure and function of tropomyosin and troponin.

Gordon, A. M., A. F. Huxley, and F. J. Julian. 1966. The variation in isometric tension with sarcomere length in vertebrate muscle fibers. *J. Physiol. (London)* 184: 170–192. Evidence at the level of the sarcomere to account for the length–tension relationships observed in whole muscles.

Heuser, J. E., and R. Cooke. 1983. Actin-myosin interactions visualized by the quick-freeze, deep-etch replica technique. *J. Mol. Biol.* 169: 97–122. Description of a new technique that provided three-dimensional surface views of myofilaments and buttressed the interpretations of muscle structure that were based on electron microscopy of thin sections.

Huxley, H. E. 1969. The mechanism of muscular contraction. *Science* 164: 1356–1366. A classic paper that set the stage for many decades of further research in muscle physiology.

Lindstedt, S. L., P. C. LaStayo, and T. E. Reich. 2001. When active muscles lengthen: Properties and consequences of eccentric contractions. *News Physiol. Sci.* 16: 256–261. An overview of the functions and energetics of lengthening contractions.

Marden, J. H. 2000. Variability in the size, composition, and function of insect flight muscles. *Annu. Rev. Physiol.* 62: 157–178. A review of the variety of types of flight muscles found in insects, and how this variation reflects functions of the whole animal.

Moller, P. 1995. *Electric Fishes: History and Behavior*. Chapman & Hall, London. A full account of the diversity of electric organs that generate electricity and the ways in which animals use these organs in behavior.

Rome, L. C., and S. L. Lindstedt. 1998. The quest for speed: Muscles built for high-frequency contractions. *News Physiol. Sci.* 13: 261–268. A review of the adaptations of rapidly contracting muscles and the trade-offs made to achieve high-speed contractions.

Rüegg, C., C. Veigel, J. E. Molloy, S. Schmitz, J. C. Sparrow, and R. H. A. Fink. 2002. Molecular motors: Force and movement generated by single myosin II molecules. *News Physiol. Sci.* 17: 213–218. A review of studies that reveal the functions of different isoforms of myosin.

Selected References

Assad, C., B. Rasnow, and P. K. Stoddard. 1999. Electric organ discharges and electric images during electrolocation. *J. Exp. Biol.* 202: 1185–1193.

Berchtold, M. W., H. Brinkmeier, and M. Müntener. 2000. Calcium ion in skeletal muscle: Its crucial role for muscle function, plasticity, and disease. *Physiol. Rev.* 80: 1215–1265.

Bernal, D., K. A. Dickson, R. E. Shadwick, and J. B. Graham. 2001. Review: Analysis of the evolutionary convergence for high performance swimming in lamnid sharks and tunas. *Comp. Biochem. Physiol. A* 129: 695–726.

Conley, K. E., and S. L. Lindstedt. 2002. Energy-saving mechanisms in muscle: The minimization strategy. *J. Exp. Biol.* 205: 2175–2181.

Dickinson, M. H., C. T. Farley, R. J. Full, M. A. R. Koehl, R. Dram, and S. Lehman. 2000. How animals move: An integrative view. *Science* 288: 100–106.

Dudley, R. 2000. *The Biomechanics of Insect Flight*. Princeton University Press, Princeton, NJ.

Gordon, A. M., M. Regnier, and E. Homsher. 2001. Skeletal and cardiac muscle contractile activation: Tropomyosin "rocks and rolls." *News Physiol. Sci.* 16: 49–55.

Gotter, A. L., M. A. Kaetzel, and J. R. Dedman. 1998. *Electrophorus electricus* as a model system for the study of membrane excitability. *Comp. Biochem. Physiol.* 119A: 225–241.

Gregorio, C. C., H. Granzier, H. Sorimachi, and S. Labeit. 1999. Muscle assembly: A titanic achievement? *Curr. Opin. Cell Biol.* 11: 18–25.

Heuser, J. E. 1983. Structure of the myosin crossbridge lattice in insect flight muscle. *J. Mol. Biol.* 169: 123–154.

Hoh, J. F. Y. 2002. "Superfast" or masticatory myosin and the evolution of jaw-closing muscles of vertebrates. *J. Exp. Biol.* 205: 2203–2210.

Johnston, I. A., D. A. Fernandez, J. Calvo, V. L. A. Vieira, A. W. North, M. Abercromby, and T. Garland, Jr. 2003. Reduction in muscle fibre number during the adaptive radiation of notothenioid fishes: A phylogenetic perspective. *J. Exp. Biol.* 206: 2595–2609.

Josephson, R. K., J. G. Malamud, and D. R. Stokes. 2000. Asynchronous muscle: A primer. *J. Exp. Biol.* 203: 2713–2722.

Kessel, R. G., and R. H. Kardon. 1979. *Tissues and Organs. A Text-Atlas of Scanning Electron Microscopy*. Freeman, San Francisco.

Marx, S. O., K. Ondrias, and A. R. Marks. 1998. Coupled gating between individual skeletal muscle Ca^{2+} release channels (ryanodine receptors). *Science* 281: 818–821.

Nybo, L., and B. Nielsen. 2001. Perceived exertion is associated with an altered brain activity during exercise with progressive hyperthermia. *J. Appl. Physiol.* 91: 2017–2023.

Parkin, J. M., M. F. Carey, S. Zhao, and M. A. Febbraio. 1999. Effect of ambient temperature on human skeletal muscle metabolism during fatiguing submaximal exercise. *J. Appl. Physiol.* 86: 902–908.

Rome, L. C., and A. A. Klimov. 2000. Superfast contractions without superfast energetics: ATP usage by SR-Ca^{2+} pumps and crossbridges in toadfish swimbladder muscle. *J. Physiol. (London)* 526: 279–286.

Rome, L. C., C. Cook, D. A. Syme, M. A. Connaugton, M. Ashley-Ross, A. Klimov, B. Tikunov, and Y. E. Goldman. 1999. Trading force for speed: Why superfast crossbridge kinetics leads to superlow forces. *Proc. Natl. Acad. Sci. U.S.A.* 96: 5826–5831.

Rome, L. C., D. A. Syme, S. Hollingworth, S. L. Lindstedt, and S. M. Baylor. 1996. The whistle and the rattle: The design of sound producing muscles. *Proc. Natl. Acad. Sci. U.S.A.* 93: 8095–8100.

Savineau, J.-P., and R. Marthan. 2000. Cytosolic calcium oscillations in smooth muscle cells. *News Physiol. Sci.* 15: 50–55.

Schmitz, H., M. C. Reedy, M. K. Reedy, R. T. Tregear, and K. A. Taylor. 1997. Tomographic three-dimensional reconstruction of insect flight muscle partially relaxed by AMPPNP and ethylene glycol. *J. Cell Biol.* 139: 695–707.

Somlyo, A. P., and A. V. Somlyo. 2003. Ca^{2+} sensitivity of smooth muscle and nonmuscle myosin II: Modulated by G proteins, kinases, and myosin phosphatase. *Physiol. Rev.* 83: 1325–1358.

Squire, J. M., and E. P. Morris. 1998. A new look at thin filament regulation in vertebrate skeletal muscle. *FASEB J.* 12: 761–771.

Strohm, E., and W. Daniels. 2003. Ultrastructure meets reproductive success: Performance of a sphecid wasp is correlated with the fine structure of the light-muscle mitochondria. *Proc. R. Soc. London [Biol.]* 270: 749–754.

Sutlive, T. G., J. R. McClung, and S. J. Goldberg. 1999. Whole-muscle and motor-unit contractile properties of the styloglossus muscle in rat. *J. Neurophysiol.* 82: 584–592.

Vale, R. D., and R. A. Milligan. 2000. The way things move: Looking under the hood of molecular motor proteins. *Science* 288: 88–95.

Vogel, S. 2001. *Prime Mover: A Natural History of Muscle*. Norton, New York.

Westerblad, K., D. G. Allen, and J. Lännergren. 2002. Muscle fatigue: Lactic acid or inorganic phosphate the major cause? *News Physiol. Sci.* 17: 17–21.

Wu, C. H. 1984. Electric fish and the discovery of animal electricity. *Amer. Sci.* 72: 598–607.

Xu, C., R. Craig, L. Tobacman, R. Horowitz, and W. Lehman. 1999. Tropomyosin positions in regulated thin filaments revealed by cryoelectron microscopy. *Biophys. J.* 77: 985–992.

Young, I. S., C. L. Harwood, and L. C. Rome. 2003. Cross-bridge blocker BTS permits direct measurement of SR Ca^{2+} pump ATP utilization in toadfish swimbladder muscle fibers. *Amer. J. Physiol. Cell Physiol.* 285: C781–C787.

Zakon, H. H., and G. A. Unguez. 1999. Development and regeneration of the electric organ. *J. Exp. Biol.* 202: 1427–1434.

18 Control of Movement: The Motor Bases of Animal Behavior

When a locust flies, eats, or mates, its behavioral performance results from the output of the nervous system to effectors. All externally observable behavior is a direct result of the activation of effectors. That is, all behavior that can be seen by an observer consists of movements (usually resulting from muscle contractions), sounds (also from muscle contractions), gland secretions, color changes, and other outputs of effectors. Thus the mechanisms by which animals generate patterns of behavior involve the control of muscles and other effectors to produce patterned movements, secretions, and so forth.

One school of psychology (behaviorism) views human behavior entirely in terms of effector activity, ignoring mental states. We will leave such issues to psychologists; for our purposes it is sufficient to assert that animal behavior results from the coordinated activation of effectors. The nervous system exerts the major and immediate control of effectors to generate behavior, although hormonal and other controls may also be important. In this chapter we will consider issues in the neural control and coordination of movement, particularly including patterns of motor behavior such as walking, swimming, and flying.

Behavioral Background: Reflexes and Fixed Action Patterns

In *Cellular Basis of Behavior*, Eric Kandel has provided a useful classification of behavior. **Complex behavior** consists of patterns or sequences of responses (e.g., locomotion or feeding) that may involve several effector organs. **Elementary behavior** consists of an isolated response or *act* of a single effector organ. Complex behavior may be made up of sequences of elementary behaviors (Table 18.1). Both the patterns of complex behavior and the acts of elementary behavior are classified as *lower-order behavior*; *higher-order behavior* consists of still more elaborate patterns or sequences of activity, such as courtship, nest building, or interactive communication. In general, higher-order behavior is too complex to be readily analyzable in terms of the neural control of effectors, and it is not considered further here.

Lower-order behavior, both elementary and complex, can be further categorized as either reflex or fixed (see Table 18.1). A **reflex** act or pattern is a response to an eliciting stimulus and by definition is graded, being greater or more complete with a stronger eliciting stimulus. A familiar example of a spinal reflex

Flight in locusts depends on patterned motor output from the central nervous system

TABLE 18.1 Classification of types of behavior

Lower-order behavior	Higher-order behavior
Elementary (acts)	Courtship
Reflex acts	Nest-building
Fixed acts	Interactive communication
Complex (patterns)	etc.
Reflex patterns	
Fixed action patterns	

Source: After Kandel 1976.

is the flexion reflex of mammals: the act of lifting (flexing) a leg in response to a sharp or painful stimulus delivered to the foot.

Reflex patterns of behavior require continuous sensory stimulation, each movement producing sensory feedback that stimulates or modifies other movements in the sequence, as we will see. Reflexive behavior can be learned and may be subject to considerable learned modification. A familiar example is provided by the classical conditioning experiments of Ivan Pavlov (1849–1936), in which dogs were trained to salivate reflexively in response to a bell by repeated pairing of the bell with a food stimulus.

The terms *fixed acts* and *fixed action patterns* stem from the behavioral studies of Konrad Lorenz (1903–1989) and Nikolaas Tinbergen (1907–1988) prior to and during the 1950s. **Fixed acts** (elementary) and **fixed action patterns** (complex) are stereotyped, species-specific, and little modified by learning. A fixed action pattern is typically triggered by a specific stimulus (the *sign stimulus*), but once initiated, it proceeds in an all-or-nothing way. For example, greylag geese nest on the ground; if an egg rolls out of the nest, the goose retrieves it by a fixed action pattern of bill movement. Once this response has been triggered, it goes to completion even if the egg is taken away.

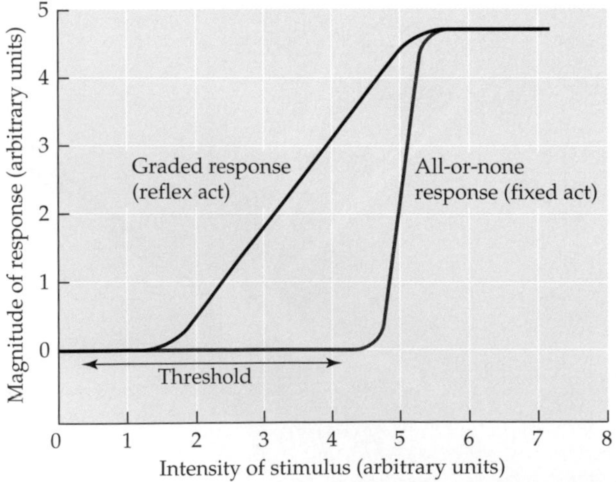

Figure 18.1 A comparison of graded reflex acts and all-or-none fixed acts The stimulus–response curve of a reflex act has a gradual slope, so a stronger stimulus elicits a greater response. In contrast, the curve for a fixed act has a steep slope. Reflex responses, therefore, are more proportional to the amount of sensory stimulation, and fixed acts are more nearly all-or-none. (After Kandel 1976.)

Fixed acts and fixed action patterns may occur in the apparent absence of a stimulus, a phenomenon known as *vacuum activity*. Single fixed acts are usually triggered by stimuli, but unlike reflex acts (which are smoothly graded responses that depend on the amount and form of the stimulus), fixed acts are all-or-none responses, the form of which is relatively independent of the amount or form of triggering stimuli (Figure 18.1). The distinction between reflex and fixed acts and action patterns is not absolute, but rather can be viewed as representing the ends of a spectrum, with intermediate forms of behavior between these extremes. Nevertheless, the distinction between stereotyped, fixed patterns of behavior and stimulus-dependent, modifiable reflex patterns remains an important organizing principle for studies of the physiological basis of behavior.

Neural Circuits Mediating Reflexes and Fixed Acts

A major question in the analysis of the neural basis of behavior is, What are the **neural circuits**—the assemblies of neurons and the patterns of synaptic interconnections between them—that produce particular patterns of behavioral movements? In this chapter we examine several neural circuits and relate their actions to the behaviors they mediate. Here we consider the circuits underlying simple reflexes and fixed acts. If reflexes are rather proportional responses and fixed acts are more all-or-none (see Figure 18.1), then their neural circuits should explain this difference.

Crayfish escape behavior is a fixed act mediated by a giant interneuron

A crayfish escapes from a predator (or from a fight) by means of a *tail flip*: a sudden abdominal flexion that propels the animal away from a tactile stimulus. The escape response has two components: a single rapid flexion of the abdomen and a subsequent swimming sequence of repeated flexions and extensions of the abdomen. We will consider only the well-studied initial single flexion. There are three different kinds of initial tail flip, each triggered by a separate pathway.

The first kind of tail flip is triggered by a sudden tap or water-jet stimulus to the abdomen, resulting in an abdominal flexion mediated by the *lateral giant interneurons* (Figure 18.2*a*). This flexion is confined to the anterior abdominal segments and pitches the crayfish upward and forward, as in a somersault. Other tail flips are triggered by a sudden stimulus to the anterior portion of the animal, causing a backward escape mediated by the *medial giant interneurons*, and a slower and less stereotyped tail flip that is mediated by separate, nongiant neurons. The two types of abdominal flexions that are mediated by giant interneurons are fixed acts that occur in an all-or-none manner.

We will focus on the neural circuit for the escape tail flip that is mediated by lateral giant interneurons (LGIs). This circuit is shown schematically in Figure 18.2*b*. A crayfish has two lateral giant interneurons, one on each side.[1] For simplicity we show the circuit of

[1] Each LGI is actually a compound series of cells electrically coupled by gap junctions (see Box 11.2).

(a) The tail flip

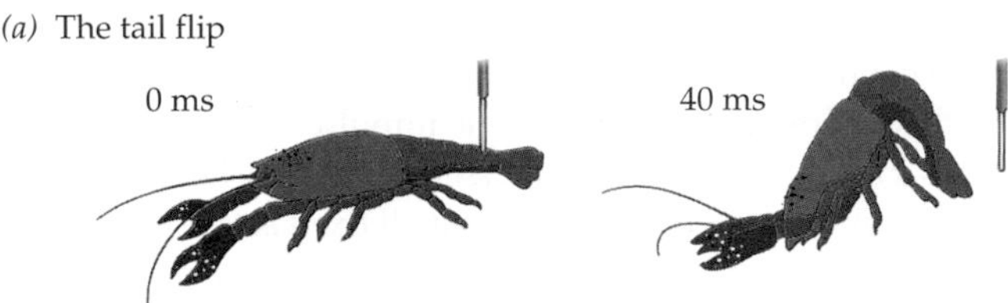

(b) The neural circuit for the tail flip

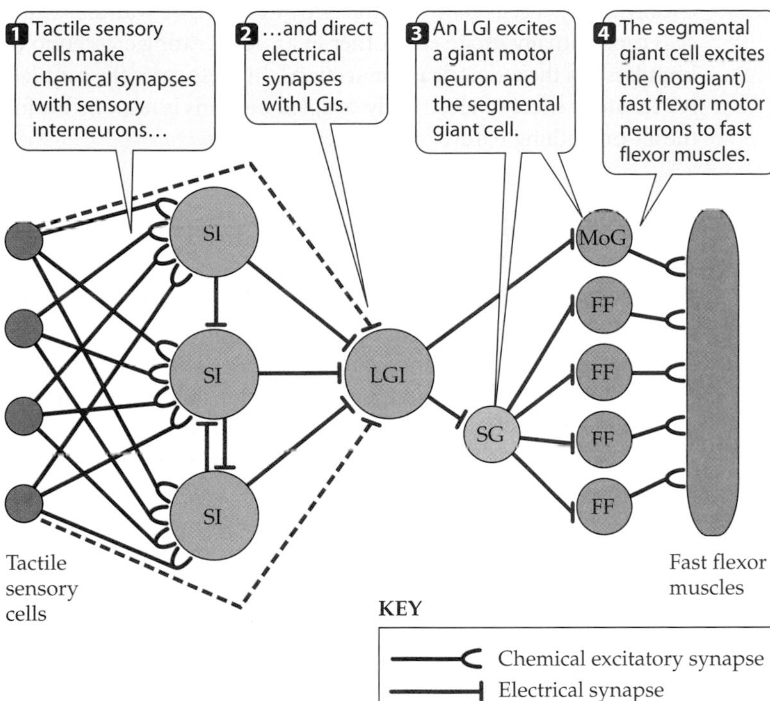

Figure 18.2 The escape response of a crayfish is a fixed act *(a)* Stimulation of the abdomen of a crayfish (*Procambarus*) evokes an abdominal flexion that moves the crayfish up and forward. This response (a fixed act) is mediated by the lateral giant interneurons (LGIs). *(b)* The neural circuit for the LGI-mediated escape tail-flip response for one side of a single abdominal segment. The circuit is unusual in that many of the synapses are electrical, including direct synapses of sensory cells onto LGI (dashed lines). This preponderance of electrical synapses makes the circuit fast (of selective advantage for an escape circuit) but doesn't affect whether it mediates a reflex or a fixed act. FF = fast flexor motor neuron; MoG = giant motor neuron; SG = segmental giant cell; SI = sensory interneuron. (After Camhi 1984.)

only one side, and we consider the LGI as a single neuron. The major sensory input to the escape circuit is from tactile sensory neurons associated with cuticular sensory hairs, which cover the surface of the animal and are excited by touch or waterborne vibration. The axons of these sensory neurons excite sensory interneurons within the abdominal portion of the central nervous system (see Figure 18.2*b*). The sensory interneurons in turn excite the LGI. In addition, many of the tactile sensory neurons make direct, weak electrical synapses onto the LGI. The LGI excites segmental giant motor neurons in different abdominal segments, each of which excites fast flexor muscles that produce the escape flexion. The LGI also excites a segmental giant neuron, which in turn excites other fast flexor motor neurons to the fast flexor muscles.

Unlike the situation in the vertebrate spinal cord (discussed shortly), most of these cells in the crayfish tail-flip circuit are uniquely identified,[2] individual neurons, rather than populations. Investigators' ability to work with such uniquely identifiable neurons, over and over in different animals, has greatly enhanced our understanding of this circuit, as well as of other circuits composed of identified neurons.

To elicit a tail flip, the excitatory synaptic input of many tactile sensory neurons must converge (largely via sensory interneurons) onto the LGI at the same time, in order for their synaptic input to be great enough to trigger an action potential in the LGI. A single LGI action potential, in contrast, powerfully excites both giant motor neurons and other fast flexor motor neurons. There is thus considerable convergence of input onto the LGI, and considerable divergence of LGI output to many motor neurons. What gives this tail-flip behavior its all-or-nothing character is precisely this convergence of sensory input onto LGI. A single action potential in LGI is sufficient to command an escape tail flip, and therefore the behavior is as all-or-none as the action potential itself.

The LGI can be considered a *command neuron,* a neuron whose activity is sufficient to command a particular element of behavior. An analogous fixed act, the rapid turn of a fish away from a vibrational stimulus, has a similar cellular basis. A pair of giant *Mauthner neurons* in the brainstem of a fish act as the decision and command neurons, commanding a flexion to the side opposite the stimulus. (The two Mauthner neurons generate flexions in opposite directions, and inhibitory circuitry prevents the two from acting together.) Mauthner neurons are the only known command neurons in a vertebrate.

Another case somewhat similar to the crayfish escape response is the escape response of cockroaches, discussed in Chapter 10 (see Figure 10.3). Cockroaches also employ giant interneurons to mediate escape, but there are several sets of giant interneurons, and their effects on motor neurons appear more additive and more variable than in the crayfish. The escape startle response of a cockroach may therefore have properties intermediate between those of a fixed act and of a reflex act.

Gill withdrawal in *Aplysia* is a reflex act

A second well-studied neural circuit controls gill withdrawal in the marine mollusc *Aplysia* (Figure 18.3). Defensive gill withdrawal is a reflex act, in contrast to the fixed act of the crayfish escape tail flip. Mechanical stimulation of the siphon or of the gill itself elicits gill withdrawal; stronger stimulation evokes proportionally more vigorous and rapid withdrawal responses.

The neural circuit mediating the gill-withdrawal reflex was described in Chapter 12 (see page 339). The circuit consists of about 24 siphon sensory neurons, 6 motor neurons to the gill, and sever-

[2]An *identified neuron* is a neuron whose structure, location, electrical activity, and/or other properties are distinctive enough that the neuron can be recognized and studied in every individual of a species.

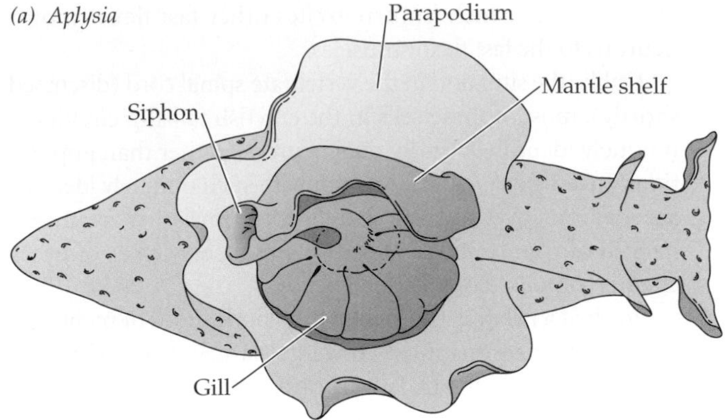

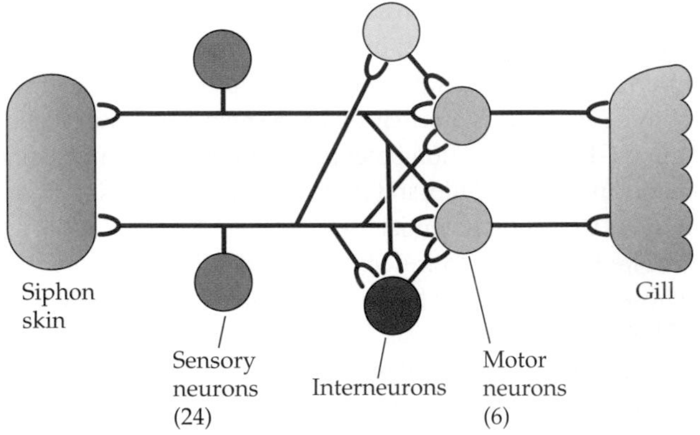

Figure 18.3 The gill-withdrawal reflex of *Aplysia* *(a)* In the marine mollusc *Aplysia*, the gill, siphon, and mantle shelf are covered by a parapodium that can be unfolded open. Tactile stimulation of the siphon reflexively evokes gill withdrawal. *(b)* The neural circuit (somewhat simplified) of the gill-withdrawal reflex. Tactile sensory neurons (24) synaptically excite six gill motor neurons, both directly and via interneurons. Some interneurons are inhibitory (magenta). Unlike crayfish LGI-mediated escape (see Figure 18.2), in *Aplysia* there is no point of convergence on a single neuron, where an action potential could make this reflexive behavior all-or-none.

al interneurons. Most of these neurons are identified, so the same neurons can be studied repeatedly in different animals. The most important synaptic connections for the gill-withdrawal reflex appear to be the direct, monosynaptic connections from sensory to motor neurons, with the interneurons playing a subsidiary role. However, optical recordings from the visceral ganglion show that hundreds of neurons in the ganglion are active during reflexive gill withdrawal.

The circuit for the gill-withdrawal reflex elucidates the neural basis for the distinction between the graded and proportional nature of reflexes and the all-or-nothing nature of a fixed act. Increasing the strength of siphon stimulation elicits more impulses in a particular sensory neuron and recruits more sensory neurons. (A weak stimulus might excite 8 sensory neurons with overlapping receptive fields on the siphon, whereas a strong or widespread stimulus might excite all 24.) The increased number and frequency of sensory impulses leads to larger EPSPs (excitatory postsynaptic potentials) and more impulses in the motor neurons, producing a larger contraction of the gill. In contrast to the crayfish escape tail-flip circuit, at no point in the gill-withdrawal circuit is a single impulse in a single neuron necessary or sufficient for complete expression of the behavior. This analysis should not imply, however, that all fixed acts critically depend on the activity of a single command neuron.

Two fixed acts in *Aplysia*—egg laying and the defensive release of ink—are all-or-nothing responses mediated by populations of electrically coupled cells. When sufficiently excited by a stimulus, such cells can mutually excite each other in an accelerating crescendo of impulses, via their electrical connections. In these cases the mutually excitatory effect of electrically coupled neurons is responsible for the all-or-nothing nature of the response.

Vertebrate spinal reflexes compensate for circumstances, as well as initiating movements

Spinal reflexes are reflex acts mediated by the neural circuits of the vertebrate spinal cord. As a result of pioneering studies by Charles Sherrington (1857–1952) and Ivan Pavlov (1849–1936) on spinal reflexes at the beginning of the twentieth century, analysis of behavior in terms of reflexes dominated studies of neural circuits until at least the 1960s. These spinal reflexes are therefore perhaps the best-known vertebrate neural circuits.

In spinal reflexes, somatosensory input (from receptors of the skin, muscles, tendons, and joints; see Chapter 13) enters the spinal cord through the dorsal roots. This sensory input, via intervening synapses in the spinal cord, excites some motor neurons and inhibits others (see Figure 12.5), leading to movements by selectively activating muscular contraction. The sensory inputs from different populations of receptors have different connections in the spinal cord and thereby initiate different reflexes. We will examine two of the many reflexes of the mammalian hindlimb that have been studied extensively in the last century: the stretch reflex and the flexion reflex.

THE STRETCH REFLEX The first spinal reflex we will consider is the **stretch reflex** (or *myotactic reflex*). A familiar example of the stretch reflex is the knee-jerk response to a tap on the patellar tendon at the knee joint, a test that is a staple of routine medical examinations. When the doctor taps you on the patellar tendon, the hit of the hammer stretches the extensor muscle in the thigh. This stretch stimulates *muscle spindles* (see Chapter 13, page 369), which contain stretch-sensitive receptor endings located in noncontractile portions of specialized *intrafusal* muscle fibers.[3] The sensory axons associated with muscle spindles are known as **1a afferent** fibers—*afferent* meaning "conducting toward the central nervous system," and *1a* because they are the largest and most rapidly conducting sensory fibers in the body. (Not all muscle-spindle sensory neurons are of the 1a class, but we will simplify the discussion by considering only the 1a sensory neurons here.)

The 1a axons from muscle spindles enter the spinal cord and make direct, excitatory synaptic contact with motor neurons to the same muscle (Figure 18.4). This direct synaptic excitation is unusual; most vertebrate sensory neurons directly synapse only onto interneurons (intrinsic neurons that do not leave the central nervous

[3] Intrafusal muscle fibers are defined on p. 371.

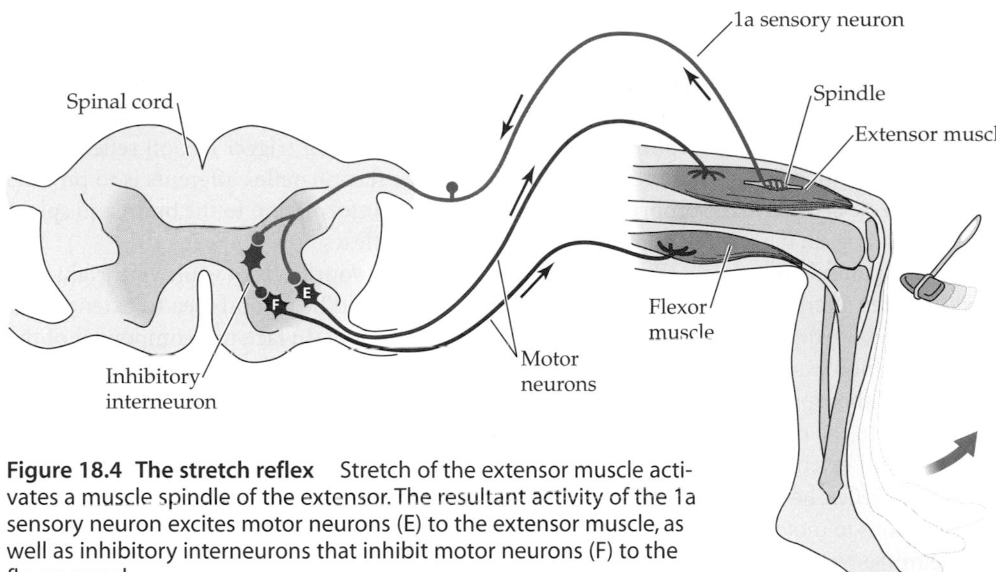

Figure 18.4 The stretch reflex Stretch of the extensor muscle activates a muscle spindle of the extensor. The resultant activity of the 1a sensory neuron excites motor neurons (E) to the extensor muscle, as well as inhibitory interneurons that inhibit motor neurons (F) to the flexor muscle.

system). The simplest manifestation of a stretch reflex, then, involves only two kinds of neurons: 1a sensory neurons and motor neurons.

When a muscle spindle is stretched, its 1a afferent neuron generates a train of nerve impulses. These impulses elicit EPSPs in motor neurons, leading to motor neuron impulses and ultimately to contraction of the stretched muscle, and the leg kicks upward. Muscle spindles are said to be *in parallel* to the other fibers of a muscle (termed **extrafusal** fibers) because they act *beside* the extrafusal fibers (see Figures 13.24 and 18.4). Because muscle spindles are in parallel to the force-producing extrafusal muscle fibers, they are sensitive to muscle *length*.

Other aspects of the neural circuit of the stretch reflex, however, complicate the picture. The 1a afferent axons also synapse upon other neurons in addition to motor neurons. Some of these connections excite interneurons that excite the motor neurons indirectly. Other synapses excite inhibitory interneurons that inhibit the motor neurons to the antagonist (opposing) muscle. Thus a tap on the patellar tendon not only excites motor neurons to the extensor muscle to produce the familiar knee jerk, but also inhibits motor neurons to the antagonist flexor muscle (see Figure 18.4).

These synaptic connections illustrate one of the most basic features of reflexes and of the organization of motor systems: the principle of **reciprocity**. Muscles tend to be arranged in antagonist pairs that oppose each other, such as the flexor that bends the knee and the extensor that straightens it. The principle of reciprocity states that any signal that activates movements, whether it is the sensory input to a reflex or a command of the central nervous system, is coordinated to contract a set of muscles that work together (the *agonists*) while relaxing the opposite (*antagonist*) set. This reciprocal control of muscles ensures that two mutually antagonistic muscles do not usually counteract each other.

Another added complexity in the stretch reflex (and in motor circuits in general) is the number of neurons involved in even the simplest behavioral act. Figure 18.4 shows one 1a sensory neuron and one extensor motor neuron, but these only represent larger populations of these neuron types. A large leg muscle contains many muscle spindles, so its stretch activates many sensory neurons. Moreover, the muscle is supplied by at least 300 motor neurons (see Chapter 17). Unlike the situations in crayfish and *Aplysia*, none of these neurons are identified; you cannot find the same extensor motor neuron from animal to animal. Each 1a sensory neuron synapses with most, and probably all, of the (α) extensor motor neurons, as well as with many interneurons of different types.

The stretch reflex illustrates the principle of **divergence** of central neural connections: Each presynaptic neuron usually contacts many postsynaptic neurons. The converse principle, **convergence**, is also true; each postsynaptic neuron is contacted by many presynaptic neurons. For example, each extensor motor neuron receives about 10,000 synapses, representing many 1a sensory neurons and many more excitatory and inhibitory interneurons. Thus the cartoon view of the circuit for a stretch reflex in Figure 18.4 is a great oversimplification.

A third aspect of the stretch reflex that increases its organizational complexity is the motor innervation of the stretch receptor organs by **gamma (γ) motor neurons.** Recall that in muscle spindles the stretch-sensitive 1a sensory neurons are associated with intrafusal muscle fibers. The intrafusal muscle fibers are innervated by a separate population of small motor neurons, the γ motor neurons. The extrafusal muscle fibers (i.e., all the fibers that are not part of muscle spindles) are innervated by alpha (α) motor neurons. (When the term *motor neuron* is used without the Greek-letter prefix, it denotes an alpha motor neuron. Hence the previous discussions of vertebrate motor neurons in Chapter 12 and in this chapter refer to alpha motor neurons.)

Activation of γ motor neurons excites the 1a sensory neurons by contracting the contractile ends of intrafusal fibers and thereby stretching the noncontractile central sensory portion of the spindle. Therefore, there are two ways to increase muscle-spindle receptor activity: by passive (external) stretch of the muscle and by γ-motor neuron activity. Note also that activity of γ motor neurons and activity of α motor neurons to the surrounding extrafusal fibers have opposite effects on muscle-spindle sensory activity.

What is the function of the stretch reflex? Surely it has not evolved to mediate a sudden extension of the leg when the knee is struck with a rubber mallet. We can illustrate one aspect of the function of stretch reflexes with the following example: Suppose that while you are standing, a large monkey or a small person jumps on your back. The added weight will cause your knees to start to buckle, stretching the extensor muscles and activating the sensory neurons of muscle spindles. This sensory activity will reflexively excite motor neurons to the extensor muscles, generating more muscle force to counteract the increased load and maintain upright posture. This scenario illustrates a functional role of the stretch reflex in postural maintenance, counteracting changes in load, muscle fatigue, or other factors.

After describing another spinal reflex—the flexion reflex—we will return to a somewhat different view of the function of a stretch reflex.

THE FLEXION REFLEX The other vertebrate spinal reflex we will examine is the **flexion reflex.** When you step on a tack, you reflexively withdraw your foot from the offending stimulus. The neural circuit mediating this flexion reflex is shown in Figure 18.5. A diverse array of sensory neurons known as *flexion-reflex afferents* have endings in the skin, muscles, and joints; some of these are sensitive to painful and noxious stimuli. The central endings of the flexion-reflex afferents make excitatory synaptic contacts on interneurons that excite flexor motor neurons, as well as inhibitory interneurons that inhibit extensor motor neurons. Thus, as in the stretch reflex (and in other spinal reflexes), synaptic interactions in the spinal cord maintain the reciprocity of action between antagonist pools of flexor and extensor motor neurons. Unlike the sensory neurons of the stretch reflex, however, flexion reflex afferents make only indirect connections to motor neurons, via at least one layer of intervening interneurons.

The obvious function of the flexion reflex is protective; the offended limb is flexed, lifted, and withdrawn from a painful and potentially damaging stimulus. The reflex circuit is relatively short, local, and rapid. Of course, flexion-reflex afferents also connect to other interneurons that ascend the spinal column to the brain, so you become aware of the painful stimulus. This slower process occurs while the reflex flexion is taking place, so in most cases the foot is lifted (or the hand is withdrawn from the hot stove) before you are aware of the stimulus triggering the withdrawal. Note that many receptors other than pain receptors can trigger flexion reflexes, and that the main function of the flexion-reflex afferents is to provide proprioceptive and cutaneous information to the brain and spinal cord, not just to elicit flexion reflexes.

If you stepped on a tack with your left foot while your right foot was lifted off the ground, it would be a good idea to extend your right foot while flexing your left foot. In fact, one component of the flexion reflex ensures this. As Figure 18.5 shows, flexion-reflex afferents synapse onto interneurons that cross the midline of the spinal cord and indirectly excite extensor motor neurons of the contralateral ("opposite side") leg. Thus the right leg is extended (by exciting extensor motor neurons and inhibiting flexor motor neurons) while the stimulated left leg is flexed (by exciting flexor motor neurons and inhibiting extensor motor neurons). The reflex extension of the contralateral leg has been given a separate name (the *crossed extension reflex*), but functionally it is an integral part of the flexion reflex, a product of the synaptic connections "wired in" to the spinal cord.

This example illustrates that reflexes do not operate in a vacuum, influencing only a single antagonist pair of muscles. Instead, reflexes may have diverse and widespread effects, and they must interact with all other synaptic influences upon motor neurons.

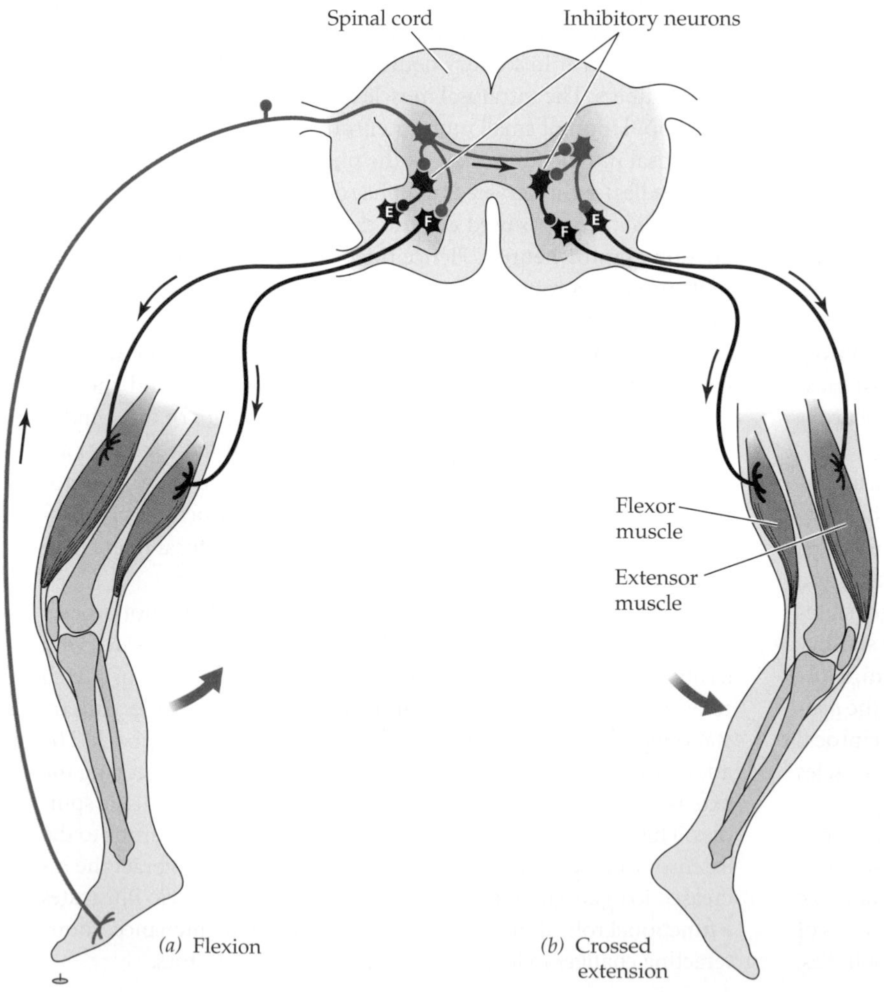

Motor neurons are activated primarily by central input rather than by spinal reflexes

Having described two spinal reflexes, we now address the more difficult question of just how fruitful such analysis is for an understanding of motor function of the spinal cord. Neurophysiologists have approached spinal circuitry from the viewpoint of reflexes for largely technical reasons that go back to Sherrington's work a century ago. This focus has led to the *reflex model* of behavior, in which behavior is viewed as stimulus-produced responses ($S \rightarrow R$), generated by elaborations of reflexes.

However, it is more accurate to consider the *primary* input to the motor circuitry of the spinal cord to be the descending input from higher centers of the central nervous system; the sensory fibers mediating spinal reflexes are a *secondary* input. This viewpoint is shown in Figure 18.6, in which the primary descending inputs enter from the top and sensory inputs enter from the bottom. Such a view implies that a major role of the sensory input to the spinal cord is to supply *sensory feedback* that can modulate or correct the responses of motor neurons to central signals.

Spinal cord
Inhibitory neurons
Flexor muscle
Extensor muscle
(*a*) Flexion
(*b*) Crossed extension

Figure 18.5 The neural circuit of the flexion reflex and the crossed extension reflex *(a)* Noxious stimulation of the skin activates pain sensory neurons that (via interneurons) excite flexor (F) motor neurons and inhibit extensor (E) motor neurons on the stimulated side. *(b)* On the opposite side, extensor motor neurons are excited and flexors are inhibited.

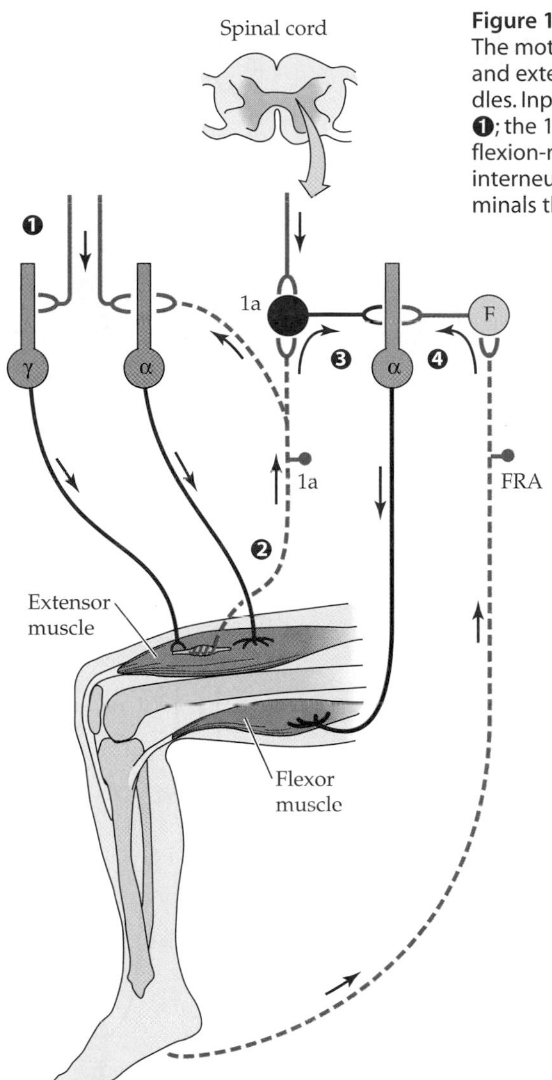

Figure 18.6 The basic circuit diagram of the ventral horn of the mammalian spinal cord The motor output neurons of the spinal cord are the alpha (α) motor neurons to flexor and extensor muscles, and gamma (γ) motor neurons to intrafusal fibers of muscle spindles. Input pathways to the ventral horn are central descending pathways from the brain ❶; the 1a afferent excitatory ❷ and inhibitory ❸ pathways from muscle spindles; and the flexion-reflex afferent (FRA) pathway ❹. Local interneurons shown are 1a inhibitory interneurons (black) and interneurons in the flexion-reflex pathway (F). Neurons and terminals that are excitatory are in colors.

To illustrate how spinal reflex circuits provide sensory feedback, next we consider how the stretch reflex compensates for a resistance or load during the execution of a centrally generated, voluntary movement. Suppose you decide to pick up a plate from the table. Because this is a voluntary movement, the CNS must program the activation of motor neurons, rather than sensory input initiating the movement. Essentially the CNS estimates the amount of force necessary to pick up the plate and sends a command to the motor neurons to generate that force. At the same time, the stretch reflex mediates **load compensation**, augmenting the contraction if there is extra weight or resistance to the intended movement.

The central command for a voluntary movement excites both α and γ motor neurons, a process termed **α–γ coactivation**. Suppose that the plate is not very heavy, and the CNS has correctly estimated its weight. (We will call this the *no-load* condition.) The coactivation of α and γ motor neurons activates contraction of both the intrafusal fibers and the extrafusal fibers of the working muscle. In the absence of a load (Figure 18.7*a*), the extrafusal muscle fibers shorten to flex the arm, allowing the intrafusal fiber associated with a muscle spindle to shorten as it contracts. The shortening of the intrafusal fiber prevents it from generating tension and activating the stretch receptor. Because shortening unloads the muscle spindle, the 1a afferent neuron of the stretch receptor doesn't generate action potentials. That is, if the muscle shortens when it is signaled to shorten, no follow-up is needed.

Now suppose that the plate is very heavy and the CNS's force estimate is insufficient (Figure 18.7*b*). (We will call this the *loaded* condition.) Coactivation excites the α and γ motor neurons as before, but now the extrafusal muscle fibers do not shorten, because of the unanticipated load. In the absence of shortening, excitation of an intrafusal fiber by γ motor neurons will activate the stretch receptor (see Figure 13.24), producing a train of action potentials. This stretch receptor activity constitutes an *error signal*, a measure of how much the muscle failed to shorten as commanded.

As Figure 18.7 shows, the stretch receptor's 1a axon makes excitatory synaptic contact with an α motor neuron that innervates the working extrafusal muscle fibers. Activity in the stretch receptor neuron (the error signal) excites proportional activity in the α motor neuron, generating additional tension in the working muscle to overcome the load. The neuron pair of the stretch receptor and α motor neuron is a reflex circuit that functions as a *load-compensating servo loop*, detecting an error (failure to shorten) and counteracting it ("more force, please") within a centrally commanded movement.

Summary: Neural Circuits Mediating Reflexes and Fixed Acts

- The pattern of motor output of a nervous system produces behavioral actions. The nervous system can generate this pattern centrally or in response to discrete stimuli. Simple, stimulus-evoked responses are often reflexive: Stronger stimulation evokes stronger responses.
- Fixed acts are rather all-or-none in their behavioral expression, with a stimulus threshold analogous to that of an action potential. In crayfish escape tail flips mediated by lateral giant interneurons (LGIs), a single LGI action potential elicits the behavioral response, explaining its all-or-none nature. Other neural circuits that produce behavioral fixed acts depend on a network of electrically coupled neurons to produce the all-or-none effect.
- Vertebrate spinal reflexes (such as the stretch reflex and the flexion reflex) have the simplest and best-understood neural circuits for a vertebrate behavior. Even the simplest vertebrate reflexes, however, have large numbers of neurons in a circuit and lack uniquely identifiable neurons.
- The primary synaptic input of spinal motor neurons is from the central nervous system; and sensory (reflex) input is secondary. Many reflexes mediate adjustments of centrally programmed movements, such as load compensation.

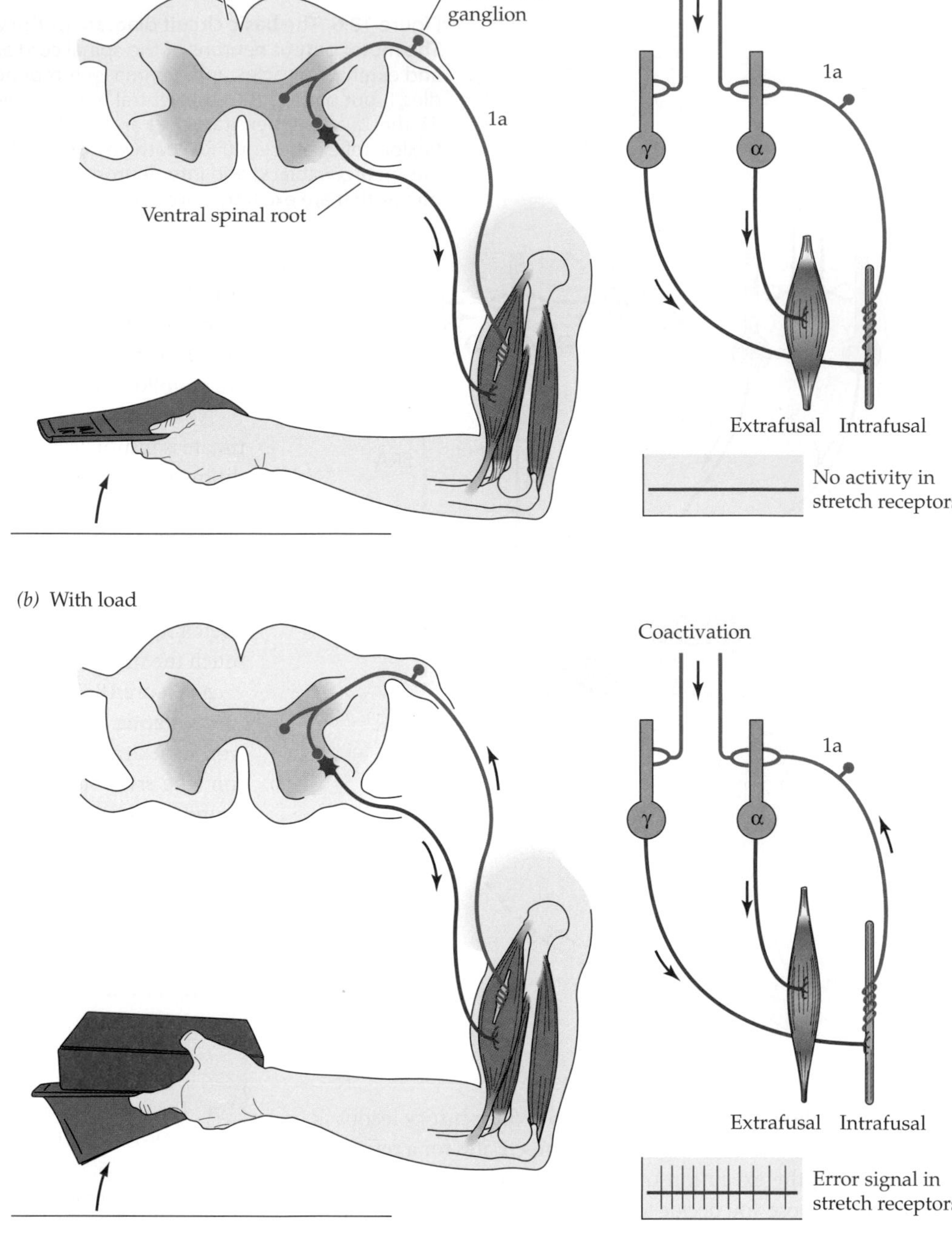

Figure 18.7 The stretch reflex mediates load compensation in a voluntary movement A descending command coactivates both α and γ motor neurons. The γ motor neurons innervate intrafusal muscle fibers in parallel to the extrafusal muscle fibers that will lift the load. *(a)* If there is no load, the extrafusal fibers shorten the whole muscle, unloading the intrafusal fiber and preventing it from activating the muscle-spindle stretch receptor. *(b)* If there is a load that prevents the muscle from shortening, intrafusal fiber contraction activates the muscle spindle 1a afferent neuron. This load-dependent excitation of the muscle spindle is an error signal that activates the α motor neuron and adds proportional excitation to the muscle to achieve the desired shortening.

Action Patterns: Neural Generation of Rhythmic Behavior

Most animal behavior consists not just of isolated single acts of the sorts that we have been discussing, but rather of action patterns: *sequences* of effector actions that result from sequences of motor output of the nervous system. These sequences of motor activity are patterned in space and time. For example, consider the activity of your nervous system that is required to pick up a pencil. First you extend your arm by contracting extensor muscles at the shoulder and upper arm, then you flex your fingers to oppose your thumb, and then you elevate and flex the arm to lift the pencil. You may alter the activity of postural muscles in your trunk or legs to compensate for the arm movements. This motor performance may involve varying amounts of visual, tactile, and proprioceptive sensory input; moreover, the temporal and spatial pattern of the sequence of contractions may differ considerably from one time to the next. Variability of this sort heightens the difficulty of study.

As a way to study the sequences of behavioral action patterns, neurophysiologists have concentrated on **rhythmic behavior**: stereotyped, repetitive sequences of movement such as walking, swimming, and flying in which the motor output is stable, repeatable, and predictable from cycle to cycle of the activity. We now examine several examples of neurophysiological analysis of rhythmic behavior, attempting to extract principles that may be of general importance in motor control systems.

Locust flight results from an interplay of central and peripheral control

Let's begin our exploration of the control of rhythmic behavior by asking, How does a locust fly? As Figure 18.8*a* shows, the movement of a single wing of a flying locust can be viewed as a simple up-and-down oscillation, generated by a set of elevator (or levator) and depressor muscles. These muscles are synchronous flight muscles (see Box 17.2). The electrical activity of these muscles can be recorded from a tethered locust flying in a wind stream. This activity consists of alternating bursts of muscle potentials—the depressors being activated when the wings are up, and the levators being activated when the wings are down. Because each muscle potential results from an action potential in a motor neuron to that muscle, it is clear that flight results from the generation in the CNS of alternating bursts of action potentials in levator and depressor motor neurons. This kind of pattern—alternating bursts of activity in motor neurons to antagonist muscles—underlies most forms of rhythmic behavior.

How are the motor neurons to antagonist muscles activated in alternation to produce a rhythmic movement such as that of a locust wing? Historically, two kinds of hypotheses have been advanced to explain the neural basis of rhythmic movements: peripheral control and central control. According to the hypothesis of **peripheral control**, each movement activates receptors that trigger the next movement in the sequence. The position of a locust wing is monitored by several proprioceptors (Figure 18.8*a*): a single wing-hinge stretch receptor that generates a train of impulses when the wing is elevated, and several other receptors that are activated when the wing is depressed.

Locust flight could (in principle) operate by peripheral control by having sensory feedback from wing receptors activate the motor neurons for the next movement (Figure 18.8*b*). Thus, elevation of the wings would excite the wing-hinge stretch receptor, which would synaptically excite depressor motor neurons, lowering the wing. The lowered wing would terminate excitation of the wing-hinge stretch receptor and would excite the depression-sensitive receptors, which would synaptically excite levator motor neurons, elevating the wing and completing the cycle. The peripheral-control hypothesis is also called the *chained-reflex* hypothesis because each movement is a reflex response to sensory feedback resulting from the last movement.

According to the hypothesis of **central control**, locust flight is sustained by a **central pattern generator**—a neural circuit in the CNS that can generate the sequential, patterned activation of motor neurons to antagonistic muscles that underlies a behavior pattern, without requiring sensory feedback to trigger the next movement. Thus, in central control of locust flight, the basic pattern of alternation of levator and depressor motor neurons would result from an intrinsic central pattern generator (CPG) rather than from a chained reflex (Figure 18.8*c*).

How would one determine whether peripheral control or central control is responsible for the patterned motor activity underlying locust flight? The obvious answer is to remove the relevant sensory input, a process termed *deafferentation* (*afferent* = "toward the CNS"). In the locust, most if not all wing sensory input can be removed by cutting of the nerves to the wing-hinge area.

(*a*) The motor pattern of locust wing muscle excitation

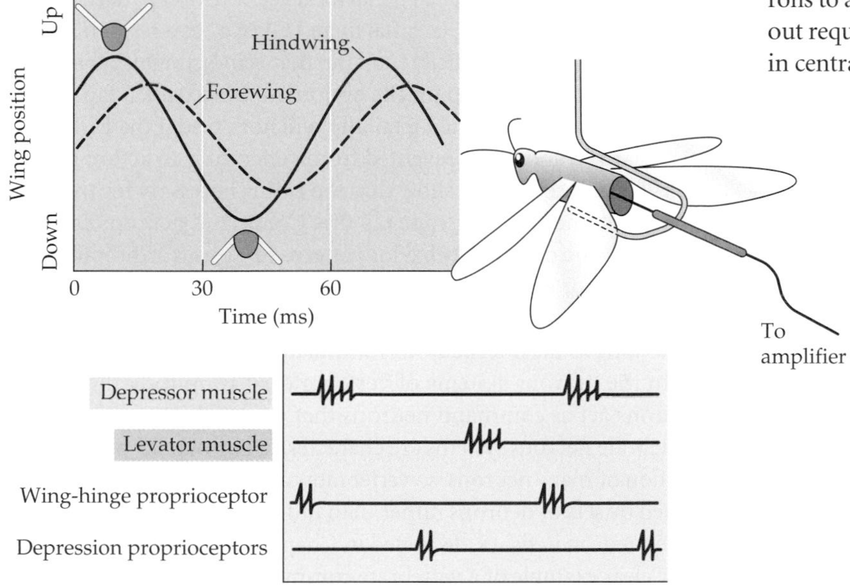

(*b*) The peripheral-control hypothesis (*c*) The central-control hypothesis

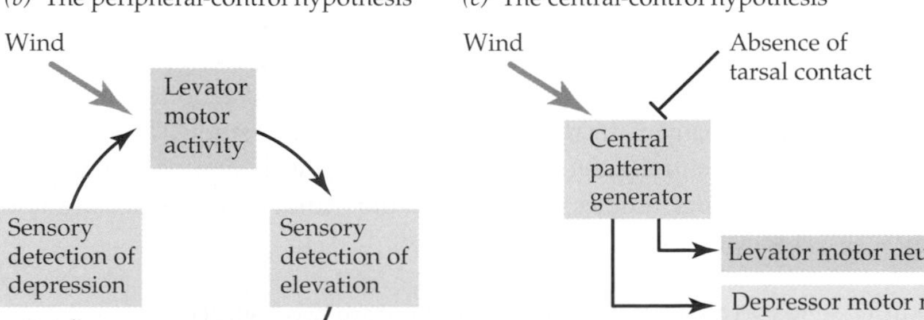

Figure 18.8 Control of flight in the locust (*a*) Cyclic wing movements and the associated temporal pattern of motor and sensory activity, recorded from a tethered locust. Two sorts of hypotheses could explain the generation of the motor pattern of wing muscle excitation: a peripheral-control hypothesis (*b*), in which sensory feedback resulting from a movement triggers the next movement; and a central-control hypothesis (*c*), in which a central pattern generator produces the motor pattern without requiring moment-to-moment sensory timing. (The central pattern generator is gated by tonic sensory input: wind on the head and the absence of tarsal contact.)

Donald Wilson, who pioneered analysis of locust flight control, found that tethered locusts could maintain flight in the absence of wing sensory feedback, although the flight frequency was slower than normal. Normal flight frequency could be restored by the provision of temporally unpatterned stimulation of the cut sensory nerve stumps or of the ventral nerve cord, a finding that suggested that the sensory input provided general excitation to the CNS but was not necessary to supply timing information for pattern generation. These experiments demonstrated the existence of a central pattern generator for locust flight.

Subsequent experiments in other animals have shown that many patterns of rhythmic behavior are under central control. These rhythmic activities include walking, swimming, breathing or ventilation (see Chapter 21), and feeding in a variety of invertebrates and vertebrates. Thus the concept of central pattern generation is a generally important aspect of the control of coordinated behavior.

The hypotheses of central control and peripheral control may appear to be logical alternatives, but they are not mutually exclusive. Therefore, the demonstration of a central pattern generator for a behavior pattern such as locust flight does not mean that sensory input is unimportant. Sensory feedback can play significant roles in a centrally controlled behavior. This statement may seem paradoxical, but consider the fact that if you were walking down the sidewalk and suddenly lost all sensation to your legs, you would probably still be able to generate the motor output sequence of walking; evidence from cats (see page 503) suggests that most mammals have a CPG for walking. Does that mean that sensory input is irrelevant? Of course not.

Sensory input may affect the quality of performance of walking and is essential for correcting the basic pattern, such as when one is walking over uneven terrain. In the locust, several functions of sensory feedback have been found. First, as already noted, sensory input has a generally stimulatory effect, speeding up the flight rhythm. Moreover, sensory feedback can provide specific timing information, adding an element of peripheral control to the system. Electrophysiological studies have shown that wing proprioceptors *do* have the synaptic effects diagrammed in Figure 18.8*b*: The stretch receptor monitoring wing elevation excites depressor motor neurons, and depression-sensitive receptors excite levator motor neurons. Thus the synaptic connections necessary for a chained reflex are present, and these reflexes operate with latencies appropriate to reinforce the flight rhythm—although they are *not necessary* for it.

Sensory feedback can *entrain* a CPG. In another experiment, one wing of a tethered, flying locust was moved up and down at a set frequency by a motor, the forced cyclic movement overriding normal flight-generated sensory feedback. When the forced movement of the one wing was at a rate close to the normal flight frequency, the flight frequency (recorded from muscles to all four wings) changed to match the driving frequency of the motor! Therefore, sensory information from the driven wing can entrain the CPG to the driven frequency.

Our conclusions at this point are these: (1) There is a central pattern generator for flight that can maintain the flight pattern in the absence of sensory timing information, and (2) sensory timing information (when present) can reset the central pattern generator, entraining it to a slightly different driven frequency. The roles of the CPG and the sensory timing information in this case are analogous to those in a circadian (about 24-hour) endogenous activity rhythm of an animal. As discussed in Chapter 10, many animals kept in constant light conditions will exhibit an activity rhythm with a period near (but not exactly) 24 hours. If a light–dark cycle (such as 12 hours light, 12 hours dark) is added, it provides timing information (the zeitgeber) that will entrain the endogenous circadian rhythm to an exactly 24-hour period (see Figure 10.16).

In the same manner, sensory timing information in locust flight can entrain the central pattern generator. Thus the original hypotheses of central and peripheral (reflex) control are not mutually exclusive. The CPG is *sufficient* to maintain flight, but this sufficiency does not rule out contributions of peripheral control. Similar interactions of CPGs with sensory entrainment have been demonstrated in other activities and animals, such as swimming in dogfish sharks. The relative contributions of central and peripheral control can be expected to differ in different cases. It is likely, however, that the two kinds of control interact in most cases.

Command neurons turn on the pattern generator

One principle exemplified by the crayfish escape flexion circuit is that of a **command neuron.** A command neuron is a neuron whose activity is *sufficient* to elicit a behavioral act or action pattern. In the crayfish, a single action potential in an LGI (e.g., as a result of direct electrical stimulation) elicits a tail flip that is indistinguishable from one in response to a tap on the abdomen. Moreover, a tap that is normally adequate to elicit a tail flip will not do so if the LGI is hyperpolarized and thus prevented from generating an action potential. These observations show that the LGI is necessary for the generation of one kind of escape tail flip. Command neurons can also evoke specific rhythmic behavior patterns in various arthropods and molluscs. For example, crayfish have command interneurons that drive rhythmic beating of the swimmerets (abdominal appendages) for as long as the interneuron is stimulated.

In the nervous systems of vertebrates, it is unlikely that single neurons act as command neurons that control behavior patterns. Vertebrate nervous systems are characterized by parallel paths of activation of many neurons, so vertebrate commands are probably mediated by sets of neurons rather than individual neurons. The Mauthner neuron in fishes, discussed in Chapter 11 (see Box 11.2), is the only clear example of a vertebrate command neuron.

There are different mechanisms of central pattern generation

How do neurons and networks of neurons in the CNS act as a central pattern generator that determines the spatiotemporal patterns of motor output to generate a rhythmic fixed action pattern? Because many of the rhythmic behavior patterns studied are oscillatory, the central pattern generators underlying them have been termed **oscillators.** Studies have clarified the neural basis of some of these oscillators.

OSCILLATOR THEORY In theory, there are two logical categories of oscillators: cellular oscillators and network oscillators. The central pattern generators that have been studied appear to employ a mix of these two kinds of oscillatory mechanisms in differing degree—

some with cellular oscillators playing a dominant role, and others apparently dominated by network properties. We will introduce the theoretical types first and then describe their roles in two real examples of central-pattern-generating neural circuits.

A **cellular oscillator** is a neuron that generates temporally patterned activity by itself, without depending on synaptic interaction with other cells. Such cells may generate endogenous bursts of action potentials (Figure 18.9*a*), or they may show oscillations of membrane potential without generating any action potentials (Figure 18.9*b*). The underlying mechanisms of oscillation may be similar for both types because some cells that generate impulse bursts will continue to oscillate after impulse generation is blocked with tetrodotoxin (TTX) (see Chapter 11, page 298). The ability of neurons to act as cellular oscillators depends on several sorts of voltage-gated ion channels, in addition to those that generate ordinary action potentials.

Cellular oscillators are thought to play a role in central pattern generation in several cases studied, including those controlling molluscan feeding, crustacean heartbeat, and crustacean scaphognathites (gill bailers) (see Figure 21.27 in Chapter 21).

A **network oscillator** is a network of neurons that interact in such a way that the output of the network is temporally patterned, although no neuron in the network functions as a cellular oscillator. Thus the oscillatory or pattern-generating property is said to be an *emergent property* of the network, resulting from cellular interactions in the network rather than from intrinsic cellular properties. The simplest model of an oscillatory network, termed a *half-center model*, is shown in Figure 18.9*c*. Two neurons (or pools of neurons) synaptically inhibit each other so that when one stops generating impulses, the other is released from inhibition and generates a train of impulses, inhibiting the first.

The half-center model appears straightforward but is actually rather unstable in its simplest form. Unless an additional time-dependent property is added to allow the first half-cycle to run down, the first neuron to reach threshold will tend to remain active and perpetually inhibit the other. There are several possible mechanisms to "fatigue" a half center, such as postinhibitory rebound, slow inactivation of the impulse-generating capability of the cells, or antifacilitation of the inhibitory synapses (see Figure 12.22 in Chapter 12).

Another possible network oscillator that is more stable contains three or more neurons in a cyclic inhibitory loop. Figure 18.9*d* shows a *closed-loop model* that would produce a stable pattern of bursts in the sequence 1–2–3–1–2–3...—without any cell possessing endogenous oscillator properties.

Neural circuits acting as network oscillators are reported to underlie several sorts of rhythmic behaviors, including the neurogenic leech heartbeat, as well as swimming in leeches, molluscs (*Tritonia* and *Clione*), lampreys, and clawed toad (*Xenopus*) tadpoles. Each described circuit is different, but each displays some of the mutual inhibition that is characteristic of half-center models, as well as closed-loop inhibitory and excitatory elements.

(*a*) Oscillating and generating impulses

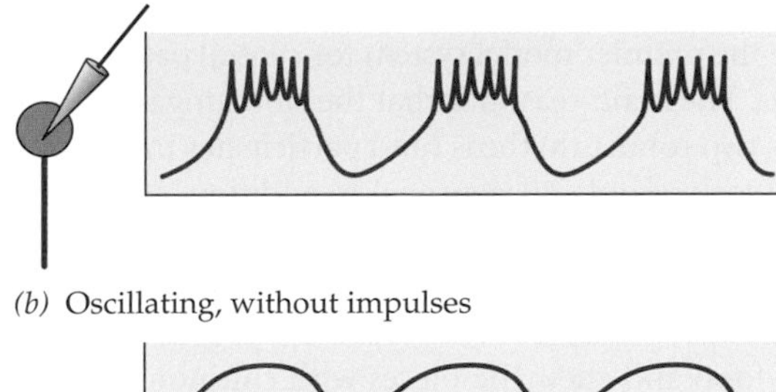

(*b*) Oscillating, without impulses

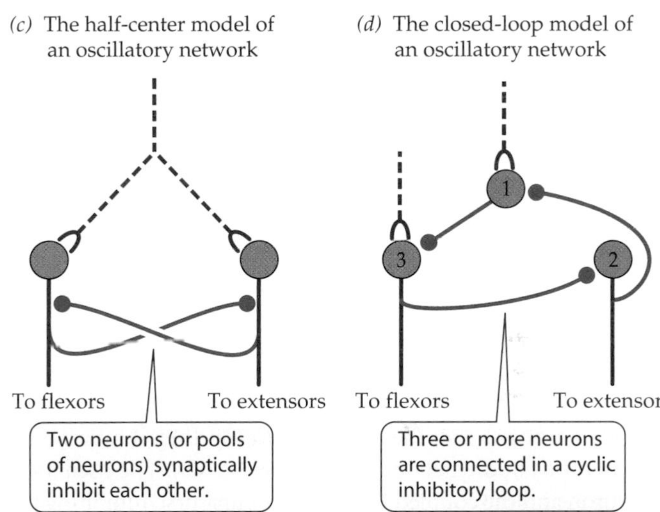

KEY

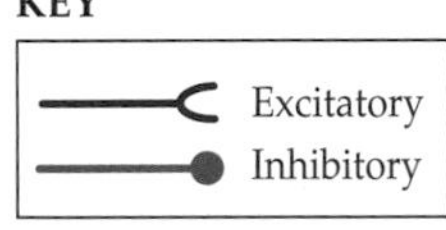

Figure 18.9 Models of oscillators underlying central pattern generators (*a*) An oscillator neuron generating bursts of impulses (e.g., in *Aplysia*). (*b*) A neuron with membrane-potential oscillation but without impulses (e.g., a neuron controlling pumping of the crustacean scaphognathite, or gill bailer). (*c*) A network oscillator composed of reciprocal inhibitory half-centers. (*d*) A network oscillator composed of closed-loop cyclic inhibition. All three cells may be spontaneously active or may receive unpatterned excitatory input (dashed lines). If cell 1 is active first, its activity inhibits cell 3, but this inhibition prevents cell 3 from inhibiting cell 2. Cell 2 can now be active, inhibiting cell 1 and thus releasing cell 3 from inhibition. Cell 3 can then be active, inhibiting cell 2 and releasing cell 1 from inhibition, and so forth.

CIRCUITS FOR CENTRAL PATTERN GENERATION: THE STOMATOGASTRIC GANGLION A central pattern generator can combine the properties of both cellular oscillators and network oscillators. Such a **hybrid oscillator** might have one or more oscillatory cells acting within a network that stabilizes and reinforces the oscillation. Most examples of CPGs will probably turn out to be hybrid oscillators with a mixture of cellular and network oscillatory properties. Here we describe the two most completely studied CPG networks, both of which are hybrid oscillatory networks in the crustacean stomatogastric ganglion. We will use these examples to introduce principles of general importance about oscillatory circuits and their control.

The **stomatogastric ganglion** sits on the external surface of the stomach of a lobster, crayfish, crab, or other decapod crustacean. It contains about 30 neurons, most of which are motor neurons controlling stomach muscles. Why is the stomatogastric ganglion important? We would not expect the control of the stomach of a crab or lobster to be the premier model system for central pattern generators, but it is. The main reason is that the stomatogastric ganglion generates two robust rhythms (and participates in two others) with a network of only 30 neurons! In addition, crustacean stomachs are ectodermal, chitin-lined, and controlled by striated muscle; thus their control is more like that of an appendage than what we would expect of a stomach. Crabs and lobsters swallow large pieces of food and chew the pieces with chitinous teeth in their stomachs, so the behavior mediated by the stomatogastric ganglion is more analogous to the control of vertebrate jaws than that of vertebrate stomachs.

As we discussed in Chapter 4 (see Figure 4.16*b*), the crustacean stomach consists of two chambers: an anterior cardiac chamber containing teeth that function as a *gastric mill* to grind and chew food, and a posterior pyloric chamber containing a sieve that serves to keep food particles from passing to the rest of the gut until the particles are small. Figure 18.10 shows a simplified neural circuit[4] and rhythmic output of one rhythm of the stomatogastric ganglion: the *pyloric rhythm* that controls the straining of food particles by the pyloric filter.

The pyloric circuit (see Figure 18.10*a*) acts as a hybrid oscillator, containing an oscillator neuron (AB) that serves as the pacemaker for the rhythm. The oscillator cell is tightly electrically coupled to two PD neurons so that these three burst together, inhibiting the other neurons in the network (see Figure 18.10*b*). The oscillatory AB neuron and the coupled PD neurons burst first, inhibiting follower cells (LP, PY, and two others not shown). At the end of the AB/PD burst, the LP cell recovers from inhibition faster than the PY cells; therefore the LP cell bursts next and prolongs PY inhibition. PY neurons then burst and inhibit LP, until the next AB burst starts a new cycle.

The pyloric circuit thus has both cellular oscillator and network oscillator properties. The generation of the pyloric rhythm depends primarily on the AB cellular oscillatory neuron, but its triphasic cycle (AB/PD → LP → PY → …) and timing depend on the strength and time course of inhibitory synapses and on intrinsic currents of the follower cells.

The other major circuit of the stomatogastric ganglion generates the *gastric rhythm*, which controls the gastric mill of teeth in the anterior stomach chamber. Originally the gastric network was thought to be a "pure" network oscillator, without oscillatory neurons. Subsequent studies showed that with modulatory input from other parts of the CNS, at least one neuron in the circuit can act as a conditional cellular oscillator, so the circuit for the gastric chewing rhythm is probably also a hybrid oscillatory circuit. Nevertheless, the expression of the gastric rhythm is much more dependent on network interactions than is the pyloric rhythm.

The rhythms and circuits of the stomatogastric ganglion exemplify another feature that may be of general importance: They are profoundly subject to *modulation*. The stomatogastric ganglion receives about 100 axons of neurons from other parts of the nervous system, many of which can secrete neuromodulators that act diffusely in the small ganglion to alter its motor output. At least 15 modulators are present, including the amines serotonin, dopamine, octopamine, and histamine; the classical transmitters acetylcholine and GABA; and several peptides, including proctolin, FMRFamide-like, and cholecystokinin (CCK)-like peptides. Many of these modulate both the pyloric and gastric rhythms.

The most common effect of a neuromodulator is to initiate and maintain rhythmic activity in a network. For example, adding serotonin, octopamine, or dopamine to a previously quiescent isolated stomatogastric ganglion induces a pyloric rhythm (Figure 18.11), although the rhythms induced by the three modulators differ in detail. In general, the stomatogastric ganglion requires permissive modulatory input from extrinsic neurons for the expression of both of the rhythms we have discussed. Many of the central and sensory neurons that provide this modulatory input are well characterized.

The neuromodulators of the stomatogastric ganglion act in two sorts of ways: They alter the intrinsic membrane properties of individual stomatogastric neurons, and they alter the strengths and dynamics of synaptic connections of the neurons. Modulatory effects on intrinsic neuronal currents can induce cellular oscillation (many stomatogastric neurons are conditional oscillators), excite or inhibit

[4] We simplify the circuit because although only six types of stomatogastric neurons participate in the pyloric rhythm, the extent to which they inhibit each other makes the whole circuit very difficult to follow and understand!

(a) The pyloric circuit (simplified)

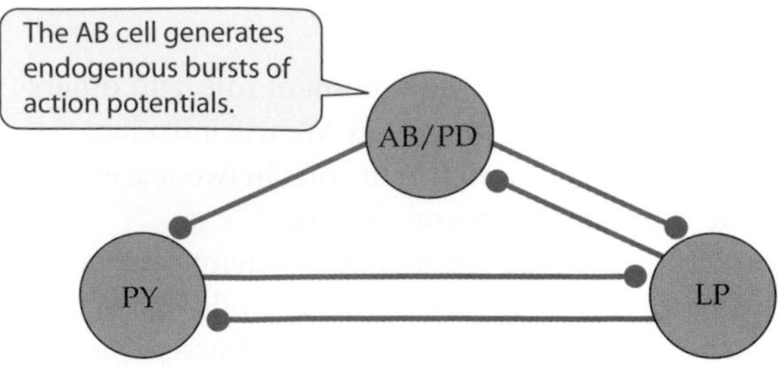

(b) A basic pyloric rhythm

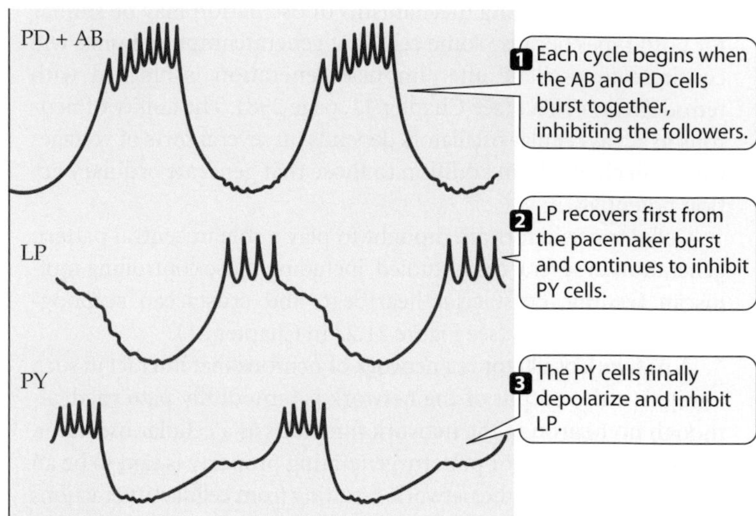

Figure 18.10 The network of neurons producing the pyloric rhythm of the crustacean stomatogastric ganglion *(a)* The pyloric circuit (simplified). AB is a strong cellular oscillator that serves as the pacemaker for the rhythm. The AB and PD cells are strongly electically coupled and represented by a single symbol. Other (follower) cells are interconnected by inhibitory chemical synapses. *(b)* A basic pyloric rhythm.

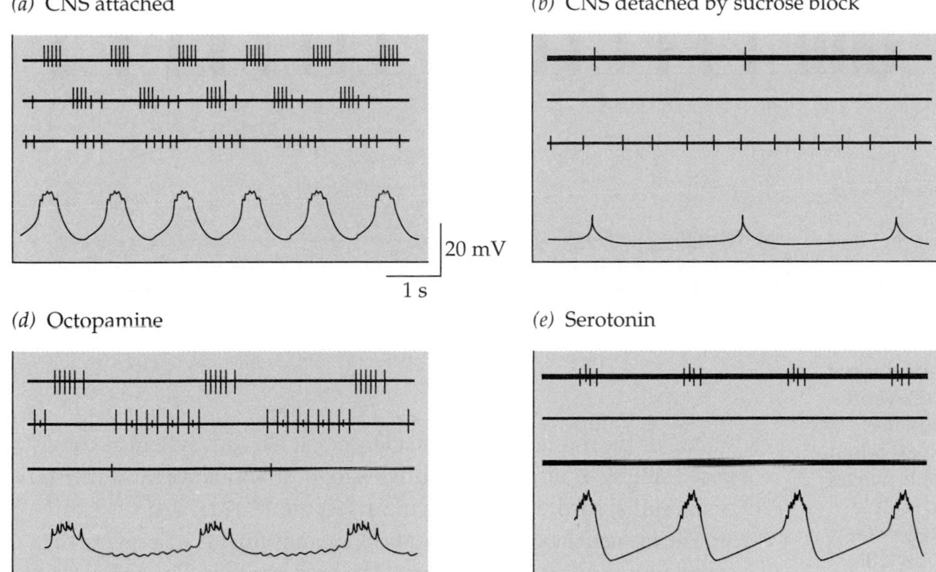

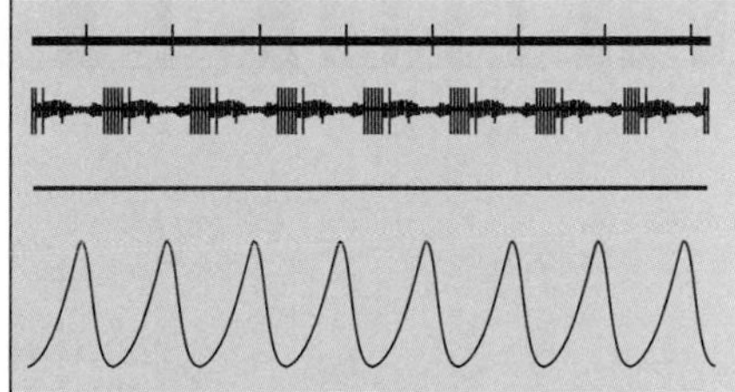

Figure 18.11 Neurotransmitters modulate the pyloric rhythm of the stomatogastric ganglion The top three traces in each part are extracellular recordings from motor nerves containing axons of different neurons; the bottom trace is an intracellular recording from cell AB. *(a)* When the stomatogastric ganglion is attached to the rest of the central nervous system, endogenous neuromodulators maintain a normal pyloric rhythm. *(b)* Separating the ganglion from the rest of the CNS (by sucrose block or by cutting the stomatogastric nerve) often leads to cessation of rhythmic activity, as shown. *(c–e)* Addition of any of the indicated neuromodulators to the bathing solution reestablishes the rhythm. Note that the pattern of the rhythm differs somewhat with different modulators. These effects result from specific, individualized actions of the modulators on both cellular ionic currents and synaptic strengths. (From Harris-Warrick and Flamm 1986.)

particular neurons, or alter other excitable properties. Moreover, modulators can make individual synapses more or less potent, changing the functional circuit connections, as well as cellular activities.

It is at once exhilarating and sobering to realize that neuronal circuits such as those of the stomatogastric ganglion are not rigidly "hard-wired," but rather are plastic and malleable—exhilarating because the ability of neuromodulation to free a circuit from the "tyranny of the wiring" may underlie adaptive plasticity of neural control of behavior, but sobering because of the realization that a circuit diagram such as that in Figure 18.10*a* is descriptive of only one state of a dynamically shifting circuit. Neurons can even shift from one functional circuit to another, firing in "gastric time" or "pyloric time" under modulatory influence, and circuit elements can combine to form new patterns of output. Growing evidence suggests that these roles of neuromodulators are of widespread importance among central pattern generators.

Central pattern generators can underlie relatively complex behavior

How elaborate a behavioral performance can we expect to depend on mechanisms of central pattern generation? It is, after all, a long way from a short-term rhythmic activity such as locust flight or crayfish swimmeret beating to complex behavior patterns such as mating behavior in three-spined stickleback fish. The courtship rituals of stickleback fish are complex. Are these complex patterns simply elaborations and chains of centrally programmed acts? For technical reasons, it has been difficult for scientists to explore the neurophysiological bases of increasingly complex fixed action patterns, but some progress has been made.

Central-pattern-generating mechanisms have been shown to be sufficient for some behavior patterns that are significantly more complex and longer-lasting than the simple cyclic patterns already described. One such long-term sequence involves gill movements of the horseshoe crab *Limulus*. Figure 21.25 in Chapter 21 shows the unusual *book gills* of these animals. These gills consist of many thin sheets of tissue termed *lamellae* (singular *lamella*) (the "pages" of the book) positioned under segmental flaps termed *gill plates*. Rhythmic beating of the gill plates ventilates the gill lamellae. This gill ventilation, however, is often intermittent: Periods of rhythmic ventilation a few minutes long may alternate with shorter periods of quiescence (apnea) or of gill cleaning. Gill cleaning is a relatively complex fixed action pattern in which the paired gill plates are brought across the midline and a part of one plate rhythmically cleans the lamellae of the opposite book gill. Animals roughly alternate between two mirror-image patterns of gill cleaning, termed *left-leading* and *right-leading*.

When the abdominal ventral nerve cord is dissected out of a *Limulus* and its unstimulated motor activity is recorded in isolation, the motor output pattern underlying all of the behavior just described persists (Figure 18.12): Periods of a ventilatory motor output rhythm alternate with periods of a gill-cleaning motor pattern. Moreover, the rough alternation between left-leading and right-leading gill cleaning can also be expressed in isolation. Thus, long, relatively elaborate sequences of behaviorally significant motor patterns can be expressed in isolated CNS tissue, without muscles, movement, or sensory feedback.

Fixed action patterns that are still more complex have been analyzed in insects, including molting activity (see Chapter 14, page 417) of crickets and moths and reproductive behavior in several insects. These stereotyped behaviors appear to have centrally patterned components but also to have stages at which appropriate sensory feedback is necessary to proceed to the next stage. The central motor programs could function as modules, like stored subroutines in a

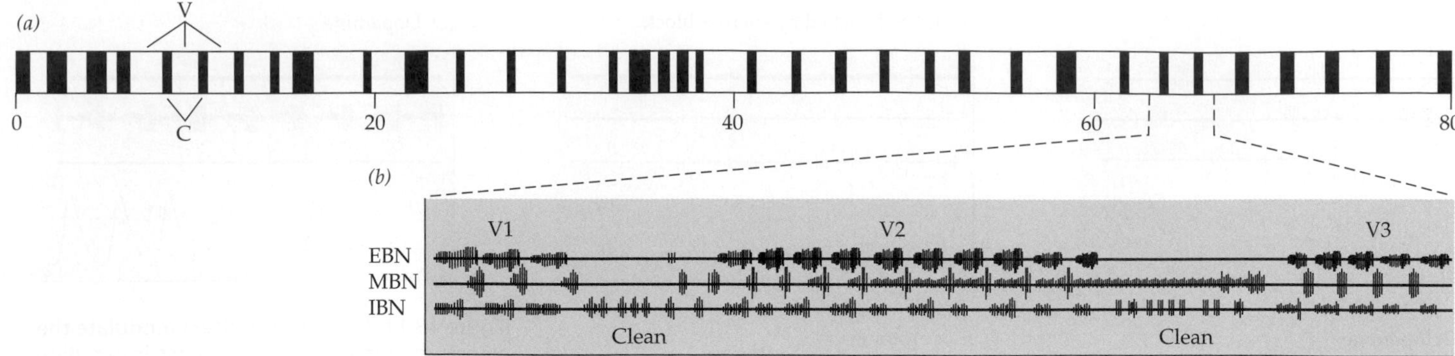

Figure 18.12 Central pattern generation of long-term behavior patterns The pattern of long-term sequential alternation of gill-ventilation (V) and gill-cleaning (C) motor rhythms persists in an isolated abdominal nerve cord of the horseshoe crab *Limulus*. *(a)* An 80-minute period of recorded activity during which gill-ventilation bouts (white) and gill-cleaning bouts (black) alternate fairly regularly. Such stable alternation is common in intact animals. *(b)* An approximately 140-s record of left branchial nerves of the first abdominal ganglion (EBN, MBN, IBN). Three bouts of gill-ventilation rhythm (V1–V3) alternate with two bouts of gill-cleaning pattern. (After Wyse, Sanes, and Watson 1980.)

computer program; when "called" by the CNS, a module would generate a particular motor pattern, after which sensory input would determine how to proceed next. It is likely that increasingly complex behavior patterns will have increasingly elaborate interactions between sensory components and central motor programs.

Summary: Action Patterns

- Most rhythmic patterns of animal behavior (walking, swimming, flying, and so on) involve a central pattern generator (CPG), which can produce the basic motor pattern without requiring sensory input at particular times in the cycle. The CPG interacts with sensory feedback from the cyclic movements, which can entrain the CPG.
- Many CPG circuits that produce rhythmic behavior patterns can be activated by commands from central neurons. Command neurons can be single neurons in arthropods or molluscs, but in vertebrates, commands involve many neurons.
- CPG circuits may depend on cellular oscillators, network oscillators, or a combination of both, as in the crustacean stomatogastric ganglion.
- CPG circuits are subject to neuromodulation, in which a neurotransmitter/neuromodulator can alter circuit function to generate or alter rhythmic output.

Control and Coordination of Vertebrate Movement

The principles of central pattern generation and the interaction of central and peripheral control of movement were first developed from invertebrate studies, principally with arthropods and molluscs. In this section we consider the degree to which these principles also apply to vertebrates. We can start with the question, How does a cat walk? For the moment, let's consider the cat nervous system as composed simply of three compartments that can influence movement: brain, spinal cord, and sensory input (Figure 18.13).

The immediate generators of walking movements in a cat are the spinal motor neurons that control the limb muscles. The spinal circuitry associated with these motor neurons was introduced earlier (see Figure 18.6). The motor neurons receive direct or indirect synaptic input from three kinds of sources: descending input from the brain, sensory input from proprioceptors and other receptors in the periphery, and local input from intrinsic spinal circuits. If the spinal motor neurons are to be activated in the correct spatiotemporal pattern to produce walking, what are the roles of these three compartments in generating this pattern?

Locomotion in cats involves a spinal central pattern generator

In the arthropod systems already discussed, the compartments of neural control could be experimentally isolated with relative ease. For example, it is technically easy to isolate the abdominal ganglia of a crayfish or a horseshoe crab, and to ask the question, What do abdominal ganglia do by themselves, without the brain and without sensory feedback? Such questions are harder to investigate experimentally in vertebrates, in which an isolated spinal cord usually dies quickly. With refinements of technique in the last 30 years, however, it has become possible to perform experiments analogous to those in invertebrates.

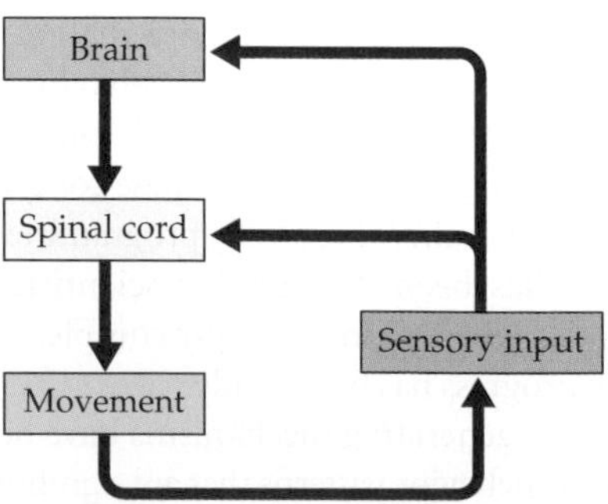

Figure 18.13 The major components of the control of movement in a vertebrate Motor neurons in the spinal cord activate muscles as a result of receiving three sorts of input: input from local spinal circuits, descending input from the brain, and sensory input.

It is possible to experimentally separate the brain, spinal cord, and sensory feedback in cats, to begin to determine their roles in walking. A cat is able to make fairly normal stepping movements on a treadmill after transection of the spinal cord to remove brain influence. First, in chronic (long-term) experiments, cats with the spinal cord transected 1 to 2 weeks after birth recover the ability to walk on a treadmill at a speed dependent on the treadmill speed. Second, in acute (short-term) experiments, spinally transected cats can walk on a treadmill if given the norepinephrine precursor dopa or the norepinephrine receptor stimulator clonidine.

These experiments show that the brain is not needed to provide timing information for walking. Noradrenergic fibers descending from the brain in intact cats presumably command or enable the expression of the walking pattern by spinal circuits, but they are not necessary for timing the stepping cycle of a limb; certainly injected dopa does not provide timing information. In other experiments, cats with brain sections (at point 1 in Figure 18.14) can walk on a treadmill when given unpatterned electrical stimulation to a mesencephalic locomotor command area. With increasing strength of stimulation, the rate of locomotion increases and the gait changes to a trot and finally to a gallop. Thus the brain may initiate locomotion and modulate it subject to conditions, but the brain is not necessary for generating the locomotor pattern.

Sensory feedback from the hindlimbs is also unnecessary for hindlimb stepping movements, as can be shown by experiments similar to those just described. Cats with or without spinal transection can make normally alternating stepping sequences following hindlimb deafferentation by cuts of the dorsal roots that contain the sensory afferent axons. (For the spinally transected cats, walking is initiated with dopa or clonidine.) These experiments indicate that the cat spinal cord contains a central pattern generator for walking movements. Similar experiments indicate that fish, salamanders, toads, and turtles also have spinal locomotor CPGs.

Sensory feedback can still have important functional roles in locomotion of intact vertebrates. Spinal reflexes stabilize and modulate the effects of centrally patterned locomotor output, but spinal reflexes themselves may also be modulated by the CPG. For example, the effect of mechanical stimulation of the top of the foot of a walking cat depends on the position of the foot in the stepping cycle. If the foot is off the ground and swinging forward, it is lifted higher when stimulated ("exaggerated flexion"). If the foot is on the ground and bearing the cat's weight, the same stimulation produces a more forceful extension. This reversal of a spinal reflex (which is clearly adaptive for stable walking) shows that the central events of the stepping cycle can strongly modulate reflex function.

The experiments described in this section demonstrate that the mechanisms of control of rhythmic locomotor movements are fundamentally similar in many invertebrates and vertebrates. For example, Pearson (see the reference on page 509) describes impressive similarities in the control of walking in cats and cockroaches. Although the cellular aspects may vary (e.g., different cellular mechanisms of central pattern generation), the functional roles of central and reflex aspects of control appear to be generally similar in many cases.

The generation of movement involves several areas in the mammalian brain

The vertebrate brain is profoundly important in the control of movement. We have discussed experiments showing that patterned locomotor movements can persist in spinally transected vertebrates and hence do not require the brain. This finding, however, does not contradict the importance of the brain in initiation, coordination, and regulation of normal movements. Next we will consider the ways in which brain areas interact with sensory input and spinal centers in movement control. The roles of various brain areas in motor control, however, are incompletely understood and are difficult to separate from sensory, motivational, and other aspects of brain function.

CEREBRAL CORTEX We begin our examination of the execution of a voluntary movement with the motor areas of the cerebral cortex. Three adjacent regions of the cerebral cortex have predominantly motor functions (Figure 18.15). The **primary motor cortex** (or simply *motor cortex*) lies just anterior to the central fissure, a prominent fold in the convoluted cortical surface of most mammals. Just ante-

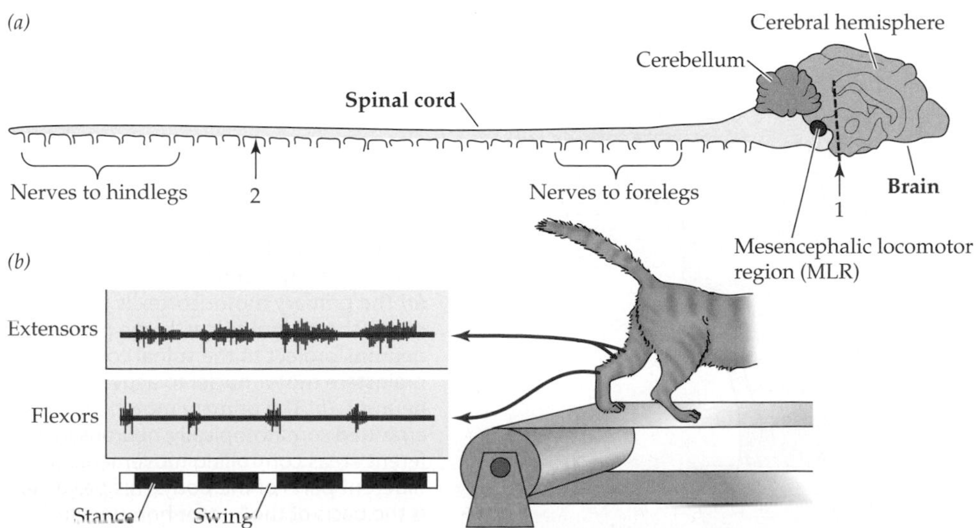

Figure 18.14 Spinal and brain control of mammalian locomotion *(a)* The spinal cord and lower brainstem of a cat isolated from cerebral hemispheres by transection at point 1. Electrical stimulation of the mesencephalic locomotor region can produce locomotion in this preparation. Transection of the spinal cord at point 2 isolates the hindlimb segments of the cord. The hindlimbs are still able to walk on a treadmill after recovery from surgery. *(b)* Locomotion on a treadmill of a cat with a spinal transection at point 2. Reciprocal bursts of electrical activity are recorded from flexors during the swing phase of walking and from extensors during the stance phase. (After Kandel et al. 1995.)

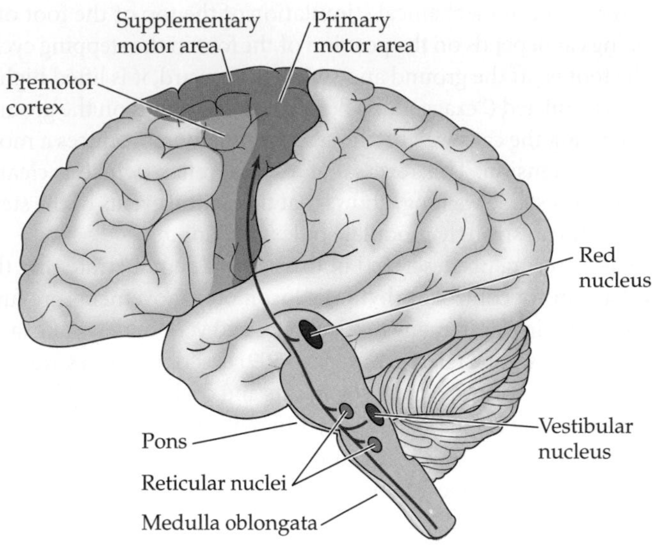

Figure 18.15 The location of major motor areas of the mammalian cerebral cortex and brainstem The primary motor cortex is just anterior to the central sulcus (which separates it from the somatosensory cortex). Anterior to the primary motor cortex are the premotor cortex and the supplementary motor area. Neurons from the motor cortex descend to activate motor nuclei of the brainstem and circuits of the spinal cord.

rior to the primary motor cortex on the lateral surface of the brain is the **premotor cortex**. The **supplementary motor area** is also just anterior to the primary motor cortex, but it extends along the medial cortical surface rather than the lateral surface (see Figure 18.15).

Early studies demonstrated that electrical stimulation of areas of the primary motor cortex elicited movements of particular parts of the body, with a point-to-point correspondence between the area stimulated and the muscles activated. Thus the body regions are represented on the surface of the primary motor cortex by a somatotopic map (Figure 18.16), which is distorted by the disproportionately large areas of cortex that serve regions such as the hand and the mouth. There are separate somatotopic maps in the other two cortical motor areas, as well as elsewhere.[5]

The neurons of the primary motor cortex that mediate motor responses to stimulation are pyramidal cells (neurons with pyramid-shaped somata), the axons of which synapse on brainstem motor nuclei and also continue down the spinal cord as major components of the corticospinal tract. (This tract is known as the *pyramidal tract* because the axons funnel through a pyramid-shaped structure on the ventral surface of the brainstem—not because the cells are pyramidal neurons.) The corticospinal axons end primarily on interneurons in the spinal cord, although in primates some also end directly on spinal motor neurons. The neurons of the primary motor cortex therefore activate spinal motor circuits directly via the corticospinal tract, and indirectly via brainstem motor nuclei: the pontine and medullary reticular nuclei, the vestibular nucleus, and the red nucleus (see Figure 18.15). The brainstem motor nuclei (except the red nucleus) are generally more important in involuntary postural control than in voluntary movements.

It seems reasonable to hypothesize that activation of pyramidal cells in the primary motor cortex initiates a voluntary movement, although definitive evidence is lacking. The corticospinal tract appears essential for voluntary movement, but subcortical areas are also important. Activity of neurons in the primary motor cortex precedes and correlates well with voluntary movements, a finding that suggests a control function but is by no means conclusive.[6] Individual neuron activities in primary motor cortex encode the force and direction of movements. For some neurons the amount of activity

[5] Somatotopic maps were discussed in Chapter 10 (see Figure 10.11).

[6] One argument against an initiating function of the motor cortex is that nonmammalian vertebrates lack a corticospinal tract and a clearly organized motor cortex. Thus it is argued that, for nonmammalian vertebrates and perhaps also for mammals, the generation of movement is controlled subcortically, with an evolutionarily recent cortical overlay in mammals that may modulate and correct the movement. This view, although not widely held, demonstrates our uncertainty about even basic issues in motor control.

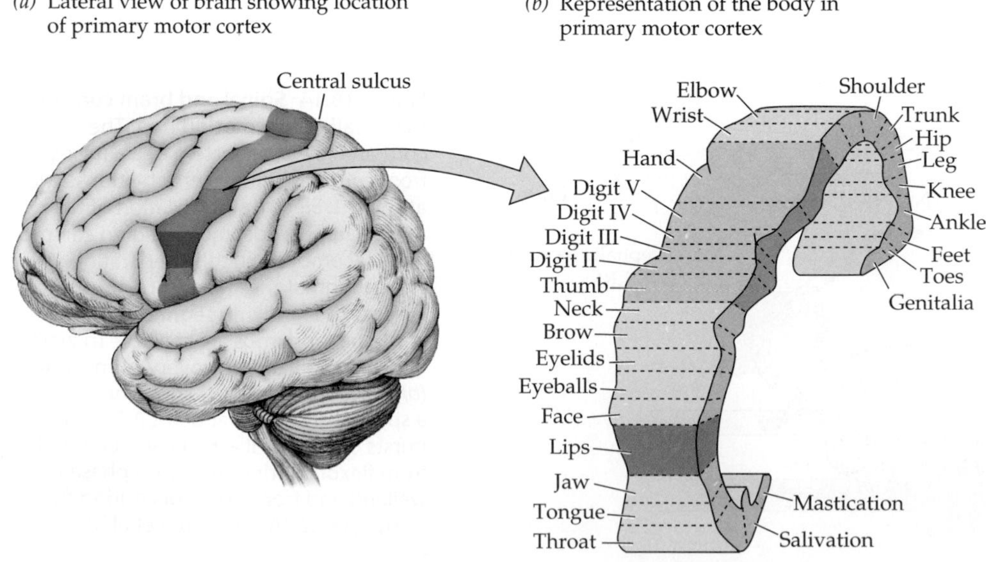

Figure 18.16 The primary motor cortex controls the generation of movements *(a)* The primary motor cortex is just anterior to the central sulcus. Large pyramidal neurons project to the spinal cord and to brainstem motor nuclei to activate motor neurons. *(b)* The primary motor cortex is arranged somatotopically, neurons in different areas controlling movements of different parts of the body. This projection is the basis of the "motor homunculus" shown in Figure 10.11*a*.

predicts the amount of force of a movement, for others it predicts the change of force, and for still others it predicts the direction of the movement independent of the muscles and forces used to achieve it. The last type of cellular activity suggests that neurons in the primary motor cortex can code a parameter of the movement more abstract than the forces and muscles that generate it.

Even if we accept the hypothesis that the primary motor cortex initiates voluntary movements, we have brushed only the tip of the iceberg. How does a decision to initiate a voluntary movement initiate the neural events that lead to the activation of motor-cortex pyramidal cells? How does the motor cortex interact with other brain areas, such as the cerebellum and basal ganglia, to produce smoothly coordinated, skilled movements without our having to expend continuous conscious effort?

For the first question we know from recordings and from imaging studies that extensive areas of the brain are active before a decision to move. For example, if we place surface electroencephalogram (EEG) electrodes on the skull of a subject and ask her to move one finger whenever she wishes, we record a consistent pattern of activity from much of the entire cortical surface: a small, widespread rising wave of electrical activity that precedes the movement by about 800 ms. This electrical activity, termed a *readiness potential,* becomes localized to the relevant portion of the primary motor cortex only in the last 50 to 80 ms preceding the movement. The decision to initiate a voluntary movement appears to involve many areas of the cortex, including so-called association areas, and to be passed to a specific motor cortical site for initiation of the movement.

The cortical areas anterior to the primary motor cortex (premotor cortex and supplementary motor area) play important roles in preparing motor systems for movement. Strong electrical stimulation in these areas leads to coordinated movements, rather than the elemental movements elicited by stimulation in the primary motor cortex. The supplementary motor area programs motor sequences. Figure 18.17 shows the results of imaging studies of regional activity of cortical areas, measured by changes in blood flow to active areas. The supplementary motor area is active during relatively complex motion sequences, and in rehearsing the sequences as well as in performing the movements themselves. Lesions in premotor and supplementary motor areas lead to deficits in the organization ("strategy") of complex movements, rather than to weakness of the movements themselves.

The premotor cortex is important in visually guided movements and controls postural adjustments associated with a movement. In general, the premotor cortex is important primarily in programming *externally* cued movements (in which, say, a color provides the cue for which movement to make), and the supplementary motor area is important primarily in *internally* cued movements (in which a learned sequence must be remembered).

The supplementary motor area and the premotor cortex project directly to the primary motor cortex, but they (and other cerebral areas) also have important projections to subcortical areas. Many studies suggest that both the initial preprogramming of a movement and its modification once initiated involve interaction of the cerebral cortex with two subcortical areas that are important in voluntary movement. Next we will consider these subcortical areas: the cerebellum and the basal ganglia.

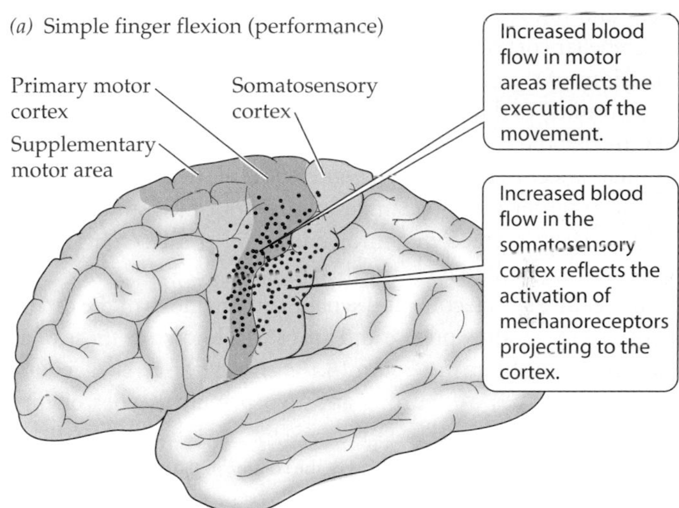

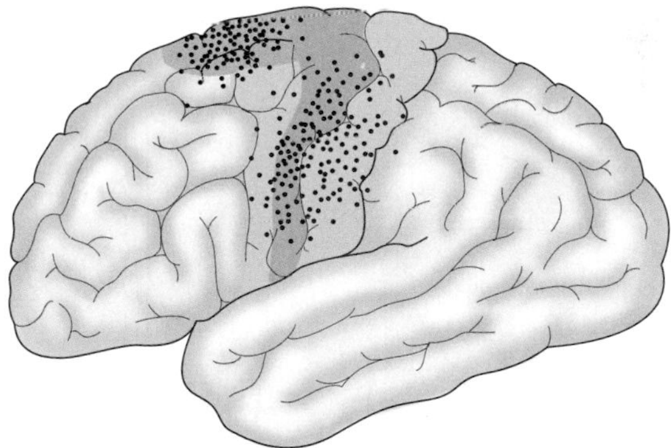

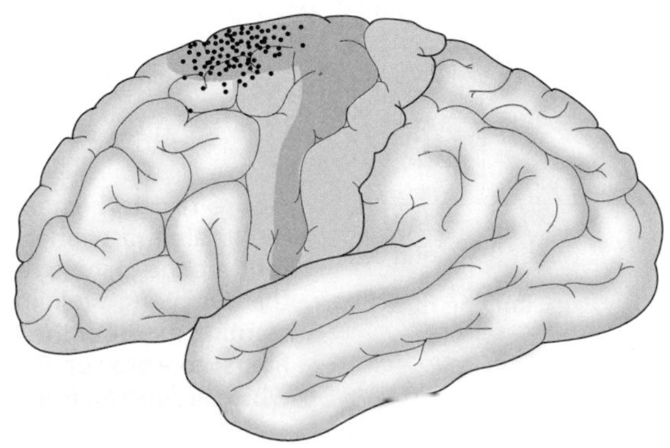

Figure 18.17 Activation of neurons in motor cortical areas during the planning and execution of conscious movements Activity increases were imaged as local increases in cerebral blood flow. *(a)* During a simple finger flexion against a spring, blood flow increased in the hand areas of primary motor cortex and somatosensory cortex (posterior to the central sulcus). *(b)* During a complex sequence of finger movements there is also increased blood flow in the supplementary motor area as well. *(c)* During mental rehearsal of the same complex sequence of finger movements, only the supplementary motor area shows an increase in blood flow. (After Kandel et al. 1995.)

CEREBELLUM The **cerebellum** is a large,[7] highly convoluted structure at the dorsal side of the hindbrain. It is present in all vertebrates. The cerebellum regulates movement indirectly, adjusting the descending motor output of other brain areas. The cerebellum is clearly involved in the coordination of movement, as demonstrated by the effects of cerebellar lesions in various animals, including humans. Voluntary movements are still possible following cerebellar lesions, but they are clumsy and disordered, lacking the smooth and effortless precision of normal movements. Movements are accompanied by tremor, and patients with cerebellar injuries report that they have to concentrate on each part of a movement, joint by joint. It is as if the cerebellum serves to compile instructions for the smooth and coordinated execution of complex movements.

The cerebellum contains two major parts: an outer **cerebellar cortex** and underlying **deep cerebellar nuclei**. The sole output of the cerebellar cortex is to the deep cerebellar nuclei. Three functional divisions of the cerebellar cortex receive inputs from and project (send outputs) to different parts of the brain via different deep cerebellar nuclei: the *vestibulocerebellum* (posterior, interacting with the vestibular system), the medial *spinocerebellum* (receiving spinal sensory information and projecting to the red nucleus), and the lateral *cerebrocerebellum* (receiving information from the cerebral cortex and projecting back to it via the thalamus).

The cellular architecture and synaptic interactions of the cerebellar cortex are elegantly precise and are as well known as those of any other area of the brain. As Figure 18.18*a* shows, the cerebellar cortex contains five types of neurons and two principal types of input fibers. The axons of Purkinje cells constitute the only output of the cerebellar cortex; these end in the deep cerebellar nuclei below the cortical surface. The major synaptic interactions of the cerebellar cortex are shown in Figure 18.18*a*. Climbing fibers make powerful excitatory 1:1 synaptic contacts with Purkinje cells. Mossy fibers, in contrast, provide divergent excitatory input to many granule cells. Axons of granule cells ascend to the surface layer of the cortex and branch in opposite directions as parallel fibers, which make excitatory synaptic contacts with the other types of cerebellar cortical cells.

The synaptic interaction of parallel fibers and Purkinje cells is especially prominent. The parallel fibers pass through the flattened, planar dendrites of Purkinje cells at right angles (Figure 18.18*b*). Each Purkinje cell receives excitatory synapses from about 100,000 parallel fibers (in addition to one climbing fiber). Thus the climbing-fiber and mossy-fiber inputs differ greatly in the degree of divergence and convergence of their synaptic effects.

With the exception of the granule cells, which exert excitatory synaptic effects, all of the other cell types of the cerebellar cortex are *inhibitory* in their effects. Basket cells, stellate cells, and Golgi cells mediate different sorts of inhibition within the cerebellar cortex. Moreover, the Purkinje axons that are the sole output of the cerebellar cortex are inhibitory in their effects on the deep cerebellar nuclei. In contrast, the cerebellar nuclei have excitatory synaptic effects on their targets in the thalamus and brainstem.

Despite extensive studies of the circuitry of the cerebellum (only superficially described here), we still do not have a clear understanding of how it modulates or coordinates movements. There are several sorts of models of cerebellar action—some stressing millisecond timing of motor discharges, others stressing corrective feed-

[7] The cerebellum accounts for about 30% of the volume of the human brain.

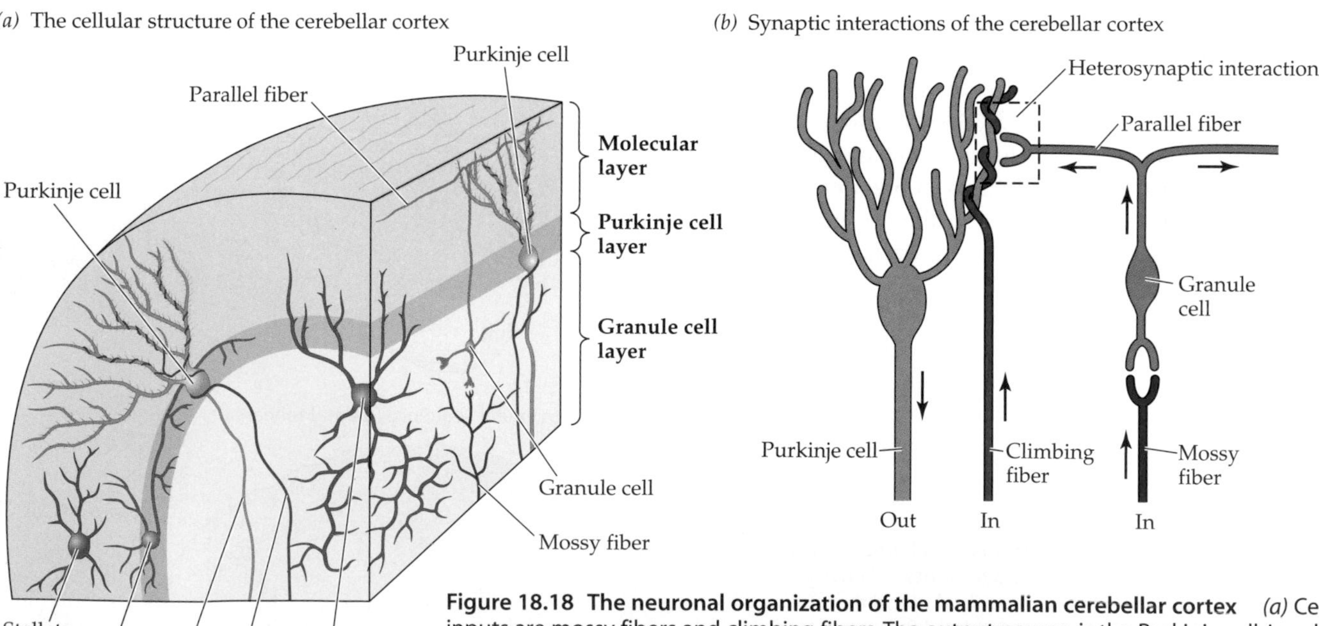

Figure 18.18 The neuronal organization of the mammalian cerebellar cortex *(a)* Cerebellar inputs are mossy fibers and climbing fibers. The output neuron is the Purkinje cell. Local interneurons are the granule cell, stellate cell, basket cell, and Golgi cell. Parallel fibers are granule cell axons that run along a folium (ridge), at right angles to the planar dendritic trees of Purkinje cells. *(b)* Motor learning model of the cerebellar cortex. The two inputs to the cerebellum (mossy fibers and climbing fibers) converge on Purkinje cells. The learning of a motor task involves heterosynaptic interaction in which a climbing fiber depresses the synaptic actions of the parallel fibers on a Purkinje cell.

back or coordination of disparate body regions and muscle groups. One popular model is that the cerebellum learns motor tasks so that they can be performed unconsciously and automatically in the appropriate sensory context. This motor learning is thought to involve changes (long-term depression, the reverse of the long-term potentiation discussed in Chapter 12; see page 340) at synapses of parallel fibers onto Purkinje cells (see Figure 18.18*b*).

BASAL GANGLIA The **basal ganglia** are a set of nuclei (clusters of brain neurons) located in the fore- and midbrain, under the cerebral hemispheres. The most important areas (in terms of motor control) are the *caudate nucleus*, the *putamen*, and the *globus pallidus*. The caudate and putamen are similar in origin and function, and together they are termed the *neostriatum* (or simply *striatum*).[8] The caudate nucleus and putamen receive excitatory input from the cerebral cortex, from both motor and association areas. The caudate nucleus and putamen send inhibitory neurons to the globus pallidus. The major output of the basal ganglia is inhibitory, neurons from the internal segment of the globus pallidus inhibiting neurons in the thalamus that excite motor areas of the cerebral cortex.

The basic circuit of the basal ganglia is therefore an example of a **loop** circuit—one in which the output of the circuit loops back to the site of the circuit's input. Loop circuits are common in the vertebrate brain, and they appear to be important for many aspects of motor control, emotions, and other brain activities. The basal ganglia also form a second loop circuit, between the striatum and the substantia nigra. Dopaminergic neurons from the substantia nigra project to the striatum and receive inhibitory feedback from it.

The basal ganglia are important in initiating movements, but it is not clear just how they act in motor control. Most evidence about their role comes from lesions and diseases that affect movement, and from pharmacological manipulations of neurotransmitter systems. Let's consider two examples of abnormal function of the basal ganglia: Parkinson's disease and Huntington's disease (Huntington's chorea). Parkinson's disease is characterized by difficulty in initiating movements (*akinesia*), so a simple task such as climbing stairs or getting up from a chair becomes almost impossible to carry out. Akinesia is often accompanied by postural rigidity and by tremors in limbs at rest. Huntington's chorea represents the opposite problem from Parkinsonism: Movements occur uncontrolledly and are difficult to stop. Both *chorea* (uncontrolled but coordinated jerky movements) and *athetosis* (slow writhing movements) are associated with damage to the striatum.

Both Parkinson's and Huntington's diseases appear to involve neuronal degeneration and altered neurotransmitter activity in the basal ganglia. In Parkinson's disease, dopaminergic neurons in the substantia nigra degenerate. The synaptic endings of these dopamine neurons are in the striatum, so degeneration of the neurons deprives the striatum of dopaminergic input. Dopamine replacement therapy (through provision of the dopamine precursor L-dopa, which crosses the blood–brain barrier) alleviates many symptoms of Parkinson's disease, at least for a while. Implants of dopaminergic tissue into the striatal area may provide longer-lasting relief from Parkinson's symptoms. In Huntington's disease, in contrast, extensive loss of neurons in the striatum decreases striatal inhibition of part of the globus pallidus.

What role, then, do the basal ganglia play in normal motor function? The implications of the studies of abnormal function cited here are suggestive but hardly conclusive. One hypothesis is that the function of the basal ganglia is to select and maintain motor behavior patterns, suppressing conflicting motor activity while reinforcing the ongoing behavior. According to this hypothesis, chorea results from a decreased inhibitory output of the basal ganglia, decreasing the ability to suppress unwanted activity. Parkinsonian symptoms, in contrast, reflect an increased inhibitory output, leading to an inability to select and change behavioral activities. In any case, the basal ganglia appear to be involved primarily in the initiation, planning, and selection of movements, in contrast to the cerebellum, which appears to "fine-tune" movement, coordinating and smoothly modifying the execution of a movement to match a command.

THE INTERACTION OF BRAIN AREAS IN MOVEMENT CONTROL We will now attempt to integrate the hypothesized roles of the cerebral cortex, cerebellum, and basal ganglia in the control of voluntary movement. As Figure 18.19 shows, the planning and programming of a movement can be viewed as separate from the execution of the movement. We can suppose that the decision to move starts in association cortex (cortex that is not linked to any particular sensory or motor system), because the readiness potentials recorded prior to a movement are not localized to a specific cerebral area. Two loop circuits from the association cortex are thought to be involved in prepro-

[8] Other basal ganglia with important functions in motor control are the *substantia nigra* and the *subthalamic nucleus*.

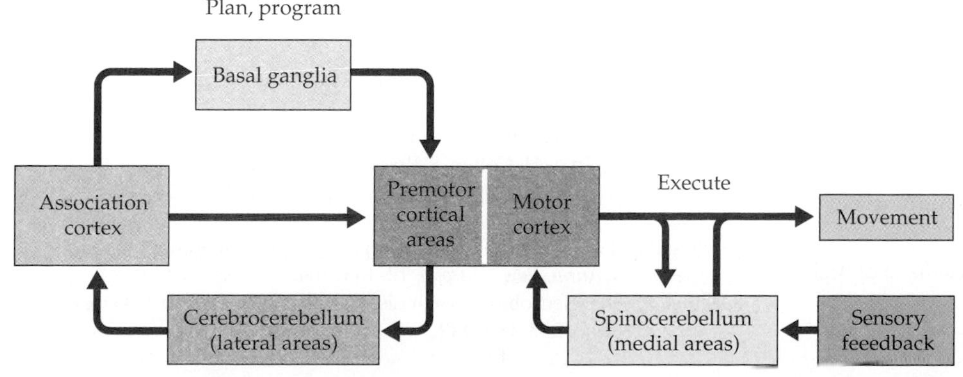

Figure 18.19 The interaction of brain areas in the planning, execution, and control of voluntary movement Sensory association cortex (as well as other cortical areas) funnels activity to premotor cortical areas, employing loops through the basal ganglia and the cerebrocerebellum in the planning and programming of the movement. Activity passes to the primary motor cortex, which is seen as executing the movement, with correction from a second cerebellar loop through the spinocerebellum. Inputs to the cerebral cortex pass through the thalamus (not shown).

gramming a movement: one loop through the basal ganglia (selection and initiation) and another through the lateral cerebellum (initial programming). Both loops feed back to the motor cortex via the ventrolateral nucleus of the thalamus. The motor cortex then generates the appropriate pattern of activity to initiate the movement.

Information about the command is sent to the spinocerebellum, via several subcortical nuclei. This process, termed *command monitoring*, "informs" the cerebellum of the intended movement. The spinocerebellum also receives ascending information—both sensory information about joint position and muscle tension, and central information from spinal and brainstem motor centers. The spinocerebellum may integrate this feedback information about the state of lower motor centers (internal feedback) and about the periphery (external feedback) with the monitored cerebral command. The cerebellar output can then modify and correct the command on a continuous basis as the movement evolves, using an integral of all relevant information (command, motor state, and sensory feedback). This continuous correction is presumably faster and smoother than, say, a correction system based on sensory feedback alone.

Even if the preceding description of the interaction of brain areas in the execution of a voluntary movement is correct, it begs other questions (such as, How is a decision to move actually made?). Nevertheless, the relations shown in Figure 18.19 illustrate some important principles to use in thinking about higher motor control in mammals.

Summary: Control and Coordination of Vertebrate Movement

- The vertebrate brain and spinal cord interact in the generation of behavior patterns such as locomotion and voluntary movements.
- Mammals have a spinal central pattern generator for stepping during walking. Descending commands from the brain can activate the CPG, and sensory feedback can modulate it.
- Several brain areas are important in generating and coordinating movements in mammals. In the cerebral cortex, the primary motor cortex directly activates spinal motor centers to generate movements, while other cortical areas (premotor cortex, supplemental motor area) are involved in planning and rehearsing movements.
- The cerebellum and the basal ganglia are connected to the cerebral cortex in looping circuits. The cerebellum is active in coordinating movements and in motor learning. The basal ganglia are involved in the initiation of movements. Parkinson's and Huntington's diseases stem from abnormalities in function of the basal ganglia.

Study Questions

1. Suppose that an arthropod such as a locust or crayfish has about 10^5 neurons, a fish 10^8, and a rat 10^{10}. Is the behavior of the fish a thousand times more complicated than that of the arthropod? Why or why not? What does the difference in number of neurons suggest about how these different animals generate patterns of behavior?
2. How do we find out whether a vertebrate such as a cat has a central pattern generator for walking?
3. A neural circuit for a central pattern generator (see Figure 18.9) may include motor neurons or may consist entirely of interneurons. How do you think the inclusion of motor neurons might restrict the flexibility of a CPG circuit? For what sorts of behavior patterns do you think a CPG circuit might include motor neurons?
4. Some movements (such as picking up a cup) are visually guided, or *steered*. Others (such as shooting a basketball) cannot be corrected once launched and are termed *ballistic*. How, and to what extent, might you expect the neural control of these movements to differ?

Key References

Carew, T. J. 2000. *Behavioral Neurobiology: The Cellular Organization of Natural Behavior*. Sinauer, Sunderland, MA. Excellent, accessible text on neuroethology, explicating the neural basis of animal behavior for a variety of well-studied cases.

Delcomyn, F. 1998. *Foundations of Neurobiology*. Freeman, New York. The most animal behavior–friendly of the neurobiology texts available.

Marder, E., and R. L. Calabrese. 1996. Principles of rhythmic motor pattern generation. *Physiol. Rev.* 76: 687–717. An excellent review of neural mechanisms generating and modulating behaviorally important rhythmic motor patterns.

Stein, P. S. G., S. Grillner, A. I. Selverston, and D. G. Stuart (eds.). 1997. *Neurons, Networks, and Motor Behavior.* MIT Press, Cambridge, MA. A good, multiauthor summary of motor control of behavior patterns in a variety of animals.

Selected references

Burrows, M. 1996. *The Neurobiology of an Insect Brain*. Oxford University Press, Oxford, UK.

Camhi, J. M. 1984. *Neuroethology*. Sinauer, Sunderland, MA.

Chesselet, M.-F., and J. M. Delfs. 1996. Basal ganglia and movement disorders: An update. *Trends Neurosci.* 19: 417–422.

Clarac, F., D. Cattaert, and D. Le Ray. 2000. Central control components of a "simple" stretch reflex. *Trends Neurosci.* 23: 199–208.

Cohen, A. H., S. Rossignol, and S. Grillner (eds.). 1988. *Neural Control of Rhythmic Movements in Vertebrates*. Wiley, New York.

Dickinson, M. H., C. T. Farley, R. J. Full, M. A. R. Koehl, R. Kram, and S. Lehman. 2000. How animals move: An integrative view. *Science* 288: 100–106.

Edwards, D. H., W. J. Heitler, and F. B. Krasne. 1999. Fifty years of a command neuron: The neurobiology of escape behavior in the crayfish. *Trends Neurosci.* 22: 153–161.

Ferrell, W. R., and U. Proske (eds.). 1995. *Neural Control of Movement.* Plenum, New York.

Gerfen, C. R. 1991. The neostriatal mosaic: Multiple levels of compartmental organization in the basal ganglia. *Annu. Rev. Neurosci.* 15: 285–320.

Georgopoulos, A. P. 1991. Higher motor control. *Annu. Rev. Neurosci.* 14: 361–377.

Georgopoulos, A. P. 1997. Neural networks and motor control. *Neuroscientist* 3: 52–60.

Getting, P. A. 1988. Comparative analysis of invertebrate central pattern generators. In A. H. Cohen, S. Rossignol, and S. Grillner (eds.), *Neural Control of Rhythmic Movements in Vertebrates*, pp. 101–128. Wiley, New York.

Getting P. A. 1989. Emerging principles governing the operation of neural networks. *Annu. Rev. Neurosci.* 12: 185–204.

Grillner, S. 1985. Neurobiological basis of rhythmic motor acts in vertebrates. *Science* 228: 143–149.

Harris-Warrick, R. M., and E. Marder. 1991. Modulation of neural networks for behavior. *Annu. Rev. Neurosci.* 14: 39–57.

Jacklett, J. W. (ed.). 1989. *Neuronal and Cellular Oscillators*. Decker, New York.

Johnson, B. R., and S. L. Hooper. 1992. Overview of the stomatogastric nervous system. In R. M. Harris-Warrick, W. Marder, A. I. Selverston, and M. Moulins (eds.), *Dynamic Biological Networks*, pp. 1–30. MIT Press, Cambridge, MA.

Kandel, E. R. 1976. *Cellular Basis of Behavior*. Freeman, San Francisco.

Kandel, E. R., J. H. Schwartz, and T. M. Jessel (eds.). 2000. *Principles of Neural Science*, 4th ed. McGraw-Hill, New York.

Orlovsky, G. N., T. G. Deliagina, and S. Grillner. 1999. *Neuronal Control of Locomotion: From Mollusc to Man*. Oxford University Press, Oxford, UK.

Pearson, K. G. 1993. Common features of motor control in vertebrates and invertebrates. *Annu. Rev. Neurosci.* 16: 265–297.

Rothwell, J. 1994. *Control of Human Voluntary Movement*, 2nd ed. Chapman and Hall, London.

Simmons, P. J., and D. Young. 1999. *Nerve Cells and Animal Behavior*, 2nd ed. Cambridge University Press, Cambridge.

Squire, L. R., F. E. Bloom, S. K. McConnell, J. L. Roberts, N. C. Spitzer, and M. J. Zigmond (eds.). 2002. *Fundamental Neuroscience*, 2nd ed. Academic Press, San Diego, CA.

Tanji, J. 2001. Sequential organization of multiple movements: Involvement of cortical motor areas. *Annu. Rev. Neurosci.* 24: 631–651.

Thatch, W. T., H. G. Goodkin, and J. G. Keating. 1991. Cerebellum and adaptive coordination of movement. *Annu. Rev. Neurosci.* 15: 403–442.

19 Muscle and Movement at Work: Muscle in Human Health and Disease

Skeletal muscle is notable for its adaptability. "Use it or lose it!," the mantra of fitness enthusiasts, is, in a nutshell, an expression of physiologists' observations that muscle is *plastic*. Muscles change, both in mass and in cellular characteristics, as a result of their use. For example, a body builder employs specific types of physical activity to build muscle mass, while someone who experiences reduced muscle activity (as when healing a fractured bone in a cast) will have a reduction in muscle mass, a condition called **atrophy**. The remarkable ability of muscle to change with use is especially interesting because muscle cells (muscle fibers) in adults are post-mitotic; that is, once a muscle forms, the cells do not divide by mitosis, and their number cannot increase. When a muscle increases in bulk, it does so by **hypertrophy**—that is, by adding structural proteins to individual cells (not by adding new cells). When a muscle is not used, it becomes smaller because the individual muscle fibers lose actin and myosin components of myofibrils. This type of change is called *disuse atrophy*. Atrophy of a muscle can also result from loss of cells, a phenomenon seen in some disease states and also in aging.

The adaptations of muscle are not limited to changes in size, however. We know from Chapters 6 and 17 that muscles typically consist of mixtures of different types of muscle fibers that are characterized by different speeds of contraction and different metabolic properties. The three main types of fibers are slow twitch fibers and two types of fast twitch fibers. Researchers have found that, to some extent, these types of fibers can be interconverted depending on a muscle's activity.

Other modifications also occur in response to different types of activity. For example, a weight lifter does resistance exercises, which stimulate the individual fibers in the exercised muscles to increase their synthesis of actin and myosin. The formation of more myofibrils leads to hypertrophy. To support increased synthetic activities and the functions of a larger volume of structural proteins, these muscle fibers incorporate nuclei from satellite cells (muscle stem cells) that lie close to the muscle fiber's cell membrane.

Different changes take place in the individual fibers in the leg muscles of a person training for long-distance running. These muscle fibers form more and larger mitochondria so that more ATP is produced by oxidative phosphorylation. Unlike the fibers in resistance-trained muscles, the fibers in the endurance-trained muscles of runners show relatively little hypertrophy. This morphology is in keeping with maintaining short diffusion distances between the blood in capillaries and the interior

Resistance exercise (left) and microgravity (right) are two examples of conditions that produce changes in skeletal muscles

of the muscle fibers. As we will see, endurance exercise also stimulates more capillaries to form around the muscle fibers. While the muscles of athletes provide extreme examples, similar but more subtle changes occur in any person's muscles over the course of day-to-day activities.

There is no shortage of examples demonstrating the plasticity of muscles, but the signals that trigger changes in muscle fibers are only beginning to be understood. What signals trigger changes in numbers and dimensions of myofibrils, populations of mitochondria, or density of capillaries? How does the nature of a muscle's mechanical activity cause individual muscle fibers to convert from one type to another? What conditions lead to atrophy? Indeed, the aim of many ongoing studies is to find ways to inhibit the muscle atrophy that occurs under conditions such as microgravity during space travel, in diseases like muscular dystrophy, and in aging. Studies on humans and laboratory mammals have contributed a great deal to our current understanding of use-related changes in muscle. In this chapter we explore several studies on muscle use and disuse, as well as disease states that shed light on muscle function and plasticity.

Exercise

Endurance exercise, such as long-distance running, cycling, or swimming, involves repetitive actions that generate relatively low forces. Slow twitch fibers that have narrow diameters and depend on aerobic metabolism play a dominant role in endurance exercise. They are ideal for maintaining isometric force economically and for carrying out repetitive isotonic contractions. *Strength* or *resistance exercise*, such as stair running or weight lifting, involves fewer repetitions of movements that generate large forces. Fast twitch fibers that depend more on anaerobic metabolism are important in strength or resistance exercises, which are often referred to as "power pursuits."

The different types of muscle fibers contain different isoforms of the myosin heavy chain. Specific genes of the myosin heavy-chain gene family encode different myosin heavy-chain isoforms. Because different isoforms hydrolyze ATP at different rates, the particular myosin isoform expressed by a muscle fiber strongly influences the fiber's functional properties.[1] The rate of hydrolysis is directly related to the rate at which cross-bridges cycle and therefore to the speed of contraction. Speed of contraction is also related to the rate at which Ca^{2+} ions are taken back into the sarcoplasmic reticulum (SR) from the cytoplasm. Slow-oxidative (SO, also called Type I)[2] twitch fibers have the slowest myosin isoform and thus the slowest rates of cross-bridge cycling. SO fibers also have a slow SR Ca^{2+}-ATPase isoform. They contain abundant mitochondria and tend to be fatigue-resistant. Fast-oxidative (FO, also called Type IIa) twitch fibers have a somewhat slower myosin isoform and an intermediate speed of contraction. Fast-glycolytic (FG, also called Type IIx) twitch fibers have the fastest myosin isoform and (like Type IIa fibers) a fast SR Ca^{2+}-ATPase isoform. Type IIx fibers have the fastest speed of contraction. Indeed, in humans, the fast Type IIx muscle fibers can contract up to ten times faster than Type I fibers. Type IIx fibers typically have the largest diameters of the three main fiber types, and they contain relatively few mitochondria (see Figures 6.10 and 17.15).

The different fiber types constitute motor units with differing physiological characteristics. The muscle fibers innervated by any one motor neuron are all of the same fiber type. The muscle fiber type is strongly influenced by the motor neuron; indeed, early experiments found that some motor units could be converted from one type to another by cross-innervation (cutting the original nerve fibers and allowing different ones to innervate the muscle fibers). During contractions, motor units are recruited in a fixed order: first Type I, then Type IIa, and finally Type IIx. The two fast fiber types are recruited to produce more powerful movements than slow fibers produce, or to supplement isometric contractions produced by slow fibers. The pattern of recruitment and the diversity of fibers with different velocities of contraction and fatigability allow a muscle to shorten at varied, appropriate speeds as it generates the forces required for different types of movement.

Power output determines a muscle's contractile performance

We know from Chapter 17 (p. 476) that the greater the cross-sectional area devoted to the contractile elements in a muscle fiber, the greater force (tension) it can generate. We also know that the speed of contraction is determined by the rate at which a given myosin isoform hydrolyzes ATP and produces cross-bridge actions. These qualities—*force* (tension) generated and *velocity* (speed) of shortening—determine the contractile performance of a muscle. Indeed, the force generated by a muscle multiplied by the velocity of shortening determines the mechanical **power** produced:

$$\text{Power} = \text{force} \times \text{shortening velocity} \qquad (19.1)$$

Figure 19.1 shows a *power curve* superimposed on the same force–velocity relationship shown in Figure 17.10. The force–velocity (load–velocity) relationship reflects the interaction between the force a muscle generates against a load and the speed at which it shortens: The velocity of shortening decreases as the load increases.

Points on the power curve are determined from the instantaneous product of force and velocity. Power is zero when the muscle shortens against no load (at V_{max}, maximum velocity) and also when the muscle contracts isometrically but does not shorten. Power output of most muscles is maximal when the muscle shortens at 20% to 40% of V_{max} and at about 30% to 40% of the load that prevents it from shortening. Power is measured in watts: one watt = 1 Newton × 1 m/s or one watt = $1 \text{ J} \cdot \text{s}^{-1}$.

We know that in isotonic contractions the force generated by a muscle equals the force of the load it moves. Thus, the x axis of the graph indicates the force exerted by the load, and it also reflects the force (tension) produced by the muscle. For a particular muscle, the cross-sectional area of the myofibrils in its muscle fibers determines the force it produces. The y axis of the graph, which indicates velocity of shortening, is related to the myosin isoforms expressed by the motor units of a particular muscle. Therefore, the force–velocity relationship—and thus the power output—of a whole muscle will be de-

[1] Some research suggests that the light chains associated with the heavy chain in the head of the myosin molecule also influence the speed of contraction.

[2] Because this chapter focuses on human muscles, we will adopt the nomenclature of Types I, IIa, and IIx used in literature describing human studies. Type IIb fibers are a very fast type of muscle fiber found in small mammals, but early studies on humans used IIb when referring to what is now known as IIx.

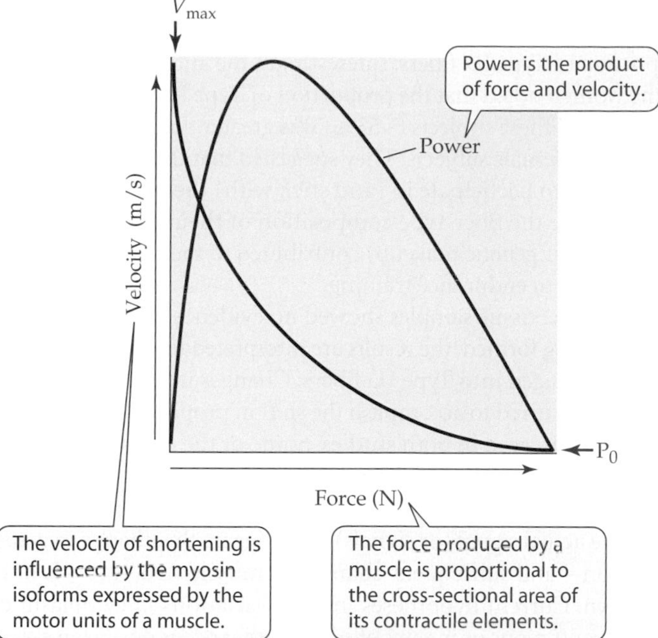

Figure 19.1 The power a muscle is capable of generating reflects its functional capabilities The blue line represents the force–velocity relationship (see Figure 17.10); the red line represents the power curve. Power is zero (P_0) when shortening velocity is maximun (no load) or zero (isometric contraction).

termined by the girth of individual muscle fibers (how many myofibrils they contain) and the different fiber types that make up the muscle. When the individual fibers of muscles change in response to use or disuse, their cross-sectional area (determined by hypertrophy or atrophy) and speed of contraction (determined by fiber type) change. It is useful to keep in mind that the *force* generated by a unit of cross-sectional area of any fiber type is basically the same, but the *power* normalized to that area is much greater for fast twitch fibers than for slow twitch fibers because of their greater speed of contraction.

How does the power output of a muscle relate to the *work* it performs (shown in Figure 17.13)? We know that work is the product of the force generated by a muscle and the distance it shortens (force × Δlength). Power is work performed *per unit of time*. Thus,

$$\begin{aligned}\text{Power} &= \text{force} \times \text{shortening velocity}\\ &= \text{force} \times (\Delta\text{length/time})\\ &= (\text{force} \times \Delta\text{length})\text{/time}\\ &= \text{work/time}\end{aligned} \tag{19.2}$$

Endurance training elicits changes in fiber type, capillary density, and mitochondrial density

In an average active person, the muscles used in locomotion typically have a mix of about half slow Type I fibers and half fast fibers; the majority of fast fibers are Type IIa. Elite athletes show distinct differences from the average person and also from one another. For example, the leg muscles of marathon runners tend to have a preponderance of Type I fibers and relatively few fast fibers, which are mainly Type IIa. By contrast, the leg muscles of sprinters tend to have a preponderance of fast fibers (a mix of Types IIa and IIx) and relatively few Type I fibers. It is not clear whether these extreme distributions of fibers expressing different myosin isoforms result mainly from genetics, or from specific training programs, or from a combination of these (and possibly other) factors. Between these extremes, humans show considerable variation in the distributions of fiber types within their muscles. It appears likely that individuals vary genetically, with some programmed to have muscles more like those of a marathoner and others to have muscles more like those of a sprinter. Still, studies of humans and experimental animals have shown that exercise training can cause some degree of interconversion between these fiber types, in particular between Type IIa and IIx fibers.

Table 19.1 presents the results of two studies of endurance training in humans. In one study, seven previously untrained women, 21 to 24 years old, did supervised cross-country running for 24 weeks, running 45 minutes a day, 3 days a week. In the other study, five previously untrained men, 20 to 23 years old, trained for a period of 8 weeks by pedaling a bicycle ergometer 40 minutes a day, 4 days a week. In both studies, samples of muscle tissue were taken before

TABLE 19.1 Fiber type distribution and capillaries around each fiber before and after training in two studies

	Before Training			After Training		
Fiber type	**Number of fibers**	**Distribution (%) mean ± S.E.M.**	**Capillaries around each fiber mean ± S.E.M.**	**Number of fibers**	**Distribution (%) mean ± S.E.M.**	**Capillaries around each fiber mean ± S.E.M.**
7 women	168–265 per subject			137–197 per subject		
Type I		58.2 ± 2.8	4.11 ± 0.15		57.5 ± 2.9	5.04 ± 0.21
Type IIa		24.9 ± 2.6	3.4 ± 0.16		31.6 ± 2.7	4.15 ± 0.21
Type IIx		11.8 ± 2.7	2.33 ± 0.19		2.7 ± 2.5	2.68 ± 0.14
Intermediate		5.2	6.9		7.9	8.0
5 men	1035 ± 126 fibers/ 5 subjects			937 ± 270 fibers/ 5 subjects		
Type I		39 ± 2.1	3.9 ± 0.18		42 ± 2.2	5.4 ± 0.32
Type IIa		36 ± 2.9	4.2 ± 0.2		42 ± 2.4	5.5 ± 0.45
Type IIx		20 ± 1.6	3.0 ± 0.22		13 ± 1.5	4.2 ± 0.5

Sources: After Ingjer 1979 (women) and Andersen and Henriksson 1977 (men).

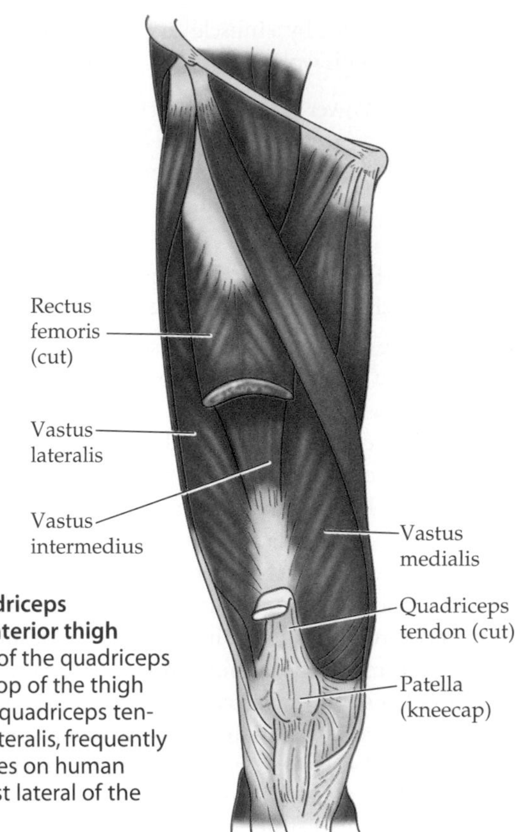

Figure 19.2 Quadriceps muscles of the anterior thigh The four muscles of the quadriceps group lie on the top of the thigh and insert on the quadriceps tendon. The vastus lateralis, frequently biopsied for studies on human muscle, is the most lateral of the four muscles.

and after training from the vastus lateralis muscle of the quadriceps group (Figure 19.2). Figure 19.3 illustrates the instruments used in a **needle biopsy**, the procedure used to obtain the samples. The vastus lateralis is a mixed muscle, with representation of all three fiber types. In an average active person, the myosin isoform profile is about 50% Type I, 40% Type IIa, and 10% Type IIx.

The proportion of Type I fibers remained unchanged as a result of training in both studies (see Table 19.1). However, endurance training caused a significant change in the proportions of the two types of fast fibers in the vastus lateralis muscle. The proportion of Type IIa fibers increased and was accompanied by a decrease in the proportion of Type IIx fibers. Interestingly, the authors of the study on the women noted that the proportion of Type I fibers in the tested muscles of these subjects (~58%) was greater than usually found in untrained female subjects. They suggested that these women may have chosen to participate in (and stick with) the nearly half-year study because the fiber-type composition of their muscles (determined by their genetic makeup) contributed to their inherent ability to respond to endurance training.

Because the tissue samples showed no evidence of cells dying or new cells being formed, the results are interpreted to mean that Type IIx fibers changed into Type IIa fibers. Changes in gene expression must have occurred to accomplish the shift in proportions of the fast fiber types observed in both studies. Some of the original Type IIx muscle fibers must have repressed their genes for the faster IIx myosin isoform and switched on their genes for the IIa myosin isoform. The actual signal (or signals) that triggers this change in gene expression—and subsequent changes in muscle structure—is not yet known. Current hypotheses include variations in cytoplasmic Ca^{2+} concentrations or in rates of ATP use that occur during an exercise program. These and other ideas are topics of active investigation. In addition to a shift in proportions of the fast fiber types, several investigators have reported observing an increase in intermediate fiber types in muscle tissues sampled after training. Individual fibers that stain as intermediate in histological preparations appear to be expressing different myosin isoforms at the same time. It would be interesting to know if these are fibers in the process of transforming from one type to another. Investigators have noted that fibers with co-expressed myosin isoforms are suggestive of the adaptive potential of muscles.

The two studies also showed an increase as a result of training in the density of capillaries in the vastus lateralis muscle (see Table 19.1). In both studies, there were increases in the number of capillaries in contact with each muscle fiber. These data indicate that endurance training caused new capillaries to grow by sprouting from existing vessels. The formation of new blood vessels is called **angiogenesis**. Many studies using a variety of tissues have demonstrated that the cytokine **VEGF** (**vascular endothelial growth factor**) is involved in angiogenesis. Although we know that muscle fibers can produce and secrete VEGF, the actual stimulus that initiates its pro-

(a)

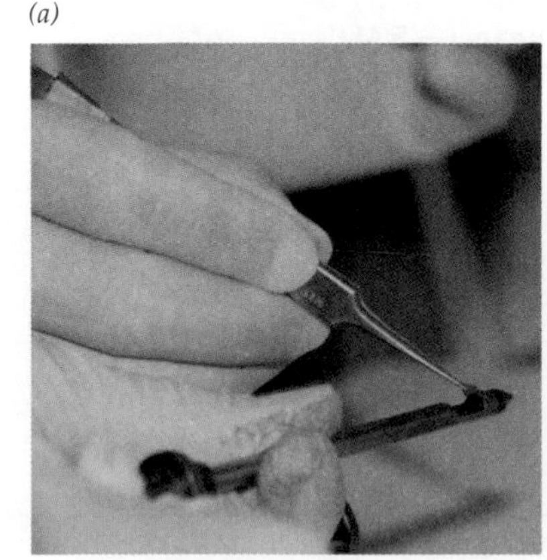

(b)

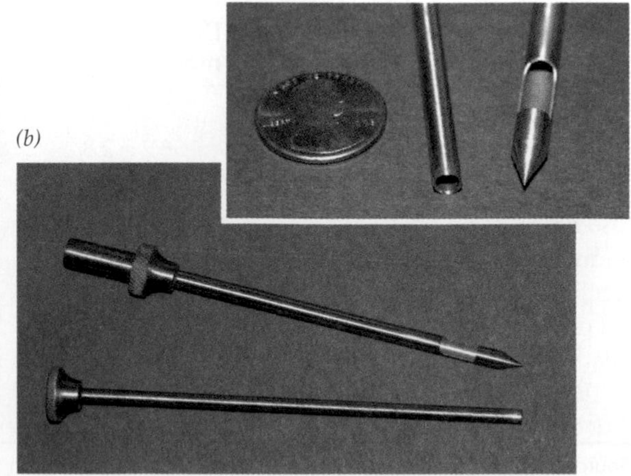

Figure 19.3 A needle biopsy is used to obtain samples of muscle tissue *(a)* The biopsy sample is a small plug of tissue removed from the muscle. *(b)* The biopsy apparatus consists of a pointed outer needle and an inner slider with a razor-sharp cutting edge. After application of local anaesthetic, the apparatus is inserted into the muscle through an incision. A small amount of surrounding muscle tissue bulges into the opening on the side of the pointed needle. The inner slider is pressed forward to sever the protruding bit of muscle tissue. The slider is withdrawn and the tissue sample removed from the pointed needle using forceps. (*a* courtesy of S. P. Scordilis, Smith College.)

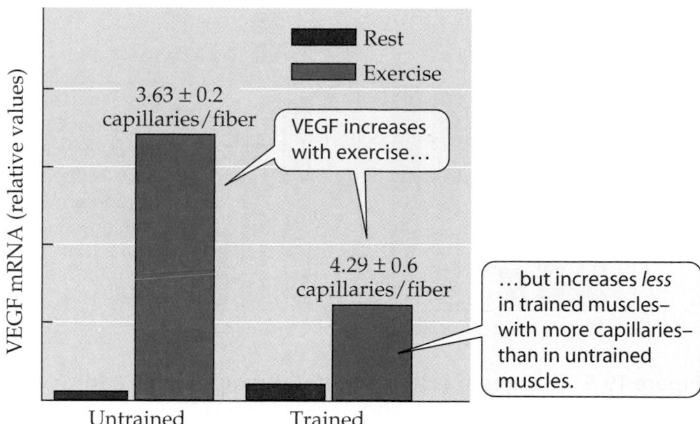

Figure 19.4 VEGF responses to a single bout of endurance exercise Vascular endothelial growth factor (VEGF) mRNA increased in the vastus lateralis muscle after a single bout of knee-extensor exercise in both untrained and trained muscles, but the increase was attenuated in the trained muscles. (After Richardson et al. 2000.)

duction is not known. Recent studies on human subjects showed that mRNA for VEGF is expressed in muscle cells within *1 hour* after an exercise bout is performed by a muscle!

Figure 19.4 shows changes in VEGF mRNA in response to exercise in six untrained human subjects. At the beginning of the study, each subject exercised only the left leg for 30 minutes on an ergometer. In this pre-training *knee-extensor exercise* bout, the subjects repeatedly contracted the quadriceps muscle group against a load to extend the leg from a "sitting angle" to a straightened position. About 1 hour after the exercise bout, biopsies were taken from the vastus lateralis muscle of the exercised leg and from the same muscle of the rested right leg. The subjects then trained the left quadriceps muscle (with varying exercises and loads to optimize training) for a period of 8 weeks. At the end of the training period, they performed the same exercise as in the pre-training bout (but at a higher load), and biopsies of the exercised vastus lateralis were taken about 1 hour afterward. Two days later, biopsies were taken from the same, now rested, muscle.

The biopsies showed low levels of VEGF mRNA at rest in both untrained and trained muscles. A single exercise bout initiated upregulation of VEGF mRNA within an hour in both untrained and trained muscles. Interestingly, the response in the trained muscles was less than that in the untrained muscles. Morphological studies of the biopsied muscle tissues showed that training also induced significant angiogenesis, with an 18% increase in the number of capillaries around each muscle fiber. Thus, exercise causes mechanical and/or metabolic perturbations in muscle cells that increase their expression of VEGF. When a regular training program of exercise upregulates VEGF repeatedly, capillaries are stimulated to proliferate. Once angiogenesis has taken place in the exercise-adapted muscle, a single bout of exercise appears to stimulate less upregulation of VEGF.

Endurance exercise increases not only the density of capillaries, but also the density of mitochondria in muscles. In one experiment, five women and five men (29 ± 5.1 years old), all previously untrained, participated in a 6-week endurance regimen using bicycle ergometers. Tissue samples were taken by biopsy from the vastus lateralis muscle before and after the training period. Measurements of structures in electron micrographs of the tissue samples showed that the total volume of mitochondria per volume of muscle fiber increased by 43% in the men and 37% in the women (Figure 19.5). In addition, the number of capillaries per muscle fiber increased by 26%. These increases in capillaries and mitochondria provide morphological correlates to account for the improved aerobic capacity of the exercised muscle and, by extension, improved athletic endurance.

Other studies using biochemical techniques have demonstrated that increases in mitochondrial enzymes accompany endurance training. Thus, the fact that exercise triggers an increase in the metabolic capacity of endurance-trained muscles has been documented by both biochemical and morphological results. How is endurance exercise transduced into signals that cause the formation of new mitochondria? Ongoing studies of the effects of exercise on experimental animals have identified transcription factors that appear to play a role in stimulating increased biogenesis of mitochondria. These transcription factors are expressed in exercised muscles within a few hours of an exercise bout.

In sum, endurance exercise triggers multifaceted responses in human skeletal muscles, including shifts in the expression of different myosin isoforms, changes in the expression of mRNA for VEGF, and increases in transcription factors that influence mitochondrial content. Different sets of genes are involved in generating these responses, and their specific and independent effects coincide to adapt the muscle to endurance exercise. Future studies will no doubt reveal additional and more detailed information about the signals that influ-

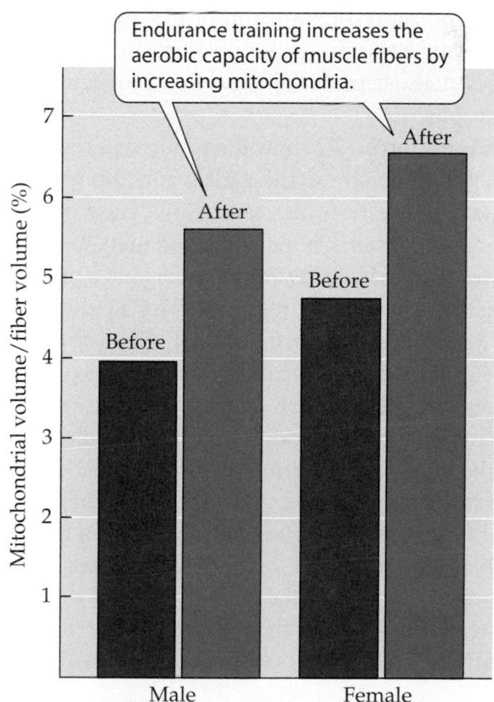

Figure 19.5 Endurance training increases the number of mitochondria (After Hoppeler et al. 1985.)

ence gene expression in muscles adapting to endurance demands. It is interesting that interconversions between the two types of fast fibers appear to occur readily, whereas transformations from Type II fibers to Type I fibers, and vice versa, do not. Researchers have observed conversions of fast fibers into slow fibers in rodent muscles treated with more aggressive techniques than possible in exercise training programs. Whether such changes could be achieved under physiological conditions over longer periods of time than so far tested in humans is still an open question. It is possible that remodeling of entire motor units, including the motor neurons, is necessary to achieve more extensive interconversions between slow and fast types of muscle fibers.

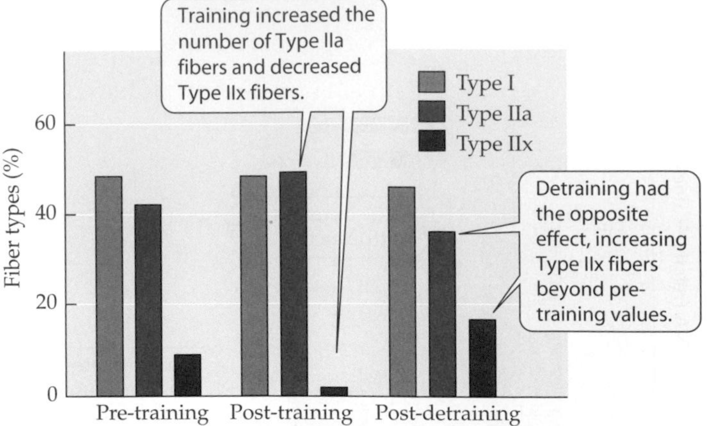

Figure 19.6 Changes in fast fiber types during training and detraining (After Andersen and Aagaard 2000.)

Resistance training causes hypertrophy and changes in fiber type

Resistance training aims at increasing muscle strength. Resistance exercise programs use repetitions of short, intensive bouts of shortening and/or lengthening contractions as well as isometric contractions. Key to resistance training is keeping exercise bouts as short bursts so that they do not stimulate the changes produced by longer-duration endurance exercise. Depending on the program of exercises, hypertrophy can be substantial. Resistance training usually does not increase the aerobic capacity of the muscle or produce an increase in capillary density. Biopsies of muscle tissue taken from previously untrained subjects who participate in resistance-training programs over several weeks typically show no change in the proportion of Type I fibers in the trained muscle. Instead, resistance training appears to cause the transformation of Type IIx fibers into Type IIa fibers. To explain this change, investigators have proposed that mechanical deformations of the muscle fiber membrane and cytoskeleton could stimulate a stretch-activated signal that influences gene expression. Such a signal could promote expression of the Type IIa gene and repress expression of the Type IIx gene. At the same time resistance training produces shifts in the proportions of fast fiber types, it also causes individual fibers to increase in diameter by hypertrophy.

Figure 19.6 and Table 19.2 show the results of a resistance-training experiment that illustrates these ideas and also reveals an additional piece of information that can be made use of by sprinters preparing for competitions. Nine untrained men, 29 ± 5 years old, volunteered for the study. Each subject underwent supervised resistance training of the legs three times a week for a period of 90 days.[3] As training proceeded, the loads against which the leg muscles worked were progressively increased. After 90 days of training, the subjects were directed to return to their previous (non-training) level of activity for another period of 90 days. Needle biopsies were taken from the vastus lateralis muscle immediately before training, at the end of the training period, and after 90 days of detraining.

Neither training nor detraining had an effect on the percentage of Type I fibers present in the vastus lateralis muscle (see Figure 19.6). As expected, after 3 months of training, all nine subjects experienced an increase in Type IIa fibers from about 42% to about 50% and a decrease in Type IIx fibers from about 9% to about 2%. Surprising changes occurred during the 90 days of detraining, however, after resistance exercises were stopped. The two types of fast fibers did *not* return to their pre-training proportions. Instead, the Type IIx fibers actually increased in abundance up to about 17%—twice their proportion in untrained muscles—during the detraining period. The increase in Type IIx fibers was accompanied by a corresponding decrease in Type IIa fibers, which declined to an average of 37%. The signals responsible for this "overshoot" of Type IIx fibers are not known, and further experiments are needed to determine how long the increased representation of Type IIx fibers persists. Despite these unknowns, this observation can be exploited by sprinters. Because large-diameter, glycolytic Type IIx fibers produce the fastest and most powerful contractions, they offer sprinters a strong competitive advantage. Thus, to increase the Type IIx fibers in their muscles, sprinters would find it useful to taper their training in advance of a major competition.

In this study, the fast Type II fibers in the nine subjects increased an average of 16% in diameter following training (see Table 19.2). Slow Type I fibers increased slightly in diameter, but not significantly. Other resistance-training experiments have elicited significant hypertrophy of Type I fibers. It will be interesting to follow further studies on human subjects designed to determine whether Type I or Type II fibers preferentially undergo hypertrophy. While variations in exercise programs produce useful information about hypertrophy in human subjects, it is difficult to design experiments that reveal the time course of changes and impossible to do invasive manipulations. Thus, researchers use animal models in conjunction with human studies to investigate the mechanisms that stimulate growth of muscle fibers.

TABLE 19.2 Average diameters of individual *vastus lateralis* muscle fibers in nine men

	Type I (μm^2)	Type II (μm^2)
Pre-resistance training	3660 ± 349	4104 ± 369
Post-resistance training	4114 ± 315	4881 ± 244 * †
Post-detraining	3993 ± 403	4521 ± 255

Source: From Andersen and Aagaard 2000.
*Significantly different from pre-resistance training ($P < 0.01$).
†Significant difference between Type I and Type II fiber size ($P < 0.01$).

[3]The training consisted of 4 to 5 sets of 6 to 15 repetitions of the following exercises: hack squat, incline leg press, knee extension, hamstring curl, and calf raise.

Stretch and force production activate protein synthesis in rabbit muscles

We know that resistance exercise programs can stimulate the addition of proteins to individual muscle fibers, but what are the actual stimuli that elicit this effect? Figure 19.7 illustrates the results of an experiment performed on the tibialis anterior (TA) muscle of adult rabbits. This muscle lies along the anterior aspect of the leg, just lateral to the sharp edge of the tibia (shinbone) (see Figure 19.8). It is a fast muscle made up mainly of Type II fibers. The TA muscle very rapidly begins to hypertrophy when it is chronically stretched or stretched and electrically stimulated. In this experiment, the TA was stretched by putting the leg in a cast that held the foot in an extended (pointed) position. Implanted electrodes stimulated the muscle with electrical pulses applied at a frequency of 5 Hz. The TA muscle of the opposite legs served as controls. The combined effects of stretch and stimulation produced a 35% increase in wet weight of the muscle within 4 days! The total RNA content (which is about 80% ribosomal RNA) of the muscle fibers increased more than 200% and peaked 2 days after the beginning of the experiment. An increase in ribosomal RNA content correlates with an increase in the rate of translation of messenger RNA into protein. Increased synthesis of actin and myosin would lead to the addition of myofibrils and hypertrophy. But what signals transduce mechanical factors such as stretch and contractions into increased protein synthesis?

Recently, researchers identified and cloned a growth factor, named **mechano growth factor (MGF)** that is expressed in muscle cells when muscles are activated by stretch or other stimulation. MGF may serve as a link between mechanical activity and the gene expression required for hypertrophy. MGF is one of several molecules derived from the *insulin-like growth factor-I (IGF-I)* gene. IGF-I is a hormone produced by the liver and is part of the growth hormone–IGF-I axis that stimulates the growth of many tissues, especially in early life. Active muscle cells also produce and secrete a muscle variant of IGF-I, which circulates in the blood and appears to promote growth specifically in muscles. Unlike IGF-Is, which travel long distances in the circulation, the molecular structure of MGF indicates that it is designed to act locally in muscle as an autocrine or paracrine signal. Minor damage to the muscle fiber membrane (*sarcolemma*) caused by mechanical activity is thought to initiate the production of MGF. Experiments suggest that an important function of MGF is to activate **satellite cells**, the muscle stem cells that lie just outside the sarcolemma. To direct the protein synthesis necessary for repair and hypertrophy, a muscle cell incorporates satellite cells, which provide additional nuclei to support its increased functions. MGF appears to stimulate satellite cells to proliferate and fuse with the muscle fibers. It is interesting that MGF is not expressed in the muscles of persons with Duchenne muscular dystrophy, the most severe form of muscular dystrophy, in which muscle fibers become atrophic and die. Aging muscles also express less MGF. We will return to these ideas when we discuss muscular dystrophies later in this chapter.

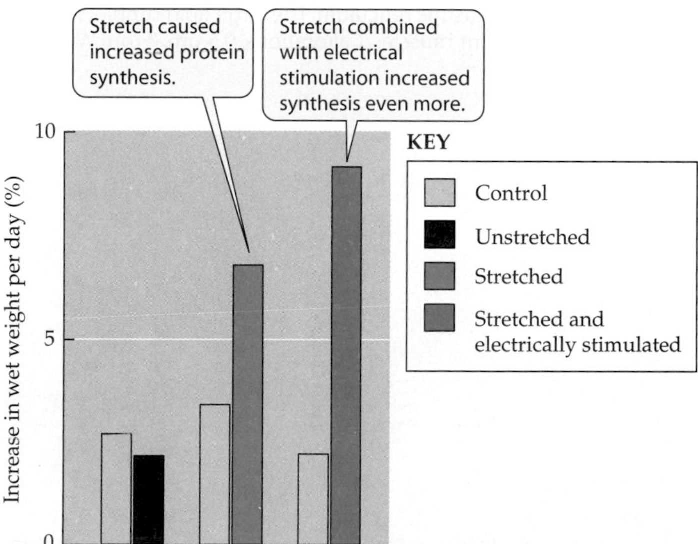

Figure 19.7 Stretch or stretch combined with electrical stimulation increased protein synthesis (After Goldspink 1999.)

Atrophy

Muscle that is not used will atrophy. When a person wears a cast on her leg, for example, that immobilizes the muscles of the leg, she can lose as much as 20% of the mass of the affected muscles in just a few weeks. The tissue wastes away, and for this reason atrophy is also referred to as *wasting*. Clinical forms of disuse atrophy arise not only from limb casting, but also from bed rest and from spinal cord injuries that cut off nerve input to skeletal muscles. Further, disuse atrophy in sedentary elderly persons adds to and compounds the inevitable atrophy that occurs with aging. Because skeletal muscles play important roles in maintaining posture and producing body movements, impairment of their functions has a strong impact both on the quality of life of individuals and on broader dimensions of public health. Prolonged disuse can lead to changes in muscle structure and function that require lengthy rehabilitation programs and major commitments of health care resources.

A nonclinical form of disuse atrophy results from weightlessness during space travel. For example, when researchers used scanning techniques to analyze the cross-sectional area (CSA) of astronauts' leg muscles after space shuttle missions of 8 to 17 days, they found that the CSA had decreased from 4% to 10% in gastrocnemius (calf) muscles and from 5% to 15% in quadriceps muscles.

When muscles atrophy, the muscle fibers decrease their uptake of amino acids from the circulation and reduce protein synthesis. However, the most important cause of atrophy appears to be the breakdown of proteins through proteolytic pathways. With the loss of actin and myosin, the myofibrils are reduced, and the diameter of the muscle fiber decreases. Additional structural changes include a reduction in numbers of both mitochondria and nuclei. Biochemical changes reflect these structural changes. For example, atrophic muscles have decreased amounts of actin mRNA, cytochrome *c* mRNA, and oxidative enzymes.

Humans experience atrophy in microgravity

The effects of microgravity on human muscles were the subject of a study on four astronauts who spent 17 days on a space shuttle mission. Forty-five days prior to launch, tissue samples were taken

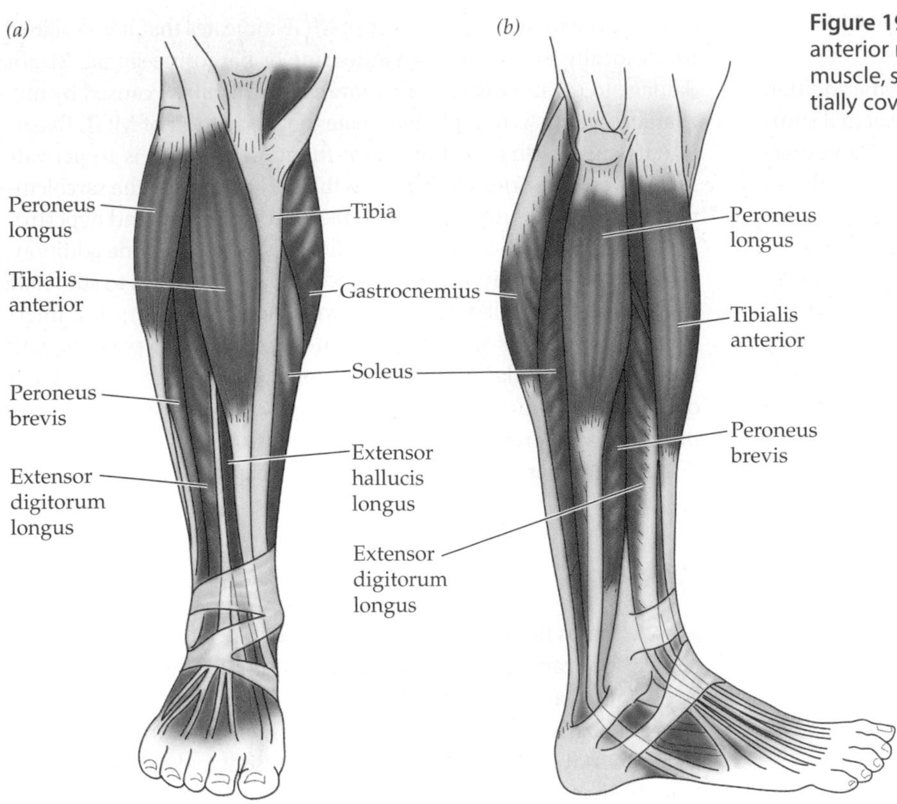

Figure 19.8 Major muscles of the lower leg *(a)* The tibialis anterior muscle lies lateral to the tibia. *(b)* The gastrocnemius muscle, seen in profile, forms the calf. The soleus muscle is partially covered by the gastrocnemius.

from the soleus muscle by needle biopsy. Within 3 hours after the shuttle landed, post-flight biopsies were taken. In humans, the soleus consists of about 70% slow Type I fibers. It is an important weight-bearing ("anti-gravity") muscle that lies just beneath the gastrocnemius muscle on the posterior of the leg (Figure 19.8). Electron micrographs of longitudinal sections of two Type I muscle fibers (Figure 19.9) provide ample evidence of atrophy that occurred in microgravity. Compared with the myofibrils in the pre-flight tissue, the diameters of the post-flight myofibrils are drastically reduced, indicating loss of both thick and thin myofilaments. Indeed, within 17 days, the average diameter of the Type I fibers measured from all four astronauts on the mission decreased from 96 ± 1 μm to 88 ± 1 μm.

Because the production of force is proportional to the cross-sectional area of a muscle, atrophy reduces both maximal force production and power output. Decreased contractile performance caused by atrophy could contribute to a variety of risks during space flights, including inability to perform emergency escape procedures and inability to carry out tasks specific to a mission. It is clear that, until the adaptation of muscle to microgravity is understood and addressed, atrophy remains a major obstacle to overcome before humans can contemplate long-term interplanetary space travel.

A variety of techniques are used to study disuse atrophy in small mammals

While information about atrophy can be gained from observational studies of human muscles, experiments using small animal models permit controlled studies that aim at revealing the underlying

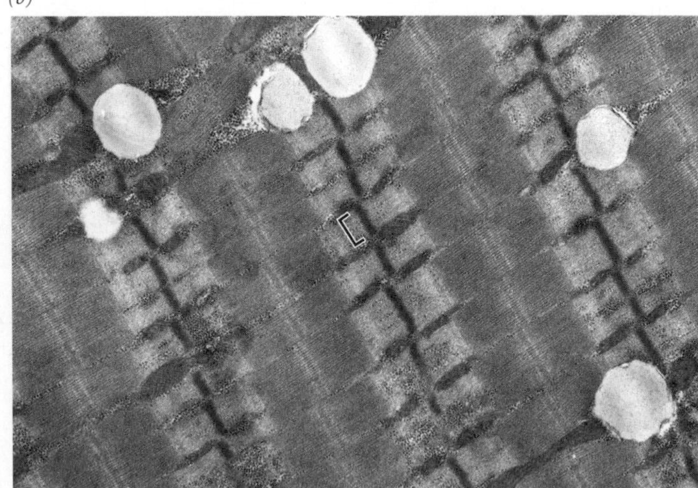

Figure 19.9 Disuse atrophy occurs during space travel These electron micrographs show longitudinal sections of Type I fibers from the soleus muscle of an astronaut 45 days prior to launch *(a)* and within 3 hours of landing *(b)* after a 17-day space shuttle mission. To minimize any effects of locomotion on the muscle fibers, the astronauts used wheelchairs after landing until the biopsy was taken. The dramatic reduction in myofibrillar diameter is indicated by the shorter Z discs (brackets) in the post-flight tissue. These slow Type I fibers have very little sarcoplasmic reticulum. The large lipid droplets are often seen in post-flight muscles. Calibrations, 0.5 μm. (From Widrick et al. 1999.)

causes of atrophy and means of preventing it. Most such studies have focused on muscles of the hindlimb because they are both active in locomotion and support the animal against gravity. Four categories of techniques are typically used to study atrophy:

1. *Denervation* caused by cutting or crushing the nerve to a muscle prevents motor neuron action potentials from signaling the muscle fibers to contract. However, it also prevents neural input to vessels supplying the muscle tissue, and therefore leaves open the possibility that a change in blood supply, in addition to denervation, could contribute to the results.
2. Cutting the tendon of a muscle (a process called *tenotomy*) eliminates the load on the muscle fibers by freeing them from their attachment to the bone. But cutting the tendon often allows the muscle to shrink to a length shorter than its normal resting length, thus potentially introducing an additional variable.
3. Pinning the joints between long bones at specific angles or using plaster casts to set limbs in fixed positions accomplishes *immobilization*, and these techniques have been used extensively. Their drawback is that muscle fibers can generate isometric contractions even though the muscles are not being used to produce movement.
4. *Suspension* techniques lift the back end of an animal so that its hindlimbs do not touch the ground. The suspension rig is attached to a framework above the cage that allows the animal to move using its forelimbs. This arrangement allows the hindlimbs a full range of voluntary movement, but not against a load, creating a situation similar to the weightless conditions of space travel.

Limited studies have been done on animals in space. One study compared the soleus muscles of rats after 1 week of weightlessness in space with those of rats after 1 week of hindlimb suspension on Earth. Muscle fiber diameters decreased in both groups, although the ground-based effects were somewhat greater than those produced in microgravity.

Disuse influences the fiber-type composition of muscles

Studies on the soleus muscles of small mammals under conditions of disuse have shown increases in Type II fibers and decreases in Type I fibers. These results were observed using different experimental techniques, such as immobilizing the hindlimb with a cast to set the muscle in a shortened position (to eliminate the possibility of its exerting force on a load or receiving any mechanical stimulus) and denervation (to eliminate neural stimulation). The soleus is normally composed of predominantly slow muscle fibers.[4] Changes in fiber-type composition in experimental animals occur very shortly after the onset of disuse. One study reported that rat soleus muscles immobilized in a shortened position for 5 days began to transcribe the fast Type IIb gene within 1 day of immobilization. In humans, some short-term studies of disuse (for example, due to imposed bed rest or exposure to microgravity) have found changes in fiber-type composition, while others have found little evidence of change. Limitations on the length as well as the stringency of controls of such studies in humans probably contribute to variations in results. However, we know that long periods of no activity at all produce profound changes in the paralyzed muscles of persons with spinal cord injuries. Individual muscle fibers of muscles paralyzed for several years contain reduced amounts of contractile proteins and are severely atrophied. In addition, they have hardly any slow oxidative (Type I) fibers. Instead, their muscles consist of about half Type IIa and half Type IIx fibers.

Taking these observations into account, researchers have proposed that the fast Type IIx gene is the "default gene" in muscle fibers. If a muscle fiber doesn't produce contractions that generate force, or isn't mechanically stretched, it expresses the Type IIx gene, produces the fastest myosin isoform, and attains a fast phenotype. Mechanical activity is thought to activate expression of the Type I and Type IIa genes, and probably to repress the Type IIx gene.

Muscles atrophy with age

Figure 19.10 illustrates the record speeds achieved by athletes of different ages in track and field events. The message from these data is that even the most physically fit individuals become slower with age. The figure reveals two additional points. First, peak performance is reached at different ages for events of different distances. Thus, sprinters in the 200-meter dash achieve their maximum performance at about 21or 22 years of age, while record marathon runners peak at about 26 years. Second, the slopes of decline in speed with age are steeper for the fast events, which depend more on power output, than for the marathon, which requires endurance. These observations suggest that power output (combined force and velocity of shortening) decreases more rapidly with age than does stamina.

What happens to muscles as they age that causes them to decrease in contractile performance? If disuse atrophy were the only change, then world-class athletes who maintain rigorous exercise programs would not show a decline. But they do. In fact, aging muscles show not only disuse atrophy of individual fibers, but also the loss of motor units. Age-related loss of muscle mass, with its related loss of strength and function, is called **sarcopenia.**

A study comparing the gastrocnemius muscles of adult (10–12 months) and old (24–26 months) rats provided information about changes in motor units with age. In this study, the gastrocnemius muscles of old rats had 34% fewer fast fatigable (FF) motor units[5] than adult rats. Further, the number of fibers in the remaining FF motor units decreased from an average of 333 fibers innervated by a single motor neuron to 286. The number of slow motor units did not change, but the number of muscle fibers in each slow motor unit increased significantly—nearly tripling from an average of 57 to 165. These results suggest that many FF fibers became denervated, atrophied, and disappeared, but that some denervated FF fibers survived by becoming reinnervated by sprouts sent out from axons innervating slow motor units. This change would account for the higher

[4]The soleus muscle of rats consists of 89% Type I fibers and 11% Type IIa fibers, while the soleus muscle of cats consists of 99% Type I fibers and only 1% Type IIa fibers. The fast Type IIb is entirely absent in both rats and cats.

[5]The authors of this study defined motor units by their contractile responses to electrical stimulation. The FF motor units can be thought of as consisting mainly of fast glycolytic Type IIb fibers, and the S motor units as consisting of slow Type I fibers.

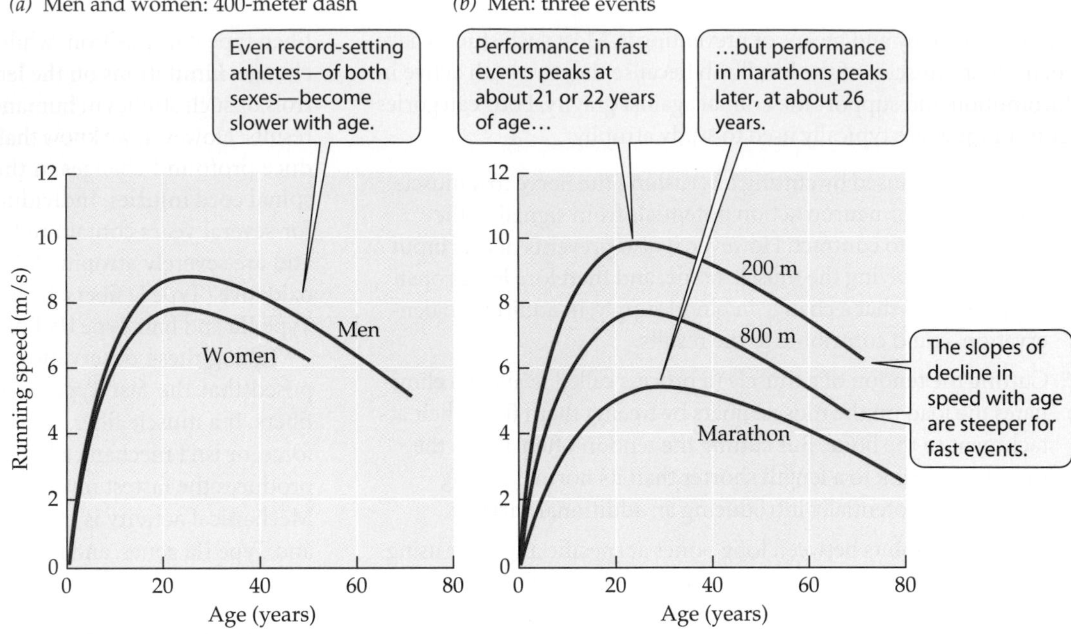

Figure 19.10 Record speeds achieved by athletes decrease with age *(a)* Record speeds for men and women running in 400-meter dash. *(b)* Record running speeds for men in events of three different lengths. The curves were fitted to data obtained from published records. (After Moore 1975.)

number of muscle fibers in the slow motor units of the old rats. Figure 19.11 illustrates this remodeling of motor units.

Physiological results complemented these morphological observations. FF motor units remaining in the muscles of old rats developed only 70% of the mean maximum force developed by adult muscles, dropping from an average force production of 206 milliNewtons (mN) to 149 mN. Slow motor units of old muscles increased their force production from an average of 19 mN in adults to 50 mN in old muscles. Slow fibers have narrower diameters than fast fibers and produce less force. This result suggests that the former fast fibers were influenced by their new (slow) motor neurons to convert to a slow phenotype. The functional result is that the increase in slow motor unit function did not make up for loss and atrophy of the FF muscle fibers. The overall force production of the whole muscle was reduced by about one-third.

Studies on muscles of elderly humans show similar changes in structure. Sarcopenia is a major cause of disability in the elderly. It is clear that loss of muscle fibers through atrophy reduces force generation. The selective loss of Type IIx fibers contributes to a disproportionately greater reduction in power. And larger and fewer motor units lead to reduced fine motor control. All of these factors contribute to an inability to adjust posture and maintain balance, which leads to the potential for falls or inappropriate movements that lead to injury.

While muscles in people of all ages experience minor injuries in normal activities, muscles in the elderly are more susceptible to injury and also take longer to repair. Ongoing animal studies are investigating the cellular mechanisms involved. For example, experiments on adult and old rats showed that old muscles stimulated by a continuous overload do not express as much MGF as those of young rats in response to the same stimulus, nor do they upregulate

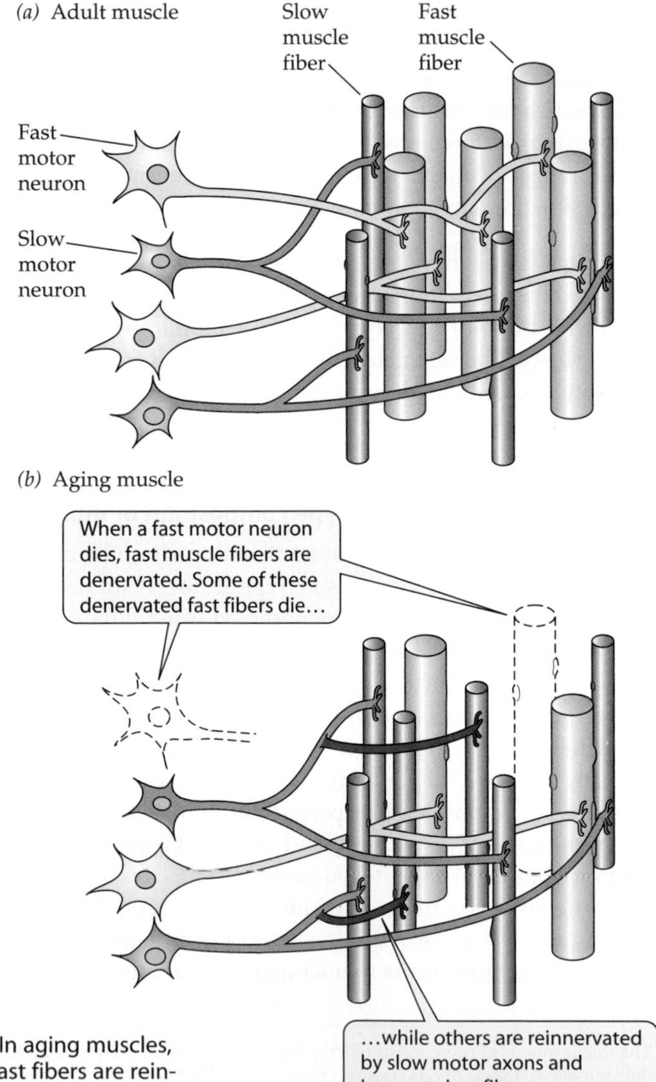

Figure 19.11 Remodeling of motor units with aging In aging muscles, some fast motor units are lost. Some of the denervated fast fibers are reinnervated by sprouts from the axons of slow motor neurons.

BOX 19.1 No Time to Lose

The Australian green-striped burrowing frog (*Cyclorana alboguttata*) can undergo months or even years of immobility without experiencing muscle atrophy. *C. alboguttata* aestivates underground within a cocoon of shed skin and mucus and emerges only during periods of heavy summer rains, which don't necessarily happen every year. During the few weeks when water is available, these frogs must breed and feed before again becoming dormant. To be successful in accomplishing these goals in a limited amount of time, they require muscles that are immediately functional. Interestingly, a study of frogs that aestivated for 3 months found no atrophy in all muscles examined, nor did isolated muscles tested in vitro show a decrease in force production compared with those of control nonaestivating frogs. The authors of this study noted that frogs typically have low metabolic rates, which decrease to extremely low levels during aestivation. Reduced metabolism would result in frog muscle tissues producing only low levels of reactive oxygen species (ROS), such as hydrogen peroxide. ROS are released from mitochondria during aerobic metabolism, and they are known to cause cellular damage. To protect against ROS, cells produce antioxidants, such as superoxide dismutase, that scavenge and inactivate ROS. It is possible that the frog's low metabolic rate is one factor that contributes to sparing its muscles when they are inactive. Low mitochondrial activity would yield very low levels of ROS and keep protein degradation and atrophy at a minimum. An additional protective measure suggested by studies on other animals would be for muscles of dormant animals to produce elevated levels of antioxidants.

Cyclorana alboguttata

receptors for the circulatory hormone IGF-I. Decreased production of MGF and responsiveness to IGF-I impair the muscle's ability to incorporate satellite cells to support repair and add myofibrils. Given this reduced capacity for repair and growth, it is essential to prevent injuries in the elderly in order to maintain muscle function. At present no mechanisms are known to stop the effects of aging on muscle. However, conditioning exercises to prevent injuries and reduce atrophy from disuse can forestall the effects of motor unit loss (which is inevitable) and contribute to ensuring mobility and independence well into old age.

Some animals experience little or no disuse atrophy

As we saw in Chapter 17, the muscles of all vertebrate animals are structurally and functionally very similar. Yet, while humans and many experimental mammals experience notable losses of muscle after remarkably short periods of disuse, other animals can spend extended periods of time hibernating or aestivating and have very little loss of muscle structure and function. Box 19.1 describes the amazing resistance to atrophy shown by an aestivating Australian frog. Black bears also show little disuse atrophy. They remain inactive in their winter dens for 5 to 7 months a year, but their muscles do not decline in contractile capability. Field biologists give heart-stopping accounts of accidentally disturbing overwintering bears and discovering that they show no loss of locomotor ability. Recently, experiments were performed to test the strength of the tibialis anterior muscles of black bears in the fall, shortly after they entered their dens, and again in the spring, shortly before they emerged. The muscles lost only 23% of their strength (measured as production of force) after 130 days of inactivity. In contrast, human muscles immobilized for that length of time are predicted to lose 90% of their strength!

In other studies of overwintering black bears, biopsies were taken from the gastrocnemius and biceps femoris muscles of the hindlimbs in the fall and again in the spring. Compared with the fall samples, the spring samples showed no obvious atrophy. The biceps femoris muscle had an approximate 10% increase in Type II fibers over the course of the winter, while the gastrocnemius showed no change in the relative distribution of fiber types. Researchers have proposed that overwintering bears (which do not eat, drink, urinate, or defecate) may preserve skeletal muscle protein by recycling nitrogen from urea into amino acids that can be used in protein synthesis. Alternatively, or in addition, they may use shivering as a form of isometric exercise to maintain a low level of muscle use. Studying the functions of animals with muscles that resist atrophy may provide insights into how to preserve structure and function in animals (including humans) whose muscles are susceptible to wasting.

Muscle Disease

Diseases of muscle are referred to as **myopathic diseases**. Myopathic diseases can be inherited or acquired. An example of an acquired myopathy is *dermatomyositis*—inflammation (*itis*) of the skin (*derma to*) and skeletal muscles (*myo*). The disease occurs in both children and adults. The main symptoms are skin rash and muscle weakness. Dermatomyositis is an autoimmune disorder and is treated with anti-inflammatory drugs and immunosuppressants. Two examples of inherited myopathies are *malignant hyperthermia* (MH) and the *muscular dystrophies.* MH is transmitted as a dominant autosomal genetic trait. People who are susceptible to MH have no symptoms except when they are exposed to general anesthesia or certain muscle relaxants, which cause them to develop life-threatening symptoms. MH is one of the main causes of death during general anesthesia. Anesthesia stimulates MH muscle cells to release excessive amounts of Ca^{2+} into the cytoplasm. The muscles contract violently and cause a dramatic rise in body temperature. This hyperthermia leads to damaged muscle tissue and other effects, such as heart and kidney failure.

There are several different forms of muscular dystrophy. They all cause gradual degeneration and weakening of muscle tissue. The mutations responsible for many of the muscular dystrophies affect either a large protein called **dystrophin** or other proteins that are associated with it. Dystrophin is a cytoskeletal protein that connects thin actin filaments to a complex of proteins at the sarcolemma. The most common and most severe of the muscular dystrophies is **Duchenne muscular dystrophy** (**DMD**). We will consider this disease in some detail.

DMD results from mutation of the *DMD* gene on the short arm of the X chromosome

The mutations responsible for most muscular dystrophies, including DMD, involve deletions and duplications of DNA of the recessive *DMD* gene.[6] *Becker muscular dystrophy* (BMD), a milder disease than DMD, also results from derangements of the *DMD* gene. A woman with a mutation of the *DMD* gene on one of her X chromosomes would have a 50% chance of passing that mutation on to a son, who would develop DMD (or BMD), and a 50% chance of passing it on to a daughter, who would be a carrier. Since the *DMD* gene was identified some 20 years ago, molecular genetic analyses have been developed for use in genetic counseling and prenatal testing.

DMD afflicts mainly boys, at a frequency of about 1 in 3500. A girl would have to inherit a mutated gene from both of her parents to develop the full-blown disease, an extremely rare event. However, about 5% to 10% of girls who carry a mutated gene develop some degree of muscle weakness. Boys show the first signs of DMD in early childhood, from 2 to 5 years of age; they have difficulties initially in running and then in climbing stairs. In the early stages of the disease, muscles show some degree of regeneration, but they lose this capacity over time. Thus they weaken progressively as muscle fibers atrophy and die and are replaced by fat and connective tissue. A wheelchair is usually required by age 12. Respiratory problems arise from a weakened diaphragm and intercostal muscles (between the ribs), and these problems, combined with heart complications, typically lead to death in the late teens or early twenties. Tissues in people afflicted with DMD do not express the protein dystrophin. In BMD, mutations of the *DMD* gene cause reduced levels of dystrophin or the production of dystrophin molecules that are only partially functional. The presence of some dystrophin may account for the later onset of BMD and the longer survival of BMD patients.

The *DMD* gene was localized to the X chromosome in the 1980s. It was subsequently isolated, cloned, and sequenced, and its product was named dystrophin. Normally, dystrophin is expressed in significant amounts in skeletal and cardiac muscles. Its absence in DMD is thought to make the sarcolemma vulnerable to damage during contractions. Somehow, the mechanical stress on the sarcolemma triggers intracellular changes that lead to cell death. Dystrophin is also normally expressed in smaller amounts in the brain. Up to one-third of boys with DMD show some mental impairment, and about 20% have an IQ of 70 or less. Investigators have found dystrophin in the postsynaptic membrane of normal neurons and have suggested that its absence in DMD may contribute to impaired cognition.

Dystrophin Connects F-Action To The Sarcolemma

Dystrophin has a molecular weight of 427 kDa and consists of 3678 amino acids. The N (amino) terminus of this rod-shaped molecule binds to F-actin of the cytoskeleton. Macromolecules of F-actin not only form the thin filaments of myofibrils, but also contribute to the cytoskeleton—the strutwork of molecules that helps to maintain the organization of intracellular structures. The C (carboxyl) terminus of dystrophin interacts with a complex of glycoprotein molecules within the sarcolemma (Figure 19.12). This membrane-spanning complex binds to extracellular glycoproteins, which in turn bind to

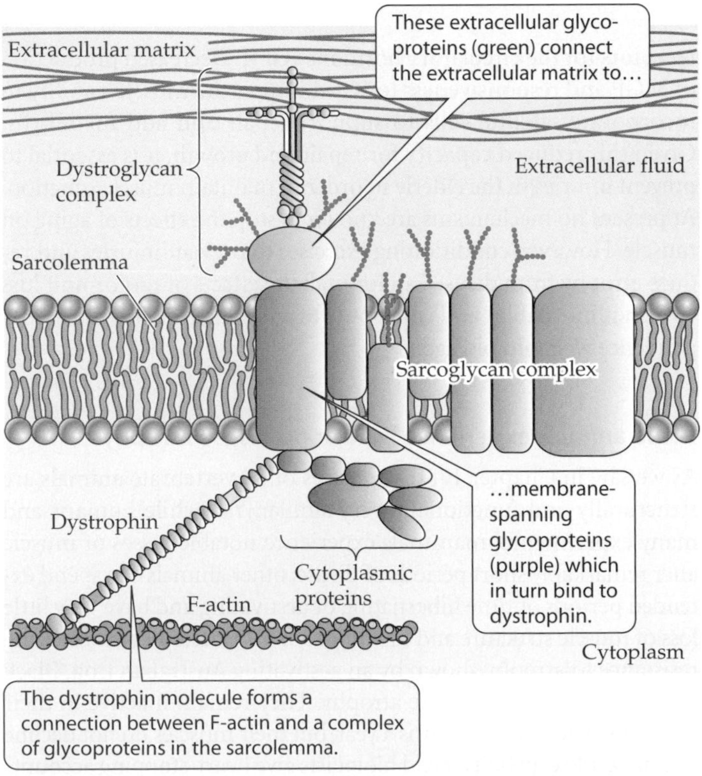

Figure 19.12 Dystrophin connects F-actin of the cytoskeleton to the sarcolemma The dystrophin–sarcolemma–ECM complex is believed to protect the sarcolemma and contribute to the transmission of force. (After Davies 2002.)

[6]The *DMD* gene is extremely large, spanning about 2.5 million base pairs on the X chromosome. The only known larger gene is that for titin, the sarcomeric protein that extends from the Z disc to the M line (see Chapter 17; page 467).

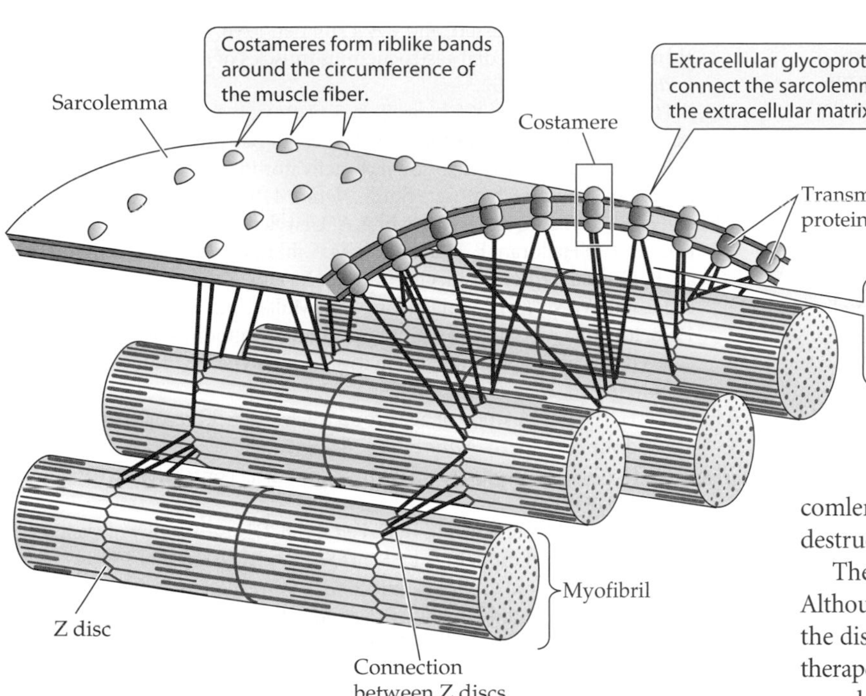

Figure 19.13 Costameres Dystrophin–sarcolemma–ECM complexes form costameres at regular positions around the circumference of the sarcolemma. (After Spence et al. 2002.)

molecules in the extracellular matrix (ECM) that surrounds each muscle fiber. The connections between cytoskeletal actin and the sarcolemma are made at the level of the Z discs.

As we know from Chapter 17, all of the myofibrils of a muscle fiber are lined up in register, so that all of the Z discs lying under the sarcolemma form a distinct visible band around the entire circumference of the cell. Dystrophin and its associated proteins connect the Z discs of the myofibrils to the sarcolemma (Figure 19.13). They contribute to complexes of proteins, called **costameres**. Costameres are defined as regions of attachment between the myofibrils, the sarcolemma, and the ECM. They form a pattern of riblike circumferential bands that are coincident with the Z discs at regular intervals along the length of the muscle fiber. These connections are thought to be important in transmitting force exerted by the muscle fiber. In addition, they support the sarcolemma and appear to be important in protecting it from damage.

In the absence of dystrophin in DMD, the complex of sarcolemmal and ECM glycoproteins is also absent. Thus links between the myofibrils, sarcolemma, and ECM are lost throughout the muscle fiber. This flaw in structure, which leaves the muscle fibers vulnerable to damage and cell death, applies to every muscle in the entire musculoskeletal system. It is believed that the absence of dystrophin and its associated structural proteins allows transient, abnormal changes in permeability of the sarcolemma produced by opening stress-activated channels or making small tears in the membrane during mechanical stresses of contractions. Ca^{2+},which enters the cell inappropriately as a result of these stress-induced changes in the permeability of the sarcolemma, is suspected as the primary cause of subsequent degeneration. Inside the cell, Ca^{2+} itself could cause sarcolemmal Ca^{2+} channels to open, which would permit further influx of Ca^{2+}. Excessive intracellular Ca^{2+} is known to cause intracellular damage in many different types of cells by turning on proteases that destroy cellular structures. Thus, stress-induced changes in the sarcomlemma permit inappropriate influx of Ca^{2+}, which leads to the destruction and death of the muscle fiber.

There are no cures for DMD or the other muscular dystrophies. Although gene therapy seems attractive as an approach to alleviating the disease, so far no effective ways have been developed to deliver therapeutic transgenes to skeletal muscles throughout the body. Pharmacological treatments that would cause overexpression of other genes in dystrophic muscle may be more promising. For example, it would be advantageous to develop treatments that would prolong the regeneration that occurs in early stages of dystrophy. Researchers recently found that overexpression of the muscle-specific IGF-I caused hypertrophy and regeneration in aging muscles. If a pharmacological agent were developed to stimulate overexpression of IGF-I, it could possibly be used to treat both aging and diseased muscles. These and many other ideas are now being intensively pursued.

Study Questions

1. Explain the reason sprinters benefit from tapered training prior to a competition. Include considerations of changes in fiber type distribution during detraining and the factors that influence power output of muscles.
2. Define/describe dystrophin and explain the significance of its absence in Duchenne muscular dystrophy.
3. Why can't resistance exercises prevent decreased power output in aging muscles?
4. List and describe three changes that occur during endurance training and explain how each improves endurance.

Key References

Andersen, J. L., P. Schjerling, and B. Saltin. 2000. Muscle, genes and athletic performance. *Sci. Amer.* 283: 48–55. An overview of the cellular effects of different types of training in human muscles.

Emery, A. E. H. 2002. The muscular dystrophies. *Lancet* 359: 687–695. A review of the clinical characteristics of several different muscular dystrophies.

Goldspink, G. 2003. Gene expression in muscle in response to exercise. *J. Muscle Res. Cell Motil.* 24: 121–126. A review of the adaptability of skeletal muscle, mechanical factors that influence gene expression and its relationship to muscle phenotype, and growth factors produced by muscle cells that support repair and hypertrophy.

Selected References

Adams, G. R., V. J. Caiozzo, and K. M. Baldwin. 2003. Skeletal muscle unweighting: Spaceflight and ground-based models. *J. Appl. Physiol.* 95: 2185–2201.

Andersen, J. L., and P. Aagaard. 2000. Myosin heavy chain IIX overshoot in human skeletal muscle. *Muscle Nerve* 23: 1095–1104.

Andersen, P., and J. Henriksson. 1977. Capillary supply of the quadriceps femoris muscle of man: Adaptive response to exercise. *J. Physiol.* 270: 677–690.

Baar, K., A. R. Wende, T. E. Jones, M. Marison, L. A. Nolte, M. Chen, D. P. Kelly, and J. O. Holloszy. 2002. Adaptations of skeletal muscle to exercise: Rapid increase in the transcriptional coactivator PGC-1. *FASEB J.* 16: 1879–1886.

Brooks, S. V. 2003. Current topics for teaching skeletal muscle physiology. *Adv. Physiol. Educ.* 27: 171–182.

Davies, K. E. 2002. Dystrophin-associated proteins and muscular dystrophies. *Sci. Med.*(Phila) 8: 308–317.

Goldspink, G. 1999. Changes in muscle mass and phenotype and the expression of autocrine and systemic growth factors by muscle in response to stretch and overload. *J. Anat.* 194: 323–334.

Goldspink, G., A. Scutt, P. T. Loughna, D. J. Wells, T. Jaenicke, and G. F. Gerlach. 1992. Gene expression in skeletal muscle in response to stretch and force generation. *Amer. J. Physiol.* 262: R356–R363.

Harlow, H. J., T. Lohuis, T. D. I. Beck, and P. A. Iaizzo. 2001. Muscle strength in overwintering bears. *Nature* 409: 997.

Hoppeler, H., and M. Flück. 2002. Normal mammalian skeletal muscle and its phenotypic plasticity. *J. Exp. Biol.* 205: 2143–2152.

Hoppeler, H., H. Howald, K. Conley, S. L. Lindstedt, H. Claassen, P. Vock, and E. R. Weibel. 1985. Endurance training in humans: Aerobic capacity and structure of skeletal muscle. *J. Appl. Physiol.* 59: 320–327.

Hudson, N. J., and C. E. Franklin. 2002. Effect of aestivation on muscle characteristics and locomotor performance in the green-striped burrowing frog, *Cyclorana alboguttata*. *J. Comp. Physiol. B* 172: 177–182.

Hudson, N. J., and C. E. Franklin. 2002. Maintaining muscle mass during extended disuse: Aestivating frogs as a model species. *J. Exp. Biol.* 205: 2297–2303.

Hurst, J. E., and R. H. Fitts. 2003. Hindlimb unloading-induced muscle atrophy and loss of function: Protective effect of isometric exercise. *J. Appl. Physiol.* 95: 1405–1417.

Ingjer, F. 1979. Effects of endurance training on muscle fibre ATP-ase activity, capillary supply and mitochondrial content in man. *J. Physiol.* 294: 419–432.

Kadhiresan, V. A., C. A. Hassett, and J. A. Faulkner. 1996. Properties of single motor units in medial gastrocnemius muscles of adult and old rats. *J. Physiol.* 493: 543–552.

Lin, J., H. Wu, P. T. Tarr, C.-Y. Zhang, Z. Wu, O. Boss, L. F. Michael, P. Pulgserver, E. Isotani, E. N. Olson, B. B. Lowell, R. Bassel-Duby, and B. M. Splegelman. 2002. Transcriptional co-activator PGC-1α drives the formation of slow-twitch muscle fibres. *Nature* 418: 797–801.

Manning, B. M., K. A. Quane, H. Ording, A. Urwyler, V. Tegazzin, M. Lehane, J. O'Halloran, E. Hartung, L. M. Giblin, P. J. Lynch, P. Vaughan, K. Censier, D. Bendixen, G. Comi, L. Heytens, K. Monsieurs, T. Fagerlund, W. Wolz, J. J. A. Heffron, C. R. Muller, and T. V. McCarthy. 1998. Identification of novel mutations in the ryanodine-receptor gene (RYR1) in malignant hyperthermia: Genotype-phenotype correlation. *Amer. J. Hum. Genet.* 62: 599–609.

Moore, D. H. II. 1975. A study of age group track and field records to relate age and running speed. *Nature* 253: 264–265.

Moyes, C. D. 2003. Controlling muscle mitochondrial content. *J. Exp. Biol.* 206: 4385–4391.

Musacchia, X. J., J. M. Steffen, and R. D. Fell. 1988. Disuse atrophy of skeletal muscle: Animal models. *Exer. Sport Sci. Rev.* 16: 61–87.

Owino, V., S. Y. Yang, and G. Goldspink. 2001. Age-related loss of skeletal muscle function and the inability to express the autocrine form of insulin-like growth factor-1 (MGF) in response to mechanical overload. *FEBS Lett.* 505: 259–263.

Pilegaard, H., B. Saltin, and P. D. Neufer. 2003. Exercise induces transient transcriptional activation of the PGC-1α gene in human skeletal muscle. *J. Physiol.* 546: 851–858.

Richardson, R. S., H. Wagner, S. R. D. Mudaliar, E. Saucedo, R. Henry, and P. D. Wagner. 2000. Exercise adaptation attenuates VEGF gene expression in human skeletal muscle. *Amer. J. Physiol.* 279: H772–H778.

Schiaffino, S., and C. Reggiani. 1996. Molecular diversity of myofibrillar proteins: Gene regulation and functional significance. *Physiol. Rev.* 76: 371–423.

Spence, H. J., Y.-J. Chen, and S. J. Winder. 2002. Muscular dystrophies, the cytoskeleton and cell adhesion. *BioEssays* 24: 542–552.

Tinker, D. B., H. J. Harlow, and T. D. I. Beck. 1998. Protein use and muscle-fiber changes in free-ranging, hibernating black bears. *Physiol. Zool.* 71: 414–424.

Widrick, J. J., S. T. Knuth, K. M. Norenberg, J. G. Romatowski, J. L. W. Bain, D. A. Riley, M. Karhanek, S. W. Trappe, T. A. Trappe, D. L. Costill, and R. H. Fitts. 1999. Effect of a 17 day spaceflight on contractile properties of human soleus muscle fibres. *J. Physiol.* 516: 915–930.

PART V

Oxygen, Carbon Dioxide, and Internal Transport

Part V Oxygen, Carbon Dioxide, and Internal Transport

Previous Page: These marine fanworms, which belong to the sabellid family of annelids, live in protective tubes. Their arrays ("fans") of tentacles, which they extend into the water when not disturbed, function as gills and food-collection organs. Blood flowing through the tentacles picks up oxygen from the seawater and carries it throughout the body.

20 Introduction to Oxygen and Carbon Dioxide Physiology

Looking at this sea turtle depositing her eggs deep in the sand of a beach, one must wonder how the young turtles in the eggs will acquire oxygen (O_2) and void carbon dioxide (CO_2) during their development. After their mother refills the hole in which they are laid, the eggs of the various species of sea turtles will be buried 0.5 to 1 meter under the surface of the sand. They will be there for 6 to 12 weeks. During that time, each fertilized ovum will develop into a small turtle, and as that happens, the rate at which each egg uses O_2 and produces CO_2 will rise. There are typically 80 to 120 eggs in each clutch. Because the eggs in a clutch develop synchronously, their collective O_2 needs will increase during development as a multiple of the O_2 needs of the individual eggs. By the time hatching is imminent, the clutch as a whole will weigh 2 to 4 kg and will have a substantial O_2 requirement. Neither the individual eggs nor the clutch as a whole has any sort of breathing tube to the surface of the sand. Nonetheless, the atmosphere above the sand is their source of O_2. How does the O_2 they consume reach them, and how does the CO_2 they produce get away? As we discuss these questions and others in this chapter, we sometimes will refer to O_2 and CO_2 as the **respiratory gases**. They are called this because they are the principal gases consumed and produced by cellular respiration (aerobic catabolism).

Of all the exchanges of materials between an animal and its environment, the exchanges of the respiratory gases are usually the most urgent. A person, for example, dies within minutes if denied O_2 but can live for days without exchanging nutrients, nitrogenous wastes, or water. The urgency of the need for O_2 arises from the role of O_2 as final electron acceptor in aerobic catabolism: As discussed in Chapter 6 (see Figure 6.2), energy cannot be transferred from bonds of food molecules to bonds of ATP by the aerobic catabolic apparatus of a cell unless O_2 is available in the cell to combine with electrons exiting the electron-transport chain. The need to void CO_2 is ordinarily not as urgent as the need to acquire O_2. Nonetheless, export of CO_2 is often a pressing concern because accumulation of

A sea turtle lays her eggs in a hole she digs on a beach After she has finished laying, she fills the hole with sand. How do the developing turtles in the eggs acquire oxygen for their metabolism as they develop, buried in the sand, over the ensuing weeks?

CO_2 in the body can rapidly acidify the body fluids and exert other harmful effects.

The respiratory gases move from place to place principally by two mechanisms. Before we discuss those mechanisms, it will be useful to have a general term for referring to gas movements, and as in Chapter 3, we shall use **transport**. That is, "gas transport" refers in an entirely general way to any and all movements of gases from place to place, regardless of mechanism. The two principal mechanisms of transport of the respiratory gases are *simple diffusion* and *convection*. These in fact are the only mechanisms of O_2 transport, because active transport of O_2 is unknown. Carbon dioxide, however, is actively transported across some cell membranes in the form of bicarbonate ions (HCO_3^-) formed by reaction with water.

An important first step for understanding respiratory gases and gas transport is to address the concept of *chemical potential* and how it is expressed in studies of gases. In Chapter 3, when we discussed the diffusion of glucose and other uncharged solutes in aqueous solutions, we concluded that (1) a solute like glucose always diffuses from regions of a solution where it is relatively high in concentration to ones where it is relatively low in concentration, and that (2) the rate of diffusion of such a solute from region to region is directly proportional to its difference in concentration between regions (see Equation 3.1).[1] Whereas these principles apply to the sorts of solutes and situations that have dominated our attention thus far in this book, they are not in fact entirely general.

A truly general statement of the principles of diffusion is worded in terms of **chemical potential**. The abstract definition of chemical potential is a topic in physical chemistry; loosely, it is the strength of the tendency of a chemical substance to undergo a physical or chemical change. A practical definition of chemical potential for our purposes here is to say that chemical potential provides the basis for a *truly general law of diffusion*: In all cases of diffusion, materials tend to move in net fashion from regions where their chemical potential is high to regions where their chemical potential is low and at a rate proportional to the difference in chemical potential. This is as true in animal bodies as in a physics laboratory.

In the study of diffusion within a single aqueous solution, the concentration of a solute (the amount of solute per unit volume) is a useful surrogate for chemical potential because, for most purposes, if one uses diffusion equations expressed in terms of the concentrations of solutes, one obtains correct answers. Concentration, however, is not a useful surrogate for chemical potential when analyzing the diffusion of materials *between* gas mixtures and aqueous solutions. Materials such as O_2 and CO_2 exist in both phases—both gas mixtures and aqueous solutions—and they diffuse between them. If one attempts to analyze their diffusion between phases on the basis of their concentrations, one often reaches erroneous conclusions. To analyze such diffusion accurately, a measure of chemical potential other than concentration is needed. Physiologists usually use *partial pressure* to express chemical potential in the study of gases. Thus, as we proceed in our study of gases, we will often encounter the concept of partial pressure, and it will be important to understand not only partial pressure itself, but also how partial pressure relates to concentration.

[1]This is not to say that the rate of diffusion depends on just the difference in concentration; as seen in Equation 3.1, there are additional factors that affect the rate.

The Properties of Gases in Gas Phases and Aqueous Solutions

Besides existing in air and other types of gas phases, gases also *dissolve* in aqueous solutions, as just suggested. When a gas dissolves in an aqueous solution, molecules of the gas become distributed among H_2O molecules in much the same way as glucose molecules or Na^+ ions are incorporated among H_2O molecules when solids dissolve. Thus, molecules of a gas disappear when they go into solution. They do not appear as tiny bubbles any more than glucose in solution appears as tiny sugar grains; bubbles of gas represent gas that is *not* in solution. In this section, we first address gases in the gas phase, then gases in aqueous solution.

Gases in the gas phase

The modern study of gases in the gas phase traces back to John Dalton (1766–1844), who articulated the *law of partial pressures*. According to this concept, the total pressure exerted by a mixture of gases (such as the atmosphere) is the sum of individual pressures exerted by each of the several component gases in the mixture (Figure 20.1). The individual pressure exerted by any particular gas in a gas mixture is termed the **partial pressure** of that gas. An important property of the partial pressure of each gas in a mixture is that it is independent of the other gases present. Moreover, in a volume of mixed gases, each component gas behaves in terms of its partial pressure as if it alone occupied the entire volume. Thus the partial pressure of each gas in a gas mixture can be calculated from the **universal gas law**,

$$PV = nRT \tag{20.1}$$

where P represents pressure, V is volume, n is the number of moles (the quantity) of gas, R is the universal gas constant, and T is the absolute temperature.[2] To use the universal gas law to calculate the partial pressure of a particular gas in a gas mixture, one sets n equal to the molar quantity of the *particular gas* of interest and V equal to the volume occupied by the *gas mixture as a whole*.

A simple procedure exists to calculate the partial pressure of a gas in a mixture if one knows the proportions of the various gases in the mixture. The proportions of gases in a mixture are expressed as *fractional concentrations*. The **mole fractional concentration** of any particular gas in a mixture is the fraction of the total moles of gas present represented by the gas in question. To illustrate, in a volume of dry atmospheric air near sea level, the number of moles of O_2 is 20.95% of the total number of moles of all gases taken together; thus, the mole fractional concentration of O_2 in dry air is 0.2095. The **volume fractional concentration** of a particular gas in a mixture is the fraction of the total volume represented by that particular gas. Because equal molar quantities of different gases occupy essentially equal volumes, the volume fractional concentration of a gas in a mixture is essentially the same as its mole fractional concentration.

[2]The partial pressure of a particular gas is symbolized by writing P with the molecular formula of the gas as a subscript. Thus, for example, the partial pressure of O_2 is symbolized P_{O_2}. In the SI, R is 8.314 J/mol·K (where K is a Kelvin and equivalent to one degree Celsius). Values for R in other systems of units are listed in standard reference books on chemistry and physics.

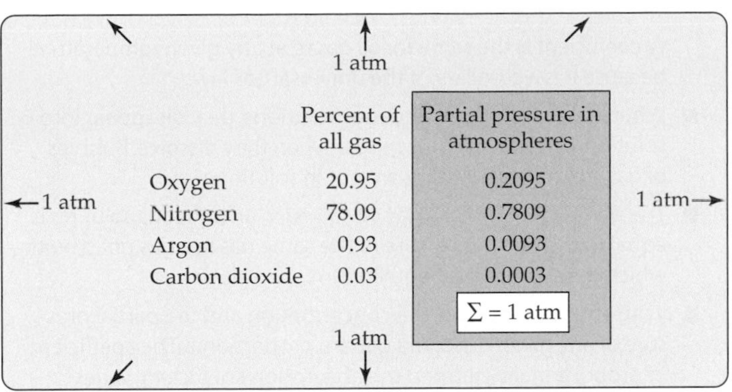

Figure 20.1 The total pressure exerted by a mixture of gases is the sum of the partial pressures exerted by the individual constituents of the mixture The diagram shows a container surrounding a body of dry atmospheric air at sea level. Data on the four most abundant constituents of dry air are shown. The air exerts a total pressure of 1 atmosphere (1 atm), which is the sum of the partial pressures. Each constituent would exert the same partial pressure even if the other constituents were absent.

This means, for example, that if we removed the O_2 from a volume of dry atmospheric air at a given temperature and pressure and then restored the residual gas to the same temperature and pressure, the volume of the air would be reduced by a quantity equal to 0.2095 (20.95%) of the original volume. It is easy to show, using the universal gas law (Equation 20.1), that the partial pressure exerted by each gas in a mixture is "its fraction" of the total pressure, where by "its fraction" we mean its mole fractional concentration or volume fractional concentration. Expressed algebraically, if P_{tot} is the total pressure of a gas mixture, P_x is the partial pressure of a particular gas (x) in the mixture, and F_x is the mole or volume fractional concentration of that gas, then

$$P_x = F_x P_{tot} \qquad (20.2)$$

To illustrate the application of Equation 20.2, let's calculate the partial pressures of oxygen, nitrogen, and carbon dioxide in dry atmospheric air. Near sea level, dry air consists chemically of 20.95% O_2, 78.09% N_2, and 0.03% CO_2. If the total pressure of the air is 1 atmosphere (1 atm), then the partial pressure of O_2 in the air is (0.2095)(1 atm) = 0.2095 atm, and the partial pressures of N_2 and CO_2 are 0.7809 atm and 0.0003 atm (see Figure 20.1). We discuss other units of measure for pressure later in this chapter.

What is the relation between the partial pressure of a gas in a gas mixture and the concentration (amount per unit volume) of the gas? If one rearranges the universal gas law (Equation 20.1), one gets $P = (n/V)RT$. If n is the molar amount of a particular gas in a gas mixture that occupies a total volume V, then n/V is the concentration of that gas, and P is its partial pressure. Moreover, RT is a constant, provided the temperature T is constant. Thus, the partial pressure and concentration are simply proportional to each other at any given T. For example, if there are two gas mixtures that are identical in temperature and if the concentration of O_2 is 5 millimole/liter in one and 10 millimole/liter in the other, the partial pressure of O_2 will be exactly twice as high in the latter.

Gases in solution

The partial pressure of a gas dissolved in an aqueous solution (or any other sort of solution) is defined to be equal to the partial pressure of the gas in a gas phase with which the solution is at equilibrium. To illustrate, consider what happens if O_2-free water is brought into contact with air containing O_2 at a partial pressure of 0.21 atm. Let's assume, specifically, that the volume of air is so great that as O_2 dissolves in the water, there is essentially no change in the O_2 concentration of the air, and thus the partial pressure of O_2 in the air remains at 0.21 atm. Oxygen will dissolve in the water until an equilibrium is reached with the air. Then the partial pressure of O_2 *in the aqueous solution* will be 0.21 atm. If this solution is later exposed to air containing O_2 at a partial pressure of 0.19 atm, the solution will lose O_2 to the air until a new equilibrium is established. The partial pressure of O_2 in the solution will then be 0.19 atm. The term *tension* is sometimes used as a synonym for partial pressure when speaking of gases in solution, as when the O_2 partial pressure is called the "O_2 tension." We do not use the tension terminology in this book.

The partial pressure and the concentration of a gas in a solution are proportional, but the nature of this proportionality is more complicated than that in gas phases. **Henry's law** is the fundamental law that relates partial pressure and concentration in solutions. There are several ways in which this law is expressed, and the various expressions employ a number of different, but related, coefficients.[3] Our approach here is to use the **absorption coefficient**, defined to be the dissolved concentration of a gas when the partial pressure of the gas in solution is 1 atm. Using P to represent the partial pressure, C, the concentration, and A, the absorption coefficient,

$$C = AP \qquad (20.3)$$

In gas phases, where C is also proportional to P, all gases have essentially identical quantitative relations between C and P (because all adhere to the universal gas law). In aqueous solutions, however, the proportionality constant A varies a great deal not only from one type of dissolved gas to another, but also from one solution to another, depending on the temperature and salinity of the water. For practical purposes, one must look up the absorption coefficient (A) that applies in any particular situation.

The absorption coefficient is a way of expressing gas *solubility*. A high absorption coefficient, for example, signifies high solubility, meaning that a lot of gas will dissolve at any given partial pressure. With this in mind, three important characteristics of gases dissolved in aqueous solutions come to light when absorption coefficients are examined:

- The solubilities of different gases are different, and specifically, CO_2 has a far higher solubility than O_2 or N_2. The absorption coefficients of CO_2, O_2, and N_2 in cold (0°C) distilled water are 77, 2.2, and 1.1 mmol/L respectively. These absorption coefficients tell you that if each of these gases is brought to a partial pressure of 1 atm in cold distilled water, the amount dissolved in each liter

[3]During the history of the study of gas solubility, an uncommonly large number of coefficients were defined by various scientists working on the subject. Many of these have survived into the present time and today are referred to by names such as *Henry's law coefficients*, the *Bunsen coefficient*, and the *Ostwald coefficient*. Although differing in detail, all describe the same principles and can be interconverted using standard equations.

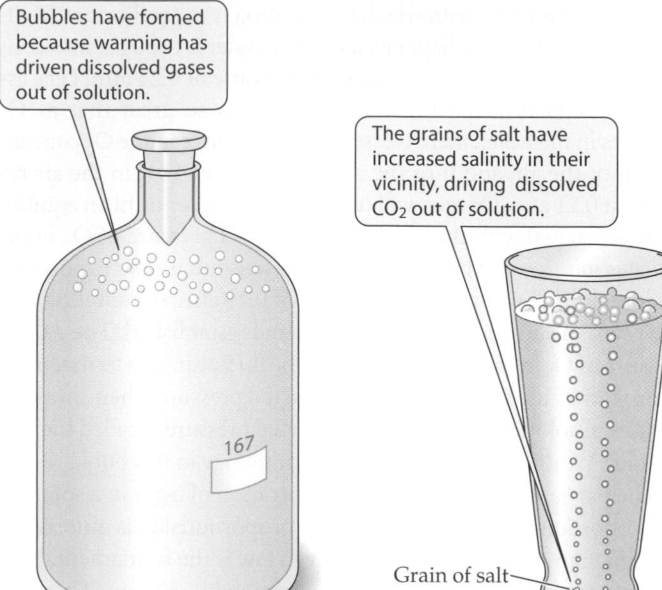

Figure 20.2 Everyday illustrations of the effects of temperature and salinity on gas solubility

of water will be 77 mmol of CO_2, but only 2.2 mmol of O_2, and 1.1 mmol of N_2.

- The solubilities of gases in aqueous solutions decrease strongly with increasing water temperature. This is true of all gases. We will use O_2 as an example. The absorption coefficients of O_2 in distilled water at 0°C, 20°C, and 40°C are 2.2, 1.4, and 1.0 mmol/L respectively. Thus, if the partial pressure of O_2 is 1 atm, a liter of distilled water dissolves 2.2 mmol of O_2 when at 0°C but less than half as much, 1.0 mmol, when at 40°C. Gases tend to come out of solution and form bubbles as water warms (Figure 20.2*a*).
- The solubilities of gases in aqueous solutions decrease with increasing salinity. For instance, the absorption coefficients of O_2 at 0°C in distilled water, 80% seawater, and full-strength seawater are 2.2, 1.8, and 1.7 mmol/L respectively. The tendency for increasing salinity to drive gases out of solution by decreasing the solubilities of the gases is termed the *salting-out effect* (Figure 20.2*b*).

Summary: The Properties of Gases in Gas Phases and Aqueous Solutions

- The total pressure of a gas mixture is the sum of the partial pressures exerted by the individual gases in the mixture. The partial pressure of each gas is independent of the other gases present.
- In gas mixtures, the concentration and the partial pressure of any given gas are simply proportional to each other. The proportionality coefficient is the same for all gases, at any given temperature, because it is a corollary of the universal gas law.
- When gases dissolve in aqueous solutions, they disappear into solution, just as sugars disappear when they dissolve. Bubbles of gas, no matter how tiny, are not in solution.
- The partial pressure of a gas dissolved in an aqueous solution is equal to the partial pressure of the same gas in a gas phase with which the solution is at equilibrium.
- In an aqueous solution, the concentration and the partial pressure of any given dissolved gas are proportional. The coefficient of proportionality, termed the absorption coefficient, varies greatly from gas to gas, as well as depending on temperature and salinity.
- Gas solubility decreases with increasing temperature and also with increasing salinity.

Diffusion of Gases

When gases diffuse from place to place, they do so by the same fundamental mechanism by which solutes diffuse through solutions (see Chapter 3, page 67). Whether gas molecules are in a gas phase or dissolved in an aqueous solution or other liquid phase, the molecules move ceaselessly at random on an atomic-molecular scale. Just by the operation of the laws of probability, when the chemical potential of a gas differs from place to place, these random movements carry more gas molecules away from regions of high chemical potential than into such regions. Macroscopically, therefore, net gas transport occurs.

The fundamental law of gas diffusion is that *gases diffuse in net fashion from areas of relatively high partial pressure to areas of relatively low partial pressure*. This is true within gas mixtures, within aqueous solutions, and across gas–water interfaces. Sometimes, but only sometimes, diffusion in the direction of the partial-pressure gradient also means diffusion in the direction of the concentration gradient. *Within* a gas mixture of *uniform temperature* or *within* an aqueous solution of *uniform temperature and salinity*, if the partial pressure of any given gas is greater in one region than in another, the concentration is also greater. Thus, within such gas mixtures and aqueous solutions, when gases diffuse from high to low partial pressure, they also diffuse from high to low concentration. These gas mixtures and solutions are merely special cases, however. In general, gas diffusion is *not* necessarily from high concentration to low. Suppose, for instance, that two masses of distilled water are separated by an O_2-permeable membrane: One is a mass of warm water at 40°C and having an O_2 partial pressure of 0.21 atm, whereas the other is a mass of cold water at 0°C and having a *lower* O_2 partial pressure, 0.17 atm. You can calculate from the absorption coefficients already given that the O_2 concentration in the warm water is 0.21 mmol/L, whereas that in the cold water is *higher*, 0.37 mmol/L (because O_2 has a higher solubility in cold water). Oxygen diffusion between these two masses of water will, of course, follow the fundamental law of gas diffusion: Oxygen will diffuse from high partial pressure to low partial pressure, from the warm water into the cold water. This diffusion, however, will carry O_2 toward the side of higher concentration! *In any system, equilibrium is attained with respect to any given gas when the partial pressure of the gas is uniform throughout.*

Although we will use the principles of gas diffusion in this book primarily to analyze O_2 and CO_2, these principles apply to all substances that exist as gases under physiological conditions. Thus, for example, partial pressures govern the diffusion of N_2, a topic that is important in understanding decompression sickness (the bends) in scuba divers and dolphins (see Chapter 24). Partial pressures also govern the diffusion of gaseous anesthetics into and out of the body. Water vapor (gaseous water) is a bit of a special case because, unlike other gases, it does not merely dissolve in aqueous solutions but can *become* liquid water or be generated from liquid water; it is given special treatment in Chapter 25.

The equation for the *rate* of gas diffusion is similar in form to the equation for the rate of solute diffusion presented in Chapter 3 (see page 68). Consider a fluid system (consisting of a gas phase, liquid phase, or both) in which a gas, such as O_2, is at a relatively high partial pressure P_1 in one region and a relatively low partial pressure P_2 in another. Think of an imaginary plane that is perpendicular to the direction of diffusion between the two regions. Let J be the rate of net movement of gas through the plane, per unit of cross-sectional area. Then

$$J = K\frac{P_1 - P_2}{X} \qquad (20.4)$$

where X is the distance separating P_1 and P_2 and K is a proportionality factor that is often called the **Krogh diffusion coefficient**. The rate of diffusion is directly proportional to the difference in partial pressures, but it is inversely proportional to the distance separating the two partial pressures. The diffusion coefficient K depends on the particular diffusing gas, the temperature, and the ease with which the gas is able to pass through the material separating the regions of different partial pressure. If the two regions of interest are separated by a layer of tissue (e.g., a gill epithelium), K is termed the **gas permeability** of the tissue.

Gases diffuse far more readily through gas phases than through aqueous solutions

The ease with which gases diffuse is far greater when they are diffusing through air than through water. The Krogh diffusion coefficient (K) for O_2, for example, is about 200,000 times higher in air than in water at 20°C! One way to see the enormous implications of this difference is to consider O_2 diffusing between two regions that have a particular difference in O_2 partial pressure [i.e., a particular value for ($P_1 - P_2$) in Equation 20.4]. If the diffusion is occurring through air rather than water, the length of the diffusion path can be 200,000 times longer and still have the same rate of O_2 transport. Because of the relative ease of diffusion through air, a buried nest of sea turtle eggs on a beach can receive the O_2 it needs by diffusion from the atmosphere, provided the beach sand is dry and not too rich with microbes that might intercept O_2 on its way through the sand. The metabolism of the eggs draws down the partial pressure of O_2 in the nest, creating a gradient of partial pressure from the atmosphere to the nest; O_2 then diffuses to the nest through the network of minute gas-filled spaces among sand grains. There are many different paths—many "angles of approach"—by which diffusion can occur from the beach surface to the nest, and when transport along all these paths is summed, the O_2 needs of the nestling turtles can be met by diffusion even though the nest is 0.5 to 1.0 m underground. Diffusion over such distances would not suffice if it had to occur through water. In fact, the usual rule of thumb is that diffusion through water can meet the O_2 requirements of living tissues only if the distances to be covered are about *1 millimeter* or less (Box 20.1)! Among other things, the dramatic difference between the ease of diffusion through air and through water explains why just a small accumulation of body fluids in the terminal air spaces of a person's lungs creates a dire medical emergency. The difference between air and water in the ease of diffusion of CO_2 is less than that for O_2, but still substantial; the Krogh diffusion coefficient for CO_2 at 20°C is about 9000 times greater in air than in water.

Gas molecules that combine chemically with other molecules cease to contribute to the gas partial pressure

Only gas molecules that exist as free gas molecules contribute to the partial pressure of a gas. This may seem an odd point to stress until you realize that, especially in body fluids, gas molecules often undergo chemical reactions. Molecules of O_2 in the blood of a person or other vertebrate, for example, combine chemically with hemoglobin, and those molecules then *do not* contribute to the partial pressure of O_2 in the blood. Only O_2 molecules that are free in solution contribute to the partial pressure. Similarly, CO_2 molecules added to blood—or to an environmental solution such as seawater—may react with H_2O to yield chemical forms such as HCO_3^- (bicarbonate ions). Only the molecules of CO_2 that are free in solution as unchanged CO_2 molecules contribute to the partial pressure of CO_2.

One noteworthy implication of these principles is that if you know the partial pressure of a gas in an aqueous solution and you use the absorption coefficient to calculate the concentration of the gas in the solution from the partial pressure, the concentration you obtain is that of gas free in solution as unchanged gas molecules. Any gas that is present in chemically combined form (or chemically altered form) is not included in the concentration calculated this way.

A second noteworthy implication is that *only the gas molecules that are free in solution as unchanged gas molecules affect the direction and rate of gas diffusion* because only those gas molecules affect the partial-pressure gradient. This point is exceedingly important in many situations, one of which is the dynamics of O_2 uptake by blood in the lungs of an animal. When O_2 diffuses from the air in a vertebrate animal's lungs into the blood, the O_2 molecules that combine with hemoglobin in the blood do not increase the blood's partial pressure of O_2. Thus, the hemoglobin-combined O_2 molecules do not interfere with further O_2 diffusion into the blood. Suppose, to elaborate, that the O_2 partial pressure in the lung air exceeds that in the blood by a certain amount. If all the O_2 that diffuses into the blood over a period of time combines with hemoglobin, the difference in O_2 partial pressure stays constant, and thus the rate of O_2 diffusion will remain undiminished even as more and more O_2 enters the blood.

BOX 20.1 Over What Distance Can Diffusion Meet the O_2 Requirements of Tissues?

August Krogh (1874–1949), one of the great names in respiratory physiology, was the first to quantify the distance over which the simple diffusion of O_2 through tissues might meet the O_2 requirements of life. Oxygen diffuses less rapidly through animal tissues than through pure water, and Krogh assumed, on the basis of the evidence available in his lifetime, a rate of one-third the rate in water. As his model, he considered a spherical cell with a rate of O_2 consumption of 0.1 mL O_2 g•h (relatively low) and an O_2 partial pressure near 0.21 atm at the surface (relatively high). He calculated that diffusion of O_2 through such a cell will meet the O_2 demands of all its parts only if the cell radius is no greater than 0.9 millimeter! Modern recalculations come up with similar results. Thus, Krogh's major conclusion still stands: Diffusion from a high-partial-pressure source through the aqueous medium of tissue can be expected to supply the ordinary O_2 requirements of aerobic catabolism over only short distances (~ 1 mm or less). Over truly minute distances, such as the distance across a cell membrane, diffusion transport is very fast. However, the rate of diffusion falls as distance increases, and in general, for O_2 transport to occur at a sufficient rate to supply the needs of tissues at distances greater than about 1 mm, some sort of convective transport needs to supplement diffusion. Convective transport is transport of O_2 by a moving fluid, such as by the circulation of blood.

Larval fish present important and interesting examples of the application of these principles. The young of many fish species are so immature at hatching that they lack an effective circulatory system. Thus, their interior tissues must receive O_2 from their body surfaces largely by diffusion. As a young larva grows, diffusion becomes less and less certain to be able to meet all its O_2 needs because the increasing thickness of its body means that its deepest tissues are positioned farther from its body surface. In a 3-week-old larval anchovy (illustrated), the gills are far from fully developed and circulatory development is still incomplete, yet the thickness of the body approximates the maximum over which diffusion can meet O_2 needs. As a larval fish develops through such a transition phase—a stage at which its convective-transport mechanisms are required but still immature—it is particularly vulnerable to dying if it finds itself in low-O_2 water. If schools of larvae die because of such effects, there can be great impacts on the sizes of adult fish populations.

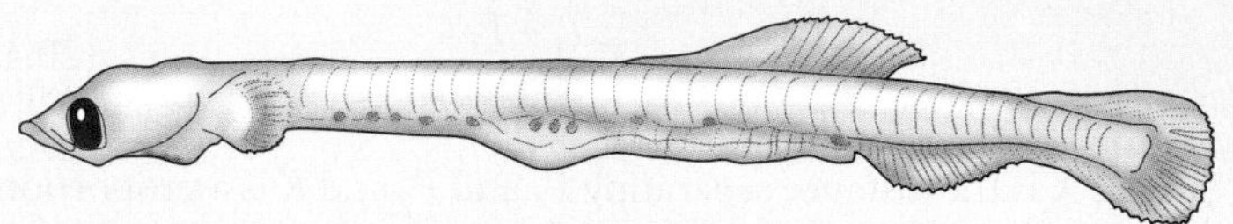

A 3-week-old larva of the anchovy *Engraulis mordax*, common along the West Coast of the United States The O_2 demand of the larva is met primarily by diffusion. The larva's average body radius is about 0.6 mm and approaching the maximum that calculations for the species indicate is compatible with uptake of O_2 by diffusion, yet the circulatory and gill-breathing systems remain immature. (Graham 1990 provides a more complete discussion.)

Summary: Diffusion of Gases

- Gases always diffuse from regions of high partial pressure to regions of low partial pressure and at a rate that is proportional to the difference in partial pressure. A gas is at equilibrium in a system when its partial pressure is uniform throughout.
- Diffusion occurs much more readily through air than through water. Specifically, the Krogh diffusion coefficient is 200,000 times greater for O_2 and 9000 times greater for CO_2 in air than in water (at 20°C).
- When gas molecules undergo chemical combination they cease to contribute to the partial pressure of the gas in question and thus no longer affect the direction or rate of diffusion of the gas.

Convective Transport of Gases

When a fluid—whether it is a gas or a liquid—flows from place to place, gas molecules in the fluid are carried along, a phenomenon called **convective gas transport.** Convective gas transport can be far faster than gas diffusion because it moves gas molecules in a deterministic, forced fashion rather than depending on random molecular movements. Breathing and the pumping of blood are the two principal processes by which animals set fluids in motion to transport gases. Both cost metabolic energy. The reward an animal gets for making the metabolic investment in breathing or blood pumping is that it is able to speed the transport of O_2 and CO_2 in advantageous ways. Convection does not always require muscular effort from an animal, however. Ambient winds and water currents often move O_2 and CO_2 from place to place in the environments of animals in ways that are helpful, as when global-scale water currents renew the O_2 supplies of the deep sea. Sometimes, ambient winds and water currents even induce internal flow within animals or animal-built structures (Box 20.2).

The precise effects of fluid flow on gas transport depend in part on the geometry of the flow. Two geometries of common importance in the circulation and breathing of animals are depicted in Figure 20.3*a*. One is **unidirectional flow** through a tube, such as the flow of blood through a blood vessel; the second is **tidal flow** (back-and-forth flow) in and out of a blind-ended cavity, such as occurs in the lungs of mammals and many other animals. To exemplify how the rate of convective gas transport can be calculated, let's use as a model the simple case of O_2 transport by the unidirectional flow of a fluid through a tube without any exchange of gases through the walls (Figure 20.3*b*). This sort of model applies, for example, to O_2 transport through the arteries of animals because arteries basically act as conduits, neither adding nor removing O_2 from the blood passing through them. The rate of gas transport is simply the product of the total concentration of gas in the fluid (the amount of gas per unit volume) and the rate of fluid flow (the volume passing through per unit time). The gas concentration used for this sort of calculation, it is important to emphasize, is the *total* concentration, including both gas that is chemically combined (e.g., O_2 combined with hemoglobin) and gas that is free

BOX 20.2 Induction of Internal Flow by Ambient Currents

Animals are sometimes able to take advantage of ambient air currents (winds) or water currents in such ways that the currents induce fluid flow through their bodies or through structures they build. The most obvious way in which this can occur is if the opening of a tube-shaped structure is positioned so that it faces partly or fully into the path of an ambient current, as in part 1 of the figure. But what if a structure is oriented at a right angle to the current, as in part 2? Is it then possible for an ambient current to induce internal flow?

Part 3 of the figure illustrates one principle by which an ambient current can induce internal flow when flowing across (rather than into) the orifices of a tubular structure. In this case, a burrow or other tubular system opens at two spots on the ground or some other surface. One of the openings is higher than the other. As a wind or water current flows along the surface, the fluid stream accelerates as it passes over the elevation, and *Bernoulli's principle* (familiar in the study of airfoils) comes into play. According to this principle, the lateral pressure exerted by a fluid stream decreases as the speed of the stream increases. Thus, the ambient pressure at the elevated orifice of the tube is lower than that at the lower orifice, and consequently, fluid is forced to flow *through* the tubular system. One of the remarkable features of this type of induction of internal flow is that, unlike the situation in part 1, it does not depend on the direction of the ambient wind or water current. Provided the ambient current is homogeneous, the acceleration of the fluid stream as it passes over the elevation will cause the pressure at the upper orifice to be lower than that at the lower one, regardless of the direction. The burrow systems of prairie dogs are believed to be ventilated by a mechanism such as this. The mounds often built at various burrow-system openings differ in height and shape, and experiments using tracer gases indicate that even light winds along the surface of the ground induce internal flow of fresh air through the burrow systems, thereby helping to refresh O_2 and purge CO_2.

Another mechanism by which ambient currents can potentially induce internal flow depends on the principle of *viscous entrainment* or *viscous suction*. When a current flows across the opening of a tube, there is a tendency for fluid to be drawn from the tube into the current because of the viscosity, or resistance to shear forces, of the fluid. Other things being equal, this force of suction is greater, the greater the velocity of the current. Thus, if a tubular structure is open at two places and the two orifices are exposed to different ambient current velocities, as in part 4, fluid will tend to flow through the tube toward the end exposed to the greater velocity. This principle can lead to flow through a tubular system in situations where the Bernoulli principle does not apply. When a fluid flows in laminar fashion along a substrate, as illustrated in part 5, there is a gradient of fluid velocity within a short distance of the substrate surface; fluid relatively far from the substrate flows more rapidly than that near the substrate. In this commonplace situation, there is no difference in pressure associated with the velocity gradient; Bernoulli's principle does not say that *any* rapidly moving stream exerts less pressure than *any* more slowly moving stream, but instead, it applies only when a *particular* fluid stream is accelerated or decelerated. As illustrated, however, the velocity gradient itself can induce flow through a tubular system because of differential viscous suction at orifices located different distances from the substrate surface. Again, the direction of the ambient current does not matter.

(1) Structure faces current

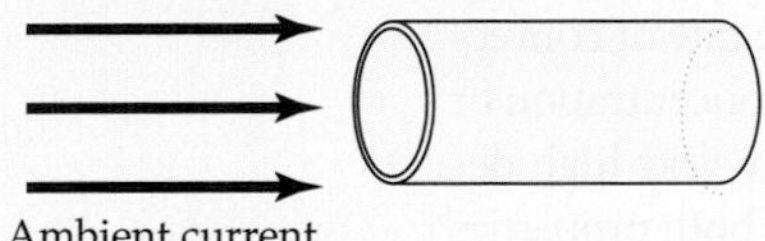

(2) Structure at right angle to current

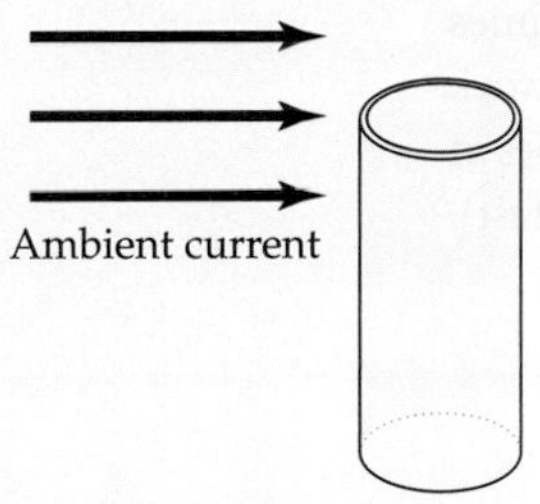

(3) Induction of flow by Bernoulli's principle

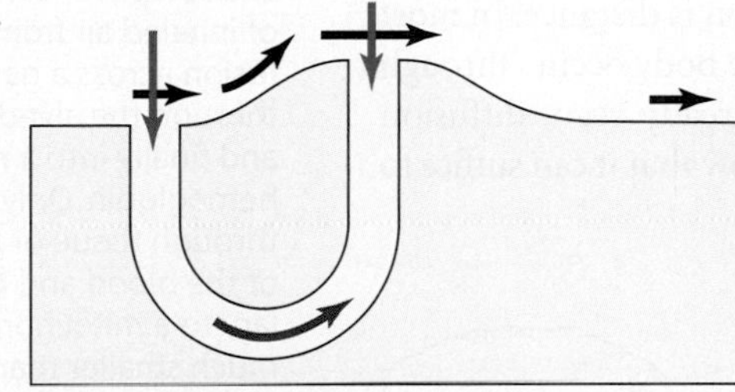

Black-tailed prairie dogs, *Cynomys indovicianus*

(4) Induction of flow by viscous suction

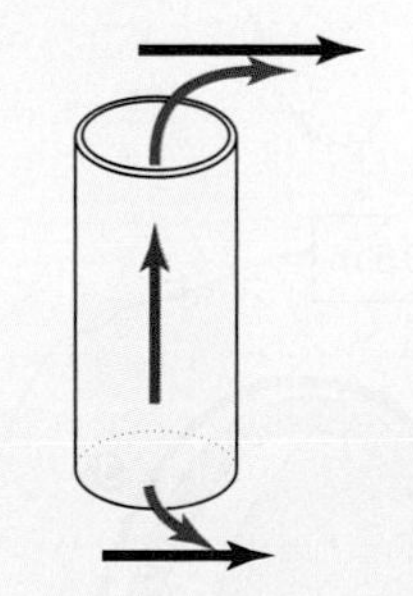

(5) Viscous suction in a laminar flow velocity gradient

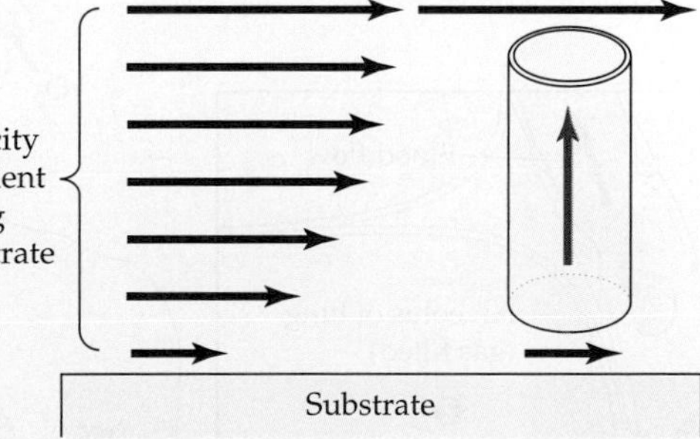

KEY

- Ambient current
- Ambient pressure
- Internal flow
- Force of viscous suction

Mechanisms by which ambient currents can induce flow through tubular structures The relative lengths of arrows symbolize relative magnitudes. For instance, long arrows representing the ambient current symbolize higher rates of current flow than short arrows. (Vogel 1994 provides a more complete discussion.)

Figure 20.3 Convective transport *(a)* Two geometries of convective transport that are common in animals: unidirectional flow through a tube and tidal flow in and out of a blind-ended cavity. Arrows depict mass flow of fluid. *(b)* Calculation of the rate of convective transport of a gas by a fluid flowing unidirectionally through a tube, assuming no gas exchange across the tube walls. *F* is used here as a general symbol for the rate of fluid flow; according to established convention in circulatory and respiratory physiology, the fluid flow rate would be symbolized by *Q* if the fluid were a liquid (e.g., blood) or *V* if it were a gas (e.g., air).

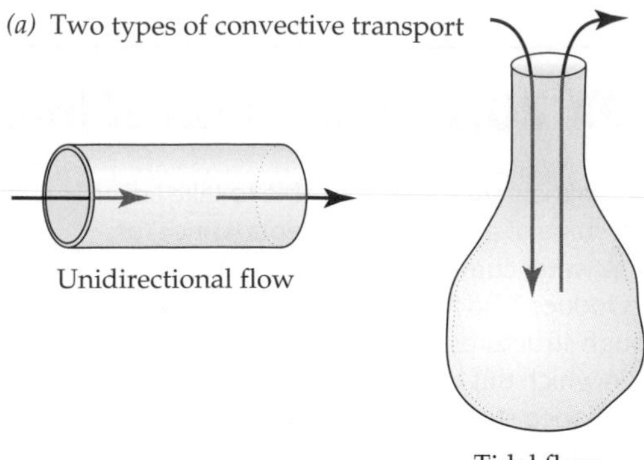

(b) Calculation of the rate of convective gas transport

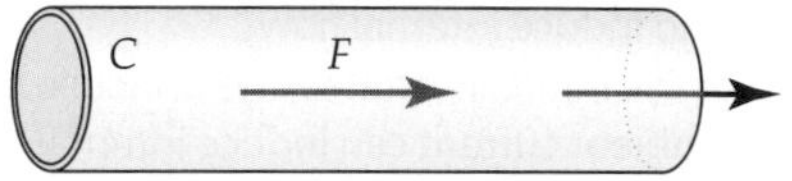

C = Total concentration of gas in flowing fluid (mol/L)

F = Flow rate of fluid (L/second)

Rate of convective gas transport = C • F

in solution. As the formula in Figure 20.3*b* shows, the rate of convective gas transport can be increased by increasing the concentration or the flow rate. Mammals and other animals that have very high demands for O_2 have evolved favorable modifications of both properties to enhance the rate of convective O_2 transport by the flow of blood through their arteries. The high concentration of hemoglobin in mammalian blood enables each volume of blood to carry 50 times more O_2 than it could carry in simple solution, and the intense work done by the mammalian heart propels blood through the arteries at high rates (e.g., 5 L/min through the systemic aorta of a person at rest).

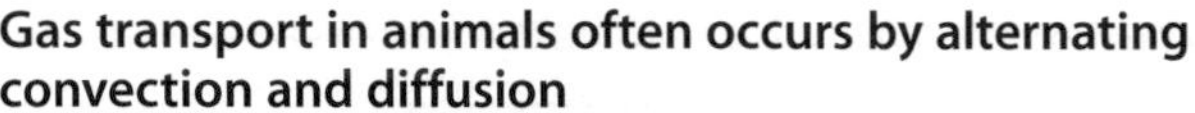

Gas transport in animals often occurs by alternating convection and diffusion

Although in small animals O_2 may be able to move as fast as needed from the environment to all parts of the body by diffusion (see Box 20.1), this is not true in animals that are larger than 2 to 3 g, or even in many that are smaller than that. The reason is distance. In most animal body plans, diffusion of O_2 within the body occurs through body fluids and tissues, and as we have already seen, diffusion through such liquid media is so inherently slow that it can suffice to

Figure 20.4 Mechanisms of oxygen transport during successive steps in the delivery of O_2 from the atmosphere to the mitochondria in a person Four steps are recognized; the distance covered in each step is shown in meters: ❶ is convective transport by movement of inhaled air from the atmosphere to the depths of the lung; ❷ is diffusion across a gas-filled alveolar end sac (including alveolus), then through the alveolar epithelium and pulmonary-capillary epithelium, and finally into a red blood cell (RBC), where O_2 combines with hemoglobin. Only the final 0.000006 m of this step requires diffusion through tissue or liquid. ❸ is convective transport by the circulation of the blood and ❹ is diffusion from a RBC in a systemic blood capillary to a mitochondrion in a muscle cell. Mitochondria are actually much smaller than shown.

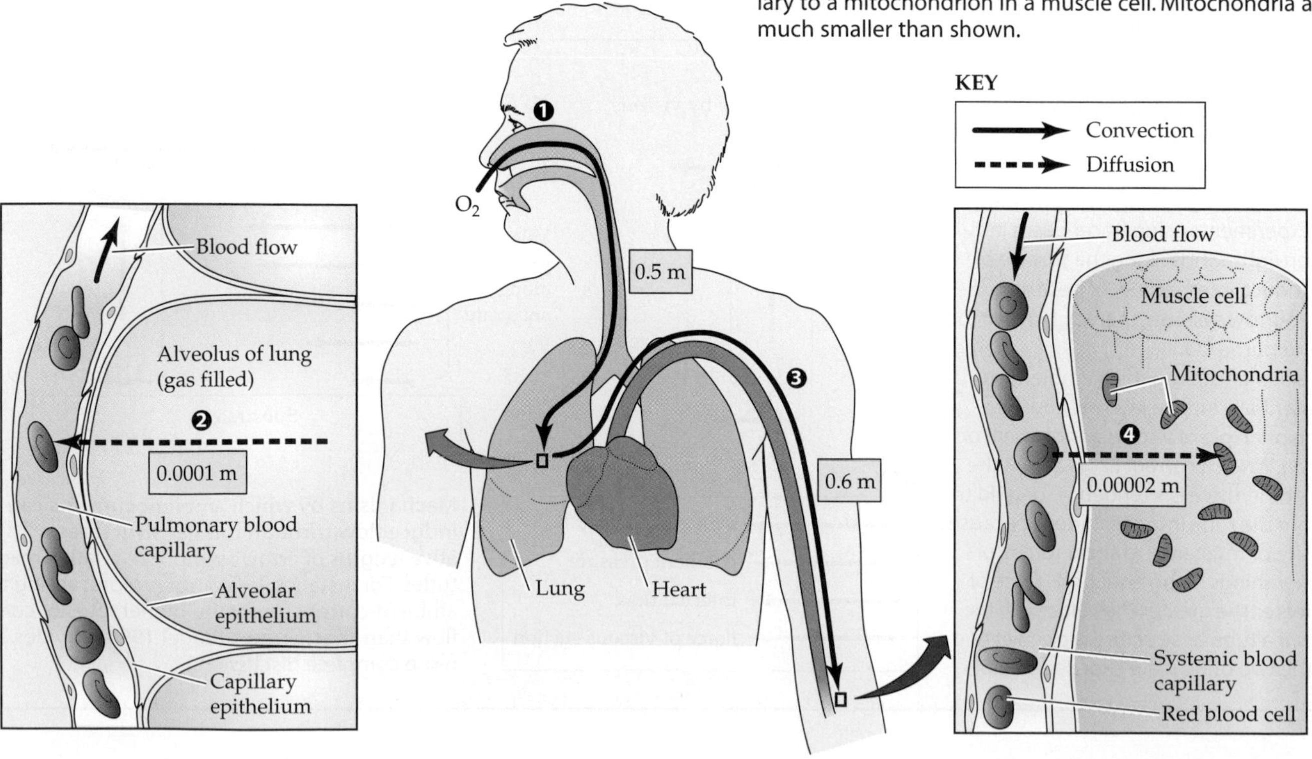

meet transport needs only over very short distances. For aquatic animals, even O_2 transport to the gills from the environment must occur through a liquid medium, and thus again, diffusion can suffice only if the distances to be covered are very short.

Animals that face these diffusion limitations typically employ convective transport as well as diffusion to exchange respiratory gases. Usually the two sorts of transport alternate as O_2 makes its way from the environment to the mitochondria. During this alternation, diffusion meets needs for gas transport over short distances, whereas, convective transport is employed for the long hauls. Oxygen transport from the atmosphere to the mitochondria in people provides an apt example (Figure 20.4). The first step in this transport is the movement of O_2 over a distance of about 0.5 m from the atmosphere to minute air passageways deep in the lungs. For this transport to occur at an adequate rate through the branching system of airways in the lungs, convection must be employed. Thus, we breathe, moving large masses of air rapidly over long distances (step 1 in Figure 20.4). Oxygen then moves by diffusion over the next 0.0001 m of distance, as it crosses the alveolar end sacs deep in the lungs and the two epithelia (alveolar and capillary) that separate the lung air from the blood (step 2). The next long haul is the movement of O_2 from the lungs to systemic tissues. Transport to a muscle in the forearm, 0.6 m from the lungs, is an example. This transport must be by convection—in this case, the circulation of the blood (step 3). Finally, O_2 must travel 0.00002 m from blood within blood capillaries into mitochondria within tissue cells (step 4). This last step occurs by diffusion.

The Oxygen Cascade

The concept of the **oxygen cascade** is a particularly informative way to summarize the transport of O_2 from the environment to the mitochondria of an animal. In everyday language, a *cascade* is a series of steep waterfalls along a stream. Suppose that a waterwheel (or turbine) is positioned at the base of a cascade, as shown in Figure 20.5*a*. The force available to turn the waterwheel is that available when the water drops from height 4 to height 5. The water starts at height 1 as it enters the cascade, but gradually, as it flows through the cascade, its height drops from 1 to 2, and then from 2 to 3, and so forth, so that the final fall of the water—which provides the actual force to turn the waterwheel—is only from 4 to 5. The oxygen cascade is analogous to such a water cascade. To construct an oxygen cascade, one plots the O_2 partial pressure at successive steps along the route between an animal's environmental source of O_2 and its mitochondria, as shown in Figure 20.5*b*. Because O_2 is not subject to active transport, the partial pressure always drops with each step, and thus the plot resembles the hydraulic model of a cascading mountain stream.

Why is the *partial pressure* plotted in an oxygen cascade? The answer to this question is at the core of understanding O_2 transport in animals. The mitochondria of an animal are where O_2 is ultimately used, and *O_2 enters the mitochondria by diffusion at a rate that depends on the difference in O_2 partial pressure between the blood in systemic capillaries and the mitochondria themselves* (Figure 20.4, step 4). Just as a waterwheel on a stream requires water of a certain height immediately upstream to function, the mitochondria in each cell require a certain partial pressure of O_2 in the capillary blood near the cell surface if they are to receive O_2 fast enough by diffusion to meet their needs. With this in mind, the whole point of the O_2-transport systems in an animal can be considered to be the maintenance of a relatively high O_2 partial pressure at cell surfaces throughout the body. With each step along the oxygen cascade, the O_2 partial pressure drops. The partial pressure at the cell surfaces must nonetheless be kept high enough for O_2 to diffuse to the mitochondria at an adequate rate.

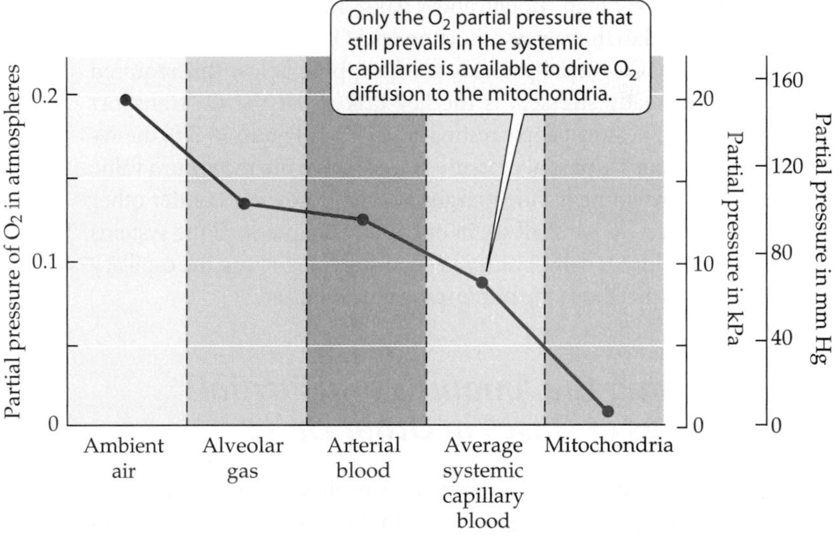

Figure 20.5 The oxygen cascade The concept of the oxygen cascade is based on an analogy with a cascade along a mountain stream. *(a)* In a cascading stream, water loses potential energy each time it falls lower. *(b)* In a person, O_2 partial pressure drops with each step in oxygen transport. The numerical values are for a healthy individual at sea level. The O_2 partial pressures depicted for alveolar gas, arterial blood, and capillary blood are measured averages. The partial pressure depicted for the mitochondria is estimated and (as would be true under the circumstances shown) is higher than the minimum required for unimpaired aerobic catabolism. The scales at the right show the O_2 partial pressures in kilopascals (kPa) and millimeters of mercury (mm Hg).

The particular oxygen cascade shown in Figure 20.5*b* is for healthy, resting human beings at sea level. The first step, the transport of O_2 from the ambient air to the gas in the alveolar end sacs in the lungs, is by convection, as we have seen. When people breathe in their normal way, the drop in O_2 partial pressure during this step is from about 0.2 atm in the ambient air[4] to about 0.13 atm in alveolar gas. The second step of the O_2 cascade, the transport of O_2 across the alveolar sacs and through the epithelia separating the alveolar gas from the blood, occurs by diffusion. Its rate therefore depends on the difference in O_2 partial pressure between the alveolar gas and the blood. In healthy lungs, a partial-pressure difference of about 0.007 atm is sufficient to cause O_2 to diffuse at the rate required. Lungs damaged from smoking or disease often require a larger partial-pressure difference, which might mean that the blood partial pressure is lower than normal. The third step of the cascade is the convective step between arterial blood and capillary blood. Whereas the arterial blood has an O_2 partial pressure near 0.12 atm, the partial pressure in the capillaries averages about 0.09 atm; this capillary partial pressure is dynamically determined by the rate at which blood flow brings O_2 to the capillaries and the rate at which diffusion into tissues removes O_2 from them.[5]

The final step of the oxygen cascade is the "payoff" of O_2 transport: the diffusion-mediated transport of O_2 from the blood in systemic capillaries to the mitochondria in surrounding cells, driven by the difference in O_2 partial pressure between the blood and the mitochondria. The reason the O_2 partial pressure in the mitochondria is lower than that in the capillary blood is that the mitochondria constantly draw down the partial pressure in their vicinity by chemically consuming O_2, converting it to water. For aerobic catabolism to be unimpaired, the *mitochondrial* partial pressure cannot be allowed to fall below about 0.001 atm. The *capillary* O_2 partial pressure must therefore be kept high enough above 0.001 atm for diffusion to occur at a rate equal to the rate of mitochondrial O_2 consumption. Keeping the capillary O_2 partial pressure from slipping below this required value, as already stressed, is the key task of all the O_2-transport processes. In healthy people resting at sea level (Figure 20.5*b*), the average capillary O_2 partial pressure is well above the minimum value required, providing a wide margin of safety. However, under other circumstances, as we shall see in upcoming chapters, all the systems responsible for O_2 delivery may need to struggle to keep the capillary partial pressure above the minimum value required.

Expressing the Amounts and Partial Pressures of Gases in Other Units

Up to now, to keep units of measure simple while we have developed the sometimes complicated principles of gas properties and gas transport, we have always expressed the amounts of gases in moles and the partial pressures in atmospheres. Most of the alternative units in use are very straightforward. Amounts, for example, are sometimes expressed in grams, and for any given gas there is simply a proportional relation between moles and grams (1 mole weighs 1 gram molecular weight). Similarly, all the units of pressure in use are simply proportional to each other; thus, understanding the units of pressure is simple, although the choice of which units to use can seem complicated because a particularly large array of pressure units exists.[6]

What can be confusing is that amounts of gases are often expressed as volumes of gases. The amount of a gas in moles is simply proportional to the volume of the gas, provided the volume is expressed at **standard conditions of temperature and pressure (STP)**, meaning a temperature of 0°C and a pressure of 1 atm (101 kPa; 760 mm Hg). In fact, this relation is so simple that it does not even vary between gases; for essentially all gases, 1 mole occupies 22.4 L at STP. The reason that volumetric expressions can get confusing is that the volume occupied by a given molar amount of gas depends on temperature and pressure (see Equation 20.1). Therefore, if a gas actually exists at a temperature and pressure different from 0°C and 1 atm, it has two volumes of potential interest. One is the volume the gas actually occupies under the conditions where it exists. The other is its volume at STP. Suppose, for example, that we are studying a person who is standing in a place where the barometric pressure is 740 mm Hg and we are concerned about 0.0446 mole of O_2 in his or her lungs. In the lungs, where this gas is at a temperature of 37°C and a pressure of 740 mm Hg, the gas occupies a volume of 1170 mL. However, this same quantity of gas occupies only 1000 mL at STP. Appendix D (Gases at Standard Temperature and Pressure) provides more detail on these calculations. When gas volume is used as a way of expressing the absolute amount of gas (the molar quantity), the volume at STP is the volume used, because it is this volume that bears a one-to-one relation to the number of moles.

Another aspect of volumes that can be confusing arises when the amount of gas dissolved in an aqueous solution is expressed as a volume of gas at STP. Suppose, to illustrate, that a liter of a solution has dissolved in it an amount of O_2 that, in gaseous form, would occupy 2 mL at STP. Biologists then often say that the concentration of O_2 in the solution is "2 mL O_2/L." Except for the units used, this expression is no different from saying that there is a certain number of moles dissolved per liter. One just has to keep in mind that when the gas is in solution, it does not occupy any volume as a gas; it occupies the stated volume only if it is removed from solution.

[4]We say "about 0.2 atm" here because in the real world, air has a variable content of water vapor (see Chapter 25). In the open atmosphere, the gas *other than water vapor* (the "dry air") is always 20.95 % O_2. However, this gas is diluted when water vapor is present, lowering the overall O_2 percentage and O_2 partial pressure.

[5]This step and the next are actually more complex than portrayed here, as discussed in Chapter 23, because O_2 diffuses out of capillaries all along their lengths. The "capillary partial pressure" is thus not a single number, but drops from a high value at the arterial ends of capillaries to a lower value at the venous ends.

[6]The units of measure for pressure can present vexing problems at present because several disparate systems of units exist and different branches of physiology have progressed to different degrees in adopting SI units. Moreover, in the United States (although not elsewhere), there has been great resistance in everyday life to abandoning old-fashioned units of measure such as millimeters of mercury and pounds per square inch (psi). The basic SI unit of pressure is the pascal (Pa)—equal to a newton per square meter (N/m^2). This is such a tiny pressure that nearly always in physiology, the kilopascal (kPa) is the SI unit used. The millimeter of mercury (mm Hg), although now being left behind, has seen extensive use in physiology for many decades; it is the pressure exerted by a column of mercury 1 mm high under standard gravitational acceleration. Another old unit used often in physiology for many decades is the torr, which is essentially identical to a millimeter of mercury. 1 kPa ≅ 7.5 mm Hg. Also, 1 kPa ≅ 0.01 atm. An atmosphere (atm) is 760 mm Hg or about 101 kPa.

The Contrasting Physical Properties of Air and Water

Air and other gas phases differ dramatically from water and aqueous solutions in many physical properties that are of critical importance in the physiology of the respiratory gases. We have already stressed that gases diffuse much more readily through air than through water. Other properties that differ dramatically between air and water are density, viscosity, and the maximal concentrations of gases found in natural environments.

Water is much more dense and viscous than air. At 20°C, for example, the density of water (about 1 g/mL) is over 800 times higher than that of sea-level air (about 0.0012 g/mL). The viscosity of water is 35 times higher than that of air at 40°C and over 100 times higher at 0°C. The greater density and viscosity of water dictate that water-breathing animals must generally expend more energy to move a given volume of fluid through their respiratory passages than air-breathing ones.

Dramatic differences typically exist between air and water in natural environments in the amount of O_2 they contain per unit of volume (Table 20.1). Ordinary atmospheric air is about 21% O_2, as we have seen. This means that if air is at sea level and at 0°C, it contains 210 mL of O_2, measured at STP, per liter. Warming the air lowers its absolute concentration of O_2 a little because gases expand and become more rarefied when they are heated, but even at 24°C, atmospheric air at sea level contains 192 mL of O_2 (measured at STP) per liter.[7] In sharp contrast, the amount of O_2 per liter in water depends on the solubility of O_2 in water, and O_2 is not particularly soluble in water. If freshwater at 0°C is equilibrated with atmospheric air at sea level so that the partial pressure of O_2 in the water is identical to that in the atmosphere (0.21 atm), the water dissolves 10.2 mL of O_2 per liter. Seawater, because of its salinity, dissolves less O_2 under the same conditions. Both freshwater and seawater dissolve less O_2 as they are warmed because the solubility of O_2 decreases as temperature increases. Overall, as seen in Table 20.1, the maximum O_2 concentration likely to occur in bodies of water is 5% or less of the concentration in air at sea level. One way to appreciate the quantitative significance of these numbers is to consider a hypothetical terrestrial or aquatic animal trying to obtain a liter of O_2 by completely extracting the O_2 from a volume of its environmental medium. At 0°C, this animal would need to process 4.8 L of air, 98 L of freshwater, and 125 L of seawater!

Water actually presents animals with a *combination* of properties that together make it a far more difficult place to acquire O_2 than air. First, each liter of water is typically more costly to pump during breathing than air because of the relatively high density and viscosity of water. Second, each liter of water has a lower O_2 reward to provide (see Table 20.1). Together, these considerations mean that a water-breathing animal must often work much harder than an air-breathing one to obtain a given quantity of O_2, which means that in the water breather, a greater fraction of the O_2 taken up must be dedicated to obtaining more O_2. Although in resting people, the cost of ventilating the lungs with air is 1% to 2% of the total metabolic rate, in resting fish, the cost of ventilating the gills with water is probably near 10% of the total metabolic rate.[8] The dramatic differences in the physical properties of air and water also undoubtedly help explain why the highest absolute metabolic rates in the animal kingdom are found in air breathers: insects, mammals, and birds. A high absolute rate of O_2 uptake is far more feasible in air than in water.

TABLE 20.1 The usual maximum concentration of O_2 in air, freshwater, and seawater at three temperatures The concentrations listed are for air at sea level and fully aerated water equilibrated with such air; in other words, the O_2 partial pressure is 0.21 atm in all cases. For the most part, actual O_2 concentrations in natural environments are either as high as shown or lower (because of O_2 depletion by organisms).

	Concentration of O_2 (mL O_2 at STP/L) at specified temperature		
	0°C	12°C	24°C
Air	210	200	192
Freshwater	10.2	7.7	6.2
Seawater[a]	8.0	6.1	4.9

[a] The values given are for full-strength seawater having a salinity of 36 g/kg.

Respiratory Environments

When we consider the actual respiratory environments of animals, biotic processes are as important as physical processes and physical properties in determining the concentrations and partial pressures of O_2 and CO_2 that prevail. The processes at work in a portion of a terrestrial or aquatic environment are diagrammed in Figure 20.6. The animals and plants living there exert strong influences on the local concentrations and partial pressures of O_2 and CO_2. During the day, with adequate sunlight, photosynthetic organisms add O_2 to the air or water and extract CO_2. Animals, bacteria, and fungi remove O_2 and add CO_2, and at night the photosynthetic organisms do the same. Any portion of the environment, as the figure shows, exchanges O_2 and CO_2 with neighboring regions by way of diffusion and convection (breezes or water currents). Diffusion always tends to equalize partial pressures across boundaries, and convection usually does (by physical mixing). Thus, in any portion of the environment, the resident organisms may collectively raise or lower the O_2 partial pressure relative to that in neighboring regions, while simultaneously, diffusion and convection tend to even out the partial pressure from place to place. If the rate of exchange among subparts of an environment by diffusion and convection is high, very little difference in partial pressure may develop from one region to another. On the other hand, if the rate of exchange is low, the O_2

[7] Air at 24°C contains 210 mL of O_2 per liter when the volume is measured *at 24°C*. However, the air is less dense than air at 0°C and thus contains less mass of O_2 per liter than air at 0°C. This effect of temperature is removed by correcting to STP. The O_2 that occupies 210 mL at 24°C will occupy only 192 mL at STP.

[8] Measurement of the percentage for fish has proved challenging, and a wide range of values have been reported in the experimental literature. The value of 10% is often cited as a "best guess" of the average.

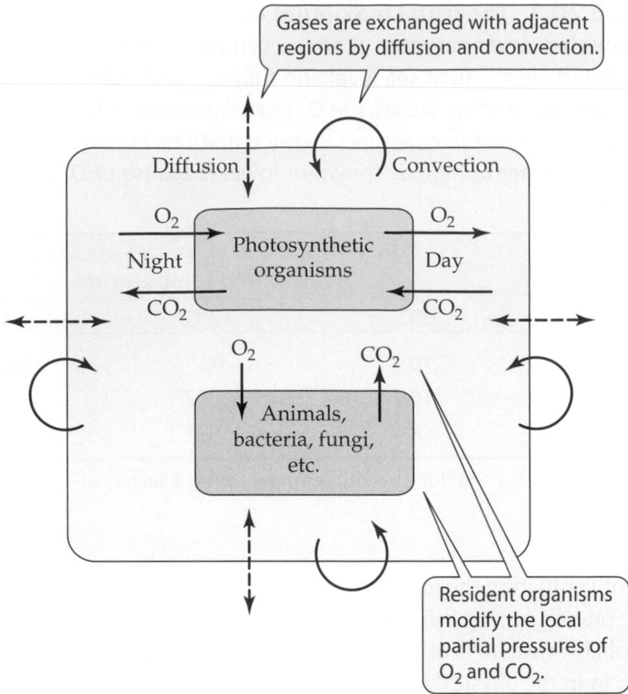

Figure 20.6 The processes that affect the partial pressures of O_2 and CO_2 within a portion of a terrestrial or aquatic environment The box symbolizes a portion of the natural world. The organisms living within it tend to modify the local partial pressures of O_2 and CO_2. Diffusion and convection, however, tend to make the local partial pressures similar to those in neighboring regions.

partial pressure in a particular region may rise well above or fall well below that in neighboring ones.

In open environments on land, because breezes are ubiquitous and diffusion occurs relatively rapidly through air, the fractional composition of dry air is virtually uniform from place to place, both at sea level and over the altitudinal range occupied by animals. Oxygen, for example, represents 20.95% of the volume of dry air just about everywhere. When the concentrations and partial pressures of gases vary from place to place in open-air environments, they do so principally because of altitude (see Figure 1.12). In human settlements at 4500 m (14,800 ft) in the Peruvian Andes, for example, where dry air consists of 21% percent O_2 just as it does at sea level, the total atmospheric pressure is only about 0.6 atm, and—because of the rarefaction of the air—the concentration and the partial pressure of O_2 are only about 60% as high as at sea level. In secluded terrestrial environments, such as underground burrows or tunnels in wood, the metabolic activities of resident organisms often draw down the local concentration and partial pressure of O_2 and increase the level of CO_2; these local changes are possible in secluded places because of restricted convective exchange with the open atmosphere.

Regional differences in the concentrations and partial pressures of O_2 and CO_2 are far more common in aquatic environments than in terrestrial ones because in water the processes that act to even out gas levels are weaker than in air. With the evening-out processes less effective, local biotic processes are readily able to modify local gas levels. Thus, even in a wide-open body of water like a lake, the concentrations and partial pressures of gases often vary considerably from place to place (see Figure 1.13). In secluded aquatic environments populated by organisms having substantial O_2 demands, it is not at all uncommon to find O_2 partial pressures approaching zero.

Study Questions

1. Explain in your own words why the oxygen cascade is presented as a cascade of partial pressures.
2. Plot the rate of O_2 consumption of a fish as a function of water temperature (see Figure 8.8 if you need help). Then plot the O_2 concentration of fully aerated water as a function of water temperature. Why do physiologists sometimes comment that high water temperatures can create a "respiratory trap" for fish, catching them between two conflicting trends?
3. A gas mixture contains 0.2 mol of N_2, 0.4 mol of O_2, and 0.1 mol of CO_2. If the mixture is at STP, what is the partial pressure of each gas, and what is the volume occupied by the mixture?
4. The absorption coefficient of O_2 in seawater at 0°C is 1.7 mmol/L. What is the concentration of O_2 if the O_2 partial pressure is 0.1 atm? Express your answer first in mmol/L, then in mL/L.
5. A team of investigators is out on a boat on a lake on a marvelous, sunny summer day, and they are taking water samples from various depths in the lake so as to construct a plot of dissolved O_2 concentration versus depth, similar to the plot in Figure 1.13. When the investigators bring up a water sample from a depth of 20 m, why is it imperative that they measure its dissolved O_2 concentration immediately, rather than letting it sit on the deck for 5 minutes?
6. In the sediments at the bottoms of ponds and lakes, there is often absolutely no O_2 at depths of 0.5 cm or more, even if the water above is rich in O_2. How is this possible?
7. Archimedes' principle states that when an object is immersed in a fluid, it is buoyed up by a force equal to the weight of the fluid displaced. Assume (slightly incorrectly) that tissue has the same density as water. Using Archimedes' principle, explain why gill filaments typically flop into a drooping mass in air while floating near neutral buoyancy in water. How do these differences between air and water affect the ability of gills to function in gas exchange?
8. For a biologist, what are the three or four most important things to know about gas solubility?
9. Compare and contrast diffusion and convection. In what way do they "alternate" in the O_2 transport system of a mammal?
10. The hemoglobin in mammalian blood is usually thought of simply as increasing the amount of O_2 that can be carried by each liter of blood. In a lecture on hemoglobin, a respiratory physiologist made the following statement: "The presence of hemoglobin in the blood also makes possible the rapid uptake of O_2 by the blood as it flows through the lungs." Explain the lecturer's point.

Selected References

Boutilier, R. G. 1990. Respiratory gas tensions in the environment. *Advances in Comparative and Environmental Physiology*, vol. 6, pp. 1–13. Springer, New York.

Clever, H. L., and R. Battino. 1975. The solubility of gases in liquids. In M. R. J. Dack (ed.), *Techniques of Chemistry*, vol. 8, part 1, pp. 379–441. Wiley, New York.

Dejours, P. 1988. *Respiration in Water and Air*. Elsevier, New York.
Denny, M. W. 1993. *Air and Water: The Biology and Physics of Life's Media*. Princeton University Press, Princeton, NJ.
Farhi, L. E., and S. M. Tenney (eds.). 1987. *The Respiratory System*, vol. 4, *Gas Exchange*. (Handbook of Physiology [Bethesda, MD], section 3). Oxford University Press, New York [Includes articles on the physics of gases.]
Fenn, W. O., and H. Rahn (eds.). 1964. *Respiration*, vol. 1. Handbook of Physiology (Bethesda, MD), section 3. Oxford University Press, New York [Includes articles on the physics of gases.]
Graham, J. B. 1990. Ecological, evolutionary, and physical factors influencing aquatic animal respiration. *Amer. Zool.* 30: 137–146.
Hlastala, M. P., and A. J. Berger. 1996. *Physiology of Respiration*. Oxford University Press, New York.
Maclean, G. S. 1981. Factors influencing the composition of respiratory gases in mammal burrows. *Comp. Biochem. Physiol. A* 69: 373–380.
Vogel, S. 1988. How organisms use flow-induced pressures. *Amer. Sci.* 76: 28–34.
Vogel, S. 1994. *Life in Moving Fluids: The Physical Biology of Flow*, 2nd ed. Princeton University Press, Princeton, NJ.
Weibel, E. R. 1984. *The Pathway for Oxygen*. Harvard University Press, Cambridge, MA.
Weiss, T. F. 1996. *Cellular Biophysics*, vol. 1. MIT Press, Cambridge, MA.

21 External Respiration: The Physiology of Breathing

Tunas represent one of the pinnacles of water breathing. If they lived on land, where they could be readily observed, they would undoubtedly be classed metaphorically with wolves, African hunting dogs, and other strong, mobile predators. Judging by the length of time spent in motion, tunas are actually more mobile than any terrestrial predator. Using their red swimming muscles, they swim *continuously*, day and night, at speeds of one to two body lengths per second; some species cover over 100 km per day during migrations.[1] Tunas thus rank with the elite endurance athletes among fish, the others being salmon, mackerel, billfish, and certain sharks. To meet the O_2 demands of their vigorous lifestyle, tunas require a respiratory system that can take up O_2 rapidly from the sea and a circulatory system that can deliver O_2 rapidly from the gills to tissues throughout the body. They are, in John Magnuson's memorable phrase, "astounding bundles of adaptations for efficient and rapid swimming."

Tunas breathe with gills, as do most fish. Their gills are hardly average, however; instead, they are exceptionally specialized for O_2 uptake, illustrating a general principle in the study of respiration: that a single type of breathing system may exhibit a wide range of evolutionary refinements in various species. For comparative purposes, the gills of rainbow trout can reasonably be considered to exemplify average fish gills. The gills of yellowfin tuna and skipjack tuna have about eight times more surface area than the gills of a rainbow trout of equal body size; if you were to mark off on a floor a square that is 1.3 meters on each side, the area within the square would correspond to the surface area of the gills in a 1-kg (2.2-lb) tuna. The gill membranes of tunas are also exceptionally thin; whereas the average distance between blood and water is 6 μm in a rainbow trout's gills, it is 0.6 μm in the gills of skipjack and yellowfin tuna. Compared to an

[1]Burst speeds, which are powered relatively anaerobically by the white swimming muscles, are twelve to fifteen body lengths per second.

Endurance athletes like tunas must be able to acquire oxygen rapidly Although water is not a particularly rich source of O_2, tunas have gills and breathing processes that enable them to live as highly active predators. These are longtail tunas (*Thunnus*).

average fish, therefore, tunas have evolved gills that present an extraordinarily large surface of extraordinarily thin membrane to the water for gas exchange.

Most fish, including rainbow trout, drive water across their gills by a pumping cycle that is powered by their buccal and opercular muscles (discussed later in this chapter). Some species are adept at alternating between this mechanism of ventilating their gills and another mechanism termed **ram ventilation**. During ram ventilation, a fish simply holds its mouth open while it swims powerfully forward, thereby "ramming" water into its buccal cavity and across its gills; in this way, the swimming muscles take over responsibility for powering the flow of water across the gills. During their evolution, tunas completely abandoned the buccal–opercular pumping mechanism and became *obligate ram ventilators*, a distinction they share with just a few other sorts of fish. As obligate ram ventilators, they have no choice regarding how much of their time they spend swimming. They must swim continuously forward, or they suffocate! Physiologists debate whether ram ventilation is intrinsically superior as a way of moving water across the gills. Less debatable is the fact that tunas achieve extraordinarily high rates of water flow using ram ventilation. During routine cruising, a 1-kg skipjack or yellowfin tuna drives about 3.6 L of water across its gills per minute; this is seven times the resting flow rate in a 1-kg rainbow trout, and twice the maximum rate of the trout.

Fundamental Concepts of External Respiration

In all animals, the systems used to exchange respiratory gases with the environment can be diagrammed as in Figure 21.1. Always, a thin layer of tissue, consisting typically of one or two simple epithelia and termed the **gas-exchange membrane** or **respiratory exchange membrane**,[2] separates the internal tissues of the animal from the environmental medium (air or water). **External respiration**, or **breathing**, consists of the processes by which O_2 is transported to the gas-exchange membrane from the environmental medium and by which CO_2 is transported away from the membrane into the environment. Often (but not always) during external respiration, air or water flows convectively between the environment and the gas-exchange membrane. In such cases, the mass flow of air or water to and from the gas-exchange membrane is termed **ventilation**.

Oxygen always crosses the gas-exchange membrane by diffusion, as stressed in Chapter 20. This means that for O_2 to enter an animal from the environment, it is *essential* that the partial pressure of O_2 on the inside of the gas-exchange membrane be lower than that on the outside. The fact that O_2 enters animals by diffusion explains why the area and thickness of the gas-exchange membrane play critical roles in O_2 acquisition. The rate of diffusion across a membrane increases in proportion to the area of the membrane. Furthermore, according to the fundamental diffusion equation (see Equation 20.4), the rate of diffusion increases as the thickness of a membrane

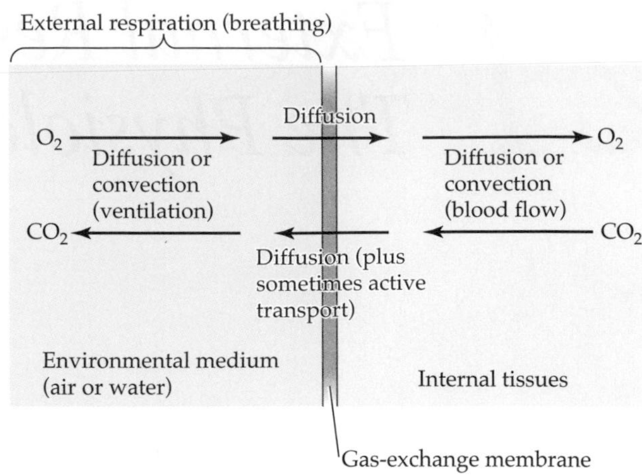

Figure 21.1 Generalized features of animal gas exchange Oxygen and carbon dioxide move between the environmental medium and the internal tissues of an animal across a gas-exchange membrane.

decreases. These physical laws explain why an expansive, thin gas-exchange membrane is a great asset for animals like tunas that must acquire O_2 at high rates. As for CO_2, diffusion is the exclusive mechanism by which it crosses the gas-exchange membrane in some animals (e.g., humans); in others, however, although diffusion is the principal CO_2 transport mechanism, active transport also occurs. Active outward transport of CO_2 (as HCO_3^-) is best documented in freshwater animals (see Figure 3.14).

Although many animals have discrete, easily identified "breathing organs," this is not always the case. The lungs of mammals and the gills of tunas stand out as "the breathing organs" of these animals because the rest of their body surface (their skin) is virtually impermeable to gases. In most adult frogs, however, the skin is rather permeable to gases, and the lungs therefore are not the sole breathing organs. Usually, just a single part of an animal is identified *by name* as a breathing organ. This naming convention, it will be clear, does not necessarily mean that that organ is the only place where significant external respiration occurs. In frogs, although only the lungs are named for their breathing function, the skin also has a breathing role. Similarly, some fish breathe with their stomachs as well as their gills, and sea stars breathe with their tube feet as well as their branchial papulae ("gills").

Body structures that are specialized for external respiration typically have gas-exchange membranes that are thrown into extensive patterns of invagination or evagination, which greatly increase the membrane surface area. For physiologists, **gills** and **lungs** are generic labels that refer to two such patterns (Figure 21.2). Gills are respiratory structures that are evaginated from the body and surrounded by the environmental medium. Lungs, in contrast, are respiratory structures that are invaginated into the body and contain the environmental medium. The adjective **branchial** refers to structures or processes associated with gills, whereas **pulmonary** refers to those associated with lungs. Although exceptions occur, water breathing is usually by gills, and air breathing is usually by lungs. The comparative method strongly indicates that lungs are adaptive for terrestrial

[2]"Membrane" is used here in an entirely different way than when speaking of a cell membrane or intracellular membrane. The gas-exchange membrane is a tissue formed by one or more layers of cells.

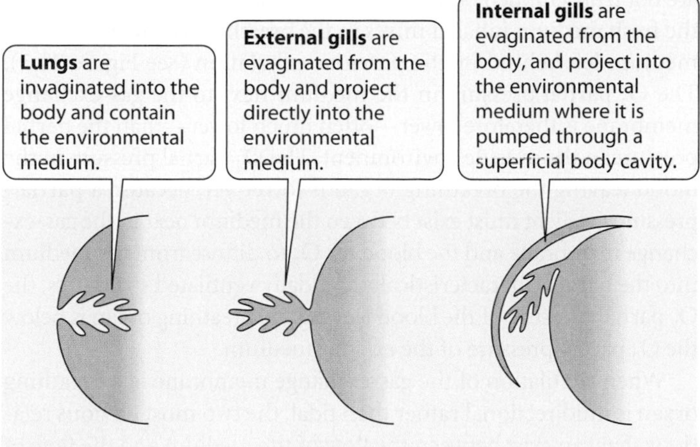

Figure 21.2 Three types of respiratory structures

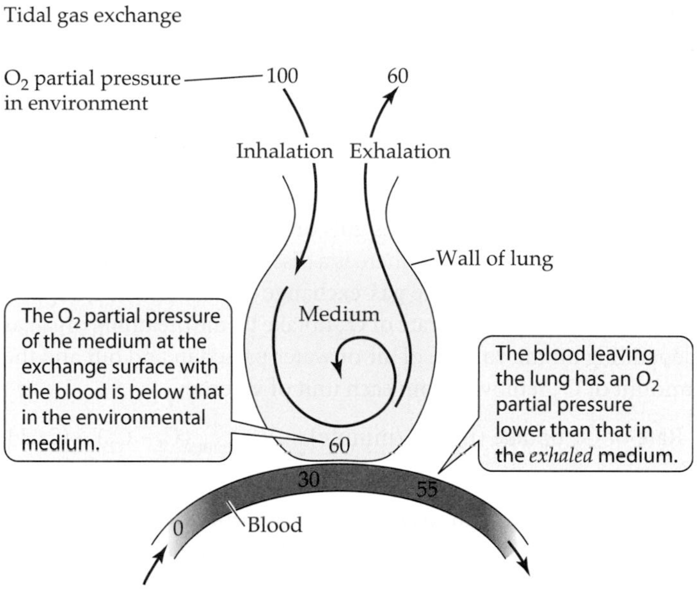

Figure 21.3 Oxygen transfer from the environmental medium to the blood in a tidally ventilated lung Numbers are O_2 partial pressures in arbitrary units; the environmental O_2 partial pressure is assumed to be 100. Because a tidally ventilated lung is never fully emptied, fresh medium mixes in the lung with "stale" medium and the partial pressure of the medium at the exchange surface with the blood is below ambient.

life (see Figure 1.17). One advantage of lungs on land is that their finely divided elements receive structural support by being embedded in the body. In the unusual cases when terrestrial animals have gills, the fine, evaginated processes of a gill on land must ordinarily be provided with skeletal support to prevent them from drooping together into clumps, and this need for skeletal support can easily be at odds with keeping the gas-exchange membrane thin.

Gills can be **external gills** or **internal gills** (see Figure 21.2). External gills are located on an exposed body surface and project directly into the surrounding environmental medium. Internal gills are enclosed within a superficial body cavity. Whereas external placement permits ambient water currents to flow over the gills, internal placement usually requires an animal to use metabolic energy to ventilate them. Internal placement has its advantages, nonetheless. When the gills are internal, the enclosing structures physically protect them and may canalize the flow of water across the gills in ways that enhance the efficiency or control of breathing.

Ventilation of lungs, gills, or other gas-exchange membranes may be **active** or **passive**. Ventilation is active if the animal creates the ventilatory currents of air or water that flow to and from the gas-exchange membrane, using forces of suction or positive pressure that it generates by use of metabolic energy (as by contracting muscles or beating cilia). On the other hand, ventilation is passive if *environmental* air or water currents directly or indirectly induce flow to and from the gas-exchange membrane (see Box 20.2). Active ventilation, although it uses an animal's energy resources, is potentially more reliable, controllable, and vigorous than passive ventilation. Active ventilation may be **unidirectional**, **tidal** (**bidirectional**), or **nondirectional**. It is unidirectional if air or water is pumped over the gas-exchange membrane in a one-way path. It is tidal if, as in mammals, air or water is brought to the gas-exchange membrane and carried away via the same path (Figure 21.3). Ventilation is nondirectional if air or water flows across the gas-exchange membrane in many directions; animals with external gills that wave the gills back and forth in the water exemplify nondirectional ventilation.

In air-breathing animals with lungs, the lungs are usually ventilated. Some lungs, however, exchange gases with the environment entirely by diffusion and are thus **diffusion lungs**. Some insects and spiders, for example, are believed to breathe with diffusion lungs. Within a diffusion lung, the air is still, and O_2 and CO_2 travel the full length of the lung passages by diffusion. Lungs in aquatic animals (which are rare) must be ventilated because diffusion through water is so slow (see page 531).

A **dual breather** or **bimodal breather** is an animal that can breathe from either air or water. Examples include certain amphibians and air-breathing fish. Dual breathers often have at least two distinct respiratory structures (e.g., both lungs and gills), which they employ when breathing from the two media.

Summary: Fundamental Concepts of External Respiration

- Oxygen always crosses the gas-exchange membrane by diffusion. This means that O_2 enters an animal only if the O_2 partial pressure on the outside of the gas-exchange membrane is higher than that on the inside.
- Breathing organs are categorized as gills if they are evaginated structures that project into the environmental medium. They are lungs if they are invaginated structures that contain the medium.
- Ventilation is the forced flow (convection) of the environmental medium to and from the gas-exchange membrane. It is categorized as active if an animal generates the forces for flow using metabolic energy. Ventilation may be unidirectional, tidal, or nondirectional.

Principles of Gas Exchange by Active Ventilation

Active ventilation is very common, and its analysis involves several specialized concepts that apply to a variety of animals. Thus there is good reason to discuss the principles of ventilation before we turn to the particulars of breathing in various animal groups.

In an animal that ventilates gills or lungs directionally—either unidirectionally or tidally—there is a discrete current of air or water that flows to and from the gas-exchange membrane. It is logical, therefore, to consider the rate of O_2 uptake by the breathing organ as depending on the volume of air or water passed in and out and the amount of O_2 removed from each unit of volume:

$$\text{Rate of } O_2 \text{ uptake (mL } O_2\text{/minute)} = V_{\text{medium}}\,(C_I - C_E) \quad (21.1)$$

where V_{medium} is the rate of flow of the air or water through the breathing organ (L/minute), C_I is the O_2 concentration of the inhaled (inspired) medium (mL O_2/L medium), and C_E is the O_2 concentration of the exhaled (expired) medium. The difference (C_I – C_E) represents the amount of O_2 removed from each unit of volume of the ventilated medium. The *percentage* of the O_2 available in the inhaled medium that is removed is $100(C_I - C_E)/C_I$. This ratio, known as the **oxygen utilization coefficient** or **oxygen extraction coefficient**, expresses how well an animal is able to use the O_2 in the air or water it pumps through its lungs or gills.

To illustrate these calculations, if the water entering the mouth of a fish contains 6 mL O_2/L, that exiting from the gills contains 4 mL O_2/L, and the rate of ventilation is 0.5 L/minute, then the fish's rate of O_2 uptake is 0.5 L/minute × 2 mL O_2/L = 1 mL O_2/minute. Moreover, its oxygen utilization coefficient is 33%. In words, the utilization coefficient says that the fish is removing 33% of the O_2 from each volume of water it pumps and is allowing the other 67% to flow out with the exhaled water. Tunas are especially efficient in using the O_2 in the water they drive over their gills. Whereas rainbow trout use 33%, yellowfin and skipjack tunas use 50% to 60%.

The O_2 partial pressure in blood leaving a breathing organ depends on the relation between the flow of the blood and the flow of the air or water

The O_2 partial pressure in the blood leaving a breathing organ can in many ways be considered the best single measure of the breathing organ's effectiveness. This is clear from the oxygen cascade concept (see Figure 20.5): An animal with an especially high O_2 partial pressure in the blood leaving its breathing organ is particularly well poised to maintain a high O_2 partial pressure in its mitochondria. Breathing organs with different designs exhibit inherent differences in the blood O_2 partial pressure they can maintain, depending on the relation between the flow of blood and the flow of the air or water. These differences are not absolute, because in all designs, the blood O_2 partial pressure is affected by additional factors besides the flow relations. Nonetheless, the flow relations between the blood and the air or water have considerable importance.

Tidally ventilated breathing organs, such as the lungs of mammals, are distinguished by the fact that the medium (air) next to the gas-exchange membrane is never fully fresh. Such breathing organs are not truly emptied between breaths. Thus, during each inhalation, the fresh medium inhaled mixes in the breathing organ with "stale" medium left behind by the previous exhalation (see Figure 21.3). The O_2 partial pressure in the medium next to the gas-exchange membrane is therefore lower—often much lower—than the partial pressure in the outside environment. The O_2 partial pressure in the blood leaving the breathing organ is lower yet, because a partial-pressure gradient must exist between the medium next to the gas-exchange membrane and the blood for O_2 to diffuse from the medium into the blood. Characteristically, in tidally ventilated structures, the O_2 partial pressure of the blood leaving the breathing organ is below the O_2 partial pressure of the *exhaled* medium.

When ventilation of the gas-exchange membrane in a breathing organ is unidirectional rather than tidal, the two most obvious relations that can exist between the flow of the medium and the flow of the blood are *concurrent* and *countercurrent*. In the concurrent arrangement (Figure 21.4*a*), the medium flows along the gas-exchange membrane in the same direction as the blood, resulting in **concurrent gas exchange.** In the countercurrent arrangement (Figure 21.4*b*), the medium and blood flow in opposite directions, and **countercurrent gas exchange** occurs.

In an organ that exhibits concurrent gas exchange, when O_2-depleted afferent[3] blood first reaches the gas-exchange membrane, it meets fresh, incoming medium, as shown at the *left* of Figure 21.4*a*. Then, as the blood and medium flow along the exchange membrane in the same direction, they gradually approach equilibrium with each other at an O_2 partial pressure that is intermediate between their respective starting partial pressures. When the blood reaches the place where it leaves the exchange membrane, its final exchange of O_2 is with medium that has a partial pressure considerably below that of the environmental medium. Concurrent gas exchange thus resembles tidal exchange in that the O_2 partial pressure of blood leaving the breathing organ cannot ordinarily rise above the partial pressure of *exhaled* medium.

In an organ that exhibits countercurrent gas exchange, when O_2-depleted afferent blood first reaches the gas-exchange membrane, it initially meets medium that has already been substantially deoxygenated, as shown at the *right* of Figure 21.4*b*. However, as the blood flows along the exchange surface in the direction opposite to the flow of medium, it steadily encounters medium of higher and higher O_2 partial pressure. Thus, even as the blood picks up O_2 and its partial pressure rises, a partial-pressure gradient favoring further uptake of O_2 is maintained. The final exchange of the blood is with fresh, incoming medium of high O_2 partial pressure. Countercurrent exchange is thus an intrinsically more effective mode of exchange than either tidal or concurrent exchange. One way to see this clearly is to note that the blood O_2 partial pressure created by countercurrent exchange is characteristically *much higher than the partial pressure in exhaled medium*; in principle, the O_2 partial pressure of the blood leaving the breathing organ may even approach the partial pressure of *inhaled* medium. In addition, if you compare Figure 21.4*a* and 21.4*b*, you will see that the O_2 partial pressure of the medium falls more—and that of the blood rises more—in 21.4*b*: Countercurrent exchange permits a more complete transfer of O_2 from

[3]Recall that *afferent* means "flowing toward;" in this case it refers to blood flowing toward the gas-exchange membrane. *Efferent* means "flowing away."

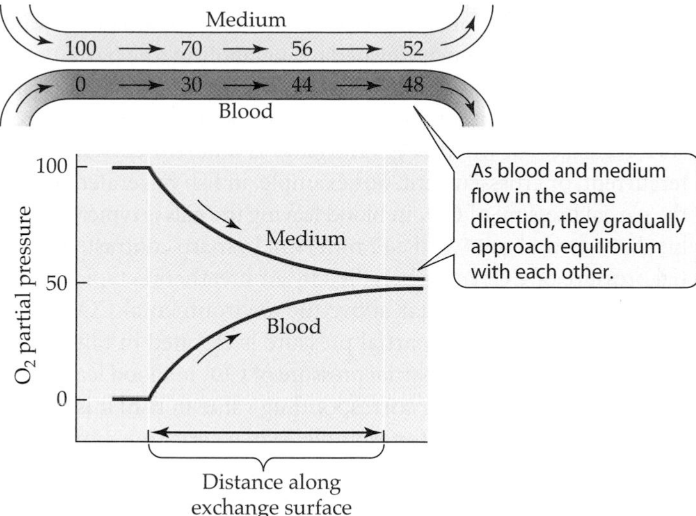

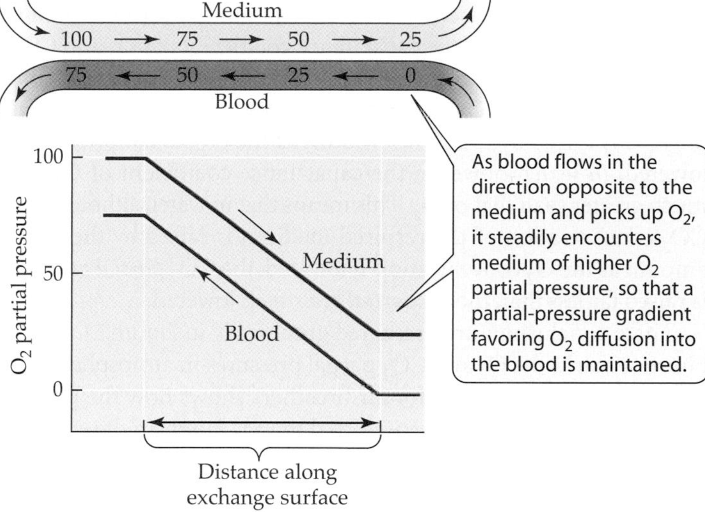

Figure 21.4 Oxygen transfer from the environmental medium to the blood by concurrent and countercurrent gas exchange The upper diagrams in each case depict the flow of medium and blood along the gas-exchange membrane. Numbers are O_2 partial pressures in arbitrary units, with the environmental partial pressure assumed to be 100. The blood reaches a higher partial pressure when the exchange is countercurrent because the blood exchanges with fresh medium just before leaving the gas-exchange membrane.

the medium to the blood than concurrent (or tidal) exchange under comparable conditions.

A third mode of gas exchange that sometimes occurs in unidirectionally ventilated breathing organs is **cross-current gas exchange**. In this mode, the blood flow breaks up into streams, each of which undergoes exchange with the medium along just part of the path followed by the medium. Therefore, some of the blood undergoes gas exchange exclusively with O_2-rich medium, and some exchanges exclusively with O_2-poor medium (Figure 21.5). Consequently, cross-current exchange permits the O_2 partial pressure of the mixed blood leaving the breathing organ to be above that of exhaled medium, but it does not permit as high a blood O_2 partial pressure as countercurrent exchange under comparable circumstances.

The modes of exchange between the blood and the medium can be ranked in terms of their inherent ability to create a high O_2 partial pressure in the blood leaving the breathing organ (corresponding to their inherent efficiency in transferring O_2 from the medium to the blood): *Countercurrent exchange is superior to cross-current exchange, and cross-current exchange is superior to concurrent or tidal exchange.* As already noted, however, this ranking, although important, is not absolute because additional factors affect how real breathing organs function in real animals.

The relative changes in the partial pressures of O_2 and CO_2 depend on whether air or water is breathed

During breathing, the extent to which the CO_2 partial pressure of the medium changes relative to the change in O_2 partial pressure differs dramatically between air breathers and water breathers. This difference arises not from the physiology of animals, but from the different physical and chemical properties of air and water.

The property of air and water that is critical for understanding relative changes in the partial pressures of O_2 and CO_2 is the **capacitance coefficient** (β), defined to be the change in total gas *concentration* per unit of change in gas *partial pressure*. In air of a given temperature, the capacitance coefficient is the same for O_2 and CO_2 because both gases adhere identically to the universal gas law (see Equation 20.1). In water of a given temperature, however, the capacitance coefficient for CO_2 is at least 23 times higher than that for O_2, in part because CO_2 is much more soluble than O_2 (see page 529).

To see the implications of these patterns in the capacitance coefficient, let's assume that in the course of an animal's metabolism, the number of moles of CO_2 produced is equal to the number of moles

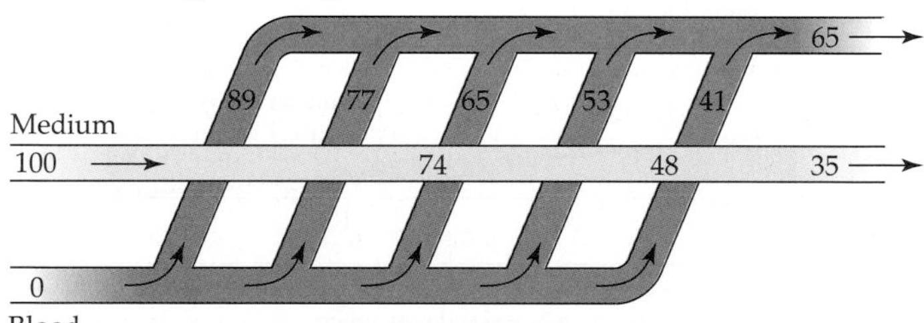

Figure 21.5 Oxygen transfer from the environmental medium to the blood by cross-current gas exchange Numbers are O_2 partial pressures in arbitrary units, with the environmental partial pressure assumed to be 100. The afferent blood vessel breaks up into smaller vessels, each of which makes exchange contact with just a limited part of the structure through which the medium is flowing unidirectionally. The smaller vessels then coalesce to form a single efferent vessel. Cross-current exchange is intermediate between concurrent and countercurrent exchange in its intrinsic gas-transfer efficiency.

of O_2 consumed.[4] Under this assumption, as air or water passes over the gas-exchange membrane of an animal, the *concentration* of CO_2 in the air or water is raised by the same amount as the *concentration* of O_2 is lowered. *In air*, the capacitance coefficients of O_2 and CO_2 are identical. Thus, because the CO_2 concentration is raised by the same amount as the O_2 concentration is lowered, the CO_2 *partial pressure* is raised by the same amount as the O_2 *partial pressure* is lowered. *In water*, however, the capacitance coefficient of CO_2 is much greater than that of O_2. This means that in water, although the CO_2 concentration of the respired medium is raised by the same amount as the O_2 concentration is lowered, the CO_2 *partial pressure* is raised far less than the O_2 *partial pressure* is lowered.

These conclusions are presented graphically in Figure 21.6. The blue dot marks the O_2 and CO_2 partial pressures in atmospheric air or aerated water. The line for air breathers shows how the partial pressures of O_2 and CO_2 are predicted to vary *jointly* in the exhaled air of an air breather. When an air breather lowers the O_2 partial pressure in the respired air to any given extent, it raises the partial pressure of CO_2 in the respired air to about the same extent. The line for water breathers shows how the partial pressures of O_2 and CO_2 are predicted to vary jointly in the exhaled water of a water breather. An animal breathing water *never* raises the partial pressure of CO_2 in the respired water by much, even if it removes *all* the O_2.

[4] As discussed in Chapter 5 (see Table 5.2), the ratio of CO_2 production to O_2 consumption can be 1.0, as we are assuming, but may be as low as 0.7, depending on the types of foodstuffs being oxidized to produce ATP.

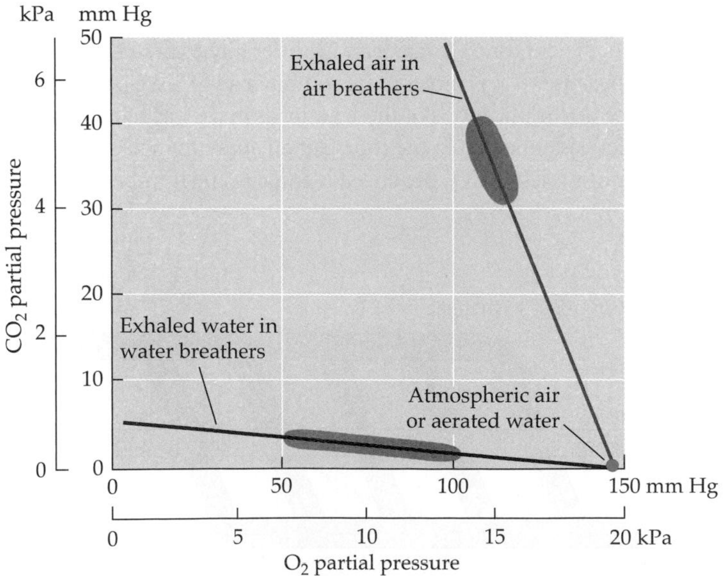

Figure 21.6 Simultaneous values for the partial pressures of CO_2 and O_2 in exhaled air or water for air and water breathers The blue dot shows values for natural fresh air or aerated water. As air or water is subjected to gas exchange, its composition shifts toward the upper left along one line or the other. The orange areas show typical compositions of arterial blood in resting vertebrates. Lines were calculated assuming the capacitance coefficient for CO_2 to be about 30 times that for O_2 in water (and equal to that for O_2 in air) and assuming a respiratory quotient of 0.9. (After Scheid and Piiper 1997.)

These contrasts have critical implications for the partial pressures of CO_2 in the arterial blood of water breathers and air breathers.[5] Because the partial pressure of CO_2 in water breathed by water breathers *cannot* be much elevated by metabolism, the partial pressure of CO_2 in the blood leaving the breathing organs of water breathers always remains near the environmental CO_2 partial pressure, regardless of whether gas exchange is tidal, concurrent, countercurrent, or cross-current. For example, in fish in aerated waters, the partial pressure of CO_2 in blood leaving the gills is typically just a bit above zero, about 0.3 kPa (2 mm Hg). In sharp contrast, the partial pressure of CO_2 in air breathed by air breathers is typically elevated by metabolism to far above the environmental CO_2 partial pressure, and the blood partial pressure is elevated in tandem. In mammals and birds, the partial pressure of CO_2 in blood leaving the lungs is over 10 times the corresponding value in fish; it is 5.3 kPa (40 mm Hg) in humans, for example! When vertebrates and invertebrates emerged onto land in the course of evolution, one of the key physiological consequences was a dramatic shift upward in the level of CO_2 in their blood and other body fluids.

[5] The orange regions in Figure 21.6 show representative arterial-blood values in water breathers and air breathers.

Summary: Principles of Gas Exchange by Active Ventilation

- The oxygen utilization coefficient is the percentage of the O_2 in inhaled medium that an animal removes before exhaling the medium.
- The four major types of gas exchange that can occur during directional ventilation can be ranked in terms of their inherent ability to establish a high O_2 partial pressure in blood exiting the breathing organ. Countercurrent gas exchange ranks highest. Cross-current gas exchange ranks second; concurrent and tidal gas exchange rank lowest.
- Because of the different capacitance coefficients of air and water, air breathers typically raise the CO_2 partial pressure in air they breathe to roughly the same extent as they lower the O_2 partial pressure. Water breathers, however, do not elevate the CO_2 partial pressure in water they breathe to any great extent. Air breathers consequently tend to have distinctly higher blood CO_2 partial pressures than water breathers.

Introduction to Vertebrate Breathing

The vertebrates living today are usually thought of, in a rough way, as representing an evolutionary sequence. In actuality, of course, today's fish were not the progenitors of today's amphibians, and today's reptiles were not the progenitors of today's mammals and birds. Thus, when we think of the sequence from fish to mammals and birds, caution is always called for in thinking of it as an evolutionary sequence. Comparisons among today's animals nonetheless often provide revealing insights into trends that occurred during evolution. Before we start our study of breathing in the various vertebrate groups, an overview of general trends in the vertebrates will help place the individual vertebrate groups in a larger context.

Among the species within any one group of phylogenetically related vertebrates, the total surface area of the gas-exchange membrane in a species is typically an allometric function of body size. The allometric relation differs, however, from one phylogenetic group to another, as seen in Figure 21.7*a*. Most groups of fish, the amphibians, and the reptiles have roughly similar allometric relations. This means, for example, that the total gill area of a fish of particular body size is roughly similar to the total lung area of an amphibian or reptile of the same size, suggesting that when vertebrates emerged onto land, there was not immediately much of a change in the area of the gas-exchange surface in their breathing organs. Mammals and birds, however, exhibit a dramatic step upward in the area of gas-exchange surface in their lungs. The mammals and the birds are independently evolved; thus, these two groups independently evolved lungs with markedly enhanced gas-exchange surface areas. Every reason exists to think that the rise in gas-exchange surface area in the lungs of mammals and birds was associated with their evolution of homeothermy. As we saw in Chapter 8, homeothermy increases an animal's metabolic rate by a factor of at least five to ten. Thus, lungs with an enhanced ability to take up O_2 and void CO_2 are required by homeothermic animals. Remarkably, the supremely active tunas have gill surface areas that approximate the lung surface areas of equal-sized mammals and birds!

The fact that mammals and birds have exceptionally large gas-exchange surface areas does not mean that they have large lungs compared to reptiles or amphibians. In fact, the opposite is often the case. For example, if a reptile and a rodent of the same body size are compared, the lung volume of the reptile is likely to be at least five times greater than that of the rodent. Yet, as we have seen (Figure 21.7*a*), the surface area of the gas-exchange membrane is likely to be ten times greater in the rodent than in the reptile! The explanation for the high gas-exchange surface area in the lungs of a mammal or bird is that the lungs of these animals are extraordinarily densely filled with branching and rebranching airways. Reptilian and amphibian lungs, in contrast, typically have parts that are simply like balloons: little more than a sheet of tissue surrounding an open central cavity. Even the parts of reptilian and amphibian lungs that are subdivided are much less elaborately subdivided than mammalian or avian lungs. Thus, whereas the lungs are large in a reptile compared to a rodent, they provide the reptile with a comparatively small area of gas-exchange membrane.

The thickness of the barrier between the blood and the environmental medium also shows significant evolutionary trends in the major vertebrate groups. In most fish, the sheet of gill tissue between blood and water is roughly 5 to 10 μm thick. Vertebrate lungs uniformly have a much reduced barrier between blood and air (Figure 21.7*b*). Among the air-breathing vertebrates, mammals tend to have a thinner sheet of tissue between blood and air than reptiles, and birds have a dramatically thinner barrier than either of those groups, roughly 0.2 μm. Tunas, which are so highly specialized for O_2 uptake compared to most fish, have a blood–water barrier in their gills that is similar in thickness to the blood–air barrier in mammals.

The skin varies widely in its role as a gas-exchange site in vertebrates (Figure 21.8). Some fish and most reptiles place little reliance on skin breathing. In contrast, the skin can be responsible for a third or more of gas exchange in other fish and reptiles and in many amphibians; in the lungless salamanders, which not only lack lungs but have an epidermis that is vascularized (highly unusual), 100% of gas

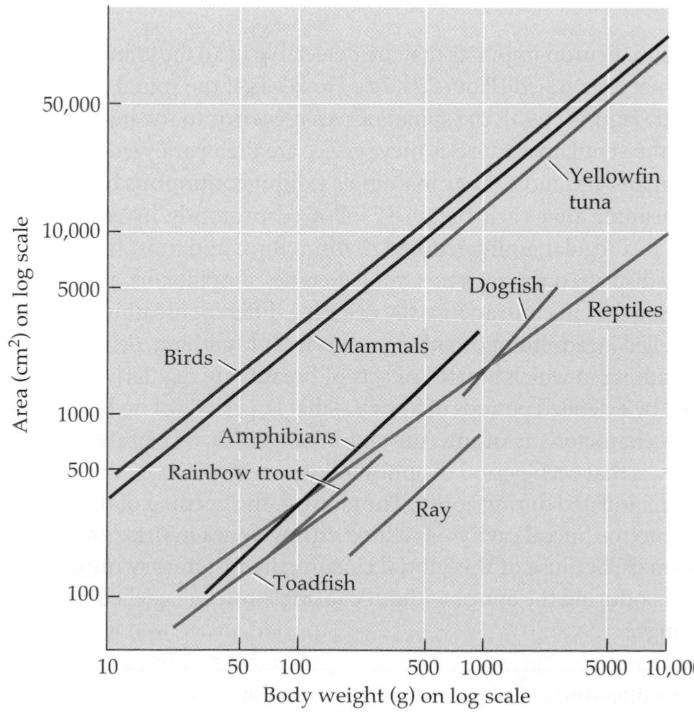

Figure 21.7 **Total area and thickness of the gas-exchange membrane in the gills or lungs of vertebrates as functions of body size** The lines for mammals and birds in both *(a)* and *(b)* and those for amphibians and reptiles in *(a)* are for many species (e.g., about 40 species of mammals, ranging in size from shrews to horses). The lines for fish in *(a)* are for various-sized individuals within single species. The thickness in *(b)* is the average distance between the blood and the water or air. (After Perry 1990.)

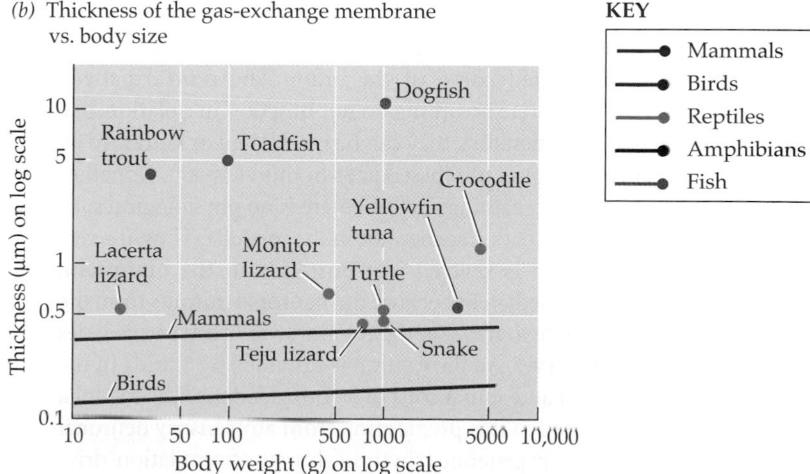

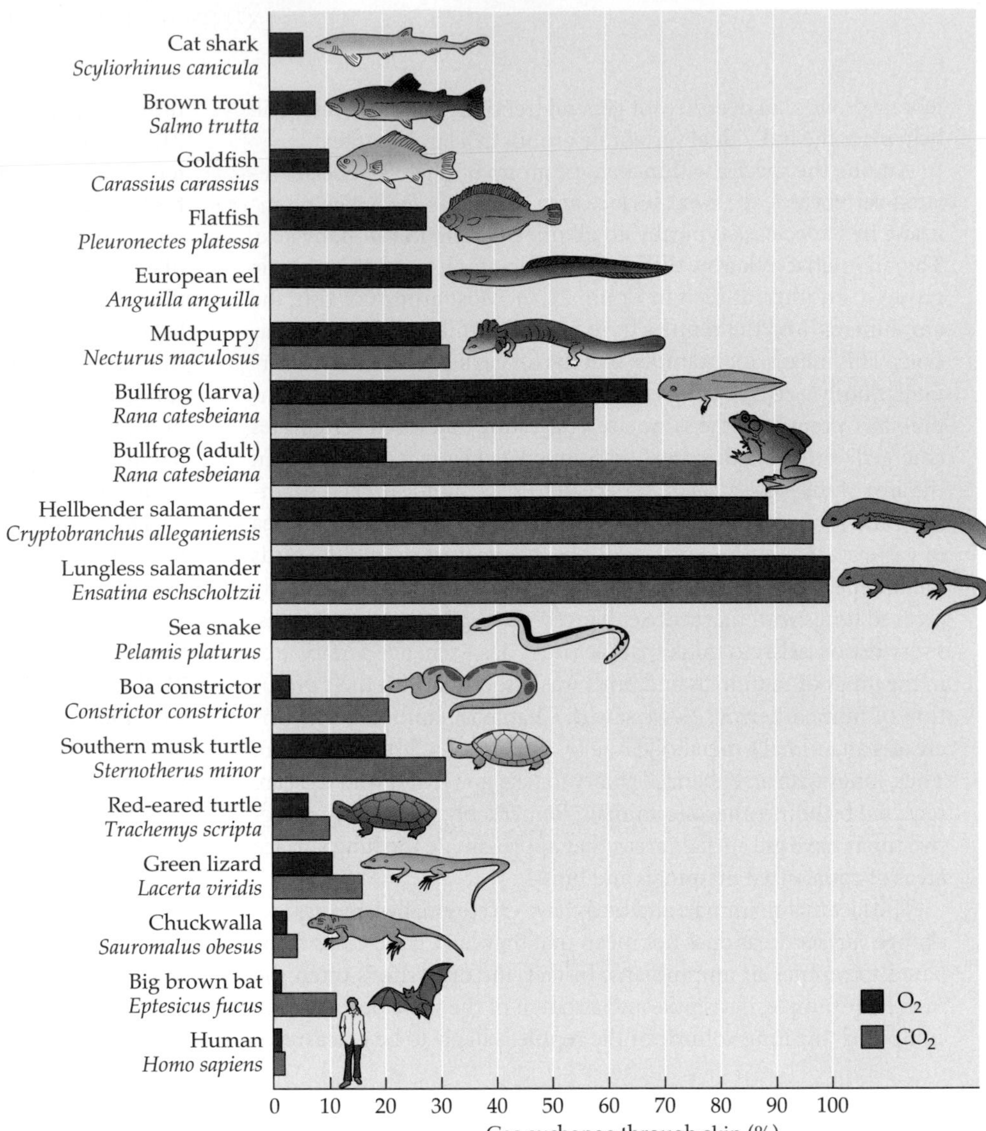

Figure 21.8 The percentage of O_2 and CO_2 exchange that occurs across the skin in vertebrates The exact extent of skin breathing within a species often depends on environmental conditions (e.g., temperature). (After Feder and Burggren 1985.)

exchange occurs across the skin. Most mammals and birds resemble humans in relying almost entirely on the lungs, and very little on the skin, for gas exchange. In terrestrial vertebrates, the role of the skin in breathing is clearly related to the skin's desiccation resistance. Groups such as mammals and birds that are well defended against water loss through their skin tend to make little use of cutaneous breathing, whereas groups such as frogs and salamanders that are poorly defended against cutaneous water loss tend to engage in significant cutaneous breathing. Modifications that render the skin poorly permeable to water also make it poorly permeable to O_2 and CO_2.

The control of the active ventilation of the gills or lungs is the final subject that deserves mention in introducing vertebrate breathing. One important topic in this regard is the control of the rhythmic contractions of muscles that produce breathing movements, such as our own inhalation and exhalation movements or the rhythmic movements of a rainbow trout's mouth and opercular flaps that pump water over its gills. In all vertebrates, the muscles responsible for breathing movements are skeletal muscles. They thus require stimulation by motor neurons for each contraction they undergo. The breathing rhythm originates in the neurons of a central pattern generator (see Chapter 18); these neurons initiate rhythmic outputs of nerve impulses (action potentials) that travel to the breathing muscles and stimulate them into rhythmic patterns of contraction. The weight of evidence suggests that the central pattern generators for breathing in all vertebrates are located within the brainstem: in the medulla, and sometimes other associated parts, of the brain. The central pattern generators for vertebrate breathing do not function in isolation; like all central pattern generators, they can be modulated or entrained by sensory neuronal inputs (discussed later in this chapter). Nonetheless, for creating the breathing rhythm, there is no physiological substitute for the central pattern generator in the medulla. Thus, if a vertebrate's spinal cord is severed just posterior to the brainstem, breathing stops immediately, because the neuronal outputs from the brainstem are unable to travel in the usual way to breathing muscles by way of spinal nerves. An illustration is provided by a study of the isolated brainstem and spinal cord of a young rodent. The medulla of the brain, isolated in this preparation from any sensory neuronal inputs, endogenously generates rhythmic bursts of ventilation-driving motor-neuron impulses that are detectable in all the cranial and spinal nerves studied (Figure 21.9*a*). However, if the spinal cord is severed (Figure 21.9*b*), the spinal nerves posterior to the injury go silent; the ventilatory muscles they service are then paralyzed.

Humans and most other mammals exhibit **continuous breathing**, meaning that each breath is followed promptly by another breath in a regular, uninterrupted rhythm. Birds and most fish also usually display continuous breathing. Reptiles, amphibians, and air-breathing fish, in contrast, usually exhibit a different breathing pattern, called **intermittent breathing** or **periodic breathing**, defined to be breathing in which breaths or sets of breaths are regularly interrupted by extended periods of **apnea**—that is, periods of no breathing.[6] A characteristic of intermittent breathing in vertebrates that show it, is that each period of apnea follows an inspiration; thus the lungs are inflated during apnea. The glottis—the opening of the airways into the buccal cavity—is closed during apnea in these groups of animals; because of this glottal closure, the inspiratory muscles can relax during the apnea without causing air to be expelled from the lungs.

[6]Apnea is defined to be the absence of breathing movements.

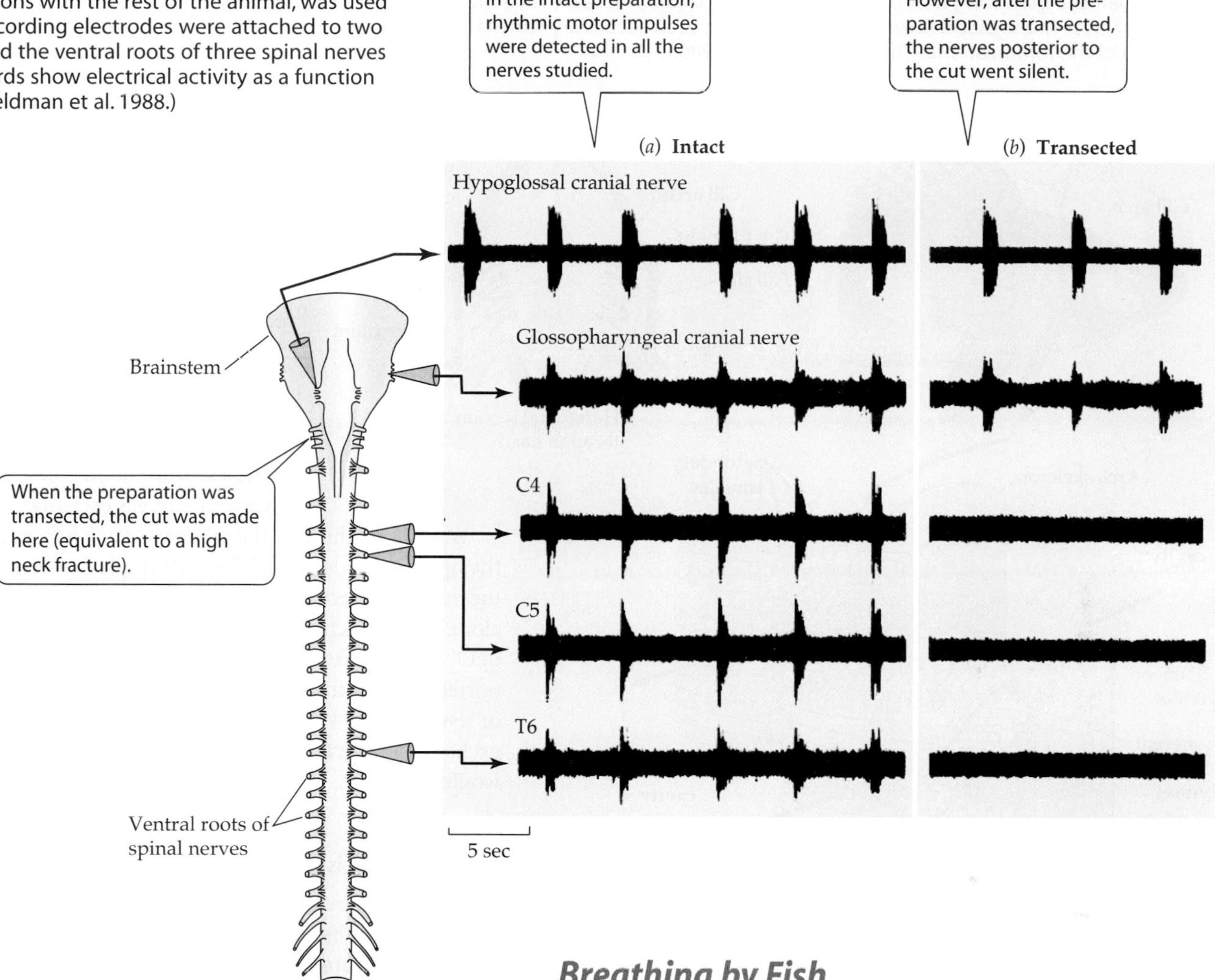

Figure 21.9 Motor nerve impulses from a central pattern generator produce the breathing rhythm The *isolated* brainstem and spinal cord of a neonatal rat, cut free from all connections with the rest of the animal, was used for this study. Recording electrodes were attached to two cranial nerves and the ventral roots of three spinal nerves (C4, C5, T6). Records show electrical activity as a function of time. (From Feldman et al. 1988.)

Summary: Introduction to Vertebrate Breathing

- The gill surface area of most fish of a given body size is similar to the lung surface area of amphibians and reptiles of the same size. Mammals and birds have much more lung surface area than amphibians and reptiles. Reptiles tend to have a thinner barrier between blood and air in their lungs than fish have between blood and water in their gills. The barrier is thinner yet in mammals and especially thin in birds.
- The skin can account for 20% or more of gas exchange in some fish and reptiles, and up to 100% in some amphibians. The skin is a minor contributor to gas exchange, however, in mammals and birds.
- The breathing muscles of vertebrates are skeletal muscles activated by motor nerve impulses. The breathing rhythm originates in a central pattern generator in the brainstem.

Breathing by Fish

Many fish start life breathing only by diffusion across their general body surfaces (see Box 20.1). It can be easy to disregard this point in the study of breathing because the animals are tiny and have no specialized breathing organs. However, mortality can be staggering at these stages, and there can be little doubt that gas-exchange insufficiencies sometimes are responsible. Thus, these early diffusion-respiration stages can be of great importance ecologically and evolutionarily. As a young fish grows, its body becomes too thick for diffusion to suffice (Box 20.1). At the same time, its gills develop, as does its circulatory system, which is required for O_2 from the gills to reach the rest of the body.

The buccal cavity of a teleost fish[7] communicates with the environment not only by way of the mouth, but also by way of lateral pharyngeal openings—the *gill slits*. The gills found in adult fish and mature young are arrayed across these lateral openings. A protective external flap, the operculum, covers the set of gills on each side of the head. Looking in detail at the structure of the gill apparatus, there are four **gill arches** that run dorsoventrally between the gill slits on

[7] The teleost fish are the principal group of bony fish.

Figure 21.10 The branchial breathing system in teleost fish The lateral view in *(a)* shows the orientation of the gill arches *under* the operculum. *(b)* and *(c)* are consecutive enlargements showing the structure of the gill sieve. *(d)* is an enlarged view of a filament and three secondary lamellae, showing that blood flow in the secondary lamellae is countercurrent to water flow across them. Blood flows in a sheet through each secondary lamella, as seen in the section of the foremost one. (After Hill and Wyse 1989.)

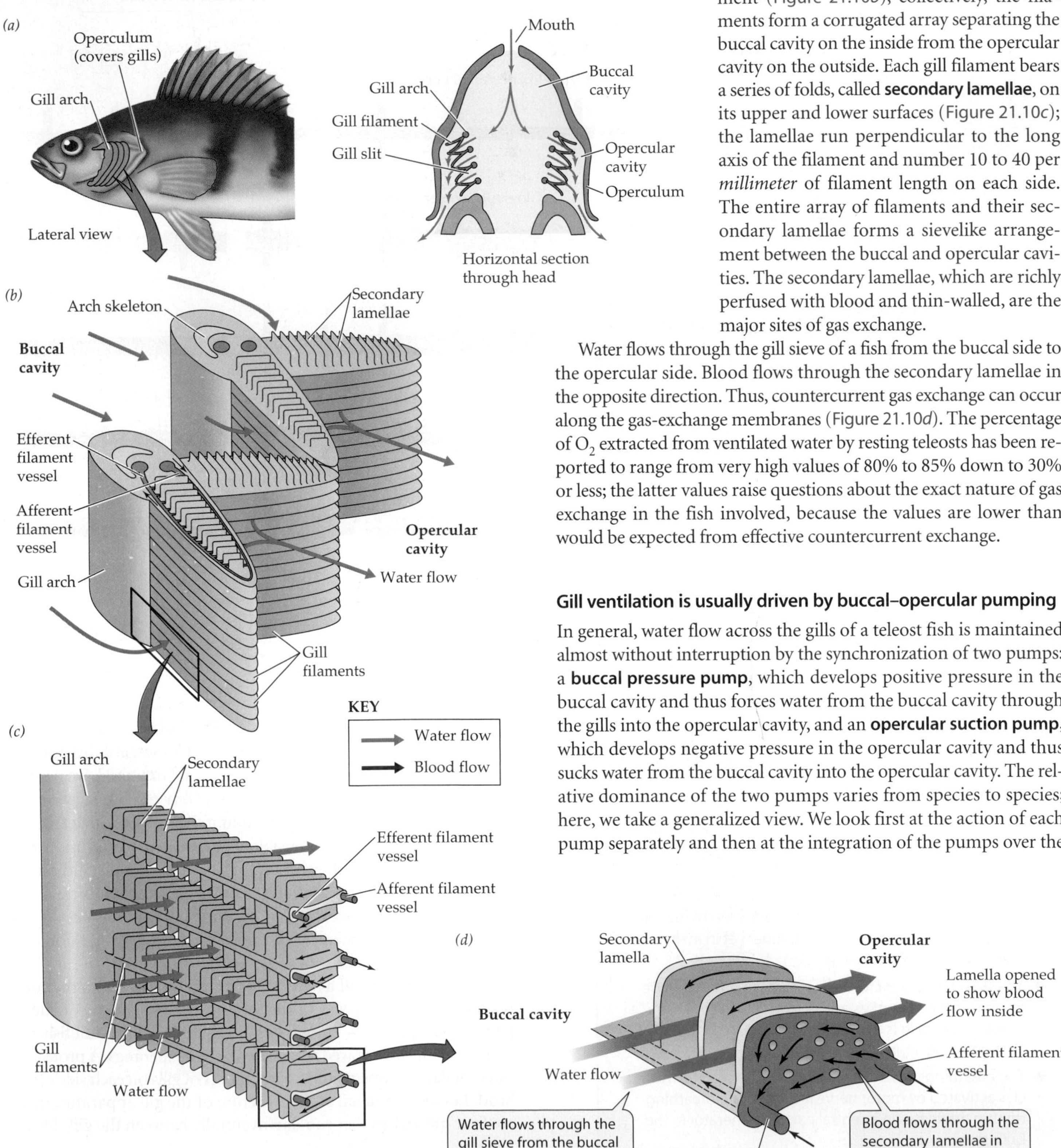

each side of the head (Figure 21.10*a*); these are reinforced with skeletal elements and provide strong supports for the gills proper. Each gill arch bears two rows of **gill filaments** splayed out laterally in a V-shaped arrangement (Figure 21.10*b*); collectively, the filaments form a corrugated array separating the buccal cavity on the inside from the opercular cavity on the outside. Each gill filament bears a series of folds, called **secondary lamellae**, on its upper and lower surfaces (Figure 21.10*c*); the lamellae run perpendicular to the long axis of the filament and number 10 to 40 per *millimeter* of filament length on each side. The entire array of filaments and their secondary lamellae forms a sievelike arrangement between the buccal and opercular cavities. The secondary lamellae, which are richly perfused with blood and thin-walled, are the major sites of gas exchange.

Water flows through the gill sieve of a fish from the buccal side to the opercular side. Blood flows through the secondary lamellae in the opposite direction. Thus, countercurrent gas exchange can occur along the gas-exchange membranes (Figure 21.10*d*). The percentage of O_2 extracted from ventilated water by resting teleosts has been reported to range from very high values of 80% to 85% down to 30% or less; the latter values raise questions about the exact nature of gas exchange in the fish involved, because the values are lower than would be expected from effective countercurrent exchange.

Gill ventilation is usually driven by buccal–opercular pumping

In general, water flow across the gills of a teleost fish is maintained almost without interruption by the synchronization of two pumps: a **buccal pressure pump**, which develops positive pressure in the buccal cavity and thus forces water from the buccal cavity through the gills into the opercular cavity, and an **opercular suction pump**, which develops negative pressure in the opercular cavity and thus sucks water from the buccal cavity into the opercular cavity. The relative dominance of the two pumps varies from species to species; here, we take a generalized view. We look first at the action of each pump separately and then at the integration of the pumps over the breathing cycle. It will be important to remember throughout this discussion that water consistently flows from regions of relatively high pressure to ones of relatively low pressure.

THE BUCCAL PRESSURE PUMP The stage is set for the buccal pressure pump to operate when a fish fills its buccal cavity with water by depressing the floor of its buccal cavity while holding its mouth open. The lowering of the buccal floor increases the volume of the buccal cavity, thereby decreasing buccal pressure below ambient pressure and causing an influx of water. The mouth is then closed, and the buccal pump enters its positive-pressure phase. The fish raises the floor of the buccal cavity during this phase. This action increases the buccal pressure above ambient pressure and drives water from the buccal cavity through the gills into the opercular cavities. Thin flaps of tissue, which act as passive valves, project across the inside of the oral opening from the upper and lower jaws. During the refilling phase of the buccal cycle, when buccal pressure is below ambient, these flap valves are pushed inward and open by the influx of water through the mouth. During the positive-pressure phase, however, the flap valves are forced against the oral opening on the inside and help to prevent water from exiting the buccal cavity through the mouth.

THE OPERCULAR SUCTION PUMP A teleost fish is able to expand and contract its opercular cavities by lateral movements of its opercula and other muscular actions. Running around the rim of each operculum is a thin sheet of tissue that acts as a passive valve, capable of sealing the slitlike opening between the opercular cavity on the inside and the ambient water on the outside. The negative-pressure or suction phase of the opercular pump occurs when the opercular cavity is expanded. At this time, the pressure in the cavity falls below the pressures in the buccal cavity and the ambient water. The negative pressure in the opercular cavity causes water to be sucked from the buccal cavity into the opercular cavity through the gill sieve. Water would also be sucked in from the environment were it not for the action of the opercular rim valve; because the pressure in the opercular cavity is lower than ambient pressure during the sucking phase, the rim valve is pushed medially against the body wall at that time, sealing the opercular opening and preventing influx of ambient water. After its sucking phase, the opercular pump enters its discharge phase. The opercular cavity is contracted, raising the pressure inside above ambient pressure; this forces the rim valve open and discharges water from the opercular cavity through the opercular opening.

INTEGRATION OF THE TWO PUMPS PRODUCES NEARLY CONTINUOUS, UNIDIRECTIONAL FLOW The temporal integration of the buccal and opercular pumps is diagrammed in Figure 21.11. In stage 1, the buccal cavity is being refilled. Expansion of the buccal cavity produces a pressure below ambient; thus, if the buccal pump were the only pump, flow of water through the gills from the buccal side would not occur at this time, and in fact, there would be backflow through the gills into the buccal cavity because of the lowered buccal pressure. It is during stage 1, however, that the opercular pump is in its sucking phase. The pressure in the opercular cavity is reduced well below buccal pressure, and water is drawn through the gills from the buccal cavity. Stage 2 is a short transition period in which the opercular pump is completing its sucking phase and the buccal pump is beginning its pressure phase. In stage 3, the opercular pump is in its discharge phase, and the pressure in the opercular cavity is elevated.

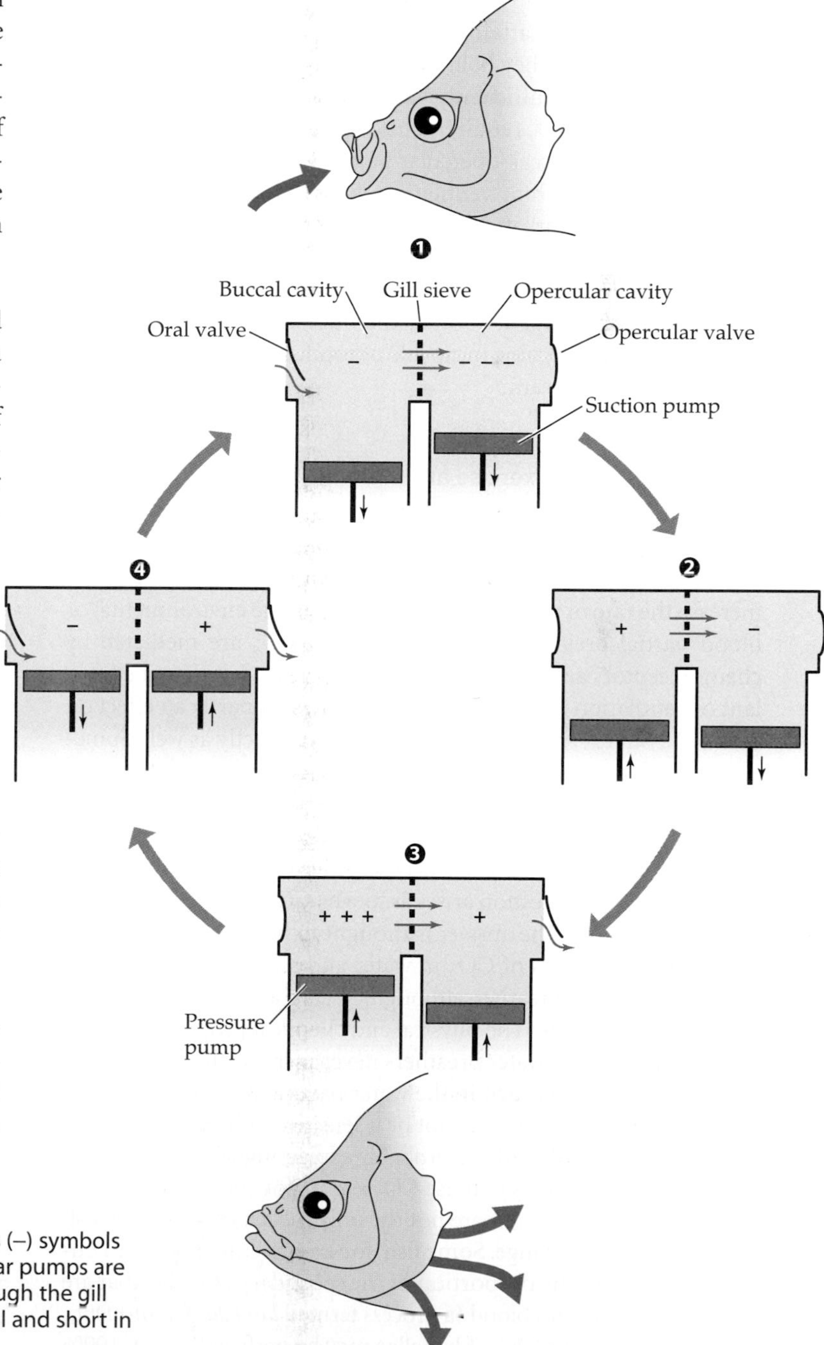

Figure 21.11 The breathing cycle in teleost fish Plus (+) and minus (–) symbols indicate pressures *relative to ambient pressure*. The buccal and opercular pumps are represented by pistons. Blue arrows represent water flow. Arrows through the gill sieve indicate water flow over the gills. Phases ❷ and ❹ are transitional and short in duration. (After Hughes 1961; Moyle 1993.)

However, because the buccal pump is simultaneously in its pressure phase, the buccal pressure is elevated to an even greater degree, and water again flows through the gills from the buccal cavity. Only in stage 4, which occupies just a short part of the breathing cycle, is the pressure gradient in the direction favoring backflow of water through the gills. In all, therefore, the two pumps are beautifully integrated to produce almost continuous, unidirectional flow across the gills: The opercular pump sucks while the buccal pump is being refilled, and the buccal pump develops positive pressure while the opercular pump is being emptied.

Many fish use ram ventilation on occasion and some use it all the time

When swimming fish attain speeds of 50 to 80 cm/s, their motion through the water, if they hold their mouths open, can itself elevate the buccal pressure sufficiently to ventilate their gills at rates adequate to meet their O_2 requirements. Many fast-swimming teleost fish, in fact, cease buccal–opercular pumping when they reach such speeds and employ ram ventilation, which in theory lowers the metabolic cost of ventilation. Tunas, mackerel, dolphinfish, bonitos, and lamnid sharks swim continuously and use ram ventilation all the time. In the tunas, at least, ram ventilation is obligatory, as we have already mentioned, because the buccal–opercular pumping mechanisms have become incapable of producing a sufficiently vigorous ventilatory stream.

Decreased O_2 and exercise are the major stimuli for increased ventilation in fish

Fish are capable of as much as 30-fold changes in their rates of gill ventilation by buccal–opercular pumping. One potent stimulus to increase the rate of ventilation is a decrease in the environmental or blood partial pressure of O_2. Responses to O_2 are mediated by chemoreceptors and neural signals. Exercise also is a potent stimulant of ventilation in fish; the effect of exercise is partly an effect on blood O_2, but exercise may affect ventilation directly as well. Sometimes, fish increase ventilation in response to increases in the external or internal partial pressure of CO_2. Carbon dioxide, however, is overall a relatively weak ventilatory stimulus in fish. Ventilation in mammals, by contrast, is extremely sensitive to elevated CO_2 (as we will see later). The question arises as to why CO_2 plays different roles in the two groups. The answer is thought to lie in the different capacitance coefficients of CO_2 in water and air (see page 545). Put loosely, for a water breather, although O_2 can be difficult to acquire, CO_2 is easy to excrete. The physical and chemical properties of CO_2 in water dictate that water breathers never increase the partial pressure of CO_2 by very much in the water passing over their gills (see Figure 21.6). Thus, CO_2 may not be a sensitive indicator of ventilatory status in water breathers, probably explaining why fish have not evolved control systems that use CO_2 as a major cue.

Changes in ventilation are not the only means employed by fish to adjust gill O_2 exchange. Some fish, for example, are known to undergo changes in the proportion of the secondary lamellae that are actively perfused with blood (a process termed *lamellar recruitment*). Whereas only about 60% of lamellae may be perfused at rest, 100% may be perfused during exercise or exposure to reduced O_2 partial pressures.

Several hundred species of bony fish are able to breathe air

Many species of bony fish have evolved mechanisms for tapping the rich O_2 resources of the air. Nearly 400 species of air-breathing fish are known, especially in freshwater. Most air-breathing fish retain functional gills and are dual breathers, acquiring O_2 from both water and air. The extent to which these fish rely on the atmosphere depends on several factors. They typically increase their use of air breathing as the level of dissolved O_2 in their aquatic habitat falls. They also tend to resort increasingly to air breathing as the temperature rises, because high temperatures elevate their O_2 needs. Notably, air-breathing fish typically void most of their CO_2 into the water—across their gills or skin—even when relying on the atmosphere for most of their O_2. The traditional view of the adaptive value of air breathing is that it arose in groups of fish living over evolutionary scales of time in O_2-poor waters, as a means of solving the problem of O_2 shortage in the water (see page 19). A new hypothesis, discussed in Chapter 23, is that it sometimes arose as a means of oxygenating the heart.

Some air-breathing fish lack marked anatomical specializations for exploiting the air. American eels (*Anguilla rostrata*) provide a case in point. Eels sometimes come out onto land in moist situations. They then meet about 60% of their O_2 requirement by uptake across the skin and 40% by buccal air gulping. Their gills, which are of quite ordinary structure, are probably the primary site of O_2 uptake from the air they gulp.

In most air-breathing fish, some part or branch of the alimentary canal has become specialized as an air-breathing organ. The specialized region is always highly vascularized, and its walls may be thrown into extensive patterns of evagination or invagination. In some species the buccal cavity has become adapted for air breathing; this is true, for example, in electric eels (*Electrophorus electricus*), which have innumerable vascular papillae on the walls of their buccal cavity and pharynx. The opercular cavities form air-breathing organs in some species; mudskippers (*Periophthalmus*), for example, breathe air using expanded gill chambers lined with vascularized, folded membranes. Many air-breathing species employ so-called *suprabranchial chambers*, situated in the dorsal head above the gills; a dramatic specialization of this sort of air-breathing organ is the presence in some catfish (*Heteropneustes*) of tubular diverticula that extend from the head into half the length of the body. Quite a few air-breathing fish employ vascularized portions of the stomach to breathe. Certain species of armored catfish (family Callichthyidae) employ the intestine; in these, half or more of the intestinal length is highly vascularized and devoted to breathing from air that is swallowed and later expelled via the anus. The swim bladder (gas bladder) is used as an air-breathing organ by many fish; in such species, the swim bladder wall may be thrown into exceedingly elaborate patterns of folding. The "tinkering" aspect of evolution (see page 8) is nowhere better illustrated than in the fantastic diversity of body parts that fish have diverted from old functions to the task of getting O_2 from the atmosphere.

The air-breathing organs of fish are most often inflated by buccal pumping. A fish takes air into its buccal cavity, then closes its mouth and compresses the buccal cavity. In this way, air is driven into its stomach, swim bladder, or other air-breathing structure.

A potential problem for air-breathing fish is that O_2 taken up from the atmosphere may be lost to O_2-poor water across their gills! This possibility probably helps explain why the gills of these animals are often reduced in their gas-exchange capability in comparison to those of other fish (the gill surface area, for instance, is often reduced). In extreme cases, the gills are so atrophied that air breathing is obligatory; this is true in electric eels, for example. In addition, many air-breathing fish have evolved specialized circulatory shunts by which oxygenated blood can bypass the gas-exchange surfaces of the gills, as a way of limiting O_2 loss across the gills.

Of all air-breathing fish, the six species of **lungfish** (dipnoans) have received perhaps the most attention because they are believed to be the modern descendants of the group of fish that gave rise to terrestrial vertebrates. The walls of the lungs of lungfish are thrown into complex patterns of ridges and septa (Figure 21.12*a*), thus resembling the walls of many amphibian lungs. The Australian lungfish (genus *Neoceratodus*) has a single lung and is a dual breather with well-developed gills. The four species of African lungfish (*Protopterus*) and the South American lungfish (*Lepidosiren*), by contrast, have bilobed lungs and in general have much-reduced gills lacking secondary lamellae. They are obligate air breathers, and thus—like electric eels—they drown if they are denied air.

The air-breathing organs of lungfish have often been viewed as "particularly homologous" to the lungs of terrestrial vertebrates. This view explains why the organs came to be called *lungs*, and why the fish themselves were named *lung*fish. Many authorities now argue that the swim bladders as well as the "lungs" of modern fish are derived from early air-breathing organs that gave rise to tetrapod lungs. Thus, if air-breathing organs were being renamed, a case could be made that air-breathing swim bladders have as much right to be called "lungs" as the air-breathing organs of the dipnoans. To a physiologist, all invaginated breathing organs are lungs (see page 542), and thus all the air-breathing organs of fish are lungs.

(*a*) The inner wall of a lungfish lung

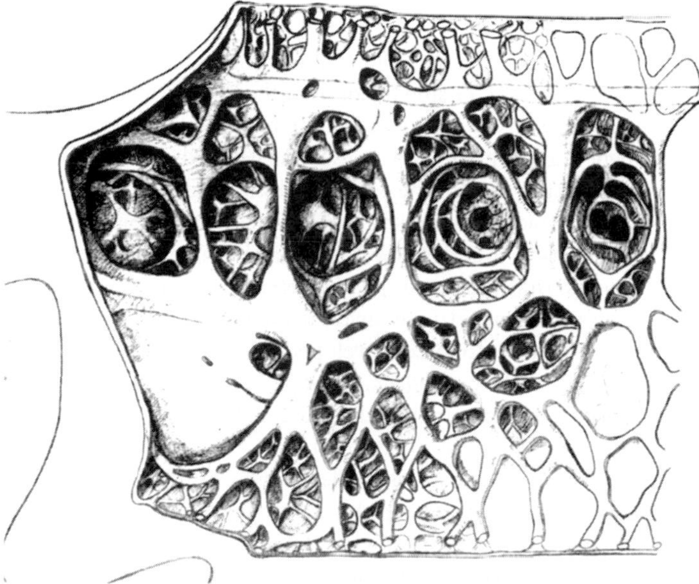

(*b*) An African lungfish in the genus *Protopterus*

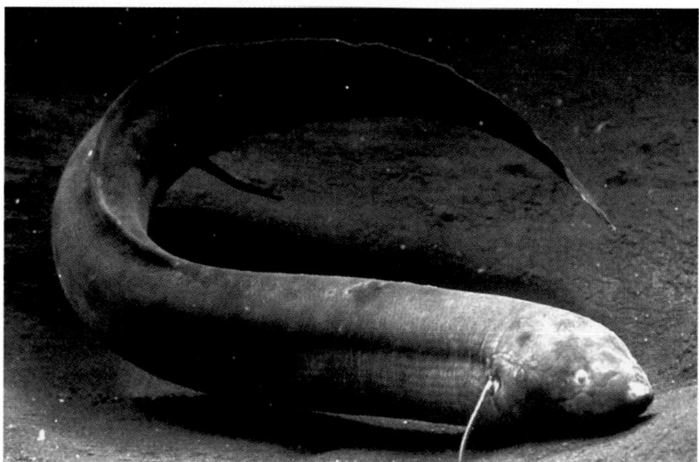

Figure 21.12 Lungfish and their lungs *(a)* The inner wall of part of a lung of *Protopterus aethiopicus*. The respiratory surface area of the lung is greatly enhanced by a complex pattern of vascularized ridges and septa. The side compartments in the wall of the lung open to a central cavity that runs the length of the lung and communicates anteriorly with a short pulmonary canal leading to the esophagus. *(b) Protopterus dolloi*. The lungfish of the genus *Protopterus*, found in Africa, are obligatory air breathers (*a* from Poll 1962.)

Summary: Breathing by Fish

- The secondary lamellae are the principal sites of gas exchange in fish gills. Countercurrent gas exchange occurs in the lamellae.
- Water flow across the gills is virtually unidirectional. It is driven by a buccal pressure pump and an opercular suction pump acting in an integrated rhythm, so that the buccal pump drives water across the gills when the opercular pump is being emptied of water and the opercular pump sucks water across the gills while the buccal pump is being refilled with water.
- Some fish turn to ram ventilation when swimming fast enough. Others, such as tunas, are obligate ram ventilators and must swim all the time to avoid suffocation.
- A lowered O_2 partial pressure in the blood is a more potent stimulus for increased ventilation in fish than an elevated CO_2 partial pressure.
- Most of the 400 or so species of air-breathing fish have an air-breathing organ that is derived from the buccal cavity, opercular cavity, stomach, or intestines—or one that originates as an out-pocketing of the foregut (e.g., a swim bladder).

Breathing by Amphibians

Of all the vertebrate groups, amphibians mix water breathing and air breathing to the greatest extent. Many move from an aquatic environment to a terrestrial one during their individual development, and many are dual breathers as adults.

The gills of aquatic amphibian larvae (tadpoles) are of different origin and structure than the gills of adult fish. They develop as

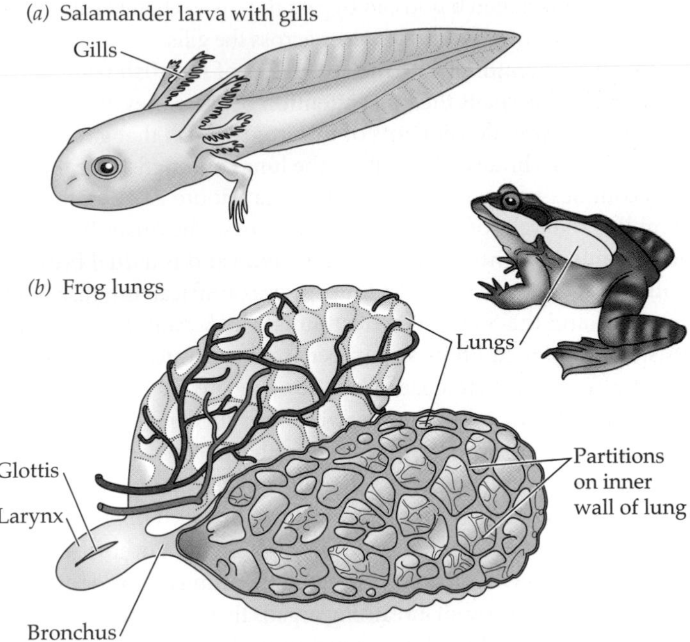

Figure 21.13 Breathing organs of amphibians *(a)* The external gills of a 3-week-old salamander larva (*Ambystoma maculatum*). *(b)* The lungs of a frog (*Rana temporaria*). One is cut open to reveal the compartmentalization of the inner wall by ridges and septa. (*b* after Poll 1962.)

outgrowths of the integument of the pharyngeal region and project into the water from the body wall (Figure 21.13*a*). The gills are external in all young amphibian larvae; in salamander larvae they remain so, but in the frogs and toads (Anura), an outgrowth of the integument, termed the operculum (different from the bony operculum of a fish), soon encloses the gills in a chamber that opens to the outside posteriorly, usually via a single aperture. Ventilation of the gills enclosed in the opercular cavity is accomplished by buccal pumping. The gills of amphibians are generally lost at metamorphosis, but external gills remain throughout life in certain aquatic salamanders, such as mudpuppies (*Necturus*). Mudpuppies ventilate their gills by waving them back and forth in the water and increase the frequency of these movements in response to decreased O_2 or increased temperature.

In most amphibians, paired lungs develop from the ventral wall of the pharynx near the time of metamorphosis. The lungs of many adult amphibians are simple, well-vascularized sacs; their internal surface area is increased little, if at all, by folding, and in this respect they are less well developed than the lungs of modern lungfish. Among frogs and toads, the inner walls of the lungs are often more elaborate; they may be thrown into a complex pattern of folds and septa, giving them a honeycombed appearance (Figure 21.13*b*). Still, the lungs retain their basic saclike form. The central cavity of each lung remains open and provides access to the various side compartments formed by the folding of the walls.

Lunged amphibians fill their lungs by buccopharyngeal pressure. This basic mechanism is presumably carried over from their piscine ancestors and, as mentioned earlier, is often employed by amphibian larvae to ventilate their gills. Most studies of pulmonary ventilation have been performed on frogs, and although several patterns differing in detail have been reported, the essentials of the buccopharyngeal pressure pump are quite uniform. Air is taken into the buccal cavity through the nares or mouth when the pressure in the cavity is reduced by lowering the floor of the cavity. When the floor of the buccal cavity is then raised with the mouth closed and the nares at least partially sealed by valves, the increase in pressure forces air into the lungs. The inflation of the lungs elevates the air pressure within them. Thus, the lungs would discharge upon opening of the mouth or nares were it not for the glottis (see Figure 21.13*b*), which is closed by muscular contraction after inhalation. A period of no breathing, with the lungs inflated, then follows (i.e., breathing is intermittent). The nares are opened during the apneic period, and a frog often pumps air in and out of its buccal cavity through its nares at that time by lowering and raising the floor of the buccal cavity, a process termed **buccopharyngeal pumping.** After a period of time, the glottis is opened, and air from the lungs is exhaled. Exhalation results in part from elastic recoil of the expanded lungs and may also be promoted by contraction of muscles in the walls of the lungs and body wall. Exhalation is thus described as having both **passive** and **active** components. In this context, *passive* refers to forces developed by simple elastic rebound (not involving contraction of muscles), whereas *active* refers to forces developed by muscular contraction.

The bullfrog (*Rana catesbeiana*) provides a well-studied specific example of the pulmonary breathing cycle. A bullfrog fills its buccal cavity in preparation for inflation of its lungs *before* it empties its lungs; with its glottis closed, it inhales air, which mostly comes to lie in a posterior depression of the buccal floor (step 1, Figure 21.14). Next (step 2), the glottis is opened, and pulmonary exhalant air passes in a coherent stream across the dorsal part of the buccopharyngeal cavity to exit through the nares. The fresh air in the depression of the buccal floor is then driven into the lungs when the buccal floor is raised with the nares closed (step 3). An important effect of the buccopharyngeal pumping between breaths is that it washes out residual pulmonary exhalant air from the buccal cavity, so that when the next pulmonary ventilatory cycle begins, the buccal cavity is filled with a relatively fresh mixture.

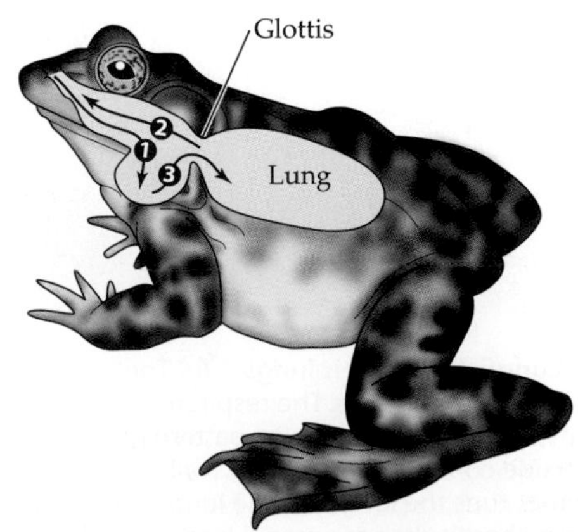

Figure 21.14 The three major steps in the ventilatory cycle of an adult bullfrog (*Rana catesbeiana*) (After Gans 1970.)

Gills, lungs, and skin are used in various combinations to achieve gas exchange

A central question in amphibian respiratory physiology is how the total gas exchange of an individual is partitioned among the available gas-exchange sites: the gills, lungs, and skin. Bullfrogs, to continue with them as an example, start their lives without lungs, and when they are living at 20°C as aquatic tadpoles, their gills and skin each account for about half of their O_2 and CO_2 exchange (Figure 21.15). As bullfrog tadpoles mature and their lungs become functional, their lungs gradually assume primary responsibility for their O_2 uptake. In adulthood, as one might guess, the lungs take up most O_2. The lungs of bullfrogs, however, do not play a large role in CO_2 exchange at any age; instead, when the gills are lost at metamorphosis, the skin increases its role in eliminating CO_2 (Figure 21.15). The pattern seen in adult bullfrogs at 20°C—that the lungs are primarily responsible for O_2 uptake whereas the skin eliminates most CO_2—is seen also in the adults of a variety of other species of frogs, toads, and lunged salamanders when at rest on land at similar temperatures.

Some species of frogs in temperate regions of the world hibernate at the bottoms of ponds and lakes during winter. All their O_2 and CO_2 exchange is then across their skin. One reason the skin can suffice under these circumstances is that the hibernating animals have relatively low needs for O_2 and CO_2 exchange because of their low body temperatures and seasonal metabolic depression. The relatively high permeability of the skin is also important; whereas it increases rates of dehydration on land, it makes underwater breathing possible.

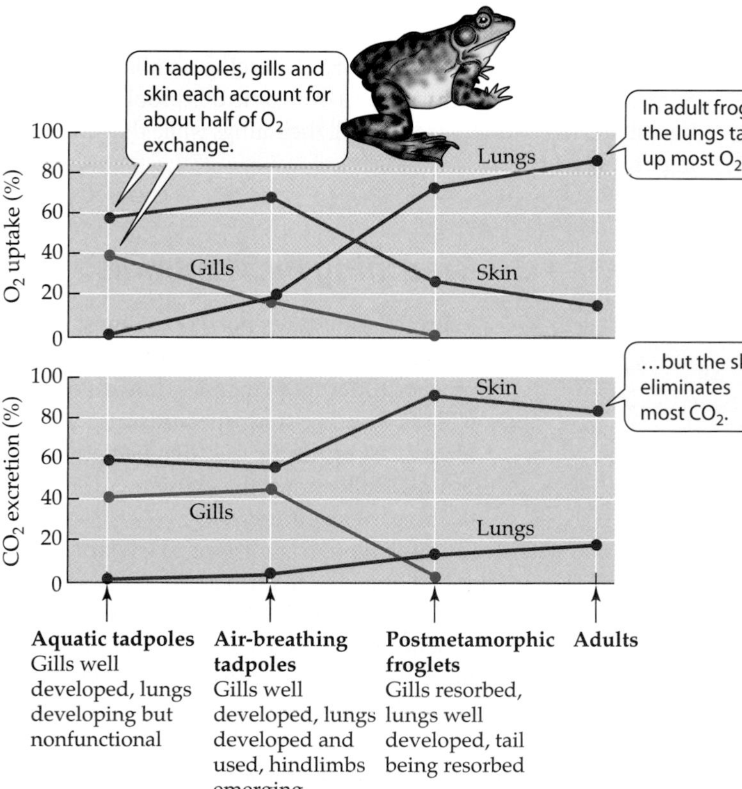

Figure 21.15 The development of external respiration in the bullfrog (*Rana catesbeiana*) Shown are the percentages of O_2 uptake and CO_2 excretion that occur across the gills, lungs, and skin of bullfrogs as they develop from tadpoles to adults. The animals studied (at 20°C) had free access to well-aerated water and air. (After Burggren and West 1982.)

Breathing by Reptiles

In most reptiles, the lungs take up essentially all O_2 and eliminate essentially all CO_2. The skin of reptiles is generally much less permeable than amphibian skin, meaning that it protects far better against evaporative dehydration, but does not readily allow the respiratory gases to pass through.

The simplest type of reptilian lung—seen in many lizards and snakes—resembles an amphibian lung in being a saclike structure with an open central cavity (Figure 21.16*a*); this type of lung is termed *unicameral* ("one-chambered"). The walls of the lung are thrown into a honeycomb-like pattern of vascularized partitions, increasing their surface area (Figure 21.16*b*). Air flows in and out of the central cavity, but gas exchange between the central cavity and the depths of the honeycomb-like cells on the walls is probably largely by diffusion.

A major evolutionary advance observed in several groups of reptiles is the subdivision of the main lung cavity into numerous smaller cavities, forming a so-called *multicameral* ("multiple-chambered") lung. This type of lung occurs in monitor lizards (Figure 21.16*c*), reptiles noted for their especially active ways of life and relatively high aerobic competence. It is also found in crocodilians and turtles. A noteworthy evolutionary development in the multicameral lung is the appearance of a cartilage-reinforced tube (*bronchus*) that runs lengthwise through the lung (Figure 21.16*c*); this tube allows air to flow to all the chambers, which are arrayed consistently in three rows. Clearly, the multicameral type of lung can provide a great deal more surface area of gas-exchange membrane per unit of lung volume than the unicameral type because the septa between lung chambers as well as the outer walls of the lung can develop elaborate gas-exchange surfaces.

In reptiles, the lungs are filled principally or exclusively by suction rather than by buccal pressure. Air is drawn into the lungs by an expansion of the lung volume, which creates a subatmospheric pressure within the lung chambers. This mode of ventilation represents a major evolutionary transition from the earlier buccal pressure–filled lungs of air-breathing fish and amphibians. It is a transition that is carried forward to the mammals and birds, which also employ suction to fill their lungs. Because suction is created by the action of thoracic and abdominal muscles, not by buccal muscles, the evolution of suction ventilation ultimately freed the buccal cavity from one of its ancient functions, allowing it to evolve in new directions without ventilatory constraints.

Suction is developed in the lungs of reptiles in two different ways. In one sort of breathing cycle, seen in at least some snakes, thoracic and abdominal expiratory muscles compress the lungs to a volume smaller than their passive relaxation volume during the exhalation of air; suction for inhalation is then developed when the lungs rebound elastically. In the other sort of breathing cycle, seen in lizards

(a) A unicameral lung in a lacertid lizard

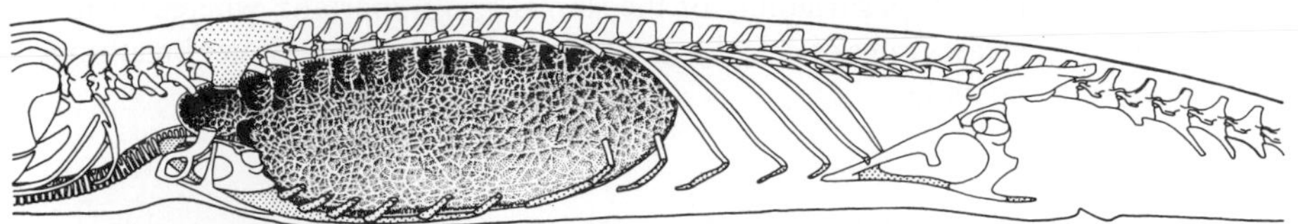

(b) Scanning electron micrograph of the wall of a tegu lizard lung

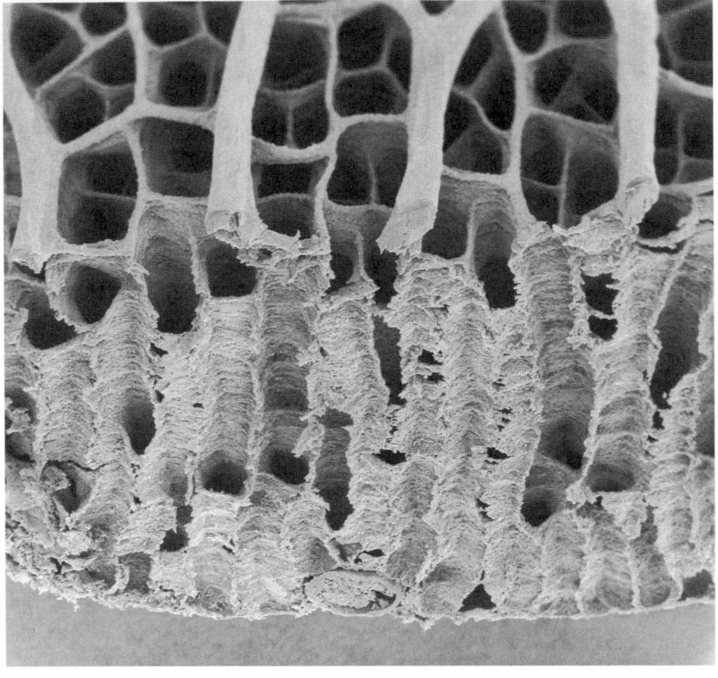

(c) A multicameral lung in a monitor lizard

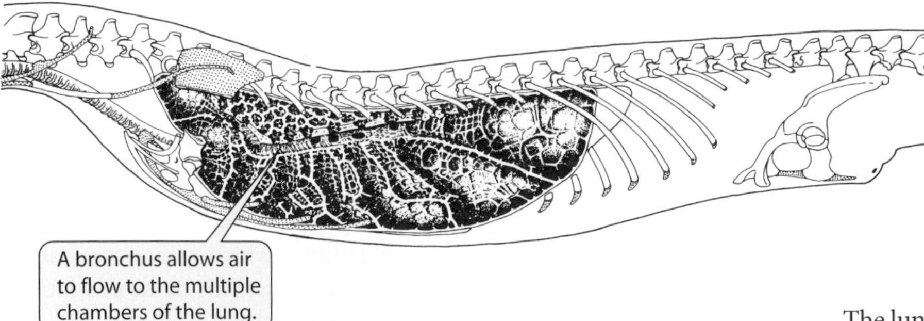

Figure 21.16 Lizard lungs *(a)* Gross internal structure of the unicameral lung of a green lizard (*Lacerta viridis*). *(b)* Scanning electron micrograph of the lung wall of a tegu lizard (*Tupinambis nigropunctatus*), showing the honeycomb-like pattern of vascularized partitions. Magnification: 34 ×. *(c)* Gross internal structure of the multicameral lung of a monitor lizard (*Varanus exanthematicus*). (*a* and *c* courtesy of Hans-Rainer Duncker, reprinted from Duncker 1978; *b* courtesy of Daniel Luchtel and Michael Hlastala.)

and crocodilians, inspiratory muscles actively create suction in the lungs during inhalation by expanding the lungs to a size larger than their passive relaxation volume; then, when the muscular activity stops, the lungs rebound elastically, becoming smaller, and the elastic rebound contributes to exhalation.

Lizards provide an instructive example to study in more detail. Running over and between the ribs on each side of a lizard's body are sheets of muscles that can expand or contract the volume enclosed by the rib cage. These **intercostal muscles** (*costa*, "rib") play a central role in lizard ventilation. When a resting lizard inhales, certain of its intercostal muscles are activated and expand the rib cage; this mechanism is sometimes called the **costal suction pump**. After the inflation of the lungs has occurred, the glottis is closed, and the inspiratory muscles relax. The inhaled air is then held in the lungs for several seconds to several minutes of apnea, while often the buccal cavity is ventilated by buccopharyngeal pumping, thought generally to aid olfaction. Exhalation then occurs, followed quickly by another inhalation. Exhalation is brought about by both passive and active processes: elastic rebound of the lungs toward their unexpanded volume, aided by the contraction of intercostal muscles and abdominal muscles that compress the thoracic cavity.

Recent research has demonstrated that when lizards are walking or running, some of their intercostal muscles help produce the back-and-forth flexions of the body that are so characteristic of lizard locomotion. This involvement of the intercostals in locomotion can interfere with their ability to develop ventilation forces. Some, but not all, species of lizards overcome this problem by using buccal pressure pumping to help fill their lungs while they are walking or running.

Breathing by Mammals

Mammals and birds possess the most elaborate lungs of all animals. Their lungs are built on very different principles, however, as we will see when we study the birds later on. The plastic cast of the airways of a person in Figure 21.17 illustrates the extreme intricacy of the lungs of a mammal.

The lung system of an adult human consists of 23 levels of airway branching. The *trachea* first branches to form a major airway, a *primary bronchus*, that enters each lung (see Figure 21.17). Each primary bronchus then branches and rebranches dendritically (as a tree branches) within the lung, giving rise to *secondary* and *higher-order bronchi* of smaller and smaller diameter, and then to ever-smaller fine tubes known as *bronchioles*. At the outer limits of this branching tree of airways, 23 branches from the trachea, the final bronchioles end blindly in **alveolar ducts** and **alveolar sacs**, the walls of which are composed of numerous semi-spherical outpocketings called **alveoli** (singular *alveolus*) (**Figure 21.18**). There are about 300 million alveoli in the lungs of a human adult; they vary in size but average about 0.25 mm in diameter. The alveoli make up a total area of gas-exchange membrane of 120 to 140 m^2! The floor of an 80-stu-

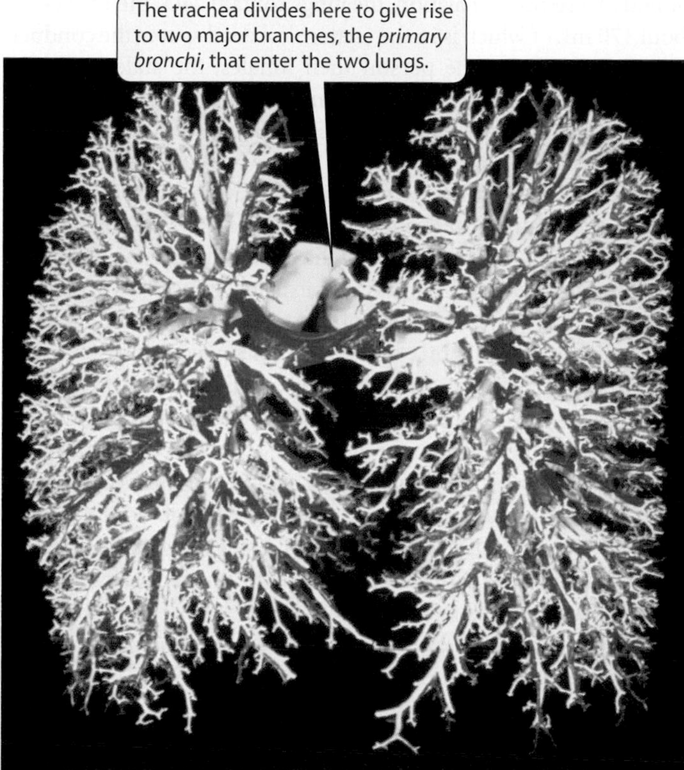

Figure 21.17 The airways in human lungs To visualize the lung passages, the airways were injected with white plastic. Similarly, the arteries were injected with red plastic. After the plastic hardened, the tissue was removed, leaving just the plastic to mark the airways and arteries. Only the airways and arteries of relatively large diameter are preserved with this technique. Thus the airway system and arterial system branch far more finely than seen here. Note that the arteries tend to branch in parallel with the airways.

(*a*) The finest airways of the mammalian lung, ending in alveoli

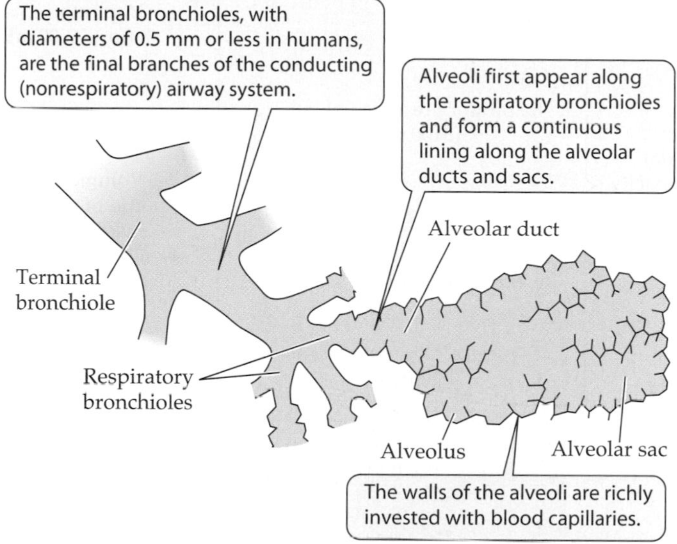

(*b*) Scanning electron micrograph of a human lung

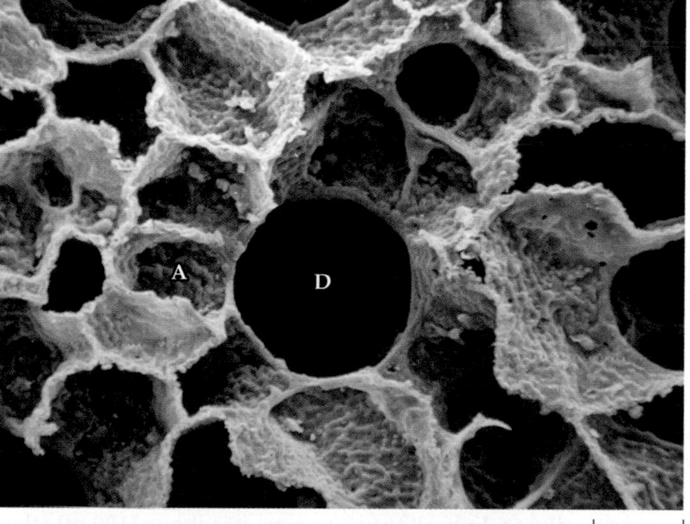

Figure 21.18 Respiratory airways of the mammalian lung *(a)* Longitudinal section of the final branches of the airways in a mammalian lung, showing respiratory bronchioles, alveolar ducts, alveolar sacs, and alveoli. *(b)* Scanning electron micrograph of a 0.4 mm^2 area of a human lung, showing alveolar ducts (D) and alveoli (A). (*a* after Hildebrandt and Young 1960; *b* courtesy of Ewald Weibel.)

dent classroom is likely to have a similar area. Thus, by virtue of elaborate branching, a highly vascularized interface between air and blood that is the size of a large classroom floor is fit within the compact volume occupied by our lungs.

The trachea and bronchi and all but the last few branches of bronchioles are not much involved in gas exchange; they are thus known as the **conducting airways** and are said to constitute the **anatomical dead space** of the lungs. They are lined with a relatively thick epithelium and do not receive a particularly rich vascular supply. Gas exchange between air and blood occurs in the **respiratory airways** of the lungs (Figure 21.18*a*), which consist of *respiratory bronchioles* (the last two or three branches of bronchioles), *alveolar ducts* (through-passages lined with alveoli), and *alveolar sacs* (end sacs also lined with alveoli). The walls of the respiratory airways are composed of a single layer of thin, highly flattened epithelial cells and are richly supplied with blood capillaries. The alveoli constitute most of the gas-exchange surface. In them, blood and air are separated by just two delicate epithelia (the alveolar and capillary epithelia) and a basement membrane in between. The total average diffusive thickness of these structures is only 0.3 to 0.6 μm.

The total lung volume is employed in different ways in different sorts of breathing

When mammals breathe at rest, they do not come close to inflating their lungs fully when they inhale, nor do they come close to deflating their lungs fully when they exhale. Thus, a wide margin exists for increasing the use of total lung volume. The **tidal volume** is the volume of air inhaled and exhaled per breath. In resting young men, the volume of the lungs at the end of inhalation is about 2900 mL, whereas that at the end of exhalation is about 2400 mL. Thus, the *resting tidal volume* is about 500 mL (Figure 21.19). No mammal can completely empty its lungs of air, but, as already stated, the lungs can

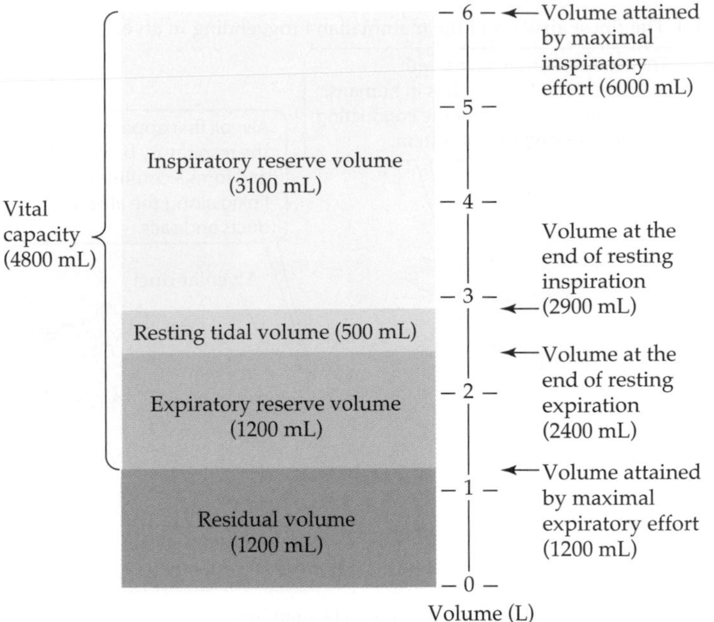

Figure 21.19 Dynamic lung volumes in healthy young adult men The lung volumes shown include the anatomical dead space as well as the respiratory airways. The inspiratory and expiratory reserve volumes shown are the reserves available during resting breathing. The residual volume is the volume remaining in the lungs after maximal expiratory effort.

be emptied more than they are during resting exhalation. The maximal volume of air that an individual can expel beyond the resting expiratory level is termed the resting **expiratory reserve volume.** In healthy young men, it is about 1200 mL; that is, of the 2400 mL of air left in the lungs at the end of a resting exhalation, 1200 mL can be exhaled by maximal expiratory effort, but 1200 mL cannot be exhaled at all (Figure 21.19). The maximal volume of air that can be inhaled beyond the resting inspiratory level is the resting **inspiratory reserve volume.** In healthy young men, it is about 3100 mL; thus the total lung volume at the end of a maximal inspiratory effort is about 6000 mL (Figure 21.19). Using the terminology developed here, when mammals increase their tidal volume above the resting level, they do so by using parts of their inspiratory and expiratory reserve volumes. The maximal possible tidal volume, which is sometimes termed the **vital capacity**, is attained by fully using both reserves and thus is the sum of the resting tidal volume and the resting inspiratory and expiratory reserve volumes: about 4800 mL in young men. The vital capacity of people tends to be increased by physical training, but advancing age and some diseases tend to decrease it.

The gas in the final airways differs from atmospheric air in composition and is motionless

The gas in the alveoli, which is the gas that undergoes exchange with the blood, differs dramatically in composition from atmospheric air. The most fundamental reason is that the alveolar sacs form the blind ends of tidally ventilated airways that are never fully emptied. To see the implications of this anatomical fact, consider a resting person. At the end of a resting exhalation, the lungs contain 2400 mL of stale air, about 170 mL of which is in the anatomical dead space (the conducting airways). When the person then inhales, the stale air in the anatomical dead space is the first to move deeper into the lungs, into the respiratory airways; thus, the entire 2400 mL of stale air is in the respiratory airways after inhalation. Of the 500 mL of fresh atmospheric air inhaled during a resting breath, about 330 mL passes through the anatomical dead space and enters the respiratory airways; the other 170 mL—the last of the air to be inhaled—simply fills the anatomical dead space and is later exhaled, unused, when exhalation occurs. Therefore, at the end of a resting inhalation, the gas in the respiratory airways is a mix of 2400 mL of stale air and 330 mL of fresh atmospheric air. Accordingly, the O_2 partial pressure in the alveoli is far below the atmospheric O_2 partial pressure, and the CO_2 partial pressure in the alveoli is far above the atmospheric partial pressure. Another significant property that arises from these quantitative realties is that the gas partial pressures in the alveoli do not change much between inhalation and exhalation. Using the values for resting people, only about 12% of the air in the respiratory airways at the end of an inhalation is fresh, whereas 88% is carried over from previous breaths. The large carry-over from breath to breath helps impart stability to the gas composition deep in the lungs.

The exact partial pressures of gases that prevail in the alveoli depend dynamically on the *rate* at which fresh air is brought into the depths of the lungs and the *rate* of gas exchange with the blood. The rate of fresh air input depends not only on the tidal volume but also on the number of breaths per minute. Ventilatory control systems (to be discussed later) ordinarily adjust these properties relative to the rate of gas exchange with the blood so that certain set-point partial pressures of O_2 and CO_2 are maintained in the alveoli. In people near sea level, for example, the alveolar partial pressure of O_2 is nearly always about 13.3 kPa (100 mm Hg), and the alveolar partial pressure of CO_2 is about 5.3 kPa (40 mm Hg).

Another important property of gas exchange in the respiratory airways also arises from the anatomical fact that the alveolar sacs form the blind ends of a tidally ventilated airway system. This additional property is that the gas occupying the final respiratory airways is essentially motionless (Figure 21.20). Therefore, O_2 and CO_2 transport through this gas must occur by diffusion. During inhalation, convective air movement carries air rapidly down the trachea and through the various bronchi. As the air flows ever deeper into the lungs, however, it slows, because the collective volume of the airways rapidly increases as the airways branch into greater and greater numbers. Moreover, the end of the airway system is a dead end. Thus, convective airflow ceases before the air reaches the alveolar sacs.

The layer of still gas deep in the lungs that separates the gas-exchange membranes from moving gas is very thin (recall, for instance, that a human alveolus averages only 0.25 mm in diameter). Thus, diffusion across the still layer of gas can readily occur fast enough for rates of O_2 and CO_2 exchange with the blood to be adequate. This happy situation is contingent, however, on the alveoli being filled with gas, not water. If the alveoli become filled with body fluid, rates of diffusion sharply plummet. That is why any disease that causes even minute accumulation of body fluid within the airways is a threat to life.

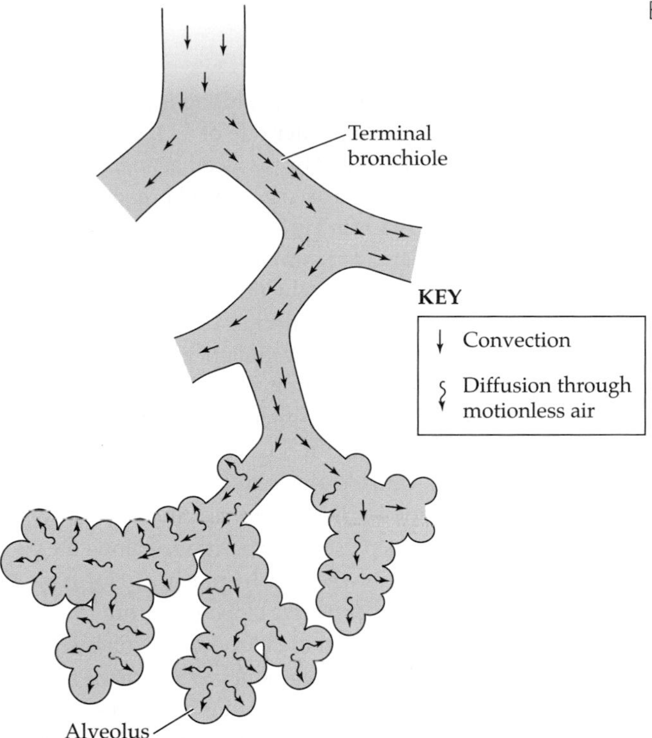

Figure 21.20 Mechanisms of gas transport in the final branches of mammalian lungs during inhalation Although gases are drawn by convection (mass flow) into the finest bronchioles, the gases in the alveolar sacs and alveoli are motionless. Therefore, O_2 must travel by diffusion across the final (minute) stretches of distance it must cover to reach the gas-exchange membrane, and CO_2 must diffuse in the opposite direction. (After Weibel 1984.)

The power for ventilation is developed by the diaphragm and the intercostal and abdominal muscles

Unlike other vertebrates, mammals have a true **diaphragm**: a sheet of muscular and connective tissue that completely separates the thoracic and abdominal cavities (see Figure 1.17). The diaphragm is dome-shaped, projecting farther into the thorax at its center than at its edges. Contraction of the diaphragm tends to flatten it, pulling the center away from the thorax toward the abdomen. This movement increases the volume of the thoracic cavity, resulting in expansion of the lungs and inflow of air by suction.

The **external** and **internal intercostal muscles** that run obliquely between each pair of adjacent ribs are also important in ventilation. Contraction of the external intercostals rotates the ribs anteriorly and outward, expanding the thoracic cavity. The contractile filaments of the internal intercostals run crossways to those of the externals, and in general, their contraction rotates the ribs posteriorly and inward, decreasing the volume of the thoracic cavity. You can easily demonstrate the action of these muscles on yourself by consciously expanding and contracting your rib cage while feeling your ribs with your fingers.

To fully understand ventilation, it is essential to recognize that the lungs and thoracic wall together form an elastic system. Much like a hollow rubber ball with a hole on one side, they assume a certain equilibrium volume, known as their **relaxation volume**, if they are free of any external forces. For the volume of the lungs to deviate from the relaxation volume, muscular effort must be exerted. In adult human males, the relaxation volume of the lungs, measured as the volume of gas they hold at relaxation, is about 2400 mL. This volume is the same as that of the lungs after a resting exhalation (see Figure 21.19), and in fact, the lungs and thorax are in their relaxed state after a resting exhalation.

Inhalation in people at rest is active (meaning that it entails muscular effort). During inhalation, the lungs are *expanded to greater than their relaxation volume* by contraction of the diaphragm, external intercostal muscles, and anterior internal intercostals. Mammals do not close the glottis during ordinary breathing, and humans are no exception. Thus, inhalation continues only as long as the inspiratory muscles contract. Exhalation in resting people is largely or completely passive (meaning that it does not entail muscular effort). When the inspiratory muscles relax at the end of inhalation, lung volume *returns elastically to its passive equilibrium state: the relaxation volume.*

When people exercise, they increase not only their tidal volume, but also their breathing frequency. Thus, additional muscular activity is required, both to amplify changes in lung volume during each breath and to hasten the inspiratory and expiratory processes. The external intercostals assume a greater role in inhalation during exercise than they play during rest. Whereas expansion of the rib cage by these muscles during quiet breathing is of relatively minor importance by comparison to the action of the diaphragm, expansion of the rib cage by the external intercostals during heavy exertion accounts for about half of the inspiratory increase in lung volume. Active forces also contribute to exhalation during exercise. The most important muscles in exhalation are the internal intercostals, which actively contract the rib cage, plus muscles of the abdominal wall, which contract the abdominal cavity, forcing the diaphragm upward into the thoracic cavity. These muscles hasten exhalation during exercise and may also compress the lungs beyond their relaxation volume, thereby enhancing tidal volume through use of some of the expiratory reserve volume.

The same basic groups of muscles are used for ventilation in other mammals, but their relative importance varies. Among large quadrupeds, for example, movements of the rib cage tend to be constrained, and the diaphragm bears especially great responsibility for ventilation.

The control of ventilation

THE VENTILATORY RHYTHM ORIGINATES IN THE PRE-BÖTZINGER COMPLEX In the past decade, investigators have obtained strong evidence that the specific neurons that *initiate* the breathing rhythm in mammals are located in a bilaterally arrayed pair of neuron clusters within the ventrolateral medulla of the brainstem, called the **pre-Bötzinger complex**. Additional clusters of neurons that are important in modulating the neural control of breathing are also found in the medulla and the nearby pons. A stunning recent discovery is that thin medullary tissue slices containing the pre-Bötzinger complex produce neural outputs that, judging from their patterns, control not only the routine breathing rhythm, but also sighs and gasps (Figure 21.21). As knowledge of the pre-Bötzinger complex unfolds, it will not be surprising to find that aberrations in this specific set of medullary neurons are at the root of pathological apneic syndromes that sometimes result in death (such as sudden infant death syndrome).

VENTILATION IS MODULATED BY CHEMOSENSATION OF CO_2, H^+, AND O_2 As briefly mentioned earlier, the partial pressures of O_2 and CO_2 in alveolar gas are ordinarily held at set-point levels under a wide range of functional states. In humans at rest, the alveolar O_2 partial pressure is 13.3 kPa (100 mm Hg), and the CO_2 partial pressure is 5.3 kPa (40 mm Hg). These partial pressures remain the same during light to moderately intense exercise. Only during heavy exertion do the gas partial pressures in the alveoli deviate more than slightly from the resting values. To explain the stability of alveolar gas composition, ventilatory controls based on chemosensation of CO_2, H^+, and O_2 are critical, although they are not the only important controls of ventilation.

Controls based on sensation of CO_2 and H^+ are the most potent of the chemosensory controls. When the concentration of CO_2 in the blood or other body fluids rises or falls, the concentration of H^+ typically increases or decreases in tandem (see page 599).[8] By means not yet fully understood, the blood concentrations of CO_2 and H^+ are independently sensed by chemosensitive neural zones near the ventral surface of the medulla. A deviation of either concentration from its normal level exerts a potent influence on breathing, and because the two tend to vary together, they often exert synergistic effects. Ventilation increases or decreases (through increases or decreases in tidal volume and breathing frequency) in such a way as to bring the concentrations back toward normal (a negative-feedback system). For example, if the CO_2 concentration of the blood is elevated, ventilation is increased, resulting in a greater rate of CO_2 exhalation. The potency of these effects is illustrated by the fact that an increase of just 0.5 kPa (4 mm Hg) in a person's arterial partial pressure of CO_2—from 5.4 to 5.9 kPa (40 to 44 mm Hg)—will cause about a *doubling* of the volume of air ventilated per minute. During high-intensity exercise, as discussed in Chapter 6, production of lactic acid by anaerobic glycolysis can increase the H^+ concentration of the blood, quite apart from the effects of CO_2 on H^+. Ventilatory drive caused by this acidification sometimes is potent enough to give athletes a sense of profound discomfort because of the intensity of their breathing.

The blood partial pressure of O_2 is sensed principally by chemoreceptive bodies outside the central nervous system, the **carotid bodies** and **aortic bodies**. The two carotid bodies (each measuring about 0.5 cm in a person) are positioned along the paired common carotid arteries, near where each branches to form internal and external carotids; the bodies receive blood flow from the carotids and relay sensory information on the blood O_2 partial pressure to the brainstem via the glossopharyngeal nerves. Humans have only carotid bodies, but dogs, cats, and many other mammals have aortic bodies as well. The aortic bodies are located along the aortic arch and relay their sensory information via the vagus nerves. Both the carotid and aortic bodies, being richly perfused with blood from major arteries, are in excellent positions to monitor arterial O_2 partial pressure.

Whereas ventilation is markedly stimulated by even small elevations of arterial CO_2 above normal, the arterial O_2 partial pressure must fall far below normal before ventilation is reliably stimulated. Thus, medullary sensation of CO_2 and H^+ is paramount in regulating ventilation under usual resting conditions. In people, if the arterial O_2 partial pressure falls below 7 or 8 kPa (50–60 mm Hg) [as compared to a normal arterial value of ~12.7 kPa (~95 mm Hg)], marked stimulation of ventilation occurs. Thus, sensation of O_2 partial pressure is important in ventilatory regulation in certain circum-

[8] CO_2 has aptly been termed a "gaseous acid" because it reacts with H_2O to form carbonic acid (H_2CO_3), which then dissociates to form H^+ and HCO_3^-. The changes in H^+ concentration are sometimes expressed as changes in pH. When H^+ concentration increases, pH decreases, and vice versa.

Figure 21.21 The fountainhead of breathing A 1-mm-thick slice of the medulla of a nestling mouse produces electrical signals that, judging from their patterns, control the rhythm of ordinary breathing and sighs. Each recording shows electrical activity as a function of time. The recordings presented here have been integrated to remove random noise, explaining why they are smoother than those in Figure 21.9. (After Lieske et al. 2000.)

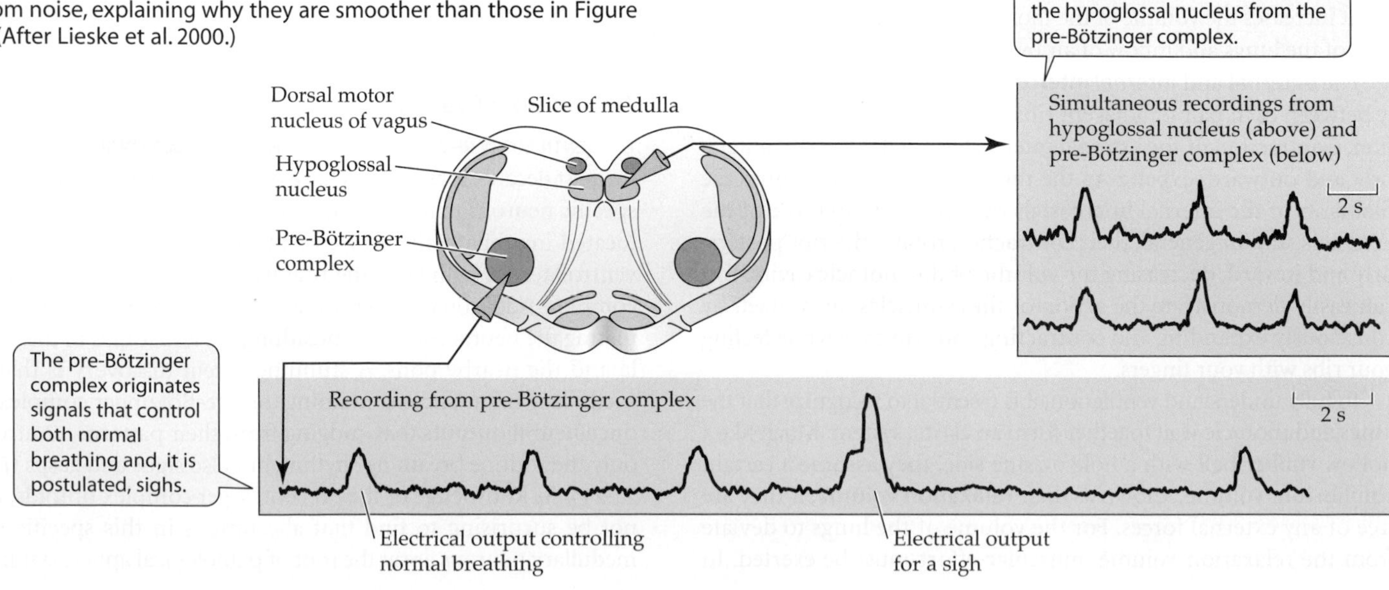

stances, as at high altitude (Box 21.1). The carotid and aortic bodies become more sensitive to lowered O_2 when the blood concentration of CO_2 or H^+ is elevated.

VENTILATION IS ALSO MODULATED IN OTHER IMPORTANT WAYS The most obvious type of modulation of ventilation in people is conscious control. We can temporarily stop breathing by choosing to stop. There are, in addition, other types of control besides the chemosensory ones already discussed.

One well-understood set of controls is based on mechanoreceptors in the lungs, which sense stretch or tension in the airways. Information from these receptors is relayed via sensory neurons to the brainstem, where signals for inhalation tend to be inhibited by lung expansion and excited by lung compression. Certain of these mechanosensory responses are known as the *Hering-Breuer reflexes*.

During exercise, there are clearly important controls operating in addition to chemosensory ones. Whereas these other controls are important, they are not well understood. As already stressed, the arteri-

BOX 21.1 Mammals at High Altitude

The environment at high montane altitudes is challenging in many respects. It can be cold, windy, dehydrating, and high in ultraviolet radiation. The most immediate challenge for a mammal at high elevation, however, is to meet the O_2 demands of its cells, because the source of O_2—the atmosphere—is rarefied. Here, we discuss several dimensions of this challenge; other aspects are discussed in Boxes 6.3 and 22.3.

Permanent human settlements occur at 4600 m (15,000 ft) and moderately higher elevations. Although the atmospheric O_2 partial pressure is only 60% as high as at sea level, the people in these settlements are far from sedentary. Peruvians living at such altitudes, for example, are miners and play soccer as a pastime. At the summit of Mt. Everest—8848 m—the atmospheric O_2 partial pressure is only about one-third that at sea level: near 7.1 kPa (53 mm Hg). As noted in Box 6.3, some people have climbed to the summit without supplementary O_2, but the achievement is at the very margin of human capability.

When a person born and reared at low altitude ascends the high mountains, his or her functional traits change over time as acclimatization occurs. Debate remains, however, over whether such an individual can ever reach the physiological status of people born and reared at high altitude. So, in analyzing people at high altitude, at least three classes need to be distinguished: the newly arrived lowlander, the acclimatized lowlander, and the native highlander. Among other mammals, many species resemble humans in being predominantly of lowland distribution, but there are also species that are limited to high altitudes.

In the figure, part (1) depicts the oxygen cascade for Peruvians living at sea level and at 4500 m. It is immediately evident that despite the large drop in *ambient* O_2 partial pressure at 4500 m, the *venous* partial pressure of the highlanders is reduced only a little. This *conservation* of the venous O_2 partial pressure results from significant reductions at high altitude in two of the partial pressure drops (steps) of the oxygen cascade. The drop in partial pressure between ambient air and alveolar gas is about 4.3 kPa (32 mm Hg) at high altitude, in contrast to 5.7 kPa (43 mm Hg) at sea level; and the drop between arterial blood and mixed venous blood is about 1.5 kPa (11 mm Hg) at high altitude, versus 7.3 kPa (55 mm Hg) at sea level. The explanation of the small arterial-to-venous drop at high altitude is a topic in blood gas transport and thus is discussed in Chapter 22 (see Box 22.3). Here, we examine the effects of *lung function* and *systemic tissue physiology*.

One of the most important defenses marshaled by humans at high altitude is **hyperventilation**, defined to be an increase in the rate of ventilation associated with any given rate of O_2 consumption. At a given O_2 demand, the ventilation rate in native highlanders at 4500 m, for example, is about 40% higher than that in people residing at sea level. When lowlanders

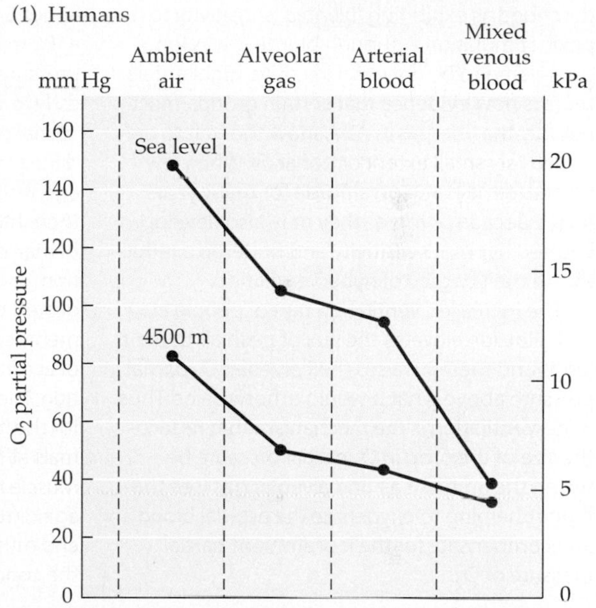

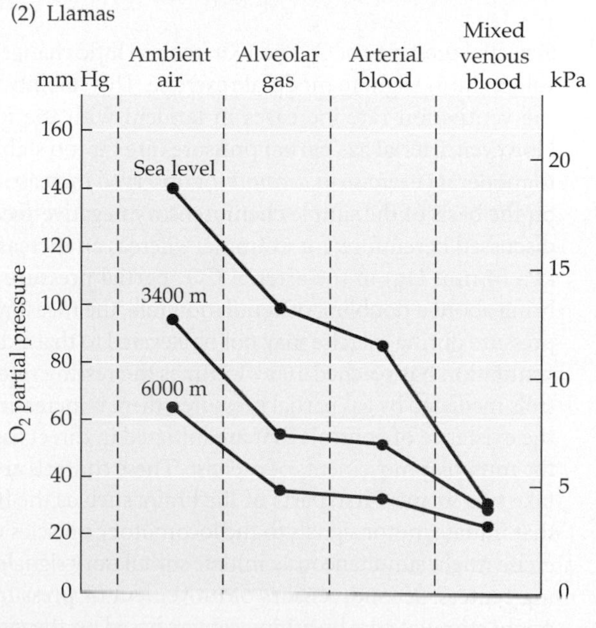

Oxygen cascades at high altitude for people and llamas (1) To obtain the data on people, two groups of native male Peruvians were studied at their altitudes of residence. (2) To obtain the llama (*Lama glama*) data, a set of animals reared at sea level was studied at sea level, at 3400 m after 10 weeks of acclimatization, and during an acute exposure to the atmospheric pressure at 6000 m. (Data from Banchero, Grover, and Will 1971; Torrance et al. 1970.)

BOX 21.1 *(Continued)*

first ascend to high altitude, a prompt (acute) increase in their rate of ventilation occurs; this increase is probably principally activated by the reduction in their arterial O_2 partial pressure, sensed by the carotid bodies. As lowlanders pass their first days at high altitude, their rate of ventilation becomes even higher, evidently because of an increasing physiological sensitivity of the breathing control mechanisms to hypoxic stimulation. After the initial acclimatization, the degree of hyperventilation remains about the same for years. Some native highlanders at a given altitude exhibit a markedly lower degree of hyperventilation than acclimatized lowlanders and are described as exhibiting "*blunted*" sensitivity to hypoxic stimulation. Although blunting was long considered to be universal in native highlanders, there is now evidence that certain groups, most notably the Sherpas in Asia, show the trait to only a relatively small extent or not at all. When lowlanders remain at high altitude for many years (e.g., a decade or more), they may also develop blunted hypoxic sensitivity and undergo a reduction in their degree of hyperventilation.

The increased ventilation rate of people at high altitude elevates the flux of fresh air to their lungs and thereby raises their alveolar O_2 partial pressure above what it would otherwise be. Thus, hyperventilation is the mechanism that reduces the size of the drop in O_2 partial pressure between the ambient air and alveolar gas (see the figure), helping to oxygenate the arterial blood and compensate for the low ambient partial pressure of O_2.

Individuals vary in their hyperventilation responses, and these differences affect their success at high altitude. In general, people who exhibit a particularly vigorous acute hyperventilation response to hypoxia are the most likely to be strong climbers at high altitudes.

Although processes such as hyperventilation help to keep O_2 partial pressures in the systemic blood capillaries from falling excessively, capillary O_2 partial pressures do in fact decline at high altitude. In people at sea level, as part (1) of the figure shows, blood enters the capillaries at an arterial partial pressure of about 12.5 kPa (94 mm Hg) and exits at a mixed venous partial pressure of about 5.2 kPa (39 mm Hg). When the altitude is 4500 m, blood enters at a much lower partial pressure, 5.9 kPa (44 mm Hg), and exits at a modestly lower one, 4.4 kPa (33 mm Hg). Thus, the O_2 partial pressure in the capillaries, which drives O_2 diffusion to the mitochondria in cells, is, on average, reduced at high altitude. For two decades, there has been a lot of interest in how the tissues of mammals might accommodate to this reduction. Investigations of various species have suggested that several types of tissue-level adjustments sometimes occur, either as a consequence of acclimatization or as a result of adaptive evolution. There is evidence, some of it controversial, for the following sorts of tissue changes in mammals at high altitude: (1) a decrease in average muscle fiber diameter, leading to a shorter average diffusion distance between blood capillaries and muscle fiber mitochondria; (2) an increase in the concentration of myoglobin, a compound that facilitates diffusion of O_2 through cells (see Chapter 22); (3) suppression of metabolism (hypometabolism) in some tissues; (4) a switch of the heart muscle toward greater use of glucose, the cardiac fuel that yields the most ATP per O_2 molecule; (5) a switch of skeletal muscle away from use of anaerobic glycolysis (because the ultimate O_2 cost per ATP is greater if lactic acid is synthesized and later metabolized than if the ATP is simply made aerobically, see Chapter 6); and (6) other changes that emphasize the use of metabolic fuels and catabolic pathways that tend to maximize the yield of ATP per O_2 molecule delivered to the tissues.

Oxygen cascades of llamas, which have a long evolutionary history of life at high altitude, are shown in part (2) of the figure. Llamas hyperventilate at high altitude; this seems to be a near-universal mammalian response. However, llamas are less reliant on this defense than humans, perhaps because their hemoglobin has a higher affinity for O_2 than ours and thus does not require as high an alveolar O_2 partial pressure to become loaded with O_2. Unlike humans, llamas do not hyperventilate at all at 3400 m (note that the size of the ambient-to-alveolar drop at 3400 m is essentially the same as that at sea level); at 6000 m, llamas hyperventilate less than humans at the same altitude. At all altitudes, llamas have a lower mixed venous partial pressure of O_2 than humans, suggesting that they may place greater reliance on tissue-level adaptations that aid function at reduced capillary O_2 partial pressures.

al partial pressures of O_2 and CO_2 remain little changed from resting values during light to moderate exercise. This stability results *because* the ventilation rate increases in tandem with the metabolic rate. However, arterial gas partial pressures are far too stable during light to moderate exercise to *account* for observed increases in ventilation on the basis of the simple chemosensory negative-feedback systems discussed heretofore; for example, whereas an increase of about 0.5 kPa (4 mm Hg) in the arterial CO_2 partial pressure is required to bring about a doubling of ventilation rate, the measured CO_2 partial pressure during exercise may not be elevated to that extent even when ventilation has reached 10 to 15 times the resting rate! Besides controls mediated by gas partial pressures, there is increasing evidence for the existence of controls that are initiated in direct association with the muscular movements of exercise. These controls are postulated to take two forms: First, parts of the brain, such as the hypothalamus, that initiate motor signals to the locomotory muscles involved in exercise might simultaneously initiate stimulatory signals to the breathing centers. Second, sensors of movement or pressure in the limbs might stimulate the breathing centers based on the vigor of the limb activity they detect. One persuasive piece of evidence for these sorts of controls is that when people suddenly begin to exercise at a moderate level, a marked increase in their ventilation rate is evident within just *one or two breaths*; this response seems far too rapid to be mediated by changes in the chemical composition of the body fluids. Another piece of evidence is that in many mammals, breathing movements and limb movements are synchronized during running.

BOTH TIDAL VOLUME AND BREATHING FREQUENCY ARE MODIFIED BY THE CONTROL SYSTEMS The overall rate of lung ventilation depends on two properties: the tidal volume, V_T, and the frequency of breaths, f, which is usually expressed as the number of breaths per minute. The product of these is the **respiratory minute volume**:

$$\text{Respiratory minute volume (mL/minute)} = V_T\,(\text{mL/breath}) \times f\,(\text{breaths/minute}) \tag{21.2}$$

To illustrate, resting people have a tidal volume of about 500 mL and breathe about 12 times per minute. Thus, their respiratory minute volume is about 6 L/minute.

During exercise or other states that increase the rate of metabolism, both tidal volume and breathing frequency are increased. People and other mammals cannot maximize both of these variables simultaneously because the time needed for one breathing cycle tends to increase as the tidal volume increases. Nonetheless, during vigorous exercise, trained athletes are able simultaneously to maintain a tidal volume of 3 L or more while breathing 30 times per minute or more. In this way, their respiratory minute volume can reach over 100 L/minute.

The **alveolar ventilation rate**, typically expressed as the **alveolar minute volume**, is defined to be the rate at which new air is brought into the respiratory airways (the respiratory bronchioles, the alveolar ducts, and the alveolar sacs with their borders of alveoli). This rate is important because air that reaches only the anatomical dead space (conducting airways) does not participate in gas exchange; it is the air that reaches the respiratory airways that is "the payoff" of ventilation. The alveolar minute volume is calculated by subtracting the volume of the anatomical dead space, V_D, from the tidal volume and multiplying by the breathing frequency:

$$\text{Alveolar minute volume} = (V_T - V_D) \times f \tag{21.3}$$

Besides this minute volume, another property of importance relating to the respiratory airways is the fraction of all inhaled air that reaches them. This fraction—calculated by dividing the alveolar minute volume (Equation 21.3) by the total minute volume (Equation 21.2)—is $(V_T - V_D)/V_T$. If V_D is presumed to be a constant, it is clear that the fraction of air reaching the respiratory airways increases as the tidal volume increases.

This conclusion helps resolve a paradox. When the overall ventilation rate is increased during exercise or other states that raise metabolism, the oxygen utilization coefficient increases. People, for example, use about 20% of the O_2 in the air they breathe when their tidal volume is 500 mL, but use about 30% when their tidal volume is 2000 mL. How is this possible if, as we have often emphasized, the control systems keep the alveolar O_2 partial pressure constant (except during the most strenuous work)? The paradox is resolved by recognizing two aspects of gas exchange in the respiratory airways. First, air that reaches the respiratory airways always gives up about the same fraction of its O_2 (accounting for the constancy of alveolar O_2 partial pressure). Second, however, a greater proportion of all the air that is breathed actually enters the respiratory airways as the tidal volume increases.

In species of different sizes, lung volume tends to be a constant proportion of body size but breathing frequency varies allometrically

If we look at the full range of mammals, ranging in size from shrews to whales, lung volume turns out to be a rather constant proportion of total body volume: Lung volume in liters averages about 6% of body weight in kilograms. The resting tidal volume tends consistently to be about a tenth of lung volume, meaning that mammals of all sizes have a resting tidal volume of roughly 6 mL/kg body weight. The resting oxygen utilization coefficient is also about the same in all mammals regardless of weight (about 20%). These facts imply that when mammals of all sizes are at rest, the amount of O_2 they obtain per breath is approximately a constant proportion of their body weight. However, as discussed in detail in Chapter 5, the resting weight-specific rate at which mammals metabolically consume O_2 increases allometrically as body size decreases. If the weight-specific O_2 demand of a small mammal is greater than that of a large mammal, and yet the small animal obtains only about the same amount of O_2 per breath per unit of weight, then the small animal must breathe more frequently. This is exactly how small and large species relate, and it explains why the breathing frequencies of mammals are allometrically related to body size. At rest, a mouse breathes 100 times per minute, whereas a person breathes 12 times per minute.

Pulmonary surfactant keeps the alveoli from collapsing

The tendency of any sort of bubble to collapse shut is an inverse function of its radius. Thus, alveoli, which are bubblelike and of very small radius, might be expected to exhibit a high inherent tendency to collapse. This tendency is reduced by the presence on their inner walls of an exceedingly thin layer of material that reduces the surface tension of the walls, thus opposing collapse. This material, known as **pulmonary surfactant**, consists of a complex mix of surface-active lipids and proteins secreted by certain alveolar cells. The surfactant layer exhibits the remarkable property that its surface tension decreases dramatically if alveolar size decreases; this decrease in surface tension helps prevent small alveoli from undergoing collapse by emptying into larger ones. If there could be any doubt about the physiological importance of lung surfactant, it is dispelled by the plight of people born with abnormally low amounts of surfactant or abnormal surfactant types. For example, the disease of premature infants known as *infant respiratory distress syndrome* is caused by surfactant abnormalities. Infants with the disease may die because they require extra effort to expand their lungs during inhalation and also suffer alveolar collapse.

Pulmonary surfactants that share basic chemical similarities have been reported from the lungs of all groups of terrestrial vertebrates, lungfish, and some other air-breathing fish. Thus, these surfactants seem to have a long evolutionary history, dating back to the origins of vertebrate air breathing.

Summary: Breathing by Mammals

- The lungs of mammals consist of dendritically branching airways that end blindly in small, thin-walled, well-vascularized outpocketings, the alveoli. The airways exhibit 23 levels of branching in the human adult lung, giving rise to 300 million alveoli. The airways in a mammalian lung are categorized as conducting airways, where little gas exchange with the blood occurs, and respiratory airways, where most gas exchange with the blood takes place.
- Because of the blind-ended structure of the mammalian lung, the gas in the alveoli has a substantially lower O_2 partial pressure and higher CO_2 partial pressure than atmospheric air.
- Contraction of the diaphragm is a principal force for inhalation in mammals, especially large quadrupeds. External intercostal muscles may contribute to inhalation; internal intercostal muscles and abdominal muscles may contribute to exhalation. Inhalation occurs by suction as the lungs are expanded by contraction of inspiratory mus-

cles. At rest, exhalation occurs passively by elastic rebound of the lungs to their relaxation volume when the inspiratory muscles relax.

- The breathing rhythm in mammals originates in a central pattern generator in the pre-Bötzinger complex in the medulla of the brainstem.
- The most potent chemosensory stimulus for increased ventilation in mammals is a rise in blood CO_2 partial pressure and/or H^+ concentration, sensed in the medulla. The blood O_2 partial pressure, ordinarily a less influential factor in controlling ventilation, is sensed by the carotid bodies along the carotid arteries (in humans) or by carotid and aortic bodies. The control of ventilation during exercise involves stimuli generated in association with limb movement as well as chemosensory controls.
- Pulmonary surfactant, a surface tension–reducing mix of lipids and proteins, makes a critical contribution to maintaining the proper microscopic conformation of the lungs in all air-breathing vertebrates.

Breathing by Birds

The lungs of birds, although they are logical derivatives of reptilian lungs, differ structurally from the lungs of all other modern vertebrates. The fundamental structural difference between avian and mammalian lungs inevitably invites comparisons. Are avian lungs *functionally* superior to mammalian lungs? Some authorities have concluded that the designs of the lungs in birds and mammals are "different but equal;" according to this view, the two sorts of lungs are equal in their gas-exchange ability. Other authorities have concluded that the lungs of birds are in fact superior organs of gas exchange. As evidence, they point to the large surface areas for gas exchange and the thin gas-exchange membranes of avian lungs (see Figure 21.7) and to experiments on gas exchange at high altitude. There are birds that can *fly*—not just mope around and survive—at the altitude of Mount Everest, and some scientists argue that their lungs are the key organs making such feats possible.

A bird's trachea bifurcates to give rise to two primary bronchi, which enter the lungs. Here the similarity to mammals ends, however. The primary bronchus that enters each lung *passes through* the lung, being known as the **mesobronchus** within the lung. Two groups of branching **secondary bronchi** arise from the mesobronchus. One group, which arises at the anterior end of the mesobronchus, branches over the ventral surface of the lung. The other group originates toward the posterior end of the mesobronchus and branches over the dorsolateral lung surface. For simplicity, we call these the *anterior* and *posterior* groups of secondary bronchi, although they are formally termed the *medioventral* and *mediodorsal* groups, respectively. Also in the spirit of simplicity, each group is represented as just a single passageway in Figure 21.22.

The anterior and posterior secondary bronchi are connected by a great many small tubes, 0.5 to 2.0 mm in internal diameter, termed **tertiary bronchi** or **parabronchi**. As depicted

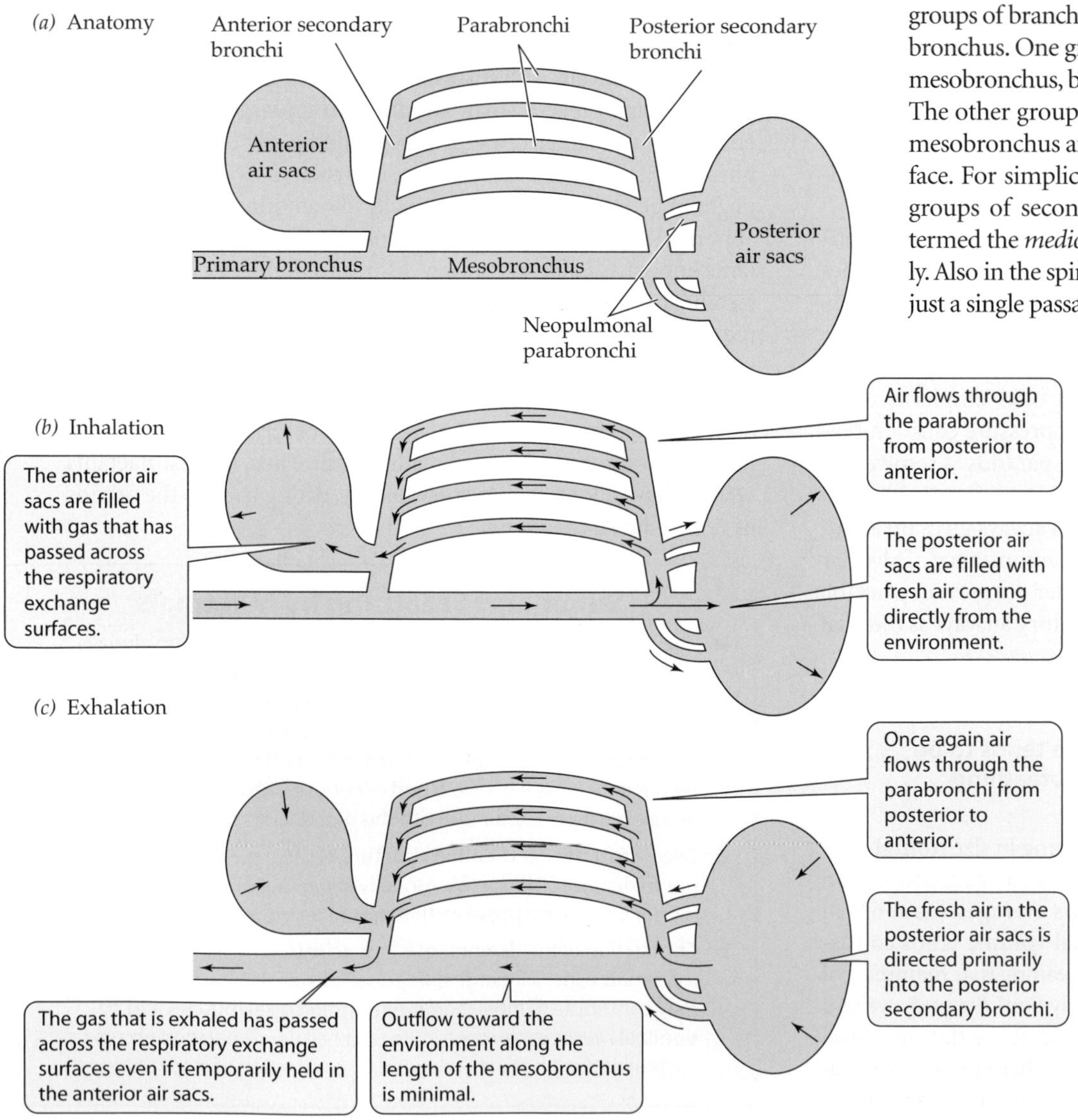

Figure 21.22 Airflow in the lungs and air sacs of birds *(a)* Basic anatomy of the avian lung and its connections with the air sacs. In this presentation, the anterior and posterior groups of secondary bronchi are each represented as a single passageway. The tubes labeled simply *parabronchi* are those of the dominant paleopulmonal system. *(b, c)* Airflow during *(b)* inhalation (when the air sacs are expanded) and *(c)* exhalation (when the air sacs are compressed).

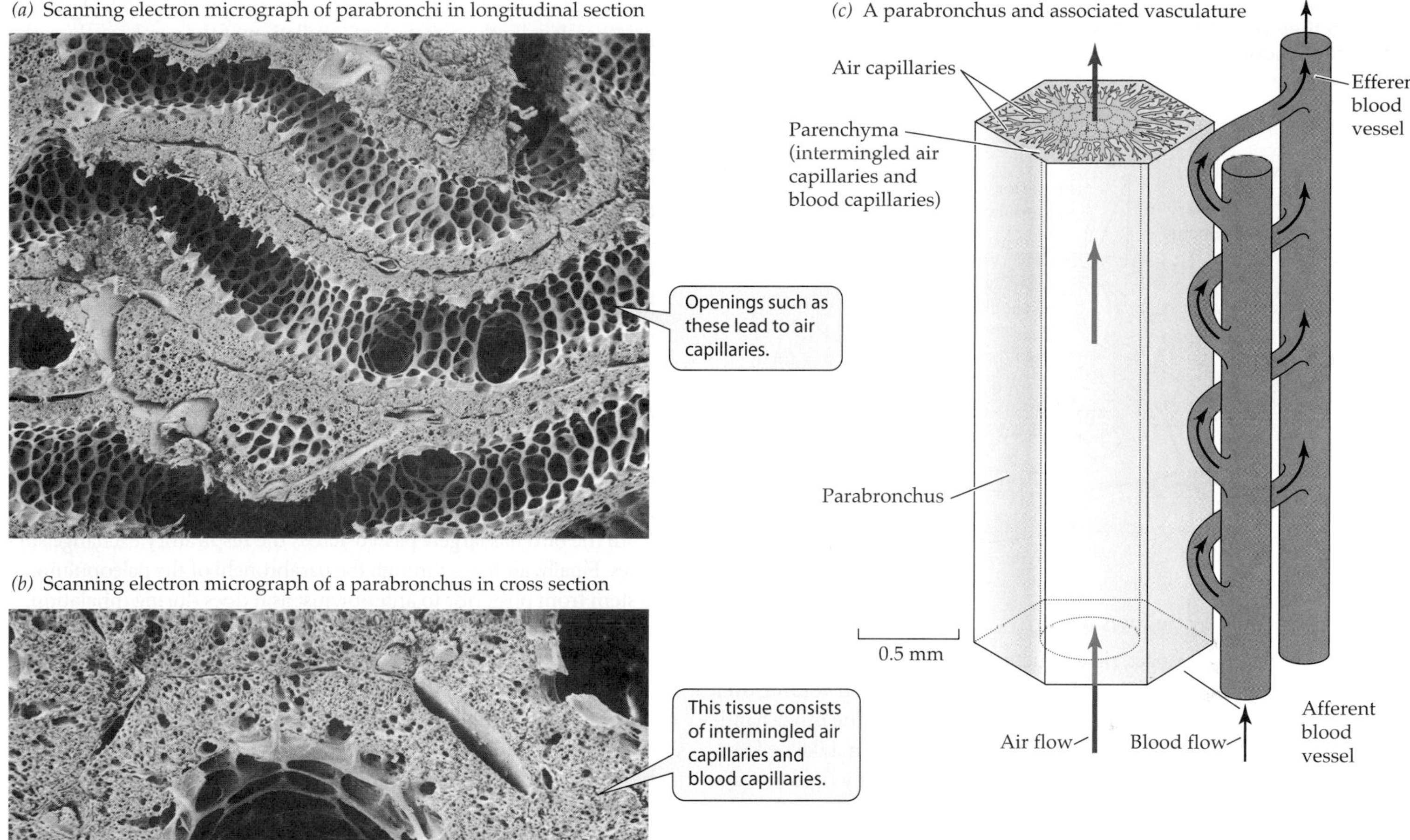

Figure 21.23 Parabronchi and air capillaries: The gas-exchange sites in avian lungs *(a)* Scanning electron micrograph of a chicken lung (*Gallus*). Magnification: 16 ×. *(b)* Scanning electron micrograph of a single parabronchus of a chicken lung in cross section. Magnification: 58 ×. *(c)* Diagram of the structure of a parabronchus and its associated vasculature. (*a* and *b* courtesy of Dave Hinds and Walter S. Tyler.)

in Figure 21.23, each parabronchus gives off radially along its length an immense number of finely branching **air capillaries**. The air capillaries are profusely surrounded by blood capillaries and are the sites of gas exchange. They are only 3 to 10 μm in diameter, and collectively, they form an enormous gas-exchange surface amounting to 200 to 300 mm^2/mm^3 of tissue in the parabronchial walls. Air flows through the parabronchi, but exchange between the central lumen of each parabronchus and the surfaces of its air capillaries is probably largely by diffusion. The parabronchi, air capillaries, and associated vasculature constitute the bulk of the lung tissue of a bird.

A bird's **air sacs**, which are part of the breathing system, are located *outside* the lungs and occupy a considerable portion of the thoracic and abdominal body cavities (Figure 21.24). Usually, there are nine air sacs, divisible into two groups. The *anterior air sacs* (two cervical sacs, two anterior thoracic sacs, and a single interclavicular sac) connect to various anterior secondary bronchi. The *posterior air sacs* (two abdominal sacs and two posterior thoracic sacs) connect to the posterior portions of the mesobronchi. (Each mesobronchus terminates at its connection with an abdominal air sac.) The air sacs are thin-walled, poorly vascularized structures that play little role in gas exchange between the air and blood. Nonetheless, as we will see, they are essential for breathing.

The structures of the lung described thus far are present in all birds, and their connections with the air sacs are similar in all birds. These pulmonary structures are collectively termed the **paleopulmonal system**, or simply **paleopulmo**. Most birds, in addition, have a more or less extensively developed system of respiratory parabronchial tubes running directly between the posterior air sacs and the posterior parts of the mesobronchi and posterior secondary bronchi (see Figure 21.22*a*)—and sometimes having even more elaborate connections. This system is called the **neopulmonal system**, or **neopulmo**. The paleopulmonal system is always dominant; the neopulmo never represents more than 20% to 25% of the total lung volume, even in its most highly developed form.

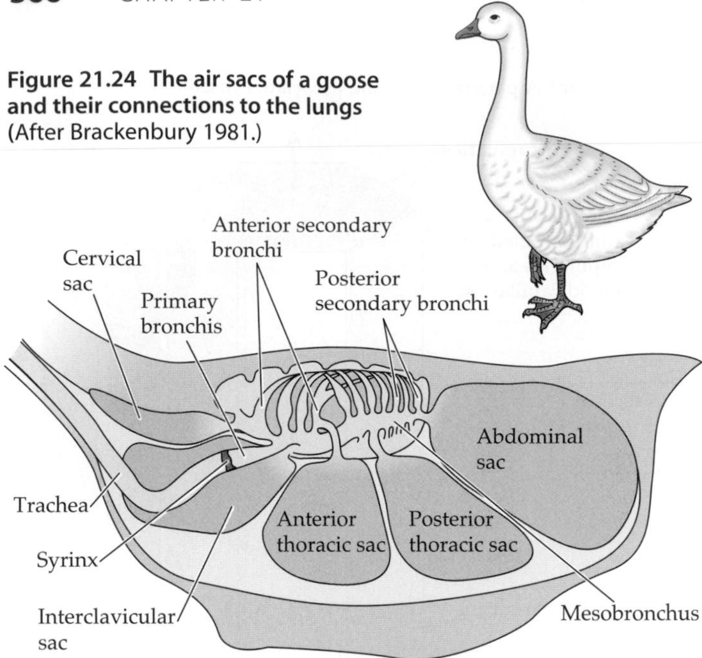

Figure 21.24 The air sacs of a goose and their connections to the lungs (After Brackenbury 1981.)

Ventilation is by bellows action

Avian lungs are compact, rigid structures. They contrast sharply with mammalian lungs in that they undergo little change in volume over the course of each breathing cycle. The air sacs, on the other hand, expand and contract substantially during ventilation. Like bellows, the air sacs suck and push gases through the relatively rigid airways of the lungs. In comparison to mammalian ventilation, this avian process turns out to be an energetically inexpensive way to move air.

The part of the rib cage surrounding the lungs themselves is relatively rigid. During inhalation, contraction of internal intercostal muscles and certain other thoracic muscles expands other parts of the rib cage (especially those posterior to the lungs), and the sternum swings downward and forward. These movements enlarge the air sacs by expanding the thoracoabdominal cavity. Some of the external intercostals and abdominal muscles compress the thoracoabdominal cavity and air sacs during exhalation. Resting birds typically breathe at only about one-half or one-third the frequency of resting mammals of equivalent body size, but the birds have greater tidal volumes.

Air flows unidirectionally through the parabronchi

Air flows unidirectionally through the parabronchi of the paleopulmonal system. To see how this occurs, we must describe the movement of air during both inhalation and exhalation. During inhalation, both the anterior and posterior sets of air sacs are expanded. Suction therefore is developed in *both* sets of air sacs, and *both* receive gas. As depicted in Figure 21.22*b*, air drawn in from the atmosphere flows through the mesobronchus of each lung to enter the posterior air sacs and posterior secondary bronchi; in turn, the air entering the posterior secondary bronchi is drawn anteriorly through the parabronchi by suction developed in the anterior air sacs. Three aspects of the events during inhalation deserve emphasis: First, the posterior air sacs are filled with relatively fresh air coming directly from the environment. Second, the anterior air sacs are filled for the most part with gas that has passed across the respiratory exchange surfaces in the parabronchi; this gas has a lowered O_2 partial pressure and an elevated CO_2 partial pressure. Finally, the direction of ventilation of the parabronchi in the paleopulmonal system is from posterior to anterior.

During exhalation, both sets of air sacs are compressed and discharge gas. As shown in Figure 21.22*c*, air exiting the posterior air sacs predominantly enters the posterior secondary bronchi to pass anteriorly through the parabronchi. This air is relatively fresh, having entered the posterior sacs more or less directly from the environment during inhalation. Gas exiting the parabronchi anteriorly, combined with gas exiting the anterior air sacs, is directed into the mesobronchus via the anterior secondary bronchi and is exhaled. Recall that the anterior air sacs were filled with gas from the parabronchi during inhalation. Thus, the exhaled gas is mostly gas that has passed across the respiratory exchange surfaces. Three aspects of the expiratory events deserve emphasis: First, the relatively fresh air of the posterior air sacs is directed mostly to the parabronchi; outflow toward the environment along the length of the mesobronchus is minimal, according to the evidence available. Second, the gas that is exhaled from the bird has largely passed across the respiratory exchange surfaces. Finally, air flows through the parabronchi of the paleopulmonal system from posterior to anterior, just as it does during inhalation.

One of the biggest remaining questions in the study of avian lungs is *how* air is directed along its elaborate and, in some ways, counterintuitive path through the air sacs and paleopulmonal system. Passive, flaplike valves appear to be entirely absent. Active, muscular valves could be present, but evidence for their existence is at best circumstantial. The weight of present evidence suggests that the complex architecture of the lung passages creates aerodynamic conditions that direct air along its inspiratory and expiratory paths without need of either passive or active valves.

Ventilation of the neopulmonal system is incompletely understood. Probably, however, airflow through many of the neopulmonal parabronchi is bidirectional (see Figure 21.22*b, c*).

As discussed in Box 21.2, the preparation of the lungs of birds to take over full responsibility for gas exchange at hatching presents unique challenges compared to other vertebrates.

The gas-exchange system is cross-current

When the unidirectional flow of air through the paleopulmonal parabronchi in the lungs of birds was first discovered, countercurrent exchange between the blood and air was quickly hypothesized. Soon, however, this hypothesis was disproved by clever experiments, which showed that the efficiency of gas exchange between air and blood is not diminished if the direction of airflow in the parabronchi is artificially reversed. Morphological and functional studies have now shown convincingly that blood flow in the respiratory exchange vessels occurs in a cross-current pattern relative to the flow of air through the parabronchi (see Figure 21.23*c*).

Summary: Breathing by Birds

- The lungs of birds are relatively compact, rigid structures consisting mostly of numerous tubes, running in parallel, termed parabronchi. Fine air capillaries, extending radially from the

BOX 21.2 Bird Development: Filling the Lungs with Air for Hatching

The lungs of both birds and mammals initially develop in a fluid-filled condition, and thus young animals of both groups face the problem of filling their lungs with air so as to be able to breathe at hatching or birth. When mammals are born, they are able to fill their lungs sufficiently to survive by inflating them suddenly with their first breath from the atmosphere. Birds, however, cannot inflate their lungs in this way. The gas-exchange elements of avian lungs, the air capillaries, cannot be inflated suddenly out of a collapsed state because the lungs are relatively rigid and the air capillaries themselves are of such extremely small diameter. Another obstacle to a sudden-inflation strategy for birds is that avian lungs probably will not work correctly unless *every single* critical airway becomes gas-filled, because the pattern of airflow through the lungs is determined by complex aerodynamic interactions among the airways. Birds have thus evolved a way to fill their lungs with air gradually before the lungs become essential for breathing.

During most of a bird's development inside an egg, its breathing organ is its highly vascular chorio-allantoic membrane, which is pressed against the eggshell on the inside. Oxygen and CO_2 pass between the atmosphere and the chorio-allantoic blood vessels by diffusion through gas-filled pores in the eggshell.

As an egg develops, it dehydrates by diffusion of water vapor outward through the eggshell pores. This dehydration results in the formation of a gas-filled space, called the *air cell*, within the egg at its blunt end. The air cell is critical for the young bird inside to fill its lungs with air, and the amount of water loss from an egg is controlled within fairly narrow limits to yield an air cell of particular size. About 2 or 3 days before a young bird hatches, it starts to breathe from the air space within its shell. During the ensuing hours until hatching occurs, the bird makes a *gradual* transition from respiration across its chorio-allantoic membrane to full-fledged pulmonary breathing. The air capillaries, in fact, undergo most of their pre-hatching development during this period. By the time hatching begins, the airways and air capillaries in the lungs are already gas-filled, poising the lungs to assume full responsibility for the life of the animal.

lumen of each parabronchus, are the principal sites of gas exchange. Air sacs, which are nonrespiratory, are integral parts of the breathing system.

- The lungs are ventilated by a bellows action generated by expansion and compression of the air sacs.
- Airflow through the parabronchi of the paleopulmonal system (the major part of the lungs) is posterior to anterior during both inhalation and exhalation. Cross-current gas exchange occurs.

Breathing by Aquatic Invertebrates and Allied Groups

Many small aquatic invertebrates, and some large ones that have low weight-specific metabolic rates, have no specialized breathing organs. They exchange gases across general body surfaces, which sometimes are ventilated by cilia- or flagella-generated water currents, or by being waved about in the water. Many larvae and some adults also lack a circulatory system. Thus, gases move about within the bodies of these animals by diffusion or by the squishing of body fluids from place to place. To the human eye, these sorts of gas-exchange systems are confining. They suffice only if the animals are tiny (see Box 20.1) or if they adhere to specialized body plans, such as the arrangement in jellyfish whereby most cells are very near a body surface regardless of how thick the gelatinous structural parts of the body may be.

For the most part, the adults of the relatively advanced phyla have gills of some sort. The gills of various major phyletic groups are often independently evolved. Thus, whereas they all are evaginated structures that project into the water (thus meeting the definition of gills), they vary widely in their structures and in how they are ventilated (Figure 21.25).

Molluscs exemplify an exceptional diversity of breathing organs built on a common plan

The phylum Mollusca nicely illustrates that within a single phyletic group, a single basic sort of breathing apparatus can undergo wide diversification. In molluscs, outfolding of the dorsal body wall produces a sheet of tissue, the **mantle**, that commonly overhangs or surrounds all or part of the rest of the body and is responsible for generating the shell. Where the mantle overhangs the rest of the body, it encloses an external body cavity, the **mantle cavity**. The gills of molluscs typically are suspended in this cavity and thus are internal gills (see Figure 21.2). Some aquatic snails provide a straightforward example. In these snails (Figure 21.26*a*), a series of modest-sized *gill leaflets* hangs in the mantle cavity and is ventilated unidirectionally by ciliary currents. At least sometimes, blood flow through the leaflets is opposite to the direction of water flow, and countercurrent gas exchange occurs.

One major modification of the gills within the phylum Mollusca is the evolution of extensive *sheetlike gills* in the clams, mussels, oysters, and other lamellibranchs. The very name of these animals—*lamellibranchs*—means "sheet-gilled." Four gill sheets, or lamellae, composed of fused or semifused filaments, hang prominently within the mantle cavity (Figure 21.26*b*). Cilia on the gill sheets drive incoming water through pores on the gill surfaces into water channels that run *within* the gill sheets; the water channels carry the water to exhalant passages. The direction of water flow within the water channels is opposite to the direction of blood flow in the major gill blood vessels, meaning that countercurrent gas exchange can again occur. In the lamellibranchs, the specialized sheetlike gills represent, in part, an adaptation for feeding: As the abundant flow of incoming water passes through the arrays of pores leading to the interior water channels of the gill sheets, food particles suspended in the water are captured for delivery to the mouth (a type of suspension feeding; see page 103). In some (not all) lamellibranch molluscs, the food-collection function has become paramount; enough respiratory gas exchange can occur across gen-

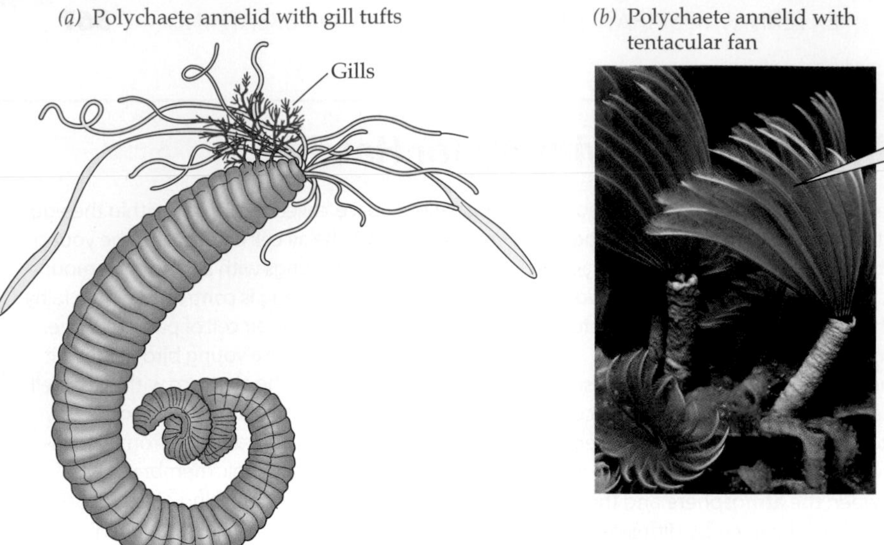

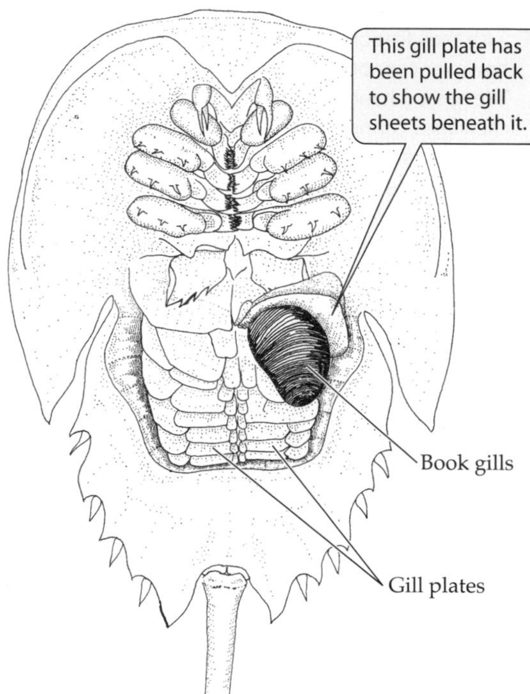

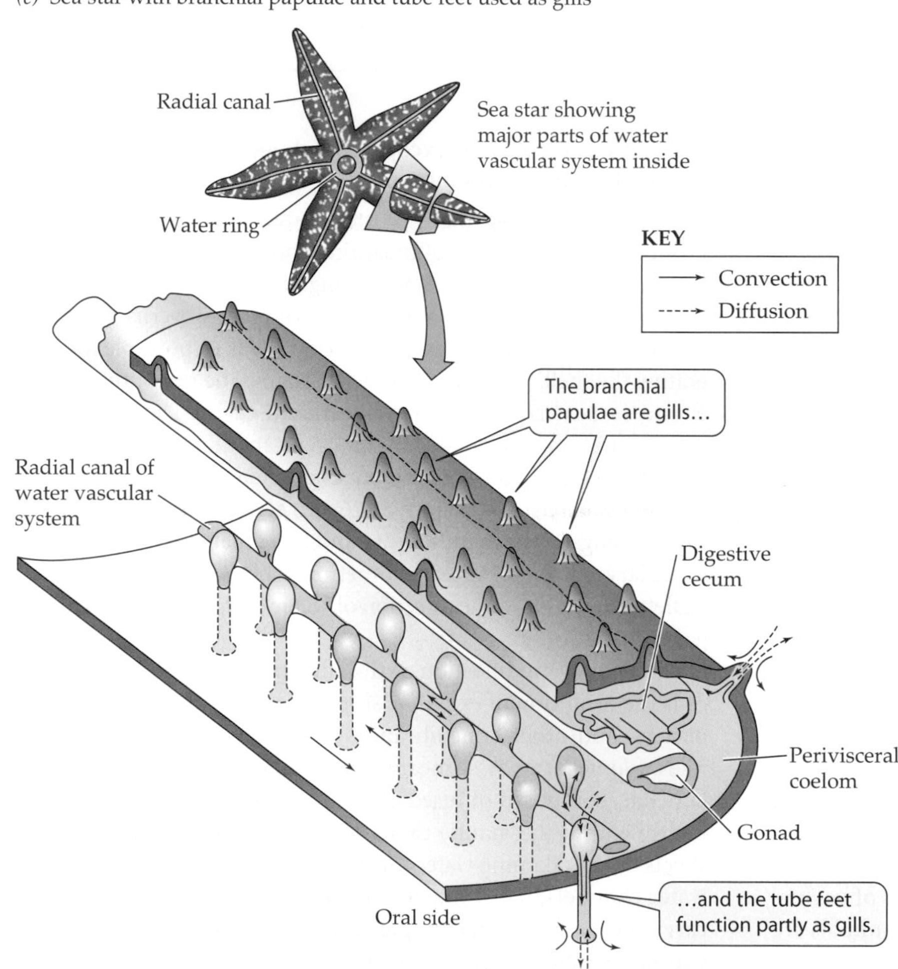

Figure 21.25 A diversity of gills in aquatic invertebrates *(a)* This terebellid worm (*Amphitrite*), a type of marine annelid, lives inside a tube it constructs and can pump water in and out of the tube. *(b)* This fanworm, another type of marine annelid, also lives in a tube, but when undisturbed, projects its well-developed array of pinnately divided tentacles into the ambient water. The tentacles are used for both feeding and respiratory gas exchange; they are ventilated by ciliary currents. *(c)* Sea stars bear many thin-walled, fingerlike projections from their coelomic cavity, termed branchial papulae ("gill processes"), on their upper body surfaces; respiratory gases pass between the coelomic fluid and ambient water by diffusion through the walls of the papulae. In much the same way, gases diffuse between the coelomic fluid and ambient water through the tube feet and associated parts of the water vascular system. Cilia accelerate these processes by circulating fluids over the inner and outer surfaces of the papulae and tube feet. *(d)* Horseshoe crabs (*Limulus*) have unique *book gills*, consisting of many thin gill sheets arranged like pages of a book. The book gills are protected under thick gill plates, which undergo rhythmic flapping motions that ventilate the gills. (*d* drawn by Ralph Russell, Jr.)

eral body surfaces to meet metabolic needs. Thus the "gills" have become primarily feeding organs.

In the cephalopod molluscs—the squids, cuttlefish, and octopuses—it is not so much the gills that are specialized, but the mechanism of ventilation. The gills are feathery structures that follow the usual molluscan plan of being positioned in the mantle cavity (Figure 21.26*c*). They are ventilated, however, by muscular contraction rather than beating of cilia. Squids and cuttlefish (and sometimes

Figure 21.26 The diversification of the breathing system in molluscs

(a) **Aquatic snail**

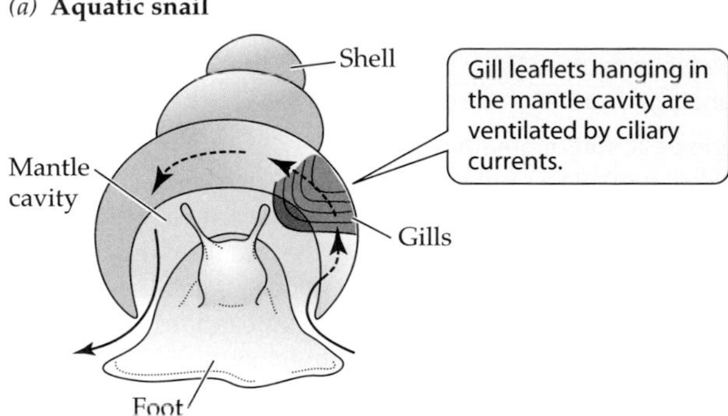

(b) **Clam (a lamellibranch mollusc)**

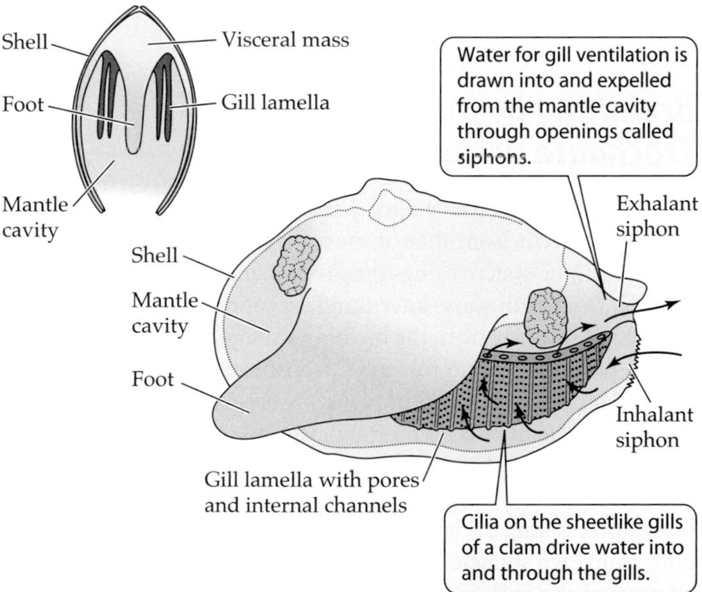

(c) **Squid (a cephalopod)**

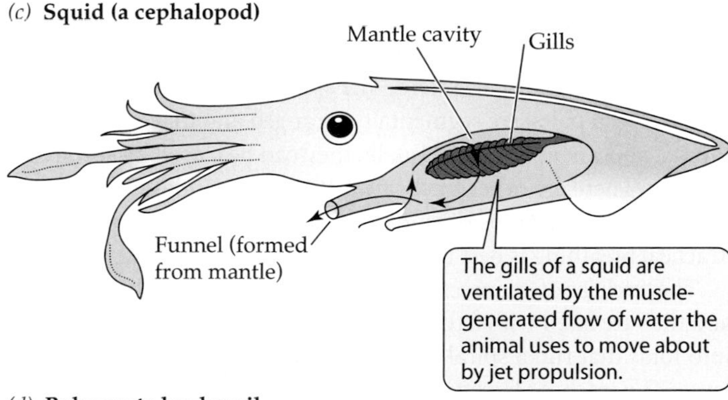

(d) **Pulmonate land snail**

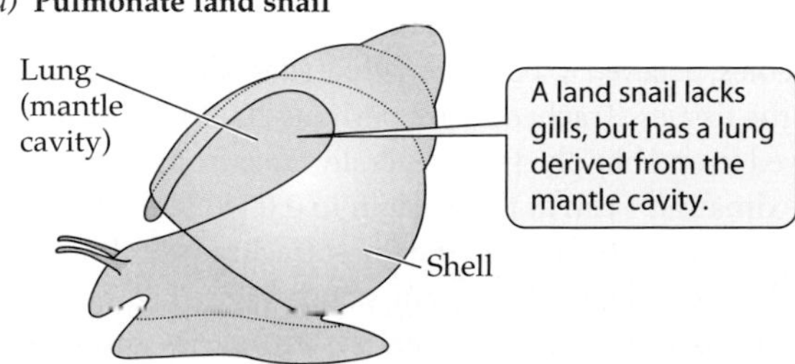

octopuses) swim by using muscular contractions of the mantle; they alternately suck water into the mantle cavity through diffuse incurrent openings and then drive it forcibly outward through a ventral funnel by mantle contraction, producing a jet-propulsive force. The gills are ventilated (in countercurrent fashion) by the vigorous flow of water used for propulsion. Some species move so much water through their mantle cavities for propulsion that they use only 5% to 10% of the O_2 in the water. They are thus very unusual in comparison to most highly active aquatic animals in that their O_2 supply from breathing is not even close to being limited by the low solubility of O_2 in the environmental water.

A final specialization worthy of note in molluscs is the evolution of the mantle cavity into a lung in the dominant group of snails and slugs that live on land, a group known aptly as the pulmonate gastropods, or Pulmonata (Figure 21.26*d*). In the terrestrial pulmonates, gills have disappeared, but the walls of the mantle cavity have become highly vascularized and well suited for gas exchange. Some species are thought to employ the mantle cavity as a diffusion lung, but others ventilate it by raising and lowering the floor of the cavity.

Decapod crustaceans include many important water breathers and some air breathers

In the decapod crustaceans—which include many ecologically and commercially important crabs, shrimps, lobsters, and crayfish—the head and thorax are covered with a continuous sheet of exoskeleton, the carapace, that overhangs the thorax laterally, fitting more or less closely around the bases of the thoracic legs. The carapace encloses two lateral branchial chambers in which the gills lie (Figure 21.27*a*). The gills arise from near the bases of the thoracic legs. Each gill consists of a central axis to which are attached many richly vascularized lamellar plates, filaments, or dendritically branching tufts. The gill surfaces—like all the external body surfaces of crustaceans—are covered with a chitinous cuticle. The cuticle on the gills is thin, however, and permeable to gases.

Ventilation in crustaceans is always accomplished by muscular contraction; like all arthropods, they lack external cilia that might otherwise produce ventilatory currents. In decapods, each branchial chamber is ventilated by a specialized appendage located toward its anterior end, known as the *scaphognathite* or *gill bailer* (Figure 21.27*b*). Beating of this appendage, under control of nerve impulses from a central pattern generator in the central nervous system, generally drives water outward through an anterior exhalant opening. Negative pressure is thus created within the branchial chamber, causing water to be drawn in at other openings, the location of which varies with species. Ventilation is unidirectional, and countercurrent exchange may occur.

Some crabs and crayfish have invaded the land, especially in the tropics. The semiterrestrial and terrestrial species retain gills, which are supported to some degree by their cuticular covering. Two trends observed in semiterrestrial and terrestrial crabs are that the gills tend to be reduced in size and number by comparison to those of aquatic crabs, and that the branchial chambers tend to be enlarged ("ballooned out"). Furthermore, the tissue[9] that lines the branchial cham-

[9]This tissue is called the *branchiostegites*.

(a) A transverse section through the thorax of a crayfish

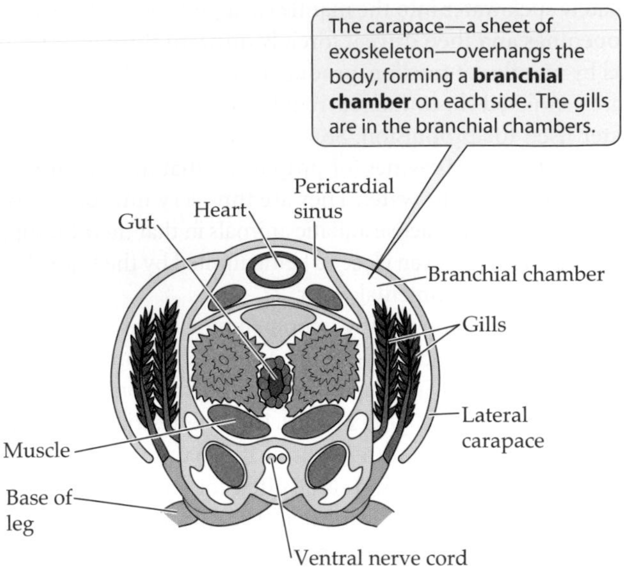

(b) A lateral view showing the gills under the carapace

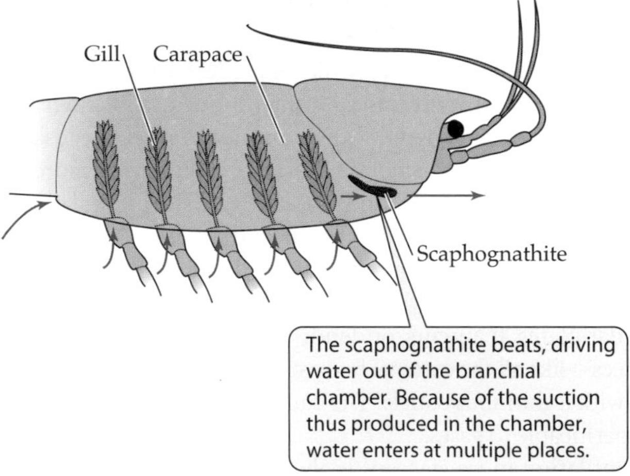

Figure 21.27 The gills and ventilation in a crayfish

bers in these crabs tends to be specialized by being well vascularized, thin, and thrown into patterns of invagination or evagination. These trends toward the reduction of the gills and the development of lunglike branchial chambers provide striking parallels to what we have seen earlier in pulmonate snails and air-breathing fish. The branchial chambers of crabs on land are typically ventilated with air by beating of the scaphognathites. In semiterrestrial (amphibious) species, the gills are kept wet by regular trips to bodies of water. Recent evidence suggests that in these crabs, O_2 is chiefly taken up by the branchial-chamber epithelium, whereas CO_2 is chiefly voided across the gills. In fully terrestrial species of crabs, however, water is not carried in the branchial chambers, and the branchial-chamber epithelium may bear chief responsibility for exchange of both O_2 and CO_2.

Summary: Aquatic Invertebrates and Allied Groups

- The gills of various groups of aquatic invertebrates are often independently evolved. Wide variation thus exists in both gill morphology and the mode of gill ventilation.
- A single basic sort of breathing apparatus can undergo wide diversification within a single phyletic group. This general principle is illustrated by the molluscs, the great majority of which have breathing organs associated with the mantle and located in the mantle cavity. Whereas both aquatic snails and lamellibranchs employ ciliary ventilation, the gills are modest-sized leaflets in snails, but expansive sheets (used partly for feeding) in the lamellibranchs. Cephalopods, such as squids, ventilate their gills by muscular effort. Most land snails lack gills and breathe with a lung derived from the mantle.

Breathing by Insects and Other Tracheate Arthropods

The insects have evolved a remarkable strategy for breathing that is entirely different from that of most metabolically active animals. Their breathing system brings the gas-exchange surface itself close to all cells. Thus, with some interesting exceptions, the cells of insects get their O_2 directly from the breathing system, and the circulatory system plays little or no role in O_2 transport. Insect blood, in fact, usually lacks any O_2 transport pigment such as hemoglobin.

The body of an insect is thoroughly invested with a system of gas-filled tubes termed **tracheae** (Figure 21.28*b*). This system opens to the atmosphere by way of pores, termed **spiracles**, located at the body surface along the lateral body wall. Tracheae penetrate into the body from each spiracle and branch repeatedly, collectively reaching all parts of the animal (only major branches are shown in Figure 21.28*b*). The tracheal trees arising from different spiracles typically join via large *longitudinal* and *transverse connectives* to form a fully interconnected tracheal system. The spiracles, which number from one to eleven pairs, are segmentally arranged and may occur on the thorax, abdomen, or both. Usually they can be closed by spiracular muscles. There are other tracheate arthropods besides insects; most notably, certain groups of spiders and ticks have tracheal systems. Tracheal breathing is best understood in insects, however.

The tracheae of an insect develop as invaginations of the epidermis and thus are lined with a thin cuticle. Typically, the cuticle is thrown into folds that run a spiral course, providing resistance against collapse. The tracheae become finer with increasing distance from the spiracles and finally give rise to very fine, thin-walled end-tubules termed **tracheoles**, believed to be the principal sites of O_2 and CO_2 exchange with the tissues. Tracheoles are perhaps 200 to 350 μm long and are believed to end blindly. They generally taper from a lumen diameter approximating 1 μm at their origin to 0.05 to 0.20 μm at the end. The walls of the tracheoles and the finest tracheae are about 0.02 to 0.2 μm thick—exceedingly thin by any standard (see Figure 21.7*b*).

Although the layout of the tracheal system varies immensely among various species of insects, the usual result is that all organs and

(a) A praying mantis, one of the largest animals to breathe with a tracheal system

(b) Major parts of the tracheal system in a flea

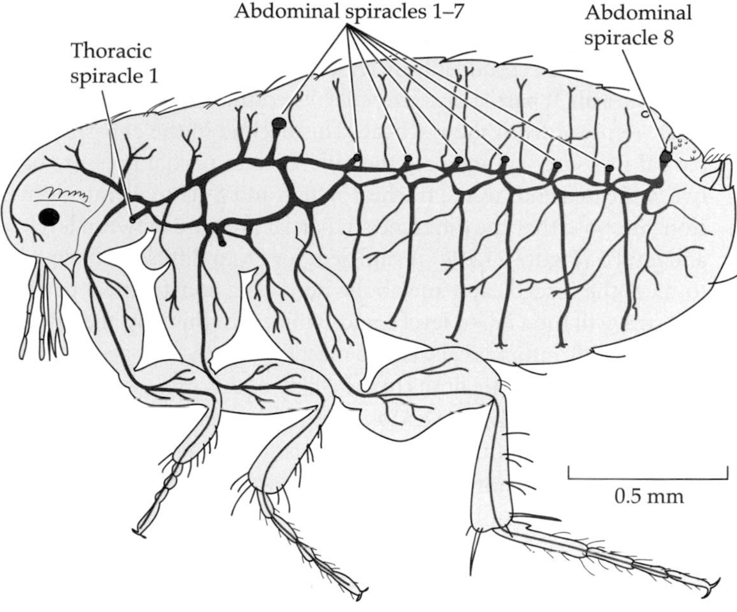

(c) Air sacs in the abdomen of a worker honeybee

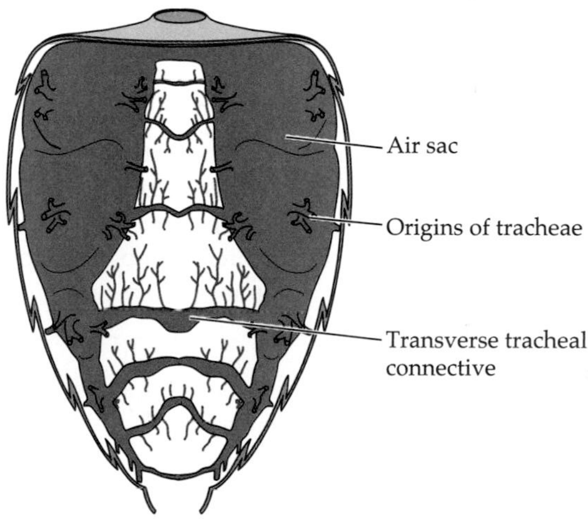

Figure 21.28 All insects breathe using a tracheal system of gas-filled tubes that ramifies to all tissues from the body surface *(a)* To look at a praying mantis, one could imagine it breathing through its mouth. Nothing could be further from the truth. Like all insects, praying mantises breathe with a tracheal system. *(b)* The principal tracheae in a flea (*Xenopsylla*), which has ten pairs of spiracles (two thoracic pairs and eight abdominal). *(c)* Air sacs and associated tracheae in the abdomen of a worker honeybee (*Apis*). Additional air sacs occur in the head and thorax. (*b* after Wigglesworth 1935.)

tissues are thoroughly invested with fine tracheae and tracheoles. The degree of tracheation of various organs and tissues tends to vary directly with their metabolic requirements. In most organs and tissues, the tracheoles typically run between the cells. In the flight muscles of many species, however, the tracheoles penetrate the muscle cells, indenting the cell membrane inward, and run among the individual myofibrils, in close proximity to the arrays of mitochondria. The average distance between adjacent tracheoles within the flight muscles of strong fliers is commonly on the order of just 3 μm. The nervous system, rectal glands, and other active tissues—including muscles besides the flight muscles—also tend to be richly supplied by the tracheal system, although intracellular penetration is not nearly as common as in flight muscles. In the epidermis of the bug *Rhodnius*, which has been carefully studied, tracheoles are much less densely distributed than in active flight muscles, but nonetheless, cells are usually within 30 μm of a tracheole. In other words, no cell is separated from a branch of the tracheal system by more than two or three other cells!

In some tissues of some species, the terminal ends of the tracheoles are filled with liquid when the animals are at rest. During exercise, or when the insects are exposed to O_2-deficient environments, the amount of liquid decreases and gas penetrates farther into the tracheoles. This more-complete filling of the tracheoles with gas facilitates the exchange of O_2 and CO_2 between the tissues and the environment because of the greater ease of diffusion in gas than in liquid.

Enlargements of the tracheal system called **air sacs** are a common feature of insect breathing systems (Figure 21.28*c*). Some air sacs take the form of swellings along tracheae, through which air can flow. Others occur as blind endings or blind, lateral diverticula of tracheae. Air sacs tend to be particularly well developed in active insects, in which they may occupy a considerable fraction of the body volume.

Diffusion is important in gas transport through the tracheal system

The traditional dogma has been that the tracheal system of most insects functions as a diffusion lung, meaning that gas transport through the system occurs solely by diffusion. This dogma is presently undergoing profound revision, as discussed in the following sections. Diffusion, nonetheless, is probably an important gas-transport mechanism in subparts of the tracheal system in most or all insects and may be the sole transport process in some. Diffusion can occur fast enough to play this role because the tracheae are gas-filled.

A question that arises is how the *rate* of diffusion can be varied to correspond to the transport needs of an insect. Although diffusion may sound like a process that is purely physical and therefore independent of animal needs, in fact its rate responds to an insect's metabolic needs because the animal's metabolism alters the partial pressures of gases. Consider an insect in which diffusion is the only mechanism of O_2 transport through the tracheal system, and suppose the insect has an adequate rate of O_2 diffusion when the atmospheric partial pressure is at the level marked 1 and the partial pressure at the inner end of its tracheal system is at the level marked 2 in Figure 21.29*a*. If the insect suddenly increases its rate of O_2 consumption, its end-tracheal O_2 partial pressure will fall because of the increased rate of O_2 removal from the tracheae. This decline in the end-tracheal partial pressure will increase the difference in partial pressure between the two ends of the tracheal system and thus accelerate diffusion. Suppose that the difference in partial pressure between level 1 and level 3 (Figure 21.29*a*) is sufficient for O_2 to diffuse fast enough to meet the insect's new metabolic need. The end-tracheal partial pressure will then fall to level 3 and stabilize. In this way, the rate of diffusion will automatically rise to meet the animal's O_2 need.

Of course, there are limits to the ability of this process to increase the rate of O_2 supply. The end-tracheal O_2 partial pressure must itself remain high enough for O_2 to diffuse from the ends of the tracheae to the mitochondria in cells. Thus, we come face to face with the insect version of the oxygen cascade between ambient air and mitochondria. If the cascade follows line 1-2-4 in Figure 21.29*b* when an insect has a low rate of cellular O_2 use, it might follow line 1-3-5 when the insect's rate of O_2 use is raised. A further increase in the rate of O_2 diffusion may not be possible while keeping the mitochondrial O_2 level adequate.

Some insects employ conspicuous ventilation

Conspicuous (macroscopic) ventilation of the tracheal system occurs in some large species of insects at rest and is common among active insects. An easily observed example of conspicuous ventilation is provided by grasshoppers and locusts, which pump their abdomens under a variety of conditions. Their abdominal pumping motions alternately expand and compress various tracheal airways, causing air to be sucked in via certain spiracles and flow through major tracheal connectives and tracheae, ultimately to be expelled via other spiracles. Air sacs, when present (as they are in grasshoppers), commonly act as bellows during such muscular pumping movements. A mechanism of conspicuous tracheal ventilation that is important in many insects during flight is **autoventilation**: ventilation of the tracheae supplying the flight muscles *driven by flight movements*. During autoventilation, as the thorax is deformed in synchrony with the wing beats, tracheae and air sacs within it are compressed and expanded, driving gas tidally in and out of the thoracic spiracles and tracheae.

Conspicuous ventilation has been known for a long time, and physiologists have generally hypothesized that it ventilates only major tracheae, with diffusion being the principal mode of gas transport through the rest of the tracheal system. If this hypothesis ultimately proves true, then the function of this ventilation would essentially be to reduce the path length for diffusion by moving air convectively to a certain depth in the tracheal system.

Microscopic ventilation is far more common than believed even a decade ago

A revolution is underway in the understanding of microscopic ventilation: ventilation that occurs on such fine scales that it is invisible without use of technology. Probably the most dramatic recent discovery is that when microscopic X-ray videos are made of various insects—such as beetles, crickets, and ants—the major tracheae in the head and thorax are observed to undergo heretofore unknown cycles of partial compression and relaxation. These tracheal pulsations occur every 1 to 2 seconds and are substantial enough that each compression reduces the volumes of the tracheae to 50% to 60% of their relaxed volumes. Insects that undergo these microscopic pulsations may well be exhaling and inhaling if their spiracles are open. Even if their spiracles are closed, these cycles of tracheal compression probably move gases convectively within the major tracheal airways.

One of the first sorts of evidence for microscopic ventilation was the discovery and analysis of **discontinuous gas exchange** in diapausing pupae[9] of moths some decades ago. The hallmark and defining feature of discontinuous gas exchange is that CO_2 is released to the atmosphere in dramatic, intermittent bursts, although the rate of O_2 uptake from the atmosphere is relatively steady. This pattern of CO_2 release arises in large part from spiracular control. In the periods between one burst of CO_2 release and the next, the spiracles are closed or partly closed, and CO_2 produced by metabolism *accumulates in body*

[9] *Diapause* is a programmed resting stage in the life cycle.

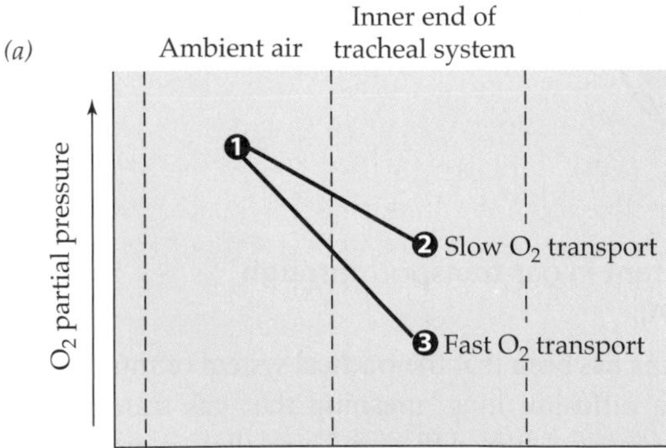

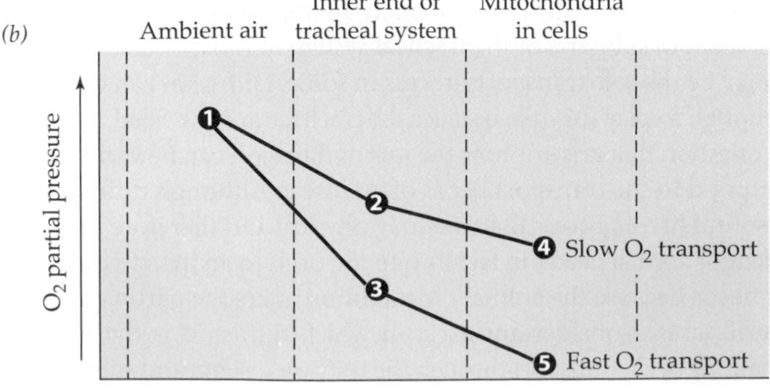

Figure 21.29 Insect oxygen cascades assuming oxygen transport by diffusion *(a)* A drop in the partial pressure at the inner end of the tracheal system from level 2 to level 3 will speed diffusion through the tracheal system by increasing the difference in partial pressure from one end to the other. *(b)* For the rate of diffusion to the mitochondria to be accelerated, the difference in partial pressure between the inner end of the tracheal system and the mitochondria (as well as that between the ambient air and the inner tracheae) must be increased.

fluids by dissolving and reacting to form bicarbonate (HCO_3^-). Because O_2 is removed from the tracheal airways by metabolism during these periods, but the CO_2 produced by metabolism temporarily accumulates in the body fluids rather than in the airways, it is possible for a partial vacuum—a negative pressure—to develop in the airways. If this occurs, then when the spiracles open or partly open, atmospheric air will be sucked in convectively on an inconspicuous, microscopic scale. No one knows the depth to which air is drawn in this way, but it travels by convection at least through the spiracles and into the major tracheae. Discontinuous gas exchange is known today to occur very widely in quiescent or resting insects, including not only butterfly and moth pupae but also adults of many species of ants, bees, beetles, and grasshoppers; it also occurs in certain ticks, mites, and spiders. Investigators have directly assessed in a number of species whether a vacuum-driven inward suction of air actually occurs, and it does in some of the species tested (but not in all).

Besides the forms of microscopic ventilation we have already discussed, several other types have been reported during the last decade. These include processes named "miniature ventilation pulses" in grasshoppers and tiny "Prague cycles" of CO_2 release in beetles. Moreover, some authorities have become convinced by calculations that continuous respiration must often involve alternating diffusion and convection phases of gas transport through the spiracles of insects.

According to the old dogma, insects that were not conspicuously pumping their abdomens or ventilating in other conspicuous ways were breathing entirely by diffusion. The evidence is now overwhelming that convective phenomena are widely employed by visibly motionless insects, but the exact interplay between convection and diffusion in the tracheal system remains a topic for future research to clarify.

Control of breathing

A vulnerability of the insect respiratory system is that it has a great potential to permit desiccating rates of evaporative water loss. The gas in the tracheal airways is humid—ordinarily saturated with water vapor—and when the spiracles are open, only a minute distance separates the humid tracheal gas from the atmosphere. Thus, outward diffusion of water vapor could readily cause rapid losses of water. Insects commonly solve this problem by keeping their spiracles partly closed or by periodically opening and closing them. If the spiracles are experimentally forced to remain fully open all the time, a resting insect's rate of evaporative water loss increases 2 to 12 times, demonstrating the importance of keeping them partly closed.

In insects using diffusion transport, constricting the spiracles makes diffusion of O_2 into the tracheal system more difficult and thus tends to lower the O_2 partial pressure at the inner ends of the tracheae. This effect, nonetheless, need not preclude a high enough inner-tracheal O_2 partial pressure for O_2 to diffuse to the mitochondria at an adequate rate to meet *resting* O_2 needs. When insects become active and have higher O_2 requirements, they may need to open their spiracles more fully to keep their inner-tracheal O_2 partial pressure adequately high. Active insects, in fact, do increase the opening of their spiracles (and pay the price of increased evaporation).

In terms of the chemosensory basis for spiracular control, the most potent stimulus for opening of the spiracles is an increase in the CO_2 partial pressure and/or H^+ concentration of the body fluids (as might occur when metabolism is raised during activity). A decrease in the O_2 partial pressure in the body fluids may also stimulate spiracular opening, but typically is far less influential. In these respects, the control of the spiracles in insects resembles the control of pulmonary ventilation in mammals. Insects that ventilate their tracheal systems by abdominal pumping or other conspicuous mechanisms during activity are well known to control their rates of ventilation in ways that parallel their control of spiracular opening.

Some insects have gills or lungs formed by tracheae

Many insect species live underwater in streams, rivers, and ponds during parts of their life cycles. Some of these life stages lack functional spiracular openings and breathe dissolved O_2 from the water using superficial arrays of fine tracheae. These insects often have a dense proliferation of fine tracheae under their general integument. Moreover—in a remarkable parallel with numerous other groups of aquatic animals—many have evolved evaginations of the body surface that are densely supplied with superficial tracheae and covered with just a thin cuticle. These evaginations, termed **tracheal gills**, may be positioned on the outer body surface or in the rectum. The tracheal system inside the bodies of these insects remains gas-filled. Thus, after O_2 has diffused into their tracheal airways from the water across the walls of their tracheal gills or other superficial tracheae, their internal tracheal system serves as the path of least resistance for the O_2 to move throughout their bodies.

Although tracheal gills have been known for decades, the tracheal "lungs" of some terrestrial insects are among the newest discoveries in insect physiology. These structures are presently known only in certain caterpillars. They consist of profuse arrays of fine tracheae suspended in the blood spaces. Oxygen reaches the tracheal lungs of a caterpillar from its spiracles through the tracheal system and can then be picked up by the blood or blood cells. Tracheal "lungs" are really lungs within a lung, because the whole tracheal system of an insect is a lung by the usual physiological definition.

Some spiders and other arachnids have unique breathing organs, called book lungs, that oxygenate their blood but, unlike caterpillar lungs, are not derived from tracheae and have structures very different from tracheae (Box 21.3).

Many aquatic insects breathe when under water through spiracles using external gas spaces

Although some insects that live in streams, rivers, and ponds have evolved tracheal gills, others have evolved alternative ways of interfacing their tracheal breathing systems with the ambient water or air, ways based on retaining functional spiracles. In an immediate sense, insects of this sort breathe from external gas spaces. There are three distinctive ways in which they do so, however. The simplest to understand is the system used by insects like mosquito larvae, which hang at the water's surface and have their functional spiracles localized to the body region that contacts the atmosphere. Underwater insects such as this breathe from the atmosphere, very much in the way that terrestrial insects do.

BOX 21.3 The Book Lungs of Arachnids

Some arachnids possess a novel type of respiratory structure, the **book lung**. Scorpions have only book lungs. Many species of spiders also have book lungs, but they may have systems of tracheae as well. The number of book lungs in an individual arachnid varies from a single pair (as in certain spiders) to four pairs (in scorpions). Book lungs are invaginations of the ventral abdomen, lined with a thin chitinous cuticle. Each book lung consists of a chamber, the atrium, which opens to the outside through a closable ventral pore, the spiracle (see figure). The dorsal or anterior surface of the atrium is thrown into many lamellar folds: the "pages of the book." Blood streams through the lamellae, whereas the spaces among the lamellae are filled with gas. The lamellae commonly number into the hundreds. The blood-to-gas distance across their walls is often less than 1 μm. Some book lungs may function as diffusion lungs, whereas others are clearly ventilated by pumping motions. They oxygenate the blood, which then carries O_2 throughout the body.

A book lung The section shows the internal structure of a book lung in a two-lunged spider. (After Comstock 1912.)

A second strategy employed by underwater insects with functional spiracles is to carry a conspicuous bubble of gas captured from the atmosphere. Many water beetles, for example, carry a conspicuous bubble either under the wings or at the tip of the abdomen. Their functional spiracles open into the bubble and exchange O_2 and CO_2 with the gas in the bubble. A remarkable attribute of a gas bubble like this is that it *acts as a gill*, not just an O_2 store. As an insect swimming under water withdraws O_2 from its bubble, the O_2 partial pressure in the bubble gas falls. Once it has fallen below the O_2 partial pressure in the surrounding water, dissolved O_2 from the water starts to diffuse into the bubble, renewing its O_2 supply. Thus, an insect is able to remove much more O_2 from a bubble than simply the amount captured from the atmosphere. The rate at which O_2 diffuses into a bubble from the water depends on the bubble surface area. A bubble of the sort we are discussing gradually shrinks, and as its surface area diminishes, it ceases to function effectively as a gill. Thus, a bubble of this sort must periodically be renewed with air from the atmosphere.

The third strategy employed by underwater insects with functional spiracles is certainly the most unexpected. It is also a type of bubble breathing, but a very different type. In some aquatic insects, parts of the body surface are covered extremely densely with fine water-repelling hairs; the bug *Aphelocheirus aestivalis*, for example, has 2 to 2.5 million of these hairs per square *millimeter*! Such densely distributed water-repelling hairs on the body surface trap among themselves a thin, almost invisible film of gas *that cannot be displaced*. This film of gas, known as a **plastron**, is incompressible and permanent. Thus, its surface area remains constant, and it can serve as a gill (an air space into which O_2 diffuses from the water) for an indefinite length of time. Some aquatic insects with plastrons remain submerged continuously for months!

Summary: Breathing by Insects and Other Tracheate Arthropods

- Insects and many arachnids breathe using a tracheal system that connects to the atmosphere by way of spiracles on the body surface and ramifies throughout the body so that gas-filled tubes bring O_2 to the vicinity of all cells.
- The modes of gas exchange through the tracheal system include diffusion, conspicuous ventilation (such as abdominal pumping and autoventilation), and several forms of microscopic ventilation.
- Aquatic insects may lack functional spiracles and breathe using superficial tracheal beds. Alternatively, they may have functional spiracles and breathe from the atmosphere, large bubbles, or plastrons.

Study Questions

1. Lungs ventilated with water occur in some animals, most notably sea cucumbers, but are rare. Why would water lungs be unlikely to be favored by natural selection?
2. In Chapter 1 we discussed François Jacob's question of whether evolution is more like tinkering or engineering. Jacob's view is that tinkering is a far better analogy than engineering. How could the evolution of air breathing in fish be used to argue for the validity of the tinkering analogy?
3. Outline the differences among the three most sophisticated lungs found in modern animals: the mammalian lung, the avian lung, and the insect tracheal system.
4. If a mammal's tidal volume is 2 L, its tracheal volume is 80 mL, its anatomical dead space volume is 350 mL, and its breathing frequency is 9 breaths/minute, what is its alveolar minute volume?
5. When researchers first discovered that airflow through a bird's paleopulmonal parabronchi is unidirectional, the question arose as to whether gas exchange is countercurrent, concurrent, or cross-current. Some ingenious investigations carried out experiments in which they measured the efficiency of gas exchange between air and blood in duck lungs when parabronchial airflow was in its normal direction and when the direction of parabronchial airflow was artificially reversed. The efficiency did not change. How is this evidence against countercurrent and concurrent gas exchange? How is it evidence for cross-current exchange?

6. What is the evidence that the breathing rhythm in mammals originates in the brainstem?
7. Explain mechanistically how hyperventilation alters the oxygen cascade of an animal.
8. Arthropods are distinguished by having an exoskeleton and lacking external cilia. How do these traits of the phylum affect the breathing structures and ventilation modes of arthropods?
9. Why do fish suffocate when taken out of the water and placed in air, whereas mammals suffocate if the air in their alveoli is replaced with water?
10. In your own words, explain how a person's oxygen utilization coefficient can increase as the person's rate of ventilation increases even though the composition of alveolar gas is held constant by negative-feedback controls.
11. A species of water beetle that carries a conspicuous bubble while under water was placed in an aquarium system in which the only gas was O_2: The atmosphere above the water was pure O_2, and the gas dissolved in the water was pure O_2. The beetles in this system had to resurface *far more frequently* than beetles in a system containing ordinary air and water. The explanation lies in the *physics of gases*. Explain why a bubble of pure O_2 cannot act as a gill under water. Then explain why the nitrogen in air is a crucial factor for an atmospheric bubble to function as a gill under water.

Key References

Duncker, H.-R. 2001.The emergence of macroscopic complexity: An outline of the history of the respiratory apparatus of vertebrates from diffusion to language production. *Zoology* 103: 240–259. Provocative and opinionated but consistently intellectually rewarding. Duncker is a meticulous pulmonary morphologist and functional morphologist who has made major contributions to the study of lungs in nonmammalian vertebrates.

Maina, J. N. 2000. What it takes to fly: The structural and functional respiratory refinements in birds and bats. *J. Exp. Biol.* 203: 3045–3064. An intriguing, comparative inquiry into the question of how the respiratory systems of bats and birds are adapted to meet the high gas-exchange requirements of flight.

Scheid, P., and J. Piiper. 1997. Vertebrate respiratory gas exchange. In W. H. Dantzler (ed.), *Comparative Physiology*, vol. 1 (Handbook of Physiology [Bethesda, MD], section 13), pp. 309–356. Oxford University Press, New York. An overview of gas exchange principles and the comparative physiology of breathing in vertebrates, written by two long-term leaders in these fields. An excellent source to master the quantitative analysis of tidal, concurrent, countercurrent, and cross-current gas exchange.

Weibel, E. R. 1984. *The Pathway for Oxygen*. Harvard University Press, Cambridge, MA. One of the all-time great treatments of its subject, the pathway for oxygen, written by a founder of the field of morphometric lung analysis. Becoming dated, unfortunately.

West, J. B. 2000. *Respiratory Physiology: The Essentials*, 6th ed. Lippincott Williams & Wilkins, Philadelphia. An articulate, compact treatment of medical respiratory physiology.

Wood, S. C. (ed.). 1989. *Comparative Pulmonary Physiology: Current Concepts*. Lung Biology in Health and Disease, vol. 39. Dekker, New York. One of the best relatively recent books in which all the vertebrate groups are covered in detail by experts within a single volume.

Selected References

Ainsworth, D. M., C. A. Smith, S. W. Eicker, N. G. Ducharme, K. S. Henderson, K. Snedden, and J. A. Dempsey. 1997. Pulmonary-locomotory interactions in exercising dogs and horses. *Respir. Physiol.* 110: 287–294.

Block, B. A., and E. D. Stevens (eds.). 2001. *Tuna: Physiology, Ecology, and Evolution*. Academic Press, New York.

Boutilier, R. G. (ed.). 1990. *Vertebrate Gas Exchange: From Environment to Cell*. Advances in Comparative and Environmental Physiology, vol. 6. Springer-Verlag, New York.

Bouverot, P. 1985. *Adaptation to Altitude-Hypoxia in Vertebrates*. Springer-Verlag, New York.

Brainerd, E. L. 1994. The evolution of lung-gill bimodal breathing and the homology of vertebrate respiratory pumps. *Amer. Zool.* 34: 289–299.

Bramble, D. M., and F. A. Jenkins, Jr. 1993. Mammalian locomotor-respiratory integration: Implications for diaphragmatic and pulmonary design. *Science* 262: 235–240.

Bridges, C. R., and P. J. Butler (eds.). 1989. *Techniques in Comparative Respiratory Physiology: An Experimental Approach*. Society for Experimental Biology Seminar Series, no. 37. Cambridge University Press, New York.

Burggren, W. W., and A. W. Pinder. 1991. Ontogeny of cardiovascular and respiratory physiology in lower vertebrates. *Annu. Rev. Physiol.* 53: 107–135.

Burggren, W. W., and N. H. West. 1982. Changing respiratory importance of gills, lungs and skin during metamorphosis in the bullfrog *Rana catesbeiana*. *Respir. Physiol.* 47: 151–164.

Burggren, W. W., K. Johansen, and B. McMahon. 1985. Respiration in phyletically ancient fishes. In R. E. Forman, A. Gorbman, J. M. Dodd, and R. Olsson (eds.), *Evolutionary Biology of Primitive Fishes*, pp. 217–252. Plenum, New York.

Cameron, J. N. 1989. *The Respiratory Physiology of Animals*. Oxford University Press, New York.

Daniels, C. B., S. Orgeig, and A. W. Smits. 1995. The evolution of the vertebrate pulmonary surfactant system. *Physiol. Zool.* 68: 539–566.

Deeming, D. C., and M. W. J. Ferguson (eds.). 1991. *Egg Incubation: Its Effects on Embryonic Development in Birds and Reptiles*. Cambridge University Press, New York.

Dejours, P. 1988. *Respiration in Water and Air*. Elsevier, New York.

Dempsey, J. A., and A. I. Pack (eds.). 1995. *Regulation of Breathing*, 2nd ed. Dekker, New York.

Denny, M. W. 1993. *Air and Water: The Biology and Physics of Life's Media*. Princeton University Press, Princeton, NJ.

Duncker, H.-R. 1974. Structure of the avian respiratory tract. *Respir. Physiol.* 22: 1–19.

Egginton, S., and H. F. Ross (eds.). 1992. *Oxygen Transport in Biological Systems: Modelling of Pathways from Environment to Cell*. Society for Experimental Biology Seminar Series, vol. 51. Cambridge University Press, New York.

Farmer, C. G. 1999. Evolution of the vertebrate cardiopulmonary system. *Annu. Rev. Physiol.* 61: 573–592.

Fedde, M. R., J. A. Orr, H. Shams, and P. Scheid. 1989. Cardiopulmonary function in exercising bar-headed geese during normoxia and hypoxia. *Respir. Physiol.* 77: 239–262.

Fincke, T., and R. Paul. 1989. Book lung function in arachnids. III. The function and control of the spiracles. *J. Comp. Physiol. B* 159: 433–441.

Floros, J., and P. Kala. 1998. Surfactant proteins: Molecular genetics of neonatal pulmonary diseases. *Annu. Rev. Physiol.* 60: 365–384.

Fregly, M. J., and C. M. Blatteis (eds.). 1996. *Environmental Physiology*, vol. 2. (Handbook of Physiology [Bethesda, MD], section 4.) Oxford University Press, New York. [See Part V: *The Terrestrial Altitude Environment.*]

Graham, J. B. 1997. *Air-Breathing Fishes*. Academic Press, New York.

Greenaway, P., and C. Farrelly. 1990. Vasculature of the gas-exchange organs in air-breathing brachyurans. *Physiol. Zool.* 63: 117–139.

Henry, R. P. 1994. Morphological, behavioral, and physiological characterization of bimodal breathing crustaceans. *Amer. Zool.* 34: 205–215.

Hlastala, M. P., and A. J. Berger. 1996. *Physiology of Respiration*. Oxford University Press, New York.

Hochachka, P. W. 1998. Mechanism and evolution of hypoxia tolerance in humans. *J. Exp. Biol.* 201: 1243–1254.

Houlihan, D. F., J. C. Rankin, and T. J. Shuttleworth (eds.). 1982. *Gills*. Cambridge University Press, Cambridge.

Howe, S., and D. L. Kilgore, Jr. 1987. Convective and diffusive gas exchange in nest cavities of the northern flicker (*Colaptes auratus*). *Physiol. Zool.* 60: 707–712.

Innes, A. J., and E. W. Taylor. 1986. The evolution of air-breathing in crustaceans: A functional analysis of branchial, cutaneous and pulmonary gas exchange. *Comp. Biochem. Physiol. A* 85: 621–637.

King, A. S., and T. McLelland (eds.). 1989. *Form and Function in Birds*, vol. 4. Academic Press, New York.

LaBarbera, M. 1990. Principles of design of fluid transport systems in zoology. *Science* 249: 992–1000.

Lenfant, C., and K. Johansen. 1968. Respiration in the African lungfish, *Protopterus aethiopicus*. *J. Exp. Biol.* 49: 437–452.

Lieske, S. P., M. Thoby-Brisson, P. Telgkamp, and J. M. Ramirez. 2000. Reconfiguration of the neural network controlling multiple breathing patterns: Eupnea, sighs, and gasps. *Nature Neurosci.* 3: 600–608.

Lighton, J. R. B. 1996. Discontinuous gas exchange in insects. *Annu. Rev. Entomol.* 41: 309–324.

Lighton, J. R. B. 1998. Notes from underground: Towards ultimate hypotheses of cyclic, discontinuous gas-exchange in tracheate arthropods. *Amer. Zool.* 38: 483–491.

Maina, J. N. 1994. Comparative pulmonary morphology and morphometry: The functional design of respiratory systems. *Adv. Comp. Environ. Physiol.* 20: 111–232.

Mill, P. J. 1997. Invertebrate respiratory systems. In W. H. Dantzler (ed.), *Comparative Physiology*, vol. 2 (Handbook of Physiology [Bethesda, MD], section 13), pp. 1009–1096. Oxford University Press, New York.

Milsom, W. K. 1991. Intermittent breathing in vertebrates. *Annu. Rev. Physiol.* 53: 87–105.

Monge, C., and F. León-Velarde. 1991. Physiological adaptation to high altitude: Oxygen transport in mammals and birds. *Physiol. Rev.* 71: 1135–1172.

Moore, L. G., F. Armaza V., M. Villena, and E. Vargas. 2000. Comparative aspects of high-altitude adaptation in human populations. In S. Lahiri, N. R. Prabhakar, and R. E. Forster II (eds.), *Oxygen Sensing: Molecule to Man*, pp. 45–62. Kluwer/Plenum, New York.

Owerkowicz, T., C. G. Farmer, J. W. Hicks, and E. L. Brainerd. 1999. Contribution of gular pumping to lung ventilation in monitor lizards. *Science* 284: 1661–1663.

Paul, R., T. Finke, and B. Linzen. 1987. Respiration in the tarantula *Eurypelma californicum*: Evidence for diffusion lungs. *J. Comp. Physiol. B* 157: 209–217.

Perry, S. F. 1983. Reptilian lungs: Functional anatomy and evolution. *Adv. Anat. Embryol. Cell Biol.* 79: 1–81.

Perry, S. F. 1989. Mainstreams in the evolution of vertebrate respiratory structures. In A. S. King and T. McLelland (eds.), *Form and Function in Birds*, vol. 4, pp. 1–67. Academic Press, New York.

Perry, S. F. 1990. *Recent advances and trends in the comparative morphometry of vertebrate gas exchange organs*. Advances in Comparative and Environmental Physiology, vol. 6: 45–71.

Piiper, J. (ed.). 1978. *Respiratory Function in Birds, Adult and Embryonic*. Springer-Verlag, New York.

Randall, D. J., W. W. Burggren, A. P. Farrell, and M. S. Haswell. 1981. *The Evolution of Air-Breathing in Vertebrates*. Cambridge University Press, New York.

Rankin, J. C., and F. B. Jensen (eds.). 1993. *Fish Ecophysiology*. Chapman & Hall, New York.

Ruben, J. A., T. D. Jones, N. R. Geist, and W. J. Hillenius. 1997. Lung structure and ventilation in theropod dinosaurs and early birds. *Science* 278: 1267–1270.

Scheid, P., and J. Piiper. 1972. Cross-current gas exchange in avian lungs: Effects of reversed parabronchial air flow in ducks. *Respir. Physiol.* 16: 304–312.

Schmitz, A., and S. F. Perry. 1999. Stereological determination of tracheal volume and diffusing capacity of the tracheal walls in the stick insect *Carausius morosus* (Phasmatodea, Lonchodidae). *Physiol. Biochem. Zool.* 72: 205–218.

Seller, T. J. (ed.). 1987. *Bird Respiration*. CRC Press, Boca Raton, FL.

Shadwick, R. W., R. K. O'Dor, and J. M. Gosline. 1990. Respiratory and cardiac function during exercise in squid. *Can. J. Zool.* 68: 792–798.

Sláma, K. 1999. Active regulation of insect respiration. *Ann. Entomol. Soc. Amer.* 92: 916–929.

Smatresk, N. J. 1994. Respiratory control in the transition from water to air breathing in vertebrates. *Amer. Zool.* 34: 264–279.

Snyder, G. K. 1987. Capillary growth and diffusion distances in muscle. *Comp. Biochem. Physiol. A* 87: 859–861.

Taylor, E. W., D. Jordan, and J. H. Coote. 1999. Central control of the cardiovascular and respiratory systems and their interactions in vertebrates. *Physiol. Rev.* 79: 855–916.

Wagner, P. D. 1996. Determinants of maximal oxygen transport and utilization. *Annu. Rev. Physiol.* 58: 21–50.

Ward, M. P., J. S. Milledge, and J. B. West (eds.). 2000. *High Altitude Medicine and Physiology*, 3rd ed. Arnold, London.

Weibel, E. R. 1987. Scaling of structural and functional variables in the respiratory system. *Annu. Rev. Physiol.* 49: 147–159.

Weibel, E. R., and C. R. Taylor (eds.). 1981. Design of the mammalian respiratory system. *Respir. Physiol.* 44: 1–86, 151–164.

West, J. B. 1984. Human physiology at extreme altitudes on Mount Everest. *Science* 223: 784–788.

Westneat, M. W., O. Betz, R. W. Blob, K. Fezzaa, W. J. Cooper, and W.-K. Lee. 2003. Tracheal respiration in insects visualized with synchrotron X-ray imaging. *Science* 299: 558–560.

See also references in Appendix A.

22 Transport of Oxygen and Carbon Dioxide in Body Fluids (With an Introduction to Acid-Base Physiology)

A developing mammalian fetus receives oxygen and voids carbon dioxide by means of a close juxtaposition of its own blood vessels with those of its mother's circulatory system in the placenta. The fetus's heart pumps blood along the umbilical cord to the placenta, where that blood picks up O_2 that its mother's blood has brought there from her lungs. The now-oxygenated fetal blood returns to the fetus along the umbilical cord and is circulated to all the parts of the fetus's body. By 10 weeks after conception, a human fetus already has hemoglobin-rich blood; the amount of hemoglobin per milliliter of its blood has reached 50% of the adult concentration and is increasing rapidly, so that it will be about 80% of the adult value at 20 weeks of age.

In the placenta, O_2 must cross from the mother's blood to the fetus's blood by diffusion through tissue separating the two circulatory systems. Exactly how this occurs remains a topic of active research. The basic options for the mode of gas transfer are those we discussed in Chapter 21 for the transfer of O_2 from water to blood in a gill-breathing animal, although in the present case, the fluids involved are the blood of the mother and the blood of the fetus: concurrent, countercurrent, or cross-current gas exchange. The mode of gas transfer in the human placenta remains uncertain for two reasons. The first is the inappropriateness of doing experiments on human fetuses. The second is that other species of placental mammals exhibit such wide variation in the physiology of their placentas that researchers do not know which animal model would reveal how the human placenta works. Enough is known about placental physiology in several mammalian species to make clear, however, that contrary to expectation, substantial impediments to O_2 transfer from maternal to fetal blood often exist in mammalian placentas. Placentas are emphatically not like lungs, in which high rates of air and blood flow and minutely thin intervening membranes result readily in dramatic blood oxygenation.

An important reason a human fetus in fact can obtain enough O_2 from its placenta is that it produces a different molecular form of hemoglobin from the one its mother produces. This is also true in many other species

This human fetus obtains oxygen by pumping blood along its umbilical cord to the placenta, where its blood picks up O_2 from its mother's blood Hemoglobin plays a major role in the acquisition and transport of O_2 by the blood of a mammalian fetus.

of placental mammals that have been studied. For reasons we explore in detail later in this chapter, fetal hemoglobin has a higher affinity for O_2 than adult hemoglobin. This greater affinity has two important, interrelated consequences. First, the *difference* in affinity between the maternal and fetal hemoglobins means that O_2 has a chemical tendency to leave the lower-affinity hemoglobin of the mother to bind with the higher-affinity hemoglobin of the fetus. Second, the *high absolute affinity* of fetal hemoglobin means that it can become well oxygenated even if the O_2 partial pressure in the fetal blood remains relatively low, as it typically does.

The hemoglobins are one of several types of **respiratory pigments** or **O_2-transport pigments** that animals have evolved. The defining property of the respiratory pigments is that they undergo reversible combination with molecular oxygen (O_2). Thus they can pick up O_2 in one place, such as the lungs of an adult or the placenta of a fetus, and release the O_2 in another place, such as the systemic tissues[1] of the adult or fetus. All the types of respiratory pigments are *metalloproteins*: proteins that contain metal atoms, exemplified by the iron in hemoglobin. All are also strongly colored at least some of the time, explaining why they are called *pigments*.

The most straightforward function of the respiratory pigments is to increase the amount of O_2 that can be carried by a unit of volume of blood. Oxygen dissolves in the blood plasma[2] just as it dissolves in any other aqueous solution (see Chapter 20). A limitation on transporting O_2 in this mode is that O_2 has a relatively low solubility in plasma, as it does in all aqueous solutions; thus the amount of O_2 that can be carried in dissolved form per unit of volume is not high. When a respiratory pigment is present in the blood, however, the blood can carry O_2 in two ways: in simple solution and in chemical combination with the pigment. Thus a respiratory pigment increases the **oxygen-carrying capacity** of blood, meaning the total amount of O_2 that can be carried by each unit of volume. In some cases, the increase is very large. For example, when the blood of an adult person leaves the lungs, it contains about 0.4 mL O_2 per 100 mL of blood in solution, but in addition, it contains about 19.6 mL O_2 per 100 mL in chemical combination with hemoglobin. Thus the blood's concentration of O_2 is increased about 50-fold by the presence of hemoglobin. This means, among other things, that the heart can work far less intensely; roughly calculated, the circulation of a liter of actual human blood delivers the same amount of O_2 as would the circulation of 50 liters of blood without hemoglobin.

There are multiple molecular forms of hemoglobin, as already exemplified by the contrast between fetal and maternal hemoglobins. Not only may one species have multiple molecular forms, but different species have different forms. Thus the word *hemoglobin* refers to a family of many compounds, not just a single compound. To emphasize this fact, we refer to these compounds as *hemoglobins* (plural), rather than just *hemoglobin* (singular).

Increasing the oxygen-carrying capacity of blood is not the only function of hemoglobins. Blood hemoglobins, for example, play important roles as buffers and participate in CO_2 transport as well as O_2 transport. Moreover, certain hemoglobins are found *within* muscle cells or neurons, where they facilitate diffusion of O_2 into those cells and may also serve as storage depots for O_2. This chapter emphasizes the role of hemoglobins in blood O_2 transport, but touches on other functions as well.

[1] The **systemic tissues** are all tissues other than the tissues of the breathing organs.

[2] The **plasma** of the blood is the aqueous solution in which the cells are suspended. Operationally, plasma is obtained by removing all cells from blood (e.g., by centrifugation).

BOX 22.1 Absorption Spectra of Respiratory Pigments

The hemoglobins and other respiratory pigments—like all pigments—differentially absorb various wavelengths of light. The pattern of absorption by a pigment when expressed as a function of wavelength, is known as an **absorption spectrum** (plural *spectra*). The absorption spectrum of a specific respiratory pigment (e.g., human hemoglobin) changes with the oxygenation or deoxygenation of the pigment, as shown in the accompanying figure. These changes are qualitatively evident to our eyes: We know, for example, that oxygenated hemoglobin (bright red) differs in color from deoxygenated hemoglobin (purple-red). By using quantitative light-absorption measurements, the percentage of heme groups that are oxygenated in blood can be determined. This is the principle behind the finger probes (oximeters) used to monitor blood oxygenation in hospital patients.

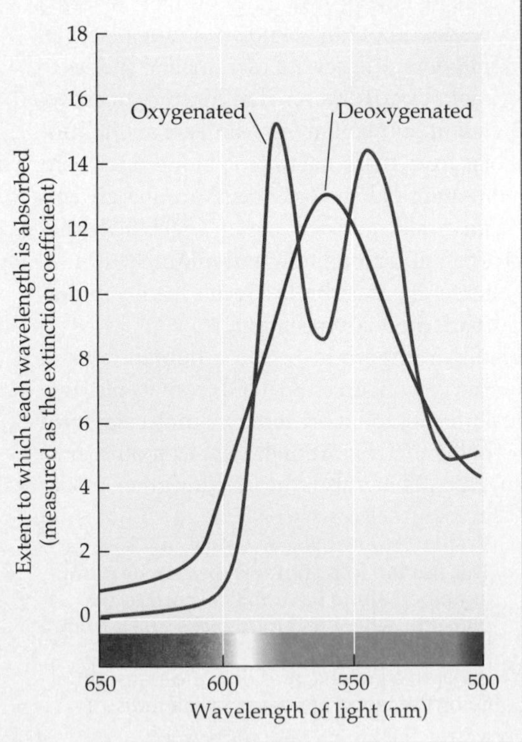

Absorption spectra for fully oxygenated and fully deoxygenated human hemoglobin To measure absorption, light of each wavelength is passed through a hemoglobin solution of defined concentration and thickness (in this case, 1 m*M* and 1 cm), and the fraction of the incoming photon energy that fails to pass through the solution is measured. From the data, one calculates the *extinction coefficient*, which is a measure of the absorption of the light by the hemoglobin (a high extinction coefficient signifies great absorption). (After Waterman 1978.)

A final introductory point worth noting is that when hemoglobins or other respiratory pigments combine with O_2, they are said to be **oxygenated**, and when they release O_2, they are **deoxygenated**. They are not said to be *oxidized* and *reduced*. The reason for these distinctions is that their process of combining with O_2 is not chemically equivalent to oxidation. During the oxygenation of a hemoglobin molecule, for example, electrons are partially transferred from iron atoms in the hemoglobin molecule to the O_2, but the transfer is not complete, as it would be in full-fledged oxidation. In fact, if a hemoglobin molecule accidentally becomes truly oxidized (so that its iron atoms are converted from their ordinary ferrous state to the ferric state), the molecule (now called *methemoglobin*) loses its ability to combine with O_2 and becomes useless for O_2 transport. The prefixes **oxy-** and **deoxy-** are used to specify the oxygenated and deoxygenated states of respiratory-pigment molecules. Hemoglobin, for example, is called **oxyhemoglobin** when it is combined with O_2 (oxygenated) and **deoxyhemoglobin** when it is not combined with O_2 (deoxygenated). Respiratory pigments change color when they are oxygenated and deoxygenated, and measures of these color changes can be used to monitor the oxygenation and deoxygenation of blood (Box 22.1)

The Chemical Properties and Distributions of the Respiratory Pigments

Four chemical categories of respiratory pigments are recognized: **hemoglobins**, **hemocyanins**, **hemerythrins**, and **chlorocruorins**. The prefix *hemo-* is from the Greek for "blood," explaining its use in the names of three of the categories of pigments. Like the hemoglobins, the other categories are groups of related compounds, not single chemical structures.

Many of the important chemical properties of the respiratory pigments resemble the properties of the enzyme proteins we studied in Chapter 2. The parallels are so great, in fact, that biochemists have occasionally dubbed the respiratory pigments "honorary enzymes." The point of mentioning these parallels is not to suggest that respiratory pigments are enzymes; they are *not*. The point, instead, is to highlight that, based on your knowledge of enzyme proteins, you will find that you already know a great deal about the molecular features of respiratory pigments.

When a hemoglobin molecule, for example, combines with O_2, it does so at *defined binding sites*, resembling the way in which enzymes combine with their substrates at defined binding sites. Moreover, the combination of the O_2-binding sites with O_2 is *highly specific* and occurs by *noncovalent, weak bonding*, just as enzyme–substrate binding is specific and noncovalent. Accordingly, O_2 is a *ligand* of hemoglobin, based on the definition of "ligand" we developed in Chapter 2. When a hemoglobin molecule combines with O_2, it undergoes a change in its molecular conformation that is analogous to the conformational change an enzyme molecule undergoes when it combines with its substrate; the ability of a hemoglobin molecule to *flex* in this way is an essential attribute of its function, just as molecular flexibility is critical for enzyme function. One of a hemoglobin molecule's most important properties is its *affinity* for O_2, meaning the ease with which it binds with the O_2 molecules it encounters; thus a hemoglobin molecule (like an enzyme) is characterized in part by *how readily* it binds with its primary ligand.

A hemoglobin molecule also has specific sites at which it combines with *ligands other than* O_2. Using the same terminology we used in Chapter 2 in connection with enzymes, such ligands are *allosteric ligands* or *allosteric modulators*, because when they bind with their specific sites on the hemoglobin molecule, they affect the ability of the hemoglobin to bind with its primary ligand, O_2. Allosteric ligands, for example, can potently affect a hemoglobin molecule's affinity for O_2. Within a hemoglobin molecule, just as in an enzyme molecule, allosteric ligands exert their effects *at a distance*; that is, the binding sites for allosteric ligands on a hemoglobin molecule are separate from the O_2-binding sites, and when allosteric ligands affect O_2 binding, they do so by modifying the conformation and flexibility of the molecule as a whole.

Hemoglobin molecules are usually *multisubunit proteins*; that is, each whole molecule consists of two or more proteins bonded together by noncovalent bonds. Multisubunit hemoglobins have an O_2-binding site on each subunit and thus have multiple O_2-binding sites. In common with multisubunit enzymes that exhibit cooperativity among substrate-binding sites, multisubunit hemoglobin molecules exhibit *cooperativity among their O_2-binding sites*, meaning that binding of O_2 to any one site on a molecule affects how readily the other sites bind O_2.[3] These interactions among O_2-binding sites themselves occur at a distance; the various O_2-binding sites on a molecule are separate and distinct, and they influence each other by effects that are relayed through the structure of the protein, rather than by direct site-to-site effects.

The points we have made using hemoglobin as an example apply to the other categories of respiratory pigments as well. Thus, in the study of all respiratory pigments it is helpful to keep them in mind.

Despite sharing many key properties with enzymes, however, the respiratory pigments differ from enzymes in a major way: They do not modify their primary ligand. After they combine with O_2, they later release O_2.

Hemoglobins contain heme and are the most widespread respiratory pigments

The chemical structures of all hemoglobin molecules share two features. First, they contain **heme** (Figure 22.1*a*), which is a particular metalloporphyrin containing iron in the ferrous state (ferrous protoporphyrin IX). Second, the heme is noncovalently bonded to a protein known as a **globin** (Figure 22.1*b*). The combination of heme with globin accounts for the name *hemoglobin*. Oxygen binds at the heme site at a ratio of one O_2 molecule per heme. In all hemoglobin molecules, the heme is identical. Thus the multiple molecular forms of hemoglobin differ in their protein (globin) structures (and in the numbers of unit molecules of hemoglobin that are linked together).

Biochemical studies reveal that small changes in the protein structure of a hemoglobin can cause highly significant alterations in the functional properties of the hemoglobin. Consider, for example, the numerous (100+) mutant forms of human hemoglobin. Each

[3] In the terminology developed in Chapter 2, this is *homotropic* cooperativity. See page 45 for more on cooperativity within multisubunit proteins.

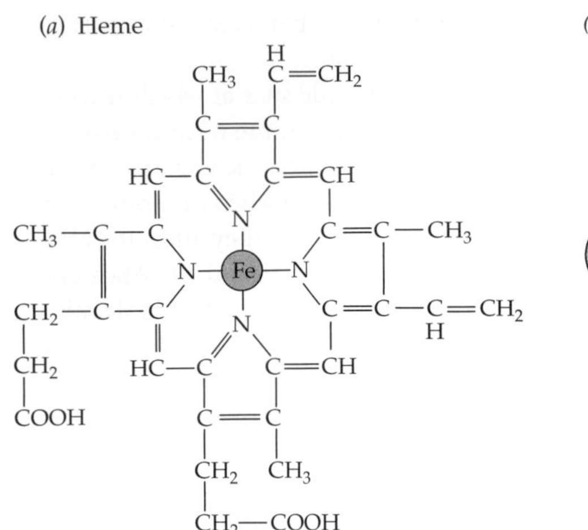

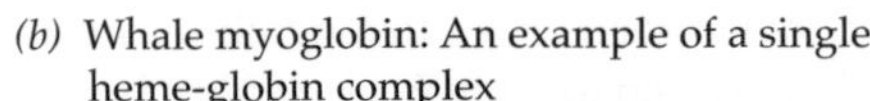

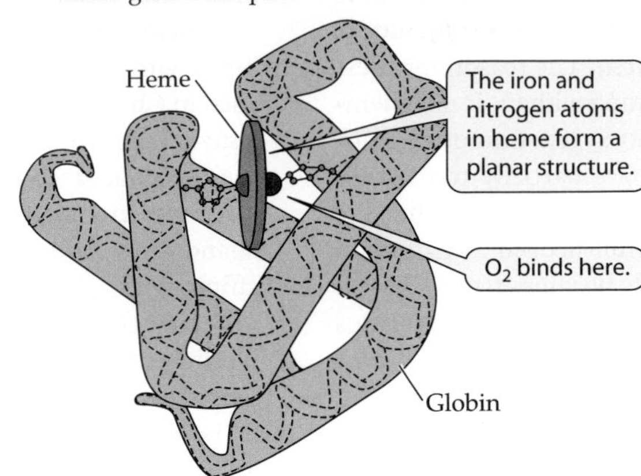

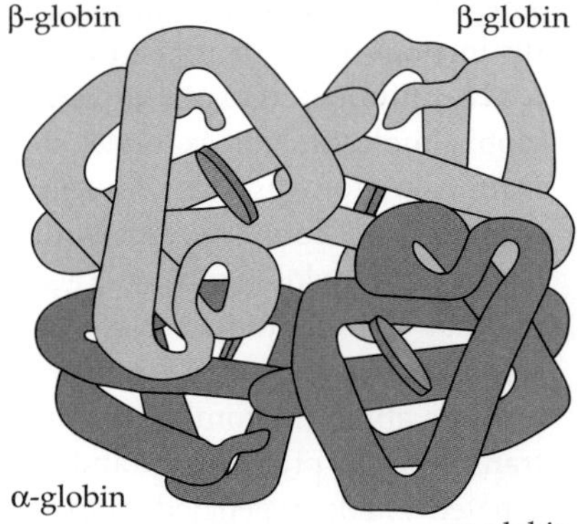

Figure 22.1 The chemical structure of hemoglobin *(a)* The structure of heme: Ferrous iron is complexed with protoporphyrin. The positions assigned to double and single bonds in the porphyrin ring are arbitrary because resonance occurs. *(b)* A single heme-globin complex. The specific molecule shown is myoglobin (muscle hemoglobin) taken from the muscle of a whale. The structure of the globin protein includes eight segments in which the amino acid backbone of the protein (seen spiraling inside the cylindrical outline) forms a helix. The outer, cylindrical part of the drawing shows the major contours of the globin structure. *(c)* A tetrameric hemoglobin molecule of the sort found in mammalian red blood cells. In adults, each tetramer consists two α-globins, two β-globins, and a total of four heme groups. (Parts *b* and *c* after Dickerson and Geis 1983.)

human globin protein consists of over 140 amino acids, and most of the mutant forms differ from the normal form in just one of those amino acids. Nonetheless, some mutant forms differ markedly from the normal form in their affinity for O_2 or in other key properties, such as their solubility or structural stability.

The blood hemoglobins of vertebrates are almost always four-unit (tetrameric) molecules (Figure 22.1*c*) that can bind a total of four O_2 molecules. The molecular weight of each unit molecule is typically about 16,000 to 17,000 daltons. Thus the four-unit blood hemoglobins have molecular weights of about 64,000 to 68,000 daltons. Two types of globins, termed α and β, are found in adult blood hemoglobins. The genes for the two types seem clearly to have originated by gene duplication, perhaps 500 million years ago. Each molecule of adult blood hemoglobin consists of two α units and two β units. The α-globin of humans contains 141 amino acids, whereas the human β-globin contains 146 amino acids. Although other species are also described as having α- and β-globins, the particular chemical structures of those globins vary from species to species.

Relatively huge hemoglobin molecules are found in some invertebrates. For example, in some annelid worms, including common earthworms (*Lumbricus*), each whole molecule of blood hemoglobin contains almost 150 unit molecules—including a like number of heme groups (O_2-binding sites)—and has a molecular weight of 4 to 5 million daltons.

Animals sometimes have hemoglobins in muscle tissues or in other tissues besides blood. When this is the case, the hemoglobins in these different locations typically differ from blood hemoglobins in their chemical structure. Vertebrate muscle hemoglobins, termed **myoglobins** (*myo-*, "muscle"), provide apt examples. The myoglobins occur within the cytoplasm of muscle fibers, tending to be especially abundant in cardiac muscle fibers and in the slow oxidative (SO) class of skeletal muscle fibers (see page 160). When present at

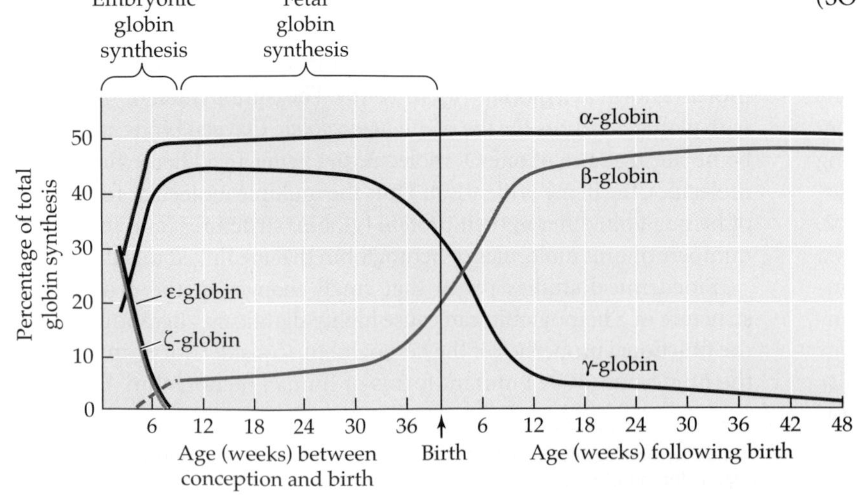

Figure 22.2 Human developmental changes in the types of globins synthesized for incorporation into blood hemoglobins Each blood-hemoglobin molecule consists of four globin units plus four O_2-binding heme groups (see Figure 22.1*c*). In early embryos, the globins synthesized are α-globin, ε-globin, and ζ-globin; *embryonic hemoglobins* are made up of these globins (e.g., one common form consists of two α-globin and two ε-globin units). By about 8 weeks after conception, α-globin and γ-globin are the principal globins synthesized, and most hemoglobin molecules are *fetal hemoglobin* molecules consisting of two α-globin and two γ-globin units. Following birth, synthesis of γ-globin ebbs, whereas that of β-globin increases, so that by 20 weeks of postnatal age, the blood hemoglobin is predominantly *adult hemoglobin* consisting of α- and β-globins. (After Wood 1976.)

high concentrations, they impart a reddish color to the tissue; "red" muscles are red because of myoglobins. Unlike blood hemoglobins, vertebrate myoglobins appear always to be single-unit (monomeric) molecules (see Figure 22.1*b*). They also have distinctive globins. In adult people, for example, whereas blood hemoglobin is a tetramer containing α- and β-globins, myoglobin is a monomer (~17,800 daltons) in which the globin is of different structure than the α- or β-globins.

Besides varying spatially within an animal, the chemical nature of hemoglobin often also changes temporally over the life cycle. For example, as we have already seen, the blood hemoglobin of fetal mammals is often different from that of adults. In humans, fetuses synthesize α-globin (as in the adult) and γ-globin, which differs from the β-globin synthesized by adults (Figure 22.2); each fetal blood-hemoglobin molecule consists of two α-globin and two γ-globin subunits. Earlier in development, still different globins are expressed in embryos.

Although the adult blood hemoglobin of many animals (e.g., humans and most other mammals) is of essentially uniform composition, in many species of poikilothermic vertebrates and invertebrates, the blood of adults normally consists of mixes of two, three, or even ten or more chemically different forms of hemoglobin. A relatively simple example is provided by the blood hemoglobin of the sucker fish *Catostomus clarkii*, which consists of about 80% of one major type of hemoglobin and 20% of another. When multiple chemical forms of hemoglobin occur in a species, the forms sometimes differ substantially in their O_2-binding characteristics. Possession of multiple blood hemoglobins may thus permit a species to maintain adequate O_2 transport over a broader range of conditions than would be possible with only a single hemoglobin type.

THE DISTRIBUTION OF HEMOGLOBINS Hemoglobins are the most widely distributed of the respiratory pigments, being found in at least ten phyla of animals (Figure 22.3) and even in some protists and plants. They are the only respiratory pigments found in vertebrates, and with a few interesting exceptions (see page 598), all vertebrates have hemoglobin in their blood. The blood hemoglobins of vertebrates are always contained in specialized cells, the **red blood cells (erythrocytes)**, discussed in Box 22.2.

Among the invertebrates, the distribution of hemoglobins is not only wide, but sporadic. Hemoglobins may occur within certain subgroups of a phylum but not others, and even within certain species but not other closely related species. Sometimes, among all the members of a large assemblage of related species, only an isolated few possess hemoglobins. The evolution of the

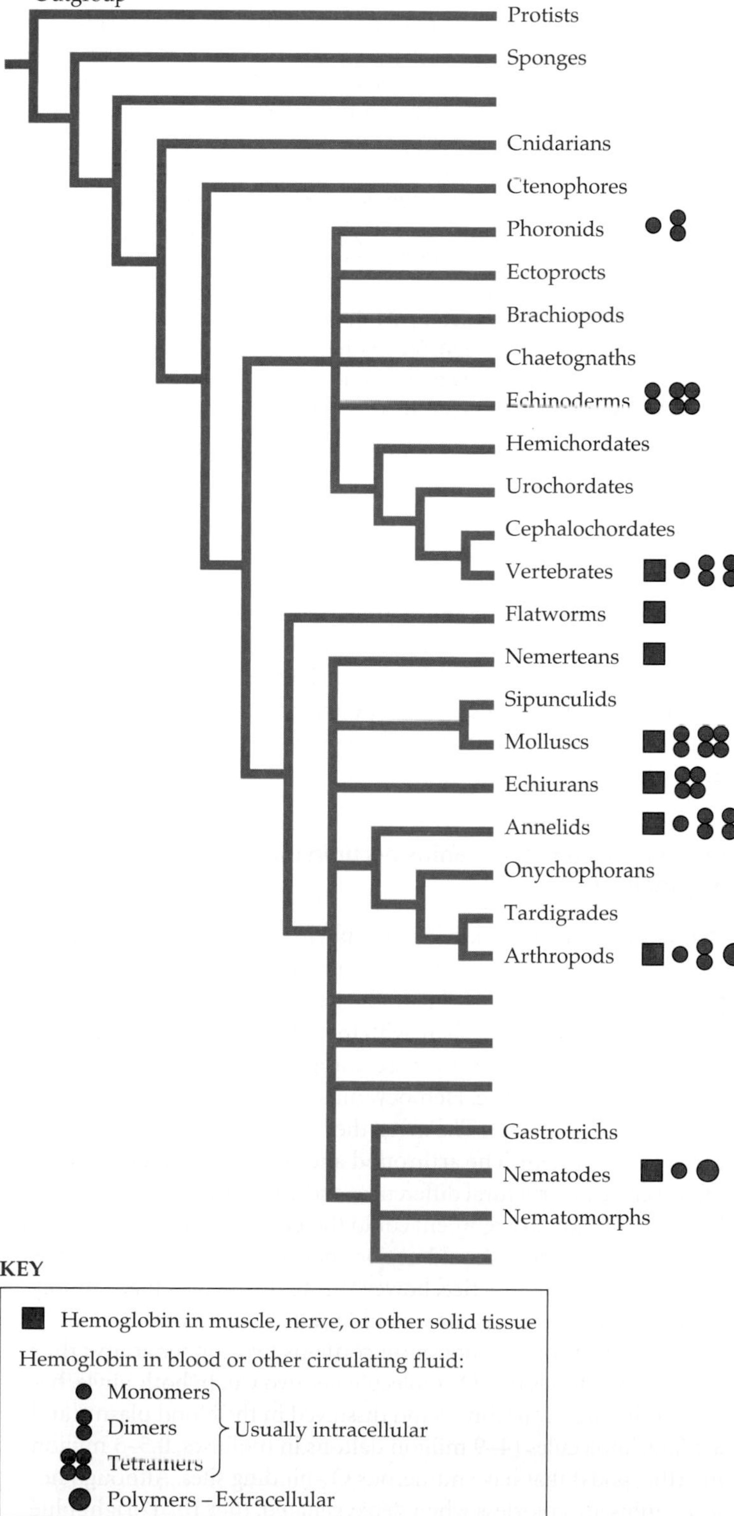

Figure 22.3 The distribution of hemoglobins in animals A square indicates that hemoglobins occur in solid tissues such as muscle or nerve tissues. Circles indicate that hemoglobins occur in circulating body fluids. The small circle corresponds to hemoglobins consisting of one unit molecule of heme plus globin: hemoglobin monomers. A pair or foursome of small circles symbolizes dimers or tetramers, respectively. A large circle indicates polymeric hemoglobins of high molecular weight, consisting of many joined unit molecules. The polymeric hemoglobins are always extracellular—dissolved in the circulating fluid. The monomeric, dimeric, and tetrameric forms, with few exceptions, are intracellular—contained within circulating cells such as erythrocytes. In groups having hemoglobin, not all species have it; nor do all the species with hemoglobin have all the chemical forms shown. Nearly all vertebrates, for example, have only tetrameric hemoglobin in their blood (the monomeric and dimeric forms occur only in cyclostome fish), and, whereas some arthropods and molluscs have hemoglobins, the majority lack them. This summary is not exhaustive. Some branches in the cladogram are simplified or unlabeled for ease of reading. (Hemoglobin data after Terwilliger 1980; cladogram after Brusca and Brusca 2003.)

BOX 22.2 Blood Cells and Their Production

The red blood cells (erythrocytes; RBCs) of vertebrates vary in size, shape, and other properties. Mammals have relatively small RBCs, usually 4–10 μm in diameter; human RBCs, for example, average 7.4 μm. Other groups of vertebrates have distinctly larger RBCs; the oval RBCs of frogs and toads, for example, average 23 × 14 μm in their major dimensions. A significant difference between the RBCs of mammals and those of other vertebrates is that the mature RBCs of mammals are essentially devoid of cell organelles; they have no nucleus, mitochondria, or ribosomes. The RBCs of all other vertebrates, and all the respiratory pigment–containing blood cells of invertebrates that have been studied, are nucleated, emphasizing how distinctive mammalian RBCs are in this regard. One way of expressing the blood's content of RBCs is as the **hematocrit**, defined to be the percentage of total blood volume occupied by the RBCs. The normal hematocrit for people at sea level is 42% to 45%.

The process by which the body makes RBCs is called **erythropoiesis**. Because RBCs have relatively short life spans, they are continually being replaced. The rate of turnover seems to be particularly rapid in mammals, possibly because mature mammalian RBCs—lacking a nucleus and ribosomes—lack any ability to repair proteins. The average human RBC lasts 4 months. We replace almost 1% of our RBCs every day, meaning that we make about 2 million new RBCs per second! The principal site of erythropoiesis in adult mammals is the soft interior of the bones, the **bone marrow**.

Erythropoiesis is under endocrine control. In mammals, the principal control is exerted by a glycoprotein hormone, **erythropoietin**. When low-O_2 conditions are detected in the body, erythropoietin is secreted in increased amounts and accelerates erythropoiesis. The control of mammalian erythropoietin secretion has become a favored system for study of the control of gene expression. The kidneys are the principal site of erythropoietin secretion in adult mammals. Molecular probes for erythropoietin messenger RNA (mRNA) have now established that erythropoietin is synthesized by secretory cells in the tissue (termed interstitial tissue) located between nephron tubules in the cortex of the kidneys. Low levels of O_2 in the kidneys cause increased expression of the erythropoietin gene in the secretory cells, thereby increasing formation of mRNA for synthesis of erythropoietin. The erythropoietin then travels in the blood to the bone marrow, where it stimulates production of RBCs.

wide but sporadic distribution of hemoglobins has attracted curiosity. An important point of background information for understanding this distribution is that metalloporphyrins are extremely ancient molecules. The cytochrome pigments of the electron-transport chain (see page 153), which are found universally in animals and undoubtedly evolved very early, consist of protein groups conjugated with heme itself or with closely similar iron porphyrins. Because all animals share such hemoglobin-like compounds, repeated hemoglobin evolution would be plausible. Some authorities conclude, nonetheless, that all hemoglobins in existing animals trace to a single common ancestor.

The circulating hemoglobins of invertebrates may be found in blood or in other moving fluids, such as coelomic fluids. Sometimes, as in vertebrates, these hemoglobins are contained within cells and thus are categorized as *intracellular*. The intracellular hemoglobins of invertebrates are always of relatively low molecular weight (~14,000–70,000 daltons); structurally, they are generally one-, two-, or four-unit molecules. Unlike those of vertebrates, the blood hemoglobins of some invertebrates are found dissolved in the blood plasma and thus are categorized as *extracellular*. The hemoglobin of earthworms (*Lumbricus*), for example, is dissolved in the blood plasma; when earthworm blood is held to the light, it is wine red and clear—quite unlike vertebrate bloods, which are opaque because of their high concentrations of red blood cells. The extracellular, dissolved hemoglobins of invertebrates are almost always large multiunit molecules, having molecular weights of 200,000 to 12 million daltons.

The concentration of blood hemoglobin in some invertebrates changes so dramatically from one environment to another that the animals change color. Water fleas (*Daphnia*), for example, have little hemoglobin and are pale when they have been living in O_2-rich waters. However, if they are placed in O_2-poor waters, they increase their levels of hemoglobin markedly within days and become bright red.

Hemoglobins—usually single-unit molecules—are found widely in solid tissues of invertebrates, not only in muscles but also in certain other tissues. Both muscle and neuronal hemoglobins occur in certain annelids and molluscs, for example; the neuronal hemoglobins sometimes impart a striking pinkish color to the ganglia or nerves. Insects usually lack respiratory pigments of any kind, but in a few, such as some of the backswimmer bugs, hemoglobins occur in tracheal organs, where they store O_2 for release to the tracheae during diving.

Copper-based hemocyanins occur in many arthropods and molluscs

Hemocyanins are found in just two phyla—the arthropods and the molluscs—but clearly rank as the second most common class of respiratory pigments. In turning to the hemocyanins, we encounter a minor problem that they share with the chlorocruorins and hemerythrins: The names given to these compounds provide no clue to their chemical structure. Hemocyanins do not contain heme, iron, or porphyrin structures. The metal they contain is copper, bound directly to the protein. The arthropod and mollusc hemocyanins exhibit consistent structural differences and are clearly of separate evolutionary origin. An argument could therefore be made that a single name, "hemocyanin," should not be applied to the molecules in the two phyla. Current practice, however, is to distinguish them simply as "arthropod hemocyanins" and "mollusc hemocyanins." Each O_2-binding site of a hemocyanin contains two copper atoms; thus the binding ratio is one O_2 molecule per two Cu. In both phyla, hemocyanins are invariably found dissolved in the blood plasma and are large molecules (4–9 million daltons in molluscs, 0.5–3 million in arthropods) that have numerous O_2-binding sites. Although hemocyanins are colorless when deoxygenated, they turn bright blue when oxygenated. Species that have high concentrations of hemocyanins are dramatically blue-blooded!

The molluscs that possess hemocyanins include the squids and octopuses (cephalopods), many chitons and gastropods (snails and slugs), and a relatively small subset of bivalves. Hemocyanins are not present in most bivalve molluscs (clams, scallops, and the like); indeed, bivalves usually lack circulating respiratory pigments of any kind. Groups of arthropods in which hemocyanins are important include the decapod crustaceans (crabs, lobsters, shrimps, and crayfish), the horseshoe crabs, and the spiders and scorpions. Hemocyanins are never found within muscle or other solid tissues. Certain molluscs that have blood hemocyanins have hemoglobins in their muscles or neurons, as already noted, or in their gills.

Chlorocruorins resemble hemoglobins and occur in certain annelids

Chlorocruorins, also sometimes called "green hemoglobins," occur in just four families of marine annelid worms, including the two families of fanworms and feather-duster worms (the Sabellidae and Serpulidae) that are so popular with aquarists. Chlorocruorins are always found extracellularly, dissolved in the blood plasma. They have close chemical similarities to the extracellular hemoglobins found dissolved in the blood plasma of many other annelids. Some authorites argue, in fact, that the chlorocruorins and the extracellular hemoglobins are so closely related from an evolutionary viewpoint that the chlorocruorins should not be recognized as a separate category of respiratory pigments. Like the extracellular hemoglobins, the chlorocruorins are large molecules, with molecular weights near 3 million daltons, composed of unit molecules consisting of iron-porphyrin groups conjugated with protein. They bind one O_2 per iron-porphyrin group. The chlorocruorins differ from hemoglobins in the type of iron porphyrin they contain.[4] This difference gives the chlorocruorins a distinctive and dramatic color. In dilute solution, they are greenish. In more concentrated solution, they are deep red by transmitted light but greenish by reflected light.

Iron-based hemerythrins do not contain heme and occur in four phyla

Hemerythrins have a distribution that is puzzling because it is simultaneously limited and far-flung, encompassing four phyla. Hemerythrins occur in all known sipunculid worms, in many brachiopods (lamp shells), in both genera of the small phylum Priapulida, and in one family of marine annelid worms (the magelonids). Despite their name, hemerythrins do not contain heme. They do contain iron (ferrous when deoxygenated), bound directly to the protein. Each O_2-binding site contains two iron atoms, and there is one such site per 13,000 to 14,000 daltons of molecular weight. In some instances, single-unit hemerythrins, known as **myohemerythrins**, occur within muscle cells. Better known are the circulating hemerythrins, which are always located intracellularly, in blood or coelomic cells, and typically have molecular weights of 40,000 to 110,000 daltons; many are octomers, having eight O_2-binding sites per molecule. Hemerythrins are colorless when deoxygenated but turn reddish-violet when oxygenated.

[4] The porphyrin differs from heme in that one of the vinyl chains (–CH=CH_2) on the periphery of the protoporphyrin ring in heme (see Figure 22.1*a*) is replaced with a formyl group (–CHO).

Summary: The Chemical Properties and Distributions of the Respiratory Pigments

- The four chemical classes of respiratory pigments are all metalloproteins. They bind reversibly with O_2 at specific O_2-binding sites associated with the metal atoms in their molecular structures.
- In hemoglobins, the unit molecule consists of heme bonded with protein (globin). The heme structure—an iron (ferrous) porphyrin—is identical in all hemoglobins. The globin, however, varies widely among species and among different molecular forms of hemoglobin within any single species.
- Hemoglobins are the most common and widespread respiratory pigments, occurring in at least ten phyla. Virtually all vertebrates have blood hemoglobin. The blood-hemoglobin molecules of vertebrates are usually tetramers consisting (in adults) of two α-globin and two β-globin unit molecules; they always occur in red blood cells. Although many invertebrates also have hemoglobins in blood cells, some invertebrates have hemoglobins dissolved in their blood plasma.
- Hemocyanins are the second most common of the respiratory pigments in animals. They contain copper and turn bright blue when oxygenated. There are two types of hemocyanins, which are of separate evolutionary origin: arthropod hemocyanins (occurring in crabs, lobsters, crayfish, horseshoe crabs, spiders, and some other arthropods) and mollusc hemocyanins (occurring in squids, octopuses, many snails, and some other molluscs). Hemocyanins are always dissolved in the blood plasma.
- Chlorocruorins, which are similar to hemoglobins, occur only in four families of marine annelid worms, always dissolved in the blood plasma.
- Hemerythrins are non-heme, iron-containing respiratory pigments that have a limited and scattered distribution, occurring in four different invertebrate phyla.

The O_2-Binding Characteristics of Respiratory Pigments

A key tool for understanding the function of a respiratory pigment is the **oxygen equilibrium curve** or **oxygen dissociation curve**. In a body fluid containing a respiratory pigment, there is a large population of O_2-binding sites. Human blood, for example, contains about 5.4×10^{20} heme groups per 100 mL. The combination of O_2 with each individual O_2-binding site is *stoichiometric*: One and only one O_2 molecule can bind, for example, with each heme group of a hemoglobin or with each pair of copper atoms in a hemocyanin. However, in blood, where there are great numbers of O_2-binding sites, all sites do not simultaneously bind with O_2 or release O_2. Instead, the fraction of the O_2-binding sites that are oxygenated varies in a *graded* manner with the O_2 partial pressure. The oxygen equilibrium curve shows the functional relation between the percentage of binding sites that are oxygenated and the O_2 partial pressure (Figure 22.4*a*).

The respiratory pigment in the blood of an animal is said to be **saturated** if the O_2 partial pressure is high enough for all O_2-binding sites to be oxygenated. At saturation, the amount of O_2 carried per volume of blood equals the blood's oxygen-carrying capacity. Ordinary human blood has an oxygen-carrying capacity of about 20 mL O_2 per 100 mL of blood and actually has that concentration of O_2 when saturated.[5] The volume of gas carried per 100 volumes of blood is often termed the **volumes percent** (**vol %**) of gas. In this system of units, the oxygen-carrying capacity of ordinary human blood is 20 vol %. If one-fourth of all O_2-binding sites in blood are oxygenated, the blood's O_2 content per unit of volume is one-fourth of its oxygen-carrying capacity, and similarly, if half are oxygenated, the O_2 content per unit of volume is half the carrying capacity. Thus, when the oxygen-carrying capacity of blood is known, one can easily convert the fraction of oxygenated binding sites at each partial pressure (Figure 22.4*a*) into a corresponding blood O_2 concentration at each partial pressure. One thus can obtain an alternative form of the oxygen equilibrium curve (Figure 22.4*b*).

In blood that contains a respiratory pigment, O_2 is actually present in two chemical forms: dissolved in the blood plasma and chemically combined with the pigment. These two portions of O_2 bear different relations to the O_2 partial pressure. The amount of dissolved O_2 per unit of volume simply follows the principles of gas solution discussed in Chapter 20 (see Equation 20.3) and is *proportional* to the O_2 partial pressure (the two properties have a straight-line relation; see Figure 22.4*b*). The amount of O_2 per unit of volume that is chemically combined with the respiratory pigment exhibits a more complex relation to O_2 partial pressure. Usually, this relation is sigmoid, and therefore the oxygen equilibrium curve for blood as a whole (including both dissolved and chemically combined O_2) is sigmoid (see Figure 22.4*b*).

As we will see later in this chapter, the O_2-binding properties of respiratory pigments are often affected by temperature, pH, and other properties of the blood chemical environment. We will also later discuss the reason for the sigmoidality we have noted. Before we consider those factors, however, it is important to understand the basic elements of respiratory pigment function in living animals and to appreciate the interpretive value of oxygen equilibrium curves. To these ends, and recognizing that refinements will later be needed, let's look at the fundamentals of O_2 transport by our own blood.

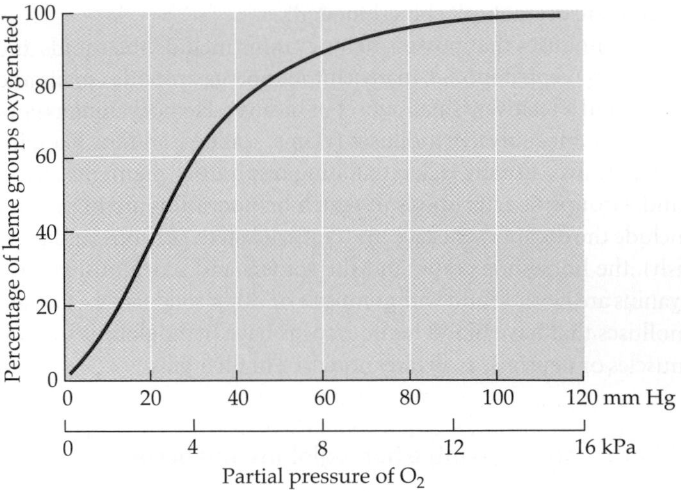

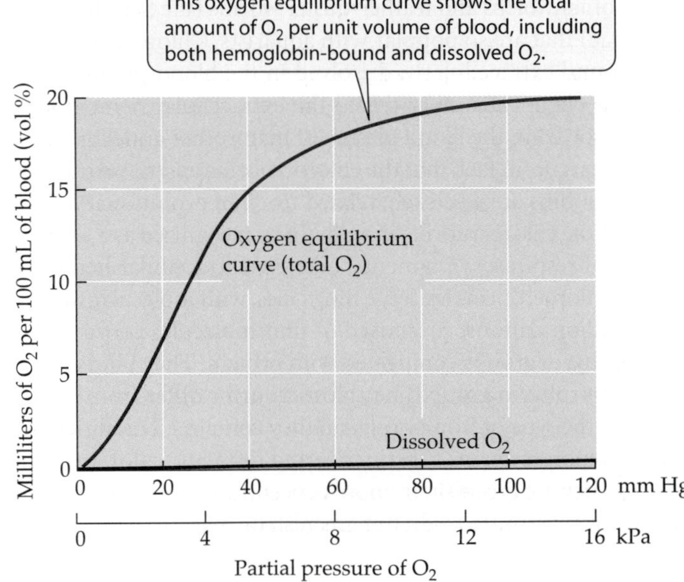

Figure 22.4 Typical oxygen equilibrium curves for human arterial blood *(a)* The percentage of heme groups oxygenated as a function of the O_2 partial pressure. *(b)* The total blood O_2 concentration—including both hemoglobin-bound and dissolved O_2—as a function of the O_2 partial pressure; the portion of the total O_2 present as dissolved O_2 is plotted at the bottom. Normal arterial values of CO_2 partial pressure, pH, and temperature are assumed. In humans, as in other animals, significant individual variation occurs. (After Roughton 1964; *b* assumes an O_2 concentration of 20 vol % at 16 kPa.)

Human O_2 transport provides an instructive case study

To understand the uptake of O_2 by the blood in a person's lungs, it is important to recall from Chapter 21 that breathing maintains the O_2 partial pressure in the alveolar gases of the lungs at about 13.3 kPa (100 mm Hg). Blood arriving at the alveoli has a lower O_2 partial pressure. Thus O_2 diffuses into the blood from the alveolar gas, raising the blood O_2 partial pressure. At each partial pressure, the hemoglobin in the blood takes up an amount of O_2 that is dictated by its oxygen equilibrium curve (see Figure 22.4).[6]

If the O_2 partial pressure of the blood in the lungs were to rise to the alveolar partial pressure, 13.3 kPa (100 mm Hg), we can see from

[5] Gas volumes are always expressed at standard conditions of temperature and pressure (see Appendix D) unless otherwise stated. Such volumes are proportional to molar quantities, as discussed in Chapter 20.

[6] The importance of hemoglobin for *speeding* the uptake of O_2 is discussed in Chapter 20 (see page 531).

Figure 22.4 that the hemoglobin in the blood would become virtually saturated with O_2. In fact, mixed blood leaving the lungs is at a somewhat lower O_2 partial pressure: 12 to 12.7 kPa (90–95 mm Hg) in a person at rest. This lower partial pressure hardly affects the blood O_2 content, however, because as shown by the oxygen equilibrium curve (Figure 22.4), there is a plateau in the relation between the blood O_2 content and the O_2 partial pressure at these high partial pressures; provided the blood O_2 partial pressure is high enough to be in the plateau region, hemoglobin will be almost saturated with O_2 regardless of the partial pressure. The alveolar O_2 partial pressure could even vary a bit, and still, because of this property of hemoglobin, blood delivered to the systemic arteries would remain almost entirely saturated. The close "matching" of the saturation partial pressure of hemoglobin and the alveolar partial pressure represents a striking evolutionary coadaptation: The hemoglobin molecule has evolved O_2-binding properties that suit it to oxygenate well at the O_2 partial pressures maintained in the lungs by the breathing system.

After leaving the lungs and flowing through the left side of the heart, the blood flows to the systemic tissues. To understand the events in the systemic tissues, it is crucial to recall that in the mitochondria, O_2 is continually being combined with electrons and protons to form H_2O. By this process, O_2 molecules are removed from solution, and the O_2 partial pressure in and around the mitochondria is lowered. Blood arriving in capillaries of systemic tissues from the lungs has a high O_2 partial pressure; O_2 thus diffuses from the blood to the mitochondria (see Figure 20.4). As dissolved O_2 leaves the blood and the O_2 partial pressure of the blood therefore falls, hemoglobin releases (unloads) O_2. The oxygen equilibrium curve (see Figure 22.4) again comes into play: The amount of O_2 released from hemoglobin at each O_2 partial pressure is dictated by the curve.

Knowing that hemoglobin leaves the lungs in a virtually saturated condition, we can calculate its yield of O_2 to the systemic tissues by obtaining a measure of its degree of saturation in the venous blood draining the tissues. The simplest way to obtain this measure is to determine the degree of saturation in blood drawn from the great veins leading back to the heart; such blood is termed **mixed venous blood** because it represents a mixture of the venous blood coming from all parts of the body. In people at rest, the O_2 partial pressure of mixed venous blood is about 5.3 kPa (40 mm Hg). The oxygen equilibrium curve (see Figure 22.4*b*) indicates that blood at this partial pressure contains about 15 mL of O_2 per 100 mL. Recalling that arterial blood contains about 20 mL of O_2/100 mL, we see that the O_2 content of the blood falls by about 5 mL of O_2/100 mL when the blood circulates through the systemic tissues in humans at rest. In other words, about 5 mL of O_2 is released from each 100 mL of blood (Figure 22.5). The release of O_2 from the blood is often expressed as the **blood oxygen utilization coefficient**, defined to be the *percentage* of arterial O_2 that is released to the systemic tissues. In people at rest, recognizing that arterial blood contains 20 mL of O_2/100 mL and that 5 mL of O_2/100 mL is released to the tissues, the oxygen utilization coefficient is about 25%. Thus only one-fourth of the O_2 brought to the systemic tissues in the arterial blood is actually used at rest.

THE SIGNIFICANCE OF MIXED VENOUS O_2 PARTIAL PRESSURE The O_2 partial pressure of mixed venous blood represents an average of the O_2 partial pressures of blood leaving the various systemic tissues. It thus allows us to gauge the blood's *overall* drop in O_2 partial pressure during circulation through all tissues combined. It does not necessarily reflect, however, the drop in partial pressure as the blood flows through any particular tissue. For example, blood entering a given tissue at a partial pressure of 12.7 kPa (95 mm Hg) might exit at a partial pressure that is either higher or lower than the mixed venous partial pressure; then the drop in blood O_2 content as the blood flows through that tissue would differ from the mean value calculated from the mixed venous partial pressure. The mixed venous partial pressure is, in fact, a *weighted* average of the O_2 partial pressures of blood leaving the various tissues. It is weighted according to the rate of blood flow through each tissue; tissues with high rates of blood flow influence the mixed venous partial pressure more than ones with low rates of flow.

THE DETERMINANTS OF A TISSUE'S VENOUS O_2 PARTIAL PRESSURE The O_2 partial pressure to which the blood falls in its passage through a particular tissue is not a static property of that tissue. Instead, it is a dynamic and changing property. It depends on the rate of blood flow

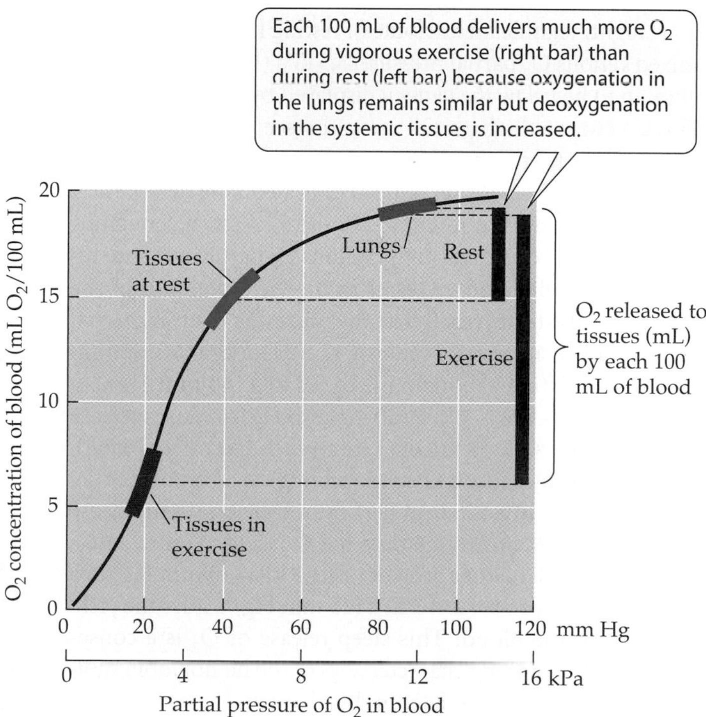

Figure 22.5 Oxygen delivery by human blood at rest and during vigorous exercise The oxygen equilibrium curve shown is that for human arterial blood (see Figure 22.4*b*). The rectangles on the curve show representative values for blood O_2 concentration and partial pressure in the lungs, in the systemic tissues during rest, and in the systemic tissues during vigorous exercise (tissue values are mixed venous blood values). The vertical purple bars to the right show how much O_2 is delivered to the tissues by each 100 mL of blood. All values are semi-quantitative; the intent of this diagram is conceptual rather than literal. Effects of pH and other variables of the blood-hemoglobin milieu are not included.

through the tissue, the arterial O_2 partial pressure, the amount of hemoglobin per unit of blood volume, and the tissue's rate of O_2 consumption. To illustrate, if the rate of blood flow through a tissue decreases while all the other relevant factors remain unchanged, the venous partial pressure will decline, because each unit of blood volume will have to yield more O_2 in its passage through the tissue. Similarly, an increase in the rate of O_2 consumption by the tissue will cause a decrease in venous partial pressure. The venous partial pressures normally seen in people at rest result from the set of conditions that ordinarily prevail at rest.

DELIVERY OF O_2 DURING EXERCISE As discussed in Chapter 21, controls on breathing tend to keep the alveolar O_2 partial pressure stable, near 13.3 kPa (100 mm Hg), as people exercise more and more intensely. During vigorous exercise, however, the *blood* O_2 partial pressure reached in the lungs tends to decline from the resting blood partial pressure. There are several potential reasons for this decline. One is that blood passes through the pulmonary circulation faster as the intensity of exercise increases, thus decreasing the time available for equilibration between the alveolar gas and blood. The relative flatness of the oxygen equilibrium curve at high O_2 partial pressures again comes to the rescue (see Figure 22.4). Even if the blood reaches a partial pressure of only 11 kPa (80 mm Hg) (which is often the case during intense exercise), the drop in the O_2 *content* of the arterial blood is modest. For simplicity, we treat the O_2 content of the arterial blood as a constant as we discuss exercise in more detail.

The modest utilization of blood O_2 at rest leaves a large margin to increase utilization during exercise. As we have seen, only about 25% of the O_2 carried by the systemic arterial blood is used when people are at rest. The remaining amount, the amount of O_2 in mixed venous blood, is called the **venous reserve.** During exercise, more O_2 is withdrawn from each unit of blood volume as the blood passes through the systemic tissues, and the venous reserve becomes smaller.

A highly significant attribute of resting function is that the mixed venous O_2 partial pressure, 5.3 kPa (40 mm Hg), is low enough to be *below the plateau* displayed by the oxygen equilibrium curve at high partial pressures (see Figure 22.5). During exercise, therefore, when the venous O_2 partial pressure declines below the resting value, it does so on the *steep* part of the oxygen equilibrium curve (Figure 22.6). Consequently, relatively small decreases in the venous O_2 partial pressure result in relatively large increases in the yield of O_2 from the blood. To illustrate, recall first that at rest, a drop in partial pressure from an arterial value of 12 to 12.7 kPa (90–95 mm Hg) to the resting venous value of 5.3 kPa (40 mm Hg)—a total drop of 6.7 to 7.4 kPa (50–55 mm Hg)—causes release of about 5 vol % O_2 (5 mL of O_2 from each 100 mL of blood). A further drop of just 2 kPa (15 mm Hg) to a venous partial pressure of 3.3 kPa (25 mm Hg) causes the blood to release another 5 vol %, thus *doubling* the O_2 yield (Figure 22.6). Moreover, a still further drop of just 1.3 kPa (10 mm Hg) to a venous partial pressure of 2 kPa (15 mm Hg) triples the yield of O_2 from the blood! This steep release of O_2 is a consequence of the binding characteristics of the hemoglobin molecule, as we will discuss later in this chapter.

Looking at actual O_2 delivery during exercise in mammals, a useful benchmark is that, over a wide range of exercise states, the O_2 partial pressure of blood leaving the working skeletal muscles is about 2.7 kPa (20 mm Hg), not only in humans but also in several other species on which measurements have been made. Thus the venous O_2 partial pressure during exercise is about 2.7 kPa (20 mm Hg) lower than the resting value. This drop in the venous partial pressure increases the amount of O_2 released from each volume of blood by a factor of about 2.5 and increases the oxygen utilization coefficient to about 65%. We have said that a partial pressure of about 2.7 kPa (20 mm Hg) prevails in the blood leaving the muscles over a wide *range* of exercise states; that is, even as the muscles work harder and demand more O_2, in this range little change occurs in their venous partial pressure or in the amount of O_2 they obtain from each unit of blood volume. This stability of venous partial pressure occurs because the *rate of blood flow* to the muscles is adjusted: As the O_2 demand of the muscles rises, their rate of blood flow increases in parallel, enabling them to draw O_2 from an enhanced volume of blood per unit of time. Of course, the rate of blood flow cannot increase indefinitely. Once it is maximized, further increases in a muscle's intensity of work result in further decreases in the venous O_2 partial pressure. Indeed, during extreme exertion, the O_2 partial pressure of blood leaving some muscles may fall close to zero, indicating virtually complete deoxygenation (100% O_2 utilization).

As the O_2 partial pressure of blood in the systemic capillaries declines, there is a risk that the rate of O_2 diffusion from the blood to the mitochondria will become too low to support aerobic catabolism. The venous O_2 partial pressure below which aerobic catabolism becomes impaired is known as the **critical venous O_2 partial pressure.** It is about 1.3 kPa (10 mm Hg) in mammalian muscles. As we

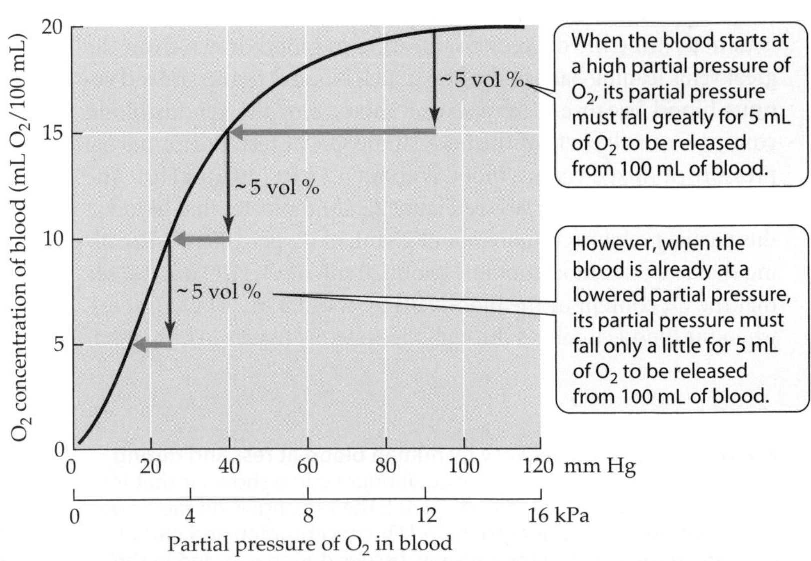

Figure 22.6 As the O_2 partial pressure of blood falls, less and less of a drop in partial pressure is required to cause unloading of 5 vol % O_2 Each green horizontal arrow depicts the drop in O_2 partial pressure required to cause the unloading depicted by the red vertical arrow to its right. The oxygen equilibrium curve shown is that for human arterial blood (see Figure 22.4*b*).

have seen, the rate of blood flow through muscles is usually increased sufficiently to maintain the venous partial pressure above this critical level over a wide range of exercise states, although the partial pressure sometimes falls below the critical value during particularly intense exertion. A noteworthy feature of human hemoglobin is that it yields about 90% of its O_2 before the venous partial pressure falls below the critical level, as can be seen from the oxygen equilibrium curve (see Figure 22.4*a*). Here again we see that the O_2-binding properties of hemoglobin are closely integrated with other physiological features.

Let's now look briefly at *whole-body* O_2 utilization and O_2 delivery during exercise. Although blood draining active muscles may be rather thoroughly deoxygenated during heavy exercise, the partial pressure of *mixed* venous blood generally does not fall below 2.1 to 2.7 kPa (16–20 mm Hg) in humans, even during strenuous work, because blood from the muscles mixes in the great veins with blood from other parts of the body in which O_2 utilization is not so great. The whole-body oxygen utilization coefficient therefore rises to a peak of about 60% to 75% during exercise—indicating that 2.5 to 3.0 times as much O_2 is extracted from each volume of blood as at rest (see Figure 22.5). In average young people, the rate of blood circulation can be increased to 4 or 4.5 times the resting level. These figures, taken together, show that the total rate of O_2 delivery by the circulatory system can increase to 10 to 13 times the resting rate. Trained athletes often achieve still higher O_2 delivery rates; a key reason is that endurance training increases the rate at which a person's heart can pump blood.

THE "MOLECULAR DESIGN" OF HUMAN HEMOGLOBIN We have seen in this section (1) that human hemoglobin is nearly saturated at the O_2 partial pressures that are maintained in the lungs by breathing; (2) that the oxygen equilibrium curve of hemoglobin is nearly flat at pulmonary O_2 partial pressures, so that high oxygenation is assured regardless of modest variation in pulmonary function; and (3) that the oxygen equilibrium curve is shaped in such a way that 90% of the O_2 bound to hemoglobin can be released for use at blood partial pressures that are compatible with full mitochondrial function. These functional properties of human hemoglobin are consequences of its chemical structure, and its normal chemical structure is but one of thousands of possible structures. Many physiologists have concluded that the human hemoglobin molecule provides a particularly convincing example of "evolutionary molecular design." Natural selection has produced a molecule with functional properties that are integrated in strikingly harmonious ways with the attributes of the organs that provide O_2 to the blood and draw O_2 from the blood.

A set of general principles helps elucidate O_2 transport by respiratory pigments

From our study of hemoglobin function in people, we can develop a set of key principles that are useful for understanding the function of blood respiratory pigments in general:

1. To determine the extent of pigment oxygenation, ask first: What are the blood O_2 partial pressures established in the breathing organs? Then examine the oxygen equilibrium curve to determine the extent of pigment oxygenation at those partial pressures.
2. To determine the extent of pigment deoxygenation in systemic tissues, start by acquiring some knowledge of blood O_2 partial pressures in those tissues. The mixed venous O_2 partial pressure is a useful and easily measured indicator; one must remember, however, that it does not necessarily provide information on O_2 release in any *particular* tissue. After a figure for tissue O_2 partial pressure has been obtained, examine the oxygen equilibrium curve to determine the extent of pigment deoxygenation in the systemic tissues.
3. To compute circulatory O_2 delivery, the rate of blood flow is as important as the yield of O_2 per unit of blood volume, because O_2 delivery is the product of flow rate and O_2 yield per unit of volume. Complexity is introduced by the fact that these two factors are not independent: The rate of blood flow helps to determine the venous O_2 partial pressure and thus the yield of O_2 per unit of volume.
4. The operational parameters of the O_2-transport system are strongly affected by exercise and possibly by other factors, such as the environmental O_2 partial pressure. Full understanding of the function of an O_2-transport system requires that animals be studied over a range of physiological and environmental conditions.

The shape of the oxygen equilibrium curve depends on O_2-binding site cooperativity

What determines the shape of the oxygen equilibrium curve? As we explore this question, vivid parallels to principles we addressed in the study of enzymes will again be evident. In Chapter 2,(see page 43) we saw that when the catalytic sites of a particular enzyme function independently of each other, a hyperbolic relation exists between enzyme activity and substrate concentration; when the sites exhibit cooperativity, however, a sigmoid relation occurs. Similarly, when the O_2-binding sites of a respiratory pigment function independently, the oxygen equilibrium curve is hyperbolic, and when they exhibit cooperativity, a sigmoid curve results.

Hyperbolic oxygen equilibrium curves are exemplified by the vertebrate myoglobins. They (and most of the invertebrate myoglobins) contain just one O_2-binding site (heme) per molecule. Thus their O_2-binding sites function independently of each other, and the chemical reaction between myoglobin and O_2 can be written simply as

$$Mb + O_2 \rightleftharpoons MbO_2 \quad (22.1)$$

where Mb is a molecule of deoxymyoglobin and MbO_2 is one of oxymyoglobin. According to the principles of mass action (see page 46), increasing the partial pressure (and thus the chemical potential) of O_2 will shift the reaction to the right, increasing myoglobin oxygenation. The principles of mass action predict that for such a simple chemical reaction, the fraction of myoglobin molecules oxygenated will be a *hyperbolic* function of the O_2 partial pressure. In fact, oxygen equilibrium curves for human myoglobin (Figure 22.7*a*) and other monomeric myoglobins do exhibit this predicted shape.

Because vertebrate blood hemoglobins have four O_2-binding sites within each molecule, the opportunity exists for cooperativity, and in fact, positive cooperativity occurs. Binding of O_2 at one or two of the sites on a molecule of blood hemoglobin alters the conformation of the molecule in ways that enhance the affinity of the remaining

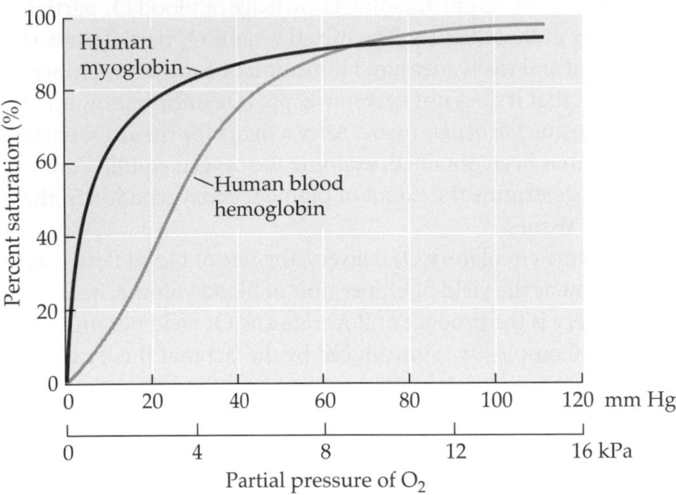

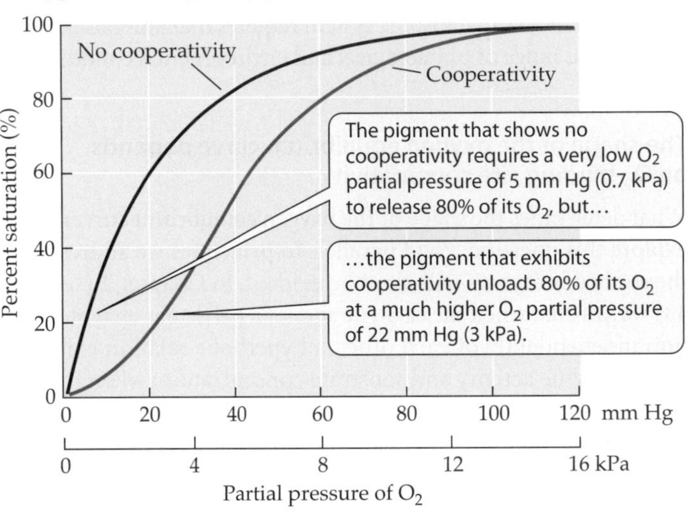

Figure 22.7 Respiratory pigments display hyperbolic or sigmoid oxygen equilibrium curves depending on whether they exhibit cooperativity in O_2 binding *(a)* The hyperbolic oxygen equilibrium curve of human myoglobin—a pigment that exhibits no cooperativity—compared with the sigmoid curve of human blood hemoglobin—a pigment that displays cooperativity. Both curves were determined under similar conditions: 38°C, pH 7.40. *(b)* Comparison of oxygen equilibrium curves for two hypothetical pigments that reach saturation at about the same O_2 partial pressure, but differ in whether they exhibit cooperativity. (*a* after Roughton 1964.)

sites for O_2, meaning that a partially oxygenated molecule is more likely than an entirely deoxygenated one to bind additional O_2.[7] The consequence is a sigmoid oxygen-equilibrium curve, exhibiting a particularly steep relation between O_2 binding and O_2 partial pressure in the mid-range of O_2 partial pressures. Figure 22.8 presents oxygen equilibrium curves for the blood of eleven animal species, including six vertebrates that have four-unit hemoglobins and five invertebrates that have high-molecular-weight hemoglobins or hemocyanins with numerous O_2-binding sites. All the curves are sigmoid to some degree, reflecting intramolecular cooperativity; however, the extent of cooperativity and sigmoidality varies. The extent of cooperativity is usually expressed using a mathematical index called the **Hill coefficient** (n), named after A. V. Hill, an important twentieth-century physiologist. The coefficient is 1.0 for pigments that show no cooperativity (e.g., myoglobins) and reaches 6 or more in some high-molecular-weight pigments with high cooperativity. Mammalian hemoglobins exhibit values of 2.4–3.0.

Cooperativity affects deoxygenation (unloading) of respiratory pigments as much as it affects their oxygenation (loading). During

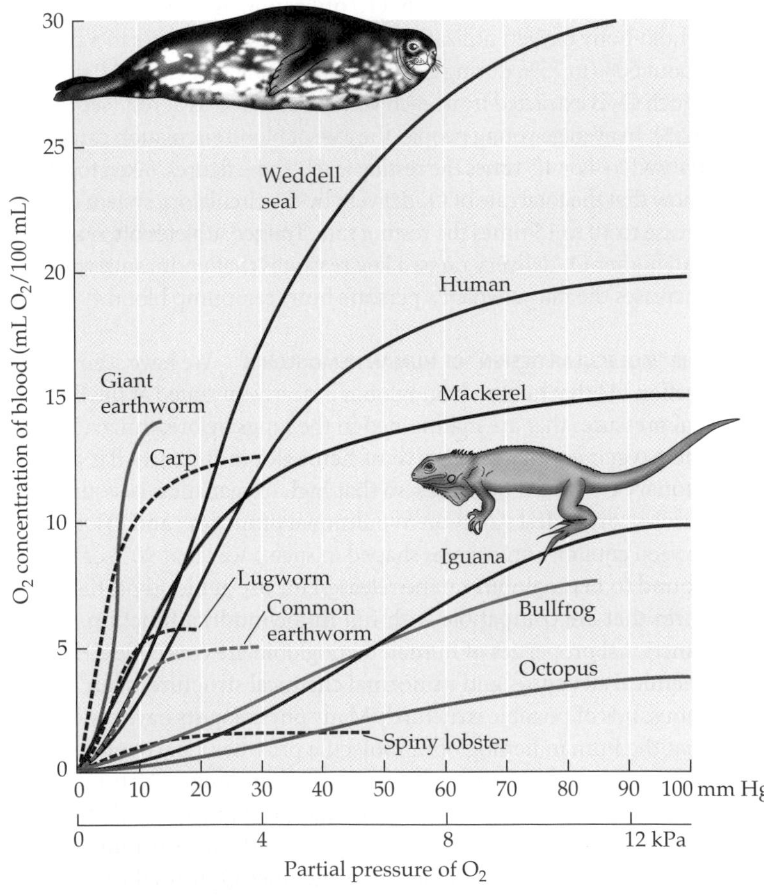

Figure 22.8 A diversity of blood oxygen equilibrium curves The blood oxygen equilibrium curves of 11 animal species vary in two ways. First, they vary in shape, a property that reflects the different molecular forms of the respiratory pigments in different species. Second, they vary in height, a property that reflects how much respiratory pigment is present per unit of blood volume (oxygen-carrying capacity). Species: Bullfrog, *Rana catesbeiana*; carp, *Cyprinus carpio*; common earthworm, the nightcrawler *Lumbricus terrestris*; giant earthworm, the 1-meter-long South American earthworm *Glossoscolex giganteus*; iguana, *Iguana iguana*; lugworm, the seacoast annelid *Arenicola* sp.; mackerel, *Scomber scombrus*; octopus, the giant octopus *Enteroctopus dofleini* of the North American Pacific coast; spiny lobster, *Panulirus interruptus*; Weddell seal, *Leptonychotes weddelli*. (From Hill and Wyse 1989.)

[7] Because the four O_2-binding sites are located within the four different protein subunits of the hemoglobin tetramer, the cooperativity displayed by the tetramer is often termed *subunit interaction*. It used to be termed *heme–heme interaction*, but this term has been dropped because the interaction between the O_2-binding sites is indirect, not directly between one heme and another.

deoxygenation of a molecule that exhibits cooperativity, removal of O_2 from some of the O_2-binding sites tends to decrease the affinity of the remaining sites for O_2, thereby promoting even further deoxygenation. Figure 22.7*b* shows the consequences, using two hypothetical pigments that are similar in the O_2 partial pressure at which they become saturated, but differ in that one exhibits cooperativity whereas the other does not. If we assume that both pigments are initially fully oxygenated and ask how they behave during deoxygenation, it is clear that in the mid-range of O_2 partial pressures, the pigment showing cooperativity deoxygenates more readily, giving up more of its O_2 at any given O_2 partial pressure. In a sentence, whether a molecule of a respiratory pigment is loading or unloading, cooperativity enhances the *responsiveness* of the process to changes in the O_2 partial pressure within the mid-range of partial pressures.

Respiratory pigments exhibit a wide range in their affinities for O_2

The respiratory pigments evolved by different species of animals vary widely in how readily they combine with O_2, a property known as their **affinity** for O_2. Pigments that require relatively high O_2 partial pressures for full loading and that conversely unload substantial amounts of O_2 at relatively high partial pressures are said to have a relatively **low affinity** for O_2. Pigments that load fully at low partial pressures and consequently also require low partial pressures for substantial unloading are said to have a relatively **high affinity** for O_2. Affinity for O_2 is an *inverse* function of the O_2 partial pressure required for loading: The *higher* the O_2 partial pressure required to load a pigment, the *lower* is the pigment's affinity for O_2. The hemoglobins of humans and carp (see Figure 22.8) provide examples of pigments that differ in their affinity for O_2. Human hemoglobin requires a far higher O_2 partial pressure to become saturated than carp hemoglobin, indicating that the human hemoglobin combines less readily with O_2 and has a lower affinity.

A convenient index of O_2 affinity is $\boldsymbol{P_{50}}$ (pronounced "P fifty"), defined as the partial pressure of O_2 at which a pigment is 50% saturated. Figure 22.9 shows how P_{50} is measured. With Figure 22.9 in mind, a glance at Figure 22.8 reveals that human hemoglobin has a much higher P_{50} ($\cong$ 3.5 kPa in arterial blood) than carp hemoglobin ($\cong$ 0.7 kPa). *Affinity and* P_{50} *are inversely related: As* P_{50} *increases,* O_2 *affinity decreases.*

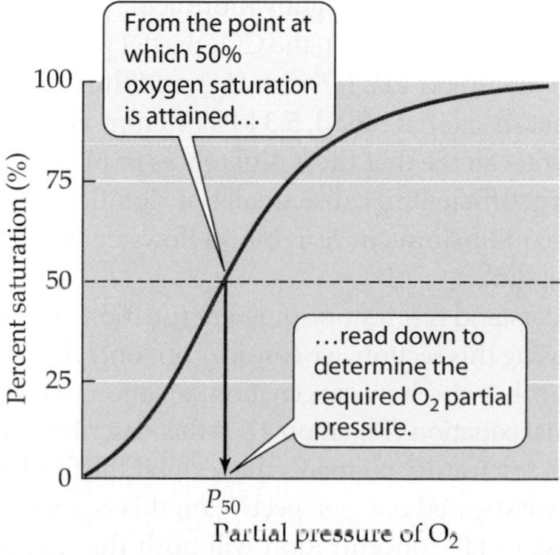

Figure 22.9 How to measure P_{50}

In the jargon of respiratory-pigment physiology, lowering the O_2 affinity is said to "shift the oxygen equilibrium curve to the right." This is because lowering the affinity of a pigment causes the oxygen equilibrium curve to extend farther to the right on the *x* axis (the axis of O_2 partial pressure) before reaching saturation. Conversely, raising the O_2 affinity is said to "shift the curve to the left."

The Bohr effect: Oxygen affinity depends on the partial pressure of CO_2 and the pH

A decrease in the pH or an increase in the CO_2 partial pressure in a body fluid or tissue containing a respiratory pigment often causes the O_2 affinity of the respiratory pigment to decrease, thus shifting the oxygen equilibrium curve to the right. This effect, illustrated for the blood hemoglobins of humans and dogs in Figure 22.10, is known as the **Bohr effect** or **Bohr shift**, in commemoration of Christian Bohr (1855–1911), the prominent Danish physiologist (and father of Nobel laureate Niels Bohr) who led the discovery of it in 1904[8]. Part of the reason that an increase in CO_2 partial pressure causes such a shift is that the pH of a solution tends to decline as its CO_2 partial pressure is raised. However, CO_2 also exerts a direct negative effect on the O_2 affinities of some respiratory pigments, such as the blood hemoglobins of humans and other mammals. Recognizing that protons (H^+ ions) and CO_2 can exert independent affinity-lowering effects, modern workers often distinguish two types of Bohr effects: a **fixed-acid Bohr effect**—which results from influences of the proton (H^+) concentration on respiratory-pigment molecules—and a **CO_2 Bohr effect**—which results from the immediate influences of increased CO_2 partial pressure. Species that show these effects vary widely in the *magnitudes* of the effects.

Protons exert their effects on O_2 affinity by combining with pigment molecules. Referring to hemoglobin (Hb) as a specific example, we can write the following, *strictly conceptual* equation to summarize the effects of protons on O_2 affinity (the equation does not reflect the true stoichiometry of the reaction):

$$HbO_2 + H^+ \rightleftharpoons HbH^+ + O_2 \qquad (22.2)$$

Increasing the H^+ concentration tends to increase the combination of Hb with H^+, thus shifting the reaction to the right and favoring dissociation of O_2. The H^+ ions bind at sites on the hemoglobin molecules different from the O_2-binding sites (e.g., at histidine residues). Thus H^+ acts as an *allosteric* modulator of O_2 binding. Carbon dioxide also combines chemically with pigment molecules and functions as an allosteric modulator in cases in which it exerts direct effects on affinity.

The Bohr effect often has adaptive consequences for O_2 delivery. Because the CO_2 partial pressure is generally higher and the pH is generally lower in the systemic tissues than in the lungs or gills, a respiratory pigment that displays a Bohr effect shifts to lower O_2 affinity each time the blood enters the systemic tissues and reverts back to higher O_2 affinity each time the blood returns to the breathing or-

[8] In unusual cases, such as some species of molluscs and spiders, Bohr effects opposite to the usual direction, termed *reverse Bohr effects*, are observed.

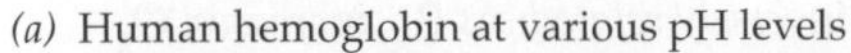

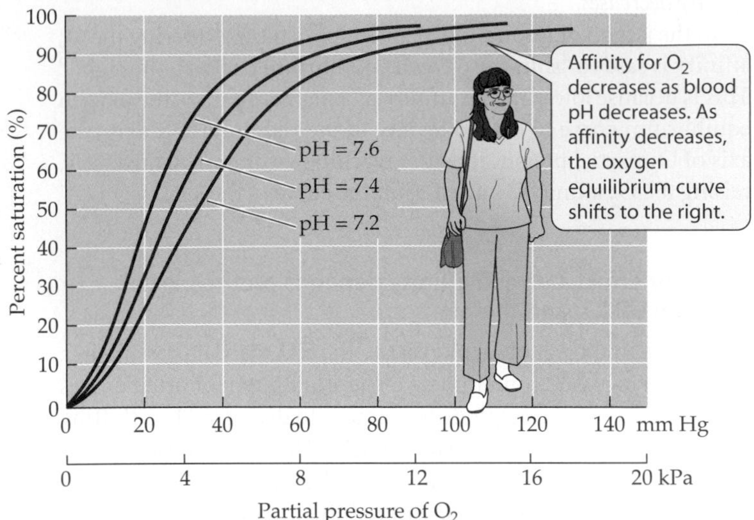

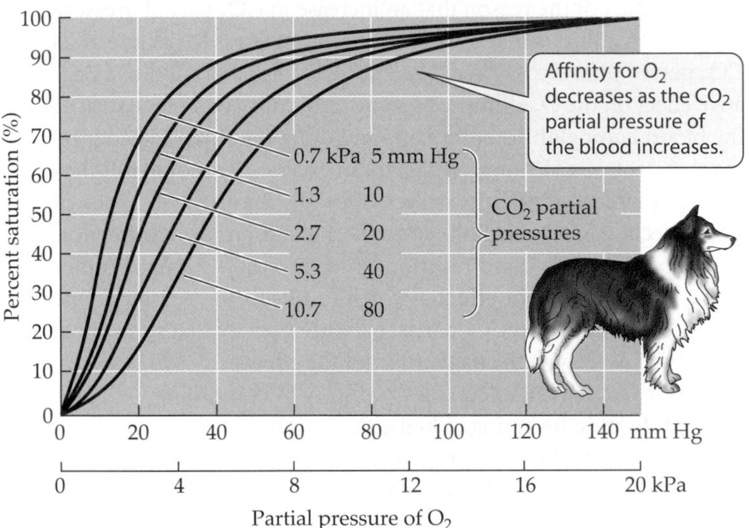

Figure 22.10 The Bohr effect: Affinity for O_2 decreases as pH decreases or CO_2 partial pressure increases *(a)* Oxygen equilibrium curves of human hemoglobin at three different pHs at 38°C. In resting people, the normal pH of arterial blood is about 7.4, whereas that of mixed venous blood is about 0.04 unit less. *(b)* Oxygen equilibrium curves of dog hemoglobin at five different CO_2 partial pressures at 38°C. The data in part *b* are from the original work of Bohr and his coworkers. (After Roughton 1964.)

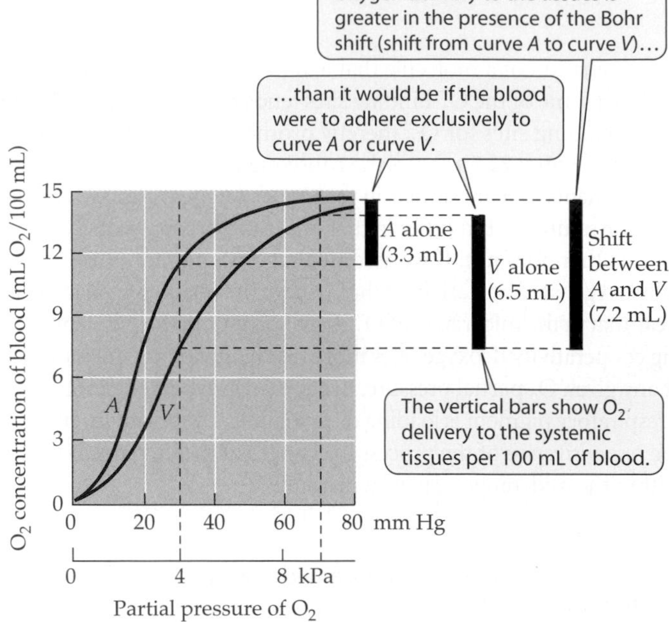

Figure 22.11 The Bohr effect typically enhances O_2 delivery in an animal The diagram shows oxygen equilibrium curves for arterial (*A*) and venous blood (*V*) in a hypothetical animal. The venous blood displays a reduced O_2 affinity because its CO_2 partial pressure and H^+ concentration are higher than those in arterial blood. In this hypothetical animal, the arterial O_2 partial pressure is assumed to be 9.3 kPa (70 mm Hg), and the blood is assumed to fall to a venous O_2 partial pressure of 4 kPa (30 mm Hg) as it flows through the systemic tissues. The magnitude of the Bohr shift shown is exaggerated for clarity.

gans. The shift to lower affinity in the systemic tissues promotes release of O_2 because it facilitates deoxygenation. Conversely, the shift back to higher affinity in the breathing organs promotes uptake of O_2 by facilitating oxygenation. Figure 22.11 illustrates the net effect of this shifting back and forth between two oxygen equilibrium curves as the blood flows between the breathing organs and systemic tissues. At any given O_2 partial pressures in the arterial and venous blood, more O_2 is delivered to the systemic tissues than would be if the pigment followed just one or the other equilibrium curve alone.

During exercise, the CO_2 partial pressure in the systemic tissues may rise above that prevailing during rest because of the increased production of CO_2. Furthermore, the tissue pH may fall below the resting pH not only because of the elevated CO_2 partial pressure, but also because of the accumulation of acid metabolites such as lactic acid. These changes may augment the Bohr shift during exercise, thereby enhancing O_2 delivery to the active tissues.

Now it will be clear why we indicated earlier in this chapter that refinements would ultimately be needed in our analysis of O_2 delivery in humans. We based our earlier analysis on the arterial oxygen equilibrium curve alone (see Figure 22.4), whereas in reality, Bohr shifts occur as blood flows between the lungs and systemic tissues. In people at rest, venous blood is slightly more acid (pH 7.36) than arterial blood (pH 7.40). Moreover, the CO_2 partial pressure in mixed venous blood is about 6.1 kPa (46 mm Hg), and thus is significantly higher than that in arterial blood, 5.3 kPa (40 mm Hg). Looking at Figure 22.10, you can see that these differences in pH and CO_2 partial pressure are sufficient to cause small but significant Bohr shifts of the oxygen equilibrium curve as blood flows between the lungs and systemic tissues.

To fully understand respiratory-pigment function, it is important that, before closing this section, we consider not only how pH can affect oxygenation, but also how oxygenation can affect pH. Let's return to the conceptual equation (Equation 22.2) that describes the reaction of H^+ ions with respiratory pigments that exhibit the fixed-acid Bohr effect. Earlier, we stressed one perspective on this equation; namely, that an increase in H^+ concentration will push the reaction to the right, decreasing the tendency of pigment molecules to bind to O_2.

Now we must also stress that the equation has a complementary and equally important property: Removal of O_2 from pigment molecules will pull the reaction to the right, causing the pigment molecules to take up H^+ from their surroundings. When blood passes through the systemic tissues, metabolism tends to increase the H^+ concentration of the blood solution. Simultaneously, however, because of the diffusion of O_2 out of the blood, respiratory-pigment molecules unload O_2 and thus bind with H^+. This removal of free H^+ from the blood, induced by the deoxygenation of the pigment molecules, limits the increase in blood H^+—and the decrease in blood pH—caused by the metabolic addition of CO_2 and H^+.

The Root effect: In unusual cases, CO_2 and pH affect the oxygen-carrying capacity of the respiratory pigment

In some types of animals, because of distinctive properties of their respiratory pigments, an increase in the CO_2 partial pressure or a decrease in the pH of the blood not only causes a Bohr effect, but also reduces the amount of O_2 the respiratory pigment binds when saturated.[9] The reduction in the amount of O_2 bound to the pigment at saturation (Figure 22.12) is termed the **Root effect**, after its discoverer. Root effects are not common. Among vertebrates, they are observed only in fish, principally teleost fish. Some molluscs are also reported to show either normal or reversed Root effects.

[9]Some modern authorities view the Root effect as an exaggerated Bohr effect.

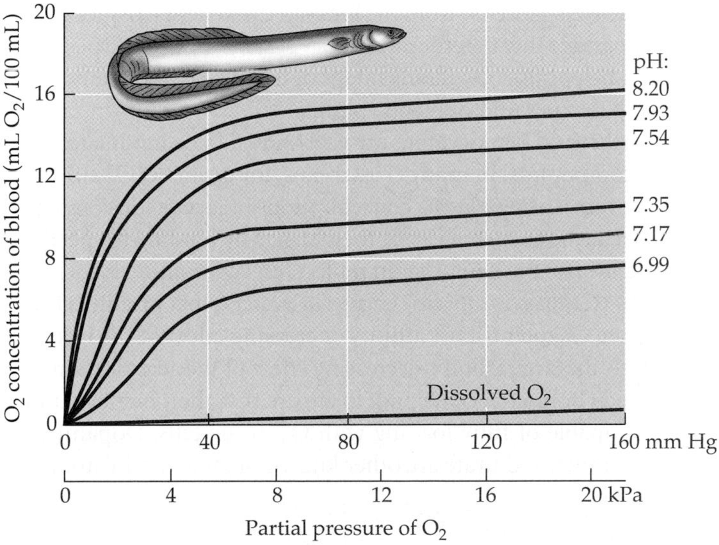

Figure 22.12 The Root effect in eels: Acidification lowers the oxygen-carrying capacity of hemoglobin Oxygen equilibrium curves are shown for the whole blood of eels (*Anguilla vulgaris*) at six different pH levels; the pH was adjusted at a constant temperature (14°C) by addition of lactic acid. Because the oxygen equilibrium curves include dissolved O_2, saturation of the hemoglobin is indicated when the slope of an oxygen equilibrium curve parallels the slope of the dissolved O_2 line. Experiments on some fish have shown that O_2 binding by hemoglobin is reduced at low pH even when the hemoglobin is exposed to an O_2 partial pressure of 140 atm (14,000 kPa)! (After Steen 1963.)

Root effects provide a mechanism by which the O_2 *partial pressure* of even well-oxygenated blood can be dramatically increased under the control of blood pH. Consider, for example, hemoglobin in the blood of eels when it is fully loaded with O_2 (see Figure 22.12). At a pH of 7.54, the hemoglobin is chemically combined with 12.6 mL of O_2 per 100 mL of blood. Acidification to a pH of 7.35 lowers the O_2-binding capacity of the hemoglobin to 9.4 mL/100 mL because of the Root effect. Thus acidification from pH 7.54 to 7.35 will cause this hemoglobin to unload 3.2 mL of O_2 into each 100 mL of blood. The O_2 released in this way, in an immediate sense, has no place to go except into blood solution. By dissolving, it dramatically elevates the blood O_2 partial pressure.

The Root effect is employed in various species of teleost fish to help create high O_2 partial pressures in two regions of the body: the swim bladder and the eyes. In both regions, the pH of well-oxygenated blood is lowered by a tissue-specific addition of lactic acid, inducing a rise in the O_2 partial pressure because of the Root effect. Furthermore, in both types of organs, this rise is magnified by a countercurrent vascular arrangement (a *rete mirabile*) that favors multiplication of the initial effect.[10] The creation of high O_2 partial pressures helps inflate the swim bladder in many fish (swim-bladder gas is often very rich in O_2). The retinas of some fish require high O_2 partial pressures (up to 2 atm) for proper function.

Thermal effects: Oxygen affinity depends on tissue temperature

The O_2 affinity of respiratory pigments is often decreased by increases in temperature, as seen in Figure 22.13 (changes in temperature rarely affect the O_2 content of blood at saturation, however). When humans or other animals exercise, if the blood temperature in their exercising muscles exceeds the temperature in their lungs, thermal shifts in affinity will enhance O_2 delivery to the muscles in a way analogous to that already described for the Bohr effect (see Figure 22.11). In total, therefore, unloading of O_2 to the exercising muscles will be promoted in a concerted manner by both temperature effects and Bohr effects, both of which independently tend to decrease the O_2 affinity of the respiratory pigment when the blood passes through the muscles.

Organic modulators often exert chronic effects on oxygen affinity

Organic compounds synthesized by metabolism often play major roles as allosteric modulators of the function of respiratory pigments. In vertebrates, the principal compounds acting in this role are organophosphate compounds within the red blood cells, which affect hemoglobin O_2 affinity. The organophosphate of chief importance in most mammals, including humans, is **2,3-bisphosphoglycerate**, which is synthesized in red blood cells from intermediates of glycolysis. For historical reasons, the abbreviated name **2,3-DPG** (standing for 2,3-diphosphoglycerate) is often used in the physiological literature to refer to this compound. The effect of 2,3-DPG is to reduce the O_2

[10]The process of countercurrent multiplication is explained, in a different context, in Chapter 27 (see page 735).

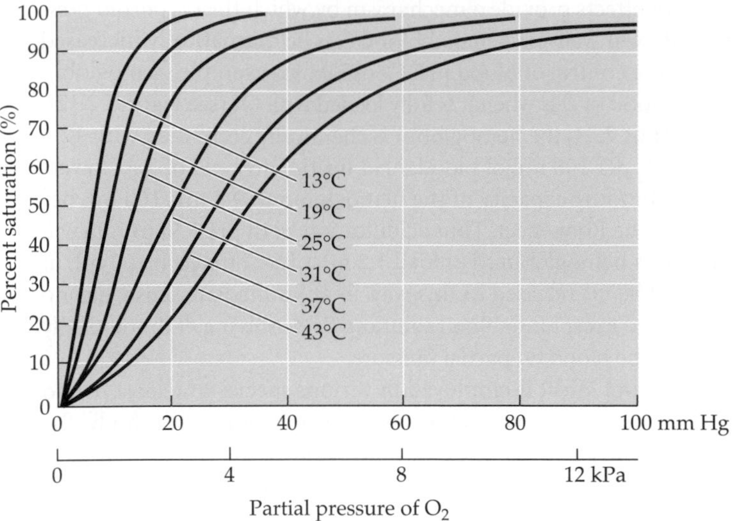

Figure 22.13 An increase in temperature typically causes a decrease in O_2 affinity Oxygen equilibrium curves are shown for human blood at six different temperatures, with pH held constant at 7.4. These results show the pure effect of changes in temperature because of the constancy maintained in pH. The results, however, tend to understate the effects of temperature in many real-life situations because when the pH is not artificially controlled, a rise of blood temperature typically induces a decrease in blood pH, as discussed later in this chapter (see Figure 22.23), meaning that the immediate effects of temperature are often reinforced by thermally induced fixed-acid Bohr effects. The CO_2 partial pressure was held constant during the studies shown. (After Reeves 1980.)

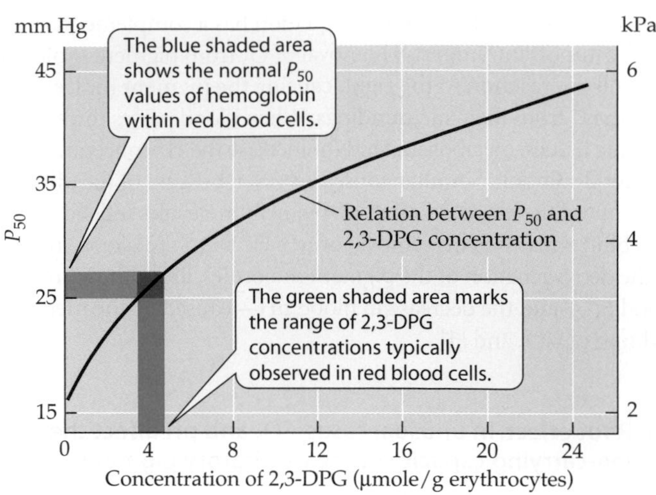

Figure 22.14 The normal P_{50} of human hemoglobin within red blood cells depends on a normal intracellular concentration of 2,3-DPG There is usually about one 2,3-DPG molecule per hemoglobin molecule in human red blood cells. The temperature (37°C), CO_2 partial pressure (5.3 kPa), and extracellular pH (7.4) were held constant during the measurements presented here. (After Duhm 1971.)

affinity (raise the P_{50}) of the hemoglobin molecules with which it binds as shown by the line in Figure 22.14. The hemoglobin of humans and most other mammals is continuously exposed to and modulated by 2,3-DPG within the red blood cells. Thus, as stressed in Figure 22.14, the "normal" O_2 affinity of human hemoglobin is in part a consequence of modulation by a "normal" 2,3-DPG concentration within the red blood cells.[11]

In nonmammalian vertebrates, modulation of hemoglobin O_2 affinity by red blood cell organophosphates is also very common, although the specific phosphate compounds that bind with and allosterically affect hemoglobin vary from group to group and usually do not include 2,3-DPG. ATP and guanosine triphosphate (GTP) are generally the principal organophosphate modulators in fish. In birds, inositol pentaphosphate (IPP) and ATP are especially important. As in mammals, when organophosphates act as modulators in these other groups, their effect is to lower O_2 affinity.

Chronic changes in the concentration of organophosphate modulators in red blood cells serve as *mechanisms of acclimation or acclimatization* in many vertebrates. People suffering from anemia, to mention one example, often exhibit a chronic increase in the concentration of 2,3-DPG in their red blood cells; the O_2 affinity of their hemoglobin is thereby lowered by comparison to the usual affinity. The resulting shift to the right in their oxygen equilibrium curve is not great enough to cause any substantial impairment of O_2 loading in their lungs, but it significantly facilitates O_2 unloading in their systemic tissues (Figure 22.15). Thus each molecule of hemoglobin, on average, delivers more O_2 from the lungs to the systemic tissues during each passage through the circulatory system. In anemic people, this effect helps offset the disadvantage of having a reduced amount of hemoglobin per unit of blood volume.

The arthropod hemocyanins are well known to be modulated by organic compounds. In many crustaceans, for example, O_2 affinity is raised by increasing plasma concentrations of lactate ions (specifically L-lactate ions), which exert their effects by binding to specific allosteric sites on the hemocyanin molecules. When animals such as blue crabs (*Callinectes sapidus*) engage in exercise that produces lactic acid (see Chapter 6), the affinity-increasing effect of the lactate ions offsets the large affinity-decreasing effect of the acidification of their blood (Bohr effect), helping to ensure that their hemocyanin remains capable of fully loading with O_2 in the gills. Dopamine, trimethylamine, and urate are other known organic modulators in crustaceans.

Inorganic ions may also act as modulators of respiratory pigments

Concentrations of specific inorganic ions in blood cells or blood plasma sometimes allosterically modulate the O_2 affinity or other attributes of respiratory pigments. Recent research, for example, has revealed that the concentration of Cl^- in the red blood cells is a critical allosteric modulator of hemoglobin function in many ruminant mammals and certain bears. The bicarbonate ion (HCO_3^-) is a modulator in crocodilians. The divalent ions Ca^{2+} and Mg^{2+} are important

[11]Some mammals have hemoglobins that are not modulated by organophosphate compounds under ordinary physiological conditions. Included are some ruminants, cats, civets, and related species. Their hemoglobins, within the red blood cells, display functionally appropriate O_2 affinities without 2,3-DPG, and the red blood cells of adults usually contain little 2,3-DPG.

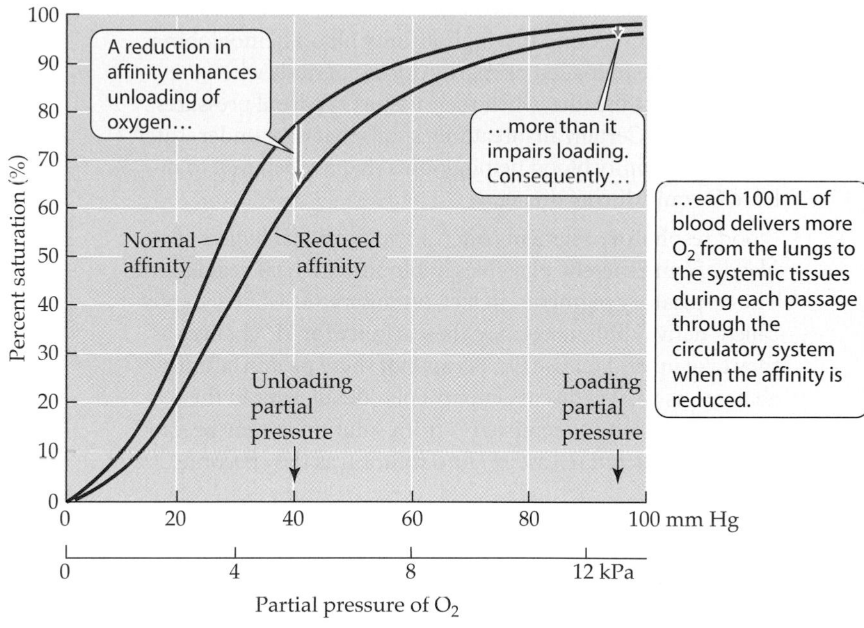

Figure 22.15 A decrease in the O_2 affinity of hemoglobin can aid O_2 delivery to the systemic tissues when the oxygen partial pressure in the lungs or gills remains high Two human oxygen equilibrium curves, representing normal and reduced O_2 affinities, are shown. The loading O_2 partial pressure in the lungs is assumed to be 12.7 kPa (95 mm Hg), and the unloading O_2 partial pressure in the systemic tissues is 5.3 kPa (40 mm Hg). The green vertical arrows show the changes in percent saturation at these two partial pressures caused by an overall shift from normal affinity to reduced affinity. (For simplicity and clarity, other effects on affinity, such as Bohr effects, are ignored, and the reduction in affinity is exaggerated.)

allosteric modulators of hemocyanin in crustaceans. Blue crabs (*Callinectes*) exposed to O_2-poor waters, for example, increase their blood Ca^{2+} concentration, which raises the O_2 affinity of their hemocyanin.

Summary: The O_2-Binding Characteristics of Respiratory Pigments

- The oxygen equilibrium curve of a respiratory pigment, which shows the relation between the extent of O_2 binding by the pigment and the O_2 partial pressure, is a key tool for interpreting respiratory-pigment function. An examination of the function of human hemoglobin allows the development of guidelines for analyzing O_2 transport (see the section titled "A set of general principles helps elucidate O_2 transport by respiratory pigments").
- The shape of the oxygen equilibrium curve depends on the degree of cooperativity among O_2-binding sites on respiratory-pigment molecules. When there is no cooperativity—as is the case when each molecule has only a single O_2-binding site—the oxygen equilibrium curve is hyperbolic. The curve is sigmoid when molecules have multiple O_2-binding sites that exhibit positive cooperativity. Hyperbolic curves are the norm for myoglobins; sigmoid curves are the norm for blood pigments.
- The Bohr effect is a reduction in O_2 affinity caused by a decrease in pH and/or an increase in CO_2 partial pressure. The Bohr effect typically enhances O_2 delivery because it promotes O_2 unloading in systemic tissues while promoting loading in the breathing organs.
- The Root effect, which occurs only rarely, is a reduction of the oxygen-carrying capacity of a respiratory pigment caused by a decrease in pH and/or an increase in CO_2 partial pressure. In teleost fish it helps inflate the swim bladder and oxygenate the retina.
- Elevated temperatures often decrease the O_2 affinity of respiratory pigments.
- Organic molecules and inorganic ions frequently serve as allosteric modulators of respiratory-pigment function. A particularly well known organic modulator is 2,3-DPG in the red blood cells of mammals, which acts to decrease the O_2 affinity of the hemoglobin in the cells.

The Functions of Respiratory Pigments in Animals

It would be hard to exaggerate the diversity of functional properties found among animal respiratory pigments. The *oxygen affinity* (P_{50}) of respiratory pigments varies from less than 0.2 kPa to more than 7 kPa. *Cooperativity* (the Hill coefficient, n) varies from 1 to more than 6. The *concentration* of the respiratory pigment in an animal's blood may be so low that the pigment merely doubles the oxygen-carrying capacity of the blood in comparison to the dissolved O_2 concentration; alternatively, a pigment may be so concentrated that it allows blood to carry 80 times more O_2 than can be dissolved. Based on its chemical properties, a respiratory pigment may or may not exhibit a *Bohr effect* or *temperature effect*. One pigment may be modulated by 2,3-DPG, another by ATP. With this diversity of properties, even when the respiratory pigments of various animals carry out a single function, they do so in a diversity of detailed ways. Moreover, several major functions are carried out by respiratory pigments. The overall range of roles assumed by respiratory pigments is thus extensive. At least seven functions carried out by respiratory pigments can be identified. They are not mutually exclusive; often a single respiratory pigment carries out two or more functions simultaneously. The seven functions are as follows:

1. Respiratory pigments in blood (or other circulating body fluids) typically aid the *routine transport of O_2 from the breathing organs to the systemic tissues.* This is the function to which we have devoted most of our attention up to this point in the chapter.
2. Respiratory pigments in the blood of some invertebrates probably function primarily as *O_2 stores*, rather than participating in routine O_2 transport. The pigments that fit this description have very high O_2 affinities. Consequently, they hold so tightly to O_2 that they probably do not unload under routine conditions. Instead, they seem to release their O_2 when animals face severe O_2

shortages. In certain species of tube-dwelling marine worms, for example, O_2 bound to a high-affinity blood hemoglobin is believed to be unloaded primarily during periods when the worms do not breathe, when their tissue O_2 partial pressures fall very low. Certain air-breathing snails that dive under water have high-affinity blood hemoglobins that are believed to release O_2 only during diving.

3. Blood respiratory pigments often serve as critical *buffers of blood pH* and thereby play roles in blood acid–base regulation. The respiratory pigments, in fact, often serve as what are aptly called "active" buffers because their affinity for H^+ changes as they take up and release O_2. Because of these proton-affinity changes, as we will discuss in particular detail later in the chapter, they tend to remove H^+ from solution as they become deoxygenated and release H^+ into solution as they become oxygenated.
4. Blood respiratory pigments often play critical roles in *CO_2 transport*, as we will also see later in the chapter.
5. Hemoglobins in the cytoplasm of muscle cells (myoglobins), or in the cells of other solid tissues, play two principal roles. First, they *facilitate* (*increase the rate of*) *O_2 diffusion through the cytoplasm of the cells*: At any given difference in O_2 partial pressure between the blood capillaries and the mitochondria of the cells, O_2 diffuses through the cytoplasm to the mitochondria faster if myoglobin is present. The second role played by hemoglobins within solid tissues is *O_2 storage for the tissues*. The myoglobin-bound O_2 store in muscles, for example, can be called upon at the start of vigorous muscular work to help sustain aerobic ATP production while circulatory O_2 delivery is accelerated.[12]
6. Respiratory pigments in blood or other circulating body fluids sometimes play *nonrespiratory transport roles*. Recent groundbreaking studies indicate, for example, that blood hemoglobin in mammals reversibly combines with *nitric oxide* (*NO*) and transports it from the lungs to systemic tissues. Because NO is a potent vasodilator, release of NO from the blood hemoglobin in the microcirculatory beds of systemic tissues helps regulate blood flow through the capillaries and helps determine blood pressure. In worms that have symbiotic sulfur-oxidizing bacteria, such as the hydrothermal vent worms seen in Figure 4.13, the blood hemoglobin has sulfide-binding sites and transports S^{2-} as well as O_2.
7. Finally, very recent studies that still need to be confirmed indicate that, at least in mammals, deoxyhemoglobin may *act as an enzyme*, not in carrying out its role in O_2 transport, but in catalyzing the local formation of NO from blood nitrite.

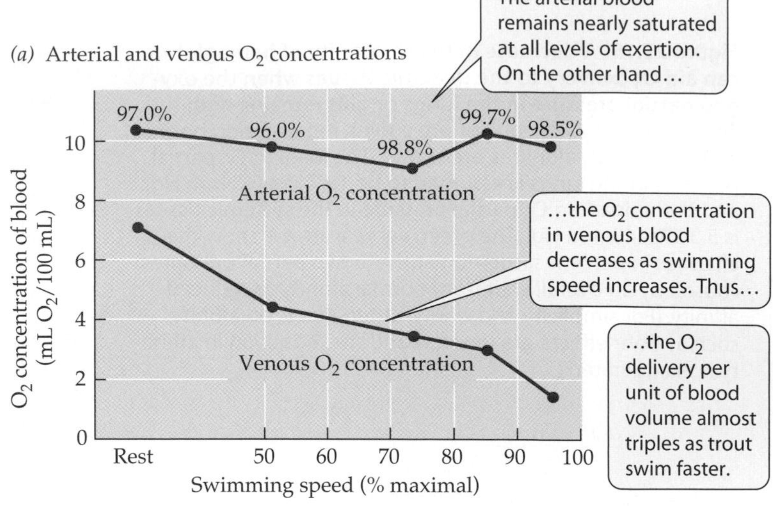

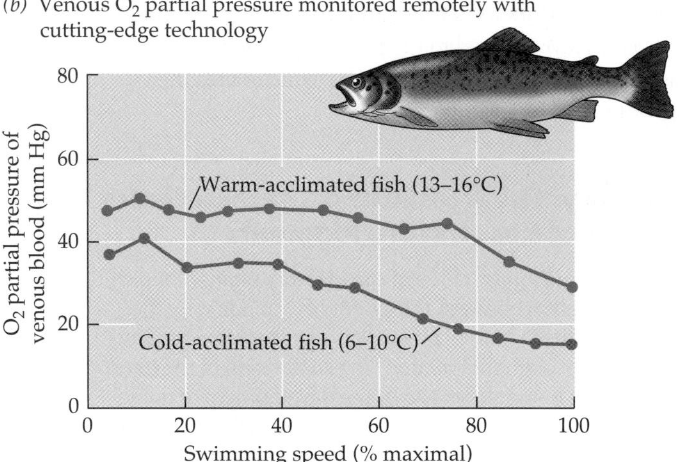

Figure 22.16 Blood O_2 transport in rainbow trout in relation to exercise *(a)* The lines show the average O_2 concentration of arterial and venous blood in trout (*Oncorhynchus mykiss*) at rest and swimming at various speeds in well-aerated water. The numbers above the arterial points show the average arterial percent saturation of the particular fish studied at each speed. As fish increase their speed from rest to maximal, they increase O_2 delivery per unit of blood volume about threefold. They also increase their rate circulation about threefold. Thus, they increase the total rate of O_2 delivery to their tissues about tenfold. *(b)* The relation between venous O_2 pressure and swimming speed in trout acclimated to two temperatures. The data were gathered using remote technology that permitted automatic and continuous measurement of the venous O_2 partial pressure in swimming fish. (*a* after Jones and Randall 1978; *b* after Farrell and Clutterham 2003.)

Patterns of circulatory O_2 transport: The mammalian model is common but not universal

Circulatory O_2 transport in most animals qualitatively follows the pattern we described earlier for mammals (see Figure 22.5). Rainbow trout, for example, adhere to this pattern (Figure 22.16) which has several major features. First, the blood respiratory pigment reaches near-saturation in the lungs or gills when the animals are living in well-aerated environments. Second, the respiratory pigment yields just a modest fraction of its O_2 to the systemic tissues during rest, meaning that venous blood in resting individuals is far from being fully deoxygenated. Third, the large resting venous O_2 reserve is used (i.e., venous blood becomes more deoxygenated) during exercise or other states of heightened metabolism. Thus increased tissue O_2 demands are met by increasing the amount of O_2 delivered per unit of blood volume, as well as by increasing the rate of blood flow.

[12] This role is discussed at length in Chapters 6 and 17.

Squids and octopuses are important examples of animals that follow a different pattern of circulatory O_2 transport, and thus illustrate that the pattern observed in mammals and fish is not universal. The squids and octopuses that have been studied have only a very small venous O_2 reserve when they are at rest: *Inactive* individuals use 80% to 90% of the O_2 available in their arterial blood (Figure 22.17). Thus, during exercise, little room exists for unloading to be enhanced, and the animals must meet their heightened O_2 demands almost entirely by increasing their circulatory rates. This aspect of their respiratory-pigment physiology places high demands on their hearts and constrains their ability to exercise, as we shall see in more detail when we discuss their circulation in Chapter 23. The inherently small venous O_2 reserve of the squids and octopuses also places limits on their ability to live in poorly aerated waters. If a squid or octopus ventures into O_2-poor waters and is therefore unable to oxygenate its arterial blood fully, it cannot compensate to any great degree (as a fish can) by enhancing the deoxygenation of its venous blood, because the venous blood is already highly deoxygenated even in aerated waters. Squids and octopuses are notoriously intolerant of low-O_2 environments.

Regardless of the exact pattern of circulatory O_2 delivery an animal displays, the oxygen carrying capacity of its blood which depends on the amount of respiratory pigment per unit of blood volume—is a key determinant of how much O_2 can be delivered to its tissues. As already seen in Figure 22.8, a wide range of oxygen-carrying capacities is observed. The range of known values in animals that have blood respiratory pigments is from about 30 to 40 vol % in some diving mammals to just 1 or 2 vol % in many crustaceans and molluscs. Among vertebrates, a rough correlation exists between metabolic intensity and the oxygen-carrying capacity of the blood; mammals and birds usually have carrying capacities of 15 to 20 vol %, whereas fish, amphibians, and reptiles usually have less hemoglobin per unit of volume and have carrying capacities of 5 to 15 vol %. Active species of fish such as tunas and lamnid sharks tend to have higher oxygen-carrying capacities than related sluggish species. Animals with hemocyanin tend to have low oxygen-carrying capacities. The highest carrying capacities known for hemocyanin-containing bloods occur in squids and octopuses and are just 2 to 5 vol % (at or below the lower end of the range for fish); animals with hemocyanin (which is always dissolved in the blood plasma rather than being contained in blood cells) probably cannot have much higher carrying capacities because the hemocyanin concentrations needed for higher capacities would make their blood too viscous to pump.

Individual animals can vary their oxygen-carrying capacity by raising or lowering the amount of respiratory pigment per unit of volume of blood. The most common responses of this sort are long-term, occurring during acclimation or acclimatization to changed environments (to be discussed shortly). Some vertebrates, however, can acutely change their carrying capacity because they can remove red blood cells from the blood, store the cells, and quickly release them back into the blood. Horses, dogs, and some seals are well known to store massive quantities of red blood cells in the spleen when at rest. When the cells are needed during exercise, they are quickly released back into the blood under control of the sympathetic nervous system; foxhounds, for example, can promptly increase their oxygen-carrying capacity from 16 vol % to 23 vol % in this way.

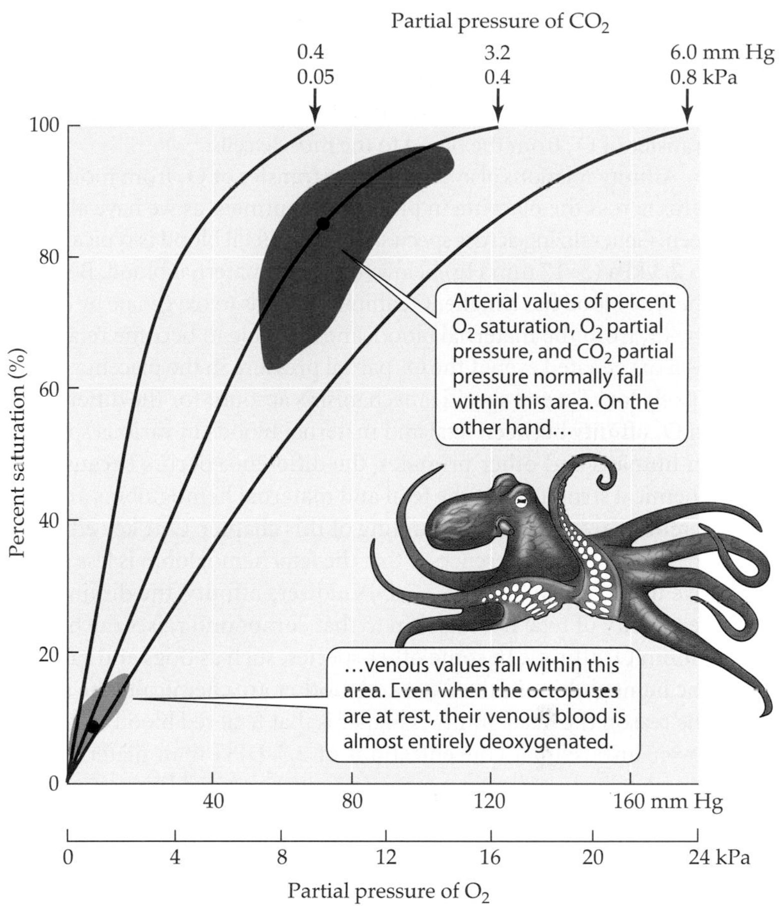

Figure 22.17 Blood O_2 delivery in an octopus Oxygen equilibrium curves are shown for the octopus *Enteroctopus dofleini* during rest or moderate activity in well-aerated water. The oxygen equilibrium curve of the hemocyanin-containing blood of this octopus depends on the CO_2 partial pressure. The two dots represent the approximate means for arterial and venous blood. (After Johansen and Lenfant 1966.)

Respiratory pigments often display differences in O_2 affinity that aid successful O_2 transport

Different respiratory pigments often differ in their O_2 affinities in ways that are critical for understanding their functions in O_2 transport. The significance of relative O_2 affinities is particularly obvious when two respiratory pigments exist within one animal and pass O_2 from one to the other. This occurs, for example, in animals that have myoglobin. In these animals, the blood respiratory pigment (hemoglobin or hemocyanin) and the myoglobin act as a sort of "O_2 bucket brigade"; the blood pigment carries O_2 from the lungs or gills to the muscles, then passes the O_2 to the myoglobin in the muscle cells. The myoglobin typically has a higher O_2 affinity—a lower P_{50}—than the blood pigment; human myoglobin, for example, has a P_{50} of about 0.8 kPa (6 mm Hg), indicating a dramatically higher O_2 affinity than that of human blood hemoglobin, about 3.5 kPa (27 mm Hg) (see Figure 22.7*a*). The higher O_2 affinity of the myoglobin means that it tends to load with O_2 at the expense of unloading of the blood hemoglobin. Thus the difference in affinity promotes transfer of O_2 from the blood to the muscle cells.

Affinity relations also promote the transfer of O_2 from mother to fetus across the placenta in placental mammals, as we have already seen. Generalizing across species, the P_{50} of fetal blood is typically 0.4 to 2.3 kPa (3–17 mm Hg) below the P_{50} of maternal blood. Because the fetal blood has a higher O_2 affinity, it tends to oxygenate by drawing O_2 from the maternal blood, and it is able to become relatively well oxygenated even if the O_2 partial pressure in the placenta is relatively low. Several specific mechanisms account for the differences in O_2 affinity between fetal and maternal bloods in various species. In humans and other primates, the difference occurs because the chemical structures of the fetal and maternal hemoglobins are different, as we saw at the beginning of this chapter. One key effect of these structural differences is that the fetal hemoglobin is less sensitive to 2,3-DPG; because 2,3-DPG lowers affinity, the diminished sensitivity of fetal hemoglobin to that compound raises the hemoglobin O_2 affinity. In some other species, such as dogs and rabbits, the hemoglobins in the fetus and mother are chemically the same; the reason the fetal affinity is higher is that fetal red blood cells have lower intracellular concentrations of 2,3-DPG than maternal red blood cells. In still other species of mammals, additional mechanisms of raising the fetal O_2 affinity are observed; ruminants, for example, have fetal forms of hemoglobin that are intrinsically higher in affinity than maternal hemoglobin, without 2,3-DPG modulation. The relatively high O_2 affinity of fetal hemoglobin is not necessarily the only factor that promotes O_2 transfer from the mother's blood to the fetus's. An extremely interesting additional factor is that often, the loss of CO_2 from the fetal blood to the maternal blood induces a synchronous rise in fetal O_2 affinity and fall in maternal O_2 affinity because of opposite-directed Bohr shifts!

Evolved differences in affinity among related species

In comparisons of blood hemoglobins among related species, a trend observed in mammals and some other vertebrate groups is that the hemoglobin O_2 affinity tends to be directly correlated with body size: Small species have relative high P_{50} values and, therefore, relatively low O_2 affinities (Figure 22.18). Natural selection is hypothesized to have favored this pattern because of the inverse relation between weight-specific metabolic rate and body size (see Chapter 5). Arterial blood oxygenates similarly in all species of aboveground mammals because the O_2 partial pressure in the lungs is high enough in all species to be on the plateaus of their oxygen equilibrium curves (where differences in affinity have little effect; see Figure 22.15). The lower-affinity hemoglobins in the smaller species unload O_2 to the tissues more readily, however. In this way, the lower affinity in the small species is hypothesized to help them meet their higher weight-specific O_2 needs.

Speaking generally, it is a complicated matter to *predict* how related species will differ in O_2 affinity. An important reason this is true is that the relative advantages of low and high O_2 affinity depend on how well oxygenated the blood becomes in the lungs or gills. If arterial O_2 partial pressures are high enough for the respiratory pigment to be well oxygenated regardless of its affinity, then the main effect of a low affinity is to promote unloading of O_2 in the systemic tissues, and low affinity tends to aid O_2 delivery.[13] On the other hand, if arterial O_2 partial pressures are so low that the respiratory pigments fall far short of being fully oxygenated, a pigment with a high O_2 affinity tends best to promote O_2 delivery to the systemic tissues by allowing the blood to become more fully loaded in the lungs or gills.

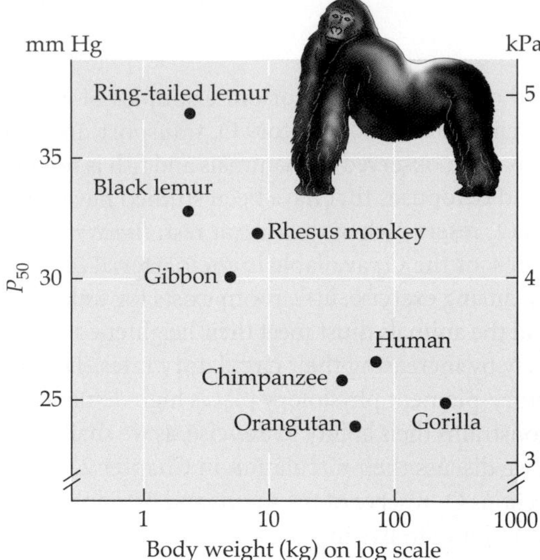

Figure 22.18 The O_2 affinity of the hemoglobin in the whole blood of primates is a regular function of body size Small-bodied species tend to exhibit lower O_2 affinity—and thus higher P_{50}—than large-bodied ones. (After Dhindsa et al. 1972.)

Many species that have long histories of existence in low-O_2 environments have evolved higher O_2 affinities than related species living in high-O_2 environments. For example, carp and catfish, which often inhabit waters low in O_2, have average P_{50} values of 0.1 to 0.7 kPa (1–5 mm Hg), whereas mackerel and rainbow trout, which live in well-aerated waters, have far higher P_{50} values of 2.1 to 2.4 kPa (16–18 mm Hg) (see Figure 22.8).[14] One reason goldfish survive the care of kindergarteners is that these members of the carp family have high-affinity hemoglobins that can load well in O_2-poor water. Mammal species that live underground typically have evolved higher O_2 affinities than aboveground species of the same body size; similarly, some species native to high altitudes have evolved higher O_2 affinities than lowland species (see Box 22.3).

The respiratory-pigment physiology of individuals undergoes acclimation and acclimatization

When individual animals are exposed chronically to reduced O_2 availability in their environments, they often respond with chronic alterations of their respiratory-pigment physiology. The most common response of this sort in both vertebrates and invertebrates is for the

[13] Figure 22.15, although it applies to changes in O_2 affinity within a species, illustrates this effect.

[14] These measurements were made at approximately the same CO_2 partial pressures and temperatures.

BOX 22.3 Blood and Circulation in Mammals at High Altitude

The study of blood O_2 transport in humans and other mammals at high altitude is, in its own particular way, one of the most intriguing chapters in the annals of evolutionary physiology. If you think back to the oxygen cascade for people in the high Andes that we discussed in Box 21.1, you will recall that when people are exposed to the reduced atmospheric partial pressure of O_2 at high altitude, they do not experience an equal reduction in their venous O_2 partial pressure. In fact, the venous partial pressure is reduced *far* less than the atmospheric partial pressure. A key reason for this conservation of venous O_2 partial pressure is blood O_2 transport. The drop in O_2 partial pressure between arterial and venous blood is much smaller at high altitude than at sea level. This reduction is important because it helps keep the venous O_2 partial pressure from falling too low.

The principal explanation for the reduced arteriovenous drop in O_2 partial pressure at high altitude does not entail any special adaptations. Instead, the reduction is simply a consequence of the shape of the mammalian oxygen equilibrium curve. Living at high altitude lowers the arterial O_2 partial pressure. Figure 22.6 illustrates the consequence: When the arterial O_2 partial pressure is moved off the plateau of the equilibrium curve, there is a sharp reduction in the arteriovenous drop in partial pressure required for the blood to yield any particular quantity of O_2.

Twenty years ago, two changes in respiratory-pigment physiology that occur in humans at high altitude were claimed to help limit the arteriovenous drop in partial pressure: (1) an increase in the number of red blood cells (RBCs) per unit of blood volume and (2) a drop in the O_2 affinity of hemoglobin. At that time, both changes were held up as shining examples of *adaptation* to high altitude. Now the significance of both (and the existence of one) is sharply questioned.

When people and some other species of *lowland* mammals go to high altitude, their oxygen-carrying capacities rise to well above sea-level values because of an erythropoietin-mediated increase in the number of RBCs per unit of blood volume: a state known as **polycythemia** ("many cells in the blood"). For example, whereas a person at sea level typically has an oxygen-carrying capacity of 20 vol %, his or her carrying capacity might well rise to 28 vol % at 4000 or 5000 m. A change such as this could well be helpful *in principle* because it increases the amount of O_2 the blood can carry. However, as we discussed in Chapter 1 (see page 24), modern evolutionary physiologists stress that a trait is not necessarily helpful or adaptive merely because it exists. Early researchers ignored the need to *demonstrate* an advantage for polycythemia and quickly ballyhooed human polycythemia as a prime example of the adaptive process. Skeptics have argued that polycythemia might actually *interfere with* O_2 delivery because an increase in RBC concentration increases the viscosity of the blood. Thus polycythemia not only places a greater demand on the heart, but also could impair blood flow through small blood vessels.

By now, sufficient comparative data have accumulated that we can make the following statement with good confidence: Species of mammals (and birds) that are *native* to high altitudes do *not* have unusually high RBC concentrations or oxygen-carrying capacities. Thus, exceptional concentrations of RBCs in the blood do not seem in any general sense to be adaptive to high altitudes. Why, then, do people develop polycythemia? We do not know. In an effort to carry out a direct test of the implications of human polycythemia, researchers have medically removed RBCs from people at high altitude. Some (but not all) studies of this sort have found that the subjects experienced either no change or an improvement in their ability to function.

With regard to O_2 affinity, there has been at least equal debate. Humans and some other *lowland* species undergo an increase in the concentration of 2,3-DPG in their RBCs at high altitude. When this change was first discovered, it was promptly claimed to help prevent tissue hypoxia by lowering the O_2 affinity of hemoglobin and thus promoting O_2 unloading into the systemic tissues. The counterarguments are complex. One concern has been whether the reasoning is even theoretically correct, because a reduction in O_2 affinity also has the potential to impair loading in the lungs, especially at high altitude. More to the point, the collection of comparative data on many additional species now permits confidence in the following conclusion: Species of mammals (and birds) *native* to high altitudes typically have either ordinary O_2 affinities or particularly *high* affinities. (In fact, recent research shows that the birds renowned for flying at altitudes of 9,000 to 11,000 m, such as the bar-headed goose, have extraordinarily high affinities.) Thus, if people at high altitude have reduced affinities, we must be wary of interpreting those reduced affinities as being beneficial.

Why, then, might human O_2 affinity be reduced at high altitude? One sobering argument is that the human increase in 2,3-DPG at altitude may in fact be a pathologically *misplaced* response. According to this reasoning, humans have had so little contact with high altitude in their evolutionary history that when they go to high altitude, they mistakenly activate a response that evolved to help them deal with *anemia* at sea level (see page 592)!

The reality of the human decrease in O_2 affinity at high altitude is itself now open to question. Researchers now recognize that the effects of any one hemoglobin modulator cannot be interpreted in isolation. If the concentration of 2,3-DPG within RBCs increases, that does not in itself necessarily mean that the affinity of hemoglobin goes down. Blood pH is typically shifted in an alkaline direction at high altitude because the hyperventilation we discussed in Box 21.1 tends to lower the CO_2 concentration of the blood. The operation of the Bohr effect could thus cancel the 2,3-DPG effect. A recent careful study of people native to the Andes revealed that their hemoglobin O_2 affinities were indistinguishable from those of people at sea level, and during the American Medical Expedition to Mount Everest in 1981—a famous research expedition—the P_{50} values of the climbers remained unaltered up to about 6000 m (then fell). Thus there is reason to believe that people may in fact resemble many native-highland mammals in having unaltered O_2 affinities at high altitude.

At first sight, it would seem logical to think that the rate of blood flow would increase at high altitude. However, cardiac output is not systematically elevated in humans or other mammals at high altitude, either at rest or at any given level of exercise. The failure to increase cardiac output at altitude may seem paradoxical, but recent theoretical analyses show that it is not. Consider a person at rest. At high altitude, as noted earlier in this box, the *arterial* O_2 partial pressure is on the *steep* part of the oxygen equilibrium curve. If the heart pumps blood faster to try to assist with the problems of high altitude, the arterial O_2 partial pressure tends to go down a bit, in part because the speeding blood has a reduced opportunity to move toward full equilibrium with alveolar air. On the steep part of the oxygen equilibrium curve, the reduction in arterial O_2 *content* because of decreased equilibration is great enough to offset the advantages of pumping more blood per unit of time. Theory indicates, therefore, that the heart is not in a position to help raise tissue partial pressures of O_2 at high altitude.

concentration of the respiratory pigment in the blood to be increased. Fish, for example, often increase the concentration of red blood cells in their blood when they live in poorly oxygenated waters.

Respiratory-pigment modulators also often play roles in the chronic responses of animals to low-O_2 environments. In principle, an animal that finds itself in an O_2-poor environment might respond by synthesizing a different molecular form of respiratory pigment, but this sort of response is almost unknown. Instead, if a change in pigment functional properties occurs, it is achieved by modulating the preexisting molecular form. Fish provide outstanding examples. When transferred from well-aerated to poorly aerated waters, fish often decrease the concentrations of ATP and GTP within their red blood cells over time. These chronic changes in the intracellular environment of hemoglobin alter its functional properties: The O_2 affinity of the hemoglobin is raised, and the pigment therefore loads better from O_2-poor waters. Figure 22.19 shows the impressive advantages that eels gain by increasing both their amount of hemoglobin per unit of blood volume and the O_2 affinity of their hemoglobin during acclimation to low-O_2 waters. Blue crabs, as earlier noted, chronically raise the concentration of Ca^{2+} in their blood when exposed to O_2-poor waters, apparently by mobilizing Ca^{2+} from the exoskeleton; the effect is to raise the O_2 affinity of their hemocyanin. Box 22.3 discusses the interesting question of how mammals respond to the low atmospheric O_2 partial pressures of high altitudes.

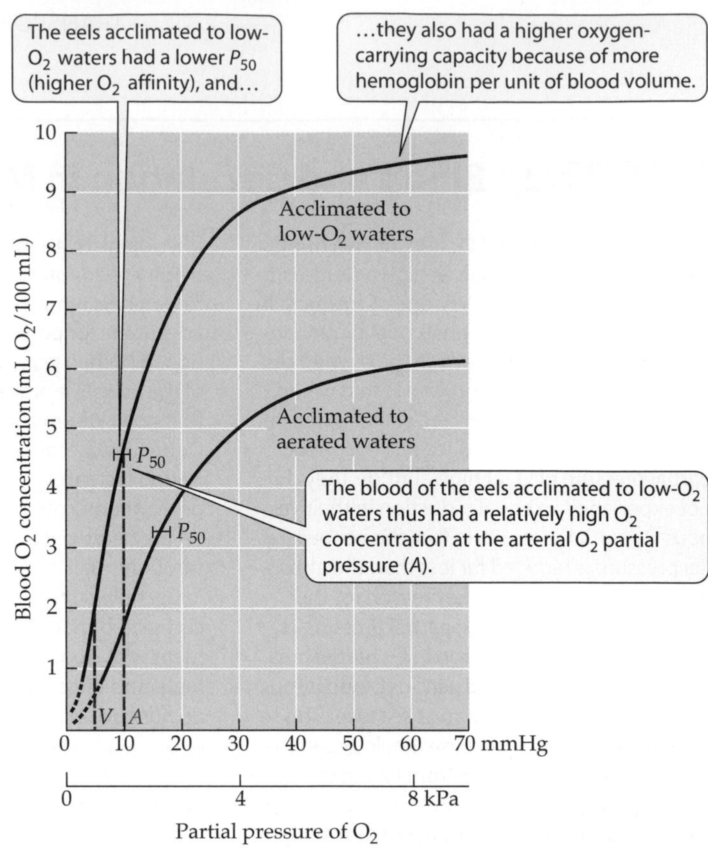

Figure 22.19 Acclimation of the respiratory-pigment physiology of eels to low-O_2 and high-O_2 environments Average oxygen equilibrium curves for two groups of eels (*Anguilla anguilla*), one of which was kept in aerated waters (19 kPa O_2 partial pressure) and the other in low-O_2 waters (2–5 kPa O_2 partial pressure) for 1–2 weeks. The horizontal bars delimit the mean $P_{50} \pm$ twice the standard error of P_{50} for each set of eels. The increase in the O_2 affinity of hemoglobin in the eels acclimated to low-O_2 waters was mediated by a decrease in the concentrations of ATP and GTP in their red blood cells. The vertical dashed lines show mean arterial (*A*) and venous (*V*) O_2 partial pressures in the eels living in low-O_2 waters. Measurements were carried out at 20°C and pH = 7.8. (After Wood and Johansen 1973.)

Icefish live without hemoglobin

We now end our discussion of the transport of O_2 by respiratory pigments by considering a group of unusual and puzzling vertebrates that represent the opposite case: animals that have no O_2 transport by hemoglobin because they have no hemoglobin. Virtually all vertebrates have substantial amounts of hemoglobin in their blood. However, the fish of the family Channichthyidae, which occupy frigid Antarctic seas (see Figure 1.9), lack blood hemoglobin (or have negligibly small amounts) and also lack tissue myoglobin. Without red blood, they are whitish and translucent, thus their common names *icefish* or *white-blooded fish*. Recent genetic research has shown that one key reason for their lack of hemoglobin is that they lack a functional gene for synthesis of β-globin. The icefish are not small, making their lack of hemoglobin all the more remarkable; one species, *Chaenocephalus aceratus*, for example, grows to over a half meter long. The oxygen-carrying capacity of icefish blood is limited to the amount that can dissolve in the blood plasma.

The habitat of the icefish is undoubtedly critical in permitting them to live without hemoglobin. The Antarctic seas tend to be consistently well aerated, and the temperature of the water typically remains near –1.9°C the year round. The solubility of O_2 in aqueous solutions increases as temperature decreases. Thus, in the Antarctic seas, the solubility of O_2 is relatively high, not only in the ambient water but also in the fish's blood. The low temperatures also depress the metabolic rates and the O_2 needs of the fish.

To elucidate the physiology of icefish, comparisons have been made between them and other Antarctic fish, such as species of the genus *Notothenia*, that possess more-usual levels of hemoglobin. The oxygen-carrying capacities of icefish (about 0.7 vol %) are only about one-tenth as high as those of *Notothenia* (6–7 vol %). A key adjustment in icefish is an exceptionally high rate of blood circulation. The hearts of icefish are dramatically larger than those of most fish of their body size and are estimated to pump at least four to ten times as much blood per stroke. Another morphological feature that aids rapid blood flow in icefish is that the blood vessels in the microcirculatory beds of their tissues are of exceptionally large diameter, thus posing exceptionally low resistance to rapid throughflow of blood.

Summary: The Functions of Respiratory Pigments in Animals

- Respiratory pigments are diverse in their functional properties. The functions they can potentially perform include O_2 transport, CO_2 transport, transport of substances other than respiratory gases, blood buffering, facilitation of O_2 diffusion through the cells of solid tissues such as muscle, and O_2 storage in blood or solid tissues.

- Blood respiratory pigments typically become well oxygenated in the breathing organs, and when animals are at rest, the respiratory pigments typically release only a modest fraction of their O_2 to the systemic tissues (25% in humans). During exercise, O_2 delivery is enhanced by increases in both the extent of pigment unloading and the rate of blood flow.
- The relative O_2 affinities of respiratory pigments are often critical for pigment function. When O_2 is transferred from one respiratory pigment to another in an individual animal—as when blood hemoglobin donates O_2 to myoglobin—it is usual for the pigment receiving the O_2 to have a higher O_2 affinity. Comparing related species, those with long evolutionary histories in O_2-poor environments often have evolved blood respiratory pigments with particularly high O_2 affinities.
- Respiratory-pigment physiology undergoes acclimation and acclimatization, as by changes in pigment amounts and O_2 affinity.

Carbon Dioxide Transport

Carbon dioxide dissolves in blood as CO_2 molecules, but usually only a small fraction of the carbon dioxide in blood is present in this chemical form (about 5% in human arterial blood). Thus the first step in understanding carbon dioxide transport is to discuss the other chemical forms in which carbon dioxide exists in blood. Because carbon dioxide can be present in multiple chemical forms, not just CO_2, we must distinguish the material from its exact chemical forms. We do this by speaking of "carbon dioxide" when we refer to the sum total of the material in all its chemical forms and by specifying the chemical form (e.g., CO_2) when we refer to a particular form.

When carbon dioxide dissolves in aqueous solutions, it undergoes a series of reactions. The first is hydration to form carbonic acid (H_2CO_3):

$$CO_2 + H_2O \rightleftharpoons H_2CO_3 \tag{22.3}$$

The second is dissociation of the carbonic acid to yield bicarbonate (HCO_3^-) and a proton:

$$H_2CO_3 \rightleftharpoons H^+ + HCO_3^- \tag{22.4}$$

Bicarbonate can then dissociate further to yield carbonate (CO_3^{2-}) and an additional proton. This final dissociation, however, occurs to only a small extent in most animals. Moreover, although carbonic acid is an important intermediate compound, it never accumulates to more than very slight concentrations. For most purposes, therefore, the reaction of CO_2 with water can be viewed simply as yielding HCO_3^- and protons:

$$CO_2 + H_2O \rightleftharpoons HCO_3^- + H^+ \tag{22.5}$$

Equation 22.5 emphasizes that carbon dioxide acts as an acid in aqueous systems because it reacts to produce H^+; as earlier mentioned, it has aptly been termed a "gaseous acid."

The extent of bicarbonate formation depends on blood buffers

Although almost no bicarbonate is generated when CO_2 is dissolved in distilled water or a simple salt (NaCl) solution, bicarbonate is typically the dominant form in which carbon dioxide exists in the bloods of animals. To understand why—and to understand why there is more bicarbonate in some animal bloods than in others—we need to examine the factors that affect bicarbonate formation.

Suppose that we bring a liter of an aqueous solution—initially devoid of carbon dioxide—into contact with a gas that acts as a source of CO_2, and that this gas remains at a constant CO_2 partial pressure regardless of how much CO_2 it donates to the solution. From Chapter 20, we know that after the solution comes to equilibrium with the gas, the concentration of carbon dioxide in solution in the form of CO_2 will be simply proportional to the CO_2 partial pressure. Thus the amount of CO_2 taken up in *dissolved* form by our liter of solution will depend simply on the principles of gas solubility. In contrast, the extent of bicarbonate formation is governed, not by the principles of solubility, but by the action of compounds that act as *buffers of pH*. In blood, these are the blood buffers.

How do blood buffers determine the amount of HCO_3^- formation? A straightforward way to see the answer is to return to the analysis of our solution and apply the principles of mass action (see page 46) to Equation 22.5. According to the principles of mass action, the following equation holds true at equilibrium:

$$\frac{[HCO_3^-][H^+]}{[CO_2]} = K \tag{22.6}$$

where the square brackets signify the concentrations of the various chemical entities, and K is a constant. Because $[CO_2]$, the concentration of CO_2, is a constant at equilibruim in our solution at a given CO_2 partial pressure, and because K is also a constant, Equation 22.6 reveals that the amount of HCO_3^- formed per unit of volume of solution depends inversely on the H^+ concentration. If $[H^+]$ is kept relatively low, $[HCO_3^-]$ will be high; that is, a lot of HCO_3^- will be formed. However, if $[H^+]$ is allowed to rise to high levels, $[HCO_3^-]$ will be low; little HCO_3^- will be formed. When carbon dioxide enters our solution from the gas and undergoes the reaction in Equation 22.5, the degree to which the H^+ made by the reaction is allowed to accumulate, driving $[H^+]$ up, is determined by the *buffers* in the solution. If the buffers are ineffective, and the H^+ produced by the reaction simply accumulates, $[H^+]$ will rapidly become high, and the whole reaction will quickly reach an end point with little uptake of carbon dioxide and little formation of HCO_3^-. However, if the buffers are highly effective, so that most H^+ is removed from solution as it is formed and $[H^+]$ stays low, a great deal of carbon dioxide will be able to undergo reaction, causing a large buildup of HCO_3^-.

Buffer reactions are represented by the general equation

$$HX \rightleftharpoons H^+ + X^- \tag{22.7}$$

where X^- is a chemical group or compound that can combine with H^+. When H^+ is added to a buffered solution, the buffer reaction is shifted to the left, removing some of the H^+ from free solution; on the other hand, if H^+ is extracted from a buffered solution, the reaction shifts to the right, releasing free H^+ from compound HX. Together, HX and X^- are termed a **buffer pair**. According to the principles of mass action, the following equation describes a buffer reaction at equilibrium:

$$\frac{[H^+][X^-]}{[HX]} = K' \qquad (22.8)$$

where K' is a constant that depends on the particular buffer reaction and the prevailing conditions, notably temperature. The negative of the common logarithm of K' is symbolized **pK'**, just as the negative of the logarithm of $[H^+]$ is called pH. The buffering effectiveness of any given buffer reaction is greatest when half of the X^- groups are combined with H^+ and half are not; that is, the change in pH caused by the addition or removal of H^+ is minimized when $[HX] = [X^-]$. From Equation 22.8, it is clear that for [HX] and $[X^-]$ to be equal, $[H^+]$ must equal K'; that is, pH must equal pK'. Thus the buffering effectiveness of any given buffer reaction is greatest when the prevailing pH matches the pK' of the reaction. Applying this principle to the blood (or any other body fluid) of an animal, we can say that the blood may contain an enormous variety of potential buffer pairs, but typically, *the buffer reactions that are important will be those with pK′ values within one pH unit of the pH prevailing in the blood.*

The blood of mammals and most other vertebrates is highly effective in buffering the H^+ generated from CO_2 because it has a high concentration of effective buffer groups. These groups are found mostly on blood protein molecules, *especially hemoglobin*! Two types of abundant chemical groups on blood proteins have appropriate pK' values to contribute in major ways to the buffering function of blood. They are the terminal amino groups of the protein chains and the imidazole groups found wherever the amino acid histidine occurs in the protein structure. *The imidazole groups are the dominant buffering groups.* The buffering of human blood is so effective that when CO_2 undergoes the reaction in Equation 22.5, forming HCO_3^- and H^+, the buffer groups remove from free solution over 99.999% of the H^+ produced! This buffering permits a great deal of HCO_3^- to be formed, meaning that the blood can take up a great deal of carbon dioxide.

Carbon dioxide transport is interpreted by use of carbon dioxide equilibrium curves

Blood equilibrium curves for carbon dioxide have interpretive value similar to that of oxygen equilibrium curves. To understand the use of carbon dioxide equilibrium curves, we must first establish the meaning of the **total carbon dioxide concentration** of the blood. Suppose that some blood is brought to equilibrium with an atmosphere containing no CO_2, so that the CO_2 partial pressure of the blood is zero. Suppose that the blood is then exposed to an atmosphere containing CO_2 at some fixed, positive partial pressure. And suppose that as the blood comes to equilibrium with the new atmosphere, we measure the total quantity of CO_2 it takes up, regardless of the chemical form assumed by the CO_2 in the blood. This quantity—the total amount of CO_2 that must enter each unit of volume of blood to raise the blood CO_2 partial pressure from zero to any particular positive CO_2 partial pressure—is termed the blood's *total carbon dioxide concentration* at that partial pressure. A plot of

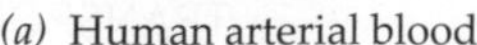

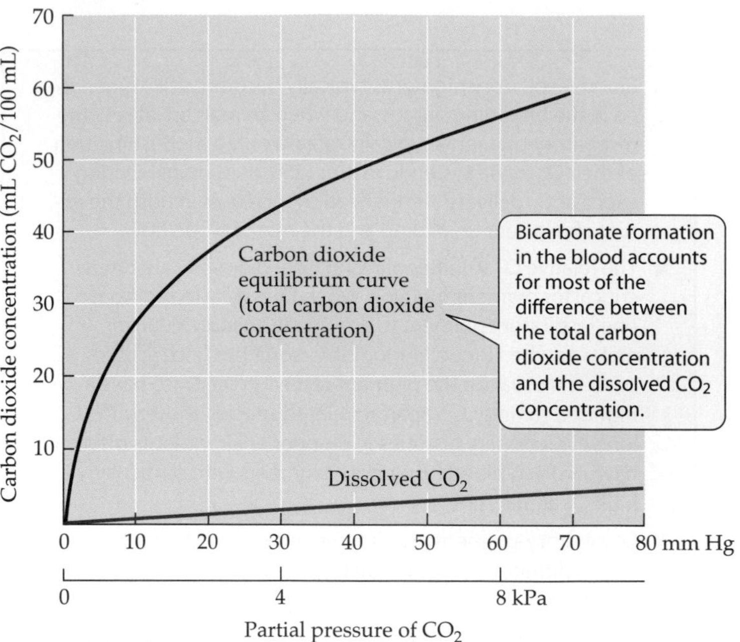

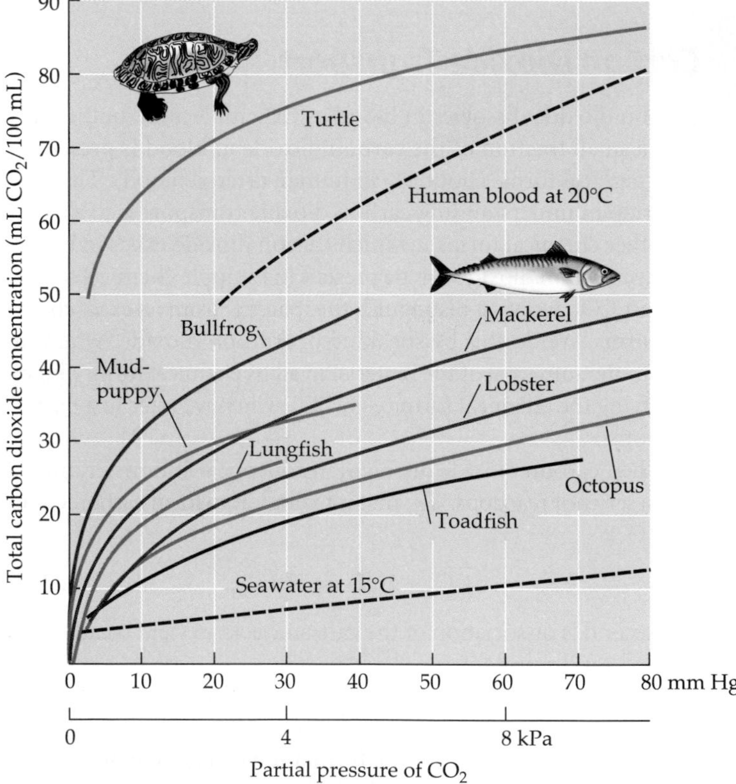

Figure 22.20 Carbon dioxide equilibrium curves *(a)* The carbon dioxide equilibrium curve of fully oxygenated human blood at normal body temperature. The portion of the total carbon dioxide concentration attributable to dissolved CO_2 is shown at the bottom. *(b)* Carbon dioxide equilibrium curves for oxygenated blood of nine species at 15°C to 25°C. Because all curves were not determined at exactly the same temperature, some of the differences among curves may arise from temperature effects. Species: bullfrog, *Rana catesbeiana*; lobster, *Panulirus vulgaris*; lungfish, *Neoceratodus forsteri*; mackerel, *Scomber scombrus*; mudpuppy, *Necturus maculosus*; octopus, *Octopus macropus*; toadfish, *Opsanus tau*; turtle, *Pseudemys floridana*. (After Hill and Wyse 1989.)

the total carbon dioxide concentration as a function of CO_2 partial pressure is known as a **carbon dioxide equilibrium curve** or **carbon dioxide dissociation curve** (Figure 22.20*a*).

What determines the *shape* of the carbon dioxide equilibrium curve? In mammals, carbon dioxide exists in blood in three chemical forms, and thus the total carbon dioxide concentration has three components. Two, as we have already discussed, are dissolved CO_2 and HCO_3^-. The third is carbon dioxide that is directly chemically combined (in a reversible manner) with amino groups on hemoglobin and other blood proteins, forming **carbamate groups** (–NH–COO$^-$) (also called *carbamino groups*). The role of carbamate formation in groups of animals other than mammals—vertebrate and invertebrate—remains debatable. Typically, in both mammals and other groups, the great preponderance of blood carbon dioxide is in the form of HCO_3^-; 90% of the carbon dioxide in human blood, for example, is in that form. The shapes of the carbon dioxide equilibrium curves of animals are thus determined largely by the kinetics of HCO_3^- formation in their bloods. This, in turn, means that the shapes depend on the blood buffer systems: the concentrations of buffer groups, their pK' values, and the extent to which they are loaded with H^+ from other acids besides CO_2.

A diversity of carbon dioxide equilibrium curves is found in the animal kingdom (Figure 22.20*b*). Air-breathing and water-breathing animals typically operate on quite different parts of their carbon dioxide equilibrium curves. This is because, as we discussed in Chapter 21 (see Figure 21.6), air breathers typically have far higher arterial CO_2 partial pressures than water breathers. Resting mammals and birds breathing atmospheric air, for example, have systemic arterial CO_2 partial pressures of 3.3 kPa (25 mm Hg) or higher, whereas gill-breathing fish in well-aerated waters commonly have systemic arterial CO_2 partial pressures of only 0.1 to 0.4 kPa (1–3 mm Hg). In air breathers, the CO_2 partial pressure of blood rises from a high arterial value to a still higher venous value as the blood circulates through the systemic tissues, meaning that the part of the carbon dioxide equilibrium curve that is used is the part at relatively high CO_2 partial pressures. In water breathers, by contrast, both the arterial and venous CO_2 partial pressures are relatively low; the part of the equilibrium curve that is used by water breathers is therefore the steep part at relatively low CO_2 partial pressures.

The Haldane effect: The carbon dioxide equilibrium curve depends on blood oxygenation

The carbon dioxide equilibrium curve of an animal's blood commonly changes with the state of oxygenation of the respiratory pigment (the O_2-transport pigment) in the blood, a phenomenon named the **Haldane effect** after one of its discoverers. When a blood exhibits the Haldane effect, *deoxygenation promotes CO_2 uptake by the blood and oxygenation promotes CO_2 unloading*. Thus the total carbon dioxide concentration at any given CO_2 partial pressure is greater when the blood is deoxygenated than when it is oxygenated (Figure 22.21). The reason for the Haldane effect is that the respiratory pigments play major roles as buffering compounds, and their buffering function depends on their degree of oxygenation. Deoxygenation of a respiratory pigment alters its buffering function in such a way that it tends to keep the blood at a more alkaline pH and take up more H^+. According to Equation 22.6, this means that when a respiratory pigment becomes deoxygenated, more HCO_3^- can form, and the blood therefore reaches a higher total carbon dioxide concentration. This phenomenon is the necessary converse of the Bohr effect, as we noted earlier.

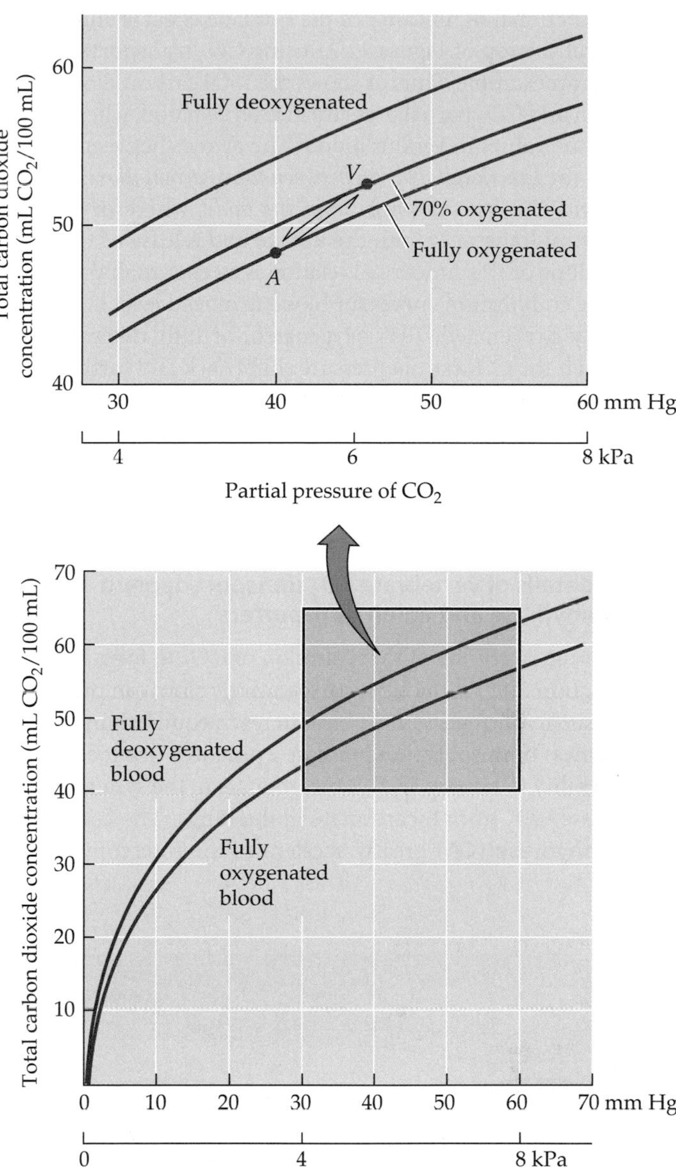

Figure 22.21 The Haldane effect and its implications for human carbon dioxide transport The principal graph (bottom) shows carbon dioxide equilibrium curves for fully oxygenated and essentially fully deoxygenated human blood, illustrating the Haldane effect. The inset at the top summarizes carbon dioxide transport in people who are approximately at rest. Point *A* represents arterial blood, which is fully oxygenated and has a CO_2 partial pressure of about 5.3 kPa (40 mm Hg). Point *V* represents mixed venous blood, which is about 70% oxygenated and has a CO_2 partial pressure of about 6.1 kPa (46 mm Hg). The arrows show the functional relation between total carbon dioxide concentration and CO_2 partial pressure as blood circulates through the body, becoming alternately arterial and venous.

The functional significance of the Haldane effect is illustrated in the inset at the top of Figure 22.21 using CO_2 transport in resting people as an example. Point *A* shows the total carbon dioxide concentration and CO_2 partial pressure in arterial blood, whereas point *V* shows the values in venous blood. The arrows between *A* and *V* represent the *functional relation between total carbon dioxide concentration and CO_2 partial pressure in the body*, where oxygenation changes simultaneously with the uptake and release of CO_2. Note that the slope of this functional relation is steeper than the slope of any of the equilibrium curves for blood at a fixed level of oxygenation (fully oxygenated, 70% oxygenated, or fully deoxygenated). Thus, when the CO_2 partial pressure shifts back and forth between the arterial value of 5.3 kPa (40 mm Hg) and the venous value of 6.1 kPa (46 mm Hg), the blood can take up and release more CO_2 than would be possible without the Haldane effect. In this way, hemoglobin function *simultaneously* aids CO_2 transport and O_2 transport!

Critical details of vertebrate CO_2 transport depend on carbonic anhydrase and anion transporters

An important attribute of the hydration of CO_2 to form bicarbonate (Equation 22.5) is that it occurs relatively slowly in the absence of catalysis (requiring a minute or so to reach equilibrium). The native slowness of this reaction presents a potential bottleneck in the blood's ability to take up CO_2 as bicarbonate in the systemic tissues and release CO_2 from bicarbonate in the lungs. The enzyme **carbonic anhydrase** (**CA**) greatly accelerates the interconversion of CO_2 and HCO_3^-, thereby preventing this reaction from acting as a bottleneck.[15] The reaction is the only one known to be catalyzed in CO_2 transport.

The morphological location of CA has important consequences for CO_2 transport. In vertebrates, CA is found within the red blood cells but almost never in the blood plasma. Despite the absence of CA from the plasma, catalysis is sometimes brought about in the plasma by membrane-associated CA that is in contact with the plasma in certain tissues. Membrane-associated CA has been found, for example, in the cell membranes of the endothelial cells that form the walls of blood capillaries in the lungs, as well as in the skeletal muscles, of some vertebrates. A key point to observe is that *CA is both essential and localized*; CA catalysis must take place for CO_2 transport to occur, but it takes place only in certain physical locations.

When CO_2 enters the blood in the capillaries of a systemic tissue (Figure 22.22), the one place where CA is always found is within the red blood cells that are present there. The CO_2 that enters the red blood cells is quickly converted to HCO_3^- and H^+. CO_2 readily diffuses into red blood cells, and if there is no plasma CA activity, virtually all the reaction of CO_2 to form HCO_3^- and H^+ occurs inside the red blood cells. Hemoglobin—the most important blood buffer—is immediately available inside the red blood cells to take up H^+ and thus play its critical role in promoting HCO_3^- formation. In fact, be-

[15] When CO_2 is hydrated to form HCO_3^- by carbonic anhydrase catalysis, H_2CO_3 is not formed as an intermediate. Instead, the reaction proceeds by a pathway not involving H_2CO_3 formation.

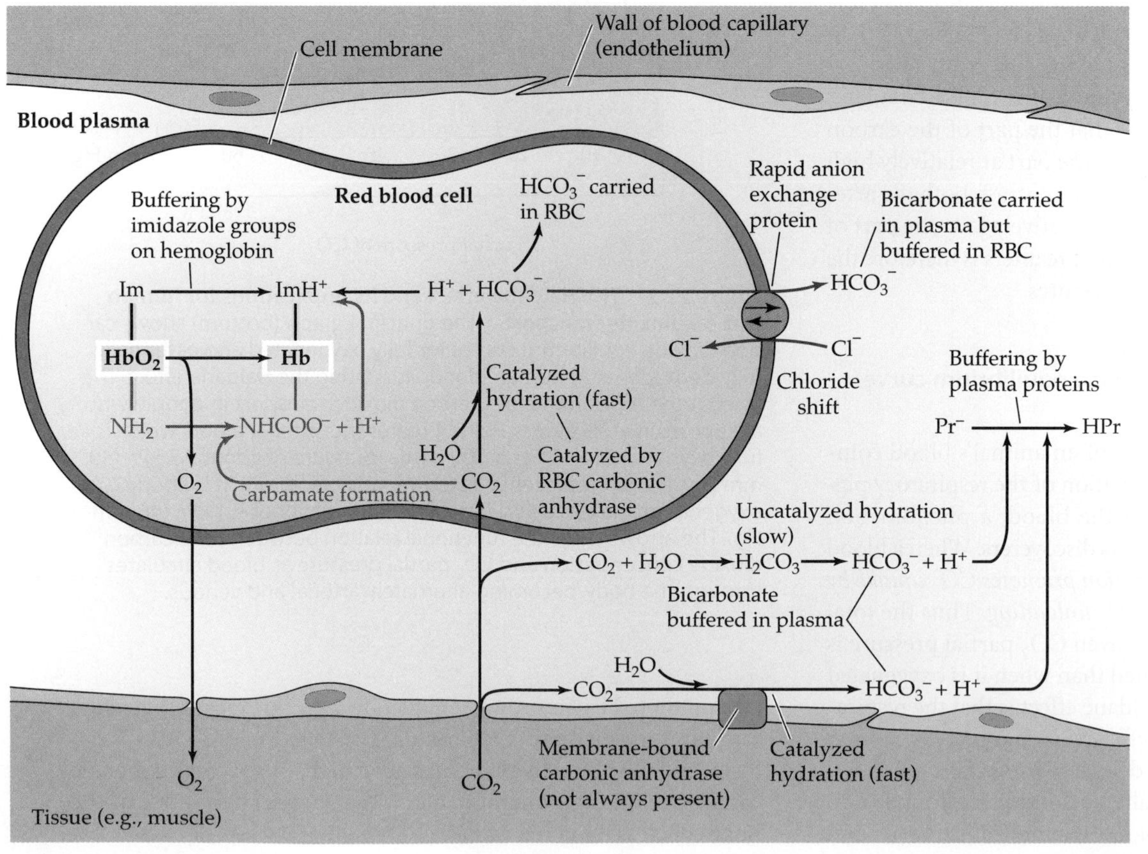

Figure 22.22 The processes of CO_2 uptake by the blood in a systemic blood capillary of a vertebrate The processes shown here occur in reverse in the lungs or gills. Where the labels refer to the site of buffering, they are specifying where the H^+ generated during bicarbonate production is taken up by buffering compounds. Carbamate formation occurs to a significant extent in mammals, but not necessarily in other vertebrates. RBC, red blood cell; Hb, hemoglobin; Im, imidazole groups; Pr, plasma proteins.

cause hemoglobin is undergoing deoxygenation as CO_2 is added to the blood, hemoglobin develops a greater affinity for H^+ just as it is needed. The red blood cell membranes of nearly all vertebrates are well endowed with a transporter protein—a rapid anion exchanger often termed the *band 3 protein*—that facilitates diffusion of HCO_3^- and Cl^- across the membranes in a 1:1 ratio. The HCO_3^- that is formed and buffered inside red blood cells thus tends to diffuse out into the plasma, so that the plasma ultimately carries most of the HCO_3^- added to the blood in the systemic capillaries. As HCO_3^- diffuses out of the red blood cells into the plasma, Cl^- diffuses into the cells from the plasma—a process called the **chloride shift**. In tissues that have CA associated with the blood-capillary endothelium, such as the skeletal muscles of at least certain vertebrates, some rapid formation of HCO_3^- and H^+ occurs in the plasma, where the HCO_3^- must be buffered by plasma proteins. All these events occur in reverse when the blood flows through the lungs or gills.

The operations of these kinetic details govern the exact ways in which CO_2 is transported under any set of conditions. For example, from recent research, we know that the operations of the kinetic details differ during exercise and rest—a difference that may ultimately prove to be critical for a full understanding of exercise physiology.

Summary: Carbon Dioxide Transport

- The carbon dioxide equilibrium curve, which shows the relation between the total carbon dioxide concentration of blood and the CO_2 partial pressure, is a key tool for analyzing carbon dioxide transport. In water breathers, the CO_2 partial pressures of both systemic arterial blood and systemic venous blood are typically low and on the steep portion of the carbon dioxide equilibrium curve. In air breathers, blood CO_2 partial pressures tend to be far higher and therefore on the flatter portion of the carbon dioxide equilibrium curve.
- Most carbon dioxide carried in blood is typically in the form of bicarbonate, HCO_3^-. The extent of HCO_3^- formation depends on blood buffers and determines the shape of the carbon dioxide equilibrium curve. Because respiratory pigments are major blood buffers, they play major roles in carbon dioxide transport.
- The Haldane effect, which is in part the necessary converse of the Bohr effect, is an increase in the total carbon dioxide concentration of the blood caused by deoxygenation of the respiratory pigment. The Haldane effect aids carbon dioxide transport by promoting CO_2 uptake by the blood in the systemic tissues and CO_2 loss from the blood in the breathing organs.
- Rapid uptake of CO_2 by the blood or loss of CO_2 from the blood requires the action of carbonic anhydrase—an enzyme localized to certain places (e.g., RBCs)—because the conversion between CO_2 and HCO_3^- is otherwise slow.

Acid–Base Physiology

The pH of the body fluids cannot vary far from normal levels without serious functional consequences. In humans, for instance, the normal pH of arterial blood at 37°C is about 7.4, and individuals are placed near death if their pH rises to just 7.7, or falls to 6.8! Abnormal H^+ concentrations inflict their adverse effects to a large extent by influencing the function of proteins. As buffer groups on a protein take up or lose H^+, the electrical charge of the whole molecule is rendered more positive or negative; beyond certain limits, these changes result in changes in molecular conformation or other properties that interfere with proper protein function. For the electrical-charge and ionization status of a protein molecule to remain within limits compatible with protein function, the pH must remain within parallel limits.

The neutral pH is defined to be the pH of pure water. By this definition, as shown in Figure 22.23, the neutral pH varies with temperature, being higher at low temperatures than at high ones. In poikilotherms, the normal *blood* pH varies with body temperature *in parallel* with the neutral pH. Specifically, within each species, the blood pH is displaced by a relatively fixed amount to the alkaline side of the neutral pH, rising and falling with body temperature to maintain this fixed displacement (see Figure 22.23). Each species is therefore said to maintain a *constant relative alkalinity* of blood pH.

In the early days of the study of acid–base physiology, studies of humans and other large mammals gave rise to the notion that pH is always regulated at a single, invariant level (e.g., 7.4 in human arterial blood). We now realize that this type of pH regulation is a special case that occurs only in animals that maintain a constant deep-body temperature. In animals as a whole, the blood pH that is maintained by acid–base regulatory mechanisms is a temperature-dependent variable.

What advantage might animals gain by increasing their blood pH as their body temperature falls? This question might well be re-

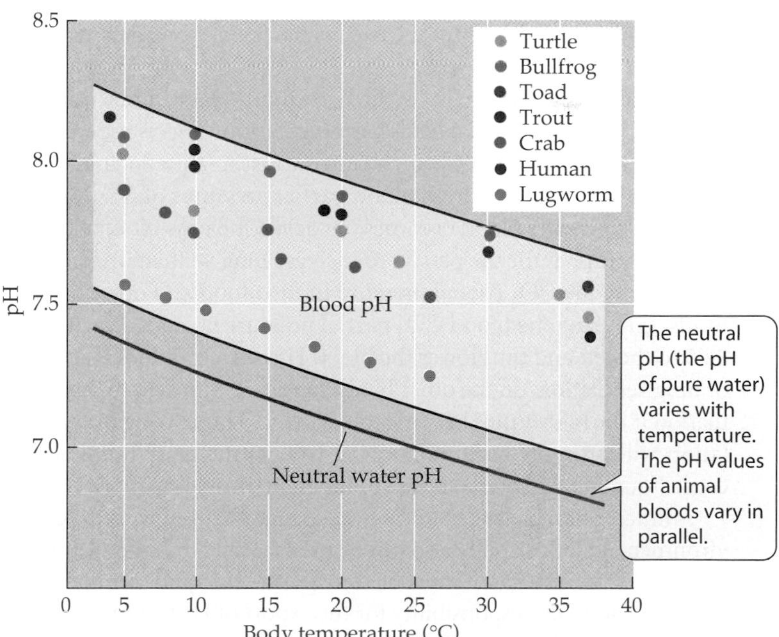

Figure 22.23 Normal blood pH is a temperature-dependent variable As the neutral pH varies with body temperature, blood pH—which is more alkaline than neutral—varies in parallel; thus the blood pH of any particular species tends to remain alkaline to a fixed extent. Data are shown for six poikilothermic species and for humans. Species differ in how alkaline their blood is relative to the neutral pH. (From Dejours 1981.)

phrased, because the pH of the intracellular fluids of cells (although different from blood pH) also increases as temperature falls. Thus, what advantage might animals gain by increasing the pH in both their extra- and intracellular fluids as temperature falls? According to the leading hypothesis, termed the **alphastat hypothesis**, the *changes* in pH are a means of maintaining a *constant* state of electrical charge on protein molecules. The reason that changes in pH are required is that changes in temperature alter the chemical behavior of the buffer groups on the molecules. Most importantly, as temperature falls, the pK' values of imidazole groups increase. This means that at reduced temperatures, imdazole groups increase their inherent tendency to combine with H^+. If this change in chemical behavior were unopposed, more of the imidazole groups on proteins would be combined with H^+ at low temperatures. When the temperature is lowered, decreasing the H^+ concentration of the body fluids serves to oppose the heightened tendency of the imidazole groups to take up H^+. Accordingly, it helps prevent the proportion of positively charged groups on proteins from changing.

Acid–base regulation involves excretion or retention of chemical forms affecting H^+ concentration

When a process occurs that tends to cause a protracted increase in the amount of acid in the body of an animal, maintenance of the animal's temperature-dependent normal pH demands that other processes be set in motion that will either export acid from the body or increase the body's content of base. Conversely, if a disturbance occurs that decreases body acid, acid–base regulation requires a compensating uptake of acid or export of base. There are two competing "worldviews" of acid–base regulation.[16] In our brief overview here, we adopt the simpler of the two, which focuses on adjustments in CO_2, H^+, and HCO_3^-.

The concentration of CO_2 in the body fluids of an animal can be raised or lowered to assist acid–base regulation. This is especially true in terrestrial animals, which (in contrast to aquatic animals) normally have relatively high blood partial pressures of CO_2. Suppose that a person's blood becomes too acid. One possible compensatory response is for the person to increase lung ventilation, thereby lowering the CO_2 partial pressure in the blood and other body fluids. Lowering the blood CO_2 partial pressure will pull Equation 22.5 to the left and thus lower the blood H^+ concentration. Slowing of lung ventilation, on the other hand, can assist with acid–base regulation if the body fluids become too alkaline. The slowing of ventilation will promote accumulation of CO_2 in the body fluids and cause Equation 22.5 to be shifted to the right, providing more H^+.

Animals often have the ability to exchange H^+ itself with the environment, and this ability also can be used for acid–base regulation. Because H^+ is not a gas, it must be transported in liquid solution. In terrestrial animals, responsibility for the export of H^+ from the body rests with the kidneys. Humans, for example, are regularly confronted with an excess of H^+ from their diet, and they void the excess in their urine; this urinary elimination of H^+ can be curtailed entirely, however, when appropriate. In aquatic animals, including both fish and crustaceans, H^+ is exchanged with the environment by the gill epithelium (see Box 3.1, for example).

Bicarbonate ions are also exchanged with the environment to assist acid–base regulation. These HCO_3^- exchanges are mediated by the kidneys in terrestrial animals but principally, it appears, by the gill epithelium in fish and crabs (see Box 3.1). Bicarbonate functions as a base. If retention of HCO_3^- in the body is increased, Equation 22.5 is shifted to the left, tending to remove H^+ from solution in the body fluids and make the body fluids more alkaline. Conversely, increased elimination of HCO_3^- tends to raise the H^+ concentration of the body fluids.[17]

Disturbances of acid–base regulation fall into respiratory and metabolic categories

Disturbances of the pH of the body fluids are categorized as **acidosis** or **alkalosis**. Acidosis occurs when the pH of the body fluids is shifted to the acid side of the normal pH at a given body temperature. Alkalosis is a shift in pH to the alkaline side of the normal pH at a given body temperature. Disturbances of pH are also classed as **respiratory** or **metabolic** according to their primary cause.

The *respiratory* disturbances of pH are ones that are brought about by an abnormal alteration of the rate of CO_2 elimination by the lungs or gills. **Respiratory alkalosis** arises when the exhalation of CO_2 is abnormally increased relative to CO_2 production, causing the CO_2 partial pressure in the body fluids to be driven below the level needed to maintain a normal pH. Panting by mammals, for example, sometimes causes this type of disturbance (see page 226). **Respiratory acidosis** occurs when exhalation of CO_2 is impaired and metabolically produced CO_2 therefore accumulates excessively in the body.

Whereas the blood property that is *initially* altered in *respiratory* disturbances of pH is the CO_2 partial pressure, *metabolic* disturbances of pH—by definition—initially alter the blood *bicarbonate* concentration. **Metabolic alkalosis** and **metabolic acidosis** both have numerous possible causes. Metabolic acidosis, for example, can result from excessive loss of HCO_3^- in gastrointestinal fluids during chronic diarrhea or from excessive loss of HCO_3^- in saliva during thermoregulatory saliva spreading by rodents (see page 227). Metabolic acidosis can also result from excessive addition of H^+ to the body fluids, as when lactic acid is accumulated during vigorous exercise; the added H^+ from lactic acid reacts with the pool of HCO_3^- in the body fluids, lowering the concentration of HCO_3^-.

Animals typically respond to disturbances of pH using their acid–base regulatory mechanisms. Lung ventilation by human athletes performing work of ever-increasing intensity provides an example. Whereas athletes increase their rate of lung ventilation in parallel with their rate of CO_2 production when they are not accumulating lactic acid, they increase their rate of ventilation more than their rate of CO_2 production when lactic acid accumulation occurs. This disproportionate increase in ventilation, termed *hyperventila-*

[16]The books by Davenport and Stewart in the Selected Readings provide readable introductions to these two worldviews. For those who become interested in the strong ion difference approach, not covered here, the reference by Johnson et al. is also worthwhile.

[17]One way to view this effect of HCO_3^- elimination is to recognize that HCO_3^- originates from H_2CO_3; when HCO_3^- is eliminated, just the H^+ of H_2CO_3 remains in the body fluids, acidifying them.

tion, causes CO_2 to be exhaled from the body faster than it is being produced. The CO_2 partial pressure in the blood and body fluids is thereby reduced, helping to limit the degree of acidosis caused by the accumulation of lactic acid.

Summary: Acid–Base Physiology

- The neutral pH varies with temperature, being higher at low temperatures than at high ones. In animals with variable body temperatures, the normal blood pH typically varies in parallel with the neutral pH, being displaced in the alkaline direction to a constant extent (constant relative alkalinity).
- Acidosis and alkalosis are categories of acid–base disturbance. They occur, respectively, when the blood pH is to the acid or alkaline side of the normal pH for the prevailing body temperature. Either sort of disturbance can be respiratory (originating because of changes in CO_2 loss by breathing) or metabolic (originating because of changes in the blood bicarbonate concentration).
- Animals can correct chronic acid–base disturbances by modulating CO_2, H^+, or HCO_3^- elimination in regulatory ways.

Study Questions

1. While touring a saltwater aquarium, suppose you see a striped bass, a hammerhead shark, an octopus, a feather-duster worm, and a lobster. What type of respiratory pigment would you expect to find in the blood of each?
2. One could say that a respiratory pigment with relatively low O_2 affinity is potentially disadvantageous for loading, but advantageous for unloading. Explain both parts of this statement.
3. In most species of mammals, the O_2 affinity of a fetus's blood hemoglobin is greater than that of its mother's blood hemoglobin. However, mammal species are not all the same in the *mechanism* that causes the affinities to be different. Specify three distinct mechanisms for the difference in affinity between fetal and maternal blood hemoglobin. Recall from Chapter 1 that François Jacob argued that evolution is analogous to tinkering rather than engineering. Would you say that the evolution of the mechanism of the fetal–maternal difference in O_2 affinity provides evidence for Jacob's argument? Explain.
4. Outline the ways in which mammalian hemoglobin *simultaneously* plays important roles in O_2 transport, CO_2 transport, and control of blood pH.
5. A fish swims from a body of cool water into a body of warm water. As its body temperature rises, its rate of O_2 consumption increases. The warm water, however, is likely to have a lower concentration of dissolved O_2 than the cool water because the solubility of O_2 in water decreases as temperature increases. These two factors taken together—an increase in the fish's rate of O_2 consumption and a decrease in the dissolved O_2 concentration of its environmental water—can make it difficult for the fish to obtain enough O_2 to meet its needs. The two factors can act as a two-pronged trap. Actually, however, the fish may face a three-pronged trap. How is the increase in temperature likely to affect the O_2 affinity of the fish's hemoglobin, and how could the effect on hemoglobin add even further to the challenge the fish faces? Does global warming pose concerns of this sort?
6. When fishing boats pull trawling nets through the water, many fish escape by vigorously swimming away. Others, after vigorous escape swimming, get caught, but later are released because they are not of legal size or are not the species desired. Fish that escape or are released sometimes die anyway. The accumulation of lactic acid from anaerobic work in such fish seems in certain cases to be a key factor in their deaths. How could a large lactic acid accumulation in a fish interfere with its ability to obtain enough O_2 to survive?
7. To study the chemical properties of the blood hemoglobin of a vertebrate, it might seem convenient to remove the hemoglobin from the red blood cells so that it is in simple aqueous solution. However, removing the hemoglobin from red blood cells often promptly alters its O_2-binding characteristics. Why?
8. Give an example of respiratory alkalosis and one of metabolic acidosis. In each case, explain how your example illustrates that type of acid–base disturbance.
9. As noted in this chapter, respiratory pigments that are dissolved in the blood plasma usually have very high molecular weights. The statement has been made that in animals with dissolved respiratory pigments, "the polymerization of unit respiratory-pigment molecules into high-molecular-weight polymers allows the blood solution to have a high oxygen-carrying capacity without having its osmotic pressure boosted to high levels by the presence of the respiratory pigment." Explain. (Hint: Review in Chapter 3 how dissolved entities affect the colligative properties of solutions.)
10. Among related species (e.g., vertebrates), there is often a positive correlation between the *oxygen*-carrying capacity of blood and the height of the *carbon dioxide* equilibrium curve (the total carbon dioxide concentration at high CO_2 partial pressures). Why?
11. Studies have shown that reindeer and musk-ox, two Arctic mammals (see Chapter 9), have evolved hemoglobins that are unusually low in their sensitivity to temperature compared to the hemoglobins of most large mammals. The researchers who made this discovery hypothesized that a particularly low thermal sensitivity is required for hemoglobin to unload O_2 to an adequate extent in the distal parts of these animals' legs, where tissue temperatures may be 25°C cooler than in the thorax (see Figure 8.26). Explain the rationale for this hypothesis. In answering, discuss how the temperature of the hemoglobin molecules changes as blood flows between the lungs and the legs in cold weather. If possible, design studies to test the hypothesis.

Key References

Cameron, J. N. 1989. *The Respiratory Physiology of Animals*. Oxford University Press, New York. Extensive discussion of blood gas transport, including a set of thorough case studies of animals such as blue crabs, trout, and humans.

Dickerson, R. E., and I. Geis. 1983. *Hemoglobin: Structure, Function, Evolution, and Pathology*. Benjamin-Cummings, Menlo Park, CA. A book one will never forget. The combination of science and art is masterful.

Hardison, R. 1999. The evolution of hemoglobin. *Amer. Sci.* 87: 126–137. A rewarding overview of hemoglobin and related compounds in animals, plants, and microbes.

Ingermann, R. L. 1997. Vertebrate hemoglobins. In W. H. Dantzler (ed.), *Comparative Physiology*, vol. 1 (Handbook of Physiology [Bethesda, MD], section 13), pp. 357–408. Oxford University Press, New York. A thorough review of the comparative physiology of vertebrate hemoglobins.

Mangum, C. P. 1997. Invertebrate blood oxygen carriers. In W. H. Dantzler (ed.), *Comparative Physiology*, vol. 2 (Handbook of Physiology [Bethesda, MD], section 13), pp. 1097–1135. Oxford University Press,

New York. A thorough review of the comparative physiology of invertebrate respiratory pigments.

Wagner, P. D. 1996. Determinants of maximal oxygen transport and utilization. *Annu. Rev. Physiol.* 58: 21–50. An accessible introduction to modern thinking on the transport of O_2 from hemoglobin to the mitochondria. The process is more complex than heretofore thought, and new details may prove to explain important human disease syndromes. The article also provides insight into high-altitude acclimatization and other issues.

Selected References

Ackers, G. K., M. L. Doyle, D. Myers, and M. A. Daugherty. 1992. Molecular code for cooperativity in hemoglobin. *Science* 255: 54–63.

Adamczewska, A. M., and S. Morris. 1998. The functioning of the haemocyanin of the terrestrial Christmas Island red crab *Gecarcoidea natalis* and roles for organic modulators. *J. Exp. Biol.* 201: 3233–3244.

Angersbach, D. 1978. Oxygen transport in the blood of the tarantula *Eurypelma californicum*: pO_2 and pH during rest, activity, and recovery. *J. Comp. Physiol.* 123: 113–125.

Arp, A. J., J. J. Childress, and R. D. Vetter. 1987. The sulphide-binding protein in the blood of the vestimentiferan tube-worm, *Riftia pachyptila*, is the extracellular haemoglobin. *J. Exp. Biol.* 128: 139–158.

Bauer, C., G. Gros, and H. Bartels (eds.). 1980. *Biophysics and Physiology of Carbon Dioxide*. Springer-Verlag, New York.

Bernal, D., K. A. Dickson, R. E. Shadwick, and J. B. Graham. 2001. Review: Analysis of the evolutionary convergence for high performance swimming in lamnid sharks and tunas. *Comp. Biochem. Physiol. A* 129: 695–726.

Booth, C. E., B. R. McMahon, and A. W. Pinder. 1982. Oxygen uptake and the potentiating effects of increased hemolymph lactate on oxygen transport during exercise in the blue crab, *Callinectes sapidus*. *J. Comp. Physiol.* 148: 111–121.

Bridges, C. R. 2001. Modulation of hemocyanin oxygen affinity: Properties and physiological implications in a changing world. *J. Exp. Biol.* 204: 1021–1032.

Brix, O., A. Bårdgard, S. Mathisen, N. Tyler, M. Nuutinen, S. G. Condo, and B. Giardina. 1990. Oxygen transport in the blood of arctic mammals: Adaptation to local heterothermia. *J. Comp. Physiol. B* 159: 655–660.

Brown, A. C., and N. B. Terwilliger. 1998. Ontogeny of hemocyanin function in the Dungeness crab *Cancer magister*: Hemolymph modulation of hemocyanin oxygen-binding. *J. Exp. Biol.* 201: 819–826.

Burggren, W., B. McMahon, and D. Powers. 1991. Respiratory functions of blood. In C. L. Prosser (ed.), *Environmental and Metabolic Animal Physiology* (*Comparative Animal Physiology*, 4th ed.), pp. 437–508. Wiley-Liss, New York.

Cocca, E., M. Ratnayake-Lecamwasam, S. K. Parker, L. Camardella, M. Ciaramella, G. di Prisco, and H. W. Detrich III. 1995. Genomic remnants of α-globin genes in the hemoglobinless antarctic icefishes. *Proc. Natl. Acad. Sci. U.S.A.* 92: 1817–1821.

Cosby, K., K. S. Partovi, J. H. Crawford, R. P. Patel, C. D. Reiter, S. Martyr, B. K. Yang, M. A. Waclawiw, G. Zalos, X. Xu, K. T. Huang, H. Shields, D. B. Kim-Shapiro, A. N. Schechter, R O. Cannon III, and M. T. Gladwin. 2003. Nitrite reduction to nitric oxide by deoxyhemoglobin vasodilates the human circulation. *Nature Med.* 9: 1498–1505.

Davenport, H. W. 1974. *The ABC of Acid–Base Chemistry*, 6th ed. University of Chicago Press, Chicago.

Edsall, J. T., and J. Wyman. 1958. *Biophysical Chemistry*, vol. 1. Academic Press, New York. [A classic reference on carbon dioxide chemistry.]

Egginton, S. 1997. A comparison of the response to induced exercise in red- and white-blooded antarctic fishes. *J. Comp. Physiol. B* 167: 129–134.

Fänge, R. 1983. Gas exchange in fish swim bladder. *Rev. Physiol. Biochem. Pharmacol.* 97: 111–158.

Farrell, A. P., and S. M. Clutterham. 2003. On-line venous oxygen tensions in rainbow trout during graded exercise at two acclimation temperatures. *J. Exp. Biol.* 206: 487–496.

Geers, C., and G. Gros. 2000. Carbon dioxide transport and carbonic anhydrase in blood and muscle. *Physiol. Rev.* 80: 681–715.

Heath, D. 1989. *High-Altitude Medicine and Pathology*. Butterworth, Boston.

Hebbel, R. P., J. W. Eaton, R. S. Kronenberg, E. D. Zanjani, L. G. Moore, and E. M. Berger. 1978. Human llamas. *J. Clin. Invest.* 62: 593–600.

Heisler, N. (ed.). 1986. *Acid–Base Regulation in Animals*. Elsevier, New York.

Heisler, N. 1990. Acid–base regulation: Interrelationships between gaseous and ionic regulation. *Adv. Comp. Environ. Physiol.* 6: 211–251.

Houlihan, D. F., G. Duthie, P. J. Smith, M. J. Wells, and J. Wells. 1986. Ventilation and circulation during exercise in *Octopus vulgaris*. *J. Comp. Physiol. B* 156: 683–689.

Johansen, K., and C. Lenfant. 1966. Gas exchange in the cephalopod, *Octopus dofleini*. *Amer. J. Physiol.* 210: 910–918.

Johansen, K., G. Lykkeboe, R. E. Weber, and G. M. O. Maloiy. 1976. Respiratory properties of blood in awake and estivating lungfish, *Protopterus amphibius*. *Respir. Physiol.* 27: 335–345.

Johnson, R. L., Jr., G. J. F. Heigenhauser, C. C. W. Hsia, N. L. Jones, and P. D. Wagner. 1996. Determinants of gas exchange and acid–base balance during exercise. In L. B. Rowell and J. T. Shepherd (eds.), *Exercise: Regulation and Integration of Multiple Systems* (Handbook of Physiology [Bethesda, MD], section 12), pp. 515–584. Oxford University Press, New York.

Kiceniuk, J. W., and D. R. Jones. 1977. The oxygen transport system in trout (*Salmo gairdneri*) during sustained exercise. *J. Exp. Biol.* 69: 247–260.

Kraus, D. W., and J. M. Colacino. 1986. Extended oxygen delivery from the nerve hemoglobin of *Tellina alternata* (Bivalvia). *Science* 232: 90–92.

Lahiri, S., N. S. Cherniack, and R. S. Fitzgerald (eds.). 1991. *Response and Adaptation to Hypoxia: Organ to Organelle*. Oxford University Press, New York.

Linzen, B. (ed.). 1986. *Invertebrate Oxygen Carriers*. Springer-Verlag, New York.

Lykkeboe, G., and K. Johansen. 1982. A cephalopod approach to rethinking about the importance of the Bohr and Haldane effects. *Pacific Sci.* 36: 305–313.

Mangum, C. P. (ed.). 1992. Blood and Tissue Oxygen Carriers. *Advances in Comparative and Environmental Physiology*, vol. 13. Springer-Verlag, New York.

Mangum, C., and D. Towle. 1977. Physiological adaptation to unstable environments. *Amer. Sci.* 65: 67–75. [For an update, see Mason, R. P., C. P. Mangum, and G. Godette. 1983. *Biol. Bull.* 164: 104–123.]

McMahon, B. R. 2001. Respiratory and circulatory compensation to hypoxia in crustaceans. *Respir. Physiol.* 128: 349–364.

McMahon, B. R., D. G. McDonald, and C. M. Wood. 1979. Ventilation, oxygen uptake and haemolymph oxygen transport, following enforced exhausting activity in the Dungeness crab *Cancer magister*. *J. Exp. Biol.* 80: 271–285.

Miller, K. I. 1994. Cephalopod haemocyanins: A review of structure and function. *Mar. Fresh. Behav. Physiol.* 25: 101–120.

Monge, C., and F. León-Velarde. 1991. Physiological adaptation to high altitude: Oxygen transport in mammals and birds. *Physiol. Rev.* 71: 1135–1172.

Morris, S., and C. R. Bridges. 1994. Properties of respiratory pigments in bimodal breathing animals: Air and water breathing by fish and crustaceans. *Amer. Zool.* 34: 216–228.

Nikinmaa, M. 1990. *Vertebrate Red Blood Cells*. Springer-Verlag, New York.

Pawloski, J. R., D. T. Hess, and J. S. Stamler. 2001. Export by red blood cells of nitric oxide bioactivity. *Nature* 409: 622–626.

Pelster, B., and D. Randall. 1998. The physiology of the Root effect. In S. F. Perry and B. L. Tufts (eds.), *Fish Respiration*, pp. 113–139. Academic Press, New York.

Perry, S. F., and B. L. Tufts (eds.). 1998. *Fish Respiration*. Academic Press, New York.

Perutz, M. F. 1978. Hemoglobin structure and respiratory transport. *Sci. Amer.* 239(6): 92–125.

Perutz, M. F. 1990. Mechanisms regulating the reactions of human hemoglobin with oxygen and carbon monoxide. *Annu. Rev. Physiol.* 52: 1–25.

Ratcliffe, P. J., K.-U. Eckardt, and C. Bauer. 1996. Hypoxia, erythropoietin gene expression, and erythropoiesis. In M. J. Fregly and C. M. Blatteis (eds.), *Environmental Physiology*, vol. 2 (Handbook of Physiology [Bethesda, MD], section 4), pp. 1125–1153. Oxford University Press, New York.

Reeves, R. B., and H. Rahn. 1979. Patterns in vertebrate acid–base regulation. In S. C. Wood and C. Lenfant (eds.), *Evolution of Respiratory Processes*, pp. 225–252. Dekker, New York.

Riggs, A. F. 1998. Self-association, cooperativity and supercooperativity of oxygen binding by hemoglobins. *J. Exp. Biol.* 201: 1073–1084.

Sanders, N. K., A. J. Arp, and J. J. Childress. 1988. Oxygen binding characteristics of the hemocyanins of two deep sea hydrothermal vent crustaceans. *Respir. Physiol.* 71: 57–68.

Schmidt-Nielsen, K., and C. R. Taylor. 1968. Red blood cells: Why or why not? *Science* 162: 274–275.

Snyder, G. K., and B. A. Sheafor. 1999. Red blood cells: Centerpiece in the evolution of the vertebrate circulatory system. *Amer. Zool.* 39: 189–198.

Snyder, L. R. G. 1985. Low P_{50} in deer mice native to high altitude. *J. Appl. Physiol.* 58: 193–199.

Snyder, L. R. G., J. P. Hayes, and M. A. Chappell. 1988. Alpha-chain hemoglobin polymorphisms are correlated with altitude in the deer mouse, *Peromyscus maniculatus*. *Evolution* 42: 689–697. [See also pp. 681–688.]

Stewart, P. A. 1978. Independent and dependent variables of acid–base control. *Respir. Physiol.* 33: 9–26

Stewart, P. A. 1981. *How to Understand Acid–Base: A Quantitative Acid–Base Primer for Biology and Medicine*. Elsevier, New York.

Stryer, L. 1995. *Biochemistry*, 4th ed. Freeman, New York.

Tetens, V., and N. J. Christensen. 1987. Beta-adrenergic control of blood oxygen affinity in acutely hypoxia exposed rainbow trout. *J. Comp. Physiol. B* 157: 667–675.

Truchot, J. P. 1987. *Comparative Aspects of Extracellular Acid–Base Balance*. Springer-Verlag, New York.

Urich, K. 1994. *Comparative Animal Biochemistry*. Springer-Verlag, New York.

Ward, M. P., J. S. Milledge, and J. B. West. 2000. *High Altitude Medicine and Physiology*, 3rd ed. Oxford University Press, New York.

Wells, R. M. G. 1990. Hemoglobin physiology in vertebrate animals: A cautionary approach to adaptationist thinking. *Advances in Comparative and Environmental Physiology*, vol. 6: 143–161.

West, J. B. 1984. Human physiology at extreme altitudes on Mount Everest. *Science* 223: 784–788.

Westen, E. A., and H. D. Prange. 2003. A reexamination of the mechanisms underlying the arteriovenous chloride shift. *Physiol. Biochem. Zool.* 76: 603–614.

Wilkening, R. B., and G. Meschia. 1992. Current topic: Comparative physiology of placental oxygen transport. *Placenta* 13: 1–15.

See also references in Appendix A.

23 Circulation

When we look at the contours of the muscles of an athletic person, we can imagine that the muscles are a force unto themselves. The muscles, however, can be only as effective as their circulation permits them to be. For each cell in a muscle, the streaming of blood in nearby capillaries is a lifeline for resupply with O_2 and other necessities. Although a muscle may have a highly developed contractile apparatus, it is able to contract only with as much endurance and power as its rate of ATP production permits, and during sustained exertion, the capacity of a muscle to produce ATP is determined by the rate at which O_2 is brought to it by the circulation of the blood (see Chapter 7). In traditional human societies, in which food is gathered, shelter is built, and battles are fought with human muscles, the rate of O_2 supply to the muscles by the circulation can help determine whether a person and his family thrive or perish. In industrialized human societies, the rate of O_2 supply to the muscles by the circulation helps determine whether a person can perform a job requiring hard manual labor, or be a successful basketball player or long-distance runner.

Besides the transport of commodities such as O_2, the circulatory system also performs other vital functions. Blood pressure is used to initiate the formation of urine. Blood pressure is also used to stiffen erectile tissues in both sexes during sexual intercourse. In men, the circulatory system of the testes plays a thermoregulatory role; the arteries and veins are arranged into a countercurrent heat exchanger (see page 223) that short-circuits heat from arterial blood flowing toward the testes into venous blood leaving them, thereby reducing heat flow into the testes and helping to keep them at the diminished temperature they require for sperm production. Vascular countercurrent heat exchangers also occur in the arms of both women and men and in the appendages of various other species of mammals. These arrays of blood vessels reduce energy costs by limiting heat losses from the body in chilly environments.

Ancient people undoubtedly were as aware as we are that no matter where the body is cut, it oozes blood. This was the first sign that blood at high pressure streams through every region of tissue. The idea that the blood makes a round trip through the body—the concept that it circulates—was first put forth by William Harvey in 1628. However, Harvey and his contemporaries could not possibly have understood the circulation as we do today, because blood capillaries were unknown in his time, and O_2 was not identified as a defined gas for another 150 years. From Harvey's

The heart circulates blood to every tissue and organ, sustaining all bodily strength The *kaduma*, the young men, of the Nuba tribe in the Sudan—one of Earth's disappearing peoples—prize strength and physical health, as well as the serenity of the music they pluck on their lyres. (From Riefenstahl 1973.)

time to the present, scientists have learned steadily more about the tasks that are accomplished by the circulation of the blood. As they have done so, the very word *circulation* has taken on a progressively different meaning. A generation after Harvey, early microscopists discovered that minute blood vessels just barely wider than red blood cells—the capillaries—weave among the cells of every tissue. Bleeding from every cut could then, for the first time, be understood, and the stage was set for biologists ultimately to realize that the blood brings needed substances to all cells. In the nineteenth century, scientists demonstrated that the circulation of the blood brings O_2 to cells (see Box 5.1). In the twentieth century, hormones, antibodies, and immune cells were discovered, and the concept of the circulation of the blood was expanded to include their transport from one region of the body to another.

To define **circulation** today in a way that is relevant to all kinds of animals, two perspectives can be taken. From the perspective of mechanics, circulation is the pressure-driven mass flow of a body fluid called **blood** through a system of tubular vessels or other passages that brings the fluid to all parts of the body. The system of vessels or other blood passages—plus the blood itself—is called the **circulatory system**. When we think of circulation, however, we usually do not think of it in just these mechanical terms. From a second perspective, circulation is defined by what it accomplishes. Thus, for us today, circulation is a pressure-driven mass flow of fluid that rapidly transports O_2, CO_2, nutrients, organic wastes, hormones, agents of the immune system, heat, and other commodities throughout the body and that often provides a source of hydraulic pressure for organ function.

The *speed* of transport by the circulation is one of its most central and defining attributes. As we first saw in discussing Table 3.1, diffusion through aqueous solutions is too slow to transport commodities at biologically significant rates over distances exceeding a millimeter or so. Only very small animals, therefore, can depend on diffusion as their sole means of internal transport (see Box 20.1). Convective transport—transport by mass flow of body fluids—is intrinsically far faster than diffusion. Thus, as stressed in Chapter 20 (see Figure 20.4), animals larger than a millimeter or so generally require blood circulation (or at least some other mass flow of body fluids) to move commodities from place to place in their bodies at adequate rates.

As we study the circulation, both of the defining perspectives we have identified will be important. For example, looking back at the man in our opening photograph, one of our two key questions must be mechanical: How do his heart and blood vascular system function to bring blood to and from the cells in his head, feet, skeletal muscles, and all other parts of his body at the rate required? Our second key question must concern the functions accomplished by his circulation.

In people and in most other types of animals, the transport of O_2 is the most pressing and urgent function performed by the circulation. Of all the commodities that tissues require to be brought to them by the circulation of the blood, O_2 is the one that the tissues typically can least afford to have brought more slowly. Conversely, the rate at which the tissues need O_2 typically determines the rate at which blood must be circulated; in most cases, the heart could pump blood much more slowly if the only commodities requiring blood transport were those other than O_2. Thus, in most kinds of animals, metabolic intensity and the capacity of the circulatory system to transport O_2 have coevolved, and they are strongly correlated in modern species.

Hearts

A logical starting point for our study of the circulation is the hearts of animals. A **heart** is a discrete, localized pumping structure. Some animals that have a circulatory system lack a heart; in many annelid worms, for example, the blood is propelled through the circulatory system entirely by peristaltic contractions of blood vessels. Hearts are very common in circulatory systems, however, and often assume principal responsibility for driving the flow of blood through the blood vessels.

In some types of animals, such as arthropods, the heart is **single-chambered**, consisting of a single muscular tube or sac. In others, such as vertebrates, the heart is composed of two or more compartments through which blood passes in sequence, and thus is classified as **multichambered**. Many types of animals, in addition to their principal heart, possess other hearts that assist with the pumping of blood through localized parts of the body. Such secondary or local hearts are called **accessory hearts** or **auxiliary hearts**.

The muscle tissue of a heart is known as the **myocardium**. It consists of **cardiac muscle**, which typically has distinctive structural and physiological properties in comparison to other types of muscle, as we discussed in Chapter 17 (see Figure 17.19 and Table 17.3).

To study the morphology of a heart in detail, there is no more appropriate example than our own (Figure 23.1). The hearts of other mammals and of birds are similar to ours. The left side of the human heart, which consists of two chambers—a weakly muscular **atrium** and a strongly muscular **ventricle**—receives freshly oxygenated blood from the lungs and pumps it to the systemic tissues of the body.[1] Blood arrives in the **left atrium** via the **pulmonary veins** that drain the lungs.[2] It leaves the **left ventricle** via a single massive artery called the **systemic aorta**, which branches to send arterial vessels to the head, arms, abdomen, and all other body regions, even the myocardium itself. Passive valves, consisting of flaps of connective tissue covered with endothelial tissue, are positioned between the atrium and ventricle (the **left atrioventricular valve**) and between the ventricle and aorta (the **aortic valve**); these valves allow blood to flow freely in the correct direction, but prevent it from flowing backward. After blood leaves the systemic aorta, it passes through the **systemic circuit**—the blood vessels that take blood to and from the systemic tissues—and ultimately returns in the great collecting veins (**venae cavae;** singular *vena cava*) to the heart, where it enters the **right atrium** and then the **right ventricle**. The function of the right heart is to pump blood through the **pulmonary circuit**—the blood vessels that take blood to and from the lungs. The right ventricle propels blood into a large vessel, the **pulmonary trunk**, which divides to form the **pulmonary arteries** to the lungs. As in the left heart, pas-

[1] The **systemic tissues** are all the tissues other than the tissues of the breathing organs.

[2] By definition, **veins** are vessels that carry blood toward the heart, and **arteries** are vessels that carry blood away from the heart.

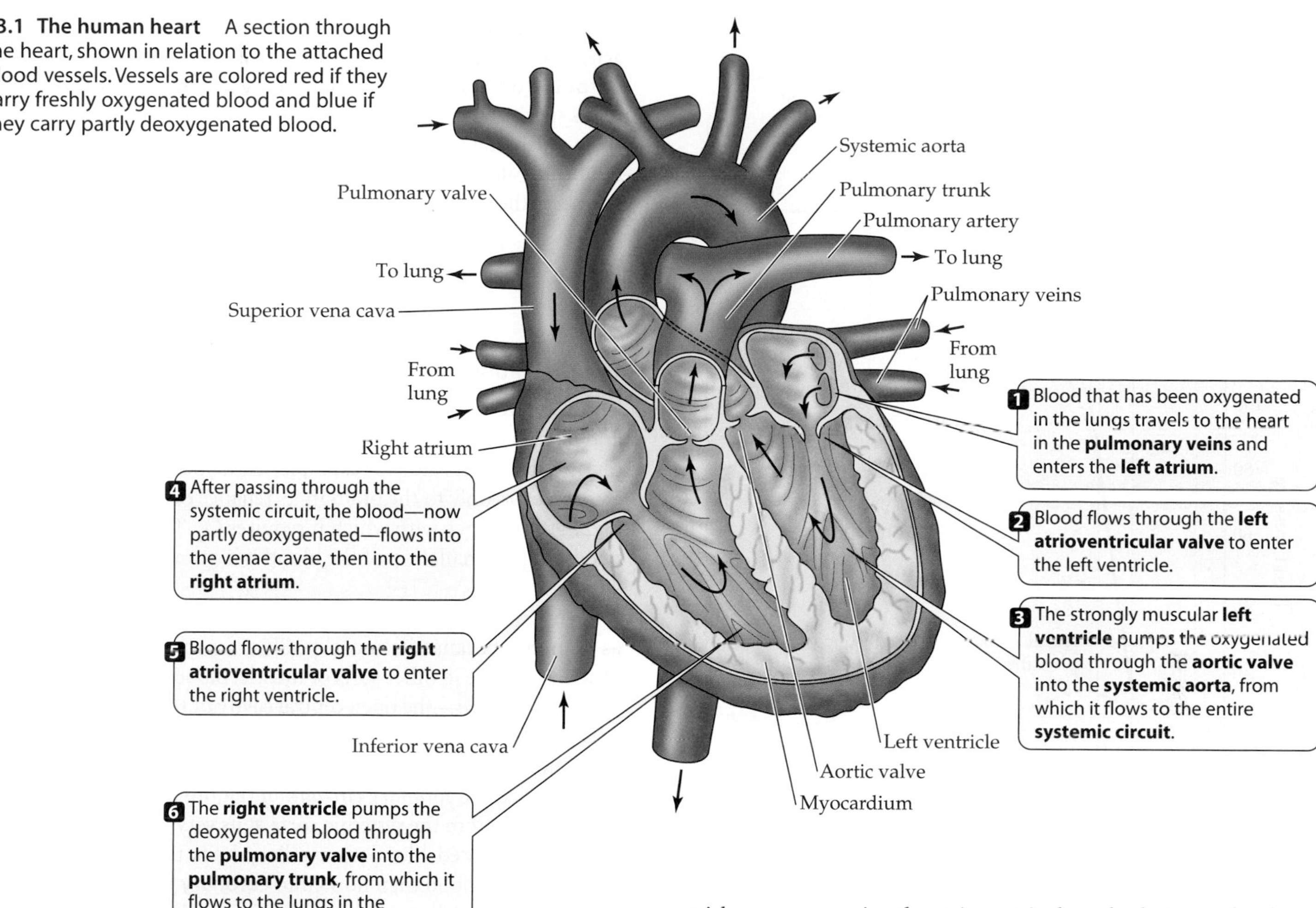

23.1 The human heart A section through the heart, shown in relation to the attached blood vessels. Vessels are colored red if they carry freshly oxygenated blood and blue if they carry partly deoxygenated blood.

sive flap valves prevent backward flow in the right heart; these valves are positioned between the atrium and ventricle (the **right atrioventricular valve**) and between the ventricle and pulmonary trunk (the **pulmonary valve**). After blood has been oxygenated in the lungs, it returns to the left atrium.

The heart as a pump: The action of a heart can be analyzed in terms of the physics of pumping

During the beating cycle of any type of heart, the period of contraction is called **systole** (pronounced with a long *e*: systol*ee*), and the period of relaxation is termed **diastole** (diastol*ee*). We can understand the workings of the heart as a pump by analyzing pressure, flow, and volume during these periods, using the sort of diagram shown in Figure 23.2. By studying the relations of pressure, flow, and volume, researchers have been able to construct such summary diagrams for many types of animal hearts. The example we study here is the human left heart (the left atrium and ventricle).

At the time marked by the arrow at the bottom of the figure, ventricular systole begins. Whereas the pressure inside the ventricle was previously lower than that inside the atrium, once the ventricle starts to contract, the ventricular pressure rises immediately to exceed the atrial pressure, causing the atrioventricular valve between the chambers to flip shut. For a brief interval of time (about 0.1 s), however, the ventricular pressure remains below the pressure in the systemic aorta, meaning that the aortic valve is not forced open. During this interval, therefore, both the inflow and outflow valves of the ventricle are shut. The volume of blood in the ventricular chamber during this time is thus constant, and the interval is called the phase of **isovolumetric contraction** ("contraction with unchanging volume") or **isometric contraction**. The contraction of the ventricle on the fixed volume of blood within causes the intraventricular pressure to rise rapidly. As soon as the ventricular pressure rises high enough to exceed the aortic pressure, the aortic valve flips open, and the blood in the ventricle accelerates extremely rapidly, gushing out into the aorta (thus increasing aortic pressure). The opening of the aortic valve marks the start of the phase of **ventricular ejection**. Toward the end of this phase, the aortic pressure comes to exceed the ventricular pressure slightly, but ejection of blood into the aorta continues for a while—at a rapidly falling rate—because of blood momentum. Ultimately, the ventricle starts to relax. The ventricular pressure then falls rapidly away from the aortic pressure, and the aortic valve shuts. A period of **isovolumetric relaxation** follows, as ventricular pressure falls with both the inflow and outflow valves shut. When the ventricular pressure drops below the atrial pressure, the atrioventricular valve is forced open, and **ventricular filling** begins. Most filling of the ventricle occurs *before* atrial systole—that is, before the atrial muscle contracts; the motive force for this filling is the pressure built up

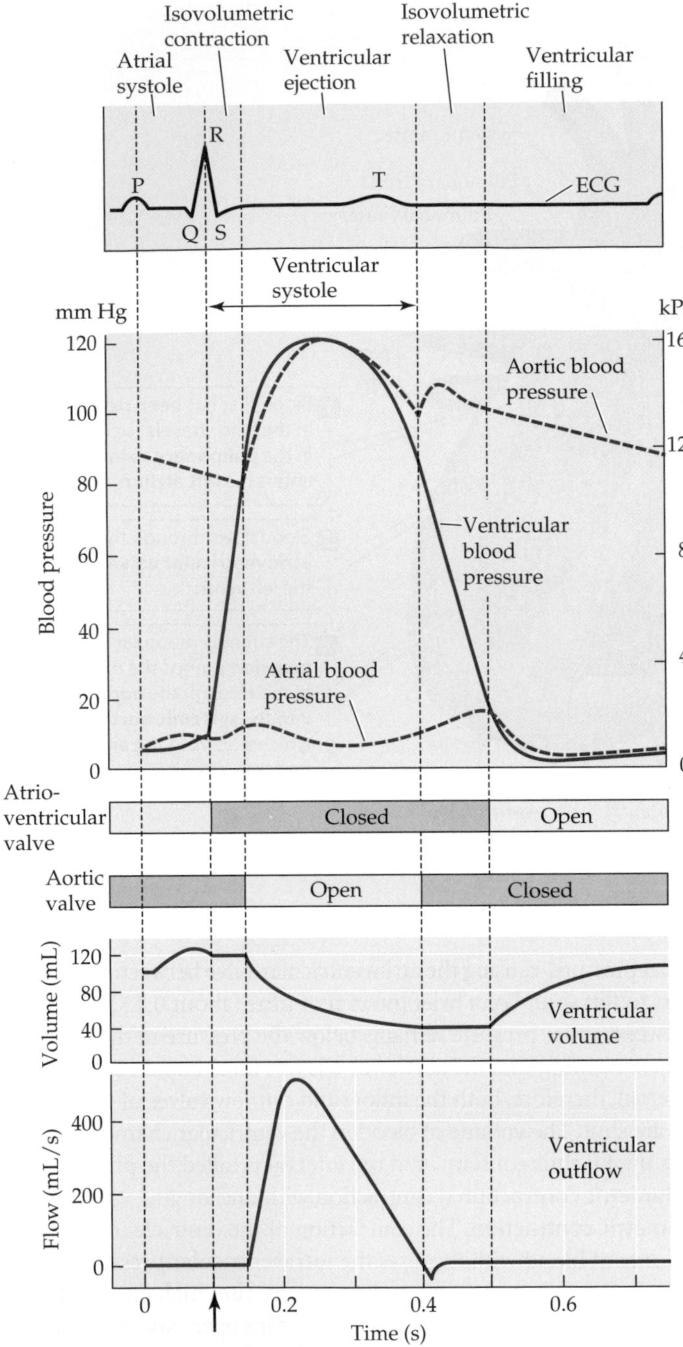

Figure 23.2 The heart as a pump: The dynamics of the left side of the human heart The heart cycle is divided into five phases, labeled at the top and demarcated by the vertical lines that run through the diagram. The diagram shows the synchronous changes that occur in left ventricular blood pressure, systemic aortic blood pressure, left atrial blood pressure, ventricular volume, the rate of blood flow out of the ventricle, and the closing and opening of the atrioventricular and aortic valves in people at rest. The arrow at the bottom marks the start of ventricular systole. The ECG is the electrocardiogram, discussed later in this chapter.

by *accumulation* of pulmonary venous blood in the atrium. When atrial systole occurs, it forces some additional blood into the ventricle just before the next ventricular systole.

The volume of blood pumped by a heart per unit of time into the artery or arteries emanating from the heart is known as the **cardiac output.** (In the case of the mammalian or avian heart, the term refers specifically to the output of the left ventricle into the systemic aorta unless stated otherwise.) The cardiac output is the product of the heart rate and the volume of blood pumped per heart cycle, known as the **stroke volume:**

$$\text{Cardiac output (mL/minute)} = \text{heart rate (beats/minute)} \times \text{stroke volume (mL/beat)} \quad (23.1)$$

The circulation must deliver O_2 to the myocardium

The myocardium of any heart performs sustained, vigorous work, and its cells therefore are especially dependent on a steady O_2 supply. In most vertebrates, for example, the myocardium of the ventricles is second only to the brain in its reliance on aerobic catabolism and in the urgency with which it requires O_2. The myocardium depends on the circulation of the blood it is pumping to renew its O_2 supplies.

In mammals and birds, the ventricular myocardium is classified as **compact** because its muscle cells are packed closely together, much as they are in other sorts of muscle. Blood passing through the *ventricular lumen*—the open central cavity of the ventricular chamber—cannot directly flow among the myocardial muscle cells. Instead, the myocardium is supplied with blood by a system of blood vessels called the *coronary circulation* (Figure 23.3*a*). **Coronary arteries** branch from the systemic aorta at its very beginning and carry freshly oxygenated blood to capillary beds throughout the myocardium; the blood then flows into **coronary veins**, which carry it out of the myocardium and into the right atrium. If a coronary artery becomes blocked, the part of the myocardium it supplies quickly deteriorates from O_2 deprivation, explaining why occlusions in the coronary arteries are extremely dangerous.

The way that the cells of the ventricular myocardium obtain O_2 is very different in most teleost (bony) fish, amphibians, and reptiles. In these animals, the ventricular myocardium is classified as **spongy** because the muscle is permeated by an anastomosing network of open spaces. The blood that is passing through the lumen of the ventricle—ventricular *luminal* blood—flows among the spaces of the spongy tissue, and the myocardial cells obtain their O_2 from this blood (Figure 23.3*b*). The coronary circulation is absent or relatively poorly developed. This arrangement for oxygenation of the myocardium is clearly the original (evolutionarily primitive) arrangement in vertebrates. As we will see later in this chapter, the blood passing through the ventricular lumen in fish and other animals with this arrangement is usually *not* fully oxygenated. Consequently, the myocardial cells are not always assured of a rich O_2 supply.

In some fish—including salmonids, tunas, and sharks—and some amphibians and reptiles, the ventricular myocardium consists of an outer, compact muscle layer surrounding a spongy layer (Figure 23.3*c*). The compact layer is supplied by a system of coronary blood vessels that receive freshly oxygenated blood. The spongy myocardium may receive branches of the coronary circulation, but also must obtain O_2 from the luminal blood, which—as in the cases discussed in the preceding paragraph—is not freshly oxygenated.

(a) Compact myocardium with coronary arteries and veins

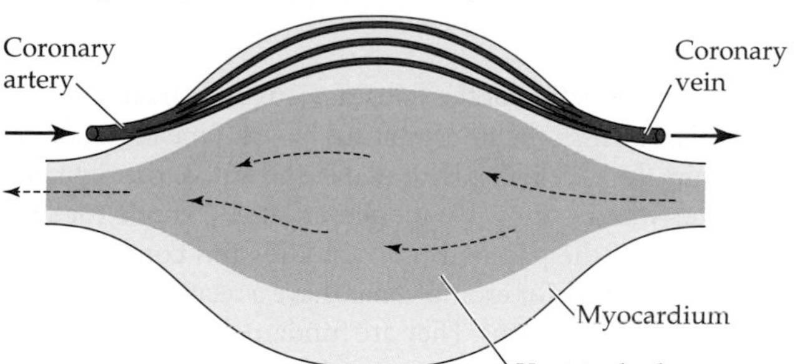

(b) Spongy myocardium with little or no development of coronary vessels

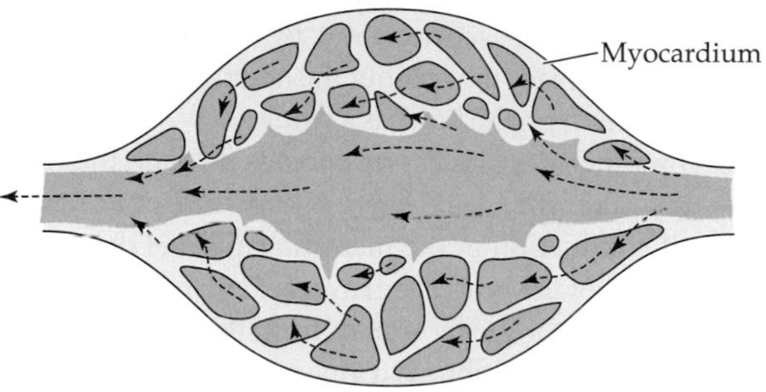

(c) Myocardium composed of outer compact tissue and inner spongy tissue

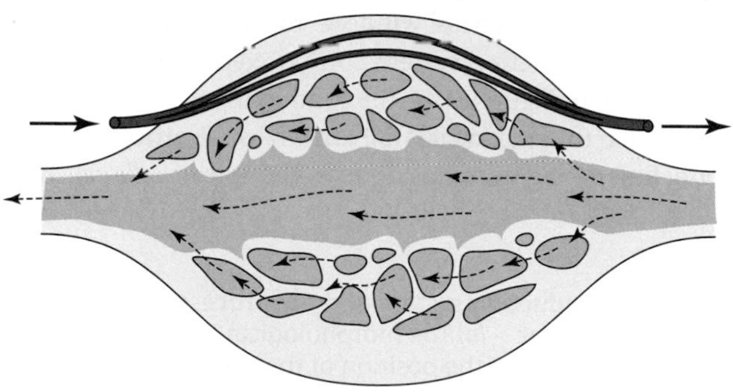

(d) Myocardium of mixed structure with blood flowing from lumen into coronary veins

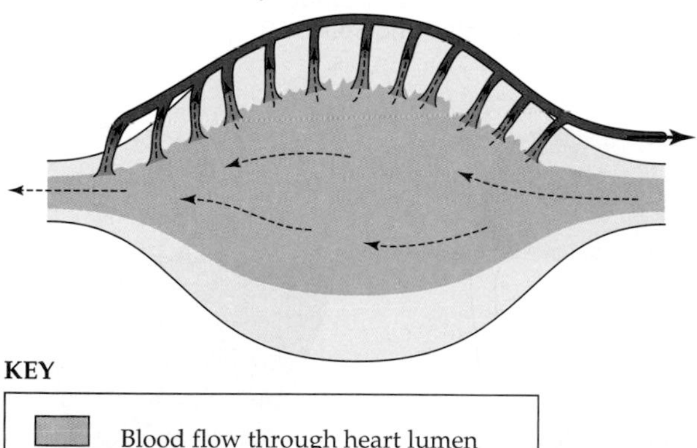

KEY

Blood flow through heart lumen

Blood flow through coronary vessels

Figure 23.3 Four systems evolved by animals to supply O_2 to the myocardium Except for flow through spongy tissue, details of circulation are shown only in the upper half of each diagram. *(a)* The compact myocardium of mammals and birds is supplied by coronary arteries and veins. *(b)* The spongy myocardium of most teleost fish shows little or no development of coronary vessels. *(c)* The ventricular myocardium of some fish, such as salmonid fish, tunas, and sharks, consists of an outer, compact muscle layer with a coronary circulation and an inner spongy layer. *(d)* In some octopuses, such as *Octopus vulgaris,* blood flows from the lumen through the myocardium and into coronary veins.

Relatively little is known about the mechanisms of O_2 supply to the myocardium in invertebrates. The systemic heart of the octopus *Octopus vulgaris*, one of the best-studied invertebrate examples, has an arrangement different from those seen in vertebrates. Based on the evidence available, some of the blood flowing through the heart lumen enters a blood capillary system that carries it through the myocardium, whereupon it is collected into coronary veins on the surface of the heart (Figure 23.3*d*). The blood flowing through the lumen of the systemic heart of an octopus is freshly oxygenated, meaning that the myocardial cells ordinarily receive a relatively rich O_2 supply. The myocardium of some other octopuses besides *O. vulgaris* may have a coronary artery system and thus be oxygenated more along the lines of the pattern shown in Figure 23.3*a*.

The electrical impulses for heart contraction may originate in muscle cells or neurons

The rhythmic contraction of a heart reflects a rhythmic depolarization of the cell membranes of its constituent muscle cells. As described in Chapter 17, the cell membranes of muscle cells are polarized electrically (inside negative, outside positive), but can become depolarized. Depolarization is the immediate stimulus for contraction. A key question about any heart is, where does the impetus for the rhythmic depolarization of the muscle cells originate? Do the muscle cells themselves spontaneously depolarize in a rhythmic manner? Or are they induced to depolarize by impulses arriving from other cells? If the latter is the case, which are the cells that spontaneously depolarize and thus originate the impulses? A heart's **pacemaker** is the cell or set of cells that spontaneously *initiates* the rhythm of depolarization in the heart.[3]

In some animals, each electrical impulse to contract originates in muscle cells or modified muscle cells; the hearts of such animals are described as **myogenic** ("beginning in muscle"). In other animals, each impulse to contract originates in neurons (nerve cells), and the heart is termed **neurogenic** ("beginning in neurons"). Let's explore some of the attributes of myogenic and neurogenic hearts by looking at a classic example of each.

MYOGENIC HEARTS The hearts of vertebrates are myogenic. In almost all cases, they are innervated, but they continue to beat even if all nervous connections are stripped away.

An important feature of vertebrate heart muscle is that adjacent muscle cells are electrically coupled. This electrical coupling occurs at gap junctions (see Figure 2.7), which in humans and other mammals occur primarily at specialized regions of intercellular contact known as *intercalated discs*.[4] Because adjacent cells are electrically

[3] See page 301 for discussion of the cellular physiology of pacemaker depolarization.
[4] Electrical coupling is discussed on page 314. Gap junctions are discussed on pages 39 and 315. Intercalated discs are illustrated in Figure 17.19.

coupled, depolarization of any one cell in the myocardium directly and quickly causes depolarization of other cells neighboring it. In turn, those cells quickly induce their neighbors to depolarize, and so on. Thus, within large regions of the heart muscle, once depolarization is initiated at any point, it rapidly spreads—from muscle cell to muscle cell—to all cells in the region, leading all to contract together, as a unit.

Most or all of the muscle cells in a vertebrate heart possess an inherent ability to undergo rhythmic depolarization and contraction. Thus, pieces of muscle cut from any part of the heart will beat. In the intact heart, of course, individual bits of heart muscle do not depolarize and contract on their own—at their own rhythms. Instead, all the cells in the myocardium are controlled by a particular group of specialized muscle cells: the pacemaker. In fish, amphibians, and reptiles, the pacemaker is located in the wall of the sinus venosus, the first heart chamber, or at the junction of the sinus venosus and atrium. In birds and mammals, in which the sinus venosus has become incorporated into the atrium, the pacemaker is located in the wall of the right atrium (Figure 23.4), and is known as the **sinoatrial (S-A) node** (or **sinus node**). The cells of the pacemaker are modified in comparison to most heart muscle cells; for example, they have a relatively poorly developed contractile apparatus. They are fundamentally muscle cells, however, meaning that the heart is myogenic. A critical attribute of the pacemaker cells is that they exhibit the highest frequency of spontaneous depolarization of all cells in the heart and therefore are normally the first to depolarize at each heartbeat. By thus *initiating* a wave of depolarization that spreads throughout the heart, they *impose* their rhythm of depolarization on the heart as a whole.

The process by which depolarization spreads through the vertebrate heart or any other myogenic heart is known as **conduction**. Critical details of conduction in the mammalian heart depend on key structural features of the heart. The myocardium of the two atria of the heart is separated, for the most part, from the myocardium of the two ventricles by a layer of fibrous connective tissue across which myocardial cells are not electrically coupled by gap junctions and through which depolarization therefore cannot pass. In the mammalian heart, the one "electrical window" through this fibrous layer is provided by a **conducting system** composed of specialized muscle cells. As shown in Figure 23.4*a*, the conducting system starts with a group of cells in the right atrial wall known as the **atrioventricular (A-V) node.** Emanating from this node is a bundle of cells called the

(*a*) The conducting system and sinoatrial node

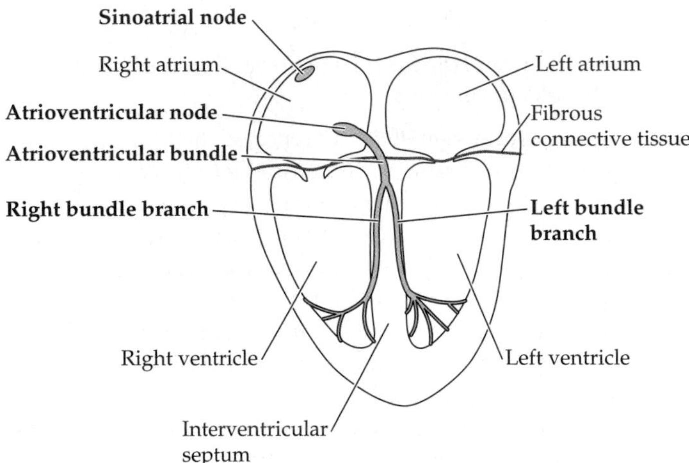

(*b*) The initiation and spread of depolarization during a heartbeat

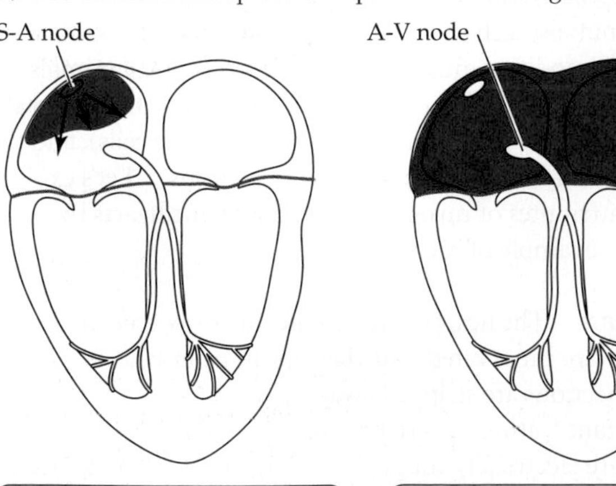

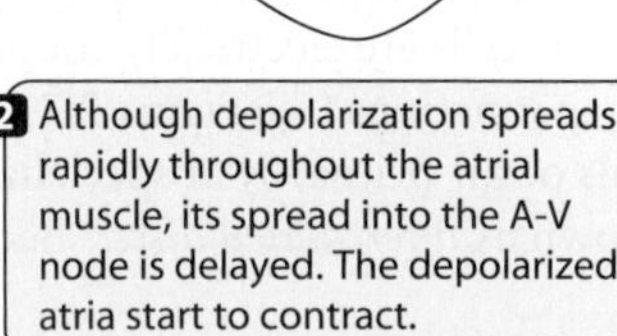

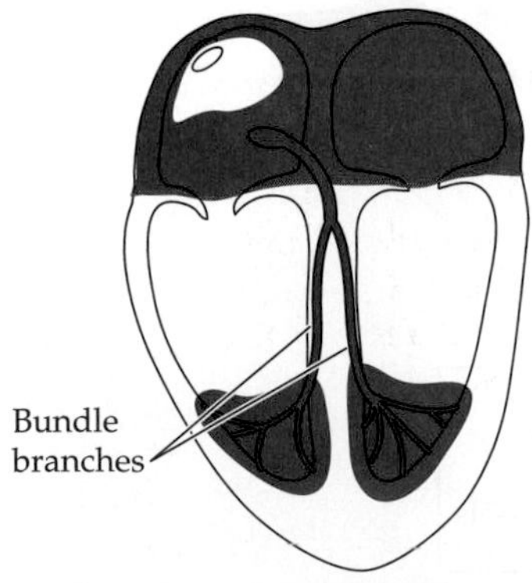

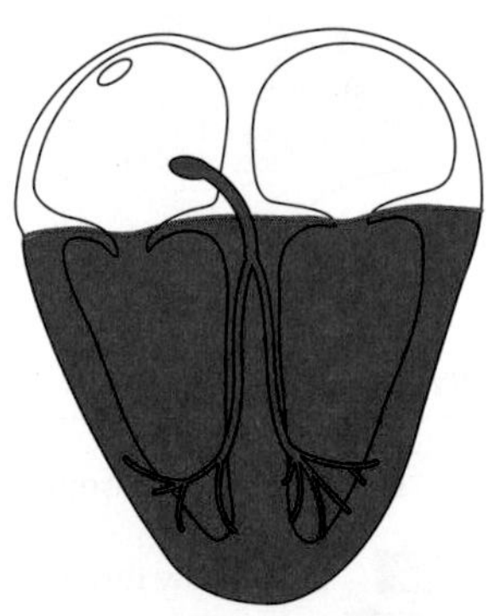

Figure 23.4 The conducting system and the process of conduction in the mammalian heart *(a)* The morphological arrangement of the conducting system and the position of the sinoatrial node. The branches of the right and left bundle branches are in fact more elaborate than shown; traveling along the inner surfaces of the ventricles and across the ventricular cavities, they run to much of the inner wall of each ventricle. *(b)* The initiation and conduction of depolarization during a heartbeat. (*a* after Scher and Spach 1979; *b* after Rushmer 1976.)

atrioventricular bundle (*common bundle, bundle of His*), which penetrates the fibrous layer and enters the *interventricular septum*—the wall of tissue that separates the right and left ventricles. Once in the septum, the atrioventricular bundle divides into right and left portions, the **bundle branches**, which travel along the right and left surfaces of the septum and connect with systems of large, distinctive cells, the **Purkinje fibers**, that branch into the ventricular myocardium on each side.

Functionally, the conducting system of the mammalian heart has two key properties: (1) depolarization enters and traverses the A-V node relatively slowly, and (2) depolarization spreads down the atrioventricular bundle, bundle branches, and systems of Purkinje fibers much more rapidly than it could travel through ordinary ventricular muscle. The implications of these properties become apparent when we consider the sequence of events during a heartbeat, shown in Figure 23.4*b*. Steps 1 and 2 show that once the sinoatrial node initiates a heartbeat by depolarizing spontaneously, the depolarization spreads rapidly throughout the muscle of both atria, leading to atrial contraction. Spread into the ventricular muscle does not occur as rapidly, however, because it is dependent on activation of the conducting system, and the spread of depolarization into and through the initial member of the conducting system—the A-V node—is relatively slow (step 2). This slowness of depolarization of the A-V node is responsible for the sequencing of contraction: atrial contraction distinctly first, ventricular contraction distinctly second. Once the A-V node is activated, depolarization sweeps rapidly down the conducting system into the ventricles (step 3), precipitating wholesale ventricular depolarization and contraction (step 4). The rapid delivery of the depolarizing wave to far-flung parts of the ventricular tissue by the conducting system ensures that all parts of the ventricular myocardium will contract approximately together.

NEUROGENIC HEARTS The defining feature of neurogenic hearts is that the rhythmic depolarization responsible for initiating the heartbeats originates in nervous tissue. The hearts of lobsters are excellent examples of neurogenic hearts. The muscle cells in a lobster heart are induced to depolarize and contract in much the same way as skeletal muscle cells: Each muscle cell is innervated and typically contracts when and only when stimulated to do so by nerve impulses. As shown in Figure 23.5, a **cardiac ganglion** consisting of nine neurons is attached to the inside dorsal surface of the heart of a lobster. The axonal processes of the five most anterior cells (numbered 1–5) innervate the heart muscle. Those of the four posterior cells (numbered 6–9) are confined to the ganglion and make synaptic contact with the five anterior cells. One of the posterior cells ordinarily assumes the role of pacemaker. This cell acts as a cellular oscillator and central pattern generator (see Chapter 18). Periodically and spontaneously, it produces a train of impulses, which excite the other posterior cells. The impulses from the posterior cells activate the five anterior cells, which in turn send trains of impulses to the muscle cells of the heart, causing the latter to contract approximately in unison. If the ganglion and heart muscle are dissected apart, the ganglion continues to produce bursts of impulses periodically, but the muscle ordinarily stops contracting! Other animals known or believed to have neurogenic hearts include other decapod crustaceans (e.g., crabs, shrimp, and crayfish), horseshoe crabs (*Limulus*), and spiders and scorpions.

A heart produces an electrical signature, the electrocardiogram

When a mass of heart muscle is *in the process* of being depolarized, such that some regions of cells are depolarized already and others await depolarization, a difference in electrical potential exists between the extracellular fluids in the depolarized regions of the mus-

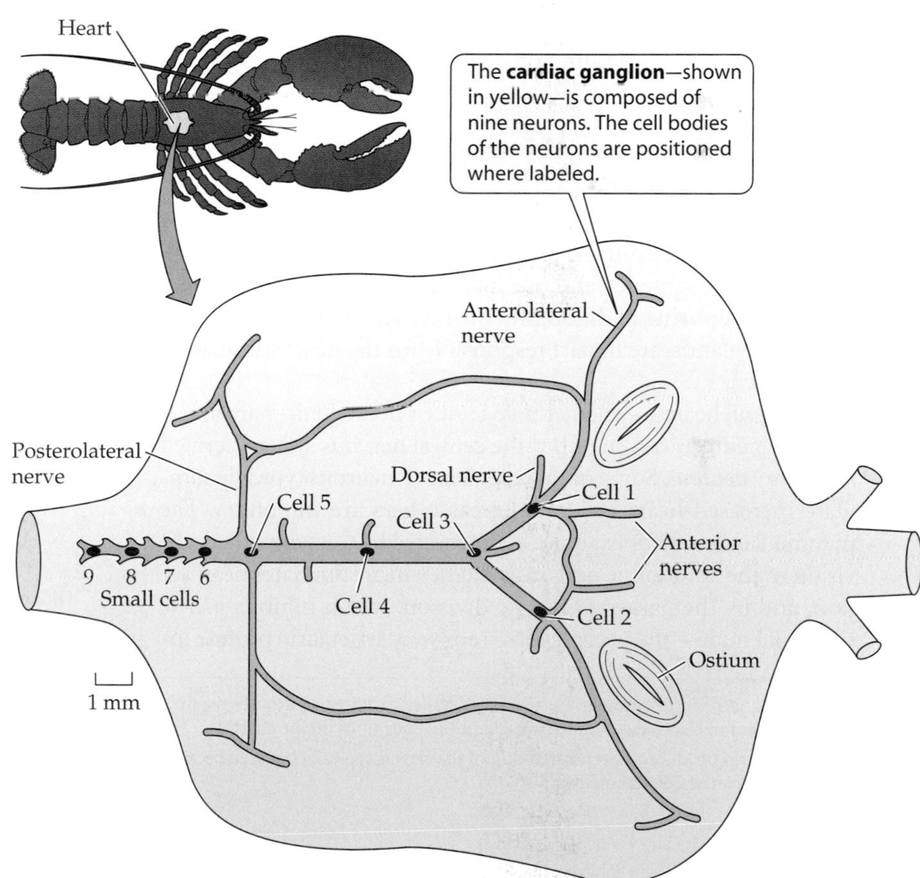

Figure 23.5 The neurogenic heart of a lobster and the cardiac ganglion that initiates and controls its contractions A dorsal view of the heart of the American lobster (*Homarus americanus*), showing the cardiac ganglion, which is positioned on the inside dorsal surface. The cell bodies of the nine neurons that compose the ganglion are marked in black. Neuronal processes go out from each cell body and together form the structure of the ganglion. The posterior four neurons (numbers 6–9) are small, whereas the anterior five are large. Neurons that carry impulses to heart muscle cells exit the ganglion through the various nerves. Regulatory neurons from the central nervous system enter the ganglion in the dorsal nerve. The function of the ostia (slitlike openings through the heart wall) is discussed later in this chapter. (After Hartline 1967.)

cle and those in the polarized regions (Figure 23.6*a*). A voltage difference of this sort within the heart muscle sets up ionic currents in the tissues and body fluids surrounding the heart. In this way, the voltage difference within the heart induces voltage differences elsewhere in the body, even between various parts of the external body surface. **Electrocardiograms** (**ECGs, EKGs**) are measurements over time of voltage differences of this sort. They are recorded using extracellular electrodes, which usually are placed on the body surface. To record the elementary ECG of a person, a physician or nurse places electrodes on the skin of the person's two arms and left leg. The electrodes detect voltage differences on the skin surface that are induced by voltage differences within the heart muscle.[5]

The ECGs of two species, human and octopus, are shown in Figure 23.6*b*. The waveforms in the human ECG are named with letters (Figure 23.6*c*). The **P wave** is produced by the depolarization of the myocardium of the two atria (= atrial contraction). The Q, R, and S waves, together known as the **QRS complex**, arise from the depolarization of the myocardium of the two ventricles (= ventricular contraction). Repolarization of the ventricles generates the **T wave**.[6] Figure 23.2 shows the relation of the ECG waveforms to the mechanical events during the heart pumping cycle.

Heart action is modulated by hormonal, nervous, and intrinsic controls

Heart action is subject to hormonal, nervous, and intrinsic controls. Possibly the controls we are most aware of are hormonal. When we are frightened and our heart pounds and races, the hormones epinephrine and norepinephrine, secreted by the adrenal medullary glands, are in part responsible for the heart stimulation we experience.

Nearly all hearts—whether myogenic or neurogenic—are innervated by neurons coming from the central nervous system, termed *regulatory neurons*. Some of these regulatory neurons typically stimulate increased heart action, whereas others are inhibitory. The mammalian heart, for example, is innervated by the sympathetic division of the autonomic nervous system, which stimulates heart action, and by the parasympathetic division, which inhibits it. The sinoatrial node—the pacemaker—receives particularly profuse innervation; sympathetic impulses delivered to the S-A node increase the frequency of spontaneous depolarization by the pacemaker cells and thus accelerate the heart rhythm, whereas parasympathetic impulses exert opposite effects. The heart muscle proper is also innervated. Sympathetic impulses delivered to cardiac-muscle cells in the myocardium markedly enhance the vigor of their contraction, whereas parasympathetic impulses suppress their vigor of contraction. Sympathetic stimulation of the heart is increased during exer-

[5] More-advanced electrocardiograms used for detailed diagnostic purposes require the attachment of electrodes at numerous additional positions on the chest.

[6] The waveform produced by repolarization of the atria is typically not seen because it is obscured by the QRS complex.

(*a*) Relative charges in myocardial extracellular fluids as the human ventricular myocardium depolarizes

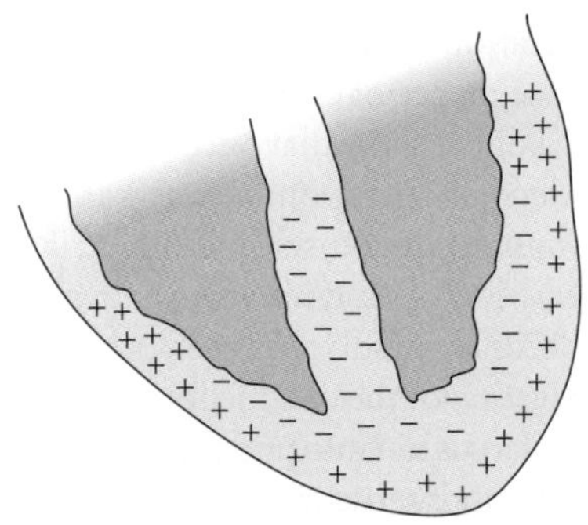

(*b*) Electrocardiograms of human and octopus

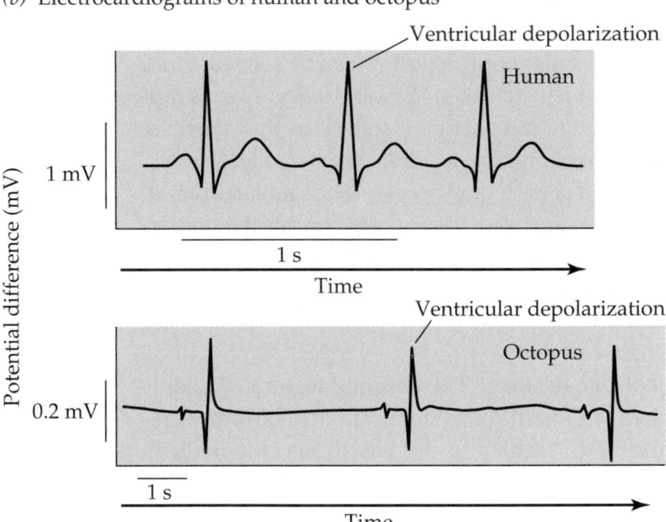

(*c*) Waveforms in the normal human electrocardiogram

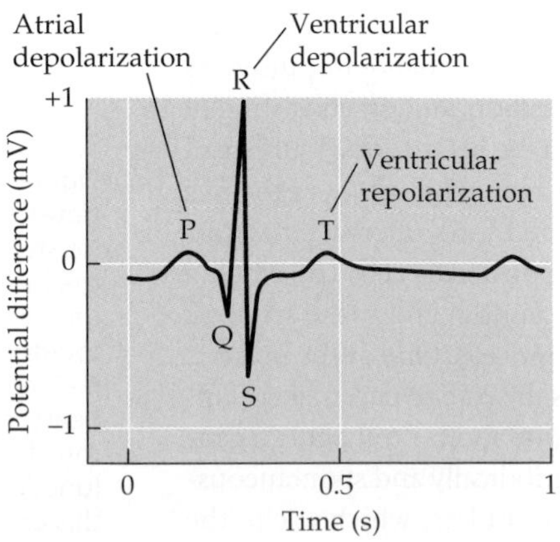

Figure 23.6 Electrocardiography (*a*) Relative electrical charges in the extracellular fluids of the human ventricular myocardium at an instant during passage of a wave of depolarization. The part of the ventricular myocardium lying nearest the ventricular chambers depolarizes first because it is the part supplied immediately by the branches of the conducting system (see Figure 23.4). (*b*) Electrocardiograms of a human and an octopus (*Eledone cirrhosa*) during three heartbeats. The human ECG was obtained using electrodes placed on the skin surface on the right arm and left leg. Electrodes attached to the surface of the ventricle (main body) of the systemic heart were used to record the octopus ECG. (*c*) A human ECG during one heartbeat with waveforms identified. (Octopus ECG after Smith 1981.)

cise, for example. In lobsters, the cardiac ganglion is innervated by both excitatory and inhibitory regulatory neurons; these neurons modulate both the frequency and intensity of the bursts of impulses generated by the ganglion and thus affect the heart rate and the vigor of heart contraction.

Intrinsic controls of heart action are controls that occur without the mediation of hormones or extrinsic neurons. The **Frank-Starling relation** describes one of the important intrinsic controls of the mammalian heart; namely, that stretching of the cardiac muscle tends to increase the vigor of its contraction. This relation plays an important role in enabling the heart to match its output of blood to its input. Consider, for example, what happens when the rate of blood flow into the heart is increased. Because the heart chambers then tend to take in more blood in the time between beats, they become more stretched (distended) between beats. The heart muscle then automatically contracts more vigorously, ejecting more blood—a response that tends to match heart output to the increased blood received by the heart. Lobster hearts function similarly in that they *intrinsically* increase both the vigor and rate of their contraction as they are stretched. The mechanism of this response in lobsters is, at least in part, quite different from that in vertebrates, however, because the cardiac ganglion is involved. Stretch induces the ganglion to fire more frequently and intensely.

Summary: Hearts

- The output of a heart, known as the cardiac output, depends on the heart rate and stroke volume.
- The cells in the heart muscle, the myocardium, must have means of receiving O_2. In some hearts the myocardium is spongy, and blood flowing through the heart chambers flows through the spongy spaces, supplying O_2 to the cells. In other hearts, including those of mammals, the myocardium is compact and is supplied with blood by means of coronary blood vessels.
- A heart is myogenic if the depolarization impulses required for heartbeats originate in muscle cells or modified muscle cells. A heart is neurogenic if the impulses originate in neurons. Vertebrate hearts are myogenic.
- In the mammalian heart, the sinoatrial node in the wall of the right atrium acts as pacemaker, initiating waves of depolarization. Conduction of a wave of depolarization from the atria to the ventricles occurs through the conducting system, which ensures both that the ventricles contract slightly later than the atria and that the whole ventricular myocardium contracts approximately at once.
- When a part of the myocardium is in the process of contracting, voltage differences in the extracellular fluids develop between regions of muscle cells that have already undergone depolarization and regions that have not. These differences can be detected on the body surface. The electrocardiogram is a recording of such differences as a function of time.
- The rate and force of heart contraction are governed by nervous, endocrine, and intrinsic controls.

Principles of Pressure, Resistance, and Flow in Vascular Systems

Having discussed the fundamental features of the hearts of animals, we now need to turn our attention to the perfusion of the vascular system. **Perfusion** refers to the forced flow of blood through blood vessels.

The **blood pressure** produced by the heart—or, in some animals, by other muscular activity—is the principal factor that causes blood to flow through the vascular system. What we mean by *blood pressure* is the amount by which the pressure of the blood exceeds the ambient pressure. Blood pressure is often expressed in *kilopascals* (*kPa*) by physiologists, but usually in *millimeters of mercury* (*mm Hg*) in medicine and related disciplines.[7] When we say that the blood pressure in a vessel in an animal is 10 kPa (= 75 mm Hg), this means that the pressure there is 10 kPa, or 75 mm Hg, higher than the pressure simultaneously present in the animal's surrounding environment.

In arteries, the blood pressure rises and falls over the heart cycle. The highest pressure attained at the time of cardiac contraction is termed the **systolic pressure**, whereas the lowest pressure reached during cardiac relaxation is called the **diastolic pressure.** In young adult people at rest, for example, the systolic pressure in the systemic aorta is usually about 16 kPa or 120 mm Hg, and the aortic diastolic pressure is about 10 kPa or 75 mm Hg. When these pressures are measured for clinical reasons, the results are often expressed as a pseudo-ratio—for example, 120/75 ("120 over 75"). The **mean pressure** in an artery is obtained by averaging the pressure over the entire cardiac cycle; it usually does not equal the average of the two extreme pressures, systolic and diastolic, because the systolic and diastolic phases are not the same in duration. In resting young adults, the mean pressure in the systemic aorta is ordinarily about 12.7 kPa or 95 mm Hg.

Besides the pressures produced dynamically by the beating of the heart, pressures resulting from *fluid-column effects* can also be important in circulatory systems. Any unobstructed vertical column of fluid exerts a pressure—termed a *hydrostatic* pressure—that increases as its height increases (Figure 23.7*a*). Because blood in the vessels of an animal forms fluid columns, fluid-column pressures are present in circulatory systems. The pressure produced by the beating of the heart is added to the fluid-column pressures that are present in arteries to determine the total arterial blood pressure. These relations are relevant for the clinical measurement of blood pressure. The pressure of interest during clinical measurement is that produced by the heart, unconfounded by fluid-column effects. The measurement device must, therefore, be in the same horizontal plane as the heart, or the measurement must be corrected to remove fluid-column effects.

In a vertical column of blood, each 13 cm of height exerts about 10 mm Hg of pressure. At levels of the body below the heart, this hydrostatic fluid-column pressure adds to the pressure contributed by the heart; thus, in a person standing up, the blood pressure in arteries in the legs is far above that in the systemic aorta (Figure 23.7*b*). At levels of the body above the heart, some of the pressure developed by

[7] Appendix B and the footnote on page 536 of Chapter 20 discuss the relations among these units.

(a) The physics of fluid-column effects

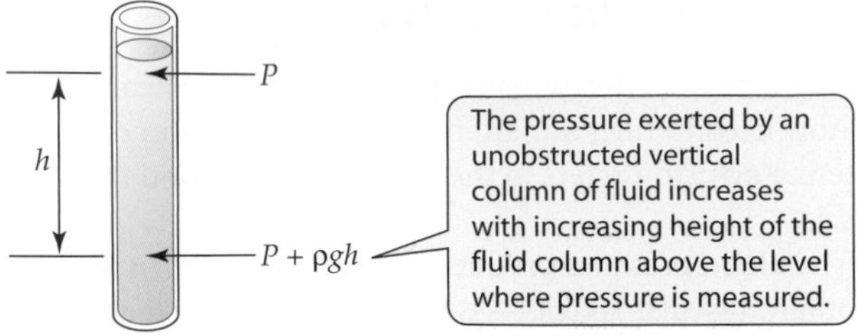

(b) Mean blood pressure in major arteries of a quietly standing person

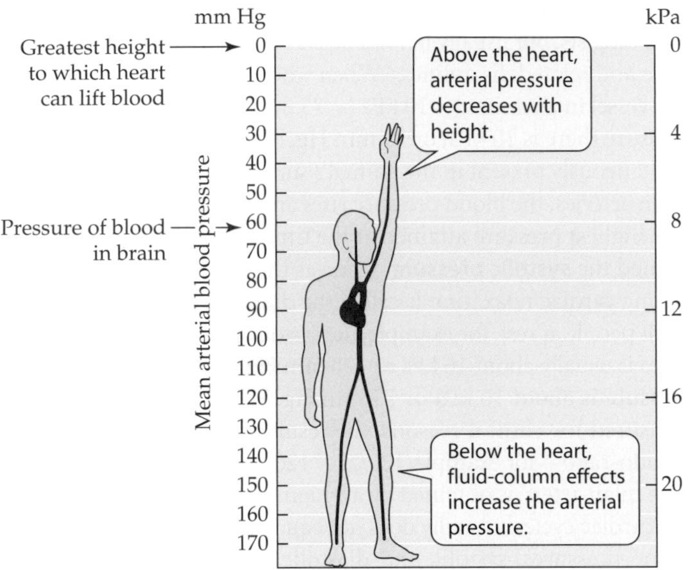

Figure 23.7 Fluid-column effects on blood pressure in the arterial vascular system *(a)* The physics of fluid-column effects. In an unobstructed, vertically positioned tube filled with a nonmoving fluid, if the pressure at one height is P, then the pressure at a height h units lower is $P + \rho gh$, where ρ is the mass density of the fluid and g is acceleration due to gravity. *(b)* Fluid-column effects on arterial blood pressure in a person standing quietly. (*b* after Rushmer 1976.)

the heart is lost in simply supporting the fluid column of blood. Thus the blood pressure in the arteries of the neck and head decreases by approximately 10 mm Hg for every 13 cm of height above the heart.

Considering all the factors that could influence blood flow, how can we predict the *direction* of flow? Throughout most of this chapter, we will make simplifying assumptions so that we will be able to analyze blood flow based just on the pressures produced by the beating of the heart. However, it is important to be aware of a more general concept, the **total fluid energy** of the blood, that allows one to analyze blood flow in any situation, without simplifying assumptions.

Blood can possess three forms of energy that affect its flow, and the *total fluid energy* of the blood is the sum of these three forms, as shown in Figure 23.8. The first form of energy (labeled 1 in Figure 23.8) is the pressure the blood is under because of the beating of the heart; this is a form of potential energy, which means that it can produce motion, but is not motion itself. The second form of energy (2) is the blood's energy of motion—its kinetic energy. The third form (3) is the potential energy the blood possesses because of its position in Earth's gravitational field. Blood always flows from where its total fluid energy is higher to where its total fluid energy is lower.

Why be concerned with this complex concept? Actually, we have already encountered two situations in which the total fluid energy must be taken into account to understand the flow of blood. First, we saw in Figure 23.2 that toward the end of ventricular systole in the human heart, blood briefly continues to flow out of the left ventricle into the systemic aorta even though the blood pressure in the ventricle (factor 1 in the total fluid energy) has fallen below the pressure in the systemic aorta. If flow were governed only by simple pressure differences, blood would not flow out of the ventricle when the pressure in the ventricle is lower than that in the aorta. Blood in fact flows out of the ventricle at the end of ventricular systole because at that time, the blood on the ventricular side of the aortic valve has a greater forward momentum—a greater kinetic energy (factor 2)—than that on the aortic side. Thus the *total fluid energy*—the true driving force for blood flow—is higher in the ventricle than in the aorta, and blood flows out of the ventricle under such circumstances.

The second case we have encountered that cannot be understood without taking account of total fluid energy is the flow of blood into our legs when we are standing up. At any given time, if one uses a simple pressure-measurement device to measure the existing pressure in the systemic aorta of a person standing up and the simultaneous pressure in the arteries in the lower legs, the pressure in the lower legs is perhaps 10 kPa (70–80 mm Hg) *higher*.[8] Thus, if pressure alone governed blood flow, blood would flow from the legs into the aorta, not the other way around. Here again, to understand the flow that actually occurs, one must analyze the total fluid energy. A simple pressure-measurement device confounds and confuses two distinct factors in the total fluid energy: the pressure produced by the beating of the heart (factor 1) and the blood's potential energy of position in Earth's gravitational field (factor 3). These factors become confounded by a simple pressure-measurement device because in parts of the body below the heart, high potential energy of position in the blood near the heart is recorded as high pressure in the blood below the heart. To analyze blood flow accurately, one must go back to the equation for total fluid energy and analyze its three components in an unconfounded way. The pressure produced by the heart (factor 1) is slightly higher in the aorta than in the major leg arteries, and furthermore, the blood in the aorta has a greater potential energy of position (factor 3) than that in the leg arteries because it is at a greater elevation (put loosely, the blood in the aorta tends to "fall" into the legs). Considering

[8] The true value depends on how tall the person is.

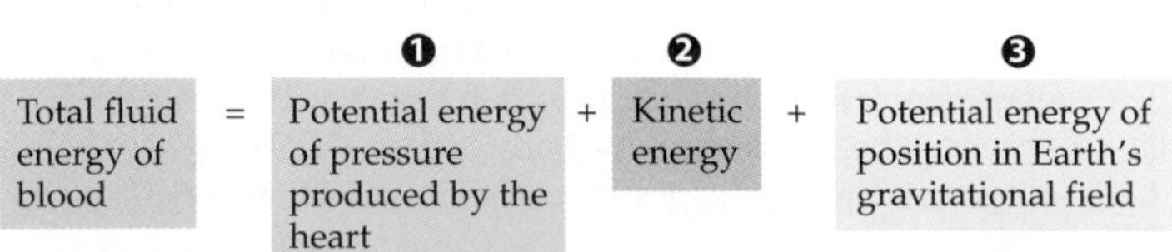

Figure 23.8 Total fluid energy: The true driving force for blood flow Kinetic energy is energy of motion. Potential energy is not motion but can produce motion.

all three factors, the *total fluid energy* is higher in the aorta, and blood therefore flows into the legs.

For many purposes, the analysis of blood flow can be usefully simplified by making two assumptions that in fact are often reasonably realistic. One of these assumptions is that the kinetic energy of the blood (factor 2) does not vary from place to place within the system being analyzed. To illustrate the application of this assumption, consider the analysis of blood flow in human arteries. Kinetic energy accounts for a very small fraction of the total fluid energy in the arteries (only 1%–3% in the systemic aorta); thus, when analyzing blood flow in the arteries, little accuracy is lost by disregarding differences in kinetic energy from place to place. The second simplifying step that can be taken is to assume that the animal under study is in a horizontal posture. For a person or other animal lying on a horizontal surface, all blood is roughly in one horizontal plane, and one can reasonably assume that potential energy of position (factor 3) is equal everywhere. The horizontal posture is in fact the one usually used for experiments or analysis, and one can demonstrate that its use usually does not detract from the generality of conclusions reached.

With these simplifying assumptions made, blood flow can be analyzed using the pressure developed by the heart as the sole driving force.[9] This is the approach we will use except in special cases in which the additional complexities of the total fluid energy must be considered.

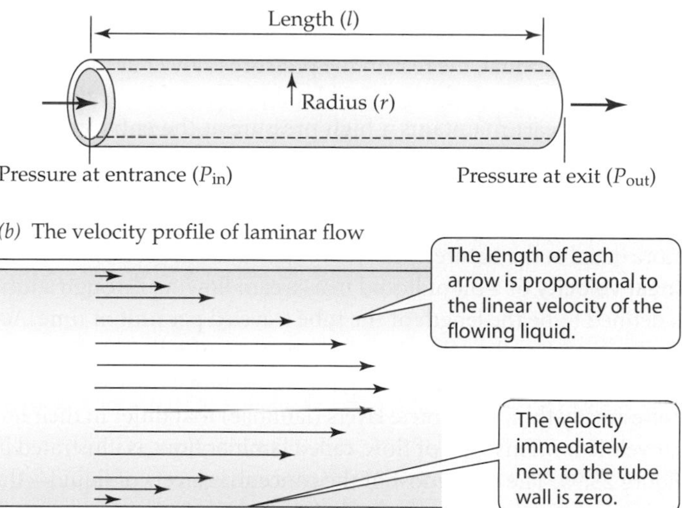

Figure 23.9 The physics of flow through tubes *(a)* Critical factors for understanding the rate of flow using the Poiseuille equation. *(b)* Linear velocity as a function of distance from the tube wall. A microscopically thin layer of liquid touching the tube wall does not move.The velocity profile seen here applies when a simple liquid such as water flows in a laminar fashion through a tube; the velocity profile for blood differs from that for a simple liquid.

The rate of blood flow depends on differences in blood pressure and on vascular resistance

Already in the nineteenth century, physiologists were seeking to understand the perfusion of blood vessels by analyzing the steady, nonturbulent flow of a simple liquid such as water through a horizontal, rigid-walled tube (**Figure 23.9*a***). Their analyses led to insights that are still considered important today. The factors that determine the rate of flow (mL/minute) from one end of a tube to the other, they discovered, are the pressure at the entry to the tube (P_{in}), the pressure at the exit (P_{out}), the radius of the lumen of the tube (r), the tube length (l), and the viscosity of the liquid (η).[10] The formula relating these quantities is named the **Poiseuille equation** or **Hagen-Poiseuille equation**, after Jean Poiseuille (1797–1869) and Gotthilf Hagen (1797–1884), who derived it:

$$\text{Flow rate} = (P_{in} - P_{out})\left(\frac{\pi}{8}\right)\left(\frac{1}{\eta}\right)\left(\frac{r^4}{l}\right) \qquad (23.2)$$

According to the Poiseuille equation, increasing the difference in pressure between the ends of a tube increases the rate of flow through the tube. Raising the viscosity of the fluid diminishes the rate of flow. The final term in the equation is an important geometric term: It shows that the rate of flow through a tube is a direct function of *the fourth power* of the radius of the lumen. Because of this relation, the rate of flow is extraordinarily sensitive to changes in the luminal radius. If water is flowing through a tube at a certain rate and the radius of the lumen is reduced to half of the starting radius, the rate of flow falls to 1/16 of the original rate!

The Poiseuille equation, as mentioned, was derived to apply to simple liquids like water flowing through unbranched, rigid-walled tubes. Blood is not a simple liquid because it contains suspended cells, and blood vessels are not unbranched or rigid-walled. Nonetheless, the Poiseuille equation often proves to be a useful approximate model for understanding the flow of blood through animal blood vessels. From the Poiseuille equation, we see that when muscles in the walls of a blood vessel change the radius of the vessel by constricting it or allowing it to dilate, the rate of flow through the vessel is profoundly affected.

Another equation that is useful for understanding the rate of steady blood flow through a horizontal system of blood vessels is

$$\text{Flow rate} = \frac{\Delta P}{R} \qquad (23.3)$$

where ΔP is the difference in blood pressure between the entry vessels of the vascular system and the exit vessels, and R is the **resistance to flow** or **vascular resistance**. This equation simply says that the rate of flow increases when the difference in pressure increases, but the rate of flow decreases when the resistance to flow increases. If we are considering a simple tubular vessel, rather than a whole system of blood vessels, an easy relation exists between Equation 23.3 and the Poiseuille equation. Because ΔP and $(P_{in} - P_{out})$ represent the same quantity in this case, the resistance R is equal to $8\eta l/\pi r^4$. From the latter expression, one can see that *resistance is inversely proportional to the fourth power of the vessel radius*. Halving the radius of a vessel increases the resistance to flow through the vessel by a factor of 16!

[9] Under these simplifying assumptions, the total fluid energy is equal to the pressure developed by the heart—factor 1 in Figure 23.8—because factors 2 and 3 do not apply.

[10] We will describe viscosity shortly. For the moment, it refers to how syruplike a liquid is.

The dissipation of energy: Pressure and flow turn to heat during circulation of the blood

When blood flows through a horizontal blood vessel or system of vessels in an animal, the reason the pressure is lower at the exit end than at the entry end is that the blood pressure falls as the blood flows. The heart maintains a high pressure at the entry end. This pressure drives the blood through the vessel or system of vessels, but as the blood passes through, the pressure becomes diminished. Why?

To answer this question accurately, we need to look briefly in more detail at the nature of flow through a tube or blood vessel. The **linear velocity** of a bit of liquid in a stream flowing through a tube is defined to be the length of the tube traveled per unit of time. According to ideal flow theory, when the flow of a liquid through a tube is steady and nonturbulent, the liquid moves in a series of infinitesimally thin, concentric layers (laminae) that differ in their linear velocities. This type of flow, called **laminar flow**, is illustrated in Figure 23.9*b*. The outermost of the concentric layers of liquid—the layer immediately next to the wall of the tube—does not move at all. Layers closer and closer to the center move faster and faster.

A crucial aspect of this sort of flow is that the adjacent layers of the liquid that are moving at different linear velocities do not slip effortlessly past each other. Instead, there is a sort of friction that must be overcome to make them move relative to each other. The total magnitude of this *internal friction* in a moving liquid depends in part on the dimensions of the tube. In addition, however, it depends on a property intrinsic to the particular liquid. This property is the **viscosity** (specifically, *dynamic viscosity*) of the liquid. Viscosity refers to *a lack of intrinsic slipperiness* between liquid layers moving at different linear velocities; liquids that are particularly low in internal slipperiness have high viscosities and exhibit syruplike properties. The internal friction within a moving liquid is very real and, like other friction, results in the degradation of kinetic energy (energy of motion) into heat. Thus, as a liquid flows through a tube, some of its kinetic energy is steadily degraded.[11]

Understanding that energy of motion is degraded to heat, we can now see the broad outlines of the energetics of blood flow through a horizontal system of blood vessels. The pressure provided at the entry end by the heart is a form of potential energy. Some of this potential energy is converted to kinetic energy: energy of motion of the blood. Then, along each millimeter of the tubular system through which the blood flows, some of the kinetic energy is lost as heat in overcoming internal friction. *Ultimately, therefore, pressure is converted to heat.* During horizontal flow, the drop in pressure from one point in a tubular system to another point downstream is in fact a good index of the heat produced in overcoming opposing viscous forces. *Thus the drop in blood pressure from place to place can be used as a measure of the energy cost of blood flow.*

To see the application of these concepts, consider blood flow through our arteries. The mean blood pressure resulting from the pumping of the heart changes by only a small amount from one end of an artery to the other; for example, the pressure in the radial artery in our wrist—0.7 m from our heart—is only about 0.5 kPa (3 mm Hg) lower than that in the systemic aorta. This small drop in pressure signifies that the cost of driving blood through the arteries is low. Shortly we will see that in some other types of vessels, the blood pressure drops steeply as blood flows through. This steep pressure drop signifies that such vessels are very costly to perfuse.

[11] See Chapter 5 for a discussion of energy and energy degradation. Sometimes, people say that pumping the blood through vessels involves overcoming friction between the blood and the vessel walls. This is not correct. In fact, there is no frictional force to be overcome between a laminarly flowing liquid and the walls of a tube because no relative motion occurs between the walls and the liquid layer next to them (that layer is still). The frictional resistance to the flow of a liquid is *entirely internal* to the liquid.

Summary: Principles of Pressure, Resistance, and Flow in Vascular Systems

- Blood pressure is measured relative to environmental pressure; it is the extent to which the pressure in the blood exceeds that in the environment.
- During steady flow of blood through horizontal vessels or systems of vessels, the rate of blood flow is directly proportional to the difference in pressure between the inlet and outlet. It is also inversely proportional to vascular resistance. According to the Poiseuille equation, vascular resistance varies inversely with the fourth power of vessel radius.
- Blood pressure declines during the flow of blood through vessels because the potential energy represented by the pressure is converted to kinetic energy, which then is converted to heat in overcoming viscous resistance to flow. During steady flow through a horizontal system, this drop in blood pressure is a measure of the energy cost of perfusion.

Circulation in Mammals and Birds

The basic circulatory plan of mammals and birds is illustrated in Figure 23.10*a*. Oxygen-depleted blood returning from the systemic tissues enters the right heart via the great veins and is pumped by the right ventricle to the lungs, where O_2 is taken up and CO_2 is released. The oxygenated blood from the lungs travels to the left heart and is then pumped by the left ventricle to the systemic aorta, which divides to supply all the systemic tissues. A key feature of the circulatory plan is that it places the lungs in *series* with the systemic tissues. This serial arrangement, emphasized in Figure 23.10*b*, maximizes the efficiency of O_2 delivery to the systemic tissues: All the blood pumped to the systemic tissues by the heart is freshly oxygenated, and the tissues receive blood at the full level of oxygenation achieved in the lungs. These features are important ways in which the circulatory system of mammals and birds is able to meet their high O_2 demands.

The circulatory system is closed

Circulatory systems are classified as **open** or **closed**, depending on whether the entire circulatory path is enclosed in discrete vessels. In an *open* system, the blood leaves discrete vessels and bathes at least some nonvascular tissues directly. In a *closed* system, there is always at least a thin vessel wall separating the blood from the other tissues. The distinction between open and closed systems is relative because there are many intergradations. Mammals and birds, as well as other vertebrates, have essentially closed circulatory systems.

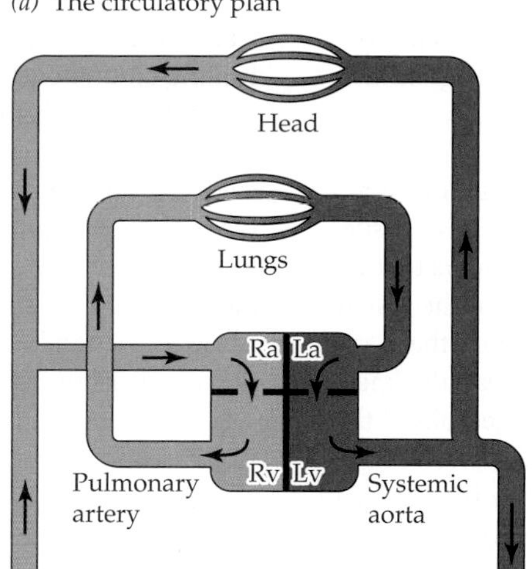

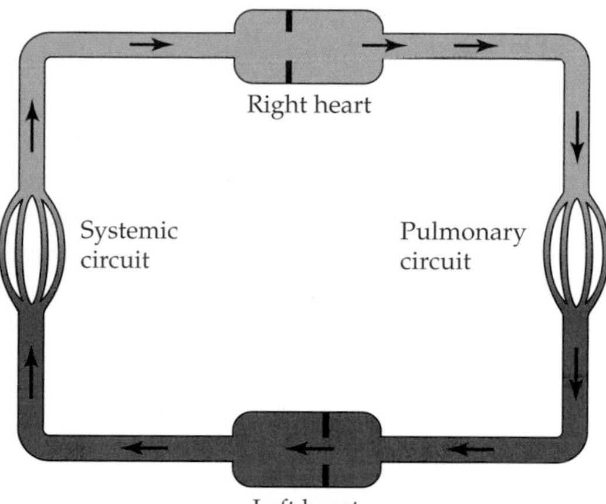

Figure 23.10 The circulatory plan in mammals and birds *(a)* The circulatory plan as it exists geometrically in the body. *(b)* The same plan, redrawn as a schematic to emphasize the arrangement of the pulmonary and systemic circuits in series with each other. Red and blue portions carry relatively oxygenated and deoxygenated blood, respectively. Ra, right atrium of the heart; La, left atrium; Rv, right ventricle; Lv, left ventricle.

Each part of the systemic vascular system has distinctive anatomical and functional features

The blood vessels at various points in the systemic vascular system differ anatomically and functionally in important ways. In this section, we discuss the major types of vessels in the order in which blood passes through them. A significant point to mention at the start is that in vertebrates, *all* types of vessels—and the chambers of the heart—are lined on the inside with a simple epithelium (i.e., a single layer of epithelial cells). For historical reasons, this epithelium that lines the vascular system is called the **vascular endothelium**. The cells of the endothelium are the cells that actually make contact with the blood. They perform many functions, which are only gradually being understood. Some endothelial cells secrete agents into the blood—such as nitric oxide, prostacyclin, and prostaglandin I_2—that affect the contraction and relaxation of vascular smooth muscle or help control clotting. Endothelial cells also sometimes synthesize active hormones from hormone precursors in the blood. Modern endocrinologists believe that the vascular endothelium may in fact be the largest endocrine structure in a vertebrate's body! In addition, endothelial cells sometimes terminate hormone action (e.g., by degrading hormones), and they participate in immune responses.

ARTERIES The great arteries have thick walls that are heavily invested with smooth muscle and with elastic and collagenous connective tissue. They thus are equipped to convey blood under considerable pressure from the heart to the peripheral parts of the circulatory system. The elasticity of the great arteries enables them to perform two important hydrodynamic functions: (1) the damping of pressure oscillations and (2) the maintenance of a pressure reservoir between systoles. If the heart were to discharge blood into rigid, inelastic tubes, the blood pressure in the great arteries would oscillate violently upward and downward with each contraction and relaxation of the heart. Instead, the arteries are elastic, and they stretch when they receive blood discharged from the heart. Some of the energy of each heart contraction is thus stored as elastic potential energy in the artery walls, and consequently the increase in arterial pressure during systole is limited to some extent. The energy stored elastically at the time of systole is released as the arteries rebound to their unstretched dimensions during diastole. In this way, some of the energy of heart contraction is used to maintain the pressure in the great arteries between contractions. The end result is that variations in arterial pressure over the cardiac cycle are reduced (the *pressure-damping effect*), and a substantial pressure is maintained in the arteries even when the heart is at rest between beats (the *pressure-reservoir effect*).

The arteries become smaller as they branch outward toward the periphery of the circulatory system. The walls of the arteries simultaneously become thinner, a fact that at first appears paradoxical when we recall that the mean blood pressure diminishes hardly at all in the arteries. The paradox is resolved in good measure by a principle identified by the great mathematician Pierre Simon, the Marquis de Laplace (1749–1827), and now known as **Laplace's law**.[12] This law deals with the relation between luminal pressure and wall tension in hollow structures. As applied to tubes, it says that when the pressure in the lumen of a tube exceeds that outside the tube by any given amount, the circumferential tension developed within the walls of the tube is directly proportional to the radius of the tube: $T = r\Delta P$, where T is wall tension, r is the luminal radius, and ΔP here represents the pressure difference across the walls. Because of this relation, even though a small artery may be exposed to the same blood pressure as a large one, the tension developed within its walls is lower than that developed within the walls of the large artery. Accordingly, the walls of small arteries need not be as well fortified to resist overexpansion as those of large arteries. The same principle explains why blood capillaries can be exceedingly thin-walled and yet resist substantial pressures.

[12] Laplace's law applies quantitatively only to simple elastic materials like rubber. It is a useful principle for analyzing complex elastic structures like arterial walls, but the detailed study of arteries involves additional considerations.

MICROCIRCULATORY BEDS The systemic arteries ultimately deliver blood to networks of microscopically tiny blood vessels in all the systemic organs and tissues. These **microcirculatory beds**, diagrammed in Figure 23.11, consist of three types of vessels: *arterioles*, *capillaries*, and *venules*.

Arterioles, although minute, have the same basic structure as arteries. Their walls consist of smooth muscle and connective tissue. The walls, in fact, are so well-endowed with these elements that on average, they are quite thick relative to the dimensions of the vessels themselves. Among the arterioles of humans, for example, the mean diameter of the lumen is about 30 μm, whereas the mean thickness of the walls is about 20 μm. The smooth muscles in the walls of the arterioles play a pivotal role in the *vasomotor control of blood distribution*. The adjective **vasomotor** refers to changes in the luminal radius of blood vessels. A decrease in the luminal radius is called **vasoconstriction**, whereas an increase is called **vasodilation**. In our earlier discussion of the Poiseuille equation (Equation 23.2), we noted that the rate of flow through a tubular vessel is extremely sensitive to the radius of the tubular lumen. By contracting and relaxing, the smooth muscles in the wall of an arteriole control the luminal radius of the vessel and thereby profoundly affect the rate of blood flow into the capillary beds that the arteriole supplies. Control of the arteriolar muscles is mediated by the sympathetic division of the autonomic nervous system, by circulating hormones, and by incompletely understood chemical and physical effects at the local tissue level. In addition to the arterioles, both precapillary sphincters (see Figure 23.11) and small terminal arteries also participate in controlling blood flow to capillary beds.

One familiar example of the control of tissue perfusion by arterioles is provided by the responses of the skin blood flow to warm and cold environments. As we discussed in Chapter 8 (see page 220), people and other mammals often maintain relatively vigorous blood flow to the surface of the skin surface in warm environments, but reduce blood flow to the surface in cold environments. Flow is reduced by vasoconstriction in arterioles that control the flow of blood through the superficial microcirculatory beds of the skin. This vasoconstriction is brought about in part by the sympathetic nervous system; direct (local) vasoconstrictive effects are also exerted by skin cooling. Modulation of arteriolar–venular anastomoses (see Figure 23.11) also plays a role in this process. Changes in the perfusion of skeletal muscles during exercise provide another important example of arteriolar control: Blood flow through a skeletal muscle can be increased by a factor of ten or more in a person during exercise. Part of the reason for the increased blood flow is arteriolar vasodilation, which appears to be mediated principally by local effects of metabolites produced in exercising muscles. A final familiar example of the control of tissue perfusion by arterioles and small arteries is the erection of the penis or clitoris. When a man is not sexually aroused, the arterioles and small terminal arteries that control blood flow to the spongy erectile regions of the penis are vasoconstricted. During sexual arousal, parasympathetic neurons and cells in the local vascular endothelium release nitric oxide, which acts as a potent signal (see Figure 2.26) for dilation of the arterioles and small arteries, so the penis inflates with blood and stiffens under blood pressure. Erection of the clitoris is believed to occur by similar mechanisms.

The heart produces pressure that is transmitted to all the microcirculatory beds in the body by way of the arteries. This driving force is always available at the entry to each microcirculatory bed, ensuring that vasodilation or vasoconstriction of arterioles will cause immediate changes in tissue perfusion. Each microcirculatory bed has its own arterioles—which determine the rate of flow into the bed—and thus is readily controlled independently of other microcirculatory beds. These important features of the circulatory system in mammals and birds permit highly sensitive temporal and spatial control of blood distribution. These features depend on the fact that the entire circulatory system is enclosed in vessels and represent major advantages of the closed type of circulatory system.

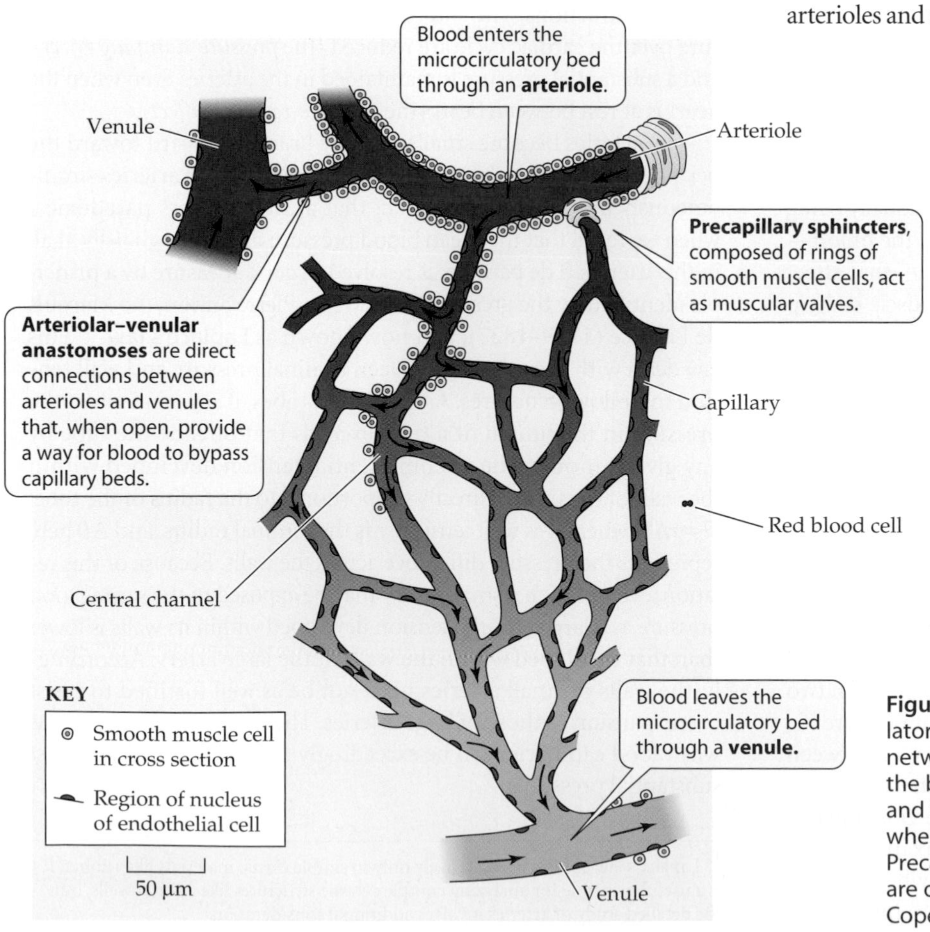

Figure 23.11 A microcirculatory bed In this microcirculatory bed of a mammal, capillaries form an anastomosing network between an arteriole at the top and a venule at the bottom. The endothelial cells that line the arterioles and form the walls of the capillaries are thin and flat, except where each cell is thickened in the region of its nucleus. Precapillary sphincters and arteriolar–venular anastomoses are opened and closed by smooth muscle cells. (After Copenhaver, Kelly, and Wood 1978.)

From the arterioles, the blood typically enters the **capillaries**. The walls of capillaries consist of *only the vascular endothelium*—a single layer of highly flattened cells resting on an outer basement membrane (see Figure 2.6*c*). Because the walls of capillaries are very thin (< 1 μm) and fenestrated (porous) to some extent, the capillaries are the *preeminent sites of exchange of O_2 and other materials between the blood and the tissues.*

Capillary beds (see Figure 23.11) consist of many capillaries that branch and anastomose among the cells of the tissues. The lumens of capillaries are often barely wide enough to allow red blood cells to pass through in single file. The density of capillaries is different in different tissues, being particularly high in skeletal muscles, the myocardium, and the brain. In tissues rich in capillaries, the exchange surface provided by the capillary beds is nothing short of amazing. A cubic centimeter of skeletal muscle or cardiac muscle may well contain 10–20 meters of capillaries! In cross sections of mammalian gastrocnemius muscles, one can see from 300 to over 600 capillaries per *square millimeter*. Within the capillary beds of skeletal muscles, only some of the capillaries are open at rest, but all may open during exercise. It is also possible for new capillaries to develop, or for old ones to disappear, during acclimation to changed environmental conditions, aging, or disease. Modern theories hold that a pathological loss of capillaries may be an invisible but major contributing factor to several chronic debilitating diseases. The capillary beds drain into **venules**, which are small vessels with thin walls (2–5 μm in humans) containing connective tissue and muscle cells.

VEINS The blood flows from the venules back to the heart through a series of veins of increasing diameter. As we will see, the blood pressure has declined precipitously by the time the blood leaves the microcirculatory beds. Thus the walls of the veins need not be capable of resisting high tensions, and in fact are quite thin compared with those of the arteries. To a far greater extent than the arterial system, the venous system contains passive one-way valves. These valves are sheets of tissue shaped so that they permit flow toward the heart, but pinch shut if flow reverses. The valves help ensure that blood will consistently move toward the heart in the veins even though the driving pressures are low and variable. The pressure developed by the heart is not entirely gone in the veins and aids the return of blood to the heart. Another important force promoting the flow of blood in the veins is the contraction of *skeletal* muscles; when the muscles contract, they squeeze the blood in nearby veins, and because the valves in the veins permit only one-way flow, the squeezing moves blood toward the heart.

Mammals and birds have a high-pressure systemic circuit

The blood pressures maintained in the great systemic arteries of mammals and birds are the highest blood pressures found in any animals (Table 23.1). The need for these high pressures arises from the particular flow requirements of mammals and birds and from their systemic vascular resistance: a classic application of Equation 23.3. Mammals and birds require high rates of blood flow because of their high O_2-transport demands. The resistance to flow through their systemic circuits is high, however. To maintain high rates of flow in the face of this high resistance, the mammals and birds require the exceptionally high pressures observed in their systemic arteries. Why is the resistance to flow so high in the systemic circuit? The Poiseuille equation tells us that vessels of small radius tend to make particularly great contributions to vascular resistance. In the closed circulatory system of a vertebrate, every tissue is densely invested with microscopically minute vessels through which the blood must be forced to flow. Collectively, these minute vessels give the systemic circuit its high resistance.

As we have already seen, the blood pumped by the heart makes its way from the systemic aorta through all the major arteries with little loss in pressure. Flow through the terminal arteries, arterioles, and capillaries is the process that dissipates most of the pressure (the potential energy) produced by the heart because those are the vessels that have very small radii.

As blood enters the terminal arteries and the microcirculatory beds, the total cross-sectional area of the vasculature increases markedly, and simultaneously, the average linear velocity of the blood slows dramatically (Figure 23.12*a*). Although individual capillaries are minute in cross section, they are so numerous that their collective cross-sectional area greatly exceeds that of the systemic aorta. The major reason the blood slows as it moves from the great arteries into the capillaries is precisely that the cross-sectional area increases; much as a woodland stream slows where it widens, the blood slows where it has a greater cross-sectional area to pass through. Capillaries are so short (< 1 mm) that the blood must have a low linear velocity when it passes through them if it is to remain in them long enough to exchange with the tissues.

TABLE 23.1 Systolic and diastolic blood pressures in the arteries leaving the heart and the cardiac outputs of some resting vertebrates[a]

Species	Blood pressure (mm Hg)[b] Systolic	Blood pressure (mm Hg)[b] Diastolic	Cardiac output (mL/kg•min)
Human (young adult male)	120	75	80–90
Bottlenose dolphin	150	121	47–105
Horse	171	103	150
Ground squirrel	139	99	313
Laboratory rat	130	91	209
Turtle	25	10	57
Iguana	48	37	58
Leopard frog	32	21	20–30
Rainbow trout	45	33	18–37
Catfish	40	30	11
Dogfish shark	30	24	25

Source: After Farrell 1991.

[a] Where both systemic and pulmonary arteries leave the heart, the data are for systemic arteries.

[b] To obtain blood pressures in units of kilopascals, divide the values given by 7.5.

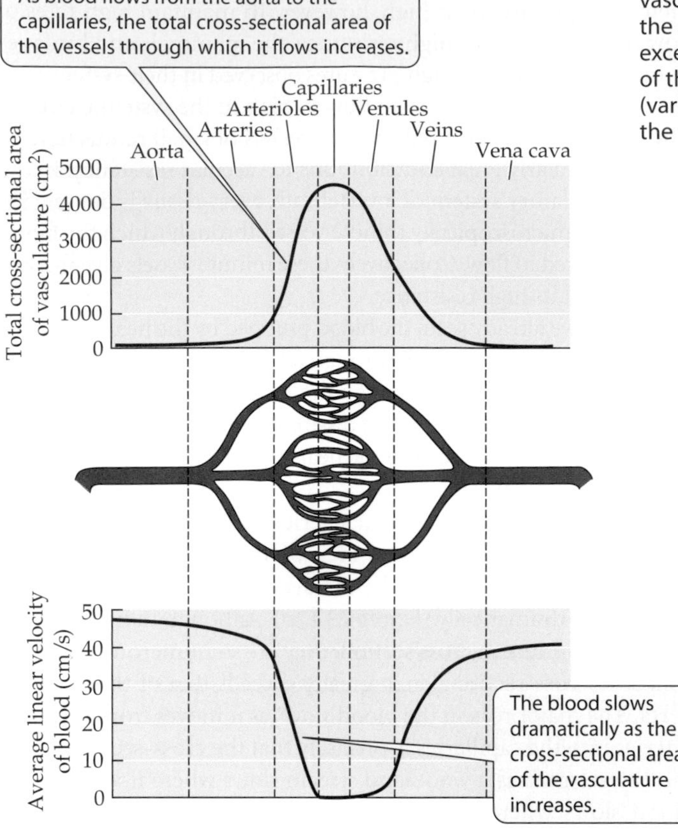

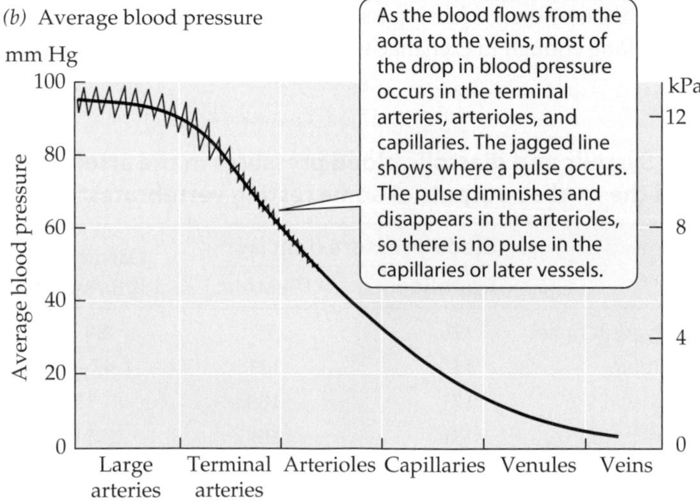

Figure 23.12 Blood flow in the human systemic vasculature *(a)* Cross-sectional area and linear velocity of blood flow in the various parts of the systemic vasculature. Although the cross-sectional area of individual capillaries is minute, the capillaries are so numerous that their collective cross-sectional area greatly exceeds that of the systemic aorta. *(b)* Mean blood pressure in the various parts of the systemic vasculature. The jagged line does not represent the actual pulse (variation of blood pressure between systole and diastole), but symbolizes where the pulse occurs. (*a* from Feigl 1974.)

Even though the blood slows as it passes through the small vessels, they are where the blood loses pressure (Figure 23.12*b*). In people at rest, the mean blood pressure falls from about 12 kPa to 8 kPa (90 mm Hg to 60 mm Hg) as blood flows through the terminal arteries. Then the mean pressure falls to about 4.4 kPa (33 mm Hg) as the blood flows through the arterioles, and to about 2 kPa (15 mm Hg) as it flows through the capillaries and initial venules. A pressure drop of only about 0.4 kPa (3 mm Hg) is sufficient to move blood at a high linear velocity from the systemic aorta to the wrist in the major arteries of the arm. However, a 15-fold greater pressure drop, about 6 kPa (45 mm Hg), is required to move blood at a low linear velocity over a distance of *a few millimeters* from the beginnings of arterioles to the initial venules in a microcirculatory bed! These values emphasize that a closed circulatory system is energetically costly.

The venous part of the systemic circuit is a low-pressure, low-resistance system. The mean pressure of blood entering the venous vasculature (the postcapillary pressure) in people at rest is only about 1.3–2 kPa (10–15 mm Hg). The pressure developed by the left ventricle is essentially entirely degraded to heat by the time the blood reaches the entry of the great veins into the right atrium.

Fluid undergoes complex patterns of exchange across the walls of systemic capillaries

A capillary has fluid on both sides of its walls: blood plasma on the inside and extracellular tissue fluid on the outside. By *extracellular tissue fluid* we mean the fluid between cells in the tissue through which the capillary passes (interstitial fluid). The mean blood pressure in systemic capillaries typically exceeds the hydrostatic pressure in the surrounding extracellular tissue fluid. This difference in *hydrostatic* pressure favors pressure-driven oozing (mass flow) of fluid out of the blood plasma across the capillary walls: a process termed **ultrafiltration** (see page 722). On the other hand, the blood plasma has a higher *osmotic* pressure than the extracellular tissue fluid. This difference in osmotic pressure favors the osmosis of water into the blood plasma across the capillary walls. The interplay of the processes favoring efflux of fluid from capillaries and those favoring influx has attracted considerable interest, in part because imbalances can give rise to serious disease states. Yet fluid exchanges in the capillaries still are not thoroughly understood.

Studies by E. H. Starling and E. M. Landis in the late nineteenth and early twentieth centuries produced a model of fluid exchange across capillary walls (Figure 23.13) that is still believed to summarize its essentials in many tissues. The osmotic pressure of the blood plasma in most mammals exceeds that of the extracellular tissue fluid by about 3.3 kPa (25 mm Hg) along the entire length of the capillaries. This difference, called the **colloid osmotic pressure** of the blood plasma, arises because the plasma is richer than the tissue fluid in dissolved proteins (e.g., albumins) that do not pass freely through capillary walls (see page 87). At the arterial ends of the capillaries, the hydrostatic pressure of the blood exceeds that of the tissue fluid by more than 4 kPa (30 mm Hg); because the hydrostatic-pressure difference favoring efflux of fluid from the capillaries is thus greater than the osmotic-pressure difference favoring influx, fluid is forced out of the capillaries in net fashion. On the other hand, as the blood flows through the capillaries, its hydrostatic pressure falls. At the ve-

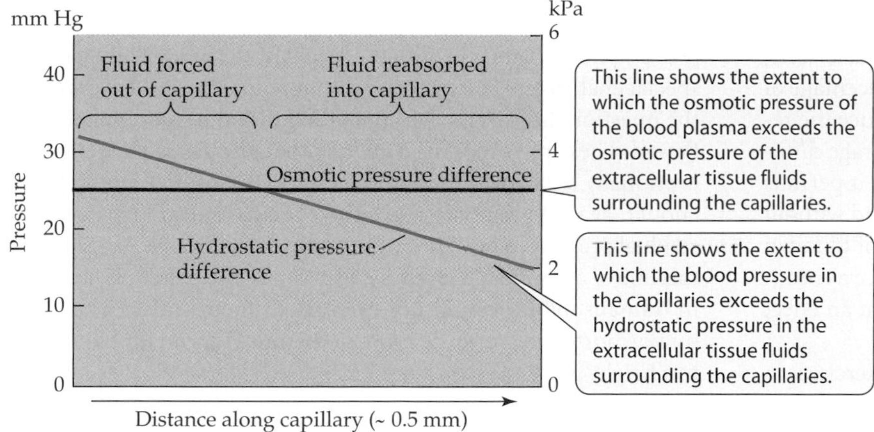

Figure 23.13 Fluid exchange across mammalian systemic capillary walls: The Starling-Landis hypothesis The blood plasma loses fluid in the early parts of capillaries, but regains much of it in the later parts. The osmotic-pressure difference can be represented as a horizontal line because the losses and gains of fluid are not great enough to change the relative osmotic pressures substantially. Values are approximate.

nous ends of the capillaries, the blood pressure exceeds the hydrostatic pressure of the tissue fluid by only about 2 kPa (15 mm Hg). Thus, at the venous ends, the osmotic-pressure difference favoring influx of fluid into the capillaries exceeds the hydrostatic-pressure difference favoring efflux, and a net influx of fluid into the capillaries occurs.

The **Starling-Landis hypothesis**, stated briefly, is that the blood plasma loses volume in the initial segments of systemic blood capillaries, but regains volume in the final segments. *The overall effect is often a net loss*: Fluid from the blood plasma tends to be transferred to the extracellular tissue fluid in net fashion. The fluid added to the tissue fluid is picked up by the *lymphatic system*, which ultimately returns it to the blood.

The pulmonary circuit is a comparatively low-pressure system

Discussion of the pulmonary circulation has been put off to this point because an understanding of the forces affecting fluid movement across capillary walls is essential to grasping a significant problem that could arise in perfusing the lungs. Efficient gas exchange in the lungs requires that the pulmonary capillaries be closely associated with the air spaces in the lungs: the alveoli of mammals or the air capillaries of birds. Rapid net loss of fluid from the blood plasma across the walls of the pulmonary capillaries would flood these air spaces with liquid and impair O_2 and CO_2 exchange.[13] Alveolar flooding—called *pulmonary edema*—is a medical emergency when it occurs because of disease. Flooding of the alveoli or air capillaries does not normally occur, however, because the pulmonary circulation is a comparatively low-resistance, low-pressure system.

The pulmonary circulatory path is much shorter than the systemic path and lacks arterioles. For these and other reasons, the vascular resistance of the pulmonary circuit is far lower than that of the systemic circuit. The volume of blood passing through the pulmonary circuit must match that passing through the systemic circuit because the two are connected in series (see Figure 23.10). However, because the resistance of the pulmonary circuit is low, the arterial pressure in the pulmonary circuit can be far lower than that in the systemic circuit (see Equation 23.3). In people at rest, the mean pressure maintained in the pulmonary arteries by the right ventricle is about 1.9 kPa (14 mm Hg), in contrast to the much higher mean pressure, about 12.7 kPa (95 mm Hg), maintained in the systemic aorta by the left ventricle. The mean blood pressure in the pulmonary capillaries is about 1 kPa (8 mm Hg), in contrast to about 3.3 kPa (25 mm Hg) in systemic capillaries. The low blood pressure in the pulmonary capillaries largely precludes any net loss of liquid from the plasma across the capillary walls by ultrafiltration. Thus the air spaces of the lungs remain air-filled, not liquid-filled.

During exercise, blood flow is increased by orchestrated changes in cardiac output and vascular resistance

The circulatory system of mammals and birds operates at a relatively leisurely pace when individuals are at rest. Exercise heightens demands on the circulatory system and thus brings out its full capabilities. A useful equation for understanding O_2 delivery in the systemic circuit is

$$\text{Rate of } O_2 \text{ delivery} = \text{cardiac output} \times (\text{arterial } O_2 \text{ concentration} - \text{venous } O_2 \text{ concentration}) \quad (23.4)$$

During exercise, if we assume that arterial blood is always fully oxygenated, the rate of O_2 delivery can be increased by increasing the rate of blood flow (i.e., increasing cardiac output) or by extracting more O_2 from each unit volume of blood that circulates (i.e., decreasing venous O_2 concentration). Both strategies of increasing O_2 delivery are simultaneously employed in mammals and birds. In people exercising vigorously, for example, we saw in Chapter 22 that venous blood becomes much more deoxygenated than in people at rest (see Figure 22.5). Simultaneously, average young adults can increase cardiac output by at least a factor of 4 over the resting level by increasing both heart rate and stroke volume. Athletes who are successful in endurance events tend to have relatively high maximal cardiac outputs; some can sustain cardiac outputs that are six to seven times their resting values.

A substantial decline in the total resistance of the systemic vasculature occurs during exercise. Were it not for this drop in systemic resistance, the aortic blood pressure would have to increase dramatically to drive blood through the vasculature at rates that are several times faster than at rest (see Equation 23.3). In fact, during whole-body exercise, the mean blood pressure in the human systemic aorta increases by only about 2.7 kPa (20 mm Hg), even when the cardiac output is far above its resting level.

Vasodilation in the vascular beds of the active muscles (including the breathing muscles) is responsible for much of the decrease in systemic vascular resistance that occurs during exercise. One of the

[13] The rate of diffusion of O_2 and CO_2 is dramatically lower through water than through air, as emphasized in Chapter 20 (see page 531).

consequences of vasodilation in the terminal arteries and arterioles, as noted earlier, is that there is a great increase in the percentage of muscle capillaries that are open. This increase helps reduce the resistance to through-flow of blood. It also reduces the average diffusion distance between capillaries and muscle fibers, and it permits the capillary beds to carry a greatly increased flow of blood without major changes in the linear velocity (or residence time) of blood in individual capillaries. Endurance training tends to produce a chronic increase in the number of capillaries per muscle fiber, an effect that clearly helps meet high muscle O_2 demand.

The response of the systemic vascular system during exercise is in fact highly coordinated and adaptive in that it preferentially distributes the increased cardiac output to the tissues that require increased perfusion. Although the skeletal muscles of people receive about 20% of the cardiac output at rest, they receive 70% to 80% or more of the much-increased cardiac output during vigorous whole-body exertion. Simultaneously, blood flow to the intestines, kidneys, liver, and inactive muscles may be substantially reduced by vasoconstriction. The brain receives about the same amount of blood regardless of exercise state.[14]

Species have evolved differences in their circulatory physiology

Comparing mammal species of various body sizes, there is a very apparent allometric relation between resting cardiac output and body size: Cardiac output per unit of body weight tends to increase as body size decreases. Thus small mammals meet their relatively high weight-specific demands for O_2 transport in part by maintaining relatively high weight-specific rates of blood flow. In relation to body size, small mammal species typically do not have larger hearts than large species. Instead, as we discussed in Chapter 5 (see Table 5.4), the principal way in which small species maintain their relatively high weight-specific cardiac outputs is by having relatively high heart rates. The mean aortic blood pressure does not vary systematically with body size. Birds tend to have larger hearts, lower heart rates, and higher aortic blood pressures than mammals of equivalent body size.

Many species face unique cardiovascular challenges. Among them, giraffes present one of the most thought-provoking cases. Full-grown giraffes can be over 5 m tall. Because of their long necks, their brains can be 1.6 m above their hearts, a distance over four times greater than in people. Moreover, because of their tall columns of blood, standing giraffes have exceptionally high blood pressures in their legs and feet. A few aspects of giraffe circulatory physiology are agreed upon by all authorities who have studied them. One is that giraffes have an exceptionally well developed left ventricle and maintain unusually high systemic aortic blood pressures. Their mean aortic pressure when they are standing and at rest is about 29 kPa (220 mm Hg), in contrast to about 13 kPa (100 mm Hg) in most other mammals.

The question of how blood is circulated to the brain in a standing giraffe is a topic of contentious debate. Some physiologists argue that the arteries and veins of the head and neck function analogously to a siphon system, in which the height of the brain would pose no special challenges. The majority of physiologists who have studied the question, however, have concluded that the siphon analogy is flawed. These physiologists conclude that giraffes require their exceptionally high aortic blood pressures to be able to lift blood high enough against gravity to reach their brains and still have enough residual pressure at head height to perfuse their brains. According to this line of argument, if giraffes had mean aortic pressures like those in humans, blood would not even reach their brains! The mean human aortic pressure (12.7 kPa or 95 mm Hg) can lift blood only to a height of 1.2 m (see Figure 23.7*b*).

Summary: Circulation in Mammals and Birds

- Mammals and birds, like virtually all other vertebrates, have closed circulatory systems, meaning that the blood always remains within blood vessels lined with vascular endothelium.
- The pulmonary and systemic circuits are connected in series. The left ventricle develops high pressures to force blood through the high-resistance systemic circuit. The right ventricle develops lower pressures to force blood through the low-resistance pulmonary circuit.
- In the systemic circuit, arteries convey blood over relatively long distances with little loss of blood pressure; they perform pressure-damping and pressure-reservoir functions because of their elasticity. Within systemic microcirculatory beds, fine spatial and temporal control of blood flow is exerted by contraction and relaxation of smooth muscles in the walls of arterioles (vasomotor controls). The capillaries are the principal sites of exchange between the blood and systemic tissues because their walls consist of just a single layer of endothelial cells and because they are densely distributed.
- As blood flows through systemic capillaries, blood pressure tends to force fluid to pass outward through the capillary walls by ultrafiltration. The colloid osmotic pressure of the blood plasma tends to cause fluid movement into the blood. The net effect of this interplay is a loss of fluid, which is picked up by the lymphatic system. The lower blood pressures in the pulmonary circuit help to prevent pulmonary flooding (edema).
- During exercise, cardiac output is increased by increases in both heart rate and stroke volume. Arterial blood pressure does not rise excessively because vascular resistance is decreased, mainly by vasodilation in active muscles.

Circulation in Fish

Fish in general resemble the vast majority of other vertebrates in having closed circulatory systems. The circulatory plan of most fish is illustrated in Figure 23.14*a*. Blood is pumped anteriorly by the heart into the **ventral aorta**, which distributes it to the afferent gill vessels.[15] The blood then passes through the blood channels of the gills and is brought by the efferent gill vessels to the **dorsal aorta**, a

[14] Responses of the circulation to high altitude are discussed in Box 22.3.

[15] As mentioned in earlier chapters, *afferent* means "going toward," and *efferent* means "going away."

(a) The circulatory plan

4 Blood is distributed to the systemic tissues by the dorsal aorta.

Dorsal aorta

Head

Ef

Ef

3 Blood is collected into efferent branchial vessels (Ef), which empty into the dorsal aorta.

Gut, liver, kidneys, body wall

Gills

Gills

2 Blood perfuses the gills.

Af

Af

Heart

Great veins

Ventral aorta

5 The great veins return deoxygenated blood from the systemic tissues to the heart.

1 The heart pumps blood anteriorly into the ventral aorta, which gives off afferent branchial vessels (Af) to the gill arches.

(b) A teleost heart

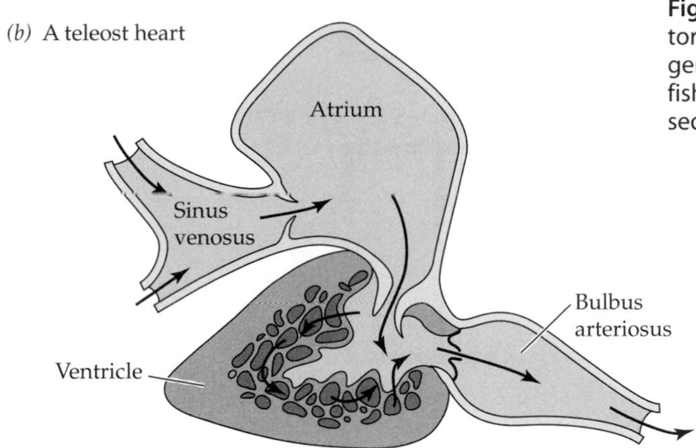

Figure 23.14 The circulatory plan in gill-breathing fish *(a)* The overall circulatory plan of fish. Red and blue portions carry relatively oxygenated and deoxygenated blood, respectively. *(b)* The heart of a teleost fish (trout). In both teleost fish and elasmobranch fish (sharks, skates, and rays), the heart consists of four sequential chambers. (*b* after Randall 1968.)

large dorsal artery that distributes the blood to the systemic tissues. After perfusing the systemic capillaries, the blood returns in the veins to the heart. As in mammals and birds, the circulatory plan places the respiratory organs in series with the systemic tissues, thus assuring efficient O_2 transport.

There are two ways in which the circulatory plan of fish poses potential problems that do not exist in the mammalian and avian plan. First, in fish, there is no heart between the breathing-organ circulation and the systemic circulation to impart fresh energy to oxygenated blood as it leaves for the systemic tissues. When the heart pumps blood into the ventral aorta, the pressure it provides must be sufficient to drive the blood through the resistances of *both* the gill vasculature *and* the systemic vasculature. Second, the primitive condition in fish is for the myocardium to be oxygenated by the blood flowing through the lumen of the heart (see Figure 23.3*b*); this blood is *relatively deoxygenated* because it has just passed through the systemic circuit.

The fish heart (Figure 23.14*b*) consists of four chambers arranged in series: a **sinus venosus** into which the great veins empty, an **atrium**, a **ventricle**, and a bulbous segment that empties into the ventral aorta. The main propulsive force is developed by the ventricle. In elasmobranch fish (sharks, skates, and rays) and lungfish (dipnoans), the bulbous segment—known as the **conus arteriosus**—is endowed with cardiac muscle and contracts in sequence with the ventricle, helping to pump the blood. In teleost fish,[16] the bulbous segment—called the **bulbus arteriosus**—consists of vascular smooth muscle and elastic tissue and does not contract in sequence with the other heart chambers; it seems to act primarily as an elastic chamber that smoothes pressure oscillations and serves as a pressure reservoir between heart contractions (see page 621).

Fish generally have much lower cardiac outputs than mammals of similar body size (see Table 23.1). Their lower cardiac outputs correlate with their lower O_2 demands: Fish have lower metabolic rates than mammals (see Chapter 8) and thus can satisfy their O_2 needs with lower rates of blood flow. Fish also maintain lower arterial pressures: The mean pressure in the ventral aorta is typically between 3 kPa (20 mm Hg) and 12 kPa (90 mm Hg)—usually toward the lower end of this range. The blood pressure drops significantly as the blood passes through the gills. Because of this drop, the mean pressure in the dorsal aorta—the pressure available to perfuse the systemic circuit—is generally only 60% to 80 % of the pressure produced by the heart.

Among fish, there is a very wide and interesting range in the size and performance of the heart, correlated with how physically active the animals are. Species that are relatively inactive and sluggish tend to have relatively small hearts and low cardiac outputs relative to their body sizes, whereas active species tend to have large hearts and high cardiac outputs. Tunas, the supremely active fish we featured at the start of Chapter 21, stand out among all fish in the performance of their circulatory systems, just as they stand out in the performance of their breathing systems. Yellowfin and skipjack tunas have weight-specific cardiac outputs at rest that are twice as high as the highest known in other fish; their ventral aortic blood pressures are also twice as high, accounting for the ability of the blood to flow through the vascular system at a high rate. The tunas are also among the fish with the highest amounts of hemoglobin per unit of blood volume. Thus

[16] Teleost fish are the principal group of fish with bony skeletons.

tunas not only employ the most sophisticated breathing systems found among fish to obtain O_2, but also exploit both major factors in Equation 23.4 to sustain exceptionally high rates of circulatory O_2 delivery to their tissues: They pump blood rapidly, and each unit of blood volume is able to deliver a large amount of O_2.

Exercise in fish is only starting to be well understood. Rainbow trout, often the model species for exercise studies, can increase their cardiac output about threefold as they swim faster by increasing both heart rate and stroke volume. Simultaneously, as we saw in Figure 22.16, they deoxygenate their blood more than they do at rest, thereby increasing the amount of O_2 delivered per unit of blood volume. This enhanced blood deoxygenation potentially has important implications for myocardial O_2 supply. The myocardium obtains much of its O_2 from the venous blood flowing through its lumen (see Figure 23.3*b,c*). As the intensity of exercise increases, the myocardium requires O_2 at a greater rate, yet the partial pressure of O_2 in the luminal blood decreases because of increased unloading of O_2 to the systemic tissues (see Figure 22.16). There is every reason to believe that at some point, the myocardial O_2 supply in many fish places limits on exercise performance, especially in O_2-poor environments.

The circulatory plans of air-breathing organs (ABOs) in fish pose unresolved questions

A few hundred species of fish are adapted to breathing air, as discussed in Chapter 21 (see page 552). In most cases, the air-breathing organs (ABOs) of these species are derived from structures such as the mouth membranes, gut, or swim bladder, that are primitively served by the *systemic* circulation. An important attribute of these structures is that venous blood leaving them typically flows into the systemic venous vasculature—not the systemic arterial vasculature—even when the structures are adapted for air breathing.

Consider, for illustration, the electric eel (*Electrophorus electricus*), which is an obligate air breather. Its gills are so reduced that they no longer contribute to O_2 uptake, and most of its O_2 uptake occurs in its mouth cavity and pharynx. The lining of the mouth cavity is so thoroughly covered with highly vascularized tufts of tissue that it resembles a red cauliflower. The circulatory plan of the electric eel is illustrated in **Figure 23.15*b***. The afferent blood vessels to its air-breathing organ (the mouth cavity) arise from vessels leading toward the gills. The efferent vessels from the air-breathing organ, carrying oxygenated blood, empty into the systemic venous vasculature. Thus, in contrast to the usual fish pattern (**Figure 23.15*a***) and to the pattern in mammals and birds, the circulation of the O_2 source is placed in *parallel* with the circulation of the systemic tissues. The circulatory circuit of the air-breathing organ is typically also in parallel with the systemic circuit in fish that breathe air using a part of the gastrointestinal tract or the swim bladder (**Figure 23.15*c,d***).

The implications of the parallel arrangement are profound. Experiments on electric eels confirm what anatomy suggests: Oxygenated blood from the air-breathing organ freely *mixes* with deoxygenated blood from the systemic tissues in the systemic veins and heart; thus the heart pumps a mixture of deoxygenated and oxygenated blood to both the air-breathing organ and the systemic arteries (see Figure 23.15*b*). The degree of O_2 saturation of hemoglobin in the blood pumped by the heart of the electric eel never

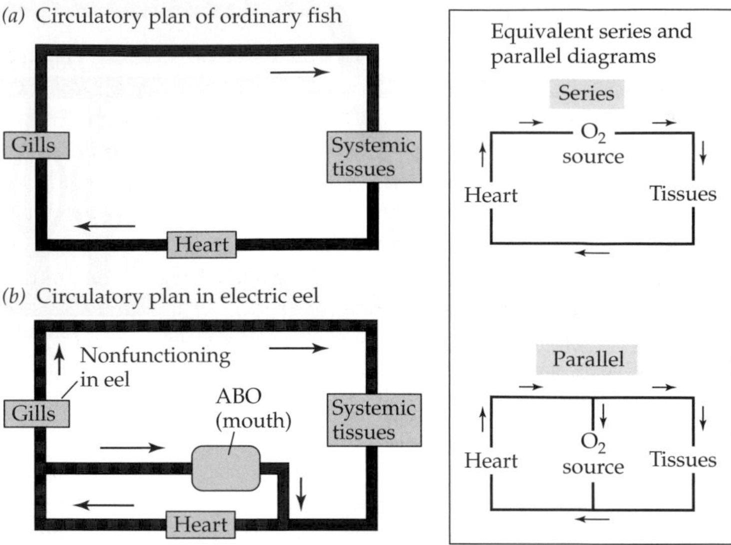

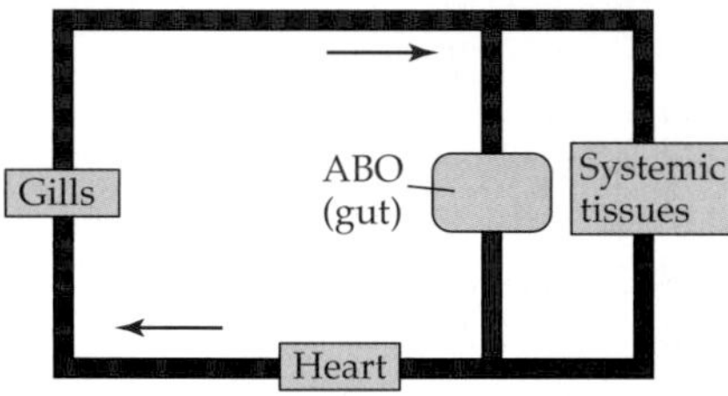

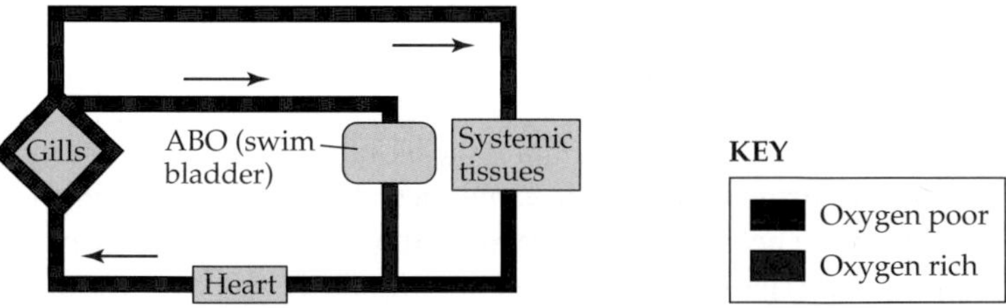

Figure 23.15 The circulatory plans of some air-breathing fish The proportions of red and blue in all of the circulatory paths represent the approximate proportions of oxygenated and deoxygenated blood carried in the vessels. *(a)* The circulatory plan in ordinary, water-breathing fish, shown as a base of comparison. *(b)* The circulatory plan in the electric eel, *Electrophorus electricus*, which has an air-breathing organ (ABO) derived from the mouth cavity; its gills are not effective in taking up O_2. Afferent vessels to the ABO are derived from the afferent gill vessels; efferent vessels from the ABO connect to systemic veins. In some of the other fish that have an ABO derived from the mouth, pharyngeal, or opercular cavities, the circulatory plan is similar to that in the electric eel, but the gills remain partly functional. *(c)* The circulatory plan in certain catfish (e.g., *Hoplosternum*, *Plecostomus*) and other fish in which part of the stomach or intestine serves as an ABO. Afferent vessels to the ABO are derived from the dorsal aorta; efferent vessels enter systemic veins. *(d)* The circulatory plan in the bowfin (*Amia*) and certain other fish that use the swim bladder as an ABO. Afferent vessels to the swim bladder are specialized, arising from efferent gill vessels. Efferent vessels from the swim bladder enter systemic veins. The gill vasculature is represented as two tracts to emphasize that the afferent vessels to the swim bladder arise from efferent vessels of only some gill arches. (After Johansen 1970.)

exceeds 60% to 65%! This is true even though the blood leaving the air-breathing organ itself may be over 90% saturated.

In considering the implications of the circulatory plan in electric eels and other fish with air-breathing organs placed in parallel with the systemic tissues, biologists have traditionally emphasized that the mixing of the oxygenated and deoxygenated bloods *reduces the efficiency of O_2 transport*. Blood that has been deoxygenated in the systemic tissues is in part recycled directly back to the systemic tissues. Moreover, blood oxygenated in the air-breathing organ is in part recycled directly back to the air-breathing organ. According to the traditional interpretation of most biologists, these consequences of the parallel circulatory plan are, quite simply, shortcomings. When these fish evolved air breathing, according to this line of thinking, they would have benefited if the organs they adopted for air breathing evolved a modified venous vasculature—a vasculature that would keep the oxygenated blood separate from deoxygenated blood. Instead, they retained a disadvantageous, primitive venous vasculature that empties oxygenated blood into ordinary veins, where it mixes with deoxygenated blood.

An entirely new perspective has been brought to this matter in the last few years. This perspective emphasizes that the myocardium in fish must get much or all of its O_2 from the blood passing through the heart chambers, as mentioned in the previous section. When a fish obtains its O_2 from the air, mixing the oxygenated blood from its air-breathing organ with its systemic venous blood could be a great advantage because such mixing ensures that the blood flowing through the heart chambers contains O_2 for the heart muscle. Only future research will resolve whether the circulatory plan in these air-breathing fish is a less-than-ideal arrangement based on primitive venous vessels or a helpful adaptation (or a compromise).

Lungfish have specializations to promote separation of oxygenated and deoxygenated blood

The lungfish (dipnoans), which are usually considered the modern fish that most closely resemble the progenitors of terrestrial vertebrates (see page 553), display circulatory plans that differ in significant ways from those of nearly all other air-breathing fish. Here we consider just the African lungfish genus *Protopterus*, the members of which are obligate air breathers. These lungfish (one of which is pictured in Figure 21.12) have evolved a circulatory system that can maintain considerable separation between the blood oxygenated in the lungs and other, deoxygenated blood.

The lungs of *Protopterus* have a venous vasculature that is dramatically different from that of most air-breathing organs in fish: The veins from the lungs lead directly into the left side of the atrium of the heart, rather than connecting with the systemic venous vasculature. This means that blood from the lungs is kept anatomically separate from systemic venous blood until both sorts of blood have entered the atrium. The sinus venosus, which now receives only systemic venous blood, connects to the right side of the atrium.

The heart of *Protopterus* and other lungfish is very different from that of most fish. For one thing, the atrium and ventricle are partly (but not completely) divided into right and left halves by septa. For another, the conus arteriosus—which takes the form of a sharply twisted tube—possesses two longitudinal ridges that project toward each other from opposite sides of its lumen, partially dividing the lumen into two channels. The ventral aorta is virtually or entirely eliminated in lungfish. Thus the four pairs of afferent branchial arteries arise immediately from the anterior end of the conus arteriosus, as do the homologous vessels of amphibians.

As shown in **Figure 23.16**, two of the four pairs of afferent branchial arteries—constituting aortic arches 3 and 4—arise from the ventral channel of the conus arteriosus and travel to anterior gill arches that lack gill lamellae; these arteries do not break up into capillaries, but instead form direct through-connections to the dorsal aorta. The other two pairs of afferent branchial arteries—constituting aortic arches 5 and 6—lead from the dorsal channel of the conus arteriosus to two pairs of posterior gill arches that retain rudimentary lamellae; these arteries break up to supply the gill lamellae, but the lamellar capillaries are of large diameter and can be bypassed to

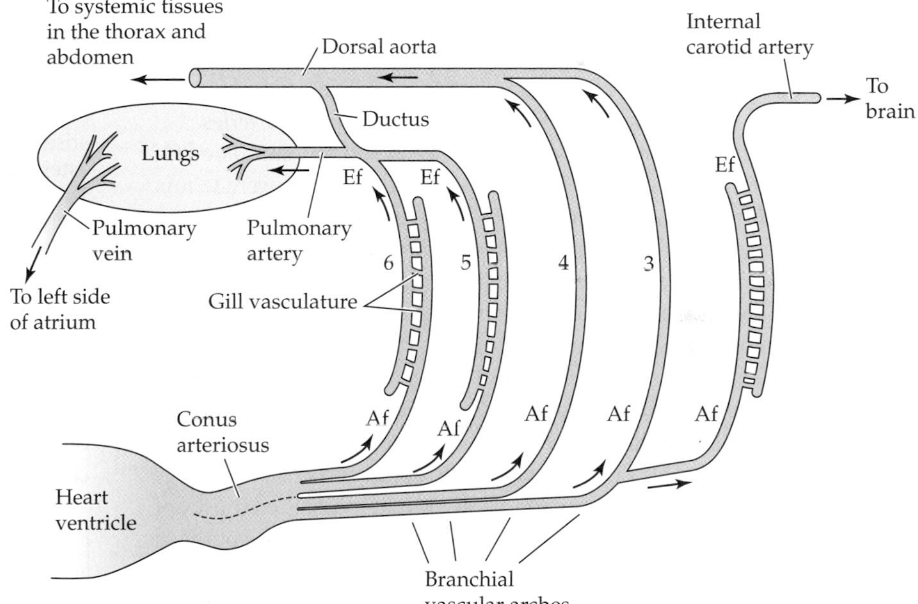

Figure 23.16 The branchial vascular arches of a lungfish and their relation to heart, lungs, and systemic tissues In this diagram of one side of *Protopterus aethiopicus*, the branchial vascular arches are numbered according to the aortic arches they represent (other numbering systems are sometimes used). Branchial arches 3 and 4 go to gill arches that lack gill lamellae; those branchial vascular arches do not break up into capillaries, but instead form direct through-connections to the dorsal aorta. Gill lamellae of a rudimentary sort are found on the gill arches supplied by branchial vascular arches 5 and 6. Within these gill arches, the afferent branchial arteries (Af) break up to supply the gill lamellae, and blood is collected from the lamellae by efferent branchial vessels (Ef). A similar arrangement is found in the most anterior gill arch, as shown. The efferent vessels of arches 5 and 6 form the pulmonary artery. The vessel called the ductus connects the pulmonary artery and dorsal aorta; flow through it is not well understood, although it is generally believed to help blood bypass the lungs during periods when ventilation of the lungs is suspended. Arrows show directions of blood flow. (After Laurent, DeLaney, and Fishman 1978.)

at least some extent by way of vascular shunts, meaning that blood often encounters only a relatively low resistance in flowing from the afferent to the efferent vessels of these gill arches. The efferent vessels give rise to the arteries that carry blood to the lungs—arteries homologous to the pulmonary arteries of tetrapod vertebrates.

To understand blood flow through the central circulation[17] of *Protopterus*, *functional* studies are obviously needed, because the anatomy by itself does not dictate particular patterns of flow. The functional studies that have been done vary in their results, but demonstrate that oxygenated pulmonary venous blood and deoxygenated systemic venous blood in fact follow substantially different paths through the central circulation.

One method of studying blood-flow patterns is to inject fluids that are opaque to X-rays into selected vessels and, using X-rays, monitor where these fluids are carried by the blood.[18] Studies using this technique show that pulmonary venous blood tends to follow a course through the left parts of the atrium and ventricle and is delivered preferentially to the ventral channel of the conus arteriosus , and then (Figure 23.17) the blood is pumped into vascular arches 3 and 4, which carry the blood directly to the dorsal aorta (compare Figure 23.16). In this way, the blood oxygenated in the lungs is directed preferentially into the systemic circuit. Systemic venous blood tends to pass through the right parts of the atrium and ventricle. That blood then appears in X-ray images either to be distributed about evenly to all four pairs of vascular arches (those numbered 3–6) or to be delivered preferentially to the dorsal channel of the conus arteriosus and into vascular arches 5 and 6, which in part supply blood to the pulmonary arteries.

Another way to study blood-flow patterns is to monitor blood O_2 levels. If oxygenated blood from the lungs of a lungfish were to mix thoroughly with systemic venous blood in the heart, then the blood pumped into all the output vessels of the heart would be the same in O_2 partial pressure. On the other hand, if the oxygenated and deoxygenated bloods are distributed selectively, the blood pumped into certain outflow vessels will have a higher O_2 partial pressure than that pumped into others. In one study of *Protopterus*, when the systemic venous O_2 partial pressure averaged 2 mm Hg and the pulmonary venous partial pressure was 46 mm Hg, the blood pumped into vascular arches 3 and 4 for direct passage to the dorsal aorta had an average O_2 partial pressure of 38 mm Hg, showing that it consisted mostly of oxygenated blood from the lungs. Simultaneously, the blood pumped to vascular arches 5 and 6 had a much lower O_2 partial pressure.

The exact mechanisms by which oxygenated and deoxygenated bloods are distributed selectively in the central circulation of *Protopterus* are not fully understood. The functional studies show clearly, however, that these lungfish are able to distribute the two types of blood in ways that promote efficiency of O_2 transport. Another efficiency-promoting capability of lungfish is that they are able to vary the *fraction* of their total cardiac output that they pump to their lungs. Because the fish are intermittent breathers, their lung air varies in its O_2 content. Lungfish increase and decrease the fraction of their cardiac output pumped to the lungs as the O_2 concentration of the lung air rises and falls. Their ability to *redistribute* their cardiac output is potentially advantageous and is made possible by their incompletely divided central circulation (Box 23.1).

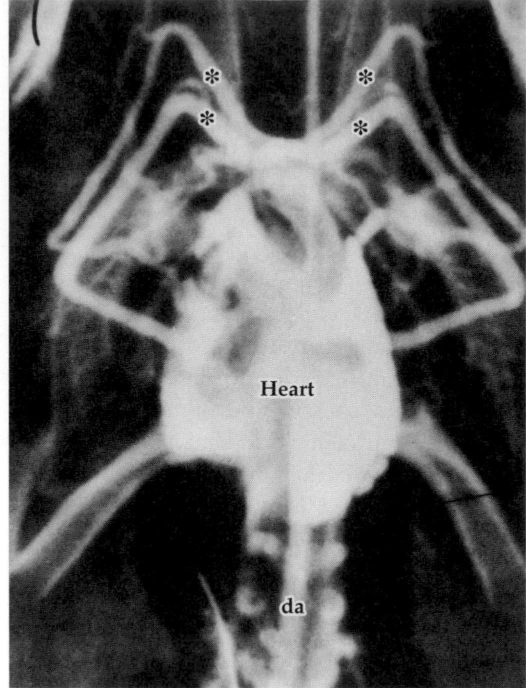

Figure 23.17 Oxygenated blood from the lungs of a lungfish is pumped selectively by the heart into certain branchial vascular arches The fish under study (*Protopterus aethiopicus*) was viewed from below using X-rays. A fluid opaque to X-rays was injected into the pulmonary vein, flowed into the heart, and was being pumped by the heart when this image was made. White-colored parts of the image show the location of the fluid at the moment the image was made. Note that of the four pairs of branchial vascular arches that arise directly from the heart, only two pairs—those labeled with asterisks—received substantial amounts of the X-ray-opaque fluid. The arches receiving the fluid were numbers 3 and 4 (compare Figure 23.16). This image shows that blood from the lungs is pumped by the heart into these arches preferentially and thus flows primarily to the dorsal aorta (da). (Courtesy of Kjell Johansen; from Johansen and Hol 1968.)

Summary: Circulation in Fish

- In most fish, the heart pumps blood to the gills, after which the blood passes through the systemic circuit before returning to the heart. The gill and systemic circuits are thus arranged in series.
- In air-breathing fish, blood leaving the air-breathing organ usually mixes with systemic venous blood. Thus the circulation of the air-breathing organ is in parallel with the systemic circuit. This arrangement decreases the efficiency of O_2 transport but may help oxygenate the myocardium.
- Lungfish have a modified central circulation in which blood from the lungs enters the left side of the atrium and the atrium, ventricle, and conus arteriosus are partly divided. Deoxygenated and oxygenated blood can be kept relatively separate and pumped selectively to the lungs and systemic circuit. Redistribution of cardiac output is possible, and occurs in synchrony with the intermittent breathing cycle.

[17]The term **central circulation** refers to the heart and the veins and arteries that immediately connect to the heart.

[18]The same sort of technique is used to visualize blood flow to organs like the brain and myocardium in medical practice.

BOX 23.1 An Incompletely Divided Central Circulation Can Be an Advantage for Intermittent Breathers

Intermittent breathing is common in lungfish, amphibians, and noncrocodilian reptiles—groups in which the heart is incompletely divided. It is also common in crocodilians, in which the heart is completely divided but the central circulation outside the heart is incompletely divided. An intermittent breather, having ventilated its lungs with air, holds its breath for a substantial time before ventilating again (see page 548). Intermittent breathing presents both distinct opportunities and challenges because the lungs vary from time to time in how effectively they are able to oxygenate the blood. Immediately after an animal has taken a series of breaths, the air in the lungs is rich in O_2, and blood flowing through the lungs can be well oxygenated. However, after a long interval of apnea (cessation of breathing), the air in the lungs may be depleted of O_2, and little opportunity may exist for blood flowing through the lungs to gain O_2.

When the lungs vary over time in their capacity to oxygenate the blood, animals can potentially benefit by varying the rate of blood flow to their lungs. Mammals and birds could not vary their lung blood flow without varying blood flow *throughout their bodies*; because their pulmonary and systemic circuits are connected in series, the rate of blood flow to the lungs must match the rate of flow to the systemic tissues. However, animals with incompletely divided hearts—or with central circulations that are incompletely divided in other ways—can potentially *redistribute* total blood flow along different paths through the central circulation so that the amount of blood pumped to the lungs can be increased or decreased relatively independently of flow to the systemic tissues For an animal with this capability, blood flow to the lungs could be increased immediately after each period of breathing—when the lung air is rich in O_2—and decreased toward the end of each period of apnea—when the lung air is O_2-depleted. In this way, the incompletely divided central circulation could have advantages. To emphasize the nature of these potential advantages, consider that an animal with an incompletely divided heart is, in principle, capable of either raising blood flow to its lungs to a high rate or bypassing its lungs completely while steadily circulating blood round and round in its systemic circuit.

A great deal of research has convincingly demonstrated that most or all air-breathing vertebrates with incompletely divided central circulations do in fact modulate blood flow to their lungs, at least under certain circumstances. The circulatory function of *Protopterus* lungfish, for example, changes markedly in synchrony with their breathing cycle. Immediately after a lungfish has come to the water's surface and ventilated its lungs with air, circulatory adjustments take place that favor rapid uptake of O_2 from the lungs and efficient distribution of the O_2 to the body. One such adjustment is that the rate of blood flow to the lungs is increased to about four times the rate observed just prior to ventilation. This increase is achieved in part by an increase in cardiac output. In addition, however, it results from the *redistribution* of the cardiac output by changes in the pattern of flow through the heart and the rest of the central circulation. The percentage of the heart's total outflow directed to the pulmonary arteries, whereas it may be as low as 20% just before ventilation, may rise to as high as 70% just after ventilation. Then, it falls back to 20% as the O_2 in the lung air is depleted.

Such adjustments represent a special case of **ventilation–perfusion matching**. An important tenet of circulatory and respiratory physiology is that, ideally, the ability of the circulatory system to *perfuse* the gas-exchange organ—and thus carry O_2 away—should be closely matched to the ability of the breathing system to *ventilate* the breathing organ with O_2 from the environment. The study of ventilation–perfusion matching is the study of whether and how the breathing and circulatory systems function together to approximate this ideal. Animals such as lungfish exhibit a type of ventilation–perfusion matching when they vary blood flow to their lungs in synchrony with their intermittent breathing cycles.

Circulation in Amphibians and Reptiles

Reptiles and lung-breathing amphibians typically have two completely separate atrial chambers. Oxygenated blood from the lungs enters the left atrium, whereas systemic venous blood enters the right atrium via the sinus venosus. Separation of oxygenated and deoxygenated blood is anatomically ensured until the blood enters the ventricle. Thereafter, however, the central circulation of all amphibians and reptiles is incompletely divided in one way or another.

In amphibians, the ventricle entirely lacks a septum. The ventricular lumen, however, is not a wide-open cavity because it is crisscrossed with strands and cords of spongy myocardium. The ventricle discharges into a contractile conus arteriosus, which in turn typically discharges into paired carotid, systemic, and pulmonary arteries. Running along the inside of the conus in most amphibians is a complexly twisted ridge of endothelial tissue (the *spiral fold*) that incompletely divides the conus lumen.

Although amphibians are diverse, it is clear that many of them—including certain frogs, toads, and salamanders—are capable of substantial selective distribution of blood in their incompletely divided central circulation. Bullfrogs (*Rana catesbeiana*) provide a dramatic example. In one study, when blood entering the right atrium of a bullfrog from the systemic tissues had an average O_2 concentration of 4.2 vol % and that entering the left atrium from the lungs had a concentration of 8.6 vol %, the blood pumped into the pulmonary arteries contained 4.4 vol % and that pumped into the systemic arteries contained 8.0 vol %. These data indicate that 91% of the pulmonary venous blood arriving in the heart was channeled into the systemic arteries, whereas 84% of the systemic venous blood was directed into the pulmonary arteries. Such a high degree of selective distribution of oxygenated and deoxygenated blood functionally places the pulmonary circuit in series with the systemic circuit, an arrangement that promotes efficiency of O_2 transport, as we noted earlier. When amphibians achieve such remarkable levels of selective blood distribution, they do so with a ventricle that entirely lacks a septum and with a conus arteriosus that is only incompletely divided! Physiologists do not know exactly how.

Because the skin is typically an important respiratory site in amphibians (see page 555), its circulation deserves note. The pulmonary arteries (also known as the *pulmocutaneous* arteries) not only send blood to the lungs, but also give rise to some of the major cutaneous arteries. Thus, if deoxygenated blood is preferentially pumped to the lungs, it is also preferentially pumped to the skin. The veins draining the skin connect with the general systemic venous vasculature—a property traditionally viewed as disadvantageous because it allows blood oxygenated in the skin to mix with deoxygenated systemic venous blood rather than staying separate. However, from the new viewpoint that emphasizes adequate O_2 for the myocardium, the arrangement of the cutaneous veins could be an advantage. The myocardium of amphibians gets much of its O_2 from the blood flowing through the heart lumen, and that blood is oxygenated when it receives input from the cutaneous veins.

In reptiles, the conus arteriosus is has disappeared, and arteries arise directly from the ventricle. In turtles, snakes, and lizards—the noncrocodilian groups—the ventricle is usually incompletely divided into three chambers by muscular ridges and partial septa. Although pulmonary and systemic venous bloods are potentially able to mix in the ventricle, numerous studies have demonstrated that these animals typically display a high degree of selective blood distribution (their pulmonary arteries, for example, receive mostly deoxygenated blood). The mechanism of this selective distribution is not fully understood and seems likely to vary from group to group.

Amphibians and noncrocodilian reptiles are often intermittent breathers. During intermittent breathing, they take advantage of the incomplete division of their hearts to modulate where the blood from the heart is pumped in an adaptive manner synchronized with the breathing cycle (see Box 23.1). Some diving turtles provide dramatic examples. When their lung air becomes O_2-depleted during a dive, they continue to circulate blood round and round in their systemic circuit while sending hardly any blood to their lungs.[19]

In the crocodilian reptiles, the ventricle is completely divided into two chambers by a septum. This does not mean, however, that the crocodilians resemble mammals and birds—far from it. Crocodilians have two systemic aortas, which arise from the left and right ventricles (Figure 23.18*a*). The two aortas are connected shortly after their exit from the ventricles by an aperture called the **foramen of Panizza**, and the vascular beds perfused by the two aortas are also connected more distally. The pulmonary artery arises from the right ventricle. Functional studies have demonstrated that the crocodilian heart can achieve virtually perfect selective distribution of deoxygenated blood to the lungs and oxygenated blood to the systemic circuit. The way this happens is that deoxygenated systemic venous blood, which flows from the right atrium into the right ventricle, can be pumped exclusively into the pulmonary artery because valves and pressure relationships prevent it from entering the systemic aorta that leaves the right ventricle (Figure 23.18*b*). This flow pattern is not guaranteed by anatomy, however, and flow may follow alternative patterns during diving or breath-holding (Figure 23.18*c*).

[19]In the professional literature on the cardiovascular physiology of animals with incompletely divided central circulations, the variable distribution of total cardiac output to the pulmonary and systemic circuits is often called *shunting*. *Right-to-left* (*R-L*) *shunting* is diversion of systemic venous blood directly back into the systemic circuit, bypassing the lungs. This is the sort of shunting displayed by the diving turtles. *Left-to-right* (*L-R*) *shunting* is diversion of pulmonary venous blood directly back to the lungs.

Concluding Comments on Vertebrates

In studying the circulatory systems of fish, amphibians, and reptiles, there is perhaps a yearning to find a linear advance toward the condition of mammals and birds. The facts, however, do not at all reflect such a linear progression. The groups of animals alive today are not in the business of striving to be mammals and birds (how could they be?).

Among air-breathing vertebrates, logic suggests that efficiency of O_2 transport is promoted by having the pulmonary and systemic circuits connected in series during periods of time when the lungs are being ventilated and the lung air is rich in O_2. In mammals and birds, which breathe continuously, the lung air is always rich in O_2, and the pulmonary and systemic circuits are connected in series in an anatomically rigid manner. This does not mean, however, that the anatomically rigid series arrangement is always ideal. For animals that fill their lungs with air only intermittently, an incomplete anatomical division of the central circulation can potentially better promote efficiency of O_2 transport, because the incompletely divid-

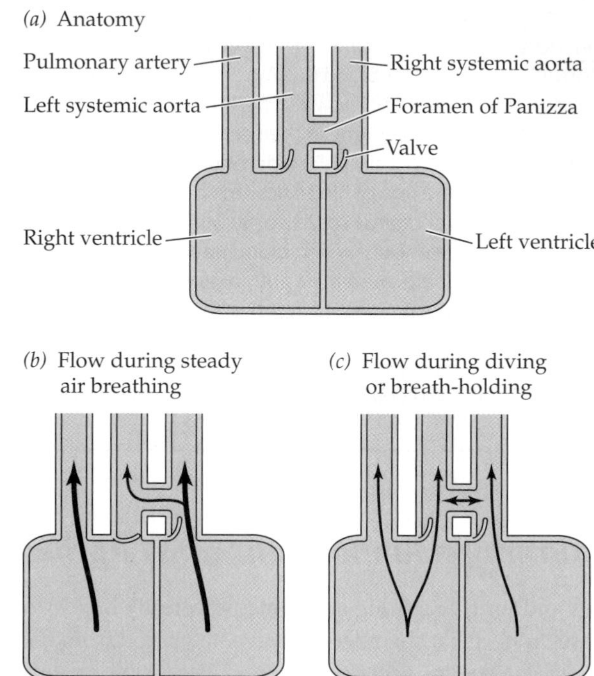

Figure 23.18 Blood flow in the heart ventricles and the systemic and pulmonary arteries of crocodilian reptiles *(a)* The basic plan of the ventricles and vessels. *(b)* During steady air breathing, blood flows readily from the right ventricle into the pulmonary artery. Consequently, the systolic pressure in the right ventricle does not rise particularly high, and in fact remains lower than that in the left systemic aorta, so the valve between the right ventricle and left systemic aorta is not forced open. Blood from the left ventricle is pumped mostly into the right systemic aorta; some may enter the left systemic aorta through the foramen of Panizza, although this does not always occur. *(c)* During diving or prolonged breath-holding, changed pressure relations may result in opening of the valve into the left systemic aorta during ventricular systole, permitting some of the systemic venous blood pumped by the right ventricle to bypass the lungs. In this way, the incompletely divided central circulation permits pulmonary blood flow to be reduced during times when lung air is O_2-depleted (see Box 23.1).

ed state permits the rate of pulmonary blood flow to be modulated independently of systemic flow and in synchrony with O_2 availability in the lungs. For animals in which the myocardium obtains its O_2 from luminal blood, an incompletely divided heart can potentially also provide other advantages because it allows the myocardium of the right side of the heart to acquire O_2 from oxygenated blood flowing through the left side. In brief, the circulatory plans of the "lower" vertebrates have their own potential virtues and are not simply defective versions of the mammalian or avian plan.

Invertebrates with Closed Circulatory Systems

Closed circulatory systems are found in only a few groups of invertebrates, notably some annelid worms and the cephalopod molluscs (the squids and octopuses and their relatives). The cephalopods, which we emphasize here, are of great interest because they are extraordinarily active animals compared to most members of their phylum (e.g., clams and snails). Some squids, in fact, jet around at such high speeds that they rank with fish as being among the most active of all aquatic animals. The giant squids are the largest invertebrates on Earth.The circulatory system of a squid or octopus consists of extensive networks of arteries and veins—both of which are muscular—joined in large part by capillary beds.

A squid or octopus has a principal heart, called the **systemic heart**, and two weaker, auxiliary **branchial hearts**. The systemic heart—which is myogenic—consists of a powerful muscular chamber, valved at its inflows and outflows. The circulatory plan is presented in Figure 23.19. Blood enters the systemic heart from the gills and is pumped from the systemic heart into major arteries (aortae) that take it to the systemic tissues. As blood returns to the vicinity of the systemic heart in the major veins, it is split into two symmetrical paths to be directed to the gills. Near the base of each gill is a bulbous branchial heart, which pumps blood into an afferent branchial vessel. The blood then passes through capillaries in the gill to arrive in the efferent branchial vessel and return to the systemic heart. This circulatory plan resembles that of mammals, birds, and most fish in that it places the respiratory organs and the systemic tissues strictly in series. The arrangement of hearts may seem a bit odd at first, but is actually quite "close to home." As seen by comparing Figures 23.10*b* and 23.19*a*, the arrangement of hearts in a squid or octopus is identical to that in a mammal, except that the respiratory pumps are anatomically separate from the systemic pump.

Physiologically, the circulatory system of squids and octopuses resembles that of vertebrates far more than it does that of the other molluscan groups (which have low-pressure, open circulatory systems). In the octopus *Enteroctopus dofleini* and the highly active squid *Loligo pealii*, the mean blood pressure produced by the systemic heart is 4 to 5 kPa (30–37 mm Hg), and by the time the blood has passed through the systemic tissues and reached the great systemic veins, its mean pressure has dropped to less than 1 kPa (7 mm Hg). The systemic pressure gradient in cephalopods thus resembles that in fish (see Table 23.1). The cardiac output of resting or mildly active *E. dofleini* is typically about 10–20 mL/kg·minute; accordingly, it is also in the range seen in fish (see Table 23.1). When the vascular resistance of the systemic circuit in cephalopods is calculated (Equation 23.3), it is relatively high, as would be expected for a closed circulatory system. Overall, therefore, the cephalopods are similar to vertebrates in that they maintain reasonably rapid rates of blood flow through a systemic circuit of relatively high resistance by maintaining relatively high pressures in the systemic arteries.

Exercise in cephalopods presents intriguing questions, which we have already started to discuss in earlier chapters. Box 23.2, besides summarizing points made in those other chapters, explains the special burden borne by the circulatory system during cephalopod exercise.

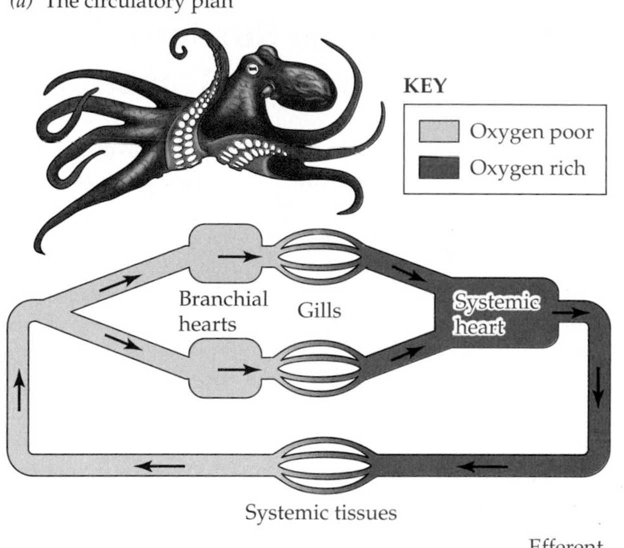

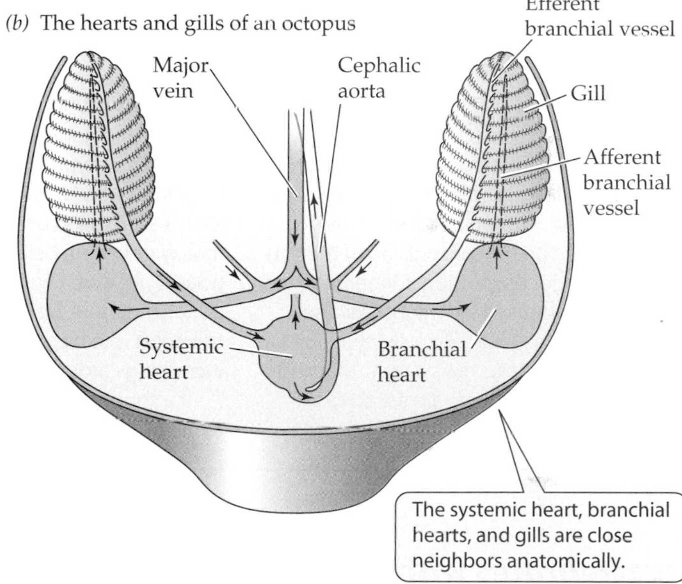

Figure 23.19 The circulatory plan of squids and octopuses *(a)* The arrangement of the hearts, gills, and systemic tissues relative to one another. Hemocyanin, the respiratory pigment of squids and octopuses, turns blue when well oxygenated, but is clear or nearly clear when deoxygenated. *(b)* A more realistic drawing of the central circulatory system of an octopus. (*b* after Johansen and Lenfant 1966.)

BOX 23.2 Bearing the Burden of Athleticism, Sort Of: A Synthesis of Cephalopod O_2 Transport

Both the vertebrate and the invertebrate worlds have produced high-speed, swimming predators. The principal vertebrates in this ecological role are fish. As improbable as it may seem when gazing at a clam or snail, the molluscs have produced the invertebrate entry: the squids. The fossil record indicates that swimming cephalopods were abundant in the oceans before the appearance of marine fish. In that early era, the cephalopods were alone in playing the role of being fast-swimming predators. The arrival of fish then produced competition. Looking back, it is difficult not to conclude that the fish, taken as a group, have come out ahead of the cephalopods as a group. Of course, some species of squids alive today have been as successful as fish. However, the squids might have dominated the oceans had fish not appeared, and today it is the fish that dominate.

To generate swimming power, the cephalopods became committed to jet propulsion early in their evolution and have never turned back. Compared to propulsion of the sort employed by fish, jet propulsion is inherently inefficient. For example, a squid might well use twice the O_2 of a like-sized fish to swim at half the speed of the fish. Thus the systems that supply O_2 to the tissues of squids are asked to assume a large burden.

The gills of cephalopods are positioned in the mantle cavity, where they have always been throughout the evolutionary history of molluscs (see Figure 21.26). Being in the mantle cavity, the gills are exposed to extreme rates of water flow during jet propulsion and have no difficulty obtaining O_2, as we saw in Chapter 21 (see page 568).

The circulatory system is where limits develop. For one thing, molluscs are committed to hemocyanin dissolved in the blood plasma as their respiratory pigment. Dissolved hemocyanin makes the blood ever more viscous as its concentration rises. Squids and octopuses have evolved the highest concentrations of hemocyanin in any animals: They seem to have raised blood viscosity as high as it can go. Even thus maximally concentrated, hemocyanin allows the blood leaving the gills to carry only about 5 vol % O_2, as we saw in Chapter 22. Hemoglobin in red blood cells allows many fish to have oxygen-carrying capacities over twice as high. As also seen in Chapter 22 (in Figure 22.17), squids and octopuses, even when *approximately at rest*, are believed to have almost no venous O_2 reserve. Thus, when they start jetting about, their sole available option to increase O_2 transport is to increase their rate of blood flow by increasing cardiac work.

If a squid and a fish were both to increase cardiac output by fourfold during exercise, the squid would increase its rate of circulatory O_2 delivery about fourfold, but the fish might increase its rate tenfold by simultaneously tapping its venous O_2 reserve to increase the amount of O_2 delivered per unit of blood volume (see Figure 22.16). This difference illustrates the implications of the resting venous O_2 reserve: If there is little resting venous reserve, O_2 delivery can be increased only by increasing the rate of blood flow.

Viewed metaphorically, the sophisticated circulatory system of the cephalopods is doomed to bearing an almost impossible burden by a series of irrevocable evolutionary commitments in other functional attributes, such as the commitment to jetting as the source of propulsive power and the commitment to hemocyanin. When a squid powers up, its O_2 cost of swimming is high compared to that of a fish, but because its blood oxygen-carrying capacity and venous O_2 reserve are comparatively low, the squid's hearts must bear almost the whole responsibility of increasing O_2 delivery. Fortunately, some squids prosper nonetheless, but the "O_2 equation" is probably a key reason why the squids as a group were outcompeted by fish over the eons in the oceans.

Invertebrates with Open Circulatory Systems

The animals with open circulatory systems—which include all arthropods and most molluscs—commonly have a well-developed central heart. The heart may discharge into a network of arteries, and the arteries may lead to capillary beds of discrete, minute vessels lined only with a single layer of endothelial cells. Ultimately, however, in an open circulatory system, the blood exits discrete vessels (**Figure 23.20**), and thereafter it flows through lacunae and sinuses. **Lacunae** are small spaces among tissue cells. A tissue with lacunae is often thoroughly permeated by an anastomosing network of lacunar spaces—a network that brings blood close to all the cells in the tissue. Lacunar networks are not dissimilar to capillary networks in this respect. However, lacunar channels are characteristically irregular in shape and are not in the form of discrete vessels lined with endothelium. **Sinuses** are larger blood spaces, commonly representing thoroughfare channels for the blood. Lacunae and sinuses in animals with open circulatory systems are sometimes bounded by a membrane of some type. However, they are believed by most physiologists to be sometimes bounded simply by ordinary tissue cells. This latter point is of key importance because it means that the blood bathes the tissue cells directly, and *there is in fact no clear distinction between blood and extracellular tissue fluid* (*interstitial fluid*). In recognition of this fact, some authors call the blood **hemolymph** ("blood lymph").

As already suggested, experts debate the exact nature of the lining of lacunae and sinuses in some cases. Regardless of the ultimate outcome of these debates, it is clear that the design of the circulatory system in animals with open circulatory systems is significantly different from that in vertebrates or cephalopods. In the latter groups, not only is blood *carried to* the depths of each tissue by discrete vessels, but it remains in vessels as it *passes through* each tissue, and it is *collected from* each tissue by discrete vessels. In animals with open circulatory systems, after blood leaves the arteries (or sometimes capillaries), it is left to follow a relatively ill-defined path through lacunae and sinuses before it is again channeled into discrete vessels.

The crustacean circulatory system provides an example of an open system

Some immature, small, or sessile crustaceans lack a heart and blood vessels. Their circulation is exclusively through sinuses and lacunae, and

(a) Shortly before arrival of dye

(b) 1 second later

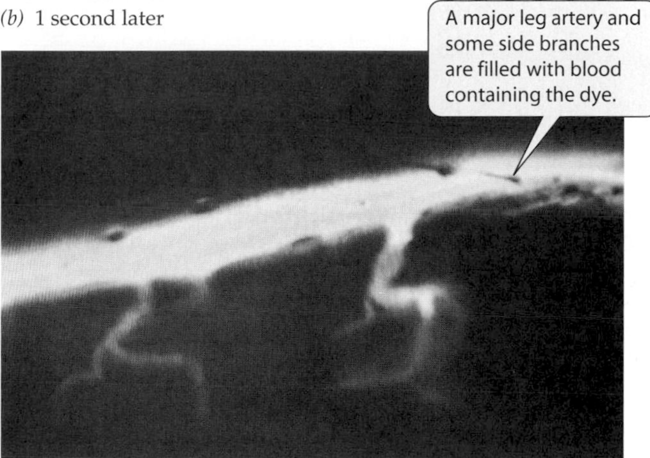

(c) Another moment later

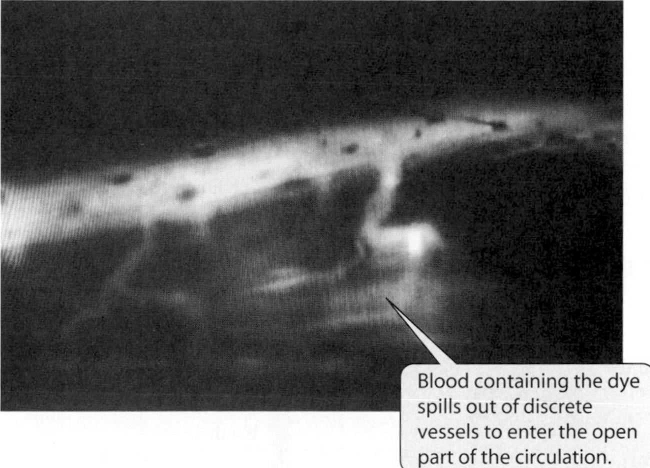

Figure 23.20 The transition of blood flow from the closed part of the circulation to the open part A fluorescent dye was injected into arteries of a spider. As the dye traveled with the flowing blood, it was imaged using fluorescent videomicroscopy. *(a)* A portion of a leg just before the arrival of blood containing dye. *(b)* The same spot 1 s later. *(c)* The same spot a moment later, showing blood and dye flowing out of discrete vessels. These are the first images ever obtained of this process. The specific animal studied was a spider (*Pholcus*) with book lungs, chosen for investigation because its integument (including the exoskeleton) is naturally transparent over much of the body. (Photos courtesy of R. J. Paul; from Paul et al. 1994.)

the propulsive force is provided by ordinary body movements. In some other crustaceans, a heart is present, but empties through holes rather than arteries. In still others, there are arteries, but they end abruptly after extending only a short distance from the heart. Here we emphasize the decapod crustaceans—crabs, crayfish, lobsters, and their relatives—which have especially elaborate open circulatory systems.

In adult decapod crustaceans, the heart is a single-chambered saclike structure—positioned in the dorsal thorax—the beat of which is initiated neurogenically (see Figure 23.5).[20] *All the vessels connected to the heart are arteries* (as is typical of crustaceans in general). The arteries are valved at their origins and leave the heart in several directions (**Figure 23.21*a***). Blood enters the heart not through vessels, but

[20]The heart is myogenic in the embryos and juveniles of at least some decapod crustaceans and is partly or wholly myogenic in adults of some other types of crustaceans.

(a) The heart and the arteries emanating from the heart

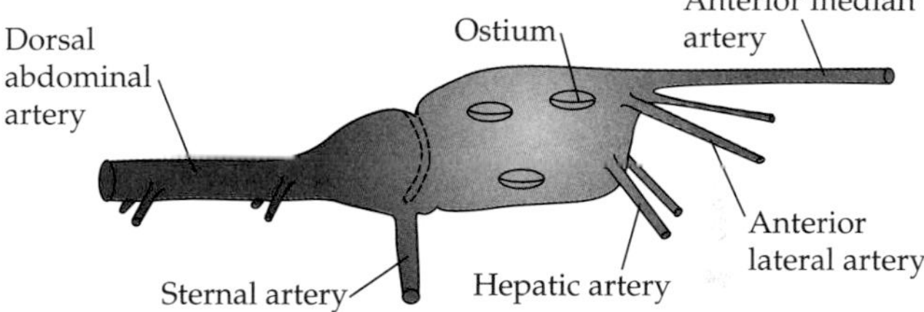

(b) The array of suspensory ligaments around the heart

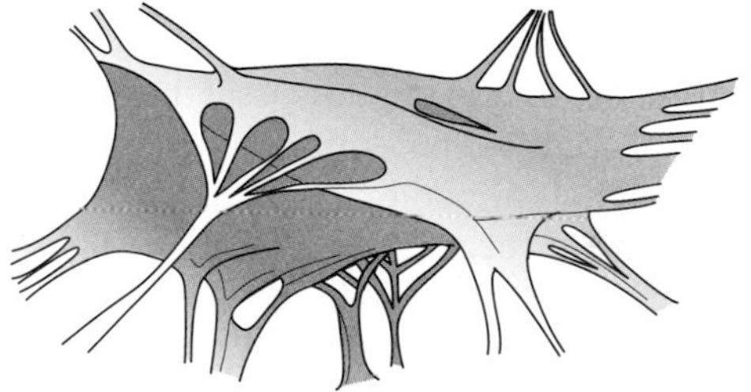

(c) Flow of blood through the central circulation

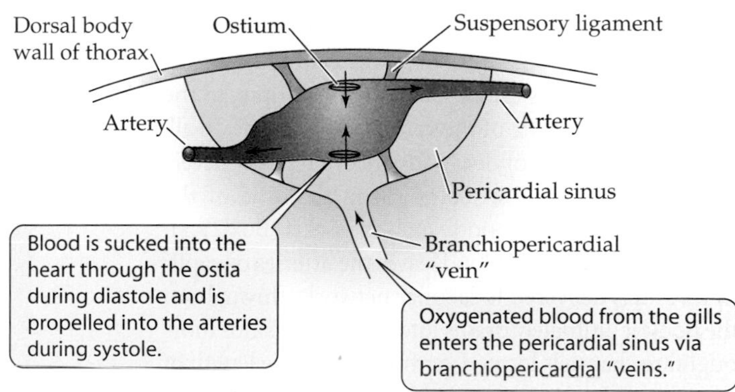

Figure 23.21 The heart of a decapod crustacean *(a)* The heart and the arteries leaving it. All the vessels connected to the heart in a crustacean are arteries. *(b)* The heart is surrounded by an array of elastic suspensory ligaments that run between the heart wall and the wall of the surrounding pericardial sinus, suspending the heart in the pericardial sinus. *(c)* The position of the heart in the pericardial sinus and the pattern of blood flow through the central circulation. (*a* after Wilkens 1999; *b* after Plateau 1880.)

through slits in the heart wall, called **ostia**, of which decapods usually have three pairs. The heart is suspended by elastic **suspensory ligaments** within a bounded sinus, the **pericardial sinus** (Figure 23.21*b,c*). The only blood channels that enter the pericardial sinus are "veins" that bring blood from the gills or other respiratory surfaces.[21]

When the heart contracts at systole, the ostia are closed by muscular tension, flap valves, or both; blood within the heart thus cannot spew out through the ostia into the pericardial sinus, and is driven into the arteries. Contraction of the heart stretches the suspensory ligaments attached to the heart wall. During diastole, elastic rebound of these ligaments expands the heart back to its presystolic volume. This elastic rebound is the primary force for refilling because as the heart is stretched open, the pressure within is reduced below that in the pericardial sinus, and blood is sucked inward from the pericardial sinus through the ostia.

The arterial walls of crustaceans, although elastic, are typically nonmuscular. Thus the arteries themselves are neither pulsatile nor capable of vasomotor control. For this reason, until about 20 years ago, physiologists assumed that crustaceans had little, if any, control over the regional distribution of cardiac output. Now it is clear, however, that blood distribution is under control of highly localized, muscular valves. The best studied are the **cardioarterial valves**, positioned at the points where the various arteries leave the heart. Each valve contains innervated muscle. Excitatory nerve impulses, by stimulating contraction of a valve, can limit blood flow into the artery guarded by that valve; simultaneously, inhibitory impulses sent to other valves might enhance blood flow into other arteries. Hormones also help control the cardioarterial valves. Other muscular valves occur elsewhere in the arterial system, but are poorly understood.

Decapods have extensive arterial systems (Figure 23.22*a*). Branching arteries lead blood from the heart to most regions of the body. Sometimes the blood is discharged from the arteries directly to lacunar networks. Sometimes the arteries lead to capillary beds, which then discharge into lacunar networks. Prominent capillary beds are observed, for example, in the brain and the ganglia of the ventral nerve cords. In all, the arterial systems of decapods are often impressively elaborate, especially in comparison to some of the other groups of arthropods such as insects (Box 23.3).

In a decapod, after blood leaves the arteries or capillaries and is delivered to lacunar networks throughout the body, it ultimately drains into a system of sinus thoroughfare channels located ventrally along the length of the animal. In these sinuses, blood from the posterior regions of the body flows anteriorly, and that from the anterior regions of the body flows posteriorly. As shown in Figure 23.22*b*, both flows converge on a ventral thoracic sinus termed the **infrabranchial sinus** (*infrabranchial*, "below the gills"). Afferent branchial channels that carry blood into the gills arise from this sinus. Blood flows into these channels, traverses systems of small sinuses or lacunae in the gill filaments or lamellae, and then exits the gills by way of efferent branchial channels. Importantly, these channels discharge into the branchiopericardial "veins," which lead directly into the pericardial sinus. Thus, after a period of seemingly poorly controlled flow following its discharge from the arterial system, the blood is channeled in a most significant way prior to its return to the heart. The gills are placed in series with the rest of the circulation, and oxygenation of the blood pumped by the heart is ensured.[22]

[21]We term these channels "veins," in quotation marks, because they are not typical venous vessels of the sort seen in vertebrates or cephalopods, but are usually called *veins* nonetheless.

[22]A small fraction of the blood returning to the heart bypasses the gills by flowing through the tissue lining the gill chambers. This circulatory path is greatly elaborated in air-breathing crabs that use the lining of the gill chambers as a breathing surface.

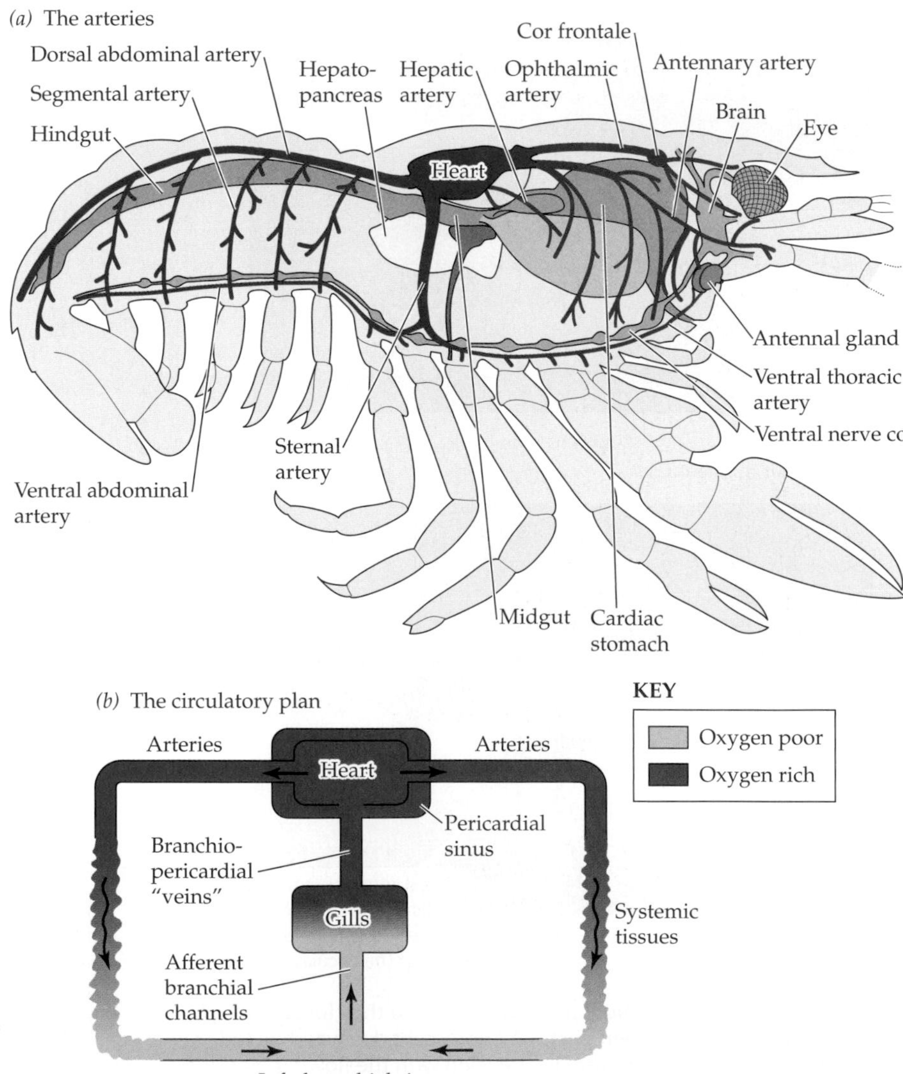

Figure 23.22 Circulation through the body of a crayfish or lobster *(a)* The heart and major arteries. The *cor frontale* is an accessory heart (powered by skeletal muscles) positioned upstream of the extensive vascular networks of the brain; it is believed to help ensure that adequate perfusion pressures are maintained in the brain. *(b)* The circulatory plan. Because the respiratory pigment of decapod crustaceans is hemocyanin, the blood turns blueish when oxygenated, but is clear when fully deoxygenated. (*a* after McLaughlin 1980.)

BOX 23.3 Circulation and O_2: Lessons from the Insect World

Like all arthropods, insects have an open circulatory system. A dorsal vessel that runs along most of the body is divisible—often indistinctly—into a posterior heart (usually restricted to the abdomen) and an anterior dorsal aorta that runs forward into the thorax and head, as shown in the figure. The heart has ostia and refills much as a crustacean heart does. It contracts in a peristaltic wave, forcing blood into the dorsal aorta, which often continues the peristaltic wave of contraction. The dorsal aorta is the only blood vessel in many species of insects. In others, segmentally arranged lateral arteries branch off from the dorsal vessel. Whether there are lateral arteries or not, the major blood vessels end abruptly with little branching, discharging blood directly to the lacunar circulation. The blood-vessel system is far less extensive than that in decapod crustaceans, and there are no capillary beds.

The modest development of the circulatory system in insects seems paradoxical at first. Many insects are very active animals with relatively high O_2 demands. During sustained flight, some insects, in fact, attain weight-specific rates of O_2 consumption that not only are many times higher than any reported for crustaceans, but are among the highest in the entire animal kingdom. Within phyla other than the arthropods (most notably chordates and molluscs), we find that groups with a capacity for relatively intense aerobic catabolism show unambiguous circulatory refinements in comparison to related groups with a lower capacity for aerobic catabolism: Mammals have far faster rates of blood flow than fish, for example, and squids have many circulatory refinements compared to clams. When we look at the arthropods, however, it is clear that the circulatory system in insects is not more advanced than that in crustaceans or other arthropods, despite the extraordinary metabolic intensity of many insects.

The paradox of the modest circulatory system of insects finds its likely answer in the tracheal breathing system. Recall that O_2 is brought to all the cells of an insect by a system of gas-filled tracheal tubes that functions *independently* of the circulatory system (see Figure 21.28). The insects have a lessen to teach: namely, that *the circulatory system can remain quite simple in even highly active animals if it need not assume the burden of O_2 transport.* The tasks the circulatory system must perform in insects (e.g., nutrient and hormone transport) are far less urgent than O_2 transport. Thus the insect circulatory system can be quite modest. The decapod crustaceans, in contrast, depend on their circulatory system to supply their tissues with O_2 at the rate required by their active ways of life. Thus the decapods require a relatively sophisticated version of the arthropod circulatory system.

Blood flow through the tissues of an insect is principally through lacunae and sinuses The blood vessels emanating from the heart are all arteries and end abruptly after little branching. Blood enters the heart through ostia (not shown). Insects often have accessory pulsatile structures that aid flow through their legs, wings, and antennae. The blood of insects lacks a respiratory pigment. (From Jones 1964.)

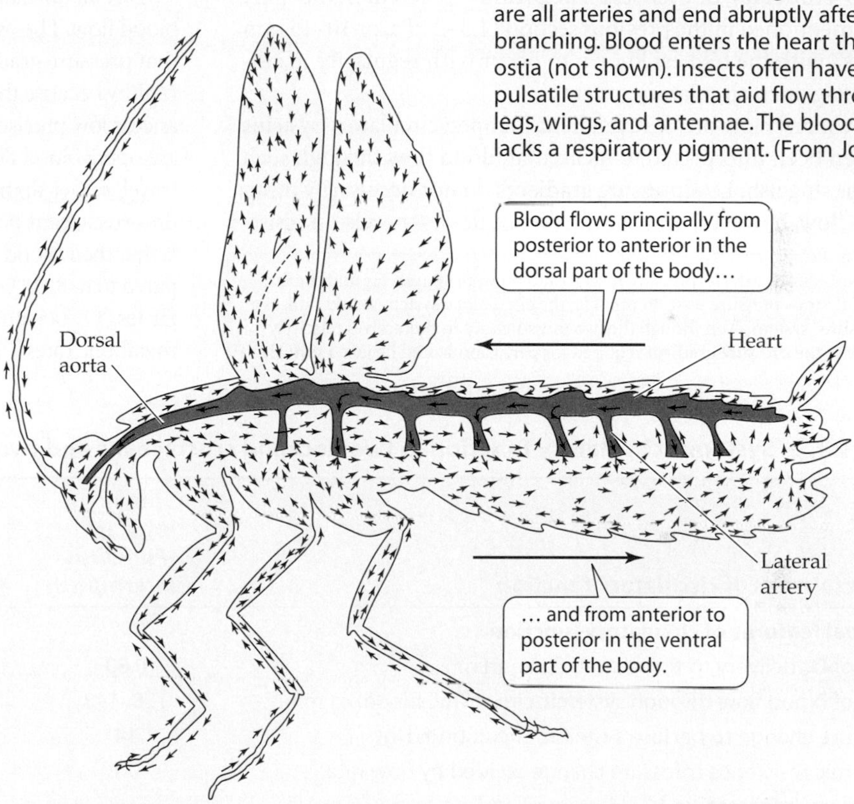

Blood flows principally from posterior to anterior in the dorsal part of the body...
Dorsal aorta
Heart
Lateral artery
... and from anterior to posterior in the ventral part of the body.

Like fish, decapod crustaceans perfuse the systemic tissues and the gills in sequence, without repumping the blood between the two. In decapods, however—unlike fish—blood leaving the heart passes through the systemic circuit first and the gill circuit second. Thus blood pumped to the systemic circuit is at the pressure the heart develops. The drop of blood pressure in the systemic circuit has been quantified in only a few species, but in general may be classified as moderately large; the difference between arterial pressure and infrabranchial-sinus pressure is 1.3 to 2 kPa (10–15 mm Hg) in spiny lobsters (*Panulirus*) and about 1 kPa (6–7 mm Hg) in American lobsters (*Homarus*). The gill circulation typically poses a low resistance to flow and requires only a small pressure difference between the afferent and efferent branchial channels—about 0.2 kPa (1–2 mm Hg)—to be perfused.

Open systems are functionally different from closed systems but may be equal in critical ways

In animals with open circulatory systems, the pressures in the blood spaces are often affected substantially by body movements. This phenomenon must be considered for the function of these systems to be properly analyzed. When body movements create pressure *differences* from one part of an animal's body to another, they aid blood flow; in American lobsters, for example, quick flexion of the abdomen—an important swimming movement (see page 149)—raises pressures in the abdomen above those in the thorax and can cause the rate of blood flow toward the thorax in the ventral abdominal sinuses to increase by a factor of 10 or more. Body movements and changes in posture, however, may also affect the *overall* level of pressure *throughout* the body; a change of posture, for instance, might compress all the blood spaces in the body and pressurize the blood everywhere. If pressure is changed to the same extent everywhere, blood flow is not immediately affected. Only *differences* in pressure cause flow.

When we speak of circulatory systems as being "high-pressure" or "low-pressure" systems, the *pressure gradient required to perfuse them*—not the overall pressure that exists everywhere—is the characteristic at issue. If, for example, a pressure gradient of 4 mm Hg (0.5 kPa) is all that is required to drive blood through an animal's circulatory system, the system is a "low-pressure" system. This remains true even if the animal is in a posture that pressurizes its whole body to a high level.[23]

Open circulatory systems are typically low-pressure or moderate-pressure systems. In bivalve molluscs and nonpulmonate snails, arterial pressures typically exceed venous pressures by just 0.1–0.5 kPa (1 to 4 mm Hg), or even less. The pressure gradient in the spiny lobsters mentioned in the previous section (1.3–2 kPa, or 10–15 mm Hg) ranks with the highest known to occur with regularity in animals having open circulatory systems.

The relatively low pressure gradients in open circulatory systems have often been interpreted to mean that blood flow through such systems is sluggish. Low pressure gradients do not necessarily mean sluggish flow, however, because the rate of flow depends on resistance as well as pressure (Equation 23.3). The first four lines of Table 23.2 compare four critical functional attributes of the open circulatory systems in lobsters and crabs with those of the closed circulatory systems of fish of similar body sizes. As the first line shows, all the species have similar metabolic rates and similar rates of circulatory O_2 delivery under the conditions of study. The crustaceans (see the second line) circulate their blood *at least three times faster* than the fish, however. The crustaceans, that is, are the exact opposite of being sluggish! Both of the crustaceans (third line) maintain lower pressure gradients than the fish; the crab in particular has a very low pressure gradient of just 3 mm Hg. Circulation is fast in the crustaceans, nonetheless, because (fourth line) the resistance to flow through the crustacean circulatory systems is much lower than the resistance in the fish circulatory systems.

Not all animals with open circulatory systems have high rates of blood flow. The overall picture in the decapod crustaceans, however, is that pressure gradients—while low—are adequate to circulate blood rapidly because the resistance to flow is low. Without doubt, the resistance is low precisely because the circulatory systems of the crustaceans are open; blood flows through lacunae and sinuses during much of its travel, rather than through minutely narrow capillary vessels. Why do the crustaceans *need* their high blood flow rates? One reason, certainly, is that their blood oxygen-carrying capacities are low (1–2 vol %, compared to 6–8 vol % in fish). The lobster and crab in Table 23.2 transport far less O_2 per unit of blood volume than the fish. To sustain similar metabolic rates, therefore, they must circulate their blood faster.

[23]For instance, if the arterial pressure in a crustacean were 40 mm Hg and the infrabranchial-sinus pressure were 36 mm Hg, the circulatory system would rank as a "low-pressure" system, even though the two pressures are in themselves relatively high, because the pressure gradient required for perfusion would be just 4 mm Hg.

TABLE 23.2 Systemic circulatory function in two decapod crustaceans and two fish of similar body sizes

Characteristics of circulatory function	Spiny lobster (*Panulirus interruptus*)	Rock crab[a] (*Cancer productus*)	Starry flounder (*Platichthys stellatus*)	Rainbow trout (*Oncorynchus mykiss*)
Principal features of circulatory function				
Rate of O_2 delivery to tissues (mL O_2/kg·min)[b]	0.80	0.60	0.46	0.65
Rate of blood flow through systemic circuit (mL blood/kg·min)[c]	128–148	125	39	18
Pressure change to perfuse systemic circuit (mm Hg)[d]	14	3	16	22
Systemic resistance (pressure change divided by flow rate)[e]	0.1	0.03	0.4	1.2
Secondary information				
Heart rate (beats/min)	65	101	35	63
Stroke volume (mL/kg·stroke)	2.1	1.2	1.2	0.3
Blood pressure in major systemic arteries (mm Hg)[d]	35	10	18	26
Blood pressure in major systemic veins or venous sinuses (mm Hg)[d]	21	7	2	4
Blood oxygen-carrying capacity (vol %)	2.0	1.3	5.7	7.8
Temperature during studies (°C)	16	12–16	8–11	9–15
Body weight (g)	515	~370	684	~210

Source: Hill and Wyse 1989; based on a compilation of data from several sources.

[a] The rock crab periodically ceases heart action; after heart action resumes, the crab's rate of O_2 consumption, cardiac output, and stroke volume gradually fall. Values given are averages measured 5 minutes or more after a cardiac pause.

[b] Measured as the animals' rates of O_2 consumption.

[c] Equals cardiac output in steady state.

[d] Pressure change is the difference between blood pressures in arteries and in veins (or venous sinuses), both given below. The arterial pressures were measured in systemic arteries, except for the crab pressure, which was measured as systolic ventricular pressure and thus probably overestimates arterial pressure. The venous pressures were measured in the systemic veins of the fish and in the infrabranchial sinus of the crustaceans.

[e] The values of systemic resistance are in units of mm Hg·kg·min/mL blood.

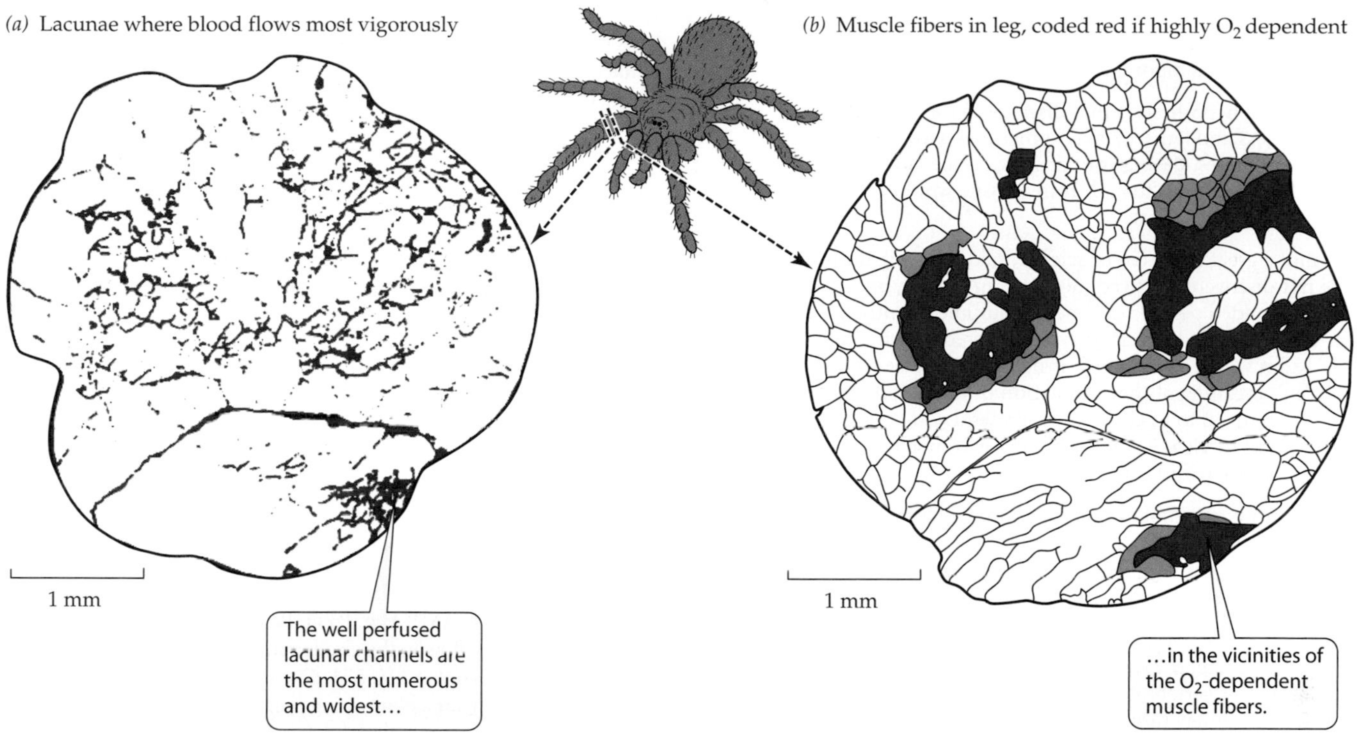

(a) Lacunae where blood flows most vigorously

(b) Muscle fibers in leg, coded red if highly O_2 dependent

1 mm

The well perfused lacunar channels are the most numerous and widest…

1 mm

…in the vicinities of the O_2-dependent muscle fibers.

Figure 23.23 The microanatomy of blood flow in a lacunar system: Blood flows most vigorously where it is needed most Adjacent cross-sections of a tarantula's leg, showing the walking muscles inside. *(a)* Locations where blood flows particularly vigorously in the lacunar networks of the muscles. In this cross-section, hemocyanin was stained immunohistochemically using monoclonal antibodies (antibodies specific to hemocyanin). The stain was then visualized. *(b)* Locations of muscle fibers that are particularly O_2-dependent. In this cross-section, fibers were assessed using their profiles of catabolic enzymes (similar to Figure 6.10). The fibers colored dark red are the most poised to produce ATP aerobically and are the most O_2-dependent. Those colored light red are also aerobically poised, but less so than those colored dark red. (From Paul et al. 1994.)

Inevitably, the question arises of whether open or closed circulatory systems are "better." Data such as those in Table 23.2 suggest that the systems are "different but equal" in certain key respects. The greatest void that remains in physiological knowledge of open circulatory systems is the question of how well the flow of blood can be directed and controlled. Scientists have tended to assume that little direction or control is possible. Increasing evidence exists, however, that—by means of cardioarterial valves or other (mostly unknown) mechanisms—open circulatory systems can direct blood flow far more exactly than scientists have assumed. Figure 23.23, for example, presents a novel set of data gathered by use of modern molecular techniques (monoclonal antibodies), showing that in the leg muscles of a spider, *lacunar* blood flow tends to be most vigorous near the particular sets of muscle cells that are most O_2-dependent. Closed circulatory systems are justly touted for their ability to exert fine spatial and temporal control over blood flow. Although open systems seem unlikely to be the equal of closed systems in this respect, the difference between the two is probably smaller than traditionally believed.

Summary: Invertebrates with Open Circulatory Systems

- Most invertebrates have open circulatory systems in which blood leaves discrete vessels and flows through systems of lacunae and sinuses, where it comes into contact with ordinary tissue cells (accounting for the fact the blood is sometimes called hemolymph).
- Animals with open circulatory systems typically have a heart, and they may have extensive systems of blood vessels, including even capillary beds. The blood ultimately leaves the vessels, however.
- Open circulatory systems tend to be characterized by relatively small changes of blood pressure across the systemic circuit, compared to those in closed circulatory systems. Resistance is low in open systems, however, because blood is not forced through capillary beds in most or all tissues. Thus the rate of blood flow may be high despite the small pressure changes.
- Little is known about the spatial and temporal control of blood flow within animals having open circulatory systems. Some control is known to be possible, however (as by cardioarterial valves in crustaceans), and control is probably more sophisticated than generally assumed in the past.

Study Questions

1. Imagine that the digestion of a meal has just provided a new supply of glucose and amino acids in your midgut (small intestine). Then imagine all the steps that will take place as your body uses the glucose and amino acids and disposes of the waste products that are

produced. List all the individual and specific functions your circulatory system will perform during this sequence of events. Arrange your list in the correct temporal order.

2. A current working hypothesis is that diseases characterized by incurable weakness and lethargy may sometimes arise in part from loss of blood capillaries. Basing your answer on the laws of diffusion, explain how transport to and from a tissue would be affected if 20% of its capillaries were lost.

3. Thomas Kuhn, a great historian of science, said that "big ideas" in science are not discovered at fixed moments in history, but are discovered over extended periods of time by multiple scientists. We often hear it said that William Harvey "discovered the circulation of the blood." Kuhn would say that this is not true because when we speak of today's meaning of the "circulation of the blood," the concept required centuries to be discovered. Pretend you are Kuhn, and argue for his point of view. Then, speaking for yourself, explain why you agree or disagree with Kuhn.

4. Figure 23.12*b* shows mean blood pressures in the large arteries, terminal arteries, arterioles, and capillaries in the systemic circuit of a resting person. Using those pressures and assuming a cardiac output of 6 L/minute, calculate the collective vascular resistance of the large arteries, the terminal arteries, the arterioles, and the capillaries.

5. Looking ahead to Chapter 27, the process that initiates urine formation in most types of vertebrates is ultrafiltration of fluid from the blood plasma into kidney tubules under the force of blood pressure. What properties of the blood plasma in the urine-forming structures of the kidney are likely to affect the *rate* of ultrafiltration into the kidney tubules? Why is failure of urine production a potential side effect of heart disease?

6. During exercise—as compared to rest—what are the aspects of blood O_2 transport that can be modified to increase O_2 delivery to the exercising muscles? Based on the information discussed in this chapter, how do squids differ from vertebrates in the specific ways in which they increase O_2 delivery?

7. List the major types of blood vessels in the systemic circuit of a mammal, and outline the *functions* of each type of vessel.

8. What are the pros and cons of closed and open circulatory systems? Consider not only the types of information in Table 23.2, but also the other relevant information discussed in this chapter.

9. Mechanistically, why is pulmonary edema a threat to life? Thinking as creatively as possible, list three distinctly different potential causes of pulmonary edema.

10. Many objects designed by engineers have "circulatory systems" of one type or another. Describe at least three or four examples. Why do engineers sometimes include systems of flowing fluids in objects they design? We often hear it said that animal systems can teach us how to design engineered systems. If you were a biologist teaching a group of engineers about the lessons of animal circulatory systems, what principles would you emphasize?

Key References

Burggren, W., A. Farrell, and H. Lillywhite. 1997. Vertebrate cardiovascular systems. In W. H. Dantzler (ed.), *Comparative Physiology*, vol. 1 (Handbook of Physiology [Bethesda, MD], section 13), pp. 215–308. Oxford University Press, New York. A thorough and articulate review of most aspects of vertebrate circulatory function.

Cameron, J. N. 1989. *The Respiratory Physiology of Animals*. Oxford University Press, New York. An attractive feature of this book is its set of case studies in which circulatory physiology is synthesized with the physiology of breathing and blood gas transport. The case studies include invertebrates and vertebrates.

Farmer, C. G. 1997. Did lungs and the intracardiac shunt evolve to oxygenate the heart in vertebrates? *Paleobiology* 23: 358–372. A provocative effort to shake up thinking about incompletely divided hearts in vertebrates. The ideas are still being sorted out and evaluated.

Farrell, A. P. 1991. Circulation of body fluids. In C. L. Prosser (ed.), *Environmental and Metabolic Animal Physiology* (*Comparative Animal Physiology*, 4th ed.), pp. 509–558. Wiley-Liss, New York. A particularly useful attempt to review the circulation in all animals within 40 pages.

Guyton, A. C., and J. E. Hall. 2001. *A Textbook of Medical Physiology*, 10th ed. Saunders, Philadelphia. A human physiology with a tradition of excellence in its treatment of circulatory physiology.

Wilkens, J. L. 1999. Evolution of the cardiovascular system in Crustacea. *Amer. Zool.* 39: 199–214. An analysis of the pros and cons of open circulatory systems in an evolutionary context.

Selected References

Agnisola, C., and D. F. Houlihan. 1994. Some aspects of cardiac dynamics in *Octopus vulgaris* (LAM). *Mar. Fresh. Behav. Physiol.* 25: 87–100.

Badeer, H. S. 1997. Is the flow in the giraffe's jugular vein a "free" fall? *Comp. Biochem. Physiol. A* 118: 573–576. [See the article by Seymour, Hargens, and Pedley for a contrasting point of view.]

Belman, B. W. 1975. Some aspects of the circulatory physiology of the spiny lobster *Panulirus interruptus*. *Mar. Biol.* 29: 295–305.

Bourne, G. B., and B. R. McMahon (eds.). 1990. The physiology of invertebrate circulatory systems [A symposium]. *Physiol. Zool.* 63: 1–189.

Boutilier, R. G., M. L. Glass, and N. Heisler. 1986. The relative distribution of pulmocutaneous blood flow in *Rana catesbeiana*: Effects of pulmonary or cutaneous hypoxia. *J. Exp. Biol.* 126: 33–39.

Brill, R. W., and P. G. Bushnell. 2001. The cardiovascular system of tunas. In B. A. Block and E. D. Stevens (eds.), *Tuna: Physiology, Ecology, and Evolution*, pp.79–120. Academic Press, New York.

Burggren, W., and K. Johansen. 1982. Ventricular haemodynamics in the monitor lizard *Varanus exanthematicus*: Pulmonary and systemic pressure separation. *J. Exp. Biol.* 96: 343–354.

Burggren, W. W. 1982. Pulmonary blood plasma filtration in reptiles: A "wet" vertebrate lung? *Science* 215: 77–78.

Burggren, W. W. 1987. Form and function in reptilian circulations. *Amer. Zool.* 27: 5–19.

Burggren, W. W. 1995. Central cardiovascular function in amphibians: Qualitative influences of phylogeny, ontogeny, and season. *Advances in Comparative and Environmental Physiology,* vol. 21: 175–197.

Burggren, W. W., and K. Johansen. 1986. Circulation and respiration in lungfishes (Dipnoi). *J. Morphol.* Suppl. 1: 217–236.

Burggren, W. W., and A. W. Pinder. 1991. Ontogeny of cardiovascular and respiratory physiology in lower vertebrates. *Annu. Rev. Physiol.* 53: 107–135.

Burggren, W. W., B. B. Keller, and C. Weinstein (eds.). 1997. *Development of Cardiovascular Systems: Molecules to Organisms*. Cambridge University Press, New York.

Bushnell, P. G., and D. R. Jones. 1994. Cardiovascular and respiratory physiology of tuna: Adaptations for support of exceptionally high metabolic rates. *Environ. Biol. Fish.* 40: 303–318.

Cardiovascular adaptation in reptiles [A symposium]. 1987. *Amer. Zool.* 27: 3–131.

Cooke, I. M. 2002. Reliable, responsive pacemaking and pattern generation with minimal cell numbers: The crustacean cardiac ganglion. *Biol. Bull.* 202: 108–136.

Farmer, C. G. 1999. Evolution of the vertebrate cardiopulmonary system. *Annu. Rev. Physiol.* 61: 573–592.

Farrell, A. P. 1991. From hagfish to tuna: A perspective on cardiac function in fish. *Physiol. Zool.* 64: 1137–1164.

Farrell, A. P., and S. M. Clutterham. 2003. On-line venous oxygen tensions in rainbow trout during graded exercise at two acclimation temperatures. *J. Exp. Biol.* 206: 487–496.

Fritsche, R., M. Axelsson, C. E. Franklin, G. G. Grigg, S. Holmgren, and S. Nilsson. 1993. Respiratory and cardiovascular responses to hypoxia in the Australian lungfish. *Respir. Physiol.* 94: 173–187.

Functional morphology of the heart of vertebrates [A symposium]. 1968. *Amer. Zool.* 8: 177–229.

Grigg, G. C., and K. Johansen. 1987. Cardiovascular dynamics in *Crocodylus porosus* breathing air and during voluntary aerobic dives. *J. Comp. Physiol. B* 157: 381–392.

Hamlett, W. C., and R. Muñoz-Chápuli (eds.). 1996. Comparative cardiovascular biology of lower vertebrates [A symposium]. *J. Exp. Zool.* 275: 71–251.

Hargens, A. R., R. W. Millard, K. Pettersson, and K. Johansen. 1987. Gravitational haemodynamics and oedema prevention in the giraffe. *Nature* 329: 59–60. [A useful commentary is provided on pp. 13–14 of the same journal issue.]

Hartline, D. K. 1979. Integrative neurophysiology of the lobster cardiac ganglion. *Amer. Zool.* 19: 53–65.

Hicks, J. W., and T. Wang. 1998. Cardiovascular regulation during anoxia in the turtle: An in vivo study. *Physiol. Zool.* 71: 1–14.

Hicks, J. W., A. Ishimatsu, S. Molloi, A. Erskin, and N. Heisler. 1996. The mechanism of cardiac shunting in reptiles: A new synthesis. *J. Exp. Biol.* 199: 1435–1446.

Hill, R. B. (ed.) 1992. Control of circulation in invertebrates [A symposium]. *Experientia* 48: 797–858.

Hoppeler, H., O. Mathieu, E. R. Weibel, R. Krauer, S. L. Lindstedt, and C. R. Taylor. 1981. Design of the mammalian respiratory system. VII. Capillaries in skeletal muscles. *Respir. Physiol.* 44: 129–150.

Johansen, K. 1970. Air breathing in fishes. In W. S. Hoar and D. J. Randall (eds.), *Fish Physiology*, vol. 4, pp. 361–411. Academic Press, New York.

Johansen, K., and W. Burggren. 1980. Cardiovascular functions in the lower vertebrates. In G. H. Bourne (ed.), *Hearts and Heart-like Organs*, vol. 1, pp. 61–117. Academic Press, New York.

Johansen, K., and W. W. Burggren (eds.). 1985. *Cardiovascular Shunts.* Munksgaard, Copenhagen.

Jones, D. R., and D. J. Randall. 1978. The respiratory and circulatory systems during exercise. In W. S. Hoar and D. J. Randall (eds.), *Fish Physiology*, vol. 7, pp. 425–501. Academic Press, New York.

Jones, H. D., and D. Peggs. 1983. Hydrostatic and osmotic pressures in the heart and pericardium of *Mya arenaria* and *Anodonta cygnea*. *Comp. Biochem. Physiol. A* 76: 381–385.

Kiceniuk, J. W., and D. R. Jones. 1977. The oxygen transport system in trout (*Salmo gairdneri*) during sustained exercise. *J. Exp. Biol.* 69: 247–260.

LaBarbera, M. 1990. Principles of design of fluid transport systems in zoology. *Science* 249: 992–1000.

Laughlin, M. H. 1987. Skeletal muscle blood flow capacity: Role of muscle pump in exercise hyperemia. *Amer. J. Physiol.* 253: H993–H1004.

Lillywhite, H. B. 1996. Gravity, blood circulation, and the adaptation of form and function in lower vertebrates. *J. Exp. Zool.* 275: 217–225.

Mantel, L. H. (ed.). 1983. *The Biology of Crustacea*, vol. 5. Academic Press, New York. [See Chapters 1 and 6.]

Martin, A. W. 1980. Some invertebrate myogenic hearts: The hearts of worms and molluscs. In G. H. Bourne (ed.), *Hearts and Heart-like Organs*, vol. 1, pp. 1–39. Academic Press, New York.

McMahon, B. R., J. L. Wilkens, and P. J. S. Smith. 1997. Invertebrate circulatory systems. In W. H. Dantzler (ed.), *Comparative Physiology*, vol. 2 (Handbook of Physiology [Bethesda, MD], section 13), pp. 931–1008. Oxford University Press, New York.

Nilsson, S. 1995. Central cardiovascular dynamics in reptiles. *Adv. Comp. Environ. Physiol.* 21: 159–173.

O'Dor, R., H. O. Pörtner, and R. E. Shadwick. 1990. Squid as elite athletes: Locomotory, respiratory, and circulatory integration. In D. L. Gilbert, W. J. Adelman, Jr., and J. M. Arnold (eds.), *Squid as Experimental Animals*, pp. 481–503. Plenum, New York.

Paul, R. J., S. Bihlmayer, M. Colmorgen, and S. Zahler. 1994. The open circulatory system of spiders (*Eurypelma californicum, Pholcus phalangioides*): A survey of functional morphology and physiology. *Physiol. Zool.* 67: 1360–1382.

Pawloski, J. R., D. T. Hess, and J. S. Stamler. 2001. Export by red blood cells of nitric oxide bioactivity. *Nature* 409: 622–626.

Randall, D. J., and C. Daxboeck. 1982. Cardiovascular changes in the rainbow trout (*Salmo gairdneri* Richardson) during exercise. *Can. J. Zool.* 60: 1135–1140.

Randall, D. J., W. W. Burggren, A. P. Farrell, and M. S. Haswell. 1981. *The Evolution of Air Breathing in Vertebrates.* Cambridge University Press, New York.

Rose, R. A., J. L. Wilkens, and R. L. Walker. 1998. The effects of walking on heart rate, ventilation rate and acid–base status in the lobster *Homarus americanus*. *J. Exp. Biol.* 201: 2601–2608.

Satchell, G. H. 1991. *Physiology and Form of Fish Circulation.* Cambridge University Press, London.

Seymour, R. S., A. R. Hargens, and T. J. Pedley. 1993. The heart works against gravity. *Amer. J. Physiol.* 265: R715–R720. [See the article by Badeer for a contrasting point of view.]

Shadwick, R. E. 1998. Elasticity in arteries. *Amer. Sci.* 86: 535–541.

Shadwick, R. E., R. K. O'Dor, and J. M. Gosline. 1990. Respiratory and cardiac function during exercise in squid. *Can. J. Zool.* 68: 792–798.

Snyder, G. K., and B. A. Sheafor. 1999. Red blood cells: Centerpiece in the evolution of the vertebrate circulatory system. *Amer. Zool.* 39: 189–198.

Taylor, E. W., D. Jordan, and J. H. Coote. 1999. Central control of the cardiovascular and respiratory systems and their interactions in vertebrates. *Physiol. Rev.* 79: 855–916.

Tazawa, H., M. Mochizuki, and J. Piiper. 1979. Respiratory gas transport by the incompletely separated double circulation in the bullfrog, *Rana catesbeiana*. *Respir. Physiol.* 36: 77–95.

Trueman, E. R., and A. C. Brown. 1985. Dynamics of burrowing and pedal extension in *Donax serra* (Mollusca: Bivalvia). *J. Zool.* 207A: 345–355.

Wagner, P. D. 1996. Determinants of maximal oxygen transport and utilization. *Annu. Rev. Physiol.* 58: 21–50.

Weibel, E. R. 1984. *The Pathway for Oxygen.* Harvard University Press, Cambridge, MA.

Wells, M. J. 1992. The cephalopod heart: The evolution of a high-performance invertebrate pump. *Experientia* 48: 800–808.

Wells, M. J., and P. J. S. Smith. 1987. The performance of the octopus circulatory system: A triumph of engineering over design. *Experientia* 43: 487–499.

Wells, M. J., G. G. Duthie, D. F. Houlihan, P. J. S. Smith, and J. Wells. 1987. Blood flow and pressure changes in exercising octopuses (*Octopus vulgaris*). *J. Exp. Biol.* 131: 175–187.

West, N. H., and D. R. Jones (eds.). 1994. The form and function of open and closed circulations [A symposium]. *Physiol. Zool.* 67: 1257–1425.

See also references in Appendix A.

24 Oxygen, Carbon Dioxide, and Internal Transport at Work: Diving by Marine Mammals

The Weddell seal (*Leptonychotes weddellii*) captures all its food underwater, yet depends entirely on the atmosphere for oxygen. It seems, when first considered, to be one of the most physiologically improbable creatures one could imagine. Not only is it incapable of breathing where it hunts, it is large (400–500 kg), homeothermic, entirely predatory, and confronted by cold throughout its life. A Weddell seal lives through all seasons in the frigid Antarctic, either hauled out on ice sheets or diving in seawater at −2°C to capture fish, crustaceans, and squids. Despite the seeming improbability of such animals, the numbers of individuals and species tell a tale of ecological success. Early polar explorers, such as the Scottish sea captain James Weddell, brought home news of stupendous populations of diving mammals. Weddell seals today number near 1 million.

Worldwide, there are over 30 species of seals and sea lions, over 70 of whales and dolphins, and several other diving species of marine mammals, such as sea otters and sea cows. All are descended from terrestrial ancestors, and all were probably lured to the sea by a bounty of food. But as they evolved methods of feeding on the ocean's riches, marine mammals did not evolve ways of gleaning O_2 from the sea, and thus they remain tied to their ancestors' mode of breathing.

Diving by marine mammals first came under serious study by physiologists around 1930. From the beginning, physiological studies have focused on two basic questions: First, how do animals that are strictly dependent on the atmosphere for O_2 meet their metabolic energy demands during long periods under water? Second, how do diving mammals cope with the high pressures found at depth? This chapter focuses on the first question, but considers the second briefly at the end.

Diving Feats and Behavior

The diving feats of seals and whales have been appreciated in a limited way for centuries. During the heroic era of whaling (chronicled most famously in *Moby-Dick*), for example, the whalers were amazed by the depths to which wounded whales could dive. Sometimes a sperm whale had to be cut loose after "sounding" so deep as to draw out two lengths of harpoon line, each over 370 m long. When physio-

Weddell seals These large seals occur only in Antarctica, where they collect all their food by diving. Weddell seals often cut and maintain holes through the ice so as to be able to access their dual habitats: water for feeding and air for breathing.

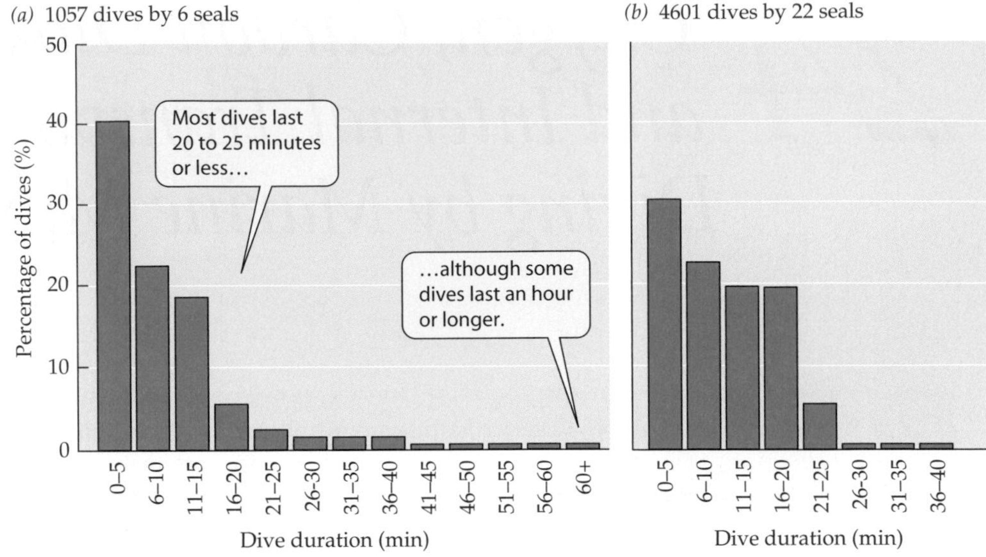

Figure 24.1 Durations of dives by wild Weddell seals These data come from two studies, which included over 5600 dives by free-living Weddell seals. Each vertical bar shows the percentage of dives falling within a particular duration category. Most dives are far shorter than the maximum duration for the species. (After Kooyman et al. 1980.)

logical research on animal diving began, few techniques existed for the study of free-ranging animals in the wild, and thus the first investigations were mostly carried out in bathtubs and swimming pools. By 1970, however, innovative scientists were starting to take advantage of the revolution in technology to monitor the diving behavior of free-living animals. They discovered that fact is indeed sometimes stranger than fiction. There are mammals that electively stay underwater—holding their breath—for 2 hours, and there are ones that go so deep that they leave the atmosphere—their source of O_2—a mile behind.

The Weddell seal proved to be a perfect subject for many of the early studies of diving behavior in the wild because in those early days, the only devices available to obtain data on the durations and depths of dives were innovative but primitive instruments that could record relatively small amounts information while attached to an animal. The instruments had to be reclaimed for the data to be acquired. Weddell seals living on ice sheets gain access to the water for feeding by cutting and maintaining holes through the ice. After a seal dives through an ice hole, it must ultimately return to that or another hole to breathe. Gerald Kooyman and his colleagues capitalized on this trait. They would attach a data recorder to a wild Weddell seal and then permit the seal to live in its ordinary way for several days, whereupon they would find the seal at one of its breathing holes to remove the device. Using this approach, they were able—for the first time with any species—to describe the durations and depths of hundreds and thousands of voluntary dives.

The most fantastic revelation of the early studies was that Weddell seals sometimes stay submerged voluntarily for over an hour (**Figure 24.1**); we know today that their dives occasionally last as long as 80 minutes! Equally important, however, was the revelation that the vast majority of dives by Weddell seals are considerably shorter than the maximum durations observed; most dives last 20 to 25 minutes or less. The pattern of diving depths resembles that of diving durations in that during most dives, seals go to depths that are considerably shallower than the maximum the species is capable of reaching (**Figure 24.2**). On rare occasions, a Weddell seal descends to nearly 600 m—over one-third of a mile down—but few dives are deeper than 400 m, and most are to depths of 200 m or less. While diving, a seal is sometimes subject to stupendous physical pressures. A useful rule of thumb is that water pressure increases by about 1 atmosphere (101 kPa) for every 10 m of depth. Thus a seal diving to 400 m voluntarily subjects itself to about 40 atm of hydrostatic pressure.

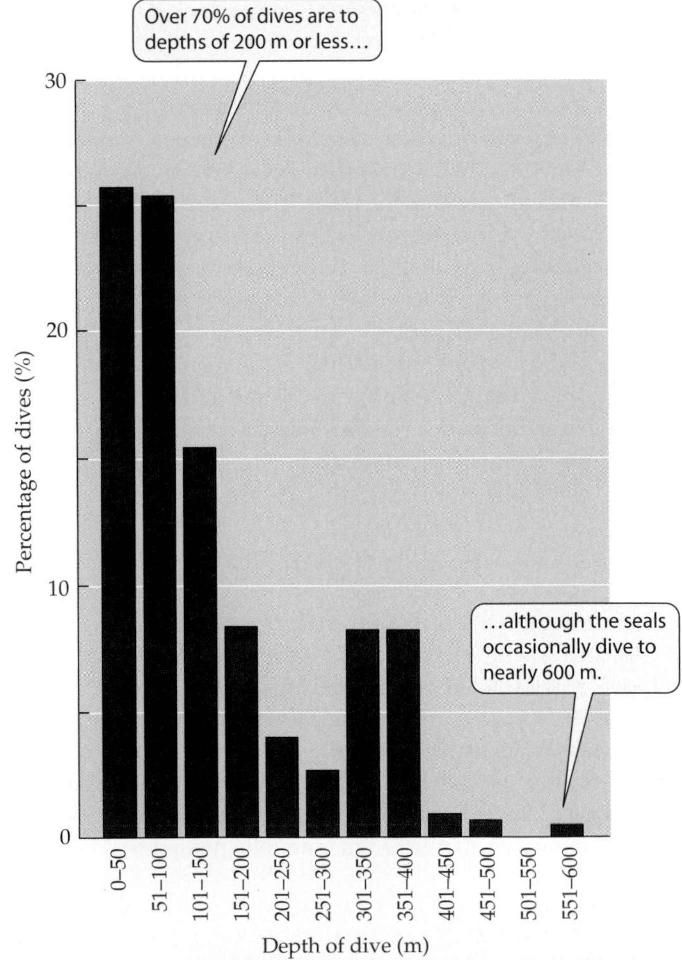

Figure 24.2 Depths of dives by wild Weddell seals These data represent over 380 dives by 27 individuals. Each vertical bar shows the percentage of dives falling within a particular depth category. Most dives are substantially shallower than the maximum depth for the species. (After Kooyman 1966.)

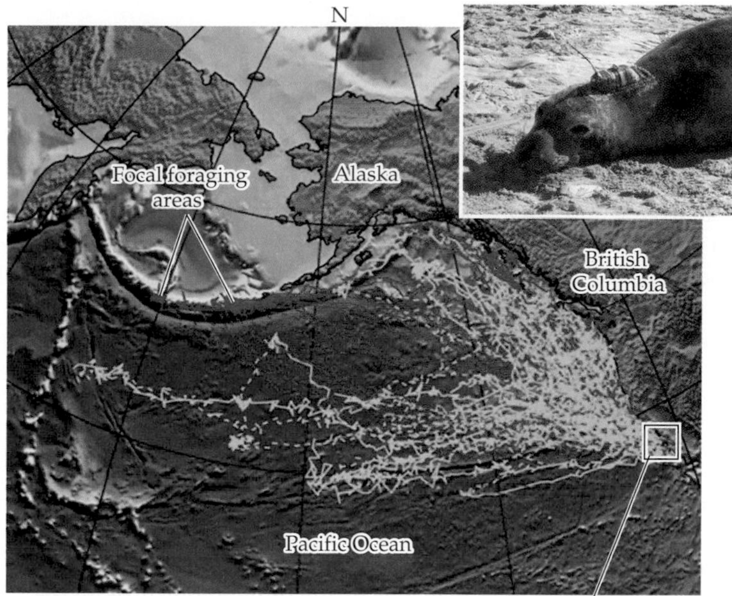

Figure 24.3 Migrations of northern elephant seals The elephant seals that breed at Año Nuevo in California get their food at distant places in the Pacific Ocean. Shown here are the migration routes of over 30 animals from the Año Nuevo rookery. They were equipped with radio transmitters that signaled their location (see inset) and followed by satellite. The "focal foraging area" of an animal, which usually looks like a tight cluster of data in these records, is the part of the ocean where it lingered for many weeks of diving and feeding following its outward journey. When animals leave Año Nuevo, they tend to travel without pause to their focal foraging areas. During successive migratory trips away from Año Nuevo, individuals often return to their specific focal foraging areas using much the same outbound and inbound travel routes on each trip. Red represents males; yellow, females. (Courtesy of Dan Costa; after Le Boeuf et al. 2000.)

Today, technology has advanced to the point that data can be radioed from free-living marine mammals and picked up on a global scale by satellite receivers. A dramatic study using this technology was recently completed on the northern elephant seals (*Mirounga angustirostris*) that breed in a famous colony at Año Nuevo in California (near San Francisco). Both sexes spend most of their lives at sea. They visit land on two occasions each year to mate, molt, and (in the case of females) rear young. Between these visits to land, they undertake long oceanic migrations, often across half the breadth of the Pacific Ocean, during which they feed intensively. The longitudes and latitudes of individual animals can be followed throughout their migrations using radio and satellite technology (Figure 24.3). For monitoring diving depths and durations, instruments that log the data rather than transmitting them are still sometimes used. Modern digital logging instruments have such prodigious memories that they can archive a seal's depth at all times for many months; thus, once an instrument attached to a seal at the start of its migration is reclaimed, all the animal's dives throughout its whole migratory trip can be correlated with its satellite-recorded longitudes and latitudes in the ocean. Figure 24.4 shows the depths of all the dives undertaken by an elephant seal called Moo while in transit between Año Nuevo and his principal foraging area near the Aleutian Islands. Moo's deepest dive was to about 750 m (approaching half a mile). Another elephant seal (a female) was once observed to dive to almost 1600 m (essentially a mile): the deepest dive ever documented for an air-breathing life form! Note, however, that most dives by northern elephant seals are to depths far less than these maxima (Figure 24.4).

The marine mammals are far from uniform in their diving capabilities. The major groups—such as the true (phocid) seals, fur seals, and whales—have different phylogenetic histories, and within each group, species have diversified. Weddell seals, elephant seals, and ribbon seals are among the most proficient divers. The longest voluntary dive ever recorded, 2 hours, was observed in a southern elephant seal. Sperm whales (*Physeter catodon*) also rank with the proficient divers. They occasionally remain submerged for 60 to 70 minutes when diving freely and have been known to dive for at least 90 minutes after being harpooned. Free-ranging sperm whales have been detected by sonar at 1140 m (7/10 of a mile down). The opposite end of the spectrum of diving proficiency is exemplified by certain of the fur seals, such as northern fur seals (*Callorhinus ursinus*).[1] Based on records of over 3000 dives, northern fur seals do not dive longer than 8 minutes or deeper than 260 m. Despite wide variation in the extreme dives of which species are capable, all species of diving mam-

[1] The fur seals and sea lions—which compose the group known as otariid seals—tend as a group to be less proficient divers than the true (phocid) seals.

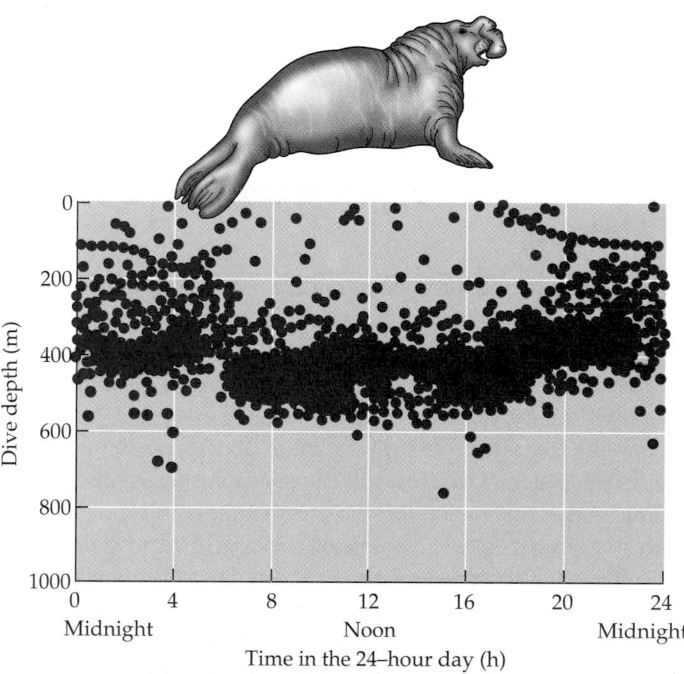

Figure 24.4 Diving by an individual elephant seal The maximum depth reached by Moo, a male northern elephant seal, during each of his dives is plotted against the time of day of the dive. Each symbol represents a single dive. This plot includes all dives that Moo undertook during several weeks while he was in transit between Año Nuevo and his focal foraging area in Alaskan waters. Even though Moo had to pay the metabolic cost of swimming 6500 km (round trip) during this migration, he increased his body weight from 1270 to 1650 kg by diving for food. Moo's dives were recorded by a time–depth logger with prodigious memory that was attached to him before his departure from Año Nuevo and recovered after his return. (After Le Boeuf et al. 2000.)

mals (and birds) seem to adhere to the two important generalizations that we have already illustrated: (1) The durations of most dives are substantially shorter than the maximum duration of which each species is capable, and (2) most dives are to depths substantially more shallow than the species-specific maximum depth. Dives of record duration and depth, although awe-inspiring and important, are uncommon.

To put the performances of marine mammals into perspective, it is instructive to look at the capabilities of a representative terrestrial mammal, *Homo sapiens*. Trained human breath-hold divers are generally limited to about 3 minutes of submergence when at rest and about 90 seconds when swimming. The *ama* of Korea and Japan earn their living by diving for shellfish and edible seaweeds. Their abilities are typical of those exhibited by other people (e.g., pearl and sponge divers) who employ breath-hold diving as a career. The ama who dive in deep waters receive assistance during descent and ascent so their time at the bottom can be maximized. They carry weights to aid descent and are pulled back to the water's surface on a rope by an assistant in a boat. They routinely remain submerged for 60 to 80 seconds and reach depths of 15 to 25 m. They rest only about a minute between dives and thus average about 30 dives per hour. The extremes recorded for human performance during breath-hold diving have been attained by competitive divers. The world-record holders reach depths of over 100 m—with assistance during descent and ascent—in dives that last about 3.5 minutes. Dives of this duration and depth can be carried out only by exceptional people and, even for them, are dangerous. However, such dives are trivial and commonplace for average individuals of many species of marine mammals.

Types of Dives and the Importance of Method

An important theme throughout science—dramatized in the study of diving physiology—is that the methods employed in experiments often affect the results and, therefore, our perception of reality. During the first decades of the modern study of diving physiology (roughly 1935–1970), scientists believed that they should "bring diving into the laboratory" to study it. In the laboratory, they could establish experimental controls and employ sophisticated instrumentation; another motivation for laboratory studies was simply that the study of diving in the wild seemed to be largely impossible. In the laboratory studies that were carried out, animals were often strapped to a movable platform and lowered underwater whenever an investigator wished to elicit diving responses. By the 1960s, research of this sort had produced such a seemingly complete picture of diving physiology that most biologists believed that the subject was fully understood and could be relegated to the history of science. Then some radical investigators began to study animals such as Weddell seals in the wild. The methods they had available to use were initially primitive compared to those available in the laboratory. Nonetheless, the early studies they carried out showed that the physiology of *voluntary* diving in the wild often differs from that of *forced* diving in laboratories. Perceptions thus changed. The initial view had been that diving physiology shows little variation from one dive to another because it is controlled by highly stereotyped physiological reflexes. The newer view—now abundantly confirmed—was that marine mammals use a spectrum of physiological strategies during different sorts of dives. Today, even as advances in technology have greatly expanded the range of physiological mechanisms that can be studied in the wild, laboratory studies continue to play a role because certain mechanisms can be studied only in laboratories even now.

Recognizing that laboratory and field studies of diving may produce different results, one of the first questions to ask about any set of data is whether it was obtained by forcing animals underwater or by studying free-living animals diving voluntarily. Another important question is whether the dives studied were long or short *relative to the species-specific maximum dive length*, because we know now that qualitatively different suites of responses may be marshaled in long and in short dives. Still another key question is whether the studied animals were active or quiet during diving. These, then, are some of the important organizing distinctions among dives in the study of diving physiology:

- Forced or voluntary?
- Short or long (relative to the maximum length for the species)?
- Quiet or active?

Physiology: The Big Picture

When diving mammals break contact with the atmosphere, they carry O_2 with them in three major internal stores: O_2 bound to blood hemoglobin, O_2 bound to muscle myoglobin, and O_2 contained in air in their lungs. These stores permit aerobic catabolism to continue to some extent during a dive. The internal stores of O_2 are adequate in principle to permit *all* the tissues of a diver's body to function aerobically throughout a relatively short dive. When we look at the physiology of actual dives, we find, in fact, that voluntary dives of relatively short duration are mostly or completely aerobic in a variety of diving species, according to available evidence.

The internal O_2 stores of diving mammals, however, are utterly inadequate to permit fully aerobic catabolism throughout an animal's body during protracted dives. How, then, are some species able to survive for 30 minutes, 60 minutes, or even longer without breathing? Our basic concept of how energy needs are met during protracted dives was first proposed by Laurence Irving (1895–1979) in 1934. He recognized that certain tissues—notably the central nervous system and heart—are strongly or exclusively dependent on aerobic catabolism for production of ATP; they need O_2 on a steady basis and are quickly damaged by O_2 insufficiency. However, other tissues—such as skeletal muscle—have a well-developed ability to meet their ATP demands anaerobically and thus are relatively tolerant of O_2 deprivation. Irving then reasoned that dives could be prolonged if animals were to "reserve" a certain portion of their O_2 supplies for the tissues that are O_2-dependent. During a dive, if all tissues have equal access to an animal's entire O_2 store, then the concentration of O_2 throughout the body will fall quickly to such a low level that the O_2-dependent tissues are impaired. Under such circumstances, a seal or whale will need to surface, even though many of its tissues could continue to function—anaerobically—for a longer time. On the other hand, if some O_2 is reserved for use by the O_2-dependent tissues during a dive, then those tissues can continue to have adequate O_2 even while other parts of the body exhaust their O_2 supplies and resort to anaerobic catabolism, thus extending the time the animal can remain submerged.

The preferential delivery of some O_2 to the O_2-dependent tissues is achieved, as Irving predicted, by adjustments of circulatory function. These adjustments were first elucidated by Irving and Per Scholander (1905–1980) in laboratory experiments involving forced dives, many of which were conducted in the Swarthmore College swimming pool. During forced dives, blood flow to the skeletal muscles of the appendages and trunk, the skin, the gut, the kidneys, and certain other tissues is curtailed by vasoconstriction in the arterial vessels that supply those tissues and organs. The skeletal muscles and other tissues that are deprived of active blood flow can make use of the hemoglobin-bound O_2 in the blood sequestered in their capillaries, and they can use their myoglobin-bound O_2. However, as those local O_2 stores are depleted, the circulation-deprived tissues turn to anaerobic catabolism, and lactic acid accumulates in them. *With the circulation to many parts of the body curtailed, the heart pumps blood primarily between itself and the lungs and head.* The O_2 stores of the circulating blood are thereby reserved primarily for the O_2-dependent tissues—the brain and heart—and whatever O_2 is extracted from the air in the lungs is likewise delivered preferentially to those tissues. Consequently, adequate O_2 partial pressures can be maintained in the O_2-dependent tissues for a long period. Lactic acid produced by the skeletal muscles and other circulation-deprived tissues tends to remain sequestered in those tissues during a dive precisely because the tissues receive little or no blood flow. When the animal surfaces, however, circulation to such tissues is restored, and there is a sudden rise of lactic acid in the circulating blood. *The observation that circulating lactic acid increases principally* after *a forced dive was one of the earliest pieces of evidence that circulatory function is radically altered during diving.*

When animals are forcibly submerged, the adjustments in the pattern of blood flow just described tend to occur rapidly, consistently, and to a profound extent. Accordingly, during the era prior to 1970, when diving was studied mainly in laboratories, these responses were labeled a *diving reflex*. Today, however, based on studies of voluntarily diving animals, physiologists recognize that the responses of the circulatory system are not nearly as inflexible and stereotyped as once was thought. Indeed, as suggested earlier, there is good reason to believe that in certain species, relatively little redistribution of blood flow occurs during voluntary dives that are short enough for all energy needs to be met by aerobic catabolism using an animal's O_2 stores. Seals and whales in the wild undergo a profound redistribution of blood flow when they dive voluntarily for long periods of time. Animals that are forced underwater probably exhibit stereotyped and "reflexive" circulatory responses because they sense that they have no control over the length of time they will be submerged, and thus they consistently marshal the responses they employ for prolonged diving in the wild.

From this overview, you can see that the O_2 stores, circulatory physiology, and metabolic physiology of marine mammals all play critical roles in understanding their diving physiology. The next three sections discuss these three elements in more detail.

The Oxygen Stores of Divers

Because a dive can last only as long as the brain is supplied with O_2, the size of a diving mammal's total O_2 store is obviously a key determinant of how long the animal can stay submerged. Besides determining how long the brain can endure, the size of the total O_2 store is also highly relevant even in determining how long the skeletal muscles and other anoxia-tolerant tissues can endure during a dive. The skeletal muscles and other such tissues can continue to function only until they reach their maximum tolerable lactic acid concentration. A large O_2 store can prolong muscle endurance during dives by postponing the onset of anaerobic catabolism and the consequent accumulation of lactic acid.

The blood O_2 store tends to be large in diving mammals

The amount of O_2 that can be stored in the blood depends on three features: (1) the oxygen-carrying capacity of the blood, (2) the total volume of blood, and (3) the degree to which the blood is fully loaded (saturated) with O_2 at the time of submergence.

Regarding the oxygen-carrying capacity of the blood, although some species of diving mammals have values that are well within the ordinary range for nondiving, terrestrial mammals, some other species have exceptionally high values. Bottlenose dolphins, northern fur seals, and Steller (northern) sea lions, to cite some examples, have oxygen-carrying capacities of 17 to 22 vol %—quite ordinary for mammals. On the other hand, species of diving mammals known to have especially high oxygen-carrying capacities include the harbor seal (26–29 vol %), pygmy sperm whale (32 vol %), Weddell seal (29–36 vol %), and ribbon seal (34 vol %). Among the seals taken as a group, a trend is evident for species that undergo long dives—such as the three true seals just mentioned—to have higher oxygen-carrying capacities than species that perform shorter dives.

Accurate determinations of blood volume have been made for only a relatively few species of diving mammals, but the evidence indicates a tendency toward high volumes. Humans, dogs, horses, and rabbits have average blood volumes of 60 to 110 mL per kilogram of body weight. By contrast, such accomplished divers as harbor and ribbon seals have blood volumes of 130 to 140 mL/kg, and the blood volumes of Weddell seals, northern elephant seals, and sperm whales are estimated to be 200 to 220 mL/kg—two or three times as high as is typical of terrestrial mammals.

An animal's maximum possible blood store of O_2 is calculated by multiplying the oxygen-carrying capacity of its blood by its blood volume. This figure is only indirectly relevant to normal physiology because the entire volume of blood is never fully oxygenated. However, the figure is useful for comparing species. The maximum possible blood O_2 store of humans and horses is about 14 to 15 mL O_2 per kilogram of body weight. In contrast, very much higher O_2 storage capacities are found in the species of seals that combine the advantages of both a high oxygen-carrying capacity and a high blood volume. Average reported values for the maximum blood O_2 store in Weddell seals, harbor seals, and ribbon seals are 40 to 70 mL O_2/kg—as much as four or five times higher than the stores of humans and horses.

Diving mammals have high myoglobin concentrations and large myoglobin-bound O_2 stores

The amount of O_2 stored as oxymyoglobin at the time of submergence depends on how much myoglobin is present in each unit of muscle tissue. *One of the most consistent features of diving species of mammals is that they have dramatically high myoglobin concentrations in their skeletal muscles compared to terrestrial species.* The skeletal muscles of some proficient divers are so rich in myoglobin that they are almost black. In humans and horses, skeletal muscles contain about 4 to 9 mg of myoglobin per gram of wet weight. By contrast, harbor seals, Weddell seals, and northern elephant seals have 50 to 55 mg/g, and ribbon seals and Fraser's dolphins have 70 to 80 mg/g!

Oxymyoglobin represents an essentially private store of O_2 for the muscles. As discussed in Chapter 22 (see page 595), myoglobin has such a high affinity for O_2 that it typically draws O_2 from blood hemoglobin rather than donating O_2 to the blood. Thus, even if muscles receive blood circulation during a dive, oxymyoglobin within them does not yield much O_2 to the blood for use elsewhere in the body. Instead, the O_2 remains bound to the myoglobin until the O_2 partial pressure in the muscles falls to a low level; then it is donated to the muscle mitochondria to permit continued aerobic ATP production in the muscles.

Diving mammals vary in their use of the lungs as an O_2 store

To understand lung function in diving mammals, one must first recognize the unusual structural flexibility of the thorax in these animals. In the deep-diving marine mammals that have been studied, the thorax is structured in ways that permit the thoracic walls to be shoved freely inward as the outside water pressure increases (Figure 24.5). Thus, over extensive ranges of depth and pressure, the *thoracic cavity and lungs are freely compressible*, meaning that their volumes respond to changes of outside pressure as would be predicted from a simple application of the universal gas law (see Equation 20.1). A tenfold increase in the pressure applied to the thorax, for example, reduces the volume of the air inside the lungs to a tenth of what it previously was. Humans and other terrestrial mammals are different in that their thoracic walls are structured less flexibly and resist compression. In the seals, which are the marine mammals for which the most-rigorous quantitative data are available, the thoracic wall offers no resistance at all to collapse over the whole range of depths and pressures the animals experience.

One might at first think that a large air store in the lungs would be of unquestioned advantage for a diving mammal. Three considerations, however, argue against this conclusion. First, a large amount of air in the lungs can strongly buoy a diving animal upward, forcing it to work hard to remain submerged. Second, the alveoli are believed to be typically the first parts of the lungs to collapse as the lung air compresses at depth. Hence, at depth, the lung air comes to be contained mostly in the *conducting* airways—the trachea, bronchi, and nonrespiratory bronchioles. Once this occurs, the O_2 in the lung air is unavailable, because little transfer of O_2 from the lung air to the blood can occur in the conducting airways. Thus, whereas diving mammals can make effective use of their pulmonary O_2 stores during shallow diving, they are not able to do so at depth. A final consideration is that a large pulmonary air store

(a) Dolphin near the surface

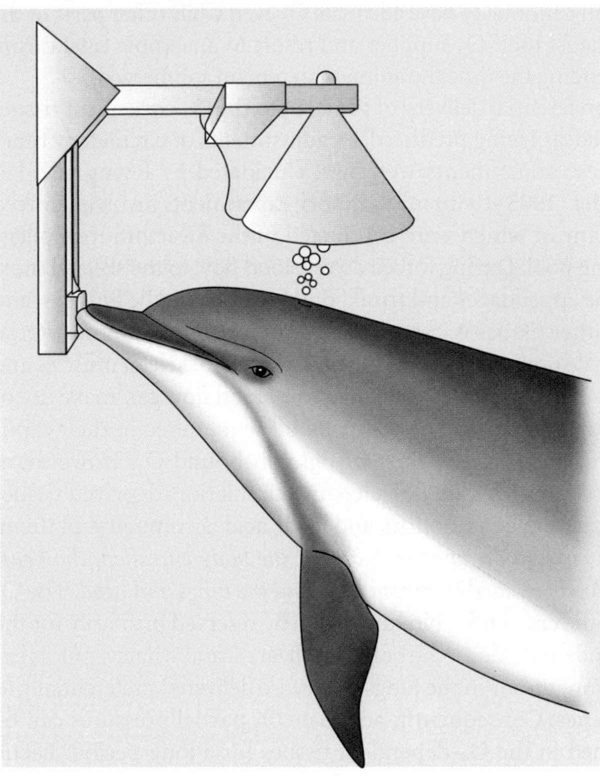

(b) Dolphin at a depth of 300 m

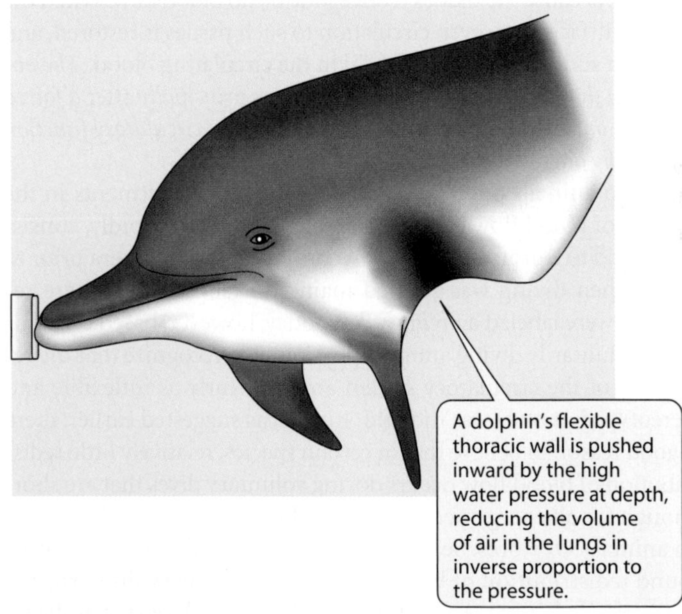

Figure 24.5 The thorax is highly compressible in marine mammals Portrayed here is a bottlenose dolphin *(Tursiops truncatus)* that had been trained to push a button to signal its presence as it traveled between the surface of the water and a depth of 300 meters. When it exhaled near the surface of the water *(a)*, its exhalant air was collected for analysis. When it pushed the button at 300 m *(b)*, a camera was activated to take its photograph. The ambient pressure at 300 m was about 30 atm. Photographs of the dolphin at that depth revealed that its thoracic wall was pushed far inward by the elevated ambient pressure. (After Ridgway, Scronce, and Kanwisher 1969.)

means not only a large store of O_2 in the lungs, but also a large store of N_2 (air is 78% N_2), which can increase the odds that an animal will develop decompression sickness (the bends), as discussed later in this chapter.

The size of the air store in a marine mammal's lungs at the start of a dive depends on two factors: the *volumetric capacity* of the lungs (the amount of air they can hold when maximally inflated) and the degree to which the animal inflates its lungs before diving. Regarding the first factor, marine mammals do not as a rule have exceptionally large lungs; their volumetric capacities per unit of body weight are generally similar to those of terrestrial mammals or just modestly larger. Moreover, many of the deep-diving seals and whales have volumetric capacities that are relatively low by comparison to those of other species of seals or whales of similar size that dive more shallowly. Regarding the second factor, the degree of lung inflation, some marine mammals dive after a vigorous inhalation and thus may carry an amount of lung air that approaches their volumetric capacity. This appears to be true, for example, of whales (including dolphins). By contrast, many deep-diving species of true (phocid) seals dive following *exhalation*; their lungs are thus filled to just 20% to 60% of their volumetric capacity. All things considered, little premium seems to be placed on having an exceptionally large pulmonary air (and O_2) store in marine mammals.

Total O_2 stores never permit dives of maximal duration to be fully aerobic

Three major points emerge from the study of the O_2 stores of marine mammals:

1. Some species of diving mammals have much greater total O_2 stores per unit of body weight than terrestrial mammals because they have high blood oxygen-carrying capacities and high blood volumes—giving them high blood O_2 stores—and they have high concentrations of myoglobin in their muscles—giving them high myoglobin-bound O_2 stores (**Figure 24.6**).
2. Among diving mammals, species such as fur seals and sea lions that dive for relatively short periods tend to have smaller O_2 stores per unit of body weight than species such as harbor and ribbon seals that are more proficient as divers and dive for longer periods (see Figure 24.6).
3. The O_2 stores of diving mammals are utterly inadequate to sustain a rate of O_2 consumption during long dives that is equivalent to the rate of O_2 consumption of these animals while they are at rest and breathing air.

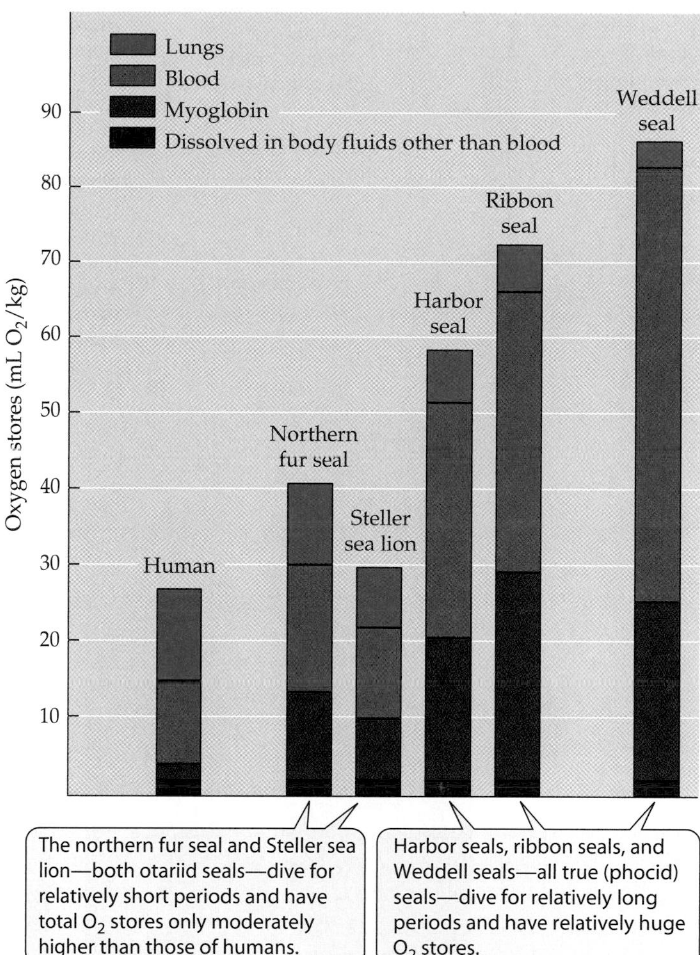

Figure 24.6 A comparison of the total O_2 stores of five species of marine mammals and humans The total O_2 available per kilogram of body weight for each species is estimated by adding the O_2 dissolved in body fluids other than blood throughout the body, the O_2 bound to myoglobin in muscles, the O_2 in blood (mostly bound to hemoglobin), and the O_2 in lung air. The lungs of humans are assumed to be fully inflated, but those of the other species are assumed to be only half inflated because many seals dive after exhaling. The data for the four species in the middle were calculated by one research team using a standardized set of procedures and assumptions (Lenfant, Johansen, and Torrance 1970), whereas the data for the Weddell seal (Ponganis, Kooyman, and Castellini 1993) and human, although approximately comparable, are from other sources.

The third point, which is very important, is illustrated by the following recent measurements carried out by Paul Ponganis and his colleagues. They estimated the total available O_2 store of a 450-kg Weddell seal to be 38.8 L.[2] The resting rate of O_2 consumption of such a seal while breathing air is 1.9 to 2.3 L O_2 per minute. Thus the seal could sustain its resting, aerial rate of O_2 consumption for 17 to 20 minutes during diving if it completely used its available O_2 store. Actually, however, Weddell seals sometimes dive for 60 to 80 minutes (see Figure 24.1*a*). Similar calculations have been performed for many other diving species, and always, dives of maximal length turn out to last from two to several times longer than would be predicted if the animal were to function aerobically at the rate seen during rest in air. Furthermore, the diving capabilities of long-duration divers are disproportionately greater than can be accounted for by the relative expansion of their O_2 stores compared to terrestrial mammals of similar body size. For example, consider harbor seals and humans, two species of roughly the same body size. Although the weight-specific O_2 stores of harbor seals are 2 to 2.5 times those of humans (see Figure 24.6), the seals can remain submerged for over 12 times as

[2] To derive this amount from Figure 24.6, multiply the size of the O_2 store per kilogram, 86 mL O_2/kg, by the seal's weight, 450 kg.

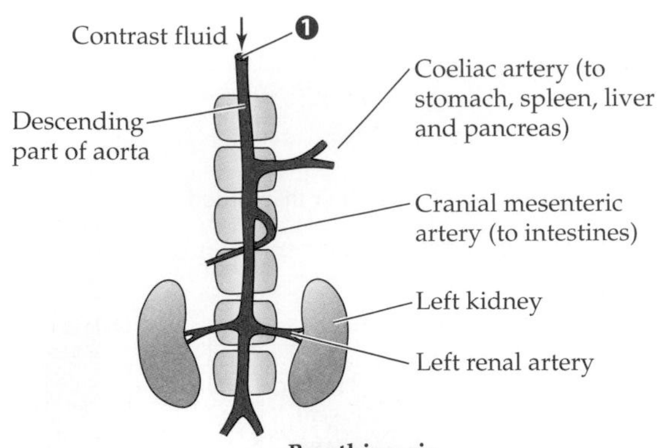

Figure 24.7 Circulatory patterns are radically changed during forced or prolonged submergence X-ray images of the visceral cavity of a harbor seal (*Phoca vitulina*), viewed from its ventral side after injection of contrast fluid during air breathing and during forced submergence. Although the blood labeled with contrast fluid flowed into all major branches of the aorta within seconds during air breathing, the blood was unable to freely flow out of the aorta when the seal was submerged. ❶ Contrast fluid was injected into the descending portion of the systemic aorta. Within 0.5 seconds after the contrast fluid was injected when the seal was breathing air, the coeliac artery ❷ and renal artery ❸ were filled with blood containing contrast fluid; and within 2 second after injection, the entire left kidney was filled with blood containing contrast fluid ❹. When the seal was submerged, however, the coeliac artery ❺ and renal artery did not readily fill with blood containing contrast fluid, and after 4 seconds, the kidney ❻ remained dark—showing that it had not received blood with contrast fluid. L2 =second lumbar vertebra; R = right side of seal. (Photos courtesy of Klaus Bron; from Bron et al. 1966.)

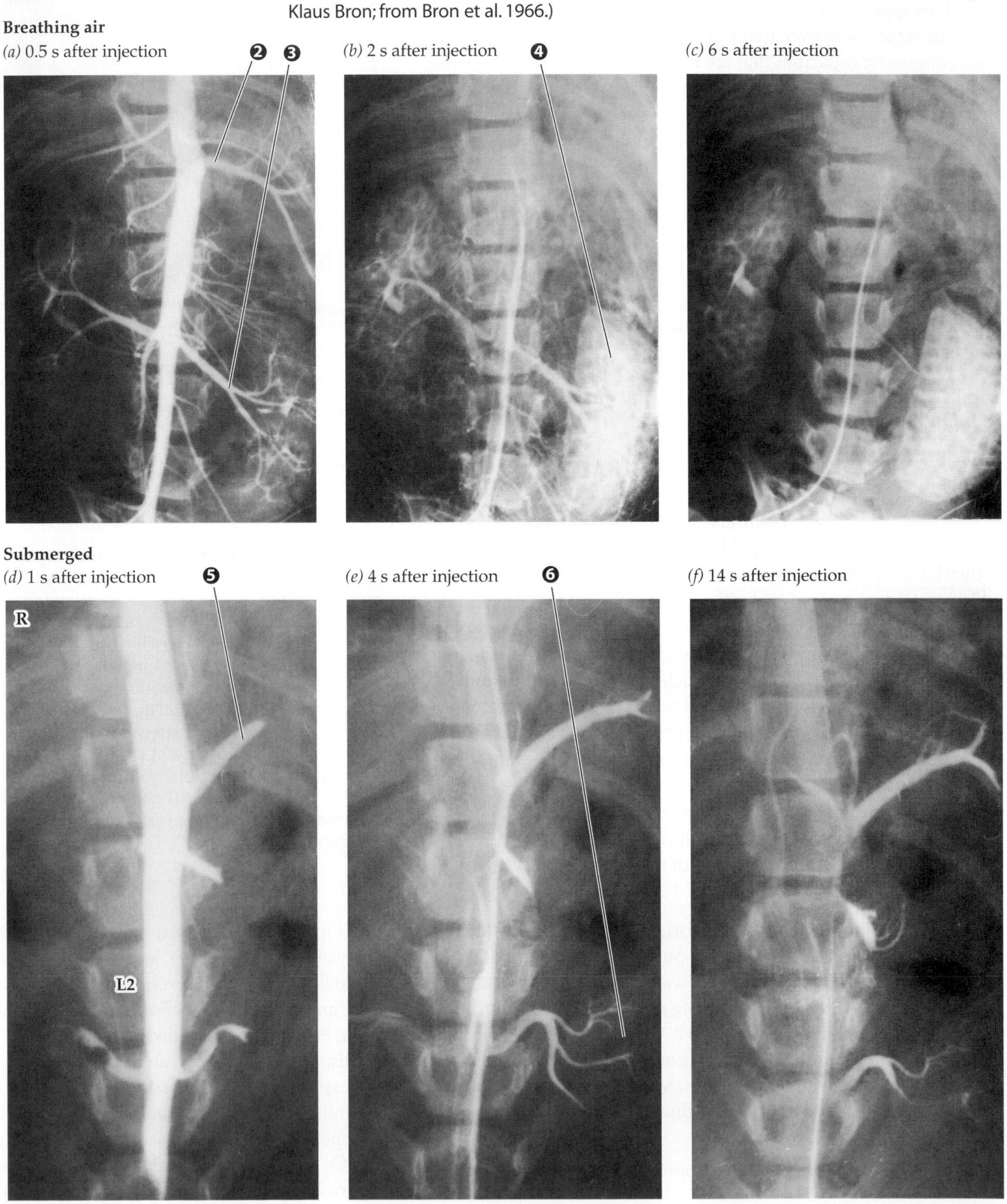

Breathing air

(a) 0.5 s after injection ❷ ❸

(b) 2 s after injection ❹

(c) 6 s after injection

Submerged

(d) 1 s after injection ❺

R

L2

(e) 4 s after injection ❻

(f) 14 s after injection

long as humans when diving under comparable conditions. Regardless of how large the O_2 stores of diving species may be, they do not in themselves explain the lengths of submergence of which the animals are capable.

Circulatory Adjustments during Dives

The circulation holds a special place in the chronicles of diving physiology because the very first physiological observations on diving were measures of heart rates. These observations began in 1870, when the French physiologist Paul Bert (1833–1886) discovered that the heart rates of ducks decreased from 100 to 14 beats per minute when he forced the ducks under water. A decrease in heart rate during diving is called **diving bradycardia** (*brady*, "slow"). Bert and others soon demonstrated that the phenomenon is a consistent feature of forced submergence in diving mammals and birds. The universality of diving bradycardia persuaded early physiologists that the slowing of the heart is important in permitting animals to stay submerged for extended periods of time. But how is it important? Diving bradycardia proved to be only the tip of a physiological iceberg, and like the tips of real icebergs, this phenomenon—considered in isolation—barely hinted at the nature of the systematic physiological response of which it is a part. In the 1930s, Irving and Scholander took the crucial step of postulating that other parts of the cardiovascular system, in addition to the heart, respond during diving. One of their earliest tests of their idea was to study bleeding from the paw skin of seals. Small cuts that bled freely when the seals were breathing air stopped bleeding suddenly and completely when the seals were pushed underwater. From simple observations like this, the modern revolution in diving physiology began, and after seven decades of knowing of the *existence* of diving bradycardia, scientists began to understand its *significance*.

Regional vasoconstriction: Much of a diving mammal's body is cut off from blood flow during forced or protracted dives

Once physiologists realized that a marine mammal's pattern of blood flow might change during dives, they employed several techniques to examine how the vascular system functions during diving. One such technique involves the use of a *contrast fluid*, a fluid that is impenetrable by X-rays. When a contrast fluid is injected into an animal's bloodstream, its flow can be observed on X-ray images. In one of the watershed studies in the history of diving physiology, a group of diagnostic radiologists headed by Klaus Bron injected a contrast fluid into the aorta of a harbor seal, then took X-ray images to observe subsequent blood flow in the seal's visceral arterial system while the animal was breathing air and while it was forcibly submerged (Figure 24.7).

While the seal breathed air, an X-ray image taken 0.5 seconds after the injection of the contrast fluid showed that blood was flowing vigorously from the aorta into abdominal arteries that branch off from the aorta, such as the coeliac and renal arteries (Figure 24.7*a*). Within 2 s after injection, the left kidney was illuminated in the X-ray image, showing that the profusely branching arterial system within the kidney had filled with blood that contained the contrast fluid (Figure 24.7*b*). After only 6 s, the coeliac and renal arteries had already faded from view (Figure 24.7*c*), demonstrating that the contrast fluid had flowed through them.

The sequence of X-ray images taken while the seal was submerged revealed a strikingly different circulatory pattern. At 1 s after injection, the X-ray image showed that the blood containing the contrast fluid had penetrated only a short distance from the aorta into the coeliac and renal arteries and stopped, leaving the branches of those arteries unfilled (Figure 24.7*d*; compare to Figure 24.7*a*). Even after 4 s, the left kidney remained dark, indicating that most of its arterial vessels had not received blood containing contrast fluid (Figure 24.7*e*). As long as 14 s after injection, the bases of the coeliac and renal arteries remained filled with contrast fluid (Figure 24.7*f*), indicating that the blood in those arteries had stagnated and was unable to flow through the vessels.

Bron's pioneering radiological study revealed that blood flow to major parts of a marine mammal's body is profoundly curtailed during a forced dive. The coeliac and renal arteries supply blood to the stomach, spleen, liver, pancreas, and kidneys. During a dive, blood is unable to flow freely (if at all) into any of those visceral organs. Subsequent research revealed that the flow of blood is blocked by vasoconstriction under control of the sympathetic nervous system. Strikingly, although arterioles are the usual sites of vasomotor control in mammals (see page 622), pronounced vasoconstriction occurs in *sizable arteries* in at least some diving species, as revealed by Figure 24.7. Arteries constrict shut, and the organs they supply are denied blood flow.

The parts of a diving mammal's body that receive little or no blood flow during forced dives commonly include the animal's limbs, the skeletal muscles of its trunk, its pectoral muscles, its skin and body wall, and such visceral organs as the gastrointestinal tract, liver, spleen, and kidneys. Simultaneously, because vasoconstriction occurs selectively, blood flows freely—or relatively freely—to the animal's brain, lungs, and myocardium, as well as possibly to some other body regions. A pundit once said that during forced dives, a seal or whale becomes a "heart-lung-brain machine," and there is truth to this remark (Figure 24.8). The unrestricted, *active* circulation in the forcibly submerged seal is as follows: Blood is pumped by

Figure 24.8 The forcibly submerged seal as a "heart-lung-brain machine" With blood flow to much of the body curtailed by vasoconstriction, active circulation is principally from the right heart to the lungs, the lungs to the left heart, the left heart to the brain, and the brain back to the right heart.

the right heart to the lungs, travels from the lungs to the left heart, then is pumped by the left heart to the head, and finally returns to the right heart to be pumped again to the lungs. During diving, the rate of blood flow to the brain tends to remain similar to the rate during air breathing. Meanwhile, as the blood travels round and round between heart and head, flow to the parts of the body posterior to the heart is severely restricted.

Diving bradycardia matches cardiac output to the circulatory task

Once experiments had revealed that blood flow to many regions of a diving mammal's body is curtailed during forced or protracted dives, diving bradycardia could at last be understood for what it is: a single part of an integrated, body-wide reorganization of cardiovascular function. During a forced or protracted dive, vasoconstriction greatly reduces the dimensions of the active circulatory system. Accordingly, less output of blood from the heart is required to maintain flow in the parts of the circulatory system where unrestricted flow continues. An analogy is provided by a faucet that sends a flow of water to six hoses; if five of the hoses are pinched off, the faucet can maintain flow to the sixth hose with only a sixth of its preexisting output. Bradycardia is a mechanism of reducing the heart's output—that is, of matching cardiac output to the dimensions of the vascular system requiring perfusion. In marine mammals, the stroke volume of the heart changes only modestly, if at all, during diving, according to studies of several species. Thus cardiac output declines during a dive roughly in proportion to the decline in heart rate (see Equation 23.1).

Studies of blood pressure demonstrate in a particularly graphic way that the drop in cardiac output is highly integrated with the reduction in the dimensions of the active circulatory system during a dive. As discussed in Chapter 23 (see Equation 23.3), the pressure in the systemic aorta and the other great systemic arteries is a function of both the rate of cardiac pumping and the resistance to blood flow posed by the vascular system. A change in either factor without a compensatory adjustment in the other can severely disturb the systemic arterial blood pressure. The vasoconstriction that occurs during a dive increases the overall resistance to blood flow posed by the vascular system. However, the blood pressure in the great systemic arteries remains unaltered or changes only modestly during dives in marine mammals. This stability of the arterial blood pressure is an elegant demonstration that the reduction of cardiac output during dives is closely matched to the increase in peripheral vascular resistance.

During diving bradycardia, the myocardium has a reduced demand for ATP and O_2 because the heart is doing less work per unit of time. Thus one might expect that the myocardium would require a lower rate of blood flow during dives than during air breathing. Studies of Weddell seals and harbor seals confirm that in these species (others may be different), myocardial blood flow is reduced markedly during forced dives, probably because of partial vasoconstriction in the coronary arteries. Weddell seals showing an 86% drop in cardiac work exhibit an 85% reduction in blood flow to their ventricular heart muscle.

Cardiovascular responses are graded in freely diving animals

The cardiovascular function of free-living marine mammals undergoing voluntary dives has proved particularly challenging to study. Today however, methods finally exist to monitor the electrocardiogram continuously in unfettered animals at sea. Physiologists employing these methods have found that the heart rates of freely diving mammals typically decrease in a *graded* manner as the animals increase the durations of their dives (**Figure 24.9**). Heart rate responses during forced dives are more of an "on–off" sort; the heart rate of a forcibly submerged harbor seal, for instance, drops to less than 10% of its pre-dive level within 10 seconds every time the animal is submerged. Thus an important distinction between voluntary and forced dives is the graded versus stereotyped nature of the heart rate response during the two sorts of dives.

Ideal methods still do not exist for studying vasoconstriction in freely diving animals. For two major reasons, however, scientists believe that the vasoconstrictor response is also graded during voluntary dives. First, as we have discussed, heart rate and vasoconstriction are believed to be integrated in a way that matches cardiac output to the dimensions of the vascular system requiring blood flow. If this

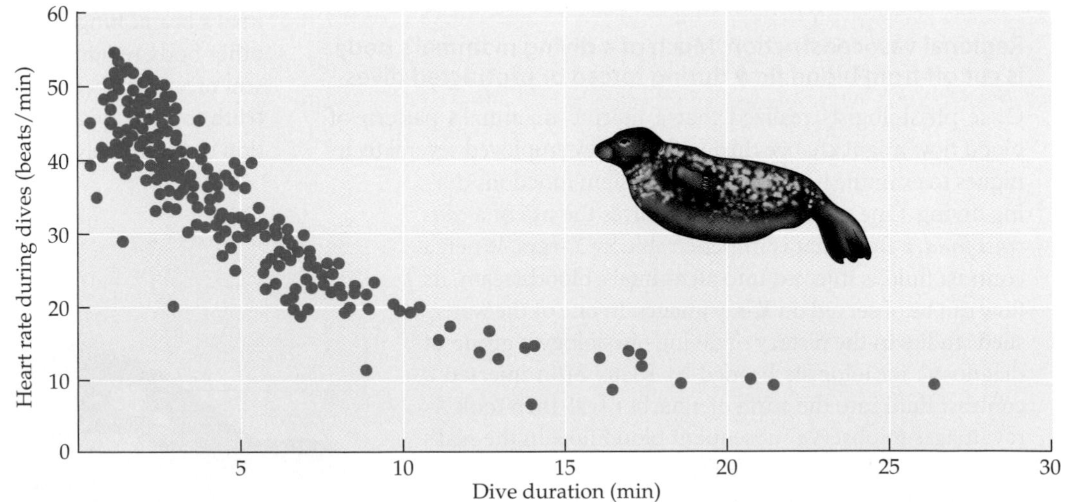

Figure 24.9 Diving heart rate varies with dive duration in a graded manner in freely diving seals Grey seals (*Halichoerus grypus*) living freely in the ocean near Scotland were monitored using radio and sonic transmitters. When not diving, their average heart rate was 119 beats per minute. During their voluntary dives, the seals always exhibited bradycardia, but they adjusted the degree of bradycardia depending on how long they spent underwater. This graded bradycardia is in contrast to the stereotyped bradycardia of forcibly submerged seals, which typically exhibit a maximal or near-maximal drop in heart rate whenever forced underwater. (After Thompson and Fedak 1993.)

integrated response consistently occurs, then little vasoconstriction would occur in animals that exhibit just a small reduction of heart rate, and progressively greater vasoconstriction would develop as bradycardia becomes more profound. A second line of evidence that corroborates this view is provided by studies of systemic organ function. Urine formation by the kidneys, for instance, requires blood flow. Thus one can learn about blood flow to the kidneys during dives by use of chemical markers that permit the extent of urine formation to be measured. Although measures of organ function do not tell a consistent story, they often indicate that vasoconstriction is graded in voluntary dives. Urine formation, for example, seems to continue during relatively short dives but stop during long dives by free-living Weddell seals.

The current working hypothesis of most physiologists who study diving is that the whole suite of cardiovascular responses to voluntary diving occurs in a graded manner. During protracted voluntary dives, free-living animals probably function much like forcibly submerged ones: They undergo profound vasoconstriction accompanied by a profound drop in heart rate, and large parts of the body are cut off from active blood flow. During relatively short voluntary dives, however, vasoconstriction is probably modest, so that only a small reduction in cardiac output is warranted, and most (or all) parts of the body continue to receive blood flow. When all parts of the body receive blood flow, all can share the blood O_2. This means that a dive cannot be of extreme length, but it also means that the animal avoids the stresses of anaerobic catabolism (discussed in Chapter 6 and later in this chapter). Interestingly, the cardiovascular responses of marine mammals during diving seem to be specializations of responses to asphyxic conditions that occur throughout the vertebrates, rather than being entirely novel evolutionary innovations (Box 24.1).

BOX 24.1 The Evolution of Vertebrate Cardiac and Vascular Responses to Asphyxia

In the decades following Scholander and Irving's seminal discoveries regarding the diving physiology of marine mammals, Scholander and others sought to learn when the diving responses had originated in the evolution of vertebrates. They were intrigued to find similar responses in fish taken out of water! Two sorts of fish that they studied were flying fish and grunion, both of which breathe with gills, but occasionally emerge into air voluntarily. Flying fish sometimes spend tens of seconds out of water as they skitter across the sea surface, and grunion sometimes spend minutes high on beaches where they slither out of the water to mate and lay eggs. The researchers found that both species exhibit profound bradycardia (as the figure shows), as well as evidence of peripheral vasoconstriction, when they are out of water. The common denominator between diving marine mammals and these fish is that both the mammals and the fish are asphyxic (unable to breathe) when they undergo bradycardia and peripheral vasoconstriction. Scholander and the other researchers thus argued that bradycardia and peripheral vasoconstriction first evolved as defenses against asphyxia in fish. By now there is a large body of evidence indicating that bradycardia is a typical response of both cartilaginous and bony fish when they are exposed to O_2-poor (hypoxic) water, reinforcing the view that bradycardia is an ancient response of vertebrates to O_2 insufficiency.

If, in fact, mammals inherited the rudiments of their cardiovascular responses to asphyxia from piscine ancestors, one would guess that those responses would be observed in many kinds of mammals, not just diving ones. In fact, the species of mammals that habitually dive are not alone in undergoing bradycardia and peripheral vasoconstriction when they are unable to obtain O_2 by breathing. Scholander himself reported evidence for bradycardia and peripheral vasoconstriction in neonatal humans and other mammals when they pass through the birth canal, and he observed bradycardia in human pearl divers. Abundant evidence exists today that adult humans and adults of at least some other terrestrial mammals routinely display bradycardia and redistribution of blood flow (e.g., restriction of blood flow to skeletal muscles) when their whole bodies are submerged, or even if they simply immerse their faces in a bowl of water. Humans, however, are not "just like" marine mammals, because human physiological responses to immersion are not as profound as those seen in marine mammals and are not coordinated in the same way (e.g., blood pressure often soars in humans during long breath-hold dives because the output of blood from the heart and the resistance to blood flow through the vascular system are mismatched). Thus, in a sense, the marine mammals seem to have "perfected" responses that all or most mammals display, and that mammals may well have inherited from fish. An important objective for future research on the comparative physiology of diving is to better understand what exactly occurred during this "perfecting" process in the course of evolution and how exactly the control mechanisms in marine mammals differ from those in terrestrial ones.

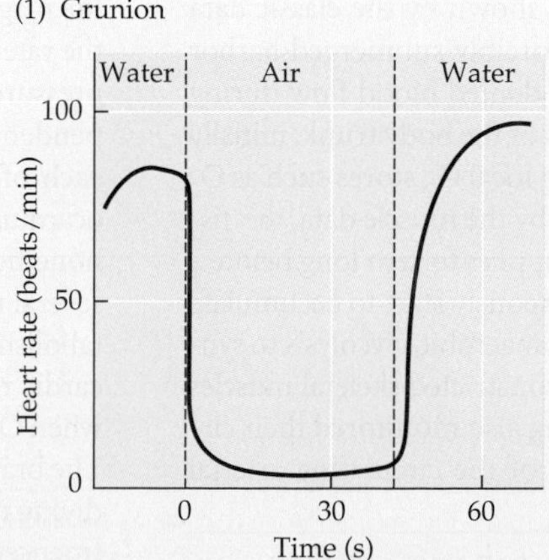

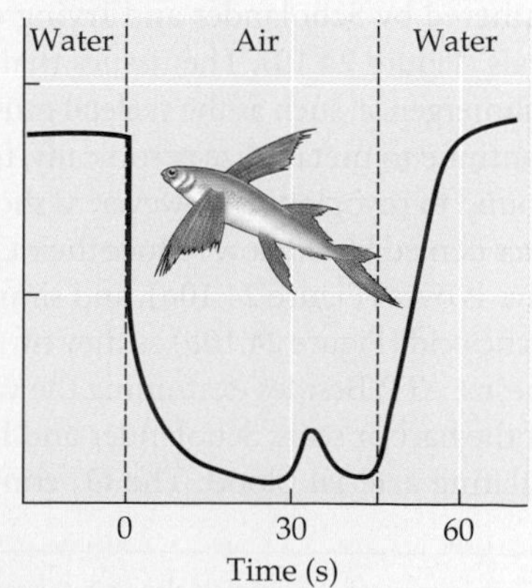

The heart rates of fish removed from water. The graphs show Scholander's original data for (1) grunion and (2) flying fish. (After Scholander 1964.)

Red blood cells are removed from the blood between dive sequences in some seals

Species of seals with large blood O_2 stores, such as Weddell seals and ribbon seals (see Figure 24.6), typically have exceptionally large concentrations of red blood cells in their blood *when they are diving.* Although an elevated red blood cell concentration enables the blood to store a large amount of O_2, it has a downside as well, because it increases the viscosity of the blood, forcing the heart to work harder to pump blood. Recent studies on several species of true (phocid) seals have shown that some red blood cells are removed from the circulating blood and stored in the spleen when the animals are resting at the water's surface or on land, then returned to the blood during diving.[3] The blood of a Weddell seal, for instance, may be 38% red blood cells (by volume) when the animal is resting in the air and become enriched to 52% red blood cells during diving. The removal of red blood cells from the blood when a seal is not diving means that the heart does not have to work exceptionally hard to pump blood during rest. Red blood cells require at least 10 to 20 minutes to move between spleen and blood. Thus they do not move in and out of the blood with each dive, but enter the blood during sequences of dives and are withdrawn during extended periods of rest.

Metabolism during Dives

The stage is now set to discuss metabolism during dives. Is it aerobic or anaerobic, or is it both? In terrestrial mammals that are resting or just modestly active, blood flows freely throughout the body—making O_2 available to all tissues—and all tissues make ATP aerobically. A similar picture seems to exist during short, voluntary dives by many marine mammals, even though they are not breathing; the circulatory system remains open to most or all body regions, and the tissues receive enough O_2 from the body's O_2 stores to function aerobically. A dramatically different picture develops during forced or prolonged dives.

The body becomes metabolically subdivided during forced or protracted dives

The intense peripheral vasoconstrictor response that occurs during forced or protracted dives effectively divides a diving animal's body into two metabolically distinct parts, as shown by the classic data gathered by Scholander and Irving on forcibly submerged harbor seals (**Figure 24.10**). The tissues that are denied blood flow during submergence, such as the skeletal muscles of the body trunk, initially continue to metabolize aerobically, using local O_2 stores such as O_2 bound to myoglobin. However, as shown by the muscle data, the tissues denied blood flow reduce their O_2 supplies to zero long before a dive is over (Figure 24.10*a*), and simultaneously start to accumulate lactic acid (Figure 24.10*b*) as they turn to anaerobic glycolysis to synthesize ATP. Besides examining the vasoconstricted skeletal muscles of the harbor seals, Scholander and Irving also monitored their circulating arterial blood. The O_2 content of the circulating arterial blood falls much more slowly than the O_2 content of the skeletal muscles (see Figure 24.10*a*); thus, long after the skeletal muscles have exhausted their O_2, the brain, myocardium, and other perfused tissues receive substantial O_2 supplies from the blood. The perfused tissues accordingly remain aerobic and do not produce lactic acid for most of the duration of a dive, as shown by the fact that little lactic acid accumulates in the circulating arterial blood during diving (see Figure 24.10*b*). All things considered, you can see that well before a long dive is over, a seal's body becomes divided into two regions, one of which remains aerobic while the other becomes O_2-depleted and dependent on anaerobic ATP production.

An important point to note is that lactic acid remains sequestered in the skeletal muscles and other vasoconstricted tissues while a dive is in progress, rather than entering the circulating blood (see Figure 24.10*b*). However, when a dive ends and the animal again starts to breathe, blood flow is promptly restored to the vasoconstricted tissues, and the lactic acid accumulated in them is washed out. Based on the favored current hypothesis, the concentration of lactic acid in the skeletal muscles falls exponentially after blood flow is restored (**Figure 24.11**). The lactic acid released from the muscles enters the circulating blood, driving the blood lactic acid concentration higher and higher for many minutes following the end of a dive. Eventually, the rate at which metabolic processes clear lactic acid from the blood (see Figure 6.4) exceeds the rate of lactic acid release from the muscles, at which time the blood lactic acid concentration starts to fall (see Figure 24.11). However, the concentration of lactic acid in muscles and blood may not return to normal, baseline levels *for many tens of minutes after a dive*, because the metabolism of lactic acid is a slow process, as emphasized in Chapter 6 (see page 159).

Metabolic limits on dive duration are determined by O_2 supplies, rates of O_2 use and lactic acid accumulation, and tissue tolerances

A diving mammal usually elects to end a dive before its metabolic limits are reached. However, a highly protracted dive could be terminated by exhaustion of O_2 in the part of the body receiving active circulation, by excessive accumulation of lactic acid in anaerobic tissues, or by other metabolic limitations.

Three factors determine the limits of endurance of the O_2-dependent tissues that receive active blood circulation during dives: (1) the magnitude of the O_2 store available to those parts of the body, (2) the rate of use of the O_2 store, and (3) the extent to which the partial pressure of O_2 can fall before impairing the function of the O_2-dependent tissues. A scattering of interesting details are known about each of these considerations. For example, even though the myocardium is believed to have a continuous requirement for O_2, it may nonetheless start to employ anaerobic glycolysis to some extent after several minutes of diving. Both this partial recourse to anaerobic catabolism and the drop in cardiac work associated with diving bradycardia reduce the heart's O_2 needs and thus help postpone the time when O_2 supplies become inadequate to sustain myocardial function. The brain, in contrast to the heart, is believed to be entirely aerobic in diving mammals, just as it is in terrestrial mammals. However, electroencephalographic signs of cerebral impairment appear at lower O_2 partial pressures in at least some seals than in terrestrial mammals.

[3] Certain terrestrial mammals also vary the red blood cell concentration of their blood by sequestering cells in the spleen, as noted on page 595.

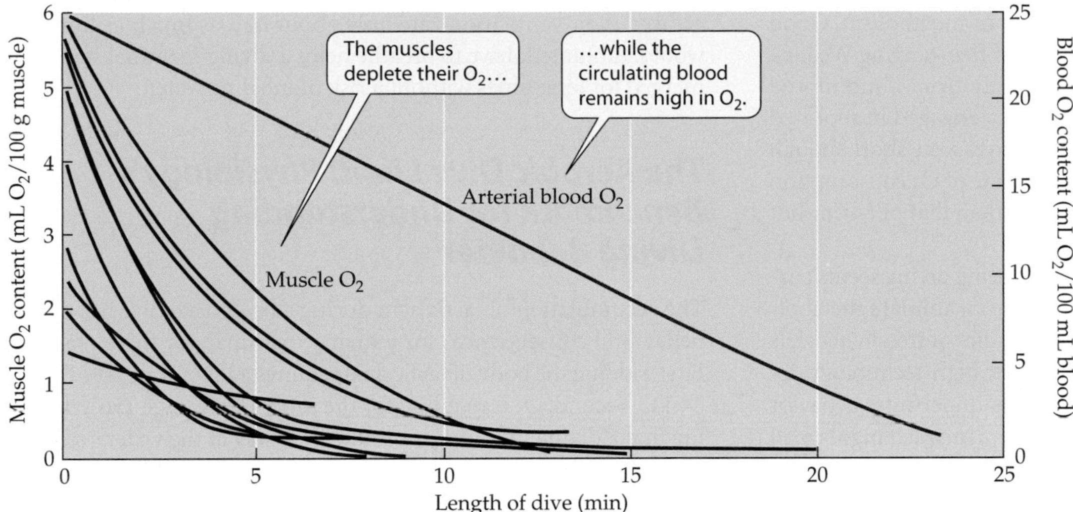

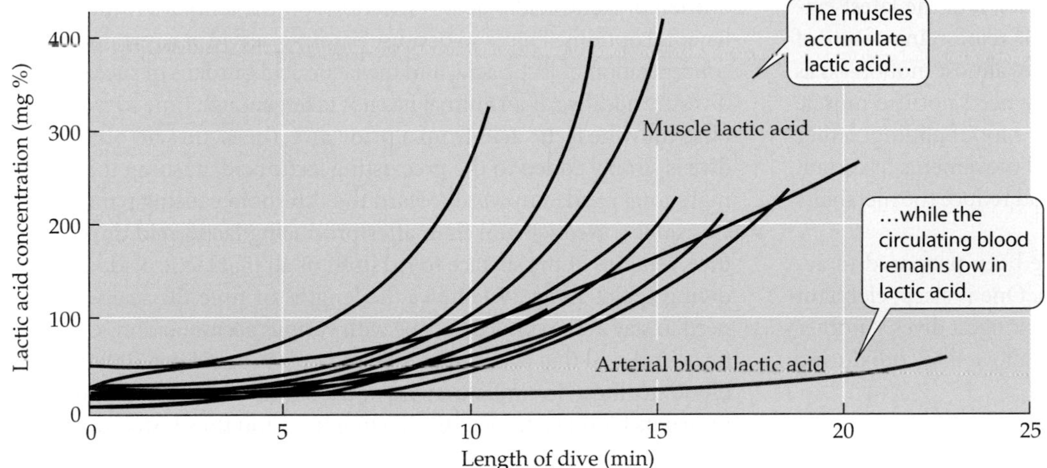

Figure 24.10 Metabolic subdivision of the body during forced submergence The O_2 content and lactic acid content of the dorsal trunk muscles and circulating arterial blood of harbor seals (*Phoca vitulina*) are shown as functions of time underwater during forced submergence. *(a)* Muscle O_2 content in each of ten seals and the average course of O_2 depletion in circulating arterial blood. *(b)* Muscle lactic acid content in each of ten seals and the average accumulation of lactic acid in circulating arterial blood. (From Scholander, Irving, and Grinnell 1942.)

This higher tolerance of neurological hypoxia in seals probably extends the time that the brain can continue to function during dives.

METABOLIC RATES DURING DIVING As you might expect, an animal's metabolic rate during diving is one of the most important factors in determining metabolic limits on dive duration. A low metabolic rate could slow both the rate of O_2 depletion and the rate of lactic acid accumulation, thereby lengthening a dive whether O_2 or lactic acid sets the limits. Scientists have sought evidence for depressed metabolism in diving mammals from the dawn of the modern study of diving. Rigorous measurements of metabolic rates during diving are not easy, however, and knowledge about this subject remains scattered and uncertain.

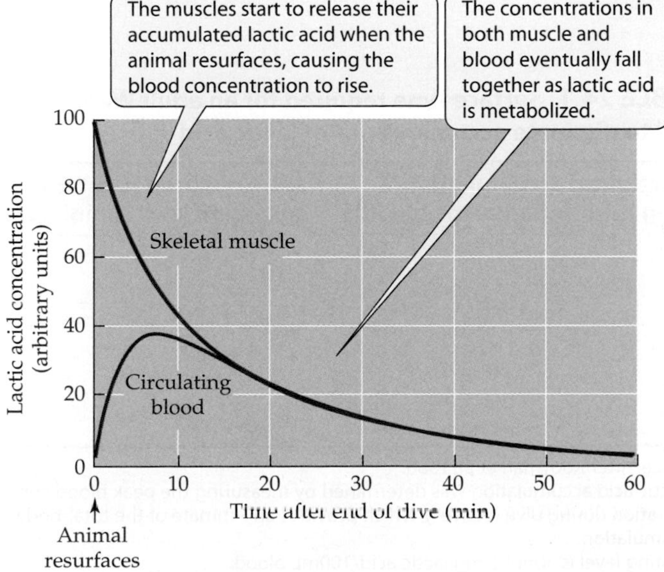

Figure 24.11 The release of lactic acid from muscles to blood after a dive The graph shows concentrations of lactic acid in skeletal muscles (blue line) and in circulating blood (red line) in the hour after a dive. Lactic acid concentrations are expressed relative to the concentration in muscle at dive's end, arbitrarily set equal to 100 units. A 43-minute dive by an adult Weddell seal is assumed. Blood values are empirical, whereas muscle values are based on the assumption of exponential kinetics. (After Butler and Jones 1997.)

The preponderance of available evidence indicates that submergence commonly brings about a depression of metabolism. Good insight is provided, for example, by studies of free-ranging Weddell seals that were trained to breathe consistently from a monitored source of air for hours at a time, even as they engaged in repeated dives involving food capture (most of these dives were short enough to be fully aerobic). Remarkably, the average rate of O_2 consumption of these *active* animals was equal to or lower than that of *resting* but nondiving seals.

Evidence for depression of metabolism during diving seems paradoxical because swimming ought to increase an animal's metabolic rate. Just in the last few years, behavioral studies of free-living seals employing miniature video cameras and other high-technology devices have provided new insights into the seals' underwater behavior that sometimes help explain the paradox. A principle taught to all scuba divers is that once a diver becomes negatively buoyant, sinking becomes a self-reinforcing process; a little sinking compresses air cavities in the diving gear, making a diver less buoyant and thus even more prone to sinking. The lungs of seals, positioned within a flexible thorax, behave as compressible air cavities, and the latest evidence indicates that seals often employ the self-reinforcing nature of negative buoyancy to sink, holding their bodies almost motionless as they descend ever quicker. In this way, they need not use muscle power to descend. Video cameras also show a lot of gliding, rather than active swimming, during the horizontal movements of certain seals and dolphins underwater; gliding should reduce the metabolic rates of the animals as they move about.

Other energy-sparing mechanisms have been proposed to account for low metabolic rates during dives. One is sleep; elephant seals might be sleeping during some of their longest dives. Another possibility is hypothermia; tissues might be allowed to cool during diving.

ADAPTATIONS TO THE ACCUMULATION OF METABOLIC END PRODUCTS Both the aerobic and anaerobic catabolic pathways produce end products that accumulate in a diving mammal: CO_2 and lactic acid, respectively. The ways in which a diver responds to the buildup of these compounds are important determinants of maximal dive duration. For one thing, both CO_2 and lactic acid tend to cause acidification of tissues and body fluids. How is the pH prevented from falling rapidly to a level that would force a dive to end? Part of the answer is that *the blood of diving species typically has a particularly high buffering capacity.*

Anybody who has ever held his or her breath as long as possible underwater knows that eventually the urge to breathe becomes impossible to resist. The buildup of blood CO_2 and the drop of blood pH during breath-holding are believed to drive the irrepressible urge to breathe because both factors are known to be potent stimulants of pulmonary ventilation in terrestrial mammals, as discussed in Chapter 21 (see page 560). *Diving species exhibit blunted (i.e., reduced) ventilatory sensitivity to changes in blood CO_2 and pH compared to terrestrial species.* An example is provided by free-ranging harbor seals exposed to elevated concentrations of CO_2 in their breathing air. As the CO_2 concentration was raised, the seals increased the amount of air they breathed per unit of time, mainly by spending a greater fraction of their time breathing rather than diving. However, for a given increase in alveolar CO_2 partial pressure, the seals increased their ventilation rate only about half as much as humans would. A blunted drive to breathe helps a diving mammal stay submerged for long times without a risk of inhaling water.

The Aerobic Dive Limit: Physiology's Benchmark for Understanding Diving Behavior

The accumulation of lactic acid during a prolonged dive has major behavioral consequences for a diving mammal, for three reasons. First, ridding the body of lactic acid requires a lot of time (see Figure 24.11). Second, O_2 is required for the process (see page 156), meaning that a diving mammal typically must stay at the water's surface where it can breathe, or return often to the surface, while it metabolizes lactic acid. Third, if a diving mammal already has a lactic acid burden to metabolize following a prolonged dive, it cannot engage immediately in a dive of maximal duration; that is, the preexisting burden of lactic acid places restrictions on the animal's future behavioral options. This is so because paralysis sets in at a certain high concentration of lactic acid, and the lactic acid burdens of successive dives are additive; if an animal has not taken enough time to metabolize fully the lactic acid from a prior dive, the lactic acid of a new dive is simply added to the preexisting lactic acid, pushing the animal's total burden upward toward the threshold causing paralysis. Consider a Weddell seal that, after producing lactic acid during a dive, remains at the surface to rid itself of all that lactic acid before diving again. Table 24.1 shows the lengths of time the seal would need to stay at the surface to cope with various accumulations of lactic acid. A seal that accumulates 80 mg of lactic acid per 100 mL of blood during a dive must remain at the surface for over an hour to return its blood lactic acid to a resting level and thus restore its full range of behavioral options.

Accumulation of lactic acid can be avoided if dives are kept short. This is illustrated by the famous set of data in Figure 24.12, gathered by Gerald Kooyman and his colleagues. To gather these data, blood was drawn from wild Weddell seals when they returned to ice holes

TABLE 24.1 Surface time required for an adult Weddell seal to dissipate accumulations of lactic acid

Lactic acid accumulation (mg lactic acid/100 mL blood)[a]	Time required to return to resting level[b] (min)
20	11
40	27
80	70
120	105
145	120

Source: After Kooyman et al. 1980.
[a] Lactic acid accumulation was determined by measuring the peak blood concentration during dive recovery, which provides an estimate of the total-body accumulation.
[b] Resting level is about 5 mg lactic acid/100mL blood.

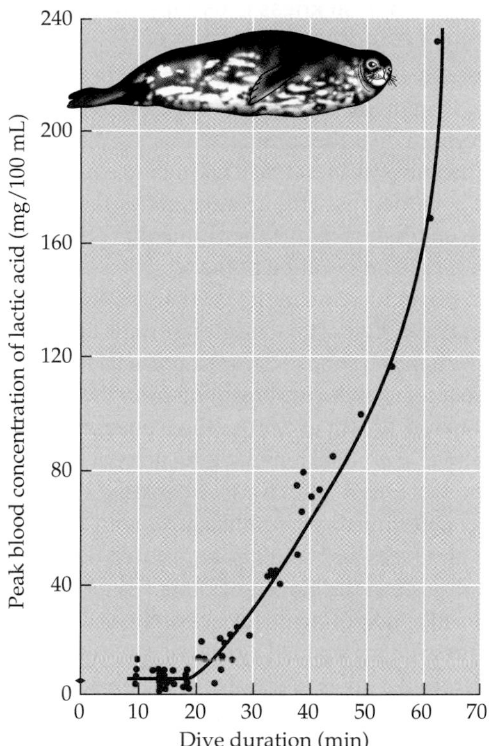

Figure 24.12 Peak concentration of lactic acid in arterial blood of freely diving Weddell seals following dives of various durations Because the circulatory system is fully open after a dive, lactic acid produced anywhere in the body during a dive, even if temporarily sequestered during submergence, appears in the arterial blood after a dive. Thus the peak concentration of lactic acid in arterial blood following a dive provides an estimate of total-body net production of lactic acid during the dive. The diamond on the *y* axis marks the average resting concentration of lactic acid in arterial blood. (After Kooyman et al. 1980.)

to breathe following voluntary dives. Recall that although lactic acid stays sequestered within skeletal muscles during dives, it is released from the muscles into the general blood circulation after an animal surfaces and starts to breathe (see Figure 24.11). If the investigators who stepped up to a seal returning from a dive simply waited long enough for the release process to drive the blood concentration to its peak level before drawing a blood sample, they could estimate the seal's total-body accumulation of lactic acid from the blood level. Figure 24.12 shows how the peak blood concentration of lactic acid depends on dive duration. Weddell seals accumulate lactic acid when they dive for 25 minutes or longer. In fact, they accumulate very high levels during lengthy dives. However, they do not accumulate lactic acid above the resting, baseline level if they dive for 20 minutes or less.

Based on the data shown in Figure 24.12, Kooyman and his colleagues defined the concept of the **aerobic dive limit** (**ADL**), which is the longest dive that can be undertaken by a species without a net accumulation of lactic acid above the resting level. Unfortunately, measuring lactic acid levels in freely diving animals of many species borders on impossible, and thus the ADL often cannot be determined as originally defined. Accordingly, the ADL is often estimated by calculations using data on a species' O_2 stores and diving metabolic rate. The assumption of this computational approach is that a dive will not cause accumulation of lactic acid if the O_2 cost to meet all the metabolic demands of the dive is less than the total available O_2 store. To determine the ADL using the computational approach, researchers measure the species' O_2 stores (see Figure 24.6) and metabolic rate during diving as accurately as possible. Then, the O_2 stores are corrected to take account of amounts that are unavailable for use during dives, and the animal's rate of O_2 use per minute during diving is estimated from the metabolic rate by assuming that all metabolism is aerobic. Finally, the total available O_2 store is divided by the rate of O_2 use to determine the maximum dive length that will not require lactic acid production. For instance, if a seal has a usable O_2 store of 2800 mL and must consume O_2 at a rate of 280 mL/minute during diving for its metabolism to be entirely aerobic, its ADL is estimated to be 10 minutes. The ADL has been calculated (or, less commonly, directly measured) for many species of marine mammals, and just as species vary widely in their diving competence, they also vary widely in ADL. The ADL for adult Weddell seals is about 20 minutes (see Figure 24.12). ADLs in fur seals and California sea lions (both otariids) are about 4 to 6 minutes. Some fairly crude estimates indicate that sperm whales and male southern elephant seals have ADLs of 40 to 50 minutes!

A central hypothesis of modern diving physiology is that it is adaptive for diving mammals to keep most dives shorter than their species-specific ADL. One reason that diving in this way is thought to be adaptive is that animals can minimize their time at the surface and maximize their time underwater by keeping their dives fully aerobic.

How do aerobic diving patterns translate into increased time spent underwater? When a dive is fully aerobic, replenishing O_2 stores is the sole task an animal must carry out to recover before diving again. Vigorous breathing can replenish body O_2 stores rapidly. Thus, based on observations on several species, marine mammals need to remain at the surface for only 1 to 4 minutes between one dive and the next *provided their dives are all shorter than the ADL.* Short surface intervals, in turn, provide lots of underwater time for foraging. Consider, for instance, a Weddell seal that undertakes a series of six 15-minute dives. Because these dives are shorter than the ADL (see Figure 24.12) and fully aerobic, the seal needs to be at the surface for only about 4 minutes between dives. Thus, counting dive time and recovery time, the six dives will take 114 minutes. Of this time, 90 minutes will be spent underwater. Thus this fully aerobic diving schedule will permit the seal to forage and feed underwater for almost 80% of its time. What would happen, in contrast, if the seal were to stay underwater for a single, continuous period that exceeds the ADL? Long dives can mean even longer surface intervals because of the slow rate at which lactic acid is metabolized. A 44-minute dive, for example, will result in a peak blood level of lactic acid of about 80 mg/100 mL (see Figure 24.12) and therefore require about 70 minutes of recovery at the surface (see Table 24.1). The 44-minute dive and recovery will together take 114 minutes, the same time as the six 15-minute dives. However, the seal diving for 44 minutes will spend less than 40% of its time underwater. In brief, a diver can typically spend a greater fraction of its time foraging and feeding if it makes many aerobic dives, each shorter than the ADL, than if it makes just a few lengthy and highly anaerobic dives.

Dives shorter than the ADL are also postulated to be adaptive because they permit homeostasis to be maintained without interruption throughout the body. As discussed earlier, investigators think that most or all organs receive a continuing blood flow and O_2 supply during dives that are relatively short.[4] Thus, such dives permit most organs to continue functioning in an approximately normal way. Protracted dives, by contrast, can force many organs away from homeostasis; the kidneys, for instance, may be forced to stop urine production during a protracted dive, and enzymes throughout the body may be forced to function at highly altered levels of pH when CO_2 and lactic acid are allowed to accumulate together.

If dives shorter than the ADL are, in fact, adaptive, then diving mammals are predicted to elect short dive lengths during their natural diving behavior. With this thought, we come full circle. We stressed at the start of this chapter that dives of maximal duration, although important and amazing, tend to be uncommon; most dives are far shorter than the species-specific maximum. Now we can be more exacting in our prediction of dive lengths: *Most dives should be shorter than the species-specific aerobic dive limit*, according to the line of thought developed here. A great deal of contemporary research is directed toward testing this prediction, and with some notable exceptions, the theory holds up to scrutiny. For instance, over 90% of the voluntary dives of wild Weddell seals are shorter than the 20-minute ADL of Weddell seals (compare Figures 24.1 and 24.12). Grey seals are calculated to have a shorter ADL, about 10 minutes, and over 90% of their dives are shorter than 10 minutes. The ADL of bottlenose dolphins is calculated to be about 4 minutes, and over 90% of their dives are shorter than 4 minutes.

Thus the aerobic dive limit—a physiological feature of marine mammals—seems to be an important behavioral benchmark. As we have seen, diving species have a sometimes-incredible ability to dive for far longer than their ADL; they may use that ability when faced with extraordinary behavioral challenges, such as avoiding danger or searching for new foraging areas. However, dives that are longer than the ADL force many organs away from homeostasis and require long surface intervals. Thus, dives are usually kept shorter than the ADL. The significance of large body O_2 stores becomes clearer in this light: Large O_2 stores give a species a high ADL value. Thus, large O_2 stores permit a greater range of diving options while an animal adheres to the "rule" that its ADL is the upper length limit for the majority of its dives.

Decompression Sickness

Many "diver's diseases" have been described in the annals of human diving, and physiologists have focused much attention on learning how these diseases are avoided by marine mammals. We consider here just one of these diseases, known as decompression sickness, caisson disease, or the bends. An informative starting point is to examine the etiology of this disease in human divers.

[4] This is not to say that blood flow is unaltered. Flow is likely to be redistributed during short dives much as it is redistributed in terrestrial animals during exercise, with some organs receiving relatively more of the total flow than they do during rest and some receiving less. Nonetheless, the current working hypothesis is that blood flow (at one rate or another) is maintained to most or all organs during short dives.

Human decompression sickness is usually caused by N_2 absorption from a compressed-air source

Unambiguous cases of decompression sickness occur in humans when they are diving with a source of compressed air (such as a scuba tank). During a dive, the compressed-air source maintains the air pressure in the lungs at a level high enough to equal the ambient water pressure at all depths. This arrangement prevents the lungs from collapsing under the force of the ambient pressure and allows continued breathing. The elevation of the total pressure in the lungs means that there is an increase in the partial pressure of each individual pulmonary gas. Of particular interest is the alveolar N_2 partial pressure, which may be boosted from a normal level of about 0.8 atm (the level that prevails during breathing from the atmosphere at sea level) to 5 or even 10 atm at depth. When a person starts a dive, his or her tissues are at equilibrium with the normal sea-level alveolar N_2 partial pressure; that is, each tissue contains a concentration of dissolved N_2 determined by equilibration with N_2 at a partial pressure of 0.8 atm (see Chapter 20, page 529). At depth, increased amounts of N_2 dissolve in the blood and other tissues because the N_2 partial pressure in the tissues is initially below the new, elevated alveolar N_2 partial pressure maintained by the compressed-air source. As N_2 is extracted from the lungs to dissolve in the blood and other tissues, the N_2 partial pressure in the lungs does not fall, because the lung air is constantly renewed from the compressed-air source. If a dive is continued for long enough, the blood and other tissues dissolve sufficient N_2 to reach a N_2 partial pressure that matches the elevated N_2 partial pressure in the lungs. Decompression sickness may then occur if the person suddenly surfaces ("decompresses"). During sudden surfacing, the N_2 partial pressure in the lungs falls abruptly to its ordinary value, and the N_2-charged blood and other tissues start to lose N_2 into the lung air. Problems can arise after a long, deep dive because the *partial pressure of dissolved* N_2 in the blood and other tissues may be many atmospheres and thus may exceed the *hydrostatic pressure* prevailing in the body at sea level. Under these circumstances, bubbles of N_2 may form within the blood and other tissues.

Bubble formation is not difficult to understand in principle. If the hydrostatic pressure in a particular tissue is 1 atm (as it is at sea level) and a microscopically tiny pocket of gas develops there, the pressure in the gas pocket will be 1 atm. If the body fluids surrounding the gas pocket contain dissolved N_2 at a high partial pressure—say, 6 atm—N_2 will diffuse from solution into the gas pocket *even if the gas pocket is filled purely with* N_2. The minute gas pocket will therefore grow into a visible bubble. An analogy is provided by a recently opened pop bottle. When the pressure inside the bottle is suddenly reduced by removing the cap, gas in solution at high partial pressure diffuses into minute gas pockets at low pressure, causing bubbles to grow to visible size.

Nitrogen bubbles formed in a human diver in the way we have described are believed to be the primary agents causing decompression sickness. The most common symptom of decompression sickness is throbbing pain in the joints and muscles of the arms and legs (the bends). In addition, an afflicted person may suffer neurological symptoms, such as paralysis, and severe breathing problems (the chokes). Exactly how the bubbles cause these symptoms remains a topic of ongoing research. Bubbles can block blood flow, press on nerve endings, and even disturb the structures of proteins because of electrical phenomena at gas-water interfaces.

The factors that determine whether *clinical* decompression sickness occurs are not completely known. One consideration is that excess dissolved gas is steadily eliminated across the lungs after a dive. If the N_2 overload is not too great, this elimination may lower the N_2 partial pressure in the blood and other tissues rapidly enough that even if bubbles start to form, their growth and proliferation are halted before clinical symptoms occur. In general, as a very rough rule of thumb, humans can surface immediately without fear of the bends if their blood and tissue N_2 partial pressure is under about 2 atm. Otherwise they must surface gradually to avoid illness (which in severe cases can be fatal).

Breath-hold dives must be repeated many times to cause decompression sickness in humans

When we consider humans undergoing *breath-hold* diving, a crucial difference from diving with compressed air is immediately apparent. A breath-hold diver descends with only the limited amount of extra N_2 contained within the lung air upon submergence; the pulmonary N_2 supply is not steadily renewed. During descent to depth, the lungs of a breath-hold diver are compressed under the force of increasing ambient pressure, and the N_2 partial pressure in the lungs increases initially to high levels, just as in diving with compressed air. This process creates a partial-pressure gradient favoring the transfer of N_2 from the lungs to the blood and other tissues. However, the *quantity* of N_2 in the lungs is limited, and thus only a limited quantity can be transferred. In humans, the amount of N_2 that is transferred to the blood and other tissues during a single breath-hold dive is far too small to cause decompression sickness.

What happens, however, if a person undergoes *many repeated* breath-hold dives? If the time between successive dives is insufficient for the tissues to release accumulated dissolved N_2 after each dive, the tissue N_2 partial pressure could conceivably be elevated *in a series of upward steps* to a threatening level. This possibility seems to be far from theoretical. A number of reports exist of people who have developed symptoms of decompression sickness after sequences of many breath-hold dives. In a classic instance described by Paulev, for example, an individual complained of decompression symptoms after diving about 60 times to depths of 15 to 20 m over a period of 5 hours.

Marine mammals avoid decompression sickness during deep dives by alveolar collapse

Marine mammals are breath-hold divers, and the first question to be asked is whether they face a potential problem of decompression sickness. Calculations and experiments indicate that if N_2 is presumed to be able to move freely from the lungs into the rest of the body, N_2 partial pressures in the blood and other tissues of a diving mammal may often rise to threatening levels following many repeated dives.

During deep dives, the risk of decompression sickness is reduced in diving mammals by structural features of their lungs and thorax that serve to keep pulmonary N_2 within the lungs, thereby limiting buildup of N_2 in the blood and other tissues. Because of the compressibility of the thorax in marine mammals (see Figure 24.5), the volume of the air held in their lungs decreases in inverse proportion to the increase in pressure as the animals descend during diving. As the lung air is compressed into a smaller volume, the *respiratory* air spaces of the lungs, such as the alveolar sacs (see Figure 21.18), are believed to collapse preferentially, meaning that the lung air gradually becomes contained entirely in the *conducting* airways (anatomical dead space) of the lungs (**Figure 24.13**). Consequently, below a certain depth, N_2 invasion from the lungs into the blood and other tissues cannot occur, because the N_2 in the lung air is safely sequestered in the parts of the lungs where gas exchange between the lung air and blood is not possible. Oxygen, as we have already stressed, is also sequestered within the conducting airways. The sequestration of O_2 might simply be a price an animal must pay to avoid the bends, or possibly—as discussed in the final section of this chapter—it may hold advantages of its own.

The depth at which the alveoli and alveolar sacs collapse completely is the depth at which the hydrostatic pressure is great enough to reduce the initial volume of lung air to the volume of the conducting airways. For example, if the volume of the conducting airways in an animal is 2 L, and if the animal carries 22 L of lung air on submergence, then alveolar collapse should be complete at about 100 m, because at that depth the hydrostatic pressure is near 11 atm—sufficient to reduce the initial volume by a factor of 11 (to 2 L). One reason that the deep-diving seals exhale before diving may be to hasten alveolar collapse by reducing the initial volume of lung air relative to the volume of the conducting airways. Studies of Weddell and northern elephant seals indicate that their alveolar exchange area is substantially reduced at depths as shallow as 30 m.

Dives too shallow for alveolar collapse present special problems

Alveolar collapse, despite its importance during deep dives, is probably not the only consideration in understanding how marine mammals avoid decompression sickness. For alveolar collapse to prevent a buildup of N_2 in the tissues, an animal must dive deep enough for effective collapse to occur. However, certain marine mammals, (e.g., some dolphins, fur seals, and sea lions) sometimes dive extensively to depths that, although sufficient to elevate the pulmonary N_2 pressure to levels of questionable safety, are too shallow to induce alveo-

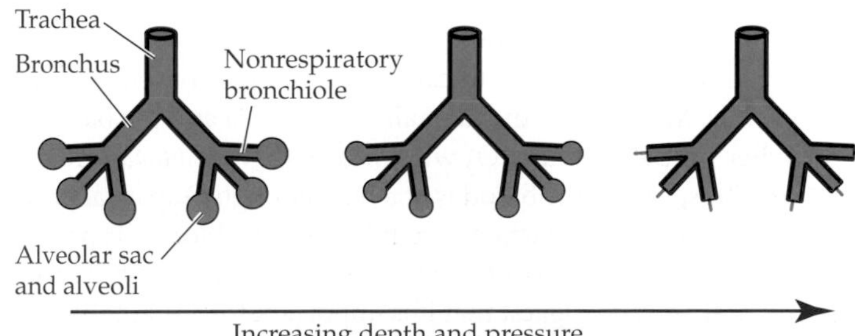

Figure 24.13 The hypothesis of preferential collapse of the alveoli and alveolar sacs at depth As the lung air is compressed to a smaller volume at depth, the alveoli and alveolar sacs collapse preferentially, and all air moves into the trachea, bronchi, and nonrespiratory bronchioles—the conducting airways (anatomical dead space)—where gas exchange with the blood is negligible.

lar collapse. During shallow dives of this sort, N_2 could invade the blood and other tissues to a great enough extent to pose a risk of decompression sickness. Experiments on trained bottlenose dolphins undergoing repeated dives to modest depths suggest that this concern might in fact be realistic. After an hour of diving, the animals developed muscle N_2 partial pressures—1.7 to 2.1 atm—that in a human would be only marginally safe. Possibly, the dolphins avoid unsafe tissue N_2 partial pressures during normal diving by behaviorally limiting their diving schedules. Possibly, they can cope with higher tissue N_2 partial pressures than humans can.

A Possible Advantage for Pulmonary O_2 Sequestration in Deep Dives

A new theory on the relation between alveolar collapse and lung O_2 provides a thought-provoking note on which to end this chapter. From the beginning of the scientific study of diving, investigators have tended to perceive the lungs as analogous to scuba tanks. Only a defective scuba tank would fail to deliver all its O_2. Thus, when lungs fail to deliver all their O_2, that failure is perceived as a flaw. Alveolar collapse in deep-diving marine mammals, in this view, is simultaneously advantageous and flawed. It is advantageous because it prevents much of the lung N_2 from dissolving in the blood and other tissues. It is flawed because it denies the animal the use of some of its O_2 store. Translated into evolutionary terms, this view presupposes that alveolar collapse was favored by natural selection as a means of preventing decompression sickness, and that the sequestration of O_2 is simply a negative side effect. The new theory postulates that in deep divers, O_2 sequestration by alveolar collapse is itself an advantage.

As a diving mammal ascends to the water's surface at the end of a deep dive, the total pressure in its lungs declines as its lungs expand. The partial pressure of O_2 in the lung air accordingly decreases as the mammal ascends. Thus, if the lungs were to pass most of their O_2 to the body during a dive, the O_2 partial pressure in the lungs could drop precipitously to near zero during the ascent at the end of the dive. To illustrate, suppose that during a dive to 400 m, the O_2 partial pressure in an animal's lung air fell to 13 kPa (100 mm Hg) as O_2 left the lungs and entered the blood. As the animal ascended from 400 m to 40 m, the decreased total pressure and the expansion of the lung air would lower the O_2 partial pressure to a tenth of that, 1.3 kPa (10 mm Hg). And the animal would still have another 40 m to ascend before it could breathe! A blood O_2 partial pressure of 1.3 kPa is low enough to cause blackout (unconsciousness), even in seals. Moreover, if the lung air is at 1.3 kPa and the blood is at a higher partial pressure, O_2 will diffuse *from the blood into the lung air*! Thus, even if the blood is at a high enough O_2 partial pressure to prevent blackout when ascent begins, the blood O_2 partial pressure could fall to blackout-inducing levels before ascent is completed. These worries suggest that it is important for the lungs to retain a sizable portion of their O_2 throughout a deep dive. Thus, in addition to its other advantages, alveolar collapse may in fact be a mechanism for preventing the risk of blackout during ascent by ensuring that enough O_2 remains in the lungs to keep the O_2 partial pressure acceptably high in the ever-expanding pulmonary air.

One reason to end on this novel note is that these thoughts emphasize the highly *interactive* nature of the challenges faced by the lungs, blood, and other tissues during all stages of diving. A second reason is to stress the importance of striving to see physiological challenges from the point of view of the animals studied. Lungs cease to look like mere scuba tanks when viewed from the perspective of a deep-diving marine mammal.

Study Questions

1. Some researchers define comparative physiology as the search for the "ideal" species for the study of each phenomenon of interest. Why was the Weddell seal the "ideal" species for the study of voluntary diving in the wild at the time such studies began?
2. Outline the pros and cons of carrying lots of lung air during a dive.
3. Based on the study of O_2 needs and stores, the aerobic dive limit for young Weddell seals weighing 140 kg is calculated to be 10 minutes, whereas that for fully grown 400-kg Weddell seals is calculated to be about 20 minutes. Why might small individuals in general be expected to have shorter aerobic dive limits than large individuals? To carry out an empirical study of the differences in aerobic dive limit between small and large individuals of a species, what experiments and measurements would you plan?
4. Helicopters have been used to move young salmon from lake to lake for purposes of aquaculture in remote areas such as Alaska. From the viewpoint of the behavior of dissolved gases, why is it important for helicopters used in this way to stay at very low altitudes? (Hint: After fish were transported at altitudes where the total ambient pressure was 0.9 atm, high percentages suffered disease or death because of gas bubble formation in various tissues.)
5. There is evidence that marine mammals practice unihemispheric sleep: sleep that occurs in only one brain hemisphere at a time, so that while one hemisphere sleeps, the other is awake. Such sleep is essentially unheard of in terrestrial mammals (although common in birds). What might be the advantages of unihemispheric sleep for a marine mammal?

Key References

Butler, P. J., and D. R. Jones. 1997. Physiology of diving of birds and mammals. *Physiol. Rev.* 77: 837–899. A thorough, intellectually rigorous, and modern review of diving behavior and physiology in birds as well as mammals, written by two of the pros in the field.

Kooyman, G. L. 1981. *Weddell Seal, Consummate Diver.* Cambridge University Press, New York. A "scientific natural history" that's hard to put down once you start reading. A well-told story about a fascinating animal written by the man who teamed up with the animal to initiate the modern renaissance in diving physiology.

Kooyman, G. L., and P. J. Ponganis. 1998. The physiological basis of diving to depth: Birds and mammals. *Annu. Rev. Physiol.* 60: 19–32. A compact review that is a compelling read both because it does a superb job of highlighting crucial observations in our current body of knowledge and because it points the way toward key questions for future research.

Kooyman, G. L., E. A. Wahrenbrock, M. A. Castellini, R. W. Davis, and E. E. Sinnett. 1980. Aerobic and anaerobic metabolism during voluntary diving in Weddell seals: Evidence of preferred pathways from blood chemistry and behavior. *J. Comp. Physiol. B* 138: 335–346. A watershed paper that reports the use of combined studies of chemistry and behavior to break new ground in the study of diving in free-living animals. The fountainhead of the concept of aerobic dive limit.

Le Boeuf, B. J., D. E. Crocker, D. P. Costa, S. B. Blackwell, P. M. Webb, and D. S. Houser. 2000. Foraging ecology of northern elephant seals. *Ecol. Monogr.* 70: 353–382. A tour de force on the migrations and foraging behavior of elephant seals as studied with modern technology. For the curious mind, this is a report that generously stimulates the generation of hypotheses for future studies of diving.

Selected References

Bennett, K. A., B. J. McConnell, and M. A. Fedak. 2001. Diurnal and seasonal variations in the duration and depth of the longest dives in southern elephant seals (*Mirounga leonina*): Possible physiological and behavioral constraints. *J. Exp. Biol.* 204: 649–662.

Bennett, P. B., and D. H. Elliott (eds.). 1993. *The Physiology and Medicine of Diving*. Saunders, Philadelphia.

Boutilier, R. G., J. Z. Reed, and M. A. Fedak. 2001. Unsteady-state gas exchange and storage in diving marine mammals: The harbor porpoise and gray seal. *Amer. J. Physiol.* 281: R490–R494.

Boyd, I. L., J. P. Y. Arnould, T. Barton, and J. P. Croxall. 1994. Foraging behaviour of Antarctic fur seals during periods of contrasting prey abundance. *J. Anim. Ecol.* 63: 703–713.

Bron, K. M., H. V. Murdaugh, Jr., J. E. Millen, R. Lenthall, P. Raskin, and E. D. Robin. 1966. Arterial constrictor response in a diving mammal. *Science* 152: 540–543. [A classic.]

Burns, J. M. 1999. The development of diving behavior in juvenile Weddell seals: Pushing physiological limits in order to survive. *Can. J. Zool.* 77: 737–747.

Comparative Physiology and Biochemistry of Cardiovascular, Respiratory, and Metabolic Responses to Hypoxia, Diving, and Hibernation (A Symposium). 1988. *Can. J. Zool.* 66: 1–190.

Costa, D. P., N. J. Gales, and M. E. Goebel. 2001. Aerobic dive limit: How often does it occur in nature? *Comp. Biochem. Physiol. A* 129: 771–783.

Davis, R. W., and S. B. Kanatous. 1999. Convective oxygen transport and tissue oxygen consumption in Weddell seals during aerobic dives. *J. Exp. Biol.* 202: 1091–1113.

Falke, K. J., R. D. Hill, J. Qvist, R. C. Schneider, M. Guppy, G. C. Liggins, P. W. Hochachka, R. E. Elliott, and W. M. Zapol. 1985. Seal lungs collapse during free diving: Evidence from arterial nitrogen tensions. *Science* 229: 556–558.

Ferretti, G. 2001. Extreme human breath-hold diving. *Eur. J. Appl. Physiol.* 84: 254–271.

Guppy, M., R. D. Hill, R. C. Schneider, J. Qvist, G. C. Liggins, W. M. Zapol, and P. W. Hochachka. 1986. Microcomputer-assisted metabolic studies of voluntary diving of Weddell seals. *Amer. J. Physiol.* 250: R175–R187.

Guyton, G. P., K. S. Stanek, R. C. Schneider, P. W. Hochachka, W. E. Hurford, D. G. Zapol, G. C. Liggins, and W. M. Zapol. 1995. Myoglobin saturation in free-diving Weddell seals. *J. Appl. Physiol.* 79: 1148–1155.

Hochachka, P. W. 2000. Pinniped diving response mechanism and evolution: A window on the paradigm of comparative biochemistry and physiology. *Comp. Biochem. Physiol. A* 126: 435–458.

Hong, S. K. 1996. The hyperbaric environment. In M. J. Fregly and C. M. Blatteis (eds.), *Environmental Physiology*, vol. 2 (Handbook of Physiology [Bethesda, MD], (Section 4), pp. 975–1056. Oxford University Press, New York.

Hong, S. K., and H. Rahn. 1967. The diving women of Korea and Japan. *Sci. Amer.* 216(5): 34–43. [Although published in 1967, still informative because of the detail included on traditional ama practices.]

Hurley, J. A., and D. P. Costa. 2001. Standard metabolic rate at the surface and during trained submersions in adult California sea lions (*Zalophus californianus*). *J. Exp. Biol.* 204: 3273–3281.

Kjekshus, J. K., A. S. Blix, R. Elsner, R. Hol, and E. Amundsen. 1982. Myocardial blood flow and metabolism in the diving seal. *Amer. J. Physiol.* 242: R97–R104.

Kooyman, G. L. 1989. *Diverse Divers: Physiology and Behavior*. Springer-Verlag, New York.

Noren, S. R., and T. M. Williams. 2000. Body size and skeletal muscle myoglobin of cetaceans: Adaptations for maximizing dive duration. *Comp. Biochem. Physiol. A* 126: 181–191.

Noren, S. R., T. M. Williams, D. A. Pabst, W. A. McLellan, and J. L. Dearolf. 2001. The development of diving in marine endotherms: Preparing the skeletal muscles of dolphins, penguins, and seals for activity during submergence. *J. Comp. Physiol. B* 171: 127–134.

Ponganis, P. J., G. L. Kooyman, and M. A. Castellini. 1993. Determinants of the aerobic dive limit of Weddell seals: Analysis of diving metabolic rates, postdive end tidal P_{O_2}'s, and blood and muscle oxygen stores. *Physiol. Zool.* 66: 732–749.

Ridgway, S. H., and R. Howard. 1979. Dolphin lung collapse and intramuscular circulation during free diving: Evidence from nitrogen washout. *Science* 206: 1182–1183.

Ridgway, S. H., B. L. Scronce, and J. Kanwisher. 1969. Respiration and deep diving in the bottlenose porpoise. *Science* 166: 1651–1654. [A classic.]

Rommel, S. A., D. A. Pabst, and W. A. McLellan. 1998. Reproductive thermoregulation in marine mammals. *Amer. Sci.* 86: 440–448.

Schreer, J. F., K. M. Kovacs, and R. J. O. Hines. 2001. Comparative diving patterns of pinnipeds and seabirds. *Ecol. Monogr.* 71: 137–162.

Snyder, G. K. 1983. Respiratory adaptations in diving mammals. *Respir. Physiol.* 54: 269–294.

Thompson, D., and M. A. Fedak. 1993. Cardiac responses of grey seals during diving at sea. *J. Exp. Biol.* 174: 139–164.

Thornton, S. J., D. M. Spielman, N. J. Pelc, W. F. Block, D. E. Crocker, D. P. Costa, B. J. Le Boeuf, and P. W. Hochachka. 2001. Effects of forced diving on the spleen and hepatic sinus in northern elephant seal pups. *Proc. Natl. Acad. Sci. U.S.A.* 98: 9413–9418.

Watkins, W. A., M. A. Daher, K. M. Fristrup, T. J. Howald, and G. N. DiSciara. 1993. Sperm whales tagged with transponders and tracked underwater by sonar. *Mar. Mamm. Sci.* 9: 55–67.

Williams, E. E., B. S. Stewart, C. A. Beuchat, G. N. Somero, and J. R. Hazel. 2001. Hydrostatic-pressure and temperature effects on the molecular order of erythrocyte membranes from deep-, shallow-, and non-diving mammals. *Can. J. Zool.* 79: 888–894.

Williams, T. M., R. W. Davis, L. A. Fuiman, J. Francis, B. L. Le Boeuf, M. Horning, J. Calambokidis, and D. A. Croll. 2000. Sink or swim: Strategies for cost-efficient diving by marine mammals. *Science* 288: 133–136.

Williams, T. M., J. E. Haun, and W. A. Friedl. 1999. The diving physiology of bottlenose dolphins (*Tursiops truncatus*). I. Balancing the demands of exercise for energy conservation at depth. *J. Exp. Biol.* 202: 2739–2748.

Zapol, W. M., G. C. Liggins, R. C. Schneider, J. Qvist, M. T. Snider, R. K. Creasy, and P. W. Hochachka. 1979. Regional blood flow during simulated diving in the conscious Weddell seal. *J. Appl. Physiol.* 47: 968–973.

PART VI

Water, Salts, and Excretion

Part VI Water, Salts, and Excretion

Previous Page: Animals gain and lose water on a regular basis. The composition and volume of the body fluids depend on the processes of water gain and loss.

25 Water and Salt Physiology: Introduction and Mechanisms

Crustaceans like this blue crab must molt to grow. Between molts, the body of a crustacean is enclosed in a tough, calcified exoskeleton. The exoskeleton has the advantage that it acts like armor, protecting the soft tissues inside from predators. It has the disadvantage, however, that it must be shed for the animal to increase in size. A blue crab must molt over 25 times in the course of its 2-year life.

The soft tissues of the blue crab in the photograph are so grossly swollen that the crab is literally breaking out of its old exoskeleton. Prior to the actual time of molting, seams in the exoskeleton are weakened under hormonal control. Then—also under hormonal control—the body of the animal takes on excess water from its surrounding environment and swells. This swelling in a nearly mature blue crab increases the weight of the animal's tissues by 50%, or sometimes even 100%. The first function of the swelling is to crack the old exoskeleton along its seams so the animal can crawl out. The swelling then serves two additional major functions. First, it gives the animal's body a degree of structural rigidity while the exoskeleton is gone. Without the swelling, the crab's muscles (which are ordinarily attached to the exoskeleton) would have no purchase, would have nothing to pull against; because of the swelling, the muscles can use the soft but taut integument as a substitute for the exoskeleton. Second, the swelling a crab undergoes enables the animal to start *very quickly* to make a new, larger exoskeleton. Because of its swelling with water, a crab's body is bigger as soon as the animal crawls out of its old exoskeleton. Synthesis of the new, bigger exoskeleton can therefore begin immediately, so the animal soon regains its armor. Later, the crab will grow into its new exoskeleton, voiding excess water as it does so. In this way, it is protected inside its new exoskeleton while most of its growth takes place.

The swelling that blue crabs and other aquatic crustaceans routinely undergo serves as a visible, dramatic reminder that watery solutions play crucial roles in the lives of animals. It

Blue crabs dramatize the dynamism of animal water and salt relations when they shed their exoskeleton to grow When a blue crab (*Callinectes sapidus*) sheds its exoskeleton, it takes on extra water in a carefully orchestrated way and swells. The swelling cracks open the old exoskeleton so the crab can exit.

also helps emphasize the *dynamism* of water exchanges in all animals: Body water comes and goes, often turning over very rapidly. Mammals the size of mice commonly turn over 25% of their water every day when living in their natural habitats. People, being larger, turn over only about 7% of their body water per day on average. Even a turnover rate of 7% per day, however, means that a person's body weight could, in principle, increase or decrease by 5 kg (11 pounds) in a single day merely because of an imbalance in water gains and losses.

Despite the dynamism of water turnover, most animals, when healthy, maintain a relatively steady body-water content by means of highly responsive mechanisms that match water gains and water losses. The routine swelling of crustaceans during molting is really just a variation on this theme of water balance. The crustaceans unbalance their water gains and losses in a controlled way to swell, but between molts they regulate their body-water content.

Two principal types of body fluids are recognized in animals: the **intracellular fluids** (**cytosol**) inside cells and the **extracellular fluids** outside cells. The extracellular fluids are divided into two subparts: the **interstitial fluids** found between cells in ordinary tissues[1] and the **blood plasma**, the part of the blood other than the blood cells (the two types of extracellular fluids are identical in animals with open circulatory systems). Each body fluid is an aqueous solution in which are dissolved a variety of inorganic ions—such as Na^+, Cl^-, and K^+—plus organic compounds obtained from food or synthesized metabolically, such as the plasma proteins. The intracellular fluids, interstitial fluids, and blood plasma are often described as the **fluid compartments** of the body.

In this chapter and the three that follow, we discuss the nature and significance of the body fluids, the challenges animals face to maintain the composition and volume of their three body-fluid compartments within viable limits, and the mechanisms of body-fluid regulation. The present chapter will be focused on concepts and mechanisms that pertain to animals in general, or at least to large subsets of animals. Later chapters discuss the specifics of particular animal groups, environments, and organs involved in water-salt physiology.

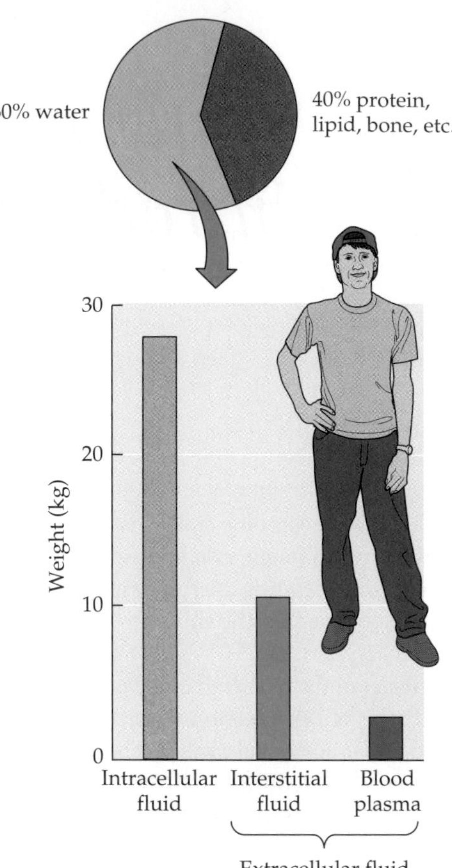

Figure 25.1 Body fluids account for 60% of the body weight of young adult people The bar chart shows the normal weights of the intracellular fluid and the two categories of extracellular fluids—interstitial fluid and blood plasma—in a 70-kg person. (See Figure 4.1 for the composition of the 40% of the body that is not body fluids.)

The Importance of Animal Body Fluids

One sign of the importance of body fluids is simply their abundance. Collectively they account for over half of an animal's body weight in most cases. They represent 60% of body weight in adult people (**Figure 25.1**) and in many other animals. In some types of animals, they account for as much as 95%.

Most of the cellular and subcellular structures in an animal's body, and most of its proteins and other molecules, are either bathed with body fluids or in contact with body fluids. The body fluids thus constitute the immediate environment for cells and molecules in the body, a fact already recognized by Claude Bernard (see page 10) when he developed the concept of the *internal environment* (*milieu intérieur*). The *compositions* of the body fluids provide the context in which cells, organelles, and molecules function.

The particular *inorganic ions*, or *salts* (often called *electrolytes* in the study of medicine), in the body fluids—and the concentrations of those ions—are important aspects of body-fluid composition for several reasons. One is that the ionic composition of body fluids affects the crucial three-dimensional molecular conformations of enzymes and other proteins; the ionic composition can have immediate effects on the conformations of proteins, or it can have indirect effects, as by influencing pH. The ion concentrations in the intracellular and interstitial fluids are also important in maintaining correct electrical gradients across cell membranes. Related to this point, ion concentrations play key roles in nerve-impulse transmission and muscle excitation (see Chapters 11 and 17).

The *water* in body fluids is also important for several reasons. The conceptually simplest reason is that the water acts as the matrix in which the ions are dissolved; the amount of water is as instrumental in determining ion concentrations as are the ions themselves. Water is also important because it affects the *volumes* of cells and tissues. In addition, osmotic movements of water affect the hydrostatic pressures that prevail in the body.

An important modern theme in the study of body fluids and other aqueous solutions is that the ions, H_2O molecules, and organic molecules within a solution relate in ways that impart *structure* to the solution. Individual ions in a solution, for example, induce the

[1] For simplicity, in earlier chapters we have usually referred to the *interstitial fluids* by their more-general name *extracellular fluids*.

H_2O molecules immediately surrounding them to orient around the ions, forming "shells of hydration." Molecules of H_2O also associate with proteins and other macromolecules as "water of hydration." In regions of free water, the H_2O molecules structure themselves into a flickering array by mutual hydrogen bonding ("flickering" because individual H_2O molecules move from place to place in the array, rather than occupying fixed positions). There can be little doubt that new insights about the roles of body fluids will emerge as the structured nature of aqueous solutions is taken into full account in future research.

All animals, wherever they live, have evolved specific, controlled relations of their body-fluid composition to the composition of the external environment in which they live. No known species has body fluids that are simply at passive equilibrium with the external environment. In some cases, the active control exerted by animals over their body-fluid composition is dramatic. Freshwater animals, for example, maintain their body fluids far from equilibrium with freshwater, such that the major inorganic ions are much more concentrated in their body fluids than they are in the water they inhabit. On the other hand, the water–salt regulatory processes in some other animals are quite subtle. Many invertebrates that live in the open ocean, for example, maintain just relatively small differences in salt composition between their blood plasma and seawater. Whether the differences between an animal's body fluids and its external environment are dramatic or subtle, actively maintained differences always exist. This observation suggests that for all animals, the effects of salts and water on protein function and other functional properties have led to fine-tuning of body-fluid composition during the course of evolution and to the evolutionary development of mechanisms that can control the water–salt composition of the body fluids in at least certain respects.

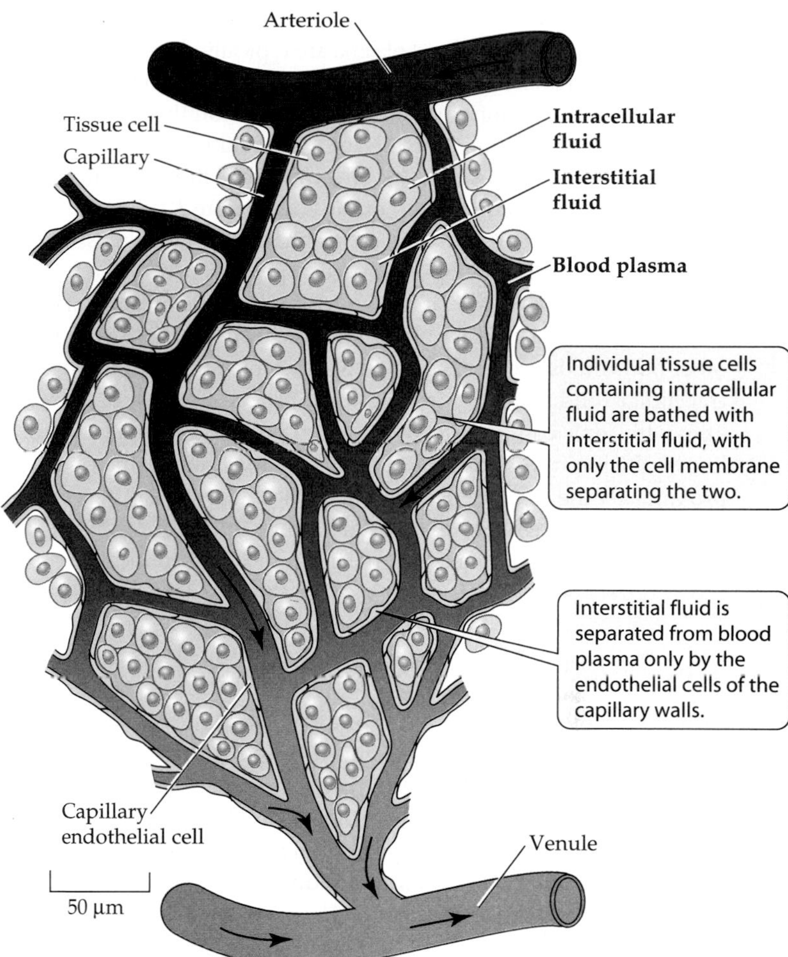

Figure 25.2 The three major types of body fluids are closely juxtaposed This diagram of a microcirculatory bed in a mammal shows the close juxtaposition of the three major types of body fluids. For simplicity, red blood cells and other blood cells are not shown. (See Figure 23.11 for more on the structure of microcirculatory beds.)

The Relations among Body Fluids

The three major body-fluid compartments of an animal—the intracellular fluids, interstitial fluids, and blood plasma—interact and affect one another. Figure 25.2 emphasizes that the three compartments are closely juxtaposed within an animal's tissues. The intracellular fluids are separated from the interstitial fluids only by the cell membranes of the tissue cells. The interstitial fluids are separated from the blood plasma only by the blood-capillary endothelium.

There is often a brisk exchange of water and ions between the intracellular fluids and the interstitial fluids across cell membranes. As discussed in Chapter 3, this is true in part because cell membranes are permeable to water and ions; the membranes have water channels and ion channels through which osmosis and diffusion occur, and water can move by osmosis directly through the membrane phospholipid bilayer. Moreover, transporter proteins in cell membranes often aid ion transport between the intracellular fluid and interstitial fluid by facilitated diffusion or active transport.

There is often also a brisk exchange of water and ions between the interstitial fluids and the blood plasma. The blood-capillary endothelium that separates the interstitial fluids and blood plasma is in most cases relatively densely perforated by minute, open pores of one sort or another. Some of these pores are formed by tiny gaps (about 4 nm in diameter) *between* cells in the endothelium; other pores are formed in some tissues by holes *through* endothelial cells.[2] Water and ions freely cross the capillary endothelium through these pores by osmosis and diffusion, bringing the fluids on each side close to equilibrium with each other. In addition, as discussed in Chapter 23 (see Figure 23.13), the elevated blood pressure inside capillaries forces a bulk flow of fluid—containing both water and ions—out of the plasma and into the interstitial fluids by ultrafiltration.

Most studies of the water–salt physiology of animals living free in their environments focus on the composition of the blood plasma. This is true because the blood can be sampled readily and therefore provides a convenient window into an animal's body-fluid physiology. Because of the preeminence of blood in body-fluid studies, we must wonder what we can deduce about the composition of the other

[2] The capillaries in various tissues of an animal's body vary in their types and densities of pores. In the human body, for example, liver capillaries are very porous. Brain capillaries, on the other hand, virtually lack pores. The capillaries in most tissues are intermediate. The pores are often called *fenestrations*.

body fluids from knowledge of the composition of the blood plasma. The interstitial fluid and blood plasma are typically closely similar in their osmotic pressures and ionic compositions because of the free exchange of water and ions by osmosis and diffusion through blood-capillary pores.[3] Although the intracellular fluid inside cells differs strongly in ionic composition from the interstitial fluid surrounding cells because of the action of ion pumps (see Figure 3.10), the intracellular fluid and interstitial fluid are ordinarily the same in osmotic pressure because cell membranes are freely permeable to water.

Summarizing, therefore, if one knows the osmotic pressure of the blood plasma, one typically also knows the osmotic pressure of the other body fluids because the blood plasma, interstitial fluid, and intracellular fluid are either isosmotic or close to isosmotic.[4] If one knows the ionic composition of the blood plasma, one typically also knows the ionic composition of the interstitial fluid (which is approximately the same); however, the intracellular fluid differs dramatically from the other two fluids in its ionic composition and thus must be studied in its own right to determine its composition.

The Types of Regulation and Conformity

Three types of regulation of the composition of the blood plasma are possible: osmotic regulation, ionic regulation, and volume regulation. Some animals—such as humans and other mammals—exhibit regulation in all these respects. Other animals, however, function as conformers in certain aspects of their water–salt physiology.[5]

Osmotic regulation (**osmoregulation**) is the maintenance of a constant or nearly constant osmotic pressure in the blood plasma. To illustrate osmotic regulation of the blood plasma, consider an aquatic animal that is placed in environmental water of various osmotic pressures. If the animal exhibits perfect osmotic regulation, the blood osmotic pressure is kept the same regardless of the osmotic pressure of the environmental water (Figure 25.3*a*). Ordinarily, osmotic regulation of the interstitial fluid and intracellular fluid goes hand in hand with that of the blood plasma because all three fluids are similar in osmotic pressure. The dashed line in Figure 25.3*a* is a line of equality between the blood osmotic pressure and the environmental osmotic pressure and thus is called the **isosmotic line** (similar to the *isothermal line* discussed in Chapter 8). A perfect osmoregulator exhibits no tendency to follow the isosmotic line. The opposite of osmotic regulation is **osmotic conformity**. In this case, the blood osmotic pressure always equals the osmotic pressure of the environmental water, and thus falls on the isosmotic line (Figure 25.3*b*).

Actual animals often display gradations or mixes of osmotic regulation and osmotic conformity. Figure 25.3*c*, for example, shows the relation between blood osmotic pressure and ambient osmotic pressure in three invertebrates. The shrimp is almost a perfect osmoreg-

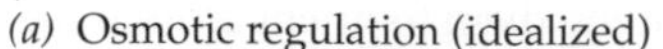

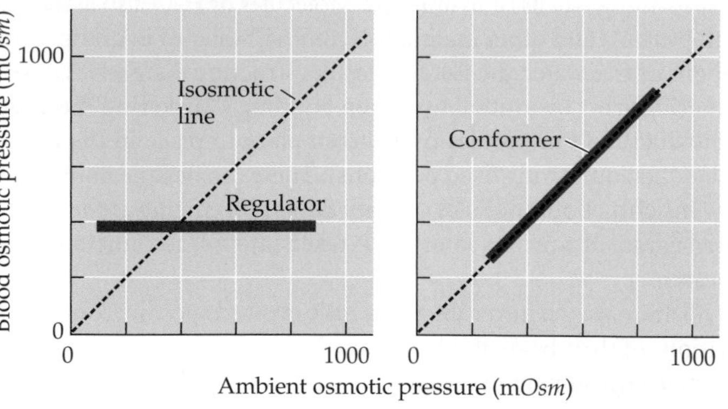

(*c*) Actual relations of three marine invertebrates

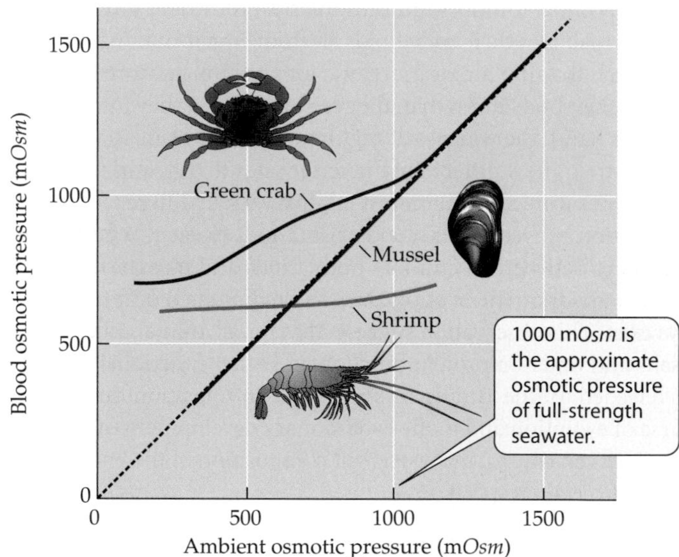

Figure 25.3 Osmotic regulation and conformity In each graph, the solid line shows the osmotic pressure of blood plasma as a function of the ambient (environmental) osmotic pressure; the dashed line is a line of equality between blood osmotic pressure and ambient osmotic pressure (an isosmotic line). *(a, b)* The osmotic pressure of the blood plasma as a function of the osmotic pressure of the ambient water in *(a)* a perfect osmotic regulator and *(b)* a perfect osmotic conformer. *(c)* Actual animals often exhibit responses that are mixed or intermediate between idealized regulation and idealized conformity. The osmotic pressure of the blood plasma as a function of the ambient osmotic pressure is shown for three species of marine invertebrates: *Mytilus edulis*, the blue mussel; *Carcinus maenas*, the green crab; and *Palaemonetes varians*, a species of grass shrimp. The mussel is a strict osmotic conformer. The crab regulates in waters more dilute than seawater but is an osmotic conformer at higher ambient osmotic pressures. The shrimp regulates over a wide range of ambient osmotic pressures. (*c* after Hill and Wyse 1989.)

[3] Blood plasma is typically more concentrated than interstitial extracellular fluid in dissolved proteins that cannot pass through the capillary endothelium. This disparity produces a difference in osmotic pressure between the two fluids that can be of critical importance for understanding certain phenomena, such as bulk-fluid exchanges across capillary walls (see Figure 23.13 and the discussion of the Starling-Landis hypothesis in Chapter 23). The *absolute* osmotic pressures of the blood plasma and interstitial fluid are quite similar, nonetheless. In human systemic tissues, as specified in Chapter 23, the blood osmotic pressure is about 3.3 kPa (25 mm Hg) higher than the interstitial-fluid osmotic pressure. In the osmolarity system of units, 3.3 kPa is equivalent to about 1.5 milliosmolar (m*Osm*). Thus, when the blood plasma has an osmotic pressure of 300 m*Osm* (as is typical of human blood), the interstitial fluid has an osmotic pressure of 298.5 m*Osm*. For most purposes, therefore, the blood plasma and interstitial fluid can be considered to be virtually identical in osmotic pressure.

[4] As we learned in Chapter 3, *isosmotic* means "of equal osmotic pressure."

[5] See page 11 for the distinction between *regulation* and *conformity*.

ulator; its blood osmotic pressure varies just a little with the osmotic pressure of its external environment. The green crab exhibits impressive but imperfect osmotic regulation in waters more dilute than 1000 m*Osm* (the approximate osmotic pressure of full-strength seawater), but it is an osmotic conformer in waters more concentrated than 1000 m*Osm*. The mussel is an osmotic conformer at all the concentrations studied.

The concepts of osmotic regulation and conformity apply to terrestrial animals as well. People, for example, maintain a remarkably stable blood-plasma osmotic pressure of about 300 m*Osm* whether they drink much water per day or little per day. Thus people are excellent osmotic regulators.

Ionic regulation of the blood plasma is the maintenance of a constant or nearly constant concentration of an inorganic ion in the plasma. There are as many potential types of ionic regulation as there are ions because each ion is subject to ion-specific physiological controls. The opposite of ionic regulation is **ionic conformity**; an animal that allows the concentration of an ion in its blood plasma to match the concentration in its external environment is a conformer for that ion. As a broad generalization, ionic regulation of the interstitial fluid goes hand in hand with that of the blood plasma; regulation of an ion in the plasma implies regulation of that ion in the interstitial fluid, and conformity in the plasma means conformity in the interstitial fluid. The intracellular fluid, however, is subject to its own regulatory processes, and ionic regulation or conformity in the intracellular fluid cannot be predicted from the physiology of the other body fluids.

Volume regulation is the regulation of the total *amount* of water in a body fluid. Volume regulation of the blood plasma, for example, is regulation of the amount of water in the plasma. Animals generally regulate their blood volume. We, for example, maintain an approximately constant amount of water in our blood plasma regardless of how much water we drink per day. The concept of volume regulation also applies to the interstitial fluid and intracellular fluid; for example, cells exhibit volume regulation if they maintain a constant amount of intracellular water.

Lack of volume regulation is illustrated by a blue crab at the specific times in its life when it swells and molts. Otherwise, blue crabs are volume regulators, as already stressed. True **volume conformity**, referring to a completely passive osmotic expansion of a body fluid, is a relatively rare phenomenon.

The integrated study of osmotic regulation, ionic regulation, and volume regulation can seem confusing at first because although the three types of regulation are in fact distinct, they are often intimately related. Consider a freshwater fish that takes on a quantity of water from its dilute environment by osmosis. This influx of water into its body will tend to have three *simultaneous* effects:

1. It will lower the osmotic pressure of the fish's blood plasma (a challenge to osmotic regulation).
2. It will dilute ions in the fish's blood plasma (a challenge to ionic regulation).
3. It will swell the amounts of water in the fish's blood plasma and other body fluids (a challenge to volume regulation).

Simultaneous effects like this can give the illusion that osmotic regulation, ionic regulation, and volume regulation are the same thing. They are in fact different, however, as we will see by example later in this chapter. *An important tool for understanding animal water–salt physiology is to maintain the conceptual distinction between the three types of regulation and analyze problems in water–salt physiology from all three perspectives.*

Regulation and conformity have pros and cons. If an animal exhibits osmotic and ionic regulation of its blood plasma, the composition of its interstitial fluid is also regulated, and therefore its *cells*—which are bathed by interstitial fluid—experience a constant osmotic–ionic environment. Animals must invest energy to maintain this constancy of the internal environment, however. If animals exhibit osmotic–ionic conformity in their blood plasma and interstitial fluid, they avoid these energy costs, but their cells must then cope with changing osmotic–ionic conditions in the interstitial fluid that bathes them.[6]

Summary: The Types of Regulation and Conformity

- Osmotic regulation is the maintenance of a steady osmotic pressure in the blood plasma.
- Ionic regulation of any particular inorganic ion is the maintenance of a steady concentration of that ion in the blood plasma.
- Volume regulation is the maintenance of a steady volume (amount) of water in the body as a whole or in particular fluid compartments. Cell-volume regulation is regulation of the amount of intracellular water.

Natural Aquatic Environments

Some aquatic animals live in environments that are quite uniform and stable in their water–salt composition. The animals that live in the open ocean are in this category. Full-strength seawater is approximately the same everywhere in its composition of major ions (Table 25.1). The *total* concentration of all salts taken together in a body of water is expressed as its **salinity**, defined to be the number of grams of dissolved inorganic matter per kilogram of the water. The salinity of seawater from the open ocean is 34–36 g/kg.[7] An important and easily remembered feature of seawater is that its osmotic pressure is almost exactly 1000 m*Osm*, as already noted.

Freshwater is usually defined as water having a salinity less than 0.5 g/kg (the worldwide average for lakes and rivers is 0.1–0.2 g/kg). Freshwater is essentially a uniform environment in terms of its osmotic pressure, which is always very low (0.5–15 m*Osm*). On the other hand, although ions are always dilute in freshwater (see Table 25.1), certain individual ions vary in concentration from one body of freshwater to another in ways that are biologically consequential. An ion of particular note in this regard is calcium (Ca^{2+}). Variations

[6] See Chapter 1 (page 13) for a more complete discussion of the pros and cons of regulation and conformity (i.e., homeostasis and lack of homeostasis).

[7] Several alternative ways exist to write this unit of measure. To refer to g/kg, some scientists use the symbol ‰; others write *parts per thousand* (*ppt*) or *parts per mille*. Salinity can also be expressed using *practical salinity units*; values in this system are dimensionless (have no units), but are numerically closely similar to values expressed in g/kg.

in its concentration can exert substantial effects on the water–salt physiology of freshwater animals by affecting their membrane permeabilities and possibly other functional properties (freshwaters high in Ca^{2+} are often called "hard").

Where ocean water mixes with freshwater along continental coasts, waters of intermediate salinity, termed **brackish waters**, are formed. Many of the places where brackish waters occur are classified as **estuaries.** An estuary is any body of water that is partially surrounded by land and that has inflows of both freshwater and seawater. Estuaries are of great importance in human affairs. They also are among the most interesting aquatic habitats physiologically, because the salinity in estuaries typically varies dramatically from place to place and time to time, thereby presenting animals with unusual challenges. Brackish waters are usually defined to have salinities between 0.5 and 30 g/kg—corresponding to osmotic pressures of 15 to 850 m*Osm*. As illustrated by the Chesapeake Bay (Figure 25.4), a single estuary may exhibit almost this entire range of salinities within a distance of less than 200 miles. Estuaries are the characteristic habitat of blue crabs, which are effective osmoregulators at brackish salinities when not molting. As a blue crab travels about into waters of various salinities in an estuary, its blood osmotic pressure remains much the same everywhere it goes (Figure 25.5).

TABLE 25.1 Concentrations of major ions in seawater and freshwater

Ion	Concentration (m*M*) Seawater[a]	Freshwater[b]
Sodium (Na^+)	470	0.35
Chloride (Cl^-)	548	0.23
Magnesium (Mg^{2+})	54	0.21
Sulfate (SO_4^{2-})	28	0.19
Calcium (Ca^{2+})	10	0.75
Potassium (K^+)	10	0.08
Bicarbonate (HCO_3^-)	2	1.72

Sources: Data for seawater from Barnes 1954; data for freshwater from Bayly and Williams 1973.

[a] Seawater values are for a salinity of 34 g/kg.

[b] Freshwater values are worldwide averages for rivers.

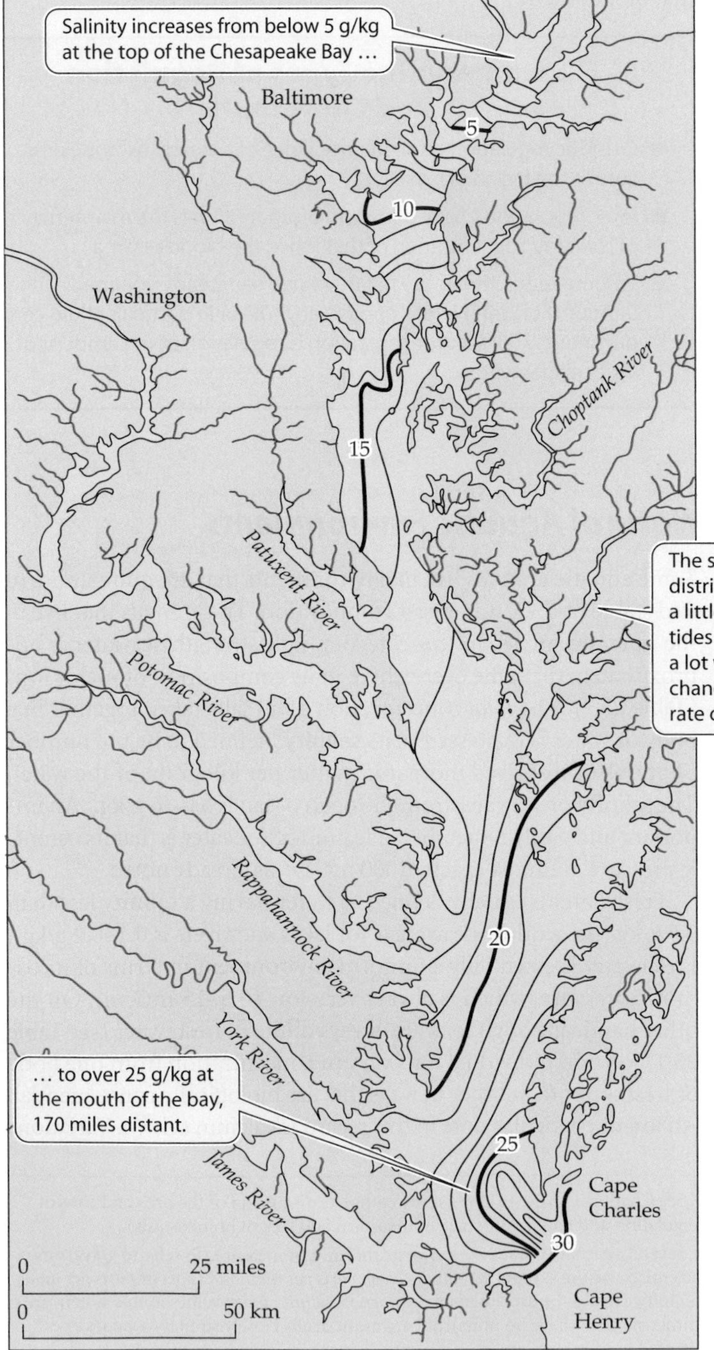

A typical shoreline of the Chesapeake Bay

Figure 25.4 Salinity trends in an estuary Salinity varies greatly along the length of the Chesapeake Bay, one of the world's most famous estuaries. The map shows a typical pattern of salinity in water near the surface of the water column. Numbers are salinities in g/kg. Each line connects all the surface waters of a particular salinity; thus, for example, the line for 20 g/kg connects all the places where the salinity of the surface water is 20 g/kg. (Salinity distribution after McHugh 1967.)

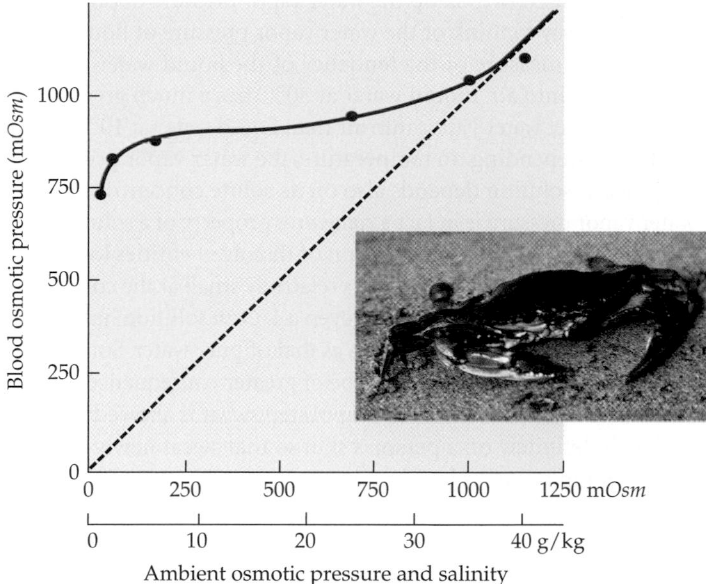

Figure 25.5 The responses of a resident osmotic regulator to variations in salinity in an estuary Blue crabs (*Callinectes sapidus*) are abundant throughout the Chesapeake Bay and other similar estuaries along the Atlantic and Gulf coasts of the United States. The graph shows the average osmotic pressure of the blood plasma of blue crabs as a function of the osmotic pressure and salinity of the ambient water during nonmolting parts of their lives, when they osmoregulate. The dashed line is the isosmotic line. (After Kirschner 1991.)

Natural Terrestrial Environments

Animals on land are surrounded by a fluid—air—that contains water only in the gaseous state and, of course, is essentially free of salts. One might guess that terrestrial animals would commonly be able to gain water from the air when the humidity of the air is high. That is not the case, however. For most terrestrial animals most of the time, the atmosphere is a sink for water—often a severe sink. Because living in air is inherently dehydrating, the study of the water–salt physiology of terrestrial animals is dominated by the study of *water*.

For scientists interested in the water relations of terrestrial animals, the world's deserts are particularly important and intriguing habitats because they present animals (including people) with the extremes of water stress. Deserts cover substantial areas of Earth's landmasses (Figure 25.6). Although deserts are often hot, they are not necessarily so, because they may occur at high altitudes or polar latitudes. They are defined by their dryness, not their temperature. A simple definition of a desert is that it receives less than about 25 cm (10 inches) of rain or precipitation per year; we discuss more-refined definitions in Chapter 28.

To understand the relations of animals to atmospheric water in deserts or other terrestrial habitats, one must study the principles of **evaporation**, the change of water from a liquid to a gas. Evaporation is a special case of gas diffusion. As discussed in Chapter 20, gas dif-

1 Sahara Desert
2 Arabian Desert
3 Gobi Desert
4 Great Victorian Desert
5 Namib Desert
6 Sonoran and Chihuahuan Deserts
7 Atacama Desert

Figure 25.6 Deserts occur on all continents Sandy-colored areas are mostly deserts. Animals that have succeeded in living in deserts have had to evolve extreme adaptations to water stress, including extreme abilities to acquire water, conserve the water they have, and tolerate dehydration.

fusion is analyzed using the *partial pressures* of gases.[8] Gases always diffuse in net fashion from regions of high partial pressure to regions of low partial pressure. Thus, if a body fluid, or any other aqueous solution, is in contact with the atmosphere, net evaporation occurs if the partial pressure of water in the solution exceeds that in the atmosphere, and the rate of evaporation increases as the difference in partial pressure increases.

What do we mean by the partial pressure of water in a solution and in the atmosphere? Let's start with the latter. Gaseous water, called **water vapor**, is simply a gas like any other gas. Thus it is a constituent of the atmosphere in the same way that other gases, such as O_2, are. The partial pressure of water—often called the **water vapor pressure**—in the atmosphere is simply the portion of the total atmospheric pressure that is exerted by the water vapor present. It can be calculated from the universal gas law (Equation 20.1) and is independent of the partial pressures of other gases in the atmosphere. **Humidity** is an informal term referring loosely to the water content of air. Although several measures of humidity are in common use, the water vapor pressure is the most useful for physiological analysis.[9]

Water vapor is distinguished from the other gases in the atmosphere by having a limit on its partial pressure. That is, the water vapor pressure can rise only to a certain maximum in air of a particular temperature. The limit on water vapor pressure is a direct consequence of the fact that (unlike the other atmospheric gases) water can exist as a liquid (not just a gas) under ordinary atmospheric conditions. Air that has reached its maximum water vapor pressure is said to be **saturated**, and its water vapor pressure is termed the **saturation water vapor pressure**. If a body of air has reached saturation with water vapor, it cannot hold any more water in the gaseous state. Thus, if water vapor is added to such air from an outside source, the excess water vapor promptly condenses out in the form of liquid water droplets (e.g., fog). The saturation water vapor pressure increases dramatically with the temperature of air, as Table 25.2 shows.

To illustrate the immediate significance of these concepts, consider that we humans exhale air at a temperature above 30°C, whereas a toad with a body temperature of 20°C breathes out air at 20°C. In either case, the air exhaled is saturated with water vapor. Warmer air holds more water vapor when saturated, however, and therefore people lose more water with each liter of air they exhale than toads lose.

Now we need to ask what is meant by the partial pressure of water in an aqueous solution. Any particular aqueous solution, if it is placed in contact with air in a closed system, will tend to establish a characteristic, equilibrium water vapor pressure in the air. That vapor pressure is the **water vapor pressure** of the aqueous solution. The water vapor pressure of pure liquid water depends on the temperature of the water. Specifically, the water vapor pressure of pure liquid water at a particular temperature is the same as the saturation water vapor pressure of air at the same temperature; Table 25.2 can therefore be used to look up the water vapor pressure of pure liquid water. One way to think of the water vapor pressure of liquid water is that it is a measure of the tendency of the liquid water to inject water vapor into air. Liquid water at 30°C has a much greater tendency to inject water vapor into air than liquid water at 10°C.

Besides depending on temperature, the water vapor pressure of an aqueous solution depends also on its solute concentration. The water vapor pressure is in fact a *colligative* property of a solution (see page 82). Raising the concentration of dissolved entities lowers the water vapor pressure. This effect is relatively small at the concentrations of most animal body fluids; even a 1-*Osm* solution has a water vapor pressure that is 98% as high as that of pure water. Sometimes, however, the effect of solutes can be of greater consequence. For instance, if salt left behind from evaporated sweat is allowed to accumulate indefinitely on a person's skin so that sweat newly secreted onto the skin becomes highly concentrated by dissolving accumulated salt, the water vapor pressure of the sweat can be reduced to only 75% of that of pure water.

Having discussed the partial pressure of water both in the atmosphere and in aqueous solutions, we can now analyze the physical laws that govern the evaporation of water from terrestrial animals. As mentioned earlier, water always diffuses from regions of relatively high water vapor pressure to regions of lower water vapor pressure. Thus water evaporates from an aqueous solution (e.g., a body fluid) provided the water vapor pressure of the solution exceeds the water vapor pressure of the air next to the solution. The *rate* of evaporation depends on the difference in water vapor pressures. Specifically, if J is the net rate of evaporation per unit of solution surface area, WVP_s is the water vapor pressure of the solution, and WVP_a is the water vapor pressure in the air, then

$$J = K\frac{WVP_s - WVP_a}{X} \qquad (25.1)$$

where X is the distance separating WVP_s and WVP_a, and K is a proportionality factor.

TABLE 25.2 The saturation water vapor pressure at selected temperatures The saturation water vapor pressure is independent of the composition of the air or other gas. The table also shows the mass of water per unit of volume when air or other gas is saturated at the stated temperature.

Temperature (°C)	Saturation water vapor pressure (mm Hg)	(kPa)	Mass of water per unit of volume at saturation (mg H_2O/L)
0	4.6	0.61	4.9
10	9.2	1.23	9.4
20	17.5	2.33	17.3
30	31.8	4.24	30.4
37[a]	47.1	6.28	43.9
40	55.3	7.37	51.1

[a] Data for 37°C are included because 37°C is the usual deep-body temperature of people and other placental mammals.

[8] If you are unfamiliar with the concept of *partial pressure*, you should review pages 528–530 and Figure 20.1.

[9] Besides being expressed simply as the water vapor pressure, the humidity of air is often expressed *relative* to the air's temperature-specific *saturation water vapor pressure*, discussed in the next paragraph. One expression of this sort is the **saturation deficit**, which is the *difference* between the actual, prevailing water vapor pressure and the saturation water vapor pressure. Another expression of humidity is the **relative humidity**, defined to be the *ratio* of the actual water vapor pressure over the saturation water vapor pressure.

From Equation 25.1, we see that the rate at which an animal loses water by evaporation in a terrestrial environment depends partly on the environmental humidity, expressed as WVP_a. Lowering the water vapor pressure of the air speeds evaporation if all other factors are held constant—a very familiar relation. Note also, however, that the rate of evaporation depends on the water vapor pressure of the body fluid from which evaporation is occurring, WVP_s. Thus, for example, evaporation occurs faster from body fluids that are warm than from ones that are cooler. When animals such as mammals, birds, and basking lizards are exposed to the open air in hot deserts, they can face exceptional risks of dehydration by evaporation because their body fluids are warm (meaning WVP_s is high; see Table 25.2) while simultaneously the air is dry (WVP_a is low).

The rate of evaporation also depends on the rate of air movement. If an animal is standing in still air, evaporation from the animal itself tends to humidify the air immediately next to its skin, creating a boundary layer of elevated WVP_a near the skin, very similar to the boundary layer depicted in Figure 3.3. The elevated WVP_a slows evaporation. A wind, however, blows water-vapor-laden air away from the skin surface, replacing it with drier air from the open atmosphere, thereby decreasing the WVP_a next to the skin and speeding evaporation. A wind in a desert can be extremely dehydrating.

In terms of Equation 25.1, the chief physiological defense of animals exposed to the open air in deserts is to have a low value of K. When evaporation is occurring through an animal's integument, K represents the permeability of the integument to water. A low permeability slows the loss of body water by evaporation.

To complete our discussion of the diffusion of water vapor in terrestrial animals, let's return to a point made at the start; namely, that animals rarely gain water by condensation. Thus animals tend to lose water to the atmosphere rather than gaining water from it. In most types of animals, condensation can occur only when the body surface is cooler than the air. Under such unusual circumstances, animals can function like glasses of iced tea.

We are all familiar with the fact that gaseous water from the atmosphere condenses into water droplets on the outside of a cold glass of iced tea on a humid summer day. The physics of the *initiation* of water-droplet formation is an advanced topic in physical chemistry. However, once microscopically tiny water droplets have been initiated, their *growth* follows the simple principles of water diffusion. Water will diffuse from the water vapor in the atmosphere into a droplet of liquid water—causing the droplet to grow—only if the water vapor pressure in the atmosphere exceeds the water vapor pressure of the water in the droplet. If a water droplet is cold, it has a relatively low water vapor pressure (see Table 25.2). If the water vapor pressure in the air is simultaneously high (as it is on a warm, humid day), water will move from the gaseous state in the air into the liquid state in the droplet. In this way, a microscopically tiny droplet will grow to be big and visible.

For the most part, the body surfaces of animals are not cooler than the air. Thus the water vapor pressure of body fluids is typically higher than that of the air, and animals lose water by evaporation rather than gaining it by condensation. Occasionally, animals in special circumstances have cool body surfaces. For example, when lizards that have spent the night in chilly underground burrows emerge into the open air in the morning, their skin can be cooler than the air, and if the air is humid enough, water droplets can form on their skin and potentially be ingested. The condensation is transitory, however, because the animals warm as time passes. As soon as they are as warm as the air, they start to lose water by evaporation.[10]

Summary: Natural Terrestrial Environments

- The water vapor pressure of air is the partial pressure of water vapor in the air and is the most useful expression of humidity for analysis of evaporation and condensation.
- The water vapor pressure of an aqueous solution is the equilibrium water vapor pressure the solution tends to create in juxtaposed air if the solution and air are sealed in a closed system.
- Water vapor diffuses from regions of high water vapor pressure to regions of low water vapor pressure. Thus evaporation occurs if the water vapor pressure of an aqueous solution exceeds that of the surrounding air; it takes place at a rate proportional to the difference in vapor pressure.

Organs of Blood Regulation

It is routine for animals living in their natural environments to experience conditions that tend to change their blood composition. A mammal exposed to the dryness of desert air, for example, tends to lose water by evaporation, and its loss of water tends to raise the osmotic pressure of its blood, concentrate ions in its blood, and decrease the volume of its blood (challenges to osmotic regulation, ionic regulation, and volume regulation). Another example would be a fish that migrates from the lower part of Chesapeake Bay, where the salinity is high, into the upper part, where the salinity approaches that of freshwater (see Figure 25.4). In the dilute waters, the fish will tend to take on extra water by osmosis, and the influx of water will tend to lower the osmotic pressure of its blood, lower the concentrations of ions in its blood, and expand the volume of its blood, as we saw earlier. Animals like mammals and fish are regulators of blood composition and respond to such challenges in negative-feedback fashion (see Box 1.1). Certain of their organs change their functions in ways that tend to reverse any changes in blood composition; in this way, their blood composition tends to be kept stable despite the environmental challenges they encounter.

The kidneys play major roles in stabilizing blood composition, not only in mammals and fish but in most other types of animals. There also may be other organs that play major roles in blood regu-

[10] Animals could, in principle, gain water at a low rate while at the *same* temperature as *saturated* air, because the solutes in body fluids lower the body-fluid water vapor pressure slightly. However, in reality, when animals are in saturated air, their metabolic heat production raises their temperature to be at least slightly above air temperature, forcing the water vapor pressure gradient to favor evaporation. Thus, without a cooling mechanism, water is not gained by diffusion from the air. Of course, evaporation of body fluids can cool an animal's body surfaces. However, thermodynamics dictates that evaporative cooling cannot be sufficient to cause simultaneous condensation (one process cannot cause evaporation and condensation simultaneously). Thus, for water to be obtained by diffusion from air because of body cooling, the cooling must be caused by some other process than simultaneous evaporation. Some insects, ticks, and other terrestrial arthropods are exceptions to the general rule in that they can gain water steadily from atmospheric water vapor; these cases are discussed in Chapter 26 (see page 712).

lation; the gills are important organs of blood regulation in fish and other aquatic animals, and salt glands, discussed in Chapter 26, are important in certain birds and reptiles. In Part VI we emphasize kidneys. Our specific goal will be to develop a set of principles that—rather than being addressed to particular types of kidneys—are of *general* usefulness for analyzing the function of kidneys as organs of blood regulation. Some of these principles can also be applied directly or indirectly to analyzing the function of gills and salt glands.

The most fundamental function of kidneys is to regulate the composition of the blood plasma by removing water, salts, and other solutes from the plasma in a controlled way. Kidneys are fluid-processing organs. They start with blood plasma and produce urine; thus, speaking conceptually, blood plasma is their input, and urine is their output. Many of the effects of kidneys on blood composition can be analyzed by comparing their output to their input—that is, by comparing the urine to the plasma. This sort of comparison is usually carried out with **U/P ratios**: ratios of urine ("U") composition over plasma ("P") composition.

The effects of kidney function on osmotic regulation depend on the osmotic U/P ratio

The **osmotic U/P ratio** is the osmotic pressure of the urine divided by the osmotic pressure of the blood plasma. For example, if an animal's urine has an osmotic pressure of 150 m*Osm* and the plasma osmotic pressure is 300 m*Osm*, the osmotic U/P ratio is 0.5. Urine may, in principle, be isosmotic, hyperosmotic, or hyposmotic to the blood plasma.[11] This **relative osmoticity** of the urine is reflected in the U/P ratio. If U/P = 1, the urine is isosmotic to the plasma. If U/P < 1, the urine is hyposmotic to the plasma. If U/P > 1, the urine is hyperosmotic to the plasma. The kidneys typically have control over the U/P ratio and can adjust it within a species-specific range. People, for example, can have an osmotic U/P ratio as high as 4 or as low as 0.1.

To explore the interpretive value of the U/P ratio, let's start by considering a freshwater fish. As discussed in Chapter 3 (see Figure 3.18), the body fluids of a freshwater fish have an osmotic pressure far higher than the osmotic pressure of freshwater: The blood plasma of a freshwater fish is strongly hyperosmotic to freshwater. Suppose that a fish takes up a quantity of water into its body fluids by osmosis from the pond or stream in which it lives. This water will dilute the fish's blood and reduce its plasma osmotic pressure. Can the fish restore its original plasma osmotic pressure (i.e., can it osmoregulate) by producing urine? A bit of reflection will reveal that the answer is "yes" *only if the fish is able to produce a urine that is more dilute than its plasma—that is, a hyposmotic urine* (U/P < 1).

A urine that is hyposmotic to the plasma will *preferentially void water*. By this we mean that in comparison to the blood plasma, urine of this sort is richer in water and poorer in dissolved solutes. Thus, when the urine is excreted, it disproportionately depletes the blood of water. Because of this preferential removal of water from the plasma—and because of the converse, the preferential retention of solutes in the plasma—voiding the urine will act to elevate the osmotic pressure of the plasma back toward its original level. One way to see this point is to contrast this outcome with what would happen if a fish's urine were always isosmotic to its plasma. A fish with a U/P ratio of 1 would merely excrete water and solutes in the same ratio at which they existed in its blood plasma. Thus, if the fish's plasma had become diluted, it would remain diluted regardless of how much isosmotic urine (U/P = 1) the fish excreted.

By analyzing the freshwater fish, we arrive at two general principles of kidney function (Figure 25.7):

1. The production of urine isosmotic to plasma (U/P = 1) cannot serve directly to change the osmotic pressure of the plasma or bring about osmotic regulation of the plasma.
2. The production of hyposmotic urine (U/P < 1) aids osmotic regulation of the blood plasma if an animal's plasma has become diluted and the plasma osmotic pressure therefore needs to be raised.

[11] As noted in Chapter 3, if A and B are two solutions and A has a higher osmotic pressure than B, then A is *hyperosmotic* to B, whereas B is *hyposmotic* to A.

Figure 25.7 The interpretive significance of the osmotic U/P ratio The terms *solute* and *solutes* refer to total numbers of osmotically effective dissolved entities.

	Implications for excretion		
U/P ratio	**Effects on water excretion**	**Effects on solute excretion**	**Effects on composition of blood plasma**
U/P = 1 (isosmotic urine)	Water is excreted in the same relation to solutes as prevails in the blood plasma.	Solutes are excreted in the same relation to water as prevails in the blood plasma.	The formation of urine leaves the ratio of solutes to water in the blood plasma unchanged, thus does not alter the plasma osmotic pressure.
U/P < 1 (hyposmotic urine)	Water is preferentially excreted. Urine contains more water relative to solutes than plasma.	Solutes are preferentially held back from excretion. Urine contains less solutes relative to water than plasma.	The ratio of solutes to water in the plasma is shifted upward. The osmotic pressure of the plasma is raised.
U/P > 1 (hyperosmotic urine)	Water is preferentially held back from excretion. Urine contains less water relative to solutes than plasma.	Solutes are preferentially excreted. Urine contains more solutes relative to water than plasma.	The ratio of solutes to water in the plasma is shifted downward. The osmotic pressure of the plasma is lowered.

As one might expect, most freshwater animals have evolved kidneys that have the capacity to make urine that is hyposmotic to their plasma. We humans, as well as many other terrestrial animals, also have that capacity, which serves us well after an evening of too much iced tea or beer.

If the plasma osmotic pressure of an animal has been raised to abnormally *high* levels, the production of urine that is isosmotic to plasma will again fail to bring about osmoregulation. Urine more concentrated than the plasma, however, will preferentially void solutes (and preferentially retain water); such urine will favorably alter the ratio of solutes to water in the plasma. Thus we arrive at a third principle of kidney function (see Figure 25.7):

3. The production of hyperosmotic urine (U/P > 1) aids osmotic regulation of the blood plasma if an animal's plasma has become overly concentrated and the plasma osmotic pressure therefore needs to be lowered.

The ability to produce urine that is hyperosmotic to the plasma is not nearly as widespread as the ability to produce hyposmotic urine. The greatest capacities to concentrate the urine are found in mammals, birds, and insects—all primarily terrestrial groups that frequently face risks of dehydration.

The effects of kidney function on volume regulation depend on the amount of urine produced

The kidneys usually help regulate the *quantity of water* in an animal's body by voiding greater or lesser *amounts* of water as required. We ourselves provide a familiar example: We make a lot of urine after drinking a lot of water, but we make little urine if we are short of water.

The kidneys, in fact, can play a critical role in *volume regulation* even when not playing any direct role in *osmotic regulation*. In this respect, kidney function illustrates that volume regulation and osmotic regulation are *distinct processes*, as earlier stressed. There are some true crabs that live in rivers, mostly in tropical and subtropical parts of the world. These freshwater crabs provide striking examples of animals in which the kidneys participate in volume regulation, but not osmotic regulation. The crabs are strongly hyperosmotic to the freshwater in which they live, and thus experience a steady osmotic flux of water into their body fluids. The crabs meet this challenge to volume regulation by producing an ample flow of urine; each day their kidneys excrete the same amount of water as the crabs gain by osmosis. However, at least in the species that have been investigated, the kidneys of freshwater crabs are unable to produce urine that is more dilute than the blood plasma. The urine is always isosmotic to the plasma (U/P = 1), and consequently, the production of urine does not alter the plasma osmotic concentration. The kidneys of freshwater crabs, therefore, help them dispose of their excess *volume* of water, but other organs must maintain the high *osmotic pressure* of their blood.

The effects of kidney function on ionic regulation depend on ionic U/P ratios

The role played by the kidneys in ionic regulation can be analyzed in terms analogous to those used to analyze osmotic regulation (see Figure 25.7). For each ion, an **ionic U/P ratio** can be computed; this ratio is the concentration of that ion in the urine divided by its concentration in the blood plasma. To see the interpretive value of an ionic U/P ratio, take the case of Na^+. The concentration Na^+ in urine over the concentration of Na^+ in plasma is the sodium U/P ratio. If this ratio is greater than 1.0, the urine contains more Na^+ per unit of water volume than the plasma; thus the excretion of urine by the kidneys lowers the plasma Na^+ concentration. On the other hand, if the sodium U/P ratio is less than 1.0, the excretion of urine acts to retain Na^+ preferentially in the body and raise the plasma Na^+ concentration.

The kidneys can play a role in *ionic regulation* even while not playing any direct role in *osmotic regulation*. In this way, the kidneys illustrate that ionic regulation is a distinct concept from osmotic regulation. Marine teleost (bony) fish provide examples of animals in which the kidneys participate in ionic regulation, but not osmotic regulation. These fish are hyposmotic to the seawater in which they live. They thus lose water osmotically to their environment, and they also gain ions by diffusion from the seawater; both of these processes tend to raise the osmotic pressure and ion concentrations of their blood plasma. The marine teleost fish produce a urine that is isosmotic to their plasma (osmotic U/P = 1); thus their urine can play no direct role in solving their *osmotic* regulatory problem. However, their urine differs greatly from their plasma in its solute *composition*. In particular, the U/P ratios for Mg^{2+}, SO_4^{2-}, and Ca^{2+} are far greater than 1, signifying that these ions are higher in concentration in the urine than in the plasma. The excretion of urine by these fish therefore serves the important *ionic* regulatory role of keeping down the internal concentrations of these ions, which the fish tend to gain from the seawater.

> **Summary: Organs of Blood Regulation**
>
> - The effects of kidney function on the composition of the blood plasma are analyzed using osmotic and ionic U/P ratios. Figure 25.7 summarizes the interpretation of U/P ratios.
> - Osmotic regulation, volume regulation, and ionic regulation are separable kidney functions in the sense that the kidneys can participate in volume regulation while simultaneously not aiding osmotic regulation, or they can carry out ionic regulation independently of osmotic regulation.

Food and Drinking Water

The specific composition of food and drinking water often has major implications for the water–salt physiology of animals living in their natural environments—illustrating once again that physiology and ecology are intimately related. To start our discussion of this topic, let's focus on the relative osmoticities of predators and their prey.

When one animal captures and eats another, the water–salt composition of the prey animal—not just its nutrient content—is significant for the predator. Consider, for example, predator–prey relations in the ocean. Marine mammals and teleost fish are strongly hyposmotic to seawater. However, most marine invertebrates are approximately isosmotic to seawater. Consider a mammal or fish that consumes a meal of invertebrates. The body fluids of the invertebrate prey are markedly more concentrated in salts than the body fluids of the predator. The predator therefore ingests fluids that are saltier than its own and must eliminate the excess salts to maintain its normal body-fluid composition. By contrast, consider a mammal or fish that consumes a meal of fish. In this case, the body fluids of the prey are similar in salt concentration to those of the predator. Thus the fish-eating predator incurs little or no excess salt load when it eats, in contrast to the large salt load incurred by the invertebrate-eating predator. In a very real sense, a fish-eating predator benefits from the work that its prey performed to keep their body fluids more dilute than seawater—an intriguing lesson in ecological energetics.

Salty drinking water may not provide H_2O

When drinking water contains salts, it may not serve as a useful source of H_2O. Whether an animal can gain H_2O by drinking salty water (e.g., seawater) depends on whether it can eliminate the salts from the salty water using less H_2O than was ingested with them. This principle, which applies to all animals, is a critical consideration when people suffering from dehydration are presented with the option of drinking salty water.

We have all heard Coleridge's famous line from *The Rime of the Ancient Mariner*, "Water, water, everywhere, nor any drop to drink." Sailors desperate for water discovered long ago that drinking ocean water was worse than drinking no water at all: Drinking the seawater paradoxically dehydrated them. We now know that a key consideration in understanding this paradox is that the maximum Cl^- concentration that the human kidney can produce in the urine is lower than the concentration of Cl^- in seawater. Thus, if people drink seawater, the Cl^- they ingest can be excreted only by voiding more H_2O than was taken in with the Cl^-. That is, for people to get rid of the Cl^- in ingested seawater, they not only must use all the H_2O ingested with the seawater, but also must draw on other bodily reserves of H_2O, thereby dehydrating their tissues. Some animals are able to excrete salts in more concentrated form than people can and thus are able to gain H_2O by drinking salty solutions like seawater (by excreting the salts in less H_2O than was ingested with them).

Plants and algae with salty tissue fluids pose challenges for herbivores

Some plants in terrestrial environments—particularly ones native to deserts—have very salty tissue fluids. If herbivores eat such plants, they receive a substantial salt load along with the food value of the plants.

The soils in certain desert regions are very saline. One reason for this condition is that salts tend to accumulate over eons of time in the places where rain settles; the evaporation of the water from rainfall leaves the salts it contains behind in the soil, and each rainfall adds to the salts left by preceding rainfalls. Plants called **halophytes** ("salt plants") root in these saline soils and often have high salt concentrations in their tissue fluids. Such plants form a major part of the diet of sand rats (*Psammomys obesus*) and are consumed in large quantities at times by dromedary camels. The total salt concentration in some halophytes exceeds that of seawater by as much as 50%. Many of the halophytes are succulent plants with juicy leaves. Animals that eat them obtain considerable water, but they obtain a large salt load as well.

Analytically, the salt levels of salty plants pose much the same problems for animals as those we discussed in considering salty drinking water. To take the case of just one ion in particular, the Na^+ concentration in a halophyte's tissues may be five times that in mammalian blood plasma. Thus, a mammal must have kidneys that can produce a sodium U/P ratio greater than 5 if it is to eat such halophytes, excrete the Na^+, and obtain a net gain of H_2O from the plants. Most mammals cannot achieve such a high sodium U/P ratio. Sand rats, however, have kidneys with legendary concentrating abilities. Thus, after scraping off and discarding the saltiest parts of leaves, sand rats are able to eat halophytes without ill effects from the salt they ingest, thereby reducing competition for food with other rodents, which avoid the salty plants.

Air-dried foods contain water in variable amounts

Many terrestrial animals consume air-dried seeds or other dry plant matter. These air-dried foods contain moisture, even though they are ostensibly dry. This fact is significant, particularly for animals that live where drinking water is difficult to find. Such foods equilibrate with air moisture, so they vary in their water content as humidity varies. "Dry" barley, for example, contains almost 4 g of water per 100 g dry weight at 10% relative humidity, and it contains five times more water at 76% relative humidity.

When dry plant material is exposed to an altered air humidity, its moisture content changes within hours. The humidity of the air tends to be higher at night than during the day, and it is often higher belowground than aboveground. Animals that get water from air-dried foods can thus increase their water intake by feeding at night or by storing food in burrows prior to ingesting it.

Protein-rich foods can be dehydrating for terrestrial animals

Because carbohydrates and lipids consist primarily of carbon, hydrogen, and oxygen, their oxidation during metabolism results mostly in formation of CO_2 and H_2O. The CO_2 is exhaled into the atmosphere, and the H_2O contributes to an animal's water resources. Proteins, in contrast, contain large amounts of nitrogen, and their catabolism results in nitrogenous wastes. These nitrogenous wastes may affect a terrestrial animal's water balance when they are excreted in solution in the urine.

In mammals, for example, the principal nitrogenous waste is urea, a highly soluble compound voided in the urine. The amount of urinary water required to void urea depends on the urea-concentrating ability of an animal's kidneys. When a mammal is producing urine with as high a urea concentration as it can, a high-protein meal may well force the animal to void more water than a low-protein meal. In an environment where drinking water is hard to find, a high-protein meal may therefore increase the animal's water stress.

Metabolic Water

When organic food molecules are aerobically catabolized, water is formed, as illustrated by the equation for glucose oxidation:

$$C_6H_{12}O_6 + 6\ O_2 \rightarrow 6\ CO_2 + 6\ H_2O \qquad (25.2)$$

TABLE 25.3 Average gross amount of metabolic water formed in the oxidation of pure foodstuffs The values in this table apply to the oxidation of materials that have been absorbed from a meal and to the oxidation of materials stored in the body. To emphasize this, the materials are called foodstuffs rather than foods. The gross amount of metabolic water formed is, by definition, simply the amount made by the oxidation reactions.

Foodstuff	Grams of H_2O formed per gram of foodstuff
Carbohydrate[a]	0.56
Lipid	1.07
Protein with urea production[b]	0.40
Protein with uric acid production[b]	0.50

Source: After Schmidt-Nielsen 1964.
[a] Starch is assumed for the specific value listed.
[b] Water yield in protein catabolism depends on the nitrogenous end product.

The water produced by catabolic reactions such as this is known as **metabolic water** (**oxidation water**), in contrast to **preformed water**, which is water taken in as H_2O from the environment. Metabolic water is produced by all animals. Table 25.3 lists the amount of water formed per gram of foodstuff oxidized.

The simple fact that aerobic catabolism produces water does not by itself mean that animals gain water in net fashion from catabolism. This is true because the catabolism of foodstuffs not only produces water, but also obligates an animal to certain water losses. The **obligatory water losses** of catabolism are the losses that must take place for catabolism to occur. To assess the *net* impact of catabolism on the water balance of an animal, the obligatory water losses must be subtracted from the gains of metabolic water. This important principle will be exemplified after we briefly discuss the *respiratory*, *urinary*, and *fecal* components of obligatory water loss. For simplicity, as we discuss them, we assume a terrestrial, air-breathing animal.

The **obligatory respiratory water loss** is defined to be the loss of water that is necessary to obtain O_2 for catabolism. Aerobic catabolism requires O_2, and when animals breathe to obtain O_2, they lose water by evaporation. Thus the aerobic catabolism of all types of food molecules causes obligatory respiratory water loss. The magnitude of the loss depends on a species' physiology of breathing and on the humidity of the ambient air.

The **obligatory urinary water loss** is the loss of water that is necessary for the urine to void products of catabolism requiring urinary excretion. Only protein catabolism can cause obligatory urinary water loss because the oxidation of carbohydrates and lipids does not produce products that must be excreted in urine. As we saw in the previous section, the oxidation of proteins yields nitrogenous wastes, such as urea, that often must be excreted in solution in the urine, thereby demanding water excretion.

The **obligatory fecal water loss** is the loss of water that must occur in feces to catabolize food; the idea here is that feces contain water, and possibly a loss of water in feces may be necessary to catabolize food. Organic materials stored in the body, such as lipid stores, produce no fecal waste when they are catabolized. Thus obligatory fecal water loss occurs only when *ingested* foods are catabolized. Ingested foods usually contain *preformed* water. If an animal must lose more water in its feces than it took in as preformed water with its ingested food, it incurs a net fecal water loss to catabolize the food; this net loss is the obligatory fecal water loss.[12]

The use of these concepts is exemplified in Box 25.1. The kangaroo rats discussed in this box are able to thrive in some of the driest places in North America's southwestern deserts. In the experiments discussed in the box, the kangaroo rats were given no water to drink and fed nothing but air-dried barley. Remarkably, they were able to gain metabolic water in net fashion by catabolizing the barley, and they did not suffer from dehydration. Note that their net water gain from catabolizing the barley is computed by first determining the immediate (gross) yield of metabolic water from the oxidation of the food molecules absorbed from a gram of ingested barley and then subtracting the obligatory respiratory, urinary, and fecal water losses required to catabolize a gram of ingested barley.

Metabolic water matters most in animals that conserve water effectively

At moderate temperatures and even at low relative humidities (as low as about 20%), kangaroo rats and some other desert rodents can live indefinitely on air-dried seeds without drinking water, meeting most of their water needs with metabolic water and the remainder with the small amounts of preformed water in the seeds. Under similar conditions, most other mammals would quickly die. These striking contrasts have given rise to a myth that desert rodents produce especially large amounts of metabolic water.

In fact, *the amount of metabolic water produced per gram of food oxidized is fixed by chemistry*; it depends simply on the stoichiometry of the aerobic catabolic pathways. Thus, for a given type of food, all animals produce the same amount of metabolic water per gram of food oxidized.[13]

If this is so, then why do some animals depend more on metabolic water than others? The answer is that some *conserve* water more effectively than others. If two animals have the same metabolic rate and oxidize the same foods, both will produce the same amount of metabolic water. If one conserves water poorly and thus has a high overall rate of water turnover, its production of metabolic water will be small relative to its total water intake and loss; thus the metabolic water it produces will represent just a small part of its water budget. If the other animal conserves water well and has a low overall rate of water turnover, its production of metabolic water will be large relative to its total water intake and loss—and will represent a large part of its water budget.

To illustrate, suppose we place kangaroo rats and laboratory rats in cages at an air temperature of 25°C and a relative humidity of 33%, and we feed both species the same food, air-dried barley. Let's assume (as is approximately true) that individuals of the two species have similar rates of metabolism. If we provide no drinking water,

[12] If the feces contain less water than was taken in with ingested food, the animal realizes a net gain of *preformed* water, which is not a factor in metabolic water calculations.

[13] The only exception is that when protein is catabolized, the amount of metabolic water produced depends to some extent on the particular nitrogenous end product made (see Table 25.3) because the chemistry of the reactions depends on the particular end product. The nitrogenous end products are discussed in detail in Chapter 27.

BOX 25.1 Net Metabolic Water Gain in Kangaroo Rats

Desert kangaroo rats (*Dipodomys*) were studied at an air temperature of 25°C and a relative humidity of 33%. They were fed air-dried barley and given no drinking water. This box shows how one takes obligatory water losses into account to calculate the kangaroo rats' *net* gain of metabolic water.

Gross Metabolic Water Production (0.54 g H_2O/g barley)

By knowing the composition of barley, the extent of absorption of barley from the gut, and the amount of water produced by oxidizing absorbed carbohydrates, lipids, and proteins (see Table 25.3), one can calculate that the food molecules absorbed by a kangaroo rat from a gram of ingested barley yield about 0.54 g of metabolic water during cellular oxidation.

Obligatory Water Losses (total: 0.47 g H_2O/g barley)

1. *Respiratory* (0.33 g H_2O/g barley): Oxidation of the food molecules absorbed from a gram of barley requires consumption of about 810 mL of O_2. To acquire this amount of O_2 by breathing, a kangaroo rat loses about 0.33 g of water by pulmonary evaporation under the conditions of the experiments.
2. *Urinary* (0.14 g H_2O/g barley): The protein absorbed from a gram of ingested barley yields about 0.03 g of urea. Kangaroo rats can concentrate urea sufficiently in their urine that this amount of urea can be excreted in about 0.14 g of water.
3. *Fecal* (0 g H_2O/g barley): About 0.1 g of preformed water is obtained with each gram of barley ingested. However, the feces resulting from the digestion of a gram of barley contain only about 0.03 g of water. Thus there is a net gain of *preformed* water in the digestion of the barley. This gain of preformed water is ignored in calculating the net gain of metabolic water.

Net Gain of Metabolic Water (0.07 g H_2O/g barley)

The total of all the obligatory water losses during the catabolism of a gram of ingested barley is 0.47 g. To calculate the net gain of metabolic water, this total must be subtracted from the gross metabolic water production per gram of barley, 0.54 g. The net gain is therefore 0.07 g of water per gram of ingested barley. This gain of metabolic water can be used to offset other water losses, or if it represents an excess, it can be excreted.

A kangaroo rat This species, *Dipodomys merriami*, is one of several species of kangaroo rats noted for their success in the deserts and semideserts of the American Southwest.

the kangaroo rats will thrive—living largely on metabolic water—but the laboratory rats will deteriorate and ultimately need to be given drinking water to save their lives. Chemical stoichiometry suggests that the two species produce similar amounts of metabolic water. Why, then, can only one survive without drinking water? The answer can be seen in Table 25.4, which presents a full accounting of the net effects of catabolism on water balance in the two species. The kangaroo rats conserve water more effectively than the laboratory rats; they have lower urinary water losses because they can concentrate urea to a greater degree in their urine, and they have lower fecal water losses, in part because they can dry their feces more thoroughly. When the obligatory water losses of the two species are subtracted from the gross amount of metabolic water produced, the kangaroo rats enjoy a *net gain* of metabolic water per gram of barley they ingest and catabolize. The laboratory rats, however, suffer a *net loss*. Note that the critical difference between the species is not in how abundantly they *produce* metabolic water. It is the exceptional *water-conservation* capability of the kangaroo rats that permits them to live on metabolic water.

TABLE 25.4 Approximate catabolic gains and losses of water in caged kangaroo rats (*Dipodomys*) and laboratory rats (*Rattus*) when eating air-dried barley and denied drinking water at 25°C and 33% relative humidity The values given are grams of H_2O per gram (dry weight) of barley ingested. Those for the kangaroo rats are from Box 25.1.

Category of water gain or loss	Kangaroo rats	Laboratory rats
Gross metabolic water produced	0.54 g/g	0.54 g/g
Obligatory water losses		
Respiratory	0.33	0.33
Urinary	0.14	0.24
Fecal	0.00	0.03
Total obligatory water losses	0.47	0.60
Net gain of metabolic water	+ 0.07	– 0.06

Summary: Metabolic Water

- Metabolic water is produced by all animals in amounts determined by the chemical stoichiometry of the oxidation of organic molecules.
- To determine the *net* effect of catabolism on water balance, obligatory respiratory, urinary, and fecal water losses must be subtracted from gross metabolic water production.
- The role played by metabolic water in the overall water budget of an animal depends on the animal's capacity to *conserve* body water.

The Water and Salt Physiology of Tissue Cells

An important attribute of cells in the body is that they have particular *volumes*. As we noted earlier, cell membranes are so permeable to water that the intracellular fluids remain isosmotic to the extracellular fluids bathing the cells (the interstitial fluids) simply by osmosis. This condition may at first sound benign, but it in fact means that any change in the osmotic pressure of the extracellular fluids is a threat to the maintenance of a constant cell volume. If the osmotic pressure of the extracellular fluids decreases, cells tend to take on water by osmosis and swell. If the extracellular osmotic pressure increases, cells tend to lose water and shrink.

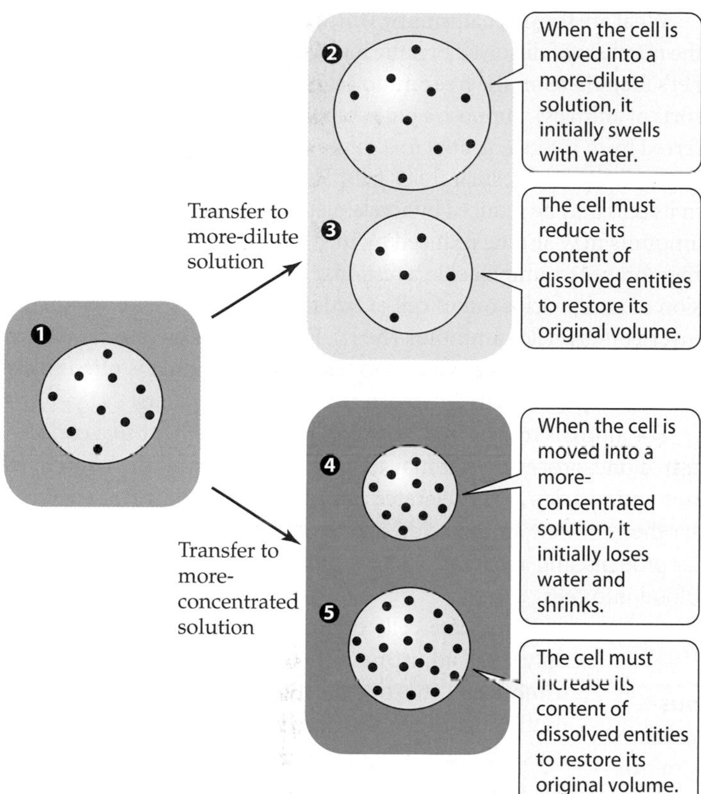

Figure 25.8 The fundamental principles of cell-volume regulation Dots represent dissolved entities in the intracellular fluid of a model cell. Shading around each cell represents the concentration of the ambient solution.

How, then, can cells maintain a constant cell volume if the osmotic pressure of the extracellular fluids changes? In other words, how can they display **cell-volume regulation**? The answer is that they must alter their total content of osmotically effective dissolved entities. To see this, let's consider a model cell placed in ambient solutions of our choosing in Figure 25.8. In step 1, the cell contains ten osmotically effective dissolved entities and is at its normal size in the initial ambient solution. If the cell is transferred into an ambient solution that has an osmotic pressure only half as great as the starting solution, its initial response (step 2) is to take on water osmotically and swell to twice its normal volume, thereby reducing the concentration of dissolved entities inside by half. If the cell remains in the diluted ambient solution, how can it restore its normal volume? The restoration of the original cell volume requires that the cell reduce its content of dissolved entities. As seen in step 3, if the cell halves the number of dissolved entities in its intracellular fluid, it will return to its original volume while remaining isosmotic to the diluted ambient solution. Suppose now that we return to the original cell (step 1) and we transfer it into an ambient solution that has an osmotic pressure double that of the starting solution. The cell's initial response (step 4) will then be to lose water osmotically and shrink to half its normal volume, doubling the concentration of dissolved entities inside. To restore its normal volume in the concentrated ambient solution, the cell will need to increase its content of dissolved entities. As seen in step 5, if the cell doubles the number of dissolved entities in its intracellular fluid, it will return to its original volume while remaining isosmotic to the concentrated ambient solution.

We see from this exercise that *if the extracellular fluids of an animal become diluted, its cells must reduce their intracellular numbers of osmotically effective dissolved entities to retain their original volumes.* Conversely, *if an animal's extracellular fluids become more concentrated, cell-volume regulation requires that its cells increase their intracellular numbers of osmotically effective dissolved entities.*

Most animals employ organic solutes for cell-volume regulation

When osmoconforming invertebrates such as blue mussels (see Figure 25.3*c*) experience changes in their blood osmotic pressure because of changes in the salinity of their environment, the cells throughout their bodies face volume-regulation challenges. If mussels are exposed to diluted environmental water, for example, the osmotic pressure of their blood falls, and all their cells become bathed with diluted extracellular fluids; without volume-regulatory processes, the cells will swell. This type of challenge is not limited to osmoconformers, because osmoregulating animals are rarely perfect regulators. Even the shrimp in Figure 25.3*c*—which are exceptionally effective osmoregulators—exhibit small changes in their blood osmotic pressure as they swim into waters with different salinities. They thus can experience significant tendencies for their cells to swell in dilute environments and shrink in concentrated environments.

In invertebrates, cells typically increase and decrease the amounts of *organic solutes* in their intracellular fluids to maintain constant cell volumes. This strategy is diagrammed in Figure 25.9, which shows a cell bathed with four different concentrations of extracellular fluids. The blue dots inside the cell symbolize inorganic ions (principally K^+, Mg^{2+}, Na^+, and Cl^-). The yellow dots symbolize dissolved organic molecules, such as free amino acids. The cells of invertebrates typically regulate their volume by decreasing their intracellular numbers of organic molecules when bathed with dilute extracellular fluids and by increasing their intracellular numbers of organic mole-

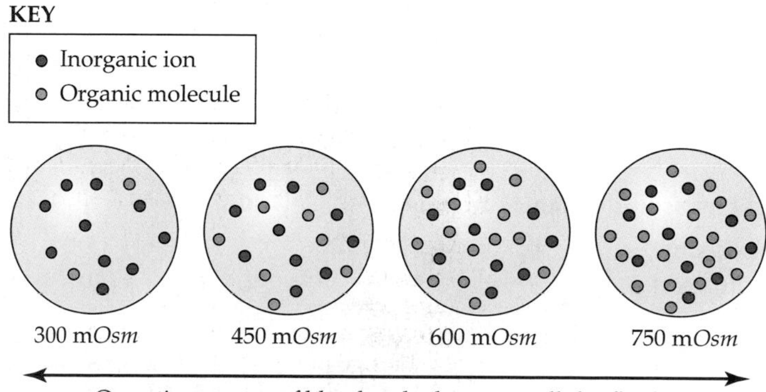

Figure 25.9 Many animal cells alter their content of organic molecules to achieve cell-volume regulation As the osmotic pressure of the blood and interstitial fluids rises, a cell adds organic solutes to its intracellular fluids, so that both the cell volume and the intracellular concentration of inorganic ions remain approximately constant.

cules when bathed with concentrated extracellular fluids. The numbers and concentrations of inorganic ions in the intracellular fluids are maintained at nearly constant levels. Thus the cells achieve ionic regulation at the same time they regulate their volume!

An actual set of data on the process of cell-volume regulation is shown in Table 25.5. Blue crabs that were not in the process of molting were first acclimated to full-strength seawater and then to 50% seawater. Their blood osmotic pressure, while regulated to a substantial degree, fell from 1100 m*Osm* in the full-strength seawater to 850 m*Osm* in the 50% seawater. Despite this change in the osmotic pressure of the extracellular fluids, the cells in the crabs' muscles maintained virtually constant volumes. The cells regulated their volumes by reducing their intracellular content of osmotically effective dissolved entities when the crabs were moved into the diluted seawater. Inorganic ions decreased slightly. Intracellular organic compounds (standard amino acids and taurine) decreased much more. Unlike the model cell in Figure 25.9, these real cells adjusted both inorganic and organic cell constituents. However, the organic constituents accounted for about 80% of the total decrease in intracellular dissolved entities.

TABLE 25.5 Intracellular osmotic contributions of solutes in the leg muscles of blue crabs (*Callinectes sapidus*) acclimated to full-strength seawater and 50% seawater Although the data are presented as *concentrations*, they mirror the *amounts* of solutes in cells because the cells exhibited almost perfect volume regulation and their water content did not change.

	Osmotic concentration (milliosmole/kg intracellular water)	
Compound	**Crabs acclimated to full-strength seawater**	**Crabs acclimated to 50% seawater**
Free standard amino acids		
Glycine	362	282
Arginine	136	95
Proline	74	49
Serine	52	5
Alanine	38	23
Glutamate	10	3
Aspartate	8	3
Methionine	6	1
Valine	5	6
Other	19	9
Taurine	69	37
Total of measured organic compounds	**780**	**514**
Inorganic ions		
Sodium (Na^+)	40	28
Potassium (K^+)	186	162
Chloride (Cl^-)	46	26
Total of measured inorganic ions	**271**	**217**
Total intracellular osmotic pressure accounted for by listed solutes	**1051**	**731**
Blood osmotic pressure (milliosmole/L)	1100	850

Source: After Gerard and Gilles 1972.

What are the mechanisms by which cells can increase and decrease the numbers of dissolved organic molecules in their intracellular fluid? Let's focus here on the free amino acids (which are adjusted by many sorts of animals). Suppose a blue crab (or other invertebrate) is transferred from a concentrated to a dilute environment and the concentration of its extracellular fluids falls. The amounts of free amino acids in its cells may be reduced by accelerated oxidation of amino acids; the amounts may also be reduced by increasing the permeability of cell membranes to amino acids, a response that permits accelerated diffusion of amino acids out of cells. Oxidation of amino acids in aquatic invertebrates yields ammonia (NH_3). Experiments on several species have shown that an increase in excretion of ammonia is often easily detected when animals are transferred into dilute waters!

For animals transferred from a dilute environment to a concentrated one, processes that may help *raise* the intracellular content of free amino acids are decelerated amino acid catabolism, accelerated synthesis of new amino acids, accelerated breakdown of intracellular proteins, and accelerated active transport of amino acids from the blood into cells.

Vertebrate cells typically are called upon to make only small adjustments in their intracellular content of osmotically effective dissolved entities because the osmotic pressure of the extracellular fluids in vertebrates is usually tightly regulated. Inorganic ions, particularly K^+, are usually the principal solutes that undergo adjustment to maintain a constant volume in vertebrate cells, at least in the short term. Organic molecules, however, are the main solutes employed for volume regulation by cells in certain critical situations. The vertebrate cells that employ principally organic molecules include the cells of the brain (where stable concentrations of inorganic ions are particularly important for nerve impulse generation) and the cells of the mammalian kidney interior (where the osmotic pressure of the extracellular fluid is high and variable) (see page 740).

Organic solutes have played major roles in the adjustment of intracellular osmotic pressure over evolutionary time

As a broad generalization, all modern-day animals have similar intracellular concentrations of inorganic ions. This is evident in the lower row of boxes in Figure 25.10. The figure illustrates that when taxonomic position and habitat are taken into account, modern-day animals fall into four major groups in terms of their body-fluid composition. Two of the groups, A and B, consist of marine animals having body fluids that are approximately isosmotic with seawater. The other two groups, C and D, consist of animals, mostly terrestrial and freshwater ones, in which the body-fluid osmotic pressure is only about one third as high as the osmotic pressure of seawater. All four groups, despite their large differences in other respects, are about the same in the concentrations of inorganic ions in their intracellular fluids, indicating that intracellular ion concentrations have been con-

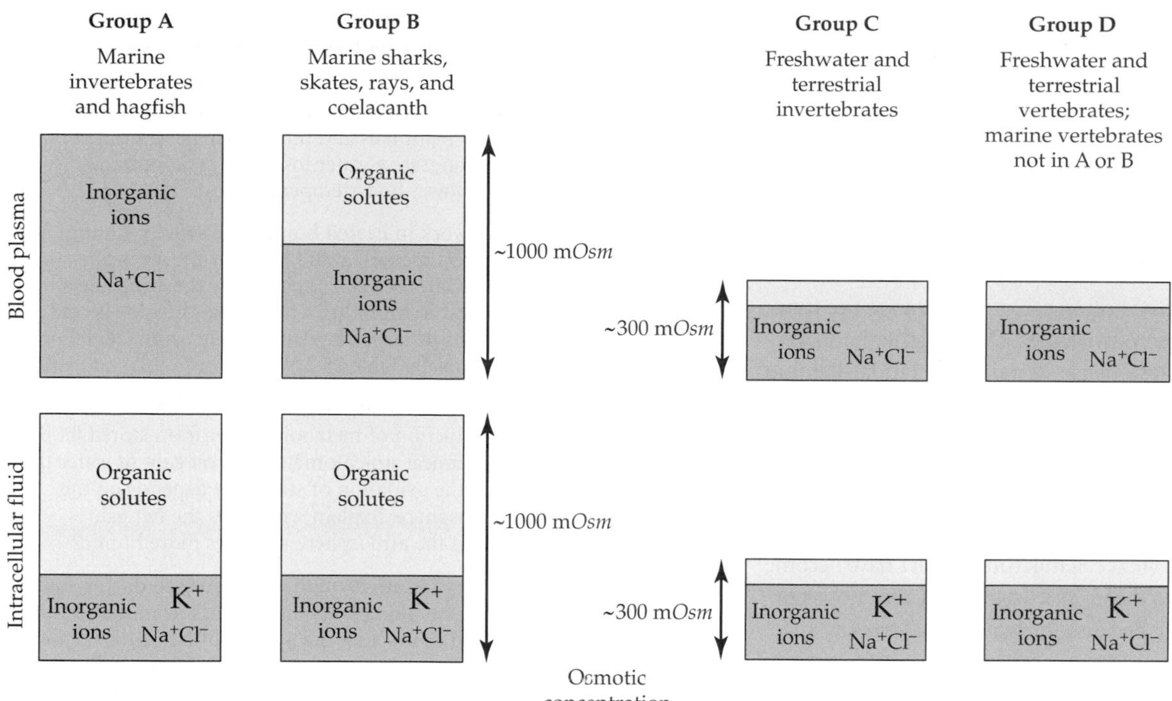

Figure 25.10 Animals tend to be similar in their intracellular concentrations of inorganic ions although some differ dramatically from others in the osmotic pressure of their body fluids and the composition of their blood plasma Modern-day animals can be categorized into four groups—A, B, C, and D—based on their taxonomic position (vertebrate or invertebrate), habitat, and body-fluid composition. The total height of each box represents the osmotic pressure of the blood plasma (upper row) or intracellular fluid (lower row). The heights of the blue and yellow areas inside each box represent the portions of the total osmotic pressure contributed by inorganic ions and organic solutes, respectively. The figure applies to unstressed animals in their normal environments. The sizes of the chemical symbols for the inorganic ions symbolize the ion composition of the body fluids; K^+ is at high concentration inside cells but at low concentration in blood plasma and other extracellular fluids. (After Gilles and Delpire 1997.)

served over the course of evolution. How is it possible for intracellular ion concentrations to have remained approximately constant while some animals have evolved body fluids as osmotically concentrated as seawater yet others have evolved body fluids only a third as concentrated? As you can see in Figure 25.10, organic solutes have played a role over the course of evolution that is analogous to the role they often play during the acclimation of individual animals to changed-salinity environments. As animals have evolved different *total* intracellular concentrations, most of the change has been in concentrations of intracellular organic solutes rather than inorganic ions.

Summary: The Water and Salt Physiology of Tissue Cells

- For a cell to maintain a constant volume, it must reduce the amounts of osmotically effective dissolved entities in its intracellular fluid when the osmotic pressure of the surrounding extracellular fluid falls, and it must increase its content of dissolved entities when the osmotic pressure of the extracellular fluid rises.
- Organic molecules such as free amino acids are the principal intracellular solutes employed for cell-volume regulation by individual cells of invertebrates and by certain types of individual cells (those of the brain and interior of the mammalian kidney) in vertebrates.
- The use of organic molecules as principal agents of cell-volume regulation permits simultaneous ionic regulation of the intracellular fluid.

From Osmolytes to Compatible Solutes: Terms and Concepts

Any solute that exerts a sufficiently large effect on the osmotic pressure of a body fluid to be of consequence for understanding water–salt physiology is termed an **osmolyte** (or **osmotic agent**), regardless of what other functions it may perform. An osmolyte that an individual animal or individual cell increases or decreases in amount to achieve osmotic regulation or cell-volume regulation is called an **osmotic effector**.

Animals, as we have seen, are selective in which solutes they employ as intracellular osmotic effectors. They generally use organic solutes, rather than inorganic ions, in this role. Why they do so is an important contemporary question in the study of water–salt physiology. The hypothesis favored by most physiologists is that the solutes employed preferentially as osmotic effectors are the ones that have minimal effects on the functions of proteins and other macromolecules.

The term "compatible solute" was first used over 30 years ago to refer to a solute that, at high concentrations, does not interfere with the ability of enzymes to function effectively. Modern definitions are less specific and more rough-and-ready, but they retain the spirit of the original. Today, a **compatible solute** is a solute that, when concentrated enough to contribute significantly to the osmotic pressure of a body fluid, has little or no effect on the structure and function of macromolecules with which the fluid is in contact.

Inorganic ions are generally the antithesis of compatible solutes. Changes in their concentrations often seriously derange the functions of enzymes or other macromolecules. On the other hand, certain types of organic molecules, such as certain free standard amino acids,[14] taurine, and glycerol, qualify as outstanding examples of compatible solutes.

According to the theory of compatible solutes, animals modulate intracellular organic solutes more than they do inorganic ions because the organic solutes they employ are compatible solutes, whereas the inorganic ions are not. In this view, cell macromolecules have become adapted by natural selection to the inorganic-ion concentrations normally found in cells. However, because inorganic ions strongly affect macromolecular properties, *changes* in inorganic-ion concentrations may derange cell function. Thus animals emphasize the use of compatible organic solutes as intracellular osmotic effectors.

The concept of *counteracting solutes* is also significant for understanding the roles of particular solutes in water–salt physiology. Unlike compatible solutes, individual counteracting solutes have strong effects on macromolecules. However, the effects of one counteracting solute are "opposite" to those of others, meaning that such solutes can offset each other's effects. **Counteracting solutes** are osmolytes that act *in teams* of two or more to modify the osmotic pressures of body fluids without greatly perturbing macromolecules, because the individual members of the teams have mutually offsetting effects on the macromolecules. The defining example of this concept is provided by the interplay of urea and methylamines in the body fluids of certain marine fish. Urea tends to have strong destabilizing and inhibiting effects on enzymes and other macromolecules. On the other hand, certain methylamine compounds—such as glycine betaine, sarcosine, and trimethylamine oxide (TMAO)—tend to stabilize and activate enzymes, and thus can counteract the effects of urea. In animals that employ urea as an osmolyte in their body fluids, one of the methylamines is usually also present in a quantity that more or less exactly "titrates away" the effects of the urea. Sharks are famous for employing this type of system, as we shall see in Chapter 26.

Study Questions

1. In your own words, explain why an animal suffering from excessively high blood osmotic pressure must be able to produce urine hyperosmotic to its blood plasma for its kidneys to help correct the problem it faces. In your answer, explain why neither urine isosmotic to the blood plasma nor urine hyposmotic to the plasma would help.

[14] Among the standard amino acids (those used in protein synthesis), glycine, arginine, proline, and serine—when present as free amino acids—have particularly small effects on macromolecules and thus stand out as superior compatible solutes.

2. How are compatible intracellular solutes employed in cell-volume regulation, and why do they have advantages as intracellular osmotic effectors?
3. Based on the physical laws of evaporation, explain why the high body temperatures of mammals and birds make them prone to having higher rates of evaporative water loss than poikilothermic terrestrial vertebrates with lower body temperatures.
4. People who live and work in heated homes and office buildings in places with cold winters are well aware how dry the air inside becomes during the winter season. Explain why the air inside a heated building is often low in humidity when the air outside is cold. Keep in mind that outside air is circulated inside by incidental or forced flow. (Hint: Consult Table 25.2.)
5. When animals oxidize stored fat, they produce metabolic water. Even though the production of metabolic water from stored fat follows principles of chemical stoichiometry, the *net* gain of water that animals realize from the oxidation of stored fat depends on the humidity of the atmosphere. Explain why. Does the net gain increase or decrease as the atmosphere becomes more humid?
6. When blue crabs living in full-strength seawater swell during molting, they take on the H_2O that bloats their bodies from the seawater in which they live. Data show that the activity of Na^+–K^+-ATPase in the gills of blue crabs increases as they start to swell during molting. One hypothesis to explain the *mechanism* of swelling is that the gills actively transport ions from the surrounding seawater into the blood of the crabs. How could this account for their uptake of H_2O?
7. Some species of animals gain physiological advantages by exploiting the specialized physiological capabilities of other species. Explain how marine mammals that prey on teleost fish rather than invertebrates illustrate this principle. Also explain how carnivorous mammals that prey on herbivorous mammals in deserts illustrate the same principle.
8. The kidney function of freshwater crabs illustrates that volume regulation can occur independently of osmotic regulation, and the kidney function of marine teleost fish illustrates that ionic regulation can occur independently of osmotic regulation. Explain both points.
9. People dying of dehydration because of lack of freshwater to drink inevitably ponder drinking their own urine. Analyzing the matter by use of the principles developed in this chapter, is a dehydrated person likely to gain H_2O by drinking his or her urine?
10. In many parts of the world, if one goes out at daybreak in autumn, it is a common sight to see a layer of fog formed just above the surface of the water of ponds and lakes. A fog consists of minute droplets of liquid water suspended in the atmosphere. When a layer of fog forms above a pond, the pond itself is the source of the water in the atmospheric water droplets. Inevitably, the temperature of the water in a pond is higher than the temperature of the air on mornings when fog layers form. Using the quantitative information in Table 25.2, explain why these fog layers form.

Key References

Frank, C. L. 1988. Diet selection by a heteromyid rodent: Role of net metabolic water production. *Ecology* 69: 1943–1951. Do desert animals select foods based on their promise to provide a net yield of metabolic water? This interesting study at the interface of physiology and behavior addresses this question.

Gilles, R., and E. Delpire. 1997. Variations in salinity, osmolarity, and water availability: Vertebrates and invertebrates. In W. H. Dantzler (ed.), *Comparative Physiology*, vol. 2 (Handbook of Physiology [Bethesda,

MD], section 13), pp. 1523–1586. Oxford University Press, New York. A detailed but well-integrated review that includes advanced information on most of the topics covered in this chapter.

Hochachka, P. W., and G. N. Somero. 2002. *Biochemical Adaptation*, 2nd ed. Oxford University Press, New York. The single best modern discussion of the biochemistry of water and solutes in the animal kingdom. Essential reading for anyone interested in the evolution of cell-volume regulation and the aspects of physiology related to such regulation.

Somero, G. N. 1992. Adapting to water stress: Convergence on common solutions. In G. N. Somero, C. B. Osmond, and C.L Bolis (eds.), *Water and Life*, pp. 3–18. Springer-Verlag, New York. Although becoming dated, this paper is still noteworthy as a compact, easy-to-read overview of the challenges of cell-volume stress and the responses of cells to such stress, emphasizing compatible solutes.

Selected References

Barnhart, M. C. 1983. Gas permeability of the epiphragm of a terrestrial snail, *Otala lactea*. *Physiol. Zool.* 56: 436–444.

Beyenbach, K. W. (ed.). 1990. *Cell Volume Regulation*. Karger, Basel.

Borowitzka, L. J. 1981. Solute accumulation and regulation of cell water activity. In L. G. Paleg and D. Aspinall (eds.), *The Physiology and Biochemistry of Drought Resistance in Plants*, pp. 97–130. Academic Press, New York.

Campbell, G. S., and J. M. Norman. 1988. *An Introduction to Environmental Biophysics*, 2nd ed. Springer-Verlag, New York. [Discusses all aspects of the biophysics of water in terrestrial systems.]

Denny, M. W. 1993. *Air and Water: The Biology and Physics of Life's Media*. Princeton University Press, Princeton.

Galcheva-Gargova, Z., B. Dérijard, I.-H. Wu, and R. J. Davis. 1994. An osmosensing signal transduction pathway in mammalian cells. *Science* 265: 806–808.

Groot, J. A., and R. Bakker. 1988. NaCl transport in the vertebrate intestine. Advances in Comparative and Environmental Physiology, vol. 1, pp. 103–152. Springer, New York.

Haond, C., L. Bonnal, R. Sandeaux, G. Charmantier, and J.-P. Trilles. 1999. Ontogeny of intracellular isosmotic regulation in the European lobster *Homarus gammarus* (L.). *Physiol. Biochem. Zool.* 72: 534–544.

Kirschner, L. B. 1991. Water and ions. In C. L. Prosser (ed.), *Environmental and Metabolic Animal Physiology* (*Comparative Animal Physiology*, 4th ed.), pp. 13–107. Wiley-Liss, New York.

Lang, F. (ed.). 1998. *Cell Volume Regulation*. Karger, Basel. [The one defect in this otherwise admirable book is that it does not notify readers that its coverage is mammals, not animals in general.]

Mangum, C. 1992. Physiological aspects of molting in the blue crab *Callinectes sapidus*. *Amer. Zool.* 32: 459–469.

Oschman, J. L. 1980. Water transport, cell junctions, and "structured water." In E. E. Bittar (ed.), *Membrane Structure and Function*, vol. 2, pp. 141–170. Wiley, New York.

Skadhauge, E. 1980. Water transport in the vertebrate intestine. In R. Gilles (ed.), *Animals and Environmental Fitness*, pp. 79–90. Pergamon, New York.

Spring, K. R. 1998. Routes and mechanism of fluid transport by epithelia. *Annu. Rev. Physiol.* 60: 105–119.

Strange, K. (ed.) 1994. *Cellular and Molecular Physiology of Cell Volume Regulation*. CRC Press, Boca Raton, FL.

Taylor, J. R. A., and W. M. Kier. 2003. Switching skeletons: Hydrostatic support in molting crabs. *Science* 301: 209–210.

Ussing, H. H., N. Bindslev, N. A. Lassen, and O. Sten-Knudsen (eds.). 1981. *Water Transport Across Epithelia*. Munksgaard, Copenhagen.

Wheatly, M. G., and A. T. Gannon. 1995. Ion regulation in crayfish: Freshwater adaptations and the problem of molting. *Amer. Zool.* 35: 49–59.

See also references in Appendix A.

26 Water and Salt Physiology of Animals in Their Environments

The body fluids of these ocean fish are far more dilute than the seawater in which they are swimming. The body fluids of marine teleost (bony) fish have osmotic pressures of about 300 to 500 m*Osm*. Seawater, in contrast, has an osmotic pressure of approximately 1000 m*Osm*. From the viewpoint of fluid composition, marine teleost fish are packets of low-salinity fluids cruising about within a high-salinity environment. Their dilute body fluids help explain why their flesh does not taste as salty as seawater and is in fact generally not saltier to the tongue than the flesh of freshwater fish.

Because the body fluids inside a marine teleost fish are not at equilibrium with the seawater surrounding the fish, passive processes occur that tend to alter the composition of the body fluids, drawing them toward equilibrium with seawater. The body fluids, having a lower osmotic pressure than seawater, lose water by osmosis, a process that tends to concentrate them. Moreover, a number of inorganic ions diffuse inward from the seawater, thereby also tending to concentrate the body fluids. Marine teleost fish must steadily expend energy to maintain their body fluids out of equilibrium with the seawater in which they swim.

When we reflect on the questions that are raised by the body-fluid composition of marine teleost fish, we quickly recognize that the questions of *mechanism* and *origin* stressed in Chapter 1 are both important. *How* do the fish keep their body fluids more dilute than seawater, and *why* do they do so? These same two questions—which are the central questions of mechanistic physiology and evolutionary physiology—arise in the study of the water–salt relations of all animals.

In terms of their water–salt physiology, animals have been versatile in adapting to an astounding range of environments on Earth—not just seawater and ordinary freshwater, but also salt lakes far more concentrated than seawater, glacial ponds almost as dilute as distilled water, estuaries with highly variable salinity, and terrestrial environments ranging from rain forests to extreme deserts. Each type of habitat poses distinct challenges, which all types of animals living there must meet. Animals of diverse types that live in a particular habitat often have converged on

These coral reef fish expend energy to keep their body fluids more dilute than seawater Major questions raised are why they do so and what mechanisms they employ.

similar mechanisms for meeting the challenges the habitat poses. Because of these considerations, it is logical to organize our detailed study of animal water–salt relations in this chapter around habitats. We start by focusing on animals in freshwater habitats, in part because we have already emphasized them in Chapter 3 and in part because the study of freshwater fish will help set the stage for better understanding ocean fish.

Animals in Freshwater

Biologists generally believe that the major animal phyla originated in the oceans. Seawater was probably somewhat different in its total salinity and salt composition in the early eras of animal evolution than it is today. Nonetheless, when the animal phyla invaded freshwater from the oceans, there can be no doubt that they encountered a drastic reduction in the concentration of their surroundings. The osmotic pressure of freshwater is typically less than 1% as high as that of seawater today, and the major ions in freshwater are very dilute compared to their concentrations in seawater (see Table 25.1).

All freshwater animals regulate their blood osmotic pressures at levels hyperosmotic to freshwater. Freshwater animals are therefore classified as **hyperosmotic regulators**. As Table 26.1 shows, the blood osmotic pressures of various types of freshwater animals span an order of magnitude, but even freshwater mussels, which are among the most dilute animals on Earth, have blood that is substantially more concentrated than freshwater; body fluids as dilute as freshwater seem to be incompatible with life. The solutes in the blood plasma of freshwater animals are mainly inorganic ions, Na^+ and Cl^- being the two principal ions. As Table 26.1 shows, each of the individual inorganic ions in the blood plasma of freshwater animals tends to be substantially more concentrated in the blood than in freshwater.

Passive water and ion exchanges: Freshwater animals tend to gain water by osmosis and lose major ions by diffusion

Being hyperosmotic to their surroundings, freshwater animals tend to gain water continuously by osmosis, and this water gain tends to dilute their body fluids. The relatively high concentrations of ions in their blood suggest that ions will tend to diffuse in net fashion from their blood into the ambient water. The analysis of ion diffusion is actually more complicated, as discussed in Chapter 3, because ion diffusion depends on electrical gradients as well as concentration gradients. Nonetheless (see page 72), when both electrical and concentration effects are taken into account, the direction of diffusion for the major ions—such as Na^+ and Cl^-—in freshwater animals is as the concentration gradients suggest: from the blood into the environmental water. This loss of major ions by diffusion tends, like osmotic water gain, to dilute the body fluids (Figure 26.1*a*).

In a broad sense, we expect a freshwater animal's energy costs for osmotic and ionic regulation to be directly dependent on the animal's rates of passive water gain and passive ion loss. The more rapidly water is taken up by osmosis, and the more rapidly ions are lost by diffusion, the more rapidly an animal will need to expend energy to counteract these processes so as to maintain a normal blood composition. Three factors determine the rates of passive exchange of water and ions: (1) the magnitudes of the osmotic and ionic gradients between the blood and environmental water, (2) the permeability of an animal's outer body covering to water and ions, and (3) the surface area for exchange. The first two of these factors deserve further discussion.

OSMOTIC AND IONIC GRADIENTS Most types of freshwater animals have much less concentrated body fluids than their marine relatives. The decapod crustaceans (e.g., crayfish, crabs, and lobsters) provide good examples of this general pattern. Although most marine decapod crustaceans are essentially isosmotic to seawater (about 1000 m*Osm*), most freshwater decapods have blood osmotic pressures of 500 m*Osm* or less (e.g., ~440 m*Osm* in the crayfish in Table 26.1). Similarly, although marine molluscs are approximately isosmotic to seawater, freshwater molluscs have far lower blood osmotic pressures (e.g., ~40 m*Osm* in the mussels in Table 26.1). The lower blood concentrations seen in freshwater animals result in smaller osmotic and ionic gradients between their blood and the freshwater environment than would otherwise be the case. For example, if decapod crustaceans and molluscs had retained their ancient blood concentrations when they invaded freshwater, the osmotic difference between their blood and freshwater would be over 950 m*Osm* (the difference

TABLE 26.1 The composition of blood plasma in some freshwater animals The worldwide average composition of river water is shown for comparison.

Animal	Osmotic pressure (milliosmole per kg of H_2O)	Ion concentrations (millimole per liter)					
		Na^+	K^+	Ca^{2+}	Mg^{2+}	Cl^-	HCO_3^-
Freshwater mussel (*Anodonta cygnaea*)	44	16	0.5	6	0.2	12	12
Snail (*Viviparus viviparus*)	76	34	1.2	5.7	< 0.5	31	11
Crayfish (*Astacus fluviatilis*)	436	212	4.1	16	1.5	199	15
Mosquito larva (*Aedes aegypti*)	266	100	4.2	—	—	51	—
Brown trout (*Salmo trutta*)	326	161	5.3	6.3	0.9	119	—
Frog (*Rana esculenta*)	237	109	2.6	2.1	1.3	78	27
River water	0.5–10	0.4	0.1	0.8	0.2	0.2	1.7

Source: Hill and Wyse 1989; river water data from Table 25.1.

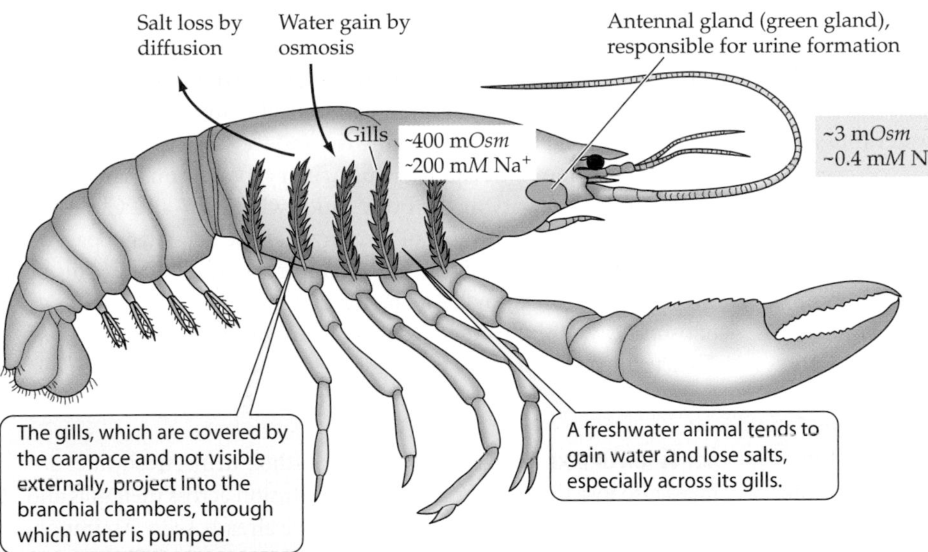

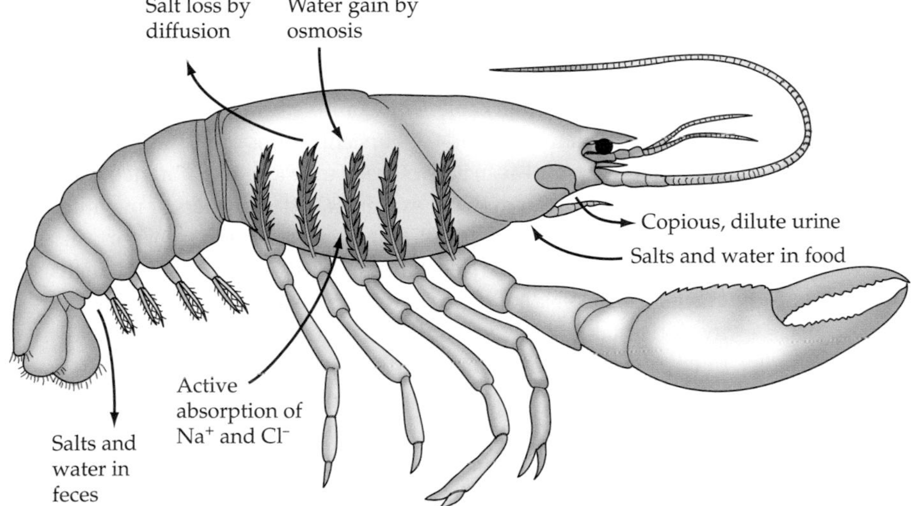

Figure 26.1 Water–salt relations in a freshwater animal *(a)* A freshwater animal, such as a crayfish, faces challenges because of passive water and salt exchange. The numbers are the approximate osmotic pressure and Na^+ concentration found in the blood of a crayfish and the ambient water. The antennal glands or green glands, which function as the kidneys of a crayfish, open at the bases of the second antennae. *(b)* A summary of all the major processes of water and salt exchange, including the processes the animal uses to maintain water–salt balance.

between about 1000 m*Osm* in the blood and perhaps 10 m*Osm* in freshwater). The actual osmotic difference between the blood and the surrounding water in freshwater decapods and molluscs is much less because their blood is less concentrated than that of their marine progenitors. In the crayfish in Table 26.1, for example, the osmotic difference between the blood and the surrounding water is less than 440 m*Osm*.

The evolution of blood of reduced concentration when animals invaded freshwater was probably an adaptation to reduce the energy costs of living in freshwater. Less-concentrated blood means smaller osmotic and ionic gradients between the blood and freshwater, and thus lower rates of water gain and ion loss by osmosis and diffusion.

PERMEABILITIES The permeability of the integument[1] of a freshwater animal to water and salts is generally relatively low. Freshwater crayfish, for example, are no more than 10% as permeable to water, Na^+, and Cl^- as marine decapod crustaceans of similar body size. The low permeabilities evolved by freshwater animals are important in reducing their rates of passive water and ion exchange and thus in reducing their energy costs for maintaining a normal blood composition. For a freshwater animal, or any other animal that maintains a difference in composition between its blood and the ambient water, a low integumentary permeability is analogous to an insulatory pelage in an Arctic mammal; the low permeability slows the processes that tend to bring the blood and ambient water to equilibrium, just as pelage insulation slows heat losses that tend to cool an Arctic mammal to ambient temperature.

If freshwater animals did not need to breathe, they might cover themselves entirely in an integument of extremely low permeability to water and ions. However, they do need to breathe, and there seems to be no way to make gills that are both highly permeable to O_2 and poorly permeable to H_2O and inorganic ions. Just as the gills of freshwater animals provide a "window" for O_2 to enter the body, they provide a window for water to enter by osmosis and for ions to leave by diffusion. Gills not only tend to exhibit high permeabilities to water and ions per unit of surface area, but also have large surface areas. In many types of freshwater animals, osmosis and diffusion occur mostly across the gills (plus possibly a few other localized body surfaces[2]).

The importance of the gills as windows for passive water and ion exchange has the interesting and significant implication that differences in whole-body permeability to water and salts among related freshwater animals are sometimes secondary effects of differences in their metabolic intensities and demands for O_2. Species with high O_2 demands often have gill systems that are particularly suited to rapid inward rates of O_2 diffusion and also, as a corollary, permit particularly rapid rates of osmotic water uptake and diffusional salt loss.

[1] *Integument* is a general term for the outer body covering. For example, the integument of a vertebrate is its skin, and the integument of an arthropod is its exoskeleton (cuticle) or shell.

[2] The membranes of the buccal and opercular cavities in fish, for example, are relatively important sites of passive water–salt exchange. Little water or ion exchange occurs across a fish's outer skin, however.

Most types of freshwater animals share similar regulatory mechanisms

Most freshwater animals share a fundamentally similar suite of mechanisms for osmotic–ionic regulation. This suite of mechanisms is found in such diverse phylogenetic groups as freshwater teleost fish, lampreys, frogs, toads, soft-shelled turtles, freshwater mussels, crayfish, earthworms, leeches, and mosquito larvae.

URINE As we have seen, freshwater animals are faced with a continuous influx of excess water by osmosis. They void this excess water by making a copious (abundant) urine. A goldfish or frog, for example, might excrete urine equivalent to one-third of its body weight per day (Table 26.2). Because urine production balances osmotic water gain, the rate of urinary water excretion provides a measure of the rate of osmotic water influx. The daily osmotic water influx of a goldfish or frog is therefore equal to one-third of its body weight![3]

The urine of freshwater animals, besides being produced in abundance, is typically strongly hyposmotic to their blood plasma and contains much lower concentrations of Na^+ and Cl^- than the plasma. That is, the U/P ratios (urine:plasma ratios) for osmotic pressure, Na^+, and Cl^- are far less than 1 in these animals (see Table 26.2). Recall from Figure 25.7 that an osmotic U/P ratio of less than 1 signifies that urine production tends to raise the plasma osmotic pressure. Similarly, when the U/P ratio for an ion is less than 1, urine production tends to raise the plasma concentration of that ion. Typically, therefore, the kidneys of a freshwater animal not only solve the animal's volume-regulation problem by voiding the animal's excess volume of water, but also aid osmotic and ionic regulation by tending to maintain high osmotic pressures and ion concentrations in the blood.

Whereas the urine of freshwater animals is generally copious and dilute, an important concept to understand is that kidneys are *regulatory* organs that characteristically adjust their function according to conditions in the body fluids. Thus the exact volume and composition of the urine vary with the circumstances. For example, if an animal experiences an increase in the rate at which it takes in water by osmosis, its kidneys ordinarily increase their rate of urine production.

Although freshwater animals can typically limit the concentrations of Na^+ and Cl^- in their urine to low levels, some loss of these ions in the urine is inevitable. This urinary loss of ions can pose a threat to the integrity of the body fluids when Na^+ and Cl^- are in short supply. The rate of loss of ions in the urine depends in part on the rate of urine production, and therefore on the rate of osmotic water flux into an animal. Any factor that increases an animal's rate of osmotic water influx tends to increase the animal's rate of ion loss. Thus *volume regulation and ionic regulation are intimately related and basically at conflict with each other in freshwater animals.*

ACTIVE ION UPTAKE Besides losing ions in their urine, freshwater animals also lose them by direct outward diffusion across their gills and other permeable body surfaces, as we have already seen. An important way that freshwater animals replace lost Na^+ and Cl^- is by actively transporting these ions into their blood directly from the pond or river water in which they live. Other ions may also be transported inward in this way. Freshwater fish, for example, are well known to take up Ca^{2+} from the ambient water by active transport.

The capacities of most freshwater animals for active uptake of Na^+ and Cl^- are remarkable. For example, some crayfish, fish, and frogs—which have Na^+ and Cl^- concentrations of 100 to 200 m*M* in their blood plasma—can actively take up Na^+ and Cl^- in net fashion from ambient waters as dilute as 0.01 m*M* (four orders of magnitude more dilute than their blood)! The site of active ion transport varies among animal groups. Teleost fish and decapod crustaceans such as crayfish absorb Na^+ and Cl^- across the gill epithelium. In frogs, active ion uptake occurs across the gills when the animals are tadpoles, but across the skin when they are adults. Active ion uptake also occurs across the general integument in earthworms and leeches.

We discussed the cellular mechanisms of active Na^+ and Cl^- uptake by certain freshwater animals in Chapter 3 (see especially Box 3.1). Although details of the Na^+ and Cl^- pumping mechanisms vary from group to group of freshwater animals, the following key points are believed to apply to all or most groups:

- The active uptake of ions from the ambient water requires ATP. Thus active ion uptake places demands on an animal's energy resources.
- The mechanisms that pump Na^+ and Cl^- from the ambient water into the blood are typically different and independent from each other.
- The Cl^- pump typically exchanges bicarbonate ions (HCO_3^-) for Cl^- ions, in this way remaining electroneutral (Figure 26.2).
- The Na^+ pump typically exchanges protons (H^+) for Na^+ ions [or possibly exchanges ammonium ions (NH_4^+) for Na^+ in some groups of animals], thereby remaining electroneutral (see Figure 26.2).

[3] If a 70-kg person had a similar weight-specific rate of water uptake, he or she would gain 6 gallons of water per day and thus would need to excrete 6 gallons of urine per day.

TABLE 26.2 Rates of urine production, and osmotic and Na^+ U/P ratios, in some freshwater animals

Animal	Rate of urine production (mL/100 g body wt·day)	Osmotic U/P ratio[a]	Na^+ U/P ratio[b]
Snail (*Viviparus viviparus*)	36–131	0.20	0.28
Crayfish (*Astacus fluviatilis*)	8	0.10	0.006–0.06
Mosquito larva (*Aedes aegypti*)	≤20	0.12	0.05
Frog (*Rana clamitans*)	32	—	—
Clawed toad (*Xenopus laevis*)	58	0.16	0.10
Goldfish (*Carassius auratus*)	33	0.14	0.10

Source: Hill and Wyse 1989.

[a] The osmotic U/P ratio is the osmotic pressure of the urine divided by the osmotic pressure of the blood plasma.

[b] The Na^+ U/P ratio is the urine Na^+ concentration divided by the plasma Na^+ concentration.

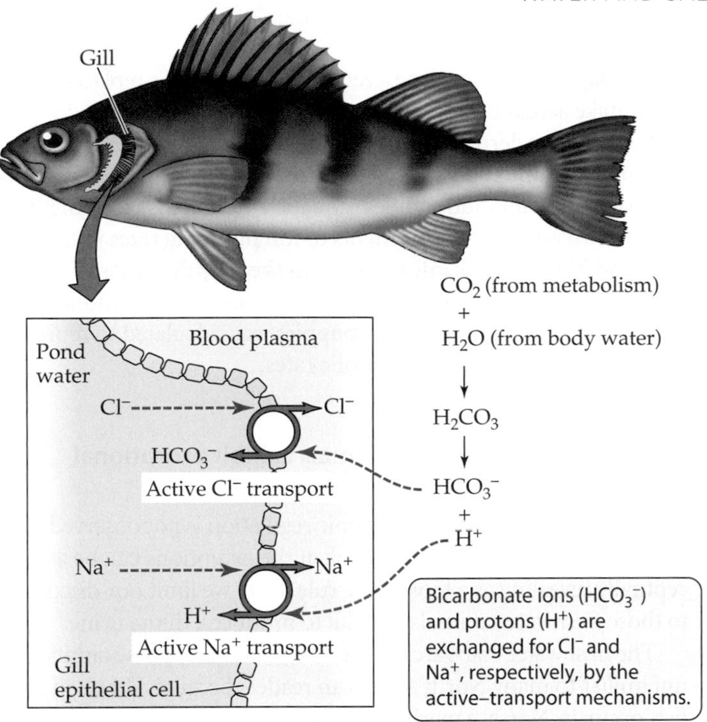

Figure 26.2 Ion exchanges mediated by active Na^+ and Cl^- transport in the gill epithelium of freshwater teleost fish The mechanisms of active transport exist within single epithelial cells. More detail is available in Chapter 3, where we first discussed this process.

- The HCO_3^- pumped from the blood into the ambient water by the Cl^- pump and the H^+ pumped outward by the Na^+ pump are produced by aerobic catabolism. The HCO_3^- and H^+ ions are formed by the reaction of metabolically produced CO_2 with H_2O (see Figure 26.2). Thus the Na^+ and Cl^- pumps participate in removal of metabolic wastes.
- Because HCO_3^- and H^+ are principal players in acid–base regulation (see page 603), the Na^+ and Cl^- pumps sometimes play critical roles in the acid–base physiology of freshwater animals.

In freshwater teleost fish—which have been particularly thoroughly studied compared to most freshwater animals—the gill epithelium consists principally of two types of cells. One type is termed **chloride cells** or **mitochondria-rich cells**;[4] the second type is termed **pavement cells** (Figure 26.3). There has been a tendency to assume that Cl^- uptake occurs across the chloride cells and Na^+ uptake occurs across both types of cells, or just the pavement cells, in freshwater fish. Modern molecular techniques indicate, however, that the localization of pumps to cells is more complicated than heretofore assumed; moreover, subtypes of cells probably occur within the two major types, and the ion pumps may be localized to certain subtypes. Oxygen uptake is believed to occur principally across the pavement cells, which usually occupy over 90% of the gill epithelium and are thinner than chloride cells.

The number of chloride cells in the gill epithelium of a freshwater fish is variable and under adaptive (partly hormonal) control. One condition that increases numbers of chloride cells is excess blood HCO_3^- (alkalosis). During alkalosis, each chloride cell also increases its synthesis of a key Cl^-/HCO_3^- countertransport protein (see Box 3.1) that exports HCO_3^- from the body fluids in exchange for Cl^-.

A second condition that leads to an increase in chloride cells is life in very "soft" water: water of exceptionally low Ca^{2+} concentration (Figure 26.4). Freshwater fish acquire most of their Ca^{2+} from the water in which they live, rather than from their food. The chloride cells, or a subset of them, are the sites of active Ca^{2+} uptake. Increasing the number of chloride cells is believed to help fish acquire enough Ca^{2+} when they are living in Ca^{2+}-poor waters. However, increasing the number of chloride cells can also interfere with uptake of O_2. Recent research on several species has shown that in fish living in very soft water, the replacement of pavement cells by chloride cells can double the average diffusion distance between blood and water in the gills because chloride cells are thicker than the pavement cells they replace (see Figure 26.4). This doubling of the diffusion distance measurably interferes with O_2 uptake. *Thus freshwater fish exhibit a trade-off between their ability to take up Ca^{2+} and their ability to take up O_2; increasing one ability decreases the other.* In the modern study of animal–environment relations, this is one of the physiological trade-offs that is best understood at a cellular level.

FOOD AND DRINKING WATER Besides acquiring ions by active transport from the ambient water, freshwater animals also gain ions in their food. The role of food in meeting their ion needs is not well understood, although the prevailing view is that inputs of ions via active transport exceed those from food in the majority of species. Besides eating food, freshwater animals also have the opportunity to drink water. But do they? These animals typically have to produce

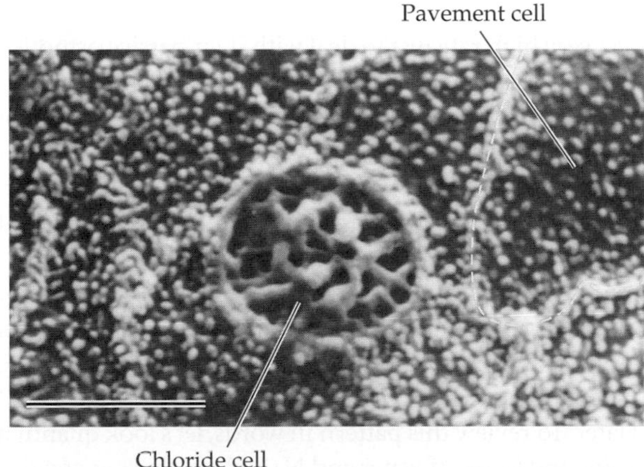

Figure 26.3 A chloride cell surrounded by pavement cells in the gill epithelium of a freshwater teleost fish The image is a scanning electron micrograph of the outer surface of the gill epithelium of a brown bullhead (*Ictalurus nebulosus*) that had been living in ordinary freshwater. One pavement cell is outlined to show its limits. The black bar at the lower left corresponds to 2 μm. (See Figure 26.6 for more detail on chloride cells.) (Photograph courtesy of Greg Goss and Steve Perry; from Goss et al. 1998.)

[4] Chloride cells are also found in the gills of freshwater crustaceans and in a number of other situations. Marine teleost fish have prominent chloride cells in their gill epithelium as later discussed, and chloride cells occur in the rectal salt glands of sharks, skates, and rays (elasmobranch fish). Chloride cells are sometimes called *ionocytes*.

(a) Fish living in ordinary fresh water

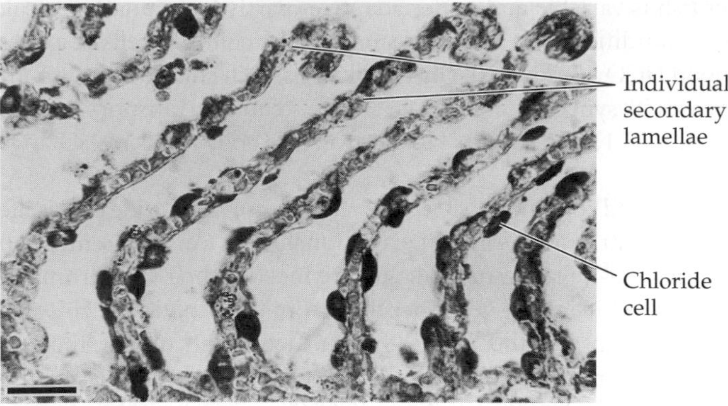

(b) Fish living in very "soft" fresh water

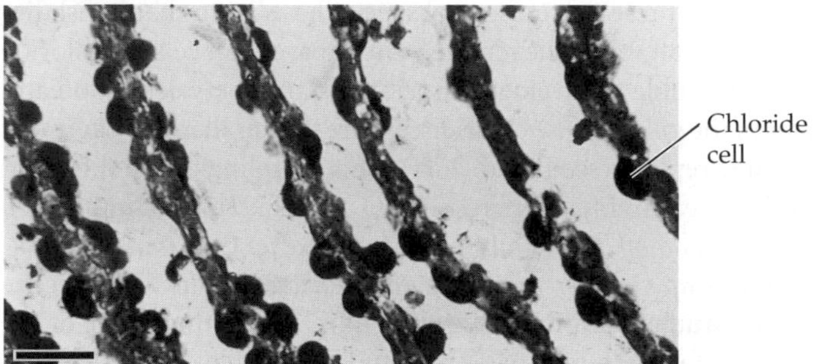

Figure 26.4 Cellular acclimation to living in two types of water in the gill epithelium of freshwater fish Sections of the secondary lamellae in the gills of rainbow trout (*Oncorhynchus mykiss*), viewed using light microscopy and stained to show chloride cells (the circular black objects), from *(a)* a fish that had been living in ordinary freshwater with a Ca^{2+} concentration of 0.4 mmol/L and *(b)* a fish that had been living for 2 weeks in very "soft" freshwater with a Ca^{2+} concentration of 0.05 mmol/L. The bar at the lower left in each photograph corresponds to 15 μm. (Photographs courtesy of Steve Perry; from Perry 1998.)

urine at a very high rate just to deal with their passive osmotic water influxes. Thus one would not expect them to drink, and usually they do not. However, recent studies on teleost fish in freshwater have revealed that some species—especially when they are larvae—do drink sufficiently to raise their total water influx by 5% to 50% over their osmotic influx alone. The reasons for and consequences of this drinking are unclear.

QUANTITATIVE EXAMPLE AND COST ESTIMATES The usual pattern of water–salt balance in freshwater animals is summarized in Figure 26.1*b* (showing a crayfish) and in Figure 26.7*a* (showing a fish) presented later. To review this pattern in words, let's look quantitatively at the gains and losses of water and Na^+ in a freshwater crayfish (*Astacus*). When fasting at 20°C, a 29-g crayfish excretes about 2.4 mL of urine per day. This amount provides a measure of the animal's rate of uptake of water by osmosis, probably principally across its gills. The crayfish's urine is very dilute (1 m*M* Na^+) relative to its blood plasma (~200 m*M* Na^+). Only about 2 or 3 μmole of Na^+ is therefore lost per day in its urine. The animal's loss of Na^+ by diffusion into the surrounding water is much greater, approximately 240 μmole/day. The Na^+ lost by excretion and diffusion is replaced by active uptake across the gills at a rate near 240 μmole/day.

Recently, modern methods were used to estimate the energy cost of osmotic–ionic regulation in two species of freshwater animals, rainbow trout and mudpuppies (aquatic amphibians). The methods used were based on measurements of ion pumping rates in the gills, skin, and kidneys, plus information on the ATP demands of the ion pumps involved. By this approach, the energy cost of osmotic–ionic regulation in the trout and mudpuppies was calculated to represent 3% to 7% of their resting metabolic rates.

A few types of freshwater animals exhibit exceptional patterns of regulation

The typical pattern of osmotic–ionic regulation is not observed in all freshwater animals. As usual, a look at the exceptions can be as conceptually revealing as a look at the rule. Here we limit our discussion to those freshwater animals that fail to produce a dilute urine.

The freshwater crabs are outstanding examples. These animals are unfamiliar to many North American readers because, although crayfish occur in North American freshwaters, crabs have traditionally not.[5] Nonetheless, freshwater crabs occur commonly on most other continents, usually at tropical or subtropical latitudes. Two species that have been studied are *Potamon niloticus*, an African crab, and *Eriocheir sinensis*, which is found in Asia and Europe. Freshwater crabs typically maintain high blood osmotic pressures in comparison to other freshwater animals: about 510 m*Osm* in *Potamon* and 640 m*Osm* in *Eriocheir*, for example (compare Table 26.1). As usual, the major blood solutes are Na^+ and Cl^-. The urine of freshwater crabs, unlike the urine of nearly all other freshwater animals, is virtually isosmotic to their blood plasma. Its ionic composition is also very similar to that of blood plasma. Thus the loss of Na^+ and Cl^- *per unit of volume* of urine in the freshwater crabs is very high compared to that in most freshwater animals. A critical feature of the freshwater crabs is that their bodies exhibit exceptionally low permeability to water. This low permeability means that the crabs experience relatively low rates of osmotic water influx. Thus the rates at which they must excrete their high-concentration urine are low. *Eriocheir*, for instance, excretes water equivalent to only about 3.6% of its body weight per day, and *Potamon* excretes less than 0.6% per day (compare Table 26.2). The unusually low urine output of these crabs—which results from their unusually low permeability to water—helps limit the rate at which they lose ions in their urine. The quantities of Na^+ and Cl^- that the crabs lose by excretion in their urine and by diffusion, although large by comparison to the quantities lost by freshwater crayfish, are "manageable," in the sense that they can be successfully replaced by active uptake (across the gills) from the ambient water.

Among teleost fish living in freshwater, there are a few known species that—like the crabs just described—excrete urine that is nearly isosmotic to their blood plasma. An example is the toadfish *Opsanus tau*, a primarily marine fish that enters freshwater creeks. Reason exists to believe that both the freshwater crabs and the toad-

[5] The freshwater crab *Eriocheir* has recently been introduced into some river systems along the west coast of the United States, and is now established there as an alien species.

fish have invaded freshwater only relatively recently in evolutionary time. Their short history in freshwater may help explain why their kidneys have not evolved the ability to make dilute urine.

Why do most freshwater animals make dilute urine?

The freshwater crabs and the toadfish prove that production of dilute urine is not a necessity for life in freshwater; the production of relatively concentrated urine increases ion losses per unit of volume, but if total urinary losses can be restrained enough that ions can be replaced, existence in freshwater is possible. Why, then, have the overwhelming majority of freshwater animals evolved the capacity to make urine that is strongly hyposmotic to their blood plasma?

The answer is probably energy savings. To produce a dilute urine, the kidneys start with a fluid that is as concentrated as blood plasma and actively extract NaCl from it, lowering its concentration and returning the ions to the blood. Every Na^+ or Cl^- ion thus removed from the urine prior to excretion is an ion that does not have to be replaced by active uptake from the ambient water. In the urine as it is formed in the kidneys, the concentrations of Na^+ and Cl^- are initially as high as in the blood plasma; only gradually—as ions are reabsorbed—do urine ion concentrations fall to low levels. By contrast, the concentrations of Na^+ and Cl^- in the ambient water are *always* very low. Two lines of argument—one based on thermodynamic principles and the other based on molecular details of transport mechanisms—indicate that active uptake of Na^+ or Cl^- from a relatively concentrated source costs less energy per ion than uptake from a dilute source. Thus removing ions from urine prior to excretion is less costly than replacing the same ions from the ambient water.

Summary: Animals in Freshwater

- All freshwater animals are hyperosmotic to the water in which they live. They tend to gain water by osmosis and lose ions by diffusion, especially across their permeable gill membranes. These passive fluxes of water and ions tend to dilute their body fluids.
- To void their excess of water, freshwater animals produce a copious urine.
- In nearly all freshwater animals, the urine is dilute compared to the blood, thereby helping to maintain the osmotic pressure of the blood and the blood concentrations of major ions at levels higher than those in the environment.
- To replace ions lost by direct diffusion into the environment and excretion in urine, freshwater animals take up Na^+, Cl^-, and some other ions by active transport. Active uptake occurs across the gill epithelium in fish and crayfish. Foods also help to replenish ions.

Animals in the Ocean

The fauna of the oceans is far more phylogenetically diverse than that of the other major habitats on Earth, probably in good part because life originated in the oceans. All phyla and most classes of animals have marine representatives. Many phylogenetic groups have invaded freshwater and the land. In turn, many freshwater and terrestrial groups have reinvaded the oceans. Thus, whereas some modern sea animals have a continuously marine ancestry, others trace their history to forms that occupied other habitats. As we will see in this section, phylogenetic history appears sometimes to have left a major imprint on the water–salt physiology of contemporary species.

Most marine invertebrates are isosmotic to seawater

Although the cuttlefish and the marine teleost fish seen swimming in seawater in Figure 26.5 may look like they would have similar blood osmotic concentrations, they actually are very different. Most marine invertebrates—unlike the hyposmotic teleosts—are isosmotic, or nearly so, to seawater. The invertebrates that are isosmotic to seawater include the molluscs, like the cuttlefish, and such other marine animals as sponges, coelenterates, annelids, echinoderms, and most arthropods. These animals do not face problems of osmotic regulation.

The solutes in the blood of these marine invertebrates are mostly inorganic ions (see Figure 25.10). The ionic composition of their blood, however, is rarely, if ever, identical to the ionic composition of seawater (Table 26.3). A particular ion often proves to be relatively concentrated in some animals but relatively dilute in others; for example, Mg^{2+} is relatively high in concentration in the blood of the squid *Loligo*, but low in that of the crab *Carcinus*. The adaptive significance of such differences in blood ionic composition is generally unknown.

Isosmotic marine invertebrates maintain the differences in ionic composition between their blood and seawater by ionic regulatory processes. These animals are typically relatively permeable to both water and ions. Ions therefore tend to diffuse between blood and seawater with ease, following their electrochemical gradients. One process these animals commonly use to maintain their blood ionic

Figure 26.5 Two marine animals with different blood osmotic pressures, an invertebrate with blood isosmotic to seawater and a teleost fish with blood dramatically hyposmotic to seawater The cuttlefish (*Sepia officinalis*; top) belongs to the group of marine molluscs known as cephalopods, which also includes squids and octopuses.

composition is active uptake of ions from seawater at the body surface or from ingested seawater in the gut. A second common process is kidney regulation of blood composition. In crustaceans, molluscs, and some other groups, although the excretory organs make a urine that is approximately isosmotic to the blood plasma, they alter its ionic composition, thus contributing to ionic regulation. In most marine decapod crustaceans, for example, urine is produced that is richer in Mg^{2+} and SO_4^{2-} than the blood (U/P = 1.1–4.2 in several species), which helps keep blood concentrations of these ions lower than seawater concentrations.

TABLE 26.3 The composition of the blood plasma or other extracellular body fluids in some marine invertebrates and hagfish All these animals are isosmotic to seawater. The ion concentrations listed are for animals living in seawater of the composition specified in the last line of the table.

	Ion concentration (mmol/kg H_2O)					
Animal and body fluid	Na^+	K^+	Ca^{2+}	Mg^{2+}	Cl^-	SO_4^{2-}
Mussel (*Mytilus*), blood plasma	474	12.0	11.9	52.6	553	28.9
Squid (*Loligo*), blood plasma	456	22.2	10.6	55.4	578	8.1
Crab (*Carcinus*), blood plasma	531	12.3	13.3	19.5	557	16.5
Sea urchin (*Echinus*), coelomic fluid	474	10.1	10.6	53.5	557	28.7
Jellyfish (*Aurelia*), mesogleal fluid	474	10.7	10.0	53.0	580	15.8
Hagfish (*Myxine*), blood plasma	537	9.1	5.9	18.0	542	6.3
Seawater	478	10.1	10.5	54.5	558	28.8

Source: After Potts and Parry 1964.

Hagfish are the only vertebrates with blood inorganic ion concentrations that make them isosmotic to seawater

The hagfish, an exclusively marine group of jawless primitive vertebrates, resemble the great majority of marine invertebrates in two key respects: Not only is their blood approximately isosmotic to seawater, but their blood solutes are principally Na^+, Cl^-, and other inorganic ions (see Table 26.3 and Figure 25.10). The ionic regulatory processes of hagfish are similar to those of marine invertebrates. Hagfish appear to be the only modern vertebrates (craniates) that have a continuously marine ancestry (see Box 26.1).

The marine teleost fish are markedly hyposmotic to seawater

As stressed at the start of this chapter, the marine teleost fish are **hyposmotic regulators**, in which blood osmotic pressures are far lower than the osmotic pressure of the seawater in which they swim. As in freshwater teleosts, Na^+, Cl^-, and other inorganic ions constitute most of the solutes in the blood plasma of marine teleosts. One of the most intriguing questions about marine teleosts concerns the *origin* of their hyposmotic state: *Why* do they have blood that is more dilute in ions and lower in osmotic pressure than seawater? As discussed in Box 26.1, the dilute body fluids of marine teleosts are thought to be an evolutionary vestige: These fish are generally believed to be descended from ancient ancestors that lived in freshwater.

The blood osmotic pressures of marine teleosts are typically 300 to 500 m*Osm*—higher than those of freshwater teleosts (about 250–350 m*Osm*), but not exceptionally so. It is hypothesized that when teleost fish invaded the oceans from freshwater in the course of their evolution, they evolved modest increases in total blood solutes. This change reduced the difference between their blood and

BOX 26.1 Where Were Vertebrates at Their Start?

The traditional hypothesis held by most biologists about the early evolution of the vertebrates is that they originated as jawless animals in the oceans about 500 million years ago. The lineage leading to modern hagfish (which are themselves jawless) then never left the oceans. Today's hagfish, in this view, are the one group of modern vertebrates or vertebrate-like animals (craniates) that have lived in the oceans throughout their evolutionary history. According to the traditional hypothesis, early jawless vertebrates entered freshwater, and jaws then originated in freshwater. Thus, according to the traditional hypothesis, all jawed vertebrates (plus modern lampreys) are descended from freshwater ancestors.

The water–salt physiology of modern vertebrates provides one of the principal arguments for the traditional hypothesis. As we saw in Figure 25.10, total blood salt concentrations are monotonously similar in all the jawed vertebrates alive today and are more like those in freshwater invertebrates than in marine invertebrates. This similarity in blood salt concentrations suggests that all the living jawed vertebrates had ancient ancestors that lived in freshwater and that the blood ion concentrations of vertebrates became relatively fixed at that time in evolution. According to this view, modern marine teleost fish have blood salt concentrations far lower than those in seawater because they are descended from freshwater ancestors.

You will notice the circularity of reasoning here. The features of the blood composition of modern animals are used as evidence for a freshwater origin of jawed vertebrates. The purported freshwater origin of jawed vertebrates is then used to explain the blood composition of modern animals. This unsettling state of affairs arises in part because the fossil record is too sketchy to provide firm independent confirmation or rejection of the freshwater-origins hypothesis for jawed vertebrates.

Some evolutionary biologists believe that the available circumstantial evidence supports a scenario different from the traditional one. One alternative view, for example, is that the earliest vertebrates were closely associated with the continental margins and experienced both marine and freshwater environments. All scenarios for the early evolution of vertebrates concur in postulating that the blood composition of modern jawed vertebrates is a consequence of early evolutionary experience with freshwater. No other interpretation seems plausible.

the seawater in their new habitat. Nonetheless, in terms of osmotic pressure, modern marine teleost fish face a difference between their blood and their environmental water that is far greater than that faced by freshwater teleosts. In freshwater teleosts, blood osmotic pressure averages about 300 m*Osm* higher than the osmotic pressure of freshwater. In marine teleosts, given that the blood osmotic pressure averages 400 m*Osm* and the osmotic pressure of seawater is near 1000 m*Osm*, the difference between the blood and environmental osmotic pressures is about 600 m*Osm*: *approximately twice as great* as the difference maintained by freshwater teleosts! This consideration in itself would tend to saddle the marine fish with a relatively high rate of osmotic water flux. Marine teleosts, however, are typically less permeable to water than freshwater teleosts, so in fact the osmotic fluxes experienced by the two groups are roughly similar (for a given body size). Of course, in the case of the marine fish, water tends to *leave* the body osmotically, rather than entering. *For a hyposmotic animal, the sea is a desiccating environment.*

The concentrations of Na^+, Cl^-, Mg^{2+}, SO_4^{2-}, and some other ions are far lower in the blood plasma of marine teleosts than in seawater, suggesting that marine teleost fish also face problems of inward diffusion of multiple ions. The concentration gradients between the blood plasma and the environment for the two major plasma ions, Na^+ and Cl^-, are large by comparison to the (oppositely directed) gradients seen in freshwater teleosts. Nonetheless, actual rates of ion diffusion depend on electrical gradients and gill permeability, not just on ion concentration gradients. When all these factors are taken into account, inward Na^+ diffusion turns out not to be much of a problem (or may not even occur in some species) because the gill epithelium is positively charged on the inside, repelling Na^+. On the other hand, some other ions—most notably Cl^-—tend to diffuse into the blood plasma from seawater at substantial rates, tending to concentrate the body fluids of the fish.

REPLACEMENT OF WATER LOSSES Marine teleost fish lose water by osmosis and, to a lesser extent, by urine production. To replace the water they lose and thus volume regulate, these fish drink seawater. Although some drink an amount of seawater that is less than 1% of their body weight per day, others drink over 50% per day, and the average is probably 10% to 20% per day. Recognizing that the major water losses of these fish are osmotic, these data on water ingestion give us an idea of the magnitude of the daily osmotic outflux of water.

In marine teleosts, as in freshwater teleosts, the process of volume regulation worsens problems of ionic regulation.[6] At first sight, drinking seawater seems to be a straightforward way to obtain water. Consider, however, that when seawater is first taken into the gut of a marine teleost, it is strongly *hyperosmotic* to the fish's blood. Thus H_2O would be expected to travel by osmosis *out of the blood into the gut fluids*, not vice versa—and that is exactly what happens. Studies on several species indicate that as ingested seawater travels through the esophagus, stomach, and (in at least some instances) anterior intestine, not only do Na^+ and Cl^- diffuse into the blood across the gut wall, but H_2O *enters the gut fluids* by osmosis. Thus, gradually, the ingested seawater in the gut becomes diluted. In other parts of the intestine, Na^+ and Cl^- are *actively transported out of the gut contents into the blood.* Here H_2O is finally taken up from the gut fluids into the blood. In these parts of the intestine, the active uptake of Na^+ and Cl^- into the blood creates conditions that favor the osmotic uptake of water. In the simplest cases, the active uptake of the ions from the gut fluids renders the gut fluids hyposmotic to the blood, but often more-complex processes based on highly localized osmotic gradients within the intestinal epithelium (termed *local osmosis*) are involved in translating ion uptake into water uptake.

In the end, about 50% to 85% of the H_2O in ingested seawater is absorbed into the blood as the seawater passes through the gastrointestinal tract of a marine teleost fish. However, a much greater proportion of the NaCl in the ingested seawater—often over 97%—is absorbed. Because H_2O cannot be absorbed without NaCl absorption, the influx of NaCl into the blood is *obligated by the process of obtaining* H_2O. The influx of NaCl into the blood increases problems of Na^+ and Cl^- regulation faced by the fish, explaining the statement made earlier that volume regulation worsens problems of ionic regulation. The gut is poorly permeable to the major divalent ions in seawater, Mg^{2+} and SO_4^{2-}.[7] Thus, although the divalent ions diffuse into the blood to a small extent as seawater passes through the gut, for the most part they remain in the gut and are expelled in the feces.

URINE We now turn to the question of how marine teleosts eliminate the excess ions that enter their body fluids from the gut or that diffuse into their body fluids from seawater across their gills or other body surfaces. For the most part, excess *divalent* ions are removed from the body fluids by excretion in the urine, and *monovalent* ions are excreted by the gills.

The kidneys of marine teleosts are unable to produce urine that is hyperosmotic to the blood plasma; the kidneys excrete urine that is about as concentrated as they can possibly produce, approximately isosmotic to the plasma. The fact that the osmotic pressure of the urine matches that of the blood plasma (osmotic $U/P \cong 1$) means that the excretion of urine cannot help the fish with their osmotic regulatory problem (see Figure 25.7) The *ionic composition* of the urine, however, differs dramatically from that of the plasma, and the kidneys are the principal organs that carry out ionic regulation of Mg^{2+} and SO_4^{2-} (and Ca^{2+}). Whereas U/P ratios for Na^+, Cl^-, and K^+ are below 1, those for Mg^{2+} and SO_4^{2-} are much greater than 1. The kidneys thus void the major divalent ions preferentially in relation to water and keep plasma concentrations of those ions from increasing.

For every milliliter of water that is first ingested and absorbed and then excreted as urine, a marine teleost fish is left with an excess of solutes because, although the water enters its body hyperosmotic to its body fluids, the water leaves its body isosmotic to its body fluids. From the viewpoint of osmoregulation, therefore, production of urine by marine teleost fish is an outright liability, and we would expect them to limit their volume of urine to the minimum necessary for excretion of solutes that are not excreted by other routes. Nitrogenous wastes and the principal ions, Na^+ and Cl^-, are voided across the gills. Thus, the role of the kidneys is largely limited to ex-

[6] Volume regulation and ionic regulation are inextricably linked in the marine and freshwater teleosts because the major solutes of the body fluids in both of these groups of fish are ions.

[7] *Divalent* ions have two charges per ion, either two positive charges or two negative charges.

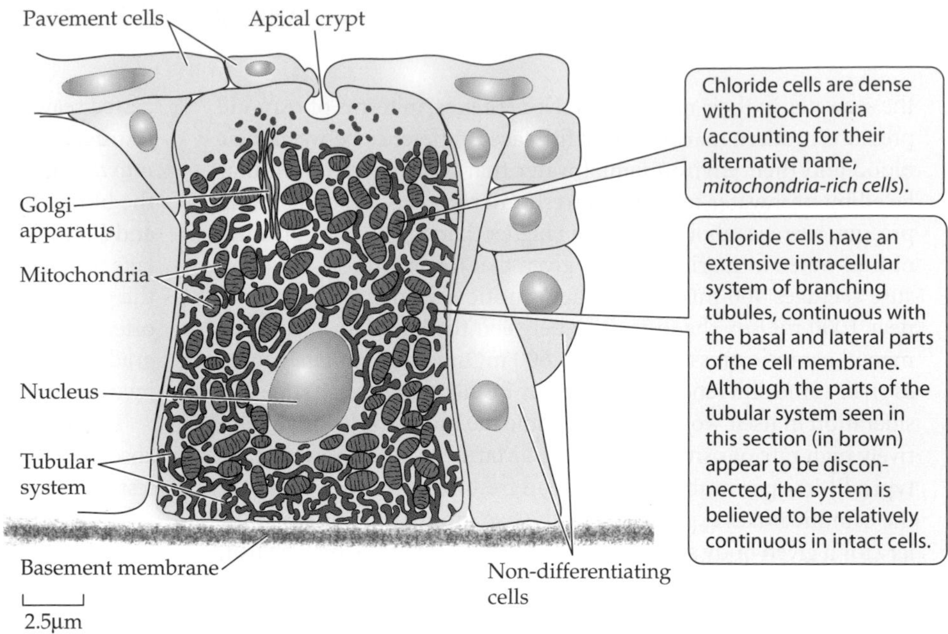

Figure 26.6 A section of a chloride cell of a marine teleost fish In some species of marine teleosts, chloride cells occur not just in the gill epithelium but also in the inner opercular epithelium, jaw epithelium, and certain other surfaces, implicating these other body parts—as well as the gills—in ion excretion. The chloride cell drawn here is from the opercular epithelium of a seawater-acclimated killifish (*Fundulus heteroclitus*). The structure of chloride cells in the gill epithelium is the same, but the cellular surroundings in the gill epithelium are different. Chloride cell morphology differs between freshwater and marine fish (see Figure 26.2); in marine fish, for example, the apical crypt is more prominent, and (in contrast to freshwater fish) the chloride cells in the gill epithelium tend to be found in multi-cell groups that include accessory cells. (After Degnan, Karnaky, and Zadunaisky 1977.)

cretion of divalent ions, and the rate of urine production can be low. The urine volumes of several species have been measured to be just 0.5% to 3.5% of body weight per day.

EXTRARENAL NaCl EXCRETION BY THE GILLS The gills assume primary responsibility for excreting excesses of the major ions, Na^+ and Cl^-, from the blood into the surrounding ocean. The excretion of Cl^- is active and is carried out by chloride cells (Figure 26.6). Box 26.2 outlines the mechanism of Cl^- pumping as it is presently understood. Although the gill epithelium is believed always to transport Na^+ as well as Cl^- out of the blood into the seawater, the excretion of Na^+ is probably active in about half the species that have been studied, but passive in the others (in which Na^+ diffuses outward, attracted by an outside-negative electrical gradient generated by active Cl^- transport). The elimination of Cl^- and Na^+ by the gills of marine teleost fish provides our first example of **extrarenal salt excretion**: excretion of ions by structures other than the kidneys.

Present evidence indicates that excretion of NaCl by the gills is accomplished without concomitant excretion of water; the material excreted is purely ions. Thus, besides voiding NaCl from the blood (ionic regulation), the process maintains the osmotic pressure of the blood at a level lower than the ambient osmotic pressure. Thus the gills are where osmotic regulation is principally accomplished.[8]

QUANTITATIVE EXAMPLE AND COST ESTIMATES The pattern of water–salt regulation in marine teleost fish is summarized in Figure 26.7*b*. Let's review it using quantitative data for one particular species, the southern flounder (*Paralichthys lethostigma*). An individual flounder that weighs about 1 kg loses water equivalent to about 7.9% of its body weight per day because of osmosis from its body fluids into the

[8] The effective osmotic pressure of the material excreted across the gills is far higher than the plasma osmotic pressure. Thus, according to the principles presented in Figure 25.7, its excretion tends to lower the plasma osmotic pressure.

(a) Freshwater teleost
Salt loss by diffusion
Water uptake by osmosis
Gills
Salts and water in food (generally do not drink)
Hyperosmotic to ambient water
Salts and water in feces
Large amounts of urine, very hyposmotic to plasma
Active uptake of Na^+ and Cl^-

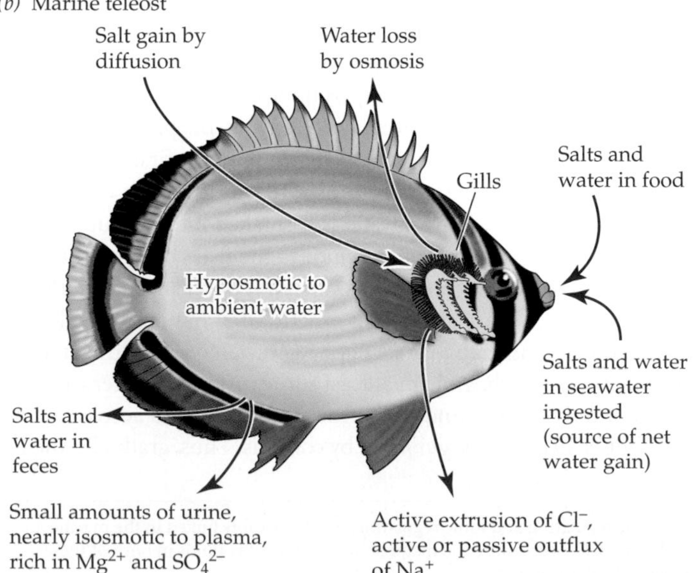

Figure 26.7 Water–salt relations in freshwater and marine teleost fish

BOX 26.2 Epithelial NaCl Secretion

NaCl secretion by the gill epithelium of marine teleost fish is believed to occur by the mechanism shown in the accompanying figure. The same model of the secretory mechanism applies to other NaCl-secreting structures to be discussed later in this chapter: the cranial salt glands of marine birds and sea turtles and the rectal salt glands of marine sharks, skates, and rays.

To understand the model, let's focus first on Cl^- secretion by the chloride cell in the figure. The Cl^- transporter protein shown in the basolateral membrane of the cell is *not* an ATPase. Thus Cl^- transport is by *secondary* rather than primary active transport (see page 76). The Cl^- transporter protein is an electroneutral cotransporter that moves one Na^+ ion, one K^+ ion, and two Cl^- ions into the cell across the basolateral membrane on each transport cycle; in the jargon of the study of transporter proteins, it is a *Na–K–2Cl cotransporter*. The energy for Cl^- transport is supplied from ATP indirectly by the action of Na^+–K^+-ATPase (see page 78), also located in the basolateral membrane. The Na^+–K^+-ATPase uses ATP-bond energy to pump Na^+ out of the cell, thereby creating a strong electrochemical gradient favoring diffusion of Na^+ from the blood into the cell. Following its electrochemical gradient, Na^+ diffuses into the cell bound to the Na–K–2Cl cotransporter, and this process brings Cl^- from the blood into the cell. The entry of Cl^- into the cell creates an electrochemical gradient favoring the diffusion of Cl^- out of the cell, and Cl^- exits the cell by way of Cl^- channel proteins in the apical cell membrane. In the case of the gill epithelium of marine teleosts, the Cl^- leaving the cell enters the ambient water; in the case of the cranial and rectal salt glands of other vertebrates (discussed later in this chapter), the Cl^- enters the secretory ducts of the glands. Potassium (K^+) ions accumulated in the cell by the actions of Na^+–K^+-ATPase and the Na–K–2Cl cotransporter simply diffuse back to the blood by way of K^+ channels.

The Cl^- and K^+ channels we have described are gated channels in at least some cases. Secretion is controlled in part by regulatory mechanisms that open and close these channels. In the cells of avian salt glands, the Cl^- and K^+ channels are known to be Ca^{2+}-activated; thus secretion is under immediate control of a Ca^{2+}-based second-messenger system.

Let's now turn to the question of how Na^+ is secreted across the epithelium. The transfer of Cl^- across the secretory epithelium by the mechanism we have described is electrogenic, and as shown in the figure, it renders the apical side of the epithelium electrically negative relative to the blood. This difference in electrical potential across the epithelium can be large enough to create an electrochemical gradient that favors *diffusion* of Na^+ from the blood outward across the epithelium (Na^+ is repelled from the positively charged inner epithelial surface and attracted to the negatively charged outer surface). Current evidence indicates that when Na^+ travels outward across the epithelium by diffusion, it follows paracellular (between-cell) pathways. In some cases, diffusion is believed to be the sole mechanism of Na^+ secretion, but in other cases there is evidence for secretion of Na^+ by active transport.

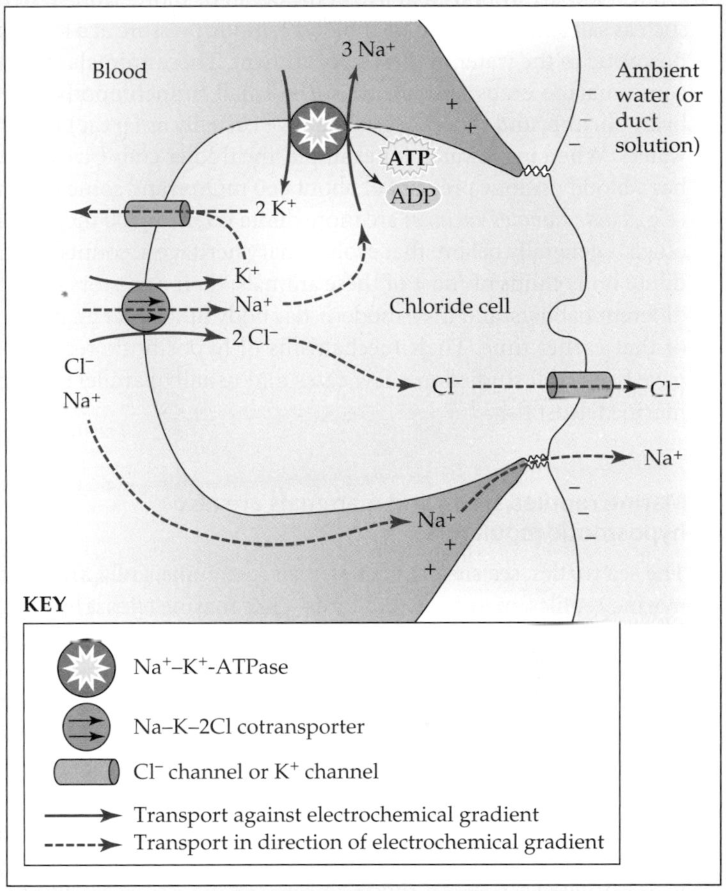

The probable mechanism of epithelial NaCl secretion The diagram shows a chloride cell flanked by two other cells (which may or may not themselves be chloride cells) in a secretory epithelium. In the gills of marine teleosts, NaCl is secreted directly into the ambient water; in the cranial salt glands of birds and sea turtles and the rectal salt glands of elasmobranch fish, NaCl is secreted into ducts, and the salty solution thus formed flows out of the animal by way of a duct system. The Na–K–2Cl cotransporter, K^+ channel, and Na^+–K^+-ATPase are proteins in the basolateral cell membrane. The Cl^- channel is a protein in the apical membrane.

surrounding seawater. To replace this water, as well as urinary water losses of 0.4% of body weight per day and fecal losses of 2.7% per day, the fish drinks seawater equivalent to 11% of its body weight per day. From the seawater it ingests, the fish absorbs 76% of the water (H_2O), but in doing so it absorbs much higher percentages of the Na^+ (99%) and Cl^- (96%). The flounder produces scanty urine having a U/P ratio for Mg^{2+} of 99 and a U/P ratio for SO_4^{2-} of 330; the urine removes the excesses of both of these divalent ions from the body fluids. The gills of the flounder excrete virtually all of the excess monovalent ions, Na^+ and Cl^-.

The energy cost of Na^+, Cl^-, and osmotic regulation in marine teleosts has been estimated to be 8% to 17% of the resting metabolic rates of the fish, based on measured ion-pumping rates and the known ATP costs of pumping. Rainbow trout—which can live in freshwater or seawater—are estimated to devote 3% of their resting energy use to Na^+, Cl^-, and osmotic regulation when living in freshwater, but 8% when living in seawater. The higher cost in seawater reflects the fact, earlier stressed, that for a teleost fish, the difference in concentration between the blood and the ambient water is about twice as great in seawater as in freshwater.

Some arthropods of saline waters are hyposmotic regulators

Quite a few arthropods that live in the ocean or more-saline waters, such as salt lakes, maintain their blood osmotic pressure at a level hyposmotic to the water in their environment. These animals include some marine crabs and shrimps, the small branchiopods called brine shrimp, and those insects that live (usually as larvae) in salty waters. When in seawater, for example, the fiddler crab *Uca pugnax* has a blood osmotic pressure of about 860 m*Osm*, and some prawns (e.g., *Palaemonetes varians*) are more dilute yet, near 700 m*Osm*. Biologists generally believe that evolutionary heritage accounts for the dilute body fluids of most of these animals; their ancestors lived in different habitats, and their modern-day body fluids bear an imprint of that earlier time. Their mechanisms of hyposmotic regulation have been well studied in a few cases and usually parallel those of marine teleost fish.

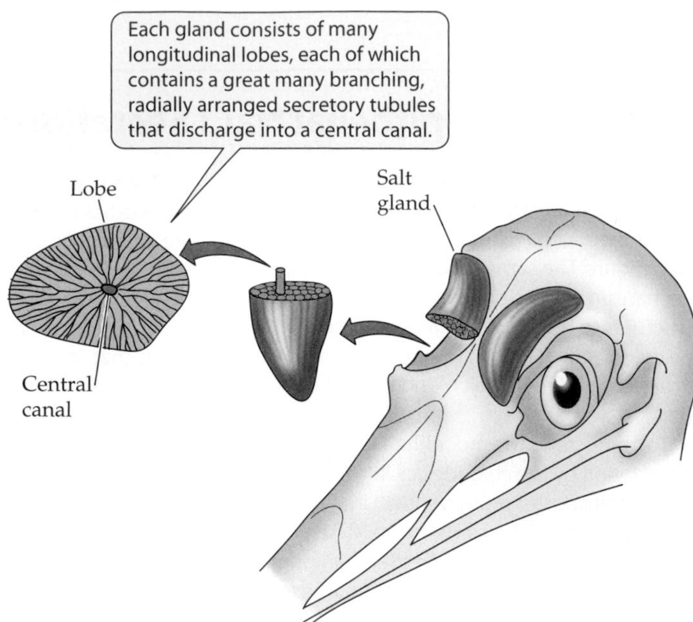

Figure 26.8 The salt glands of a herring gull Each salt gland lies in a shallow depression in the skull above the eye. The glands empty into the nasal passages via ducts, and their secretion flows out through the external nares (nostrils). (After Schmidt-Nielsen 1960; for exhaustive detail on morphology, consult the article by Gerstberger and Gray in the Selected References.)

Marine reptiles, birds, and mammals are also hyposmotic regulators

The sea turtles, sea snakes, whales, seals, penguins, gulls, and other marine reptiles, mammals, and birds—like marine teleost fish—are markedly hyposmotic to seawater. All are descended from terrestrial ancestors, and their blood compositions are clearly carryovers from their ancestors. The blood osmotic pressures of all these marine vertebrates tend to be about 400 m*Osm*: just modestly higher than the values seen in modern-day terrestrial and freshwater vertebrates.

Because the marine reptiles, birds, and mammals are air breathers, they do not expose permeable respiratory membranes to seawater. Another advantage of their terrestrial heritage is that they have inherited integuments that were originally adapted to limiting water losses in the dehydrating terrestrial environment, so they tend to exhibit low integumentary permeabilities. These animals nonetheless confront problems of water loss and salt loading. They lose water, for example, by pulmonary evaporation during breathing; they also lose water to some extent across their skin, not only when they are immersed in seawater, but also when they are exposed to the air. They often gain excess salts from the foods they eat; for example, when they prey on marine plants or invertebrates that are isosmotic to seawater, they ingest body fluids that have far higher salt concentrations than their own (see Figure 25.10). In addition, they probably often take in quantities of seawater with the foods they eat, although for the most part, they are thought not to drink seawater.

MARINE REPTILES AND BIRDS The marine reptiles—like other reptiles—are generally unable to produce urine hyperosmotic to their blood. The urine-concentrating capabilities of marine birds are incompletely understood, but for most, the maximum urine concentration appears to be isosmotic to the blood or only modestly hyperosmotic ($U/P \leq 2$). Because of these facts, the kidneys of marine reptiles and birds are in general unable—by themselves—to maintain the blood of the animals hyposmotic to seawater (see Figure 25.7).

Organs of extrarenal salt excretion called **salt glands** play a central role in the osmotic–ionic regulation of marine reptiles and most marine birds.[9] The salt glands are located in the head (Figure 26.8). They produce concentrated salt solutions, which are discharged into the nasal passages in birds and lizards, and the mouth in sea snakes. In sea turtles, the secretions (referred to as *orbital* secretions) are emitted like tears. Not only are the secretions of the salt glands strongly hyperosmotic to the blood (by a factor of 4 or 5 in many species), but they also, as indicated in Table 26.4, contain concentrations of Na^+ and Cl^- (and K^+ as well) that exceed those in seawater. Thus, reptiles and birds with salt glands are, in principle, able to extract H_2O from seawater; they could drink seawater and void the major monovalent ions in less H_2O than they ingested, retaining the excess H_2O in their bodies. Their kidneys alone do not provide such a capability. The cellular mechanism of salt secretion in the salt glands of birds and sea turtles follows the model presented in Box 26.2.

The ingestion of a salt load by an animal with salt glands is followed promptly by an increase in the rate of secretion by the glands. Control of this response, at least in birds, is mediated principally by the parasympathetic division of the autonomic nervous system. When osmoreceptors located in or near the heart and brain detect high blood osmotic pressures, the parasympathetic nervous system releases acetylcholine in the salt glands; this chemical message induces the gated Cl^- and K^+ channels in the chloride cells (seen in the figure in Box 26.2) to open, activating secretion. Besides these acute responses, salt glands also undergo chronic responses (acclimatization). If a bird experiences a chronic increase in salt ingestion—as it would after migrating from a freshwater habitat to an ocean habitat—its salt glands typically increase in size, concentrating ability, and peak secretory rate. These changes are reversible.

[9] Salt glands have been reported in 14 orders of birds. However, they have not been reported in the passerine ("perching") birds. A few species of passerines that live in association with the ocean, such as certain salt marsh sparrows, have evolved kidneys that—based on published reports (now disputed by some authorities) —have unusually great concentrating abilities compared to most birds.

TABLE 26.4 Sodium concentrations in the salt-gland secretions of marine reptiles and birds Chloride concentrations are typically about the same as sodium concentrations. All examples listed are nasal secretions except those of the sea turtles, which are orbital. All data are for adults except those for the green sea turtles.

Animal	Sodium concentration (m*M*)
Standard seawater	470
Marine iguana (*Amblyrhynchus cristatus*)	840
Loggerhead sea turtle (*Caretta caretta*)	730–880
Newly hatched green sea turtle (*Chelonia mydas*)	460–830
Herring gull (*Larus argentatus*)	600–800
Brown pelican (*Pelecanus occidentalis*)	600–750
Humboldt penguin (*Speniscus humboldti*)	725–850
Leach's storm-petrel (*Oceanodroma leucorhoa*)	900–1100

Source: Marshall and Cooper 1988; Schmidt-Nielsen 1960; Schmidt-Nielsen and Fange 1958.

The tears observed flowing down the faces of sea turtles when they emerge onto beaches to lay eggs are of some renown. We now understand that they are secretions of salt glands, not tears of emotion. If you carefully watch a herring gull standing by the ocean, you will see salt-gland secretions dripping slowly out of its nostrils and off the end of its bill, or perhaps the gull will flick the drops away with a shake of its head.

MARINE MAMMALS Mammals, as a group, are capable of producing the most concentrated urine of all vertebrates. This ability is important on land, and for marine mammals, it is a key to hyposmotic regulation. Salt glands or other mechanisms of extrarenal salt excretion are not known in mammals.

As important as the kidneys are in seals and whales, the urine-concentrating abilities of these animals are not exceptional by mammalian standards, according to available data. For example, in harbor seals (*Phoca vitulina*), which are typical of the marine species that are superior in urine-concentrating ability, the maximum osmotic U/P ratio is about 6, and the maximum urinary Na^+ and Cl^- concentrations that have been measured are slightly above 500 m*M*. Although these renal concentrating abilities are high in comparison to those of reptiles and most birds, they are not particularly high in comparison to those of nonmarine mammals of similar body size.

The overall patterns of water and salt balance in seals and whales remain open to debate. Existence on a diet of teleost fish poses no great challenges. To date, however, the data available leave unclear whether or how most species could exist while eating only invertebrates, which have substantially saltier body fluids than fish, as noted earlier. Although seals and whales are thought generally not to drink seawater, research in the past decade has revealed that some species of both groups do drink under certain circumstances; some otariid (e.g., fur) seals, for instance, drink seawater when hauled out on land for weeks in hot climates during their breeding season. Physiologists are debating the potential advantages of drinking for animals that in general cannot concentrate salts in their urine to levels higher than the concentrations seen in seawater.

Marine elasmobranch fish are hyperosmotic but hypoionic to seawater

The marine sharks, skates, and rays—collectively known as the *elasmobranch fish*—have evolved a novel solution to the osmotic problems of living in the sea. As illustrated by the example of the dogfish shark in Table 26.5, their blood concentrations of *inorganic ions* are similar to those of marine teleost fish and well below those in seawater. The osmotic pressure of their blood, however, is slightly *higher* than that of seawater. These fish are able to be hyperosmotic to seawater—even though their blood has far lower concentrations of inorganic ions than seawater—because they have high concentrations in their body fluids of two *organic solutes*: urea and, to a lesser extent, trimethylamine oxide (TMAO).[10] Because their blood is hyperosmotic to seawater, the marine elasmobranchs experience a small *osmotic influx* of water, in sharp contrast to the marine teleosts, which suffer from osmotic desiccation. The hyperosmoticity of the elasmobranchs—caused by their high blood concentrations of urea and TMAO—is, in effect, a mechanism for obtaining water.

Urea (Box 26.3) is the principal nitrogenous end product of protein catabolism in elasmobranch fish. The gills of marine elasmo-

[10] For the chemical structures of urea and TMAO, see Figure 27.23.

TABLE 26.5 Composition of the blood plasma and excretory fluids of the dogfish shark and coelacanth For each species, the composition of the seawater in which the fish were living when studied is also given. TMAO = trimethylamine oxide.

	Osmotic pressure (m*Osm*)	Solute concentration (m*M*)			
		Na^+	Cl^-	Urea	TMAO
Dogfish shark (*Squalus acanthias*)					
Blood plasma	1018	286	246	351	71
Urine	780	337	203	72	6
Rectal-gland secretion	1018	540	533	15	—
Seawater	930	440	496	0	0
Coelacanth (*Latimeria chalumnae*)					
Blood plasma	931	197	187	377	122
Urine	961	184	15	388	94
Seawater	1035	470	548	0	0

Source: Hill and Wyse 1989.

BOX 26.3 The Evolution of Urea Synthesis in Vertebrates

Proteins are 16% nitrogen. Thus, when proteins are catabolized, the disposition of their nitrogen is a significant matter. Simple deamination of amino acids during protein breakdown leads to formation of ammonia (NH_3) as the nitrogen-containing end product of catabolism. One of the major alternative nitrogenous end products is urea, shown at the top of the figure. The biochemical pathway for the synthesis of urea from protein nitrogen in vertebrates is known as the *ornithine–urea cycle*, discussed further in Chapter 27. For present purposes, a point to be stressed is that the synthesis of urea costs energy: Each urea molecule requires the energy from four or five ATP molecules (four or five high-energy phosphate bonds) for its synthesis. This cost is an "extra cost" that is avoided if ammonia is made instead of urea.

In the past 15 years, a consensus has emerged that the ornithine–urea cycle existed in the earliest vertebrates. Because the cycle is actually a set of enzyme-catalyzed reactions, the preceding sentence can be reworded to state that the earliest vertebrates are believed to have had genes coding for all the ornithine–urea cycle enzymes. The figure presents the evolutionary relationships of the modern vertebrates and shows which groups of modern vertebrates actually synthesize urea from protein nitrogen. Some groups that do not do so might have suffered deletion of one or more required genes. A current working hypothesis, however, is that, in general, all modern-day vertebrates have the genes for the ornithine–urea cycle, but one or more of the genes are not expressed in those vertebrates that fail to synthesize urea from protein nitrogen. To the extent that all vertebrates have the genes, all are *poised* to synthesize urea: The genes need only be expressed for urea synthesis to occur.

Two principal advantages of urea synthesis seem to account for the cases in which aquatic vertebrates invest extra energy to make urea rather than ammonia from their waste protein nitrogen. First, urea may be employed as an osmolyte to raise the osmotic pressure of the blood (e.g., to render the blood of marine animals hyperosmotic to seawater); in such cases, the urea is typically paired with the counteracting solute TMAO. This particular use of urea is seen in the elasmobranch fish (sharks, skates, and rays), holocephalan fish, and the coelacanth. The cladogram in the figure suggests, therefore, that this use *evolved at least twice independently*. Second, urea may be employed as a detoxification compound for waste nitrogen in animals that are temporarily unable to excrete wastes; compared to ammonia, urea is far less toxic and is therefore more suited to being accumulated in the body.

Recent studies have shown that the molecular forms of two key enzymes of the ornithine–urea cycle are fundamentally different in the organisms above and below the dashed line in the figure. The phylogenetic perspective permits us to pinpoint the spot marked with an asterisk as the likely point in the evolutionary history of the vertebrates when the enzymes changed.

A cladogram of the vertebrates, showing which employ the ornithine–urea cycle to synthesize urea from protein nitrogen The animals included in this cladogram are ones that have been tested for activity of the ornithine–urea cycle. Two enzymes of the ornithine–urea cycle are carbamoyl phosphate synthase (CPS) and arginase. Animals below the dashed line synthesize a molecular form of CPS named CPS I and cytosolic arginase; animals above the line predominantly synthesize CPS III (or synthesize a mix of CPS III and a CPS I-like enzyme) and mitochondrial arginase. Thus the enzyme forms changed at the point in evolutionary history represented by the asterisk. (After Mommsen and Walsh 1989.)

branchs are much less permeable to urea than those of marine teleosts, and elasmobranchs reabsorb urea from their urine as the urine forms in their kidneys; these two features are important adaptations for maintaining a high concentration of urea in their body fluids. Urea in high concentrations alters the structure of many proteins, and its concentration is kept low in most vertebrates (about 2–7 m*M* in human plasma). The levels in marine elasmobranchs—over 300 m*M* (see Table 26.5)—are "out of sight" by comparison. Some enzymes and other macromolecules in elasmobranchs have evolved exceptional resistance to urea's destabilizing effects. Some elasmobranch organs, such as the heart, have in fact become dependent on urea for proper function. Recent studies have revealed, however, that many elasmobranch proteins are just as sensitive to urea's destabilizing effects as homologous proteins in other vertebrates. How can this be? A key part of the answer is that TMAO serves as a *counteracting solute* (see page 682). In the amounts present, TMAO offsets the effects of urea.

In most aquatic animals, the blood osmotic pressure is attributable primarily to inorganic ions dissolved in the blood plasma (see Figure 25.10). Because of this, problems of osmotic and ionic regulation are related in particular ways: If an animal tends to gain water by osmosis, it tends to lose ions by diffusion, and vice versa. These relations

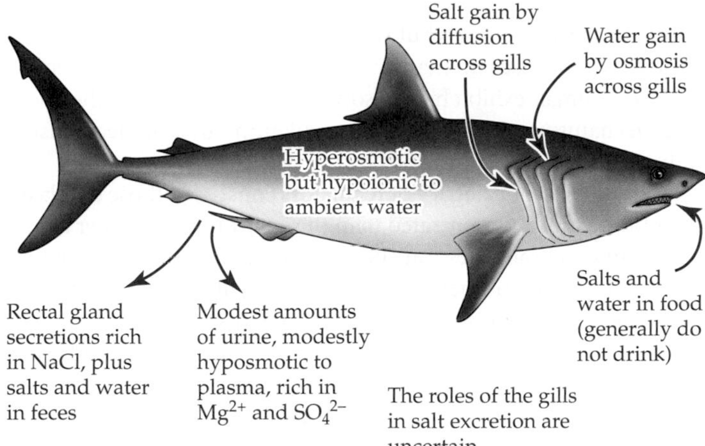

Figure 26.9 Water–salt relations in a marine shark

are uncoupled in the marine elasmobranch fish. Being slightly hyperosmotic to seawater, the elasmobranchs tend to gain water by osmosis, but because their blood ion concentrations are below those in seawater, they also tend to gain excess ions by diffusion from seawater. (They also gain ions by consuming foods that are as salty as seawater.) Because of their osmotic inputs of water, the elasmobranchs, unlike marine teleosts, need not drink seawater to meet their water needs. Thus they do not incur a NaCl load by drinking.

Excess salts are removed from the body fluids of elasmobranchs by the kidneys and, extrarenally, by *rectal salt glands*. These salt glands, consisting of thousands of secretory tubules, void into the rectum a secretion (see Table 26.5) that is isosmotic to the blood, but contains only traces of urea and approximates or exceeds seawater in its concentrations of Na^+ and Cl^-. The mechanism of NaCl secretion is as described in Box 26.2. Excess divalent ions are voided largely by the kidneys in elasmobranchs, along with significant quantities of NaCl (see Table 26.5). It remains unclear whether active ion excretion occurs across the gills of elasmobranchs.

Ever since the "elasmobranch strategy" of water–salt regulation in the sea (Figure 26.9) was discovered, biologists have speculated about its possible advantages over the "teleost strategy." Until recently, the usual conclusion was that the elasmobranch strategy costs less energy than the strategy used by marine teleosts because elasmobranchs obtain H_2O by "cost-free" osmosis, whereas the teleosts must drink seawater and pump NaCl out of it to get H_2O. The error in this view, we now recognize, is that the osmosis of water into an elasmobranch is *not* "cost-free." To keep its blood hyperosmotic to seawater, an elasmobranch must synthesize urea, which costs more ATP than merely making ammonia from waste nitrogen (as most teleost fish do).[11] The elasmobranch might also need to pay ATP costs to recover urea from its urine. In a careful analysis, Leonard Kirschner concluded that the costs of the elasmobranch and teleost strategies are essentially the same. At least in terms of energy, the strategies seem to be "different but equal."

The modern marine elasmobranchs are believed to be descended from freshwater ancestors; the concentration of salts in their blood supports this viewpoint (see Figure 25.10). Their retention of urea and TMAO is believed to be an adaptation to colonization of the sea. Many elasmobranch species are found today in freshwater. Some are permanent residents of freshwater; these species resemble freshwater teleosts in their water–salt biology. However, most species of elasmobranchs found in freshwater also occur in the ocean; when these animals venture into freshwater, they retain somewhat elevated blood urea concentrations, even though doing so promotes osmotic uptake of excess water in the freshwater environment.

Besides marine elasmobranchs, two other types of marine fish maintain high blood concentrations of urea and TMAO: the coelacanth and the holocephalans (chimaeras). The coelacanth (see Table 26.5) is a particularly interesting case because it is the only living example of the crossopterygian fish, presumed ancestors of the terrestrial vertebrates.

[11] The ATP costs of producing urea and ammonia are discussed more in Chapter 27.

Summary: Animals in the Ocean

- Most marine invertebrates are approximately isosmotic to seawater, but their blood differs from seawater in ionic composition. They exhibit ionic regulation, but little or no osmotic regulation. Hagfish display the same pattern.
- Marine teleost fish are hyposmotic to seawater, apparently because they are descended from freshwater or coastal ancestors.
- Being hyposmotic to seawater, marine teleosts tend to lose water by osmosis and gain ions by diffusion. To replace water, they drink; however, to absorb H_2O from the seawater in their gut, they must actively take up NaCl, increasing their problem of salt loading. Their kidneys make urine that is approximately isosmotic to their blood, but rich in divalent ions, thereby assuming chief responsibility for divalent ion regulation. Monovalent ions are excreted across their gills; although Cl^- is secreted into the ambient water actively by chloride cells, Na^+ secretion may be secondary to Cl^- secretion.
- Marine reptiles and birds have cranial salt glands that permit them to excrete ions at higher concentrations than possible in their urine.
- Marine mammals lack salt glands, but have kidneys that can produce more-concentrated urine than those of reptiles or birds can. Their urinary concentrating abilities are not exceptional compared to those of other mammals, however, and their water–salt balance is not entirely understood.
- Marine elasmobranch fish, although they have blood ion concentrations far lower than those of seawater, are slightly hyperosmotic to seawater because of high concentrations of two counteracting organic solutes, urea and trimethylamine oxide. Unlike teleosts, therefore, elasmobranchs need not drink and need not incur an extra NaCl load to gain H_2O from ingested seawater.

Animals that Face Changes in Salinity

Many aquatic animals that live near the margins of the continents—and some that venture far from the coastlines—can face large changes in the salinity of the waters they occupy during their lifetimes. Brackish waters, as mentioned in Chapter 25, occur in estuaries, salt marshes, and other coastal settings. Ocean animals that venture into brackish coastal waters encounter lower salinities than they experience when living in the open ocean. Freshwater animals face elevated salinities when they enter brackish waters. Some species live *within* estuaries for most or all of their lives; they face changes in salinity as they move from place to place (see Figure 25.4) or as tides or other water movements shift the waters around them. Fish like salmon that migrate between the ocean and rivers—and other animals that migrate between seawater and freshwater—provide perhaps the most dramatic examples of animals that face changes in salinity during their lives.

In their relations to changing salinities, animals are often categorized as *stenohaline* or *euryhaline*. **Stenohaline** species are able to survive only over narrow ranges of ambient salinity. **Euryhaline** species, on the other hand, can survive broad ranges of salinity.

Animals are also classified as *osmoconformers* or *osmoregulators* (see Figure 25.3). **Osmoconformers**—sometimes called **poikilosmotic animals**—permit their blood osmotic pressure to match the ambient osmotic pressure. **Osmoregulators**—also called **homeosmotic animals**—maintain a relatively stable blood osmotic pressure over a range of ambient osmotic pressures.

Most species of invertebrates that occur in the ocean are stenohaline osmoconformers; when they are placed in brackish waters, their blood osmotic pressure falls, and because they cannot tolerate blood osmotic pressures much lower than those they have in seawater, they do not prosper. Certain osmoconformers are exceptional, however, in that they are euryhaline. Oysters and mussels provide outstanding examples; despite being osmoconformers, some species thrive over wide ranges of salinity, from seawater itself to waters less than 20% as concentrated as seawater. For osmoconformers to be so euryhaline, their cells must have remarkable abilities to function over wide ranges of blood osmotic pressure. The cells of euryhaline osmoconformers are noted for having dramatic powers of cell-volume regulation.

An intriguing and commercially important illustration of how water–salt physiology can feature in the lives of euryhaline osmoconformers is provided by the story of MSX, a debilitating protistan parasite of the commercial oyster (*Crassostrea virginica*) of the Atlantic seaboard of the United States. These oysters live in estuaries, where the ambient salinity varies from place to place. Because their blood osmotic pressure matches the ambient osmotic pressure, their blood osmotic pressure also varies from place to place. The MSX parasite cannot survive at blood osmotic pressures below about 400 m*Osm*. For the oysters, therefore, ambient waters with osmotic pressures lower than 400 m*Osm* are safe havens from the parasite. In the Chesapeake Bay (see Figure 25.4), serious spread of MSX has occurred during droughts. When there is little rain, rivers bring less freshwater into the bay, and the salinity rises above 400 m*Osm* in places where it is ordinarily lower. The oysters living in such places experience a rise in blood osmotic pressure and become vulnerable to the parasite.

Among animals that are osmoregulators, regulation is often limited to certain ranges of ambient osmotic pressure. Thus different categories of regulators are recognized. In one common pattern, called **hyper-isosmotic regulation**, a species keeps its blood more concentrated than the environmental water at low environmental salinities, but allows its blood osmotic pressure to match the ambient osmotic pressure at higher salinities. Species that are predominantly freshwater animals but venture into brackish waters typically show this pattern, as do many coastal marine invertebrates (Figure 26.10*a*). Animals exhibit hyper-isosmotic regulation when they possess mechanisms of hyperosmotic regulation but lack mechanisms of hyposmotic regulation.

A second major category of regulators consists of those that keep their blood more concentrated than the environmental water at low environmental salinities but more dilute than the environmental water at high environmental salinities. This pattern is called **hyper-hyposmotic regulation** and requires mechanisms of both hyperosmotic and hyposmotic regulation. It is observed in salmon and other migratory fish and in a variety of crustaceans (Figure 26.10*b*).

Among the species of crustaceans that live in the ocean, some are osmoregulators of one sort or another, but others are osmoconformers. Both osmoregulating and osmoconforming species also occur in some other groups of marine invertebrates, such as annelids. In these phylogenetic groups, the most-euryhaline species are typically those that osmoregulate. Thus success in dilute waters in these groups has been achieved by protecting the cells of the body from exposure to low blood osmotic pressures.

The migratory fish are dramatic examples of hyper-hyposmotic regulators

The fish that migrate between freshwaters and the oceans typically breed in one habitat and undergo much of their growth and maturation in the other. Some species—termed **anadromous** ("running upward")—ascend rivers and streams from the oceans to breed; these fish include salmon and certain smelts, shad, and lampreys. Other species—termed **catadromous** ("running downward")—grow in freshwaters and descend to the oceans for breeding; they include the freshwater eels of North America and Europe.

The migratory fish are superb osmoregulators. They function as hyperosmotic regulators when in freshwater and as hyposmotic regulators when in seawater, and they are so effective in both settings that their blood osmotic pressure generally changes only a little between the two habitats. The chinook salmon (*Oncorhynchus tshawytscha*), for example, has a plasma osmotic pressure averaging about 410 m*Osm* when in the ocean and about 360 m*Osm* when at its freshwater spawning grounds.

The mechanisms of regulation employed by migratory fish in seawater and in freshwater are the same as those we earlier discussed for marine and freshwater teleosts (see Figure 26.7). Thus, when the fish move from seawater to freshwater, they reverse the direction of active NaCl transport across their gills (outward transport in seawater, inward in freshwater); they decrease the rate at which they drink, or stop drinking altogether; they increase their rate of urine production; and they switch from producing urine that is approximately isosmotic to their blood plasma to producing urine that is markedly hyposmotic to the plasma. The activity of the NaCl-uptake mechanisms in the intestine also decreases when they enter freshwater. Some crustaceans that exhibit hyper-hyposmotic regulation, such as fiddler crabs (*Uca*), are known to display quite similar and equally dramatic shifts in their regulatory mechanisms as they move between salinities.

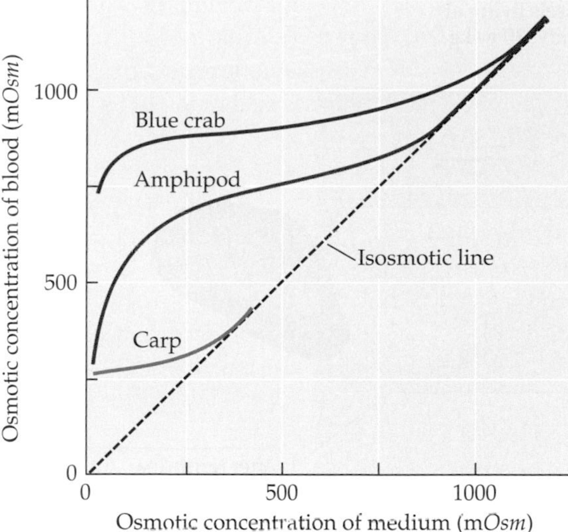

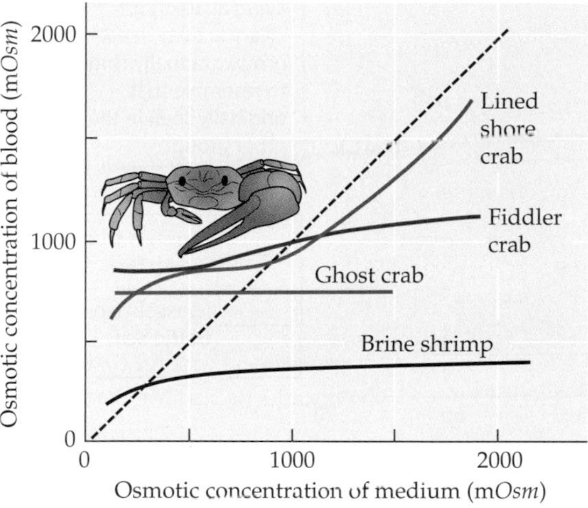

Figure 26.10 Types of osmotic regulation For each animal shown, blood osmotic pressure is plotted as a function of the osmotic pressure of the ambient water. Each dashed line is a line of equality between blood osmotic pressure and ambient osmotic pressure (an isosmotic line). *(a)* Three species of hyper-isosmotic regulators. Such regulation is typical of freshwater animals that enter brackish waters (e.g., the carp); it occurs also in many shore or estuarine crabs (e.g., the blue crab) and in some euryhaline amphipods and annelids. *(b)* Four species of hyper-hyposmotic regulators. Such regulation occurs in many shore crabs (e.g., the fiddler crab shown), semiterrestrial crabs (e.g., the ghost crab), coastal shrimps, and animals adapted to inland saline environments (e.g., the brine shrimp), as well as euryhaline and migratory fish. The species included are: amphipod, *Gammarus oceanicus*; blue crab, *Callinectes sapidus*; brine shrimp, *Artemia salina*; fiddler crab, *Uca pugilator*; ghost crab, *Ocypode cursor*; and lined shore crab, *Pachygrapsus crassipes*. (After Hill and Wyse 1989; D'Orazio and Holliday 1985; Greenaway 1988; Kirschner 1991.)

In the migratory fish, many of the physiological changes that occur when they move between seawater and freshwater are known to be under hormonal control, although the specific controls are not always well understood. In many species, the adenohypophyseal hormone prolactin is vital for existence in freshwater; it reduces the permeability of the gills to Na^+, and augments urine flow by effects on the kidneys and urinary bladder. Aldosterone, cortisol, epinephrine, natriuretic peptides, and hormones of the urophysis and neurohypophysis also have roles in modulating gill or kidney function.

Animals undergo change in all time frames in their relations to ambient salinity

The relations of animals to salinity may change in all the time frames we highlighted in Chapter 1 (see Table 1.2). Up to now, we have emphasized acute responses: the responses that individuals undergo soon after the salinity of their environment becomes altered. Another time frame in which individuals respond to changes in their environment is the chronic time frame (i.e., acclimation or acclimatization). An instructive example is provided by studies of groups of blue mussels (*Mytilus edulis*) collected from the North Sea—where the salinity of the ambient water was 30 g/kg—and from the brackish Baltic Sea—where the salinity was 15 g/kg. Figure 26.11*a* shows the ranges of salinities at which the two groups of mussels were able to maintain ciliary activity (rated on a scale of 0 to 3 on the *y* axis) at the time of collection. The groups were then switched in the salinities at which they lived. As each group acclimated to its new salinity, the range of salinities over which it could maintain ciliary activity gradually shifted. After 30 days (Figure 26.11*e*), the North Sea animals living at a salinity of 15 g/kg displayed normal ciliary activity over approximately the same salinity range as originally seen in the Baltic animals; the Baltic animals living at a salinity of 30 g/kg also acclimated, coming to resemble the original North Sea animals. These results show acclimation in action and suggest that the original difference between the North Sea and Baltic mussels was itself largely a consequence of acclimation.

Evidence also suggests that populations of animals *evolve* differences in their water–salt physiology when living in different environments over multiple generations. An intriguing illustration is provided by populations of lampreys (*Petromyzon marinus*) that have become landlocked in North American freshwater lakes. Lampreys, which are anadromous like salmon, ordinarily migrate to the ocean when they are young adults. The populations that are landlocked in lakes have had no experience with the sea for many generations, however. Adults from some landlocked populations exhibit osmoregulatory difficulties when they are placed in water that is half or more saline as seawater salinity. However, adults from migratory populations—when tested—can osmoregulate at the full salinity of seawater *even before* they have migrated and had actual experience with salty waters. These observations strongly suggest genetic divergence between the landlocked and migratory populations.

Many animals are also known to undergo *developmental changes* in their water–salt relations. Young stages of crustaceans, for example, are often more stenohaline than adults of their species. The eggs of blue crabs provide a specific case. Whereas adults can live in water that is virtually fresh, their eggs require salinities above 23 g/kg to develop. In places like the Chesapeake Bay (see Figure 25.4), adult females that are living in relatively dilute parts of the estuary when they mate, must undertake lengthy migrations to high-salinity waters before they lay their eggs.

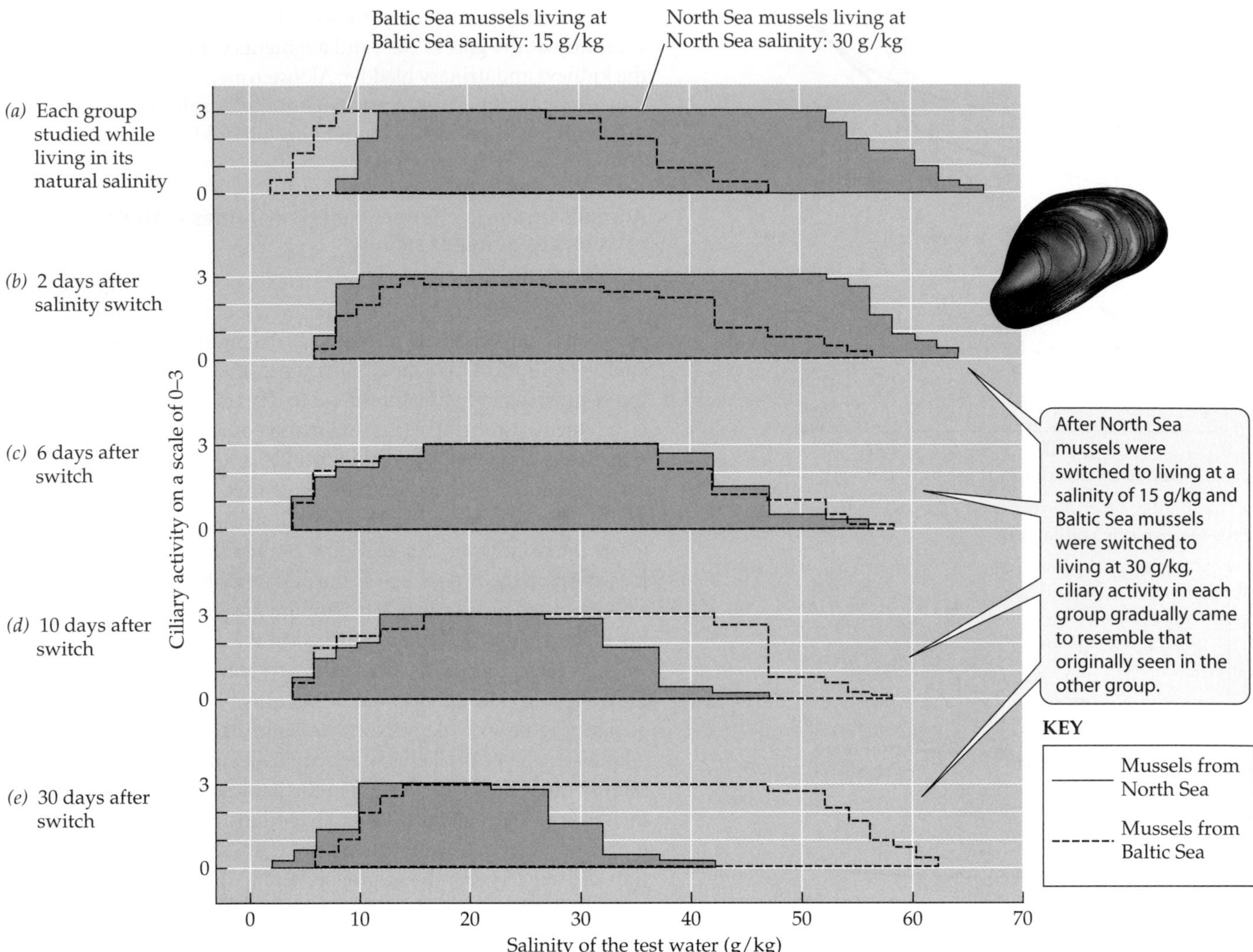

Figure 26.11 Acclimation of mussels to changed salinity Blue mussels (*Mytilus edulis*) were collected from the North Sea, where the ordinary ambient salinity is about 30 g/kg, and from the Baltic Sea, where the ordinary ambient salinity is about 15 g/kg. After initial testing *(a)*, each group was switched to living in water of the opposite salinity and periodically retested *(b–e)*. The aspect of their physiology studied was the activity of their gill cilia—important for pumping water through the body so food and O_2 can be collected. Ciliary activity was scored on a scale of 0 (low) to 3 (high) as a function of salinity (*x* axis). (After Theede 1965.)

Summary: Animals that Face Changes in Salinity

- Some groups of marine invertebrates, such as molluscs, are uniformly osmoconformers. The euryhaline species in these groups are tolerant of wide ranges of blood osmotic pressure.
- Other groups of marine invertebrates, such as crustaceans, include osmoconforming and osmoregulating species. In general, the euryhaline species in these groups are osmoregulators.
- Animals that are hyper-isosmotic regulators have mechanisms for hyperosmotic regulation but not hyposmotic regulation. Hyper-hyposmotic regulators have mechanisms for both types of regulation. Euryhaline fish, such as species that migrate between seawater and freshwater, are excellent hyper-hyposmotic regulators.

Responses to Drying of the Habitat in Aquatic Animals

Residents of puddles, small ponds, intermittent streams, and the like are often confronted with drying of their habitat. The lungfish (dipnoan fish), which have lungs and fleshy fins, are of particular interest in this regard because of their relation to the fish that gave rise to amphibians. All lungfish occur in transient bodies of freshwater. If an African lungfish (*Protopterus aethiopicus*) is confronted with drying of its habitat, it digs a chamber in the bed of the lake or stream where it lives. In the chamber, the fish curls up and secretes mucus that hardens into a cocoon opening only to its mouth. The cocoon acts as a barrier to evaporative water loss. The fish then undergoes metabolic depression (see page 166); its metabolic rate ultimately drops to about 10% of the ordinary resting level. This hypometabolism reduces its rates of respiratory water loss and use of stored energy. The lungfish's kidneys virtually stop making urine. Water is thus saved, but wastes must accumulate in the body. As its nitrogenous end product, the fish switches from producing ammonia to producing urea, a far less toxic compound (see Box 26.3), and urea may accumulate in its blood to levels ap-

BOX 26.4 Life as Nothing More than a Morphological State

Survival while dried as fully as possible by desiccation in air is called **anhydrobiosis**: "remaining alive without water." It occurs in a great variety of small freshwater animals. Some of these animals, such as many nematodes and bdelloid rotifers—and freshwater as well as terrestrial tardigrades ("water bears")—are able to enter anhydrobiosis at any stage of their life cycle. In some other types of animals, anhydrobiosis is possible only during certain life stages, such as the eggs of some water fleas (cladocerans) and other crustaceans, the larvae of some chironomid flies, the embryonic cysts of coelenterates, and the gemmules of sponges. The encysted embryos (gastrulae) of brine shrimp (*Artemia*; often marketed as aquarium food under the strange name "sea monkeys") provide some of the best-studied examples of anhydrobiosis.

Being dried in air does not mean that every molecule of water is gone. Proteins and other materials tend to hold on to adhesively and cohesively bound water, more so when the humidity of the air is high than when it is low. Nevertheless, anhydrobiotic forms are very dry, typically having less than 2% as much water as they have when hydrated. Some require certain minimal air humidities to survive. Some, however, can survive drying to the ultimate possible extent; brine shrimp embryos, for example, can survive drying in a vacuum, which reduces their water content to 0.007 gram per gram of dry weight!

Many anhydrobiotic forms, when dried to their maximal tolerable extent, cease to exhibit any of the usual signs of metabolism; for example, they neither consume O_2 nor accumulate wastes. Thus, as stated by one authority, they become "nothing more than a morphological state." They are described as *ametabolic*. More specifically, they are **reversibly ametabolic** because, although their metabolism stops, it can be restarted. These ametabolic forms raise interesting questions about how animal life is to be defined, inasmuch as metabolism is often included in the list of properties that distinguish living systems from nonliving ones.

The disaccharide trehalose often accumulates in animal forms entering a state of anhydrobiosis. In ways that remain debated, the trehalose prevents the structures of macromolecules, cell membranes, and intracellular membranes from being permanently destabilized by the loss of water.

Many anhydrobiotic stages are tolerant of a variety of environmental extremes, not just dryness, and can endure their dried state for many years. They frequently blow about in the wind, springing magically back to an animated life when they land in water.

proaching those of marine elasmobranch fish. The lungfish can survive in this dormant condition for more than a year!

Many other freshwater animals burrow into the substrate—often encasing themselves in mucoid coverings—and enter a resting condition during times of drought. Included are some leeches, snails, water mites, and amphibians. Some small animals from freshwater, saline, or terrestrial habitats are tolerant of being fully dried in air. They represent the extreme in animal desiccation tolerance, as described in Box 26.4.

Animals on Land: Fundamental Physiological Principles

Animal life originated and spent much of its early evolutionary history in water. The land and its plant life represented a vast ecological resource within which animals would ultimately establish themselves in an immense variety of ecological niches. The earliest animals that ventured to spend time on land, to consume the productivity of the land, and, ultimately, to develop on land were able to escape competitors and predators in their primordial aquatic habitat. For this reason, positive selective pressure for terrestriality must have been great. But early animal life was adapted to living in an abundance of water, and evaporative losses of water on land constituted a physiological problem of paramount importance for all stages of the life cycle. In our discussion of animals on land, we will focus on water. Although terrestrial animals sometimes face problems of salt balance, water balance usually presents their most pressing challenge.

The distinction between *humidic* and *xeric* animals provides a useful organizing principle for the study of water relations in terrestrial animals. The **humidic animals** are those that, while living on land, are restricted to humid, water-rich microenvironments. The **xeric animals** are those that are capable of living in dry, water-poor places.[12]

The humidic animals include earthworms, slugs, centipedes, most amphibians, and most terrestrial crabs. Some live underground. Others live in leaf litter or under logs or rocks. The majority of frogs and toads stay in or near bodies of water, and when they venture away from water, they remain in protected microenvironments, such as the tall grass frequented by leopard frogs. Some humidic animals, such as most amphibians and all terrestrial crabs, still resemble their aquatic progenitors in that they require standing water to breed.

The major groups of xeric animals are the mammals, birds, reptiles, insects, and arachnids (e.g., spiders and ticks). Although xeric animals often seek protected, humid microenvironments, they are not stringently tied to such environments, as humidic animals are. The xeric animals can live successfully in the open air, and many of them routinely expose themselves to the full drying power of the terrestrial environment. Some thrive in deserts and other equally dry environments, such as grain stores.

In many ways, the physiological difference between humidic and xeric animals is a distinction in how *rapidly* they get into trouble by loss of water in desiccating environments. The humidic animals dehydrate rapidly in dry environments. Thus they cannot remain long in such places, as xeric animals can.

[12] The term *xeric* has a standardized meaning and is widely used. However, there is no standardized term to describe the animals restricted to moist habitats; although we use *humidic*, alternative terms are used in other books and articles. The term *mesic* is sometimes applied to animals intermediate between those that are xeric and those that are humidic.

A low integumentary permeability to water is a key to reducing evaporative water loss on land

Evaporation is one of the chief modes by which terrestrial animals lose water. In this section we begin our analysis of the physiological principles of living on land by discussing evaporation across the integument of the body. In the following section we will turn to respiratory evaporation. As discussed in Chapter 25 (see Equation 25.1), the rate of evaporation through an animal's integument depends on the difference in water vapor pressure between the animal's body fluids and the air, and it also depends on the permeability of the integument to water (K in Equation 25.1).

A high integumentary permeability to water is one of the specific characteristics that is most important in restricting humidic animals to their protective microhabitats. The skin of an earthworm, the skin of most amphibians, and most of the fleshy surfaces of a snail or slug, for example, have high permeabilities and provide little barrier to water loss: These animals often lose water *through their integuments* at rates that are 50% to 100% as great as rates of evaporation from *open dishes of water* of equivalent surface area! With such a high integumentary permeability, a humidic animal can restrict its integumentary rate of evaporation only by limiting the difference in water vapor pressure that exists across its integument. From the viewpoint of physics, this explains why humidic animals are tied to humid habitats, where the air has a high water vapor pressure.

The xeric animals have integuments with a low permeability to water. Indeed, the evolution of a low integumentary permeability to water is one of the single most important steps toward a xeric existence. For a long time, the physical basis for the low integumentary permeability of the vertebrate xeric groups was not known. Research has started to resolve the issue, however, so that now it seems correct to say that in all the major xeric groups—vertebrate and invertebrate alike—*very thin layers of lipids* are responsible for low integumentary permeability. In mammals, and probably in birds and reptiles, these lipids—such as glycolipids—are laid down in the spaces between cells in the thin, superficial horny layer of the skin. In insects and arachnids, the lipids—such as long-chain hydrocarbons and wax esters—are contained in the outermost layers of the exoskeleton (cuticle); these layers, termed the *epicuticle*, are only 1–2 μm thick. Interestingly, because most of the resistance to water loss across an insect's integument resides in its thin, lipoid epicuticle and not in the thicker, chitinous part of its exoskeleton, soft-bodied insects can have permeabilities as low as hard-bodied ones. The importance of the epicuticular lipids is emphasized by the fact that arthropods such as many millipedes and centipedes—which either lack epicuticular lipids or possess lipids of different types than insects—often are quite permeable to water despite the superficial similarity of their exoskeletons to those of insects.

A great diversity of lipids is found in the epicuticles of insects. The lipid composition differs among species. It may also differ among populations within a species and give rise to significant differences between populations in their physiology of water balance (Figure 26.12).

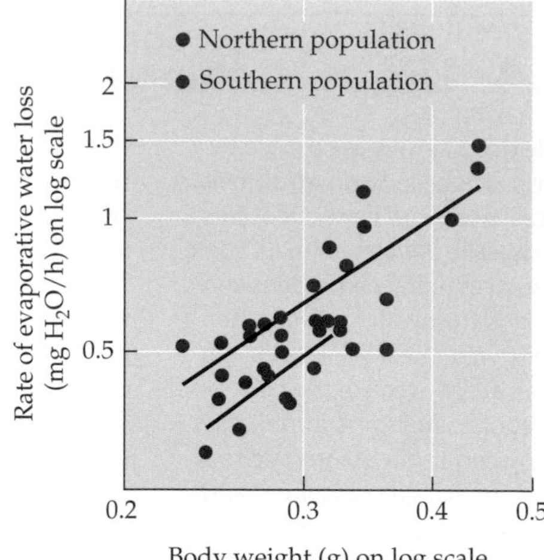

Figure 26.12 Differentiation between populations of one species in protection against evaporative water loss The graph shows the rate of evaporative water loss at 25°C as a function of body weight in live grasshoppers of a single species (*Melanoplus sanguinipes*) from two geographically separate populations in northern and southern California. According to current evidence, these populations exhibit genetically controlled differences in the lipid composition of their epicuticular layer. These differences contribute to the greater resistance of the southern animals to water loss. In each population, the rate of water loss is an allometric function of body weight, so the relation plots as a straight line on log–log coordinates (see Appendix G: Exponential and Allometric Equations). (After Gibbs 1998.)

The permeability of the integument to water in insects and arachnids depends on the cuticular temperature as well as the epicuticular lipid composition. In many cases, if the temperature is gradually raised, the permeability increases just slightly up to a certain temperature—termed the **transition temperature**—and then increases dramatically (Figure 26.13). The marked increase in permeability at the transition temperature is a consequence of lipid melting. Often the transition temperature is so high that it would not be experienced by insects in nature. This is not always the case, however. The common cockroach *Periplaneta americana*, for example, experiences a marked increase in its permeability starting at 25°C to 30°C; thus it might naturally encounter temperatures high enough to degrade its protection against evaporative water loss significantly.

Within phylogenetic groups of xeric animals, if integumentary permeability is measured in a variety of species under similar conditions, species from relatively arid habitats are often found to have relatively low permeabilities. An illustration is provided by the two iguanid lizards, the chuckwalla and the common iguana, included in Table 26.6. The chuckwalla, which occurs in deserts, loses water through each square centimeter of its skin at a rate that is only about 30% as great as that of the iguana, which lives in moister environments.

Respiratory evaporative water loss depends on the function of the breathing organs and the rate of metabolism

Some groups of humidic animals have respiratory surfaces that are directly exposed to the air. Earthworms, some isopods, and amphibians, for instance, breathe substantially or entirely across their general integuments. This arrangement is a disadvantage from the viewpoint of evaporative water loss because movement of air across exposed respiratory surfaces can greatly exceed that necessary for ex-

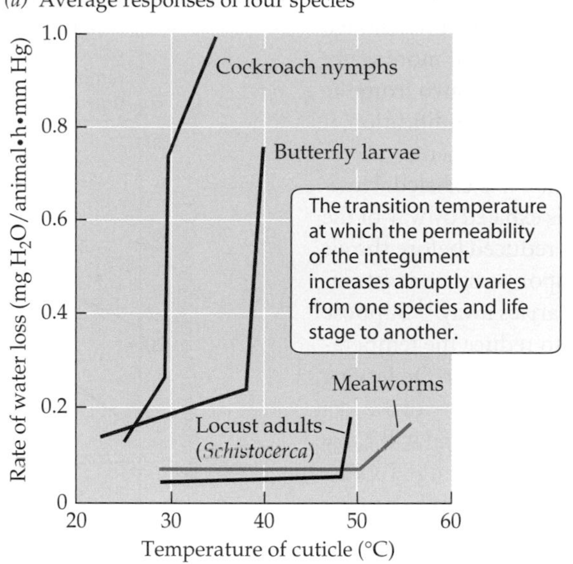

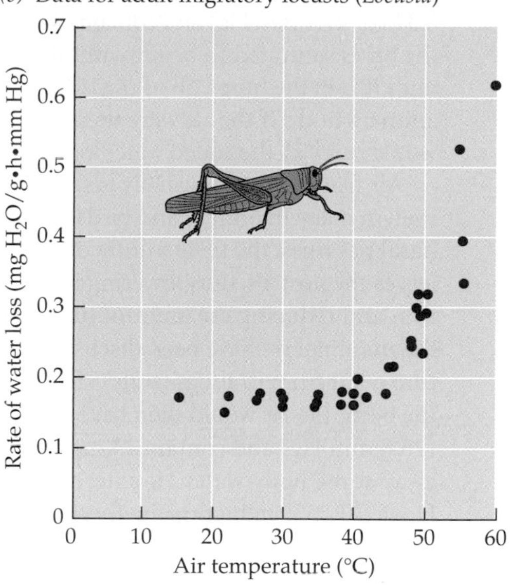

Figure 26.13 The rate of evaporative water loss of insects often increases abruptly at a transition temperature *(a)* Average relations between rate of evaporative water loss and temperature in four insect species: cockroach nymphs, *Periplaneta americana*; butterfly larvae, *Pieris brassicae*; locusts, *Schistocerca gregaria*; and mealworms, larval *Tenebrio molitor*. *(b)* Actual data for the African migratory locust (*Locusta migratoria*). The insects used in all these studies were dead. The rates of evaporative water loss are expressed "per mm Hg," referring to the difference (expressed in millimeters of mercury) between the actual water vapor pressure of the air during the measurements and the saturation vapor pressure at the various temperatures; measured and expressed in this way, changes in the rate of evaporative water loss within a species reflect changes in the permeability of the integument. (*a* from Hill and Wyse 1989; *b* after Loveridge 1968.)

change of O_2 and CO_2, so water loss can be much greater than the minimum required for respiratory gas exchange.

Most terrestrial animals have evolved invaginated respiratory structures (see Figure 1.17). In the xeric groups, breathing is carried out exclusively by such invaginated structures, and the general integument is virtually impermeable to O_2 and CO_2. The enormous advantage of this arrangement is that access of air to the thin, moist respiratory membranes can be closely controlled and thus limited to the levels required for exchange of O_2 and CO_2. The mammals, birds, and reptiles control access of air to their lungs by regulating their breathing movements. Insects close and open the spiracles of their tracheal system (see Figure 21.28).

WATER CONSERVATION BY COOLING OF EXHALANT AIR A physical law of great consequence for warm-bodied air breathers is that when air is saturated with water vapor, its content of water per unit of volume depends strongly on its temperature, approximately doubling with every 11°C increase in temperature (see Table 25.2). When a mammal or bird inhales air into its lungs, the temperature of the air is raised to the deep-body temperature of the animal, and the air becomes saturated with water vapor at the elevated temperature. Depending on conditions, a substantial amount of water may be added to the air. For example, consider a mammal inhaling *saturated* air at 20°C. Such air has a water vapor pressure of 2.3 kPa (17.5 mm Hg) and contains about 17 mg H_2O/L. By the time the air reaches the

TABLE 26.6 Rates of cutaneous and respiratory evaporative water loss in four reptile species The species are listed, from top to bottom, in order of decreasing aridity of their habitats, the chuckwalla being desert-dwelling and the slider turtle being amphibious. Note that resistance to evaporative water loss decreases in the same order. Based on their appearances, the integuments of xeric animals sometimes give the impression that they are so impervious that the animals' integumentary water losses would be trivial. The data shown here, however, demonstrate that in reptiles, the cutaneous losses often constitute well over half of total resting evaporative water losses. When studied, the animals were resting at 23°C in dry air.

Reptile species	Cutaneous water loss per unit of body surface area (mg $H_2O/cm^2 \cdot$day)	Respiratory water loss per unit of O_2 consumed (mg H_2O/mL O_2)	Total rate of evaporative water loss (% body weight/day)	Cutaneous loss as percentage of total
Chuckwalla lizard (*Sauromalus obesus*)	1.3	0.5	0.3	66
Iguana (*Iguana iguana*)	4.8	0.9	0.8	72
Eastern box turtle (*Terrapene carolina*)	5.3	4.2	0.9	76
Slider turtle (*Trachemys scripta*)	12.2	4.2	2.0	78

Source: After Bentley and Schmidt-Nielsen 1966.

lungs, it is saturated at 37°C; it therefore has a water vapor pressure of 6.3 kPa (47.1 mm Hg) and contains 44 mg H_2O/L. Thus, even though the air is saturated to begin with, it contains 27 mg/L more water once it is in the lungs, all of this added water being drawn from the animal's body. If the air were then exhaled without modification, it would carry all the added water away into the environment.

Air from the lungs usually is saturated when it is exhaled. However, in many mammals and birds, when air is exhaled by way of the nasal passages, the *temperature* of the air is reduced before the air leaves the nostrils, thus lowering its water vapor pressure at saturation and reducing the amount of water it carries away.[13] Suppose the mammal we have been discussing were to reduce the temperature of air from its lungs to 25°C before exhaling the air. On leaving the body, the air would then have a water vapor pressure of 3.2 kPa (23.8 mm Hg) and contain 23 mg H_2O/L. The air would still carry away some body water (it entered the body with 17 mg H_2O/L). However, the reduction in the temperature of the exhalant air would cause 78% of the water added during inhalation to be recovered before exhalation.

In those mammals and birds that reduce the temperature of air before it is exhaled, the air is cooled by a countercurrent mechanism in the nasal passages. To understand the process, let's first look in more detail at what happens during inhalation, using our example of a mammal breathing 20°C air. As inhaled ambient air travels up the nasal passages, it is progressively warmed to about 37°C, and it takes up water vapor as its temperature is elevated. The heat that warms the air and the latent heat of vaporization for the added water vapor are drawn from the walls of the nasal passages.[14] Thus the walls of the nasal passages are cooled by the process of inhalation. The outer ends of the nasal passages are typically cooled most and the inner ends, least. During the ensuing exhalation, air coming up from the lungs arrives at the interior ends of the nasal passages at a temperature of 37°C and saturated. However, as the air moves down through the nasal passages toward the nostrils, it encounters the increasingly cooler surfaces created by the previous inhalation. Thus the air being exhaled is cooled as it travels toward the nostrils. This cooling lowers the saturation water vapor pressure of the air (Table 25.2), causing water to condense out of the air onto the nasal-passage walls. The overall process is considered a *countercurrent* process because it depends on flow of air in opposite directions.

If the cooling of nasal exhalant air seems unfamiliar, it may be because the cooling that occurs in humans is quite modest. In contrast, the cooling of nasal exhalant air in small mammals is dramatic, as illustrated by the data plotted as black dots in Figure 26.14; a small mammal that inhales air at 20°C and warms it to 37°C in its lungs might well exhale the air at 22°C or 23°C. Even in medium-sized and large mammals, nasal cooling can be substantial in species with long snouts (and thus long nasal passages), such as dogs.

AN ALGEBRAIC EXPRESSION FOR RESPIRATORY EVAPORATIVE WATER LOSS A useful way to think about the rate of respiratory evaporative water loss is to recognize that it depends on an animal's rate of O_2 consumption and the amount of water lost per unit of O_2 consumed:

$$\text{Rate of water loss (mg H}_2\text{O/hour)} = \text{rate of O}_2 \text{ consumption (mL O}_2\text{/hour)} \times \text{water loss per unit of O}_2 \text{ consumed (mg H}_2\text{O/mL O}_2) \quad (26.1)$$

This equation emphasizes that an animal's rate of metabolism is a major determinant of its rate of evaporative water loss. Mammals and birds, as we stressed in Chapters 5 and 8 (see Figure 5.9), typically have rates of O_2 consumption that are far higher than those of reptiles of similar body size. Among terrestrial vertebrates, therefore, the rate of respiratory evaporative water loss tends to be higher in mammals and birds than in reptiles.[15]

The amount of water lost per unit of O_2 consumed is affected by several factors. One is the efficiency of the breathing organs in removing O_2 from inhaled air. Marine mammals, for instance, are noted for having a relatively high efficiency; they breathe relatively slowly and deeply, extracting a comparatively large fraction of the O_2 from the air they breathe, an effect that reduces their rates of evaporative water

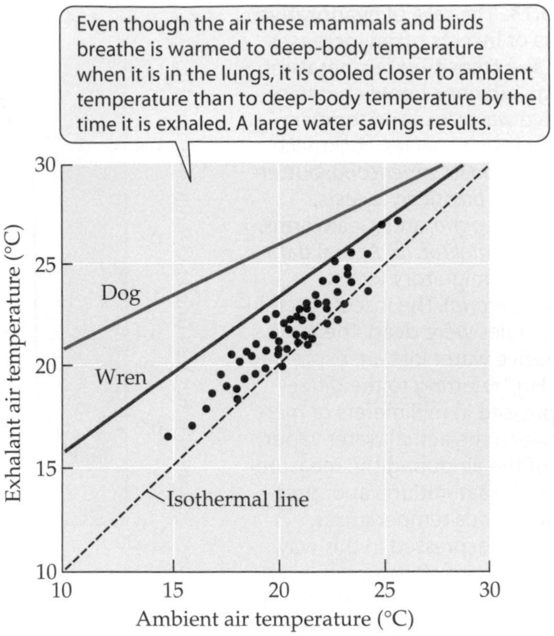

Even though the air these mammals and birds breathe is warmed to deep-body temperature when it is in the lungs, it is cooled closer to ambient temperature than to deep-body temperature by the time it is exhaled. A large water savings results.

Exhalant air temperature (°C)

Ambient air temperature (°C)

Dog

Wren

Isothermal line

Figure 26.14 The temperature of air exhaled from the nostrils as a function of ambient air temperature in mammals and birds The black dots are individual data points gathered on animals representing 18 species of small mammals, including shrews, bats, mice, squirrels, and rabbits. The solid lines are for mongrel domestic dogs (blue) and cactus wrens (*Campylorhynchus brunneicapillum*) (red). The dashed line is a line of equality between exhalant air temperature and ambient air temperature (an isothermal line). The data for dogs and wrens were gathered at low humidities; those for the small mammals were gathered at unspecified humidities. (From Hill and Wyse 1989.)

[13] Cooling of nasal exhalant air also takes place in some lizards when they are maintaining high, behaviorally regulated body temperatures.

[14] The nasal passages are not simple tubes. Their walls are often thrown into elaborate surface-enhancing structures (*turbinates*) over which air flows.

[15] A compensation, which will be mentioned again later, is that animals with relatively high rates of O_2 consumption also have relatively high rates of metabolic water production.

loss. Cooling of nasal exhalant air is also, in essence, a mechanism of reducing the amount of water lost per unit of O_2 consumed.

The total rate of evaporative water loss depends on an animal's body size and phylogenetic group

An animal's total rate of evaporative water loss (EWL) is the sum of its rates of integumentary and respiratory water loss. The broad statistical trend within sets of phylogenetically related species is for the total rate of EWL measured under particular conditions to vary allometrically with body size. This important pattern is exemplified using data for birds at moderate temperatures in Figure 26.15. Small-bodied species tend to have higher weight-specific rates of EWL than large-bodied ones. A qualitatively similar pattern been found in virtually every phylogenetic group examined.

There are two reasons why small-bodied species tend to have relatively high weight-specific rates of EWL. First, small animals tend to have relatively high weight-specific rates of water loss *through the integument* because they have a high surface-to-volume ratio (see Equation 5.6) and thus a greater integumentary surface area per unit of weight than related large animals. Second, small animals also tend to have relatively high weight-specific rates of *respiratory* water loss because they have higher metabolic rates per unit of body weight than related large animals (see Equation 26.1; see also Figure 5.9).

As we have stressed, there are also consistent differences *between phylogenetic groups* in their total rates of EWL. Animals in humidic phylogenetic groups, such as amphibians, have highly permeable integuments and other properties that give them high total rates of EWL in comparison to similar-sized animals in xeric groups. Some of these differences in total rates of EWL are brought to light in Figure 26.16. All the vertebrates in the figure are roughly similar in body size; thus, comparing them brings out differences between the phylogenetic groups. Reptiles have very low total rates of EWL compared to amphibians because the reptiles have low-permeability integuments and enclosed breathing systems. Mammals and birds share these properties of reptiles, but have higher total rates of EWL because they have higher metabolic rates than reptiles. The two species of semiterrestrial crabs in the figure are similar in size to the vertebrates; semiterrestrial crabs, as can be seen, are similar to amphibians in their rates of EWL and are humidic. The isopods and insects in the figure are tiny compared to the other animals shown. The isopods combine the disadvantages of small size and poor defenses against EWL. Their total weight-specific rates of EWL are staggering. One can only marvel at the insects, especially the tsetse flies (weighing 0.03 g), which have such excellent defenses against water loss that they have relatively low total rates of EWL per unit of weight, despite being very small.

Excretory water loss depends on the concentrating ability of the excretory organs and the amount of solute that needs to be excreted

Besides evaporative water loss, excretion in urine is another major way that terrestrial animals lose water. Like aquatic animals, terrestrial ones modulate the concentration, composition, and volume of their urine to serve changing requirements for osmotic, ionic, and volume regulation. We humans, for example, when dehydrated, produce a limited volume of urine that is hyperosmotic to our blood (our osmotic U/P ratio can reach 4), but after we have consumed large amounts of water, we void a copious urine that is hyposmotic to our blood (our osmotic U/P ratio can be as low as 0.1–0.2). While recognizing the fundamental *regulatory* role of urine excretion, a key question in the study of water balance in terrestrial animals is how effectively the animals can *minimize* their urinary water losses. There are two basic ways to reduce the amount of water lost in urine. One is to concentrate the urine, thereby decreasing the amount of water

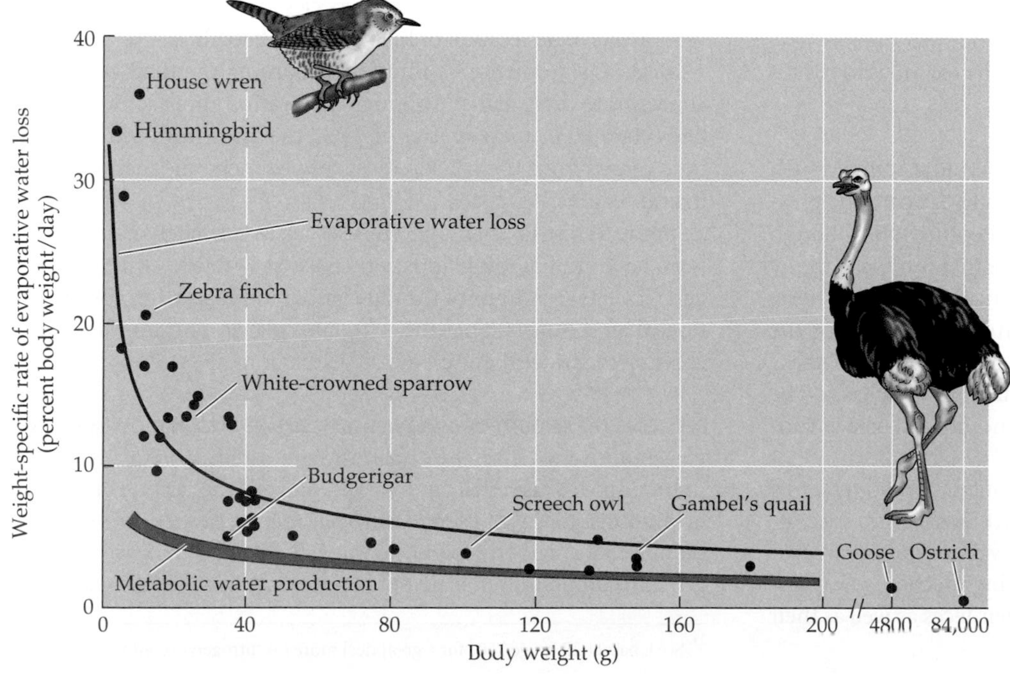

Figure 26.15 Within a phylogenetic group, the total rate of evaporative water loss is an allometric function of body size This trend is illustrated here using data for birds resting at 23°C to 25°C in relatively dry air. Evaporative water loss is expressed in weight-specific units (percent of body weight per day = mg H_2O lost per 100 mg body weight per day). The solid line is a regression line fitted to the data points. The graph also shows the gross rate of metabolic water production (as a percentage of body weight per day) to be expected on the basis of standard relations between resting metabolic rate and weight (pink area); there is a range of values for metabolic water production at each weight because production depends on the type of food oxidized. (From Hill and Wyse 1989.)

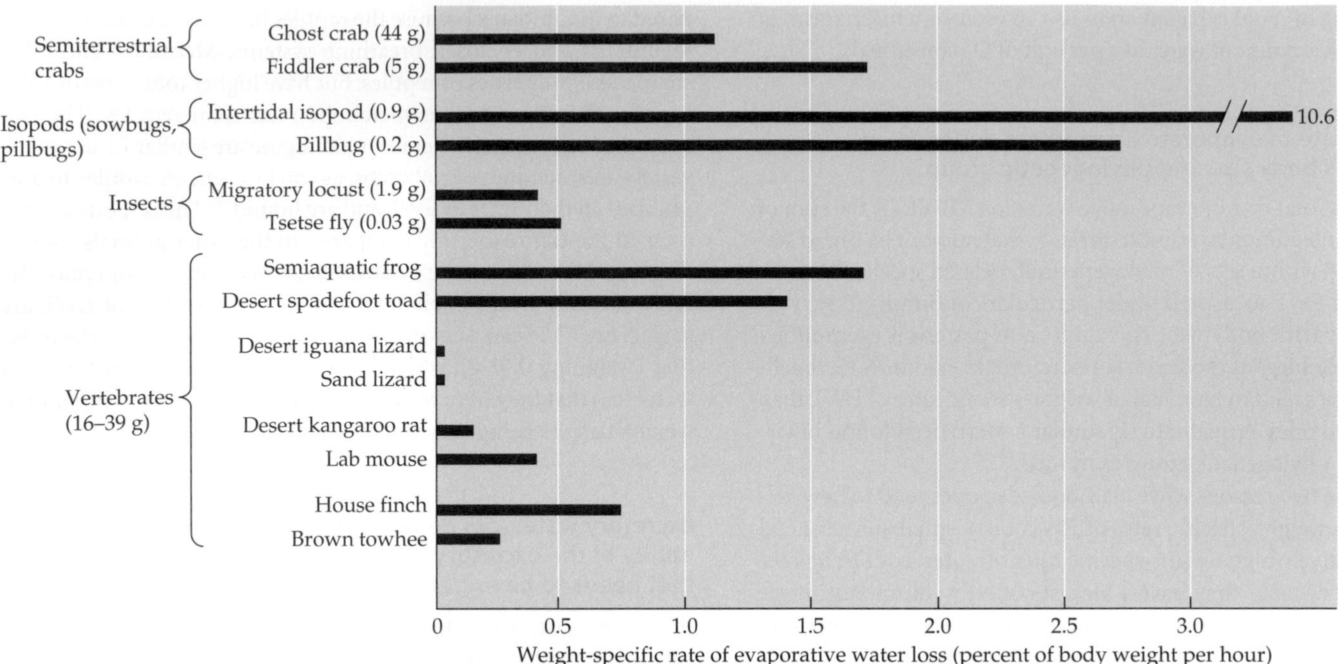

Figure 26.16 The total rate of evaporative water loss varies greatly among different types of vertebrates and arthropods All the animals shown were studied in dry or relatively dry air between 25°C and 32°C. The vertebrates are all of about the same body size (16–39 g). Body weights are given for the individual arthropod species. Evaporative water loss is expressed in weight-specific units (percent of body weight per hour = mg H_2O lost per 100 mg body weight per hour). Species, listed from top to bottom: semiterrestrial crabs, *Ocypode quadrata* and *Uca annulipes*; isopods, *Ligia oceanica* and *Armadillidium vulgare*; insects, *Locusta migratoria* and *Glossina morsitans*; amphibians, *Rana temporaria* and *Scaphiopus couchii*; lizards, *Dipsosaurus dorsalis* and *Uma notata*; mammals, *Dipodomys merriami* and *Mus domesticus*; birds, *Carpodacus mexicanus* and *Pipilo fuscus*. (After Hill and Wyse 1989.)

required to excrete a given amount of solute. The second is to reduce the amount of solute excreted in the urine.

URINARY CONCENTRATING ABILITY Most of the humidic animals, such earthworms and amphibians, are unable to raise the osmotic pressure of their urine above that of their blood. Reptiles too, although xeric, are generally incapable of making urine that is hyperosmotic to their blood plasma. By contrast, certain of the major xeric groups—notably insects, mammals, and birds—have evolved the ability to make hyperosmotic urine. Probably all three of these groups evolved the ability independently—a fact that emphasizes the selective advantages of being able to make concentrated urine in animals confronted with desiccation stresses.

Among terrestrial insects, maximum osmotic U/P ratios of 2 to 4 have been observed in certain blowflies, desert locusts, and stick insects. Mealworms (*Tenebrio*), which live in dry grain stores, can produce urine with a U/P of 8. The concentrating mechanisms of insects reside in the rectum and thus act on their feces as well as their urine (see Chapter 27).

In mammals, an enormous range of urinary concentrating abilities is found. The U/P ratios achieved by many species are well above those seen in other groups of animals. To illustrate both of these points, the maximum reported osmotic U/P ratio is about 3 for muskrats, 4 for humans, 8 for dromedary camels, 9 for laboratory rats, 14 for Merriam's kangaroo rats and Mongolian gerbils (*Meriones*), and—at the highest extreme—about 26 for certain species of Australian desert hopping mice (e.g., *Notomys alexis*)! A significant and unexpected feature of mammalian kidney function is that the maximum concentrating ability tends to decrease with body size. In Figure 26.17, which shows this relation, note that there is a lot of scatter at any given body size. Much of this scatter correlates with habitat; the highest values at a given body size tend to represent species that confront relatively severe threats of desiccation or high dietary salt loads.

Birds as a group are far inferior to mammals in their abilities to concentrate their urine. Although U/P ratios approaching 6 have been claimed in a few species of birds, the maximum osmotic U/P ratio ranges from 1.5 to 2.5 in most species, including many that live in arid regions.

Some terrestrial lizards and birds, such as ostriches and roadrunners, have cranial salt glands, which assist with secreting Na^+, K^+, and Cl^-. The secretions of the salt glands have higher ion concentrations than the urine. Thus the salt glands play important roles in decreasing the water required to void excess ions.

REDUCING THE AMOUNT OF DISSOLVED MATTER EXCRETED IN THE URINE Nitrogenous wastes from the catabolism of proteins are usually voided in the urine by terrestrial animals.[16] Many of the xeric groups have evolved the ability to excrete nitrogen mostly in the form of compounds that are poorly soluble in water. By incorporating waste nitrogen into compounds that are not in solution, these animals reduce the

[16] See Chapter 27 (page 750) for a great deal more on nitrogen excretion.

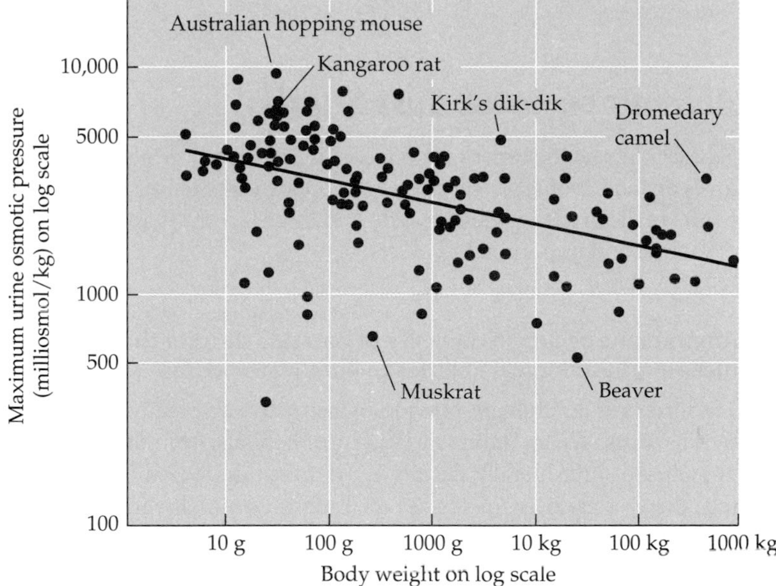

Figure 26.17 The maximum concentration of urine mammals can produce is in part a function of their body size Data for 146 species of mammals are included. The species chosen for labeling were deliberately selected to represent extremes. Because the overall relation is allometric, it plots as a straight line on log–log coordinates (see Appendix G: Exponential and Allometric Equations). Because all mammals have approximately the same plasma osmotic pressure, a plot of the maximum osmotic U/P ratio as a function of body size would resemble this plot. (After Beuchat 1990).

amount of solute in their urine and thus reduce urinary water loss. It is a testimony to the advantages of this type of nitrogen excretion that it has evolved independently many times: Insects, arachnids, some terrestrial snails, a few xeric frogs, reptiles, and birds all produce poorly soluble nitrogenous wastes. The principal poorly soluble compounds employed are uric acid, urate salts, allantoin, and guanine. Urate salts such as sodium and potassium urate—excreted by birds, reptiles, and some other groups—have the advantage that they carry away not only nitrogen, but also inorganic cations in precipitated form.

By no means do all terrestrial animals exclude waste nitrogen from being excreted in solution, however. Many of the humidic animals—including earthworms, isopods (e.g., pillbugs), some terrestrial molluscs, and most amphibians—excrete nitrogen principally as urea or ammonia, both of which are highly soluble. Mammals, paradoxically, also do; they produce mostly urea. If highly soluble nitrogenous wastes are voided in the urine, they demand water for their excretion. Some of the groups that produce such highly soluble compounds have evolved means of reducing the water demands of nitrogen excretion. Some isopods and snails void ammonia as a gas, for example, and mammals possess world-record abilities to concentrate urea in their urine (see Box 24.4 on page 753).

Terrestrial animals sometimes enter dormancy or tolerate wide departures from homeostasis to cope with water stress

Many terrestrial animals, both humidic and xeric, enter dormancy—often called **estivation**—in response to immediate or predictable water stress. Mammals and birds that enter such dormancy—including desert ground squirrels and poorwills—often become hypothermic (see page 228). Metabolism is depressed during dormancy in poikilotherms as well as in mammals and birds.[17] Metabolic depression has several advantages: First, an animal in metabolic depression can live on body fat or other stored foods for a long time, and thus it can remain continuously in a protective microenvironment. Second, the animal's requirement for O_2 is reduced, thereby reducing its respiratory evaporation. Finally, nitrogenous and other wastes are produced at a low rate, which may be vital in permitting protracted existence with little or no urine output.

Besides dormancy, another common "change of status" that animals—particularly poikilotherms—commonly undergo during water stress is to allow large changes in their body-fluid composition while continuing to be active. Species that live in places where they are prone to dehydration often are especially able to tolerate large losses of water from their body fluids and profound concentration of solutes in their body fluids. When Claude Bernard spoke of the *internal environment* in his seminal studies that led to the concept of homeostasis, he referred to the body fluids. The ability to remain active and functional despite profound alteration of the body fluids has been termed **anhomeostasis** and can be a key to existence during water stress.

The total rates of water turnover of free living terrestrial animals follow allometric patterns

A logical conclusion for our introductory discussion of animals on land is to focus on the total rates of *water turnover* (water lost and gained per day) of such animals when they are living freely in their natural habitats. The rate of water turnover can be measured by use of isotopically labeled water (e.g., heavy water). An animal living in the wild is trapped, injected with a known amount of labeled water, and turned loose. Days later, it is recaptured, and from the amount of labeled water remaining in its body, its rate of water turnover while free can be calculated. For an animal in water balance, losses of water by evaporation, urination, and other processes each day are replaced by gains. The rate of water turnover measures the rate of these mutually balancing processes. A high rate of turnover means that an animal loses and replaces a lot of water per day. Such an animal is particularly vulnerable to dehydration if it faces problems of finding replacement water.

Figure 26.18 shows general water-turnover patterns in the vertebrates. As might be guessed from what we have said before in this chapter, the total rate of water turnover (T) of animals in any particular phylogenetic group tends to be an allometric function of body weight (W):

$$T = aW^b \tag{26.2}$$

(see Appendix G: Exponential and Allometric Equations). The exponent b of this allometric function is typically 0.6 to 0.8. This means that if we consider the relation between the *weight-specific* rate of water turnover (T/W) and weight, the exponent ($b - 1$) is negative: –0.2 to –0.4. The rate of water turnover *per gram of body weight* thus decreases as animals get bigger; big species within any given phylogenetic group tend to turn over a *smaller fraction* of their total water per day than little ones. As Figure 26.18 shows, mammals and birds of any given body size turn over much more water per day than reptiles of the same size.

[17] Anhydrobiosis, involving *suspension* of metabolism, is discussed in Box 26.4.

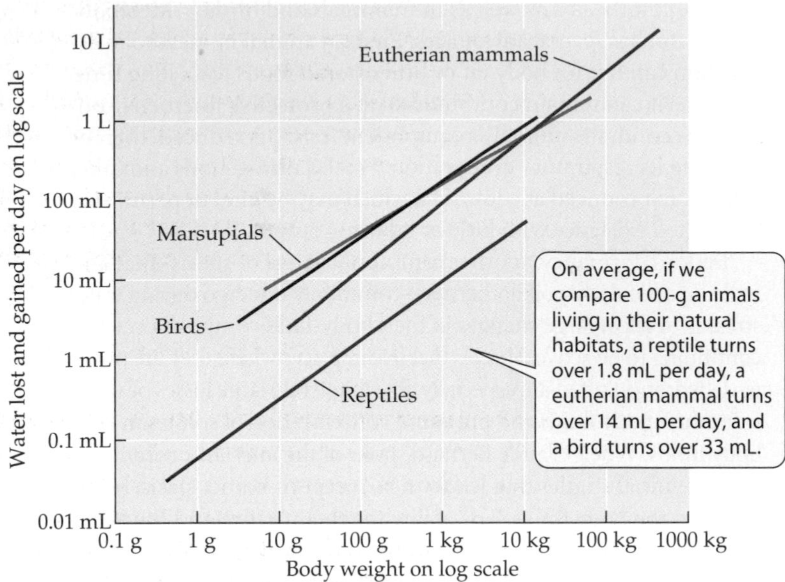

Figure 26.18 Water-turnover rates of free-living terrestrial vertebrates as a function of body size The water turnover is the amount of water lost and gained per day when animals are in water balance. Each line is a plot of Equation 26.2 for the animal group specified. Because the relation for each group is allometric, it plots as a straight line on log–log coordinates. (After Nagy and Peterson 1988.)

Summary: Animals on Land: Fundamental Physiological Principles

- Humidic terrestrial animals are restricted to humid, water-rich microenvironments. Xeric terrestrial animals are those that are capable of a fully exposed existence in the open air.
- A low integumentary permeability to water—which reduces integumentary evaporative water loss—is required for animals to be xeric. All the major xeric groups—insects, arachnids, reptiles, birds, and mammals—have low permeability because of integumentary lipids.
- Respiratory evaporative water loss depends directly on an animal's rate of O_2 consumption (its metabolic rate) and the amount of H_2O lost per unit of O_2 consumed. One way to reduce the latter in mammals and birds is countercurrent cooling of nasal exhalant air.
- The animals with the lowest total rates of evaporative water loss (EWL) are those, like lizards, that combine the advantages of low integumentary permeability to water, tightly controlled access of air to breathing organs, and low metabolic rates.
- Water loss in urine can be reduced by producing concentrated urine (which reduces the amount of water needed to void soluble wastes) or by producing poorly soluble nitrogenous end products such as uric acid (which remove waste nitrogen from solution). Only three groups of animals can make urine hyperosmotic to their blood: insects, birds, and mammals.
- Within groups of related species, water dynamism tends to vary allometrically with body size. Weight-specific EWL and weight-specific total water turnover tend to decrease as size increases.

Animals on Land: Case Studies

Now let's look at the water relations of some particular groups of animals on land. Doing so will provide an opportunity to integrate the points made in the previous section and discuss some new features of interest and importance.

Amphibians occupy a diversity of habitats despite their meager physiological abilities to limit water losses

The terrestrial amphibians provide an instructive case study because they have invaded an impressive variety of habitats, from the shores of ponds to, quite literally, deserts. Yet, despite this diversity of habitats, the vast majority of species are humidic animals that are remarkably similar to one another in their physiological water-balance characteristics. Their diversification into a wide range of habitats has depended to a substantial extent on the evolution of *protective behaviors* and *advantageous patterns of seasonality*.

Most species of amphibians, including most that live in deserts, share several attributes that significantly circumscribe their physiological abilities to limit water losses. First and foremost, they have an integument that poses little barrier to evaporative water loss. They incorporate waste nitrogen mostly into urea, a highly soluble compound requiring considerable amounts of water for its excretion. Moreover, although amphibians are notably adept at simply shutting off urine outflow when faced with dehydration, they are unable, when they do excrete urine, to produce a urine any more concentrated in total solutes than their blood plasma.

Amphibians have the same basic sources of water as most other animals: preformed water in food, preformed water taken in as "drink," and metabolic water. For the most part, adult amphibians are carnivores. Their food is therefore succulent, but it yields a lot of urea, which they cannot excrete in concentrated form. Significantly, if an amphibian is eating insects, the total amount of water it gets from its food (preformed and metabolic) is likely to be no more than about 15% of the amount it needs just to excrete the urea it produces from its food (not to mention having water to meet evaporative and other water losses). This calculation emphasizes the overwhelming importance of "drink" as a water source for the majority of amphibians.

Most amphibians do not in fact drink, but instead absorb water across their skin. This absorption does not necessarily require immersion in water. Many species can gain water at substantial rates merely by pressing their ventral skin against moist soil, moss, or other substrates.[18] A region of the ventral skin at the posterior of the abdomen and extending onto the thighs—called the *pelvic patch* or *seat patch*—is often specialized for rapid water uptake. A medium-sized and well-hydrated leopard frog (*Rana pipiens*) sitting on wet soil in its native habitat might well absorb 6 to 10 g of water per hour from the soil across its ventral surfaces, while simultaneously losing a like amount into the air by evaporation across its dorsal surfaces!

Most terrestrial amphibians are able to ward off dehydration for a time when they are away from sources of water by using their blad-

[18] Amphibians are far from being the only animals that take up water from moist substrates. The phenomenon has been documented in certain isopods, millipedes, insects, spiders, scorpions, land crabs, and snails.

der as a canteen. If an animal starts to dehydrate, the walls of its bladder are rendered permeable to water. If the urine inside is hyposmotic to the blood plasma, water then leaves the bladder contents by osmosis. NaCl is actively transported out of the bladder, thereby promoting osmotic outflux of water. The capacity of the bladder to hold fluid in terrestrial frogs and toads is remarkable: The water contained in the filled bladder is equal to 20% to 50% of an animal's bladder-empty weight. By contrast, in strictly aquatic amphibians, the bladder is usually tiny.

HORMONAL CONTROL OF RESPONSES TO DEHYDRATION In terrestrial amphibians, the neurohypophyseal hormone arginine vasotocin (see page 399), called *antidiuretic hormone* (ADH), activates a suite of coordinated responses that collectively retard or reverse the process of dehydration. Release of ADH is stimulated if the volume of the body fluids is decreased or if their osmotic pressure is increased. An amphibian's overall reaction to ADH has appropriately been called the amphibian *water-balance response*. In its complete form (not shown by all species), this response involves changes at three sites in the body: the kidneys, bladder, and skin. First, ADH causes the kidneys to reduce their rate of urine production and elevate the urine concentration toward isosmoticity with the blood (see page 729). Second, ADH stimulates the bladder to increase its permeability to water and its rate of NaCl reabsorption, responses that augment return of water from the bladder to the blood. Finally, ADH causes the ventral skin through which water absorption occurs to increase its apparent permeability to water influx, thereby facilitating rehydration.

ADH is by no means the only hormone active in water–salt physiology. Recently, for example, *angiotensin II* (see page 412) has been shown to be a principal controller of "cutaneous drinking" in frogs, stimulating the animals to press their ventral skin against moist substrates.

HOW DO DEHYDRATION-PRONE AMPHIBIANS LIVE IN DESERTS? As already mentioned, some species of frogs and toads, such as *Bufo cognatus* and the spadefoot *Scaphiopus couchii* in North America, live successfully in deserts or other arid habitats, even though their skin provides no more protection against evaporative water loss than that of semiaquatic frogs like leopard frogs (see Figure 26.16). Such desert species are in fact remarkably similar to the majority of terrestrial amphibians in all physiological respects, although some species show modest quantitative improvements over amphibians that live in moist habitats, such as by having a larger bladder, a somewhat greater tolerance of dehydration, or an accelerated pace of rehydration.

Behavior and seasonal dormancy are critical keys to the success of these desert amphibians. Dehydration can kill them *in a few hours* if they are exposed in the desert. Stringent behavioral control of water loss is therefore a prime requirement of life. These desert amphibians spend much of their time in protective microhabitats, especially in burrows underground, and are largely nocturnal. They also employ seasonal dormancy to simply "retire from the scene" and protect their water status during dry seasons. Spadefoot toads (*S. couchii*), for example, are known to spend many months of each year in dormancy. Overall, these desert amphibians are reclusive animals, holed up in secluded places during much of their lives. Nonetheless, they are able to survive in deserts despite the high permeability of their skin and other vulnerabilities.

"RADICAL" PHYSIOLOGICAL SPECIALIZATIONS OCCUR IN SOME ARBOREAL, ARID-ZONE FROGS For an amphibian to exist in arid places without being restricted to a secluded life, it must evolve superior physiological mechanisms to cope with the challenges of dehydration stress. Biologists are gradually learning more about such mechanisms in several types of unusual arboreal frogs that live exposed lives in arid or semiarid habitats. Frogs of this sort in two genera—*Phyllomedusa* of South and Central America and *Chiromantis* of Africa—have been studied for several decades and are known to have physiological abilities to conserve water that are extraordinarily different from those of most amphibians. One distinctive trait of these frogs is that their skin is exceptionally poorly permeable to water; their rates of evaporative water loss are consequently little different from those of some lizards of similar size. Unusual cutaneous lipids are responsible for this low skin permeability in both genera. *Phyllomedusa*, for example, secretes lipids (mainly waxy esters) from skin glands and spreads them on its skin surface (Figure 26.19). A second highly distinctive trait of these genera is that they excrete much of their nitrogenous waste in the form of poorly soluble uric acid or urates; *P. sauvagei*, for example, excretes 80% of its waste nitrogen in these forms.

The tiny, dramatically colorful reed frogs (*Hyperolius*) of the African savannas represent another remarkable group of arboreal frogs—a group that is only now starting to be well understood. At least some species have exceptionally low skin permeabilities. They do not, however, routinely produce uric acid or other related compounds like the frogs just discussed. During the dry season, reed frogs remain in exposed locations on the branches of bushes and

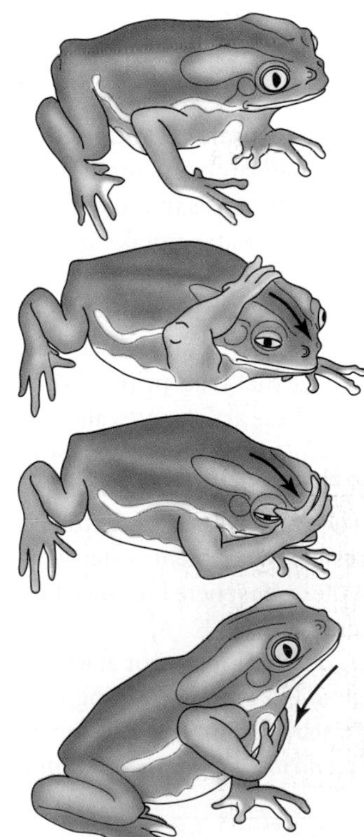

Figure 26.19 Arboreal frogs of the genus *Phyllomedusa* spread protective lipids secreted by integumentary glands over their skin surface The lipids sharply reduce their rate of evaporative water loss across their skin. The spreading is carried out by a series of stereotyped limb movements, as shown. (After Blaylock, Ruibal, and Platt-Aloia 1976.)

trees even as they undergo profound dehydration. They stop producing urine at such times, and much of their waste nitrogen accumulates in their body fluids as urea. However, as they dehydrate, they start to synthesize guanine from waste nitrogen retained in their bodies; in *H. viridiflavus*, 25% of waste nitrogen becomes guanine. Guanine is a low-solubility purine like uric acid (see Figure 27.23). Remarkably, the frogs deposit much of the guanine they synthesize in skin cells (iridophores), and the little animals turn bright white from its presence. The formation of guanine lengthens the time the frogs can store waste nitrogen (because it keeps the nitrogen out of solution). Guanine deposition in the skin also helps reduce solar heating by increasing the reflectance of the skin to incoming radiation!

Xeric invertebrates: Because of exquisite water conservation, some insects and arachnids have only small water needs

The suite of characteristics that permits many species of insects and arachnids (e.g., spiders and scorpions) to prosper in dry habitats includes several physiological attributes that promote highly effective water conservation: low integumentary permeability to water, provided chiefly by epicuticular lipids; stringent limitation of respiratory water losses by control of the opening of the spiracles; excretion of waste nitrogen in poorly soluble forms; and an ability (at least in many insects) to concentrate dissolved wastes in the urine to such an extent that the urinary osmotic pressure is two to eight times higher than blood osmotic pressure.

Some insects, ticks, and mites, in addition, have a way to obtain water that is unique in the animal kingdom; namely, they are able to gain water from the gaseous water vapor in the air in a *steady* manner while they are at the same temperature as the air. For example, the desert cockroach *Arenivaga investigata* can gain water from the air steadily even when the ambient relative humidity is as low as 79% to 83%. The mealworms (*Tenebrio*) that people often use as food for pets can gain water down to 88% relative humidity, and firebrats (*Thermobia*) can do so down to 45%. At high ambient humidities, a water gain of about 10% of body weight per day is the rule if these insects are dehydrated. The mechanism of water uptake in these arthropods in many cases remains subject to debate. The site of uptake in the desert cockroach and some ticks is known to be the mouth. Mealworms and firebrats, on the other hand, absorb water via the rectum. Some investigators believe that true, primary active transport of water sometimes occurs in these arthropods, although most disagree. A mechanism that is known to operate in some species is the production—at the mouth or rectum—of *localized*, superficial pockets of body fluids with such high solute concentrations that they have water vapor pressures below ambient water vapor pressure;[19] water vapor diffuses from the atmosphere into such body fluids, following the vapor pressure gradient.

Few water budgets have been worked out for insects or arachnids, but it is clear that certain species are so effective in limiting water losses that they can maintain water balance at moderate temperatures and low humidities while having *no drinking water* and *eating only air-dried foods*; common examples include grain beetles and clothes moths. Besides the small amounts of preformed water in air-dried foods, the only sources of water for such insects are metabolic water and whatever water they may gain from atmospheric water vapor. Mealworms can maintain their body fluids at nearly constant levels for at least a month while fasting in dry air (from which water vapor uptake is impossible)! Their ability to conserve water is apparently so extraordinary that their water losses can be replaced by metabolic water production from the oxidation of stored organic materials in their bodies.

Because of their short generation times, insects are among the most useful animals for studies of physiological evolution using laboratory populations. Experiments using insects demonstrate that the resistance of animals to water stress can respond dramatically to evolutionary selective pressures (Box 26.5).

Xeric vertebrates: Studies of lizards and small mammals help clarify the complexities of desert existence

When one treks to the driest deserts, one sees only a few sorts of animals leading active lives. Besides insects and arachnids, lizards and small mammals[20] are two groups that especially stand out. A fact that draws interest to both is that they often have no chance of finding drinking water for months on end. Birds also occur in severe deserts, but if they need to drink, they can fly to watering places at some distance. However, when there is no water that could be drunk (or absorbed) in the immediate locale of a lizard or small mammal, the animal must do without. Desert lizards characteristically eat insects or living plant tissues, foods that supply significant amounts of preformed water. Many species of small mammals in deserts also receive substantial quantities of preformed water in their food because they eat primarily insects or plants; in North America, these animals include pack rats, grasshopper mice, and ground squirrels. In all the major deserts of the world, however, there have evolved extraordinary species of small mammals that live away from drinking water and eat principally air-dried seeds or other air-dried plant matter. The kangaroo rats (*Dipodomys*) of North America's southwestern deserts provide classic examples. Others animals of this sort are the kangaroo mice and pocket mice of North America, the gerbils and jerboas of Old World deserts, and the hopping mice of Australian deserts.

A significant, intriguing trait of the lizards that prosper away from drinking water is that, like other lizards, they are diurnal. The key traits that permit the existence of lizards as diurnal animals in the driest places on Earth include, first, their relatively low metabolic rates (see Figure 5.9). As we have seen several times in this chapter, a low metabolic rate can greatly reduce water losses; it also reduces food needs, an asset in habitats where populations of food organisms are themselves stressed and relatively unproductive. Second, the desert lizards excrete their nitrogenous wastes as water-sparing uric acid or urates. They also use behavior to avoid those stresses that are avoidable, as by moving into underground burrows or shadows during the heat of the day. Some species employ salt glands. Finally, desert lizards tend to be remarkably tolerant of large

[19] Recall from Chapter 3 that the water vapor pressure of a solution is a colligative property and decreases as solute concentration increases.

[20] See Chapter 28 for a discussion of large mammals in deserts.

BOX 26.5 The Study of Physiological Evolution by Artificial Selection

One way to learn how the physiological features of organisms respond to different selective pressures in different environments is to expose animals to artificial (human-mediated) selection. For example, the evolutionary physiology of desiccation resistance can be studied by selecting for animals that are particularly resistant to desiccation and then by determining how such animals differ physiologically from control animals. Experiments of this sort have been carried out on fruit flies (*Drosophila melanogaster*). Starting with genetically identical fly populations, investigators exposed some populations—termed the D populations—to selection for desiccation resistance for over 150 generations, whereas other populations—termed C populations—were maintained for the same number of generations without desiccation experience. Shortly after their emergence as adults, the flies of each generation of a D population were exposed to dry air until 80% died, only the 20% that survived were then bred to create the next generation. The flies of a C population were not subjected to this selection regime.

After the 150+ generations of selection, flies from the D and C populations were bred for two generations in identical, nonstressful environments to ensure that differences of acclimatization were not present during testing. Then, D and C flies were placed in dry air to determine average survival times. The figure shows the dramatic results: D flies survived desiccation stress for over three times longer than C flies! Thus fewer than 200 generations were required to cause shifts in gene frequencies that strongly favored dehydration tolerance. Current research is focused on identifying the specific physiological features that have evolved to give the D flies their superior resistance to the stresses of desiccation. One feature is that the D flies have more body water than the C flies (1.0 versus 0.8 mg of water per fly). Another is that the D flies conserve water more effectively; whereas the D flies lose 26 nL H_2O/hour in a dehydration test, the C flies lose 45 nL/hour.

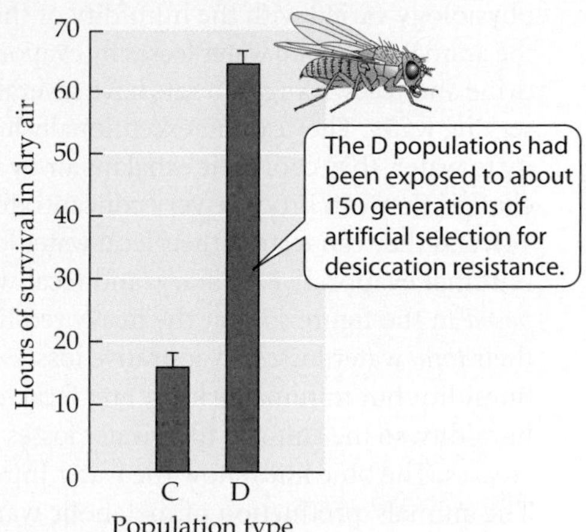

Survival times of fruit flies in C and D populations when held in dry air without food or water The height of each bar shows the average survival time. The T-shaped line segment at the top of each bar shows one standard error (SE) of the mean. (After Bradley, Williams, and Rose 1999.)

shifts in their body-fluid composition, such as high solute concentrations during dehydration. They can survive perturbations of their body fluids—often for long periods—that would spell a quick, sure death for a mammal or bird.

The small mammals of deserts operate on a different, higher scale of metabolic intensity than the lizards. High metabolic intensity is in itself a vulnerability in deserts; it raises the rate of respiratory evaporative water loss, as we have seen, and it can contribute enough endogenously produced heat to add significantly to heat stress. Small mammals that live in deserts have, in general, evolved lower than average basal metabolic rates for their size. Some species, moreover, undergo daily torpor or estivation when they are short of food or dehydrated. Still, when small mammals are active, their metabolic rates are far higher than those of lizards.

The species of small desert mammals that, by far, have attracted the most curiosity are the ones that eat primarily air-dried plant matter, such as kangaroo rats. In classic studies 50 years ago that contributed to the genesis of modern animal physiology, researchers produced the water-balance summary for kangaroo rats (*D. merriami*) shown in Figure 26.20. As we discuss this information, it will be

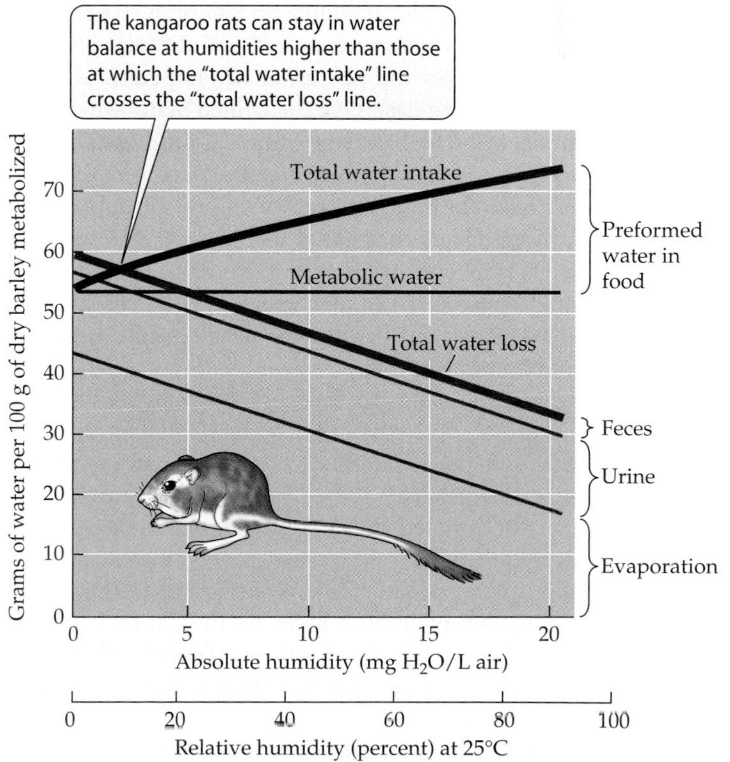

Figure 26.20 A kangaroo rat water budget For this study, carried out at 25°C, kangaroo rats (*Dipodomys merriami*) were fed husked barley seeds (pearled barley) at equilibrium with atmospheric moisture and provided no drinking water (they usually do not drink even if water is available). The water losses shown are the minimal possible water losses; in actuality, if water intake exceeds minimal losses, the animals increase their losses (as by excreting more-dilute urine) so that losses match inputs. The amounts of water graphed on the *y* axis are those gained or lost each time 100 g (dry weight) of barley is consumed. Under the conditions of study, the animals normally consume 100 g of barley in about a month. (After Schmidt-Nielson and Schmidt-Nielsen 1951.)

important to keep in mind that the animals were studied at 25°C. They had no drinking water and were fed only barley seed. Each of the five lines in the figure shows how a key attribute of their water physiology varied with the humidity of the air. The red lines show the animals' *minimal* water losses by evaporation and elimination of urine and feces. Kangaroo rats have several specializations for conserving water. They exhibit exceptionally low cutaneous permeability to water; they cool their exhalant air by nasal countercurrent exchange; they can produce very concentrated urine (osmotic U/P = 14); and they can restrict their fecal water losses exceptionally. Their minimal evaporative, urinary, and fecal water losses are *superimposed* in the figure so that the heavy red line at the top represents their *total* water losses. Evaporative losses decrease with increasing humidity, but minimal urinary and fecal losses are independent of humidity, so the animals' total water losses decrease as humidity increases. The blue lines show the water inputs of the kangaroo rats. The animals' production of metabolic water (determined by their metabolic rates) is the same at all humidities. Their input of preformed water increases as humidity increases, however, because seeds come to equilibrium with the water vapor in the air and contain more water when the humidity is high. The animals' inputs of metabolic and preformed water are superimposed so that the heavy blue line represents their total water intake.

Kangaroo rats can be in water balance if their total water inputs equal or exceed their total, minimal water losses. Based on Figure 26.20, therefore, the animals can be in water balance while eating air-dried seeds and drinking nothing if the relative humidity is above about 10%. Most of their water input is metabolic water. As stressed in Chapter 25 (see page 677), this is not because they produce exceptional amounts of metabolic water. It is because they conserve water so well that metabolic water can meet most of their needs.

Now we need to recall that the studies in Figure 26.20 were done at 25°C. The researchers who carried out the studies also obtained data on soil temperatures and nocturnal air temperatures in the parts of the southwestern deserts of the United States where they worked.[21] They concluded that temperatures at the depths of kangaroo rat burrows and at night on the desert surface were generally about 25°C or less. The kangaroo rats, being nocturnal, lived in their cool burrows during the heat of the day, and they emerged onto the desert surface only in the cool of the night. Thus, not only were free-living kangaroo rats never heat-stressed, but also—based on Figure 26.20—they could stay in water balance while eating air-dried seeds and drinking nothing.

How is the water physiology of kangaroo rats affected if the ambient temperature is cooler than 25°C? The *dominant* modes of water gain and loss in these animals are metabolic water production and evaporative water loss (EWL) (see Figure 26.20). A straightforward (although not complete) way to gain insight into the effects of temperature on the animals' water balance is to examine how the rate of metabolic water production and the rate of EWL relate to each other as functions of ambient temperature. The rate of metabolic water production—because it varies in parallel with the animals' metabolic rates—*increases with decreasing ambient temperature* at temperatures below the thermoneutral zone (see Figure 8.22). The rate of EWL by rodents *typically decreases with decreasing ambient temperature*. The **water-balance index** is the ratio of the rate of water *gain* by metabolic water production over the rate of water *loss* by evaporation. Because decreasing the ambient temperature causes the rate of metabolic water production to increase but the rate of EWL to decrease, this ratio becomes dramatically more favorable as the ambient temperature falls, as Figure 26.21*a* shows.

Based on the analysis in Figure 26.21*a*, we would expect kangaroo rats to be under far less water stress during the cool seasons of the year than the warm seasons. In fact, studies of kangaroo rats (*D. merriami*) in the wild match this expectation. The urinary osmotic pressure of the animals in a particular population averaged about 1000 m*Osm* (U/P = 3) in midwinter, but rose to about 4000 m*Osm* (U/P = 11) in midsummer. Low winter temperatures apparently placed the animals in such a favorable situation that they had a water surplus, which they excreted in relatively dilute urine. In summer they produced concentrated urine that conserved water.

Ominously, Figure 26.21*a* suggests that if kangaroo rats were to live at ambient temperatures of 30°C or higher, where their rates of evaporative water loss would be greatly heightened, they might potentially fall into severe negative water balance if eating only seeds. Until very recently, researchers assumed that kangaroo rats and other "simple seed eaters" of the world's deserts could in fact exist on a diet of seeds alone because they avoid temperatures higher than about 25°C by staying in benign burrows during the day and venturing out only during mild times of night. Desert climates vary in their severity, however, and past studies on such rodents were not conducted in the hottest places where the animals exist. What has happened recently to revise our view of these animals is that some brave investigators have, for the first time, studied kangaroo rats in the hottest parts of the Sonoran Desert (see Figure 25.6). The animals, it turns out, are even more remarkable than recognized before.

During the summer months in the hot parts of the Sonoran Desert, the soil at the depths of kangaroo rat burrows remains steadily at a temperature of about 35°C (95°F)! Moreover, the air temperature at the desert surface during the times of night when the animals are particularly active is between 35°C and 37°C. The animals living in these places are emphatically *not* exempt from heat stress. Recognition of that reality has led to the new discovery that the kangaroo rats in at least some populations are physiologically specialized to tolerate body temperatures of 41°C to 42°C—exceptionally high for mammals. Under the conditions in the hottest parts of the Sonoran Desert, because the animals have far higher rates of evaporation than under the more moderate conditions in Figure 26.20, the sum of the water they gain from metabolism and from preformed water in seeds is *less than half of their total water need.* There is no water to drink. Fortunately, insects are able to exist and accumulate body fluids in this severe environment. Current evidence indicates that the kangaroo rats achieve water balance in the hottest desert regions where they live by adding insects and green-plant parts to their usual seed diet. They also show some evidence of specialization in their water-balance physiology (Figure 26.21*b*). The unfolding of this intriguing story is helping physiologists realize that in certain places, probably all small desert animals meet challenges that we have not appreciated or understood before.

[21] For soil temperatures they in fact used the information in Figure 1.15.

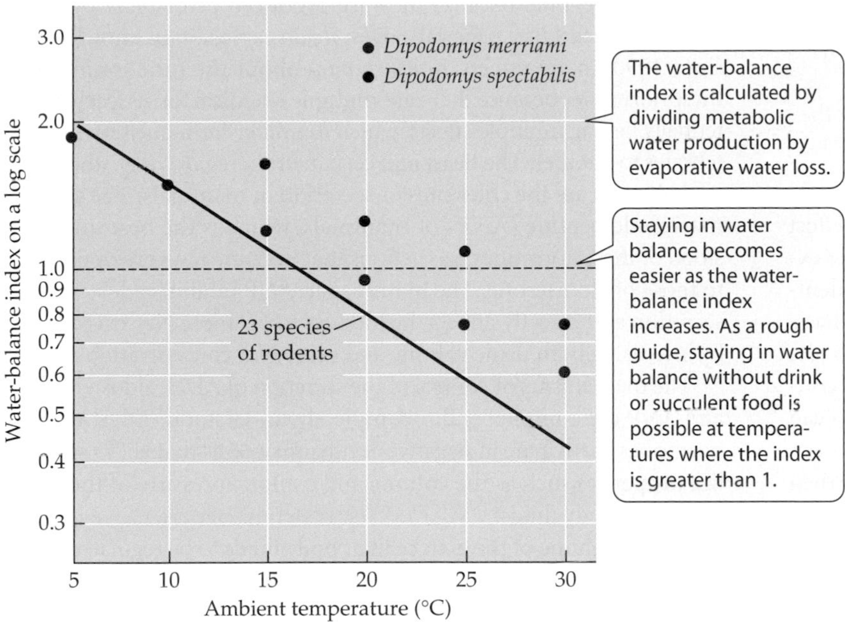

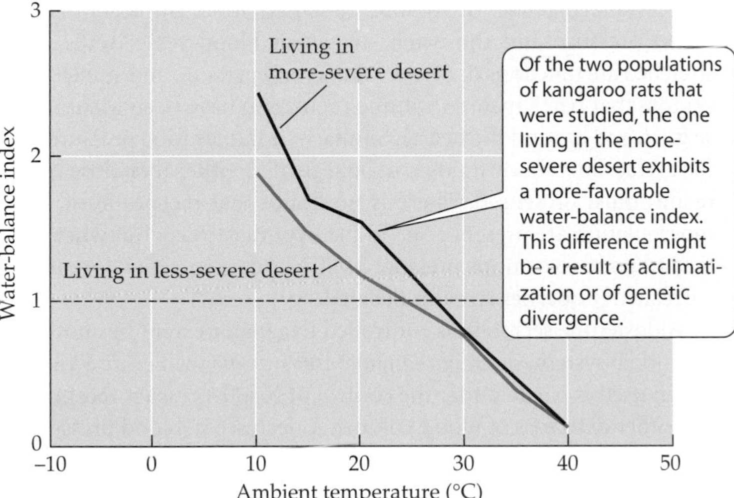

Figure 26.21 Effects of temperature on water balance in kangaroo rats In desert rodents that eat dry foods and drink no water, the rate of metabolic water production and the rate of evaporative water loss dominate the water budget (see Figure 26.20). Thus the ratio of the one over the other—termed the *water-balance index*—serves as a useful (although not complete) measure of the animals' ability to stay in water balance. *(a)* The line plots the average relation of the water-balance index to ambient temperature in 23 desert and nondesert species of rodents studied in dry or nearly dry air. The symbols show the relation for two species of kangaroo rats (*Dipodomys*), illustrating that particular species can differ from the average. *(b)* The relation between the water-balance index and ambient temperature in two populations of one kangaroo rat species (*D. merriami*). The two populations, which were studied by a single team of investigators, lived in two regions of desert that differed in the severity of water stress. The difference between the populations is statistically significant. (The *D. merriami* populations studied in *b* differed from the one studied in *a*.) (*a* after MacMillen and Grubbs 1976; *b* after Tracy and Walsberg 2001.)

Summary: Animals on Land: Case Studies

- Most terrestrial amphibians have meager physiological abilities to limit water loss because their skin is highly permeable to water and they cannot make urine that is hyperosmotic to their body fluids. Stringent behavioral control of water balance and seasonal dormancy are essential for their success in arid places. A few types of arboreal amphibians that live in arid areas have unusual adaptations such as cutaneous lipids that protect against rapid evaporation.
- Insects and lizards are among the animals that are most physiologically capable of living in the driest places on Earth. Their key traits for existence in extreme places include very low integumentary permeability to water, relatively low metabolic rates, excretion of poorly soluble nitrogenous wastes, and—at least in lizards—tolerance of profound changes in body-fluid composition. Insects can produce hyperosmotic urine and sometimes gain water from water vapor, but the fact that they are small is in itself a disadvantage.
- Some seed-eating mammals can live in deserts without needing to drink. Besides having highly evolved physiological mechanisms of water conservation, these animals depend on behavioral selection of benign microhabitats to maintain water balance. Recent evidence indicates that in the hottest places, these animals probably need to supplement their seed diet with water-rich foods like insects.

Control of Water and Salt Balance in Terrestrial Animals

The kidneys and other organs responsible for water and salt regulation in terrestrial animals are generally under hormonal control in both vertebrates and invertebrates, although the nervous system is sometimes a principal player, as in the control of avian salt glands. Some of the most important hormones involved are those called the *antidiuretic* and *diuretic* hormones. **Diuresis** is the production of an abundant (and usually dilute) urine. A **diuretic hormone** promotes diuresis. An **antidiuretic hormone** opposes diuresis or, in other words, modulates the excretory organs so that a relatively low volume of (usually concentrated) urine is produced.

Vertebrates are considered to produce only antidiuretic hormones. In insects, on the other hand, both diuretic and antidiuretic hormones have been described. Some species of insects, such as stick insects (*Carausius*), are known to produce both a diuretic hormone and an antidiuretic hormone. In other species, only one or the other type of hormone is presently known. Certain blood-sucking insects, for example, have a diuretic hormone, which is secreted after a blood meal, promoting rapid excretion of much of the water in the blood and thereby concentrating the nutritious part of the meal (e.g., proteins) in the gut.

In vertebrates, which will be our focus in this brief discussion, three hormones or types of hormones play particularly important roles in the regulation of water–salt physiology:

1. Antidiuretic hormone (ADH), produced by the neurohypophysis (see page 410 and Table 14.3)
2. Mineralocorticoids, most notably **aldosterone**, produced by the adrenal cortex or homologous interrenal tissue (see page 402)
3. **Natriuretic hormones** (see page 412)

The principal effect of ADH in mammals, and its principal effect on the kidneys in other terrestrial vertebrates, is to control the excretion of *pure water* (*osmotically free water*) relatively independently of solute excretion. Consider, for example, that a mammal has a certain *quantity* of urea, salts, and other solutes that it must excrete per day. If the solutes collectively are excreted at the maximal concentration the species can achieve, the accompanying water loss can be considered to be strictly obligated by solute excretion. However, if the solutes are excreted at less than maximal concentration, then the urine contains some additional water that is not obligated by solute excretion; in essence, the urine is diluted by the addition of pure water above and beyond the amount needed to void solutes, and the additional water represents a specific excretion of water itself. The urine can therefore be considered to consist of two components: (1) the solutes and their associated water and (2) a quantity of additional pure water. The magnitude of the latter component is controlled by ADH. If a person, for instance, has a constant daily solute output, but consumes little water on one day and a lot of water on another day, secretion of ADH on the first day will restrict the amount of water excreted with the solutes; however, inhibition of ADH secretion on the second day will permit diuresis and thus the excretion of a great deal of water with the solutes.[22]

The principal effect of aldosterone is to cause the kidneys to hold back Na^+ from excretion while promoting the excretion of K^+ in the urine. These actions of aldosterone most obviously affect the quantities of Na^+ and K^+ in the body. Less obviously, the action of aldosterone is one of the most important elements in the routine regulation of extracellular-fluid volume (including blood plasma volume). To understand this latter role of aldosterone, consider that Na^+ is largely excluded from intracellular fluids by being actively transported out of cells. Thus increases or decreases in the amount of Na^+ in the body lead to quantitatively similar increases and decreases in the amount of Na^+ (and accompanying anions, notably Cl^-) in the extracellular fluids. When Na^+ is retained in the body by being held back from the urine—and thus accumulates in the extracellular fluids—the systems controlling the osmotic pressure of the extracellular fluids cause water to be retained as well, so that the osmotic pressure of the extracellular fluids remains normal. In that way, the extracellular volume is expanded. Actually, although aldosterone has its major effects on the kidneys, it often functions as an all-purpose Na^+-retention hormone. In mammals, for example, it stimulates the salivary glands, sweat glands, and intestines to increase reabsorption of Na^+. Moreover, aldosterone stimulates salt appetite.

The natriuretic hormones—often termed *natriuretic peptides* because of their structure—promote addition of Na^+ to the urine, as their name indicates (*natri*, "sodium"; *uretic*, "having to do with excretion in the urine"). Certain of these hormones are called *atrial natriuretic peptides* because they are produced principally by the atria of the heart (see page 412). Despite an explosion of knowledge since 1990, many uncertainties remain about the functioning of these hormones because there are multiple chemical forms, each potentially having multiple effects, which can differ from one set of vertebrates to another. The heart and certain brain regions (e.g., the hypothalamus) are the chief sites of secretion in mammals. The atrial natriuretic peptide (ANP) of mammals, which is the best understood of these hormones, has actions that in many ways are opposite to those of aldosterone. The mammalian ANP inhibits aldosterone secretion and directly affects the kidneys to promote Na^+ excretion by increasing both urine volume and urine Na^+ concentration.

The mechanisms of control of the secretion of ADH, aldosterone, and ANP are complex and incompletely understood. All of these hormones participate in negative-feedback systems that act to stabilize properties such as the volume and osmotic pressure of the extracellular body fluids.

For the volume of the extracellular body fluids to be regulated, either the volume itself or reliable correlates of volume must be sensed, so that the regulatory systems will "know" whether to promote an increase or a decrease in volume at any particular moment. Probably volume itself is not sensed. However, there is good evidence that certain correlates of volume are sensed; for example, both the blood pressure and the extent to which blood-vessel walls are stretched are functions of blood volume, and pressure and stretch receptors that participate in volume regulation have been identified (e.g., in and around the heart). Similarly, if the osmotic pressure of the extracellular body fluids is to be regulated, either it or close correlates must be sensed. There is no doubt that receptors for osmoregulation are present (e.g., in the hypothalamus), but whether they respond to osmotic pressure itself, Na^+ concentration, or other correlated properties remains debated.

Aldosterone secretion is controlled to a major extent by another hormonal system, the *renin–angiotensin system* (see Figure 14.14), which itself is partly under the control of blood-pressure receptors and other detectors of blood volume. A decrease in blood pressure, signifying a reduction in blood volume, activates secretion by the kidneys of the hormone *renin* (pronounced reenin), which in turn causes formation in the blood of *angiotensin II*. The angiotensin stimulates the adrenal glands to secrete aldosterone, which induces increased Na^+ reabsorption from the urine, tending to expand extracellular-fluid volume and raise blood pressure.

Secretion of ADH is partly responsive to changes in blood volume; pressure sensors and other sensors of volume affect ADH secretion by way of nervous inputs to the hypothalamus and also via the renin–angiotensin–aldosterone system. Decreases in pressure activate ADH secretion, a response favoring fluid retention. Secretion of ADH is also under the control of osmoreceptors or other detectors of the concentration of the body fluids. Increases in the osmotic pressure of the body fluids induce increased ADH secretion; the ADH then favors the specific retention of water by the renal tubules, thus tending to lower the osmotic pressure of the body fluids.

Secretion of ANP is stimulated by expansion of extracellular-fluid volume, which is detected by stretching of the walls of the atria in the

[22] The action of ADH is discussed in more detail in Chapters 14 and 27.

heart. ANP then promotes loss of extracellular fluid. One of ANP's principal overall effects is a decrease in blood pressure.

Summary: Control of Water and Salt Balance in Terrestrial Animals

- The control of body-fluid volume, composition, and osmotic pressure is mediated mostly by hormones that are themselves secreted under control of negative-feedback systems. Stretch or pressure receptors provide information on blood volume, and osmoreceptors provide information on blood osmotic pressure.
- In vertebrates, antidiuretic hormone (ADH) regulates the amount of pure, osmotically free water that is excreted; it does so by controlling whether a more-than-minimal amount of water is excreted with solutes.
- Aldosterone and natriuretic hormones in vertebrates act to favor Na^+ retention or Na^+ excretion, respectively. The control of body Na^+ content by these hormones helps to control extracellular-fluid volume because body Na^+ is present mostly in those fluids.

Study Questions

1. It has been said that in terms of water balance, behavior can compensate for physiology, or physiology can compensate for behavior. Do you agree? Give examples to support your answer.
2. When a salmon or other teleost fish migrates from seawater into freshwater, what are all the changes that take place in its patterns of water–salt physiology?
3. Theory predicts that when both terrestrial and freshwater animals evolve higher metabolic rates, they can be expected to evolve greater challenges to maintaining water balance. Explain why this is so for both terrestrial and freshwater animals.
4. Although dogs inhale and exhale exclusively through the nose when not under heat stress, they exhale orally to some (variable) extent when panting. Air exhaled by way of the mouth remains nearly at deep-body temperature. We noted in Chapter 8 that dogs are believed to benefit when they pant by breathing at a fixed resonant frequency. Explain how a panting dog could *vary its rate of evaporative cooling*—even while *breathing at a fixed frequency*—by modulating how much it exhales by way of its nose or mouth.
5. Animals often face trade-offs, and one of the goals of modern physiology is to understand mechanistically why the improvement of performance in one way may necessarily degrade performance in other ways. In freshwater fish, explain why O_2 uptake and Ca^{2+} uptake can probably not be simultaneously maximized. In other words, why is there a trade-off between the ability to take up O_2 and the ability to take up Ca^{2+}?
6. Being as specific as you can, discuss how global warming can be expected to alter the water physiology of animals and what the ecological consequences might be. Figure 26.21 might provide a useful starting point, but do not limit yourself to it. Parks set aside to protect animals have defined geographical boundaries. How might the relation between global warming and water physiology affect the effectiveness of the park program?
7. Walter Cannon, the Nobel laureate who coined the term *homeostasis*, argued that lizards and amphibians are less highly evolved than mammals and birds because they are not as fully homeostatic as mammals and birds. Considering water–salt relations and any other aspects of physiology you find to be pertinent, explain in detail why you agree or disagree.
8. Green crabs (*Carcinus maenas*) are hyperosmotic regulators in brackish waters. When water salinity is lowered, a green crab responds by increasing its rate of urine production; its urine output, on average, rises from 4% to 30% of its body weight per day when the ambient salinity is reduced from 35 g/kg to 14 g/kg. Explain the value of this response. If one arranges to keep a green crab in full-strength seawater and bathe just its antennules with an alternative water source, urine production increases as the salinity of the water bathing the antennules is lowered. What can one conclude from this result?
9. Related species of terrestrial animals typically display allometric relations between body-water dynamism and body size. For example, the weight-specific rate of evaporative water loss tends to decrease allometrically as body size increases. What are the mechanistic reasons for this relationship?
10. We discussed the fact in Chapter 21 (see Figure 21.10) that the efficiency of O_2 exchange across the gills of teleost fish is enhanced by countercurrent exchange between the water pumped over the gills and the blood flowing through the secondary lamellae of the gills. A recent review article makes the point that the countercurrent arrangement of blood flow and water flow—which has positive consequences for O_2 exchange—has disadvantageous side effects in that it enhances osmotic water uptake by freshwater fish, osmotic water loss by marine teleosts, loss of NaCl by diffusion in freshwater fish, and NaCl gain by diffusion in seawater teleosts. Do you agree or disagree? Explain fully.
11. Probably the most complete account of water and salt balance in a marine mammal is not for a seal or whale, but for a remarkable bat, *Pizonyx vivesi*, that lives on desert islands in the Gulf of California, where it subsists on fish and crustaceans. It gets much of its water as preformed and metabolic water from its food. However, it needs more water than its food provides, and the amount of additional water it needs increases as it flies more, because flight greatly increases its rate of evaporative water loss. The bat can concentrate NaCl to about 620 m*M* in its urine. Based on this information, could it maintain water balance by drinking increasing amounts of seawater as it flies more? What additional information would you need to be certain?

Key References

NOTE: This list of Key References is short because many of the publications in this discipline are focused exclusively on particular animal groups. Scan the Selected References to find excellent "key" readings on most major animal groups.

Kirschner, L. B. 1991. Water and ions. In C. L. Prosser (ed.), *Environmental and Metabolic Animal Physiology* (*Comparative Animal Physiology*, 4th ed.), pp. 13–107. Wiley-Liss, New York. This is one of the best reviews of manageable length ever produced on water–salt physiology throughout the animal kingdom. Becoming dated, but still valuable.

Kirschner, L. B. 1997. Extrarenal mechanisms in hydromineral and acid–base regulation of aquatic vertebrates. In W. H. Dantzler (ed.), *Comparative Physiology*, vol. 1 (Handbook of Physiology [Bethesda, MD], section 13), pp. 577–622. Oxford University Press, New York. This review focuses on transport mechanisms in organs other than kidneys. It includes summary treatments of Kirschner's innovative attempts to estimate the energy costs of water–salt regulation.

Walsberg, G. E. 2000. Small mammals in hot deserts: Some generalizations revisited. *BioScience* 50: 109–120. Although this article focuses exclusively on small mammals in hot deserts, it is interesting reading for anyone interested in the physiology of free-living animals because of its commentaries on the use and abuse of model systems.

Selected References

Ahearn, G. A., J. M. Duerr, Z. Zhuang, R. J. Brown, A. Aslamkhan, and D. A. Killebrew. 1999. Ion transport processes of crustacean epithelial cells. *Physiol. Biochem. Zool.* 72: 1–18.

Ar, A., and H. Rahn. 1980. Water in the avian egg: Overall budget of incubation. *Amer. Zool.* 20: 373–384.

Baldwin, G. F., and L. B. Kirschner. 1976. Sodium and chloride regulation in *Uca*. *Physiol. Zool.* 49: 158–171, 172–180.

Beuchat, C. 1990. Body size, medullary thickness, and urine concentrating ability in mammals. *Amer. J. Physiol.* 258: R298–R308.

Bradley, T. J. 1987. Physiology of osmoregulation in mosquitoes. *Annu. Rev. Entomol.* 32: 439–462.

Bradley, T. J., A. E. Williams, and M. R. Rose. 1999. Physiological responses to selection for desiccation resistance in *Drosophila melanogaster*. *Amer. Zool.* 39: 337–345.

Bradshaw, S. D. 1997. *Homeostasis in Desert Reptiles*. Springer, New York.

Bryan, G. W. 1960. Sodium regulation in the crayfish *Astacus fluviatilis*. *J. Exp. Biol.* 37: 83–112.

Calder, W. A. III, and E. J. Braun. 1983. Scaling of osmotic regulation in mammals and birds. *Amer. J. Physiol.* 244: R601–R606.

Carey, C. 1986. Tolerance of variation in eggshell conductance, water loss, and water content by red-winged blackbird embryos. *Physiol. Zool.* 59: 109–122.

Carpenter, R. E. 1968. Salt and water metabolism in the marine fish-eating bat, *Pizonyx vivesi*. *Comp. Biochem. Physiol.* 24: 951–964.

Charmantier, G., C. Haond, J.-H. Lignot, and M. Charmantier-Daures. 2001. Ecophysiological adaptation to salinity throughout a life cycle: A review in homarid lobsters. *J. Exp. Biol.* 204: 967–977.

Che Mat, C. R. B., and W. T. W. Potts. 1985. Water balance in *Crangon vulgaris*. *Comp. Biochem. Physiol. A* 82: 705–710.

Coast, G. M. 1998. Insect diuretic peptides: Structures, evolution, and actions. *Amer. Zool.* 38: 442–449.

Crowe, J. H., F. A. Hoekstra, and L. M. Crowe. 1992. Anhydrobiosis. *Annu. Rev. Physiol.* 54: 579–599.

Crowe, J. H., A. E. Oliver, and F. Tablin. 2002. Is there a single biochemical adaptation to anhydrobiosis? *Integr. Comp. Biol.* 42: 497–503.

Danulat, E. 1995. Biochemical-physiological adaptations of teleosts to highly alkaline, saline lakes. In P. W. Hochachka and T. P. Mommsen (eds.), *Biochemistry and Molecular Biology of Fishes*, vol. 5, *Environmental and Ecological Biochemistry*, pp. 229–249. Elsevier, New York.

Davenport, J. 1985. Osmotic control in marine animals. *Symp. Soc. Exp. Biol.* 39: 207–244.

Diehl, W. J. 1986. Osmoregulation in echinoderms. *Comp. Biochem. Physiol. A* 84: 199–205.

Evans, D. H. (ed.). 1998. *The Physiology of Fishes*, 2nd ed. CRC Press, Boca Raton, FL.

Evans, D. H., P. M. Piermarini, and W. T. W. Potts. 1999. Ionic transport in the fish gill epithelium. *J. Exp. Zool.* 283: 641–652.

Feder, M. E., and W. W. Burggren (eds.). 1992. *Environmental Physiology of the Amphibians*. University of Chicago Press, Chicago.

Frank, C. L. 1988. Diet selection by a heteromyid rodent: Role of net metabolic water production. *Ecology* 69: 1943–1951.

Gerstberger, R., and D. A. Gray. 1993. Fine structure, innervation, and functional control of avian salt glands. *Int. Rev. Cytol.* 144: 129–215.

Gibbs, A. G. 1998. Water-proofing properties of cuticular lipids. *Amer. Zool.* 38: 471–482.

Gilles, R., and E. Delpire. 1997. Variations in salinity, osmolarity, and water availability: Vertebrates and invertebrates. In W. H. Dantzler (ed.), *Comparative Physiology*, vol. 2 (Handbook of Physiology [Bethesda, MD], section 13), pp. 1523–1586. Oxford University Press, New York.

Goldstein, D. L., J. B. Williams, and E. J. Braun. 1990. Osmoregulation in the field by salt-marsh savannah sparrows, *Passerculus sandwichensis beldingi*. *Physiol. Zool.* 63: 669–682.

Goss, G. G., S. F. Perry, J. N. Fryer, and P. Laurent. 1998. Gill morphology and acid–base regulation in freshwater fishes. *Comp. Biochem. Physiol. A* 119: 107–115.

Greenaway, P. 1985. Calcium balance and moulting in the Crustacea. *Biol. Rev.* 60: 425–454.

Greenaway, P. 1988. Ion and water balance. In W. W. Burggren and B. R. McMahon (eds.), *Biology of the Land Crabs*, pp. 211–248. Cambridge University Press, New York.

Gregor, R. (ed.). 1988. *NaCl Transport in Epithelia*. Advances in Comparative and Environmental Physiology, vol. 1. Springer, New York.

Griffith, R. W. 1980. Chemistry of the body fluids of the coelacanth, *Latimeria chalumnae*. *Proc. Roy. Soc. London* [*Biol.*] 208: 329–347.

Griffith, R. W. 1994. The life of the first vertebrates. *BioScience* 44: 408–417.

Hadley, N. F. 1994. *Water Relations of Terrestrial Arthropods*. Academic Press, New York.

Hand, S. C. 1991. Metabolic dormancy in aquatic invertebrates. Advances in Comparative and Environmental Physiology, vol. 8, pp. 1–50. Springer, New York.

Hand, S. C. 1992. Water content and metabolic organization in anhydrobiotic animals. In G. N. Somero, C. B. Osmond, and C.L Bolis (eds.), *Water and Life*, pp. 104–127. Springer, New York.

Hand, S. C., and I. Hardewig. 1996. Downregulation of cellular metabolism during environmental stress: Mechanisms and implications. *Annu. Rev. Physiol.* 58: 539–563.

Hazon, N., F. B. Eddy, and G. Flik (eds.). 1997. *Ionic Regulation in Animals: A Tribute to Professor W. T. W. Potts*. Springer, New York.

Heisler, N (ed.). 1995. *Mechanisms of Systemic Regulation: Acid–Base Regulation, Ion-Transfer and Metabolism*. Advances in Comparative and Environmental Physiology, vol. 22. Springer, New York.

Hickman, C. P., Jr. 1968. Ingestion, intestinal absorption, and elimination of seawater and salts in the southern flounder, *Paralichthys lethostigma*. *Can. J. Zool.* 46: 457–466.

Hillyard, S. D. 1999. Behavioral, molecular and integrative mechanisms of amphibian osmoregulation. *J. Exp. Zool.* 283: 662–674.

Hochachka, P. W., and T. P. Mommsen (eds.). 1995. *Biochemistry and Molecular Biology of Fishes*, vol. 5, *Environmental and Ecological Biochemistry*. Elsevier, New York.

Holmes, W. N., and J. G. Phillips. 1985. The avian salt gland. *Biol. Rev.* 60: 213–256.

Jönsson, K. I. 2003. Causes and consequences of excess resistance in cryptobiotic metazoans. *Physiol. Biochem. Zool.* 76: 429–435.

Kinne, R. K. H. (ed.). 1990. *Basic Principles in Transport*. Karger, Basel.

Kirschner, L. B. 1961. Thermodynamics and osmoregulation. *Nature* 191: 815–816.

Kirschner, L. B. 1993. The energetics of osmotic regulation in ureotelic and hypoosmotic fishes. *J. Exp. Zool.* 267: 19–26.

Kirschner, L. B. 1995. Energetics of osmoregulation in fresh water vertebrates. *J. Exp. Zool.* 271: 243–252.

Koehn, R. K. 1983. Biochemical genetics and adaptation in molluscs. In K. M. Wilbur (ed.), *The Mollusca*, vol. 2, pp. 305–330. Academic Press, New York.

Koehn, R. K., and T. J. Hilbish. 1987. The adaptive importance of genetic variation. *Amer. Sci.* 75: 134–141.

Land, S. C., and N. Bernier. 1995. Estivation: Mechanisms and control of metabolic suppression. In P. W. Hochachka and T. P. Mommsen (eds.), *Biochemistry and Molecular Biology of Fishes*, vol. 5, *Environmental and Ecological Biochemistry*, pp. 381–412. Elsevier, New York.

MacMillen, R. E., and D. S. Hinds. 1983. Water regulatory efficiency in heteromyid rodents: A model and its applications. *Ecology* 64: 152–164.

Marshall, A. T., and P. D. Cooper. 1988. Secretory capacity of the lachrymal salt gland of hatchling sea turtles, *Chelonia mydas*. *J. Comp. Physiol. B* 157: 821–827.

McClanahan, L., Jr. 1972. Changes in body fluids of burrowed spadefoot toads as a function of soil water potential. *Copeia* 1972: 209–216.

Mommsen, T. P., and P. J. Walsh. 1989. Evolution of urea synthesis in vertebrates: The piscine connection. *Science* 243: 72–75.

Nagy, K. A., and C. C. Peterson. 1988. *Scaling of Water Flux Rate in Animals*. University of California Publications in Zoology, vol. 120. University of California Press, Berkeley.

Nielsen, B. 1984. The effect of dehydration on circulation and temperature regulation during exercise. *J. Thermal Biol.* 9: 107–112.

O'Donnell, M. J. 1997. Mechanisms of excretion and ion transport in invertebrates. In W. H. Dantzler (ed.), *Comparative Physiology*, vol. 2 (Handbook of Physiology [Bethesda, MD], section 13), pp. 1207–1289. Oxford University Press, New York.

Oglesby, L. C. 1981. Volume regulation in aquatic invertebrates. *J. Exp. Zool.* 215: 289–301.

Oglesby, L. C. 1982. Salt and water balance in the sipunculan *Phascolopsis gouldi*: Is any animal a "simple osmometer"? *Comp. Biochem. Physiol. A* 71: 363–368.

Onken, H., and S. Riestenpatt. 1998. NaCl absorption across split gill lamellae of hyperregulating crabs: Transport mechanisms and their regulation. *Comp. Biochem. Physiol. A* 119: 883–893.

Ortiz, R. M. 2001. Osmoregulation in marine mammals. *J. Exp. Biol.* 204: 1831–1844.

Perry, S. F. 1997. The chloride cell: Structure and function in the gills of freshwater fishes. *Annu. Rev. Physiol.* 59: 325–347.

Perry, S. F. 1998. Relationships between branchial chloride cells and gas transfer in freshwater fish. *Comp. Biochem. Physiol. A* 119: 9–16.

Perry, S. F., A. Shahsavarani, T. Georgalis, M. Bayaa, M. Furimsky, and S. L. Y. Thomas. 2003. Channels, pumps, and exchangers in the gill and kidney of freshwater fishes: Their role in ionic and acid–base regulation. *J. Exp. Zool. A* 300: 53–62.

Peterson, C. C. 1996. Anhomeostatis: Seasonal water and solute relations in two populations of the desert tortoise (*Gopherus agassizii*) during chronic drought. *Physiol. Zool.* 69: 1324–1358.

Podrabsky, J. E., and S. C. Hand. 1999. The bioenergetics of embryonic diapause in an annual killifish, *Austrofundulus limnaeus*. *J. Exp. Biol.* 202: 2567–2580.

Rahn, H., C. V. Paganelli, I. C. T. Nisbet, and G. C. Whittow. 1976. Regulation of incubation water loss in eggs of seven species of terns. *Physiol. Zool.* 49: 245–259.

Remane, A., and C. Schlieper. 1971. Biology of brackish water. *Die Binnengewässer* 25: 1–372.

Schmidt-Nielsen, K. 1964. *Desert Animals*. Oxford University Press, London.

Schmuck, R., and K. E. Linsenmair. 1988. Adaptations of the reed frog *Hyperolius viridiflavus* (Amphibia, Anura, Hyperoliidae) to its arid environment. III. Aspects of nitrogen metabolism and osmoregulation in the reed frog, *Hyperolius viridiflavus taeniatus*, with special reference to the role of iridophores. *Oecologia* 75: 354–361.

Schmuck, R., and K. E. Linsenmair. 1997. Regulation of body water balance in reedfrogs (superspecies *Hyperolius viridiflavus* and *Hyperolius marmoratus*: Amphibia, Anura, Hyperoliidae) living in unpredictably varying savannah environments. *Comp. Biochem. Physiol. A* 118: 1335–1352.

Schroter, R. C., and N. V. Watkins. 1989. Respiratory heat exchange in mammals. *Respir. Physiol.* 78: 357–368.

Shaw, J. 1959. Salt and water balance in the East African fresh-water crab, *Potamon niloticus* (M. Edw.). *J. Exp. Biol.* 36: 157–176.

Shuttleworth, T. J., and J.-P. Hildebrandt. 1999. Vertebrate salt glands: Short- and long-term regulation of function. *J. Exp. Zool.* 283: 689–701.

Smedley, S. R., and T. Eisner. 1995. Sodium uptake by puddling in a moth. *Science* 270: 1816–1818.

Somero, G. N., C. B. Osmond, and C.L Bolis (eds.). 1992. *Water and Life*. Springer, New York.

Tracy, C. R. 1976. A model of the dynamic exchanges of water and energy between a terrestrial amphibian and its environment. *Ecol. Monogr.* 46: 293–326.

Tracy, R. L., and G. E. Walsberg. 2002. Kangaroo rats revisited: Re-evaluating a classic case of desert survival. *Oecologia* 133: 449–457.

Warburg, M. R. 1997. *Ecophysiology of Amphibians Inhabiting Xeric Environments*. Springer, New York.

Westin, L., and A. Nissling. 1991. Effects of salinity on spermatozoa motility, percentage of fertilized eggs and egg development of Baltic cod (*Gadus morhua*), and implications for cod stock fluctuations in the Baltic. *Mar. Biol.* 108: 5–9.

Wheatly, M. 1993. Physiological adaptations in decapodan crustaceans for life in fresh water. Advances in Comparative and Environmental Physiology, vol. 15, pp. 77–132. Springer, New York.

Wheatly, M. G., and A. T. Gannon. 1995. Ion regulation in crayfish: Freshwater adaptations and the problem of molting. *Amer. Zool.* 35: 49–59.

Williams, J. B., and B. I. Tieleman. 2001. Physiological ecology and behavior of desert birds. *Current Ornithology* 16: 299–353.

Withers, P. C., and M. Guppy. 1996. Do Australian desert frogs co-accumulate counteracting solutes with urea during aestivation? *J. Exp. Biol.* 199: 1809–1816.

Wright, J. L., P. Westh, and H. Ramløv. 1992. Cryptobiosis in Tardigrada. *Biol. Rev.* 67: 1–29.

Wright, P. A., and P. M. Anderson. 2001. *Nitrogen Excretion*. Fish Physiology, vol. 20. Academic Press, New York.

Wright, P. A., and M. D. Land. 1998. Urea production and transport in teleost fishes. *Comp. Biochem. Physiol. A* 119: 47–54.

See also references in Appendix A.

27 Kidneys and Excretion (With Notes on Nitrogen Excretion)

An animal's body fluids are dynamic, continuously gaining and losing water, inorganic ions, and organic solutes. When this kangaroo rat is successfully maintaining water balance in the desert in moderate summer weather, about 15% of the water in its body fluids is lost and replaced each day. This rate of turnover is low for a mammal of its body size; species of mice and rats living in moist temperate habitats often turn over 35% or more of their water per day. Humans, being larger, have lower percentage turnover rates, but even in people going about their ordinary daily lives, about 7% of the water in the body fluids is lost and replaced every day. Organic solutes and inorganic ions in the body fluids also undergo incessant turnover, as already mentioned. Ions such as Na^+ and Cl^-, for example, are added from foods and excreted in urine.

The great dynamism of the body fluids means that their composition is continuously in danger of being shifted away from normal. As water, for example, leaves the body fluids and is replaced each day, any mismatch between the rates of loss and replacement may make the body fluids become excessively dilute or concentrated.

For the blood and other body fluids to be maintained at a normal composition, an animal requires organs that are capable of correcting any departures from normal that develop during the complex fluxes of materials in and out. In aquatic animals such as fish and crayfish, the kidneys and gills are primarily responsible for this task. In terrestrial animals such as kangaroo rats, the kidneys bear the primary responsibility alone.

Kangaroo rats and other desert rodents have a status in the physiology of kidney function not unlike the status of cheetahs in the physiology of running. The kidneys of these desert rodents represent the ultimate product of evolution in their ability to concentrate urine. Most types of animals cannot concentrate their urine at all, in the sense that they cannot produce urine with an osmotic pressure higher than their blood osmotic pressure. Some birds and insects can make urine that is up to 6 to 8 times higher in osmotic pressure than their blood plasma. Certain species of kangaroo rats, however, can make urine that is 14 times higher in osmotic pressure than their blood plasma, and some of the desert hopping mice of Australia can make urine over 20 times higher.

The advantage of the high concentrating ability of the kidneys of desert rodents is that the kidneys have an

Small mammals that have evolved in deserts are, among all animals, the extreme performers in their ability to concentrate their urine Excretion of dissolved wastes in as little water as possible is a key to maintaining water balance in water-poor environments. The kidneys of some kangaroo rats can produce urine 14 times more concentrated than the blood plasma. Shown here is *Dipodomys deserti*.

extraordinary capacity to correct a type of error in blood composition that desert rodents are particularly likely to face. When a kidney can produce urine with a higher osmotic pressure than the blood, it can dilute the blood (see Figure 25.7). During the dynamic daily flux of water and solutes in and out of the body, the blood of desert rodents tends often to be shifted toward higher concentrations. The kidneys of these rodents, with their unique abilities to produce urine hyperosmotic to the blood, are exceptionally suited to correcting this sort of problem.

Considering animals in general, what are **kidneys**? We will see in this chapter that the kidneys of various types of animals are very diverse in morphology as well as in details of their physiology. All kidneys, however, have three features in common: First, they are all tubular structures that communicate directly or indirectly with the outside world. Second, they all produce and eliminate aqueous solutions derived from the blood or other extracellular body fluids. Third, *their function is the regulation of the composition and volume of the blood and other extracellular body fluids by means of* controlled excretion *of solutes and water.*

Urine, the product of the kidneys, is typically a complex solution containing multiple inorganic and organic solutes. All the constituents of the urine are drawn from the blood, and the urine concentration of each affects the blood concentration according to the principles we discussed in Chapter 25 (see page 675 and Figure 25.7). The urine often contains nitrogenous wastes, but the role of the urine is much broader than merely excreting waste nitrogen. The urine of a mammal, for example—while it contains urea (the nitrogenous end product)—also contains Na^+, Cl^-, K^+, PO_4^{3-}, SO_4^{2-}, creatinine, and numerous other components. The kidneys excrete each of these in greater or lesser amounts day by day, closely regulating the concentration of each in the blood. The kidneys also excrete greater or lesser amounts of H^+ in the urine, thereby helping to maintain a steady blood pH. Moreover, the kidneys regulate the osmotic pressure of the blood through the controlled excretion of water relative to total solutes. It seems almost impossible—but is true—that the kidneys perform all these functions *simultaneously* by structuring the composition of a *single* fluid output: the urine.

Basic Mechanisms of Kidney Function

Urine formation can usually be conceptualized as occurring in two steps, although these "steps" may sometimes be partly contemporaneous. First, an aqueous solution, called **primary urine**, is introduced into the kidney tubules. Second, this solution is modified as it moves through the kidney tubules and other excretory passages, ultimately becoming the **definitive urine** that is eliminated.

Primary urine is introduced into kidney tubules by ultrafiltration or secretion

One widespread mechanism by which fluid is introduced into kidney tubules is **ultrafiltration**. This is the mechanism used in most vertebrates and in many invertebrates, such as molluscs and decapod crustaceans (e.g., crayfish and crabs). Ultrafiltration into a kidney tubule occurs when the hydrostatic pressure is higher outside the tubule than inside the tubule lumen[1] at a place where the tubule wall is structured in a specialized, minutely porous way that permits fluid to pass through the wall. The difference in hydrostatic pressure then forces fluid to enter the tubule through the wall by means of pressure-driven mass flow, or streaming. It is this flow that is termed *ultrafiltration*. The process is literally a form of *filtration* because solutes of large molecular size typically are unable to pass through the wall of the tubule. Thus the fluid introduced into the tubule lumen—which is termed a **filtrate** or **ultrafiltrate**—consists only of water and the subset of solutes that are able to stream through with the water. The blood plasma is the source of the water and solutes that stream through. Although there are exceptions in some groups of invertebrates, the blood pressure produced by the heart typically is the pressure that drives ultrafiltration, explaining why heart weakening can interfere with urine formation.

To understand the formation of primary urine by ultrafiltration more fully, let's examine the process in the vertebrate kidney. Each kidney consists of many tubules, called **nephrons**, the walls of which consist of a single layer of epithelial cells (see Figure 2.6*b*). As diagrammed in Figure 27.1*a*, each nephron *begins blindly* with its walls thrown into a hemispherical, invaginated structure termed a **Bowman's capsule**, named after William Bowman (1816–1892), who first described it. Tucked inside each Bowman's capsule is an anastomosing cluster of blood capillaries, termed a **glomerulus** (Figure 27.1*b*), which is supplied with blood at relatively high pressure by branches of the renal artery. A Bowman's capsule and its glomerulus together constitute a **renal corpuscle**.[2] The glomerular capillaries are intimately juxtaposed to the walls of the Bowman's capsule. The wall of each capillary consists of a single layer of epithelial cells (the capillary endothelium), just as the wall of the Bowman's capsule consists of a single layer of cells. Thus the lumen of the capillaries is separated from the lumen of the Bowman's capsule by only two layers of cells and a basement membrane between the cell layers. These intervening structures have a specialized morphology and act as a filter. Fluid is driven through this filter from the blood plasma into the lumen of the Bowman's capsule by the hydrostatic pressure of the blood. Although a critical determinant of whether a solute will pass through is its molecular size, molecular charge and shape can also be significant; thus the filter has complex properties, which include, but are not limited to, simple physical pores or slits.

Inorganic ions and small organic molecules such as glucose, urea, and amino acids move freely with the fluid as it passes from the blood plasma into the lumen of a Bowman's capsule. Thus the concentrations of these solutes are virtually the same in the capsular fluid as in the blood plasma. At the other extreme, solutes with molecular weights of about 10,000 daltons or more—such as albumins and other plasma proteins—are almost entirely excluded from passage. The filtrate introduced into the Bowman's capsules—the primary urine—thus closely resembles the blood plasma in its composition of inorganic ions and low-molecular-weight organic solutes, but differs from the plasma in being almost devoid of high-molecular-weight organic solutes such as proteins.

[1] The *lumen* of a hollow structure such as a kidney tubule is the open central cavity.

[2] Another name for *renal corpuscle* is *Malpighian corpuscle*. Sometimes the entire renal corpuscle is called a *glomerulus*.

(a) The general form of a vertebrate nephron at the end where primary urine is formed

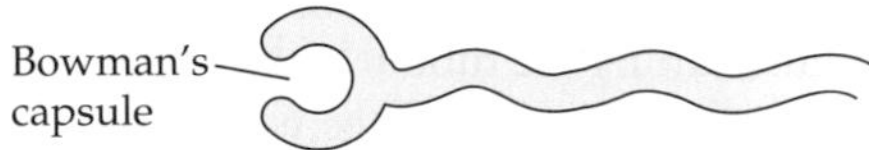

(b) A human glomerulus positioned in a Bowman's capsule

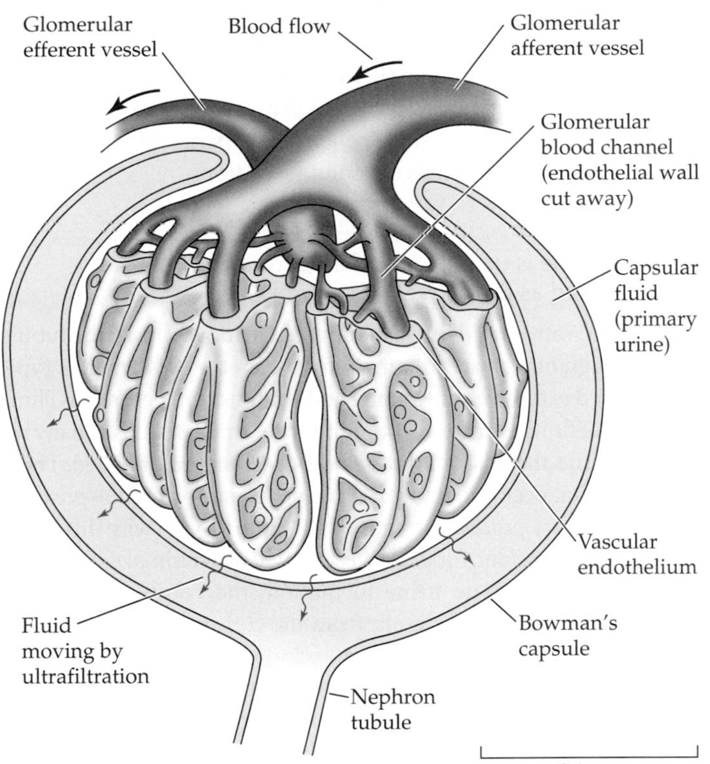

(c) Forces that favor and impede filtration

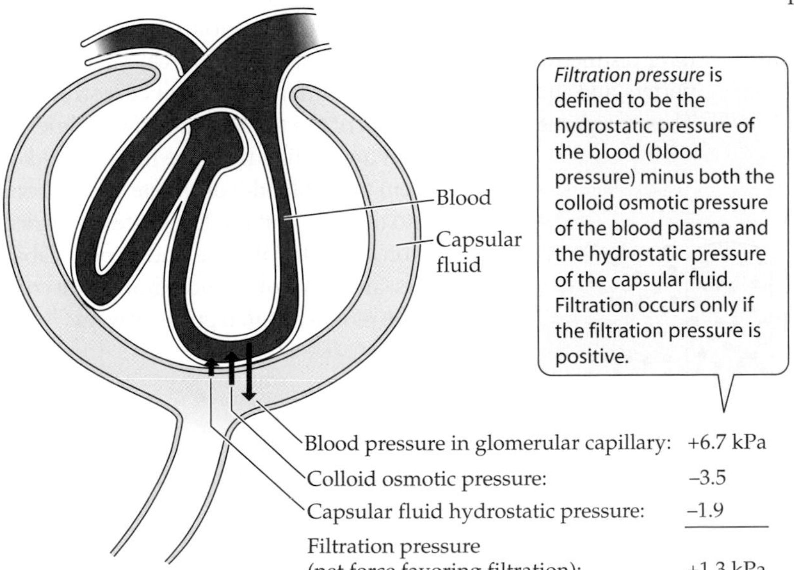

Figure 27.1 The structural and functional basis for formation of primary urine by ultrafiltration in the vertebrate kidney *(a)* The blind end of a vertebrate nephron, where ultrafiltration occurs. *(b)* A human renal corpuscle, consisting of glomerulus and Bowman's capsule. The vascular endothelium that forms the walls of the glomerular capillaries has been cut away at the top of the drawing, so that only the blood channels are shown there. The Bowman's capsule is drawn diagrammatically; the inner membrane of the capsule actually interdigitates with the sheets of vascular endothelium so that there is intimate juxtaposition of all blood capillaries and the capsular membrane. *(c)* The forces of hydrostatic pressure and colloid osmotic pressure that affect the rate of filtration: The relative lengths of the arrows in this part symbolize the relative magnitudes of these forces. (*b* after Elias et al. 1960.)

Because proteins remain more concentrated in the blood plasma than in the filtrate, the osmotic pressure of the blood plasma is higher than the osmotic pressure of the filtrate (see page 87). This difference in osmotic pressure is called the *colloid osmotic pressure* of the blood. Taking the colloid osmotic pressure into account, there are two processes that tend to cause water (H_2O) to move between the blood plasma and the fluid in the Bowman's capsule. The first is the *difference in osmotic pressure*, which tends to cause osmosis of water from the capsular fluid into the blood plasma. The second is the *difference in hydrostatic pressure*, which tends to cause streaming of water from the blood plasma into the capsular fluid. *Net filtration* of fluid into the capsular lumen will occur only if the difference in hydrostatic pressure is greater than the difference in osmotic pressure. In the renal corpuscles of mammals that are used as model systems for research, the blood pressure (hydrostatic pressure of the blood) is about 6.7 kPa, and the opposing hydrostatic pressure in the capsular fluid is about 1.9 kPa, meaning that the difference in hydrostatic pressure is about 4.8 kPa. The colloid osmotic pressure averages about 3.5 kPa. Thus the net force favoring filtration (termed *filtration pressure*) is about 1.3 kPa (**Figure 27.1*c***). The blood pressure in the glomerular capillaries is significantly higher than that in most capillaries in mammals, helping to promote filtration and formation of primary urine. Part of the reason for the high capillary blood pressure is that the arterioles leading to the glomeruli are relatively large in diameter and thus offer a relatively low resistance to blood flow.

The rate of primary-urine formation by all of an animal's kidney tubules taken together is called the **filtration rate.** In vertebrates, it is termed specifically the **glomerular filtration rate**, or **GFR**. Adult humans, for example, have a GFR of about 120 mL/min. At this rate, *the equivalent of all the plasma water in a person's body is filtered about every half hour*! As this example suggests, the GFR in a vertebrate greatly exceeds the rate of excretion of definitive urine. Most of the filtered water is ultimately reabsorbed back into the blood, rather than being excreted. The sheer magnitude of the rate of filtration means, however, that the nephrons have very intimate access to the blood plasma to carry out their function of regulating plasma composition.

The rate of production of definitive urine by an individual vertebrate animal can, in principle, be controlled in part by regulating the GFR. This mode of controlling urine flow is employed to some degree by mammals, but to a greater extent by other types of vertebrates. One way to adjust the GFR is to vary the rate of filtration into all the nephrons of the kidney collectively. An alternative way is to modulate the proportion of nephrons that are functioning as filtration units. The latter strategy is the norm in nonmammalian vertebrates. The rate of filtration into an individual nephron depends on its glomerular blood pressure, which can be modulated by vasomotor changes in the diameter (and hence flow resistance) of the glomerular afferent vessel. Vasomotor changes of this sort are under the control of the autonomic nervous system and circulating hormones. Although variation in the GFR can be important, it is not—as we will see later—the only mechanism of controlling the rate of definitive-urine production. Modulation of the *rate of reabsorption of filtered fluid* by the nephrons is also important; in mammals it is the preeminent process of urine volume control.

Besides ultrafiltration, **active solute secretion** is the second mechanism by which water and solutes can be moved into kidney tubules to form the primary urine. This is the mechanism employed, for example, by insects and some marine fish.

To see how urine formation can be initiated by secretion, consider Figure 27.2, which presents the essentials of a secretory system in conceptual, stepwise fashion. For simplicity, only two uncharged solutes are assumed to be present. Moreover, the fluid outside the tubule is assumed to be abundant, so that over short periods of time, movements of solutes and water into the tubule do not greatly modify its composition. At the start, which is labeled step 1, the osmotic pressure and the concentrations of both solutes are equal on the inside and outside of the kidney tubule. In step 2, an active-transport pump uses energy from ATP to secrete a quantity of solute X into the lumen of the kidney tubule, increasing the inside concentration of X and also increasing the inside osmotic pressure. In step 3, water moves inward by osmosis, following the osmotic gradient set up by secretion of solute X, and the volume of fluid in the tubule increases. Because of this increase in volume, the inside concentration of solute Y, initially the same as the outside concentration, is reduced so that it is now lower than the outside concentration. In step 4, solute Y diffuses inward following its concentration gradient. Although simplified and artificial, this model system demonstrates that *active secretion of even just a single solute into a kidney tubule can lead to passive influx of water and other solutes.* Thus a complex solution of many solutes can be introduced into the lumen of a kidney tubule from the body fluids bathing the tubule by a secretory mechanism. During the operation of a secretory system, the epithelium of the kidney tubule acts as something of a filter. The permeability of the epithelium to the various solutes that *might* passively diffuse into the tubular lumen determines which solutes *do*, in fact, enter.

Whether the process of primary-urine formation is ultrafiltration or secretion, energy is required. In ultrafiltration systems, energy is expended in maintaining a suitably high blood pressure to cause net filtration. In secretory systems, energy is expended by the active-transport pump that carries out solute secretion.

Primary urine is usually modified extensively before it is excreted as definitive urine

As the fluid introduced into a kidney tubule moves down the tubule and through other parts of an animal's excretory system, it is typically altered extensively in volume and composition before it is eliminated as definitive urine. Most of the water in the fluid is usually reabsorbed into the blood. Solutes may be reabsorbed or added. These processes that occur *after* primary-urine formation are the *predominant regulatory processes* in kidney function. To see why this statement is true, consider, for example, that when an animal employs ultrafiltration to initiate urine formation, the composition of its primary urine is little different from the composition of its blood plasma (except that proteins are excluded from the urine). The processes taking place *after* filtration are the ones that adjust the urine composition to be different from plasma composition.

As urine flows through a kidney tubule, it is separated from blood capillaries or blood spaces by the epithelium that forms the wall of the tubule. This epithelium is typically differentiated into distinct regions along the length of the tubule. Within each region, the epithelial cells have distinctive membrane proteins, such as ion channels, transporters, and aquaporins; and the cells may have a distinctive structure. These properties give each region of the tubule distinctive abilities to reabsorb water and solutes from the tubular fluid—returning them to the blood—or secrete solutes from the blood into the tubular fluid. The processes carried out by a region of a kidney tubule, and its permeability properties, are commonly under hormonal control and can be modulated in regulatory ways.

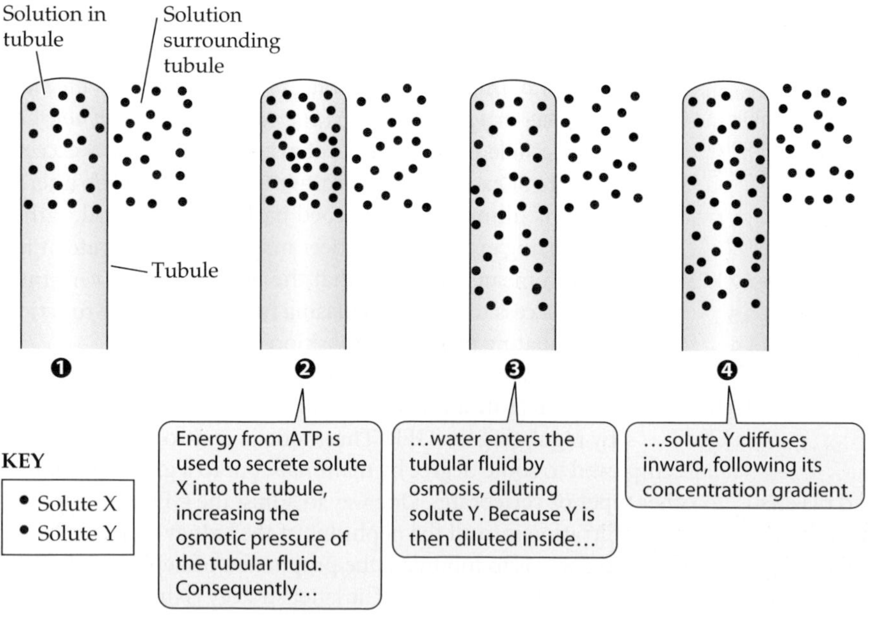

Figure 27.2 Formation of primary urine by active solute secretion In this model system, there are two uncharged solutes. Although the renal tubule is completely surrounded by the outside solution, only a small sample of the outside solution is shown at the upper right of the tubule in each step. For simplicity, the outside solution is assumed to stay constant in volume and composition. Movement of water into the tubule is represented by an increase in the length of the tubule filled with solution.

The processes that adjust the volume and composition of the urine are not necessarily limited to the kidney tubules. In mammals, the volume and composition of the urine are finalized once the urine leaves the kidneys. In other types of animals, however, the urine is often modified by the urinary bladder, cloaca, or other postrenal ("after kidney") structures.

> **Summary: Basic Mechanisms of Kidney Function**
>
> - Primary urine is formed by ultrafiltration or by active solute secretion.
> - During ultrafiltration, fluid is driven by elevated hydrostatic pressure from the blood into the kidney tubules through intervening epithelia and basement membranes that act as a filter. The filtrate, which is the primary urine, is almost identical to blood plasma in its composition, except that it lacks high-molecular-weight solutes such as plasma proteins.
> - In cases in which primary urine is formed by active solute secretion, the process that initiates and drives primary-urine formation is the active transport of one or more solutes into the kidney tubules. Water then follows by osmosis, and other solutes enter by diffusion, following electrochemical gradients set up by the active solute transport and osmosis.
> - As primary urine flows through the kidney tubules, its volume and composition are modified by active or passive transport of solutes and by osmosis of water across the epithelial walls of the tubules. These processes are the *predominant regulatory processes in the kidney tubules* because they determine the volume and composition of the urine that is excreted from the kidneys. The urine produced by the kidneys is sometimes (as in mammals) the definitive urine, but in many animals, the volume and composition are further adjusted by postrenal processing.

Urine Formation in Amphibians

The amphibians provide an excellent starting point for the study of vertebrate nephron function. Much is known about amphibian nephrons because of practical considerations that make them relatively easy to study. Furthermore, the amphibian nephron can reasonably be considered a "generalized" vertebrate nephron. Our purpose in this section is not only to describe how amphibians form urine, but also to bring out many additional general principles of vertebrate kidney function by example.

Each nephron of an amphibian (Figure 27.3*a*,*b*) consists of (1) a Bowman's capsule; (2) a convoluted segment known as the **proximal convoluted tubule**; (3) a short, relatively straight segment of small diameter, the **intermediate segment**; (4) a second convoluted segment known as the **distal convoluted tubule**; and (5) a relatively straight

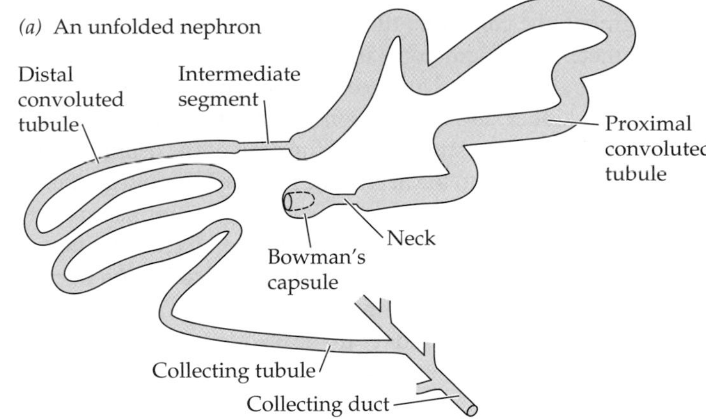

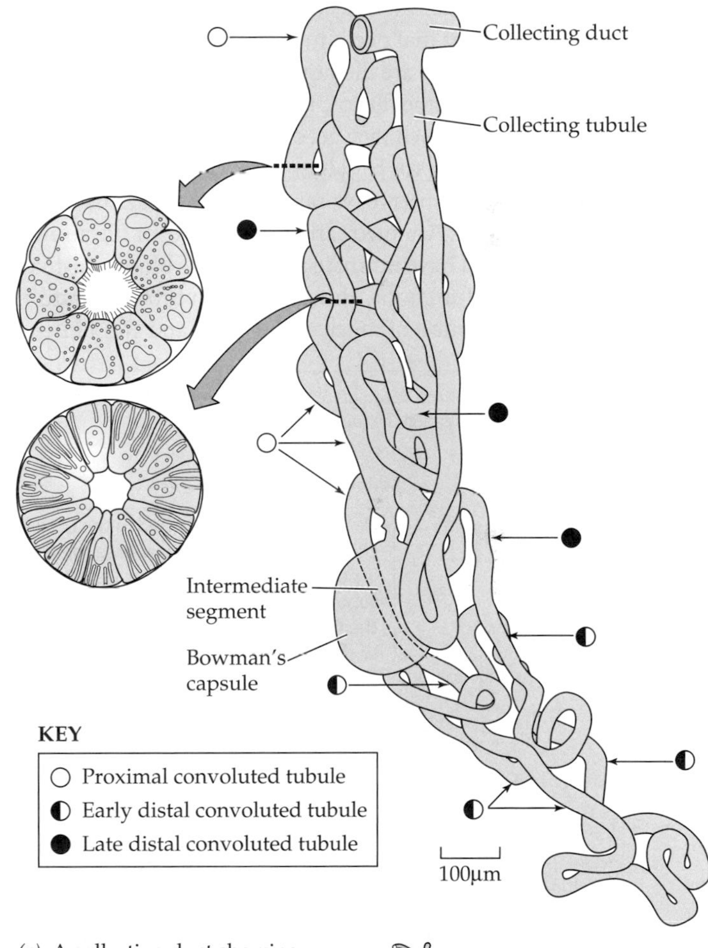

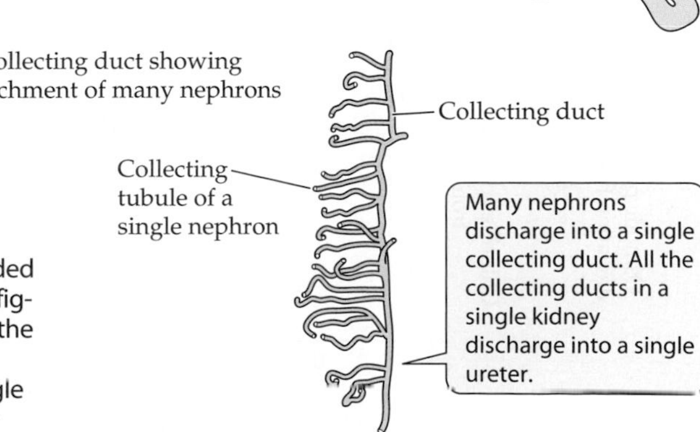

Figure 27.3 Amphibian nephrons and their connections to collecting ducts *(a)* An unfolded amphibian nephron. *(b)* A nephron of the toad *Bufo bufo*, shown realistically in its natural configuration. Symbols help to trace the nephron along the course of its intricate geometry. Two of the segments—part of the proximal convoluted tubule and part of the late distal convoluted tubule—are shown in cross section at greater magnification than the main drawing. *(c)* A single collecting duct, showing the connections of the collecting tubules of many nephrons. (*b* after Møbjerg, Larsen, and Jespersen 1998; *c* after Huber 1932.)

segment, the **collecting tubule.**[3] The nephrons are microscopic in diameter but macroscopic in length; in an average-sized toad, each might be 1 cm long. Hundreds or thousands of nephrons are found in each kidney, and the nephrons constitute much of the bulk of the kidney tissue. The structure and function of the nephron epithelium change from one to the next of the major nephron segments we have described. They also often change *within* a major nephron segment (e.g., within the distal convoluted tubule). In each kidney, the collecting tubules of all the nephrons feed into **collecting ducts** (Figure 27.3c), and all the collecting ducts connect to a single **ureter**, which carries fluid from the kidney to the bladder.

The proximal convoluted tubule reabsorbs much of the filtrate without changing osmotic pressure

The amount of water filtered each day by the kidneys typically far exceeds the amount that needs to be excreted. The same can be said of Na^+ and Cl^-, which are the principal solutes in the blood plasma and therefore also in the filtrate. They also are filtered at rates that far exceed the rates at which they need to be excreted; Na^+ and Cl^- enter the Bowman's capsules as briskly as water does, yet amphibians often need ultimately to conserve Na^+ and Cl^- to the maximal possible extent (see page 688). As noted earlier, a high rate of filtration ensures that the nephrons have intimate access to the blood plasma to perform regulatory functions. However, a high filtration rate also necessitates reabsorption of much of the water and NaCl filtered.

[3] The names of the nephron segments are not standardized. For example, the *collecting tubule* is sometimes called the *initial collecting duct.*

The reabsorption begins in the proximal convoluted tubule. Sodium is actively reabsorbed across the walls of the proximal tubule. Chloride may also be reabsorbed actively in some species, but in general its reabsorption is passive, induced by the electrical gradient set up by active Na^+ reabsorption. Although the quantities of Na^+ and Cl^- reabsorbed in the proximal tubule are substantial, the osmotic pressure of the tubular fluid does not fall in the proximal tubule. Instead, the fluid—which is isosmotic to the blood plasma when introduced into the Bowman's capsule by ultrafiltration[4]—remains isosmotic to the plasma as it flows through the proximal tubule. Its osmotic pressure remains unchanged because as NaCl is reabsorbed, there is also a proportional reabsorption of water from the tubular fluid. The epithelial walls of the proximal tubule are freely permeable to water. Thus water moves out of the tubular fluid by osmosis rapidly enough to prevent the active reabsorption of NaCl from producing a lower osmotic pressure in the tubular fluid than in the blood. In those species of amphibians that have been studied, 20% to 40% of the filtered NaCl and water are reabsorbed in the proximal tubule. Even as these large *amounts* of NaCl and water are reabsorbed, the *concentrations* of ions and water in the tubular fluid—as shown in Figure 27.4—remain unaltered because ions and water are removed in proportion to each other (Box 27.1).

[4] The difference in osmotic pressure that actually exists between the filtrate and the blood plasma—the colloid osmotic pressure—is large enough to affect filtration, as earlier discussed. However, not only in amphibians but also in other vertebrates, the difference in osmotic pressure is less than 1% of the *absolute* osmotic pressure of either the filtrate or plasma. Thus, for most purposes, the filtrate and the plasma can be considered isosmotic.

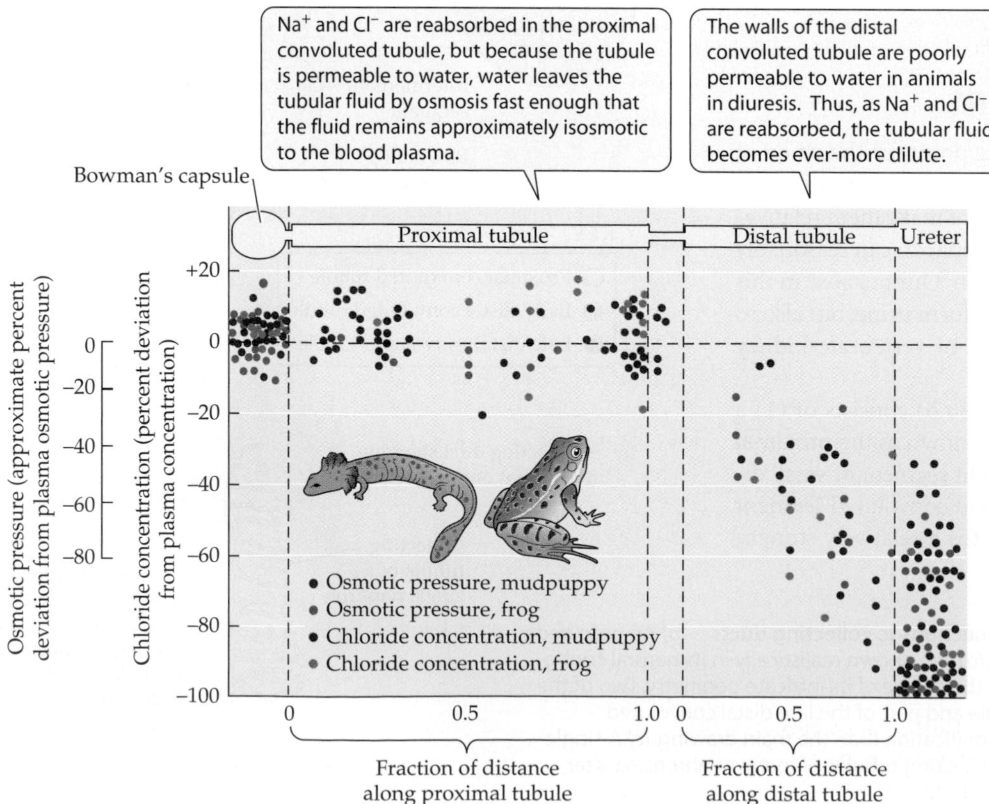

Figure 27.4 Urine formation in amphibians during diuresis The osmotic pressure and Cl^- concentration of urine as it flows through the nephrons of two species of amphibians—the semiterrestrial leopard frog (*Rana pipiens*) and the aquatic mudpuppy (*Necturus maculosus*)—when the animals are in diuresis. Fluid was sampled for analysis by use of minute pipettes inserted into the nephrons (see Box 27.2). Concentrations are expressed as percentage deviations from plasma concentrations; for example, a value of –40 indicates that the concentration in the tubular fluid was below the plasma concentration by an amount equal to 40% of the plasma concentration. (After Walker et al. 1937.)

BOX 27.1 Quantity versus Concentration

When analyzing kidney function, it is important to maintain a clear distinction between measures of *quantity* (or *mass*) and measures of *concentration*. The importance of this distinction is illustrated nicely by the events in the proximal tubule of amphibians. As shown in Figure 27.4, the *concentrations* of Na^+, Cl^-, and water in the tubular fluid remain, on average, unchanged. Yet, the *quantities* of these substances exiting the proximal tubule are much lower than those entering.

Measures of quantity and concentration are each informative, although in different ways. Quantity is an absolute measure, whereas concentration is a relative measure (quantity of solute relative to quantity of water). As a general principle, measures of *quantity* provide the most direct insight into questions of salt and water *balance*. For instance, to determine whether an animal is in Na^+ balance, you would measure the quantity of Na^+ gained per day and the quantity lost per day (including the quantity lost in urine) and compare them. Urine *concentrations* are not directly useful for balance calculations. However, as discussed in Chapter 25 (see Figure 25.7), urine *concentrations* provide the most direct insight into the effects of urine production on blood composition. For instance, if you wanted to know whether the kidneys are lowering the Na^+ concentration of the blood plasma, you would examine the urine Na^+ concentration relative to the plasma Na^+ concentration. Urine production is lowering the plasma Na^+ concentration if the Na^+ U/P ratio exceeds 1.0.

Another important process that takes place in the proximal tubule is the reabsorption of glucose. Glucose is a valuable metabolite that—because of its small molecular size—cannot be withheld from the primary urine during ultrafiltration. However, in amphibians and other vertebrates, it is promptly reclaimed and returned to the blood. Glucose is reabsorbed by secondary active transport, driven by the primary active transport of Na^+—a mechanism similar to that diagrammed in Figure 3.12. Amino acids are also valuable, small organic molecules that are freely carried into the Bowman's capsules by ultrafiltration. Their reabsorption is known to begin in the proximal tubules. Box 27.2 discusses some of the methods used to study kidney function: methods that have played important roles in creating the knowledge discussed here—in our review of amphibian proximal tubule function—and throughout the chapter.

The distal convoluted tubule can differentially reabsorb water and solutes, thus controlling water excretion

Active reabsorption of NaCl from the tubular fluid continues in the distal convoluted tubule. In this way, the quantity of NaCl destined for excretion from the body is gradually lowered toward the level that is appropriate for maintenance of NaCl balance.

A major function of the distal convoluted tubule in many amphibians—a function that may be shared by the collecting ducts and urinary bladder—is control of the excretion of *pure, osmotically free water*. Recall from Chapter 26 (see page 716) that the water in urine may be considered to consist of two parts: (1) water that is *required* to accompany excreted solutes and (2) additional water that may be excreted, but is not required for solute excretion. The second component may be considered to represent an excretion of pure, or "free," water precisely because it is not required for solute excretion.

In amphibians, the amount of water that is *required* to be excreted with solutes is determined by the fact that the maximum possible osmotic urine:plasma ratio (U/P ratio) is 1.0. The urine osmotic pressure cannot exceed the plasma osmotic pressure, meaning that *at least* enough water must be excreted with solutes in the urine to create a solution that is isosmotic to the blood plasma. If the urine osmotic pressure of an amphibian in fact equals the animal's plasma osmotic pressure, the urine contains *only* water that is required for solute excretion. That is, the urine contains no water of the second kind: no pure, osmotically free water. On the other hand, if the urine osmotic pressure of an amphibian is less than the animal's plasma osmotic pressure (osmotic U/P < 1), the urine carries an "extra" quantity of water, an amount not strictly required by solute excretion. This extra quantity represents an excretion of pure water. This excretion of pure water can be varied: The more dilute the urine, the more free water it contains. Thus *an animal can control its excretion of water independently of its excretion of solutes by varying the osmotic pressure of its urine.*

The extent of pure-water excretion is controlled in the distal convoluted tubule by varying the degree to which osmotic water reabsorption keeps pace with solute reabsorption there. The extent of water reabsorption is itself controlled by modulating the permeability of the walls of the tubule to water. This control of permeability is exercised at least partly by **antidiuretic hormone** (**ADH**) secreted by the neurohypophysis (posterior pituitary).[5] As we now explain these points, refer to Figure 27.5 for a visual summary.

When ADH levels are low (Figure 27.5*a*), the permeability of the wall of the distal convoluted tubule to water is low. Consequently, NaCl and water reabsorption from the tubular fluid are significantly uncoupled. The active reabsorption of NaCl tends to dilute the tubular fluid and thus create an osmotic gradient that favors water reabsorption by outward osmosis. However, the low permeability of the tubule wall to water impedes osmosis. This limitation of water reabsorption has two important and complementary consequences. First, NaCl reabsorption makes the tubular fluid more dilute than the blood plasma, both in osmotic pressure and in ion concentrations. This dilution (see Figure 27.4) is progressive: As fluid flows through the distal tubule, the fluid becomes ever-more dilute as ever-more ions are reabsorbed from it. The active reabsorption of solutes from the urine across tubule walls that are poorly permeable to water is the mechanism by which a urine hyposmotic to the blood plasma is made in amphibians and, in its fundamentals, is the *universal mechanism by which animals make hyposmotic urine.* The second principal consequence of low permeability to water in the amphibian distal convoluted tubule is that a high proportion of the

[5] The antidiuretic hormone of amphibians, reptiles, and birds is *arginine vasotocin*; see Table 14.3 on page 399.

BOX 27.2 Renal Clearance and Other Methods of Study of Kidney Function

Some of the methods used to study kidney function, although technically difficult, are intuitively easy to understand. A technique of this sort that has revolutionized renal physiology is **micropuncture**. Fine micropipettes are inserted into individual nephrons at identified points, permitting samples of tubular fluid to be withdrawn for analysis of composition. Such samples from amphibian nephrons reveal, for example, that the glucose concentration falls virtually to zero by the end of the proximal convoluted tubule. This is how we know that the proximal tubule is the site of glucose reabsorption.

A method that is not so intuitively simple to understand—but important in both physiological research and medical practice—is the study of renal *clearance*. **Plasma clearance**, the most fundamental clearance concept, is defined to be the volume of blood plasma that would need to completely depleted—or "cleared"—of a solute to yield the quantity of the solute excreted in the urine over a specified period of time. The plasma clearance can be different for different solutes and thus is solute-specific.

Before we examine how the concept of plasma clearance can be used to study kidney function, let's establish how it is calculated. Suppose a toad excretes 100 mL of urine per day, and that urine contains 2 μmol Na^+/mL. The animal's rate of excretion of Na^+ would then be $2 \times 100 = 200$ μmol/day. Suppose we measure the animal's plasma concentration of Na^+ and find that it is 100 μmol/mL. We would then know that to obtain the quantity of Na^+ excreted in a day, 2 mL of plasma would have to be completely cleared of Na^+. The plasma clearance of Na^+ would therefore be 2 mL/day. In algebraic form, if V is the volume of urine excreted per unit of time, U is the urinary concentration of the substance in question, and P is the plasma concentration, then the plasma clearance of the substance per unit of time, C, is computed as

$$C = \frac{UV}{P}$$

The way that plasma clearance is calculated should not be taken to imply that the kidneys obtain the quantity of a solute they excrete by completely clearing part of the plasma while leaving the rest untouched. Actually, of course, the quantity excreted is obtained by incompletely clearing the plasma at large. Although the plasma clearance is in this sense an artificial concept, it is also a powerful concept in the analysis of renal function.

One way in which the plasma clearance is used is to measure the glomerular filtration rate. To use the plasma clearance for this purpose, an artificial solute is introduced into the blood plasma (e.g., by injection). A solute useful for this purpose must meet three criteria: First, it must enter the nephrons by glomerular filtration and not be introduced into the urine by any other mechanism. Second, during ultrafiltration, it must be freely filtered, so that its concentration in the filtrate is the same as its concentration in the blood plasma. Third, it must not be reabsorbed in the nephrons or elsewhere, so that whatever amount enters the nephrons by filtration is fully excreted. The most commonly used substance meeting these criteria is **inulin**, a fructose polysaccharide derived from Jerusalem artichokes. It is known to meet the required criteria in mammals, amphibians, and some (but not all) other vertebrates. *In these animals, the glomerular filtration rate equals the plasma clearance of inulin.* We can use algebra to demonstrate this equality. Suppose inulin has been introduced into the blood plasma of an animal at concentration P. If F is the animal's glomerular filtration rate (volume of filtrate produced per unit of time), then the amount of inulin filtered per unit of time is FP, because the concentration of inulin in the filtrate equals the inulin concentration in the plasma. The amount of inulin excreted in urine per unit of time is UV (the concentration of inulin in the urine times the volume of urine excreted per unit of time). For a substance like inulin that, following filtration, is neither added to the urine nor reabsorbed, the amount filtered and the amount excreted must be the same. Thus $FP = UV$. By rearrangement,

$$F = \frac{UV}{P}$$

Thus, as earlier stated, the glomerular filtration rate equals the plasma clearance.

Most native plasma solutes that enter the nephrons by filtration (e.g., Na^+ and glucose) undergo postfiltration exchange with the blood; a solute, for example, might be secreted into the urine across the walls of the nephrons following primary-urine formation by filtration, or it might be reabsorbed from the urine between filtration and excretion. *Whether a solute is added or removed in net fashion following filtration can be determined by comparing its clearance with the GFR.* For example, if the clearance of a solute exceeds the GFR, one knows that the volume of plasma being cleared of that solute per unit of time exceeds the volume of plasma filtered per unit of time. A difference of this sort indicates that the solute is added to the urine following filtration. To illustrate, pioneering studies of bullfrogs showed that their urea clearance can be several times higher than their GFR; this finding provided evidence that in bullfrogs, urea is transferred from the plasma into the urine by secretion across the walls of the nephons in addition to being introduced by filtration at the Bowman's capsules. Conversely, if the clearance of a freely filtered solute is less than the GFR, the difference indicates that the solute is reabsorbed after filtration. Consider, for example, results for toads (*Bufo marinus*) living in distilled water: Their GFR (inulin clearance) was about 6 mL/hour, but their Na^+ clearance was far lower, about 0.04 mL/hour. Because we know that Na^+ is freely filtered, these data tell us that, every hour, the amount of Na^+ contained in 6 mL of plasma entered the nephrons by filtration, yet only the amount contained in 0.04 mL of plasma was excreted. Most of the filtered Na^+ must have been reabsorbed.

A useful quantitative measure is the **relative clearance** of a substance, defined to be the ratio of that substance's plasma clearance to the GFR (e.g., inulin clearance). For a freely filtered substance, this ratio is the fraction of the amount filtered that is excreted. To illustrate, the relative Na^+ clearance for the toads just discussed was the Na^+ clearance (0.04 mL/hour) divided by the GFR (6 mL/hour), or 0.007 (0.7%). This value tells us that the kidneys of the toads were excreting only 0.7% of the amount of Na^+ they filtered. When the toads were transferred to saline water, their relative Na^+ clearance rose to 53%, indicating that under those conditions, their kidneys reduced their reabsorption of Na^+ and excreted 53% of the amount they filtered.

Although substances like inulin were introduced to the study of kidney physiology because their plasma clearance permits measurement of the GFR, such substances also have important applications in micropuncture studies. To illustrate, suppose that a frog is injected with inulin prior to a micropuncture study of its proximal tubular fluids. What will be found is that the inulin concentration of the tubular fluid increases as the fluid flows through the proximal tubule. This increase provides direct evidence that water is reabsorbed in the proximal tubule; because inulin enters nephrons only by glomerular filtration, the increase of its concentration from the beginning to the end of proximal tubule must result from water removal, not inulin addition. In bullfrogs, micropuncture studies show that as tubular fluid flows through the proximal tubule, its concentration of urea increases even more than its concentration of inulin. This result is evidence that the proximal tubule is a site of urea secretion into the urine.

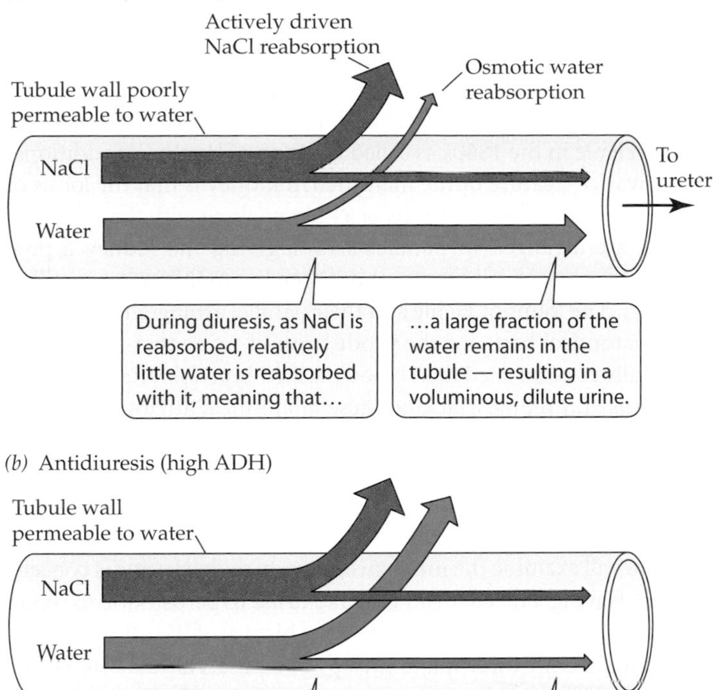

Figure 27.5 Major solute and water fluxes in the late distal convoluted tubule during diuresis and antidiuresis The pathways followed by NaCl and water are symbolized by the relative sizes of the arrows. *(a)* In diuresis, low permeability of the tubule walls to water impedes osmotic water reabsorption; thus a relatively large fraction of the water remains in the tubule, and the ratio of solute to water (the osmotic pressure) in the tubular fluid is dramatically reduced. *(b)* In antidiuresis, the tubule walls are more permeable to water; thus a larger fraction of the water is reabsorbed, and the ratio of solute to water in the tubular fluid is affected relatively little. This conceptual diagram is simplified in two ways: first, in that it assumes NaCl and water reabsorption occur in the same parts of the tubule, and second, in that solutes other than NaCl (e.g., urea) are ignored.

water that enters the distal tubule passes through to be excreted in the urine. Considering both consequences together, one can see that in the presence of low levels of ADH, the urine is *dilute* and *voluminous*. It carries away far more water than is necessary just to excrete solutes; thus it has a high content of pure, osmotically free water.

When ADH levels are high (Figure 27.5*b*), the presence of ADH induces the wall of the late distal convoluted tubule (the half or so of the distal tubule closest to the collecting tubule) to become relatively permeable to water, and the distal tubule then functions more like the proximal tubule. Osmotic water reabsorption is promoted. Thus, as NaCl is reabsorbed, more water is reabsorbed than when ADH levels are low. Again, there are two complementary consequences. First, the tubular fluid stays more nearly isosmotic to the blood plasma than when ADH levels are low. Second, a smaller quantity of the water that enters the distal tubule passes through to be excreted. In the presence of high levels of ADH, therefore, the urine is relatively *concentrated* and *scanty*. It carries away little or no pure, osmotically free water.

You can see now why ADH has the name it does. Recall from Chapter 26 that **diuresis** is production of abundant urine. High levels of ADH promote the opposite: **antidiuresis**.

ADH is believed to control the permeability of the amphibian distal convoluted tubule to water by controlling the insertion and retrieval of aquaporin proteins (see page 741) in cell membranes in parts of the tubular epithelium. When the level of ADH is high, aquaporins are inserted into the cell membranes, and water can pass through the epithelium relatively readily by osmosis. When the level of ADH is low, aquaporins are retrieved from the cell membranes, and osmosis through the epithelium is impeded. More will be said of aquaporin function later in this chapter, when we discuss mammals.

Active H^+ secretion into the tubular fluid is an additional function that is known to occur in the distal convoluted tubule. The amount of H^+ added is adjusted to maintain a normal pH in the body fluids.

ADH exerts an elaborate pattern of control over nephron function

In amphibians—and also in reptiles and birds—ADH not only increases the permeability of parts of the distal convoluted tubules to water, but also decreases the glomerular filtration rate. Specifically, ADH reduces the GFR in these vertebrate groups by reducing the numbers of actively filtering nephrons, an effect mediated by inducing vasoconstriction in glomerular afferent blood vessels. The decrease in GFR tends to reduce urine flow, thus complementing the increase in water reabsorption induced by ADH in the distal tubules.

ADH has also been shown in some frogs and toads to increase the rate of active NaCl reabsorption from the renal tubules. This effect, like the others we have mentioned, tends to reduce urine volume because it promotes solute-driven water reabsorption and decreases the solute load of the urine.

Clearly, ADH mediates an elaborate *pattern* of control over nephron function. If an amphibian experiences excess water influx—as will occur during immersion in freshwater—secretion of ADH is restricted. Then the GFR is relatively high and distal-tubule reabsorption of water is relatively low, and a voluminous, dilute urine results. If dehydration sets in, ADH is secreted from the neurohypophysis, apparently under the control of osmoreceptors (which detect an increase in body-fluid osmolarity) and of pressure or stretch receptors (which signal a decrease in blood volume). The ADH induces a reduction in GFR, an increase in distal-tubule water reabsorption, and an increase in NaCl reabsorption, thus promoting production of a scanty, concentrated urine. The renal responses to ADH are not as well developed in some amphibians from consistently moist or wet habitats as they are in species that are more terrestrial and thus more likely to experience dehydration (see page 711).

The bladder functions in urine formation in amphibians

In many species of amphibians, the bladder not only stores urine, but also plays a substantial role in adjusting the volume and composition of the urine. In these species, the function of the bladder can be described very much in the way we have described that of the distal convoluted tubules. The bladder wall is poorly permeable to water when ADH levels are low, but becomes quite permeable to water when ADH levels are high; the participation of aquaporins in these changes of permeability in the amphibian bladder has been directly demonstrated. NaCl is actively reabsorbed across the bladder wall, and this reabsorption is stimulated by ADH.

The amphibian excretory system has mechanisms to promote excretion of urea

Urea is the principal compound used to excrete waste nitrogen in most adult amphibians. The nephrons, bladder, and other excretory passages of adults seem generally to be poorly permeable to urea; thus urea introduced into the tubular fluid tends to be retained in it and be excreted. Filtration is one process by which urea enters the nephrons, and in many amphibians it is probably the sole process. However, in at least some ranid frogs (e.g., bullfrogs), urea is also actively secreted into the tubular fluid across the nephron walls. As water is reabsorbed from the tubular fluid, urea in the fluid tends to be concentrated.

Summary: Urine Formation in Amphibians

- A primary function of the proximal convoluted tubule of the amphibian nephron is the isosmotic reduction of urine volume. NaCl is actively reabsorbed from the tubular fluid. Because the epithelial wall of the proximal tubule is permeable to water, water exits the tubular fluid by osmosis, keeping the tubular fluid isosmotic to the blood plasma.
- Glucose and amino acids are actively reabsorbed from the tubular fluid in the proximal tubule.
- The distal convoluted tubule controls the volume and concentration of the definitive urine produced by the kidney. An important mechanism by which this control is exercised is that the epithelial wall of the distal convoluted tubule can have high or low permeability to water, depending on blood levels of antidiuretic hormone (ADH) secreted by the neurohypophysis (posterior pituitary).
- When ADH levels are low, the distal tubule is poorly permeable to water. Active reabsorption of NaCl dilutes the tubular fluid. However, because water cannot readily move out of the tubular fluid by osmosis, the volume of the tubular fluid remains high, and both the osmotic pressure and the NaCl concentration of the fluid become progressively lower.
- When ADH levels are high, aquaporins are inserted into cell membranes in the epithelial wall of the distal tubule, causing the permeability of the wall to water to become high. As active reabsorption of NaCl takes place, osmosis carries water out of the tubular fluid. Thus the volume of the tubular fluid is reduced, and the fluid remains similar to the blood plasma in its osmotic pressure and NaCl concentration.

Urine Formation in Mammals

The nephrons of amphibians, as earlier noted, may reasonably be considered to represent the generalized vertebrate condition. Reptilian nephrons resemble them. Mammalian nephrons differ, however. Compared to an amphibian nephron, each nephron of a mammal has an added, long segment of tubule, positioned between the proximal and distal convoluted tubules. This added segment is arranged in the shape of a hairpin loop and, having first been described by Jacob Henle in the 1860s, is called the **loop of Henle.** An additional "innovative" feature of the mammalian kidney is that the loops of Henle of the various nephrons in a kidney, along with the collecting ducts, are arranged in parallel arrays, giving the kidney a pronounced macroscopic structure not seen in amphibian or reptilian kidneys. The loops of Henle and their parallel arrangement provide the anatomical basis for the production of urine that is more osmotically concentrated than blood plasma: *hyperosmotic urine*. Amphibians and reptiles, lacking these anatomical attributes, cannot produce hyperosmotic urine.

We have seen in Chapter 26 that the ability of mammals to concentrate their urine is one of their most dramatic and important adaptations for life on land. Now, as we study the mammalian kidney, we will examine the mechanism by which their urine is concentrated. Having a mechanism to make urine hyperosmotic to blood might at first seem trivial; however, the history of life offers a different verdict. Only three major groups of animals have mastered the task: mammals, birds, and insects. In each case, the ability to concentrate the urine has opened up new habitats and ways of life—as by aiding small mammals such as the kangaroo rats described at the start of this chapter to survive as seed eaters in deserts. Thus, as we examine the mammalian mechanism of concentrating urine, we focus on a physiological attribute of enormous ecological and evolutionary significance.

The nephrons, singly and collectively, give the mammalian kidney a distinctive structure

The loop of Henle in a mammalian nephron consists of two parallel limbs connected by a hairpin bend: The **descending limb** leads from the proximal convoluted tubule to the bend, and the **ascending limb** runs from the bend to the distal convoluted tubule (Figure 27.6*b*). The descending limb begins with a **thick segment**, and the ascending limb terminates with a thick segment. Interposed between these thick segments, at various positions and for various lengths, is a segment of very small diameter, the **thin segment**. The thin segment differs cytologically from the intermediate segment discussed earlier and occurs only in mammals and birds. The loop of Henle varies considerably in length among species of mammals and among the nephrons within the kidneys of one species.

As can be seen in Figure 27.6*b*, the Bowman's capsules and convoluted tubules of the nephrons in each kidney of a mammal are aggregated toward the outer surface of the kidney, whereas the loops of Henle and collecting ducts project inward, toward the **renal pelvis**, a tubular structure that represents the expanded inner end of the ureter that drains the kidney (see Figure 27.7*a*). Because of this highly ordered arrangement of the renal tubules, histologically distinct zones are apparent in the kidney tissue. In cross section, the tissue of each kidney consists of an outer layer, the **cortex**, which surrounds an inner body of tissue, the **medulla** (Figure 27.6*a*). The cortex (see Figure 27.6*b*) consists of Bowman's capsules, convoluted tubules, the beginnings of collecting ducts, and associated vasculature. The medulla consists of loops of Henle and collecting ducts, as well as

(a) Kidney in cross section

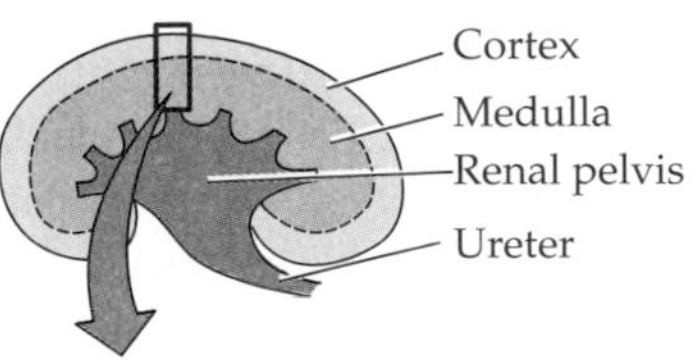

(b) Detail of nephrons and blood supply

Nephron with outer cortical capsule and short loop
Nephron with juxtamedullary capsule and long loop
Artery
Proximal convoluted tubule
Bowman's capsule
Peritubular capillary
Distal convoluted tubule
Thick segment of descending limb of loop of Henle
Thick segment of ascending limb of loop of Henle
Cortex
Thin segment
Vein
Outer zone of medulla
Collecting duct
Medulla
Vasa recta
Collecting duct
Thin segment of loop of Henle
Inner zone of medulla
Collecting duct discharges to pelvis

Figure 27.6 Human nephrons and their circulatory supply in the kidney *(a)* A mammalian kidney in cross section, showing the gross structure. *(b)* A more detailed look at nephrons and their blood supply. The intricate shape of the kidney and renal pelvis seen in *(a)* is not observed in all mammals; in many species, each kidney has a shape like that in Figure 27.7*a*. The structures of the nephrons and vasculature seen in *(b)* are similar in all mammals, however. (*b* after Smith 1951.)

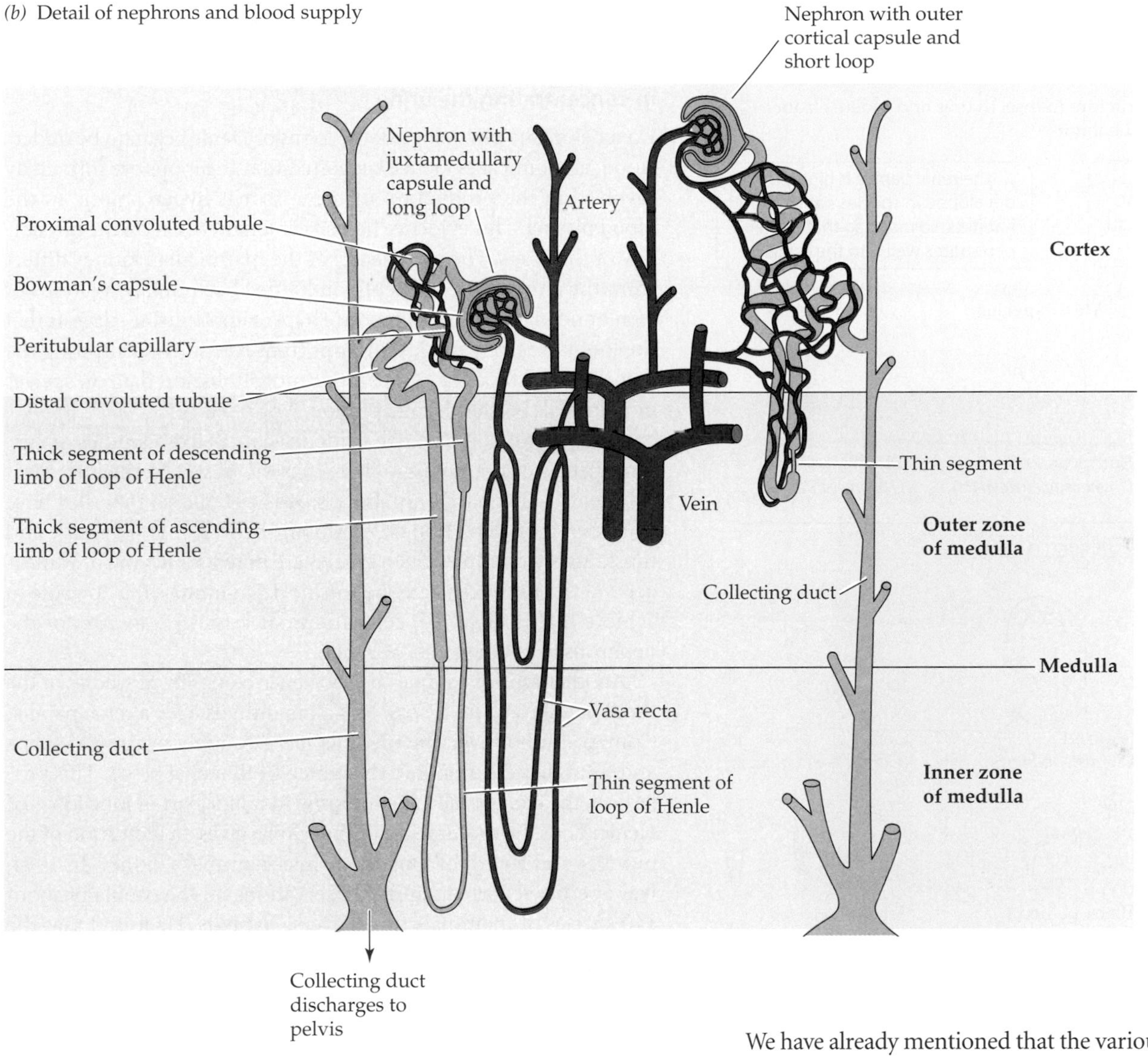

their associated vasculature. Within the medulla, the loops of Henle and collecting ducts run in parallel to one another.

To get oriented to fluid flow patterns in the mammalian kidney, let's now trace the path of fluid through a nephron, focusing on the nephron to the left in Figure 27.6*b*. After filtration into the Bowman's capsule, fluid moves first through the proximal convoluted tubule and then descends into the medulla in the loop of Henle. After rounding the bend of the loop, the fluid returns to the cortex, passes through the distal convoluted tubule, and leaves the nephron to enter a collecting duct. The fluid then again passes through the medulla, this time in the collecting duct. After the fluid is discharged from the collecting duct into the renal pelvis, it flows into the ureter to be excreted. A convention worthy of note is that when fluid flows from the cortex toward the medulla, it is said to move *deeper* into the kidney.

We have already mentioned that the various nephrons in the kidney of a species may have loops of Henle of dramatically different lengths. Nephrons differing in this regard are positioned differently within the kidney, a fact that contributes to gross kidney structure. As can be seen at the left side of Figure 27.6*b*, there is a region deep in the medulla—termed the **inner zone** of the medulla—that is defined by the fact that the only *loop* elements present are *thin* descending and ascending segments of *relatively long* loops of Henle. The surrounding, more superficial part of the medulla is the **outer zone.** Loops of Henle that project into the inner zone are termed **long loops.** Loops that turn back within the outer zone of the medulla or within the cortex are called **short loops.** The thin segments of long and short loops of Henle differ cytologically. Bowman's capsules may be positioned near the outer cortical surface, at mid-depth in the cortex, or within the cortical tissue next to the medulla; the last location is termed the *juxtamedullary* ("near the

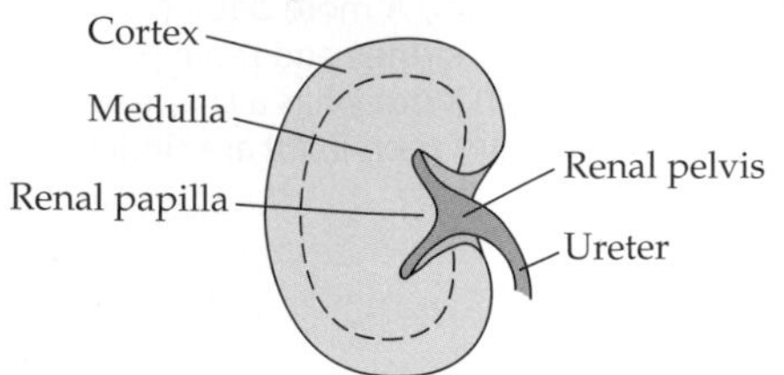

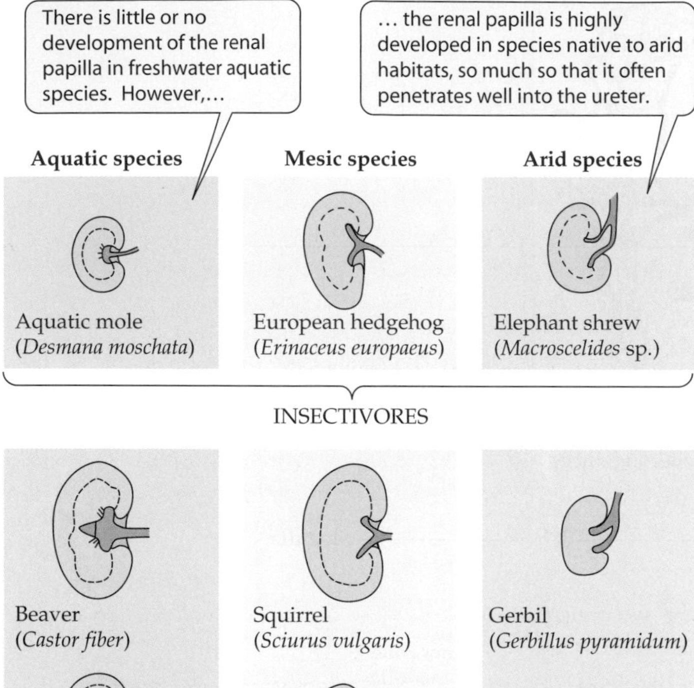

Figure 27.7 Evolutionary development of the renal papilla in mammals native to different habitats *(a)* Cross section of a generalized mammalian kidney, showing the location of the renal papilla. The papilla, when present, is the part of the medulla that tends to project into the lumen of the renal pelvis. *(b)* Kidney structures of insectivores and rodents from aquatic, mesic, and arid habitats. (After Sperber 1944.)

medulla") position. As depicted in Figure 27.6*b*, nephrons with short loops tend to have their capsules positioned toward the outer cortex, whereas those having long loops tend to have midcortical or juxtamedullary capsules. Laboratory rats have about 30,000 nephrons of all types combined in each kidney. Domestic dogs have about 400,000, and humans have between 0.4 and 1.2 million.

A final morphological point worth making is that the thick ascending segment of each nephron, near its upper end, passes next to the Bowman's capsule of the very same nephron.[6] At this point, the wall of the thick ascending segment is modified, forming the **macula densa**. The macula densa and other associated cells form a structure called the **juxtaglomerular apparatus**.[7] Specialized vascular endothelial cells in this apparatus are the source of renin, which controls aldosterone secretion (see page 411 and Figure 14.14).

Comparative anatomy points to a role for the loops of Henle in concentrating the urine

Even before the physiology of the loops of Henle began to be understood, anatomical evidence suggested that the loops are intimately involved in the production of urine that is hyperosmotic to the blood plasma. This evidence helped center attention on the physiology of the loops. The mere fact that the mammalian kidney differs from the amphibian and reptilian kidney both in having loops of Henle and in being able to produce hyperosmotic urine suggests that the loops are likely to play an important role in concentrating the urine. Several types of comparative morphological data on species of mammals bolster this impression.

One type of comparative evidence comes from studies of certain species of mammals—characteristic of wet freshwater environments—that have only short loops of Henle, so that they have no inner medulla. Hippos, mountain beavers (*Aplodontia*), and muskrats are examples. Such species are noted for having only meager abilities to concentrate their urine. In mammals that are able to achieve high urinary concentrations, at least 15% to 20% of the nephrons have long loops of Henle.

Another type of comparative evidence comes from studies of the **renal papilla** (Figure 27.7*a*). Not all mammals have a renal papilla. Commonly, however, the medulla has a roughly pyramidal shape and forms a projection into the lumen of the renal pelvis. This projection, the renal papilla, is composed in major part of long loops of Henle. Thus the prominence of the papilla gives an indication of the number and length of long loops in a mammal's kidney. In 1944, Ivar Sperber reported seminal observations on the papilla in about 140 species of mammals from diverse habitats. He found that the papilla was uniformly lacking or poorly developed in species inhabiting wet or aquatic freshwater habitats. The papilla was present in species from mesic (moderately moist) habitats and was most developed in species from arid habitats (Figure 27.7*b*). Insofar as habitat may be taken as an indicator of demand for urinary concentration, Sperber's results indicated that there is a greater evolutionary development of the long loops of Henle in species that produce relatively concentrated urine.

Inspired by Sperber's work, comparative studies have since been conducted on **medullary thickness**. The thickness of the medulla provides a measure of the lengths of the longest loops of Henle. A problem that needs to be addressed in such comparative studies is that medullary thickness depends on the body size of mammals; humans, for example, have a very thick medulla compared to all species of mice merely because humans are over 1000 times bigger than mice. To re-

[6] Figure 27.6*b* is drawn to emphasize other features and does not show this.

[7] Be certain not to confuse juxtamedullary capsules and the juxtaglomerular apparatus.

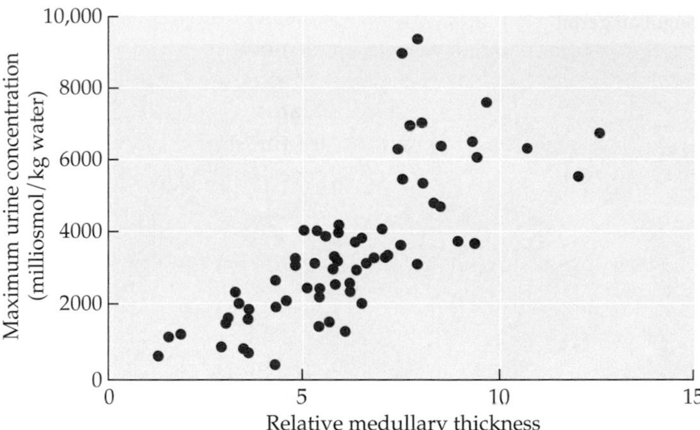

Figure 27.8 Maximum urine concentration correlates with the relative thickness of the medulla The relation is plotted for 68 species of mammals, each point representing a different species. The relative medullary thickness is a dimensionless number. To calculate it, an index of kidney size is first calculated by taking the cube root of the product of the three principal linear dimensions (length, width, thickness) of the kidney. Medullary thickness is then expressed as a ratio of the index of kidney size to obtain relative medullary thickness. (After Beuchat 1990.)

move the effects of absolute kidney size, medullary thickness is expressed as a ratio of kidney size. This ratio is called **relative medullary thickness**. A high relative medullary thickness means that the longest loops of Henle are long relative to the overall dimensions of the kidney. By now, data are available on many species, and as Figure 27.8 shows, urinary concentrating ability is strongly correlated with relative medullary thickness: Species with high relative medullary thickness tend to be able to produce especially concentrated urine.

The latest incarnation of Sperber's work is shown in Figure 27.9, where relative medullary thickness is plotted as a function of body size for mammals from several types of habitats. This modern analysis reveals that the relative thickness of the medulla tends to decrease allometrically with body size (just as concentrating ability tends to decrease with body size, as seen in Figure 26.17). Habitat, however, is a significant factor. At any given body size, mammals from arid habitats tend to have the thickest medullas and longest loops of Henle, whereas those from freshwater aquatic habitats have the thinnest and shortest, and those from intermediate mesic habitats are in between.

A stunning morphological comparison of the kidneys of three species of rodents of roughly similar body size is seen in Figure 27.10. Two of the species, the Mongolian gerbil and sand rat, evolved in deserts. Both have far more-prominent renal papillas and thicker medullas than the laboratory rat (Figure 27.10*a–c*). Moreover, the sand rat has a thicker, longer papilla than the gerbil. Detailed studies of the sand rat reveal that its renal medulla is particularly elaborately organized; in comparison to most mammals, an especially high proportion of the long loops of Henle in the sand rat kidney extend far into the papilla (Figure 27.10*d*), rather than turning back only a fraction of the way toward the tip. The sand rat, when living in its natural habitat, experiences far higher dietary salt loads than the gerbil because it subsists largely on succulent plants of very high salt content (see page 676), whereas the gerbil is a seed eater. The sand rat can produce a bit more-concentrated urine than the gerbil (6300 m*Osm* versus 5000 m*Osm*). What is more striking, however, is that the sand rat produces far greater *volumes* of highly concentrated urine than the gerbil. The longer, thicker papilla in the sand rat kidney correlates with the species' ability to produce an abundance of concentrated urine.

Countercurrent multiplication is the key to producing concentrated urine

When renal physiologists finally figured out how mammals make urine hyperosmotic to their blood plasma, they were guided to the loops of Henle by Sperber's studies of comparative kidney morphology. We will soon return to the loops, but first we need to dis-

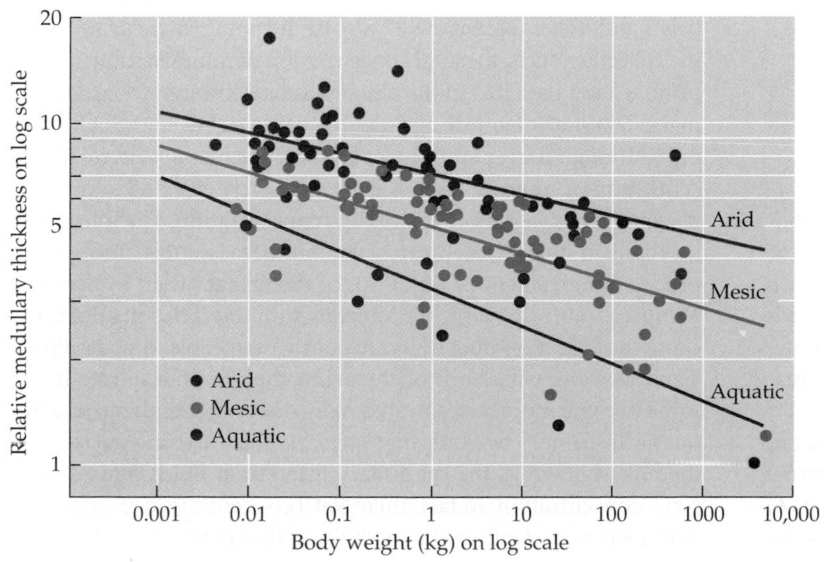

Figure 27.9 The relation between relative medullary thickness and body size depends on whether mammals are native to arid, mesic, or aquatic habitats Each point represents a different mammal species. The three lines are fitted statistically through the points for the arid, mesic, and freshwater aquatic species. The straight lines on this log–log plot indicate that the relations are allometric (see Appendix G: Exponential and Allometric Equations). To interpret this plot, keep in mind that logarithmic scales tend to cause visual compression of data. The differences in relative medullary thickness among the arid, mesic, and aquatic mammals of a particular body size are substantial; for example, the medullary thickness of a representative 1-kg arid species would be over twice that of a 1-kg aquatic species. See the legend of Figure 27.8 for a description of how relative medullary thickness was calculated. (After Beuchat 1996.)

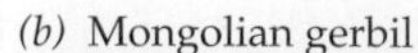

(*a*) Laboratory rat

(*b*) Mongolian gerbil

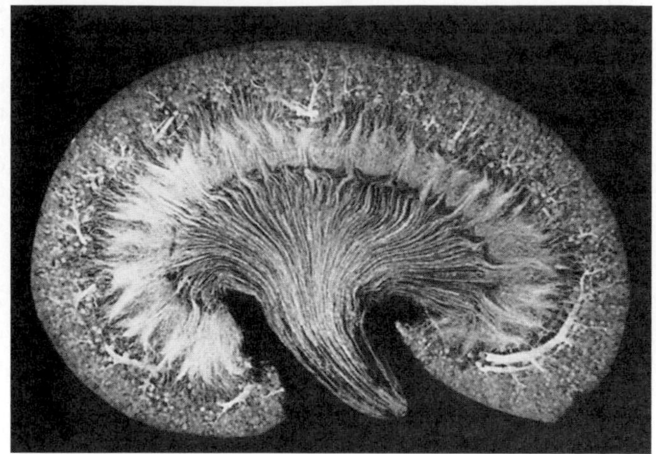

(*c*) Sand rat

(*d*) A long-looped nephron in the sand rat kidney

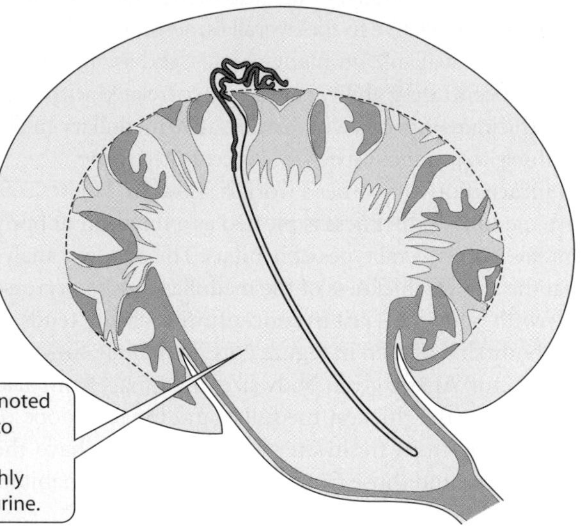

Figure 27.10 Kidney structure visualized by injection of the microvasculature Parts *a*, *b*, and *c* are midsagittal sections of the kidneys of three species of rodents of similar adult body size, in which the microscopic blood vessels of the kidneys have been injected with rubber for visualization. *(d)* is a drawing of a sand rat kidney showing a nephron with a long loop of Henle. The gerbil in *b* is *Meriones shawii*; the sand rat in *c* and *d* is *Psammomys obesus*. (Photographs in parts *a–c* courtesy of Lise Bankir [see Bankir and de Rouffignac 1985]; *d* after Kaissling et al. 1975.)

tinguish *urea* and *nonurea solutes* and discuss the immediate concentrating process for the latter. The nonurea solutes are simply the solutes other than urea. They consist mostly of inorganic ions such as Na^+, K^+, Cl^-, and SO_4^{2-}. An important operational parameter is the osmotically effective concentration of all the nonurea solutes taken together, termed the **total concentration of nonurea solutes.**

THE IMMEDIATE CONCENTRATING PROCESS FOR NONUREA SOLUTES The immediate concentrating process for the nonurea solutes is the removal of water from the urine as it flows through the collecting ducts to leave the kidney. Recall that on its way out of the kidney, urine is discharged from the nephrons into the collecting ducts, and then flows down the collecting ducts—passing first through the renal cortex and then the medulla—prior to being discharged into the renal pelvis and ureter (see Figure 27.6*b*). At the point where urine enters the collecting ducts, the total concentration of nonurea solutes in the urine is *lower* than that in the blood plasma. However, when a mammal is in a state of antidiuresis, as the urine passes in the collecting ducts through deeper and deeper layers of the medulla, its total concentration of nonurea solutes is progressively elevated, ultimately reaching levels well above the plasma concentration. The immediate mechanism that concentrates the nonurea solutes during this process is movement of water out of the urinary fluid by osmosis. Nonurea solutes are largely trapped within the collecting ducts because the collecting-duct walls are poorly permeable to such solutes. Thus, as water passes out of the urine, the nonurea solutes in the urine become more concentrated. Why does water undergo osmosis out of the urine? The fluids that *surround* the collecting ducts in the medulla, known as the **medullary interstitial fluids**, have a high NaCl concentration. In fact, their NaCl concentration rises steadily with increasing depth in the medulla, so that in the deepest parts of the medulla the osmotic pressure attributable to NaCl is far above plasma osmotic pressure. During antidiuresis, the walls of the collecting ducts are freely permeable to water. As urine inside the collecting ducts flows deeper into the medulla and encounters ever-more-concentrated interstitial fluids just on the other side of the collecting-duct walls, water progressively moves by osmosis out of the urine into the medullary interstitial fluids.

An important attribute of these processes is that *a high NaCl concentration on the outside of the collecting ducts serves to concentrate not only NaCl, but also many other nonurea solutes, on the inside.* This happens because the solutes involved cannot readily cross the walls of the collecting ducts, yet the duct walls are freely permeable to water. Because of this difference between permeability to solutes and to water, when high interstitial NaCl concentrations are encountered deep in the medulla, the primary process of equilibration between the urine and the interstitial fluid is osmosis. As this osmosis occurs, nonurea solutes in the urine are concentrated *indiscriminately* until their total osmotic concentration matches the osmotic concentration of NaCl in the medullary interstitial fluid.

A SINGLE EFFECT BASED ON ACTIVE NaCl TRANSPORT Now we must consider how the gradient of NaCl concentration in the medullary interstitial fluids is created. The loops of Henle are responsible. The first step in understanding how the loops of Henle produce the NaCl gradient is to study a phenomenon, often termed the **single effect**, that is well documented in the outer zone of the medulla, where the thick segments of the ascending limbs of the loops of Henle occur.

The cells in the walls of the ascending thick segment of a loop of Henle actively transport NaCl from the fluid inside the loop into the adjacent medullary interstitial fluid. The consequences of this NaCl transport, illustrated in Figure 27.11, depend on the permeability characteristics of the ascending limb and the adjacent descending limb of the loop of Henle. The walls of the ascending limb are essentially impermeable to water. Thus the active transport of NaCl out of the fluid inside the ascending limb creates a difference in osmotic pressure between that fluid and the adjacent interstitial fluid, as well as decreasing the NaCl concentration of the fluid inside the ascending limb and increasing the NaCl concentration of the interstitial fluid. The permeability characteristics of the descending limb appear to vary from species to species. Nonetheless, by one set of passive processes or another, the fluid inside the descending limb readily approaches equilibrium or near-equilibrium with the interstitial fluid in terms of osmotic pressure and ion concentration.

In a few words, the active transport of NaCl out of the ascending limb lowers the NaCl concentration and osmotic pressure of the ascending-limb fluid and raises the NaCl concentration and osmotic pressure of both the adjacent interstitial fluid and adjacent descending-limb fluid. These differences between the ascending-limb fluid and the *adjacent* interstitial and descending-limb fluid represent the *single effect* of the active-transport mechanism.

(a) Initial condition

Interstitial fluid

Descending limb Ascending limb

The walls of the ascending limb are essentially impermeable to water and are drawn thick to symbolize that.

(b) Processes that generate the single effect

H_2O

Na^+

Cl^-

NaCl

Active transport of NaCl out of the fluid in the ascending limb dilutes that fluid and concentrates the interstitial fluid. The walls of the descending limb are permeable to water, so the fluid inside becomes concentrated by losing water osmotically to the interstitial fluid and sometimes by gaining Na^+ and Cl^- by diffusion.

(c) The single effect

The **single effect** is the difference in osmotic pressure and NaCl concentration between (i) the ascending-limb fluid and (ii) the adjacent interstitial fluid and descending-limb fluid.

KEY

Active transport

Osmosis or diffusion

Na^+

Cl^-

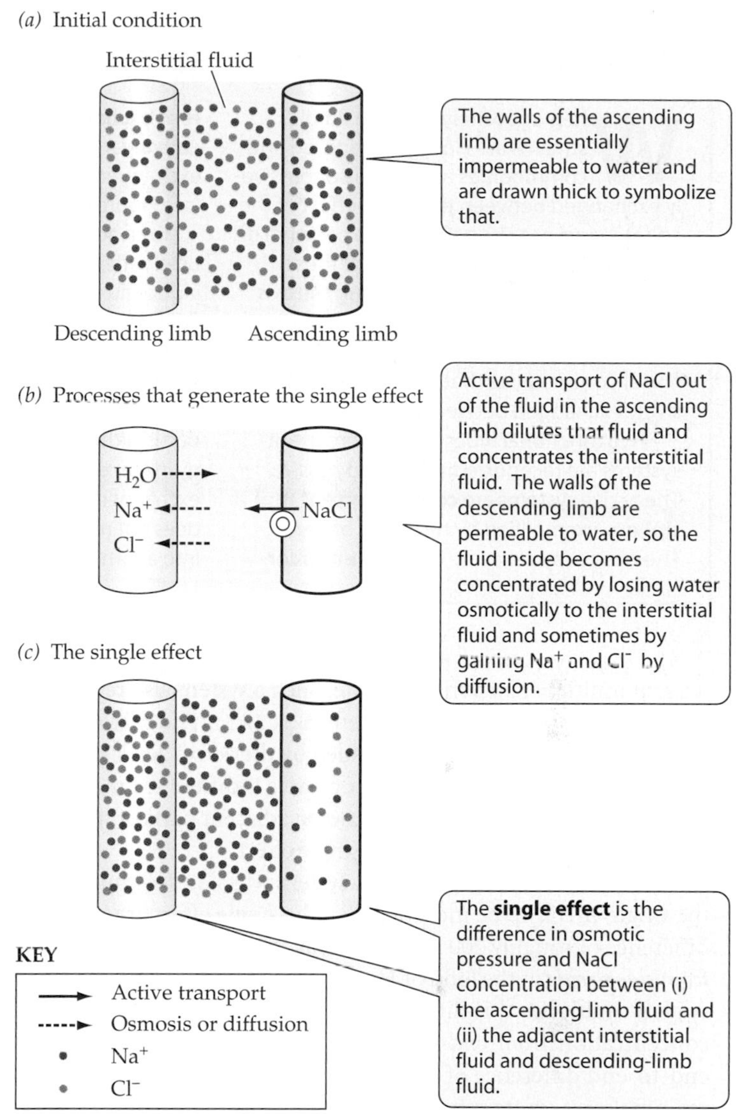

Figure 27.11 Generation of the single effect Shown here are the ascending limb of the loop of Henle and adjacent descending limb in the outer zone of the medulla, where the thick segments of the ascending limbs occur. As a thought exercise, the diagrams show how the single effect could be generated from scratch. *(a)* In the initial condition, all the fluids are identical in their osmotic pressures and ion concentrations. *(b)* The processes that generate the single effect. *(c)* The single effect that is produced. The osmotic pressure and the concentrations of ions in the ascending-limb fluid are lowered from their original levels, whereas the osmotic pressure and the concentrations of ions in the interstitial and descending-limb fluids are raised.

COUNTERCURRENT MULTIPLICATION The major hurdle in understanding how mammals produce concentrated urine was crossed in the 1940s and 1950s when Werner Kuhn (1899–1968), Heinrich Wirz (1914–1993), and several other investigators demonstrated that the concept of **countercurrent multiplication** applies to the loops of Henle. In the classic model of countercurrent multiplication generated by their work, it was assumed that all parts of each ascending limb actively transport NaCl in the manner just described. Here, we develop that classic model. Later, we discuss complexities introduced by more-recent research.

The hairpin shape of a loop of Henle sets up two fluid streams that are oppositely directed (countercurrent), intimately juxtaposed, and connected. These properties are all requirements for a counter-

BOX 27.3 Countercurrent Multipliers versus Countercurrent Exchangers

When oppositely directed fluid streams are closely juxtaposed and commodities are actively or passively exchanged between them, the effect of the countercurrent arrangement is to preserve or magnify differences in the levels of those commodities from *end to end* along the axis of fluid flow. The countercurrent arrangement has this effect because it *impedes end-to-end flux* of commodities that are actively or passively exchanged between the fluid streams.

Two functional types of countercurrent systems are recognized: *active* and *passive*. The active systems are **countercurrent multipliers**, exemplified by the loops of Henle. The passive systems are called **countercurrent exchangers** (or *countercurrent diffusion exchangers*) and are exemplified by the heat exchangers in the appendages of mammals (see Figure 8.29*b*).

In an active system, metabolic energy is used *within the countercurrent system itself* to induce flux of commodities into or out of the fluid streams; within the loop of Henle, for example, energy is used to transport NaCl out of the ascending limb. In a passive system, fluxes of commodities into or out of the fluid streams occur without expenditure of metabolic energy *in the countercurrent system itself*. In the heat exchanger in Figure 8.29*b*, for example, heat does not move out of one blood vessel and into another because of any metabolic energy expenditure *within* the countercurrent system; instead, heat follows temperature gradients that exist because energy expenditure *elsewhere* in the body has caused the body core to be warmer than the environment.

Active countercurrent systems *create* differences in levels of commodities from end to end along their axis of flow. Note, for instance, that if the loops of Henle were turned off, the gradient of osmotic pressure and NaCl concentration from the outer to the inner end of the loops would disappear. Passive systems, by contrast, do not create end-to-end differences, but instead *preserve* or *accentuate* end-to-end differences that already exist for other reasons.

current multiplier system to operate. Such a system also requires an energy investment within the system (Box 27.3). The energy investment in the loop of Henle is provided by the active NaCl transport we have already discussed, which creates a difference in osmotic pressure and ion concentration between adjacent parts of the oppositely directed fluid streams—the single effect.

The countercurrent multiplier system translates—multiplies—the single effect. To be more specific, the single effect amounts to a difference of roughly 200 m*Osm* and, as already stressed, is oriented *from side to side* in the loop of Henle. The countercurrent multiplier system multiplies this difference into a much-larger difference in concentration *from end to end* in the loop (Figure 27.12*a*). An end-to-end difference of 600 m*Osm* would not be unusual. Many mammals can create substantially greater end-to-end differences.

The mechanism of countercurrent multiplication is diagrammed in Figure 27.12*b*. Although osmotic pressures are shown in the figure and the following discussion is phrased in those terms, it will be important to remember that differences in osmotic pressure in the loop of Henle are paralleled by differences in NaCl concentration. In step 1 of Figure 27.12*b*, the entire loop of Henle and the interstitial space are filled with fluid of the same osmotic pressure as that exiting the proximal convoluted tubule—approximately isosmotic with the blood plasma (300 m*Osm*). In step 2, active transport establishes a single-effect osmotic gradient of 200 m*Osm* all along the loop. In step 3, fluid moves through the loop in countercurrent fashion. Fluid that was concentrated in the descending limb during step 2 is thus brought around into the ascending limb and now lies opposite to the descending limb, so that both limbs and the interstitial space are filled with concentrated fluid at the inner end of the loop. Now when, in step 4, the single-effect osmotic gradient is again established, the interstitial fluid is elevated to 500 m*Osm* at the inner end, rather than the 400 m*Osm* developed in step 2, and the fluid in the descending limb also reaches this higher osmotic concentration. Steps 5 and 6, and steps 7 and 8, repeat this process. *Fluid concentrated in the descending limb moves around into the ascending limb, setting the stage for the single effect to produce an ever-increasing osmotic concentration in the interstitial fluid and descending limb at the inner end of the loop.* Meanwhile, the steady influx of 300-m*Osm* fluid into the beginning of the descending limb—and the dilution of the ascending-limb fluid as it flows from deep in the medulla to the top of the ascending limb—combine to keep the osmotic pressure of the interstitial fluid at the outer (cortical) end of the loop near 300 m*Osm*. Thus the difference in osmotic pressure between the two ends of the loop becomes greater and greater, so much so that it greatly exceeds the single effect (see Figure 27.12*a*).

As earlier noted, during the early years when the countercurrent multiplication concept was initially applied to understanding mammalian kidney function, the single effect was postulated to be created along the entire length of a loop of Henle by active NaCl transport out of the ascending limb. However, by 1970, research had established that the thin segment of the ascending limb deep in the medulla is unlikely to be carrying out such active transport. That discovery started a saga that continues unended today. Active transport of NaCl out of the tubular fluid occurs in the thick segment of the ascending limb, and a consensus exists that the single effect is created according to the classic model in the outer region of a loop of Henle where the thick ascending segment occurs. On the other hand, the single effect is now assumed to be created by some other mechanism in the inner region of a loop of Henle where the ascending limb is thin. The question of the mechanism in the inner region has proved to be very complex and remains unresolved.

That said, it is important to return to the big picture: Countercurrent multiplication of a single effect along much or all of the length of a loop of Henle creates a large gradient of osmotic pressure from one end of the loop to the other. In the medulla, there are thousands of loops of Henle, all aligned in parallel. We would expect that, by their combined action, they would create in the medullary tissue as a whole a dramatic gradient of increasing osmotic pressure from its outer to its inner side. The classic data that originally confirmed this expectation are shown in Figure 27.13.

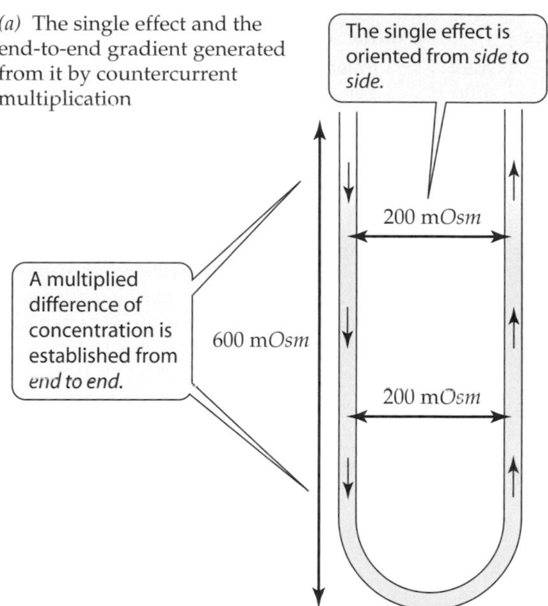

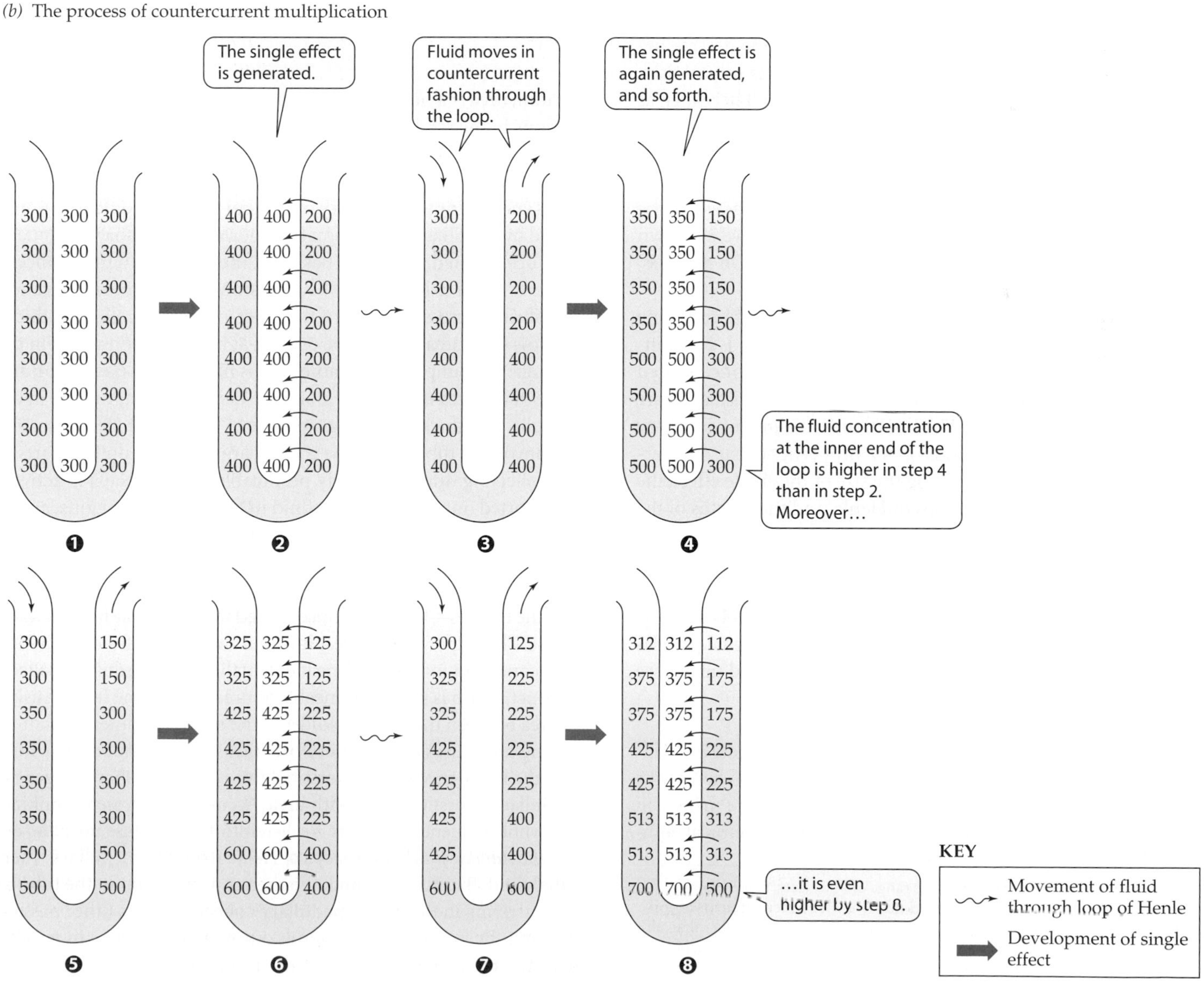

(b) The process of countercurrent multiplication

The single effect is generated.

Fluid moves in countercurrent fashion through the loop.

The single effect is again generated, and so forth.

❶

300	300	300
300	300	300
300	300	300
300	300	300
300	300	300
300	300	300
300	300	300
300	300	300

❷

400	400	200
400	400	200
400	400	200
400	400	200
400	400	200
400	400	200
400	400	200
400	400	200

❸

300	200
300	200
300	200
300	200
400	400
400	400
400	400
400	400

❹

350	350	150
350	350	150
350	350	150
350	350	150
500	500	300
500	500	300
500	500	300
500	500	300

The fluid concentration at the inner end of the loop is higher in step 4 than in step 2. Moreover…

❺

300	150
300	150
350	300
350	300
350	300
350	300
500	500
500	500

❻

325	325	125
325	325	125
425	425	225
425	425	225
425	425	225
425	425	225
600	600	400
600	600	400

❼

300	125
325	225
325	225
425	225
425	225
425	400
425	400
600	600

❽

312	312	112
375	375	175
375	375	175
425	425	225
425	425	225
513	513	313
513	513	313
700	700	500

…it is even higher by step 8.

KEY
Movement of fluid through loop of Henle
Development of single effect

Figure 27.12 Countercurrent multiplication in the loop of Henle *(a)* The distinction between the side-to-side (transverse) difference in osmotic pressure and the end-to-end (axial) difference in the loop of Henle. The side-to-side difference is the single effect. The end-to-end difference is generated from the side-to-side difference by countercurrent multiplication. *(b)* The *process* by which countercurrent multiplication occurs. The numbers are osmotic pressures in units of milliosmolarity (m*Osm*). The operation of the multiplier is presented conceptually as a series of alternating steps. In ❶, the entire system is at 300 m*Osm*. In ❷, a single-effect osmotic gradient of 200 m*Osm* is developed all along the loop, and in ❸, fluid flows through the loop. These steps are repeated in ❹ through ❽. The amount of fluid movement through the loop decreases progressively in ❸, ❺, and ❼. Fluid entering the descending limb is always at 300 m*Osm*, creating a tendency for the osmotic pressure at the cortical end of the descending limb and interstitial space to remain near 300 m*Osm*. Although both *(a)* and *(b)* are presented in terms of osmotic pressures, the differences in osmotic pressure are paralleled by differences in NaCl concentration (*b* after Pitts 1974.)

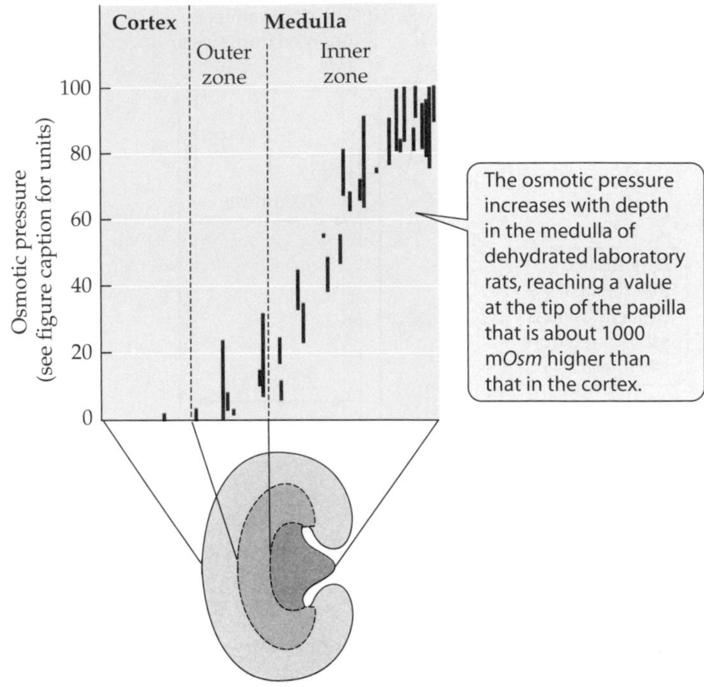

Figure 27.13 Osmotic pressure increases with depth in the medulla Each vertical red line shows the range of osmotic pressures measured at a particular depth in the cortex or medulla of kidneys taken from five dehydrated laboratory rats. All the rats, besides being sampled at various places, were deliberately sampled at the tip of the papilla, accounting for the cluster of data there. On the *y* axis, 0 represents an osmotic pressure equal to that of the blood plasma in the general circulation, whereas 100 represents the highest osmotic pressure measured (about 1000 m*Osm* greater). Intermediate osmotic pressures are scaled relative to the two extremes; specifically, any particular measured osmotic pressure (OP) is expressed as 100 × (measured OP – plasma OP)/(maximal OP – plasma OP). Throughout the cortex, the osmotic pressure is equivalent to the osmotic pressure of plasma in the general circulation. The increase in osmotic pressure with depth in the medulla is attributable both to an increase in NaCl concentration and to an increase in urea concentration. (After Wirz, Hargitay, and Kuhn 1951.)

CONCLUDING POINTS ON THE MECHANISM OF CONCENTRATING NONUREA SOLUTES Figure 27.14*a* summarizes the changes in the total concentration of nonurea solutes in the tubular fluid of nephrons and collecting ducts when the kidney of a mammal is producing concentrated urine. As fluid in a nephron travels down the descending limb of the loop of Henle, its concentration of nonurea solutes rises, reaching a high level at the hairpin bend of the loop. Thereafter, as the fluid comes back out of the medulla in the ascending limb of the loop of Henle, its concentration of nonurea solutes falls, so that by the time the fluid exits the loop, it is actually *more dilute* than when it started and more dilute than the blood plasma. Then, however, the fluid makes a final, crucial pass through the medulla, traveling down a collecting duct to be discharged into the renal pelvis. On this pass, final concentration of the nonurea solutes occurs.

The total concentration of nonurea solutes in the definitive urine depends on the NaCl concentration of the interstitial fluids of the innermost medulla, because in a kidney producing concentrated urine, the urine osmotically equilibrates with those fluids just before leaving the kidney (see Figure 27.14*a*). In turn, the inner-medullary NaCl concentration itself depends on the properties of the countercurrent multiplier system, including the size of the single effect, the rate of fluid flow through the loops of Henle, and the lengths of the loops. Lengthening of the loops tends to increase the end-to-end gradient of NaCl concentration that can be maintained by the loops and thus tends to raise the inner-medullary NaCl concentration. This explains why, among related species of similar body size, the species with relatively thick medullas and prominent renal papillas tend to be capable of producing relatively concentrated urine (see Figures 27.7, 27.8).[8]

CONCENTRATION OF UREA The mechanisms that concentrate urea differ from those that concentrate the nonurea solutes. Whereas the walls of the collecting ducts prevent most solutes in the urine and medullary interstitial fluid from diffusing to electrochemical equilibrium, urea is a notable exception. Urea is present at high concentrations in the medullary interstitial fluid, and when mammals are in an antidiuretic state, the walls of the collecting ducts *in the inner medulla* permit free diffusion of urea between the urine inside the ducts and the inner-medullary interstitial fluid (this diffusion is mediated by a facilitated-diffusion transporter protein that is dramatically upregulated by ADH). Basically, therefore, high urea concentrations in the urine reflect the diffusion of urea to concentration equilibrium across the walls of the inner-medullary collecting ducts.

How does urea come to be present at high concentrations in the medulla? Put simply, much more urea is filtered than is excreted, and some of the urea reabsorbed along the nephrons accumulates in the medulla. The thick ascending segment of the loop of Henle, the distal convoluted tubule, and the cortical and outer-medullary parts of the collecting duct are poorly permeable to urea. NaCl is actively transported out of the tubular fluid in these tubular regions, and at least in the cortical and outer-medullary collecting duct, water leaves the tubular fluid by osmosis during antidiuresis. Because the permeability to urea in these tubular regions is low, urea—trapped inside the tubules—becomes concentrated in the tubular fluid as water is lost. The important net result is that the tubular fluid has a high urea concentration by the time it enters the inner-medullary collecting duct, which is highly permeable to urea, permitting free diffusion of urea between the tubular fluid and the medullary interstitial fluid, during antidiuresis. Urea therefore diffuses from the tubular fluid into the inner-medullary interstitial fluid, a process that charges the medullary interstitial fluid with urea. According to present thinking, this whole sequence of events is self-reinforcing because urea *also enters the tubular fluid in the loops of Henle* from the medullary interstitial fluid. By such recycling, the urea concentration in the tubular fluid arriving in the inner-medullary collecting ducts (after passage through the loops of Henle) tends automatically to rise in parallel with the urea concentration of the interstitial fluid. Thus, with a

[8] When comparing species that cover a wide range of body sizes, *relative* loop length—estimated as relative medullary thickness—is a far better predictor of concentrating ability than *absolute* loop length (see Figure 27.8). Factors other than absolute length thus clearly play major roles in kidney concentrating function, but these additional factors are not yet understood.

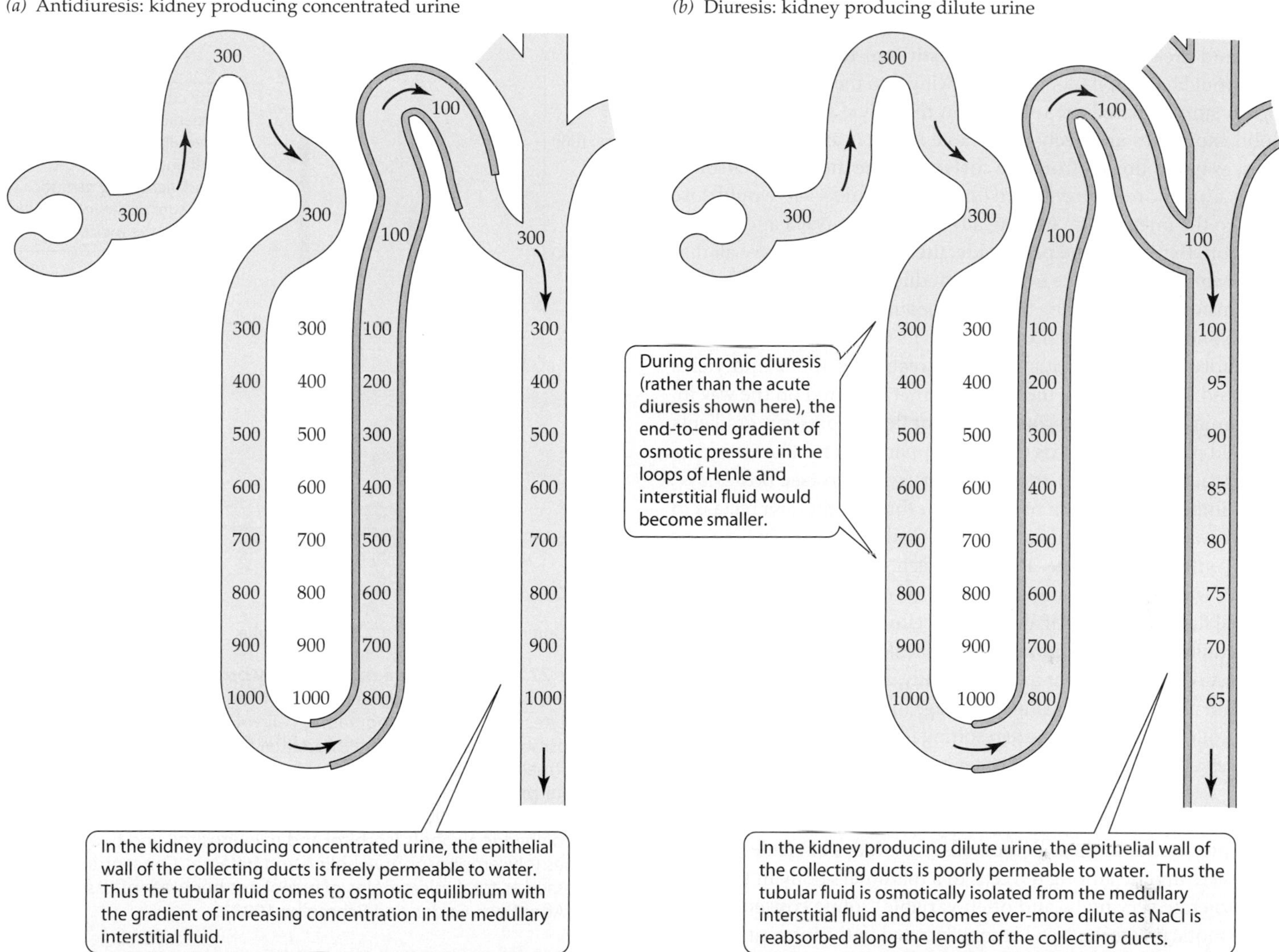

Figure 27.14 Osmotic pressures attributable to nonurea solutes in the nephrons and collecting ducts during antidiuresis and diuresis The thickness of the walls symbolizes permeability to water; parts of the tubules that are poorly permeable to water are thick, whereas those permeable to water are thin. The numbers, expressed in units of milliosmolarity, are approximate and intended only to illustrate general trends.

steady influx of new urea from filtration, a gradient favorable for diffusion of urea *into* the interstitial fluid from the collecting ducts is maintained, even though the interstitial-fluid concentration rises to a high level. High concentrations of urea in the interstitial fluid promote high urinary concentrations because diffusive outflux of urea from the collecting-duct fluid continues only to the point of concentration equilibrium with the medullary interstitial fluid.

How does the process of urea concentration relate to the process by which nonurea solutes are concentrated? This is actually an intricate question because of the interacting dynamism of the two processes. However, two important global points can be made fairly simply. First, because urea and nonurea solutes are concentrated by rather separate mechanisms, a high urea concentration in the urine does not in any simple mathematical fashion displace nonurea solutes or reduce the concentration of nonurea solutes that is possible. The urine of a mammal can simultaneously contain high concentrations of both urea and nonurea solutes. The second point to be made is that the osmotic reabsorption of water from the urine in the inner medulla is controlled by the processing of the *nonurea* solutes. Because urea diffuses to concentration equilibrium across the walls of the inner-medullary collecting ducts, it does not (except transiently) make a direct contribution to the *difference* in osmotic pressure between the collecting-duct fluid and the interstitial fluid. The *difference* in osmotic pressure—which governs the osmotic reabsorption of water—is a consequence of different concentrations of the nonurea solutes.

THE BLOOD SUPPLY OF THE MEDULLA: THE VASA RECTA The blood capillaries of the medulla form hairpin loops—known as **vasa recta**—that parallel the loops of Henle. This arrangement, diagrammed in Figure 27.6*b*, is vividly evident in Figures 27.10*a–c*, in which the structures visualized are the blood vessels.

The looped shape of the vasa recta prevents the circulation of blood to the medulla from destroying the concentration gradients of NaCl and urea in the medullary interstitium. To see this, consider what would happen if blood, after flowing into the medulla from the cortex, simply exited the medulla on the pelvic side. The walls of blood capillaries are freely permeable to water and small solutes. Thus, as blood flowed from the cortex into the medulla—encountering ever-more-concentrated interstitial fluids—it would lose water to the interstitium osmotically and take up NaCl and urea by diffusion. Exiting on the pelvic side, the blood would leave all that water behind and take the solutes away, diluting the medulla in both ways. Instead, after entering the inner medulla, the blood reverses direction and flows back to the cortex. On its way out, as it encounters ever-more-dilute interstitial fluids, it reabsorbs water and yields NaCl and urea, reversing the processes that occurred on the way in. The familiar tendency of countercurrent flow to preserve gradients oriented parallel to the axis of flow is apparent here. The vasa recta are described as *countercurrent diffusion exchangers* (see Box 27.3).

An important function of blood flow through the vasa recta is to remove water from the medullary interstitial fluids. The final process of concentrating the urine, as we have seen, involves osmotic movement of water from the collecting ducts into the inner-medullary interstitial fluid. This water, if allowed to accumulate, would itself dilute the medullary concentration gradient. The flow of blood through the vasa recta carries the water away. Evidently, as the blood dynamically loses and regains water during its passage into and out of the medulla, the colloid osmotic pressure resulting from the blood proteins introduces a bias for the gains of water by the blood to exceed losses.

CELL VOLUME REGULATION IN THE MEDULLA The cells in the medulla of the kidney—such as those in the walls of the loops of Henle and vasa recta—are unique among the cells in a mammal's body in that they must tolerate very high (and often variable) solute concentrations and osmotic pressures. NaCl is concentrated in the interstitial fluids bathing these cells. For the cells to maintain normal cell volumes, they must contain adequate levels of intracellular solutes to balance the interstitial NaCl (see Figure 25.8), but a high intracellular level of NaCl would not be acceptable because such a concentration of NaCl would disrupt cell macromolecules. Instead, the cells contain high intracellular concentrations of a set of organic osmolytes of metabolic origin, notably polyhydric alcohols and methylamines (Figure 27.15). These organic compounds serve as *compatible solutes* (see page 682): They balance the high extracellular NaCl concentration while having relatively small effects on cell macromolecules.

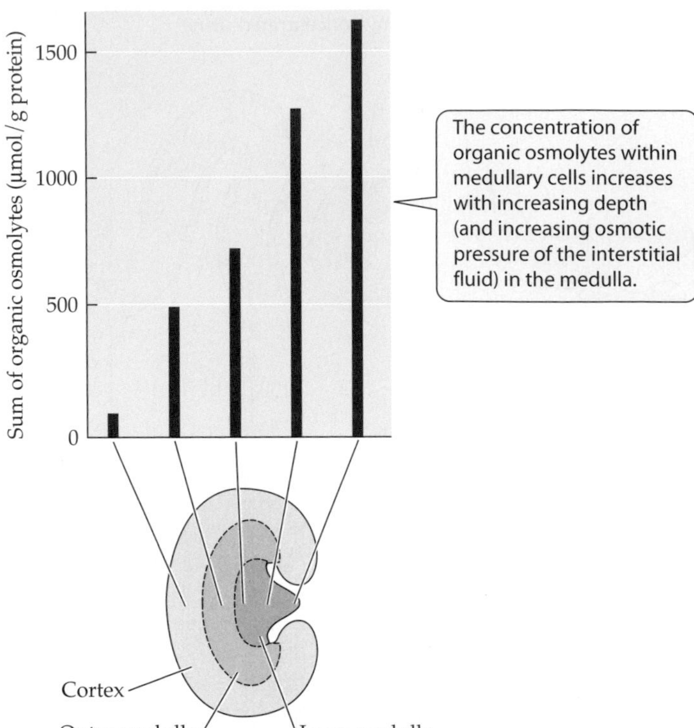

Figure 27.15 Cell volume regulation by organic osmolytes The cells in the renal medulla have high concentrations of organic osmolytes as a means of regulating cell volume in the face of the high osmotic pressures in the interstitial fluids. The data shown are for normally hydrated laboratory rats. Each bar represents the sum of the four principal organic osmolytes at the designated anatomical location. The osmolytes summed together for presentation here are two polyhydric alcohols (sorbitol and *myo*-inositol) and two trimethylamines (glycerophosphorylcholine and betaine). Concentration is expressed as total micromoles of osmolytes per gram of tissue protein. (After Beck, Guder, and Schmolke 1998.)

The kidney tubules carry out many processes to produce definitive urine

Thus far we have focused on how the mammalian kidney can produce urine more concentrated than the blood plasma. The mammalian kidney resembles other kidneys, however, in that it carries out many functions simultaneously as it forms definitive urine. In this section we take more of an overview of how the nephrons and collecting ducts function in mammals. A useful way to approach this task is, first, to discuss how multiple solutes and water are processed when the kidney is producing a concentrated urine, then discuss how the kidney switches between producing concentrated and dilute urine.

AN OVERVIEW OF EVENTS IN THE CONCENTRATING KIDNEY Glomerular filtration is, of course, the first step in forming urine. Compared to other vertebrates, mammals—with some known exceptions (e.g., dromedary camels)—tend to maintain relatively stable GFRs and adjust their rate of urine production principally by adjusting the fraction of filtered fluid that they ultimately reabsorb prior to excretion. The fluid introduced into the Bowman's capsules by filtration is approximately isosmotic to the blood plasma and contains similar concentrations of inorganic ions, glucose, and amino acids. A major function of the proximal convoluted tubule is reabsorption of NaCl and water; in fact, 60% to 80% of the filtered amounts of NaCl and water are reabsorbed by the time the tubular fluid reaches the beginning of the loop of Henle. The walls of the proximal tubule are freely permeable to water, so water exits osmotically as NaCl and other solutes are reabsorbed, and the tubular fluid stays isosmotic to the blood plasma. Glucose, many amino acids, and HCO_3^- (bicarbonate ion) are almost completely reabsorbed in the proximal tubule.

A major contemporary area for research is establishing the molecular basis for the function of the proximal convoluted tubule and all other regions of the kidney tubules. The ultimate goal of this research is to understand every aspect of reabsorption and secretion along all parts of the renal tubules in terms of the specific transporter proteins, channel proteins, and other molecules that mediate the processes. The pursuit of this goal will provide insight into disease processes and will clarify ways in which the kidneys of various species differ in their functional properties. An illustration of the sort of insight obtained is provided by the summary of the major solute processes in the epithelial wall of the early proximal convoluted tubule in Figure 27.16*a*. As shown, Na^+ reabsorption from the urine in this part of the proximal tubule is driven by primary active transport carried out by Na^+–K^+-ATPase in the basolateral membrane. Glucose reabsorption, however, occurs by secondary active transport (see page 76). Until the relatively recent discovery of aquaporins, the molecular basis of the high permeability of the epithelial wall of the proximal tubule to water was a mystery. In fact, aquaporins are responsible. The aquaporins in the proximal tubule are classified as *constituitive* because they are integral members of the cell membranes, not much subject to modulation by external agents.

After fluid leaves the proximal convoluted tubule, its next step is to travel through the loop of Henle. Although the tubular fluid enters the loop isosmotic to plasma (~300 m*Osm*), it exits the loop hyposmotic to plasma (perhaps 100–150 m*Osm*), as we have seen (see Figure 27.14*a*). In the ascending thick segment of the loop of Henle, a key process that both creates the single effect for countercurrent multiplication and ultimately renders the tubular fluid hyposmotic to the blood plasma is the transport of NaCl out of the tubular fluid. A current model of the molecular biology of this process is shown in Figure 27.16*b*. Illustrating the value of molecular knowledge for the practice of medicine, a set of modern medications used to treat hypertension (high blood pressure)—known as **loop diuretics**—is targeted at the Na–K–2Cl cotransporter protein in the thick ascending limb. These medications inhibit NaCl transport out of the tubular fluid by inhibiting the cotransporter, resulting in increased Na^+ excretion, which tends to decrease the volume of the blood plasma (see page 716).

After exiting the loop of Henle, the tubular fluid passes through the distal convoluted tubule. The walls of much or all of this tubule are poorly permeable to water and actively transport NaCl out. Thus the tubular fluid remains strongly hyposmotic to plasma (see Figure 27.14*a*). Potassium is added to the tubular fluid (partly passively, partly actively) in the distal convoluted tubule and cortical collecting duct. Because most K^+ from filtration has been reabsorbed, this addition of K^+ controls the amount of K^+ eliminated in the urine.

Perhaps 5% or less of the originally filtered volume reaches the collecting duct. In the concentrating kidney, the collecting duct is permeable to water.[9] Thus dilute tubular fluid arriving there promptly comes to isosmoticity with the cortical interstitial fluid (~300 m*Osm*) by osmotic outflux of water (see Figure 27.14*a*). This tubular fluid then descends from the cortical to the pelvic end of the collecting duct. As it does so, it encounters ever-higher interstitial NaCl concentrations and attains higher concentrations of urea and nonurea solutes by the mechanisms we have discussed. Water is reabsorbed osmotically. Especially in the cortical part of the collecting duct, but also in the inner-medullary part, NaCl is actively reabsorbed (Figure 27.16*c*). This reabsorption of NaCl in the collecting duct sets the final amount of NaCl excreted. It also plays a key role in controlling urine volume, because by reducing the amount of nonurea solute in the urine, it enhances osmotic loss of water from the urine. In the end, mammals in antidiuresis typically excrete only 1% or less of the filtered NaCl and water.

[9] The terminal distal tubule may also be permeable to water.

THE DILUTING KIDNEY AND THE MODULATION OF URINE CONCENTRATION AND VOLUME Individual mammals are typically capable of adjusting the concentration and volume of their urine over broad ranges. Concentration and volume tend to undergo reciprocal change. A person in antidiuresis might produce urine that is as concentrated as about 1200 m*Osm* (U/P = 4), but limited in volume to less than 1% of the filtered amount. In diuresis, by contrast, that person might lower the concentration of the urine to about 50 m*Osm* (U/P = 0.2) and increase the volume to about 15% of the filtered amount.

The principal agent of control of these urine properties is antidiuretic hormone (ADH). The ADH of most mammals is *arginine vasopressin*; thus ADH is often called **vasopressin** in books on mammalian physiology and medicine. As knowledge advances, the known effects of ADH become more extensive and complex. The action of ADH that is of most central importance is that it modulates the permeability to water of the collecting ducts.

The effect of ADH on the permeability of the collecting duct epithelium is an exciting area for research at present because after decades of mystery, the mechanism is being revealed (see Figure 14.13). It entails a specific type of aquaporin molecule, termed aquaporin-2 (AQP-2), that is inserted into and retrieved from the apical cell membranes of the epithelial cells. The presence of ADH causes insertion of aquaporin molecules into the cell membranes and an increase in permeability to water (Figure 27.17). When ADH levels fall, the aquaporins are retrieved from the cell membranes and permeability to water falls.

In the concentrating kidney, although we have not said so heretofore, high blood levels of ADH are responsible for the high permeability of the collecting ducts to water. This high permeability permits water to leave the collecting ducts by following the osmotic gradient between the collecting-duct fluid and medullary interstitial fluid. The osmotic exit of water, as previously stressed, accounts for both the high concentration and the low volume of the urine produced during antidiuresis.

When blood levels of ADH are low and aquaporins are retrieved from the apical cell membranes of the collecting-duct epithelium, the collecting ducts are poorly permeable to water. The distal convoluted tubules are poorly permeable also. Thus, during diuresis, from the time the tubular fluid exits the loops of Henle to the time it is discharged into the renal pelvis, it is blocked from coming freely to osmotic equilibrium with the surrounding cortical and medullary interstitial fluids. Recall that the tubular fluid is hyposmotic to plasma when it exits the loops; thus it would lose water osmotically if it could. However, the low permeability of the walls of the distal tubules and collecting ducts impedes such water loss when ADH levels are low. In fact, because of the low permeability, as NaCl is actively reabsorbed in the distal tubules and collecting ducts, the tubu-

In all three parts of the kidney tubule, Na^+ diffuses into the epithelial cells from the tubular fluid because there is an electrochemical gradient favoring such diffusion.

In the early proximal tubule, the tubular fluid is rich in glucose and amino acids, and much of the Na^+ entry into a cell occurs by means of cotransporters that bring about the secondary active transport of glucose and amino acids into the cell. Only the Na-glucose cotransporter is shown here.

(a) **Early proximal convoluted tubule**

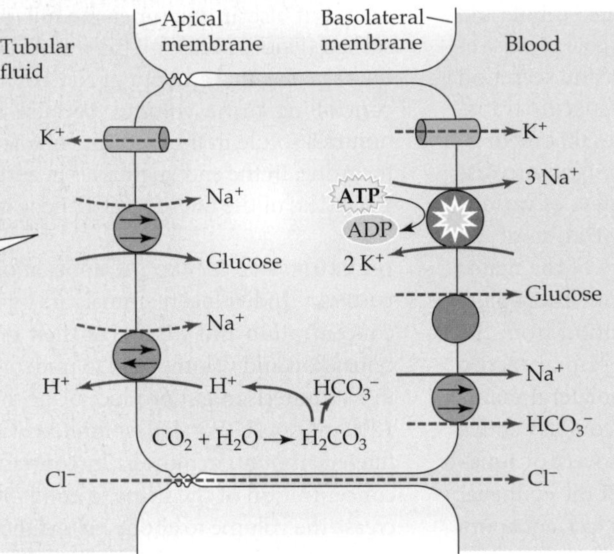

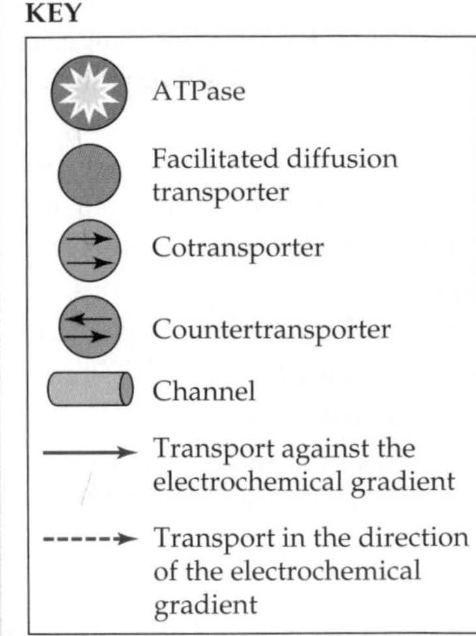

In all three parts of the kidney tubule, energy for Na^+ reabsorption comes from ATP used for primary active transport by Na^+–K^+-ATPase. The ATPase removes Na^+ from each type of epithelial cell across the basolateral cell membrane. Na^+ enters each cell across the apical cell membrane by diffusion down the electrochemical gradient generated by Na^+–K^+-ATPase. The membrane proteins involved in Na^+ entry are different in all three types of epithelial cells, however.

(b) **Thick ascending limb of loop of Henle**

In the thick ascending limb, much of the Na^+ entry into a cell occurs by means of a Na–K–2Cl cotransporter that carries K^+ and Cl^- inward by secondary active transport. **Loop diuretics** inhibit this cotransporter.

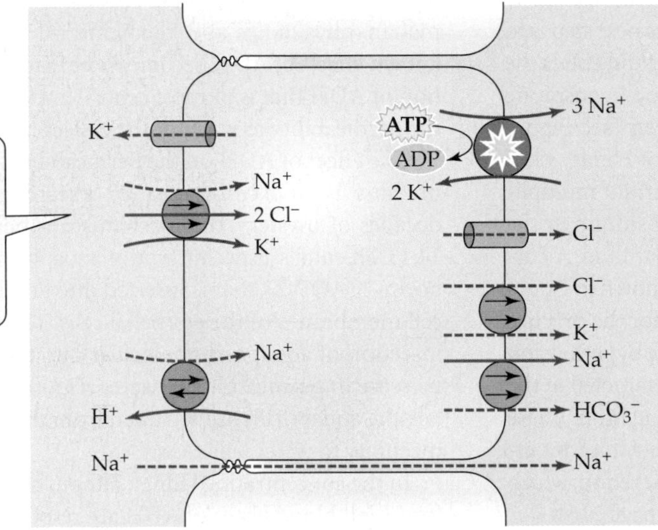

(c) **Collecting duct Na^+-reabsorbing cell (principal cell)**

In the collecting duct, Na^+ enters the principal cells by a channel. **Diuretic drugs such as amiloride** block or inhibit this channel. The principal cells are the *main targets of aldosterone*, which promotes Na^+ reabsorption by increasing synthesis of the Na^+ channel protein and the Na^+–K^+ -ATPase, as well as other actions.

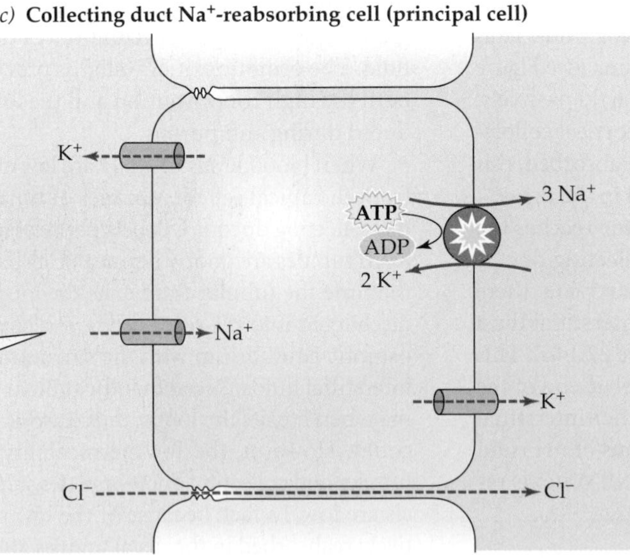

Figure 27.16 Major molecular mechanisms of NaCl reabsorption and associated processes in three parts of the kidney tubule Each drawing shows a representative epithelial cell from a different part of a mammalian kidney tubule. *(a)* The early proximal convoluted tubule. Cl^- is removed from the tubular fluid at this site by simple diffusion, largely via paracellular pathways, following an electrochemical gradient created by Na^+–K^+-ATPase. The Na–H countertransporter in the apical membrane moves H^+ into the tubular fluid, where the H^+ combines with tubular bicarbonate (HCO_3^-), forming CO_2, which enters the cell, supplying CO_2 to the intracellular reaction shown. *(b)* The thick segment of the ascending limb of the loop of Henle. Here, Cl^- is reabsorbed through the cells, rather than by the paracellular route, in a process mediated by a Na–K–2Cl cotransporter in the apical membrane. *(c)* In the collecting ducts, different cells carry out Na^+ and Cl^- reabsorption. The drawing shows a Na^+-reabsorbing cell, known as a *principal cell*. In all three segments, most K^+ brought into cells by Na^+–K^+-ATPase diffuses out via channels.

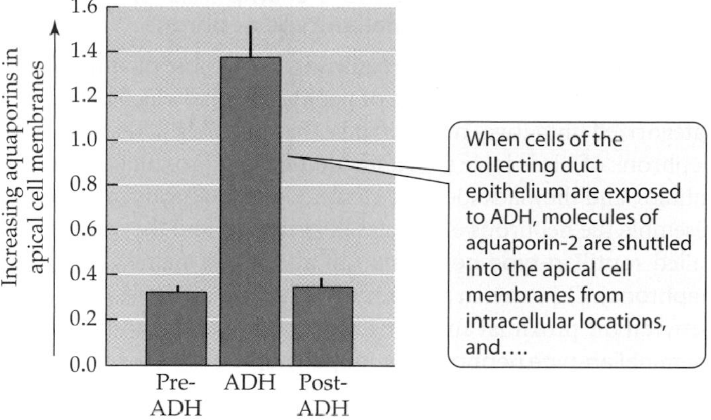

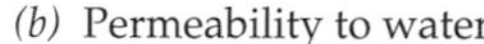

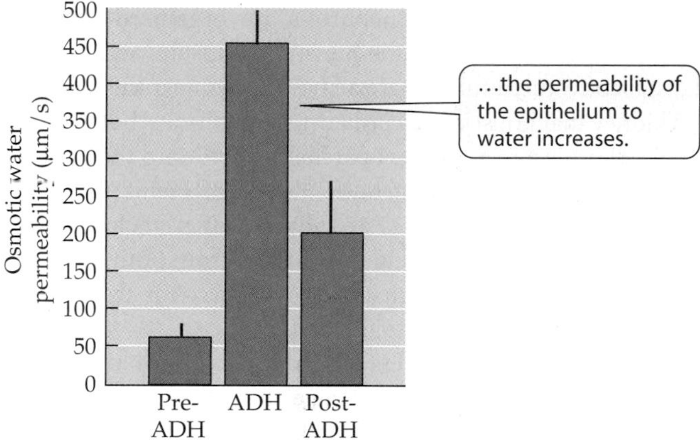

Figure 27.17 The intracellular position of aquaporin-2 and permeability to water Studies were carried out on collecting ducts from the inner medulla. The distribution of aquaporin-2 (AQP-2) molecules was determined by visualizing and directly counting the molecules by means of immunological reactions and electron microscopy. According to the shuttle hypothesis, the AQP-2 molecules in each epithelial cell are shuttled back and forth between the apical cell membrane and intracellular vesicular membranes. *(a)* The number of AQP-2 molecules in apical cell membranes as a ratio of the number in intracellular vesicular membranes. *(b)* Permeability of the collecting-duct epithelium to water. (After Knepper et al. 1996.)

lar fluid becomes ever-more hyposmotic to the plasma (Figure 27.14*b*). An abundant, dilute urine results.

Note how fundamentally similar the action of ADH is in mammals and in amphibians. In both groups—and indeed, in all groups of tetrapod vertebrates—the primary effect of ADH on the renal tubules is to increase the permeability to water of tubular epithelia that otherwise are poorly permeable. This increase in permeability to water has the important consequence that it allows the tubular fluids to come to osmotic equilibrium with the fluids surrounding the tubules. In amphibians, the fluids surrounding the distal tubules and collecting ducts are osmotically similar to the blood; thus the presence of ADH causes production of urine that approaches isosmoticity with the blood. In mammals, the collecting ducts are surrounded by fluids that are hyperosmotic to the blood. Thus ADH causes production of a hyperosmotic urine.

The excretion of *water* is controlled by ADH in mammals, as it is in amphibians.[10] The amount of each nonurea *solute* excreted is adjusted by solute-specific tubular mechanisms (e.g., active reabsorption and secretion). The concentration of ADH helps determine the amount of water excreted with the nonurea solutes.

When an individual mammal switches between chronic antidiuresis and chronic diuresis, an additional change besides the permeability adjustments occurs: The magnitude of the osmotic gradient in the medullary interstitial fluids—the gradient between the cortex and inner medulla—dimishes. For example, in a dog shifted from antidiuresis to chronic profound diuresis, the osmotic pressure of the inner-medullary interstitial fluid might fall from 2400 m*Osm* to 500 m*Osm*. Periods of hours or days are required for such changes to be fully realized.

ADH is not the only hormone that controls kidney function. As discussed in Chapter 26 (see page 716), aldosterone and natriuretic hormones help control the reabsorption and secretion of Na^+ and K^+. In addition, calcitonin affects renal function, and the kidneys themselves employ paracrines, such as eicosanoids and kinins, as *local* chemical messengers.

[10] See Chapter 26 (page 716) for a full explanation of this point and Figure 27.5 for a diagram of how changes in tubular permeability to water can alter the amount of water excreted in the urine.

Summary: Urine Formation in Mammals

- The loops of Henle, collecting ducts, and vasa recta form parallel arrays in the medulla of the mammalian kidney, creating the structural basis for the ability to form urine hyperosmotic to the blood plasma. Among species of mammals of a particular body size, the species with long loops of Henle tend to be able to produce more-concentrated urine than those with shorter loops.
- The proximal convoluted tubule reabsorbs much of the NaCl and water from the filtrate by processes that do not alter the osmotic pressure of the tubular fluid. It also fully reabsorbs glucose and amino acids.
- After the tubular fluid passes through the loop of Henle, it is less concentrated than when it entered. Nonetheless, processes in the loop of Henle create the gradient of osmotic pressure and NaCl concentration in the medullary interstitial fluid that is responsible for the ultimate concentration of the urine. In the part of the loop where the ascending limb is thick, active NaCl transport creates a single-effect difference in osmotic pressure and NaCl concentration between adjacent parts of the ascending and descending limbs. By acting as a countercurrent multiplication system, the loop generates a difference in osmotic pressure and NaCl concentration from end to end that is much larger than the single effect.
- During antidiuresis, as tubular fluid makes its last pass through the medulla in the collecting ducts, nonurea solutes are concentrated because the collecting-duct walls are permeable to water, permitting osmotic equilibration between the tubular fluid and the medullary interstitial fluid. The permeability of the collecting-duct walls to water results from insertion of aquaporin-2 molecules into cell membranes in response to ADH (vasopressin).
- During diuresis, the collecting-duct walls are poorly permeable to water, so tubular fluid is osmotically isolated from the medullary interstitial fluid and can be diluted by solute reabsorption.

Urine Formation in Other Vertebrates

Freshwater and marine teleost fish differ in nephron structure and function

The sort of nephrons we described in amphibians apparently evolved in their freshwater progenitors, because the nephrons of nearly all freshwater teleost (bony) fish are structurally similar to those of amphibians. In freshwater fish, as in amphibians, the distal convoluted tubule plays a key role in diluting the urine. NaCl is reabsorbed there, but the walls of the tubule are nearly impermeable to water, so water remains behind in the tubule, and a dilute urine is produced.

Marine teleost fish commonly lack the distal convoluted tubule. If they are descended from freshwater ancestors, as is often argued (see Box 26.1), the absence of the distal tubule probably represents a secondary loss rather than a primitive condition. The reason for the loss seems relatively straightforward: Marine teleosts are hyposmotic to their environment and thereby face continuous osmotic desiccation. They have no need of a nephron segment specialized for the production of a dilute urine rich in osmotically free water.

Besides differing in the presence or absence of the distal convoluted tubule, freshwater and marine teleosts differ in other ways. Freshwater teleosts typically have relatively large numbers of nephrons and well-developed glomeruli. Their GFRs are relatively high, as suits animals that have excesses of water that must be voided in urine. In contrast, marine teleosts tend to have relatively few nephrons and small glomeruli. They have low GFRs, a condition that seems logical for animals that face desiccation and produce relatively little urine.

Current evidence suggests that many marine teleosts form their primary urine not entirely by ultrafiltration, but partly by *secretion* into the proximal tubules. The mechanism of secretion is that ions—including Na^+, Cl^-, Mg^{2+}, and SO_4^{2-}—are transported into the proximal tubules, and water and other solutes follow (see Figure 27.2). In a small number of marine teleost species—which are described as being **aglomerular**—the trend toward small glomeruli in the marine environment is carried to its logical extreme, and the nephrons lack glomeruli. The aglomerular species form their primary urine entirely by secretion.

Some of the most interesting fish from the viewpoint of kidney function are the euryhaline teleost species that can live in either freshwater or seawater. The control of kidney function in teleosts has in fact been most thoroughly studied in some of these species. When a euryhaline fish is transferred from seawater to freshwater, it typically undergoes a large increase in GFR, mediated for the most part by a rise in the number of filtering nephrons. Active secretion of Mg^{2+} and SO_4^{2-} into the urine is curtailed. Moreover, the nephrons—and sometimes other excretory structures (e.g., the bladder)—undergo decreases in their overall permeability to water, a change that favors water excretion. Prolactin, arginine vasotocin (the "ADH" of fish; see Table 14.3), and angiotensin II are implicated in controlling these changes, but the controls are not well understood.

Reptiles have nephrons like those of amphibians, but birds have some mammalian-type nephrons

The nephrons of reptiles are broadly similar to those of amphibians. Birds, by contrast, have a range of nephron forms, which are usually categorized into two major types (**Figure 27.18**). Some of the nephrons of birds have short, uncomplicated proximal and distal tubules, and they lack loops of Henle. These nephrons superficially resemble the nephrons of reptiles in structure, and they are usually called **reptilian-type nephrons** (an alternative name is **loopless nephrons**). Other avian nephrons have a loop of Henle interposed between the proximal and distal convoluted tubules and are called **mammalian-type nephrons** (or **looped nephrons**). These nephrons have relatively large glomeruli and elaborate proximal tubules. Approximately 10% to 30% of the nephrons in a bird's kidney are typically of the mammalian type; the remainder are of the reptilian type. The mammalian-type nephrons are organized into sets. Among the nephrons of a set, the Bowman's capsules and proximal and distal convoluted tubules are all positioned near the same part of the kidney that houses the reptilian-type nephrons, but the loops of Henle all project in a compact parallel array toward the direction of the ureter. Each parallel array of loops of Henle is called a **medullary cone** (see Figure 27.18). Each kidney includes many cones. Collecting ducts carrying the outflow from both the reptilian-type and the mammalian-type nephrons run through the medullary cones on their way to the ureter.

Neither the nephrons of a reptile's kidneys nor the reptilian-type nephrons in a bird's kidneys can produce urine that is hyperosmotic to blood. However, the loops of Henle of the mammalian-type nephrons in a bird's kidney carry out countercurrent multiplication

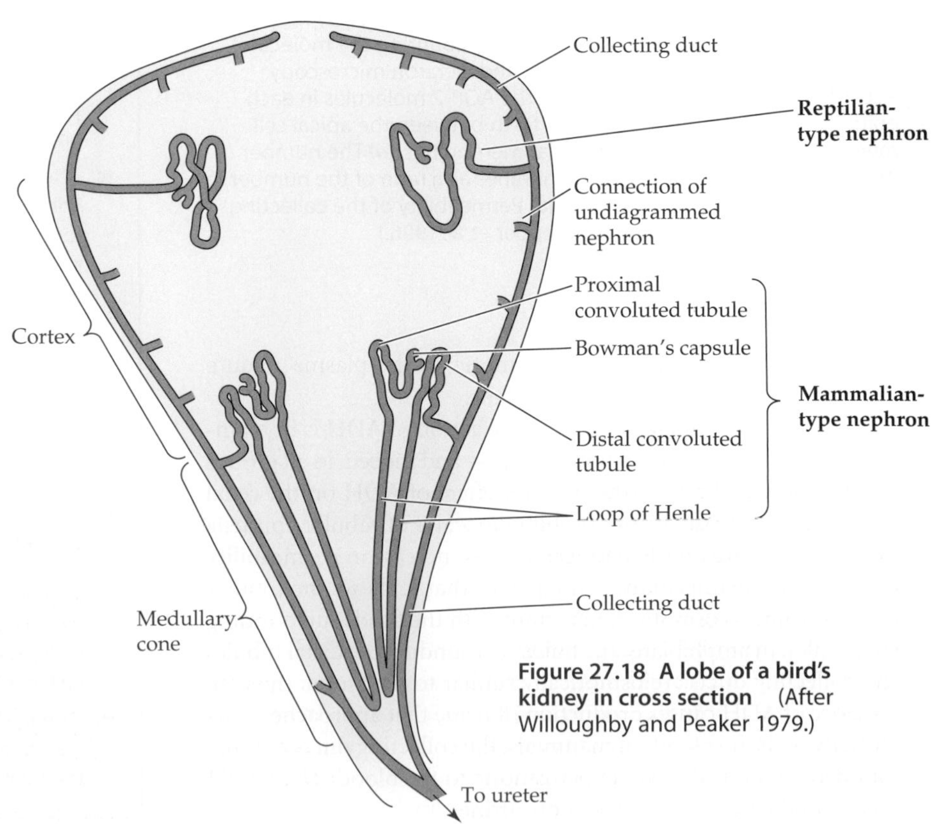

Figure 27.18 A lobe of a bird's kidney in cross section (After Willoughby and Peaker 1979.)

and can raise the urine osmotic pressure above the blood osmotic pressure. Details of the countercurrent mechanism in birds are probably different from those in mammals and are being elucidated; among other things, the ascending limbs of the loops of Henle in birds are entirely thick. For most birds, the maximum urine osmotic pressure is no more than about 2.5 times blood osmotic pressure. Among related species of birds, there tends to be a direct correlation between the maximal concentrating ability of the kidneys and the numbers of mammalian-type nephrons and medullary cones per unit of volume of renal tissue.

Uric acid, the principal nitrogenous end product of birds and most reptiles, is introduced into the nephrons by filtration. In addition, both birds and reptiles actively secrete uric acid into the urine as the urine flows through the nephrons. Secretion accounts for the greater part of the excreted amount.

In both birds and reptiles, the ureters discharge into the cloaca. From the cloaca, the urine is not uncommonly moved by reverse peristalsis into the lower intestine. Both the cloaca and intestine may modify the composition and volume of the urine before excretion. When the urine enters the cloaca from the ureters, the uric acid and urates in it are often present largely in the form of supersaturated colloidal suspensions stabilized by special proteins. The uric acid and urates are then precipitated into solid, particulate form in the cloaca–intestine. Precipitation *after* the urine has left the ureters helps prevent clogging of the renal tubules with the precipitate.

Summary: Urine Formation in Other Vertebrates

- Freshwater teleost fish have nephrons structurally similar to amphibian nephrons. Marine teleost fish, however, usually lack the distal convoluted tubule and have a relatively poorly developed glomerular filtration apparatus that seems often to be supplemented by active solute secretion. A few marine fish are aglomerular and depend entirely on secretion.
- Reptiles and birds have nephrons structurally similar to amphibian nephrons. Birds, in addition, have mammalian-type nephrons (with loops of Henle) organized into parallel arrays—the medullary cones—in which urine hyperosmotic to blood plasma can be made.

Urine Formation in Decapod Crustaceans

An adult crayfish, crab, lobster, or other decapod crustacean has two renal organs, known as **antennal glands** or **green glands**, which are located in its head and open to the outside independently near the bases of its second antennae (Figure 27.19*a*). Each antennal gland is basically a single tube, sometimes loosely described as resembling "a single giant nephron." In a freshwater crayfish (Figure 27.19*b*), each antennal gland begins with a closed **end sac** or **coelomosac** lying to the side of the esophagus. Following the end sac is the **labyrinth** (or *green body*), a sheet of spongy tissue consisting of a channel that branches and anastomoses extensively along its length. The **nephridial canal**, which also has a spongy internal morphology, leads from the labyrinth to the expanded **bladder**, and the bladder empties to the outside. The nephridial canal is found only in certain freshwater decapod crustaceans and is generally considered to be a freshwater "invention." The morphology of the antennal gland in a lobster or marine crab is like that described for the crayfish, except that the labyrinth connects relatively directly with the bladder.

The walls of the end sac are thin, and arteries from the heart supply a network of small vessels or lacunae on the outer surface of the end sac. This morphological evidence has long suggested that fluid enters the end sac by filtration under the force of blood pressure. Additional morphological evidence for filtration at the end sac is provided by the presence of podocyte-like cells in the end sac epithelium. **Podocytes** are cells of distinctive morphology that are often viewed as being diagnostic of sites of ultrafiltration. The physiological evidence that is available supports the hypothesis that primary urine is formed in the end sac by filtration from the blood under force of blood pressure. Some authorities argue, however, that

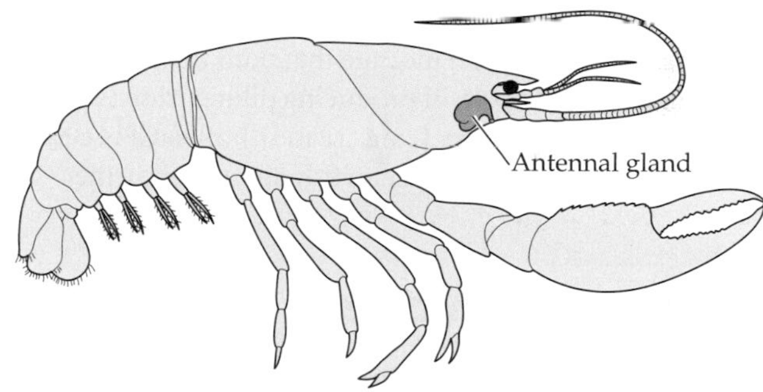

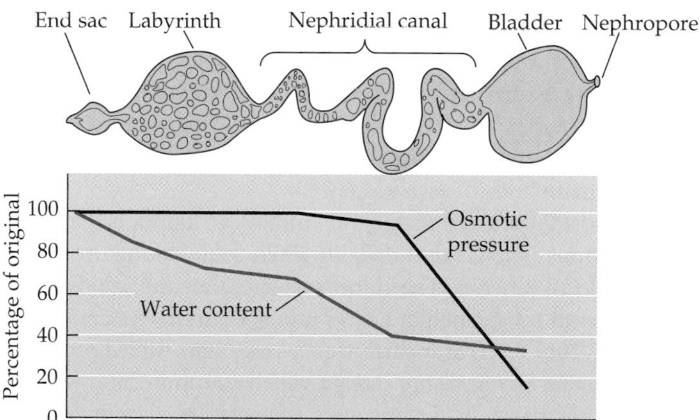

Figure 27.19 The antennal gland and urine formation in a freshwater crayfish *(a)* The position of the antennal gland on the right side of a crayfish's body. *(b)* Although the nephridial canal is in fact tightly convoluted and partly enveloped by the sheetlike labyrinth, the antennal gland can be stretched out to reveal its parts. The graph shows measured changes in the osmotic pressure and water content of the urine as it passes through the parts of the antennal gland in crayfish (*Australopotamobius pallipes* and *Orconectes virilis*) living in freshwater. Values are plotted immediately below the anatomical locations where they were measured and are expressed as percentages of the values in the end sac. (After Riegel 1977.)

whereas filtration forms the primary urine, other ill-defined processes besides blood pressure drive the filtration.

Evidence indicates that the composition of the urine is modified in all the structures through which it flows as it passes through the antennal gland. The labyrinth of American lobsters, for example, is known to reclaim glucose, and probably reclaims amino acids, from the filtrate. In marine crabs, the bladder is an important site of glucose reabsorption from the urine and Mg^{2+} secretion into the urine. Unfortunately, a truly synthetic understanding of the handling of solutes by all parts of an antennal gland is not yet available for any species.

Modifications of the osmotic pressure of the urine have been a major focus of study in decapods. The labyrinth is by all accounts incapable of rendering the urine hyposmotic to the blood. There is a much-emphasized correlation between the presence of a nephridial canal and the ability to produce urine hyposmotic to the blood. Freshwater crayfish, which can make dilute urine, have a nephridial canal; marine decapods and freshwater crabs such as *Eriocheir sinensis* (see page 690), which are unable to make dilute urine, lack the nephridial canal. Thus the nephridial canal has often been described as being analogous to the vertebrate distal convoluted tubule. The results of some studies on crayfish indicate that both the bladder and the nephridial canal participate in producing dilute urine (see Figure 27.19*b*). Active reabsorption (conservation) of NaCl is clearly one of the major processes by which crayfish dilute their urine.

Urine Formation in Molluscs

The renal organs of molluscs are tubular or saccular structures, called **nephridia** or **kidneys**, that empty into the mantle cavity or directly to the outside. Bivalves, most cephalopods (octopuses and squids), and some gastropods have two kidneys, but most gastropods have only one. In an arrangement that initially strikes most people as truly bizarre, each kidney typically opens into the *pericardial cavity* that surrounds the heart, and the fluid in the pericardial cavity—the pericardial fluid—is believed to be the primary urine. A canal called the **renopericardial canal** leads from the pericardial cavity to the main body of each kidney.

The most thoroughly understood molluscan kidney is that of the giant octopus *Enteroctopus dofleini*, often found along the Pacific coast of North America. The kidneys of octopuses and squids are associated with the branchial hearts rather than the systemic heart (Figure 27.20*a*). Each branchial heart bears a thin-walled protuberance, the *branchial heart appendage*, which communicates with the lumen of the heart. In *Enteroctopus*, the pericardial cavity of each branchial heart encloses only the side of the heart bearing the heart appendage as seen in Figure 27.20*a*. A kidney connects to each pericardial cavity. Compelling evidence exists in *Enteroctopus* and certain other cephalopods that the pericardial fluid is an ultrafiltrate of the blood, forced into the pericardial cavity across the branchial heart appendage under the force of pressure developed in the heart. In each kidney, this filtrate flows through a long renopericardial canal and then an enlarged **renal sac** before being discharged into the mantle cavity. Studies have shown that the renopericardial canal alters the composition of the urine. Glucose and amino acids are promptly reabsorbed, for example—a process reminiscent of their prompt reclamation in vertebrate nephrons.

The kidneys of bivalve and gastropod molluscs, such as clams and snails, open into the pericardial cavity surrounding the systemic heart (Figure 27.20*b*).[11] Tubular structures called **pericardial glands** are found in the walls of the atria of the heart or in association with the membranes surrounding the pericardial cavity of these animals. Podocytes (suggestive of filtration) are characteristic cells in the pericardial glands. The incomplete physiological evidence that exists indicates that although the pressures developed by the hearts of bi-

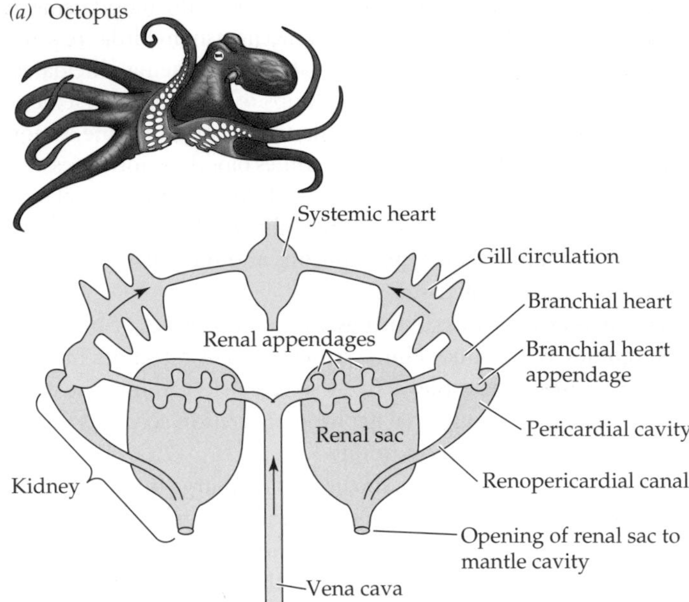

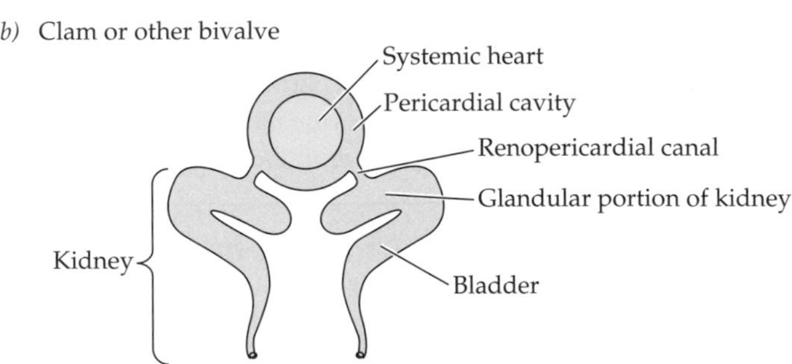

Figure 27.20 The kidneys of molluscs and their relations to the circulatory system *(a)* The kidneys and associated parts of the circulatory system in the giant octopus *Enteroctopus*. After the principal vein returning blood from the systemic tissues, the vena cava, branches, each branch passes by one of the renal sacs, and there it bears many glandular diverticula, called *renal appendages*, which are closely juxtaposed to the walls of the sac. In *Enteroctopus*, ammonia is believed to be secreted into the renal sacs across the renal appendages before blood reaches the branchial hearts, where ultrafiltration occurs. *(b)* A stylized cross section through the systemic heart in a clam, mussel, or other bivalve, showing the relation of the heart and pericardial cavity to the kidneys. (After Martin and Harrison 1966.)

[11] Bivalves and gastropods lack branchial hearts.

valves and gastropods are usually low, the pericardial fluid (which is the primary urine) is formed by filtration, and this filtration occurs in the pericardial glands. The primary urine is altered as it flows through the kidneys, but little is known about the process. In freshwater mussels, the osmotic pressure of the urine is reduced to about half that of the blood by the time the urine arrives in the bladder.

Urine Formation in Insects

The formation of urine has been much more thoroughly studied in insects than in any other group of invertebrates. Most insects possess Malpighian tubules,[12] which are often loosely referred to as the "excretory tubules." A point to be stressed from the outset, however, is that the hindgut is as important as the Malpighian tubules in the formation of urine.

The **Malpighian tubules** are long, slender, blind-ended structures that typically arise from the junction of the midgut and hindgut (Figure 27.21). They number from 2 to over 200, depending on the species. Projecting into the hemocoel, they are bathed by the blood (hemolymph). The walls of the tubules consist of a single layer of epithelial cells, surrounded on the outside by a thin basement membrane. Although the tubules exhibit little histological differentiation along their length in some species, they are differentiated into two to six (or possibly more) distinct regions in numerous others. In many species, the various tubules within an individual are morphologically similar, but in others, two or more types of tubules are present.

[12] The tubules are named after Marcello Malpighi (1628–1694), one of the great early microscopists, who was the first to describe the blood capillaries and renal corpuscles of vertebrates as well as the Malpighian tubules and tracheae of insects.

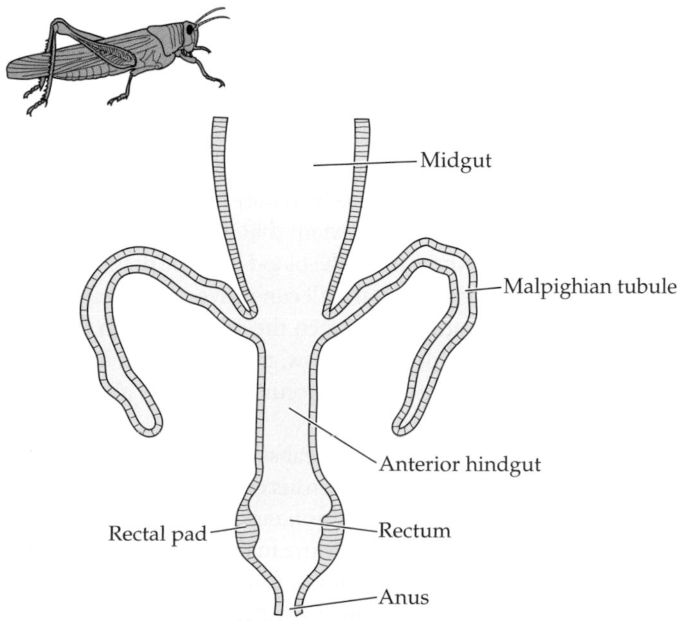

Figure 27.21 The posterior gut and Malpighian tubules of an insect The Malpighian tubules empty into the gut at the junction of the midgut and hindgut.

The hindgut, which is lined with cuticle (see Figure 4.16), typically consists of a relatively small-diameter **anterior hindgut** (*ileum* or *intestine*) and an expanded posterior part, the **rectum** (see Figure 27.21). The walls of the anterior hindgut usually consist of a single layer of cuboidal or squamous epithelial cells. The walls of the rectum are similar to those of the anterior hindgut in some insects, but in many other species, parts of the rectal wall consist of thick columnar epithelial cells, sometimes associated with secondary cell layers. These thickened parts of the rectal wall are termed **rectal pads** or **rectal papillae**, depending on their gross morphology.

The Malpighian tubules form and sometimes modify the primary urine

The function shared by the Malpighian tubules of all insects studied is the formation of the primary urine. The tubules are not supplied with blood vessels, and filtration does not occur. Instead, the primary urine is formed by a secretory mechanism (see Figure 27.2). In the most common scenario, potassium chloride (KCl) is secreted vigorously by the tubular epithelium from the blood into the lumen of a Malpighian tubule—so vigorously that the K^+ concentration in the tubular fluid is 6 to 30 times higher than that in the blood. According to theories that have rapidly matured in the past 15 years, the K^+ secretion is by secondary active transport, with the energy being drawn most commonly from electrochemical gradients maintained by the action of a H^+-ATPase. The secretion of K^+ into a Malpighian tubule is electrogenic, and Cl^- accompanies the K^+ passively by following the electrical gradient set up by K^+ secretion. The result is that KCl is secreted at a cost of ATP. The flux of KCl into the lumen of a Malpighian tubule drives osmotic entry of water, which typically occurs briskly enough that the tubular fluid remains approximately isosmotic to the blood. Many additional solutes then enter the tubular fluid passively, as by following concentration gradients set up by water influx; these solutes include amino acids, sugars, diverse organic wastes and toxins, and several inorganic ions. Proteins are largely excluded because of their size. Certain organic compounds—notably some detoxification products and plant toxins (e.g., alkaloids)—are actively secreted into the primary urine by some insects.

Although KCl is most commonly the principal salt secreted to initiate primary-urine formation, NaCl plays this role in some species. There are also species in which KCl predominates under some conditions, whereas NaCl does under other conditions.

As primary urine passes down the Malpighian tubules and flows through the parts of the tubules nearest the gut, the tubular epithelium may also be involved in reabsorption of salts, water, or other molecules such as glucose. In the end, the fluid passed into the hindgut by the Malpighian tubules contains numerous solutes. It has a solute composition quite unlike that of the blood, typically being far richer in KCl than the blood. It usually resembles the blood in osmotic pressure, however.

The rate at which primary urine is formed and flows toward the hindgut in an insect's Malpighian tubules can be quite high relative to the total volume of the insect's body fluids. Current estimates, for example, indicate that in a female yellow-fever mosquito (*Aedes aegypti*), during an ordinary 24-hour day when she has *not* taken a blood meal, the Malpighian tubules produce primary urine equivalent to 12 times her total body volume of extracellular fluid! If this sounds strange, it is really not; recall the very high rate at which human kidneys produce primary urine—also equivalent to about 12 times the entire extracellular fluid volume every day. In insects, as in vertebrates, most primary urine is reabsorbed rather than being excreted. The overall process—a high rate of primary-urine formation followed by a high rate of reabsorption—gives the excretory system intimate access to the blood to carry out its regulatory functions, as already stressed. An aspect of the reabsorption process in insects that is particularly noteworthy is that it reclaims KCl. KCl must be secreted at a high rate to drive the production of primary urine, but it could not be lost from the body at that rate; instead, most KCl is reabsorbed back into the blood in the hindgut and recycled to produce more primary urine.

TABLE 27.1 Average composition of the rectal fluid and other body fluids in locusts provided with tap water or a saline solution to drink The animals (*Schistocerca gregaria*) were deprived of food.

Experimental treatment	Fluid	Osmotic pressure (*Osm*)	Ion Concentration (m*M*)		
			Cl^-	Na^+	K^+
Water-fed	Rectal fluid	0.82[a]	5	1	22
	Anterior-hindgut fluid	0.42	93	20	139
	Blood	0.40	115	108	11
Saline-fed	Rectal fluid	1.87	569	405	241
	Anterior-hindgut fluid	—	192	67	186
	Blood	0.52	163	158	19

Source: After Phillips 1964.

[a] The high osmotic pressure in the scanty rectal fluid of water-fed animals is presumed to be caused by organic solutes.

The hindgut modulates urine composition, concentration, and volume

Although the reabsorption of KCl and water from the primary urine sometimes starts in the lower parts of the Malpighian tubules, it occurs predominantly in the hindgut, especially in the rectum. After urine is discharged from the Malpighian tubules, it flows with the feces through the hindgut, where its composition, concentration, and volume are modified to produce the definitive urine, which is excreted. The rectum (which is far better understood than the anterior hindgut) not only reabsorbs most of the water, K^+, Na^+, and Cl^- introduced into the hindgut by the Malpighian tubules, but also often reabsorbs amino acids, acetate, and phosphate. The rectum also has some secretory functions. For example, acidification of the urine by secretion of H^+ occurs in the rectum and contributes to the precipitation of uric acid and urates there.

Research has made it increasingly clear that the insect rectum has impressive *regulatory* abilities. It adjusts the osmotic pressure of the urine by varying the relative rates of reabsorption of water and total solutes. The ionic composition of the urine is also adjusted in the rectum. In one set of experiments, for example, fasting locusts were permitted to drink either tap water or a saline solution containing K^+, Na^+, Cl^-, and other ions. The rectum in the water-fed locusts *lowered* ion concentrations in the urine, compared to those in the anterior-hindgut fluid; but the rectum in the saline-fed locusts *raised* ion concentrations (Table 27.1). The saline-fed locusts also accumulated greater volumes of urine in the rectum, so that overall, the quantities of ions excreted in their urine exceeded those excreted by the water-fed animals by hundreds of times.

The study of the hormonal control of urine production in insects is a burgeoning field at present, in part because of the expectation that the next generation of controls for insect pests might include procedures that defeat vital control mechanisms. Numerous diuretic and antidiuretic neurohormones—which affect both Malpighian-tubule and rectal function—have been identified in various species. The interactions of such hormones within single species are currently an active area for research.

PRODUCTION OF URINE HYPEROSMOTIC TO THE BLOOD When insects produce urine that is hyperosmotic to their blood (see page 708), the process of concentration usually occurs in the rectum. The mechanisms that concentrate the urine have attracted particular interest, in part because the ability to produce hyperosmotic urine is so unusual among animals. At least three different mechanisms of concentrating the urine have evolved in insects.

In insects that have rectal pads or rectal papillae—such as cockroaches (*Periplaneta*), desert locusts (*Schistocerca*), and blowflies (*Calliphora*)—the urine in the rectal lumen is concentrated by *water reabsorption in excess of solute reabsorption*. This water reabsorption is highly intriguing because it can continue even when the osmotic pressure of the rectal contents has risen to be two or more times higher than the osmotic pressure of the blood bathing the rectum! The existence of this seemingly paradoxical process has been demonstrated in several ways. Perhaps the most compelling evidence comes from experiments in which the rectum has been filled with a pure solution of a solute (e.g., trehalose) that is neither reabsorbed nor secreted across the rectal wall. The *amount* of such a solute in the rectum is fixed during the course of an experiment. When locusts (*Schistocerca*) were treated in this way, they reabsorbed water from their rectal contents to such an extent that the rectal osmotic pressure rose to be nearly three times the blood osmotic pressure. Results of this sort show that the rectal wall can move water against large, opposing osmotic gradients between the rectal fluid on the inside and blood on the outside. They show, moreover, that in the short term, this water reabsorption can occur even in the absence of simultaneous solute reabsorption.

The mechanism of such water reabsorption was mystifying for many years. Now, however, a consensus exists that it is a case of osmosis on a microscopic scale: *local osmosis*. The mechanism depends in part on a complex microarchitecture in the rectal pads or papillae. The details of structure and possibly of function vary from species to species. Here, we focus on the blowfly (*Calliphora*) as an example.

In the rectal papilla of a blowfly (Figure 27.22*a*), adjacent cells of the columnar epithelium are tightly joined on the side facing the rectal lumen and on the opposite (basal) side, but in between, the cells are separated by an elaborate network of minute channels and spaces, here termed the **intercellular spaces** (*intercellular*, "between cells"). The network of intercellular spaces communicates at the apex of the papilla with subepithelial spaces—here called **infundibular channels**—that are positioned under the basal side of the epithelial cell layer and connect with general blood spaces. It is believed that the epithelial cells actively secrete solutes into the intercellular spaces, thereby rendering the fluid in the intercellular spaces strongly hyperosmotic both to the blood and to the fluid in the rectal lumen (Figure 27.22*b*). Osmosis then carries water out of the rectal lumen into the intercellular spaces; that is, because of the *locally* high osmotic pressure in the intercellular spaces, water is osmotically withdrawn from the rectal fluid even though the latter is thereby made increasingly hyperosmotic to the blood. Entry of water into the intercellular spaces causes fluid to flow in streams through the intercellular spaces toward the apex of the papilla and then through the infundibular channels toward the main blood cavity of the body (hemocoel). The fluid exiting the intercellular spaces is highly concentrated, but as it flows under the epithelial cells in the infundibular channels, solutes are believed to be actively or passively reabsorbed from the fluid into the cells across membranes poorly permeable to water, with two highly significant consequences. First, the fluid flowing through the infundibular channels is diluted, so in the end a fluid rich in water, rather than in solutes, is introduced into the blood. Second, solutes are returned to the epithelial cells and thus can again be secreted into the intercellular spaces, permitting continued osmotic water absorption from the rectal fluid without great need for new solutes from any source. The nature of the solutes involved remains under investigation. The inorganic ions Na^+, K^+, and Cl^- are strongly implicated; some organic solutes also play roles.

A second type of concentrating mechanism has been described in insects that have a **cryptonephridial complex.** These insects include mealworms (larval *Tenebrio molitor*), certain larval and adult coleopterans (beetles), and certain larval lepidopterans (butterflies and moths). Mealworms can produce pellets of excrement (feces and urine combined) that are particularly concentrated relative to the excrement of other insects, and ostensibly dry. The structural basis for concentrating and drying the excrement is a specialized association between the Malpighian tubules and rectum: the *cryptonephridial complex*. In the complex, the distal parts of the Malpighian tubules (the parts nearest the blind ends)—which float freely in the hemocoel of most insects—are closely associated with the outer rectal wall, and these parts of the tubules and the rectum are together enclosed by a **perinephric membrane**, which separates them from the hemocoel. KCl and NaCl are actively transported from the blood into the lumen of each of these cryptonephric Malpighian tubules. In a marked departure from the usual condition in insects, however, water is prevented from entering the tu-

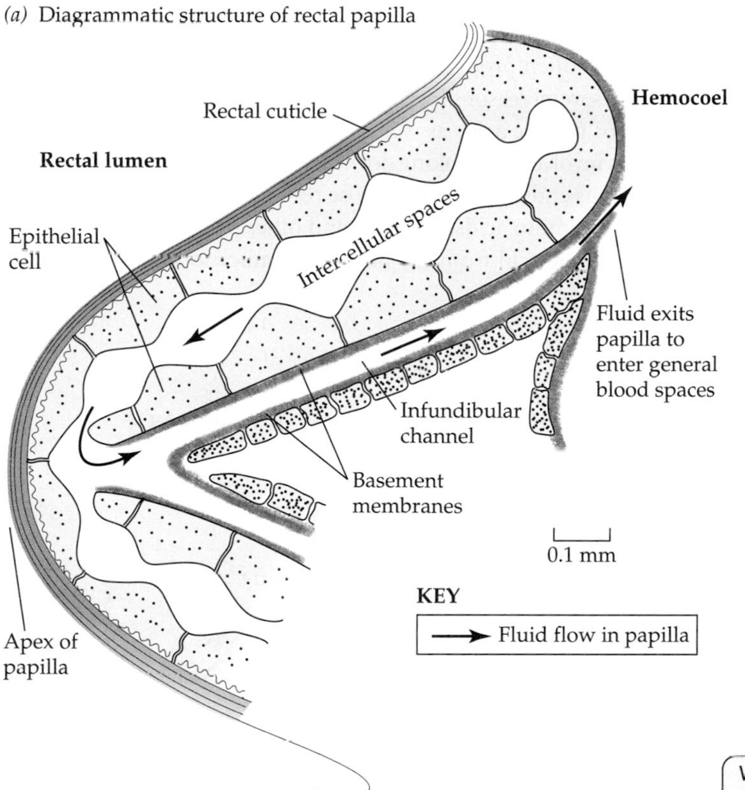

Figure 27.22 The structure and function of the blowfly rectal papilla Each blowfly (*Calliphora erythrocephala*) has four rectal papillae. *(a)* The structure of a papilla, shown highly diagrammatically. The intercellular spaces, depicted for simplicity as a single broad cavity running through the epithelial cells, actually consist of a complex, interconnecting network of small channels and spaces *between* the epithelial cells. The papilla, being in the rectum, is covered with a thin cuticle; although this material is part of the exoskeleton, it is highly permeable to water and solutes of small ionic or molecular size. *(b)* The proposed mechanism of water absorption from the rectal lumen. In the short term, water absorption can occur without solute absorption from the rectum. (*a* after Gupta and Berridge 1966.)

bular fluid from the blood (probably by water-impermeability of the perinephric membrane). The fluid in the cryptonephric Malpighian tubules thus is made strongly hyperosmotic to the blood, and a gradient favoring osmotic reabsorption of water from the closely juxtaposed rectal lumen is created. Mealworms and some of the other insects with a cryptonephridial complex are known to absorb water vapor from the atmosphere across their rectal linings; the high concentrations of KCl and NaCl in the cryptonephric Malpighian tubules appear to be responsible (see page 712).

A third mechanism of concentrating urine is known in saline-water mosquito larvae (*Aedes*). These larvae can live in waters that are markedly hyperosmotic to their blood, and under such circumstances they drink to replace the water that they lose osmotically across their body surfaces. Unlike the insects discussed up to now, their main challenge is not water conservation; instead, it is to excrete ions at higher concentrations than in their drinking water, so that they can realize a net gain of H_2O from drinking and excretion. They concentrate their urine not by reabsorbing water to a greater degree than solutes from the rectal fluid, but by secreting ions into the rectal fluid.

Summary: Urine Formation in Insects

- Primary urine is introduced into the Malpighian tubules by a secretory process usually based on active transport of KCl into the tubular fluid. As the primary urine flows down the Malpighian tubules, it may be modified by reabsorption or secretion, but typically remains isosmotic to the blood.
- The Malpighian tubules empty into the hindgut at the junction of the midgut and hindgut.
- The rectum modifies the volume, composition, and osmotic pressure of the urine in regulatory ways. The production of hyposmotic urine occurs by reabsorption of solutes in excess of water. Two of the known mechanisms of producing hyperosmotic urine, on the other hand, involve reabsorption of water in excess of solutes. Some of the insects that produce hyperosmotic urine in this way do so by local osmosis and solute recycling in rectal pads or papillae; others do so with a cryptonephridial complex. Saline-water insects may form hyperosmotic urine by secretion of solutes into the rectum.

Nitrogen Disposition and Excretion

When animals catabolize organic molecules to release chemical energy, the atoms of the molecules appear in a variety of catabolic end products. During aerobic catabolism, the three most abundant atoms—carbon, hydrogen, and oxygen—appear in CO_2 and H_2O. The CO_2 is typically voided promptly into the environment across lungs, gills, or skin. The H_2O (*metabolic water*) simply becomes part of an animal's body water resources. The fourth most abundant atom is nitrogen, which is a characteristic constituent of proteins and nucleic acids. The disposition of nitrogen atoms from catabolism is not so simple as that of carbon, hydrogen, and oxygen.

Some of the compounds into which animals incorporate nitrogen during catabolism are shown in Figure 27.23. Each of these nitrogenous end products has advantages and disadvantages for the animals that synthesize it. Some of the compounds are relatively cheap to make, for example, whereas others are low in toxicity. There is no single end product that is ideal in all ways. Thus animals have evolved a variety of strategies for dealing with the nitrogen atoms released from organic molecules by catabolism.

The relation between nitrogen excretion and kidney function varies from one group of animals to another. Mammals, birds, and reptiles exemplify one end of the spectrum: They excrete nitrogenous end products entirely in their urine. At the other end of the spectrum, there are some aquatic animals in which nitrogenous end products are excreted mainly across the gills or skin, and the kidneys play little or no role.

Animals often produce two or more nitrogenous end products. There are several reasons for this. One is that nitrogen is a major constituent of nucleic acids as well as proteins, and often the catabolic pathways involved in breaking down nucleic acids produce a different nitrogenous end product than those responsible for breaking down proteins or amino acids. Humans and other primates, for example, synthesize uric acid from the nitrogen of nucleic-acid purines, but

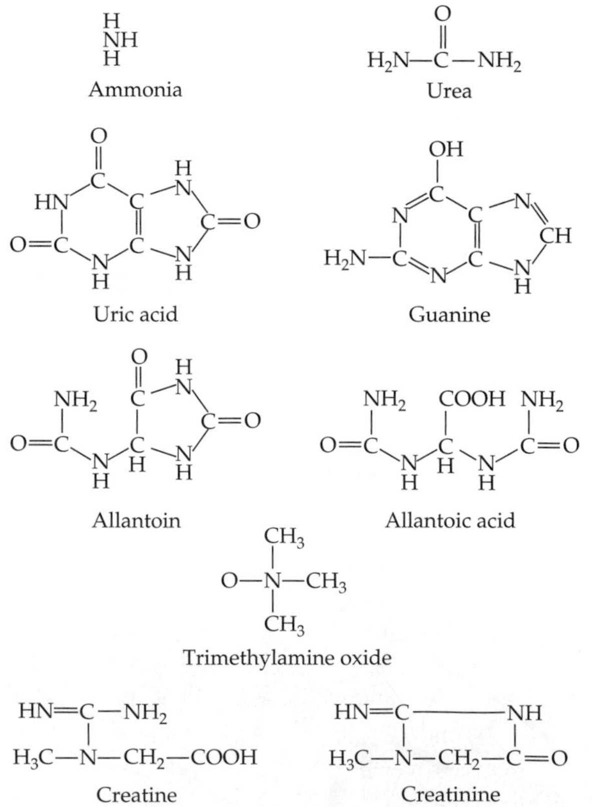

Figure 27.23 Some nitrogenous compounds excreted by animals Uric acid and guanine are purines. Allantoin and allantoic acid are poorly soluble breakdown products of uric acid. Trimethylamine oxide and its precursor, trimethylamine, are found in a variety of marine animals, but do not occur in freshwater animals; both are highly soluble. Creatine and its internal anhydride, creatinine, occur as relatively minor excretory compounds in many vertebrates and some invertebrates. Some animals, mostly invertebrates, also lose significant amounts of amino acids to the environment.

synthesize principally urea from protein nitrogen.[13] Protein catabolism dominates as a source of nitrogen—accounting for an average of about 95% of catabolic nitrogen in animals. Thus in humans and other primates, there is a little uric acid, and a lot of urea synthesized.

When ammonia (NH_3) or the ammonium ion (NH_4^+) is the principal nitrogenous end product of an animal, the animal is described as **ammonotelic.** If urea is the principal nitrogenous end product, an animal is termed **ureotelic.** If uric acid is the principal end product, an animal is **uricotelic.**

Ammonotelism is the primitive state

Ammonia (NH_3) reacts with hydrogen ions to form the ammonium ion (NH_4^+). At the ordinary pH values of animal body fluids and tissues, this reaction is shifted strongly toward the formation of NH_4^+. For simplicity, we use the word *ammonia* here to refer to either chemical form.

Ammonia is clearly the primitive nitrogenous end product of animals. Strong evidence for this view comes from the fact that the vast majority of today's marine invertebrates are ammonotelic.

Ammonia is quite toxic. The reasons for its toxicity are numerous and only starting to be well understood. Blood concentrations are ordinarily kept low: usually under 0.3 m*M* in vertebrates and under 7 m*M* even in the most tolerant of invertebrates.

Because of its toxicity, ammonia cannot ordinarily be allowed to accumulate in an animal's body. Thus, for an animal to be ammonotelic, it must have a means of unfailingly voiding ammonia as rapidly as it is formed by catabolism. Aquatic animals can meet this challenge because of the abundance of water in which they live; often, much of the ammonia they produce is voided directly into the ambient water across their gills or other external body surfaces. Not only marine invertebrates, but also other water-breathing aquatic animals, are ammonotelic. Both freshwater and marine teleost fish are typically ammonotelic, and both aquatic tadpoles of amphibians and adult aquatic amphibians (e.g., mudpuppies) are ammonotelic. Ammonotelism, however, is unusual on land. A terrestrial animal is likely to depend on excretion in urine to rid itself of nitrogenous wastes. Urinary excretion of ammonia requires the excretion of a great deal of water.[14] Because of this and because it is urgent for an animal to prevent accumulation of ammonia in its body, excretion of ammonia in the urine might sometimes require a terrestrial animal to void large amounts of water when water itself is in short supply.

Although ammonotelism is unusual on land, some *humidic* terrestrial invertebrates are either ammonotelic or at least produce substantial quantities of ammonia. Terrestrial isopod crustaceans (e.g., pillbugs), for example, are ammonotelic. They void much of their ammonia into the atmosphere as NH_3 gas, an expedient that prevents them from needing to void water to get rid of ammonia. Many terrestrial snails, although not ammonotelic, void substantial NH_3 gas as well. In both isopods and snails, the fundamental reason for ammonia production may be that ammonia plays a role in the process of deposition of calcium carbonate in their exoskeleton or shell.

[13] The affliction known as gout results from abnormal uric acid metabolism.

[14] The high water requirement of urinary ammonia excretion is compared to the lower water requirement of urinary urea excretion in the next section.

Ammonia is the cheapest nitrogenous end product to produce. This probably explains why ammonotelism is so common among aquatic animals, which can easily avoid problems with ammonia's toxicity. Ammonia is generally formed during the catabolism of proteins by way of reactions that have no ATP cost; some of these, termed *transamination* reactions, move amino groups to particular amino acids for which there are deamination enzymes, and then the latter amino acids undergo *deamination* (see Figure 4.2*b*).

Urea is more costly to synthesize but less toxic than ammonia

Urea is highly soluble and generally diffuses readily across membranes. Although hardly benign in its effects on macromolecules, it is far less toxic than ammonia. In humans, blood concentrations are normally in the range of 3 m*M* to 7 m*M*, and much higher concentrations, although abnormal, can be tolerated. As seen in Chapter 26, very high urea concentrations can be found in marine elasmobranch fish ($\geq$ 300 m*M*) and some other animals.

If urea is less toxic than ammonia, why are so many animals ammonotelic? Two reasons seem to be of principal importance. First, ammonotelism is the primitive condition, and some groups of animals may never have evolved the enzymes required to synthesize urea. Second, urea is more costly to make than ammonia: The synthesis of each urea molecule requires the energy from four or five high-energy phosphate bonds (equivalent to that released by converting four or five ATP molecules to ADP).[15] As is so often the case in biology, animals face trade-offs. Ammonia is toxic but cheap; urea is less toxic but more costly.

Most of the animals that routinely employ urea as their principal nitrogeneous end product are terrestrial vertebrates. Adult terrestrial amphibians are predominantly ureotelic. All mammals are ureotelic, as are some turtles. Terrestrial invertebrates have more often evolved uricotelism; only some flatworms and earthworms are at times ureotelic.

There is persuasive evidence that vertebrates adopted ureotelism when they emerged onto land. One line of supportive evidence is the fact that modern terrestrial amphibians are ureotelic, in contrast to freshwater fish, which are nearly always ammonotelic. Another line of evidence is that the tadpoles of amphibians are usually ammonotelic, but increasingly express the enzymes for urea synthesis as they go through metamorphosis, eventually becoming ureotelic adults (Figure 27.24). For terrestrial animals that excrete nitrogenous wastes in their urine, the advantage of ureotelism is that urea excretion requires less water than ammonia excretion. This is a direct consequence of the lower toxicity of urea. Because of its lower toxicity, steady-state blood concentrations of urea in ureotelic vertebrates—expressed in molar terms—are typically *at least* 20 times higher than steady-state blood concentrations of ammonia in ammonotelic vertebrates. Thus, if a urea-producing and an ammonia-producing species were both to excrete all waste nitrogen in their urine at identical urine:plasma ratios for their respective nitrogenous end products, the urea-producing species could excrete any particular amount of nitrogen in 2.5% or less of the water required by the am-

[15] Authorities differ in whether they estimate the cost to be four or five per molecule.

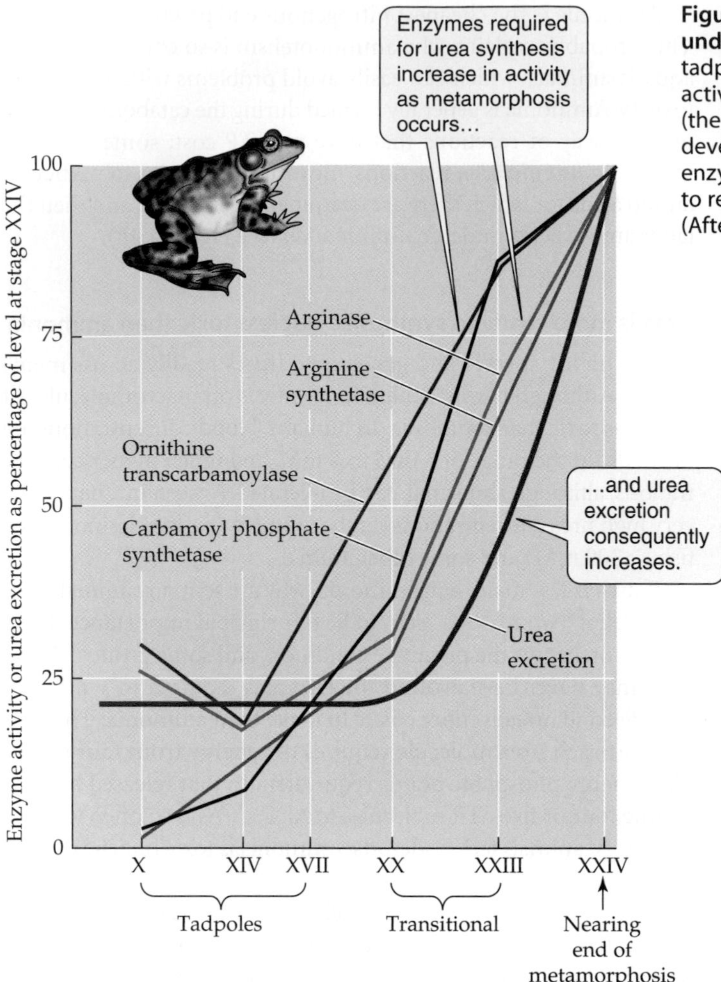

Figure 27.24 Bullfrogs shift from ammonotelism to ureotelism as they undergo metamorphosis Bullfrogs (*Rana catesbeiana*) start life as aquatic tadpoles, but metamorphose into semiterrestrial adults. The graph shows the activities of four enzymes of the metabolic pathway that synthesizes urea (the ornithine–urea cycle) and the excretion of urea as functions of stage of development; all are expressed relative to their levels at stage XXIV. The enzymes were measured in the liver. ("Arginine synthetase" is now recognized to represent the activity of two enzymes that together produce arginine.) (After Brown and Cohen 1958.)

monia-producing one.[16] These considerations have no significance for aquatic animals that void nitrogen across general body surfaces. They have great significance, however, for animals on land that excrete nitrogen in urine.

Some animals employ ureotelism for functions other than the simple, routine excretion of nitrogen. One category of such animals consists of the marine fish that routinely accumulate urea as an osmolyte to aid them with osmoregulation in the ocean—notably marine elasmobranchs, holocephalans, and the coelacanth *Latimeria* (see page 697). A second category consists of ammonotelic vertebrates that switch to producing urea during water crises. Certain fish, such as lungfish of the genus *Protopterus* (see page 702) and some freshwater teleost species from stillwater environments, switch from being ammonotelic to being ureotelic when confronted with drying of their habitat or other stresses. They then can stop voiding urine and allow their nitrogenous waste to accumulate in their body fluids, a strategy that would be impossible if they were synthesizing highly toxic ammonia.

The most thoroughly known biochemical mechanism for the synthesis of urea from protein nitrogen—the mechanism used by all vertebrates and some invertebrates—is the **ornithine–urea cycle**, a set of biochemical reactions requiring five enzymes.[17] In the ornithine–urea cycle, one of the nitrogen atoms incorporated into urea originates from free ammonia derived from deamination reactions, especially deamination of glutamic acid. The second nitrogen incorporated into urea comes from the amino group of aspartic acid. Amino groups from most amino acids can make their way to glutamic acid or aspartic acid by transamination reactions. In vertebrates that synthesize urea, the liver is the one tissue that expresses the full suite of enzymes of the ornithine–urea cycle.

As discussed in Box 26.3, many authorities now believe that the ornithine–urea cycle appeared early in the evolution of vertebrates. A current working hypothesis is that all vertebrates have the genes for all the required enzymes, and thus that the existence of a functioning ornithine–urea cycle in a species depends on whether the species expresses the genes. If the ability to synthesize urea is such a "readily available" option, it would help explain why ureotelism occurs in such a scattered and wide variety of vertebrates (see the cladogram in Box 26.3).

Uric acid and related compounds remove nitrogen from solution

An animal is classified as uricotelic if its primary nitrogenous end product is uric acid, the dihydrate of uric acid, urate salts, or a mix of these compounds, all of which are purines. These compounds all have low toxicity and solubility. Uric acid itself is poorly soluble in water (0.4 mmole—65 mg—dissolves in a liter at 37°C). The urate salts, although more soluble than uric acid, also have very low solubility in comparison to urea or ammonia. Because of their low solubility, uric acid and urates remove nitrogen from solution, reducing the water costs of excretion; they can be excreted as semisolid pastes or even as dry pellets or powders.

Experiments have shown that a variety of cations—including Na^+, K^+, NH_4^+, Ca^{2+}, and Mg^{2+}—can be incorporated into uric acid excrement in a poorly soluble state. There is uncertainty over the chemical form taken by these ions; they can be present as urate salts, but apparently they also can be bound in some manner to precipitated uric acid. Regardless, the cations are removed from solution, and this state can appreciably reduce the water demands of cation excretion. Calculations indicate, for example, that because of precipitation of K^+ in the excrement, desert iguanas (*Dipsosaurus*) can void as much as 5000 milliequivalents of K^+ per liter of water—an effective concentration that is well above the highest K^+ concentration achieved by reptilian salt glands!

[16] This is true because the urinary molar concentration of urea would be at least 20 times higher than that of ammonia, and each molecule of urea contains two atoms of nitrogen, rather than just one.

[17] The reactions are presented in any biochemistry text.

BOX 27.4 Why Are Mammals Not Uricotelic?

If birds, lizards, and snakes are uricotelic, why aren't mammals? An informed biologist knowing the patterns of nitrogen excretion in other animal groups, but knowing nothing about nitrogen excretion in mammals, would surely predict uricotelism in mammals, at least in desert species. Yet, without known exception, all mammals are ureotelic. We do not know the answer to this riddle. However, a point worth stressing is that mammals are able to do things with urea that are quite unique.

The mammalian kidney is in a class by itself in its ability to concentrate urea. The maximum U/P ratio for urea in mammalian urine is typically higher than that for any other solute and can greatly exceed the maximum osmotic U/P ratio—signifying that urea can be concentrated to a much greater extent than solutes as a whole. In humans, for example, the maximal osmotic U/P ratio is about 4.2, and the maximal Cl^- U/P ratio is about 3.5, but the maximal urea U/P ratio is about 170. Many desert rodents can achieve urinary urea concentrations of 2.5 *M* to 5.0 *M*, corresponding to 70 to 140 g of nitrogen per liter! Because the urinary urea concentrations achieved by mammals are much greater than those attained by other animals, the water losses obligated by nitrogen excretion in mammals are exceptionally low in comparison to other ureotelic groups. In fact, urinary nitrogen:water ratios attainable by desert rodents can equal or exceed those observed in some of the uricotelic vertebrates that void their uric acid in a relatively fluid mix (e.g., certain birds). On the other hand, some birds and reptiles that void uric acid in the form of relatively dry pellets achieve nitrogen:water ratios that are several times higher than the highest mammalian values.

Some people have argued that mammals have remained ureotelic because the elaborately developed countercurrent multiplication system of the mammalian kidney provides such great potential for concentrating urea that the selective advantage of uricotelism has been blunted. The reverse argument is that mammals for some reason were unable to evolve the biochemical and physiological attributes required for uricotelism. Being tied to ureotelism, so this argument goes, the mammals experienced great selective pressures to evolve exceptional urea-concentrating abilities.

Focusing again on nitrogen, uric acid and urates not only permit nitrogen to be excreted with little water, but also have great advantages in times of water crisis when urine production is curtailed or stopped. If a ureotelic animal in water crisis stops producing urine, the urea concentration in its body fluids steadily rises because urea is so soluble that its solubility limits are never reached. This buildup cannot continue indefinitely because at very high concentrations, urea becomes toxic. By contrast, if a uricotelic animal stops producing urine, uric acid and urates are deposited as precipitates within its body. Because the solubilities of these compounds are low, their concentrations in the body fluids cannot increase above low levels regardless of the amounts stored. Uric acid and urates therefore are suited to indefinite storage.

Besides uric acid and urates, other purines or compounds derived from purines are sometimes employed as nitrogenous end products. The purine *guanine*—which is even less soluble than uric acid—is a primary nitrogenous end product in some animals. *Allantoin* and *allantoic acid*—compounds formed by the partial breakdown of uric acid—may also be primary nitrogenous end products; although more soluble than uric acid, they have low solubilities.

Uric acid probably requires considerably more energy per nitrogen atom for its synthesis than urea.[18] Nonetheless, uricotelism—or the production of other purines as principal nitrogenous end products—is the most common state in terrestrial animals. Birds, lizards, and snakes are uricotelic (Box 27.4). (The white matter in bird droppings is uric acid.) Turtles that inhabit dry terrestrial habitats tend toward uricotelism. Among terrestrial invertebrates, the great majority that live in the open air employ purines or purine derivatives as their primary nitrogenous end products. In most terrestrial insects, the principal nitrogenous excretion is uric acid, allantoin, or allantoic acid (the question of why some insect species excrete mainly one of these compounds and other species excrete another is interesting and unanswered). Spiders, scorpions, and certain ticks excrete mostly guanine. Among most terrestrial snails, uric acid is the predominant nitrogenous end product, but other purines—guanine and xanthine—are often produced in quantity as well. Temporary or permanent storage of purines has been observed in certain land crabs and in many insects and snails.

The biochemical pathways employed for the synthesis of uric acid or related compounds from protein nitrogen are complicated.[19] However, they in fact are only relatively small modifications of very ancient and universal pathways for the synthesis of the purine constituents of compounds like DNA—helping to explain how uricotelism (or "purinotelism") could have evolved independently in several phyla on land. The convergence of several terrestrial phyla on uricotelism testifies to the advantages of poorly soluble nitrogenous end products for terrestrial existence.

[18] Although some biochemists calculate that the synthesis of uric acid from protein nitrogen costs about the same amount of ATP-bond energy per nitrogen atom as urea synthesis (2–2.5 high-energy phosphate bonds per atom), others calculate that each nitrogen atom costs as much as 6 high-energy phosphate bonds to be incorporated into uric acid. Regardless of that matter, uric acid probably has a higher overall cost than urea in at least some animals because of extra processes that must be carried out to prevent it from precipitating prematurely.

[19] These pathways are reviewed in biochemistry texts.

Summary: Nitrogen Disposition and Excretion

- Animals that synthesize ammonia or urea as their primary nitrogenous end product are termed, respectively, ammonotelic or ureotelic. Animals that synthesize mainly uric acid or urates are uricotelic.
- Ammonotelism is the primitive condition and is seen in most water-breathing aquatic animals. Ammonia has the advantage of costing no extra ATP to produce. It is toxic, however, and thus, for an animal to be ammonotelic, the animal must have a means to void ammonia reliably as fast as it is produced so that blood levels are kept low. Aquatic animals void ammonia into the ambient water across their gills or general body surfaces.

- Ureotelism is more costly than ammonotelism because producing urea has an ATP cost. Urea is far less toxic than ammonia, however. Ureotelism has evolved principally in certain groups of vertebrates, in which it serves one or more of three possible functions: reducing the water requirement of routine nitrogen excretion (e.g., in terrestrial amphibians and mammals), adjusting the blood osmotic pressure in advantageous ways (e.g., in elasmobranch fish), and detoxification of waste nitrogen during periods when water-stressed animals cease urine production.
- Uricotelism is even more costly per nitrogen atom that ureotelism, but uric acid and related compounds have the advantage that they are so poorly soluble that they are low in toxicity, can be excreted in little water, and can be accumulated in the body indefinitely. Most groups of terrestrial animals, including invertebrates (e.g., insects) and vertebrates (e.g., birds and many reptiles), are uricotelic or produce predominantly other purines (e.g., guanine) or purine derivatives.

Study Questions

1. Considering the distal convoluted tubule of the amphibian nephron, explain how changes in the permeability of the tubule wall to water affect the amount of pure, osmotically free water excreted in the urine. Be sure to define what is meant by pure, osmotically free water.
2. Outline how the orientation of nephrons relative to each other imparts gross structure to the kidneys of mammals and birds.
3. If you were attempting to tell whether an animal produces its primary urine by ultrafiltration or secretion, what measurements would you make *on the primary urine*? If your measurements on the primary urine indicated that ultrafiltration might be occurring, what other types of measurements would you make to determine whether physical and physicochemical conditions favorable to ultrafiltration exist? Explain.
4. When researchers first proposed the countercurrent multiplication hypothesis for concentration of urine in the mammalian kidney, there was great resistance to its acceptance in certain quarters. The anatomist Ivar Sperber, whose comparative morphological studies originally helped draw attention to the loops of Henle, pointed out that there were certain rodents in which the anatomy of the kidney should make it relatively simple to sample blood from the hairpin bends of the vasa recta deep in the medulla. Samples of such blood were obtained, and the osmotic pressure of this blood proved to be far higher than the osmotic pressure of blood in the general circulation. This research convinced doubters of the validity of the countercurrent multiplication process. Why does blood at the hairpin bends of the vasa recta have a high osmotic pressure, and why would knowing its osmotic pressure provide strong support for the countercurrent multiplication hypothesis?
5. Production of any sort of nitrogenous waste besides ammonia costs energy. Name at least three distinctly different advantages an animal might gain by investing in urea production or uric acid production.
6. Explain how primary urine is introduced into the Malpighian tubules of an insect.
7. The immediate effect of ADH on the renal tubules of frogs and mice is the same, yet when ADH is secreted, frogs produce urine that is approximately isosmotic to their blood plasma, whereas mice produce urine far more concentrated than their blood plasma. Explain this difference in terms of the factors affecting osmosis in the kidneys of frogs and mice.
8. Drugs that increase urine flow (diuretic drugs) are often employed in the treatment of hypertension (high blood pressure) or other disease states. Three physiological categories of such drugs are ones that (1) function as loop diuretics, (2) inhibit the action of aldosterone, and (3) block Na^+ channels in the collecting ducts. Explain why each of these categories would be expected to increase Na^+ excretion and urine flow. (Hint: Rereading the section on hormones at the end of Chapter 26 might prove helpful.)
9. In some rats studied in a state of antidiuresis, the urea concentration and osmotic pressure in the inner-medullary collecting-duct fluid were about the same as the urea concentration and osmotic pressure in the inner-medullary interstitial fluid. However, the concentration of Na^+ in the collecting-duct fluid was only about 1/8 of the Na^+ concentration in the interstitial fluid, and the collecting-duct fluid was rich in K^+ even though the interstitial fluid contained hardly any K^+. How are these results to be explained by the difference in concentrating mechanisms for urea and for inorganic ions?
10. In mammals, the kidneys are the only organs that regulate routine excretion of water, salts, and nitrogenous wastes from the blood. As logical as this may sound to us, it is unique among vertebrates. For each of the other groups of vertebrates, describe the functions of as many organs as you can—in addition to the kidneys—that participate in these processes.
11. Whenever the concentrating ability of mammalian kidneys has been studied in relation to the lengths of the loops of Henle in various species, a clear correlation between the two has been found—indicating that loop length matters—but in addition, there has been a great deal of scatter in the data (e.g., see Figure 27.8). Fifty years from now, physiologists will probably understand the mechanistic reasons *why* loop length is not a perfect predictor of concentrating ability. Suppose a government agency has decided to give you all the resources you need to study whatever you desire. As a brainstorming exercise, what specific aspects of mammalian kidney function besides loop length would you investigate to try to account better for differences among species in concentrating ability?

Key References

Beuchat, C. A. 1996. Structure and concentrating ability of the mammalian kidney: Correlations with habitat. *Amer. J. Physiol.* 271: R157–R179. A good example of modern comparative biology that takes full advantage of the massive amounts of data available in our "information era" to review what is known and generate hypotheses for future research.

Beyenbach, K. W. 2003. Transport mechanisms of diuresis in Malpighian tubules of insects. *J. Exp. Biol.* 206: 3845–3856. An accessible, well illustrated, and definitive review of Malpighian tubule function in insects, written by one of the leaders in the field. Emphasizes molecular mechanisms but includes an enlightening overview of organ function.

Braun, E. J., and W. H. Dantzler. 1997. Vertebrate renal system. In W. H. Dantzler (ed.), *Comparative Physiology*, vol. 1 (Handbook of Physiology [Bethesda, MD], section 13), pp. 481–576. Oxford University Press, New York. A modern review of comparative kidney function in the vertebrates, emphasizing cellular and molecular physiology.

Masilamani, S., M. A. Knepper, and M. B. Burg. 2000. Urine concentration and dilution. In B. M. Brenner (ed.), *Brenner and Rector's The Kidney*, 6th ed., vol. 1, pp. 595–635. Saunders, Philadelphia. A unusually articulate and clear review of concentration and dilution in the mammalian kidney, including thorough discussion of models for the single effect in the inner medulla. Many enlightening diagrams.

Vander, A. J. 1995. *Renal Physiology*, 5th ed. McGraw-Hill, New York. A compact summary of medical renal physiology justly respected for its excellent pedagogy.

Selected References

Abramson, R. G., and M. S. Lipkowitz. 1990. Evolution of the uric acid transport mechanisms in vertebrate kidney. In R. K. H. Kinne (ed.), *Basic Principles in Transport*, pp. 115–153. Karger, Basel.

Atkinson, D. E. 1992. Functional roles of urea synthesis in vertebrates. *Physiol. Zool.* 65: 243–267. [Emphasizes a novel perspective on the evolution of ureo- and uricotelism.]

Bankir, L., and C. de Rouffignac. 1985. Urinary concentrating ability: Insights from comparative anatomy. *Amer. J. Physiol.* 249: R643–R666.

Beuchat, C. A. 1990. Body size, medullary thickness, and urine concentrating ability in mammals. *Amer. J. Physiol.* 258: R298–R308.

Bidigare, R. R. 1983. Nitrogen excretion by marine zooplankton. In E. J. Carpenter and D. G. Capone (eds.), *Nitrogen in the Marine Environment*, pp. 385–409. Academic Press, New York.

Bradley, T. J. 1987. Physiology of osmoregulation in mosquitoes. *Annu. Rev. Entomol.* 32: 439–462.

Braun, E. J. 1999. Integration of organ systems in avian osmoregulation. *J. Exp. Zool.* 283: 702–707.

Brenner, B. M. (ed.). 2000. *Brenner and Rector's The Kidney*, 6th ed. Saunders, Philadelphia.

Brown, J. A., R. J. Balment, and J. C. Rankin (ed.). 1993. *New Insights in Vertebrate Kidney Function*. Cambridge University Press, New York.

Campbell, J. W. 1991. Excretory nitrogen metabolism. In C. L. Prosser (ed.), *Environmental and Metabolic Animal Physiology* (*Comparative Animal Physiology*, 4th ed.), pp. 277–324. Saunders, Philadelphia.

Campbell, J. W., J. E. Vorhaben, and D. D. Smith, Jr. 1987. Uricoteley: Its nature and origin during the evolution of tetrapod vertebrates. *J. Exp. Zool.* 243: 349–363.

Casotti, G., K. C. Richardson, and J. S. Bradley. 1993. Ecomorphological constraints imposed by the kidney component measurements in honeyeater birds inhabiting different environments. *J. Zool. (London)* 231: 611–625.

Christiansen, J. S., R. A. Dalmo, and K. Ingebrigtsen. 1996. Xenobiotic excretion in fish with aglomerular kidneys. *Mar. Ecol. Progr. Ser.* 136: 303–304.

Coast, G. M., I. Orchard, J. E. Phillips, and D. A. Schooley. 2002. Insect diuretic and antidiuretic hormones. *Adv. Insect Physiol.* 29: 279–409.

Dantzler, W. H. 1982. Renal adaptations of desert vertebrates. *BioScience* 32: 108–113.

Dantzler, W. H. 1985. Comparative aspects of renal function. In D. W. Seldon and G. Giebisch (eds.), *The Kidney: Physiology and Pathophysiology*, pp. 333–364. Raven, New York.

Dobbs, G. H. III, Y. Lin, and A. L. DeVries. 1974. Aglomerularism in antarctic fish. *Science* 185: 793–794.

Dow, J. A. T., and S. A. Davies. 2003. Integrative physiology and functional genomics of epithelial function in a genetic model organism. *Physiol. Rev.* 83: 687–729.

Evans, D. H., and J. N. Cameron. 1986. Gill ammonia transport. *J. Exp. Zool.* 239: 17–23.

Greenwald, L. 1989. The significance of renal relative medullary thickness. *Physiol. Zool.* 62: 1005–1014.

Hadley, N. F. 1994. *Water Relations of Terrestrial Arthropods*. Academic Press, New York. [Includes a good discussion of excretory structure and function.]

Heisler, N. 1995. Ammonia vs. ammonium: Elimination pathways of nitrogenous wastes in ammonotelic fishes. Advances in Comparative and Environmental Physiology, vol. 22, pp. 63–87. Springer, New York.

Hevert, F. 1984. Urine formation in the lamellibranchs: Evidence for ultrafiltration and quantitative description. *J. Exp. Biol.* 111: 1–12.

Jones, H. D., and D. Peggs. 1983. Hydrostatic and osmotic pressures in the heart and pericardium of *Mya arenaria* and *Anodonta cygnea*. *Comp. Biochem. Physiol. A* 76: 381–385.

Karnaky, K. J., Jr. 1998. Osmotic and ionic regulation. In D. H. Evans (ed.), *The Physiology of Fishes*, 2nd ed. CRC Press, Boca Raton, FL.

Kinne, R. K. H. (ed.). 1989. *Structure and Function of the Kidney*. Karger, Basel.

Kinne, R. K. H. (ed.). 1990. *Basic Principles in Transport*. Karger, Basel.

Kinne, R. K. H. (ed.). 1990. *Urinary Concentrating Mechanisms*. Karger, Basel.

Lang, F. (ed.). 1998. *Cell Volume Regulation*. Karger, Basel. [Includes background chapters and a specific chapter on the kidney.]

Little, C. 1965. The formation of urine by the prosobranch gastropod mollusc *Viviparus viviparus* Linn. *J. Exp. Biol.* 43: 39–54.

Long, W. S. 1973. Renal handling of urea in *Rana catesbeiana*. *Amer. J. Physiol.* 224: 482–490.

Lote, C. J. (ed.). 1986. *Advances in Renal Physiology*. Croom Helm, London.

Maddrell, S. H. P. 1981. Functional design of the insect excretory system. *J. Exp. Biol.* 90: 1–15.

O'Donnell, M. J. 1997. Mechanisms of excretion and ion transport in invertebrates. In W. H. Dantzler (ed.), *Comparative Physiology*, vol. 2 (Handbook of Physiology [Bethesda, MD], section 13), pp. 1207–1289. Oxford University Press, New York.

Regnault, M. 1987. Nitrogen excretion in marine and fresh-water Crustacea. *Biol. Rev.* 62: 1–24.

Reilly, R. F., and D. H. Ellison. 2000. Mammalian distal tubule: Physiology, pathophysiology, and molecular anatomy. *Physiol. Rev.* 80: 277–313.

Robinson, W. E., and M. P. Morse. 1994. Biochemical constituents of the blood plasma and pericardial fluid of several marine bivalve molluscs: Implications for ultrafiltration. *Comp. Biochem. Physiol. B* 107: 117–123.

Rupert, E. E. 1994. Evolutionary origins of the vertebrate nephron. *Amer. Zool.* 34: 542–553.

Rupert, E. E., and P. R. Smith. 1988. The functional organization of filtration nephridia. *Biol. Rev.* 63: 231–258.

Sasaki, S., K. Ishibashi, and F. Marumo. 1998. Aquaporin-2 and -3: Representatives of two subgroups of the aquaporin family colocalized in the kidney collecting duct. *Annu. Rev. Physiol.* 60: 199–220.

Schipp, R., and F. Hevert. 1981. Ultrafiltration in the branchial heart appendage of dibranchiate cephalopods: A comparative ultrastructural and physiological study. *J. Exp. Biol.* 92: 23–35.

Schnermann, J. B., and S. I. Sayegh. 1998. *Kidney Physiology*. Lippincott-Raven, Philadelphia.

Tyler-Jones, R., and E. W. Taylor. 1986. Urine flow and the role of the antennal glands in water balance during aerial exposure in the crayfish, *Austropotamobius pallipes* (Lereboullet). *J. Comp. Physiol. B* 156: 529–535.

Urich, K. 1994. *Comparative Animal Biochemistry*. Springer, Verlag, New York.

Ussing, H. H., N. Bindslev, N. A. Lassen, and O. Sten-Knudsen (eds.). 1981. *Water Transport across Epithelia*. Munksgaard, Copenhagen.

Walsh, P. J. 1997. Evolution and regulation of urea synthesis and ureotely in (Batrachoidid) fishes. *Annu. Rev. Physiol.* 59: 299–323.

Walsh, P. J., and P. Wright (eds.). 1995. *Nitrogen Metabolism and Excretion.* CRC Press, Boca Raton, FL.

Wenning, A. 1996. Managing high salt loads: From neuron to urine in the leech. *Physiol. Zool.* 69: 719–745.

Wessing, A. (ed.). 1975. Excretion. *Fortschr. Zool.* 23: 1–362. [Includes articles on many animal groups.]

Wright, P. A., and P. M. Anderson (eds.). 2001. *Nitrogen Excretion.* Fish Physiology, vol. 20. Academic Press, New York.

Yancey, P. H. 1992. Compatible and counteracting aspects of organic osmolytes in mammalian kidney cells in vivo and in vitro. In G. N. Somero, C. B. Osmond, and C.L Bolis (eds.), *Water and Life*, pp. 19–32. Springer-Verlag, New York.

See also references in Appendix A.

28 Water, Salts, and Excretion at Work: Mammals of Deserts and Dry Savannas

In the desert, large herbivores such as the oryxes in this photograph are usually the equivalent of nomads, moving about to find moisture to survive. Similarly, in the dry savannas—the grassland plains—large herbivores often function as nomads or migrants. Conditions vary so much from time to time and from place to place in these arid zones that living in an invariant locale is often not a viable strategy. The same can be said for people in a state of nature; most of the traditionial cultures of the deserts and dry savannas were nomadic.

Our goal in this chapter is to explore how the oryxes and other species of sizable mammals—5 kg or larger in body weight—can live in hot deserts or dry-savanna ecosystems. One way to understand these mammals is to compare them with small species, so we will sometimes mention the kangaroo rats, gerbils, and other small mammals that coexist in arid habitats with the large species. However, the small mammals were discussed in Chapter 26, and we will not return to them here in any detail. The mammals of Africa will receive greatest emphasis because they form the most diverse and abundant set of large, arid-land mammals in the world today, and their diversity has drawn the interest of physiologists for decades.

Desert and Dry-Savanna Environments

Deserts and semideserts by most accountings cover about a third of the land on Earth. There is no uncomplicated way to define a desert. One of the simplest definitions is that a desert receives less than about 25 cm (10 inches) of precipitation per year, but this is an imperfect standard because there are regions that receive much more than that but are undoubtedly deserts. Following the lead of Imanuel

Oryxes are extreme examples of large mammals that can survive indefinitely in deserts without drinking water Physiologists are still learning how these animals orchestrate their water losses and gains to stay in water balance in such water-poor places. The animals shown are gemsbok oryxes (*Oryx gazella*), often called simply gemsbok.

Noy-Meir, most biologists prefer a definition that emphasizes two attributes of deserts that are of extreme importance for plants and animals. First, a desert is a place where precipitation is so low that *availability of water exerts a dominant controlling effect on biological processes.* Second, when precipitation occurs in deserts, *it comes in infrequent, largely unpredictable events.* Rains in deserts are highly irregular and unreliable in both time and space. One year may bring five times (or even twenty times) the rain of another. One 30-km^2 area may be drenched during a storm while an adjacent area of similar size receives little or nothing, because rains in deserts are usually produced by isolated storms rather than broad fronts. In deserts, large herbivores must often conduct their lives in ways that—more than anything—are opportunistic, taking advantage of rain or moisture whenever and wherever it occurs.

The classic look of the dry savannas is of endless plains of grass with trees dotted here and there. Dry savannas and deserts are in many cases neighboring environments. Gradients of moisture often occur on continental scales of space, with moist forests in some regions, deserts in others, and dry savannas in between, where there is more moisture than in the deserts, but far less than in the forests. In dry savannas—as in deserts—water is a dominant controlling factor for biological processes, but in a somewhat different way. Savannas usually have discrete rainy seasons interrupted by discrete rainless seasons. Thus, although on an annual basis, savannas receive more rain than deserts and receive it more predictably, the plants and animals living in savannas must often endure profound drought for long periods each year. The Serengeti plains of East Africa and other areas nearby are examples of this second sort of water-controlled ecosystem. During the long dry season each year, virtually no rain may fall for 4 to 6 months; thus, streams dry up, and soils become so parched that the grasses turn crisp and brown.

Deserts and dry savannas exist for several reasons. The single most important cause of desert and semidesert conditions on a planetary scale is the global pattern of air movements whereby air warmed at the equator rises to high altitudes, displacing air at those altitudes in such a way that high-altitude air descends to Earth's surface somewhere else. Air at high altitudes, partly because it is cold (see Table 25.2), contains little moisture, even when saturated with water vapor. Thus, in regions where high-altitude air tends consistently to fall to low altitude, the land can become parched. The present global pattern is for high-altitude air to descend in two bands encircling the globe at latitudes about 30° north and 30° south of the equator. Most of the world's great deserts—including the largest, the Sahara—are products of this process. Another common but more localized cause of desert or savanna conditions is rain-shadowing caused by highlands. If a region's only reliable source of moisture is winds blowing in from the ocean, and if intervening highlands force the winds to rise—so that the air cools and its moisture condenses to form rain or snow—the winds may have little moisture left by the time they blow into the region of interest. Near Los Angeles, for example, the coastal mountains force the prevailing winds blowing east from the Pacific Ocean to rise, causing mountain rain. The land still farther east, the Mojave Desert, looks much like a dry, sandy-colored shadow cast by the mountains.

The existence and extent of deserts and dry savannas, we see, often depend on nothing more permanent than the gossamer movements of air. Accordingly, over the long reaches of geological history, deserts and savannas have shifted their positions as patterns of air movement have changed. Despite shifting about, deserts and savannas have probably always existed somewhere. Thus, they have long presented plants and animals with special challenges, and they have long presented opportunities for species that are able to make do with little water and tap the meager water resources available.

The Relations of Animals to Water

Large mammals have advantages and disadvantages compared to small ones in hot deserts and dry savannas. Many of the most apparent disadvantages of large size are behavioral. Large mammals, for example, cannot take shelter underground by digging burrows, as kangaroo rats, gerbils, and other desert rodents do. Moreover, large animals are less likely than small species to find adequate shade or other protective microhabitats on the surface of the ground. In contrast to the *behavioral* limitations that they face, however, large mammals often find themselves in a distinctly more favorable *physiological* position than small mammals.

In terms of water costs, large body size is a physiological advantage

For a mammal to thermoregulate in a hot place, it must evaporate water to void from its body not only the *exogenous* heat that enters its body from the environment but also the *endogenous* heat produced by its own metabolism. Large size is an advantage from both of these perspectives because it tends to reduce an animal's weight-specific rate of heat gain from both exogenous and endogenous sources.

Speaking of exogenous heat inputs, large mammals enjoy an advantage because they have less body-surface area per unit of weight than small mammals (their surface-to-volume ratios are lower). When the air temperature is higher than body temperature and the sun is beating down, environmental heat enters an animal's body across its body surfaces. Because large mammals have relatively little body-surface area per unit of weight, they tend to experience less heat entry per unit of weight over a period of time than small mammals. Thus, their water costs to void the incoming heat are lower per unit of weight.

Speaking of endogenous heat inputs, large mammals are in a more favorable position than small ones because of the allometric relation between metabolic rate and body size. Large mammals tend to have far lower weight-specific metabolic rates than small mammals (see Figure 5.9). Thus, their rates of endogenous heat production per unit of body weight are relatively low, and their weight-specific water costs to void endogenously produced heat are also low.

Quantitatively, the physiological advantages of large size can be quite dramatic. In the 1930s, D. B. Dill (1891-1986)—a pioneer in the study of exercise and heat—undertook his famous "walks in the desert" that provided the very first quantitative information on the water costs of mammalian exercise under hot conditions. He himself was one of the human subjects, and he took along with him both dogs and burros. Together, man and beast trekked along under the searing sun in the Nevada desert, sweating or panting to thermoregulate; and as they walked, Dill measured their rates of dehydration.

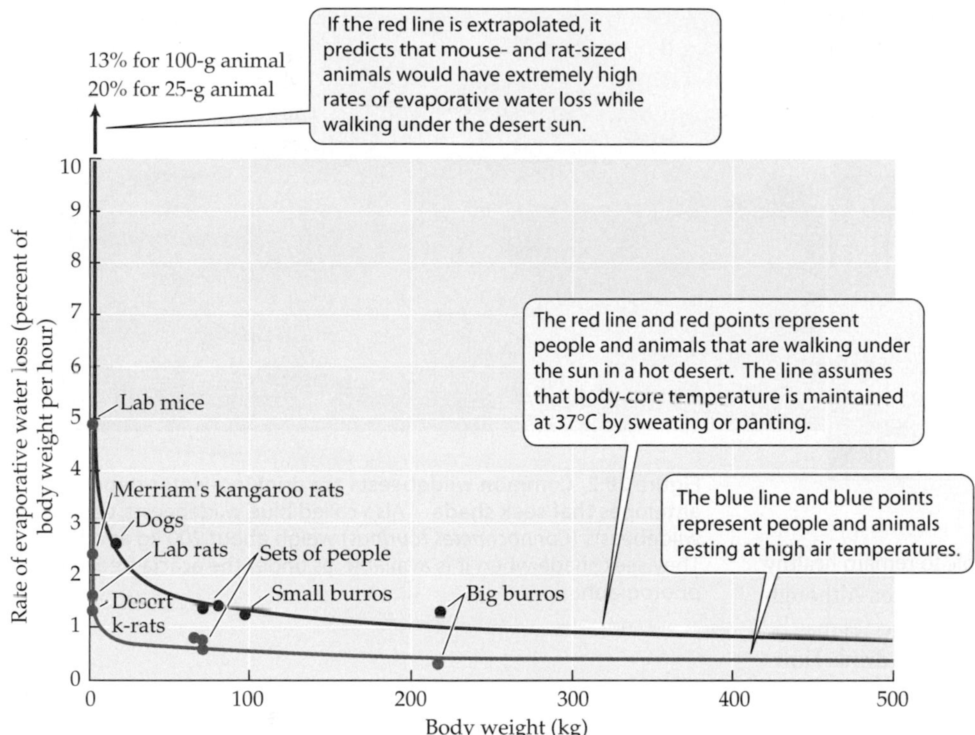

Figure 28.1 The rate of evaporative water loss experienced by mammals exposed to heat stress depends strongly on body size The graph shows weight-specific rates of evaporative water loss for exercising animals (red) and resting animals (blue) under hot conditions. The resting people and burros were studied while resting in the desert; the resting mice and rats were studied in a laboratory chamber maintained at an air temperature of 40°C (104°F). The equation for the red line is $E = 6.03W^{-0.33}$, where E is rate of evaporative water loss in percentage of body weight per hour and W is body weight in kilograms. The equation for the blue line is $E = 1.38W^{-0.21}$. The two species of kangaroo rats included in the resting studies are Merriam's kangaroo rats (*Dipodomys merriami*) and desert kangaroo rats (abbreviated "desert k-rats"; *D. deserti*), weighing 34 g and 114 g, respectively. (After Schmidt-Nielsen 1954; Soholt, Dill, and Oddershede 1977.)

Knut Schmidt-Nielsen (b. 1915), another pioneer in desert research, used fundamental principles and Dill's data to construct the classic line shown in red in Figure 28.1. Later, Dill and collaborators gathered more data of a similar sort on burros and high school students in the Colorado desert, accounting for the total of five data points plotted in red on the figure.

The red line shows that there is a very strong relation between body size and the water cost of being active in the daytime desert. The line applies specifically to mammals that thermoregulate at a body temperature of about 37°C as they exercise.[1] To interpret the line, it is important to recognize that the rate of evaporation is plotted in weight-specific terms on the y axis. Specifically, the rate of evaporation is expressed as grams of water lost from the body per 100 grams of body weight during an hour, that is, *percentage of body weight* lost per hour. As you can see, the water cost of being active in the daytime desert is predicted by the red line to *soar* at small body sizes.

According to the red line, if a 100-g rat were to walk along under the desert sun and thermoregulate by evaporation of water, its water cost would be almost 13% of its body weight *per hour*. If a 25-g mouse were to do so, its cost would be about 20% per hour. Dehydration ordinarily becomes lethal under hot conditions when 10% to 15% of body weight has been lost. Thus, the rat walking along under the sun would die of dehydration after about an hour, and the mouse would die after a half hour! In other words, from what we know, it would be *impossible* for mammals of these sizes to lead fully exposed, active lives in the daytime desert. Of course, rats and mice have never been actually tested to see if these predictions hold true for animals walking under the desert sun. However, the blue line in Figure 28.1 shows actual data, not only for burros and people, but also for four species of rats and mice that were studied while resting under milder but thermally stressful conditions. These data verify that water costs rise steeply at body sizes as small as those of rats and mice. From this perspective, it is indeed fortunate that small mammals can readily escape the full intensity of the daytime desert heat by burrowing and other behavioral means. Physiologically, such small mammals are ill-equipped to face the full brunt of desert conditions.

Facing the full brunt of desert conditions, however, is exactly what large desert mammals must often do. And fortunately, as Figure 28.1 shows, large mammals are in a far better physiological position to do so than small mammals. Based on the red line, a 100-kg mammal, such as a small adult burro, can thermoregulate while exercising in the heat of the desert day at a water cost of just 1.3% of its body weight per hour, and a 400-kg mammal, such as an adult dromedary camel, can do so at a water cost of only 0.8% per hour. The predicted water costs of large mammals in the desert are still lower when the animals are not exercising, as shown by the blue line. Large size, in brief, is a great advantage physiologically when animals live under hot conditions, quite apart from any specific adaptations that individual species may have evolved.

[1] The humans, dogs, and burros used to construct the line in fact thermoregulate in approximately this way, although not all species do.

Coexisting species are diverse in their relations to drinking water

One might imagine that all large species of mammals that have long histories in arid environments would have evolved minimal water requirements. That is not the case, however. Among the species that are observed commonly in dry, hot places, some are classified as

Figure 28.2 Common wildebeests are drinking-water-dependent antelopes that seek shade Also called blue wildebeests, common wildebeests (*Connochaetes taurinus*) weigh about 200 kg as adults. They seek shade when it is available, as under the acacia tree in the photograph.

drinking-water-independent because they are able to remain healthy for many days or weeks without access to drinking water. Although most of these animals must drink on occasion, some species may be literally independent of drinking water and *never* need to drink. However, other coexisting species are classified as **drinking-water-dependent** because they must drink each day, or at least every other day.

Whether a particular species falls into the drinking-water-independent or -dependent category depends in part on the severity of conditions. A species that is independent of drinking water in the cool seasons of the year might, for example, become dependent on it in the hot seasons. Despite such ambiguities, the distinction between drinking-water-independent and -dependent species is a useful organizing principle for discussing large desert and dry-savanna herbivores.

An outstanding example of coexisting drinking-water-independent and -dependent species is provided by the fabled herds of herbivores of the Serengeti ecosystem and other similar ecosystems in East Africa and South Africa. In the Serengeti ecosystem, wildebeests and zebras are drinking-water-dependent.[2] During the hot seasons of the year, they must drink every day or every other day. This means, in effect, that they cannot wander more than about 25 km (15 miles) from standing water. Species that often coexist with wildebeests and zebras but are drinking-water-independent include Grant's gazelle, the common eland, and the dik-dik (dwarf antelope). These animals are able to travel far from standing water (because they need not return to it every day), and in that way they gain access to food resources that are unavailable to the drinking-water-dependent species. On even just modestly hot days in areas where trees are found, wildebeests seek shade in the heat of the day (Figure 28.2), whereas Grant's gazelles on such days are indifferent to sun or shade (Figure 28.3). This contrast in shade-seeking behavior correlates with the water physiology of the two species: The species that dehydrates

[2] The species discussed in this paragraph are the common or blue wildebeest (*Connochaetes taurinus*), the plains or Burchell's zebra (*Equus burchelli*), the common eland (*Tragelaphus oryx*), Grant's gazelle (*Gazella granti*), and Kirk's dik-dik (*Madoqua kirkii*).

Figure 28.3 Grant's gazelles are drinking-water-independent antelopes noted for their indifference to sun or shade These gazelles (*Gazella granti*) weigh about 50 kg as adults. Grant's gazelles are one of the species often reputed by naturalists to be able to live indefinitely without drinking water in the dry savannas and deserts.

more readily in the absence of drinking water also exploits shade more readily when shade is available.

The most dramatic manifestation of the difference between drinking-water-independent and -dependent species in the Serengeti ecosystem is the differential participation of the two categories of animals in the annual migration. The annual migration in the Serengeti is itself perhaps the most dramatic illustration in today's world of the way water can exert a dominant controlling effect on biological processes in an ecosystem. Similar migrations occurred elsewhere in Africa a century ago, but the Serengeti is the last remaining, truly large ecosystem in which they occur.

In the Serengeti, there is a gradient of decreasing moisture availability (increasing aridity) from the northwest to the southeast because of the positions of mountains and highlands, the directions of winds, and the courses followed by rivers; total annual rainfall in the northwest is about twice that in the southeast. In the rainy season, enough rain falls everywhere for lush growth of grasses. However, during the long dry season between June and November, the southeastern plains typically receive almost no rain. As the dry season progresses, streams and rivers in the southeast first stop flowing, becoming nothing but a series of puddles and ponds. Then even the puddles and ponds dry up. The soil becomes parched, and the grasses turn brown and brittle. These seasonal patterns set the stage for the annual migration.

Not all species of large herbivores participate in the migration. In fact, most of the drinking-water-independent species do not. Grant's gazelles, elands, dik-diks, and most other drinking-water-independent species simply "stay put" even as the thronging herds of migratory animals move through. The most abundant migratory species is the wildebeest, which in recent times has numbered about 1 million. Another highly visible migratory species is the zebra. Both the wildebeest and zebra are drinking-water-dependent. As the rainy season starts in November or December, they move on a broad front from the northwest toward the southeastern plains (Figure 28.4*b*). While the rains fall in the southeast (Figure 28.4*c*), they find water and lush grass there. However, after the rains stop in the southeast during the following May or June, they must get out of the area. First they move west (Figure 28.4*d*), and then—as the landscape becomes ever drier—they move in thundering herds back to the far northwest (Figure 28.4*e*), where they have the best chance of finding the water they require during the season of little rain. As we discussed at the start of Chapter 4, scientists have only hypotheses—not certain knowledge—about why wildebeests and zebras move from the northwest to the southeast when the rainy season arrives. However, there is no doubt why these animals leave the southeast when the rains stop: They need to drink each day, and doing so in the southeast becomes impossible. Meanwhile, drinking-water-independent species are free to stay put or make just relatively subtle adjustments to the strong seasonality of water availability.

As earlier noted, migrations of drinking-water-dependent large herbivores used to occur in some African ecosystems where they no longer do. Thus, although the Serengeti seems almost unique today, it was not unique primordially; migratory herds evolved in other settings as well. For example, the wildebeests in the large Kalahari ecosystem at the border of Botswana and South Africa used to migrate in much the same pattern as the those in the Serengeti. However, this migration was disrupted in the twentieth century by human intervention, first by a fence constructed (in theory) to protect cattle from the spread of disease by migrating wildebeests and later by the installation of wells, which the wildebeests congregate around rather than migrating to seek water.

Carnivorous mammals such as lions and cheetahs, it is worth noting, are generally assumed (on the basis of little evidence) to achieve water balance without having to drink much, if at all, because of the high water content of their foods. Some in the Serengeti are indifferent to the migration, whereas others follow the prey animals.

Water conflicts threaten animals and people

Water conflicts seem to be almost inevitable, as well as intrinsically ominous, in ecosystems where water exerts dominant controlling effects on biological processes. The conflicts that presently exist are worth noting because they threaten the future existence of the species and systems of life we are discussing.

One sort of water conflict is between the water needs of desert and dry-savanna ecosystems and the water claims of outside human enterprises. In the United States, burgeoning cities in arid parts of the West and equally burgeoning demands for agricultural irrigation water have led to the draining and diversion of many of the rivers that once flowed through the Western deserts and savanna-like prairies. The most dramatic example is probably provided by the Colorado River, which is tapped for so many human uses as it flows from the Rocky Mountains to the Sea of Cortez (Gulf of California) that in many years it now runs out of water before reaching its ancient destination and dries up in the desert. Near the northern boundary of the Serengeti ecosystem, there are several present-day activities or proposals for activities that threaten water conflicts with wildlife. As we have seen, the drinking-water-dependent species rely on a dependable supply of drinking water in the northwestern part of the ecosystem to survive the dry season. The sufficiency of this supply is already known to be less than fully certain; during drought years, 20% to 30% of the wildebeest population has been known to die. The principal reliable water source is the Mara River, which brings water into the ecosystem from forested highlands farther north. Deforestation of the highlands, diversion of river flow for irrigation uses, and diversion for hydroelectric power production are all possibilities at present. Such human activities would create water conflicts with the wild animals in their dry-season refuge, potentially disrupting the migration and the whole Serengeti ecosystem.

Another major sort of water conflict arises from the relationship between indigenous peoples and the land within desert and dry-savanna ecosystems. Throughout the world, the traditional way of life for hundreds of cultural groups in such places has been nomadism. Surviving on very small amounts of water compared to modern urban water usage (Table 28.1), indigenous people lived in deserts and dry savannas by moving with the rains. The notions of private land ownership that have spread around the world in the last few centuries do not readily accommodate nomadism; a nomad today would be spending time on one piece of private property after another. Thus, as we look at a modern-day Maasai (Figure 28.5) or modern-day representative of any other nomadic group, we see two things. First we are reminded of the peoples—not just the wild ani-

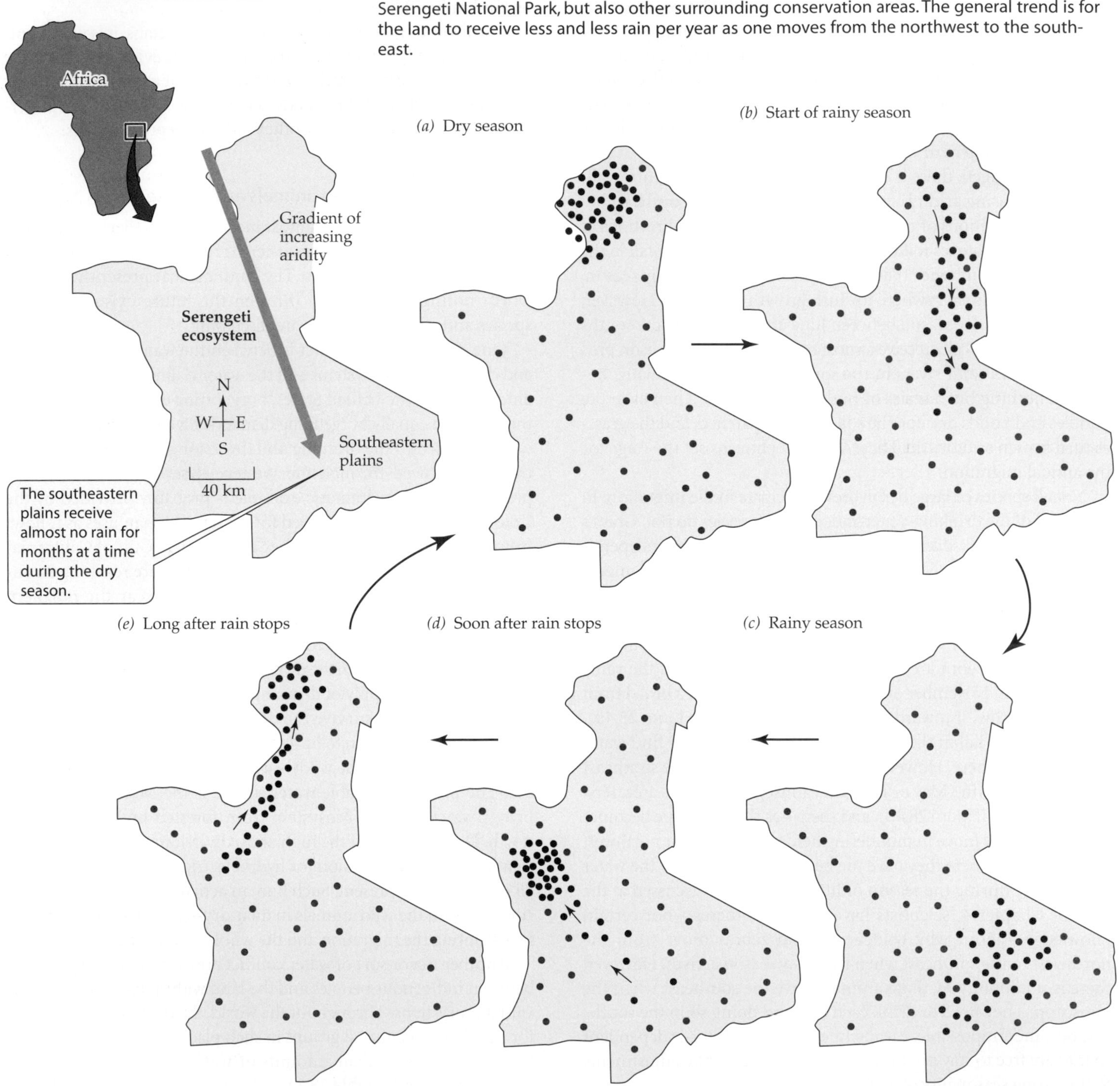

Figure 28.4 The annual migration in the Serengeti ecosystem Driven by seasonal changes in water availability, the drinking-water-dependent wildebeests and zebras migrate from the northwest in the dry season to the southeast in the rainy season and back again, whereas most drinking-water-independent species adjust only in much more subtle ways to the seasonal rhythm of dry and rainy seasons. The Serengeti ecosystem as mapped here includes not just Serengeti National Park, but also other surrounding conservation areas. The general trend is for the land to receive less and less rain per year as one moves from the northwest to the southeast.

mals—that lived for millennia in the deserts and dry savannas. Second, we see the troubled ending of a way of life as modern governments impose changed relationships to the land and, importantly, to water. Many nomadic peoples have been settled on their own private parcels of land, but in the deserts and dry savannas, a single parcel is unlikely to provide adequate water through all years and all seasons; the people used to be nomads precisely because movement was necessary.

A debate that today has reached its time of ultimate resolution is whether some nomadic cultures should be given a chance to continue by protecting large tracts of land for nomadic use or even by assimilating them into parks now reserved exclusively for wild ani-

TABLE 28.1 Water use by people in different sorts of communities in Arabia People in indigenous desert settlements use one-tenth the water of people in modern towns. The figures are for all domestic water use, including drinking, washing, bathing, and other water demands.

Type of community	Domestic water use per person (L/day)
Modern Arabian town without major industry[a]	240
Traditional agricultural village	120
Small desert settlement with supply by government water truck	80
Small desert settlement with traditional water supply	28

Source: After Goudie and Wilkinson 1977.
[a] New York City has a similar usage rate.

mals. The challenges are particularly acute for the herding cultures, whose herds of cattle or goats traditionally provided them with a way to glean the essentials for life from the arid land, but today greatly magnify the water conflicts with other uses of the land. Protected parks for wild animals in deserts and dry savannas have nearly always had their borders set to include the most reliable water sources. When the herds kept by nomadic peoples are banned from the parks, the herds are kept from the most dependable places to drink the water they need during dry seasons, but when the herds are let into the parks, they compete with the wild animals for water.

Figure 28.5 A Maasai youth of college age He is a member of the current generation of a people who until recent times were nomadic in the dry savannas of East Africa.

All species of large herbivores require considerable amounts of preformed water

Let's now take a more quantitative approach to understanding the water physiology of oryxes, gazelles, wildebeests, and the other large wild herbivores of deserts and dry savannas. Recall that under many conditions, kangaroo rats and other small desert rodents get well over half their water as metabolic water; at 25°C, for example, 80% to 90% of a kangaroo rat's water needs may be met by metabolic water (see Figure 26.20). In sharp contrast, *all the species of large herbivores of deserts and dry savannas acquire most of their water as preformed water.*

To see this more clearly, let's consider the water budgets of two very different mammals of similar body size: the highly drinking-water-dependent common wildebeest and the highly drinking-water-independent beisa oryx (**Figure 28.6**). These animals were studied under two sets of conditions. The first set of conditions was designed to be about as mild as the animals would ever encounter in nature; the environmental temperature was kept constantly at about 22°C. The second set of conditions was intended to simulate more closely the high water stresses the animals sometimes encounter; during each 24-hour day, the temperature was raised to 40°C (104°F) (simulating daytime heat) for 12 hours and lowered to 22°C (simulating night) for the other 12 hours. Under both sets of conditions, the animals were gradually given less and less preformed water per day until they reached the minimum they required to maintain health. They were then maintained on that minimum for 2 weeks so that their water budgets could be studied.

The results (see Figure 28.6) reveal that the drinking-water-dependent wildebeest has a far greater total water need, and a far greater need for preformed water, than the drinking-water-independent oryx when both species are studied side-by-side under identical conditions. Thus there is a dramatically measurable physiological difference between the two species.

A second major insight revealed by the results, however, is that even the oryx—which is famous for being one of the most drinking-water-independent of all large mammals—requires considerable quantities of preformed water to stay in water balance. Metabolic water met only about 20% of the oryx's total water need under both sets of conditions. Thus preformed water had to meet about 80% of its need. How is it possible for a species to be drinking-water-independent, yet require preformed water to meet much of its daily water requirement? The answer is that the oryx must meet much of its daily water need from preformed water *in its foods.*

The water budgets of the wildebeest and oryx are quite typical of large herbivores in deserts and dry savannas. With them in mind, we can make the following key points about these animals:

- *All* species depend more on preformed water than on metabolic water to meet their water needs.
- Drinking-water-dependent species cannot get all the preformed water they need from their foods, either because they need more preformed water than available foods could ever provide or because they are ineffective in selecting and processing foods to maximize their acquisition of preformed water. Therefore, they have to drink each day for their total inputs of preformed water to be adequate to keep them healthy.

(a) Water losses

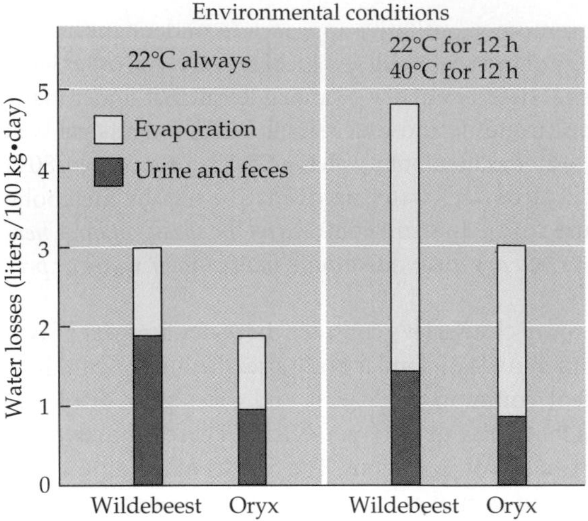

(b) Water gains

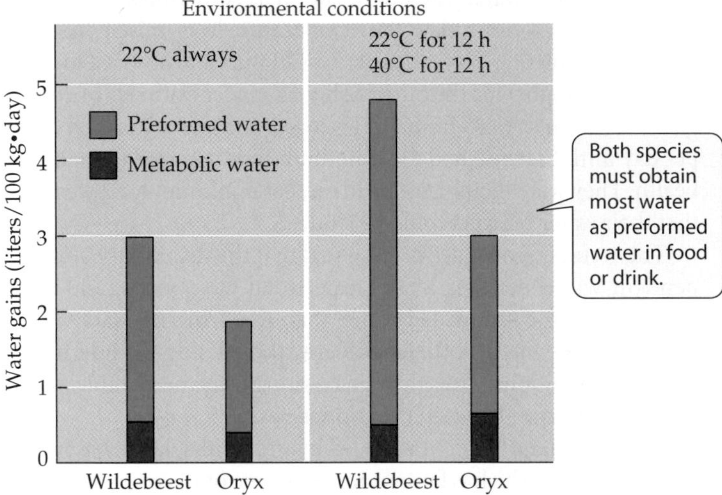

Figure 28.6 Water budgets of wildebeests and oryxes The animals of each species studied were roughly similar in body weight; the wildebeests (*Connochaetes taurinus*) averaged 160 kg, whereas the oryxes (*Oryx beisa*) averaged 100 kg. They were studied in identical chambers that permitted their water losses *(a)* and gains *(b)* in urine, feces, food, and drink to be quantified. During some experiments (left) the air temperature was a constant 22°C. During others (right) the air temperature was 22°C for half of each 24-hour day and 40°C for the other half. The animals had been acclimated to water restriction before the start of these measurements and were maintained on water restriction throughout the study. (After Taylor 1968, 1970.)

- Drinking-water-independent species, in contrast, can get all the preformed water they need to stay healthy—for many days in a row—from their foods. Whereas doing this may sound simple, it often is far from simple, as we will soon see, because foods in deserts and dry savannas are often parched.
- Most drinking-water-independent species in fact need to drink occasionally during the hot–dry seasons of the year. Although the preformed water they get from their foods is *almost* enough to meet their total need for preformed water, it is not *fully* adequate. Thus, when they do not drink on a given day, they become a bit dehydrated. They become a bit more dehydrated the next day, and so forth, until ultimately—after a week or two (or even three)—they need to find drinking water.
- A few drinking-water-independent species are believed to be able to meet their entire need for preformed water from the preformed water in their foods even in the hot–dry seasons of the year, meaning that they need never drink. The oryx and eland are often cited as examples. Field biologists report that these species are never observed to drink in water-poor regions, and physiological data (which we will discuss shortly) make it plausible to believe that these animals acquire enough preformed water from their foods to not ever require drinking water.

Water and food resources in the deserts and dry savannas are often complex

Now that we have addressed the needs of large herbivores for preformed water in drink or food, a next logical step is to consider the nature of the drinking-water and food resources they have available. These resources are often far less straightforward than meets the eye.

One problem in both deserts and dry savannas is that *when standing water is found, it is very common for it to be salty water*. Recent measurements have revealed, for example, that in the central and southeastern Serengeti ecosystem, it is common to find salinities of 5 to 15 g/kg in the headwaters of major rivers and 20 to 30 g/kg (approaching seawater salinity) in stagnant pools or landlocked lakes. Not all waters are this salty; some have salinities lower than 1 g/kg. Nonetheless, it is impressive how salty the waters can be in this benign-looking grassland ecosystem, and there is some possibility that the immediate stimulus for the migrating herds to leave the southeast at the end of the rainy season is the increasing salinity of the drinking water as the landscape dries.

There are several reasons why waters in deserts and dry savannas are often salty. Briefly speaking, soils in arid regions are commonly salty, and the waters become salty by dissolving salts out of the soils. One key reason the soils are salty is that they get so little rain that they are rarely thoroughly flushed with water. Each rain may bring a low concentration of salt with it (tiny amounts of aerosolized seawater can be detected in the atmosphere hundreds of miles inland from the oceans, for example, and are picked up by rain). When the rain hits the ground in arid regions, it soaks in to only a shallow depth and evaporates. Rain after rain causes salts to accumulate in the upper layers of the soil. These salts then dissolve into any pools of water that form on top of the soil, making the water salty until it itself evaporates, leaving the salts behind once more. This process is aggravated in low places where rain tends to settle in pools or flow in temporary streams. In such places, especially large amounts of water accumulate after each rain and leave their salts behind after evaporating.

When animals find salty water to drink, their kidneys must be able to excrete each ion in the water at a higher concentration than in the water itself if the animals are to gain H_2O by drinking the water (see page 676). Thus the evolution of kidneys with exceptional concentrating abilities has been important in two ways for desert and savanna mammals. First, a high concentrating ability allows ordinary soluble wastes to be excreted with relatively little water. Second, it permits animals to gain H_2O from saltier water sources than would otherwise be possible. Without question, when drinking waters approach half the concentration of seawater, they cease being useful H_2O sources for some of the species of mammals that are native to the deserts and dry savannas.

A second highly consequential complexity of water and food resources in deserts and dry savannas is that *in dry seasons, the leaves of plants often undergo large and rapid swings in their content of preformed water, because of equilibration with the humidity of the air.* When this phenomenon was first reported about 35 years ago, it seemed that it might be merely a footnote in the biology of a few species. By now, however, desert biologists recognize the phenomenon as often being of extreme importance in the water biology of large herbivores. Figure 28.7 shows, for example, how a perennial grass that is a favored food source for animals varies in its moisture content between midday, evening, and the dark hours before dawn in the Namib Desert. The measurements were made in the dry season, when the blades of the grass were dead and brown. Although ostensibly "dry" all the time, the grass blades in fact quickly equilibrate with changes in atmospheric humidity and have a far higher water content at the end of the night than at midday or evening. Animals thus can increase their input of preformed water by eating preferentially in the hours around dawn, and research has shown that many species do.

A final aspect of the complexity of water and food resources has both obvious and less obvious aspects. In deserts and dry savannas, *both the growth and the nutritional composition of plants are often highly correlated with changes in water availability.* The effects of rainfall on plant growth are obvious: In deserts, where rainfall occurs irregularly and unpredictably, plant growth is likewise irregular and unpredictable. In savannas, plant growth waxes in the wet season and wanes in the dry season. What is not obvious is that some plants (though not all) vary markedly in their content of nutritionally important materials as rains come and go. Certain perennial grasses in the Kalahari Desert ecosystem, for example, are known to vary in their protein content between the wet and dry seasons (Figure 28.8). The leaves and seed pods of coexisting trees do not show such strong seasonal changes.

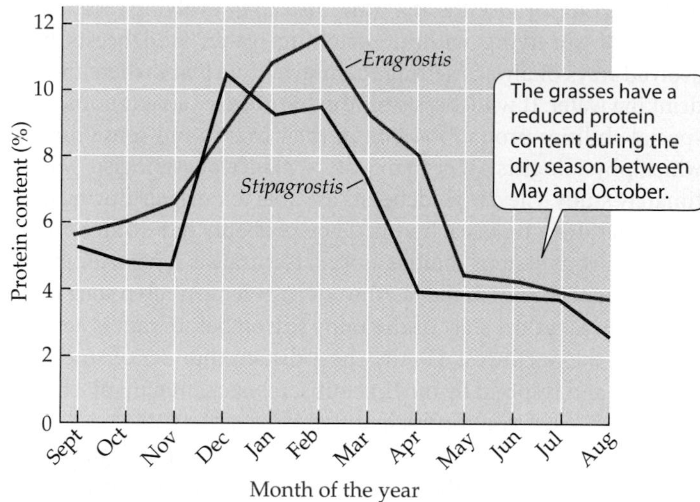

Figure 28.8 The nutritional content of some plants varies with rainy and dry seasons The percentage of protein in two species of perennial grasses (*Eragrostis lehmanniana* and *Stipagrostis obtusa*) eaten by wildebeests and oryxes was measured throughout the year in the southern Kalahari Desert ecosystem. (After Lovegrove 1993.)

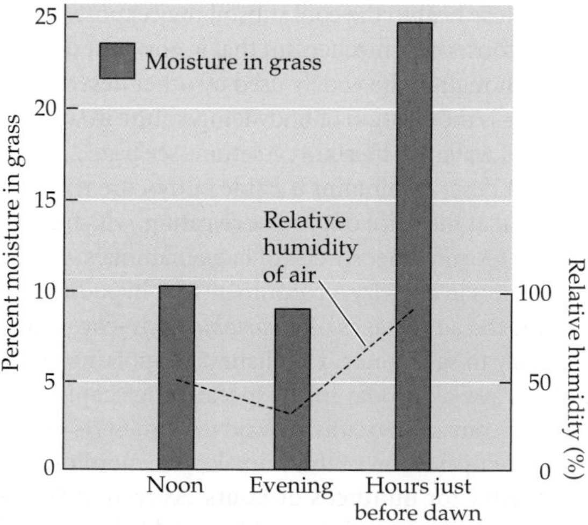

Figure 28.7 The moisture content of "dry" grass varies with time of day or night Data were gathered on dead, brown blades of the perennial desert grass *Stipagrostis uniplumus* in the Namib Desert. The moisture content of the grass is particularly high in the hours just before dawn, when the relative humidity is highest. (After Louw 1972.)

The Dramatic Adaptations of Particular Species

For the most part, the physiologists who have made the effort to study the large mammals of deserts and dry savannas have been water-balance specialists. It's a commentary on human nature, rather than animal nature, that scientists—like all people—have personal proclivities that lead them to attach greater importance to some things they observe than to others. Water-balance physiologists tend to attach greatest importance to adaptations that permit animals to live with as little water as possible. Thus, with water-balance physiologists organizing most physiological research, wildebeests have not seemed to amount to much.

Before we turn to some of the species that attract interest because of their extreme water economy, it is worth remembering how remarkably successful wildebeests are. In the Serengeti, for example, they outnumber all the other species of large mammals combined! This is true even though they are noted more for needing water than for conserving it. Faced with life in an environment that dries out for months at a time, instead of evolving exceptional modes of surviving without drinking water, wildebeests have evolved ways of always getting themselves to a place where there is drinking water. If wildebeests have extraordinary adaptations, their special abilities probably reside in their brains and sense organs, which physiologists other than water-balance physiologists will ultimately study. Clearly wildebeests are able to *navigate* during their long migratory treks each year. They probably have other exceptional nervous system abilities as well. Naturalists report, for example, that wildebeests in the northwestern Serengeti often start moving southeast at the start of the rainy season before rain is actually falling in the northwest. Do they hear distant thunder or smell distant rain and respond by moving out across a mental map? The answers cannot help but be fascinating.

Let's now discuss three case studies of the sort that intrigue water-balance physiologists: the cases of the oryxes, the Thomson's and Grant's gazelles, and the dromedary (one-humped) camel. We saw earlier that large body size in itself is a physiological advantage in hot–dry environments (see Figure 28.1). All of these species enjoy the advantages of large size. In addition, all have evolved exceptional species-specific adaptations for maintaining water balance in places where water exerts a dominant controlling effect on life.

Figure 28.9 Oryxes in deserts often exist on dead, dry grasses and the leaves of water-stressed bushes and trees Oryxes are able to stay in water balance with their only preformed water coming from such sources because they conserve water exceptionally well by mechanisms that are only partly understood. The animal shown is a gemsbok (*Oryx gazella*).

Oryxes represent the pinnacle of desert survival

Among the truly wild large mammals, oryxes (Figure 28.9) represent the pinnacle of evolution in their ability to survive in deserts. Dromedaries might be their equal physiologically, but wild dromedaries went extinct before historical times, and today's dromedaries are either domesticated or escaped from domestication. Four types of oryxes are recognized, but whether some are species or subspecies is debated. The three types that have featured in physiological research are the gemsbok oryx (*Oryx gazella*) and beisa oryx (*O. beisa*)—which closely resemble each other—and the Arabian oryx (*O. leucoryx*). The Arabian oryx was driven to extinction in the wild for a time, but has been reestablished. Oryxes are known to live in some of the most inhospitable places on Earth, including so-called *hyperarid deserts*, where there is no drinking water except immediately after rains and the rains themselves are so infrequent that average precipitation is less than 5 cm (2 inches) per year.

Just recently, researchers measured the water turnover rates of free-living Arabian oryxes by use of isotopically labeled water (see page 709). They found that the water turnover rates of the oryxes are only one-quarter to one-half as high as expected for mammals of their body size. These recent results confirm the conclusion from the earlier laboratory studies that oryxes have evolved unusually low water needs (see Figure 28.6).

The *mechanisms* by which oryxes achieve exceptional water conservation and low water-turnover rates are only starting to be understood. Their kidneys can concentrate their urine to an osmotic urine:plasma (U/P) ratio of about 8 (Table 28.2), which is a high value for their body size (see Figure 26.17). Although oryxes are famous for being indifferent to shade under many circumstances, they methodically stay in shade—if they can find it—during the heat of the day when under severe water stress. At least two species sometimes dig shallow depressions in which they settle down; this behavior is thought, without evidence, to shield them from the sun or allow them to lose heat to the cool subsoil they expose.

The water-conserving mechanism that is now best understood in oryxes—and thought to be widely used by other desert and dry-savanna species—is modulation of body temperature in ways that minimize the use of water for thermoregulation (see page 224). A person placed in a hot desert maintains a stable body-core temperature of about 37°C, but at the price of profuse sweating, which entails rapid water loss. Oryxes and other species of large mammals with long evolutionary histories in hot–dry environments are hypothesized in general to exploit the advantages of a *variable* body-core temperature when necessary to save water. The firmest supporting evidence for this hypothesis available today in any species is in fact provided by recent studies of Arabian oryxes living wild in the deserts of Saudi Arabia. The body temperatures of the animals were monitored remotely by radiotelemetry for hundreds of hours. As seen in Figure 28.10, their average body-core temperature rose and fell to a small extent over the 24-hour day during winter, when the ambient air temperature varied from an average low of 13°C at night to an average high of 27°C during the day. In summer, however, when the ambient air temperature varied from a nighttime low of 29°C to a daytime high of 44°C (111°F), the amplitude of the oryxes' cycle of body-core tem-

TABLE 28.2 Maximal urine concentrations and minimal fecal water concentrations in selected African mammals All values are from dehydrated subjects.

Species	Urine osmotic pressure (mosmol/kg H_2O)	Osmotic U/P ratio	Fecal water content (g H_2O/100 g)
African buffalo (*Syncerus caffer*)	1120	4	
Hereford cow (*Bos taurus*)	1160	4	75
Zebu cow (*Bos indicus*)	1300	4	
Somali donkey (*Equus asinus*)	1680	5	61
Common wildebeest (*Connochaetes taurinus*)	1830	6	
Thomson's gazelle (*Gazella thomsoni*)	2640	7	
Grant's gazelle (*Gazella granti*)	2790	8	
Oryx (*Oryx beisa*)	3100	8	
Dromedary camel (*Camelus dromedarius*)	3200	8	44
Dik-dik (*Madoqua kirkii*)	4760	~12	44

Source: After Maloiy, MacFarlane, and Shkolnik 1979; dik-dik urine data updated from Beuchat 1996.

perature became much larger. On average, the oryxes let their body temperatures fall to 36.5°C overnight, but then rise to 40.5°C during the day: an average amplitude of about 4°C. In one extreme case, an oryx allowed its body-core temperature to increase by 7.5°C between its nighttime low and its daytime high.

The oryxes' high-amplitude cycles of body temperature are believed to save water in two ways (see page 224). First, when the animals let their body temperatures rise during the heat of the day, they are in effect *storing* heat instead of panting or sweating to get rid of it. They then exploit the coolness of the nighttime environment to lose the heat by nonevaporative means (convection and thermal radiation). The water savings per day from allowing the body temperature to cycle—rather than holding it constant by panting or sweating—is about 0.3 liters for an oryx of average size. The elevation of body temperature during the day also has the advantage that it reduces the difference in temperature between an oryx's tissues and the hot environment, thereby slowing the rate at which heat enters its body.

Remote-monitoring techniques have only recently started to be used to study the physiology of desert and dry-savanna mammals, and only two studies have been completed on the body temperatures of free-living oryxes (none has been completed on free-living camels or gazelles). One of the studies is the one we have discussed (Figure 28.10). The second study did not find large cycles of body temperature in the oryxes monitored (*O. gazella*). This may mean only that the animals in the second study were not experiencing a tendency to dehydrate. Many laboratory investigations have demonstrated that desert and dry-savanna mammals are in general far more likely to allow large swings in body temperature when they are tending to dehydrate than otherwise. A goal for physiologists as they make greater use of remote-monitoring techniques will be to understand better how free-living animals modulate their actual use of the water-conserving mechanisms they possess.

Up to now we have discussed just one side of the water-balance equation in oryxes: their ability to limit water losses. What about water gains? Studies on free-living oryxes in the wild confirm that, as laboratory data suggest (see Figure 28.6), they must meet most of their water needs with preformed water. Metabolic water can account for only one-quarter or less of their water inputs in the wild.

Although oryxes drink when they find water and are even known at times to dig for water in river beds, there can be no doubt that they often live for months or even years without drinking. They

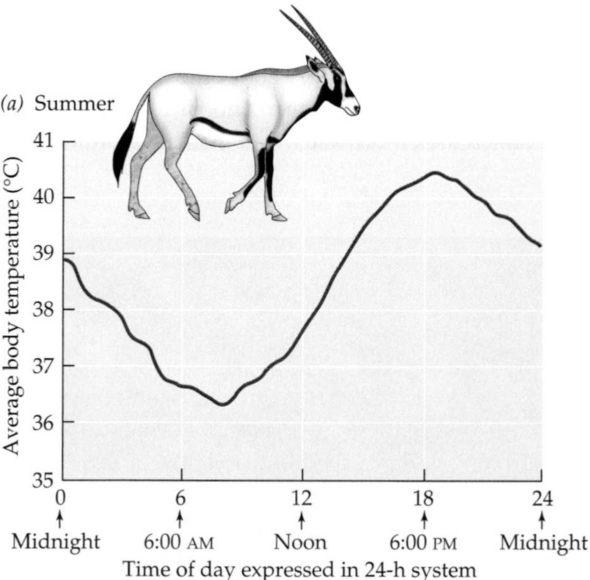

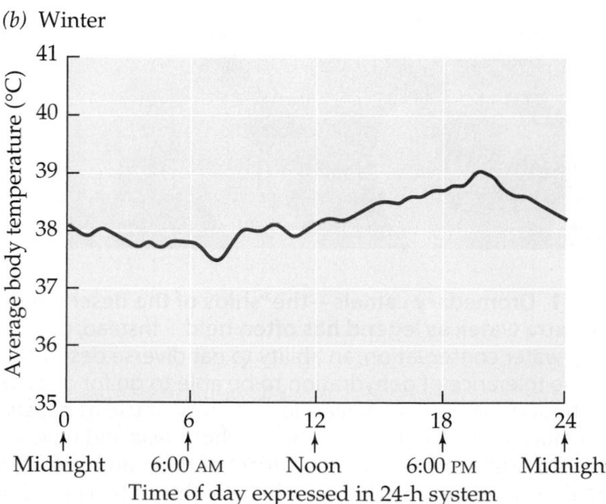

Figure 28.10 Average body-core temperatures of free-living Arabian oryxes over the 24-hour day during summer and winter The six oryxes (*Oryx leucoryx*) studied were living in their natural environment in Saudi Arabia and monitored using implanted radiotelemeters. They exhibited high-amplitude cycles of body temperature in the summer. (After Ostrowski, Williams, and Ismael 2003.)

then get the preformed water they need from their foods—an astounding proposition, because the foods available to them often seem to the eye to be distinctly unpromising. Oryxes eat the leaves of desert trees, such as acacia trees. These leaves, although they are firm and not ostensibly juicy, turn out, when analyzed, to be at least 50% to 60% water, even in the midst of severe droughts. Oryxes also eat dried grasses and other dead, dry plants (see Figure 28.9), and when they do so, they time their feeding to occur principally in the predawn hours, when the plants are richest in water (see Figure 28.7). Oryxes become nomads in the deserts, seeking out places where fog, dew, or rain has allowed plants to build up more water than average. They have uncanny abilities to find and excavate the underground dormancy organs of desert plants, structures that can be 1 meter deep but contain 50% to 70% water. All these strategies used by oryxes to obtain preformed water from food seem so marginal that they might not work to sustain life, and in fact there is no quantitative proof that they do work. However, they apparently must work, because the oryxes survive. One reason life is possible on such meager water resources is water conservation. Surviving on the water in the leaves of water-stressed or dead plants would be impossible without mechanisms that can profoundly limit the amount of water an animal needs.

Oryxes have been recorded to survive rainless droughts in the desert for at least 34 months. Sometimes their body condition declines under such extreme duress, and some die. Even after 3 years without drinking, however, the problem that undermines them may not be insufficient water, but rather the deteriorating nutritional quality (e.g., declining protein content) of the plants in their environment.

Grant's and Thomson's gazelles differ in their relations to water

Grant's gazelle (*Gazella granti*) and Thomson's gazelle (*G. thomsoni*) are two look-alike antelopes, weighing 10 to 50 kg, that often coexist in abundance in the savannas of East Africa (see Figure 28.3). Both are drinking-water-independent and often stand in the open all day long, indifferent to shade. For four decades the contrasts between these two species have highlighted that straightforward studies of water input and water output do not always tell the whole story about the water relations of savanna animals.

When Grant's and Thomson's gazelles are subjected to standard water-balance analyses in laboratory settings, they prove to be similar. If anything, Thomson's gazelle seems more adept at conserving water during such tests. However, in their natural environment, the species clearly have different relations to water, and of the two, Thomson's gazelle paradoxically seems to be the inferior one in coping with water stress. In the Serengeti, Thomson's gazelle is a participant in the annual migration; despite being drinking-water-independent, it treks along in great numbers with the drinking-water-dependent wildebeests and zebras. Where savannas give way to deserts, Grant's gazelle occupies the deserts as well as the savannas, but Thomson's gazelle stays largely in the moister savannas.

The study of physiology has yet to explain the divergence of the water relations of these gazelles in the wild. Some insight is probably provided by their different responses to extreme ambient temperatures in laboratory tests. Exposed to an air temperature of 45°C, Grant's gazelle allows its body temperature to rise that high or higher; some individuals reach a rectal temperature of 46.5°C (116°F), among the highest ever recorded in a vertebrate. Thomson's gazelle, however, keeps its body temperature below 43°C, a process that means more panting and more use of water for thermoregulation. The two species may also differ in food selection, behavioral thermoregulation, or other traits that only careful field studies will bring fully to light.

The dromedary camel does not store water, but conserves it and tolerates profound dehydration

The modern era of physiological research on large desert mammals began with studies on dromedary camels (Figure 28.11) by Knut and Bodil Schmidt-Nielsen in the mid-twentieth century. More has steadily been learned since then. Dromedaries drink, in part because they are domesticated and their human handlers water them. Nonetheless, they are capable of extreme performance in some of the most inhospitable places. During the cool seasons of the year, for example, dromedaries are well known to be able to travel for several weeks and cover hundreds of desert miles—serving as beasts of burden—without drinking along the way.

From the time of ancient Rome until the Schmidt-Nielsens' pioneering studies, the reigning theory about the water independence of camels was that they had a canteen onboard. That is, people assumed that camels took more water with them than ordinary mam-

Figure 28.11 Dromedary camels—the "ships of the desert"—do not carry extra water, as legend has often held Instead, they depend on water conservation, an ability to eat diverse desert foods, and extreme tolerance of dehydration to be able to go for days or weeks without drinking. They have a long history of use as beasts of burden in much of northern Africa and southern Asia and have been imported into Australia. They are the largest of the mammals adapted to desert life, sometimes growing to adult weights of 500 or 600 kg.

mals do when they started a trek. For centuries, debate therefore focused on the location of the canteen, with most commentators supposing that the rumen or parts of the rumen played this role. After early biochemists established that water is made by metabolism, even the hump became a proposed site of canteen function. Not that the hump contains water: Scientists have long recognized that the hump is filled with fat. However, fat yields more than its own weight in H_2O when it is oxidized (see Table 25.3), and thus it seemed only logical that the fat in the hump was a lightweight way to carry H_2O. The Schmidt-Nielsens and later investigators systematically demolished all these myths, some of which had "provided the answer to the camel question" for 2000 years. A camel's rumen contains no more water than any other ruminant's rumen. When the total amount of water in a camel's body is measured, it is no more than the average for all mammals. The hump fat actually costs H_2O to oxidize because the H_2O lost in breathing to obtain O_2 for the oxidation of the fat exceeds the metabolic water produced (see page 676). In the end, scientists established beyond doubt that camels drink only to make up for prior water losses, never to store water for future water needs.

There are two principal keys to the extraordinary ability of camels to survive in deserts for long periods without drinking. One of these is by now very familiar: Camels conserve water extraordinarily well. Based on studies of penned animals, dromedaries dramatically exploit the water-conservation benefits of large daily changes in body temperature. When dehydrated, they may allow their body temperature to rise as much as 6°C during the day and fall as much in the cool of night. One reason they can permit their body temperature to rise as high as it does is that they keep the brain cooler than the rest of the body (see page 225)—a trait also reported in oryxes and gazelles. Dromedaries also employ several other water conservation strategies: (1) they can produce exceptionally dry feces and concentrated urine (see Table 28.2); (2) they curtail their urine production relatively rapidly and profoundly when faced with dehydration; (3) they have thick, sometimes glossy fur that acts as a heat shield, helping to slow influx of heat from the fur surface and reflect or reradiate incoming solar radiation (see page 224); and (4) they minimize heat influx behaviorally, as by steadily facing the sun when allowed to rest during the day, a practice by which they consistently present a narrow body profile to the sun's direct rays.

The second known key to the water independence of dromedary camels is their unusual tolerance of dehydration. Species of mammals without long evolutionary histories in deserts, such as dogs or horses, are in danger of death if they dehydrate by 10% to 15% of their body weight while under heat stress. Dromedaries, in contrast, can tolerate at least twice this extent of dehydration. There are known cases of dromedaries being alert and functional in the desert heat while dehydrated by 30% to 40% of their body weight.

Thus, when a dromedary sets off on a desert trek, it has the same amount of water, in relation to its body size, as you or I. However, it dehydrates by a smaller percentage of its body weight per day because it conserves water exceptionally well, and it can tolerate a level of total dehydration at least twice what we can tolerate. Along the way it munches desert plants, ranging from green leaves to dried-out thorn bushes and saltbushes, replacing some of its water losses from preformed water in its food. Thus it can forestall its need to drink for many days in the summer and many weeks in the cool seasons.

Study Questions

1. What are the pros and cons of studying water physiology in the laboratory and in the field? Could an animal's water physiology be fully understood by field studies alone? By laboratory studies alone? Explain.
2. In Chapter 1 we noted that an animal's body size is one of its most important properties. Considering all the mammals that live in deserts, explain why their body sizes matter. Include discussion of physiological mechanisms where appropriate.
3. The effects of water loss in milk on the water budgets of lactating female mammals in arid environments are not well understood. Specify three hypotheses regarding mechanisms by which females might prevent water losses in their milk from forcing them into negative water balance (which could lead to fatal dehydration). For each hypothesis, design a study to test or evaluate the hypothesis.

Key References

Louw, G., and M. Seely. 1982. *Ecology of Desert Organisms*. Longman, New York. A fine introduction to the ecology of the world's deserts written by two eminent desert biologists.

Lovegrove, B. 1993. *The Living Deserts of Southern Africa*. Fernwood Press, Vlaeberg, South Africa. It's hard to imagine a more delightful and astute book on deserts and desert organisms than this one, written by a physiologist for the general public. Illustrated with a profusion of outstanding color photographs. Although the book is limited to southern Africa in its coverage, it will provide insight into desert processes in general.

Ostrowski, S., J. B. Williams, and K. Ismael. 2003. Heterothermy and the water economy of free-living Arabian oryx (*Oryx leucoryx*). *J. Exp. Biol.* 206: 1471–1478. This paper not only provides an important new set of data, but offers a good review of the controversy swirling around the concept of adaptive heterothermy in desert and dry-savanna mammals, including references to many relevant research reports.

Schmidt-Nielsen, K. 1964. *Desert Animals*. Oxford University Press, London. One of the classic monographs in animal physiology. The book still inspires fascination about the relations of desert animals to water and stands as a model of clear thinking about its subject, although much of its detailed content is dated by now.

Selected References

Evenari, M. 1985. The desert environment. In M. Evenari, I. Noy-Meir, and D. W. Goodall (eds.), *Ecosystems of the World*, vol. 12A, *Hot Deserts and Arid Shrublands*, pp. 1–22. Elsevier, New York.

Gauthier-Pilters, H., and A. I. Dagg. 1981. *The Camel: Its Evolution, Ecology, Behavior, and Relationship to Man*. University of Chicago Press, Chicago.

Gereta, E., E. Wolanski, M. Borner, and S. Serneels. 2002. Use of an ecohydrology model to predict the impact on the Serengeti ecosystem of deforestation, irrigation and the proposed Amala Weir Water Diversion Project in Kenya. *Ecohydrol. Hydrobiol.* 2: 135–142.

Goudie, A., and J. Wilkinson. 1977. *The Warm Desert Environment*. Cambridge University Press, New York.

Maloiy, G. M. O. (ed.). 1972. *Comparative Physiology of Desert Animals*. Symposia of the Zoological Society of London, no. 31. Academic Press, New York.

Maloney, S. K., A. Fuller, G. Mitchell, and D. Mitchell. 2002. Brain and arterial blood temperatures of free-ranging oryx (*Oryx gazella*). *Pflügers Arch.* 443: 437–445.

Mitchell, D., S. K. Maloney, C. Jessen, H. P. Laburn, P. R. Kamerman, G. Mitchell, and A. Fuller. 2002. Adaptive heterothermy and selective brain cooling in arid-zone mammals. *Comp. Biochem. Physiol. B* 131:

571–585. [A disputatious challenge to much of conventional wisdom about the ways large arid-zone mammals reduce water losses.]

Monbiot, G. 1994. *No Man's Land*. Macmillan, London. [An exploration into the fate of the nomadic peoples of the East African arid lands.]

Nix, H. A. 1983. Climate of tropical savannas. In F. Bourlière (ed.), *Ecosystems of the World*, vol. 13, *Tropical Savannas*, pp. 37–62. Elsevier, New York.

Reisner, M. 1993. *Cadillac Desert: The American West and Its Disappearing Water*, rev. ed. Penguin Books, New York.

Schroter, R. C., D. Robertshaw, and R. Z. Filadi. 1989. Brain cooling and respiratory heat exchange in camels during rest and exercise. *Respir. Physiol.* 78: 95–105.

Taylor, C. R. 1968. The minimum water requirements of some East African bovids. Symposia of the Zoological Society of London, no. 21, pp. 195–206. Academic Press, London.

Taylor, C. R. 1970. Dehydration and heat: Effects on temperature regulation of East African ungulates. *Amer. J. Physiol.* 219: 1136–1139.

Taylor, C. R. 1970. Strategies of temperature regulation: Effect on evaporation in East African ungulates. *Amer. J. Physiol.* 219: 1131–1135.

Williams, J. B., S. Ostrowski, E. Bedin, and K. Ismail. 2001. Seasonal variation in energy expenditure, water flux and food consumption of Arabian oryx *Oryx leucoryx*. *J. Exp. Biol.* 204: 2301–2311.

Williamson, D. T. 1987. Plant underground storage organs as a source of moisture for Kalahari wildlife. *Afr. J. Ecol.* 25: 63–64.

Wilson, R. T. 1989. *Ecophysiology of the Camelidae and Desert Ruminants*. Springer-Verlag, New York.

Wolanski, E., and E. Gereta. 2001. Water quantity and quality as the factors driving the Serengeti ecosystem, Tanzania. *Hydrobiologia* 458: 169–180.

Yagil, R. 1985. *The Desert Camel: Comparative Physiological Adaptation*. Karger, New York.

APPENDIX A References

Each of the works listed here covers a variety of aspects of physiology. Many are excellent reference sources. They have generally not been listed with individual chapters because many of them would have to be listed in virtually all chapters, such is their breadth of coverage.

General Works

Barrington, E. J. W. 1979. *Invertebrate Structure and Function*, 2nd ed. Nelson, Sudbury-on-Thames.

Dantzler, W. H. (ed.). 1997. *Comparative Physiology* (Handbook of Physiology [Bethesda, MD], section 13). 2 vols. Oxford University Press, New York.

Florkin, M., and B. T. Scheer (eds.). 1967–1979. *Chemical Zoology*. 13 vols. Academic Press, New York.

Fregly, M. J., and C. M. Blatteis (eds.). 1997. *Environmental Physiology* (Handbook of Physiology [Bethesda, MD], section 4). 2 vols. Oxford University Press, New York.

Fretter, V., and A. Graham. 1976. *A Functional Anatomy of Invertebrates*. Academic Press, New York.

Hochachka, P. W., and G. N. Somero. 2002. *Biochemical Adaptation. Mechanism and Process in Physiological Evolution*. Oxford University Press, New York.

McNab, B. K. 2002. *The Physiological Ecology of Vertebrates: A View from Energetics*. Cornell University Press, Ithaca, NY.

Michal, G. (ed.). 1999. *Biochemical Pathways: An Atlas of Biochemistry and Molecular Biology*. Wiley, New York.

Prosser, C. L. 1986. *Adaptational Biology: Molecules to Organisms*. Wiley, New York.

Prosser, C. L. (ed.). 1991. *Environmental and Metabolic Animal Physiology* (*Comparative Animal Physiology*, 4th ed.). Wiley-Liss, New York.

Prosser, C. L. (ed.). 1991. *Neural and Integrative Animal Physiology* (*Comparative Animal Physiology*, 4th ed.). Wiley-Liss, New York.

Seymour, R. S. (ed.). 1983. *Respiration and Metabolism of Embryonic Vertebrates*. Junk, Dordrecht.

Urich, K. 1994. *Comparative Animal Biochemistry*. Springer-Verlag, New York.

Withers, P. C. 1992. *Comparative Animal Physiology*. Saunders, Philadelphia.

Works Dealing with Specific Groups of Invertebrates

Anderson, O. R. 1988. *Comparative Protozoology: Ecology, Physiology, Life History*. Springer-Verlag, New York.

Bergquist, P. R. 1978. *Sponges*. University of California Press, Berkeley.

Bliss, D. E. (ed.). 1982–1985. *The Biology of Crustacea*. 10 vols. Academic Press, New York.

Blum, M. S. (ed.). 1985. *Fundamentals of Insect Physiology*. Wiley, New York.

Burggren, W. W., and B. R. McMahon (eds.). 1988. *Biology of the Land Crabs*. Cambridge University Press, New York.

Holdich, D.M. 2002. Biology of Freshwater Crayfish. Blackwell Science, Oxford, UK.

Hughes, R. N. 1986. *A Functional Biology of Marine Gastropods*. Johns Hopkins University Press, Baltimore, MD.

Kerkut, G. A., and L. I. Gilbert (eds.). 1985. *Comprehensive Insect Physiology, Biochemistry and Pharmacology*. 13 vols. Pergamon, New York.

Lawrence, J. 1987. *A Functional Biology of Echinoderms*. Johns Hopkins University Press, Baltimore, MD.

Nentwig, W. (ed.). 1987. *Ecophysiology of Spiders*. Springer-Verlag, New York.

Obenchain, F. D., and R. Galun (eds.). 1982. *Physiology of Ticks*. Pergamon, New York.

Sawyer, R. T. 1986. *Leech Biology and Behaviour*. 3 vols. Clarendon, Oxford.

Schram, F. R. 1986. *Crustacea*. Oxford University Press, New York.

Sonenshine, D. E. 1991–1993. *Biology of Ticks*. 2 vols. Oxford University Press, New York.

Sutton, S. L., and D. M. Holdrich (eds.). 1984. *The Biology of Terrestrial Isopods*. Symposia of the Zoological Society of London, no. 53. Clarendon, Oxford.

Wilbur, K. M. (ed.). 1983–1988. *The Mollusca*. 12 vols. Academic Press, New York.

Works Dealing with Specific Groups of Vertebrates

Bemis, W. E., W. W. Burggren, and N. E. Kemp (eds.). 1987. *The Biology and Evolution of Lungfishes*. Liss, New York.

Evans, D. H. (ed.). 1998. *The Physiology of Fishes*, 2nd ed. CRC Press, Boca Raton, FL.

Farner, D. S., J. R. King, and K. C. Parkes (eds.). 1982–1985. *Avian Biology*, vols. 6–8. Academic Press, New York.

Feder, M.E., and W.W. Burggren (eds.). 1992. Environmental Physiology of the Amphibians. University of Chicago Press, Chicago, IL.

Foreman, R. E., A. Gorbman, J. M. Dodd, and R. Olsson (eds.). 1985. *Evolutionary Biology of Primitive Fishes*. Plenum, New York.

Gans, C. (ed.). 1969–1998. *Biology of the Reptilia*. 19 vols. Academic Press, New York.

Handbook of Physiology. 1960–2003. American Physiological Society, Bethesda. MD. (A comprehensive mammalian physiology; numerous volumes and editors.)

Hoar, W. S., D. J. Randall and A. P. Farrell (eds.). 1969–2001. *Fish Physiology*. 20 vols. Academic Press, New York.

Hochachka, P. W., and T. P. Mommsen (eds.). 1991–1995. *Biochemistry and Molecular Biology of Fishes*. 5 vols. Elsevier Science, New York.

King, A. S., and J. McLelland (eds.). 1979–1989. *Form and Function in Birds*. 4 vols. Academic Press, New York.

Rankin, J. C., and F. B. Jensen (eds.). 1993. *Fish Ecophysiology*. Chapman & Hall, New York.

Stevens, L. 1996. *Avian Biochemistry and Molecular Biology*. Cambridge University Press, New York.

Tytler, P., and P. Calow (eds.). 1985. *Fish Energetics: New Perspectives*. Johns Hopkins University Press, Baltimore, MD.

Vander, A. J., J. H. Sherman, and D. S. Luciano. 2001. *Human Physiology: The Mechanisms of Body Function*, 8th ed. McGraw-Hill, Boston.

Whittow, G. C. (ed.). 2000. *Sturkie's Avian Physiology*, 5th ed. Academic Press, New York.

APPENDIX B The Système International and Other Units of Measure

Physiology is in transition to full use of a system of units of measure called the **Système International (SI)**. Some subdisciplines of physiology today use almost entirely SI units, whereas other subdisciplines continue to use older systems of measure along with SI units. The SI recognizes seven base units of measure, listed in the first section of the accompanying table: the meter, kilogram, degree Celsius, second, ampere, mole, and candela. All other units in the SI are to be derived from these seven; thus, for example, the SI unit for velocity (distance per time) is the meter/second. Many derived SI units are listed in the second section of the table. Some derived units are given special names; for example, the unit of force (mass × acceleration) is technically a (kilogram × meter)/second^2, which is more commonly known as a newton. The unit for energy (force × distance) is a newton × meter, called a joule. Note that when names of people are used for units of measure, the names are written entirely in lowercase (although abbreviations are uppercase; e.g., 1 newton is abbreviated 1 N).

Prefixes indicating orders of magnitude greater or smaller are acceptable in the SI. For instance, the prefix kilo- can be used to indicate "three orders of magnitude greater" [a kilojoule (kJ) is 1000 J]. See Appendix C for definitions of prefixes. Note that in scientific writing, abbreviations for symbols are not followed by a period (unless by coincidence, as at the end of a sentence). For example, "kg"—not "kg."—is the correct abbreviation for "kilogram."

Two sorts of notation are used for derived units: *numerator/denominator* notation and *exponential* notation. In numerator/denominator notation the unit of velocity, for example, is the meter/second or m/s; in exponential notation the unit is written $m \cdot s^{-1}$. In this text we use numerator/denominator notation because we consider it more intuitive and easier to understand, but many students are also familiar with exponential notation and find it easier for complex calculations. In this appendix we list units in both numerator/denominator notation and exponential notation.

Relations between SI units and selected traditional units are shown in the table, in the column labeled "Relations among units." Some relations between different sets of traditional units are also listed. To obtain the relations converse to those shown, divide both sides of the relevant equation by the number to the right of the equals sign. For instance, the table states, "1 m = 3.28 ft." Dividing both sides by 3.28 yields, "1 ft = 0.305 m."

More information on the SI and traditional units of measure can be obtained in *Handbook of Chemistry and Physics* (CRC Press, Boca Raton, FL) or *Lange's Handbook of Chemistry* (McGraw-Hill, New York).

Quantity	SI Unit		Relations among units
Base SI Units			
Length	meter (m)		1 m = 3.28 feet (ft)
			1 inch (in) = 25.4 millimeter (mm)
			1 statute mile (mi) = 1609.3 m
Mass	kilogram (kg)		1 kg = 2.20 pound, avoirdupois (lb)
			1 ounce, avoirdupois (oz) = 28.3 gram (g)
Temperature	1 degree Celsius (°C) = 1 kelvin (K)		A difference of 1°C = a difference of 1.8 degree Fahrenheit (°F)
Time	second (s)		—
Electric current	ampere (A)		—
Amount of substance	mole (mol)		—
Luminous intensity	candela (cd)		1 candela ≅ 1 candle (pentane)
	SI Unit		
	Numerator/ denominator notation	**Exponential notation**	
Derived SI Units			
Area	m^2	m^2	1 m^2 = 10,000 square centimeters (cm^2)
			1 m^2 = 10.8 ft^2
Volume	m^3	m^3	1 m^3 = 1×10^6 cm^3
			1 m^3 = 1000 liter (L)
			1 cm^3 = 1 milliliter (mL)
			1 U.S. gallon = 3.785 L
			1 U.S. fluid ounce = 29.6 mL
Density	kg/m^3	$kg{\cdot}m^{-3}$	1 kg/m^3 = 0.001 g/mL
Velocity	m/s	$m{\cdot}s^{-1}$	1 m/s = 3.28 ft/s
			1 m/s = 2.24 statute mile/hour
Acceleration	m/s^2	$m{\cdot}s^{-2}$	1 m/s^2 = 3.28 ft/s^2
Force	$kg{\cdot}m/s^2$ = 1 newton (N)	$kg{\cdot}m{\cdot}s^{-2}$ = 1 N	1 N = 0.102 kilogram of force
			1 N = 0.225 pound of force
			1 N = 1×10^5 dyne
Energy, work	$kg{\cdot}m^2/s^2$ = 1 N·m = 1 joule (J)	$kg{\cdot}m^2{\cdot}s^{-2}$ = 1 N·m = 1 J	1 J = 0.239 calorie (cal)
			1 J = 1×10^7 erg
			1 J = 0.000948 British thermal unit
			1 J = 0.738 foot-pound
Power	$kg{\cdot}m^2/s^3$ = 1 J/s = 1 watt (W)	$kg{\cdot}m^2{\cdot}s^{-3}$ = 1 $J{\cdot}s^{-1}$ = 1 W	1 W = 0.239 cal/s
			1 W = 0.0013 horsepower
			1 W = 3.41 British thermal unit/h
Pressure	$kg/(s^2{\cdot}m)$ = 1 N/m^2 = 1 pascal (Pa)	$kg{\cdot}s^{-2}{\cdot}m^{-1}$ = 1 $N{\cdot}m^{-2}$ = 1 Pa	1 Pa = 0.0075 mm of mercury (mm Hg)
			1 kilopascal (kPa) = 1000 Pa
			1 atmosphere = 101.3 kPa
			1 atmosphere = 760 mm Hg
			1 mm Hg = 1 torr
			1 lb/in^2 = 6.89 kPa
Frequency	1/s = 1 hertz (Hz)	s^{-1}= 1 Hz	1 Hz = 1 cycle/s
Electric potential	$kg{\cdot}m^2/(s^3{\cdot}A)$ = 1 W/A = 1 volt (V)	$kg{\cdot}m^2{\cdot}s^{-3}{\cdot}A^{-1}$ = 1 $W{\cdot}A^{-1}$ =1 V	—
Electric resistance	$kg{\cdot}m^2/(s^3{\cdot}A^2)$ = 1 V/A = 1 ohm (Ω)	$kg{\cdot}m^2{\cdot}s^{-3}{\cdot}A^{-2}$= 1 $V{\cdot}A^{-1}$ =1 Ω	—
Electric charge	s·A = 1 coulomb (C)	s·A = 1 C	1 C = 0.00028 ampere-hour

APPENDIX C Prefixes Indicating Orders of Magnitude

You undoubtedly know that kilo- means "three orders of magnitude greater."
This and a series of less familiar prefixes are summarized for convenience in the following table.

Prefix	Abbreviation (placed before unit abbreviation)	Meaning in words	Meaning in powers of 10
tera-	T	12 orders of magnitude greater	$\times 10^{12}$
giga-	G	9 orders of magnitude greater	$\times 10^{9}$
mega-	M	6 orders of magnitude greater	$\times 10^{6}$
kilo-	k	3 orders of magnitude greater	$\times 10^{3}$
deci-	d	1 order of magnitude smaller	$\times 10^{-1}$
centi-	c	2 orders of magnitude smaller	$\times 10^{-2}$
milli-	m	3 orders of magnitude smaller	$\times 10^{-3}$
micro-	μ	6 orders of magnitude smaller	$\times 10^{-6}$
nano-	n	9 orders of magnitude smaller	$\times 10^{-9}$
pico-	p	12 orders of magnitude smaller	$\times 10^{-12}$
femto-	f	15 orders of magnitude smaller	$\times 10^{-15}$

APPENDIX D Gases at Standard Temperature and Pressure

Temperature and pressure exert such great effects on gas volume that they must be specified for a volume to have meaning. Physiologists usually convert all measured gas volumes to volumes at **standard temperature and pressure** (**STP**)—0°C and 1 atmosphere—so that the volumes can be compared. Reported values are assumed to be at STP unless otherwise stated. The equation for making the conversion is

$$\text{volume at STP} = \text{volume measured} \times \frac{P_{meas}}{1 \text{ atm}} \times \frac{273 \text{ K}}{T_{meas}} \quad \text{(D.1)}$$

where P_{meas} is the pressure prevailing during measurement, in atmospheres, and T_{meas} is the temperature prevailing during measurement, in absolute (Kelvin) degrees.

Suppose, to illustrate, that you study a bird in a chamber where the pressure is 0.92 atm and the temperature is 20°C. On the Kelvin scale, the temperature is 293 K (273 plus its value on the Celsius scale). Suppose the animal uses 50 mL of O_2, measured under the *prevailing* conditions, in an hour. Expressed at STP, the volume used is

$$\text{volume at STP} = 50 \text{ mL} \times \frac{0.92 \text{ atm}}{1 \text{ atm}} \times \frac{273 \text{ K}}{293 \text{ K}} = 42.9 \text{ mL} \quad \text{(D.2)}$$

Thus, the animal's rate of O_2 consumption *at STP*, a measure of its metabolic rate, is 42.9 mL O_2/hour.

A significant property of gases is that identical molar amounts of different gases occupy almost identical volumes when at the same pressure and temperature. A mole of gas occupies about 22.4 L at STP, whether the gas is O_2, N_2, or some other gas.

APPENDIX E Fitting Lines to Data

Suppose you are interested in how two traits of an animal are related to each other. The traits might, for example, be resting heart rate and age, but here we will call them Y and X for generality. If you make measurements on ten individuals, you will have ten pairs of values that you can plot as in Figure A1 to display visually the relation between the traits. Each point is for one of the studied individuals and represents that particular individual's measures for traits Y and X.

The points plotted seem to fall along a straight line. You could simply sketch a line through them, but that would be subjective. A better approach is to use a statistical procedure to determine which line best fits the data according to objective criteria. The procedure most commonly used to do this is called **linear least squares regression**. This procedure starts by assuming that a linear equation applies. That is, it assumes that the equation for the line of best fit will take the form $Y = a + bX$, where a and b are constants. An investigator must decide first whether that assumption is reasonable. If one doubts the suitability of the linear model, there are advanced statistical procedures that will help one to decide. Once the decision is made to use the equation $Y = a + bX$, the procedure of linear least squares regression provides an objective way to find the values for a (intercept) and b (slope) that position the line most appropriately relative to the data.

A proposed line relating Y to X is drawn in blue in Figure A2. The red lines in the figure show the vertical distance between each data point and the proposed line. The central proposition (axiom) of least squares regression, approximately stated, is that the best line relating Y to X is the one that minimizes the sum of the lengths of the red lines. Exactly stated, the procedure squares the length of each red line, adds the squares of all the red lines, and determines the best blue line as the one that minimizes this sum of squares (thus, "least squares"). You might guess from appearances that the blue line in Figure A2 is not the best line relating Y to X, by this standard. It isn't, but you need not try one blue line after another to find the best. Instead, the calculus of minimization (i.e., minimization of the sum of squares) provides algorithms for carrying out the procedure, and these algorithms are simple enough to be programmed into even cheap calculators today. Figure A3 shows the line that best fits the data by the least squares standard; the line is plotted, and its equation is given in the lower right.

An important reason for understanding the procedure as described is that you will see that the best line is strictly a function of the data (Y, X values) used for the computation. Different data invariably lead to different values for a and b. If one investigator measures Y and X on ten individuals and another measures them on ten different individuals, the investigators will obtain different values for a and b, and different "best" lines. Even the addition of one individual to a data set will change a and b; in Figure A4, data for an eleventh individual are added to the ten data points in Figure A3, resulting in a new line of best fit by the least squares procedure. These facts do not mean that the statistically fitted lines are "just guesses." The facts merely reflect the reality that a fitted line can be no more certain than the data on which the line is based. Ten data points typically leave much room for uncertainty. If you had a hundred data points, the uncertainty would be diminished, and a and b would probably be altered only slightly by adding one more data point. Statistical procedures permit one to calculate a numerical measure of the level of uncertainty of a fitted line. Mastery of statistics is in fact crucial for the choice of line-fitting procedures, because there are several nuances involved in applying the procedures correctly.

Figure A Challenges and procedures of statistical line fitting (1) Ten *X,Y* data points measured on ten individuals. (2) A subjectively placed, proposed line (blue) to fit the data points. The red lines show the vertical deviations between the data points and line. (3) The line of best statistical fit to the ten points, based on the linear least squares model. (4) An additional data point has been added, and the line of best statistical fit has been recalculated to take account of the information provided by all eleven data points.

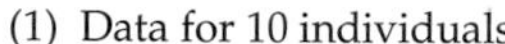

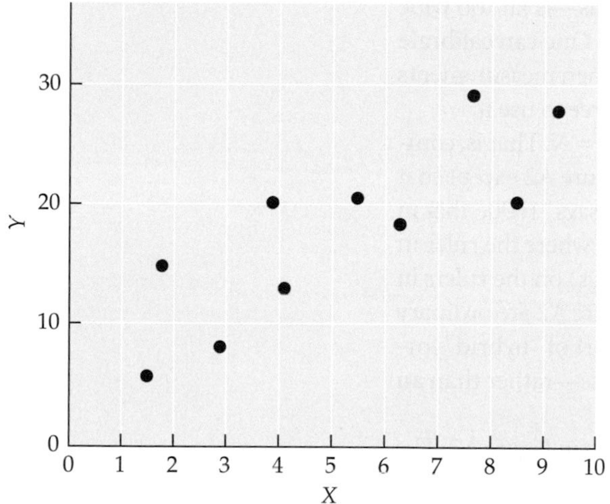

(2) A rough line through the points with deviations marked

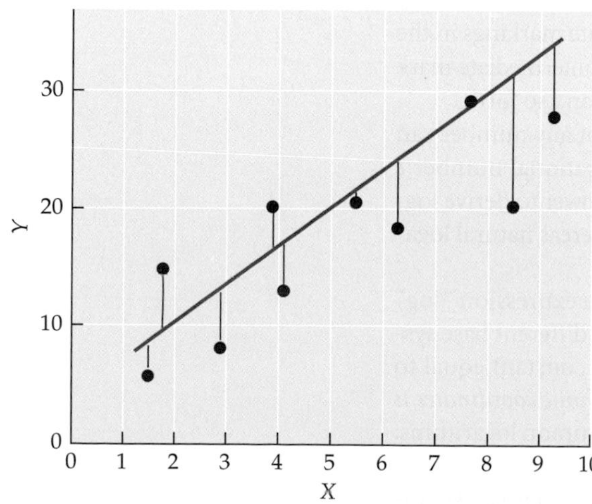

(3) Least squares fit to the data in (1)

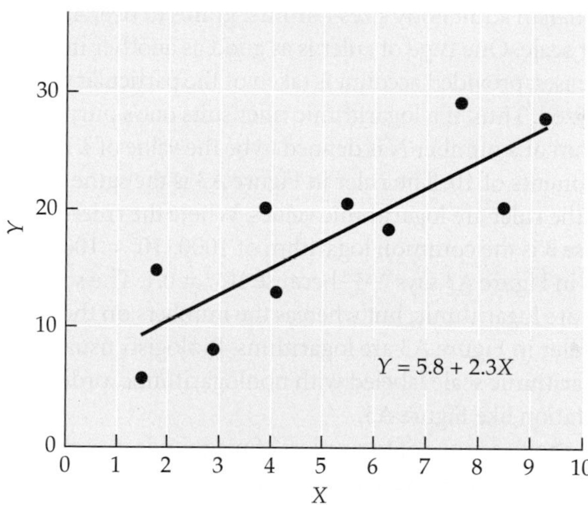

(4) Least squares fit to the data in (1) plus one more point

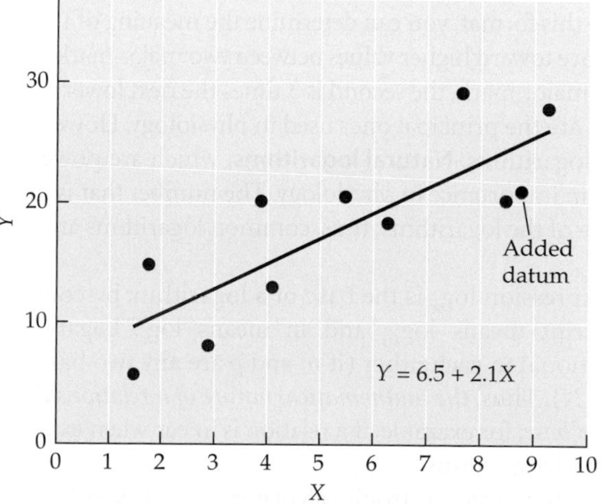

APPENDIX F Logarithms

The ruler in Figure A1 is calibrated in a linear scale of numbers of the sort we use in everyday life; each major marking is 1 unit greater than the preceding one. The ruler in Figure A2 is calibrated in a nonlinear, *logarithmic* scale; each marking on the scale is an *order of magnitude* greater than the preceding one. Note that the logarithmic ruler covers a far wider range. That ruler runs between 0.01 and 100,000; for the linear ruler to do the same, it would need to be extremely long, or the markings would need to be extremely close. One reason logarithmic scales are sometimes used by biologists is their wide range. For instance, a physiologist comparing metabolic rates in all mammals would find a logarithmic scale useful because the range of mammalian adult body sizes—from 2 grams to over 4,000,000 grams—is far too wide to fit on any single linear scale. One type of ruler is as good as another in the abstract. One can calibrate a ruler in any way one pleases, provided account is taken of the particular calibration when measurements are interpreted and analyzed. Thus, if a logarithmic ruler suits one's purposes, one is free to use it.

The **common logarithm** of a number N is defined to be the value of L for which $10^L = N$. That is, common logarithms are exponents of 10. The ruler in Figure A3 is the same as that in Figure A2 except that the numbers printed on the ruler are logarithmic values. Where the ruler in Figure A2 says "1000," that in Figure A3 says "3" because 3 is the common logarithm of 1000 ($10^3 = 1000$). Similarly, where the ruler in Figure A2 says "0.1," that in Figure A3 says "–1" because $10^{-1} = 0.1$. The scales (markings) on the rulers in Figure A2 and Figure A3 are logarithmic, but whereas the numbers on the ruler in Figure A2 are ordinary numbers, those on the ruler in Figure A3 are logarithms. Biologists usually use the sort of "hybrid" format in Figure A2—a logarithmic scale labeled with nonlogarithmic, ordinary numbers—rather than an "all-logarithmic" presentation like Figure A3.

The ruler in Figure A4 shows a range of two orders of magnitude in greater detail than in Figure A2. The presentation is hybrid; that is, whereas the markings are logarithmic, the numbers are ordinary. Note that numbers that would be evenly spaced on a linear ruler—such as the numbers 2, 3, 4, and 5—are not evenly spaced on the ruler in Figure A4 because of the logarithmic scaling. Very commonly, logarithmic scales are printed as on the ruler in Figure A5, with the major markings labeled but the intermediate markings unlabeled. When you see this format, you can determine the meaning of the intermediate markings in the following way: As you move toward higher values between two major markings, the first intermediate mark is 2 times the next lower major mark, the second is 3 times the next lower major mark, and so forth.

Common logarithms are the principal ones used in physiology. However, powers of any number can in principle be used as logarithms. **Natural logarithms**, which are powers of the irrational number e (2.71828...), are second in importance in physiology. The number that is raised to a power to derive logarithms is called the **base** of the logarithms; thus, common logarithms are base 10, whereas natural logarithms are base e.

The value x in the expression $\log_x$ is the base of a logarithm; by convention, the expression "log" (written without a subscript) means "$\log_{10}$" and "ln" means "$\log_e$." Logarithms in two different base systems are simply proportional to each other (if m and p are any two bases and c is a constant equal to $\log_p m$, $\log_p N = c \times \log_m N$). Thus, *the mathematical nature of a relationship in logarithmic coordinates is not altered by the choice of base*; for example, if a relation is linear when expressed in common logarithms, it remains linear in natural logarithms.

Before the invention of the cheap electronic calculator around 1970, logarithms were essential as computational tools. Today, logarithms are rarely used that way and have two principal applications: graphical and mathematical/statistical, both mentioned in Appendix G.

Figure A Five rulers illustrating properties and styles of labeling of linear and logarithmic scales of measurement

(1) An ordinary ruler (linear scale)

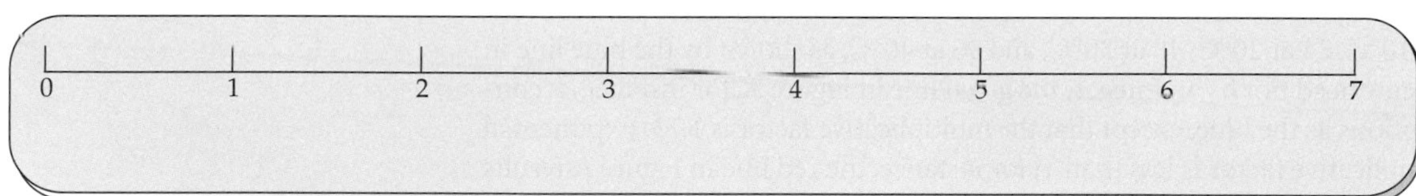

(2) A ruler scaled logarithmically

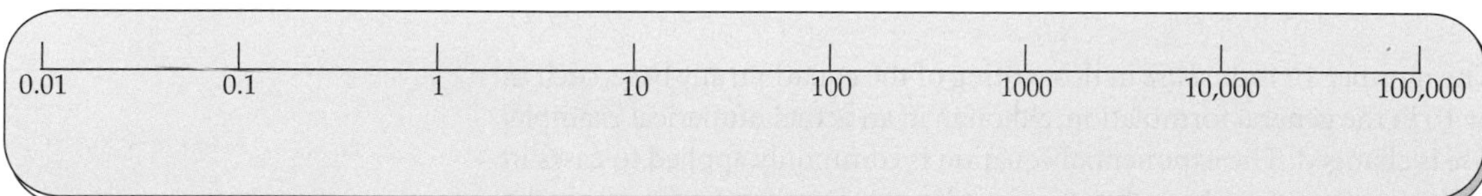

(3) The same ruler as in (2) with values expressed in logarithmic units

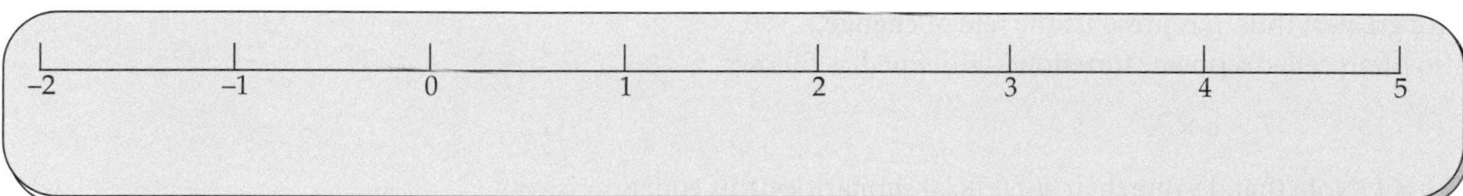

(4) The section of (2) between 1 and 100 with minor marks added

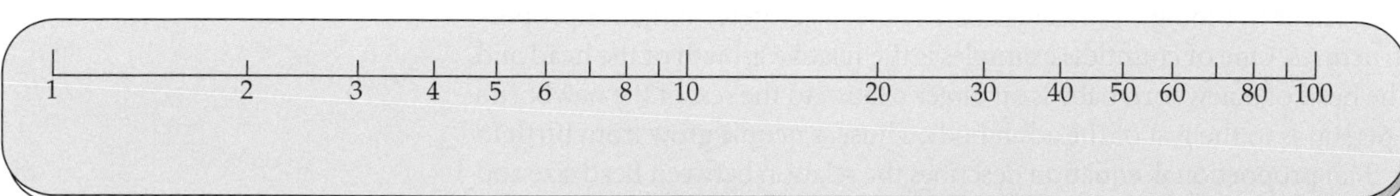

(5) Five orders of magnitude on a logarithmic ruler with only major marks labeled, showing interpretation of three minor marks

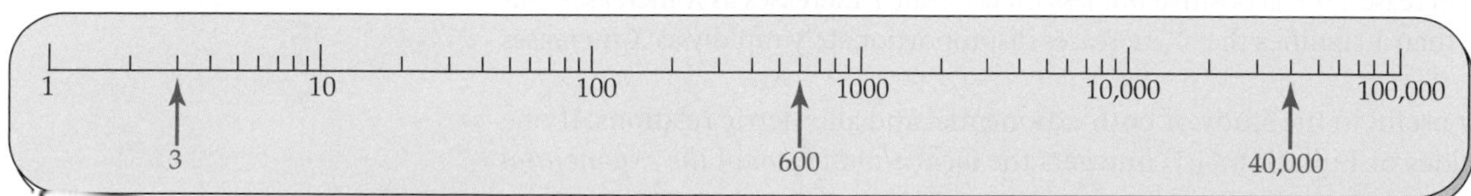

APPENDIX G Exponential and Allometric Equations

The two most common types of nonlinear equations in the study of physiology are exponential and allometric equations. A variable Y is an **exponential function** of a variable X if Y changes by a fixed *multiplicative factor* every time X changes by a fixed *additive increment*. An example of exponential increase would be a rate that doubles every time the temperature increases 10°C. In this instance, the multiplicative factor would be 2, and the additive increment would be 10°C. If the actual rate at 0°C is 6 (in arbitrary units), the rate would be 12 at 10°C, 24 at 20°C, 48 at 30°C, and 96 at 40°C, as shown by the blue line in Figure A. The multiplicative factor need not be an integer; the green line in Figure A, for instance, is constructed from the same assumptions as the blue, except that the multiplicative factor is 1.73. Exponential decrease occurs when the multiplicative factor is less than 1; for instance, the red line in Figure A results when the factor is 0.5.

The general mathematical form of an exponential function is

$$Y = m \times 10^{n \times X} \tag{G.1}$$

where m and n are constants. The number 10 is the *base* in this writing of the equation; any base, such as e, could be freely substituted for 10 in the general formulation, although in an actual numerical example, the value of n changes if the base is changed. The exponential equation is commonly applied to cases in which Y is time-dependent and X represents time. In such cases, since $Y = m$ when $X = 0$, m is simply the value of Y at the start (time zero). If n is positive, exponential increase occurs, whereas if n is negative, exponential decrease occurs. Whether increase or decrease is occurring, the rate of change becomes greater as the absolute value of n becomes greater; thus, n represents the rate of change.

The **allometric function**—also often called a **power function**—is defined as follows:

$$Y = a \times X^b \tag{G.2}$$

where a and b are constants and $b \neq 1$. Note that despite their superficial similarities, this equation is very different from the exponential function. Whereas the independent variable X is in the exponent in the exponential equation, it is raised to a power (thus, *power* function) in the allometric equation. The allometric equation first became prominent in biology as a mathematical means of describing nonproportional relative growth of body structures. One of countless examples is the relative growth of the head and the rest of the body in people. The head of a newborn baby is far larger relative to the rest of the newborn's body than the head of an adult person is to the rest of the adult body. Thus, as people grow from birth to adulthood, an allometric, rather than proportional, equation describes the relation between head size and body size. Today, the allometric equation is extremely important in physiology, developmental biology, paleontology, and other fields that focus on the *relative scaling of animal traits*. If b is negative in Equation G.2, trait Y decreases as trait X increases. If b is positive but less than 1, trait Y increases as X increases, but not as fast. A value of b greater than 1 signifies that Y increases disproportionately rapidly as X increases (Figure B).

Logarithmic scales are very useful in the study of both exponential and allometric relations. If one takes the logarithm of both sides of Equation G.1, one gets the *logarithmic form of the exponential equation*:

$$\log Y = \log m + n \times X \tag{G.3}$$

This equation tells you that for an exponential relation, if you plot the logarithm of Y on one axis of a graph and X itself on the other axis—a plot called *semilogarithmic* because one axis is logarithmic and the other is not—you will get a straight line with slope n and intercept log m.

The *logarithmic form of the allometric equation* is obtained by taking the logarithm of both sides of Equation G.2:

$$\log Y = \log a + b \times \log X \qquad \text{(G.4)}$$

This equation reveals that for an allometric relation, plotting the logarithm of Y on one axis and the logarithm of X on the other—a graph termed *log–log* because both axes are logarithmic—yields a straight line with b as slope and log a as intercept.

The constants in the exponential and allometric equations (n, m and a, b) are predicted from theory in some applications. More usually, however, they are determined empirically: Actual data are gathered on the relation between Y and X, and the data are used to estimate the values of the constants. For this approach, the usual practice is to plot the raw data (X,Y values) on the sort of graph that produces a straight line (semilogarithmic if exponential, log–log if allometric). Linear least squares regression (see Appendix E) is then used to fit a line through the data. Finally, the slope and intercept of the line are used to estimate the equation parameters. R. H. Peters (*The Ecological Implications of Body Size*, Cambridge University Press, Cambridge, 1983) thoughtfully discusses the theoretical details and limitations of this approach.

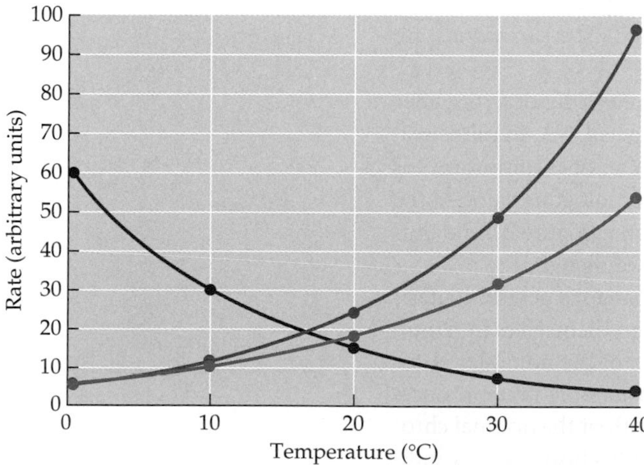

Figure A Three exponential equations Each point plotted along the blue line, at 10°C steps from left to right, is 2 times higher than the point preceding. Each point along the green line is 1.73 times higher. These examples of exponential increase illustrate the self-reinforcing nature of exponential change. Each point along the red line is 0.5 times the height of the preceding point, illustrating exponential decrease.

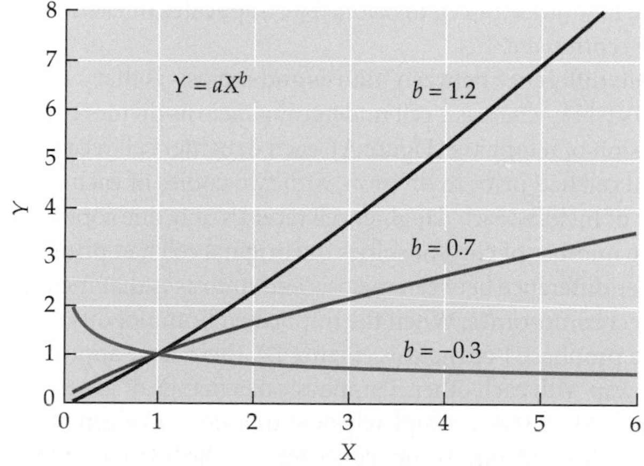

Figure B Three allometric equations The parameter a in the allometric equation (Equation G.2) is set equal to 1 in all three lines shown. The three lines differ in the value of b, the exponent in the allometric equation.

APPENDIX H Mitosis and Meiosis

All cells use the information encoded in the genes of their DNA to orchestrate their functions. When cells divide, it is critical that the DNA be duplicated and distributed properly to the daughter cells so that they receive a complete set of operational instructions. To accomplish this goal, DNA forms chromosomes that are inherited when the cell replicates itself in cell division. One chromosome includes one complete molecule of DNA, which incorporates many genes. The majority of cells in the body, the **somatic cells**, contain two copies of each DNA molecule, one inherited from each parent. They are referred to as *diploid*, which means having two identical sets of chromosomes. A set of two copies of the same chromosome is called a *homologous pair*. By contrast, **germ cells**—sperm or ova—have only one copy of each DNA molecule, so they are *haploid*, which means having a single set of chromosomes. When a sperm cell and an ovum unite in fertilization, each contributes one member of each homologous pair, and the resulting zygote is diploid.

Animals use two types of cell division. A diploid somatic cell uses **mitosis** to produce two diploid daughter cells. A specialized diploid cell in the testis or ovary uses **meiosis** to produce haploid daughter cells that develop into germ cells. The life cycle of a cell is separated into two main segments: the time during which it is not dividing, which is called **interphase**, and the time during which it is actively dividing. Some cells, once differentiated, don't divide at all. A motor neuron in the vertebrate spinal cord is a good example of such a cell. It remains in interphase continuously. Others, such as the epithelial cells lining the gut, divide frequently by mitosis. The figures in this appendix outline the events that occur during mitosis (Figure A) and meiosis (Figure B). In both types of cell division, the DNA is present as *chromatin* during interphase. It is incorporated into visible chromosomes only when the cell is dividing. In both types of division, the DNA is replicated (doubled) near the end of interphase, and duplicated chromosomes form in the first phase of cell division. The duplicates of each chromosome, called *sister chromatids*, are joined by a *centromere*.

The main difference between mitosis and meiosis is that a cell undergoing mitosis divides once after doubling its DNA, whereas a cell undergoing meiosis divides twice after doubling its DNA. In mitosis, at the conclusion of telophase (Figure A), each daughter cell receives the same number of chromosomes as the original cell had prior to division, with two copies of each chromosome—homologous pairs. At the conclusion of meiosis, each daughter cell receives only one copy of each chromosome (Figure B), and thus has half the number of chromosomes the original cell had prior to the initiation of meiosis.

A further difference between meiosis and mitosis is that meiosis permits exchange of genetic material between chromosomes. When the duplicated homologous chromosomes align themselves in *tetrads* during late prophase I of meiosis (Figure B), their close apposition allows parts of the nonsister chromatids to swap with each other. This physical exchange of genetic material, called *crossing-over*, produces chromosomes that have a completely new, unique mix of genes different from both of the original chromosomes. If these chromosomes contribute to the formation of a zygote in fertilization, the new individual will be unique, not only because it contains chromosomes from two different parents, but also because the genetic material of the contributing parents' chromosomes was reshuffled in the process of crossing-over. Biologists have observed that genetic diversity among the members of a species appears to contribute to the endurance of that species over long evolutionary periods. The reasons underlying this adaptive advantage of genetic diversity are being actively investigated.

Figure A Mitosis

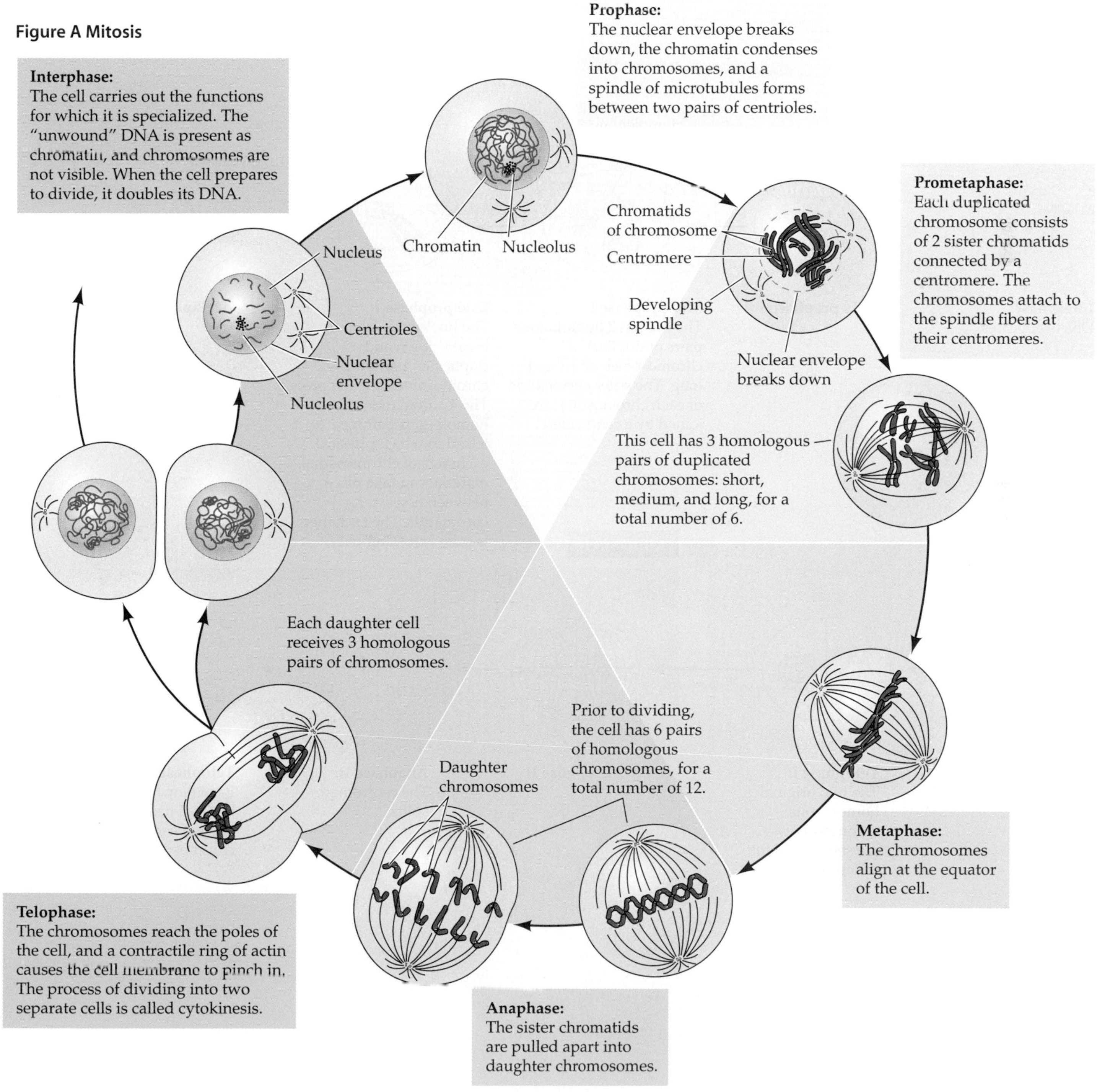

Interphase:
The cell carries out the functions for which it is specialized. The "unwound" DNA is present as chromatin, and chromosomes are not visible. When the cell prepares to divide, it doubles its DNA.

Nucleus
Centrioles
Nuclear envelope
Nucleolus

Prophase:
The nuclear envelope breaks down, the chromatin condenses into chromosomes, and a spindle of microtubules forms between two pairs of centrioles.

Chromatin
Nucleolus

Prometaphase:
Each duplicated chromosome consists of 2 sister chromatids connected by a centromere. The chromosomes attach to the spindle fibers at their centromeres.

Chromatids of chromosome
Centromere
Developing spindle
Nuclear envelope breaks down

This cell has 3 homologous pairs of duplicated chromosomes: short, medium, and long, for a total number of 6.

Metaphase:
The chromosomes align at the equator of the cell.

Prior to dividing, the cell has 6 pairs of homologous chromosomes, for a total number of 12.

Daughter chromosomes

Anaphase:
The sister chromatids are pulled apart into daughter chromosomes.

Each daughter cell receives 3 homologous pairs of chromosomes.

Telophase:
The chromosomes reach the poles of the cell, and a contractile ring of actin causes the cell membrane to pinch in. The process of dividing into two separate cells is called cytokinesis.

Figure B Meiosis

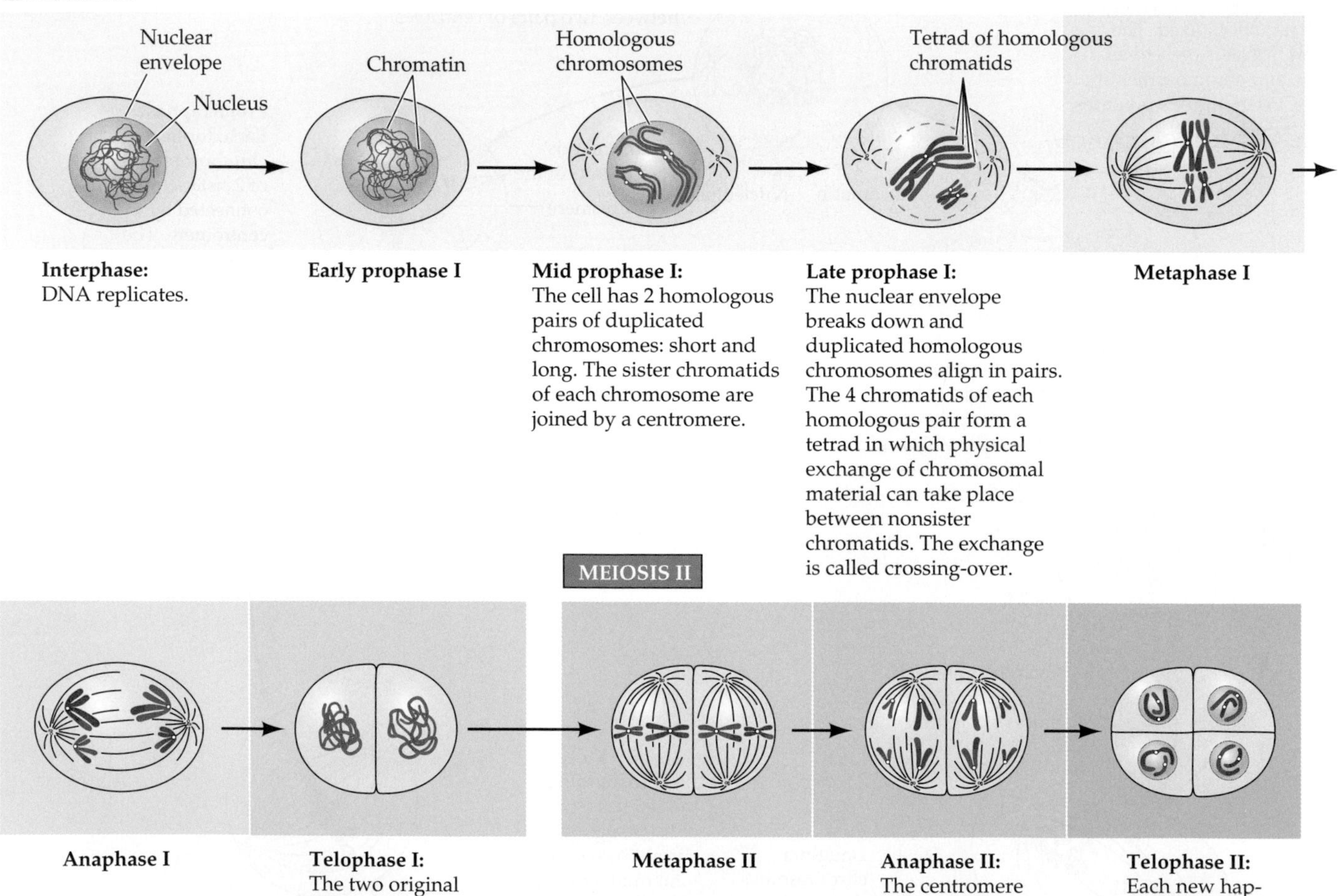

APPENDIX I The Standard Amino Acids

The standard amino acids are those employed in the synthesis of most proteins and polypeptides. The 20 most common standard amino acids are listed here, along with their usual abbreviations. There are two systems of abbreviation. In the three-letter system, each amino acid is represented by a three-letter code (the first letter of which is uppercase). In the one-letter system, each is represented by a single uppercase letter. Figure 4.2 shows the chemical structures of five of the standard amino acids.

Names and abbreviations of the 20 most common standard amino acids

Amino acid	Three-letter code	One-letter code
Alanine	Ala	A
Arginine	Arg	R
Asparagine	Asn	N
Aspartic acid	Asp	D
Cysteine	Cys	C
Glutamic acid	Glu	E
Glutamine	Gln	Q
Glycine	Gly	G
Histidine	His	H
Isoleucine	Ile	I
Leucine	Leu	L
Lysine	Lys	K
Methionine	Met	M
Phenylalanine	Phe	F
Proline	Pro	P
Serine	Ser	S
Threonine	Thr	T
Tryptophan	Trp	W
Tyrosine	Tyr	Y
Valine	Val	V

APPENDIX J Basic Physics Terms

This appendix defines and interrelates basic terms in mechanics and the physics of electricity, and refers back to some places in the text where these are discussed in a physiological context. It is deliberately selective rather than comprehensive in both these respects. Symbols for units are roman (upright) and symbols for variables are italic (slanted); for example, m is meters and m is mass. Note that Appendix B treats units of measure and Appendix D discusses gases at standard temperature and pressure. Physical principles not treated here are discussed elsewhere in the text; for example, the physics of gas transfer is discussed in Chapter 20 and supplemented with treatment of water vapor in Chapter 25.

Mass (m) The amount of material in an object. The basic unit is the *kilogram* (kg), which is a unit of mass, not of weight.

Velocity Rate of movement; $v = \mathrm{d}x/\mathrm{d}t$, where x is distance and t is time. Velocity is similar to speed, except that velocity is a vector quantity that considers direction as well as speed; speed is scalar and is independent of direction.

Acceleration Rate of change of velocity; $a = \mathrm{d}v/\mathrm{d}t$.

Force (F) A push or pull on a body; what it takes to make mass accelerate. $F = ma$ and is measured in *newtons*; $1\mathrm{N} = \mathrm{kg} \cdot \mathrm{m/s^2}$.

Weight The force exerted on a mass by Earth's gravitational acceleration; $w = mg$, where g is the acceleration due to gravity.

Pressure Force per unit of area. The standard unit of pressure is the *pascal* (Pa; $1\ \mathrm{Pa} = 1\mathrm{N/m^2}$). Pressure differences determine how materials flow, whether in blood circulation (Chapter 23), breathing (Chapter 21), respiratory gas exchange (Chapter 22), or filtration of water (Chapters 23, 27).

Work (W) The mechanical definition of work is the product of force and distance; $W = Fd$, measured in *joules*. The mechanical definition of work does not apply in all contexts of work in physiology. Physiological work is discussed in Chapter 5, pages 127–129. The concept of work applied specifically to muscle is discussed in Chapter 17, page 476, particularly Figure 17.13, and in Chapter 19, page 513, including Equation 19.2.

Energy The capacity to do work. In mechanics, this is divided into kinetic energy and potential energy. Energy is measured in joules (the same units as work).

Kinetic energy Energy inherent in the motion of mass; it increases with increasing mass and velocity; $K = 1/2mv^2$.

Potential energy Stored energy that can be released to do work. An object that gravity can move downhill, a coiled spring, a battery, and the chemical bond energy in molecules all represent examples of potential energy.

Power Work per unit of time, in *watts* ($1\mathrm{W} = 1\mathrm{J/s}$). The concept of power applied specifically to muscle is discussed in Chapter 19, pp. 512–513, particularly Figure 19.1 and Equations 19.1 and 19.2.

Light Electromagnetic radiation with a wavelength of about 400–700 nm; shorter wavelengths are ultraviolet radiation and longer wavelengths are infrared radiation.

Sound Waves of compression of air or water, which propagate away from a vibrating source. See Chapter 13, page 373.

Temperature The atoms and molecules within any substance undergo constant random motions on an atomic–molecular scale. The temperature of a substance is a measure of the speed—or intensity—of these random motions, and is independent of the macroscopic mass of the substance (although it is dependent on molecular mass). Chapter 8 discusses the physics of heat transfer, and of temperature and heat in general.

Heat Heat (unlike temperature) is a form of energy: the energy that a substance possesses by virtue of the random motions of its atomic–molecular constituents. The amount of heat in a substance depends on both the number and kinds of atoms and molecules in the substance and on the speed of each. A copper block with many copper atoms moving at a given average speed contains proportionally more heat energy than one with fewer atoms moving at the same speed, although they are at the same temperature.

Electricity The flow of electricity through electrical circuits is analogous to the hydraulic flow of water or any liquid through pipes. Similar relationships and equations describe both systems, and the hydraulic analogy helps to develop an intuitive understanding of electrical concepts. The section in Chapter 11 titled "The Ionic Basis of Membrane Potentials" discusses electrical concepts in a physiological perspective.

Electrical charge (Q) All matter is made up of charged particles (protons and electrons). Atoms or molecules that gain or lose electrons have unequal numbers of positive and negative charges and are called ions. Ions of unlike charge attract each other, and ions of like charge repel each other. Charge can be defined as the source of this electrical force of attraction/repulsion. The unit of charge is the *coulomb* (C), which is the charge carried by 6.24×10^{18} protons. Charge relates to mass through Faraday's constant (96,500 C/mol).

Current (I) Electric current is the flow of charge, measured in *amperes* (1 A = 1 C/s). In biological systems, ions moving in solution carry essentially all currents.

Voltage (V or E) Potential difference; the electromotive force or electrical potential. (In common usage these terms mean the same thing.) The potential difference in an electrical circuit is analogous to the pressure difference in a hydraulic circuit. No current will flow between two points at the same pressure, or the same voltage. In a hydraulic circuit, a pump or a difference in height supplies the pressure difference (the latter because water flows downhill due to the force of gravity). In an electrical circuit, a battery separates positive and negative charges to provide the potential difference (acting as a voltage source). Potential difference is measured in *volts* (V); 1 V is the energy required to move 1 C of charge a distance of 1 m against a force of 1 N. Chapter 11, pp. 288–291 describes how membrane permeability to ions generates a membrane potential difference.

Resistance (R) The property that hinders the flow of current. In an analogous hydraulic system, for a given pressure difference, more water will flow through a pipe with a large cross-sectional area than through a narrower pipe. The narrower pipe has a higher resistance to current flow. Similarly, in an electrical circuit, the lower the resistance of an element of the circuit, the more current will flow for a given voltage. The unit of resistance is the *ohm* (Ω).

Conductance (g) The inverse of resistance; the ease of current flow for a given voltage, measured in *siemens* (S).

Ohm's law $E = IR$. The voltage is the product of the current and the resistance in a circuit, or by rearrangement, the current equals the voltage over the resistance ($I = E/R$).

Capacitance (C) Measure of the ability of a nonconductor to store charge. Capacitance is the amount of charge stored per unit of voltage ($C = Q/V$) and is measured in *farads* (F). A capacitor has two conducting plates separated by an insulator; capacitance increases with increased size of the plates and with decreased distance between them. When a capacitor is first connected to a battery, positive charges flow onto one plate of the capacitor and negative charges onto the other. Although no charges cross from one plate to the other, a transient capacitive current flows because the charges on plate will repel like charges and attract unlike charges on the other plate. See Figures 11.5 and 11.22.

Glossary

A band The region of the sarcomere that spans the length of the thick filaments. It includes the H zone and regions on both ends where thick and thin filaments overlap.

Absolute refractory period The time during and immediately after an action potential in which the voltage threshold is infinite. Thus, no depolarization can exceed threshold and no new action potentials can be initiated.

Absorbed energy In the study of nutritional physiology, the chemical-bond energy of the compounds that an animal absorbs (assimilates) from its digestive tract. Also called *assimilated energy*.

Absorption In the study of nutritional physiology, the entry of organic molecules into the living tissues of an animal from outside those tissues. Absorption includes the entry of materials from the lumen of the gut inasmuch as the lumen is continuous with the outside environment and therefore outside the animal. Also called *assimilation*.

Absorption coefficient The dissolved concentration of a gas when partial pressure of that gas in solution is 1 atm. See also *Henry's law*.

Absorption efficiency In the study of nutritional physiology, the fraction of molecules or the fraction of chemical-bond energy absorbed from the digestive tract expressed in relation to the amount ingested. Also called *assimilation efficiency*.

Absorption spectrum The absorption of electromagnetic energy by a molecule or other object as a function of the wavelength of the energy.

Accessory heart In an animal with two or more hearts, a heart other than the primary heart. Also called *auxiliary heart*.

Acclimation A chronic response of an individual to a changed environment in cases in which the old and new environments differ in just one or two highly defined ways. Acclimation is a laboratory phenomenon.

Acclimatization A chronic response of an individual to a changed environment in cases in which the new and old environments are natural environments that can differ in numerous ways, such as winter and summer environments, or low and high altitudes.

Acidosis A state in which the pH of the body fluids is excessively acid.

Acoustico-lateralis system A vertebrate sensory system in which the sensory receptors are hair cells and their derivatives. Includes auditory receptors, organs of balance and gravity detection, and the lateral line system of fish and amphibians.

Acrosomal reaction The release of enzymes from the acrosome in the head of a sperm; occurs when the cell membrane of the sperm head binds to species-specific receptor sites on the zona pellucida surrounding the oocyte.

Actin One of the contractile proteins of muscle cells. Globular G-actin monomers polymerize to form the filamentous F-actin of the thin myofilament. Actin also contributes to motility in many other kinds of cells.

Action potential A brief electrical signal of about 100 mV across the cell membrane of a neuron or other excitable cell. It is initiated by a depolarization above threshold and is propagated to the end of the axon or cell. Also called a *nerve impulse*.

Activation energy The minimal amount of energy a molecule must gain to enter its transition state during a chemical reaction. Without entering its transition state, a molecule cannot react. Enzymes speed reactions by lowering the activation energy, making achievement of the transition state more likely.

Active evaporative cooling Evaporative cooling that occurs because evaporation is accelerated by some physiological process other than the simple production of heat, such as panting or sweating. See also *evaporative cooling*.

Active site A specific region of an enzyme molecule, at or near the surface of the molecule, where the enzyme bonds with its substrate and where the conversion of substrate to product is catalyzed. Also called a *substrate-binding site*.

Active transport The transport of a solute across a membrane by a mechanism that is capable of using metabolic energy to cause solute molecules to move across the membrane. An active-transport mechanism is capable of making a solute move away from equilibrium (against its electrochemical gradient). Also called *uphill transport*.

Acute response A response exhibited by an animal during the first minutes or hours after an environmental change.

Adaptation A genetically controlled trait that, through the process of natural selection, has come to be present at high frequency in a population because it confers a greater probability of survival and successful reproduction in the prevailing environment than available alternative states.

Adaptation, sensory See *receptor adaptation*.

Adenohypophysis The nonneural endocrine portion of the vertebrate pituitary gland. It is commonly called the anterior pituitary and includes three parts with variable representation in different species: pars tuberalis, pars intermedia, and pars distalis

Adequate stimulus The kind of stimulus energy to which a receptor is most sensitive, or the kind to which it normally responds.

Adipocyte An animal cell specialized for the storage of fats or oils.

Adrenergic neuron A neuron that synthesizes and releases norepinenphrine (noradrenaline) or epinephrine (adrenaline) as a neurotransmitter.

Aerobic Requiring O_2.

Aerobic capacity A synonym for $\dot{V}_{O_2max}$

Aerobic dive limit In diving mammals and birds, the length of the longest dive that can be undertaken without net accumulation of lactic acid above the resting level.

Aerobic expansibility An animal's peak rate of O_2 consumption during locomotion expressed as a ratio of its resting rate of O_2 consumption.

Aerobic scope for activity The difference between an animal's peak rate of O_2 consumption during locomotion and its resting rate of O_2 consumption.

Afferent Going toward. Thus, for example, an afferent blood vessel carries blood toward an organ of interest. Contrast with *efferent*.

Afferent neuron A neuron, normally sensory, that conducts signals from the periphery into the central nervous system.

Affinity See *enzyme–substrate affinity, oxygen affinity*.

Aglomerular Lacking a glomerulus.

Air capillaries Minute, gas-filled, blind-ended channels that branch off from the principal respiratory tubules, the parabronchi, within the lungs of a bird and that act as the primary sites of respiratory gas exchange between the air and blood.

Alkalosis A state in which the pH of the body fluids is excessively alkaline.

All-or-none Occurring fully or not at all; not graded.

Allometric equation Two variables, X and Y, are related in an allometric manner when $Y = aX^b$ ($b \neq 1$). See Appendix G. Also called the *power equation*.

Allosteric modulation In relation to enzymes, modulation of the catalytic properties of an enzyme by the binding of nonsubstrate ligands to specific nonsubstrate-binding sites, called regulatory sites or allosteric sites; a type of *cooperativity*, commonly important in the regulation of enzyme-catalyzed reactions and pathways. In relation to respiratory pigments, modulation of O_2 binding by the binding of ligands other than O_2. See also *cooperativity*.

Allosteric modulator In relation to enzymes, a nonsubstrate ligand of an enzyme that modulates the catalytic activity of the enzyme by binding to a specific regulatory site on the enzyme molecule. In relation to respiratory pigments, a ligand other than O_2 that modulates binding with O_2.

Alpha (α) motor neuron A relatively large motor neuron in a vertebrate spinal cord that innervates extrafusal muscle fibers.

Alphastat hypothesis The hypothesis that in poikilotherms the pH of body fluids is typically regulated at a relatively fixed difference from neutral pH—even as the body temperature changes and the neutral pH therefore also changes—as a way of maintaining a relatively constant state of electrical charge on proteins (particularly on protein imidazole groups).

Alveoli (singular **alveolus**) In the lungs of vertebrates, the blind-ended terminations of multiple branchings of the respiratory tract that form the surface for exchange of gases between the air and the blood. In the mammary glands of mammals, the hollow glandular structures that secrete milk.

Amacrine cell A retinal neuron that mediates lateral antagonistic effects and is sensitive to visual movements.

Ambient Relating to the surroundings of an organism. Synonymous with *environmental*; for instance, the ambient temperature is the environmental temperature.

Ammonotelic Incorporating most nitrogen from the catabolism of nitrogenous compounds into ammonia. "Most" is defined differently by different authorities; a common approach is to categorize an animal as ammonotelic if 50% or more of the nitrogen released by catabolism is incorporated into ammonia.

Amphipathic molecule A molecule that consists of a polar subpart and a nonpolar subpart.

Amplification In a sequence of enzyme-catalyzed biochemical reactions, an increase in the number of reacting molecules resulting from the fact that one enzyme molecule can catalyze the formation of more than one product molecule.

Anabolism Constructive metabolism; the set of metabolic processes that build relatively large molecules from smaller molecular building blocks using energy.

Anadromous Relating to an aquatic animal that undergoes most of its growth in seawater but enters freshwater to breed.

Anaerobic Capable of functioning without O_2.

Anaerobic glycolysis The reactions that convert glucose to lactic acid.

Anastomose To form an interconnected network of tubules, vessels, or similar structures by patterns of branching, reconnection, and rebranching.

Anatomical dead space In the lungs of an animal, the sum total of the conducting airways; that is, that portion of the lungs that does not participate in the exchange of respiratory gases between air and blood.

Androgen A masculinizing hormone such as testosterone.

Angiogenesis The formation of new blood vessels by sprouting of branches from existing vessels.

Angiotensin converting enzyme (**ACE**) A membrane-bound enzyme produced by vascular endothelial cells that converts inactive angiotensin I to angiotensin II.

Anhydrobiosis Survival while dried as fully as possible by desiccation in air.

Anion A negatively charged ion.

Anoxia In relation to the tissues of an animal, the state of being devoid of O_2.

Antagonism A type of influence of one substance such as a hormone over another in which one opposes the action of the other on a target tissue.

Antagonist An opponent. In biochemical reactions, a substance that opposes the action of another substance. In muscles, a muscle that performs the opposite action of the muscle or group of muscles with which it is paired.

Antennal gland The urine-producing structure of a decapod crustacean. Also called a *green gland*.

Anterior pituitary The nonneural endocrine portion of the vertebrate pituitary gland. It includes three parts with variable representation in different species: the pars tuberalis, pars intermedia, and pars distalis. Also called the *adenohypophysis*.

Antidiuresis The production of urine at a low rate. Such urine is usually concentrated as well as scanty.

Antidiuretic Opposing diuresis; promoting antidiuresis.

Antidiuretic hormone (**ADH**) A hormone that promotes antidiuresis. In mammals, also called vasopressin.

Antifreeze compound A metabolically synthesized compound, added to a body fluid, that has the principal function of lowering the freezing point of that body fluid.

Anti-Müllerian hormone See *Müllerian inhibitory substance*.

Antiporter See *countertransporter*.

Apical surface In an epithelium or epithelial cell, the surface that faces toward the lumen or open space lined by the epithelium. Also called the *mucosal surface*.

Apnea The absence of breathing.

Aquaporin A chemically defined type of water channel, often highly specific for water as opposed to solutes. See also *water channel*.

Arachnid A spider, tick, scorpion, or mite.

Archimedes' principle The principle that states that a body immersed in a fluid is buoyed up by a force equal to the weight of the fluid displaced.

Aromatase The enzyme that converts androgens such as testosterone to estrogens such as estradiol.

Arousal The emergence of an animal from hibernation, involving the rewarming of tissues to ordinary homeothermic temperatures.

Arteriole A microscopically fine blood vessel with muscular walls that carries blood from arteries to capillaries in a microcirculatory bed of a vertebrate. Contraction and relaxation of the muscular walls controls the rate of blood flow to the capillaries supplied by the arteriole.

Artery A macroscopic blood vessel that carries blood away from the heart.

Asexual reproduction The formation of new individuals without the union of gametes from two different parents. The offspring are genetically identical to their parent.

Assimilation In the study of nutritional physiology, synonymous with *absorption*.

Atmosphere A unit of measure for pressure. An *atmosphere* or more technically speaking, a *normal atmosphere* is the average pressure exerted by Earth's atmosphere at sea level. It is quantitatively specified by international agreement to be 1.01325×10^5 pascal at 15°C. It is also specified to be equal to the pressure exerted by a column of mercury 760 mm high under standard gravitational acceleration.

ATPase An enzyme capable of hydrolyzing adenosine triphosphate (ATP) and thereby releasing energy from ATP.

Atrial natriuretic peptide A member of a family of peptide hormones secreted by specialized cells in the atria of the heart that stimulate the excretion of water and sodium.

Atrophy A decrease in the mass of a tissue or organ by the loss of cells or of intracellular components of cells. Also called *wasting*.

Audition Hearing; sound detection and perception.

Autocrine A locally acting chemical signal that binds to receptors and exerts a regulatory effect on the same cell that secreted it.

Autonomic effector An effector other than skeletal muscle; includes smooth and cardiac muscles and tissues of the viscera and exocrine glands.

Autonomic nervous system (ANS) The division of the nervous system that innervates and controls autonomic effectors and conveys sensory information from internal organs.

Autoreceptor In synaptic processes, a receptor protein on a presynaptic terminal that is stimulated by neurotransmitter released by the same neuron.

Autotroph An organism that can obtain the energy it needs to stay alive from sources other than the chemical bonds of organic compounds. Some autotrophs (*photoautotrophs*) use photon energy; others (*chemoautotrophs*) use energy from inorganic chemical reactions. Contrast with *heterotroph*. See also *primary production*.

Auxiliary heart See *accessory heart*.

Average daily metabolic rate (ADMR) An animal's rate of metabolism averaged over all hours of the 24-hour day.

Avogadro's number The number of molecules in a mole; also, the number of independent dissolved entities in an osmole. Equal to 6.022×10^{23}.

Axis In the study of endocrinology, two or more hormone-secreting tissues that together form an *hierarchical* control system.

Axon A process of a neuron specialized for conveying action potentials (usually) away from the cell body. An axon may be myelinated or unmyelinated.

Axonal transport The transport of materials within the cytoplasm of an axon; may be either anterograde (away from the cell body) or retrograde (toward the cell body).

Baculum A rodlike structure of cartilage or bone that gives stiffness to the shaft of the penis of some species of mammals.

Basal lamina *See basement membrane.*

Basal metabolic rate (BMR) The metabolic rate of a homeothermic animal when it is in its thermoneutral zone, resting, and fasting.

Basal surface In an epithelium or epithelial cell, the surface that faces toward the underlying tissue on which the epithelium rests or to which the epithelium is attached. Also called the *serosal surface*.

Basement membrane A thin, permeable, noncellular, and nonliving sheet of matrix material on which an epithelium rests. It is composed of glycoproteins and particular types of collagen, and is secreted mostly by the cells of the overlying epithelium, but cells of the underlying tissue also help secrete it. Also called the *basal lamina*.

Basilar membrane A membranous tissue within the cochlea of the vertebrate ear that contains the auditory sensory hair cells and is vibrated by sound waves.

Behavioral thermoregulation In poikilotherms, the maintenance of a relatively constant body temperature by behavioral means. In homeotherms, the use of behaviors to assist in the maintenance of a relatively constant body temperature.

Bends See *decompression sickness*.

Bile salt A type of emulsifying compound, produced by the biliary system in the liver of a vertebrate, that plays a key role in lipid digestion and absorption because of its ability to emulsify lipids within the gut lumen.

Bimodal breather See *dual breather*.

Biological clock A physiological mechanism that gives an organism an endogenous capability to keep track of the passage of time.

Bipolar cell A type of neuron in the vertebrate retina that mediates the direct (straight through) pathway connecting rods and cones to ganglion cells.

Bivalve mollusc A clam, mussel, scallop, oyster, or related animal.

Black-body temperature See *radiant temperature*.

Blastocyst A hollow sphere of cells formed in early mammalian development that implants in the uterus. Its inner cell mass will form the embryo and extraembryonic structures, and its outer trophoblast cells will contribute to formation of the placenta.

Blood The fluid (including suspended cells) that is circulated within the circulatory system.

Blood plasma The part of the blood that remains after blood cells are removed; the part of the blood other than cells.

Blood pressure The extent to which the pressure of the blood exceeds the ambient pressure.

Bohr effect A decrease in the O_2 affinity of a respiratory pigment (a shift of the oxygen equilibrium curve to the right) caused by a decrease in pH or an increase in CO_2 partial pressure.

Bomb calorimeter A device used to measure the energy content of organic materials.

Book gills Unusual gill structures consisting of many sheets of tissue arrayed like pages of a book; found in horseshoe crabs.

Book lungs Lungs within which sheets of gas-exchange tissue alternate with sheet-like air spaces, like pages of a book slightly separated by air. Found in spiders, scorpions, and some other arachnids.

Brackish water Water that is intermediate in salinity between seawater and freshwater. One criterion sometimes used is that water is brackish if its salinity is between 0.5 and 30 g/kg.

Bradycardia A heart rate that is unusually low.

Brain The anterior enlargement of the central nervous system in an animal with a cephalized nervous system.

Branchial Relating to gills.

Brown adipose tissue (BAT) A metabolically active form of adipose tissue, known only in mammals, that serves as the primary site of mammalian nonshivering thermogenesis and expresses a distinctive mitochondrial protein, *uncoupling protein 1* (UCP1). Also called *brown fat*. Contrast with *white adipose tissue*.

Brush border An apical epithelial surface bearing *microvilli*.

Buccal pressure pump The development of positive pressure within the buccal cavity of a vertebrate, used to force air into lungs or water across gills.

Buffer reaction A chemical reaction in a solution that tends to stabilize the pH of the solution by removing H^+ from the solution when H^+ is added by some external process and by adding H^+ to the solution when H^+ is removed by an external process.

Bulk solution Solution that is not immediately next to a membrane or other surface.

Burst exercise Sudden, intense exercise.

Caching behavior The storage of food or other material for later use.

Caisson disease See *decompression sickness.*

Calorie (cal) A unit of energy equal to the amount of heat required to raise the temperature of 1 g of water by 1°C, from 14.5 to 15.5°C. Some people outside science distinguish the calorie from the kilocalorie (kcal; 1000 calories) by writing "calorie" (small "c") to represent the calorie and "Calorie" (large "c") to represent the kilocalorie. Sometimes, in fact, in writing outside science, "calorie" (small "c") is used to refer to the kilocalorie; thus one must be cautious in interpreting uses of "calorie" outside science.

Camera eye An eye that optically resembles a camera, with a lens that focuses an image on a retina of light-sensitive cells.

cAMP See *cyclic adenosine monophosphate.*

Capacitance The ability of an insulating material, such as the cell membrane, to store electrical charges.

Capacitance coefficient (β) In the study of respiratory gas exchange, the change in total gas *concentration* per unit of change in gas *partial pressure* in air, water, or a body fluid like blood. The capacitance coefficient sometimes differs from simple gas-solubility coefficients such as the absorption coefficient because in the case of the capacitance coefficient, the gas concentration includes bound gases, such as O_2 bound to hemoglobin.

Capacitation The final maturation of sperm that takes place in the female reproductive tract to make the sperm capable of rapid forward swimming and fertilization.

Capillary A microscopically fine blood vessel, the wall of which consists of only a single layer of epithelial cells. Capillaries are the principal sites of exchange between blood and tissues in a closed circulatory system.

Carbon dioxide dissociation curve See *carbon dioxide equilibrium curve.*

Carbon dioxide equilibrium curve In relation to the CO_2-carrying properties of blood, a graph of the *total carbon dioxide concentration* as a function of the CO_2 partial pressure of the blood. Also called the *carbon dioxide dissociation curve.*

Carbonic anhydrase (CA) An enzyme that accelerates the conversion of CO_2 and H_2O to form HCO_3^- and H^+, or the reverse reaction.

Cardiac muscle Muscle that forms the wall of the heart. In vertebrates, heart muscle consists of branched, generally uninucleate, striated muscle cells that are connected by intercalated discs. Their contractions are produced by endogenously generated myogenic action potentials and may be modified by neural and hormonal factors. The hearts of invertebrates may consist of striated or nonstriated cells that may be endogenously active (myogenic) or controlled by neural input (neurogenic).

Cardiac output The volume of blood pumped by a heart per unit of time, calculated as the stroke volume multiplied by the number of beats per unit of time (heart rate). In the case of the mammalian or avian heart, the cardiac output is specifically the output of the left ventricle into the systemic aorta unless stated otherwise.

Carrier See *transporter.*

Carrier-mediated transport Solute transport across a membrane that requires the reversible, noncovalent binding of solute molecules with a protein (called a *transporter* or *carrier*) in the membrane. There are two kinds of carrier-mediated transport: *active transport* (which employs metabolic energy) and *facilitated diffusion* (which does not).

Catabolism Destructive metabolism; the set of metabolic processes by which complex chemical compounds are broken down to release energy, create smaller chemical building blocks, or prepare chemical constituents for elimination.

Catadromous Relating to an aquatic animal that undergoes most of its growth in freshwater but enters seawater to breed.

Catalytic rate constant (k_{cat}) The number of substrate molecules a unit of enzyme is capable of converting to product per unit of time when the enzyme is saturated.

Catalytic vacuole The part of an enzyme molecule that provides a suitable setting for catalysis during the conversion of substrate to product; includes, but is not limited to, the *active site.*

Cation A positively charged ion.

Cell body The portion of a neuron that contains the cell nucleus; also called the *soma* or *perikaryon.*

Cell membrane The membrane that encloses an animal cell and forms the outer boundary of the cell. Also called the *plasma membrane.*

Cell signal transduction The processes by which cells alter their intracellular activities in response to extracellular signals.

Cell theory The theory, developed in the nineteenth century, that organisms are composed of cells, which act as structural, functional, and developmental units of organization.

Cellular oscillator A neuron that can generate a rhythmic change in membrane potential or activity without needing synaptic input to do so; one possible basis of a central pattern generator.

Central circulation The heart and the veins and arteries that connect immediately to the heart.

Central nervous system (CNS) The consolidated integrative part of an animal's nervous system; in vertebrates, consists of the brain and spinal cord.

Central pattern generator A neural circuit that generates a behaviorally significant pattern of motor output in space and time without requiring temporally patterned sensory input.

Centralization The evolutionary tendency of animal groups to concentrate integrative neural functions into a central nervous system.

Cephalization The concentration of structures of the nervous system toward the anterior end of an animal, a trend underlying the evolution of anterior brains in many animal groups.

Cephalopod mollusc An octopus, squid, cuttlefish, or related animal.

Cerebellar cortex The outer layer of the cerebellum of the vertebrate hindbrain; involved in motor coordination and learning.

Cerebral cortex The outer part of the cerebral hemispheres of the vertebrate forebrain; greatly enlarged in birds and mammals.

cGMP See *cyclic guanosine monophosphate.*

cGMP phosphodiesterase An enzyme that cleaves cGMP into 5′-GMP.

Channel A membrane protein that aids the passive transport of a solute across a membrane without undergoing any sort of chemical binding with that solute. Channels participate particularly in passive transport of inorganic ions across membranes. See also *water channel.*

Chaperones See *molecular chaperones.*

Chemical energy (**chemical-bond energy**) Energy that is liberated or required when atoms are rearranged into new configurations. Animals obtain the energy they need to stay alive by reconfiguring atoms in food molecules, thereby liberating chemical energy.

Chemical potential Qualitatively speaking, the strength of the tendency of a chemical substance to undergo a physical or chemical change. Measures of chemical potential are useful for predicting the direction of change because chemical substances tend to go from high chemical potential to low chemical potential.

Chemical synapse A synapse that employs a chemical neurotransmitter that is released presynaptically and acts on postsynaptic neurotransmitter receptors.

Chemiosmotic theory A theory concerning the biochemical relationship between electron transport and oxidative phosphorylation in mitochondria. According to the theory, electron transport pumps protons into the mitochondrial intermembrane space, and the back-diffusion of the protons through ATP synthase results in ATP formation.

Chemoautotroph An organism that can obtain the energy it needs to stay alive from energy-yielding inorganic chemical reactions.

Chloride cell A type of epithelial cell specialized for ion transport, found particularly in the gill epithelia of fish but also in some other epithelia of aquatic animals. Characterized by an abundance of mitochondria and other signs of secretory or absorptive activity. Also called a *mitochondria-rich cell.*

Chloride shift In vertebrate blood, diffusion of Cl^- from the blood plasma into the red blood cells in exchange for diffusion of HCO_3^- out of the cells, or the reverse process. Mediated by a membrane countertransporter often called the band 3 protein.

Chlorocruorin A type of respiratory pigment found in marine annelid worms, formed by the combination of hemelike structures with protein. Undergoes reversible combination with O_2 at the hemelike loci.

Cholinergic neuron A neuron that synthesizes and releases acetylcholine as a neurotransmitter.

Chorionic gonadotropin A hormone secreted by the embryonic placenta in horses and primates that maintains the function of the corpus luteum beyond the time when it would degenerate if pregnancy did not occur.

Chronic response A response expressed by an animal following days, weeks, or other prolonged exposure to new environmental conditions.

Ciliary photoreceptor A photoreceptor cell in which the light-sensitive part is a modified cilium; characteristic of vertebrates.

Circadian oscillator The biological clock that times a circadian rhythm.

Circadian rhythm An endogenous rhythm with a period of about a day (ca. 24 hours).

Circannual rhythm An endogenous rhythm with a period of about a year.

Circatidal rhythm An endogenous rhythm with a period approximating a tidal cycle (ca. 12.4 hours).

Circulation The pressure-driven mass flow of blood through a system of tubular vessels or other passages that brings it to all parts of the body.

Circulatory system The blood and the system of vessels or other passages through which it circulates.

Classical conditioning A form of associative learning in which an animal learns the association between two stimuli.

Clearance See *plasma clearance.*

Clone A new individual produced by asexual reproduction that is genetically identical to its parent, or a group of such individuals. Used as a verb, *to clone* means to make an exact genetic copy.

Closed circulatory system A circulatory system in which the blood is enclosed within blood vessels throughout and is therefore distinct from the interstitial fluids. Contrast with *open circulatory system.*

Cochlea A part of the inner ear of many vertebrates, coiled in mammals, that contains the auditory sensory hair cells.

Colligative properties The properties of an aqueous solution that depend simply on the number of dissolved entities per unit of volume, rather than on the chemical nature of the dissolved entities. The three principal colligative properties in animal physiology are osmotic pressure, freezing point, and water vapor pressure.

Colloid osmotic pressure The difference in osmotic pressure that arises between two solutions on either side of a cell membrane or epithelium because the two solutions differ in their concentrations of nonpermeating protein solutes. Because the solutes responsible for colloid osmotic pressure cannot cross the membrane or epithelium, the colloid osmotic pressure represents a "fixed" difference in osmotic pressure between the solutions. Also called *oncotic pressure.*

Commissure A bundle of axons that connects the two sides of a bilaterally symmetrical central ganglion or bilateral regions of a central nervous system.

Comparative method A method of analysis that seeks to identify adaptive traits or adaptive evolutionary trends by comparing how a particular function is carried out by related and unrelated species in similar and dissimilar environments.

Compass direction North, south, east, west, or a combination of these; the direction of a movement or orientation as it could be described using a compass.

Compatible solute A solute that, when concentrated enough to contribute significantly to the osmotic pressure of a fluid, has little or no effect on the structure and function of macromolecules with which the fluid is in contact.

Compensation In the study of the acclimation or acclimatization, the return of a physiological property toward its value that existed prior to an environmental change even though the animal remains in the changed environment. Compensation is *partial* if the physiological property returns only partly toward its preceding value. It is *complete* if the physiological property returns to the value that existed prior to the environmental change.

Complete compensation See *compensation.*

Compound eye A multifaceted eye characteristic of arthropods, composed of many individual optical units called ommatidia.

Concentration gradient Technically, the difference in the concentration of a solute between two places divided by the distance separating those two places. Often used more loosely to refer simply to a difference in concentration.

Condensation A change in the physical state of a compound from a gas to a liquid.

Conductance A measure of how easily current will flow through a conductive pathway. Contrast with electrical *resistance*. See also *thermal conductance*.

Conducting airways In the lungs of an animal, the airways that do not participate in the exchange of respiratory gases between air and blood, but rather simply conduct air from one place to another.

Conducting system A system of specialized muscle cells by which depolarization spreads throughout the vertebrate heart.

Conduction In the study of heat, the transfer of heat by intermolecular collisions through a material that is macroscopically motionless. In the study of heart physiology, the process by which depolarization spreads through the vertebrate heart or any other myogenic heart. In the study of electrical phenomena, the transmission of electrical currents or signals through a conductive pathway.

Cone A type of photoreceptor in the vertebrate retina. Cones are smaller and less light sensitive than rods and are used for diurnal vision and color vision.

Conformity A state in which an animal's internal conditions match the external environmental conditions.

Connective A bundle of neuronal axons in the central nervous system that connects central ganglia; found in the ganglionic nervous systems of arthropods, annelids, and molluscs.

Connexin The protein that makes up connexons.

Connexon The protein channel of a gap junction at which adjacent cells join so there is cytoplasmic continuity between them. Two connexons in the membranes of adjacent cells form a channel that allows electrical current flow between electrically coupled cells.

Constitutive enzyme An enzyme (or other protein) that is always expressed in a tissue. Contrast with *inducible enzyme*.

Consumption, of energy Conversion of chemical-bond energy to heat or external work.

Contact chemoreceptor A chemosensory cell (of a terrestrial animal) that is normally stimulated by chemicals that contact it in a liquid. Contrast *distance chemoreceptor*.

Contraction The condition in which a skeletal muscle is activated to produce force. Muscle contraction can be shortening (isotonic), isometric (remaining the same length), or lengthening (isotonic).

Convection Always refers to mass flow, but has different specific meanings depending on context. In the study of heat transfer, *convection* is the transfer of heat by the mass flow of a material substance (e.g., wind). In the study of material transport, *convection* refers to (1) the flow of a fluid from place to place and (2) the transport of molecules in the fluid from place to place by the fluid flow. Blood flow, for example, is a type of convection, and when the blood carries a material such as O_2 from place to place, the material is said to undergo convective transport.

Convergence Coming together. In neurophysiology, a pattern in which signals from many presynaptic neurons come together to affect a particular postsynaptic neuron. Contrast with *divergence*.

Cooperativity A type of chemical behavior that occurs in protein molecules that have multiple ligand-binding sites, in which the binding of any one site to its ligand may facilitate or inhibit the binding of other sites on the same molecule to their ligands. The binding sites do not interact directly; instead, binding at one site induces protein conformational shifts that affect other sites at a distance.

Coronary artery An artery that carries blood into the myocardium of a heart.

Coronary vein A vein that carries blood out of the myocardium of a heart.

Corpus allatum (plural **corpora allata**) One of a bilateral pair of organs in insects that serve both as neurohemal organs, where prothoracotropic hormone (PTTH) is released from axon terminals of neurosecretory cells in the brain, and as nonneural endocrine tissue, which secretes juvenile hormone.

Corpus luteum The endocrine structure in the mammalian ovary that is formed by the reorganization of the cells of an ovarian follicle that has undergone ovulation. It secretes progesterone, estrogen, and inhibin. If fertilization occurs, it remains active during pregnancy; if not, it degenerates.

Cost of transport For an animal that is undergoing directional locomotion, the energy cost of covering a unit of distance.

Costal suction pump The development of negative pressure within the thorax of an air breathing vertebrate by action of the costal muscles between the ribs, used to suck air into the lungs.

Costamere In muscle fibers, a complex of protein molecules that serves as a region of attachment between the myofibrils, sarcolemma, and extracellular matrix. Costameres are organized into many parallel bands that circumscribe each muscle fiber at regular intervals coincident with the Z discs.

Cotransmitter In neurons that synthesize and release more than one kind of neurotransmitter molecule, the second kind of neurotransmitter.

Cotransporter A transporter that obligatorily carries two different solutes in the same direction simultaneously. Through the participation of a cotransporter, the diffusion of one solute in the direction of its electrochemical gradient can cause the other solute to move away from equilibrium. Also called a *symporter*.

Counteracting solutes Osmolytes that act in teams of two or more to modify the osmotic pressures of body fluids without greatly perturbing macromolecules because the individual members of the teams have mutually offsetting effects on the macromolecules.

Countercurrent exchange Exchange of heat, O_2, or other substances by passive-transport processes (e.g., diffusion) between two closely juxtaposed fluid streams flowing in opposite directions (e.g., blood flowing in opposite directions in two blood vessels).

Countercurrent multiplication A process that occurs in a system consisting of two juxtaposed fluid streams flowing in opposite directions, in which metabolic processes produce a difference (termed the *single effect*) between adjacent parts of the two streams. Because of the dynamics of such a system, the difference produced between adjacent parts of the two streams is multiplied to create a much larger difference between the two ends of the system.

Countertransporter A transporter that obligatorily carries two different solutes in opposite directions simultaneously. Through the participation of a countertransporter, the diffusion of one solute in the direction of its electrochemical gradient can cause the other solute to move away from equilibrium. Also called an *antiporter*.

Covalent bond A chemical bond in which atoms fully share electrons. Covalent bonds are strong and, for the most part in biological systems, require enzyme catalysis to be made or broken.

Covalent modulation Modulation of the catalytic properties of an enzyme or other protein by chemical reactions that make or break covalent bonds between a modulator and the enzyme or other modulated protein. The most common mode of covalent modulation is phosphorylation and dephosphorylation catalyzed by protein kinases and protein phosphatases. Also called *covalent modification.*

Cranial nerves Peripheral nerves that connect to the brain.

Cross-bridge The head of a myosin molecule that interacts with actin molecules to produce muscle contraction by repeated oarlike power strokes. Each myosin molecule has two heads, and each head has an actin-binding site and an enzymatic site that binds and hydrolyzes ATP to produce energy to fuel each power stroke.

Cross-current exchange A type of exchange between two fluid streams, most commonly found in breathing organs in which O_2 and CO_2 are exchanged between air and blood. During cross-current exchange of this sort, the total blood flow is broken up into multiple separate streams, each of which "crosses" the flow of air at a particular point, thereby exchanging respiratory gases with the air at just one point along the path of airflow. The streams then coalesce to reestablish a unified total flow.

Cryptobiosis Latent life. A resting state in which there is little or no metabolism, so that there are few, if any, signs that the organism is still alive.

Cryptonephridial complex A specialized, close association between the Malpighian tubules and rectum that occurs in certain types of insects (e.g., mealworms) and plays a role in the production of concentrated urine and the absorption of water vapor from the atmosphere.

Current, electric The flow of electrical charge.

Cutaneous Relating to the skin.

Cuticle The exoskeleton of an arthropod, or the material of which the exoskeleton is composed.

Cyclic adenosine monophosphate (cyclic AMP, cAMP) A second messenger produced intracellularly in response to several neurotransmitters and hormones.

Cyclic guanosine monophosphate (cyclic GMP, cGMP) An intracellular messenger in rod and cone photoreceptors and a second messenger in some neurotransmitter actions.

Cytokines Peptide regulatory molecules that are involved in cell development and differentiation and in the immune response.

Cytoskeleton Intracellular structural support elements (e.g., microtubules, intermediate filaments, actin microfilaments) composed primarily of fibrous protein polymers.

Cytosol See *intracellular fluids.*

$D_2{}^{18}O$ method See *doubly labeled water method.*

Daily torpor In mammals and birds, a form of controlled hypothermia in which the body temperature is able to approximate ambient temperature for part (but only part) of each 24-hour day, generally on many consecutive days.

Dale's principle The idea that a neuron releases the same kind of neurotransmitter at all its endings. With the discovery of cotransmitters, Dale's principle may be revised to say "the same kind(s) of neurotransmitters."

Dalton (Da) A unit of measure for atomic and molecular mass, equal to 1⁄12 the mass of an atom of the most abundant carbon isotope, ^{12}C. Thus, the mass of a ^{12}C atom is 12 daltons.

Dark current The ionic current, carried mainly by Na^+ ions, that flows into the outer segments of vertebrate photoreceptors in the dark. Light absorption leads to closing of the Na^+ channels, turning off the dark current and hyperpolarizing the photoreceptor.

Data logger A small microcomputer with a large amount of memory that can be placed in or on an animal to collect and store time-labeled information on physiological or behavioral variables. After the data logger is recovered from the animal, the information it has stored is off-loaded to a computer.

Deamination Removal of nitrogen-containing amino groups from amino acids or proteins.

Decapod crustacean A lobster, crayfish, crab, shrimp, or related animal.

Decompression sickness A pathological state that arises after diving when bubbles are formed within body fluids because the reduction in pressure during surfacing allows gases (especially N_2) present at high dissolved partial pressures to come out of solution. Also called *the bends, caisson disease.*

Decremental spread Spread of a signal in such a way that signal strength decreases with distance. In electrophysiology, passive voltage changes (synaptic potentials, receptor potentials, etc.) spread decrementally, in contrast to propagated action potentials. Also called *electrotonic conduction, passive spread.*

Definitive urine The final urine that is excreted by the kidneys. Contrast with *primary urine.*

Degradation of energy Conversion of energy from a high-grade form to heat.

Denaturation A change in the tertiary (three-dimensional) structure of an intact protein that renders the protein nonfunctional. The primary structure is not altered during denaturation.

Dendrite The receptive element of most neurons, which receives synaptic input from other neurons. Most neurons have many, multiply branching dendrites, in contrast to one sparsely branching axon.

Dendritic Branching in a way that resembles the branching of a tree.

Deoxygenation The release of O_2 from a combined state with a respiratory pigment such as hemoglobin. Not equivalent to *reduction.*

Depolarization A decrease in amplitude of the inside-negative potential of a cell membrane toward zero. More generally, any increase in the inside positivity of a cell membrane, even if it exceeds zero.

Desmosome A "spot weld" type of junction between two adjacent cells in which protein filaments intermingle across the space between the cells, thereby strengthening and stabilizing the morphological arrangement of the cells.

Diaphragm A sheet of muscular and connective tissue that completely separates the thoracic and abdominal cavities, found only in mammals.

Diastole The period of relaxation during the beating cycle of a heart.

Diet-induced thermogenesis (DIT) A chronic increase in metabolic rate induced in certain types of animals by chronic overeating.

Diffusion See *simple diffusion.*

Diffusion lung A lung within which the air is still, so that O_2 and CO_2 must travel the full length of the lung passages by diffusion.

Digestion The process of splitting up ingested food molecules into smaller chemical components that an animal is capable of distributing to the tissues of its body. Most commonly carried out by hydrolytic enzymes.

Dihydropyridine receptor (DHPR) The voltage-sensitive molecule in the transverse tubules of vertebrate striated muscle fibers that links excitation to the release of Ca^{2+} ions from the sarcoplasmic reticulum. It serves as a Ca^{2+} channel in cardiac muscle and as a voltage sensor in skeletal muscle.

Dipnoan A lungfish. A member of the sarcopterygian class of bony fish, believed to be among the closest living relatives of the ancestors of amphibians. There are three genera of dipnoans: *Protopterus* in Africa, *Neoceratodus* in Australia, and *Lepidosiren* in South America.

Direct calorimetry Measurement of metabolic rates by quantifying heat and external work.

Direct measurement A measurement procedure that quantifies a property by measuring exactly what the *definition* of the property specifies. Contrast with *indirect measurement*.

Distance chemoreceptor A chemosensory cell (of a terrestrial animal) that is normally stimulated by chemicals that are airborne over a considerable distance. Distance chemoreceptors are usually more sensitive than contact chemoreceptors. Contrast with *contact chemoreceptor*.

Diuresis The production of urine at a high rate. Such urine is usually dilute as well as abundant.

Diuretic Promoting diuresis.

Diurnal Active in the daytime.

Divergence Spreading apart; a pattern in which signals from a particular presynaptic neuron synaptically excite or inhibit many postsynaptic neurons. Contrast with *convergence*.

Diving bradycardia Slowing of the heart rate during diving.

Donnan equilibrium A complex multi-ionic equilibrium state that tends to be reached by the interacting diffusion of multiple permeating ions and water across a cell membrane or epithelium when there is a set of nonpermeating ions (ions that cannot cross the membrane or epithelium) that are more abundant on one side than on the other. Because of their content of nonpermeating anionic proteins and nucleic acids, animal cells would approach Donnan equilibrium if it were not for the fact that living processes hold them away from any sort of equilibrium.

Dorsal root ganglion A peripheral ganglion at the dorsal root of a spinal nerve, containing cell bodies of the sensory neurons in that nerve.

Doubly labeled water method A method used to measure the metabolic rate of a free-living animal in which water labeled with unusual isotopes of hydrogen and oxygen is injected into the animal. It estimates the animal's rate of CO_2 production. Also called the *$D_2{}^{18}O$ method*.

Downregulation A downward shift in the catalytic activity of an enzyme, the rate of functioning of a biochemical pathway, or the rate of some other similar process brought about in a controlled manner by a regulatory system.

Dry heat transfer Heat transfer by conduction, convection, or thermal radiation; heat transfer that does not involve evaporation or condensation of water.

Dual breather An animal that simultaneously possesses the ability to breathe from air and from water. Also called *bimodal breather*.

Duchenne muscular dystrophy The most common form of muscular dystrophy caused by mutations of the *DMD* gene on the X chromosome. It is characterized by the lack of the cytoskeletal protein dystrophin, which leads to progressive degeneration and weakening of muscle tissue.

Dynamic viscosity See *viscosity*.

Dystrophin A cytoskeletal protein in muscle fibers that connects actin filaments of the cytoskeleton to a complex of proteins in the sarcolemma.

Ecdysis The process of shedding the outer body covering. In arthropods, the shedding of the old cuticle (exoskeleton) from one instar as the animal makes the transition to the next instar. Also called *molting*.

Ecdysone The steroid prohormone secreted by the thoracic glands of arthropods that stimulates ecdysis, or molting. Also called *molting hormone*.

Ectotherm See *poikilotherm*.

Effector A tissue, organ, or cell that carries out functions under the direction of the nervous system.

Efferent Going away. Thus, for example, an efferent blood vessel carries blood away from an organ of interest. Contrast with *afferent*.

Efferent neuron A neuron that conveys signals from the central nervous system to the periphery, usually exerting motor control.

Efficiency of energy transformation In any process that transforms high-grade energy from one form to another, the output of high-grade energy expressed as a ratio of the input of high-grade energy. See also *absorption efficiency*, gross *growth efficiency*, *net growth efficiency*.

Elasmobranch fish Sharks, skates, and rays.

Electrical energy Energy that a system possesses by virtue of the separation of positive and negative electrical charges.

Electrical gradient Technically, the difference in electrical potential (voltage) between two places divided by the distance separating those two places. Often used more loosely to refer simply to a difference in electrical potential.

Electrical synapse A synapse at which current spreads directly from cell to cell through a low-resistance gap junction.

Electrocardiogram (EKG, ECG) A recording as a function of time of differences in electrical potential set up in extracellular body fluids by the depolarization and repolarization of the myocardium during heart contraction and relaxation. These differences can be detected on the surface of the body and recorded from there.

Electrochemical equilibrium A term that is synonymous with *equilibrium*, but emphasizes that the equilibrium state for an ion or other charged solute depends on both electrical and chemical effects. A charged solute is at electrochemical equilibrium across a membrane when the effect of concentration on its diffusion and the effect of the electrical potential on its diffusion sum to zero.

Electroencephalogram A record of gross electrical activity in the brain, usually recorded using multiple electrodes placed on the skin of the head or the brain surface.

Electrogenic pump An active-transport process that pumps net charge across a membrane, acting to generate an electric current across the membrane and to produce a voltage difference across the membrane. Contrast with *electroneutral pump*.

Electrolyte An inorganic ion in a body fluid.

Electroneutral pump An active-transport process that pumps charges across a membrane such that no difference of charge is created across the membrane, and so is not a current source. Contrast with *electrogenic pump*.

Electroreceptor A sensory receptor cell that responds to environmental electrical stimuli (e.g., in weakly electric fish).

Electrotonic conduction See *decremental spread*.

Empirical Based on data rather than merely reasoning.

Endocrine cell A nonneural epithelial cell or a neuron that secretes a hormone or neurohormone.

Endocrine gland A gland or tissue without ducts that secretes a hormone. Nonneural endocrine cells may form glands that are discrete, (with all cells grouped together), or diffuse, (with cells scattered within other tissues), or intermediate between discrete and diffuse.

Endogenous rhythm A rhythmic pattern of activity in space and time that is an intrinsic property of the nervous system and does not require external timing information.

Endothelium The epithelium that lines the heart and the lumen of blood vessels in vertebrates.

Endotherm An animal in which the body temperature is elevated by metabolically produced heat.

Energy The ability to maintain or increase order in a system.

Energy degradation See *degradation of energy*.

Energy metabolism The set of processes by which energy is acquired, transformed, channeled into useful functions, and dissipated by cells or organisms.

Enteric division One of three divisions of the autonomic nervous system; exerts largely autonomous control over the gut.

Entrainment The process of synchronizing an endogenous rhythm to an environmental stimulus.

Environmental See *ambient*.

Enzyme A protein that catalyzes a chemical reaction in which covalent bonds are made or broken.

Enzyme–substrate affinity The proclivity of an enzyme to form a complex with its substrate when the enzyme and substrate meet.

Epicuticle The outermost layer of the exoskeleton of an arthropod, where lipids that protect against desiccation are deposited in insects and arachnids.

Epithelium A sheet of cells that lines a cavity or covers an organ or body surface, thereby forming a boundary between functionally different regions of the body or between an animal and its external environment.

Equilibrium The state toward which an *isolated system* changes; that is, the state toward which a system moves—internally—when it has no inputs or outputs of energy or matter. A system is at equilibrium when internal changes have brought it to an internally stable state from which further net change is impossible without system inputs or outputs. The state of equilibrium is a state of minimal capacity to do work under locally prevailing conditions. See also *electrochemical equilibrium*.

Equilibrium potential The membrane potential at which an ion species is at electrochemical equilibrium, with concentration-diffusion forces offset by electrical forces so that there is no net flux of that ion species across the membrane.

Erythrocyte See *red blood cell*.

Erythropoiesis Production of red blood cells.

Essential amino acid A standard amino acid that an animal cannot synthesize and which thus must be obtained from food, microbial symbionts, or other sources other than animal biosynthesis.

Estivation In mammals and birds, a form of controlled hypothermia in which the body temperature is able to approximate ambient temperature continuously for two or more consecutive days during summer. In other animals, a nonspecific term referring to a resting or dormant condition during the summer or during drought.

Estrogen A feminizing hormone such as estradiol. Estrogens, which may be secreted by the ovary, placenta, testes, and possibly the adrenal cortex, are essential for female secondary sexual characteristics and reproduction.

Estrous cycle The cycle in which oocytes mature and are ovulated periodically in most female mammals.

Estrus A stage of the estrous cycle around the time of ovulation during which a female uses behaviors to indicate that she is fertile. Also called *heat*.

Estuary A body of water along a seacoast that is partially enclosed by land and that receives inputs of both freshwater and seawater, so that it is intermediate in salinity.

Euryhaline Able to live over a wide range of environmental salinities. Contrast with *stenohaline*.

Eurythermal Referring to poikilotherms, able to live over a broad range of body temperatures. Contrast with *stenothermal*.

Evaporation A change in the physical state of a compound from a liquid to a gas; most commonly used to refer to water. Evaporation can in principle occur at any temperature.

Evaporative cooling Removal of heat by the evaporation of water. The evaporative cooling caused by the evaporation of a gram of water is determined by the *latent heat of vaporization* of water per gram.

Evolution A change in gene frequencies over time in a population. Evolution can result in *adaptation*, or it can be *nonadaptive*.

Excess postexercise oxygen consumption (EPOC) An elevation of the actual O_2 uptake by breathing above the theoretical O_2 requirement of rest when an animal is resting immediately after exercise; "breathing hard" after exercise. Also called *oxygen debt*.

Excitable cells Cells that can generate action potentials because their membranes contain voltage-gated channels.

Excitation-contraction coupling Events at the transverse tubules and sarcoplasmic reticulum of a muscle cell that permit electrical activity conducted along the cell membrane to initiate contractile activity.

Excitatory Resulting in an increase in activity or probability of activity. In a neuron, depolarization is excitatory because it increases the likelihood of generation of action potentials. Contrast with *inhibitory*.

Excitatory postsynaptic potential (EPSP) A voltage change in a postsynaptic cell—normally a depolarization—that excites the cell.

Exponential Referring to a type of relationship between two variables in which the dependent variable goes up in multiplicative steps as the independent variable goes up in additive steps. See *Appendix G*.

External respiration Breathing.

External work Mechanical work that involves applying forces to objects outside an animal's body. Locomotion is the principal example; other examples would be processes such as a squirrel chewing into a nut, or a mole pushing soil aside.

Exteroceptor A sensory receptor cell that is activated by stimuli from outside the body.

Extracellular fluids The aqueous solutions outside cells. In animals with closed circulatory systems, subdivided into *blood plasma* and *interstitial fluids*.

Extrafusal muscle fiber An "ordinary" muscle fiber that is not associated with a muscle-spindle stretch receptor.

Extrarenal salt excretion Excretion of inorganic ions by structures other than the kidneys, such as the gills in marine teleost fish and salt glands in marine birds.

Facilitated diffusion Passive transport of a solute across a membrane mediated by the noncovalent and reversible binding of solute molecules to a solute-specific transporter (carrier) protein in the membrane. Facilitated diffusion is the principal mode of passive transport of polar organic solutes, such as glucose and amino acids, across membranes.

Fast-glycolytic (FG) muscle fibers Muscle fibers that are poised to make ATP principally by anaerobic catabolism, develop contractile tension rapidly, have relatively high peak power outputs, and fatigue relatively rapidly.

Fermentation Enzyme-catalyzed reactions that occur without O_2, such as reactions that accomplish the anaerobic breakdown of compounds to liberate energy for metabolic use.

Fick equation An equation used to predict the rate of diffusion. See Equation 3.1.

Field metabolic rate (**FMR**) The *average daily metabolic rate* of an animal when living free in its natural environment.

Filter feeding Feeding on objects suspended in water that are very small by comparison to the feeding animal in which the mechanism of food collection is some sort of sieving. If the mechanism is of another sort or is unknown, the feeding should be termed *suspension feeding*.

Filtrate See *ultrafiltrate*.

Fixed act A simple behavioral response to a stimulus that is all-or-none. Contrast with *reflex*.

Flexion reflex A reflex response that flexes or withdraws a limb from a painful or noxious stimulus.

Fluid compartment A defined subpart of the body fluids, often distributed throughout the body rather than occupying a discrete physical location. The simplest subdivision of the body fluids recognizes three fluid compartments: blood plasma, interstitial fluids, and intracellular fluids.

Fluid mosaic model A theory of the nature of cell membranes. According to this theory, a cell membrane consists of a mosaic of protein and lipid molecules, all of which move about in directions parallel to the membrane faces because of the fluid state of the lipid matrix.

Fluidity In reference to the phospholipids in a cell membrane, the ability of individual phospholipid molecules to diffuse through the population of all such molecules in a membrane leaflet because the molecules are not covalently bonded to one another. Fluidity is quantitatively variable. Diffusion is relatively rapid in some membranes, which are said to exhibit high fluidity.

Follicle In secretory tissues such as the thyroid gland, a globe-shaped, hollow structure enclosed by an epithelium. In the ovary, an oocyte and the layers of somatic cells surrounding it.

Follicular phase The ovarian phase of the menstrual or estrous cycle prior to ovulation, during which primary oocytes, each enclosed in a follicle, develop, mature, and secrete hormones.

Foodstuff A material in the body, such as carbohydrate or lipid, that an animal uses as a source of energy or chemical building blocks. Foodstuffs are often derived directly from foods but differ from foods in being already in the body.

Foregut fermenter An animal that has a specialized foregut chamber housing communities of fermenting microbes that assist with the breakdown of food materials and that often provide biosynthetic capabilities the animal lacks. The microbial communities commonly include bacteria, protists, yeasts, and fungi.

Fossorial Living underground. Used usually to distinguish species that live underground but belong to phylogenetic groups that generally live aboveground.

Free-running rhythm An endogenous rhythm that is not entrained by an environmental stimulus.

Freezing-point depression (**DFP**) The difference (sign ignored) between the freezing point of a solution and the freezing point of pure water. For example, if the freezing point of a solution is –1.6°C, its freezing-point depression is 1.6°C. A colligative property.

G protein A protein involved in signal transduction that is activated by binding with guanosine triphosphate (GTP). Some G proteins occur in cell membranes and are typically trimers; others occur intracellularly and are typically monomers.

G protein–coupled receptor A membrane receptor protein that, when it binds to its specific extracellular signal ligand, relays a signal into the cell by activating G proteins in the cell membrane.

Gamete A reproductive cell, also called a germ cell; an egg (ovum) produced by an ovary or a sperm produced by a testis.

Gametogenesis The formation of haploid eggs or sperm (gametes) through the process of meiosis. It is called specifically *oogenesis* when referring processes in the ovary and *spermatogenesis* in the testis.

Gamma (γ) motor neuron A small motor neuron in a vertebrate spinal cord that innervates an intrafusal muscle fiber.

Ganglion (plural **ganglia**) A discrete collection of neuronal cell bodies. In arthropod nervous systems, most ganglia are segmental components of the central nervous system; in vertebrates, ganglia are components of the peripheral nervous system.

Ganglion cell An output cell of the vertebrate retina, with an axon extending in the optic nerve to visual processing areas of the brain.

Gap junction A region of adjacent cell membranes that includes shared channels (formed by adjoining connexons in vertebrates) that permit cytoplasmic continuity between the cells.

Gas tension See *partial pressure*.

Gated channel A channel that "opens" and "closes" to facilitate or inhibit solute passage. Some gated channels "open" and "close" in response to changes in ligand binding; others do so in response to voltage changes or other changes.

Gene lineage In the study of evolution, a set of genes that are related by descent from a common ancestral gene; also a representation of the family tree of such a set of genes.

Genetic drift Changes in gene frequencies within a population over time resulting primarily from chance.

Germ cell See *gamete*.

Gills In the most general sense used by physiologists, structures specialized for external respiration that project from the body into the ambient medium and are thereby surrounded by the environmental medium. Contrast with *lungs*.

Glial cells Cells in the neural tissue other than neurons. Glial cells are considered support cells, only secondarily involved in signaling and integration. Also called *neuroglia*.

Glomerular filtration rate (**GFR**) The rate at which all the nephrons in the kidneys of a vertebrate produce primary urine by ultrafiltration.

Glomerulus A minute anastomosing cluster of blood capillaries associated with a nephron in the kidney of a vertebrate, serving as the site of formation of primary urine by ultrafiltration. The term is also sometimes used to refer not only to such a cluster of capillaries, but also the Bowman's capsule with which it is associated.

Glucocorticoids Steroid hormones, such as cortisol and corticosterone, that are released from the cortex of the adrenal gland and regulate carbohydrate, lipid, and protein metabolism.

Glycolysis The reactions that convert glucose to pyruvic acid.

Goldman equation An equation that describes membrane potential in terms of the concentrations of and membrane permeabilities to more than one ion species.

Gonadotropin A hormone that stimulates the gonads (ovaries or testes) to produce gametes and secrete hormones, and also supports and maintains the gonadal tissue.

Graded potential A voltage change that is variable in amplitude, rather than all-or-none (like an action potential). Examples include synaptic potentials and receptor potentials.

Granular cells See *juxtaglomerular cells*.

Granulosa cells Somatic cells surrounding the primary oocyte of an ovarian follicle.

Gray matter A histological region of a vertebrate central nervous system that contains neuronal cell bodies, dendrites, and synapses as well as axons.

Green gland See *antennal gland.*

Gross growth efficiency In a growing animal, the chemical-bond energy of new biomass added by growth expressed as a ratio of the animal's ingested energy over the same time period.

Gular fluttering Rapid up-and-down oscillation of the floor of the mouth cavity of a bird or reptile to enhance the rate of evaporative cooling by increasing air flow over moist membranes in the mouth.

H zone A region at the center of the sarcomere that contains only thick filaments; shortens during contraction. Also called *H band.*

Habituation A simple, nonassociative form of learning; a learned decrease in a behavioral response with repeated presentations of a nonthreatening stimulus.

Hagen-Poiseuille equation See *Poiseuille equation.*

Hair cell A sensory epithelial cell in a vertebrate acoustico-lateralis system that transduces displacement of its apical stereocilia into an electrical signal.

Haldane effect A shift of the carbon dioxide equilibrium curve of the blood caused by the oxygenation and deoxygenation of the blood respiratory pigment. In blood that exhibits a Haldane effect, the total carbon dioxide concentration is higher when the blood is deoxygenated than when it is oxygenated.

Half-life The time required to reduce something by one-half; applies to the concentration of substances such as hormones or to radioactivity.

Halophyte A plant that roots in saline soils, often characterized by high salt concentrations in its tissue fluids.

Heart A discrete, localized structure specialized for pumping blood. It may be neurogenic or myogenic and associated with an open or a closed circulatory system.

Heat The energy that matter possesses by virtue of the ceaseless, random motions that all of the atoms and molecules of which it is composed undergo on an atomic-molecular scale of distance. Also called *molecular kinetic energy.*

Heat-shock proteins An evolutionarily related group of proteins that are principally inducible—being expressed in the aftermath of heat stress or other stress—and that function as molecular chaperones. Heat-shock proteins use ATP to assist in the repair of stress-damaged proteins by preventing those proteins from aggregating with one another and by promoting molecular folding patterns that restore them to correct three-dimensional conformations.

Hemerythrin A type of respiratory pigment found in scattered groups of animals in four phyla, consisting of an iron-based metalloprotein. Undergoes reversible combination with O_2 at iron-containing loci.

Hemimetabolous insect An insect that undergoes gradual metamorphosis. Its juvenile stages (nymphs) appear similar to the adult, except that they are smaller and have immature wings and external genitalia. Contrast with *holometabolous insect.*

Hemocyanin A type of respiratory pigment found in arthropods and molluscs, consisting of a copper-based metalloprotein. Undergoes reversible combination with O_2 at copper-containing loci.

Hemoglobin A type of respiratory pigment formed by the combination of heme (ferrous protoporphyrin IX) with a globin protein. Undergoes reversible combination with O_2 at the heme loci.

Hemolymph A synonym for *blood* in an animal that has an open circulatory system. The term emphasizes that the blood in such animals includes all extracellular fluids, and thus that there is no distinction between the fluid that is in the blood vessels at any one time and the interstitial fluid between tissue cells.

Henry's law A law that relates partial pressure and concentration for gases dissolved in an aqueous solution (or in another liquid solution). One way to express this law is that $C = AP$, where C is dissolved concentration, P is partial pressure in atmospheres, and A is the *absorption coefficient* (dissolved concentration when partial pressure is 1 atm).

Hepatopancreas An organ in crustaceans that connects with the stomach and functions in secretion of digestive enzymes; absorption of nutrient molecules; storage of lipids, glycogen, and Ca^{2+}; and sequestration of foreign compounds.

Hermaphrodite An individual that possesses both ovaries and testes. An animal may be a simultaneous hermaphrodite, in which both types of gonads produce gametes at the same time, or a sequential hermaphrodite, in which only one type of gonad produces gametes at a time.

Heterothermy The property of exhibiting different thermal relations from time to time or place to place. There are two types of heterothermy: In *regional heterothermy*, some regions of an individual animal's body exhibit different thermal relations than other regions at the same time. In *temporal heterothermy*, an individual exhibits one type of thermal relation at certain times and another type of thermal relation at other times (e.g., hibernation at some times and homeothermy at others).

Heterotroph An organism that obtains the energy it needs to stay alive by breaking up organic compounds that it obtains from other organisms, thereby releasing the chemical-bond energy of those organic compounds. Contrast with *autotroph.*

Hibernaculum The place where an animal resides while in hibernation.

Hibernation In mammals and birds, a form of controlled hypothermia in which the body temperature is able to approximate ambient temperature continuously for two or more consecutive days during winter. In other animals, a nonspecific term referring to a resting or dormant condition during winter.

High-grade energy Energy in a form that can do physiological work. Chemical, electrical, and mechanical energy are high-grade forms of energy.

Hill coefficient A measure of the degree of cooperatively among O_2-binding sites in a molecule of a respiratory pigment.

Hindgut fermenter An animal that has a specialized hindgut chamber housing communities of fermenting microbes that assist with the breakdown of food materials and that often provide biosynthetic capabilities the animal lacks.

Hodgkin cycle The cycle that explains the rising phase of an action potential: Depolarization opens voltage-gated Na^+ channels, increasing membrane permeability to Na^+. The resulting inflow of Na^+ further depolarizes the membrane, opening more Na^+ channels.

Holometabolous insect An insect that undergoes complete metamorphosis, in which the larval and pupal stages do not resemble the adult. Contrast with *hemimetabolous.*

Homeosmotic animal Synonym for *osmoregulator.*

Homeostasis Internal constancy and the physiological regulatory systems that automatically make adjustments to maintain it. In the words of Walter Cannon, who coined the term, "the coordinated physiological processes which maintain most of the [constant] states in the organism."

Homeotherm An animal that thermoregulates by physiological means (rather than simply by behavior).

Homeoviscous adaptation The maintenance of a relatively constant lipid fluidity regardless of tissue temperature.

Homing The ability of an animal to return to its home site after being displaced.

Homologous Relating to features of organisms that are similar because of common evolutionary descent.

Horizontal cell A neuron in the vertebrate retina that is part of the lateral pathway, mediating center–surround antagonistic effects in retinal neuron receptive fields.

Hormone A chemical substance, released by nonneural endocrine cells or by neurons, that is carried in the blood to distant target cells, where it exerts regulatory influences on their function. There are three main chemical classes of hormones: steroids, peptides or proteins, and amines.

Hormonal axis A hormonal sequence in which one hormone stimulates the secretion of a second hormone that may in turn stimulate secretion of a third hormone. Known hormonal axes consist of two or three hormones in sequence.

Hourglass timer See *interval timer*.

Humidic Restricted to humid, water-rich terrestrial microenvironments; unable to live steadily in the open air.

Humidity A general term referring loosely to the concentration of gaseous water in a gas. See also *water vapor pressure*, *relative humidity*, and *saturation deficit*—all of which represent ways to express humidity.

Hydrophilic Dissolving readily in water ("water loving").

Hydrophobic Not dissolving readily in water ("water hating"); typically lipid-soluble.

Hydrostatic pressure The sort of pressure that is developed in a fluid (gas or liquid) when forces are applied that tend to increase the amount of matter per unit of volume. Hydrostatic pressure is what is meant by everyday scientific uses of the word *pressure*; A tire pump, for instance, produces a hydrostatic pressure in a tire.

Hyperbolic kinetics A type of saturation kinetics in which the velocity of a chemical reaction increases in a smooth, strictly asymptotic way toward its maximum.

Hyper-hyposmotic regulator An aquatic animal that maintains a blood osmotic pressure that is higher than the osmotic pressure of the water in which it lives when the ambient osmotic pressure is low, but lower than the osmotic pressure of the water in which it lives when the ambient osmotic pressure is high. Such an animal may have a stable blood osmotic pressure over a wide range of ambient osmotic pressures.

Hyper-isosmotic regulator An aquatic animal that maintains a blood osmotic pressure that is higher than the osmotic pressure of the water in which it lives when the ambient osmotic pressure is low, but the same as the osmotic pressure of the water in which it lives when the ambient osmotic pressure is high.

Hyperosmotic Having a higher osmotic pressure. Said of a solution in comparison to another, specific solution. *Hyperosmotic* is a relative term that is meaningless unless the comparison solution is specified.

Hyperosmotic regulator An aquatic animal that maintains a blood osmotic pressure higher than the osmotic pressure of the water in which it lives.

Hyperpolarization A voltage change that makes a cell membrane potential more inside-negative (normally moves it further from zero).

Hyperthermia The state of having a body temperature that is higher than the temperature considered to be normal or usual.

Hypertrophy The addition of structural components to cells in a way that increases the size of a tissue or organ. In muscle, the addition of contractile proteins to skeletal muscle fibers.

Hyposmotic Having a lower osmotic pressure. Said of a solution in comparison to another, specific solution. *Hyposmotic* is a relative term that is meaningless unless the comparison solution is specified. The word is a contraction of *hypo-osmotic*.

Hyposmotic regulator An aquatic animal that maintains a blood osmotic pressure lower than the osmotic pressure of the water in which it lives.

Hypothalamo-hypophyseal portal system A system of blood vessels that connects capillaries in the hypothalamus to capillaries in the anterior pituitary, providing a direct pathway by which hypothalamic hormones can reach specific populations of cells in the adenohypophysis (anterior pituitary).

Hypothermia The state of having a body temperature that is lower than the temperature considered to be normal or usual.

Hypoxia Referring to the tissues of an animal, the state of having an unusually low level of O_2.

I band A region of two adjacent sarcomeres that includes only thin filaments and is bisected by the Z disc; shortens during contraction.

Ice-nucleating agent A dissolved or undissolved substance that promotes freezing (i.e., limits supercooling).

Ideal gas law See *universal gas law*.

Imidazole group A type of chemical group, found on the amino acid histidine and some other compounds, that is often extremely important in buffering body fluids because of its particular chemical buffering properties.

Indirect calorimetry Measurement of metabolic rates by quantifying respiratory gas exchange or some other property besides heat and external work.

Indirect measurement A measurement procedure that quantifies a property by measuring something other than what the definition of the property specifies. Contrast with *direct measurement*.

Inducible enzyme An enzyme (or other protein) that is expressed only when "induced" by the presence of a molecule that serves as an inducing agent. An inducible enzyme disappears from a cell or tissue when its inducing agent is absent, but is expressed when the cell or tissue is exposed to its inducing agent. Contrast with *constitutive enzyme*.

Ingested energy The energy present in the chemical bonds of an animal's food.

Inhibitory Resulting in a decrease in activity or probability of activity. In a neuron, hyperpolarization is inhibitory because it decreases the likelihood of generation of action potentials. Contrast with *excitatory*.

Inhibitory postsynaptic potential (IPSP) A voltage change in a postsynaptic cell—normally a hyperpolarization—that inhibits the cell.

Inka cells Endocrine cells on the tracheae of insects that secrete two peptides: pre-ecdysis triggering hormone and ecdysis triggering hormone.

Innate behavior Behavior that has a strong genetic basis or results from genetic preprogramming.

Inner ear The cochlea and the canals of the vestibular system.

Inner hair cells In the organ of Corti of the cochlea, a single row of hair cells that transduce sound vibrations into voltage changes that excite auditory sensory neurons.

Innervate To provide neural input.

Inorganic ion A non-carbon atom (e.g., Na^+) or group of atoms (e.g., SO_4^{-2}) that bears a net negative or positive charge.

Insensible water loss In humans, water loss across the skin without sweating; a synonym for *transpirational water loss* in people.

Insertion The movement of channel or transporter proteins from inactive intracellular locations into the membrane where they are active.

Instar A period between ecdyses, or molts, in the arthropod life cycle.

Insulation The resistance to dry heat transfer through a material or between an animal and its environment. See also *resistance to heat transfer*.

Integral membrane protein A protein that is part of a cell membrane and cannot be removed without extraction procedures that take the membrane apart.

Integration The summing and coordination of input signals to provide a harmonious control of output. Cellular integration refers to the integration of signals within a cell, and physiological integration refers to the integration of sensory, central nervous system, and endocrine signals for harmonious control of effectors in the body.

Integument The outer body covering of an animal, such as the skin of a vertebrate or the exoskeleton of an arthropod.

Intercalated disc An intercellular contact between adjacent cardiac muscle fibers of vertebrates that contains desmosomes and gap junctions.

Intercellular fluids See *interstitial fluids*.

Intercostal muscles Sheetlike muscles that run between adjacent ribs, the contraction of which expands or contracts the volume of the rib cage.

Intermittent breathing Breathing in which breaths or sets of breaths are regularly interrupted by extended periods of apnea.

Internal environment The set of conditions—temperature, pH, Na^+ concentration, and so forth—existing in the body fluids of an animal and therefore experienced by cells within an animal's body. Also called the *milieu intérieur*.

Internal work Mechanical work performed inside the body, such as the beating of the heart or peristalsis in the gut.

Interneuron A neuron that is confined to the central nervous system; and is therefore neither a sensory neuron nor a motor neuron.

Internode The region of a myelinated axon that lies between two nodes of Ranvier and is covered by a myelin sheath.

Interoceptor A sensory receptor cell that is activated by stimuli within the body and thus monitors some aspect of the internal state.

Interspecific enzyme homologs Different molecular forms of a single enzyme occurring in two or more related species; called *homologs* because they are assumed to be related by evolutionary decent.

Interstitial cells See *Leydig cells*.

Interstitial fluids The fluids between cells in tissues. More specifically, in animals with closed circulatory systems, the fluids between cells in tissues other than blood; that is, the interstitial fluids are the extracellular fluids other than the blood plasma. Sometimes called *intercellular fluids* or *tissue fluids*.

Interval timer A biological clock that times an interval shorter than a day but appears to be noncyclic, having to be restarted each time, like a kitchen timer or an hourglass. Also called an *hourglass timer*.

Intracellular fluids The aqueous solutions inside cells. Also called the *cytosol*.

Intracrine A peptide signaling molecule that regulates intracellular functions and is known to act as a hormone, paracrine, or autocrine in different contexts. It may be retained within the cell that synthesized it or internalized from the extracellular space.

Intrafusal muscle fiber A specialized muscle fiber associated with a muscle spindle stretch receptor. Muscle stretching or activation of the intrafusal muscle fiber by a gamma motor neuron can activate the stretch receptor.

Intraluminal digestive enzyme A digestive enzyme that is secreted into the lumen of the gut or another body cavity, where it mixes with food materials and digests them.

Inulin A polysaccharide widely used in studies of plasma clearance. See *plasma clearance*.

Ion An atom or group of atoms that bears a net negative or positive charge.

Ionic conformity A state in which the concentration of an inorganic ion in the blood plasma matches, and varies with, the concentration of that ion in the external environment.

Ionic regulation The maintenance of a constant or nearly constant concentration of an inorganic ion in the blood plasma regardless of the concentration of that ion in the external environment.

Ionotropic receptor A neurotransmitter receptor molecule that changes the membrane permeability of the postsynaptic cell to particular ions when it binds neurotransmitter molecules; usually a *ligand-gated channel*.

Islets of Langerhans Clusters of endocrine cells distributed among the exocrine-gland tissue of the pancreas. They contain specific endocrine cells that secrete specific hormones: B cells secrete insulin, A cells glucagon, and D cells somatostatin.

Isoform In relation to a protein that exists on multiple molecular forms, any one of those molecular forms.

Isolated system A defined part of the material universe that (at least as a thought exercise) cannot exchange either matter or energy with its surroundings.

Isometric contraction A contraction in which a muscle does not shorten significantly as it exerts force (tension) against a load it cannot move.

Isosmotic Having the same osmotic pressure. Said of a solution in comparison to another, specific solution. *Isosmotic* is a relative term that is meaningless unless the comparison solution is specified.

Isotonic contraction A contraction in which a muscle changes its length as it exerts force (tension) against a load. The tension during contraction remains constant and equal to that of the load.

Isozymes Various molecular forms of a single kind of enzyme synthesized by one species. Also called *isoenzymes*.

Juvenile hormone In insects, a lipid-soluble hormone secreted by the nonneural endocrine cells of the corpora allata that stimulates its target tissues to maintain immature characteristics.

Juxtaglomerular cells Specialized smooth muscle cells in the arterioles associated with the glomerulus of the vertebrate nephron that secrete renin in response to low blood pressure. Also called *granular cells*.

Kairomone A chemical signal released by a member of one species in the course of its activities that is detected and exploited by a member of another species. Its detection confers a benefit on the detecting species that is detrimental to the releasing species.

k_{cat} See *catalytic rate constant*.

Kidney An organ that regulates of the composition and volume of the blood and other extracellular body fluids by producing and eliminating from the body an aqueous solution (*urine*) derived from the blood or other extracellular fluids.

Knockout animal An animal that has been genetically engineered to lack functional copies of a gene of interest.

Labeled lines The principle that sensory receptors encode the modality or quality of a sensory stimulus by having different sensory receptor cells respond to different kinds of stimuli, so that the CNS can decode the stimulus by monitoring which axons ("lines") produce action potentials.

Laminar flow Flow of a fluid without turbulence. When a fluid flows without turbulence over a surface or through a tube, the flow can be envisioned as occurring in a series of thin layers (laminae) of fluid at progressively greater distances from the surface or tube walls. The layer juxtaposed to the surface or tube wall does not flow at all, and layers at increasing distances from it flow faster and faster.

Laplace's law The tension developed within the walls of a hollow structure exposed to a particular difference in pressure between inside and outside is directly related to the radius of the structure.

Larva A free-living developmental stage that is very different in appearance from the adult stage.

Latch state A characteristic state of smooth muscle in which the dephosphorylated myosin head remains attached to actin and maintains tension for long periods of time.

Latent heat of vaporization The heat that must be provided to convert a material from a liquid to a gas at constant temperature (called *latent* because although heat is provided, the temperature of the material does not rise). It is expressed per unit of mass of the material under study.

Lateral geniculate nucleus (**LGN**) A region of the thalamus in the diencephalon of the vertebrate brain that receives axons of retinal ganglion cells and relays visual information to the primary visual cortex.

Leaflet One of the two phospholipid layers within a cell membrane.

Leak channel A channel in a cell membrane that is normally open and mediates the membrane permeability (primarily to K^+ ions) that underlies the resting membrane potential.

Left-to-right shunting In the study of blood flow through the central circulation of an air-breathing fish, amphibian, or reptile, diversion of pulmonary venous blood directly back to the lungs, bypassing the systemic circuit.

Length constant The distance along a cell over which a change in passively spreading electrical potential decays in amplitude to 37% of its amplitude at the origin.

Length–tension relationship The relationship between the length of a whole muscle or a sarcomere prior to stimulation and the tension developed during isometric contraction. Maximum tension develops when there is optimal overlap of thick and thin filaments to allow effective cross-bridge action.

Lengthening contraction A contraction in which the muscle lengthens as it is activated and exerts force.

Leydig cells Testosterone-secreting cells located in the connective tissue between the seminiferous tubules of the testis; also called *interstitial cells.*

Ligand Any molecule that selectively binds noncovalently to a structurally and chemically complementary site on a specific protein. The substrate of an enzyme, for example, is a ligand of that enzyme.

Ligand-gated channel A gated channel that opens to allow diffusion of a solute as a result of binding by a neurotransmitter (or other specific signaling molecule) to a receptor site on the channel protein.

Linear heat transfer equation An equation that relates heat loss from an animal to the difference between body temperature and ambient temperature by means of a proportionality coefficient. See Equation 8.9.

Load The force against which a contracting muscle exerts an opposing force, referred to as tension.

Load–velocity relationship The principle that velocity of shortening during isotonic contraction decreases as load increases.

Loading Oxygenation of a respiratory pigment.

Long-term potentiation (**LTP**) A stable, long-lasting increase in the amplitude of the response of a neuron after it has been stimulated repeatedly by presynaptic input at a high frequency.

Loop of Henle A portion of a nephron tubule shaped like a hairpin; found in the nephrons of mammals and some of the nephrons of birds, that is the site of countercurrent multiplication, the process responsible for production of urine hyperosmotic to the blood plasma.

Low-grade energy Energy in a form that cannot do any physiological work (i.e., heat).

Lower critical temperature In a homeotherm, the ambient temperature that represents the lower limit of the thermoneutral zone. See also *thermoneutral zone.*

Lumen The open central cavity or core of a "hollow" organ or tissue; for example, the open central core of a blood vessel.

Luminal Related to the lumen.

Lungfish See *dipnoan.*

Lungs In the most general sense used by physiologists, structures specialized for external respiration that are invaginated into the body and thereby contain the environmental medium. Contrast with *gills.*

Luteal phase The ovarian phase of the menstrual or estrous cycle that follows ovulation, during which the corpus luteum is functional and then degenerates.

Lymph In an animal with a closed circulatory system, the interstitial fluids. *Lymph* is often used in a more restrictive sense to refer specifically to excess volumes of interstitial fluid that are collected from the tissues and returned to the blood by the lymphatic vascular system.

Lymphatic vascular system An elaborate system of vessels that approximately parallels the blood vascular system in vertebrates and serves to remove excess interstitial fluids (lymph) from tissues throughout the body, returning the fluids ultimately to the blood plasma.

M line In muscle fibers, a web of accessory proteins at the center of the sarcomere that anchors the thick filaments and titin.

Magnetic compass A mechanism by which an animal uses Earth's magnetic field to determine compass direction in navigation.

Magnetoreceptor A sensory receptor cell that is stimulated by Earth's magnetic field.

Malpighian tubules Fine tubules that initiate urine formation in insects. They empty their product into the gut at the junction of the midgut and hindgut.

Mantle cavity In molluscs, an external body cavity formed where a sheetlike outfolding of the dorsal body wall, the *mantle*, overhangs or surrounds all or part of the rest of the body. The gills typically are suspended in the mantle cavity.

Map-and-compass navigation Navigation in which an animal has information about where it is in the world (the "map") and about the direction it should take to reach its destination (the "compass").

Mass action The inherent tendency for the reactants and products of a chemical reaction to shift in their concentrations, by way of the reaction, until an equilibrium state—defined by particular concentration ratios—is achieved. Mass action impels reactions toward equilibrium, although other processes may operate against equilibrium and therefore prevent an equilibrium from actually being established.

Maximal aerobic power A synonym for $\dot{V}_{O_2max}$.

Maximal exercise Exercise that requires an individual's maximal rate of O_2 consumption.

Maximum reaction velocity (V_{max}) The greatest rate at which an enzyme-catalyzed reaction can convert substrate to product with the number of active enzyme molecules that are present.

Mechanical energy Energy of *organized* motion in which many molecules move simultaneously in the same direction. The energy of motion of a moving arm provides an example. Mechanical energy and heat are the two forms of kinetic energy.

Mechano growth factor A locally acting peptide, produced by muscle fibers when they are stimulated to produce mechanical activity, that stimulates muscle satellite cells to proliferate and become incorporated into the muscle fiber.

Membrane-associated digestive enzyme A digestive enzyme that is positioned in the apical membranes of epithelial cells lining the gut or another body cavity. Food materials in the gut lumen must make physical contact with the gut epithelium to be digested by this sort of enzyme.

Membrane fluidity See *fluidity*.

Membrane potential The potential difference (voltage) across a cell membrane or other selectively permeable membrane.

Menstrual cycle The cycle in which oocytes mature and are ovulated periodically in some female primates; characterized by menstruation, the shedding of the uterine lining in a blood-tinged discharge from the vagina.

Mesic Moderately moist. "Mesic" animals are terrestrial animals that live in moderately moist environments.

Metabolic acidosis or **alkalosis** Acidosis or alkalosis that is caused by abnormal excretion or retention of bicarbonate (HCO_3^-) in the body fluids. Contrast with *respiratory acidosis* or *alkalosis*.

Metabolic depression A reduction in the ATP needs of an animal (or a specific tissue) to below the level ordinarily associated with rest in a way that does not present an immediate physiological threat to life.

Metabolic rate An animal's rate of energy consumption; the rate at which it converts chemical-bond energy to heat and external work.

Metabolic water Water that is formed by chemical reaction within the body. For example, when glucose is oxidized, one of the products is H_2O that did not previously exist. Also called *oxidation water*. Contrast with *preformed water*. See also *net metabolic water production*.

Metabolism The set of processes by which cells and organisms acquire, rearrange, and void commodities in ways that sustain life.

Metabotropic receptor A neurotransmitter receptor that alters a metabolic function of the postsynaptic cell, often by stimulating production of a second messenger.

Metamorphosis The process of changing from one form to another, such as changing from a juvenile form to an adult form.

Metalloprotein A protein that includes one or more metal ions in its structure.

Michaelis constant (K_m) The half-saturation constant of an enzyme-catalyzed reaction that exhibits hyperbolic kinetics—that is, the concentration of substrate at which the reaction velocity is half of the maximal velocity. This constant is a measure of enzyme–substrate affinity.

Michaelis-Menten equation An equation that describes the relation between reaction velocity and substrate concentration in an enzyme-catalyzed reaction that exhibits hyperbolic kinetics. See Equation 2.2

Microcirculatory bed In a closed circulatory system, one of the systems of microscopically fine blood vessels that connect arteries and veins, weaving among ordinary tissue cells. In vertebrate systemic tissues, microcirculatory beds consist of arterioles, capillaries, and venules.

Microclimate The set of climatic conditions (temperature, humidity, wind speed, and so forth) prevailing in a subpart of a larger environmental system.

Microenvironment A place within a larger environment in which the physical and chemical conditions differ significantly from the average conditions characterizing the larger environment.

Microvilli Microscopic finger-shaped projections from the apical membranes of certain types of epithelial cells. Microvilli occur most commonly in epithelia that are active in secretion or reabsorption.

Middle ear The portion of the vertebrate ear between the tympanic membrane and the cochlea.

Migration A seasonal or other periodic movement of animals from one geographic region to another.

Milieu intérieur See *internal environment*.

Miniature EPSP (**mEPSP**) A small excitatory postsynaptic potential at a neuromuscular junction or postsynaptic neuron produced by presynaptic release of a single quantal packet of neurotransmitter.

Minute volume The amount of air, water, or blood pumped per minute in a breathing system or circulatory system.

Mitochondria-rich cell See *chloride cell*.

Mixed venous blood The blood in the great veins leading back to the heart, formed by mixing of the venous blood coming from the various regions and organs of the body.

Modality The subjective sensation of a particular sense, such as sight, taste, or hearing.

Molal A unit of measure of chemical concentration; specifically, the molal concentration is the number of moles mixed with a kilogram of water to make a solution.

Molar A unit of measure of chemical concentration; specifically, the molar concentration is the number of moles dissolved in a liter of solution.

Mole (**mol**) A set of 6.022×10^{23} items (i.e., an Avogadro's number of items). For example, a mole of a chemical compound contains 6.022×10^{23} molecules of the compound. The mass in grams of a mole of a chemical compound is identical to the formula mass of the compound. Thus, if you have 18 g of water, a compound that has a formula mass of 18, you have one mole of water, or 6.022×10^{23} water molecules. See also *Avogadro's number, osmole*.

Molecular chaperones Proteins that use ATP to guide the folding of other proteins into correct three-dimensional configurations. Molecular chaperones ensure that target proteins assume the correct tertiary structures during initial synthesis and sometimes are able to guide reversibly denatured proteins back to functional conformations, thereby preventing permanent denaturation.

Molecular kinetic energy See *heat*.

Molting See *ecdysis*.

Molting hormone See *ecdysone*.

Monounsaturated fatty acid See *unsaturated fatty acid*.

Motility Any sort of muscular activity by the gut, such as peristalsis or segmentation.

Motor neuron A neuron that conveys motor signals from the central nervous system to the periphery to control an effector such as skeletal muscle.

Motor unit A motor neuron and all the muscle fibers it innervates.

Mucosal surface See *apical surface*.

Müllerian inhibitory substance A glycoprotein hormone secreted by Sertoli cells in the developing testes of genetic males that causes degeneration of the Müllerian ducts, which would otherwise give rise to the uterus and oviducts. Also called *anti-Müllerian hormone*.

Multiterminal innervation A pattern of innervation in which a single axon branches near its end to make many synaptic contacts along the length of a muscle fiber.

Muscle A group of muscle cells and associated tissues. It may be smooth, skeletal, or cardiac.

Muscle fiber A term typically applied to skeletal or cardiac muscle cells, but sometimes also applied to smooth muscle cells. This term was initially used to emphasize the long, fibrous appearance of certain muscle cells, such as vertebrate skeletal muscle cells (fibers), that are multinucleate because they are formed during development by the fusion of two or more embryonic myoblasts.

Muscle spindle A stretch receptor that is arranged in parallel with the tension producing fibers of a muscle and sends action potentials to the CNS when the muscle is stretched.

Myelin An insulating sheath around an axon, composed of multiple wrappings of glial cell membranes, that increases the velocity of propagation of action potentials.

Myocardium The muscle tissue of a heart.

Myofibril A longitudinal component of a striated muscle cell that consists of a series of sarcomeres and extends the length of the cell. Bundles of myofibrils, each surrounded by a sleeve of sarcoplasmic reticulum, constitute the girth of the muscle fiber.

Myofilament One of two types of longitudinal components of sarcomeres. Thick myofilaments consist of polymerized myosin molecules and thin myofilaments consist of polymerized actin molecules.

Myogenic heart A heart in which the electrical impulse to contract during each beating cycle originates in muscle cells or modified muscle cells. Contrast with *neurogenic heart*.

Myoglobin Any type of hemoglobin found in the cytoplasm of muscle.

Myometrium The layer of the uterus made of smooth muscle.

Myosin A molecular motor found in many types of cells that converts chemical energy of ATP into mechanical energy of motion. In muscle cells, it functions as a contractile protein. Myosin monomers polymerize to form thick myofilaments.

Natriuretic Promoting loss of sodium in the urine.

Navigation The act of moving on a particular course or toward a specific destination, using sensory cues to determine direction and position.

Nebulin In muscle fibers, a large inelastic protein that extends along the thin filament from the Z disc to the margin of the H zone.

Negative feedback A process by which the deviations of a property from a specific set-point level are opposed, thereby tending to keep the property at the set-point level.

Nephridium A term used to refer to the kidneys in certain types of animals such as molluscs.

Nephron One of the tubules that forms urine in the kidney of a vertebrate.

Nernst equation An equation used to determine the equilibrium potential for a particular ion species, given its concentrations on both sides of a membrane.

Nerve A collection of axons in a peripheral nervous system.

Nerve impulse See *action potential*.

Nerve net A simple, uncentralized, and unpolarized network of neurons, found in cnidarians and locally in many other groups and considered to be a primitive stage in the evolution of nervous systems.

Nervous system An organized constellation of neurons and glial cells specialized for repeated conduction of electrical signals (action potentials) within and between cells. These signals pass from sensory receptors and neurons to other neurons and effectors. Nervous systems integrate the signals of convergent neurons, generate new signals, and modify the properties of neurons based on their interactions.

Net growth efficiency In a growing animal, the chemical-bond energy of new biomass added by growth expressed as a ratio of the animal's absorbed energy over the same time period.

Net metabolic water production The production of metabolic water by a process minus the losses of water that are obligatory for that process to take place.

Network oscillator A neural circuit that acts as a central pattern generator as a result of the synaptic interaction of the neurons, rather than as a result of cellular oscillator activity.

Neurogenic heart A heart in which the electrical impulse to contract during each beating cycle originates in neurons. Contrast with *myogenic heart*.

Neuroglia See *glial cells*.

Neurohemal organ An organ made up of axon terminals of neurosecretory cells in association with a well-developed bed of capillaries or other circulatory specializations, in which the axon terminals store neurohormones and secrete them into the blood.

Neurohormone A hormone secreted by a neuron (also called a neuroendocrine or neurosecretory cell) into the blood.

Neurohypophysis The neuroendocrine portion of the vertebrate pituitary gland from which neurohormones are released from axon terminals. It is an extension of the brain, commonly called the posterior pituitary. See *posterior pituitary*.

Neuron A nerve cell; the fundamental signaling unit of the nervous system, composed of a cell body and elongated processes—dendrites and axon—that carry electrical signals.

Neuron doctrine The theory that the nervous system, like other organ systems, is composed of discrete cellular elements (neurons) that are its fundamental signaling elements.

Neuronal integration The process by which a postsynaptic neuron sums the inputs from several presynaptic neurons to control its generation of action potentials.

Neurosecretory cell A neuron that synthesizes and releases hormones.

Neurotransmitter A molecule that is used as a chemical signal in synaptic transmission.

Neutral pH The pH of pure water. The neutral pH varies with temperature; at any given temperature, pH values less that the neutral pH represent an acid solution, whereas pH values above the neutral pH represent an alkaline solution.

Newton's law of cooling A common, although not necessarily historically defensible, synonym for the *linear heat transfer equation*.

Nocturnal Active in the nighttime.

Node of Ranvier An interruption in the myelin sheath around an axon that allows active propagation of an action potential by saltatory conduction.

Nonadaptive evolution Evolution that occurs by processes other than natural selection and which therefore can produce traits that are not adaptations. See also *evolution*, *genetic drift*, *pleiotropy*.

Noncovalent bond A chemical bond that does not involve covalent bonding. Because they are not covalent, noncovalent bonds are flexible rather than rigid, and they can be made and broken with relative ease, without enzyme catalysis. There are four types: hydrogen bonds, ionic bonds, van der Waals interactions, and hydrophobic bonds. Also called *weak bonds.*

Nonpolar molecule A molecule in which electrons are evenly distributed, the various regions of which are all therefore similar in charge.

Nonshivering thermogenesis (NST) In mammals and some birds, elevation of heat production for thermoregulation by means other than shivering. The same mechanisms potentially function in body weight regulation by serving to get rid of excess organic food molecules in the form of heat and inorganic chemical products.

Nonspiking neuron A neuron that transmits information without generating action potentials.

Nucleating agent See *ice-nucleating agent.*

Nucleus (plural **nuclei**) A cluster of functionally related neuronal cell bodies in a vertebrate central nervous system.

Nutrition The study of the chemical components of animal bodies and how animals are able to synthesize those chemical components from the chemical materials they collect from their environments.

Olfaction The sense of smell; chemoreception of molecules released at a distance away from the animal.

Omega 3 and **omega 6 fatty acids** Fatty acids characterized by double bonds at particular points in their molecular structures. If the first double bond encountered when scanning a fatty acid molecule from its methyl ($-CH_3$) end is the third bond in the carbon-chain backbone of the molecule, the fatty acid is an omega 3 fatty acid. If the first double bond occurs at the sixth position, the fatty acid is of the omega 6 type.

Oncotic pressure Synonym for *colloid osmotic pressure.*

Oogenesis The formation of haploid eggs through the process of meiosis in the ovary.

Oogonia (singular **oogonium**) Diploid cells in the vertebrate ovary that proliferate by mitosis, then go through meiosis to give rise to gametes (ova). In elasmobranchs, birds, and most mammals, oogonia proliferate only during fetal life and remain suspended in an early stage of meiosis until the female becomes reproductively mature.

Open circulatory system A circulatory system in which the blood leaves discrete vessels and bathes at least some nonvascular tissues directly, meaning that blood and interstitial fluid are the same. Contrast with *closed circulatory system.*

Open system A defined part of the material universe that is not isolated, meaning that it is capable of exchanging matter, energy, or both with its surroundings.

Opsin The protein part of the photopigment rhodopsin, which is a G protein–coupled receptor molecule.

Orientation The way an organism positions itself in relation to environmental cues.

Ornithine–urea cycle A cyclic metabolic pathway that produces the nitrogenous end product urea.

Osmoconformer An aquatic animal that allows its blood osmotic pressure to match and vary with the environmental osmotic pressure. Also called a *poikilosmotic animal.*

Osmoconformity See *osmotic conformity.*

Osmolar (*Osm*) A unit of measure of osmotic pressure. A 1-osmolar solution is defined to be a solution that behaves osmotically as if it has one Avogadro's number of independent dissolved entities per liter.

Osmole An Avogadro's number of osmotically effective dissolved entities; a set of 6.022×10^{23} osmotically effective dissolved entities.

Osmolyte Any solute that exerts a sufficiently large effect on the osmotic pressure of a body fluid to be of consequence for understanding water–salt physiology.

Osmoregulation See *osmotic regulation.*

Osmoregulator An animal that maintains an approximately constant blood osmotic pressure even as the osmotic pressure of its environment varies.

Osmosis The passive transport of water across a membrane.

Osmotic conformity A state in which the osmotic pressure of the body fluids matches, and varies with, the osmotic pressure in the external environment.

Osmotic effector An osmolyte that an individual animal or individual cell increases or decreases in amount to achieve the regulation of the osmotic pressure of a body fluid or of cell volume. Also called an *osmoticum.*

Osmotic pressure The property of a solution that allows one to predict whether the solution will gain or lose water by osmosis when it undergoes exchange with another solution; osmosis is always from lower osmotic pressure to higher osmotic pressure. An alternative, measurement-oriented definition is that the osmotic pressure of a solution is the difference in hydrostatic pressure that must be created between the solution and pure water to prevent any net osmotic movement of water when the solution and the pure water are separated by a semipermeable membrane. A colligative property.

Osmotic regulation The maintenance of a constant or nearly constant osmotic pressure in body fluids regardless of the osmotic pressure in the external environment.

Osmotically free water Water that is excreted in the urine above and beyond whatever water is absolutely required for solute excretion.

Osmoticum See *osmotic effector.*

Outer hair cells In the Organ of Corti of the cochlea, the three rows of hair cells that amplify the sound-produced local movements that stimulate the inner hair cells to activate cochlear sensory neurons.

Ovulation The process of releasing an egg (ovum) from the ovary.

Ovum A haploid gamete (egg) produced by oogenesis in the ovary.

Oxidation Removal of electrons or hydrogen atoms from a molecule.

Oxidation water Synonym for *metabolic water.*

Oxidative phosphorylation The formation of ATP using energy released by the transport of electrons through the electron-transport chain.

Oxygen affinity The readiness with which a respiratory pigment such as hemoglobin combines with O_2. When oxygen affinity is high, only a low O_2 partial pressure is required to cause extensive oxygenation.

Oxygen-carrying capacity In a body fluid (e.g., blood) that contains a respiratory pigment, the amount of O_2 per unit of fluid volume when the respiratory pigment is saturated with O_2. More generally, the maximal amount of O_2 per unit of volume in a fluid under ordinary physiological conditions.

Oxygen cascade The sequential drop in the partial pressure of O_2 from one step to the next in the series of steps by which O_2 is transported from the environment of an animal to the animal's mitochondria.

Oxygen conformity A response in which an animal exposed to decreasing O_2 concentrations in its environment allows its rate of O_2 consumption to decrease in parallel.

Oxygen debt An older term for *excess postexercise oxygen consumption.*

Oxygen deficit A difference between actual O_2 uptake by breathing and the theoretical O_2 requirement of exercise during the first minutes at the start of exercise. During the oxygen deficit phase, aerobic catabolism based on O_2 uptake by breathing is unable to meet fully the ATP requirement of exercise, and other ATP-producing mechanisms (such as phosphagens) must contribute ATP.

Oxygen dissociation curve See *oxygen equilibrium curve.*

Oxygen equilibrium curve Referring to the O_2-carrying properties of blood, a graph of the amount of O_2 per unit of blood volume as a function of the O_2 partial pressure of the blood. Also called the *oxygen dissociation curve.*

Oxygen regulation A response in which an animal exposed to a decreasing O_2 concentration in its environment maintains a stable rate of O_2 consumption.

Oxygen-transport pigment See *respiratory pigment.*

Oxygen utilization coefficient In breathing, the fraction (or percentage) of the total O_2 in respired air or water that is removed by the breathing process. In circulation, the fraction of the total O_2 carried by blood that is removed from the blood as it passes around the body.

Oxygenation Referring to water, the dissolution of O_2 in that water. Referring to respiratory pigments, the combination of those pigments with O_2; oxygenation of this sort is reversible and not equivalent to oxidation.

Oxytocin A neurohormone synthesized by neurosecretory cells that have their cell bodies in the hypothalamus and their axon terminals in the pars nervosa of the posterior pituitary. Its major functions are to stimulate contraction of myoepithelial cells of the mammary glands to cause milk ejection and contraction of the myometrium during parturition.

P450 enzymes A set of inducible enzymes involved in the detoxification of foreign compounds.

P_{50} The partial pressure of O_2 that causes a blood or respiratory pigment to become 50% saturated with O_2. A measure of O_2 affinity; a high P_{50} signifies a low O_2 affinity.

Pacemaker The cell or set of cells that spontaneously initiates a rhythm. In a heart, for example, the pacemaker is the cell or set of cells that spontaneously initiates the rhythmic contractions of the heart muscle.

Panting An increase in the rate of breathing; often serves to increase the rate of evaporative cooling by increasing air flow over moist surfaces of the airways of the breathing system.

Paracrine A locally acting chemical signal that binds to receptors and exerts a regulatory effect on cells in the neighborhood of the cell that released it.

Parallel In reference to the arrangement of parts in an electrical circuit, vascular system, or other analogous system in which substances flow from place to place, the parts of the system are in *parallel* if one path of flow branches to give rise to two or more paths in which the parts reside, so only a fraction of the total flow occurs through any one of the parts. Contrast with *series.*

Parasympathetic division A division of the autonomic nervous system that is connected to the CNS via cranial and sacral nerves; the parasympathetic and sympathetic divisions tend to exert opposing controls on autonomic effectors.

Parthenogenesis A form of asexual reproduction in which eggs produced by females develop into genetically identical clones of their female parent.

Partial compensation See *compensation.*

Partial pressure The pressure exerted by a particular gas within a mixture of gases. In terms of the pressure it exerts (its partial pressure), each gas in a mixture behaves as if it alone occupies the entire volume occupied by the mixture; thus each gas's partial pressure can be calculated from the universal gas law. The concept of partial pressure is also applied to gases dissolved in aqueous solution: The partial pressure of a gas in aqueous solution is equal to the partial pressure of the same gas in a gas phase with which the solution is at equilibrium. *Gas tension* is synonymous with *partial pressure* for a gas in solution. See also *universal gas law.*

Passive electrical properties The electrical properties of a cell that do not involve a change in membrane ion permeability, and thus involve no change in electrical resistance.

Passive spread See *decremental spread.*

Passive transport The transport of a material by a mechanism that is capable of carrying the material only in the direction of equilibrium.

Patch clamping A method of measuring single-channel currents by sealing a glass capillary microelectrode to a patch of membrane.

Path integration An animal's summation of the distances and directions of its past movements, so that the animal knows where it is (relative to home) without using landmarks.

Pay-as-you-go phase A phase during exercise when aerobic catabolism using O_2 taken up by breathing is meeting the full ATP requirement of the exercise.

Perfusion The forced flow of blood through blood vessels.

Perikaryon See *cell body.*

Periodic breathing See *intermittent breathing.*

Peripheral activation Conversion of a hormone after secretion to a more physiologically potent form.

Peripheral membrane protein A protein that is associated with a cell membrane, typically on one side or the other, but that can be removed without destroying the membrane.

Peripheral nervous system (PNS) The portion of a nervous system outside of the central nervous system, consisting of afferent and efferent nerves that connect the central nervous system to various parts of the body.

Peristalsis One of the gut's principal modes of muscular activity, in which constriction of the gut at one point along its length initiates constriction at a neighboring point farther along the gut, producing a "wave" of constriction that moves progressively along the gut, propelling food material before it.

Permeability In reference to a cell membrane or epithelium, the ease with which a particular solute can move through it by diffusion, or the ease with which water can move through it by osmosis.

Permeating In reference to a solute, able to pass through a cell membrane or epithelium.

Permissiveness A type of relationship between hormones in which one hormone must be present to allow the other hormone to exert its effect.

pH The negative of the common logarithm of the concentration of H^+. The pH is inversely related to the concentration of H^+ (acidity). A low pH signifies acid conditions, whereas a high pH signifies alkaline conditions. See also *neutral pH.*

Phasic receptor See *rapidly adapting receptor.*

Phasing factor An environmental cue that can entrain a biological clock, synchronizing it to environmental changes (such as the daily light-dark cycle). Also called a *zeitgeber.*

Phenotypic plasticity The ability of an individual animal to express two or more genetically controlled phenotypes.

Pheromone A chemical signal that conveys information between two or more individuals that are members of the same species. It typically signals the sexual readiness or social status of the releasing animal and triggers stereotyped behaviors or physiological changes in the detecting animal.

Phosphagens Compounds that can donate high-energy phosphate bonds to ADP to make ATP; thus, compounds that act as stores of high-energy phosphate bonds. The two most common are creatine phosphate and arginine phosphate.

Phospholipid A lipid compound in which a phosphate group or groups occur. A typical membrane phospholipid consists of two hydrocarbon tails linked by a phosphate group to a compound such as choline.

Photoautotroph An organism that can obtain the energy it needs to stay alive from photons; an organism capable of photosynthesis.

Photoperiod Day length; the number of hours of daylight in a 24-hour day.

Phyletic See *phylogenetic.*

Phylogenetic Having to do with the evolutionary relationships of organisms; the patterns of relationship that organisms exhibit by virtue of common evolutionary descent. Also called *phyletic.*

Phylogenetic reconstruction A reconstruction of the family tree (the ancestry) of groups of related species, often using molecular genetic data.

Physiological work Any process carried out by an animal that increases order and requires energy. For example, an animal does physiological work when it synthesizes proteins, generates electrical or chemical gradients by actively transporting solutes across cell membranes, or contracts muscles to move materials inside or outside its body.

Pilomotor Related to the erection or compression of the hairs in the fur (pelage) of a mammal under the control of muscles attached to the bases of the hairs.

Place theory The theory that the pitch or frequency of a sound is encoded by the place of maximal vibration and maximal hair cell stimulation along the length of the basilar membrane.

Placenta An organ formed jointly by the embryo and the mother in marsupial and eutherian mammals. It allows intimate association (but not mixing) of the maternal and embryonic or fetal bloodstreams for the exchange of materials.

Plasma See *blood plasma.*

Plasma clearance The volume of blood plasma that would have to be completely cleared of a solute to obtain the amount of that solute excreted in urine over a specified period of time.

Plasma membrane See *cell membrane.*

Pleiotropy The control of two or more distinct and seemingly unrelated traits by an allele of a single gene.

Poikilosmotic animal See *osmoconformer.*

Poikilotherm An animal in which the body temperature is determined by equilibration of the body with the thermal conditions in the environment. Also called an *ectotherm.*

Poiseuille equation An equation that describes the quantitative relation between the rate of flow of fluid through a horizontal tube and factors such as pressure, luminal radius, and length. The equation states that flow rate depends directly on the fourth power of the luminal radius. See Equation 23.2. Also called the *Hagen-Poiseuille equation.*

Polar body A small cell formed by the first or second meiotic division of oogenesis. In these uneven divisions, most of the cytoplasm is retained in the oocyte. Polar bodies receive chromosomes but very little cytoplasm, and they degenerate.

Polar molecule A molecule in which electrons are unevenly distributed, so that some regions of the molecule are relatively negative while others are relatively positive.

Polycythemia A state of having an unusually high concentration of red blood cells in the blood; seen, for example, in humans and some other lowland mammals when they acclimate to high altitudes.

Polyneuronal innervation A pattern of innervation in which a single muscle fiber receives synaptic contacts from more than one motor neuron.

Polysaccharide A carbohydrate molecule that consists of many simple-sugar (monosaccharide) molecules polymerized together. Starch, cellulose, and chitin are important examples.

Polyunsaturated fatty acid See *unsaturated fatty acid.*

Postganglionic neuron In the autonomic nervous system, a neuron that extends from an autonomic ganglion to an effector.

Postsynaptic cell A neuron or effector that receives a signal (chemical or electrical) from a presynaptic cell at a synapse.

Postsynaptic potential See *Synaptic potential.*

Power curve In muscles, the relationship between the velocity of shortening and the force exerted against a load.

Power equation See *allometric equation.*

Posterior pituitary The neuroendocrine portion of the vertebrate pituitary gland in which neurohormones are released from axon terminals. The posterior pituitary is an extension of the brain. Also called the *neurohypophysis.*

Pre-Bötzinger complex A bilaterally arrayed pair of neuron clusters within the medulla of the brainstem of a mammal, believed to be the source of the breathing rhythm.

Preferred body temperature The body temperature that is maintained in a poikilotherm by behavioral thermoregulation.

Preflight warm-up In an insect, warming of the flight muscles prior to flight. Preflight warm-up is often essential for flight because the flight muscles must be warm to generate enough power for flight. It can occur by behavioral means (e.g., sun basking) or by physiological means (shivering).

Preformed water Water that enters the body in the form of H_2O. Contrast with *metabolic water.*

Preganglionic neuron In the autonomic nervous system, a neuron that extends from the central nervous system to an autonomic ganglion.

Pressure The force a fluid (liquid or gas) exerts in a perpendicular direction on solid surfaces with which it is in contact. Expressed per unit of surface area.

Presynaptic cell A neuron or other cell that transmits a signal to a postsynaptic cell at a synapse.

Presynaptic inhibition Inhibition of a neuron by decreasing the amount of neurotransmitter released by an excitatory presynaptic neuron.

Primary active transport Active transport driven by a mechanism that draws energy directly from ATP. Contrast with *secondary active transport.*

Primary follicle A primary oocyte surrounded by a single layer of somatic granulosa cells.

Primary oocyte A cell that forms in the vertebrate ovary from a diploid oogonium that has replicated its DNA. In elasmobranchs, birds, and most mammals, it remains arrested in this state until the female becomes reproductively mature.

Primary production The production of organic matter from inorganic chemical precursors. The most common type of primary production is photosynthesis.

Primary spermatocyte A cell that forms in the testis of reproductively mature male vertebrates from a diploid spermatogonium that has replicated its DNA.

Primary structure The sequence of amino acids in a protein molecule.

Primary urine The fluid initially introduced into the tubules of a kidney. It is processed as it flows through the kidney tubules, ultimately becoming the *definitive urine* that is excreted from the body.

Progesterone A sex steroid hormone secreted by the corpus luteum of most vertebrates and the placenta of eutherian mammals.

Prolactin A hormone secreted by cells in the anterior pituitary that stimulates the production of milk in mammals and performs a variety of other functions in vertebrates related to reproduction, water and mineral balance, and caring for the young.

Propagation Spatial transmission of a signal such as an action potential without any decrease in amplitude with distance.

Proprioceptor A sensory receptor that provides an animal with information about the relative position or movement of parts of its body.

Protein kinase An enzyme that covalently bonds a phosphate group to a protein using ATP as the phosphate donor.

Prothoracic glands The paired glands located in the thorax of insects that synthesize and secrete ecdysone in response to prothoracotropic hormone.

Prothoracotropic hormone A tropic protein neurohormone secreted by neuroendocrine cells with cell bodies in the brain and axon terminals in the neurohemal region of the corpora allata; stimulates the prothoracic glands to secrete ecdysone.

Ptilomotor Related to the erection or compression of the feathers in the plumage of a bird under the control of muscles attached to the bases of the feathers.

Pulmonary Related to the lungs.

Pulmonary circuit In the circulatory system of an air-breathing animal, the blood vessels that take blood to and from the lungs.

Pump In the study of cellular-molecular biology, a mechanism of *active transport*.

Pupa The stage in holometabolous insects in which the larval tissues are destroyed and replaced by adult tissues. The pupa metamorphoses into the adult.

Purinotelic Incorporating most nitrogen from the catabolism of nitrogenous compounds into purines, such as uric acid, guanine, and xanthine.

Q_{10} See *temperature coefficient*.

Quantal release The release of neurotransmitter molecules in multimolecular packets (quanta) corresponding to exocytosis of synaptic vesicles.

Quaternary protein structure In a protein molecule that is composed of two or more separate proteins, the three-dimensional arrangement of the protein subunits relative to one another.

Radiant heat transfer Transfer of heat by electromagnetic radiation between two surfaces that are not in contact. Also called *thermal radiation heat transfer*.

Radiant temperature The surface temperature of an object as judged by its rate of emission of electromagnetic energy. The radiant temperature is the surface temperature that would have to prevail for the object to emit energy at the rate it actually does if its emissivity were exactly 1 (i.e., if it were a true black body). Radiant temperature is calculated by measuring the intensity of radiation and entering it in the Stefan-Boltzmann equation with the emissivity set equal to 1. Also called the *blackbody temperature*.

Radiotelemetry The use of a radio transmitter placed in or on an animal to transmit data on physiological or behavioral variables.

Radular apparatus A feeding apparatus found in snails, slugs, chitons, squids, and some other molluscs. The radula itself is a band of connective tissue, studded with teeth, that is pulled back and forth to create grinding or scraping action.

Ram ventilation A type of gill ventilation observed in certain types of fish (e.g., tunas) in which the fish holds its mouth open as it swims forward, thereby using its swimming motions to drive water over its gills.

Rapidly adapting receptor A sensory receptor cell that exhibits a rapidly decreasing response to a maintained stimulus. Also called a *phasic receptor*.

Receptive field In sensory systems, the region of a sensory surface within which stimulation changes the activity of a particular neuron.

Receptor A protein that binds noncovalently with specific molecules and, as a consequence of this binding, initiates a change in membrane permeability or cell metabolism. Receptors mediate the response of a cell to chemical messages (signals) arriving from outside the cell. Although most receptors are in the cell membrane, some are intracellular. See also *sensory receptor*.

Receptor adaptation A decrease in the response of a sensory receptor cell to a prolonged or repeated stimulus.

Receptor potential The graded change in membrane potential that occurs in a sensory receptor cell when it is stimulated.

Receptor, sensory See *sensory receptor*.

Reciprocity A principle of motor control in which signals that activate motor neurons and muscles also inhibit activation of antagonist motor neurons and muscles.

Red blood cell A hemoglobin-containing cell in the blood of an animal. Also called an *erythrocyte*.

Red muscle In general, a loose term referring to a vertebrate muscle that is rich in myoglobin and thus reddish in color. In fish, a large mass of muscle that consists almost entirely of myoglobin-rich muscle fibers that make ATP mostly by aerobic catabolism; routine cruising by fish is powered by red muscle.

Redox balance (reduction–oxidation balance) A state in which a cell has the capability to remove electrons from a compound that undergoes reversible reduction and oxidation as fast as they are added to the compound.

Reduction Addition of electrons or hydrogen atoms to a molecule.

Reflex A simple, relatively stereotyped, but graded behavioral response to a specific stimulus.

Regional heterothermy See *heterothermy*.

Regulation The maintenance of internal conditions at an approximately constant level while external conditions vary.

Relative humidity In air of a particular temperature, the existing water vapor pressure divided by the saturation water vapor pressure characteristic of that temperature.

Relative refractory period The brief period following an action potential during which the voltage threshold is elevated and it is difficult to generate another action potential.

Release-inhibiting hormone A hormone secreted by neuroendocrine cells in the hypothalamus that travels to the anterior pituitary through the hypothalamo-hypophyseal portal system and inhibits the secretion of a hormone by a specific population of anterior pituitary endocrine cells. All release-inhibiting hormones are peptides except the catecholamine dopamine, which is known to inhibit the secretion of prolactin.

Releasing hormone A peptide hormone secreted by neuroendocrine cells in the hypothalamus that travels to the anterior pituitary through the hypothalamo-hypophyseal portal system and stimulates the secretion of a hormone by a specific population of anterior pituitary endocrine cells; also exerts a tropic action to maintain and support those anterior pituitary cells.

Renal Related to kidneys.

Renal corpuscle In the kidney of a vertebrate, a glomerulus and its associated Bowman's capsule. See also *glomerulus.*

Renin A substance secreted into the blood by juxtaglomerular cells of the vertebrate nephron in response to low blood pressure; converts angiotensinogen to angiotensin I.

Renin–angiotensin–aldosterone system A hormonal complex that ensures adequate arterial blood pressure. Set in motion by renin, it produces angiotensin II, which stimulates secretion of aldosterone (important in renal conservation of sodium) and vasopressin, promotes thirst, and stimulates vasoconstriction.

Residue In biochemistry, the modified form assumed by a small molecule when it is incorporated by covalent bonding into a larger molecule. For example, when an amino acid is incorporated into a protein, it is technically an amino acid *residue* because its structure is no longer the full structure of the free amino acid.

Resistance, electrical The property that hinders the flow of electric current, measured in ohms (Ω). Contrast with *conductance.*

Resistance to heat transfer A measure of the difficulty of dry heat transfer either through a material or between an animal and its environment. In relation to an animal, resistance to heat transfer is defined to be the ratio of the difference between body temperature and ambient temperature over the rate of dry heat transfer between the animal and the environment. Contrast with *thermal conductance.*

Resistance, vascular See *vascular resistance.*

Resonant frequency In any elastic system, the frequency at which the system oscillates when left alone following activation by a pulse of energy; the natural frequency. The energy cost of energy-driven oscillation tends to be lowest if the oscillation is at the resonant frequency of the system involved.

Respiratory acidosis or **alkalosis** Acidosis or alkalosis that is caused by an abnormally rapid or slow rate of removal of CO_2 from the body fluids by breathing. Contrast with *metabolic acidosis* or *alkalosis.*

Respiratory gases Oxygen (O_2) and carbon dioxide (CO_2).

Respiratory pigments Any of the metalloprotein pigments that undergo reversible combination with O_2 and thus are able to pick up O_2 in certain places in an animal's body (e.g., the breathing organs) and release it in other places (e.g., systemic tissues). Respiratory pigments include hemoglobin, hemocyanin, hemerythrin, and chlorocruorin. Also called *O_2-transport pigments.*

Respiratory quotient (**RQ**) The moles of CO_2 produced by a cell expressed as a ratio of the moles of O_2 simultaneously consumed. Because the RQ changes with the type of foodstuff being oxidized, it can be used to assess what foodstuffs a cell is using in aerobic catabolism.

Resting membrane potential The normal electrical potential across the cell membrane of a cell at rest.

Rete mirabile A Latin expression, meaning literally "wonderful net." A morphological term referring to any intricately complex vascular system composed of closely juxtaposed, small-diameter arterial and venous blood vessels. The term *rete* is a shorthand synonym.

Retina The layer of photoreceptor cells and other neurons that line the inside of a vertebrate eye.

Retinal The aldehyde of vitamin A, one of two components of the photopigment rhodopsin.

Retrieval The movement of channel or transporter protein molecules out of the membrane where they are active to intracellular locations where they are inactive.

Retrograde messenger A chemical signal thought to be released by a postsynaptic cell that alters the synaptic properties (such as neurotransmitter release) of the presynaptic cell.

Reversal potential (**E_{EPSP}**) The membrane potential at which the amplitude of a voltage response (such as a postsynaptic potential or receptor potential) is zero because there is no net driving force for ion flow.

Rhodopsin The light-absorbing pigment of photoreceptors that initiates the visual response to light; composed of retinal and the protein opsin.

Right-to-left shunting In the study of blood flow through the central circulation of an air-breathing fish, amphibian, or reptile, diversion of systemic venous blood directly back into the systemic circuit, bypassing the lungs.

Rod A type of photoreceptor in the vertebrate retina. Rods are larger than cones, respond at lower light levels, and are used for nocturnal vision.

Root effect A decrease in the amount of O_2 a respiratory pigment can bind at saturation—and thus a decrease in the oxygen-carrying capacity of blood—caused by a decrease in pH or an increase in CO_2 partial pressure. Unusual; observed only in certain fish and molluscs.

Ryanodine receptor The calcium channel of the sarcoplasmic reticulum in striated muscles.

Salinity The sum total concentration of inorganic dissolved matter in water, usually expressed as grams of dissolved matter per kilogram of water.

Salt A synonym for *inorganic ion,* or a compound formed by inorganic ions.

Salt glands Organs other than kidneys that excrete concentrated solutions of inorganic ions. Examples include the cranial salt glands of marine birds and the rectal salt glands of marine elasmobranch fish.

Saltatory conduction Propagation of action potentials in a spatially discontinuous manner in a myelinated axon by jumping from one node of Ranvier to another.

Sarcolemma The cell membrane of a muscle fiber.

Sarcomere The contractile unit of striated muscle that consists of contractile, regulatory, and cytoskeletal proteins. Many sarcomeres in series, delineated by Z discs, constitute a myofibril.

Sarcopenia The loss of skeletal muscle mass as a result of aging. It involves both the loss of contractile proteins from individual muscle fibers and the loss of complete fibers by cell death.

Sarcoplasmic reticulum A system of internal compartments of a muscle cell that envelops myofibrils and stores Ca^{2+} ions. The sarcoplasmic reticulum has calcium ATPase pumps that transport Ca^{2+} ions into its lumen and calcium channels that open in response to excitation along the associated transverse tubules.

Satellite cells Muscle stem cells that lie immediately outside the sarcolemma.

Saturated fatty acid A fatty acid in which all the bonds between carbon atoms in the carbon-chain backbone of the molecule are single bonds. Also called a *saturate.*

Saturated respiratory pigment A respiratory pigment that has combined with as much O_2 as it can possibly hold.

Saturation deficit In air of a particular temperature, the difference between the saturation water vapor pressure characteristic of that temperature and the existing water vapor pressure.

Saturation kinetics The kinetics characteristic of a chemical reaction or other chemical process that is limited to a maximum velocity by a limited supply of some type of molecule with which other molecules must reversibly combine for the reaction or process to take place.

Saturation water vapor pressure The maximum possible water vapor pressure that can stably exist in a gas of a particular temperature. If the water vapor pressure rises above the saturation water vapor pressure, condensation occurs. The saturation water vapor pressure varies strongly with temperature.

Second law of thermodynamics A law, believed to apply without exception in the biosphere, stating that if an isolated system undergoes internal change, the net effect of the change in the system as a whole is always to increase disorder (entropy).

Second messenger An intracellular signaling molecule that is produced inside a cell in response to the arrival of a chemically different extracellular signaling compound.

Secondary active transport Active transport driven by a mechanism that does not draw energy directly from ATP, but instead obtains it from the potential energy inherent in an electrochemical gradient. During secondary active transport, an electrochemical gradient of a solute (e.g., Na^+) is created using ATP-bond energy, and the active-transport mechanism obtains its energy from that gradient. Contrast with *primary active transport*.

Secondary lamellae In fish gills, the microscopically fine folds of tissue on the surfaces of the gill filaments that serve as the primary sites of exchange of respiratory gases between the ambient water and blood.

Secondary oocyte The cell formed at the time of ovulation by the primary oocyte when it concludes its first meiotic division and extrudes the first polar body.

Secondary spermatocyte One of two cells formed from a primary spermatocyte as a result of the first meiotic division during spermatogenesis.

Secondary structure The arrangement of the amino acids within subregions of a protein molecule into highly regular geometric shapes. The two most common types of such highly ordered arrays of amino acids are the α-helix and the β-sheet (pleated sheet).

Segmentation One of the gut's principal modes of muscular activity, in which circular muscles contract and relax in patterns that push the gut contents back and forth.

Selective permeability The state of having a high permeability to some solutes but a low permeability to others. See also *permeability*.

Semipermeable membrane A membrane that is permeable only to water. All true semipermeable membranes are human-made, as there are no natural biological membranes that are strictly semipermeable.

Sense organ A complex multicellular structure specialized to detect a particular type of sensory stimulus.

Sensillum A sensory hair of arthropods; not related to vertebrate hair, but rather a hollow chitinous projection of the exoskeleton that is associated with sensory receptor neurons.

Sensitization Enhancement of a learned behavioral response to a harmless stimulus after exposure to a strong or harmful stimulus.

Sensory adaptation See *receptor adaptation*.

Sensory neuron A sensory receptor that is a neuron, or a peripheral neuron that is excited by a non-neural sensory receptor cell.

Sensory receptor A sensory cell that is specialized to respond to a particular kind of environmental stimulus.

Sensory transduction The process by which the energy of a physical stimulus is converted into an electrical signal in a sensory receptor cell.

Septate junction A type of junction between epithelial cells that differs in fine structure from a tight junction but otherwise has similar properties. Found in invertebrates.

Series In relation to the arrangement of parts in an electrical circuit, vascular system, or other analogous system in which substances flow from place to place; the parts of the system are in *series* if they occur sequentially along a single path of flow, so that all flow must occur sequentially through all parts. Contrast with *parallel*.

Serosal surface See *basal surface*.

Sertoli cells Epithelial cells that form the walls of the seminiferous tubules of the testes of vertebrates and function to support spermatogenesis. Also called *sustentacular cells*.

Sexual reproduction The formation of a new, genetically unique individual from the union of male and female gametes.

Shivering In a mammal or bird, the unsynchronized contraction and relaxation of motor units in skeletal muscles in high-frequency rhythms, producing heat rather than organized motion as the primary product. In an insect, contraction of the flight muscles in a nonflying mode to generate heat rather than flight.

Short-chain fatty acids Fatty acids that consist of relatively few carbon atoms, including acetic acid (2 carbons), propionic acid (3), and butyric acid (4). They are produced, for example, by symbiotic fermenting microbes and are readily absorbed and metabolized by animals. Also called *volatile fatty acids*.

Signal transduction In the study of cell signaling, the translation of a signal from one chemical form (e.g., a hormone) into another chemical form (e.g., a second messenger inside a cell).

Simple diffusion Transport of solutes, water, gases, or other materials that arises from the molecular agitation that exists in all systems above absolute zero and from the simple statistical tendency for such agitation to carry more molecules out of regions of relatively high concentration than into such regions.

Single effect The difference produced by use of metabolic energy between adjacent parts of the two oppositely flowing fluid streams in a countercurrent multiplier system. See also *countercurrent multiplication*.

Skeletal muscle Muscle that produces locomotory movements or other external movements of the body. In vertebrates, skeletal muscle is attached to the endoskeleton. In invertebrates, it is attached to the exoskeleton, shell, other other external covering, or it may constitute the hydrostatic skeleton. Vertebrate skeletal muscle consists of large, cylindrical, multinucleate striated cells (*muscle fibers*). Depending on the species, invertebrate skeletal muscles may consist of striated or smooth muscle cells.

Sliding-filament theory The well-documented theory that muscle contraction results from the active interaction between thick and thin myofilaments, which causes them to slide past each other.

Slow-oxidative (SO) muscle fibers Muscle fibers that are poised to make ATP principally by aerobic catabolism, develop contractile tension slowly, have relatively low peak power outputs, and are relatively resistant to fatigue.

Slowly adapting receptor A sensory receptor cell that responds to a maintained stimulus in a way that decreases slowly and incompletely. Also called a *tonic receptor*.

Smooth muscle Muscle that consists of small, spindle-shaped, uninucleate cells without striations. Thin actin filaments and thick myosin filaments are present, but not organized into sarcomeres. In vertebrates, smooth muscle is found in hollow and tubular internal organs, such as certain blood vessels or the gut.

Sociobiology The study of the social relations and behaviors of animals from an evolutionary perspective.

Solute An entity, notably a molecule or ion, that is in solution.

Soma (plural **somata**) See *cell body*.

Somatic nervous system The part of the vertebrate peripheral nervous system that controls skeletal muscles and provides afferent information from sensory receptors not associated with internal organs.

Somatotopic map The topographic representation or mapping of the body surface onto a region of the brain—for example, in the vertebrate somatosensory cortex.

Spatial summation The summation of postsynaptic potentials that result from presynaptic action potentials at different synapses onto the same postsynaptic cell.

Specific dynamic action (**SDA**) A rise in the metabolic rate of an animal caused by the processing of ingested food.

Spermatid A haploid cell formed from a secondary spermatocyte by the second meiotic division of spermatogenesis. It will mature into a sperm cell.

Spermatogenesis The formation of haploid sperm through the process of meiosis in the testis.

Spermatogonia (singular **spermatogonium**) Diploid cells formed by mitosis in the seminiferous tubules of the testes of male vertebrates.

Spermatozoa Haploid gametes, produced by spermatogenesis in the testis. Also called *sperm cells* or *sperm*.

Sphincter A circular muscle, located between two chambers, that can contract tightly and steadily (tonically) for long periods of time, thus preventing exchange between the chambers.

Spinal nerves Segmental nerves of the peripheral nervous system that attach to the spinal cord.

Spinal reflex A reflex mediated by neural circuits of the vertebrate spinal cord.

Spiracle In a terrestrial arthropod (e.g., insect), a porelike aperture on the surface of the body that opens into the breathing system (e.g., tracheal system).

Standard amino acid One of the 20–22 amino acids that are employed by organisms to synthesize proteins.

Standard metabolic rate (**SMR**) The metabolic rate of a poikilothermic (ectothermic) animal when it is resting and fasting. The SMR is specific to the body temperature prevailing during measurement.

Standard temperature and pressure (**STP**) A temperature of 0°C and a pressure of 1 atm (101 kPa; 760 mm Hg). Gas volumes are often expressed under these conditions as a way of standardizing the effects of temperature and pressure on volume.

Star compass The mechanism by which an animal uses the positions of stars and constellations (and an internal clock) to determine compass direction in nocturnal navigation.

Statocyst A sense organ detecting acceleration and the direction of gravitational force.

Statolith A stony mineral concretion in a statocyst that is denser than the medium in which it sits, so that it is pulled downward by gravity and stimulates mechanoreceptors.

Stefan-Boltzmann equation An equation that relates the temperature of a surface and the rate at which the surface emits electromagnetic energy.

Stenohaline Able to live only within a narrow range of environmental salinities. Contrast with *euryhaline*.

Stenothermal Referring to poikilotherms, able to live only within a narrow range of body temperatures. Contrast with *eurythermal*.

Stimulus A change in the external or internal environment that can be detected by an animal.

Stoichiometry The existence of fixed ratios in chemical reactions. For instance, the fact that hydrogen and oxygen atoms react in a 2-to-1 ratio to form water is an example of stoichiometry.

Stress response The response of an animal to a threatening situation. In vertebrates, it typically involves functions of the autonomic nervous system and the hypothalamus–pituitary–adrenal cortex axis.

Stretch reflex A spinal reflex in which muscle stretching activates a muscle spindle stretch receptor, generating nerve impulses in 1a afferent axons that excite motor neurons innervating the same muscle to oppose the stretch.

Striated muscle Muscle that consists of cells in which the thick myosin and thin actin filaments are arranged in sarcomeres. The sarcomeres are aligned in register to form stripes or striations at right angles to the long axis of the cell. In vertebrates, skeletal and cardiac muscles are striated.

Stroke volume In reference to a heart or other organ that pumps fluid by rhythmic cycles of contraction, the volume of fluid pumped per cycle.

Submaximal exercise Exercise that requires less than an individual's maximal rate of O_2 consumption.

Substrate One of the initial reactants of an enzyme-catalyzed reaction.

Substrate-binding site See *active site*.

Summation In excitable cells, the addition of graded subthreshold potentials (electrical events). In muscle fibers, the addition of twitches (mechanical events) produced by high frequencies of action potentials.

Sun compass The mechanism by which an animal uses the sun's position and an internal clock to determine compass direction in navigation.

Supercooling Cooling of a solution to below its freezing point without freezing.

Supercooling point The highest temperature at which freezing is almost certain to occur promptly in a supercooled solution. (Freezing of a supercooled solution is probabilistic. At some temperatures freezing is unlikely. The temperature needs to be lowered to the supercooling point for prompt freezing to be likely.)

Supramaximal exercise Exercise that requires ATP at a greater rate than it can be made aerobically even when an individual's rate of O_2 consumption is maximized.

Surface-to-volume ratio The ratio of the total area of the outer surface of a three-dimensional object over the volume of that object.

Suspension feeding Feeding on objects suspended in water that are very small by comparison to the feeding animal. See also *filter feeding*.

Sustentacular cells See *Sertoli cells*.

Sweating The secretion onto the skin surface of a low-salinity aqueous solution (*sweat*) by specialized *sweat glands* to increase the rate of evaporative cooling. Occurs only in some groups of mammals.

Symmorphosis A hypothesis about the evolution of multiple organ systems in a species, which posits that the performance limits of those systems remain roughly matched because it would make no sense for any one system to have evolved capabilities that could never be used because of more-restrictive limits in other systems.

Sympathetic division A division of the autonomic nervous system that is connected to the CNS via thoracic and lumbar spinal nerves; the sympathetic and parasympathetic divisions tend to exert opposing controls over autonomic effectors.

Symporter See *cotransporter.*

Synapse A specialized site of communication between two neurons, between a neuron and an effector, or between a non-neural sensory cell and a neuron.

Synaptic antifacilitation A decrease in the amplitude of postsynaptic responses to repeated presynaptic action potentials. Also called *synaptic depression.*

Synaptic cleft The extracellular gap between presynaptic and postsynaptic cells at a synapse, typically 20–40 nm.

Synaptic facilitation An increase in the amplitude of postsynaptic responses that occurs after repeated presynaptic action potentials.

Synaptic plasticity Change in properties of synapses or strength of synaptic interactions with time or circumstance. Developmental changes and changes with learning are thought to reflect synaptic plasticity.

Synaptic potential A graded change in a postsynaptic cell's membrane potential produced by synaptic input. Also called a *postsynaptic potential.*

Synaptic transmission The process whereby one neuron influences the excitability of another neuron or effector. Synaptic transmission can be either chemical or electrical.

Synergism Interactions between two or more agents (e.g., hormones) whereby they have a greater effect acting together than the simple sum of their individual effects.

Systemic circuit In a circulatory system, the blood vessels that take blood to and from the systemic tissues.

Systemic tissues All tissues other than those of the breathing organs.

Systole The period of contraction during the beating cycle of a heart.

Tachycardia A heart rate that is unusually high.

Target cell A cell that responds to a chemical signaling molecule such as a paracrine or hormone because it expresses specific receptors for that molecule. It may express receptors for more than one signaling molecule. Its sensitivity depends on the number of receptors present, and can be changed by upregulation or downregulation of the receptors.

Taste bud A collection of epithelial taste receptor cells and support cells on the tongue or, in fish, on the skin surface.

Telemetry See *radiotelemetry.*

Teleost fish The principal group of fish having bony skeletons.

Temperature A measure of the speed or intensity of the ceaseless random motions that all the atoms and molecules of any substance undergo on an atomic-molecular scale. More exactly, the temperature of a substance is proportional to the product of the mean square speed of random molecular motions and the molecular mass.

Temperature coefficient The ratio of a rate at one body temperature over that rate at a body temperature 10°C lower. Symbolized Q_{10}.

Temperature gradient Technically, the difference in temperature between two places divided by the distance separating those two places. Often used more loosely to refer simply to a difference in temperature. Also called a *thermal gradient.*

Temporal heterothermy See *heterothermy.*

Temporal summation The summation of synaptic potentials in response to repeated presynaptic action potentials at the same synapse.

Tension, mulscular The force produced by cross-bridge action in a contracting muscle.

Tension, gas See *partial pressure.*

Tertiary structure The natural arrangement of an entire protein molecule in three dimensions, including its secondary structure and the other patterns of folding that give the molecule its particular conformation. Tertiary structure is flexible because it is stabilized by noncovalent bonds.

Testosterone A sex steroid hormone, produced by the Leydig cells of the testes, that is essential for male secondary sexual characteristics and reproduction. An androgen.

Tetanus Summed twitches of skeletal muscles produced by trains of motor action potentials. Fused tetanus is a smooth rise in tension produced by a high-frequency train of action potentials. Unfused tetanus is produced by a lower-frequency train of action potentials and shows some relaxation of each twitch between action potentials.

Theca cells Somatic connective tissue cells that form the outer layer of a developing ovarian follicle.

Thermal conductance A measure of the ease of dry heat transfer between an animal and its environment. Contrast with *resistance to heat transfer.*

Thermal gradient See *temperature gradient.*

Thermal hysteresis protein (**THP**) A protein or glycoprotein that acts as an antifreeze by chemically interfering with the formation or growth of ice crystals. Thermal hysteresis proteins lower the freezing point more than the melting point, thus their name.

Thermal radiation heat transfer See *radiant heat transfer.*

Thermogenin An older term referring to uncoupling protein 1. See *uncoupling protein.*

Thermoneutral zone (**TNZ**) In a homeotherm, the range of ambient temperatures over which the metabolic rate is constant regardless of ambient temperature.

Thermoregulation The maintenance of a relatively constant body temperature.

Tidal flow Flow that occurs alternately in and out through a single set of passageways.

Tidal volume In an animal that exhibits tidal breathing, the amount of air inhaled and exhaled per breath.

Tight junction A place where the cell membranes of adjacent cells in an epithelium are tightly joined so that there is no extracellular space between the cells.

Time constant The time required for an exponential process to reach 63% of completion. In neurophysiology, it is a measure of the time needed to change membrane potential and is proportional to the product of resistance and capacitance.

Time-energy budget A method used to estimate an animal's average daily metabolic rate (ADMR). In this method, the time per day spent in each type of activity is measured and multiplied by an estimate of the energy cost of the activity per unit of time to get the total daily energy cost of the activity. The total costs of all activities are then added to get the ADMR. See also *average daily metabolic rate.*

Tissue A group of similar cells organized into a functional unit.

Tissue fluids See *interstitial fluids.*

Titin A giant elastic protein molecule that spans an entire half-sarcomere from Z-disc to M-line.

Tonic receptor See *slowly adapting receptor.*

Total carbon dioxide concentration The amount of CO_2 a solution or body fluid takes up per unit of volume to reach a particular CO_2 partial pressure, regardless of the chemical form the CO_2 assumes when in the solution or body fluid.

Totipotent Capable of all things. For animals, chemical energy is the only form of *totipotent* energy.

Trabeculae Strands of muscle tissue that run through the open central cavity of a heart chamber, crisscrossing from one part of the chamber wall to another.

Trachea A principal tube in the breathing system of a terrestrial animal, typically with reinforced walls. In vertebrates, the trachea is the initial, large airway that carries air from the buccal cavity to the lungs. In insects, a trachea is any of multitudinous airways that ramify throughout the body to form the tracheal respiratory system. See also *tracheole.*

Tracheole A very fine, thin-walled end-tubule at the innermost reaches of the tracheal breathing system of an insect or other tracheate arthropod. The tracheoles are believed to be the principal sites of O_2 and CO_2 exchange with the tissues.

Tract A bundle of axons within a vertebrate central nervous system.

Transducin A G protein that is activated by rhodopsin in photoreceptors, leading to a receptor potential.

Transduction In the study of energy, the transformation of one form of energy into another. In the study of cell signaling, the translation of a signal from one chemical form to another chemical form. In neurophysiology, the conversion of stimulus energy into an electrical signal in sensory receptor cells; the electrical signal is usually a receptor potential.

Transpirational water loss Water loss across the integument of an animal that occurs without sweating or any other active mechanism of transporting water across the integument; passive water loss across the integument.

Transport An entirely general term referring to any and all movements of solutes, water, gases, or other materials from place to place, regardless of the mechanisms of movement.

Transporter A membrane protein that mediates the transport of solute molecules across a membrane and must undergo reversible, noncovalent bonding with the solute molecules in order to do so. Transporters participate in *facilitated diffusion* and *active transport.* Also sometimes called a *carrier.*

Transverse tubules Fingerlike indentations of the cell membrane at regular intervals over the entire surface of a muscle cell. Transverse tubules conduct electrical excitation into the interior of the cell and are intimately associated with the sarcoplasmic reticulum.

Triacylglycerol A fat or oil; a lipid composed of glycerol esterified with three fatty acids. Also called a *triglyceride.*

Triglyceride See *triacylglycerol.*

Trophoblast The outer layer of cells of a blastocyst that will form the fetal portion of the placenta.

Tropic action An action performed by a hormone acting on a target endocrine gland to stimulate secretion of hormone by the target gland and also to maintain the structure and function of the target gland.

Tropomyosin In muscles, a coiled protein molecule that spans seven actin monomers on the thin filament and is associated with one troponin molecule.

Troponin In muscles, a protein consisting of three subunits that is associated with actin and tropomyosin on the thin filaments.

Twitch A single contraction and relaxation of a skeletal muscle fiber produced by an action potential that triggers release of Ca^{2+} ions from the sarcoplasmic reticulum. A single twitch is produced by many repeated cross-bridge power strokes that draw the thin filaments toward the center of each sarcomere.

Tympanal organ An organ of hearing in which sound vibrates a tympanal membrane ("eardrum") to activate auditory receptor cells. The term is usually used for insect hearing organs, although the vertebrate ear is also a tympanal organ.

UCP1 An uncoupling protein that is expressed in the mitochondria of *brown adipose tissue* in mammals. UCP1 permits protons from the space between the inner and outer mitochondrial membranes to move to the core of a mitochondrion without driving the production of ATP; thus it promotes uncoupling of oxidative phosphorylation and production of heat from the energy released by electron transport.

U/P ratio The concentration of urine expressed as a ratio of the concentration of blood plasma. The *osmotic* U/P ratio is the ratio of urine osmotic pressure over plasma osmotic pressure. The U/P ratio for a particular *ion* is the ratio of the urine concentration of that ion over the plasma concentration of that ion.

Ultrafiltrate An aqueous solution produced by ultrafiltration. Also called *filtrate.*

Ultrafiltration Pressure-driven mass flow (oozing, streaming) of fluid out of the blood plasma across the walls of blood capillaries (or sometimes through other barriers), considered a form of filtration because solutes of high molecular weight are left behind while ones of low molecular weight travel with the fluid. In ordinary tissues, fluid that leaves the blood plasma by ultrafiltration enters the tissue interstitial fluids. In kidneys that form primary urine by ultrafiltration, the fluid that leaves the blood plasma enters the kidney tubules (e.g., nephrons).

Uncoupling of oxidative phosphorylation The state of making little or no ATP from the energy that is released by the transport of electrons through the electron-transport chain.

Uncoupling protein (UCP) Unless otherwise stated, refers to *UCP1.*

Universal gas law In the study of gases, an equation that relates pressure (P), volume (V), molar quantity (n), and absolute temperature (T): $PV = nRT$, where R is the universal gas constant (8.314 J/mol·K, where K is a Kelvin and equivalent to one degree Celsius). Also called the *ideal gas law.*

Unloading Deoxygenation of a respiratory pigment.

Unsaturated fatty acid A fatty acid in which one or more of the bonds between carbon atoms in the carbon-chain backbone of the molecule are double bonds; also called an *unsaturate. Monounsaturated fatty acids* (*monounsaturates*) have a single double bond in their structure. *Polyunsaturated fatty acids* (*polyunsaturates*) have two or more double bonds in their structure.

Uphill transport See *active transport.*

Upper critical temperature In a homeotherm, the ambient temperature that represents the upper limit of the thermoneutral zone. See also *thermoneutral zone.*

Upregulation An upward shift in the catalytic activity of an enzyme, the rate of functioning of a biochemical pathway, or the rate of some other similar process brought about in a controlled manner by a regulatory system.

Ureotelic Incorporating most nitrogen from the catabolism of nitrogenous compounds into urea. "Most" is defined differently by different authorities; a common approach is to categorize an animal as *ureotelic* if 50% or more of the nitrogen released by catabolism is incorporated into urea.

Uricotelic Incorporating most nitrogen from the catabolism of nitrogenous compounds into uric acid or closely similar compounds such as urate salts. "Most" is defined differently by different authorities; a common approach is to categorize an animal as *uricotelic* if 50% or more of the nitrogen released by catabolism is incorporated into uric acid.

Urine The fluid excreted by a *kidney.*

Vaporization A change in the physical state of a material from a liquid to a gas.

Vasa recta Blood vessels of minute diameter that are arranged in long hairpin shapes and that constitute the principal vasculature of the medulla of the mammalian kidney. They function as countercurrent diffusion exchangers because blood flow is in opposite directions in the two limbs of each hairpin.

Vascular endothelial growth factor (VEGF) A locally acting cytokine that stimulates angiogenesis.

Vascular resistance The resistance to blood flow through a blood vessel or system of blood vessels, calculated as the pressure drop divided by the flow rate.

Vasomotor Related to changes in the inside (luminal) diameters of blood vessels mediated by contraction and relaxation of smooth muscles in the blood vessel walls.

Vasopressin A neurohormone of mammals, consisting of nine amino acids, that stimulates the reabsorption of water by the collecting ducts in the kidneys. Also called *antidiuretic hormone.* Its molecular structure is similar to that of the neurohormone vasotocin that controls water conservation in many nonmammalian vertebrates. Both are synthesized by neurosecretory cells that have their cell bodies in the hypothalamus and their axon terminals in the pars nervosa of the posterior pituitary.

Vein A macroscopic blood vessel that carries blood toward the heart.

Venous reserve In the study of O_2 transport, the amount of O_2 remaining in venous blood after the blood has passed through the systemic tissues.

Ventilation Forced flow (convection) of air or water into and out of structures used for external respiration or over body surfaces used for external respiration.

Ventilation-perfusion matching In a breathing organ, matching of the rate of blood flow and the rate of ventilation so that efficiency is promoted in the transfer of O_2 from air or water to the blood.

Vesicle A small, membrane-bound, spherical organelle in the cytoplasm of a cell.

Viscosity Internal friction in a moving fluid; a lack of intrinsic slipperiness between fluid layers that are moving at different linear velocities. Fluids that are particularly high in internal friction—low in internal slipperiness—have high viscosities and exhibit syruplike properties. Viscosity can also be thought of as resistance to shear forces within a moving fluid.

Vitamin An organic compound that an animal must obtain from food, symbiotic microbes, or another source other than animal biosynthesis because the animal cannot synthesize it, yet requires it in small amounts.

$\dot{V}_{O_2max}$ An animal's maximal rate of O_2 consumption. It is usually elicited by exercising animals to peak sustained exercise intensity.

Volatile fatty acids See *short-chain fatty acids.*

Voltage clamping An experimental method to measure ionic current flow by imposing a selected membrane potential on a cell and monitoring the current necessary to hold it at that voltage.

Voltage-gated channel An ion channel that opens in response to membrane depolarization.

Voltage threshold The critical value of depolarization that is just enough to trigger an action potential.

Volume regulation The maintenance of a constant or nearly constant volume (amount) of body fluid. The term can be applied to cells, in which case it refers to a constant volume of intracellular fluid. It can also be applied to whole animals, in which case it refers to a constant volume of all body fluids.

Volumes percent (vol %) Milliliters (at STP) of gas dissolved or chemically combined within a fluid per 100 milliliters of the fluid.

Vomeronasal organ An accessory olfactory organ of vertebrates that mediates many (but not all) sensory responses to pheromones.

Wasting See *atrophy.*

Water channel A membrane protein that aids passive water transport—osmosis—through a cell membrane.

Water vapor Water in the gaseous state.

Water vapor pressure In reference to a gas phase, the prevailing partial pressure of gaseous water. In reference to an aqueous solution, the partial pressure of gaseous water that the solution will create by evaporation in a gas phase with which it is in contact if permitted to come to equilibrium with the gas phase. A colligative property.

Water-vapor-pressure depression The difference between the water vapor pressure of a solution and that of pure water under the same conditions. A colligative property.

Weak bonds See *noncovalent bonds.*

White adipose tissue The ordinary fat-storage tissue of vertebrates, exemplified by the "fat" we speak of in poultry or beef prepared for food. Also called *white fat.* Contrast with *brown adipose tissue.*

White matter A histological region of the vertebrate central nervous system that consists largely of tracts of neuronal axons. The abundance of myelin imparts a glistening white sheen to the tissue.

White muscle In general, a loose term referring to a vertebrate muscle that is poor in myoglobin and thus whitish (rather than reddish) in color. In fish, a large mass of muscle that consists almost entirely of myoglobin-poor muscle fibers that make ATP mostly by anaerobic catabolism; white muscle powers burst exercise in fish.

Work See *external work, internal work, physiological work.*

Xeric Able to live steadily in the open air and thus face the full drying power of the terrestrial environment.

Z disc A web of accessory proteins at each end of a sarcomere that anchors the proteins titin and nebulin and the actin thin filaments. Also called *Z line, Z band.*

Zeitgeber See *phasing factor.*

Zona pellucida An extracellular layer of glycoproteins secreted by the primary oocyte.

Zygote A genetically unique, diploid cell produced by the union of an egg (ovum) and a sperm in sexual reproduction.

Photo Credits

Part Openers
Part 1: Craig Lovell/Corbis.
Part 2: Jose B. Ruiz/Nature Picture Library.
Part 3: Constantinos Petrinos/Nature Picture Library.
Part 4: Steve Turner/Oxford Scientific Films.
Part 5: Don Kreuter/Painet, Inc.
Part 6: Doug Allan/Nature Picture Library.

Table of Contents *p. ix*: Craig Lovell/Corbis. *p. x*: Data from PDB file 1OED, Miyazawa et al, 2003. *p. xi*: Jose B. Ruiz/Nature Picture Library. *p. xii*: Stephen Frink Collection/Alamy Images. *p. xiv*: Constantinos Petrinos/Nature Picture Library. *p. xvii*: Steve Turner/Oxford Scientific Films. *p. xviii*: SCPhotos/Alamy Images. *p. xix*: Don Kreuter/Painet, Inc. *p. xxi*: Steve Bloom/Alamy Images. *p. xxii*: Doug Allan/Nature Picture Library. *p. xxiii*: Painet, Inc.

Chapter 1 *Opener*: François Gohier/Photo Researchers, Inc. 1.4*a*: Larry Jon Friesen. 1.4*b*: Paul Kay/Oxford Scientific Films. 1.9: Norbert Wu/www.norbertwu.com. 1.10: A. H. Rider/Photo Researchers, Inc. 1.11: G. C. Kelley/Photo Researchers, Inc. 1.12: François Carrel/Montagne Magazine/Corbis. 1.14: Juan Carlos Calvin/AGE Fotostock.

Chapter 2 *Opener, snail*: Alex Kerstitch/Visuals Unlimited. *adder*: Mark Boulton/Photo Researchers, Inc. 2.10*a*: Rod Planck/Photo Researchers, Inc. 2.10*b*: Suzanne L. Collins/Photo Researchers, Inc. 2.24: Alex Kerstitch/Visuals Unlimited.

Chapter 3 *Opener*: Aflo Foto Agency.

Chapter 4 *Opener*: Peter Blackwell/Nature Picture Library. 4.6*a*: Ken Lucas/Visuals Unlimited. 4.6*b*: Alexis Rosenfeld/SPL/Photo Researchers, Inc. 4.8*a*: Larry Jon Friesen. 4.8*c*: Harry Rogers/Photo Researchers, Inc. 4.8*d*: Volker Steger/SPL/Photo Researchers, Inc. 4.10: Courtesy of Richard Hill. 4.12*a*: Randy Harwood/Painet, Inc. 4.12*b*: A. Flowers & L. Newman/Photo Researchers, Inc. 4.13: Dudley Foster/Woods Hole Oceanographic Institution. 4.21: Tim David/Photo Researchers, Inc.

Chapter 5 *Opener*: Gordon Wiltsie/National Geographic Image Collection. 5.3: Mark Hanauer/Corbis. 5.11: Charles V. Angelo/Photo Researchers, Inc.

Chapter 6 *Opener*: Richard Herrmann/Visuals Unlimited. Box 6.3: Jurgen Ankenbrand/Painet, Inc.

Chapter 7 *Opener*: Royalty-Free/Corbis. *inset*: Robert Barber/Painet, Inc. 7.2: Courtesy of COSMED Srl.

Chapter 8 *Opener*: Courtesy of Bernd Heinrich. 8.24: M. H. Sharp/Photo Researchers, Inc. 8.27, 8.28: Thermal maps courtesy of Richard Hill.

Chapter 9 *Opener*: Bryan & Cherry Alexander Photography.

Chapter 10 *Opener*: Jacana/Photo Researchers, Inc. 10.9: Bettmann/Corbis.

Chapter 11 *Opener*: Georgette Douwma/SPL/Photo Researchers, Inc. 11.19*a*: Data from PDB file 1J95, Morais-Cabral et al, 2001.

Chapter 12 *Opener*: Courtesy the Office of Communications, Princeton University. 12.3: Dennis Kunkel Microscopy, Inc.

Chapter 13 *Opener*: Jeff Lepore/Photo Researchers, Inc. Box 13.1: Dietmar Nill/Nature Picture Library.

Chapter 14 *Opener*: John and Barbara Gerlach/Visuals Unlimited. Box 14.1: Courtesy of Lisa Davis and Lowell Getz.

Chapter 15 *Opener*: Eric & David Hosking/Photo Researchers, Inc. 15.1*a*: LogicStock/Painet Inc. 15.1*b*: John Walsh/SPL/Photo Researchers, Inc. 15.1*c*: Colin Milkins/Oxford Scientific Films. 15.1*d*: Joseph L. Collins/Photo Researchers, Inc.

Chapter 16 *Opener*: David Westphalen/Painet, Inc. 16.1*d*: Jeff Foott/Nature Picture Library. 16.4: Reprinted with permission from the cover of Science, 287 (5454). 16.9: Charles Walcott.

Chapter 17 *Opener*: Stephen Dalton/Photo Researchers, Inc. 17.15: Biophoto Associates/Photo Researchers, Inc. 17.19: Dr. Richard Kessel/Visuals Unlimited.

Chapter 18 *Opener*: Stephen Dalton/Photo Researchers, Inc.

Chapter 19 *Opener, body builder*: Clark Brooks/Painet, Inc. *astronaut*: Courtesy of NASA. 19.3*b*: David McIntyre. 19.9: Courtesy of Danny A. Riley and James L. W. Bain, Medical College of Wisconsin, Milwaukee, WI. Box 19.1: Courtesy of Ant Backer.

Chapter 20 *Opener*: Gregory G. Dimijian/Photo Researchers, Inc. Box 20.2: Jerry L. Ferrara/Photo Researchers, Inc.

Chapter 21 *Opener*: Peter Scoones/SPL/Photo Researchers, Inc. 21.12: John G. Shedd Aquarium/Visuals Unlimited. 21.17: SPL/Photo Researchers, Inc. 21.25*b*: Tom Stack/Painet, Inc. 21.28: Pete Oxford/ Nature Picture Library. Box 21.1: Jurgen Ankenbrand/Painet, Inc.

Chapter 22 *Opener*: Dr Derek Bromhall/Oxford Scientific Films.

Chapter 23 *Opener*: Leni Riefenstahl.

Chapter 24 *Opener*: Doug Allan/Nature Picture Library. 24.3 *inset*: Daniel Costa.

Chapter 25 *Opener*, 25.4a, and 25.5: Courtesy of Richard Hill. 25.6: NASA Goddard Space Flight Center Image by Reto Stöckli; enhancements by Robert Simmon.

Chapter 26 *Opener*: Jeff Rotman/Nature Picture Library. 26.5: Larry Jon Friesen. 26.19: Jack Milchanowski/AGE Fotostock.

Chapter 27 *Opener*: Dr. Lloyd Glenn Ingles/California Academy of Sciences.

Chapter 28 *Opener*: M. Harvey/DRK Photo. 28.2: Courtesy of Richard Hill. 28.3: Photodisc Blue/Getty Images. 28.3 *inset*: Rob Simpson/Painet, Inc. 28.5: Courtesy of Richard Hill. 28.9: Frans Lemmens/Masterfile. 28.11: Doug Allan/Nature Picture Library.

Figure and Table Citations

Chapter 1

Coulianos, C.-C., and A. G. Johnels. 1963. Note on the subnivean environment of small mammals. *Ark. Zool.* 15: 363–370.

Lloyd, J. E. 1966. *Studies on the Flash Communication System in* Photinus *Fireflies*. Miscellaneous Publications (University of Michigan. Museum of Zoology), no. 130. Museum of Zoology, University of Michigan, Ann Arbor.

Lynch, C. B. 1992. Clinal variation in cold adaptation in *Mus domesticus*: Verification of predictions from laboratory populations. *Amer. Nat.* 139: 1219–1236.

Misonne, X. 1959. Analyse Zoogéographique des Mammifères de l'Iran. *Memoires, 2me Série, no. 59 Institut Royal Sciences Naturelles de Belgique*, Brussells.

Owen-Smith, R. N. 1988. *Megaherbivores*. Cambridge University Press, Cambridge.

Pandolf, K. B., and A. J. Young. 1992. Environmental extremes and endurance performance. In R. J. Shephard and P.-O. Åstrand (eds.), *Endurance in Sport*, pp. 270–282. Blackwell, London.

Scriber, J. M. 1973. Latitudinal gradients in larval feeding specialization of the world Papilionidae (Lepidoptera). *Psyche* 80: 355–373.

Videler, J. J. 1993. *Fish Swimming*. Chapman & Hall, New York.

Walls, G. L. 1942. *The Vertebrate Eye and Its Adaptive Radiation*. Cranbrook Instititute of Science, Bloomfield Hills, MI.

Wells, M. J. 1966. Cephalopod sense organs. In K. M. Wilbur and C. M. Yonge (eds.), *Physiology of Mollusca*, vol. 2, pp. 523–545. Academic Press, New York.

Young, J. Z. 1971. *The Anatomy of the Nervous System of* Octopus vulgaris. Oxford University Press, London.

Chapter 2

Logue, J. A., A. L. DeVries, E. Fodor, and A. R. Cossins. 2000. Lipid compositional correlates of temperature-adaptive interspecific differences in membrane physical structure. *J. Exp. Biol.* 203. 2105–2115.

Powers, D. A., M. Smith, I. Gonzalez-Villasenor, L. DiMichele, D. Crawford, G. Bernardi, and T. Lauerman. 1993. A multidisciplinary approach to the selectionist/neutralist controversy using the model teleost, *Fundulus heteroclitus. Oxford Surv. Evol. Biol.* 9: 43–107.

Stock, D. W., J. M. Quattro, G. S. Whitt, and D. A. Powers. 1997. Lactate dehydrogenase (LDH) gene duplication during chordate evolution: The cDNA sequence of the LDH of the tunicate *Styela licata. Mol. Biol. Evol.* 14: 1273–1284.

Walker, R. 1983. *The Molecular Biology of Enzyme Synthesis*. Wiley, New York.

Chapter 3

Weiss, T. F. 1996. *Cellular Biophysics*. MIT Press, Cambridge, MA.

Chapter 4

Arp, A. J., J. J. Childress, and C. R. Fisher, Jr. 1985. Blood gas transportation in *Riftia pachyptila. Bull. Biol. Soc. Wash.* no. 6: 289–300.

Buller, H. A., and R. J. Grand. 1990. Lactose intolerance. *Annu. Rev. Med.* 41: 141–148.

Burton, B. T., and W. R. Foster. 1988. *Human Nutrition*. McGraw-Hill, New York.

Hofmann, R. R. 1989. Evolutionary steps of ecophysiological adaptation and diversification of ruminants: A comparative view of their digestive system. *Oecologia* 78: 443–457.

Hyman. L. H. 1967. *The Invertebrates: Mollusca I*. McGraw-Hill, New York.

Morris, J. G. 1991. Nutrition. In C. L. Prosser (ed.), *Environmental and Metabolic Animal Physiology*. (*Comparative Animal Physiology*, 4th ed.), pp. 231–276. Wiley-Liss, New York.

Pengelley, E. T., and S. J. Asmundson. 1969. Free-running periods of endogenous circannian rhythms in the golden-mantled ground squirrel, *Citellus lateralis. Comp. Biochem. Physiol.* 30: 177–183.

Slijper, E. J. 1979. *Whales*. Cornell University Press, Ithaca, NY.

Stevens, C. E. 1977. Comparative physiology of the digestive system. In M. J. Swenson (ed.), *Duke's Physiology of Domestic Animals*, 9th ed., pp. 216–232. Cornell University Press, Ithaca, NY.

Chapter 5

Brown, A. C., and G. Brengelmann. 1965. Energy metabolism. In T. C. Ruch and H. D. Patton (eds.), *Physiology and Biophysics*, 19th ed., pp. 1030–1049. Saunders, Philadelphia.

Darveau, C.-A., R. K. Suarez, R. D. Andrews, and P. W. Hochachka. 2002. Allometric cascade as a unifying principle of body mass effects on metabolism. *Nature* 417: 166–170.

Golley, F. B. 1960. Energy dynamics of a food chain of an old-field community. *Ecol. Monogr.* 30: 187–206.

Hayssen, V., and R. C. Lacy. 1985. Basal metabolic rates in mammals: Taxonomic differences in the allometry of BMR and body mass. *Comp. Biochem. Physiol. A* 81: 741–754.

Jobling, M. 1993. Bioenergetics: Feed intake and energy partitioning. In J. C. Rankin and F. B. Jensen (eds.), *Fish Ecophysiology*, pp. 1–44. Chapman & Hall, New York.

Kleiber, M. 1975. *The Fire of Life*, 2nd ed. Krieger, Huntington, NY.

Lasiewski, R. C., and W. R. Dawson. 1967. A re-examination of the relation between standard metabolic rate and body weight in birds. *Condor* 69: 13–23.

Lasker, R. 1970. Utilization of zooplankton energy by a Pacific sardine population in the California current. In J. H. Steele (ed.), *Marine Food Chains*, pp. 265–284. University of California Press, Berkeley.

Lavoisier, A. L. 1862. *Oeuvres de Lavoisier*, vol. 2. Imprimerie Impériale, Paris.

Mandelbrot, B. B. 1983. *The Fractal Geometry of Nature*. Freeman, San Francisco.

McNab, B. K. 1986. The influence of food habits on the energetics of eutherian mammals. *Ecol. Monogr.* 56: 1–19.

Owen-Smith, R. N. 1988. *Megaherbivores: The Influence of Very Large Body Size on Ecology*. Cambridge University Press, New York.

Roberts, J. L. 1957. Thermal acclimation of metabolism in the crab *Pachygrapsus crassipes* Randall. I. The influence of body size, starvation, and molting. *Physiol. Zool.* 30: 232–242.

Saunders, J. B. de C. M., and C. D. O'Malley. 1950. *The Illustrations from the Works of Andreas Vesalius of Brussels*. World, Cleveland, OH.

Seymour, R. S., and A. J. Blaylock. 2000. The principle of Laplace and scaling of ventricular wall stress and blood pressure in mammals and birds. *Physiol. Biochem. Zool.* 73: 389–405.

Templeton, J. R. 1970. Reptiles. In G. C. Whittow (ed.), *Comparative Physiology of Thermoregulation*, vol. 1, pp. 167–221. Academic Press, New York.

Whitford, W. G. 1973. The effects of temperature on respiration in the Amphibia. *Amer. Zool.* 13: 505–512.

Chapter 6

Åstrand, P.-O., and K. Rodahl. 1986. *Textbook of Work Physiology: Physiological Bases of Exercise*, 3rd ed. McGraw-Hill, New York.

Billeter, R., and H. Hoppeler. 1992. Muscular basis of strength. In P. V. Komi (ed.), *Strength and Power in Sport*, pp. 39–63. Blackwell, Oxford, UK.

Coyle, E. F. 1991. Carbohydrate metabolism and fatigue. In G. Atlan, L. Beliveau, and P. Bouissou (eds.), *Muscle Fatigue: Biochemical and Physiological Aspects*, pp. 153–164. Masson, Paris.

Famme, P., and J. Knudsen. 1984. Total heat balance study of anaerobiosis in *Tubifex tubifex* (Müller). *J. Comp. Physiol. B* 154: 587–591.

Fulco, C. S., P. B. Rock, and A. Cymerman. 1998. Maximal and submaximal exercise performance at altitude. *Aviat. Space Environ. Med.* 69: 793–801.

Hand, S. C., and E. Gnaiger. 1988. Anaerobic dormancy quantified in *Artemia* embryos: A calorimetric test of the control mechanism. *Science* 239: 1425–1427.

Hochachka, P. W., and G. N. Somero. 2002. *Biochemical Adaptation*. Oxford University Press, New York.

Newsholme, E. A., E. Blomstrand, N. McAndrew, and M. Parry-Billings. 1992. Biochemical causes of fatigue and overtraining. In R. J. Shephard and P.-O. Åstrand (eds.), *Endurance in Sport*, pp. 351–364. Blackwell, Oxford, UK.

Saraste, M. 1999. Oxidative phosphorylation at the *fin de siècle*. *Science* 283: 1488–1493.

Ultsch, G. R., H. Borschung, and M. J. Ross. 1978. Metabolism, critical oxygen tension, and habitat selection in darters (*Etheostoma*). *Ecology* 59: 99–107.

Wine, J. J., and F. B. Krasne. 1972. The organization of escape behaviour in the crayfish. *J. Exp. Biol.* 56: 1–18.

Chapter 7

Alerstam, T., and A. Hedenström. 1998. The development of bird migration theory. *J. Avian Biol.* 29: 343–369.

Åstrand. P.-O., and K. Rodahl. 1986. *Textbook of Work Physiology*, 3rd ed. McGraw-Hill, New York.

Brett, J. R. 1964. The respiratory metabolism and swimming performance of young sockeye salmon. *J. Fish. Res. Board Canada* 21: 1183–1226.

Dal Monte, A., M. Faina, and C. Menchinelli. 1992. Sport-specific ergonomic equipment. In R. J. Shephard and P.-O. Åstrand (eds.), *Endurance in Sport*, pp. 210. Blackwell, Oxford, UK.

Dial, K. P., A. A. Biewener, B. W. Tobalske, and D. R. Warrick. 1997. Mechanical power output of bird flight. *Nature* 390: 67–70.

Hammond, K. A., and J. Diamond. 1997. Maximal sustained energy budgets in humans and animals. *Nature* 386: 457–462.

Henriksson, J., and J. S. Reitman. 1977. Time course of changes in human skeletal muscle succinate dehydrogenase and cytochrome oxidase activities and maximal oxygen uptake with physical activity and inactivity. *Acta Physiol. Scand.* 99: 91–97.

Hill, R. W., and G. A. Wyse. 1989. *Animal Physiology*, 2nd ed. HarperCollins, New York.

Lindstedt, S. L., J. F. Hokanson, D. J. Wells, S. D. Swain, H. Hoppeler, and V. Navarro. 1991. Running energetics in the pronghorn antelope. *Nature* 353: 748–750.

Nagy, K. A., W. R. Siegfried, and R. P. Wilson. 1984. Energy utilization by free-ranging jackass penguins, *Speniscus demersus*. *Ecology* 65: 1648–1655.

Peterson, C. C., K. A. Nagy, and J. Diamond. 1990. Sustained metabolic scope. *Proc. Natl. Acad. Sci. U.S.A.* 87: 2324–2328.

Taylor, C. R., K. Schmidt-Nielsen, and J. L. Raab. 1970. Scaling of energetic cost of running to body size in mammals. *Amer. J. Physiol.* 219: 1104–1107.

Tucker, V. A. 1968. Respiratory exchange and evaporative water loss in the flying budgerigar. *J. Exp. Biol.* 48: 67–87.

Tucker, V. A. 1969. The energetics of bird flight. *Sci. Amer.* 220(5): 70–78.

Tucker, V. A. 1975. The energetic cost of moving about. *Amer. Sci.* 63: 413–419.

Videler, J. J. 1993. *Fish Swimming*. Chapman & Hall, London.

Chapter 8

Armitage, K. B., B. C. Woods, and C. M. Salsbury. 2000. Energetics of hibernation in woodchucks (*Marmota monax*). In G. Heldmaier and M. Klingenspor (eds.), *Life in the Cold*, pp. 73–80. Springer, New York.

Bauwens, D., P. E. Hertz, and A. M. Castilla. 1996. Thermoregulation in a lacertid lizard: The relative contributions of distinct behavioral mechanisms. *Ecology* 77: 1818–1830.

Block, B. A., and J. R. Finnerty. 1994. Endothermy in fishes: A phylogenetic analysis of constraints, predispositions, and selection pressures. *Environ. Biol. Fishes* 40: 283–302.

Brown, J. H., and G. A. Bartholomew. 1969. Periodicity and energetics of torpor in the kangaroo mouse, *Microdipodops pallidus*. *Ecology* 50: 705–709.

Bullock, T. H. 1955. Compensation for temperature in the metabolism and activity of poikilotherms. *Biol. Rev. (Cambridge)* 30: 311–342.

Carey, F. G., and J. M. Teal. 1969. Regulation of body temperature by the bluefin tuna. *Comp. Biochem. Physiol.* 28: 205–213.

Dawkins, M. J. R., and D. Hull. 1964. Brown adipose tissue and the response of new-born rabbits to cold. *J. Physiol. (London)* 172: 216–238.

Dawson, W. R., and G. A. Bartholomew. 1956. Relation of oxygen consumption to body weight, temperature, and temperature acclimation in lizards *Uca stansburiana* and *Sceloporus occidenalis*. *Physiol. Zool.* 29: 40–51.

Dizon, A. E., and R. W. Brill. 1979. Thermoregulation in tunas. *Amer. Zool.* 19: 249–265.

Fletcher, G. L., S. V. Goddard, P. L. Davies, Z. Gong, K. V. Ewart, and C. L. Hew. 1998. New insights into fish antifreeze proteins: Physiological significance and molecular regulation. In H. O. Pörtner and R. C. Playle (eds.), *Cold Ocean Physiology*, pp. 239–265. Cambridge University Press, New York.

Harrison, J. F., J. H. Fewell, S. P. Roberts, and H. G. Hall. 1996. Achievement of thermal stability by varying metabolic heat production in flying honeybees. *Science* 274: 88–90.

Hart, J. S. 1957. Climatic and temperature induced changes in the energetics of homeotherms. *Rev. Can. Biol.* 16: 133–174.

Heinrich, B. 1971. Temperature regulation of the sphinx moth, *Manduca sexta*. I. Flight energetics and body temperature during free and tethered flight. *J. Exp. Biol.* 54: 141–152.

Heinrich, B. 1974. Thermoregulation in bumblebees. I. Brood incubation by *Bombus vosnesenskii* queens. *J. Comp. Physiol.* 88: 129–140.

Hill, R. W. 1975. Daily torpor in *Peromyscus leucopus* on an adequate diet. *Comp. Biochem. Physiol. A* 51: 413–423.

Hill, R. W., D. P. Christian, and J. H. Veghte. 1980. Pinna temperature in exercising jackrabbits, *Lepus californicus*. *J. Mamm.* 61: 30–38.

Hochachka, P. W., and G. N. Somero. 2002. *Biochemical Adaptation: Mechanism and Process in Physiological Evolution*. Oxford University Press, New York.

Hudson, J. W., W. R. Dawson, and R. W. Hill. 1974. Growth and development of temperature regulation in nestling cattle egrets. *Comp. Biochem. Physiol. A* 49: 717–741.

Huey, R. B., and J. G. Kingsolver. 1993. Evolution of resistance to high temperature in ectotherms. *Amer. Nat.* 142: S21–S46.

Irving, L., and J. Krog. 1955. Temperature of skin in the arctic as a regulator of heat. *J. Appl. Physiol.* 7: 355–364.

Johnson, R. E. 1968. Temperature regulation in the white-tailed ptarmigan, *Lagopus leucurus. Comp. Biochem. Physiol.* 24: 1003–1014.

Maloney, S. K., and G. Mitchell. 1997. Selective brain cooling: Role of angularis oculi vein and nasal thermoreception. *Amer. J. Physiol.* 273: R1108–R1116.

Root, T. 1988. Energy constraints on avian distributions and abundances. *Ecology* 69: 330–339.

Rubinsky, B., S. T. S. Wong, J.-S. Hong, J. Gilbert, M. Roos, and K. B. Storey. 1994. ^{1}H magnetic resonance imaging of freezing and thawing in freeze-tolerant frogs. *Amer. J. Physiol.* 266: R1771–R1777.

Scholander, P. F., W. Flagg, V. Walters, and L. Irving. 1953. Climatic adaptation in arctic and tropical poikilotherms. *Physiol. Zool.* 26: 67–92.

Scholander, P. F., R. Hock, V. Walters, F. Johnson, and L. Irving. 1950. Heat regulation in some arctic and tropical mammals and birds. *Biol. Bull.* 99: 237–258.

Somero, G. N. 1997. Temperature relationships: From molecules to biogeography. In W. H. Dantzler (ed.), *Comparative Physiology*, vol. 2 (Handbook of Physiology [Bethesda, MD], section 13), pp. 1391–1444. Oxford University Press, New York.

Chapter 9

Arnold, W. 1988. Social thermoregulation during hibernation in alpine marmots (*Marmota marmota*). *J. Comp. Physiol. B* 158: 151–156.

Barnes, B. M., and C. L. Buck. 2000. Hibernation in the extreme: Burrow and body temperatures, metabolism, and limits to torpor bout length in arctic ground squirrels. In G. Heldmaier and M. Klingenspor (eds.), *Life in the Cold*, pp. 65–72. Springer, New York.

Boyer, B. B., and B. M. Barnes. 1999. Molecular and metabolic aspects of mammalian hibernation. *BioScience* 49: 713–724.

Dawkins, M. J. R., and D. Hull. 1965. The production of heat by fat. *Sci. Amer.* 213: 62–67.

Geiser, F., and G. J. Kenagy. 1987. Polyunsaturated lipid diet lengthens torpor and reduces body temperature in a hibernator. *Amer. J. Physiol.* 252: R897–R901.

Heldmaier, G. 1993. Seasonal acclimatization of small mammals. *Verh. Dtsch. Zool. Ges.* 86(2): 67–77.

Heldmaier, G., and T. Ruf. 1992. Body temperature and metabolic rate during natural hypothermia in endotherms. *J. Comp. Physiol. B* 162: 696–706.

Hull, D. 1973. Thermoregulation in young mammals. In G. C. Whittow (ed.), Comparative Physiology of Thermoregulation, vol. 3, pp. 167–200. Academic Press, NY.

Johnsen, H. K., A. Rognmo, K. J. Nilssen, and A. S. Blix. 1985. Seasonal changes in the relative importance of different avenues of heat loss in resting and running reindeer. *Acta Physiol. Scand.* 123: 73–79.

Markussen, K. A., A. Rognmo, and A. S. Blix. 1985. Some aspects of thermoregulation in newborn reindeer calves (*Rangifer tarandus tarandus*). *Acta Physiol. Scand.* 123: 215–220.

Meng, M., G. C. West, and L. Irving. 1969. Fatty acid composition of caribou bone marrow. *Comp. Biochem. Physiol.* 30: 187–191.

Nieminen, M. 1980. Nutritional and seasonal effects on the haematology and blood chemistry in reindeer (*Rangifer tarandus tarandus* L.). *Comp. Biochem. Physiol. A* 66: 399–413.

Nilssen, K. J., J. A. Sundsfjord, and A. S. Blix. 1984. Regulation of metabolic rate in Svalbard and Norwegian reindeer. *Amer. J. Physiol.* 247: R837–R841.

Orpin, C. G., S. D. Mathiesen, Y. Greenwood, and A. S. Blix. 1985. Seasonal changes in the ruminal microflora of the high-arctic Svalbard reindeer (*Rangifer tarandus platyrhynchus*). *Appl. Environ. Microbiol.* 50: 144–151.

Power, G. G., T. R. Gunn, B. M. Johnston, and P. D. Gluckman. 1987. Oxygen supply and the placenta limit thermogenic responses in fetal sheep. *J. Appl. Physiol.* 63: 1896–1901.

Ruf, T., and W. Arnold. 2000. Mechanisms of social thermoregulation in hibernating alpine marmots (*Marmota marmota*). In G. Heldmaier and M. Klingenspor (eds.), *Life in the Cold*, pp. 81–94. Springer, New York.

Soppela, P., M. Nieminen, S. Saarela, and R. Hissa. 1986. The influence of ambient temperature on metabolism and body temperature of newborn and growing reindeer calves (*Rangifer tarandus tarandus* L.). *Comp. Biochem. Physiol. A* 83: 371–386.

Chapter 10

Camhi, J. M. 1984. *Neuroethology: Nerve Cells and the Natural Behavior of Animals.* Sinauer, Sunderland, MA.

Campbell, N. A., J. B. Reece, and L. G. Mitchell. 1999. *Biology*, 5th ed. Benjamin Cummings, Menlo Park, CA.

Chen, W., X. H. Zhu, K. R. Thulborn, and K. Ugurbil. 1999. Retinotopic mapping of lateral geniculate nucleus in humans using functional magnetic resonance imaging. *Proc. Natl. Acad. Sci. U.S.A.* 96:2430–2434.

Kandel, E. R., J. H. Schwartz, and T. M. Jessel (eds.). 2000. *Principles of Neural Science*, 4th ed. McGraw-Hill, New York.

Nicholls, J. G., A. R. Martin, B. G. Wallace, and P. A. Fuchs 2001. *From Neuron to Brain*, 4th ed. Sinauer, Sunderland, MA.

Purves, W. K., G. H. Orians, and H. C. Heller. 1992. *Life: The Science of Biology*, 3rd ed. Sinauer, Sunderland, MA.

Rosenzweig, M. R., S. M. Breedlove, and A. L. Leiman. 2002. *Biological Psychology: An Introduction to Behavioral, Cognitive, and Clinical Neuroscience*, 3rd ed. Sinauer, Sunderland, MA.

Vander, A., J. Sherman, and D. Luciano 1998. *Human Physiology: The Mechanisms of Body Function*, 7th ed. WCB McGraw-Hill, Boston.

Chapter 11

Bear, M. F., B. W. Conners, and M. A. Paradiso. 2001. *Neuroscience: Exploring the Brain*, 2nd ed. Lippincott Williams & Wilkins, Baltimore, MD.

Bullock, T. H., and G. A. Horridge. 1965. *Structure and Function in the Nervous Systems of Invertebrates.* Freeman, San Francisco.

Kandel, E. R., J. A. Schwartz, and T. M. Jessell (eds.). 1995. *Essentials of Neural Science and Behavior.* Appleton & Lange, Norwalk, CT.

Kandel, E. R., J. H. Schwartz, and T. M. Jessell (eds.). 2000. *Principles of Neural Science and Behavior*, 4th ed. McGraw-Hill, New York.

McCormick, D. A 1999. Membrane potential and action potential. In M. J. Zigmond, F. E. Bloom, S. C. Landis, J. L. Roberts, and L. R. Squire (eds.), *Fundamental Neuroscience*, pp. 129–154. Academic Press, San Diego, CA.

Miller, C. 2001. See potassium run. *Nature* 414: 23–24.

Morais-Cabral, J. H., Y. Zhou, and R. MacKinnon. 2001. Energetic optimization of ion conduction rate by the K^+ selectivity filter. *Nature* 414: 37–42.

Schmidt, R. F. (ed.). 1985. *Fundamentals of Neurophysiology*, 3rd ed. Springer, New York.

Zhou, Y., J. H. Morais-Cabral, A. Kaufman, and R. MacKennon. 2001. Chemistry of ion coordination and hydration revealed by a K^+ channel–Fab complex at 2.0 Å resolution. *Nature* 414: 43–48.

Chapter 12

Kandel, E. R., J. A. Schwartz, and T. M. Jessell (eds.). 1995. *Essentials of Neural Science and Behavior.* Appleton & Lange, Norwalk, CT.

Kandel, E. R., J. H. Schwartz, and T. M. Jessell (eds.). 2000. *Principles of Neural Science*, 4th ed. McGraw-Hill, New York.

Purves, D., G. J. Augustione, D. Fitzpatrick, L. C. Katz, A.-S. LaMantia, J. O. McNamara, and S. M. Williams (eds.). *Neuroscience*, 2nd ed. Sinauer, Sunderland, MA.

Chapter 13

Delcomyn, F. 1998. *Foundations of Neurobiology.* Freeman, New York.

Eakin, R. M. 1968. In T. Dobzhansky, M. K. Hecht, and W. C. Steere (eds.), *Evolutionary Biology*, vol. 2, pp. 194–242. Appleton-Century-Crofts, New York.

Kandel, E. R., J. H. Schwartz, and T. M. Jessel (eds.). 1995. *Essentials of Neural Science and Behavior.* Appleton & Lange, Stamford, CT.

Knudsen, E. I., and M. Konishi. 1978. Centersurround organization of auditory receptive fields in the owl. *Science.* 202: 778–780.

Kuffler, S. W., J. G. Nicholls, and A. R. Martin, 1984. *From Neuron to Brain*, 2nd ed. Sinauer, Sunderland, MA.

Masland, R.H. 1986. The functional architecture of the retina. *Sci. Am.* 255(6): 102–111.

Strassmaier, M., and P. G. Gillespie. 2002. The hair cell's transduction channel. *Curr. Opin. Neurobiol.* 12: 380–386.

Wersäll, J., and D. Bagger-Sjöbäck. 1974. Morphology of the vestibular sense organ. In H. H. Kornhuber (ed.), *Vestibular System* (Handbook of Sensory Physiology, vol. 6, part 1), pp. 123–170. Springer, New York.

Wolken, J. J. 1975. *Photoprocesses, Photoreceptors, and Evolution.* Academic Press, New York.

Chapter 14

Bentley, P. J. 1998. *Comparative Vertebrate Endocrinology*, 3rd ed. Cambridge University Press, Cambridge.

Goodman, H. M. 2001. Endocrinology concepts for medical students. *Adv. Physiol. Educ.* 25: 213–224.

Henderson, I. W. 2000. Endocrine system in vertebrates. In *Nature Encyclopedia of Life Sciences.* Nature Publishing Group, London. http//www.els.net/[doi:10.1038/npg.els.000 1845]

Müller, W. A., G. R. Faloona, E. Aguilar-Parada, and R. H. Unger. 1970. Abnormal alpha-cell function in diabetes: Response to carbohydrate and protein ingestion. *New Eng. J. Med.* 283: 109–115.

O'Brien, M. A., E. J. Katahira, T. R. Flanagan, L. W. Arnold, G. Haughton, and W. E. Bollenbacher. 1988. A monoclonal antibody to the insect prothoracicotropic hormone. *J. Neurosci.* 8: 3247–3257.

Randall, D., W. Burggren, and K. French. 2002. *Eckert Animal Physiology: Mechanisms and Adaptations*, 5th ed. W. H. Freeman, New York.

Sacca, L., N. Eigler, P. E. Cryer, and R. S. Sherwin. 1979. Insulin antagonistic effects of epinephrine and glucagon in the dog. *Amer. J. Physiol.* 237: E487–E492.

Schmidt-Nielsen, K. 1997. *Animal Physiology: Adaptation and Environment*, 5th ed. Cambridge University Press, New York.

Sherwood, L., 2004. *Human Physiology: From Cells to Systems*, 5th ed. Brooks/Cole—Thompson Learning, Belmont, CA.

Silverthorn, D. U., 2004. *Human Physiology: An Integrated Approach*, 3rd ed. Benjamin Cummings, San Francisco.

Watson, R. D., E. Spaziani, and W. E. Bollenbacher. 1989. Regulation of ecdysone biosynthesis in insects and crustaceans: a comparison. In J. Koolman (ed.), *Ecdysone: From Chemistry to Mode of Action*, pp. 188–203. Thieme, Stuttgart.

Widmaier, E. P., H. Raff, and K. T. Strang. 2004. *Vander, Sherman, & Luciano's Human Physiology.* McGraw-Hill, New York.

Žitňan, D., I. Žitňanová, I. Spalovská, P. Takáč, Y. Park, and M. E. Adams. 2003. Conservation of ecdysis-triggering hormone signaling in insects. *J. Exp. Biol.* 206: 1275–1289.

Chapter 15

Allen, W. R. 2001. Equine Pregnancy. *Reproduction.* 121: 513–527.

Primakoff, P. and D. G. Myles. 2002. Penetration, adhesion, and fusion in mammalian sperm-egg interaction. *Science* 296: 2138–2185.

Chapter 16

Brower, L. P. 1996. Monarch butterfly orientation: Missing pieces of a magnificent puzzle. *J. Exp. Biol.* 199: 93–103.

Collett, T. S., and J. Zeil. 1998. Places and landmarks: An arthropod perspective. In S. Healy (ed.), *Spatial Representation in Animals*, pp. 18–53. Oxford University Press, Oxford, UK.

Dittman, A. H., and T. P. Quinn. 1996. Homing in Pacific salmon: Mechanisms and ecological basis. *J. Exp. Biol.* 199: 83–91.

Drickamer, L. C., S. H. Vessey, and E. M. Jakob (eds.). 2002. *Animal Behavior: Mechanisms, Ecology, Evolution*, 5th ed. McGraw Hill, Boston.

Goodenough, J., B. McGuire and R. A. Wallace. 2001. *Perspectives on Animal Behavior*, 2nd ed. Wiley, New York.

Helbig, A. J. 1996. Genetic basis, mode of inheritance, and evolutionary changes in migratory directions in palearctic warblers (Aves: Sylviide). *J. Exp. Biol.* 199: 49–55.

Kandel, E. R., J. H. Schwartz, and T. M. Jessel (eds.). 2000. *Principles of Neural Science*, 4th ed. McGraw Hill, New York.

Lohmann, K. J., S. D. Cain, S. A. Dodge, and C. M. F. Lohmann. 2001. Regional magnetic fields as navigational markers for sea turtles. *Science* 294: 364–366.

Papi, F. (ed.). 1992. *Animal Homing.* Chapman & Hall, London.

Perdick, A. C. 1958. Two types of orientation in migrating starlings, *Sturmus vulgaris* L. and chaffinches, *Fringilla coelebs* L., as revealed by displacement experiments. *Ardea* 46: 1–37.

Srinivasan, M. V., S. W. Zhang, M. Altwein, and J. Tautz. 2000. Honeybee navigation: Nature and calibration of the "odometer." *Science* 287: 851–853.

Walcott, C. and R. P. Green. 1974. Orientation of homing pigeons altered by a change in the direction of an applied magnetic field. *Science* 184: 180–182.

Wehner, R. 1997. The ant's celestial compass system": Spectral and polarization channels. In Lehrer M. (ed.), *Orientation and Communication in Arthropods,* pp. 145–185. Birkhäuser, Basel, Switzerland.

Wiltschko, R., and W. Wiltschko. 1995. *Magnetic Orientation in Animals.* Zoophysiology, vol. 33. Springer, Berlin.

Wiltschko, W., and R. Wiltschko. 1996. Magnetic orientation in birds. *J. Exp. Biol.* 199: 29–38.

Chapter 17

Gordon, A. M., A. F. Huxley, and F. J. Julian. 1966. The variation in isometric tension with sarcomere length in vertebrate muscle fibers. *J. Physiol. (London)* 184: 170–192.

Gotter, A. L., M. A. Kaetzel, and J. R. Dedman. 1998. *Electrophorus electricus* as a model system for the study of membrane excitability. *Comp. Biochem. Physiol. A* 119: 225–241.

Pringle, J.W.S. 1975. *Insect Flight.* Oxford Biology Reader No. 52 (J.J. Head, ed.), Oxford University Press, Oxford.

Randall, D., W. Burggren, and K. French. 2002. *Eckert Animal Physiology: Mechanisms and Adaptations,* 5th ed. Freeman, New York.

Sherwood, L. 2004. *Human Physiology: From Cells to Systems*, 5th ed. Thomson/Brooks/Cole, Belmont, CA.

Silverthorn, D. U. 2004. *Human Physiology: An Integrated Approach*, 3rd ed. Benjamin Cummings, San Francisco.

Chapter 18

Camhi, J. M. 1984. *Neuroethology.* Sinauer, Sunderland, MA.

Harris-Warrick, R. M., and R. E. Flamm. 1986. Chemical modulation of a small central pattern generator circuit. *Trends Neurosci.* 9: 432–437.

Stomatogastric Nervous System, pp. 1–30. MIT Press, Cambridge, MA.

Kandel, E. R. 1976. *Cellular Basis of Behavior: An Introduction to Behavioral Neurobiology.* Freeman, San Francisco.

Kandel, E. R., J. A. Schwartz, and T. M. Jessell (eds.). 1995. *Essentials of Neural Science and Behavior.* Appleton & Lange, Norwalk, CT.

Wyse, G. A., D. H. Sanes, and W. H. Watson. 1980. Central neural motor programs underlying sthort- and long-term patterns of *Limulus* respiratory activity. *J. Comp. Physiol.* 141: 87–92.

Chapter 19

Andersen, J. L., and P. Aagaard. 2000. Myosin heavy chain IIX overshoot in human skeletal muscle. *Muscle Nerve* 23: 1095–1104.

Andersen, P., and J. Henriksson. 1977. Capillary supply of the quadriceps femoris muscle of man: Adaptive response to exercise. *J. Physiol.* 270: 677–690.

Goldspink, G. 1999. Changes in muscle mass and phenotype and the expression of autocrine and systemic growth factors by muscle in response to stretch and overload. *J. Anat.* 194: 323–334.

Hoppeler, H., H. Howald, K. Conley, S. L. Lindstedt, H. Claassen, P. Vock, and E. R. Weibel. 1985. Endurance training in humans: Aerobic capacity and structure of skeletal muscle. *J. Appl. Physiol.* 59: 320–327.

Ingjer, F. 1979. Effects of endurance training on muscle fibre ATP-ase activity, capillary supply and mitochondrial content in man. *J. Physiol.* 294: 419–432.

Moore, D. H. II. 1975. A study of age group track and field records to relate age and running speed. *Nature* 253: 264–265.

Owino, V., S. Y. Yang, and G. Goldspink. 2001. Age-related loss of skeletal muscle function and the inability to express the autocrine form of insulin-like growth factor-1 (MGF) in response to mechanical overload. *FEBS Lett.* 505: 259–263.

Richardson, R. S., H. Wagner, S. R. D. Mudaliar, E. Saucedo, R. Henry, and P. D. Wagner. 2000. Exercise adaptation attenuates VEGF gene expression in human skeletal muscle. *Amer. J. Physiol.* 279: H772–H778.

Widrick, J. J., S. T. Knuth, K. M. Norenberg, J. G. Romatowski, J. L. W. Bain, D. A. Riley, M. Karhanek, S. W. Trappe, T. A. Trappe, D. L. Costill, and R. H. Fitts. 1999. Effect of a 17 day spaceflight on contractile properties of human soleus muscle fibres. *J. Physiol.* 915–930.

Chapter 21

Banchero, N., R. F. Grover, and J. A. Will. 1971. Oxygen transport in the llama (*Llama glama*). *Respir. Physiol.* 13: 102–115.

Brackenbury, J. H. 1981. Airflow and respired gases within the lung-air-sac system of birds. *Comp. Biochem. Physiol. A* 68: 1–8.

Burggren, W. W., and N. H. West. 1982. Changing respiratory importance of gills, lungs and skin during metamorphosis in the bullfrog *Rana catesbeiana*. *Respir. Physiol.* 47: 151–164.

Comstock, J. H. *The Spider Book*. Doubleday, Garden City, N. J.

Duncker, H. R. 1978. General morphological principles of amniotic lungs. In J. Piiper (ed.), *Respiratory Function in Birds, Adult and Embryonic*, pp. 2–15. Springer-Verlag, New York.

Feder, M. E., and W. W. Burggren. 1985. Skin breathing in vertebrates. *Sci. Amer.* 253(5): 126–142.

Feldman, J. L., J. C. Smith, D. R. McCrimmon, H. H. Ellenberger, and D. F. Speck. 1988. Generation of respiratory pattern in mammals. In A. H. Cohen, S. Rossignol, and S. Grillner (eds.), *Neural Control of Rhythmic Movements in Vertebrates*, pp. 73–100. Wiley, New York.

Gans, C. 1970. Respiration in early tetrapods—The frog is a red herring. *Evolution* 24: 723–734.

Hildebrandt, J. and A. C. Young. 1960. Anatomy and physics of respiration. In T. C, Ruch and H. D. Patton (eds.), *Physiology and Biophysics*, 19th ed., pp. 733–760. Saunders, Philadelphia.

Hill, R. W., and G. A. Wyse. 1989. *Animal Physiology*, 2nd ed. Harper & Row, New York.

Hughes, G. M. 1961. How a fish extracts oxygen from water. *New Sci.* 11: 346–348.

Lieske, S. P., M. Thoby-Brisson, P. Telgkamp, and J. M. Ramirez. 2000. Reconfiguration of the neural network controlling multiple breathing patterns: Eupnea, sighs, and gasps. *Nature Neurosci.* 3: 600–608.

Moyle, P. B. 1993. *Fish: An Enthusiast's Guide*. University of California Press, Berkeley, CA.

Perry, S. F. 1990. Recent advances and trends in the comparative morphometry of vertebrate gas exchange organs. *Adv. Comp. Environ. Physiol.* 6: 45–71.

Poll, M. 1962. Etude sur la structure adulte et la formation de sacs pulmonaires des Protoptères. *Ann. Mus. R. Afr. Centr. (Ser. 8)* 108: 129–172.

Scheid, P., and J. Piiper. 1997. Vertebrate respiratory gas exchange. In W. H. Dantzler (ed.), *Comparative Physiology*, vol. 1 (Handbook of Physiology [Bethesda, MD], section 13), pp. 309–356. Oxford University Press, New York.

Torrance, J. D., C. Lenfant, J. Cruz, and E. Marticorena. 1970. Oxygen transport mechanisms in residents at high altitude. *Respir. Physiol.* 11: 1–15.

Weibel, E. R. 1979. Oxygen demand and the size of the respiratory structures in mammals. In S. C. Wood and C. Lenfant (eds.), *Evolution of Respiratory Processes*, pp. 289–316. Dekker, New York.

Weibel, E. R. 1984. *The Pathway for Oxygen*. Harvard University Press, Cambridge, MA.

Wigglesworth, V. B. 1935. The regulation of respiration in the flea, *Xenopsylla cheopis*, Roths. (Pulicidae) *Proc. Roy. Soc. London [Biol.]* 118: 397–419.

Chapter 22

Brusca, R. C., and G. J. Brusca. 2003. *Invertebrates*, 2nd ed. Sinauer, Sunderland, MA.

Dejours, P. 1981. *Principles of Comparative Respiratory Physiology*, 2nd ed. Elsevier/North-Holland, New York.

Dhindsa, D. S., J. Metcalfe, and A. S. Hoversland. 1972. Comparative studies of the respiratory functions of mammalian blood. IX. Ring-tailed lemur (*Lemur catta*) and black lemur (*Lemur macaco*). *Respir. Physiol.* 15: 331–342.

Dickerson, R. E., and I. Geis. 1983. *Hemoglobin: Structure, Function, Evolution, and Pathology*. Benjamin Cummings, Menlo Park, CA.

Duhm, J. 1971. Effects of 2,3-diphosphoglycerate and other organic phosphate compounds on oxygen affinity and intracellular pH of human erythrocytes. *Pflügers Arch.* 326: 341–356.

Farrell, A. P., and S. M. Clutterham. 2003. On-line venous oxygen tensions in rainbow trout during graded exercise at two acclimation temperatures. *J. Exp. Biol.* 206: 487–496.

Hill, R. W., and G. A. Wyse. 1989. *Animal Physiology*, 2nd ed. Harper & Row, New York.

Johansen, K., and C. Lenfant. 1966. Gas exchange in the cephalopod, *Octopus dofleini*. *Amer. J. Physiol.* 210: 910–918.

Jones, D. R., and D. J. Randall. 1978. The respiratory and circulatory systems during exercise. In W. S. Hoar and D. J. Randall (eds.), *Fish Physiology*, vol. 7, pp. 425–501. Academic Press, New York.

Reeves, R. B. 1980. The effect of temperature on the oxygen equilibrium curve of human blood. *Respir. Physiol.* 42: 317–328.

Roughton, F. J. W. 1964. Transport of oxygen and carbon dioxide. In W. O. Fenn and H. Rahn (eds.), *Respiration*, vol. 1 (Handbook of Physiology [Bethesda, MD], section 3), pp. 767–825. American Physiological Society, Washington, DC.

Steen, J. B. 1963. The physiology of the swimbladder of the eel *Angullia vulgaris*. I. The solubility of the gases and the buffer capacity of the blood. *Acta Physiol. Scand.* 58: 124–137.

Terwilliger, R. C. 1980. Structures of invertebrate hemoglobins. *Amer. Zool.* 20: 53–67.

Waterman, M. R. 1978. Spectral characterization of human hemoglobin and its derivatives. In S. Fleischer and L. Packer (eds.), *Methods in Enzymology*, vol. 52, pp. 456–463. Academic Press, New York.

Wood, S. C., and K. Johansen. 1973. Blood oxygen transport and acid–base balance in eels during hypoxia. *Amer. J. Physiol.* 225: 849–851.

Wood, W. G. 1976. Haemoglobin synthesis during human fetal development. *Brit. Med. Bull.* 32: 282–287.

Chapter 23

Copenhaver, W. M., D. E. Kelly, and R. L. Wood. 1978. *Bailey's Textbook of Histology*, 17th ed. Williams & Wilkins, Baltimore.

Farrell, A. P. 1991. Circulation of body fluids. In C. L. Prosser (ed.), *Environmental and Metabolic Animal Physiology* (*Comparative Animal Physiology*, 4th ed.), pp. 509–558. Wiley-Liss, New York.

Feigl, E. O. 1974. Physics of the cardiovascular system. In T. C. Ruch and H. D. Patton (eds.), *Physiology and Biophysics*, 20th ed., vol. 2, pp. 10–22. Saunders, Philadelphia.

Hartline, D. K. 1967. Impulse identification and axon mapping of the nine neurons in the cardiac ganglion of the lobster, *Homarus americanus*. *J. Exp. Biol.* 47: 327–346.

Hartline, D. K. 1979. Integrative neurophysiology of the lobster cardiac ganglion. *Amer. Zool.* 19: 53–65.

Hill, R. W., and G. A. Wyse. 1989. *Animal Physiology*, 2nd ed. Harper & Row, New York.

Johansen, K. 1970. Air breathing in fishes. In W. S. Hoar and D. J. Randall (eds.), *Fish Physiology*, vol. 4, pp. 361–411. Academic Press, New York.

Johansen, K., and R. Hol. 1968. A radiological study of the central circulation in the lungfish, *Protopterus aethiopicus*. *J. Morphol.* 126: 333–348.

Johansen, K., and C. Lenfant. 1966. Gas exchange in the cephalopod, *Octopus dofleini*. *Amer. J. Physiol.* 210: 910–918.

Jones, J. C. 1964. The circulatory system of insects. In M. Rockstein (ed.), *The Physiology of Insecta*, vol. 3, pp. 1–107. Academic Press, New York.

Laurent, P., R. G. DeLaney, and A. P. Fishman. 1978. The vasculature of the gills in the aquatic and aestivating lungfish (*Protopterus aethiopicus*). *J. Morphol.* 156: 173–208.

McLaughlin, P. A. 1980. *Comparative Morphology of Recent Crustacea*. W. H. Freeman, San Francisco.

Paul, R. J., S. Bihlmayer, M. Colmorgen, and S. Zahler. 1994. The open circulatory system of spiders (*Eurypelma californicum, Pholcus phalangioides*): A survey of functional morphology and physiology. *Physiol. Zool.* 67: 1360–1382.

Plateau, F. 1880. Le coeur des crustacés décapodes. *Arch. Biol.* 1: 595–695.

Randall, D. J. 1968. Functional morphology of the heart in fishes. *Amer. Zool.* 8: 179–189.

Riefenstahl, L. 1973. *The Last of the Nuba*. Harper & Row, New York.

Rushmer, R. F. 1976. *Cardiovascular Dynamics*, 4th ed. Saunders, Philadelphia.

Scher, A. M., and M. S. Spach. 1979. Cardiac depolarization and repolarization and the electrocardiogram. In R. M. Berne (ed.), *The Cardiovascular System*, vol. 1 (Handbook of Physiology [Bethesda, MD], section 2), pp. 357–392. Oxford University Press, New York.

Smith, P. J. S. 1981. The octopod ventricular cardiogram. *Comp. Biochem. Physiol. A* 70: 103–105.

Wilkens, J. L. 1999. Evolution of the cardiovascular system in Crustacea. *Amer. Zool.* 39: 199–214.

Chapter 24

Bron, K. M., H. V. Murdaugh, Jr., J. E. Millen, R. Lenthall, P. Raskin, and E. D. Robin. 1966. Arterial constrictor response in a diving mammal. *Science* 152: 540–543.

Butler, P. J., and D. R. Jones. 1997. Physiology of diving of birds and mammals. *Physiol. Rev.* 77: 837–899.

Kooyman, G. L. 1966. Maximum diving capabilities of the Weddell seal, *Leptonychotes weddelli*. *Science* 151: 1553–1554.

Kooyman, G. L., E. A. Wahrenbrock, M. A. Castellini, R. W. Davis, and E. E. Sinnett. 1980. Aerobic and anaerobic metabolism during voluntary diving in Weddell seals: Evidence of preferred pathways from blood chemistry and behavior. *J. Comp. Physiol. B* 138: 335–346.

Le Boeuf, B. J., D. E. Crocker, D. P. Costa, S. B. Blackwell, P. M. Webb, and D. S. Houser. 2000. Foraging ecology of northern elephant seals. *Ecol. Monogr.* 70: 353–382.

Lenfant, C., K. Johansen, and J. D. Torrance. 1970. Gas transport and oxygen storage capacity in some pinnipeds and the sea otter. *Respir. Physiol.* 9: 277–286.

Ponganis, P. J., G. L. Kooyman, and M. A. Castellini. 1993. Determinants of the aerobic dive limit of Weddell seals: Analysis of diving metabolic rates, postdive end tidal P_{O_2}'s, and blood and muscle oxygen stores. *Physiol. Zool.* 66: 732–749.

Ridgway, S. H., B. L. Scronce, and J. Kanwisher. 1969. Respiration and deep diving in the bottlenose porpoise. *Science* 166: 1651–1654.

Scholander, P. F. 1964. Animals in aquatic environments: Diving mammals and birds. In D. B. Dill (ed.), *Adaptation to the Environment* (Handbook of Physiology [Bethesda, MD], section 4), pp. 729–739. Oxford University Press, New York.

Scholander, P. F., L. Irving, and S. W. Grinnell. 1942. Aerobic and anaerobic changes in seal muscles during diving. *J. Biol. Chem.* 142: 431–440.

Thompson, D., and M. A. Fedak. 1993. Cardiac responses of grey seals during diving at sea. *J. Exp. Biol.* 174: 139–164.

Chapter 25

Barnes, H. 1954. Some tables for the ionic composition of seawater. *J. Exp. Biol.* 31: 582–588.

Bayly, I. A. E., and W. D. Williams. 1973. *Inland Waters and Their Ecology*. Longman, Camberwell, Australia.

Gerard, J. F., and R. Gilles. 1972. The free amino-acid pool in *Callinectes sapidus* (Rathbun) tissues and its rôle in the osmotic intracellular regulation. *J. Exp. Mar. Biol. Ecol.* 10: 125–136.

Gilles, R., and E. Delpire. 1997. Variations in salinity, osmolarity, and water availability: Vertebrates and invertebrates. In W. H. Dantzler (ed.), *Comparative Physiology*, vol. 2 (Handbook of Physiology [Bethesda, MD], section 13), pp. 1523–1586. Oxford University Press, New York.

Hill, R. W., and G. A. Wyse. 1989. *Animal Physiology*, 2nd ed. HarperCollins, New York.

Kirschner, L. B. 1991. Water and Ions. In C. L., Prosser (ed.), *Environmental and Metabolic Animal Physiology* (*Comparative Animal Physiology*, 4th ed.), pp. 13–107. Wiley-Liss Press, New York.

Schmidt-Nielsen, K. 1964. Terrestrial animals in dry heat: Desert rodents. In D. B. Dill (ed.), *Adaptation to the Environment* (Handbook of Physiology [Bethesda, MD], section 4), pp. 493–507. Oxford University Press, New York.

McHugh, J. L. 1967. Estuarine nekton. In G. H. Lauff (ed.), *Estuaries*, pp. 581–620. American Association for the Advancement of Science, Washington, DC.

Chapter 26

Bentley, P. J., and K. Schmidt-Nielsen. 1966. Cutaneous water loss in reptiles. *Science* 151: 1547–1549.

Beuchat, C. A. 1990. Body size, medullary thickness, and urine concentrating ability in mammals. *Amer. J. Physiol.* 258: R298–R308.

Blaylock, L. A. R. Ruibal, and K. Platt-Aloia. 1976. Skin structure and wiping behavior of phyllomedusine frogs. *Copeia* 1976: 283–295.

Bradley, T. J., A. E. Williams, and M. R. Rose. 1999. Physiological responses to selection for desiccation resistance in *Drosophila melanogaster*. *Amer. Zool.* 39: 337–345.

Degnan, K. J., K. J. Karnaky, Jr., and J. A. Zadunaisky. 1977. Active chloride transport in the in vitro opercular skin of a teleost (*Fundulus heteroclitus*), a gill-like epithelium rich in chloride cells. *J. Physiol.* 271: 155–191.

D'Orazio, S. E., and C. W. Holliday. 1985. Gill Na, K-ATPase and osmoregulation in the sand fiddler crab, *Uca pugilator*. *Physiol. Zool.* 58: 364–373.

Gibbs, A. G. 1998. Water-proofing properties of cuticular lipids. *Amer. Zool.* 38: 471–482.

Goss, G. G., S. F. Perry, J. N. Fryer, and P. Laurent. 1998. Gill morphology and acid–base regulation in freshwater fishes. *Comp. Biochem. Physiol. A* 119: 107–115.

Greenaway, P. 1988. Ion and water balance. In W. W. Burggren and B. R. McMahon (eds.), *Biology of the Land Crabs*, pp. 211–248. Cambridge University Press, New York.

Hill, R. W., and G. A. Wyse. 1989. *Animal Physiology*, 2nd ed. Harper & Row, New York.

Kirschner, L. B. 1991. Water and ions. In C. L. Prosser (ed.), *Environmental and Metabolic Animal Physiology* (Comparative Animal Physiology, 4th ed.), pp. 13–107. Wiley-Liss, New York.

Loveridge, J. P. 1968. The control of water loss in *Locusta migratoria migratorioides* R & F. I: Cuticular water loss. *J. Exp. Biol.* 49: 1–13.
MacMillen, R. E., and D. E. Grubbs. 1976. Water metabolism in rodents. In H. D. Johnson (ed.), *Progress in Animal Biometeorology*, vol. 1, pp. 67–73. Swets & Zeitlinger, Lisse, The Netherlands.
Marshall, A. T., and P. D. Cooper. 1988. Secretory capacity of the lachrymal salt gland of hatchling sea turtles, *Chelonia mydas*. *J. Comp. Physiol. B* 157: 821–827.
Mommsen, T. P., and P. J. Walsh. 1989. Evolution of urea synthesis in vertebrates: The piscine connection. *Science* 243: 72–75.
Nagy, K. A., and C. C. Peterson. 1988. *Scaling of Water Flux Rate in Animals*. University of California Publications in Zoology, vol. 120. University of California Press, Berkeley.
Perry, S. F. 1998. Relationships between branchial chloride cells and gas transfer in freshwater fish. *Comp. Biochem. Physiol. A* 119: 9–16.
Potts, W. T. W., and G. Parry. 1964. *Osmotic and Ionic Regulation in Animals*. Pergamon, Oxford.
Schmidt-Nielson, B., and K. Schmidt-Nielsen. 1951. A complete account of the water metabolism in kangaroo rats and an experimental verification. *J. Cell. Comp. Physiol.* 38: 165–181.
Schmidt-Nielsen, K. 1960. The salt-secreting gland of marine birds. *Circulation* 21: 955–967.
Schmidt-Nielsen, K., and R. Fange. 1958. Salt glands in marine reptiles. *Nature* 182: 783–785.
Theede, H. 1965. Vergleichende experimentelle Untersuchungen über die zelluläre Gefrierresistenz mariner Muscheln. *Kieler Meeresforschungen* 21: 153–166.
Tracy, R. L., and G. E. Walsberg. 2001. Intraspecific variation in water loss in a desert rodent, *Dipodomys merriami*. *Ecology* 82: 1130–1137.

Chapter 27

Bankir, L., and C. de Rouffignac. 1985. Urinary concentrating ability: Insights from comparative anatomy. *Amer. J. Physiol.* 249: R643–R666.
Beck, F.-X., W. G. Guder, and M. Schmolke. 1998. Cellular osmoregulation in kidney medulla. In F. Lang (ed.), *Cell Volume Regulation*, pp. 169–184. Karger, Basel.
Beuchat, C. A. 1990. Body size, medullary thickness, and urine concentrating ability in mammals. *Amer. J. Physiol.* 258: R298–R308.
Beuchat, C. A. 1996. Structure and concentrating ability of the mammalian kidney: Correlations with habitat. *Amer. J. Physiol.* 271: R157–R179.
Brown, G. W., Jr., and P. P. Cohen. 1958. Biosynthesis of urea in metamorphosing tadpoles. In W. D. McElroy and B. Glass (eds.), *A Symposium on the Chemical Basis of Development*, pp. 495–513. Johns Hopkins Press, Baltimore, MD.
Elias, H., A. Hossman, I. B. Barth, and A. Solmor. 1960. Blood flow in the renal glomerulus. *J. Urol.* 83: 790–798.
Gupta, B. L., and M. J. Berridge. 1966. Fine structural organization of the rectum in the blowfly, *Calliphora erythrocephala* (Meig.) with special reference to connective tissue, tracheae and neurosecretory innervation in the rectal papillae. *J. Morphol.* 120: 23–82.
Huber, G. C. 1932. Renal tubules. In E. V. Cowdry (ed.), *Special Cytology*, 2nd ed., vol. 2, pp. 933–977. Paul B. Hoeber, New York.
Kaissling, B., C. de Rouffignac, J. M. Barrett, and W. Kriz. 1975. The structural organization of the kidney of the desert rodent *Psammomys obesus*. *Anat. Embryol.* 148: 121–143.
Knepper, M. A., J. B. Wade, J. Terris, C. A. Ecelbarger, D. Marples, B. Mandon, C.-L. Chou, B. K. Kishore, and S. Nielsen. 1996. Renal aquaporins. *Kidney International* 49: 1712–1717.
Martin, A. W., and F. M. Harrison. 1966. Excretion. In K. M. Wilbur and C. M. Yonge (eds.), *Physiology of Mollusca*, vol. 2, pp. 275–308. Academic Press, New York.
Møbjerg, N., E. H. Larsen, and Å. Jespersen. 1998. Morphology of the nephron in the mesonephros of *Bufo bufo* (Amphibia, Anura, Bufonidae). *Acta Zool.* 79: 31–50.
Phillips, J. E. 1964. Rectal absorption in the desert locust, *Schistocerca gregaria* Forskål. III. The nature of the excretory process. *J. Exp. Biol.* 41: 69–80.
Pitts, R. F. 1974. *Physiology of the Kidney and Body Fluids*, 3rd ed. Year Book Medical, Chicago.
Riegel, J. A. 1977. Fluid movement through the crayfish antennal gland. In B. L. Gupta, R. B. Moreton, J. L. Oschman, and B. J. Wall (eds.), *Transport of Ions and Water in Animals*, pp. 613–631. Academic Press, New York.
Smith, H. W. 1951. *The Kidney: Structure and Function in Health and Disease*. Oxford University Press, New York.
Sperber, I. 1944. Studies on the mammalian kidney. *Zool. Bidr. Uppsala* 22: 249–432.
Walker, A. M., C. L. Hudson, T. Findley, Jr., and A. N. Richards. 1937. The total molecular concentration and the chloride concentration of fluid from different segments of the renal tubule of Amphibia: The site of chloride reabsorption. *Amer. J. Physiol.* 118: 121–129.
Willoughby, E. J., and M. Peaker. 1979. Birds. In G. M. O. Maloiy (ed.), *Comparative Physiology of Osmoregulation in Animals*, vol. 2, pp. 1–55. Academic Press, New York.
Wirz, H., B. Hargitay, and W. Kuhn. 1951. Lokalisation des Konzentrierungsprozesses in der Niere durch direkte Kryoskopie. *Helv. Physiol. Pharmacol. Acta* 9: 196–207.

Chapter 28

Beuchat, C. A. 1996. Structure and concentrating ability of the mammalian kidney: Correlations with habitat. *Amer. J. Physiol.* 271: R157–R179.
Goudie, A., and J. Wilkinson. 1977. *The Warm Desert Environment*. Cambridge University Press, New York.
Louw, G. N. 1972. The role of advective fog in the water economy of certain Namib desert animals. Symposia of the Zoological Society of London, no. 31, pp. 297–314. Academic Press, New York.
Lovegrove, B. 1993. *The Living Deserts of Southern Africa*. Fernwood Press, Vlaeberg, South Africa.
Maloiy, G. M. O., W. V. MacFarlane, and A. Shkolnik. 1979. Mammalian herbivores. In G. M. O. Maloiy (ed.), *Comparative Physiology of Osmoregulation in Animals*, vol. 2, pp. 185–209. Academic Press, New York.
Ostrowski, S., J. B. Williams, and K. Ismael. 2003. Heterothermy and the water economy of free-living Arabian oryx (*Oryx leucoryx*). *J. Exp. Biol.* 206: 1471–1478.
Schmidt-Nielsen, K. 1954. Heat regulation in small and large desert animals. In J. L. Cloudsley-Thompson (ed.), *Biology of Deserts*, pp. 182–187. Institute of Biology, London.
Soholt, L. F., D. B. Dill, and I. Oddershede. 1977. Evaporative cooling in desert heat: Sun and shade; rest and exercise. *Comp. Biochem. Physiol. A* 57: 369–371.
Taylor, C. R. 1968. The minimum water requirements of some East African bovids. Symposia of the Zoological Society of London, no. 21, pp. 195–206. Academic Press, London.
Taylor, C. R. 1970. Strategies of temperature regulation: Effect on evaporation in East African ungulates. *Amer. J. Physiol.* 219: 1131–1135.

Index